Food Composition and Nutrition Tables

Die Zusammensetzung der Lebensmittel Nährwert-Tabellen

La composition des aliments Tableaux des valeurs nutritives

Founded by
Begründet von
Fondés par

S. W. Souci, W. Fachmann, H. Kraut

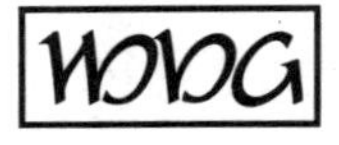

Wissenschaftliche Verlagsgesellschaft mbH Stuttgart

Food Composition and Nutrition Tables 1989/90

Die Zusammensetzung der Lebensmittel Nährwert-Tabellen 1989/90

La composition des aliments Tableaux des valeurs nutritives 1989/90

on behalf of the
im Auftrage des
par ordre du

Bundesministerium für Ernährung, Landwirtschaft und Forsten, Bonn

edited by
herausgegeben von
publié par

Deutsche Forschungsanstalt für Lebensmittelchemie, Garching b. München

4th revised and completed edition
4., revidierte und ergänzte Auflage
4ème édition, revue et complétée

compiled by
bearbeitet von
revue et complétée par

Heimo Scherz und Friedrich Senser

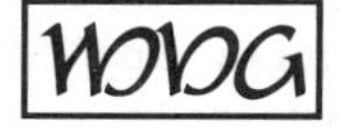

Wissenschaftliche Verlagsgesellschaft mbH Stuttgart 1989

CIP-Kurztitelaufnahme der Deutschen Bibliothek:
Food composition and nutrition tables ... =
Die Zusammensetzung der Lebensmittel, Nährwert-Tabellen ... / on behalf of the Bundesministerium für Ernährung, Landwirtschaft u. Forsten, Bonn, ed. by Dt. Forschungsanst. für Lebensmittelchemie, Garching b. München Founded by S. W. Souci ...
- Stuttgart: Wissenschaftliche Verlagsgesellschaft ISSN 0721-6912
Erscheint ca. zwei- bis dreijährl.
1. Aufl. Losebl.-Ausg. u.d.T.: Die Zusammensetzung der Lebensmittel
Ed. 2. 1981/82 (1981) ff.
NE: Souci, Siegfried W. [Begr.]; PT

Ed. 3. 1986/87 (1986). ISBN 3-8047-0833-1

Druck: Zechnersche Buchdruckerei, Speyer

Preface for the fourth edition

After two years the third issue of the table has been sold out. For the present edition, the material has been carefully revised and up dated using all the informational sources available to us (original publications, monographs, reviews, unpublished data etc.).

For the shaping of the individual tables as well as for the entire book, the approved form has been kept essentially.

The tables have been up dated by the introduction of new foods in the following food categories:

- Milk and milk products (e.g. some kinds of yoghurt)
- Meat, deer and fowl (e.g. pheasant, quail)
- Seafish
- Oilseed
- Wild berries (e.g. blackthorn)

The number of the tables has not been increased but reduced from 734 to 693 by the following alterations:

- Some tables have been eliminated, the contents being not actual any more (e.g. whole half bodies of animals for slaughter, not actual kinds of flour and bread etc.).
- In the group "meat" the data for each meat portions of animals for slaughter are not presented as individual tables but in a shortened form, as a comprehesive table in the appendix.

New groups of nutritional constituents are purines (e.g. adenine, hypoxanthine) and phospholipides (e.g. phosphatidylcholine, phosphatidylethanolamine). In the group "organic acids", salicylic acid has been introduced.

We would like to gratefully thank

Vorwort zur vierten Auflage

Nach zwei Jahren ist die dritte Auflage der Tabelle bereits vergriffen. In der vorliegenden vierten Auflage wurden alle Daten kritisch überprüft und anhand der uns zur Verfügung stehenden Informationen (Originalarbeiten, Übersichtsartikel, Monographien, unpublizierte Daten etc.) auf den neuesten Stand gebracht.

Für die Gestaltung der einzelnen Tabellen wie auch des gesamten Tabellenwerkes wurde die bewährte Form im wesentlichen beibehalten.

Folgende Lebensmittelgruppen wurden durch Aufnahme neuer Produkte erweitert:

- Milch und Milchprodukte (z. B. Joghurtarten)
- Fleisch, Wild und Geflügel (z. B. Wildgeflügel wie Fasan, Wachtel)
- Seefische
- Ölsamen
- Wildfrüchte (z. B. Schlehe)

Die Zahl der Tabellenblätter hat sich dadurch aber nicht erhöht, sondern durch folgende Änderungen sogar von 734 auf 693 reduziert:

- Einige nicht mehr aktuelle Tabellenblätter (u.a. ganze Tierhälften bei Schlachttieren, nicht mehr gängige Mehl- und Brotsorten etc.) wurden gestrichen.
- Die Daten für Teilstücke von Schlachttieren wurden nicht mehr in den Einzeltabellen aufgeführt, sondern im Appendix unter Berücksichtigung der einzelnen Handelsklassen in abgekürzter Form zusammengestellt.

Neue Inhaltsstoffgruppen sind Purine (z. B. Adenin, Hypoxanthin) und Phospholipide (z. B. Phosphatidylcholin, Phosphatidylethanolamin). In der

Préface de la quatriéme édition

Deux ans après son apparition, la 3e édition des tableaux est déjà épuisée.

Dans cette 4e édition toutes les données ont été revues de façon critique et mises à jour sur la base de la documentation accessible (monographies, travaux originaux, données non encore publiées etc.).

La forme et l'organisation des tableaux et de l'ouvrage n'ont pas été changées.

Depuis la troisième édition les classes alimentaires suivantes ont été élargies:

- lait et produits laitiers (p. ex. différentes sortes de yogourt)
- viande, gibier et volaille (p. ex. faisan et caille)
- poissons maritimes
- graines oléagineuses
- fruits sauvages (p. ex. prunelle)

Quand même le nombre des tableaux a pu être réduit de 734 à 693 par les mesures suivantes:

- Quelques tableaux manquant d'actualité ont été omis (p. ex. moitiés de bétail, certaines sortes de farine et de pain).
- Les valeurs pour les pièces d'animaux de boucherie ne se trouvent plus dans les tableaux particuliers, mais furent rassemblées dans l'appendice sous une forme réduite en tenant compte des différentes classes de marchandises.

Nous avons ajouté des groupes de constituants alimentaires comme les purines (p. ex. adénine, hypoxanthine) et les phospholipides (p. ex. phospatidylcholine, phosphatidyléthanolamine). Le groupe des acides organiques comprend maintenant l'acide salicylique.

Prof. S. Braun and Prof. G. Schmidt (Hochschule der Bundeswehr) for their excellent support. We would also like to thank Mrs. I. Ströhlein (Technische Universität München) for her help and the Leibniz-Rechenzentrum der Bayerischen Akademie der Wissenschaft, and especially Dr. P. Sarreither, for the support as well as for the care and the control. Further thanks are due to Mrs. A. Klein and Mrs. Ch. Hertlein for the excellent technical assistance.

Several institutions had provided us with data for this edition (see page XXVIII).

For the critical reviews of the data and for the intense discussions, we would like to express our thanks to Dr. J. Oehlenschläger (Bundesforschungsanstalt für Fischerei, Hamburg), Dr. E. Rabe (Bundesforschungsanstalt für Getreide- und Kartoffelverarbeitung, Detmold), Prof. Dr. K. O. Honikel and Dr. K. Hofmann (Bundesanstalt für Fleischforschung, Kulmbach) and Prof. Dr. Dr. H. Steinhart (Institut für Lebensmittelchemie, Universität Hamburg).

Garching, July 1989

H.-D. Belitz
W. Grosch

Gruppe der organischen Säuren wurde Salizylsäure aufgenommen.

Unser Dank gilt den Herren Professoren Dr. S. Braun und Dr. G. Schmidt (beide Hochschule der Bundeswehr, München), die uns beim Betrieb des Datenbanksystems LINDAS in bewährter Weise unterstützt haben. Dank gilt auch Frau I. Ströhlein (Technische Universität München) für ihre tatkräftige Mitarbeit und dem Leibniz-Rechenzentrum der Bayerischen Akademie der Wissenschaften, hier vor allem Herrn Dr. P. Sarreither, für die gewährte Unterstützung und Betreuung. Frau A. Klein und Frau Ch. Hertlein danken wir für die ausgezeichnete Mitarbeit.

Die Arbeit an den Tabellen wurde dankenswerterweise von den auf S. XXVIII angegebenen Institutionen unterstützt.

Für die Durchsicht von Daten wie auch für intensive Diskussionen sind wir den Herren Dr. J. Oehlenschläger (Bundesforschungsanstalt für Fischerei, Hamburg), Dr. E. Rabe (Bundesforschungsanstalt für Getreide und Kartoffelverarbeitung, Detmold), Prof. Dr. K. O. Honikel, Dr. K. Hofmann (Bundesanstalt für Fleischforschung, Kulmbach) und Prof. Dr. Dr. H. Steinhart (Institut für Lebensmittelchemie, Universität Hamburg) zu besonderem Dank verpflichtet.

Garching, Juli 1989

H.-D. Belitz
W. Grosch

Nous remercions vivement M. le Professeur Dr. S. Braun et M. le Professeur Dr. G. Schmidt (Hochschule der Bundeswehr, Munich), qui nous ont conseillés dans toutes les questions relatives au fonctionnement de la banque de données LINDAS. Nous remercions également Madame I. Ströhlein (Technische Universität Munich) et le Leibniz Rechenzentrum der Bayerischen Akademie der Wissenschaften á Munich, et là surtout Monsieur le Dr. P. Sarreither, de l'appui ainsi que Madame A. Klein et Madame Ch. Hertlein de leur travail excellent.

Nous avons pu bénéficier de nouveau de l'appui des instituts mentionnés page XXVIII: ils ont mis notre disposition un trés grand nombre de résultats d'analyses.

M. le Dr. J. Oehlenschläger (Bundesforschungsanstalt für Fischerei, Hamburg), M. le Dr. E. Rabe (Bundesanstalt für Getreide- und Kartoffelverarbeitung, Detmold), M. le Professeur Dr. K. O. Honikel et M. le Dr. K. Hofmann (Bundesanstalt für Fleischforschung, Kulmbach) et M. le Professeur Dr. Dr. H. Steinhart (Institut für Lebensmittelchemie, Hamburg) ont eu l'amabilité de revoir et de corriger des données. Que tous ceux qui ont ainsi contribué au succés de cet ouvrage trouvent ici l'expression de notre gratitude.

Garching, Juillet 1989

H.-D. Belitz
W. Grosch

From the prefaces of preceding editions

Aus Vorworten früherer Auflagen

Préfaces des éditions précédentes: extraits

2nd edition

In 1979 the 4th supplement to the nutrition tables founded by S. W. Souci, W. Fachmann, and H. Kraut, "The Composition of Foods", was published. As the continuation of this reliable tabular study in its previous form no longer takes account of the possibilities of data aquisition and processing offered today, we have therefore developed a data bank, LINDAS (= Food-Constituent-Data-System), by means of which a considerable volume of data can be stored, processed and printed out in a very flexible form. The publication of updated editions of the entire tables or special tables for specific purposes, too, will thus be possible at suitable intervals in the form of computer printouts.

The present nutrition tables are on the one hand tailored to the needs of dietetics and nutritional guidance; on the other, they are intended to offer those employed in administration, science and industry, in as much as they are concerned with the production, marketing and control of food products or with questions of nutrition, ready acces to a comprehensive source of information regarding the composition of food. The wide range of data provided caters for the varying information needs and takes account of the circumstance that the nutritive evaluation of constituents is constantly changing.

2. Auflage

Im Jahre 1979 wurde die IV. Ergänzungslieferung der von S. W. Souci, W. Fachmann und H. Kraut begründeten Nährwert-Tabellen „Die Zusammensetzung der Lebensmittel" veröffentlicht. Da die Weiterführung dieses bewährten Tabellenwerkes in der bisherigen Form nicht mehr den heute gegebenen Möglichkeiten der Datenerfassung und -verarbeitung entspricht, haben wir eine Datenbank LINDAS (Lebensmittel-Inhaltsstoff-Daten-System) entwickelt, mit der größere Mengen an Datenmaterial gespeichert, verarbeitet und in sehr flexibler Form ausgedruckt werden können. Die Herausgabe jeweils aktualisierter Ausgaben des gesamten Tabellenwerkes oder auch von Spezialtabellen für bestimmte Zwecke wird damit in Form von Computerausdrucken in angemessenen Abständen möglich.

Die vorliegenden Nährwerttabellen sind einerseits auf die Bedürfnisse der Diätetik und Ernährungsberatung abgestimmt. Andererseits sollen sie auch den in Administration, Wissenschaft und Wirtschaft Tätigen, soweit sie mit der Erzeugung, Vermarktung und Überwachung von Lebensmitteln bzw. mit Fragen der Ernährung befaßt sind, eine schnelle und umfassende Information über die Zusammensetzung der Lebensmittel bieten. Die Breite des vorgelegten Datenmaterials trägt den unterschiedlichen Informationsbedürfnissen und dem Umstand Rechnung, daß die ernährungsphysiologische Bewertung von Inhaltsstoffen sich ständig ändert.

2ᵉ édition

Nous avons publié en 1979 la IVᵉ édition des tableaux des valeurs nutritives de S. W. Souci, W. Fachmann et H. Kraut. Comme l'édition de ces tables sous leur forme traditionnelle ne répond plus aux possibilités actuelles de collecte et de traitement des données, nous avons développé une banque de données LINDAS qui permet de stokker et de traiter un plus grand nombre d'information et de les imprimer d'une manière très flexible. Ainsi il nous sera possible de publier, selon les besoins, des éditions complètes de l'ouvrage, mises à jour, ainsi que des tableaux spéciaux, sous forme de fiches d'ordinateur.

Les présents tableaux des valeurs nutritives sont destinés d'une part à l'usage des diététiciens et des conseillers alimentaires; d'autre part ils doivent fournir à tous ceux qui, dans l'administration, dans les sciences et dans l'économie, ont à s'occuper de la production, de la commercialisation et du contrôle des denrées alimentaires ou sont confrontés à des problèmes de l'alimentation, des informations rapidement accessibles et complètes sur la composition des aliments. L'étendue des données présentées ici tient compte et du fait que les divers utilisateurs ont des besoins d'information différents, et de celui que l'évaluation physiologique des nutriments est constamment soumise à des changements.

3rd edition

The good acceptance of the second edition in many countries showed that the new design of the tables had been appropriate.

The changes of the nutritional information on "available carbohydrates" and on "total dietary fibre" are important additional alterations.

The new column "nutrient density" in the tables enables better declarations of the nutritional values.

3. Auflage

Die gute Aufnahme der 2. Auflage im In- und Ausland hat gezeigt, daß mit der völligen Neugestaltung der Tabelle der richtige Weg eingeschlagen wurde.

Bei den Inhaltsstoffen sind als wichtige Änderungen besonders die Angaben der „verwertbaren Kohlenhydrate" und der „Gesamtballaststoffe" zu erwähnen.

Zur leichteren Beurteilung des Nährwertes wurde eine Spalte „Nährstoffdichte" aufgenommen.

3e edition

L'écho positif qu'a trouvé la 2e édition dans beaucoup de pays nous a montré, qu'on avait bien fait de réorganiser les tableaux.

Pour les constituants alimentaires nous avons ajouté des données sur le contenu en hydrates de carbone utilisables et sur le contenu total en substances de lest.

La nouvelle rubrique densité nutritive servira à faciliter la déclaration de la valeur nutritive.

Contents

Inhaltsverzeichnis

Table des matières

Introduction

Einleitung

Introduction

I. Structure of the tables

The nutrition tables, obtained from the data bank LINDAS (Lebensmittel-Inhaltsstoff-Datensystem), consist of the following parts:
1.1 Head of the table; the name of the food is in german, english and french; in some cases, the scientific latin name has also been included.
1.2 Energy values: listed are the total energy value as well as the energy values corresponding to the content of protein, fat and carbohydrates in Kilojoule (kJ) and Kilocalories (kcal) per 100 g edible portion, with and without consideration of the digestibility.
1.3 Quotation of the waste
1.4 Concentration of the food constituents*); the data are grouped as follows:
Main constituents (water, protein, fat, available carbohydrates, total dietary fiber, minerals), individual minerals and trace elements, vitamins, amino acids, fatty acids, individual carbohydrates (mono-, oligo- and polysaccharides), hydroxycarbonic acids (fruit acids and phenolic acids), sterols, biogene amines, purines, phospholipides and other special constituents. The amounts of the nutrients are both quoted per 100 g edible portion (mean value and variation) and per 100 g product as purchased (mean value). Additionally the nutrition density (nutrition content/energy value) and in

*) The present data collection is a so-called "open table", meaning that the quotation of the individual nutrient can be varied according to the needs.

I. Aufbau der Tabellen

Die aus der Datenbank LINDAS (Lebensmittel-Inhaltsstoff-Daten-System) für jedes Lebensmittel erhaltene Tabelle gliedert sich in die folgenden Abschnitte:
1.1 Tabellenkopf mit dem Namen des Lebensmittels in deutscher, englischer und französischer Sprache. Gegebenenfalls ist auch der wissenschaftliche Name angegeben.
1.2 Angabe des Nährwerts (Energiegehalt):
Gesamtwerte sowie Werte für Eiweiß, Fett und Kohlenhydrate in Kilojoule (kJ) und Kilocalorien (kcal) pro 100 g eßbarem Anteil ohne und mit Berücksichtigung der Resorbierbarkeit.
1.3 Angabe des Abfalls.
1.4 Angaben über Konzentrationen an Inhaltsstoffen,*) gegliedert in folgende Gruppen:
Hauptbestandteile (Wasser, Protein, Fett, verwertbare Kohlenhydrate, Gesamtballaststoffe, Mineralstoffe), einzelne Mineralstoffe und essentielle Spurenelemente, Vitamine, Aminosäuren, Fettsäuren, einzelne Kohlenhydrate (Mono-, Oligo- und Polysaccharide), Carbonsäuren (Fruchtsäuren bzw. phenolische Säuren), Sterine, biogene Amine, Purine, Phospholipide und sonstige Bestandteile. Angegeben ist die Menge pro 100 g eßbarem Anteil (Mittelwert und Schwankungsbreite), die Menge pro 100 g käuflicher

*) Es handelt sich um eine offene Tabelle, d. h. die angeführten Inhaltsstoffe können je nach den Erfordernissen variieren.

I. Structure des tableaux

La banque des données LINDAS (Lebensmittel-Inhaltsstoff-Datensystem) fournit pour chaque aliment un tableau structuré de manière suivante:
1.1 Entête du tableau avec le nom des aliments en allemand, anglais et en français et si nécessaire, le nom scientifique latin.
1.2 Valeur nutritive (valeurs caloriques): contenu total ainsi que valeurs particulières pour les protéines, les matières grasses et les hydrates de carbone en kilojoule (kJ) et en kilocalorie (kcal) pour 100 g de matière comestible en tenant compte du degré de résorption.
1.3 Données sur les déchèts.
1.4 Données sur la concentration des différents constituants alimentaires.*) Ceux-ci sont classifiés de la manière suivante: composants principaux (eau, protéines, lipides, hydrates de carbone utilisables, substances de lest, substances minérales); substances minérales et substances de trace individuelles, vitamines, acides aminés, acides gras, hydrates de carbone particuliers (oligo-, mono- et polysaccharides) acides carboniques (acides de fruits et acides phénoliques), stérols, amines biogènes, purines, phospholipides etc.

Les valeurs se rapportent à 100 g de matière comestible (valeur moyenne et écart moyen), et à 100 g de produit brut tel qu'il se présente à l'achat (va-

*) Il s'agit d'un tableau ouvert: la liste des constituants varie suivant les besoins.

the case of protein rich foods, the amino acid composition in mole percent is given.

1.5 Footnote
Additional remarks to 1.1–1.4

Rohware (Mittelwert) sowie die Nährstoffdichte (Nährstoffgehalt/Brennwert). Bei proteinhaltigen Lebensmitteln ist die Aminosäurezusammensetzung zusätzlich in Molprozenten angegeben.

1.5 Fußnoten:
Ergänzende Angaben zu den in 1.1–1.4 enthaltenen Daten.

leur moyenne). En outre on y trouve la densité nutritive (quantité du nutriment divisé par la valeur calorique). Pour les aliments contenant des protéines, la composition en aminoacides est donnée en moles pour cent.

1.5 Remarques supplémentaires en bas de la page, concernant les rubriques 1.1 jusqu'à 1.4.

II. General Remarks

2.1 Raw product – edible portion – waste

All the values refer to the "edible portion" of the particular food, and are calculated as follows: raw product as purchased, minus waste. The waste is given in the tables as the percentage of the raw product as purchased.

"Raw product as purchased" denotes the food product as it is when it reaches the kitchen. Many food products in this state contain portions which have to be removed before the food is prepared. The remaining portion is the "edible portion". The waste can vary considerably for a number of reasons (e.g. maturity, type and duration of transport and storage). The mean values were used for the calculations.

2.2 Quotation of the concentrations

The amounts are given in gram (Abbreviation: GRAM), milligram (MILLI), microgram (MICRO) or nanogram (NANO) per 100 g edible portion (average value: abreviation AV; variation) and per 100 g raw product as purchased (average value raw product: abreviation AVR).

Besides the average value calculated from individual data, the variation, consisting in the corresponding highest and lowest values known to us, was entered in a separate column headed "variation". In some cases, the data available varied so widely, that only

II. Allgemeine Anmerkungen

2.1 Rohware – Eßbarer Anteil – Abfall

Alle Werte beziehen sich auf den „eßbaren Anteil" des Lebensmittels in seiner handels- bzw. verzehrsüblichen Form, der sich wie folgt ergibt: Käufliche Rohware minus Abfall = Eßbarer Anteil. Der Abfall ist in den Tabellen in Prozent der käuflichen Rohware angegeben. Unter „käuflicher Rohware" ist das Lebensmittel zu verstehen, wie es in die Küche gelangt. Viele Lebensmittel enthalten in diesem Zustand Anteile, die vor der Zubereitung entfernt werden müssen. Der verbleibende Rest ist der „eßbare Anteil". Der Abfall kann auch beim gleichen Lebensmittel erheblichen Schwankungen unterworfen sein, deren Ursachen mannigfaltig sein können (z. B. Reifegrad, Art und Dauer des Transports und der Lagerung, Art der Verarbeitung und Zubereitung). Für die Rechnung wurden mittlere Werte benutzt.

2.2 Konzentrationsangaben

Die Konzentrationsangabe erfolgt in Gramm (Abkürzung in der englischen Fassung: GRAM), Milligramm (MILLI), Mikrogramm (MICRO) oder Nanogramm (NANO) pro 100 g eßbarem Anteil (Mittelwert AV: englische Abkürzung für „average value"; Variation: Schwankungsbreite) und pro 100 g käuflicher Rohware (Mittelwert AVR: englische Abkürzungen für „average value raw product").

II. Remarques generales

2.1 Produit brut – partie comestible – déchets

Les données se rapportent à la partie comestible des aliments tel qu'elle se présente au marché. Elle est calculée de la manière suivante: produit brut moins déchets = partie comestible. Le déchet est indiqué en pourcent du produit brut. On entend par produit brut l'aliment tel qu'il parvient a la cuisine. A ce stade beaucoup d'aliments comprennent des parties qui doivent être enlevées avant la préparation; la partie résiduelle est dite partie comestible. L'importance du déchet est sujette à de fortes variations dont les causes peuvent être multiples (p. ex. degré de maturité, conditions et durée du transport et du magasinage). Pour le calcul on a pris les valuers moyennes.

2.2 Concentration

Les concentrations sont données en grammes (abbréviation: GRAM), milligrammes (MILLI), microgrammes (MICRO) ou en nanogrammes (NANO) pour 100 g de matière comestible (AV: anglais: average value, valeur moyenne; Variation: écart moyen) ainsi que pour 100 g de matière brute (AVR: valeur moyenne).

A côté de la valeur moyenne, les tables donnent parfois les valeurs numériques disponibles diffèrent tellement que la valeur moyenne ne nous a pas semblé licite. Dans ce cas nous donnons seulement les valeurs maximales

the variation is given without calculating an average value.

Average values were taken from the literature, without quoting variation nor individual values. Nutrients, which do not occur in a particular food, are quoted 0.000 in the table. Nutrients for which no data are available, are not listet.

In the preceding issues of these tables, the amount of carbohydrates was calculated by difference. The total of the main constituents was in general 100%. In the present table, the carbohydrate content is derived from the experimental data, when available (see below under 3.2 and 3.3); therefore the sum of the main constituents can diverge of 2 to 5% from 100% in the corresponding tables.

2.3 Energy value

The energy values quoted are the physiological energy values, with and without consideration of the individual digestibility. The values for proteins and fats were calculated using the factors given by Merrill and Watt (1). For the carbohydrate constituents, an average factor of 4 kcal/g was used (2).

In the case of alcoholic beverages for ethanol a factor was used of 7.1 kcal/g (1) and for extracts of 4 kcal/g (2) (see the footnotes in the corresponding tables).

2.4 Nutrition density

The nutrition density listed in a separate column, is the ratio between the amount of a food constituent (g, mg, µg) and the total energy value of the digestible constituents.

Neben dem arithmetischen Mittelwert, gebildet aus den vorliegenden Einzeldaten, ist in einer getrennten Spalte die Schwankungsbreite in Form der jeweiligen höchsten und niedrigsten Werte angegeben. In manchen Fällen weichen die vorliegenden Daten so sehr voneinander ab, daß eine Mittelwertbildung unzulässig erschien. Es wurden dann nur die Grenzwerte angegeben. Mittelwerte aus der Literatur, für die weder Einzelwerte noch Schwankungsbreiten vorliegen, sind als solche aufgeführt. Mit der Angabe 0,000 sind Inhaltsstoffe in den Tabellen aufgeführt, die nachweislich nicht vorkommen. Inhaltsstoffe für die keine Daten vorliegen, werden dagegen nicht aufgeführt.

In früheren Auflagen wurde der Kohlenhydratgehalt in der herkömmlichen Weise als Differenz berechnet, so daß sich als Summe der Hauptbestandteile rechnerisch 100% ergab. In der vorliegenden Ausgabe ist, soweit Daten verfügbar waren, der Kohlenhydratanteil als Summe der Analysendaten für Einzelkomponenten angegeben (vgl. 3.2 und 3.3). In entsprechenden Tabellenblättern kann die Summe der Hauptbestandteile deshalb geringfügig von 100 abweichen.

2.3 Energieinhalt

Als Energieinhalt ist der physiologische Brennwert angegeben, korrigiert durch Resorptionsfaktoren, die der unterschiedlichen Ausnutzung Rechnung tragen. Die Daten für Proteine und Fette wurden nach Merrill und Watt (1) berechnet. Die verwertbaren Kohlenhydrate wurden einheitlich mit 4 kcal/g bewertet (2). Bei alkoholischen Getränken wurde für Ethanol mit 7,1 kcal/g (1), für den Extrakt mit 4 kcal/g (2) gerechnet (vgl. die Fußnoten in den entsprechenden Tabellen).

2.4 Nährstoffdichte

Die in einer gesonderten Spalte angegebene Nährstoffdichte ist der Quotient aus der jeweiligen Inhaltsstoffmenge (g, mg, µg) und dem Gesamtenergiegehalt des verdaulichen Anteils (Megajoule).

et minimales et pas de valeur moyenne. Dans d'autres cas, nous ne donnons que les valeurs moyennes, mais pas d'intervalle de variation; c'est le cas quand nous avons utilisé des sources ne contenant que des valeurs moyennes. La valeur 0,000 correspond à des constituants pour lesquels il a été prouvé qu'ils ne sont pas présents dans l'aliment en question. Par contre on a omis les substances pour lesquelles on n'a pas trouvé des données.

Dans des éditions précédentes le contenu en hydrates de carbone a été calculé par différence. De cette façon la somme des constituants principaux donnait normalement 100%. Dans cette édition le contenu en hydrates de carbone a été, si possible, basé sur des analyses concrètes (voir: 3.2 et 3.3). De cette façon la somme des constituants peut différer un peu sur 100% principaux.

2.3 Valeurs caloriques

Les valeurs caloriques données sont des valeurs physiologiques. Pour les lipides et les protéines elles ont été corrigées par des facteurs de résorption selon Merrill et Watt (1). Pour les hydrates de carbone utilisables on a pris toujours 4 kcal/g (2). L'ethanol des boissons alcooliques entre dans ce calcul avec 7.1 kcal/g (1) l'exrait avec 4 kcal/g (2) (voir les remarques en bas des tableaux).

2.4 Densité nutritive

La densité nutritive, que nous donnons dans une rubrique séparée, est la quantité du constituant respectif (g, mg, µg) divisée par le contenu énergétique de la partie digérable (en Megajoule).

III. Comments on individual nutrient

3.1 Protein

Quoted are the values for raw protein calculated from the total N-content by multiplication with the following factors (3): milk and milk products, 6.38; cereals and cereal products, 5.80; soy and soy products, 5.71; oil seeds and nuts, 5.3; mushrooms, 4.17 and for all the other foods (e.g. meat und meat products, fish, fruits, vegetables and related products), 6.25.

The raw protein includes also low molecular N-compounds like free amino acids and peptides (fruits, vegetables, fish) as well as urea (mushroom).

3.2 Available carbohydrates

For the majority of foods, the available carbohydrates are quoted as the total of the individual constituents such as glucose, fructose, sucrose maltose, lactose, starch, dextrin (for mono-, oligo- and polysaccharides) for sorbitol and xylitol (for sugar alcohols) and citric acid, malic acid and lactic acid (for fruit acids).

For foods, where no data or uncomplete data were present, the total amount of available carbohydrates was calculated by difference, according to the following equation:

available carbohydrates = 100 – (water + protein + fat + minerals + total dietary fibre)

For such foods, where neither values of individual carbohydrates nor values of total dietary fibre were available, the amount of available carbohydrates was calculated by difference (available carbohydrates = 100 – (water + protein + fat + minerals)).

The values for available carbohydrates calculated by difference, are marked by footnotes.

III. Anmerkungen zu einzelnen Inhaltsstoffen

3.1 Protein

Angegeben ist Rohprotein, berechnet aus dem Gesamt-N-Gehalt durch Multiplikation mit folgenden Faktoren (3): Milch und Milchprodukte 6,38, Getreide und Getreideprodukte 5,80, Soja und Sojaerzeugnisse 5,71, Ölsamen, Schalenobst 5,30, Pilze 4,17, alle anderen Lebensmittel (z.B. Fleisch und Fleischwaren, Fisch, Obst, Gemüse und deren Verarbeitungsprodukte) 6,25.

Das Rohprotein umfaßt auch niedermolekulare N-Verbindungen, bei denen es sich vorwiegend um freie Aminosäuren und Peptide (Obst, Gemüse, Fische), aber z.B. auch um Harnstoff (Pilze) handelt.

3.2 Verwertbare Kohlenhydrate

Bei der Mehrzahl der Lebensmittel ist die Summe der Einzeldaten angegeben für Mono-, Oligo- und Polysaccharide (Glucose, Fructose, Saccharose, Maltose, Lactose, Stärke, Dextrin), Zuckeralkohole (Sorbit, Xylit) und Fruchtsäuren (Zitronensäure, Apfelsäure, Milchsäure).

Bei Lebensmitteln, für die keine oder nur unvollständige Daten für die Kohlenhydratkomponenten vorlagen, wurde die Gesamtmenge an verwertbaren Kohlenhydraten durch ein Differenzverfahren unter Berücksichtigung der Gesamtballaststoffe nach folgender Formel berechnet:

Verwertbare Kohlenhydrate = 100 – (Wasser + Protein + Fett + Asche + Gesamtballaststoffe).

Bei einigen Lebensmitteln, für die weder Einzeldaten der verwertbaren Kohlenhydrate noch Gesamtballaststoffwerte vorlagen, wurden die verwertbaren Kohlenhydrate in der früher üblichen Weise als Differenz 100 – (Wasser + Protein + Fett + Asche) ermittelt.

III. Remarques concernant quelques constituants particuliers

3.1 Protéines

Nous avons indiqué la quantité brute, calculée à partir du contenu total en azote, multiplié par les facteurs suivants (3): lait et produits de lait 6,38, céréales et produits de céréales 5,80, soja et produits de soja 5,71, graines oleagineuses et fruits a coques 5,30, champignons 4,17, les autres aliments (p.ex. viande et produits de viande, poissons, fruits et légumes) 6,25.

La protéine brute comprend en outre des substances azotées comme les aminoacides libres et les peptides (fruits, légumes, poissons) mais aussi urée (champignon).

3.2 Hydrates de carbone utilisables

Pour la plupart des aliments c'est la somme des valeurs pour les mono-, oligo et polysaccharides (glucose, fructose, galactose, saccharose, maltose, lactose, amidon, dextrines), les alcools de sucre (sorbite, xylite) et les acides de fruits (acide citrique, malique, lactique). Si on n'a pas pu trouver de valeurs pour les hydrates de carbone, on si les données etaient incomplètes on a calculé le contenu en hydrates de carbone utilisables de la façon suivante:

hydrates de carbone utilisables = 100 – (eau + protéines + lipides + substances minérales + total en substances de lest).

Si en surplus il manquaient les concentrations pour les substances de lest ce calcul est le suivant: 100 – (eau + protéines + lipides + substances minerales). Le valeurs calculées de cette façon sont marquées en bas de la page.

3.3 Substances de lest

Cette catégorie de substances comprend tout ce qui ne peut être clivé par les enzymes des organes digestifs

3.3 Dietary fibres

This group contains all the higher molecular weight substances which are not split by the enzymes of the human intestine. Basically, the compounds are some polysaccharides (e.g. cellulose, hemicellulose) and some polymeric phenol type of compounds (e.g. lignin). A distinction must be made between water soluble and water insoluble dietary fibre. The total of both fractions, the "total dietary fibre" replaces the foregoing term "crude fibre".

Different analytical methods used for the determination of the dietary fibre content may give different results. For these tables, only the values from enzymatic methods (4–10) have been accepted as remarked in the footnotes.

For some foods, where a quotation of the dietary fibres was not available, (e.g. coffee, cocoa), it seemed to be appropiate to give approximate values, calculated in the following manner:

total dietary fibre = 100 − (water + protein + fat + available carbohydrates + minerals)

These values are also marked with special footnotes.

3.4 Trace elements

The amount of trace elements in a food is strongly dependent on its origin. The values in the tables can only be utilized for orientation.

3.5 Vitamins

Vitamins exist in foods in free and bound forms. The values in the tables are the total value for all forms.

Vitamin A and carotene:

Besides Vitamin A, the total amount of provitamin A carotenoids is expressed as "carotene" (6 µg carotene is 1 µg Vitamin A).

Vitamin E:

listed are:

- the individual tocopherols (α, β, γ and δ), and tocotrienols, and their total amount,
- the total Vitamin E activity (ex-

Alle aufgrund von Differenzrechnungen erhaltenen Werte für verwertbare Kohlenhydrate sind in den jeweiligen Tabellen durch entsprechende Fußnoten gekennzeichnet.

3.3 Ballaststoffe

Die Gruppe umfaßt alle hochmolekularen Stoffe, die von den Enzymen des menschlichen Verdauungsapparates nicht gespalten werden. Es handelt sich dabei in der Regel um bestimmte Polysaccharide (z.B. Cellulose, Hemicellulosen etc.) und um polymere phenolische Verbindungen (Lignin). Man unterscheidet zwischen wasserlöslichen und -unlöslichen Ballaststoffen. Die Summe beider Fraktionen, die Gesamtballaststoffe, ersetzen den früher gebräuchlichen Begriff Rohfaser.

Für die Ballaststoffanalyse existieren verschiedene Methoden, die zu unterschiedlichen Resultaten führen können. In die Tabelle wurden nur die mit enzymatischen Methoden (4–10) erhaltenen Werte aufgenommen, unter Angabe der zur Bestimmung jeweils verwendeten Methode in der Fußnote.

Bei einigen Lebensmitteln mit fehlenden Ballaststoffdaten (z.B. bei Kaffee, Kakao) erschien es angebracht, Orientierungswerte anzugeben. Diese wurden in folgender Weise berechnet:

Gesamtballaststoffe = 100 − (Wasser + Protein + Fett + Asche + verwertbare Kohlenhydrate). Diese berechneten Ballaststoffwerte sind durch entsprechende Fußnoten gekennzeichnet.

3.4 Spurenelemente

Je nach Provenienz des Lebensmittels können große Schwankungen im Spurenelementgehalt auftreten. Die Werte haben deshalb nur orientierenden Charakter.

3.5 Vitamine

Vitamine können in freier und gebundener Form vorkommen. Die Tabellenwerte entsprechen immer der Summe aller möglichen Formen.

humains, par exemple certains polysaccharides comme la cellulose et l'hemicellulose ou les polymères phénoliques comme la lignine.

On distingue les substances de lest solubles et insolubles dans l'eau. La somme de ces deux fractions est dite contenu total en substances de lest ce qui remplace l'expression: fibres naturelles. Il existe différentes méthodes pour l'analyse des substances de lest qui peuvent fournir des résultats divergents. Les tableaux ne contiennent en général que les resultats des analyses enzymatiques (4–10). Les méthodes utilisées sont marquées au bas du tableau.

Pour certains aliments on n'a pas trouvé le données pour les substances de lest (p.ex. café, cacao). Il nous a semblé utile de calculer ces valeurs de la maniere suivante. Total substance de lest = 100 − (eau + protéines + lipides + substances minerales + hydrates de carbone utilisables).

Dans ce cas il se trouve une remarque au bas de la page.

3.4 Substances de trace

Les valeurs peuvent différer fortement d'une region à l'autre et ne peuvent servir que d'orientation.

3.5 Vitamines

On trouve les vitamines sous forme libre et liée. Les tableaux en indiquent la somme.

Vitamine A et carotine:

A coté de la vitamine A on a indiqué sous le nom de «carotine» la somme des provitamines A (6 µg de carotine représentent 1 µg de vitamine A).

Vitamine E:

On donne le contenu en tocophérols différents (α, β, γ, δ) et en tocotriénols (α, β, γ) ainsi que la somme de ces substances et l'activité totale de la vitamine E, (exprimée en mg, resp. µg en α-tocophérol) calculée avec les coefficients de Mc Laughlin et Weihrauch (11).

pressed as mg or µg α-tocopherol) calculated by means of the activity coefficients, according to Mc-Laughlin and Weihrauch (11).

Vitamin B_6:
Quotation of the total of pyridoxine, pyridoxale and pyridoxamine.

Folic acid:
Quotation of the total amount of free and bound folic acid. A value for each form is not possible because of the presence of γ-glutamylhydrolase (conjugase) in many foods and also because of their different manners of production and preparation.

Vitamin C:
Quotation of the total amount of ascorbic acid and dehydroascorbic acid.

Food processing leads to different degrees of vitamin losses. The following table serves as an example of the extent of such losses.

Percentage of vitamin loss due to food processing in the household.
Vitamin loss (%)

	I	II	III	IV	V	VI
Vitamin A	40	–	–	0	–	–
Carotene	30	–	–	–	–	–
Vitamin D	40	–	–	0-20	–	–
Vitamin E	55	–	–	–	–	–
Vitamin K	5	–	–	–	–	–
Vitamin B_1	80	65-70	15-50	10-30	20-40	5-40
Vitamin B_2	75	25-40	10-70	0-10	10-20	5-30
Nicotinamide	70	30-70	30-40	0	10-30	–
Pantothenic acid	50	30-50	30	0	20	–
Vitamin B_6	50	30-60	40	0-50	0-40	–
Biotin	60	–	–	0	–	–
Folic acid	100	–	20-50	0-50	–	–
Vitamin B_{12}	–	–	–	50	–	–
Vitamin C	100	–	10-75	10-70	–	20-50

I: Maximal loss due to cooking (12)
II: Meat cooking (solid matter) (13, 14)
III: Vegetable cooking (solid matter) (13, 14, 16)
IV: Heating of milk (different heating methods) (15)
V: Meat roasting and frying (solid matter) (15)
VI: Vegetable stewing and steaming (solid matter) (13)

3.6 Amino acids

Listed is the total amount of the free and bound amino acids. For fruits and fruit juices the free amino acid content

Vitamin A und Carotin:
Außer Vitamin A ist die Summe aller Provitamin-A-Carotinoide als „Carotin" angegeben (6 µg „Carotin" entsprechen 1 µg Vitamin A).

Vitamin E:
Angegeben sind
- die einzelnen Tocopherole (α, β, γ, δ) und Tocotrienole (α, β, γ), sowie deren Summe.
- die Gesamtaktivität an Vitamin E (ausgedrückt als mg oder µg α-Tocopherol) berechnet mit den Aktivitätskoeffizienten nach Mc Laughlin und Weihrauch (11).

Vitamin B_6:
Angegeben ist die Summe von Pyridoxin, Pyridoxal und Pyridoxamin.

Folsäure:
Angegeben ist die Summe an freier und gebundener Folsäure. Generelle Aussagen über den Anteil der beiden Formen sind wegen des Vorkommens der γ-Glutamyl-Hydrolase (Konjugase) in vielen Lebensmitteln und unterschiedlicher Verarbeitungs- und Zubereitungsverfahren nicht möglich.

Vitamin C:
Angegeben ist die Summe von Ascorbinsäure und Dehydroascorbinsäure.

Bei der Lebensmittelverarbeitung treten unterschiedliche Vitaminverluste auf. Die nachstehende Tabelle orientiert über die Größenordnungen.

3.6 Aminosäuren

Angegeben ist die Summe der freien und gebundenen Aminosäuren. Bei Früchten und Fruchtsäften sind im Anhang in einer Spezialtabelle die freien Aminosäuren gesondert angeführt.

3.7 Einzelne Kohlenhydrate

In diesem Abschnitt sind Monosaccharide, Oligosaccharide, Polysaccharide und Zuckeralkohole zusammengefaßt.

Bei den Polysacchariden werden neben Stärke und Cellulose die Anteile

Vitamin B_6:
On indique la somme pyridoxine + pyridoxal + pyridoxamine.

Acide folique:
Vu la présence de la γ-glutamyl-hydrolase dans beaucoup d'aliments, il n'est pas possible de différencier parmi l'acide folique associé ou libre. Les tableaux en présentent la somme.

Vitamine C:
On a indiqué la somme en acide ascorbique et en acide dehydroascorbique.

La préparation des aliments peut causer des pertes en vitamines. Le tableau suivant tent d'en montrer l'ordre de grandeur.

Perte en vitamines (%) lors de la préparation des aliments

	I	II	III	IV	V	VI
Vitamine A	40	–	–	0	–	–
Carotine	30	–	–	–	–	–
Vitamine D	40	–	–	0-20	–	–
Vitamine E	55	–	–	–	–	–
Vitamine K	5	–	–	–	–	–
Vitamine B_1	80	65-70	15-50	10-30	20-40	5-40
Vitamine B_2	75	25-40	10-70	0-10	10-20	5-30
Nicotinamide	70	30-70	30-40	0	10-30	–
Acid pantoténique	50	30-50	30	0	20	–
Vitamine B_6	50	30-60	40	0-50	0-40	–
Biotine	60	–	–	0	–	–
Acide folique	100	–	20-50	0-50	–	–
Vitamine B_{12}	–	–	–	50	–	–
Vitamine C	100	–	10-75	10-70	–	20-50

I: perte maximale lors de la cuisson (12)
II: viande cuite (matière solide) (13, 14)
III: legumes cuits (matière solide) (13, 14, 16)
IV: lait chauffé selon differentes méthodes (15)
V: viande grillée et rôtie (matière solide) (13)
VI: légumes etuvés (matière solide) (13)

3.6 Aminoacides

On a indiqué la somme en aminoacides libres et liés. Pour les fruits et les jus de fruits on il se trouve un tableau special contenant les aminoacides libres dans l'appendice.

3.7 Hydrates de carbone

On présente le contenu en mono-, oligo- et polysaccharides ainsi qu'en alcools de sucre. Dans la rubrique poly-

is listed in a special table at the appendix.

3.7 Individual carbohydrates

In this group, the amounts of monosaccharides, oligosaccharides, polysaccharides and sugar alcohols are quoted.

Besides starch and cellulose the fractions of pentoses, hexoses and uronic acid in the cellulose free total dietary fibre is given under the terms "pentosan", "hexosan" and "polyuronic acid".

3.8 Phenolic acids

Quotation of the total amount of free and bound compounds.

3.9 Biogenic amines

The data consist of the physiologically active aromatic and heterocyclic monoamines (e.g. tyramine, histamine, dopamine) and of the aliphatic mono- and diamines (e.g. trimethylamine, putrescine, cadaverine).

The amount of these compounds in foods is dependent on different parameters (e.g. maturity, freshness etc.) and therefore, in many cases only the variations are quoted.

3.10 Sterols

Quotation of the total amount of the free and bound compounds.

3.11 Purines

The total amount of free and bound compound is given for each component. The line "total purines" contains the total of all the individual components and is calculated as uric acid.

3.12 Phospholipides

The value listed includes also the lysophospholipides (e.g. phosphatidylcholine and lysophosphatidylcholine). The line "total phospholipids" gives the total of the individual compounds.

Vitaminverluste (in %) bei der haushaltsmäßigen Lebensmittelverarbeitung

	I	II	III	IV	V	VI
Vitamin A	40	–	–	0	–	–
Carotin	30	–	–	–	–	–
Vitamin D	40	–	–	0–20	–	–
Vitamin E	55	–	–	–	–	–
Vitamin K	5	–	–	–	–	–
Vitamin B_1	80	65–70	15–50	10–30	20–40	5–40
Vitamin B_2	75	25–40	10–70	0–10	10–20	5–30
Nicotinamid	70	30–70	30–40	0	10–30	–
Pantothensäure	50	30–50	30	0	20	–
Vitamin B_6	50	30–60	40	0–50	0–40	–
Biotin	60	–	–	0	–	–
Folsäure	100	–	20–50	0–50	–	–
Vitamin B_{12}	–	–	–	50	–	–
Vitamin C	100	–	10–75	10–70	–	20–50

I: Maximaler Kochverlust (12)
II: Garen von Fleisch (Gargut) (13, 14)
III: Garen von Gemüse (Gargut) (13, 14, 16)
IV: Erhitzen von Milch nach verschiedenen Verfahren (15)
V: Grillen und Braten von Fleisch (Gargut) (13)
VI: Dünsten und Dämpfen von Gemüse (Gargut) (13)

der Pentosen, Hexosen und Uronsäuren am cellulosefreien Gesamtballaststoff als „Pentosan", „Hexosan" und „Polyuronic acid" angegeben.

3.8 Phenolische Säuren

Angegeben ist die Summe der freien und gebundenen Verbindungen.

3.9 Biogene Amine

Die Angaben umfassen die physiologisch wirksamen aromatischen und heterocyclischen Monoamine (z.B. Tyramin, Histamin, Dopamin) wie auch die aliphatischen Mono- und Diamine (u.a. Trimethylamin, Putrescin, Cadaverin). Der Gehalt an diesen Verbindungen in Lebensmitteln kann in Abhängigkeit von verschiedenen Parametern (z.B. Reifezustand, Frischezustand) stark schwanken. In vielen Fällen sind deshalb keine Mittelwerte angegeben.

3.10 Sterine

Angegeben ist die Summe der freien und gebundenen Verbindungen.

3.11 Purine

Bei den Einzelverbindungen ist die Summe der freien und gebundenen saccharides on a indiqué a côte de l'amidon et de la cellulose le contenu en pentoses, hexoses et en acides uroniques des substances de lest non cellulosiques sous le nom «Pentosan», «Hexosan» et «Polyuronic acid».

3.8 Acides phénoliques

Somme en substances libres et liées.

3.9 Amines biogènes

On donne les monoamines aromatiques et heterocycliques avec leur activité physiologique (p. ex. tyramine, histamine, dopamine) ainsi que les monoamines et les diamines aliphatiques (p.ex. trimethylamine, putrescine, cadavérine). Le contenu peut varier fortement p.ex. avec le degré de maturité ou de fraîcheur du produit. Dans ce cas on n'a pas pris la moyenne.

3.10 Sterols

Somme des substances libres et liées.

3.11 Purines

On a indiqué pour les différents composés la somme en substances libres et licés. La rubrique «total purines» contient la somme de toutes ces substances exprimée en acide urique.

3.12 Phospholipides

Les valeurs impliquent des lysophosphatides (p. ex. phosphatidylcholine, lysophosphatidylcholine). La somme des substances particulière est indiquée dans la rubrique «total phospholipids».

Literature

(1) A. Merill, B.K. Watt: Energy value of food, Agriculture Handbook No. 74 [1955], Agriculture Research Service, United States Department of Agriculture.
(2) BGBl. 1977 I pp. 2595.
(3) Lebensmittelchem. Gerichtl. Chem. **39** (1985) 59.
(4) T. Schweizer, P. Würsch, J. Sci. Food Agric. **30** (1979) 613.
(5) D.A.T. Southgate: J. Sci. Food Agric. **20** (1969) 331.
(6) D.A.T. Southgate, G.J. Hudson u. H. Englyst: J. Sci. Food Agric. **29** (1978) 979.

References

(1) A. Merill, B. K. Watt: Energy value of food, Agriculture Handbook No. 74 [1955], Agriculture Research Service, United States Department of Agriculture.
(2) BGBl. 1977 I pp. 2595.
(3) Lebensmittelchem. Gerichtl. Chem. **39** (1985) 59.
(4) T. Schweizer, P. Würsch, J. Sci. Food Agric. **30** (1979) 613.
(5) D. A. T. Southgate: J. Sci. Food Agric. **20** (1969) 331.
(6) D. A. T. Southgate, G. J. Hudson u. H. Englyst: J. Sci. Food Agric. **29** (1978) 979.
(7) F. Meuser, P. Suckow, W. Kulikowski, Z. Lebensm. Unters. Forsch. **181** (1985) 101.
(8) H. N. Englyst, S. A. Bingham, S. A. Runswick, E. Collison u. J. H. Cummings, Journal of Human Nutrition and Dietetics **1** (1988) 247.
(9) Amtliche Sammlung von Untersuchungsverfahren nach § 35 LMBG Band I/3 L 00.00-18 (1988); Beuth-Verlag Berlin, Köln.
(10) E. Rabe: Getreide Mehl Brot **41** (1987) 302.
(11) J. Mc Laughlin, J. C. Weihrauch, J. Amer. Diet. Assoc. **75** (1979) 647.
(12) F. Lee, Basic Food Chemistry, pp. 218 AVI Publ. Comp., Westport, Connecticut (1983).
(13) A. Bognar, AID-Verbraucherdienst **28** (1983) 179, 201.
(14) A. A. Paul, D. A. T. Southgate, Mc-Cance u. Widdowson's, Composition of Food, pp. 18–25, Elsevier North-Holland Biomedical Press, Amsterdam, New York, Oxford 1978.
(15) M. Rechcigl, Handbook of Nutritive Value of Processed Food, pp. 232, 387, CRC-Press., Boca-Raton-Florida, 1982.
(16) B. Klein, Y. Kuo, G. Boyd: J. Food Sci. **46** (1981) 640.

Verbindungen angegeben. Die Spalte „Gesamtpurine" enthält die Summe aller Einzelverbindungen als Harnsäure berechnet.

3.12 Phospholipide

Die Werte umfassen auch die Lyso-Verbindungen, also z. B. Phosphatidylcholin und Lysophosphatidylcholin. Die Spalte „Gesamtphospholipid" enthält die Summe der Einzelverbindungen.

Literatur

(1) A. Merill, B. K. Watt: Energy value of food, Agriculture Handbook No. 74 [1955], Agriculture Research Service, United States Department of Agriculture.
(2) BGBl. 1977 I S. 2595.
(3) Lebensmittelchem. Gerichtl. Chem. **39** (1985) 59.
(4) T. Schweizer, P. Würsch, J. Sci. Food Agric. **30** (1979) 613.
(5) D. A. T. Southgate: J. Sci. Food Agric. **20** (1969) 331.
(6) D. A. T. Southgate, G. J. Hudson u. H. Englyst: J. Sci. Food Agric. **29** (1978) 979.
(7) F. Meuser, P. Suckow, W. Kulikowski, Z. Lebensm. Unters. Forsch. **181** (1985) 101.
(8) H. N. Englyst, S. A. Bingham, S. A. Runswick, E. Collison u. J. H. Cummings, Journal of Human Nutrition and Dietetics **1** (1988) 247.
(9) Amtliche Sammlung von Untersuchungsverfahren nach § 35 LMBG Band I/3 L 00.00-18 (1988); Beuth-Verlag Berlin, Köln.
(10) E. Rabe: Getreide Mehl Brot **41** (1987) 302.
(11) J. Mc Laughlin, J. C. Weihrauch, J. Amer. Diet. Assoc. **75** (1979) 647.
(12) F. Lee, Basic Food Chemistry, S. 218 AVI Publ. Comp., Westport, Connecticut (1983).
(13) A. Bognar, AID-Verbraucherdienst **28** (1983) 179, 201.
(14) A. A. Paul, D. A. T. Southgate, Mc-Cance u. Widdowson's, Composition of Food, S. 18–25, Elsevier North-Holland Biomedical Press, Amsterdam, New York, Oxford 1978.
(15) M. Rechcigl, Handbook of Nutritive Value of Processed Food, S. 232, 387, CRC-Press., Boca-Raton-Florida, 1982.
(16) B. Klein, Y. Kuo, G. Boyd: J. Food Sci. **46** (1981) 640.

(7) F. Meuser, P. Suckow, W. Kulikowski, Z. Lebensm. Unters. Forsch. **181** (1985) 101.
(8) H. N. Englyst, S. A. Bingham, S. A. Runswick, E. Collison u. J. H. Cummings, Journal of Human Nutrition and Dietetics **1** (1988) 247.
(9) Amtliche Sammlung von Untersuchungsverfahren nach § 35 LMBG Band I/3 L 00.00-18 (1988); Beuth-Verlag Berlin, Köln.
(10) E. Rabe: Getreide Mehl Brot **41** (1987) 302.
(11) J. Mc Laughlin, J. C. Weihrauch, J. Amer. Diet. Assoc. **75** (1979) 647.
(12) F. Lee, Basic Food Chemistry, pp. 218 AVI Publ. Comp., Westport, Connecticut (1983).
(13) A. Bognar, AID-Verbraucherdienst **28** (1983) 179, 201.
(14) A. A. Paul, D. A. T. Southgate, Mc-Cance u. Widdowson's, Composition of Food, pp. 18–25, Elsevier North-Holland Biomedical Press, Amsterdam, New York, Oxford 1978.
(15) M. Rechcigl, Handbook of Nutritive Value of Processed Food, pp. 232, 387, CRC-Press., Boca-Raton-Florida, 1982.
(16) B. Klein, Y. Kuo, G. Boyd: J. Food Sci. **46** (1981) 640.

Glossary of the food constituents / Glossar der Lebensmittelinhaltsstoffe / Glossair des composants des aliments

Glossary of the food constituents	Glossar der Lebensmittelinhaltsstoffe	Glossair des composants des aliments
Acetic acid	Essigsäure	Acide acétique
Adenine	Adenin	Adénine
Agmatine	Agmatin	Agmatine
Alanine	Alanin	Alanine
Albumen	Albumin	Albumines
Allylisocyanat	Allylisocyanat	Isocyanate d'allyle
Alpha-Ketoglutaric acid	Alpha-Ketoglutarsäure	Acide alpha-ketoglutarique
Alpha-Tocopherol	Alpha-Tocopherol	Alpha tocophérol
Alpha-Tocotrienol	Alpha-Tocotrienol	Alpha tocotriénol
Aluminium	Aluminium	Aluminium
Amino acids	Aminosäuren	Acides aminés
Arabinose	Arabinose	Arabinose
Arachidic acid	Arachinsäure	Acide arachidique
Arachidonic acid	Arachidonsäure	Acide arachidonique
Arginine	Arginin	Arginine
Asparagine	Asparagin	Asparagine
Aspartic acid	Asparaginsäure	Acide aspartique
Available carbohydrates	Verwertbare Kohlenhydrate	Hydrates de carbone utilisables
Avenasterol-5	Avenasterin-5	Avenastérol-5
Avenasterol-7	Avenasterin-7	Avenastérol-7
Behenic acid	Behensäure	Acide béhénique
Benzoic acid	Benzoesäure	Acide benzoique
Beta-Phenylethylamine	Beta-Phenylethylamin	Beta-phenylethylamine
Beta-Sitosterol	Beta-Sitosterin	Beta-Sitostérine
Beta-Tocopherol	Beta-Tocopherol	Beta-tocophérol
Beta-Tocotrienol	Beta-Tocotrienol	Beta-tocotriénol
Biotin	Biotin	Biotine
Boron	Bor	Bore
Brassicastereol	Brassicasterin	Brassicastérol
Bromine	Brom	Brome
Butyleneglycol	Butylenglykol	Glycol de Butyléne
Butyric acid	Buttersäure	Acide butyrique
$C_{14:1}$ Fatty acid	$C_{14:1}$ Fettsäure	$C_{14:1}$ acide gras
$C_{15:0}$ Fatty acid	$C_{15:0}$ Fettsäure	$C_{15:0}$ acide gras
$C_{22:1}$ Fatty acid	$C_{22:1}$ Fettsäure	$C_{22:1}$ acide gras
$C_{18:4}$ $(n-3)$ Fatty acid	$C_{18:4}$ $(n-3)$ Fettsäure	$C_{18:4}$ $(n-3)$ acide gras
$C_{20:5}$ $(n-3)$ Fatty acid	$C_{20:5}$ $(n-3)$ Fettsäure	$C_{20:5}$ $(n-3)$ acide gras
$C_{22:5}$ $(n-3)$ Fatty acid	$C_{22:5}$ $(n-3)$ Fettsäure	$C_{22:5}$ $(n-3)$ acide gras
$C_{22:6}$ $(n-3)$ Fatty acid	$C_{22:6}$ $(n-3)$ Fettsäure	$C_{22:6}$ $(n-3)$ acide gras
Cadaverine	Cadaverin	Cadavérine

Caffeic acid	Kaffeesäure	Acide caféique
Calcium	Calcium	Calcium
Campesterol	Campesterin	Campestérol
Capric acid	Caprinsäure	Acide caprique
Caproic acid	Capronsäure	Acide caproique
Caprylic acid	Caprylsäure	Acide caprylique
Carotene	Carotin	Carotine
Carotenoids total	Gesamtcarotinoide	Carotinoides, contenu total
Casein	Casein	Caséine
Cellulose	Cellulose	Cellulose
Chloride	Chlorid	Chlorure
Chlorogenic acid	Chlorogensäure	Acide chlorogénique
Cholesterol	Cholesterin	Cholestérol
Choline	Cholin	Choline
Chromium	Chrom	Chrome
Citric acid	Zitronensäure	Acide citrique
Cobalt	Kobalt	Cobalt
Coffeine	Coffein	Caféine
Connective tissue protein	Bindegewebseiweiß	Protéines du tissu conjonctif
Copper	Kupfer	Cuivre
Creatinine	Kreatinin	Créatinine
Cysteamine	Cysteamin	Cysteamine
Cystine	Cystin	Cystine
Delta-Tocopherol	Delta-Tocopherol	Delta-tocophérol
Desmosterol	Desmosterin	Desmostérol
Dextrin	Dextrin	Dextrine
Dietary Fibre, water soluble	Ballaststoffe, wasserlöslich	Substances de lest soluble dans l'eau
Dietary Fibre, water insoluble	Ballaststoffe, wasserunlöslich	Substances de lest insoluble dans l'eau
Dopamine	Dopamin	Dopamine
Eicosenic acid	Eicosensäure	Acide eicosénique
Erucic acid	Erucasäure	Acide erucique
Ethanol	Ethanol	Ethanol
Ethanolamine	Ethanolamin	Ethanolamine
Ethylamine	Ethylamin	Ethylamine
Extract	Extrakt	Extrait
Extract, sugar free	Extrakt, zuckerfrei	Extrait, sans sucre
Extract, water soluble	Extrakt, wasserlöslich	Extrait, soluble dans l'eau
Fat	Fett	Matiéres grasses
Fatty acids total	Gesamtfettsäuren	Acides gras, contenu total
Ferulic acid	Ferulasäure	Acide férulique
Feruloylputrescine	Feruloylputrescin	Feruloylputrescine
Fluoride	Fluorid	Fluorure
Folic acid	Folsäure	Acide folique
Formic acid	Ameisensäure	Acide formique
Fructose	Fructose	Fructose
Fumaric acid	Fumarsäure	Acide fumarique
Galactitol	Galactit	Galactite
Galactose	Galactose	Galactose
Galacturonic acid	Galacturonsäure	Acide galacturonique
Gallic acid	Gallussäure	Acide gallique

Gamma-Aminobutyric acid	Gamma-Aminobuttersäure	Acide-gamma-aminobutyrique
Gamma-Tocopherol	Gamma-Tocopherol	Gamma-tocophérol
Gamma-Tocotrienol	Gamma-Tocotrienol	Gamma-tocotriénol
Globulin	Globulin	Globulines
Glucodifructose	Glucodifructose	Glucodifructose
Gluconic acid	Gluconsäure	Acide gluconique
Glucose	Glucose	Glucose
Glutamic acid	Glutaminsäure	Acide glutamique
Glutamine	Glutamin	Glutamine
Glutaric acid	Glutarsäure	Acide glutarique
Glyceric acid	Glycerinsäure	Acide glycérique
Glycerol	Glycerin	Glycérine
Glycine	Glycin	Glycine
Glycogen	Glycogen	Glycogène
Glycolic acid	Glykolsäure	Acide glycolique
Guanine	Guanin	Guanine
Hemicellulose	Hemicellulose	Hémicellulose
Hexosan	Hexosan	Hexosane
Hippuric acid	Hippursäure	Acide hippurique
Histamine	Histamin	Histamine
Histidine	Histidin	Histidine
Hydroxyproline	Hydroxyprolin	Hydroxyproline
Hypoxanthine	Hypoxanthin	Hypoxanthine
Inositol	Inosit	Inosite
Invert sugar	Invertzucker	Sucre inverti
Iron	Eisen	Fer
Isocitric acid	Isozitronensäure	Acide isocitrique
Isoleucine	Isoleucin	Isoleucine
Iodide	Jodid	Iodure
Lactic acid	Milchsäure	Acide lactique
Lactose	Lactose	Lactose
Lauric acid	Laurinsäure	Acide laurique
Leucine	Leucin	Leucine
Levulinic acid	Lävulinsäure	Acide lévulique
Lignin	Lignin	Lignine
Lignoceric acid	Lignocerinsäure	Acide lignocérique
Linoleic acid	Linolsäure	Acide linolique
Linolenic acid	Linolensäure	Acide linolénique
Lysine	Lysin	Lysine
Magnesium	Magnesium	Magnésium
Malic acid	Apfelsäure	Acide malique
Malonic acid	Malonsäure	Acide malonique
Maltose	Maltose	Maltose
Manganese	Mangan	Manganése
Mannitol	Mannit	Mannite
Margaric acid	Margarinsäure	Acide margarique
Methanol	Methanol	Méthanol
Methionine	Methionin	Méthionine
Methylamine	Methylamin	Méthylamine
Methylenecholesterol-24	Methylencholesterin-24	Méthylénecholestérol-24

Milk protein	Milcheiweiß	Protéines du lait
Minerals	Mineralstoffe	Substance minerales
Molybdenum	Molybdän	Molybdène
Myoinositol	Myoinosit	Myoinosite
Myristic acid	Myristinsäure	Acide myristique
Nickel	Nickel	Nickel
Nicotinamide	Nicotinamid	Nicotinamide
Nitrate	Nitrat	Nitrate
Nitrite	Nitrit	Nitrite
Nitrogen compounds	Stickstoff-Verbindungen	Composés azotés
Nonreducing sugar	Nichtreduzierender Zucker	Sucre non réducteur
Nonvolatile acid	Nichtflüchtige Säure	Acide non volatile
Noradrenaline	Noradrenalin	Noradrénaline
Oleic acid	Ölsäure	Acide oléique
Original wort	Stammwürze	Moût de bière
Oxalic acid, soluble	Oxalsäure, löslich	Acide oxalique, soluble
Oxalic acid, total	Oxalsäure, gesamt	Acide oxalique, contenu total
Oxaloacetic acid	Oxalessigsäure	Acide oxalacétique
Palmitic acid	Palmitinsäure	Acide palmitique
Palmitoleic acid	Palmitoleinsäure	Acide palmitoléique
Pantothenic acid	Pantothensäure	Acide pantothenique
Para-Coumaric acid	Para-Cumarsäure	Acide para-cumarique
Para-Hydroxybenzoic acid	Para-Hydroxybenzoesäure	Acide para-hydroxybenzoique
Parasorbic acid	Parasorbinsäure	Acide parasorbique
Pectin	Pectin	Pectine
Pentosan	Pentosan	Pentosane
Phenylalanine	Phenylalanin	Phénylalanine
Phosphate	Phosphat	Phosphate
Phosphatidylcholine	Phosphatidylcholin	Phosphatidylcholine
Phosphatidylethanolamine	Phosphatidylethanolamin	Phosphatidyléthanolamine
Phosphatidylglycerol	Phosphatidylglycerin	Phosphatidylglycérine
Phosphatidylinositol	Phosphatidylinosit	Phosphatidylinosite
Phosphatidylserine	Phosphatidylserin	Phosphatidylserine
Phospholipides total	Phospholipide, gesamt	Phospholipides contenu total
Phosphorus	Phosphor	Phosphore
Phytic acid	Phytinsäure	Acide phytique
Polyuronic acid	Polyuronsäure	Acide polyuronique
Potassium	Kalium	Potassium
Proline	Prolin	Proline
Propionic acid	Propionsäure	Acide propionique
Protein	Eiweiß	Protéine
Protocatechuic acid	Protocatechusäure	Acide protocatéchique
Purin-(base)-nitrogen	Purin-(basen)-stickstoff	Azote des (bases) purines
Purines total	Gesamtpurine	Purines, contenu total
Putrescine	Putrescin	Putrescine
Pyrrolidonecarbonic acid	Pyrrolidoncarbonsäure	Acide carboxylique de pyrrolidone
Pyruvic acid	Brenztraubensäure	Acide pyroracémique
Quinic acid	Chinasäure	Acide quinique
Raffinose	Raffinose	Raffinose
Reducing sugar	Reduzierender Zucker	Sucre réducteur

Saturated fatty acids	Gesättigte Fettsäuren	Acides gras saturés
Selenium	Selen	Sélénium
Serine	Serin	Sérine
Serotonine	Serotonin	Sérotonine
Shikimic acid	Shikimisäure	Acide shikimique
Silicon	Silicium	Silice
Sinapic acid	Sinapinsäure	Acide sinapique
Sodium	Natrium	Sodium
Sorbitol	Sorbit	Sorbite
Spermidine	Spermidin	Spermidine
Spermine	Spermin	Spermine
Sphingomyelin	Sphingomyelin	Sphingomyelin
Stachyose	Stachyose	Stachyose
Starch	Stärke	Amidon
Stearic acid	Stearinsäure	Acide stéarique
Sterols, free	Sterine, frei	Stérols non liés
Sterols, total	Sterine, gesamt	Stérols contenu total
Stigmasterol	Stigmasterin	Stigmastérol
Succinic acid	Bernsteinsäure	Acide succinique
Sucrose	Saccharose	Saccharose
Sulphate	Sulfat	Sulfate
Synephrine	Synephrin	Synephrine
Tannic acid	Gerbstoff	Acide tannique
Tartaric acid	Weinsäure	Acide tartrique
Taurine	Taurin	Taurine
Theobromine	Theobromin	Théobromine
Threonine	Threonin	Thréonine
Tin	Zinn	Etain
Total acid	Gesamtsäure	Acide contenu total
Total Dietary Fibre	Gesamt-Ballaststoffe	Substances de lest contenu total
Total sterols	Gesamtsterine	Contenu total en stérols
Total sugar content	Gesamtzucker	Contenu total en sucre
Trimethylamine	Trimethylamin	Triméthylamine
Trehalose	Trehalose	Trehalose
Tryptamine	Tryptamin	Tryptamine
Tryptophan	Tryptophan	Tryptophane
Tyramine	Tyramin	Tyramine
Tyrosine	Tyrosin	Tyrosine
Unsaturated fatty acids	Ungesättigte Fettsäuren	Acides gras insaturés
Uric acid	Harnsäure	Acide urique
Valine	Valin	Valine
Vanadium	Vanadium	Vanadium
Vanillic acid	Vanillinsäure	Acide vanillique
Verbascose	Verbascose	Verbascose
Vitamin A	Vitamin A	Vitamine A
Vitamin B_1	Vitamin B_1	Vitamine B_1
Vitamin B_{12}	Vitamin B_{12}	Vitamine B_{12}
Vitamin B_2	Vitamin B_2	Vitamine B_2
Vitamin B_6	Vitamin B_6	Vitamine B_6
Vitamin C	Vitamin C	Vitamine C

Vitamin D	Vitamin D	Vitamine D
Vitamin E Activity	Vitamin E Aktivität	Vitamine E activité
Vitamin K	Vitamin K	Vitamine K
Volatile acid	Flüchtige Säure	Acide volatil
Water	Wasser	Eau
Whey protein	Molkenprotein	Protéines de petit lait
Xanthine	Xanthin	Xanthine
Xylitol	Xylit	Xylite
Xylose	Xylose	Xylose
Zinc	Zink	Zinc

Glossary of the table	Glossar der Tabelle	Glossair des tableaux
Energy value (average) per 100 g edible portion	Brennwert (Mittel) pro 100 g eßbare Ware	Valeur calorifique (moyenne) dans 100 g de matière comestible
Amount of digestible constituents per 100 g	Aus 100 g eßbarer Ware resorbiert	Partie digérable dans 100 g de matière comestible
Energy value (average) of the digestible fraction per 100 g edible portion	Brennwert (Mittel) des resorbierten Anteils pro 100 g eßbarer Ware	Valeur calorifique (moyenne) de la partie digérable de 100 g de matière comestible
Protein	Eiweiß	protéines
Fat	Fett	matière grasse
available Carbohydrates	verwertbare Kohlenhydrate	hydrates de carbone utilisables
waste	Abfall	déchets
percentage	Prozent	pourcentage
average:	Mittel:	moyenne
minimum:	Minimum:	minimum:
maximum:	Maximum:	maximum:
constituents	Bestandteile	constituants
variation	Schwankungsbreite	variation
Abbrevations: DIM: dimension (g, mg, µg) AV: average value/100 g edible portion AVR: average value in edible portion/100 g raw product (as purchased) NUTR. DENS.: Nutrient density = amount of constituent (g, mg, µg)/total energy* of digestible nutrients (in Megajoule) MOLPERC. Molpercent	Abkürzungen: DIM: Dimension (g, mg, µg) AV: Mittelwert/100 g eßbaren Anteil AVR: Mittelwert im eßbaren Anteil/100 g käufliche Rohware NUTR. DENS.: Nährstoffdichte = Inhaltsstoffmenge (g, mg, µg)/Gesamtenergie* der resorbierten Nährstoffe (in Megajoule) MOLPERC. Molprozent	Abréviations: DIM: Dimension (g, mg, µg) AV: Valeur moyenne/100 g de matière comestible AVR: Valeur moyenne de matière comestible/100 g de produit brut NUTR. DENS.: Densité nutritive = contenu du constituant (g, mg, µg)/énergie totale* de la matière digérable (en mégajoule) MOLPERC. Mole pourcent
* before rounding up or down	* Vor Auf- bzw. Abrundung	* non arrondie

Acknowledgement

The following institutions listed below have provided us with analytical data for the present tables.

Danksagung

Von den nachfolgend genannten Institutionen wurden uns Analysenergebnisse für die vorliegenden Tabellen zur Verfügung gestellt.

Remerciements

Les instituts mentionnés ci-aprés ont eu l'amabilité de mettre à notre disposition leurs résultats d'analyses.

Alfa-Institut für hauswirtschaftliche Produkt- und Verfahrensentwicklung GmbH, Eltville a. Rhein

Bahlsens Keksfabrik KG, Hannover

Biologische Bundesanstalt, Berlin und Bernkastel

Bundesanstalt für Fettforschung, Münster

Bundesanstalt für Fleischforschung, Kulmbach

Bundesanstalt für Milchforschung, Kiel

Bundesanstalt für alpenländische Milchwirtschaft, Rotholz/Tirol, Österreich

Bundesforschungsanstalt für Ernährung, Karlsruhe

Bundesforschungsanstalt für Fischerei, Hamburg

Bundesforschungsanstalt für Getreide- und Kartoffelverarbeitung, Detmold

Bundesforschungsanstalt für Pflanzenzüchtung, Ahrensburg

Bundesforschungsanstalt für Rebenzüchtung, Geilweilerhof, Siebeldingen/Pfalz

Centrale Marketinggesellschaft der deutschen Agrarwirtschaft mbH, Bonn Bad Godesberg

Chemisches Landesuntersuchungsamt, Stuttgart

Chemisches Untersuchungsamt, Oberhausen

Department of Food Chemistry and Technology, Universität Helsinki, Finnland

Deutsche Gesellschaft für Ernährung e. V., Frankfurt a. M.

Eidgenössische Forschungsanstalt für Milchwirtschaft, Liebefeld/Bern, Schweiz

Entwicklungsbüro für Getreideverarbeitung, Düsseldorf

Ferrero – offene Handelsgesellschaft mbH, Frankfurt a. M.

Forschungsinstitut für Ernährungswissenschaft, Wien, Österreich

Institut für Agrikulturchemie, Universität Göttingen

Institut für Ernährungswissenschaft, Universität Gießen

Institut für Humanernährung und Lebensmittelkunde, Universität Kiel

Institut für Konserventechnologie, Braunschweig

Institut für Lebensmittelchemie, Universität Hamburg

Institut für Lebensmittelchemie, Universität Münster

Institut für Lebensmitteltechnologie, Getreidetechnologie, Technische Universität Berlin

Institut für Nutzpflanzenforschung, Technische Universität Berlin

Institut für physikalische Chemie, Kernforschungsanlage Jülich

Landesuntersuchungsamt für das Gesundheitswesen Nordbayern, Würzburg

Landesuntersuchungsamt für das Gesundheitswesen Südbayern, München

Lebensmittelchemisches Institut des Bundesverbandes der Deutschen Süßwarenindustrie e. V., Köln

Lehrstuhl für experimentelle Zahnheilkunde, Universität Würzburg

MRC – Dunn Clinical Nutrition Centre, Cambridge, Großbritannien

Maizena GmbH, Heilbronn

Margarineinstitut für gesunde Ernährung, Hamburg

Milchwirtschaftliche Untersuchungs- und Versuchsanstalt, Kempten

National Marine Fisheries Service, Charleston, USA

Nestlé Deutschland AG, Frankfurt a. M.

Pfanni-Werk Otto Eckart KG, München

Staatl. Veterinäruntersuchungsamt, Frankfurt a. M.

Staatl. Veterinäruntersuchungsamt, Gießen

Zentrale Erfassungs- und Bewertungsstelle für Umweltchemikalien (ZEBS), Berlin

Milch

Milk
Lait

Milch

MILK

LAIT

BÜFFELMILCH BUFFALO MILK LAIT DE BUFFLONNE

		PROTEIN	FAT	CARBOHYDRATES	TOTAL
ENERGY VALUE (AVERAGE) PER 100 G EDIBLE PORTION	KJOULE	74	308	85	467
	(KCAL)	18	74	20	112
AMOUNT OF DIGESTIBLE CONSTITUENTS PER 100 G	GRAM	3.88	7.57	5.05	
ENERGY VALUE (AVERAGE) OF THE DIGESTIBLE FRACTION PER 100 G EDIBLE PORTION	KJOULE	72	293	85	449
	(KCAL)	17	70	20	107

WASTE PERCENTAGE AVERAGE 0.00

CONSTITUENTS	DIM	AV	VARIATION		AVR	NUTR. DENS.		MOLPERC.
WATER	GRAM	81.10	75.30	83.60	81.10	GRAM/MJ	180.56	
PROTEIN	GRAM	4.01	3.78	4.32	4.01	GRAM/MJ	8.93	
FAT	GRAM	7.97	6.37	13.70	7.97	GRAM/MJ	17.74	
AVAILABLE CARBOHYDR.	GRAM	5.05	-	-	5.05	GRAM/MJ	11.24	
MINERALS	GRAM	0.74	0.70	0.83	0.74	GRAM/MJ	1.65	
SODIUM	MILLI	40.00	-	-	40.00	MILLI/MJ	89.06	
POTASSIUM	MILLI	100.00	-	-	100.00	MILLI/MJ	222.64	
CALCIUM	MILLI	195.00	190.00	200.00	195.00	MILLI/MJ	434.15	
MANGANESE	MICRO	5.00	2.00	8.00	5.00	MICRO/MJ	11.13	
IRON	MILLI	0.15	0.06	0.19	0.15	MILLI/MJ	0.33	
COPPER	MICRO	25.00	15.00	34.00	25.00	MICRO/MJ	55.66	
ZINC	MILLI	0.60	0.28	0.96	0.60	MILLI/MJ	1.34	
PHOSPHORUS	MILLI	130.00	-	-	130.00	MILLI/MJ	289.43	
CHLORIDE	MILLI	62.00	61.00	62.00	62.00	MILLI/MJ	138.04	
VITAMIN A	MICRO	64.00	-	-	64.00	MICRO/MJ	142.49	
VITAMIN B1	MICRO	50.00	-	-	50.00	MICRO/MJ	111.32	
VITAMIN B2	MILLI	0.10	-	-	0.10	MILLI/MJ	0.22	
NICOTINAMIDE	MICRO	80.00	-	-	80.00	MICRO/MJ	178.11	
PANTOTHENIC ACID	MILLI	0.37	-	-	0.37	MILLI/MJ	0.82	
VITAMIN B6	MICRO	25.00	-	-	25.00	MICRO/MJ	55.66	
BIOTIN	MICRO	11.00	-	-	11.00	MICRO/MJ	24.49	
VITAMIN B12	MICRO	0.30	-	-	0.30	MICRO/MJ	0.67	
VITAMIN C	MILLI	2.50	-	-	2.50	MILLI/MJ	5.57	
ALANINE	GRAM	0.13	-	-	0.13	GRAM/MJ	0.29	
ARGININE	GRAM	0.11	-	-	0.11	GRAM/MJ	0.24	

CONSTITUENTS	DIM	AV	VARIATION			AVR	NUTR. DENS.		MOLPERC.
ASPARTIC ACID	GRAM	0.31	-	-	-	0.31	GRAM/MJ	0.69	
CYSTINE	MILLI	48.00	-	-	-	48.00	MILLI/MJ	106.87	
GLUTAMIC ACID	GRAM	0.48	-	-	-	0.48	GRAM/MJ	1.07	
HISTIDINE	MILLI	78.00	-	-	-	78.00	MILLI/MJ	173.66	
ISOLEUCINE	GRAM	0.20	-	-	-	0.20	GRAM/MJ	0.45	
LEUCINE	GRAM	0.37	-	-	-	0.37	GRAM/MJ	0.82	
METHIONINE	MILLI	97.00	-	-	-	97.00	MILLI/MJ	215.96	
PHENYLALANINE	MILLI	162.00	-	-	-	162.00	MILLI/MJ	360.68	
SERINE	GRAM	0.23	-	-	-	0.23	GRAM/MJ	0.51	
THREONINE	GRAM	0.18	-	-	-	0.18	GRAM/MJ	0.40	
TRYPTOPHAN	MILLI	53.00	-	-	-	53.00	MILLI/MJ	118.00	
TYROSINE	MILLI	183.00	-	-	-	183.00	MILLI/MJ	407.43	
VALINE	GRAM	0.22	-	-	-	0.22	GRAM/MJ	0.49	
CITRIC ACID	MILLI	150.00	130.00	-	180.00	150.00	MILLI/MJ	333.96	
LACTOSE	GRAM	4.90	-	-	-	4.90	GRAM/MJ	10.91	
BUTYRIC ACID	GRAM	0.32	0.18	-	0.47	0.32	GRAM/MJ	0.71	
CAPROIC ACID	GRAM	0.17	0.08	-	0.22	0.17	GRAM/MJ	0.38	
CAPRYLIC ACID	MILLI	80.00	40.00	-	130.00	80.00	MILLI/MJ	178.11	
CAPRIC ACID	GRAM	0.16	0.08	-	0.25	0.16	GRAM/MJ	0.36	
LAURIC ACID	GRAM	0.19	0.11	-	0.28	0.19	GRAM/MJ	0.42	
MYRISTIC ACID	GRAM	0.81	0.58	-	1.05	0.81	GRAM/MJ	1.80	
C15 : 0 FATTY ACID	GRAM	0.10	0.06	-	0.17	0.10	GRAM/MJ	0.22	
PALMITIC ACID	GRAM	2.31	1.90	-	3.10	2.31	GRAM/MJ	5.14	
C17 : 0 FATTY ACID	MILLI	40.00	30.00	-	110.00	40.00	MILLI/MJ	89.06	
STEARIC ACID	GRAM	0.79	0.79	-	1.30	0.79	GRAM/MJ	1.76	
ARACHIDIC ACID	MILLI	20.00	10.00	-	50.00	20.00	MILLI/MJ	44.53	
PALMITOLEIC ACID	GRAM	0.16	0.11	-	0.27	0.16	GRAM/MJ	0.36	
OLEIC ACID	GRAM	1.81	1.04	-	2.70	1.81	GRAM/MJ	4.03	
C14 : 1 FATTY ACID	MILLI	50.00	20.00	-	90.00	50.00	MILLI/MJ	111.32	
LINOLEIC ACID	MILLI	80.00	50.00	-	170.00	80.00	MILLI/MJ	178.11	
LINOLENIC ACID	MILLI	90.00	20.00	-	110.00	90.00	MILLI/MJ	200.38	
TOTAL PHOSPHOLIPIDS	MILLI	34.00	-	-	-	34.00	MILLI/MJ	75.70	
PHOSPHATIDYLCHOLINE	MILLI	9.00	-	-	-	9.00	MILLI/MJ	20.04	
PHOSPHATIDYLETHANOLAMINE	MILLI	9.00	-	-	-	9.00	MILLI/MJ	20.04	
PHOSPHATIDYLSERINE	MILLI	1.00	-	-	-	1.00	MILLI/MJ	2.23	
PHOSPHATIDYLINOSITOL	MILLI	1.00	-	-	-	1.00	MILLI/MJ	2.23	
SPHINGOMYELIN	MILLI	12.00	-	-	-	12.00	MILLI/MJ	26.72	
CASEIN	GRAM	3.50	-	-	-	3.50	GRAM/MJ	7.79	

ESELSMILCH — DONKEY MILK (ASSES MILK) — LAIT D'ÂNE

		PROTEIN	FAT	CARBOHYDRATES	TOTAL
ENERGY VALUE (AVERAGE) PER 100 G EDIBLE PORTION	KJOULE	37	39	102	178
	(KCAL)	8.8	9.4	24	43
AMOUNT OF DIGESTIBLE CONSTITUENTS PER 100 G	GRAM	1.94	0.95	6.10	
ENERGY VALUE (AVERAGE) OF THE DIGESTIBLE FRACTION PER 100 G EDIBLE PORTION	KJOULE	36	37	102	175
	(KCAL)	8.5	8.9	24	42

WASTE PERCENTAGE AVERAGE 0.00

CONSTITUENTS	DIM	AV	VARIATION	AVR	NUTR. DENS.		MOLPERC.
WATER	GRAM	91.00	90.10 - 92.20	91.00	GRAM/MJ	519.59	
PROTEIN	GRAM	2.00	1.90 - 2.10	2.00	GRAM/MJ	11.42	
FAT	GRAM	1.01	0.54 - 1.50	1.01	GRAM/MJ	5.77	
AVAILABLE CARBOHYDR.	GRAM	6.10	- - -	6.10	GRAM/MJ	34.83	
MINERALS	GRAM	0.47	- - -	0.47	GRAM/MJ	2.68	
CALCIUM	MILLI	110.00	80.00 - 140.00	110.00	MILLI/MJ	628.07	
IRON	MICRO	10.00	- - -	10.00	MICRO/MJ	57.10	
PHOSPHORUS	MILLI	61.00	- - -	61.00	MILLI/MJ	348.29	
VITAMIN B1	MICRO	41.00	21.00 - 60.00	41.00	MICRO/MJ	234.10	
VITAMIN B2	MICRO	64.00	30.00 - 97.00	64.00	MICRO/MJ	365.42	
NICOTINAMIDE	MICRO	74.00	57.00 - 90.00	74.00	MICRO/MJ	422.52	
VITAMIN B12	MICRO	0.11	- - -	0.11	MICRO/MJ	0.63	
VITAMIN C	MILLI	2.00	- - -	2.00	MILLI/MJ	11.42	
LACTOSE	GRAM	6.10	- - -	6.10	GRAM/MJ	34.83	
ALBUMEN	GRAM	0.60	- - -	0.60	GRAM/MJ	3.43	
CASEIN	GRAM	0.70	- - -	0.70	GRAM/MJ	4.00	

FRAUENMILCH (MUTTERMILCH) 1	HUMAN MILK (MOTHER'S MILK) 1	LAIT DE FEMME (LAIT MATERNEL) 1

		PROTEIN	FAT	CARBOHYDRATES	TOTAL
ENERGY VALUE (AVERAGE) PER 100 G EDIBLE PORTION	KJOULE	21	156	119	295
	(KCAL)	5.0	37	28	71
AMOUNT OF DIGESTIBLE CONSTITUENTS PER 100 G	GRAM	1.09	3.82	7.09	
ENERGY VALUE (AVERAGE) OF THE DIGESTIBLE FRACTION PER 100 G EDIBLE PORTION	KJOULE	20	148	119	287
	(KCAL)	4.8	35	28	69

WASTE PERCENTAGE AVERAGE 0.00

CONSTITUENTS	DIM	AV	VARIATION			AVR	NUTR.	DENS.	MOLPERC.
WATER	GRAM	87.50	85.20	-	89.70	87.50	GRAM/MJ	304.87	
PROTEIN	GRAM	1.13 2	1.03	-	1.43	1.13	GRAM/MJ	3.94	
FAT	GRAM	4.03	3.50	-	4.62	4.03	GRAM/MJ	14.04	
AVAILABLE CARBOHYDR.	GRAM	7.09	-	-	-	7.09	GRAM/MJ	24.70	
MINERALS	GRAM	0.21	0.19	-	0.23	0.21	GRAM/MJ	0.73	
SODIUM	MILLI	16.00	14.00	-	19.00	16.00	MILLI/MJ	55.75	
POTASSIUM	MILLI	53.00	50.00	-	64.00	53.00	MILLI/MJ	184.66	
MAGNESIUM	MILLI	3.80	2.90	-	5.00	3.80	MILLI/MJ	13.24	
CALCIUM	MILLI	31.00	25.00	-	41.00	31.00	MILLI/MJ	108.01	
MANGANESE	MICRO	1.40	-	-	-	1.40	MICRO/MJ	4.88	
IRON	MICRO	29.00	26.00	-	50.00	29.00	MICRO/MJ	101.04	
COBALT	MICRO	1.40	0.63	-	2.70	1.40	MICRO/MJ	4.88	
COPPER	MICRO	35.00	24.00	-	52.00	35.00	MICRO/MJ	121.95	
ZINC	MILLI	0.22	0.12	-	0.39	0.22	MILLI/MJ	0.77	
NICKEL	MICRO	1.00	0.40	-	1.00	1.00	MICRO/MJ	3.48	
CHROMIUM	MICRO	67.00	53.00	-	80.00	67.00	MICRO/MJ	233.44	
MOLYBDENUM	MICRO	1.00	-	-	-	1.00	MICRO/MJ	3.48	
VANADIUM	MICRO	0.50	0.00	-	1.00	0.50	MICRO/MJ	1.74	
PHOSPHORUS	MILLI	15.00	12.00	-	17.00	15.00	MILLI/MJ	52.26	
CHLORIDE	MILLI	40.00	32.00	-	49.00	40.00	MILLI/MJ	139.37	
FLUORIDE	MICRO	17.00	13.00	-	25.00	17.00	MICRO/MJ	59.23	
IODIDE	MICRO	6.30	4.30	-	9.00	6.30	MICRO/MJ	21.95	
SELENIUM	MICRO	3.30	1.30	-	5.30	3.30	MICRO/MJ	11.50	
BROMINE	MILLI	0.10	-	-	-	0.10	MILLI/MJ	0.35	
VITAMIN A	MICRO	54.00	37.00	-	75.00	54.00	MICRO/MJ	188.15	
CAROTENE	MICRO	24.00	16.00	-	32.00	24.00	MICRO/MJ	83.62	

1 MATURE MILK FROM 10TH DAY POST PARTUM

2 THE AMOUNT IS VALID FOR RAW PROTEIN: THIS CONSISTS OF 0,81 G/100G PURE PROTEIN , BESIDE FREE AMINO ACID, PEPTIDES AND UREA

CONSTITUENTS	DIM	AV	VARIATION		AVR	NUTR. DENS.		MOLPERC.
VITAMIN D	NANO	50.00	10.00	- 120.00	50.00	NANO/MJ	174.21	
VITAMIN E ACTIVITY	MILLI	0.52	0.28	- 0.68	0.52	MILLI/MJ	1.81	
ALPHA-TOCOPHEROL	MILLI	0.52	-	- -	0.52	MILLI/MJ	1.81	
VITAMIN K	MICRO	3.00	2.00	- 4.00	3.00	MICRO/MJ	10.45	
VITAMIN B1	MICRO	15.00	13.00	- 17.00	15.00	MICRO/MJ	52.26	
VITAMIN B2	MICRO	38.00	30.00	- 44.00	38.00	MICRO/MJ	132.40	
NICOTINAMIDE	MILLI	0.17	0.13	- 0.20	0.17	MILLI/MJ	0.59	
PANTOTHENIC ACID	MILLI	0.21	0.16	- 0.26	0.21	MILLI/MJ	0.73	
VITAMIN B6	MICRO	13.00	10.00	- 18.00	13.00	MICRO/MJ	45.29	
BIOTIN	MICRO	0.58	0.40	- 1.00	0.58	MICRO/MJ	2.02	
FOLIC ACID	MICRO	5.00	-	- -	5.00	MICRO/MJ	17.42	
VITAMIN B12	NANO	50.00	30.00	- 100.00	50.00	NANO/MJ	174.21	
VITAMIN C	MILLI	4.40	3.50	- 5.50	4.40	MILLI/MJ	15.33	
ALANINE	MILLI	56.00	51.00	- 58.00	56.00	MILLI/MJ	195.12	
ARGININE	MILLI	51.00	40.00	- 65.00	51.00	MILLI/MJ	177.70	
ASPARTIC ACID	GRAM	0.12	0.11	- 0.12	0.12	GRAM/MJ	0.42	
CYSTINE	MILLI	24.00	17.00	- 29.00	24.00	MILLI/MJ	83.62	
GLUTAMIC ACID	GRAM	0.22	0.20	- 0.24	0.22	GRAM/MJ	0.77	
GLYCINE	MILLI	36.00	26.00	- 39.00	36.00	MILLI/MJ	125.43	
HISTIDINE	MILLI	31.00	21.00	- 41.00	31.00	MILLI/MJ	108.01	
ISOLEUCINE	MILLI	77.00	54.00	- 93.00	77.00	MILLI/MJ	268.29	
LEUCINE	GRAM	0.13	0.10	- 0.18	0.13	GRAM/MJ	0.45	
LYSINE	MILLI	86.00	77.00	- 96.00	86.00	MILLI/MJ	299.64	
METHIONINE	MILLI	24.00	18.00	- 36.00	24.00	MILLI/MJ	83.62	
PHENYLALANINE	MILLI	54.00	43.00	- 74.00	54.00	MILLI/MJ	188.15	
PROLINE	GRAM	0.12	0.10	- 0.13	0.12	GRAM/MJ	0.42	
SERINE	MILLI	59.00	57.00	- 65.00	59.00	MILLI/MJ	205.57	
THREONINE	MILLI	63.00	57.00	- 79.00	63.00	MILLI/MJ	219.51	
TRYPTOPHAN	MILLI	22.00	20.00	- 23.00	22.00	MILLI/MJ	76.65	
TYROSINE	MILLI	56.00	35.00	- 72.00	56.00	MILLI/MJ	195.12	
VALINE	MILLI	81.00	61.00	- 97.00	81.00	MILLI/MJ	282.22	
CITRIC ACID	MILLI	85.00	78.00	- 92.00	85.00	MILLI/MJ	296.16	
LACTOSE	GRAM	7.00	4.90	- 9.50	7.00	GRAM/MJ	24.39	
CAPRIC ACID	MILLI	61.00	30.00	- 86.00	61.00	MILLI/MJ	212.54	
LAURIC ACID	GRAM	0.21	0.15	- 0.34	0.21	GRAM/MJ	0.74	
MYRISTIC ACID	GRAM	0.34	0.24	- 0.66	0.34	GRAM/MJ	1.19	
PALMITIC ACID	GRAM	0.96	0.79	- 1.49	0.96	GRAM/MJ	3.36	
STEARIC ACID	GRAM	0.29	0.21	- 0.47	0.29	GRAM/MJ	1.02	
ARACHIDIC ACID	MILLI	46.00	42.00	- 49.00	46.00	MILLI/MJ	160.27	
PALMITOLEIC ACID	GRAM	0.12	0.08	- 0.19	0.13	GRAM/MJ	0.45	
OLEIC ACID	GRAM	1.34	0.99	- 1.54	1.34	GRAM/MJ	4.67	
LINOLEIC ACID	GRAM	0.38	0.29	- 0.54	0.38	GRAM/MJ	1.32	
LINOLENIC ACID	MILLI	22.00	15.00	- 28.00	22.00	MILLI/MJ	76.65	
ARACHIDONIC ACID	MILLI	4.20	1.50	- 11.40	4.20	MILLI/MJ	14.63	
CHOLESTEROL	MILLI	25.00	19.00	- 35.00	25.00	MILLI/MJ	87.11	
CASEIN	GRAM	0.36	0.31	- 0.39	0.36	GRAM/MJ	1.25	

VORTRANSITORISCHE FRAUENMILCH
2.-3. TAG POST PARTUM

HUMAN MILK PRETRANSITIONAL
2nd–3rd Day POST PARTUM

LAIT DE FEMME LAIT PRÉTRANSITOIRE
2E AU 3E JOUR APRÈS

		PROTEIN	FAT	CARBOHYDRATES	TOTAL
ENERGY VALUE (AVERAGE)	KJOULE	47	112	82	242
PER 100 G	(KCAL)	11	27	20	58
EDIBLE PORTION					
AMOUNT OF DIGESTIBLE	GRAM	2.49	2.75	4.90	
CONSTITUENTS PER 100 G					
ENERGY VALUE (AVERAGE)	KJOULE	46	107	82	235
OF THE DIGESTIBLE	(KCAL)	11	25	20	56
FRACTION PER 100 G					
EDIBLE PORTION					

WASTE PERCENTAGE AVERAGE 0.00

CONSTITUENTS	DIM	AV	VARIATION			AVR	NUTR. DENS.		MOLPERC.
WATER	GRAM	89.30	-	-	-	89.30	GRAM/MJ	380.77	
PROTEIN	GRAM	2.57	2.02	-	2.99	2.57	GRAM/MJ	10.96	
FAT	GRAM	2.90	-	-	-	2.90	GRAM/MJ	12.37	
AVAILABLE CARBOHYDR.	GRAM	4.90 1	4.40	-	5.40	4.90	GRAM/MJ	20.89	
MINERALS	GRAM	0.36	0.32	-	0.39	0.36	GRAM/MJ	1.54	
SODIUM	MILLI	54.00	-	-	-	54.00	MILLI/MJ	230.25	
POTASSIUM	MILLI	64.00	-	-	-	64.00	MILLI/MJ	272.89	
MAGNESIUM	MILLI	3.30	2.90	-	3.80	3.30	MILLI/MJ	14.07	
CALCIUM	MILLI	29.00	23.00	-	35.00	29.00	MILLI/MJ	123.65	
MANGANESE	MICRO	1.10	0.60	-	1.60	1.10	MICRO/MJ	4.69	
IRON	MICRO	48.00	36.00	-	60.00	48.00	MICRO/MJ	204.67	
COPPER	MICRO	46.00	28.00	-	64.00	46.00	MICRO/MJ	196.14	
VITAMIN B1	MICRO	10.00	-	-	-	10.00	MICRO/MJ	42.64	

1 ESTIMATED BY THE DIFFERENCE METHOD
100 - (WATER + PROTEIN + FAT + MINERALS)

TRANSITORISCHE FRAUENMILCH
(ÜBERGANGSMILCH)
6.–10. TAG POST PARTUM

HUMAN MILK TRANSITIONAL
6–10th DAY POST PARTUM

LAIT DE FEMME LAIT TRANSITOIRE
6E AU 10E JOUR APRÈS

		PROTEIN	FAT	CARBOHYDRATES	TOTAL
ENERGY VALUE (AVERAGE) PER 100 G EDIBLE PORTION	KJOULE	30	136	110	277
	(KCAL)	7.2	33	26	66
AMOUNT OF DIGESTIBLE CONSTITUENTS PER 100 G	GRAM	1.58	3.34	6.60	
ENERGY VALUE (AVERAGE) OF THE DIGESTIBLE FRACTION PER 100 G EDIBLE PORTION	KJOULE	29	129	110	269
	(KCAL)	7.0	31	26	64

WASTE PERCENTAGE AVERAGE 0.00

CONSTITUENTS	DIM	AV	VARIATION		AVR	NUTR. DENS.		MOLPERC.
WATER	GRAM	86.40	84.70	- 88.30	86.40	GRAM/MJ	321.21	
PROTEIN	GRAM	1.63	1.47	- 1.86	1.63	GRAM/MJ	6.06	
FAT	GRAM	3.52	3.10	- 3.87	3.52	GRAM/MJ	13.09	
AVAILABLE CARBOHYDR.	GRAM	6.60	5.70	- 7.40	6.60	GRAM/MJ	24.54	
MINERALS	GRAM	0.27	0.24	- 0.32	0.27	GRAM/MJ	1.00	
SODIUM	MILLI	29.00	19.00	- 54.00	29.00	MILLI/MJ	107.81	
POTASSIUM	MILLI	64.00	53.00	- 77.00	64.00	MILLI/MJ	237.93	
MAGNESIUM	MILLI	3.50	2.60	- 5.40	3.50	MILLI/MJ	13.01	
CALCIUM	MILLI	40.00	34.00	- 46.00	40.00	MILLI/MJ	148.71	
IRON	MICRO	40.00	20.00	- 50.00	40.00	MICRO/MJ	148.71	
COPPER	MICRO	50.00	40.00	- 70.00	50.00	MICRO/MJ	185.88	
ZINC	MILLI	0.38	0.04	- 0.59	0.38	MILLI/MJ	1.41	
PHOSPHORUS	MILLI	18.00	10.00	- 32.00	18.00	MILLI/MJ	66.92	
CHLORIDE	MILLI	50.00	46.00	- 54.00	50.00	MILLI/MJ	185.88	
IODIDE	MICRO	2.40	-	- -	2.40	MICRO/MJ	8.92	
VITAMIN A	MICRO	88.00	58.00	- 180.00	88.00	MICRO/MJ	327.16	
CAROTENE	MICRO	26.00	-	- -	26.00	MICRO/MJ	96.66	
VITAMIN E ACTIVITY	MILLI	1.32	0.48	- 3.00	1.32	MILLI/MJ	4.91	
ALPHA-TOCOPHEROL	MILLI	1.32	0.48	- 3.00	1.32	MILLI/MJ	4.91	
VITAMIN B1	MICRO	20.00	1.00	- 33.00	20.00	MICRO/MJ	74.35	
VITAMIN B2	MICRO	4.00	1.00	- 7.00	4.00	MICRO/MJ	14.87	
NICOTINAMIDE	MILLI	0.18	0.06	- 0.36	0.18	MILLI/MJ	0.67	
PANTOTHENIC ACID	MILLI	0.29	0.14	- 0.41	0.29	MILLI/MJ	1.08	
BIOTIN	MICRO	0.40	0.00	- 1.80	0.40	MICRO/MJ	1.49	
FOLIC ACID	MICRO	0.50	0.20	- 0.80	0.50	MICRO/MJ	1.86	
VITAMIN B12	NANO	36.00	3.00	- 70.00	36.00	NANO/MJ	133.84	

CONSTITUENTS	DIM	AV	VARIATION			AVR	NUTR. DENS.		MOLPERC.
VITAMIN C	MILLI	5.50	3.90	-	7.10	5.50	MILLI/MJ	20.45	
ALANINE	MILLI	84.00	-	-	-	84.00	MILLI/MJ	312.29	6.5
ARGININE	MILLI	72.00	55.00	-	102.00	72.00	MILLI/MJ	267.67	2.9
ASPARTIC ACID	GRAM	0.17	-	-	-	0.17	GRAM/MJ	0.63	8.7
CYSTINE	MILLI	34.00	-	-	-	34.00	MILLI/MJ	126.40	1.0
GLUTAMIC ACID	GRAM	0.32	-	-	-	0.32	GRAM/MJ	1.19	15.0
GLYCINE	MILLI	53.00	-	-	-	53.00	MILLI/MJ	197.04	4.9
HISTIDINE	MILLI	44.00	35.00	-	51.00	44.00	MILLI/MJ	163.58	2.0
ISOLEUCINE	GRAM	0.12	0.09	-	0.15	0.12	GRAM/MJ	0.45	6.2
LEUCINE	GRAM	0.18	0.13	-	0.24	0.18	GRAM/MJ	0.67	9.6
LYSINE	GRAM	0.13	0.10	-	0.17	0.13	GRAM/MJ	0.48	6.0
METHIONINE	MILLI	28.00	19.00	-	41.00	28.00	MILLI/MJ	104.10	1.3
PHENYLALANINE	MILLI	74.00	56.00	-	86.00	74.00	MILLI/MJ	275.11	3.1
PROLINE	GRAM	0.18	-	-	-	0.18	GRAM/MJ	0.67	10.9
SERINE	MILLI	83.00	-	-	-	83.00	MILLI/MJ	308.57	5.4
THREONINE	MILLI	95.00	75.00	-	118.00	95.00	MILLI/MJ	353.18	5.5
TRYPTOPHAN	MILLI	32.00	26.00	-	41.00	32.00	MILLI/MJ	118.97	1.1
TYROSINE	MILLI	53.00	-	-	-	53.00	MILLI/MJ	197.04	2.0
VALINE	GRAM	0.14	0.10	-	0.17	0.14	GRAM/MJ	0.52	8.0
LACTOSE	GRAM	6.60	5.70	-	7.40	6.60	GRAM/MJ	24.54	
CHOLESTEROL	MILLI	29.00	14.00	-	44.00	29.00	MILLI/MJ	107.81	

KAMELMILCH[1] CAMEL MILK[1] LAIT DE CHAMELLE[1]

		PROTEIN	FAT	CARBOHYDRATES	TOTAL
ENERGY VALUE (AVERAGE) PER 100 G EDIBLE PORTION	KJOULE	94	159	80	333
	(KCAL)	22	38	19	80
AMOUNT OF DIGESTIBLE CONSTITUENTS PER 100 G	GRAM	4.94	3.89	4.80	
ENERGY VALUE (AVERAGE) OF THE DIGESTIBLE FRACTION PER 100 G EDIBLE PORTION	KJOULE	91	151	80	322
	(KCAL)	22	36	19	77

WASTE PERCENTAGE AVERAGE 0.00

CONSTITUENTS	DIM	AV	VARIATION			AVR	NUTR. DENS.		MOLPERC.
WATER	GRAM	85.90	-	-	-	85.90	GRAM/MJ	266.65	
PROTEIN	GRAM	5.10	4.60	-	5.70	5.10	GRAM/MJ	15.83	
FAT	GRAM	4.10	3.90	-	4.30	4.10	GRAM/MJ	12.73	
AVAILABLE CARBOHYDR.	GRAM	4.80	-	-	-	4.80	GRAM/MJ	14.90	
MINERALS	GRAM	0.58	0.50	-	0.63	0.58	GRAM/MJ	1.80	
SODIUM	MILLI	30.00	23.00	-	37.00	30.00	MILLI/MJ	93.12	
POTASSIUM	MILLI	144.00	120.00	-	156.00	144.00	MILLI/MJ	447.00	
MAGNESIUM	MILLI	10.00	-	-	-	10.00	MILLI/MJ	31.04	
CALCIUM	MILLI	132.00	-	-	-	132.00	MILLI/MJ	409.75	
PHOSPHORUS	MILLI	22.00	15.00	-	35.00	22.00	MILLI/MJ	68.29	
CHLORIDE	MILLI	140.00	-	-	-	140.00	MILLI/MJ	434.58	
LACTOSE	GRAM	4.80	-	-	-	4.80	GRAM/MJ	14.90	

1 IN THE PERIOD OF THIRSTING OF THE ANIMAL, THE CAMEL-MILK HAS THE FOLLOWING COMPOSITION PER 100 G: WATER: 90,1 G; PROTEIN: 2,9 G; FAT: 1,9 G; LACTOSE: 3,1 G; ASH: 0,39 G; NA: 39 MG; K: 188 MG; CA: 111 MG; MG: 7,9 MG; CL: 265 MG; UREA: 38 MG.

KUHMILCH	**COW'S MILK**	**LAIT DE VACHE**
VOLLMILCH	WHOLE MILK	LAIT ENTIER
(ROHMILCH, VORZUGSMILCH)	(RAW MILK)	(L. CRU, L. AMÉLIORÉ)

		PROTEIN	FAT	CARBOHYDRATES	TOTAL
ENERGY VALUE (AVERAGE)	KJOULE	61	146	79	287
PER 100 G	(KCAL)	15	35	19	69
EDIBLE PORTION					
AMOUNT OF DIGESTIBLE	GRAM	3.23	3.59	4.75	
CONSTITUENTS PER 100 G					
ENERGY VALUE (AVERAGE)	KJOULE	59	139	79	278
OF THE DIGESTIBLE	(KCAL)	14	33	19	66
FRACTION PER 100 G					
EDIBLE PORTION					

WASTE PERCENTAGE AVERAGE 0.00

CONSTITUENTS	DIM	AV	VARIATION			AVR	NUTR.	DENS.	MOLPERC.
WATER	GRAM	87.50	86.80	-	88.30	87.50	GRAM/MJ	314.82	
PROTEIN	GRAM	3.33	3.08	-	3.70	3.33	GRAM/MJ	11.98	
FAT	GRAM	3.78 1	3.60	-	3.88	3.78	GRAM/MJ	13.60	
AVAILABLE CARBOHYDR.	GRAM	4.75	-	-	-	4.75	GRAM/MJ	17.09	
MINERALS	GRAM	0.74	0.67	-	0.81	0.74	GRAM/MJ	2.66	
SODIUM	MILLI	48.00	40.00	-	58.00	48.00	MILLI/MJ	172.70	
POTASSIUM	MILLI	157.00	144.00	-	178.00	157.00	MILLI/MJ	564.87	
MAGNESIUM	MILLI	12.00	9.00	-	16.00	12.00	MILLI/MJ	43.17	
CALCIUM	MILLI	120.00	107.00	-	133.00	120.00	MILLI/MJ	431.75	
MANGANESE	MICRO	2.50	1.30	-	4.00	2.50	MICRO/MJ	8.99	
IRON	MICRO	46.00	30.00	-	70.00	46.00	MICRO/MJ	165.50	
COBALT	NANO	80.00	50.00	-	130.00	80.00	NANO/MJ	287.83	
COPPER	MICRO	10.00	2.00	-	30.00	10.00	MICRO/MJ	35.98	
ZINC	MILLI	0.38	0.21	-	0.55	0.38	MILLI/MJ	1.37	
NICKEL	MICRO	2.50	0.40	-	6.00	2.50	MICRO/MJ	8.99	
CHROMIUM	MICRO	2.50	1.00	-	4.00	2.50	MICRO/MJ	8.99	
MOLYBDENUM	MICRO	4.20	2.40	-	6.00	4.20	MICRO/MJ	15.11	
VANADIUM	MICRO	-	0.00	-	10.00				
ALUMINIUM	MICRO	46.00	-	-	-	46.00	MICRO/MJ	165.50	
PHOSPHORUS	MILLI	92.00	63.00	-	102.00	92.00	MILLI/MJ	331.01	
CHLORIDE	MILLI	102.00	90.00	-	106.00	102.00	MILLI/MJ	366.99	
FLUORIDE	MICRO	17.00 2	11.00	-	21.00	17.00	MICRO/MJ	61.16	
IODIDE	MICRO	3.30 3	2.00	-	6.00	3.30	MICRO/MJ	11.87	
BORON	MICRO	27.00	19.00	-	95.00	27.00	MICRO/MJ	97.14	
SELENIUM	MICRO	9.00	5.00	-	13.00	9.00	MICRO/MJ	32.38	
BROMINE	MICRO	224.00	154.00	-	293.00	224.00	MICRO/MJ	805.93	
NITRATE	MILLI	0.08	0.02	-	1.24	0.08	MILLI/MJ	0.29	
VITAMIN A	MICRO	30.00 4	27.00	-	34.00	30.00	MICRO/MJ	107.94	
CAROTENE	MICRO	18.00	14.00	-	22.00	18.00	MICRO/MJ	64.76	

1 FEW ENGLISH BREEDS (GUERNSEY; JERSEY) CONTAIN TILL 5 % FAT.
COMPOSITION OF THE LIPID: TRIGLYCERIDES: 95 - 98 %; PHOSPHORLIPIDES 0,2 - 1,0 %.

2 IN THE ENVIROMENT OF AL-PLANTS, THE CONTENT CAN RAISE TILL TO 60 UG/100G.

3 THE AMOUNT IS DEPENDENT ON THE CONCENTRATION OF IODINE IN FEED AND WATER

4 YEAR-ROUND AVERAGE: JANUARY-APRIL: VITAMIN A: 22 (19-27) UG/100G CAROTIN: 16 (13-19) UG/100G;
MAI-DECEMBER: VITAMIN A: 34 (29-38) UG/100G, CAROTIN: 27 (16-25) UG/100G.

CONSTITUENTS	DIM	AV	VARIATION		AVR	NUTR. DENS.		MOLPERC.
VITAMIN D	NANO	63.00	20.00	- 90.00	63.00	NANO/MJ	226.67	
VITAMIN E ACTIVITY	MICRO	88.00	41.00	- 100.00	88.00	MICRO/MJ	316.62	
ALPHA-TOCOPHEROL	MICRO	88.00	41.00	- 100.00	88.00	MICRO/MJ	316.62	
VITAMIN K	MICRO	4.00	0.00	- 33.00	4.00	MICRO/MJ	14.39	
VITAMIN B1	MICRO	37.00	30.00	- 55.00	37.00	MICRO/MJ	133.12	
VITAMIN B2	MILLI	0.18	0.14	- 0.22	0.18	MILLI/MJ	0.65	
NICOTINAMIDE	MICRO	90.00	70.00	- 110.00	90.00	MICRO/MJ	323.81	
PANTOTHENIC ACID	MILLI	0.35	0.28	- 0.42	0.35	MILLI/MJ	1.26	
VITAMIN B6	MICRO	46.00	22.00	- 70.00	46.00	MICRO/MJ	165.50	
BIOTIN	MICRO	3.50	2.00	- 5.00	3.50	MICRO/MJ	12.59	
FOLIC ACID	MICRO	5.90	3.80	- 9.00	5.90	MICRO/MJ	21.23	
VITAMIN B12	MICRO	0.42	0.30	- 0.76	0.42	MICRO/MJ	1.51	
VITAMIN C	MILLI	1.70	1.00	- 2.40	1.70	MILLI/MJ	6.12	
ALANINE	GRAM	0.13	0.13	- 0.14	0.13	GRAM/MJ	0.47	4.9
ARGININE	GRAM	0.13	0.12	- 0.15	0.13	GRAM/MJ	0.47	2.5
ASPARTIC ACID	GRAM	0.29	0.26	- 0.31	0.29	GRAM/MJ	1.04	7.5
CYSTINE	MILLI	28.00	22.00	- 32.00	28.00	MILLI/MJ	100.74	0.4
GLUTAMIC ACID	GRAM	0.79	0.73	- 0.82	0.79	GRAM/MJ	2.84	18.5
GLYCINE	MILLI	76.00	71.00	- 84.00	76.00	MILLI/MJ	273.44	3.5
HISTIDINE	MILLI	95.00	84.00	- 106.00	95.00	MILLI/MJ	341.80	2.1
ISOLEUCINE	GRAM	0.22	0.19	- 0.26	0.22	GRAM/MJ	0.79	5.7
LEUCINE	GRAM	0.36	0.34	- 0.38	0.36	GRAM/MJ	1.30	9.6
LYSINE	GRAM	0.28	0.25	- 0.30	0.28	GRAM/MJ	1.01	6.5
METHIONINE	MILLI	90.00	84.00	- 99.00	90.00	MILLI/MJ	323.81	2.1
PHENYLALANINE	GRAM	0.18	0.17	- 0.19	0.18	GRAM/MJ	0.65	3.7
PROLINE	GRAM	0.34	0.33	- 0.37	0.34	GRAM/MJ	1.22	10.2
SERINE	GRAM	0.21	0.20	- 0.22	0.21	GRAM/MJ	0.76	6.9
THREONINE	GRAM	0.16	0.15	- 0.18	0.16	GRAM/MJ	0.58	4.6
TRYPTOPHAN	MILLI	49.00	45.00	- 55.00	49.00	MILLI/MJ	176.30	0.8
TYROSINE	GRAM	0.18	0.17	- 0.21	0.18	GRAM/MJ	0.65	3.4
VALINE	GRAM	0.24	0.21	- 0.27	0.24	GRAM/MJ	0.86	7.1
CITRIC ACID	GRAM	0.21	0.17	- 0.28	0.21	GRAM/MJ	0.76	
LACTOSE	GRAM	4.54	4.35	- 4.80	4.54	GRAM/MJ	16.33	
BUTYRIC ACID	GRAM	0.12	0.04	- 0.20	0.13	GRAM/MJ	0.46	
CAPROIC ACID	MILLI	82.00	29.00	- 157.00	82.00	MILLI/MJ	295.03	
CAPRYLIC ACID	MILLI	46.00	18.00	- 93.00	46.00	MILLI/MJ	165.50	
CAPRIC ACID	MILLI	96.00	43.00	- 179.00	96.00	MILLI/MJ	345.40	
LAURIC ACID	GRAM	0.12	0.04	- 0.38	0.12	GRAM/MJ	0.44	
MYRISTIC ACID	GRAM	0.38	0.19	- 0.64	0.38	GRAM/MJ	1.37	
C15 : 0 FATTY ACID	MILLI	46.00	29.00	- 82.00	46.00	MILLI/MJ	165.50	
PALMITIC ACID	GRAM	0.96	0.66	- 1.50	0.96	GRAM/MJ	3.46	
C17 : 0 FATTY ACID	MILLI	36.00	4.00	- 85.00	36.00	MILLI/MJ	129.52	
STEARIC ACID	GRAM	0.36	0.14	- 0.54	0.36	GRAM/MJ	1.30	
ARACHIDIC ACID	MILLI	25.00	10.00	- 40.00	25.00	MILLI/MJ	89.95	
BEHENIC ACID	MILLI	2.10	1.40	- 3.90	2.10	MILLI/MJ	7.56	
PALMITOLEIC ACID	GRAM	0.11	0.05	- 0.21	0.11	GRAM/MJ	0.41	
OLEIC ACID	GRAM	0.94	0.45	- 1.43	0.94	GRAM/MJ	3.38	
C14 : 1 FATTY ACID	MILLI	79.00	21.00	- 146.00	79.00	MILLI/MJ	284.23	
LINOLEIC ACID	MILLI	89.00	11.00	- 257.00	89.00	MILLI/MJ	320.21	
LINOLENIC ACID	MILLI	61.00	4.00	- 225.00	61.00	MILLI/MJ	219.47	
CHOLESTEROL	MILLI	12.30	10.00	- 14.80	12.30	MILLI/MJ	44.25	
TOTAL PHOSPHOLIPIDS	MILLI	35.00	-	- -	35.00	MILLI/MJ	125.93	
PHOSPHATIDYLCHOLINE	MILLI	12.00	-	- -	12.00	MILLI/MJ	43.17	
PHOSPHATIDYLETHANOLAMINE	MILLI	10.00	-	- -	10.00	MILLI/MJ	35.98	
PHOSPHATIDYLSERINE	MILLI	1.00	-	- -	1.00	MILLI/MJ	3.60	
PHOSPHATIDYLINOSITOL	MILLI	0.20	-	- -	0.20	MILLI/MJ	0.72	
SPHINGOMYELIN	MILLI	9.00	-	- -	9.00	MILLI/MJ	32.38	
ALBUMEN AND GLOBULIN	GRAM	0.51	0.45	- 0.68	0.51	GRAM/MJ	1.83	
CASEIN	GRAM	2.66	2.45	- 3.00	2.66	GRAM/MJ	9.57	

KUHMILCH	**COW'S MILK**	**LAIT DE VACHE**
KONSUMMILCH	CONSUMER MILK	LAIT DE CONSOMMATION
MIND. 3,5% FETT	MIN. 3.5% FAT CONTENT	MIN. 3,5% MAT. GR.

		PROTEIN	FAT	CARBOHYDRATES	TOTAL
ENERGY VALUE (AVERAGE)	KJOULE	61	138	80	279
PER 100 G	(KCAL)	15	33	19	67
EDIBLE PORTION					
AMOUNT OF DIGESTIBLE	GRAM	3.23	3.39	4.76	
CONSTITUENTS PER 100 G					
ENERGY VALUE (AVERAGE)	KJOULE	60	131	80	271
OF THE DIGESTIBLE	(KCAL)	14	31	19	65
FRACTION PER 100 G					
EDIBLE PORTION					

WASTE PERCENTAGE AVERAGE 0.00

CONSTITUENTS	DIM	AV	VARIATION			AVR	NUTR. DENS.		MOLPERC.
WATER	GRAM	87.70	87.30	-	88.40	87.70	GRAM/MJ	324.14	
PROTEIN	GRAM	3.34	3.08	-	3.70	3.34	GRAM/MJ	12.34	
FAT	GRAM	3.57	3.50	-	3.62	3.57	GRAM/MJ	13.19	
AVAILABLE CARBOHYDR.	GRAM	4.76	-	-	-	4.76	GRAM/MJ	17.59	
MINERALS	GRAM	0.74	0.67	-	0.81	0.74	GRAM/MJ	2.74	
SODIUM	MILLI	48.00	40.00	-	58.00	48.00	MILLI/MJ	177.41	
POTASSIUM	MILLI	157.00	144.00	-	178.00	157.00	MILLI/MJ	580.27	
MAGNESIUM	MILLI	12.00	9.00	-	16.00	12.00	MILLI/MJ	44.35	
CALCIUM	MILLI	120.00	107.00	-	133.00	120.00	MILLI/MJ	443.52	
MANGANESE	MICRO	2.50	1.30	-	4.00	2.50	MICRO/MJ	9.24	
IRON	MICRO	46.00	30.00	-	73.00	46.00	MICRO/MJ	170.01	
COBALT	NANO	80.00	50.00	-	130.00	80.00	NANO/MJ	295.68	
COPPER	MICRO	17.00	2.00	-	30.00	17.00	MICRO/MJ	62.83	
ZINC	MILLI	0.38	0.21	-	0.55	0.38	MILLI/MJ	1.40	
NICKEL	MICRO	1.00	-	-	-	1.00	MICRO/MJ	3.70	
CHROMIUM	MICRO	1.70	-	-	-	1.70	MICRO/MJ	6.28	
PHOSPHORUS	MILLI	92.00	63.00	-	102.00	92.00	MILLI/MJ	340.03	
CHLORIDE	MILLI	102.00	90.00	-	106.00	102.00	MILLI/MJ	376.99	
FLUORIDE	MICRO	17.00	11.00	-	21.00	17.00	MICRO/MJ	62.83	
IODIDE	MICRO	3.30	2.00	-	6.00	3.30	MICRO/MJ	12.20	
BROMINE	MILLI	0.10	-	-	-	0.10	MILLI/MJ	0.37	
NITRATE	MILLI	0.08	0.02	-	1.24	0.08	MILLI/MJ	0.30	
VITAMIN A	MICRO	28.00	25.00	-	32.00	28.00	MICRO/MJ	103.49	
CAROTENE	MICRO	17.00	13.00	-	21.00	17.00	MICRO/MJ	62.83	
VITAMIN D	NANO	60.00	19.00	-	85.00	60.00	NANO/MJ	221.76	
VITAMIN E ACTIVITY	MICRO	84.00	39.00	-	94.00	84.00	MICRO/MJ	310.46	

CONSTITUENTS	DIM	AV	VARIATION			AVR	NUTR. DENS.		MOLPERC.
ALPHA-TOCOPHEROL	MICRO	84.00	39.00	-	94.00	84.00	MICRO/MJ	310.46	
VITAMIN K	MICRO	4.00	0.00	-	33.00	4.00	MICRO/MJ	14.78	
VITAMIN B1	MICRO	37.00	30.00	-	55.00	37.00	MICRO/MJ	136.75	
VITAMIN B2	MILLI	0.18	0.14	-	0.22	0.18	MILLI/MJ	0.67	
NICOTINAMIDE	MICRO	90.00	70.00	-	110.00	90.00	MICRO/MJ	332.64	
PANTOTHENIC ACID	MILLI	0.35	0.28	-	0.42	0.35	MILLI/MJ	1.29	
VITAMIN B6	MICRO	46.00	22.00	-	70.00	46.00	MICRO/MJ	170.01	
BIOTIN	MICRO	3.50	2.00	-	5.00	3.50	MICRO/MJ	12.94	
FOLIC ACID	MICRO	5.90	3.80	-	9.00	5.90	MICRO/MJ	21.81	
VITAMIN B12	MICRO	0.42	0.30	-	0.76	0.42	MICRO/MJ	1.55	
VITAMIN C	MILLI	1.70	1.00	-	2.40	1.70	MILLI/MJ	6.28	
ALANINE	GRAM	0.12	-	-	-	0.13	GRAM/MJ	0.47	5.1
ARGININE	GRAM	0.12	0.11	-	0.14	0.12	GRAM/MJ	0.44	2.5
ASPARTIC ACID	GRAM	0.28	-	-	-	0.28	GRAM/MJ	1.03	7.5
CYSTINE	MILLI	26.00	20.00	-	30.00	26.00	MILLI/MJ	96.10	0.4
GLUTAMIC ACID	GRAM	0.75	-	-	-	0.75	GRAM/MJ	2.77	18.2
GLYCINE	MILLI	80.00	-	-	-	80.00	MILLI/MJ	295.68	3.8
HISTIDINE	MILLI	89.00	80.00	-	100.00	89.00	MILLI/MJ	328.94	2.0
ISOLEUCINE	GRAM	0.21	0.18	-	0.25	0.21	GRAM/MJ	0.78	5.7
LEUCINE	GRAM	0.35	0.33	-	0.36	0.35	GRAM/MJ	1.29	9.5
LYSINE	GRAM	0.26	0.24	-	0.28	0.26	GRAM/MJ	0.96	6.3
METHIONINE	MILLI	84.00	80.00	-	90.00	84.00	MILLI/MJ	310.46	2.0
PHENYLALANINE	GRAM	0.17	0.16	-	0.18	0.17	GRAM/MJ	0.63	3.7
PROLINE	GRAM	0.35	-	-	-	0.35	GRAM/MJ	1.31	11.0
SERINE	GRAM	0.19	-	-	-	0.19	GRAM/MJ	0.72	6.6
THREONINE	GRAM	0.15	0.14	-	0.17	0.15	GRAM/MJ	0.55	4.5
TRYPTOPHAN	MILLI	46.00	43.00	-	50.00	46.00	MILLI/MJ	170.01	0.8
TYROSINE	GRAM	0.17	0.16	-	0.19	0.17	GRAM/MJ	0.63	3.3
VALINE	GRAM	0.23	0.22	-	0.26	0.23	GRAM/MJ	0.85	7.0
HISTAMINE	MICRO	50.00	30.00	-	70.00	50.00	MICRO/MJ	184.80	
CITRIC ACID	GRAM	0.21	0.17	-	0.28	0.21	GRAM/MJ	0.78	
LACTOSE	GRAM	4.55	4.36	-	4.82	4.55	GRAM/MJ	16.82	
BUTYRIC ACID	GRAM	0.12	-	-	-	0.12	GRAM/MJ	0.44	
CAPROIC ACID	MILLI	60.00	-	-	-	60.00	MILLI/MJ	221.76	
CAPRYLIC ACID	MILLI	40.00	-	-	-	40.00	MILLI/MJ	147.84	
CAPRIC ACID	MILLI	80.00	-	-	-	80.00	MILLI/MJ	295.68	
LAURIC ACID	GRAM	0.10	-	-	-	0.10	GRAM/MJ	0.37	
MYRISTIC ACID	GRAM	0.36	-	-	-	0.36	GRAM/MJ	1.33	
PALMITIC ACID	GRAM	0.93	-	-	-	0.93	GRAM/MJ	3.44	
STEARIC ACID	GRAM	0.40	-	-	-	0.40	GRAM/MJ	1.48	
PALMITOLEIC ACID	MILLI	80.00	-	-	-	80.00	MILLI/MJ	295.68	
OLEIC ACID	GRAM	0.89	-	-	-	0.89	GRAM/MJ	3.29	
LINOLEIC ACID	MILLI	92.00	74.00	-	110.00	92.00	MILLI/MJ	340.03	
LINOLENIC ACID	MILLI	24.00	20.00	-	30.00	24.00	MILLI/MJ	88.70	
CHOLESTEROL	MILLI	11.70	9.50	-	14.10	11.70	MILLI/MJ	43.24	
TOTAL PHOSPHOLIPIDS	MILLI	33.00	-	-	-	33.00	MILLI/MJ	121.97	
PHOSPHATIDYLCHOLINE	MILLI	11.00	-	-	-	11.00	MILLI/MJ	40.66	
PHOSPHATIDYLETHANOLAMINE	MILLI	9.00	-	-	-	9.00	MILLI/MJ	33.26	
PHOSPHATIDYLSERINE	MILLI	1.00	-	-	-	1.00	MILLI/MJ	3.70	
PHOSPHATIDYLINOSITOL	MILLI	2.00	-	-	-	2.00	MILLI/MJ	7.39	
SPHINGOMYELIN	MILLI	9.00	-	-	-	9.00	MILLI/MJ	33.26	
ALBUMEN AND GLOBULIN	GRAM	0.51	0.45	-	0.68	0.51	GRAM/MJ	1.88	
CASEIN	GRAM	2.66	2.45	-	3.00	2.66	GRAM/MJ	9.83	

KUHMILCH
MIND. 1,5%, HÖCHST. 1,8% FETT

COW'S MILK
MIN. 1.5%, MAX. 1.8% FAT CONTENT

LAIT DE VACHE
MIN. 1,5%, MAX. 1,8% MAT. GR.

		PROTEIN	FAT	CARBOHYDRATES	TOTAL
ENERGY VALUE (AVERAGE) PER 100 G EDIBLE PORTION	KJOULE	62	62	80	204
	(KCAL)	15	15	19	49
AMOUNT OF DIGESTIBLE CONSTITUENTS PER 100 G	GRAM	3.24	1.55	4.78	
ENERGY VALUE (AVERAGE) OF THE DIGESTIBLE FRACTION PER 100 G EDIBLE PORTION	KJOULE	60	60	80	200
	(KCAL)	14	14	19	48

WASTE PERCENTAGE AVERAGE 0.00

CONSTITUENTS	DIM	AV	VARIATION			AVR	NUTR.	DENS.	MOLPERC.
WATER	GRAM	89.60	89.00	-	90.30	89.60	GRAM/MJ	448.26	
PROTEIN	GRAM	3.35	3.05	-	3.84	3.35	GRAM/MJ	16.76	
FAT	GRAM	1.60	1.50	-	1.80	1.60	GRAM/MJ	8.00	
AVAILABLE CARBOHYDR.	GRAM	4.78	-	-	-	4.78	GRAM/MJ	23.91	
MINERALS	GRAM	0.73	0.66	-	0.76	0.73	GRAM/MJ	3.65	
SODIUM	MILLI	47.00	39.00	-	57.00	47.00	MILLI/MJ	235.13	
POTASSIUM	MILLI	155.00	142.00	-	176.00	155.00	MILLI/MJ	775.44	
MAGNESIUM	MILLI	12.00	9.00	-	16.00	12.00	MILLI/MJ	60.03	
CALCIUM	MILLI	118.00	106.00	-	131.00	118.00	MILLI/MJ	590.34	
MANGANESE	MICRO	2.50	1.30	-	3.90	2.50	MICRO/MJ	12.51	
IRON	MICRO	45.00	30.00	-	69.00	45.00	MICRO/MJ	225.13	
COBALT	NANO	80.00	50.00	-	130.00	80.00	NANO/MJ	400.23	
COPPER	MICRO	10.00	2.00	-	30.00	10.00	MICRO/MJ	50.03	
ZINC	MILLI	0.37	0.21	-	0.54	0.37	MILLI/MJ	1.85	
PHOSPHORUS	MILLI	91.00	62.00	-	101.00	91.00	MILLI/MJ	455.26	
CHLORIDE	MILLI	101.00	89.00	-	105.00	101.00	MILLI/MJ	505.29	
FLUORIDE	MICRO	17.00	11.00	-	21.00	17.00	MICRO/MJ	85.05	
IODIDE	MICRO	3.30	2.00	-	6.00	3.30	MICRO/MJ	16.51	
VITAMIN A	MICRO	13.00	11.00	-	14.00	13.00	MICRO/MJ	65.04	
CAROTENE	MICRO	8.00	6.00	-	9.00	8.00	MICRO/MJ	40.02	
VITAMIN D	NANO	28.00	8.00	-	38.00	28.00	NANO/MJ	140.08	
VITAMIN E ACTIVITY	MICRO	37.00	17.00	-	42.00	37.00	MICRO/MJ	185.11	
ALPHA-TOCOPHEROL	MICRO	37.00	17.00	-	42.00	37.00	MICRO/MJ	185.11	
VITAMIN K	MICRO	2.00	0.00	-	16.00	2.00	MICRO/MJ	10.01	
VITAMIN B1	MICRO	37.00	30.00	-	55.00	37.00	MICRO/MJ	185.11	
VITAMIN B2	MILLI	0.18	0.14	-	0.22	0.18	MILLI/MJ	0.90	

CONSTITUENTS	DIM	AV	VARIATION			AVR	NUTR. DENS.		MOLPERC.
NICOTINAMIDE	MICRO	90.00	70.00	-	110.00	90.00	MICRO/MJ	450.26	
PANTOTHENIC ACID	MILLI	0.35	0.28	-	0.42	0.35	MILLI/MJ	1.75	
VITAMIN B6	MICRO	46.00	22.00	-	70.00	46.00	MICRO/MJ	230.13	
BIOTIN	MICRO	3.50	2.00	-	5.00	3.50	MICRO/MJ	17.51	
FOLIC ACID	MICRO	-	0.29	-	6.80				
VITAMIN B12	MICRO	0.42	0.30	-	0.76	0.42	MICRO/MJ	2.10	
VITAMIN C	MILLI	1.70	1.00	-	2.40	1.70	MILLI/MJ	8.50	
ALANINE	GRAM	0.13	-	-	-	0.13	GRAM/MJ	0.65	4.9
ARGININE	GRAM	0.13	-	-	-	0.13	GRAM/MJ	0.65	2.5
ASPARTIC ACID	GRAM	0.29	-	-	-	0.29	GRAM/MJ	1.45	7.5
CYSTINE	MILLI	28.00	-	-	-	28.00	MILLI/MJ	140.08	0.4
GLUTAMIC ACID	GRAM	0.79	-	-	-	0.79	GRAM/MJ	3.95	18.5
GLYCINE	MILLI	76.00	-	-	-	76.00	MILLI/MJ	380.22	3.5
HISTIDINE	MILLI	95.00	-	-	-	95.00	MILLI/MJ	475.27	2.1
ISOLEUCINE	GRAM	0.22	-	-	-	0.22	GRAM/MJ	1.10	5.7
LEUCINE	GRAM	0.36	-	-	-	0.36	GRAM/MJ	1.80	9.6
LYSINE	GRAM	0.28	-	-	-	0.28	GRAM/MJ	1.40	6.5
METHIONINE	MILLI	90.00	-	-	-	90.00	MILLI/MJ	450.26	2.1
PHENYLALANINE	GRAM	0.18	-	-	-	0.18	GRAM/MJ	0.90	3.7
PROLINE	GRAM	0.34	-	-	-	0.34	GRAM/MJ	1.70	10.2
SERINE	GRAM	0.21	-	-	-	0.21	GRAM/MJ	1.05	6.9
THREONINE	GRAM	0.16	-	-	-	0.16	GRAM/MJ	0.80	4.6
TRYPTOPHAN	MILLI	49.00	-	-	-	49.00	MILLI/MJ	245.14	0.8
TYROSINE	GRAM	0.18	-	-	-	0.18	GRAM/MJ	0.90	3.4
VALINE	GRAM	0.24	-	-	-	0.24	GRAM/MJ	1.20	7.1
CITRIC ACID	GRAM	0.21	0.17	-	0.28	0.21	GRAM/MJ	1.05	
LACTOSE	GRAM	4.57	4.40	-	4.85	4.57	GRAM/MJ	22.86	
BUTYRIC ACID	MILLI	50.00	-	-	-	50.00	MILLI/MJ	250.14	
CAPROIC ACID	MILLI	30.00	-	-	-	30.00	MILLI/MJ	150.09	
CAPRYLIC ACID	MILLI	20.00	-	-	-	20.00	MILLI/MJ	100.06	
CAPRIC ACID	MILLI	40.00	-	-	-	40.00	MILLI/MJ	200.11	
LAURIC ACID	MILLI	40.00	-	-	-	40.00	MILLI/MJ	200.11	
MYRISTIC ACID	GRAM	0.15	-	-	-	0.15	GRAM/MJ	0.75	
PALMITIC ACID	GRAM	0.40	-	-	-	0.40	GRAM/MJ	2.00	
STEARIC ACID	GRAM	0.18	-	-	-	0.18	GRAM/MJ	0.90	
PALMITOLEIC ACID	MILLI	30.00	-	-	-	30.00	MILLI/MJ	150.09	
OLEIC ACID	GRAM	0.38	-	-	-	0.38	GRAM/MJ	1.90	
LINOLEIC ACID	MILLI	60.00	-	-	-	60.00	MILLI/MJ	300.17	
LINOLENIC ACID	MILLI	30.00	-	-	-	30.00	MILLI/MJ	150.09	
CHOLESTEROL	MILLI	5.20	4.20	-	6.30	5.20	MILLI/MJ	26.01	
ALBUMEN AND GLOBULIN	GRAM	0.52	0.46	-	0.69	0.52	GRAM/MJ	2.60	
CASEIN	GRAM	2.71	2.49	-	3.05	2.71	GRAM/MJ	13.56	

KUHMILCH	**COW'S MILK**	**LAIT DE VACHE**
STERILMILCH	STERILIZED	STERILISÉ
1	1	1

		PROTEIN	FAT	CARBOHYDRATES	TOTAL
ENERGY VALUE (AVERAGE) PER 100 G EDIBLE PORTION	KJOULE	61	146	78	285
	(KCAL)	15	35	19	68
AMOUNT OF DIGESTIBLE CONSTITUENTS PER 100 G	GRAM	3.23	3.59	0.00	
ENERGY VALUE (AVERAGE) OF THE DIGESTIBLE FRACTION PER 100 G EDIBLE PORTION	KJOULE	59	139	0.0	198
	(KCAL)	14	33	0.0	47

WASTE PERCENTAGE AVERAGE 0.00

CONSTITUENTS	DIM	AV	VARIATION		AVR	NUTR. DENS.		MOLPERC.
WATER	GRAM	87.50	86.80	88.30	87.50	GRAM/MJ	440.93	
PROTEIN	GRAM	3.33	3.08	3.70	3.33	GRAM/MJ	16.78	
FAT	GRAM	3.78	3.60	3.88	3.78	GRAM/MJ	19.05	
AVAILABLE CARBOHYDR.	GRAM	4.65 2	-	-	4.65	GRAM/MJ	23.43	
MINERALS	GRAM	0.74	0.67	0.81	0.74	GRAM/MJ	3.73	
SODIUM	MILLI	48.00	40.00	58.00	48.00	MILLI/MJ	241.88	
POTASSIUM	MILLI	157.00	144.00	178.00	157.00	MILLI/MJ	791.16	
MAGNESIUM	MILLI	12.00	9.00	16.00	12.00	MILLI/MJ	60.47	
CALCIUM	MILLI	120.00	107.00	133.00	120.00	MILLI/MJ	604.71	
PHOSPHORUS	MILLI	92.00	63.00	102.00	92.00	MILLI/MJ	463.61	
CHLORIDE	MILLI	102.00	90.00	106.00	102.00	MILLI/MJ	514.00	
IODIDE	MICRO	3.30	2.00	6.00	3.30	MICRO/MJ	16.63	
VITAMIN A	MICRO	30.00	27.00	34.00	30.00	MICRO/MJ	151.18	
CAROTENE	MICRO	18.00	10.00	120.00	18.00	MICRO/MJ	90.71	
VITAMIN D	NANO	20.00	20.00	60.00	20.00	NANO/MJ	100.78	
VITAMIN E ACTIVITY	MICRO	88.00	40.00	100.00	88.00	MICRO/MJ	443.45	
ALPHA-TOCOPHEROL	MICRO	88.00	40.00	100.00	88.00	MICRO/MJ	443.45	
VITAMIN B1	MICRO	24.00	20.00	39.00	24.00	MICRO/MJ	120.94	
VITAMIN B2	MILLI	0.18	0.14	0.22	0.18	MILLI/MJ	0.91	
NICOTINAMIDE	MICRO	90.00	70.00	110.00	90.00	MICRO/MJ	453.53	
PANTOTHENIC ACID	MILLI	0.35	0.28	0.42	0.35	MILLI/MJ	1.76	
VITAMIN B6	MICRO	23.00	10.00	35.00	23.00	MICRO/MJ	115.90	
BIOTIN	MICRO	3.50	2.00	5.00	3.50	MICRO/MJ	17.64	
FOLIC ACID	MICRO	2.90	1.90	4.50	2.90	MICRO/MJ	14.61	
VITAMIN B12	NANO	40.00	30.00	80.00	40.00	NANO/MJ	201.57	
VITAMIN C	MILLI	0.17	0.10	0.24	0.17	MILLI/MJ	0.86	

1 HEATING OF THE MILK IN CLOSED VESSELS FOR 20 - 40 MIN AT 110 - 120 GRAD C.

2 ESTIMATED BY THE DIFFERENCE METHOD 100 - (WATER + PROTEIN + FAT + MINERALS)

KUHMILCH
UHT (ULTRAHOCHERHITZT)

COW'S MILK
UHT
(ULTRA HIGH TEMPERATURE HEATED)

LAIT DE VACHE
UHT
(ULTRA HAUT TEMPERATURE ÉCHAUFFÉ)

		PROTEIN	FAT	CARBOHYDRATES	TOTAL
ENERGY VALUE (AVERAGE) PER 100 G EDIBLE PORTION	KJOULE	61	146	78	285
	(KCAL)	15	35	19	68
AMOUNT OF DIGESTIBLE CONSTITUENTS PER 100 G	GRAM	3.23	3.59	4.65	
ENERGY VALUE (AVERAGE) OF THE DIGESTIBLE FRACTION PER 100 G EDIBLE PORTION	KJOULE	59	139	78	276
	(KCAL)	14	33	19	66

WASTE PERCENTAGE AVERAGE 0.00

CONSTITUENTS	DIM	AV	VARIATION			AVR	NUTR. DENS.		MOLPERC.
WATER	GRAM	87.50	86.80	-	88.30	87.50	GRAM/MJ	316.72	
PROTEIN	GRAM	3.33	3.08	-	3.70	3.33	GRAM/MJ	12.05	
FAT	GRAM	3.78	3.60	-	3.88	3.78	GRAM/MJ	13.68	
AVAILABLE CARBOHYDR.	GRAM	4.65 2	-	-	-	4.65	GRAM/MJ	16.83	
MINERALS	GRAM	0.74	0.67	-	0.81	0.74	GRAM/MJ	2.68	
SODIUM	MILLI	48.00	40.00	-	58.00	48.00	MILLI/MJ	173.75	
POTASSIUM	MILLI	157.00	144.00	-	178.00	157.00	MILLI/MJ	568.29	
MAGNESIUM	MILLI	12.00	9.00	-	16.00	12.00	MILLI/MJ	43.44	
CALCIUM	MILLI	120.00	107.00	-	133.00	120.00	MILLI/MJ	434.36	
PHOSPHORUS	MILLI	92.00	63.00	-	102.00	92.00	MILLI/MJ	333.01	
CHLORIDE	MILLI	102.00	90.00	-	106.00	102.00	MILLI/MJ	369.21	
IODIDE	MICRO	3.30	2.00	-	6.00	3.30	MICRO/MJ	11.95	
VITAMIN A	MICRO	30.00	27.00	-	34.00	30.00	MICRO/MJ	108.59	
CAROTENE	MICRO	18.00	10.00	-	20.00	18.00	MICRO/MJ	65.15	
VITAMIN D	NANO	20.00	20.00	-	60.00	20.00	NANO/MJ	72.39	
VITAMIN E ACTIVITY	MICRO	88.00	40.00	-	100.00	88.00	MICRO/MJ	318.53	
ALPHA-TOCOPHEROL	MICRO	88.00	40.00	-	100.00	88.00	MICRO/MJ	318.53	
VITAMIN B1	MICRO	33.00	27.00	-	36.00	33.00	MICRO/MJ	119.45	
VITAMIN B2	MILLI	0.18	0.14	-	0.22	0.18	MILLI/MJ	0.65	
NICOTINAMIDE	MICRO	90.00	70.00	-	110.00	90.00	MICRO/MJ	325.77	
PANTOTHENIC ACID	MILLI	0.35	0.28	-	0.42	0.35	MILLI/MJ	1.27	
VITAMIN B6	MICRO	41.00	18.00	-	63.00	41.00	MICRO/MJ	148.41	
BIOTIN	MICRO	3.50	2.00	-	5.00	3.50	MICRO/MJ	12.67	
FOLIC ACID	MICRO	5.30	3.40	-	8.10	5.30	MICRO/MJ	19.18	
VITAMIN B12	MICRO	0.38	0.27	-	0.68	0.38	MICRO/MJ	1.38	
VITAMIN C	MILLI	1.28	0.75	-	1.80	1.28	MILLI/MJ	4.63	

1 DIRECT VAPOR INJECTION: 1 - 4 SEC.; 130 - 150 GRAD C.

2 ESTIMATED BY THE DIFFERENCE METHOD
100 - (WATER + PROTEIN + FAT + MINERALS)

KUHMILCH
ABGEKOCHT

COW'S MILK
BOILED

LAIT DE VACHE
BOUILLI

		PROTEIN	FAT	CARBOHYDRATES	TOTAL
ENERGY VALUE (AVERAGE) PER 100 G EDIBLE PORTION	KJOULE	50	114	79	243
	(KCAL)	12	27	19	58
AMOUNT OF DIGESTIBLE CONSTITUENTS PER 100 G	GRAM	2.61	2.80	4.71	
ENERGY VALUE (AVERAGE) OF THE DIGESTIBLE FRACTION PER 100 G EDIBLE PORTION	KJOULE	48	108	79	236
	(KCAL)	12	26	19	56

WASTE PERCENTAGE AVERAGE 0.00

CONSTITUENTS	DIM	AV	VARIATION			AVR	NUTR.	DENS.	MOLPERC.
WATER	GRAM	88.90	-	-	-	88.90	GRAM/MJ	377.49	
PROTEIN	GRAM	2.70	-	-	-	2.70	GRAM/MJ	11.46	
FAT	GRAM	2.95	-	-	-	2.95	GRAM/MJ	12.53	
AVAILABLE CARBOHYDR.	GRAM	4.71 1	-	-	-	4.71	GRAM/MJ	20.00	
MINERALS	GRAM	0.74	0.67	-	0.81	0.74	GRAM/MJ	3.14	
SODIUM	MILLI	48.00	40.00	-	58.00	48.00	MILLI/MJ	203.82	
POTASSIUM	MILLI	157.00	144.00	-	178.00	157.00	MILLI/MJ	666.66	
MAGNESIUM	MILLI	12.00	9.00	-	16.00	12.00	MILLI/MJ	50.95	
CALCIUM	MILLI	120.00	107.00	-	133.00	120.00	MILLI/MJ	509.55	
PHOSPHORUS	MILLI	92.00	63.00	-	102.00	92.00	MILLI/MJ	390.65	
CHLORIDE	MILLI	102.00	90.00	-	106.00	102.00	MILLI/MJ	433.11	
IODIDE	MICRO	3.10	2.00	-	6.00	3.10	MICRO/MJ	13.16	
VITAMIN A	MICRO	30.00	27.00	-	34.00	30.00	MICRO/MJ	127.39	
CAROTENE	MICRO	18.00	10.00	-	20.00	18.00	MICRO/MJ	76.43	
VITAMIN D	NANO	20.00	20.00	-	60.00	20.00	NANO/MJ	84.92	
VITAMIN B1	MICRO	26.00	21.00	-	28.00	26.00	MICRO/MJ	110.40	
VITAMIN B2	MILLI	0.16	0.13	-	0.18	0.16	MILLI/MJ	0.68	
NICOTINAMIDE	MICRO	90.00	70.00	-	110.00	90.00	MICRO/MJ	382.16	
PANTOTHENIC ACID	MILLI	0.35	0.28	-	0.42	0.35	MILLI/MJ	1.49	
VITAMIN B6	MICRO	41.00	18.00	-	63.00	41.00	MICRO/MJ	174.09	
BIOTIN	MICRO	3.50	2.00	-	5.00	3.50	MICRO/MJ	14.86	
FOLIC ACID	MICRO	4.70	3.00	-	7.20	4.70	MICRO/MJ	19.96	
VITAMIN B12	MICRO	0.21	0.13	-	0.39	0.21	MICRO/MJ	0.89	
VITAMIN C	MILLI	0.85	0.50	-	1.20	0.85	MILLI/MJ	3.61	

1 ESTIMATED BY THE DIFFERENCE METHOD
100 - (WATER + PROTEIN + FAT + MINERALS)

KUHMILCH
MAGERMILCH
(ENTRAHMTE MILCH)

COW'S MILK
SKIMMED MILK

LAIT DE VACHE
LAIT MAIGRE
(LAIT ÉCRÉMÉ)

		PROTEIN	FAT	CARBOHYDRATES	TOTAL
ENERGY VALUE (AVERAGE)	KJOULE	64	2.7	84	151
PER 100 G	(KCAL)	15	0.6	20	36
EDIBLE PORTION					
AMOUNT OF DIGESTIBLE	GRAM	3.39	0.06	5.02	
CONSTITUENTS PER 100 G					
ENERGY VALUE (AVERAGE)	KJOULE	63	2.6	84	149
OF THE DIGESTIBLE	(KCAL)	15	0.6	20	36
FRACTION PER 100 G					
EDIBLE PORTION					

WASTE PERCENTAGE AVERAGE 0.00

CONSTITUENTS	DIM	AV	VARIATION			AVR	NUTR. DENS.		MOLPERC.
WATER	GRAM	90.90	90.60	-	91.20	90.90	GRAM/MJ	609.70	
PROTEIN	GRAM	3.50	2.50	-	5.50	3.50	GRAM/MJ	23.48	
FAT	MILLI	70.00	20.00	-	120.00	70.00	MILLI/MJ	469.52	
AVAILABLE CARBOHYDR.	GRAM	5.02	-	-	-	5.02	GRAM/MJ	33.67	
MINERALS	GRAM	0.75	-	-	-	0.75	GRAM/MJ	5.03	
SODIUM	MILLI	53.00	20.00	-	89.00	53.00	MILLI/MJ	355.49	
POTASSIUM	MILLI	150.00	95.00	-	195.00	150.00	GRAM/MJ	1.01	
MAGNESIUM	MILLI	14.00	-	-	-	14.00	MILLI/MJ	93.90	
CALCIUM	MILLI	123.00	100.00	-	180.00	123.00	MILLI/MJ	825.01	
MANGANESE	MICRO	-	0.45	-	6.75				
IRON	MILLI	0.12	0.06	-	0.25	0.12	MILLI/MJ	0.80	
COBALT	MICRO	-	0.00	-	0.23				
COPPER	MICRO	2.30	0.00	-	22.00	2.30	MICRO/MJ	15.43	
ZINC	MILLI	0.40	-	-	-	0.40	MILLI/MJ	2.68	
NICKEL	MICRO	1.00	-	-	-	1.00	MICRO/MJ	6.71	
CHROMIUM	MICRO	1.10	0.50	-	1.70	1.10	MICRO/MJ	7.38	
PHOSPHORUS	MILLI	97.00	80.00	-	120.00	97.00	MILLI/MJ	650.62	
CHLORIDE	MILLI	100.00	-	-	-	100.00	MILLI/MJ	670.74	
IODIDE	MICRO	3.40	2.10	-	6.20	3.40	MICRO/MJ	22.81	
SELENIUM	MICRO	4.75	4.50	-	5.00	4.75	MICRO/MJ	31.86	
BROMINE	MILLI	0.20	-	-	-	0.20	MILLI/MJ	1.34	
VITAMIN A	MICRO	2.40	1.80	-	3.00	2.40	MICRO/MJ	16.10	
VITAMIN D	-	0.00	-	-	-	0.00			
ALPHA-TOCOPHEROL	-	TRACES	-	-	-				
VITAMIN B1	MICRO	38.00	35.00	-	40.00	38.00	MICRO/MJ	254.88	
VITAMIN B2	MILLI	0.17	0.16	-	0.18	0.17	MILLI/MJ	1.14	

CONSTITUENTS	DIM	AV	VARIATION			AVR	NUTR. DENS.		MOLPERC.
NICOTINAMIDE	MICRO	95.00	90.00	-	100.00	95.00	MICRO/MJ	637.20	
PANTOTHENIC ACID	MILLI	0.28	-	-	-	0.28	MILLI/MJ	1.88	
VITAMIN B6	MICRO	50.00	36.00	-	54.00	50.00	MICRO/MJ	335.37	
BIOTIN	MICRO	1.50	-	-	-	1.50	MICRO/MJ	10.06	
FOLIC ACID	MICRO	5.00	-	-	-	5.00	MICRO/MJ	33.54	
VITAMIN B12	MICRO	0.30	-	-	-	0.30	MICRO/MJ	2.01	
VITAMIN C	MILLI	-	0.23	-	2.00				
ALANINE	GRAM	0.14	-	-	-	0.14	GRAM/MJ	0.94	5.1
ARGININE	GRAM	0.13	0.09	-	0.17	0.13	GRAM/MJ	0.87	2.5
ASPARTIC ACID	GRAM	0.29	-	-	-	0.29	GRAM/MJ	1.95	7.5
CYSTINE	MILLI	31.00	13.00	-	55.00	31.00	MILLI/MJ	207.93	0.4
GLUTAMIC ACID	GRAM	0.82	-	-	-	0.82	GRAM/MJ	5.50	18.2
GLYCINE	MILLI	80.00	-	-	-	80.00	MILLI/MJ	536.59	3.8
HISTIDINE	MILLI	92.00	42.00	-	170.00	92.00	MILLI/MJ	617.08	2.0
ISOLEUCINE	GRAM	0.22	0.15	-	0.29	0.22	GRAM/MJ	1.48	5.7
LEUCINE	GRAM	0.34	0.28	-	0.43	0.34	GRAM/MJ	2.28	9.5
LYSINE	GRAM	0.27	0.18	-	0.33	0.27	GRAM/MJ	1.81	6.3
METHIONINE	MILLI	86.00	56.00	-	130.00	86.00	MILLI/MJ	576.84	2.0
PHENYLALANINE	GRAM	0.17	0.13	-	0.23	0.17	GRAM/MJ	1.14	3.7
PROLINE	GRAM	0.38	-	-	-	0.38	GRAM/MJ	2.55	11.0
SERINE	GRAM	0.21	-	-	-	0.21	GRAM/MJ	1.41	6.6
THREONINE	GRAM	0.16	0.12	-	0.21	0.16	GRAM/MJ	1.07	4.5
TRYPTOPHAN	MILLI	49.00	34.00	-	64.00	49.00	MILLI/MJ	328.66	0.8
TYROSINE	GRAM	0.18	0.14	-	0.22	0.18	GRAM/MJ	1.21	3.3
VALINE	GRAM	0.24	0.15	-	0.30	0.24	GRAM/MJ	1.61	7.0
CITRIC ACID	GRAM	0.22	-	-	-	0.22	GRAM/MJ	1.48	
LACTOSE	GRAM	4.80	-	-	-	4.80	GRAM/MJ	32.20	
BUTYRIC ACID	MILLI	2.40	-	-	-	2.40	MILLI/MJ	16.10	
CAPROIC ACID	MILLI	1.50	-	-	-	1.50	MILLI/MJ	10.06	
CAPRYLIC ACID	MILLI	0.90	-	-	-	0.90	MILLI/MJ	6.04	
CAPRIC ACID	MILLI	1.70	-	-	-	1.70	MILLI/MJ	11.40	
LAURIC ACID	MILLI	2.20	-	-	-	2.20	MILLI/MJ	14.76	
MYRISTIC ACID	MILLI	7.10	-	-	-	7.10	MILLI/MJ	47.62	
PALMITIC ACID	MILLI	17.80	-	-	-	17.80	MILLI/MJ	119.39	
STEARIC ACID	MILLI	6.70	-	-	-	6.70	MILLI/MJ	44.94	
PALMITOLEIC ACID	MILLI	2.10	-	-	-	2.10	MILLI/MJ	14.09	
OLEIC ACID	MILLI	17.40	-	-	-	17.40	MILLI/MJ	116.71	
LINOLEIC ACID	MILLI	0.20	-	-	-	0.20	MILLI/MJ	1.34	
LINOLENIC ACID	MILLI	0.10	-	-	-	0.10	MILLI/MJ	0.67	
CHOLESTEROL	MILLI	3.00	-	-	-	3.00	MILLI/MJ	20.12	

SCHAFMILCH | EWE'S MILK (SHEEP MILK) | LAIT DE BREBIS

		PROTEIN	FAT	CARBOHYDRATES	TOTAL
ENERGY VALUE (AVERAGE)	KJOULE	97	242	78	417
PER 100 G	(KCAL)	23	58	19	100
EDIBLE PORTION					
AMOUNT OF DIGESTIBLE	GRAM	5.11	5.94	4.67	
CONSTITUENTS PER 100 G					
ENERGY VALUE (AVERAGE)	KJOULE	94	230	78	402
OF THE DIGESTIBLE	(KCAL)	22	55	19	96
FRACTION PER 100 G					
EDIBLE PORTION					

WASTE PERCENTAGE AVERAGE 0.00

CONSTITUENTS	DIM	AV	VARIATION			AVR	NUTR. DENS.		MOLPERC.
WATER	GRAM	82.70	81.50	-	84.10	82.70	GRAM/MJ	205.50	
PROTEIN	GRAM	5.27	4.95	-	11.60	5.27	GRAM/MJ	13.10	
FAT	GRAM	6.26	2.00	-	13.00	6.26	GRAM/MJ	15.56	
AVAILABLE CARBOHYDR.	GRAM	4.67	-	-	-	4.67	GRAM/MJ	11.60	
MINERALS	GRAM	0.86	0.80	-	0.90	0.86	GRAM/MJ	2.14	
SODIUM	MILLI	30.00	28.50	-	31.00	30.00	MILLI/MJ	74.55	
POTASSIUM	MILLI	182.00	174.00	-	190.00	182.00	MILLI/MJ	452.26	
MAGNESIUM	MILLI	11.50	7.50	-	18.90	11.50	MILLI/MJ	28.58	
CALCIUM	MILLI	183.00	136.00	-	200.00	183.00	MILLI/MJ	454.74	
MANGANESE	MICRO	13.00	-	-	-	13.00	MICRO/MJ	32.30	
IRON	MILLI	0.10	-	-	-	0.10	MILLI/MJ	0.25	
COPPER	MICRO	88.00	-	-	-	88.00	MICRO/MJ	218.67	
ZINC	MILLI	0.47	-	-	-	0.47	MILLI/MJ	1.17	
NICKEL	MICRO	23.00	-	-	-	23.00	MICRO/MJ	57.15	
PHOSPHORUS	MILLI	115.00	80.00	-	145.00	115.00	MILLI/MJ	285.77	
CHLORIDE	MILLI	76.00	71.00	-	92.00	76.00	MILLI/MJ	188.85	
VITAMIN A	MICRO	50.00	-	-	-	50.00	MICRO/MJ	124.25	
CAROTENE	MICRO	5.00	2.00	-	7.00	5.00	MICRO/MJ	12.42	
VITAMIN B1	MICRO	48.00	28.00	-	70.00	48.00	MICRO/MJ	119.28	
VITAMIN B2	MILLI	0.23	0.16	-	0.30	0.23	MILLI/MJ	0.57	
NICOTINAMIDE	MILLI	0.45	0.40	-	0.50	0.45	MILLI/MJ	1.12	
PANTOTHENIC ACID	MILLI	0.35	-	-	-	0.35	MILLI/MJ	0.87	
BIOTIN	MICRO	9.00	-	-	-	9.00	MICRO/MJ	22.36	
VITAMIN B12	MICRO	0.51	0.30	-	0.71	0.51	MICRO/MJ	1.27	
VITAMIN C	MILLI	4.25	3.00	-	6.00	4.25	MILLI/MJ	10.56	

CONSTITUENTS	DIM	AV	VARIATION			AVR	NUTR. DENS.		MOLPERC.
ALANINE	GRAM	0.22	-	-	-	0.22	GRAM/MJ	0.55	5.9
ARGININE	GRAM	0.18	-	-	-	0.18	GRAM/MJ	0.45	2.0
ASPARTIC ACID	GRAM	0.46	-	-	-	0.46	GRAM/MJ	1.14	8.2
CYSTINE	MILLI	60.00	-	-	-	60.00	MILLI/MJ	149.10	0.6
GLUTAMIC ACID	GRAM	1.07	-	-	-	1.07	GRAM/MJ	2.66	17.4
GLYCINE	GRAM	0.11	-	-	-	0.11	GRAM/MJ	0.27	3.5
HISTIDINE	GRAM	0.13	-	-	-	0.13	GRAM/MJ	0.32	1.4
ISOLEUCINE	GRAM	0.28	-	-	-	0.28	GRAM/MJ	0.70	5.1
LEUCINE	GRAM	0.53	-	-	-	0.53	GRAM/MJ	1.32	9.6
LYSINE	GRAM	0.46	-	-	-	0.46	GRAM/MJ	1.14	6.0
METHIONINE	GRAM	0.14	-	-	-	0.14	GRAM/MJ	0.35	2.2
PHENYLALANINE	GRAM	0.26	-	-	-	0.26	GRAM/MJ	0.65	3.8
PROLINE	GRAM	0.55	-	-	-	0.55	GRAM/MJ	1.37	11.4
SERINE	GRAM	0.32	-	-	-	0.32	GRAM/MJ	0.80	7.3
THREONINE	GRAM	0.24	-	-	-	0.24	GRAM/MJ	0.60	4.8
TRYPTOPHAN	MILLI	70.00	-	-	-	70.00	MILLI/MJ	173.95	0.8
TYROSINE	GRAM	0.26	-	-	-	0.26	GRAM/MJ	0.65	3.4
VALINE	GRAM	0.32	-	-	-	0.32	GRAM/MJ	0.80	6.5
CITRIC ACID	GRAM	0.12	0.07	-	0.17	0.12	GRAM/MJ	0.30	
LACTOSE	GRAM	4.55	4.25	-	5.20	4.55	GRAM/MJ	11.31	
BUTYRIC ACID	GRAM	0.18	-	-	-	0.18	GRAM/MJ	0.45	
CAPROIC ACID	GRAM	0.12	-	-	-	0.12	GRAM/MJ	0.30	
CAPRYLIC ACID	GRAM	0.12	-	-	-	0.12	GRAM/MJ	0.30	
CAPRIC ACID	GRAM	0.36	-	-	-	0.36	GRAM/MJ	0.89	
LAURIC ACID	GRAM	0.21	-	-	-	0.21	GRAM/MJ	0.52	
MYRISTIC ACID	GRAM	0.59	-	-	-	0.59	GRAM/MJ	1.47	
PALMITIC ACID	GRAM	1.44	-	-	-	1.44	GRAM/MJ	3.58	
STEARIC ACID	GRAM	0.80	-	-	-	0.80	GRAM/MJ	1.99	
PALMITOLEIC ACID	GRAM	0.12	-	-	-	0.12	GRAM/MJ	0.30	
OLEIC ACID	GRAM	1.39	-	-	-	1.39	GRAM/MJ	3.45	
LINOLEIC ACID	GRAM	0.16	-	-	-	0.16	GRAM/MJ	0.40	
LINOLENIC ACID	GRAM	0.12	-	-	-	0.12	GRAM/MJ	0.30	
TOTAL PHOSPHOLIPIDS	MILLI	46.00	-	-	-	46.00	MILLI/MJ	114.31	
PHOSPHATIDYLCHOLINE	MILLI	13.00	-	-	-	13.00	MILLI/MJ	32.30	
PHOSPHATIDYLETHANOLAMINE	MILLI	16.00	-	-	-	16.00	MILLI/MJ	39.76	
PHOSPHATIDYLSERINE	MILLI	2.00	-	-	-	2.00	MILLI/MJ	4.97	
PHOSPHATIDYLINOSITOL	MILLI	2.00	-	-	-	2.00	MILLI/MJ	4.97	
SPHINGOMYELIN	MILLI	13.00	-	-	-	13.00	MILLI/MJ	32.30	
CASEIN	GRAM	4.46	4.29	-	4.60	4.46	GRAM/MJ	11.08	

STUTENMILCH — MARE'S MILK (HORSE MILK) — LAIT DE JUMENT

		PROTEIN	FAT	CARBOHYDRATES	TOTAL
ENERGY VALUE (AVERAGE) PER 100 G EDIBLE PORTION	KJOULE	41	58	105	204
	(KCAL)	9.7	14	25	49
AMOUNT OF DIGESTIBLE CONSTITUENTS PER 100 G	GRAM	2.13	1.42	6.28	
ENERGY VALUE (AVERAGE) OF THE DIGESTIBLE FRACTION PER 100 G EDIBLE PORTION	KJOULE	39	55	105	200
	(KCAL)	9.4	13	25	48

WASTE PERCENTAGE AVERAGE 0.00

CONSTITUENTS	DIM	AV	VARIATION		AVR	NUTR. DENS.		MOLPERC.
WATER	GRAM	89.70	89.20	90.70	89.70	GRAM/MJ	448.87	
PROTEIN	GRAM	2.20	2.10	2.30	2.20	GRAM/MJ	11.01	
FAT	GRAM	1.50	1.30	2.00	1.50	GRAM/MJ	7.51	
AVAILABLE CARBOHYDR.	GRAM	6.28	-	-	6.28	GRAM/MJ	31.43	
MINERALS	GRAM	0.36	0.34	0.40	0.36	GRAM/MJ	1.80	
POTASSIUM	MILLI	64.00	-	-	64.00	MILLI/MJ	320.26	
MAGNESIUM	MILLI	9.00	-	-	9.00	MILLI/MJ	45.04	
CALCIUM	MILLI	110.00	90.00	120.00	110.00	MILLI/MJ	550.45	
IRON	MICRO	65.00	30.00	100.00	65.00	MICRO/MJ	325.27	
COPPER	MICRO	30.00	20.00	40.00	30.00	MICRO/MJ	150.12	
ZINC	MICRO	0.15	0.10	0.20	0.15	MICRO/MJ	0.75	
PHOSPHORUS	MILLI	54.00	45.00	63.00	54.00	MILLI/MJ	270.22	
VITAMIN A	MICRO	12.00	-	-	12.00	MICRO/MJ	60.05	
CAROTENE	MICRO	32.00	18.00	45.00	32.00	MICRO/MJ	160.13	
VITAMIN B1	MICRO	30.00	-	-	30.00	MICRO/MJ	150.12	
VITAMIN B2	MICRO	30.00	-	-	30.00	MICRO/MJ	150.12	
NICOTINAMIDE	MILLI	0.14	-	-	0.14	MILLI/MJ	0.70	
PANTOTHENIC ACID	MILLI	0.30	-	-	0.30	MILLI/MJ	1.50	
VITAMIN B6	MICRO	30.00	-	-	30.00	MICRO/MJ	150.12	
VITAMIN B12	MICRO	0.30	-	-	0.30	MICRO/MJ	1.50	
VITAMIN C	MILLI	15.00	-	-	15.00	MILLI/MJ	75.06	
CITRIC ACID	MILLI	80.00	7.00	170.00	80.00	MILLI/MJ	400.33	
LACTOSE	GRAM	6.20	-	-	6.20	GRAM/MJ	31.03	
CASEIN	GRAM	1.20	-	-	1.20	GRAM/MJ	6.00	

ZIEGENMILCH GOAT'S MILK LAIT DE CHÈVRE

		PROTEIN	FAT	CARBOHYDRATES	TOTAL
ENERGY VALUE (AVERAGE) PER 100 G EDIBLE PORTION	KJOULE	68	152	72	292
	(KCAL)	16	36	17	70
AMOUNT OF DIGESTIBLE CONSTITUENTS PER 100 G	GRAM	3.57	3.72	4.33	
ENERGY VALUE (AVERAGE) OF THE DIGESTIBLE FRACTION PER 100 G EDIBLE PORTION	KJOULE	66	144	72	282
	(KCAL)	16	34	17	68

WASTE PERCENTAGE AVERAGE 0.00

CONSTITUENTS	DIM	AV		VARIATION			AVR	NUTR. DENS.		MOLPERC.
WATER	GRAM	86.60		85.80	-	87.40	86.60	GRAM/MJ	306.56	
PROTEIN	GRAM	3.69		2.90	-	4.70	3.69	GRAM/MJ	13.06	
FAT	GRAM	3.92	1	3.40	-	5.10	3.92	GRAM/MJ	13.88	
AVAILABLE CARBOHYDR.	GRAM	4.33		-	-	-	4.33	GRAM/MJ	15.33	
MINERALS	GRAM	0.79		0.70	-	0.85	0.79	GRAM/MJ	2.80	
SODIUM	MILLI	42.00		34.00	-	50.00	42.00	MILLI/MJ	148.68	
POTASSIUM	MILLI	181.00		135.00	-	235.00	181.00	MILLI/MJ	640.74	
MAGNESIUM	MILLI	14.00		10.00	-	21.00	14.00	MILLI/MJ	49.56	
CALCIUM	MILLI	127.00		106.00	-	192.00	127.00	MILLI/MJ	449.58	
MANGANESE	MICRO	13.00		5.00	-	18.00	13.00	MICRO/MJ	46.02	
IRON	MICRO	50.00		42.00	-	75.00	50.00	MICRO/MJ	177.00	
COPPER	MICRO	18.00		13.00	-	75.00	18.00	MICRO/MJ	63.72	
ZINC	MILLI	0.26		0.18	-	0.73	0.26	MILLI/MJ	0.92	
NICKEL	MICRO	19.00		-	-	-	19.00	MICRO/MJ	67.26	
CHROMIUM	MICRO	13.00		10.00	-	15.00	13.00	MICRO/MJ	46.02	
MOLYBDENUM	MICRO	1.80		1.20	-	2.50	1.80	MICRO/MJ	6.37	
PHOSPHORUS	MILLI	109.00		92.00	-	148.00	109.00	MILLI/MJ	385.86	
CHLORIDE	MILLI	142.00		100.00	-	198.00	142.00	MILLI/MJ	502.68	
IODIDE	MICRO	4.10		2.10	-	11.00	4.10	MICRO/MJ	14.51	
BROMINE	MICRO	457.00		411.00	-	503.00	457.00	MILLI/MJ	1.62	
VITAMIN A	MICRO	68.00		-	-	-	68.00	MICRO/MJ	240.72	
CAROTENE	MICRO	35.00		-	-	-	35.00	MICRO/MJ	123.90	
VITAMIN D	MICRO	0.25		-	-	-	0.25	MICRO/MJ	0.88	
VITAMIN B1	MICRO	49.00		40.00	-	61.00	49.00	MICRO/MJ	173.46	
VITAMIN B2	MILLI	0.15		0.11	-	0.18	0.15	MILLI/MJ	0.53	
NICOTINAMIDE	MILLI	0.32		0.30	-	0.37	0.32	MILLI/MJ	1.13	

1 AMERICAN AND AFRICAN MOUNTAIN GOATS: PER 100G: 7,3 G FAT, 5,7 G LACTOSE.

CONSTITUENTS	DIM	AV	VARIATION			AVR	NUTR. DENS.		MOLPERC.
PANTOTHENIC ACID	MILLI	0.31	-	-	-	0.31	MILLI/MJ	1.10	
VITAMIN B6	MICRO	27.00	7.00	-	48.00	27.00	MICRO/MJ	95.58	
BIOTIN	MICRO	3.90	-	-	-	3.90	MICRO/MJ	13.81	
FOLIC ACID	MICRO	0.80	-	-	-	0.80	MICRO/MJ	2.83	
VITAMIN B12	NANO	70.00	-	-	-	70.00	NANO/MJ	247.80	
VITAMIN C	MILLI	2.00	1.00	-	3.00	2.00	MILLI/MJ	7.08	
ALANINE	GRAM	0.14	0.13	-	0.15	0.14	GRAM/MJ	0.50	4.6
ARGININE	GRAM	0.13	0.12	-	0.15	0.13	GRAM/MJ	0.46	2.2
ASPARTIC ACID	GRAM	0.30	0.26	-	0.33	0.30	GRAM/MJ	1.06	6.4
CYSTINE	MILLI	83.00	53.00	-	102.00	83.00	MILLI/MJ	293.82	1.0
GLUTAMIC ACID	GRAM	0.78	0.76	-	0.80	0.78	GRAM/MJ	2.76	15.3
GLYCINE	MILLI	74.00	68.00	-	84.00	74.00	MILLI/MJ	261.96	2.8
HISTIDINE	MILLI	79.00	56.00	-	102.00	79.00	MILLI/MJ	279.66	1.5
ISOLEUCINE	GRAM	0.23	0.20	-	0.25	0.23	GRAM/MJ	0.81	4.9
LEUCINE	GRAM	0.39	0.36	-	0.40	0.39	GRAM/MJ	1.38	8.6
LYSINE	GRAM	0.34	0.31	-	0.39	0.34	GRAM/MJ	1.20	6.7
METHIONINE	MILLI	94.00	83.00	-	105.00	94.00	MILLI/MJ	332.76	1.8
PHENYLALANINE	GRAM	0.18	0.15	-	0.24	0.18	GRAM/MJ	0.64	3.2
PROLINE	GRAM	0.72	0.36	-	0.47	0.72	GRAM/MJ	2.55	18.2
SERINE	GRAM	0.21	0.19	-	0.23	0.21	GRAM/MJ	0.74	5.8
THREONINE	GRAM	0.23	0.19	-	0.29	0.23	GRAM/MJ	0.81	5.5
TRYPTOPHAN	MILLI	50.00	48.00	-	52.00	50.00	MILLI/MJ	177.00	0.7
TYROSINE	GRAM	0.24	0.22	-	0.26	0.24	GRAM/MJ	0.85	3.8
VALINE	GRAM	0.28	0.27	-	0.30	0.28	GRAM/MJ	0.99	6.9
CITRIC ACID	MILLI	130.00	95.00	-	170.00	130.00	MILLI/MJ	460.20	
LACTOSE	GRAM	4.20	4.00	-	4.90	4.20	GRAM/MJ	14.87	
BUTYRIC ACID	GRAM	0.14	0.12	-	0.18	0.14	GRAM/MJ	0.50	
CAPROIC ACID	MILLI	80.00	60.00	-	110.00	80.00	MILLI/MJ	283.20	
CAPRYLIC ACID	MILLI	80.00	50.00	-	130.00	80.00	MILLI/MJ	283.20	
CAPRIC ACID	GRAM	0.29	0.24	-	0.41	0.29	GRAM/MJ	1.03	
LAURIC ACID	GRAM	0.12	0.09	-	0.19	0.12	GRAM/MJ	0.42	
MYRISTIC ACID	GRAM	0.38	0.31	-	0.41	0.38	GRAM/MJ	1.35	
C15 : 0 FATTY ACID	MILLI	40.00	30.00	-	50.00	40.00	MILLI/MJ	141.60	
PALMITIC ACID	GRAM	1.18	0.92	-	1.42	1.18	GRAM/MJ	4.18	
C17 : 0 FATTY ACID	MILLI	30.00	10.00	-	30.00	30.00	MILLI/MJ	106.20	
STEARIC ACID	GRAM	0.37	0.22	-	0.55	0.37	GRAM/MJ	1.31	
ARACHIDIC ACID	MILLI	10.00	4.00	-	15.00	10.00	MILLI/MJ	35.40	
PALMITOLEIC ACID	MILLI	40.00	20.00	-	60.00	40.00	MILLI/MJ	141.60	
OLEIC ACID	GRAM	0.71	0.56	-	1.04	0.71	GRAM/MJ	2.51	
C14 : 1 FATTY ACID	MILLI	2.00	1.00	-	7.00	2.00	MILLI/MJ	7.08	
LINOLEIC ACID	MILLI	90.00	70.00	-	150.00	90.00	MILLI/MJ	318.60	
LINOLENIC ACID	MILLI	20.00	10.00	-	40.00	20.00	MILLI/MJ	70.80	
CHOLESTEROL	MILLI	11.00	-	-	-	11.00	MILLI/MJ	38.94	
CASEIN	GRAM	2.90	2.85	-	3.00	2.90	GRAM/MJ	10.27	

Milchprodukte (ohne Käse)

DAIRY PRODUCTS (EXCEPT CHEESE)

PRODUITS LAITIERS (À L'EXCEPTION DU FROMAGE)

KONDENSMILCH
MIND. 7,5% FETT

CONDENSED MILK
MIN. 7.5% FAT CONTENT

LAIT CONDENSÉ
MIN. 7,5% MAT. GR.

		PROTEIN	FAT	CARBOHYDRATES	TOTAL
ENERGY VALUE (AVERAGE) PER 100 G EDIBLE PORTION	KJOULE	119	293	159	572
	(KCAL)	29	70	38	137
AMOUNT OF DIGESTIBLE CONSTITUENTS PER 100 G	GRAM	6.29	7.19	9.53	
ENERGY VALUE (AVERAGE) OF THE DIGESTIBLE FRACTION PER 100 G EDIBLE PORTION	KJOULE	116	278	159	554
	(KCAL)	28	67	38	132

WASTE PERCENTAGE AVERAGE 0.00

CONSTITUENTS	DIM	AV	VARIATION			AVR	NUTR.	DENS.	MOLPERC.
WATER	GRAM	74.70	74.30	-	75.00	74.70	GRAM/MJ	134.91	
PROTEIN	GRAM	6.49	6.44	-	6.57	6.49	GRAM/MJ	11.72	
FAT	GRAM	7.57	7.50	-	7.76	7.57	GRAM/MJ	13.67	
AVAILABLE CARBOHYDR.	GRAM	9.53	-	-	-	9.53	GRAM/MJ	17.21	
MINERALS	GRAM	1.52	1.44	-	1.59	1.52	GRAM/MJ	2.75	
SODIUM	MILLI	98.00	82.00	-	150.00	98.00	MILLI/MJ	176.99	
POTASSIUM	MILLI	322.00	295.00	-	357.00	322.00	MILLI/MJ	581.53	
MAGNESIUM	MILLI	27.00	18.00	-	33.00	27.00	MILLI/MJ	48.76	
CALCIUM	MILLI	242.00	219.00	-	262.00	242.00	MILLI/MJ	437.05	
MANGANESE	MICRO	5.10	2.70	-	8.20	5.10	MICRO/MJ	9.21	
IRON	MICRO	94.00	62.00	-	140.00	94.00	MICRO/MJ	169.76	
COBALT	MICRO	0.16	0.10	-	0.27	0.16	MICRO/MJ	0.29	
COPPER	MICRO	21.00	4.00	-	62.00	21.00	MICRO/MJ	37.93	
ZINC	MILLI	0.78	0.43	-	1.10	0.78	MILLI/MJ	1.41	
TIN	MILLI	-	0.70	-	10.50 1				
PHOSPHORUS	MILLI	189.00	150.00	-	209.00	189.00	MILLI/MJ	341.33	
CHLORIDE	MILLI	209.00	185.00	-	217.00	209.00	MILLI/MJ	377.45	
FLUORIDE	MICRO	35.00	23.00	-	43.00	35.00	MICRO/MJ	63.21	
IODIDE	MICRO	6.70	4.10	-	12.20	6.70	MICRO/MJ	12.10	
SELENIUM	MICRO	1.20	-	-	-	1.20	MICRO/MJ	2.17	

1 SN-CONTENT DEPENDENT ON THE TIME OF STORAGE.
4 MONTHS: 2MG/100G; 9 MONTHS: 9 MG/100G

CONSTITUENTS	DIM	AV	VARIATION			AVR	NUTR. DENS.		MOLPERC.
VITAMIN A	MICRO	48.00	46.00	-	54.00	48.00	MICRO/MJ	86.69	
CAROTENE	MICRO	34.00	28.00	-	38.00	34.00	MICRO/MJ	61.40	
VITAMIN D	MICRO	0.10	0.04	-	0.18	0.10	MICRO/MJ	0.18	
VITAMIN E ACTIVITY	MILLI	0.17	0.08	-	0.20	0.17	MILLI/MJ	0.31	
ALPHA-TOCOPHEROL	MILLI	0.17	0.08	-	0.20	0.17	MILLI/MJ	0.31	
VITAMIN B1	MICRO	67.00	50.00	-	90.00	67.00	MICRO/MJ	121.00	
VITAMIN B2	MILLI	0.37	0.32	-	0.44	0.37	MILLI/MJ	0.67	
NICOTINAMIDE	MILLI	0.20	0.18	-	0.24	0.20	MILLI/MJ	0.36	
PANTOTHENIC ACID	MILLI	0.64	0.58	-	0.70	0.64	MILLI/MJ	1.16	
VITAMIN B6	MICRO	59.00	30.00	-	90.00	59.00	MICRO/MJ	106.55	
BIOTIN	MICRO	6.30	3.00	-	9.10	6.30	MICRO/MJ	11.38	
FOLIC ACID	MICRO	6.00	-	-	-	6.00	MICRO/MJ	10.84	
VITAMIN B12	MICRO	0.41	0.30	-	0.63	0.41	MICRO/MJ	0.74	
VITAMIN C	MILLI	2.10	1.40	-	3.00	2.10	MILLI/MJ	3.79	
ALANINE	GRAM	0.27	-	-	-	0.27	GRAM/MJ	0.49	5.0
ARGININE	GRAM	0.23	0.21	-	0.25	0.23	GRAM/MJ	0.42	2.5
ASPARTIC ACID	GRAM	0.55	-	-	-	0.55	GRAM/MJ	0.99	7.5
CYSTINE	MILLI	66.00	28.00	-	115.00	66.00	MILLI/MJ	119.20	0.4
GLUTAMIC ACID	GRAM	1.49	-	-	-	1.50	GRAM/MJ	2.71	18.5
GLYCINE	GRAM	0.18	-	-	-	0.18	GRAM/MJ	0.33	3.5
HISTIDINE	GRAM	0.17	0.15	-	0.19	0.17	GRAM/MJ	0.31	2.1
ISOLEUCINE	GRAM	0.45	0.38	-	0.51	0.45	GRAM/MJ	0.81	5.6
LEUCINE	GRAM	0.72	0.68	-	0.75	0.72	GRAM/MJ	1.30	9.6
LYSINE	GRAM	0.52	0.47	-	0.55	0.52	GRAM/MJ	0.94	6.5
METHIONINE	GRAM	0.17	0.13	-	0.20	0.17	GRAM/MJ	0.31	2.1
PHENYLALANINE	GRAM	0.34	0.32	-	0.37	0.34	GRAM/MJ	0.61	3.7
PROLINE	GRAM	0.72	-	-	-	0.72	GRAM/MJ	1.30	10.2
SERINE	GRAM	0.41	-	-	-	0.41	GRAM/MJ	0.74	7.0
THREONINE	GRAM	0.33	0.31	-	0.35	0.33	GRAM/MJ	0.60	4.6
TRYPTOPHAN	MILLI	88.00	69.00	-	97.00	88.00	MILLI/MJ	158.93	0.8
TYROSINE	GRAM	0.32	0.29	-	0.37	0.32	GRAM/MJ	0.58	3.4
VALINE	GRAM	0.48	0.39	-	0.54	0.48	GRAM/MJ	0.87	7.0
CITRIC ACID	GRAM	0.34	0.31	-	0.40	0.34	GRAM/MJ	0.61	
LACTOSE	GRAM	9.19	-	-	-	9.19	GRAM/MJ	16.60	
BUTYRIC ACID	GRAM	0.20	-	-	-	0.20	GRAM/MJ	0.36	
CAPROIC ACID	GRAM	0.13	-	-	-	0.13	GRAM/MJ	0.23	
CAPRYLIC ACID	MILLI	50.00	-	-	-	50.00	MILLI/MJ	90.30	
CAPRIC ACID	GRAM	0.11	-	-	-	0.11	GRAM/MJ	0.20	
LAURIC ACID	GRAM	0.16	-	-	-	0.16	GRAM/MJ	0.29	
MYRISTIC ACID	GRAM	0.73	-	-	-	0.73	GRAM/MJ	1.32	
PALMITIC ACID	GRAM	2.03	-	-	-	2.03	GRAM/MJ	3.67	
STEARIC ACID	GRAM	0.92	-	-	-	0.92	GRAM/MJ	1.66	
PALMITOLEIC ACID	GRAM	0.16	-	-	-	0.16	GRAM/MJ	0.29	
OLEIC ACID	GRAM	2.10	-	-	-	2.10	GRAM/MJ	3.79	
LINOLEIC ACID	GRAM	0.19	0.16	-	0.24	0.19	GRAM/MJ	0.34	
LINOLENIC ACID	MILLI	50.00	42.00	-	64.00	50.00	MILLI/MJ	90.30	
CHOLESTEROL	MILLI	24.80	20.10	-	29.80	24.80	MILLI/MJ	44.79	
CASEIN	GRAM	5.18	4.78	-	5.85	5.18	GRAM/MJ	9.36	

KONDENSMILCH
MIND. 10% FETT

CONDENSED MILK
MIN. 10% FAT CONTENT

LAIT CONDENSÉ
MIN. 10% MAT. GR.

		PROTEIN	FAT	CARBOHYDRATES	TOTAL
ENERGY VALUE (AVERAGE) PER 100 G EDIBLE PORTION	KJOULE	161	391	209	761
	(KCAL)	39	93	50	182
AMOUNT OF DIGESTIBLE CONSTITUENTS PER 100 G	GRAM	8.49	9.59	12.50	
ENERGY VALUE (AVERAGE) OF THE DIGESTIBLE FRACTION PER 100 G EDIBLE PORTION	KJOULE	156	371	209	737
	(KCAL)	37	89	50	176

WASTE PERCENTAGE AVERAGE 0.00

CONSTITUENTS	DIM	AV	VARIATION		AVR	NUTR. DENS.		MOLPERC.
WATER	GRAM	66.70	66.30	- 67.00	66.70	GRAM/MJ	90.51	
PROTEIN	GRAM	8.76	8.48	- 8.84	8.76	GRAM/MJ	11.89	
FAT	GRAM	10.10	10.00	- 10.30	10.10	GRAM/MJ	13.70	
AVAILABLE CARBOHYDR.	GRAM	12.50	12.00	- 13.20	12.50	GRAM/MJ	16.96	
MINERALS	GRAM	1.98	1.90	- 2.15	1.98	GRAM/MJ	2.69	
SODIUM	MILLI	128.00	107.00	- 195.00	128.00	MILLI/MJ	173.68	
POTASSIUM	MILLI	420.00	384.00	- 465.00	420.00	MILLI/MJ	569.90	
MAGNESIUM	MILLI	35.00	23.00	- 43.00	35.00	MILLI/MJ	47.49	
CALCIUM	MILLI	315.00	285.00	- 341.00	315.00	MILLI/MJ	427.42	
MANGANESE	MICRO	6.60	3.50	- 10.70	6.60	MICRO/MJ	8.96	
IRON	MILLI	0.12	0.08	- 0.18	0.12	MILLI/MJ	0.16	
COBALT	MICRO	0.21	0.13	- 0.35	0.21	MICRO/MJ	0.28	
COPPER	MICRO	27.00	5.00	- 81.00	27.00	MICRO/MJ	36.64	
ZINC	MILLI	1.00	0.56	- 1.40	1.00	MILLI/MJ	1.36	
TIN	MILLI	-	0.70	- 10.50 1				
PHOSPHORUS	MILLI	246.00	195.00	- 272.00	246.00	MILLI/MJ	333.80	
CHLORIDE	MILLI	272.00	241.00	- 283.00	272.00	MILLI/MJ	369.08	
FLUORIDE	MICRO	46.00	30.00	- 56.00	46.00	MICRO/MJ	62.42	
IODIDE	MICRO	8.80	5.30	- 16.00	8.80	MICRO/MJ	11.94	
SELENIUM	MICRO	1.20	-	- -	1.20	MICRO/MJ	1.63	
VITAMIN A	MICRO	64.00	61.00	- 72.00	64.00	MICRO/MJ	86.84	
CAROTENE	MICRO	45.00	37.00	- 51.00	45.00	MICRO/MJ	61.06	
VITAMIN D	MICRO	0.13	0.01	- 0.24	0.13	MICRO/MJ	0.18	
VITAMIN E ACTIVITY	MILLI	0.23	0.11	- 0.27	0.23	MILLI/MJ	0.31	
ALPHA-TOCOPHEROL	MILLI	0.23	0.11	- 0.27	0.23	MILLI/MJ	0.31	
VITAMIN B1	MICRO	88.00	65.00	- 120.00	88.00	MICRO/MJ	119.41	

1 SN-CONTENT DEPENDENT ON THE TIME OF STORAGE.
4 MONTHS: 2MG/100G; 9 MONTHS: 9 MG/100G

CONSTITUENTS	DIM	AV	VARIATION			AVR	NUTR. DENS.		MOLPERC.
VITAMIN B2	MILLI	0.48	0.42	-	0.58	0.48	MILLI/MJ	0.65	
NICOTINAMIDE	MILLI	0.26	0.24	-	0.31	0.26	MILLI/MJ	0.35	
PANTOTHENIC ACID	MILLI	0.84	0.76	-	0.92	0.84	MILLI/MJ	1.14	
VITAMIN B6	MICRO	77.00	39.00	-	120.00	77.00	MICRO/MJ	104.48	
BIOTIN	MICRO	8.20	3.90	-	11.90	8.20	MICRO/MJ	11.13	
FOLIC ACID	MICRO	8.00	-	-	-	8.00	MICRO/MJ	10.86	
VITAMIN B12	MICRO	0.54	0.39	-	0.82	0.54	MICRO/MJ	0.73	
VITAMIN C	MILLI	2.70	1.80	-	3.90	2.70	MILLI/MJ	3.66	
ALANINE	GRAM	0.34	-	-	-	0.34	GRAM/MJ	0.46	5.0
ARGININE	GRAM	0.33	-	-	-	0.33	GRAM/MJ	0.45	2.5
ASPARTIC ACID	GRAM	0.76	-	-	-	0.76	GRAM/MJ	1.03	7.5
CYSTINE	MILLI	70.00	-	-	-	70.00	MILLI/MJ	94.98	0.4
GLUTAMIC ACID	GRAM	2.08	-	-	-	2.08	GRAM/MJ	2.82	18.5
GLYCINE	GRAM	0.20	-	-	-	0.20	GRAM/MJ	0.27	3.5
HISTIDINE	GRAM	0.25	-	-	-	0.25	GRAM/MJ	0.34	2.1
ISOLEUCINE	GRAM	0.56	-	-	-	0.56	GRAM/MJ	0.76	5.6
LEUCINE	GRAM	0.96	-	-	-	0.96	GRAM/MJ	1.30	9.6
LYSINE	GRAM	0.73	-	-	-	0.73	GRAM/MJ	0.99	6.5
METHIONINE	GRAM	0.24	-	-	-	0.24	GRAM/MJ	0.33	2.1
PHENYLALANINE	GRAM	0.47	-	-	-	0.47	GRAM/MJ	0.64	3.7
PROLINE	GRAM	0.90	-	-	-	0.90	GRAM/MJ	1.22	10.2
SERINE	GRAM	0.56	-	-	-	0.56	GRAM/MJ	0.76	7.0
THREONINE	GRAM	0.42	-	-	-	0.42	GRAM/MJ	0.57	4.6
TRYPTOPHAN	GRAM	0.13	-	-	-	0.13	GRAM/MJ	0.18	0.8
TYROSINE	GRAM	0.47	-	-	-	0.47	GRAM/MJ	0.64	3.4
VALINE	GRAM	0.63	-	-	-	0.63	GRAM/MJ	0.85	7.0
CITRIC ACID	GRAM	0.44	0.41	-	0.52	0.44	GRAM/MJ	0.60	
LACTOSE	GRAM	12.54	-	-	-	12.54	GRAM/MJ	17.02	
LINOLEIC ACID	GRAM	0.25	0.21	-	0.32	0.25	GRAM/MJ	0.34	
LINOLENIC ACID	MILLI	67.00	56.00	-	85.00	67.00	MILLI/MJ	90.91	
CHOLESTEROL	MILLI	33.10	26.80	-	39.80	33.10	MILLI/MJ	44.91	
CASEIN	GRAM	6.99	6.45	-	7.90	6.99	GRAM/MJ	9.49	

KONDENSMILCH
GEZUCKERT

CONDENSED MILK
SWEETENED

LAIT CONDENSÉ
SUCRÉ

		PROTEIN	FAT	CARBOHYDRATES	TOTAL
ENERGY VALUE (AVERAGE)	KJOULE	151	341	869	1360
PER 100 G	(KCAL)	36	81	208	325
EDIBLE PORTION					
AMOUNT OF DIGESTIBLE	GRAM	7.95	8.36	51.90	
CONSTITUENTS PER 100 G					
ENERGY VALUE (AVERAGE)	KJOULE	146	324	869	1339
OF THE DIGESTIBLE	(KCAL)	35	77	208	320
FRACTION PER 100 G					
EDIBLE PORTION					

WASTE PERCENTAGE AVERAGE 0.00

CONSTITUENTS	DIM	AV	VARIATION			AVR	NUTR. DENS.		MOLPERC.
WATER	GRAM	26.10	25.00	-	27.00	26.10	GRAM/MJ	19.50	
PROTEIN	GRAM	8.20	8.10	-	8.30	8.20	GRAM/MJ	6.13	
FAT	GRAM	8.80	8.40	-	9.00	8.80	GRAM/MJ	6.57	
AVAILABLE CARBOHYDR.	GRAM	51.90	-	-	-	51.90	GRAM/MJ	38.77	
MINERALS	GRAM	1.80	1.70	-	1.90	1.80	GRAM/MJ	1.34	
SODIUM	MILLI	88.00	84.00	-	92.00	88.00	MILLI/MJ	65.74	
POTASSIUM	MILLI	360.00	340.00	-	376.00	360.00	MILLI/MJ	268.94	
MAGNESIUM	MILLI	27.00	-	-	-	27.00	MILLI/MJ	20.17	
CALCIUM	MILLI	238.00	223.00	-	252.00	238.00	MILLI/MJ	177.80	
IRON	MILLI	0.25	0.20	-	0.30	0.25	MILLI/MJ	0.19	
COPPER	MICRO	40.00	-	-	-	40.00	MICRO/MJ	29.88	
ZINC	MILLI	1.00	-	-	-	1.00	MILLI/MJ	0.75	
PHOSPHORUS	MILLI	236.00	228.00	-	250.00	236.00	MILLI/MJ	176.31	
BORON	MILLI	0.11	-	-	-	0.12	MILLI/MJ	0.09	
VITAMIN A	MICRO	81.00	-	-	-	81.00	MICRO/MJ	60.51	
VITAMIN D	MICRO	0.13	-	-	-	0.13	MICRO/MJ	0.10	
VITAMIN E ACTIVITY	MILLI	0.20	0.10	-	0.42	0.20	MILLI/MJ	0.15	
ALPHA-TOCOPHEROL	MILLI	0.20	0.10	-	0.42	0.20	MILLI/MJ	0.15	
VITAMIN B1	MICRO	94.00	90.00	-	97.00	94.00	MICRO/MJ	70.22	
VITAMIN B2	MILLI	0.39	0.35	-	0.44	0.39	MILLI/MJ	0.29	
NICOTINAMIDE	MILLI	0.24	-	-	-	0.24	MILLI/MJ	0.18	
PANTOTHENIC ACID	MILLI	0.82	-	-	-	0.82	MILLI/MJ	0.61	
VITAMIN B6	MICRO	59.00	-	-	-	59.00	MICRO/MJ	44.08	
BIOTIN	MICRO	3.20	-	-	-	3.20	MICRO/MJ	2.39	
FOLIC ACID	MICRO	10.00	-	-	-	10.00	MICRO/MJ	7.47	
VITAMIN B12	MICRO	0.50	-	-	-	0.50	MICRO/MJ	0.37	
VITAMIN C	MILLI	3.80	-	-	-	3.80	MILLI/MJ	2.84	

CONSTITUENTS	DIM	AV	VARIATION			AVR	NUTR. DENS.		MOLPERC.
ALANINE	GRAM	0.30	0.28	-	0.31	0.30	GRAM/MJ	0.22	5.0
ARGININE	GRAM	0.32	0.30	-	0.33	0.32	GRAM/MJ	0.24	2.5
ASPARTIC ACID	GRAM	0.66	0.62	-	0.69	0.66	GRAM/MJ	0.49	7.5
CYSTINE	MILLI	76.00	31.00	-	130.00	76.00	MILLI/MJ	56.78	0.4
GLUTAMIC ACID	GRAM	1.80	1.72	-	1.87	1.80	GRAM/MJ	1.34	18.5
GLYCINE	GRAM	0.18	0.17	-	0.18	0.18	GRAM/MJ	0.13	3.5
HISTIDINE	GRAM	0.23	0.10	-	0.40	0.23	GRAM/MJ	0.17	2.1
ISOLEUCINE	GRAM	0.49	0.35	-	0.68	0.49	GRAM/MJ	0.37	5.6
LEUCINE	GRAM	0.81	0.66	-	1.00	0.81	GRAM/MJ	0.61	9.6
LYSINE	GRAM	0.65	0.42	-	0.78	0.65	GRAM/MJ	0.49	6.5
METHIONINE	GRAM	0.21	0.13	-	0.31	0.21	GRAM/MJ	0.16	2.1
PHENYLALANINE	GRAM	0.41	0.30	-	0.54	0.41	GRAM/MJ	0.31	3.7
PROLINE	GRAM	0.78	0.77	-	0.79	0.78	GRAM/MJ	0.58	10.2
SERINE	GRAM	0.47	0.45	-	0.48	0.47	GRAM/MJ	0.35	7.0
THREONINE	GRAM	0.38	0.29	-	0.50	0.38	GRAM/MJ	0.28	4.6
TRYPTOPHAN	GRAM	0.12	0.08	-	0.15	0.12	GRAM/MJ	0.09	0.8
TYROSINE	GRAM	0.39	0.33	-	0.51	0.39	GRAM/MJ	0.29	3.4
VALINE	GRAM	0.56	0.36	-	0.69	0.56	GRAM/MJ	0.42	7.0
SUCROSE	GRAM	41.70	41.00	-	42.70	41.70	GRAM/MJ	31.15	
LACTOSE	GRAM	10.20	-	-	-	10.20	GRAM/MJ	7.62	
BUTYRIC ACID	GRAM	0.28	-	-	-	0.28	GRAM/MJ	0.21	
CAPROIC ACID	GRAM	0.17	-	-	-	0.17	GRAM/MJ	0.13	
CAPRYLIC ACID	GRAM	0.10	-	-	-	0.10	GRAM/MJ	0.07	
CAPRIC ACID	MILLI	70.00	-	-	-	70.00	MILLI/MJ	52.29	
LAURIC ACID	GRAM	0.18	-	-	-	0.18	GRAM/MJ	0.13	
MYRISTIC ACID	GRAM	0.78	-	-	-	0.78	GRAM/MJ	0.58	
PALMITIC ACID	GRAM	2.40	-	-	-	2.40	GRAM/MJ	1.79	
STEARIC ACID	GRAM	1.21	-	-	-	1.21	GRAM/MJ	0.90	
PALMITOLEIC ACID	GRAM	0.14	-	-	-	0.14	GRAM/MJ	0.10	
OLEIC ACID	GRAM	2.41	1.62	-	2.80	2.41	GRAM/MJ	1.80	
LINOLEIC ACID	GRAM	0.18	0.08	-	0.31	0.18	GRAM/MJ	0.13	
LINOLENIC ACID	MILLI	59.00	49.00	-	75.00	59.00	MILLI/MJ	44.08	
CHOLESTEROL	MILLI	29.00	-	-	-	29.00	MILLI/MJ	21.66	

KONDENSMAGER-MILCH GEZUCKERT

CONDENSED SKIMMED MILK SWEETENED

LAIT CONDENSÉ MAIGRE SUCRÉ

		PROTEIN	FAT	CARBOHYDRATES	TOTAL
ENERGY VALUE (AVERAGE) PER 100 G EDIBLE PORTION	KJOULE	184	7.7	949	1141
	(KCAL)	44	1.9	227	273
AMOUNT OF DIGESTIBLE CONSTITUENTS PER 100 G	GRAM	9.70	0.19	56.70	
ENERGY VALUE (AVERAGE) OF THE DIGESTIBLE FRACTION PER 100 G EDIBLE PORTION	KJOULE	179	7.4	949	1135
	(KCAL)	43	1.8	227	271

WASTE PERCENTAGE AVERAGE 0.00

CONSTITUENTS	DIM	AV	VARIATION			AVR	NUTR. DENS.		MOLPERC.
WATER	GRAM	29.00	-	-	-	29.00	GRAM/MJ	25.55	
PROTEIN	GRAM	10.00	9.38	-	10.60	10.00	GRAM/MJ	8.81	
FAT	GRAM	0.20	0.10	-	0.30	0.20	GRAM/MJ	0.18	
AVAILABLE CARBOHYDR.	GRAM	56.70	-	-	-	56.70	GRAM/MJ	49.96	
MINERALS	GRAM	2.30	-	-	-	2.30	GRAM/MJ	2.03	
SODIUM	MILLI	180.00	-	-	-	180.00	MILLI/MJ	158.61	
POTASSIUM	MILLI	500.00	-	-	-	500.00	MILLI/MJ	440.58	
MAGNESIUM	MILLI	38.00	-	-	-	38.00	MILLI/MJ	33.48	
CALCIUM	MILLI	340.00	-	-	-	340.00	MILLI/MJ	299.60	
IRON	MILLI	0.29	-	-	-	0.29	MILLI/MJ	0.26	
COPPER	MICRO	30.00	-	-	-	30.00	MICRO/MJ	26.44	
ZINC	MILLI	1.20	-	-	-	1.20	MILLI/MJ	1.06	
PHOSPHORUS	MILLI	270.00	-	-	-	270.00	MILLI/MJ	237.92	
CHLORIDE	MILLI	310.00	-	-	-	310.00	MILLI/MJ	273.16	
VITAMIN A	MICRO	3.60	-	-	-	3.60	MICRO/MJ	3.17	
VITAMIN D	-	0.00	-	-	-	0.00			
VITAMIN B1	MILLI	0.12	-	-	-	0.12	MILLI/MJ	0.11	
VITAMIN B2	MILLI	0.41	-	-	-	0.41	MILLI/MJ	0.36	
NICOTINAMIDE	MILLI	0.29	-	-	-	0.29	MILLI/MJ	0.26	
PANTOTHENIC ACID	MILLI	1.00	-	-	-	1.00	MILLI/MJ	0.88	
VITAMIN B6	MICRO	73.00	-	-	-	73.00	MICRO/MJ	64.33	
BIOTIN	MICRO	3.80	-	-	-	3.80	MICRO/MJ	3.35	
FOLIC ACID	MICRO	10.00	-	-	-	10.00	MICRO/MJ	8.81	
VITAMIN B12	MICRO	0.60	-	-	-	0.60	MICRO/MJ	0.53	
VITAMIN C	MILLI	2.40	-	-	-	2.40	MILLI/MJ	2.11	

CONSTITUENTS	DIM	AV	VARIATION			AVR	NUTR. DENS.		MOLPERC.
ALANINE	GRAM	0.38	-	-	-	0.38	GRAM/MJ	0.33	4.9
ARGININE	GRAM	0.38	-	-	-	0.38	GRAM/MJ	0.33	2.1
ASPARTIC ACID	GRAM	0.86	-	-	-	0.86	GRAM/MJ	0.76	7.4
CYSTINE	MILLI	80.00	-	-	-	80.00	MILLI/MJ	70.49	0.4
GLUTAMIC ACID	GRAM	2.37	-	-	-	2.37	GRAM/MJ	2.09	18.6
GLYCINE	GRAM	0.23	-	-	-	0.23	GRAM/MJ	0.20	3.5
HISTIDINE	GRAM	0.29	-	-	-	0.29	GRAM/MJ	0.26	1.5
ISOLEUCINE	GRAM	0.65	-	-	-	0.65	GRAM/MJ	0.57	5.7
LEUCINE	GRAM	1.09	-	-	-	1.09	GRAM/MJ	0.96	9.6
LYSINE	GRAM	0.83	-	-	-	0.83	GRAM/MJ	0.73	5.2
METHIONINE	GRAM	0.27	-	-	-	0.27	GRAM/MJ	0.24	2.1
PHENYLALANINE	GRAM	0.53	-	-	-	0.53	GRAM/MJ	0.47	3.7
PROLINE	GRAM	1.03	-	-	-	1.03	GRAM/MJ	0.91	10.3
SERINE	GRAM	0.64	-	-	-	0.64	GRAM/MJ	0.56	7.0
THREONINE	GRAM	0.48	-	-	-	0.48	GRAM/MJ	0.42	4.6
TRYPTOPHAN	GRAM	0.15	-	-	-	0.15	GRAM/MJ	0.13	0.8
TYROSINE	GRAM	0.54	-	-	-	0.54	GRAM/MJ	0.48	3.4
VALINE	GRAM	0.92	-	-	-	0.92	GRAM/MJ	0.81	9.0
SUCROSE	GRAM	43.90	-	-	-	43.90	GRAM/MJ	38.68	
LACTOSE	GRAM	12.80	-	-	-	12.80	GRAM/MJ	11.28	
CHOLESTEROL	MILLI	0.70	-	-	-	0.70	MILLI/MJ	0.62	

TROCKENVOLLMILCH (VOLLMILCHPULVER) 1

DRIED MILK WHOLE MILK POWDER 1

LAIT ENTIER EN POUDRE 1

		PROTEIN	FAT	CARBOHYDRATES	TOTAL
ENERGY VALUE (AVERAGE)	KJOULE	464	1014	615	2092
PER 100 G EDIBLE PORTION	(KCAL)	111	242	147	500
AMOUNT OF DIGESTIBLE CONSTITUENTS PER 100 G	GRAM	24.44	24.89	36.72	
ENERGY VALUE (AVERAGE)	KJOULE	450	963	615	2028
OF THE DIGESTIBLE FRACTION PER 100 G EDIBLE PORTION	(KCAL)	108	230	147	485

WASTE PERCENTAGE AVERAGE 0.00

CONSTITUENTS	DIM	AV	VARIATION			AVR	NUTR. DENS.		MOLPERC.
WATER	GRAM	3.50	1.91	-	4.04	3.50	GRAM/MJ	1.73	
PROTEIN	GRAM	25.20	23.70	-	26.50	25.20	GRAM/MJ	12.43	
FAT	GRAM	26.20	24.60	-	26.80	26.20	GRAM/MJ	12.92	
AVAILABLE CARBOHYDR.	GRAM	36.72 2	-	-	-	36.72	GRAM/MJ	18.11	
MINERALS	GRAM	7.00	-	-	-	7.00	GRAM/MJ	3.45	

1 SPRAY-DRIED

2 CALCULATED FROM THE RAW PRODUCT

CONSTITUENTS	DIM	AV	VARIATION			AVR	NUTR. DENS.		MOLPERC.
SODIUM	MILLI	371.00	302.00	-	410.00	371.00	MILLI/MJ	182.95	
POTASSIUM	GRAM	1.16	1.10	-	1.20	1.16	GRAM/MJ	0.57	
MAGNESIUM	MILLI	110.00	-	-	-	110.00	MILLI/MJ	54.24	
CALCIUM	GRAM	0.92	0.80	-	1.00	0.92	GRAM/MJ	0.45	
MANGANESE	MICRO	70.00	60.00	-	80.00	70.00	MICRO/MJ	34.52	
IRON	MILLI	0.70	0.20	-	3.70	0.70	MILLI/MJ	0.35	
COPPER	MILLI	0.16	0.14	-	0.30	0.17	MILLI/MJ	0.08	
ZINC	MILLI	2.10	0.95	-	2.50	2.10	MILLI/MJ	1.04	
NICKEL	MICRO	15.00	-	-	-	15.00	MICRO/MJ	7.40	
CHROMIUM	MICRO	36.00	7.00	-	65.00	36.00	MICRO/MJ	17.75	
MOLYBDENUM	MICRO	50.00	-	-	-	50.00	MICRO/MJ	24.66	
VANADIUM	MICRO	41.00	2.00	-	80.00	41.00	MICRO/MJ	20.22	
ALUMINIUM	MILLI	0.13	-	-	-	0.14	MILLI/MJ	0.07	
PHOSPHORUS	MILLI	714.00	700.00	-	728.00	714.00	MILLI/MJ	352.10	
CHLORIDE	MILLI	810.00	-	-	-	810.00	MILLI/MJ	399.44	
FLUORIDE	MILLI	0.12	0.05	-	0.19	0.12	MILLI/MJ	0.06	
IODIDE	MICRO	26.60	16.10	-	48.30	26.60	MICRO/MJ	13.12	
BORON	MILLI	0.33	-	-	-	0.33	MILLI/MJ	0.16	
SELENIUM	MICRO	14.00	-	-	-	14.00	MICRO/MJ	6.90	
BROMINE	MILLI	1.50	0.50	-	2.40	1.50	MILLI/MJ	0.74	
SILICON	MILLI	1.00	0.50	-	1.00	1.00	MILLI/MJ	0.49	
VITAMIN A	MILLI	0.23	0.22	-	0.24	0.23	MILLI/MJ	0.11	
CAROTENE	MILLI	0.14	-	-	-	0.14	MILLI/MJ	0.07	
VITAMIN D	MICRO	0.46	-	-	-	0.46	MICRO/MJ	0.23	
VITAMIN E ACTIVITY	MILLI	0.75	0.50	-	1.00	0.75	MILLI/MJ	0.37	
ALPHA-TOCOPHEROL	MILLI	0.75	0.50	-	1.00	0.75	MILLI/MJ	0.37	
VITAMIN K	MILLI	-	0.01	-	0.10				
VITAMIN B1	MILLI	0.27	0.19	-	0.30	0.27	MILLI/MJ	0.13	
VITAMIN B2	MILLI	1.40	0.70	-	1.46	1.40	MILLI/MJ	0.69	
NICOTINAMIDE	MILLI	0.70	0.60	-	0.76	0.70	MILLI/MJ	0.35	
PANTOTHENIC ACID	MILLI	2.70	-	-	-	2.70	MILLI/MJ	1.33	
VITAMIN B6	MILLI	0.20	0.19	-	0.21	0.20	MILLI/MJ	0.10	
BIOTIN	MICRO	24.00	-	-	-	24.00	MICRO/MJ	11.84	
FOLIC ACID	MICRO	40.00	-	-	-	40.00	MICRO/MJ	19.73	
VITAMIN B12	MICRO	2.40	1.60	-	5.90	2.40	MICRO/MJ	1.18	
VITAMIN C	MILLI	11.00	7.00	-	17.00	11.00	MILLI/MJ	5.42	
ALANINE	GRAM	0.91	-	-	-	0.91	GRAM/MJ	0.45	5.0
ARGININE	GRAM	0.92	0.68	-	1.24	0.92	GRAM/MJ	0.45	2.6
ASPARTIC ACID	GRAM	1.99	-	-	-	1.99	GRAM/MJ	0.98	7.3
CYSTINE	GRAM	0.23	0.10	-	0.39	0.23	GRAM/MJ	0.11	0.5
GLUTAMIC ACID	GRAM	5.51	-	-	-	5.51	GRAM/MJ	2.72	18.2
GLYCINE	GRAM	0.56	-	-	-	0.56	GRAM/MJ	0.28	3.6
HISTIDINE	GRAM	0.66	0.31	-	1.22	0.66	GRAM/MJ	0.33	2.1
ISOLEUCINE	GRAM	1.61	1.09	-	2.09	1.61	GRAM/MJ	0.79	6.0
LEUCINE	GRAM	2.47	2.03	-	3.08	2.47	GRAM/MJ	1.22	9.2
LYSINE	GRAM	1.96	1.28	-	2.41	1.96	GRAM/MJ	0.97	6.5
METHIONINE	GRAM	0.62	0.40	-	0.97	0.62	GRAM/MJ	0.31	2.0
PHENYLALANINE	GRAM	1.22	0.93	-	1.67	1.22	GRAM/MJ	0.60	3.6
PROLINE	GRAM	2.55	-	-	-	2.55	GRAM/MJ	1.26	10.8
SERINE	GRAM	1.43	-	-	-	1.43	GRAM/MJ	0.71	6.6
THREONINE	GRAM	1.16	0.89	-	1.52	1.16	GRAM/MJ	0.57	4.7
TRYPTOPHAN	GRAM	0.35	0.24	-	0.46	0.35	GRAM/MJ	0.17	0.8
TYROSINE	GRAM	1.28	1.01	-	1.58	1.28	GRAM/MJ	0.63	3.4
VALINE	GRAM	1.73	1.11	-	2.13	1.73	GRAM/MJ	0.85	7.2

CONSTITUENTS	DIM	AV	VARIATION			AVR	NUTR. DENS.		MOLPERC.
CITRIC ACID	GRAM	1.62	-	-	-	1.62	GRAM/MJ	0.80	
LACTOSE	GRAM	35.10	-	-	-	35.10	GRAM/MJ	17.31	
BUTYRIC ACID	GRAM	0.87	-	-	-	0.87	GRAM/MJ	0.43	
CAPROIC ACID	GRAM	0.24	-	-	-	0.24	GRAM/MJ	0.12	
CAPRYLIC ACID	GRAM	0.27	-	-	-	0.27	GRAM/MJ	0.13	
CAPRIC ACID	GRAM	0.60	-	-	-	0.60	GRAM/MJ	0.30	
LAURIC ACID	GRAM	0.61	-	-	-	0.61	GRAM/MJ	0.30	
MYRISTIC ACID	GRAM	2.82	-	-	-	2.82	GRAM/MJ	1.39	
PALMITIC ACID	GRAM	7.52	-	-	-	7.52	GRAM/MJ	3.71	
STEARIC ACID	GRAM	2.85	-	-	-	2.85	GRAM/MJ	1.41	
PALMITOLEIC ACID	GRAM	1.20	-	-	-	1.20	GRAM/MJ	0.59	
OLEIC ACID	GRAM	7.18	4.84	-	8.36	7.18	GRAM/MJ	3.54	
LINOLEIC ACID	GRAM	0.55	0.22	-	0.91	0.55	GRAM/MJ	0.27	
LINOLENIC ACID	GRAM	0.17	0.15	-	0.22	0.17	GRAM/MJ	0.08	
CHOLESTEROL	MILLI	97.00	-	-	-	97.00	MILLI/MJ	47.83	

TROCKENMAGER-MILCH (MAGERMILCHPULVER)

DRIED SKIMMED MILK (SKIMMED MILK POWDER)

LAIT ÈCRÉMÉ EN POUDRE

		PROTEIN	FAT	CARBOHYDRATES	TOTAL
ENERGY VALUE (AVERAGE) PER 100 G EDIBLE PORTION	KJOULE	644	38	882	1564
	(KCAL)	154	9.0	211	374
AMOUNT OF DIGESTIBLE CONSTITUENTS PER 100 G	GRAM	33.95	0.92	52.71	
ENERGY VALUE (AVERAGE) OF THE DIGESTIBLE FRACTION PER 100 G EDIBLE PORTION	KJOULE	625	36	882	1543
	(KCAL)	149	8.5	211	369

WASTE PERCENTAGE AVERAGE 0.00

CONSTITUENTS	DIM	AV	VARIATION			AVR	NUTR. DENS.		MOLPERC.
WATER	GRAM	4.30	2.87	-	5.85	4.30	GRAM/MJ	2.79	
PROTEIN	GRAM	35.00	33.90	-	35.60	35.00	GRAM/MJ	22.69	
FAT	GRAM	0.97	0.50	-	1.50	0.97	GRAM/MJ	0.63	
AVAILABLE CARBOHYDR.	GRAM	52.71 1	-	-	-	52.71	GRAM/MJ	34.16	
MINERALS	GRAM	7.80	7.49	-	8.00	7.80	GRAM/MJ	5.06	

CONSTITUENTS	DIM	AV	VARIATION		AVR	NUTR. DENS.		MOLPERC.
SODIUM	MILLI	557.00	210.00	935.00	557.00	MILLI/MJ	361.03	
POTASSIUM	GRAM	1.58	1.00	2.05	1.58	GRAM/MJ	1.02	
MAGNESIUM	MILLI	110.00	-	-	110.00	MILLI/MJ	71.30	
CALCIUM	GRAM	1.29	0.96	1.89	1.29	GRAM/MJ	0.84	
MANGANESE	MICRO	-	5.00	700.00				
IRON	MILLI	0.80	0.60	1.00	0.80	MILLI/MJ	0.52	
COBALT	MICRO	-	0.00	2.40				
COPPER	MICRO	29.00	0.00	450.00	29.00	MICRO/MJ	18.80	
ZINC	MILLI	4.10	-	-	4.10	MILLI/MJ	2.66	
CHROMIUM	MICRO	13.00	-	-	13.00	MICRO/MJ	8.43	
MOLYBDENUM	MICRO	29.00	-	-	29.00	MICRO/MJ	18.80	
PHOSPHORUS	GRAM	1.02	0.90	1.89	1.02	GRAM/MJ	0.66	
IODIDE	MICRO	35.80	21.70	65.20	35.80	MICRO/MJ	23.20	
BORON	MILLI	0.43	-	-	0.44	MILLI/MJ	0.28	
SELENIUM	MICRO	5.00	-	-	5.00	MICRO/MJ	3.24	
VITAMIN A	MICRO	12.00	-	-	12.00	MICRO/MJ	7.78	
CAROTENE	-	0.00	-	-	0.00			
VITAMIN D	NANO	25.00	-	-	25.00	NANO/MJ	16.20	
VITAMIN E ACTIVITY	MICRO	40.00	-	-	40.00	MICRO/MJ	25.93	
ALPHA-TOCOPHEROL	MICRO	40.00	-	-	40.00	MICRO/MJ	25.93	
VITAMIN B1	MILLI	0.34	0.23	0.45	0.34	MILLI/MJ	0.22	
VITAMIN B2	MILLI	2.18	1.80	2.63	2.18	MILLI/MJ	1.41	
NICOTINAMIDE	MILLI	-	1.00	1.20				
PANTOTHENIC ACID	MILLI	3.45	3.39	3.50	3.45	MILLI/MJ	2.24	
VITAMIN B6	MILLI	0.28	-	-	0.28	MILLI/MJ	0.18	
BIOTIN	MICRO	14.00	-	-	14.00	MICRO/MJ	9.07	
FOLIC ACID	MICRO	21.00	-	-	21.00	MICRO/MJ	13.61	
VITAMIN B12	MICRO	2.20	-	-	2.20	MICRO/MJ	1.43	
VITAMIN C	MILLI	2.00	-	-	2.00	MILLI/MJ	1.30	
ALANINE	GRAM	1.25	-	-	1.25	GRAM/MJ	0.81	5.0
ARGININE	GRAM	1.28	0.94	1.72	1.28	GRAM/MJ	0.83	2.6
ASPARTIC ACID	GRAM	2.74	-	-	2.74	GRAM/MJ	1.78	7.3
CYSTINE	GRAM	0.31	0.13	0.55	0.31	GRAM/MJ	0.20	0.5
GLUTAMIC ACID	GRAM	7.57	-	-	7.57	GRAM/MJ	4.91	18.2
GLYCINE	GRAM	0.77	-	-	0.77	GRAM/MJ	0.50	3.6
HISTIDINE	GRAM	0.92	0.43	1.70	0.92	GRAM/MJ	0.60	2.1
ISOLEUCINE	GRAM	2.24	1.51	2.91	2.24	GRAM/MJ	1.45	6.0
LEUCINE	GRAM	3.43	2.82	4.27	3.43	GRAM/MJ	2.22	9.2
LYSINE	GRAM	2.72	1.78	3.34	2.72	GRAM/MJ	1.76	6.5
METHIONINE	GRAM	0.86	0.56	1.34	0.86	GRAM/MJ	0.56	2.0
PHENYLALANINE	GRAM	1.70	1.30	2.31	1.70	GRAM/MJ	1.10	3.6
PROLINE	GRAM	3.50	-	-	3.50	GRAM/MJ	2.27	10.8
SERINE	GRAM	1.97	-	-	1.97	GRAM/MJ	1.28	6.6
THREONINE	GRAM	1.61	1.23	2.11	1.61	GRAM/MJ	1.04	4.7
TRYPTOPHAN	GRAM	0.49	0.34	0.64	0.49	GRAM/MJ	0.32	0.8
TYROSINE	GRAM	1.78	1.41	2.19	1.78	GRAM/MJ	1.15	3.4
VALINE	GRAM	2.40	1.54	2.96	2.40	GRAM/MJ	1.56	7.2
CITRIC ACID	GRAM	2.21 [1]	-	-	2.21	GRAM/MJ	1.43	
LACTOSE	GRAM	50.50 [1]	-	-	50.50	GRAM/MJ	32.73	
CHOLESTEROL	MILLI	3.00	-	-	3.00	MILLI/MJ	1.94	

1 CALCULATED FROM THE RAW PRODUCT

SAHNE / CREAM / CRÈME

MIND. 10% FETT — MIN. 10% FAT CONTENT — MIN. 10% MAT. GR.

		PROTEIN	FAT	CARBOHYDRATES	TOTAL
ENERGY VALUE (AVERAGE)	KJOULE	57	406	68	531
PER 100 G EDIBLE PORTION	(KCAL)	14	97	16	127
AMOUNT OF DIGESTIBLE CONSTITUENTS PER 100 G	GRAM	3.00	9.97	4.05	
ENERGY VALUE (AVERAGE)	KJOULE	55	386	68	509
OF THE DIGESTIBLE FRACTION PER 100 G EDIBLE PORTION	(KCAL)	13	92	16	122

WASTE PERCENTAGE AVERAGE 0.00

CONSTITUENTS	DIM	AV	VARIATION			AVR	NUTR. DENS.		MOLPERC.
WATER	GRAM	81.70	81.60	-	81.90	81.70	GRAM/MJ	160.45	
PROTEIN	GRAM	3.10	3.06	-	3.13	3.10	GRAM/MJ	6.09	
FAT	GRAM	10.50	10.10	-	11.20	10.50	GRAM/MJ	20.62	
AVAILABLE CARBOHYDR.	GRAM	4.05	2.50	-	4.60	4.05	GRAM/MJ	7.95	
MINERALS	GRAM	0.61	0.60	-	0.62	0.61	GRAM/MJ	1.20	
SODIUM	MILLI	40.00	35.00	-	46.00	40.00	MILLI/MJ	78.56	
POTASSIUM	MILLI	132.00	129.00	-	136.00	132.00	MILLI/MJ	259.23	
MAGNESIUM	MILLI	11.20	7.90	-	13.00	11.20	MILLI/MJ	22.00	
CALCIUM	MILLI	101.00	87.00	-	109.00	101.00	MILLI/MJ	198.35	
MANGANESE	MICRO	2.60	1.70	-	3.40	2.60	MICRO/MJ	5.11	
IRON	MILLI	0.11	0.10	-	0.12	0.11	MILLI/MJ	0.22	
COBALT	NANO	61.00	0.00	-	200.00	61.00	NANO/MJ	119.80	
COPPER	MICRO	22.00	17.00	-	26.00	22.00	MICRO/MJ	43.21	
ZINC	MILLI	0.30	0.09	-	0.43	0.30	MILLI/MJ	0.59	
PHOSPHORUS	MILLI	85.00	-	-	-	85.00	MILLI/MJ	166.93	
CHLORIDE	MILLI	76.50	74.60	-	79.00	76.50	MILLI/MJ	150.24	
FLUORIDE	MICRO	17.00	5.00	-	28.00	17.00	MICRO/MJ	33.39	
IODIDE	MICRO	2.90	1.80	-	5.40	2.90	MICRO/MJ	5.70	
VITAMIN A	MICRO	66.00	51.00	-	80.00	66.00	MICRO/MJ	129.62	
CAROTENE	MICRO	50.00	40.00	-	60.00	50.00	MICRO/MJ	98.19	
VITAMIN D	MICRO	0.82	0.38	-	1.25	0.82	MICRO/MJ	1.61	
VITAMIN B1	MICRO	30.00	10.00	-	40.00	30.00	MICRO/MJ	58.92	
VITAMIN B2	MILLI	0.16	0.14	-	0.17	0.16	MILLI/MJ	0.31	
NICOTINAMIDE	MILLI	0.10	-	-	-	0.10	MILLI/MJ	0.20	
PANTOTHENIC ACID	MICRO	80.00	80.00	-	90.00	80.00	MICRO/MJ	157.11	
VITAMIN B6	MICRO	35.00	-	-	-	35.00	MICRO/MJ	68.74	

CONSTITUENTS	DIM	AV	VARIATION			AVR	NUTR. DENS.		MOLPERC.
BIOTIN	MICRO	3.40	-	-	-	3.40	MICRO/MJ	6.68	
FOLIC ACID	MICRO	2.00	-	-	-	2.00	MICRO/MJ	3.93	
VITAMIN B12	MICRO	0.40	-	-	-	0.40	MICRO/MJ	0.79	
VITAMIN C	MILLI	1.00	0.00	-	1.00	1.00	MILLI/MJ	1.96	
ALANINE	GRAM	0.12	-	-	-	0.12	GRAM/MJ	0.24	5.0
ARGININE	GRAM	0.11	0.08	-	0.15	0.11	GRAM/MJ	0.22	2.6
ASPARTIC ACID	GRAM	0.25	0.24	-	0.26	0.25	GRAM/MJ	0.49	7.6
CYSTINE	MILLI	28.00	12.00	-	49.00	28.00	MILLI/MJ	54.99	0.5
GLUTAMIC ACID	GRAM	0.68	0.65	-	0.71	0.68	GRAM/MJ	1.34	18.1
GLYCINE	MILLI	68.00	60.00	-	70.00	68.00	MILLI/MJ	133.55	3.6
HISTIDINE	MILLI	81.00	38.00	-	150.00	81.00	MILLI/MJ	159.08	2.2
ISOLEUCINE	GRAM	0.20	0.14	-	0.26	0.20	GRAM/MJ	0.39	5.7
LEUCINE	GRAM	0.31	0.25	-	0.38	0.31	GRAM/MJ	0.61	9.5
LYSINE	GRAM	0.24	0.16	-	0.30	0.24	GRAM/MJ	0.47	6.5
METHIONINE	MILLI	80.00	50.00	-	120.00	80.00	MILLI/MJ	157.11	2.2
PHENYLALANINE	GRAM	0.15	0.12	-	0.21	0.15	GRAM/MJ	0.29	3.8
PROLINE	GRAM	0.29	0.28	-	0.30	0.29	GRAM/MJ	0.57	10.2
SERINE	GRAM	0.17	0.16	-	0.18	0.17	GRAM/MJ	0.33	6.6
THREONINE	GRAM	0.14	0.11	-	0.19	0.14	GRAM/MJ	0.27	4.8
TRYPTOPHAN	MILLI	44.00	30.00	-	56.00	44.00	MILLI/MJ	86.41	0.8
TYROSINE	GRAM	0.15	0.13	-	0.20	0.15	GRAM/MJ	0.29	3.2
VALINE	GRAM	0.22	0.14	-	0.27	0.22	GRAM/MJ	0.43	7.3
LACTOSE	GRAM	4.05	-	-	-	4.05	GRAM/MJ	7.95	
BUTYRIC ACID	GRAM	0.37	-	-	-	0.37	GRAM/MJ	0.73	
CAPROIC ACID	GRAM	0.22	-	-	-	0.22	GRAM/MJ	0.43	
CAPRYLIC ACID	GRAM	0.13	-	-	-	0.13	GRAM/MJ	0.26	
CAPRIC ACID	GRAM	0.29	-	-	-	0.29	GRAM/MJ	0.57	
LAURIC ACID	GRAM	0.32	-	-	-	0.32	GRAM/MJ	0.63	
MYRISTIC ACID	GRAM	1.16	-	-	-	1.16	GRAM/MJ	2.28	
PALMITIC ACID	GRAM	3.02	-	-	-	3.02	GRAM/MJ	5.93	
STEARIC ACID	GRAM	1.39	-	-	-	1.39	GRAM/MJ	2.73	
PALMITOLEIC ACID	GRAM	0.26	-	-	-	0.26	GRAM/MJ	0.51	
OLEIC ACID	GRAM	2.89	-	-	-	2.89	GRAM/MJ	5.68	
LINOLEIC ACID	GRAM	0.26	-	-	-	0.26	GRAM/MJ	0.51	
LINOLENIC ACID	GRAM	0.17	-	-	-	0.17	GRAM/MJ	0.33	
CHOLESTEROL	MILLI	34.00	-	-	-	34.00	MILLI/MJ	66.77	

SAHNE
MIND. 30% FETT
(SCHLAGSAHNE, SCHLAG-RAHM)

CREAM
MIN. 30% FAT CONTENT
(WHIPPING CREAM)

CRÈME
MIN. 30% MAT. GR.
(CRÈME FOUETTÉE)

		PROTEIN	FAT	CARBOHYDRATES	TOTAL
ENERGY VALUE (AVERAGE)	KJOULE	43	1227	55	1325
PER 100 G	(KCAL)	10	293	13	317
EDIBLE PORTION					
AMOUNT OF DIGESTIBLE	GRAM	2.28	30.11	3.27	
CONSTITUENTS PER 100 G					
ENERGY VALUE (AVERAGE)	KJOULE	42	1166	55	1262
OF THE DIGESTIBLE	(KCAL)	10	279	13	302
FRACTION PER 100 G					
EDIBLE PORTION					

WASTE PERCENTAGE AVERAGE 0.00

CONSTITUENTS	DIM	AV	VARIATION			AVR	NUTR. DENS.		MOLPERC.
WATER	GRAM	62.00	59.70	-	64.10	62.00	GRAM/MJ	49.11	
PROTEIN	GRAM	2.36	2.18	-	2.60	2.36	GRAM/MJ	1.87	
FAT	GRAM	31.70	30.00	-	33.00	31.70	GRAM/MJ	25.11	
AVAILABLE CARBOHYDR.	GRAM	3.27	-	-	-	3.27	GRAM/MJ	2.59	
MINERALS	GRAM	0.50	0.48	-	0.53	0.50	GRAM/MJ	0.40	
SODIUM	MILLI	34.00	28.00	-	35.00	34.00	MILLI/MJ	26.93	
POTASSIUM	MILLI	112.00	99.00	-	148.00	112.00	MILLI/MJ	88.72	
MAGNESIUM	MILLI	10.00	8.00	-	12.00	10.00	MILLI/MJ	7.92	
CALCIUM	MILLI	80.00	75.00	-	85.00	80.00	MILLI/MJ	63.37	
MANGANESE	MICRO	1.70	0.90	-	2.70	1.70	MICRO/MJ	1.35	
IRON	MICRO	30.00	20.00	-	50.00	30.00	MICRO/MJ	23.76	
COBALT	NANO	50.00	30.00	-	90.00	50.00	NANO/MJ	39.61	
COPPER	MICRO	6.20	3.00	-	9.80	6.20	MICRO/MJ	4.91	
ZINC	MILLI	0.26	0.14	-	0.37	0.26	MILLI/MJ	0.21	
PHOSPHORUS	MILLI	63.00	43.00	-	69.00	63.00	MILLI/MJ	49.91	
CHLORIDE	MILLI	69.00	61.00	-	72.00	69.00	MILLI/MJ	54.66	
FLUORIDE	MICRO	12.00	7.00	-	14.00	12.00	MICRO/MJ	9.51	
IODIDE	MICRO	2.40	1.40	-	4.30	2.40	MICRO/MJ	1.90	
SELENIUM	MICRO	0.55	0.50	-	0.60	0.55	MICRO/MJ	0.44	
NITRATE	MILLI	0.19	0.00	-	0.99	0.19	MILLI/MJ	0.15	
VITAMIN A	MILLI	0.25	0.23	-	0.29	0.25	MILLI/MJ	0.20	
CAROTENE	MILLI	0.15	0.12	-	0.18	0.15	MILLI/MJ	0.12	
VITAMIN D	MICRO	1.10	0.73	-	2.00	1.10	MICRO/MJ	0.87	
VITAMIN E ACTIVITY	MILLI	0.77	0.55	-	1.00	0.77	MILLI/MJ	0.61	
ALPHA-TOCOPHEROL	MILLI	0.77	0.55	-	1.00	0.77	MILLI/MJ	0.61	
VITAMIN B1	MICRO	25.00	20.00	-	30.00	25.00	MICRO/MJ	19.80	

CONSTITUENTS	DIM	AV	VARIATION			AVR	NUTR. DENS.		MOLPERC.
VITAMIN B2	MILLI	0.15	0.12	-	0.17	0.15	MILLI/MJ	0.12	
NICOTINAMIDE	MICRO	80.00	70.00	-	100.00	80.00	MICRO/MJ	63.37	
PANTOTHENIC ACID	MILLI	0.30	0.26	-	0.34	0.30	MILLI/MJ	0.24	
VITAMIN B6	MICRO	36.00	35.00	-	40.00	36.00	MICRO/MJ	28.52	
BIOTIN	MICRO	3.40	1.90	-	4.80	3.40	MICRO/MJ	2.69	
FOLIC ACID	MICRO	4.00	-	-	-	4.00	MICRO/MJ	3.17	
VITAMIN B12	MICRO	0.40	0.29	-	0.72	0.40	MICRO/MJ	0.32	
VITAMIN C	MILLI	1.00	-	-	-	1.00	MILLI/MJ	0.79	
ALANINE	MILLI	86.00	82.00	-	89.00	86.00	MILLI/MJ	68.13	5.0
ARGININE	MILLI	86.00	70.00	-	100.00	86.00	MILLI/MJ	68.13	2.6
ASPARTIC ACID	GRAM	0.20	0.18	-	0.21	0.20	GRAM/MJ	0.16	7.6
CYSTINE	MILLI	21.00	14.00	-	22.00	21.00	MILLI/MJ	16.64	0.5
GLUTAMIC ACID	GRAM	0.51	0.49	-	0.53	0.51	GRAM/MJ	0.40	18.1
GLYCINE	MILLI	52.00	50.00	-	52.00	52.00	MILLI/MJ	41.19	3.6
HISTIDINE	MILLI	66.00	57.00	-	74.00	66.00	MILLI/MJ	52.28	2.2
ISOLEUCINE	GRAM	0.14	0.14	-	0.15	0.14	GRAM/MJ	0.11	5.7
LEUCINE	GRAM	0.24	0.23	-	0.26	0.24	GRAM/MJ	0.19	9.5
LYSINE	GRAM	0.18	0.17	-	0.20	0.18	GRAM/MJ	0.14	6.5
METHIONINE	MILLI	62.00	57.00	-	67.00	62.00	MILLI/MJ	49.11	2.2
PHENYLALANINE	GRAM	0.12	0.11	-	0.13	0.12	GRAM/MJ	0.10	3.8
PROLINE	GRAM	0.23	0.22	-	0.23	0.23	GRAM/MJ	0.18	10.2
SERINE	GRAM	0.13	0.13	-	0.14	0.13	GRAM/MJ	0.10	6.6
THREONINE	GRAM	0.11	0.10	-	0.12	0.11	GRAM/MJ	0.09	4.8
TRYPTOPHAN	MILLI	33.00	30.00	-	35.00	33.00	MILLI/MJ	26.14	0.8
TYROSINE	GRAM	0.11	0.10	-	0.13	0.11	GRAM/MJ	0.09	3.2
VALINE	GRAM	0.16	0.15	-	0.18	0.16	GRAM/MJ	0.13	7.3
HISTAMINE	MICRO	10.00	0.00	-	90.00	10.00	MICRO/MJ	7.92	
TYRAMINE	MILLI	0.17	0.10	-	0.75	0.17	MILLI/MJ	0.13	
LACTOSE	GRAM	3.27	2.70	-	3.46	3.27	GRAM/MJ	2.59	
BUTYRIC ACID	GRAM	1.08	-	-	-	1.08	GRAM/MJ	0.86	
CAPROIC ACID	GRAM	0.30	-	-	-	0.30	GRAM/MJ	0.24	
CAPRYLIC ACID	GRAM	0.31	-	-	-	0.31	GRAM/MJ	0.25	
CAPRIC ACID	GRAM	0.63	-	-	-	0.63	GRAM/MJ	0.50	
LAURIC ACID	GRAM	0.37	-	-	-	0.37	GRAM/MJ	0.29	
MYRISTIC ACID	GRAM	3.29	-	-	-	3.29	GRAM/MJ	2.61	
PALMITIC ACID	GRAM	8.84	-	-	-	8.84	GRAM/MJ	7.00	
STEARIC ACID	GRAM	3.37	-	-	-	3.37	GRAM/MJ	2.67	
PALMITOLEIC ACID	GRAM	1.61	-	-	-	1.61	GRAM/MJ	1.28	
OLEIC ACID	GRAM	7.66	-	-	-	7.66	GRAM/MJ	6.07	
LINOLEIC ACID	GRAM	0.81	0.65	-	1.00	0.81	GRAM/MJ	0.64	
LINOLENIC ACID	GRAM	0.21	0.18	-	0.27	0.21	GRAM/MJ	0.17	
CHOLESTEROL	MILLI	109.00	97.00	-	122.00	109.00	MILLI/MJ	86.34	
TOTAL PHOSPHOLIPIDS	MILLI	131.00	-	-	-	131.00	MILLI/MJ	103.77	

SAHNE, SAUER
SAUERRAHM

SOUR CREAM

CRÈME AIGRE

		PROTEIN	FAT	CARBOHYDRATES	TOTAL
ENERGY VALUE (AVERAGE)	KJOULE	52	697	70	818
PER 100 G	(KCAL)	12	167	17	196
EDIBLE PORTION					
AMOUNT OF DIGESTIBLE	GRAM	2.71	17.10	4.20	
CONSTITUENTS PER 100 G					
ENERGY VALUE (AVERAGE)	KJOULE	50	662	70	782
OF THE DIGESTIBLE	(KCAL)	12	158	17	187
FRACTION PER 100 G					
EDIBLE PORTION					

WASTE PERCENTAGE AVERAGE 0.00

CONSTITUENTS	DIM	AV	VARIATION			AVR	NUTR. DENS.		MOLPERC.
WATER	GRAM	74.50	-	-	-	74.50	GRAM/MJ	95.26	
PROTEIN	GRAM	2.80	-	-	-	2.80	GRAM/MJ	3.58	
FAT	GRAM	18.00	-	-	-	18.00	GRAM/MJ	23.02	
AVAILABLE CARBOHYDR.	GRAM	4.20 1	-	-	-	4.20	GRAM/MJ	5.37	
MINERALS	GRAM	0.50	-	-	-	0.50	GRAM/MJ	0.64	
SODIUM	MILLI	53.00	-	-	-	53.00	MILLI/MJ	67.77	
POTASSIUM	MILLI	144.00	-	-	-	144.00	MILLI/MJ	184.12	
MAGNESIUM	MILLI	11.00	-	-	-	11.00	MILLI/MJ	14.06	
CALCIUM	MILLI	100.00	-	-	-	100.00	MILLI/MJ	127.86	
IRON	MICRO	60.00	-	-	-	60.00	MICRO/MJ	76.72	
PHOSPHORUS	MILLI	80.00	-	-	-	80.00	MILLI/MJ	102.29	
IODIDE	MICRO	2.80	1.70	-	5.20	2.80	MICRO/MJ	3.58	
NITRATE	MILLI	0.17	0.44	-	1.33	0.17	MILLI/MJ	0.22	
VITAMIN B1	MICRO	35.00	-	-	-	35.00	MICRO/MJ	44.75	
VITAMIN B2	MILLI	0.15	-	-	-	0.15	MILLI/MJ	0.19	
NICOTINAMIDE	MICRO	67.00	-	-	-	67.00	MICRO/MJ	85.67	
PANTOTHENIC ACID	MILLI	0.34	-	-	-	0.34	MILLI/MJ	0.43	
VITAMIN B6	MICRO	17.00	-	-	-	17.00	MICRO/MJ	21.74	
BIOTIN	MICRO	3.00	-	-	-	3.00	MICRO/MJ	3.84	
FOLIC ACID	MICRO	7.00	3.00	-	10.00	7.00	MICRO/MJ	8.95	
VITAMIN B12	MICRO	0.30	-	-	-	0.30	MICRO/MJ	0.38	
VITAMIN C	MILLI	0.90	-	-	-	0.90	MILLI/MJ	1.15	
HISTAMINE	MICRO	10.00	0.00	-	50.00	10.00	MICRO/MJ	12.79	
TYRAMINE	MILLI	0.14	0.10	-	1.10	0.14	MILLI/MJ	0.18	

1 ESTIMATED BY THE DIFFERENCE METHOD
100 - (WATER + PROTEIN + FAT + MINERALS)

CONSTITUENTS	DIM	AV	VARIATION			AVR	NUTR. DENS.		MOLPERC.
LACTIC ACID	GRAM	0.73	0.60	-	0.84	0.73	GRAM/MJ	0.93	
BUTYRIC ACID	GRAM	0.60	-	-	-	0.60	GRAM/MJ	0.77	
CAPROIC ACID	GRAM	0.35	-	-	-	0.35	GRAM/MJ	0.45	
CAPRYLIC ACID	GRAM	0.21	-	-	-	0.21	GRAM/MJ	0.27	
CAPRIC ACID	GRAM	0.46	-	-	-	0.46	GRAM/MJ	0.59	
LAURIC ACID	GRAM	0.52	-	-	-	0.52	GRAM/MJ	0.66	
MYRISTIC ACID	GRAM	1.85	-	-	-	1.85	GRAM/MJ	2.37	
PALMITIC ACID	GRAM	4.84	-	-	-	4.84	GRAM/MJ	6.19	
STEARIC ACID	GRAM	2.23	-	-	-	2.23	GRAM/MJ	2.85	
PALMITOLEIC ACID	GRAM	0.42	-	-	-	0.42	GRAM/MJ	0.54	
OLEIC ACID	GRAM	4.63	-	-	-	4.63	GRAM/MJ	5.92	
LINOLEIC ACID	GRAM	0.43	-	-	-	0.43	GRAM/MJ	0.55	
LINOLENIC ACID	GRAM	0.27	-	-	-	0.27	GRAM/MJ	0.35	
CHOLESTEROL	MILLI	59.00	-	-	-	59.00	MILLI/MJ	75.44	

BUTTERMILCH BUTTERMILK LAIT DE BEURRE

		PROTEIN	FAT	CARBOHYDRATES	TOTAL
ENERGY VALUE (AVERAGE) PER 100 G EDIBLE PORTION	KJOULE	64	20	81	165
	(KCAL)	15	4.7	19	39
AMOUNT OF DIGESTIBLE CONSTITUENTS PER 100 G	GRAM	3.39	0.48	4.81	
ENERGY VALUE (AVERAGE) OF THE DIGESTIBLE FRACTION PER 100 G EDIBLE PORTION	KJOULE	63	19	81	162
	(KCAL)	15	4.5	19	39

WASTE PERCENTAGE AVERAGE 0.00

CONSTITUENTS	DIM	AV	VARIATION			AVR	NUTR. DENS.		MOLPERC.
WATER	GRAM	91.20	90.20	-	92.00	91.20	GRAM/MJ	563.83	
PROTEIN	GRAM	3.50	3.00	-	3.70	3.50	GRAM/MJ	21.64	
FAT	GRAM	0.51	0.28	-	0.74	0.51	GRAM/MJ	3.15	
AVAILABLE CARBOHYDR.	GRAM	4.81	-	-	-	4.81	GRAM/MJ	29.74	
MINERALS	GRAM	0.75	0.70	-	0.80	0.75	GRAM/MJ	4.64	
SODIUM	MILLI	57.00	50.00	-	64.00	57.00	MILLI/MJ	352.39	
POTASSIUM	MILLI	147.00	140.00	-	150.00	147.00	MILLI/MJ	908.80	
MAGNESIUM	MILLI	16.00	-	-	-	16.00	MILLI/MJ	98.92	
CALCIUM	MILLI	109.00	100.00	-	118.00	109.00	MILLI/MJ	673.87	
IRON	MILLI	0.10	-	-	-	0.10	MILLI/MJ	0.62	
COBALT	MICRO	0.10	-	-	-	0.10	MICRO/MJ	0.62	
PHOSPHORUS	MILLI	90.00	80.00	-	96.00	90.00	MILLI/MJ	556.41	
CHLORIDE	MILLI	100.00	-	-	-	100.00	MILLI/MJ	618.23	
FLUORIDE	MICRO	10.00	-	-	-	10.00	MICRO/MJ	61.82	
BORON	MICRO	27.00	-	-	-	27.00	MICRO/MJ	166.92	
NITRATE	MILLI	0.28	0.06	-	0.84	0.28	MILLI/MJ	1.73	
VITAMIN A	MICRO	7.50	-	-	-	7.50	MICRO/MJ	46.37	
CAROTENE	MICRO	9.00	5.00	-	12.00	9.00	MICRO/MJ	55.64	
ALPHA-TOCOPHEROL	-	TRACES	-	-	-				
VITAMIN B1	MICRO	34.00	30.00	-	40.00	34.00	MICRO/MJ	210.20	
VITAMIN B2	MILLI	0.16	0.15	-	0.20	0.16	MILLI/MJ	0.99	
NICOTINAMIDE	MILLI	0.10	0.09	-	0.10	0.10	MILLI/MJ	0.62	
PANTOTHENIC ACID	MILLI	0.30	-	-	-	0.30	MILLI/MJ	1.85	
VITAMIN B6	MICRO	40.00	-	-	-	40.00	MICRO/MJ	247.29	
BIOTIN	MICRO	1.50	-	-	-	1.50	MICRO/MJ	9.27	
FOLIC ACID	MICRO	5.00	-	-	-	5.00	MICRO/MJ	30.91	

CONSTITUENTS	DIM	AV	VARIATION			AVR	NUTR. DENS.		MOLPERC.
VITAMIN B12	MICRO	0.20	-	-	-	0.20	MICRO/MJ	1.24	
VITAMIN C	MILLI	0.60	0.20	-	1.00	0.60	MILLI/MJ	3.71	
ALANINE	GRAM	0.13	-	-	-	0.13	GRAM/MJ	0.80	5.3
ARGININE	GRAM	0.17	0.11	-	0.35	0.17	GRAM/MJ	1.05	2.9
ASPARTIC ACID	GRAM	0.28	-	-	-	0.28	GRAM/MJ	1.73	7.6
CYSTINE	MILLI	33.00	-	-	-	33.00	MILLI/MJ	204.02	0.5
GLUTAMIC ACID	GRAM	0.68	-	-	-	0.68	GRAM/MJ	4.20	16.7
GLYCINE	MILLI	77.00	-	-	-	77.00	MILLI/MJ	476.04	3.7
HISTIDINE	GRAM	0.10	0.08	-	0.12	0.10	GRAM/MJ	0.62	1.6
ISOLEUCINE	GRAM	0.22	0.16	-	0.26	0.22	GRAM/MJ	1.36	6.1
LEUCINE	GRAM	0.35	0.28	-	0.43	0.35	GRAM/MJ	2.16	9.7
LYSINE	GRAM	0.29	0.33	-	0.34	0.29	GRAM/MJ	1.79	5.8
METHIONINE	MILLI	84.00	67.00	-	100.00	84.00	MILLI/MJ	519.31	2.0
PHENYLALANINE	GRAM	0.19	0.14	-	0.19	0.19	GRAM/MJ	1.17	4.2
PROLINE	GRAM	0.35	-	-	-	0.35	GRAM/MJ	2.16	11.0
SERINE	GRAM	0.18	-	-	-	0.18	GRAM/MJ	1.11	6.2
THREONINE	GRAM	0.17	0.14	-	0.20	0.17	GRAM/MJ	1.05	5.2
TRYPTOPHAN	MILLI	38.00	18.00	-	53.00	38.00	MILLI/MJ	234.93	0.7
TYROSINE	GRAM	0.14	0.05	-	0.18	0.14	GRAM/MJ	0.87	2.8
VALINE	GRAM	0.26	0.24	-	0.28	0.26	GRAM/MJ	1.61	8.0
HISTAMINE	MICRO	10.00	0.00	-	70.00	10.00	MICRO/MJ	61.82	
TYRAMINE	MILLI	0.22	0.10	-	0.60	0.22	MILLI/MJ	1.36	
CITRIC ACID	MILLI	6.00	-	-	-	6.00	MILLI/MJ	37.09	
LACTIC ACID	GRAM	0.80	0.59	-	0.89	0.80	GRAM/MJ	4.97	
FORMIC ACID	MILLI	11.00	-	-	-	11.00	MILLI/MJ	68.01	
ACETIC ACID	MILLI	85.00	-	-	-	85.00	MILLI/MJ	525.50	
LACTOSE	GRAM	4.01	3.50	-	4.40	4.01	GRAM/MJ	24.79	
CHOLESTEROL	MILLI	4.00	-	-	-	4.00	MILLI/MJ	24.73	

BUTTERMILCH-PULVER
(TROCKENBUTTERMILCH)

BUTTERMILK POWDER

LAIT DE BEURRE EN POUDRE

		PROTEIN	FAT	CARBOHYDRATES	TOTAL
ENERGY VALUE (AVERAGE)	KJOULE	615	88	888	1591
PER 100 G EDIBLE PORTION	(KCAL)	147	21	212	380
AMOUNT OF DIGESTIBLE CONSTITUENTS PER 100 G	GRAM	32.39	2.16	53.04	
ENERGY VALUE (AVERAGE)	KJOULE	596	84	888	1568
OF THE DIGESTIBLE FRACTION PER 100 G EDIBLE PORTION	(KCAL)	143	20	212	375

WASTE PERCENTAGE AVERAGE 0.00

CONSTITUENTS	DIM	AV	VARIATION		AVR	NUTR.	DENS.	MOLPERC.
WATER	GRAM	3.10	-	-	3.10	GRAM/MJ	1.98	
PROTEIN	GRAM	33.40	33.00	33.80	33.40	GRAM/MJ	21.30	
FAT	GRAM	2.28	1.55	3.00	2.28	GRAM/MJ	1.45	
AVAILABLE CARBOHYDR.	GRAM	53.04 [1]	-	-	53.04	GRAM/MJ	33.83	
MINERALS	GRAM	6.50	-	-	6.50	GRAM/MJ	4.15	
SODIUM	MILLI	354.00	332.00	376.00	354.00	MILLI/MJ	225.77	
POTASSIUM	GRAM	1.30	1.24	1.36	1.30	GRAM/MJ	0.83	
MAGNESIUM	MILLI	110.00	-	-	110.00	MILLI/MJ	70.16	
CALCIUM	MILLI	894.00	863.00	924.00	894.00	MILLI/MJ	570.17	
IRON	MILLI	0.30	-	-	0.30	MILLI/MJ	0.19	
ZINC	MILLI	4.00	-	-	4.00	MILLI/MJ	2.55	
PHOSPHORUS	MILLI	915.00	900.00	930.00	915.00	MILLI/MJ	583.57	
IODIDE	MICRO	35.80	21.70	65.10	35.80	MICRO/MJ	22.83	
VITAMIN A	MICRO	66.00	-	-	66.00	MICRO/MJ	42.09	
VITAMIN B1	MILLI	0.35	-	-	0.35	MILLI/MJ	0.22	
VITAMIN B2	MILLI	1.50	-	-	1.50	MILLI/MJ	0.96	
NICOTINAMIDE	MILLI	1.00	-	-	1.00	MILLI/MJ	0.64	
PANTOTHENIC ACID	MILLI	3.20	-	-	3.20	MILLI/MJ	2.04	
VITAMIN B6	MILLI	0.34	-	-	0.34	MILLI/MJ	0.22	
FOLIC ACID	MICRO	47.00	-	-	47.00	MICRO/MJ	29.98	
VITAMIN B12	MICRO	4.00	-	-	4.00	MICRO/MJ	2.55	
VITAMIN C	MILLI	5.00	-	-	5.00	MILLI/MJ	3.19	
ALANINE	GRAM	1.18	-	-	1.18	GRAM/MJ	0.75	5.0
ARGININE	GRAM	1.17	0.90	1.60	1.17	GRAM/MJ	0.75	2.1
ASPARTIC ACID	GRAM	2.60	-	-	2.60	GRAM/MJ	1.66	7.4

1 CALCULATED FROM THE RAW PRODUCT

CONSTITUENTS	DIM	AV	VARIATION			AVR	NUTR. DENS.		MOLPERC.
CYSTINE	GRAM	0.30	0.13	-	0.52	0.30	GRAM/MJ	0.19	0.5
GLUTAMIC ACID	GRAM	7.18	-	-	-	7.18	GRAM/MJ	4.58	18.5
GLYCINE	GRAM	0.73	-	-	-	0.73	GRAM/MJ	0.47	3.7
HISTIDINE	GRAM	0.88	0.41	-	1.60	0.88	GRAM/MJ	0.56	1.5
ISOLEUCINE	GRAM	2.13	1.49	-	2.77	2.13	GRAM/MJ	1.36	6.2
LEUCINE	GRAM	3.31	2.67	-	4.05	3.31	GRAM/MJ	2.11	9.6
LYSINE	GRAM	2.56	1.71	-	3.20	2.56	GRAM/MJ	1.63	5.3
METHIONINE	GRAM	0.82	0.53	-	1.28	0.82	GRAM/MJ	0.52	2.1
PHENYLALANINE	GRAM	1.60	1.28	-	2.24	1.60	GRAM/MJ	1.02	3.7
PROLINE	GRAM	3.32	-	-	-	3.32	GRAM/MJ	2.12	10.9
SERINE	GRAM	1.87	-	-	-	1.87	GRAM/MJ	1.19	6.7
THREONINE	GRAM	1.49	1.17	-	2.03	1.49	GRAM/MJ	0.95	4.7
TRYPTOPHAN	GRAM	0.47	0.32	-	0.61	0.47	GRAM/MJ	0.30	0.9
TYROSINE	GRAM	1.71	1.39	-	2.13	1.71	GRAM/MJ	1.09	3.6
VALINE	GRAM	2.35	1.49	-	2.88	2.35	GRAM/MJ	1.50	7.6
CITRIC ACID	MILLI	70.00	-	-	-	70.00	MILLI/MJ	44.64	
LACTIC ACID	GRAM	8.81	-	-	-	8.81	GRAM/MJ	5.62	
LACTOSE	GRAM	44.16	-	-	-	44.16	GRAM/MJ	28.16	
CHOLESTEROL	MILLI	18.00	-	-	-	18.00	MILLI/MJ	11.48	

MOLKE
SÜSS

WHEY
SWEET

SÈRUM
(PETIT LAIT)
DOUX

		PROTEIN	FAT	CARBOHYDRATES	TOTAL
ENERGY VALUE (AVERAGE)	KJOULE	15	9.3	84	108
PER 100 G	(KCAL)	3.6	2.2	20	26
EDIBLE PORTION					
AMOUNT OF DIGESTIBLE	GRAM	0.79	0.22	4.99	
CONSTITUENTS PER 100 G					
ENERGY VALUE (AVERAGE)	KJOULE	15	8.8	84	107
OF THE DIGESTIBLE	(KCAL)	3.5	2.1	20	26
FRACTION PER 100 G					
EDIBLE PORTION					

WASTE PERCENTAGE AVERAGE 0.00

CONSTITUENTS	DIM	AV	VARIATION			AVR	NUTR. DENS.		MOLPERC.
WATER	GRAM	93.60	93.00	-	95.00	93.60	GRAM/MJ	874.93	
PROTEIN	GRAM	0.82	0.50	-	1.10	0.82	GRAM/MJ	7.67	
FAT	GRAM	0.24	0.06	-	0.80	0.24	GRAM/MJ	2.24	
AVAILABLE CARBOHYDR.	GRAM	4.99	-	-	-	4.99	GRAM/MJ	46.64	
MINERALS	GRAM	0.58	0.40	-	0.80	0.58	GRAM/MJ	5.42	

CONSTITUENTS	DIM	AV	VARIATION			AVR	NUTR. DENS.		MOLPERC.
SODIUM	MILLI	45.00	-	-	-	45.00	MILLI/MJ	420.64	
POTASSIUM	MILLI	129.00	110.00	-	147.00	129.00	GRAM/MJ	1.21	
MAGNESIUM	MILLI	1.00	-	-	-	1.00	MILLI/MJ	9.35	
CALCIUM	MILLI	67.90	51.00	-	99.00	67.90	MILLI/MJ	634.70	
IRON	MILLI	0.10	-	-	-	0.10	MILLI/MJ	0.93	
ZINC	MICRO	50.00	50.00	-	200.00	50.00	MICRO/MJ	467.38	
NICKEL	MICRO	4.00	-	-	-	4.00	MICRO/MJ	37.39	
MOLYBDENUM	MICRO	34.00	-	-	-	34.00	MICRO/MJ	317.82	
PHOSPHORUS	MILLI	43.00	33.00	-	53.00	43.00	MILLI/MJ	401.95	
CHLORIDE	MILLI	67.00	-	-	-	67.00	MILLI/MJ	626.29	
VITAMIN A	MICRO	3.00	-	-	-	3.00	MICRO/MJ	28.04	
VITAMIN B1	MICRO	37.00	35.00	-	40.00	37.00	MICRO/MJ	345.86	
VITAMIN B2	MILLI	0.15	0.10	-	0.20	0.15	MILLI/MJ	1.40	
NICOTINAMIDE	MILLI	0.19	0.10	-	0.25	0.19	MILLI/MJ	1.78	
PANTOTHENIC ACID	MILLI	0.34	0.21	-	0.41	0.34	MILLI/MJ	3.18	
VITAMIN B6	MICRO	42.00	21.00	-	77.00	42.00	MICRO/MJ	392.60	
BIOTIN	MICRO	1.40	1.30	-	1.50	1.40	MICRO/MJ	13.09	
FOLIC ACID	MICRO	1.00	-	-	-	1.00	MICRO/MJ	9.35	
VITAMIN B12	MICRO	0.20	0.15	-	0.24	0.20	MICRO/MJ	1.87	
VITAMIN C	MILLI	0.89	0.47	-	1.30	0.89	MILLI/MJ	8.32	
ALANINE	MILLI	42.00	-	-	-	42.00	MILLI/MJ	392.60	6.7
ARGININE	MILLI	27.00	-	-	-	27.00	MILLI/MJ	252.38	2.2
ASPARTIC ACID	GRAM	0.10	-	-	-	0.10	GRAM/MJ	0.93	10.9
CYSTINE	MILLI	12.00	-	-	-	12.00	MILLI/MJ	112.17	0.7
GLUTAMIC ACID	GRAM	0.16	-	-	-	0.16	GRAM/MJ	1.50	15.9
GLYCINE	MILLI	20.00	-	-	-	20.00	MILLI/MJ	186.95	3.8
HISTIDINE	MILLI	20.00	-	-	-	20.00	MILLI/MJ	186.95	1.8
ISOLEUCINE	MILLI	58.00	-	-	-	58.00	MILLI/MJ	542.16	6.3
LEUCINE	MILLI	96.00	-	-	-	96.00	MILLI/MJ	897.37	10.4
LYSINE	MILLI	79.00	-	-	-	79.00	MILLI/MJ	738.46	7.7
METHIONINE	MILLI	16.00	-	-	-	16.00	MILLI/MJ	149.56	1.5
PHENYLALANINE	MILLI	34.00	-	-	-	34.00	MILLI/MJ	317.82	2.9
PROLINE	MILLI	39.00	-	-	-	39.00	MILLI/MJ	364.56	3.6
SERINE	MILLI	44.00	-	-	-	44.00	MILLI/MJ	411.29	5.9
THREONINE	MILLI	70.00	-	-	-	70.00	MILLI/MJ	654.33	8.4
TRYPTOPHAN	MILLI	17.00	-	-	-	17.00	MILLI/MJ	158.91	1.2
TYROSINE	MILLI	32.00	-	-	-	32.00	MILLI/MJ	299.12	2.5
VALINE	MILLI	62.00	-	-	-	62.00	MILLI/MJ	579.55	7.6
CITRIC ACID	GRAM	0.16	0.14	-	0.17	0.16	GRAM/MJ	1.50	
LACTIC ACID	GRAM	0.13	0.05	-	0.20	0.13	GRAM/MJ	1.22	
LACTOSE	GRAM	4.70	4.40	-	5.20	4.70	GRAM/MJ	43.93	

TROCKENMOLKE (MOLKENPULVER)	**DRIED WHEY** (WHEYPOWDER)	**POUDRE DE SÈRUM** (POUDRE DE PETIT LAIT)

		PROTEIN	FAT	CARBOHYDRATES	TOTAL
ENERGY VALUE (AVERAGE) PER 100 G EDIBLE PORTION	KJOULE	221	46	1212	1480
	(KCAL)	53	11	290	354
AMOUNT OF DIGESTIBLE CONSTITUENTS PER 100 G	GRAM	11.64	1.14	72.43	
ENERGY VALUE (AVERAGE) OF THE DIGESTIBLE FRACTION PER 100 G EDIBLE PORTION	KJOULE	214	44	1212	1471
	(KCAL)	51	11	290	351

WASTE PERCENTAGE AVERAGE 0.00

CONSTITUENTS	DIM	AV	VARIATION			AVR	NUTR. DENS.		MOLPERC.
WATER	GRAM	7.10	6.20	-	8.00	7.10	GRAM/MJ	4.83	
PROTEIN	GRAM	12.00	9.90	-	13.50	12.00	GRAM/MJ	8.16	
FAT	GRAM	1.20	-	-	-	1.20	GRAM/MJ	0.82	
AVAILABLE CARBOHYDR.	GRAM	72.43 [1]	-	-	-	72.43	GRAM/MJ	49.25	
MINERALS	GRAM	8.20	7.70	-	8.70	8.20	GRAM/MJ	5.58	
SODIUM	GRAM	1.29	0.53	-	2.91	1.29	GRAM/MJ	0.88	
POTASSIUM	GRAM	1.86	1.46	-	2.52	1.86	GRAM/MJ	1.26	
MAGNESIUM	MILLI	180.00	130.00	-	260.00	180.00	MILLI/MJ	122.40	
CALCIUM	GRAM	0.89	0.68	-	1.10	0.89	GRAM/MJ	0.61	
IRON	MILLI	0.90	0.60	-	1.40	0.90	MILLI/MJ	0.61	
COPPER	MILLI	0.30	0.10	-	0.60	0.30	MILLI/MJ	0.20	
ZINC	MILLI	2.10	0.20	-	4.20	2.10	MILLI/MJ	1.43	
NICKEL	MICRO	80.00	-	-	-	80.00	MICRO/MJ	54.40	
PHOSPHORUS	MILLI	576.00	-	-	-	576.00	MILLI/MJ	391.68	
SELENIUM	MICRO	6.60	5.50	-	8.10	6.60	MICRO/MJ	4.49	
VITAMIN A	MICRO	15.00	-	-	-	15.00	MICRO/MJ	10.20	
VITAMIN E ACTIVITY	MICRO	60.00	14.00	-	250.00	60.00	MICRO/MJ	40.80	
ALPHA-TOCOPHEROL	MICRO	60.00	14.00	-	250.00	60.00	MICRO/MJ	40.80	
VITAMIN B1	MILLI	0.49	-	-	-	0.49	MILLI/MJ	0.33	
VITAMIN B2	MILLI	2.50	-	-	-	2.50	MILLI/MJ	1.70	
NICOTINAMIDE	MILLI	0.80	-	-	-	0.80	MILLI/MJ	0.54	
PANTOTHENIC ACID	MILLI	11.50	8.20	-	15.00	11.50	MILLI/MJ	7.82	
VITAMIN B6	MILLI	0.60	0.36	-	0.77	0.60	MILLI/MJ	0.41	
BIOTIN	MICRO	43.00	28.00	-	115.00	43.00	MICRO/MJ	29.24	
FOLIC ACID	MICRO	12.00	4.00	-	300.00	12.00	MICRO/MJ	8.16	
VITAMIN B12	MICRO	2.40	0.90	-	3.70	2.40	MICRO/MJ	1.63	

1 CALCULATED FROM THE RAW PRODUCT

CONSTITUENTS	DIM	AV	VARIATION			AVR	NUTR. DENS.		MOLPERC.
VITAMIN C	MILLI	1.41	0.00	-	9.10	1.41	MILLI/MJ	0.96	
ALANINE	GRAM	0.61	-	-	-	0.61	GRAM/MJ	0.41	6.7
ARGININE	GRAM	0.39	-	-	-	0.39	GRAM/MJ	0.27	2.2
ASPARTIC ACID	GRAM	1.49	-	-	-	1.49	GRAM/MJ	1.01	10.9
CYSTINE	GRAM	0.17	-	-	-	0.17	GRAM/MJ	0.12	0.7
GLUTAMIC ACID	GRAM	2.40	-	-	-	2.40	GRAM/MJ	1.63	15.9
GLYCINE	GRAM	0.29	-	-	-	0.29	GRAM/MJ	0.20	3.8
HISTIDINE	GRAM	0.29	-	-	-	0.29	GRAM/MJ	0.20	1.8
ISOLEUCINE	GRAM	0.85	-	-	-	0.85	GRAM/MJ	0.58	6.3
LEUCINE	GRAM	1.40	-	-	-	1.40	GRAM/MJ	0.95	10.4
LYSINE	GRAM	1.15	-	-	-	1.15	GRAM/MJ	0.78	7.7
METHIONINE	GRAM	0.23	-	-	-	0.23	GRAM/MJ	0.16	1.5
PHENYLALANINE	GRAM	0.49	-	-	-	0.49	GRAM/MJ	0.33	2.9
PROLINE	GRAM	0.43	-	-	-	0.43	GRAM/MJ	0.29	3.6
SERINE	GRAM	0.64	-	-	-	0.64	GRAM/MJ	0.44	5.9
THREONINE	GRAM	1.02	-	-	-	1.02	GRAM/MJ	0.69	8.4
TRYPTOPHAN	GRAM	0.25	-	-	-	0.25	GRAM/MJ	0.17	1.2
TYROSINE	GRAM	0.47	-	-	-	0.47	GRAM/MJ	0.32	2.5
VALINE	GRAM	0.91	-	-	-	0.91	GRAM/MJ	0.62	7.6
CITRIC ACID	GRAM	2.32	-	-	-	2.32	GRAM/MJ	1.58	
LACTIC ACID	GRAM	1.89	-	-	-	1.89	GRAM/MJ	1.29	
LACTOSE	GRAM	68.22	-	-	-	68.22	GRAM/MJ	46.39	

JOGHURT
MIND. 3,5% FETT

YOGHURT
MIN. 3.5% FAT CONTENT

YAOURT
MIN. 3,5% MAT. GR.

		PROTEIN	FAT	CARBOHYDRATES	TOTAL
ENERGY VALUE (AVERAGE) PER 100 G EDIBLE PORTION	KJOULE	71	145	91	307
	(KCAL)	17	35	22	73
AMOUNT OF DIGESTIBLE CONSTITUENTS PER 100 G	GRAM	3.76	3.56	5.42	
ENERGY VALUE (AVERAGE) OF THE DIGESTIBLE FRACTION PER 100 G EDIBLE PORTION	KJOULE	69	138	91	298
	(KCAL)	17	33	22	71

WASTE PERCENTAGE AVERAGE 0.00

CONSTITUENTS	DIM	AV	VARIATION			AVR	NUTR. DENS.		MOLPERC.
WATER	GRAM	87.00	86.30	-	88.10	87.00	GRAM/MJ	292.07	
PROTEIN	GRAM	3.88 [1]	3.30	-	4.20	3.88	GRAM/MJ	13.03	
FAT	GRAM	3.75	3.50	-	4.12	3.75	GRAM/MJ	12.59	
AVAILABLE CARBOHYDR.	GRAM	5.42	-	-	-	5.42	GRAM/MJ	18.20	
MINERALS	GRAM	0.74	0.67	-	0.81	0.74	GRAM/MJ	2.48	

1 VALID FOR UNCONCENTRATED PRODUCTS; CONCENTRATION IN VACCUO OR THE ADDITION OF SKIM MILK POWDER BEFORE THE FERMENTATION ENHANCES THE AMOUNT OF PROTEINS TO 4,6 G/100G

CONSTITUENTS	DIM	AV	VARIATION	AVR	NUTR. DENS.		MOLPERC.
SODIUM	MILLI	48.00	40.00 - 58.00	48.00	MILLI/MJ	161.14	
POTASSIUM	MILLI	157.00	144.00 - 178.00	157.00	MILLI/MJ	527.07	
MAGNESIUM	MILLI	12.00	9.00 - 16.00	12.00	MILLI/MJ	40.29	
CALCIUM	MILLI	120.00	107.00 - 133.00	120.00	MILLI/MJ	402.86	
MANGANESE	MICRO	2.50	1.30 - 4.00	2.50	MICRO/MJ	8.39	
IRON	MICRO	46.00	30.00 - 340.00	46.00	MICRO/MJ	154.43	
COBALT	NANO	80.00	50.00 - 130.00	80.00	NANO/MJ	268.57	
COPPER	MICRO	10.00	0.00 - 55.00	10.00	MICRO/MJ	33.57	
ZINC	MILLI	0.38	0.21 - 0.82	0.38	MILLI/MJ	1.28	
NICKEL	MICRO	1.00	- - -	1.00	MICRO/MJ	3.36	
CHROMIUM	MICRO	4.00	0.50 - 7.40	4.00	MICRO/MJ	13.43	
MOLYBDENUM	MICRO	-	0.00 - 5.00				
PHOSPHORUS	MILLI	92.00	63.00 - 102.00	92.00	MILLI/MJ	308.86	
CHLORIDE	MILLI	102.00	96.00 - 106.00	102.00	MILLI/MJ	342.43	
FLUORIDE	MICRO	17.00	11.00 - 21.00	17.00	MICRO/MJ	57.07	
IODIDE	MICRO	3.50	2.10 - 6.40	3.50	MICRO/MJ	11.75	
BORON	MICRO	60.00	35.00 - 75.00	60.00	MICRO/MJ	201.43	
SELENIUM	MICRO	0.30	- - -	0.30	MICRO/MJ	1.01	
NITRATE	MILLI	0.24	0.02 - 0.84	0.24	MILLI/MJ	0.81	
VITAMIN A	MICRO	29.00	26.00 - 33.00	29.00	MICRO/MJ	97.36	
CAROTENE	MICRO	18.00	14.00 - 22.00	18.00	MICRO/MJ	60.43	
VITAMIN D	NANO	62.00	20.00 - 88.00	62.00	NANO/MJ	208.14	
VITAMIN E ACTIVITY	MICRO	87.00	41.00 - 98.00	87.00	MICRO/MJ	292.07	
ALPHA-TOCOPHEROL	MICRO	87.00	41.00 - 98.00	87.00	MICRO/MJ	292.07	
VITAMIN B1	MICRO	37.00	30.00 - 55.00	37.00	MICRO/MJ	124.21	
VITAMIN B2	MILLI	0.18	0.14 - 0.22	0.18	MILLI/MJ	0.60	
NICOTINAMIDE	MICRO	90.00	70.00 - 110.00	90.00	MICRO/MJ	302.14	
PANTOTHENIC ACID	MILLI	0.35	0.28 - 0.42	0.35	MILLI/MJ	1.18	
VITAMIN B6	MICRO	46.00	22.00 - 70.00	46.00	MICRO/MJ	154.43	
BIOTIN	MICRO	3.50	2.00 - 5.00	3.50	MICRO/MJ	11.75	
FOLIC ACID	MICRO	10.00	- - -	10.00	MICRO/MJ	33.57	
VITAMIN B12	NANO	90.00	80.00 - 100.00	90.00	NANO/MJ	302.14	
VITAMIN C	MILLI	1.00	0.50 - 1.50	1.00	MILLI/MJ	3.36	
ALANINE	GRAM	0.17	- - -	0.17	GRAM/MJ	0.57	5.8
ARGININE	GRAM	0.14	0.12 - 0.16	0.14	GRAM/MJ	0.47	2.4
ASPARTIC ACID	GRAM	0.31	- - -	0.31	GRAM/MJ	1.04	7.1
CYSTINE	MILLI	30.00	23.00 - 35.00	30.00	MILLI/MJ	100.71	0.4
GLUTAMIC ACID	GRAM	0.76	- - -	0.76	GRAM/MJ	2.55	15.7
GLYCINE	MILLI	94.00	- - -	94.00	MILLI/MJ	315.57	3.8
HISTIDINE	GRAM	0.10	0.09 - 0.12	0.10	GRAM/MJ	0.34	2.0
ISOLEUCINE	GRAM	0.24	0.21 - 0.29	0.24	GRAM/MJ	0.81	5.6
LEUCINE	GRAM	0.41	0.38 - 0.42	0.41	GRAM/MJ	1.38	9.5
LYSINE	GRAM	0.31	0.28 - 0.35	0.31	GRAM/MJ	1.04	6.5
METHIONINE	GRAM	0.10	0.09 - 0.11	0.10	GRAM/MJ	0.34	2.0
PHENYLALANINE	GRAM	0.21	- - -	0.21	GRAM/MJ	0.71	3.9
PROLINE	GRAM	0.46	- - -	0.46	GRAM/MJ	1.54	12.2
SERINE	GRAM	0.24	- - -	0.24	GRAM/MJ	0.81	6.9
THREONINE	GRAM	0.17	0.16 - 0.20	0.17	GRAM/MJ	0.57	4.3
TRYPTOPHAN	MILLI	45.00	22.00 - 58.00	45.00	MILLI/MJ	151.07	0.7
TYROSINE	GRAM	0.20	0.19 - 0.21	0.20	GRAM/MJ	0.67	3.4
VALINE	GRAM	0.30	0.26 - 0.32	0.30	GRAM/MJ	1.01	7.8
HISTAMINE	MICRO	20.00	0.00 - 210.00	20.00	MICRO/MJ	67.14	
TYRAMINE	MILLI	0.13	0.10 - 0.90	0.13	MILLI/MJ	0.44	
LACTIC ACID	GRAM	1.05	0.79 - 1.46	1.05	GRAM/MJ	3.54	

CONSTITUENTS	DIM	AV	VARIATION			AVR	NUTR. DENS.		MOLPERC.
GLUCOSE	MILLI	30.00	30.00	-	30.00	30.00	MILLI/MJ	100.71	
LACTOSE	GRAM	3.19	3.19	-	3.19	3.19	GRAM/MJ	10.71	
GALACTOSE	GRAM	1.15	1.15	-	1.15	1.15	GRAM/MJ	3.86	
BUTYRIC ACID	GRAM	0.13	-	-	-	0.13	GRAM/MJ	0.44	
CAPROIC ACID	MILLI	80.00	-	-	-	80.00	MILLI/MJ	268.57	
CAPRYLIC ACID	MILLI	50.00	-	-	-	50.00	MILLI/MJ	167.86	
CAPRIC ACID	GRAM	0.10	-	-	-	0.10	GRAM/MJ	0.34	
LAURIC ACID	GRAM	0.12	-	-	-	0.12	GRAM/MJ	0.40	
MYRISTIC ACID	GRAM	0.38	-	-	-	0.38	GRAM/MJ	1.28	
PALMITIC ACID	GRAM	0.95	-	-	-	0.95	GRAM/MJ	3.19	
STEARIC ACID	GRAM	0.36	-	-	-	0.36	GRAM/MJ	1.21	
PALMITOLEIC ACID	GRAM	0.11	-	-	-	0.11	GRAM/MJ	0.37	
OLEIC ACID	GRAM	0.93	-	-	-	0.93	GRAM/MJ	3.12	
LINOLEIC ACID	MILLI	90.00	77.00	-	119.00	90.00	MILLI/MJ	302.14	
LINOLENIC ACID	MILLI	60.00	-	-	-	60.00	MILLI/MJ	201.43	
CHOLESTEROL	MILLI	12.20	9.90	-	14.70	12.20	MILLI/MJ	40.96	

JOGHURT
MIND. 1,5%, HÖCHST. 1,8% FETT

YOGHURT
MIN. 1.5, MAX. 1.8% FAT CONTENT

YAOURT
MIN. 1,5%, MAX. 1,8% MAT. GR.

		PROTEIN	FAT	CARBOHYDRATES	TOTAL
ENERGY VALUE (AVERAGE)	KJOULE	65	62	94	221
PER 100 G EDIBLE PORTION	(KCAL)	16	15	22	53
AMOUNT OF DIGESTIBLE CONSTITUENTS PER 100 G	GRAM	3.44	1.52	5.60	
ENERGY VALUE (AVERAGE)	KJOULE	63	59	94	216
OF THE DIGESTIBLE FRACTION PER 100 G EDIBLE PORTION	(KCAL)	15	14	22	52

WASTE PERCENTAGE AVERAGE 0.00

CONSTITUENTS	DIM	AV	VARIATION			AVR	NUTR. DENS.		MOLPERC.
WATER	GRAM	88.90	88.50	-	89.90	88.90	GRAM/MJ	411.68	
PROTEIN	GRAM	3.55	3.31	-	4.14	3.55	GRAM/MJ	16.44	
FAT	GRAM	1.60	1.50	-	1.80	1.60	GRAM/MJ	7.41	
AVAILABLE CARBOHYDR.	GRAM	5.60	-	-	-	5.60	GRAM/MJ	25.93	
MINERALS	GRAM	0.75	0.69	-	0.87	0.75	GRAM/MJ	3.47	

CONSTITUENTS	DIM	AV	VARIATION			AVR	NUTR.	DENS.	MOLPERC.
SODIUM	MILLI	45.00	38.00	-	55.00	45.00	MILLI/MJ	208.39	
POTASSIUM	MILLI	149.00	136.00	-	168.00	149.00	MILLI/MJ	690.00	
MAGNESIUM	MILLI	11.00	9.00	-	15.00	11.00	MILLI/MJ	50.94	
CALCIUM	MILLI	114.00	101.00	-	126.00	114.00	MILLI/MJ	527.92	
MANGANESE	MICRO	2.40	1.20	-	3.80	2.40	MICRO/MJ	11.11	
IRON	MICRO	44.00	28.00	-	66.00	44.00	MICRO/MJ	203.76	
COBALT	NANO	80.00	50.00	-	120.00	80.00	NANO/MJ	370.47	
COPPER	MICRO	9.00	2.00	-	28.00	9.00	MICRO/MJ	41.68	
ZINC	MILLI	0.36	0.20	-	0.52	0.36	MILLI/MJ	1.67	
PHOSPHORUS	MILLI	87.00	60.00	-	96.00	87.00	MILLI/MJ	402.89	
CHLORIDE	MILLI	96.00	85.00	-	100.00	96.00	MILLI/MJ	444.56	
FLUORIDE	MICRO	16.00	10.00	-	20.00	16.00	MICRO/MJ	74.09	
IODIDE	MICRO	3.60	2.20	-	6.50	3.60	MICRO/MJ	16.67	
VITAMIN A	MICRO	13.00	11.00	-	14.00	13.00	MICRO/MJ	60.20	
CAROTENE	MICRO	8.00	6.00	-	9.00	8.00	MICRO/MJ	37.05	
VITAMIN D	NANO	28.00	8.00	-	38.00	28.00	NANO/MJ	129.66	
VITAMIN E ACTIVITY	MICRO	37.00	17.00	-	43.00	37.00	MICRO/MJ	171.34	
ALPHA-TOCOPHEROL	MICRO	37.00	17.00	-	43.00	37.00	MICRO/MJ	171.34	
VITAMIN B1	MICRO	35.00	29.00	-	52.00	35.00	MICRO/MJ	162.08	
VITAMIN B2	MILLI	0.17	0.13	-	0.21	0.17	MILLI/MJ	0.79	
NICOTINAMIDE	MICRO	86.00	67.00	-	105.00	86.00	MICRO/MJ	398.25	
PANTOTHENIC ACID	MILLI	0.33	0.27	-	0.40	0.33	MILLI/MJ	1.53	
VITAMIN B6	MICRO	44.00	21.00	-	67.00	44.00	MICRO/MJ	203.76	
BIOTIN	MICRO	3.30	1.90	-	4.80	3.30	MICRO/MJ	15.28	
FOLIC ACID	MICRO	-	0.28	-	6.50				
VITAMIN B12	MICRO	0.40	0.29	-	0.72	0.40	MICRO/MJ	1.85	
VITAMIN C	MILLI	1.60	1.00	-	2.30	1.60	MILLI/MJ	7.41	
ALANINE	GRAM	0.16	-	-	-	0.16	GRAM/MJ	0.74	5.8
ARGININE	GRAM	0.13	-	-	-	0.13	GRAM/MJ	0.60	2.4
ASPARTIC ACID	GRAM	0.28	-	-	-	0.28	GRAM/MJ	1.30	7.1
CYSTINE	MILLI	27.00	-	-	-	27.00	MILLI/MJ	125.03	0.4
GLUTAMIC ACID	GRAM	0.70	-	-	-	0.70	GRAM/MJ	3.24	15.7
GLYCINE	MILLI	86.00	-	-	-	86.00	MILLI/MJ	398.25	3.8
HISTIDINE	MILLI	95.00	-	-	-	95.00	MILLI/MJ	439.93	2.0
ISOLEUCINE	GRAM	0.22	-	-	-	0.22	GRAM/MJ	1.02	5.6
LEUCINE	GRAM	0.38	-	-	-	0.38	GRAM/MJ	1.76	9.5
LYSINE	GRAM	0.28	-	-	-	0.28	GRAM/MJ	1.30	6.5
METHIONINE	MILLI	92.00	-	-	-	92.00	MILLI/MJ	426.04	2.1
PHENYLALANINE	GRAM	0.19	-	-	-	0.19	GRAM/MJ	0.88	3.9
PROLINE	GRAM	0.42	-	-	-	0.42	GRAM/MJ	1.94	12.2
SERINE	GRAM	0.22	-	-	-	0.22	GRAM/MJ	1.02	7.0
THREONINE	GRAM	0.16	-	-	-	0.16	GRAM/MJ	0.74	4.4
TRYPTOPHAN	MILLI	42.00	-	-	-	42.00	MILLI/MJ	194.50	0.7
TYROSINE	GRAM	0.18	-	-	-	0.18	GRAM/MJ	0.83	3.4
VALINE	GRAM	0.27	-	-	-	0.27	GRAM/MJ	1.25	7.8
LACTIC ACID	GRAM	1.11	1.11	-	1.11	1.11	GRAM/MJ	5.14	
GLUCOSE	MILLI	30.00	30.00	-	30.00	30.00	MILLI/MJ	138.93	
LACTOSE	GRAM	3.28	3.28	-	3.28	3.28	GRAM/MJ	15.19	
GALACTOSE	GRAM	1.18	1.18	-	1.18	1.18	GRAM/MJ	5.46	
BUTYRIC ACID	MILLI	50.00	-	-	-	50.00	MILLI/MJ	231.54	
CAPROIC ACID	MILLI	30.00	-	-	-	30.00	MILLI/MJ	138.93	
CAPRYLIC ACID	MILLI	20.00	-	-	-	20.00	MILLI/MJ	92.62	
CAPRIC ACID	MILLI	40.00	-	-	-	40.00	MILLI/MJ	185.23	
LAURIC ACID	MILLI	50.00	-	-	-	50.00	MILLI/MJ	231.54	
MYRISTIC ACID	GRAM	0.16	-	-	-	0.16	GRAM/MJ	0.74	
PALMITIC ACID	GRAM	0.42	-	-	-	0.42	GRAM/MJ	1.94	
STEARIC ACID	GRAM	0.15	-	-	-	0.15	GRAM/MJ	0.69	
PALMITOLEIC ACID	MILLI	30.00	-	-	-	30.00	MILLI/MJ	138.93	
OLEIC ACID	GRAM	0.35	-	-	-	0.35	GRAM/MJ	1.62	
LINOLEIC ACID	MILLI	41.00	33.00	-	51.00	41.00	MILLI/MJ	189.87	
LINOLENIC ACID	MILLI	11.00	9.00	-	14.00	11.00	MILLI/MJ	50.94	
CHOLESTEROL	MILLI	5.20	4.20	-	6.30	5.20	MILLI/MJ	24.08	

JOGHURT
MAGER
HÖCHST. 0,3% FETT

YOGHURT
LOW FAT
MAX. 0.3% FAT CONTENT

YAOURT
MAIGRE
MAX. 0,3% MAT. GR.

		PROTEIN	FAT	CARBOHYDRATES	TOTAL
ENERGY VALUE (AVERAGE)	KJOULE	80	3.9	81	165
PER 100 G	(KCAL)	19	0.9	19	39
EDIBLE PORTION					
AMOUNT OF DIGESTIBLE	GRAM	4.21	0.09	4.85	
CONSTITUENTS PER 100 G					
ENERGY VALUE (AVERAGE)	KJOULE	78	3.7	81	163
OF THE DIGESTIBLE	(KCAL)	19	0.9	19	39
FRACTION PER 100 G					
EDIBLE PORTION					

WASTE PERCENTAGE AVERAGE 0.00

CONSTITUENTS	DIM	AV	VARIATION			AVR	NUTR. DENS.		MOLPERC.
WATER	GRAM	89.80	87.40	-	90.50	89.80	GRAM/MJ	552.53	
PROTEIN	GRAM	4.35	3.47	-	4.80	4.35	GRAM/MJ	26.77	
FAT	GRAM	0.10	0.04	-	0.30	0.10	GRAM/MJ	0.62	
AVAILABLE CARBOHYDR.	GRAM	4.85 1	-	-	-	4.85	GRAM/MJ	29.84	
MINERALS	GRAM	0.90	0.80	-	0.97	0.90	GRAM/MJ	5.54	
SODIUM	MILLI	57.00	48.00	-	69.00	57.00	MILLI/MJ	350.71	
POTASSIUM	MILLI	187.00	171.00	-	212.00	187.00	GRAM/MJ	1.15	
MAGNESIUM	MILLI	14.00	11.00	-	19.00	14.00	MILLI/MJ	86.14	
CALCIUM	MILLI	143.00	127.00	-	158.00	143.00	MILLI/MJ	879.86	
MANGANESE	MICRO	3.00	1.50	-	4.80	3.00	MICRO/MJ	18.46	
IRON	MICRO	55.00	36.00	-	83.00	55.00	MICRO/MJ	338.41	
COBALT	MICRO	0.10	0.06	-	0.15	0.10	MICRO/MJ	0.62	
COPPER	MICRO	12.00	2.00	-	36.00	12.00	MICRO/MJ	73.83	
ZINC	MILLI	0.45	0.25	-	0.65	0.45	MILLI/MJ	2.77	
PHOSPHORUS	MILLI	109.00	75.00	-	121.00	109.00	MILLI/MJ	670.66	
CHLORIDE	MILLI	121.00	107.00	-	126.00	121.00	MILLI/MJ	744.50	
FLUORIDE	MICRO	20.00	13.00	-	25.00	20.00	MICRO/MJ	123.06	
IODIDE	MICRO	3.80	2.30	-	7.60	3.80	MICRO/MJ	23.38	
VITAMIN A	MICRO	0.80	0.70	-	0.90	0.80	MICRO/MJ	4.92	
CAROTENE	MICRO	0.50	0.40	-	0.60	0.50	MICRO/MJ	3.08	
VITAMIN D	-	TRACES	-	-	-				
VITAMIN E ACTIVITY	MICRO	2.30	1.10	-	2.70	2.30	MICRO/MJ	14.15	
ALPHA-TOCOPHEROL	MICRO	2.30	1.10	-	2.70	2.30	MICRO/MJ	14.15	
VITAMIN K	MICRO	0.50	0.00	-	0.90	0.50	MICRO/MJ	3.08	
VITAMIN B1	MICRO	38.00	31.00	-	56.00	38.00	MICRO/MJ	233.81	
VITAMIN B2	MILLI	0.18	0.14	-	0.23	0.18	MILLI/MJ	1.11	

1 ESTIMATED BY THE DIFFERENCE METHOD
100 - (WATER + PROTEIN + FAT + MINERALS)

CONSTITUENTS	DIM	AV	VARIATION			AVR	NUTR.	DENS.	MOLPERC.
NICOTINAMIDE	MICRO	92.00	72.00	-	113.00	92.00	MICRO/MJ	566.06	
PANTOTHENIC ACID	MILLI	0.36	0.29	-	0.43	0.36	MILLI/MJ	2.22	
VITAMIN B6	MICRO	47.00	23.00	-	72.00	47.00	MICRO/MJ	289.19	
BIOTIN	MICRO	3.60	2.00	-	5.10	3.60	MICRO/MJ	22.15	
FOLIC ACID	MICRO	12.00	-	-	-	12.00	MICRO/MJ	73.83	
VITAMIN B12	MICRO	0.43	0.30	-	0.78	0.43	MICRO/MJ	2.65	
VITAMIN C	MILLI	1.70	1.00	-	2.50	1.70	MILLI/MJ	10.46	
ALANINE	GRAM	0.19	-	-	-	0.19	GRAM/MJ	1.17	5.8
ARGININE	GRAM	0.16	-	-	-	0.16	GRAM/MJ	0.98	2.4
ASPARTIC ACID	GRAM	0.35	-	-	-	0.35	GRAM/MJ	2.15	7.0
CYSTINE	MILLI	34.00	-	-	-	34.00	MILLI/MJ	209.20	0.4
GLUTAMIC ACID	GRAM	0.85	-	-	-	0.85	GRAM/MJ	5.23	15.6
GLYCINE	GRAM	0.11	-	-	-	0.11	GRAM/MJ	0.68	3.8
HISTIDINE	GRAM	0.12	-	-	-	0.12	GRAM/MJ	0.74	2.0
ISOLEUCINE	GRAM	0.33	-	-	-	0.33	GRAM/MJ	2.03	6.7
LEUCINE	GRAM	0.46	-	-	-	0.46	GRAM/MJ	2.83	9.4
LYSINE	GRAM	0.35	-	-	-	0.35	GRAM/MJ	2.15	6.4
METHIONINE	GRAM	0.11	-	-	-	0.11	GRAM/MJ	0.68	2.0
PHENYLALANINE	GRAM	0.24	-	-	-	0.24	GRAM/MJ	1.48	3.8
PROLINE	GRAM	0.52	-	-	-	0.52	GRAM/MJ	3.20	12.1
SERINE	GRAM	0.27	-	-	-	0.27	GRAM/MJ	1.66	6.9
THREONINE	GRAM	0.18	-	-	-	0.18	GRAM/MJ	1.11	4.0
TRYPTOPHAN	MILLI	50.00	-	-	-	50.00	MILLI/MJ	307.64	0.7
TYROSINE	GRAM	0.22	-	-	-	0.22	GRAM/MJ	1.35	3.3
VALINE	GRAM	0.34	-	-	-	0.34	GRAM/MJ	2.09	7.7
LACTIC ACID	GRAM	1.21	1.21	-	1.21	1.21	GRAM/MJ	7.44	
LINOLEIC ACID	MILLI	2.60	2.10	-	3.20	2.60	MILLI/MJ	16.00	
LINOLENIC ACID	MILLI	0.66	0.56	-	0.85	0.66	MILLI/MJ	4.06	
CHOLESTEROL	MILLI	0.33	0.27	-	0.39	0.33	MILLI/MJ	2.03	

FRUCHTJOGHURT
VOLLFETT

YOGHURT WITH FRUITS
FULL FAT

YAOURT AVEC FRUITS
GRAS

		PROTEIN	FAT	CARBOHYDRATES	TOTAL
ENERGY VALUE (AVERAGE)	KJOULE	72	101	259	432
PER 100 G	(KCAL)	17	24	62	103
EDIBLE PORTION					
AMOUNT OF DIGESTIBLE	GRAM	3.78	2.48	15.48	
CONSTITUENTS PER 100 G					
ENERGY VALUE (AVERAGE)	KJOULE	70	96	259	425
OF THE DIGESTIBLE	(KCAL)	17	23	62	102
FRACTION PER 100 G					
EDIBLE PORTION					

WASTE PERCENTAGE AVERAGE 0.00

CONSTITUENTS	DIM	AV	VARIATION	AVR	NUTR. DENS.		MOLPERC.
WATER	GRAM	74.40	74.40 - 74.40	74.40	GRAM/MJ	175.04	
PROTEIN	GRAM	3.90	3.90 - 3.90	3.90	GRAM/MJ	9.18	
FAT	GRAM	2.62	2.62 - 2.62	2.62	GRAM/MJ	6.16	
AVAILABLE CARBOHYDR.	GRAM	15.48	15.48 - 15.48	15.48	GRAM/MJ	36.42	
CALCIUM	MILLI	127.00	127.00 - 127.00	127.00	MILLI/MJ	298.79	
PHOSPHORUS	MILLI	95.60	95.60 - 95.60	95.60	MILLI/MJ	224.92	
TOTAL SUGAR	GRAM	15.48	15.48 - 15.48	15.48	GRAM/MJ	36.42	
LACTOSE	GRAM	3.08	3.08 - 3.08	3.08	GRAM/MJ	7.25	

FRUCHTJOGHURT
FETTARM

YOGHURT WITH FRUITS
REDUCED FAT

YAOURT AVEC FRUITS
GRAISSE REDUIT

		PROTEIN	FAT	CARBOHYDRATES	TOTAL
ENERGY VALUE (AVERAGE) PER 100 G EDIBLE PORTION	KJOULE	66	51	226	343
	(KCAL)	16	12	54	82
AMOUNT OF DIGESTIBLE CONSTITUENTS PER 100 G	GRAM	3.47	1.26	13.51	
ENERGY VALUE (AVERAGE) OF THE DIGESTIBLE FRACTION PER 100 G EDIBLE PORTION	KJOULE	64	49	226	339
	(KCAL)	15	12	54	81

WASTE PERCENTAGE AVERAGE 0.00

CONSTITUENTS	DIM	AV	VARIATION	AVR	NUTR.	DENS.	MOLPERC.
WATER	GRAM	78.90	78.90 - 78.90	78.90	GRAM/MJ	232.79	
PROTEIN	GRAM	3.58	3.58 - 3.58	3.58	GRAM/MJ	10.56	
FAT	GRAM	1.33	1.33 - 1.33	1.33	GRAM/MJ	3.92	
AVAILABLE CARBOHYDR.	GRAM	13.51	13.51 - 13.51	13.51	GRAM/MJ	39.86	
CALCIUM	MILLI	114.00	114.00 - 114.00	114.00	MILLI/MJ	336.35	
PHOSPHORUS	MILLI	91.90	91.90 - 91.90	91.90	MILLI/MJ	271.15	
TOTAL SUGAR	GRAM	13.51	13.51 - 13.51	13.51	GRAM/MJ	39.86	
LACTOSE	GRAM	3.11	3.11 - 3.11	3.11	GRAM/MJ	9.18	

FRUCHTJOGHURT MAGER

YOGHURT WITH FRUITS SKIMMED

YAOURT AVEC FRUITS MAIGRE

		PROTEIN	FAT	CARBOHYDRATES	TOTAL
ENERGY VALUE (AVERAGE) PER 100 G EDIBLE PORTION	KJOULE	70	4.3	215	289
	(KCAL)	17	1.0	51	69
AMOUNT OF DIGESTIBLE CONSTITUENTS PER 100 G	GRAM	3.68	0.10	12.83	
ENERGY VALUE (AVERAGE) OF THE DIGESTIBLE FRACTION PER 100 G EDIBLE PORTION	KJOULE	68	4.0	215	287
	(KCAL)	16	1.0	51	69

WASTE PERCENTAGE AVERAGE 0.00

CONSTITUENTS	DIM	AV	VARIATION	AVR	NUTR. DENS.		MOLPERC.
WATER	GRAM	81.40	81.40 - 81.40	81.40	GRAM/MJ	283.99	
PROTEIN	GRAM	3.80	3.80 - 3.80	3.80	GRAM/MJ	13.26	
FAT	GRAM	0.11	0.11 - 0.11	0.11	GRAM/MJ	0.38	
AVAILABLE CARBOHYDR.	GRAM	12.83	12.83 - 12.83	12.83	GRAM/MJ	44.76	
CALCIUM	MILLI	128.00	128.00 - 128.00	128.00	MILLI/MJ	446.58	
PHOSPHORUS	MILLI	95.40	95.40 - 95.40	95.40	MILLI/MJ	332.84	
TOTAL SUGAR	GRAM	12.83	12.83 - 12.83	12.83	GRAM/MJ	44.76	
LACTOSE	GRAM	2.98	2.98 - 2.98	2.98	GRAM/MJ	10.40	

KEFIR KEFIR KÉFIR

		PROTEIN	FAT	CARBOHYDRATES	TOTAL
ENERGY VALUE (AVERAGE)	KJOULE	61	135	80	277
PER 100 G	(KCAL)	15	32	19	66
EDIBLE PORTION					
AMOUNT OF DIGESTIBLE CONSTITUENTS PER 100 G	GRAM	3.20	3.32	4.80	
ENERGY VALUE (AVERAGE)	KJOULE	59	129	80	268
OF THE DIGESTIBLE	(KCAL)	14	31	19	64
FRACTION PER 100 G EDIBLE PORTION					

WASTE PERCENTAGE AVERAGE 0.00

CONSTITUENTS	DIM	AV	VARIATION			AVR	NUTR.	DENS.	MOLPERC.
WATER	GRAM	87.60	-	-	-	87.60	GRAM/MJ	326.93	
PROTEIN	GRAM	3.30	-	-	-	3.30	GRAM/MJ	12.32	
FAT	GRAM	3.50	-	-	-	3.50	GRAM/MJ	13.06	
AVAILABLE CARBOHYDR.	GRAM	4.80 [1]	-	-	-	4.80	GRAM/MJ	17.91	
MINERALS	GRAM	0.80	-	-	-	0.80	GRAM/MJ	2.99	
SODIUM	MILLI	46.00	-	-	-	46.00	MILLI/MJ	171.68	
POTASSIUM	MILLI	160.00	-	-	-	160.00	MILLI/MJ	597.14	
MAGNESIUM	MILLI	14.00	-	-	-	14.00	MILLI/MJ	52.25	
IRON	MILLI	0.13	-	-	-	0.13	MILLI/MJ	0.49	
PHOSPHORUS	MILLI	90.00	-	-	-	90.00	MILLI/MJ	335.89	
LACTIC ACID	GRAM	0.70	-	-	-	0.70	GRAM/MJ	2.61	
ETHANOL	GRAM	0.50	0.20	-	0.80	0.50	GRAM/MJ	1.87	

1 ESTIMATED BY THE DIFFERENCE METHOD
100 - (WATER + PROTEIN + FAT + MINERALS)

Käse

CHEESE FROMAGE

BEL PAESEKÄSE BEL PAESE CHEESE FROMAGE BEL PAESE

		PROTEIN	FAT	CARBOHYDRATES	TOTAL
ENERGY VALUE (AVERAGE) PER 100 G EDIBLE PORTION	KJOULE	468	1169	0.0	1636
	(KCAL)	112	279	0.0	391
AMOUNT OF DIGESTIBLE CONSTITUENTS PER 100 G	GRAM	24.63	28.69	0.00	
ENERGY VALUE (AVERAGE) OF THE DIGESTIBLE FRACTION PER 100 G EDIBLE PORTION	KJOULE	454	1110	0.0	1564
	(KCAL)	108	265	0.0	374

WASTE PERCENTAGE AVERAGE 0.00

CONSTITUENTS	DIM	AV	VARIATION			AVR	NUTR. DENS.		MOLPERC.
WATER	GRAM	38.90	-	-	-	38.90	GRAM/MJ	24.87	
PROTEIN	GRAM	25.40	-	-	-	25.40	GRAM/MJ	16.24	
FAT	GRAM	30.20	-	-	-	30.20	GRAM/MJ	19.31	
CALCIUM	MILLI	604.00	-	-	-	604.00	MILLI/MJ	386.21	
PHOSPHORUS	MILLI	480.00	-	-	-	480.00	MILLI/MJ	306.92	
VITAMIN B1	MICRO	30.00	-	-	-	30.00	MICRO/MJ	19.18	
VITAMIN B2	MILLI	0.22	-	-	-	0.22	MILLI/MJ	0.14	
NICOTINAMIDE	MILLI	0.26	-	-	-	0.26	MILLI/MJ	0.17	
BUTYRIC ACID	GRAM	1.73	-	-	-	1.73	GRAM/MJ	1.11	
CAPROIC ACID	GRAM	0.87	-	-	-	0.87	GRAM/MJ	0.56	
CAPRYLIC ACID	GRAM	0.55	-	-	-	0.55	GRAM/MJ	0.35	
CAPRIC ACID	GRAM	1.03	-	-	-	1.03	GRAM/MJ	0.66	
LAURIC ACID	GRAM	1.13	-	-	-	1.13	GRAM/MJ	0.72	
MYRISTIC ACID	GRAM	2.86	-	-	-	2.86	GRAM/MJ	1.83	
PALMITIC ACID	GRAM	7.79	-	-	-	7.79	GRAM/MJ	4.98	
STEARIC ACID	GRAM	2.89	-	-	-	2.89	GRAM/MJ	1.85	
PALMITOLEIC ACID	GRAM	0.67	-	-	-	0.67	GRAM/MJ	0.43	
OLEIC ACID	GRAM	7.89	-	-	-	7.89	GRAM/MJ	5.04	
LINOLEIC ACID	GRAM	0.26	-	-	-	0.26	GRAM/MJ	0.17	
LINOLENIC ACID	GRAM	0.31	-	-	-	0.31	GRAM/MJ	0.20	

BRIEKÄSE
50% FETT I. TR.

BRIE CHEESE
50% FAT CONTENT IN DRY MATTER

FROMAGE BRIE
50% DE MAT. GR. S. SEC

		PROTEIN	FAT	CARBOHYDRATES	TOTAL
ENERGY VALUE (AVERAGE) PER 100 G EDIBLE PORTION	KJOULE	416	1080	17	1512
	(KCAL)	99	258	4.0	361
AMOUNT OF DIGESTIBLE CONSTITUENTS PER 100 G	GRAM	21.92	26.50	0.99	
ENERGY VALUE (AVERAGE) OF THE DIGESTIBLE FRACTION PER 100 G EDIBLE PORTION	KJOULE	404	1026	17	1446
	(KCAL)	96	245	4.0	346

WASTE PERCENTAGE AVERAGE 0.00

CONSTITUENTS	DIM	AV	VARIATION			AVR	NUTR.	DENS.	MOLPERC.
WATER	GRAM	45.50	39.60	-	54.00	45.50	GRAM/MJ	31.47	
PROTEIN	GRAM	22.60	20.80	-	24.30	22.60	GRAM/MJ	15.63	
FAT	GRAM	27.90	23.00	-	32.00	27.90	GRAM/MJ	19.30	
AVAILABLE CARBOHYDR.	GRAM	0.99 1	-	-	-	0.99	GRAM/MJ	0.68	
MINERALS	GRAM	4.00	-	-	-	4.00	GRAM/MJ	2.77	
SODIUM	GRAM	1.17	-	-	-	1.17	GRAM/MJ	0.81	
POTASSIUM	MILLI	152.00	-	-	-	152.00	MILLI/MJ	105.12	
CALCIUM	MILLI	400.00	-	-	-	400.00	MILLI/MJ	276.64	
IRON	MILLI	0.50	-	-	-	0.50	MILLI/MJ	0.35	
PHOSPHORUS	MILLI	188.00	-	-	-	188.00	MILLI/MJ	130.02	
VITAMIN A	MILLI	0.14	0.14	-	0.15	0.14	MILLI/MJ	0.10	
CAROTENE	MILLI	0.10	0.09	-	0.10	0.10	MILLI/MJ	0.07	
VITAMIN B1	MICRO	50.00	-	-	-	50.00	MICRO/MJ	34.58	
VITAMIN B2	MILLI	0.34	0.33	-	0.37	0.34	MILLI/MJ	0.24	
NICOTINAMIDE	MILLI	1.13	1.00	-	1.24	1.13	MILLI/MJ	0.78	
PANTOTHENIC ACID	MILLI	0.69	-	-	-	0.69	MILLI/MJ	0.48	
VITAMIN B6	MILLI	0.23	-	-	-	0.23	MILLI/MJ	0.16	
BIOTIN	MICRO	6.20	4.50	-	8.00	6.20	MICRO/MJ	4.29	
FOLIC ACID	MICRO	65.00	-	-	-	65.00	MICRO/MJ	44.95	
VITAMIN B12	MICRO	1.70	-	-	-	1.70	MICRO/MJ	1.18	
ALANINE	GRAM	0.91	-	-	-	0.91	GRAM/MJ	0.63	5.7
ARGININE	GRAM	0.78	-	-	-	0.78	GRAM/MJ	0.54	2.5
ASPARTIC ACID	GRAM	1.43	-	-	-	1.43	GRAM/MJ	0.99	6.0
CYSTINE	GRAM	0.12	-	-	-	0.12	GRAM/MJ	0.08	0.3
GLUTAMIC ACID	GRAM	4.65	-	-	-	4.65	GRAM/MJ	3.22	17.7

1 INCLUDES THE AMOUNTS OF LACTOSE AND LACTIC ACID

CONSTITUENTS	DIM	AV	VARIATION			AVR	NUTR. DENS.		MOLPERC.
GLYCINE	GRAM	0.41	-	-	-	0.41	GRAM/MJ	0.28	3.1
HISTIDINE	GRAM	0.76	-	-	-	0.76	GRAM/MJ	0.53	2.7
ISOLEUCINE	GRAM	1.08	-	-	-	1.08	GRAM/MJ	0.75	4.6
LEUCINE	GRAM	2.05	-	-	-	2.05	GRAM/MJ	1.42	8.7
LYSINE	GRAM	1.96	-	-	-	1.96	GRAM/MJ	1.36	7.5
METHIONINE	GRAM	0.63	-	-	-	0.63	GRAM/MJ	0.44	2.4
PHENYLALANINE	GRAM	1.23	-	-	-	1.23	GRAM/MJ	0.85	4.2
PROLINE	GRAM	2.61	-	-	-	2.61	GRAM/MJ	1.81	12.7
SERINE	GRAM	1.24	-	-	-	1.24	GRAM/MJ	0.86	6.6
THREONINE	GRAM	0.80	-	-	-	0.80	GRAM/MJ	0.55	3.8
TRYPTOPHAN	GRAM	0.34	-	-	-	0.34	GRAM/MJ	0.24	0.9
TYROSINE	GRAM	1.27	-	-	-	1.27	GRAM/MJ	0.88	3.9
VALINE	GRAM	1.42	-	-	-	1.42	GRAM/MJ	0.98	6.8
TYRAMINE	MILLI	-	0.00	-	26.00				
LACTIC ACID	GRAM	0.89	0.89	-	0.89	0.89	GRAM/MJ	0.62	
LACTOSE	MILLI	97.00	97.00	-	97.00	97.00	MILLI/MJ	67.08	
CHOLESTEROL	MILLI	100.00	-	-	-	100.00	MILLI/MJ	69.16	

BUTTERKÄSE
50% FETT I. TR.

„BUTTERKAESE“
50% FAT CONTENT IN DRY MATTER

FROMAGE „BUTTERKAESE“
50% DE MAT. GR. S. SEC

		PROTEIN	FAT	CARBOHYDRATES	TOTAL
ENERGY VALUE (AVERAGE)	KJOULE	388	1115	0.0	1503
PER 100 G	(KCAL)	93	266	0.0	359
EDIBLE PORTION					
AMOUNT OF DIGESTIBLE	GRAM	20.46	27.36	0.00	
CONSTITUENTS PER 100 G					
ENERGY VALUE (AVERAGE)	KJOULE	377	1059	0.0	1436
OF THE DIGESTIBLE	(KCAL)	90	253	0.0	343
FRACTION PER 100 G					
EDIBLE PORTION					

WASTE PERCENTAGE AVERAGE 0.00

CONSTITUENTS	DIM	AV	VARIATION			AVR	NUTR. DENS.		MOLPERC.
WATER	GRAM	46.20	43.90	-	50.00	46.20	GRAM/MJ	32.18	
PROTEIN	GRAM	21.10	20.10	-	22.70	21.10	GRAM/MJ	14.70	
FAT	GRAM	28.80	25.00	-	29.70	28.80	GRAM/MJ	20.06	
MINERALS	GRAM	3.35	2.97	-	3.64	3.35	GRAM/MJ	2.33	
SODIUM	GRAM	0.86	0.66	-	1.07	0.87	GRAM/MJ	0.60	
POTASSIUM	MILLI	78.00	77.00	-	80.00	78.00	MILLI/MJ	54.33	
MAGNESIUM	MILLI	53.00	50.00	-	59.00	53.00	MILLI/MJ	36.92	
CALCIUM	MILLI	694.00	569.00	-	818.00	694.00	MILLI/MJ	483.40	
IRON	MILLI	0.57	0.49	-	0.71	0.57	MILLI/MJ	0.40	
PHOSPHORUS	MILLI	417.00	356.00	-	461.00	417.00	MILLI/MJ	290.46	

CAMEMBERTKÄSE
30% FETT I. TR.

CAMEMBERT CHEESE
30% FAT CONTENT IN DRY MATTER

FROMAGE CAMEMBERT
30% DE MAT. GR. S. SEC

		PROTEIN	FAT	CARBOHYDRATES	TOTAL
ENERGY VALUE (AVERAGE)	KJOULE	433	522	0.0	955
PER 100 G	(KCAL)	103	125	0.0	228
EDIBLE PORTION					
AMOUNT OF DIGESTIBLE	GRAM	22.79	12.82	0.00	
CONSTITUENTS PER 100 G					
ENERGY VALUE (AVERAGE)	KJOULE	420	496	0.0	916
OF THE DIGESTIBLE	(KCAL)	100	119	0.0	219
FRACTION PER 100 G					
EDIBLE PORTION					

WASTE PERCENTAGE AVERAGE 0.00

CONSTITUENTS	DIM	AV	VARIATION	AVR	NUTR. DENS.		MOLPERC.
WATER	GRAM	58.20	56.50 - 62.00	58.20	GRAM/MJ	63.54	
PROTEIN	GRAM	23.50	21.90 - 25.30	23.50	GRAM/MJ	25.66	
FAT	GRAM	13.50	11.40 - 14.00	13.50	GRAM/MJ	14.74	
MINERALS	GRAM	4.00	3.80 - 4.10	4.00	GRAM/MJ	4.37	
SODIUM	GRAM	0.90	0.81 - 1.43	0.90	GRAM/MJ	0.98	
POTASSIUM	MILLI	120.00	100.00 - 140.00	120.00	MILLI/MJ	131.00	
MAGNESIUM	MILLI	19.00	13.00 - 22.00	19.00	MILLI/MJ	20.74	
CALCIUM	MILLI	600.00	- - -	600.00	MILLI/MJ	655.02	
IRON	MILLI	0.17	0.10 - 0.22	0.17	MILLI/MJ	0.19	
COPPER	MICRO	80.00	70.00 - 90.00	80.00	MICRO/MJ	87.34	
ZINC	MILLI	3.40	2.00 - 4.60	3.40	MILLI/MJ	3.71	
PHOSPHORUS	MILLI	540.00	- - -	540.00	MILLI/MJ	589.52	
CHLORIDE	GRAM	1.75	1.30 - 2.20	1.75	GRAM/MJ	1.91	
FLUORIDE	MICRO	28.00	- - -	28.00	MICRO/MJ	30.57	
BORON	MICRO	65.00	- - -	65.00	MICRO/MJ	70.96	
SELENIUM	MICRO	6.00	- - -	6.00	MICRO/MJ	6.55	
NITRATE	MILLI	0.20	0.10 - 1.10	0.20	MILLI/MJ	0.22	
VITAMIN A	MILLI	0.20	0.14 - 0.25	0.20	MILLI/MJ	0.22	
CAROTENE	MILLI	0.10	0.08 - 0.15	0.10	MILLI/MJ	0.11	
VITAMIN D	MICRO	0.17	0.12 - 0.21	0.17	MICRO/MJ	0.19	
VITAMIN E ACTIVITY	MILLI	0.30	0.14 - 0.35	0.30	MILLI/MJ	0.33	
ALPHA-TOCOPHEROL	MILLI	0.30	0.14 - 0.35	0.30	MILLI/MJ	0.33	
VITAMIN B1	MICRO	50.00	45.00 - 56.00	50.00	MICRO/MJ	54.59	
VITAMIN B2	MILLI	0.67	0.50 - 0.84	0.67	MILLI/MJ	0.73	
NICOTINAMIDE	MILLI	1.20	0.62 - 2.00	1.20	MILLI/MJ	1.31	
PANTOTHENIC ACID	MILLI	0.90	0.50 - 1.45	0.90	MILLI/MJ	0.98	
VITAMIN B6	MILLI	0.28	0.19 - 0.38	0.28	MILLI/MJ	0.31	
BIOTIN	MICRO	5.00	3.90 - 6.00	5.00	MICRO/MJ	5.46	
FOLIC ACID	MICRO	66.00	24.00 - 110.00	66.00	MICRO/MJ	72.05	
VITAMIN B12	MICRO	3.10	2.90 - 3.50	3.10	MICRO/MJ	3.38	

CONSTITUENTS	DIM	AV	VARIATION			AVR	NUTR. DENS.		MOLPERC.
ALANINE	GRAM	0.93	0.89	-	0.98	0.93	GRAM/MJ	1.02	
ARGININE	GRAM	0.90	0.85	-	0.94	0.90	GRAM/MJ	0.98	
ASPARTIC ACID	GRAM	1.75	1.53	-	1.97	1.75	GRAM/MJ	1.91	
CYSTINE	GRAM	0.11	0.09	-	0.12	0.11	GRAM/MJ	0.12	
GLUTAMIC ACID	GRAM	5.18	4.99	-	5.34	5.18	GRAM/MJ	5.66	
GLYCINE	GRAM	0.48	0.45	-	0.52	0.48	GRAM/MJ	0.52	
HISTIDINE	GRAM	0.78	0.76	-	0.82	0.78	GRAM/MJ	0.85	
ISOLEUCINE	GRAM	1.34	1.21	-	1.48	1.34	GRAM/MJ	1.46	
LEUCINE	GRAM	2.25	2.17	-	2.36	2.25	GRAM/MJ	2.46	
LYSINE	GRAM	1.90	1.71	-	2.00	1.90	GRAM/MJ	2.07	
METHIONINE	GRAM	0.69	0.63	-	0.75	0.69	GRAM/MJ	0.75	
PHENYLALANINE	GRAM	1.31	1.25	-	1.42	1.31	GRAM/MJ	1.43	
PROLINE	GRAM	2.48	2.19	-	2.78	2.48	GRAM/MJ	2.71	
SERINE	GRAM	1.35	1.32	-	1.38	1.35	GRAM/MJ	1.47	
THREONINE	GRAM	0.85	0.82	-	0.88	0.85	GRAM/MJ	0.93	
TRYPTOPHAN	GRAM	0.37	0.31	-	0.41	0.37	GRAM/MJ	0.40	
TYROSINE	GRAM	1.31	1.11	-	1.49	1.31	GRAM/MJ	1.43	
VALINE	GRAM	1.62	1.57	-	1.68	1.62	GRAM/MJ	1.77	
CADAVERINE	MILLI	2.40	1.20	-	3.70	2.40	MILLI/MJ	2.62	
HISTAMINE	MILLI	0.26	0.00	-	48.00	0.26	MILLI/MJ	0.28	
PUTRESCINE	MILLI	1.50	0.70	-	3.30	1.50	MILLI/MJ	1.64	
TYRAMINE	MILLI	3.70	2.00	-	200.00	3.70	MILLI/MJ	4.04	
SALICYLIC ACID	MICRO	10.00	10.00	-	10.00	10.00	MICRO/MJ	10.92	
BUTYRIC ACID	GRAM	0.27	-	-	-	0.27	GRAM/MJ	0.29	
CAPROIC ACID	GRAM	0.16	-	-	-	0.16	GRAM/MJ	0.17	
CAPRYLIC ACID	GRAM	0.15	-	-	-	0.15	GRAM/MJ	0.16	
CAPRIC ACID	GRAM	0.33	-	-	-	0.33	GRAM/MJ	0.36	
LAURIC ACID	GRAM	0.25	-	-	-	0.25	GRAM/MJ	0.27	
MYRISTIC ACID	GRAM	1.49	-	-	-	1.49	GRAM/MJ	1.63	
PALMITIC ACID	GRAM	4.02	-	-	-	4.02	GRAM/MJ	4.39	
STEARIC ACID	GRAM	1.40	-	-	-	1.40	GRAM/MJ	1.53	
PALMITOLEIC ACID	GRAM	0.49	-	-	-	0.49	GRAM/MJ	0.53	
OLEIC ACID	GRAM	3.19	-	-	-	3.19	GRAM/MJ	3.48	
LINOLEIC ACID	GRAM	0.25	0.23	-	0.27	0.25	GRAM/MJ	0.27	
LINOLENIC ACID	GRAM	0.14	0.15	-	0.16	0.14	GRAM/MJ	0.15	
CHOLESTEROL	MILLI	38.00	35.00	-	40.00	38.00	MILLI/MJ	41.48	

CAMEMBERTKÄSE
40% FETT I. TR.

CAMEMBERT CHEESE
40% FAT CONTENT IN DRY MATTER

FROMAGE CAMEMBERT
40% DE MAT. GR. S. SEC

		PROTEIN	FAT	CARBOHYDRATES	TOTAL
ENERGY VALUE (AVERAGE)	KJOULE	414	793	0.0	1208
PER 100 G EDIBLE PORTION	(KCAL)	99	190	0.0	289
AMOUNT OF DIGESTIBLE CONSTITUENTS PER 100 G	GRAM	21.82	19.47	0.00	
ENERGY VALUE (AVERAGE)	KJOULE	402	754	0.0	1156
OF THE DIGESTIBLE FRACTION PER 100 G EDIBLE PORTION	(KCAL)	96	180	0.0	276

WASTE PERCENTAGE AVERAGE 0.00

CONSTITUENTS	DIM	AV	VARIATION			AVR	NUTR. DENS.		MOLPERC.
WATER	GRAM	51.70	50.40	-	58.00	51.70	GRAM/MJ	44.74	
PROTEIN	GRAM	22.50	-	-	-	22.50	GRAM/MJ	19.47	
FAT	GRAM	20.50	16.80	-	21.60	20.50	GRAM/MJ	17.74	
MINERALS	GRAM	3.51	-	-	-	3.51	GRAM/MJ	3.04	
SODIUM	MILLI	830.00	-	-	-	830.00	MILLI/MJ	718.30	
POTASSIUM	MILLI	100.00	-	-	-	100.00	MILLI/MJ	86.54	
CALCIUM	MILLI	570.00	-	-	-	570.00	MILLI/MJ	493.29	
IRON	MILLI	0.46	-	-	-	0.46	MILLI/MJ	0.40	
ZINC	MILLI	2.90	1.70	-	3.80	2.90	MILLI/MJ	2.51	
PHOSPHORUS	MILLI	350.00	-	-	-	350.00	MILLI/MJ	302.90	
BORON	MICRO	65.00	-	-	-	65.00	MICRO/MJ	56.25	
SELENIUM	MICRO	6.00	-	-	-	6.00	MICRO/MJ	5.19	
NITRATE	MILLI	0.20	0.10	-	1.10	0.20	MILLI/MJ	0.17	
VITAMIN A	MILLI	0.30	-	-	-	0.30	MILLI/MJ	0.26	
CAROTENE	MILLI	0.17	0.13	-	0.23	0.17	MILLI/MJ	0.15	
VITAMIN E ACTIVITY	MILLI	0.44	-	-	-	0.44	MILLI/MJ	0.38	
ALPHA-TOCOPHEROL	MILLI	0.44	-	-	-	0.44	MILLI/MJ	0.38	
VITAMIN B1	MICRO	49.00	27.00	-	65.00	49.00	MICRO/MJ	42.41	
VITAMIN B2	MILLI	0.49	0.38	-	0.62	0.49	MILLI/MJ	0.42	
NICOTINAMIDE	MILLI	1.57	1.20	-	2.37	1.57	MILLI/MJ	1.36	
PANTOTHENIC ACID	MILLI	0.90	0.50	-	1.40	0.90	MILLI/MJ	0.78	
VITAMIN B6	MILLI	0.27	0.18	-	0.37	0.27	MILLI/MJ	0.23	
BIOTIN	MICRO	3.80	2.20	-	5.40	3.80	MICRO/MJ	3.29	
ALANINE	GRAM	0.89	0.85	-	0.94	0.89	GRAM/MJ	0.77	
ARGININE	GRAM	0.82	-	-	-	0.82	GRAM/MJ	0.71	
ASPARTIC ACID	GRAM	1.67	1.46	-	1.88	1.67	GRAM/MJ	1.45	
CYSTINE	GRAM	0.13	-	-	-	0.13	GRAM/MJ	0.11	
GLUTAMIC ACID	GRAM	4.95	4.77	-	5.10	4.95	GRAM/MJ	4.28	
GLYCINE	GRAM	0.46	0.43	-	0.50	0.46	GRAM/MJ	0.40	
HISTIDINE	GRAM	0.73	-	-	-	0.73	GRAM/MJ	0.63	
ISOLEUCINE	GRAM	1.52	-	-	-	1.52	GRAM/MJ	1.32	
LEUCINE	GRAM	2.19	-	-	-	2.19	GRAM/MJ	1.90	
LYSINE	GRAM	1.65	-	-	-	1.65	GRAM/MJ	1.43	
METHIONINE	GRAM	0.59	-	-	-	0.59	GRAM/MJ	0.51	
PHENYLALANINE	GRAM	1.20	-	-	-	1.20	GRAM/MJ	1.04	
PROLINE	GRAM	2.38	2.09	-	2.66	2.38	GRAM/MJ	2.06	
SERINE	GRAM	1.29	1.27	-	1.32	1.29	GRAM/MJ	1.12	
THREONINE	GRAM	0.84	-	-	-	0.84	GRAM/MJ	0.73	
TRYPTOPHAN	GRAM	0.31	-	-	-	0.31	GRAM/MJ	0.27	
TYROSINE	GRAM	1.08	-	-	-	1.08	GRAM/MJ	0.93	
VALINE	GRAM	1.61	-	-	-	1.61	GRAM/MJ	1.39	
CADAVERINE	MILLI	2.40	1.20	-	3.70	2.40	MILLI/MJ	2.08	
HISTAMINE	MILLI	0.26	0.00	-	48.00	0.26	MILLI/MJ	0.23	
PUTRESCINE	MILLI	1.50	0.70	-	3.30	1.50	MILLI/MJ	1.30	
TYRAMINE	MILLI	3.70	2.00	-	200.00	3.70	MILLI/MJ	3.20	
SALICYLIC ACID	MICRO	10.00	10.00	-	10.00	10.00	MICRO/MJ	8.65	
BUTYRIC ACID	GRAM	0.41	-	-	-	0.41	GRAM/MJ	0.35	
CAPROIC ACID	GRAM	0.24	-	-	-	0.24	GRAM/MJ	0.21	
CAPRYLIC ACID	GRAM	0.22	-	-	-	0.22	GRAM/MJ	0.19	
CAPRIC ACID	GRAM	0.50	-	-	-	0.50	GRAM/MJ	0.43	
LAURIC ACID	GRAM	0.37	-	-	-	0.37	GRAM/MJ	0.32	
MYRISTIC ACID	GRAM	2.27	-	-	-	2.27	GRAM/MJ	1.96	
PALMITIC ACID	GRAM	6.11	-	-	-	6.11	GRAM/MJ	5.29	
STEARIC ACID	GRAM	2.13	-	-	-	2.13	GRAM/MJ	1.84	
PALMITOLEIC ACID	GRAM	0.74	-	-	-	0.74	GRAM/MJ	0.64	
OLEIC ACID	GRAM	4.86	-	-	-	4.86	GRAM/MJ	4.21	
LINOLEIC ACID	GRAM	0.38	-	-	-	0.38	GRAM/MJ	0.33	
LINOLENIC ACID	GRAM	0.23	-	-	-	0.23	GRAM/MJ	0.20	

CAMEMBERTKÄSE
45% FETT I. TR.

CAMEMBERT CHEESE
45% FAT CONTENT IN DRY MATTER

FROMAGE CAMEMBERT
45% DE MAT. GR. S. SEC

		PROTEIN	FAT	CARBOHYDRATES	TOTAL
ENERGY VALUE (AVERAGE)	KJOULE	387	863	3.2	1253
PER 100 G EDIBLE PORTION	(KCAL)	92	206	0.8	299
AMOUNT OF DIGESTIBLE CONSTITUENTS PER 100 G	GRAM	20.37	21.18	0.19	
ENERGY VALUE (AVERAGE)	KJOULE	375	820	3.2	1198
OF THE DIGESTIBLE FRACTION PER 100 G EDIBLE PORTION	(KCAL)	90	196	0.8	286

WASTE PERCENTAGE AVERAGE 0.00

CONSTITUENTS	DIM	AV	VARIATION		AVR	NUTR. DENS.		MOLPERC.
WATER	GRAM	52.00	47.50	56.00	52.00	GRAM/MJ	43.40	
PROTEIN	GRAM	21.00	19.40	22.00	21.00	GRAM/MJ	17.53	
FAT	GRAM	22.30	19.80	23.30	22.30	GRAM/MJ	18.61	
AVAILABLE CARBOHYDR.	GRAM	0.19 [1]	-	-	0.19	GRAM/MJ	0.16	
MINERALS	GRAM	3.90	3.57	4.35	3.90	GRAM/MJ	3.26	
SODIUM	GRAM	0.97	0.65	1.35	0.98	GRAM/MJ	0.81	
POTASSIUM	MILLI	110.00	92.00	128.00	110.00	MILLI/MJ	91.81	
MAGNESIUM	MILLI	17.00	12.00	20.00	17.00	MILLI/MJ	14.19	
CALCIUM	MILLI	570.00	550.00	600.00	570.00	MILLI/MJ	475.76	
IRON	MILLI	0.15	0.09	0.20	0.15	MILLI/MJ	0.13	
COPPER	MICRO	70.00	60.00	80.00	70.00	MICRO/MJ	58.43	
ZINC	MILLI	3.10	1.80	4.10	3.10	MILLI/MJ	2.59	
PHOSPHORUS	MILLI	350.00	-	-	350.00	MILLI/MJ	292.13	
CHLORIDE	GRAM	1.44	1.07	1.81	1.44	GRAM/MJ	1.20	
FLUORIDE	MICRO	25.00	-	-	25.00	MICRO/MJ	20.87	
BORON	MICRO	65.00	-	-	65.00	MICRO/MJ	54.25	
SELENIUM	MICRO	6.00	-	-	6.00	MICRO/MJ	5.01	
NITRATE	MILLI	0.20	0.10	1.10	0.20	MILLI/MJ	0.17	
VITAMIN A	MILLI	0.33	0.24	0.42	0.33	MILLI/MJ	0.28	
CAROTENE	MILLI	0.19	0.14	0.25	0.19	MILLI/MJ	0.16	
VITAMIN D	MICRO	0.28	0.20	0.35	0.28	MICRO/MJ	0.23	
VITAMIN E ACTIVITY	MILLI	0.50	0.24	0.59	0.50	MILLI/MJ	0.42	
ALPHA-TOCOPHEROL	MILLI	0.50	0.24	0.59	0.50	MILLI/MJ	0.42	
VITAMIN B1	MICRO	45.00	40.00	50.00	45.00	MICRO/MJ	37.56	
VITAMIN B2	MILLI	0.60	0.45	0.75	0.60	MILLI/MJ	0.50	
NICOTINAMIDE	MILLI	1.10	0.55	1.80	1.10	MILLI/MJ	0.92	
PANTOTHENIC ACID	MILLI	0.80	0.45	1.30	0.80	MILLI/MJ	0.67	
VITAMIN B6	MILLI	0.25	0.17	0.34	0.25	MILLI/MJ	0.21	
BIOTIN	MICRO	4.50	3.50	5.40	4.50	MICRO/MJ	3.76	
FOLIC ACID	MICRO	59.00	21.00	97.00	59.00	MICRO/MJ	49.25	
VITAMIN B12	MICRO	2.80	2.60	3.10	2.80	MICRO/MJ	2.34	
VITAMIN C	-	TRACES	-	-				

1 INCLUDES THE AMOUNTS OF LACTOSE AND LACTIC ACID

CONSTITUENTS	DIM	AV	VARIATION			AVR	NUTR. DENS.		MOLPERC.
ALANINE	GRAM	0.83	0.79	-	0.87	0.83	GRAM/MJ	0.69	
ARGININE	GRAM	0.80	0.76	-	0.84	0.80	GRAM/MJ	0.67	
ASPARTIC ACID	GRAM	1.55	1.36	-	1.74	1.55	GRAM/MJ	1.29	
CYSTINE	GRAM	0.10	0.08	-	0.11	0.10	GRAM/MJ	0.08	
GLUTAMIC ACID	GRAM	4.59	4.43	-	4.73	4.59	GRAM/MJ	3.83	
GLYCINE	GRAM	0.43	0.40	-	0.46	0.43	GRAM/MJ	0.36	
HISTIDINE	GRAM	0.70	0.68	-	0.73	0.70	GRAM/MJ	0.58	
ISOLEUCINE	GRAM	1.20	1.08	-	1.32	1.20	GRAM/MJ	1.00	
LEUCINE	GRAM	2.01	1.94	-	2.11	2.01	GRAM/MJ	1.68	
LYSINE	GRAM	1.70	1.53	-	1.79	1.70	GRAM/MJ	1.42	
METHIONINE	GRAM	0.62	0.56	-	0.67	0.62	GRAM/MJ	0.52	
PHENYLALANINE	GRAM	1.17	1.12	-	1.27	1.17	GRAM/MJ	0.98	
PROLINE	GRAM	2.20	1.94	-	2.47	2.20	GRAM/MJ	1.84	
SERINE	GRAM	1.19	1.17	-	1.22	1.19	GRAM/MJ	0.99	
THREONINE	GRAM	0.76	0.73	-	0.79	0.76	GRAM/MJ	0.63	
TRYPTOPHAN	GRAM	0.33	0.28	-	0.37	0.33	GRAM/MJ	0.28	
TYROSINE	GRAM	1.17	0.99	-	1.33	1.17	GRAM/MJ	0.98	
VALINE	GRAM	1.45	1.40	-	1.50	1.45	GRAM/MJ	1.21	
CADAVERINE	MILLI	2.40	1.20	-	3.70	2.40	MILLI/MJ	2.00	
HISTAMINE	MILLI	0.26	0.00	-	48.00	0.26	MILLI/MJ	0.22	
PUTRESCINE	MILLI	1.50	0.70	-	3.30	1.50	MILLI/MJ	1.25	
TYRAMINE	MILLI	3.70	2.00	-	200.00	3.70	MILLI/MJ	3.09	
LACTIC ACID	MILLI	88.00	31.00	-	210.00	88.00	MILLI/MJ	73.45	
SALICYLIC ACID	MICRO	10.00	10.00	-	10.00	10.00	MICRO/MJ	8.35	
LACTOSE	MILLI	102.00	102.00	-	102.00	102.00	MILLI/MJ	85.14	
BUTYRIC ACID	GRAM	0.45	-	-	-	0.45	GRAM/MJ	0.38	
CAPROIC ACID	GRAM	0.26	-	-	-	0.26	GRAM/MJ	0.22	
CAPRYLIC ACID	GRAM	0.24	-	-	-	0.24	GRAM/MJ	0.20	
CAPRIC ACID	GRAM	0.55	-	-	-	0.55	GRAM/MJ	0.46	
LAURIC ACID	GRAM	0.41	-	-	-	0.41	GRAM/MJ	0.34	
MYRISTIC ACID	GRAM	2.47	-	-	-	2.47	GRAM/MJ	2.06	
STEARIC ACID	GRAM	2.32	-	-	-	2.32	GRAM/MJ	1.94	
PALMITOLEIC ACID	GRAM	0.81	-	-	-	0.81	GRAM/MJ	0.68	
OLEIC ACID	GRAM	5.28	-	-	-	5.28	GRAM/MJ	4.41	
LINOLEIC ACID	GRAM	0.41	0.38	-	0.45	0.41	GRAM/MJ	0.34	
LINOLENIC ACID	GRAM	0.25	0.23	-	0.27	0.25	GRAM/MJ	0.21	
CHOLESTEROL	MILLI	62.00	58.00	-	66.00	62.00	MILLI/MJ	51.75	

CAMEMBERTKÄSE
50% FETT I. TR.
(RAHMCAMEMBERT)

CAMEMBERT CHEESE
50% FAT CONTENT IN
DRY MATTER

FROMAGE CAMEMBERT
50% DE MAT. GR. S. SEC

		PROTEIN	FAT	CARBOHYDRATES	TOTAL
ENERGY VALUE (AVERAGE)	KJOULE	377	995	3.7	1376
PER 100 G	(KCAL)	90	238	0.9	329
EDIBLE PORTION					
AMOUNT OF DIGESTIBLE	GRAM	19.88	24.41	0.22	
CONSTITUENTS PER 100 G					
ENERGY VALUE (AVERAGE)	KJOULE	366	945	3.7	1315
OF THE DIGESTIBLE	(KCAL)	87	226	0.9	314
FRACTION PER 100 G					
EDIBLE PORTION					

WASTE PERCENTAGE AVERAGE 0.00

CONSTITUENTS	DIM	AV	VARIATION			AVR	NUTR. DENS.		MOLPERC.
WATER	GRAM	50.00	47.80	-	54.00	50.00	GRAM/MJ	38.03	
PROTEIN	GRAM	20.50	18.10	-	21.70	20.50	GRAM/MJ	15.59	
FAT	GRAM	25.70	23.00	-	27.10	25.70	GRAM/MJ	19.55	
AVAILABLE CARBOHYDR.	GRAM	0.22 1	-	-	-	0.22	GRAM/MJ	0.17	
MINERALS	GRAM	3.60	3.20	-	3.80	3.60	GRAM/MJ	2.74	
SODIUM	GRAM	0.90	0.81	-	1.15	0.90	GRAM/MJ	0.68	
POTASSIUM	MILLI	96.00	80.00	-	110.00	96.00	MILLI/MJ	73.02	
MAGNESIUM	MILLI	15.00	10.00	-	17.00	15.00	MILLI/MJ	11.41	
CALCIUM	MILLI	510.00	-	-	-	510.00	MILLI/MJ	387.93	
IRON	MILLI	0.13	0.08	-	0.17	0.13	MILLI/MJ	0.10	
COPPER	MICRO	60.00	50.00	-	70.00	60.00	MICRO/MJ	45.64	
ZINC	MILLI	2.70	1.60	-	3.60	2.70	MILLI/MJ	2.05	
PHOSPHORUS	MILLI	390.00	-	-	-	390.00	MILLI/MJ	296.65	
CHLORIDE	GRAM	1.24	1.00	-	1.40	1.24	GRAM/MJ	0.94	
FLUORIDE	MICRO	22.00	-	-	-	22.00	MICRO/MJ	16.73	
SELENIUM	MICRO	6.00	-	-	-	6.00	MICRO/MJ	4.56	
NITRATE	MILLI	0.20	0.10	-	1.10	0.20	MILLI/MJ	0.15	
VITAMIN A	MILLI	0.38	0.28	-	0.48	0.38	MILLI/MJ	0.29	
CAROTENE	MILLI	0.22	0.16	-	0.29	0.22	MILLI/MJ	0.17	
VITAMIN D	MICRO	0.32	0.23	-	0.40	0.32	MICRO/MJ	0.24	
VITAMIN E ACTIVITY	MILLI	0.58	0.28	-	0.68	0.58	MILLI/MJ	0.44	
ALPHA-TOCOPHEROL	MILLI	0.58	-	-	-	0.58	MILLI/MJ	0.44	
VITAMIN B1	MICRO	43.00	38.00	-	48.00	43.00	MICRO/MJ	32.71	
VITAMIN B2	MILLI	0.57	0.43	-	0.71	0.57	MILLI/MJ	0.43	
NICOTINAMIDE	MILLI	1.00	0.53	-	1.70	1.00	MILLI/MJ	0.76	
PANTOTHENIC ACID	MILLI	0.76	0.43	-	1.24	0.76	MILLI/MJ	0.58	
VITAMIN B6	MILLI	0.24	0.16	-	0.32	0.24	MILLI/MJ	0.18	
BIOTIN	MICRO	4.30	3.30	-	5.10	4.30	MICRO/MJ	3.27	
FOLIC ACID	MICRO	56.00	20.00	-	92.00	56.00	MICRO/MJ	42.60	
VITAMIN B12	MICRO	2.60	2.40	-	2.90	2.60	MICRO/MJ	1.98	
ALANINE	GRAM	0.81	0.77	-	0.85	0.81	GRAM/MJ	0.62	5.4
ARGININE	GRAM	0.76	0.72	-	0.81	0.76	GRAM/MJ	0.58	2.6
ASPARTIC ACID	GRAM	1.52	1.33	-	1.71	1.52	GRAM/MJ	1.16	6.8
CYSTINE	GRAM	0.18	0.13	-	0.20	0.18	GRAM/MJ	0.14	0.4
GLUTAMIC ACID	GRAM	4.50	4.34	-	4.64	4.50	GRAM/MJ	3.42	18.3
GLYCINE	GRAM	0.42	0.39	-	0.45	0.42	GRAM/MJ	0.32	3.3
HISTIDINE	GRAM	0.66	0.61	-	0.71	0.66	GRAM/MJ	0.50	2.5
ISOLEUCINE	GRAM	1.08	1.00	-	1.24	1.08	GRAM/MJ	0.82	4.9
LEUCINE	GRAM	1.96	1.83	-	2.06	1.96	GRAM/MJ	1.49	8.9
LYSINE	GRAM	1.69	1.44	-	1.83	1.69	GRAM/MJ	1.29	6.9
METHIONINE	GRAM	0.58	0.53	-	0.63	0.58	GRAM/MJ	0.44	2.3
PHENYLALANINE	GRAM	1.11	1.06	-	1.20	1.11	GRAM/MJ	0.84	4.0
PROLINE	GRAM	2.16	1.90	-	2.42	2.16	GRAM/MJ	1.64	11.2
SERINE	GRAM	1.17	1.15	-	1.20	1.17	GRAM/MJ	0.89	6.6
THREONINE	GRAM	0.82	0.69	-	1.00	0.82	GRAM/MJ	0.62	4.1
TRYPTOPHAN	GRAM	0.31	0.26	-	0.35	0.31	GRAM/MJ	0.24	0.9
TYROSINE	GRAM	1.06	0.90	-	1.25	1.06	GRAM/MJ	0.81	3.5
VALINE	GRAM	1.39	1.32	-	1.41	1.39	GRAM/MJ	1.06	7.1
CADAVERINE	MILLI	2.40	1.20	-	3.70	2.40	MILLI/MJ	1.83	
HISTAMINE	MILLI	0.26	0.00	-	48.00	0.26	MILLI/MJ	0.20	
PUTRESCINE	MILLI	1.50	0.70	-	3.30	1.50	MILLI/MJ	1.14	
TYRAMINE	MILLI	3.70	2.00	-	200.00	3.70	MILLI/MJ	2.81	

1 INCLUDES THE AMOUNTS OF LACTOSE AND LACTIC ACID

CONSTITUENTS	DIM	AV	VARIATION			AVR	NUTR. DENS.		MOLPERC.
LACTIC ACID	MILLI	107.00	27.00	-	240.00	107.00	MILLI/MJ	81.39	
SALICYLIC ACID	MICRO	10.00	10.00	-	10.00	10.00	MICRO/MJ	7.61	
LACTOSE	MILLI	101.00	101.00	-	101.00	101.00	MILLI/MJ	76.83	
BUTYRIC ACID	GRAM	0.52	-	-	-	0.52	GRAM/MJ	0.40	
CAPROIC ACID	GRAM	0.30	-	-	-	0.30	GRAM/MJ	0.23	
CAPRYLIC ACID	GRAM	0.28	-	-	-	0.28	GRAM/MJ	0.21	
CAPRIC ACID	GRAM	0.63	-	-	-	0.63	GRAM/MJ	0.48	
LAURIC ACID	GRAM	0.47	-	-	-	0.47	GRAM/MJ	0.36	
MYRISTIC ACID	GRAM	2.85	-	-	-	2.85	GRAM/MJ	2.17	
PALMITIC ACID	GRAM	7.66	-	-	-	7.66	GRAM/MJ	5.83	
STEARIC ACID	GRAM	2.67	-	-	-	2.67	GRAM/MJ	2.03	
PALMITOLEIC ACID	GRAM	0.93	-	-	-	0.93	GRAM/MJ	0.71	
OLEIC ACID	GRAM	6.09	-	-	-	6.09	GRAM/MJ	4.63	
LINOLEIC ACID	GRAM	0.48	-	-	-	0.48	GRAM/MJ	0.37	
LINOLENIC ACID	GRAM	0.29	-	-	-	0.29	GRAM/MJ	0.22	
CHOLESTEROL	MILLI	71.00	66.00	-	75.00	71.00	MILLI/MJ	54.01	

CAMEMBERTKÄSE
60% FETT I. TR.

CAMEMBERT CHEESE
60% FAT CONTENT IN DRY MATTER

FROMAGE CAMEMBERT
60% DE MAT. GR. S. SEC

		PROTEIN	FAT	CARBOHYDRATES	TOTAL
ENERGY VALUE (AVERAGE) PER 100 G EDIBLE PORTION	KJOULE	330	1316	0.0	1645
	(KCAL)	79	315	0.0	393
AMOUNT OF DIGESTIBLE CONSTITUENTS PER 100 G	GRAM	17.36	32.30	0.00	
ENERGY VALUE (AVERAGE) OF THE DIGESTIBLE FRACTION PER 100 G EDIBLE PORTION	KJOULE	320	1250	0.0	1570
	(KCAL)	76	299	0.0	375

WASTE PERCENTAGE AVERAGE 0.00

CONSTITUENTS	DIM	AV	VARIATION			AVR	NUTR. DENS.		MOLPERC.
WATER	GRAM	43.90	42.80	-	48.00	43.90	GRAM/MJ	27.97	
PROTEIN	GRAM	17.90	16.80	-	19.00	17.90	GRAM/MJ	11.40	
FAT	GRAM	34.00	31.20	-	35.30	34.00	GRAM/MJ	21.66	
MINERALS	GRAM	3.32	3.14	-	3.50	3.32	GRAM/MJ	2.12	
SODIUM	MILLI	944.00	-	-	-	944.00	MILLI/MJ	601.38	
POTASSIUM	MILLI	105.00	93.00	-	112.00	105.00	MILLI/MJ	66.89	
MAGNESIUM	MILLI	29.00	25.00	-	36.00	29.00	MILLI/MJ	18.47	
CALCIUM	MILLI	400.00	-	-	-	400.00	MILLI/MJ	254.82	
IRON	MILLI	0.58	0.34	-	0.95	0.58	MILLI/MJ	0.37	
ZINC	MILLI	2.70	1.60	-	3.60	2.70	MILLI/MJ	1.72	
PHOSPHORUS	MILLI	310.00	274.00	-	322.00	310.00	MILLI/MJ	197.49	
BORON	MICRO	65.00	-	-	-	65.00	MICRO/MJ	41.41	
SELENIUM	MICRO	6.00	-	-	-	6.00	MICRO/MJ	3.82	
NITRATE	MILLI	0.20	0.10	-	1.10	0.20	MILLI/MJ	0.13	

CONSTITUENTS	DIM	AV	VARIATION			AVR	NUTR. DENS.		MOLPERC.
VITAMIN A	MILLI	0.63	-	-	-	0.63	MILLI/MJ	0.40	
VITAMIN E ACTIVITY	MILLI	0.77	-	-	-	0.77	MILLI/MJ	0.49	
ALPHA-TOCOPHEROL	MILLI	0.77	-	-	-	0.77	MILLI/MJ	0.49	
VITAMIN B1	MICRO	37.00	20.00	-	49.00	37.00	MICRO/MJ	23.57	
VITAMIN B2	MILLI	0.37	0.29	-	0.46	0.37	MILLI/MJ	0.24	
NICOTINAMIDE	MILLI	1.18	0.90	-	1.78	1.18	MILLI/MJ	0.75	
PANTOTHENIC ACID	MILLI	0.70	0.40	-	1.10	0.70	MILLI/MJ	0.45	
VITAMIN B6	MILLI	0.20	0.14	-	0.28	0.20	MILLI/MJ	0.13	
BIOTIN	MICRO	2.80	1.60	-	4.10	2.80	MICRO/MJ	1.78	
ALANINE	GRAM	0.70	0.67	-	0.74	0.70	GRAM/MJ	0.45	
ARGININE	GRAM	0.65	-	-	-	0.65	GRAM/MJ	0.41	
ASPARTIC ACID	GRAM	1.32	1.16	-	1.49	1.32	GRAM/MJ	0.84	
CYSTINE	GRAM	0.11	-	-	-	0.11	GRAM/MJ	0.07	
GLUTAMIC ACID	GRAM	3.92	3.78	-	4.04	3.92	GRAM/MJ	2.50	
GLYCINE	GRAM	0.37	0.34	-	0.39	0.37	GRAM/MJ	0.24	
HISTIDINE	GRAM	0.58	-	-	-	0.58	GRAM/MJ	0.37	
ISOLEUCINE	GRAM	1.21	-	-	-	1.21	GRAM/MJ	0.77	
LEUCINE	GRAM	1.74	-	-	-	1.74	GRAM/MJ	1.11	
LYSINE	GRAM	1.31	-	-	-	1.31	GRAM/MJ	0.83	
METHIONINE	GRAM	0.47	-	-	-	0.47	GRAM/MJ	0.30	
PHENYLALANINE	GRAM	0.96	-	-	-	0.96	GRAM/MJ	0.61	
PROLINE	GRAM	1.88	1.65	-	2.11	1.88	GRAM/MJ	1.20	
SERINE	GRAM	1.02	1.00	-	1.04	1.02	GRAM/MJ	0.65	
THREONINE	GRAM	0.67	-	-	-	0.67	GRAM/MJ	0.43	
TRYPTOPHAN	GRAM	0.25	-	-	-	0.25	GRAM/MJ	0.16	
TYROSINE	GRAM	0.86	-	-	-	0.86	GRAM/MJ	0.55	
VALINE	GRAM	1.28	-	-	-	1.28	GRAM/MJ	0.82	
CADAVERINE	MILLI	2.40	1.20	-	3.70	2.40	MILLI/MJ	1.53	
HISTAMINE	MILLI	0.26	0.00	-	48.00	0.26	MILLI/MJ	0.17	
PUTRESCINE	MILLI	1.50	0.70	-	3.30	1.50	MILLI/MJ	0.96	
TYRAMINE	MILLI	3.70	2.00	-	200.00	3.70	MILLI/MJ	2.36	
SALICYLIC ACID	MICRO	10.00	10.00	-	10.00	10.00	MICRO/MJ	6.37	
BUTYRIC ACID	GRAM	0.69	-	-	-	0.69	GRAM/MJ	0.44	
CAPROIC ACID	GRAM	0.39	-	-	-	0.39	GRAM/MJ	0.25	
CAPRYLIC ACID	GRAM	0.37	-	-	-	0.37	GRAM/MJ	0.24	
CAPRIC ACID	GRAM	0.83	-	-	-	0.83	GRAM/MJ	0.53	
LAURIC ACID	GRAM	0.62	-	-	-	0.62	GRAM/MJ	0.39	
MYRISTIC ACID	GRAM	3.77	-	-	-	3.77	GRAM/MJ	2.40	
PALMITIC ACID	GRAM	10.13	-	-	-	10.13	GRAM/MJ	6.45	
STEARIC ACID	GRAM	3.53	-	-	-	3.53	GRAM/MJ	2.25	
PALMITOLEIC ACID	GRAM	1.23	-	-	-	1.23	GRAM/MJ	0.78	
OLEIC ACID	GRAM	8.06	-	-	-	8.06	GRAM/MJ	5.13	
LINOLEIC ACID	GRAM	0.63	-	-	-	0.63	GRAM/MJ	0.40	
LINOLENIC ACID	GRAM	0.38	-	-	-	0.38	GRAM/MJ	0.24	

CHESTERKÄSE
(CHEDDARKÄSE)
50% FETT I. TR.

CHEDDAR CHEESE
(CHESTER CHEESE)
50% FAT CONTENT IN DRY MATTER

FROMAGE CHESTER
(FROMAGE CHEDDAR)
50% DE MAT. GR. S. SEC

		PROTEIN	FAT	CARBOHYDRATES	TOTAL
ENERGY VALUE (AVERAGE) PER 100 G EDIBLE PORTION	KJOULE	468	1246	28	1742
	(KCAL)	112	298	6.8	416
AMOUNT OF DIGESTIBLE CONSTITUENTS PER 100 G	GRAM	24.63	30.59	1.69	
ENERGY VALUE (AVERAGE) OF THE DIGESTIBLE FRACTION PER 100 G EDIBLE PORTION	KJOULE	454	1184	28	1666
	(KCAL)	108	283	6.8	398

WASTE PERCENTAGE AVERAGE 0.00

CONSTITUENTS	DIM	AV	VARIATION			AVR	NUTR. DENS.		MOLPERC.
WATER	GRAM	36.30	31.30	-	38.00	36.30	GRAM/MJ	21.79	
PROTEIN	GRAM	25.40	23.70	-	28.00	25.40	GRAM/MJ	15.25	
FAT	GRAM	32.20	31.00	-	34.50	32.20	GRAM/MJ	19.33	
AVAILABLE CARBOHYDR.	GRAM	1.69 1	-	-	-	1.69	GRAM/MJ	1.01	
MINERALS	GRAM	4.41	3.70	-	5.50	4.41	GRAM/MJ	2.65	
SODIUM	MILLI	675.00	565.00	-	769.00	675.00	MILLI/MJ	405.22	
POTASSIUM	MILLI	102.00	82.00	-	116.00	102.00	MILLI/MJ	61.23	
MAGNESIUM	MILLI	37.00	25.00	-	55.00	37.00	MILLI/MJ	22.21	
CALCIUM	MILLI	810.00	741.00	-	984.00	810.00	MILLI/MJ	486.27	
MANGANESE	MICRO	17.00	13.00	-	19.00	17.00	MICRO/MJ	10.21	
IRON	MILLI	0.59	0.44	-	0.74	0.59	MILLI/MJ	0.35	
COPPER	MICRO	80.00	30.00	-	110.00	80.00	MICRO/MJ	48.03	
ZINC	MILLI	3.90	3.80	-	4.50	3.90	MILLI/MJ	2.34	
NICKEL	MICRO	1.40	-	-	-	1.40	MICRO/MJ	0.84	
CHROMIUM	MICRO	12.00	-	-	-	12.00	MICRO/MJ	7.20	
PHOSPHORUS	MILLI	530.00	478.00	-	610.00	530.00	MILLI/MJ	318.17	
CHLORIDE	GRAM	1.10	0.80	-	1.30	1.10	GRAM/MJ	0.66	
SELENIUM	MICRO	11.00	-	-	-	11.00	MICRO/MJ	6.60	
VITAMIN A	MILLI	0.39	0.25	-	0.52	0.39	MILLI/MJ	0.23	
CAROTENE	MILLI	0.30	-	-	-	0.30	MILLI/MJ	0.18	
VITAMIN D	MICRO	0.34	0.33	-	0.35	0.34	MICRO/MJ	0.20	
VITAMIN E ACTIVITY	MILLI	1.00	-	-	-	1.00	MILLI/MJ	0.60	
ALPHA-TOCOPHEROL	MILLI	1.00	-	-	-	1.00	MILLI/MJ	0.60	
VITAMIN B1	MICRO	37.00	27.00	-	50.00	37.00	MICRO/MJ	22.21	
VITAMIN B2	MILLI	0.44	0.32	-	0.50	0.44	MILLI/MJ	0.26	
NICOTINAMIDE	MILLI	0.11	0.09	-	0.17	0.11	MILLI/MJ	0.07	

1 INCLUDES THE AMOUNTS OF GLUCOSE, GALACTOSE, LACTOSE AND LACTIC ACID

CONSTITUENTS	DIM	AV	VARIATION			AVR	NUTR. DENS.		MOLPERC.
PANTOTHENIC ACID	MILLI	0.29	0.18	-	0.41	0.29	MILLI/MJ	0.17	
VITAMIN B6	MICRO	55.00	33.00	-	98.00	55.00	MICRO/MJ	33.02	
BIOTIN	MICRO	1.90	1.70	-	2.00	1.90	MICRO/MJ	1.14	
FOLIC ACID	MICRO	19.00	12.00	-	30.00	19.00	MICRO/MJ	11.41	
VITAMIN B12	MICRO	1.10	0.52	-	2.00	1.10	MICRO/MJ	0.66	
ALANINE	GRAM	0.76	-	-	-	0.76	GRAM/MJ	0.46	3.9
ARGININE	GRAM	0.90	0.79	-	0.97	0.90	GRAM/MJ	0.54	2.4
ASPARTIC ACID	GRAM	1.81	-	-	-	1.81	GRAM/MJ	1.09	6.2
CYSTINE	GRAM	0.21	0.10	-	0.40	0.21	GRAM/MJ	0.13	0.4
GLUTAMIC ACID	GRAM	6.62	-	-	-	6.62	GRAM/MJ	3.97	20.5
GLYCINE	GRAM	0.47	-	-	-	0.47	GRAM/MJ	0.28	2.9
HISTIDINE	GRAM	0.80	0.76	-	0.83	0.80	GRAM/MJ	0.48	2.3
ISOLEUCINE	GRAM	1.81	1.60	-	2.12	1.81	GRAM/MJ	1.09	6.3
LEUCINE	GRAM	2.52	1.85	-	3.06	2.52	GRAM/MJ	1.51	8.8
LYSINE	GRAM	2.07	1.86	-	2.33	2.07	GRAM/MJ	1.24	6.5
METHIONINE	GRAM	0.77	0.66	-	1.00	0.77	GRAM/MJ	0.46	2.4
PHENYLALANINE	GRAM	1.45	1.30	-	1.63	1.45	GRAM/MJ	0.87	4.0
PROLINE	GRAM	3.05	-	-	-	3.05	GRAM/MJ	1.83	12.1
SERINE	GRAM	1.58	-	-	-	1.58	GRAM/MJ	0.95	6.8
THREONINE	GRAM	0.98	0.88	-	1.07	0.98	GRAM/MJ	0.59	3.7
TRYPTOPHAN	GRAM	0.29	0.26	-	0.34	0.29	GRAM/MJ	0.17	0.6
TYROSINE	GRAM	1.30	0.21	-	1.30	1.30	GRAM/MJ	0.78	3.3
VALINE	GRAM	1.81	1.62	-	1.96	1.81	GRAM/MJ	1.09	7.0
CADAVERINE	MILLI	-	2.00	-	36.00				
HISTAMINE	MILLI	3.40	1.20	-	5.80	3.40	MILLI/MJ	2.04	
BETA-PHENYLETHYLAMINE	MILLI	12.00	0.00	-	30.00	12.00	MILLI/MJ	7.20	
PUTRESCINE	MILLI	-	1.00	-	70.00				
TRYPTAMINE	MILLI	0.10	0.00	-	0.20	0.10	MILLI/MJ	0.06	
TYRAMINE	MILLI	35.00	5.70	-	112.00	35.00	MILLI/MJ	21.01	
LACTIC ACID	GRAM	1.33	1.03	-	1.62	1.33	GRAM/MJ	0.80	
SUCCINIC ACID	MILLI	6.40	0.00	-	18.90	6.40	MILLI/MJ	3.84	
GLUCOSE	MILLI	3.60	0.40	-	10.70	3.60	MILLI/MJ	2.16	
LACTOSE	MILLI	298.00	72.00	-	479.00	298.00	MILLI/MJ	178.90	
GALACTOSE	MILLI	55.00	2.00	-	147.00	55.00	MILLI/MJ	33.02	
BUTYRIC ACID	GRAM	1.05	-	-	-	1.05	GRAM/MJ	0.63	
CAPROIC ACID	GRAM	0.49	-	-	-	0.49	GRAM/MJ	0.29	
CAPRYLIC ACID	GRAM	0.27	-	-	-	0.27	GRAM/MJ	0.16	
CAPRIC ACID	GRAM	0.58	-	-	-	0.58	GRAM/MJ	0.35	
LAURIC ACID	GRAM	0.52	-	-	-	0.52	GRAM/MJ	0.31	
MYRISTIC ACID	GRAM	3.23	-	-	-	3.23	GRAM/MJ	1.94	
PALMITIC ACID	GRAM	9.52	-	-	-	9.52	GRAM/MJ	5.72	
STEARIC ACID	GRAM	3.89	-	-	-	3.89	GRAM/MJ	2.34	
PALMITOLEIC ACID	GRAM	0.97	-	-	-	0.97	GRAM/MJ	0.58	
OLEIC ACID	GRAM	7.67	-	-	-	7.67	GRAM/MJ	4.60	
LINOLEIC ACID	GRAM	0.56	-	-	-	0.56	GRAM/MJ	0.34	
LINOLENIC ACID	GRAM	0.38	-	-	-	0.38	GRAM/MJ	0.23	
CHOLESTEROL	MILLI	100.00	95.00	-	105.00	100.00	MILLI/MJ	60.03	

COTTAGEKÄSE (HÜTTENKÄSE) | COTTAGE CHEESE | FROMAGE COTTAGE

		PROTEIN	FAT	CARBOHYDRATES	TOTAL
ENERGY VALUE (AVERAGE) PER 100 G EDIBLE PORTION	KJOULE	226	166	60	453
	(KCAL)	54	40	14	108
AMOUNT OF DIGESTIBLE CONSTITUENTS PER 100 G	GRAM	11.93	4.08	3.60	
ENERGY VALUE (AVERAGE) OF THE DIGESTIBLE FRACTION PER 100 G EDIBLE PORTION	KJOULE	220	158	60	438
	(KCAL)	52	38	14	105

WASTE PERCENTAGE AVERAGE 0.00

CONSTITUENTS	DIM	AV	VARIATION			AVR	NUTR. DENS.		MOLPERC.
WATER	GRAM	78.50	-	-	-	78.50	GRAM/MJ	179.23	
PROTEIN	GRAM	12.30	-	-	-	12.30	GRAM/MJ	28.08	
FAT	GRAM	4.30	-	-	-	4.30	GRAM/MJ	9.82	
AVAILABLE CARBOHYDR.	GRAM	3.60 1	-	-	-	3.60	GRAM/MJ	8.22	
SODIUM	MILLI	230.00	-	-	-	230.00	MILLI/MJ	525.12	
POTASSIUM	MILLI	88.00	-	-	-	88.00	MILLI/MJ	200.92	
CALCIUM	MILLI	95.00	-	-	-	95.00	MILLI/MJ	216.90	
IRON	MILLI	0.30	-	-	-	0.30	MILLI/MJ	0.68	
PHOSPHORUS	MILLI	150.00	-	-	-	150.00	MILLI/MJ	342.47	
VITAMIN A	MICRO	17.00	-	-	-	17.00	MICRO/MJ	38.81	
VITAMIN B1	MICRO	29.00	-	-	-	29.00	MICRO/MJ	66.21	
VITAMIN B2	MILLI	0.25	-	-	-	0.25	MILLI/MJ	0.57	
NICOTINAMIDE	MILLI	0.11	-	-	-	0.11	MILLI/MJ	0.25	
VITAMIN B12	MICRO	2.00	-	-	-	2.00	MICRO/MJ	4.57	
LACTIC ACID	GRAM	0.30	-	-	-	0.30	GRAM/MJ	0.68	
LACTOSE	GRAM	3.30	-	-	-	3.30	GRAM/MJ	7.53	

1 INCLUDES THE AMOUNTS OF LACTOSE AND LACTIC ACID

EDAMERKÄSE
30% FETT I. TR.

EDAM CHEESE
30% FAT CONTENT IN DRY MATTER

FROMAGE EDAM
30% DE MAT. GR. S. SEC

		PROTEIN	FAT	CARBOHYDRATES	TOTAL
ENERGY VALUE (AVERAGE)	KJOULE	486	627	0.0	1113
PER 100 G EDIBLE PORTION	(KCAL)	116	150	0.0	266
AMOUNT OF DIGESTIBLE CONSTITUENTS PER 100 G	GRAM	25.60	15.39	0.00	
ENERGY VALUE (AVERAGE)	KJOULE	471	596	0.0	1067
OF THE DIGESTIBLE FRACTION PER 100 G EDIBLE PORTION	(KCAL)	113	142	0.0	255

WASTE PERCENTAGE AVERAGE 0.00

CONSTITUENTS	DIM	AV	VARIATION	AVR	NUTR.	DENS.	MOLPERC.
WATER	GRAM	49.10	45.00 - 51.00	46.65	GRAM/MJ	46.01	
PROTEIN	GRAM	26.40	23.50 - 30.70	25.08	GRAM/MJ	24.74	
FAT	GRAM	16.20	14.70 - 18.70	15.39	GRAM/MJ	15.18	
MINERALS	GRAM	4.80	4.40 - 5.49	4.56	GRAM/MJ	4.50	
SODIUM	GRAM	0.80	0.77 - 1.33	0.76	GRAM/MJ	0.75	
POTASSIUM	MILLI	95.00	89.00 - 102.00	90.25	MILLI/MJ	89.03	
MAGNESIUM	MILLI	59.00	51.00 - 66.00	56.05	MILLI/MJ	55.29	
CALCIUM	MILLI	800.00	- - -	760.00	MILLI/MJ	749.73	
MANGANESE	MILLI	0.18	- - -	0.17	MILLI/MJ	0.17	
IRON	MILLI	0.60	0.50 - 1.30	0.57	MILLI/MJ	0.56	
COPPER	MILLI	0.78	0.00 - 1.50	0.74	MILLI/MJ	0.73	
ZINC	MILLI	-	1.10 - 10.60				
NICKEL	MICRO	89.00	33.00 - 144.00	84.55	MICRO/MJ	83.41	
CHROMIUM	MICRO	95.00	34.00 - 156.00	90.25	MICRO/MJ	89.03	
PHOSPHORUS	MILLI	570.00	- - -	541.50	MILLI/MJ	534.18	
CHLORIDE	GRAM	1.25	1.22 - 1.28	1.19	GRAM/MJ	1.17	
FLUORIDE	MILLI	-	0.04 - 0.17				
IODIDE	MICRO	5.30	- - -	5.04	MICRO/MJ	4.97	
NITRATE	MILLI	-	0.10 - 3.30				
VITAMIN A	MILLI	0.15	0.10 - 0.19	0.14	MILLI/MJ	0.14	
VITAMIN E ACTIVITY	MILLI	0.23	0.22 - 0.24	0.22	MILLI/MJ	0.22	
ALPHA-TOCOPHEROL	MILLI	0.23	- - -	0.22	MILLI/MJ	0.22	
VITAMIN B1	MICRO	60.00	32.00 - 78.00	57.00	MICRO/MJ	56.23	
VITAMIN B2	MILLI	0.35	0.17 - 0.55	0.33	MILLI/MJ	0.33	
NICOTINAMIDE	MILLI	0.10	0.04 - 0.11	0.10	MILLI/MJ	0.09	

CONSTITUENTS	DIM	AV	VARIATION			AVR	NUTR.	DENS.	MOLPERC.
ALANINE	GRAM	0.90	0.81	-	0.99	0.86	GRAM/MJ	0.84	4.5
ARGININE	GRAM	1.03	1.02	-	1.04	0.98	GRAM/MJ	0.97	2.6
ASPARTIC ACID	GRAM	2.02	1.84	-	2.19	1.92	GRAM/MJ	1.89	6.7
CYSTINE	GRAM	0.25	-	-	-	0.24	GRAM/MJ	0.23	0.5
GLUTAMIC ACID	GRAM	6.21	5.94	-	6.48	5.90	GRAM/MJ	5.82	18.7
GLYCINE	GRAM	0.56	0.52	-	0.59	0.53	GRAM/MJ	0.52	3.3
HISTIDINE	GRAM	0.94	0.79	-	1.09	0.89	GRAM/MJ	0.88	2.7
ISOLEUCINE	GRAM	1.49	1.38	-	1.76	1.42	GRAM/MJ	1.40	5.0
LEUCINE	GRAM	2.68	2.49	-	2.83	2.55	GRAM/MJ	2.51	9.0
LYSINE	GRAM	2.39	2.02	-	2.81	2.27	GRAM/MJ	2.24	7.2
METHIONINE	GRAM	0.79	0.76	-	0.90	0.75	GRAM/MJ	0.74	2.3
PHENYLALANINE	GRAM	1.45	1.35	-	1.54	1.38	GRAM/MJ	1.36	3.9
PROLINE	GRAM	2.94	2.43	-	3.43	2.79	GRAM/MJ	2.76	11.3
SERINE	GRAM	1.59	1.53	-	1.64	1.51	GRAM/MJ	1.49	6.7
THREONINE	GRAM	1.14	0.98	-	1.27	1.08	GRAM/MJ	1.07	4.2
TRYPTOPHAN	GRAM	0.40	0.36	-	0.47	0.38	GRAM/MJ	0.37	0.9
TYROSINE	GRAM	1.34	1.15	-	1.54	1.27	GRAM/MJ	1.26	3.3
VALINE	GRAM	1.90	1.78	-	2.01	1.81	GRAM/MJ	1.78	7.2
TYRAMINE	MILLI	31.00	-	-	-	29.45	MILLI/MJ	29.05	
BUTYRIC ACID	GRAM	0.59	-	-	-	0.56	GRAM/MJ	0.55	
CAPROIC ACID	GRAM	0.27	-	-	-	0.26	GRAM/MJ	0.25	
CAPRYLIC ACID	GRAM	0.17	-	-	-	0.16	GRAM/MJ	0.16	
CAPRIC ACID	GRAM	0.35	-	-	-	0.33	GRAM/MJ	0.33	
LAURIC ACID	GRAM	0.29	-	-	-	0.28	GRAM/MJ	0.27	
MYRISTIC ACID	GRAM	1.74	-	-	-	1.65	GRAM/MJ	1.63	
PALMITIC ACID	GRAM	4.78	-	-	-	4.54	GRAM/MJ	4.48	
STEARIC ACID	GRAM	1.76	-	-	-	1.67	GRAM/MJ	1.65	
PALMITOLEIC ACID	GRAM	0.39	-	-	-	0.37	GRAM/MJ	0.37	
OLEIC ACID	GRAM	4.09	-	-	-	3.89	GRAM/MJ	3.83	
LINOLEIC ACID	GRAM	0.25	-	-	-	0.24	GRAM/MJ	0.23	
LINOLENIC ACID	GRAM	0.15	-	-	-	0.14	GRAM/MJ	0.14	

EDAMERKÄSE
40% FETT I. TR.

EDAM CHEESE
40% FAT CONTENT IN DRY MATTER

FROMAGE EDAM
40% DE MAT. GR. S. SEC

		PROTEIN	FAT	CARBOHYDRATES	TOTAL
ENERGY VALUE (AVERAGE) PER 100 G EDIBLE PORTION	KJOULE	480	906	0.0	1386
	(KCAL)	115	216	0.0	331
AMOUNT OF DIGESTIBLE CONSTITUENTS PER 100 G	GRAM	25.31	22.23	0.00	
ENERGY VALUE (AVERAGE) OF THE DIGESTIBLE FRACTION PER 100 G EDIBLE PORTION	KJOULE	466	860	0.0	1326
	(KCAL)	111	206	0.0	317

WASTE PERCENTAGE AVERAGE 0.00

CONSTITUENTS	DIM	AV	VARIATION			AVR	NUTR.	DENS.	MOLPERC.
WATER	GRAM	44.80	40.40	-	47.00	42.56	GRAM/MJ	33.78	
PROTEIN	GRAM	26.10	24.40	-	27.70	24.80	GRAM/MJ	19.68	
FAT	GRAM	23.40	21.20	-	25.40	22.23	GRAM/MJ	17.64	
MINERALS	GRAM	4.68	4.07	-	5.19	4.45	GRAM/MJ	3.53	

CONSTITUENTS	DIM	AV	VARIATION			AVR	NUTR. DENS.		MOLPERC.
SODIUM	GRAM	0.90	0.74	-	1.07	0.86	GRAM/MJ	0.68	
POTASSIUM	MILLI	105.00	76.00	-	159.00	99.75	MILLI/MJ	79.16	
MAGNESIUM	MILLI	31.00	28.00	-	40.00	29.45	MILLI/MJ	23.37	
CALCIUM	MILLI	793.00	675.00	-	908.00	753.35	MILLI/MJ	597.85	
MANGANESE	MICRO	36.00	20.00	-	53.00	34.20	MICRO/MJ	27.14	
IRON	MILLI	0.27	0.19	-	0.50	0.26	MILLI/MJ	0.20	
COBALT	MICRO	1.00	1.00	-	2.00	0.95	MICRO/MJ	0.75	
COPPER	MICRO	49.00	30.00	-	70.00	46.55	MICRO/MJ	36.94	
ZINC	MILLI	4.90	4.20	-	5.30	4.66	MILLI/MJ	3.69	
NICKEL	MICRO	89.00	33.00	-	144.00	84.55	MICRO/MJ	67.10	
CHROMIUM	MICRO	95.00	34.00	-	156.00	90.25	MICRO/MJ	71.62	
MOLYBDENUM	MICRO	6.00	-	-	-	5.70	MICRO/MJ	4.52	
PHOSPHORUS	MILLI	498.00	387.00	-	550.00	473.10	MILLI/MJ	375.45	
CHLORIDE	GRAM	1.37	1.18	-	1.62	1.30	GRAM/MJ	1.03	
FLUORIDE	MICRO	70.00	40.00	-	160.00	66.50	MICRO/MJ	52.77	
IODIDE	MICRO	5.00	-	-	-	4.75	MICRO/MJ	3.77	
SELENIUM	MICRO	4.00	3.00	-	5.00	3.80	MICRO/MJ	3.02	
BROMINE	MILLI	0.10	-	-	-	0.10	MILLI/MJ	0.08	
SILICON	MILLI	0.50	0.50	-	1.00	0.48	MILLI/MJ	0.38	
NITRATE	MILLI	-	0.10	-	3.30				
VITAMIN A	MILLI	0.23	0.10	-	0.31	0.22	MILLI/MJ	0.17	
CAROTENE	MICRO	60.00	-	-	-	57.00	MICRO/MJ	45.23	
VITAMIN D	MICRO	0.29	0.23	-	0.35	0.28	MICRO/MJ	0.22	
VITAMIN E ACTIVITY	MILLI	0.34	0.33	-	0.36	0.32	MILLI/MJ	0.26	
ALPHA-TOCOPHEROL	MILLI	0.34	-	-	-	0.32	MILLI/MJ	0.26	
VITAMIN B1	MICRO	48.00	30.00	-	67.00	45.60	MICRO/MJ	36.19	
VITAMIN B2	MILLI	0.37	0.20	-	0.52	0.35	MILLI/MJ	0.28	
NICOTINAMIDE	MICRO	70.00	38.00	-	100.00	66.50	MICRO/MJ	52.77	
PANTOTHENIC ACID	MILLI	0.35	0.32	-	0.39	0.33	MILLI/MJ	0.26	
VITAMIN B6	MICRO	73.00	60.00	-	84.00	69.35	MICRO/MJ	55.04	
BIOTIN	MICRO	1.50	-	-	-	1.43	MICRO/MJ	1.13	
FOLIC ACID	MICRO	20.00	-	-	-	19.00	MICRO/MJ	15.08	
VITAMIN B12	MICRO	1.90	1.20	-	2.60	1.81	MICRO/MJ	1.43	
ALANINE	GRAM	0.89	0.80	-	0.98	0.85	GRAM/MJ	0.67	4.5
ARGININE	GRAM	1.02	1.01	-	1.03	0.97	GRAM/MJ	0.77	2.6
ASPARTIC ACID	GRAM	2.00	1.82	-	2.17	1.90	GRAM/MJ	1.51	6.7
CYSTINE	GRAM	0.25	-	-	-	0.24	GRAM/MJ	0.19	0.5
GLUTAMIC ACID	GRAM	6.15	5.88	-	6.42	5.84	GRAM/MJ	4.64	18.7
GLYCINE	GRAM	0.55	0.51	-	0.58	0.52	GRAM/MJ	0.41	3.3
HISTIDINE	GRAM	0.93	0.78	-	1.08	0.88	GRAM/MJ	0.70	2.7
ISOLEUCINE	GRAM	1.48	1.37	-	1.74	1.41	GRAM/MJ	1.12	5.0
LEUCINE	GRAM	2.65	2.47	-	2.80	2.52	GRAM/MJ	2.00	9.0
LYSINE	GRAM	2.37	2.00	-	2.78	2.25	GRAM/MJ	1.79	7.2
METHIONINE	GRAM	0.78	0.75	-	0.89	0.74	GRAM/MJ	0.59	2.3
PHENYLALANINE	GRAM	1.44	1.34	-	1.52	1.37	GRAM/MJ	1.09	3.9
PROLINE	GRAM	2.91	2.41	-	3.40	2.76	GRAM/MJ	2.19	11.3
SERINE	GRAM	1.57	1.51	-	1.62	1.49	GRAM/MJ	1.18	6.7
THREONINE	GRAM	1.13	0.97	-	1.26	1.07	GRAM/MJ	0.85	4.2
TRYPTOPHAN	GRAM	0.40	0.36	-	0.47	0.38	GRAM/MJ	0.30	0.9
TYROSINE	GRAM	1.33	1.14	-	1.52	1.26	GRAM/MJ	1.00	3.3
VALINE	GRAM	1.88	1.76	-	1.99	1.79	GRAM/MJ	1.42	7.2

CONSTITUENTS	DIM	AV	VARIATION			AVR	NUTR. DENS.		MOLPERC.
TYRAMINE	MILLI	31.00	-	-	-	29.45	MILLI/MJ	23.37	
BUTYRIC ACID	GRAM	0.86	-	-	-	0.82	GRAM/MJ	0.65	
CAPROIC ACID	GRAM	0.39	-	-	-	0.37	GRAM/MJ	0.29	
CAPRYLIC ACID	GRAM	0.26	-	-	-	0.25	GRAM/MJ	0.20	
CAPRIC ACID	GRAM	0.50	-	-	-	0.48	GRAM/MJ	0.38	
LAURIC ACID	GRAM	0.43	-	-	-	0.41	GRAM/MJ	0.32	
MYRISTIC ACID	GRAM	2.52	-	-	-	2.39	GRAM/MJ	1.90	
PALMITIC ACID	GRAM	6.92	-	-	-	6.57	GRAM/MJ	5.22	
STEARIC ACID	GRAM	2.55	-	-	-	2.42	GRAM/MJ	1.92	
PALMITOLEIC ACID	GRAM	0.69	-	-	-	0.66	GRAM/MJ	0.52	
OLEIC ACID	GRAM	5.92	-	-	-	5.62	GRAM/MJ	4.46	
LINOLEIC ACID	GRAM	0.36	-	-	-	0.34	GRAM/MJ	0.27	
LINOLENIC ACID	GRAM	0.21	-	-	-	0.20	GRAM/MJ	0.16	
CHOLESTEROL	MILLI	71.00	67.00	-	77.00	67.45	MILLI/MJ	53.53	

EDAMERKÄSE
45% FETT I. TR.

EDAM CHEESE
45% FAT CONTENT IN DRY MATTER

FROMAGE EDAM
45% DE MAT. GR. S. SEC

		PROTEIN	FAT	CARBOHYDRATES	TOTAL
ENERGY VALUE (AVERAGE) PER 100 G EDIBLE PORTION	KJOULE	457	1095	0.0	1552
	(KCAL)	109	262	0.0	371
AMOUNT OF DIGESTIBLE CONSTITUENTS PER 100 G	GRAM	24.05	26.88	0.00	
ENERGY VALUE (AVERAGE) OF THE DIGESTIBLE FRACTION PER 100 G EDIBLE PORTION	KJOULE	443	1041	0.0	1483
	(KCAL)	106	249	0.0	355

WASTE PERCENTAGE AVERAGE 0.00

CONSTITUENTS	DIM	AV	VARIATION			AVR	NUTR. DENS.		MOLPERC.
WATER	GRAM	41.90	38.60	-	45.00	39.81	GRAM/MJ	28.25	
PROTEIN	GRAM	24.80	-	-	-	23.56	GRAM/MJ	16.72	
FAT	GRAM	28.30	24.80	-	28.80	26.89	GRAM/MJ	19.08	
MINERALS	GRAM	4.00	-	-	-	3.80	GRAM/MJ	2.70	

CONSTITUENTS	DIM	AV	VARIATION			AVR	NUTR. DENS.		MOLPERC.
SODIUM	MILLI	654.00	484.00	-	832.00	621.30	MILLI/MJ	440.89	
POTASSIUM	MILLI	67.00	39.00	-	84.00	63.65	MILLI/MJ	45.17	
CALCIUM	MILLI	678.00	598.00	-	784.00	644.10	MILLI/MJ	457.07	
MANGANESE	MILLI	0.15	-	-	-	0.14	MILLI/MJ	0.10	
IRON	MILLI	0.62	0.18	-	1.22	0.59	MILLI/MJ	0.42	
COPPER	MILLI	0.65	0.00	-	1.20	0.62	MILLI/MJ	0.44	
ZINC	MILLI	-	0.89	-	8.87				
NICKEL	MICRO	89.00	33.00	-	144.00	84.55	MICRO/MJ	60.00	
CHROMIUM	MICRO	95.00	34.00	-	156.00	90.25	MICRO/MJ	64.04	
PHOSPHORUS	MILLI	403.00	343.00	-	464.00	382.85	MILLI/MJ	271.68	
CHLORIDE	GRAM	1.05	1.02	-	1.07	1.00	GRAM/MJ	0.71	
FLUORIDE	MILLI	-	0.04	-	0.14				
IODIDE	MICRO	4.00	-	-	-	3.80	MICRO/MJ	2.70	
NITRATE	MILLI	-	0.10	-	3.30				
VITAMIN A	MILLI	0.22	0.12	-	0.34	0.21	MILLI/MJ	0.15	
VITAMIN E ACTIVITY	MILLI	0.41	0.40	-	0.43	0.39	MILLI/MJ	0.28	
ALPHA-TOCOPHEROL	MILLI	0.42	-	-	-	0.40	MILLI/MJ	0.28	
VITAMIN B1	MICRO	56.00	30.00	-	73.00	53.20	MICRO/MJ	37.75	
VITAMIN B2	MILLI	0.35	0.16	-	0.52	0.33	MILLI/MJ	0.24	
NICOTINAMIDE	MICRO	69.00	39.00	-	100.00	65.55	MICRO/MJ	46.52	
VITAMIN B12	MICRO	2.10	1.90	-	2.30	2.00	MICRO/MJ	1.42	
ALANINE	GRAM	0.85	0.76	-	0.93	0.81	GRAM/MJ	0.57	4.5
ARGININE	GRAM	0.97	0.96	-	0.98	0.92	GRAM/MJ	0.65	2.6
ASPARTIC ACID	GRAM	1.90	1.73	-	2.06	1.81	GRAM/MJ	1.28	6.7
CYSTINE	GRAM	0.24	-	-	-	0.23	GRAM/MJ	0.16	0.5
GLUTAMIC ACID	GRAM	5.84	5.59	-	6.10	5.55	GRAM/MJ	3.94	18.7
GLYCINE	GRAM	0.52	0.48	-	0.55	0.49	GRAM/MJ	0.35	3.3
HISTIDINE	GRAM	0.88	0.74	-	1.03	0.84	GRAM/MJ	0.59	2.7
ISOLEUCINE	GRAM	1.41	1.30	-	1.65	1.34	GRAM/MJ	0.95	5.0
LEUCINE	GRAM	2.52	2.35	-	2.66	2.39	GRAM/MJ	1.70	9.0
LYSINE	GRAM	2.25	1.90	-	2.64	2.14	GRAM/MJ	1.52	7.2
METHIONINE	GRAM	0.74	0.71	-	0.85	0.70	GRAM/MJ	0.50	2.3
PHENYLALANINE	GRAM	1.37	1.27	-	1.44	1.30	GRAM/MJ	0.92	3.9
PROLINE	GRAM	2.76	2.29	-	3.23	2.62	GRAM/MJ	1.86	11.3
SERINE	GRAM	1.49	1.43	-	1.54	1.42	GRAM/MJ	1.00	6.7
THREONINE	GRAM	1.07	0.92	-	1.20	1.02	GRAM/MJ	0.72	4.2
TRYPTOPHAN	GRAM	0.38	0.34	-	0.45	0.36	GRAM/MJ	0.26	0.9
TYROSINE	GRAM	1.26	1.08	-	1.44	1.20	GRAM/MJ	0.85	3.3
VALINE	GRAM	1.79	1.67	-	1.89	1.70	GRAM/MJ	1.21	7.2
TYRAMINE	MILLI	31.00	-	-	-	29.45	MILLI/MJ	20.90	
BUTYRIC ACID	GRAM	1.04	-	-	-	0.99	GRAM/MJ	0.70	
CAPROIC ACID	GRAM	0.48	-	-	-	0.46	GRAM/MJ	0.32	
CAPRYLIC ACID	GRAM	0.31	-	-	-	0.29	GRAM/MJ	0.21	
CAPRIC ACID	GRAM	0.61	-	-	-	0.58	GRAM/MJ	0.41	
LAURIC ACID	GRAM	0.52	-	-	-	0.49	GRAM/MJ	0.35	
MYRISTIC ACID	GRAM	3.05	-	-	-	2.90	GRAM/MJ	2.06	
PALMITIC ACID	GRAM	8.37	-	-	-	7.95	GRAM/MJ	5.64	
STEARIC ACID	GRAM	3.09	-	-	-	2.94	GRAM/MJ	2.08	
PALMITOLEIC ACID	GRAM	0.84	-	-	-	0.80	GRAM/MJ	0.57	
OLEIC ACID	GRAM	7.16	-	-	-	6.80	GRAM/MJ	4.83	
LINOLEIC ACID	GRAM	0.44	-	-	-	0.42	GRAM/MJ	0.30	
LINOLENIC ACID	GRAM	0.26	-	-	-	0.25	GRAM/MJ	0.18	

EDELPILZKÄSE
50% FETT I. TR.

BLUE CHEESE
50% FAT CONTENT IN DRY MATTER

FROMAGE BLEU
50% DE MAT. GR. S. SEC

		PROTEIN	FAT	CARBOHYDRATES	TOTAL
ENERGY VALUE (AVERAGE)	KJOULE	388	1153	12	1554
PER 100 G	(KCAL)	93	276	2.9	371
EDIBLE PORTION					
AMOUNT OF DIGESTIBLE	GRAM	20.46	28.31	0.72	
CONSTITUENTS PER 100 G					
ENERGY VALUE (AVERAGE)	KJOULE	377	1096	12	1484
OF THE DIGESTIBLE	(KCAL)	90	262	2.9	355
FRACTION PER 100 G					
EDIBLE PORTION					

WASTE PERCENTAGE AVERAGE 0.00

CONSTITUENTS	DIM	AV	VARIATION			AVR	NUTR. DENS.		MOLPERC.
WATER	GRAM	42.80	40.00	-	50.00	42.80	GRAM/MJ	28.83	
PROTEIN	GRAM	21.10	19.40	-	23.00	21.10	GRAM/MJ	14.21	
FAT	GRAM	29.80	25.50	-	32.00	29.80	GRAM/MJ	20.07	
AVAILABLE CARBOHYDR.	GRAM	0.72 1	-	-	-	0.72	GRAM/MJ	0.49	
MINERALS	GRAM	5.45	4.40	-	6.60	5.45	GRAM/MJ	3.67	
SODIUM	GRAM	1.45	1.23	-	1.79	1.45	GRAM/MJ	0.98	
POTASSIUM	MILLI	138.00	100.00	-	186.00	138.00	MILLI/MJ	92.96	
MAGNESIUM	MILLI	39.00	23.00	-	47.00	39.00	MILLI/MJ	26.27	
CALCIUM	MILLI	526.00	390.00	-	720.00	526.00	MILLI/MJ	354.33	
MANGANESE	MILLI	0.19	-	-	-	0.19	MILLI/MJ	0.13	
IRON	MILLI	0.19	0.17	-	0.20	0.19	MILLI/MJ	0.13	
COBALT	MICRO	1.00	-	-	-	1.00	MICRO/MJ	0.67	
COPPER	MILLI	0.14	0.09	-	0.17	0.14	MILLI/MJ	0.09	
ZINC	MILLI	3.10	-	-	-	3.10	MILLI/MJ	2.09	
CHROMIUM	MICRO	1.00	-	-	-	1.00	MICRO/MJ	0.67	
PHOSPHORUS	MILLI	362.00	330.00	-	425.00	362.00	MILLI/MJ	243.85	
CHLORIDE	GRAM	2.50	1.80	-	3.00	2.50	GRAM/MJ	1.68	
FLUORIDE	MICRO	50.00	-	-	-	50.00	MICRO/MJ	33.68	
SELENIUM	MICRO	2.00	-	-	-	2.00	MICRO/MJ	1.35	
BROMINE	MILLI	0.30	-	-	-	0.30	MILLI/MJ	0.20	
VITAMIN A	MILLI	0.26	0.16	-	0.33	0.26	MILLI/MJ	0.18	
CAROTENE	MILLI	0.18	0.12	-	0.27	0.18	MILLI/MJ	0.12	
VITAMIN E ACTIVITY	MILLI	0.77	0.60	-	1.00	0.77	MILLI/MJ	0.52	
ALPHA-TOCOPHEROL	MILLI	0.77	-	-	-	0.77	MILLI/MJ	0.52	
VITAMIN B1	MICRO	43.00	30.00	-	60.00	43.00	MICRO/MJ	28.97	
VITAMIN B2	MILLI	0.50	0.30	-	0.61	0.50	MILLI/MJ	0.34	

1 INCLUDES ONLY THE AMOUNT OF LACTIC ACID

CONSTITUENTS	DIM	AV	VARIATION			AVR	NUTR. DENS.		MOLPERC.
NICOTINAMIDE	MILLI	0.81	0.36	-	1.24	0.81	MILLI/MJ	0.55	
PANTOTHENIC ACID	MILLI	2.00	-	-	-	2.00	MILLI/MJ	1.35	
VITAMIN B6	MILLI	0.18	0.13	-	0.23	0.18	MILLI/MJ	0.12	
BIOTIN	MICRO	3.00	-	-	-	3.00	MICRO/MJ	2.02	
FOLIC ACID	MICRO	40.00	-	-	-	40.00	MICRO/MJ	26.95	
VITAMIN B12	MICRO	0.59	-	-	-	0.59	MICRO/MJ	0.40	
ALANINE	GRAM	0.73	0.68	-	0.80	0.73	GRAM/MJ	0.49	4.5
ARGININE	GRAM	1.65	1.55	-	1.72	1.65	GRAM/MJ	1.11	5.2
ASPARTIC ACID	GRAM	0.83	0.58	-	0.97	0.83	GRAM/MJ	0.56	3.4
CYSTINE	GRAM	0.12	0.10	-	0.17	0.12	GRAM/MJ	0.08	0.3
GLUTAMIC ACID	GRAM	4.97	4.51	-	5.16	4.97	GRAM/MJ	3.35	18.4
GLYCINE	GRAM	0.51	0.49	-	0.56	0.51	GRAM/MJ	0.34	3.7
HISTIDINE	GRAM	0.99	0.87	-	1.05	0.99	GRAM/MJ	0.67	3.5
ISOLEUCINE	GRAM	1.19	1.14	-	1.31	1.19	GRAM/MJ	0.80	4.9
LEUCINE	GRAM	2.14	2.01	-	2.23	2.14	GRAM/MJ	1.44	8.9
LYSINE	GRAM	2.38	2.23	-	2.45	2.38	GRAM/MJ	1.60	8.9
METHIONINE	GRAM	0.52	0.45	-	0.59	0.52	GRAM/MJ	0.35	1.9
PHENYLALANINE	GRAM	1.22	1.14	-	1.26	1.22	GRAM/MJ	0.82	4.0
PROLINE	GRAM	2.35	2.21	-	2.50	2.35	GRAM/MJ	1.58	11.1
SERINE	GRAM	1.33	1.24	-	1.48	1.33	GRAM/MJ	0.90	6.9
THREONINE	GRAM	0.92	0.90	-	0.99	0.92	GRAM/MJ	0.62	4.2
TRYPTOPHAN	GRAM	0.21	0.17	-	0.25	0.21	GRAM/MJ	0.14	0.6
TYROSINE	GRAM	1.02	0.77	-	1.11	1.02	GRAM/MJ	0.69	3.1
VALINE	GRAM	1.46	1.36	-	1.58	1.46	GRAM/MJ	0.98	6.8
LACTIC ACID	GRAM	0.72	0.11	-	1.33	0.72	GRAM/MJ	0.49	
BUTYRIC ACID	GRAM	0.68	-	-	-	0.68	GRAM/MJ	0.46	
CAPROIC ACID	GRAM	0.37	-	-	-	0.37	GRAM/MJ	0.25	
CAPRYLIC ACID	GRAM	0.26	-	-	-	0.26	GRAM/MJ	0.18	
CAPRIC ACID	GRAM	0.62	-	-	-	0.62	GRAM/MJ	0.42	
LAURIC ACID	GRAM	0.51	-	-	-	0.51	GRAM/MJ	0.34	
MYRISTIC ACID	GRAM	3.42	-	-	-	3.42	GRAM/MJ	2.30	
PALMITIC ACID	GRAM	9.49	-	-	-	9.49	GRAM/MJ	6.39	
STEARIC ACID	GRAM	3.36	-	-	-	3.36	GRAM/MJ	2.26	
PALMITOLEIC ACID	GRAM	0.85	-	-	-	0.85	GRAM/MJ	0.57	
OLEIC ACID	GRAM	6.87	-	-	-	6.87	GRAM/MJ	4.63	
LINOLEIC ACID	GRAM	0.56	-	-	-	0.56	GRAM/MJ	0.38	
LINOLENIC ACID	GRAM	0.27	-	-	-	0.27	GRAM/MJ	0.18	
CHOLESTEROL	MILLI	88.00	78.00	-	98.00	88.00	MILLI/MJ	59.28	

EMMENTALERKÄSE
45% FETT I. TR.

EMMENTAL CHEESE
45% FAT CONTENT IN DRY MATTER

FROMAGE EMMENTALER
45% DE MAT. GR. S. SEC

		PROTEIN	FAT	CARBOHYDRATES	TOTAL
ENERGY VALUE (AVERAGE) PER 100 G EDIBLE PORTION	KJOULE	528	1149	7.5	1685
	(KCAL)	126	275	1.8	403
AMOUNT OF DIGESTIBLE CONSTITUENTS PER 100 G	GRAM	27.83	28.21	0.45	
ENERGY VALUE (AVERAGE) OF THE DIGESTIBLE FRACTION PER 100 G EDIBLE PORTION	KJOULE	513	1092	7.5	1612
	(KCAL)	122	261	1.8	385

WASTE PERCENTAGE AVERAGE[1] 6.0 MINIMUM 3.0 MAXIMUM 10

CONSTITUENTS	DIM	AV	VARIATION		AVR	NUTR.	DENS.	MOLPERC.
WATER	GRAM	35.70	33.60	- 38.00	33.56	GRAM/MJ	22.15	
PROTEIN	GRAM	28.70	27.40	- 30.20	26.98	GRAM/MJ	17.80	
FAT	GRAM	29.70	27.90	- 32.00	27.92	GRAM/MJ	18.42	
AVAILABLE CARBOHYDR.	GRAM	0.45 [2]	-	- -	0.42	GRAM/MJ	0.28	
MINERALS	GRAM	3.88	3.53	- 4.40	3.65	GRAM/MJ	2.41	
SODIUM	MILLI	450.00	270.00	- 600.00	423.00	MILLI/MJ	279.15	
POTASSIUM	MILLI	107.00	100.00	- 119.00	100.58	MILLI/MJ	66.38	
MAGNESIUM	MILLI	35.00	21.00	- 48.00	32.90	MILLI/MJ	21.71	
CALCIUM	GRAM	1.02	0.89	- 1.18	0.96	GRAM/MJ	0.63	
MANGANESE	MICRO	27.00	20.00	- 44.00	25.38	MICRO/MJ	16.75	
IRON	MILLI	0.31	0.18	- 0.50	0.29	MILLI/MJ	0.19	
COBALT	MICRO	1.00	-	- -	0.94	MICRO/MJ	0.62	
COPPER	MILLI	1.17	0.53	- 1.80	1.10	MILLI/MJ	0.73	
ZINC	MILLI	4.63	4.30	- 5.00	4.35	MILLI/MJ	2.87	
NICKEL	MICRO	20.00	-	- -	18.80	MICRO/MJ	12.41	
CHROMIUM	MICRO	5.00	1.00	- 156.00	4.70	MICRO/MJ	3.10	
MOLYBDENUM	MICRO	10.00	-	- -	9.40	MICRO/MJ	6.20	
PHOSPHORUS	MILLI	636.00	540.00	- 860.00	597.84	MILLI/MJ	394.54	
CHLORIDE	MILLI	370.00	280.00	- 500.00	347.80	MILLI/MJ	229.53	
FLUORIDE	MICRO	60.00	50.00	- 60.00	56.40	MICRO/MJ	37.22	
BORON	MILLI	0.13	-	- -	0.12	MILLI/MJ	0.08	
SELENIUM	MICRO	11.00	-	- -	10.34	MICRO/MJ	6.82	
BROMINE	MILLI	0.40	0.10	- 1.00	0.38	MILLI/MJ	0.25	
SILICON	MILLI	1.00	0.50	- 2.00	0.94	MILLI/MJ	0.62	
VITAMIN A	MILLI	0.32	0.10	- 0.42	0.30	MILLI/MJ	0.20	
CAROTENE	MILLI	0.14	0.11	- 0.17	0.13	MILLI/MJ	0.09	

1 RIND

2 INCLUDES ONLY THE AMOUNT OF LACTIC ACID

CONSTITUENTS	DIM	AV	VARIATION			AVR	NUTR. DENS.		MOLPERC.
VITAMIN D	MICRO	1.10	0.35	-	3.10	1.03	MICRO/MJ	0.68	
VITAMIN E ACTIVITY	MILLI	0.35	0.30	-	0.40	0.33	MILLI/MJ	0.22	
ALPHA-TOCOPHEROL	MILLI	0.35	-	-	-	0.33	MILLI/MJ	0.22	
VITAMIN B1	MICRO	50.00	22.00	-	74.00	47.00	MICRO/MJ	31.02	
VITAMIN B2	MILLI	0.34	0.22	-	0.53	0.32	MILLI/MJ	0.21	
NICOTINAMIDE	MILLI	0.18	0.10	-	0.30	0.17	MILLI/MJ	0.11	
PANTOTHENIC ACID	MILLI	0.40	0.26	-	0.55	0.38	MILLI/MJ	0.25	
VITAMIN B6	MICRO	65.00	54.00	-	90.00	61.10	MICRO/MJ	40.32	
BIOTIN	MICRO	3.00	2.00	-	4.00	2.82	MICRO/MJ	1.86	
FOLIC ACID	MICRO	4.30	3.50	-	6.00	4.04	MICRO/MJ	2.67	
VITAMIN B12	MICRO	2.20	1.20	-	3.60	2.07	MICRO/MJ	1.36	
VITAMIN C	MILLI	0.50	0.00	-	1.00	0.47	MILLI/MJ	0.31	
ALANINE	GRAM	0.92	-	-	-	0.86	GRAM/MJ	0.57	4.4
ARGININE	GRAM	1.00	0.92	-	1.04	0.94	GRAM/MJ	0.62	2.4
ASPARTIC ACID	GRAM	1.58	-	-	-	1.49	GRAM/MJ	0.98	5.0
CYSTINE	GRAM	0.17	0.14	-	0.29	0.16	GRAM/MJ	0.11	0.3
GLUTAMIC ACID	GRAM	5.76	-	-	-	5.41	GRAM/MJ	3.57	16.6
GLYCINE	GRAM	0.51	-	-	-	0.48	GRAM/MJ	0.32	2.9
HISTIDINE	GRAM	1.02	0.93	-	1.08	0.96	GRAM/MJ	0.63	2.8
ISOLEUCINE	GRAM	1.73	1.55	-	1.93	1.63	GRAM/MJ	1.07	5.6
LEUCINE	GRAM	2.99	2.79	-	3.24	2.81	GRAM/MJ	1.85	9.7
LYSINE	GRAM	2.39	2.10	-	2.61	2.25	GRAM/MJ	1.48	6.9
METHIONINE	GRAM	0.79	0.74	-	0.92	0.74	GRAM/MJ	0.49	2.2
PHENYLALANINE	GRAM	1.61	1.53	-	1.68	1.51	GRAM/MJ	1.00	4.1
PROLINE	GRAM	3.73	-	-	-	3.51	GRAM/MJ	2.31	13.8
SERINE	GRAM	1.66	-	-	-	1.56	GRAM/MJ	1.03	6.7
THREONINE	GRAM	1.14	1.06	-	1.30	1.07	GRAM/MJ	0.71	4.1
TRYPTOPHAN	GRAM	0.43	0.37	-	0.52	0.40	GRAM/MJ	0.27	0.9
TYROSINE	GRAM	1.61	1.37	-	1.84	1.51	GRAM/MJ	1.00	3.8
VALINE	GRAM	2.12	2.07	-	2.30	1.99	GRAM/MJ	1.32	7.7
CADAVERINE	MILLI	0.50	0.01	-	8.00	0.47	MILLI/MJ	0.31	
HISTAMINE	MILLI	4.10	0.40	-	250.00	3.85	MILLI/MJ	2.54	
BETA-PHENYLETHYLAMINE	MILLI	1.60	0.00	-	7.80	1.50	MILLI/MJ	0.99	
PUTRESCINE	MILLI	0.40	0.01	-	15.00	0.38	MILLI/MJ	0.25	
TRYPTAMINE	-	0.00	-	-	-	0.00			
TYRAMINE	MILLI	8.40	0.50	-	28.00	7.90	MILLI/MJ	5.21	
LACTIC ACID	GRAM	0.45	0.16	-	0.67	0.42	GRAM/MJ	0.28	
BUTYRIC ACID	GRAM	1.15	-	-	-	1.08	GRAM/MJ	0.71	
CAPROIC ACID	GRAM	0.51	-	-	-	0.48	GRAM/MJ	0.32	
CAPRYLIC ACID	GRAM	0.30	-	-	-	0.28	GRAM/MJ	0.19	
CAPRIC ACID	GRAM	0.65	-	-	-	0.61	GRAM/MJ	0.40	
LAURIC ACID	GRAM	0.54	-	-	-	0.51	GRAM/MJ	0.33	
MYRISTIC ACID	GRAM	3.19	-	-	-	3.00	GRAM/MJ	1.98	
PALMITIC ACID	GRAM	8.14	-	-	-	7.65	GRAM/MJ	5.05	
STEARIC ACID	GRAM	3.39	-	-	-	3.19	GRAM/MJ	2.10	
PALMITOLEIC ACID	GRAM	0.92	-	-	-	0.86	GRAM/MJ	0.57	
OLEIC ACID	GRAM	6.29	-	-	-	5.91	GRAM/MJ	3.90	
LINOLEIC ACID	GRAM	0.65	-	-	-	0.61	GRAM/MJ	0.40	
LINOLENIC ACID	GRAM	0.37	-	-	-	0.35	GRAM/MJ	0.23	
ARACHIDONIC ACID	MILLI	28.00	-	-	-	26.32	MILLI/MJ	17.37	
CHOLESTEROL	MILLI	92.00	-	-	-	86.48	MILLI/MJ	57.07	

FETAKÄSE	FETA CHEESE	FROMAGE FETA
45% FETT I. TR.	45% FAT CONTENT IN DRY MATTER	45% DE MAT. GR. S. SEC.

		PROTEIN	FAT	CARBOHYDRATES	TOTAL
ENERGY VALUE (AVERAGE) PER 100 G EDIBLE PORTION	KJOULE	313	701	33	1046
	(KCAL)	75	167	7.8	250
AMOUNT OF DIGESTIBLE CONSTITUENTS PER 100 G	GRAM	16.49	17.19	1.95	
ENERGY VALUE (AVERAGE) OF THE DIGESTIBLE FRACTION PER 100 G EDIBLE PORTION	KJOULE	304	665	33	1002
	(KCAL)	73	159	7.8	239

WASTE PERCENTAGE AVERAGE 0.00

CONSTITUENTS	DIM	AV	VARIATION	AVR	NUTR.	DENS.	MOLPERC.
WATER	GRAM	59.10	- - -	59.10	GRAM/MJ	59.00	
PROTEIN	GRAM	17.00	- - -	17.00	GRAM/MJ	16.97	
FAT	GRAM	18.10	- - -	18.10	GRAM/MJ	18.07	
AVAILABLE CARBOHYDR.	GRAM	1.95 1	- - -	1.95	GRAM/MJ	1.95	
MINERALS	GRAM	5.20	- - -	5.20	GRAM/MJ	5.19	
MAGNESIUM	MILLI	19.00	- - -	19.00	MILLI/MJ	18.97	
CALCIUM	MILLI	429.00	- - -	429.00	MILLI/MJ	428.28	
IRON	MILLI	0.65	- - -	0.65	MILLI/MJ	0.65	
ZINC	MILLI	2.90	- - -	2.90	MILLI/MJ	2.90	
PHOSPHORUS	MILLI	337.00	- - -	337.00	MILLI/MJ	336.43	
LACTIC ACID	GRAM	1.42	1.21 - 1.57	1.42	GRAM/MJ	1.42	
LACTOSE	MILLI	527.00	527.00 - 527.00	527.00	MILLI/MJ	526.11	
BUTYRIC ACID	GRAM	0.77	- - -	0.77	GRAM/MJ	0.77	
CAPROIC ACID	GRAM	0.57	- - -	0.57	GRAM/MJ	0.57	
CAPRYLIC ACID	GRAM	0.55	- - -	0.55	GRAM/MJ	0.55	
CAPRIC ACID	GRAM	1.96	- - -	1.96	GRAM/MJ	1.96	
LAURIC ACID	GRAM	1.16	- - -	1.16	GRAM/MJ	1.16	
MYRISTIC ACID	GRAM	2.76	- - -	2.76	GRAM/MJ	2.76	
PALMITIC ACID	GRAM	5.15	- - -	5.15	GRAM/MJ	5.14	
STEARIC ACID	GRAM	1.48	- - -	1.48	GRAM/MJ	1.48	
PALMITOLEIC ACID	GRAM	0.38	- - -	0.38	GRAM/MJ	0.38	
OLEIC ACID	GRAM	3.97	- - -	3.97	GRAM/MJ	3.96	
LINOLEIC ACID	GRAM	0.33	- - -	0.33	GRAM/MJ	0.33	
LINOLENIC ACID	GRAM	0.26	- - -	0.26	GRAM/MJ	0.26	

1 INCLUDES THE AMOUNTS OF LACTOSE AND LACTIC ACID

FRISCHKÄSE
DOPPELRAHM-
MIND. 60%, HÖCHST. 85% FETT I. TR.

FRESH CHEESE
MIN. 60%, MAX. 85% FAT CONTENT IN DRY MATTER

FROMAGE BLANC, FROMAGE FRAIS
MIN. 60%, MAX. 85% DE MAT. GR. S. SEC

		PROTEIN	FAT	CARBOHYDRATES	TOTAL
ENERGY VALUE (AVERAGE)	KJOULE	208	1219	50	1477
PER 100 G EDIBLE PORTION	(KCAL)	50	291	12	353
AMOUNT OF DIGESTIBLE CONSTITUENTS PER 100 G	GRAM	10.96	29.92	3.00	
ENERGY VALUE (AVERAGE)	KJOULE	202	1158	50	1410
OF THE DIGESTIBLE FRACTION PER 100 G EDIBLE PORTION	(KCAL)	48	277	12	337

WASTE PERCENTAGE AVERAGE 0.00

CONSTITUENTS	DIM	AV	VARIATION			AVR	NUTR. DENS.		MOLPERC.
WATER	GRAM	52.80	50.30	-	55.10	52.80	GRAM/MJ	37.44	
PROTEIN	GRAM	11.30	8.40	-	14.60	11.30	GRAM/MJ	8.01	
FAT	GRAM	31.50	27.40	-	35.20	31.50	GRAM/MJ	22.34	
AVAILABLE CARBOHYDR.	GRAM	3.00 1	-	-	-	3.00	GRAM/MJ	2.13	
MINERALS	GRAM	1.40	1.22	-	1.56	1.40	GRAM/MJ	0.99	
SODIUM	MILLI	375.00	285.00	-	450.00	375.00	MILLI/MJ	265.93	
POTASSIUM	MILLI	95.00	70.00	-	130.00	95.00	MILLI/MJ	67.37	
MAGNESIUM	MILLI	7.20	6.00	-	8.90	7.20	MILLI/MJ	5.11	
CALCIUM	MILLI	79.00	68.00	-	91.00	79.00	MILLI/MJ	56.02	
IRON	MILLI	0.55	0.45	-	0.67	0.55	MILLI/MJ	0.39	
PHOSPHORUS	MILLI	137.00	106.00	-	189.00	137.00	MILLI/MJ	97.15	
VITAMIN A	MILLI	0.30	0.25	-	0.40	0.30	MILLI/MJ	0.21	
CAROTENE	MILLI	0.15	0.10	-	0.20	0.15	MILLI/MJ	0.11	
VITAMIN E ACTIVITY	MILLI	0.70	0.34	-	0.83	0.70	MILLI/MJ	0.50	
ALPHA-TOCOPHEROL	MILLI	0.70	0.34	-	0.83	0.70	MILLI/MJ	0.50	
VITAMIN B1	MICRO	45.00	38.00	-	70.00	45.00	MICRO/MJ	31.91	
VITAMIN B2	MILLI	0.23	0.18	-	0.33	0.23	MILLI/MJ	0.16	
NICOTINAMIDE	MILLI	0.11	0.09	-	0.17	0.11	MILLI/MJ	0.08	
PANTOTHENIC ACID	MILLI	0.44	0.35	-	0.53	0.44	MILLI/MJ	0.31	
VITAMIN B6	MICRO	60.00	20.00	-	90.00	60.00	MICRO/MJ	42.55	
BIOTIN	MICRO	4.40	2.50	-	6.30	4.40	MICRO/MJ	3.12	
VITAMIN B12	MICRO	0.53	0.38	-	0.96	0.53	MICRO/MJ	0.38	
VITAMIN C	-	0.00	-	-	-	0.00			
ALANINE	GRAM	0.40	0.33	-	0.43	0.40	GRAM/MJ	0.28	4.5
ARGININE	GRAM	0.52	0.51	-	0.54	0.52	GRAM/MJ	0.37	3.0

1 INCLUDES THE AMOUNTS OF LACTOSE AND LACTIC ACID

CONSTITUENTS	DIM	AV	VARIATION			AVR	NUTR. DENS.		MOLPERC.
ASPARTIC ACID	GRAM	1.12	1.08	-	1.18	1.12	GRAM/MJ	0.79	8.4
CYSTINE	GRAM	0.10	0.08	-	0.11	0.10	GRAM/MJ	0.07	0.4
GLUTAMIC ACID	GRAM	2.50	2.45	-	2.54	2.50	GRAM/MJ	1.77	17.0
GLYCINE	GRAM	0.27	0.26	-	0.31	0.27	GRAM/MJ	0.19	3.6
HISTIDINE	GRAM	0.52	0.50	-	0.59	0.52	GRAM/MJ	0.37	3.3
ISOLEUCINE	GRAM	0.70	0.65	-	0.73	0.70	GRAM/MJ	0.50	5.3
LEUCINE	GRAM	1.18	1.16	-	1.19	1.18	GRAM/MJ	0.84	9.0
LYSINE	GRAM	1.20	1.19	-	1.22	1.20	GRAM/MJ	0.85	8.2
METHIONINE	GRAM	0.34	0.29	-	0.38	0.34	GRAM/MJ	0.24	2.3
PHENYLALANINE	GRAM	0.59	0.52	-	0.69	0.59	GRAM/MJ	0.42	3.6
PROLINE	GRAM	1.10	1.05	-	1.13	1.10	GRAM/MJ	0.78	9.5
SERINE	GRAM	0.75	0.70	-	0.93	0.75	GRAM/MJ	0.53	7.1
THREONINE	GRAM	0.54	0.51	-	0.57	0.54	GRAM/MJ	0.38	4.5
TRYPTOPHAN	GRAM	0.15	0.14	-	0.16	0.15	GRAM/MJ	0.11	0.7
TYROSINE	GRAM	0.59	0.54	-	0.63	0.59	GRAM/MJ	0.42	3.3
VALINE	GRAM	0.73	0.67	-	0.76	0.73	GRAM/MJ	0.52	6.2
LACTIC ACID	GRAM	0.44	0.31	-	0.57	0.44	GRAM/MJ	0.31	
LACTOSE	GRAM	2.56	-	-	-	2.56	GRAM/MJ	1.82	
BUTYRIC ACID	GRAM	0.90	-	-	-	0.90	GRAM/MJ	0.64	
CAPROIC ACID	GRAM	0.26	-	-	-	0.26	GRAM/MJ	0.18	
CAPRYLIC ACID	GRAM	0.31	-	-	-	0.31	GRAM/MJ	0.22	
CAPRIC ACID	GRAM	0.61	-	-	-	0.61	GRAM/MJ	0.43	
LAURIC ACID	GRAM	0.42	-	-	-	0.42	GRAM/MJ	0.30	
MYRISTIC ACID	GRAM	3.25	-	-	-	3.25	GRAM/MJ	2.30	
PALMITIC ACID	GRAM	9.52	-	-	-	9.52	GRAM/MJ	6.75	
STEARIC ACID	GRAM	3.65	-	-	-	3.65	GRAM/MJ	2.59	
PALMITOLEIC ACID	GRAM	0.86	-	-	-	0.86	GRAM/MJ	0.61	
OLEIC ACID	GRAM	7.57	-	-	-	7.57	GRAM/MJ	5.37	
LINOLEIC ACID	GRAM	0.80	0.65	-	1.00	0.80	GRAM/MJ	0.57	
LINOLENIC ACID	GRAM	0.20	0.17	-	0.27	0.20	GRAM/MJ	0.14	
CHOLESTEROL	MILLI	103.00	99.00	-	109.00	103.00	MILLI/MJ	73.04	

FRISCHKÄSE
RAHM-
50% FETT I. TR.

FRESH CHEESE
50% FAT CONTENT IN
DRY MATTER

FROMAGE BLANC, FROMAGE FRAIS
50% DE MAT. GR. S. SEC

		PROTEIN	FAT	CARBOHYDRATES	TOTAL
ENERGY VALUE (AVERAGE)	KJOULE	254	913	72	1239
PER 100 G	(KCAL)	61	218	17	296
EDIBLE PORTION					
AMOUNT OF DIGESTIBLE	GRAM	13.38	22.42	4.30	
CONSTITUENTS PER 100 G					
ENERGY VALUE (AVERAGE)	KJOULE	246	868	72	1186
OF THE DIGESTIBLE	(KCAL)	59	207	17	283
FRACTION PER 100 G					
EDIBLE PORTION					

WASTE PERCENTAGE AVERAGE 0.00

CONSTITUENTS	DIM	AV	VARIATION			AVR	NUTR. DENS.		MOLPERC.
WATER	GRAM	57.00	55.80	-	57.80	57.00	GRAM/MJ	48.06	
PROTEIN	GRAM	13.80	13.40	-	14.30	13.80	GRAM/MJ	11.63	
FAT	GRAM	23.60	22.50	-	24.60	23.60	GRAM/MJ	19.90	
AVAILABLE CARBOHYDR.	GRAM	4.30 1	-	-	-	4.30	GRAM/MJ	3.63	
MINERALS	GRAM	1.30	0.90	-	1.70	1.30	GRAM/MJ	1.10	
POTASSIUM	MILLI	118.00	87.00	-	160.00	118.00	MILLI/MJ	99.49	
MAGNESIUM	MILLI	9.00	7.00	-	11.00	9.00	MILLI/MJ	7.59	
CALCIUM	MILLI	98.00	84.00	-	113.00	98.00	MILLI/MJ	82.62	
IRON	MILLI	0.68	0.56	-	0.83	0.68	MILLI/MJ	0.57	
PHOSPHORUS	MILLI	170.00	131.00	-	234.00	170.00	MILLI/MJ	143.33	
VITAMIN A	MILLI	0.23	0.19	-	0.30	0.23	MILLI/MJ	0.19	
CAROTENE	MILLI	0.11	0.08	-	0.15	0.11	MILLI/MJ	0.09	
ALPHA-TOCOPHEROL	MILLI	0.70	0.34	-	0.83	0.70	MILLI/MJ	0.59	
VITAMIN B1	MICRO	56.00	47.00	-	87.00	56.00	MICRO/MJ	47.21	
VITAMIN B2	MILLI	0.28	0.22	-	0.41	0.29	MILLI/MJ	0.24	
NICOTINAMIDE	MILLI	0.14	0.11	-	0.21	0.14	MILLI/MJ	0.12	
PANTOTHENIC ACID	MILLI	0.54	0.43	-	0.66	0.55	MILLI/MJ	0.46	
VITAMIN B6	MICRO	74.00	25.00	-	112.00	74.00	MICRO/MJ	62.39	
BIOTIN	MICRO	5.50	3.10	-	7.80	5.50	MICRO/MJ	4.64	
VITAMIN B12	MICRO	0.66	0.47	-	1.20	0.66	MICRO/MJ	0.56	
VITAMIN C	-	0.00	-	-	-	0.00			
ALANINE	GRAM	0.49	0.40	-	0.52	0.49	GRAM/MJ	0.41	4.5
ARGININE	GRAM	0.52	0.48	-	0.55	0.52	GRAM/MJ	0.44	3.0
ASPARTIC ACID	GRAM	1.37	1.32	-	1.44	1.37	GRAM/MJ	1.16	8.4
CYSTINE	GRAM	0.12	0.10	-	0.13	0.12	GRAM/MJ	0.10	0.4
GLUTAMIC ACID	GRAM	3.05	2.99	-	3.10	3.05	GRAM/MJ	2.57	17.0
GLYCINE	GRAM	0.33	0.32	-	0.38	0.33	GRAM/MJ	0.28	3.6
HISTIDINE	GRAM	0.45	0.41	-	0.55	0.45	GRAM/MJ	0.38	3.3
ISOLEUCINE	GRAM	0.85	0.79	-	0.89	0.85	GRAM/MJ	0.72	5.3
LEUCINE	GRAM	1.40	1.37	-	1.42	1.40	GRAM/MJ	1.18	9.0
LYSINE	GRAM	1.21	1.11	-	1.30	1.21	GRAM/MJ	1.02	8.2
METHIONINE	GRAM	0.41	0.35	-	0.46	0.41	GRAM/MJ	0.35	2.3
PHENYLALANINE	GRAM	0.72	0.63	-	0.84	0.72	GRAM/MJ	0.61	3.6
PROLINE	GRAM	1.34	1.28	-	1.38	1.34	GRAM/MJ	1.13	9.5
SERINE	GRAM	0.92	0.85	-	1.13	0.92	GRAM/MJ	0.78	7.1
THREONINE	GRAM	0.66	0.62	-	0.70	0.66	GRAM/MJ	0.56	4.5
TRYPTOPHAN	GRAM	0.15	0.12	-	0.17	0.15	GRAM/MJ	0.13	0.7
TYROSINE	GRAM	0.66	0.62	-	0.72	0.66	GRAM/MJ	0.56	3.3
VALINE	GRAM	0.89	0.82	-	0.93	0.89	GRAM/MJ	0.75	6.2
LACTIC ACID	GRAM	0.89	0.82	-	0.93	0.89	GRAM/MJ	0.75	
LACTOSE	GRAM	3.41	-	-	-	3.41	GRAM/MJ	2.87	
BUTYRIC ACID	GRAM	0.67	-	-	-	0.67	GRAM/MJ	0.56	
CAPROIC ACID	GRAM	0.19	-	-	-	0.19	GRAM/MJ	0.16	
CAPRYLIC ACID	GRAM	0.23	-	-	-	0.23	GRAM/MJ	0.19	
CAPRIC ACID	GRAM	0.45	-	-	-	0.45	GRAM/MJ	0.38	
LAURIC ACID	GRAM	0.31	-	-	-	0.31	GRAM/MJ	0.26	
MYRISTIC ACID	GRAM	2.43	-	-	-	2.43	GRAM/MJ	2.05	
PALMITIC ACID	GRAM	7.13	-	-	-	7.13	GRAM/MJ	6.01	
STEARIC ACID	GRAM	2.73	-	-	-	2.73	GRAM/MJ	2.30	
PALMITOLEIC ACID	GRAM	0.64	-	-	-	0.64	GRAM/MJ	0.54	
OLEIC ACID	GRAM	5.67	-	-	-	5.67	GRAM/MJ	4.78	
LINOLEIC ACID	GRAM	0.60	0.49	-	0.75	0.60	GRAM/MJ	0.51	
LINOLENIC ACID	GRAM	0.15	0.13	-	0.20	0.15	GRAM/MJ	0.13	
CHOLESTEROL	MILLI	77.00	74.00	-	82.00	77.00	MILLI/MJ	64.92	

1 INCLUDES THE AMOUNTS OF LACTOSE AND LACTIC ACID

GORGONZOLAKÄSE GORGONZOLA CHEESE FROMAGE GORGONZOLA

		PROTEIN	FAT	CARBOHYDRATES	TOTAL
ENERGY VALUE (AVERAGE)	KJOULE	357	1208	10	1575
PER 100 G	(KCAL)	85	289	2.5	376
EDIBLE PORTION					
AMOUNT OF DIGESTIBLE	GRAM	18.81	29.64	0.62	
CONSTITUENTS PER 100 G					
ENERGY VALUE (AVERAGE)	KJOULE	346	1147	10	1504
OF THE DIGESTIBLE	(KCAL)	83	274	2.5	359
FRACTION PER 100 G					
EDIBLE PORTION					

WASTE PERCENTAGE AVERAGE 0.00

CONSTITUENTS	DIM	AV	VARIATION			AVR	NUTR. DENS.		MOLPERC.
WATER	GRAM	42.40	-	-	-	42.40	GRAM/MJ	28.19	
PROTEIN	GRAM	19.40	-	-	-	19.40	GRAM/MJ	12.90	
FAT	GRAM	31.20	-	-	-	31.20	GRAM/MJ	20.75	
AVAILABLE CARBOHYDR.	GRAM	0.62 1	-	-	-	0.62	GRAM/MJ	0.41	
CALCIUM	MILLI	612.00	-	-	-	612.00	MILLI/MJ	406.93	
IRON	MILLI	0.30	-	-	-	0.30	MILLI/MJ	0.20	
PHOSPHORUS	MILLI	356.00	-	-	-	356.00	MILLI/MJ	236.71	
VITAMIN B1	MICRO	50.00	-	-	-	50.00	MICRO/MJ	33.25	
VITAMIN B2	MILLI	0.43	-	-	-	0.43	MILLI/MJ	0.29	
NICOTINAMIDE	MILLI	0.32	-	-	-	0.32	MILLI/MJ	0.21	
VITAMIN B6	MILLI	0.11	-	-	-	0.11	MILLI/MJ	0.07	
BIOTIN	MICRO	2.00	-	-	-	2.00	MICRO/MJ	1.33	
FOLIC ACID	MICRO	31.00	-	-	-	31.00	MICRO/MJ	20.61	
VITAMIN B12	MICRO	1.20	-	-	-	1.20	MICRO/MJ	0.80	
LACTIC ACID	GRAM	0.62	0.62	-	0.62	0.62	GRAM/MJ	0.41	

1 INCLUDES ONLY THE AMOUNT OF LACTIC ACID

GOUDAKÄSE
45% FETT I. TR.

GOUDA CHEESE
45% FAT CONTENT IN DRY MATTER

FROMAGE GOUDA
45% DE MAT. GR. S. SEC

		PROTEIN	FAT	CARBOHYDRATES	TOTAL
ENERGY VALUE (AVERAGE)	KJOULE	469	1130	0.0	1600
PER 100 G	(KCAL)	112	270	0.0	382
EDIBLE PORTION					
AMOUNT OF DIGESTIBLE	GRAM	24.73	27.74	0.00	
CONSTITUENTS PER 100 G					
ENERGY VALUE (AVERAGE)	KJOULE	455	1074	0.0	1529
OF THE DIGESTIBLE	(KCAL)	109	257	0.0	365
FRACTION PER 100 G					
EDIBLE PORTION					

WASTE PERCENTAGE AVERAGE 0.00

CONSTITUENTS	DIM	AV	VARIATION		AVR	NUTR. DENS.		MOLPERC.
WATER	GRAM	36.40	34.90	38.20	34.58	GRAM/MJ	23.81	
PROTEIN	GRAM	25.50	22.70	27.40	24.23	GRAM/MJ	16.68	
FAT	GRAM	29.20	28.60	30.40	27.74	GRAM/MJ	19.10	
MINERALS	GRAM	4.20	-	-	3.99	GRAM/MJ	2.75	
SODIUM	MILLI	869.00	812.00	926.00	825.55	MILLI/MJ	568.36	
POTASSIUM	MILLI	76.00	75.00	77.00	72.20	MILLI/MJ	49.71	
MAGNESIUM	MILLI	28.00	-	-	26.60	MILLI/MJ	18.31	
CALCIUM	MILLI	820.00	760.00	880.00	779.00	MILLI/MJ	536.31	
IRON	MILLI	0.50	-	-	0.48	MILLI/MJ	0.33	
COPPER	MICRO	70.00	-	-	66.50	MICRO/MJ	45.78	
ZINC	MILLI	3.90	-	-	3.71	MILLI/MJ	2.55	
NICKEL	MICRO	89.00	33.00	144.00	84.55	MICRO/MJ	58.21	
CHROMIUM	MICRO	95.00	34.00	156.00	90.25	MICRO/MJ	62.13	
PHOSPHORUS	MILLI	443.00	350.00	535.00	420.85	MILLI/MJ	289.74	
NITRATE	MILLI	-	0.40	4.00				
VITAMIN A	MILLI	0.26	-	-	0.25	MILLI/MJ	0.17	
VITAMIN D	MICRO	1.25	-	-	1.19	MICRO/MJ	0.82	
VITAMIN B1	MICRO	30.00	-	-	28.50	MICRO/MJ	19.62	
VITAMIN B2	MILLI	0.20	-	-	0.19	MILLI/MJ	0.13	
NICOTINAMIDE	MILLI	0.10	-	-	0.10	MILLI/MJ	0.07	
PANTOTHENIC ACID	MILLI	0.34	-	-	0.32	MILLI/MJ	0.22	
VITAMIN B6	MICRO	80.00	-	-	76.00	MICRO/MJ	52.32	
FOLIC ACID	MICRO	21.00	-	-	19.95	MICRO/MJ	13.73	
VITAMIN C	MILLI	1.00	-	-	0.95	MILLI/MJ	0.65	
ALANINE	GRAM	0.78	-	-	0.74	GRAM/MJ	0.51	

CONSTITUENTS	DIM	AV	VARIATION			AVR	NUTR. DENS.		MOLPERC.
ARGININE	GRAM	0.98	-	-	-	0.93	GRAM/MJ	0.64	
ASPARTIC ACID	GRAM	1.78	-	-	-	1.69	GRAM/MJ	1.16	
GLUTAMIC ACID	GRAM	6.28	-	-	-	5.97	GRAM/MJ	4.11	
GLYCINE	GRAM	0.50	-	-	-	0.48	GRAM/MJ	0.33	
HISTIDINE	GRAM	1.05	-	-	-	1.00	GRAM/MJ	0.69	
ISOLEUCINE	GRAM	1.34	-	-	-	1.27	GRAM/MJ	0.88	
LEUCINE	GRAM	2.62	-	-	-	2.49	GRAM/MJ	1.71	
LYSINE	GRAM	2.79	-	-	-	2.65	GRAM/MJ	1.82	
METHIONINE	GRAM	0.74	-	-	-	0.70	GRAM/MJ	0.48	
PHENYLALANINE	GRAM	1.46	-	-	-	1.39	GRAM/MJ	0.95	
PROLINE	GRAM	0.34	-	-	-	0.32	GRAM/MJ	0.22	
SERINE	GRAM	1.57	-	-	-	1.49	GRAM/MJ	1.03	
THREONINE	GRAM	0.95	-	-	-	0.90	GRAM/MJ	0.62	
TYROSINE	GRAM	1.48	-	-	-	1.41	GRAM/MJ	0.97	
VALINE	GRAM	1.85	-	-	-	1.76	GRAM/MJ	1.21	
CADAVERINE	MILLI	2.50	-	-	-	2.38	MILLI/MJ	1.64	
HISTAMINE	MILLI	4.60	3.50	-	18.00	4.37	MILLI/MJ	3.01	
PUTRESCINE	MILLI	-	2.00	-	20.00				
TYRAMINE	MILLI	-	2.00	-	67.00				
BUTYRIC ACID	GRAM	1.06	-	-	-	1.01	GRAM/MJ	0.69	
CAPROIC ACID	GRAM	0.68	-	-	-	0.65	GRAM/MJ	0.44	
CAPRYLIC ACID	GRAM	0.46	-	-	-	0.44	GRAM/MJ	0.30	
CAPRIC ACID	GRAM	0.98	-	-	-	0.93	GRAM/MJ	0.64	
LAURIC ACID	GRAM	1.29	-	-	-	1.23	GRAM/MJ	0.84	
MYRISTIC ACID	GRAM	3.23	-	-	-	3.07	GRAM/MJ	2.11	
PALMITIC ACID	GRAM	7.29	-	-	-	6.93	GRAM/MJ	4.77	
STEARIC ACID	GRAM	3.11	-	-	-	2.95	GRAM/MJ	2.03	
PALMITOLEIC ACID	GRAM	0.95	-	-	-	0.90	GRAM/MJ	0.62	
OLEIC ACID	GRAM	6.80	-	-	-	6.46	GRAM/MJ	4.45	
LINOLEIC ACID	GRAM	0.28	-	-	-	0.27	GRAM/MJ	0.18	
LINOLENIC ACID	GRAM	0.42	-	-	-	0.40	GRAM/MJ	0.27	
CHOLESTEROL	MILLI	114.00	-	-	-	108.30	MILLI/MJ	74.56	

GRUYEREKÄSE GRUYERE CHEESE FROMAGE GRUYÈRE

		PROTEIN	FAT	CARBOHYDRATES	TOTAL
ENERGY VALUE (AVERAGE)	KJOULE	549	1250	14	1813
PER 100 G	(KCAL)	131	299	3.4	433
EDIBLE PORTION					
AMOUNT OF DIGESTIBLE	GRAM	28.90	30.68	0.85	
CONSTITUENTS PER 100 G					
ENERGY VALUE (AVERAGE)	KJOULE	532	1188	14	1734
OF THE DIGESTIBLE	(KCAL)	127	284	3.4	414
FRACTION PER 100 G					
EDIBLE PORTION					

WASTE PERCENTAGE AVERAGE 0.00

CONSTITUENTS	DIM	AV		VARIATION			AVR	NUTR. DENS.		MOLPERC.
WATER	GRAM	33.20		-	-	-	33.20	GRAM/MJ	19.15	
PROTEIN	GRAM	29.80		-	-	-	29.80	GRAM/MJ	17.19	
FAT	GRAM	32.30		-	-	-	32.30	GRAM/MJ	18.63	
AVAILABLE CARBOHYDR.	GRAM	0.85	1	-	-	-	0.85	GRAM/MJ	0.49	
MINERALS	GRAM	4.30		-	-	-	4.30	GRAM/MJ	2.48	
SODIUM	MILLI	336.00		-	-	-	336.00	MILLI/MJ	193.78	
POTASSIUM	MILLI	81.00		-	-	-	81.00	MILLI/MJ	46.71	
CALCIUM	GRAM	1.00		-	-	-	1.00	GRAM/MJ	0.58	
PHOSPHORUS	MILLI	605.00		-	-	-	605.00	MILLI/MJ	348.92	
VITAMIN B1	MICRO	50.00		40.00	-	60.00	50.00	MICRO/MJ	28.84	
VITAMIN B2	MILLI	0.30		-	-	-	0.30	MILLI/MJ	0.17	
NICOTINAMIDE	MILLI	0.14		-	-	-	0.14	MILLI/MJ	0.08	
PANTOTHENIC ACID	MILLI	0.52		-	-	-	0.52	MILLI/MJ	0.30	
VITAMIN B6	MICRO	80.00		-	-	-	80.00	MICRO/MJ	46.14	
BIOTIN	MICRO	1.30		-	-	-	1.30	MICRO/MJ	0.75	
FOLIC ACID	MICRO	10.00		-	-	-	10.00	MICRO/MJ	5.77	
VITAMIN B12	MICRO	2.00		-	-	-	2.00	MICRO/MJ	1.15	
ALANINE	GRAM	0.96		-	-	-	0.96	GRAM/MJ	0.55	4.4
ARGININE	GRAM	0.98		-	-	-	0.98	GRAM/MJ	0.57	2.3
ASPARTIC ACID	GRAM	1.65		-	-	-	1.65	GRAM/MJ	0.95	5.0
CYSTINE	GRAM	0.30		-	-	-	0.30	GRAM/MJ	0.17	0.5
GLUTAMIC ACID	GRAM	5.98		-	-	-	5.98	GRAM/MJ	3.45	16.6
GLYCINE	GRAM	0.53		-	-	-	0.53	GRAM/MJ	0.31	2.9
HISTIDINE	GRAM	1.11		-	-	-	1.11	GRAM/MJ	0.64	2.9
ISOLEUCINE	GRAM	1.61		-	-	-	1.61	GRAM/MJ	0.93	5.0
LEUCINE	GRAM	3.10		-	-	-	3.10	GRAM/MJ	1.79	9.6
LYSINE	GRAM	2.71		-	-	-	2.71	GRAM/MJ	1.56	7.5
METHIONINE	GRAM	0.82		-	-	-	0.82	GRAM/MJ	0.47	2.2
PHENYLALANINE	GRAM	1.74		-	-	-	1.74	GRAM/MJ	1.00	4.3
PROLINE	GRAM	3.87		-	-	-	3.87	GRAM/MJ	2.23	13.7
SERINE	GRAM	1.72		-	-	-	1.72	GRAM/MJ	0.99	6.7
THREONINE	GRAM	1.09		-	-	-	1.09	GRAM/MJ	0.63	3.7
TRYPTOPHAN	GRAM	0.42		-	-	-	0.42	GRAM/MJ	0.24	0.8
TYROSINE	GRAM	1.78		-	-	-	1.78	GRAM/MJ	1.03	4.0
VALINE	GRAM	2.24		-	-	-	2.24	GRAM/MJ	1.29	7.8
LACTIC ACID	GRAM	0.85		0.85	-	0.85	0.85	GRAM/MJ	0.49	
BUTYRIC ACID	GRAM	1.05		-	-	-	1.05	GRAM/MJ	0.61	
CAPROIC ACID	GRAM	0.62		-	-	-	0.62	GRAM/MJ	0.36	
CAPRYLIC ACID	GRAM	0.35		-	-	-	0.35	GRAM/MJ	0.20	
CAPRIC ACID	GRAM	0.75		-	-	-	0.75	GRAM/MJ	0.43	
LAURIC ACID	GRAM	0.91		-	-	-	0.91	GRAM/MJ	0.52	
MYRISTIC ACID	GRAM	3.36		-	-	-	3.36	GRAM/MJ	1.94	
PALMITIC ACID	GRAM	7.76		-	-	-	7.76	GRAM/MJ	4.48	
STEARIC ACID	GRAM	2.32		-	-	-	2.32	GRAM/MJ	1.34	
PALMITOLEIC ACID	GRAM	0.73		-	-	-	0.73	GRAM/MJ	0.42	
OLEIC ACID	GRAM	8.57		-	-	-	8.57	GRAM/MJ	4.94	
LINOLEIC ACID	GRAM	1.30		-	-	-	1.30	GRAM/MJ	0.75	
LINOLENIC ACID	GRAM	0.43		-	-	-	0.43	GRAM/MJ	0.25	

1 INCLUDES ONLY THE AMOUNT OF LACTIC ACID

LIMBURGERKÄSE	LIMBURGER CHEESE	FROMAGE LIMBOURG
20% FETT I. TR.	20% FAT CONTENT IN DRY MATTER	20% DE MAT. GR. S. SEC

		PROTEIN	FAT	CARBOHYDRATES	TOTAL
ENERGY VALUE (AVERAGE) PER 100 G EDIBLE PORTION	KJOULE	486	332	1.0	819
	(KCAL)	116	79	0.2	196
AMOUNT OF DIGESTIBLE CONSTITUENTS PER 100 G	GRAM	25.60	8.14	0.06	
ENERGY VALUE (AVERAGE) OF THE DIGESTIBLE FRACTION PER 100 G EDIBLE PORTION	KJOULE	471	315	1.0	788
	(KCAL)	113	75	0.2	188

WASTE PERCENTAGE AVERAGE 0.00

CONSTITUENTS	DIM	AV	VARIATION			AVR	NUTR.	DENS.	MOLPERC.
WATER	GRAM	58.50	58.00	-	65.00	58.50	GRAM/MJ	74.28	
PROTEIN	GRAM	26.40	25.50	-	27.50	26.40	GRAM/MJ	33.52	
FAT	GRAM	8.57	7.00	-	9.32	8.57	GRAM/MJ	10.88	
AVAILABLE CARBOHYDR.	MILLI	60.00 1	-	-	-	60.00	MILLI/MJ	76.19	
MINERALS	GRAM	4.33	4.04	-	4.61	4.33	GRAM/MJ	5.50	
SODIUM	GRAM	1.28	1.08	-	1.45	1.28	GRAM/MJ	1.63	
POTASSIUM	MILLI	116.00	103.00	-	128.00	116.00	MILLI/MJ	147.30	
MAGNESIUM	MILLI	39.00	32.00	-	45.00	39.00	MILLI/MJ	49.52	
CALCIUM	MILLI	510.00	430.00	-	639.00	510.00	MILLI/MJ	647.59	
IRON	MILLI	0.40	0.30	-	0.40	0.40	MILLI/MJ	0.51	
PHOSPHORUS	MILLI	285.00	198.00	-	373.00	285.00	MILLI/MJ	361.89	
NITRATE	MILLI	0.20	0.10	-	0.40	0.20	MILLI/MJ	0.25	
VITAMIN A	MICRO	37.00	-	-	-	37.00	MICRO/MJ	46.98	
CAROTENE	MICRO	20.00	-	-	-	20.00	MICRO/MJ	25.40	
VITAMIN B1	MICRO	35.00	34.00	-	36.00	35.00	MICRO/MJ	44.44	
ARGININE	GRAM	0.96	-	-	-	0.96	GRAM/MJ	1.22	
ASPARTIC ACID	GRAM	1.75	-	-	-	1.75	GRAM/MJ	2.22	
CYSTINE	GRAM	0.15	-	-	-	0.15	GRAM/MJ	0.19	
GLUTAMIC ACID	GRAM	5.32	-	-	-	5.32	GRAM/MJ	6.76	
GLYCINE	GRAM	0.48	-	-	-	0.48	GRAM/MJ	0.61	
HISTIDINE	GRAM	0.86	-	-	-	0.86	GRAM/MJ	1.09	
ISOLEUCINE	GRAM	1.79	-	-	-	1.79	GRAM/MJ	2.27	
LEUCINE	GRAM	2.57	-	-	-	2.57	GRAM/MJ	3.26	
LYSINE	GRAM	1.94	-	-	-	1.94	GRAM/MJ	2.46	
METHIONINE	GRAM	0.69	-	-	-	0.69	GRAM/MJ	0.88	

1 INCLUDES ONLY THE AMOUNT OF LACTIC ACID

CONSTITUENTS	DIM	AV	VARIATION			AVR	NUTR. DENS.		MOLPERC.
PHENYLALANINE	GRAM	1.41	-	-	-	1.41	GRAM/MJ	1.79	
PROLINE	GRAM	2.88	-	-	-	2.88	GRAM/MJ	3.66	
SERINE	GRAM	1.35	-	-	-	1.35	GRAM/MJ	1.71	
THREONINE	GRAM	0.98	-	-	-	0.98	GRAM/MJ	1.24	
TRYPTOPHAN	GRAM	0.36	-	-	-	0.36	GRAM/MJ	0.46	
TYROSINE	GRAM	1.26	-	-	-	1.26	GRAM/MJ	1.60	
VALINE	GRAM	1.89	-	-	-	1.89	GRAM/MJ	2.40	
LACTIC ACID	MILLI	60.00	60.00	-	60.00	60.00	MILLI/MJ	76.19	
BUTYRIC ACID	GRAM	0.25	-	-	-	0.25	GRAM/MJ	0.32	
CAPROIC ACID	GRAM	0.15	-	-	-	0.15	GRAM/MJ	0.19	
CAPRYLIC ACID	MILLI	90.00	-	-	-	90.00	MILLI/MJ	114.28	
CAPRIC ACID	GRAM	0.15	-	-	-	0.15	GRAM/MJ	0.19	
LAURIC ACID	GRAM	0.27	-	-	-	0.27	GRAM/MJ	0.34	
MYRISTIC ACID	GRAM	0.88	-	-	-	0.88	GRAM/MJ	1.12	
PALMITIC ACID	GRAM	2.33	-	-	-	2.33	GRAM/MJ	2.96	
STEARIC ACID	GRAM	0.96	-	-	-	0.96	GRAM/MJ	1.22	
PALMITOLEIC ACID	GRAM	0.32	-	-	-	0.32	GRAM/MJ	0.41	
OLEIC ACID	GRAM	2.26	-	-	-	2.26	GRAM/MJ	2.87	
LINOLEIC ACID	GRAM	0.11	-	-	-	0.11	GRAM/MJ	0.14	
LINOLENIC ACID	MILLI	50.00	-	-	-	50.00	MILLI/MJ	63.49	

LIMBURGERKÄSE
40% FETT I. TR.

LIMBURGER CHEESE
40% FAT CONTENT IN DRY MATTER

FROMAGE LIMBOURG
40% DE MAT. GR. S. SEC

		PROTEIN	FAT	CARBOHYDRATES	TOTAL
ENERGY VALUE (AVERAGE)	KJOULE	412	762	1.2	1176
PER 100 G	(KCAL)	99	182	0.3	281
EDIBLE PORTION					
AMOUNT OF DIGESTIBLE	GRAM	21.72	18.71	0.07	
CONSTITUENTS PER 100 G					
ENERGY VALUE (AVERAGE)	KJOULE	400	724	1.2	1125
OF THE DIGESTIBLE	(KCAL)	96	173	0.3	269
FRACTION PER 100 G					
EDIBLE PORTION					

WASTE PERCENTAGE AVERAGE 0.00

CONSTITUENTS	DIM	AV	VARIATION			AVR	NUTR. DENS.		MOLPERC.
WATER	GRAM	51.70	50.50	-	58.00	51.70	GRAM/MJ	45.94	
PROTEIN	GRAM	22.40	-	-	-	22.40	GRAM/MJ	19.90	
FAT	GRAM	19.70	16.80	-	20.80	19.70	GRAM/MJ	17.50	
AVAILABLE CARBOHYDR.	MILLI	70.00 1	-	-	-	70.00	MILLI/MJ	62.20	
MINERALS	GRAM	4.15	-	-	-	4.15	GRAM/MJ	3.69	

1 INCLUDES ONLY THE AMOUNT OF LACTIC ACID

CONSTITUENTS	DIM	AV	VARIATION			AVR	NUTR. DENS.		MOLPERC.
SODIUM	GRAM	1.30	-	-	-	1.30	GRAM/MJ	1.16	
POTASSIUM	MILLI	128.00	-	-	-	128.00	MILLI/MJ	113.73	
MAGNESIUM	MILLI	21.00	-	-	-	21.00	MILLI/MJ	18.66	
CALCIUM	MILLI	534.00	-	-	-	534.00	MILLI/MJ	474.46	
IRON	MILLI	0.60	-	-	-	0.60	MILLI/MJ	0.53	
COPPER	MILLI	0.11	-	-	-	0.11	MILLI/MJ	0.10	
ZINC	MILLI	2.10	-	-	-	2.10	MILLI/MJ	1.87	
PHOSPHORUS	MILLI	256.00	-	-	-	256.00	MILLI/MJ	227.46	
NITRATE	MILLI	0.20	0.10	-	0.40	0.20	MILLI/MJ	0.18	
VITAMIN A	MILLI	0.38	-	-	-	0.38	MILLI/MJ	0.34	
VITAMIN B1	MICRO	50.00	-	-	-	50.00	MICRO/MJ	44.43	
VITAMIN B2	MILLI	0.35	-	-	-	0.35	MILLI/MJ	0.31	
NICOTINAMIDE	MICRO	60.00	-	-	-	60.00	MICRO/MJ	53.31	
PANTOTHENIC ACID	MILLI	1.18	-	-	-	1.18	MILLI/MJ	1.05	
VITAMIN B6	MICRO	86.00	-	-	-	86.00	MICRO/MJ	76.41	
BIOTIN	MICRO	8.60	2.00	-	20.00	8.60	MICRO/MJ	7.64	
FOLIC ACID	MICRO	58.00	-	-	-	58.00	MICRO/MJ	51.53	
VITAMIN C	-	TRACES	-	-	-				
ARGININE	GRAM	0.82	-	-	-	0.82	GRAM/MJ	0.73	
ASPARTIC ACID	GRAM	1.48	-	-	-	1.48	GRAM/MJ	1.31	
CYSTINE	GRAM	0.13	-	-	-	0.13	GRAM/MJ	0.12	
GLUTAMIC ACID	GRAM	4.51	-	-	-	4.51	GRAM/MJ	4.01	
GLYCINE	GRAM	0.41	-	-	-	0.41	GRAM/MJ	0.36	
HISTIDINE	GRAM	0.73	-	-	-	0.73	GRAM/MJ	0.65	
ISOLEUCINE	GRAM	1.52	-	-	-	1.52	GRAM/MJ	1.35	
LEUCINE	GRAM	2.19	-	-	-	2.19	GRAM/MJ	1.95	
LYSINE	GRAM	1.65	-	-	-	1.65	GRAM/MJ	1.47	
METHIONINE	GRAM	0.59	-	-	-	0.59	GRAM/MJ	0.52	
PHENYLALANINE	GRAM	1.20	-	-	-	1.20	GRAM/MJ	1.07	
PROLINE	GRAM	2.44	-	-	-	2.44	GRAM/MJ	2.17	
SERINE	GRAM	1.14	-	-	-	1.14	GRAM/MJ	1.01	
THREONINE	GRAM	0.84	-	-	-	0.84	GRAM/MJ	0.75	
TRYPTOPHAN	GRAM	0.31	-	-	-	0.31	GRAM/MJ	0.28	
TYROSINE	GRAM	1.08	-	-	-	1.08	GRAM/MJ	0.96	
VALINE	GRAM	1.61	-	-	-	1.61	GRAM/MJ	1.43	
LACTIC ACID	MILLI	70.00	70.00	-	70.00	70.00	MILLI/MJ	62.20	
BUTYRIC ACID	GRAM	0.58	-	-	-	0.58	GRAM/MJ	0.52	
CAPROIC ACID	GRAM	0.35	-	-	-	0.35	GRAM/MJ	0.31	
CAPRYLIC ACID	GRAM	0.20	-	-	-	0.20	GRAM/MJ	0.18	
CAPRIC ACID	GRAM	0.35	-	-	-	0.35	GRAM/MJ	0.31	
LAURIC ACID	GRAM	0.63	-	-	-	0.63	GRAM/MJ	0.56	
MYRISTIC ACID	GRAM	2.02	-	-	-	2.02	GRAM/MJ	1.79	
PALMITIC ACID	GRAM	5.35	-	-	-	5.35	GRAM/MJ	4.75	
STEARIC ACID	GRAM	2.20	-	-	-	2.20	GRAM/MJ	1.95	
PALMITOLEIC ACID	GRAM	0.73	-	-	-	0.73	GRAM/MJ	0.65	
OLEIC ACID	GRAM	5.19	-	-	-	5.19	GRAM/MJ	4.61	
LINOLEIC ACID	GRAM	0.25	-	-	-	0.25	GRAM/MJ	0.22	
LINOLENIC ACID	GRAM	0.12	-	-	-	0.12	GRAM/MJ	0.11	
CHOLESTEROL	MILLI	90.00	-	-	-	90.00	MILLI/MJ	79.97	

MOZZARELLAKÄSE | MOZZARELLA CHEESE | FROMAGE MOZZARELLA

		PROTEIN	FAT	CARBOHYDRATES	TOTAL
ENERGY VALUE (AVERAGE) PER 100 G EDIBLE PORTION	KJOULE (KCAL)	366 88	623 149	0.0 0.0	989 236
AMOUNT OF DIGESTIBLE CONSTITUENTS PER 100 G	GRAM	19.30	15.29	0.00	
ENERGY VALUE (AVERAGE) OF THE DIGESTIBLE FRACTION PER 100 G EDIBLE PORTION	KJOULE (KCAL)	355 85	592 141	0.0 0.0	947 226

WASTE PERCENTAGE AVERAGE 0.00

CONSTITUENTS	DIM	AV	VARIATION	AVR	NUTR. DENS.		MOLPERC.
WATER	GRAM	60.10	- - -	60.10	GRAM/MJ	63.44	
PROTEIN	GRAM	19.90	- - -	19.90	GRAM/MJ	21.01	
FAT	GRAM	16.10	- - -	16.10	GRAM/MJ	17.00	
CALCIUM	MILLI	403.00	- - -	403.00	MILLI/MJ	425.42	
IRON	MILLI	0.20	- - -	0.20	MILLI/MJ	0.21	
PHOSPHORUS	MILLI	239.00	- - -	239.00	MILLI/MJ	252.29	
VITAMIN B1	MICRO	30.00	- - -	30.00	MICRO/MJ	31.67	
VITAMIN B2	MILLI	0.27	- - -	0.27	MILLI/MJ	0.29	
NICOTINAMIDE	MILLI	0.14	- - -	0.14	MILLI/MJ	0.15	
VITAMIN B6	MICRO	60.00	- - -	60.00	MICRO/MJ	63.34	
BIOTIN	MICRO	2.00	- - -	2.00	MICRO/MJ	2.11	
FOLIC ACID	MICRO	10.00	- - -	10.00	MICRO/MJ	10.56	
SALICYLIC ACID	MICRO	20.00	20.00 - 20.00	20.00	MICRO/MJ	21.11	
BUTYRIC ACID	GRAM	0.54	- - -	0.54	GRAM/MJ	0.57	
CAPROIC ACID	GRAM	0.10	- - -	0.10	GRAM/MJ	0.11	
CAPRYLIC ACID	GRAM	0.12	- - -	0.12	GRAM/MJ	0.13	
CAPRIC ACID	GRAM	0.26	- - -	0.26	GRAM/MJ	0.27	
LAURIC ACID	GRAM	0.17	- - -	0.17	GRAM/MJ	0.18	
MYRISTIC ACID	GRAM	1.67	- - -	1.67	GRAM/MJ	1.76	
PALMITIC ACID	GRAM	5.08	- - -	5.08	GRAM/MJ	5.36	
STEARIC ACID	GRAM	2.03	- - -	2.03	GRAM/MJ	2.14	
PALMITOLEIC ACID	GRAM	0.46	- - -	0.46	GRAM/MJ	0.49	
OLEIC ACID	GRAM	4.05	- - -	4.05	GRAM/MJ	4.28	
LINOLEIC ACID	GRAM	0.35	- - -	0.35	GRAM/MJ	0.37	
LINOLENIC ACID	GRAM	0.14	- - -	0.14	GRAM/MJ	0.15	

MÜNSTERKÄSE
45% FETT I. TR.

MUENSTER CHEESE
45% FAT CONTENT IN DRY MATTER

FROMAGE MUNSTER
45% DE MAT. GR. S. SEC

		PROTEIN	FAT	CARBOHYDRATES	TOTAL
ENERGY VALUE (AVERAGE)	KJOULE	398	875	1.2	1273
PER 100 G EDIBLE PORTION	(KCAL)	95	209	0.3	304
AMOUNT OF DIGESTIBLE CONSTITUENTS PER 100 G	GRAM	20.95	21.47	0.07	
ENERGY VALUE (AVERAGE)	KJOULE	386	831	1.2	1218
OF THE DIGESTIBLE FRACTION PER 100 G EDIBLE PORTION	(KCAL)	92	199	0.3	291

WASTE PERCENTAGE AVERAGE 0.00

CONSTITUENTS	DIM	AV	VARIATION			AVR	NUTR. DENS.		MOLPERC.
WATER	GRAM	52.10	50.00	-	54.00	52.10	GRAM/MJ	42.78	
PROTEIN	GRAM	21.60	-	-	-	21.60	GRAM/MJ	17.74	
FAT	GRAM	22.60	22.00	-	23.50	22.60	GRAM/MJ	18.56	
AVAILABLE CARBOHYDR.	MILLI	70.00 1	-	-	-	70.00	MILLI/MJ	57.48	
MINERALS	GRAM	3.70	-	-	-	3.70	GRAM/MJ	3.04	
SODIUM	GRAM	1.02	-	-	-	1.02	GRAM/MJ	0.84	
POTASSIUM	MILLI	134.00	-	-	-	134.00	MILLI/MJ	110.03	
MAGNESIUM	MILLI	27.00	-	-	-	27.00	MILLI/MJ	22.17	
CALCIUM	MILLI	310.00	210.00	-	410.00	310.00	MILLI/MJ	254.55	
IRON	MILLI	0.41	-	-	-	0.41	MILLI/MJ	0.34	
ZINC	MILLI	2.81	-	-	-	2.81	MILLI/MJ	2.31	
PHOSPHORUS	MILLI	240.00	200.00	-	300.00	240.00	MILLI/MJ	197.07	
VITAMIN A	MILLI	0.34	-	-	-	0.34	MILLI/MJ	0.28	
VITAMIN B1	MICRO	10.00	-	-	-	10.00	MICRO/MJ	8.21	
VITAMIN B2	MILLI	0.32	-	-	-	0.32	MILLI/MJ	0.26	
NICOTINAMIDE	MILLI	0.10	-	-	-	0.10	MILLI/MJ	0.08	
PANTOTHENIC ACID	MILLI	0.19	-	-	-	0.19	MILLI/MJ	0.16	
VITAMIN B6	MICRO	56.00	-	-	-	56.00	MICRO/MJ	45.98	
FOLIC ACID	MICRO	12.00	-	-	-	12.00	MICRO/MJ	9.85	
VITAMIN B12	MICRO	1.40	-	-	-	1.40	MICRO/MJ	1.15	
ALANINE	GRAM	0.63	-	-	-	0.63	GRAM/MJ	0.52	4.1
ARGININE	GRAM	0.81	-	-	-	0.81	GRAM/MJ	0.67	2.7
ASPARTIC ACID	GRAM	1.48	-	-	-	1.48	GRAM/MJ	1.22	6.4
CYSTINE	GRAM	0.12	-	-	-	0.12	GRAM/MJ	0.10	0.3
GLUTAMIC ACID	GRAM	5.13	-	-	-	5.13	GRAM/MJ	4.21	20.0

1 INCLUDES ONLY THE AMOUNT OF LACTIC ACID

CONSTITUENTS	DIM	AV	VARIATION			AVR	NUTR. DENS.		MOLPERC.
GLYCINE	GRAM	0.41	-	-	-	0.41	GRAM/MJ	0.34	3.1
HISTIDINE	GRAM	0.77	-	-	-	0.77	GRAM/MJ	0.63	2.8
ISOLEUCINE	GRAM	1.06	-	-	-	1.06	GRAM/MJ	0.87	4.6
LEUCINE	GRAM	2.09	-	-	-	2.09	GRAM/MJ	1.72	9.1
LYSINE	GRAM	1.98	-	-	-	1.98	GRAM/MJ	1.63	7.8
METHIONINE	GRAM	0.53	-	-	-	0.53	GRAM/MJ	0.44	2.0
PHENYLALANINE	GRAM	1.14	-	-	-	1.14	GRAM/MJ	0.94	4.0
PROLINE	GRAM	2.39	-	-	-	2.39	GRAM/MJ	1.96	11.9
SERINE	GRAM	1.20	-	-	-	1.20	GRAM/MJ	0.99	6.5
THREONINE	GRAM	0.82	-	-	-	0.82	GRAM/MJ	0.67	3.9
TRYPTOPHAN	GRAM	0.30	-	-	-	0.30	GRAM/MJ	0.25	0.8
TYROSINE	GRAM	1.03	-	-	-	1.03	GRAM/MJ	0.85	3.3
VALINE	GRAM	1.37	-	-	-	1.37	GRAM/MJ	1.12	6.7
LACTIC ACID	MILLI	70.00	70.00	-	70.00	70.00	MILLI/MJ	57.48	
BUTYRIC ACID	GRAM	0.78	-	-	-	0.78	GRAM/MJ	0.64	
CAPROIC ACID	GRAM	0.18	-	-	-	0.18	GRAM/MJ	0.15	
CAPRYLIC ACID	GRAM	0.21	-	-	-	0.21	GRAM/MJ	0.17	
CAPRIC ACID	GRAM	0.44	-	-	-	0.44	GRAM/MJ	0.36	
LAURIC ACID	GRAM	0.28	-	-	-	0.28	GRAM/MJ	0.23	
MYRISTIC ACID	GRAM	2.31	-	-	-	2.31	GRAM/MJ	1.90	
PALMITIC ACID	GRAM	6.95	-	-	-	6.95	GRAM/MJ	5.71	
STEARIC ACID	GRAM	2.69	-	-	-	2.69	GRAM/MJ	2.21	
PALMITOLEIC ACID	GRAM	0.73	-	-	-	0.73	GRAM/MJ	0.60	
OLEIC ACID	GRAM	5.53	-	-	-	5.53	GRAM/MJ	4.54	
LINOLEIC ACID	GRAM	0.32	-	-	-	0.32	GRAM/MJ	0.26	
LINOLENIC ACID	GRAM	0.17	-	-	-	0.17	GRAM/MJ	0.14	
CHOLESTEROL	MILLI	96.00	-	-	-	96.00	MILLI/MJ	78.83	

MÜNSTERKÄSE
50% FETT I. TR.

MUENSTER CHEESE
50% FAT CONTENT IN DRY MATTER

FROMAGE MUNSTER
50% DE MAT. GR. S. SEC

		PROTEIN	FAT	CARBOHYDRATES	TOTAL
ENERGY VALUE (AVERAGE)	KJOULE	385	1018	1.2	1404
PER 100 G EDIBLE PORTION	(KCAL)	92	243	0.3	336
AMOUNT OF DIGESTIBLE CONSTITUENTS PER 100 G	GRAM	20.27	24.98	0.07	
ENERGY VALUE (AVERAGE)	KJOULE	373	967	1.2	1341
OF THE DIGESTIBLE FRACTION PER 100 G EDIBLE PORTION	(KCAL)	89	231	0.3	321

WASTE PERCENTAGE AVERAGE 0.00

CONSTITUENTS	DIM	AV	VARIATION			AVR	NUTR. DENS.		MOLPERC.
WATER	GRAM	49.10	47.10	-	52.90	49.10	GRAM/MJ	36.60	
PROTEIN	GRAM	20.90	-	-	-	20.90	GRAM/MJ	15.58	
FAT	GRAM	26.30	24.40	-	27.30	26.30	GRAM/MJ	19.61	
AVAILABLE CARBOHYDR.	MILLI	70.00 1	-	-	-	70.00	MILLI/MJ	52.19	
MINERALS	GRAM	3.70	-	-	-	3.70	GRAM/MJ	2.76	

1 INCLUDES ONLY THE AMOUNT OF LACTIC ACID

CONSTITUENTS	DIM	AV	VARIATION		AVR	NUTR. DENS.		MOLPERC.
SODIUM	MILLI	900.00	-	-	900.00	MILLI/MJ	670.96	
POTASSIUM	MILLI	134.00	-	-	134.00	MILLI/MJ	99.90	
MAGNESIUM	MILLI	27.00	-	-	27.00	MILLI/MJ	20.13	
CALCIUM	MILLI	230.00	220.00	- 240.00	230.00	MILLI/MJ	171.47	
IRON	MILLI	0.41	-	-	0.41	MILLI/MJ	0.31	
ZINC	MILLI	2.81	-	-	2.81	MILLI/MJ	2.09	
PHOSPHORUS	MILLI	170.00	100.00	- 200.00	170.00	MILLI/MJ	126.74	
VITAMIN A	MILLI	0.34	-	-	0.34	MILLI/MJ	0.25	
VITAMIN B1	MICRO	10.00	-	-	10.00	MICRO/MJ	7.46	
VITAMIN B2	MILLI	0.32	-	-	0.32	MILLI/MJ	0.24	
NICOTINAMIDE	MILLI	0.10	-	-	0.10	MILLI/MJ	0.08	
PANTOTHENIC ACID	MILLI	0.19	-	-	0.19	MILLI/MJ	0.14	
VITAMIN B6	MICRO	56.00	-	-	56.00	MICRO/MJ	41.75	
FOLIC ACID	MICRO	12.00	-	-	12.00	MICRO/MJ	8.95	
VITAMIN B12	MICRO	1.40	-	-	1.40	MICRO/MJ	1.04	
ALANINE	GRAM	0.61	-	-	0.61	GRAM/MJ	0.45	4.1
ARGININE	GRAM	0.79	-	-	0.79	GRAM/MJ	0.59	2.7
ASPARTIC ACID	GRAM	1.43	-	-	1.43	GRAM/MJ	1.07	6.4
CYSTINE	GRAM	0.12	-	-	0.12	GRAM/MJ	0.09	0.3
GLUTAMIC ACID	GRAM	4.97	-	-	4.97	GRAM/MJ	3.71	20.1
GLYCINE	GRAM	0.39	-	-	0.39	GRAM/MJ	0.29	3.1
HISTIDINE	GRAM	0.74	-	-	0.74	GRAM/MJ	0.55	2.8
ISOLEUCINE	GRAM	1.03	-	-	1.03	GRAM/MJ	0.77	4.7
LEUCINE	GRAM	2.02	-	-	2.02	GRAM/MJ	1.51	9.2
LYSINE	GRAM	1.77	-	-	1.77	GRAM/MJ	1.32	7.2
METHIONINE	GRAM	0.51	-	-	0.51	GRAM/MJ	0.38	2.0
PHENYLALANINE	GRAM	1.11	-	-	1.11	GRAM/MJ	0.83	4.0
PROLINE	GRAM	2.31	-	-	2.31	GRAM/MJ	1.72	12.0
SERINE	GRAM	1.16	-	-	1.16	GRAM/MJ	0.86	6.6
THREONINE	GRAM	0.79	-	-	0.79	GRAM/MJ	0.59	4.0
TRYPTOPHAN	GRAM	0.29	-	-	0.29	GRAM/MJ	0.22	0.8
TYROSINE	GRAM	1.00	-	-	1.00	GRAM/MJ	0.75	3.3
VALINE	GRAM	1.32	-	-	1.32	GRAM/MJ	0.98	6.7
LACTIC ACID	MILLI	70.00	70.00	- 70.00	70.00	MILLI/MJ	52.19	
BUTYRIC ACID	GRAM	0.91	-	-	0.91	GRAM/MJ	0.68	
CAPROIC ACID	GRAM	0.21	-	-	0.21	GRAM/MJ	0.16	
CAPRYLIC ACID	GRAM	0.24	-	-	0.24	GRAM/MJ	0.18	
CAPRIC ACID	GRAM	0.52	-	-	0.52	GRAM/MJ	0.39	
LAURIC ACID	GRAM	0.32	-	-	0.32	GRAM/MJ	0.24	
MYRISTIC ACID	GRAM	2.69	-	-	2.69	GRAM/MJ	2.01	
PALMITIC ACID	GRAM	8.08	-	-	8.08	GRAM/MJ	6.02	
STEARIC ACID	GRAM	3.13	-	-	3.13	GRAM/MJ	2.33	
PALMITOLEIC ACID	GRAM	0.85	-	-	0.85	GRAM/MJ	0.63	
OLEIC ACID	GRAM	6.43	-	-	6.43	GRAM/MJ	4.79	
LINOLEIC ACID	GRAM	0.38	-	-	0.38	GRAM/MJ	0.28	
LINOLENIC ACID	GRAM	0.20	-	-	0.20	GRAM/MJ	0.15	
CHOLESTEROL	MILLI	96.00	-	-	96.00	MILLI/MJ	71.57	

PARMESANKÄSE — PARMESAN CHEESE — FROMAGE PARMESAN

		PROTEIN	FAT	CARBOHYDRATES	TOTAL
ENERGY VALUE (AVERAGE)	KJOULE	655	999	1.0	1655
PER 100 G	(KCAL)	157	239	0.2	396
EDIBLE PORTION					
AMOUNT OF DIGESTIBLE	GRAM	34.53	24.51	0.06	
CONSTITUENTS PER 100 G					
ENERGY VALUE (AVERAGE)	KJOULE	636	949	1.0	1585
OF THE DIGESTIBLE	(KCAL)	152	227	0.2	379
FRACTION PER 100 G					
EDIBLE PORTION					

WASTE PERCENTAGE AVERAGE 5.0

CONSTITUENTS	DIM	AV	VARIATION			AVR	NUTR. DENS.		MOLPERC.
WATER	GRAM	29.60	27.60	-	32.00	28.12	GRAM/MJ	18.67	
PROTEIN	GRAM	35.60	32.20	-	37.00	33.82	GRAM/MJ	22.46	
FAT	GRAM	25.80	22.00	-	29.70	24.51	GRAM/MJ	16.27	
AVAILABLE CARBOHYDR.	MILLI	60.00 1	-	-	-	57.00	MILLI/MJ	37.85	
MINERALS	GRAM	5.53	4.55	-	6.70	5.25	GRAM/MJ	3.49	
SODIUM	MILLI	704.00	601.00	-	880.00	668.80	MILLI/MJ	444.08	
POTASSIUM	MILLI	131.00	81.00	-	153.00	124.45	MILLI/MJ	82.63	
MAGNESIUM	MILLI	44.50	40.40	-	49.60	42.28	MILLI/MJ	28.07	
CALCIUM	GRAM	1.29	0.95	-	1.60	1.23	GRAM/MJ	0.81	
IRON	MILLI	1.02	0.37	-	1.90	0.97	MILLI/MJ	0.64	
COPPER	MILLI	0.36	-	-	-	0.34	MILLI/MJ	0.23	
ZINC	MILLI	3.00	1.00	-	4.00	2.85	MILLI/MJ	1.89	
PHOSPHORUS	GRAM	0.84	0.77	-	1.00	0.81	GRAM/MJ	0.53	
CHLORIDE	GRAM	0.95	0.63	-	1.35	0.91	GRAM/MJ	0.60	
VITAMIN A	MILLI	0.34	0.29	-	0.41	0.32	MILLI/MJ	0.21	
VITAMIN D	MICRO	0.65	0.30	-	1.00	0.62	MICRO/MJ	0.41	
VITAMIN B1	MICRO	20.00	20.00	-	30.00	19.00	MICRO/MJ	12.62	
VITAMIN B2	MILLI	0.62	0.53	-	0.73	0.59	MILLI/MJ	0.39	
NICOTINAMIDE	MILLI	0.17	0.11	-	0.20	0.16	MILLI/MJ	0.11	
PANTOTHENIC ACID	MILLI	0.53	-	-	-	0.50	MILLI/MJ	0.33	
VITAMIN B6	MICRO	96.00	-	-	-	91.20	MICRO/MJ	60.56	
BIOTIN	MICRO	3.00	1.70	-	4.30	2.85	MICRO/MJ	1.89	
FOLIC ACID	MICRO	20.00	-	-	-	19.00	MICRO/MJ	12.62	
ALANINE	GRAM	1.19	1.04	-	1.34	1.13	GRAM/MJ	0.75	4.6
ARGININE	GRAM	1.33	1.31	-	1.40	1.26	GRAM/MJ	0.84	2.6

1 INCLUDES THE AMOUNTS OF LACTOSE

CONSTITUENTS	DIM	AV	VARIATION			AVR	NUTR. DENS.		MOLPERC.
ASPARTIC ACID	GRAM	2.60	2.23	-	2.96	2.47	GRAM/MJ	1.64	6.7
CYSTINE	GRAM	0.25	0.23	-	0.33	0.24	GRAM/MJ	0.16	0.4
GLUTAMIC ACID	GRAM	8.10	8.03	-	8.17	7.70	GRAM/MJ	5.11	18.8
GLYCINE	GRAM	0.70	0.62	-	0.78	0.67	GRAM/MJ	0.44	3.2
HISTIDINE	GRAM	1.20	1.06	-	1.38	1.14	GRAM/MJ	0.76	2.6
ISOLEUCINE	GRAM	1.93	1.89	-	1.96	1.83	GRAM/MJ	1.22	5.0
LEUCINE	GRAM	3.50	3.44	-	3.57	3.33	GRAM/MJ	2.21	9.1
LYSINE	GRAM	2.98	2.61	-	3.29	2.83	GRAM/MJ	1.88	7.0
METHIONINE	GRAM	0.96	0.93	-	1.00	0.91	GRAM/MJ	0.61	2.2
PHENYLALANINE	GRAM	1.91	-	-	-	1.81	GRAM/MJ	1.20	4.0
PROLINE	GRAM	3.88	3.30	-	4.18	3.69	GRAM/MJ	2.45	11.5
SERINE	GRAM	2.07	-	-	-	1.97	GRAM/MJ	1.31	6.7
THREONINE	GRAM	1.45	1.31	-	1.73	1.38	GRAM/MJ	0.91	4.2
TRYPTOPHAN	GRAM	0.49	0.48	-	0.51	0.47	GRAM/MJ	0.31	0.8
TYROSINE	GRAM	1.75	1.70	-	1.99	1.66	GRAM/MJ	1.10	3.3
VALINE	GRAM	2.52	2.44	-	2.57	2.39	GRAM/MJ	1.59	7.4
HISTAMINE	MILLI	-	0.00	-	58.00				
TYRAMINE	MILLI	-	0.40	-	29.00				
LACTOSE	MILLI	60.00	60.00	-	60.00	57.00	MILLI/MJ	37.85	
BUTYRIC ACID	GRAM	1.30	-	-	-	1.24	GRAM/MJ	0.82	
CAPROIC ACID	GRAM	0.49	-	-	-	0.47	GRAM/MJ	0.31	
CAPRYLIC ACID	GRAM	0.26	-	-	-	0.25	GRAM/MJ	0.16	
CAPRIC ACID	GRAM	0.65	-	-	-	0.62	GRAM/MJ	0.41	
LAURIC ACID	GRAM	0.87	-	-	-	0.83	GRAM/MJ	0.55	
MYRISTIC ACID	GRAM	2.91	-	-	-	2.76	GRAM/MJ	1.84	
PALMITIC ACID	GRAM	6.97	-	-	-	6.62	GRAM/MJ	4.40	
STEARIC ACID	GRAM	2.30	-	-	-	2.19	GRAM/MJ	1.45	
PALMITOLEIC ACID	GRAM	0.39	-	-	-	0.37	GRAM/MJ	0.25	
OLEIC ACID	GRAM	6.66	-	-	-	6.33	GRAM/MJ	4.20	
LINOLEIC ACID	GRAM	0.27	-	-	-	0.26	GRAM/MJ	0.17	
LINOLENIC ACID	GRAM	0.30	-	-	-	0.29	GRAM/MJ	0.19	
CHOLESTEROL	MILLI	68.00	-	-	-	64.60	MILLI/MJ	42.89	

PROVOLONEKÄSE PROVOLONE CHEESE FROMAGE PROVOLONE

		PROTEIN	FAT	CARBOHYDRATES	TOTAL
ENERGY VALUE (AVERAGE)	KJOULE	484	1118	0.0	1603
PER 100 G	(KCAL)	116	267	0.0	383
EDIBLE PORTION					
AMOUNT OF DIGESTIBLE	GRAM	25.51	27.45	0.00	
CONSTITUENTS PER 100 G					
ENERGY VALUE (AVERAGE)	KJOULE	470	1063	0.0	1532
OF THE DIGESTIBLE	(KCAL)	112	254	0.0	366
FRACTION PER 100 G					
EDIBLE PORTION					

WASTE PERCENTAGE AVERAGE 0.00

CONSTITUENTS	DIM	AV	VARIATION			AVR	NUTR. DENS.		MOLPERC.
WATER	GRAM	39.60	-	-	-	39.60	GRAM/MJ	25.85	
PROTEIN	GRAM	26.30	-	-	-	26.30	GRAM/MJ	17.16	
FAT	GRAM	28.90	-	-	-	28.90	GRAM/MJ	18.86	
MINERALS	GRAM	5.87	-	-	-	5.87	GRAM/MJ	3.83	
MAGNESIUM	MILLI	31.00	-	-	-	31.00	MILLI/MJ	20.23	
CALCIUM	MILLI	881.00	-	-	-	881.00	MILLI/MJ	574.99	
IRON	MILLI	0.50	-	-	-	0.50	MILLI/MJ	0.33	
COPPER	MILLI	0.24	-	-	-	0.24	MILLI/MJ	0.16	
ZINC	MILLI	3.90	-	-	-	3.90	MILLI/MJ	2.55	
PHOSPHORUS	MILLI	576.00	-	-	-	576.00	MILLI/MJ	375.93	
VITAMIN B1	MICRO	19.00	-	-	-	19.00	MICRO/MJ	12.40	
VITAMIN B2	MILLI	0.32	-	-	-	0.32	MILLI/MJ	0.21	
NICOTINAMIDE	MILLI	0.16	-	-	-	0.16	MILLI/MJ	0.10	
CADAVERINE	MILLI	-	2.00	-	20.00				
PUTRESCINE	MILLI	-	1.00	-	20.00				
BUTYRIC ACID	GRAM	1.06	-	-	-	1.06	GRAM/MJ	0.69	
CAPROIC ACID	GRAM	0.40	-	-	-	0.40	GRAM/MJ	0.26	
CAPRYLIC ACID	GRAM	0.30	-	-	-	0.30	GRAM/MJ	0.20	
CAPRIC ACID	GRAM	0.63	-	-	-	0.63	GRAM/MJ	0.41	
LAURIC ACID	GRAM	0.63	-	-	-	0.63	GRAM/MJ	0.41	
MYRISTIC ACID	GRAM	3.06	-	-	-	3.06	GRAM/MJ	2.00	
PALMITIC ACID	GRAM	8.67	-	-	-	8.67	GRAM/MJ	5.66	
STEARIC ACID	GRAM	2.97	-	-	-	2.97	GRAM/MJ	1.94	
PALMITOLEIC ACID	GRAM	0.76	-	-	-	0.76	GRAM/MJ	0.50	
OLEIC ACID	GRAM	6.96	-	-	-	6.96	GRAM/MJ	4.54	
LINOLEIC ACID	GRAM	0.57	-	-	-	0.57	GRAM/MJ	0.37	
LINOLENIC ACID	GRAM	0.30	-	-	-	0.30	GRAM/MJ	0.20	

RICOTTAKÄSE RICOTTA CHEESE FROMAGE RICOTTA

		PROTEIN	FAT	CARBOHYDRATES	TOTAL
ENERGY VALUE (AVERAGE) PER 100 G EDIBLE PORTION	KJOULE	175	581	5.5	761
	(KCAL)	42	139	1.3	182
AMOUNT OF DIGESTIBLE CONSTITUENTS PER 100 G	GRAM	9.21	14.25	0.33	
ENERGY VALUE (AVERAGE) OF THE DIGESTIBLE FRACTION PER 100 G EDIBLE PORTION	KJOULE	170	552	5.5	727
	(KCAL)	41	132	1.3	174

WASTE PERCENTAGE AVERAGE 0.00

CONSTITUENTS	DIM	AV	VARIATION	AVR	NUTR. DENS.	MOLPERC.
WATER	GRAM	75.00	- - -	75.00	GRAM/MJ 103.21	
PROTEIN	GRAM	9.50	- - -	9.50	GRAM/MJ 13.07	
FAT	GRAM	15.00	- - -	15.00	GRAM/MJ 20.64	
AVAILABLE CARBOHYDR.	GRAM	0.33 [1]	- - -	0.33	GRAM/MJ 0.45	
CALCIUM	MILLI	274.00	- - -	274.00	MILLI/MJ 377.06	
PHOSPHORUS	MILLI	270.00	- - -	270.00	MILLI/MJ 371.56	
LACTOSE	GRAM	0.33	0.33 - 0.33	0.33	GRAM/MJ 0.45	
BUTYRIC ACID	GRAM	0.66	- - -	0.66	GRAM/MJ 0.91	
CAPROIC ACID	GRAM	0.10	- - -	0.10	GRAM/MJ 0.14	
CAPRYLIC ACID	GRAM	0.13	- - -	0.13	GRAM/MJ 0.18	
CAPRIC ACID	GRAM	0.27	- - -	0.27	GRAM/MJ 0.37	
LAURIC ACID	GRAM	0.18	- - -	0.18	GRAM/MJ 0.25	
MYRISTIC ACID	GRAM	1.57	- - -	1.57	GRAM/MJ 2.16	
PALMITIC ACID	GRAM	4.67	- - -	4.67	GRAM/MJ 6.43	
STEARIC ACID	GRAM	1.49	- - -	1.49	GRAM/MJ 2.05	
PALMITOLEIC ACID	GRAM	0.62	- - -	0.62	GRAM/MJ 0.85	
OLEIC ACID	GRAM	3.31	- - -	3.31	GRAM/MJ 4.56	
LINOLEIC ACID	GRAM	0.32	- - -	0.32	GRAM/MJ 0.44	
LINOLENIC ACID	GRAM	0.13	- - -	0.13	GRAM/MJ 0.18	

1 INCLUDES THE AMOUNT OF LACTOSE

ROMADURKÄSE
20% FETT I. TR.

ROMADUR CHEESE
20% FAT CONTENT IN DRY MATTER

FROMAGE ROMADOUR
20% DE MAT. GR. S. SEC

		PROTEIN	FAT	CARBOHYDRATES	TOTAL
ENERGY VALUE (AVERAGE)	KJOULE	440	355	0.0	794
PER 100 G	(KCAL)	105	85	0.0	190
EDIBLE PORTION					
AMOUNT OF DIGESTIBLE	GRAM	23.18	8.70	0.00	
CONSTITUENTS PER 100 G					
ENERGY VALUE (AVERAGE)	KJOULE	427	337	0.0	764
OF THE DIGESTIBLE	(KCAL)	102	80	0.0	182
FRACTION PER 100 G					
EDIBLE PORTION					

WASTE PERCENTAGE AVERAGE 0.00

CONSTITUENTS	DIM	AV	VARIATION			AVR	NUTR. DENS.		MOLPERC.
WATER	GRAM	60.30	54.00	-	65.00	60.30	GRAM/MJ	78.97	
PROTEIN	GRAM	23.90	21.40	-	26.40	23.90	GRAM/MJ	31.30	
FAT	GRAM	9.16	7.00	-	11.40	9.16	GRAM/MJ	12.00	
MINERALS	GRAM	5.33	-	-	-	5.33	GRAM/MJ	6.98	
CALCIUM	MILLI	448.00	347.00	-	555.00	448.00	MILLI/MJ	586.71	
PHOSPHORUS	MILLI	325.00	218.00	-	472.00	325.00	MILLI/MJ	425.63	
NITRATE	MILLI	0.15	0.10	-	0.20	0.15	MILLI/MJ	0.20	

ROMADURKÄSE
30% FETT I. TR.

ROMADUR CHEESE
30% FAT CONTENT IN DRY MATTER

FROMAGE ROMADOUR
30% DE MAT. GR. S. SEC

		PROTEIN	FAT	CARBOHYDRATES	TOTAL
ENERGY VALUE (AVERAGE)	KJOULE	436	530	0.0	967
PER 100 G	(KCAL)	104	127	0.0	231
EDIBLE PORTION					
AMOUNT OF DIGESTIBLE	GRAM	22.98	13.01	0.00	
CONSTITUENTS PER 100 G					
ENERGY VALUE (AVERAGE)	KJOULE	423	504	0.0	927
OF THE DIGESTIBLE	(KCAL)	101	120	0.0	222
FRACTION PER 100 G					
EDIBLE PORTION					

WASTE PERCENTAGE AVERAGE 0.00

CONSTITUENTS	DIM	AV	VARIATION			AVR	NUTR. DENS.		MOLPERC.
WATER	GRAM	57.20	55.70	-	62.00	57.20	GRAM/MJ	61.71	
PROTEIN	GRAM	23.70	22.40	-	25.00	23.70	GRAM/MJ	25.57	
FAT	GRAM	13.70	11.40	-	14.30	13.70	GRAM/MJ	14.78	
MINERALS	GRAM	4.61	3.50	-	5.71	4.61	GRAM/MJ	4.97	
SODIUM	GRAM	1.23	-	-	-	1.23	GRAM/MJ	1.33	
POTASSIUM	MILLI	117.00	-	-	-	117.00	MILLI/MJ	126.22	
CALCIUM	MILLI	374.00	304.00	-	493.00	374.00	MILLI/MJ	403.48	
PHOSPHORUS	MILLI	316.00	243.00	-	398.00	316.00	MILLI/MJ	340.91	
NITRATE	MILLI	0.15	0.10	-	0.20	0.15	MILLI/MJ	0.16	

ROMADURKÄSE
40% FETT I. TR.

ROMADUR CHEESE
40% FAT CONTENT IN DRY MATTER

FROMAGE ROMADOUR
40% DE MAT. GR. S. SEC

		PROTEIN	FAT	CARBOHYDRATES	TOTAL
ENERGY VALUE (AVERAGE)	KJOULE	425	778	5.5	1209
PER 100 G EDIBLE PORTION	(KCAL)	102	186	1.3	289
AMOUNT OF DIGESTIBLE CONSTITUENTS PER 100 G	GRAM	22.40	19.09	0.33	
ENERGY VALUE (AVERAGE)	KJOULE	413	739	5.5	1157
OF THE DIGESTIBLE FRACTION PER 100 G EDIBLE PORTION	(KCAL)	99	177	1.3	277

WASTE PERCENTAGE AVERAGE 0.00

CONSTITUENTS	DIM	AV	VARIATION			AVR	NUTR. DENS.		MOLPERC.
WATER	GRAM	51.80	51.40	-	58.00	51.80	GRAM/MJ	44.77	
PROTEIN	GRAM	23.10	-	-	-	23.10	GRAM/MJ	19.96	
FAT	GRAM	20.10	16.80	-	20.60	20.10	GRAM/MJ	17.37	
AVAILABLE CARBOHYDR.	GRAM	0.33 1	-	-	-	0.33	GRAM/MJ	0.29	
MINERALS	MILLI	4.48	-	-	-	4.48	MILLI/MJ	3.87	
CALCIUM	MILLI	403.00	347.00	-	480.00	403.00	MILLI/MJ	348.30	
PHOSPHORUS	MILLI	326.00	277.00	-	350.00	326.00	MILLI/MJ	281.75	
NITRATE	MILLI	0.15	0.10	-	0.20	0.15	MILLI/MJ	0.13	
LACTIC ACID	GRAM	0.33	-	-	-	0.33	GRAM/MJ	0.29	

1 INCLUDES ONLY THE AMOUNT OF LACTIC ACID

ROMADURKÄSE
45% FETT I. TR.

ROMADUR CHEESE
45% FAT CONTENT IN DRY MATTER

FROMAGE ROMADOUR
45% DE MAT. GR. S. SEC

		PROTEIN	FAT	CARBOHYDRATES	TOTAL
ENERGY VALUE (AVERAGE)	KJOULE	390	894	0.0	1284
PER 100 G	(KCAL)	93	214	0.0	307
EDIBLE PORTION					
AMOUNT OF DIGESTIBLE	GRAM	20.56	21.94	0.00	
CONSTITUENTS PER 100 G					
ENERGY VALUE (AVERAGE)	KJOULE	379	849	0.0	1228
OF THE DIGESTIBLE	(KCAL)	90	203	0.0	293
FRACTION PER 100 G					
EDIBLE PORTION					

WASTE PERCENTAGE AVERAGE 0.00

CONSTITUENTS	DIM	AV	VARIATION			AVR	NUTR. DENS.		MOLPERC.
WATER	GRAM	51.00	50.40	-	56.00	51.00	GRAM/MJ	41.53	
PROTEIN	GRAM	21.20	-	-	-	21.20	GRAM/MJ	17.27	
FAT	GRAM	23.10	19.80	-	23.90	23.10	GRAM/MJ	18.81	
MINERALS	GRAM	4.26	-	-	-	4.26	GRAM/MJ	3.47	
CALCIUM	MILLI	273.00	255.00	-	317.00	273.00	MILLI/MJ	222.33	
PHOSPHORUS	MILLI	212.00	199.00	-	268.00	212.00	MILLI/MJ	172.65	
NITRATE	MILLI	0.15	0.10	-	0.20	0.15	MILLI/MJ	0.12	

ROMADURKÄSE
50% FETT I. TR.

ROMADUR CHEESE
50% FAT CONTENT IN DRY MATTER

FROMAGE ROMADOUR
50% DE MAT. GR. S. SEC

		PROTEIN	FAT	CARBOHYDRATES	TOTAL
ENERGY VALUE (AVERAGE)	KJOULE	368	995	0.0	1363
PER 100 G	(KCAL)	88	238	0.0	326
EDIBLE PORTION					
AMOUNT OF DIGESTIBLE	GRAM	19.40	24.41	0.00	
CONSTITUENTS PER 100 G					
ENERGY VALUE (AVERAGE)	KJOULE	357	945	0.0	1302
OF THE DIGESTIBLE	(KCAL)	85	226	0.0	311
FRACTION PER 100 G					
EDIBLE PORTION					

WASTE PERCENTAGE AVERAGE 0.00

CONSTITUENTS	DIM	AV	VARIATION			AVR	NUTR. DENS.		MOLPERC.
WATER	GRAM	50.00	49.10	-	54.00	50.00	GRAM/MJ	38.40	
PROTEIN	GRAM	20.00	-	-	-	20.00	GRAM/MJ	15.36	
FAT	GRAM	25.70	23.00	-	26.20	25.70	GRAM/MJ	19.74	
MINERALS	GRAM	4.15	-	-	-	4.15	GRAM/MJ	3.19	
CALCIUM	MILLI	264.00	202.00	-	328.00	264.00	MILLI/MJ	202.76	
PHOSPHORUS	MILLI	235.00	192.00	-	254.00	235.00	MILLI/MJ	180.48	
NITRATE	MILLI	0.15	0.10	-	0.20	0.15	MILLI/MJ	0.12	

ROQUEFORTKÄSE ROQUEFORT CHEESE FROMAGE ROQUEFORT

		PROTEIN	FAT	CARBOHYDRATES	TOTAL
ENERGY VALUE (AVERAGE)	KJOULE	396	1184	3.0	1583
PER 100 G EDIBLE PORTION	(KCAL)	95	283	0.7	378
AMOUNT OF DIGESTIBLE CONSTITUENTS PER 100 G	GRAM	20.85	29.07	0.18	
ENERGY VALUE (AVERAGE)	KJOULE	384	1125	3.0	1512
OF THE DIGESTIBLE FRACTION PER 100 G EDIBLE PORTION	(KCAL)	92	269	0.7	361

WASTE PERCENTAGE AVERAGE 0.00

CONSTITUENTS	DIM	AV	VARIATION			AVR	NUTR. DENS.		MOLPERC.
WATER	GRAM	39.40	-	-	-	39.40	GRAM/MJ	26.06	
PROTEIN	GRAM	21.50	-	-	-	21.50	GRAM/MJ	14.22	
FAT	GRAM	30.60	-	-	-	30.60	GRAM/MJ	20.24	
AVAILABLE CARBOHYDR.	GRAM	0.18 1	-	-	-	0.18	GRAM/MJ	0.12	
MINERALS	GRAM	6.44	-	-	-	6.44	GRAM/MJ	4.26	
SODIUM	GRAM	1.81	-	-	-	1.81	GRAM/MJ	1.20	
POTASSIUM	MILLI	91.00	-	-	-	91.00	MILLI/MJ	60.18	
MAGNESIUM	MILLI	30.00	-	-	-	30.00	MILLI/MJ	19.84	
CALCIUM	MILLI	662.00	-	-	-	662.00	MILLI/MJ	437.83	
IRON	MILLI	0.60	-	-	-	0.60	MILLI/MJ	0.40	
PHOSPHORUS	MILLI	392.00	-	-	-	392.00	MILLI/MJ	259.26	
VITAMIN A	MILLI	0.31	-	-	-	0.31	MILLI/MJ	0.21	
VITAMIN B1	MICRO	40.00	-	-	-	40.00	MICRO/MJ	26.45	
VITAMIN B2	MILLI	0.59	-	-	-	0.59	MILLI/MJ	0.39	
NICOTINAMIDE	MILLI	0.73	-	-	-	0.73	MILLI/MJ	0.48	
PANTOTHENIC ACID	MILLI	1.73	-	-	-	1.73	MILLI/MJ	1.14	
VITAMIN B6	MILLI	0.12	-	-	-	0.12	MILLI/MJ	0.08	
FOLIC ACID	MICRO	49.00	-	-	-	49.00	MICRO/MJ	32.41	
VITAMIN B12	MICRO	0.60	-	-	-	0.60	MICRO/MJ	0.40	

1 INCLUDES ONLY THE AMOUNT OF LACTIC ACID

CONSTITUENTS	DIM	AV	VARIATION			AVR	NUTR. DENS.		MOLPERC.
CADAVERINE	MILLI	8.40	7.10	-	9.30	8.40	MILLI/MJ	5.56	
HISTAMINE	MILLI	6.50	1.00	-	16.80	6.50	MILLI/MJ	4.30	
PUTRESCINE	MILLI	2.40	1.50	-	3.30	2.40	MILLI/MJ	1.59	
TYRAMINE	MILLI	-	2.70	-	110.00				
LACTIC ACID	GRAM	0.18	0.18	-	0.18	0.18	GRAM/MJ	0.12	
BUTYRIC ACID	GRAM	0.88	-	-	-	0.88	GRAM/MJ	0.58	
CAPROIC ACID	GRAM	0.66	-	-	-	0.66	GRAM/MJ	0.44	
CAPRYLIC ACID	GRAM	0.67	-	-	-	0.67	GRAM/MJ	0.44	
CAPRIC ACID	GRAM	2.16	-	-	-	2.16	GRAM/MJ	1.43	
LAURIC ACID	GRAM	1.30	-	-	-	1.30	GRAM/MJ	0.86	
MYRISTIC ACID	GRAM	3.25	-	-	-	3.25	GRAM/MJ	2.15	
PALMITIC ACID	GRAM	6.57	-	-	-	6.57	GRAM/MJ	4.35	
STEARIC ACID	GRAM	3.14	-	-	-	3.14	GRAM/MJ	2.08	
PALMITOLEIC ACID	GRAM	0.73	-	-	-	0.73	GRAM/MJ	0.48	
OLEIC ACID	GRAM	7.46	-	-	-	7.46	GRAM/MJ	4.93	
LINOLEIC ACID	GRAM	0.62	-	-	-	0.62	GRAM/MJ	0.41	
LINOLENIC ACID	GRAM	0.70	-	-	-	0.70	GRAM/MJ	0.46	

SAUERMILCHKÄSE
(HARZER-, MAINZER-, HAND-, STANGENKÄSE)
HÖCHST. 10% FETT I. TR.

ACID CURD CHEESE
MAX. 10% FAT CONTENT IN DRY MATTER

FROMAGE DE LAIT CAILLÉ
MOINS DE 10% DE MAT. GR. S. SEC

		PROTEIN	FAT	CARBOHYDRATES	TOTAL
ENERGY VALUE (AVERAGE)	KJOULE	552	27	5.4	585
PER 100 G EDIBLE PORTION	(KCAL)	132	6.5	1.3	140
AMOUNT OF DIGESTIBLE CONSTITUENTS PER 100 G	GRAM	29.10	0.66	0.32	
ENERGY VALUE (AVERAGE)	KJOULE	536	26	5.4	567
OF THE DIGESTIBLE FRACTION PER 100 G EDIBLE PORTION	(KCAL)	128	6.2	1.3	135

WASTE PERCENTAGE AVERAGE 0.00

CONSTITUENTS	DIM	AV	VARIATION			AVR	NUTR. DENS.		MOLPERC.
WATER	GRAM	64.00	60.60	-	67.60	64.00	GRAM/MJ	112.91	
PROTEIN	GRAM	30.00	26.00	-	36.30	30.00	GRAM/MJ	52.93	
FAT	GRAM	0.70	0.35	-	1.30	0.70	GRAM/MJ	1.23	
AVAILABLE CARBOHYDR.	GRAM	0.32 [1]	-	-	-	0.32	GRAM/MJ	0.56	
MINERALS	GRAM	4.70	3.90	-	6.03	4.70	GRAM/MJ	8.29	

1 INCLUDES ONLY THE AMOUNT OF LACTIC ACID

CONSTITUENTS	DIM	AV	VARIATION			AVR	NUTR.	DENS.	MOLPERC.
SODIUM	GRAM	1.52	1.22	-	2.14	1.52	GRAM/MJ	2.68	
POTASSIUM	MILLI	106.00	103.00	-	109.00	106.00	MILLI/MJ	187.01	
MAGNESIUM	MILLI	13.00	10.00	-	17.00	13.00	MILLI/MJ	22.94	
CALCIUM	MILLI	125.00	70.00	-	195.00	125.00	MILLI/MJ	220.53	
IRON	MILLI	0.29	0.19	-	0.44	0.29	MILLI/MJ	0.51	
COPPER	MICRO	80.00	20.00	-	240.00	80.00	MICRO/MJ	141.14	
ZINC	MILLI	2.00	1.10	-	2.90	2.00	MILLI/MJ	3.53	
PHOSPHORUS	MILLI	266.00	173.00	-	336.00	266.00	MILLI/MJ	469.29	
CHLORIDE	GRAM	2.10	1.70	-	2.70	2.10	GRAM/MJ	3.70	
VITAMIN B1	MICRO	30.00	-	-	-	30.00	MICRO/MJ	52.93	
VITAMIN B2	MILLI	0.36	-	-	-	0.36	MILLI/MJ	0.64	
NICOTINAMIDE	MILLI	0.70	-	-	-	0.70	MILLI/MJ	1.23	
ALANINE	GRAM	1.18	1.04	-	1.45	1.18	GRAM/MJ	2.08	5.1
ARGININE	GRAM	1.28	1.11	-	1.42	1.28	GRAM/MJ	2.26	2.8
ASPARTIC ACID	GRAM	2.45	2.39	-	2.52	2.45	GRAM/MJ	4.32	7.1
CYSTINE	GRAM	0.20	0.15	-	0.24	0.20	GRAM/MJ	0.35	0.3
GLUTAMIC ACID	GRAM	6.99	6.85	-	7.16	6.99	GRAM/MJ	12.33	18.2
GLYCINE	GRAM	0.73	0.65	-	0.83	0.73	GRAM/MJ	1.29	3.7
HISTIDINE	GRAM	1.31	1.25	-	1.35	1.31	GRAM/MJ	2.31	3.2
ISOLEUCINE	GRAM	1.66	1.59	-	1.69	1.66	GRAM/MJ	2.93	4.9
LEUCINE	GRAM	3.04	2.87	-	3.11	3.04	GRAM/MJ	5.36	8.9
LYSINE	GRAM	2.94	2.80	-	3.25	2.94	GRAM/MJ	5.19	7.7
METHIONINE	GRAM	0.52	0.21	-	0.93	0.52	GRAM/MJ	0.92	1.3
PHENYLALANINE	GRAM	1.66	1.59	-	1.70	1.66	GRAM/MJ	2.93	3.9
PROLINE	GRAM	3.11	2.94	-	3.22	3.11	GRAM/MJ	5.49	10.4
SERINE	GRAM	1.85	1.79	-	1.96	1.85	GRAM/MJ	3.26	6.8
THREONINE	GRAM	1.42	1.28	-	1.45	1.42	GRAM/MJ	2.51	4.6
TRYPTOPHAN	GRAM	0.28	0.21	-	0.31	0.28	GRAM/MJ	0.49	0.5
TYROSINE	GRAM	1.80	1.49	-	2.08	1.80	GRAM/MJ	3.18	3.8
VALINE	GRAM	2.08	1.94	-	2.22	2.08	GRAM/MJ	3.67	6.8
HISTAMINE	MILLI	39.00	-	-	-	39.00	MILLI/MJ	68.81	
LACTIC ACID	GRAM	0.32	0.10	-	0.95	0.32	GRAM/MJ	0.56	

SCHICHTKÄSE
10% FETT I. TR.

LAYERED CHEESE
10% FAT CONTENT IN DRY MATTER

FROMAGE STRATIFORME
10% DE MAT. GR. S. SEC

		PROTEIN	FAT	CARBOHYDRATES	TOTAL
ENERGY VALUE (AVERAGE)	KJOULE	234	94	16	343
PER 100 G	(KCAL)	56	22	3.8	82
EDIBLE PORTION					
AMOUNT OF DIGESTIBLE	GRAM	12.31	2.29	0.94	
CONSTITUENTS PER 100 G					
ENERGY VALUE (AVERAGE)	KJOULE	227	89	16	331
OF THE DIGESTIBLE	(KCAL)	54	21	3.8	79
FRACTION PER 100 G					
EDIBLE PORTION					

WASTE PERCENTAGE AVERAGE 0.00

CONSTITUENTS	DIM	AV		VARIATION			AVR	NUTR. DENS.		MOLPERC.
WATER	GRAM	78.60		76.60	-	79.80	78.60	GRAM/MJ	237.11	
PROTEIN	GRAM	12.70		-	-	-	12.70	GRAM/MJ	38.31	
FAT	GRAM	2.42		2.25	-	2.62	2.42	GRAM/MJ	7.30	
AVAILABLE CARBOHYDR.	GRAM	0.94	1	-	-	-	0.94	GRAM/MJ	2.84	
MINERALS	GRAM	1.00		-	-	-	1.00	GRAM/MJ	3.02	
SODIUM	MILLI	39.00		29.00	-	46.00	39.00	MILLI/MJ	117.65	
POTASSIUM	MILLI	125.00		120.00	-	130.00	125.00	MILLI/MJ	377.08	
CALCIUM	MILLI	77.00		72.00	-	82.00	77.00	MILLI/MJ	232.28	
PHOSPHORUS	MILLI	163.00		-	-	-	163.00	MILLI/MJ	491.71	
LACTIC ACID	GRAM	0.94		-	-	-	0.94	GRAM/MJ	2.84	

1 INCLUDES ONLY THE AMOUNT OF LACTIC ACID

SCHICHTKÄSE
20% FETT I. TR.

LAYERED CHEESE
20% FAT CONTENT IN DRY MATTER

FROMAGE STRATIFORME
20% DE MAT. GR. S. SEC

		PROTEIN	FAT	CARBOHYDRATES	TOTAL
ENERGY VALUE (AVERAGE)	KJOULE	219	192	15	426
PER 100 G EDIBLE PORTION	(KCAL)	52	46	3.6	102
AMOUNT OF DIGESTIBLE CONSTITUENTS PER 100 G	GRAM	11.54	4.71	0.89	
ENERGY VALUE (AVERAGE)	KJOULE	213	182	15	410
OF THE DIGESTIBLE FRACTION PER 100 G EDIBLE PORTION	(KCAL)	51	44	3.6	98

WASTE PERCENTAGE AVERAGE 0.00

CONSTITUENTS	DIM	AV		VARIATION			AVR	NUTR. DENS.		MOLPERC.
WATER	GRAM	78.00		77.50	-	78.50	78.00	GRAM/MJ	190.35	
PROTEIN	GRAM	11.90		-	-	-	11.90	GRAM/MJ	29.04	
FAT	GRAM	4.96		4.33	-	5.46	4.96	GRAM/MJ	12.10	
AVAILABLE CARBOHYDR.	GRAM	0.89	1	-	-	-	0.89	GRAM/MJ	2.17	
MINERALS	GRAM	0.80		-	-	-	0.80	GRAM/MJ	1.95	
SODIUM	MILLI	35.00		30.00	-	39.00	35.00	MILLI/MJ	85.42	
POTASSIUM	MILLI	120.00		112.00	-	127.00	120.00	MILLI/MJ	292.85	
CALCIUM	MILLI	79.00		78.00	-	80.00	79.00	MILLI/MJ	192.80	
LACTIC ACID	GRAM	0.89		0.79	-	0.99	0.89	GRAM/MJ	2.17	

1 INCLUDES ONLY THE AMOUNT OF LACTIC ACID

SCHICHTKÄSE
40% FETT I. TR.

LAYERED CHEESE
40% FAT CONTENT IN DRY MATTER

FROMAGE STRATIFORME
40% DE MAT. GR. S. SEC

		PROTEIN	FAT	CARBOHYDRATES	TOTAL
ENERGY VALUE (AVERAGE) PER 100 G EDIBLE PORTION	KJOULE	199	437	12	648
	(KCAL)	48	105	2.8	155
AMOUNT OF DIGESTIBLE CONSTITUENTS PER 100 G	GRAM	10.47	10.73	0.71	
ENERGY VALUE (AVERAGE) OF THE DIGESTIBLE FRACTION PER 100 G EDIBLE PORTION	KJOULE	193	415	12	620
	(KCAL)	46	99	2.8	148

WASTE PERCENTAGE AVERAGE 0.00

CONSTITUENTS	DIM	AV	VARIATION	AVR	NUTR. DENS.		MOLPERC.
WATER	GRAM	75.00	73.00 - 76.90	75.00	GRAM/MJ	120.93	
PROTEIN	GRAM	10.80	9.53 - 12.00	10.80	GRAM/MJ	17.41	
FAT	GRAM	11.30	10.50 - 12.10	11.30	GRAM/MJ	18.22	
AVAILABLE CARBOHYDR.	GRAM	0.71 1	- - -	0.71	GRAM/MJ	1.14	
MINERALS	GRAM	0.73	- - -	0.73	GRAM/MJ	1.18	
SODIUM	MILLI	42.00	35.00 - 48.00	42.00	MILLI/MJ	67.72	
POTASSIUM	MILLI	118.00	115.00 - 120.00	118.00	MILLI/MJ	190.26	
CALCIUM	MILLI	82.00	77.00 - 87.00	82.00	MILLI/MJ	132.21	
LACTIC ACID	GRAM	0.71	0.58 - 0.84	0.71	GRAM/MJ	1.14	

1 INCLUDES ONLY THE AMOUNT OF LACTIC ACID

SCHMELZKÄSE
45% FETT I. TR.

PROCESSED CHEESE
45% FAT CONTENT IN DRY MATTER

FROMAGE FONDU
45% DE MAT. GR. S. SEC

		PROTEIN	FAT	CARBOHYDRATES	TOTAL
ENERGY VALUE (AVERAGE) PER 100 G EDIBLE PORTION	KJOULE	265	913	0.0	1178
	(KCAL)	63	218	0.0	282
AMOUNT OF DIGESTIBLE CONSTITUENTS PER 100 G	GRAM	13.96	22.42	0.00	
ENERGY VALUE (AVERAGE) OF THE DIGESTIBLE FRACTION PER 100 G EDIBLE PORTION	KJOULE	257	868	0.0	1125
	(KCAL)	61	207	0.0	269

WASTE PERCENTAGE AVERAGE 0.00

CONSTITUENTS	DIM	AV	VARIATION			AVR	NUTR. DENS.		MOLPERC.
WATER	GRAM	51.30	51.10	-	51.50	51.30	GRAM/MJ	45.61	
PROTEIN	GRAM	14.40	14.00	-	15.20	14.40	GRAM/MJ	12.80	
FAT	GRAM	23.60	22.60	-	24.70	23.60	GRAM/MJ	20.98	
MINERALS	GRAM	4.60	4.00	-	4.90	4.60	GRAM/MJ	4.09	
SODIUM	GRAM	1.26	0.79	-	1.47	1.26	GRAM/MJ	1.12	
POTASSIUM	MILLI	65.00	40.00	-	73.00	65.00	MILLI/MJ	57.79	
CALCIUM	MILLI	547.00	300.00	-	661.00	547.00	MILLI/MJ	486.29	
IRON	MILLI	1.00	-	-	-	1.00	MILLI/MJ	0.89	
COPPER	MILLI	0.46	-	-	-	0.46	MILLI/MJ	0.41	
PHOSPHORUS	MILLI	944.00	-	-	-	944.00	MILLI/MJ	839.23	
CHLORIDE	MILLI	725.00	643.00	-	807.00	725.00	MILLI/MJ	644.53	
VITAMIN A	MILLI	0.30	0.15	-	0.42	0.30	MILLI/MJ	0.27	
VITAMIN D	MICRO	3.13	1.25	-	5.00	3.13	MICRO/MJ	2.78	
VITAMIN B1	MICRO	34.00	10.00	-	50.00	34.00	MICRO/MJ	30.23	
VITAMIN B2	MILLI	0.38	0.30	-	0.45	0.38	MILLI/MJ	0.34	
NICOTINAMIDE	MILLI	0.22	0.07	-	0.60	0.22	MILLI/MJ	0.20	
PANTOTHENIC ACID	MILLI	0.52	0.15	-	0.95	0.52	MILLI/MJ	0.46	
VITAMIN B6	MICRO	70.00	60.00	-	80.00	70.00	MICRO/MJ	62.23	
BIOTIN	MICRO	3.60	1.00	-	7.00	3.60	MICRO/MJ	3.20	
FOLIC ACID	MICRO	3.46	2.00	-	5.00	3.46	MICRO/MJ	3.08	
VITAMIN B12	MICRO	0.25	-	-	-	0.25	MICRO/MJ	0.22	

SCHMELZKÄSE
60% FETT I. TR.
(KÄSECREME)

PROCESSED CHEESE
60% FAT CONTENT IN
DRY MATTER

PRÉP. DE FROMAGE FONDU
60% DE MAT. GR. S. SEC

		PROTEIN	FAT	CARBOHYDRATES	TOTAL
ENERGY VALUE (AVERAGE)	KJOULE	243	1177	0.0	1420
PER 100 G EDIBLE PORTION	(KCAL)	58	281	0.0	339
AMOUNT OF DIGESTIBLE CONSTITUENTS PER 100 G	GRAM	12.80	28.88	0.00	
ENERGY VALUE (AVERAGE)	KJOULE	236	1118	0.0	1353
OF THE DIGESTIBLE FRACTION PER 100 G EDIBLE PORTION	(KCAL)	56	267	0.0	323

WASTE PERCENTAGE AVERAGE 0.00

CONSTITUENTS	DIM	AV	VARIATION			AVR	NUTR. DENS.		MOLPERC.
WATER	GRAM	50.60	45.70	-	53.00	50.60	GRAM/MJ	37.39	
PROTEIN	GRAM	13.20	11.60	-	13.70	13.20	GRAM/MJ	9.75	
FAT	GRAM	30.40	28.20	-	33.60	30.40	GRAM/MJ	22.46	
MINERALS	GRAM	4.10	3.90	-	4.40	4.10	GRAM/MJ	3.03	

CONSTITUENTS	DIM	AV	VARIATION			AVR	NUTR. DENS.		MOLPERC.
SODIUM	GRAM	1.01	0.95	-	1.05	1.01	GRAM/MJ	0.75	
POTASSIUM	MILLI	108.00	53.00	-	165.00	108.00	MILLI/MJ	79.80	
MAGNESIUM	MILLI	48.00	47.00	-	50.00	48.00	MILLI/MJ	35.47	
CALCIUM	MILLI	355.00	340.00	-	370.00	355.00	MILLI/MJ	262.30	
IRON	MILLI	1.40	1.10	-	1.60	1.40	MILLI/MJ	1.03	
ZINC	MILLI	-	1.00	-	4.00				
PHOSPHORUS	MILLI	795.00	728.00	-	871.00	795.00	MILLI/MJ	587.40	
VITAMIN B1	MICRO	40.00	40.00	-	50.00	40.00	MICRO/MJ	29.55	
VITAMIN B2	MILLI	0.35	0.32	-	0.41	0.35	MILLI/MJ	0.26	
NICOTINAMIDE	MICRO	70.00	40.00	-	110.00	70.00	MICRO/MJ	51.72	
PANTOTHENIC ACID	MILLI	0.47	0.36	-	0.57	0.47	MILLI/MJ	0.35	
VITAMIN B6	MICRO	80.00	70.00	-	100.00	80.00	MICRO/MJ	59.11	
BIOTIN	MICRO	2.80	2.10	-	3.40	2.80	MICRO/MJ	2.07	
FOLIC ACID	MICRO	3.40	2.60	-	4.60	3.40	MICRO/MJ	2.51	
VITAMIN B12	MICRO	0.25	0.16	-	0.37	0.25	MICRO/MJ	0.18	
VITAMIN C	-	TRACES	-	-	-				

SPEISEQUARK
MAGER

QUARK, FRESH CHEESE
FROM SKIMMILK

FROMAGE BLANC
MAIGRE

		PROTEIN	FAT	CARBOHYDRATES	TOTAL
ENERGY VALUE (AVERAGE)	KJOULE	249	9.7	66	325
PER 100 G	(KCAL)	59	2.3	16	78
EDIBLE PORTION					
AMOUNT OF DIGESTIBLE	GRAM	13.09	0.23	3.97	
CONSTITUENTS PER 100 G					
ENERGY VALUE (AVERAGE)	KJOULE	241	9.2	66	317
OF THE DIGESTIBLE	(KCAL)	58	2.2	16	76
FRACTION PER 100 G					
EDIBLE PORTION					

WASTE PERCENTAGE AVERAGE 0.00

CONSTITUENTS	DIM	AV	VARIATION			AVR	NUTR. DENS.		MOLPERC.
WATER	GRAM	81.30	79.50	-	83.00	81.30	GRAM/MJ	256.70	
PROTEIN	GRAM	13.50	12.00	-	14.60	13.50	GRAM/MJ	42.63	
FAT	GRAM	0.25	0.15	-	0.50	0.25	GRAM/MJ	0.79	
AVAILABLE CARBOHYDR.	GRAM	3.97 [1]	-	-	-	3.97	GRAM/MJ	12.54	
MINERALS	GRAM	0.85	0.81	-	1.00	0.85	GRAM/MJ	2.68	

1 INCLUDES THE AMOUNTS OF LACTOSE AND LACTIC ACID

CONSTITUENTS	DIM	AV	VARIATION		AVR	NUTR. DENS.		MOLPERC.
SODIUM	MILLI	40.00	25.00	- 52.00	40.00	MILLI/MJ	126.30	
POTASSIUM	MILLI	95.00	70.00	- 117.00	95.00	MILLI/MJ	299.96	
MAGNESIUM	MILLI	12.00	6.00	- 19.00	12.00	MILLI/MJ	37.89	
CALCIUM	MILLI	92.00	71.00	- 104.00	92.00	MILLI/MJ	290.49	
MANGANESE	MICRO	70.00	40.00	- 110.00	70.00	MICRO/MJ	221.02	
IRON	MILLI	0.40	0.30	- 0.50	0.40	MILLI/MJ	1.26	
COBALT	MICRO	0.12	0.08	- 0.19	0.12	MICRO/MJ	0.38	
COPPER	MICRO	15.00	3.00	- 50.00	15.00	MICRO/MJ	47.36	
ZINC	MILLI	0.57	0.32	- 0.83	0.57	MILLI/MJ	1.80	
PHOSPHORUS	MILLI	160.00	138.00	- 175.00	160.00	MILLI/MJ	505.20	
FLUORIDE	MICRO	25.00	17.00	- 32.00	25.00	MICRO/MJ	78.94	
IODIDE	MICRO	4.00	0.63	- 7.20	4.00	MICRO/MJ	12.63	
NITRATE	MILLI	0.14	0.00	- 0.62	0.14	MILLI/MJ	0.44	
VITAMIN A	MICRO	2.10	1.90	- 2.40	2.10	MICRO/MJ	6.63	
CAROTENE	MICRO	1.30	1.00	- 1.50	1.30	MICRO/MJ	4.10	
VITAMIN E ACTIVITY	MICRO	6.20	2.90	- 7.00	6.20	MICRO/MJ	19.58	
ALPHA-TOCOPHEROL	MICRO	6.20	2.90	- 7.00	6.20	MICRO/MJ	19.58	
VITAMIN K	MICRO	1.20	0.00	- 2.30	1.20	MICRO/MJ	3.79	
VITAMIN B1	MICRO	43.00	20.00	- 80.00	43.00	MICRO/MJ	135.77	
VITAMIN B2	MILLI	0.30	0.23	- 0.35	0.30	MILLI/MJ	0.95	
NICOTINAMIDE	MILLI	0.15	0.10	- 0.20	0.15	MILLI/MJ	0.47	
PANTOTHENIC ACID	MILLI	0.74	0.59	- 0.89	0.74	MILLI/MJ	2.34	
VITAMIN B6	MILLI	0.10	0.05	- 0.15	0.10	MILLI/MJ	0.32	
BIOTIN	MICRO	7.00	4.00	- 11.00	7.00	MICRO/MJ	22.10	
FOLIC ACID	MICRO	18.00	17.00	- 19.00	18.00	MICRO/MJ	56.83	
VITAMIN B12	MICRO	0.88	0.75	- 1.00	0.88	MICRO/MJ	2.78	
VITAMIN C	MILLI	0.70	0.00	- 1.00	0.70	MILLI/MJ	2.21	
ARGININE	GRAM	0.54	0.50	- 0.64	0.54	GRAM/MJ	1.71	
CYSTINE	GRAM	0.12	0.09	- 0.14	0.12	GRAM/MJ	0.38	
HISTIDINE	GRAM	0.41	0.36	- 0.44	0.41	GRAM/MJ	1.29	
ISOLEUCINE	GRAM	0.85	0.78	- 0.88	0.85	GRAM/MJ	2.68	
LEUCINE	GRAM	1.39	1.25	- 1.45	1.39	GRAM/MJ	4.39	
LYSINE	GRAM	1.13	1.05	- 1.28	1.13	GRAM/MJ	3.57	
METHIONINE	GRAM	0.42	0.38	- 0.46	0.42	GRAM/MJ	1.33	
PHENYLALANINE	GRAM	0.70	0.68	- 0.73	0.70	GRAM/MJ	2.21	
THREONINE	GRAM	0.63	0.58	- 0.68	0.63	GRAM/MJ	1.99	
TRYPTOPHAN	GRAM	0.17	0.14	- 0.19	0.17	GRAM/MJ	0.54	
TYROSINE	GRAM	0.70	0.67	- 0.73	0.70	GRAM/MJ	2.21	
VALINE	GRAM	0.90	0.77	- 0.99	0.90	GRAM/MJ	2.84	
LACTIC ACID	GRAM	0.77	0.72	- 0.82	0.77	GRAM/MJ	2.43	
LACTOSE	GRAM	3.20	3.00	- 3.50	3.20	GRAM/MJ	10.10	
CHOLESTEROL	MILLI	0.80	-	- -	0.80	MILLI/MJ	2.53	

SPEISEQUARK	QUARK, FRESH CHEESE	**FROMAGE BLANC**
20% FETT I. TR.	20% FAT CONTENT IN DRY MATTER	20% DE MAT. GR. S. SEC

		PROTEIN	FAT	CARBOHYDRATES	TOTAL
ENERGY VALUE (AVERAGE)	KJOULE	230	197	58	485
PER 100 G	(KCAL)	55	47	14	116
EDIBLE PORTION					
AMOUNT OF DIGESTIBLE	GRAM	12.12	4.84	3.44	
CONSTITUENTS PER 100 G					
ENERGY VALUE (AVERAGE)	KJOULE	223	188	58	468
OF THE DIGESTIBLE	(KCAL)	53	45	14	112
FRACTION PER 100 G					
EDIBLE PORTION					

WASTE PERCENTAGE AVERAGE 0.00

CONSTITUENTS	DIM	AV	VARIATION	AVR	NUTR.	DENS.	MOLPERC.
WATER	GRAM	78.00	76.10 - 79.30	78.00	GRAM/MJ	166.56	
PROTEIN	GRAM	12.50	11.20 - 13.50	12.50	GRAM/MJ	26.69	
FAT	GRAM	5.10	4.70 - 5.50	5.10	GRAM/MJ	10.89	
AVAILABLE CARBOHYDR.	GRAM	3.44 [1]	- - -	3.44	GRAM/MJ	7.35	
MINERALS	GRAM	0.80	0.72 - 0.91	0.80	GRAM/MJ	1.71	
SODIUM	MILLI	35.00	23.00 - 48.00	35.00	MILLI/MJ	74.74	
POTASSIUM	MILLI	87.00	64.00 - 108.00	87.00	MILLI/MJ	185.78	
MAGNESIUM	MILLI	11.00	6.00 - 17.00	11.00	MILLI/MJ	23.49	
CALCIUM	MILLI	85.00	76.00 - 110.00	85.00	MILLI/MJ	181.51	
MANGANESE	MICRO	60.00	40.00 - 100.00	60.00	MICRO/MJ	128.12	
IRON	MILLI	0.37	0.28 - 0.46	0.37	MILLI/MJ	0.79	
COBALT	MICRO	0.12	0.07 - 0.18	0.12	MICRO/MJ	0.26	
COPPER	MICRO	14.00	3.00 - 37.00	14.00	MICRO/MJ	29.90	
ZINC	MILLI	0.50	0.30 - 0.80	0.50	MILLI/MJ	1.07	
NICKEL	MICRO	1.00	- - -	1.00	MICRO/MJ	2.14	
CHROMIUM	MICRO	1.00	- - -	1.00	MICRO/MJ	2.14	
MOLYBDENUM	MICRO	7.00	- - -	7.00	MICRO/MJ	14.95	
PHOSPHORUS	MILLI	165.00	135.00 - 190.00	165.00	MILLI/MJ	352.34	
CHLORIDE	MILLI	130.00	100.00 - 175.00	130.00	MILLI/MJ	277.60	
FLUORIDE	MICRO	23.00	14.00 - 28.00	23.00	MICRO/MJ	49.11	
IODIDE	MICRO	3.70	0.60 - 6.60	3.70	MICRO/MJ	7.90	
SELENIUM	MICRO	5.00	1.00 - 9.00	5.00	MICRO/MJ	10.68	
SILICON	MILLI	0.40	- - -	0.40	MILLI/MJ	0.85	
VITAMIN A	MICRO	40.00	36.00 - 46.00	40.00	MICRO/MJ	85.42	
CAROTENE	MICRO	24.00	19.00 - 30.00	24.00	MICRO/MJ	51.25	
VITAMIN D	NANO	85.00	27.00 - 120.00	85.00	NANO/MJ	181.51	

1 INCLUDES THE AMOUNTS OF LACTOSE AND LACTIC ACID

CONSTITUENTS	DIM	AV	VARIATION			AVR	NUTR. DENS.		MOLPERC.
VITAMIN E ACTIVITY	MILLI	0.12	0.05	-	0.14	0.12	MILLI/MJ	0.26	
ALPHA-TOCOPHEROL	MILLI	0.12	0.05	-	0.14	0.12	MILLI/MJ	0.26	
VITAMIN K	MICRO	23.00	0.00	-	45.00	23.00	MICRO/MJ	49.11	
VITAMIN B1	MICRO	37.00	18.00	-	73.00	37.00	MICRO/MJ	79.01	
VITAMIN B2	MILLI	0.27	0.21	-	0.32	0.27	MILLI/MJ	0.58	
NICOTINAMIDE	MILLI	0.14	0.09	-	0.18	0.14	MILLI/MJ	0.30	
PANTOTHENIC ACID	MILLI	0.68	0.54	-	0.82	0.68	MILLI/MJ	1.45	
VITAMIN B6	MICRO	90.00	50.00	-	140.00	90.00	MICRO/MJ	192.18	
BIOTIN	MICRO	6.40	3.70	-	10.00	6.40	MICRO/MJ	13.67	
FOLIC ACID	MICRO	16.00	15.00	-	17.00	16.00	MICRO/MJ	34.17	
VITAMIN B12	MICRO	0.81	0.69	-	0.92	0.81	MICRO/MJ	1.73	
VITAMIN C	MILLI	0.60	0.00	-	0.90	0.60	MILLI/MJ	1.28	
ARGININE	GRAM	0.50	0.46	-	0.59	0.50	GRAM/MJ	1.07	
CYSTINE	GRAM	0.11	0.08	-	0.13	0.11	GRAM/MJ	0.23	
HISTIDINE	GRAM	0.38	0.33	-	0.41	0.38	GRAM/MJ	0.81	
ISOLEUCINE	GRAM	0.79	0.72	-	0.81	0.79	GRAM/MJ	1.69	
LEUCINE	GRAM	1.29	1.16	-	1.34	1.29	GRAM/MJ	2.75	
LYSINE	GRAM	1.05	0.97	-	1.19	1.05	GRAM/MJ	2.24	
METHIONINE	GRAM	0.39	0.35	-	0.43	0.39	GRAM/MJ	0.83	
PHENYLALANINE	GRAM	0.65	0.63	-	0.68	0.65	GRAM/MJ	1.39	
THREONINE	GRAM	0.58	0.54	-	0.63	0.58	GRAM/MJ	1.24	
TRYPTOPHAN	GRAM	0.16	0.13	-	0.18	0.16	GRAM/MJ	0.34	
TYROSINE	GRAM	0.65	0.62	-	0.68	0.65	GRAM/MJ	1.39	
VALINE	GRAM	0.83	0.71	-	0.92	0.83	GRAM/MJ	1.77	
HISTAMINE	MICRO	20.00	0.00	-	90.00	20.00	MICRO/MJ	42.71	
TYRAMINE	MILLI	0.24	0.10	-	0.80	0.24	MILLI/MJ	0.51	
LACTIC ACID	GRAM	0.74	0.69	-	0.79	0.74	GRAM/MJ	1.58	
LACTOSE	GRAM	2.70	2.00	-	3.30	2.70	GRAM/MJ	5.77	
BUTYRIC ACID	GRAM	0.14	-	-	-	0.14	GRAM/MJ	0.30	
CAPROIC ACID	MILLI	40.00	-	-	-	40.00	MILLI/MJ	85.42	
CAPRYLIC ACID	MILLI	50.00	-	-	-	50.00	MILLI/MJ	106.77	
CAPRIC ACID	GRAM	0.10	-	-	-	0.10	GRAM/MJ	0.21	
LAURIC ACID	MILLI	70.00	-	-	-	70.00	MILLI/MJ	149.48	
MYRISTIC ACID	GRAM	0.53	-	-	-	0.53	GRAM/MJ	1.13	
PALMITIC ACID	GRAM	1.54	-	-	-	1.54	GRAM/MJ	3.29	
STEARIC ACID	GRAM	0.59	-	-	-	0.59	GRAM/MJ	1.26	
PALMITOLEIC ACID	GRAM	0.14	-	-	-	0.14	GRAM/MJ	0.30	
OLEIC ACID	GRAM	1.22	-	-	-	1.22	GRAM/MJ	2.61	
LINOLEIC ACID	GRAM	0.13	0.11	-	0.16	0.13	GRAM/MJ	0.28	
LINOLENIC ACID	MILLI	34.00	28.00	-	43.00	34.00	MILLI/MJ	72.60	
CHOLESTEROL	MILLI	17.00	-	-	-	17.00	MILLI/MJ	36.30	

SPEISEQUARK	QUARK, FRESH CHEESE	FROMAGE BLANC
40% FETT I. TR. (SPEISEQUARK MIT SAHNE)	40% FAT CONTENT IN DRY MATTER	40% DE MAT. GR. S. SEC

		PROTEIN	FAT	CARBOHYDRATES	TOTAL
ENERGY VALUE (AVERAGE)	KJOULE	204	441	55	701
PER 100 G	(KCAL)	49	105	13	167
EDIBLE PORTION					
AMOUNT OF DIGESTIBLE	GRAM	10.76	10.83	3.30	
CONSTITUENTS PER 100 G					
ENERGY VALUE (AVERAGE)	KJOULE	198	419	55	673
OF THE DIGESTIBLE	(KCAL)	47	100	13	161
FRACTION PER 100 G					
EDIBLE PORTION					

WASTE PERCENTAGE AVERAGE 0.00

CONSTITUENTS	DIM	AV		VARIATION		AVR	NUTR.	DENS.	MOLPERC.
WATER	GRAM	73.50	72.60	-	74.60	73.50	GRAM/MJ	109.28	
PROTEIN	GRAM	11.10	9.50	-	12.00	11.10	GRAM/MJ	16.50	
FAT	GRAM	11.40	10.30	-	12.50	11.40	GRAM/MJ	16.95	
AVAILABLE CARBOHYDR.	GRAM	3.30 1	-	-	-	3.30	GRAM/MJ	4.91	
MINERALS	GRAM	0.75	0.73	-	0.78	0.75	GRAM/MJ	1.12	
SODIUM	MILLI	34.00	22.00	-	45.00	34.00	MILLI/MJ	50.55	
POTASSIUM	MILLI	82.00	60.00	-	101.00	82.00	MILLI/MJ	121.92	
MAGNESIUM	MILLI	10.00	5.00	-	16.00	10.00	MILLI/MJ	14.87	
CALCIUM	MILLI	95.00	75.00	-	110.00	95.00	MILLI/MJ	141.25	
MANGANESE	MICRO	60.00	30.00	-	90.00	60.00	MICRO/MJ	89.21	
IRON	MILLI	0.34	0.26	-	0.43	0.34	MILLI/MJ	0.51	
COBALT	MICRO	0.11	0.07	-	0.17	0.11	MICRO/MJ	0.16	
COPPER	MICRO	13.00	3.00	-	34.00	13.00	MICRO/MJ	19.33	
ZINC	MILLI	0.50	0.30	-	0.80	0.50	MILLI/MJ	0.74	
NICKEL	MICRO	-	5.00	-	26.00				
CHROMIUM	MICRO	-	5.00	-	13.00				
MOLYBDENUM	MILLI	-	0.01	-	0.02				
PHOSPHORUS	MILLI	187.00	179.00	-	195.00	187.00	MILLI/MJ	278.03	
CHLORIDE	MILLI	130.00	110.00	-	160.00	130.00	MILLI/MJ	193.28	
FLUORIDE	MICRO	22.00	13.00	-	26.00	22.00	MICRO/MJ	32.71	
IODIDE	MICRO	3.40	0.50	-	6.20	3.40	MICRO/MJ	5.06	
VITAMIN A	MICRO	90.00	81.00	-	103.00	90.00	MICRO/MJ	133.81	
CAROTENE	MICRO	54.00	43.00	-	67.00	54.00	MICRO/MJ	80.29	
VITAMIN D	MICRO	0.19	0.06	-	0.27	0.19	MICRO/MJ	0.28	
VITAMIN E ACTIVITY	MILLI	0.27	0.11	-	0.31	0.27	MILLI/MJ	0.40	
ALPHA-TOCOPHEROL	MILLI	0.27	0.11	-	0.31	0.27	MILLI/MJ	0.40	

1 INCLUDES THE AMOUNTS OF LACTOSE AND LACTIC ACID

CONSTITUENTS	DIM	AV	VARIATION			AVR	NUTR. DENS.		MOLPERC.
VITAMIN K	MICRO	50.00	0.00	-	100.00	50.00	MICRO/MJ	74.34	
VITAMIN B1	MICRO	33.00	16.00	-	65.00	33.00	MICRO/MJ	49.06	
VITAMIN B2	MILLI	0.24	0.19	-	0.28	0.24	MILLI/MJ	0.36	
NICOTINAMIDE	MILLI	0.12	0.08	-	0.16	0.12	MILLI/MJ	0.18	
PANTOTHENIC ACID	MILLI	0.61	0.48	-	0.73	0.61	MILLI/MJ	0.91	
VITAMIN B6	MICRO	80.00	40.00	-	120.00	80.00	MICRO/MJ	118.94	
BIOTIN	MICRO	6.00	3.30	-	8.90	6.00	MICRO/MJ	8.92	
FOLIC ACID	MILLI	-	0.01	-	0.02				
VITAMIN B12	MICRO	0.72	0.61	-	0.82	0.72	MICRO/MJ	1.07	
VITAMIN C	MILLI	0.50	0.00	-	0.80	0.50	MILLI/MJ	0.74	
ARGININE	GRAM	0.44	0.41	-	0.52	0.44	GRAM/MJ	0.65	
CYSTINE	GRAM	0.10	0.07	-	0.11	0.10	GRAM/MJ	0.15	
HISTIDINE	GRAM	0.33	0.30	-	0.36	0.33	GRAM/MJ	0.49	
ISOLEUCINE	GRAM	0.70	0.64	-	0.73	0.70	GRAM/MJ	1.04	
LEUCINE	GRAM	1.14	1.10	-	1.19	1.14	GRAM/MJ	1.69	
LYSINE	GRAM	0.93	0.87	-	1.05	0.93	GRAM/MJ	1.38	
METHIONINE	GRAM	0.34	0.31	-	0.38	0.34	GRAM/MJ	0.51	
PHENYLALANINE	GRAM	0.58	0.56	-	0.60	0.58	GRAM/MJ	0.86	
THREONINE	GRAM	0.52	0.47	-	0.56	0.52	GRAM/MJ	0.77	
TRYPTOPHAN	GRAM	0.14	0.12	-	0.16	0.14	GRAM/MJ	0.21	
TYROSINE	GRAM	0.58	0.55	-	0.60	0.58	GRAM/MJ	0.86	
VALINE	GRAM	0.74	0.64	-	0.81	0.74	GRAM/MJ	1.10	
HISTAMINE	MICRO	20.00	0.00	-	90.00	20.00	MICRO/MJ	29.74	
TYRAMINE	MILLI	0.24	0.10	-	0.80	0.24	MILLI/MJ	0.36	
LACTIC ACID	GRAM	0.70	0.65	-	0.75	0.70	GRAM/MJ	1.04	
LACTOSE	GRAM	2.60	2.00	-	3.10	2.60	GRAM/MJ	3.87	
BUTYRIC ACID	GRAM	0.32	-	-	-	0.32	GRAM/MJ	0.48	
CAPROIC ACID	MILLI	90.00	-	-	-	90.00	MILLI/MJ	133.81	
CAPRYLIC ACID	GRAM	0.11	-	-	-	0.11	GRAM/MJ	0.16	
CAPRIC ACID	GRAM	0.22	-	-	-	0.22	GRAM/MJ	0.33	
LAURIC ACID	GRAM	0.15	-	-	-	0.15	GRAM/MJ	0.22	
MYRISTIC ACID	GRAM	1.18	-	-	-	1.18	GRAM/MJ	1.75	
PALMITIC ACID	GRAM	3.46	-	-	-	3.46	GRAM/MJ	5.14	
STEARIC ACID	GRAM	1.32	-	-	-	1.32	GRAM/MJ	1.96	
PALMITOLEIC ACID	GRAM	0.31	-	-	-	0.31	GRAM/MJ	0.46	
OLEIC ACID	GRAM	2.74	-	-	-	2.74	GRAM/MJ	4.07	
LINOLEIC ACID	GRAM	0.29	0.24	-	0.36	0.29	GRAM/MJ	0.44	
LINOLENIC ACID	MILLI	76.00	63.00	-	96.00	76.00	MILLI/MJ	113.00	
CHOLESTEROL	MILLI	37.00	-	-	-	37.00	MILLI/MJ	55.01	

TILSITERKÄSE
30% FETT I. TR.

TILSIT CHEESE
30% FAT CONTENT IN DRY MATTER

FROMAGE TILSIT
30% DE MAT. GR. S. SEC

		PROTEIN	FAT	CARBOHYDRATES	TOTAL
ENERGY VALUE (AVERAGE)	KJOULE	528	666	0.0	1194
PER 100 G	(KCAL)	126	159	0.0	285
EDIBLE PORTION					
AMOUNT OF DIGESTIBLE	GRAM	27.83	16.34	0.00	
CONSTITUENTS PER 100 G					
ENERGY VALUE (AVERAGE)	KJOULE	513	632	0.0	1145
OF THE DIGESTIBLE	(KCAL)	122	151	0.0	274
FRACTION PER 100 G					
EDIBLE PORTION					

WASTE PERCENTAGE AVERAGE 0.00

CONSTITUENTS	DIM	AV	VARIATION			AVR	NUTR.	DENS.	MOLPERC.
WATER	GRAM	46.20	41.80	-	51.00	46.20	GRAM/MJ	40.35	
PROTEIN	GRAM	28.70	-	-	-	28.70	GRAM/MJ	25.07	
FAT	GRAM	17.20	14.70	-	17.50	17.20	GRAM/MJ	15.02	
MINERALS	GRAM	4.83	-	-	-	4.83	GRAM/MJ	4.22	
SODIUM	GRAM	1.00	-	-	-	1.00	GRAM/MJ	0.87	
CALCIUM	MILLI	830.00	-	-	-	830.00	MILLI/MJ	724.96	
PHOSPHORUS	MILLI	580.00	-	-	-	580.00	MILLI/MJ	506.60	
CHLORIDE	GRAM	1.50	-	-	-	1.50	GRAM/MJ	1.31	
NITRATE	MILLI	0.70	0.00	-	3.00	0.70	MILLI/MJ	0.61	
VITAMIN A	MILLI	0.12	-	-	-	0.12	MILLI/MJ	0.10	
ALANINE	GRAM	1.02	-	-	-	1.02	GRAM/MJ	0.89	
ARGININE	GRAM	0.98	-	-	-	0.98	GRAM/MJ	0.86	
ASPARTIC ACID	GRAM	2.08	-	-	-	2.08	GRAM/MJ	1.82	
GLUTAMIC ACID	GRAM	6.38	-	-	-	6.38	GRAM/MJ	5.57	
GLYCINE	GRAM	0.58	-	-	-	0.58	GRAM/MJ	0.51	
HISTIDINE	GRAM	0.82	-	-	-	0.82	GRAM/MJ	0.72	
ISOLEUCINE	GRAM	1.72	-	-	-	1.72	GRAM/MJ	1.50	
LEUCINE	GRAM	2.95	-	-	-	2.95	GRAM/MJ	2.58	
LYSINE	GRAM	2.37	-	-	-	2.37	GRAM/MJ	2.07	
METHIONINE	GRAM	0.87	-	-	-	0.87	GRAM/MJ	0.76	
PHENYLALANINE	GRAM	1.58	-	-	-	1.58	GRAM/MJ	1.38	
PROLINE	GRAM	3.44	-	-	-	3.44	GRAM/MJ	3.00	
SERINE	GRAM	1.64	-	-	-	1.64	GRAM/MJ	1.43	
THREONINE	GRAM	1.05	-	-	-	1.05	GRAM/MJ	0.92	
TRYPTOPHAN	GRAM	0.40	-	-	-	0.40	GRAM/MJ	0.35	

CONSTITUENTS	DIM	AV	VARIATION			AVR	NUTR. DENS.		MOLPERC.
TYROSINE	GRAM	1.69	-	-	-	1.69	GRAM/MJ	1.48	
VALINE	GRAM	2.04	-	-	-	2.04	GRAM/MJ	1.78	
BUTYRIC ACID	GRAM	0.58	-	-	-	0.58	GRAM/MJ	0.51	
CAPROIC ACID	GRAM	0.36	-	-	-	0.36	GRAM/MJ	0.31	
CAPRYLIC ACID	GRAM	0.25	-	-	-	0.25	GRAM/MJ	0.22	
CAPRIC ACID	GRAM	0.49	-	-	-	0.49	GRAM/MJ	0.43	
LAURIC ACID	GRAM	0.71	-	-	-	0.71	GRAM/MJ	0.62	
MYRISTIC ACID	GRAM	1.72	-	-	-	1.72	GRAM/MJ	1.50	
PALMITIC ACID	GRAM	5.09	-	-	-	5.09	GRAM/MJ	4.45	
STEARIC ACID	GRAM	1.45	-	-	-	1.45	GRAM/MJ	1.27	
PALMITOLEIC ACID	GRAM	0.44	-	-	-	0.44	GRAM/MJ	0.38	
OLEIC ACID	GRAM	4.02	-	-	-	4.02	GRAM/MJ	3.51	
LINOLEIC ACID	GRAM	0.26	-	-	-	0.26	GRAM/MJ	0.23	
LINOLENIC ACID	GRAM	0.21	-	-	-	0.21	GRAM/MJ	0.18	

TILSITERKÄSE
45% FETT I. TR.

TILSIT CHEESE
45% FAT CONTENT IN DRY MATTER

FROMAGE TILSIT
45% DE MAT. GR. S. SEC

		PROTEIN	FAT	CARBOHYDRATES	TOTAL
ENERGY VALUE (AVERAGE)	KJOULE	484	1072	0.0	1556
PER 100 G	(KCAL)	116	256	0.0	372
EDIBLE PORTION					
AMOUNT OF DIGESTIBLE	GRAM	25.51	26.31	0.00	
CONSTITUENTS PER 100 G					
ENERGY VALUE (AVERAGE)	KJOULE	470	1018	0.0	1488
OF THE DIGESTIBLE	(KCAL)	112	243	0.0	356
FRACTION PER 100 G					
EDIBLE PORTION					

WASTE PERCENTAGE AVERAGE 0.00

CONSTITUENTS	DIM	AV	VARIATION			AVR	NUTR. DENS.		MOLPERC.
WATER	GRAM	40.60	39.10	-	45.00	40.60	GRAM/MJ	27.28	
PROTEIN	GRAM	26.30	26.00	-	26.80	26.30	GRAM/MJ	17.67	
FAT	GRAM	27.70	24.80	-	29.40	27.70	GRAM/MJ	18.61	
MINERALS	GRAM	4.85	4.00	-	5.75	4.85	GRAM/MJ	3.26	
SODIUM	MILLI	773.00	641.00	-	865.00	773.00	MILLI/MJ	519.46	
POTASSIUM	MILLI	60.00	51.00	-	78.00	60.00	MILLI/MJ	40.32	
MAGNESIUM	MILLI	31.00	-	-	-	31.00	MILLI/MJ	20.83	
CALCIUM	MILLI	858.00	740.00	-	976.00	858.00	MILLI/MJ	576.58	
IRON	MILLI	0.23	-	-	-	0.23	MILLI/MJ	0.15	
COPPER	MICRO	70.00	-	-	-	70.00	MICRO/MJ	47.04	
ZINC	MILLI	3.50	-	-	-	3.50	MILLI/MJ	2.35	
PHOSPHORUS	MILLI	522.00	429.00	-	555.00	522.00	MILLI/MJ	350.79	
CHLORIDE	GRAM	1.40	-	-	-	1.40	GRAM/MJ	0.94	
NITRATE	MILLI	0.70	0.00	-	3.00	0.70	MILLI/MJ	0.47	

CONSTITUENTS	DIM	AV	VARIATION			AVR	NUTR.	DENS.	MOLPERC.
VITAMIN A	MILLI	0.12	-	-	-	0.12	MILLI/MJ	0.08	
VITAMIN B1	MICRO	61.00	-	-	-	61.00	MICRO/MJ	40.99	
VITAMIN B2	MILLI	0.36	-	-	-	0.36	MILLI/MJ	0.24	
NICOTINAMIDE	MILLI	0.21	-	-	-	0.21	MILLI/MJ	0.14	
PANTOTHENIC ACID	MILLI	0.35	-	-	-	0.35	MILLI/MJ	0.24	
VITAMIN B12	MICRO	2.20	2.00	-	2.50	2.20	MICRO/MJ	1.48	
VITAMIN C	MILLI	1.00	-	-	-	1.00	MILLI/MJ	0.67	
ALANINE	GRAM	0.94	-	-	-	0.94	GRAM/MJ	0.63	
ARGININE	GRAM	0.90	-	-	-	0.90	GRAM/MJ	0.60	
ASPARTIC ACID	GRAM	1.91	-	-	-	1.91	GRAM/MJ	1.28	
GLUTAMIC ACID	GRAM	5.85	-	-	-	5.85	GRAM/MJ	3.93	
GLYCINE	GRAM	0.53	-	-	-	0.53	GRAM/MJ	0.36	
HISTIDINE	GRAM	0.75	-	-	-	0.75	GRAM/MJ	0.50	
ISOLEUCINE	GRAM	1.58	-	-	-	1.58	GRAM/MJ	1.06	
LEUCINE	GRAM	2.71	-	-	-	2.71	GRAM/MJ	1.82	
LYSINE	GRAM	2.17	-	-	-	2.17	GRAM/MJ	1.46	
METHIONINE	GRAM	0.80	-	-	-	0.80	GRAM/MJ	0.54	
PHENYLALANINE	GRAM	1.45	-	-	-	1.45	GRAM/MJ	0.97	
PROLINE	GRAM	3.16	-	-	-	3.16	GRAM/MJ	2.12	
SERINE	GRAM	1.50	-	-	-	1.50	GRAM/MJ	1.01	
THREONINE	GRAM	0.96	-	-	-	0.96	GRAM/MJ	0.65	
TRYPTOPHAN	GRAM	0.37	-	-	-	0.37	GRAM/MJ	0.25	
TYROSINE	GRAM	1.55	-	-	-	1.55	GRAM/MJ	1.04	
VALINE	GRAM	1.87	-	-	-	1.87	GRAM/MJ	1.26	
HISTAMINE	MILLI	5.50	-	-	-	5.50	MILLI/MJ	3.70	
BUTYRIC ACID	GRAM	0.94	-	-	-	0.94	GRAM/MJ	0.63	
CAPROIC ACID	GRAM	0.58	-	-	-	0.58	GRAM/MJ	0.39	
CAPRYLIC ACID	GRAM	0.41	-	-	-	0.41	GRAM/MJ	0.28	
CAPRIC ACID	GRAM	0.80	-	-	-	0.80	GRAM/MJ	0.54	
LAURIC ACID	GRAM	1.14	-	-	-	1.14	GRAM/MJ	0.77	
MYRISTIC ACID	GRAM	2.78	-	-	-	2.78	GRAM/MJ	1.87	
PALMITIC ACID	GRAM	8.19	-	-	-	8.19	GRAM/MJ	5.50	
STEARIC ACID	GRAM	2.33	-	-	-	2.33	GRAM/MJ	1.57	
PALMITOLEIC ACID	GRAM	0.71	-	-	-	0.71	GRAM/MJ	0.48	
OLEIC ACID	GRAM	6.47	-	-	-	6.47	GRAM/MJ	4.35	
LINOLEIC ACID	GRAM	0.43	-	-	-	0.43	GRAM/MJ	0.29	
LINOLENIC ACID	GRAM	0.34	-	-	-	0.34	GRAM/MJ	0.23	

Eier und Eiprodukte

Eggs and egg products
Oeufs et produits à base d'oeufs

Eier und Eiprodukte

EGGS AND EGG PRODUCTS — OEUFS ET PRODUITS À BASE D'OEUFS

ENTENEI — DUCK EGG — OEUF DE CANE

GESAMTEI-INHALT — WHOLE EGG CONTENT — OEUF ENTIER

		PROTEIN	FAT	CARBOHYDRATES	TOTAL
ENERGY VALUE (AVERAGE) PER 100 G EDIBLE PORTION	KJOULE	245	572	12	829
	(KCAL)	59	137	2.8	198
AMOUNT OF DIGESTIBLE CONSTITUENTS PER 100 G	GRAM	12.61	13.68	0.70	
ENERGY VALUE (AVERAGE) OF THE DIGESTIBLE FRACTION PER 100 G EDIBLE PORTION	KJOULE	237	544	12	793
	(KCAL)	57	130	2.8	190

WASTE PERCENTAGE AVERAGE 11

CONSTITUENTS	DIM	AV	VARIATION		AVR	NUTR. DENS.		MOLPERC.
WATER	GRAM	70.90	70.10 -	72.00	63.10	GRAM/MJ	89.42	
PROTEIN	GRAM	13.00	12.70 -	13.30	11.57	GRAM/MJ	16.40	
FAT	GRAM	14.40	14.00 -	15.10	12.82	GRAM/MJ	18.16	
AVAILABLE CARBOHYDR.	GRAM	0.70	- -	-	0.62	GRAM/MJ	0.88	
MINERALS	GRAM	1.00	- -	-	0.89	GRAM/MJ	1.26	
SODIUM	MILLI	100.00	- -	-	89.00	MILLI/MJ	126.12	
POTASSIUM	MILLI	150.00	- -	-	133.50	MILLI/MJ	189.18	
CALCIUM	MILLI	63.00	56.00 -	70.00	56.07	MILLI/MJ	79.46	
IRON	MILLI	2.70	2.50 -	2.80	2.40	MILLI/MJ	3.41	
PHOSPHORUS	MILLI	178.00	160.00 -	195.00	158.42	MILLI/MJ	224.50	
VITAMIN A	MILLI	0.54	- -	-	0.48	MILLI/MJ	0.68	
CAROTENE	MILLI	1.20	- -	-	1.07	MILLI/MJ	1.51	
VITAMIN D	MICRO	5.00	- -	-	4.45	MICRO/MJ	6.31	
VITAMIN B1	MILLI	0.16	0.10 -	0.25	0.14	MILLI/MJ	0.20	
VITAMIN B2	MILLI	0.53	0.30 -	0.80	0.47	MILLI/MJ	0.67	
NICOTINAMIDE	MILLI	0.13	0.10 -	0.20	0.12	MILLI/MJ	0.16	
VITAMIN C	MILLI	-	0.00 -	0.50				
ALANINE	GRAM	0.79	- -	-	0.70	GRAM/MJ	1.00	7.7
ARGININE	GRAM	0.84	- -	-	0.75	GRAM/MJ	1.06	4.2
ASPARTIC ACID	GRAM	0.90	- -	-	0.80	GRAM/MJ	1.14	5.9
CYSTINE	GRAM	0.31	- -	-	0.28	GRAM/MJ	0.39	1.1
GLUTAMIC ACID	GRAM	2.04	- -	-	1.82	GRAM/MJ	2.57	12.1
GLYCINE	GRAM	0.55	- -	-	0.49	GRAM/MJ	0.69	6.4
HISTIDINE	GRAM	0.35	- -	-	0.31	GRAM/MJ	0.44	2.0
ISOLEUCINE	GRAM	0.70	- -	-	0.62	GRAM/MJ	0.88	4.6

CONSTITUENTS	DIM	AV	VARIATION			AVR	NUTR. DENS.		MOLPERC.
LEUCINE	GRAM	1.26	-	-	-	1.12	GRAM/MJ	1.59	8.4
LYSINE	GRAM	1.06	-	-	-	0.94	GRAM/MJ	1.34	6.3
METHIONINE	GRAM	0.65	-	-	-	0.58	GRAM/MJ	0.82	3.8
PHENYLALANINE	GRAM	0.94	-	-	-	0.84	GRAM/MJ	1.19	5.0
PROLINE	GRAM	0.57	-	-	-	0.51	GRAM/MJ	0.72	4.3
SERINE	GRAM	1.17	-	-	-	1.04	GRAM/MJ	1.48	9.7
THREONINE	GRAM	0.87	-	-	-	0.77	GRAM/MJ	1.10	6.4
TRYPTOPHAN	GRAM	0.29	-	-	-	0.26	GRAM/MJ	0.37	1.2
TYROSINE	GRAM	0.68	-	-	-	0.61	GRAM/MJ	0.86	3.3
VALINE	GRAM	1.05	-	-	-	0.93	GRAM/MJ	1.32	7.8
TOTAL PHOSPHOLIPIDS	GRAM	3.83	-	-	-	3.41	GRAM/MJ	4.83	
PHOSPHATIDYLCHOLINE	GRAM	2.90	-	-	-	2.58	GRAM/MJ	3.66	
PHOSPHATIDYLETHANOLAMINE	MILLI	632.00	-	-	-	562.48	MILLI/MJ	797.09	
SPHINGOMYELIN	MILLI	94.00	-	-	-	83.66	MILLI/MJ	118.55	

HÜHNEREI
GESAMTEI-INHALT

CHICKEN EGG
WHOLE EGG CONTENT

OEUF DE POULE
OEUF ENTIER

		PROTEIN	FAT	CARBOHYDRATES	TOTAL
ENERGY VALUE (AVERAGE)	KJOULE	243	445	12	700
PER 100 G EDIBLE PORTION	(KCAL)	58	106	2.8	167
AMOUNT OF DIGESTIBLE CONSTITUENTS PER 100 G	GRAM	12.51	10.64	0.70	
ENERGY VALUE (AVERAGE)	KJOULE	236	423	12	670
OF THE DIGESTIBLE FRACTION PER 100 G EDIBLE PORTION	(KCAL)	56	101	2.8	160

WASTE PERCENTAGE AVERAGE 12 MINIMUM 7.0 MAXIMUM 15

CONSTITUENTS	DIM	AV	VARIATION			AVR	NUTR. DENS.		MOLPERC.
WATER	GRAM	74.10	72.50	-	75.00	65.21	GRAM/MJ	110.56	
PROTEIN	GRAM	12.90	12.50	-	13.30	11.35	GRAM/MJ	19.25	
FAT	GRAM	11.20	10.70	-	11.60	9.86	GRAM/MJ	16.71	
AVAILABLE CARBOHYDR.	GRAM	0.70	-	-	-	0.62	GRAM/MJ	1.04	
MINERALS	GRAM	1.10	1.00	-	1.10	0.97	GRAM/MJ	1.64	

CONSTITUENTS	DIM	AV	VARIATION			AVR	NUTR. DENS.		MOLPERC.
SODIUM	MILLI	144.00	122.00	-	156.00	126.72	MILLI/MJ	214.85	
POTASSIUM	MILLI	147.00	128.00	-	155.00	129.36	MILLI/MJ	219.33	
MAGNESIUM	MILLI	12.00	11.00	-	13.00	10.56	MILLI/MJ	17.90	
CALCIUM	MILLI	56.00	54.00	-	60.00	49.28	MILLI/MJ	83.55	
MANGANESE	MICRO	30.00	0.00	-	50.00	26.40	MICRO/MJ	44.76	
IRON	MILLI	2.10	1.54	-	2.70	1.85	MILLI/MJ	3.13	
COBALT	MICRO	0.50	-	-	-	0.44	MICRO/MJ	0.75	
COPPER	MILLI	-	0.05	-	0.23				
ZINC	MILLI	1.35	0.80	-	2.00	1.19	MILLI/MJ	2.01	
NICKEL	MICRO	24.00	8.00	-	40.00	21.12	MICRO/MJ	35.81	
CHROMIUM	MICRO	-	0.50	-	50.00				
MOLYBDENUM	MICRO	49.00	22.00	-	49.00	43.12	MICRO/MJ	73.11	
PHOSPHORUS	MILLI	216.00	206.00	-	255.00	190.08	MILLI/MJ	322.28	
CHLORIDE	MILLI	180.00	170.00	-	180.00	158.40	MILLI/MJ	268.57	
FLUORIDE	MILLI	0.11	0.01	-	0.12	0.10	MILLI/MJ	0.16	
IODIDE	MICRO	9.70	1.00	-	40.00	8.54	MICRO/MJ	14.47	
BORON	MICRO	75.00	63.00	-	94.00	66.00	MICRO/MJ	111.90	
SELENIUM	MICRO	10.40	5.60	-	50.20	9.15	MICRO/MJ	15.52	
SILICON	MILLI	0.30	-	-	-	0.26	MILLI/MJ	0.45	
VITAMIN A	MILLI	0.22	0.14	-	0.50	0.19	MILLI/MJ	0.33	
CAROTENE	-	TRACES	-	-	-				
VITAMIN D	MICRO	1.80	1.30	-	5.00	1.58	MICRO/MJ	2.69	
VITAMIN E ACTIVITY	MILLI	0.74	-	-	-	0.65	MILLI/MJ	1.10	
TOTAL TOCOPHEROLS	MILLI	1.05	-	-	-	0.92	MILLI/MJ	1.57	
ALPHA-TOCOPHEROL	MILLI	0.70	-	-	-	0.62	MILLI/MJ	1.04	
GAMMA-TOCOPHEROL	MILLI	0.35	-	-	-	0.31	MILLI/MJ	0.52	
VITAMIN K	MICRO	45.00	-	-	-	39.60	MICRO/MJ	67.14	
VITAMIN B1	MILLI	0.10	0.07	-	0.14	0.09	MILLI/MJ	0.15	
VITAMIN B2	MILLI	0.31	0.29	-	0.62	0.27	MILLI/MJ	0.46	
NICOTINAMIDE	MICRO	83.00	50.00	-	100.00	73.04	MICRO/MJ	123.84	
PANTOTHENIC ACID	MILLI	1.60	1.10	-	1.80	1.41	MILLI/MJ	2.39	
VITAMIN B6	MILLI	0.12	0.09	-	0.18	0.11	MILLI/MJ	0.18	
BIOTIN	MICRO	25.00	-	-	-	22.00	MICRO/MJ	37.30	
FOLIC ACID	MICRO	65.00	-	-	-	57.20	MICRO/MJ	96.98	
VITAMIN B12	MICRO	-	0.84	-	3.13				
VITAMIN C	-	0.00	-	-	-	0.00			
ALANINE	GRAM	0.89	0.36	-	0.93	0.78	GRAM/MJ	1.33	8.7
ARGININE	GRAM	0.89	0.78	-	1.34	0.78	GRAM/MJ	1.33	4.4
ASPARTIC ACID	GRAM	1.46	1.33	-	1.57	1.28	GRAM/MJ	2.18	9.5
CYSTINE	GRAM	0.31	0.18	-	0.40	0.27	GRAM/MJ	0.46	1.1
GLUTAMIC ACID	GRAM	1.81	1.74	-	1.86	1.59	GRAM/MJ	2.70	10.7
GLYCINE	GRAM	0.53	0.50	-	0.55	0.47	GRAM/MJ	0.79	6.1
HISTIDINE	GRAM	0.33	0.20	-	0.63	0.29	GRAM/MJ	0.49	1.8
ISOLEUCINE	GRAM	0.93	0.75	-	1.23	0.82	GRAM/MJ	1.39	6.1
LEUCINE	GRAM	1.26	1.16	-	1.43	1.11	GRAM/MJ	1.88	8.3
LYSINE	GRAM	0.89	0.71	-	1.11	0.78	GRAM/MJ	1.33	5.3
METHIONINE	GRAM	0.45	0.20	-	0.66	0.40	GRAM/MJ	0.67	2.6
PHENYLALANINE	GRAM	0.80	0.64	-	1.14	0.70	GRAM/MJ	1.19	4.2
PROLINE	GRAM	0.59	0.57	-	0.60	0.52	GRAM/MJ	0.88	4.4
SERINE	GRAM	1.15	1.09	-	1.20	1.01	GRAM/MJ	1.72	9.5
THREONINE	GRAM	0.71	0.57	-	1.10	0.62	GRAM/MJ	1.06	5.2
TRYPTOPHAN	GRAM	0.23	0.16	-	0.30	0.20	GRAM/MJ	0.34	1.0
TYROSINE	GRAM	0.59	0.38	-	0.83	0.52	GRAM/MJ	0.88	2.8
VALINE	GRAM	1.12	0.99	-	1.31	0.99	GRAM/MJ	1.67	8.3

CONSTITUENTS	DIM	AV	VARIATION		AVR	NUTR.	DENS.	MOLPERC.
GLUCOSE	GRAM	0.34	-	-	0.30	GRAM/MJ	0.51	
MYRISTIC ACID	MILLI	30.00	-	-	26.40	MILLI/MJ	44.76	
PALMITIC ACID	GRAM	2.30	2.10	- 2.50	2.02	GRAM/MJ	3.43	
STEARIC ACID	GRAM	0.78	0.71	- 0.87	0.69	GRAM/MJ	1.16	
PALMITOLEIC ACID	GRAM	0.53	0.39	- 0.68	0.47	GRAM/MJ	0.79	
OLEIC ACID	GRAM	4.20	4.10	- 4.20	3.70	GRAM/MJ	6.27	
LINOLEIC ACID	GRAM	1.35	1.20	- 1.45	1.19	GRAM/MJ	2.01	
LINOLENIC ACID	MILLI	70.00	30.00	- 120.00	61.60	MILLI/MJ	104.44	
ARACHIDONIC ACID	MILLI	70.00	40.00	- 90.00	61.60	MILLI/MJ	104.44	
CHOLESTEROL	MILLI	396.00	372.00	- 444.00	348.48	MILLI/MJ	590.84	
TOTAL PHOSPHOLIPIDS	GRAM	3.51	-	-	3.09	GRAM/MJ	5.24	
PHOSPHATIDYLCHOLINE	GRAM	2.70	-	-	2.38	GRAM/MJ	4.03	
PHOSPHATIDYLETHANOLAMINE	MILLI	581.00	-	-	511.28	MILLI/MJ	866.87	
SPHINGOMYELIN	MILLI	82.00	-	-	72.16	MILLI/MJ	122.35	

HÜHNEREIGELB
FLÜSSIGEIGELB

CHICKEN EGG, YOLK
EGG YOLK, LIQUID

JAUNE D'OEUF
LIQUIDE

		PROTEIN	FAT	CARBOHYDRATES	TOTAL
ENERGY VALUE (AVERAGE)	KJOULE	303	1268	5.0	1576
PER 100 G EDIBLE PORTION	(KCAL)	72	303	1.2	377
AMOUNT OF DIGESTIBLE CONSTITUENTS PER 100 G	GRAM	15.61	30.30	0.30	
ENERGY VALUE (AVERAGE)	KJOULE	294	1205	5.0	1504
OF THE DIGESTIBLE FRACTION PER 100 G EDIBLE PORTION	(KCAL)	70	288	1.2	359

WASTE PERCENTAGE AVERAGE 0.00

CONSTITUENTS	DIM	AV	VARIATION	AVR	NUTR. DENS.		MOLPERC.
WATER	GRAM	50.00	49.40 - 50.90	50.00	GRAM/MJ	33.25	
PROTEIN	GRAM	16.10	16.00 - 16.30	16.10	GRAM/MJ	10.71	
FAT	GRAM	31.90	31.00 - 33.00	31.90	GRAM/MJ	21.22	
AVAILABLE CARBOHYDR.	GRAM	0.30	0.20 - 0.70	0.30	GRAM/MJ	0.20	
MINERALS	GRAM	1.70	- - -	1.70	GRAM/MJ	1.13	
SODIUM	MILLI	51.00	26.00 - 86.00	51.00	MILLI/MJ	33.92	
POTASSIUM	MILLI	138.00	108.00 - 155.00	138.00	MILLI/MJ	91.78	
MAGNESIUM	MILLI	16.00	15.00 - 17.00	16.00	MILLI/MJ	10.64	
CALCIUM	MILLI	140.00	131.00 - 147.00	140.00	MILLI/MJ	93.11	
MANGANESE	MILLI	-	0.05 - 0.20				
IRON	MILLI	7.20	5.10 - 12.20	7.20	MILLI/MJ	4.79	
COBALT	MICRO	35.00	20.00 - 50.00	35.00	MICRO/MJ	23.28	
COPPER	MILLI	0.35	- - -	0.35	MILLI/MJ	0.23	
ZINC	MILLI	3.80	2.60 - 4.00	3.80	MILLI/MJ	2.53	
CHROMIUM	MICRO	20.00	- - -	20.00	MICRO/MJ	13.30	
MOLYBDENUM	MICRO	17.00	- - -	17.00	MICRO/MJ	11.31	
VANADIUM	MICRO	68.00	- - -	68.00	MICRO/MJ	45.22	
PHOSPHORUS	MILLI	590.00	568.00 - 607.00	590.00	MILLI/MJ	392.39	
CHLORIDE	MILLI	180.00	180.00 - 190.00	180.00	MILLI/MJ	119.71	
FLUORIDE	MICRO	30.00	- - -	30.00	MICRO/MJ	19.95	
IODIDE	MICRO	-	7.50 - 15.80				
BORON	MICRO	66.00	- - -	66.00	MICRO/MJ	43.89	
SELENIUM	MICRO	30.00	17.00 - 40.00	30.00	MICRO/MJ	19.95	
VITAMIN A	MILLI	0.55	0.30 - 2.25	0.55	MILLI/MJ	0.37	
CAROTENE	-	TRACES	- - -				
VITAMIN D	MICRO	-	2.50 - 12.00				

CONSTITUENTS	DIM	AV	VARIATION			AVR	NUTR. DENS.		MOLPERC.
VITAMIN E ACTIVITY	MILLI	2.10	-	-	-	2.10	MILLI/MJ	1.40	
TOTAL TOCOPHEROLS	MILLI	3.10	-	-	-	3.10	MILLI/MJ	2.06	
ALPHA-TOCOPHEROL	MILLI	2.00	-	-	-	2.00	MILLI/MJ	1.33	
GAMMA-TOCOPHEROL	MILLI	1.00	-	-	-	1.00	MILLI/MJ	0.67	
VITAMIN B1	MILLI	0.29	0.17	-	0.35	0.29	MILLI/MJ	0.19	
VITAMIN B2	MILLI	0.40	0.24	-	0.54	0.40	MILLI/MJ	0.27	
NICOTINAMIDE	MICRO	65.00	30.00	-	100.00	65.00	MICRO/MJ	43.23	
PANTOTHENIC ACID	MILLI	3.72	2.96	-	5.04	3.72	MILLI/MJ	2.47	
VITAMIN B6	MILLI	0.30	0.26	-	0.36	0.30	MILLI/MJ	0.20	
BIOTIN	MICRO	-	35.00	-	70.00				
FOLIC ACID	MILLI	0.15	-	-	-	0.15	MILLI/MJ	0.10	
VITAMIN B12	MICRO	2.00	0.80	-	3.30	2.00	MICRO/MJ	1.33	
VITAMIN C	-	0.00	-	-	-	0.00			
ALANINE	GRAM	1.03	-	-	-	1.03	GRAM/MJ	0.69	8.0
ARGININE	GRAM	1.28	1.20	-	1.30	1.28	GRAM/MJ	0.85	5.1
ASPARTIC ACID	GRAM	1.76	1.59	-	2.02	1.76	GRAM/MJ	1.17	9.2
CYSTINE	GRAM	0.31	0.26	-	0.56	0.31	GRAM/MJ	0.21	0.9
GLUTAMIC ACID	GRAM	2.20	2.05	-	2.31	2.20	GRAM/MJ	1.46	10.4
GLYCINE	GRAM	0.62	0.59	-	0.66	0.62	GRAM/MJ	0.41	5.7
HISTIDINE	GRAM	0.44	0.27	-	0.52	0.44	GRAM/MJ	0.29	2.0
ISOLEUCINE	GRAM	1.09	0.78	-	1.31	1.09	GRAM/MJ	0.72	5.8
LEUCINE	GRAM	1.63	1.60	-	1.66	1.63	GRAM/MJ	1.08	8.6
LYSINE	GRAM	1.30	0.91	-	1.61	1.30	GRAM/MJ	0.86	6.2
METHIONINE	GRAM	0.47	0.36	-	0.68	0.47	GRAM/MJ	0.31	2.2
PHENYLALANINE	GRAM	0.79	0.70	-	0.85	0.79	GRAM/MJ	0.53	3.3
PROLINE	GRAM	0.78	0.70	-	0.82	0.78	GRAM/MJ	0.52	4.7
SERINE	GRAM	1.62	1.47	-	1.73	1.62	GRAM/MJ	1.08	10.7
THREONINE	GRAM	1.01	0.68	-	1.21	1.01	GRAM/MJ	0.67	5.9
TRYPTOPHAN	GRAM	0.29	0.19	-	0.46	0.29	GRAM/MJ	0.19	1.0
TYROSINE	GRAM	0.78	0.73	-	0.88	0.78	GRAM/MJ	0.52	3.0
VALINE	GRAM	1.24	1.06	-	1.44	1.24	GRAM/MJ	0.82	7.4
GLUCOSE	GRAM	0.21	0.14	-	0.32	0.21	GRAM/MJ	0.14	
MYRISTIC ACID	MILLI	80.00	-	-	-	80.00	MILLI/MJ	53.20	
PALMITIC ACID	GRAM	6.50	5.90	-	7.00	6.50	GRAM/MJ	4.32	
STEARIC ACID	GRAM	2.20	2.00	-	2.45	2.20	GRAM/MJ	1.46	
ARACHIDIC ACID	GRAM	0.40	-	-	-	0.40	GRAM/MJ	0.27	
PALMITOLEIC ACID	GRAM	1.50	1.10	-	1.90	1.50	GRAM/MJ	1.00	
OLEIC ACID	GRAM	11.70	11.60	-	11.80	11.70	GRAM/MJ	7.78	
LINOLEIC ACID	GRAM	3.80	3.50	-	4.10	3.80	GRAM/MJ	2.53	
LINOLENIC ACID	GRAM	0.22	0.09	-	0.34	0.22	GRAM/MJ	0.15	
ARACHIDONIC ACID	GRAM	0.21	0.13	-	0.28	0.21	GRAM/MJ	0.14	
C22 : 5 N-3 FATTY ACID	MILLI	60.00	-	-	-	60.00	MILLI/MJ	39.90	
C22 : 6 N-3 FATTY ACID	GRAM	0.18	-	-	-	0.18	GRAM/MJ	0.12	
CHOLESTEROL	GRAM	1.26	1.07	-	1.36	1.26	GRAM/MJ	0.84	
TOTAL PHOSPHOLIPIDS	GRAM	10.34	-	-	-	10.34	GRAM/MJ	6.88	
PHOSPHATIDYLCHOLINE	GRAM	6.79	-	-	-	6.79	GRAM/MJ	4.52	
PHOSPHATIDYLETHANOLAMINE	GRAM	1.93	-	-	-	1.93	GRAM/MJ	1.28	
PHOSPHATIDYLINOSITOL	MILLI	64.00	-	-	-	64.00	MILLI/MJ	42.56	
SPHINGOMYELIN	MILLI	488.00	-	-	-	488.00	MILLI/MJ	324.55	

HÜHNEREIWEISS / CHICKEN EGG, WHITE / BLANC D'OEUF

FLÜSSIGEIWEISS, EIKLAR / EGG WHITE, LIQUID / LIQUIDE

		PROTEIN	FAT	CARBOHYDRATES	TOTAL
ENERGY VALUE (AVERAGE)	KJOULE	209	7.9	12	229
PER 100 G EDIBLE PORTION	(KCAL)	50	1.9	2.8	55
AMOUNT OF DIGESTIBLE CONSTITUENTS PER 100 G	GRAM	10.76	0.19	0.70	
ENERGY VALUE (AVERAGE)	KJOULE	203	7.6	12	222
OF THE DIGESTIBLE FRACTION PER 100 G EDIBLE PORTION	(KCAL)	48	1.8	2.8	53

WASTE PERCENTAGE AVERAGE 0.00

CONSTITUENTS	DIM	AV	VARIATION			AVR	NUTR. DENS.		MOLPERC.
WATER	GRAM	87.30	85.60	-	88.00	87.30	GRAM/MJ	393.26	
PROTEIN	GRAM	11.10	10.80	-	12.30	11.10	GRAM/MJ	50.00	
FAT	GRAM	0.20	0.20	-	0.30	0.20	GRAM/MJ	0.90	
AVAILABLE CARBOHYDR.	GRAM	0.70	0.70	-	0.80	0.70	GRAM/MJ	3.15	
MINERALS	GRAM	0.70	0.60	-	0.95	0.70	GRAM/MJ	3.15	
SODIUM	MILLI	170.00	155.00	-	200.00	170.00	MILLI/MJ	765.81	
POTASSIUM	MILLI	154.00	136.00	-	174.00	154.00	MILLI/MJ	693.73	
MAGNESIUM	MILLI	12.00	11.00	-	12.00	12.00	MILLI/MJ	54.06	
CALCIUM	MILLI	11.00	5.30	-	20.00	11.00	MILLI/MJ	49.55	
MANGANESE	MICRO	40.00	-	-	-	40.00	MICRO/MJ	180.19	
IRON	MILLI	0.20	0.10	-	0.20	0.20	MILLI/MJ	0.90	
COBALT	MICRO	6.00	-	-	-	6.00	MICRO/MJ	27.03	
COPPER	MILLI	0.13	-	-	-	0.13	MILLI/MJ	0.59	
ZINC	MICRO	20.00	-	-	-	20.00	MICRO/MJ	90.09	
MOLYBDENUM	MICRO	12.00	-	-	-	12.00	MICRO/MJ	54.06	
VANADIUM	MICRO	37.00	-	-	-	37.00	MICRO/MJ	166.68	
PHOSPHORUS	MILLI	21.00	15.00	-	30.00	21.00	MILLI/MJ	94.60	
CHLORIDE	MILLI	0.17	-	-	-	0.17	MILLI/MJ	0.77	
FLUORIDE	MILLI	-	0.01	-	0.15				
IODIDE	MICRO	6.80	-	-	-	6.80	MICRO/MJ	30.63	
BORON	MICRO	27.00	20.00	-	30.00	27.00	MICRO/MJ	121.63	
SELENIUM	MICRO	-	4.30	-	10.00				
VITAMIN A	-	TRACES	-	-	-				
CAROTENE	-	0.00	-	-	-	0.00			
VITAMIN D	-	0.00	-	-	-	0.00			
VITAMIN E ACTIVITY	-	0.00	-	-	-	0.00			

CONSTITUENTS	DIM	AV	VARIATION			AVR	NUTR. DENS.		MOLPERC.
VITAMIN B1	MICRO	22.00	15.00	-	30.00	22.00	MICRO/MJ	99.10	
VITAMIN B2	MILLI	0.32	0.25	-	0.45	0.32	MILLI/MJ	1.44	
NICOTINAMIDE	MICRO	90.00	80.00	-	100.00	90.00	MICRO/MJ	405.43	
PANTOTHENIC ACID	MILLI	0.14	0.13	-	0.16	0.14	MILLI/MJ	0.63	
VITAMIN B6	MICRO	12.00	3.00	-	20.00	12.00	MICRO/MJ	54.06	
BIOTIN	MICRO	7.00	-	-	-	7.00	MICRO/MJ	31.53	
FOLIC ACID	MICRO	16.00	-	-	-	16.00	MICRO/MJ	72.08	
VITAMIN B12	MICRO	0.10	-	-	-	0.10	MICRO/MJ	0.45	
VITAMIN C	MILLI	0.30	-	-	-	0.30	MILLI/MJ	1.35	
ALANINE	GRAM	0.83	0.79	-	0.89	0.83	GRAM/MJ	3.74	9.3
ARGININE	GRAM	0.68	0.50	-	0.79	0.68	GRAM/MJ	3.06	3.9
ASPARTIC ACID	GRAM	1.23	1.11	-	1.38	1.23	GRAM/MJ	5.54	9.3
CYSTINE	GRAM	0.29	0.11	-	0.61	0.29	GRAM/MJ	1.31	1.2
GLUTAMIC ACID	GRAM	1.64	1.50	-	1.73	1.64	GRAM/MJ	7.39	11.2
GLYCINE	GRAM	0.50	0.46	-	0.53	0.50	GRAM/MJ	2.25	6.7
HISTIDINE	GRAM	0.28	0.14	-	0.33	0.28	GRAM/MJ	1.26	1.8
ISOLEUCINE	GRAM	0.74	0.67	-	0.82	0.74	GRAM/MJ	3.33	5.7
LEUCINE	GRAM	1.08	0.96	-	1.10	1.08	GRAM/MJ	4.87	8.2
LYSINE	GRAM	0.74	0.70	-	0.81	0.74	GRAM/MJ	3.33	5.1
METHIONINE	GRAM	0.47	0.33	-	0.68	0.47	GRAM/MJ	2.12	3.2
PHENYLALANINE	GRAM	0.76	0.65	-	0.99	0.76	GRAM/MJ	3.42	4.6
PROLINE	GRAM	0.50	0.49	-	0.51	0.50	GRAM/MJ	2.25	4.4
SERINE	GRAM	0.92	0.82	-	0.98	0.92	GRAM/MJ	4.14	8.8
THREONINE	GRAM	0.58	0.45	-	0.67	0.58	GRAM/MJ	2.61	4.9
TRYPTOPHAN	GRAM	0.20	0.13	-	0.27	0.20	GRAM/MJ	0.90	1.0
TYROSINE	GRAM	0.46	0.37	-	0.65	0.46	GRAM/MJ	2.07	2.5
VALINE	GRAM	0.98	0.89	-	1.03	0.98	GRAM/MJ	4.41	8.4
GLUCOSE	GRAM	0.41	0.32	-	0.50	0.41	GRAM/MJ	1.85	
CHOLESTEROL	-	0.00	-	-	-	0.00			

HÜHNEREI
GESAMTEI-INHALT
GETROCKNET
(TROCKENVOLLEI)

CHICKEN EGG
WHOLE EGG CONTENT
DRIED
(DRIED WHOLE EGG)

OEUF ENTIER DE POULE
EN POUDRE

		PROTEIN	FAT	CARBOHYDRATES	TOTAL
ENERGY VALUE (AVERAGE)	KJOULE	866	1661	40	2568
PER 100 G EDIBLE PORTION	(KCAL)	207	397	9.6	614
AMOUNT OF DIGESTIBLE CONSTITUENTS PER 100 G	GRAM	44.62	39.71	2.40	
ENERGY VALUE (AVERAGE)	KJOULE	840	1578	40	2459
OF THE DIGESTIBLE FRACTION PER 100 G EDIBLE PORTION	(KCAL)	201	377	9.6	588

WASTE PERCENTAGE AVERAGE 0.00

CONSTITUENTS	DIM	AV	VARIATION	AVR	NUTR. DENS.		MOLPERC.
WATER	GRAM	6.13	5.00 - 7.00	6.13	GRAM/MJ	2.49	
PROTEIN	GRAM	46.00	43.20 - 47.90	46.00	GRAM/MJ	18.71	
FAT	GRAM	41.80	40.90 - 42.60	41.80	GRAM/MJ	17.00	
AVAILABLE CARBOHYDR.	GRAM	2.40	2.00 - 2.50	2.40	GRAM/MJ	0.98	
MINERALS	GRAM	3.67	3.60 - 3.81	3.67	GRAM/MJ	1.49	
SODIUM	MILLI	521.00	- - -	521.00	MILLI/MJ	211.90	
POTASSIUM	MILLI	490.00	- - -	490.00	MILLI/MJ	199.30	
MAGNESIUM	MILLI	46.00	- - -	46.00	MILLI/MJ	18.71	
CALCIUM	MILLI	190.00	- - -	190.00	MILLI/MJ	77.28	
IRON	MILLI	8.80	- - -	8.80	MILLI/MJ	3.58	
ZINC	MILLI	5.00	- - -	5.00	MILLI/MJ	2.03	
PHOSPHORUS	MILLI	757.00	746.00 - 767.00	757.00	MILLI/MJ	307.89	
SELENIUM	MILLI	0.10	- - -	0.10	MILLI/MJ	0.04	
VITAMIN A	MILLI	0.80	- - -	0.80	MILLI/MJ	0.33	
VITAMIN D	MICRO	5.00	- - -	5.00	MICRO/MJ	2.03	
VITAMIN E ACTIVITY	MILLI	2.70	- - -	2.70	MILLI/MJ	1.10	
TOTAL TOCOPHEROLS	MILLI	3.90	- - -	3.90	MILLI/MJ	1.59	
ALPHA-TOCOPHEROL	MILLI	2.60	- - -	2.60	MILLI/MJ	1.06	
GAMMA-TOCOPHEROL	MILLI	1.30	- - -	1.30	MILLI/MJ	0.53	
VITAMIN B1	MILLI	0.44	0.34 - 0.54	0.44	MILLI/MJ	0.18	
VITAMIN B2	MILLI	1.38	1.06 - 1.70	1.38	MILLI/MJ	0.56	
NICOTINAMIDE	MILLI	0.24	0.20 - 0.28	0.24	MILLI/MJ	0.10	
PANTOTHENIC ACID	MILLI	7.35	6.70 - 10.00	7.35	MILLI/MJ	2.99	
VITAMIN B6	MICRO	80.00	- - -	80.00	MICRO/MJ	32.54	
BIOTIN	MICRO	84.00	- - -	84.00	MICRO/MJ	34.16	
FOLIC ACID	MILLI	0.18	- - -	0.18	MILLI/MJ	0.07	

CONSTITUENTS	DIM	AV	VARIATION			AVR	NUTR.	DENS.	MOLPERC.
VITAMIN B12	MICRO	9.57	-	-	-	9.57	MICRO/MJ	3.89	
VITAMIN C	-	0.00	-	-	-	0.00			
ALANINE	GRAM	3.21	3.05	-	3.30	3.21	GRAM/MJ	1.31	8.8
ARGININE	GRAM	3.13	2.75	-	4.76	3.13	GRAM/MJ	1.27	4.4
ASPARTIC ACID	GRAM	5.29	5.14	-	5.57	5.29	GRAM/MJ	2.15	9.7
CYSTINE	GRAM	1.08	0.65	-	1.43	1.08	GRAM/MJ	0.44	1.1
GLUTAMIC ACID	GRAM	6.39	6.15	-	6.53	6.39	GRAM/MJ	2.60	10.6
GLYCINE	GRAM	1.89	1.79	-	1.97	1.89	GRAM/MJ	0.77	6.1
HISTIDINE	GRAM	1.19	0.68	-	1.79	1.19	GRAM/MJ	0.48	1.9
ISOLEUCINE	GRAM	3.22	2.67	-	4.37	3.22	GRAM/MJ	1.31	6.0
LEUCINE	GRAM	4.50	4.07	-	5.05	4.50	GRAM/MJ	1.83	8.3
LYSINE	GRAM	3.25	2.53	-	3.89	3.25	GRAM/MJ	1.32	5.4
METHIONINE	GRAM	1.64	0.73	-	2.35	1.64	GRAM/MJ	0.67	2.7
PHENYLALANINE	GRAM	2.79	2.24	-	4.03	2.79	GRAM/MJ	1.13	4.1
PROLINE	GRAM	2.09	2.04	-	2.12	2.09	GRAM/MJ	0.85	4.4
SERINE	GRAM	4.08	4.03	-	4.18	4.08	GRAM/MJ	1.66	9.4
THREONINE	GRAM	2.63	2.00	-	3.88	2.63	GRAM/MJ	1.07	5.4
TRYPTOPHAN	GRAM	0.81	0.56	-	1.02	0.81	GRAM/MJ	0.33	1.0
TYROSINE	GRAM	2.08	1.30	-	2.97	2.08	GRAM/MJ	0.85	2.8
VALINE	GRAM	3.90	3.52	-	4.67	3.90	GRAM/MJ	1.59	8.1
MYRISTIC ACID	GRAM	0.12	-	-	-	0.12	GRAM/MJ	0.05	
PALMITIC ACID	GRAM	8.60	7.80	-	9.30	8.60	GRAM/MJ	3.50	
STEARIC ACID	GRAM	2.90	2.60	-	3.30	2.90	GRAM/MJ	1.18	
PALMITOLEIC ACID	GRAM	2.00	1.45	-	2.50	2.00	GRAM/MJ	0.81	
OLEIC ACID	GRAM	15.50	15.40	-	15.70	15.50	GRAM/MJ	6.30	
LINOLEIC ACID	GRAM	5.00	4.60	-	5.40	5.00	GRAM/MJ	2.03	
LINOLENIC ACID	GRAM	0.29	0.11	-	0.45	0.29	GRAM/MJ	0.12	
ARACHIDONIC ACID	GRAM	0.28	0.17	-	0.37	0.28	GRAM/MJ	0.11	
CHOLESTEROL	GRAM	1.44	1.35	-	1.61	1.44	GRAM/MJ	0.59	

HÜHNEREIGELB
GETROCKNET
(TROCKENEIGELB)

CHICKEN EGG, YOLK
DRIED
(DRIED EGG YOLK)

JAUNE D'OEUF
EN POUDRE

		PROTEIN	FAT	CARBOHYDRATES	TOTAL
ENERGY VALUE (AVERAGE)	KJOULE	597	2357	35	2989
PER 100 G	(KCAL)	143	563	8.4	714
EDIBLE PORTION					
AMOUNT OF DIGESTIBLE	GRAM	30.74	56.33	2.10	
CONSTITUENTS PER 100 G					
ENERGY VALUE (AVERAGE)	KJOULE	579	2239	35	2853
OF THE DIGESTIBLE	(KCAL)	138	535	8.4	682
FRACTION PER 100 G					
EDIBLE PORTION					

WASTE PERCENTAGE AVERAGE 0.00

CONSTITUENTS	DIM	AV	VARIATION			AVR	NUTR. DENS.		MOLPERC.
WATER	GRAM	3.40	3.00	-	3.80	3.40	GRAM/MJ	1.19	
PROTEIN	GRAM	31.70	31.20	-	32.20	31.70	GRAM/MJ	11.11	
FAT	GRAM	59.30	56.30	-	61.20	59.30	GRAM/MJ	20.78	
AVAILABLE CARBOHYDR.	GRAM	2.10	1.30	-	3.10	2.10	GRAM/MJ	0.74	
MINERALS	GRAM	3.50	3.30	-	3.60	3.50	GRAM/MJ	1.23	
SODIUM	MILLI	91.00	-	-	-	91.00	MILLI/MJ	31.89	
POTASSIUM	MILLI	168.00	-	-	-	168.00	MILLI/MJ	58.88	
MAGNESIUM	MILLI	28.00	-	-	-	28.00	MILLI/MJ	9.81	
CALCIUM	MILLI	282.00	-	-	-	282.00	MILLI/MJ	98.83	
IRON	MILLI	13.80	-	-	-	13.80	MILLI/MJ	4.84	
ZINC	MILLI	6.15	-	-	-	6.15	MILLI/MJ	2.16	
PHOSPHORUS	GRAM	1.12	-	-	-	1.12	GRAM/MJ	0.39	
VITAMIN A	MILLI	1.06	-	-	-	1.06	MILLI/MJ	0.37	
VITAMIN E ACTIVITY	MILLI	3.90	-	-	-	3.90	MILLI/MJ	1.37	
TOTAL TOCOPHEROLS	MILLI	5.55	-	-	-	5.55	MILLI/MJ	1.95	
ALPHA-TOCOPHEROL	MILLI	3.70	-	-	-	3.70	MILLI/MJ	1.30	
GAMMA-TOCOPHEROL	MILLI	1.82	-	-	-	1.82	MILLI/MJ	0.64	
VITAMIN B1	MILLI	0.50	-	-	-	0.50	MILLI/MJ	0.18	
VITAMIN B2	MILLI	0.66	-	-	-	0.66	MILLI/MJ	0.23	
NICOTINAMIDE	MILLI	0.10	-	-	-	0.10	MILLI/MJ	0.04	
PANTOTHENIC ACID	MILLI	8.24	-	-	-	8.24	MILLI/MJ	2.89	
VITAMIN B6	MILLI	0.58	-	-	-	0.58	MILLI/MJ	0.20	
BIOTIN	MILLI	0.11	-	-	-	0.11	MILLI/MJ	0.04	
FOLIC ACID	MILLI	0.21	-	-	-	0.21	MILLI/MJ	0.07	
VITAMIN B12	MICRO	7.08	-	-	-	7.08	MICRO/MJ	2.48	

CONSTITUENTS	DIM	AV	VARIATION			AVR	NUTR. DENS.		MOLPERC.
ALANINE	GRAM	2.04	-	-	-	2.04	GRAM/MJ	0.71	8.1
ARGININE	GRAM	2.46	2.37	-	2.55	2.46	GRAM/MJ	0.86	5.0
ASPARTIC ACID	GRAM	3.13	-	-	-	3.13	GRAM/MJ	1.10	8.3
CYSTINE	GRAM	0.58	0.50	-	1.10	0.58	GRAM/MJ	0.20	0.8
GLUTAMIC ACID	GRAM	4.52	-	-	-	4.52	GRAM/MJ	1.58	10.8
GLYCINE	GRAM	1.29	-	-	-	1.29	GRAM/MJ	0.45	6.0
HISTIDINE	GRAM	0.81	0.53	-	1.02	0.81	GRAM/MJ	0.28	1.8
ISOLEUCINE	GRAM	2.28	1.51	-	2.60	2.28	GRAM/MJ	0.80	6.1
LEUCINE	GRAM	3.19	3.12	-	3.20	3.19	GRAM/MJ	1.12	8.5
LYSINE	GRAM	2.41	1.80	-	3.15	2.41	GRAM/MJ	0.84	5.8
METHIONINE	GRAM	0.94	0.70	-	1.34	0.94	GRAM/MJ	0.33	2.2
PHENYLALANINE	GRAM	1.60	1.36	-	1.67	1.60	GRAM/MJ	0.56	3.4
PROLINE	GRAM	1.60	-	-	-	1.60	GRAM/MJ	0.56	4.9
SERINE	GRAM	3.25	-	-	-	3.25	GRAM/MJ	1.14	10.9
THREONINE	GRAM	2.00	1.33	-	2.37	2.00	GRAM/MJ	0.70	5.9
TRYPTOPHAN	GRAM	0.52	0.38	-	0.91	0.52	GRAM/MJ	0.18	0.9
TYROSINE	GRAM	1.60	1.54	-	1.72	1.60	GRAM/MJ	0.56	3.1
VALINE	GRAM	2.48	2.25	-	2.80	2.48	GRAM/MJ	0.87	7.4
MYRISTIC ACID	GRAM	0.17	-	-	-	0.17	GRAM/MJ	0.06	
PALMITIC ACID	GRAM	12.20	11.10	-	13.20	12.20	GRAM/MJ	4.28	
STEARIC ACID	GRAM	4.10	3.80	-	4.60	4.10	GRAM/MJ	1.44	
ARACHIDIC ACID	GRAM	0.70	-	-	-	0.70	GRAM/MJ	0.25	
PALMITOLEIC ACID	GRAM	2.80	2.10	-	3.60	2.80	GRAM/MJ	0.98	
OLEIC ACID	GRAM	22.00	21.80	-	22.20	22.00	GRAM/MJ	7.71	
LINOLEIC ACID	GRAM	7.10	6.60	-	7.70	7.10	GRAM/MJ	2.49	
LINOLENIC ACID	GRAM	0.41	0.16	-	0.64	0.41	GRAM/MJ	0.14	
ARACHIDONIC ACID	GRAM	0.39	0.24	-	0.52	0.39	GRAM/MJ	0.14	
C22 : 5 N-3 FATTY ACID	GRAM	0.10	-	-	-	0.10	GRAM/MJ	0.04	
C22 : 6 N-3 FATTY ACID	GRAM	0.30	-	-	-	0.30	GRAM/MJ	0.11	
CHOLESTEROL	GRAM	2.43	2.07	-	2.63	2.43	GRAM/MJ	0.85	

HÜHNEREIWEISS GETROCKNET (TROCKENEIWEISS)

CHICKEN EGG, WHITE DRIED (DRIED EGG WHITE)

BLANC D'OEUF EN POUDRE

		PROTEIN	FAT	CARBOHYDRATES	TOTAL
ENERGY VALUE (AVERAGE) PER 100 G EDIBLE PORTION	KJOULE	455	4.8	136	1596
	(KCAL)	348	1.1	32	381
AMOUNT OF DIGESTIBLE CONSTITUENTS PER 100 G	GRAM	74.98	0.11	8.10	
ENERGY VALUE (AVERAGE) OF THE DIGESTIBLE FRACTION PER 100 G EDIBLE PORTION	KJOULE	412	4.5	136	1552
	(KCAL)	337	1.1	32	371

WASTE PERCENTAGE AVERAGE 0.00

CONSTITUENTS	DIM	AV	VARIATION			AVR	NUTR. DENS.		MOLPERC.
WATER	GRAM	9.02	3.00	-	13.50	9.02	GRAM/MJ	5.81	
PROTEIN	GRAM	77.30	67.60	-	85.90	77.30	GRAM/MJ	49.81	
FAT	GRAM	0.12	0.02	-	0.30	0.12	GRAM/MJ	0.08	
AVAILABLE CARBOHYDR.	GRAM	8.10	5.44	-	11.90	8.10	GRAM/MJ	5.22	
MINERALS	GRAM	5.46	4.50	-	6.86	5.46	GRAM/MJ	3.52	
SODIUM	GRAM	1.42	1.27	-	1.68	1.42	GRAM/MJ	0.92	
POTASSIUM	GRAM	1.07	1.00	-	1.12	1.07	GRAM/MJ	0.69	
MAGNESIUM	MILLI	72.00	-	-	-	72.00	MILLI/MJ	46.40	
CALCIUM	MILLI	84.00	48.00	-	120.00	84.00	MILLI/MJ	54.13	
IRON	MILLI	0.90	0.20	-	1.60	0.90	MILLI/MJ	0.58	
ZINC	MILLI	0.15	-	-	-	0.15	MILLI/MJ	0.10	
PHOSPHORUS	MILLI	110.00	85.00	-	140.00	110.00	MILLI/MJ	70.88	
VITAMIN A	-	0.00	-	-	-	0.00			
VITAMIN D	-	0.00	-	-	-	0.00			
VITAMIN B1	MICRO	37.00	-	-	-	37.00	MICRO/MJ	23.84	
VITAMIN B2	MILLI	2.05	-	-	-	2.05	MILLI/MJ	1.32	
NICOTINAMIDE	MILLI	0.70	-	-	-	0.70	MILLI/MJ	0.45	
PANTOTHENIC ACID	MILLI	1.96	-	-	-	1.96	MILLI/MJ	1.26	
VITAMIN B6	MICRO	24.00	-	-	-	24.00	MICRO/MJ	15.47	
BIOTIN	MICRO	57.00	-	-	-	57.00	MICRO/MJ	36.73	
FOLIC ACID	MICRO	96.00	-	-	-	96.00	MICRO/MJ	61.86	
VITAMIN B12	MICRO	0.53	-	-	-	0.53	MICRO/MJ	0.34	
VITAMIN C	-	0.00	-	-	-	0.00			
ALANINE	GRAM	6.01	-	-	-	6.01	GRAM/MJ	3.87	9.7
ARGININE	GRAM	4.79	-	-	-	4.79	GRAM/MJ	3.09	4.0

CONSTITUENTS	DIM	AV	VARIATION			AVR	NUTR.	DENS.	MOLPERC.
ASPARTIC ACID	GRAM	7.62	-	-	-	7.62	GRAM/MJ	4.91	8.2
CYSTINE	GRAM	1.98	-	-	-	1.98	GRAM/MJ	1.28	1.2
GLUTAMIC ACID	GRAM	11.84	-	-	-	11.84	GRAM/MJ	7.63	11.6
GLYCINE	GRAM	3.66	-	-	-	3.66	GRAM/MJ	2.36	7.0
HISTIDINE	GRAM	1.81	-	-	-	1.81	GRAM/MJ	1.17	1.7
ISOLEUCINE	GRAM	5.41	-	-	-	5.41	GRAM/MJ	3.49	5.9
LEUCINE	GRAM	7.56	-	-	-	7.56	GRAM/MJ	4.87	8.3
LYSINE	GRAM	5.02	-	-	-	5.02	GRAM/MJ	3.23	4.9
METHIONINE	GRAM	3.30	-	-	-	3.30	GRAM/MJ	2.13	3.2
PHENYLALANINE	GRAM	5.30	-	-	-	5.30	GRAM/MJ	3.42	4.6
PROLINE	GRAM	3.32	-	-	-	3.32	GRAM/MJ	2.14	4.1
SERINE	GRAM	6.63	-	-	-	6.63	GRAM/MJ	4.27	9.1
THREONINE	GRAM	3.91	-	-	-	3.91	GRAM/MJ	2.52	4.7
TRYPTOPHAN	GRAM	1.26	-	-	-	1.26	GRAM/MJ	0.81	0.9
TYROSINE	GRAM	3.39	-	-	-	3.39	GRAM/MJ	2.18	2.7
VALINE	GRAM	6.74	-	-	-	6.74	GRAM/MJ	4.34	8.3

Fette

Fats
Graisses

Tierische Fette und Öle

ANIMAL FATS AND OILS

GRAISSES ET HUILES ANIMALES

BUTTER (SÜSS- UND SAUERRAHM-BUTTER)

BUTTER (FROM CREAM AND SOUR CREAM)

BEURRE (BEURRE DE CRÈME) (DOUCE ET CAILLÉE)

		PROTEIN	FAT	CARBOHYDRATES	TOTAL
ENERGY VALUE (AVERAGE)	KJOULE	12	3220	0.0	3232
PER 100 G EDIBLE PORTION	(KCAL)	2.9	770	0.0	773
AMOUNT OF DIGESTIBLE CONSTITUENTS PER 100 G	GRAM	0.64	79.04	0.00	
ENERGY VALUE (AVERAGE)	KJOULE	12	3059	0.0	3071
OF THE DIGESTIBLE FRACTION PER 100 G EDIBLE PORTION	(KCAL)	2.9	731	0.0	734

WASTE PERCENTAGE AVERAGE 0.00

CONSTITUENTS	DIM	AV	VARIATION		AVR	NUTR. DENS.		MOLPERC.
WATER	GRAM	15.30	14.10	- 16.00	15.30	GRAM/MJ	4.98	
PROTEIN	GRAM	0.67	0.50	- 0.82	0.67	GRAM/MJ	0.22	
FAT	GRAM	83.20	82.00	- 83.70	83.20	GRAM/MJ	27.09	
MINERALS	GRAM	0.11	0.10	- 0.13	0.11	GRAM/MJ	0.04	
SODIUM	MILLI	5.10	3.10	- 8.00	5.10	MILLI/MJ	1.66	
POTASSIUM	MILLI	16.00	10.00	- 240.00	16.00	MILLI/MJ	5.21	
MAGNESIUM	MILLI	3.00	1.80	- 4.20	3.00	MILLI/MJ	0.98	
CALCIUM	MILLI	13.00	11.00	- 19.00	13.00	MILLI/MJ	4.23	
MANGANESE	MICRO	-	2.50	- 40.00				
IRON	MILLI	-	0.03	- 0.20				
COBALT	MICRO	1.00	-	- -	1.00	MICRO/MJ	0.33	
COPPER	MICRO	-	2.00	- 15.00				
ZINC	MILLI	0.23	-	- -	0.23	MILLI/MJ	0.07	
NICKEL	MICRO	10.00	3.00	- 20.00	10.00	MICRO/MJ	3.26	
CHROMIUM	MICRO	6.00	1.00	- 17.00	6.00	MICRO/MJ	1.95	
MOLYBDENUM	MILLI	-	0.00	- 0.02				
PHOSPHORUS	MILLI	21.00	18.00	- 27.00	21.00	MILLI/MJ	6.84	
CHLORIDE	MILLI	-	17.00	- 29.00				
FLUORIDE	MILLI	0.13	0.12	- 0.15	0.13	MILLI/MJ	0.04	
IODIDE	MICRO	4.40	-	- -	4.40	MICRO/MJ	1.43	
BORON	MICRO	75.00	50.00	- 95.00	75.00	MICRO/MJ	24.42	
SELENIUM	MICRO	0.30	-	- -	0.30	MICRO/MJ	0.10	
VITAMIN A	MILLI	0.59	0.52	- 0.67	0.59	MILLI/MJ	0.19	
CAROTENE	MILLI	0.38	0.30	- 0.46	0.38	MILLI/MJ	0.12	
VITAMIN D	MICRO	1.30	0.30	- 2.50	1.30	MICRO/MJ	0.42	
VITAMIN E ACTIVITY	MILLI	2.20	1.80	- 2.60	2.20	MILLI/MJ	0.72	

CONSTITUENTS	DIM	AV	VARIATION			AVR	NUTR. DENS.		MOLPERC.
ALPHA-TOCOPHEROL	MILLI	2.20	1.80	-	2.60	2.20	MILLI/MJ	0.72	
VITAMIN K	MICRO	60.00	-	-	-	60.00	MICRO/MJ	19.54	
VITAMIN B1	MICRO	5.00	4.00	-	7.00	5.00	MICRO/MJ	1.63	
VITAMIN B2	MICRO	22.00	10.00	-	35.00	22.00	MICRO/MJ	7.16	
NICOTINAMIDE	MICRO	34.00	10.00	-	100.00	34.00	MICRO/MJ	11.07	
PANTOTHENIC ACID	MICRO	47.00	38.00	-	57.00	47.00	MICRO/MJ	15.30	
VITAMIN B6	MICRO	5.00	4.00	-	6.00	5.00	MICRO/MJ	1.63	
BIOTIN	-	TRACES	-	-	-				
FOLIC ACID	-	TRACES	-	-	-				
VITAMIN B12	-	TRACES	-	-	-				
VITAMIN C	MILLI	0.20	0.10	-	1.00	0.20	MILLI/MJ	0.07	
ALANINE	MILLI	28.00	-	-	-	28.00	MILLI/MJ	9.12	
ARGININE	MILLI	25.00	23.00	-	28.00	25.00	MILLI/MJ	8.14	
ASPARTIC ACID	MILLI	57.00	54.00	-	59.00	57.00	MILLI/MJ	18.56	
CYSTINE	MILLI	5.00	4.00	-	6.00	5.00	MILLI/MJ	1.63	
GLYCINE	MILLI	17.00	-	-	-	17.00	MILLI/MJ	5.54	
HISTIDINE	MILLI	19.00	16.00	-	30.00	19.00	MILLI/MJ	6.19	
ISOLEUCINE	MILLI	45.00	39.00	-	54.00	45.00	MILLI/MJ	14.65	
LEUCINE	MILLI	74.00	72.00	-	79.00	74.00	MILLI/MJ	24.10	
LYSINE	MILLI	54.00	50.00	-	59.00	54.00	MILLI/MJ	17.58	
METHIONINE	MILLI	19.00	17.00	-	20.00	19.00	MILLI/MJ	6.19	
PHENYLALANINE	MILLI	36.00	34.00	-	38.00	36.00	MILLI/MJ	11.72	
PROLINE	MILLI	69.00	66.00	-	72.00	69.00	MILLI/MJ	22.47	
SERINE	MILLI	42.00	41.00	-	43.00	42.00	MILLI/MJ	13.68	
THREONINE	MILLI	34.00	31.00	-	38.00	34.00	MILLI/MJ	11.07	
TRYPTOPHAN	MILLI	9.00	9.00	-	10.00	9.00	MILLI/MJ	2.93	
TYROSINE	MILLI	36.00	33.00	-	40.00	36.00	MILLI/MJ	11.72	
VALINE	MILLI	51.00	46.00	-	58.00	51.00	MILLI/MJ	16.61	
LACTOSE	GRAM	0.57	0.57	-	0.57	0.57	GRAM/MJ	0.19	
BUTYRIC ACID	GRAM	2.60	-	-	-	2.60	GRAM/MJ	0.85	
CAPROIC ACID	GRAM	1.50	-	-	-	1.50	GRAM/MJ	0.49	
CAPRYLIC ACID	GRAM	0.90	-	-	-	0.90	GRAM/MJ	0.29	
CAPRIC ACID	GRAM	2.00	-	-	-	2.00	GRAM/MJ	0.65	
LAURIC ACID	GRAM	2.20	-	-	-	2.20	GRAM/MJ	0.72	
MYRISTIC ACID	GRAM	8.10	-	-	-	8.10	GRAM/MJ	2.64	
PALMITIC ACID	GRAM	21.10	-	-	-	21.10	GRAM/MJ	6.87	
STEARIC ACID	GRAM	9.70	-	-	-	9.70	GRAM/MJ	3.16	
PALMITOLEIC ACID	GRAM	1.80	-	-	-	1.80	GRAM/MJ	0.59	
OLEIC ACID	GRAM	20.10	-	-	-	20.10	GRAM/MJ	6.55	
C14 : 1 FATTY ACID	GRAM	1.20	-	-	-	1.20	GRAM/MJ	0.39	
LINOLEIC ACID	GRAM	1.80	-	-	-	1.80	GRAM/MJ	0.59	
LINOLENIC ACID	GRAM	1.20	-	-	-	1.20	GRAM/MJ	0.39	
CHOLESTEROL	MILLI	240.00	220.00	-	270.00	240.00	MILLI/MJ	78.15	
TOTAL PHOSPHOLIPIDS	MILLI	200.00	-	-	-	200.00	MILLI/MJ	65.13	

BUTTERSCHMALZ BUTTER FAT BEURRE FONDUE

		PROTEIN	FAT	CARBOHYDRATES	TOTAL
ENERGY VALUE (AVERAGE)	KJOULE	4.6	3851	0.0	3855
PER 100 G	(KCAL)	1.1	920	0.0	921
EDIBLE PORTION					
AMOUNT OF DIGESTIBLE	GRAM	0.24	94.52	0.00	
CONSTITUENTS PER 100 G					
ENERGY VALUE (AVERAGE)	KJOULE	4.5	3658	0.0	3663
OF THE DIGESTIBLE	(KCAL)	1.1	874	0.0	875
FRACTION PER 100 G					
EDIBLE PORTION					

WASTE PERCENTAGE AVERAGE 0.00

CONSTITUENTS	DIM	AV	VARIATION			AVR	NUTR. DENS.		MOLPERC.
WATER	GRAM	0.25	0.20	-	0.32	0.25	GRAM/MJ	0.07	
PROTEIN	GRAM	0.25	0.10	-	0.30	0.25	GRAM/MJ	0.07	
FAT	GRAM	99.50	-	-	-	99.50	GRAM/MJ	27.17	
AVAILABLE CARBOHYDR.	-	0.00	-	-	-	0.00			
MINERALS	-	0.00	-	-	-	0.00			
VITAMIN A	MILLI	0.85	-	-	-	0.85	MILLI/MJ	0.23	
CAROTENE	MILLI	0.20	-	-	-	0.20	MILLI/MJ	0.05	
VITAMIN E ACTIVITY	MILLI	3.60	2.30	-	5.00	3.60	MILLI/MJ	0.98	
ALPHA-TOCOPHEROL	MILLI	3.60	2.30	-	5.00	3.60	MILLI/MJ	0.98	
VITAMIN C	-	0.00	-	-	-	0.00			
BUTYRIC ACID	GRAM	3.20	-	-	-	3.20	GRAM/MJ	0.87	
CAPROIC ACID	GRAM	1.90	-	-	-	1.90	GRAM/MJ	0.52	
CAPRYLIC ACID	GRAM	1.10	-	-	-	1.10	GRAM/MJ	0.30	
CAPRIC ACID	GRAM	2.50	-	-	-	2.50	GRAM/MJ	0.68	
LAURIC ACID	GRAM	2.80	-	-	-	2.80	GRAM/MJ	0.76	
MYRISTIC ACID	GRAM	10.00	-	-	-	10.00	GRAM/MJ	2.73	
PALMITIC ACID	GRAM	26.20	-	-	-	26.20	GRAM/MJ	7.15	
STEARIC ACID	GRAM	12.10	-	-	-	12.10	GRAM/MJ	3.30	
ARACHIDIC ACID	GRAM	0.95	-	-	-	0.95	GRAM/MJ	0.26	
PALMITOLEIC ACID	GRAM	2.30	-	-	-	2.30	GRAM/MJ	0.63	
OLEIC ACID	GRAM	25.10	-	-	-	25.10	GRAM/MJ	6.85	
C14 : 1 FATTY ACID	GRAM	1.50	-	-	-	1.50	GRAM/MJ	0.41	
LINOLEIC ACID	GRAM	2.30	-	-	-	2.30	GRAM/MJ	0.63	
LINOLENIC ACID	GRAM	1.40	-	-	-	1.40	GRAM/MJ	0.38	

GÄNSEFETT GOOSE FAT GRAISSE D'OIE

		PROTEIN	FAT	CARBOHYDRATES	TOTAL
ENERGY VALUE (AVERAGE) PER 100 G EDIBLE PORTION	KJOULE	0.0	3955	0.0	3955
	(KCAL)	0.0	945	0.0	945
AMOUNT OF DIGESTIBLE CONSTITUENTS PER 100 G	GRAM	0.00	94.52	0.00	
ENERGY VALUE (AVERAGE) OF THE DIGESTIBLE FRACTION PER 100 G EDIBLE PORTION	KJOULE	0.0	3757	0.0	3757
	(KCAL)	0.0	898	0.0	898

WASTE PERCENTAGE AVERAGE 0.00

CONSTITUENTS	DIM	AV	VARIATION			AVR	NUTR. DENS.		MOLPERC.
FAT	GRAM	99.50	-	-	-	99.50	GRAM/MJ	26.48	
MYRISTIC ACID	GRAM	0.48	-	-	-	0.48	GRAM/MJ	0.13	
PALMITIC ACID	GRAM	20.00	-	-	-	20.00	GRAM/MJ	5.32	
STEARIC ACID	GRAM	6.20	-	-	-	6.20	GRAM/MJ	1.65	
PALMITOLEIC ACID	GRAM	2.40	-	-	-	2.40	GRAM/MJ	0.64	
OLEIC ACID	GRAM	55.00	-	-	-	55.00	GRAM/MJ	14.64	
LINOLEIC ACID	GRAM	9.00	-	-	-	9.00	GRAM/MJ	2.40	
LINOLENIC ACID	GRAM	1.90	-	-	-	1.90	GRAM/MJ	0.51	

HAIFISCHÖL SHARK OIL HUILE DE SQUALE

		PROTEIN	FAT	CARBOHYDRATES	TOTAL
ENERGY VALUE (AVERAGE)	KJOULE	0.0	3955	0.0	3955
PER 100 G	(KCAL)	0.0	945	0.0	945
EDIBLE PORTION					
AMOUNT OF DIGESTIBLE	GRAM	0.00	94.52	0.00	
CONSTITUENTS PER 100 G					
ENERGY VALUE (AVERAGE)	KJOULE	0.0	3757	0.0	3757
OF THE DIGESTIBLE	(KCAL)	0.0	898	0.0	898
FRACTION PER 100 G					
EDIBLE PORTION					

WASTE PERCENTAGE AVERAGE 0.00

CONSTITUENTS	DIM	AV	VARIATION			AVR	NUTR. DENS.		MOLPERC.
FAT	GRAM	99.50	-	-	-	99.50	GRAM/MJ	26.48	
MYRISTIC ACID	GRAM	6.12	-	-	-	6.12	GRAM/MJ	1.63	
PALMITIC ACID	GRAM	27.75	-	-	-	27.75	GRAM/MJ	7.39	
STEARIC ACID	GRAM	4.20	-	-	-	4.20	GRAM/MJ	1.12	
PALMITOLEIC ACID	GRAM	5.86	-	-	-	5.86	GRAM/MJ	1.56	
OLEIC ACID	GRAM	16.18	-	-	-	16.18	GRAM/MJ	4.31	
EICOSENOIC ACID	GRAM	2.89	-	-	-	2.89	GRAM/MJ	0.77	
C14 : 1 FATTY ACID	GRAM	0.44	-	-	-	0.44	GRAM/MJ	0.12	
LINOLEIC ACID	GRAM	0.27	-	-	-	0.27	GRAM/MJ	0.07	
ARACHIDONIC ACID	GRAM	5.08	-	-	-	5.08	GRAM/MJ	1.35	
C20 : 5 N-3 FATTY ACID	GRAM	3.50	-	-	-	3.50	GRAM/MJ	0.93	
C22 : 6 N-3 FATTY ACID	GRAM	16.36	-	-	-	16.36	GRAM/MJ	4.35	

HAMMELTALG
(HAMMELFETT)
NICHT AUSGELASSEN

MUTTON FAT
NOT RENDERED

SUIF DE MOUTON
NON FONDU

		PROTEIN	FAT	CARBOHYDRATES	TOTAL
ENERGY VALUE (AVERAGE)	KJOULE	71	3232	0.0	3302
PER 100 G	(KCAL)	17	772	0.0	789
EDIBLE PORTION					
AMOUNT OF DIGESTIBLE	GRAM	3.73	77.23	0.00	
CONSTITUENTS PER 100 G					
ENERGY VALUE (AVERAGE)	KJOULE	69	3070	0.0	3139
OF THE DIGESTIBLE	(KCAL)	16	734	0.0	750
FRACTION PER 100 G					
EDIBLE PORTION					

WASTE PERCENTAGE AVERAGE 0.00

CONSTITUENTS	DIM	AV	VARIATION			AVR	NUTR. DENS.		MOLPERC.
WATER	GRAM	14.70	11.10	-	18.70	14.70	GRAM/MJ	4.68	
PROTEIN	GRAM	3.85	2.29	-	5.42	3.85	GRAM/MJ	1.23	
FAT	GRAM	81.30	75.70	-	86.80	81.30	GRAM/MJ	25.90	
AVAILABLE CARBOHYDR.	-	0.00	-	-	-	0.00			
MINERALS	GRAM	0.10	0.05	-	0.14	0.10	GRAM/MJ	0.03	
SODIUM	MILLI	2.00	-	-	-	2.00	MILLI/MJ	0.64	
POTASSIUM	MILLI	4.00	-	-	-	4.00	MILLI/MJ	1.27	
CALCIUM	-	0.00	-	-	-	0.00			
MYRISTIC ACID	GRAM	1.90	0.78	-	3.10	1.90	GRAM/MJ	0.61	
PALMITIC ACID	GRAM	18.70	15.50	-	21.80	18.70	GRAM/MJ	5.96	
STEARIC ACID	GRAM	22.20	19.40	-	25.00	22.20	GRAM/MJ	7.07	
OLEIC ACID	GRAM	32.30	28.00	-	36.50	32.30	GRAM/MJ	10.29	
LINOLEIC ACID	GRAM	3.34	-	-	-	3.34	GRAM/MJ	1.06	

HERINGSÖL HERRING OIL HUILE DE HARENG

		PROTEIN	FAT	CARBOHYDRATES	TOTAL
ENERGY VALUE (AVERAGE)	KJOULE	0.0	3955	0.0	3955
PER 100 G	(KCAL)	0.0	945	0.0	945
EDIBLE PORTION					
AMOUNT OF DIGESTIBLE	GRAM	0.00	94.52	0.00	
CONSTITUENTS PER 100 G					
ENERGY VALUE (AVERAGE)	KJOULE	0.0	3757	0.0	3757
OF THE DIGESTIBLE	(KCAL)	0.0	898	0.0	898
FRACTION PER 100 G					
EDIBLE PORTION					

WASTE PERCENTAGE AVERAGE 0.00

CONSTITUENTS	DIM	AV	VARIATION			AVR	NUTR. DENS.		MOLPERC.
FAT	GRAM	99.50	-	-	-	99.50	GRAM/MJ	26.48	
IODIDE	MILLI	0.12	-	-	-	0.12	MILLI/MJ	0.03	
SELENIUM	MICRO	13.50	-	-	-	13.50	MICRO/MJ	3.59	
BROMINE	MILLI	0.66	-	-	-	0.66	MILLI/MJ	0.18	
VITAMIN E ACTIVITY	MILLI	5.90	-	-	-	5.90	MILLI/MJ	1.57	
MYRISTIC ACID	GRAM	6.50	-	-	-	6.50	GRAM/MJ	1.73	
PALMITIC ACID	GRAM	12.90	-	-	-	12.90	GRAM/MJ	3.43	
STEARIC ACID	GRAM	1.40	-	-	-	1.40	GRAM/MJ	0.37	
PALMITOLEIC ACID	GRAM	6.50	-	-	-	6.50	GRAM/MJ	1.73	
OLEIC ACID	GRAM	14.80	-	-	-	14.80	GRAM/MJ	3.94	
EICOSENOIC ACID	GRAM	14.00	-	-	-	14.00	GRAM/MJ	3.73	
C22 : 1 FATTY ACID	GRAM	18.20	-	-	-	18.20	GRAM/MJ	4.84	
LINOLEIC ACID	GRAM	2.40	-	-	-	2.40	GRAM/MJ	0.64	
LINOLENIC ACID	GRAM	0.90	-	-	-	0.90	GRAM/MJ	0.24	

HÜHNERFETT POULTRY FAT GRAISSE DE POULET

		PROTEIN	FAT	CARBOHYDRATES	TOTAL
ENERGY VALUE (AVERAGE) PER 100 G EDIBLE PORTION	KJOULE	0.0	3955	0.0	3955
	(KCAL)	0.0	945	0.0	945
AMOUNT OF DIGESTIBLE CONSTITUENTS PER 100 G	GRAM	0.00	94.52	0.00	
ENERGY VALUE (AVERAGE) OF THE DIGESTIBLE FRACTION PER 100 G EDIBLE PORTION	KJOULE	0.0	3757	0.0	3757
	(KCAL)	0.0	898	0.0	898

WASTE PERCENTAGE AVERAGE 0.00

CONSTITUENTS	DIM	AV	VARIATION			AVR	NUTR. DENS.		MOLPERC.
FAT	GRAM	99.50	-	-	-	99.50	GRAM/MJ	26.48	
VITAMIN E ACTIVITY	MILLI	2.73	-	-	-	2.73	MILLI/MJ	0.73	
MYRISTIC ACID	GRAM	0.50	-	-	-	0.50	GRAM/MJ	0.13	
PALMITIC ACID	GRAM	19.00	-	-	-	19.00	GRAM/MJ	5.06	
STEARIC ACID	GRAM	7.50	-	-	-	7.50	GRAM/MJ	2.00	
PALMITOLEIC ACID	GRAM	3.00	-	-	-	3.00	GRAM/MJ	0.80	
OLEIC ACID	GRAM	47.00	-	-	-	47.00	GRAM/MJ	12.51	
LINOLEIC ACID	GRAM	21.50	-	-	-	21.50	GRAM/MJ	5.72	
LINOLENIC ACID	GRAM	1.50	-	-	-	1.50	GRAM/MJ	0.40	

POTTWALÖL (WALÖL) | WHALE OIL | HUILE DE BALEINE

		PROTEIN	FAT	CARBOHYDRATES	TOTAL
ENERGY VALUE (AVERAGE)	KJOULE	0.0	3955	0.0	3955
PER 100 G	(KCAL)	0.0	945	0.0	945
EDIBLE PORTION					
AMOUNT OF DIGESTIBLE	GRAM	0.00	94.52	0.00	
CONSTITUENTS PER 100 G					
ENERGY VALUE (AVERAGE)	KJOULE	0.0	3757	0.0	3757
OF THE DIGESTIBLE	(KCAL)	0.0	898	0.0	898
FRACTION PER 100 G					
EDIBLE PORTION					

WASTE PERCENTAGE AVERAGE 0.00

CONSTITUENTS	DIM	AV	VARIATION	AVR	NUTR. DENS.		MOLPERC.
FAT	GRAM	99.50	- - -	99.50	GRAM/MJ	26.48	
MYRISTIC ACID	GRAM	8.10	- - -	8.10	GRAM/MJ	2.16	
C15 : 0 FATTY ACID	GRAM	0.95	- - -	0.95	GRAM/MJ	0.25	
PALMITIC ACID	GRAM	11.00	- - -	11.00	GRAM/MJ	2.93	
STEARIC ACID	GRAM	0.95	- - -	0.95	GRAM/MJ	0.25	
PALMITOLEIC ACID	GRAM	17.60	- - -	17.60	GRAM/MJ	4.68	
OLEIC ACID	GRAM	31.00	- - -	31.00	GRAM/MJ	8.25	
EICOSENOIC ACID	GRAM	16.00	- - -	16.00	GRAM/MJ	4.26	
C22 : 1 FATTY ACID	GRAM	8.10	- - -	8.10	GRAM/MJ	2.16	
LINOLEIC ACID	GRAM	0.48	- - -	0.48	GRAM/MJ	0.13	

RINDERTALG　BEEF-SUET TALLOW　SUIF DE BOEUF

		PROTEIN	FAT	CARBOHYDRATES	TOTAL
ENERGY VALUE (AVERAGE) PER 100 G EDIBLE PORTION	KJOULE	15	3836	0.0	3850
	(KCAL)	3.5	917	0.0	920
AMOUNT OF DIGESTIBLE CONSTITUENTS PER 100 G	GRAM	0.77	91.67	0.00	
ENERGY VALUE (AVERAGE) OF THE DIGESTIBLE FRACTION PER 100 G EDIBLE PORTION	KJOULE	14	3644	0.0	3658
	(KCAL)	3.4	871	0.0	874

WASTE PERCENTAGE AVERAGE 0.00

CONSTITUENTS	DIM	AV	VARIATION			AVR	NUTR.	DENS.	MOLPERC.
WATER	GRAM	1.97	0.30	-	6.00	1.97	GRAM/MJ	0.54	
PROTEIN	GRAM	0.80	0.15	-	2.00	0.80	GRAM/MJ	0.22	
FAT	GRAM	96.50	90.00	-	99.10	96.50	GRAM/MJ	26.38	
AVAILABLE CARBOHYDR.	-	0.00	-	-	-	0.00			
MINERALS	MILLI	70.00	-	-	-	70.00	MILLI/MJ	19.14	
SODIUM	MILLI	10.50	1.00	-	20.00	10.50	MILLI/MJ	2.87	
POTASSIUM	MILLI	6.00	2.00	-	10.00	6.00	MILLI/MJ	1.64	
MAGNESIUM	MILLI	-	0.00	-	6.00				
CALCIUM	-	0.00	-	-	-	0.00			
MANGANESE	MICRO	1.00	-	-	-	1.00	MICRO/MJ	0.27	
IRON	MILLI	0.32	0.10	-	0.50	0.32	MILLI/MJ	0.09	
COPPER	MICRO	80.00	40.00	-	110.00	80.00	MICRO/MJ	21.87	
MOLYBDENUM	-	0.00	-	-	-	0.00			
PHOSPHORUS	MILLI	7.00	-	-	-	7.00	MILLI/MJ	1.91	
VITAMIN A	MILLI	0.22	0.05	-	0.54	0.22	MILLI/MJ	0.06	
CAROTENE	MILLI	0.22	0.19	-	0.36	0.22	MILLI/MJ	0.06	
VITAMIN E ACTIVITY	MILLI	1.30	0.00	-	2.50	1.30	MILLI/MJ	0.36	
ALPHA-TOCOPHEROL	MILLI	1.30	0.00	-	2.50	1.30	MILLI/MJ	0.36	
LAURIC ACID	MILLI	90.00	-	-	-	90.00	MILLI/MJ	24.60	
MYRISTIC ACID	GRAM	2.90	0.96	-	7.60	2.90	GRAM/MJ	0.79	
PALMITIC ACID	GRAM	24.80	19.10	-	35.30	24.80	GRAM/MJ	6.78	
STEARIC ACID	GRAM	18.60	5.70	-	38.20	18.60	GRAM/MJ	5.08	
ARACHIDIC ACID	GRAM	-	0.00	-	1.00				
PALMITOLEIC ACID	GRAM	3.30	0.70	-	8.40	3.30	GRAM/MJ	0.90	
OLEIC ACID	GRAM	38.20	24.90	-	52.60	38.20	GRAM/MJ	10.44	
C14 : 1 FATTY ACID	GRAM	0.50	0.50	-	1.40	0.50	GRAM/MJ	0.14	
LINOLEIC ACID	GRAM	4.30	0.50	-	4.80	4.30	GRAM/MJ	1.18	
LINOLENIC ACID	GRAM	0.47	-	-	-	0.47	GRAM/MJ	0.13	
ARACHIDONIC ACID	GRAM	0.23	-	-	-	0.23	GRAM/MJ	0.06	
CHOLESTEROL	MILLI	100.00	70.00	-	140.00	100.00	MILLI/MJ	27.34	

SCHWEINESCHMALZ LARD SAINDOUX

		PROTEIN	FAT	CARBOHYDRATES	TOTAL
ENERGY VALUE (AVERAGE) PER 100 G EDIBLE PORTION	KJOULE	1.8	3963	0.0	3965
	(KCAL)	0.4	947	0.0	948
AMOUNT OF DIGESTIBLE CONSTITUENTS PER 100 G	GRAM	0.09	94.71	0.00	
ENERGY VALUE (AVERAGE) OF THE DIGESTIBLE FRACTION PER 100 G EDIBLE PORTION	KJOULE	1.8	3765	0.0	3767
	(KCAL)	0.4	900	0.0	900

WASTE PERCENTAGE AVERAGE 0.00

CONSTITUENTS	DIM	AV	VARIATION		AVR	NUTR. DENS.		MOLPERC.
WATER	GRAM	0.16	0.15	- 0.50	0.16	GRAM/MJ	0.04	
PROTEIN	GRAM	0.10	-	- -	0.10	GRAM/MJ	0.03	
FAT	GRAM	99.70	99.00	- 99.80	99.70	GRAM/MJ	26.47	
AVAILABLE CARBOHYDR.	-	0.00	-	- -	0.00			
SODIUM	MILLI	1.00	1.00	- 2.00	1.00	MILLI/MJ	0.27	
POTASSIUM	MILLI	1.00	1.00	- 1.50	1.00	MILLI/MJ	0.27	
MAGNESIUM	MILLI	-	0.00	- 1.30				
CALCIUM	MILLI	-	0.00	- 0.80				
IRON	MILLI	-	0.00	- 0.10				
COPPER	MILLI	-	0.00	- 0.02				
PHOSPHORUS	MILLI	-	0.00	- 3.00				
CHLORIDE	MILLI	4.00	-	- -	4.00	MILLI/MJ	1.06	
IODIDE	MICRO	9.70	-	- -	9.70	MICRO/MJ	2.58	
VITAMIN E ACTIVITY	MILLI	1.60	1.20	- 2.00	1.60	MILLI/MJ	0.42	
TOTAL TOCOPHEROLS	MILLI	1.80	1.30	- 2.20	1.80	MILLI/MJ	0.48	
ALPHA-TOCOPHEROL	MILLI	1.60	1.20	- 2.00	1.60	MILLI/MJ	0.42	
GAMMA-TOCOPHEROL	MICRO	90.00	70.00	- 110.00	90.00	MICRO/MJ	23.89	
MYRISTIC ACID	GRAM	0.90	0.50	- 2.70	0.90	GRAM/MJ	0.24	
PALMITIC ACID	GRAM	22.90	19.10	- 30.50	22.90	GRAM/MJ	6.08	
STEARIC ACID	GRAM	13.30	4.80	- 22.90	13.30	GRAM/MJ	3.53	
ARACHIDIC ACID	GRAM	-	0.10	- 1.00				
PALMITOLEIC ACID	GRAM	3.80	1.60	- 4.80	3.80	GRAM/MJ	1.01	
OLEIC ACID	GRAM	41.10	19.30	- 59.30	41.10	GRAM/MJ	10.91	
C14 : 1 FATTY ACID	GRAM	0.50	0.00	- 0.50	0.50	GRAM/MJ	0.13	
LINOLEIC ACID	GRAM	8.60	2.80	- 15.40	8.60	GRAM/MJ	2.28	
LINOLENIC ACID	GRAM	1.00	0.40	- 1.90	1.00	GRAM/MJ	0.27	
ARACHIDONIC ACID	GRAM	1.70	0.40	- 3.00	1.70	GRAM/MJ	0.45	
CHOLESTEROL	MILLI	86.00	50.00	- 122.00	86.00	MILLI/MJ	22.83	

Pflanzliche Fette und Öle

PLANT FATS AND OILS **GRAISSES ET HUILES VÉGÉTALES**

BAUMWOLLSAMENÖL (BAUMWOLLSAATÖL, COTTONÖL) GEREINIGT

COTTON SEED OIL REFINED

HUILE DE COTON ÉPURÉE

		PROTEIN	FAT	CARBOHYDRATES	TOTAL
ENERGY VALUE (AVERAGE)	KJOULE	0.0	3879	0.0	3879
PER 100 G	(KCAL)	0.0	927	0.0	927
EDIBLE PORTION					
AMOUNT OF DIGESTIBLE	GRAM	0.00	94.71	0.00	
CONSTITUENTS PER 100 G					
ENERGY VALUE (AVERAGE)	KJOULE	0.0	3685	0.0	3685
OF THE DIGESTIBLE	(KCAL)	0.0	881	0.0	881
FRACTION PER 100 G					
EDIBLE PORTION					

WASTE PERCENTAGE AVERAGE 0.00

CONSTITUENTS	DIM	AV	VARIATION			AVR	NUTR.	DENS.	MOLPERC.
WATER	GRAM	0.30	-	-	-	0.30	GRAM/MJ	0.08	
FAT	GRAM	99.70	-	-	-	99.70	GRAM/MJ	27.05	
MINERALS	-	TRACES	-	-	-				
SODIUM	MILLI	1.00	-	-	-	1.00	MILLI/MJ	0.27	
POTASSIUM	-	0.00	-	-	-	0.00			
MANGANESE	MICRO	13.00	-	-	-	13.00	MICRO/MJ	3.53	
IRON	MICRO	30.00	10.00	-	50.00	30.00	MICRO/MJ	8.14	
COPPER	MICRO	5.00	0.20	-	10.00	5.00	MICRO/MJ	1.36	
VITAMIN A	-	0.00	-	-	-	0.00			
VITAMIN E ACTIVITY	MILLI	38.30	-	-	-	38.30	MILLI/MJ	10.39	
TOTAL TOCOPHEROLS	MILLI	65.20	-	-	-	65.20	MILLI/MJ	17.69	
ALPHA-TOCOPHEROL	MILLI	35.30	-	-	-	35.30	MILLI/MJ	9.58	
GAMMA-TOCOPHEROL	MILLI	30.00	-	-	-	30.00	MILLI/MJ	8.14	
MYRISTIC ACID	GRAM	1.40	0.40	-	[illegible].90	1.40	GRAM/MJ	0.38	
PALMITIC ACID	GRAM	21.00	16.20	-	27.70	21.00	GRAM/MJ	5.70	
STEARIC ACID	GRAM	4.80	-	-	-	4.80	GRAM/MJ	1.30	
PALMITOLEIC ACID	GRAM	1.40	0.00	-	1.31	1.40	GRAM/MJ	0.38	
OLEIC ACID	GRAM	18.20	12.40	-	22.90	18.20	GRAM/MJ	4.94	
LINOLEIC ACID	GRAM	47.80	38.20	-	55.40	47.80	GRAM/MJ	12.97	
LINOLENIC ACID	GRAM	1.00	0.50	-	1.40	1.00	GRAM/MJ	0.27	
TOTAL STEROLS	MILLI	327.00	-	-	-	327.00	MILLI/MJ	88.73	
5-AVENASTEROL	MILLI	5.00	-	-	-	5.00	MILLI/MJ	1.36	
BRASSICASTEROL	-	TRACES	-	-	-				
CAMPESTEROL	MILLI	20.00	-	-	-	20.00	MILLI/MJ	5.43	
BETA-SITOSTEROL	MILLI	303.00	-	-	-	303.00	MILLI/MJ	82.21	
STIGMASTEROL	-	TRACES	-	-	-				

ERDNUSSÖL (ARACHISÖL) GEREINIGT	PEANUT OIL (ARACHIS OIL) REFINED	HUILE D'ARACHIDE ÉPURÉE

		PROTEIN	FAT	CARBOHYDRATES	TOTAL
ENERGY VALUE (AVERAGE)	KJOULE	0.0	3868	0.0	3868
PER 100 G	(KCAL)	0.0	924	0.0	924
EDIBLE PORTION					
AMOUNT OF DIGESTIBLE	GRAM	0.00	94.43	0.00	
CONSTITUENTS PER 100 G					
ENERGY VALUE (AVERAGE)	KJOULE	0.0	3674	0.0	3674
OF THE DIGESTIBLE	(KCAL)	0.0	878	0.0	878
FRACTION PER 100 G					
EDIBLE PORTION					

WASTE PERCENTAGE AVERAGE 0.00

CONSTITUENTS	DIM	AV	VARIATION			AVR	NUTR. DENS.		MOLPERC.
WATER	GRAM	0.40	-	-	-	0.40	GRAM/MJ	0.11	
PROTEIN	-	0.00	-	-	-	0.00			
FAT	GRAM	99.40	-	-	-	99.40	GRAM/MJ	27.05	
SODIUM	-	0.00	-	-	-	0.00			
POTASSIUM	-	0.00	-	-	-	0.00			
CALCIUM	-	0.00	-	-	-	0.00			
MANGANESE	MICRO	2.00	-	-	-	2.00	MICRO/MJ	0.54	
IRON	MICRO	64.00	TRACES	-	100.00	64.00	MICRO/MJ	17.42	
COPPER	MICRO	3.00	1.00	-	5.00	3.00	MICRO/MJ	0.82	
VITAMIN A	-	0.00	-	-	-	0.00			
VITAMIN E ACTIVITY	MILLI	17.20	-	-	-	17.20	MILLI/MJ	4.68	
TOTAL TOCOPHEROLS	MILLI	33.10	-	-	-	33.10	MILLI/MJ	9.01	
ALPHA-TOCOPHEROL	MILLI	15.60	-	-	-	15.60	MILLI/MJ	4.25	
GAMMA-TOCOPHEROL	MILLI	15.60	-	-	-	15.60	MILLI/MJ	4.25	
DELTA-TOCOPHEROL	MILLI	1.90	-	-	-	1.90	MILLI/MJ	0.52	
MYRISTIC ACID	GRAM	0.23	0.00	-	0.47	0.23	GRAM/MJ	0.06	
PALMITIC ACID	GRAM	10.00	5.70	-	16.40	10.00	GRAM/MJ	2.72	
STEARIC ACID	GRAM	2.80	1.10	-	6.20	2.80	GRAM/MJ	0.76	
ARACHIDIC ACID	GRAM	3.40	2.10	-	4.70	3.40	GRAM/MJ	0.93	
PALMITOLEIC ACID	GRAM	-	0.00	-	1.00				
OLEIC ACID	GRAM	52.50 1	31.20	-	64.20	52.50	GRAM/MJ	14.29	
EICOSENOIC ACID	GRAM	1.40	0.30	-	2.70	1.40	GRAM/MJ	0.38	
LINOLEIC ACID	GRAM	23.90 1	11.70	-	45.90	23.90	GRAM/MJ	6.50	
LINOLENIC ACID	GRAM	-	0.00	-	1.30				
TOTAL STEROLS	MILLI	240.00	-	-	-	240.00	MILLI/MJ	65.32	
FREE STEROLS	MILLI	150.00	-	-	-	150.00	MILLI/MJ	40.82	
BRASSICASTEROL	MILLI	-	0.00	-	1.90				
CAMPESTEROL	MILLI	29.00	17.00	-	54.50	29.00	MILLI/MJ	7.89	
CHOLESTEROL	MILLI	1.20	0.00	-	2.40	1.20	MILLI/MJ	0.33	
BETA-SITOSTEROL	MILLI	179.00	-	-	-	179.00	MILLI/MJ	48.72	
STIGMASTEROL	MILLI	25.90	-	-	-	25.90	MILLI/MJ	7.05	

1 VALID FOR AFRICAN PEANUT OIL. ORIGIN FROM ARGENTINA:
35,5 G OLEIC ACID/100G; 39,1 G LINOLEIC ACID/100G

ILLIPEFETT ILLIPE FAT BEURRE DE ILLIPE

		PROTEIN	FAT	CARBOHYDRATES	TOTAL
ENERGY VALUE (AVERAGE)	KJOULE	0	3872	0.0	3888
PER 100 G	(KCAL)	0.0	925	0.0	929
EDIBLE PORTION					
AMOUNT OF DIGESTIBLE	GRAM	0.00	94.52	0.00	
CONSTITUENTS PER 100 G					
ENERGY VALUE (AVERAGE)	KJOULE	0.0	3678	0.0	3678
OF THE DIGESTIBLE	(KCAL)	0.0	879	0.0	879
FRACTION PER 100 G					
EDIBLE PORTION					

WASTE PERCENTAGE AVERAGE 0.00

CONSTITUENTS	DIM	AV	VARIATION			AVR	NUTR. DENS.		MOLPERC.
FAT	GRAM	99.50	-	-	-	99.50	GRAM/MJ	27.05	
CAPRYLIC ACID	GRAM	0.20	-	-	-	0.20	GRAM/MJ	0.05	
CAPRIC ACID	GRAM	0.20	-	-	-	0.20	GRAM/MJ	0.05	
LAURIC ACID	GRAM	0.30	-	-	-	0.30	GRAM/MJ	0.08	
PALMITIC ACID	GRAM	16.30	-	-	-	16.30	GRAM/MJ	4.43	
STEARIC ACID	GRAM	43.30	-	-	-	43.30	GRAM/MJ	11.77	
ARACHIDIC ACID	GRAM	1.60	-	-	-	1.60	GRAM/MJ	0.44	
PALMITOLEIC ACID	GRAM	0.20	-	-	-	0.20	GRAM/MJ	0.05	
OLEIC ACID	GRAM	30.90	-	-	-	30.90	GRAM/MJ	8.40	
LINOLEIC ACID	GRAM	1.50	-	-	-	1.50	GRAM/MJ	0.41	
LINOLENIC ACID	GRAM	0.10	-	-	-	0.10	GRAM/MJ	0.03	

KAKAOBUTTER COCOA BUTTER GRAISSE DE CACAO

		PROTEIN	FAT	CARBOHYDRATES	TOTAL
ENERGY VALUE (AVERAGE)	KJOULE	0.0	3872	0.0	3872
PER 100 G	(KCAL)	0.0	925	0.0	925
EDIBLE PORTION					
AMOUNT OF DIGESTIBLE	GRAM	0.00	94.52	0.00	
CONSTITUENTS PER 100 G					
ENERGY VALUE (AVERAGE)	KJOULE	0.0	3678	0.0	3678
OF THE DIGESTIBLE	(KCAL)	0.0	879	0.0	879
FRACTION PER 100 G					
EDIBLE PORTION					

WASTE PERCENTAGE AVERAGE 0.00

CONSTITUENTS	DIM	AV	VARIATION			AVR	NUTR. DENS.		MOLPERC.
FAT	GRAM	99.50	-	-	-	99.50	GRAM/MJ	27.05	
VITAMIN E ACTIVITY	MILLI	1.04	-	-	-	1.04	MILLI/MJ	0.28	
TOTAL TOCOPHEROLS	MILLI	5.60	-	-	-	5.60	MILLI/MJ	1.52	
ALPHA-TOCOPHEROL	MILLI	0.50	-	-	-	0.50	MILLI/MJ	0.14	
BETA-TOCOPHEROL	MILLI	0.16	-	-	-	0.16	MILLI/MJ	0.04	
GAMMA-TOCOPHEROL	MILLI	4.90	-	-	-	4.90	MILLI/MJ	1.33	
PALMITIC ACID	GRAM	24.80	22.60	-	30.40	24.80	GRAM/MJ	6.74	
STEARIC ACID	GRAM	33.50	30.20	-	36.00	33.50	GRAM/MJ	9.11	
ARACHIDIC ACID	GRAM	1.00	0.00	-	1.20	1.00	GRAM/MJ	0.27	
PALMITOLEIC ACID	GRAM	0.50	0.10	-	0.50	0.50	GRAM/MJ	0.14	
OLEIC ACID	GRAM	32.60	29.20	-	36.40	32.60	GRAM/MJ	8.86	
LINOLEIC ACID	GRAM	1.30	1.30	-	4.00	1.30	GRAM/MJ	0.35	
LINOLENIC ACID	GRAM	0.40	TRACES	-	0.50	0.40	GRAM/MJ	0.11	
TOTAL STEROLS	MILLI	226.00	201.00	-	250.00	226.00	MILLI/MJ	61.45	
BRASSICASTEROL	-	TRACES	-	-	-				
CAMPESTEROL	MILLI	22.00	19.00	-	25.00	22.00	MILLI/MJ	5.98	
CHOLESTEROL	MILLI	2.70	1.40	-	4.50	2.70	MILLI/MJ	0.73	
BETA-SITOSTEROL	MILLI	138.00	134.00	-	141.00	138.00	MILLI/MJ	37.52	
STIGMASTEROL	MILLI	61.00	56.00	-	69.00	61.00	MILLI/MJ	16.58	

KOKOSFETT
GEREINIGT

COCONUT OIL
REFINED

GRAISSE DE COCO
ÉPURÉE

		PROTEIN	FAT	CARBOHYDRATES	TOTAL
ENERGY VALUE (AVERAGE)	KJOULE	15	3852	0.0	3867
PER 100 G	(KCAL)	3.5	921	0.0	924
EDIBLE PORTION					
AMOUNT OF DIGESTIBLE	GRAM	0.77	94.05	0.00	
CONSTITUENTS PER 100 G					
ENERGY VALUE (AVERAGE)	KJOULE	14	3660	0.0	3674
OF THE DIGESTIBLE	(KCAL)	3.4	875	0.0	878
FRACTION PER 100 G					
EDIBLE PORTION					

WASTE PERCENTAGE AVERAGE 0.00

CONSTITUENTS	DIM	AV	VARIATION			AVR	NUTR.	DENS.	MOLPERC.
WATER	MILLI	90.00	40.00	-	160.00	90.00	MILLI/MJ	24.50	
PROTEIN	GRAM	0.80	0.00	-	1.70	0.80	GRAM/MJ	0.22	
FAT	GRAM	99.00	97.00	-	99.90	99.00	GRAM/MJ	26.95	
AVAILABLE CARBOHYDR.	MILLI	10.00	-	-	-	10.00	MILLI/MJ	2.72	
MINERALS	GRAM	0.10	-	-	-	0.10	GRAM/MJ	0.03	
SODIUM	MILLI	-	0.00	-	4.00				
POTASSIUM	MILLI	-	0.00	-	4.00				
MAGNESIUM	MILLI	0.20	-	-	-	0.20	MILLI/MJ	0.05	
CALCIUM	MILLI	-	0.00	-	3.00				
IRON	MICRO	20.00	-	-	-	20.00	MICRO/MJ	5.44	
COPPER	MICRO	2.00	-	-	-	2.00	MICRO/MJ	0.54	
PHOSPHORUS	MILLI	0.90	0.87	-	0.92	0.90	MILLI/MJ	0.24	
CHLORIDE	MILLI	0.10	-	-	-	0.10	MILLI/MJ	0.03	
VITAMIN A	-	0.00	-	-	-	0.00			
CAROTENE	-	TRACES	-	-	-				
VITAMIN E ACTIVITY	MILLI	0.80	-	-	-	0.80	MILLI/MJ	0.22	
TOTAL TOCOPHEROLS	MILLI	3.20	-	-	-	3.20	MILLI/MJ	0.87	
ALPHA-TOCOPHEROL	MILLI	0.35	-	-	-	0.35	MILLI/MJ	0.10	
GAMMA-TOCOPHEROL	MILLI	0.17	-	-	-	0.17	MILLI/MJ	0.05	
ALPHA-TOCOTRIENOL	MILLI	1.30	-	-	-	1.30	MILLI/MJ	0.35	
BETA-TOCOTRIENOL	MILLI	0.10	-	-	-	0.10	MILLI/MJ	0.03	
GAMMA-TOCOTRIENOL	MILLI	1.30	-	-	-	1.30	MILLI/MJ	0.35	
CAPRYLIC ACID	GRAM	7.60	5.70	-	10.50	7.60	GRAM/MJ	2.07	
CAPRIC ACID	GRAM	5.70	5.70	-	6.90	5.70	GRAM/MJ	1.55	
LAURIC ACID	GRAM	45.00	40.30	-	49.00	45.00	GRAM/MJ	12.25	
MYRISTIC ACID	GRAM	17.20	12.30	-	18.20	17.20	GRAM/MJ	4.68	
PALMITIC ACID	GRAM	8.60	6.70	-	10.50	8.60	GRAM/MJ	2.34	
STEARIC ACID	GRAM	2.40	0.80	-	5.70	2.40	GRAM/MJ	0.65	
OLEIC ACID	GRAM	6.70	5.20	-	6.70	6.70	GRAM/MJ	1.82	
LINOLEIC ACID	GRAM	1.40	1.00	-	1.90	1.40	GRAM/MJ	0.38	
TOTAL STEROLS	MILLI	100.00	-	-	-	100.00	MILLI/MJ	27.22	
FREE STEROLS	MILLI	60.00	-	-	-	60.00	MILLI/MJ	16.33	
BRASSICASTEROL	-	TRACES	-	-	-				
CAMPESTEROL	MILLI	7.00	-	-	-	7.00	MILLI/MJ	1.91	
CHOLESTEROL	MILLI	0.57	-	-	-	0.57	MILLI/MJ	0.16	
BETA-SITOSTEROL	MILLI	66.00	-	-	-	66.00	MILLI/MJ	17.96	
STIGMASTEROL	MILLI	15.00	-	-	-	15.00	MILLI/MJ	4.08	

KÜRBISKERNÖL — PUMPKIN SEED OIL — HUILE DE NOYAU A CITROUILLE

		PROTEIN	FAT	CARBOHYDRATES	TOTAL
ENERGY VALUE (AVERAGE)	KJOULE	0.0	3872	0.0	3872
PER 100 G	(KCAL)	0.0	925	0.0	925
EDIBLE PORTION					
AMOUNT OF DIGESTIBLE	GRAM	0.00	94.52	0.00	
CONSTITUENTS PER 100 G					
ENERGY VALUE (AVERAGE)	KJOULE	0.0	3678	0.0	3678
OF THE DIGESTIBLE	(KCAL)	0.0	879	0.0	879
FRACTION PER 100 G					
EDIBLE PORTION					

WASTE PERCENTAGE AVERAGE 0.00

CONSTITUENTS	DIM	AV	VARIATION			AVR	NUTR. DENS.		MOLPERC.
FAT	GRAM	99.50	-	-	-	99.50	GRAM/MJ	27.05	
PALMITIC ACID	GRAM	15.00	-	-	-	15.00	GRAM/MJ	4.08	
STEARIC ACID	GRAM	4.80	-	-	-	4.80	GRAM/MJ	1.31	
PALMITOLEIC ACID	GRAM	0.48	-	-	-	0.48	GRAM/MJ	0.13	
OLEIC ACID	GRAM	23.00	-	-	-	23.00	GRAM/MJ	6.25	
LINOLEIC ACID	GRAM	51.00	-	-	-	51.00	GRAM/MJ	13.87	
LINOLENIC ACID	GRAM	0.48	-	-	-	0.48	GRAM/MJ	0.13	
TOTAL STEROLS	MILLI	523.00	-	-	-	523.00	MILLI/MJ	142.19	
CAMPESTEROL	MILLI	25.00	-	-	-	25.00	MILLI/MJ	6.80	
CHOLESTEROL	-	TRACES	-	-	-				
STIGMASTEROL	MILLI	8.00	-	-	-	8.00	MILLI/MJ	2.18	

LEINÖL | LINSEED OIL | HUILE DE LIN

		PROTEIN	FAT	CARBOHYDRATES	TOTAL
ENERGY VALUE (AVERAGE)	KJOULE	0.0	3872	0.0	3872
PER 100 G	(KCAL)	0.0	925	0.0	925
EDIBLE PORTION					
AMOUNT OF DIGESTIBLE	GRAM	0.00	94.52	0.00	
CONSTITUENTS PER 100 G					
ENERGY VALUE (AVERAGE)	KJOULE	0.0	3678	0.0	3678
OF THE DIGESTIBLE	(KCAL)	0.0	879	0.0	879
FRACTION PER 100 G					
EDIBLE PORTION					

WASTE PERCENTAGE AVERAGE 0.00

CONSTITUENTS	DIM	AV	VARIATION			AVR	NUTR. DENS.		MOLPERC.
FAT	GRAM	99.50	-	-	-	99.50	GRAM/MJ	27.05	
VITAMIN E ACTIVITY	MILLI	2.10	-	-	-	2.10	MILLI/MJ	0.57	
TOTAL TOCOPHEROLS	MILLI	19.00	-	-	-	19.00	MILLI/MJ	5.17	
ALPHA-TOCOPHEROL	MILLI	0.30	-	-	-	0.30	MILLI/MJ	0.08	
GAMMA-TOCOPHEROL	MILLI	18.40	-	-	-	18.40	MILLI/MJ	5.00	
PALMITIC ACID	GRAM	6.20	-	-	-	6.20	GRAM/MJ	1.69	
STEARIC ACID	GRAM	3.40	-	-	-	3.40	GRAM/MJ	0.92	
OLEIC ACID	GRAM	17.20	-	-	-	17.20	GRAM/MJ	4.68	
LINOLEIC ACID	GRAM	13.40	-	-	-	13.40	GRAM/MJ	3.64	
LINOLENIC ACID	GRAM	55.30	-	-	-	55.30	GRAM/MJ	15.04	
TOTAL STEROLS	GRAM	0.43	-	-	-	0.43	GRAM/MJ	0.12	
BRASSICASTEROL	MILLI	4.30	-	-	-	4.30	MILLI/MJ	1.17	
CAMPESTEROL	GRAM	0.12	-	-	-	0.12	GRAM/MJ	0.03	
CHOLESTEROL	MILLI	6.50	-	-	-	6.50	MILLI/MJ	1.77	
BETA-SITOSTEROL	GRAM	0.21	-	-	-	0.21	GRAM/MJ	0.06	
STIGMASTEROL	MILLI	38.70	-	-	-	38.70	MILLI/MJ	10.52	

MAISÖL (MAISKEIMÖL) GEREINIGT	CORN OIL REFINED	HUILE DE MAÏS ÉPURÉE

		PROTEIN	FAT	CARBOHYDRATES	TOTAL
ENERGY VALUE (AVERAGE)	KJOULE	0.0	3891	0.0	3891
PER 100 G EDIBLE PORTION	(KCAL)	0.0	930	0.0	930
AMOUNT OF DIGESTIBLE CONSTITUENTS PER 100 G	GRAM	0.00	95.00	0.00	
ENERGY VALUE (AVERAGE)	KJOULE	0.0	3697	0.0	3697
OF THE DIGESTIBLE FRACTION PER 100 G EDIBLE PORTION	(KCAL)	0.0	884	0.0	884

WASTE PERCENTAGE AVERAGE 0.00

CONSTITUENTS	DIM	AV	VARIATION			AVR	NUTR. DENS.		MOLPERC.
WATER	-	0.00	-	-	-	0.00			
PROTEIN	-	0.00	-	-	-	0.00			
FAT	GRAM	100.00	-	-	-	100.00	GRAM/MJ	27.05	
AVAILABLE CARBOHYDR.	-	0.00	-	-	-	0.00			
SODIUM	MILLI	1.00	-	-	-	1.00	MILLI/MJ	0.27	
POTASSIUM	MILLI	1.00	0.00	-	2.00	1.00	MILLI/MJ	0.27	
CALCIUM	MILLI	15.00	14.00	-	16.00	15.00	MILLI/MJ	4.06	
IRON	MILLI	1.30	-	-	-	1.30	MILLI/MJ	0.35	
COPPER	MICRO	50.00	-	-	-	50.00	MICRO/MJ	13.53	
CAROTENE	MILLI	0.14	0.12	-	0.15	0.14	MILLI/MJ	0.04	
VITAMIN E ACTIVITY	MILLI	30.90	-	-	-	30.90	MILLI/MJ	8.36	
TOTAL TOCOPHEROLS	MILLI	84.00	-	-	-	84.00	MILLI/MJ	22.72	
ALPHA-TOCOPHEROL	MILLI	25.10	-	-	-	25.10	MILLI/MJ	6.79	
BETA-TOCOPHEROL	MILLI	0.65	-	-	-	0.65	MILLI/MJ	0.18	
GAMMA-TOCOPHEROL	MILLI	55.80	-	-	-	55.80	MILLI/MJ	15.10	
ALPHA-TOCOTRIENOL	MILLI	0.69	-	-	-	0.69	MILLI/MJ	0.19	
PALMITIC ACID	GRAM	10.00	7.30	-	18.20	10.00	GRAM/MJ	2.71	
STEARIC ACID	GRAM	2.40	0.40	-	3.80	2.40	GRAM/MJ	0.65	
ARACHIDIC ACID	GRAM	0.50	0.00	-	2.30	0.50	GRAM/MJ	0.14	
PALMITOLEIC ACID	GRAM	0.50	0.00	-	0.50	0.50	GRAM/MJ	0.14	
OLEIC ACID	GRAM	31.10	16.20	-	54.00	31.10	GRAM/MJ	8.41	
LINOLEIC ACID	GRAM	50.00	20.60	-	65.20	50.00	GRAM/MJ	13.53	
LINOLENIC ACID	GRAM	0.90	0.00	-	3.10	0.90	GRAM/MJ	0.24	
TOTAL STEROLS	MILLI	850.00	-	-	-	850.00	MILLI/MJ	229.94	
FREE STEROLS	MILLI	250.00	-	-	-	250.00	MILLI/MJ	67.63	
5-AVENASTEROL	MILLI	20.00	-	-	-	20.00	MILLI/MJ	5.41	
BRASSICASTEROL	MILLI	-	0.00	-	4.20				
CAMPESTEROL	MILLI	179.00	139.00	-	195.00	179.00	MILLI/MJ	48.42	
CHOLESTEROL	MILLI	-	TRACES	-	3.50				
BETA-SITOSTEROL	MILLI	595.00	560.00	-	630.00	595.00	MILLI/MJ	160.96	
STIGMASTEROL	MILLI	51.00	51.00	-	68.00	51.00	MILLI/MJ	13.80	

MOHNÖL · POPPY SEED OIL · HUILE DE PAVOT

		PROTEIN	FAT	CARBOHYDRATES	TOTAL
ENERGY VALUE (AVERAGE)	KJOULE	0.0	3872	0.0	3872
PER 100 G EDIBLE PORTION	(KCAL)	0.0	925	0.0	925
AMOUNT OF DIGESTIBLE CONSTITUENTS PER 100 G	GRAM	0.00	94.52	0.00	
ENERGY VALUE (AVERAGE)	KJOULE	0.0	3678	0.0	3678
OF THE DIGESTIBLE FRACTION PER 100 G EDIBLE PORTION	(KCAL)	0.0	879	0.0	879

WASTE PERCENTAGE AVERAGE 0.00

CONSTITUENTS	DIM	AV	VARIATION			AVR	NUTR. DENS.		MOLPERC.
FAT	GRAM	99.50	-	-	-	99.50	GRAM/MJ	27.05	
PALMITIC ACID	GRAM	9.50	8.70	-	10.60	9.50	GRAM/MJ	2.58	
STEARIC ACID	GRAM	1.90	1.60	-	2.00	1.90	GRAM/MJ	0.52	
PALMITOLEIC ACID	-	0.00	0.00	-	300.00	0.00			
OLEIC ACID	GRAM	10.50	-	-	-	10.50	GRAM/MJ	2.85	
LINOLEIC ACID	GRAM	72.40	-	-	-	72.40	GRAM/MJ	19.68	
LINOLENIC ACID	GRAM	1.00	0.40	-	4.80	1.00	GRAM/MJ	0.27	
TOTAL STEROLS	MILLI	263.00	250.00	-	276.00	263.00	MILLI/MJ	71.50	
BRASSICASTEROL	-	TRACES	-	-	-				
CAMPESTEROL	MILLI	58.00	51.00	-	58.00	58.00	MILLI/MJ	15.77	
CHOLESTEROL	-	TRACES	-	-	-				
BETA-SITOSTEROL	MILLI	173.00	167.00	-	179.00	173.00	MILLI/MJ	47.04	
STIGMASTEROL	MILLI	16.00	7.30	-	24.00	16.00	MILLI/MJ	4.35	

OLIVENÖL | OLIVE OIL | HUILE D'OLIVE

		PROTEIN	FAT	CARBOHYDRATES	TOTAL
ENERGY VALUE (AVERAGE)	KJOULE	0.0	3876	0.0	3876
PER 100 G	(KCAL)	0.0	926	0.0	926
EDIBLE PORTION					
AMOUNT OF DIGESTIBLE	GRAM	0.00	94.62	0.00	
CONSTITUENTS PER 100 G					
ENERGY VALUE (AVERAGE)	KJOULE	0.0	3682	0.0	3682
OF THE DIGESTIBLE	(KCAL)	0.0	880	0.0	880
FRACTION PER 100 G					
EDIBLE PORTION					

WASTE PERCENTAGE AVERAGE 0.00

CONSTITUENTS	DIM	AV	VARIATION			AVR	NUTR. DENS.		MOLPERC.
WATER	GRAM	0.20	0.00	-	0.40	0.20	GRAM/MJ	0.05	
PROTEIN	-	0.00	-	-	-	0.00			
FAT	GRAM	99.60	99.40	-	99.90	99.60	GRAM/MJ	27.05	
AVAILABLE CARBOHYDR.	GRAM	0.20	-	-	-	0.20	GRAM/MJ	0.05	
SODIUM	MILLI	1.00	0.10	-	2.00	1.00	MILLI/MJ	0.27	
POTASSIUM	MILLI	-	0.00	-	0.20				
CHLORIDE	MILLI	100.00	-	-	-	100.00	MILLI/MJ	27.16	
IODIDE	MICRO	5.00	-	-	-	5.00	MICRO/MJ	1.36	
VITAMIN A	MILLI	0.12	-	-	-	0.12	MILLI/MJ	0.03	
VITAMIN E ACTIVITY	MILLI	12.00	-	-	-	12.00	MILLI/MJ	3.26	
TOTAL TOCOPHEROLS	MILLI	13.70	-	-	-	13.70	MILLI/MJ	3.72	
ALPHA-TOCOPHEROL	MILLI	11.90	-	-	-	11.90	MILLI/MJ	3.23	
BETA-TOCOPHEROL	MILLI	0.10	-	-	-	0.10	MILLI/MJ	0.03	
GAMMA-TOCOPHEROL	MILLI	0.76	-	-	-	0.76	MILLI/MJ	0.21	
PALMITIC ACID	GRAM	10.80	6.50	-	14.90	10.80	GRAM/MJ	2.93	
STEARIC ACID	GRAM	2.40	1.30	-	3.10	2.40	GRAM/MJ	0.65	
ARACHIDIC ACID	GRAM	0.41	-	-	-	0.41	GRAM/MJ	0.11	
PALMITOLEIC ACID	GRAM	1.50	1.42	-	3.20	1.50	GRAM/MJ	0.41	
OLEIC ACID	GRAM	71.70	60.00	-	79.40	71.70	GRAM/MJ	19.47	
LINOLEIC ACID	GRAM	8.00	3.47	-	12.70	8.00	GRAM/MJ	2.17	
LINOLENIC ACID	GRAM	0.95	0.57	-	0.95	0.95	GRAM/MJ	0.26	
TOTAL STEROLS	MILLI	110.00	-	-	-	110.00	MILLI/MJ	29.88	
FREE STEROLS	MILLI	60.00	-	-	-	60.00	MILLI/MJ	16.30	
5-AVENASTEROL	MILLI	9.00	-	-	-	9.00	MILLI/MJ	2.44	
CAMPESTEROL	MILLI	2.20	-	-	-	2.20	MILLI/MJ	0.60	
BETA-SITOSTEROL	MILLI	102.00	-	-	-	102.00	MILLI/MJ	27.70	
STIGMASTEROL	MILLI	0.60	-	-	-	0.60	MILLI/MJ	0.16	

PALMKERNFETT
GEREINIGT

PALM KERNEL OIL
REFINED

HUILE DE PEPIN A PALME
ÉPURÉE

		PROTEIN	FAT	CARBOHYDRATES	TOTAL
ENERGY VALUE (AVERAGE)	KJOULE	0.0	3864	0.0	3864
PER 100 G	(KCAL)	0.0	923	0.0	923
EDIBLE PORTION					
AMOUNT OF DIGESTIBLE	GRAM	0.00	94.33	0.00	
CONSTITUENTS PER 100 G					
ENERGY VALUE (AVERAGE)	KJOULE	0.0	3671	0.0	3671
OF THE DIGESTIBLE	(KCAL)	0.0	877	0.0	877
FRACTION PER 100 G					
EDIBLE PORTION					

WASTE PERCENTAGE AVERAGE 0.00

CONSTITUENTS	DIM	AV	VARIATION			AVR	NUTR. DENS.		MOLPERC.
FAT	GRAM	99.30	-	-	-	99.30	GRAM/MJ	27.05	
TOTAL TOCOPHEROLS	MILLI	6.20	-	-	-	6.20	MILLI/MJ	1.69	
CAPRYLIC ACID	GRAM	5.70	2.90	-	5.70	5.70	GRAM/MJ	1.55	
CAPRIC ACID	GRAM	3.80	3.80	-	5.70	3.80	GRAM/MJ	1.04	
LAURIC ACID	GRAM	45.00	42.00	-	51.00	45.00	GRAM/MJ	12.26	
MYRISTIC ACID	GRAM	15.30	12.40	-	16.20	15.30	GRAM/MJ	4.17	
PALMITIC ACID	GRAM	7.60	7.60	-	8.60	7.60	GRAM/MJ	2.07	
STEARIC ACID	GRAM	2.40	1.00	-	2.90	2.40	GRAM/MJ	0.65	
OLEIC ACID	GRAM	13.40	10.10	-	17.70	13.40	GRAM/MJ	3.65	
LINOLEIC ACID	GRAM	2.40	0.70	-	5.80	2.40	GRAM/MJ	0.65	
TOTAL STEROLS	MILLI	80.00	-	-	-	80.00	MILLI/MJ	21.79	
BRASSICASTEROL	MILLI	3.60	-	-	-	3.60	MILLI/MJ	0.98	
CAMPESTEROL	MILLI	8.00	7.20	-	9.10	8.00	MILLI/MJ	2.18	
CHOLESTEROL	MILLI	2.00	0.40	-	2.40	2.00	MILLI/MJ	0.54	
BETA-SITOSTEROL	MILLI	57.00	-	-	-	57.00	MILLI/MJ	15.53	
STIGMASTEROL	MILLI	9.60	9.00	-	12.70	9.60	MILLI/MJ	2.62	

PALMÖL	PALM OIL	HUILE DE PALME
GEREINIGT	REFINED	ÉPURÉE

		PROTEIN	FAT	CARBOHYDRATES	TOTAL
ENERGY VALUE (AVERAGE)	KJOULE	0.0	3923	0.0	3923
PER 100 G	(KCAL)	0.0	938	0.0	938
EDIBLE PORTION					
AMOUNT OF DIGESTIBLE	GRAM	0.00	93.76	0.00	
CONSTITUENTS PER 100 G					
ENERGY VALUE (AVERAGE)	KJOULE	0.0	3727	0.0	3727
OF THE DIGESTIBLE	(KCAL)	0.0	891	0.0	891
FRACTION PER 100 G					
EDIBLE PORTION					

WASTE PERCENTAGE AVERAGE 0.00

CONSTITUENTS	DIM	AV	VARIATION	AVR	NUTR. DENS.		MOLPERC.
FAT	GRAM	98.70	- - -	98.70	GRAM/MJ	26.48	
CAROTENE	MILLI	56.50	3.00 - 110.00	56.50	MILLI/MJ	15.16	
VITAMIN E ACTIVITY	MILLI	21.80	- - -	21.80	MILLI/MJ	5.85	
TOTAL TOCOPHEROLS	MILLI	35.50	- - -	35.50	MILLI/MJ	9.53	
ALPHA-TOCOPHEROL	MILLI	18.30	- - -	18.30	MILLI/MJ	4.91	
ALPHA-TOCOTRIENOL	MILLI	11.49	- - -	11.49	MILLI/MJ	3.08	
GAMMA-TOCOTRIENOL	MILLI	5.75	- - -	5.75	MILLI/MJ	1.54	
MYRISTIC ACID	GRAM	0.95	0.40 - 5.60	0.95	GRAM/MJ	0.25	
PALMITIC ACID	GRAM	40.50	30.50 - 48.80	40.50	GRAM/MJ	10.87	
STEARIC ACID	GRAM	4.60	1.40 - 7.60	4.60	GRAM/MJ	1.23	
ARACHIDIC ACID	GRAM	0.50	TRACES - 0.50	0.50	GRAM/MJ	0.13	
PALMITOLEIC ACID	GRAM	0.50	0.10 - 0.50	0.50	GRAM/MJ	0.13	
OLEIC ACID	GRAM	37.40	34.20 - 49.20	37.40	GRAM/MJ	10.03	
LINOLEIC ACID	GRAM	10.50	4.80 - 12.00	10.50	GRAM/MJ	2.82	
LINOLENIC ACID	GRAM	0.50	0.10 - 0.50	0.50	GRAM/MJ	0.13	
TOTAL STEROLS	MILLI	40.00	30.00 - 60.00	40.00	MILLI/MJ	10.73	
FREE STEROLS	MILLI	30.00	- - -	30.00	MILLI/MJ	8.05	
CAMPESTEROL	MILLI	6.80	4.80 - 9.30	6.80	MILLI/MJ	1.82	
CHOLESTEROL	MILLI	0.80	0.10 - 1.60	0.80	MILLI/MJ	0.21	
BETA-SITOSTEROL	MILLI	28.00	23.00 - 30.00	28.00	MILLI/MJ	7.51	
STIGMASTEROL	MILLI	4.00	3.20 - 5.50	4.00	MILLI/MJ	1.07	

RÜBÖL	**RAPE SEED OIL**	**HUILE DE NAVETTE**
GEREINIGT	REFINED	ÉPURÉE

		PROTEIN	FAT	CARBOHYDRATES	TOTAL
ENERGY VALUE (AVERAGE)	KJOULE	0.0	3852	0.0	3852
PER 100 G	(KCAL)	0.0	921	0.0	921
EDIBLE PORTION					
AMOUNT OF DIGESTIBLE	GRAM	0.00	94.05	0.00	
CONSTITUENTS PER 100 G					
ENERGY VALUE (AVERAGE)	KJOULE	0.0	3660	0.0	3660
OF THE DIGESTIBLE	(KCAL)	0.0	875	0.0	875
FRACTION PER 100 G					
EDIBLE PORTION					

WASTE PERCENTAGE AVERAGE 0.00

CONSTITUENTS	DIM	AV	VARIATION			AVR	NUTR. DENS.		MOLPERC.
FAT	GRAM	99.00	-	-	-	99.00	GRAM/MJ	27.05	
SELENIUM	MICRO	3.50	-	-	-	3.50	MICRO/MJ	0.96	
CAROTENE	MILLI	3.30	-	-	-	3.30	MILLI/MJ	0.90	
VITAMIN E ACTIVITY	MILLI	15.30	-	-	-	15.30	MILLI/MJ	4.18	
TOTAL TOCOPHEROLS	MILLI	39.60	-	-	-	39.60	MILLI/MJ	10.82	
ALPHA-TOCOPHEROL	MILLI	12.80	-	-	-	12.80	MILLI/MJ	3.50	
GAMMA-TOCOPHEROL	MILLI	25.10	-	-	-	25.10	MILLI/MJ	6.86	
DELTA-TOCOPHEROL	MILLI	1.10	-	-	-	1.10	MILLI/MJ	0.30	
MYRISTIC ACID	GRAM	-	0.00	-	0.02				
PALMITIC ACID	GRAM	3.80	3.10	-	5.90	3.80	GRAM/MJ	1.04	
STEARIC ACID	GRAM	1.40	1.00	-	2.40	1.40	GRAM/MJ	0.38	
ARACHIDIC ACID	GRAM	0.50	0.50	-	1.90	0.50	GRAM/MJ	0.14	
BEHENIC ACID	GRAM	-	0.00	-	0.40				
LIGNOCERIC ACID	GRAM	0.60	0.47	-	0.76	0.60	GRAM/MJ	0.16	
PALMITOLEIC ACID	GRAM	0.50	0.20	-	0.50	0.50	GRAM/MJ	0.14	
OLEIC ACID	GRAM	60.10	50.10	-	60.20	60.10	GRAM/MJ	16.42	
EICOSENOIC ACID	GRAM	7.40	3.80	-	10.90	7.40	GRAM/MJ	2.02	
C22 : 1 FATTY ACID	GRAM	0.50	0.00	-	1.60	0.50	GRAM/MJ	0.14	
LINOLEIC ACID	GRAM	19.10	17.80	-	26.10	19.10	GRAM/MJ	5.22	
LINOLENIC ACID	GRAM	8.60	8.00	-	14.80	8.60	GRAM/MJ	2.35	
TOTAL STEROLS	MILLI	250.00	-	-	-	250.00	MILLI/MJ	68.31	
BRASSICASTEROL	MILLI	23.00	12.00	-	28.00	23.00	MILLI/MJ	6.28	
CAMPESTEROL	MILLI	94.80	93.20	-	96.30	94.80	MILLI/MJ	25.90	
CHOLESTEROL	MILLI	1.80	1.60	-	2.00	1.80	MILLI/MJ	0.49	
BETA-SITOSTEROL	MILLI	129.00	127.00	-	131.00	129.00	MILLI/MJ	35.25	
STIGMASTEROL	MILLI	3.90	3.00	-	4.70	3.90	MILLI/MJ	1.07	

SAFLORÖL GEREINIGT | SAFFLOWER OIL REFINED | HUILE DE CARTHAME ÉPURÉE

		PROTEIN	FAT	CARBOHYDRATES	TOTAL
ENERGY VALUE (AVERAGE)	KJOULE	0.0	3872	0.0	3872
PER 100 G	(KCAL)	0.0	925	0.0	925
EDIBLE PORTION					
AMOUNT OF DIGESTIBLE	GRAM	0.00	94.52	0.00	
CONSTITUENTS PER 100 G					
ENERGY VALUE (AVERAGE)	KJOULE	0.0	3678	0.0	3678
OF THE DIGESTIBLE	(KCAL)	0.0	879	0.0	879
FRACTION PER 100 G					
EDIBLE PORTION					

WASTE PERCENTAGE AVERAGE 0.00

CONSTITUENTS	DIM	AV	VARIATION			AVR	NUTR. DENS.		MOLPERC.
FAT	GRAM	99.50	-	-	-	99.50	GRAM/MJ	27.05	
VITAMIN E ACTIVITY	MILLI	34.50	-	-	-	34.50	MILLI/MJ	9.38	
TOTAL TOCOPHEROLS	MILLI	38.10	-	-	-	38.10	MILLI/MJ	10.36	
ALPHA-TOCOPHEROL	MILLI	34.10	-	-	-	34.10	MILLI/MJ	9.27	
GAMMA-TOCOPHEROL	MILLI	3.50	-	-	-	3.50	MILLI/MJ	0.95	
DELTA-TOCOPHEROL	MILLI	0.49	-	-	-	0.49	MILLI/MJ	0.13	
PALMITIC ACID	GRAM	5.70	-	-	-	5.70	GRAM/MJ	1.55	
STEARIC ACID	GRAM	2.40	-	-	-	2.40	GRAM/MJ	0.65	
ARACHIDIC ACID	GRAM	0.47	-	-	-	0.47	GRAM/MJ	0.13	
OLEIC ACID	GRAM	11.40	-	-	-	11.40	GRAM/MJ	3.10	
EICOSENOIC ACID	GRAM	0.47	-	-	-	0.47	GRAM/MJ	0.13	
LINOLEIC ACID	GRAM	74.00	-	-	-	74.00	GRAM/MJ	20.12	
LINOLENIC ACID	GRAM	0.47	-	-	-	0.47	GRAM/MJ	0.13	
TOTAL STEROLS	MILLI	444.00	-	-	-	444.00	MILLI/MJ	120.72	
5-AVENASTEROL	MILLI	5.00	-	-	-	5.00	MILLI/MJ	1.36	
CAMPESTEROL	MILLI	49.00	-	-	-	49.00	MILLI/MJ	13.32	
BETA-SITOSTEROL	MILLI	231.00	-	-	-	231.00	MILLI/MJ	62.80	
STIGMASTEROL	MILLI	40.00	-	-	-	40.00	MILLI/MJ	10.88	

SESAMÖL
GEREINIGT

SESAME SEED OIL
REFINED

HUILE DE SESAME
ÉPURÉE

		PROTEIN	FAT	CARBOHYDRATES	TOTAL
ENERGY VALUE (AVERAGE) PER 100 G EDIBLE PORTION	KJOULE	0.0	3872	0.0	3872
	(KCAL)	0.0	925	0.0	925
AMOUNT OF DIGESTIBLE CONSTITUENTS PER 100 G	GRAM	0.00	94.52	0.00	
ENERGY VALUE (AVERAGE) OF THE DIGESTIBLE FRACTION PER 100 G EDIBLE PORTION	KJOULE	0.0	3678	0.0	3678
	(KCAL)	0.0	879	0.0	879

WASTE PERCENTAGE AVERAGE 0.00

CONSTITUENTS	DIM	AV	VARIATION			AVR	NUTR. DENS.		MOLPERC.
FAT	GRAM	99.50	-	-	-	99.50	GRAM/MJ	27.05	
VITAMIN E ACTIVITY	MILLI	4.10	-	-	-	4.10	MILLI/MJ	1.11	
TOTAL TOCOPHEROLS	MILLI	29.10	-	-	-	29.10	MILLI/MJ	7.91	
ALPHA-TOCOPHEROL	MILLI	0.37	-	-	-	0.37	MILLI/MJ	0.10	
BETA-TOCOPHEROL	MILLI	1.40	-	-	-	1.40	MILLI/MJ	0.38	
GAMMA-TOCOPHEROL	MILLI	25.20	-	-	-	25.20	MILLI/MJ	6.85	
DELTA-TOCOPHEROL	MILLI	2.10	-	-	-	2.10	MILLI/MJ	0.57	
PALMITIC ACID	GRAM	8.10	6.70	-	14.30	8.10	GRAM/MJ	2.20	
STEARIC ACID	GRAM	4.30	TRACES	-	6.80	4.30	GRAM/MJ	1.17	
ARACHIDIC ACID	GRAM	0.50	-	-	-	0.50	GRAM/MJ	0.14	
OLEIC ACID	GRAM	40.10	31.40	-	51.40	40.10	GRAM/MJ	10.90	
LINOLEIC ACID	GRAM	42.50	32.00	-	56.00	42.50	GRAM/MJ	11.55	
LINOLENIC ACID	GRAM	-	0.00	-	1.90				
TOTAL STEROLS	MILLI	732.00	600.00	-	865.00	732.00	MILLI/MJ	199.02	
CAMPESTEROL	MILLI	164.00	128.00	-	202.00	164.00	MILLI/MJ	44.59	
CHOLESTEROL	MILLI	1.40	TRACES	-	5.90	1.40	MILLI/MJ	0.38	
BETA-SITOSTEROL	MILLI	430.00	364.00	-	526.00	430.00	MILLI/MJ	116.91	
STIGMASTEROL	MILLI	60.00	41.00	-	88.00	60.00	MILLI/MJ	16.31	

SHEAFETT (SHEABUTTER) | SHEA NUT OIL (SHEABUTTER) | HUILE DE SHEA NOIX

		PROTEIN	FAT	CARBOHYDRATES	TOTAL
ENERGY VALUE (AVERAGE)	KJOULE	0.0	3891	0.0	3891
PER 100 G EDIBLE PORTION	(KCAL)	0.0	930	0.0	930
AMOUNT OF DIGESTIBLE CONSTITUENTS PER 100 G	GRAM	0.00	95.00	0.00	
ENERGY VALUE (AVERAGE)	KJOULE	0.0	3697	0.0	3697
OF THE DIGESTIBLE FRACTION PER 100 G EDIBLE PORTION	(KCAL)	0.0	884	0.0	884

WASTE PERCENTAGE AVERAGE 0.00

CONSTITUENTS	DIM	AV	VARIATION	AVR	NUTR. DENS.		MOLPERC.
FAT	GRAM	100.00	- - -	100.00	GRAM/MJ	27.05	
CAPRYLIC ACID	GRAM	0.18	- - -	0.18	GRAM/MJ	0.05	
CAPRIC ACID	GRAM	0.18	- - -	0.18	GRAM/MJ	0.05	
LAURIC ACID	GRAM	1.40	- - -	1.40	GRAM/MJ	0.38	
MYRISTIC ACID	GRAM	0.50	- - -	0.50	GRAM/MJ	0.14	
PALMITIC ACID	GRAM	3.80	- - -	3.80	GRAM/MJ	1.03	
STEARIC ACID	GRAM	36.20	- - -	36.20	GRAM/MJ	9.79	
ARACHIDIC ACID	GRAM	1.00	- - -	1.00	GRAM/MJ	0.27	
PALMITOLEIC ACID	GRAM	0.10	- - -	0.10	GRAM/MJ	0.03	
OLEIC ACID	GRAM	45.20	- - -	45.20	GRAM/MJ	12.23	
LINOLEIC ACID	GRAM	6.70	- - -	6.70	GRAM/MJ	1.81	
LINOLENIC ACID	GRAM	0.30	- - -	0.30	GRAM/MJ	0.08	
TOTAL STEROLS	MILLI	357.00	- - -	357.00	MILLI/MJ	96.58	
24-METH.-CHOLESTEROL	MILLI	22.00	- - -	22.00	MILLI/MJ	5.95	

SOJAÖL
GEREINIGT

SOYBEAN OIL
REFINED

HUILE DE SOYA
ÉPURÉE

		PROTEIN	FAT	CARBOHYDRATES	TOTAL
ENERGY VALUE (AVERAGE) PER 100 G EDIBLE PORTION	KJOULE	0.0	3837	0.0	3837
	(KCAL)	0.0	917	0.0	917
AMOUNT OF DIGESTIBLE CONSTITUENTS PER 100 G	GRAM	0.00	93.67	0.00	
ENERGY VALUE (AVERAGE) OF THE DIGESTIBLE FRACTION PER 100 G EDIBLE PORTION	KJOULE	0.0	3645	0.0	3645
	(KCAL)	0.0	871	0.0	871

WASTE PERCENTAGE AVERAGE 0.00

CONSTITUENTS	DIM	AV	VARIATION			AVR	NUTR. DENS.		MOLPERC.
FAT	GRAM	98.60	-	-	-	98.60	GRAM/MJ	27.05	
ZINC	MILLI	0.20	-	-	-	0.20	MILLI/MJ	0.05	
CAROTENE	MILLI	3.50	-	-	-	3.50	MILLI/MJ	0.96	
VITAMIN E ACTIVITY	MILLI	14.60	-	-	-	14.60	MILLI/MJ	4.01	
TOTAL TOCOPHEROLS	MILLI	93.70	-	-	-	93.70	MILLI/MJ	25.71	
ALPHA-TOCOPHEROL	MILLI	7.50	-	-	-	7.50	MILLI/MJ	2.06	
BETA-TOCOPHEROL	MILLI	0.90	-	-	-	0.90	MILLI/MJ	0.25	
GAMMA-TOCOPHEROL	MILLI	66.20	-	-	-	66.20	MILLI/MJ	18.16	
DELTA-TOCOPHEROL	MILLI	19.10	-	-	-	19.10	MILLI/MJ	5.24	
PALMITIC ACID	GRAM	9.50	6.70	-	14.50	9.50	GRAM/MJ	2.61	
STEARIC ACID	GRAM	3.40	0.50	-	8.90	3.40	GRAM/MJ	0.93	
ARACHIDIC ACID	GRAM	0.50	0.10	-	0.90	0.50	GRAM/MJ	0.14	
PALMITOLEIC ACID	GRAM	0.50	-	-	-	0.50	GRAM/MJ	0.14	
OLEIC ACID	GRAM	20.10	14.30	-	28.70	20.10	GRAM/MJ	5.51	
LINOLEIC ACID	GRAM	53.40	36.50	-	57.80	53.40	GRAM/MJ	14.65	
LINOLENIC ACID	GRAM	7.60	1.90	-	14.70	7.60	GRAM/MJ	2.09	
TOTAL STEROLS	MILLI	340.00	-	-	-	340.00	MILLI/MJ	93.28	
FREE STEROLS	MILLI	320.00	-	-	-	320.00	MILLI/MJ	87.80	
BRASSICASTEROL	MILLI	-	0.00	-	1.70				
CAMPESTEROL	MILLI	64.60	44.20	-	81.80	64.60	MILLI/MJ	17.72	
CHOLESTEROL	MILLI	2.40	1.00	-	3.70	2.40	MILLI/MJ	0.66	
BETA-SITOSTEROL	MILLI	194.00	173.00	-	230.00	194.00	MILLI/MJ	53.23	
STIGMASTEROL	MILLI	71.40	38.30	-	81.60	71.40	MILLI/MJ	19.59	

SONNENBLUMENÖL GEREINIGT | SUNFLOWER OIL REFINED | HUILE DE TOURNESOL ÉPURÉE

		PROTEIN	FAT	CARBOHYDRATES	TOTAL
ENERGY VALUE (AVERAGE)	KJOULE	0.0	3883	0.0	3883
PER 100 G	(KCAL)	0.0	928	0.0	928
EDIBLE PORTION					
AMOUNT OF DIGESTIBLE	GRAM	0.00	94.81	0.00	
CONSTITUENTS PER 100 G					
ENERGY VALUE (AVERAGE)	KJOULE	0.0	3689	0.0	3689
OF THE DIGESTIBLE	(KCAL)	0.0	882	0.0	882
FRACTION PER 100 G					
EDIBLE PORTION					

WASTE PERCENTAGE AVERAGE 0.00

CONSTITUENTS	DIM	AV	VARIATION			AVR	NUTR. DENS.		MOLPERC.
WATER	GRAM	0.20	-	-	-	0.20	GRAM/MJ	0.05	
PROTEIN	-	0.00	-	-	-	0.00			
FAT	GRAM	99.80	-	-	-	99.80	GRAM/MJ	27.05	
IRON	MICRO	30.00	-	-	-	30.00	MICRO/MJ	8.13	
COPPER	MICRO	0.70	0.40	-	1.00	0.70	MICRO/MJ	0.19	
VITAMIN A	-	0.00	-	-	-	0.00			
CAROTENE	MICRO	26.00	3.00	-	150.00	26.00	MICRO/MJ	7.05	
VITAMIN E ACTIVITY	MILLI	55.80	-	-	-	55.80	MILLI/MJ	15.13	
TOTAL TOCOPHEROLS	MILLI	59.50	-	-	-	59.50	MILLI/MJ	16.13	
ALPHA-TOCOPHEROL	MILLI	55.00	-	-	-	55.00	MILLI/MJ	14.91	
BETA-TOCOPHEROL	MILLI	1.30	-	-	-	1.30	MILLI/MJ	0.35	
GAMMA-TOCOPHEROL	MILLI	3.20	-	-	-	3.20	MILLI/MJ	0.87	
PALMITIC ACID	GRAM	6.20	2.90	-	13.30	6.20	GRAM/MJ	1.68	
STEARIC ACID	GRAM	4.80	1.00	-	9.60	4.80	GRAM/MJ	1.30	
ARACHIDIC ACID	GRAM	0.50	0.10	-	0.80	0.50	GRAM/MJ	0.14	
PALMITOLEIC ACID	GRAM	0.50	0.00	-	0.50	0.50	GRAM/MJ	0.14	
OLEIC ACID	GRAM	21.90	13.30	-	68.80	21.90	GRAM/MJ	5.94	
LINOLEIC ACID	GRAM	60.20	19.10	-	74.80	60.20	GRAM/MJ	16.32	
LINOLENIC ACID	GRAM	0.50	0.06	-	1.90	0.50	GRAM/MJ	0.14	
TOTAL STEROLS	MILLI	350.00	-	-	-	350.00	MILLI/MJ	94.87	
FREE STEROLS	MILLI	160.00	-	-	-	160.00	MILLI/MJ	43.37	
5-AVENASTEROL	MILLI	14.00	-	-	-	14.00	MILLI/MJ	3.79	
BRASSICASTEROL	-	TRACES	-	-	-				
CAMPESTEROL	MILLI	31.50	28.00	-	47.60	31.50	MILLI/MJ	8.54	
CHOLESTEROL	MILLI	-	TRACES	-	3.20				
BETA-SITOSTEROL	MILLI	210.00	200.00	-	235.00	210.00	MILLI/MJ	56.92	
STIGMASTEROL	MILLI	34.50	28.00	-	41.00	34.50	MILLI/MJ	9.35	

TRAUBENKERNÖL GRAPE SEED OIL HUILE DE NOYAU DE RAISIN

		PROTEIN	FAT	CARBOHYDRATES	TOTAL
ENERGY VALUE (AVERAGE) PER 100 G EDIBLE PORTION	KJOULE	0.0	3872	0.0	3872
	(KCAL)	0.0	925	0.0	925
AMOUNT OF DIGESTIBLE CONSTITUENTS PER 100 G	GRAM	0.00	94.52	0.00	
ENERGY VALUE (AVERAGE) OF THE DIGESTIBLE FRACTION PER 100 G EDIBLE PORTION	KJOULE	0.0	3678	0.0	3678
	(KCAL)	0.0	879	0.0	879

WASTE PERCENTAGE AVERAGE 0.00

CONSTITUENTS	DIM	AV	VARIATION			AVR	NUTR.	DENS.	MOLPERC.
FAT	GRAM	99.50	99.00	-	100.00	99.50	GRAM/MJ	27.05	
VITAMIN E ACTIVITY	MILLI	31.90	-	-	-	31.90	MILLI/MJ	8.67	
TOTAL TOCOPHEROLS	MILLI	61.60	-	-	-	61.60	MILLI/MJ	16.75	
ALPHA-TOCOPHEROL	MILLI	28.80	-	-	-	28.80	MILLI/MJ	7.83	
GAMMA-TOCOPHEROL	MILLI	30.80	-	-	-	30.80	MILLI/MJ	8.37	
DELTA-TOCOPHEROL	MILLI	2.00	-	-	-	2.00	MILLI/MJ	0.54	
PALMITIC ACID	GRAM	5.90	2.10	-	6.70	5.90	GRAM/MJ	1.60	
STEARIC ACID	GRAM	3.00	2.10	-	3.80	3.00	GRAM/MJ	0.82	
PALMITOLEIC ACID	GRAM	0.48	-	-	-	0.48	GRAM/MJ	0.13	
OLEIC ACID	GRAM	16.20	15.20	-	29.50	16.20	GRAM/MJ	4.40	
LINOLEIC ACID	GRAM	65.60	40.90	-	67.50	65.60	GRAM/MJ	17.84	
LINOLENIC ACID	GRAM	0.48	-	-	-	0.48	GRAM/MJ	0.13	
TOTAL STEROLS	MILLI	130.00	-	-	-	130.00	MILLI/MJ	35.34	
CAMPESTEROL	MILLI	13.00	-	-	-	13.00	MILLI/MJ	3.53	
CHOLESTEROL	-	TRACES	-	-	-				
BETA-SITOSTEROL	MILLI	98.00	-	-	-	98.00	MILLI/MJ	26.64	
STIGMASTEROL	MILLI	15.00	-	-	-	15.00	MILLI/MJ	4.08	

WALNUSSÖL WALNUT OIL HUILE DE NOIX

		PROTEIN	FAT	CARBOHYDRATES	TOTAL
ENERGY VALUE (AVERAGE)	KJOULE	0.0	3872	0.0	3872
PER 100 G	(KCAL)	0.0	925	0.0	925
EDIBLE PORTION					
AMOUNT OF DIGESTIBLE	GRAM	0.00	94.52	0.00	
CONSTITUENTS PER 100 G					
ENERGY VALUE (AVERAGE)	KJOULE	0.0	3678	0.0	3678
OF THE DIGESTIBLE	(KCAL)	0.0	879	0.0	879
FRACTION PER 100 G					
EDIBLE PORTION					

WASTE PERCENTAGE AVERAGE 0.00

CONSTITUENTS	DIM	AV	VARIATION		AVR	NUTR. DENS.		MOLPERC.
FAT	GRAM	99.50	-	-	99.50	GRAM/MJ	27.05	
VITAMIN E ACTIVITY	MILLI	3.20	-	-	3.20	MILLI/MJ	0.87	
TOTAL TOCOPHEROLS	MILLI	32.04	-	-	32.04	MILLI/MJ	8.71	
ALPHA-TOCOPHEROL	MILLI	0.44	-	-	0.44	MILLI/MJ	0.12	
GAMMA-TOCOPHEROL	MILLI	27.80	-	-	27.80	MILLI/MJ	7.56	
DELTA-TOCOPHEROL	MILLI	3.80	-	-	3.80	MILLI/MJ	1.03	
PALMITIC ACID	GRAM	6.70	5.10	- 10.50	6.70	GRAM/MJ	1.82	
STEARIC ACID	GRAM	1.90	0.00	- 4.80	1.90	GRAM/MJ	0.52	
OLEIC ACID	GRAM	15.70	9.90	- 37.50	15.70	GRAM/MJ	4.27	
LINOLEIC ACID	GRAM	57.50	40.60	- 72.60	57.50	GRAM/MJ	15.63	
LINOLENIC ACID	GRAM	13.40	4.80	- 15.40	13.40	GRAM/MJ	3.64	
TOTAL STEROLS	MILLI	176.00	-	-	176.00	MILLI/MJ	47.85	
CAMPESTEROL	MILLI	9.00	7.00	- 11.00	9.00	MILLI/MJ	2.45	
CHOLESTEROL	MILLI	-	0.70	- 1.90				
BETA-SITOSTEROL	MILLI	155.00	131.00	- 164.00	155.00	MILLI/MJ	42.14	

WEIZENKEIMÖL — WHEAT GERM OIL — HUILE DE GERMES DE BLÉ

		PROTEIN	FAT	CARBOHYDRATES	TOTAL
ENERGY VALUE (AVERAGE)	KJOULE	0.0	3872	0.0	3872
PER 100 G	(KCAL)	0.0	925	0.0	925
EDIBLE PORTION					
AMOUNT OF DIGESTIBLE	GRAM	0.00	94.52	0.00	
CONSTITUENTS PER 100 G					
ENERGY VALUE (AVERAGE)	KJOULE	0.0	3678	0.0	3678
OF THE DIGESTIBLE	(KCAL)	0.0	879	0.0	879
FRACTION PER 100 G					
EDIBLE PORTION					

WASTE PERCENTAGE AVERAGE 0.00

CONSTITUENTS	DIM	AV	VARIATION			AVR	NUTR.	DENS.	MOLPERC.
FAT	GRAM	99.50	-	-	-	99.50	GRAM/MJ	27.05	
VITAMIN E ACTIVITY	MILLI	215.40	-	-	-	215.40	MILLI/MJ	58.56	
TOTAL TOCOPHEROLS	MILLI	280.00	-	-	-	280.00	MILLI/MJ	76.13	
ALPHA-TOCOPHEROL	MILLI	192.00	-	-	-	192.00	MILLI/MJ	52.20	
BETA-TOCOPHEROL	MILLI	50.80	-	-	-	50.80	MILLI/MJ	13.81	
GAMMA-TOCOPHEROL	MILLI	30.40	-	-	-	30.40	MILLI/MJ	8.27	
DELTA-TOCOPHEROL	MILLI	4.10	-	-	-	4.10	MILLI/MJ	1.11	
PALMITIC ACID	GRAM	16.60	15.70	-	19.00	16.60	GRAM/MJ	4.51	
STEARIC ACID	GRAM	0.60	0.30	-	0.80	0.60	GRAM/MJ	0.16	
OLEIC ACID	GRAM	14.70	11.70	-	15.90	14.70	GRAM/MJ	4.00	
EICOSENOIC ACID	GRAM	0.95	-	-	-	0.95	GRAM/MJ	0.26	
LINOLEIC ACID	GRAM	55.80	48.40	-	55.80	55.80	GRAM/MJ	15.17	
LINOLENIC ACID	GRAM	8.90	6.70	-	11.70	8.90	GRAM/MJ	2.42	
TOTAL STEROLS	MILLI	553.00	-	-	-	553.00	MILLI/MJ	150.35	
5-AVENASTEROL	MILLI	33.00	-	-	-	33.00	MILLI/MJ	8.97	
CAMPESTEROL	MILLI	122.00	-	-	-	122.00	MILLI/MJ	33.17	
CHOLESTEROL	-	TRACES	-	-	-				
BETA-SITOSTEROL	MILLI	370.00	-	-	-	370.00	MILLI/MJ	100.60	
STIGMASTEROL	-	TRACES	-	-	-				

Margarine

MARGARINES **MARGARINE**

MARGARINE
STANDARDMARGARINE
1, 2

MARGARINE
MARGARINE, STANDARD
1, 2

MARGARINE
„STANDARDMARGARINE"
1, 2

		PROTEIN	FAT	CARBOHYDRATES	TOTAL
ENERGY VALUE (AVERAGE)	KJOULE	3.7	3113	6.7	3123
PER 100 G	(KCAL)	0.9	744	1.6	746
EDIBLE PORTION					
AMOUNT OF DIGESTIBLE	GRAM	0.19	76.00	0.40	
CONSTITUENTS PER 100 G					
ENERGY VALUE (AVERAGE)	KJOULE	3.6	2957	6.7	2968
OF THE DIGESTIBLE	(KCAL)	0.9	707	1.6	709
FRACTION PER 100 G					
EDIBLE PORTION					

WASTE PERCENTAGE AVERAGE 0.00

CONSTITUENTS	DIM	AV	VARIATION			AVR	NUTR. DENS.		MOLPERC.
WATER	GRAM	19.15	-	-	-	19.15	GRAM/MJ	6.45	
PROTEIN	GRAM	0.20	-	-	-	0.20	GRAM/MJ	0.07	
FAT	GRAM	80.00	-	-	-	80.00	GRAM/MJ	26.96	
AVAILABLE CARBOHYDR.	GRAM	0.40	-	-	-	0.40	GRAM/MJ	0.13	
MINERALS	GRAM	0.25	0.19	-	0.38	0.25	GRAM/MJ	0.08	
SODIUM	MILLI	101.00	-	-	-	101.00	MILLI/MJ	34.04	
CHLORIDE	MILLI	158.00	-	-	-	158.00	MILLI/MJ	53.24	
VITAMIN A	MILLI	0.53	0.42	-	0.75	0.53	MILLI/MJ	0.18	
CAROTENE	MILLI	0.65	-	-	-	0.65	MILLI/MJ	0.22	
VITAMIN D	MICRO	2.50	-	-	-	2.50	MICRO/MJ	0.84	
VITAMIN E ACTIVITY	MILLI	10.00	-	-	-	10.00	MILLI/MJ	3.37	
CAPRYLIC ACID	MILLI	80.00	-	-	-	80.00	MILLI/MJ	26.96	
CAPRIC ACID	MILLI	80.00	-	-	-	80.00	MILLI/MJ	26.96	
LAURIC ACID	GRAM	0.46	-	-	-	0.46	GRAM/MJ	0.16	
MYRISTIC ACID	GRAM	2.00	-	-	-	2.00	GRAM/MJ	0.67	
PALMITIC ACID	GRAM	12.20	-	-	-	12.20	GRAM/MJ	4.11	
STEARIC ACID	GRAM	7.80	-	-	-	7.80	GRAM/MJ	2.63	
ARACHIDIC ACID	GRAM	0.76	-	-	-	0.76	GRAM/MJ	0.26	
BEHENIC ACID	GRAM	0.30	-	-	-	0.30	GRAM/MJ	0.10	
LIGNOCERIC ACID	MILLI	80.00	-	-	-	80.00	MILLI/MJ	26.96	
PALMITOLEIC ACID	GRAM	2.70	-	-	-	2.70	GRAM/MJ	0.91	
OLEIC ACID	GRAM	26.80	-	-	-	26.80	GRAM/MJ	9.03	
EICOSENOIC ACID	GRAM	1.50	-	-	-	1.50	GRAM/MJ	0.51	
C22 : 1 FATTY ACID	GRAM	1.50	-	-	-	1.50	GRAM/MJ	0.51	
LINOLEIC ACID	GRAM	17.60	9.50	-	26.50	17.60	GRAM/MJ	5.93	

1 COOKING-MARGARINE USED IN PUBLIC KITCHENS

2 COMMERCIAL PRODUCTS IN THE FEDERAL REPUBLIC OF GERMANY

CONSTITUENTS	DIM	AV	VARIATION		AVR	NUTR. DENS.		MOLPERC.
LINOLENIC ACID	GRAM	1.90	TRACES	- 5.20	1.90	GRAM/MJ	0.64	
TOTAL STEROLS	MILLI	320.00	160.00	- 400.00	320.00	MILLI/MJ	107.83	
BRASSICASTEROL	MILLI	8.00	0.00	- 12.00	8.00	MILLI/MJ	2.70	
CAMPESTEROL	MILLI	51.00	21.00	- 72.00	51.00	MILLI/MJ	17.19	
CHOLESTEROL	MILLI	115.00	84.00	- 146.00	115.00	MILLI/MJ	38.75	
BETA-SITOSTEROL	MILLI	118.00	112.00	- 127.00	118.00	MILLI/MJ	39.76	
STIGMASTEROL	MILLI	13.00	12.00	- 15.00	13.00	MILLI/MJ	4.38	

MARGARINE
PFLANZENMARGARINE

1

MARGARINE
VEGETABLE OIL MARGARINE

1

MARGARINE
MARGARINE D'ORIGINE VÉGÉTALE

1

		PROTEIN	FAT	CARBOHYDRATES	TOTAL
ENERGY VALUE (AVERAGE) PER 100 G EDIBLE PORTION	KJOULE	3.7	3113	6.7	3123
	(KCAL)	0.9	744	1.6	746
AMOUNT OF DIGESTIBLE CONSTITUENTS PER 100 G	GRAM	0.19	76.00	0.40	
ENERGY VALUE (AVERAGE) OF THE DIGESTIBLE FRACTION PER 100 G EDIBLE PORTION	KJOULE	3.6	2957	6.7	2968
	(KCAL)	0.9	707	1.6	709

WASTE PERCENTAGE AVERAGE 0.00

CONSTITUENTS	DIM	AV	VARIATION	AVR	NUTR. DENS.		MOLPERC.
WATER	GRAM	19.10	- - -	19.10	GRAM/MJ	6.44	
PROTEIN	GRAM	0.20	- - -	0.20	GRAM/MJ	0.07	
FAT	GRAM	80.00	- - -	80.00	GRAM/MJ	26.96	
AVAILABLE CARBOHYDR.	GRAM	0.40	- - -	0.40	GRAM/MJ	0.13	
MINERALS	GRAM	0.26	- - -	0.26	GRAM/MJ	0.09	
SODIUM	MILLI	101.00	- - -	101.00	MILLI/MJ	34.04	
CHLORIDE	MILLI	158.00	- - -	158.00	MILLI/MJ	53.24	
VITAMIN A	MILLI	0.50	- - -	0.50	MILLI/MJ	0.17	
CAROTENE	MILLI	0.65	- - -	0.65	MILLI/MJ	0.22	
VITAMIN D	MICRO	2.50	- - -	2.50	MICRO/MJ	0.84	
VITAMIN E ACTIVITY	MILLI	16.00	- - -	16.00	MILLI/MJ	5.39	
CAPRYLIC ACID	GRAM	0.31	- - -	0.31	GRAM/MJ	0.10	
CAPRIC ACID	GRAM	0.38	- - -	0.38	GRAM/MJ	0.13	
LAURIC ACID	GRAM	3.40	- - -	3.40	GRAM/MJ	1.15	
MYRISTIC ACID	GRAM	1.80	- - -	1.80	GRAM/MJ	0.61	
PALMITIC ACID	GRAM	10.70	- - -	10.70	GRAM/MJ	3.61	
STEARIC ACID	GRAM	4.90	- - -	4.90	GRAM/MJ	1.65	
ARACHIDIC ACID	GRAM	0.23	- - -	0.23	GRAM/MJ	0.08	
OLEIC ACID	GRAM	27.50	- - -	27.50	GRAM/MJ	9.27	
EICOSENOIC ACID	GRAM	0.46	- - -	0.46	GRAM/MJ	0.16	
C22 : 1 FATTY ACID	GRAM	0.99	- - -	0.99	GRAM/MJ	0.33	
LINOLEIC ACID	GRAM	23.10	- - -	23.10	GRAM/MJ	7.78	
LINOLENIC ACID	GRAM	2.40	- - -	2.40	GRAM/MJ	0.81	
TOTAL STEROLS	MILLI	310.00	230.00 - 390.00	310.00	MILLI/MJ	104.46	
BRASSICASTEROL	MILLI	14.00	3.10 - 23.00	14.00	MILLI/MJ	4.72	
CAMPESTEROL	MILLI	74.00	59.00 - 82.00	74.00	MILLI/MJ	24.94	
CHOLESTEROL	MILLI	7.40	4.70 - 16.00	7.40	MILLI/MJ	2.49	
BETA-SITOSTEROL	MILLI	173.00	166.00 - 183.00	173.00	MILLI/MJ	58.30	
STIGMASTEROL	MILLI	26.00	17.00 - 43.00	26.00	MILLI/MJ	8.76	

1 COMMERCIAL PRODUCTS IN THE FEDERAL REPUBLIC OF GERMANY

MARGARINE DIÄTMARGARINE 1	MARGARINE DIETETIC MARGARINE 1	MARGARINE MARGARINE DIETETIQUE 1

		PROTEIN	FAT	CARBOHYDRATES	TOTAL
ENERGY VALUE (AVERAGE) PER 100 G EDIBLE PORTION	KJOULE	3.7	3113	3.3	3120
	(KCAL)	0.9	744	0.8	746
AMOUNT OF DIGESTIBLE CONSTITUENTS PER 100 G	GRAM	0.19	76.00	0.20	
ENERGY VALUE (AVERAGE) OF THE DIGESTIBLE FRACTION PER 100 G EDIBLE PORTION	KJOULE	3.6	2957	3.3	2964
	(KCAL)	0.9	707	0.8	708

WASTE PERCENTAGE AVERAGE 0.00

CONSTITUENTS	DIM	AV	VARIATION			AVR	NUTR.	DENS.	MOLPERC.
WATER	GRAM	19.10	-	-	-	19.10	GRAM/MJ	6.44	
PROTEIN	GRAM	0.20	-	-	-	0.20	GRAM/MJ	0.07	
FAT	GRAM	80.00	-	-	-	80.00	GRAM/MJ	26.99	
AVAILABLE CARBOHYDR.	GRAM	0.20	-	-	-	0.20	GRAM/MJ	0.07	
SODIUM	MILLI	39.00	-	-	-	39.00	MILLI/MJ	13.16	
CHLORIDE	MILLI	62.00	-	-	-	62.00	MILLI/MJ	20.92	
VITAMIN A	MILLI	0.50	-	-	-	0.50	MILLI/MJ	0.17	
CAROTENE	MILLI	0.20	-	-	-	0.20	MILLI/MJ	0.07	
VITAMIN D	MICRO	2.50	-	-	-	2.50	MICRO/MJ	0.84	
VITAMIN E ACTIVITY	MILLI	67.00	-	-	-	67.00	MILLI/MJ	22.60	
CAPRYLIC ACID	GRAM	0.15	-	-	-	0.15	GRAM/MJ	0.05	
CAPRIC ACID	GRAM	0.15	-	-	-	0.15	GRAM/MJ	0.05	
LAURIC ACID	GRAM	2.20	-	-	-	2.20	GRAM/MJ	0.74	
PALMITIC ACID	GRAM	6.70	-	-	-	6.70	GRAM/MJ	2.26	
STEARIC ACID	GRAM	7.50	-	-	-	7.50	GRAM/MJ	2.53	
ARACHIDIC ACID	GRAM	0.45	-	-	-	0.45	GRAM/MJ	0.15	
BEHENIC ACID	GRAM	0.61	-	-	-	0.61	GRAM/MJ	0.21	
PALMITOLEIC ACID	MILLI	80.00	-	-	-	80.00	MILLI/MJ	26.99	
OLEIC ACID	GRAM	10.90	-	-	-	10.90	GRAM/MJ	3.68	
LINOLEIC ACID	GRAM	46.30	-	-	-	46.30	GRAM/MJ	15.62	
LINOLENIC ACID	GRAM	0.40	-	-	-	0.40	GRAM/MJ	0.13	
TOTAL STEROLS	MILLI	380.00	-	-	-	380.00	MILLI/MJ	128.20	
BRASSICASTEROL	MILLI	0.40	-	-	-	0.40	MILLI/MJ	0.13	
CAMPESTEROL	MILLI	13.00	-	-	-	13.00	MILLI/MJ	4.39	
CHOLESTEROL	MILLI	1.40	-	-	-	1.40	MILLI/MJ	0.47	
BETA-SITOSTEROL	MILLI	247.00	-	-	-	247.00	MILLI/MJ	83.33	
STIGMASTEROL	MILLI	30.00	-	-	-	30.00	MILLI/MJ	10.12	

1 COMMERCIAL PRODUCTS IN THE FEDERAL REPUBLIC OF GERMANY

MARGARINE	MARGARINE	MARGARINE
HALBFETTMARGARINE	MARGARINE SEMI-FAT	MARGARINE DEMI-GRASSE
1	1	1

		PROTEIN	FAT	CARBOHYDRATES	TOTAL
ENERGY VALUE (AVERAGE)	KJOULE	29	1556	6.7	1593
PER 100 G	(KCAL)	7.0	372	1.6	381
EDIBLE PORTION					
AMOUNT OF DIGESTIBLE	GRAM	1.55	38.00	0.40	
CONSTITUENTS PER 100 G					
ENERGY VALUE (AVERAGE)	KJOULE	29	1479	6.7	1514
OF THE DIGESTIBLE	(KCAL)	6.8	353	1.6	362
FRACTION PER 100 G					
EDIBLE PORTION					

WASTE PERCENTAGE AVERAGE 0.00

CONSTITUENTS	DIM	AV	VARIATION			AVR	NUTR. DENS.		MOLPERC.
WATER	GRAM	57.90	-	-	-	57.90	GRAM/MJ	38.25	
PROTEIN	GRAM	1.60	-	-	-	1.60	GRAM/MJ	1.06	
FAT	GRAM	40.00	-	-	-	40.00	GRAM/MJ	26.42	
AVAILABLE CARBOHYDR.	GRAM	0.40	-	-	-	0.40	GRAM/MJ	0.26	
MINERALS	GRAM	1.15	-	-	-	1.15	GRAM/MJ	0.76	
SODIUM	MILLI	390.00	-	-	-	390.00	MILLI/MJ	257.61	
POTASSIUM	MILLI	7.00	-	-	-	7.00	MILLI/MJ	4.62	
MAGNESIUM	MILLI	0.50	-	-	-	0.50	MILLI/MJ	0.33	
CALCIUM	MILLI	12.00	10.00	-	15.00	12.00	MILLI/MJ	7.93	
MANGANESE	-	TRACES	-	-	-				
IRON	MICRO	30.00	-	-	-	30.00	MICRO/MJ	19.82	
COPPER	MICRO	3.00	-	-	-	3.00	MICRO/MJ	1.98	
PHOSPHORUS	MILLI	8.00	-	-	-	8.00	MILLI/MJ	5.28	
CHLORIDE	MILLI	615.00	-	-	-	615.00	MILLI/MJ	406.24	
VITAMIN A	MILLI	0.50	-	-	-	0.50	MILLI/MJ	0.33	
CAROTENE	MILLI	0.50	-	-	-	0.50	MILLI/MJ	0.33	
VITAMIN D	MICRO	2.50	-	-	-	2.50	MICRO/MJ	1.65	
VITAMIN E ACTIVITY	MILLI	6.00	-	-	-	6.00	MILLI/MJ	3.96	
CAPRYLIC ACID	MILLI	40.00	-	-	-	40.00	MILLI/MJ	26.42	
CAPRIC ACID	MILLI	40.00	-	-	-	40.00	MILLI/MJ	26.42	
LAURIC ACID	GRAM	0.80	-	-	-	0.80	GRAM/MJ	0.53	
MYRISTIC ACID	GRAM	0.42	-	-	-	0.42	GRAM/MJ	0.28	
PALMITIC ACID	GRAM	5.60	-	-	-	5.60	GRAM/MJ	3.70	
STEARIC ACID	GRAM	3.20	-	-	-	3.20	GRAM/MJ	2.11	
ARACHIDIC ACID	GRAM	0.19	-	-	-	0.19	GRAM/MJ	0.13	
BEHENIC ACID	GRAM	0.15	-	-	-	0.15	GRAM/MJ	0.10	
OLEIC ACID	GRAM	10.10	-	-	-	10.10	GRAM/MJ	6.67	
LINOLEIC ACID	GRAM	15.30	-	-	-	15.30	GRAM/MJ	10.11	
LINOLENIC ACID	GRAM	2.20	-	-	-	2.20	GRAM/MJ	1.45	
TOTAL STEROLS	MILLI	140.00	-	-	-	140.00	MILLI/MJ	92.48	
BRASSICASTEROL	MILLI	1.00	-	-	-	1.00	MILLI/MJ	0.66	
CAMPESTEROL	MILLI	27.00	-	-	-	27.00	MILLI/MJ	17.83	
CHOLESTEROL	MILLI	4.20	-	-	-	4.20	MILLI/MJ	2.77	
BETA-SITOSTEROL	MILLI	78.00	-	-	-	78.00	MILLI/MJ	51.52	
STIGMASTEROL	MILLI	18.00	-	-	-	18.00	MILLI/MJ	11.89	

1 COMMERCIAL PRODUCTS IN THE FEDERAL REPUBLIC OF GERMANY

Fleisch

Meat
Viande

Fleisch und Innereien von Schlachttieren

MEAT AND ORGANS OF SLAUGHTERED ANIMALS

VIANDE ET ABATS À ANIMAUX DE BOUCHERIE

HAMMELFLEISCH
MUSKELFLEISCH (FILET)

MUTTON
MUSCLE ONLY (FILLET)

VIANDE DE MOUTON
VIANDE DE MUSCLE (FILET)

		PROTEIN	FAT	CARBOHYDRATES	TOTAL
ENERGY VALUE (AVERAGE) PER 100 G EDIBLE PORTION	KJOULE	376	136	0.0	511
	(KCAL)	90	32	0.0	122
AMOUNT OF DIGESTIBLE CONSTITUENTS PER 100 G	GRAM	19.78	3.23	0.00	
ENERGY VALUE (AVERAGE) OF THE DIGESTIBLE FRACTION PER 100 G EDIBLE PORTION	KJOULE	364	129	0.0	493
	(KCAL)	87	31	0.0	118

WASTE PERCENTAGE AVERAGE 0.00

CONSTITUENTS	DIM	AV	VARIATION			AVR	NUTR. DENS.		MOLPERC.
WATER	GRAM	75.00	74.70	-	75.40	72.00	GRAM/MJ	152.11	
PROTEIN	GRAM	20.40	19.50	-	21.30	19.58	GRAM/MJ	41.37	
FAT	GRAM	3.41	1.92	-	4.90	3.27	GRAM/MJ	6.92	
MINERALS	GRAM	1.13	1.05	-	1.21	1.08	GRAM/MJ	2.29	
SODIUM	MILLI	94.00	93.00	-	94.00	90.24	MILLI/MJ	190.65	
POTASSIUM	MILLI	289.00	258.00	-	319.00	277.44	MILLI/MJ	586.14	
MAGNESIUM	MILLI	19.00	10.00	-	26.00	18.24	MILLI/MJ	38.54	
CALCIUM	MILLI	12.00	-	-	-	11.52	MILLI/MJ	24.34	
MANGANESE	MICRO	13.00	4.00	-	20.00	12.48	MICRO/MJ	26.37	
IRON	MILLI	1.80	1.50	-	2.30	1.73	MILLI/MJ	3.65	
COBALT	MICRO	0.70	0.50	-	1.00	0.67	MICRO/MJ	1.42	
COPPER	MICRO	90.00	50.00	-	130.00	86.40	MICRO/MJ	182.54	
ZINC	MILLI	2.30	1.50	-	3.10	2.21	MILLI/MJ	4.66	
NICKEL	MICRO	2.00	-	-	-	1.92	MICRO/MJ	4.06	
CHROMIUM	MICRO	4.00	2.00	-	10.00	3.84	MICRO/MJ	8.11	
PHOSPHORUS	MILLI	162.00	100.00	-	190.00	155.52	MILLI/MJ	328.57	
FLUORIDE	MICRO	20.00	10.00	-	20.00	19.20	MICRO/MJ	40.56	
SELENIUM	MICRO	1.00	0.50	-	2.00	0.96	MICRO/MJ	2.03	
SILICON	MILLI	1.00	0.50	-	1.00	0.96	MILLI/MJ	2.03	
VITAMIN E ACTIVITY	MILLI	0.43	-	-	-	0.41	MILLI/MJ	0.87	
TOTAL TOCOPHEROLS	MILLI	0.46	-	-	-	0.44	MILLI/MJ	0.93	
ALPHA-TOCOPHEROL	MILLI	0.43	-	-	-	0.41	MILLI/MJ	0.87	
GAMMA-TOCOPHEROL	MICRO	30.00	-	-	-	28.80	MICRO/MJ	60.85	
VITAMIN B1	MILLI	0.18	-	-	-	0.17	MILLI/MJ	0.37	
VITAMIN B2	MILLI	0.25	-	-	-	0.24	MILLI/MJ	0.51	
NICOTINAMIDE	MILLI	5.80	-	-	-	5.57	MILLI/MJ	11.76	

CONSTITUENTS	DIM	AV	VARIATION			AVR	NUTR. DENS.		MOLPERC.
FOLIC ACID	MICRO	3.00	-	-	-	2.88	MICRO/MJ	6.08	
ALANINE	GRAM	1.44	1.43	-	1.46	1.38	GRAM/MJ	2.92	8.8
ARGININE	GRAM	1.44	1.32	-	1.68	1.38	GRAM/MJ	2.92	4.5
ASPARTIC ACID	GRAM	2.15	2.13	-	2.16	2.06	GRAM/MJ	4.36	8.8
CYSTINE	GRAM	0.29	0.17	-	0.69	0.28	GRAM/MJ	0.59	0.7
GLUTAMIC ACID	GRAM	4.30	3.86	-	4.77	4.13	GRAM/MJ	8.72	15.9
GLYCINE	GRAM	1.43	1.31	-	1.55	1.37	GRAM/MJ	2.90	10.4
HISTIDINE	GRAM	0.63	0.50	-	0.74	0.60	GRAM/MJ	1.28	2.2
ISOLEUCINE	GRAM	1.21	0.99	-	1.47	1.16	GRAM/MJ	2.45	5.0
LEUCINE	GRAM	1.80	1.58	-	2.02	1.73	GRAM/MJ	3.65	7.5
LYSINE	GRAM	2.00	1.48	-	2.12	1.92	GRAM/MJ	4.06	7.4
METHIONINE	GRAM	0.56	0.49	-	0.59	0.54	GRAM/MJ	1.14	2.0
PHENYLALANINE	GRAM	0.92	0.83	-	1.10	0.88	GRAM/MJ	1.87	3.0
PROLINE	GRAM	1.02	0.84	-	1.12	0.98	GRAM/MJ	2.07	4.8
SERINE	GRAM	1.04	0.97	-	1.10	1.00	GRAM/MJ	2.11	5.4
THREONINE	GRAM	1.09	0.92	-	1.25	1.05	GRAM/MJ	2.21	5.0
TRYPTOPHAN	GRAM	0.29	0.24	-	0.36	0.28	GRAM/MJ	0.59	0.8
TYROSINE	GRAM	0.77	0.64	-	1.10	0.74	GRAM/MJ	1.56	2.3
VALINE	GRAM	1.18	1.08	-	1.32	1.13	GRAM/MJ	2.39	5.5
LINOLEIC ACID	MILLI	60.00	-	-	-	57.60	MILLI/MJ	121.69	
LINOLENIC ACID	MILLI	20.00	-	-	-	19.20	MILLI/MJ	40.56	
CHOLESTEROL	MILLI	70.00	-	-	-	67.20	MILLI/MJ	141.97	

HAMMELHERZ SHEEP'S HEART COEUR DE MOUTON

		PROTEIN	FAT	CARBOHYDRATES	TOTAL
ENERGY VALUE (AVERAGE)	KJOULE	309	397	0.0	707
PER 100 G	(KCAL)	74	95	0.0	169
EDIBLE PORTION					
AMOUNT OF DIGESTIBLE	GRAM	16.29	9.50	0.00	
CONSTITUENTS PER 100 G					
ENERGY VALUE (AVERAGE)	KJOULE	300	378	0.0	678
OF THE DIGESTIBLE	(KCAL)	72	90	0.0	162
FRACTION PER 100 G					
EDIBLE PORTION					

WASTE PERCENTAGE AVERAGE 22 MINIMUM 18 MAXIMUM 27

CONSTITUENTS	DIM	AV	VARIATION			AVR	NUTR. DENS.		MOLPERC.
WATER	GRAM	72.00	-	-	-	56.16	GRAM/MJ	106.26	
PROTEIN	GRAM	16.80	-	-	-	13.10	GRAM/MJ	24.79	
FAT	GRAM	10.00	-	-	-	7.80	GRAM/MJ	14.76	
MINERALS	GRAM	1.00	-	-	-	0.78	GRAM/MJ	1.48	
SODIUM	MILLI	118.00	113.00	-	123.00	92.04	MILLI/MJ	174.14	
POTASSIUM	MILLI	248.00	221.00	-	265.00	193.44	MILLI/MJ	365.99	
MAGNESIUM	MILLI	16.00	14.00	-	17.00	12.48	MILLI/MJ	23.61	
CALCIUM	MILLI	4.00	1.00	-	6.00	3.12	MILLI/MJ	5.90	
MANGANESE	MICRO	50.00	-	-	-	39.00	MICRO/MJ	73.79	
IRON	MILLI	6.10	-	-	-	4.76	MILLI/MJ	9.00	
COBALT	MICRO	2.00	1.00	-	3.00	1.56	MICRO/MJ	2.95	
COPPER	MILLI	0.45	-	-	-	0.35	MILLI/MJ	0.66	
ZINC	MILLI	2.12	0.98	-	2.80	1.65	MILLI/MJ	3.13	
NICKEL	MICRO	1.50	-	-	-	1.17	MICRO/MJ	2.21	
CHROMIUM	MICRO	0.75	0.50	-	1.00	0.59	MICRO/MJ	1.11	
PHOSPHORUS	MILLI	160.00	150.00	-	170.00	124.80	MILLI/MJ	236.12	
FLUORIDE	MICRO	15.00	10.00	-	20.00	11.70	MICRO/MJ	22.14	
SELENIUM	MICRO	3.00	2.00	-	6.00	2.34	MICRO/MJ	4.43	
SILICON	MICRO	1.00	0.50	-	1.00	0.78	MICRO/MJ	1.48	
VITAMIN B1	MILLI	0.31	-	-	-	0.24	MILLI/MJ	0.46	
VITAMIN B2	MILLI	0.86	-	-	-	0.67	MILLI/MJ	1.27	
NICOTINAMIDE	MILLI	4.60	-	-	-	3.59	MILLI/MJ	6.79	
PANTOTHENIC ACID	MILLI	3.00	-	-	-	2.34	MILLI/MJ	4.43	
VITAMIN B12	MICRO	5.20	-	-	-	4.06	MICRO/MJ	7.67	
HISTIDINE	GRAM	0.42	-	-	-	0.33	GRAM/MJ	0.62	
ISOLEUCINE	GRAM	0.84	-	-	-	0.66	GRAM/MJ	1.24	
LEUCINE	GRAM	1.53	-	-	-	1.19	GRAM/MJ	2.26	
LYSINE	GRAM	1.43	-	-	-	1.12	GRAM/MJ	2.11	
METHIONINE	GRAM	0.37	-	-	-	0.29	GRAM/MJ	0.55	
PHENYLALANINE	GRAM	0.79	-	-	-	0.62	GRAM/MJ	1.17	
THREONINE	GRAM	0.76	-	-	-	0.59	GRAM/MJ	1.12	
TRYPTOPHAN	GRAM	0.20	-	-	-	0.16	GRAM/MJ	0.30	
VALINE	GRAM	0.92	-	-	-	0.72	GRAM/MJ	1.36	
OLEIC ACID	GRAM	4.20	3.73	-	4.65	3.28	GRAM/MJ	6.20	
LINOLEIC ACID	GRAM	0.35	0.16	-	0.54	0.27	GRAM/MJ	0.52	
LINOLENIC ACID	MILLI	55.00	38.00	-	86.00	42.90	MILLI/MJ	81.17	
TOTAL PURINES	MILLI	241.00 [1]	241.00	-	241.00	187.98	MILLI/MJ	355.66	
ADENINE	MILLI	82.00	82.00	-	82.00	63.96	MILLI/MJ	121.01	
GUANINE	MILLI	43.00	43.00	-	43.00	33.54	MILLI/MJ	63.46	
XANTHINE	MILLI	3.00	3.00	-	3.00	2.34	MILLI/MJ	4.43	
HYPOXANTHINE	MILLI	71.00	71.00	-	71.00	55.38	MILLI/MJ	104.78	

1 VALUE OF TOTAL PURINES IS EXPRESSSED IN MG URIC ACID/100G

HAMMELHIRN SHEEP'S BRAIN CERVELLE DE MOUTON

		PROTEIN	FAT	CARBOHYDRATES	TOTAL
ENERGY VALUE (AVERAGE)	KJOULE	201	362	0.0	562
PER 100 G	(KCAL)	48	86	0.0	134
EDIBLE PORTION					
AMOUNT OF DIGESTIBLE	GRAM	10.57	8.64	0.00	
CONSTITUENTS PER 100 G					
ENERGY VALUE (AVERAGE)	KJOULE	195	344	0.0	538
OF THE DIGESTIBLE	(KCAL)	47	82	0.0	129
FRACTION PER 100 G					
EDIBLE PORTION					

WASTE PERCENTAGE AVERAGE 2.0

CONSTITUENTS	DIM	AV	VARIATION		AVR	NUTR. DENS.		MOLPERC.
WATER	GRAM	78.00	76.00	- 79.10	76.44	GRAM/MJ	144.91	
PROTEIN	GRAM	10.90	10.00	- 11.80	10.68	GRAM/MJ	20.25	
FAT	GRAM	9.10	8.00	- 11.40	8.92	GRAM/MJ	16.91	
MINERALS	GRAM	1.40	1.30	- 1.60	1.37	GRAM/MJ	2.60	
MAGNESIUM	MILLI	14.60	14.40	- 14.80	14.31	MILLI/MJ	27.12	
CALCIUM	MILLI	5.40	2.10	- 7.10	5.29	MILLI/MJ	10.03	
IRON	MILLI	3.80	2.00	- 6.70	3.72	MILLI/MJ	7.06	
PHOSPHORUS	MILLI	305.00	252.00	- 358.00	298.90	MILLI/MJ	566.63	
VITAMIN B1	MILLI	0.24	0.14	- 0.34	0.24	MILLI/MJ	0.45	
VITAMIN B2	MILLI	0.25	0.23	- 0.27	0.25	MILLI/MJ	0.46	
NICOTINAMIDE	MILLI	3.20	2.90	- 3.50	3.14	MILLI/MJ	5.95	
PANTOTHENIC ACID	MILLI	2.60	-	- -	2.55	MILLI/MJ	4.83	
VITAMIN B12	MICRO	7.30	-	- -	7.15	MICRO/MJ	13.56	
VITAMIN C	MILLI	15.20	11.00	- 19.40	14.90	MILLI/MJ	28.24	
ARGININE	GRAM	0.61	-	- -	0.60	GRAM/MJ	1.13	
HISTIDINE	GRAM	0.31	-	- -	0.30	GRAM/MJ	0.58	
ISOLEUCINE	GRAM	0.57	-	- -	0.56	GRAM/MJ	1.06	
LEUCINE	GRAM	1.00	-	- -	0.98	GRAM/MJ	1.86	
LYSINE	GRAM	1.01	-	- -	0.99	GRAM/MJ	1.88	
METHIONINE	GRAM	0.27	-	- -	0.26	GRAM/MJ	0.50	
PHENYLALANINE	GRAM	0.58	-	- -	0.57	GRAM/MJ	1.08	
THREONINE	GRAM	0.52	-	- -	0.51	GRAM/MJ	0.97	
TRYPTOPHAN	GRAM	0.15	-	- -	0.15	GRAM/MJ	0.28	
VALINE	GRAM	0.60	-	- -	0.59	GRAM/MJ	1.11	
CHOLESTEROL	GRAM	2.20	-	- -	2.16	GRAM/MJ	4.09	

HAMMELLEBER SHEEP'S LIVER FOIE DE MOUTON

		PROTEIN	FAT	CARBOHYDRATES	TOTAL
ENERGY VALUE (AVERAGE)	KJOULE	390	157	0.0	547
PER 100 G	(KCAL)	93	38	0.0	131
EDIBLE PORTION					
AMOUNT OF DIGESTIBLE	GRAM	20.56	3.75	0.00	
CONSTITUENTS PER 100 G					
ENERGY VALUE (AVERAGE)	KJOULE	379	149	0.0	528
OF THE DIGESTIBLE	(KCAL)	90	36	0.0	126
FRACTION PER 100 G					
EDIBLE PORTION					

WASTE PERCENTAGE AVERAGE 6.0 MINIMUM 4.0 MAXIMUM 7.0

CONSTITUENTS	DIM	AV	VARIATION			AVR	NUTR. DENS.		MOLPERC.
WATER	GRAM	70.40	69.50	-	71.00	66.18	GRAM/MJ	133.40	
PROTEIN	GRAM	21.20	21.00	-	21.70	19.93	GRAM/MJ	40.17	
FAT	GRAM	3.95	3.90	-	4.00	3.71	GRAM/MJ	7.48	
MINERALS	GRAM	1.43	1.40	-	1.50	1.34	GRAM/MJ	2.71	
SODIUM	MILLI	95.00	89.00	-	105.00	89.30	MILLI/MJ	180.02	
POTASSIUM	MILLI	282.00	262.00	-	303.00	265.08	MILLI/MJ	534.36	
CALCIUM	MILLI	4.30	0.60	-	8.00	4.04	MILLI/MJ	8.15	
MANGANESE	MILLI	0.33	-	-	-	0.31	MILLI/MJ	0.63	
IRON	MILLI	12.40	11.90	-	12.60	11.66	MILLI/MJ	23.50	
COPPER	MILLI	7.64	4.50	-	10.80	7.18	MILLI/MJ	14.48	
ZINC	MILLI	4.35	-	-	-	4.09	MILLI/MJ	8.24	
NICKEL	MICRO	26.00	-	-	-	24.44	MICRO/MJ	49.27	
PHOSPHORUS	MILLI	364.00	-	-	-	342.16	MILLI/MJ	689.75	
IODIDE	MICRO	3.30	-	-	-	3.10	MICRO/MJ	6.25	
BORON	MILLI	0.12	0.11	-	0.18	0.12	MILLI/MJ	0.24	
VITAMIN A	MILLI	9.50	5.40	-	15.20	8.93	MILLI/MJ	18.00	
VITAMIN D	MICRO	2.00	-	-	-	1.88	MICRO/MJ	3.79	
VITAMIN B1	MILLI	0.36	0.29	-	0.40	0.34	MILLI/MJ	0.68	
VITAMIN B2	MILLI	3.33	2.80	-	3.90	3.13	MILLI/MJ	6.31	
NICOTINAMIDE	MILLI	15.30	12.10	-	17.00	14.38	MILLI/MJ	28.99	
PANTOTHENIC ACID	MILLI	7.60	7.10	-	8.10	7.14	MILLI/MJ	14.40	
VITAMIN B6	MILLI	0.37	-	-	-	0.35	MILLI/MJ	0.70	
BIOTIN	MILLI	0.13	-	-	-	0.12	MILLI/MJ	0.25	
FOLIC ACID	MILLI	0.28	0.27	-	0.28	0.26	MILLI/MJ	0.53	
VITAMIN B12	MICRO	35.00	-	-	-	32.90	MICRO/MJ	66.32	
VITAMIN C	MILLI	30.70	20.00	-	36.60	28.86	MILLI/MJ	58.17	

CONSTITUENTS	DIM	AV	VARIATION			AVR	NUTR. DENS.		MOLPERC.
ALANINE	GRAM	1.12	-	-	-	1.05	GRAM/MJ	2.12	8.1
ARGININE	GRAM	1.12	-	-	-	1.05	GRAM/MJ	2.12	4.2
ASPARTIC ACID	GRAM	1.83	-	-	-	1.72	GRAM/MJ	3.47	8.9
CYSTINE	GRAM	0.31	-	-	-	0.29	GRAM/MJ	0.59	0.8
GLUTAMIC ACID	GRAM	2.58	-	-	-	2.43	GRAM/MJ	4.89	11.4
GLYCINE	GRAM	1.05	-	-	-	0.99	GRAM/MJ	1.99	9.1
HISTIDINE	GRAM	0.78	-	-	-	0.73	GRAM/MJ	1.48	3.3
ISOLEUCINE	GRAM	0.92	-	-	-	0.86	GRAM/MJ	1.74	4.5
LEUCINE	GRAM	1.66	-	-	-	1.56	GRAM/MJ	3.15	8.2
LYSINE	GRAM	1.80	-	-	-	1.69	GRAM/MJ	3.41	8.0
METHIONINE	GRAM	0.51	-	-	-	0.48	GRAM/MJ	0.97	2.2
PHENYLALANINE	GRAM	1.05	-	-	-	0.99	GRAM/MJ	1.99	4.1
PROLINE	GRAM	1.12	-	-	-	1.05	GRAM/MJ	2.12	6.3
SERINE	GRAM	0.98	-	-	-	0.92	GRAM/MJ	1.86	6.0
THREONINE	GRAM	0.92	-	-	-	0.86	GRAM/MJ	1.74	5.0
TRYPTOPHAN	GRAM	0.27	-	-	-	0.25	GRAM/MJ	0.51	0.9
TYROSINE	GRAM	0.64	-	-	-	0.60	GRAM/MJ	1.21	2.3
VALINE	GRAM	1.22	-	-	-	1.15	GRAM/MJ	2.31	6.7
CHOLESTEROL	MILLI	300.00	-	-	-	282.00	MILLI/MJ	568.47	

HAMMELLUNGE — SHEEP'S LUNGS (LIGHTS) — MOU DE MOUTON

		PROTEIN	FAT	CARBOHYDRATES	TOTAL
ENERGY VALUE (AVERAGE)	KJOULE	339	91	0.0	430
PER 100 G EDIBLE PORTION	(KCAL)	81	22	0.0	103
AMOUNT OF DIGESTIBLE CONSTITUENTS PER 100 G	GRAM	17.84	2.18	0.00	
ENERGY VALUE (AVERAGE)	KJOULE	329	87	0.0	415
OF THE DIGESTIBLE FRACTION PER 100 G EDIBLE PORTION	(KCAL)	79	21	0.0	99

WASTE PERCENTAGE AVERAGE 24 MINIMUM 15 MAXIMUM 34

CONSTITUENTS	DIM	AV	VARIATION			AVR	NUTR. DENS.		MOLPERC.
WATER	GRAM	78.00	76.00	-	79.70	59.28	GRAM/MJ	187.76	
PROTEIN	GRAM	18.40	16.70	-	21.20	13.98	GRAM/MJ	44.29	
FAT	GRAM	2.30	2.00	-	2.80	1.75	GRAM/MJ	5.54	
MINERALS	GRAM	1.10	1.10	-	1.20	0.84	GRAM/MJ	2.65	
SODIUM	MILLI	205.00	177.00	-	218.00	155.80	MILLI/MJ	493.47	
POTASSIUM	MILLI	292.00	269.00	-	318.00	221.92	MILLI/MJ	702.90	
CALCIUM	MILLI	17.00	-	-	-	12.92	MILLI/MJ	40.92	
IRON	MILLI	6.40	-	-	-	4.86	MILLI/MJ	15.41	
PHOSPHORUS	MILLI	66.00	-	-	-	50.16	MILLI/MJ	158.87	
BORON	MILLI	0.18	0.18	-	0.24	0.14	MILLI/MJ	0.45	
VITAMIN A	MICRO	27.00	18.00	-	35.00	20.52	MICRO/MJ	64.99	
VITAMIN B1	MILLI	0.11	-	-	-	0.08	MILLI/MJ	0.26	
VITAMIN B2	MILLI	0.47	-	-	-	0.36	MILLI/MJ	1.13	
NICOTINAMIDE	MILLI	4.70	-	-	-	3.57	MILLI/MJ	11.31	
PANTOTHENIC ACID	MILLI	1.20	-	-	-	0.91	MILLI/MJ	2.89	
VITAMIN B12	MICRO	5.00	-	-	-	3.80	MICRO/MJ	12.04	
VITAMIN C	MILLI	31.00	-	-	-	23.56	MILLI/MJ	74.62	
ARGININE	GRAM	1.14	-	-	-	0.87	GRAM/MJ	2.74	
HISTIDINE	GRAM	0.46	-	-	-	0.35	GRAM/MJ	1.11	
ISOLEUCINE	GRAM	0.88	-	-	-	0.67	GRAM/MJ	2.12	
LEUCINE	GRAM	1.67	-	-	-	1.27	GRAM/MJ	4.02	
LYSINE	GRAM	1.36	-	-	-	1.03	GRAM/MJ	3.27	
METHIONINE	GRAM	0.37	-	-	-	0.28	GRAM/MJ	0.89	
PHENYLALANINE	GRAM	0.92	-	-	-	0.70	GRAM/MJ	2.21	
THREONINE	GRAM	0.88	-	-	-	0.67	GRAM/MJ	2.12	
TRYPTOPHAN	GRAM	0.17	-	-	-	0.13	GRAM/MJ	0.41	
VALINE	GRAM	1.05	-	-	-	0.80	GRAM/MJ	2.53	

HAMMELMILZ SHEEP'S SPLEEN RATE DE MOUTON

		PROTEIN	FAT	CARBOHYDRATES	TOTAL
ENERGY VALUE (AVERAGE)	KJOULE	331	159	0.0	490
PER 100 G	(KCAL)	79	38	0.0	117
EDIBLE PORTION					
AMOUNT OF DIGESTIBLE	GRAM	17.46	3.80	0.00	
CONSTITUENTS PER 100 G					
ENERGY VALUE (AVERAGE)	KJOULE	321	151	0.0	472
OF THE DIGESTIBLE	(KCAL)	77	36	0.0	113
FRACTION PER 100 G					
EDIBLE PORTION					

WASTE PERCENTAGE AVERAGE 0.00

CONSTITUENTS	DIM	AV	VARIATION			AVR	NUTR.	DENS.	MOLPERC.
WATER	GRAM	76.20	74.00	-	78.40	76.20	GRAM/MJ	161.28	
PROTEIN	GRAM	18.00	17.20	-	18.80	18.00	GRAM/MJ	38.10	
FAT	GRAM	4.00	-	-	-	4.00	GRAM/MJ	8.47	
MINERALS	GRAM	1.45	1.30	-	1.60	1.45	GRAM/MJ	3.07	
VITAMIN B1	MICRO	90.00	-	-	-	90.00	MICRO/MJ	190.49	
VITAMIN B2	MILLI	0.27	-	-	-	0.27	MILLI/MJ	0.57	
NICOTINAMIDE	MILLI	4.70	-	-	-	4.70	MILLI/MJ	9.95	
PANTOTHENIC ACID	MILLI	1.50	-	-	-	1.50	MILLI/MJ	3.17	
VITAMIN B12	MICRO	6.70	-	-	-	6.70	MICRO/MJ	14.18	
VITAMIN C	MILLI	23.20	-	-	-	23.20	MILLI/MJ	49.10	
ARGININE	GRAM	1.04	-	-	-	1.04	GRAM/MJ	2.20	
HISTIDINE	GRAM	0.49	-	-	-	0.49	GRAM/MJ	1.04	
ISOLEUCINE	GRAM	0.94	-	-	-	0.94	GRAM/MJ	1.99	
LEUCINE	GRAM	1.69	-	-	-	1.69	GRAM/MJ	3.58	
LYSINE	GRAM	1.57	-	-	-	1.57	GRAM/MJ	3.32	
METHIONINE	GRAM	0.34	-	-	-	0.34	GRAM/MJ	0.72	
PHENYLALANINE	GRAM	0.88	-	-	-	0.88	GRAM/MJ	1.86	
THREONINE	GRAM	0.90	-	-	-	0.90	GRAM/MJ	1.90	
TRYPTOPHAN	GRAM	0.25	-	-	-	0.25	GRAM/MJ	0.53	
VALINE	GRAM	1.03	-	-	-	1.03	GRAM/MJ	2.18	
TOTAL PURINES	MILLI	773.00 [1]	773.00	-	773.00	773.00	GRAM/MJ	1.64	
ADENINE	MILLI	267.00	267.00	-	267.00	267.00	MILLI/MJ	565.11	
GUANINE	MILLI	267.00	267.00	-	267.00	267.00	MILLI/MJ	565.11	
XANTHINE	MILLI	14.00	14.00	-	14.00	14.00	MILLI/MJ	29.63	
HYPOXANTHINE	MILLI	104.00	104.00	-	104.00	104.00	MILLI/MJ	220.12	

1 VALUE OF TOTAL PURINES IS EXPRESSSED IN MG URIC ACID/100G

HAMMELNIERE SHEEP'S KIDNEY ROGNON DE MOUTON

		PROTEIN	FAT	CARBOHYDRATES	TOTAL
ENERGY VALUE (AVERAGE)	KJOULE	304	120	0.0	424
PER 100 G	(KCAL)	73	29	0.0	101
EDIBLE PORTION					
AMOUNT OF DIGESTIBLE	GRAM	16.00	2.87	0.00	
CONSTITUENTS PER 100 G					
ENERGY VALUE (AVERAGE)	KJOULE	295	114	0.0	409
OF THE DIGESTIBLE	(KCAL)	70	27	0.0	98
FRACTION PER 100 G					
EDIBLE PORTION					

WASTE PERCENTAGE AVERAGE 3.0

CONSTITUENTS	DIM	AV	VARIATION		AVR	NUTR. DENS.		MOLPERC.
WATER	GRAM	78.50	76.70	- 80.20	76.15	GRAM/MJ	191.90	
PROTEIN	GRAM	16.50	15.70	- 18.00	16.01	GRAM/MJ	40.34	
FAT	GRAM	3.03	2.20	- 3.30	2.94	GRAM/MJ	7.41	
MINERALS	GRAM	1.27	1.20	- 1.30	1.23	GRAM/MJ	3.10	
SODIUM	MILLI	239.00	227.00	- 250.00	231.83	MILLI/MJ	584.27	
POTASSIUM	MILLI	252.00	249.00	- 254.00	244.44	MILLI/MJ	616.05	
CALCIUM	MILLI	13.00	-	- -	12.61	MILLI/MJ	31.78	
MANGANESE	MILLI	0.12	-	- -	0.12	MILLI/MJ	0.29	
IRON	MILLI	7.50	4.10	- 9.20	7.28	MILLI/MJ	18.33	
COPPER	MILLI	0.40	-	- -	0.39	MILLI/MJ	0.98	
ZINC	MILLI	2.80	-	- -	2.72	MILLI/MJ	6.84	
PHOSPHORUS	MILLI	262.00	237.00	- 279.00	254.14	MILLI/MJ	640.49	
BORON	MICRO	56.00	29.00	- 104.00	54.32	MICRO/MJ	136.90	
VITAMIN A	MILLI	0.33	0.30	- 0.35	0.32	MILLI/MJ	0.81	
VITAMIN E ACTIVITY	MILLI	0.41	-	- -	0.40	MILLI/MJ	1.00	
ALPHA-TOCOPHEROL	MILLI	0.41	-	- -	0.40	MILLI/MJ	1.00	
VITAMIN B1	MILLI	0.49	0.38	- 0.57	0.48	MILLI/MJ	1.20	
VITAMIN B2	MILLI	2.35	1.78	- 3.00	2.28	MILLI/MJ	5.74	
NICOTINAMIDE	MILLI	7.13	6.10	- 8.20	6.92	MILLI/MJ	17.43	
PANTOTHENIC ACID	MILLI	4.52	4.30	- 4.74	4.38	MILLI/MJ	11.05	
VITAMIN B12	MICRO	63.00	-	- -	61.11	MICRO/MJ	154.01	
VITAMIN C	MILLI	12.00	7.00	- 18.00	11.64	MILLI/MJ	29.34	
ARGININE	GRAM	1.01	0.99	- 1.03	0.98	GRAM/MJ	2.47	
CYSTINE	GRAM	0.20	-	- -	0.19	GRAM/MJ	0.49	
HISTIDINE	GRAM	0.42	0.41	- 0.43	0.41	GRAM/MJ	1.03	

CONSTITUENTS	DIM	AV	VARIATION			AVR	NUTR. DENS.		MOLPERC.
ISOLEUCINE	GRAM	0.87	0.80	-	0.94	0.84	GRAM/MJ	2.13	
LEUCINE	GRAM	1.46	1.43	-	1.49	1.42	GRAM/MJ	3.57	
LYSINE	GRAM	1.30	1.20	-	1.39	1.26	GRAM/MJ	3.18	
METHIONINE	GRAM	0.34	0.33	-	0.34	0.33	GRAM/MJ	0.83	
PHENYLALANINE	GRAM	0.76	0.73	-	0.78	0.74	GRAM/MJ	1.86	
THREONINE	GRAM	0.76	0.73	-	0.78	0.74	GRAM/MJ	1.86	
TRYPTOPHAN	GRAM	0.21	0.17	-	1.39	0.20	GRAM/MJ	0.51	
TYROSINE	GRAM	0.61	-	-	-	0.59	GRAM/MJ	1.49	
VALINE	GRAM	0.98	0.96	-	0.99	0.95	GRAM/MJ	2.40	
CHOLESTEROL	MILLI	375.00	-	-	-	363.75	MILLI/MJ	916.74	

HAMMELZUNGE SHEEP'S TONGUE LANGUE DE MOUTON

		PROTEIN	FAT	CARBOHYDRATES	TOTAL
ENERGY VALUE (AVERAGE)	KJOULE	249	588	0.0	837
PER 100 G	(KCAL)	59	141	0.0	200
EDIBLE PORTION					
AMOUNT OF DIGESTIBLE	GRAM	13.09	14.06	0.00	
CONSTITUENTS PER 100 G					
ENERGY VALUE (AVERAGE)	KJOULE	241	559	0.0	800
OF THE DIGESTIBLE	(KCAL)	58	134	0.0	191
FRACTION PER 100 G					
EDIBLE PORTION					

WASTE PERCENTAGE AVERAGE 32

CONSTITUENTS	DIM	AV	VARIATION			AVR	NUTR. DENS.		MOLPERC.
WATER	GRAM	69.20	65.70	-	72.00	47.06	GRAM/MJ	86.51	
PROTEIN	GRAM	13.50	12.00	-	14.60	9.18	GRAM/MJ	16.88	
FAT	GRAM	14.80	14.50	-	15.00	10.06	GRAM/MJ	18.50	
MINERALS	GRAM	0.77	0.70	-	0.80	0.52	GRAM/MJ	0.96	
SODIUM	MILLI	105.00	98.00	-	111.00	71.40	MILLI/MJ	131.26	
POTASSIUM	MILLI	277.00	272.00	-	284.00	188.36	MILLI/MJ	346.28	
CALCIUM	MILLI	19.00	-	-	-	12.92	MILLI/MJ	23.75	
IRON	MILLI	3.10	-	-	-	2.11	MILLI/MJ	3.88	
PHOSPHORUS	MILLI	119.00	-	-	-	80.92	MILLI/MJ	148.76	
VITAMIN A	-	TRACES	-	-	-				
VITAMIN B1	MICRO	80.00	-	-	-	54.40	MICRO/MJ	100.01	
VITAMIN B2	MILLI	0.28	-	-	-	0.19	MILLI/MJ	0.35	
NICOTINAMIDE	MILLI	4.20	-	-	-	2.86	MILLI/MJ	5.25	
VITAMIN C	MILLI	6.80	-	-	-	4.62	MILLI/MJ	8.50	

LAMMFLEISCH — LAMB — VIANDE D'AGNEAU

REINES MUSKELFLEISCH — MUSCELS ONLY — VIANDE DE MUSCLE

		PROTEIN	FAT	CARBOHYDRATES	TOTAL
ENERGY VALUE (AVERAGE)	KJOULE	383	147	0.0	530
PER 100 G EDIBLE PORTION	(KCAL)	92	35	0.0	127
AMOUNT OF DIGESTIBLE CONSTITUENTS PER 100 G	GRAM	20.17	3.51	0.00	
ENERGY VALUE (AVERAGE)	KJOULE	371	140	0.0	511
OF THE DIGESTIBLE FRACTION PER 100 G EDIBLE PORTION	(KCAL)	89	33	0.0	122

WASTE PERCENTAGE AVERAGE 0.00

CONSTITUENTS	DIM	AV	VARIATION	AVR	NUTR. DENS.		MOLPERC.
WATER	GRAM	74.70	- - -	74.70	GRAM/MJ	146.14	
PROTEIN	GRAM	20.80	- - -	20.80	GRAM/MJ	40.69	
FAT	GRAM	3.70	3.50 - 4.00	3.70	GRAM/MJ	7.24	
MINERALS	GRAM	1.20	- - -	1.20	GRAM/MJ	2.35	
SODIUM	MILLI	67.00	64.00 - 71.00	67.00	MILLI/MJ	131.08	
POTASSIUM	MILLI	289.00	267.00 - 302.00	289.00	MILLI/MJ	565.40	
MAGNESIUM	MILLI	22.00	21.00 - 23.00	22.00	MILLI/MJ	43.04	
CALCIUM	MILLI	3.00	- - -	3.00	MILLI/MJ	5.87	
IRON	MILLI	1.60	1.20 - 1.90	1.60	MILLI/MJ	3.13	
COPPER	MILLI	0.17	0.13 - 0.24	0.17	MILLI/MJ	0.33	
ZINC	MILLI	2.90	2.50 - 3.20	2.90	MILLI/MJ	5.67	
TOTAL PURINES	MILLI	182.00 [1]	182.00 - 182.00	182.00	MILLI/MJ	356.06	
ADENINE	MILLI	24.00	24.00 - 24.00	24.00	MILLI/MJ	46.95	
GUANINE	MILLI	25.00	25.00 - 25.00	25.00	MILLI/MJ	48.91	
XANTHINE	MILLI	1.00	1.00 - 1.00	1.00	MILLI/MJ	1.96	
HYPOXANTHINE	MILLI	99.00	99.00 - 99.00	99.00	MILLI/MJ	193.68	

1 VALUE OF TOTAL PURINES IS EXPRESSSED IN MG URIC ACID/100G

LAMMFLEISCH	**LAMB**	**VIANDE DE AGNEAU**
INTERMUSCULARES FETTGEWEBE	INTERMUSCULARE ADIPOSE TISSUE	TISSU ADIPEUX INTERMUSCULAIRE

		PROTEIN	FAT	CARBOHYDRATES	TOTAL
ENERGY VALUE (AVERAGE)	KJOULE	101	2715	0.0	2816
PER 100 G	(KCAL)	24	649	0.0	673
EDIBLE PORTION					
AMOUNT OF DIGESTIBLE	GRAM	5.32	64.88	0.00	
CONSTITUENTS PER 100 G					
ENERGY VALUE (AVERAGE)	KJOULE	98	2579	0.0	2677
OF THE DIGESTIBLE	(KCAL)	23	616	0.0	640
FRACTION PER 100 G					
EDIBLE PORTION					

WASTE PERCENTAGE AVERAGE 0.00

CONSTITUENTS	DIM	AV	VARIATION			AVR	NUTR. DENS.		MOLPERC.
WATER	GRAM	25.80	25.80	-	25.80	25.80	GRAM/MJ	9.64	
PROTEIN	GRAM	5.49	5.49	-	5.49	5.49	GRAM/MJ	2.05	
FAT	GRAM	68.30	68.30	-	68.30	68.30	GRAM/MJ	25.51	
MINERALS	GRAM	0.44	0.44	-	0.44	0.44	GRAM/MJ	0.16	
CHOLESTEROL	MILLI	75.00	75.00	-	75.00	75.00	MILLI/MJ	28.02	

LAMMFLEISCH
SUBCUTANES FETTGEWEBE

LAMB
SUBCUTANEOUS ADIPOSE TISSUE

VIANDE DE AGNEAU
TISSU ADIPEUX SOUSCUTANE

		PROTEIN	FAT	CARBOHYDRATES	TOTAL
ENERGY VALUE (AVERAGE)	KJOULE	96	3057	0.0	3153
PER 100 G	(KCAL)	23	731	0.0	754
EDIBLE PORTION					
AMOUNT OF DIGESTIBLE	GRAM	5.06	73.05	0.00	
CONSTITUENTS PER 100 G					
ENERGY VALUE (AVERAGE)	KJOULE	93	2904	0.0	2997
OF THE DIGESTIBLE	(KCAL)	22	694	0.0	716
FRACTION PER 100 G					
EDIBLE PORTION					

WASTE PERCENTAGE AVERAGE 0.00

CONSTITUENTS	DIM	AV	VARIATION			AVR	NUTR. DENS.		MOLPERC.
WATER	GRAM	17.80	17.80	-	17.80	17.80	GRAM/MJ	5.94	
PROTEIN	GRAM	5.22	5.22	-	5.22	5.22	GRAM/MJ	1.74	
FAT	GRAM	76.90	76.90	-	76.90	76.90	GRAM/MJ	25.66	
MINERALS	GRAM	0.34	0.34	-	0.34	0.34	GRAM/MJ	0.11	
CHOLESTEROL	MILLI	75.00	75.00	-	75.00	75.00	MILLI/MJ	25.02	

SCHAFFLEISCH / SHEEP'S MEAT / VIANDE DE BREBIS

DURCHSCHNITT / AVERAGE / EN MOYENNE

		PROTEIN	FAT	CARBOHYDRATES	TOTAL
ENERGY VALUE (AVERAGE) PER 100 G EDIBLE PORTION	KJOULE	324	660	0.0	984
	(KCAL)	77	158	0.0	235
AMOUNT OF DIGESTIBLE CONSTITUENTS PER 100 G	GRAM	17.07	15.77	0.00	
ENERGY VALUE (AVERAGE) OF THE DIGESTIBLE FRACTION PER 100 G EDIBLE PORTION	KJOULE	314	627	0.0	941
	(KCAL)	75	150	0.0	225

WASTE PERCENTAGE AVERAGE 19 MINIMUM 12 MAXIMUM 26

CONSTITUENTS	DIM	AV	VARIATION			AVR	NUTR. DENS.		MOLPERC.
WATER	GRAM	63.90	51.00	-	72.60	51.76	GRAM/MJ	67.90	
PROTEIN	GRAM	17.60	14.40	-	19.90	14.26	GRAM/MJ	18.70	
FAT	GRAM	16.60	5.80	-	33.50	13.45	GRAM/MJ	17.64	
MINERALS	GRAM	1.08	0.90	-	1.30	0.87	GRAM/MJ	1.15	
SODIUM	MILLI	84.00	-	-	-	68.04	MILLI/MJ	89.26	
POTASSIUM	MILLI	301.00	-	-	-	243.81	MILLI/MJ	319.83	
MAGNESIUM	MILLI	24.00	-	-	-	19.44	MILLI/MJ	25.50	
CALCIUM	MILLI	10.00	9.00	-	10.00	8.10	MILLI/MJ	10.63	
IRON	MILLI	2.30	1.90	-	2.70	1.86	MILLI/MJ	2.44	
MOLYBDENUM	MILLI	-	0.01	-	0.04				
PHOSPHORUS	MILLI	194.00	-	-	-	157.14	MILLI/MJ	206.14	
CHLORIDE	MILLI	85.00	-	-	-	68.85	MILLI/MJ	90.32	
BORON	MICRO	36.00	25.00	-	48.00	29.16	MICRO/MJ	38.25	
SELENIUM	MILLI	-	0.02	-	0.03				
VITAMIN A	MICRO	45.00	0.00	-	45.00	36.45	MICRO/MJ	47.82	
VITAMIN B1	MILLI	0.17	0.14	-	0.21	0.14	MILLI/MJ	0.18	
VITAMIN B2	MILLI	0.27	0.25	-	0.30	0.22	MILLI/MJ	0.29	
NICOTINAMIDE	MILLI	5.42	4.30	-	6.90	4.39	MILLI/MJ	5.76	
VITAMIN B6	MILLI	0.20	0.14	-	0.24	0.16	MILLI/MJ	0.21	
VITAMIN B12	MICRO	5.00	-	-	-	4.05	MICRO/MJ	5.31	
VITAMIN C	MILLI	1.00	0.00	-	1.00	0.81	MILLI/MJ	1.06	
ARGININE	GRAM	1.15	1.05	-	1.34	0.93	GRAM/MJ	1.22	
CYSTINE	GRAM	0.23	0.14	-	0.55	0.19	GRAM/MJ	0.24	
HISTIDINE	GRAM	0.49	0.40	-	0.60	0.40	GRAM/MJ	0.52	
ISOLEUCINE	GRAM	0.91	0.75	-	1.11	0.74	GRAM/MJ	0.97	
LEUCINE	GRAM	1.36	1.19	-	1.51	1.10	GRAM/MJ	1.45	
LYSINE	GRAM	1.43	1.16	-	1.65	1.16	GRAM/MJ	1.52	
METHIONINE	GRAM	0.42	0.39	-	0.46	0.34	GRAM/MJ	0.45	
PHENYLALANINE	GRAM	0.72	0.65	-	0.86	0.58	GRAM/MJ	0.77	
THREONINE	GRAM	0.81	0.69	-	0.93	0.66	GRAM/MJ	0.86	
TRYPTOPHAN	GRAM	0.23	0.18	-	0.28	0.19	GRAM/MJ	0.24	
TYROSINE	GRAM	0.61	0.50	-	0.86	0.49	GRAM/MJ	0.65	
VALINE	GRAM	0.86	0.81	-	0.98	0.70	GRAM/MJ	0.91	

KALBFLEISCH
REINES MUSKELFLEISCH

VEAL
MUSCELS ONLY

VIANDE DE VEAU
VIANDE DE MUSCLE

		PROTEIN	FAT	CARBOHYDRATES	TOTAL
ENERGY VALUE (AVERAGE)	KJOULE	392	32	0.0	424
PER 100 G	(KCAL)	94	7.7	0.0	101
EDIBLE PORTION					
AMOUNT OF DIGESTIBLE	GRAM	20.66	0.76	0.00	
CONSTITUENTS PER 100 G					
ENERGY VALUE (AVERAGE)	KJOULE	380	31	0.0	411
OF THE DIGESTIBLE	(KCAL)	91	7.3	0.0	98
FRACTION PER 100 G					
EDIBLE PORTION					

WASTE PERCENTAGE AVERAGE 0.00

CONSTITUENTS	DIM	AV	VARIATION		AVR	NUTR. DENS.		MOLPERC.
WATER	GRAM	76.40	-	-	76.40	GRAM/MJ	185.91	
PROTEIN	GRAM	21.30	-	-	21.30	GRAM/MJ	51.83	
FAT	GRAM	0.81	-	-	0.81	GRAM/MJ	1.97	
MINERALS	GRAM	1.19	-	-	1.19	GRAM/MJ	2.90	
SODIUM	MILLI	94.00	75.00	110.00	94.00	MILLI/MJ	228.74	
POTASSIUM	MILLI	358.00	314.00	398.00	358.00	MILLI/MJ	871.16	
CALCIUM	MILLI	13.00	11.00	14.00	13.00	MILLI/MJ	31.63	
MANGANESE	MICRO	30.00	20.00	40.00	30.00	MICRO/MJ	73.00	
IRON	MILLI	2.10	1.50	2.60	2.10	MILLI/MJ	5.11	
COPPER	MILLI	0.16	0.09	0.24	0.16	MILLI/MJ	0.39	
ZINC	MICRO	3.00	2.50	3.50	3.00	MICRO/MJ	7.30	
PHOSPHORUS	MILLI	198.00	187.00	203.00	198.00	MILLI/MJ	481.81	
CHLORIDE	MILLI	73.00	-	-	73.00	MILLI/MJ	177.64	
FLUORIDE	MICRO	20.00	-	-	20.00	MICRO/MJ	48.67	
IODIDE	MICRO	2.80	-	-	2.80	MICRO/MJ	6.81	
VITAMIN A	-	TRACES	-	-				
VITAMIN B1	MILLI	0.14	-	-	0.14	MILLI/MJ	0.34	
VITAMIN B2	MILLI	0.27	-	-	0.27	MILLI/MJ	0.66	
NICOTINAMIDE	MILLI	6.50	5.10	7.40	6.50	MILLI/MJ	15.82	
PANTOTHENIC ACID	MILLI	0.85	-	-	0.85	MILLI/MJ	2.07	
VITAMIN B6	MILLI	0.40	0.20	0.65	0.40	MILLI/MJ	0.97	
FOLIC ACID	MICRO	5.00	4.00	7.00	5.00	MICRO/MJ	12.17	
VITAMIN B12	MICRO	2.00	-	-	2.00	MICRO/MJ	4.87	
ALANINE	GRAM	1.64	-	-	1.64	GRAM/MJ	3.99	9.5
ARGININE	GRAM	1.54	1.49	1.59	1.54	GRAM/MJ	3.75	4.6

CONSTITUENTS	DIM	AV	VARIATION			AVR	NUTR. DENS.		MOLPERC.
ASPARTIC ACID	GRAM	2.40	2.26	-	2.53	2.40	GRAM/MJ	5.84	9.4
CYSTINE	GRAM	0.28	-	-	-	0.28	GRAM/MJ	0.68	0.6
GLUTAMIC ACID	GRAM	3.97	-	-	-	3.97	GRAM/MJ	9.66	14.0
GLYCINE	GRAM	1.34	1.18	-	1.49	1.34	GRAM/MJ	3.26	9.3
HISTIDINE	GRAM	0.80	0.77	-	0.82	0.80	GRAM/MJ	1.95	2.7
ISOLEUCINE	GRAM	1.29	1.21	-	1.36	1.29	GRAM/MJ	3.14	5.1
LEUCINE	GRAM	1.89	-	-	-	1.89	GRAM/MJ	4.60	7.5
LYSINE	GRAM	2.05	-	-	-	2.05	GRAM/MJ	4.99	7.3
METHIONINE	GRAM	0.60	0.58	-	0.63	0.60	GRAM/MJ	1.46	2.1
PHENYLALANINE	GRAM	1.02	-	-	-	1.02	GRAM/MJ	2.48	3.2
PROLINE	GRAM	1.15	1.06	-	1.23	1.15	GRAM/MJ	2.80	5.2
SERINE	GRAM	1.15	1.09	-	1.18	1.15	GRAM/MJ	2.80	5.7
THREONINE	GRAM	1.13	-	-	-	1.13	GRAM/MJ	2.75	4.9
TRYPTOPHAN	GRAM	0.30	0.28	-	0.34	0.30	GRAM/MJ	0.73	0.8
TYROSINE	GRAM	0.88	-	-	-	0.88	GRAM/MJ	2.14	2.5
VALINE	GRAM	1.31	1.26	-	1.36	1.31	GRAM/MJ	3.19	5.8
LAURIC ACID	MILLI	3.00	3.00	-	3.00	3.00	MILLI/MJ	7.30	
MYRISTIC ACID	MILLI	23.00	23.00	-	23.00	23.00	MILLI/MJ	55.97	
PALMITIC ACID	MILLI	166.00	166.00	-	166.00	166.00	MILLI/MJ	403.95	
STEARIC ACID	MILLI	106.00	106.00	-	106.00	106.00	MILLI/MJ	257.94	
ARACHIDIC ACID	MILLI	1.90	1.90	-	1.90	1.90	MILLI/MJ	4.62	
PALMITOLEIC ACID	MILLI	11.00	11.00	-	11.00	11.00	MILLI/MJ	26.77	
OLEIC ACID	MILLI	179.00	179.00	-	179.00	179.00	MILLI/MJ	435.58	
LINOLEIC ACID	MILLI	197.00	197.00	-	197.00	197.00	MILLI/MJ	479.38	
LINOLENIC ACID	MILLI	9.10	9.10	-	9.10	9.10	MILLI/MJ	22.14	
ARACHIDONIC ACID	MILLI	53.00	53.00	-	53.00	53.00	MILLI/MJ	128.97	
CHOLESTEROL	MILLI	70.00	-	-	-	70.00	MILLI/MJ	170.34	
TOTAL PURINES	MILLI	158.00 1	-	-	-	158.00	MILLI/MJ	384.48	
TOTAL PHOSPHOLIPIDS	MILLI	666.00	-	-	-	666.00	GRAM/MJ	1.62	
PHOSPHATIDYLCHOLINE	MILLI	254.00	-	-	-	254.00	MILLI/MJ	618.08	
PHOSPHATIDYLETHANOLAMINE	MILLI	143.00	-	-	-	143.00	MILLI/MJ	347.98	
PHOSPHATIDYLSERINE	MILLI	72.00	-	-	-	72.00	MILLI/MJ	175.21	
PHOSPHATIDYLINOSITOL	MILLI	37.00	-	-	-	37.00	MILLI/MJ	90.04	
SPHINGOMYELIN	MILLI	40.00	-	-	-	40.00	MILLI/MJ	97.34	

1 VALUE EXPRESSED IN MG URIC ACID/100G

KALBSBRIES (THYMUSDRÜSE) | CALF NECK SWEET BREAD | RIZ DE VEAU

		PROTEIN	FAT	CARBOHYDRATES	TOTAL
ENERGY VALUE (AVERAGE)	KJOULE	317	135	0.0	452
PER 100 G	(KCAL)	76	32	0.0	108
EDIBLE PORTION					
AMOUNT OF DIGESTIBLE	GRAM	16.68	3.23	0.00	
CONSTITUENTS PER 100 G					
ENERGY VALUE (AVERAGE)	KJOULE	307	128	0.0	436
OF THE DIGESTIBLE	(KCAL)	73	31	0.0	104
FRACTION PER 100 G					
EDIBLE PORTION					

WASTE PERCENTAGE AVERAGE 3.0 MINIMUM 2.0 MAXIMUM 3.0

CONSTITUENTS	DIM	AV	VARIATION		AVR	NUTR. DENS.		MOLPERC.
WATER	GRAM	77.70	75.00	- 79.30	75.37	GRAM/MJ	178.40	
PROTEIN	GRAM	17.20	14.00	- 19.60	16.68	GRAM/MJ	39.49	
FAT	GRAM	3.40	2.90	- 4.29	3.30	GRAM/MJ	7.81	
MINERALS	GRAM	1.93	1.90	- 2.00	1.87	GRAM/MJ	4.43	
SODIUM	MILLI	87.00	56.00	- 118.00	84.39	MILLI/MJ	199.76	
POTASSIUM	MILLI	386.00	385.00	- 387.00	374.42	MILLI/MJ	886.27	
CALCIUM	MILLI	0.50	0.00	- 1.00	0.49	MILLI/MJ	1.15	
IRON	MILLI	2.00	-	- -	1.94	MILLI/MJ	4.59	
CHLORIDE	MILLI	120.00	-	- -	116.40	MILLI/MJ	275.53	
VITAMIN C	MILLI	56.00	-	- -	54.32	MILLI/MJ	128.58	
TOTAL PURINES	GRAM	1.26 1	1.26	- 1.26	1.22	GRAM/MJ	2.89	
ADENINE	MILLI	491.00	491.00	- 491.00	476.27	GRAM/MJ	1.13	
GUANINE	MILLI	434.00	434.00	- 434.00	420.98	MILLI/MJ	996.48	
XANTHINE	MILLI	51.00	51.00	- 51.00	49.47	MILLI/MJ	117.10	
HYPOXANTHINE	MILLI	91.00	91.00	- 91.00	88.27	MILLI/MJ	208.94	

1 VALUE OF TOTAL PURINES IS EXPRESSSED IN MG URIC ACID/100G

KALBSGEKRÖSE
(KALBSKUTTELN, KALDAUNEN)

CALF'S TRIPE
(OFFAL PLUCK)

FRAISE DE VEAU
(TRIPES)

		PROTEIN	FAT	CARBOHYDRATES	TOTAL
ENERGY VALUE (AVERAGE)	KJOULE	272	330	0.0	602
PER 100 G	(KCAL)	65	79	0.0	144
EDIBLE PORTION					
AMOUNT OF DIGESTIBLE	GRAM	14.35	7.88	0.00	
CONSTITUENTS PER 100 G					
ENERGY VALUE (AVERAGE)	KJOULE	264	313	0.0	578
OF THE DIGESTIBLE	(KCAL)	63	75	0.0	138
FRACTION PER 100 G					
EDIBLE PORTION					

WASTE PERCENTAGE AVERAGE 0.00

CONSTITUENTS	DIM	AV	VARIATION		AVR	NUTR.	DENS.	MOLPERC.
WATER	GRAM	75.40	74.40	- 76.40	75.40	GRAM/MJ	130.52	
PROTEIN	GRAM	14.80	12.70	- 16.90	14.80	GRAM/MJ	25.62	
FAT	GRAM	8.30	6.20	- 10.40	8.30	GRAM/MJ	14.37	
MINERALS	GRAM	0.55	0.50	- 0.60	0.55	GRAM/MJ	0.95	
IRON	MILLI	10.00	-	- -	10.00	MILLI/MJ	17.31	
VITAMIN A	-	TRACES	-	- -				
VITAMIN B1	MICRO	25.00	-	- -	25.00	MICRO/MJ	43.27	
VITAMIN B2	MICRO	70.00	-	- -	70.00	MICRO/MJ	121.17	
NICOTINAMIDE	MILLI	5.70	-	- -	5.70	MILLI/MJ	9.87	
VITAMIN C	MILLI	9.40	-	- -	9.40	MILLI/MJ	16.27	
ARGININE	GRAM	0.73	-	- -	0.73	GRAM/MJ	1.26	
CYSTINE	GRAM	0.13	-	- -	0.13	GRAM/MJ	0.23	
HISTIDINE	GRAM	0.40	-	- -	0.40	GRAM/MJ	0.69	
LEUCINE	GRAM	1.72	-	- -	1.72	GRAM/MJ	2.98	
LYSINE	GRAM	1.17	-	- -	1.17	GRAM/MJ	2.03	
METHIONINE	GRAM	0.33	-	- -	0.33	GRAM/MJ	0.57	
PHENYLALANINE	GRAM	0.46	-	- -	0.46	GRAM/MJ	0.80	
THREONINE	GRAM	0.90	-	- -	0.90	GRAM/MJ	1.56	
TRYPTOPHAN	GRAM	0.15	-	- -	0.15	GRAM/MJ	0.26	
TYROSINE	GRAM	0.62	-	- -	0.62	GRAM/MJ	1.07	
VALINE	GRAM	0.73	-	- -	0.73	GRAM/MJ	1.26	

KALBSHERZ CALF'S HEART COEUR DE VEAU

		PROTEIN	FAT	CARBOHYDRATES	TOTAL
ENERGY VALUE (AVERAGE)	KJOULE	293	201	0.0	494
PER 100 G	(KCAL)	70	48	0.0	118
EDIBLE PORTION					
AMOUNT OF DIGESTIBLE	GRAM	15.42	4.80	0.00	
CONSTITUENTS PER 100 G					
ENERGY VALUE (AVERAGE)	KJOULE	284	191	0.0	475
OF THE DIGESTIBLE	(KCAL)	68	46	0.0	114
FRACTION PER 100 G					
EDIBLE PORTION					

WASTE PERCENTAGE AVERAGE 20 MINIMUM 15 MAXIMUM 25

CONSTITUENTS	DIM	AV	VARIATION			AVR	NUTR. DENS.		MOLPERC.
WATER	GRAM	77.00	75.60	-	79.00	61.60	GRAM/MJ	162.11	
PROTEIN	GRAM	15.90	15.00	-	17.20	12.72	GRAM/MJ	33.47	
FAT	GRAM	5.06	2.90	-	7.10	4.05	GRAM/MJ	10.65	
MINERALS	GRAM	1.08	1.00	-	1.20	0.86	GRAM/MJ	2.27	
SODIUM	MILLI	104.00	90.00	-	120.00	83.20	MILLI/MJ	218.95	
POTASSIUM	MILLI	265.00	208.00	-	308.00	212.00	MILLI/MJ	557.89	
MAGNESIUM	MILLI	25.00	16.00	-	37.00	20.00	MILLI/MJ	52.63	
CALCIUM	MILLI	16.00	-	-	-	12.80	MILLI/MJ	33.68	
MANGANESE	MICRO	30.00	-	-	-	24.00	MICRO/MJ	63.16	
IRON	MILLI	3.70	3.00	-	4.30	2.96	MILLI/MJ	7.79	
COPPER	MILLI	0.32	0.29	-	0.34	0.26	MILLI/MJ	0.67	
ZINC	MILLI	0.20	-	-	-	0.16	MILLI/MJ	0.42	
PHOSPHORUS	MILLI	180.00	160.00	-	200.00	144.00	MILLI/MJ	378.95	
VITAMIN A	MICRO	6.00	0.00	-	9.00	4.80	MICRO/MJ	12.63	
VITAMIN E ACTIVITY	MILLI	0.38	0.32	-	0.44	0.30	MILLI/MJ	0.80	
ALPHA-TOCOPHEROL	MILLI	0.38	0.32	-	0.44	0.30	MILLI/MJ	0.80	
VITAMIN B1	MILLI	0.55	0.40	-	0.64	0.44	MILLI/MJ	1.16	
VITAMIN B2	MILLI	1.00	0.80	-	1.10	0.80	MILLI/MJ	2.11	
NICOTINAMIDE	MILLI	6.60	6.00	-	8.10	5.28	MILLI/MJ	13.89	
PANTOTHENIC ACID	MILLI	2.78	2.75	-	2.81	2.22	MILLI/MJ	5.85	
VITAMIN B6	MILLI	0.29	-	-	-	0.23	MILLI/MJ	0.61	
BIOTIN	MICRO	7.30	-	-	-	5.84	MICRO/MJ	15.37	
FOLIC ACID	MICRO	3.10	1.80	-	4.20	2.48	MICRO/MJ	6.53	
VITAMIN B12	MICRO	11.00	-	-	-	8.80	MICRO/MJ	23.16	
VITAMIN C	MILLI	5.40	4.00	-	7.90	4.32	MILLI/MJ	11.37	

CONSTITUENTS	DIM	AV	VARIATION			AVR	NUTR. DENS.		MOLPERC.
ARGININE	GRAM	1.00	0.85	-	1.18	0.80	GRAM/MJ	2.11	
CYSTINE	GRAM	0.16	0.14	-	0.21	0.13	GRAM/MJ	0.34	
HISTIDINE	GRAM	0.40	0.27	-	0.48	0.32	GRAM/MJ	0.84	
ISOLEUCINE	GRAM	0.81	0.70	-	1.04	0.65	GRAM/MJ	1.71	
LEUCINE	GRAM	1.42	1.26	-	1.76	1.14	GRAM/MJ	2.99	
LYSINE	GRAM	1.30	1.11	-	1.69	1.04	GRAM/MJ	2.74	
METHIONINE	GRAM	0.38	0.33	-	0.49	0.30	GRAM/MJ	0.80	
PHENYLALANINE	GRAM	0.72	0.65	-	0.81	0.58	GRAM/MJ	1.52	
THREONINE	GRAM	0.73	0.61	-	1.00	0.58	GRAM/MJ	1.54	
TRYPTOPHAN	GRAM	0.21	0.14	-	0.22	0.17	GRAM/MJ	0.44	
TYROSINE	GRAM	0.59	0.47	-	0.70	0.47	GRAM/MJ	1.24	
VALINE	GRAM	0.91	0.78	-	1.12	0.73	GRAM/MJ	1.92	
LINOLEIC ACID	GRAM	0.33	-	-	-	0.26	GRAM/MJ	0.69	
LINOLENIC ACID	-	TRACES	-	-	-				

KALBSHIRN CALF'S BRAIN CERVELLE DE VEAU

		PROTEIN	FAT	CARBOHYDRATES	TOTAL
ENERGY VALUE (AVERAGE) PER 100 G EDIBLE PORTION	KJOULE	186	302	0.0	488
	(KCAL)	44	72	0.0	117
AMOUNT OF DIGESTIBLE CONSTITUENTS PER 100 G	GRAM	9.79	7.22	0.00	
ENERGY VALUE (AVERAGE) OF THE DIGESTIBLE FRACTION PER 100 G EDIBLE PORTION	KJOULE	180	287	0.0	467
	(KCAL)	43	69	0.0	112

WASTE PERCENTAGE AVERAGE 0.00

CONSTITUENTS	DIM	AV	VARIATION			AVR	NUTR. DENS.		MOLPERC.
WATER	GRAM	80.40	78.80	-	81.40	77.99	GRAM/MJ	172.04	
PROTEIN	GRAM	10.10	9.00	-	10.80	9.80	GRAM/MJ	21.61	
FAT	GRAM	7.60	6.40	-	8.60	7.37	GRAM/MJ	16.26	
MINERALS	GRAM	1.40	1.30	-	1.50	1.36	GRAM/MJ	3.00	

CONSTITUENTS	DIM	AV	VARIATION			AVR	NUTR. DENS.		MOLPERC.
SODIUM	MILLI	158.00	125.00	-	190.00	153.26	MILLI/MJ	338.08	
POTASSIUM	MILLI	280.00	220.00	-	345.00	271.60	MILLI/MJ	599.14	
MAGNESIUM	MILLI	15.00	14.00	-	17.00	14.55	MILLI/MJ	32.10	
CALCIUM	MILLI	12.00	8.00	-	160.00	11.64	MILLI/MJ	25.68	
MANGANESE	MICRO	40.00	-	-	-	38.80	MICRO/MJ	85.59	
IRON	MILLI	2.50	2.30	-	2.90	2.43	MILLI/MJ	5.35	
COBALT	MICRO	1.00	-	-	-	0.97	MICRO/MJ	2.14	
COPPER	MILLI	0.14	-	-	-	0.14	MILLI/MJ	0.30	
ZINC	MILLI	1.28	0.90	-	1.70	1.24	MILLI/MJ	2.74	
PHOSPHORUS	MILLI	350.00	310.00	-	380.00	339.50	MILLI/MJ	748.92	
VITAMIN A	-	0.00	-	-	-	0.00			
VITAMIN B1	MILLI	0.16	0.12	-	0.23	0.16	MILLI/MJ	0.34	
VITAMIN B2	MILLI	0.26	0.22	-	0.30	0.25	MILLI/MJ	0.56	
NICOTINAMIDE	MILLI	3.60	3.00	-	4.40	3.49	MILLI/MJ	7.70	
PANTOTHENIC ACID	MILLI	2.50	-	-	-	2.43	MILLI/MJ	5.35	
VITAMIN B6	MILLI	0.16	-	-	-	0.16	MILLI/MJ	0.34	
BIOTIN	MICRO	6.10	-	-	-	5.92	MICRO/MJ	13.05	
FOLIC ACID	MICRO	13.00	11.00	-	21.00	12.61	MICRO/MJ	27.82	
VITAMIN B12	MICRO	5.70	4.70	-	7.80	5.53	MICRO/MJ	12.20	
VITAMIN C	MILLI	23.00	18.00	-	26.00	22.31	MILLI/MJ	49.21	
ARGININE	GRAM	0.60	0.52	-	0.67	0.58	GRAM/MJ	1.28	
CYSTINE	GRAM	0.13	0.12	-	0.17	0.13	GRAM/MJ	0.28	
HISTIDINE	GRAM	0.28	0.27	-	0.30	0.27	GRAM/MJ	0.60	
ISOLEUCINE	GRAM	0.53	0.51	-	0.55	0.51	GRAM/MJ	1.13	
LEUCINE	GRAM	0.89	0.85	-	0.93	0.86	GRAM/MJ	1.90	
LYSINE	GRAM	0.78	0.71	-	0.85	0.76	GRAM/MJ	1.67	
METHIONINE	GRAM	0.25	0.20	-	0.36	0.24	GRAM/MJ	0.53	
PHENYLALANINE	GRAM	0.48	0.45	-	0.54	0.47	GRAM/MJ	1.03	
THREONINE	GRAM	0.48	0.47	-	0.51	0.47	GRAM/MJ	1.03	
TRYPTOPHAN	GRAM	0.14	0.13	-	0.15	0.14	GRAM/MJ	0.30	
TYROSINE	GRAM	0.46	0.38	-	0.51	0.45	GRAM/MJ	0.98	
VALINE	GRAM	0.56	0.54	-	0.58	0.54	GRAM/MJ	1.20	
CHOLESTEROL	GRAM	2.00	-	-	-	1.94	GRAM/MJ	4.28	

KALBSLEBER CALF'S LIVER FOIE DE VEAU

		PROTEIN	FAT	CARBOHYDRATES	TOTAL
ENERGY VALUE (AVERAGE) PER 100 G EDIBLE PORTION	KJOULE	353	165	0.0	518
	(KCAL)	84	39	0.0	124
AMOUNT OF DIGESTIBLE CONSTITUENTS PER 100 G	GRAM	18.62	3.93	0.00	
ENERGY VALUE (AVERAGE) OF THE DIGESTIBLE FRACTION PER 100 G EDIBLE PORTION	KJOULE	343	156	0.0	499
	(KCAL)	82	37	0.0	119

WASTE PERCENTAGE AVERAGE 0.00

CONSTITUENTS	DIM	AV	VARIATION			AVR	NUTR. DENS.		MOLPERC.
WATER	GRAM	71.20	69.50	-	73.00	69.06	GRAM/MJ	142.63	
PROTEIN	GRAM	19.20	17.70	-	20.60	18.62	GRAM/MJ	38.46	
FAT	GRAM	4.14	2.80	-	6.50	4.02	GRAM/MJ	8.29	
MINERALS	GRAM	1.37	1.30	-	1.60	1.33	GRAM/MJ	2.74	
SODIUM	MILLI	87.00	73.00	-	119.00	84.39	MILLI/MJ	174.28	
POTASSIUM	MILLI	316.00	281.00	-	367.00	306.52	MILLI/MJ	633.03	
MAGNESIUM	MILLI	19.00	15.00	-	30.00	18.43	MILLI/MJ	38.06	
CALCIUM	MILLI	8.70	6.00	-	12.80	8.44	MILLI/MJ	17.43	
MANGANESE	MILLI	0.28	0.23	-	0.34	0.27	MILLI/MJ	0.56	
IRON	MILLI	7.90	5.70	-	9.30	7.66	MILLI/MJ	15.83	
COBALT	MICRO	47.00	-	-	-	45.59	MICRO/MJ	94.15	
COPPER	MILLI	5.50	3.50	-	7.90	5.34	MILLI/MJ	11.02	
ZINC	MILLI	8.40	6.60	-	10.10	8.15	MILLI/MJ	16.83	
VANADIUM	MICRO	0.50	0.20	-	1.00	0.49	MICRO/MJ	1.00	
PHOSPHORUS	MILLI	306.00	279.00	-	333.00	296.82	MILLI/MJ	612.99	
CHLORIDE	MILLI	89.00	-	-	-	86.33	MILLI/MJ	178.29	
FLUORIDE	MICRO	19.00	-	-	-	18.43	MICRO/MJ	38.06	
BORON	MICRO	40.00	-	-	-	38.80	MICRO/MJ	80.13	
SELENIUM	MICRO	40.00	-	-	-	38.80	MICRO/MJ	80.13	
VITAMIN A	MILLI	21.90	8.30	-	39.80	21.24	MILLI/MJ	43.87	
VITAMIN D	MICRO	0.33	0.20	-	0.50	0.32	MICRO/MJ	0.66	
VITAMIN E ACTIVITY	MILLI	0.24	-	-	-	0.23	MILLI/MJ	0.48	
ALPHA-TOCOPHEROL	MILLI	0.24	-	-	-	0.23	MILLI/MJ	0.48	
VITAMIN K	MILLI	0.15	-	-	-	0.15	MILLI/MJ	0.30	
VITAMIN B1	MILLI	0.28	0.20	-	0.37	0.27	MILLI/MJ	0.56	
VITAMIN B2	MILLI	2.61	2.05	-	3.41	2.53	MILLI/MJ	5.23	

CONSTITUENTS	DIM	AV	VARIATION			AVR	NUTR. DENS.		MOLPERC.
NICOTINAMIDE	MILLI	15.00	11.40	-	17.00	14.55	MILLI/MJ	30.05	
PANTOTHENIC ACID	MILLI	7.90	6.00	-	9.70	7.66	MILLI/MJ	15.83	
VITAMIN B6	MILLI	0.90	0.72	-	0.99	0.87	MILLI/MJ	1.80	
BIOTIN	MICRO	75.00	-	-	-	72.75	MICRO/MJ	150.24	
FOLIC ACID	MILLI	0.24	-	-	-	0.23	MILLI/MJ	0.48	
VITAMIN B12	MICRO	60.00	38.00	-	82.00	58.20	MICRO/MJ	120.19	
VITAMIN C	MILLI	35.00	26.00	-	45.00	33.95	MILLI/MJ	70.11	
ALANINE	GRAM	1.43	1.29	-	1.57	1.39	GRAM/MJ	2.86	9.1
ARGININE	GRAM	1.21	1.13	-	1.30	1.17	GRAM/MJ	2.42	3.9
ASPARTIC ACID	GRAM	2.14	1.95	-	2.34	2.08	GRAM/MJ	4.29	9.1
CYSTINE	GRAM	0.28	0.26	-	0.30	0.27	GRAM/MJ	0.56	0.7
GLUTAMIC ACID	GRAM	2.87	2.70	-	3.03	2.78	GRAM/MJ	5.75	11.1
GLYCINE	GRAM	1.42	1.28	-	1.56	1.38	GRAM/MJ	2.84	10.7
HISTIDINE	GRAM	0.68	0.57	-	0.78	0.66	GRAM/MJ	1.36	2.5
ISOLEUCINE	GRAM	1.09	0.98	-	1.18	1.06	GRAM/MJ	2.18	4.7
LEUCINE	GRAM	1.94	1.78	-	2.09	1.88	GRAM/MJ	3.89	8.4
LYSINE	GRAM	1.74	1.63	-	1.84	1.69	GRAM/MJ	3.49	6.8
METHIONINE	GRAM	0.53	-	-	-	0.51	GRAM/MJ	1.06	2.0
PHENYLALANINE	GRAM	1.10	1.09	-	1.11	1.07	GRAM/MJ	2.20	3.8
PROLINE	GRAM	1.21	1.20	-	1.22	1.17	GRAM/MJ	2.42	6.0
SERINE	GRAM	1.17	1.10	-	1.23	1.13	GRAM/MJ	2.34	6.3
THREONINE	GRAM	1.05	1.00	-	1.10	1.02	GRAM/MJ	2.10	5.0
TRYPTOPHAN	GRAM	0.31	0.26	-	0.33	0.30	GRAM/MJ	0.62	0.9
TYROSINE	GRAM	0.73	0.65	-	0.81	0.71	GRAM/MJ	1.46	2.3
VALINE	GRAM	1.39	1.34	-	1.45	1.35	GRAM/MJ	2.78	6.7
CHOLESTEROL	MILLI	360.00	-	-	-	349.20	MILLI/MJ	721.17	
TOTAL PURINES	MILLI	460.00 1	460.00	-	460.00	446.20	MILLI/MJ	921.49	
ADENINE	MILLI	128.00	128.00	-	128.00	124.16	MILLI/MJ	256.42	
GUANINE	MILLI	146.00	146.00	-	146.00	141.62	MILLI/MJ	292.47	
XANTHINE	MILLI	65.00	65.00	-	65.00	63.05	MILLI/MJ	130.21	
HYPOXANTHINE	MILLI	54.00	54.00	-	54.00	52.38	MILLI/MJ	108.18	

1 VALUE OF TOTAL PURINES IS EXPRESSSED IN MG URIC ACID/100G

1 DEPENDENT ON THE METHOD OF EXTRACTION; WITH PRECEDING PROCEDURES VALUES OF 250 - 350 MG URIC ACID/100G HAD BEEN OBTAINED

KALBSLUNGE — CALF'S LUNGS (LIGHTS) — MOU DE VEAU

		PROTEIN	FAT	CARBOHYDRATES	TOTAL
ENERGY VALUE (AVERAGE) PER 100 G EDIBLE PORTION	KJOULE	322	86	0.0	408
	(KCAL)	77	21	0.0	98
AMOUNT OF DIGESTIBLE CONSTITUENTS PER 100 G	GRAM	16.97	2.06	0.00	
ENERGY VALUE (AVERAGE) OF THE DIGESTIBLE FRACTION PER 100 G EDIBLE PORTION	KJOULE	313	82	0.0	394
	(KCAL)	75	20	0.0	94

WASTE PERCENTAGE AVERAGE 24 MINIMUM 15 MAXIMUM 34

CONSTITUENTS	DIM	AV	VARIATION			AVR	NUTR.	DENS.	MOLPERC.
WATER	GRAM	79.20	79.00	-	79.60	60.19	GRAM/MJ	200.79	
PROTEIN	GRAM	17.50	16.60	-	18.30	13.30	GRAM/MJ	44.37	
FAT	GRAM	2.17	2.00	-	2.33	1.65	GRAM/MJ	5.50	
MINERALS	GRAM	1.14	1.00	-	1.28	0.87	GRAM/MJ	2.89	
SODIUM	MILLI	154.00	135.00	-	173.00	117.04	MILLI/MJ	390.42	
POTASSIUM	MILLI	303.00	297.00	-	308.00	230.28	MILLI/MJ	768.17	
CALCIUM	MILLI	5.00	4.00	-	5.00	3.80	MILLI/MJ	12.68	
IRON	MILLI	5.00	-	-	-	3.80	MILLI/MJ	12.68	
VITAMIN B1	MILLI	0.11	-	-	-	0.08	MILLI/MJ	0.28	
VITAMIN B2	MILLI	0.36	-	-	-	0.27	MILLI/MJ	0.91	
NICOTINAMIDE	MILLI	4.00	-	-	-	3.04	MILLI/MJ	10.14	
PANTOTHENIC ACID	MILLI	1.00	-	-	-	0.76	MILLI/MJ	2.54	
VITAMIN B6	MICRO	70.00	-	-	-	53.20	MICRO/MJ	177.47	
BIOTIN	MICRO	5.90	-	-	-	4.48	MICRO/MJ	14.96	
VITAMIN B12	MICRO	3.30	-	-	-	2.51	MICRO/MJ	8.37	
VITAMIN C	MILLI	39.30	-	-	-	29.87	MILLI/MJ	99.63	
ARGININE	GRAM	1.07	-	-	-	0.81	GRAM/MJ	2.71	
HISTIDINE	GRAM	0.40	-	-	-	0.30	GRAM/MJ	1.01	
ISOLEUCINE	GRAM	0.77	-	-	-	0.59	GRAM/MJ	1.95	
LEUCINE	GRAM	1.45	-	-	-	1.10	GRAM/MJ	3.68	
LYSINE	GRAM	1.35	-	-	-	1.03	GRAM/MJ	3.42	
METHIONINE	GRAM	0.32	-	-	-	0.24	GRAM/MJ	0.81	
PHENYLALANINE	GRAM	0.79	-	-	-	0.60	GRAM/MJ	2.00	
THREONINE	GRAM	0.72	-	-	-	0.55	GRAM/MJ	1.83	
TRYPTOPHAN	GRAM	0.18	-	-	-	0.14	GRAM/MJ	0.46	
VALINE	GRAM	1.03	-	-	-	0.78	GRAM/MJ	2.61	

KALBSMILZ CALF'S SPLEEN RATE DE VEAU

		PROTEIN	FAT	CARBOHYDRATES	TOTAL
ENERGY VALUE (AVERAGE)	KJOULE	335	119	0.0	454
PER 100 G	(KCAL)	80	29	0.0	109
EDIBLE PORTION					
AMOUNT OF DIGESTIBLE	GRAM	17.65	2.85	0.00	
CONSTITUENTS PER 100 G					
ENERGY VALUE (AVERAGE)	KJOULE	325	113	0.0	438
OF THE DIGESTIBLE	(KCAL)	78	27	0.0	105
FRACTION PER 100 G					
EDIBLE PORTION					

WASTE PERCENTAGE AVERAGE 0.00

CONSTITUENTS	DIM	AV	VARIATION			AVR	NUTR. DENS.		MOLPERC.
WATER	GRAM	77.30	76.90	-	78.30	77.30	GRAM/MJ	176.37	
PROTEIN	GRAM	18.20	18.10	-	18.30	18.20	GRAM/MJ	41.53	
FAT	GRAM	3.00	-	-	-	3.00	GRAM/MJ	6.84	
MINERALS	GRAM	1.35	1.30	-	1.40	1.35	GRAM/MJ	3.08	
SODIUM	MILLI	114.00	-	-	-	114.00	MILLI/MJ	260.10	
POTASSIUM	MILLI	400.00	-	-	-	400.00	MILLI/MJ	912.65	
CALCIUM	MILLI	3.00	-	-	-	3.00	MILLI/MJ	6.84	
IRON	MILLI	9.70	8.70	-	10.60	9.70	MILLI/MJ	22.13	
PHOSPHORUS	MILLI	272.00	-	-	-	272.00	MILLI/MJ	620.60	
VITAMIN B2	MILLI	0.37	-	-	-	0.37	MILLI/MJ	0.84	
NICOTINAMIDE	MILLI	8.20	-	-	-	8.20	MILLI/MJ	18.71	
VITAMIN C	MILLI	41.00	-	-	-	41.00	MILLI/MJ	93.55	

KALBSNIERE CALF'S KIDNEY ROGNON DE VEAU

		PROTEIN	FAT	CARBOHYDRATES	TOTAL
ENERGY VALUE (AVERAGE) PER 100 G EDIBLE PORTION	KJOULE	307	253	0.0	561
	(KCAL)	73	61	0.0	134
AMOUNT OF DIGESTIBLE CONSTITUENTS PER 100 G	GRAM	16.19	6.05	0.00	
ENERGY VALUE (AVERAGE) OF THE DIGESTIBLE FRACTION PER 100 G EDIBLE PORTION	KJOULE	298	241	0.0	539
	(KCAL)	71	57	0.0	129

WASTE PERCENTAGE AVERAGE 12 MINIMUM 10 MAXIMUM 15

CONSTITUENTS	DIM	AV	VARIATION			AVR	NUTR. DENS.		MOLPERC.
WATER	GRAM	75.00	-	-	-	66.00	GRAM/MJ	139.21	
PROTEIN	GRAM	16.70	15.00	-	18.00	14.70	GRAM/MJ	31.00	
FAT	GRAM	6.37	5.00	-	8.10	5.61	GRAM/MJ	11.82	
MINERALS	GRAM	1.10	-	-	-	0.97	GRAM/MJ	2.04	
SODIUM	MILLI	200.00	180.00	-	250.00	176.00	MILLI/MJ	371.23	
POTASSIUM	MILLI	290.00	270.00	-	300.00	255.20	MILLI/MJ	538.28	
MAGNESIUM	MILLI	18.00	-	-	-	15.84	MILLI/MJ	33.41	
CALCIUM	MILLI	10.00	-	-	-	8.80	MILLI/MJ	18.56	
MANGANESE	MICRO	50.00	-	-	-	44.00	MICRO/MJ	92.81	
IRON	MILLI	11.50	7.90	-	15.00	10.12	MILLI/MJ	21.35	
COBALT	MICRO	10.00	-	-	-	8.80	MICRO/MJ	18.56	
COPPER	MICRO	0.37	-	-	-	0.33	MICRO/MJ	0.69	
ZINC	MILLI	1.80	-	-	-	1.58	MILLI/MJ	3.34	
PHOSPHORUS	MILLI	260.00	220.00	-	300.00	228.80	MILLI/MJ	482.60	
FLUORIDE	MILLI	0.20	-	-	-	0.18	MILLI/MJ	0.37	
SELENIUM	MILLI	0.26	0.11	-	0.34	0.23	MILLI/MJ	0.48	
VITAMIN A	MILLI	0.21	0.05	-	0.35	0.18	MILLI/MJ	0.39	
VITAMIN B1	MILLI	0.37	0.31	-	0.41	0.33	MILLI/MJ	0.69	
VITAMIN B2	MILLI	2.48	2.40	-	2.55	2.18	MILLI/MJ	4.60	
NICOTINAMIDE	MILLI	6.47	5.20	-	7.00	5.69	MILLI/MJ	12.01	
PANTOTHENIC ACID	MILLI	4.00	-	-	-	3.52	MILLI/MJ	7.42	
VITAMIN B6	MILLI	0.50	0.38	-	0.62	0.44	MILLI/MJ	0.93	
BIOTIN	MICRO	80.00	-	-	-	70.40	MICRO/MJ	148.49	
FOLIC ACID	MICRO	63.00	-	-	-	55.44	MICRO/MJ	116.94	
VITAMIN B12	MICRO	25.00	14.00	-	40.00	22.00	MICRO/MJ	46.40	
VITAMIN C	MILLI	12.70	10.00	-	20.00	11.18	MILLI/MJ	23.57	

CONSTITUENTS	DIM	AV	VARIATION			AVR	NUTR. DENS.		MOLPERC.
ARGININE	GRAM	1.04	0.97	-	1.16	0.92	GRAM/MJ	1.93	
CYSTINE	GRAM	0.20	0.17	-	0.24	0.18	GRAM/MJ	0.37	
HISTIDINE	GRAM	0.42	0.36	-	0.50	0.37	GRAM/MJ	0.78	
ISOLEUCINE	GRAM	0.81	0.72	-	1.04	0.71	GRAM/MJ	1.50	
LEUCINE	GRAM	1.45	1.29	-	1.67	1.28	GRAM/MJ	2.69	
LYSINE	GRAM	1.21	1.04	-	1.40	1.06	GRAM/MJ	2.25	
METHIONINE	GRAM	0.34	0.28	-	0.45	0.30	GRAM/MJ	0.63	
PHENYLALANINE	GRAM	0.79	0.72	-	0.91	0.70	GRAM/MJ	1.47	
THREONINE	GRAM	0.74	0.59	-	0.82	0.65	GRAM/MJ	1.37	
TRYPTOPHAN	GRAM	0.25	0.17	-	0.35	0.22	GRAM/MJ	0.46	
TYROSINE	GRAM	0.62	0.44	-	0.77	0.55	GRAM/MJ	1.15	
VALINE	GRAM	0.98	0.87	-	1.11	0.86	GRAM/MJ	1.82	
MYRISTIC ACID	MILLI	183.00	-	-	-	161.04	MILLI/MJ	339.67	
PALMITIC ACID	GRAM	2.16	-	-	-	1.90	GRAM/MJ	4.01	
STEARIC ACID	MILLI	579.00	-	-	-	509.52	GRAM/MJ	1.07	
PALMITOLEIC ACID	MILLI	365.00	-	-	-	321.20	MILLI/MJ	677.49	
OLEIC ACID	GRAM	2.56	-	-	-	2.25	GRAM/MJ	4.75	
C14 : 1 FATTY ACID	MILLI	91.00	-	-	-	80.08	MILLI/MJ	168.91	
LINOLEIC ACID	MILLI	61.00	-	-	-	53.68	MILLI/MJ	113.22	
LINOLENIC ACID	MILLI	61.00	-	-	-	53.68	MILLI/MJ	113.22	
CHOLESTEROL	MILLI	380.00	-	-	-	334.40	MILLI/MJ	705.33	

KALBSZUNGE   CALF'S TONGUE   LANGUE DE VEAU

		PROTEIN	FAT	CARBOHYDRATES	TOTAL
ENERGY VALUE (AVERAGE)	KJOULE	315	246	0.0	561
PER 100 G	(KCAL)	75	59	0.0	134
EDIBLE PORTION					
AMOUNT OF DIGESTIBLE	GRAM	16.58	5.89	0.00	
CONSTITUENTS PER 100 G					
ENERGY VALUE (AVERAGE)	KJOULE	305	234	0.0	539
OF THE DIGESTIBLE	(KCAL)	73	56	0.0	129
FRACTION PER 100 G					
EDIBLE PORTION					

WASTE PERCENTAGE AVERAGE 23

CONSTITUENTS	DIM	AV	VARIATION			AVR	NUTR.	DENS.	MOLPERC.
WATER	GRAM	77.40	74.90	-	80.00	59.60	GRAM/MJ	143.47	
PROTEIN	GRAM	17.10	16.30	-	17.90	13.17	GRAM/MJ	31.70	
FAT	GRAM	6.20	-	-	-	4.77	GRAM/MJ	11.49	
MINERALS	GRAM	0.91	-	-	-	0.70	GRAM/MJ	1.69	
SODIUM	MILLI	84.00	71.00	-	96.00	64.68	MILLI/MJ	155.71	
POTASSIUM	MILLI	200.00	189.00	-	212.00	154.00	MILLI/MJ	370.73	
CALCIUM	MILLI	9.00	-	-	-	6.93	MILLI/MJ	16.68	
IRON	MILLI	3.00	2.80	-	3.10	2.31	MILLI/MJ	5.56	
PHOSPHORUS	MILLI	190.00	-	-	-	146.30	MILLI/MJ	352.19	
VITAMIN B1	MILLI	0.15	0.12	-	0.16	0.12	MILLI/MJ	0.28	
VITAMIN B2	MILLI	0.29	0.28	-	0.30	0.22	MILLI/MJ	0.54	
NICOTINAMIDE	MILLI	3.70	3.00	-	5.00	2.85	MILLI/MJ	6.86	
VITAMIN B6	MILLI	0.13	-	-	-	0.10	MILLI/MJ	0.24	
BIOTIN	MICRO	3.30	-	-	-	2.54	MICRO/MJ	6.12	
ARGININE	GRAM	1.11	-	-	-	0.85	GRAM/MJ	2.06	
HISTIDINE	GRAM	0.45	-	-	-	0.35	GRAM/MJ	0.83	
ISOLEUCINE	GRAM	0.72	-	-	-	0.55	GRAM/MJ	1.33	
LEUCINE	GRAM	1.45	-	-	-	1.12	GRAM/MJ	2.69	
LYSINE	GRAM	1.49	-	-	-	1.15	GRAM/MJ	2.76	
METHIONINE	GRAM	0.39	-	-	-	0.30	GRAM/MJ	0.72	
PHENYLALANINE	GRAM	0.72	-	-	-	0.55	GRAM/MJ	1.33	
THREONINE	GRAM	0.75	-	-	-	0.58	GRAM/MJ	1.39	
TRYPTOPHAN	GRAM	0.14	-	-	-	0.11	GRAM/MJ	0.26	
VALINE	GRAM	0.91	-	-	-	0.70	GRAM/MJ	1.69	

RINDFLEISCH	**BEEF**	**VIANDE DE BOEUF**
REINES MUSKELFLEISCH	MUSCLES ONLY	VIANDE DE MUSCLE

		PROTEIN	FAT	CARBOHYDRATES	TOTAL
ENERGY VALUE (AVERAGE)	KJOULE	405	76	0.0	481
PER 100 G EDIBLE PORTION	(KCAL)	97	18	0.0	115
AMOUNT OF DIGESTIBLE CONSTITUENTS PER 100 G	GRAM	21.34	1.80	0.00	
ENERGY VALUE (AVERAGE)	KJOULE	393	72	0.0	465
OF THE DIGESTIBLE FRACTION PER 100 G EDIBLE PORTION	(KCAL)	94	17	0.0	111

WASTE PERCENTAGE AVERAGE 0.00

CONSTITUENTS	DIM	AV	VARIATION		AVR	NUTR. DENS.		MOLPERC.
WATER	GRAM	75.10	74.40	- 77.00	75.10	GRAM/MJ	161.64	
PROTEIN	GRAM	22.00	20.60	- 22.70	22.00	GRAM/MJ	47.35	
FAT	GRAM	1.90	1.55	- 3.00	1.90	GRAM/MJ	4.09	
AVAILABLE CARBOHYDR.	GRAM	1.05	-	- -	1.05	GRAM/MJ	2.26	
MINERALS	GRAM	1.23	1.20	- 1.27	1.23	GRAM/MJ	2.65	
SODIUM	MILLI	57.00	48.00	- 60.00	57.00	MILLI/MJ	122.68	
POTASSIUM	MILLI	370.00	325.00	- 440.00	370.00	MILLI/MJ	796.37	
MAGNESIUM	MILLI	21.00	14.00	- 26.00	21.00	MILLI/MJ	45.20	
CALCIUM	MILLI	3.50	3.00	- 4.50	3.50	MILLI/MJ	7.53	
MANGANESE	MICRO	22.00	-	- -	22.00	MICRO/MJ	47.35	
IRON	MILLI	1.90	1.40	- 2.10	1.90	MILLI/MJ	4.09	
COBALT	MICRO	0.27	-	- -	0.27	MICRO/MJ	0.59	
COPPER	MICRO	65.00	50.00	- 90.00	65.00	MICRO/MJ	139.90	
ZINC	MILLI	4.20	2.40	- 6.10	4.20	MILLI/MJ	9.04	
NICKEL	MICRO	1.00	-	- -	1.00	MICRO/MJ	2.15	
CHROMIUM	MICRO	14.00	-	- -	14.00	MICRO/MJ	30.13	
MOLYBDENUM	MICRO	28.00	10.00	- 45.00	28.00	MICRO/MJ	60.27	
ALUMINIUM	MICRO	-	5.00	- 170.00				
PHOSPHORUS	MILLI	194.00	171.00	- 217.00	194.00	MILLI/MJ	417.56	
CHLORIDE	MILLI	52.00	34.00	- 91.00	52.00	MILLI/MJ	111.92	
FLUORIDE	MILLI	-	0.01	- 0.10				
BORON	MICRO	45.00	10.00	- 144.00	45.00	MICRO/MJ	96.86	
SELENIUM	MICRO	3.00	-	- -	3.00	MICRO/MJ	6.46	
VITAMIN A	MICRO	20.00	-	- -	20.00	MICRO/MJ	43.05	
VITAMIN E ACTIVITY	MILLI	0.54	0.20	- 0.65	0.54	MILLI/MJ	1.16	
ALPHA-TOCOPHEROL	MILLI	0.54	0.20	- 0.65	0.54	MILLI/MJ	1.16	

CONSTITUENTS	DIM	AV	VARIATION			AVR	NUTR. DENS.		MOLPERC.
VITAMIN K	MICRO	21.00	-	-	-	21.00	MICRO/MJ	45.20	
VITAMIN B1	MILLI	0.23	-	-	-	0.23	MILLI/MJ	0.50	
VITAMIN B2	MILLI	0.26	-	-	-	0.26	MILLI/MJ	0.56	
NICOTINAMIDE	MILLI	7.50	-	-	-	7.50	MILLI/MJ	16.14	
PANTOTHENIC ACID	MILLI	0.60	-	-	-	0.60	MILLI/MJ	1.29	
VITAMIN B6	MILLI	0.40	-	-	-	0.40	MILLI/MJ	0.86	
BIOTIN	MICRO	3.00	-	-	-	3.00	MICRO/MJ	6.46	
FOLIC ACID	MICRO	15.30	-	-	-	15.30	MICRO/MJ	32.93	
VITAMIN B12	MICRO	5.00	1.00	-	8.00	5.00	MICRO/MJ	10.76	
ALANINE	GRAM	1.69	-	-	-	1.69	GRAM/MJ	3.64	9.4
ARGININE	GRAM	1.54	-	-	-	1.54	GRAM/MJ	3.31	4.4
ASPARTIC ACID	GRAM	2.34	-	-	-	2.34	GRAM/MJ	5.04	8.7
CYSTINE	GRAM	0.28	-	-	-	0.28	GRAM/MJ	0.60	0.6
GLUTAMIC ACID	GRAM	4.13	-	-	-	4.13	GRAM/MJ	8.89	14.0
GLYCINE	GRAM	1.56	-	-	-	1.56	GRAM/MJ	3.36	10.3
HISTIDINE	GRAM	0.85	-	-	-	0.85	GRAM/MJ	1.83	2.7
ISOLEUCINE	GRAM	1.25	-	-	-	1.25	GRAM/MJ	2.69	4.7
LEUCINE	GRAM	1.95	-	-	-	1.95	GRAM/MJ	4.20	7.4
LYSINE	GRAM	2.31	-	-	-	2.31	GRAM/MJ	4.97	7.9
METHIONINE	GRAM	0.65	-	-	-	0.65	GRAM/MJ	1.40	2.2
PHENYLALANINE	GRAM	1.06	-	-	-	1.06	GRAM/MJ	2.28	3.2
PROLINE	GRAM	1.28	-	-	-	1.28	GRAM/MJ	2.76	5.5
SERINE	GRAM	1.14	-	-	-	1.14	GRAM/MJ	2.45	5.4
THREONINE	GRAM	1.15	-	-	-	1.15	GRAM/MJ	2.48	4.8
TRYPTOPHAN	GRAM	0.29	-	-	-	0.29	GRAM/MJ	0.62	0.7
TYROSINE	GRAM	0.89	-	-	-	0.89	GRAM/MJ	1.92	2.5
VALINE	GRAM	1.32	-	-	-	1.32	GRAM/MJ	2.84	5.6
LACTIC ACID	MILLI	500.00	200.00	-	800.00	500.00	GRAM/MJ	1.08	
SUCCINIC ACID	MILLI	50.00	-	-	-	50.00	MILLI/MJ	107.62	
GLYCOLIC ACID	MILLI	100.00	-	-	-	100.00	MILLI/MJ	215.24	
GLUCOSE	MILLI	27.50	5.00	-	50.00	27.50	MILLI/MJ	59.19	
FRUCTOSE	MILLI	16.50	3.00	-	30.00	16.50	MILLI/MJ	35.51	
GLYCOGENE	GRAM	0.51	0.02	-	1.00	0.51	GRAM/MJ	1.10	
MYOINOSITOL	MILLI	15.00	10.00	-	20.00	15.00	MILLI/MJ	32.29	
CAPRIC ACID	MILLI	1.70	-	-	-	1.70	MILLI/MJ	3.66	
LAURIC ACID	MILLI	1.70	-	-	-	1.70	MILLI/MJ	3.66	
MYRISTIC ACID	MILLI	56.00	-	-	-	56.00	MILLI/MJ	120.53	
C15 : 0 FATTY ACID	MILLI	9.00	-	-	-	9.00	MILLI/MJ	19.37	
PALMITIC ACID	GRAM	0.42	-	-	-	0.42	GRAM/MJ	0.90	
C17 : 0 FATTY ACID	MILLI	27.00	-	-	-	27.00	MILLI/MJ	58.11	
STEARIC ACID	GRAM	0.25	-	-	-	0.25	GRAM/MJ	0.54	
PALMITOLEIC ACID	MILLI	85.00	-	-	-	85.00	MILLI/MJ	182.95	
OLEIC ACID	GRAM	0.72	-	-	-	0.72	GRAM/MJ	1.55	
C14 : 1 FATTY ACID	MILLI	15.00	-	-	-	15.00	MILLI/MJ	32.29	
LINOLEIC ACID	MILLI	80.00	-	-	-	80.00	MILLI/MJ	172.19	
CHOLESTEROL	MILLI	60.00	-	-	-	60.00	MILLI/MJ	129.14	
TOTAL PURINES	MILLI	133.00 [1]	133.00	-	133.00	133.00	MILLI/MJ	286.26	
ADENINE	MILLI	15.00	15.00	-	15.00	15.00	MILLI/MJ	32.29	
GUANINE	MILLI	16.00	16.00	-	16.00	16.00	MILLI/MJ	34.44	
XANTHINE	MILLI	3.00	3.00	-	3.00	3.00	MILLI/MJ	6.46	
HYPOXANTHINE	MILLI	75.00	75.00	-	75.00	75.00	MILLI/MJ	161.43	
TOTAL PHOSPHOLIPIDS	MILLI	827.00	-	-	-	827.00	GRAM/MJ	1.78	
PHOSPHATIDYLCHOLINE	MILLI	429.00	-	-	-	429.00	MILLI/MJ	923.36	
PHOSPHATIDYLETHANOLAMINE	MILLI	197.00	-	-	-	197.00	MILLI/MJ	424.02	
PHOSPHATIDYLSERINE	MILLI	69.00	-	-	-	69.00	MILLI/MJ	148.51	

1 VALUE OF TOTAL PURINES IS EXPRESSSED IN MG URIC ACID/100G

RINDFLEISCH INTERMUSKULARES FETTGEWEBE

BEEF INTERMUSCULARE ADIPOSE TISSUE

VIANDE DE BOEUF TISSU ADIPEUX INTERMUSCULAIRE

		PROTEIN	FAT	CARBOHYDRATES	TOTAL
ENERGY VALUE (AVERAGE) PER 100 G EDIBLE PORTION	KJOULE	151	2818	0.0	2969
	(KCAL)	36	674	0.0	710
AMOUNT OF DIGESTIBLE CONSTITUENTS PER 100 G	GRAM	7.95	67.35	0.00	
ENERGY VALUE (AVERAGE) OF THE DIGESTIBLE FRACTION PER 100 G EDIBLE PORTION	KJOULE	146	2677	0.0	2824
	(KCAL)	35	640	0.0	675

WASTE PERCENTAGE AVERAGE 0.00

CONSTITUENTS	DIM	AV	VARIATION	AVR	NUTR. DENS.		MOLPERC.
WATER	GRAM	20.20	20.20 - 20.20	20.20	GRAM/MJ	7.15	
PROTEIN	GRAM	8.20	8.20 - 8.20	8.20	GRAM/MJ	2.90	
FAT	GRAM	70.90	70.90 - 70.90	70.90	GRAM/MJ	25.11	
MINERALS	GRAM	0.30	0.30 - 0.30	0.30	GRAM/MJ	0.11	
SODIUM	MILLI	29.00	29.00 - 29.00	29.00	MILLI/MJ	10.27	
POTASSIUM	MILLI	66.00	66.00 - 66.00	66.00	MILLI/MJ	23.37	
IRON	MILLI	0.75	0.75 - 0.75	0.75	MILLI/MJ	0.27	
ZINC	MILLI	1.00	1.00 - 1.00	1.00	MILLI/MJ	0.35	
VITAMIN B1	MICRO	40.00	40.00 - 40.00	40.00	MICRO/MJ	14.17	
VITAMIN B2	MICRO	70.00	70.00 - 70.00	70.00	MICRO/MJ	24.79	
NICOTINAMIDE	MILLI	1.42	1.42 - 1.42	1.42	MILLI/MJ	0.50	
VITAMIN B12	MICRO	1.46	1.46 - 1.46	1.46	MICRO/MJ	0.52	
CHOLESTEROL	MILLI	99.00	99.00 - 99.00	99.00	MILLI/MJ	35.06	

RINDFLEISCH
SUBCUTANES FETTGEWEBE

BEEF
SUBCUTANEOUS ADIPOSE TISSUE

VIANDE DE BOEUF
TISSU ADIPEUX SOUSCUTAN

		PROTEIN	FAT	CARBOHYDRATES	TOTAL
ENERGY VALUE (AVERAGE)	KJOULE	28	3736	0.0	3764
PER 100 G	(KCAL)	6.6	893	0.0	900
EDIBLE PORTION					
AMOUNT OF DIGESTIBLE	GRAM	1.45	89.30	0.00	
CONSTITUENTS PER 100 G					
ENERGY VALUE (AVERAGE)	KJOULE	27	3549	0.0	3576
OF THE DIGESTIBLE	(KCAL)	6.4	848	0.0	855
FRACTION PER 100 G					
EDIBLE PORTION					

WASTE PERCENTAGE AVERAGE 0.00

CONSTITUENTS	DIM	AV	VARIATION			AVR	NUTR. DENS.		MOLPERC.
WATER	GRAM	4.00	4.00	-	4.00	4.00	GRAM/MJ	1.12	
PROTEIN	GRAM	1.50	1.50	-	1.50	1.50	GRAM/MJ	0.42	
FAT	GRAM	94.00	94.00	-	94.00	94.00	GRAM/MJ	26.28	
MINERALS	GRAM	0.10	0.10	-	0.10	0.10	GRAM/MJ	0.03	
CHOLESTEROL	MILLI	68.00	68.00	-	68.00	68.00	MILLI/MJ	19.01	

RINDERBLUT OX BLOOD SANG DE BOEUF

		PROTEIN	FAT	CARBOHYDRATES	TOTAL
ENERGY VALUE (AVERAGE)	KJOULE	328	5.2	0.0	333
PER 100 G	(KCAL)	78	1.2	0.0	80
EDIBLE PORTION					
AMOUNT OF DIGESTIBLE	GRAM	17.26	0.12	0.00	
CONSTITUENTS PER 100 G					
ENERGY VALUE (AVERAGE)	KJOULE	318	4.9	0.0	323
OF THE DIGESTIBLE	(KCAL)	76	1.2	0.0	77
FRACTION PER 100 G					
EDIBLE PORTION					

WASTE PERCENTAGE AVERAGE 0.00

CONSTITUENTS	DIM	AV	VARIATION			AVR	NUTR. DENS.		MOLPERC.
WATER	GRAM	80.50	80.00	-	80.80	80.50	GRAM/MJ	249.40	
PROTEIN	GRAM	17.80	17.00	-	18.10	17.80	GRAM/MJ	55.15	
FAT	GRAM	0.13	0.06	-	0.18	0.13	GRAM/MJ	0.40	
MINERALS	GRAM	0.85	-	-	-	0.85	GRAM/MJ	2.63	
SODIUM	MILLI	330.00	300.00	-	360.00	330.00	GRAM/MJ	1.02	
POTASSIUM	MILLI	44.00	34.00	-	54.00	44.00	MILLI/MJ	136.32	
MAGNESIUM	MILLI	5.00	4.00	-	5.00	5.00	MILLI/MJ	15.49	
CALCIUM	MILLI	6.00	5.00	-	7.00	6.00	MILLI/MJ	18.59	
IRON	MILLI	49.00	38.00	-	60.00	49.00	MILLI/MJ	151.81	
ZINC	MILLI	0.20	-	-	-	0.20	MILLI/MJ	0.62	
PHOSPHORUS	MILLI	19.00	18.00	-	20.00	19.00	MILLI/MJ	58.87	
CHLORIDE	MILLI	304.00	290.00	-	318.00	304.00	MILLI/MJ	941.85	
VITAMIN A	MICRO	21.00	16.00	-	30.00	21.00	MICRO/MJ	65.06	
VITAMIN B1	MICRO	90.00	-	-	-	90.00	MICRO/MJ	278.84	
VITAMIN B2	MICRO	30.00	-	-	-	30.00	MICRO/MJ	92.95	
NICOTINAMIDE	MILLI	0.85	-	-	-	0.85	MILLI/MJ	2.63	
ARGININE	GRAM	1.25	-	-	-	1.25	GRAM/MJ	3.87	
CYSTINE	GRAM	0.18	-	-	-	0.18	GRAM/MJ	0.56	
HISTIDINE	GRAM	1.92	-	-	-	1.92	GRAM/MJ	5.95	
ISOLEUCINE	GRAM	0.40	-	-	-	0.40	GRAM/MJ	1.24	
LEUCINE	GRAM	2.80	-	-	-	2.80	GRAM/MJ	8.67	
LYSINE	GRAM	3.27	-	-	-	3.27	GRAM/MJ	10.13	
METHIONINE	GRAM	0.78	-	-	-	0.78	GRAM/MJ	2.42	
PHENYLALANINE	GRAM	1.44	-	-	-	1.44	GRAM/MJ	4.46	
THREONINE	GRAM	1.28	-	-	-	1.28	GRAM/MJ	3.97	
TYROSINE	GRAM	0.77	-	-	-	0.77	GRAM/MJ	2.39	
VALINE	GRAM	2.13	-	-	-	2.13	GRAM/MJ	6.60	
CHOLESTEROL	MILLI	190.00	-	-	-	190.00	MILLI/MJ	588.66	

RINDERHERZ — OX HEART — COEUR DE BOEUF

		PROTEIN	FAT	CARBOHYDRATES	TOTAL
ENERGY VALUE (AVERAGE) PER 100 G EDIBLE PORTION	KJOULE	309	238	0.0	548
	(KCAL)	74	57	0.0	131
AMOUNT OF DIGESTIBLE CONSTITUENTS PER 100 G	GRAM	16.29	5.70	0.00	
ENERGY VALUE (AVERAGE) OF THE DIGESTIBLE FRACTION PER 100 G EDIBLE PORTION	KJOULE	300	227	0.0	527
	(KCAL)	72	54	0.0	126

WASTE PERCENTAGE AVERAGE 21

CONSTITUENTS	DIM	AV	VARIATION			AVR	NUTR. DENS.		MOLPERC.
WATER	GRAM	75.50	70.90	-	78.00	59.65	GRAM/MJ	143.38	
PROTEIN	GRAM	16.80	16.50	-	16.90	13.27	GRAM/MJ	31.90	
FAT	GRAM	6.00	3.70	-	10.10	4.74	GRAM/MJ	11.39	
MINERALS	GRAM	1.10	-	-	-	0.87	GRAM/MJ	2.09	
SODIUM	MILLI	108.00	90.00	-	124.00	85.32	MILLI/MJ	205.10	
POTASSIUM	MILLI	286.00	257.00	-	308.00	225.94	MILLI/MJ	543.14	
MAGNESIUM	MILLI	25.00	-	-	-	19.75	MILLI/MJ	47.48	
CALCIUM	MILLI	9.30	9.00	-	10.00	7.35	MILLI/MJ	17.66	
MANGANESE	MICRO	59.00	-	-	-	46.61	MICRO/MJ	112.05	
IRON	MILLI	5.13	4.60	-	6.20	4.05	MILLI/MJ	9.74	
COPPER	MILLI	0.41	-	-	-	0.32	MILLI/MJ	0.78	
ZINC	MILLI	2.00	-	-	-	1.58	MILLI/MJ	3.80	
MOLYBDENUM	MICRO	20.00	-	-	-	15.80	MICRO/MJ	37.98	
PHOSPHORUS	MILLI	214.00	203.00	-	236.00	169.06	MILLI/MJ	406.41	
FLUORIDE	MICRO	60.00	-	-	-	47.40	MICRO/MJ	113.95	
IODIDE	MICRO	30.00	-	-	-	23.70	MICRO/MJ	56.97	
BORON	MILLI	0.12	0.09	-	0.15	0.10	MILLI/MJ	0.23	
SELENIUM	MICRO	47.00	27.00	-	68.00	37.13	MICRO/MJ	89.26	
VITAMIN A	MICRO	-	0.00	-	9.00				
VITAMIN E ACTIVITY	MILLI	0.60	-	-	-	0.47	MILLI/MJ	1.14	
ALPHA-TOCOPHEROL	MILLI	0.60	-	-	-	0.47	MILLI/MJ	1.14	
VITAMIN B1	MILLI	0.51	0.24	-	0.67	0.40	MILLI/MJ	0.97	
VITAMIN B2	MILLI	0.91	0.84	-	1.00	0.72	MILLI/MJ	1.73	
NICOTINAMIDE	MILLI	7.18	6.60	-	7.80	5.67	MILLI/MJ	13.64	
PANTOTHENIC ACID	MILLI	2.78	2.75	-	2.81	2.20	MILLI/MJ	5.28	
VITAMIN B6	MILLI	0.28	0.26	-	0.29	0.22	MILLI/MJ	0.53	

CONSTITUENTS	DIM	AV	VARIATION			AVR	NUTR. DENS.		MOLPERC.
BIOTIN	MICRO	7.30	-	-	-	5.77	MICRO/MJ	13.86	
VITAMIN B12	MICRO	9.90	9.70	-	10.00	7.82	MICRO/MJ	18.80	
VITAMIN C	MILLI	5.50	4.50	-	6.00	4.35	MILLI/MJ	10.45	
ALANINE	GRAM	1.28	1.20	-	1.34	1.01	GRAM/MJ	2.43	9.4
ARGININE	GRAM	1.20	1.19	-	1.21	0.95	GRAM/MJ	2.28	4.5
ASPARTIC ACID	GRAM	1.76	1.69	-	1.81	1.39	GRAM/MJ	3.34	8.6
CYSTINE	GRAM	0.25	0.18	-	0.30	0.20	GRAM/MJ	0.47	0.7
GLUTAMIC ACID	GRAM	2.45	1.91	-	2.98	1.94	GRAM/MJ	4.65	10.9
GLYCINE	GRAM	1.21	1.04	-	1.37	0.96	GRAM/MJ	2.30	10.5
HISTIDINE	GRAM	0.50	0.48	-	0.53	0.40	GRAM/MJ	0.95	2.1
ISOLEUCINE	GRAM	1.16	1.07	-	1.48	0.92	GRAM/MJ	2.20	5.8
LEUCINE	GRAM	1.83	1.83	-	1.85	1.45	GRAM/MJ	3.48	9.1
LYSINE	GRAM	1.75	1.65	-	1.90	1.38	GRAM/MJ	3.32	7.8
METHIONINE	GRAM	0.51	0.43	-	0.68	0.40	GRAM/MJ	0.97	2.2
PHENYLALANINE	GRAM	0.88	0.84	-	0.90	0.70	GRAM/MJ	1.67	3.5
PROLINE	GRAM	0.81	0.72	-	0.87	0.64	GRAM/MJ	1.54	4.6
SERINE	GRAM	1.00	0.91	-	1.10	0.79	GRAM/MJ	1.90	6.2
THREONINE	GRAM	0.92	0.88	-	0.94	0.73	GRAM/MJ	1.75	5.0
TRYPTOPHAN	GRAM	0.22	0.19	-	0.23	0.17	GRAM/MJ	0.42	0.7
TYROSINE	GRAM	0.65	0.64	-	0.67	0.51	GRAM/MJ	1.23	2.3
VALINE	GRAM	1.11	0.98	-	1.15	0.88	GRAM/MJ	2.11	6.2
LINOLEIC ACID	GRAM	0.10	0.09	-	0.12	0.08	GRAM/MJ	0.19	
LINOLENIC ACID	MILLI	60.00	-	-	-	47.40	MILLI/MJ	113.95	
ARACHIDONIC ACID	MILLI	48.00	-	-	-	37.92	MILLI/MJ	91.16	
CHOLESTEROL	MILLI	150.00	-	-	-	118.50	MILLI/MJ	284.86	
TOTAL PURINES	MILLI	256.00 1	256.00	-	256.00	202.24	MILLI/MJ	486.17	
ADENINE	MILLI	47.00	47.00	-	47.00	37.13	MILLI/MJ	89.26	
GUANINE	MILLI	31.00	31.00	-	31.00	24.49	MILLI/MJ	58.87	
XANTHINE	MILLI	4.00	4.00	-	4.00	3.16	MILLI/MJ	7.60	
HYPOXANTHINE	MILLI	129.00	129.00	-	129.00	101.91	MILLI/MJ	244.98	

1 VALUE OF TOTAL PURINES IS EXPRESSSED IN MG URIC ACID/100G

RINDERHIRN | OX BRAIN | CERVELLE DE BOEUF

		PROTEIN	FAT	CARBOHYDRATES	TOTAL
ENERGY VALUE (AVERAGE)	KJOULE	191	383	0.0	574
PER 100 G	(KCAL)	46	91	0.0	137
EDIBLE PORTION					
AMOUNT OF DIGESTIBLE	GRAM	10.08	9.14	0.00	
CONSTITUENTS PER 100 G					
ENERGY VALUE (AVERAGE)	KJOULE	186	364	0.0	549
OF THE DIGESTIBLE	(KCAL)	44	87	0.0	131
FRACTION PER 100 G					
EDIBLE PORTION					

WASTE PERCENTAGE AVERAGE 0.00

CONSTITUENTS	DIM	AV	VARIATION			AVR	NUTR.	DENS.	MOLPERC.
WATER	GRAM	78.10	77.00	-	78.90	76.54	GRAM/MJ	142.17	
PROTEIN	GRAM	10.40	10.20	-	10.70	10.19	GRAM/MJ	18.93	
FAT	GRAM	9.63	8.90	-	11.00	9.44	GRAM/MJ	17.53	
MINERALS	GRAM	1.43	1.40	-	1.50	1.40	GRAM/MJ	2.60	
SODIUM	MILLI	167.00	155.00	-	179.00	163.66	MILLI/MJ	304.00	
POTASSIUM	MILLI	281.00	265.00	-	296.00	275.38	MILLI/MJ	511.51	
CALCIUM	MILLI	10.00	7.00	-	13.00	9.80	MILLI/MJ	18.20	
IRON	MILLI	2.53	2.10	-	3.20	2.48	MILLI/MJ	4.61	
PHOSPHORUS	MILLI	366.00	351.00	-	380.00	358.68	MILLI/MJ	666.24	
VITAMIN B1	MILLI	0.13	0.12	-	0.15	0.13	MILLI/MJ	0.24	
VITAMIN B2	MILLI	0.24	0.21	-	0.30	0.24	MILLI/MJ	0.44	
NICOTINAMIDE	MILLI	3.50	3.10	-	3.80	3.43	MILLI/MJ	6.37	
PANTOTHENIC ACID	MILLI	2.50	-	-	-	2.45	MILLI/MJ	4.55	
VITAMIN B6	MILLI	0.16	-	-	-	0.16	MILLI/MJ	0.29	
BIOTIN	MICRO	6.10	-	-	-	5.98	MICRO/MJ	11.10	
FOLIC ACID	MICRO	12.00	-	-	-	11.76	MICRO/MJ	21.84	
VITAMIN B12	MICRO	3.42	2.13	-	4.70	3.35	MICRO/MJ	6.23	
VITAMIN C	MILLI	17.00	14.00	-	19.00	16.66	MILLI/MJ	30.95	
ARGININE	GRAM	0.64	-	-	-	0.63	GRAM/MJ	1.17	
CYSTINE	GRAM	0.13	-	-	-	0.13	GRAM/MJ	0.24	
HISTIDINE	GRAM	0.31	-	-	-	0.30	GRAM/MJ	0.56	
ISOLEUCINE	GRAM	0.56	-	-	-	0.55	GRAM/MJ	1.02	
LEUCINE	GRAM	0.97	-	-	-	0.95	GRAM/MJ	1.77	
LYSINE	GRAM	0.87	-	-	-	0.85	GRAM/MJ	1.58	
METHIONINE	GRAM	0.22	-	-	-	0.22	GRAM/MJ	0.40	
PHENYLALANINE	GRAM	0.55	-	-	-	0.54	GRAM/MJ	1.00	
THREONINE	GRAM	0.52	-	-	-	0.51	GRAM/MJ	0.95	
TRYPTOPHAN	GRAM	0.16	-	-	-	0.16	GRAM/MJ	0.29	
VALINE	GRAM	0.59	-	-	-	0.58	GRAM/MJ	1.07	
CHOLESTEROL	GRAM	2.00	-	-	-	1.96	GRAM/MJ	3.64	
TOTAL PHOSPHOLIPIDS	GRAM	4.44	-	-	-	4.35	GRAM/MJ	8.08	
PHOSPHATIDYLCHOLINE	GRAM	1.07	-	-	-	1.05	GRAM/MJ	1.95	
PHOSPHATIDYLETHANOLAMINE	GRAM	1.60	-	-	-	1.57	GRAM/MJ	2.91	
PHOSPHATIDYLSERINE	MILLI	713.00	-	-	-	698.74	GRAM/MJ	1.30	
PHOSPHATIDYLINOSITOL	MILLI	198.00	-	-	-	194.04	MILLI/MJ	360.43	
SPHINGOMYELIN	MILLI	772.00	-	-	-	756.56	GRAM/MJ	1.41	

RINDERLEBER OX LIVER FOIE DE BOEUF

		PROTEIN	FAT	CARBOHYDRATES	TOTAL
ENERGY VALUE (AVERAGE)	KJOULE	363	123	28	514
PER 100 G	(KCAL)	87	29	6.6	123
EDIBLE PORTION					
AMOUNT OF DIGESTIBLE	GRAM	19.10	2.94	1.65	
CONSTITUENTS PER 100 G					
ENERGY VALUE (AVERAGE)	KJOULE	352	117	28	496
OF THE DIGESTIBLE	(KCAL)	84	28	6.6	119
FRACTION PER 100 G					
EDIBLE PORTION					

WASTE PERCENTAGE AVERAGE 7.0 MINIMUM 2.0 MAXIMUM 11

CONSTITUENTS	DIM	AV	VARIATION			AVR	NUTR. DENS.		MOLPERC.
WATER	GRAM	69.90	69.70	-	70.00	65.01	GRAM/MJ	140.80	
PROTEIN	GRAM	19.70	19.70	-	20.60	18.32	GRAM/MJ	39.68	
FAT	GRAM	3.10	3.00	-	3.20	2.88	GRAM/MJ	6.24	
AVAILABLE CARBOHYDR.	GRAM	1.65	-	-	-	1.53	GRAM/MJ	3.32	
MINERALS	GRAM	1.40	-	-	-	1.30	GRAM/MJ	2.82	
SODIUM	MILLI	116.00	77.00	-	155.00	107.88	MILLI/MJ	233.65	
POTASSIUM	MILLI	292.00	235.00	-	318.00	271.56	MILLI/MJ	588.16	
MAGNESIUM	MILLI	17.10	-	-	-	15.90	MILLI/MJ	34.44	
CALCIUM	MILLI	7.00	0.00	-	26.00	6.51	MILLI/MJ	14.10	
MANGANESE	MILLI	0.25	-	-	-	0.23	MILLI/MJ	0.50	
IRON	MILLI	7.10	6.60	-	8.30	6.60	MILLI/MJ	14.30	
COBALT	MICRO	10.50	0.00	-	45.00	9.77	MICRO/MJ	21.15	
COPPER	MILLI	3.60	2.02	-	7.94	3.35	MILLI/MJ	7.25	
ZINC	MILLI	5.10	3.00	-	8.50	4.74	MILLI/MJ	10.27	
NICKEL	MILLI	-	0.00	-	0.48				
CHROMIUM	MICRO	1.20	0.00	-	10.00	1.12	MICRO/MJ	2.42	
MOLYBDENUM	MILLI	-	0.00	-	0.24				
ALUMINIUM	MICRO	-	7.00	-	180.00				
PHOSPHORUS	MILLI	358.00	-	-	-	332.94	MILLI/MJ	721.10	
CHLORIDE	MILLI	68.10	-	-	-	63.33	MILLI/MJ	137.17	
FLUORIDE	MILLI	0.13	0.10	-	0.16	0.12	MILLI/MJ	0.26	
IODIDE	MICRO	14.00	-	-	-	13.02	MICRO/MJ	28.20	
BORON	MILLI	0.10	0.05	-	0.16	0.10	MILLI/MJ	0.21	
SELENIUM	MICRO	35.00	27.00	-	45.00	32.55	MICRO/MJ	70.50	
VITAMIN A	MILLI	15.30	2.90	-	37.90	14.23	MILLI/MJ	30.82	
VITAMIN D	MICRO	1.70	-	-	-	1.58	MICRO/MJ	3.42	

CONSTITUENTS	DIM	AV	VARIATION			AVR	NUTR. DENS.		MOLPERC.
VITAMIN E ACTIVITY	MILLI	0.67	-	-	-	0.62	MILLI/MJ	1.35	
ALPHA-TOCOPHEROL	MILLI	0.67	-	-	-	0.62	MILLI/MJ	1.35	
VITAMIN K	MICRO	45.00	-	-	-	41.85	MICRO/MJ	90.64	
VITAMIN B1	MILLI	0.30	0.23	-	0.40	0.28	MILLI/MJ	0.60	
VITAMIN B2	MILLI	2.88	2.08	-	3.33	2.68	MILLI/MJ	5.80	
NICOTINAMIDE	MILLI	14.70	13.50	-	17.00	13.67	MILLI/MJ	29.61	
PANTOTHENIC ACID	MILLI	7.30	-	-	-	6.79	MILLI/MJ	14.70	
VITAMIN B6	MILLI	0.71	-	-	-	0.66	MILLI/MJ	1.43	
BIOTIN	MILLI	0.10	-	-	-	0.09	MILLI/MJ	0.20	
FOLIC ACID	MILLI	0.22	-	-	-	0.20	MILLI/MJ	0.44	
VITAMIN B12	MICRO	65.00	-	-	-	60.45	MICRO/MJ	130.93	
VITAMIN C	MILLI	30.00	20.00	-	35.00	27.90	MILLI/MJ	60.43	
ALANINE	GRAM	1.44	1.30	-	1.58	1.34	GRAM/MJ	2.90	9.0
ARGININE	GRAM	1.30	1.14	-	1.45	1.21	GRAM/MJ	2.62	4.1
ASPARTIC ACID	GRAM	2.14	1.97	-	2.35	1.99	GRAM/MJ	4.31	8.9
CYSTINE	GRAM	0.28	0.26	-	0.30	0.26	GRAM/MJ	0.56	0.6
GLUTAMIC ACID	GRAM	2.98	2.72	-	3.16	2.77	GRAM/MJ	6.00	11.2
GLYCINE	GRAM	1.49	1.29	-	1.60	1.39	GRAM/MJ	3.00	11.0
HISTIDINE	GRAM	0.66	0.63	-	0.79	0.61	GRAM/MJ	1.33	2.4
ISOLEUCINE	GRAM	1.15	0.97	-	1.32	1.07	GRAM/MJ	2.32	4.9
LEUCINE	GRAM	1.99	1.75	-	2.14	1.85	GRAM/MJ	4.01	8.4
LYSINE	GRAM	1.75	1.64	-	1.85	1.63	GRAM/MJ	3.52	6.6
METHIONINE	GRAM	0.60	0.52	-	0.75	0.56	GRAM/MJ	1.21	2.2
PHENYLALANINE	GRAM	1.17	1.10	-	1.30	1.09	GRAM/MJ	2.36	3.9
PROLINE	GRAM	1.23	1.16	-	1.23	1.14	GRAM/MJ	2.48	5.9
SERINE	GRAM	1.11	0.99	-	1.24	1.03	GRAM/MJ	2.24	5.9
THREONINE	GRAM	1.01	0.92	-	1.00	0.94	GRAM/MJ	2.03	4.7
TRYPTOPHAN	GRAM	0.31	0.27	-	0.33	0.29	GRAM/MJ	0.62	0.8
TYROSINE	GRAM	0.77	0.67	-	0.82	0.72	GRAM/MJ	1.55	2.4
VALINE	GRAM	1.47	1.34	-	1.58	1.37	GRAM/MJ	2.96	7.0
TYRAMINE	MILLI	27.40	-	-	-	25.48	MILLI/MJ	55.19	
GLYCOGENE	GRAM	1.65	1.37	-	1.94	1.53	GRAM/MJ	3.32	
CHOLESTEROL	MILLI	265.00	-	-	-	246.45	MILLI/MJ	533.78	
TOTAL PURINES	MILLI	554.00 1	554.00	-	554.00	515.22	GRAM/MJ	1.12	
ADENINE	MILLI	147.00	147.00	-	147.00	136.71	MILLI/MJ	296.10	
GUANINE	MILLI	151.00	151.00	-	151.00	140.43	MILLI/MJ	304.15	
XANTHINE	MILLI	57.00	57.00	-	57.00	53.01	MILLI/MJ	114.81	
HYPOXANTHINE	MILLI	114.00	114.00	-	114.00	106.02	MILLI/MJ	229.63	

1 VALUE OF TOTAL PURINES IS EXPRESSSED IN MG URIC ACID/100G

1 DEPENDENT ON THE METHOD OF EXTRACTION; WITH PRECEDING PROCEDURES VALUES OF 250 - 350 MG URIC ACID/100G HAD BEEN OBTAINED

RINDERLUNGE | OX LUNGS (LIGHTS) | MOU DE BOEUF

		PROTEIN	FAT	CARBOHYDRATES	TOTAL
ENERGY VALUE (AVERAGE) PER 100 G EDIBLE PORTION	KJOULE	333	115	0.0	448
	(KCAL)	80	28	0.0	107
AMOUNT OF DIGESTIBLE CONSTITUENTS PER 100 G	GRAM	17.55	2.75	0.00	
ENERGY VALUE (AVERAGE) OF THE DIGESTIBLE FRACTION PER 100 G EDIBLE PORTION	KJOULE	323	110	0.0	433
	(KCAL)	77	26	0.0	103

WASTE PERCENTAGE AVERAGE 24 MINIMUM 15 MAXIMUM 34

CONSTITUENTS	DIM	AV	VARIATION		AVR	NUTR. DENS.		MOLPERC.
WATER	GRAM	77.50	76.10	- 80.10	58.90	GRAM/MJ	179.10	
PROTEIN	GRAM	18.10	15.60	- 20.10	13.76	GRAM/MJ	41.83	
FAT	GRAM	2.90	2.00	- 4.50	2.20	GRAM/MJ	6.70	
MINERALS	GRAM	1.02	-	- -	0.78	GRAM/MJ	2.36	
SODIUM	MILLI	198.00	178.00	- 216.00	150.48	MILLI/MJ	457.57	
POTASSIUM	MILLI	228.00	223.00	- 231.00	173.28	MILLI/MJ	526.90	
CALCIUM	MILLI	13.00	12.00	- 13.00	9.88	MILLI/MJ	30.04	
IRON	MILLI	7.50	6.60	- 8.40	5.70	MILLI/MJ	17.33	
MOLYBDENUM	MILLI	-	0.02	- 0.02				
PHOSPHORUS	MILLI	224.00	196.00	- 250.00	170.24	MILLI/MJ	517.65	
BORON	MICRO	86.00	56.00	- 152.00	65.36	MICRO/MJ	198.74	
SELENIUM	MICRO	30.00	20.00	- 40.00	22.80	MICRO/MJ	69.33	
VITAMIN A	MICRO	55.00	-	- -	41.80	MICRO/MJ	127.10	
VITAMIN B1	MICRO	90.00	30.00	- 110.00	68.40	MICRO/MJ	207.99	
VITAMIN B2	MILLI	0.34	0.30	- 0.36	0.26	MILLI/MJ	0.79	
NICOTINAMIDE	MILLI	4.27	3.10	- 6.00	3.25	MILLI/MJ	9.87	
PANTOTHENIC ACID	MILLI	1.00	-	- -	0.76	MILLI/MJ	2.31	
VITAMIN B6	MICRO	70.00	-	- -	53.20	MICRO/MJ	161.77	
BIOTIN	MICRO	5.90	-	- -	4.48	MICRO/MJ	13.63	
FOLIC ACID	MICRO	11.00	-	- -	8.36	MICRO/MJ	25.42	
VITAMIN B12	MICRO	3.30	-	- -	2.51	MICRO/MJ	7.63	
VITAMIN C	MILLI	39.00	-	- -	29.64	MILLI/MJ	90.13	
ARGININE	GRAM	1.01	0.92	- 1.10	0.77	GRAM/MJ	2.33	
CYSTINE	GRAM	0.16	-	- -	0.12	GRAM/MJ	0.37	
HISTIDINE	GRAM	0.47	0.42	- 0.51	0.36	GRAM/MJ	1.09	

CONSTITUENTS	DIM	AV	VARIATION			AVR	NUTR. DENS.		MOLPERC.
ISOLEUCINE	GRAM	0.80	-	-	-	0.61	GRAM/MJ	1.85	
LEUCINE	GRAM	1.91	1.50	-	2.32	1.45	GRAM/MJ	4.41	
LYSINE	GRAM	1.35	1.30	-	1.39	1.03	GRAM/MJ	3.12	
METHIONINE	GRAM	0.37	0.33	-	0.40	0.28	GRAM/MJ	0.86	
PHENYLALANINE	GRAM	0.84	0.81	-	0.87	0.64	GRAM/MJ	1.94	
THREONINE	GRAM	0.75	0.74	-	0.76	0.57	GRAM/MJ	1.73	
TRYPTOPHAN	GRAM	0.17	0.16	-	0.18	0.13	GRAM/MJ	0.39	
TYROSINE	GRAM	0.60	-	-	-	0.46	GRAM/MJ	1.39	
VALINE	GRAM	1.01	0.94	-	1.07	0.77	GRAM/MJ	2.33	
TOTAL PURINES	MILLI	399.00 1	399.00	-	399.00	303.24	MILLI/MJ	922.07	
ADENINE	MILLI	184.00	184.00	-	184.00	139.84	MILLI/MJ	425.21	
GUANINE	MILLI	105.00	105.00	-	105.00	79.80	MILLI/MJ	242.65	
XANTHINE	MILLI	9.00	9.00	-	9.00	6.84	MILLI/MJ	20.80	
HYPOXANTHINE	MILLI	35.00	35.00	-	35.00	26.60	MILLI/MJ	80.88	

1 VALUE OF TOTAL PURINES IS EXPRESSSED IN MG URIC ACID/100G

RINDERMILZ　　OX SPLEEN　　RATE DE BOEUF

		PROTEIN	FAT	CARBOHYDRATES	TOTAL
ENERGY VALUE (AVERAGE)	KJOULE	341	115	0.0	456
PER 100 G	(KCAL)	81	28	0.0	109
EDIBLE PORTION					
AMOUNT OF DIGESTIBLE	GRAM	17.94	2.75	0.00	
CONSTITUENTS PER 100 G					
ENERGY VALUE (AVERAGE)	KJOULE	330	110	0.0	440
OF THE DIGESTIBLE	(KCAL)	79	26	0.0	105
FRACTION PER 100 G					
EDIBLE PORTION					

WASTE PERCENTAGE AVERAGE 0.00

CONSTITUENTS	DIM	AV	VARIATION			AVR	NUTR. DENS.		MOLPERC.
WATER	GRAM	76.70	73.30	-	78.40	76.70	GRAM/MJ	174.37	
PROTEIN	GRAM	18.50	17.40	-	21.90	18.50	GRAM/MJ	42.06	
FAT	GRAM	2.90	1.70	-	4.10	2.90	GRAM/MJ	6.59	
MINERALS	GRAM	1.38	1.30	-	1.40	1.38	GRAM/MJ	3.14	
SODIUM	MILLI	99.00	94.00	-	104.00	99.00	MILLI/MJ	225.07	
POTASSIUM	MILLI	379.00	365.00	-	392.00	379.00	MILLI/MJ	861.63	
CALCIUM	MILLI	6.50	3.00	-	10.00	6.50	MILLI/MJ	14.78	
IRON	MILLI	8.90	-	-	-	8.90	MILLI/MJ	20.23	
MOLYBDENUM	MICRO	60.00	-	-	-	60.00	MICRO/MJ	136.41	
PHOSPHORUS	MILLI	236.00	-	-	-	236.00	MILLI/MJ	536.53	
BORON	MICRO	86.00	43.00	-	123.00	86.00	MICRO/MJ	195.51	
SELENIUM	MICRO	31.00	27.00	-	41.00	31.00	MICRO/MJ	70.48	
VITAMIN A	MICRO	95.00	-	-	-	95.00	MICRO/MJ	215.97	
VITAMIN B1	MILLI	0.13	0.12	-	0.13	0.13	MILLI/MJ	0.30	
VITAMIN B2	MILLI	0.30	0.28	-	0.34	0.30	MILLI/MJ	0.68	
NICOTINAMIDE	MILLI	3.90	3.30	-	4.20	3.90	MILLI/MJ	8.87	
PANTOTHENIC ACID	MILLI	1.20	-	-	-	1.20	MILLI/MJ	2.73	
VITAMIN B6	MILLI	0.12	-	-	-	0.12	MILLI/MJ	0.27	
BIOTIN	MICRO	5.70	-	-	-	5.70	MICRO/MJ	12.96	
VITAMIN B12	MICRO	5.10	-	-	-	5.10	MICRO/MJ	11.59	
VITAMIN C	MILLI	-	6.00	-	46.00				
ARGININE	GRAM	1.05	-	-	-	1.05	GRAM/MJ	2.39	
HISTIDINE	GRAM	0.44	-	-	-	0.44	GRAM/MJ	1.00	
ISOLEUCINE	GRAM	0.85	-	-	-	0.85	GRAM/MJ	1.93	
LEUCINE	GRAM	1.57	-	-	-	1.57	GRAM/MJ	3.57	

CONSTITUENTS	DIM	AV	VARIATION	AVR	NUTR. DENS.		MOLPERC.
LYSINE	GRAM	1.44	- - -	1.44	GRAM/MJ	3.27	
METHIONINE	GRAM	0.37	- - -	0.37	GRAM/MJ	0.84	
PHENYLALANINE	GRAM	0.83	- - -	0.83	GRAM/MJ	1.89	
THREONINE	GRAM	0.81	- - -	0.81	GRAM/MJ	1.84	
TRYPTOPHAN	GRAM	0.20	- - -	0.20	GRAM/MJ	0.45	
VALINE	GRAM	1.11	- - -	1.11	GRAM/MJ	2.52	
TOTAL PURINES	MILLI	444.00 [1]	444.00 - 444.00	444.00	GRAM/MJ	1.01	
ADENINE	MILLI	145.00	145.00 - 145.00	145.00	MILLI/MJ	329.65	
GUANINE	MILLI	160.00	160.00 - 160.00	160.00	MILLI/MJ	363.75	
XANTHINE	MILLI	17.00	17.00 - 17.00	17.00	MILLI/MJ	38.65	
HYPOXANTHINE	MILLI	54.00	54.00 - 54.00	54.00	MILLI/MJ	122.76	

1 VALUE OF TOTAL PURINES IS EXPRESSSED IN MG URIC ACID/100G

RINDERNIERE — OX KIDNEY — ROGNON DE BOEUF

		PROTEIN	FAT	CARBOHYDRATES	TOTAL
ENERGY VALUE (AVERAGE)	KJOULE	306	204	0.0	510
PER 100 G	(KCAL)	73	49	0.0	122
EDIBLE PORTION					
AMOUNT OF DIGESTIBLE	GRAM	16.10	4.88	0.00	
CONSTITUENTS PER 100 G					
ENERGY VALUE (AVERAGE)	KJOULE	296	194	0.0	491
OF THE DIGESTIBLE	(KCAL)	71	46	0.0	117
FRACTION PER 100 G					
EDIBLE PORTION					

WASTE PERCENTAGE AVERAGE 13 MINIMUM 12 MAXIMUM 13

CONSTITUENTS	DIM	AV	VARIATION			AVR	NUTR. DENS.		MOLPERC.
WATER	GRAM	76.10	75.00	-	77.70	66.21	GRAM/MJ	155.14	
PROTEIN	GRAM	16.60	15.00	-	17.70	14.44	GRAM/MJ	33.84	
FAT	GRAM	5.14	1.20	-	8.10	4.47	GRAM/MJ	10.48	
MINERALS	GRAM	1.17	1.10	-	1.20	1.02	GRAM/MJ	2.39	
SODIUM	MILLI	235.00	214.00	-	246.00	204.45	MILLI/MJ	479.08	
POTASSIUM	MILLI	245.00	231.00	-	266.00	213.15	MILLI/MJ	499.47	
MAGNESIUM	MILLI	20.00	18.00	-	21.00	17.40	MILLI/MJ	40.77	
CALCIUM	MILLI	11.00	9.00	-	14.00	9.57	MILLI/MJ	22.43	
MANGANESE	MILLI	0.11	0.06	-	0.17	0.10	MILLI/MJ	0.22	
IRON	MILLI	9.50	6.50	-	15.00	8.27	MILLI/MJ	19.37	
COBALT	MICRO	3.00	0.00	-	25.00	2.61	MICRO/MJ	6.12	
COPPER	MILLI	0.39	0.18	-	0.92	0.34	MILLI/MJ	0.80	
ZINC	MILLI	1.90	1.30	-	3.20	1.65	MILLI/MJ	3.87	
NICKEL	MICRO	1.50	1.50	-	48.00	1.31	MICRO/MJ	3.06	
CHROMIUM	MICRO	3.00	0.00	-	22.00	2.61	MICRO/MJ	6.12	
MOLYBDENUM	MICRO	31.00	0.00	-	190.00	26.97	MICRO/MJ	63.20	
ALUMINIUM	MICRO	6.00	-	-	-	5.22	MICRO/MJ	12.23	
PHOSPHORUS	MILLI	248.00	221.00	-	262.00	215.76	MILLI/MJ	505.59	
CHLORIDE	MILLI	251.00	246.00	-	256.00	218.37	MILLI/MJ	511.70	
FLUORIDE	MILLI	0.20	-	-	-	0.17	MILLI/MJ	0.41	
BORON	MICRO	86.00	20.00	-	155.00	74.82	MICRO/MJ	175.32	
SELENIUM	MILLI	-	0.16	-	0.55				
VITAMIN A	MILLI	0.33	0.30	-	0.35	0.29	MILLI/MJ	0.67	
VITAMIN E ACTIVITY	MILLI	0.18	-	-	-	0.16	MILLI/MJ	0.37	
ALPHA-TOCOPHEROL	MILLI	0.18	-	-	-	0.16	MILLI/MJ	0.37	
VITAMIN B1	MILLI	0.30	0.25	-	0.38	0.26	MILLI/MJ	0.61	

CONSTITUENTS	DIM	AV	VARIATION	AVR	NUTR. DENS.		MOLPERC.
VITAMIN B2	MILLI	2.26	1.90 - 2.78	1.97	MILLI/MJ	4.61	
NICOTINAMIDE	MILLI	6.17	5.30 - 7.40	5.37	MILLI/MJ	12.58	
PANTOTHENIC ACID	MILLI	3.85	3.40 - 4.05	3.35	MILLI/MJ	7.85	
VITAMIN B6	MILLI	0.39	- - -	0.34	MILLI/MJ	0.80	
BIOTIN	MILLI	-	0.02 - 0.09				
FOLIC ACID	MICRO	50.00	41.00 - 58.00	43.50	MICRO/MJ	101.93	
VITAMIN B12	MICRO	33.40	28.00 - 38.80	29.06	MICRO/MJ	68.09	
VITAMIN C	MILLI	11.00	11.00 - 13.00	9.57	MILLI/MJ	22.43	
ALANINE	GRAM	1.18	1.08 - 1.30	1.03	GRAM/MJ	2.41	8.7
ARGININE	GRAM	1.10	0.99 - 1.16	0.96	GRAM/MJ	2.24	4.1
ASPARTIC ACID	GRAM	1.73	1.50 - 2.03	1.51	GRAM/MJ	3.53	8.5
CYSTINE	GRAM	0.27	0.23 - 0.34	0.23	GRAM/MJ	0.55	0.7
GLUTAMIC ACID	GRAM	2.54	2.27 - 2.80	2.21	GRAM/MJ	5.18	11.3
GLYCINE	GRAM	1.35	1.36 - 1.60	1.17	GRAM/MJ	2.75	11.8
HISTIDINE	GRAM	0.54	0.47 - 0.62	0.47	GRAM/MJ	1.10	2.3
ISOLEUCINE	GRAM	1.02	0.78 - 1.30	0.89	GRAM/MJ	2.08	5.1
LEUCINE	GRAM	1.72	1.48 - 1.92	1.50	GRAM/MJ	3.51	8.6
LYSINE	GRAM	1.41	1.38 - 1.46	1.23	GRAM/MJ	2.87	6.3
METHIONINE	GRAM	0.44	0.39 - 0.57	0.38	GRAM/MJ	0.90	1.9
PHENYLALANINE	GRAM	0.94	0.91 - 1.00	0.82	GRAM/MJ	1.92	3.7
PROLINE	GRAM	1.05	1.02 - 1.08	0.91	GRAM/MJ	2.14	6.0
SERINE	GRAM	1.03	0.93 - 1.17	0.90	GRAM/MJ	2.10	6.4
THREONINE	GRAM	0.86	0.82 - 0.90	0.75	GRAM/MJ	1.75	4.7
TRYPTOPHAN	GRAM	0.24	0.21 - 0.27	0.21	GRAM/MJ	0.49	0.8
TYROSINE	GRAM	0.67	0.58 - 0.71	0.58	GRAM/MJ	1.37	2.4
VALINE	GRAM	1.15	1.08 - 1.19	1.00	GRAM/MJ	2.34	6.4
LAURIC ACID	MILLI	15.00	- - -	13.05	MILLI/MJ	30.58	
MYRISTIC ACID	GRAM	0.18	- - -	0.16	GRAM/MJ	0.37	
C15 : 0 FATTY ACID	MILLI	64.00	- - -	55.68	MILLI/MJ	130.47	
PALMITIC ACID	GRAM	1.49	- - -	1.30	GRAM/MJ	3.04	
C17 : 0 FATTY ACID	MILLI	74.00	- - -	64.38	MILLI/MJ	150.86	
STEARIC ACID	GRAM	1.59	- - -	1.38	GRAM/MJ	3.24	
PALMITOLEIC ACID	GRAM	0.22	- - -	0.19	GRAM/MJ	0.45	
OLEIC ACID	GRAM	1.08	- - -	0.94	GRAM/MJ	2.20	
EICOSENOIC ACID	MILLI	29.00	- - -	25.23	MILLI/MJ	59.12	
C14 : 1 FATTY ACID	MILLI	44.00	- - -	38.28	MILLI/MJ	89.70	
LINOLEIC ACID	MILLI	59.00	- - -	51.33	MILLI/MJ	120.28	
LINOLENIC ACID	MILLI	25.00	- - -	21.75	MILLI/MJ	50.97	
C16 : 2 FATTY ACID	MILLI	29.00	- - -	25.23	MILLI/MJ	59.12	
CHOLESTEROL	MILLI	375.00	- - -	326.25	MILLI/MJ	764.49	
TOTAL PURINES	MILLI	269.00 [1]	269.00 - 269.00	234.03	MILLI/MJ	548.40	
ADENINE	MILLI	71.00	71.00 - 71.00	61.77	MILLI/MJ	144.74	
GUANINE	MILLI	74.00	74.00 - 74.00	64.38	MILLI/MJ	150.86	
XANTHINE	MILLI	10.00	10.00 - 10.00	8.70	MILLI/MJ	20.39	
HYPOXANTHINE	MILLI	71.00	71.00 - 71.00	61.77	MILLI/MJ	144.74	

1 VALUE OF TOTAL PURINES IS EXPRESSSED IN MG URIC ACID/100G

RINDERZUNGE OX TONGUE LANGUE DE BOEUF

		PROTEIN	FAT	CARBOHYDRATES	TOTAL
ENERGY VALUE (AVERAGE) PER 100 G EDIBLE PORTION	KJOULE	295	632	0.0	927
	(KCAL)	70	151	0.0	221
AMOUNT OF DIGESTIBLE CONSTITUENTS PER 100 G	GRAM	15.52	15.10	0.00	
ENERGY VALUE (AVERAGE) OF THE DIGESTIBLE FRACTION PER 100 G EDIBLE PORTION	KJOULE	286	600	0.0	886
	(KCAL)	68	143	0.0	212

WASTE PERCENTAGE AVERAGE 26

CONSTITUENTS	DIM	AV	VARIATION			AVR	NUTR. DENS.		MOLPERC.
WATER	GRAM	66.80	65.60	-	68.00	49.43	GRAM/MJ	75.39	
PROTEIN	GRAM	16.00	15.70	-	16.40	11.84	GRAM/MJ	18.06	
FAT	GRAM	15.90	15.00	-	17.60	11.77	GRAM/MJ	17.94	
MINERALS	GRAM	0.95	0.90	-	1.00	0.70	GRAM/MJ	1.07	
SODIUM	MILLI	100.00	-	-	-	74.00	MILLI/MJ	112.85	
POTASSIUM	MILLI	255.00	250.00	-	260.00	188.70	MILLI/MJ	287.77	
MAGNESIUM	MILLI	10.00	-	-	-	7.40	MILLI/MJ	11.29	
CALCIUM	MILLI	9.50	9.00	-	10.00	7.03	MILLI/MJ	10.72	
IRON	MILLI	3.00	2.80	-	3.00	2.22	MILLI/MJ	3.39	
PHOSPHORUS	MILLI	229.00	187.00	-	262.00	169.46	MILLI/MJ	258.43	
CHLORIDE	MILLI	13.00	-	-	-	9.62	MILLI/MJ	14.67	
VITAMIN A	MICRO	4.00	0.00	-	5.00	2.96	MICRO/MJ	4.51	
VITAMIN B1	MILLI	0.14	0.12	-	0.17	0.10	MILLI/MJ	0.16	
VITAMIN B2	MILLI	0.29	0.25	-	0.33	0.21	MILLI/MJ	0.33	
NICOTINAMIDE	MILLI	4.60	2.50	-	5.00	3.40	MILLI/MJ	5.19	
PANTOTHENIC ACID	MILLI	2.00	-	-	-	1.48	MILLI/MJ	2.26	
VITAMIN B6	MILLI	0.13	-	-	-	0.10	MILLI/MJ	0.15	
BIOTIN	MICRO	3.30	-	-	-	2.44	MICRO/MJ	3.72	
ARGININE	GRAM	1.04	0.93	-	1.17	0.77	GRAM/MJ	1.17	
CYSTINE	GRAM	0.20	0.17	-	0.24	0.15	GRAM/MJ	0.23	
HISTIDINE	GRAM	0.40	0.31	-	0.46	0.30	GRAM/MJ	0.45	
ISOLEUCINE	GRAM	0.77	0.66	-	0.92	0.57	GRAM/MJ	0.87	
LEUCINE	GRAM	1.25	1.01	-	1.40	0.93	GRAM/MJ	1.41	
LYSINE	GRAM	1.33	1.14	-	1.62	0.98	GRAM/MJ	1.50	
METHIONINE	GRAM	0.35	0.30	-	0.40	0.26	GRAM/MJ	0.39	
PHENYLALANINE	GRAM	0.65	0.52	-	0.74	0.48	GRAM/MJ	0.73	
THREONINE	GRAM	0.69	0.51	-	0.78	0.51	GRAM/MJ	0.78	
TRYPTOPHAN	GRAM	0.19	0.17	-	0.23	0.14	GRAM/MJ	0.21	
TYROSINE	GRAM	0.54	0.49	-	0.58	0.40	GRAM/MJ	0.61	
VALINE	GRAM	0.82	0.73	-	0.90	0.61	GRAM/MJ	0.93	
CHOLESTEROL	MILLI	108.00	-	-	-	79.92	MILLI/MJ	121.88	

SCHWEINEFLEISCH
REINES MUSKELFLEISCH

PORK
MUSCLES ONLY

VIANDE DE PORC
VIANDE DE MUSCLE

		PROTEIN	FAT	CARBOHYDRATES	TOTAL
ENERGY VALUE (AVERAGE)	KJOULE	405	74	0.0	479
PER 100 G	(KCAL)	97	18	0.0	114
EDIBLE PORTION					
AMOUNT OF DIGESTIBLE	GRAM	21.34	1.76	0.00	
CONSTITUENTS PER 100 G					
ENERGY VALUE (AVERAGE)	KJOULE	393	70	0.0	463
OF THE DIGESTIBLE	(KCAL)	94	17	0.0	111
FRACTION PER 100 G					
EDIBLE PORTION					

WASTE PERCENTAGE AVERAGE 0.00

CONSTITUENTS	DIM	AV	VARIATION		AVR	NUTR. DENS.		MOLPERC.
WATER	GRAM	74.70	73.00	- 75.30	74.70	GRAM/MJ	161.31	
PROTEIN	GRAM	22.00	19.50	- 24.00	22.00	GRAM/MJ	47.51	
FAT	GRAM	1.86 1	1.00	- 2.80	1.86	GRAM/MJ	4.02	
MINERALS	GRAM	1.05	1.00	- 1.17	1.05	GRAM/MJ	2.27	
SODIUM	MILLI	56.00	39.00	- 80.00	56.00	MILLI/MJ	120.93	
POTASSIUM	MILLI	418.00	384.00	- 441.00	418.00	MILLI/MJ	902.62	
MAGNESIUM	MILLI	27.00	25.00	- 29.00	27.00	MILLI/MJ	58.30	
CALCIUM	MILLI	3.20	1.90	- 5.20	3.20	MILLI/MJ	6.91	
MANGANESE	MICRO	80.00	60.00	- 120.00	80.00	MICRO/MJ	172.75	
IRON	MILLI	1.00	0.50	- 4.70	1.00	MILLI/MJ	2.16	
COBALT	MICRO	0.10	-	- -	0.10	MICRO/MJ	0.22	
COPPER	MICRO	50.00	40.00	- 70.00	50.00	MICRO/MJ	107.97	
ZINC	MILLI	1.90 2	1.40	- 6.20	1.90	MILLI/MJ	4.10	
NICKEL	MICRO	1.00	-	- -	1.00	MICRO/MJ	2.16	
CHROMIUM	MICRO	1.40	-	- -	1.40	MICRO/MJ	3.02	
MOLYBDENUM	MICRO	27.00	13.00	- 41.00	27.00	MICRO/MJ	58.30	
ALUMINIUM	MICRO	3.00	-	- -	3.00	MICRO/MJ	6.48	
PHOSPHORUS	MILLI	204.00	197.00	- 210.00	204.00	MILLI/MJ	440.51	
BORON	MICRO	46.00	13.00	- 91.00	46.00	MICRO/MJ	99.33	
SELENIUM	MICRO	1.00	-	- -	1.00	MICRO/MJ	2.16	
VITAMIN A	MICRO	6.00	-	- -	6.00	MICRO/MJ	12.96	
VITAMIN E ACTIVITY	MICRO	80.00	-	- -	80.00	MICRO/MJ	172.75	
TOTAL TOCOPHEROLS	MILLI	0.10	-	- -	0.10	MILLI/MJ	0.22	
ALPHA-TOCOPHEROL	MICRO	80.00	-	- -	80.00	MICRO/MJ	172.75	
GAMMA-TOCOPHEROL	MICRO	20.00	-	- -	20.00	MICRO/MJ	43.19	
VITAMIN K	MICRO	18.00	-	- -	18.00	MICRO/MJ	38.87	

1 INTRA MUSCULAR FAT

THE FAT CONTENT IS DEPENDENT ON THE TYPE OF MUSCLE

2 THE ZN-CONTENT IS DEPENDENT ON THE TYPE OF MUSCLE; PSOAS MAJOR, LONGISSIMUS DORSI 1,5-1,9 MG/100G; SEMIMEMBRANOSUS: 2,3-3,1 MG/100G; TRAPEZIUS, SERRATUS VENTRALIS: 5,1 - 6,2 MG/100G

CONSTITUENTS	DIM	AV	VARIATION			AVR	NUTR. DENS.		MOLPERC.
VITAMIN B1	MILLI	0.90	0.80	-	1.00	0.90	MILLI/MJ	1.94	
VITAMIN B2	MILLI	0.23	-	-	-	0.23	MILLI/MJ	0.50	
NICOTINAMIDE	MILLI	5.00 3	4.50	-	5.30	5.00	MILLI/MJ	10.80	
PANTOTHENIC ACID	MILLI	0.70	-	-	-	0.70	MILLI/MJ	1.51	
VITAMIN B6	MILLI	0.50	-	-	-	0.50	MILLI/MJ	1.08	
BIOTIN	MICRO	5.00	-	-	-	5.00	MICRO/MJ	10.80	
FOLIC ACID	MICRO	6.00	-	-	-	6.00	MICRO/MJ	12.96	
VITAMIN B12	MICRO	5.00	1.00	-	8.00	5.00	MICRO/MJ	10.80	
VITAMIN C	MILLI	2.00	-	-	-	2.00	MILLI/MJ	4.32	
ALANINE	GRAM	1.53	1.34	-	1.78	1.53	GRAM/MJ	3.30	8.7
ARGININE	GRAM	1.53	1.04	-	1.65	1.53	GRAM/MJ	3.30	4.4
ASPARTIC ACID	GRAM	2.43	2.23	-	2.38	2.43	GRAM/MJ	5.25	9.2
CYSTINE	GRAM	0.31	0.16	-	0.36	0.31	GRAM/MJ	0.67	0.7
GLUTAMIC ACID	GRAM	3.91	3.61	-	4.14	3.91	GRAM/MJ	8.44	13.4
GLYCINE	GRAM	1.42	1.14	-	1.58	1.42	GRAM/MJ	3.07	9.6
HISTIDINE	GRAM	0.99	0.51	-	1.21	0.99	GRAM/MJ	2.14	3.2
ISOLEUCINE	GRAM	1.27	0.96	-	1.55	1.27	GRAM/MJ	2.74	4.9
LEUCINE	GRAM	1.92	1.69	-	2.11	1.92	GRAM/MJ	4.15	7.4
LYSINE	GRAM	2.20	1.73	-	2.43	2.20	GRAM/MJ	4.75	7.6
METHIONINE	GRAM	0.72	0.42	-	0.91	0.72	GRAM/MJ	1.55	2.4
PHENYLALANINE	GRAM	0.98	0.81	-	1.06	0.98	GRAM/MJ	2.12	3.0
PROLINE	GRAM	1.21	1.14	-	1.29	1.21	GRAM/MJ	2.61	5.3
SERINE	GRAM	1.12	1.07	-	1.16	1.12	GRAM/MJ	2.42	5.4
THREONINE	GRAM	1.25	0.86	-	1.45	1.25	GRAM/MJ	2.70	5.3
TRYPTOPHAN	GRAM	0.31	0.18	-	0.33	0.31	GRAM/MJ	0.67	0.8
TYROSINE	GRAM	0.91	0.58	-	1.04	0.91	GRAM/MJ	1.97	2.5
VALINE	GRAM	1.42	1.21	-	1.74	1.42	GRAM/MJ	3.07	6.1
MYRISTIC ACID	MILLI	25.00	-	-	-	25.00	MILLI/MJ	53.98	
PALMITIC ACID	GRAM	0.42	-	-	-	0.42	GRAM/MJ	0.91	
STEARIC ACID	GRAM	0.21	-	-	-	0.21	GRAM/MJ	0.45	
PALMITOLEIC ACID	MILLI	80.00	-	-	-	80.00	MILLI/MJ	172.75	
OLEIC ACID	GRAM	0.81	-	-	-	0.81	GRAM/MJ	1.75	
LINOLEIC ACID	GRAM	0.11	-	-	-	0.11	GRAM/MJ	0.24	
LINOLENIC ACID	MILLI	25.00	-	-	-	25.00	MILLI/MJ	53.98	
CHOLESTEROL	MILLI	65.00	-	-	-	65.00	MILLI/MJ	140.36	
TOTAL PURINES	MILLI	166.00 4	-	-	-	166.00	MILLI/MJ	358.46	
ADENINE	MILLI	22.00	22.00	-	22.00	22.00	MILLI/MJ	47.51	
GUANINE	MILLI	19.00	19.00	-	19.00	19.00	MILLI/MJ	41.03	
XANTHINE	MILLI	2.00	2.00	-	2.00	2.00	MILLI/MJ	4.32	
HYPOXANTHINE	MILLI	93.00	-	-	-	93.00	MILLI/MJ	200.82	
TOTAL PHOSPHOLIPIDS	MILLI	594.00	-	-	-	594.00	GRAM/MJ	1.28	
PHOSPHATIDYLCHOLINE	MILLI	327.00	-	-	-	327.00	MILLI/MJ	706.12	
PHOSPHATIDYLETHANOLAMINE	MILLI	171.00	-	-	-	171.00	MILLI/MJ	369.25	
SPHINGOMYELIN	MILLI	49.00	-	-	-	49.00	MILLI/MJ	105.81	

3 PSE-MEAT: THE MUSCLES CONTAIN MORE NICOTINAMID (8,7 MG/100G) THAN THOSE OF NORMAL MEAT (5 MG/100G). VITAMINES B1 UND B2 SHOW NO DIFFERENCES.

4 VALUE OF TOTAL PURINES IS EXPRESSSED IN MG URIC ACID/100G

SCHWEINEFLEISCH
INTERMUSKULARES FETTGEWEBE

PORK
INTERMUSCULARE ADIPOSE TISSUE

VIANDE DE PORC
TISSU ADIPEUX INTERMUSCULAIRE

		PROTEIN	FAT	CARBOHYDRATES	TOTAL
ENERGY VALUE (AVERAGE) PER 100 G EDIBLE PORTION	KJOULE	87	3049	0.0	3135
	(KCAL)	21	729	0.0	749
AMOUNT OF DIGESTIBLE CONSTITUENTS PER 100 G	GRAM	4.55	72.86	0.00	
ENERGY VALUE (AVERAGE) OF THE DIGESTIBLE FRACTION PER 100 G EDIBLE PORTION	KJOULE	84	2896	0.0	2980
	(KCAL)	20	692	0.0	712

WASTE PERCENTAGE AVERAGE 0.00

CONSTITUENTS	DIM	AV	VARIATION		AVR	NUTR. DENS.		MOLPERC.
WATER	GRAM	18.00	18.00	- 18.00	18.00	GRAM/MJ	6.04	
PROTEIN	GRAM	4.70	4.70	- 4.70	4.70	GRAM/MJ	1.58	
FAT	GRAM	76.70	76.70	- 76.70	76.70	GRAM/MJ	25.74	
MINERALS	GRAM	0.30	0.30	- 0.30	0.30	GRAM/MJ	0.10	
SODIUM	MILLI	18.00	18.00	- 18.00	18.00	MILLI/MJ	6.04	
POTASSIUM	MILLI	94.00	94.00	- 94.00	94.00	MILLI/MJ	31.54	
IRON	MILLI	0.29	0.29	- 0.29	0.29	MILLI/MJ	0.10	
ZINC	MILLI	0.53	0.53	- 0.53	0.53	MILLI/MJ	0.18	
VITAMIN B1	MILLI	0.13	0.13	- 0.13	0.13	MILLI/MJ	0.04	
VITAMIN B2	MICRO	80.00	80.00	- 80.00	80.00	MICRO/MJ	26.84	
NICOTINAMIDE	MILLI	1.58	1.58	- 1.58	1.58	MILLI/MJ	0.53	
VITAMIN B12	MICRO	0.29	0.29	- 0.29	0.29	MICRO/MJ	0.10	
CHOLESTEROL	MILLI	93.00	93.00	- 93.00	93.00	MILLI/MJ	31.21	

SCHWEINEFLEISCH
SUBKUTANES FETTGEWEBE

PORK
SUBCUTANEOUS ADIPOSE TISSUE

VIANDE DE PORC
TISSU ADIPEUX SOUSCUTAN

		PROTEIN	FAT	CARBOHYDRATES	TOTAL
ENERGY VALUE (AVERAGE)	KJOULE	53	3526	0.0	3579
PER 100 G	(KCAL)	13	843	0.0	855
EDIBLE PORTION					
AMOUNT OF DIGESTIBLE	GRAM	2.81	84.26	0.00	
CONSTITUENTS PER 100 G					
ENERGY VALUE (AVERAGE)	KJOULE	52	3349	0.0	3401
OF THE DIGESTIBLE	(KCAL)	12	801	0.0	813
FRACTION PER 100 G					
EDIBLE PORTION					

WASTE PERCENTAGE AVERAGE 0.00

CONSTITUENTS	DIM	AV	VARIATION			AVR	NUTR. DENS.		MOLPERC.
WATER	GRAM	7.70	7.70	-	7.70	7.70	GRAM/MJ	2.26	
PROTEIN	GRAM	2.90	2.90	-	2.90	2.90	GRAM/MJ	0.85	
FAT	GRAM	88.70	88.70	-	88.70	88.70	GRAM/MJ	26.08	
MINERALS	GRAM	0.70	0.70	-	0.70	0.70	GRAM/MJ	0.21	
SODIUM	MILLI	11.00	11.00	-	11.00	11.00	MILLI/MJ	3.23	
POTASSIUM	MILLI	65.00	65.00	-	65.00	65.00	MILLI/MJ	19.11	
IRON	MILLI	0.18	0.18	-	0.18	0.18	MILLI/MJ	0.05	
ZINC	MILLI	0.37	0.37	-	0.37	0.37	MILLI/MJ	0.11	
VITAMIN B1	MICRO	80.00	80.00	-	80.00	80.00	MICRO/MJ	23.52	
VITAMIN B2	MICRO	50.00	50.00	-	50.00	50.00	MICRO/MJ	14.70	
NICOTINAMIDE	MILLI	0.99	0.99	-	0.99	0.99	MILLI/MJ	0.29	
VITAMIN B12	MICRO	0.18	0.18	-	0.18	0.18	MICRO/MJ	0.05	
CHOLESTEROL	MILLI	57.00	57.00	-	57.00	57.00	MILLI/MJ	16.76	

SCHWEINEBLUT PIG'S BLOOD SANG DE PORC

		PROTEIN	FAT	CARBOHYDRATES	TOTAL
ENERGY VALUE (AVERAGE) PER 100 G EDIBLE PORTION	KJOULE	341	4.4	1.0	346
	(KCAL)	81	1.0	0.2	83
AMOUNT OF DIGESTIBLE CONSTITUENTS PER 100 G	GRAM	17.94	0.10	0.06	
ENERGY VALUE (AVERAGE) OF THE DIGESTIBLE FRACTION PER 100 G EDIBLE PORTION	KJOULE	330	4.2	1.0	336
	(KCAL)	79	1.0	0.2	80

WASTE PERCENTAGE AVERAGE 0.00

CONSTITUENTS	DIM	AV	VARIATION		AVR	NUTR. DENS.		MOLPERC.
WATER	GRAM	79.20	79.00	- 79.70	79.20	GRAM/MJ	236.05	
PROTEIN	GRAM	18.50	18.00	- 19.20	18.50	GRAM/MJ	55.14	
FAT	GRAM	0.11	0.10	- 0.12	0.11	GRAM/MJ	0.33	
AVAILABLE CARBOHYDR.	MILLI	60.00	60.00	- 70.00	60.00	MILLI/MJ	178.83	
MINERALS	GRAM	0.97	-	- -	0.97	GRAM/MJ	2.89	
SODIUM	MILLI	207.00	180.00	- 267.00	207.00	MILLI/MJ	616.96	
POTASSIUM	MILLI	185.00	157.00	- 206.00	185.00	MILLI/MJ	551.39	
MAGNESIUM	MILLI	20.00	-	- -	20.00	MILLI/MJ	59.61	
CALCIUM	MILLI	5.00	-	- -	5.00	MILLI/MJ	14.90	
IRON	MILLI	6.60	3.30	- 15.00	6.60	MILLI/MJ	19.67	
PHOSPHORUS	MILLI	49.00	44.00	- 52.00	49.00	MILLI/MJ	146.04	
CHLORIDE	MILLI	262.00	256.00	- 270.00	262.00	MILLI/MJ	780.88	
VITAMIN A	MICRO	30.00	-	- -	30.00	MICRO/MJ	89.41	
VITAMIN B1	MICRO	90.00	-	- -	90.00	MICRO/MJ	268.24	
VITAMIN B2	MICRO	30.00	-	- -	30.00	MICRO/MJ	89.41	
NICOTINAMIDE	MILLI	0.55	-	- -	0.55	MILLI/MJ	1.64	
VITAMIN C	-	0.00	-	- -	0.00			
CHOLESTEROL	MILLI	40.00	-	- -	40.00	MILLI/MJ	119.22	

SCHWEINEHERZ PIG'S HEART COEUR DE PORC

		PROTEIN	FAT	CARBOHYDRATES	TOTAL
ENERGY VALUE (AVERAGE)	KJOULE	311	143	0.0	454
PER 100 G	(KCAL)	74	34	0.0	109
EDIBLE PORTION					
AMOUNT OF DIGESTIBLE	GRAM	16.39	3.42	0.00	
CONSTITUENTS PER 100 G					
ENERGY VALUE (AVERAGE)	KJOULE	302	136	0.0	438
OF THE DIGESTIBLE	(KCAL)	72	32	0.0	105
FRACTION PER 100 G					
EDIBLE PORTION					

WASTE PERCENTAGE AVERAGE 17 MINIMUM 12 MAXIMUM 20

CONSTITUENTS	DIM	AV	VARIATION			AVR	NUTR. DENS.		MOLPERC.
WATER	GRAM	76.80	-	-	-	63.74	GRAM/MJ	175.45	
PROTEIN	GRAM	16.90	-	-	-	14.03	GRAM/MJ	38.61	
FAT	GRAM	3.60	2.40	-	4.80	2.99	GRAM/MJ	8.22	
MINERALS	GRAM	1.10	-	-	-	0.91	GRAM/MJ	2.51	
SODIUM	MILLI	80.00	-	-	-	66.40	MILLI/MJ	182.76	
POTASSIUM	MILLI	257.00	-	-	-	213.31	MILLI/MJ	587.12	
MAGNESIUM	MILLI	20.00	-	-	-	16.60	MILLI/MJ	45.69	
CALCIUM	MILLI	20.00	6.00	-	35.00	16.60	MILLI/MJ	45.69	
MANGANESE	MICRO	24.00	9.00	-	38.00	19.92	MICRO/MJ	54.83	
IRON	MILLI	4.30	2.70	-	5.80	3.57	MILLI/MJ	9.82	
COBALT	MICRO	2.00	-	-	-	1.66	MICRO/MJ	4.57	
COPPER	MILLI	0.41	-	-	-	0.34	MILLI/MJ	0.94	
ZINC	MILLI	2.20	-	-	-	1.83	MILLI/MJ	5.03	
CHROMIUM	MICRO	1.00	0.30	-	4.00	0.83	MICRO/MJ	2.28	
MOLYBDENUM	MICRO	19.00	-	-	-	15.77	MICRO/MJ	43.41	
PHOSPHORUS	MILLI	176.00	134.00	-	240.00	146.08	MILLI/MJ	402.08	
FLUORIDE	MICRO	60.00	-	-	-	49.80	MICRO/MJ	137.07	
IODIDE	MICRO	3.00	-	-	-	2.49	MICRO/MJ	6.85	
BORON	MILLI	0.10	-	-	-	0.09	MILLI/MJ	0.24	
SELENIUM	MICRO	88.00	40.00	-	126.00	73.04	MICRO/MJ	201.04	
SILICON	MILLI	3.00	1.00	-	5.00	2.49	MILLI/MJ	6.85	
VITAMIN A	MICRO	10.00	9.00	-	11.00	8.30	MICRO/MJ	22.85	
VITAMIN B1	MILLI	0.46	0.31	-	0.63	0.38	MILLI/MJ	1.05	
VITAMIN B2	MILLI	1.06	0.81	-	1.24	0.88	MILLI/MJ	2.42	
NICOTINAMIDE	MILLI	6.60	5.60	-	8.00	5.48	MILLI/MJ	15.08	
PANTOTHENIC ACID	MILLI	2.50	-	-	-	2.08	MILLI/MJ	5.71	

CONSTITUENTS	DIM	AV	VARIATION			AVR	NUTR. DENS.		MOLPERC.
VITAMIN B6	MILLI	0.43	0.35	-	0.51	0.36	MILLI/MJ	0.98	
BIOTIN	MICRO	4.00	-	-	-	3.32	MICRO/MJ	9.14	
FOLIC ACID	MICRO	4.00	-	-	-	3.32	MICRO/MJ	9.14	
VITAMIN B12	MICRO	2.70	1.80	-	4.40	2.24	MICRO/MJ	6.17	
VITAMIN C	MILLI	5.30	4.00	-	6.00	4.40	MILLI/MJ	12.11	
ALANINE	GRAM	1.25	1.22	-	1.29	1.04	GRAM/MJ	2.86	9.2
ARGININE	GRAM	1.13	1.08	-	1.16	0.94	GRAM/MJ	2.58	4.3
ASPARTIC ACID	GRAM	1.85	-	-	-	1.54	GRAM/MJ	4.23	9.1
CYSTINE	GRAM	0.24	-	-	-	0.20	GRAM/MJ	0.55	0.7
GLUTAMIC ACID	GRAM	3.34	3.26	-	3.43	2.77	GRAM/MJ	7.63	14.9
GLYCINE	GRAM	1.12	0.99	-	1.23	0.93	GRAM/MJ	2.56	9.8
HISTIDINE	GRAM	0.74	0.08	-	0.83	0.61	GRAM/MJ	1.69	3.1
ISOLEUCINE	GRAM	0.93	-	-	-	0.77	GRAM/MJ	2.12	4.7
LEUCINE	GRAM	1.55	1.47	-	1.64	1.29	GRAM/MJ	3.54	7.8
LYSINE	GRAM	1.76	1.61	-	1.91	1.46	GRAM/MJ	4.02	7.9
METHIONINE	GRAM	0.46	-	-	-	0.38	GRAM/MJ	1.05	2.0
PHENYLALANINE	GRAM	0.75	0.71	-	0.77	0.62	GRAM/MJ	1.71	3.0
PROLINE	GRAM	0.87	0.71	-	1.02	0.72	GRAM/MJ	1.99	5.0
SERINE	GRAM	0.77	0.65	-	0.90	0.64	GRAM/MJ	1.76	4.8
THREONINE	GRAM	0.85	0.81	-	0.90	0.71	GRAM/MJ	1.94	4.7
TRYPTOPHAN	GRAM	0.22	-	-	-	0.18	GRAM/MJ	0.50	0.7
TYROSINE	GRAM	0.71	0.69	-	0.74	0.59	GRAM/MJ	1.62	2.6
VALINE	GRAM	1.04	1.01	-	1.06	0.86	GRAM/MJ	2.38	5.8
MYRISTIC ACID	MILLI	10.00	-	-	-	8.30	MILLI/MJ	22.85	
PALMITIC ACID	GRAM	0.58	-	-	-	0.48	GRAM/MJ	1.33	
C17 : 0 FATTY ACID	MILLI	10.00	-	-	-	8.30	MILLI/MJ	22.85	
STEARIC ACID	GRAM	0.42	-	-	-	0.35	GRAM/MJ	0.96	
ARACHIDIC ACID	MILLI	20.00	-	-	-	16.60	MILLI/MJ	45.69	
PALMITOLEIC ACID	MILLI	30.00	-	-	-	24.90	MILLI/MJ	68.54	
OLEIC ACID	GRAM	0.63	-	-	-	0.52	GRAM/MJ	1.44	
LINOLEIC ACID	GRAM	1.16	-	-	-	0.96	GRAM/MJ	2.65	
LINOLENIC ACID	MILLI	30.00	-	-	-	24.90	MILLI/MJ	68.54	
ARACHIDONIC ACID	GRAM	0.33	-	-	-	0.27	GRAM/MJ	0.75	
CHOLESTEROL	MILLI	150.00	-	-	-	124.50	MILLI/MJ	342.68	
TOTAL PURINES	MILLI	530.00 1	530.00	-	530.00	439.90	GRAM/MJ	1.21	
ADENINE	MILLI	58.00	58.00	-	58.00	48.14	MILLI/MJ	132.50	
GUANINE	MILLI	50.00	50.00	-	50.00	41.50	MILLI/MJ	114.23	
HYPOXANTHINE	MILLI	367.00	367.00	-	367.00	304.61	MILLI/MJ	838.42	

1 VALUE OF TOTAL PURINES IS EXPRESSSED IN MG URIC ACID/100G

SCHWEINEHIRN PIG'S BRAIN CERVELLE DE PORC

		PROTEIN	FAT	CARBOHYDRATES	TOTAL
ENERGY VALUE (AVERAGE)	KJOULE	195	358	0.0	553
PER 100 G	(KCAL)	47	86	0.0	132
EDIBLE PORTION					
AMOUNT OF DIGESTIBLE	GRAM	10.28	8.55	0.00	
CONSTITUENTS PER 100 G					
ENERGY VALUE (AVERAGE)	KJOULE	189	340	0.0	529
OF THE DIGESTIBLE	(KCAL)	45	81	0.0	126
FRACTION PER 100 G					
EDIBLE PORTION					

WASTE PERCENTAGE AVERAGE 0.00

CONSTITUENTS	DIM	AV	VARIATION			AVR	NUTR. DENS.		MOLPERC.
WATER	GRAM	78.00	-	-	-	76.44	GRAM/MJ	147.41	
PROTEIN	GRAM	10.60	-	-	-	10.39	GRAM/MJ	20.03	
FAT	GRAM	9.00	-	-	-	8.82	GRAM/MJ	17.01	
MINERALS	GRAM	1.50	-	-	-	1.47	GRAM/MJ	2.83	
SODIUM	MILLI	153.00	138.00	-	187.00	149.94	MILLI/MJ	289.15	
POTASSIUM	MILLI	312.00	283.00	-	340.00	305.76	MILLI/MJ	589.64	
MAGNESIUM	MILLI	20.00	17.00	-	21.00	19.60	MILLI/MJ	37.80	
CALCIUM	MILLI	10.00	4.00	-	16.00	9.80	MILLI/MJ	18.90	
MANGANESE	MICRO	48.00	40.00	-	50.00	47.04	MICRO/MJ	90.71	
IRON	MILLI	3.60	-	-	-	3.53	MILLI/MJ	6.80	
COBALT	MICRO	0.70	0.30	-	1.00	0.69	MICRO/MJ	1.32	
COPPER	MILLI	0.54	-	-	-	0.53	MILLI/MJ	1.02	
ZINC	MILLI	1.60	1.50	-	1.80	1.57	MILLI/MJ	3.02	
PHOSPHORUS	MILLI	400.00	330.00	-	460.00	392.00	MILLI/MJ	755.95	
SELENIUM	MICRO	17.00	14.00	-	21.00	16.66	MICRO/MJ	32.13	
VITAMIN A	MICRO	9.00	-	-	-	8.82	MICRO/MJ	17.01	
VITAMIN B1	MILLI	0.16	-	-	-	0.16	MILLI/MJ	0.30	
VITAMIN B2	MILLI	0.28	-	-	-	0.27	MILLI/MJ	0.53	
NICOTINAMIDE	MILLI	4.30	-	-	-	4.21	MILLI/MJ	8.13	
PANTOTHENIC ACID	MILLI	2.80	-	-	-	2.74	MILLI/MJ	5.29	
VITAMIN B12	MICRO	2.80	-	-	-	2.74	MICRO/MJ	5.29	
VITAMIN C	MILLI	18.00	-	-	-	17.64	MILLI/MJ	34.02	
ARGININE	GRAM	0.56	-	-	-	0.55	GRAM/MJ	1.06	
HISTIDINE	GRAM	0.29	-	-	-	0.28	GRAM/MJ	0.55	
ISOLEUCINE	GRAM	0.56	-	-	-	0.55	GRAM/MJ	1.06	
LEUCINE	GRAM	0.93	-	-	-	0.91	GRAM/MJ	1.76	
LYSINE	GRAM	0.92	-	-	-	0.90	GRAM/MJ	1.74	
METHIONINE	GRAM	0.22	-	-	-	0.22	GRAM/MJ	0.42	
PHENYLALANINE	GRAM	0.55	-	-	-	0.54	GRAM/MJ	1.04	
THREONINE	GRAM	0.53	-	-	-	0.52	GRAM/MJ	1.00	
TRYPTOPHAN	GRAM	0.16	-	-	-	0.16	GRAM/MJ	0.30	
VALINE	GRAM	0.65	-	-	-	0.64	GRAM/MJ	1.23	
PALMITOLEIC ACID	MILLI	30.00	-	-	-	29.40	MILLI/MJ	56.70	
CHOLESTEROL	GRAM	2.00	-	-	-	1.96	GRAM/MJ	3.78	

SCHWEINELEBER PIG'S LIVER FOIE DE PORC

		PROTEIN	FAT	CARBOHYDRATES	TOTAL
ENERGY VALUE (AVERAGE)	KJOULE	370	227	8.2	605
PER 100 G	(KCAL)	88	54	2.0	145
EDIBLE PORTION					
AMOUNT OF DIGESTIBLE	GRAM	19.49	5.42	0.49	
CONSTITUENTS PER 100 G					
ENERGY VALUE (AVERAGE)	KJOULE	359	216	8.2	583
OF THE DIGESTIBLE	(KCAL)	86	52	2.0	139
FRACTION PER 100 G					
EDIBLE PORTION					

WASTE PERCENTAGE AVERAGE 7.0 MINIMUM 5.0 MAXIMUM 9.0

CONSTITUENTS	DIM	AV	VARIATION			AVR	NUTR. DENS.		MOLPERC.
WATER	GRAM	71.80	71.10	-	72.30	66.77	GRAM/MJ	123.21	
PROTEIN	GRAM	20.10	19.50	-	21.30	18.69	GRAM/MJ	34.49	
FAT	GRAM	5.71	4.80	-	7.32	5.31	GRAM/MJ	9.80	
AVAILABLE CARBOHYDR.	GRAM	0.49	-	-	-	0.46	GRAM/MJ	0.84	
MINERALS	GRAM	1.25	1.00	-	1.50	1.16	GRAM/MJ	2.15	
SODIUM	MILLI	77.00	-	-	-	71.61	MILLI/MJ	132.13	
POTASSIUM	MILLI	350.00	-	-	-	325.50	MILLI/MJ	600.61	
MAGNESIUM	MILLI	21.00	-	-	-	19.53	MILLI/MJ	36.04	
CALCIUM	MILLI	10.00	-	-	-	9.30	MILLI/MJ	17.16	
MANGANESE	MILLI	0.36	-	-	-	0.33	MILLI/MJ	0.62	
IRON	MILLI	22.10	18.00	-	30.30	20.55	MILLI/MJ	37.92	
COBALT	MICRO	1.53	-	-	-	1.42	MICRO/MJ	2.63	
COPPER	MILLI	5.48	2.58	-	8.38	5.10	MILLI/MJ	9.40	
ZINC	MILLI	5.90	-	-	-	5.49	MILLI/MJ	10.12	
NICKEL	MICRO	1.10	0.00	-	27.00	1.02	MICRO/MJ	1.89	
CHROMIUM	MICRO	-	0.00	-	10.00				
MOLYBDENUM	MILLI	0.30	-	-	-	0.28	MILLI/MJ	0.51	
ALUMINIUM	MICRO	-	2.80	-	210.00				
PHOSPHORUS	MILLI	362.00	-	-	-	336.66	MILLI/MJ	621.20	
CHLORIDE	MILLI	67.80	-	-	-	63.05	MILLI/MJ	116.35	
FLUORIDE	MILLI	0.29	-	-	-	0.27	MILLI/MJ	0.50	
IODIDE	MICRO	14.00	-	-	-	13.02	MICRO/MJ	24.02	
BORON	MICRO	73.00	56.00	-	93.00	67.89	MICRO/MJ	125.27	
SELENIUM	MICRO	58.00	39.00	-	89.00	53.94	MICRO/MJ	99.53	
VITAMIN A	MILLI	39.10	5.60	-	56.00	36.36	MILLI/MJ	67.10	
VITAMIN D	MICRO	-	0.00	-	0.25				
VITAMIN E ACTIVITY	MILLI	0.17	-	-	-	0.16	MILLI/MJ	0.29	
ALPHA-TOCOPHEROL	MILLI	0.17	-	-	-	0.16	MILLI/MJ	0.29	
VITAMIN K	MICRO	24.00	-	-	-	22.32	MICRO/MJ	41.18	
VITAMIN B1	MILLI	0.31	0.25	-	0.40	0.29	MILLI/MJ	0.53	
VITAMIN B2	MILLI	3.17	2.98	-	3.70	2.95	MILLI/MJ	5.44	
NICOTINAMIDE	MILLI	15.70	13.90	-	16.70	14.60	MILLI/MJ	26.94	
PANTOTHENIC ACID	MILLI	6.80	6.60	-	7.00	6.32	MILLI/MJ	11.67	

CONSTITUENTS	DIM	AV	VARIATION			AVR	NUTR. DENS.		MOLPERC.
VITAMIN B6	MILLI	0.59	0.33	-	0.85	0.55	MILLI/MJ	1.01	
BIOTIN	MICRO	27.00	-	-	-	25.11	MICRO/MJ	46.33	
FOLIC ACID	MILLI	0.22	-	-	-	0.20	MILLI/MJ	0.38	
VITAMIN B12	MICRO	39.00	23.00	-	55.00	36.27	MICRO/MJ	66.92	
VITAMIN C	MILLI	23.00	20.00	-	25.00	21.39	MILLI/MJ	39.47	
ALANINE	GRAM	1.46	1.34	-	1.67	1.36	GRAM/MJ	2.51	8.9
ARGININE	GRAM	1.36	1.21	-	1.48	1.26	GRAM/MJ	2.33	4.2
ASPARTIC ACID	GRAM	2.13	1.83	-	2.47	1.98	GRAM/MJ	3.66	8.7
CYSTINE	GRAM	0.32	0.28	-	0.36	0.30	GRAM/MJ	0.55	0.7
GLUTAMIC ACID	GRAM	2.77	2.20	-	3.21	2.58	GRAM/MJ	4.75	10.2
GLYCINE	GRAM	1.46	1.35	-	1.66	1.36	GRAM/MJ	2.51	10.6
HISTIDINE	GRAM	0.68	0.60	-	0.83	0.63	GRAM/MJ	1.17	2.4
ISOLEUCINE	GRAM	1.34	1.04	-	1.73	1.25	GRAM/MJ	2.30	5.6
LEUCINE	GRAM	2.12	1.89	-	2.24	1.97	GRAM/MJ	3.64	8.8
LYSINE	GRAM	1.83	1.72	-	1.96	1.70	GRAM/MJ	3.14	6.8
METHIONINE	GRAM	0.63	0.56	-	0.75	0.59	GRAM/MJ	1.08	2.3
PHENYLALANINE	GRAM	1.13	1.08	-	1.17	1.05	GRAM/MJ	1.94	3.7
PROLINE	GRAM	1.32	1.26	-	1.35	1.23	GRAM/MJ	2.27	6.2
SERINE	GRAM	1.18	1.07	-	1.30	1.10	GRAM/MJ	2.02	6.1
THREONINE	GRAM	1.07	0.98	-	1.16	1.00	GRAM/MJ	1.84	4.9
TRYPTOPHAN	GRAM	0.31	0.29	-	0.34	0.29	GRAM/MJ	0.53	0.8
TYROSINE	GRAM	0.78	0.70	-	0.85	0.73	GRAM/MJ	1.34	2.3
VALINE	GRAM	1.45	1.39	-	1.54	1.35	GRAM/MJ	2.49	6.7
GLYCOGENE	GRAM	0.49	-	-	-	0.46	GRAM/MJ	0.84	
MYRISTIC ACID	MILLI	20.00	-	-	-	18.60	MILLI/MJ	34.32	
PALMITIC ACID	GRAM	0.82	0.71	-	0.92	0.76	GRAM/MJ	1.41	
STEARIC ACID	GRAM	1.22	0.98	-	1.46	1.13	GRAM/MJ	2.09	
ARACHIDIC ACID	MILLI	10.00	-	-	-	9.30	MILLI/MJ	17.16	
PALMITOLEIC ACID	MILLI	90.00	-	-	-	83.70	MILLI/MJ	154.44	
OLEIC ACID	GRAM	1.22	0.97	-	1.47	1.13	GRAM/MJ	2.09	
EICOSENOIC ACID	MILLI	10.00	-	-	-	9.30	MILLI/MJ	17.16	
LINOLEIC ACID	GRAM	0.49	0.27	-	0.72	0.46	GRAM/MJ	0.84	
ARACHIDONIC ACID	GRAM	0.87	-	-	-	0.81	GRAM/MJ	1.49	
CHOLESTEROL	MILLI	340.00	-	-	-	316.20	MILLI/MJ	583.44	
TOTAL PURINES	MILLI	515.00 [1]	515.00	-	515.00	478.95	MILLI/MJ	883.75	
ADENINE	MILLI	139.00	139.00	-	139.00	129.27	MILLI/MJ	238.53	
GUANINE	MILLI	166.00	166.00	-	166.00	154.38	MILLI/MJ	284.86	
XANTHINE	MILLI	32.00	32.00	-	32.00	29.76	MILLI/MJ	54.91	
HYPOXANTHINE	MILLI	100.00	100.00	-	100.00	93.00	MILLI/MJ	171.60	
TOTAL PHOSPHOLIPIDS	GRAM	4.48	4.48	-	4.48	4.17	GRAM/MJ	7.69	
PHOSPHATIDYLCHOLINE	GRAM	2.70	2.70	-	2.70	2.51	GRAM/MJ	4.63	
PHOSPHATIDYLETHANOLAMINE	GRAM	0.95	0.95	-	0.95	0.88	GRAM/MJ	1.63	
PHOSPHATIDYLSERINE	MILLI	60.00	60.00	-	60.00	55.80	MILLI/MJ	102.96	
PHOSPHATIDYLINOSITOL	GRAM	0.32	0.32	-	0.32	0.30	GRAM/MJ	0.55	
SPHINGOMYELIN	GRAM	0.20	0.20	-	0.20	0.19	GRAM/MJ	0.34	

1 VALUE OF TOTAL PURINES IS EXPRESSSED IN MG URIC ACID/100G

DEPENDENT ON THE METHOD OF EXTRACTION; WITH PRECEDING PROCEDURES
VALUES OF 250 - 350 MG URIC ACID/100G HAD BEEN OBTAINED

SCHWEINELUNGE / PIG'S LUNGS (LIGHTS) / MOU DE PORC

		PROTEIN	FAT	CARBOHYDRATES	TOTAL
ENERGY VALUE (AVERAGE)	KJOULE	249	265	0.0	514
PER 100 G	(KCAL)	59	63	0.0	123
EDIBLE PORTION					
AMOUNT OF DIGESTIBLE	GRAM	13.09	6.33	0.00	
CONSTITUENTS PER 100 G					
ENERGY VALUE (AVERAGE)	KJOULE	241	252	0.0	493
OF THE DIGESTIBLE	(KCAL)	58	60	0.0	118
FRACTION PER 100 G					
EDIBLE PORTION					

WASTE PERCENTAGE AVERAGE 13

CONSTITUENTS	DIM	AV	VARIATION			AVR	NUTR.	DENS.	MOLPERC.
WATER	GRAM	79.10	72.20	-	83.00	68.82	GRAM/MJ	160.47	
PROTEIN	GRAM	13.50	11.90	-	14.90	11.75	GRAM/MJ	27.39	
FAT	GRAM	6.67	2.00	-	13.60	5.80	GRAM/MJ	13.53	
MINERALS	GRAM	0.93	0.80	-	10.06	0.81	GRAM/MJ	1.89	
SODIUM	MILLI	151.00	123.00	-	178.00	131.37	MILLI/MJ	306.33	
POTASSIUM	MILLI	243.00	234.00	-	252.00	211.41	MILLI/MJ	492.96	
CALCIUM	MILLI	3.00	2.00	-	4.00	2.61	MILLI/MJ	6.09	
IRON	MILLI	5.00	-	-	-	4.35	MILLI/MJ	10.14	
MOLYBDENUM	MILLI	-	0.02	-	0.20				
PHOSPHORUS	MILLI	230.00	-	-	-	200.10	MILLI/MJ	466.59	
BORON	MICRO	36.00	12.00	-	60.00	31.32	MICRO/MJ	73.03	
SELENIUM	MICRO	19.00	11.00	-	29.00	16.53	MICRO/MJ	38.54	
VITAMIN B1	MICRO	60.00	20.00	-	90.00	52.20	MICRO/MJ	121.72	
VITAMIN B2	MILLI	0.21	0.14	-	0.27	0.18	MILLI/MJ	0.43	
NICOTINAMIDE	MILLI	3.40	-	-	-	2.96	MILLI/MJ	6.90	
PANTOTHENIC ACID	MILLI	0.90	-	-	-	0.78	MILLI/MJ	1.83	
VITAMIN C	MILLI	13.10	-	-	-	11.40	MILLI/MJ	26.58	
ARGININE	GRAM	0.69	-	-	-	0.60	GRAM/MJ	1.40	
HISTIDINE	GRAM	0.31	-	-	-	0.27	GRAM/MJ	0.63	
ISOLEUCINE	GRAM	0.60	-	-	-	0.52	GRAM/MJ	1.22	
LEUCINE	GRAM	1.00	-	-	-	0.87	GRAM/MJ	2.03	
LYSINE	GRAM	0.10	-	-	-	0.09	GRAM/MJ	0.20	
METHIONINE	GRAM	0.23	-	-	-	0.20	GRAM/MJ	0.47	
PHENYLALANINE	GRAM	0.55	-	-	-	0.48	GRAM/MJ	1.12	
THREONINE	GRAM	0.51	-	-	-	0.44	GRAM/MJ	1.03	
TRYPTOPHAN	GRAM	0.12	-	-	-	0.10	GRAM/MJ	0.24	
VALINE	GRAM	0.79	-	-	-	0.69	GRAM/MJ	1.60	
TOTAL PURINES	MILLI	434.00 [1]	434.00	-	434.00	377.58	MILLI/MJ	880.44	
ADENINE	MILLI	133.00	133.00	-	133.00	115.71	MILLI/MJ	269.81	
GUANINE	MILLI	150.00	150.00	-	150.00	130.50	MILLI/MJ	304.30	
XANTHINE	MILLI	4.00	4.00	-	4.00	3.48	MILLI/MJ	8.11	
HYPOXANTHINE	MILLI	79.00	79.00	-	79.00	68.73	MILLI/MJ	160.26	

1 VALUE OF TOTAL PURINES IS EXPRESSSED IN MG URIC ACID/100G

SCHWEINEMILZ PIG'S SPLEEN RATE DE PORC

		PROTEIN	FAT	CARBOHYDRATES	TOTAL
ENERGY VALUE (AVERAGE)	KJOULE	317	145	0.0	461
PER 100 G	(KCAL)	76	35	0.0	110
EDIBLE PORTION					
AMOUNT OF DIGESTIBLE	GRAM	16.68	3.45	0.00	
CONSTITUENTS PER 100 G					
ENERGY VALUE (AVERAGE)	KJOULE	307	137	0.0	445
OF THE DIGESTIBLE	(KCAL)	73	33	0.0	106
FRACTION PER 100 G					
EDIBLE PORTION					

WASTE PERCENTAGE AVERAGE 0.00

CONSTITUENTS	DIM	AV	VARIATION			AVR	NUTR. DENS.		MOLPERC.
WATER	GRAM	77.40	77.00	-	79.30	77.40	GRAM/MJ	174.09	
PROTEIN	GRAM	17.20	17.10	-	17.60	17.20	GRAM/MJ	38.69	
FAT	GRAM	3.64	1.90	-	5.02	3.64	GRAM/MJ	8.19	
MINERALS	GRAM	1.44	1.40	-	1.57	1.44	GRAM/MJ	3.24	
SODIUM	MILLI	100.00	98.00	-	102.00	100.00	MILLI/MJ	224.92	
POTASSIUM	MILLI	380.00	377.00	-	382.00	380.00	MILLI/MJ	854.71	
CALCIUM	MILLI	7.00	4.00	-	10.00	7.00	MILLI/MJ	15.74	
IRON	MILLI	19.40	-	-	-	19.40	MILLI/MJ	43.64	
PHOSPHORUS	MILLI	173.00	-	-	-	173.00	MILLI/MJ	389.12	
BORON	MICRO	32.00	27.00	-	36.00	32.00	MICRO/MJ	71.98	
SELENIUM	MICRO	35.00	-	-	-	35.00	MICRO/MJ	78.72	
VITAMIN B1	MILLI	0.13	0.12	-	0.13	0.13	MILLI/MJ	0.29	
VITAMIN B2	MILLI	0.32	0.30	-	0.34	0.32	MILLI/MJ	0.72	
NICOTINAMIDE	MILLI	3.35	2.40	-	4.30	3.35	MILLI/MJ	7.53	
PANTOTHENIC ACID	MILLI	1.10	-	-	-	1.10	MILLI/MJ	2.47	
VITAMIN B12	MICRO	4.10	-	-	-	4.10	MICRO/MJ	9.22	
VITAMIN C	MILLI	30.00	-	-	-	30.00	MILLI/MJ	67.48	
ARGININE	GRAM	0.96	-	-	-	0.96	GRAM/MJ	2.16	
HISTIDINE	GRAM	0.43	-	-	-	0.43	GRAM/MJ	0.97	
ISOLEUCINE	GRAM	0.93	-	-	-	0.93	GRAM/MJ	2.09	
LEUCINE	GRAM	1.50	-	-	-	1.50	GRAM/MJ	3.37	
LYSINE	GRAM	1.39	-	-	-	1.39	GRAM/MJ	3.13	
METHIONINE	GRAM	0.34	-	-	-	0.34	GRAM/MJ	0.76	
PHENYLALANINE	GRAM	0.79	-	-	-	0.79	GRAM/MJ	1.78	
THREONINE	GRAM	0.74	-	-	-	0.74	GRAM/MJ	1.66	

CONSTITUENTS	DIM	AV	VARIATION		AVR	NUTR. DENS.		MOLPERC.
TRYPTOPHAN	GRAM	0.19	-	-	0.19	GRAM/MJ	0.43	
VALINE	GRAM	1.01	-	-	1.01	GRAM/MJ	2.27	
TOTAL PURINES	MILLI	516.00 1	516.00	- 516.00	516.00	GRAM/MJ	1.16	
ADENINE	MILLI	178.00	178.00	- 178.00	178.00	MILLI/MJ	400.36	
GUANINE	MILLI	179.00	179.00	- 179.00	179.00	MILLI/MJ	402.61	
XANTHINE	MILLI	3.00	3.00	- 3.00	3.00	MILLI/MJ	6.75	
HYPOXANTHINE	MILLI	75.00	75.00	- 75.00	75.00	MILLI/MJ	168.69	
TOTAL PHOSPHOLIPIDS	GRAM	1.84	-	-	1.84	GRAM/MJ	4.14	
PHOSPHATIDYLCHOLINE	MILLI	608.00	-	-	608.00	GRAM/MJ	1.37	
PHOSPHATIDYLETHANOLAMINE	MILLI	259.00	-	-	259.00	MILLI/MJ	582.55	
PHOSPHATIDYLSERINE	MILLI	239.00	-	-	239.00	MILLI/MJ	537.57	
PHOSPHATIDYLINOSITOL	MILLI	37.00	-	-	37.00	MILLI/MJ	83.22	
SPHINGOMYELIN	MILLI	351.00	-	-	351.00	MILLI/MJ	789.48	

1 VALUE OF TOTAL PURINES IS EXPRESSSED IN MG URIC ACID/100G

SCHWEINENIERE PIG'S KIDNEY ROGNON DE PORC

		PROTEIN	FAT	CARBOHYDRATES	TOTAL
ENERGY VALUE (AVERAGE)	KJOULE	304	207	0.0	510
PER 100 G	(KCAL)	73	49	0.0	122
EDIBLE PORTION					
AMOUNT OF DIGESTIBLE	GRAM	16.00	4.94	0.00	
CONSTITUENTS PER 100 G					
ENERGY VALUE (AVERAGE)	KJOULE	295	196	0.0	491
OF THE DIGESTIBLE	(KCAL)	70	47	0.0	117
FRACTION PER 100 G					
EDIBLE PORTION					

WASTE PERCENTAGE AVERAGE 13 MINIMUM 10 MAXIMUM 15

CONSTITUENTS	DIM	AV	VARIATION			AVR	NUTR. DENS.		MOLPERC.
WATER	GRAM	76.30	75.00	-	77.10	66.38	GRAM/MJ	155.40	
PROTEIN	GRAM	16.50	16.30	-	17.00	14.36	GRAM/MJ	33.60	
FAT	GRAM	5.20	4.60	-	6.00	4.52	GRAM/MJ	10.59	
MINERALS	GRAM	1.20	-	-	-	1.04	GRAM/MJ	2.44	
SODIUM	MILLI	173.00	159.00	-	198.00	150.51	MILLI/MJ	352.34	
POTASSIUM	MILLI	242.00	206.00	-	262.00	210.54	MILLI/MJ	492.87	
CALCIUM	MILLI	7.00	2.00	-	11.00	6.09	MILLI/MJ	14.26	
MANGANESE	MICRO	60.00	-	-	-	52.20	MICRO/MJ	122.20	
IRON	MILLI	10.00	8.00	-	15.00	8.70	MILLI/MJ	20.37	
COBALT	MICRO	0.40	-	-	-	0.35	MICRO/MJ	0.81	
COPPER	MILLI	0.17	-	-	-	0.15	MILLI/MJ	0.35	
ZINC	MILLI	0.37	-	-	-	0.32	MILLI/MJ	0.75	
NICKEL	MICRO	1.10	-	-	-	0.96	MICRO/MJ	2.24	
CHROMIUM	MICRO	1.00	-	-	-	0.87	MICRO/MJ	2.04	
MOLYBDENUM	MICRO	17.00	-	-	-	14.79	MICRO/MJ	34.62	
ALUMINIUM	MICRO	2.40	-	-	-	2.09	MICRO/MJ	4.89	
PHOSPHORUS	MILLI	260.00	250.00	-	300.00	226.20	MILLI/MJ	529.53	
CHLORIDE	MILLI	190.00	-	-	-	165.30	MILLI/MJ	386.96	
BORON	MICRO	72.00	36.00	-	114.00	62.64	MICRO/MJ	146.64	
SELENIUM	MILLI	-	0.19	-	0.42				
VITAMIN A	MICRO	60.00	40.00	-	230.00	52.20	MICRO/MJ	122.20	
VITAMIN D	-	0.00	-	-	-	0.00			
VITAMIN B1	MILLI	0.34	0.17	-	0.58	0.30	MILLI/MJ	0.69	
VITAMIN B2	MILLI	1.80	1.50	-	2.28	1.57	MILLI/MJ	3.67	
NICOTINAMIDE	MILLI	8.35	5.90	-	9.80	7.26	MILLI/MJ	17.01	
PANTOTHENIC ACID	MILLI	3.10	-	-	-	2.70	MILLI/MJ	6.31	

CONSTITUENTS	DIM	AV	VARIATION			AVR	NUTR. DENS.		MOLPERC.
VITAMIN B6	MILLI	0.55	0.48	-	0.70	0.48	MILLI/MJ	1.12	
BIOTIN	MILLI	-	0.03	-	0.13				
VITAMIN B12	MICRO	15.00	13.00	-	18.00	13.05	MICRO/MJ	30.55	
VITAMIN C	MILLI	16.00	13.00	-	18.00	13.92	MILLI/MJ	32.59	
ARGININE	GRAM	0.97	-	-	-	0.84	GRAM/MJ	1.98	
CYSTINE	GRAM	0.20	-	-	-	0.17	GRAM/MJ	0.41	
HISTIDINE	GRAM	0.43	-	-	-	0.37	GRAM/MJ	0.88	
ISOLEUCINE	GRAM	0.91	-	-	-	0.79	GRAM/MJ	1.85	
LEUCINE	GRAM	1.44	-	-	-	1.25	GRAM/MJ	2.93	
LYSINE	GRAM	1.25	-	-	-	1.09	GRAM/MJ	2.55	
METHIONINE	GRAM	0.36	-	-	-	0.31	GRAM/MJ	0.73	
PHENYLALANINE	GRAM	0.81	-	-	-	0.70	GRAM/MJ	1.65	
THREONINE	GRAM	0.74	-	-	-	0.64	GRAM/MJ	1.51	
TRYPTOPHAN	GRAM	0.25	-	-	-	0.22	GRAM/MJ	0.51	
TYROSINE	GRAM	0.61	-	-	-	0.53	GRAM/MJ	1.24	
VALINE	GRAM	1.01	-	-	-	0.88	GRAM/MJ	2.06	
MYRISTIC ACID	MILLI	10.00	0.00	-	20.00	8.70	MILLI/MJ	20.37	
PALMITIC ACID	GRAM	1.16	1.09	-	1.23	1.01	GRAM/MJ	2.36	
C17 : 0 FATTY ACID	MILLI	50.00	-	-	-	43.50	MILLI/MJ	101.83	
STEARIC ACID	GRAM	1.17	0.88	-	1.46	1.02	GRAM/MJ	2.38	
PALMITOLEIC ACID	GRAM	0.10	-	-	-	0.09	GRAM/MJ	0.20	
OLEIC ACID	GRAM	1.29	0.97	-	1.60	1.12	GRAM/MJ	2.63	
LINOLEIC ACID	GRAM	0.65	0.58	-	0.72	0.57	GRAM/MJ	1.32	
LINOLENIC ACID	MILLI	30.00	-	-	-	26.10	MILLI/MJ	61.10	
ARACHIDONIC ACID	GRAM	0.60	0.33	-	0.87	0.52	GRAM/MJ	1.22	
CHOLESTEROL	MILLI	365.00	-	-	-	317.55	MILLI/MJ	743.38	
TOTAL PURINES	MILLI	334.00 [1]	334.00	-	334.00	290.58	MILLI/MJ	680.24	
ADENINE	MILLI	86.00	86.00	-	86.00	74.82	MILLI/MJ	175.15	
GUANINE	MILLI	110.00	110.00	-	110.00	95.70	MILLI/MJ	224.03	
XANTHINE	MILLI	8.00	8.00	-	8.00	6.96	MILLI/MJ	16.29	
HYPOXANTHINE	MILLI	78.00	78.00	-	78.00	67.86	MILLI/MJ	158.86	
TOTAL PHOSPHOLIPIDS	GRAM	4.19	-	-	-	3.65	GRAM/MJ	8.53	
PHOSPHATIDYLCHOLINE	GRAM	1.51	-	-	-	1.31	GRAM/MJ	3.08	
PHOSPHATIDYLETHANOLAMINE	MILLI	713.00	-	-	-	620.31	GRAM/MJ	1.45	
PHOSPHATIDYLSERINE	MILLI	294.00	-	-	-	255.78	MILLI/MJ	598.78	
PHOSPHATIDYLINOSITOL	MILLI	126.00	-	-	-	109.62	MILLI/MJ	256.62	
SPHINGOMYELIN	MILLI	588.00	-	-	-	511.56	GRAM/MJ	1.20	

1 VALUE OF TOTAL PURINES IS EXPRESSSED IN MG URIC ACID/100G

SCHWEINEZUNGE PIG'S TONGUE LANGUE DE PORC

		PROTEIN	FAT	CARBOHYDRATES	TOTAL
ENERGY VALUE (AVERAGE) PER 100 G EDIBLE PORTION	KJOULE	278	727	0.0	1005
	(KCAL)	66	174	0.0	240
AMOUNT OF DIGESTIBLE CONSTITUENTS PER 100 G	GRAM	14.64	17.38	0.00	
ENERGY VALUE (AVERAGE) OF THE DIGESTIBLE FRACTION PER 100 G EDIBLE PORTION	KJOULE	270	691	0.0	961
	(KCAL)	64	165	0.0	230

WASTE PERCENTAGE AVERAGE 16

CONSTITUENTS	DIM	AV	VARIATION			AVR	NUTR. DENS.		MOLPERC.
WATER	GRAM	65.90	61.80	-	68.00	55.36	GRAM/MJ	68.60	
PROTEIN	GRAM	15.10	12.90	-	16.40	12.68	GRAM/MJ	15.72	
FAT	GRAM	18.30	15.00	-	23.60	15.37	GRAM/MJ	19.05	
MINERALS	GRAM	0.90	0.79	-	0.90	0.76	GRAM/MJ	0.94	
SODIUM	MILLI	93.00	86.00	-	100.00	78.12	MILLI/MJ	96.81	
POTASSIUM	MILLI	234.00	207.00	-	260.00	196.56	MILLI/MJ	243.58	
CALCIUM	MILLI	9.00	-	-	-	7.56	MILLI/MJ	9.37	
IRON	MILLI	3.25	2.80	-	3.70	2.73	MILLI/MJ	3.38	
PHOSPHORUS	MILLI	187.00	-	-	-	157.08	MILLI/MJ	194.66	
VITAMIN B1	MILLI	0.49	0.28	-	0.97	0.41	MILLI/MJ	0.51	
VITAMIN B2	MILLI	0.50	0.43	-	0.54	0.42	MILLI/MJ	0.52	
NICOTINAMIDE	MILLI	5.30	4.90	-	6.10	4.45	MILLI/MJ	5.52	
VITAMIN B6	MILLI	0.35	0.34	-	0.36	0.29	MILLI/MJ	0.36	
VITAMIN B12	MICRO	0.80	0.50	-	1.10	0.67	MICRO/MJ	0.83	
VITAMIN C	MILLI	4.40	-	-	-	3.70	MILLI/MJ	4.58	
ALANINE	GRAM	1.11	1.08	-	1.14	0.93	GRAM/MJ	1.16	9.2
ARGININE	GRAM	1.00	0.96	-	1.03	0.84	GRAM/MJ	1.04	4.2
ASPARTIC ACID	GRAM	1.64	-	-	-	1.38	GRAM/MJ	1.71	9.1
CYSTINE	GRAM	0.20	-	-	-	0.17	GRAM/MJ	0.21	0.6
GLUTAMIC ACID	GRAM	2.97	2.90	-	3.05	2.49	GRAM/MJ	3.09	14.9
GLYCINE	GRAM	0.99	0.88	-	1.09	0.83	GRAM/MJ	1.03	9.8
HISTIDINE	GRAM	0.66	0.07	-	0.74	0.55	GRAM/MJ	0.69	3.1
ISOLEUCINE	GRAM	0.83	-	-	-	0.70	GRAM/MJ	0.86	4.7
LEUCINE	GRAM	1.38	1.30	-	1.45	1.16	GRAM/MJ	1.44	7.8
LYSINE	GRAM	1.56	1.43	-	1.70	1.31	GRAM/MJ	1.62	7.9
METHIONINE	GRAM	0.41	-	-	-	0.34	GRAM/MJ	0.43	2.0
PHENYLALANINE	GRAM	0.66	0.63	-	0.68	0.55	GRAM/MJ	0.69	3.0
PROLINE	GRAM	0.77	0.63	-	0.91	0.65	GRAM/MJ	0.08	5.0
SERINE	GRAM	0.68	0.58	-	0.80	0.57	GRAM/MJ	0.71	4.8
THREONINE	GRAM	0.76	0.72	-	0.80	0.64	GRAM/MJ	0.79	4.7
TRYPTOPHAN	GRAM	0.19	-	-	-	0.16	GRAM/MJ	0.20	0.7
TYROSINE	GRAM	0.63	0.61	-	0.66	0.53	GRAM/MJ	0.66	2.6
VALINE	GRAM	0.93	0.90	-	0.94	0.78	GRAM/MJ	0.97	5.9

SCHWEINESPECK	**BACON**	**LARD DE PORC**
BAUCHSPECK	ABDOMINAL	LARD DE VENTRE
FRISCH	FRESH	FRAIS (VERT)

		PROTEIN	FAT	CARBOHYDRATES	TOTAL
ENERGY VALUE (AVERAGE)	KJOULE	50	3538	0.0	3587
PER 100 G	(KCAL)	12	846	0.0	857
EDIBLE PORTION					
AMOUNT OF DIGESTIBLE	GRAM	2.60	84.55	0.00	
CONSTITUENTS PER 100 G					
ENERGY VALUE (AVERAGE)	KJOULE	48	3361	0.0	3409
OF THE DIGESTIBLE	(KCAL)	11	803	0.0	815
FRACTION PER 100 G					
EDIBLE PORTION					

WASTE PERCENTAGE AVERAGE 10

CONSTITUENTS	DIM	AV	VARIATION			AVR	NUTR. DENS.		MOLPERC.
WATER	GRAM	8.06	7.12	-	9.09	7.25	GRAM/MJ	2.36	
PROTEIN	GRAM	2.69	1.56	-	4.49	2.42	GRAM/MJ	0.79	
FAT	GRAM	89.00	87.90	-	92.50	80.10	GRAM/MJ	26.11	
SODIUM	MILLI	17.00	-	-	-	15.30	MILLI/MJ	4.99	
POTASSIUM	MILLI	8.00	-	-	-	7.20	MILLI/MJ	2.35	
CALCIUM	MILLI	8.00	-	-	-	7.20	MILLI/MJ	2.35	
PHOSPHORUS	MILLI	25.00	-	-	-	22.50	MILLI/MJ	7.33	
CHOLESTEROL	MILLI	62.00	-	-	-	55.80	MILLI/MJ	18.19	

SCHWEINESPECK	BACON	LARD DE PORC
RÜCKENSPECK	DORSAL	LARD DE SELLE
FRISCH	FRESH	FRAIS (VERT)

		PROTEIN	FAT	CARBOHYDRATES	TOTAL
ENERGY VALUE (AVERAGE)	KJOULE	75	3279	0.0	3355
PER 100 G	(KCAL)	18	784	0.0	802
EDIBLE PORTION					
AMOUNT OF DIGESTIBLE	GRAM	3.97	78.37	0.00	
CONSTITUENTS PER 100 G					
ENERGY VALUE (AVERAGE)	KJOULE	73	3115	0.0	3188
OF THE DIGESTIBLE	(KCAL)	17	745	0.0	762
FRACTION PER 100 G					
EDIBLE PORTION					

WASTE PERCENTAGE AVERAGE 10 MINIMUM 9.0 MAXIMUM 11

CONSTITUENTS	DIM	AV	VARIATION			AVR	NUTR. DENS.		MOLPERC.
WATER	GRAM	13.10	12.80	-	13.40	11.79	GRAM/MJ	4.11	
PROTEIN	GRAM	4.10	-	-	-	3.69	GRAM/MJ	1.29	
FAT	GRAM	82.50	-	-	-	74.25	GRAM/MJ	25.87	
MINERALS	GRAM	0.30	-	-	-	0.27	GRAM/MJ	0.09	
SODIUM	MILLI	21.00	17.00	-	29.00	18.90	MILLI/MJ	6.59	
POTASSIUM	MILLI	14.00	8.00	-	20.00	12.60	MILLI/MJ	4.39	
CALCIUM	MILLI	2.00	1.00	-	5.00	1.80	MILLI/MJ	0.63	
IRON	MILLI	0.30	-	-	-	0.27	MILLI/MJ	0.09	
PHOSPHORUS	MILLI	13.00	11.00	-	14.00	11.70	MILLI/MJ	4.08	
VITAMIN A	-	0.00	-	-	-	0.00			
VITAMIN B1	MILLI	0.10	0.08	-	0.12	0.09	MILLI/MJ	0.03	
VITAMIN B2	MICRO	20.00	20.00	-	30.00	18.00	MICRO/MJ	6.27	
NICOTINAMIDE	MILLI	0.50	0.40	-	0.60	0.45	MILLI/MJ	0.16	
ARGININE	GRAM	0.22	-	-	-	0.20	GRAM/MJ	0.07	
CYSTINE	MILLI	30.00	-	-	-	27.00	MILLI/MJ	9.41	
HISTIDINE	MILLI	20.00	-	-	-	18.00	MILLI/MJ	6.27	
ISOLEUCINE	MILLI	60.00	-	-	-	54.00	MILLI/MJ	18.82	
LEUCINE	GRAM	0.22	-	-	-	0.20	GRAM/MJ	0.07	
LYSINE	GRAM	0.19	-	-	-	0.17	GRAM/MJ	0.06	
METHIONINE	MILLI	30.00	-	-	-	27.00	MILLI/MJ	9.41	
THREONINE	MILLI	80.00	-	-	-	72.00	MILLI/MJ	25.09	
TRYPTOPHAN	MILLI	4.00	-	-	-	3.60	MILLI/MJ	1.25	
TYROSINE	MILLI	30.00	-	-	-	27.00	MILLI/MJ	9.41	
VALINE	GRAM	0.10	-	-	-	0.09	GRAM/MJ	0.03	

KANINCHENFLEISCH — RABBIT MEAT — VIANDE DE LAPIN

DURCHSCHNITT — AVERAGE — MOYENNE

		PROTEIN	FAT	CARBOHYDRATES	TOTAL
ENERGY VALUE (AVERAGE)	KJOULE	383	303	0.0	686
PER 100 G	(KCAL)	92	72	0.0	164
EDIBLE PORTION					
AMOUNT OF DIGESTIBLE	GRAM	20.17	7.23	0.00	
CONSTITUENTS PER 100 G					
ENERGY VALUE (AVERAGE)	KJOULE	371	288	0.0	659
OF THE DIGESTIBLE	(KCAL)	89	69	0.0	158
FRACTION PER 100 G					
EDIBLE PORTION					

WASTE PERCENTAGE AVERAGE 21 MINIMUM 16 MAXIMUM 25

CONSTITUENTS	DIM	AV	VARIATION			AVR	NUTR.	DENS.	MOLPERC.
WATER	GRAM	69.60	68.00	-	71.70	54.98	GRAM/MJ	105.59	
PROTEIN	GRAM	20.80	20.30	-	21.10	16.43	GRAM/MJ	31.55	
FAT	GRAM	7.62	5.53	-	10.00	6.02	GRAM/MJ	11.56	
MINERALS	GRAM	1.08	1.05	-	1.11	0.85	GRAM/MJ	1.64	
SODIUM	MILLI	47.00	40.00	-	50.00	37.13	MILLI/MJ	71.30	
POTASSIUM	MILLI	382.00	350.00	-	400.00	301.78	MILLI/MJ	579.52	
MAGNESIUM	MILLI	29.00	-	-	-	22.91	MILLI/MJ	43.99	
CALCIUM	MILLI	14.00	10.00	-	17.90	11.06	MILLI/MJ	21.24	
IRON	MILLI	3.50	2.00	-	6.00	2.77	MILLI/MJ	5.31	
PHOSPHORUS	MILLI	224.00	200.00	-	253.00	176.96	MILLI/MJ	339.82	
CHLORIDE	MILLI	51.00	-	-	-	40.29	MILLI/MJ	77.37	
VITAMIN A	MICRO	0.30	-	-	-	0.24	MICRO/MJ	0.46	
VITAMIN B1	MILLI	0.11	0.05	-	0.20	0.09	MILLI/MJ	0.17	
VITAMIN B2	MICRO	66.00	50.00	-	100.00	52.14	MICRO/MJ	100.13	
NICOTINAMIDE	MILLI	8.60	5.20	-	12.70	6.79	MILLI/MJ	13.05	
VITAMIN B6	MILLI	0.30	-	-	-	0.24	MILLI/MJ	0.46	
ARGININE	GRAM	1.17	-	-	-	0.92	GRAM/MJ	1.77	
HISTIDINE	GRAM	0.47	-	-	-	0.37	GRAM/MJ	0.71	
ISOLEUCINE	GRAM	1.08	-	-	-	0.85	GRAM/MJ	1.64	
LEUCINE	GRAM	1.63	-	-	-	1.29	GRAM/MJ	2.47	
LYSINE	GRAM	1.81	-	-	-	1.43	GRAM/MJ	2.75	
METHIONINE	GRAM	0.54	-	-	-	0.43	GRAM/MJ	0.82	
PHENYLALANINE	GRAM	0.79	-	-	-	0.62	GRAM/MJ	1.20	
THREONINE	GRAM	1.02	-	-	-	0.81	GRAM/MJ	1.55	
VALINE	GRAM	1.02	-	-	-	0.81	GRAM/MJ	1.55	
TOTAL PURINES	MILLI	95.00 [1]	-	-	-	75.05	MILLI/MJ	144.12	

1 VALUE EXPRESSED IN MG URIC ACID/100G

PFERDEFLEISCH / HORSE MEAT / VIANDE DE CHEVAL

DURCHSCHNITT / AVERAGE / MOYENNE

		PROTEIN	FAT	CARBOHYDRATES	TOTAL
ENERGY VALUE (AVERAGE)	KJOULE	379	106	0.0	485
PER 100 G	(KCAL)	91	25	0.0	116
EDIBLE PORTION					
AMOUNT OF DIGESTIBLE	GRAM	20.18	2.53	0.00	
CONSTITUENTS PER 100 G					
ENERGY VALUE (AVERAGE)	KJOULE	372	101	0.0	472
OF THE DIGESTIBLE	(KCAL)	89	24	0.0	113
FRACTION PER 100 G					
EDIBLE PORTION					

WASTE PERCENTAGE AVERAGE 25

CONSTITUENTS	DIM	AV	VARIATION		AVR	NUTR. DENS.		MOLPERC.
WATER	GRAM	75.20	73.00	79.00	56.40	GRAM/MJ	159.16	
PROTEIN	GRAM	20.60	18.10	21.70	15.45	GRAM/MJ	43.60	
FAT	GRAM	2.67	1.20	4.10	2.00	GRAM/MJ	5.65	
MINERALS	GRAM	0.98	0.90	1.06	0.74	GRAM/MJ	2.07	
SODIUM	MILLI	44.00	-	-	33.00	MILLI/MJ	93.13	
POTASSIUM	MILLI	332.00	330.00	332.00	249.00	MILLI/MJ	702.68	
MAGNESIUM	MILLI	23.00	-	-	17.25	MILLI/MJ	48.68	
CALCIUM	MILLI	12.60	10.00	14.30	9.45	MILLI/MJ	26.67	
IRON	MILLI	4.70	2.00	7.00	3.53	MILLI/MJ	9.95	
PHOSPHORUS	MILLI	185.00	150.00	205.00	138.75	MILLI/MJ	391.56	
CHLORIDE	MILLI	9.00	-	-	6.75	MILLI/MJ	19.05	
BORON	MICRO	45.00	16.00	73.00	33.75	MICRO/MJ	95.24	
VITAMIN A	MICRO	21.00	12.00	27.00	15.75	MICRO/MJ	44.45	
VITAMIN B1	MILLI	0.11	0.07	0.14	0.08	MILLI/MJ	0.23	
VITAMIN B2	MILLI	0.15	0.10	0.27	0.11	MILLI/MJ	0.32	
NICOTINAMIDE	MILLI	4.60	4.30	4.70	3.45	MILLI/MJ	9.74	
VITAMIN B6	MILLI	0.50	0.34	0.81	0.38	MILLI/MJ	1.06	
VITAMIN B12	MICRO	3.00	2.20	4.50	2.25	MICRO/MJ	6.35	
ARGININE	GRAM	1.79	-	-	1.34	GRAM/MJ	3.79	
HISTIDINE	GRAM	0.87	-	-	0.65	GRAM/MJ	1.84	
ISOLEUCINE	GRAM	1.05	-	-	0.79	GRAM/MJ	2.22	
LEUCINE	GRAM	1.61	-	-	1.21	GRAM/MJ	3.41	
LYSINE	GRAM	1.57	-	-	1.18	GRAM/MJ	3.32	
METHIONINE	GRAM	1.28	-	-	0.96	GRAM/MJ	2.71	
PHENYLALANINE	GRAM	0.72	-	-	0.54	GRAM/MJ	1.52	
THREONINE	GRAM	0.91	-	-	0.68	GRAM/MJ	1.93	
TRYPTOPHAN	GRAM	0.12	-	-	0.09	GRAM/MJ	0.25	
VALINE	GRAM	1.09	-	-	0.82	GRAM/MJ	2.31	

CONSTITUENTS	DIM	AV	VARIATION			AVR	NUTR. DENS.		MOLPERC.
MYRISTIC ACID	GRAM	0.11	0.11	-	0.11	0.08	GRAM/MJ	0.23	
PALMITIC ACID	GRAM	0.75	0.75	-	0.75	0.56	GRAM/MJ	1.59	
C17 : 0 FATTY ACID	MILLI	8.00	8.00	-	8.00	6.00	MILLI/MJ	16.93	
STEARIC ACID	GRAM	0.11	0.11	-	0.11	0.08	GRAM/MJ	0.23	
PALMITOLEIC ACID	GRAM	0.21	0.21	-	0.21	0.16	GRAM/MJ	0.44	
OLEIC ACID	GRAM	0.92	0.92	-	0.92	0.69	GRAM/MJ	1.95	
EICOSENOIC ACID	MILLI	2.00	2.00	-	2.00	1.50	MILLI/MJ	4.23	
LINOLEIC ACID	GRAM	0.16	0.16	-	0.16	0.12	GRAM/MJ	0.34	
LINOLENIC ACID	GRAM	0.26	0.26	-	0.26	0.20	GRAM/MJ	0.55	
TOTAL PURINES	MILLI	200.00 1	-	-	-	150.00	MILLI/MJ	423.30	

1 VALUE EXPRESSED IN MG URIC ACID/100G

ZIEGENFLEISCH
DURCHSCHNITT

GOAT MEAT
AVERAGE

VIANDE DE CHEVRE
MOYENNE

		PROTEIN	FAT	CARBOHYDRATES	TOTAL
ENERGY VALUE (AVERAGE)	KJOULE	359	313	0.0	672
PER 100 G	(KCAL)	86	75	0.0	161
EDIBLE PORTION					
AMOUNT OF DIGESTIBLE	GRAM	18.91	7.48	0.00	
CONSTITUENTS PER 100 G					
ENERGY VALUE (AVERAGE)	KJOULE	348	298	0.0	646
OF THE DIGESTIBLE	(KCAL)	83	71	0.0	154
FRACTION PER 100 G					
EDIBLE PORTION					

WASTE PERCENTAGE AVERAGE 19 MINIMUM 17 MAXIMUM 22

CONSTITUENTS	DIM	AV	VARIATION			AVR	NUTR. DENS.		MOLPERC.
WATER	GRAM	70.00	66.70	-	73.40	56.70	GRAM/MJ	108.40	
PROTEIN	GRAM	19.50	18.00	-	20.70	15.80	GRAM/MJ	30.20	
FAT	GRAM	7.88	4.30	-	14.20	6.38	GRAM/MJ	12.20	
MINERALS	GRAM	1.03	0.90	-	1.25	0.83	GRAM/MJ	1.59	
CALCIUM	MILLI	9.50	8.00	-	11.00	7.70	MILLI/MJ	14.71	
IRON	MILLI	1.95	1.70	-	2.20	1.58	MILLI/MJ	3.02	
VITAMIN A	MICRO	36.00	-	-	-	29.16	MICRO/MJ	55.75	
VITAMIN B1	MILLI	0.15	0.13	-	0.17	0.12	MILLI/MJ	0.23	
VITAMIN B2	MILLI	0.28	0.24	-	0.32	0.23	MILLI/MJ	0.43	
NICOTINAMIDE	MILLI	4.90	4.20	-	5.60	3.97	MILLI/MJ	7.59	
VITAMIN B6	MILLI	0.30	-	-	-	0.24	MILLI/MJ	0.46	
VITAMIN C	-	0.00	-	-	-	0.00			

Fleischerzeugnisse (ohne Würste und Pasteten)

MEAT PRODUCTS (EXCEPT SAUSAGES AND PASTRIES)

PRODUITS DE VIANDE (À L'EXCEPTION DE CHARCUTERIE ET PÂTÉS)

BÜNDNER FLEISCH (BINDEN-FLEISCH) | BINDEN MEAT | VIANDE SÉCHÉE (BÜNDNER)

		PROTEIN	FAT	CARBOHYDRATES	TOTAL
ENERGY VALUE (AVERAGE)	KJOULE	718	378	0.0	1096
PER 100 G	(KCAL)	172	90	0.0	262
EDIBLE PORTION					
AMOUNT OF DIGESTIBLE	GRAM	37.83	9.02	0.00	
CONSTITUENTS PER 100 G					
ENERGY VALUE (AVERAGE)	KJOULE	696	359	0.0	1055
OF THE DIGESTIBLE	(KCAL)	166	86	0.0	252
FRACTION PER 100 G					
EDIBLE PORTION					

WASTE PERCENTAGE AVERAGE 0.00

CONSTITUENTS	DIM	AV	VARIATION			AVR	NUTR. DENS.		MOLPERC.
WATER	GRAM	45.00	39.20	-	48.00	45.00	GRAM/MJ	42.65	
PROTEIN	GRAM	39.00	29.60	-	47.60	39.00	GRAM/MJ	36.96	
FAT	GRAM	9.50	5.20	-	17.00	9.50	GRAM/MJ	9.00	
MINERALS	GRAM	6.00	-	-	-	6.00	GRAM/MJ	5.69	
SODIUM	GRAM	2.10	-	-	-	2.10	GRAM/MJ	1.99	
CALCIUM	MILLI	48.00	-	-	-	48.00	MILLI/MJ	45.49	
IRON	MILLI	9.80	-	-	-	9.80	MILLI/MJ	9.29	
CHLORIDE	GRAM	3.20	-	-	-	3.20	GRAM/MJ	3.03	
VITAMIN A	-	TRACES	-	-	-				
VITAMIN B1	MILLI	0.16	-	-	-	0.16	MILLI/MJ	0.15	
VITAMIN B2	MILLI	0.82	-	-	-	0.82	MILLI/MJ	0.78	
NICOTINAMIDE	MILLI	9.80	-	-	-	9.80	MILLI/MJ	9.29	

CORNED BEEF
AMERIKANISCH

CORNED BEEF
AMERICAN

CORNED BEEF
AMÉRICAIN

		PROTEIN	FAT	CARBOHYDRATES	TOTAL
ENERGY VALUE (AVERAGE)	KJOULE	466	477	0.0	943
PER 100 G	(KCAL)	111	114	0.0	225
EDIBLE PORTION					
AMOUNT OF DIGESTIBLE	GRAM	24.54	11.40	0.00	
CONSTITUENTS PER 100 G					
ENERGY VALUE (AVERAGE)	KJOULE	452	453	0.0	905
OF THE DIGESTIBLE	(KCAL)	108	108	0.0	216
FRACTION PER 100 G					
EDIBLE PORTION					

WASTE PERCENTAGE AVERAGE 0.00

CONSTITUENTS	DIM	AV	VARIATION			AVR	NUTR. DENS.		MOLPERC.
WATER	GRAM	59.30	-	-	-	59.30	GRAM/MJ	65.53	
PROTEIN	GRAM	25.30	-	-	-	25.30	GRAM/MJ	27.96	
FAT	GRAM	12.00	-	-	-	12.00	GRAM/MJ	13.26	
MINERALS	GRAM	3.40	-	-	-	3.40	GRAM/MJ	3.76	
SODIUM	MILLI	950.00	-	-	-	950.00	GRAM/MJ	1.05	
POTASSIUM	MILLI	140.00	-	-	-	140.00	MILLI/MJ	154.71	
MAGNESIUM	MILLI	15.00	-	-	-	15.00	MILLI/MJ	16.58	
CALCIUM	MILLI	14.00	-	-	-	14.00	MILLI/MJ	15.47	
IRON	MILLI	4.10	3.90	-	4.30	4.10	MILLI/MJ	4.53	
COPPER	MILLI	0.24	-	-	-	0.24	MILLI/MJ	0.27	
ZINC	MILLI	5.60	-	-	-	5.60	MILLI/MJ	6.19	
PHOSPHORUS	MILLI	150.00	-	-	-	150.00	MILLI/MJ	165.76	
CHLORIDE	GRAM	1.43	-	-	-	1.43	GRAM/MJ	1.58	
BORON	MICRO	28.00	25.00	-	30.00	28.00	MICRO/MJ	30.94	
VITAMIN B1	MICRO	20.00	-	-	-	20.00	MICRO/MJ	22.10	
VITAMIN B2	MILLI	0.23	0.22	-	0.24	0.23	MILLI/MJ	0.25	
NICOTINAMIDE	MILLI	3.20	3.00	-	3.40	3.20	MILLI/MJ	3.54	
PANTOTHENIC ACID	MILLI	0.40	-	-	-	0.40	MILLI/MJ	0.44	
VITAMIN B6	MICRO	60.00	-	-	-	60.00	MICRO/MJ	66.30	
BIOTIN	MICRO	2.00	-	-	-	2.00	MICRO/MJ	2.21	
FOLIC ACID	MICRO	2.00	-	-	-	2.00	MICRO/MJ	2.21	
VITAMIN B12	MICRO	2.00	-	-	-	2.00	MICRO/MJ	2.21	
VITAMIN C	-	0.00	-	-	-	0.00			
ARGININE	GRAM	1.63	-	-	-	1.63	GRAM/MJ	1.80	
CYSTINE	GRAM	0.32	-	-	-	0.32	GRAM/MJ	0.35	
HISTIDINE	GRAM	0.89	-	-	-	0.89	GRAM/MJ	0.98	
ISOLEUCINE	GRAM	1.24	-	-	-	1.24	GRAM/MJ	1.37	
LEUCINE	GRAM	1.92	-	-	-	1.92	GRAM/MJ	2.12	
LYSINE	GRAM	2.13	-	-	-	2.13	GRAM/MJ	2.35	
METHIONINE	GRAM	0.69	-	-	-	0.69	GRAM/MJ	0.76	
PHENYLALANINE	GRAM	0.91	-	-	-	0.91	GRAM/MJ	1.01	
THREONINE	GRAM	1.11	-	-	-	1.11	GRAM/MJ	1.23	
TRYPTOPHAN	GRAM	0.23	-	-	-	0.23	GRAM/MJ	0.25	
TYROSINE	GRAM	0.85	-	-	-	0.85	GRAM/MJ	0.94	
VALINE	GRAM	1.27	-	-	-	1.27	GRAM/MJ	1.40	

CORNED BEEF
DEUTSCH

CORNED BEEF
GERMAN

CORNED BEEF
ALLEMAND

		PROTEIN	FAT	CARBOHYDRATES	TOTAL
ENERGY VALUE (AVERAGE)	KJOULE	399	238	0.0	638
PER 100 G	(KCAL)	95	57	0.0	152
EDIBLE PORTION					
AMOUNT OF DIGESTIBLE	GRAM	21.04	5.70	0.00	
CONSTITUENTS PER 100 G					
ENERGY VALUE (AVERAGE)	KJOULE	388	227	0.0	614
OF THE DIGESTIBLE	(KCAL)	93	54	0.0	147
FRACTION PER 100 G					
EDIBLE PORTION					

WASTE PERCENTAGE AVERAGE 0.00

CONSTITUENTS	DIM	AV	VARIATION		AVR	NUTR. DENS.		MOLPERC.
WATER	GRAM	69.80	65.30	- 75.20	69.80	GRAM/MJ	113.67	
PROTEIN	GRAM	21.70	17.50	- 25.10	21.70	GRAM/MJ	35.34	
FAT	GRAM	6.00	2.85	- 11.20	6.00	GRAM/MJ	9.77	
MINERALS	GRAM	2.54	1.92	- 3.33	2.54	GRAM/MJ	4.14	
SODIUM	MILLI	833.00	711.00	- 943.00	833.00	GRAM/MJ	1.36	
POTASSIUM	MILLI	131.00	-	- -	131.00	MILLI/MJ	213.33	
CALCIUM	MILLI	33.00	-	- -	33.00	MILLI/MJ	53.74	
PHOSPHORUS	MILLI	128.00	-	- -	128.00	MILLI/MJ	208.45	
BORON	MICRO	28.00	25.00	- 30.00	28.00	MICRO/MJ	45.60	
VITAMIN B1	MICRO	30.00	-	- -	30.00	MICRO/MJ	48.85	
VITAMIN B2	MILLI	0.10	-	- -	0.10	MILLI/MJ	0.16	
NICOTINAMIDE	MILLI	3.10	-	- -	3.10	MILLI/MJ	5.05	
HISTIDINE	GRAM	0.76	-	- -	0.76	GRAM/MJ	1.24	
ISOLEUCINE	GRAM	1.06	-	- -	1.06	GRAM/MJ	1.73	
LEUCINE	GRAM	1.65	-	- -	1.65	GRAM/MJ	2.69	
LYSINE	GRAM	1.82	-	- -	1.82	GRAM/MJ	2.96	
METHIONINE	GRAM	0.56	-	- -	0.56	GRAM/MJ	0.91	
PHENYLALANINE	GRAM	0.78	-	- -	0.78	GRAM/MJ	1.27	
THREONINE	GRAM	0.96	-	- -	0.96	GRAM/MJ	1.56	
TRYPTOPHAN	GRAM	0.19	-	- -	0.19	GRAM/MJ	0.31	
VALINE	GRAM	1.09	-	- -	1.09	GRAM/MJ	1.78	

FLEISCHEXTRAKT MEAT EXTRACT EXTRAIT DE VIANDE

		PROTEIN	FAT	CARBOHYDRATES	TOTAL
ENERGY VALUE (AVERAGE)	KJOULE	42	34	51	1127
PER 100 G	(KCAL)	249	8.2	12	269
EDIBLE PORTION					
AMOUNT OF DIGESTIBLE	GRAM	54.90	0.81	3.04	
CONSTITUENTS PER 100 G					
ENERGY VALUE (AVERAGE)	KJOULE	11	32	51	1094
OF THE DIGESTIBLE	(KCAL)	242	7.8	12	261
FRACTION PER 100 G					
EDIBLE PORTION					

WASTE PERCENTAGE AVERAGE 0.00

CONSTITUENTS	DIM	AV	VARIATION		AVR	NUTR. DENS.	MOLPERC.
WATER	GRAM	19.80	16.90	- 21.00	19.80	GRAM/MJ	18.10
PROTEIN	GRAM	56.60	53.10	- 60.00	56.60	GRAM/MJ	51.73
FAT	GRAM	0.86	0.03	- 1.50	0.86	GRAM/MJ	0.79
AVAILABLE CARBOHYDR.	GRAM	3.04	-	- -	3.04	GRAM/MJ	2.78
MINERALS	GRAM	19.70	18.00	- 20.30	19.70	GRAM/MJ	18.01
SODIUM	GRAM	1.76	1.66	- 1.86	1.76	GRAM/MJ	1.61
POTASSIUM	GRAM	7.20	6.92	- 7.49	7.20	GRAM/MJ	6.58
MAGNESIUM	MILLI	374.00	264.00	- 552.00	374.00	MILLI/MJ	341.84
CALCIUM	MILLI	163.00	87.00	- 238.00	163.00	MILLI/MJ	148.98
IRON	MILLI	39.00	8.00	- 106.00	39.00	MILLI/MJ	35.65
PHOSPHORUS	GRAM	2.38	1.05	- 3.46	2.38	GRAM/MJ	2.18
CHLORIDE	GRAM	1.90	1.38	- 2.80	1.90	GRAM/MJ	1.74

LUNCHEON MEAT (FRÜHSTÜCKSFLEISCH) | LUNCHEON MEAT | LUNCHEON MEAT (VIANDE DE PETIT DÉJEUNER)

		PROTEIN	FAT	CARBOHYDRATES	TOTAL
ENERGY VALUE (AVERAGE)	KJOULE	271	1010	0.0	1280
PER 100 G	(KCAL)	65	241	0.0	306
EDIBLE PORTION					
AMOUNT OF DIGESTIBLE	GRAM	14.25	24.13	0.00	
CONSTITUENTS PER 100 G					
ENERGY VALUE (AVERAGE)	KJOULE	263	959	0.0	1222
OF THE DIGESTIBLE	(KCAL)	63	229	0.0	292
FRACTION PER 100 G					
EDIBLE PORTION					

WASTE PERCENTAGE AVERAGE 0.00

CONSTITUENTS	DIM	AV	VARIATION			AVR	NUTR. DENS.		MOLPERC.
WATER	GRAM	55.30	48.20	-	66.20	55.30	GRAM/MJ	45.27	
PROTEIN	GRAM	14.70	13.80	-	15.30	14.70	GRAM/MJ	12.03	
FAT	GRAM	25.40	12.80	-	34.50	25.40	GRAM/MJ	20.79	
MINERALS	GRAM	3.00	2.84	-	3.15	3.00	GRAM/MJ	2.46	
SODIUM	GRAM	1.06	0.83	-	1.31	1.06	GRAM/MJ	0.87	
POTASSIUM	MILLI	212.00	106.00	-	273.00	212.00	MILLI/MJ	173.54	
MAGNESIUM	MILLI	59.00	-	-	-	59.00	MILLI/MJ	48.30	
CALCIUM	MILLI	12.00	4.00	-	47.00	12.00	MILLI/MJ	9.82	
MANGANESE	MICRO	70.00	-	-	-	70.00	MICRO/MJ	57.30	
IRON	MILLI	2.20	1.60	-	3.10	2.20	MILLI/MJ	1.80	
COPPER	MICRO	50.00	-	-	-	50.00	MICRO/MJ	40.93	
PHOSPHORUS	MILLI	220.00	170.00	-	250.00	220.00	MILLI/MJ	180.09	
CHLORIDE	GRAM	1.16	-	-	-	1.16	GRAM/MJ	0.95	
VITAMIN B1	MICRO	50.00	20.00	-	90.00	50.00	MICRO/MJ	40.93	
VITAMIN B2	MILLI	0.19	0.16	-	0.27	0.19	MILLI/MJ	0.16	
NICOTINAMIDE	MILLI	4.70	3.20	-	6.00	4.70	MILLI/MJ	3.85	
PANTOTHENIC ACID	MILLI	0.20	-	-	-	0.20	MILLI/MJ	0.16	
FOLIC ACID	MICRO	1.00	-	-	-	1.00	MICRO/MJ	0.82	
VITAMIN B12	MICRO	2.70	-	-	-	2.70	MICRO/MJ	2.21	
VITAMIN C	MILLI	1.00	1.00	-	1.60	1.00	MILLI/MJ	0.82	
ARGININE	GRAM	0.79	-	-	-	0.79	GRAM/MJ	0.65	
CYSTINE	GRAM	0.12	-	-	-	0.12	GRAM/MJ	0.10	
HISTIDINE	GRAM	0.36	-	-	-	0.36	GRAM/MJ	0.29	
ISOLEUCINE	GRAM	0.56	-	-	-	0.56	GRAM/MJ	0.46	
LEUCINE	GRAM	1.03	-	-	-	1.03	GRAM/MJ	0.84	
LYSINE	GRAM	1.01	-	-	-	1.01	GRAM/MJ	0.83	
METHIONINE	GRAM	0.33	-	-	-	0.33	GRAM/MJ	0.27	
PHENYLALANINE	GRAM	0.54	-	-	-	0.54	GRAM/MJ	0.44	
THREONINE	GRAM	0.55	-	-	-	0.55	GRAM/MJ	0.45	
TRYPTOPHAN	GRAM	0.12	-	-	-	0.12	GRAM/MJ	0.10	
TYROSINE	GRAM	0.46	-	-	-	0.46	GRAM/MJ	0.38	
VALINE	GRAM	0.71	-	-	-	0.71	GRAM/MJ	0.58	
CONNECT.TISSUE PROTEIN	GRAM	1.80	-	-	-	1.80	GRAM/MJ	1.47	

GELATINE (SPEISEGELATINE)

GELATINE

GELATINE COMESTIBLE

		PROTEIN	FAT	CARBOHYDRATES	TOTAL
ENERGY VALUE (AVERAGE) PER 100 G EDIBLE PORTION	KJOULE	416	4.0	0.0	1420
	(KCAL)	338	1.0	0.0	339
AMOUNT OF DIGESTIBLE CONSTITUENTS PER 100 G	GRAM	81.67	0.09	0.00	
ENERGY VALUE (AVERAGE) OF THE DIGESTIBLE FRACTION PER 100 G EDIBLE PORTION	KJOULE	374	3.8	0.0	1378
	(KCAL)	328	0.9	0.0	329

WASTE PERCENTAGE AVERAGE 0.00

CONSTITUENTS	DIM	AV	VARIATION			AVR	NUTR. DENS.		MOLPERC.
WATER	GRAM	14.00	13.00	-	17.00	14.00	GRAM/MJ	10.16	
PROTEIN	GRAM	84.20	82.10	-	87.10	84.20	GRAM/MJ	61.12	
FAT	GRAM	0.10	-	-	-	0.10	GRAM/MJ	0.07	
MINERALS	GRAM	1.70	1.00	-	2.70	1.70	GRAM/MJ	1.23	
SODIUM	MILLI	32.00	27.00	-	36.00	32.00	MILLI/MJ	23.23	
POTASSIUM	MILLI	22.00	-	-	-	22.00	MILLI/MJ	15.97	
MAGNESIUM	MILLI	11.00	11.00	-	11.00	11.00	MILLI/MJ	7.99	
CALCIUM	MILLI	11.00	-	-	-	11.00	MILLI/MJ	7.99	
PHOSPHORUS	-	0.00	-	-	-	0.00			
IODIDE	MICRO	6.00	6.00	-	6.00	6.00	MICRO/MJ	4.36	
SELENIUM	MICRO	19.00	-	-	-	19.00	MICRO/MJ	13.79	
VITAMIN A	-	0.00	-	-	-	0.00			
VITAMIN B1	-	0.00	-	-	-	0.00			
VITAMIN B2	-	0.00	-	-	-	0.00			
NICOTINAMIDE	-	0.00	-	-	-	0.00			
VITAMIN B6	MICRO	5.80	5.80	-	5.80	5.80	MICRO/MJ	4.21	
VITAMIN C	-	0.00	-	-	-	0.00			
ALANINE	GRAM	9.27	-	-	-	9.27	GRAM/MJ	6.73	
ARGININE	GRAM	7.45	-	-	-	7.45	GRAM/MJ	5.41	
ASPARTIC ACID	GRAM	5.63	-	-	-	5.63	GRAM/MJ	4.09	
CYSTINE	-	TRACES	-	-	-				
GLUTAMIC ACID	GRAM	9.58	-	-	-	9.58	GRAM/MJ	6.95	
GLYCINE	GRAM	22.96	-	-	-	22.96	GRAM/MJ	16.67	
HISTIDINE	GRAM	0.61	-	-	-	0.61	GRAM/MJ	0.44	
ISOLEUCINE	GRAM	1.37	-	-	-	1.37	GRAM/MJ	0.99	
LEUCINE	GRAM	2.74	-	-	-	2.74	GRAM/MJ	1.99	
LYSINE	GRAM	3.80	-	-	-	3.80	GRAM/MJ	2.76	
METHIONINE	GRAM	0.76	-	-	-	0.76	GRAM/MJ	0.55	
PHENYLALANINE	GRAM	1.98	-	-	-	1.98	GRAM/MJ	1.44	
PROLINE	GRAM	13.04	-	-	-	13.04	GRAM/MJ	9.47	
SERINE	GRAM	3.50	-	-	-	3.50	GRAM/MJ	2.54	
THREONINE	GRAM	1.82	-	-	-	1.82	GRAM/MJ	1.32	
TRYPTOPHAN	MILLI	6.00	-	-	-	6.00	MILLI/MJ	4.36	
TYROSINE	GRAM	0.30	-	-	-	0.30	GRAM/MJ	0.22	
VALINE	GRAM	2.13	-	-	-	2.13	GRAM/MJ	1.55	
HYDROXY-PROLINE	GRAM	11.10	-	-	-	11.10	GRAM/MJ	8.06	

RINDERHACKFLEISCH BEEF MINCED MEAT VIANDE DE BOEUF HACHÉE

		PROTEIN	FAT	CARBOHYDRATES	TOTAL
ENERGY VALUE (AVERAGE)	KJOULE	422	556	0.0	978
PER 100 G	(KCAL)	101	133	0.0	234
EDIBLE PORTION					
AMOUNT OF DIGESTIBLE	GRAM	22.21	13.30	0.00	
CONSTITUENTS PER 100 G					
ENERGY VALUE (AVERAGE)	KJOULE	409	529	0.0	938
OF THE DIGESTIBLE	(KCAL)	98	126	0.0	224
FRACTION PER 100 G					
EDIBLE PORTION					

WASTE PERCENTAGE AVERAGE 0.00

CONSTITUENTS	DIM	AV	VARIATION	AVR	NUTR. DENS.		MOLPERC.
WATER	GRAM	60.60	- - -	60.60	GRAM/MJ	64.63	
PROTEIN	GRAM	22.90	- - -	22.90	GRAM/MJ	24.42	
FAT	GRAM	14.00	12.00 - 16.00	14.00	GRAM/MJ	14.93	
MINERALS	GRAM	2.50	- - -	2.50	GRAM/MJ	2.67	
CONNECT.TISSUE PROTEIN	GRAM	3.50	- - -	3.50	GRAM/MJ	3.73	

RINDFLEISCH IN DOSEN | BEEF CANNED | VIANDE DE BOEUF EN BOÎTES

		PROTEIN	FAT	CARBOHYDRATES	TOTAL
ENERGY VALUE (AVERAGE) PER 100 G EDIBLE PORTION	KJOULE	341	541	0.0	881
	(KCAL)	81	129	0.0	211
AMOUNT OF DIGESTIBLE CONSTITUENTS PER 100 G	GRAM	17.94	12.92	0.00	
ENERGY VALUE (AVERAGE) OF THE DIGESTIBLE FRACTION PER 100 G EDIBLE PORTION	KJOULE	330	514	0.0	844
	(KCAL)	79	123	0.0	202

WASTE PERCENTAGE AVERAGE 0.00

CONSTITUENTS	DIM	AV	VARIATION			AVR	NUTR.	DENS.	MOLPERC.
WATER	GRAM	65.30	61.00	-	67.90	65.30	GRAM/MJ	77.38	
PROTEIN	GRAM	18.50	17.80	-	19.20	18.50	GRAM/MJ	21.92	
FAT	GRAM	13.60	8.70	-	19.10	13.60	GRAM/MJ	16.12	
MINERALS	GRAM	2.32	2.18	-	2.44	2.32	GRAM/MJ	2.75	
CAROTENE	MILLI	0.13	-	-	-	0.13	MILLI/MJ	0.15	
VITAMIN B1	MICRO	20.00	10.00	-	20.00	20.00	MICRO/MJ	23.70	
VITAMIN B2	MILLI	0.15	-	-	-	0.15	MILLI/MJ	0.18	
NICOTINAMIDE	MILLI	4.60	4.20	-	4.90	4.60	MILLI/MJ	5.45	

SCHABEFLEISCH (TARTAR)	MINCED MEAT (STEAK TARTARE)	VIANDE DE BOEUF HACHÉE CRUE

		PROTEIN	FAT	CARBOHYDRATES	TOTAL
ENERGY VALUE (AVERAGE) PER 100 G EDIBLE PORTION	KJOULE	394	119	0.0	513
	(KCAL)	94	29	0.0	123
AMOUNT OF DIGESTIBLE CONSTITUENTS PER 100 G	GRAM	20.75	2.85	0.00	
ENERGY VALUE (AVERAGE) OF THE DIGESTIBLE FRACTION PER 100 G EDIBLE PORTION	KJOULE	382	113	0.0	495
	(KCAL)	91	27	0.0	118

WASTE PERCENTAGE AVERAGE 0.00

CONSTITUENTS	DIM	AV	VARIATION	AVR	NUTR. DENS.		MOLPERC.
WATER	GRAM	74.20	71.60 - 76.40	74.20	GRAM/MJ	149.77	
PROTEIN	GRAM	21.40	20.50 - 21.80	21.40	GRAM/MJ	43.19	
FAT	GRAM	3.00	2.00 - 5.00	3.00	GRAM/MJ	6.06	
MINERALS	GRAM	1.20	1.10 - 1.20	1.20	GRAM/MJ	2.42	

SCHWEINEFLEISCH IN DOSEN (SCHWEINEFLEISCH IM EIGENEN SAFT)

PORK CANNED WITH JUICE

VIANDE DE PORC EN BOÎTES DANS SON JUS

		PROTEIN	FAT	CARBOHYDRATES	TOTAL
ENERGY VALUE (AVERAGE)	KJOULE	287	1272	0.0	1559
PER 100 G	(KCAL)	69	304	0.0	373
EDIBLE PORTION					
AMOUNT OF DIGESTIBLE	GRAM	15.13	30.40	0.00	
CONSTITUENTS PER 100 G					
ENERGY VALUE (AVERAGE)	KJOULE	279	1208	0.0	1487
OF THE DIGESTIBLE	(KCAL)	67	289	0.0	355
FRACTION PER 100 G					
EDIBLE PORTION					

WASTE PERCENTAGE AVERAGE 0.00

CONSTITUENTS	DIM	AV	VARIATION			AVR	NUTR.	DENS.	MOLPERC.
WATER	GRAM	50.20	41.10	-	59.90	50.20	GRAM/MJ	33.76	
PROTEIN	GRAM	15.60	13.70	-	18.60	15.60	GRAM/MJ	10.49	
FAT	GRAM	32.00	17.20	-	57.00	32.00	GRAM/MJ	21.52	
MINERALS	GRAM	2.21	1.40	-	3.35	2.21	GRAM/MJ	1.49	
SODIUM	MILLI	789.00	-	-	-	789.00	MILLI/MJ	530.63	
POTASSIUM	MILLI	68.00	-	-	-	68.00	MILLI/MJ	45.73	
CALCIUM	MILLI	25.00	-	-	-	25.00	MILLI/MJ	16.81	
VITAMIN B1	MILLI	0.16	0.15	-	0.17	0.16	MILLI/MJ	0.11	
VITAMIN B2	MILLI	0.19	0.18	-	0.19	0.19	MILLI/MJ	0.13	
NICOTINAMIDE	MILLI	3.50	3.20	-	3.80	3.50	MILLI/MJ	2.35	
VITAMIN C	-	0.00	-	-	-	0.00			
ALANINE	GRAM	1.16	1.13	-	1.19	1.16	GRAM/MJ	0.78	9.2
ARGININE	GRAM	1.04	1.00	-	1.07	1.04	GRAM/MJ	0.70	4.2
ASPARTIC ACID	GRAM	1.71	-	-	-	1.71	GRAM/MJ	1.15	9.1
CYSTINE	GRAM	0.23	-	-	-	0.23	GRAM/MJ	0.15	0.7
GLUTAMIC ACID	GRAM	3.09	3.02	-	3.17	3.09	GRAM/MJ	2.08	14.9
GLYCINE	GRAM	1.04	0.92	-	1.14	1.04	GRAM/MJ	0.70	9.8
HISTIDINE	GRAM	0.68	0.08	-	0.77	0.68	GRAM/MJ	0.46	3.1
ISOLEUCINE	GRAM	0.86	-	-	-	0.86	GRAM/MJ	0.58	4.7
LEUCINE	GRAM	1.43	1.36	-	1.52	1.43	GRAM/MJ	0.96	7.7
LYSINE	GRAM	1.63	1.49	-	1.77	1.63	GRAM/MJ	1.10	7.9
METHIONINE	GRAM	0.43	-	-	-	0.43	GRAM/MJ	0.29	2.0
PHENYLALANINE	GRAM	0.69	0.66	-	0.71	0.69	GRAM/MJ	0.46	3.0
PROLINE	GRAM	0.80	0.66	-	0.95	0.80	GRAM/MJ	0.54	4.9
SERINE	GRAM	0.71	0.60	-	0.83	0.71	GRAM/MJ	0.48	4.8
THREONINE	GRAM	0.79	0.75	-	0.83	0.79	GRAM/MJ	0.53	4.7
TRYPTOPHAN	GRAM	0.20	-	-	-	0.20	GRAM/MJ	0.13	0.7
TYROSINE	GRAM	0.66	0.64	-	0.68	0.66	GRAM/MJ	0.44	2.6
VALINE	GRAM	0.97	0.94	-	0.98	0.97	GRAM/MJ	0.65	5.9

SCHWEINEFLEISCH
IN DOSEN
(SCHMALZFLEISCH)

PORK
CANNED

VIANDE DE PORC
EN BOÎTES
VIANDE À SAINDOUX

		PROTEIN	FAT	CARBOHYDRATES	TOTAL
ENERGY VALUE (AVERAGE)	KJOULE	203	2039	0.0	2242
PER 100 G	(KCAL)	48	487	0.0	536
EDIBLE PORTION					
AMOUNT OF DIGESTIBLE	GRAM	10.67	48.73	0.00	
CONSTITUENTS PER 100 G					
ENERGY VALUE (AVERAGE)	KJOULE	196	1937	0.0	2134
OF THE DIGESTIBLE	(KCAL)	47	463	0.0	510
FRACTION PER 100 G					
EDIBLE PORTION					

WASTE PERCENTAGE AVERAGE 0.00

CONSTITUENTS	DIM	AV	VARIATION			AVR	NUTR. DENS.		MOLPERC.
WATER	GRAM	35.90	34.20	-	40.90	35.90	GRAM/MJ	16.83	
PROTEIN	GRAM	11.00	8.80	-	12.50	11.00	GRAM/MJ	5.16	
FAT	GRAM	51.30	45.20	-	58.00	51.30	GRAM/MJ	24.04	
MINERALS	GRAM	1.73	1.61	-	1.90	1.73	GRAM/MJ	0.81	
VITAMIN A	-	0.00	-	-	-	0.00			
VITAMIN B1	MICRO	60.00	-	-	-	60.00	MICRO/MJ	28.12	
VITAMIN B2	MILLI	0.16	-	-	-	0.16	MILLI/MJ	0.07	
NICOTINAMIDE	MILLI	2.50	-	-	-	2.50	MILLI/MJ	1.17	
VITAMIN C	-	0.00	-	-	-	0.00			
ALANINE	GRAM	0.82	0.80	-	0.84	0.82	GRAM/MJ	0.38	9.2
ARGININE	GRAM	0.74	0.70	-	0.76	0.74	GRAM/MJ	0.35	4.3
ASPARTIC ACID	GRAM	1.21	-	-	-	1.21	GRAM/MJ	0.57	9.1
CYSTINE	GRAM	0.16	-	-	-	0.16	GRAM/MJ	0.07	0.7
GLUTAMIC ACID	GRAM	2.18	2.14	-	2.24	2.18	GRAM/MJ	1.02	14.9
GLYCINE	GRAM	0.73	0.65	-	0.81	0.73	GRAM/MJ	0.34	9.8
HISTIDINE	GRAM	0.48	0.05	-	0.55	0.48	GRAM/MJ	0.22	3.1
ISOLEUCINE	GRAM	0.61	-	-	-	0.61	GRAM/MJ	0.29	4.7
LEUCINE	GRAM	1.01	0.96	-	1.07	1.01	GRAM/MJ	0.47	7.7
LYSINE	GRAM	1.15	1.05	-	1.25	1.15	GRAM/MJ	0.54	7.9
METHIONINE	GRAM	0.30	-	-	-	0.30	GRAM/MJ	0.14	2.0
PHENYLALANINE	GRAM	0.49	0.47	-	0.50	0.49	GRAM/MJ	0.23	3.0
PROLINE	GRAM	0.57	0.47	-	0.67	0.57	GRAM/MJ	0.27	5.0
SERINE	GRAM	0.50	0.42	-	0.59	0.50	GRAM/MJ	0.23	4.8
THREONINE	GRAM	0.56	0.53	-	0.59	0.56	GRAM/MJ	0.26	4.7
TRYPTOPHAN	GRAM	0.14	-	-	-	0.14	GRAM/MJ	0.07	0.7
TYROSINE	GRAM	0.47	0.47	-	0.48	0.47	GRAM/MJ	0.22	2.6
VALINE	GRAM	0.68	0.66	-	0.69	0.68	GRAM/MJ	0.32	5.8
CHOLESTEROL	MILLI	80.00	-	-	-	80.00	MILLI/MJ	37.50	

SCHWEINEFLEISCH KASSELER | PORK RIB (KASSEL) | VIANDE DE PORC (DE KASSEL)

		PROTEIN	FAT	CARBOHYDRATES	TOTAL
ENERGY VALUE (AVERAGE) PER 100 G EDIBLE PORTION	KJOULE	385	676	0.0	1060
	(KCAL)	92	162	0.0	253
AMOUNT OF DIGESTIBLE CONSTITUENTS PER 100 G	GRAM	20.27	16.49	0.00	
ENERGY VALUE (AVERAGE) OF THE DIGESTIBLE FRACTION PER 100 G EDIBLE PORTION	KJOULE	373	655	0.0	1029
	(KCAL)	89	157	0.0	246

WASTE PERCENTAGE AVERAGE 17 MINIMUM 10 MAXIMUM 21

CONSTITUENTS	DIM	AV	VARIATION			AVR	NUTR.	DENS.	MOLPERC.
WATER	GRAM	58.70	56.00	-	72.20	48.72	GRAM/MJ	57.06	
PROTEIN	GRAM	20.90	18.50	-	22.20	17.35	GRAM/MJ	20.32	
FAT	GRAM	17.00	15.00	-	20.00	14.11	GRAM/MJ	16.53	
MINERALS	GRAM	3.23	2.31	-	3.90	2.68	GRAM/MJ	3.14	
SODIUM	GRAM	0.95	0.67	-	1.25	0.80	GRAM/MJ	0.93	
POTASSIUM	MILLI	324.00	209.00	-	438.00	268.92	MILLI/MJ	314.97	
CALCIUM	MILLI	6.00	-	-	-	4.98	MILLI/MJ	5.83	
IRON	MILLI	2.50	-	-	-	2.08	MILLI/MJ	2.43	
PHOSPHORUS	MILLI	160.00	-	-	-	132.80	MILLI/MJ	155.54	
ALANINE	GRAM	1.54	1.50	-	1.59	1.28	GRAM/MJ	1.50	9.2
ARGININE	GRAM	1.39	1.33	-	1.43	1.15	GRAM/MJ	1.35	4.2
ASPARTIC ACID	GRAM	2.28	-	-	-	1.89	GRAM/MJ	2.22	9.1
CYSTINE	GRAM	0.30	-	-	-	0.25	GRAM/MJ	0.29	0.7
GLUTAMIC ACID	GRAM	4.12	4.03	-	4.23	3.42	GRAM/MJ	4.01	14.9
GLYCINE	GRAM	1.38	1.22	-	1.52	1.15	GRAM/MJ	1.34	9.8
HISTIDINE	GRAM	0.91	0.10	-	1.03	0.76	GRAM/MJ	0.88	3.1
ISOLEUCINE	GRAM	1.15	-	-	-	0.95	GRAM/MJ	1.12	4.7
LEUCINE	GRAM	1.91	1.81	-	2.02	1.59	GRAM/MJ	1.86	7.8
LYSINE	GRAM	2.17	1.99	-	2.36	1.80	GRAM/MJ	2.11	7.9
METHIONINE	GRAM	0.57	-	-	-	0.47	GRAM/MJ	0.55	2.0
PHENYLALANINE	GRAM	0.92	0.88	-	0.95	0.76	GRAM/MJ	0.89	3.0
PROLINE	GRAM	1.07	0.88	-	1.26	0.89	GRAM/MJ	1.04	5.0
SERINE	GRAM	0.95	0.80	-	1.11	0.79	GRAM/MJ	0.92	4.8
THREONINE	GRAM	1.05	1.00	-	1.11	0.87	GRAM/MJ	1.02	4.7
TRYPTOPHAN	GRAM	0.27	-	-	-	0.22	GRAM/MJ	0.26	0.7
TYROSINE	GRAM	0.88	0.85	-	0.91	0.73	GRAM/MJ	0.86	2.6
VALINE	GRAM	1.29	1.25	-	1.31	1.07	GRAM/MJ	1.25	5.9

SCHWEINEHACK-FLEISCH | PORK MINCED MEAT | VIANDE DE PORC HACHÉE

		PROTEIN	FAT	CARBOHYDRATES	TOTAL
ENERGY VALUE (AVERAGE)	KJOULE	412	994	0.0	1406
PER 100 G	(KCAL)	99	238	0.0	336
EDIBLE PORTION					
AMOUNT OF DIGESTIBLE	GRAM	21.72	23.75	0.00	
CONSTITUENTS PER 100 G					
ENERGY VALUE (AVERAGE)	KJOULE	400	944	0.0	1344
OF THE DIGESTIBLE	(KCAL)	96	226	0.0	321
FRACTION PER 100 G					
EDIBLE PORTION					

WASTE PERCENTAGE AVERAGE 0.00

CONSTITUENTS	DIM	AV	VARIATION			AVR	NUTR. DENS.		MOLPERC.
WATER	GRAM	50.10	-	-	-	50.10	GRAM/MJ	37.28	
PROTEIN	GRAM	22.40	-	-	-	22.40	GRAM/MJ	16.67	
FAT	GRAM	25.00	15.00	-	34.10	25.00	GRAM/MJ	18.60	
MINERALS	GRAM	2.50	-	-	-	2.50	GRAM/MJ	1.86	
ALANINE	GRAM	1.65	1.61	-	1.70	1.65	GRAM/MJ	1.23	9.2
ARGININE	GRAM	1.49	1.42	-	1.53	1.49	GRAM/MJ	1.11	4.2
ASPARTIC ACID	GRAM	2.44	-	-	-	2.44	GRAM/MJ	1.82	9.1
CYSTINE	GRAM	0.32	-	-	-	0.32	GRAM/MJ	0.24	0.7
GLUTAMIC ACID	GRAM	4.41	4.31	-	4.53	4.41	GRAM/MJ	3.28	14.9
GLYCINE	GRAM	1.48	1.31	-	1.63	1.48	GRAM/MJ	1.10	9.8
HISTIDINE	GRAM	0.97	0.11	-	1.10	0.97	GRAM/MJ	0.72	3.1
ISOLEUCINE	GRAM	1.23	-	-	-	1.23	GRAM/MJ	0.92	4.7
LEUCINE	GRAM	2.04	1.94	-	2.16	2.04	GRAM/MJ	1.52	7.8
LYSINE	GRAM	2.32	2.13	-	2.53	2.32	GRAM/MJ	1.73	7.9
METHIONINE	GRAM	0.61	-	-	-	0.61	GRAM/MJ	0.45	2.0
PHENYLALANINE	GRAM	0.98	0.94	-	1.02	0.98	GRAM/MJ	0.73	3.0
PROLINE	GRAM	1.14	0.94	-	1.35	1.14	GRAM/MJ	0.85	5.0
SERINE	GRAM	1.02	0.86	-	1.19	1.02	GRAM/MJ	0.76	4.8
THREONINE	GRAM	1.12	1.07	-	1.19	1.12	GRAM/MJ	0.83	4.7
TRYPTOPHAN	GRAM	0.29	-	-	-	0.29	GRAM/MJ	0.22	0.7
TYROSINE	GRAM	0.94	0.91	-	0.97	0.94	GRAM/MJ	0.70	2.6
VALINE	GRAM	1.38	1.34	-	1.40	1.38	GRAM/MJ	1.03	5.9
CONNECT.TISSUE PROTEIN	GRAM	2.10	-	-	-	2.10	GRAM/MJ	1.56	

SCHWEINESCHINKEN GEKOCHT (KOCHSCHINKEN) | HAM COOKED | JAMBON DE PORC CUIT

		PROTEIN	FAT	CARBOHYDRATES	TOTAL
ENERGY VALUE (AVERAGE)	KJOULE	394	509	0.0	903
PER 100 G	(KCAL)	94	122	0.0	216
EDIBLE PORTION					
AMOUNT OF DIGESTIBLE	GRAM	20.75	12.16	0.00	
CONSTITUENTS PER 100 G					
ENERGY VALUE (AVERAGE)	KJOULE	382	483	0.0	865
OF THE DIGESTIBLE	(KCAL)	91	116	0.0	207
FRACTION PER 100 G					
EDIBLE PORTION					

WASTE PERCENTAGE AVERAGE 3.0

CONSTITUENTS	DIM	AV	VARIATION			AVR	NUTR. DENS.		MOLPERC.
WATER	GRAM	62.00	57.50	-	68.00	60.14	GRAM/MJ	71.64	
PROTEIN	GRAM	21.40	18.00	-	25.00	20.76	GRAM/MJ	24.73	
FAT	GRAM	12.80	5.60	-	23.00	12.42	GRAM/MJ	14.79	
MINERALS	GRAM	3.10	2.46	-	3.80	3.01	GRAM/MJ	3.58	
SODIUM	GRAM	0.96	0.74	-	1.18	0.94	GRAM/MJ	1.11	
POTASSIUM	MILLI	270.00	190.00	-	350.00	261.90	MILLI/MJ	311.96	
MAGNESIUM	MILLI	24.00	23.00	-	25.00	23.28	MILLI/MJ	27.73	
CALCIUM	MILLI	15.00	10.00	-	24.00	14.55	MILLI/MJ	17.33	
IRON	MILLI	2.30	1.70	-	2.90	2.23	MILLI/MJ	2.66	
PHOSPHORUS	MILLI	136.00	92.00	-	166.00	131.92	MILLI/MJ	157.14	
CHLORIDE	GRAM	1.70	1.40	-	2.10	1.65	GRAM/MJ	1.96	
VITAMIN D	-	0.00	-	-	-	0.00			
VITAMIN B1	MILLI	0.61	0.41	-	0.82	0.59	MILLI/MJ	0.70	
VITAMIN B2	MILLI	0.21	0.13	-	0.27	0.20	MILLI/MJ	0.24	
NICOTINAMIDE	MILLI	3.70	3.00	-	4.20	3.59	MILLI/MJ	4.28	
PANTOTHENIC ACID	MILLI	0.58	0.52	-	0.62	0.56	MILLI/MJ	0.67	
VITAMIN B6	MILLI	0.36	0.32	-	0.40	0.35	MILLI/MJ	0.42	
FOLIC ACID	MICRO	5.10	4.00	-	6.10	4.95	MICRO/MJ	5.89	
VITAMIN B12	MICRO	0.59	0.57	-	0.61	0.57	MICRO/MJ	0.68	
VITAMIN C	-	0.00	-	-	-	0.00			
ALANINE	GRAM	1.57	1.53	-	1.62	1.52	GRAM/MJ	1.81	9.2
ARGININE	GRAM	1.42	1.36	-	1.46	1.38	GRAM/MJ	1.64	4.3
ASPARTIC ACID	GRAM	2.33	-	-	-	2.26	GRAM/MJ	2.69	9.1
CYSTINE	GRAM	0.31	-	-	-	0.30	GRAM/MJ	0.36	0.7
GLUTAMIC ACID	GRAM	4.20	4.11	-	4.31	4.07	GRAM/MJ	4.85	14.9

CONSTITUENTS	DIM	AV	VARIATION			AVR	NUTR. DENS.		MOLPERC.
GLYCINE	GRAM	1.41	1.24	-	1.55	1.37	GRAM/MJ	1.63	9.8
HISTIDINE	GRAM	0.93	0.10	-	1.05	0.90	GRAM/MJ	1.07	3.1
ISOLEUCINE	GRAM	1.17	-	-	-	1.13	GRAM/MJ	1.35	4.7
LEUCINE	GRAM	1.95	1.85	-	2.06	1.89	GRAM/MJ	2.25	7.8
LYSINE	GRAM	2.21	2.03	-	2.41	2.14	GRAM/MJ	2.55	7.9
METHIONINE	GRAM	0.58	-	-	-	0.56	GRAM/MJ	0.67	2.0
PHENYLALANINE	GRAM	0.94	0.90	-	0.97	0.91	GRAM/MJ	1.09	3.0
PROLINE	GRAM	1.09	0.90	-	1.29	1.06	GRAM/MJ	1.26	4.9
SERINE	GRAM	0.97	0.82	-	1.13	0.94	GRAM/MJ	1.12	4.8
THREONINE	GRAM	1.07	1.02	-	1.13	1.04	GRAM/MJ	1.24	4.7
TRYPTOPHAN	GRAM	0.28	-	-	-	0.27	GRAM/MJ	0.32	0.7
TYROSINE	GRAM	0.90	0.87	-	0.93	0.87	GRAM/MJ	1.04	2.6
VALINE	GRAM	1.32	1.28	-	1.34	1.28	GRAM/MJ	1.53	5.9
LACTIC ACID	GRAM	0.67	0.39	-	0.95	0.65	GRAM/MJ	0.77	
LINOLEIC ACID	GRAM	1.10	0.96	-	1.20	1.07	GRAM/MJ	1.27	
LINOLENIC ACID	MILLI	70.00	60.00	-	80.00	67.90	MILLI/MJ	80.88	
ARACHIDONIC ACID	MILLI	50.00	-	-	-	48.50	MILLI/MJ	57.77	
CHOLESTEROL	MILLI	85.00	70.00	-	100.00	82.45	MILLI/MJ	98.21	
CONNECT.TISSUE PROTEIN	GRAM	1.20	0.56	-	1.90	1.16	GRAM/MJ	1.39	

SCHWEINESCHINKEN IN DOSEN — HAM CANNED — JAMBON DE PORC EN BOÎTES

		PROTEIN	FAT	CARBOHYDRATES	TOTAL
ENERGY VALUE (AVERAGE) PER 100 G EDIBLE PORTION	KJOULE	372	425	0.0	797
	(KCAL)	89	102	0.0	191
AMOUNT OF DIGESTIBLE CONSTITUENTS PER 100 G	GRAM	19.59	10.16	0.00	
ENERGY VALUE (AVERAGE) OF THE DIGESTIBLE FRACTION PER 100 G EDIBLE PORTION	KJOULE	361	404	0.0	765
	(KCAL)	86	97	0.0	183

WASTE PERCENTAGE AVERAGE 0.00

CONSTITUENTS	DIM	AV	VARIATION			AVR	NUTR.	DENS.	MOLPERC.
WATER	GRAM	66.00	65.50	-	67.00	66.00	GRAM/MJ	86.30	
PROTEIN	GRAM	20.20	19.90	-	20.50	20.20	GRAM/MJ	26.41	
FAT	GRAM	10.70	10.60	-	10.80	10.70	GRAM/MJ	13.99	
MINERALS	GRAM	3.07	2.47	-	3.67	3.07	GRAM/MJ	4.01	
SODIUM	GRAM	1.20	-	-	-	1.20	GRAM/MJ	1.57	
POTASSIUM	MILLI	350.00	-	-	-	350.00	MILLI/MJ	457.66	
CALCIUM	MILLI	10.00	-	-	-	10.00	MILLI/MJ	13.08	
IRON	MILLI	2.50	-	-	-	2.50	MILLI/MJ	3.27	
PHOSPHORUS	MILLI	161.00	-	-	-	161.00	MILLI/MJ	210.52	
VITAMIN B1	MILLI	0.41	0.35	-	0.46	0.41	MILLI/MJ	0.54	
VITAMIN B2	MILLI	0.16	0.15	-	0.16	0.16	MILLI/MJ	0.21	
NICOTINAMIDE	MILLI	3.00	2.50	-	3.40	3.00	MILLI/MJ	3.92	
VITAMIN C	-	0.00	-	-	-	0.00			
ALANINE	GRAM	1.49	1.46	-	1.54	1.49	GRAM/MJ	1.95	9.2
ARGININE	GRAM	1.35	1.29	-	1.39	1.35	GRAM/MJ	1.77	4.3
ASPARTIC ACID	GRAM	2.21	-	-	-	2.21	GRAM/MJ	2.89	9.1
CYSTINE	GRAM	0.29	-	-	-	0.29	GRAM/MJ	0.38	0.7
GLUTAMIC ACID	GRAM	4.00	3.91	-	4.10	4.00	GRAM/MJ	5.23	14.9
GLYCINE	GRAM	1.34	1.18	-	1.47	1.34	GRAM/MJ	1.75	9.8
HISTIDINE	GRAM	0.88	0.97	-	1.00	0.88	GRAM/MJ	1.15	3.1
ISOLEUCINE	GRAM	1.12	-	-	-	1.12	GRAM/MJ	1.46	4.7
LEUCINE	GRAM	1.85	1.76	-	1.96	1.85	GRAM/MJ	2.42	7.7
LYSINE	GRAM	2.10	1.93	-	2.29	2.10	GRAM/MJ	2.75	7.9
METHIONINE	GRAM	0.55	-	-	-	0.55	GRAM/MJ	0.72	2.0
PHENYLALANINE	GRAM	0.89	0.85	-	0.92	0.89	GRAM/MJ	1.16	3.0
PROLINE	GRAM	1.04	0.85	-	1.22	1.04	GRAM/MJ	1.36	5.0
SERINE	GRAM	0.92	0.78	-	1.08	0.92	GRAM/MJ	1.20	4.8
THREONINE	GRAM	1.02	0.97	-	1.08	1.02	GRAM/MJ	1.33	4.7
TRYPTOPHAN	GRAM	0.26	-	-	-	0.26	GRAM/MJ	0.34	0.7
TYROSINE	GRAM	0.85	0.82	-	0.88	0.85	GRAM/MJ	1.11	2.6
VALINE	GRAM	1.25	1.21	-	1.27	1.25	GRAM/MJ	1.63	5.9

SCHWEINESCHINKEN ROH, GERÄUCHERT — HAM RAW, SMOKE DRIED — JAMBON DE PORC CRU, FUMÉ

		PROTEIN	FAT	CARBOHYDRATES	TOTAL
ENERGY VALUE (AVERAGE)	KJOULE	331	1324	0.0	1655
PER 100 G	(KCAL)	79	316	0.0	396
EDIBLE PORTION					
AMOUNT OF DIGESTIBLE	GRAM	17.46	31.63	0.00	
CONSTITUENTS PER 100 G					
ENERGY VALUE (AVERAGE)	KJOULE	321	1257	0.0	1579
OF THE DIGESTIBLE	(KCAL)	77	ʼ301	0.0	377
FRACTION PER 100 G					
EDIBLE PORTION					

WASTE PERCENTAGE AVERAGE 13 MINIMUM 11 MAXIMUM 20

CONSTITUENTS	DIM	AV	VARIATION		AVR	NUTR. DENS.		MOLPERC.
WATER	GRAM	43.30	42.00	- 44.60	37.67	GRAM/MJ	27.42	
PROTEIN	GRAM	18.00	12.00	- 25.00	15.66	GRAM/MJ	11.40	
FAT	GRAM	33.30	25.00	- 40.00	28.97	GRAM/MJ	21.09	
MINERALS	GRAM	-	5.40	- 10.50				
SODIUM	GRAM	1.40	-	- -	1.22	GRAM/MJ	0.89	
POTASSIUM	MILLI	248.00	131.00	- 354.00	215.76	MILLI/MJ	157.08	
MAGNESIUM	MILLI	20.00	-	- -	17.40	MILLI/MJ	12.67	
CALCIUM	MILLI	10.00	-	- -	8.70	MILLI/MJ	6.33	
IRON	MILLI	2.25	2.00	- 2.50	1.96	MILLI/MJ	1.43	
PHOSPHORUS	MILLI	207.00	134.00	- 350.00	180.09	MILLI/MJ	131.11	
CHLORIDE	GRAM	2.10	-	- -	1.83	GRAM/MJ	1.33	
VITAMIN B1	MILLI	0.55	0.24	- 0.70	0.48	MILLI/MJ	0.35	
VITAMIN B2	MILLI	0.21	0.14	- 0.25	0.18	MILLI/MJ	0.13	
NICOTINAMIDE	MILLI	3.50	2.06	- 4.66	3.05	MILLI/MJ	2.22	
VITAMIN B6	MILLI	0.40	0.33	- 0.52	0.35	MILLI/MJ	0.25	
VITAMIN B12	MICRO	0.10	-	- -	0.09	MICRO/MJ	0.06	
VITAMIN C	-	0.00	-	- -	0.00			
ALANINE	GRAM	1.32	1.29	- 1.37	1.15	GRAM/MJ	0.84	9.2
ARGININE	GRAM	1.20	1.14	- 1.23	1.04	GRAM/MJ	0.76	4.3
ASPARTIC ACID	GRAM	1.96	-	- -	1.71	GRAM/MJ	1.24	9.1
CYSTINE	GRAM	0.26	-	- -	0.23	GRAM/MJ	0.16	0.7
GLUTAMIC ACID	GRAM	3.54	3.47	- 3.64	3.08	GRAM/MJ	2.24	14.9
GLYCINE	GRAM	1.19	1.05	- 1.31	1.04	GRAM/MJ	0.75	9.8
HISTIDINE	GRAM	0.78	0.09	- 0.89	0.68	GRAM/MJ	0.49	3.1
ISOLEUCINE	GRAM	0.99	-	- -	0.86	GRAM/MJ	0.63	4.7

CONSTITUENTS	DIM	AV	VARIATION			AVR	NUTR. DENS.		MOLPERC.
LEUCINE	GRAM	1.64	1.56	-	1.74	1.43	GRAM/MJ	1.04	7.7
LYSINE	GRAM	1.87	1.71	-	2.03	1.63	GRAM/MJ	1.18	7.9
METHIONINE	GRAM	0.49	-	-	-	0.43	GRAM/MJ	0.31	2.0
PHENYLALANINE	GRAM	0.79	0.76	-	0.82	0.69	GRAM/MJ	0.50	3.0
PROLINE	GRAM	0.92	0.76	-	1.08	0.80	GRAM/MJ	0.58	4.9
SERINE	GRAM	0.82	0.69	-	0.95	0.71	GRAM/MJ	0.52	4.8
THREONINE	GRAM	0.90	0.86	-	0.95	0.78	GRAM/MJ	0.57	4.7
TRYPTOPHAN	GRAM	0.23	-	-	-	0.20	GRAM/MJ	0.15	0.7
TYROSINE	GRAM	0.76	0.73	-	0.78	0.66	GRAM/MJ	0.48	2.6
VALINE	GRAM	1.11	1.08	-	1.13	0.97	GRAM/MJ	0.70	5.9
LINOLEIC ACID	GRAM	2.48	-	-	-	2.16	GRAM/MJ	1.57	
LINOLENIC ACID	GRAM	0.16	-	-	-	0.14	GRAM/MJ	0.10	
ARACHIDONIC ACID	GRAM	0.13	-	-	-	0.11	GRAM/MJ	0.08	

SCHWEINESPECK
DURCHWACHSEN
(FRÜHSTÜCKSSPECK, WAMMERL)

BACON
STREAKY

LARD
ENTRELARDÉ
(LARD DE PETIT DÉJEUNER)

		PROTEIN	FAT	CARBOHYDRATES	TOTAL
ENERGY VALUE (AVERAGE)	KJOULE	168	2584	0.0	2751
PER 100 G	(KCAL)	40	618	0.0	658
EDIBLE PORTION					
AMOUNT OF DIGESTIBLE	GRAM	8.82	61.75	0.00	
CONSTITUENTS PER 100 G					
ENERGY VALUE (AVERAGE)	KJOULE	163	2454	0.0	2617
OF THE DIGESTIBLE	(KCAL)	39	587	0.0	625
FRACTION PER 100 G					
EDIBLE PORTION					

WASTE PERCENTAGE AVERAGE 8.0 MINIMUM 6.0 MAXIMUM 12

CONSTITUENTS	DIM	AV	VARIATION			AVR	NUTR. DENS.		MOLPERC.
WATER	GRAM	20.00	-	-	-	18.40	GRAM/MJ	7.64	
PROTEIN	GRAM	9.10	-	-	-	8.37	GRAM/MJ	3.48	
FAT	GRAM	65.00	-	-	-	59.80	GRAM/MJ	24.84	
MINERALS	GRAM	5.00	-	-	-	4.60	GRAM/MJ	1.91	
SODIUM	GRAM	1.77	-	-	-	1.63	GRAM/MJ	0.68	
POTASSIUM	MILLI	225.00	-	-	-	207.00	MILLI/MJ	85.98	
CALCIUM	MILLI	9.00	-	-	-	8.28	MILLI/MJ	3.44	
IRON	MILLI	0.80	-	-	-	0.74	MILLI/MJ	0.31	
PHOSPHORUS	MILLI	108.00	-	-	-	99.36	MILLI/MJ	41.27	
VITAMIN B1	MILLI	0.43	0.38	-	0.50	0.40	MILLI/MJ	0.16	
VITAMIN B2	MILLI	0.14	0.10	-	0.20	0.13	MILLI/MJ	0.05	
NICOTINAMIDE	MILLI	2.30	1.90	-	3.00	2.12	MILLI/MJ	0.88	
VITAMIN B6	MILLI	0.35	-	-	-	0.32	MILLI/MJ	0.13	
VITAMIN B12	MICRO	0.70	0.30	-	1.20	0.64	MICRO/MJ	0.27	
ALANINE	GRAM	0.68	0.66	-	0.70	0.63	GRAM/MJ	0.26	9.2
ARGININE	GRAM	0.61	0.59	-	0.63	0.56	GRAM/MJ	0.23	4.2
ASPARTIC ACID	GRAM	1.00	-	-	-	0.92	GRAM/MJ	0.38	9.1
CYSTINE	GRAM	0.13	-	-	-	0.12	GRAM/MJ	0.05	0.7
GLUTAMIC ACID	GRAM	1.81	1.77	-	1.86	1.67	GRAM/MJ	0.69	14.9
GLYCINE	GRAM	0.61	0.54	-	0.67	0.56	GRAM/MJ	0.23	9.8
HISTIDINE	GRAM	0.40	0.04	-	0.45	0.37	GRAM/MJ	0.15	3.1
ISOLEUCINE	GRAM	0.51	-	-	-	0.47	GRAM/MJ	0.19	4.7
LEUCINE	GRAM	0.84	0.80	-	0.89	0.77	GRAM/MJ	0.32	7.7
LYSINE	GRAM	0.95	0.88	-	1.04	0.87	GRAM/MJ	0.36	7.9
METHIONINE	GRAM	0.25	-	-	-	0.23	GRAM/MJ	0.10	2.0
PHENYLALANINE	GRAM	0.40	0.39	-	0.42	0.37	GRAM/MJ	0.15	2.9
PROLINE	GRAM	0.47	0.39	-	0.55	0.43	GRAM/MJ	0.18	4.9
SERINE	GRAM	0.42	0.35	-	0.49	0.39	GRAM/MJ	0.16	4.8
THREONINE	GRAM	0.46	0.44	-	0.49	0.42	GRAM/MJ	0.18	4.7
TRYPTOPHAN	GRAM	0.12	-	-	-	0.11	GRAM/MJ	0.05	0.7
TYROSINE	GRAM	0.39	0.37	-	0.40	0.36	GRAM/MJ	0.15	2.6
VALINE	GRAM	0.57	0.55	-	0.58	0.52	GRAM/MJ	0.22	5.9
LINOLEIC ACID	GRAM	6.08	-	-	-	5.59	GRAM/MJ	2.32	
LINOLENIC ACID	GRAM	0.25	-	-	-	0.23	GRAM/MJ	0.10	
ARACHIDONIC ACID	GRAM	0.25	-	-	-	0.23	GRAM/MJ	0.10	

Würste und Pasteten

SAUSAGES AND PASTRIES **CHARCUTERIE ET PÂTÉS**

BIERSCHINKEN

SAUSAGE
BIERSCHINKEN

JAMBON DE BIÈRE
SAUCISSON "BIERSCHINKEN"

		PROTEIN	FAT	CARBOHYDRATES	TOTAL
ENERGY VALUE (AVERAGE)	KJOULE	285	763	0.0	1049
PER 100 G	(KCAL)	68	182	0.0	251
EDIBLE PORTION					
AMOUNT OF DIGESTIBLE CONSTITUENTS PER 100 G	GRAM	15.03	18.24	0.00	
ENERGY VALUE (AVERAGE)	KJOULE	277	725	0.0	1002
OF THE DIGESTIBLE	(KCAL)	66	173	0.0	239
FRACTION PER 100 G EDIBLE PORTION					

WASTE PERCENTAGE AVERAGE 2.0 MINIMUM 1.0 MAXIMUM 3.0

CONSTITUENTS	DIM	AV	VARIATION		AVR	NUTR.	DENS.	MOLPERC.
WATER	GRAM	62.80	57.20	- 64.70	61.54	GRAM/MJ	62.69	
PROTEIN	GRAM	15.50	14.10	- 16.90	15.19	GRAM/MJ	15.47	
FAT	GRAM	19.20	15.40	- 24.80	18.82	GRAM/MJ	19.17	
MINERALS	GRAM	2.53	2.23	- 2.73	2.48	GRAM/MJ	2.53	
SODIUM	MILLI	753.00	529.00	- 856.00	737.94	MILLI/MJ	751.65	
POTASSIUM	MILLI	261.00	242.00	- 284.00	255.78	MILLI/MJ	260.53	
MAGNESIUM	MILLI	18.00	16.00	- 20.00	17.64	MILLI/MJ	17.97	
CALCIUM	MILLI	15.00	13.00	- 17.00	14.70	MILLI/MJ	14.97	
IRON	MILLI	1.53	1.42	- 1.63	1.50	MILLI/MJ	1.53	
PHOSPHORUS	MILLI	152.00	135.00	- 177.00	148.96	MILLI/MJ	151.73	
NICOTINAMIDE	MILLI	3.80	-	- -	3.72	MILLI/MJ	3.79	
CONNECT.TISSUE PROTEIN	GRAM	1.60	-	- -	1.57	GRAM/MJ	1.60	

BLUTWURST (ROTWURST) — BLACK PUDDING — BOUDIN (NOIR)

		PROTEIN	FAT	CARBOHYDRATES	TOTAL
ENERGY VALUE (AVERAGE)	KJOULE	245	1530	0.0	1775
PER 100 G EDIBLE PORTION	(KCAL)	59	366	0.0	424
AMOUNT OF DIGESTIBLE CONSTITUENTS PER 100 G	GRAM	12.90	36.57	0.00	
ENERGY VALUE (AVERAGE)	KJOULE	238	1454	0.0	1691
OF THE DIGESTIBLE FRACTION PER 100 G EDIBLE PORTION	(KCAL)	57	347	0.0	404

WASTE PERCENTAGE AVERAGE 4.0 MINIMUM 3.0 MAXIMUM 5.0

CONSTITUENTS	DIM	AV	VARIATION		AVR	NUTR. DENS.		MOLPERC.
WATER	GRAM	45.50	40.60	47.00	43.68	GRAM/MJ	26.90	
PROTEIN	GRAM	13.30	10.40	16.00	12.77	GRAM/MJ	7.86	
FAT	GRAM	38.50	34.10	42.90	36.96	GRAM/MJ	22.76	
MINERALS	GRAM	2.37	1.77	3.05	2.28	GRAM/MJ	1.40	
SODIUM	MILLI	680.00	-	-	652.80	MILLI/MJ	402.06	
POTASSIUM	MILLI	38.00	34.00	42.00	36.48	MILLI/MJ	22.47	
MAGNESIUM	MILLI	7.60	7.10	8.00	7.30	MILLI/MJ	4.49	
CALCIUM	MILLI	6.50	4.30	8.80	6.24	MILLI/MJ	3.84	
IRON	MILLI	6.40	6.40	6.50	6.14	MILLI/MJ	3.78	
PHOSPHORUS	MILLI	22.00	21.00	23.00	21.12	MILLI/MJ	13.01	
CHLORIDE	GRAM	1.10	-	-	1.06	GRAM/MJ	0.65	
VITAMIN B1	MICRO	73.00	44.00	92.00	70.08	MICRO/MJ	43.16	
VITAMIN B2	MILLI	0.13	-	-	0.12	MILLI/MJ	0.08	
NICOTINAMIDE	MILLI	1.20	0.90	1.80	1.15	MILLI/MJ	0.71	
TOTAL PURINES	MILLI	90.00 [1]	-	-	86.40	MILLI/MJ	53.21	
CONNECT.TISSUE PROTEIN	GRAM	4.10	-	-	3.94	GRAM/MJ	2.42	

1 VALUE EXPRESSED IN MG URIC ACID/100G

BOCKWURST

SAUSAGE "BOCKWURST"

SAUCISSON "BOCKWURST"

		PROTEIN	FAT	CARBOHYDRATES	TOTAL
ENERGY VALUE (AVERAGE) PER 100 G EDIBLE PORTION	KJOULE	226	1006	0.0	1232
	(KCAL)	54	240	0.0	294
AMOUNT OF DIGESTIBLE CONSTITUENTS PER 100 G	GRAM	11.93	24.03	0.00	
ENERGY VALUE (AVERAGE) OF THE DIGESTIBLE FRACTION PER 100 G EDIBLE PORTION	KJOULE	220	955	0.0	1175
	(KCAL)	52	228	0.0	281

WASTE PERCENTAGE AVERAGE 0.00

CONSTITUENTS	DIM	AV	VARIATION		AVR	NUTR. DENS.		MOLPERC.
WATER	GRAM	59.10	56.60	- 61.50	59.10	GRAM/MJ	50.30	
PROTEIN	GRAM	12.30	-	- -	12.30	GRAM/MJ	10.47	
FAT	GRAM	25.30	23.90	- 26.70	25.30	GRAM/MJ	21.53	
MINERALS	GRAM	2.74	2.40	- 3.10	2.74	GRAM/MJ	2.33	
SODIUM	MILLI	700.00	-	- -	700.00	MILLI/MJ	595.75	
PHOSPHORUS	MILLI	67.00	-	- -	67.00	MILLI/MJ	57.02	
CHLORIDE	GRAM	1.10	-	- -	1.10	GRAM/MJ	0.94	
CONNECT.TISSUE PROTEIN	GRAM	2.30	-	- -	2.30	GRAM/MJ	1.96	

CERVELATWURST SAUSAGE "CERVELAT" SAUCISSON "CERVELAT"

		PROTEIN	FAT	CARBOHYDRATES	TOTAL
ENERGY VALUE (AVERAGE)	KJOULE	311	1717	0.0	2028
PER 100 G	(KCAL)	74	410	0.0	485
EDIBLE PORTION					
AMOUNT OF DIGESTIBLE	GRAM	16.39	41.04	0.00	
CONSTITUENTS PER 100 G					
ENERGY VALUE (AVERAGE)	KJOULE	302	1631	0.0	1933
OF THE DIGESTIBLE	(KCAL)	72	390	0.0	462
FRACTION PER 100 G					
EDIBLE PORTION					

WASTE PERCENTAGE AVERAGE 2.0

CONSTITUENTS	DIM	AV	VARIATION			AVR	NUTR. DENS.		MOLPERC.
WATER	GRAM	34.80	24.00	-	44.00	34.10	GRAM/MJ	18.00	
PROTEIN	GRAM	16.90	14.00	-	21.00	16.56	GRAM/MJ	8.74	
FAT	GRAM	43.20	34.00	-	50.00	42.34	GRAM/MJ	22.35	
SODIUM	GRAM	1.26	0.75	-	1.50	1.23	GRAM/MJ	0.65	
POTASSIUM	MILLI	300.00	204.00	-	473.00	294.00	MILLI/MJ	155.20	
MAGNESIUM	MILLI	11.00	-	-	-	10.78	MILLI/MJ	5.69	
CALCIUM	MILLI	24.00	13.00	-	35.00	23.52	MILLI/MJ	12.42	
IRON	MILLI	1.70	1.10	-	2.30	1.67	MILLI/MJ	0.88	
PHOSPHORUS	MILLI	155.00	-	-	-	151.90	MILLI/MJ	80.18	
VITAMIN B1	MILLI	0.10	-	-	-	0.10	MILLI/MJ	0.05	
VITAMIN B2	MILLI	0.20	-	-	-	0.20	MILLI/MJ	0.10	
NICOTINAMIDE	MILLI	4.00	-	-	-	3.92	MILLI/MJ	2.07	
VITAMIN C	-	0.00	-	-	-	0.00			

DOSENWÜRSTCHEN (BRÜHWÜRSTE) | SAUSAGES CANNED | SAUCISSES EN BOÎTES

		PROTEIN	FAT	CARBOHYDRATES	TOTAL
ENERGY VALUE (AVERAGE)	KJOULE	239	779	0.0	1018
PER 100 G	(KCAL)	57	186	0.0	243
EDIBLE PORTION					
AMOUNT OF DIGESTIBLE	GRAM	12.61	18.62	0.00	
CONSTITUENTS PER 100 G					
ENERGY VALUE (AVERAGE)	KJOULE	232	740	0.0	972
OF THE DIGESTIBLE	(KCAL)	55	177	0.0	232
FRACTION PER 100 G					
EDIBLE PORTION					

WASTE PERCENTAGE AVERAGE 0.00

CONSTITUENTS	DIM	AV	VARIATION			AVR	NUTR. DENS.		MOLPERC.
WATER	GRAM	65.70	60.50	-	70.00	65.70	GRAM/MJ	67.58	
PROTEIN	GRAM	13.00	11.90	-	14.10	13.00	GRAM/MJ	13.37	
FAT	GRAM	19.60	17.40	-	21.80	19.60	GRAM/MJ	20.16	
MINERALS	GRAM	2.14	1.89	-	2.41	2.14	GRAM/MJ	2.20	
SODIUM	MILLI	711.00	632.00	-	790.00	711.00	MILLI/MJ	731.29	
CALCIUM	MILLI	9.50	9.00	-	10.00	9.50	MILLI/MJ	9.77	
IRON	MILLI	2.70	2.40	-	3.00	2.70	MILLI/MJ	2.78	
PHOSPHORUS	MILLI	185.00	170.00	-	200.00	185.00	MILLI/MJ	190.28	
CHLORIDE	GRAM	1.10	-	-	-	1.10	GRAM/MJ	1.13	
VITAMIN B1	MICRO	30.00	-	-	-	30.00	MICRO/MJ	30.86	
VITAMIN B2	MICRO	80.00	-	-	-	80.00	MICRO/MJ	82.28	
NICOTINAMIDE	MILLI	3.05	3.00	-	3.10	3.05	MILLI/MJ	3.14	
ARGININE	GRAM	0.90	-	-	-	0.90	GRAM/MJ	0.93	
CYSTINE	GRAM	0.22	-	-	-	0.22	GRAM/MJ	0.23	
HISTIDINE	GRAM	0.35	-	-	-	0.35	GRAM/MJ	0.36	
ISOLEUCINE	GRAM	0.70	-	-	-	0.70	GRAM/MJ	0.72	
LEUCINE	GRAM	1.00	-	-	-	1.00	GRAM/MJ	1.03	
LYSINE	GRAM	1.00	-	-	-	1.00	GRAM/MJ	1.03	
METHIONINE	GRAM	0.34	-	-	-	0.34	GRAM/MJ	0.35	
PHENYLALANINE	GRAM	0.53	-	-	-	0.53	GRAM/MJ	0.55	
THREONINE	GRAM	0.45	-	-	-	0.45	GRAM/MJ	0.46	
TRYPTOPHAN	GRAM	0.13	-	-	-	0.13	GRAM/MJ	0.13	
TYROSINE	GRAM	0.43	-	-	-	0.43	GRAM/MJ	0.44	
VALINE	GRAM	0.73	-	-	-	0.73	GRAM/MJ	0.75	

FLEISCHKÄSE (LEBERKÄSE) | MEAT LOAF | FROMAGE DE VIANDE

		PROTEIN	FAT	CARBOHYDRATES	TOTAL
ENERGY VALUE (AVERAGE)	KJOULE	212	1208	0.0	1420
PER 100 G	(KCAL)	51	289	0.0	339
EDIBLE PORTION					
AMOUNT OF DIGESTIBLE	GRAM	11.15	28.88	0.00	
CONSTITUENTS PER 100 G					
ENERGY VALUE (AVERAGE)	KJOULE	205	1148	0.0	1353
OF THE DIGESTIBLE	(KCAL)	49	274	0.0	323
FRACTION PER 100 G					
EDIBLE PORTION					

WASTE PERCENTAGE AVERAGE 0.00

CONSTITUENTS	DIM	AV	VARIATION		AVR	NUTR. DENS.		MOLPERC.
WATER	GRAM	54.40	50.00	- 57.00	54.40	GRAM/MJ	40.20	
PROTEIN	GRAM	11.50	10.40	- 12.30	11.50	GRAM/MJ	8.50	
FAT	GRAM	30.40	28.80	- 32.40	30.40	GRAM/MJ	22.46	
MINERALS	GRAM	2.60	2.36	- 2.72	2.60	GRAM/MJ	1.92	
SODIUM	MILLI	599.00	324.00	- 873.00	599.00	MILLI/MJ	442.63	
POTASSIUM	MILLI	299.00	217.00	- 381.00	299.00	MILLI/MJ	220.94	
CALCIUM	MILLI	4.00	3.00	- 4.00	4.00	MILLI/MJ	2.96	
VITAMIN B1	MICRO	48.00	46.00	- 50.00	48.00	MICRO/MJ	35.47	
VITAMIN B2	MILLI	0.15	-	- -	0.15	MILLI/MJ	0.11	
NICOTINAMIDE	MILLI	2.40	-	- -	2.40	MILLI/MJ	1.77	
CONNECT.TISSUE PROTEIN	GRAM	1.93	1.10	- 2.60	1.93	GRAM/MJ	1.43	

FLEISCHKÄSE
NACH STUTTGARTER ART
(STUTTGARTER LEBERKÄSE)

MEAT LOAF
VAR. STUTTGARTER

FROMAGE DE VIANDE DE STUTTGART

		PROTEIN	FAT	CARBOHYDRATES	TOTAL
ENERGY VALUE (AVERAGE) PER 100 G EDIBLE PORTION	KJOULE	258	1268	0.0	1526
	(KCAL)	62	303	0.0	365
AMOUNT OF DIGESTIBLE CONSTITUENTS PER 100 G	GRAM	13.58	30.30	0.00	
ENERGY VALUE (AVERAGE) OF THE DIGESTIBLE FRACTION PER 100 G EDIBLE PORTION	KJOULE	250	1205	0.0	1455
	(KCAL)	60	288	0.0	348

WASTE PERCENTAGE AVERAGE 0.00

CONSTITUENTS	DIM	AV	VARIATION		AVR	NUTR. DENS.		MOLPERC.
WATER	GRAM	50.70	50.20	- 51.90	50.70	GRAM/MJ	34.86	
PROTEIN	GRAM	14.00	13.60	- 14.30	14.00	GRAM/MJ	9.62	
FAT	GRAM	31.90	30.40	- 32.40	31.90	GRAM/MJ	21.93	
MINERALS	GRAM	2.88	2.75	- 3.00	2.88	GRAM/MJ	1.98	
SODIUM	MILLI	775.00	749.00	- 801.00	775.00	MILLI/MJ	532.81	
POTASSIUM	MILLI	228.00	227.00	- 228.00	228.00	MILLI/MJ	156.75	
MAGNESIUM	MILLI	17.00	15.00	- 18.00	17.00	MILLI/MJ	11.69	
CALCIUM	MILLI	15.00	14.00	- 15.00	15.00	MILLI/MJ	10.31	
IRON	MILLI	1.24	0.96	- 1.51	1.24	MILLI/MJ	0.85	
PHOSPHORUS	MILLI	123.00	116.00	- 131.00	123.00	MILLI/MJ	84.56	

FLEISCHWURST

SAUSAGE "FLEISCHWURST"

SAUCISSON "FLEISCHWURST"

		PROTEIN	FAT	CARBOHYDRATES	TOTAL
ENERGY VALUE (AVERAGE)	KJOULE	243	1077	0.0	1320
PER 100 G	(KCAL)	58	257	0.0	316
EDIBLE PORTION					
AMOUNT OF DIGESTIBLE	GRAM	12.80	25.74	0.00	
CONSTITUENTS PER 100 G					
ENERGY VALUE (AVERAGE)	KJOULE	236	1023	0.0	1259
OF THE DIGESTIBLE	(KCAL)	56	245	0.0	301
FRACTION PER 100 G					
EDIBLE PORTION					

WASTE PERCENTAGE AVERAGE 1.0 MINIMUM 1.0 MAXIMUM 2.0

CONSTITUENTS	DIM	AV	VARIATION			AVR	NUTR. DENS.		MOLPERC.
WATER	GRAM	57.30	52.00	-	65.70	56.73	GRAM/MJ	45.51	
PROTEIN	GRAM	13.20	11.20	-	16.30	13.07	GRAM/MJ	10.48	
FAT	GRAM	27.10	18.50	-	33.00	26.83	GRAM/MJ	21.52	
MINERALS	GRAM	2.42	2.28	-	2.70	2.40	GRAM/MJ	1.92	
SODIUM	MILLI	829.00	809.00	-	848.00	820.71	MILLI/MJ	658.44	
POTASSIUM	MILLI	199.00	188.00	-	209.00	197.01	MILLI/MJ	158.06	
MAGNESIUM	MILLI	13.00	12.00	-	14.00	12.87	MILLI/MJ	10.33	
CALCIUM	MILLI	14.00	13.00	-	16.00	13.86	MILLI/MJ	11.12	
IRON	MILLI	1.70	1.40	-	2.10	1.68	MILLI/MJ	1.35	
PHOSPHORUS	MILLI	129.00	111.00	-	146.00	127.71	MILLI/MJ	102.46	
CONNECT.TISSUE PROTEIN	GRAM	2.50	-	-	-	2.48	GRAM/MJ	1.99	

FRANKFURTER WÜRSTCHEN / FRANKFURTER SAUSAGES / SAUCISSE DE FRANCFORT

		PROTEIN	FAT	CARBOHYDRATES	TOTAL
ENERGY VALUE (AVERAGE)	KJOULE	228	970	0.0	1198
PER 100 G	(KCAL)	55	232	0.0	286
EDIBLE PORTION					
AMOUNT OF DIGESTIBLE	GRAM	12.02	23.18	0.00	
CONSTITUENTS PER 100 G					
ENERGY VALUE (AVERAGE)	KJOULE	221	921	0.0	1143
OF THE DIGESTIBLE	(KCAL)	53	220	0.0	273
FRACTION PER 100 G					
EDIBLE PORTION					

WASTE PERCENTAGE AVERAGE 0.00

CONSTITUENTS	DIM	AV	VARIATION			AVR	NUTR. DENS.		MOLPERC.
WATER	GRAM	58.40	55.00	-	63.30	58.40	GRAM/MJ	51.10	
PROTEIN	GRAM	12.40	12.00	-	15.20	12.40	GRAM/MJ	10.85	
FAT	GRAM	24.40	-	-	-	24.40	GRAM/MJ	21.35	
MINERALS	GRAM	2.58	-	-	-	2.58	GRAM/MJ	2.26	
SODIUM	MILLI	778.00	-	-	-	778.00	MILLI/MJ	680.79	
POTASSIUM	MILLI	180.00	140.00	-	220.00	180.00	MILLI/MJ	157.51	
CALCIUM	MILLI	8.00	6.00	-	9.00	8.00	MILLI/MJ	7.00	
IRON	MILLI	1.75	1.20	-	2.30	1.75	MILLI/MJ	1.53	
PHOSPHORUS	MILLI	107.00	49.00	-	164.00	107.00	MILLI/MJ	93.63	
CHLORIDE	GRAM	1.20	-	-	-	1.20	GRAM/MJ	1.05	
VITAMIN A	MICRO	-	0.00	-	6.00				
VITAMIN B1	MILLI	0.18	0.09	-	0.23	0.18	MILLI/MJ	0.16	
VITAMIN B2	MILLI	0.19	0.10	-	0.24	0.19	MILLI/MJ	0.17	
NICOTINAMIDE	MILLI	2.30	1.65	-	2.70	2.30	MILLI/MJ	2.01	
PANTOTHENIC ACID	MILLI	0.43	0.41	-	0.44	0.43	MILLI/MJ	0.38	
VITAMIN B6	MILLI	0.14	0.11	-	0.23	0.14	MILLI/MJ	0.12	
ARGININE	GRAM	0.91	-	-	-	0.91	GRAM/MJ	0.80	
CYSTINE	GRAM	0.16	-	-	-	0.16	GRAM/MJ	0.14	
HISTIDINE	GRAM	0.35	-	-	-	0.35	GRAM/MJ	0.31	
ISOLEUCINE	GRAM	0.64	-	-	-	0.64	GRAM/MJ	0.56	
LEUCINE	GRAM	0.94	-	-	-	0.94	GRAM/MJ	0.82	
LYSINE	GRAM	1.05	-	-	-	1.05	GRAM/MJ	0.92	
METHIONINE	GRAM	0.28	-	-	-	0.28	GRAM/MJ	0.25	
PHENYLALANINE	GRAM	0.48	-	-	-	0.48	GRAM/MJ	0.42	
THREONINE	GRAM	0.54	-	-	-	0.54	GRAM/MJ	0.47	
TRYPTOPHAN	GRAM	0.11	-	-	-	0.11	GRAM/MJ	0.10	
TYROSINE	GRAM	0.43	-	-	-	0.43	GRAM/MJ	0.38	
VALINE	GRAM	0.66	-	-	-	0.66	GRAM/MJ	0.58	
CHOLESTEROL	MILLI	65.00	-	-	-	65.00	MILLI/MJ	56.88	
TOTAL PURINES	MILLI	130.00 [1]	-	-	-	130.00	MILLI/MJ	113.76	
CONNECT.TISSUE PROTEIN	GRAM	2.10	-	-	-	2.10	GRAM/MJ	1.84	

1 VALUE EXPRESSED IN MG URIC ACID/100G

GELBWURST
(HIRNWURST)

SAUSAGE "GELBWURST"

SAUCISSON "GELBWURST"

		PROTEIN	FAT	CARBOHYDRATES	TOTAL
ENERGY VALUE (AVERAGE)	KJOULE	217	1300	0.0	1517
PER 100 G	(KCAL)	52	311	0.0	363
EDIBLE PORTION					
AMOUNT OF DIGESTIBLE	GRAM	11.44	31.06	0.00	
CONSTITUENTS PER 100 G					
ENERGY VALUE (AVERAGE)	KJOULE	211	1235	0.0	1445
OF THE DIGESTIBLE	(KCAL)	50	295	0.0	345
FRACTION PER 100 G					
EDIBLE PORTION					

WASTE PERCENTAGE AVERAGE 4.0

CONSTITUENTS	DIM	AV	VARIATION			AVR	NUTR.	DENS.	MOLPERC.
WATER	GRAM	53.10	44.40	-	58.00	50.98	GRAM/MJ	36.74	
PROTEIN	GRAM	11.80	10.40	-	13.50	11.33	GRAM/MJ	8.16	
FAT	GRAM	32.70	28.90	-	42.90	31.39	GRAM/MJ	22.62	
MINERALS	GRAM	2.22	2.16	-	2.42	2.13	GRAM/MJ	1.54	
SODIUM	MILLI	640.00	540.00	-	740.00	614.40	MILLI/MJ	442.76	
POTASSIUM	MILLI	285.00	160.00	-	410.00	273.60	MILLI/MJ	197.17	
CHLORIDE	GRAM	1.06	1.00	-	1.17	1.02	GRAM/MJ	0.73	
VITAMIN B1	MILLI	-	0.05	-	0.31				
VITAMIN B2	MILLI	0.12	-	-	-	0.12	MILLI/MJ	0.08	
NICOTINAMIDE	MILLI	2.32	2.24	-	2.40	2.23	MILLI/MJ	1.60	
CONNECT.TISSUE PROTEIN	GRAM	2.10	-	-	-	2.02	GRAM/MJ	1.45	

GÖTTINGER (BLASENWURST)

SAUSAGE "GOETTINGER"

SAUCISSON "GOETTINGER"

		PROTEIN	FAT	CARBOHYDRATES	TOTAL
ENERGY VALUE (AVERAGE)	KJOULE	267	1391	0.0	1658
PER 100 G	(KCAL)	64	333	0.0	396
EDIBLE PORTION					
AMOUNT OF DIGESTIBLE	GRAM	14.06	33.25	0.00	
CONSTITUENTS PER 100 G					
ENERGY VALUE (AVERAGE)	KJOULE	259	1322	0.0	1581
OF THE DIGESTIBLE	(KCAL)	62	316	0.0	378
FRACTION PER 100 G					
EDIBLE PORTION					

WASTE PERCENTAGE AVERAGE 2.0 MINIMUM 1.0 MAXIMUM 2.0

CONSTITUENTS	DIM	AV	VARIATION			AVR	NUTR. DENS.		MOLPERC.
WATER	GRAM	47.50	38.90	-	50.00	46.55	GRAM/MJ	30.05	
PROTEIN	GRAM	14.50	12.70	-	15.00	14.21	GRAM/MJ	9.17	
FAT	GRAM	35.00	32.00	-	40.00	34.30	GRAM/MJ	22.14	
MINERALS	GRAM	3.00	-	-	-	2.94	GRAM/MJ	1.90	

JAGDWURST CHASE SAUSAGE SAUCISSE DE CHASSE

		PROTEIN	FAT	CARBOHYDRATES	TOTAL
ENERGY VALUE (AVERAGE)	KJOULE	228	1304	0.0	1532
PER 100 G	(KCAL)	55	312	0.0	366
EDIBLE PORTION					
AMOUNT OF DIGESTIBLE	GRAM	12.02	31.16	0.00	
CONSTITUENTS PER 100 G					
ENERGY VALUE (AVERAGE)	KJOULE	221	1239	0.0	1460
OF THE DIGESTIBLE	(KCAL)	53	296	0.0	349
FRACTION PER 100 G					
EDIBLE PORTION					

WASTE PERCENTAGE AVERAGE 2.0

CONSTITUENTS	DIM	AV	VARIATION		AVR	NUTR. DENS.		MOLPERC.
WATER	GRAM	52.30	50.40	- 57.40	51.25	GRAM/MJ	35.82	
PROTEIN	GRAM	12.40	11.40	- 13.80	12.15	GRAM/MJ	8.49	
FAT	GRAM	32.80	25.00	- 35.80	32.14	GRAM/MJ	22.47	
MINERALS	GRAM	2.70	2.29	- 3.10	2.65	GRAM/MJ	1.85	
SODIUM	MILLI	818.00	812.00	- 824.00	801.64	MILLI/MJ	560.28	
POTASSIUM	MILLI	260.00	256.00	- 264.00	254.80	MILLI/MJ	178.08	
MAGNESIUM	MILLI	19.00	-	- -	18.62	MILLI/MJ	13.01	
CALCIUM	MILLI	14.00	13.00	- 14.00	13.72	MILLI/MJ	9.59	
IRON	MILLI	2.90	2.70	- 3.00	2.84	MILLI/MJ	1.99	
PHOSPHORUS	MILLI	144.00	140.00	- 148.00	141.12	MILLI/MJ	98.63	
VITAMIN B1	MILLI	0.11	-	- -	0.11	MILLI/MJ	0.08	
VITAMIN B2	MILLI	0.12	-	- -	0.12	MILLI/MJ	0.08	
NICOTINAMIDE	MILLI	4.20	-	- -	4.12	MILLI/MJ	2.88	
VITAMIN C	-	0.00	-	- -	0.00			
TOTAL PURINES	MILLI	130.00 [1]	-	- -	127.40	MILLI/MJ	89.04	
CONNECT.TISSUE PROTEIN	GRAM	2.20	-	- -	2.16	GRAM/MJ	1.51	

1 VALUE EXPRESSED IN MG URIC ACID/100G

KALBSBRATWURST | FRYING SAUSAGE FROM VEAL | SAUCISSE À GRILLER DE VIANDE DE VEAU

		PROTEIN	FAT	CARBOHYDRATES	TOTAL
ENERGY VALUE (AVERAGE) PER 100 G EDIBLE PORTION	KJOULE	208	994	0.0	1202
	(KCAL)	50	238	0.0	287
AMOUNT OF DIGESTIBLE CONSTITUENTS PER 100 G	GRAM	10.96	23.75	0.00	
ENERGY VALUE (AVERAGE) OF THE DIGESTIBLE FRACTION PER 100 G EDIBLE PORTION	KJOULE	202	944	0.0	1146
	(KCAL)	48	226	0.0	274

WASTE PERCENTAGE AVERAGE 0.00

CONSTITUENTS	DIM	AV	VARIATION		AVR	NUTR. DENS.		MOLPERC.
WATER	GRAM	62.00	59.40	- 64.90	62.00	GRAM/MJ	54.11	
PROTEIN	GRAM	11.30	10.30	- 12.00	11.30	GRAM/MJ	9.86	
FAT	GRAM	25.00	-	- -	25.00	GRAM/MJ	21.82	
MINERALS	GRAM	1.60	1.40	- 1.80	1.60	GRAM/MJ	1.40	

KALBSKÄSE | **MEAT LOAF FROM VEAL** | **FROMAGE DE VEAU**

		PROTEIN	FAT	CARBOHYDRATES	TOTAL
ENERGY VALUE (AVERAGE) PER 100 G EDIBLE PORTION	KJOULE	234	1181	0.0	1414
	(KCAL)	56	282	0.0	338
AMOUNT OF DIGESTIBLE CONSTITUENTS PER 100 G	GRAM	12.31	28.21	0.00	
ENERGY VALUE (AVERAGE) OF THE DIGESTIBLE FRACTION PER 100 G EDIBLE PORTION	KJOULE	227	1121	0.0	1348
	(KCAL)	54	268	0.0	322

WASTE PERCENTAGE AVERAGE 0.00

CONSTITUENTS	DIM	AV	VARIATION	AVR	NUTR. DENS.		MOLPERC.
WATER	GRAM	55.40	54.00 - 58.60	55.40	GRAM/MJ	41.09	
PROTEIN	GRAM	12.70	10.50 - 14.90	12.70	GRAM/MJ	9.42	
FAT	GRAM	29.70	26.90 - 31.40	29.70	GRAM/MJ	22.03	
MINERALS	GRAM	2.31	2.10 - 2.44	2.31	GRAM/MJ	1.71	
CONNECT.TISSUE PROTEIN	GRAM	2.00	- - -	2.00	GRAM/MJ	1.48	

KNACKWURST | SAUSAGE "KNACKWURST" | SAUCISSON "KNACKWURST"

		PROTEIN	FAT	CARBOHYDRATES	TOTAL
ENERGY VALUE (AVERAGE)	KJOULE	219	1340	0.0	1559
PER 100 G	(KCAL)	52	320	0.0	373
EDIBLE PORTION					
AMOUNT OF DIGESTIBLE	GRAM	11.54	32.01	0.00	
CONSTITUENTS PER 100 G					
ENERGY VALUE (AVERAGE)	KJOULE	213	1273	0.0	1485
OF THE DIGESTIBLE	(KCAL)	51	304	0.0	355
FRACTION PER 100 G					
EDIBLE PORTION					

WASTE PERCENTAGE AVERAGE 8.0

CONSTITUENTS	DIM	AV	VARIATION			AVR	NUTR.	DENS.	MOLPERC.
WATER	GRAM	50.10	41.60	-	54.30	46.09	GRAM/MJ	33.74	
PROTEIN	GRAM	11.90	10.70	-	14.80	10.95	GRAM/MJ	8.01	
FAT	GRAM	33.70	25.50	-	35.00	31.00	GRAM/MJ	22.69	
MINERALS	GRAM	4.00	-	-	-	3.68	GRAM/MJ	2.69	
SODIUM	GRAM	1.19	-	-	-	1.09	GRAM/MJ	0.80	
POTASSIUM	MILLI	195.00	-	-	-	179.40	MILLI/MJ	131.31	
CALCIUM	MILLI	28.00	-	-	-	25.76	MILLI/MJ	18.85	
PHOSPHORUS	MILLI	144.00	-	-	-	132.48	MILLI/MJ	96.97	
VITAMIN A	MICRO	15.00	-	-	-	13.80	MICRO/MJ	10.10	

LEBERPASTETE (BRÜHWURSTART) — LIVER PATE — PÂTÉ DE FOIE

		PROTEIN	FAT	CARBOHYDRATES	TOTAL
ENERGY VALUE (AVERAGE)	KJOULE	261	1137	0.0	1398
PER 100 G	(KCAL)	62	272	0.0	334
EDIBLE PORTION					
AMOUNT OF DIGESTIBLE	GRAM	13.77	27.17	0.00	
CONSTITUENTS PER 100 G					
ENERGY VALUE (AVERAGE)	KJOULE	254	1080	0.0	1334
OF THE DIGESTIBLE	(KCAL)	61	258	0.0	319
FRACTION PER 100 G					
EDIBLE PORTION					

WASTE PERCENTAGE AVERAGE 0.00

CONSTITUENTS	DIM	AV	VARIATION		AVR	NUTR. DENS.		MOLPERC.
WATER	GRAM	53.90	50.50	- 56.80	53.90	GRAM/MJ	40.42	
PROTEIN	GRAM	14.20	12.80	- 15.20	14.20	GRAM/MJ	10.65	
FAT	GRAM	28.60	24.10	- 35.00	28.60	GRAM/MJ	21.45	
MINERALS	GRAM	2.66	2.07	- 3.43	2.66	GRAM/MJ	1.99	
SODIUM	MILLI	738.00	697.00	- 779.00	738.00	MILLI/MJ	553.42	
POTASSIUM	MILLI	173.00	138.00	- 207.00	173.00	MILLI/MJ	129.73	
MAGNESIUM	MILLI	15.00	13.00	- 16.00	15.00	MILLI/MJ	11.25	
CALCIUM	MILLI	10.00	6.00	- 14.00	10.00	MILLI/MJ	7.50	
MANGANESE	MILLI	0.12	-	- -	0.12	MILLI/MJ	0.09	
IRON	MILLI	6.40	5.50	- 7.20	6.40	MILLI/MJ	4.80	
COPPER	MILLI	0.40	-	- -	0.40	MILLI/MJ	0.30	
PHOSPHORUS	MILLI	191.00	181.00	- 200.00	191.00	MILLI/MJ	143.23	
VITAMIN B1	MICRO	30.00	10.00	- 50.00	30.00	MICRO/MJ	22.50	
VITAMIN B2	MILLI	0.60	0.40	- 1.00	0.60	MILLI/MJ	0.45	
NICOTINAMIDE	MILLI	3.30	1.80	- 4.30	3.30	MILLI/MJ	2.47	
PANTOTHENIC ACID	MILLI	1.20	-	- -	1.20	MILLI/MJ	0.90	
FOLIC ACID	MICRO	60.00	-	- -	60.00	MICRO/MJ	44.99	
VITAMIN B12	MICRO	3.20	-	- -	3.20	MICRO/MJ	2.40	
VITAMIN C	MILLI	2.00	1.00	- 3.00	2.00	MILLI/MJ	1.50	
ARGININE	GRAM	0.89	-	- -	0.89	GRAM/MJ	0.67	
CYSTINE	GRAM	0.17	-	- -	0.17	GRAM/MJ	0.13	
HISTIDINE	GRAM	0.30	-	- -	0.30	GRAM/MJ	0.22	
ISOLEUCINE	GRAM	0.36	-	- -	0.36	GRAM/MJ	0.27	
LEUCINE	GRAM	1.05	-	- -	1.05	GRAM/MJ	0.79	
LYSINE	GRAM	0.84	-	- -	0.84	GRAM/MJ	0.63	
METHIONINE	GRAM	0.28	-	- -	0.28	GRAM/MJ	0.21	
PHENYLALANINE	GRAM	0.58	-	- -	0.58	GRAM/MJ	0.43	
THREONINE	GRAM	0.56	-	- -	0.56	GRAM/MJ	0.42	
TRYPTOPHAN	GRAM	0.15	-	- -	0.15	GRAM/MJ	0.11	
TYROSINE	GRAM	0.45	-	- -	0.45	GRAM/MJ	0.34	
VALINE	GRAM	0.77	-	- -	0.77	GRAM/MJ	0.58	

LEBERPRESSACK LIVER BRAWN "LEBERPRESSACK"

		PROTEIN	FAT	CARBOHYDRATES	TOTAL
ENERGY VALUE (AVERAGE)	KJOULE	296	1276	0.0	1572
PER 100 G	(KCAL)	71	305	0.0	376
EDIBLE PORTION					
AMOUNT OF DIGESTIBLE	GRAM	15.61	30.49	0.00	
CONSTITUENTS PER 100 G					
ENERGY VALUE (AVERAGE)	KJOULE	288	1212	0.0	1500
OF THE DIGESTIBLE	(KCAL)	69	290	0.0	358
FRACTION PER 100 G					
EDIBLE PORTION					

WASTE PERCENTAGE AVERAGE 4.0

CONSTITUENTS	DIM	AV	VARIATION			AVR	NUTR. DENS.		MOLPERC.
WATER	GRAM	49.70	40.10	-	60.00	47.71	GRAM/MJ	33.14	
PROTEIN	GRAM	16.10	-	-	-	15.46	GRAM/MJ	10.74	
FAT	GRAM	32.10	25.10	-	44.80	30.82	GRAM/MJ	21.41	
MINERALS	GRAM	1.92	1.87	-	1.97	1.84	GRAM/MJ	1.28	

LEBERWURST — LIVER SAUSAGE "LIVERWURST" — SAUCISSE DE FOIE

		PROTEIN	FAT	CARBOHYDRATES	TOTAL
ENERGY VALUE (AVERAGE) PER 100 G EDIBLE PORTION	KJOULE	228	1638	0.0	1866
	(KCAL)	55	391	0.0	446
AMOUNT OF DIGESTIBLE CONSTITUENTS PER 100 G	GRAM	12.02	39.14	0.00	
ENERGY VALUE (AVERAGE) OF THE DIGESTIBLE FRACTION PER 100 G EDIBLE PORTION	KJOULE	221	1556	0.0	1777
	(KCAL)	53	372	0.0	425

WASTE PERCENTAGE AVERAGE 2.0

CONSTITUENTS	DIM	AV	VARIATION			AVR	NUTR. DENS.		MOLPERC.
WATER	GRAM	42.90	39.70	-	46.20	42.04	GRAM/MJ	24.14	
PROTEIN	GRAM	12.40	10.90	-	13.80	12.15	GRAM/MJ	6.98	
FAT	GRAM	41.20	36.50	-	45.90	40.38	GRAM/MJ	23.18	
MINERALS	GRAM	2.32	2.01	-	2.62	2.27	GRAM/MJ	1.31	
SODIUM	GRAM	0.81	0.61	-	1.06	0.79	GRAM/MJ	0.46	
POTASSIUM	MILLI	143.00	105.00	-	177.00	140.14	MILLI/MJ	80.47	
CALCIUM	MILLI	41.00	0.00	-	105.00	40.18	MILLI/MJ	23.07	
IRON	MILLI	5.30	5.00	-	5.40	5.19	MILLI/MJ	2.98	
PHOSPHORUS	MILLI	154.00	-	-	-	150.92	MILLI/MJ	86.65	
VITAMIN A	MILLI	8.30	2.35	-	14.70	8.13	MILLI/MJ	4.67	
VITAMIN B1	MILLI	0.21	0.14	-	0.28	0.21	MILLI/MJ	0.12	
VITAMIN B2	MILLI	0.92	-	-	-	0.90	MILLI/MJ	0.52	
NICOTINAMIDE	MILLI	3.60	-	-	-	3.53	MILLI/MJ	2.03	
ARGININE	GRAM	0.77	-	-	-	0.75	GRAM/MJ	0.43	
CYSTINE	GRAM	0.15	-	-	-	0.15	GRAM/MJ	0.08	
HISTIDINE	GRAM	0.37	-	-	-	0.36	GRAM/MJ	0.21	
ISOLEUCINE	GRAM	0.60	-	-	-	0.59	GRAM/MJ	0.34	
LEUCINE	GRAM	1.04	-	-	-	1.02	GRAM/MJ	0.59	
LYSINE	GRAM	0.97	-	-	-	0.95	GRAM/MJ	0.55	
METHIONINE	GRAM	0.26	-	-	-	0.25	GRAM/MJ	0.15	
PHENYLALANINE	GRAM	0.56	-	-	-	0.55	GRAM/MJ	0.32	
THREONINE	GRAM	0.54	-	-	-	0.53	GRAM/MJ	0.30	
TRYPTOPHAN	GRAM	0.14	-	-	-	0.14	GRAM/MJ	0.08	
TYROSINE	GRAM	0.38	-	-	-	0.37	GRAM/MJ	0.21	
VALINE	GRAM	0.77	-	-	-	0.75	GRAM/MJ	0.43	
MYRISTIC ACID	GRAM	0.54	0.44	-	0.66	0.53	GRAM/MJ	0.30	
PALMITIC ACID	GRAM	9.40	8.40	-	10.40	9.21	GRAM/MJ	5.29	
C17 : 0 FATTY ACID	GRAM	0.19	0.16	-	0.24	0.19	GRAM/MJ	0.11	
STEARIC ACID	GRAM	4.90	3.40	-	7.20	4.80	GRAM/MJ	2.76	
ARACHIDIC ACID	GRAM	0.64	0.17	-	0.79	0.63	GRAM/MJ	0.36	
PALMITOLEIC ACID	GRAM	1.25	0.99	-	1.50	1.23	GRAM/MJ	0.70	
OLEIC ACID	GRAM	20.50	17.00	-	23.30	20.09	GRAM/MJ	11.54	
LINOLEIC ACID	GRAM	1.50	0.66	-	1.90	1.47	GRAM/MJ	0.84	
ARACHIDONIC ACID	GRAM	0.23	-	-	-	0.23	GRAM/MJ	0.13	

CONSTITUENTS	DIM	AV	VARIATION			AVR	NUTR. DENS.		MOLPERC.
TOTAL PURINES	MILLI	165.00 1	165.00	-	165.00	161.70	MILLI/MJ	92.84	
ADENINE	MILLI	25.00	25.00	-	25.00	24.50	MILLI/MJ	14.07	
GUANINE	MILLI	42.00	42.00	-	42.00	41.16	MILLI/MJ	23.63	
XANTHINE	MILLI	12.00	12.00	-	12.00	11.76	MILLI/MJ	6.75	
HYPOXANTHINE	MILLI	61.00	61.00	-	61.00	59.78	MILLI/MJ	34.32	
CONNECT.TISSUE PROTEIN	GRAM	2.90	-	-	-	2.84	GRAM/MJ	1.63	

1 VALUE OF TOTAL PURINES IS EXPRESSSED IN MG URIC ACID/100G

LYONER — SAUSAGE "LYONER" — SAUCISSE "LYONER"

		PROTEIN	FAT	CARBOHYDRATES	TOTAL
ENERGY VALUE (AVERAGE)	KJOULE	232	1145	0.0	1377
PER 100 G	(KCAL)	55	274	0.0	329
EDIBLE PORTION					
AMOUNT OF DIGESTIBLE	GRAM	12.22	27.36	0.00	
CONSTITUENTS PER 100 G					
ENERGY VALUE (AVERAGE)	KJOULE	225	1088	0.0	1313
OF THE DIGESTIBLE	(KCAL)	54	260	0.0	314
FRACTION PER 100 G					
EDIBLE PORTION					

WASTE PERCENTAGE AVERAGE 2.0

CONSTITUENTS	DIM	AV	VARIATION			AVR	NUTR. DENS.		MOLPERC.
WATER	GRAM	55.90	53.00	-	62.80	54.78	GRAM/MJ	42.59	
PROTEIN	GRAM	12.60	11.10	-	15.50	12.35	GRAM/MJ	9.60	
FAT	GRAM	28.80	20.80	-	35.00	28.22	GRAM/MJ	21.94	
MINERALS	GRAM	2.68	2.56	-	2.75	2.63	GRAM/MJ	2.04	
CONNECT.TISSUE PROTEIN	GRAM	2.00	1.30	-	2.50	1.96	GRAM/MJ	1.52	

METTWURST (BRAUNSCHWEIGER METTWURST) — GERMAN SAUSAGE (METTWURST) — ANDOUILLE

		PROTEIN	FAT	CARBOHYDRATES	TOTAL
ENERGY VALUE (AVERAGE) PER 100 G EDIBLE PORTION	KJOULE	232	1789	0.0	2021
	(KCAL)	55	428	0.0	483
AMOUNT OF DIGESTIBLE CONSTITUENTS PER 100 G	GRAM	12.22	42.75	0.00	
ENERGY VALUE (AVERAGE) OF THE DIGESTIBLE FRACTION PER 100 G EDIBLE PORTION	KJOULE	225	1699	0.0	1924
	(KCAL)	54	406	0.0	460

WASTE PERCENTAGE AVERAGE 2.0

CONSTITUENTS	DIM	AV	VARIATION	AVR	NUTR. DENS.		MOLPERC.
WATER	GRAM	37.80	34.90 - 40.80	37.04	GRAM/MJ	19.64	
PROTEIN	GRAM	12.60	12.20 - 15.10	12.35	GRAM/MJ	6.55	
FAT	GRAM	45.00	43.10 - 55.00	44.10	GRAM/MJ	23.39	
MINERALS	GRAM	3.30	2.95 - 3.65	3.23	GRAM/MJ	1.71	
SODIUM	GRAM	1.09	0.94 - 1.24	1.07	GRAM/MJ	0.57	
POTASSIUM	MILLI	213.00	207.00 - 217.00	208.74	MILLI/MJ	110.69	
CALCIUM	MILLI	13.00	0.00 - 21.00	12.74	MILLI/MJ	6.76	
IRON	MILLI	1.60	- - -	1.57	MILLI/MJ	0.83	
PHOSPHORUS	MILLI	160.00	- - -	156.80	MILLI/MJ	83.15	
VITAMIN B1	MILLI	0.20	- - -	0.20	MILLI/MJ	0.10	
VITAMIN B2	MILLI	0.15	- - -	0.15	MILLI/MJ	0.08	
NICOTINAMIDE	MILLI	0.25	- - -	0.25	MILLI/MJ	0.13	
ARGININE	GRAM	0.89	- - -	0.87	GRAM/MJ	0.46	
HISTIDINE	GRAM	0.30	- - -	0.29	GRAM/MJ	0.16	
ISOLEUCINE	GRAM	0.61	- - -	0.60	GRAM/MJ	0.32	
LEUCINE	GRAM	1.00	- - -	0.98	GRAM/MJ	0.52	
LYSINE	GRAM	0.71	- - -	0.70	GRAM/MJ	0.37	
METHIONINE	GRAM	0.25	- - -	0.25	GRAM/MJ	0.13	
PHENYLALANINE	GRAM	0.57	- - -	0.56	GRAM/MJ	0.30	
THREONINE	GRAM	0.43	- - -	0.42	GRAM/MJ	0.22	
TRYPTOPHAN	GRAM	0.16	- - -	0.16	GRAM/MJ	0.08	
VALINE	GRAM	0.74	- - -	0.73	GRAM/MJ	0.38	
CONNECT.TISSUE PROTEIN	GRAM	2.27	1.24 - 2.90	2.22	GRAM/MJ	1.18	

MORTADELLA | SAUSAGE "MORTADELLA" | SAUCISSON "MORTADELLA"

		PROTEIN	FAT	CARBOHYDRATES	TOTAL
ENERGY VALUE (AVERAGE) PER 100 G EDIBLE PORTION	KJOULE	228	1304	0.0	1532
	(KCAL)	55	312	0.0	366
AMOUNT OF DIGESTIBLE CONSTITUENTS PER 100 G	GRAM	12.02	31.16	0.00	
ENERGY VALUE (AVERAGE) OF THE DIGESTIBLE FRACTION PER 100 G EDIBLE PORTION	KJOULE	221	1239	0.0	1460
	(KCAL)	53	296	0.0	349

WASTE PERCENTAGE AVERAGE 2.0

CONSTITUENTS	DIM	AV	VARIATION		AVR	NUTR. DENS.		MOLPERC.
WATER	GRAM	52.30	50.40	- 54.80	51.25	GRAM/MJ	35.82	
PROTEIN	GRAM	12.40	11.40	- 13.80	12.15	GRAM/MJ	8.49	
FAT	GRAM	32.80	29.70	- 35.80	32.14	GRAM/MJ	22.47	
MINERALS	GRAM	2.60	1.98	- 3.71	2.55	GRAM/MJ	1.78	
SODIUM	MILLI	668.00	590.00	- 795.00	654.64	MILLI/MJ	457.54	
POTASSIUM	MILLI	207.00	160.00	- 261.00	202.86	MILLI/MJ	141.78	
CALCIUM	MILLI	42.00	33.00	- 50.00	41.16	MILLI/MJ	28.77	
PHOSPHORUS	MILLI	143.00	136.00	- 150.00	140.14	MILLI/MJ	97.95	
CHLORIDE	MILLI	920.00	-	- -	901.60	MILLI/MJ	630.15	
VITAMIN B1	MILLI	0.10	-	- -	0.10	MILLI/MJ	0.07	
VITAMIN B2	MILLI	0.15	-	- -	0.15	MILLI/MJ	0.10	
NICOTINAMIDE	MILLI	3.10	-	- -	3.04	MILLI/MJ	2.12	
VITAMIN C	-	0.00	-	- -	0.00			
TOTAL PURINES	MILLI	130.00 1	-	- -	127.40	MILLI/MJ	89.04	
CONNECT.TISSUE PROTEIN	GRAM	1.95	-	- -	1.91	GRAM/MJ	1.34	

1 VALUE EXPRESSED IN MG URIC ACID/100G

MÜNCHNER WEISSWURST / SAUSAGE „MUNICH WEISSWURST" / BOUDIN BLANC DE MUNICH

		PROTEIN	FAT	CARBOHYDRATES	TOTAL
ENERGY VALUE (AVERAGE)	KJOULE	204	1073	0.0	1278
PER 100 G	(KCAL)	49	257	0.0	305
EDIBLE PORTION					
AMOUNT OF DIGESTIBLE	GRAM	10.76	25.65	0.00	
CONSTITUENTS PER 100 G					
ENERGY VALUE (AVERAGE)	KJOULE	198	1020	0.0	1218
OF THE DIGESTIBLE	(KCAL)	47	244	0.0	291
FRACTION PER 100 G					
EDIBLE PORTION					

WASTE PERCENTAGE AVERAGE 7.0

CONSTITUENTS	DIM	AV	VARIATION			AVR	NUTR. DENS.		MOLPERC.
WATER	GRAM	59.90	57.50	-	65.00	55.71	GRAM/MJ	49.19	
PROTEIN	GRAM	11.10	10.80	-	11.40	10.32	GRAM/MJ	9.12	
FAT	GRAM	27.00	20.00	-	30.00	25.11	GRAM/MJ	22.17	
MINERALS	GRAM	2.00	1.94	-	2.05	1.86	GRAM/MJ	1.64	
SODIUM	MILLI	620.00	-	-	-	576.60	MILLI/MJ	509.13	
POTASSIUM	MILLI	122.00	-	-	-	113.46	MILLI/MJ	100.18	
CALCIUM	MILLI	25.00	-	-	-	23.25	MILLI/MJ	20.53	
VITAMIN B1	MICRO	43.00	28.00	-	59.00	39.99	MICRO/MJ	35.31	
VITAMIN B2	MILLI	0.13	-	-	-	0.12	MILLI/MJ	0.11	
NICOTINAMIDE	MILLI	2.38	-	-	-	2.21	MILLI/MJ	1.95	
CONNECT.TISSUE PROTEIN	GRAM	1.80	-	-	-	1.67	GRAM/MJ	1.48	

PLOCKWURST — SAUSAGE "PLOCKWURST" — SAUCISSON "PLOCKWURST"

		PROTEIN	FAT	CARBOHYDRATES	TOTAL
ENERGY VALUE (AVERAGE) PER 100 G EDIBLE PORTION	KJOULE	355	1789	0.0	2144
	(KCAL)	85	428	0.0	512
AMOUNT OF DIGESTIBLE CONSTITUENTS PER 100 G	GRAM	18.72	42.75	0.00	
ENERGY VALUE (AVERAGE) OF THE DIGESTIBLE FRACTION PER 100 G EDIBLE PORTION	KJOULE	345	1699	0.0	2044
	(KCAL)	82	406	0.0	488

WASTE PERCENTAGE AVERAGE 4.0 MINIMUM 2.0 MAXIMUM 6.0

CONSTITUENTS	DIM	AV	VARIATION			AVR	NUTR. DENS.		MOLPERC.
WATER	GRAM	31.10	20.50	-	34.00	29.86	GRAM/MJ	15.22	
PROTEIN	GRAM	19.30	14.00	-	23.60	18.53	GRAM/MJ	9.44	
FAT	GRAM	45.00	42.00	-	50.00	43.20	GRAM/MJ	22.02	
MINERALS	GRAM	4.64	4.12	-	5.36	4.45	GRAM/MJ	2.27	

REGENSBURGER

SAUSAGE "REGENSBURGER"

SAUCISSON "REGENSBURGER"

		PROTEIN	FAT	CARBOHYDRATES	TOTAL
ENERGY VALUE (AVERAGE)	KJOULE	245	1236	0.0	1481
PER 100 G	(KCAL)	59	295	0.0	354
EDIBLE PORTION					
AMOUNT OF DIGESTIBLE	GRAM	12.90	29.54	0.00	
CONSTITUENTS PER 100 G					
ENERGY VALUE (AVERAGE)	KJOULE	238	1174	0.0	1412
OF THE DIGESTIBLE	(KCAL)	57	281	0.0	337
FRACTION PER 100 G					
EDIBLE PORTION					

WASTE PERCENTAGE AVERAGE 3.0

CONSTITUENTS	DIM	AV	VARIATION			AVR	NUTR. DENS.		MOLPERC.
WATER	GRAM	53.10	49.00	-	58.20	51.51	GRAM/MJ	37.61	
PROTEIN	GRAM	13.30	10.60	-	15.50	12.90	GRAM/MJ	9.42	
FAT	GRAM	31.10	22.60	-	35.00	30.17	GRAM/MJ	22.03	
MINERALS	GRAM	2.53	2.35	-	2.71	2.45	GRAM/MJ	1.79	

SALAMI
DEUTSCHE

SAUSAGE "SALAMI"
GERMAN

SAUCISSON "SALAMI"
ALLEMAND

		PROTEIN	FAT	CARBOHYDRATES	TOTAL
ENERGY VALUE (AVERAGE) PER 100 G EDIBLE PORTION	KJOULE	328	1975	0.0	2303
	(KCAL)	78	472	0.0	550
AMOUNT OF DIGESTIBLE CONSTITUENTS PER 100 G	GRAM	17.26	47.21	0.00	
ENERGY VALUE (AVERAGE) OF THE DIGESTIBLE FRACTION PER 100 G EDIBLE PORTION	KJOULE	318	1877	0.0	2195
	(KCAL)	76	449	0.0	525

WASTE PERCENTAGE AVERAGE 5.0 MINIMUM 2.0 MAXIMUM 8.0

CONSTITUENTS	DIM	AV	VARIATION			AVR	NUTR.	DENS.	MOLPERC.
WATER	GRAM	27.70	-	-	-	26.32	GRAM/MJ	12.62	
PROTEIN	GRAM	17.80	-	-	-	16.91	GRAM/MJ	8.11	
FAT	GRAM	49.70	-	-	-	47.22	GRAM/MJ	22.65	
MINERALS	GRAM	4.59	-	-	-	4.36	GRAM/MJ	2.09	
SODIUM	GRAM	1.26	0.75	-	1.50	1.20	GRAM/MJ	0.57	
POTASSIUM	MILLI	302.00	204.00	-	473.00	286.90	MILLI/MJ	137.61	
CALCIUM	MILLI	35.00	22.00	-	45.00	33.25	MILLI/MJ	15.95	
PHOSPHORUS	MILLI	167.00	-	-	-	158.65	MILLI/MJ	76.10	
CHLORIDE	GRAM	2.39	-	-	-	2.27	GRAM/MJ	1.09	
VITAMIN B1	MILLI	0.18	0.14	-	0.21	0.17	MILLI/MJ	0.08	
VITAMIN B2	MILLI	0.20	0.18	-	0.21	0.19	MILLI/MJ	0.09	
NICOTINAMIDE	MILLI	2.60	2.30	-	2.90	2.47	MILLI/MJ	1.18	
VITAMIN B12	MICRO	1.40	-	-	-	1.33	MICRO/MJ	0.64	
ARGININE	GRAM	1.24	-	-	-	1.18	GRAM/MJ	0.57	
CYSTINE	GRAM	0.22	-	-	-	0.21	GRAM/MJ	0.10	
HISTIDINE	GRAM	0.48	-	-	-	0.46	GRAM/MJ	0.22	
ISOLEUCINE	GRAM	0.86	-	-	-	0.82	GRAM/MJ	0.39	
LEUCINE	GRAM	1.28	-	-	-	1.22	GRAM/MJ	0.58	
LYSINE	GRAM	1.43	-	-	-	1.36	GRAM/MJ	0.65	
METHIONINE	GRAM	0.38	-	-	-	0.36	GRAM/MJ	0.17	
PHENYLALANINE	GRAM	0.65	-	-	-	0.62	GRAM/MJ	0.30	
THREONINE	GRAM	0.73	-	-	-	0.69	GRAM/MJ	0.33	
TRYPTOPHAN	GRAM	0.15	-	-	-	0.14	GRAM/MJ	0.07	
TYROSINE	GRAM	0.58	-	-	-	0.55	GRAM/MJ	0.26	
VALINE	GRAM	0.89	-	-	-	0.85	GRAM/MJ	0.41	

SCHWEINSBRATWURST | FRYING SAUSAGE FROM PORK | S. À GRILLER DE VIANDE DE PORC

		PROTEIN	FAT	CARBOHYDRATES	TOTAL
ENERGY VALUE (AVERAGE) PER 100 G EDIBLE PORTION	KJOULE	234	1288	0.0	1522
	(KCAL)	56	308	0.0	364
AMOUNT OF DIGESTIBLE CONSTITUENTS PER 100 G	GRAM	12.31	30.78	0.00	
ENERGY VALUE (AVERAGE) OF THE DIGESTIBLE FRACTION PER 100 G EDIBLE PORTION	KJOULE	227	1223	0.0	1450
	(KCAL)	54	292	0.0	347

WASTE PERCENTAGE AVERAGE 0.00

CONSTITUENTS	DIM	AV	VARIATION	AVR	NUTR. DENS.		MOLPERC.
WATER	GRAM	52.70	48.80 - 60.30	52.70	GRAM/MJ	36.34	
PROTEIN	GRAM	12.70	11.40 - 15.00	12.70	GRAM/MJ	8.76	
FAT	GRAM	32.40	25.60 - 35.90	32.40	GRAM/MJ	22.34	
MINERALS	GRAM	2.20	1.90 - 2.46	2.20	GRAM/MJ	1.52	
SODIUM	MILLI	520.00	- - -	520.00	MILLI/MJ	358.56	
PHOSPHORUS	MILLI	190.00	- - -	190.00	MILLI/MJ	131.01	
CHLORIDE	MILLI	790.00	- - -	790.00	MILLI/MJ	544.74	
VITAMIN B1	MILLI	0.28	0.27 - 0.29	0.28	MILLI/MJ	0.19	
VITAMIN B2	MILLI	0.22	0.21 - 0.23	0.22	MILLI/MJ	0.15	
NICOTINAMIDE	MILLI	3.20	2.90 - 3.50	3.20	MILLI/MJ	2.21	
CONNECT.TISSUE PROTEIN	GRAM	2.10	- - -	2.10	GRAM/MJ	1.45	

WIENER WÜRSTCHEN VIENNESE SAUSAGES SAUCISSE VIENNOISE

		PROTEIN	FAT	CARBOHYDRATES	TOTAL
ENERGY VALUE (AVERAGE)	KJOULE	274	970	0.0	1244
PER 100 G	(KCAL)	66	232	0.0	297
EDIBLE PORTION					
AMOUNT OF DIGESTIBLE	GRAM	14.45	23.18	0.00	
CONSTITUENTS PER 100 G					
ENERGY VALUE (AVERAGE)	KJOULE	266	921	0.0	1187
OF THE DIGESTIBLE	(KCAL)	64	220	0.0	284
FRACTION PER 100 G					
EDIBLE PORTION					

WASTE PERCENTAGE AVERAGE 1.0

CONSTITUENTS	DIM	AV	VARIATION		AVR	NUTR.	DENS.	MOLPERC.
WATER	GRAM	58.40	57.50	- 64.60	57.82	GRAM/MJ	49.18	
PROTEIN	GRAM	14.90	12.40	- 16.80	14.75	GRAM/MJ	12.55	
FAT	GRAM	24.40	20.80	- 25.00	24.16	GRAM/MJ	20.55	
MINERALS	GRAM	2.67	2.15	- 3.14	2.64	GRAM/MJ	2.25	
SODIUM	MILLI	941.00	881.00	- 950.00	931.59	MILLI/MJ	792.47	
POTASSIUM	MILLI	204.00	187.00	- 220.00	201.96	MILLI/MJ	171.80	
CALCIUM	MILLI	13.00	9.00	- 17.00	12.87	MILLI/MJ	10.95	
IRON	MILLI	2.40	-	- -	2.38	MILLI/MJ	2.02	
PHOSPHORUS	MILLI	170.00	-	- -	168.30	MILLI/MJ	143.17	
VITAMIN B1	MILLI	0.10	-	- -	0.10	MILLI/MJ	0.08	
VITAMIN B2	MILLI	0.12	-	- -	0.12	MILLI/MJ	0.10	
NICOTINAMIDE	MILLI	3.10	-	- -	3.07	MILLI/MJ	2.61	
ARGININE	GRAM	1.03	-	- -	1.02	GRAM/MJ	0.87	
CYSTINE	GRAM	0.25	-	- -	0.25	GRAM/MJ	0.21	
HISTIDINE	GRAM	0.40	-	- -	0.40	GRAM/MJ	0.34	
ISOLEUCINE	GRAM	0.81	-	- -	0.80	GRAM/MJ	0.68	
LEUCINE	GRAM	1.15	-	- -	1.14	GRAM/MJ	0.97	
LYSINE	GRAM	1.15	-	- -	1.14	GRAM/MJ	0.97	
METHIONINE	GRAM	0.39	-	- -	0.39	GRAM/MJ	0.33	
PHENYLALANINE	GRAM	0.61	-	- -	0.60	GRAM/MJ	0.51	
THREONINE	GRAM	0.52	-	- -	0.51	GRAM/MJ	0.44	
TRYPTOPHAN	GRAM	0.15	-	- -	0.15	GRAM/MJ	0.13	
TYROSINE	GRAM	0.49	-	- -	0.49	GRAM/MJ	0.41	
VALINE	GRAM	0.83	-	- -	0.82	GRAM/MJ	0.70	
CONNECT.TISSUE PROTEIN	GRAM	2.00	-	- -	1.98	GRAM/MJ	1.68	

Wild

GAME GIBIER

HASE / HARE / LIÈVRE

HASE	HARE	LIÈVRE
DURCHSCHNITT	AVERAGE	MOYENNE
LEPUS EUROPAEUS PALLAS		

		PROTEIN	FAT	CARBOHYDRATES	TOTAL
ENERGY VALUE (AVERAGE)	KJOULE	398	120	0.0	517
PER 100 G	(KCAL)	95	29	0.0	124
EDIBLE PORTION					
AMOUNT OF DIGESTIBLE	GRAM	20.95	2.85	0.00	
CONSTITUENTS PER 100 G					
ENERGY VALUE (AVERAGE)	KJOULE	386	114	0.0	499
OF THE DIGESTIBLE	(KCAL)	92	27	0.0	119
FRACTION PER 100 G					
EDIBLE PORTION					

WASTE PERCENTAGE AVERAGE 20

CONSTITUENTS	DIM	AV	VARIATION			AVR	NUTR. DENS.		MOLPERC.
WATER	GRAM	73.30	73.00	-	74.20	58.64	GRAM/MJ	146.78	
PROTEIN	GRAM	21.60	20.00	-	23.00	17.28	GRAM/MJ	43.25	
FAT	GRAM	3.01	0.90	-	5.00	2.41	GRAM/MJ	6.03	
MINERALS	GRAM	1.18	-	-	-	0.94	GRAM/MJ	2.36	
SODIUM	MILLI	50.00	-	-	-	40.00	MILLI/MJ	100.12	
POTASSIUM	MILLI	400.00	-	-	-	320.00	MILLI/MJ	801.00	
CALCIUM	MILLI	9.05	7.15	-	10.00	7.24	MILLI/MJ	18.12	
IRON	MILLI	2.40	2.00	-	3.20	1.92	MILLI/MJ	4.81	
PHOSPHORUS	MILLI	220.00	200.00	-	239.00	176.00	MILLI/MJ	440.55	
VITAMIN B1	MICRO	90.00	50.00	-	120.00	72.00	MICRO/MJ	180.22	
VITAMIN B2	MICRO	60.00	-	-	-	48.00	MICRO/MJ	120.15	
NICOTINAMIDE	MILLI	8.07	5.00	-	12.70	6.46	MILLI/MJ	16.16	
PANTOTHENIC ACID	MILLI	0.80	0.80	-	0.80	0.64	MILLI/MJ	1.60	
VITAMIN B6	MILLI	0.30	0.30	-	0.30	0.24	MILLI/MJ	0.60	
BIOTIN	-	TRACES	TRACES	-					
FOLIC ACID	MICRO	5.00	5.00	-	5.00	4.00	MICRO/MJ	10.01	
VITAMIN B12	MICRO	1.00	1.00	-	1.00	0.80	MICRO/MJ	2.00	
TRYPTOPHAN	GRAM	0.24	0.24	-	0.24	0.19	GRAM/MJ	0.48	
CHOLESTEROL	MILLI	65.00	-	-	-	52.00	MILLI/MJ	130.16	
TOTAL PURINES	MILLI	105.00 [1]	-	-	-	84.00	MILLI/MJ	210.26	

1 VALUE EXPRESSED IN MG URIC ACID/100G

HIRSCHFLEISCH
DURCHSCHNITT
CERVUS ELAPHUS L.

VENISON
AVERAGE

VIANDE DE CERF
VERNAISON
MOYENNE

		PROTEIN	FAT	CARBOHYDRATES	TOTAL
ENERGY VALUE (AVERAGE)	KJOULE	379	133	0.0	512
PER 100 G	(KCAL)	91	32	0.0	122
EDIBLE PORTION					
AMOUNT OF DIGESTIBLE	GRAM	19.98	3.17	0.00	
CONSTITUENTS PER 100 G					
ENERGY VALUE (AVERAGE)	KJOULE	368	126	0.0	494
OF THE DIGESTIBLE	(KCAL)	88	30	0.0	118
FRACTION PER 100 G					
EDIBLE PORTION					

WASTE PERCENTAGE AVERAGE 21

CONSTITUENTS	DIM	AV	VARIATION		AVR	NUTR. DENS.		MOLPERC.
WATER	GRAM	74.70	72.90	- 75.30	59.01	GRAM/MJ	151.22	
PROTEIN	GRAM	20.60	18.10	- 22.20	16.27	GRAM/MJ	41.70	
FAT	GRAM	3.34	1.32	- 5.40	2.64	GRAM/MJ	6.76	
MINERALS	GRAM	1.02	0.91	- 1.12	0.81	GRAM/MJ	2.06	
SODIUM	MILLI	61.00	54.00	- 70.00	48.19	MILLI/MJ	123.49	
POTASSIUM	MILLI	330.00	296.00	- 360.00	260.70	MILLI/MJ	668.04	
MAGNESIUM	MILLI	29.00	-	- -	22.91	MILLI/MJ	58.71	
CALCIUM	MILLI	7.00	5.00	- 9.00	5.53	MILLI/MJ	14.17	
PHOSPHORUS	MILLI	249.00	-	- -	196.71	MILLI/MJ	504.07	
CHLORIDE	MILLI	40.00	-	- -	31.60	MILLI/MJ	80.97	
VITAMIN B2	MILLI	0.25	-	- -	0.20	MILLI/MJ	0.51	

REHFLEISCH
KEULE
(SCHLEGEL)

VENISON
HAUNCH
(LEG)

VIANDE DE CHEVREUIL
CUISSOT

		PROTEIN	FAT	CARBOHYDRATES	TOTAL
ENERGY VALUE (AVERAGE) PER 100 G EDIBLE PORTION	KJOULE	394	50	0.0	444
	(KCAL)	94	12	0.0	106
AMOUNT OF DIGESTIBLE CONSTITUENTS PER 100 G	GRAM	20.75	1.18	0.00	
ENERGY VALUE (AVERAGE) OF THE DIGESTIBLE FRACTION PER 100 G EDIBLE PORTION	KJOULE	382	47	0.0	429
	(KCAL)	91	11	0.0	103

WASTE PERCENTAGE AVERAGE 18 MINIMUM 16 MAXIMUM 21

CONSTITUENTS	DIM	AV	VARIATION			AVR	NUTR. DENS.		MOLPERC.
WATER	GRAM	75.70	75.00	-	76.40	62.07	GRAM/MJ	176.31	
PROTEIN	GRAM	21.40	21.20	-	21.70	17.55	GRAM/MJ	49.84	
FAT	GRAM	1.25	1.16	-	1.34	1.03	GRAM/MJ	2.91	
MINERALS	GRAM	1.01	0.95	-	1.06	0.83	GRAM/MJ	2.35	
SODIUM	MILLI	60.00	57.00	-	63.00	49.20	MILLI/MJ	139.75	
POTASSIUM	MILLI	309.00	283.00	-	335.00	253.38	MILLI/MJ	719.70	
CALCIUM	MILLI	5.00	1.00	-	11.00	4.10	MILLI/MJ	11.65	
IRON	MILLI	3.00	-	-	-	2.46	MILLI/MJ	6.99	
PHOSPHORUS	MILLI	220.00	-	-	-	180.40	MILLI/MJ	512.41	
CHLORIDE	MILLI	40.00	-	-	-	32.80	MILLI/MJ	93.16	
VITAMIN B2	MILLI	0.25	-	-	-	0.21	MILLI/MJ	0.58	
TOTAL PURINES	MILLI	105.00 [1]	-	-	-	86.10	MILLI/MJ	244.56	

1 VALUE EXPRESSED IN MG URIC ACID/100G

REHFLEISCH
RÜCKEN

VENISON
BACK

VIANDE DE CHEVREUIL
SELLE

		PROTEIN	FAT	CARBOHYDRATES	TOTAL
ENERGY VALUE (AVERAGE) PER 100 G EDIBLE PORTION	KJOULE	412	141	0.0	553
	(KCAL)	99	34	0.0	132
AMOUNT OF DIGESTIBLE CONSTITUENTS PER 100 G	GRAM	21.72	3.37	0.00	
ENERGY VALUE (AVERAGE) OF THE DIGESTIBLE FRACTION PER 100 G EDIBLE PORTION	KJOULE	400	134	0.0	534
	(KCAL)	96	32	0.0	128

WASTE PERCENTAGE AVERAGE 30

CONSTITUENTS	DIM	AV	VARIATION			AVR	NUTR.	DENS.	MOLPERC.
WATER	GRAM	72.20	70.30	-	74.20	50.54	GRAM/MJ	135.19	
PROTEIN	GRAM	22.40	21.70	-	23.40	15.68	GRAM/MJ	41.94	
FAT	GRAM	3.55	0.66	-	6.43	2.49	GRAM/MJ	6.65	
MINERALS	GRAM	1.19	1.11	-	1.27	0.83	GRAM/MJ	2.23	
SODIUM	MILLI	84.00	71.00	-	103.00	58.80	MILLI/MJ	157.29	
POTASSIUM	MILLI	342.00	312.00	-	378.00	239.40	MILLI/MJ	640.38	
CALCIUM	MILLI	25.00	-	-	-	17.50	MILLI/MJ	46.81	
IRON	MILLI	3.00	-	-	-	2.10	MILLI/MJ	5.62	
PHOSPHORUS	MILLI	220.00	-	-	-	154.00	MILLI/MJ	411.94	
CHLORIDE	MILLI	40.00	-	-	-	28.00	MILLI/MJ	74.90	
VITAMIN B2	MILLI	0.25	-	-	-	0.18	MILLI/MJ	0.47	
TOTAL PURINES	MILLI	105.00 [1]	-	-	-	73.50	MILLI/MJ	196.61	

1 VALUE EXPRESSED IN MG URIC ACID/100G

WILDSCHWEIN-FLEISCH
DURCHSCHNITT

WILD BOAR MEAT
AVERAGE

SANGLIER VIANDE
MOYENNE

		PROTEIN	FAT	CARBOHYDRATES	TOTAL
ENERGY VALUE (AVERAGE)	KJOULE	359	134	0.0	493
PER 100 G	(KCAL)	86	32	0.0	118
EDIBLE PORTION					
AMOUNT OF DIGESTIBLE	GRAM	18.91	3.21	0.00	
CONSTITUENTS PER 100 G					
ENERGY VALUE (AVERAGE)	KJOULE	348	128	0.0	476
OF THE DIGESTIBLE	(KCAL)	83	31	0.0	114
FRACTION PER 100 G					
EDIBLE PORTION					

WASTE PERCENTAGE AVERAGE 0.00

CONSTITUENTS	DIM	AV	VARIATION			AVR	NUTR. DENS.		MOLPERC.
WATER	GRAM	74.70	74.70	-	74.70	74.70	GRAM/MJ	156.98	
PROTEIN	GRAM	19.50	19.50	-	19.50	19.50	GRAM/MJ	40.98	
FAT	GRAM	3.38	3.38	-	3.38	3.38	GRAM/MJ	7.10	
MINERALS	GRAM	2.39	2.39	-	2.39	2.39	GRAM/MJ	5.02	
MYRISTIC ACID	MILLI	50.00	50.00	-	50.00	50.00	MILLI/MJ	105.08	
PALMITIC ACID	GRAM	0.79	0.79	-	0.79	0.79	GRAM/MJ	1.66	
STEARIC ACID	GRAM	0.34	0.34	-	0.34	0.34	GRAM/MJ	0.71	
PALMITOLEIC ACID	GRAM	0.71	0.71	-	0.71	0.71	GRAM/MJ	1.49	
OLEIC ACID	GRAM	1.57	1.57	-	1.57	1.57	GRAM/MJ	3.30	
LINOLEIC ACID	GRAM	0.26	0.26	-	0.26	0.26	GRAM/MJ	0.55	
LINOLENIC ACID	-	TRACES	TRACES	-					
CONNECT.TISSUE PROTEIN	GRAM	2.39	2.39	-	2.39	2.39	GRAM/MJ	5.02	

Geflügel

POULTRY **VOLAILLE**

ENTE / DUCK / CANARD

ENTE	DUCK	CANARD
DURCHSCHNITT	AVERAGE	MOYENNE
ANAS BOSCHAS L.		

		PROTEIN	FAT	CARBOHYDRATES	TOTAL
ENERGY VALUE (AVERAGE)	KJOULE	333	684	0.0	1017
PER 100 G EDIBLE PORTION	(KCAL)	80	163	0.0	243
AMOUNT OF DIGESTIBLE CONSTITUENTS PER 100 G	GRAM	17.55	16.34	0.00	
ENERGY VALUE (AVERAGE)	KJOULE	323	649	0.0	973
OF THE DIGESTIBLE FRACTION PER 100 G EDIBLE PORTION	(KCAL)	77	155	0.0	232

WASTE PERCENTAGE AVERAGE 20 MINIMUM 14 MAXIMUM 25

CONSTITUENTS	DIM	AV	VARIATION			AVR	NUTR.	DENS.	MOLPERC.
WATER	GRAM	63.70	54.30	-	73.00	50.96	GRAM/MJ	65.49	
PROTEIN	GRAM	18.10	16.00	-	20.80	14.48	GRAM/MJ	18.61	
FAT	GRAM	17.20	6.00	-	28.60	13.76	GRAM/MJ	17.68	
MINERALS	GRAM	1.00	-	-	-	0.80	GRAM/MJ	1.03	
SODIUM	MILLI	-	80.00	-	200.00				
POTASSIUM	MILLI	292.00	285.00	-	300.00	233.60	MILLI/MJ	300.20	
CALCIUM	MILLI	11.00	9.00	-	15.00	8.80	MILLI/MJ	11.31	
MANGANESE	MICRO	30.00	-	-	-	24.00	MICRO/MJ	30.84	
IRON	MILLI	2.10	1.80	-	2.40	1.68	MILLI/MJ	2.16	
COPPER	MILLI	0.45	0.41	-	0.50	0.36	MILLI/MJ	0.46	
PHOSPHORUS	MILLI	187.00	172.00	-	200.00	149.60	MILLI/MJ	192.25	
CHLORIDE	MILLI	85.00	-	-	-	68.00	MILLI/MJ	87.39	
VITAMIN B1	MILLI	0.30	-	-	-	0.24	MILLI/MJ	0.31	
VITAMIN B2	MILLI	0.20	-	-	-	0.16	MILLI/MJ	0.21	
NICOTINAMIDE	MILLI	3.50	-	-	-	2.80	MILLI/MJ	3.60	
ARGININE	GRAM	1.10	1.09	-	1.10	0.88	GRAM/MJ	1.13	
HISTIDINE	GRAM	0.41	-	-	-	0.33	GRAM/MJ	0.42	
ISOLEUCINE	GRAM	0.94	0.93	-	0.95	0.75	GRAM/MJ	0.97	
LEUCINE	GRAM	1.40	-	-	-	1.12	GRAM/MJ	1.44	
LYSINE	GRAM	1.56	1.53	-	1.60	1.25	GRAM/MJ	1.60	
METHIONINE	GRAM	0.45	-	-	-	0.36	GRAM/MJ	0.46	
PHENYLALANINE	GRAM	0.71	0.71	-	0.72	0.57	GRAM/MJ	0.73	
THREONINE	GRAM	0.79	0.78	-	0.80	0.63	GRAM/MJ	0.81	
VALINE	GRAM	0.87	0.85	-	0.89	0.70	GRAM/MJ	0.89	
TOTAL PURINES	MILLI	110.00 [1]	-	-	-	88.00	MILLI/MJ	113.09	

1 VALUE EXPRESSED IN MG URIC ACID/100G

FASAN
DURCHSCHNITT
PHASIANUS VERSICOLOR

PHEASANT
AVERAGE

FAISAN
MOYENNE

		PROTEIN	FAT	CARBOHYDRATES	TOTAL
ENERGY VALUE (AVERAGE) PER 100 G EDIBLE PORTION	KJOULE	437	260	0.0	698
	(KCAL)	105	62	0.0	167
AMOUNT OF DIGESTIBLE CONSTITUENTS PER 100 G	GRAM	23.03	6.22	0.00	
ENERGY VALUE (AVERAGE) OF THE DIGESTIBLE FRACTION PER 100 G EDIBLE PORTION	KJOULE	424	247	0.0	671
	(KCAL)	101	59	0.0	160

WASTE PERCENTAGE AVERAGE 14

CONSTITUENTS	DIM	AV	VARIATION		AVR	NUTR. DENS.		MOLPERC.
WATER	GRAM	68.90	67.80	- 70.00	59.25	GRAM/MJ	102.61	
PROTEIN	GRAM	23.75	22.70	- 24.80	20.43	GRAM/MJ	35.37	
FAT	GRAM	6.55	3.80	- 9.30	5.63	GRAM/MJ	9.76	
MINERALS	GRAM	1.33	1.27	- 1.40	1.15	GRAM/MJ	1.99	
SODIUM	MILLI	40.00	40.00	- 40.00	34.40	MILLI/MJ	59.57	
POTASSIUM	MILLI	243.00	243.00	- 243.00	208.98	MILLI/MJ	361.91	
MAGNESIUM	MILLI	20.00	20.00	- 20.00	17.20	MILLI/MJ	29.79	
CALCIUM	MILLI	11.00	10.00	- 12.00	9.46	MILLI/MJ	16.38	
MANGANESE	MICRO	17.00	17.00	- 17.00	14.62	MICRO/MJ	25.32	
IRON	MILLI	0.40	0.40	- 0.40	0.34	MILLI/MJ	0.60	
COPPER	MICRO	65.00	65.00	- 65.00	55.90	MICRO/MJ	96.81	
ZINC	MILLI	0.96	0.96	- 0.96	0.83	MILLI/MJ	1.43	
PHOSPHORUS	MILLI	250.50	214.00	- 287.00	215.43	MILLI/MJ	373.08	
VITAMIN B1	MICRO	85.00	70.00	- 100.00	73.10	MICRO/MJ	126.59	
VITAMIN B2	MILLI	0.13	0.13	- 0.14	0.12	MILLI/MJ	0.20	
NICOTINAMIDE	MILLI	5.05	3.70	- 6.40	4.34	MILLI/MJ	7.52	
PANTOTHENIC ACID	MILLI	0.93	0.93	- 0.93	0.80	MILLI/MJ	1.39	
VITAMIN B6	MILLI	0.66	0.66	- 0.66	0.57	MILLI/MJ	0.98	
VITAMIN B12	MICRO	0.80	0.80	- 0.80	0.69	MICRO/MJ	1.19	

GANS
DURCHSCHNITT
ANSER ANSER L.

GOOSE
AVERAGE

OIE
MOYENNE

		PROTEIN	FAT	CARBOHYDRATES	TOTAL
ENERGY VALUE (AVERAGE)	KJOULE	289	1232	0.0	1521
PER 100 G EDIBLE PORTION	(KCAL)	69	295	0.0	364
AMOUNT OF DIGESTIBLE CONSTITUENTS PER 100 G	GRAM	15.22	29.45	0.00	
ENERGY VALUE (AVERAGE)	KJOULE	280	1171	0.0	1451
OF THE DIGESTIBLE FRACTION PER 100 G EDIBLE PORTION	(KCAL)	67	280	0.0	347

WASTE PERCENTAGE AVERAGE 37 MINIMUM 33 MAXIMUM 41

CONSTITUENTS	DIM	AV	VARIATION			AVR	NUTR.	DENS.	MOLPERC.
WATER	GRAM	52.40	51.00	-	53.40	33.01	GRAM/MJ	36.11	
PROTEIN	GRAM	15.70	14.90	-	16.40	9.89	GRAM/MJ	10.82	
FAT	GRAM	31.00	26.10	-	31.50	19.53	GRAM/MJ	21.37	
MINERALS	GRAM	0.90	-	-	-	0.57	GRAM/MJ	0.62	
SODIUM	MILLI	86.00	76.00	-	96.00	54.18	MILLI/MJ	59.27	
POTASSIUM	MILLI	420.00	-	-	-	264.60	MILLI/MJ	289.47	
MAGNESIUM	MILLI	23.00	23.00	-	23.00	14.49	MILLI/MJ	15.85	
CALCIUM	MILLI	12.00	9.00	-	15.00	7.56	MILLI/MJ	8.27	
MANGANESE	MICRO	50.00	-	-	-	31.50	MICRO/MJ	34.46	
IRON	MILLI	1.90	1.80	-	2.00	1.20	MILLI/MJ	1.31	
COPPER	MILLI	0.33	-	-	-	0.21	MILLI/MJ	0.23	
ZINC	MILLI	1.30	1.30	-	1.30	0.82	MILLI/MJ	0.90	
PHOSPHORUS	MILLI	184.00	180.00	-	188.00	115.92	MILLI/MJ	126.81	
CHLORIDE	MILLI	120.00	-	-	-	75.60	MILLI/MJ	82.71	
IODIDE	MICRO	4.00	-	-	-	2.52	MICRO/MJ	2.76	
SELENIUM	MICRO	23.00	23.00	-	23.00	14.49	MICRO/MJ	15.85	
VITAMIN A	MICRO	65.00	44.00	-	97.00	40.95	MICRO/MJ	44.80	
VITAMIN B1	MILLI	0.12	0.08	-	0.16	0.08	MILLI/MJ	0.08	
VITAMIN B2	MILLI	0.26	0.19	-	0.35	0.16	MILLI/MJ	0.18	
NICOTINAMIDE	MILLI	6.40	5.60	-	8.00	4.03	MILLI/MJ	4.41	
VITAMIN B6	MILLI	0.58	-	-	-	0.37	MILLI/MJ	0.40	
FOLIC ACID	MICRO	4.00	4.00	-	4.00	2.52	MICRO/MJ	2.76	
ALANINE	GRAM	0.97	0.97	-	0.97	0.61	GRAM/MJ	0.67	
ARGININE	GRAM	0.98	0.98	-	0.98	0.62	GRAM/MJ	0.68	
ASPARTIC ACID	GRAM	1.41	1.41	-	1.41	0.89	GRAM/MJ	0.97	
GLUTAMIC ACID	GRAM	2.34	2.34	-	2.34	1.47	GRAM/MJ	1.61	
GLYCINE	GRAM	0.99	0.99	-	0.99	0.62	GRAM/MJ	0.68	
HISTIDINE	GRAM	0.44	0.44	-	0.44	0.28	GRAM/MJ	0.30	
ISOLEUCINE	GRAM	0.74	0.74	-	0.74	0.47	GRAM/MJ	0.51	
LEUCINE	GRAM	1.32	1.32	-	1.32	0.83	GRAM/MJ	0.91	
LYSINE	GRAM	1.24	1.24	-	1.24	0.78	GRAM/MJ	0.85	

CONSTITUENTS	DIM	AV	VARIATION			AVR	NUTR. DENS.		MOLPERC.
METHIONINE	GRAM	0.38	0.38	-	0.38	0.24	GRAM/MJ	0.26	
PHENYLALANINE	GRAM	0.66	0.66	-	0.66	0.42	GRAM/MJ	0.45	
PROLINE	GRAM	0.76	0.76	-	0.76	0.48	GRAM/MJ	0.52	
SERINE	GRAM	0.62	0.62	-	0.62	0.39	GRAM/MJ	0.43	
THREONINE	GRAM	0.70	0.70	-	0.70	0.44	GRAM/MJ	0.48	
TRYPTOPHAN	GRAM	0.20	0.20	-	0.20	0.13	GRAM/MJ	0.14	
TYROSINE	GRAM	0.51	0.51	-	0.51	0.32	GRAM/MJ	0.35	
VALINE	GRAM	0.97	0.97	-	0.97	0.61	GRAM/MJ	0.67	
CHOLESTEROL	MILLI	86.00	86.00	-	86.00	54.18	MILLI/MJ	59.27	
TOTAL PURINES	MILLI	165.00 1	-	-	-	103.95	MILLI/MJ	113.72	

1 VALUE EXPRESSED IN MG URIC ACID/100G

HUHN / CHICKEN / POULE

HUHN
(BRATHUHN)
DURCHSCHNITT
GALLUS DOMESTICUS L.

CHICKEN
CHICKEN FOR ROASTING
AVERAGE

POULE
POULET POUR RÔTIR
MOYENNE

		PROTEIN	FAT	CARBOHYDRATES	TOTAL
ENERGY VALUE (AVERAGE) PER 100 G EDIBLE PORTION	KJOULE	379	223	0.0	602
	(KCAL)	91	53	0.0	144
AMOUNT OF DIGESTIBLE CONSTITUENTS PER 100 G	GRAM	19.98	5.32	0.00	
ENERGY VALUE (AVERAGE) OF THE DIGESTIBLE FRACTION PER 100 G EDIBLE PORTION	KJOULE	368	211	0.0	579
	(KCAL)	88	51	0.0	138

WASTE PERCENTAGE AVERAGE 26 MINIMUM 23 MAXIMUM 30

CONSTITUENTS	DIM	AV	VARIATION			AVR	NUTR. DENS.		MOLPERC.
WATER	GRAM	72.70	71.20	-	74.00	53.80	GRAM/MJ	125.49	
PROTEIN	GRAM	20.60	-	-	-	15.24	GRAM/MJ	35.56	
FAT	GRAM	5.60	4.50	-	7.20	4.14	GRAM/MJ	9.67	
MINERALS	GRAM	1.10	-	-	-	0.81	GRAM/MJ	1.90	

CONSTITUENTS	DIM	AV	VARIATION			AVR	NUTR. DENS.		MOLPERC.
SODIUM	MILLI	82.50	75.00	-	100.00	61.05	MILLI/MJ	142.41	
POTASSIUM	MILLI	359.00	343.00	-	470.00	265.66	MILLI/MJ	619.69	
MAGNESIUM	MILLI	37.00	-	-	-	27.38	MILLI/MJ	63.87	
CALCIUM	MILLI	12.00	10.00	-	16.00	8.88	MILLI/MJ	20.71	
MANGANESE	MICRO	20.00	-	-	-	14.80	MICRO/MJ	34.52	
IRON	MILLI	1.80	1.50	-	2.00	1.33	MILLI/MJ	3.11	
COPPER	MILLI	0.30	-	-	-	0.22	MILLI/MJ	0.52	
ZINC	MILLI	0.85	-	-	-	0.63	MILLI/MJ	1.47	
MOLYBDENUM	MICRO	40.00	20.00	-	60.00	29.60	MICRO/MJ	69.05	
PHOSPHORUS	MILLI	200.00	-	-	-	148.00	MILLI/MJ	345.23	
CHLORIDE	MILLI	85.00	-	-	-	62.90	MILLI/MJ	146.72	
FLUORIDE	MICRO	33.00	-	-	-	24.42	MICRO/MJ	56.96	
BORON	MICRO	19.00	14.00	-	24.00	14.06	MICRO/MJ	32.80	
VITAMIN A	MICRO	39.00	10.00	-	44.00	28.86	MICRO/MJ	67.32	
VITAMIN E ACTIVITY	MILLI	0.10	-	-	-	0.07	MILLI/MJ	0.17	
ALPHA-TOCOPHEROL	MILLI	0.10	-	-	-	0.07	MILLI/MJ	0.17	
VITAMIN B1	MICRO	83.00	50.00	-	130.00	61.42	MICRO/MJ	143.27	
VITAMIN B2	MILLI	0.16	0.11	-	0.25	0.12	MILLI/MJ	0.28	
NICOTINAMIDE	MILLI	6.80	3.70	-	10.20	5.03	MILLI/MJ	11.74	
PANTOTHENIC ACID	MILLI	0.96	0.82	-	1.24	0.71	MILLI/MJ	1.66	
VITAMIN B6	MILLI	0.50	0.35	-	0.85	0.37	MILLI/MJ	0.86	
BIOTIN	MICRO	2.00	-	-	-	1.48	MICRO/MJ	3.45	
FOLIC ACID	MICRO	9.00	6.00	-	11.00	6.66	MICRO/MJ	15.54	
VITAMIN B12	MICRO	0.50	0.20	-	0.60	0.37	MICRO/MJ	0.86	
VITAMIN C	MILLI	2.50	-	-	-	1.85	MILLI/MJ	4.32	
ALANINE	GRAM	1.49	1.44	-	1.53	1.10	GRAM/MJ	2.57	9.0
ARGININE	GRAM	1.44	1.33	-	1.63	1.07	GRAM/MJ	2.49	4.5
ASPARTIC ACID	GRAM	2.35	2.23	-	2.43	1.74	GRAM/MJ	4.06	9.5
CYSTINE	GRAM	0.31	0.29	-	0.34	0.23	GRAM/MJ	0.54	0.7
GLUTAMIC ACID	GRAM	3.82	3.49	-	4.04	2.83	GRAM/MJ	6.59	14.0
GLYCINE	GRAM	1.45	1.10	-	1.87	1.07	GRAM/MJ	2.50	10.4
HISTIDINE	GRAM	0.63	0.49	-	0.86	0.47	GRAM/MJ	1.09	2.2
ISOLEUCINE	GRAM	1.33	1.14	-	1.54	0.98	GRAM/MJ	2.30	5.5
LEUCINE	GRAM	1.84	1.70	-	1.99	1.36	GRAM/MJ	3.18	7.6
LYSINE	GRAM	2.11	1.76	-	2.34	1.56	GRAM/MJ	3.64	7.8
METHIONINE	GRAM	0.66	0.52	-	0.78	0.49	GRAM/MJ	1.14	2.4
PHENYLALANINE	GRAM	0.94	0.87	-	1.07	0.70	GRAM/MJ	1.62	3.1
PROLINE	GRAM	1.09	1.03	-	1.16	0.81	GRAM/MJ	1.88	5.1
SERINE	GRAM	0.95	0.87	-	1.03	0.70	GRAM/MJ	1.64	4.9
THREONINE	GRAM	1.04	0.91	-	1.18	0.77	GRAM/MJ	1.80	4.7
TRYPTOPHAN	GRAM	0.29	0.22	-	0.88	0.21	GRAM/MJ	0.50	0.8
TYROSINE	GRAM	0.79	0.69	-	0.87	0.58	GRAM/MJ	1.36	2.4
VALINE	GRAM	1.22	1.15	-	1.35	0.90	GRAM/MJ	2.11	5.6
OLEIC ACID	GRAM	1.61	-	-	-	1.19	GRAM/MJ	2.78	
LINOLEIC ACID	GRAM	1.16	1.00	-	1.28	0.86	GRAM/MJ	2.00	
LINOLENIC ACID	MILLI	86.00	-	-	-	63.64	MILLI/MJ	148.45	
CHOLESTEROL	MILLI	81.00	-	-	-	59.94	MILLI/MJ	139.82	
TOTAL PURINES	MILLI	115.00 1	-	-	-	85.10	MILLI/MJ	198.51	

1 VALUE EXPRESSED IN MG URIC ACID/100G

HUHN (SUPPENHUHN) DURCHSCHNITT

CHICKEN (BOILING FOWL) AVERAGE

POULE AU POT MOYENNE

		PROTEIN	FAT	CARBOHYDRATES	TOTAL
ENERGY VALUE (AVERAGE)	KJOULE	341	807	0.0	1147
PER 100 G	(KCAL)	81	193	0.0	274
EDIBLE PORTION					
AMOUNT OF DIGESTIBLE	GRAM	17.94	19.28	0.00	
CONSTITUENTS PER 100 G					
ENERGY VALUE (AVERAGE)	KJOULE	330	767	0.0	1097
OF THE DIGESTIBLE	(KCAL)	79	183	0.0	262
FRACTION PER 100 G					
EDIBLE PORTION					

WASTE PERCENTAGE AVERAGE 27

CONSTITUENTS	DIM	AV	VARIATION		AVR	NUTR. DENS.		MOLPERC.
WATER	GRAM	60.00	56.90	- 61.70	43.80	GRAM/MJ	54.70	
PROTEIN	GRAM	18.50	17.40	- 19.00	13.51	GRAM/MJ	16.87	
FAT	GRAM	20.30	11.30	- 24.80	14.82	GRAM/MJ	18.51	
MINERALS	GRAM	0.93	0.90	- 1.00	0.68	GRAM/MJ	0.85	
CALCIUM	MILLI	11.00	10.00	- 11.00	8.03	MILLI/MJ	10.03	
IRON	MILLI	1.40	1.30	- 1.50	1.02	MILLI/MJ	1.28	
PHOSPHORUS	MILLI	178.00	167.00	- 185.00	129.94	MILLI/MJ	162.28	
VITAMIN A	MICRO	32.00	-	- -	23.36	MICRO/MJ	29.17	
VITAMIN B1	MICRO	60.00	60.00	- 70.00	43.80	MICRO/MJ	54.70	
VITAMIN B2	MILLI	0.17	0.13	- 0.20	0.12	MILLI/MJ	0.15	
NICOTINAMIDE	MILLI	8.80	8.20	- 9.20	6.42	MILLI/MJ	8.02	
ARGININE	GRAM	1.17	-	- -	0.85	GRAM/MJ	1.07	
CYSTINE	GRAM	0.25	-	- -	0.18	GRAM/MJ	0.23	
HISTIDINE	GRAM	0.53	-	- -	0.39	GRAM/MJ	0.48	
ISOLEUCINE	GRAM	0.98	-	- -	0.72	GRAM/MJ	0.89	
LEUCINE	GRAM	1.34	-	- -	0.98	GRAM/MJ	1.22	
LYSINE	GRAM	1.63	-	- -	1.19	GRAM/MJ	1.49	
METHIONINE	GRAM	0.48	-	- -	0.35	GRAM/MJ	0.44	
PHENYLALANINE	GRAM	0.73	-	- -	0.53	GRAM/MJ	0.67	
THREONINE	GRAM	0.79	-	- -	0.58	GRAM/MJ	0.72	
TRYPTOPHAN	GRAM	0.23	-	- -	0.17	GRAM/MJ	0.21	
TYROSINE	GRAM	0.65	-	- -	0.47	GRAM/MJ	0.59	
VALINE	GRAM	0.91	-	- -	0.66	GRAM/MJ	0.83	

HUHN
BRUST

CHICKEN
BREAST

POULE
POITRINE

		PROTEIN	FAT	CARBOHYDRATES	TOTAL
ENERGY VALUE (AVERAGE) PER 100 G EDIBLE PORTION	KJOULE	420	38	0.0	457
	(KCAL)	100	9.0	0.0	109
AMOUNT OF DIGESTIBLE CONSTITUENTS PER 100 G	GRAM	22.11	0.90	0.00	
ENERGY VALUE (AVERAGE) OF THE DIGESTIBLE FRACTION PER 100 G EDIBLE PORTION	KJOULE	407	36	0.0	443
	(KCAL)	97	8.6	0.0	106

WASTE PERCENTAGE AVERAGE 28 MINIMUM 27 MAXIMUM 29

CONSTITUENTS	DIM	AV	VARIATION			AVR	NUTR.	DENS.	MOLPERC.
WATER	GRAM	75.00	74.90	-	75.10	54.00	GRAM/MJ	169.29	
PROTEIN	GRAM	22.80	22.40	-	23.30	16.42	GRAM/MJ	51.47	
FAT	GRAM	0.95	0.50	-	1.30	0.68	GRAM/MJ	2.14	
MINERALS	GRAM	1.20	1.10	-	1.20	0.86	GRAM/MJ	2.71	
SODIUM	MILLI	66.00	54.00	-	78.00	47.52	MILLI/MJ	148.98	
POTASSIUM	MILLI	264.00	208.00	-	320.00	190.08	MILLI/MJ	595.91	
CALCIUM	MILLI	14.00	-	-	-	10.08	MILLI/MJ	31.60	
IRON	MILLI	1.10	-	-	-	0.79	MILLI/MJ	2.48	
MOLYBDENUM	-	0.00	-	-	-	0.00			
PHOSPHORUS	MILLI	212.00	-	-	-	152.64	MILLI/MJ	478.53	
SELENIUM	MICRO	11.60	10.60	-	12.50	8.35	MICRO/MJ	26.18	
ALPHA-TOCOPHEROL	MILLI	0.25	0.10	-	0.40	0.18	MILLI/MJ	0.56	
VITAMIN B1	MICRO	70.00	-	-	-	50.40	MICRO/MJ	158.01	
VITAMIN B2	MICRO	90.00	-	-	-	64.80	MICRO/MJ	203.15	
NICOTINAMIDE	MILLI	10.50	-	-	-	7.56	MILLI/MJ	23.70	
PANTOTHENIC ACID	MILLI	0.84	0.82	-	0.86	0.60	MILLI/MJ	1.90	
FOLIC ACID	MICRO	9.00	6.00	-	11.00	6.48	MICRO/MJ	20.32	
VITAMIN B12	MICRO	0.40	-	-	-	0.29	MICRO/MJ	0.90	
VITAMIN C	-	0.00	-	-	-	0.00			
ARGININE	GRAM	1.35	-	-	-	0.97	GRAM/MJ	3.05	
HISTIDINE	GRAM	0.84	-	-	-	0.60	GRAM/MJ	1.90	
ISOLEUCINE	GRAM	1.21	-	-	-	0.87	GRAM/MJ	2.73	
LEUCINE	GRAM	1.57	-	-	-	1.13	GRAM/MJ	3.54	
LYSINE	GRAM	1.71	-	-	-	1.23	GRAM/MJ	3.86	
METHIONINE	GRAM	0.62	-	-	-	0.45	GRAM/MJ	1.40	
PHENYLALANINE	GRAM	0.87	-	-	-	0.63	GRAM/MJ	1.96	
THREONINE	GRAM	0.89	-	-	-	0.64	GRAM/MJ	2.01	
TRYPTOPHAN	GRAM	0.27	-	-	-	0.19	GRAM/MJ	0.61	
VALINE	GRAM	1.07	-	-	-	0.77	GRAM/MJ	2.42	

CONSTITUENTS	DIM	AV	VARIATION			AVR	NUTR. DENS.		MOLPERC.
MYRISTIC ACID	MILLI	2.30	2.30	-	2.30	1.66	MILLI/MJ	5.19	
PALMITIC ACID	MILLI	160.00	160.00	-	160.00	115.20	MILLI/MJ	361.16	
STEARIC ACID	MILLI	82.00	82.00	-	82.00	59.04	MILLI/MJ	185.09	
ARACHIDIC ACID	MILLI	1.80	1.80	-	1.80	1.30	MILLI/MJ	4.06	
BEHENIC ACID	MILLI	8.10	8.10	-	8.10	5.83	MILLI/MJ	18.28	
PALMITOLEIC ACID	MILLI	12.00	12.00	-	12.00	8.64	MILLI/MJ	27.09	
OLEIC ACID	MILLI	165.00	165.00	-	165.00	118.80	MILLI/MJ	372.44	
C14 : 1 FATTY ACID	MILLI	0.70	0.70	-	0.70	0.50	MILLI/MJ	1.58	
LINOLEIC ACID	MILLI	98.00	98.00	-	98.00	70.56	MILLI/MJ	221.21	
LINOLENIC ACID	MILLI	2.70	2.70	-	2.70	1.94	MILLI/MJ	6.09	
ARACHIDONIC ACID	MILLI	112.00	112.00	-	112.00	80.64	MILLI/MJ	252.81	
C22 : 5 N-3 FATTY ACID	MILLI	13.00	13.00	-	13.00	9.36	MILLI/MJ	29.34	
C22 : 6 N-3 FATTY ACID	MILLI	15.00	15.00	-	15.00	10.80	MILLI/MJ	33.86	
CHOLESTEROL	MILLI	60.00	-	-	-	43.20	MILLI/MJ	135.43	
TOTAL PURINES	MILLI	175.00 1	-	-	-	126.00	MILLI/MJ	395.02	
TOTAL PHOSPHOLIPIDS	MILLI	663.00	-	-	-	477.36	GRAM/MJ	1.50	
PHOSPHATIDYLCHOLINE	MILLI	332.00	-	-	-	239.04	MILLI/MJ	749.40	
PHOSPHATIDYLETHANOLAMINE	MILLI	159.00	-	-	-	114.48	MILLI/MJ	358.90	
PHOSPHATIDYLSERINE	MILLI	85.00	-	-	-	61.20	MILLI/MJ	191.87	
SPHINGOMYELIN	MILLI	48.00	-	-	-	34.56	MILLI/MJ	108.35	

1 VALUE EXPRESSED IN MG URIC ACID/100G

HUHN SCHLEGEL — CHICKEN LEG — POULE CUISSE

		PROTEIN	FAT	CARBOHYDRATES	TOTAL
ENERGY VALUE (AVERAGE)	KJOULE	379	93	0.0	473
PER 100 G	(KCAL)	91	22	0.0	113
EDIBLE PORTION					
AMOUNT OF DIGESTIBLE	GRAM	19.98	2.23	0.00	
CONSTITUENTS PER 100 G					
ENERGY VALUE (AVERAGE)	KJOULE	368	89	0.0	457
OF THE DIGESTIBLE	(KCAL)	88	21	0.0	109
FRACTION PER 100 G					
EDIBLE PORTION					

WASTE PERCENTAGE AVERAGE 25 MINIMUM 23 MAXIMUM 28

CONSTITUENTS	DIM	AV	VARIATION			AVR	NUTR. DENS.		MOLPERC.
WATER	GRAM	74.70	-	-	-	56.03	GRAM/MJ	163.60	
PROTEIN	GRAM	20.60	20.50	-	20.70	15.45	GRAM/MJ	45.12	
FAT	GRAM	2.35	2.10	-	3.10	1.76	GRAM/MJ	5.15	
MINERALS	GRAM	1.10	-	-	-	0.83	GRAM/MJ	2.41	

CONSTITUENTS	DIM	AV	VARIATION			AVR	NUTR. DENS.		MOLPERC.
SODIUM	MILLI	95.00	80.00	-	110.00	71.25	MILLI/MJ	208.06	
POTASSIUM	MILLI	250.00	-	-	-	187.50	MILLI/MJ	547.53	
CALCIUM	MILLI	15.00	-	-	-	11.25	MILLI/MJ	32.85	
IRON	MILLI	1.80	-	-	-	1.35	MILLI/MJ	3.94	
PHOSPHORUS	MILLI	188.00	-	-	-	141.00	MILLI/MJ	411.74	
SELENIUM	MICRO	14.00	-	-	-	10.50	MICRO/MJ	30.66	
VITAMIN B1	MILLI	0.10	-	-	-	0.08	MILLI/MJ	0.22	
VITAMIN B2	MILLI	0.24	-	-	-	0.18	MILLI/MJ	0.53	
NICOTINAMIDE	MILLI	5.60	-	-	-	4.20	MILLI/MJ	12.26	
PANTOTHENIC ACID	MILLI	0.84	0.82	-	0.86	0.63	MILLI/MJ	1.84	
FOLIC ACID	MICRO	28.00	22.00	-	34.00	21.00	MICRO/MJ	61.32	
ARGININE	GRAM	1.26	-	-	-	0.95	GRAM/MJ	2.76	
HISTIDINE	GRAM	0.60	-	-	-	0.45	GRAM/MJ	1.31	
ISOLEUCINE	GRAM	1.17	-	-	-	0.88	GRAM/MJ	2.56	
LEUCINE	GRAM	1.48	-	-	-	1.11	GRAM/MJ	3.24	
LYSINE	GRAM	1.79	-	-	-	1.34	GRAM/MJ	3.92	
METHIONINE	GRAM	0.58	-	-	-	0.44	GRAM/MJ	1.27	
PHENYLALANINE	GRAM	0.82	-	-	-	0.62	GRAM/MJ	1.80	
THREONINE	GRAM	0.78	-	-	-	0.59	GRAM/MJ	1.71	
TRYPTOPHAN	GRAM	0.19	-	-	-	0.14	GRAM/MJ	0.42	
VALINE	GRAM	0.95	-	-	-	0.71	GRAM/MJ	2.08	
MYRISTIC ACID	MILLI	10.00	10.00	-	10.00	7.50	MILLI/MJ	21.90	
PALMITIC ACID	GRAM	0.43	0.43	-	0.43	0.32	GRAM/MJ	0.94	
STEARIC ACID	GRAM	0.24	0.24	-	0.24	0.18	GRAM/MJ	0.53	
ARACHIDIC ACID	MILLI	10.00	10.00	-	10.00	7.50	MILLI/MJ	21.90	
BEHENIC ACID	MILLI	10.00	10.00	-	10.00	7.50	MILLI/MJ	21.90	
LIGNOCERIC ACID	MILLI	10.00	10.00	-	10.00	7.50	MILLI/MJ	21.90	
PALMITOLEIC ACID	MILLI	70.00	70.00	-	70.00	52.50	MILLI/MJ	153.31	
OLEIC ACID	GRAM	0.52	0.52	-	0.52	0.39	GRAM/MJ	1.14	
LINOLEIC ACID	GRAM	0.37	0.37	-	0.37	0.28	GRAM/MJ	0.81	
LINOLENIC ACID	MILLI	10.00	10.00	-	10.00	7.50	MILLI/MJ	21.90	
ARACHIDONIC ACID	GRAM	0.19	0.19	-	0.19	0.14	GRAM/MJ	0.42	
C22 : 5 N-3 FATTY ACID	MILLI	10.00	10.00	-	10.00	7.50	MILLI/MJ	21.90	
C22 : 6 N-3 FATTY ACID	MILLI	10.00	10.00	-	10.00	7.50	MILLI/MJ	21.90	
CHOLESTEROL	MILLI	-	67.00	-	80.00				
TOTAL PURINES	MILLI	110.00 1	-	-	-	82.50	MILLI/MJ	240.91	
TOTAL PHOSPHOLIPIDS	GRAM	1.00	-	-	-	0.75	GRAM/MJ	2.19	
PHOSPHATIDYLCHOLINE	MILLI	477.00	-	-	-	357.75	GRAM/MJ	1.04	
PHOSPHATIDYLETHANOLAMINE	MILLI	254.00	-	-	-	190.50	MILLI/MJ	556.29	
PHOSPHATIDYLSERINE	MILLI	134.00	-	-	-	100.50	MILLI/MJ	293.47	
SPHINGOMYELIN	MILLI	73.00	-	-	-	54.75	MILLI/MJ	159.88	

1 VALUE EXPRESSED IN MG URIC ACID/100G

HUHN
HERZ

CHICKEN
HEART

COEUR DE POULE

		PROTEIN	FAT	CARBOHYDRATES	TOTAL
ENERGY VALUE (AVERAGE) PER 100 G EDIBLE PORTION	KJOULE	318	232	0.0	550
	(KCAL)	76	55	0.0	132
AMOUNT OF DIGESTIBLE CONSTITUENTS PER 100 G	GRAM	16.78	5.53	0.00	
ENERGY VALUE (AVERAGE) OF THE DIGESTIBLE FRACTION PER 100 G EDIBLE PORTION	KJOULE	309	220	0.0	529
	(KCAL)	74	53	0.0	126

WASTE PERCENTAGE AVERAGE 0.00

CONSTITUENTS	DIM	AV	VARIATION			AVR	NUTR. DENS.		MOLPERC.
WATER	GRAM	74.30	69.60	-	77.30	74.30	GRAM/MJ	140.43	
PROTEIN	GRAM	17.30	15.40	-	20.50	17.30	GRAM/MJ	32.70	
FAT	GRAM	5.83	5.00	-	7.00	5.83	GRAM/MJ	11.02	
MINERALS	GRAM	1.03	0.80	-	1.30	1.03	GRAM/MJ	1.95	
SODIUM	MILLI	111.00	97.00	-	125.00	111.00	MILLI/MJ	209.80	
POTASSIUM	MILLI	262.00	222.00	-	283.00	262.00	MILLI/MJ	495.20	
CALCIUM	MILLI	22.00	21.00	-	23.00	22.00	MILLI/MJ	41.58	
IRON	MILLI	1.70	-	-	-	1.70	MILLI/MJ	3.21	
ZINC	MILLI	3.10	-	-	-	3.10	MILLI/MJ	5.86	
PHOSPHORUS	MILLI	164.00	142.00	-	185.00	164.00	MILLI/MJ	309.97	
VITAMIN A	MICRO	9.00	-	-	-	9.00	MICRO/MJ	17.01	
VITAMIN E ACTIVITY	MILLI	1.20	-	-	-	1.20	MILLI/MJ	2.27	
ALPHA-TOCOPHEROL	MILLI	1.20	-	-	-	1.20	MILLI/MJ	2.27	
VITAMIN K	MILLI	0.72	-	-	-	0.72	MILLI/MJ	1.36	
VITAMIN B1	MILLI	0.43	-	-	-	0.43	MILLI/MJ	0.81	
VITAMIN B2	MILLI	1.24	-	-	-	1.24	MILLI/MJ	2.34	
NICOTINAMIDE	MILLI	6.00	-	-	-	6.00	MILLI/MJ	11.34	
PANTOTHENIC ACID	MILLI	2.56	2.55	-	2.57	2.56	MILLI/MJ	4.84	
VITAMIN B12	MICRO	4.24	-	-	-	4.24	MICRO/MJ	8.01	
VITAMIN C	MILLI	6.00	-	-	-	6.00	MILLI/MJ	11.34	
ARGININE	GRAM	1.09	0.92	-	1.29	1.09	GRAM/MJ	2.06	
CYSTINE	GRAM	0.17	0.15	-	0.23	0.17	GRAM/MJ	0.32	
HISTIDINE	GRAM	0.44	0.30	-	0.53	0.44	GRAM/MJ	0.83	
ISOLEUCINE	GRAM	0.88	0.76	-	1.13	0.88	GRAM/MJ	1.66	
LEUCINE	GRAM	1.55	1.38	-	1.91	1.55	GRAM/MJ	2.93	
LYSINE	GRAM	1.42	1.20	-	1.84	1.42	GRAM/MJ	2.68	
METHIONINE	GRAM	0.41	0.35	-	0.53	0.41	GRAM/MJ	0.77	
PHENYLALANINE	GRAM	0.78	0.68	-	0.88	0.78	GRAM/MJ	1.47	
THREONINE	GRAM	0.79	0.67	-	1.10	0.79	GRAM/MJ	1.49	
TRYPTOPHAN	GRAM	0.22	0.16	-	0.25	0.22	GRAM/MJ	0.42	
TYROSINE	GRAM	0.64	0.50	-	0.76	0.64	GRAM/MJ	1.21	
VALINE	GRAM	1.00	0.85	-	1.22	1.00	GRAM/MJ	1.89	

CONSTITUENTS	DIM	AV	VARIATION			AVR	NUTR. DENS.		MOLPERC.
CHOLESTEROL	MILLI	170.00	-	-	-	170.00	MILLI/MJ	321.32	
TOTAL PHOSPHOLIPIDS	GRAM	3.13	-	-	-	3.13	GRAM/MJ	5.92	
PHOSPHATIDYLCHOLINE	GRAM	1.23	-	-	-	1.23	GRAM/MJ	2.32	
PHOSPHATIDYLETHANOLAMINE	MILLI	927.00	-	-	-	927.00	GRAM/MJ	1.75	
PHOSPHATIDYLSERINE	MILLI	414.00	-	-	-	414.00	MILLI/MJ	782.50	
SPHINGOMYELIN	MILLI	355.00	-	-	-	355.00	MILLI/MJ	670.98	

HUHN LEBER — CHICKEN LIVER — FOIE DE POULE

		PROTEIN	FAT	CARBOHYDRATES	TOTAL
ENERGY VALUE (AVERAGE) PER 100 G EDIBLE PORTION	KJOULE	407	187	0.0	594
	(KCAL)	97	45	0.0	142
AMOUNT OF DIGESTIBLE CONSTITUENTS PER 100 G	GRAM	21.43	4.46	0.00	
ENERGY VALUE (AVERAGE) OF THE DIGESTIBLE FRACTION PER 100 G EDIBLE PORTION	KJOULE	395	177	0.0	572
	(KCAL)	94	42	0.0	137

WASTE PERCENTAGE AVERAGE 0.00

CONSTITUENTS	DIM	AV	VARIATION			AVR	NUTR. DENS.		MOLPERC.
WATER	GRAM	70.30	69.60	-	71.50	70.30	GRAM/MJ	122.88	
PROTEIN	GRAM	22.10	-	-	-	22.10	GRAM/MJ	38.63	
FAT	GRAM	4.70	4.00	-	6.20	4.70	GRAM/MJ	8.22	
MINERALS	GRAM	1.70	-	-	-	1.70	GRAM/MJ	2.97	

CONSTITUENTS	DIM	AV	VARIATION			AVR	NUTR. DENS.		MOLPERC.
SODIUM	MILLI	68.00	51.00	-	84.00	68.00	MILLI/MJ	118.86	
POTASSIUM	MILLI	218.00	160.00	-	276.00	218.00	MILLI/MJ	381.04	
MAGNESIUM	MILLI	13.00	-	-	-	13.00	MILLI/MJ	22.72	
CALCIUM	MILLI	18.00	16.00	-	20.00	18.00	MILLI/MJ	31.46	
MANGANESE	MICRO	77.00	-	-	-	77.00	MICRO/MJ	134.59	
IRON	MILLI	7.40	-	-	-	7.40	MILLI/MJ	12.93	
COPPER	MILLI	0.30	-	-	-	0.30	MILLI/MJ	0.52	
ZINC	MILLI	3.20	-	-	-	3.20	MILLI/MJ	5.59	
PHOSPHORUS	MILLI	240.00	-	-	-	240.00	MILLI/MJ	419.49	
SELENIUM	MICRO	66.00	-	-	-	66.00	MICRO/MJ	115.36	
VITAMIN A	MILLI	12.80	5.70	-	20.10	12.80	MILLI/MJ	22.37	
VITAMIN D	MICRO	1.30	-	-	-	1.30	MICRO/MJ	2.27	
VITAMIN E ACTIVITY	MILLI	0.40	-	-	-	0.40	MILLI/MJ	0.70	
ALPHA-TOCOPHEROL	MILLI	0.40	-	-	-	0.40	MILLI/MJ	0.70	
VITAMIN K	MICRO	80.00	80.00	-	570.00	80.00	MICRO/MJ	139.83	
VITAMIN B1	MILLI	0.32	0.18	-	0.50	0.32	MILLI/MJ	0.56	
VITAMIN B2	MILLI	2.49	2.46	-	2.50	2.49	MILLI/MJ	4.35	
NICOTINAMIDE	MILLI	11.60	8.50	-	14.60	11.60	MILLI/MJ	20.28	
PANTOTHENIC ACID	MILLI	7.16	6.73	-	7.65	7.16	MILLI/MJ	12.51	
VITAMIN B6	MILLI	0.80	0.67	-	1.07	0.80	MILLI/MJ	1.40	
FOLIC ACID	MILLI	0.38	-	-	-	0.38	MILLI/MJ	0.66	
VITAMIN B12	MICRO	23.00	22.00	-	25.00	23.00	MICRO/MJ	40.20	
VITAMIN C	MILLI	28.00	20.00	-	35.00	28.00	MILLI/MJ	48.94	
ALANINE	GRAM	1.64	1.53	-	1.84	1.64	GRAM/MJ	2.87	9.1
ARGININE	GRAM	1.42	1.33	-	1.54	1.42	GRAM/MJ	2.48	4.0
ASPARTIC ACID	GRAM	2.33	2.09	-	2.68	2.33	GRAM/MJ	4.07	8.7
CYSTINE	GRAM	0.31	0.29	-	0.36	0.31	GRAM/MJ	0.54	0.6
GLUTAMIC ACID	GRAM	3.45	3.20	-	3.58	3.45	GRAM/MJ	6.03	11.6
GLYCINE	GRAM	1.56	1.35	-	1.84	1.56	GRAM/MJ	2.73	10.3
HISTIDINE	GRAM	0.70	0.46	-	0.93	0.70	GRAM/MJ	1.22	2.2
ISOLEUCINE	GRAM	1.51	1.15	-	1.95	1.51	GRAM/MJ	2.64	5.7
LEUCINE	GRAM	2.34	2.10	-	2.47	2.34	GRAM/MJ	4.09	8.8
LYSINE	GRAM	1.96	1.79	-	2.16	1.96	GRAM/MJ	3.43	6.6
METHIONINE	GRAM	0.64	0.61	-	0.68	0.64	GRAM/MJ	1.12	2.1
PHENYLALANINE	GRAM	1.23	1.11	-	1.30	1.23	GRAM/MJ	2.15	3.7
PROLINE	GRAM	1.34	1.15	-	1.45	1.34	GRAM/MJ	2.34	5.8
SERINE	GRAM	1.29	1.11	-	1.46	1.29	GRAM/MJ	2.25	6.1
THREONINE	GRAM	1.17	1.04	-	1.30	1.17	GRAM/MJ	2.05	4.9
TRYPTOPHAN	GRAM	0.32	0.25	-	0.37	0.32	GRAM/MJ	0.56	0.8
TYROSINE	GRAM	0.83	0.77	-	1.08	0.83	GRAM/MJ	1.45	2.3
VALINE	GRAM	1.55	1.30	-	1.64	1.55	GRAM/MJ	2.71	6.6
TYRAMINE	MILLI	10.00	-	-	-	10.00	MILLI/MJ	17.48	
CHOLESTEROL	MILLI	555.00	-	-	-	555.00	MILLI/MJ	970.07	
TOTAL PURINES	MILLI	243.00 1	-	-	-	243.00	MILLI/MJ	424.74	
TOTAL PHOSPHOLIPIDS	GRAM	2.13	-	-	-	2.13	GRAM/MJ	3.72	
PHOSPHATIDYLCHOLINE	MILLI	940.00	-	-	-	940.00	GRAM/MJ	1.64	
PHOSPHATIDYLETHANOLAMINE	MILLI	696.00	-	-	-	696.00	GRAM/MJ	1.22	
PHOSPHATIDYLSERINE	MILLI	123.00	-	-	-	123.00	MILLI/MJ	214.99	
SPHINGOMYELIN	MILLI	244.00	-	-	-	244.00	MILLI/MJ	426.48	

1 VALUE EXPRESSED IN MG URIC ACID/100G

TAUBE / PIGEON / PIGEON
DURCHSCHNITT / AVERAGE / MOYENNE

		PROTEIN	FAT	CARBOHYDRATES	TOTAL
ENERGY VALUE (AVERAGE)	KJOULE	385	378	0.0	762
PER 100 G	(KCAL)	92	90	0.0	182
EDIBLE PORTION					
AMOUNT OF DIGESTIBLE	GRAM	20.27	9.02	0.00	
CONSTITUENTS PER 100 G					
ENERGY VALUE (AVERAGE)	KJOULE	373	359	0.0	732
OF THE DIGESTIBLE	(KCAL)	89	86	0.0	175
FRACTION PER 100 G					
EDIBLE PORTION					

WASTE PERCENTAGE AVERAGE 53

CONSTITUENTS	DIM	AV	VARIATION	AVR	NUTR. DENS.		MOLPERC.
WATER	GRAM	68.40	68.40 - 68.40	32.15	GRAM/MJ	93.45	
PROTEIN	GRAM	20.90	20.90 - 20.90	9.82	GRAM/MJ	28.55	
FAT	GRAM	9.50	9.50 - 9.50	4.47	GRAM/MJ	12.98	
MINERALS	GRAM	1.20	1.20 - 1.20	0.56	GRAM/MJ	1.64	
SODIUM	MILLI	90.00	90.00 - 90.00	42.30	MILLI/MJ	122.96	
POTASSIUM	MILLI	330.00	330.00 - 330.00	155.10	MILLI/MJ	450.85	
CALCIUM	MILLI	45.00	45.00 - 45.00	21.15	MILLI/MJ	61.48	
PHOSPHORUS	MILLI	217.00	217.00 - 217.00	101.99	MILLI/MJ	296.47	
VITAMIN B1	MILLI	0.10	0.10 - 0.10	0.05	MILLI/MJ	0.14	
VITAMIN B2	MILLI	0.28	0.28 - 0.28	0.13	MILLI/MJ	0.38	
NICOTINAMIDE	MILLI	5.30	5.30 - 5.30	2.49	MILLI/MJ	7.24	

TRUTHAHN	TURKEY	DINDON
AUSGEWACHSENES TIER	ADULT ANIMAL	MOYENNE
DURCHSCHNITT	AVERAGE	
MELEAGRIS GALLOPAVO L.		

		PROTEIN	FAT	CARBOHYDRATES	TOTAL
ENERGY VALUE (AVERAGE)	KJOULE	372	596	0.0	968
PER 100 G	(KCAL)	89	143	0.0	231
EDIBLE PORTION					
AMOUNT OF DIGESTIBLE	GRAM	19.59	14.25	0.00	
CONSTITUENTS PER 100 G					
ENERGY VALUE (AVERAGE)	KJOULE	361	566	0.0	927
OF THE DIGESTIBLE	(KCAL)	86	135	0.0	222
FRACTION PER 100 G					
EDIBLE PORTION					

WASTE PERCENTAGE AVERAGE 27 MINIMUM 22 MAXIMUM 35

CONSTITUENTS	DIM	AV	VARIATION			AVR	NUTR. DENS.		MOLPERC.
WATER	GRAM	63.50	57.80	-	64.00	46.36	GRAM/MJ	68.49	
PROTEIN	GRAM	20.20	18.40	-	21.00	14.75	GRAM/MJ	21.79	
FAT	GRAM	15.00	-	-	-	10.95	GRAM/MJ	16.18	
MINERALS	GRAM	0.95	0.90	-	1.00	0.69	GRAM/MJ	1.02	
SODIUM	MILLI	63.00	48.00	-	77.00	45.99	MILLI/MJ	67.95	
POTASSIUM	MILLI	300.00	295.00	-	304.00	219.00	MILLI/MJ	323.58	
MAGNESIUM	MILLI	27.00	-	-	-	19.71	MILLI/MJ	29.12	
CALCIUM	MILLI	25.00	22.00	-	29.00	18.25	MILLI/MJ	26.97	
MANGANESE	MICRO	35.00	29.00	-	48.00	25.55	MICRO/MJ	37.75	
IRON	MILLI	1.40	0.95	-	1.90	1.02	MILLI/MJ	1.51	
COBALT	MICRO	1.90	1.40	-	2.40	1.39	MICRO/MJ	2.05	
COPPER	MILLI	0.10	0.04	-	0.17	0.07	MILLI/MJ	0.11	
ZINC	MILLI	2.00	1.70	-	2.30	1.46	MILLI/MJ	2.16	
PHOSPHORUS	MILLI	226.00	174.00	-	304.00	164.98	MILLI/MJ	243.76	
CHLORIDE	MILLI	106.00	95.00	-	117.00	77.38	MILLI/MJ	114.33	
BORON	MICRO	24.00	-	-	-	17.52	MICRO/MJ	25.89	
VITAMIN A	MICRO	13.00	-	-	-	9.49	MICRO/MJ	14.02	
VITAMIN B1	MILLI	0.10	-	-	-	0.07	MILLI/MJ	0.11	
VITAMIN B2	MILLI	0.18	-	-	-	0.13	MILLI/MJ	0.19	
NICOTINAMIDE	MILLI	10.50	-	-	-	7.67	MILLI/MJ	11.33	
PANTOTHENIC ACID	MILLI	1.10	-	-	-	0.80	MILLI/MJ	1.19	
FOLIC ACID	MICRO	16.00	-	-	-	11.68	MICRO/MJ	17.26	
ARGININE	GRAM	1.21	1.11	-	1.35	0.88	GRAM/MJ	1.31	
CYSTINE	GRAM	0.27	0.24	-	0.29	0.20	GRAM/MJ	0.29	
HISTIDINE	GRAM	0.52	0.41	-	0.74	0.38	GRAM/MJ	0.56	
ISOLEUCINE	GRAM	1.01	0.95	-	1.06	0.74	GRAM/MJ	1.09	
LEUCINE	GRAM	1.47	1.41	-	1.53	1.07	GRAM/MJ	1.59	
LYSINE	GRAM	1.74	1.62	-	1.85	1.27	GRAM/MJ	1.88	
METHIONINE	GRAM	0.53	0.51	-	0.57	0.39	GRAM/MJ	0.57	
PHENYLALANINE	GRAM	0.77	0.72	-	0.84	0.56	GRAM/MJ	0.83	
THREONINE	GRAM	0.81	0.66	-	0.90	0.59	GRAM/MJ	0.87	
TRYPTOPHAN	GRAM	0.16	0.15	-	0.17	0.12	GRAM/MJ	0.17	
TYROSINE	GRAM	0.28	0.27	-	0.29	0.20	GRAM/MJ	0.30	
VALINE	GRAM	0.95	0.89	-	1.02	0.69	GRAM/MJ	1.02	
LINOLEIC ACID	GRAM	4.19	3.43	-	5.32	3.06	GRAM/MJ	4.52	
LINOLENIC ACID	GRAM	0.23	0.20	-	0.26	0.17	GRAM/MJ	0.25	
CHOLESTEROL	MILLI	74.00	-	-	-	54.02	MILLI/MJ	79.82	

TRUTHAHN	**TURKEY**	**DINDONNEAU**
JUNGTIER	YOUNG ANIMAL	MOYENNE
DURCHSCHNITT	AVERAGE	
MELEAGRIS GALLOPAVO L.		
1	1	1

		PROTEIN	FAT	CARBOHYDRATES	TOTAL
ENERGY VALUE (AVERAGE)	KJOULE	412	270	0.0	682
PER 100 G	(KCAL)	99	65	0.0	163
EDIBLE PORTION					
AMOUNT OF DIGESTIBLE	GRAM	21.72	6.45	0.00	
CONSTITUENTS PER 100 G					
ENERGY VALUE (AVERAGE)	KJOULE	400	256	0.0	656
OF THE DIGESTIBLE	(KCAL)	96	61	0.0	157
FRACTION PER 100 G					
EDIBLE PORTION					

WASTE PERCENTAGE AVERAGE 27 MINIMUM 22 MAXIMUM 35

CONSTITUENTS	DIM	AV	VARIATION		AVR	NUTR. DENS.		MOLPERC.
WATER	GRAM	69.70	65.80	- 73.10	50.88	GRAM/MJ	106.19	
PROTEIN	GRAM	22.40	21.40	- 23.30	16.35	GRAM/MJ	34.13	
FAT	GRAM	6.79	5.30	- 8.20	4.96	GRAM/MJ	10.34	
MINERALS	GRAM	1.00	0.90	- 1.10	0.73	GRAM/MJ	1.52	
SODIUM	MILLI	66.00	51.00	- 81.00	48.18	MILLI/MJ	100.55	
POTASSIUM	MILLI	315.00	310.00	- 320.00	229.95	MILLI/MJ	479.89	
MAGNESIUM	MILLI	28.00	-	- -	20.44	MILLI/MJ	42.66	
CALCIUM	MILLI	26.00	23.00	- 30.00	18.98	MILLI/MJ	39.61	
MANGANESE	MICRO	37.00	30.00	- 50.00	27.01	MICRO/MJ	56.37	
IRON	MILLI	1.50	1.00	- 2.00	1.10	MILLI/MJ	2.29	
COBALT	MICRO	2.00	1.50	- 2.50	1.46	MICRO/MJ	3.05	
COPPER	MILLI	0.11	0.04	- 0.18	0.08	MILLI/MJ	0.17	
ZINC	MILLI	2.10	1.80	- 2.40	1.53	MILLI/MJ	3.20	
PHOSPHORUS	MILLI	238.00	183.00	- 320.00	173.74	MILLI/MJ	362.58	
CHLORIDE	MILLI	112.00	100.00	- 123.00	81.76	MILLI/MJ	170.63	
VITAMIN B1	MICRO	80.00	-	- -	58.40	MICRO/MJ	121.88	
VITAMIN B2	MILLI	0.14	-	- -	0.10	MILLI/MJ	0.21	
NICOTINAMIDE	MILLI	8.00	-	- -	5.84	MILLI/MJ	12.19	
PANTOTHENIC ACID	MILLI	0.84	-	- -	0.61	MILLI/MJ	1.28	
FOLIC ACID	MICRO	12.00	-	- -	8.76	MICRO/MJ	18.28	
ARGININE	GRAM	1.41	1.29	- 1.58	1.03	GRAM/MJ	2.15	
CYSTINE	GRAM	0.31	0.28	- 0.34	0.23	GRAM/MJ	0.47	
HISTIDINE	GRAM	0.61	0.48	- 0.86	0.45	GRAM/MJ	0.93	
ISOLEUCINE	GRAM	1.18	1.11	- 1.24	0.86	GRAM/MJ	1.80	
LEUCINE	GRAM	1.71	1.64	- 1.79	1.25	GRAM/MJ	2.61	

1 UP TO 24 WEEKS

CONSTITUENTS	DIM	AV	VARIATION			AVR	NUTR. DENS.		MOLPERC.
LYSINE	GRAM	2.03	1.89	-	2.16	1.48	GRAM/MJ	3.09	
METHIONINE	GRAM	0.62	0.60	-	0.66	0.45	GRAM/MJ	0.94	
PHENYLALANINE	GRAM	0.90	0.84	-	0.98	0.66	GRAM/MJ	1.37	
THREONINE	GRAM	0.95	0.77	-	1.05	0.69	GRAM/MJ	1.45	
TRYPTOPHAN	GRAM	0.19	0.18	-	0.20	0.14	GRAM/MJ	0.29	
TYROSINE	GRAM	0.33	0.31	-	0.34	0.24	GRAM/MJ	0.50	
VALINE	GRAM	1.11	1.04	-	1.19	0.81	GRAM/MJ	1.69	
LINOLEIC ACID	GRAM	1.49	1.22	-	1.89	1.09	GRAM/MJ	2.27	
LINOLENIC ACID	MILLI	81.00	71.00	-	91.00	59.13	MILLI/MJ	123.40	
TOTAL PURINES	MILLI	150.00 2	-	-	-	109.50	MILLI/MJ	228.52	

2 VALUE EXPRESSED IN MG URIC ACID/100G

TRUTHAHN BRUST · TURKEY BREAST · DINDON POITRINE

		PROTEIN	FAT	CARBOHYDRATES	TOTAL
ENERGY VALUE (AVERAGE) PER 100 G EDIBLE PORTION	KJOULE	444	39	0.0	483
	(KCAL)	106	9.4	0.0	115
AMOUNT OF DIGESTIBLE CONSTITUENTS PER 100 G	GRAM	23.37	0.94	0.00	
ENERGY VALUE (AVERAGE) OF THE DIGESTIBLE FRACTION PER 100 G EDIBLE PORTION	KJOULE	430	37	0.0	468
	(KCAL)	103	8.9	0.0	112

WASTE PERCENTAGE AVERAGE 11 MINIMUM 9.0 MAXIMUM 13

CONSTITUENTS	DIM	AV	VARIATION			AVR	NUTR. DENS.		MOLPERC.
WATER	GRAM	73.70	72.80	-	75.40	65.59	GRAM/MJ	157.56	
PROTEIN	GRAM	24.10	22.40	-	25.20	21.45	GRAM/MJ	51.52	
FAT	GRAM	0.99	0.30	-	1.79	0.88	GRAM/MJ	2.12	
MINERALS	GRAM	1.21	1.10	-	1.30	1.08	GRAM/MJ	2.59	

CONSTITUENTS	DIM	AV	VARIATION			AVR	NUTR. DENS.		MOLPERC.
SODIUM	MILLI	46.00	40.00	-	51.00	40.94	MILLI/MJ	98.34	
POTASSIUM	MILLI	333.00	320.00	-	359.00	296.37	MILLI/MJ	711.93	
MAGNESIUM	MILLI	20.00	-	-	-	17.80	MILLI/MJ	42.76	
MANGANESE	MICRO	30.00	-	-	-	26.70	MICRO/MJ	64.14	
IRON	MILLI	1.00	-	-	-	0.89	MILLI/MJ	2.14	
COPPER	MILLI	0.13	0.10	-	0.15	0.12	MILLI/MJ	0.28	
ZINC	MILLI	1.80	-	-	-	1.60	MILLI/MJ	3.85	
VITAMIN B1	MICRO	47.00	33.00	-	60.00	41.83	MICRO/MJ	100.48	
VITAMIN B2	MICRO	81.00	51.00	-	110.00	72.09	MICRO/MJ	173.17	
NICOTINAMIDE	MILLI	11.30	-	-	-	10.06	MILLI/MJ	24.16	
PANTOTHENIC ACID	MILLI	0.59	0.54	-	0.64	0.53	MILLI/MJ	1.26	
VITAMIN B6	MILLI	0.46	0.30	-	0.59	0.41	MILLI/MJ	0.98	
FOLIC ACID	MICRO	7.00	-	-	-	6.23	MICRO/MJ	14.97	
VITAMIN B12	MICRO	0.52	-	-	-	0.46	MICRO/MJ	1.11	
ARGININE	GRAM	1.57	-	-	-	1.40	GRAM/MJ	3.36	
CYSTINE	GRAM	0.26	0.22	-	0.31	0.23	GRAM/MJ	0.56	
HISTIDINE	GRAM	0.84	-	-	-	0.75	GRAM/MJ	1.80	
ISOLEUCINE	GRAM	1.22	1.21	-	1.23	1.09	GRAM/MJ	2.61	
LEUCINE	GRAM	1.85	1.83	-	1.86	1.65	GRAM/MJ	3.96	
LYSINE	GRAM	2.11	2.02	-	2.19	1.88	GRAM/MJ	4.51	
METHIONINE	GRAM	0.63	-	-	-	0.56	GRAM/MJ	1.35	
PHENYLALANINE	GRAM	0.88	0.87	-	0.89	0.78	GRAM/MJ	1.88	
THREONINE	GRAM	0.97	0.94	-	0.99	0.86	GRAM/MJ	2.07	
TRYPTOPHAN	GRAM	0.22	-	-	-	0.20	GRAM/MJ	0.47	
TYROSINE	GRAM	0.36	-	-	-	0.32	GRAM/MJ	0.77	
VALINE	GRAM	1.24	1.23	-	1.25	1.10	GRAM/MJ	2.65	
PALMITIC ACID	GRAM	0.22	-	-	-	0.20	GRAM/MJ	0.47	
STEARIC ACID	GRAM	0.12	-	-	-	0.11	GRAM/MJ	0.26	
PALMITOLEIC ACID	MILLI	10.00	-	-	-	8.90	MILLI/MJ	21.38	
OLEIC ACID	GRAM	0.21	-	-	-	0.19	GRAM/MJ	0.45	
LINOLEIC ACID	GRAM	0.18	-	-	-	0.16	GRAM/MJ	0.38	
ARACHIDONIC ACID	MILLI	50.00	-	-	-	44.50	MILLI/MJ	106.90	
CHOLESTEROL	MILLI	60.00	-	-	-	53.40	MILLI/MJ	128.28	
TOTAL PHOSPHOLIPIDS	MILLI	567.00	-	-	-	504.63	GRAM/MJ	1.21	
PHOSPHATIDYLCHOLINE	MILLI	313.00	-	-	-	278.57	MILLI/MJ	669.17	
PHOSPHATIDYLETHANOLAMINE	MILLI	125.00	-	-	-	111.25	MILLI/MJ	267.24	
PHOSPHATIDYLSERINE	MILLI	45.00	-	-	-	40.05	MILLI/MJ	96.21	
SPHINGOMYELIN	MILLI	58.00	-	-	-	51.62	MILLI/MJ	124.00	

TRUTHAHN / TURKEY / DINDON

KEULE (SCHLEGEL) / LEG / CUISSE

		PROTEIN	FAT	CARBOHYDRATES	TOTAL
ENERGY VALUE (AVERAGE) PER 100 G EDIBLE PORTION	KJOULE	377	143	0.0	521
	(KCAL)	90	34	0.0	124
AMOUNT OF DIGESTIBLE CONSTITUENTS PER 100 G	GRAM	19.88	3.42	0.00	
ENERGY VALUE (AVERAGE) OF THE DIGESTIBLE FRACTION PER 100 G EDIBLE PORTION	KJOULE	366	136	0.0	502
	(KCAL)	87	33	0.0	120

WASTE PERCENTAGE AVERAGE 19 MINIMUM 14 MAXIMUM 24

CONSTITUENTS	DIM	AV	VARIATION			AVR	NUTR.	DENS.	MOLPERC.
WATER	GRAM	74.70	73.10	-	76.40	60.51	GRAM/MJ	148.69	
PROTEIN	GRAM	20.50	19.80	-	21.10	16.61	GRAM/MJ	40.80	
FAT	GRAM	3.61	2.20	-	5.60	2.92	GRAM/MJ	7.19	
MINERALS	GRAM	1.14	1.10	-	1.20	0.92	GRAM/MJ	2.27	
SODIUM	MILLI	86.00	81.00	-	92.00	69.66	MILLI/MJ	171.18	
POTASSIUM	MILLI	289.00	246.00	-	310.00	234.09	MILLI/MJ	575.25	
MAGNESIUM	MILLI	17.00	-	-	-	13.77	MILLI/MJ	33.84	
MANGANESE	MICRO	50.00	-	-	-	40.50	MICRO/MJ	99.52	
IRON	MILLI	2.00	-	-	-	1.62	MILLI/MJ	3.98	
COPPER	MILLI	0.16	0.12	-	0.20	0.13	MILLI/MJ	0.32	
ZINC	MILLI	2.40	-	-	-	1.94	MILLI/MJ	4.78	
VITAMIN B1	MICRO	90.00	-	-	-	72.90	MICRO/MJ	179.14	
VITAMIN B2	MILLI	0.18	-	-	-	0.15	MILLI/MJ	0.36	
NICOTINAMIDE	MILLI	4.70	-	-	-	3.81	MILLI/MJ	9.36	
PANTOTHENIC ACID	MILLI	1.13	1.03	-	1.22	0.92	MILLI/MJ	2.25	
ARGININE	GRAM	1.33	1.31	-	1.35	1.08	GRAM/MJ	2.65	
CYSTINE	GRAM	0.21	-	-	-	0.17	GRAM/MJ	0.42	
HISTIDINE	GRAM	0.52	0.51	-	0.53	0.42	GRAM/MJ	1.04	
ISOLEUCINE	GRAM	0.99	0.98	-	1.00	0.80	GRAM/MJ	1.97	
LEUCINE	GRAM	1.56	1.54	-	1.58	1.26	GRAM/MJ	3.11	
LYSINE	GRAM	1.85	1.82	-	1.87	1.50	GRAM/MJ	3.68	
METHIONINE	GRAM	0.53	-	-	-	0.43	GRAM/MJ	1.05	
PHENYLALANINE	GRAM	0.77	0.76	-	0.78	0.62	GRAM/MJ	1.53	
THREONINE	GRAM	0.82	-	-	-	0.66	GRAM/MJ	1.63	
TRYPTOPHAN	GRAM	0.16	-	-	-	0.13	GRAM/MJ	0.32	

CONSTITUENTS	DIM	AV	VARIATION			AVR	NUTR. DENS.		MOLPERC.
TYROSINE	GRAM	0.29	-	-	-	0.23	GRAM/MJ	0.58	
VALINE	GRAM	1.02	0.98	-	1.05	0.83	GRAM/MJ	2.03	
PALMITIC ACID	GRAM	0.84	-	-	-	0.68	GRAM/MJ	1.67	
STEARIC ACID	GRAM	0.47	-	-	-	0.38	GRAM/MJ	0.94	
PALMITOLEIC ACID	MILLI	60.00	-	-	-	48.60	MILLI/MJ	119.43	
OLEIC ACID	GRAM	0.72	-	-	-	0.58	GRAM/MJ	1.43	
LINOLEIC ACID	GRAM	0.75	-	-	-	0.61	GRAM/MJ	1.49	
ARACHIDONIC ACID	GRAM	0.15	-	-	-	0.12	GRAM/MJ	0.30	
CHOLESTEROL	MILLI	75.00	-	-	-	60.75	MILLI/MJ	149.29	
TOTAL PHOSPHOLIPIDS	MILLI	766.00	-	-	-	620.46	GRAM/MJ	1.52	
PHOSPHATIDYLCHOLINE	MILLI	411.00	-	-	-	332.91	MILLI/MJ	818.09	
PHOSPHATIDYLETHANOLAMINE	MILLI	199.00	-	-	-	161.19	MILLI/MJ	396.11	
PHOSPHATIDYLSERINE	MILLI	50.00	-	-	-	40.50	MILLI/MJ	99.52	
SPHINGOMYELIN	MILLI	77.00	-	-	-	62.37	MILLI/MJ	153.27	

WACHTEL
DURCHSCHNITT
COLINUS VIRIGIANUS

QUAIL
AVERAGE

CAILLE
MOYENNE

		PROTEIN	FAT	CARBOHYDRATES	TOTAL
ENERGY VALUE (AVERAGE)	KJOULE	412	92	0.0	504
PER 100 G	(KCAL)	98	22	0.0	120
EDIBLE PORTION					
AMOUNT OF DIGESTIBLE	GRAM	21.70	2.19	0.00	
CONSTITUENTS PER 100 G					
ENERGY VALUE (AVERAGE)	KJOULE	400	87	0.0	487
OF THE DIGESTIBLE	(KCAL)	95	21	0.0	116
FRACTION PER 100 G					
EDIBLE PORTION					

WASTE PERCENTAGE AVERAGE 30

CONSTITUENTS	DIM	AV	VARIATION	AVR	NUTR. DENS.		MOLPERC.
WATER	GRAM	75.89	75.52 - 76.26	53.12	GRAM/MJ	155.84	
PROTEIN	GRAM	22.37	22.15 - 22.60	15.66	GRAM/MJ	45.95	
FAT	GRAM	2.31	2.26 - 2.37	1.62	GRAM/MJ	4.75	
MINERALS	GRAM	1.13	1.12 - 1.15	0.79	GRAM/MJ	2.33	
SODIUM	MILLI	46.50	45.00 - 48.00	32.55	MILLI/MJ	95.49	
POTASSIUM	MILLI	280.50	257.00 - 304.00	196.35	MILLI/MJ	576.01	
MAGNESIUM	MILLI	31.00	31.00 - 31.00	21.70	MILLI/MJ	63.66	
CALCIUM	MILLI	14.50	13.00 - 16.00	10.15	MILLI/MJ	29.78	
PHOSPHORUS	MILLI	179.00	175.00 - 183.00	125.30	MILLI/MJ	367.58	
VITAMIN B1	MILLI	0.13	0.10 - 0.17	0.09	MILLI/MJ	0.28	
VITAMIN B2	MILLI	0.17	0.16 - 0.19	0.12	MILLI/MJ	0.36	
NICOTINAMIDE	MILLI	9.95	9.60 - 10.30	6.97	MILLI/MJ	20.43	
PANTOTHENIC ACID	MILLI	0.66	0.66 - 0.66	0.46	MILLI/MJ	1.36	
VITAMIN B6	MILLI	0.67	0.66 - 0.68	0.47	MILLI/MJ	1.38	
MYRISTIC ACID	MILLI	35.00	30.00 - 40.00	24.50	MILLI/MJ	71.87	
PALMITIC ACID	GRAM	0.44	0.39 - 0.49	0.31	GRAM/MJ	0.90	
STEARIC ACID	GRAM	0.29	0.27 - 0.32	0.21	GRAM/MJ	0.61	
PALMITOLEIC ACID	MILLI	40.00	20.00 - 60.00	28.00	MILLI/MJ	82.14	
OLEIC ACID	GRAM	0.55	0.47 - 0.64	0.39	GRAM/MJ	1.14	
LINOLEIC ACID	GRAM	0.53	0.53 - 0.53	0.37	GRAM/MJ	1.09	
LINOLENIC ACID	MILLI	20.00	20.00 - 20.00	14.00	MILLI/MJ	41.07	
CHOLESTEROL	MILLI	43.75	37.60 - 49.90	30.63	MILLI/MJ	89.84	

Fische

Fish
Poissons

Seefische und Wal

SALT-WATER FISH AND WHALE — POISSONS MARITIMES ET BALEINE

ANGLERFISCH (SEETEUFEL) — ANGLERFISH — BAUDROIE

LOPHIUS PISCATORIUS (L.)

		PROTEIN	FAT	CARBOHYDRATES	TOTAL
ENERGY VALUE (AVERAGE)	KJOULE	274	27	0.0	301
PER 100 G	(KCAL)	66	6.4	0.0	72
EDIBLE PORTION					
AMOUNT OF DIGESTIBLE	GRAM	14.45	0.63	0.00	
CONSTITUENTS PER 100 G					
ENERGY VALUE (AVERAGE)	KJOULE	266	25	0.0	291
OF THE DIGESTIBLE	(KCAL)	64	6.0	0.0	70
FRACTION PER 100 G					
EDIBLE PORTION					

WASTE PERCENTAGE AVERAGE 27 MINIMUM 23 MAXIMUM 33

CONSTITUENTS	DIM	AV	VARIATION			AVR	NUTR. DENS.		MOLPERC.
WATER	GRAM	83.50	-	-	-	60.96	GRAM/MJ	286.57	
PROTEIN	GRAM	14.90	-	-	-	10.88	GRAM/MJ	51.14	
FAT	GRAM	0.67	0.63	-	0.73	0.49	GRAM/MJ	2.30	
MINERALS	GRAM	0.72	0.50	-	0.94	0.53	GRAM/MJ	2.47	
SODIUM	MILLI	109.00	101.00	-	117.00	79.57	MILLI/MJ	374.09	
POTASSIUM	MILLI	235.00	211.00	-	258.00	171.55	MILLI/MJ	806.52	

BLAULENG | BLUE LING | LINGUE BLEUE

MOLVA DYPTERYGIA

		PROTEIN	FAT	CARBOHYDRATES	TOTAL
ENERGY VALUE (AVERAGE)	KJOULE	320	28	0.0	348
PER 100 G	(KCAL)	77	6.7	0.0	83
EDIBLE PORTION					
AMOUNT OF DIGESTIBLE	GRAM	16.87	0.66	0.00	
CONSTITUENTS PER 100 G					
ENERGY VALUE (AVERAGE)	KJOULE	311	26	0.0	337
OF THE DIGESTIBLE	(KCAL)	74	6.3	0.0	81
FRACTION PER 100 G					
EDIBLE PORTION					

WASTE PERCENTAGE AVERAGE 0.00

CONSTITUENTS	DIM	AV	VARIATION	AVR	NUTR. DENS.		MOLPERC.
WATER	GRAM	81.60	81.60 - 81.60	81.60	GRAM/MJ	242.03	
PROTEIN	GRAM	17.40	17.40 - 17.40	17.40	GRAM/MJ	51.61	
FAT	GRAM	0.70	0.70 - 0.70	0.70	GRAM/MJ	2.08	
MINERALS	GRAM	1.00	1.00 - 1.00	1.00	GRAM/MJ	2.97	

BROSME (LUMB) | TUSK | „BROSME“

BROSME BROSME (ASCANIUS)

		PROTEIN	FAT	CARBOHYDRATES	TOTAL
ENERGY VALUE (AVERAGE)	KJOULE	318	13	0.0	332
PER 100 G	(KCAL)	76	3.1	0.0	79
EDIBLE PORTION					
AMOUNT OF DIGESTIBLE	GRAM	16.78	0.31	0.00	
CONSTITUENTS PER 100 G					
ENERGY VALUE (AVERAGE)	KJOULE	309	12	0.0	321
OF THE DIGESTIBLE	(KCAL)	74	3.0	0.0	77
FRACTION PER 100 G					
EDIBLE PORTION					

WASTE PERCENTAGE AVERAGE 48 MINIMUM 42 MAXIMUM 58

CONSTITUENTS	DIM	AV	VARIATION			AVR	NUTR. DENS.		MOLPERC.
WATER	GRAM	81.30	-	-	-	42.28	GRAM/MJ	252.96	
PROTEIN	GRAM	17.30	-	-	-	9.00	GRAM/MJ	53.83	
FAT	GRAM	0.33	0.20	-	0.55	0.17	GRAM/MJ	1.03	
MINERALS	GRAM	0.99	0.90	-	1.07	0.51	GRAM/MJ	3.08	
SODIUM	MILLI	113.00	94.00	-	129.00	58.76	MILLI/MJ	351.60	
POTASSIUM	MILLI	328.00	311.00	-	341.00	170.56	GRAM/MJ	1.02	
MAGNESIUM	MILLI	56.00	-	-	-	29.12	MILLI/MJ	174.24	
CALCIUM	MILLI	17.00	15.00	-	18.00	8.84	MILLI/MJ	52.89	
PHOSPHORUS	MILLI	177.00	-	-	-	92.04	MILLI/MJ	550.73	
VITAMIN B1	MICRO	30.00	-	-	-	15.60	MICRO/MJ	93.34	
VITAMIN B2	MICRO	60.00	30.00	-	80.00	31.20	MICRO/MJ	186.69	
NICOTINAMIDE	MILLI	2.25	2.20	-	2.30	1.17	MILLI/MJ	7.00	
PANTOTHENIC ACID	MILLI	0.12	-	-	-	0.06	MILLI/MJ	0.37	
BIOTIN	MICRO	0.12	-	-	-	0.06	MICRO/MJ	0.37	
FOLIC ACID	MICRO	2.00	-	-	-	1.04	MICRO/MJ	6.22	
VITAMIN B12	MICRO	0.30	-	-	-	0.16	MICRO/MJ	0.93	

DORNHAI
(DORNFISCH)

SQUALUS ACANTHIAS

GRAYFISH
(PICKED DOGFISH)
(SPINY DOGFISH)

AIGUILLAT COMMUN

		PROTEIN	FAT	CARBOHYDRATES	TOTAL
ENERGY VALUE (AVERAGE)	KJOULE	232	576	0.0	808
PER 100 G	(KCAL)	55	138	0.0	193
EDIBLE PORTION					
AMOUNT OF DIGESTIBLE	GRAM	12.22	13.77	0.00	
CONSTITUENTS PER 100 G					
ENERGY VALUE (AVERAGE)	KJOULE	225	548	0.0	773
OF THE DIGESTIBLE	(KCAL)	54	131	0.0	185
FRACTION PER 100 G					
EDIBLE PORTION					

WASTE PERCENTAGE AVERAGE 53

CONSTITUENTS	DIM	AV	VARIATION			AVR	NUTR.	DENS.	MOLPERC.
WATER	GRAM	71.50	69.00	-	74.00	33.61	GRAM/MJ	92.55	
PROTEIN	GRAM	12.60	11.20	-	14.00	5.92	GRAM/MJ	16.31	
FAT	GRAM	14.50	12.30	-	16.70	6.82	GRAM/MJ	18.77	
MINERALS	GRAM	0.90	0.80	-	1.00	0.42	GRAM/MJ	1.17	
SODIUM	MILLI	13.60	14.00	-	19.00	6.39	MILLI/MJ	17.60	
MAGNESIUM	MILLI	23.00	16.00	-	30.00	10.81	MILLI/MJ	29.77	
CALCIUM	MILLI	5.00	4.00	-	6.00	2.35	MILLI/MJ	6.47	
MANGANESE	MICRO	87.00	71.00	-	103.00	40.89	MICRO/MJ	112.62	
IRON	MILLI	0.50	0.36	-	0.66	0.24	MILLI/MJ	0.66	
COPPER	MICRO	57.00	43.00	-	71.00	26.79	MICRO/MJ	73.78	
MYRISTIC ACID	GRAM	0.21	0.13	-	0.26	0.10	GRAM/MJ	0.27	
PALMITIC ACID	GRAM	3.15	2.71	-	4.25	1.48	GRAM/MJ	4.08	
STEARIC ACID	GRAM	0.63	0.34	-	1.52	0.30	GRAM/MJ	0.82	
PALMITOLEIC ACID	GRAM	0.69	0.40	-	0.82	0.32	GRAM/MJ	0.89	
OLEIC ACID	GRAM	3.10	1.99	-	3.51	1.46	GRAM/MJ	4.02	
EICOSENOIC ACID	GRAM	0.80	0.45	-	1.28	0.38	GRAM/MJ	1.04	
C22 : 1 FATTY ACID	GRAM	0.77	0.53	-	1.07	0.37	GRAM/MJ	1.01	
LINOLEIC ACID	GRAM	0.20	0.16	-	0.29	0.09	GRAM/MJ	0.26	
LINOLENIC ACID	MILLI	80.00	80.00	-	80.00	37.60	MILLI/MJ	103.56	
C18 : 4 FATTY ACID	GRAM	0.35	0.24	-	0.44	0.16	GRAM/MJ	0.45	
C20 : 5 N-3 FATTY ACID	GRAM	0.78	0.51	-	1.01	0.37	GRAM/MJ	1.01	
C22 : 5 N-3 FATTY ACID	GRAM	0.29	0.29	-	0.29	0.14	GRAM/MJ	0.38	
C22 : 6 N-3 FATTY ACID	GRAM	1.60	1.37	-	2.07	0.75	GRAM/MJ	2.08	
CHOLESTEROL	MILLI	74.00	-	-	-	34.78	MILLI/MJ	95.79	

FLUNDER FLOUNDER FLET

PLATICHTHYS FLESUS (L.)

		PROTEIN	FAT	CARBOHYDRATES	TOTAL
ENERGY VALUE (AVERAGE)	KJOULE	304	28	0.0	332
PER 100 G	(KCAL)	73	6.7	0.0	79
EDIBLE PORTION					
AMOUNT OF DIGESTIBLE	GRAM	16.00	0.66	0.00	
CONSTITUENTS PER 100 G					
ENERGY VALUE (AVERAGE)	KJOULE	295	26	0.0	321
OF THE DIGESTIBLE	(KCAL)	70	6.3	0.0	77
FRACTION PER 100 G					
EDIBLE PORTION					

WASTE PERCENTAGE AVERAGE 55 MINIMUM 50 MAXIMUM 61

CONSTITUENTS	DIM	AV	VARIATION			AVR	NUTR. DENS.		MOLPERC.
WATER	GRAM	81.40	80.00	-	82.70	36.63	GRAM/MJ	253.52	
PROTEIN	GRAM	16.50	14.90	-	18.00	7.43	GRAM/MJ	51.39	
FAT	GRAM	0.70	0.50	-	1.00	0.32	GRAM/MJ	2.18	
MINERALS	GRAM	1.30	-	-	-	0.59	GRAM/MJ	4.05	
SODIUM	MILLI	92.00	68.00	-	115.00	41.40	MILLI/MJ	286.53	
POTASSIUM	MILLI	332.00	219.00	-	415.00	149.40	GRAM/MJ	1.03	
MAGNESIUM	MILLI	24.00	19.00	-	29.00	10.80	MILLI/MJ	74.75	
CALCIUM	MILLI	27.00	17.00	-	36.00	12.15	MILLI/MJ	84.09	
MANGANESE	MICRO	39.00	-	-	-	17.55	MICRO/MJ	121.47	
IRON	MILLI	0.54	0.28	-	0.80	0.24	MILLI/MJ	1.68	
COBALT	MICRO	6.00	2.00	-	10.00	2.70	MICRO/MJ	18.69	
COPPER	MICRO	47.00	-	-	-	21.15	MICRO/MJ	146.38	
ZINC	MILLI	0.50	-	-	-	0.23	MILLI/MJ	1.56	
NICKEL	MICRO	3.00	1.00	-	6.00	1.35	MICRO/MJ	9.34	
CHROMIUM	MICRO	2.00	0.40	-	30.00	0.90	MICRO/MJ	6.23	
PHOSPHORUS	MILLI	200.00	195.00	-	205.00	90.00	MILLI/MJ	622.90	
CHLORIDE	MILLI	127.00	-	-	-	57.15	MILLI/MJ	395.54	
FLUORIDE	MICRO	30.00	-	-	-	13.50	MICRO/MJ	93.44	
IODIDE	MICRO	29.00	-	-	-	13.05	MICRO/MJ	90.32	
SELENIUM	MICRO	28.00	20.00	-	30.00	12.60	MICRO/MJ	87.21	
VITAMIN A	MICRO	9.90	-	-	-	4.46	MICRO/MJ	30.83	
ALPHA-TOCOPHEROL	MILLI	0.36	-	-	-	0.16	MILLI/MJ	1.12	
VITAMIN B1	MILLI	0.22	0.03	-	0.66	0.10	MILLI/MJ	0.69	
VITAMIN B2	MILLI	0.21	0.11	-	0.46	0.09	MILLI/MJ	0.65	
NICOTINAMIDE	MILLI	3.40	2.09	-	4.35	1.53	MILLI/MJ	10.59	
VITAMIN B6	MILLI	0.25	0.21	-	0.30	0.11	MILLI/MJ	0.78	
FOLIC ACID	MICRO	11.00	-	-	-	4.95	MICRO/MJ	34.26	
VITAMIN B12	MICRO	1.00	-	-	-	0.45	MICRO/MJ	3.11	

CONSTITUENTS	DIM	AV	VARIATION			AVR	NUTR. DENS.		MOLPERC.
ALANINE	GRAM	1.24	1.12	-	1.35	0.56	GRAM/MJ	3.86	9.4
ARGININE	GRAM	1.16	1.04	-	1.21	0.52	GRAM/MJ	3.61	4.5
ASPARTIC ACID	GRAM	2.07	1.81	-	2.25	0.93	GRAM/MJ	6.45	10.5
CYSTINE	GRAM	0.15	0.09	-	0.26	0.07	GRAM/MJ	0.47	0.4
GLUTAMIC ACID	GRAM	3.18	2.70	-	3.47	1.43	GRAM/MJ	9.90	14.6
GLYCINE	GRAM	0.93	0.83	-	1.03	0.42	GRAM/MJ	2.90	8.4
HISTIDINE	GRAM	0.45	0.42	-	0.48	0.20	GRAM/MJ	1.40	2.0
ISOLEUCINE	GRAM	0.92	0.75	-	1.01	0.41	GRAM/MJ	2.87	4.7
LEUCINE	GRAM	1.60	1.41	-	1.70	0.72	GRAM/MJ	4.98	8.2
LYSINE	GRAM	1.82	1.59	-	1.96	0.82	GRAM/MJ	5.67	8.4
METHIONINE	GRAM	0.58	0.40	-	0.78	0.26	GRAM/MJ	1.81	2.6
PHENYLALANINE	GRAM	0.70	0.61	-	0.77	0.32	GRAM/MJ	2.18	2.9
PROLINE	GRAM	0.60	0.57	-	0.65	0.27	GRAM/MJ	1.87	3.5
SERINE	GRAM	0.86	0.75	-	0.92	0.39	GRAM/MJ	2.68	5.5
THREONINE	GRAM	0.92	0.81	-	0.95	0.41	GRAM/MJ	2.87	5.2
TRYPTOPHAN	GRAM	0.21	0.14	-	0.30	0.09	GRAM/MJ	0.65	0.7
TYROSINE	GRAM	0.64	0.60	-	0.70	0.29	GRAM/MJ	1.99	2.4
VALINE	GRAM	1.06	0.86	-	1.17	0.48	GRAM/MJ	3.30	6.1
CHOLESTEROL	MILLI	48.00	46.00	-	50.00	21.60	MILLI/MJ	149.50	

GRENADIER

MACROURUS BERGLAX

SOLDIER
(GRENADIER)

GRENADIER

		PROTEIN	FAT	CARBOHYDRATES	TOTAL
ENERGY VALUE (AVERAGE)	KJOULE	322	28	0.0	350
PER 100 G	(KCAL)	77	6.7	0.0	84
EDIBLE PORTION					
AMOUNT OF DIGESTIBLE	GRAM	16.97	0.66	0.00	
CONSTITUENTS PER 100 G					
ENERGY VALUE (AVERAGE)	KJOULE	313	26	0.0	339
OF THE DIGESTIBLE	(KCAL)	75	6.3	0.0	81
FRACTION PER 100 G					
EDIBLE PORTION					

WASTE PERCENTAGE AVERAGE 0.00

CONSTITUENTS	DIM	AV	VARIATION			AVR	NUTR. DENS.		MOLPERC.
WATER	GRAM	81.40	81.40	-	81.40	81.40	GRAM/MJ	240.16	
PROTEIN	GRAM	17.50	17.50	-	17.50	17.50	GRAM/MJ	51.63	
FAT	GRAM	0.70	0.70	-	0.70	0.70	GRAM/MJ	2.07	
MINERALS	GRAM	1.20	1.20	-	1.20	1.20	GRAM/MJ	3.54	
MAGNESIUM	MILLI	83.00	83.00	-	83.00	83.00	MILLI/MJ	244.88	
CALCIUM	MILLI	20.00	20.00	-	20.00	20.00	MILLI/MJ	59.01	
MANGANESE	MICRO	80.00	80.00	-	80.00	80.00	MICRO/MJ	236.03	
COPPER	MILLI	0.17	0.17	-	0.17	0.17	MILLI/MJ	0.50	
ZINC	MILLI	1.90	1.90	-	1.90	1.90	MILLI/MJ	5.61	

HEILBUTT (WEISSER HEILBUTT) HIPPOGLOSSUS HIPPOGLOSSUS (L.) — HALIBUT — FLÉTAN

		PROTEIN	FAT	CARBOHYDRATES	TOTAL
ENERGY VALUE (AVERAGE)	KJOULE	370	91	0.0	461
PER 100 G EDIBLE PORTION	(KCAL)	88	22	0.0	110
AMOUNT OF DIGESTIBLE CONSTITUENTS PER 100 G	GRAM	19.49	2.17	0.00	
ENERGY VALUE (AVERAGE)	KJOULE	359	86	0.0	445
OF THE DIGESTIBLE FRACTION PER 100 G EDIBLE PORTION	(KCAL)	86	21	0.0	106

WASTE PERCENTAGE AVERAGE 20 MINIMUM 11 MAXIMUM 25

CONSTITUENTS	DIM	AV	VARIATION		AVR	NUTR. DENS.		MOLPERC.
WATER	GRAM	76.10	74.50	- 77.80	60.88	GRAM/MJ	170.86	
PROTEIN	GRAM	20.10	18.60	- 21.80	16.08	GRAM/MJ	45.13	
FAT	GRAM	2.29	1.20	- 4.60	1.83	GRAM/MJ	5.14	
MINERALS	GRAM	1.26	1.16	- 1.39	1.01	GRAM/MJ	2.83	
SODIUM	MILLI	67.00	54.00	- 82.00	53.60	MILLI/MJ	150.43	
POTASSIUM	MILLI	446.00	384.00	- 487.00	356.80	GRAM/MJ	1.00	
MAGNESIUM	MILLI	28.00	-	- -	22.40	MILLI/MJ	62.86	
CALCIUM	MILLI	14.00	10.00	- 16.00	11.20	MILLI/MJ	31.43	
MANGANESE	MICRO	12.00	-	- -	9.60	MICRO/MJ	26.94	
IRON	MILLI	0.55	0.39	- 0.70	0.44	MILLI/MJ	1.23	
COBALT	MICRO	4.30	-	- -	3.44	MICRO/MJ	9.65	
COPPER	MILLI	0.20	0.16	- 0.23	0.16	MILLI/MJ	0.45	
CHROMIUM	MICRO	1.00	-	- -	0.80	MICRO/MJ	2.25	
PHOSPHORUS	MILLI	202.00	192.00	- 211.00	161.60	MILLI/MJ	453.52	
IODIDE	MICRO	52.00	-	- -	41.60	MICRO/MJ	116.75	
VITAMIN A	MICRO	32.00	15.00	- 47.00	25.60	MICRO/MJ	71.84	
VITAMIN D	MICRO	5.00	-	- -	4.00	MICRO/MJ	11.23	
VITAMIN E ACTIVITY	MILLI	0.85	0.40	- 1.30	0.68	MILLI/MJ	1.91	
ALPHA-TOCOPHEROL	MILLI	0.85	0.40	- 1.30	0.68	MILLI/MJ	1.91	
VITAMIN B1	MICRO	78.00	70.00	- 90.00	62.40	MICRO/MJ	175.12	
VITAMIN B2	MICRO	70.00	45.00	- 94.00	56.00	MICRO/MJ	157.16	
NICOTINAMIDE	MILLI	5.90	3.00	- 8.30	4.72	MILLI/MJ	13.25	
PANTOTHENIC ACID	MILLI	0.30	-	- -	0.24	MILLI/MJ	0.67	
VITAMIN B6	MILLI	0.42	-	- -	0.34	MILLI/MJ	0.94	
BIOTIN	MICRO	1.90	-	- -	1.52	MICRO/MJ	4.27	
FOLIC ACID	MICRO	12.00	-	- -	9.60	MICRO/MJ	26.94	

CONSTITUENTS	DIM	AV	VARIATION			AVR	NUTR. DENS.		MOLPERC.
VITAMIN B12	MICRO	1.00	0.70	-	1.30	0.80	MICRO/MJ	2.25	
ALANINE	GRAM	1.77	-	-	-	1.42	GRAM/MJ	3.97	11.6
ARGININE	GRAM	1.37	1.20	-	1.68	1.10	GRAM/MJ	3.08	4.6
ASPARTIC ACID	GRAM	2.15	-	-	-	1.72	GRAM/MJ	4.83	9.4
CYSTINE	GRAM	0.24	0.23	-	0.28	0.19	GRAM/MJ	0.54	0.6
GLUTAMIC ACID	GRAM	3.01	-	-	-	2.41	GRAM/MJ	6.76	11.9
GLYCINE	GRAM	1.15	-	-	-	0.92	GRAM/MJ	2.58	8.9
HISTIDINE	GRAM	0.48	0.34	-	0.74	0.38	GRAM/MJ	1.08	1.8
ISOLEUCINE	GRAM	1.27	1.25	-	1.32	1.02	GRAM/MJ	2.85	5.6
LEUCINE	GRAM	1.94	1.72	-	2.17	1.55	GRAM/MJ	4.36	8.6
LYSINE	GRAM	1.56	1.36	-	1.90	1.25	GRAM/MJ	3.50	6.2
METHIONINE	GRAM	0.80	0.58	-	0.93	0.64	GRAM/MJ	1.80	3.1
PHENYLALANINE	GRAM	0.68	0.62	-	0.79	0.54	GRAM/MJ	1.53	2.4
PROLINE	GRAM	0.81	-	-	-	0.65	GRAM/MJ	1.82	4.1
SERINE	GRAM	1.27	-	-	-	1.02	GRAM/MJ	2.85	7.0
THREONINE	GRAM	0.99	-	-	-	0.79	GRAM/MJ	2.22	4.8
TRYPTOPHAN	GRAM	0.26	0.21	-	0.32	0.21	GRAM/MJ	0.58	0.7
TYROSINE	GRAM	0.68	-	-	-	0.54	GRAM/MJ	1.53	2.2
VALINE	GRAM	1.30	-	-	-	1.04	GRAM/MJ	2.92	6.5
CADAVERINE	MILLI	-	0.00	-	0.10				
PUTRESCINE	MILLI	-	0.00	-	0.10				
MYRISTIC ACID	MILLI	32.00	20.00	-	44.00	25.60	MILLI/MJ	71.84	
PALMITIC ACID	GRAM	0.26	0.26	-	0.27	0.21	GRAM/MJ	0.59	
STEARIC ACID	GRAM	0.10	0.08	-	0.12	0.08	GRAM/MJ	0.23	
PALMITOLEIC ACID	MILLI	92.50	60.00	-	125.00	74.00	MILLI/MJ	207.68	
OLEIC ACID	GRAM	0.27	0.21	-	0.35	0.22	GRAM/MJ	0.63	
EICOSENOIC ACID	GRAM	0.36	0.11	-	0.62	0.29	GRAM/MJ	0.82	
C22 : 1 FATTY ACID	MILLI	42.00	40.00	-	44.00	33.60	MILLI/MJ	94.30	
LINOLEIC ACID	MILLI	24.00	24.00	-	24.00	19.20	MILLI/MJ	53.88	
LINOLENIC ACID	MILLI	34.00	34.00	-	34.00	27.20	MILLI/MJ	76.34	
ARACHIDONIC ACID	MILLI	57.00	40.00	-	74.00	45.60	MILLI/MJ	127.97	
C18 : 4 FATTY ACID	MILLI	58.00	58.00	-	58.00	46.40	MILLI/MJ	130.22	
C20 : 5 N-3 FATTY ACID	GRAM	0.19	0.17	-	0.21	0.15	GRAM/MJ	0.43	
C22 : 5 N-3 FATTY ACID	MILLI	36.00	20.00	-	52.00	28.80	MILLI/MJ	80.83	
C22 : 6 N-3 FATTY ACID	GRAM	0.50	0.37	-	0.63	0.40	GRAM/MJ	1.12	
CHOLESTEROL	MILLI	41.00	32.00	-	50.00	32.80	MILLI/MJ	92.05	
TOTAL PURINES	MILLI	178.00 [1]	178.00	-	178.00	142.40	MILLI/MJ	399.64	
ADENINE	MILLI	17.00	17.00	-	17.00	13.60	MILLI/MJ	38.17	
GUANINE	MILLI	25.00	25.00	-	25.00	20.00	MILLI/MJ	56.13	
HYPOXANTHINE	MILLI	104.00	104.00	-	104.00	83.20	MILLI/MJ	233.50	

1 VALUE OF TOTAL PURINES IS EXPRESSSED IN MG URIC ACID/100G

SCHWARZER HEILBUTT
(GRÖNLAND HEILBUTT)
RHEINHARDTIUS HIPPOGLOSSOIDES

GREENLAND HALIBUT

FLÉTAN NOIR

		PROTEIN	FAT	CARBOHYDRATES	TOTAL
ENERGY VALUE (AVERAGE)	KJOULE	243	389	0.0	632
PER 100 G EDIBLE PORTION	(KCAL)	58	93	0.0	151
AMOUNT OF DIGESTIBLE CONSTITUENTS PER 100 G	GRAM	12.82	9.29	0.00	
ENERGY VALUE (AVERAGE)	KJOULE	236	369	0.0	605
OF THE DIGESTIBLE FRACTION PER 100 G EDIBLE PORTION	(KCAL)	56	88	0.0	145

WASTE PERCENTAGE AVERAGE 20

CONSTITUENTS	DIM	AV	VARIATION	AVR	NUTR. DENS.		MOLPERC.
WATER	GRAM	75.70	70.20 - 76.30	60.57	GRAM/MJ	125.06	
PROTEIN	GRAM	13.21	12.40 - 13.30	10.57	GRAM/MJ	21.83	
FAT	GRAM	9.78	9.40 - 15.60	7.83	GRAM/MJ	16.16	
MINERALS	GRAM	1.00	0.90 - 1.10	0.80	GRAM/MJ	1.65	
SODIUM	MILLI	86.00	86.00 - 86.00	68.80	MILLI/MJ	142.06	
POTASSIUM	MILLI	344.80	306.00 - 500.00	275.84	MILLI/MJ	569.55	
MAGNESIUM	MILLI	22.00	22.00 - 22.00	17.60	MILLI/MJ	36.34	
CALCIUM	MILLI	18.50	12.00 - 25.00	14.80	MILLI/MJ	30.56	
IRON	MILLI	0.40	0.25 - 0.56	0.32	MILLI/MJ	0.67	
COPPER	MICRO	80.00	80.00 - 80.00	64.00	MICRO/MJ	132.15	
PHOSPHORUS	MILLI	147.09	146.00 - 158.00	117.67	MILLI/MJ	242.97	
IODIDE	MICRO	20.00	20.00 - 20.00	16.00	MICRO/MJ	33.04	
VITAMIN A	MICRO	31.00	12.00 - 50.00	24.80	MICRO/MJ	51.21	
VITAMIN D	MICRO	15.00	15.00 - 15.00	12.00	MICRO/MJ	24.78	
VITAMIN E ACTIVITY	MILLI	0.85	0.85 - 0.85	0.68	MILLI/MJ	1.40	
ALPHA-TOCOPHEROL	MILLI	0.85	0.85 - 0.85	0.68	MILLI/MJ	1.40	
VITAMIN B1	MICRO	65.00	60.00 - 70.00	52.00	MICRO/MJ	107.37	
VITAMIN B2	MICRO	70.00	60.00 - 80.00	56.00	MICRO/MJ	115.63	
NICOTINAMIDE	MILLI	1.30	1.10 - 1.50	1.04	MILLI/MJ	2.15	
VITAMIN B6	MILLI	0.43	0.43 - 0.43	0.34	MILLI/MJ	0.71	
FOLIC ACID	MICRO	12.00	12.00 - 12.00	9.60	MICRO/MJ	19.82	
VITAMIN B12	MICRO	1.00	1.00 - 1.00	0.80	MICRO/MJ	1.65	

HERING HERRING HARENG

CLUPEA HARENGUS (L.)

		PROTEIN	FAT	CARBOHYDRATES	TOTAL
ENERGY VALUE (AVERAGE)	KJOULE	335	708	0.0	1043
PER 100 G	(KCAL)	80	169	0.0	249
EDIBLE PORTION					
AMOUNT OF DIGESTIBLE	GRAM	17.65	16.91	0.00	
CONSTITUENTS PER 100 G					
ENERGY VALUE (AVERAGE)	KJOULE	325	672	0.0	997
OF THE DIGESTIBLE	(KCAL)	78	161	0.0	238
FRACTION PER 100 G					
EDIBLE PORTION					

WASTE PERCENTAGE AVERAGE 30 MINIMUM 24 MAXIMUM 36

CONSTITUENTS	DIM	AV	VARIATION			AVR	NUTR.	DENS.	MOLPERC.
WATER	GRAM	65.30	60.40	-	71.50	45.71	GRAM/MJ	65.49	
PROTEIN	GRAM	18.20	17.30	-	19.60	12.74	GRAM/MJ	18.25	
FAT	GRAM	17.80 1	9.90	-	19.40	12.46	GRAM/MJ	17.85	
MINERALS	GRAM	1.27	1.17	-	1.38	0.89	GRAM/MJ	1.27	
SODIUM	MILLI	117.00	68.00	-	200.00	81.90	MILLI/MJ	117.34	
POTASSIUM	MILLI	360.00	317.00	-	422.00	252.00	MILLI/MJ	361.03	
MAGNESIUM	MILLI	31.00	26.00	-	36.00	21.70	MILLI/MJ	31.09	
CALCIUM	MILLI	34.00	29.00	-	46.00	23.80	MILLI/MJ	34.10	
MANGANESE	MILLI	-	0.02	-	0.12				
IRON	MILLI	1.10	0.93	-	1.28	0.77	MILLI/MJ	1.10	
COBALT	MICRO	1.70	-	-	-	1.19	MICRO/MJ	1.70	
COPPER	MILLI	0.32	0.25	-	0.44	0.22	MILLI/MJ	0.32	
NICKEL	MICRO	30.00	-	-	-	21.00	MICRO/MJ	30.09	
PHOSPHORUS	MILLI	250.00	235.00	-	260.00	175.00	MILLI/MJ	250.72	
CHLORIDE	MILLI	170.00	40.00	-	300.00	119.00	MILLI/MJ	170.49	
IODIDE	MICRO	52.00	40.00	-	65.00	36.40	MICRO/MJ	52.15	
SELENIUM	MILLI	0.14	-	-	-	0.10	MILLI/MJ	0.14	
BROMINE	MICRO	472.00	464.00	-	480.00	330.40	MICRO/MJ	473.35	
VITAMIN A	MICRO	38.00	20.00	-	64.00	26.60	MICRO/MJ	38.11	
VITAMIN D	MICRO	31.00	25.00	-	38.00	21.70	MICRO/MJ	31.09	
VITAMIN E ACTIVITY	MILLI	1.50	1.10	-	1.80	1.05	MILLI/MJ	1.50	
ALPHA-TOCOPHEROL	MILLI	1.50	-	-	-	1.05	MILLI/MJ	1.50	
VITAMIN B1	MICRO	40.00	16.00	-	60.00	28.00	MICRO/MJ	40.11	
VITAMIN B2	MILLI	0.22	0.10	-	0.28	0.15	MILLI/MJ	0.22	
NICOTINAMIDE	MILLI	3.80	3.10	-	4.50	2.66	MILLI/MJ	3.81	
PANTOTHENIC ACID	MILLI	0.94	-	-	-	0.66	MILLI/MJ	0.94	

1 AMOUNTS ARE STRONGLY DEPENDENT ON THE AGE OF THE FISH

CONSTITUENTS	DIM	AV	VARIATION			AVR	NUTR. DENS.		MOLPERC.
VITAMIN B6	MILLI	0.45	0.35	-	0.51	0.32	MILLI/MJ	0.45	
BIOTIN	MICRO	4.50	3.00	-	6.00	3.15	MICRO/MJ	4.51	
FOLIC ACID	MICRO	5.00	-	-	-	3.50	MICRO/MJ	5.01	
VITAMIN B12	MICRO	8.50	5.00	-	12.00	5.95	MICRO/MJ	8.52	
ALANINE	GRAM	1.52	1.44	-	1.67	1.06	GRAM/MJ	1.52	10.3
ARGININE	GRAM	1.18	0.96	-	1.38	0.83	GRAM/MJ	1.18	4.1
ASPARTIC ACID	GRAM	2.31	2.18	-	2.38	1.62	GRAM/MJ	2.32	10.5
CYSTINE	GRAM	0.24	0.21	-	0.27	0.17	GRAM/MJ	0.24	0.6
GLUTAMIC ACID	GRAM	3.23	2.94	-	3.45	2.26	GRAM/MJ	3.24	13.2
GLYCINE	GRAM	1.13	1.08	-	1.20	0.79	GRAM/MJ	1.13	9.1
HISTIDINE	GRAM	0.52	0.50	-	0.58	0.36	GRAM/MJ	0.52	2.0
ISOLEUCINE	GRAM	1.04	0.93	-	1.24	0.73	GRAM/MJ	1.04	4.8
LEUCINE	GRAM	1.75	1.59	-	1.82	1.23	GRAM/MJ	1.76	8.0
LYSINE	GRAM	1.75	1.65	-	1.90	1.23	GRAM/MJ	1.76	7.2
METHIONINE	GRAM	0.66	0.58	-	0.74	0.46	GRAM/MJ	0.66	2.7
PHENYLALANINE	GRAM	0.75	0.64	-	0.90	0.53	GRAM/MJ	0.75	2.7
PROLINE	GRAM	0.84	0.78	-	0.90	0.59	GRAM/MJ	0.84	4.4
SERINE	GRAM	1.05	0.98	-	1.17	0.74	GRAM/MJ	1.05	6.0
THREONINE	GRAM	1.04	0.97	-	1.15	0.73	GRAM/MJ	1.04	5.3
TRYPTOPHAN	GRAM	0.21	0.19	-	0.23	0.15	GRAM/MJ	0.21	0.6
TYROSINE	GRAM	0.67	0.59	-	0.73	0.47	GRAM/MJ	0.67	2.2
VALINE	GRAM	1.21	1.13	-	1.25	0.85	GRAM/MJ	1.21	6.2
LAURIC ACID	MILLI	20.00	20.00	-	20.00	14.00	MILLI/MJ	20.06	
MYRISTIC ACID	GRAM	0.92	0.76	-	1.41	0.65	GRAM/MJ	0.92	
PALMITIC ACID	GRAM	1.56	1.29	-	2.07	1.10	GRAM/MJ	1.57	
STEARIC ACID	GRAM	0.15	0.13	-	0.25	0.11	GRAM/MJ	0.16	
ARACHIDIC ACID	MILLI	1.00	1.00	-	1.00	0.70	MILLI/MJ	1.00	
PALMITOLEIC ACID	GRAM	1.18	0.80	-	1.82	0.83	GRAM/MJ	1.19	
OLEIC ACID	GRAM	1.70	1.50	-	2.09	1.19	GRAM/MJ	1.70	
EICOSENOIC ACID	GRAM	2.37	1.82	-	2.48	1.66	GRAM/MJ	2.38	
C22 : 1 FATTY ACID	GRAM	3.64	2.63	-	4.06	2.55	GRAM/MJ	3.66	
LINOLEIC ACID	GRAM	0.15	0.11	-	0.34	0.11	GRAM/MJ	0.15	
LINOLENIC ACID	MILLI	61.66	50.00	-	90.00	43.17	MILLI/MJ	61.84	
ARACHIDONIC ACID	MILLI	36.66	20.00	-	90.00	25.67	MILLI/MJ	36.77	
C18 : 4 FATTY ACID	GRAM	1.15	1.15	-	1.15	0.81	GRAM/MJ	1.15	
C20 : 5 N-3 FATTY ACID	GRAM	2.68	0.44	-	4.08	1.88	GRAM/MJ	2.70	
C22 : 5 N-3 FATTY ACID	GRAM	0.10	0.07	-	0.26	0.08	GRAM/MJ	0.11	
C22 : 6 N-3 FATTY ACID	GRAM	0.45	0.31	-	0.91	0.32	GRAM/MJ	0.45	
CHOLESTEROL	MILLI	91.00	85.00	-	99.00	63.70	MILLI/MJ	91.26	
TOTAL PURINES	MILLI	210.00 2	-	-	-	147.00	MILLI/MJ	210.60	
TOTAL PHOSPHOLIPIDS	GRAM	2.58	-	-	-	1.81	GRAM/MJ	2.59	
PHOSPHATIDYLCHOLINE	GRAM	1.38	-	-	-	0.97	GRAM/MJ	1.38	
PHOSPHATIDYLETHANOLAMINE	MILLI	686.00	-	-	-	480.20	MILLI/MJ	687.97	
PHOSPHATIDYLSERINE	MILLI	360.00	-	-	-	252.00	MILLI/MJ	361.03	
SPHINGOMYELIN	MILLI	115.00	-	-	-	80.50	MILLI/MJ	115.33	

2 VALUE EXPRESSED IN MG URIC ACID/100G

HERING
OSTSEEHERING

CLUPEA HARENGUS MEMBRAS (L.)

HERRING
BALTIC SEA HERRING

HARENG BALTIQUE

		PROTEIN	FAT	CARBOHYDRATES	TOTAL
ENERGY VALUE (AVERAGE)	KJOULE	333	364	0.0	697
PER 100 G EDIBLE PORTION	(KCAL)	80	87	0.0	167
AMOUNT OF DIGESTIBLE CONSTITUENTS PER 100 G	GRAM	17.55	8.69	0.00	
ENERGY VALUE (AVERAGE)	KJOULE	323	346	0.0	669
OF THE DIGESTIBLE FRACTION PER 100 G EDIBLE PORTION	(KCAL)	77	83	0.0	160

WASTE PERCENTAGE AVERAGE 35 MINIMUM 33 MAXIMUM 37

CONSTITUENTS	DIM	AV	VARIATION			AVR	NUTR.	DENS.	MOLPERC.
WATER	GRAM	71.20	67.50	-	78.70	46.28	GRAM/MJ	106.47	
PROTEIN	GRAM	18.10	17.40	-	19.00	11.77	GRAM/MJ	27.07	
FAT	GRAM	9.15	3.60	-	14.10	5.95	GRAM/MJ	13.68	
MINERALS	GRAM	1.27	1.18	-	1.35	0.83	GRAM/MJ	1.90	
SODIUM	MILLI	74.00	65.00	-	88.00	48.10	MILLI/MJ	110.66	
POTASSIUM	MILLI	370.00	356.00	-	383.00	240.50	MILLI/MJ	553.29	
CALCIUM	MILLI	60.00	-	-	-	39.00	MILLI/MJ	89.72	
MANGANESE	MICRO	20.00	-	-	-	13.00	MICRO/MJ	29.91	
IRON	MILLI	1.20	-	-	-	0.78	MILLI/MJ	1.79	
COPPER	MILLI	0.30	-	-	-	0.20	MILLI/MJ	0.45	
PHOSPHORUS	MILLI	240.00	-	-	-	156.00	MILLI/MJ	358.89	
IODIDE	MICRO	50.00	-	-	-	32.50	MICRO/MJ	74.77	
VITAMIN A	MICRO	20.00	-	-	-	13.00	MICRO/MJ	29.91	
VITAMIN D	MICRO	7.80	-	-	-	5.07	MICRO/MJ	11.66	
VITAMIN B1	MICRO	55.00	-	-	-	35.75	MICRO/MJ	82.25	
VITAMIN B2	MILLI	0.24	-	-	-	0.16	MILLI/MJ	0.36	
NICOTINAMIDE	MILLI	4.30	-	-	-	2.80	MILLI/MJ	6.43	
PANTOTHENIC ACID	MILLI	9.30	-	-	-	6.05	MILLI/MJ	13.91	
VITAMIN B12	MICRO	11.00	-	-	-	7.15	MICRO/MJ	16.45	
ARGININE	GRAM	1.07	-	-	-	0.70	GRAM/MJ	1.60	
HISTIDINE	GRAM	0.36	-	-	-	0.23	GRAM/MJ	0.54	
ISOLEUCINE	GRAM	1.15	-	-	-	0.75	GRAM/MJ	1.72	
LEUCINE	GRAM	1.45	-	-	-	0.94	GRAM/MJ	2.17	
LYSINE	GRAM	1.48	-	-	-	0.96	GRAM/MJ	2.21	
METHIONINE	GRAM	0.40	-	-	-	0.26	GRAM/MJ	0.60	
PHENYLALANINE	GRAM	0.81	-	-	-	0.53	GRAM/MJ	1.21	
THREONINE	GRAM	0.81	-	-	-	0.53	GRAM/MJ	1.21	
TRYPTOPHAN	GRAM	0.13	-	-	-	0.08	GRAM/MJ	0.19	
TYROSINE	GRAM	0.40	-	-	-	0.26	GRAM/MJ	0.60	
VALINE	GRAM	0.92	-	-	-	0.60	GRAM/MJ	1.38	
CHOLESTEROL	MILLI	44.00	31.00	-	53.00	28.60	MILLI/MJ	65.80	

HERINGSMILCH	SOFT ROE OF HERRING (MILT OF HERRING)	LAITANCE DE HARENG

		PROTEIN	FAT	CARBOHYDRATES	TOTAL
ENERGY VALUE (AVERAGE)	KJOULE	385	112	0.0	497
PER 100 G EDIBLE PORTION	(KCAL)	92	27	0.0	119
AMOUNT OF DIGESTIBLE CONSTITUENTS PER 100 G	GRAM	20.27	2.68	0.00	
ENERGY VALUE (AVERAGE)	KJOULE	373	107	0.0	480
OF THE DIGESTIBLE FRACTION PER 100 G EDIBLE PORTION	(KCAL)	89	26	0.0	115

WASTE PERCENTAGE AVERAGE 0.00

CONSTITUENTS	DIM	AV	VARIATION			AVR	NUTR. DENS.		MOLPERC.
WATER	GRAM	75.40	74.00	-	76.70	75.40	GRAM/MJ	157.06	
PROTEIN	GRAM	20.90	19.80	-	22.80	20.90	GRAM/MJ	43.53	
FAT	GRAM	2.83	2.00	-	3.65	2.83	GRAM/MJ	5.89	
MINERALS	GRAM	2.18	-	-	-	2.18	GRAM/MJ	4.54	
SODIUM	MILLI	118.00	-	-	-	118.00	MILLI/MJ	245.79	
POTASSIUM	MILLI	364.00	-	-	-	364.00	MILLI/MJ	758.21	
VITAMIN B1	MICRO	36.00	-	-	-	36.00	MICRO/MJ	74.99	
VITAMIN B2	MILLI	0.60	-	-	-	0.60	MILLI/MJ	1.25	
ARGININE	GRAM	2.00	-	-	-	2.00	GRAM/MJ	4.17	
CYSTINE	GRAM	0.12	-	-	-	0.12	GRAM/MJ	0.25	
HISTIDINE	GRAM	0.79	-	-	-	0.79	GRAM/MJ	1.65	
LYSINE	GRAM	0.98	-	-	-	0.98	GRAM/MJ	2.04	
METHIONINE	GRAM	0.57	-	-	-	0.57	GRAM/MJ	1.19	
PHENYLALANINE	GRAM	0.79	-	-	-	0.79	GRAM/MJ	1.65	
THREONINE	GRAM	1.14	-	-	-	1.14	GRAM/MJ	2.37	
TRYPTOPHAN	GRAM	0.22	-	-	-	0.22	GRAM/MJ	0.46	
TYROSINE	GRAM	0.49	-	-	-	0.49	GRAM/MJ	1.02	
VALINE	GRAM	1.12	-	-	-	1.12	GRAM/MJ	2.33	

HERINGSROGEN HERRING ROE ROGUE DE HARENG

		PROTEIN	FAT	CARBOHYDRATES	TOTAL
ENERGY VALUE (AVERAGE)	KJOULE	479	121	0.0	600
PER 100 G	(KCAL)	114	29	0.0	143
EDIBLE PORTION					
AMOUNT OF DIGESTIBLE	GRAM	25.22	2.89	0.00	
CONSTITUENTS PER 100 G					
ENERGY VALUE (AVERAGE)	KJOULE	464	115	0.0	579
OF THE DIGESTIBLE	(KCAL)	111	28	0.0	138
FRACTION PER 100 G					
EDIBLE PORTION					

WASTE PERCENTAGE AVERAGE 0.00

CONSTITUENTS	DIM	AV	VARIATION		AVR	NUTR.	DENS.	MOLPERC.
WATER	GRAM	68.00	66.00	- 69.20	68.00	GRAM/MJ	117.35	
PROTEIN	GRAM	26.00	23.60	- 28.00	26.00	GRAM/MJ	44.87	
FAT	GRAM	3.05	1.00	- 4.95	3.05	GRAM/MJ	5.26	
MINERALS	GRAM	1.40	1.38	- 1.41	1.40	GRAM/MJ	2.42	
SODIUM	MILLI	91.00	-	- -	91.00	MILLI/MJ	157.04	
POTASSIUM	MILLI	221.00	-	- -	221.00	MILLI/MJ	381.39	
VITAMIN B1	MICRO	65.00	-	- -	65.00	MICRO/MJ	112.17	
VITAMIN B2	MILLI	0.39	-	- -	0.39	MILLI/MJ	0.67	
ARGININE	GRAM	2.41	-	- -	2.41	GRAM/MJ	4.16	
CYSTINE	GRAM	0.25	-	- -	0.25	GRAM/MJ	0.43	
HISTIDINE	GRAM	0.84	-	- -	0.84	GRAM/MJ	1.45	
LYSINE	GRAM	1.68	-	- -	1.68	GRAM/MJ	2.90	
METHIONINE	GRAM	0.79	-	- -	0.79	GRAM/MJ	1.36	
PHENYLALANINE	GRAM	1.04	-	- -	1.04	GRAM/MJ	1.79	
THREONINE	GRAM	1.57	-	- -	1.57	GRAM/MJ	2.71	
TRYPTOPHAN	GRAM	0.33	-	- -	0.33	GRAM/MJ	0.57	
TYROSINE	GRAM	0.80	-	- -	0.80	GRAM/MJ	1.38	
VALINE	GRAM	1.57	-	- -	1.57	GRAM/MJ	2.71	
TOTAL PURINES	MILLI	190.00 1	190.00	- 190.00	190.00	MILLI/MJ	327.89	
ADENINE	MILLI	31.00	31.00	- 31.00	31.00	MILLI/MJ	53.50	
GUANINE	MILLI	57.00	57.00	- 57.00	57.00	MILLI/MJ	98.37	
XANTHINE	MILLI	7.00	7.00	- 7.00	7.00	MILLI/MJ	12.08	
HYPOXANTHINE	MILLI	65.00	65.00	- 65.00	65.00	MILLI/MJ	112.17	

1 VALUE OF TOTAL PURINES IS EXPRESSSED IN MG URIC ACID/100G

KABELJAU (DORSCH) — COD — CABILLAUD

GADUS MORHUA (L.)

		PROTEIN	FAT	CARBOHYDRATES	TOTAL
ENERGY VALUE (AVERAGE)	KJOULE	326	16	0.0	342
PER 100 G	(KCAL)	78	3.8	0.0	82
EDIBLE PORTION					
AMOUNT OF DIGESTIBLE	GRAM	17.16	0.38	0.00	
CONSTITUENTS PER 100 G					
ENERGY VALUE (AVERAGE)	KJOULE	316	15	0.0	331
OF THE DIGESTIBLE	(KCAL)	76	3.6	0.0	79
FRACTION PER 100 G					
EDIBLE PORTION					

WASTE PERCENTAGE AVERAGE 25 MINIMUM 20 MAXIMUM 30

CONSTITUENTS	DIM	AV	VARIATION			AVR	NUTR. DENS.		MOLPERC.
WATER	GRAM	80.80	79.40	-	83.00	60.60	GRAM/MJ	243.98	
PROTEIN	GRAM	17.70	16.00	-	19.00	13.28	GRAM/MJ	53.45	
FAT	GRAM	0.40	0.15	-	0.65	0.30	GRAM/MJ	1.21	
MINERALS	GRAM	1.10	0.90	-	1.20	0.83	GRAM/MJ	3.32	
SODIUM	MILLI	72.00	55.00	-	84.00	54.00	MILLI/MJ	217.41	
POTASSIUM	MILLI	356.00	258.00	-	395.00	267.00	GRAM/MJ	1.07	
MAGNESIUM	MILLI	25.00	22.00	-	28.00	18.75	MILLI/MJ	75.49	
CALCIUM	MILLI	24.00	20.00	-	30.00	18.00	MILLI/MJ	72.47	
MANGANESE	MICRO	17.00	10.00	-	25.00	12.75	MICRO/MJ	51.33	
IRON	MILLI	0.44	0.40	-	0.52	0.33	MILLI/MJ	1.33	
COBALT	MICRO	0.40	-	-	-	0.30	MICRO/MJ	1.21	
COPPER	MILLI	0.23	0.04	-	0.47	0.17	MILLI/MJ	0.69	
ZINC	MILLI	0.50	-	-	-	0.38	MILLI/MJ	1.51	
NICKEL	MICRO	2.00	0.50	-	5.00	1.50	MICRO/MJ	6.04	
CHROMIUM	MICRO	1.00	0.20	-	1.00	0.75	MICRO/MJ	3.02	
VANADIUM	MICRO	19.00	-	-	-	14.25	MICRO/MJ	57.37	
PHOSPHORUS	MILLI	184.00	140.00	-	200.00	138.00	MILLI/MJ	555.59	
CHLORIDE	MILLI	228.00	55.00	-	400.00	171.00	MILLI/MJ	688.45	
FLUORIDE	MICRO	28.00	-	-	-	21.00	MICRO/MJ	84.55	
IODIDE	MILLI	0.12	0.10	-	0.15	0.09	MILLI/MJ	0.36	
BORON	MILLI	-	0.06	-	0.20				
SELENIUM	MICRO	27.00	10.00	-	120.00	20.25	MICRO/MJ	81.53	
BROMINE	MICRO	824.00	-	-	-	618.00	MILLI/MJ	2.49	
VITAMIN A	MICRO	10.00	8.00	-	12.00	7.50	MICRO/MJ	30.20	
CAROTENE	-	0.00	-	-	-	0.00			
VITAMIN D	MICRO	1.30	-	-	-	0.98	MICRO/MJ	3.93	

CONSTITUENTS	DIM	AV	VARIATION			AVR	NUTR. DENS.		MOLPERC.
VITAMIN E ACTIVITY	MILLI	0.26	0.15	-	0.43	0.20	MILLI/MJ	0.79	
VITAMIN B1	MICRO	55.00	40.00	-	75.00	41.25	MICRO/MJ	166.07	
VITAMIN B2	MICRO	46.00	31.00	-	70.00	34.50	MICRO/MJ	138.90	
NICOTINAMIDE	MILLI	2.30	2.00	-	3.10	1.73	MILLI/MJ	6.94	
PANTOTHENIC ACID	MILLI	0.12	-	-	-	0.09	MILLI/MJ	0.36	
VITAMIN B6	MILLI	0.20	0.12	-	0.28	0.15	MILLI/MJ	0.60	
BIOTIN	MICRO	-	0.20	-	3.00				
FOLIC ACID	MICRO	12.00	-	-	-	9.00	MICRO/MJ	36.23	
VITAMIN B12	MICRO	0.53	0.45	-	0.60	0.40	MICRO/MJ	1.60	
VITAMIN C	MILLI	2.00	-	-	-	1.50	MILLI/MJ	6.04	
ALANINE	GRAM	1.42	1.33	-	1.50	1.07	GRAM/MJ	4.29	10.0
ARGININE	GRAM	1.21	1.16	-	1.26	0.91	GRAM/MJ	3.65	4.4
ASPARTIC ACID	GRAM	2.01	1.98	-	2.03	1.51	GRAM/MJ	6.07	9.5
CYSTINE	GRAM	0.25	0.23	-	0.29	0.19	GRAM/MJ	0.75	0.7
GLUTAMIC ACID	GRAM	3.13	3.05	-	3.21	2.35	GRAM/MJ	9.45	13.4
GLYCINE	GRAM	0.94	0.90	-	0.99	0.71	GRAM/MJ	2.84	7.9
HISTIDINE	GRAM	0.52	0.44	-	0.61	0.39	GRAM/MJ	1.57	2.1
ISOLEUCINE	GRAM	0.99	0.94	-	1.07	0.74	GRAM/MJ	2.99	4.8
LEUCINE	GRAM	1.69	1.53	-	1.88	1.27	GRAM/MJ	5.10	8.1
LYSINE	GRAM	2.05	1.83	-	2.16	1.54	GRAM/MJ	6.19	8.8
METHIONINE	GRAM	0.60	0.41	-	0.67	0.45	GRAM/MJ	1.81	2.5
PHENYLALANINE	GRAM	0.84	0.72	-	0.99	0.63	GRAM/MJ	2.54	3.2
PROLINE	GRAM	0.82	0.78	-	0.85	0.62	GRAM/MJ	2.48	4.5
SERINE	GRAM	0.99	0.94	-	1.06	0.74	GRAM/MJ	2.99	5.9
THREONINE	GRAM	0.97	0.92	-	1.06	0.73	GRAM/MJ	2.93	5.1
TRYPTOPHAN	GRAM	0.24	0.20	-	0.25	0.18	GRAM/MJ	0.72	0.7
TYROSINE	GRAM	0.71	0.68	-	0.77	0.53	GRAM/MJ	2.14	2.5
VALINE	GRAM	1.09	1.04	-	1.18	0.82	GRAM/MJ	3.29	5.9
CADAVERINE	MILLI	-	0.00	-	0.10				
PUTRESCINE	MILLI	-	0.00	-	0.10				
TRIMETHYLAMINE	MILLI	-	0.00	-	5.00				
STEARIC ACID	MILLI	13.00	13.00	-	13.00	9.75	MILLI/MJ	39.25	
PALMITOLEIC ACID	MILLI	18.00	18.00	-	18.00	13.50	MILLI/MJ	54.35	
OLEIC ACID	MILLI	48.00	48.00	-	48.00	36.00	MILLI/MJ	144.94	
EICOSENOIC ACID	MILLI	22.00	22.00	-	22.00	16.50	MILLI/MJ	66.43	
LINOLEIC ACID	MILLI	4.00	4.00	-	4.00	3.00	MILLI/MJ	12.08	
LINOLENIC ACID	MILLI	2.00	2.00	-	2.00	1.50	MILLI/MJ	6.04	
ARACHIDONIC ACID	MILLI	3.00	3.00	-	3.00	2.25	MILLI/MJ	9.06	
C20 : 5 N-3 FATTY ACID	MILLI	35.00	35.00	-	35.00	26.25	MILLI/MJ	105.68	
C22 : 5 N-3 FATTY ACID	MILLI	11.00	11.00	-	11.00	8.25	MILLI/MJ	33.21	
C22 : 6 N-3 FATTY ACID	MILLI	56.00	56.00	-	56.00	42.00	MILLI/MJ	169.09	
CHOLESTEROL	MILLI	47.00	30.00	-	70.00	35.25	MILLI/MJ	141.92	
TOTAL PURINES	MILLI	109.00 1	109.00	-	109.00	81.75	MILLI/MJ	329.13	
ADENINE	MILLI	12.00	12.00	-	12.00	9.00	MILLI/MJ	36.23	
GUANINE	MILLI	10.00	10.00	-	10.00	7.50	MILLI/MJ	30.20	
HYPOXANTHINE	MILLI	66.00	66.00	-	66.00	49.50	MILLI/MJ	199.29	
TOTAL PHOSPHOLIPIDS	MILLI	353.00	-	-	-	264.75	GRAM/MJ	1.07	
PHOSPHATIDYLCHOLINE	MILLI	243.00	-	-	-	182.25	MILLI/MJ	733.74	
PHOSPHATIDYLETHANOLAMINE	MILLI	67.00	-	-	-	50.25	MILLI/MJ	202.31	
PHOSPHATIDYLSERINE	MILLI	18.00	-	-	-	13.50	MILLI/MJ	54.35	

1 VALUE OF TOTAL PURINES IS EXPRESSSED IN MG URIC ACID/100G

KATFISCH (STEINBEISSER) | CATFISH | LOUP

ANARHICHAS LUPUS (L.)
ANARHICHAS MINOR (OLAFSEN)

		PROTEIN	FAT	CARBOHYDRATES	TOTAL
ENERGY VALUE (AVERAGE)	KJOULE	291	111	0.0	402
PER 100 G	(KCAL)	70	27	0.0	96
EDIBLE PORTION					
AMOUNT OF DIGESTIBLE	GRAM	15.32	2.65	0.00	
CONSTITUENTS PER 100 G					
ENERGY VALUE (AVERAGE)	KJOULE	282	105	0.0	387
OF THE DIGESTIBLE	(KCAL)	67	25	0.0	93
FRACTION PER 100 G					
EDIBLE PORTION					

WASTE PERCENTAGE AVERAGE 48 MINIMUM 44 MAXIMUM 54

CONSTITUENTS	DIM	AV	VARIATION			AVR	NUTR. DENS.		MOLPERC.
WATER	GRAM	80.30	78.00	-	83.80	41.76	GRAM/MJ	207.23	
PROTEIN	GRAM	15.80	14.00	-	17.60	8.22	GRAM/MJ	40.77	
FAT	GRAM	2.79	1.58	-	4.30	1.45	GRAM/MJ	7.20	
MINERALS	GRAM	1.10	0.98	-	1.20	0.57	GRAM/MJ	2.84	
SODIUM	MILLI	105.00	98.00	-	116.00	54.60	MILLI/MJ	270.97	
POTASSIUM	MILLI	282.00	250.00	-	316.00	146.64	MILLI/MJ	727.75	
MAGNESIUM	MILLI	27.00	-	-	-	14.04	MILLI/MJ	69.68	
CALCIUM	MILLI	20.00	15.00	-	26.00	10.40	MILLI/MJ	51.61	
IRON	MILLI	1.00	-	-	-	0.52	MILLI/MJ	2.58	
PHOSPHORUS	MILLI	179.00	146.00	-	210.00	93.08	MILLI/MJ	461.94	
CHLORIDE	MILLI	110.00	-	-	-	57.20	MILLI/MJ	283.87	
FLUORIDE	MICRO	6.00	-	-	-	3.12	MICRO/MJ	15.48	
VITAMIN A	MICRO	18.00	6.00	-	29.00	9.36	MICRO/MJ	46.45	
VITAMIN D	MICRO	0.50	-	-	-	0.26	MICRO/MJ	1.29	
VITAMIN B1	MILLI	0.20	0.18	-	0.22	0.10	MILLI/MJ	0.52	
VITAMIN B2	MICRO	60.00	50.00	-	60.00	31.20	MICRO/MJ	154.84	
NICOTINAMIDE	MILLI	2.40	2.30	-	2.50	1.25	MILLI/MJ	6.19	
ALANINE	GRAM	1.54	-	-	-	0.80	GRAM/MJ	3.97	11.8
ARGININE	GRAM	1.01	-	-	-	0.53	GRAM/MJ	2.61	4.0
ASPARTIC ACID	GRAM	1.90	-	-	-	0.99	GRAM/MJ	4.90	9.8
CYSTINE	GRAM	0.18	-	-	-	0.09	GRAM/MJ	0.46	0.5
GLUTAMIC ACID	GRAM	2.62	-	-	-	1.36	GRAM/MJ	6.76	12.2
GLYCINE	GRAM	1.04	-	-	-	0.54	GRAM/MJ	2.68	9.5
HISTIDINE	GRAM	0.36	-	-	-	0.19	GRAM/MJ	0.93	1.6
ISOLEUCINE	GRAM	1.11	-	-	-	0.58	GRAM/MJ	2.86	5.8
LEUCINE	GRAM	1.43	-	-	-	0.74	GRAM/MJ	3.69	7.4
LYSINE	GRAM	1.56	-	-	-	0.81	GRAM/MJ	4.03	7.3
METHIONINE	GRAM	0.52	-	-	-	0.27	GRAM/MJ	1.34	2.4
PHENYLALANINE	GRAM	0.71	-	-	-	0.37	GRAM/MJ	1.83	2.9
PROLINE	GRAM	0.69	-	-	-	0.36	GRAM/MJ	1.78	4.1
SERINE	GRAM	1.01	-	-	-	0.53	GRAM/MJ	2.61	6.6
THREONINE	GRAM	0.93	-	-	-	0.48	GRAM/MJ	2.40	5.3
TRYPTOPHAN	GRAM	0.15	-	-	-	0.08	GRAM/MJ	0.39	0.5
TYROSINE	GRAM	0.54	-	-	-	0.28	GRAM/MJ	1.39	2.0
VALINE	GRAM	1.09	-	-	-	0.57	GRAM/MJ	2.81	6.4

KLIESCHE DAB LIMANDE

SCHARBE

LIMANDA LIMANDA

		PROTEIN	FAT	CARBOHYDRATES	TOTAL
ENERGY VALUE (AVERAGE)	KJOULE	318	62	0.0	380
PER 100 G	(KCAL)	76	15	0.0	91
EDIBLE PORTION					
AMOUNT OF DIGESTIBLE	GRAM	16.75	1.47	0.00	
CONSTITUENTS PER 100 G					
ENERGY VALUE (AVERAGE)	KJOULE	308	59	0.0	367
OF THE DIGESTIBLE	(KCAL)	74	14	0.0	88
FRACTION PER 100 G					
EDIBLE PORTION					

WASTE PERCENTAGE AVERAGE 35

CONSTITUENTS	DIM	AV	VARIATION		AVR	NUTR. DENS.		MOLPERC.
WATER	GRAM	81.26	81.10	- 81.90	52.82	GRAM/MJ	221.41	
PROTEIN	GRAM	17.27	15.10	- 18.00	11.23	GRAM/MJ	47.07	
FAT	GRAM	1.55	1.00	- 2.10	1.01	GRAM/MJ	4.22	
MINERALS	GRAM	1.10	1.10	- 1.10	0.72	GRAM/MJ	3.00	
SODIUM	MILLI	77.25	75.00	- 84.00	50.21	MILLI/MJ	210.48	
POTASSIUM	MILLI	350.00	350.00	- 350.00	227.50	MILLI/MJ	953.64	
MAGNESIUM	MILLI	24.00	24.00	- 24.00	15.60	MILLI/MJ	65.39	
CALCIUM	MILLI	24.28	8.00	- 27.00	15.79	MILLI/MJ	66.17	
MANGANESE	MICRO	40.00	40.00	- 40.00	26.00	MICRO/MJ	108.99	
IRON	MILLI	0.29	0.28	- 0.30	0.19	MILLI/MJ	0.80	
COPPER	MILLI	21.00	0.00	- 190.00	13.65	MILLI/MJ	57.22	
ZINC	MILLI	0.54	0.46	- 0.94	0.36	MILLI/MJ	1.49	
IODIDE	MICRO	30.00	30.00	- 30.00	19.50	MICRO/MJ	81.74	
SELENIUM	MICRO	70.00	70.00	- 70.00	45.50	MICRO/MJ	190.73	
VITAMIN A	MICRO	14.00	14.00	- 14.00	9.10	MICRO/MJ	38.15	
VITAMIN D	MICRO	1.50	1.50	- 1.50	0.98	MICRO/MJ	4.09	
VITAMIN E ACTIVITY	MILLI	0.40	0.40	- 0.40	0.26	MILLI/MJ	1.09	
ALPHA-TOCOPHEROL	MILLI	0.40	0.40	- 0.40	0.26	MILLI/MJ	1.09	
VITAMIN B1	MILLI	0.10	0.10	- 0.10	0.07	MILLI/MJ	0.27	
VITAMIN B2	MICRO	80.00	80.00	- 80.00	52.00	MICRO/MJ	217.98	
NICOTINAMIDE	MILLI	2.30	2.30	- 2.30	1.50	MILLI/MJ	6.27	
PANTOTHENIC ACID	MILLI	0.86	0.86	- 0.86	0.56	MILLI/MJ	2.34	
VITAMIN B6	MILLI	0.19	0.19	- 0.19	0.12	MILLI/MJ	0.52	
BIOTIN	MICRO	1.20	1.20	- 1.20	0.78	MICRO/MJ	3.27	
FOLIC ACID	MICRO	5.00	5.00	- 5.00	3.25	MICRO/MJ	13.62	
VITAMIN B12	MICRO	1.50	1.50	- 1.50	0.98	MICRO/MJ	4.09	

KÖHLER (SEELACHS) | SAITHE (COALFISH) | LIEU NOIR

POLLACHIUS VIRENS (L.)
POLLACHIUS POLLACHIUS (L.)

		PROTEIN	FAT	CARBOHYDRATES	TOTAL
ENERGY VALUE (AVERAGE)	KJOULE	337	31	0.0	368
PER 100 G	(KCAL)	81	7.4	0.0	88
EDIBLE PORTION					
AMOUNT OF DIGESTIBLE	GRAM	17.75	0.74	0.00	
CONSTITUENTS PER 100 G					
ENERGY VALUE (AVERAGE)	KJOULE	327	29	0.0	356
OF THE DIGESTIBLE	(KCAL)	78	7.0	0.0	85
FRACTION PER 100 G					
EDIBLE PORTION					

WASTE PERCENTAGE AVERAGE 35 MINIMUM 34 MAXIMUM 35

CONSTITUENTS	DIM	AV	VARIATION			AVR	NUTR. DENS.		MOLPERC.
WATER	GRAM	80.20	79.70	-	81.00	52.13	GRAM/MJ	225.13	
PROTEIN	GRAM	18.30	17.90	-	19.10	11.90	GRAM/MJ	51.37	
FAT	GRAM	0.78	0.24	-	1.00	0.51	GRAM/MJ	2.19	
MINERALS	GRAM	1.20	1.17	-	1.23	0.78	GRAM/MJ	3.37	
SODIUM	MILLI	81.30	75.50	-	87.00	52.85	MILLI/MJ	228.22	
POTASSIUM	MILLI	374.00	349.00	-	398.00	243.10	GRAM/MJ	1.05	
CALCIUM	MILLI	14.00	8.00	-	22.00	9.10	MILLI/MJ	39.30	
IRON	MILLI	1.00	-	-	-	0.65	MILLI/MJ	2.81	
PHOSPHORUS	MILLI	300.00	-	-	-	195.00	MILLI/MJ	842.12	
IODIDE	MICRO	200.00	142.00	-	260.00	130.00	MICRO/MJ	561.42	
BROMINE	MICRO	930.00	895.00	-	965.00	604.50	MILLI/MJ	2.61	
VITAMIN A	MICRO	10.50	9.00	-	12.00	6.83	MICRO/MJ	29.47	
VITAMIN B1	MICRO	88.00	85.00	-	90.00	57.20	MICRO/MJ	247.02	
VITAMIN B2	MILLI	0.35	-	-	-	0.23	MILLI/MJ	0.98	
NICOTINAMIDE	MILLI	4.00	-	-	-	2.60	MILLI/MJ	11.23	
VITAMIN B12	MICRO	3.50	-	-	-	2.28	MICRO/MJ	9.82	
ALANINE	GRAM	1.59	1.43	-	1.74	1.03	GRAM/MJ	4.46	10.7
ARGININE	GRAM	1.18	1.16	-	1.25	0.77	GRAM/MJ	3.31	4.1
ASPARTIC ACID	GRAM	2.23	2.07	-	2.42	1.45	GRAM/MJ	6.26	10.0
CYSTINE	GRAM	0.18	-	-	-	0.12	GRAM/MJ	0.51	0.4
GLUTAMIC ACID	GRAM	3.16	2.89	-	3.43	2.05	GRAM/MJ	8.87	12.9
GLYCINE	GRAM	1.10	1.00	-	1.20	0.72	GRAM/MJ	3.09	8.8
HISTIDINE	GRAM	0.46	0.39	-	0.51	0.30	GRAM/MJ	1.29	1.8
ISOLEUCINE	GRAM	1.14	1.03	-	1.25	0.74	GRAM/MJ	3.20	5.2
LEUCINE	GRAM	1.75	1.65	-	1.84	1.14	GRAM/MJ	4.91	8.0
LYSINE	GRAM	2.05	1.64	-	2.27	1.33	GRAM/MJ	5.75	8.4
METHIONINE	GRAM	0.66	0.60	-	0.72	0.43	GRAM/MJ	1.85	2.6
PHENYLALANINE	GRAM	0.78	0.74	-	0.86	0.51	GRAM/MJ	2.19	2.8
PROLINE	GRAM	0.78	0.71	-	0.88	0.51	GRAM/MJ	2.19	4.1
SERINE	GRAM	1.06	0.98	-	1.27	0.69	GRAM/MJ	2.98	6.0
THREONINE	GRAM	1.02	0.92	-	1.18	0.66	GRAM/MJ	2.86	5.1
TRYPTOPHAN	GRAM	0.19	-	-	-	0.12	GRAM/MJ	0.53	0.6
TYROSINE	GRAM	0.68	0.64	-	0.74	0.44	GRAM/MJ	1.91	2.2
VALINE	GRAM	1.24	1.20	-	1.26	0.81	GRAM/MJ	3.48	6.3

CONSTITUENTS	DIM	AV	VARIATION			AVR	NUTR. DENS.		MOLPERC.
CHOLESTEROL	MILLI	71.00	-	-	-	46.15	MILLI/MJ	199.30	
TOTAL PURINES	MILLI	163.00 1	163.00	-	163.00	105.95	MILLI/MJ	457.55	
ADENINE	MILLI	15.00	15.00	-	15.00	9.75	MILLI/MJ	42.11	
GUANINE	MILLI	11.00	11.00	-	11.00	7.15	MILLI/MJ	30.88	
XANTHINE	MILLI	3.00	3.00	-	3.00	1.95	MILLI/MJ	8.42	
HYPOXANTHINE	MILLI	104.00	104.00	-	104.00	67.60	MILLI/MJ	291.94	

1 VALUE OF TOTAL PURINES IS EXPRESSSED IN MG URIC ACID/100G

LENGFISCH LING LINGUE

MOLVA MOLVA (L.)

		PROTEIN	FAT	CARBOHYDRATES	TOTAL
ENERGY VALUE (AVERAGE)	KJOULE	350	23	0.0	372
PER 100 G EDIBLE PORTION	(KCAL)	84	5.4	0.0	89
AMOUNT OF DIGESTIBLE CONSTITUENTS PER 100 G	GRAM	18.43	0.54	0.00	
ENERGY VALUE (AVERAGE)	KJOULE	339	22	0.0	361
OF THE DIGESTIBLE FRACTION PER 100 G EDIBLE PORTION	(KCAL)	81	5.1	0.0	86

WASTE PERCENTAGE AVERAGE 32

CONSTITUENTS	DIM	AV	VARIATION			AVR	NUTR. DENS.		MOLPERC.
WATER	GRAM	79.40	77.70	-	82.00	53.99	GRAM/MJ	220.06	
PROTEIN	GRAM	19.00	18.00	-	20.50	12.92	GRAM/MJ	52.66	
FAT	GRAM	0.57	0.30	-	0.92	0.39	GRAM/MJ	1.58	
MINERALS	GRAM	1.03	0.68	-	1.50	0.70	GRAM/MJ	2.85	
SODIUM	MILLI	107.00	90.00	-	119.00	72.76	MILLI/MJ	296.55	
POTASSIUM	MILLI	329.00	289.00	-	378.00	223.72	MILLI/MJ	911.83	
MAGNESIUM	MILLI	62.00	61.00	-	62.00	42.16	MILLI/MJ	171.83	
CALCIUM	MILLI	15.00	12.00	-	20.00	10.20	MILLI/MJ	41.57	
IRON	MILLI	0.70	-	-	-	0.48	MILLI/MJ	1.94	
PHOSPHORUS	MILLI	215.00	183.00	-	263.00	146.20	MILLI/MJ	595.88	
SELENIUM	MICRO	34.00	31.00	-	36.00	23.12	MICRO/MJ	94.23	
BROMINE	MICRO	653.00	-	-	-	444.04	MILLI/MJ	1.81	
VITAMIN E ACTIVITY	MILLI	0.30	-	-	-	0.20	MILLI/MJ	0.83	
ALPHA-TOCOPHEROL	MILLI	0.30	-	-	-	0.20	MILLI/MJ	0.83	
VITAMIN B2	MICRO	80.00	-	-	-	54.40	MICRO/MJ	221.72	
NICOTINAMIDE	MILLI	2.30	-	-	-	1.56	MILLI/MJ	6.37	
PANTOTHENIC ACID	MILLI	0.32	-	-	-	0.22	MILLI/MJ	0.89	
VITAMIN B12	MICRO	0.55	0.50	-	0.60	0.37	MICRO/MJ	1.52	

CONSTITUENTS	DIM	AV	VARIATION			AVR	NUTR. DENS.		MOLPERC.
ALANINE	GRAM	1.58	1.40	-	1.79	1.07	GRAM/MJ	4.38	10.5
ARGININE	GRAM	1.31	1.20	-	1.40	0.89	GRAM/MJ	3.63	4.5
ASPARTIC ACID	GRAM	1.85	1.50	-	2.22	1.26	GRAM/MJ	5.13	8.2
CYSTINE	GRAM	0.27	0.22	-	0.36	0.18	GRAM/MJ	0.75	0.7
GLUTAMIC ACID	GRAM	3.94	3.27	-	4.62	2.68	GRAM/MJ	10.92	15.9
GLYCINE	GRAM	0.82	0.51	-	1.03	0.56	GRAM/MJ	2.27	6.5
HISTIDINE	GRAM	0.32	0.28	-	0.38	0.22	GRAM/MJ	0.89	1.2
ISOLEUCINE	GRAM	1.32	-	-	-	0.90	GRAM/MJ	3.66	6.0
LEUCINE	GRAM	2.09	1.95	-	2.25	1.42	GRAM/MJ	5.79	9.4
LYSINE	GRAM	2.15	1.96	-	2.36	1.46	GRAM/MJ	5.96	8.7
METHIONINE	GRAM	0.76	0.66	-	0.84	0.52	GRAM/MJ	2.11	3.0
PHENYLALANINE	GRAM	1.00	0.85	-	1.17	0.68	GRAM/MJ	2.77	3.6
PROLINE	GRAM	0.73	-	-	-	0.50	GRAM/MJ	2.02	3.8
SERINE	GRAM	0.99	0.83	-	1.15	0.67	GRAM/MJ	2.74	5.6
THREONINE	GRAM	1.02	-	-	-	0.69	GRAM/MJ	2.83	5.1
TRYPTOPHAN	GRAM	0.19	-	-	-	0.13	GRAM/MJ	0.53	0.6
TYROSINE	GRAM	0.71	0.68	-	0.73	0.48	GRAM/MJ	1.97	2.3
VALINE	GRAM	0.92	0.51	-	1.35	0.63	GRAM/MJ	2.55	4.7

LIMANDE — LEMON SOLE — LIMANDE SOLE

MICROSTOMUS KITT (WALB.)

		PROTEIN	FAT	CARBOHYDRATES	TOTAL
ENERGY VALUE (AVERAGE)	KJOULE	320	34	0.0	354
PER 100 G EDIBLE PORTION	(KCAL)	77	8.1	0.0	85
AMOUNT OF DIGESTIBLE CONSTITUENTS PER 100 G	GRAM	16.87	0.80	0.00	
ENERGY VALUE (AVERAGE)	KJOULE	311	32	0.0	343
OF THE DIGESTIBLE FRACTION PER 100 G EDIBLE PORTION	(KCAL)	74	7.7	0.0	82

WASTE PERCENTAGE AVERAGE 41

CONSTITUENTS	DIM	AV	VARIATION			AVR	NUTR. DENS.		MOLPERC.
WATER	GRAM	81.00	-	-	-	47.79	GRAM/MJ	236.28	
PROTEIN	GRAM	17.40	-	-	-	10.27	GRAM/MJ	50.76	
FAT	GRAM	0.85	0.77	-	0.92	0.50	GRAM/MJ	2.48	
MINERALS	GRAM	0.75	-	-	-	0.44	GRAM/MJ	2.19	

CONSTITUENTS	DIM	AV	VARIATION			AVR	NUTR.	DENS.	MOLPERC.
SODIUM	MILLI	80.00	70.00	-	89.00	47.20	MILLI/MJ	233.36	
POTASSIUM	MILLI	298.00	208.00	-	387.00	175.82	MILLI/MJ	869.28	
ALANINE	GRAM	1.47	-	-	-	0.87	GRAM/MJ	4.29	10.4
ARGININE	GRAM	1.19	-	-	-	0.70	GRAM/MJ	3.47	4.3
ASPARTIC ACID	GRAM	2.05	-	-	-	1.21	GRAM/MJ	5.98	9.7
CYSTINE	GRAM	0.21	-	-	-	0.12	GRAM/MJ	0.61	0.6
GLUTAMIC ACID	GRAM	2.95	-	-	-	1.74	GRAM/MJ	8.61	12.7
GLYCINE	GRAM	1.04	-	-	-	0.61	GRAM/MJ	3.03	8.7
HISTIDINE	GRAM	0.53	-	-	-	0.31	GRAM/MJ	1.55	2.2
ISOLEUCINE	GRAM	1.04	-	-	-	0.61	GRAM/MJ	3.03	5.0
LEUCINE	GRAM	1.68	-	-	-	0.99	GRAM/MJ	4.90	8.1
LYSINE	GRAM	1.84	-	-	-	1.09	GRAM/MJ	5.37	8.0
METHIONINE	GRAM	0.56	-	-	-	0.33	GRAM/MJ	1.63	2.4
PHENYLALANINE	GRAM	0.79	-	-	-	0.47	GRAM/MJ	2.30	3.0
PROLINE	GRAM	0.83	-	-	-	0.49	GRAM/MJ	2.42	4.6
SERINE	GRAM	1.01	-	-	-	0.60	GRAM/MJ	2.95	6.1
THREONINE	GRAM	0.95	-	-	-	0.56	GRAM/MJ	2.77	5.0
TRYPTOPHAN	GRAM	0.21	-	-	-	0.12	GRAM/MJ	0.61	0.6
TYROSINE	GRAM	0.67	-	-	-	0.40	GRAM/MJ	1.95	2.3
VALINE	GRAM	1.17	-	-	-	0.69	GRAM/MJ	3.41	6.3

MAKRELE MACKEREL MAQUEREAU

SCOMBER SCOMBRUS (L.)

		PROTEIN	FAT	CARBOHYDRATES	TOTAL
ENERGY VALUE (AVERAGE)	KJOULE	344	473	0.0	817
PER 100 G	(KCAL)	82	113	0.0	195
EDIBLE PORTION					
AMOUNT OF DIGESTIBLE	GRAM	18.13	11.30	0.00	
CONSTITUENTS PER 100 G					
ENERGY VALUE (AVERAGE)	KJOULE	334	449	0.0	783
OF THE DIGESTIBLE	(KCAL)	80	107	0.0	187
FRACTION PER 100 G					
EDIBLE PORTION					

WASTE PERCENTAGE AVERAGE 35 MINIMUM 28 MAXIMUM 44

CONSTITUENTS	DIM	AV	VARIATION			AVR	NUTR. DENS.		MOLPERC.
WATER	GRAM	68.00	62.20	-	75.70	44.20	GRAM/MJ	86.81	
PROTEIN	GRAM	18.70	17.20	-	20.10	12.16	GRAM/MJ	23.87	
FAT	GRAM	11.90	5.20	-	20.20	7.74	GRAM/MJ	15.19	
MINERALS	GRAM	1.40	1.20	-	1.60	0.91	GRAM/MJ	1.79	
SODIUM	MILLI	95.00	82.00	-	110.00	61.75	MILLI/MJ	121.28	
POTASSIUM	MILLI	396.00	358.00	-	425.00	257.40	MILLI/MJ	505.56	
MAGNESIUM	MILLI	30.00	25.00	-	40.00	19.50	MILLI/MJ	38.30	
CALCIUM	MILLI	12.00	5.00	-	20.00	7.80	MILLI/MJ	15.32	
MANGANESE	MICRO	37.00	20.00	-	56.00	24.05	MICRO/MJ	47.24	
IRON	MILLI	1.00	0.80	-	1.40	0.65	MILLI/MJ	1.28	
COPPER	MILLI	0.16	0.07	-	0.20	0.10	MILLI/MJ	0.20	
VANADIUM	MICRO	0.30	0.20	-	0.40	0.20	MICRO/MJ	0.38	
PHOSPHORUS	MILLI	244.00	208.00	-	276.00	158.60	MILLI/MJ	311.51	
CHLORIDE	MILLI	131.00	35.00	-	170.00	85.15	MILLI/MJ	167.24	
FLUORIDE	MICRO	30.00	-	-	-	19.50	MICRO/MJ	38.30	
IODIDE	MICRO	74.00	40.00	-	106.00	48.10	MICRO/MJ	94.47	
BORON	MICRO	75.00	-	-	-	48.75	MICRO/MJ	95.75	
SELENIUM	MICRO	35.00	22.00	-	130.00	22.75	MICRO/MJ	44.68	
BROMINE	MICRO	449.00	447.00	-	452.00	291.85	MICRO/MJ	573.23	
VITAMIN A	MILLI	0.10 [1]	0.05	-	0.14	0.07	MILLI/MJ	0.13	
VITAMIN D	MICRO	1.00	0.50	-	1.40	0.65	MICRO/MJ	1.28	
VITAMIN E ACTIVITY	MILLI	1.25	-	-	-	0.81	MILLI/MJ	1.60	
TOTAL TOCOPHEROLS	MILLI	-	0.70	-	1.60				
ALPHA-TOCOPHEROL	MILLI	1.25	-	-	-	0.81	MILLI/MJ	1.60	
VITAMIN B1	MILLI	0.13	0.09	-	0.15	0.08	MILLI/MJ	0.17	
VITAMIN B2	MILLI	0.36	0.25	-	0.50	0.23	MILLI/MJ	0.46	

1 VALID FOR MUSCLE ONLY; THE LIVER CONTAINS CONSIDERABLY HIGHER AMOUNTS

CONSTITUENTS	DIM	AV	VARIATION			AVR	NUTR. DENS.		MOLPERC.
NICOTINAMIDE	MILLI	7.50	5.50	-	9.40	4.88	MILLI/MJ	9.58	
PANTOTHENIC ACID	MILLI	0.46	-	-	-	0.30	MILLI/MJ	0.59	
VITAMIN B6	MILLI	0.63	0.56	-	0.70	0.41	MILLI/MJ	0.80	
BIOTIN	MICRO	-	1.50	-	7.00				
FOLIC ACID	MICRO	1.24	-	-	-	0.81	MICRO/MJ	1.58	
VITAMIN B12	MICRO	9.00	5.00	-	14.00	5.85	MICRO/MJ	11.49	
VITAMIN C	MILLI	-	0.00	-	0.80				
ALANINE	GRAM	1.88	1.45	-	2.48	1.22	GRAM/MJ	2.40	12.2
ARGININE	GRAM	1.16	0.77	-	1.32	0.75	GRAM/MJ	1.48	3.9
ASPARTIC ACID	GRAM	2.12	1.81	-	2.64	1.38	GRAM/MJ	2.71	9.2
CYSTINE	GRAM	0.23	0.18	-	0.29	0.15	GRAM/MJ	0.29	0.6
GLUTAMIC ACID	GRAM	3.17	2.87	-	3.52	2.06	GRAM/MJ	4.05	12.5
GLYCINE	GRAM	1.41	1.17	-	1.77	0.92	GRAM/MJ	1.80	10.9
HISTIDINE	GRAM	0.84	0.23	-	1.29	0.55	GRAM/MJ	1.07	3.1
ISOLEUCINE	GRAM	1.09	0.78	-	1.46	0.71	GRAM/MJ	1.39	4.8
LEUCINE	GRAM	1.80	1.46	-	2.16	1.17	GRAM/MJ	2.30	8.0
LYSINE	GRAM	1.73	1.31	-	1.99	1.12	GRAM/MJ	2.21	6.9
METHIONINE	GRAM	0.64	0.45	-	0.74	0.42	GRAM/MJ	0.82	2.5
PHENYLALANINE	GRAM	0.84	0.75	-	0.92	0.55	GRAM/MJ	1.07	3.0
PROLINE	GRAM	0.79	0.47	-	0.95	0.51	GRAM/MJ	1.01	4.0
SERINE	GRAM	0.90	0.56	-	1.24	0.59	GRAM/MJ	1.15	5.0
THREONINE	GRAM	0.97	0.57	-	1.17	0.63	GRAM/MJ	1.24	4.7
TRYPTOPHAN	GRAM	0.27	0.21	-	0.34	0.18	GRAM/MJ	0.34	0.8
TYROSINE	GRAM	0.64	0.53	-	0.77	0.42	GRAM/MJ	0.82	2.0
VALINE	GRAM	1.21	0.86	-	1.51	0.79	GRAM/MJ	1.54	6.0
HISTAMINE	MICRO	32.00	12.00	-	51.00	20.80	MICRO/MJ	40.85	
MYRISTIC ACID	GRAM	0.84	0.56	-	0.98	0.55	GRAM/MJ	1.07	
PALMITIC ACID	GRAM	2.32	1.98	-	2.50	1.51	GRAM/MJ	2.97	
STEARIC ACID	GRAM	0.68	0.44	-	0.80	0.44	GRAM/MJ	0.87	
PALMITOLEIC ACID	GRAM	1.05	0.57	-	1.30	0.69	GRAM/MJ	1.35	
OLEIC ACID	GRAM	1.72	1.60	-	1.98	1.12	GRAM/MJ	2.20	
EICOSENOIC ACID	GRAM	0.53	0.41	-	0.78	0.35	GRAM/MJ	0.68	
C22 : 1 FATTY ACID	GRAM	0.63	0.43	-	1.04	0.41	GRAM/MJ	0.81	
C18 : 4 FATTY ACID	GRAM	0.16	0.16	-	0.16	0.10	GRAM/MJ	0.20	
C20 : 5 N-3 FATTY ACID	GRAM	0.69	0.65	-	0.79	0.45	GRAM/MJ	0.89	
C22 : 5 N-3 FATTY ACID	GRAM	0.12	0.10	-	0.18	0.08	GRAM/MJ	0.16	
C22 : 6 N-3 FATTY ACID	GRAM	1.30	1.25	-	1.42	0.85	GRAM/MJ	1.67	
CHOLESTEROL	MILLI	69.00	68.00	-	70.00	44.85	MILLI/MJ	88.09	
TOTAL PURINES	MILLI	145.00 [2]	95.00	-	194.00	94.25	MILLI/MJ	185.12	
TOTAL PHOSPHOLIPIDS	GRAM	2.37	-	-	-	1.54	GRAM/MJ	3.03	
PHOSPHATIDYLCHOLINE	MILLI	443.00	-	-	-	287.95	MILLI/MJ	565.57	
PHOSPHATIDYLETHANOLAMINE	GRAM	1.38	-	-	-	0.90	GRAM/MJ	1.76	
PHOSPHATIDYLSERINE	MILLI	480.00	-	-	-	312.00	MILLI/MJ	612.81	
SPHINGOMYELIN	MILLI	572.00	-	-	-	371.80	MILLI/MJ	730.26	

2 VALUE EXPRESSED IN MG URIC ACID/100G

MEERÄSCHE MULLET MUGE

MUGILIDAE SPP.

		PROTEIN	FAT	CARBOHYDRATES	TOTAL
ENERGY VALUE (AVERAGE)	KJOULE	376	171	0.0	546
PER 100 G	(KCAL)	90	41	0.0	131
EDIBLE PORTION					
AMOUNT OF DIGESTIBLE	GRAM	19.78	4.08	0.00	
CONSTITUENTS PER 100 G					
ENERGY VALUE (AVERAGE)	KJOULE	364	162	0.0	527
OF THE DIGESTIBLE	(KCAL)	87	39	0.0	126
FRACTION PER 100 G					
EDIBLE PORTION					

WASTE PERCENTAGE AVERAGE 48

CONSTITUENTS	DIM	AV	VARIATION		AVR	NUTR. DENS.		MOLPERC.
WATER	GRAM	74.60	69.70	- 79.30	38.79	GRAM/MJ	141.65	
PROTEIN	GRAM	20.40	17.70	- 22.60	10.61	GRAM/MJ	38.73	
FAT	GRAM	4.30	3.90	- 4.90	2.24	GRAM/MJ	8.16	
MINERALS	GRAM	1.20	1.00	- 1.40	0.62	GRAM/MJ	2.28	
SODIUM	MILLI	69.00	60.00	- 81.00	35.88	MILLI/MJ	131.01	
POTASSIUM	MILLI	404.00	323.00	- 485.00	210.08	MILLI/MJ	767.10	
MAGNESIUM	MILLI	29.00	25.00	- 32.00	15.08	MILLI/MJ	55.06	
CALCIUM	MILLI	53.00	51.00	- 56.00	27.56	MILLI/MJ	100.63	
MANGANESE	MICRO	50.00	-	- -	26.00	MICRO/MJ	94.94	
IRON	MILLI	1.50	1.00	- 5.00	0.78	MILLI/MJ	2.85	
COPPER	MICRO	7.50	0.30	- 21.00	3.90	MICRO/MJ	14.24	
ZINC	MILLI	0.60	0.08	- 1.30	0.31	MILLI/MJ	1.14	
PHOSPHORUS	MILLI	217.00	203.00	- 231.00	112.84	MILLI/MJ	412.03	
IODIDE	MILLI	0.33	0.16	- 0.49	0.17	MILLI/MJ	0.63	
VITAMIN A	MICRO	47.00	45.00	- 50.00	24.44	MICRO/MJ	89.24	
VITAMIN B1	MICRO	60.00	10.00	- 87.00	31.20	MICRO/MJ	113.93	
VITAMIN B2	MILLI	0.15	0.10	- 0.23	0.08	MILLI/MJ	0.28	
NICOTINAMIDE	MILLI	3.80	3.00	- 4.60	1.98	MILLI/MJ	7.22	
ALANINE	GRAM	1.55	1.13	- 1.74	0.81	GRAM/MJ	2.94	9.6
ARGININE	GRAM	1.36	0.84	- 1.50	0.71	GRAM/MJ	2.58	4.3
ASPARTIC ACID	GRAM	2.90	1.86	- 3.41	1.51	GRAM/MJ	5.51	12.0
CYSTINE	GRAM	0.31	0.29	- 0.31	0.16	GRAM/MJ	0.59	0.7
GLUTAMIC ACID	GRAM	3.47	2.38	- 3.86	1.80	GRAM/MJ	6.59	13.0
GLYCINE	GRAM	1.38	1.21	- 1.52	0.72	GRAM/MJ	2.62	10.1
HISTIDINE	GRAM	0.61	0.27	- 0.84	0.32	GRAM/MJ	1.16	2.2

CONSTITUENTS	DIM	AV	VARIATION			AVR	NUTR. DENS.		MOLPERC.
ISOLEUCINE	GRAM	1.09	0.87	-	1.22	0.57	GRAM/MJ	2.07	4.6
LEUCINE	GRAM	1.95	1.43	-	2.13	1.01	GRAM/MJ	3.70	8.2
LYSINE	GRAM	2.04	0.53	-	2.48	1.06	GRAM/MJ	3.87	7.7
METHIONINE	GRAM	0.66	0.55	-	0.76	0.34	GRAM/MJ	1.25	2.4
PHENYLALANINE	GRAM	0.86	0.70	-	0.98	0.45	GRAM/MJ	1.63	2.9
PROLINE	GRAM	0.79	0.66	-	0.94	0.41	GRAM/MJ	1.50	3.8
SERINE	GRAM	0.95	0.68	-	1.05	0.49	GRAM/MJ	1.80	5.0
THREONINE	GRAM	1.02	0.79	-	1.11	0.53	GRAM/MJ	1.94	4.7
TRYPTOPHAN	GRAM	0.32	0.30	-	0.32	0.17	GRAM/MJ	0.61	0.9
TYROSINE	GRAM	0.74	0.58	-	0.88	0.38	GRAM/MJ	1.41	2.2
VALINE	GRAM	1.24	1.11	-	1.37	0.64	GRAM/MJ	2.35	5.8
MYRISTIC ACID	GRAM	0.23	0.22	-	0.27	0.12	GRAM/MJ	0.45	
C15 : 0 FATTY ACID	GRAM	0.18	0.18	-	0.18	0.09	GRAM/MJ	0.34	
PALMITIC ACID	GRAM	0.86	0.75	-	1.10	0.45	GRAM/MJ	1.65	
STEARIC ACID	GRAM	0.20	0.14	-	0.33	0.11	GRAM/MJ	0.39	
PALMITOLEIC ACID	GRAM	0.60	0.41	-	0.70	0.31	GRAM/MJ	1.14	
OLEIC ACID	GRAM	0.33	0.33	-	0.34	0.17	GRAM/MJ	0.63	
EICOSENOIC ACID	MILLI	30.00	30.00	-	30.00	15.60	MILLI/MJ	56.96	
C22 : 1 FATTY ACID	GRAM	0.48	0.48	-	0.48	0.25	GRAM/MJ	0.91	
LINOLEIC ACID	MILLI	60.00	60.00	-	60.00	31.20	MILLI/MJ	113.93	
LINOLENIC ACID	MILLI	26.66	10.00	-	60.00	13.87	MILLI/MJ	50.63	
ARACHIDONIC ACID	GRAM	0.21	0.21	-	0.21	0.11	GRAM/MJ	0.40	
C18 : 4 FATTY ACID	MILLI	40.00	40.00	-	40.00	20.80	MILLI/MJ	75.95	
C20 : 5 N-3 FATTY ACID	MILLI	40.00	40.00	-	40.00	20.80	MILLI/MJ	75.95	
C22 : 5 N-3 FATTY ACID	GRAM	0.11	0.07	-	0.14	0.06	GRAM/MJ	0.22	
C22 : 6 N-3 FATTY ACID	GRAM	0.35	0.34	-	0.36	0.18	GRAM/MJ	0.67	
CHOLESTEROL	MILLI	34.00	-	-	-	17.68	MILLI/MJ	64.56	
TOTAL PHOSPHOLIPIDS	MILLI	452.00	-	-	-	235.04	MILLI/MJ	858.24	
PHOSPHATIDYLCHOLINE	MILLI	194.00	-	-	-	100.88	MILLI/MJ	368.36	
PHOSPHATIDYLETHANOLAMINE	MILLI	54.00	-	-	-	28.08	MILLI/MJ	102.53	
PHOSPHATIDYLSERINE	MILLI	76.00	-	-	-	39.52	MILLI/MJ	144.31	
SPHINGOMYELIN	MILLI	97.00	-	-	-	50.44	MILLI/MJ	184.18	

MEERBARBE GOATFISH ROUGET

MULLIDAE SPP.

		PROTEIN	FAT	CARBOHYDRATES	TOTAL
ENERGY VALUE (AVERAGE) PER 100 G EDIBLE PORTION	KJOULE	370	79	0.0	450
	(KCAL)	88	19	0.0	107
AMOUNT OF DIGESTIBLE CONSTITUENTS PER 100 G	GRAM	19.49	1.90	0.00	
ENERGY VALUE (AVERAGE) OF THE DIGESTIBLE FRACTION PER 100 G EDIBLE PORTION	KJOULE	359	76	0.0	434
	(KCAL)	86	18	0.0	104

WASTE PERCENTAGE AVERAGE 59

CONSTITUENTS	DIM	AV	VARIATION	AVR	NUTR. DENS.		MOLPERC.
WATER	GRAM	76.30	74.50 - 78.10	31.28	GRAM/MJ	175.62	
PROTEIN	GRAM	20.10	16.90 - 22.90	8.24	GRAM/MJ	46.27	
FAT	GRAM	2.00	0.40 - 4.70	0.82	GRAM/MJ	4.60	
MINERALS	GRAM	1.70	0.60 - 4.00	0.70	GRAM/MJ	3.91	

ROCHEN SKATE (RAY) RAIE

RAJA SPP.

		PROTEIN	FAT	CARBOHYDRATES	TOTAL
ENERGY VALUE (AVERAGE) PER 100 G EDIBLE PORTION	KJOULE	378	50	0.0	428
	(KCAL)	90	12	0.0	102
AMOUNT OF DIGESTIBLE CONSTITUENTS PER 100 G	GRAM	19.92	1.19	0.00	
ENERGY VALUE (AVERAGE) OF THE DIGESTIBLE FRACTION PER 100 G EDIBLE PORTION	KJOULE	367	47	0.0	414
	(KCAL)	88	11	0.0	99

WASTE PERCENTAGE AVERAGE 0.00 [1]

1 VALID FOR THE PURE EDIBLE PORTION; DATA ABOUT THE WHOLE WASTE WERE NOT AVAILABLE

CONSTITUENTS	DIM	AV	VARIATION			AVR	NUTR. DENS.		MOLPERC.
WATER	GRAM	79.27	78.00	-	79.40	79.27	GRAM/MJ	191.36	
PROTEIN	GRAM	20.54	19.00	-	20.70	20.55	GRAM/MJ	49.59	
FAT	GRAM	1.25	0.80	-	1.30	1.26	GRAM/MJ	3.03	
MINERALS	GRAM	1.31	1.30	-	1.40	1.31	GRAM/MJ	3.17	
SODIUM	MILLI	90.00	90.00	-	90.00	90.00	MILLI/MJ	217.25	
POTASSIUM	MILLI	250.00	250.00	-	250.00	250.00	MILLI/MJ	603.47	
CALCIUM	MILLI	38.00	38.00	-	38.00	38.00	MILLI/MJ	91.73	
IRON	MILLI	0.75	0.75	-	0.75	0.75	MILLI/MJ	1.81	
ZINC	MILLI	0.98	0.98	-	0.98	0.98	MILLI/MJ	2.37	
PHOSPHORUS	MILLI	155.00	155.00	-	155.00	155.00	MILLI/MJ	374.15	
VITAMIN A	MICRO	3.00	3.00	-	3.00	3.00	MICRO/MJ	7.24	
VITAMIN B1	MICRO	20.00	20.00	-	20.00	20.00	MICRO/MJ	48.28	
VITAMIN B2	MICRO	10.00	10.00	-	10.00	10.00	MICRO/MJ	24.14	
NICOTINAMIDE	MILLI	2.00	2.00	-	2.00	2.00	MILLI/MJ	4.83	
ALANINE	GRAM	1.88	1.88	-	1.88	1.88	GRAM/MJ	4.54	10.6
ARGININE	GRAM	1.83	1.83	-	1.83	1.83	GRAM/MJ	4.42	5.3
ASPARTIC ACID	GRAM	2.33	2.33	-	2.33	2.33	GRAM/MJ	5.62	8.8
CYSTINE	GRAM	0.67	0.67	-	0.67	0.67	GRAM/MJ	1.62	1.4
GLUTAMIC ACID	GRAM	3.72	3.72	-	3.72	3.72	GRAM/MJ	8.98	12.7
GLYCINE	GRAM	1.22	1.22	-	1.22	1.22	GRAM/MJ	2.94	8.2
HISTIDINE	GRAM	0.72	0.72	-	0.72	0.72	GRAM/MJ	1.74	2.3
ISOLEUCINE	GRAM	1.55	1.55	-	1.55	1.55	GRAM/MJ	3.74	5.9
LEUCINE	GRAM	2.10	2.10	-	2.10	2.10	GRAM/MJ	5.07	8.1
LYSINE	GRAM	2.23	2.23	-	2.23	2.23	GRAM/MJ	5.38	7.7
METHIONINE	GRAM	0.70	0.70	-	0.70	0.70	GRAM/MJ	1.69	2.4
PHENYLALANINE	GRAM	1.04	1.04	-	1.04	1.04	GRAM/MJ	2.51	3.2
PROLINE	GRAM	0.86	0.86	-	0.86	0.86	GRAM/MJ	2.08	3.8
SERINE	GRAM	1.12	1.12	-	1.12	1.12	GRAM/MJ	2.70	5.4
THREONINE	GRAM	1.21	1.21	-	1.21	1.21	GRAM/MJ	2.92	5.1
TRYPTOPHAN	GRAM	0.29	0.29	-	0.29	0.29	GRAM/MJ	0.70	0.7
TYROSINE	GRAM	0.94	0.94	-	0.94	0.94	GRAM/MJ	2.27	2.6
VALINE	GRAM	1.38	1.38	-	1.38	1.38	GRAM/MJ	3.33	5.9

ROTBARSCH (GOLDBARSCH) — REDFISH (RED PERCH) — SÉBASTE

SEBASTES MARINUS (L.)
SEBASTES MENTELLA (TRAVIN)

		PROTEIN	FAT	CARBOHYDRATES	TOTAL
ENERGY VALUE (AVERAGE) PER 100 G EDIBLE PORTION	KJOULE	335	143	0.0	479
	(KCAL)	80	34	0.0	114
AMOUNT OF DIGESTIBLE CONSTITUENTS PER 100 G	GRAM	17.65	3.42	0.00	
ENERGY VALUE (AVERAGE) OF THE DIGESTIBLE FRACTION PER 100 G EDIBLE PORTION	KJOULE	325	136	0.0	461
	(KCAL)	78	33	0.0	110

WASTE PERCENTAGE AVERAGE 52 MINIMUM 46 MAXIMUM 64

CONSTITUENTS	DIM	AV	VARIATION		AVR	NUTR. DENS.		MOLPERC.
WATER	GRAM	76.90	74.40	- 78.50	36.91	GRAM/MJ	166.70	
PROTEIN	GRAM	18.20	17.80	- 19.00	8.74	GRAM/MJ	39.45	
FAT	GRAM	3.61	2.25	- 4.80	1.73	GRAM/MJ	7.83	
MINERALS	GRAM	1.14	0.92	- 1.40	0.55	GRAM/MJ	2.47	
SODIUM	MILLI	80.00	54.00	- 109.00	38.40	MILLI/MJ	173.42	
POTASSIUM	MILLI	308.00	269.00	- 350.00	147.84	MILLI/MJ	667.65	
MAGNESIUM	MILLI	29.00	22.00	- 34.00	13.92	MILLI/MJ	62.86	
CALCIUM	MILLI	22.00	17.00	- 31.00	10.56	MILLI/MJ	47.69	
IRON	MILLI	0.69	0.53	- 1.00	0.33	MILLI/MJ	1.50	
PHOSPHORUS	MILLI	201.00	184.00	- 212.00	96.48	MILLI/MJ	435.71	
IODIDE	MICRO	99.00	74.00	- 124.00	47.52	MICRO/MJ	214.60	
SELENIUM	MICRO	44.00	26.00	- 196.00	21.12	MICRO/MJ	95.38	
BROMINE	MILLI	1.10	-	- -	0.53	MILLI/MJ	2.38	
VITAMIN A	MICRO	12.00	9.00	- 15.00	5.76	MICRO/MJ	26.01	
VITAMIN D	MICRO	2.30	1.80	- 2.70	1.10	MICRO/MJ	4.99	
VITAMIN E ACTIVITY	MILLI	1.25	-	- -	0.60	MILLI/MJ	2.71	
ALPHA-TOCOPHEROL	MILLI	1.25	-	- -	0.60	MILLI/MJ	2.71	
VITAMIN B1	MILLI	0.11	0.09	- 0.15	0.05	MILLI/MJ	0.24	
VITAMIN B2	MICRO	80.00	50.00	- 110.00	38.40	MICRO/MJ	173.42	
NICOTINAMIDE	MILLI	2.50	1.90	- 3.50	1.20	MILLI/MJ	5.42	
VITAMIN B12	MICRO	3.80	-	- -	1.82	MICRO/MJ	8.24	
VITAMIN C	MILLI	0.80	0.40	- 1.30	0.38	MILLI/MJ	1.73	
ALANINE	GRAM	1.68	1.35	- 1.98	0.81	GRAM/MJ	3.64	11.4
ARGININE	GRAM	1.19	1.15	- 1.24	0.57	GRAM/MJ	2.58	4.1
ASPARTIC ACID	GRAM	2.28	-	- -	1.09	GRAM/MJ	4.94	10.3

CONSTITUENTS	DIM	AV	VARIATION			AVR	NUTR. DENS.		MOLPERC.
CYSTINE	GRAM	0.22	-	-	-	0.11	GRAM/MJ	0.48	0.6
GLUTAMIC ACID	GRAM	3.30	3.26	-	3.35	1.58	GRAM/MJ	7.15	13.5
GLYCINE	GRAM	1.10	1.08	-	1.13	0.53	GRAM/MJ	2.38	8.8
HISTIDINE	GRAM	0.42	0.40	-	0.45	0.20	GRAM/MJ	0.91	1.6
ISOLEUCINE	GRAM	1.14	0.93	-	1.37	0.55	GRAM/MJ	2.47	5.2
LEUCINE	GRAM	1.78	1.73	-	1.85	0.85	GRAM/MJ	3.86	8.2
LYSINE	GRAM	1.90	-	-	-	0.91	GRAM/MJ	4.12	7.8
METHIONINE	GRAM	0.64	-	-	-	0.31	GRAM/MJ	1.39	2.6
PHENYLALANINE	GRAM	0.84	-	-	-	0.40	GRAM/MJ	1.82	3.1
PROLINE	GRAM	0.75	-	-	-	0.36	GRAM/MJ	1.63	3.9
SERINE	GRAM	1.04	0.93	-	1.17	0.50	GRAM/MJ	2.25	6.0
THREONINE	GRAM	1.01	0.99	-	1.05	0.48	GRAM/MJ	2.19	5.1
TRYPTOPHAN	GRAM	0.20	-	-	-	0.10	GRAM/MJ	0.43	0.6
TYROSINE	GRAM	0.56	0.53	-	0.58	0.27	GRAM/MJ	1.21	1.9
VALINE	GRAM	1.04	0.91	-	1.15	0.50	GRAM/MJ	2.25	5.4
TRIMETHYLAMINE	MILLI	-	0.00	-	5.00				
LAURIC ACID	MILLI	17.69	10.00	-	20.00	8.49	MILLI/MJ	38.35	
MYRISTIC ACID	GRAM	0.13	0.13	-	0.16	0.06	GRAM/MJ	0.29	
PALMITIC ACID	GRAM	0.38	0.37	-	0.39	0.18	GRAM/MJ	0.83	
STEARIC ACID	GRAM	0.13	0.08	-	0.15	0.06	GRAM/MJ	0.28	
ARACHIDIC ACID	MILLI	10.00	10.00	-	10.00	4.80	MILLI/MJ	21.68	
PALMITOLEIC ACID	GRAM	0.28	0.22	-	0.40	0.14	GRAM/MJ	0.62	
OLEIC ACID	GRAM	0.60	0.50	-	0.64	0.29	GRAM/MJ	1.32	
EICOSENOIC ACID	GRAM	0.53	0.23	-	0.56	0.25	GRAM/MJ	1.15	
C22 : 1 FATTY ACID	GRAM	0.49	0.20	-	0.59	0.24	GRAM/MJ	1.07	
LINOLEIC ACID	GRAM	0.10	0.03	-	0.13	0.05	GRAM/MJ	0.22	
LINOLENIC ACID	MILLI	45.00	20.00	-	70.00	21.60	MILLI/MJ	97.55	
ARACHIDONIC ACID	GRAM	0.24	0.01	-	0.32	0.12	GRAM/MJ	0.53	
C18 : 4 FATTY ACID	MILLI	60.00	60.00	-	60.00	28.80	MILLI/MJ	130.06	
C20 : 5 N-3 FATTY ACID	GRAM	0.27	0.23	-	0.29	0.13	GRAM/MJ	0.60	
C22 : 5 N-3 FATTY ACID	MILLI	21.42	20.00	-	40.00	10.29	MILLI/MJ	46.45	
C22 : 6 N-3 FATTY ACID	GRAM	0.13	0.08	-	0.38	0.06	GRAM/MJ	0.28	
TOTAL PURINES	MILLI	241.00 [1]	241.00	-	241.00	115.68	MILLI/MJ	522.42	
ADENINE	MILLI	21.00	21.00	-	21.00	10.08	MILLI/MJ	45.52	
GUANINE	MILLI	29.00	29.00	-	29.00	13.92	MILLI/MJ	62.86	
HYPOXANTHINE	MILLI	149.00	149.00	-	149.00	71.52	MILLI/MJ	322.99	

1 VALUE OF TOTAL PURINES IS EXPRESSSED IN MG URIC ACID/100G

ROTZUNGE
HUNDSZUNGE

GLYPTOCEPHALUS
CYNOGLOSSUS (L.)

WITCH

PLIE GRISE

		PROTEIN	FAT	CARBOHYDRATES	TOTAL
ENERGY VALUE (AVERAGE)	KJOULE	285	45	0.0	330
PER 100 G	(KCAL)	68	11	0.0	79
EDIBLE PORTION					
AMOUNT OF DIGESTIBLE	GRAM	15.03	1.07	0.00	
CONSTITUENTS PER 100 G					
ENERGY VALUE (AVERAGE)	KJOULE	277	43	0.0	319
OF THE DIGESTIBLE	(KCAL)	66	10	0.0	76
FRACTION PER 100 G					
EDIBLE PORTION					

WASTE PERCENTAGE AVERAGE 31 MINIMUM 29 MAXIMUM 32

CONSTITUENTS	DIM	AV	VARIATION			AVR	NUTR. DENS.		MOLPERC.
WATER	GRAM	82.10	-	-	-	56.65	GRAM/MJ	257.00	
PROTEIN	GRAM	15.50	-	-	-	10.70	GRAM/MJ	48.52	
FAT	GRAM	1.13	0.50	-	1.89	0.78	GRAM/MJ	3.54	
MINERALS	GRAM	1.00	-	-	-	0.69	GRAM/MJ	3.13	
SODIUM	MILLI	121.00	98.00	-	136.00	83.49	MILLI/MJ	378.77	
POTASSIUM	MILLI	280.00	227.00	-	314.00	193.20	MILLI/MJ	876.49	
MAGNESIUM	MILLI	20.00	-	-	-	13.80	MILLI/MJ	62.61	
CALCIUM	MILLI	29.00	18.00	-	39.00	20.01	MILLI/MJ	90.78	
IRON	MILLI	0.70	-	-	-	0.48	MILLI/MJ	2.19	
PHOSPHORUS	MILLI	153.00	-	-	-	105.57	MILLI/MJ	478.94	
CHLORIDE	MILLI	190.00	-	-	-	131.10	MILLI/MJ	594.76	

SARDELLE ANCHOVY ANCHOIS

ENGRAULIS ENCRASICHOLUS (L).

		PROTEIN	FAT	CARBOHYDRATES	TOTAL
ENERGY VALUE (AVERAGE)	KJOULE	370	91	0.0	461
PER 100 G	(KCAL)	88	22	0.0	110
EDIBLE PORTION					
AMOUNT OF DIGESTIBLE	GRAM	19.49	2.18	0.00	
CONSTITUENTS PER 100 G					
ENERGY VALUE (AVERAGE)	KJOULE	359	87	0.0	446
OF THE DIGESTIBLE	(KCAL)	86	21	0.0	107
FRACTION PER 100 G					
EDIBLE PORTION					

WASTE PERCENTAGE AVERAGE 27

CONSTITUENTS	DIM	AV	VARIATION			AVR	NUTR.	DENS.	MOLPERC.
WATER	GRAM	75.30	72.40	-	79.90	54.97	GRAM/MJ	168.92	
PROTEIN	GRAM	20.10	15.30	-	23.50	14.67	GRAM/MJ	45.09	
FAT	GRAM	2.30	1.70	-	3.60	1.68	GRAM/MJ	5.16	
MINERALS	GRAM	1.90	1.10	-	2.10	1.39	GRAM/MJ	4.26	
POTASSIUM	MILLI	278.00	-	-	-	202.94	MILLI/MJ	623.62	
CALCIUM	MILLI	82.00	-	-	-	59.86	MILLI/MJ	183.95	
IRON	MILLI	4.90	-	-	-	3.58	MILLI/MJ	10.99	
ZINC	MILLI	1.40	-	-	-	1.02	MILLI/MJ	3.14	
PHOSPHORUS	MILLI	233.00	-	-	-	170.09	MILLI/MJ	522.68	
CHLORIDE	MILLI	30.00	-	-	-	21.90	MILLI/MJ	67.30	
VITAMIN B1	MICRO	70.00	-	-	-	51.10	MICRO/MJ	157.03	
VITAMIN B2	MILLI	0.27	-	-	-	0.20	MILLI/MJ	0.61	
NICOTINAMIDE	MILLI	20.00	-	-	-	14.60	MILLI/MJ	44.87	
MYRISTIC ACID	GRAM	0.16	0.15	-	0.17	0.12	GRAM/MJ	0.36	
PALMITIC ACID	GRAM	0.36	0.28	-	0.45	0.27	GRAM/MJ	0.82	
STEARIC ACID	GRAM	0.12	0.10	-	0.15	0.09	GRAM/MJ	0.28	
PALMITOLEIC ACID	GRAM	0.16	0.16	-	0.16	0.12	GRAM/MJ	0.36	
OLEIC ACID	GRAM	0.25	0.24	-	0.27	0.19	GRAM/MJ	0.57	
EICOSENOIC ACID	MILLI	30.00	30.00	-	30.00	21.90	MILLI/MJ	67.30	
C22 : 1 FATTY ACID	MILLI	40.00	40.00	-	40.00	29.20	MILLI/MJ	89.73	
LINOLEIC ACID	MILLI	50.00	40.00	-	60.00	36.50	MILLI/MJ	112.16	
LINOLENIC ACID	MILLI	30.00	30.00	-	30.00	21.90	MILLI/MJ	67.30	
ARACHIDONIC ACID	MILLI	10.00	10.00	-	10.00	7.30	MILLI/MJ	22.43	
C18 : 4 FATTY ACID	MILLI	30.00	30.00	-	30.00	21.90	MILLI/MJ	67.30	
C20 : 5 N-3 FATTY ACID	GRAM	0.21	0.21	-	0.21	0.15	GRAM/MJ	0.47	

CONSTITUENTS	DIM	AV	VARIATION			AVR	NUTR. DENS.		MOLPERC.
C22 : 5 N-3 FATTY ACID	MILLI	15.00	10.00	-	20.00	10.95	MILLI/MJ	33.65	
C22 : 6 N-3 FATTY ACID	GRAM	0.29	0.22	-	0.36	0.21	GRAM/MJ	0.65	
TOTAL PURINES	MILLI	239.00 1	239.00	-	239.00	174.47	MILLI/MJ	536.14	
ADENINE	MILLI	4.00	4.00	-	4.00	2.92	MILLI/MJ	8.97	
GUANINE	MILLI	113.00	113.00	-	113.00	82.49	MILLI/MJ	253.49	
XANTHINE	MILLI	6.00	6.00	-	6.00	4.38	MILLI/MJ	13.46	
HYPOXANTHINE	MILLI	82.00	82.00	-	82.00	59.86	MILLI/MJ	183.95	
TOTAL PHOSPHOLIPIDS	MILLI	621.00	-	-	-	453.33	GRAM/MJ	1.39	
PHOSPHATIDYLCHOLINE	MILLI	391.00	-	-	-	285.43	MILLI/MJ	877.11	
PHOSPHATIDYLETHANOLAMINE	MILLI	118.00	-	-	-	86.14	MILLI/MJ	264.70	
PHOSPHATIDYLSERINE	MILLI	25.00	-	-	-	18.25	MILLI/MJ	56.08	
PHOSPHATIDYLINOSITOL	MILLI	31.00	-	-	-	22.63	MILLI/MJ	69.54	
SPHINGOMYELIN	MILLI	8.00	-	-	-	5.84	MILLI/MJ	17.95	

1 VALUE OF TOTAL PURINES IS EXPRESSSED IN MG URIC ACID/100G

SARDINE SARDINE SARDINE

SARDINA PILCHARDUS
(WALB.)

		PROTEIN	FAT	CARBOHYDRATES	TOTAL
ENERGY VALUE (AVERAGE)	KJOULE	357	206	0.0	563
PER 100 G	(KCAL)	85	49	0.0	135
EDIBLE PORTION					
AMOUNT OF DIGESTIBLE	GRAM	18.81	4.92	0.00	
CONSTITUENTS PER 100 G					
ENERGY VALUE (AVERAGE)	KJOULE	346	196	0.0	542
OF THE DIGESTIBLE	(KCAL)	83	47	0.0	130
FRACTION PER 100 G					
EDIBLE PORTION					

WASTE PERCENTAGE AVERAGE 41 MINIMUM 36 MAXIMUM 44

CONSTITUENTS	DIM	AV	VARIATION			AVR	NUTR. DENS.		MOLPERC.
WATER	GRAM	73.80	66.70	-	75.70	43.54	GRAM/MJ	136.15	
PROTEIN	GRAM	19.40	16.40	-	21.20	11.45	GRAM/MJ	35.79	
FAT	GRAM	5.18	1.20	-	9.81	3.06	GRAM/MJ	9.56	
MINERALS	GRAM	1.62	1.28	-	1.89	0.96	GRAM/MJ	2.99	

CONSTITUENTS	DIM	AV	VARIATION			AVR	NUTR. DENS.		MOLPERC.
SODIUM	MILLI	100.00	-	-	-	59.00	MILLI/MJ	184.49	
MAGNESIUM	MILLI	24.00	-	-	-	14.16	MILLI/MJ	44.28	
CALCIUM	MILLI	85.00	42.00	-	127.00	50.15	MILLI/MJ	156.82	
IRON	MILLI	2.40	1.30	-	3.00	1.42	MILLI/MJ	4.43	
COPPER	MILLI	0.17	-	-	-	0.10	MILLI/MJ	0.31	
NICKEL	MICRO	21.00	-	-	-	12.39	MICRO/MJ	38.74	
MOLYBDENUM	MICRO	3.00	-	-	-	1.77	MICRO/MJ	5.53	
PHOSPHORUS	MILLI	258.00	212.00	-	303.00	152.22	MILLI/MJ	475.99	
IODIDE	MICRO	32.00	13.00	-	54.00	18.88	MICRO/MJ	59.04	
BORON	MICRO	60.00	-	-	-	35.40	MICRO/MJ	110.69	
SELENIUM	MICRO	85.00	-	-	-	50.15	MICRO/MJ	156.82	
VITAMIN A	MICRO	20.00	-	-	-	11.80	MICRO/MJ	36.90	
VITAMIN D	MICRO	7.50	-	-	-	4.43	MICRO/MJ	13.84	
VITAMIN B1	MICRO	20.00	10.00	-	110.00	11.80	MICRO/MJ	36.90	
VITAMIN B2	MILLI	0.25	0.15	-	0.35	0.15	MILLI/MJ	0.46	
NICOTINAMIDE	MILLI	9.70	-	-	-	5.72	MILLI/MJ	17.90	
VITAMIN B6	MILLI	0.96	-	-	-	0.57	MILLI/MJ	1.77	
VITAMIN B12	MICRO	0.14	-	-	-	0.08	MICRO/MJ	0.26	
ALANINE	GRAM	1.60	1.54	-	1.64	0.94	GRAM/MJ	2.95	10.1
ARGININE	GRAM	1.31	1.26	-	1.36	0.77	GRAM/MJ	2.42	4.2
ASPARTIC ACID	GRAM	2.32	2.30	-	3.36	1.37	GRAM/MJ	4.28	9.8
CYSTINE	GRAM	0.22	0.21	-	0.23	0.13	GRAM/MJ	0.41	0.5
GLUTAMIC ACID	GRAM	3.04	2.98	-	3.11	1.79	GRAM/MJ	5.61	11.6
GLYCINE	GRAM	1.24	1.09	-	1.33	0.73	GRAM/MJ	2.29	9.3
HISTIDINE	GRAM	0.46	0.37	-	0.49	0.27	GRAM/MJ	0.85	1.7
ISOLEUCINE	GRAM	1.19	1.09	-	1.24	0.70	GRAM/MJ	2.20	5.1
LEUCINE	GRAM	1.87	1.77	-	1.94	1.10	GRAM/MJ	3.45	8.0
LYSINE	GRAM	2.28	2.11	-	2.38	1.35	GRAM/MJ	4.21	8.8
METHIONINE	GRAM	0.64	0.59	-	0.73	0.38	GRAM/MJ	1.18	2.4
PHENYLALANINE	GRAM	0.91	0.84	-	0.94	0.54	GRAM/MJ	1.68	3.1
PROLINE	GRAM	0.85	0.83	-	0.89	0.50	GRAM/MJ	1.57	4.1
SERINE	GRAM	1.10	1.07	-	1.11	0.65	GRAM/MJ	2.03	5.9
THREONINE	GRAM	1.12	1.01	-	1.19	0.66	GRAM/MJ	2.07	5.3
TRYPTOPHAN	GRAM	0.24	0.22	-	0.25	0.14	GRAM/MJ	0.44	0.7
TYROSINE	GRAM	0.81	0.69	-	0.88	0.48	GRAM/MJ	1.49	2.5
VALINE	GRAM	1.45	1.23	-	1.57	0.86	GRAM/MJ	2.68	7.0
MYRISTIC ACID	GRAM	0.31	0.31	-	0.31	0.18	GRAM/MJ	0.57	
PALMITIC ACID	GRAM	0.73	0.73	-	0.73	0.43	GRAM/MJ	1.35	
STEARIC ACID	GRAM	0.28	0.28	-	0.28	0.17	GRAM/MJ	0.52	
PALMITOLEIC ACID	GRAM	0.44	0.44	-	0.44	0.26	GRAM/MJ	0.81	
OLEIC ACID	GRAM	0.64	0.64	-	0.64	0.38	GRAM/MJ	1.18	
EICOSENOIC ACID	MILLI	60.00	60.00	-	60.00	35.40	MILLI/MJ	110.69	
C22 : 1 FATTY ACID	GRAM	0.11	0.11	-	0.11	0.06	GRAM/MJ	0.20	
LINOLEIC ACID	GRAM	0.10	0.10	-	0.10	0.06	GRAM/MJ	0.18	
LINOLENIC ACID	MILLI	50.00	50.00	-	50.00	29.50	MILLI/MJ	92.25	
ARACHIDONIC ACID	MILLI	10.00	10.00	-	10.00	5.90	MILLI/MJ	18.45	
C18 : 4 FATTY ACID	MILLI	50.00	50.00	-	50.00	29.50	MILLI/MJ	92.25	
C20 : 5 N-3 FATTY ACID	GRAM	0.66	0.66	-	0.66	0.39	GRAM/MJ	1.22	
C22 : 5 N-3 FATTY ACID	GRAM	0.66	0.66	-	0.66	0.39	GRAM/MJ	1.22	
C22 : 6 N-3 FATTY ACID	GRAM	0.93	0.93	-	0.93	0.55	GRAM/MJ	1.72	
TOTAL PURINES	MILLI	345.00 [1]	-	-	-	203.55	MILLI/MJ	636.49	
TOTAL PHOSPHOLIPIDS	MILLI	945.00	-	-	-	557.55	GRAM/MJ	1.74	
PHOSPHATIDYLCHOLINE	MILLI	500.00	-	-	-	295.00	MILLI/MJ	922.46	
PHOSPHATIDYLETHANOLAMINE	MILLI	236.00	-	-	-	139.24	MILLI/MJ	435.40	
PHOSPHATIDYLSERINE	MILLI	19.00	-	-	-	11.21	MILLI/MJ	35.05	
PHOSPHATIDYLINOSITOL	MILLI	38.00	-	-	-	22.42	MILLI/MJ	70.11	
SPHINGOMYELIN	MILLI	57.00	-	-	-	33.63	MILLI/MJ	105.16	

1 VALUE EXPRESSED IN MG URIC ACID/100G

SCHELLFISCH HADDOCK AIGLEFIN

MELANOGRAMMUS
AEGLEFINUS (L.)

		PROTEIN	FAT	CARBOHYDRATES	TOTAL
ENERGY VALUE (AVERAGE)	KJOULE	330	4.8	0.0	334
PER 100 G	(KCAL)	79	1.1	0.0	80
EDIBLE PORTION					
AMOUNT OF DIGESTIBLE	GRAM	17.36	0.11	0.00	
CONSTITUENTS PER 100 G					
ENERGY VALUE (AVERAGE)	KJOULE	320	4.5	0.0	324
OF THE DIGESTIBLE	(KCAL)	76	1.1	0.0	77
FRACTION PER 100 G					
EDIBLE PORTION					

WASTE PERCENTAGE AVERAGE 43 MINIMUM 38 MAXIMUM 54

CONSTITUENTS	DIM	AV	VARIATION			AVR	NUTR. DENS.		MOLPERC.
WATER	GRAM	80.80	80.60	-	81.20	46.06	GRAM/MJ	249.25	
PROTEIN	GRAM	17.90	17.70	-	18.20	10.20	GRAM/MJ	55.22	
FAT	GRAM	0.12	0.10	-	0.14	0.07	GRAM/MJ	0.37	
MINERALS	GRAM	1.20	-	-	-	0.68	GRAM/MJ	3.70	
SODIUM	MILLI	116.00	80.00	-	145.00	66.12	MILLI/MJ	357.83	
POTASSIUM	MILLI	301.00	131.00	-	395.00	171.57	MILLI/MJ	928.50	
MAGNESIUM	MILLI	24.00	-	-	-	13.68	MILLI/MJ	74.03	
CALCIUM	MILLI	18.00	6.00	-	32.00	10.26	MILLI/MJ	55.53	
MANGANESE	MICRO	20.00	-	-	-	11.40	MICRO/MJ	61.69	
IRON	MILLI	0.61	0.52	-	0.70	0.35	MILLI/MJ	1.88	
COPPER	MILLI	0.23	-	-	-	0.13	MILLI/MJ	0.71	
ZINC	MILLI	0.30	-	-	-	0.17	MILLI/MJ	0.93	
PHOSPHORUS	MILLI	176.00	173.00	-	179.00	100.32	MILLI/MJ	542.91	
FLUORIDE	MICRO	35.00	20.00	-	50.00	19.95	MICRO/MJ	107.97	
IODIDE	MICRO	243.00	60.00	-	510.00	138.51	MICRO/MJ	749.59	
BORON	MICRO	35.00	21.00	-	50.00	19.95	MICRO/MJ	107.97	
SELENIUM	MICRO	20.00	11.00	-	25.00	11.40	MICRO/MJ	61.69	
BROMINE	MICRO	636.00	596.00	-	676.00	362.52	MILLI/MJ	1.96	
VITAMIN A	MICRO	17.00	-	-	-	9.69	MICRO/MJ	52.44	
VITAMIN E ACTIVITY	MILLI	0.39	-	-	-	0.22	MILLI/MJ	1.20	
ALPHA-TOCOPHEROL	MILLI	0.39	-	-	-	0.22	MILLI/MJ	1.20	
VITAMIN B1	MICRO	50.00	-	-	-	28.50	MICRO/MJ	154.24	
VITAMIN B2	MILLI	0.17	0.08	-	0.25	0.10	MILLI/MJ	0.52	
NICOTINAMIDE	MILLI	3.10	2.40	-	4.00	1.77	MILLI/MJ	9.56	
PANTOTHENIC ACID	MILLI	0.14	-	-	-	0.08	MILLI/MJ	0.43	
BIOTIN	MICRO	1.20	0.50	-	1.80	0.68	MICRO/MJ	3.70	

CONSTITUENTS	DIM	AV	VARIATION			AVR	NUTR. DENS.		MOLPERC.
FOLIC ACID	MICRO	10.00	-	-	-	5.70	MICRO/MJ	30.85	
VITAMIN B12	MICRO	-	0.53	-	1.80				
ALANINE	GRAM	2.16	-	-	-	1.23	GRAM/MJ	6.66	14.7
ARGININE	GRAM	1.30	1.10	-	1.87	0.74	GRAM/MJ	4.01	4.5
ASPARTIC ACID	GRAM	1.75	0.90	-	2.49	1.00	GRAM/MJ	5.40	8.0
CYSTINE	GRAM	0.21	0.08	-	0.28	0.12	GRAM/MJ	0.65	0.5
GLUTAMIC ACID	GRAM	3.06	2.35	-	3.67	1.74	GRAM/MJ	9.44	12.6
GLYCINE	GRAM	1.05	0.92	-	1.26	0.60	GRAM/MJ	3.24	8.5
HISTIDINE	GRAM	0.42	0.24	-	0.74	0.24	GRAM/MJ	1.30	1.6
ISOLEUCINE	GRAM	1.09	0.90	-	1.33	0.62	GRAM/MJ	3.36	5.0
LEUCINE	GRAM	1.66	0.81	-	1.99	0.95	GRAM/MJ	5.12	7.7
LYSINE	GRAM	1.93	1.32	-	2.26	1.10	GRAM/MJ	5.95	8.0
METHIONINE	GRAM	0.65	0.50	-	0.86	0.37	GRAM/MJ	2.01	2.6
PHENYLALANINE	GRAM	0.79	0.63	-	0.98	0.45	GRAM/MJ	2.44	2.9
PROLINE	GRAM	0.68	0.58	-	0.90	0.39	GRAM/MJ	2.10	3.6
SERINE	GRAM	1.06	0.74	-	1.36	0.60	GRAM/MJ	3.27	6.1
THREONINE	GRAM	0.92	0.67	-	1.26	0.52	GRAM/MJ	2.84	4.7
TRYPTOPHAN	GRAM	0.24	0.15	-	0.29	0.14	GRAM/MJ	0.74	0.7
TYROSINE	GRAM	0.62	0.58	-	0.78	0.35	GRAM/MJ	1.91	2.1
VALINE	GRAM	1.16	0.99	-	1.37	0.66	GRAM/MJ	3.58	6.0
MYRISTIC ACID	MILLI	2.00	2.00	-	2.00	1.14	MILLI/MJ	6.17	
PALMITIC ACID	MILLI	15.37	14.50	-	17.00	8.76	MILLI/MJ	47.43	
STEARIC ACID	MILLI	4.66	4.00	-	6.00	2.66	MILLI/MJ	14.40	
PALMITOLEIC ACID	MILLI	3.66	3.00	-	4.00	2.09	MILLI/MJ	11.31	
OLEIC ACID	MILLI	11.00	11.00	-	11.00	6.27	MILLI/MJ	33.93	
EICOSENOIC ACID	MILLI	2.00	2.00	-	2.00	1.14	MILLI/MJ	6.17	
LINOLEIC ACID	MILLI	2.00	2.00	-	2.00	1.14	MILLI/MJ	6.17	
LINOLENIC ACID	MILLI	1.00	1.00	-	1.00	0.57	MILLI/MJ	3.08	
ARACHIDONIC ACID	MILLI	2.00	2.00	-	2.00	1.14	MILLI/MJ	6.17	
C20 : 5 N-3 FATTY ACID	MILLI	8.66	8.00	-	9.00	4.94	MILLI/MJ	26.74	
C22 : 5 N-3 FATTY ACID	MILLI	2.33	2.00	-	3.00	1.33	MILLI/MJ	7.20	
C22 : 6 N-3 FATTY ACID	MILLI	21.33	18.00	-	28.00	12.16	MILLI/MJ	65.81	
CHOLESTEROL	MILLI	62.00	60.00	-	63.00	35.34	MILLI/MJ	191.25	
TOTAL PURINES	MILLI	139.00	1 139.00	-	139.00	79.23	MILLI/MJ	428.78	
ADENINE	MILLI	11.00	11.00	-	11.00	6.27	MILLI/MJ	33.93	
GUANINE	MILLI	8.00	8.00	-	8.00	4.56	MILLI/MJ	24.68	
HYPOXANTHINE	MILLI	95.00	95.00	-	95.00	54.15	MILLI/MJ	293.05	

1 VALUE OF TOTAL PURINES IS EXPRESSSED IN MG URIC ACID/100G

SCHOLLE | PLAICE (PLAT FISH) | PLIE

PLEURONECTES PLATESSA (L.)

		PROTEIN	FAT	CARBOHYDRATES	TOTAL
ENERGY VALUE (AVERAGE)	KJOULE	315	32	0.0	347
PER 100 G EDIBLE PORTION	(KCAL)	75	7.6	0.0	83
AMOUNT OF DIGESTIBLE CONSTITUENTS PER 100 G	GRAM	16.58	0.76	0.00	
ENERGY VALUE (AVERAGE)	KJOULE	305	30	0.0	336
OF THE DIGESTIBLE FRACTION PER 100 G EDIBLE PORTION	(KCAL)	73	7.2	0.0	80

WASTE PERCENTAGE AVERAGE 44 MINIMUM 36 MAXIMUM 54

CONSTITUENTS	DIM	AV	VARIATION			AVR	NUTR. DENS.		MOLPERC.
WATER	GRAM	80.70	79.00	-	81.50	45.19	GRAM/MJ	240.49	
PROTEIN	GRAM	17.10	16.00	-	18.00	9.58	GRAM/MJ	50.96	
FAT	GRAM	0.80	0.40	-	1.50	0.45	GRAM/MJ	2.38	
MINERALS	GRAM	1.40	-	-	-	0.78	GRAM/MJ	4.17	
SODIUM	MILLI	104.00	68.00	-	139.00	58.24	MILLI/MJ	309.92	
POTASSIUM	MILLI	311.00	243.00	-	398.00	174.16	MILLI/MJ	926.79	
MAGNESIUM	MILLI	22.00	-	-	-	12.32	MILLI/MJ	65.56	
CALCIUM	MILLI	61.00	-	-	-	34.16	MILLI/MJ	181.78	
MANGANESE	MICRO	30.00	10.00	-	50.00	16.80	MICRO/MJ	89.40	
IRON	MILLI	0.90	0.80	-	1.00	0.50	MILLI/MJ	2.68	
COPPER	MILLI	-	0.14	-	0.55				
PHOSPHORUS	MILLI	198.00	195.00	-	200.00	110.88	MILLI/MJ	590.04	
CHLORIDE	MILLI	127.00	-	-	-	71.12	MILLI/MJ	378.46	
IODIDE	MICRO	190.00	28.00	-	240.00	106.40	MICRO/MJ	566.20	
SELENIUM	MICRO	65.00	26.00	-	104.00	36.40	MICRO/MJ	193.70	
BROMINE	MICRO	559.00	542.00	-	575.00	313.04	MILLI/MJ	1.67	
VITAMIN A	MICRO	-	0.00	-	6.00				
VITAMIN B1	MILLI	0.21	-	-	-	0.12	MILLI/MJ	0.63	
VITAMIN B2	MILLI	0.22	0.09	-	0.33	0.12	MILLI/MJ	0.66	
NICOTINAMIDE	MILLI	4.00	2.15	-	10.60	2.24	MILLI/MJ	11.92	
PANTOTHENIC ACID	MILLI	0.80	-	-	-	0.45	MILLI/MJ	2.38	
VITAMIN B6	MILLI	0.22	0.16	-	0.29	0.12	MILLI/MJ	0.66	
FOLIC ACID	MICRO	11.00	-	-	-	6.16	MICRO/MJ	32.78	
VITAMIN B12	MICRO	1.45	0.50	-	2.20	0.81	MICRO/MJ	4.32	
VITAMIN C	MILLI	1.50	-	-	-	0.84	MILLI/MJ	4.47	

CONSTITUENTS	DIM	AV	VARIATION			AVR	NUTR.	DENS.	MOLPERC.
ALANINE	GRAM	1.69	-	-	-	0.95	GRAM/MJ	5.04	12.0
ARGININE	GRAM	1.15	-	-	-	0.64	GRAM/MJ	3.43	4.2
ASPARTIC ACID	GRAM	1.91	-	-	-	1.07	GRAM/MJ	5.69	9.1
CYSTINE	GRAM	0.21	-	-	-	0.12	GRAM/MJ	0.63	0.6
GLUTAMIC ACID	GRAM	2.86	-	-	-	1.60	GRAM/MJ	8.52	12.3
GLYCINE	GRAM	1.18	-	-	-	0.66	GRAM/MJ	3.52	9.9
HISTIDINE	GRAM	0.38	-	-	-	0.21	GRAM/MJ	1.13	1.5
ISOLEUCINE	GRAM	1.14	-	-	-	0.64	GRAM/MJ	3.40	5.5
LEUCINE	GRAM	1.60	-	-	-	0.90	GRAM/MJ	4.77	7.7
LYSINE	GRAM	1.77	-	-	-	0.99	GRAM/MJ	5.27	7.6
METHIONINE	GRAM	0.53	-	-	-	0.30	GRAM/MJ	1.58	2.2
PHENYLALANINE	GRAM	0.73	-	-	-	0.41	GRAM/MJ	2.18	2.8
PROLINE	GRAM	0.72	-	-	-	0.40	GRAM/MJ	2.15	3.9
SERINE	GRAM	1.21	-	-	-	0.68	GRAM/MJ	3.61	7.3
THREONINE	GRAM	0.91	-	-	-	0.51	GRAM/MJ	2.71	4.8
TRYPTOPHAN	GRAM	0.19	-	-	-	0.11	GRAM/MJ	0.57	0.6
TYROSINE	GRAM	0.59	-	-	-	0.33	GRAM/MJ	1.76	2.1
VALINE	GRAM	1.11	-	-	-	0.62	GRAM/MJ	3.31	6.0
TRIMETHYLAMINE	MILLI	-	0.00	-	3.00				
CHOLESTEROL	MILLI	63.00	55.00	-	70.00	35.28	MILLI/MJ	187.74	
TOTAL PURINES	MILLI	93.00 1	93.00	-	93.00	52.08	MILLI/MJ	277.14	
ADENINE	MILLI	8.00	8.00	-	8.00	4.48	MILLI/MJ	23.84	
GUANINE	MILLI	18.00	18.00	-	18.00	10.08	MILLI/MJ	53.64	
HYPOXANTHINE	MILLI	51.00	51.00	-	51.00	28.56	MILLI/MJ	151.98	

1 VALUE OF TOTAL PURINES IS EXPRESSSED IN MG URIC ACID/100G

SCHWERTFISCH SWORDFISH ESPADON

XIPHIAS GLADIUS (L.)

		PROTEIN	FAT	CARBOHYDRATES	TOTAL
ENERGY VALUE (AVERAGE)	KJOULE	357	175	0.0	532
PER 100 G	(KCAL)	85	42	0.0	127
EDIBLE PORTION					
AMOUNT OF DIGESTIBLE	GRAM	18.81	4.18	0.00	
CONSTITUENTS PER 100 G					
ENERGY VALUE (AVERAGE)	KJOULE	346	166	0.0	513
OF THE DIGESTIBLE	(KCAL)	83	40	0.0	123
FRACTION PER 100 G					
EDIBLE PORTION					

WASTE PERCENTAGE AVERAGE 35

CONSTITUENTS	DIM	AV	VARIATION			AVR	NUTR. DENS.		MOLPERC.
WATER	GRAM	74.50	67.60	-	77.70	48.43	GRAM/MJ	145.34	
PROTEIN	GRAM	19.40	16.90	-	21.20	12.61	GRAM/MJ	37.85	
FAT	GRAM	4.40	2.00	-	6.70	2.86	GRAM/MJ	8.58	
MINERALS	GRAM	1.40	1.00	-	2.50	0.91	GRAM/MJ	2.73	
SODIUM	MILLI	102.00	-	-	-	66.30	MILLI/MJ	198.99	
POTASSIUM	MILLI	342.00	-	-	-	222.30	MILLI/MJ	667.21	
CALCIUM	MILLI	10.00	7.00	-	12.00	6.50	MILLI/MJ	19.51	
PHOSPHORUS	MILLI	506.00	228.00	-	745.00	328.90	MILLI/MJ	987.17	
CHLORIDE	MILLI	130.00	-	-	-	84.50	MILLI/MJ	253.62	
VITAMIN A	MICRO	20.00	-	-	-	13.00	MICRO/MJ	39.02	
VITAMIN B1	MICRO	50.00	40.00	-	70.00	32.50	MICRO/MJ	97.55	
VITAMIN B2	MICRO	80.00	50.00	-	110.00	52.00	MICRO/MJ	156.07	
NICOTINAMIDE	MILLI	7.60	6.20	-	9.10	4.94	MILLI/MJ	14.83	
FOLIC ACID	MICRO	2.00	-	-	-	1.30	MICRO/MJ	3.90	
VITAMIN B12	MICRO	0.60	-	-	-	0.39	MICRO/MJ	1.17	
PALMITOLEIC ACID	GRAM	0.31	0.31	-	0.31	0.20	GRAM/MJ	0.60	
OLEIC ACID	GRAM	1.35	1.35	-	1.35	0.88	GRAM/MJ	2.63	
EICOSENOIC ACID	GRAM	0.16	0.16	-	0.16	0.10	GRAM/MJ	0.31	
C22 : 1 FATTY ACID	GRAM	0.11	0.11	-	0.11	0.07	GRAM/MJ	0.21	
LINOLEIC ACID	MILLI	40.00	40.00	-	40.00	26.00	MILLI/MJ	78.04	
LINOLENIC ACID	GRAM	0.23	0.23	-	0.23	0.15	GRAM/MJ	0.45	
ARACHIDONIC ACID	MILLI	90.00	90.00	-	90.00	58.50	MILLI/MJ	175.58	
C20 : 5 N-3 FATTY ACID	GRAM	0.13	0.13	-	0.13	0.08	GRAM/MJ	0.25	
C22 : 6 N-3 FATTY ACID	GRAM	0.66	0.66	-	0.66	0.43	GRAM/MJ	1.29	
CHOLESTEROL	MILLI	39.00	-	-	-	25.35	MILLI/MJ	76.09	

SEEHECHT (HECHTDORSCH) — **HAKE** — **MERLU**

MERLUCCIUS MERLUCCIUS (L.)

		PROTEIN	FAT	CARBOHYDRATES	TOTAL
ENERGY VALUE (AVERAGE)	KJOULE	317	34	0.0	350
PER 100 G	(KCAL)	76	8.1	0.0	84
EDIBLE PORTION					
AMOUNT OF DIGESTIBLE	GRAM	16.68	0.80	0.00	
CONSTITUENTS PER 100 G					
ENERGY VALUE (AVERAGE)	KJOULE	307	32	0.0	339
OF THE DIGESTIBLE	(KCAL)	73	7.7	0.0	81
FRACTION PER 100 G					
EDIBLE PORTION					

WASTE PERCENTAGE AVERAGE 42 MINIMUM 33 MAXIMUM 57

CONSTITUENTS	DIM	AV	VARIATION			AVR	NUTR.	DENS.	MOLPERC.
WATER	GRAM	80.80	-	-	-	46.86	GRAM/MJ	238.18	
PROTEIN	GRAM	17.20	16.50	-	17.70	9.98	GRAM/MJ	50.70	
FAT	GRAM	0.85	0.40	-	1.20	0.49	GRAM/MJ	2.51	
MINERALS	GRAM	1.07	0.83	-	1.30	0.62	GRAM/MJ	3.15	
SODIUM	MILLI	101.00	74.00	-	119.00	58.58	MILLI/MJ	297.72	
POTASSIUM	MILLI	294.00	251.00	-	363.00	170.52	MILLI/MJ	866.64	
CALCIUM	MILLI	41.00	-	-	-	23.78	MILLI/MJ	120.86	
PHOSPHORUS	MILLI	142.00	-	-	-	82.36	MILLI/MJ	418.58	
VITAMIN B1	MILLI	0.10	-	-	-	0.06	MILLI/MJ	0.29	
VITAMIN B2	MILLI	0.20	-	-	-	0.12	MILLI/MJ	0.59	
ALANINE	GRAM	1.14	-	-	-	0.66	GRAM/MJ	3.36	8.5
ARGININE	GRAM	1.07	1.05	-	1.09	0.62	GRAM/MJ	3.15	4.1
ASPARTIC ACID	GRAM	1.66	-	-	-	0.96	GRAM/MJ	4.89	8.2
CYSTINE	GRAM	0.16	-	-	-	0.09	GRAM/MJ	0.47	0.4
GLUTAMIC ACID	GRAM	2.47	-	-	-	1.43	GRAM/MJ	7.28	11.1
GLYCINE	GRAM	1.74	-	-	-	1.01	GRAM/MJ	5.13	15.3
HISTIDINE	GRAM	0.52	0.49	-	0.58	0.30	GRAM/MJ	1.53	2.2
ISOLEUCINE	GRAM	1.27	0.88	-	1.66	0.74	GRAM/MJ	3.74	6.4
LEUCINE	GRAM	1.44	1.41	-	1.47	0.84	GRAM/MJ	4.24	7.3
LYSINE	GRAM	1.56	1.54	-	1.56	0.90	GRAM/MJ	4.60	7.1
METHIONINE	GRAM	0.48	0.44	-	0.52	0.28	GRAM/MJ	1.41	2.1
PHENYLALANINE	GRAM	0.65	0.58	-	0.73	0.38	GRAM/MJ	1.92	2.6
PROLINE	GRAM	0.60	-	-	-	0.35	GRAM/MJ	1.77	3.4
SERINE	GRAM	0.95	-	-	-	0.55	GRAM/MJ	2.80	6.0
THREONINE	GRAM	0.98	-	-	-	0.57	GRAM/MJ	2.89	5.4
TRYPTOPHAN	GRAM	0.23	0.21	-	0.23	0.13	GRAM/MJ	0.68	0.7
TYROSINE	GRAM	0.80	-	-	-	0.46	GRAM/MJ	2.36	2.9
VALINE	GRAM	1.10	-	-	-	0.64	GRAM/MJ	3.24	6.2

CONSTITUENTS	DIM	AV	VARIATION			AVR	NUTR. DENS.		MOLPERC.
MYRISTIC ACID	MILLI	26.33	19.00	-	30.00	15.27	MILLI/MJ	77.62	
PALMITIC ACID	GRAM	0.13	0.11	-	0.15	0.08	GRAM/MJ	0.38	
STEARIC ACID	MILLI	26.25	20.00	-	36.00	15.23	MILLI/MJ	77.38	
PALMITOLEIC ACID	MILLI	43.75	29.00	-	50.00	25.38	MILLI/MJ	128.96	
OLEIC ACID	GRAM	0.12	0.12	-	0.15	0.07	GRAM/MJ	0.38	
EICOSENOIC ACID	MILLI	27.66	23.00	-	30.00	16.05	MILLI/MJ	81.56	
C22 : 1 FATTY ACID	MILLI	18.00	18.00	-	18.00	10.44	MILLI/MJ	53.06	
LINOLEIC ACID	MILLI	14.00	7.00	-	20.00	8.12	MILLI/MJ	41.27	
LINOLENIC ACID	MILLI	17.50	3.00	-	30.00	10.15	MILLI/MJ	51.59	
ARACHIDONIC ACID	MILLI	29.00	3.00	-	50.00	16.82	MILLI/MJ	85.48	
C18 : 4 FATTY ACID	MILLI	20.00	20.00	-	20.00	11.60	MILLI/MJ	58.95	
C20 : 5 N-3 FATTY ACID	MILLI	61.25	36.00	-	80.00	35.53	MILLI/MJ	180.55	
C22 : 5 N-3 FATTY ACID	MILLI	15.33	10.00	-	26.00	8.89	MILLI/MJ	45.20	
C22 : 6 N-3 FATTY ACID	GRAM	0.11	0.03	-	0.16	0.06	GRAM/MJ	0.33	
TOTAL PHOSPHOLIPIDS	MILLI	400.00	-	-	-	232.00	GRAM/MJ	1.18	
PHOSPHATIDYLCHOLINE	MILLI	252.00	-	-	-	146.16	MILLI/MJ	742.83	
PHOSPHATIDYLETHANOLAMINE	MILLI	84.00	-	-	-	48.72	MILLI/MJ	247.61	
PHOSPHATIDYLSERINE	MILLI	12.00	-	-	-	6.96	MILLI/MJ	35.37	
PHOSPHATIDYLINOSITOL	MILLI	24.00	-	-	-	13.92	MILLI/MJ	70.75	
SPHINGOMYELIN	MILLI	16.00	-	-	-	9.28	MILLI/MJ	47.16	

ALASKA SEELACHS

THERAGRA CHALCOGRAMMA

ALASKA POLLACK

(POLLACK, WALLEYE)

MORUE DU PACIFIQUE OCCIDENTAL

		PROTEIN	FAT	CARBOHYDRATES	TOTAL
ENERGY VALUE (AVERAGE)	KJOULE	307	32	0.0	339
PER 100 G	(KCAL)	73	7.6	0.0	81
EDIBLE PORTION					
AMOUNT OF DIGESTIBLE	GRAM	16.19	0.76	0.00	
CONSTITUENTS PER 100 G					
ENERGY VALUE (AVERAGE)	KJOULE	298	30	0.0	328
OF THE DIGESTIBLE	(KCAL)	71	7.2	0.0	78
FRACTION PER 100 G					
EDIBLE PORTION					

WASTE PERCENTAGE AVERAGE 0.00

CONSTITUENTS	DIM	AV	VARIATION			AVR	NUTR. DENS.		MOLPERC.
WATER	GRAM	81.20	81.20	-	81.20	81.20	GRAM/MJ	247.24	
PROTEIN	GRAM	16.70	16.70	-	16.70	16.70	GRAM/MJ	50.85	
FAT	GRAM	0.80	0.80	-	0.80	0.80	GRAM/MJ	2.44	
MINERALS	GRAM	1.10	1.10	-	1.10	1.10	GRAM/MJ	3.35	

CONSTITUENTS	DIM	AV	VARIATION			AVR	NUTR. DENS.		MOLPERC.
POTASSIUM	MILLI	428.00	428.00	-	428.00	428.00	GRAM/MJ	1.30	
MAGNESIUM	MILLI	57.00	57.00	-	57.00	57.00	MILLI/MJ	173.56	
CALCIUM	MILLI	8.00	8.00	-	8.00	8.00	MILLI/MJ	24.36	
PHOSPHORUS	MILLI	376.00	376.00	-	376.00	376.00	GRAM/MJ	1.14	
VITAMIN B1	MILLI	0.17	0.17	-	0.17	0.17	MILLI/MJ	0.52	
VITAMIN B2	MILLI	0.17	0.17	-	0.17	0.17	MILLI/MJ	0.52	
FOLIC ACID	MICRO	3.10	3.10	-	3.10	3.10	MICRO/MJ	9.44	
VITAMIN B12	MICRO	0.30	0.30	-	0.30	0.30	MICRO/MJ	0.91	
ALANINE	GRAM	1.23	1.23	-	1.23	1.23	GRAM/MJ	3.75	
ARGININE	GRAM	1.30	1.30	-	1.30	1.30	GRAM/MJ	3.96	
ASPARTIC ACID	GRAM	1.22	1.22	-	1.22	1.22	GRAM/MJ	3.71	
GLUTAMIC ACID	GRAM	3.02	3.02	-	3.02	3.02	GRAM/MJ	9.20	
GLYCINE	GRAM	1.10	1.10	-	1.10	1.10	GRAM/MJ	3.35	
HISTIDINE	GRAM	0.45	0.45	-	0.45	0.45	GRAM/MJ	1.37	
ISOLEUCINE	GRAM	1.39	1.39	-	1.39	1.39	GRAM/MJ	4.23	
LEUCINE	GRAM	1.61	1.61	-	1.61	1.61	GRAM/MJ	4.90	
LYSINE	GRAM	2.05	2.05	-	2.05	2.05	GRAM/MJ	6.24	
METHIONINE	GRAM	0.68	0.68	-	0.68	0.68	GRAM/MJ	2.07	
PHENYLALANINE	GRAM	0.81	0.81	-	0.81	0.81	GRAM/MJ	2.47	
PROLINE	GRAM	0.71	0.71	-	0.71	0.71	GRAM/MJ	2.16	
SERINE	GRAM	1.07	1.07	-	1.07	1.07	GRAM/MJ	3.26	
THREONINE	GRAM	1.14	1.14	-	1.14	1.14	GRAM/MJ	3.47	
TRYPTOPHAN	GRAM	0.22	0.22	-	0.22	0.22	GRAM/MJ	0.67	
TYROSINE	GRAM	0.72	0.72	-	0.72	0.72	GRAM/MJ	2.19	
VALINE	GRAM	1.13	1.13	-	1.13	1.13	GRAM/MJ	3.44	

SEEZUNGE SOLE SOLE

SOLEA VULGARIS (QUENSEL)

		PROTEIN	FAT	CARBOHYDRATES	TOTAL
ENERGY VALUE (AVERAGE)	KJOULE	322	54	0.0	377
PER 100 G	(KCAL)	77	13	0.0	90
EDIBLE PORTION					
AMOUNT OF DIGESTIBLE	GRAM	16.97	1.30	0.00	
CONSTITUENTS PER 100 G					
ENERGY VALUE (AVERAGE)	KJOULE	313	52	0.0	364
OF THE DIGESTIBLE	(KCAL)	75	12	0.0	87
FRACTION PER 100 G					
EDIBLE PORTION					

WASTE PERCENTAGE AVERAGE 29 MINIMUM 25 MAXIMUM 34

CONSTITUENTS	DIM	AV	VARIATION			AVR	NUTR. DENS.		MOLPERC.
WATER	GRAM	80.00	79.30	-	85.30	56.80	GRAM/MJ	219.64	
PROTEIN	GRAM	17.50	15.20	-	18.30	12.43	GRAM/MJ	48.05	
FAT	GRAM	1.37	0.53	-	1.94	0.97	GRAM/MJ	3.76	
MINERALS	GRAM	1.06	0.73	-	1.22	0.75	GRAM/MJ	2.91	

CONSTITUENTS	DIM	AV	VARIATION	AVR	NUTR. DENS.		MOLPERC.
SODIUM	MILLI	100.00	72.00 - 202.00	71.00	MILLI/MJ	274.55	
POTASSIUM	MILLI	309.00	218.00 - 349.00	219.39	MILLI/MJ	848.35	
MAGNESIUM	MILLI	49.00	25.00 - 73.00	34.79	MILLI/MJ	134.53	
CALCIUM	MILLI	29.00	18.00 - 27.01	20.59	MILLI/MJ	79.62	
IRON	MILLI	0.80	- - -	0.57	MILLI/MJ	2.20	
PHOSPHORUS	MILLI	195.00	- - -	138.45	MILLI/MJ	535.37	
CHLORIDE	MILLI	142.00	80.00 - 205.00	100.82	MILLI/MJ	389.86	
IODIDE	MICRO	17.00	- - -	12.07	MICRO/MJ	46.67	
BORON	MICRO	62.00	- - -	44.02	MICRO/MJ	170.22	
SELENIUM	MICRO	24.00	15.00 - 29.00	17.04	MICRO/MJ	65.89	
VITAMIN A	-	TRACES	- - -				
VITAMIN D	-	0.00	- - -	0.00			
VITAMIN B1	MICRO	60.00	- - -	42.60	MICRO/MJ	164.73	
VITAMIN B2	MILLI	0.10	- - -	0.07	MILLI/MJ	0.27	
NICOTINAMIDE	MILLI	3.00	- - -	2.13	MILLI/MJ	8.24	
VITAMIN C	-	0.00	- - -	0.00			
ALANINE	GRAM	1.34	1.14 - 1.48	0.95	GRAM/MJ	3.68	9.5
ARGININE	GRAM	1.14	0.87 - 1.32	0.81	GRAM/MJ	3.13	4.1
ASPARTIC ACID	GRAM	2.09	1.90 - 2.34	1.48	GRAM/MJ	5.74	9.9
CYSTINE	GRAM	0.42	0.39 - 0.44	0.30	GRAM/MJ	1.15	1.1
GLUTAMIC ACID	GRAM	3.17	2.79 - 3.62	2.25	GRAM/MJ	8.70	13.6
GLYCINE	GRAM	1.14	0.92 - 1.41	0.81	GRAM/MJ	3.13	9.6
HISTIDINE	GRAM	0.47	0.37 - 0.66	0.33	GRAM/MJ	1.29	1.9
ISOLEUCINE	GRAM	0.92	0.81 - 1.08	0.65	GRAM/MJ	2.53	4.4
LEUCINE	GRAM	1.66	1.55 - 1.82	1.18	GRAM/MJ	4.56	8.0
LYSINE	GRAM	1.86	1.55 - 2.01	1.32	GRAM/MJ	5.11	8.0
METHIONINE	GRAM	0.59	0.51 - 0.66	0.42	GRAM/MJ	1.62	2.5
PHENYLALANINE	GRAM	0.76	0.66 - 0.92	0.54	GRAM/MJ	2.09	2.9
PROLINE	GRAM	0.89	0.62 - 0.95	0.63	GRAM/MJ	2.44	4.9
SERINE	GRAM	0.95	0.76 - 1.26	0.67	GRAM/MJ	2.61	5.7
THREONINE	GRAM	1.10	0.78 - 1.14	0.78	GRAM/MJ	3.02	5.8
TRYPTOPHAN	GRAM	0.13	0.12 - 0.13	0.09	GRAM/MJ	0.36	0.4
TYROSINE	GRAM	0.62	0.46 - 0.78	0.44	GRAM/MJ	1.70	2.2
VALINE	GRAM	1.03	0.93 - 1.17	0.73	GRAM/MJ	2.83	5.5
MYRISTIC ACID	MILLI	50.00	30.00 - 70.00	35.50	MILLI/MJ	137.27	
PALMITIC ACID	GRAM	0.26	0.21 - 0.30	0.18	GRAM/MJ	0.71	
STEARIC ACID	MILLI	76.66	60.00 - 90.00	54.43	MILLI/MJ	210.49	
PALMITOLEIC ACID	GRAM	0.12	0.02 - 0.22	0.09	GRAM/MJ	0.35	
OLEIC ACID	GRAM	0.21	0.16 - 0.29	0.15	GRAM/MJ	0.58	
EICOSENOIC ACID	MILLI	40.00	40.00 - 40.00	28.40	MILLI/MJ	109.82	
C22 : 1 FATTY ACID	MILLI	63.33	50.00 - 80.00	44.97	MILLI/MJ	173.88	
LINOLEIC ACID	MILLI	47.50	10.00 - 160.00	33.73	MILLI/MJ	130.41	
LINOLENIC ACID	MILLI	10.00	10.00 - 10.00	7.10	MILLI/MJ	27.45	
ARACHIDONIC ACID	MILLI	23.33	10.00 - 40.00	16.57	MILLI/MJ	64.06	
C20 : 5 N-3 FATTY ACID	MILLI	33.33	10.00 - 80.00	23.67	MILLI/MJ	91.52	
C22 : 5 N-3 FATTY ACID	MILLI	72.50	50.00 - 90.00	51.48	MILLI/MJ	199.05	
C22 : 6 N-3 FATTY ACID	GRAM	0.16	0.10 - 0.31	0.12	GRAM/MJ	0.44	
CHOLESTEROL	MILLI	50.00	- - -	35.50	MILLI/MJ	137.27	
TOTAL PURINES	MILLI	131.00 [1]	125.00 - 137.00	93.01	MILLI/MJ	359.66	

1 VALUE EXPRESSED IN MG URIC ACID/100G

SPROTTE — SPRAT — SPRAT

SPRATTUS SPRATTUS (L.)

		PROTEIN	FAT	CARBOHYDRATES	TOTAL
ENERGY VALUE (AVERAGE) PER 100 G EDIBLE PORTION	KJOULE	307	660	0.0	967
	(KCAL)	73	158	0.0	231
AMOUNT OF DIGESTIBLE CONSTITUENTS PER 100 G	GRAM	16.19	15.77	0.00	
ENERGY VALUE (AVERAGE) OF THE DIGESTIBLE FRACTION PER 100 G EDIBLE PORTION	KJOULE	298	627	0.0	925
	(KCAL)	71	150	0.0	221

WASTE PERCENTAGE AVERAGE 41

CONSTITUENTS	DIM	AV	VARIATION			AVR	NUTR.	DENS.	MOLPERC.
WATER	GRAM	65.90	61.30	-	69.70	38.88	GRAM/MJ	71.24	
PROTEIN	GRAM	16.70	16.50	-	17.10	9.85	GRAM/MJ	18.05	
FAT	GRAM	16.60	3.60	-	20.60	9.79	GRAM/MJ	17.95	
MINERALS	GRAM	1.80	1.50	-	2.00	1.06	GRAM/MJ	1.95	
MANGANESE	MILLI	0.14	-	-	-	0.08	MILLI/MJ	0.15	
IRON	MILLI	1.90	-	-	-	1.12	MILLI/MJ	2.05	
COPPER	MICRO	80.00	-	-	-	47.20	MICRO/MJ	86.48	
ZINC	MILLI	1.50	-	-	-	0.89	MILLI/MJ	1.62	
VITAMIN B1	MICRO	40.00	-	-	-	23.60	MICRO/MJ	43.24	
MYRISTIC ACID	GRAM	1.07	0.89	-	1.17	0.64	GRAM/MJ	1.16	
PALMITIC ACID	GRAM	3.13	3.10	-	3.21	1.85	GRAM/MJ	3.39	
STEARIC ACID	GRAM	0.30	0.30	-	0.30	0.18	GRAM/MJ	0.32	
PALMITOLEIC ACID	GRAM	1.13	0.79	-	2.03	0.67	GRAM/MJ	1.23	
OLEIC ACID	GRAM	2.49	1.57	-	2.99	1.47	GRAM/MJ	2.70	
EICOSENOIC ACID	GRAM	1.25	1.05	-	1.35	0.74	GRAM/MJ	1.35	
C22 : 1 FATTY ACID	GRAM	1.84	1.39	-	2.09	1.09	GRAM/MJ	1.99	
LINOLEIC ACID	GRAM	0.23	0.21	-	0.27	0.14	GRAM/MJ	0.25	
LINOLENIC ACID	GRAM	0.16	0.16	-	0.16	0.09	GRAM/MJ	0.17	
ARACHIDONIC ACID	MILLI	70.00	70.00	-	70.00	41.30	MILLI/MJ	75.67	
C18 : 4 FATTY ACID	GRAM	0.23	0.14	-	0.28	0.14	GRAM/MJ	0.25	
C20 : 5 N-3 FATTY ACID	GRAM	1.14	0.89	-	1.57	0.67	GRAM/MJ	1.23	
C22 : 5 N-3 FATTY ACID	GRAM	0.17	0.17	-	0.17	0.10	GRAM/MJ	0.18	
C22 : 6 N-3 FATTY ACID	GRAM	1.62	1.14	-	2.62	0.96	GRAM/MJ	1.76	
CHOLESTEROL	MILLI	109.00	-	-	-	64.31	MILLI/MJ	117.83	

ST. PETERSFISCH (HERINGSKÖNIG) | **JOHN DORY** | **SAINT PIERRE**

ZEUS FABER

		PROTEIN	FAT	CARBOHYDRATES	TOTAL
ENERGY VALUE (AVERAGE)	KJOULE	334	56	0.0	390
PER 100 G	(KCAL)	80	13	0.0	93
EDIBLE PORTION					
AMOUNT OF DIGESTIBLE	GRAM	17.61	1.33	0.00	
CONSTITUENTS PER 100 G					
ENERGY VALUE (AVERAGE)	KJOULE	324	53	0.0	377
OF THE DIGESTIBLE	(KCAL)	77	13	0.0	90
FRACTION PER 100 G					
EDIBLE PORTION					

WASTE PERCENTAGE AVERAGE 66

CONSTITUENTS	DIM	AV	VARIATION			AVR	NUTR. DENS.		MOLPERC.
WATER	GRAM	78.08	78.00	-	78.30	26.55	GRAM/MJ	207.07	
PROTEIN	GRAM	18.15	17.80	-	18.30	6.17	GRAM/MJ	48.15	
FAT	GRAM	1.40	1.40	-	1.40	0.48	GRAM/MJ	3.71	
MINERALS	GRAM	1.40	1.40	-	1.40	0.48	GRAM/MJ	3.71	
POTASSIUM	MILLI	151.00	151.00	-	151.00	51.34	MILLI/MJ	400.42	
PHOSPHORUS	MILLI	230.09	230.00	-	231.00	78.23	MILLI/MJ	610.16	
CHLORIDE	MILLI	70.00	70.00	-	70.00	23.80	MILLI/MJ	185.63	

STEINBUTT TURBOT TURBOT

PSETTA MAXIMA (L.)

		PROTEIN	FAT	CARBOHYDRATES	TOTAL
ENERGY VALUE (AVERAGE) PER 100 G EDIBLE PORTION	KJOULE	307	69	0.0	376
	(KCAL)	73	16	0.0	90
AMOUNT OF DIGESTIBLE CONSTITUENTS PER 100 G	GRAM	16.19	1.64	0.00	
ENERGY VALUE (AVERAGE) OF THE DIGESTIBLE FRACTION PER 100 G EDIBLE PORTION	KJOULE	298	65	0.0	364
	(KCAL)	71	16	0.0	87

WASTE PERCENTAGE AVERAGE 54 MINIMUM 52 MAXIMUM 56

CONSTITUENTS	DIM	AV	VARIATION			AVR	NUTR. DENS.		MOLPERC.
WATER	GRAM	80.40	-	-	-	36.98	GRAM/MJ	221.16	
PROTEIN	GRAM	16.70	16.00	-	18.10	7.68	GRAM/MJ	45.94	
FAT	GRAM	1.73	1.00	-	2.28	0.80	GRAM/MJ	4.76	
MINERALS	GRAM	0.74	-	-	-	0.34	GRAM/MJ	2.04	
SODIUM	MILLI	114.00	90.00	-	150.00	52.44	MILLI/MJ	313.58	
POTASSIUM	MILLI	290.00	255.00	-	320.00	133.40	MILLI/MJ	797.70	
MAGNESIUM	MILLI	45.00	24.00	-	59.00	20.70	MILLI/MJ	123.78	
CALCIUM	MILLI	17.00	12.00	-	21.00	7.82	MILLI/MJ	46.76	
IRON	MILLI	0.50	-	-	-	0.23	MILLI/MJ	1.38	
PHOSPHORUS	MILLI	159.00	129.00	-	188.00	73.14	MILLI/MJ	437.36	
CHLORIDE	MILLI	140.00	-	-	-	64.40	MILLI/MJ	385.10	
VITAMIN A	-	TRACES	-	-	-				
VITAMIN B1	MICRO	20.00	-	-	-	9.20	MICRO/MJ	55.01	
VITAMIN B2	MILLI	0.15	0.14	-	0.15	0.07	MILLI/MJ	0.41	
NICOTINAMIDE	MILLI	3.00	-	-	-	1.38	MILLI/MJ	8.25	
VITAMIN C	-	TRACES	-	-	-				

STINT SMELT ÉPERLAN

OSMERUS EPERLANUS (L.)

		PROTEIN	FAT	CARBOHYDRATES	TOTAL
ENERGY VALUE (AVERAGE)	KJOULE	320	68	0.0	388
PER 100 G	(KCAL)	77	16	0.0	93
EDIBLE PORTION					
AMOUNT OF DIGESTIBLE	GRAM	16.87	1.61	0.00	
CONSTITUENTS PER 100 G					
ENERGY VALUE (AVERAGE)	KJOULE	311	64	0.0	375
OF THE DIGESTIBLE	(KCAL)	74	15	0.0	90
FRACTION PER 100 G					
EDIBLE PORTION					

WASTE PERCENTAGE AVERAGE 52

CONSTITUENTS	DIM	AV	VARIATION		AVR	NUTR. DENS.		MOLPERC.
WATER	GRAM	80.20	79.30	81.50	38.50	GRAM/MJ	213.92	
PROTEIN	GRAM	17.40	15.60	19.90	8.35	GRAM/MJ	46.41	
FAT	GRAM	1.70	1.00	2.40	0.82	GRAM/MJ	4.53	
MINERALS	GRAM	0.91	0.76	1.06	0.44	GRAM/MJ	2.43	
SODIUM	MILLI	156.00	80.00	214.00	74.88	MILLI/MJ	416.10	
POTASSIUM	MILLI	357.00	-	-	171.36	MILLI/MJ	952.23	
MAGNESIUM	MILLI	24.00	-	-	11.52	MILLI/MJ	64.02	
PHOSPHORUS	MILLI	245.00	190.00	680.00	117.60	MILLI/MJ	653.49	
VITAMIN B2	MILLI	0.12	-	-	0.06	MILLI/MJ	0.32	
NICOTINAMIDE	MILLI	1.45	-	-	0.70	MILLI/MJ	3.87	
PANTOTHENIC ACID	MILLI	0.64	-	-	0.31	MILLI/MJ	1.71	
BIOTIN	MICRO	30.00	-	-	14.40	MICRO/MJ	80.02	
FOLIC ACID	MICRO	37.00	-	-	17.76	MICRO/MJ	98.69	
VITAMIN B12	MICRO	3.44	-	-	1.65	MICRO/MJ	9.18	
CHOLESTEROL	MILLI	71.00	-	-	34.08	MILLI/MJ	189.38	
TOTAL PHOSPHOLIPIDS	MILLI	460.00	-	-	220.80	GRAM/MJ	1.23	
PHOSPHATIDYLCHOLINE	MILLI	239.00	-	-	114.72	MILLI/MJ	637.49	
PHOSPHATIDYLETHANOLAMINE	MILLI	152.00	-	-	72.96	MILLI/MJ	405.43	
SPHINGOMYELIN	MILLI	23.00	-	-	11.04	MILLI/MJ	61.35	

STÖCKER — JACK MAKEREL (MAAS BANKER) — CHINCHARD

TRACHURUS TRACHURUS

		PROTEIN	FAT	CARBOHYDRATES	TOTAL
ENERGY VALUE (AVERAGE)	KJOULE	364	153	0.0	517
PER 100 G	(KCAL)	87	37	0.0	123
EDIBLE PORTION					
AMOUNT OF DIGESTIBLE	GRAM	19.15	3.65	0.00	
CONSTITUENTS PER 100 G					
ENERGY VALUE (AVERAGE)	KJOULE	353	145	0.0	498
OF THE DIGESTIBLE	(KCAL)	84	35	0.0	119
FRACTION PER 100 G					
EDIBLE PORTION					

WASTE PERCENTAGE AVERAGE 60

CONSTITUENTS	DIM	AV	VARIATION			AVR	NUTR. DENS.		MOLPERC.
WATER	GRAM	75.30	74.60	-	76.00	30.12	GRAM/MJ	151.19	
PROTEIN	GRAM	19.75	19.50	-	20.00	7.90	GRAM/MJ	39.65	
FAT	GRAM	3.85	2.50	-	5.20	1.54	GRAM/MJ	7.73	
MINERALS	GRAM	1.40	1.20	-	1.60	0.56	GRAM/MJ	2.81	
SODIUM	MILLI	64.00	64.00	-	64.00	25.60	MILLI/MJ	128.50	
POTASSIUM	MILLI	360.00	360.00	-	360.00	144.00	MILLI/MJ	722.80	
CALCIUM	MILLI	65.00	65.00	-	65.00	26.00	MILLI/MJ	130.51	
COPPER	MILLI	0.19	0.19	-	0.19	0.08	MILLI/MJ	0.38	
ZINC	MILLI	0.66	0.66	-	0.66	0.26	MILLI/MJ	1.33	
PHOSPHORUS	MILLI	239.00	218.00	-	260.00	95.60	MILLI/MJ	479.86	
CHLORIDE	MILLI	120.00	120.00	-	120.00	48.00	MILLI/MJ	240.93	
IODIDE	MICRO	48.00	48.00	-	48.00	19.20	MICRO/MJ	96.37	
VITAMIN A	MICRO	12.00	12.00	-	12.00	4.80	MICRO/MJ	24.09	
VITAMIN B1	MILLI	0.14	0.14	-	0.14	0.06	MILLI/MJ	0.28	
VITAMIN B2	MILLI	0.14	0.14	-	0.14	0.06	MILLI/MJ	0.28	
NICOTINAMIDE	MILLI	3.40	3.40	-	3.40	1.36	MILLI/MJ	6.83	
VITAMIN B12	MICRO	6.00	6.00	-	6.00	2.40	MICRO/MJ	12.05	
ALANINE	GRAM	1.56	1.56	-	1.56	0.62	GRAM/MJ	3.13	10.0
ARGININE	GRAM	1.32	1.32	-	1.32	0.53	GRAM/MJ	2.65	4.3
ASPARTIC ACID	GRAM	2.01	2.01	-	2.01	0.80	GRAM/MJ	4.04	8.7
CYSTINE	GRAM	0.41	0.41	-	0.41	0.16	GRAM/MJ	0.82	1.0
GLUTAMIC ACID	GRAM	2.83	2.83	-	2.83	1.13	GRAM/MJ	5.68	11.0
GLYCINE	GRAM	1.94	1.94	-	1.94	0.78	GRAM/MJ	3.90	14.8
HISTIDINE	GRAM	0.67	0.67	-	0.67	0.27	GRAM/MJ	1.35	2.5
ISOLEUCINE	GRAM	1.05	1.05	-	1.05	0.42	GRAM/MJ	2.11	4.6
LEUCINE	GRAM	1.58	1.58	-	1.58	0.63	GRAM/MJ	3.17	6.9
LYSINE	GRAM	1.80	1.80	-	1.80	0.72	GRAM/MJ	3.61	7.1
METHIONINE	GRAM	0.58	0.58	-	0.58	0.23	GRAM/MJ	1.16	2.2
PHENYLALANINE	GRAM	0.73	0.73	-	0.73	0.29	GRAM/MJ	1.47	2.5
PROLINE	GRAM	1.15	1.15	-	1.15	0.46	GRAM/MJ	2.31	5.7
SERINE	GRAM	0.93	0.93	-	0.93	0.37	GRAM/MJ	1.87	5.1
THREONINE	GRAM	1.00	1.00	-	1.00	0.40	GRAM/MJ	2.01	4.8
TRYPTOPHAN	GRAM	0.20	0.20	-	0.20	0.08	GRAM/MJ	0.40	0.6
TYROSINE	GRAM	0.79	0.79	-	0.79	0.32	GRAM/MJ	1.59	2.5
VALINE	GRAM	1.17	1.17	-	1.17	0.47	GRAM/MJ	2.35	5.7

STÖR | **STURGEON** | **ESTURGEON**

ACIPESERIDAEN SPP.

		PROTEIN	FAT	CARBOHYDRATES	TOTAL
ENERGY VALUE (AVERAGE)	KJOULE	333	76	0.0	409
PER 100 G	(KCAL)	80	18	0.0	98
EDIBLE PORTION					
AMOUNT OF DIGESTIBLE	GRAM	17.55	1.80	0.00	
CONSTITUENTS PER 100 G					
ENERGY VALUE (AVERAGE)	KJOULE	323	72	0.0	395
OF THE DIGESTIBLE	(KCAL)	77	17	0.0	94
FRACTION PER 100 G					
EDIBLE PORTION					

WASTE PERCENTAGE AVERAGE 0.00

CONSTITUENTS	DIM	AV	VARIATION	AVR	NUTR. DENS.		MOLPERC.
WATER	GRAM	78.70	78.70 - 78.70	78.70	GRAM/MJ	199.26	
PROTEIN	GRAM	18.10	18.10 - 18.10	18.10	GRAM/MJ	45.83	
FAT	GRAM	1.90	1.90 - 1.90	1.90	GRAM/MJ	4.81	
MINERALS	GRAM	1.40	1.40 - 1.40	1.40	GRAM/MJ	3.54	

THUNFISCH | TUNA | THON

THUNNUS SPP.

		PROTEIN	FAT	CARBOHYDRATES	TOTAL
ENERGY VALUE (AVERAGE)	KJOULE	396	616	0.0	1012
PER 100 G	(KCAL)	95	147	0.0	242
EDIBLE PORTION					
AMOUNT OF DIGESTIBLE	GRAM	20.85	14.72	0.00	
CONSTITUENTS PER 100 G					
ENERGY VALUE (AVERAGE)	KJOULE	384	585	0.0	969
OF THE DIGESTIBLE	(KCAL)	92	140	0.0	232
FRACTION PER 100 G					
EDIBLE PORTION					

WASTE PERCENTAGE AVERAGE 39 MINIMUM 36 MAXIMUM 44

CONSTITUENTS	DIM	AV	VARIATION			AVR	NUTR.	DENS.	MOLPERC.
WATER	GRAM	61.50	57.00	-	66.00	37.52	GRAM/MJ	63.45	
PROTEIN	GRAM	21.50	18.00	-	24.00	13.12	GRAM/MJ	22.18	
FAT	GRAM	15.50	4.17	-	24.00	9.46	GRAM/MJ	15.99	
MINERALS	GRAM	1.10	-	-	-	0.67	GRAM/MJ	1.13	
SODIUM	MILLI	43.00	41.00	-	45.00	26.23	MILLI/MJ	44.37	
CALCIUM	MILLI	40.00	-	-	-	24.40	MILLI/MJ	41.27	
IRON	MILLI	1.00	-	-	-	0.61	MILLI/MJ	1.03	
PHOSPHORUS	MILLI	200.00	-	-	-	122.00	MILLI/MJ	206.35	
FLUORIDE	MICRO	28.00	-	-	-	17.08	MICRO/MJ	28.89	
IODIDE	MICRO	50.00	40.00	-	50.00	30.50	MICRO/MJ	51.59	
BORON	MICRO	45.00	35.00	-	54.00	27.45	MICRO/MJ	46.43	
SELENIUM	MILLI	0.13	-	-	-	0.08	MILLI/MJ	0.13	
VITAMIN A	MILLI	0.45	0.08	-	0.83	0.27	MILLI/MJ	0.46	
VITAMIN D	MICRO	5.38	2.50	-	8.25	3.28	MICRO/MJ	5.55	
VITAMIN B1	MILLI	0.16	-	-	-	0.10	MILLI/MJ	0.17	
VITAMIN B2	MILLI	0.16	-	-	-	0.10	MILLI/MJ	0.17	
NICOTINAMIDE	MILLI	8.50	7.70	-	8.80	5.19	MILLI/MJ	8.77	
PANTOTHENIC ACID	MILLI	0.66	-	-	-	0.40	MILLI/MJ	0.68	
VITAMIN B6	MILLI	0.46	0.37	-	0.54	0.28	MILLI/MJ	0.47	
FOLIC ACID	MICRO	15.00	-	-	-	9.15	MICRO/MJ	15.48	
VITAMIN B12	MICRO	4.25	3.70	-	4.80	2.59	MICRO/MJ	4.38	
ALANINE	GRAM	1.61	1.59	-	1.63	0.98	GRAM/MJ	1.66	9.4
ARGININE	GRAM	1.25	1.22	-	1.27	0.76	GRAM/MJ	1.29	3.7
ASPARTIC ACID	GRAM	2.88	-	-	-	1.76	GRAM/MJ	2.97	11.2
CYSTINE	GRAM	0.29	-	-	-	0.18	GRAM/MJ	0.30	0.6

CONSTITUENTS	DIM	AV	VARIATION			AVR	NUTR. DENS.		MOLPERC.
GLUTAMIC ACID	GRAM	3.52	-	-	-	2.15	GRAM/MJ	3.63	12.4
GLYCINE	GRAM	1.17	-	-	-	0.71	GRAM/MJ	1.21	8.1
HISTIDINE	GRAM	1.09	1.02	-	1.13	0.66	GRAM/MJ	1.12	3.6
ISOLEUCINE	GRAM	1.21	-	-	-	0.74	GRAM/MJ	1.25	4.8
LEUCINE	GRAM	2.17	2.12	-	2.19	1.32	GRAM/MJ	2.24	8.6
LYSINE	GRAM	2.21	2.14	-	2.28	1.35	GRAM/MJ	2.28	7.9
METHIONINE	GRAM	0.61	-	-	-	0.37	GRAM/MJ	0.63	2.1
PHENYLALANINE	GRAM	1.05	-	-	-	0.64	GRAM/MJ	1.08	3.3
PROLINE	GRAM	0.88	-	-	-	0.54	GRAM/MJ	0.91	4.0
SERINE	GRAM	1.05	0.95	-	1.16	0.64	GRAM/MJ	1.08	5.2
THREONINE	GRAM	1.18	-	-	-	0.72	GRAM/MJ	1.22	5.1
TRYPTOPHAN	GRAM	0.30	-	-	-	0.18	GRAM/MJ	0.31	0.8
TYROSINE	GRAM	0.97	-	-	-	0.59	GRAM/MJ	1.00	2.8
VALINE	GRAM	1.42	1.37	-	1.44	0.87	GRAM/MJ	1.47	6.3
CADAVERINE	MILLI	0.30	-	-	-	0.18	MILLI/MJ	0.31	
PUTRESCINE	MILLI	-	0.50	-	1.00				
MYRISTIC ACID	GRAM	0.40	0.37	-	0.43	0.24	GRAM/MJ	0.41	
PALMITIC ACID	GRAM	2.97	2.85	-	3.10	1.81	GRAM/MJ	3.07	
STEARIC ACID	GRAM	0.75	0.64	-	0.87	0.46	GRAM/MJ	0.78	
PALMITOLEIC ACID	GRAM	0.66	0.57	-	0.76	0.41	GRAM/MJ	0.69	
OLEIC ACID	GRAM	2.65	2.60	-	2.71	1.62	GRAM/MJ	2.74	
EICOSENOIC ACID	GRAM	0.47	0.43	-	0.51	0.29	GRAM/MJ	0.48	
C22 : 1 FATTY ACID	GRAM	0.26	0.23	-	0.29	0.16	GRAM/MJ	0.27	
LINOLEIC ACID	GRAM	0.26	0.22	-	0.31	0.16	GRAM/MJ	0.27	
LINOLENIC ACID	GRAM	0.27	0.18	-	0.37	0.17	GRAM/MJ	0.28	
ARACHIDONIC ACID	GRAM	0.28	0.27	-	0.29	0.17	GRAM/MJ	0.29	
C18 : 4 FATTY ACID	GRAM	0.23	0.20	-	0.27	0.14	GRAM/MJ	0.24	
C20 : 5 N-3 FATTY ACID	GRAM	1.07	0.93	-	1.22	0.66	GRAM/MJ	1.11	
C22 : 5 N-3 FATTY ACID	GRAM	0.18	0.12	-	0.25	0.11	GRAM/MJ	0.19	
C22 : 6 N-3 FATTY ACID	GRAM	2.28	1.36	-	3.21	1.39	GRAM/MJ	2.36	
TOTAL PURINES	MILLI	257.00 1	257.00	-	257.00	156.77	MILLI/MJ	265.16	
ADENINE	MILLI	14.00	14.00	-	14.00	8.54	MILLI/MJ	14.44	
GUANINE	MILLI	19.00	19.00	-	19.00	11.59	MILLI/MJ	19.60	
HYPOXANTHINE	MILLI	178.00	178.00	-	178.00	108.58	MILLI/MJ	183.65	
TOTAL PHOSPHOLIPIDS	GRAM	1.94	-	-	-	1.18	GRAM/MJ	2.00	
PHOSPHATIDYLCHOLINE	MILLI	641.00	-	-	-	391.01	MILLI/MJ	661.36	
PHOSPHATIDYLETHANOLAMI	MILLI	503.00	-	-	-	306.83	MILLI/MJ	518.97	
PHOSPHATIDYLSERINE	MILLI	194.00	-	-	-	118.34	MILLI/MJ	200.16	
SPHINGOMYELIN	MILLI	153.00	-	-	-	93.33	MILLI/MJ	157.86	

1 VALUE OF TOTAL PURINES IS EXPRESSSED IN MG URIC ACID/100G

WALFLEISCH WHALE MEAT VIANDE DE BALEINE

BALEANOPTERA MUSCULUS (L.)

		PROTEIN	FAT	CARBOHYDRATES	TOTAL
ENERGY VALUE (AVERAGE)	KJOULE	427	134	0.0	561
PER 100 G	(KCAL)	102	32	0.0	134
EDIBLE PORTION					
AMOUNT OF DIGESTIBLE	GRAM	22.50	3.21	0.00	
CONSTITUENTS PER 100 G					
ENERGY VALUE (AVERAGE)	KJOULE	414	128	0.0	542
OF THE DIGESTIBLE	(KCAL)	99	31	0.0	130
FRACTION PER 100 G					
EDIBLE PORTION					

WASTE PERCENTAGE AVERAGE 0.00

CONSTITUENTS	DIM	AV	VARIATION		AVR	NUTR.	DENS.	MOLPERC.
WATER	GRAM	70.10	65.20	- 76.00	70.10	GRAM/MJ	129.35	
PROTEIN	GRAM	23.20	20.00	- 27.20	23.20	GRAM/MJ	42.81	
FAT	GRAM	3.38	2.03	- 5.00	3.38	GRAM/MJ	6.24	
MINERALS	GRAM	1.10	0.90	- 1.30	1.10	GRAM/MJ	2.03	
SODIUM	MILLI	100.00	-	- -	100.00	MILLI/MJ	184.53	
POTASSIUM	MILLI	300.00	-	- -	300.00	MILLI/MJ	553.59	
CALCIUM	MILLI	12.20	10.00	- 14.60	12.20	MILLI/MJ	22.51	
IRON	MILLI	4.22	2.40	- 6.03	4.22	MILLI/MJ	7.79	
PHOSPHORUS	MILLI	153.00	105.00	- 200.00	153.00	MILLI/MJ	282.33	
IODIDE	MILLI	0.40	0.30	- 0.40	0.40	MILLI/MJ	0.74	
SELENIUM	MICRO	50.00	-	- -	50.00	MICRO/MJ	92.26	
VITAMIN A	MICRO	24.00	-	- -	24.00	MICRO/MJ	44.29	
VITAMIN B1	MICRO	42.00	21.00	- 74.00	42.00	MICRO/MJ	77.50	
VITAMIN B2	MILLI	0.23	0.10	- 0.39	0.23	MILLI/MJ	0.42	
NICOTINAMIDE	MILLI	4.71	3.79	- 5.94	4.71	MILLI/MJ	8.69	
ALANINE	GRAM	1.63	1.53	- 1.73	1.63	GRAM/MJ	3.01	8.7
ARGININE	GRAM	1.46	1.26	- 1.65	1.46	GRAM/MJ	2.69	4.0
ASPARTIC ACID	GRAM	2.44	2.34	- 2.55	2.44	GRAM/MJ	4.50	8.7
CYSTINE	GRAM	0.43	0.32	- 0.51	0.43	GRAM/MJ	0.79	0.8
GLUTAMIC ACID	GRAM	3.74	3.21	- 4.32	3.74	GRAM/MJ	6.90	12.0
GLYCINE	GRAM	1.76	1.17	- 2.80	1.76	GRAM/MJ	3.25	11.1
HISTIDINE	GRAM	0.93	0.67	- 1.13	0.93	GRAM/MJ	1.72	2.8
ISOLEUCINE	GRAM	1.53	1.38	- 1.53	1.53	GRAM/MJ	2.82	5.5
LEUCINE	GRAM	2.07	1.99	- 2.17	2.07	GRAM/MJ	3.82	7.5
LYSINE	GRAM	2.36	2.26	- 2.51	2.36	GRAM/MJ	4.35	7.6
METHIONINE	GRAM	0.62	0.53	- 0.68	0.62	GRAM/MJ	1.14	2.0
PHENYLALANINE	GRAM	1.29	1.17	- 1.39	1.29	GRAM/MJ	2.38	3.7
PROLINE	GRAM	1.33	1.01	- 1.43	1.33	GRAM/MJ	2.45	5.5
SERINE	GRAM	1.32	1.23	- 1.39	1.32	GRAM/MJ	2.44	5.9
THREONINE	GRAM	1.28	1.22	- 1.32	1.28	GRAM/MJ	2.36	5.1
TRYPTOPHAN	GRAM	0.33	0.27	- 0.37	0.33	GRAM/MJ	0.61	0.8
TYROSINE	GRAM	0.88	0.75	- 0.92	0.88	GRAM/MJ	1.62	2.3
VALINE	GRAM	1.47	1.29	- 1.68	1.47	GRAM/MJ	2.71	5.9
TOTAL PURINES	MILLI	88.00 [1]	-	- -	88.00	MILLI/MJ	162.39	

1 VALUE EXPRESSED IN MG URIC ACID/100G

Süßwasserfische

FRESHWATER FISH **POISSONS D'EAU DOUCE**

AAL EEL ANGUILLE

ANGUILLA ANGUILLA (L.)

		PROTEIN	FAT	CARBOHYDRATES	TOTAL
ENERGY VALUE (AVERAGE)	KJOULE	276	974	0.0	1250
PER 100 G	(KCAL)	66	233	0.0	299
EDIBLE PORTION					
AMOUNT OF DIGESTIBLE	GRAM	14.55	23.27	0.00	
CONSTITUENTS PER 100 G					
ENERGY VALUE (AVERAGE)	KJOULE	268	925	0.0	1193
OF THE DIGESTIBLE	(KCAL)	64	221	0.0	285
FRACTION PER 100 G					
EDIBLE PORTION					

WASTE PERCENTAGE AVERAGE 30 MINIMUM 24 MAXIMUM 35

CONSTITUENTS	DIM	AV	VARIATION			AVR	NUTR. DENS.		MOLPERC.
WATER	GRAM	59.30	53.50	-	64.60	41.51	GRAM/MJ	49.71	
PROTEIN	GRAM	15.00	14.00	-	15.90	10.50	GRAM/MJ	12.57	
FAT	GRAM	24.50	18.30	-	27.80	17.15	GRAM/MJ	20.54	
MINERALS	GRAM	0.94	0.87	-	1.00	0.66	GRAM/MJ	0.79	
SODIUM	MILLI	65.00	40.00	-	78.00	45.50	MILLI/MJ	54.48	
POTASSIUM	MILLI	217.00	166.00	-	250.00	151.90	MILLI/MJ	181.90	
MAGNESIUM	MILLI	21.00	14.00	-	31.00	14.70	MILLI/MJ	17.60	
CALCIUM	MILLI	17.00	13.00	-	20.00	11.90	MILLI/MJ	14.25	
MANGANESE	MICRO	25.00	-	-	-	17.50	MICRO/MJ	20.96	
IRON	MILLI	0.60	0.50	-	0.80	0.42	MILLI/MJ	0.50	
COBALT	MICRO	1.00	-	-	-	0.70	MICRO/MJ	0.84	
COPPER	MICRO	80.00	30.00	-	170.00	56.00	MICRO/MJ	67.06	
ZINC	MILLI	1.20	0.20	-	2.70	0.84	MILLI/MJ	1.01	
NICKEL	MICRO	3.00	1.00	-	4.00	2.10	MICRO/MJ	2.51	
CHROMIUM	MICRO	14.00	4.00	-	23.00	9.80	MICRO/MJ	11.74	
PHOSPHORUS	MILLI	223.00	192.00	-	297.00	156.10	MILLI/MJ	186.92	
CHLORIDE	MILLI	52.00	35.00	-	69.00	36.40	MILLI/MJ	43.59	
FLUORIDE	MICRO	30.00	-	-	-	21.00	MICRO/MJ	25.15	
IODIDE	MICRO	4.00	-	-	-	2.80	MICRO/MJ	3.35	
BORON	MICRO	30.00	-	-	-	21.00	MICRO/MJ	25.15	
SELENIUM	MICRO	47.00	35.00	-	60.00	32.90	MICRO/MJ	39.40	

CONSTITUENTS	DIM	AV	VARIATION			AVR	NUTR. DENS.		MOLPERC.
VITAMIN A	MILLI	0.98	0.48	-	1.80	0.69	MILLI/MJ	0.82	
VITAMIN D	MICRO	13.00	-	-	-	9.10	MICRO/MJ	10.90	
VITAMIN B1	MILLI	0.18	0.15	-	0.22	0.13	MILLI/MJ	0.15	
VITAMIN B2	MILLI	0.32	0.25	-	0.37	0.22	MILLI/MJ	0.27	
NICOTINAMIDE	MILLI	2.60	1.40	-	3.80	1.82	MILLI/MJ	2.18	
VITAMIN B6	MILLI	0.28	0.20	-	0.36	0.20	MILLI/MJ	0.23	
FOLIC ACID	MICRO	13.00	-	-	-	9.10	MICRO/MJ	10.90	
VITAMIN B12	MICRO	1.00	-	-	-	0.70	MICRO/MJ	0.84	
VITAMIN C	MILLI	1.80	1.70	-	2.00	1.26	MILLI/MJ	1.51	
ALANINE	GRAM	1.40	-	-	-	0.98	GRAM/MJ	1.17	10.2
ARGININE	GRAM	1.10	-	-	-	0.77	GRAM/MJ	0.92	4.1
ASPARTIC ACID	GRAM	2.14	-	-	-	1.50	GRAM/MJ	1.79	10.4
CYSTINE	GRAM	0.13	-	-	-	0.09	GRAM/MJ	0.11	0.4
GLUTAMIC ACID	GRAM	3.13	-	-	-	2.19	GRAM/MJ	2.62	13.8
GLYCINE	GRAM	1.29	-	-	-	0.90	GRAM/MJ	1.08	11.1
HISTIDINE	GRAM	0.46	-	-	-	0.32	GRAM/MJ	0.39	1.9
ISOLEUCINE	GRAM	0.92	-	-	-	0.64	GRAM/MJ	0.77	4.5
LEUCINE	GRAM	1.67	-	-	-	1.17	GRAM/MJ	1.40	8.2
LYSINE	GRAM	1.36	-	-	-	0.95	GRAM/MJ	1.14	6.0
METHIONINE	GRAM	0.53	-	-	-	0.37	GRAM/MJ	0.44	2.3
PHENYLALANINE	GRAM	0.73	-	-	-	0.51	GRAM/MJ	0.61	2.9
PROLINE	GRAM	0.79	-	-	-	0.55	GRAM/MJ	0.66	4.4
SERINE	GRAM	0.81	-	-	-	0.57	GRAM/MJ	0.68	5.0
THREONINE	GRAM	0.87	-	-	-	0.61	GRAM/MJ	0.73	4.7
TRYPTOPHAN	GRAM	0.18	-	-	-	0.13	GRAM/MJ	0.15	0.6
TYROSINE	GRAM	0.82	-	-	-	0.57	GRAM/MJ	0.69	2.9
VALINE	GRAM	1.19	-	-	-	0.83	GRAM/MJ	1.00	6.6
MYRISTIC ACID	GRAM	1.24	1.08	-	1.30	0.87	GRAM/MJ	1.04	
PALMITIC ACID	GRAM	3.76	3.09	-	4.96	2.63	GRAM/MJ	3.15	
STEARIC ACID	GRAM	0.64	0.37	-	1.19	0.45	GRAM/MJ	0.54	
PALMITOLEIC ACID	GRAM	2.63	2.02	-	2.90	1.85	GRAM/MJ	2.21	
OLEIC ACID	GRAM	6.89	6.13	-	7.25	4.83	GRAM/MJ	5.78	
EICOSENOIC ACID	GRAM	1.76	1.76	-	1.76	1.23	GRAM/MJ	1.48	
C22 : 1 FATTY ACID	GRAM	0.23	0.23	-	0.23	0.16	GRAM/MJ	0.19	
ARACHIDONIC ACID	GRAM	0.12	0.11	-	0.13	0.09	GRAM/MJ	0.10	
C20 : 5 N-3 FATTY ACID	GRAM	0.26	0.18	-	0.30	0.18	GRAM/MJ	0.22	
C22 : 5 N-3 FATTY ACID	GRAM	0.37	0.15	-	0.48	0.26	GRAM/MJ	0.31	
C22 : 6 N-3 FATTY ACID	GRAM	0.56	0.46	-	0.62	0.40	GRAM/MJ	0.48	
CHOLESTEROL	MILLI	142.00	-	-	-	99.40	MILLI/MJ	119.03	
TOTAL PHOSPHOLIPIDS	GRAM	1.94	-	-	-	1.36	GRAM/MJ	1.63	
PHOSPHATIDYLCHOLINE	MILLI	867.00	-	-	-	606.90	MILLI/MJ	726.74	
PHOSPHATIDYLETHANOLAMI	MILLI	176.00	-	-	-	123.20	MILLI/MJ	147.53	
PHOSPHATIDYLSERINE	MILLI	335.00	-	-	-	234.50	MILLI/MJ	280.81	
SPHINGOMYELIN	MILLI	407.00	-	-	-	284.90	MILLI/MJ	341.16	

BARSCH (FLUSSBARSCH) | PERCH (RIVER PERCH) | PERCHE

PERCA FLUVIATILIS (L.)

		PROTEIN	FAT	CARBOHYDRATES	TOTAL
ENERGY VALUE (AVERAGE)	KJOULE	339	32	0.0	371
PER 100 G	(KCAL)	81	7.6	0.0	89
EDIBLE PORTION					
AMOUNT OF DIGESTIBLE	GRAM	17.84	0.76	0.00	
CONSTITUENTS PER 100 G					
ENERGY VALUE (AVERAGE)	KJOULE	329	30	0.0	359
OF THE DIGESTIBLE	(KCAL)	79	7.2	0.0	86
FRACTION PER 100 G					
EDIBLE PORTION					

WASTE PERCENTAGE AVERAGE 60 MINIMUM 47 MAXIMUM 64

CONSTITUENTS	DIM	AV	VARIATION			AVR	NUTR. DENS.		MOLPERC.
WATER	GRAM	79.50	79.10	-	81.10	31.80	GRAM/MJ	221.58	
PROTEIN	GRAM	18.40	18.20	-	18.80	7.36	GRAM/MJ	51.28	
FAT	GRAM	0.80	0.70	-	0.90	0.32	GRAM/MJ	2.23	
MINERALS	GRAM	1.24	1.10	-	1.30	0.50	GRAM/MJ	3.46	
SODIUM	MILLI	47.00	37.00	-	69.00	18.80	MILLI/MJ	131.00	
POTASSIUM	MILLI	330.00	220.00	-	390.00	132.00	MILLI/MJ	919.78	
MAGNESIUM	MILLI	20.00	-	-	-	8.00	MILLI/MJ	55.74	
CALCIUM	MILLI	20.00	-	-	-	8.00	MILLI/MJ	55.74	
IRON	MILLI	1.00	-	-	-	0.40	MILLI/MJ	2.79	
PHOSPHORUS	MILLI	198.00	180.00	-	215.00	79.20	MILLI/MJ	551.87	
IODIDE	MICRO	4.00	-	-	-	1.60	MICRO/MJ	11.15	
SELENIUM	MICRO	24.00	12.00	-	66.00	9.60	MICRO/MJ	66.89	
VITAMIN A	MICRO	7.00	2.00	-	12.00	2.80	MICRO/MJ	19.51	
VITAMIN B1	MICRO	75.00	60.00	-	90.00	30.00	MICRO/MJ	209.04	
VITAMIN B2	MILLI	0.12	0.07	-	0.17	0.05	MILLI/MJ	0.33	
NICOTINAMIDE	MILLI	1.74	1.70	-	1.78	0.70	MILLI/MJ	4.85	
ALANINE	GRAM	1.25	-	-	-	0.50	GRAM/MJ	3.48	8.6
ARGININE	GRAM	1.24	-	-	-	0.50	GRAM/MJ	3.46	4.4
ASPARTIC ACID	GRAM	2.38	-	-	-	0.95	GRAM/MJ	6.63	11.0
CYSTINE	GRAM	0.26	-	-	-	0.10	GRAM/MJ	0.72	0.7
GLUTAMIC ACID	GRAM	2.80	-	-	-	1.12	GRAM/MJ	7.80	11.7
GLYCINE	GRAM	1.01	-	-	-	0.40	GRAM/MJ	2.82	8.2
HISTIDINE	GRAM	0.71	-	-	-	0.28	GRAM/MJ	1.98	2.8
ISOLEUCINE	GRAM	1.18	1.10	-	1.26	0.47	GRAM/MJ	3.29	5.5
LEUCINE	GRAM	1.80	1.74	-	1.88	0.72	GRAM/MJ	5.02	8.4

CONSTITUENTS	DIM	AV	VARIATION			AVR	NUTR. DENS.		MOLPERC.
LYSINE	GRAM	2.19	1.83	-	2.56	0.88	GRAM/MJ	6.10	9.2
METHIONINE	GRAM	0.63	0.49	-	0.76	0.25	GRAM/MJ	1.76	2.6
PHENYLALANINE	GRAM	0.86	0.81	-	0.91	0.34	GRAM/MJ	2.40	3.2
PROLINE	GRAM	0.68	-	-	-	0.27	GRAM/MJ	1.90	3.6
SERINE	GRAM	0.89	-	-	-	0.36	GRAM/MJ	2.48	5.2
THREONINE	GRAM	1.06	-	-	-	0.42	GRAM/MJ	2.95	5.5
TRYPTOPHAN	GRAM	0.22	-	-	-	0.09	GRAM/MJ	0.61	0.7
TYROSINE	GRAM	0.74	-	-	-	0.30	GRAM/MJ	2.06	2.5
VALINE	GRAM	1.22	1.19	-	1.23	0.49	GRAM/MJ	3.40	6.4
PALMITIC ACID	MILLI	105.00	105.00	-	105.00	42.00	MILLI/MJ	292.66	
STEARIC ACID	MILLI	22.00	22.00	-	22.00	8.80	MILLI/MJ	61.32	
PALMITOLEIC ACID	MILLI	23.00	23.00	-	23.00	9.20	MILLI/MJ	64.11	
OLEIC ACID	MILLI	70.00	70.00	-	70.00	28.00	MILLI/MJ	195.10	
LINOLEIC ACID	MILLI	11.00	11.00	-	11.00	4.40	MILLI/MJ	30.66	
C20 : 5 N-3 FATTY ACID	MILLI	53.00	53.00	-	53.00	21.20	MILLI/MJ	147.72	
C22 : 5 N-3 FATTY ACID	MILLI	8.00	8.00	-	8.00	3.20	MILLI/MJ	22.30	
CHOLESTEROL	MILLI	72.00	68.00	-	76.00	28.80	MILLI/MJ	200.68	

BRASSEN
(BRACHSEN, BLEI)

ABRAMIS BRAMA (L.)

BREAM

BRÈME

		PROTEIN	FAT	CARBOHYDRATES	TOTAL
ENERGY VALUE (AVERAGE)	KJOULE	306	218	0.0	524
PER 100 G	(KCAL)	73	52	0.0	125
EDIBLE PORTION					
AMOUNT OF DIGESTIBLE	GRAM	16.10	5.21	0.00	
CONSTITUENTS PER 100 G					
ENERGY VALUE (AVERAGE)	KJOULE	296	207	0.0	504
OF THE DIGESTIBLE	(KCAL)	71	50	0.0	120
FRACTION PER 100 G					
EDIBLE PORTION					

WASTE PERCENTAGE AVERAGE 44 MINIMUM 43 MAXIMUM 45

CONSTITUENTS	DIM	AV	VARIATION		AVR	NUTR. DENS.		MOLPERC.
WATER	GRAM	76.70	74.00	- 79.30	42.95	GRAM/MJ	152.26	
PROTEIN	GRAM	16.60	13.80	- 19.00	9.30	GRAM/MJ	32.95	
FAT	GRAM	5.49	4.80	- 11.70	3.07	GRAM/MJ	10.90	
MINERALS	GRAM	1.17	1.00	- 1.32	0.66	GRAM/MJ	2.32	
SODIUM	MILLI	23.00	20.00	- 28.00	12.88	MILLI/MJ	45.66	
POTASSIUM	MILLI	310.00	133.00	- 422.00	173.60	MILLI/MJ	615.40	
CALCIUM	MILLI	89.00	48.00	- 130.00	49.84	MILLI/MJ	176.68	
SELENIUM	MILLI	0.16	-	- -	0.09	MILLI/MJ	0.32	
CYSTINE	MILLI	80.00	-	- -	44.80	MILLI/MJ	158.81	
HISTIDINE	GRAM	0.33	-	- -	0.18	GRAM/MJ	0.66	
METHIONINE	GRAM	0.40	-	- -	0.22	GRAM/MJ	0.79	
TRYPTOPHAN	GRAM	0.13	-	- -	0.07	GRAM/MJ	0.26	
TYROSINE	GRAM	0.51	-	- -	0.29	GRAM/MJ	1.01	
MYRISTIC ACID	GRAM	0.54	0.07	- 0.66	0.30	GRAM/MJ	1.08	
PALMITIC ACID	GRAM	0.82	0.59	- 1.74	0.46	GRAM/MJ	1.63	
STEARIC ACID	GRAM	0.23	0.16	- 0.52	0.13	GRAM/MJ	0.46	
PALMITOLEIC ACID	MILLI	92.00	50.00	- 260.00	51.52	MILLI/MJ	182.63	
OLEIC ACID	GRAM	0.91	0.68	- 0.97	0.51	GRAM/MJ	1.81	
EICOSENOIC ACID	GRAM	0.10	0.10	- 0.10	0.06	GRAM/MJ	0.20	
C22 : 1 FATTY ACID	GRAM	0.11	0.11	- 0.11	0.06	GRAM/MJ	0.22	
LINOLEIC ACID	MILLI	76.00	20.00	- 90.00	42.56	MILLI/MJ	150.87	
LINOLENIC ACID	MILLI	90.00	10.00	- 110.00	50.40	MILLI/MJ	178.66	
C20 : 5 N-3 FATTY ACID	GRAM	0.45	0.05	- 0.56	0.26	GRAM/MJ	0.91	
C22 : 5 N-3 FATTY ACID	MILLI	60.00	60.00	- 60.00	33.60	MILLI/MJ	119.11	
C22 : 6 N-3 FATTY ACID	GRAM	0.86	0.86	- 0.89	0.48	GRAM/MJ	1.72	

FORELLE (BACHFORELLE, REGENBOGENFORELLE) — TROUT — TRUITE

SALMO GAIRDNERI

		PROTEIN	FAT	CARBOHYDRATES	TOTAL
ENERGY VALUE (AVERAGE)	KJOULE	359	109	0.0	467
PER 100 G	(KCAL)	86	26	0.0	112
EDIBLE PORTION					
AMOUNT OF DIGESTIBLE	GRAM	18.91	2.59	0.00	
CONSTITUENTS PER 100 G					
ENERGY VALUE (AVERAGE)	KJOULE	348	103	0.0	451
OF THE DIGESTIBLE	(KCAL)	83	25	0.0	108
FRACTION PER 100 G					
EDIBLE PORTION					

WASTE PERCENTAGE AVERAGE 48 MINIMUM 44 MAXIMUM 52

CONSTITUENTS	DIM	AV	VARIATION		AVR	NUTR. DENS.		MOLPERC.
WATER	GRAM	76.30	74.80	77.70	39.68	GRAM/MJ	169.07	
PROTEIN	GRAM	19.50	18.00	20.20	10.14	GRAM/MJ	43.21	
FAT	GRAM	2.73	1.90	4.55	1.42	GRAM/MJ	6.05	
MINERALS	GRAM	1.32	1.21	1.44	0.69	GRAM/MJ	2.92	
SODIUM	MILLI	40.00	31.00	54.00	20.80	MILLI/MJ	88.63	
POTASSIUM	MILLI	465.00	393.00	516.00	241.80	GRAM/MJ	1.03	
MAGNESIUM	MILLI	27.00	25.00	31.00	14.04	MILLI/MJ	59.83	
CALCIUM	MILLI	18.00	14.00	21.00	9.36	MILLI/MJ	39.88	
MANGANESE	MICRO	30.00	-	-	15.60	MICRO/MJ	66.47	
IRON	MILLI	0.69	0.38	1.00	0.36	MILLI/MJ	1.53	
COBALT	MICRO	1.00	0.30	1.00	0.52	MICRO/MJ	2.22	
COPPER	MILLI	0.15	0.01	0.33	0.08	MILLI/MJ	0.33	
ZINC	MILLI	0.48	0.13	0.65	0.25	MILLI/MJ	1.06	
NICKEL	MICRO	2.00	1.00	3.00	1.04	MICRO/MJ	4.43	
CHROMIUM	MICRO	1.00	0.30	3.00	0.52	MICRO/MJ	2.22	
PHOSPHORUS	MILLI	242.00	220.00	266.00	125.84	MILLI/MJ	536.22	
FLUORIDE	MICRO	30.00	10.00	70.00	15.60	MICRO/MJ	66.47	
IODIDE	MICRO	3.20	-	-	1.66	MICRO/MJ	7.09	
SELENIUM	MILLI	-	0.02	0.14				
BROMINE	MICRO	330.00	-	-	171.60	MICRO/MJ	731.21	
VITAMIN A	MICRO	45.00	-	-	23.40	MICRO/MJ	99.71	
VITAMIN B1	MICRO	84.00	50.00	100.00	43.68	MICRO/MJ	186.13	
VITAMIN B2	MICRO	76.00	50.00	110.00	39.52	MICRO/MJ	168.40	
NICOTINAMIDE	MILLI	3.41	3.29	3.50	1.77	MILLI/MJ	7.56	
ALANINE	GRAM	1.55	1.43	1.65	0.81	GRAM/MJ	3.43	9.8

CONSTITUENTS	DIM	AV		VARIATION		AVR	NUTR. DENS.		MOLPERC.
ARGININE	GRAM	1.40		1.33 -	1.44	0.73	GRAM/MJ	3.10	4.5
ASPARTIC ACID	GRAM	2.36		2.15 -	2.59	1.23	GRAM/MJ	5.23	10.0
CYSTINE	GRAM	0.17		- -	-	0.09	GRAM/MJ	0.38	0.4
GLUTAMIC ACID	GRAM	3.33		2.80 -	3.85	1.73	GRAM/MJ	7.38	12.7
GLYCINE	GRAM	1.47		1.39 -	1.55	0.76	GRAM/MJ	3.26	11.0
HISTIDINE	GRAM	0.57		0.48 -	0.68	0.30	GRAM/MJ	1.26	2.1
ISOLEUCINE	GRAM	1.27		0.97 -	1.19	0.66	GRAM/MJ	2.81	5.4
LEUCINE	GRAM	1.78		1.67 -	2.00	0.93	GRAM/MJ	3.94	7.6
LYSINE	GRAM	2.02		1.62 -	2.46	1.05	GRAM/MJ	4.48	7.8
METHIONINE	GRAM	0.66		0.59 -	0.73	0.34	GRAM/MJ	1.46	2.5
PHENYLALANINE	GRAM	0.92		0.85 -	0.96	0.48	GRAM/MJ	2.04	3.1
PROLINE	GRAM	0.85		0.85 -	0.88	0.44	GRAM/MJ	1.88	4.1
SERINE	GRAM	0.97		0.94 -	1.01	0.50	GRAM/MJ	2.15	5.2
THREONINE	GRAM	1.08		0.99 -	1.19	0.56	GRAM/MJ	2.39	5.1
TRYPTOPHAN	GRAM	0.24		- -	-	0.12	GRAM/MJ	0.53	0.7
TYROSINE	GRAM	0.68		0.62 -	0.73	0.35	GRAM/MJ	1.51	2.1
VALINE	GRAM	1.25		1.16 -	1.35	0.65	GRAM/MJ	2.77	6.0
MYRISTIC ACID	GRAM	0.17		0.06 -	0.23	0.09	GRAM/MJ	0.38	
PALMITIC ACID	GRAM	0.26		0.18 -	0.44	0.14	GRAM/MJ	0.59	
STEARIC ACID	MILLI	90.00		70.00 -	130.00	46.80	MILLI/MJ	199.42	
PALMITOLEIC ACID	GRAM	0.11		0.07 -	0.19	0.06	GRAM/MJ	0.24	
OLEIC ACID	GRAM	0.47		0.35 -	0.73	0.25	GRAM/MJ	1.06	
EICOSENOIC ACID	MILLI	90.00		90.00 -	90.00	46.80	MILLI/MJ	199.42	
C22 : 1 FATTY ACID	MILLI	10.00		10.00 -	10.00	5.20	MILLI/MJ	22.16	
LINOLEIC ACID	MILLI	74.00		60.00 -	450.00	38.48	MILLI/MJ	163.97	
ARACHIDONIC ACID	MILLI	30.00		30.00 -	30.00	15.60	MILLI/MJ	66.47	
C18 : 4 FATTY ACID	MILLI	40.00		40.00 -	40.00	20.80	MILLI/MJ	88.63	
C20 : 5 N-3 FATTY ACID	GRAM	0.15		0.15 -	0.15	0.08	GRAM/MJ	0.33	
C22 : 5 N-3 FATTY ACID	MILLI	63.33		50.00 -	70.00	32.93	MILLI/MJ	140.33	
C22 : 6 N-3 FATTY ACID	GRAM	0.43		0.25 -	0.53	0.23	GRAM/MJ	0.97	
CHOLESTEROL	MILLI	56.00		55.00 -	57.00	29.12	MILLI/MJ	124.08	
TOTAL PURINES	MILLI	297.00	1	297.00 -	297.00	154.44	MILLI/MJ	658.09	
ADENINE	MILLI	19.00		19.00 -	19.00	9.88	MILLI/MJ	42.10	
GUANINE	MILLI	104.00		104.00 -	104.00	54.08	MILLI/MJ	230.44	
XANTHINE	MILLI	11.00		11.00 -	11.00	5.72	MILLI/MJ	24.37	
HYPOXANTHINE	MILLI	117.00		117.00 -	117.00	60.84	MILLI/MJ	259.25	
TOTAL PHOSPHOLIPIDS	MILLI	330.00		- -	-	171.60	MILLI/MJ	731.21	
PHOSPHATIDYLCHOLINE	MILLI	220.00		- -	-	114.40	MILLI/MJ	487.48	
PHOSPHATIDYLETHANOLAMINE	MILLI	70.00		- -	-	36.40	MILLI/MJ	155.11	
PHOSPHATIDYLSERINE	MILLI	14.00		- -	-	7.28	MILLI/MJ	31.02	
PHOSPHATIDYLINOSITOL	MILLI	6.00		- -	-	3.12	MILLI/MJ	13.29	
SPHINGOMYELIN	MILLI	6.00		- -	-	3.12	MILLI/MJ	13.29	

1 VALUE OF TOTAL PURINES IS EXPRESSSED IN MG URIC ACID/100G

HECHT PIKE BROCHET

ESOX LUCIUS (L.)

		PROTEIN	FAT	CARBOHYDRATES	TOTAL
ENERGY VALUE (AVERAGE) PER 100 G EDIBLE PORTION	KJOULE (KCAL)	339 81	34 8.1	0.0 0.0	373 89
AMOUNT OF DIGESTIBLE CONSTITUENTS PER 100 G	GRAM	17.84	0.80	0.00	
ENERGY VALUE (AVERAGE) OF THE DIGESTIBLE FRACTION PER 100 G EDIBLE PORTION	KJOULE (KCAL)	329 79	32 7.7	0.0 0.0	361 86

WASTE PERCENTAGE AVERAGE 45 MINIMUM 43 MAXIMUM 52

CONSTITUENTS	DIM	AV	VARIATION			AVR	NUTR. DENS.		MOLPERC.
WATER	GRAM	79.60	79.50	-	80.00	43.78	GRAM/MJ	220.70	
PROTEIN	GRAM	18.40	17.90	-	18.80	10.12	GRAM/MJ	51.02	
FAT	GRAM	0.85	0.43	-	2.00	0.47	GRAM/MJ	2.36	
MINERALS	GRAM	1.05	0.96	-	1.13	0.58	GRAM/MJ	2.91	
SODIUM	MILLI	63.00	50.00	-	75.00	34.65	MILLI/MJ	174.67	
POTASSIUM	MILLI	250.00	191.00	-	350.00	137.50	MILLI/MJ	693.15	
MAGNESIUM	MILLI	25.00	22.00	-	29.00	13.75	MILLI/MJ	69.32	
CALCIUM	MILLI	20.00	-	-	-	11.00	MILLI/MJ	55.45	
MANGANESE	MICRO	35.00	30.00	-	40.00	19.25	MICRO/MJ	97.04	
IRON	MILLI	1.05	-	-	-	0.58	MILLI/MJ	2.91	
COBALT	MICRO	1.00	0.20	-	4.00	0.55	MICRO/MJ	2.77	
COPPER	MICRO	47.00	30.00	-	100.00	25.85	MICRO/MJ	130.31	
ZINC	MILLI	1.10	0.74	-	1.60	0.61	MILLI/MJ	3.05	
NICKEL	MICRO	50.00	0.60	-	30.01	27.50	MICRO/MJ	138.63	
CHROMIUM	MICRO	6.00	2.00	-	10.00	3.30	MICRO/MJ	16.64	
PHOSPHORUS	MILLI	192.00	160.00	-	220.00	105.60	MILLI/MJ	532.34	
FLUORIDE	MICRO	80.00	60.00	-	120.00	44.00	MICRO/MJ	221.81	
SELENIUM	MICRO	13.00	8.00	-	76.00	7.15	MICRO/MJ	36.04	
VITAMIN A	MICRO	15.00 1	-	-	-	8.25	MICRO/MJ	41.59	
VITAMIN E ACTIVITY	MILLI	0.20	-	-	-	0.11	MILLI/MJ	0.55	
VITAMIN B1	MICRO	85.00	80.00	-	90.00	46.75	MICRO/MJ	235.67	
VITAMIN B2	MICRO	55.00	40.00	-	70.00	30.25	MICRO/MJ	152.49	
NICOTINAMIDE	MILLI	1.60	1.50	-	1.70	0.88	MILLI/MJ	4.44	
VITAMIN B6	MILLI	0.15	0.09	-	0.19	0.08	MILLI/MJ	0.42	
ALANINE	GRAM	1.54	-	-	-	0.85	GRAM/MJ	4.27	

1 VALID ONLY FOR MUSCLES; THE VITAMIN A CONTENT OF THE LIVER IS 4,8 - 26,1 MG/100G

CONSTITUENTS	DIM	AV	VARIATION			AVR	NUTR. DENS.		MOLPERC.
ARGININE	GRAM	1.48	1.39	-	1.55	0.81	GRAM/MJ	4.10	
ASPARTIC ACID	GRAM	1.86	-	-	-	1.02	GRAM/MJ	5.16	
CYSTINE	GRAM	0.22	0.14	-	0.29	0.12	GRAM/MJ	0.61	
GLUTAMIC ACID	GRAM	4.58	-	-	-	2.52	GRAM/MJ	12.70	
GLYCINE	GRAM	0.62	-	-	-	0.34	GRAM/MJ	1.72	
HISTIDINE	GRAM	0.35	0.26	-	0.42	0.19	GRAM/MJ	0.97	
ISOLEUCINE	GRAM	0.94	-	-	-	0.52	GRAM/MJ	2.61	
LEUCINE	GRAM	2.39	1.53	-	3.22	1.31	GRAM/MJ	6.63	
LYSINE	GRAM	2.15	1.44	-	2.18	1.18	GRAM/MJ	5.96	
METHIONINE	GRAM	0.66	0.57	-	0.74	0.36	GRAM/MJ	1.83	
PHENYLALANINE	GRAM	0.82	0.77	-	0.88	0.45	GRAM/MJ	2.27	
PROLINE	-	0.00	-	-	-	0.00			
SERINE	GRAM	0.86	-	-	-	0.47	GRAM/MJ	2.38	
THREONINE	GRAM	0.95	-	-	-	0.52	GRAM/MJ	2.63	
TRYPTOPHAN	GRAM	0.16	-	-	-	0.09	GRAM/MJ	0.44	
TYROSINE	GRAM	0.69	-	-	-	0.38	GRAM/MJ	1.91	
VALINE	GRAM	0.93	0.82	-	1.11	0.51	GRAM/MJ	2.58	
MYRISTIC ACID	MILLI	29.00	27.00	-	30.00	15.95	MILLI/MJ	80.41	
PALMITIC ACID	MILLI	37.66	10.00	-	93.00	20.72	MILLI/MJ	104.44	
STEARIC ACID	MILLI	16.14	14.50	-	20.00	8.88	MILLI/MJ	44.76	
PALMITOLEIC ACID	MILLI	90.00	90.00	-	90.00	49.50	MILLI/MJ	249.53	
OLEIC ACID	MILLI	59.00	46.00	-	80.00	32.45	MILLI/MJ	163.58	
EICOSENOIC ACID	MILLI	20.00	20.00	-	20.00	11.00	MILLI/MJ	55.45	
C22 : 1 FATTY ACID	MILLI	7.00	1.00	-	10.00	3.85	MILLI/MJ	19.41	
LINOLEIC ACID	MILLI	27.57	12.00	-	50.00	15.16	MILLI/MJ	76.44	
LINOLENIC ACID	MILLI	47.33	42.00	-	50.00	26.03	MILLI/MJ	131.24	
ARACHIDONIC ACID	MILLI	50.00	41.00	-	50.00	27.50	MILLI/MJ	138.63	
C20 : 5 N-3 FATTY ACID	MILLI	64.57	52.00	-	70.00	35.51	MILLI/MJ	179.03	
C22 : 5 N-3 FATTY ACID	MILLI	10.33	10.00	-	11.00	5.68	MILLI/MJ	28.65	
C22 : 6 N-3 FATTY ACID	MILLI	175.00	73.00	-	248.00	96.25	MILLI/MJ	485.21	
CHOLESTEROL	MILLI	63.00	39.00	-	86.00	34.65	MILLI/MJ	174.67	
TOTAL PURINES	MILLI	140.00 2	-	-	-	77.00	MILLI/MJ	388.17	

2 VALUE EXPRESSED IN MG URIC ACID/100G

KARPFEN CARP CARPE

CYPRINUS CARPIO (L.)

		PROTEIN	FAT	CARBOHYDRATES	TOTAL
ENERGY VALUE (AVERAGE)	KJOULE	331	191	0.0	522
PER 100 G EDIBLE PORTION	(KCAL)	79	46	0.0	125
AMOUNT OF DIGESTIBLE CONSTITUENTS PER 100 G	GRAM	17.46	4.56	0.00	
ENERGY VALUE (AVERAGE)	KJOULE	321	181	0.0	503
OF THE DIGESTIBLE FRACTION PER 100 G EDIBLE PORTION	(KCAL)	77	43	0.0	120

WASTE PERCENTAGE AVERAGE 48 MINIMUM 34 MAXIMUM 61

CONSTITUENTS	DIM	AV	VARIATION			AVR	NUTR. DENS.		MOLPERC.
WATER	GRAM	75.80	72.00	-	78.00	39.42	GRAM/MJ	150.79	
PROTEIN	GRAM	18.00	16.70	-	19.30	9.36	GRAM/MJ	35.81	
FAT	GRAM	4.80	2.00	-	7.10	2.50	GRAM/MJ	9.55	
MINERALS	GRAM	1.17	1.10	-	1.30	0.61	GRAM/MJ	2.33	
SODIUM	MILLI	46.00	37.00	-	51.00	23.92	MILLI/MJ	91.51	
POTASSIUM	MILLI	306.00	284.00	-	385.00	159.12	MILLI/MJ	608.73	
MAGNESIUM	MILLI	30.00	15.00	-	45.00	15.60	MILLI/MJ	59.68	
CALCIUM	MILLI	52.00	12.00	-	182.00	27.04	MILLI/MJ	103.45	
IRON	MILLI	1.10	0.90	-	1.30	0.57	MILLI/MJ	2.19	
MOLYBDENUM	MICRO	53.00	-	-	-	27.56	MICRO/MJ	105.43	
PHOSPHORUS	MILLI	216.00	176.00	-	253.00	112.32	MILLI/MJ	429.69	
CHLORIDE	MILLI	50.00	30.00	-	62.00	26.00	MILLI/MJ	99.47	
FLUORIDE	MICRO	32.00	-	-	-	16.64	MICRO/MJ	63.66	
IODIDE	MICRO	1.70	-	-	-	0.88	MICRO/MJ	3.38	
SELENIUM	MICRO	-	7.00	-	130.00				
VITAMIN A	MICRO	44.00 1	10.00	-	140.00	22.88	MICRO/MJ	87.53	
VITAMIN E ACTIVITY	MILLI	0.50	-	-	-	0.26	MILLI/MJ	0.99	
VITAMIN B1	MICRO	68.00	50.00	-	80.00	35.36	MICRO/MJ	135.27	
VITAMIN B2	MICRO	53.00	40.00	-	80.00	27.56	MICRO/MJ	105.43	
NICOTINAMIDE	MILLI	1.90	1.50	-	2.80	0.99	MILLI/MJ	3.78	
VITAMIN B6	MILLI	0.15	0.10	-	0.20	0.08	MILLI/MJ	0.30	
VITAMIN C	MILLI	1.00	-	-	-	0.52	MILLI/MJ	1.99	
ALANINE	GRAM	1.45	1.34	-	1.52	0.75	GRAM/MJ	2.88	10.1
ARGININE	GRAM	1.27	1.22	-	1.42	0.66	GRAM/MJ	2.53	4.5
ASPARTIC ACID	GRAM	2.31	2.27	-	2.71	1.20	GRAM/MJ	4.60	10.8

1 VALID ONLY FOR MUSCLES; THE VITAMIN A CONTENT OF THE LIVER IS 14 (8,5 - 19) MG/100G

CONSTITUENTS	DIM	AV	VARIATION			AVR	NUTR. DENS.		MOLPERC.
CYSTINE	GRAM	0.18	-	-	-	0.09	GRAM/MJ	0.36	0.5
GLUTAMIC ACID	GRAM	3.19	3.09	-	3.41	1.66	GRAM/MJ	6.35	13.4
GLYCINE	GRAM	1.00	0.84	-	1.16	0.52	GRAM/MJ	1.99	8.3
HISTIDINE	GRAM	0.42	0.10	-	0.60	0.22	GRAM/MJ	0.84	1.7
ISOLEUCINE	GRAM	1.00	0.88	-	1.02	0.52	GRAM/MJ	1.99	4.7
LEUCINE	GRAM	1.68	1.34	-	1.84	0.87	GRAM/MJ	3.34	7.9
LYSINE	GRAM	2.11	1.89	-	2.22	1.10	GRAM/MJ	4.20	8.9
METHIONINE	GRAM	0.59	0.53	-	0.64	0.31	GRAM/MJ	1.17	2.5
PHENYLALANINE	GRAM	0.89	0.70	-	1.00	0.46	GRAM/MJ	1.77	3.3
PROLINE	GRAM	0.64	0.46	-	0.77	0.33	GRAM/MJ	1.27	3.4
SERINE	GRAM	0.99	0.88	-	1.09	0.51	GRAM/MJ	1.97	5.8
THREONINE	GRAM	1.04	0.75	-	1.27	0.54	GRAM/MJ	2.07	5.4
TRYPTOPHAN	GRAM	0.21	-	-	-	0.11	GRAM/MJ	0.42	0.6
TYROSINE	GRAM	0.74	0.72	-	0.77	0.38	GRAM/MJ	1.47	2.5
VALINE	GRAM	1.05	0.87	-	1.27	0.55	GRAM/MJ	2.09	5.6
MYRISTIC ACID	MILLI	83.00	30.00	-	160.00	43.16	MILLI/MJ	165.11	
PALMITIC ACID	GRAM	0.68	0.63	-	0.76	0.36	GRAM/MJ	1.37	
STEARIC ACID	GRAM	0.15	0.04	-	0.22	0.08	GRAM/MJ	0.31	
PALMITOLEIC ACID	GRAM	0.60	0.39	-	1.00	0.31	GRAM/MJ	1.19	
OLEIC ACID	GRAM	1.47	1.24	-	1.94	0.77	GRAM/MJ	2.94	
EICOSENOIC ACID	GRAM	0.10	0.01	-	0.20	0.05	GRAM/MJ	0.21	
C22 : 1 FATTY ACID	MILLI	80.00	80.00	-	80.00	41.60	MILLI/MJ	159.15	
LINOLEIC ACID	GRAM	0.41	0.10	-	0.66	0.21	GRAM/MJ	0.82	
LINOLENIC ACID	GRAM	0.14	0.02	-	0.35	0.07	GRAM/MJ	0.28	
ARACHIDONIC ACID	GRAM	0.19	0.18	-	0.20	0.10	GRAM/MJ	0.38	
C20 : 5 N-3 FATTY ACID	GRAM	0.21	0.08	-	0.38	0.11	GRAM/MJ	0.43	
C22 : 5 N-3 FATTY ACID	MILLI	70.00	70.00	-	70.00	36.40	MILLI/MJ	139.25	
C22 : 6 N-3 FATTY ACID	MILLI	83.00	30.00	-	140.00	43.16	MILLI/MJ	165.11	
CHOLESTEROL	MILLI	67.00	-	-	-	34.84	MILLI/MJ	133.28	
TOTAL PURINES	MILLI	160.00 2	-	-	-	83.20	MILLI/MJ	318.29	

2 VALUE EXPRESSED IN MG URIC ACID/100G

Fische
Fish
Poissons

LACHS (SALM) — SALMON — SAUMON

SALMO SALAR (L.)

		PROTEIN	FAT	CARBOHYDRATES	TOTAL
ENERGY VALUE (AVERAGE)	KJOULE	366	541	0.0	907
PER 100 G	(KCAL)	88	129	0.0	217
EDIBLE PORTION					
AMOUNT OF DIGESTIBLE	GRAM	19.30	12.92	0.00	
CONSTITUENTS PER 100 G					
ENERGY VALUE (AVERAGE)	KJOULE	355	514	0.0	869
OF THE DIGESTIBLE	(KCAL)	85	123	0.0	208
FRACTION PER 100 G					
EDIBLE PORTION					

WASTE PERCENTAGE AVERAGE 36 MINIMUM 31 MAXIMUM 40

CONSTITUENTS	DIM	AV	VARIATION			AVR	NUTR.	DENS.	MOLPERC.
WATER	GRAM	65.50	63.40	-	67.50	41.92	GRAM/MJ	75.38	
PROTEIN	GRAM	19.90	17.40	-	21.10	12.74	GRAM/MJ	22.90	
FAT	GRAM	13.60	12.50	-	16.50	8.70	GRAM/MJ	15.65	
MINERALS	GRAM	1.00	-	-	-	0.64	GRAM/MJ	1.15	
SODIUM	MILLI	51.00	48.00	-	53.00	32.64	MILLI/MJ	58.69	
POTASSIUM	MILLI	371.00	331.00	-	410.00	237.44	MILLI/MJ	426.97	
MAGNESIUM	MILLI	29.00	-	-	-	18.56	MILLI/MJ	33.38	
CALCIUM	MILLI	13.00	-	-	-	8.32	MILLI/MJ	14.96	
MANGANESE	MICRO	14.00	10.00	-	50.00	8.96	MICRO/MJ	16.11	
IRON	MILLI	1.00	0.95	-	1.50	0.64	MILLI/MJ	1.15	
COBALT	MICRO	2.80	0.30	-	7.00	1.79	MICRO/MJ	3.22	
COPPER	MILLI	0.20	-	-	-	0.13	MILLI/MJ	0.23	
ZINC	MILLI	0.80	-	-	-	0.51	MILLI/MJ	0.92	
NICKEL	MICRO	2.00	1.00	-	4.00	1.28	MICRO/MJ	2.30	
PHOSPHORUS	MILLI	266.00	242.00	-	289.00	170.24	MILLI/MJ	306.13	
FLUORIDE	MICRO	30.00	-	-	-	19.20	MICRO/MJ	34.53	
IODIDE	MICRO	34.00	-	-	-	21.76	MICRO/MJ	39.13	
BORON	MICRO	40.00	15.00	-	64.00	25.60	MICRO/MJ	46.03	
SELENIUM	MICRO	26.00	20.00	-	34.00	16.64	MICRO/MJ	29.92	
SILICON	MILLI	0.50	-	-	-	0.32	MILLI/MJ	0.58	
VITAMIN A	MICRO	65.00	-	-	-	41.60	MICRO/MJ	74.81	
VITAMIN D	MICRO	16.30	5.00	-	20.00	10.43	MICRO/MJ	18.76	
VITAMIN B1	MILLI	0.17	0.12	-	0.23	0.11	MILLI/MJ	0.20	
VITAMIN B2	MILLI	0.17	0.09	-	0.20	0.11	MILLI/MJ	0.20	
NICOTINAMIDE	MILLI	7.50	6.00	-	10.80	4.80	MILLI/MJ	8.63	
PANTOTHENIC ACID	MILLI	0.75	-	-	-	0.48	MILLI/MJ	0.86	

CONSTITUENTS	DIM	AV	VARIATION			AVR	NUTR. DENS.		MOLPERC.
VITAMIN B6	MILLI	0.98	-	-	-	0.63	MILLI/MJ	1.13	
BIOTIN	MICRO	0.85	-	-	-	0.54	MICRO/MJ	0.98	
FOLIC ACID	MICRO	20.00	-	-	-	12.80	MICRO/MJ	23.02	
VITAMIN B12	MICRO	2.89	-	-	-	1.85	MICRO/MJ	3.33	
VITAMIN C	MILLI	-	0.80	-	1.20				
ALANINE	GRAM	1.67	1.53	-	1.77	1.07	GRAM/MJ	1.92	10.2
ARGININE	GRAM	1.33	1.30	-	1.36	0.85	GRAM/MJ	1.53	4.2
ASPARTIC ACID	GRAM	2.22	2.16	-	2.27	1.42	GRAM/MJ	2.55	9.1
CYSTINE	GRAM	0.29	0.24	-	0.34	0.19	GRAM/MJ	0.33	0.7
GLUTAMIC ACID	GRAM	3.23	3.02	-	3.41	2.07	GRAM/MJ	3.72	12.0
GLYCINE	GRAM	1.63	1.49	-	1.75	1.04	GRAM/MJ	1.88	11.8
HISTIDINE	GRAM	0.66	0.62	-	0.72	0.42	GRAM/MJ	0.76	2.3
ISOLEUCINE	GRAM	1.16	1.02	-	1.27	0.74	GRAM/MJ	1.34	4.8
LEUCINE	GRAM	1.77	1.74	-	1.82	1.13	GRAM/MJ	2.04	7.4
LYSINE	GRAM	2.02	1.89	-	2.15	1.29	GRAM/MJ	2.32	7.5
METHIONINE	GRAM	0.70	0.65	-	0.74	0.45	GRAM/MJ	0.81	2.6
PHENYLALANINE	GRAM	0.91	0.83	-	0.95	0.58	GRAM/MJ	1.05	3.0
PROLINE	GRAM	1.00	0.96	-	1.05	0.64	GRAM/MJ	1.15	4.7
SERINE	GRAM	1.01	1.00	-	1.03	0.65	GRAM/MJ	1.16	5.2
THREONINE	GRAM	1.11	1.08	-	1.14	0.71	GRAM/MJ	1.28	5.1
TRYPTOPHAN	GRAM	0.26	0.23	-	0.31	0.17	GRAM/MJ	0.30	0.7
TYROSINE	GRAM	0.72	0.65	-	0.75	0.46	GRAM/MJ	0.83	2.2
VALINE	GRAM	1.39	1.37	-	1.42	0.89	GRAM/MJ	1.60	6.5
MYRISTIC ACID	GRAM	0.82	0.36	-	1.87	0.53	GRAM/MJ	0.95	
PALMITIC ACID	GRAM	1.74	1.31	-	1.87	1.12	GRAM/MJ	2.01	
STEARIC ACID	GRAM	0.35	0.31	-	0.47	0.23	GRAM/MJ	0.41	
PALMITOLEIC ACID	GRAM	0.80	0.59	-	0.97	0.52	GRAM/MJ	0.93	
OLEIC ACID	GRAM	2.96	2.74	-	3.32	1.90	GRAM/MJ	3.41	
EICOSENOIC ACID	GRAM	0.65	0.12	-	0.98	0.42	GRAM/MJ	0.75	
C22 : 1 FATTY ACID	GRAM	1.09	0.70	-	1.29	0.70	GRAM/MJ	1.25	
LINOLEIC ACID	GRAM	0.44	0.18	-	0.64	0.29	GRAM/MJ	0.51	
LINOLENIC ACID	GRAM	0.55	0.11	-	0.71	0.35	GRAM/MJ	0.63	
ARACHIDONIC ACID	GRAM	0.30	0.06	-	0.65	0.19	GRAM/MJ	0.35	
C18 : 4 FATTY ACID	GRAM	0.19	0.18	-	0.20	0.12	GRAM/MJ	0.22	
C20 : 5 N-3 FATTY ACID	GRAM	0.70	0.42	-	0.88	0.45	GRAM/MJ	0.82	
C22 : 5 N-3 FATTY ACID	GRAM	0.43	0.09	-	0.68	0.28	GRAM/MJ	0.50	
C22 : 6 N-3 FATTY ACID	GRAM	2.14	0.30	-	2.96	1.37	GRAM/MJ	2.47	
CHOLESTEROL	MILLI	35.00	-	-	-	22.40	MILLI/MJ	40.28	
TOTAL PURINES	MILLI	170.00 [1]	110.00	-	250.00	108.80	MILLI/MJ	195.65	

1 VALUE EXPRESSED IN MG URIC ACID/100G

RENKE (MARÄNE, FELCHEN) — POLLAN — CORÉGONE

COREGONUS SPP.

		PROTEIN	FAT	CARBOHYDRATES	TOTAL
ENERGY VALUE (AVERAGE) PER 100 G EDIBLE PORTION	KJOULE	328	128	0.0	455
	(KCAL)	78	30	0.0	109
AMOUNT OF DIGESTIBLE CONSTITUENTS PER 100 G	GRAM	17.26	3.04	0.00	
ENERGY VALUE (AVERAGE) OF THE DIGESTIBLE FRACTION PER 100 G EDIBLE PORTION	KJOULE	318	121	0.0	439
	(KCAL)	76	29	0.0	105

WASTE PERCENTAGE AVERAGE 32 MINIMUM 26 MAXIMUM 47

CONSTITUENTS	DIM	AV	VARIATION			AVR	NUTR.	DENS.	MOLPERC.
WATER	GRAM	77.70	73.40	-	82.70	52.84	GRAM/MJ	176.96	
PROTEIN	GRAM	17.80	14.60	-	19.10	12.10	GRAM/MJ	40.54	
FAT	GRAM	3.21	1.51	-	6.90	2.18	GRAM/MJ	7.31	
MINERALS	GRAM	1.10	0.91	-	1.26	0.75	GRAM/MJ	2.51	
SODIUM	MILLI	36.00	28.00	-	47.00	24.48	MILLI/MJ	81.99	
POTASSIUM	MILLI	318.00	255.00	-	402.00	216.24	MILLI/MJ	724.26	
MAGNESIUM	MILLI	30.00	-	-	-	20.40	MILLI/MJ	68.33	
CALCIUM	MILLI	60.00	18.00	-	130.00	40.80	MILLI/MJ	136.65	
MANGANESE	MICRO	49.00	23.00	-	130.00	33.32	MICRO/MJ	111.60	
IRON	MILLI	0.48	0.34	-	0.75	0.33	MILLI/MJ	1.09	
COBALT	MICRO	1.00	0.30	-	3.00	0.68	MICRO/MJ	2.28	
COPPER	MICRO	50.00	40.00	-	60.00	34.00	MICRO/MJ	113.88	
ZINC	MILLI	1.20	0.50	-	3.10	0.82	MILLI/MJ	2.73	
NICKEL	MICRO	2.00	0.80	-	3.00	1.36	MICRO/MJ	4.56	
CHROMIUM	MICRO	1.00	0.30	-	2.00	0.68	MICRO/MJ	2.28	
PHOSPHORUS	MILLI	290.00	260.00	-	340.00	197.20	MILLI/MJ	660.48	
FLUORIDE	MILLI	0.10	0.04	-	0.23	0.07	MILLI/MJ	0.23	
SELENIUM	MICRO	37.00	10.00	-	60.00	25.16	MICRO/MJ	84.27	
MYRISTIC ACID	GRAM	0.21	0.21	-	0.21	0.14	GRAM/MJ	0.48	
PALMITIC ACID	GRAM	0.47	0.47	-	0.47	0.32	GRAM/MJ	1.07	
STEARIC ACID	MILLI	10.00	10.00	-	10.00	6.80	MILLI/MJ	22.78	
PALMITOLEIC ACID	GRAM	0.24	0.18	-	0.31	0.17	GRAM/MJ	0.56	
OLEIC ACID	GRAM	0.56	0.31	-	0.81	0.38	GRAM/MJ	1.28	
EICOSENOIC ACID	GRAM	0.62	0.62	-	0.62	0.42	GRAM/MJ	1.41	
C22 : 1 FATTY ACID	MILLI	20.00	20.00	-	20.00	13.60	MILLI/MJ	45.55	
LINOLEIC ACID	GRAM	0.12	0.11	-	0.13	0.08	GRAM/MJ	0.27	
LINOLENIC ACID	GRAM	0.14	0.14	-	0.14	0.10	GRAM/MJ	0.32	
ARACHIDONIC ACID	GRAM	0.13	0.13	-	0.13	0.09	GRAM/MJ	0.30	
C18 : 4 FATTY ACID	MILLI	80.00	30.00	-	130.00	54.40	MILLI/MJ	182.20	
C20 : 5 N-3 FATTY ACID	GRAM	0.20	0.19	-	0.22	0.14	GRAM/MJ	0.47	
C22 : 5 N-3 FATTY ACID	MILLI	80.00	60.00	-	100.00	54.40	MILLI/MJ	182.20	
C22 : 6 N-3 FATTY ACID	GRAM	0.23	0.19	-	0.27	0.16	GRAM/MJ	0.52	

SCHLEIE TENCH TANCHE

TINCA TINCA (L.)

		PROTEIN	FAT	CARBOHYDRATES	TOTAL
ENERGY VALUE (AVERAGE)	KJOULE	326	29	0.0	355
PER 100 G	(KCAL)	78	7.0	0.0	85
EDIBLE PORTION					
AMOUNT OF DIGESTIBLE	GRAM	17.16	0.70	0.00	
CONSTITUENTS PER 100 G					
ENERGY VALUE (AVERAGE)	KJOULE	316	28	0.0	344
OF THE DIGESTIBLE	(KCAL)	76	6.7	0.0	82
FRACTION PER 100 G					
EDIBLE PORTION					

WASTE PERCENTAGE AVERAGE 60

CONSTITUENTS	DIM	AV	VARIATION			AVR	NUTR. DENS.		MOLPERC.
WATER	GRAM	76.50	64.70	-	80.00	30.60	GRAM/MJ	222.37	
PROTEIN	GRAM	17.70	13.70	-	20.00	7.08	GRAM/MJ	51.45	
FAT	GRAM	0.74	0.30	-	2.11	0.30	GRAM/MJ	2.15	
MINERALS	GRAM	1.75	1.66	-	1.83	0.70	GRAM/MJ	5.09	
SODIUM	MILLI	80.00	72.00	-	88.00	32.00	MILLI/MJ	232.55	
POTASSIUM	MILLI	245.00	220.00	-	270.00	98.00	MILLI/MJ	712.17	
MAGNESIUM	MILLI	18.00	16.00	-	19.00	7.20	MILLI/MJ	52.32	
CALCIUM	MILLI	31.00	28.00	-	34.00	12.40	MILLI/MJ	90.11	
IRON	MILLI	0.80	0.70	-	0.80	0.32	MILLI/MJ	2.33	
PHOSPHORUS	MILLI	156.00	140.00	-	175.00	62.40	MILLI/MJ	453.47	
CHLORIDE	MILLI	50.00	-	-	-	20.00	MILLI/MJ	145.34	
VITAMIN A	MICRO	1.00	-	-	-	0.40	MICRO/MJ	2.91	
VITAMIN B1	MICRO	75.00	50.00	-	100.00	30.00	MICRO/MJ	218.01	
VITAMIN B2	MILLI	0.18	0.16	-	0.20	0.07	MILLI/MJ	0.52	
NICOTINAMIDE	MILLI	4.00	-	-	-	1.60	MILLI/MJ	11.63	
VITAMIN C	MILLI	1.00	-	-	-	0.40	MILLI/MJ	2.91	
TOTAL PURINES	MILLI	80.00 [1]	-	-	-	32.00	MILLI/MJ	232.55	

1 VALUE EXPRESSED IN MG URIC ACID/100G

WALLER (WELS)

SILURUS GLANIS (L.)

FRESHWATER CATFISH

ESPÈCES D'EAU DOUCE

		PROTEIN	FAT	CARBOHYDRATES	TOTAL
ENERGY VALUE (AVERAGE)	KJOULE	282	449	0.0	731
PER 100 G	(KCAL)	67	107	0.0	175
EDIBLE PORTION					
AMOUNT OF DIGESTIBLE	GRAM	14.84	10.73	0.00	
CONSTITUENTS PER 100 G					
ENERGY VALUE (AVERAGE)	KJOULE	273	427	0.0	700
OF THE DIGESTIBLE	(KCAL)	65	102	0.0	167
FRACTION PER 100 G					
EDIBLE PORTION					

WASTE PERCENTAGE AVERAGE 40

CONSTITUENTS	DIM	AV	VARIATION			AVR	NUTR.	DENS.	MOLPERC.
WATER	GRAM	72.10	65.80	-	77.70	43.26	GRAM/MJ	103.01	
PROTEIN	GRAM	15.30	13.40	-	16.50	9.18	GRAM/MJ	21.86	
FAT	GRAM	11.30	3.40	-	19.90	6.78	GRAM/MJ	16.14	
MINERALS	GRAM	0.98	0.95	-	1.00	0.59	GRAM/MJ	1.40	
SODIUM	MILLI	33.00	-	-	-	19.80	MILLI/MJ	47.15	
POTASSIUM	MILLI	307.00	-	-	-	184.20	MILLI/MJ	438.63	
CALCIUM	MILLI	40.00	-	-	-	24.00	MILLI/MJ	57.15	
PHOSPHORUS	MILLI	100.00	-	-	-	60.00	MILLI/MJ	142.88	
MYRISTIC ACID	GRAM	1.89	0.19	-	3.60	1.14	GRAM/MJ	2.71	
PALMITIC ACID	GRAM	1.07	0.62	-	1.53	0.65	GRAM/MJ	1.54	
STEARIC ACID	GRAM	0.26	0.20	-	0.32	0.16	GRAM/MJ	0.37	
PALMITOLEIC ACID	GRAM	0.23	0.13	-	0.34	0.14	GRAM/MJ	0.34	
OLEIC ACID	GRAM	2.49	1.00	-	3.98	1.49	GRAM/MJ	3.56	
EICOSENOIC ACID	MILLI	20.00	20.00	-	20.00	12.00	MILLI/MJ	28.58	
LINOLEIC ACID	GRAM	1.19	0.22	-	2.16	0.71	GRAM/MJ	1.70	
LINOLENIC ACID	GRAM	0.17	0.02	-	0.32	0.10	GRAM/MJ	0.24	
ARACHIDONIC ACID	GRAM	0.12	0.08	-	0.17	0.08	GRAM/MJ	0.18	
C18 : 4 FATTY ACID	MILLI	20.00	20.00	-	20.00	12.00	MILLI/MJ	28.58	
C20 : 5 N-3 FATTY ACID	GRAM	0.15	0.15	-	0.15	0.09	GRAM/MJ	0.21	
C22 : 5 N-3 FATTY ACID	GRAM	0.11	0.07	-	0.15	0.07	GRAM/MJ	0.16	
C22 : 6 N-3 FATTY ACID	GRAM	0.39	0.34	-	0.45	0.24	GRAM/MJ	0.56	
CHOLESTEROL	MILLI	152.00	-	-	-	91.20	MILLI/MJ	217.17	

ZANDER PIKE-PERCH SANDRE

STIZOSTEDION LUCIOPERCA (L.)

		PROTEIN	FAT	CARBOHYDRATES	TOTAL
ENERGY VALUE (AVERAGE)	KJOULE	353	29	0.0	382
PER 100 G	(KCAL)	84	6.9	0.0	91
EDIBLE PORTION					
AMOUNT OF DIGESTIBLE	GRAM	18.62	0.69	0.00	
CONSTITUENTS PER 100 G					
ENERGY VALUE (AVERAGE)	KJOULE	343	28	0.0	370
OF THE DIGESTIBLE	(KCAL)	82	6.6	0.0	89
FRACTION PER 100 G					
EDIBLE PORTION					

WASTE PERCENTAGE AVERAGE 50

CONSTITUENTS	DIM	AV	VARIATION			AVR	NUTR. DENS.		MOLPERC.
WATER	GRAM	78.40	76.60	-	79.60	39.20	GRAM/MJ	211.65	
PROTEIN	GRAM	19.20	18.00	-	22.10	9.60	GRAM/MJ	51.83	
FAT	GRAM	0.73	0.30	-	1.60	0.37	GRAM/MJ	1.97	
MINERALS	GRAM	1.22	0.94	-	1.50	0.61	GRAM/MJ	3.29	
SODIUM	MILLI	81.00	-	-	-	40.50	MILLI/MJ	218.67	
POTASSIUM	MILLI	237.00	233.00	-	240.00	118.50	MILLI/MJ	639.80	
MAGNESIUM	MILLI	18.00	-	-	-	9.00	MILLI/MJ	48.59	
CALCIUM	MILLI	27.00	26.00	-	29.00	13.50	MILLI/MJ	72.89	
IRON	MILLI	1.40	-	-	-	0.70	MILLI/MJ	3.78	
PHOSPHORUS	MILLI	194.00	157.00	-	230.00	97.00	MILLI/MJ	523.72	
CHLORIDE	MILLI	41.00	-	-	-	20.50	MILLI/MJ	110.68	
SELENIUM	MICRO	26.00	21.00	-	31.00	13.00	MICRO/MJ	70.19	
VITAMIN B1	MILLI	0.16	-	-	-	0.08	MILLI/MJ	0.43	
VITAMIN B2	MILLI	0.25	-	-	-	0.13	MILLI/MJ	0.67	
NICOTINAMIDE	MILLI	2.31	-	-	-	1.16	MILLI/MJ	6.24	
VITAMIN C	MILLI	1.00	-	-	-	0.50	MILLI/MJ	2.70	
ISOLEUCINE	GRAM	0.86	-	-	-	0.43	GRAM/MJ	2.32	
LEUCINE	GRAM	1.41	-	-	-	0.71	GRAM/MJ	3.81	
LYSINE	GRAM	2.08	-	-	-	1.04	GRAM/MJ	5.62	
METHIONINE	GRAM	0.54	-	-	-	0.27	GRAM/MJ	1.46	
PHENYLALANINE	GRAM	0.72	-	-	-	0.36	GRAM/MJ	1.94	
THREONINE	GRAM	0.83	-	-	-	0.42	GRAM/MJ	2.24	
VALINE	GRAM	0.96	-	-	-	0.48	GRAM/MJ	2.59	

CONSTITUENTS	DIM	AV	VARIATION		AVR	NUTR. DENS.		MOLPERC.
MYRISTIC ACID	MILLI	44.00	36.00	- 48.00	22.00	MILLI/MJ	118.78	
PALMITIC ACID	MILLI	86.00	38.00	- 160.00	43.00	MILLI/MJ	232.17	
STEARIC ACID	MILLI	12.85	10.00	- 15.00	6.43	MILLI/MJ	34.71	
PALMITOLEIC ACID	MILLI	49.00	14.00	- 100.00	24.50	MILLI/MJ	132.28	
OLEIC ACID	MILLI	87.00	74.00	- 110.00	43.50	MILLI/MJ	234.86	
EICOSENOIC ACID	MILLI	8.50	7.00	- 10.00	4.25	MILLI/MJ	22.95	
C22 : 1 FATTY ACID	MILLI	10.00	10.00	- 10.00	5.00	MILLI/MJ	27.00	
LINOLEIC ACID	MILLI	15.16	13.00	- 20.00	7.58	MILLI/MJ	40.94	
LINOLENIC ACID	MILLI	12.33	10.00	- 20.00	6.17	MILLI/MJ	33.29	
ARACHIDONIC ACID	MILLI	15.16	10.00	- 30.00	7.58	MILLI/MJ	40.94	
C20 : 5 N-3 FATTY ACID	MILLI	102.50	21.00	- 141.00	51.25	MILLI/MJ	276.71	
C22 : 5 N-3 FATTY ACID	MILLI	8.00	6.00	- 10.00	4.00	MILLI/MJ	21.60	
C22 : 6 N-3 FATTY ACID	MILLI	54.16	29.00	- 64.00	27.08	MILLI/MJ	146.23	
TOTAL PURINES	MILLI	110.00 [1]	-	- -	55.00	MILLI/MJ	296.96	

1 VALUE EXPRESSED IN MG URIC ACID/100G

Krustentiere und Weichtiere

CRUSTACEANS AND MOLLUSCS — CRUSTACÉS ET MOLLUSQUES

AUSTER — OYSTER — HUÎTRE

OSTREA EDULIS (L.)

		PROTEIN	FAT	CARBOHYDRATES	TOTAL
ENERGY VALUE (AVERAGE)	KJOULE	166	48	80	294
PER 100 G	(KCAL)	40	11	19	70
EDIBLE PORTION					
AMOUNT OF DIGESTIBLE	GRAM	8.73	1.14	4.80	
CONSTITUENTS PER 100 G					
ENERGY VALUE (AVERAGE)	KJOULE	161	45	80	286
OF THE DIGESTIBLE	(KCAL)	38	11	19	68
FRACTION PER 100 G					
EDIBLE PORTION					

WASTE PERCENTAGE AVERAGE 90 MINIMUM 87 MAXIMUM 95

CONSTITUENTS	DIM	AV	VARIATION			AVR	NUTR. DENS.		MOLPERC.
WATER	GRAM	83.00	80.50	-	87.50	8.30	GRAM/MJ	289.84	
PROTEIN	GRAM	9.00	6.00	-	10.70	0.90	GRAM/MJ	31.43	
FAT	GRAM	1.20	0.79	-	2.10	0.12	GRAM/MJ	4.19	
AVAILABLE CARBOHYDR.	GRAM	4.80	4.80	-	5.60	0.48	GRAM/MJ	16.76	
MINERALS	GRAM	2.00	-	-	-	0.20	GRAM/MJ	6.98	
SODIUM	MILLI	-	73.00	-	505.00				
POTASSIUM	MILLI	184.00	110.00	-	258.00	18.40	MILLI/MJ	642.54	
CALCIUM	MILLI	82.00	70.00	-	94.00	8.20	MILLI/MJ	286.35	
MANGANESE	MILLI	0.60	0.20	-	1.00	0.06	MILLI/MJ	2.10	
IRON	MILLI	5.80	3.10	-	7.50	0.58	MILLI/MJ	20.25	
COPPER	MILLI	2.50	1.20	-	3.70	0.25	MILLI/MJ	8.73	
ZINC	MILLI	-	6.50	-	160.00				
NICKEL	MICRO	18.50	8.60	-	37.00	1.85	MICRO/MJ	64.60	
CHROMIUM	MICRO	9.00	-	-	-	0.90	MICRO/MJ	31.43	
VANADIUM	MICRO	11.00	-	-	-	1.10	MICRO/MJ	38.41	
PHOSPHORUS	MILLI	157.00	143.00	-	170.00	15.70	MILLI/MJ	548.26	
FLUORIDE	MILLI	0.12	0.07	-	0.16	0.01	MILLI/MJ	0.42	
IODIDE	MICRO	58.00	-	-	-	5.80	MICRO/MJ	202.54	
SELENIUM	MICRO	60.00	49.00	-	66.00	6.00	MICRO/MJ	209.53	
VITAMIN A	MICRO	93.00	90.00	-	96.00	9.30	MICRO/MJ	324.76	
VITAMIN D	MICRO	8.00	-	-	-	0.80	MICRO/MJ	27.94	
VITAMIN E ACTIVITY	MILLI	0.85	-	-	-	0.09	MILLI/MJ	2.97	
ALPHA-TOCOPHEROL	MILLI	0.85	-	-	-	0.09	MILLI/MJ	2.97	
VITAMIN B1	MILLI	0.16	0.11	-	0.26	0.02	MILLI/MJ	0.56	
VITAMIN B2	MILLI	0.20	0.11	-	0.30	0.02	MILLI/MJ	0.70	
NICOTINAMIDE	MILLI	2.17	1.00	-	4.76	0.22	MILLI/MJ	7.58	

CONSTITUENTS	DIM	AV	VARIATION			AVR	NUTR. DENS.		MOLPERC.
PANTOTHENIC ACID	MILLI	0.32	0.18	-	0.46	0.03	MILLI/MJ	1.12	
VITAMIN B6	MILLI	0.22	0.11	-	0.32	0.02	MILLI/MJ	0.77	
BIOTIN	MICRO	10.00	-	-	-	1.00	MICRO/MJ	34.92	
FOLIC ACID	MICRO	7.00	4.00	-	10.00	0.70	MICRO/MJ	24.44	
VITAMIN B12	MICRO	14.60	-	-	-	1.46	MICRO/MJ	50.98	
ALANINE	GRAM	0.75	0.56	-	0.96	0.08	GRAM/MJ	2.62	10.1
ARGININE	GRAM	0.68	0.58	-	0.73	0.07	GRAM/MJ	2.37	4.7
ASPARTIC ACID	GRAM	1.03	0.82	-	1.17	0.10	GRAM/MJ	3.60	9.3
CYSTINE	GRAM	0.15	-	-	-	0.02	GRAM/MJ	0.52	0.7
GLUTAMIC ACID	GRAM	1.58	1.47	-	1.74	0.16	GRAM/MJ	5.52	12.9
GLYCINE	GRAM	0.80	0.53	-	1.08	0.08	GRAM/MJ	2.79	12.8
HISTIDINE	GRAM	0.21	0.15	-	0.29	0.02	GRAM/MJ	0.73	1.6
ISOLEUCINE	GRAM	0.54	0.39	-	0.76	0.05	GRAM/MJ	1.89	4.9
LEUCINE	GRAM	0.82	0.69	-	0.94	0.08	GRAM/MJ	2.86	7.5
LYSINE	GRAM	0.72	0.53	-	0.99	0.07	GRAM/MJ	2.51	5.9
METHIONINE	GRAM	0.26	0.10	-	0.35	0.03	GRAM/MJ	0.91	2.1
PHENYLALANINE	GRAM	0.42	0.34	-	0.47	0.04	GRAM/MJ	1.47	3.0
PROLINE	GRAM	0.57	0.31	-	0.62	0.06	GRAM/MJ	1.99	5.9
SERINE	GRAM	0.51	0.45	-	0.57	0.05	GRAM/MJ	1.78	5.8
THREONINE	GRAM	0.48	0.44	-	0.54	0.05	GRAM/MJ	1.68	4.8
TRYPTOPHAN	MILLI	80.00	-	-	-	8.00	MILLI/MJ	279.37	0.5
TYROSINE	GRAM	0.33	0.22	-	0.40	0.03	GRAM/MJ	1.15	2.2
VALINE	GRAM	0.52	0.45	-	0.63	0.05	GRAM/MJ	1.82	5.3
GLYCOGENE	GRAM	1.09	0.80	-	1.30	0.11	GRAM/MJ	3.81	
MYRISTIC ACID	MILLI	80.00	80.00	-	80.00	8.00	MILLI/MJ	279.37	
PALMITIC ACID	GRAM	0.29	0.29	-	0.29	0.03	GRAM/MJ	1.01	
STEARIC ACID	MILLI	80.00	80.00	-	80.00	8.00	MILLI/MJ	279.37	
PALMITOLEIC ACID	MILLI	50.00	50.00	-	50.00	5.00	MILLI/MJ	174.60	
OLEIC ACID	MILLI	60.00	60.00	-	60.00	6.00	MILLI/MJ	209.53	
EICOSENOIC ACID	MILLI	30.00	30.00	-	30.00	3.00	MILLI/MJ	104.76	
LINOLEIC ACID	MILLI	10.00	10.00	-	10.00	1.00	MILLI/MJ	34.92	
LINOLENIC ACID	MILLI	40.00	40.00	-	40.00	4.00	MILLI/MJ	139.68	
ARACHIDONIC ACID	MILLI	10.00	10.00	-	10.00	1.00	MILLI/MJ	34.92	
C18 : 4 FATTY ACID	MILLI	10.00	10.00	-	10.00	1.00	MILLI/MJ	34.92	
C20 : 5 N-3 FATTY ACID	MILLI	40.00	40.00	-	40.00	4.00	MILLI/MJ	139.68	
C22 : 6 N-3 FATTY ACID	MILLI	10.00	10.00	-	10.00	1.00	MILLI/MJ	34.92	
TOTAL STEROLS	MILLI	277.00	277.00	-	277.00	27.70	MILLI/MJ	967.31	
CHOLESTEROL	MILLI	123.00	123.00	-	123.00	12.30	MILLI/MJ	429.53	
DESMOSTEROL	MILLI	86.00	86.00	-	86.00	8.60	MILLI/MJ	300.32	
24-METH.-CHOLESTEROL	MILLI	48.00	48.00	-	48.00	4.80	MILLI/MJ	167.62	
TOTAL PURINES	MILLI	90.00 [1]	-	-	-	9.00	MILLI/MJ	314.29	

1 VALUE EXPRESSED IN MG URIC ACID/100G

GARNELE (NORDSEEGARNELE) — SHRIMP — CREVETTE

CRANGON GRANGON (L.)
UND VERW. ARTEN

		PROTEIN	FAT	CARBOHYDRATES	TOTAL
ENERGY VALUE (AVERAGE)	KJOULE	342	57	0.0	400
PER 100 G	(KCAL)	82	14	0.0	96
EDIBLE PORTION					
AMOUNT OF DIGESTIBLE	GRAM	18.04	1.36	0.00	
CONSTITUENTS PER 100 G					
ENERGY VALUE (AVERAGE)	KJOULE	332	54	0.0	387
OF THE DIGESTIBLE	(KCAL)	79	13	0.0	92
FRACTION PER 100 G					
EDIBLE PORTION					

WASTE PERCENTAGE AVERAGE 57 MINIMUM 49 MAXIMUM 66

CONSTITUENTS	DIM	AV	VARIATION			AVR	NUTR. DENS.		MOLPERC.
WATER	GRAM	78.40	76.70	-	79.70	33.71	GRAM/MJ	202.83	
PROTEIN	GRAM	18.60	16.40	-	21.30	8.00	GRAM/MJ	48.12	
FAT	GRAM	1.44	0.80	-	2.30	0.62	GRAM/MJ	3.73	
MINERALS	GRAM	1.38	1.20	-	1.71	0.59	GRAM/MJ	3.57	
SODIUM	MILLI	146.00	140.00	-	150.00	62.78	MILLI/MJ	377.73	
POTASSIUM	MILLI	266.00	220.00	-	312.00	114.38	MILLI/MJ	688.19	
MAGNESIUM	MILLI	67.00	42.00	-	92.00	28.81	MILLI/MJ	173.34	
CALCIUM	MILLI	92.00	63.00	-	115.00	39.56	MILLI/MJ	238.02	
MANGANESE	MICRO	30.00	0.00	-	50.00	12.90	MICRO/MJ	77.62	
IRON	MILLI	1.76	1.30	-	2.15	0.76	MILLI/MJ	4.55	
COBALT	MICRO	12.00	-	-	-	5.16	MICRO/MJ	31.05	
COPPER	MILLI	0.24	0.08	-	0.43	0.10	MILLI/MJ	0.62	
ZINC	MILLI	2.31	1.75	-	3.10	0.99	MILLI/MJ	5.98	
NICKEL	MICRO	3.00	-	-	-	1.29	MICRO/MJ	7.76	
MOLYBDENUM	MICRO	3.00	-	-	-	1.29	MICRO/MJ	7.76	
PHOSPHORUS	MILLI	224.00	166.00	-	300.00	96.32	MILLI/MJ	579.53	
FLUORIDE	MILLI	0.16	0.10	-	0.25	0.07	MILLI/MJ	0.41	
IODIDE	MILLI	0.13	-	-	-	0.06	MILLI/MJ	0.34	
BORON	MICRO	10.00	-	-	-	4.30	MICRO/MJ	25.87	
SELENIUM	MICRO	41.00	28.00	-	51.00	17.63	MICRO/MJ	106.07	
VITAMIN A	-	TRACES	-	-	-				
VITAMIN B1	MICRO	51.00	20.00	-	70.00	21.93	MICRO/MJ	131.95	
VITAMIN B2	MICRO	34.00	27.00	-	45.00	14.62	MICRO/MJ	87.96	
NICOTINAMIDE	MILLI	2.43	1.50	-	3.20	1.04	MILLI/MJ	6.29	
PANTOTHENIC ACID	MILLI	0.37	-	-	-	0.16	MILLI/MJ	0.96	
VITAMIN B6	MILLI	0.13	0.10	-	0.15	0.06	MILLI/MJ	0.34	

CONSTITUENTS	DIM	AV	VARIATION			AVR	NUTR. DENS.		MOLPERC.
BIOTIN	MICRO	1.00	-	-	-	0.43	MICRO/MJ	2.59	
FOLIC ACID	MICRO	7.40	-	-	-	3.18	MICRO/MJ	19.15	
VITAMIN B12	MICRO	0.83	0.73	-	0.93	0.36	MICRO/MJ	2.15	
VITAMIN C	MILLI	1.90	-	-	-	0.82	MILLI/MJ	4.92	
ALANINE	GRAM	1.40	1.17	-	1.55	0.60	GRAM/MJ	3.62	9.3
ARGININE	GRAM	1.74	1.22	-	2.68	0.75	GRAM/MJ	4.50	5.9
ASPARTIC ACID	GRAM	2.05	1.56	-	2.39	0.88	GRAM/MJ	5.30	9.1
CYSTINE	GRAM	0.31	0.22	-	0.43	0.13	GRAM/MJ	0.80	0.8
GLUTAMIC ACID	GRAM	3.25	3.03	-	3.50	1.40	GRAM/MJ	8.41	13.1
GLYCINE	GRAM	1.54	0.89	-	2.21	0.66	GRAM/MJ	3.98	12.1
HISTIDINE	GRAM	0.41	0.34	-	0.79	0.18	GRAM/MJ	1.06	1.6
ISOLEUCINE	GRAM	1.00	0.76	-	1.23	0.43	GRAM/MJ	2.59	4.5
LEUCINE	GRAM	1.97	1.48	-	3.15	0.85	GRAM/MJ	5.10	8.9
LYSINE	GRAM	2.02	1.40	-	4.17	0.87	GRAM/MJ	5.23	8.2
METHIONINE	GRAM	0.67	0.50	-	1.00	0.29	GRAM/MJ	1.73	2.7
PHENYLALANINE	GRAM	0.88	0.72	-	1.09	0.38	GRAM/MJ	2.28	3.2
PROLINE	GRAM	0.87	0.60	-	1.00	0.37	GRAM/MJ	2.25	4.5
SERINE	GRAM	0.75	0.62	-	0.90	0.32	GRAM/MJ	1.94	4.2
THREONINE	GRAM	0.85	0.72	-	1.05	0.37	GRAM/MJ	2.20	4.2
TRYPTOPHAN	GRAM	0.21	0.08	-	0.32	0.09	GRAM/MJ	0.54	0.6
TYROSINE	GRAM	0.65	0.21	-	1.03	0.28	GRAM/MJ	1.68	2.1
VALINE	GRAM	0.99	0.76	-	1.38	0.43	GRAM/MJ	2.56	5.0
CADAVERINE	MILLI	0.30	-	-	-	0.13	MILLI/MJ	0.78	
PUTRESCINE	MILLI	-	0.50	-	1.00				
CHOLESTEROL	MILLI	138.00	125.00	-	150.00	59.34	MILLI/MJ	357.03	
TOTAL PURINES	MILLI	147.00 [1]	60.00	-	234.00	63.21	MILLI/MJ	380.32	

1 VALUE EXPRESSED IN MG URIC ACID/100G

HUMMER LOBSTER HOMARD

HOMARUS VULGARIS (EDW.)

		PROTEIN	FAT	CARBOHYDRATES	TOTAL
ENERGY VALUE (AVERAGE)	KJOULE	293	76	0.0	368
PER 100 G	(KCAL)	70	18	0.0	88
EDIBLE PORTION					
AMOUNT OF DIGESTIBLE	GRAM	15.42	1.80	0.00	
CONSTITUENTS PER 100 G					
ENERGY VALUE (AVERAGE)	KJOULE	284	72	0.0	356
OF THE DIGESTIBLE	(KCAL)	68	17	0.0	85
FRACTION PER 100 G					
EDIBLE PORTION					

WASTE PERCENTAGE AVERAGE 64

CONSTITUENTS	DIM	AV	VARIATION			AVR	NUTR. DENS.		MOLPERC.
WATER	GRAM	79.80	77.50	-	85.20	28.73	GRAM/MJ	224.36	
PROTEIN	GRAM	15.90	14.00	-	18.80	5.72	GRAM/MJ	44.70	
FAT	GRAM	1.90	1.80	-	1.90	0.68	GRAM/MJ	5.34	
MINERALS	GRAM	2.10	1.71	-	2.22	0.76	GRAM/MJ	5.90	
SODIUM	MILLI	270.00	210.00	-	325.00	97.20	MILLI/MJ	759.12	
POTASSIUM	MILLI	220.00	180.00	-	260.00	79.20	MILLI/MJ	618.54	
MAGNESIUM	MILLI	24.00	24.00	-	24.00	8.64	MILLI/MJ	67.48	
CALCIUM	MILLI	61.00	61.00	-	62.00	21.96	MILLI/MJ	171.50	
MANGANESE	MICRO	34.00	34.00	-	34.00	12.24	MICRO/MJ	95.59	
IRON	MILLI	1.00	0.95	-	1.30	0.36	MILLI/MJ	2.81	
COPPER	MILLI	0.70	-	-	-	0.25	MILLI/MJ	1.97	
ZINC	MILLI	1.60	-	-	-	0.58	MILLI/MJ	4.50	
NICKEL	MICRO	66.00	-	-	-	23.76	MICRO/MJ	185.56	
MOLYBDENUM	MICRO	23.00	-	-	-	8.28	MICRO/MJ	64.67	
VANADIUM	MICRO	4.00	3.00	-	5.00	1.44	MICRO/MJ	11.25	
PHOSPHORUS	MILLI	234.00	184.00	-	283.00	84.24	MILLI/MJ	657.90	
CHLORIDE	MILLI	61.00	-	-	-	21.96	MILLI/MJ	171.50	
FLUORIDE	MILLI	0.21	-	-	-	0.08	MILLI/MJ	0.59	
IODIDE	MILLI	0.10	-	-	-	0.04	MILLI/MJ	0.28	
SELENIUM	MILLI	0.13	0.10	-	0.15	0.05	MILLI/MJ	0.37	
VITAMIN A	-	0.00	-	-	-	0.00			
VITAMIN E ACTIVITY	MILLI	1.47	-	-	-	0.53	MILLI/MJ	4.13	
ALPHA-TOCOPHEROL	MILLI	1.47	-	-	-	0.53	MILLI/MJ	4.13	
VITAMIN B1	MILLI	0.13	0.09	-	0.16	0.05	MILLI/MJ	0.37	
VITAMIN B2	MICRO	88.00	50.00	-	180.00	31.68	MICRO/MJ	247.42	
NICOTINAMIDE	MILLI	1.82	1.46	-	1.90	0.66	MILLI/MJ	5.12	

CONSTITUENTS	DIM	AV	VARIATION			AVR	NUTR. DENS.		MOLPERC.
PANTOTHENIC ACID	MILLI	1.67	1.35	-	1.99	0.60	MILLI/MJ	4.70	
VITAMIN B6	MILLI	1.18	1.18	-	1.18	0.42	MILLI/MJ	3.32	
BIOTIN	MICRO	5.00	4.80	-	5.20	1.80	MICRO/MJ	14.06	
FOLIC ACID	MICRO	17.00	-	-	-	6.12	MICRO/MJ	47.80	
VITAMIN B12	MICRO	0.48	0.46	-	0.49	0.17	MICRO/MJ	1.35	
ALANINE	GRAM	1.12	1.00	-	1.24	0.40	GRAM/MJ	3.15	8.7
ARGININE	GRAM	1.32	1.24	-	1.41	0.48	GRAM/MJ	3.71	5.3
ASPARTIC ACID	GRAM	2.00	1.64	-	2.39	0.72	GRAM/MJ	5.62	10.4
CYSTINE	GRAM	0.22	0.16	-	0.27	0.08	GRAM/MJ	0.62	0.6
GLUTAMIC ACID	GRAM	2.73	2.18	-	3.23	0.98	GRAM/MJ	7.68	12.9
GLYCINE	GRAM	1.12	1.01	-	1.17	0.40	GRAM/MJ	3.15	10.3
HISTIDINE	GRAM	0.35	0.22	-	0.40	0.13	GRAM/MJ	0.98	1.6
ISOLEUCINE	GRAM	0.79	0.73	-	0.83	0.28	GRAM/MJ	2.22	4.2
LEUCINE	GRAM	1.65	1.22	-	2.20	0.59	GRAM/MJ	4.64	8.7
LYSINE	GRAM	1.99	1.11	-	3.28	0.72	GRAM/MJ	5.59	9.4
METHIONINE	GRAM	0.55	0.39	-	0.64	0.20	GRAM/MJ	1.55	2.6
PHENYLALANINE	GRAM	0.67	0.50	-	0.89	0.24	GRAM/MJ	1.88	2.8
PROLINE	GRAM	0.82	0.68	-	0.95	0.30	GRAM/MJ	2.31	4.9
SERINE	GRAM	0.80	0.69	-	0.99	0.29	GRAM/MJ	2.25	5.3
THREONINE	GRAM	0.85	0.71	-	1.05	0.31	GRAM/MJ	2.39	4.9
TRYPTOPHAN	GRAM	0.12	0.03	-	0.16	0.04	GRAM/MJ	0.34	0.4
TYROSINE	GRAM	0.57	0.15	-	0.77	0.21	GRAM/MJ	1.60	2.2
VALINE	GRAM	0.80	0.57	-	0.91	0.29	GRAM/MJ	2.25	4.7
CHOLESTEROL	MILLI	135.00	95.00	-	182.00	48.60	MILLI/MJ	379.56	
TOTAL PURINES	MILLI	118.00 [1]	60.00	-	175.00	42.48	MILLI/MJ	331.76	

1 VALUE EXPRESSED IN MG URIC ACID/100G

KREBS (FLUSSKREBS) | CRAYFISH | ECREVISSE

ASTACUS FLUVIATILIS (F.)

		PROTEIN	FAT	CARBOHYDRATES	TOTAL
ENERGY VALUE (AVERAGE)	KJOULE	276	19	0.0	295
PER 100 G	(KCAL)	66	4.5	0.0	70
EDIBLE PORTION					
AMOUNT OF DIGESTIBLE	GRAM	14.55	0.44	0.00	
CONSTITUENTS PER 100 G					
ENERGY VALUE (AVERAGE)	KJOULE	268	18	0.0	286
OF THE DIGESTIBLE	(KCAL)	64	4.2	0.0	68
FRACTION PER 100 G					
EDIBLE PORTION					

WASTE PERCENTAGE AVERAGE 77 MINIMUM 70 MAXIMUM 85

CONSTITUENTS	DIM	AV	VARIATION		AVR	NUTR. DENS.		MOLPERC.
WATER	GRAM	83.10	81.20	- 86.00	19.11	GRAM/MJ	290.96	
PROTEIN	GRAM	15.00	13.90	- 16.00	3.45	GRAM/MJ	52.52	
FAT	GRAM	0.47	0.35	- 0.59	0.11	GRAM/MJ	1.65	
MINERALS	GRAM	1.26	1.20	- 1.31	0.29	GRAM/MJ	4.41	
SODIUM	MILLI	253.00	205.00	- 300.00	58.19	MILLI/MJ	885.83	
POTASSIUM	MILLI	254.00	250.00	- 258.00	58.42	MILLI/MJ	889.33	
CALCIUM	MILLI	43.00	25.00	- 60.00	9.89	MILLI/MJ	150.56	
IRON	MILLI	2.00	1.00	- 3.00	0.46	MILLI/MJ	7.00	
PHOSPHORUS	MILLI	224.00	200.00	- 248.00	51.52	MILLI/MJ	784.29	
VITAMIN B1	MILLI	0.15	-	- -	0.03	MILLI/MJ	0.53	
VITAMIN B2	MILLI	0.10	-	- -	0.02	MILLI/MJ	0.35	
NICOTINAMIDE	MILLI	2.00	-	- -	0.46	MILLI/MJ	7.00	
ARGININE	GRAM	0.57	0.53	- 0.62	0.13	GRAM/MJ	2.00	
CYSTINE	GRAM	0.31	0.30	- 0.31	0.07	GRAM/MJ	1.09	
HISTIDINE	GRAM	0.48	0.44	- 0.52	0.11	GRAM/MJ	1.68	
LYSINE	GRAM	0.85	0.84	- 0.86	0.20	GRAM/MJ	2.98	
METHIONINE	GRAM	1.02	0.96	- 1.08	0.23	GRAM/MJ	3.57	
PHENYLALANINE	GRAM	0.82	0.81	- 0.84	0.19	GRAM/MJ	2.87	
THREONINE	GRAM	0.88	0.86	- 0.90	0.20	GRAM/MJ	3.08	
TYROSINE	GRAM	0.80	0.75	- 0.84	0.18	GRAM/MJ	2.80	
VALINE	GRAM	0.56	0.55	- 0.57	0.13	GRAM/MJ	1.96	
CHOLESTEROL	MILLI	158.00	-	- -	36.34	MILLI/MJ	553.21	
TOTAL PURINES	MILLI	60.00 [1]	-	- -	13.80	MILLI/MJ	210.08	
TOTAL PHOSPHOLIPIDS	MILLI	376.00	-	- -	86.48	GRAM/MJ	1.32	
PHOSPHATIDYLCHOLINE	MILLI	205.00	-	- -	47.15	MILLI/MJ	717.77	
PHOSPHATIDYLETHANOLAMINE	MILLI	99.00	-	- -	22.77	MILLI/MJ	346.63	
PHOSPHATIDYLSERINE	MILLI	40.00	-	- -	9.20	MILLI/MJ	140.05	
SPHINGOMYELIN	MILLI	18.00	-	- -	4.14	MILLI/MJ	63.02	

1 VALUE EXPRESSED IN MG URIC ACID/100G

KRILL KRILL KRILL

EUPHAUSIA SUPERBA

		PROTEIN	FAT	CARBOHYDRATES	TOTAL
ENERGY VALUE (AVERAGE) PER 100 G EDIBLE PORTION	KJOULE	276	135	0.0	411
	(KCAL)	66	32	0.0	98
AMOUNT OF DIGESTIBLE CONSTITUENTS PER 100 G	GRAM	14.55	3.23	0.00	
ENERGY VALUE (AVERAGE) OF THE DIGESTIBLE FRACTION PER 100 G EDIBLE PORTION	KJOULE	268	128	0.0	396
	(KCAL)	64	31	0.0	95

WASTE PERCENTAGE AVERAGE 70

CONSTITUENTS	DIM	AV	VARIATION			AVR	NUTR.	DENS.	MOLPERC.
WATER	GRAM	78.00	-	-	-	23.40	GRAM/MJ	196.85	
PROTEIN	GRAM	15.00	-	-	-	4.50	GRAM/MJ	37.86	
FAT	GRAM	3.40	1.30	-	6.20	1.02	GRAM/MJ	8.58	
MINERALS	GRAM	2.80	-	-	-	0.84	GRAM/MJ	7.07	
SODIUM	MILLI	320.00	-	-	-	96.00	MILLI/MJ	807.58	
POTASSIUM	MILLI	260.00	-	-	-	78.00	MILLI/MJ	656.16	
MAGNESIUM	MILLI	20.00	-	-	-	6.00	MILLI/MJ	50.47	
CALCIUM	MILLI	160.00	-	-	-	48.00	MILLI/MJ	403.79	
IRON	MILLI	2.70	-	-	-	0.81	MILLI/MJ	6.81	
COPPER	MILLI	2.00	-	-	-	0.60	MILLI/MJ	5.05	
PHOSPHORUS	MILLI	430.00	-	-	-	129.00	GRAM/MJ	1.09	
FLUORIDE	MILLI	34.00	30.00	-	38.00	10.20	MILLI/MJ	85.81	
VITAMIN A	MILLI	0.15	-	-	-	0.05	MILLI/MJ	0.38	
VITAMIN B1	MICRO	30.00	-	-	-	9.00	MICRO/MJ	75.71	
VITAMIN B2	MILLI	0.10	-	-	-	0.03	MILLI/MJ	0.25	
NICOTINAMIDE	MILLI	7.00	-	-	-	2.10	MILLI/MJ	17.67	
VITAMIN B6	MILLI	0.10	-	-	-	0.03	MILLI/MJ	0.25	
VITAMIN B12	MICRO	18.00	-	-	-	5.40	MICRO/MJ	45.43	
LAURIC ACID	MILLI	10.00	10.00	-	10.00	3.00	MILLI/MJ	25.24	
MYRISTIC ACID	GRAM	0.36	0.36	-	0.36	0.11	GRAM/MJ	0.91	
PALMITIC ACID	GRAM	0.88	0.88	-	0.88	0.26	GRAM/MJ	2.22	
STEARIC ACID	MILLI	60.00	60.00	-	60.00	18.00	MILLI/MJ	151.42	
OLEIC ACID	GRAM	0.32	0.32	-	0.32	0.10	GRAM/MJ	0.81	
EICOSENOIC ACID	MILLI	30.00	30.00	-	30.00	9.00	MILLI/MJ	75.71	
LINOLEIC ACID	MILLI	40.00	40.00	-	40.00	12.00	MILLI/MJ	100.95	
LINOLENIC ACID	MILLI	10.00	10.00	-	10.00	3.00	MILLI/MJ	25.24	
C20 : 5 N-3 FATTY ACID	GRAM	0.32	0.32	-	0.32	0.10	GRAM/MJ	0.81	
C22 : 5 N-3 FATTY ACID	-	TRACES	TRACES	-					
C22 : 6 N-3 FATTY ACID	GRAM	0.12	0.12	-	0.12	0.04	GRAM/MJ	0.30	

LANGUSTE | CRAWFISH | LANGOUSTE

SPINY LOBSTER

PALINURUS SPP.; PANULIRUS SPP.

		PROTEIN	FAT	CARBOHYDRATES	TOTAL
ENERGY VALUE (AVERAGE)	KJOULE	317	44	22	382
PER 100 G	(KCAL)	76	10	5.2	91
EDIBLE PORTION					
AMOUNT OF DIGESTIBLE	GRAM	16.68	1.04	1.30	
CONSTITUENTS PER 100 G					
ENERGY VALUE (AVERAGE)	KJOULE	307	42	22	370
OF THE DIGESTIBLE	(KCAL)	73	9.9	5.2	89
FRACTION PER 100 G					
EDIBLE PORTION					

WASTE PERCENTAGE AVERAGE 53

CONSTITUENTS	DIM	AV	VARIATION		AVR	NUTR. DENS.		MOLPERC.
WATER	GRAM	79.10	76.00	83.00	37.18	GRAM/MJ	213.53	
PROTEIN	GRAM	17.20	15.00	18.70	8.08	GRAM/MJ	46.43	
FAT	GRAM	1.10	0.50	1.40	0.52	GRAM/MJ	2.97	
AVAILABLE CARBOHYDR.	GRAM	1.30	-	-	0.61	GRAM/MJ	3.51	
MINERALS	GRAM	1.30	1.20	1.40	0.61	GRAM/MJ	3.51	
SODIUM	MILLI	182.00	-	-	85.54	MILLI/MJ	491.31	
POTASSIUM	MILLI	500.00	-	-	235.00	GRAM/MJ	1.35	
CALCIUM	MILLI	68.00	58.00	77.00	31.96	MILLI/MJ	183.57	
IRON	MILLI	1.30	1.00	1.50	0.61	MILLI/MJ	3.51	
PHOSPHORUS	MILLI	215.00	201.00	230.00	101.05	MILLI/MJ	580.39	
VITAMIN A	MICRO	25.00	-	-	11.75	MICRO/MJ	67.49	
VITAMIN B1	MICRO	10.00	-	-	4.70	MICRO/MJ	26.99	
VITAMIN B2	MICRO	80.00	-	-	37.60	MICRO/MJ	215.96	
NICOTINAMIDE	MILLI	3.00	-	-	1.41	MILLI/MJ	8.10	
VITAMIN C	MILLI	2.00	-	-	0.94	MILLI/MJ	5.40	
ALANINE	GRAM	1.18	-	-	0.55	GRAM/MJ	3.19	9.1
ARGININE	GRAM	1.30	-	-	0.61	GRAM/MJ	3.51	5.1
ASPARTIC ACID	GRAM	2.27	-	-	1.07	GRAM/MJ	6.13	11.7
CYSTINE	GRAM	0.23	-	-	0.11	GRAM/MJ	0.62	0.7
GLUTAMIC ACID	GRAM	3.10	-	-	1.46	GRAM/MJ	8.37	14.4
GLYCINE	GRAM	0.97	-	-	0.46	GRAM/MJ	2.62	8.9
HISTIDINE	GRAM	0.38	-	-	0.18	GRAM/MJ	1.03	1.7
ISOLEUCINE	GRAM	0.76	-	-	0.36	GRAM/MJ	2.05	4.0
LEUCINE	GRAM	0.88	-	-	0.41	GRAM/MJ	2.38	4.6
LYSINE	GRAM	1.69	-	-	0.79	GRAM/MJ	4.56	7.9

CONSTITUENTS	DIM	AV	VARIATION			AVR	NUTR.	DENS.	MOLPERC.
METHIONINE	GRAM	0.58	-	-	-	0.27	GRAM/MJ	1.57	2.7
PHENYLALANINE	GRAM	0.74	-	-	-	0.35	GRAM/MJ	2.00	3.1
PROLINE	GRAM	0.64	-	-	-	0.30	GRAM/MJ	1.73	3.8
SERINE	GRAM	0.95	-	-	-	0.45	GRAM/MJ	2.56	6.2
THREONINE	GRAM	0.84	-	-	-	0.39	GRAM/MJ	2.27	4.8
TRYPTOPHAN	GRAM	0.19	-	-	-	0.09	GRAM/MJ	0.51	0.6
TYROSINE	GRAM	1.52	-	-	-	0.71	GRAM/MJ	4.10	5.7
VALINE	GRAM	0.86	-	-	-	0.40	GRAM/MJ	2.32	5.0
MYRISTIC ACID	MILLI	10.00	10.00	-	10.00	4.70	MILLI/MJ	26.99	
PALMITIC ACID	MILLI	60.00	60.00	-	60.00	28.20	MILLI/MJ	161.97	
STEARIC ACID	MILLI	60.00	60.00	-	60.00	28.20	MILLI/MJ	161.97	
PALMITOLEIC ACID	MILLI	40.00	40.00	-	40.00	18.80	MILLI/MJ	107.98	
OLEIC ACID	MILLI	80.00	80.00	-	80.00	37.60	MILLI/MJ	215.96	
EICOSENOIC ACID	MILLI	10.00	10.00	-	10.00	4.70	MILLI/MJ	26.99	
LINOLEIC ACID	MILLI	30.00	30.00	-	30.00	14.10	MILLI/MJ	80.98	
LINOLENIC ACID	MILLI	10.00	10.00	-	10.00	4.70	MILLI/MJ	26.99	
ARACHIDONIC ACID	GRAM	0.19	0.19	-	0.19	0.09	GRAM/MJ	0.51	
C18 : 4 FATTY ACID	MILLI	10.00	10.00	-	10.00	4.70	MILLI/MJ	26.99	
C20 : 5 N-3 FATTY ACID	GRAM	0.17	0.17	-	0.17	0.08	GRAM/MJ	0.46	
C22 : 5 N-3 FATTY ACID	MILLI	60.00	60.00	-	60.00	28.20	MILLI/MJ	161.97	
C22 : 6 N-3 FATTY ACID	MILLI	80.00	80.00	-	80.00	37.60	MILLI/MJ	215.96	
CHOLESTEROL	MILLI	140.00	-	-	-	65.80	MILLI/MJ	377.93	

MIESMUSCHEL MUSSEL MOULE

(BLAU- ODER PFAHLMUSCHEL)

MYTILUS EDULIS (L.)

		PROTEIN	FAT	CARBOHYDRATES	TOTAL
ENERGY VALUE (AVERAGE)	KJOULE	181	53	0.0	234
PER 100 G	(KCAL)	43	13	0.0	56
EDIBLE PORTION					
AMOUNT OF DIGESTIBLE	GRAM	9.54	1.27	0.00	
CONSTITUENTS PER 100 G					
ENERGY VALUE (AVERAGE)	KJOULE	176	51	0.0	226
OF THE DIGESTIBLE	(KCAL)	42	12	0.0	54
FRACTION PER 100 G					
EDIBLE PORTION					

WASTE PERCENTAGE AVERAGE 82 MINIMUM 80 MAXIMUM 83

CONSTITUENTS	DIM	AV	VARIATION			AVR	NUTR. DENS.		MOLPERC.
WATER	GRAM	83.20	80.40	-	86.70	14.98	GRAM/MJ	367.63	
PROTEIN	GRAM	9.84	8.00	-	11.70	1.77	GRAM/MJ	43.48	
FAT	GRAM	1.34	0.80	-	1.90	0.24	GRAM/MJ	5.92	
MINERALS	GRAM	1.70	1.30	-	2.40	0.31	GRAM/MJ	7.51	
SODIUM	MILLI	296.00	230.00	-	431.00	53.28	GRAM/MJ	1.31	
POTASSIUM	MILLI	277.00	216.00	-	315.00	49.86	GRAM/MJ	1.22	
MAGNESIUM	MILLI	36.00	22.70	-	49.30	6.48	MILLI/MJ	159.07	
CALCIUM	MILLI	27.00	18.00	-	42.00	4.86	MILLI/MJ	119.30	
MANGANESE	MILLI	0.18	0.18	-	0.18	0.03	MILLI/MJ	0.80	
IRON	MILLI	5.12	3.57	-	6.00	0.92	MILLI/MJ	22.62	
COPPER	MILLI	0.17	0.17	-	0.17	0.03	MILLI/MJ	0.75	
ZINC	MILLI	2.70	2.70	-	2.70	0.49	MILLI/MJ	11.93	
NICKEL	MILLI	-	0.03	-	0.12				
MOLYBDENUM	MICRO	56.00	-	-	-	10.08	MICRO/MJ	247.44	
PHOSPHORUS	MILLI	246.00	236.00	-	250.00	44.28	GRAM/MJ	1.09	
CHLORIDE	MILLI	463.00	-	-	-	83.34	GRAM/MJ	2.05	
IODIDE	MILLI	0.13	0.11	-	0.15	0.02	MILLI/MJ	0.57	
BORON	MILLI	0.29	0.15	-	0.45	0.05	MILLI/MJ	1.30	
SELENIUM	MICRO	48.00	46.00	-	390.00	8.64	MICRO/MJ	212.09	
VITAMIN A	MICRO	54.00	-	-	-	9.72	MICRO/MJ	238.61	
VITAMIN E ACTIVITY	MILLI	0.75	-	-	-	0.14	MILLI/MJ	3.31	
ALPHA-TOCOPHEROL	MILLI	0.75	-	-	-	0.14	MILLI/MJ	3.31	
VITAMIN B1	MILLI	0.16	-	-	-	0.03	MILLI/MJ	0.71	
VITAMIN B2	MILLI	0.22	0.18	-	0.25	0.04	MILLI/MJ	0.97	
NICOTINAMIDE	MILLI	1.60	-	-	-	0.29	MILLI/MJ	7.07	
VITAMIN B6	MICRO	76.00	76.00	-	76.00	13.68	MICRO/MJ	335.81	

CONSTITUENTS	DIM	AV	VARIATION			AVR	NUTR. DENS.		MOLPERC.
FOLIC ACID	MICRO	33.00	33.00	-	33.00	5.94	MICRO/MJ	145.81	
VITAMIN B12	MICRO	8.00	8.00	-	8.00	1.44	MICRO/MJ	35.35	
VITAMIN C	MILLI	3.20	3.20	-	3.20	0.58	MILLI/MJ	14.14	
ALANINE	GRAM	0.55	0.55	-	0.55	0.10	GRAM/MJ	2.43	8.3
ARGININE	GRAM	0.73	0.73	-	0.73	0.13	GRAM/MJ	3.23	5.6
ASPARTIC ACID	GRAM	1.10	1.10	-	1.10	0.20	GRAM/MJ	4.86	11.1
CYSTINE	GRAM	0.16	0.16	-	0.16	0.03	GRAM/MJ	0.71	0.9
GLUTAMIC ACID	GRAM	1.37	1.37	-	1.37	0.25	GRAM/MJ	6.05	12.5
GLYCINE	GRAM	0.50	0.50	-	0.50	0.09	GRAM/MJ	2.21	9.0
HISTIDINE	GRAM	0.24	0.24	-	0.24	0.04	GRAM/MJ	1.06	2.1
ISOLEUCINE	GRAM	0.47	0.47	-	0.47	0.08	GRAM/MJ	2.08	4.8
LEUCINE	GRAM	0.76	0.76	-	0.76	0.14	GRAM/MJ	3.36	7.8
LYSINE	GRAM	0.78	0.78	-	0.78	0.14	GRAM/MJ	3.45	7.2
METHIONINE	MILLI	0.27	0.27	-	0.27	0.05	MILLI/MJ	1.19	0.0
PHENYLALANINE	GRAM	0.41	0.41	-	0.41	0.07	GRAM/MJ	1.81	3.3
PROLINE	GRAM	0.41	0.41	-	0.41	0.07	GRAM/MJ	1.81	4.8
SERINE	GRAM	0.50	0.50	-	0.50	0.09	GRAM/MJ	2.21	6.4
THREONINE	GRAM	0.46	0.46	-	0.46	0.08	GRAM/MJ	2.03	5.2
TRYPTOPHAN	GRAM	0.12	0.12	-	0.12	0.02	GRAM/MJ	0.53	0.8
TYROSINE	GRAM	0.41	0.41	-	0.41	0.07	GRAM/MJ	1.81	3.0
VALINE	GRAM	0.61	0.61	-	0.61	0.11	GRAM/MJ	2.70	7.0
MYRISTIC ACID	GRAM	0.11	0.11	-	0.11	0.02	GRAM/MJ	0.49	
PALMITIC ACID	GRAM	0.36	0.36	-	0.36	0.06	GRAM/MJ	1.59	
STEARIC ACID	MILLI	80.00	80.00	-	80.00	14.40	MILLI/MJ	353.49	
PALMITOLEIC ACID	GRAM	0.11	0.11	-	0.11	0.02	GRAM/MJ	0.49	
OLEIC ACID	MILLI	90.00	90.00	-	90.00	16.20	MILLI/MJ	397.68	
LINOLEIC ACID	MILLI	60.00	60.00	-	60.00	10.80	MILLI/MJ	265.12	
LINOLENIC ACID	MILLI	10.00	10.00	-	10.00	1.80	MILLI/MJ	44.19	
ARACHIDONIC ACID	MILLI	40.00	40.00	-	40.00	7.20	MILLI/MJ	176.74	
C18 : 4 FATTY ACID	MILLI	70.00	70.00	-	70.00	12.60	MILLI/MJ	309.30	
C20 : 5 N-3 FATTY ACID	MILLI	50.00	50.00	-	50.00	9.00	MILLI/MJ	220.93	
C22 : 6 N-3 FATTY ACID	GRAM	0.10	0.10	-	0.10	0.02	GRAM/MJ	0.44	
TOTAL STEROLS	MILLI	260.00	260.00	-	260.00	46.80	GRAM/MJ	1.15	
CHOLESTEROL	MILLI	126.00	126.00	-	126.00	22.68	MILLI/MJ	556.75	
DESMOSTEROL	MILLI	78.00	78.00	-	78.00	14.04	MILLI/MJ	344.65	
24-METH.-CHOLESTEROL	MILLI	38.00	38.00	-	38.00	6.84	MILLI/MJ	167.91	
TOTAL PURINES	MILLI	112.00 1	112.00	-	112.00	20.16	MILLI/MJ	494.89	
ADENINE	MILLI	38.00	38.00	-	38.00	6.84	MILLI/MJ	167.91	
GUANINE	MILLI	28.00	28.00	-	28.00	5.04	MILLI/MJ	123.72	
XANTHINE	MILLI	2.00	2.00	-	2.00	0.36	MILLI/MJ	8.84	
HYPOXANTHINE	MILLI	25.00	25.00	-	25.00	4.50	MILLI/MJ	110.47	

1 VALUE OF TOTAL PURINES IS EXPRESSSED IN MG URIC ACID/100G

PILGERMUSCHEL SCALLOP COQUILLE ST. JACQUES

PECTENS JACOBAEUS (L.)

		PROTEIN	FAT	CARBOHYDRATES	TOTAL
ENERGY VALUE (AVERAGE)	KJOULE	287	4.0	0.0	291
PER 100 G	(KCAL)	69	1.0	0.0	70
EDIBLE PORTION					
AMOUNT OF DIGESTIBLE	GRAM	15.13	0.09	0.00	
CONSTITUENTS PER 100 G					
ENERGY VALUE (AVERAGE)	KJOULE	279	3.8	0.0	282
OF THE DIGESTIBLE	(KCAL)	67	0.9	0.0	67
FRACTION PER 100 G					
EDIBLE PORTION					

WASTE PERCENTAGE AVERAGE 56

CONSTITUENTS	DIM	AV	VARIATION			AVR	NUTR. DENS.		MOLPERC.
WATER	GRAM	80.00	79.70	-	80.30	35.20	GRAM/MJ	283.34	
PROTEIN	GRAM	15.60	14.80	-	16.40	6.86	GRAM/MJ	55.25	
FAT	GRAM	0.10	-	-	-	0.04	GRAM/MJ	0.35	
MINERALS	GRAM	1.40	-	-	-	0.62	GRAM/MJ	4.96	
CALCIUM	MILLI	26.00	-	-	-	11.44	MILLI/MJ	92.08	
IRON	MILLI	1.80	-	-	-	0.79	MILLI/MJ	6.38	
NICKEL	MILLI	0.34	0.07	-	0.89	0.15	MILLI/MJ	1.20	
PHOSPHORUS	MILLI	208.00	-	-	-	91.52	MILLI/MJ	736.67	
VITAMIN A	-	0.00	-	-	-	0.00			
VITAMIN B1	MICRO	40.00	-	-	-	17.60	MICRO/MJ	141.67	
VITAMIN B2	MICRO	83.00	65.00	-	100.00	36.52	MICRO/MJ	293.96	
NICOTINAMIDE	MILLI	1.28	1.14	-	1.40	0.56	MILLI/MJ	4.53	
PANTOTHENIC ACID	MILLI	0.14	-	-	-	0.06	MILLI/MJ	0.50	
BIOTIN	MICRO	0.32	-	-	-	0.14	MICRO/MJ	1.13	
FOLIC ACID	MICRO	17.00	-	-	-	7.48	MICRO/MJ	60.21	
VITAMIN B12	MICRO	1.34	-	-	-	0.59	MICRO/MJ	4.75	
TOTAL STEROLS	MILLI	188.00	188.00	-	188.00	82.72	MILLI/MJ	665.84	
CHOLESTEROL	MILLI	104.00	104.00	-	104.00	45.76	MILLI/MJ	368.34	
DESMOSTEROL	MILLI	40.00	40.00	-	40.00	17.60	MILLI/MJ	141.67	
24-METH.-CHOLESTEROL	MILLI	37.00	37.00	-	37.00	16.28	MILLI/MJ	131.04	
TOTAL PURINES	MILLI	136.00 [1]	-	-	-	59.84	MILLI/MJ	481.67	

1 VALUE EXPRESSED IN MG URIC ACID/100G

SEEOHR

HALIOTIS GIGANTES SPP.

ABALONE
(SEA EAR, ORMER)

ORMEAU
(OREILLE DE MER)

		PROTEIN	FAT	CARBOHYDRATES	TOTAL
ENERGY VALUE (AVERAGE)	KJOULE	359	179	0.0	538
PER 100 G	(KCAL)	86	43	0.0	129
EDIBLE PORTION					
AMOUNT OF DIGESTIBLE	GRAM	18.91	4.27	0.00	
CONSTITUENTS PER 100 G					
ENERGY VALUE (AVERAGE)	KJOULE	348	170	0.0	518
OF THE DIGESTIBLE	(KCAL)	83	41	0.0	124
FRACTION PER 100 G					
EDIBLE PORTION					

WASTE PERCENTAGE AVERAGE 52

CONSTITUENTS	DIM	AV	VARIATION			AVR	NUTR. DENS.		MOLPERC.
WATER	GRAM	76.00	-	-	-	36.48	GRAM/MJ	146.68	
PROTEIN	GRAM	19.50	19.00	-	20.00	9.36	GRAM/MJ	37.63	
FAT	GRAM	4.50	0.40	-	5.00	2.16	GRAM/MJ	8.68	
MINERALS	GRAM	1.55	1.50	-	1.60	0.74	GRAM/MJ	2.99	
CALCIUM	MILLI	35.00	34.00	-	37.00	16.80	MILLI/MJ	67.55	
IRON	MILLI	2.70	2.40	-	2.90	1.30	MILLI/MJ	5.21	
PHOSPHORUS	MILLI	180.00	169.00	-	191.00	86.40	MILLI/MJ	347.40	
VITAMIN A	MICRO	5.00	-	-	-	2.40	MICRO/MJ	9.65	
VITAMIN B1	MILLI	0.24	-	-	-	0.12	MILLI/MJ	0.46	
VITAMIN B2	MICRO	60.00	-	-	-	28.80	MICRO/MJ	115.80	
NICOTINAMIDE	MILLI	1.60	-	-	-	0.77	MILLI/MJ	3.09	

SCHILDKRÖTE TURTLE TORTUE

CHELONIA MYDAS (L.)

		PROTEIN	FAT	CARBOHYDRATES	TOTAL
ENERGY VALUE (AVERAGE)	KJOULE	322	32	0.0	354
PER 100 G	(KCAL)	77	7.6	0.0	85
EDIBLE PORTION					
AMOUNT OF DIGESTIBLE	GRAM	16.97	0.76	0.00	
CONSTITUENTS PER 100 G					
ENERGY VALUE (AVERAGE)	KJOULE	313	30	0.0	343
OF THE DIGESTIBLE	(KCAL)	75	7.2	0.0	82
FRACTION PER 100 G					
EDIBLE PORTION					

WASTE PERCENTAGE AVERAGE 69

CONSTITUENTS	DIM	AV	VARIATION			AVR	NUTR. DENS.		MOLPERC.
WATER	GRAM	80.90	-	-	-	25.08	GRAM/MJ	236.06	
PROTEIN	GRAM	17.50	-	-	-	5.43	GRAM/MJ	51.06	
FAT	GRAM	0.80	-	-	-	0.25	GRAM/MJ	2.33	
MINERALS	GRAM	0.80	-	-	-	0.25	GRAM/MJ	2.33	
POTASSIUM	MILLI	235.00	-	-	-	72.85	MILLI/MJ	685.71	
CALCIUM	MILLI	107.00	-	-	-	33.17	MILLI/MJ	312.22	
IRON	MILLI	1.50	-	-	-	0.47	MILLI/MJ	4.38	
PHOSPHORUS	MILLI	146.00	-	-	-	45.26	MILLI/MJ	426.01	
VITAMIN B1	MILLI	0.25	-	-	-	0.08	MILLI/MJ	0.73	
VITAMIN B2	MILLI	0.50	-	-	-	0.16	MILLI/MJ	1.46	
NICOTINAMIDE	MILLI	2.60	-	-	-	0.81	MILLI/MJ	7.59	

STECKMUSCHEL (KLAFFMUSCHEL, PIEP-MUSCHEL, SANDAUSTER) — SOFT CLAM — MYE

MYA ARENARIA (L.)

		PROTEIN	FAT	CARBOHYDRATES	TOTAL
ENERGY VALUE (AVERAGE) PER 100 G EDIBLE PORTION	KJOULE	193	52	0.0	245
	(KCAL)	46	12	0.0	59
AMOUNT OF DIGESTIBLE CONSTITUENTS PER 100 G	GRAM	10.18	1.24	0.00	
ENERGY VALUE (AVERAGE) OF THE DIGESTIBLE FRACTION PER 100 G EDIBLE PORTION	KJOULE	188	49	0.0	237
	(KCAL)	45	12	0.0	57

WASTE PERCENTAGE AVERAGE 65 MINIMUM 61 MAXIMUM 71

CONSTITUENTS	DIM	AV	VARIATION			AVR	NUTR.	DENS.	MOLPERC.
WATER	GRAM	83.10	80.30	-	87.90	29.09	GRAM/MJ	350.68	
PROTEIN	GRAM	10.50	8.50	-	12.80	3.68	GRAM/MJ	44.31	
FAT	GRAM	1.31	1.21	-	1.40	0.46	GRAM/MJ	5.53	
MINERALS	GRAM	2.04	1.98	-	2.10	0.71	GRAM/MJ	8.61	
SODIUM	MILLI	121.00	-	-	-	42.35	MILLI/MJ	510.62	
POTASSIUM	MILLI	800.00	-	-	-	280.00	GRAM/MJ	3.38	
MAGNESIUM	MILLI	62.80	-	-	-	21.98	MILLI/MJ	265.01	
CALCIUM	MILLI	11.60	-	-	-	4.06	MILLI/MJ	48.95	
IRON	MILLI	0.57	-	-	-	0.20	MILLI/MJ	2.41	
NICKEL	MILLI	-	0.06	-	0.15				
CHROMIUM	MICRO	11.00	-	-	-	3.85	MICRO/MJ	46.42	
PHOSPHORUS	MILLI	310.00	-	-	-	108.50	GRAM/MJ	1.31	
CHLORIDE	MILLI	91.80	-	-	-	32.13	MILLI/MJ	387.39	
IODIDE	MILLI	0.12	-	-	-	0.04	MILLI/MJ	0.51	
VITAMIN A	MICRO	33.00	-	-	-	11.55	MICRO/MJ	139.26	
VITAMIN B1	MILLI	0.10	-	-	-	0.04	MILLI/MJ	0.42	
VITAMIN B2	MILLI	0.19	0.18	-	0.19	0.07	MILLI/MJ	0.80	
NICOTINAMIDE	MILLI	1.45	1.26	-	1.60	0.51	MILLI/MJ	6.12	
PANTOTHENIC ACID	MILLI	0.62	-	-	-	0.22	MILLI/MJ	2.62	
BIOTIN	MICRO	2.34	-	-	-	0.82	MICRO/MJ	9.87	
FOLIC ACID	MICRO	2.65	-	-	-	0.93	MICRO/MJ	11.18	
VITAMIN B12	MICRO	62.00	-	-	-	21.70	MICRO/MJ	261.64	
ALANINE	GRAM	1.08	0.89	-	1.22	0.38	GRAM/MJ	4.56	12.3
ARGININE	GRAM	0.87	0.71	-	0.96	0.30	GRAM/MJ	3.67	5.1
ASPARTIC ACID	GRAM	1.18	0.98	-	1.35	0.41	GRAM/MJ	4.98	9.0

CONSTITUENTS	DIM	AV	VARIATION			AVR	NUTR. DENS.		MOLPERC.
CYSTINE	GRAM	0.14	0.11	-	0.15	0.05	GRAM/MJ	0.59	0.6
GLUTAMIC ACID	GRAM	1.70	1.48	-	1.85	0.60	GRAM/MJ	7.17	11.7
GLYCINE	GRAM	1.05	0.70	-	1.20	0.37	GRAM/MJ	4.43	14.2
HISTIDINE	GRAM	0.23	0.17	-	0.26	0.08	GRAM/MJ	0.97	1.5
ISOLEUCINE	GRAM	0.58	0.52	-	0.74	0.20	GRAM/MJ	2.45	4.5
LEUCINE	GRAM	0.76	0.46	-	0.99	0.27	GRAM/MJ	3.21	5.9
LYSINE	GRAM	0.98	0.87	-	1.14	0.34	GRAM/MJ	4.14	6.8
METHIONINE	GRAM	0.31	0.28	-	0.33	0.11	GRAM/MJ	1.31	2.1
PHENYLALANINE	GRAM	0.42	0.34	-	0.48	0.15	GRAM/MJ	1.77	2.6
PROLINE	GRAM	0.47	0.40	-	0.51	0.16	GRAM/MJ	1.98	4.1
SERINE	GRAM	0.60	0.51	-	0.66	0.21	GRAM/MJ	2.53	5.8
THREONINE	GRAM	0.58	0.49	-	0.66	0.20	GRAM/MJ	2.45	4.9
TRYPTOPHAN	GRAM	0.16	-	-	-	0.06	GRAM/MJ	0.68	0.8
TYROSINE	GRAM	0.53	0.33	-	0.87	0.19	GRAM/MJ	2.24	3.0
VALINE	GRAM	0.59	0.53	-	0.64	0.21	GRAM/MJ	2.49	5.1
TOTAL STEROLS	MILLI	113.00	113.00	-	113.00	39.55	MILLI/MJ	476.86	
CHOLESTEROL	MILLI	66.00	66.00	-	66.00	23.10	MILLI/MJ	278.52	
DESMOSTEROL	MILLI	15.00	15.00	-	15.00	5.25	MILLI/MJ	63.30	
24-METH.-CHOLESTEROL	MILLI	24.00	24.00	-	24.00	8.40	MILLI/MJ	101.28	
TOTAL PHOSPHOLIPIDS	MILLI	481.00	-	-	-	168.35	GRAM/MJ	2.03	
PHOSPHATIDYLCHOLINE	MILLI	196.00	-	-	-	68.60	MILLI/MJ	827.12	
PHOSPHATIDYLETHANOLAMINE	MILLI	15.00	-	-	-	5.25	MILLI/MJ	63.30	
PHOSPHATIDYLSERINE	MILLI	87.00	-	-	-	30.45	MILLI/MJ	367.14	
SPHINGOMYELIN	MILLI	117.00	-	-	-	40.95	MILLI/MJ	493.74	

TINTENFISCH

SEPIA SPP.

CUTTLE FISH

SÈCHE
(SEICHE)

		PROTEIN	FAT	CARBOHYDRATES	TOTAL
ENERGY VALUE (AVERAGE) PER 100 G EDIBLE PORTION	KJOULE	296	36	0.0	332
	(KCAL)	71	8.6	0.0	79
AMOUNT OF DIGESTIBLE CONSTITUENTS PER 100 G	GRAM	15.61	0.85	0.00	
ENERGY VALUE (AVERAGE) OF THE DIGESTIBLE FRACTION PER 100 G EDIBLE PORTION	KJOULE	288	34	0.0	321
	(KCAL)	69	8.1	0.0	77

WASTE PERCENTAGE AVERAGE 21

CONSTITUENTS	DIM	AV	VARIATION			AVR	NUTR.	DENS.	MOLPERC.
WATER	GRAM	81.00	-	-	-	63.99	GRAM/MJ	251.95	
PROTEIN	GRAM	16.10	-	-	-	12.72	GRAM/MJ	50.08	
FAT	GRAM	0.90	-	-	-	0.71	GRAM/MJ	2.80	
MINERALS	GRAM	1.00	-	-	-	0.79	GRAM/MJ	3.11	
SODIUM	MILLI	387.00	387.00	-	387.00	305.73	GRAM/MJ	1.20	
POTASSIUM	MILLI	273.00	-	-	-	215.67	MILLI/MJ	849.18	
CALCIUM	MILLI	27.00	-	-	-	21.33	MILLI/MJ	83.98	
MANGANESE	MICRO	11.00	11.00	-	11.00	8.69	MICRO/MJ	34.22	
IRON	MILLI	0.80	-	-	-	0.63	MILLI/MJ	2.49	
COBALT	MICRO	1.30	-	-	-	1.03	MICRO/MJ	4.04	
ZINC	MILLI	0.70	-	-	-	0.55	MILLI/MJ	2.18	
PHOSPHORUS	MILLI	143.00	-	-	-	112.97	MILLI/MJ	444.81	
VITAMIN E ACTIVITY	MILLI	2.40	2.40	-	2.40	1.90	MILLI/MJ	7.47	
ALPHA-TOCOPHEROL	MILLI	2.40	2.40	-	2.40	1.90	MILLI/MJ	7.47	
VITAMIN B1	MICRO	70.00	-	-	-	55.30	MICRO/MJ	217.74	
VITAMIN B2	MICRO	50.00	-	-	-	39.50	MICRO/MJ	155.53	
NICOTINAMIDE	MILLI	2.60	-	-	-	2.05	MILLI/MJ	8.09	
VITAMIN B6	MILLI	0.39	0.39	-	0.39	0.31	MILLI/MJ	1.21	
TOTAL PHOSPHOLIPIDS	MILLI	704.00	-	-	-	556.16	GRAM/MJ	2.19	
PHOSPHATIDYLCHOLINE	MILLI	296.00	-	-	-	233.84	MILLI/MJ	920.72	
PHOSPHATIDYLETHANOLAMINE	MILLI	205.00	-	-	-	161.95	MILLI/MJ	637.66	
PHOSPHATIDYLSERINE	MILLI	31.00	-	-	-	24.49	MILLI/MJ	96.43	
PHOSPHATIDYLINOSITOL	MILLI	24.00	-	-	-	18.96	MILLI/MJ	74.65	
SPHINGOMYELIN	MILLI	19.00	-	-	-	15.01	MILLI/MJ	59.10	

WEINBERGSCHNECKE SNAIL EDIBLE ESCARGOT

HELIX POMATIA (L.)

		PROTEIN	FAT	CARBOHYDRATES	TOTAL
ENERGY VALUE (AVERAGE)	KJOULE	295	40	0.0	334
PER 100 G	(KCAL)	70	9.5	0.0	80
EDIBLE PORTION					
AMOUNT OF DIGESTIBLE	GRAM	15.52	0.95	0.00	
CONSTITUENTS PER 100 G					
ENERGY VALUE (AVERAGE)	KJOULE	286	38	0.0	323
OF THE DIGESTIBLE	(KCAL)	68	9.0	0.0	77
FRACTION PER 100 G					
EDIBLE PORTION					

WASTE PERCENTAGE AVERAGE 60

CONSTITUENTS	DIM	AV	VARIATION			AVR	NUTR. DENS.		MOLPERC.
WATER	GRAM	79.00	-	-	-	31.60	GRAM/MJ	244.22	
PROTEIN	GRAM	16.00	-	-	-	6.40	GRAM/MJ	49.46	
FAT	GRAM	1.00	-	-	-	0.40	GRAM/MJ	3.09	
MINERALS	GRAM	1.00	-	-	-	0.40	GRAM/MJ	3.09	

Fischerzeugnisse

FISH PRODUCTS **PRODUITS DE POISSONS**

AAL / EEL / ANGUILLE

AAL GERÄUCHERT

EEL SMOKED

ANGUILLE FUMÉ

		PROTEIN	FAT	CARBOHYDRATES	TOTAL
ENERGY VALUE (AVERAGE)	KJOULE	330	1137	0.0	1466
PER 100 G	(KCAL)	79	272	0.0	350
EDIBLE PORTION					
AMOUNT OF DIGESTIBLE	GRAM	17.36	27.17	0.00	
CONSTITUENTS PER 100 G					
ENERGY VALUE (AVERAGE)	KJOULE	320	1080	0.0	1400
OF THE DIGESTIBLE	(KCAL)	76	258	0.0	335
FRACTION PER 100 G					
EDIBLE PORTION					

WASTE PERCENTAGE AVERAGE 24 MINIMUM 19 MAXIMUM 30

CONSTITUENTS	DIM	AV	VARIATION			AVR	NUTR. DENS.		MOLPERC.
WATER	GRAM	51.10	48.20	-	55.00	38.84	GRAM/MJ	36.51	
PROTEIN	GRAM	17.90	16.10	-	19.00	13.60	GRAM/MJ	12.79	
FAT	GRAM	28.60	24.00	-	33.90	21.74	GRAM/MJ	20.43	
MINERALS	GRAM	2.16	1.92	-	2.40	1.64	GRAM/MJ	1.54	
SODIUM	MILLI	500.00	444.00	-	798.00	380.00	MILLI/MJ	357.25	
POTASSIUM	MILLI	243.00	186.00	-	280.00	184.68	MILLI/MJ	173.62	
MAGNESIUM	MILLI	18.00	16.00	-	20.00	13.68	MILLI/MJ	12.86	
CALCIUM	MILLI	19.00	15.00	-	22.00	14.44	MILLI/MJ	13.58	
MANGANESE	MICRO	28.00	-	-	-	21.28	MICRO/MJ	20.01	
IRON	MILLI	0.67	0.56	-	0.90	0.51	MILLI/MJ	0.48	
COPPER	MICRO	90.00	30.00	-	190.00	68.40	MICRO/MJ	64.30	
PHOSPHORUS	MILLI	250.00	215.00	-	333.00	190.00	MILLI/MJ	178.62	
CHLORIDE	GRAM	0.72	0.64	-	1.18	0.55	GRAM/MJ	0.52	
FLUORIDE	MILLI	0.18	-	-	-	0.14	MILLI/MJ	0.13	
IODIDE	MICRO	4.50	-	-	-	3.42	MICRO/MJ	3.22	
VITAMIN A	MILLI	0.94	0.27	-	1.80	0.71	MILLI/MJ	0.67	
VITAMIN D	MICRO	90.00	36.00	-	140.00	68.40	MICRO/MJ	64.30	
VITAMIN B1	MILLI	0.19	0.15	-	0.22	0.14	MILLI/MJ	0.14	
VITAMIN B2	MILLI	0.37	0.27	-	0.47	0.28	MILLI/MJ	0.26	
NICOTINAMIDE	MILLI	3.50	2.29	-	5.20	2.66	MILLI/MJ	2.50	
VITAMIN B6	MILLI	0.16	0.13	-	0.21	0.12	MILLI/MJ	0.11	
VITAMIN B12	MICRO	1.00	-	-	-	0.76	MICRO/MJ	0.71	
ARGININE	GRAM	0.97	0.92	-	1.01	0.74	GRAM/MJ	0.69	
CYSTINE	GRAM	0.20	0.17	-	0.24	0.15	GRAM/MJ	0.14	
HISTIDINE	GRAM	0.38	-	-	-	0.29	GRAM/MJ	0.27	

CONSTITUENTS	DIM	AV	VARIATION			AVR	NUTR. DENS.		MOLPERC.
ISOLEUCINE	GRAM	0.86	0.81	-	0.91	0.65	GRAM/MJ	0.61	
LEUCINE	GRAM	1.41	1.35	-	1.46	1.07	GRAM/MJ	1.01	
LYSINE	GRAM	1.34	1.10	-	1.57	1.02	GRAM/MJ	0.96	
METHIONINE	GRAM	0.51	0.49	-	0.52	0.39	GRAM/MJ	0.36	
PHENYLALANINE	GRAM	0.67	-	-	-	0.51	GRAM/MJ	0.48	
THREONINE	GRAM	0.82	0.76	-	0.86	0.62	GRAM/MJ	0.59	
TRYPTOPHAN	GRAM	0.18	0.17	-	0.18	0.14	GRAM/MJ	0.13	
TYROSINE	GRAM	0.62	0.49	-	0.74	0.47	GRAM/MJ	0.44	
VALINE	GRAM	0.99	0.95	-	1.03	0.75	GRAM/MJ	0.71	
MYRISTIC ACID	GRAM	1.88	1.88	-	1.88	1.43	GRAM/MJ	1.34	
PALMITIC ACID	GRAM	5.91	5.91	-	5.91	4.49	GRAM/MJ	4.22	
STEARIC ACID	GRAM	0.82	0.82	-	0.82	0.62	GRAM/MJ	0.59	
PALMITOLEIC ACID	GRAM	4.19	4.19	-	4.19	3.18	GRAM/MJ	2.99	
OLEIC ACID	GRAM	9.10	9.10	-	9.10	6.92	GRAM/MJ	6.50	
LINOLEIC ACID	GRAM	1.00	1.00	-	1.00	0.76	GRAM/MJ	0.71	
LINOLENIC ACID	GRAM	1.52	1.52	-	1.52	1.16	GRAM/MJ	1.09	
ARACHIDONIC ACID	GRAM	0.49	0.49	-	0.49	0.37	GRAM/MJ	0.35	
C22 : 6 N-3 FATTY ACID	GRAM	0.26	0.26	-	0.26	0.20	GRAM/MJ	0.19	
TOTAL PURINES	MILLI	78.00 1	45.00	-	110.00	59.28	MILLI/MJ	55.73	

1 VALUE EXPRESSED IN MG URIC ACID/100G

BRATHERING / HERRING, FRIED / HARENG FRIT ET MARINÉ

		PROTEIN	FAT	CARBOHYDRATES	TOTAL
ENERGY VALUE (AVERAGE)	KJOULE	309	604	0.0	913
PER 100 G	(KCAL)	74	144	0.0	218
EDIBLE PORTION					
AMOUNT OF DIGESTIBLE	GRAM	16.29	14.44	0.00	
CONSTITUENTS PER 100 G					
ENERGY VALUE (AVERAGE)	KJOULE	300	574	0.0	874
OF THE DIGESTIBLE	(KCAL)	72	137	0.0	209
FRACTION PER 100 G					
EDIBLE PORTION					

WASTE PERCENTAGE AVERAGE 8.0 MINIMUM 3.0 MAXIMUM 16

CONSTITUENTS	DIM	AV	VARIATION		AVR	NUTR. DENS.		MOLPERC.
WATER	GRAM	62.00	56.80	- 65.50	57.04	GRAM/MJ	70.94	
PROTEIN	GRAM	16.80	12.80	- 21.80	15.46	GRAM/MJ	19.22	
FAT	GRAM	15.20	11.10	- 21.00	13.98	GRAM/MJ	17.39	
MINERALS	GRAM	2.23	1.58	- 2.91	2.05	GRAM/MJ	2.55	
SODIUM	MILLI	569.00	335.00	- 884.00	523.48	MILLI/MJ	651.06	
POTASSIUM	MILLI	182.00	158.00	- 206.00	167.44	MILLI/MJ	208.25	
CALCIUM	MILLI	36.00	1.00	- 70.00	33.12	MILLI/MJ	41.19	
IRON	MILLI	1.10	-	- -	1.01	MILLI/MJ	1.26	
PHOSPHORUS	MILLI	240.00	-	- -	220.80	MILLI/MJ	274.61	
IODIDE	MILLI	0.13	-	- -	0.12	MILLI/MJ	0.15	
VITAMIN A	MICRO	20.00	-	- -	18.40	MICRO/MJ	22.88	
VITAMIN B1	MICRO	10.00	-	- -	9.20	MICRO/MJ	11.44	
VITAMIN B2	MILLI	0.13	-	- -	0.12	MILLI/MJ	0.15	
NICOTINAMIDE	MILLI	3.90	-	- -	3.59	MILLI/MJ	4.46	
VITAMIN C	-	0.00	-	- -	0.00			
ARGININE	GRAM	0.94	-	- -	0.86	GRAM/MJ	1.08	
CYSTINE	GRAM	0.22	-	- -	0.20	GRAM/MJ	0.25	
HISTIDINE	GRAM	0.41	-	- -	0.38	GRAM/MJ	0.47	
ISOLEUCINE	GRAM	0.85	-	- -	0.78	GRAM/MJ	0.97	
LEUCINE	GRAM	1.27	-	- -	1.17	GRAM/MJ	1.45	
LYSINE	GRAM	1.47	-	- -	1.35	GRAM/MJ	1.68	
METHIONINE	GRAM	0.49	-	- -	0.45	GRAM/MJ	0.56	
PHENYLALANINE	GRAM	0.62	-	- -	0.57	GRAM/MJ	0.71	
THREONINE	GRAM	0.73	-	- -	0.67	GRAM/MJ	0.84	
TRYPTOPHAN	GRAM	0.17	-	- -	0.16	GRAM/MJ	0.19	
TYROSINE	GRAM	0.46	-	- -	0.42	GRAM/MJ	0.53	
VALINE	GRAM	0.89	-	- -	0.82	GRAM/MJ	1.02	
CHOLESTEROL	MILLI	86.60	48.60	- 104.00	79.67	MILLI/MJ	99.09	

BÜCKLING "BUECKLING" "BUECKLING"

		PROTEIN	FAT	CARBOHYDRATES	TOTAL
ENERGY VALUE (AVERAGE)	KJOULE	390	616	0.0	1006
PER 100 G	(KCAL)	93	147	0.0	241
EDIBLE PORTION					
AMOUNT OF DIGESTIBLE	GRAM	20.56	14.72	0.00	
CONSTITUENTS PER 100 G					
ENERGY VALUE (AVERAGE)	KJOULE	379	585	0.0	964
OF THE DIGESTIBLE	(KCAL)	90	140	0.0	230
FRACTION PER 100 G					
EDIBLE PORTION					

WASTE PERCENTAGE AVERAGE 29 MINIMUM 26 MAXIMUM 32

CONSTITUENTS	DIM	AV	VARIATION			AVR	NUTR. DENS.		MOLPERC.
WATER	GRAM	62.00	57.40	-	67.50	44.02	GRAM/MJ	64.32	
PROTEIN	GRAM	21.20	20.00	-	22.60	15.05	GRAM/MJ	21.99	
FAT	GRAM	15.50	9.60	-	20.80	11.01	GRAM/MJ	16.08	
MINERALS	GRAM	1.30	1.20	-	1.46	0.92	GRAM/MJ	1.35	
SODIUM	MILLI	156.00	100.00	-	210.00	110.76	MILLI/MJ	161.85	
POTASSIUM	MILLI	320.00	295.00	-	350.00	227.20	MILLI/MJ	332.00	
MAGNESIUM	MILLI	32.00	27.00	-	37.00	22.72	MILLI/MJ	33.20	
CALCIUM	MILLI	35.00	30.00	-	45.00	24.85	MILLI/MJ	36.31	
IRON	MILLI	1.10	0.95	-	1.30	0.78	MILLI/MJ	1.14	
COBALT	MICRO	1.70	-	-	-	1.21	MICRO/MJ	1.76	
COPPER	MILLI	0.33	0.26	-	0.45	0.23	MILLI/MJ	0.34	
NICKEL	MILLI	0.17	-	-	-	0.12	MILLI/MJ	0.18	
PHOSPHORUS	MILLI	256.00	240.00	-	265.00	181.76	MILLI/MJ	265.60	
CHLORIDE	MILLI	175.00	110.00	-	236.00	124.25	MILLI/MJ	181.56	
FLUORIDE	MILLI	0.36	0.16	-	0.55	0.26	MILLI/MJ	0.37	
IODIDE	MICRO	53.00	41.00	-	67.00	37.63	MICRO/MJ	54.99	
SELENIUM	MILLI	0.14	-	-	-	0.10	MILLI/MJ	0.15	
VITAMIN A	MICRO	28.00	12.00	-	65.00	19.88	MICRO/MJ	29.05	
VITAMIN D	MICRO	30.00	20.00	-	40.00	21.30	MICRO/MJ	31.12	
VITAMIN E ACTIVITY	MILLI	1.60	1.10	-	1.90	1.14	MILLI/MJ	1.66	
ALPHA-TOCOPHEROL	MILLI	1.60	1.10	-	1.90	1.14	MILLI/MJ	1.66	
VITAMIN B1	MICRO	40.00	18.00	-	68.00	28.40	MICRO/MJ	41.50	
VITAMIN B2	MILLI	0.25	0.10	-	0.32	0.18	MILLI/MJ	0.26	
NICOTINAMIDE	MILLI	4.30	3.50	-	5.10	3.05	MILLI/MJ	4.46	
PANTOTHENIC ACID	MILLI	1.00	-	-	-	0.71	MILLI/MJ	1.04	
VITAMIN B6	MILLI	0.50	0.40	-	0.58	0.36	MILLI/MJ	0.52	

CONSTITUENTS	DIM	AV	VARIATION			AVR	NUTR. DENS.		MOLPERC.
BIOTIN	MICRO	5.10	3.40	-	6.80	3.62	MICRO/MJ	5.29	
VITAMIN B12	MICRO	9.70	5.70	-	13.60	6.89	MICRO/MJ	10.06	
VITAMIN C	-	0.00	-	-	-	0.00			
ARGININE	GRAM	1.32	1.16	-	1.55	0.94	GRAM/MJ	1.37	
CYSTINE	GRAM	0.26	0.22	-	0.30	0.18	GRAM/MJ	0.27	
HISTIDINE	GRAM	0.57	0.51	-	0.65	0.40	GRAM/MJ	0.59	
ISOLEUCINE	GRAM	1.06	0.99	-	1.12	0.75	GRAM/MJ	1.10	
LEUCINE	GRAM	1.69	1.50	-	1.90	1.20	GRAM/MJ	1.75	
LYSINE	GRAM	1.94	1.65	-	2.11	1.38	GRAM/MJ	2.01	
METHIONINE	GRAM	0.63	0.57	-	0.75	0.45	GRAM/MJ	0.65	
PHENYLALANINE	GRAM	0.85	0.72	-	1.00	0.60	GRAM/MJ	0.88	
THREONINE	GRAM	1.01	0.92	-	1.13	0.72	GRAM/MJ	1.05	
TRYPTOPHAN	GRAM	0.22	0.17	-	0.30	0.16	GRAM/MJ	0.23	
TYROSINE	GRAM	0.65	0.57	-	0.78	0.46	GRAM/MJ	0.67	
VALINE	GRAM	1.18	1.06	-	1.30	0.84	GRAM/MJ	1.22	
MYRISTIC ACID	GRAM	0.38	0.38	-	0.38	0.27	GRAM/MJ	0.39	
PALMITIC ACID	GRAM	1.72	1.72	-	1.72	1.22	GRAM/MJ	1.78	
STEARIC ACID	GRAM	0.17	0.17	-	0.17	0.12	GRAM/MJ	0.18	
PALMITOLEIC ACID	GRAM	1.12	1.12	-	1.12	0.80	GRAM/MJ	1.16	
OLEIC ACID	GRAM	1.48	1.48	-	1.48	1.05	GRAM/MJ	1.54	
EICOSENOIC ACID	GRAM	2.27	2.27	-	2.27	1.61	GRAM/MJ	2.36	
C22 : 1 FATTY ACID	GRAM	2.54	2.54	-	2.54	1.80	GRAM/MJ	2.64	
LINOLEIC ACID	GRAM	1.48	0.27	-	3.50	1.05	GRAM/MJ	1.54	
LINOLENIC ACID	GRAM	0.24	0.24	-	0.24	0.17	GRAM/MJ	0.25	
ARACHIDONIC ACID	MILLI	70.00	70.00	-	70.00	49.70	MILLI/MJ	72.62	
C20 : 5 N-3 FATTY ACID	GRAM	1.12	1.12	-	1.12	0.80	GRAM/MJ	1.16	
C22 : 5 N-3 FATTY ACID	GRAM	0.10	0.10	-	0.10	0.07	GRAM/MJ	0.10	
C22 : 6 N-3 FATTY ACID	GRAM	0.49	0.49	-	0.49	0.35	GRAM/MJ	0.51	
CHOLESTEROL	MILLI	90.00	-	-	-	63.90	MILLI/MJ	93.37	

FLUNDER
GERÄUCHERT

FLOUNDER
SMOKED

FLET
FUMÉ

		PROTEIN	FAT	CARBOHYDRATES	TOTAL
ENERGY VALUE (AVERAGE)	KJOULE	429	76	0.0	505
PER 100 G	(KCAL)	103	18	0.0	121
EDIBLE PORTION					
AMOUNT OF DIGESTIBLE	GRAM	22.60	1.82	0.00	
CONSTITUENTS PER 100 G					
ENERGY VALUE (AVERAGE)	KJOULE	416	73	0.0	489
OF THE DIGESTIBLE	(KCAL)	99	17	0.0	117
FRACTION PER 100 G					
EDIBLE PORTION					

WASTE PERCENTAGE AVERAGE 22 MINIMUM 18 MAXIMUM 25

CONSTITUENTS	DIM	AV	VARIATION	AVR	NUTR. DENS.		MOLPERC.
WATER	GRAM	71.90	71.70 - 72.10	56.08	GRAM/MJ	147.16	
PROTEIN	GRAM	23.30	21.90 - 25.30	18.17	GRAM/MJ	47.69	
FAT	GRAM	1.92	1.29 - 2.54	1.50	GRAM/MJ	3.93	
MINERALS	GRAM	2.81	2.41 - 3.37	2.19	GRAM/MJ	5.75	
SODIUM	MILLI	481.00	300.00 - 661.00	375.18	MILLI/MJ	984.49	
POTASSIUM	MILLI	410.00	361.00 - 458.00	319.80	MILLI/MJ	839.17	
CALCIUM	MILLI	22.00	18.00 - 26.00	17.16	MILLI/MJ	45.03	

HEILBUTT	**HALIBUT**	**FLÉTAN**
SCHWARZER HEILBUTT	GREENLAND HALIBUT	FLÉTAN NOIR
GERÄUCHERT	SMOKED	FUMÉ

		PROTEIN	FAT	CARBOHYDRATES	TOTAL
ENERGY VALUE (AVERAGE)	KJOULE	318	680	0.0	998
PER 100 G	(KCAL)	76	162	0.0	239
EDIBLE PORTION					
AMOUNT OF DIGESTIBLE	GRAM	16.78	16.24	0.00	
CONSTITUENTS PER 100 G					
ENERGY VALUE (AVERAGE)	KJOULE	309	646	0.0	955
OF THE DIGESTIBLE	(KCAL)	74	154	0.0	228
FRACTION PER 100 G					
EDIBLE PORTION					

WASTE PERCENTAGE AVERAGE 24 MINIMUM 15 MAXIMUM 28

CONSTITUENTS	DIM	AV	VARIATION		AVR	NUTR.	DENS.	MOLPERC.
WATER	GRAM	63.80	63.00	- 65.70	48.49	GRAM/MJ	66.83	
PROTEIN	GRAM	17.30	15.60	- 20.00	13.15	GRAM/MJ	18.12	
FAT	GRAM	17.10	15.50	- 18.70	13.00	GRAM/MJ	17.91	
MINERALS	GRAM	1.73	1.56	- 1.85	1.31	GRAM/MJ	1.81	
SODIUM	MILLI	406.00	378.00	- 433.00	308.56	MILLI/MJ	425.29	
POTASSIUM	MILLI	280.00	234.00	- 325.00	212.80	MILLI/MJ	293.30	
CALCIUM	MILLI	18.00	-	- -	13.68	MILLI/MJ	18.86	
IRON	MILLI	0.90	-	- -	0.68	MILLI/MJ	0.94	
PHOSPHORUS	MILLI	300.00	-	- -	228.00	MILLI/MJ	314.26	
VITAMIN A	MICRO	33.00	-	- -	25.08	MICRO/MJ	34.57	
VITAMIN B1	MICRO	60.00	-	- -	45.60	MICRO/MJ	62.85	
VITAMIN B2	MICRO	40.00	-	- -	30.40	MICRO/MJ	41.90	
NICOTINAMIDE	MILLI	6.00	-	- -	4.56	MILLI/MJ	6.29	
ARGININE	GRAM	0.98	-	- -	0.74	GRAM/MJ	1.03	
CYSTINE	GRAM	0.23	-	- -	0.17	GRAM/MJ	0.24	
ISOLEUCINE	GRAM	0.87	-	- -	0.66	GRAM/MJ	0.91	
LEUCINE	GRAM	1.31	-	- -	1.00	GRAM/MJ	1.37	
LYSINE	GRAM	1.52	-	- -	1.16	GRAM/MJ	1.59	
METHIONINE	GRAM	0.50	-	- -	0.38	GRAM/MJ	0.52	
PHENYLALANINE	GRAM	0.64	-	- -	0.49	GRAM/MJ	0.67	
THREONINE	GRAM	0.75	-	- -	0.57	GRAM/MJ	0.79	
TRYPTOPHAN	GRAM	0.18	-	- -	0.14	GRAM/MJ	0.19	
TYROSINE	GRAM	0.47	-	- -	0.36	GRAM/MJ	0.49	
VALINE	GRAM	0.92	-	- -	0.70	GRAM/MJ	0.96	

HERING IN GELEE

HERRING IN JELLY

HARENG EN GELÉE

		PROTEIN	FAT	CARBOHYDRATES	TOTAL
ENERGY VALUE (AVERAGE)	KJOULE	234	501	0.0	735
PER 100 G	(KCAL)	56	120	0.0	176
EDIBLE PORTION					
AMOUNT OF DIGESTIBLE	GRAM	12.31	11.97	0.00	
CONSTITUENTS PER 100 G					
ENERGY VALUE (AVERAGE)	KJOULE	227	476	0.0	703
OF THE DIGESTIBLE	(KCAL)	54	114	0.0	168
FRACTION PER 100 G					
EDIBLE PORTION					

WASTE PERCENTAGE AVERAGE 0.00

CONSTITUENTS	DIM	AV	VARIATION	AVR	NUTR. DENS.		MOLPERC.
WATER	GRAM	72.40	- - -	72.40	GRAM/MJ	103.05	
PROTEIN	GRAM	12.70	- - -	12.70	GRAM/MJ	18.08	
FAT	GRAM	12.60	- - -	12.60	GRAM/MJ	17.93	
MINERALS	GRAM	2.02	- - -	2.02	GRAM/MJ	2.88	
SODIUM	MILLI	594.00	575.00 - 613.00	594.00	MILLI/MJ	845.47	
POTASSIUM	MILLI	159.00	157.00 - 160.00	159.00	MILLI/MJ	226.31	

HERING MARINIERT (BISMARCKHERING)	HERRING VINEGAR CURED (BISMARCKHERING)	HARENG MARINÉ (HARENG BISMARCK)

		PROTEIN	FAT	CARBOHYDRATES	TOTAL
ENERGY VALUE (AVERAGE) PER 100 G EDIBLE PORTION	KJOULE	304	636	0.0	940
	(KCAL)	73	152	0.0	225
AMOUNT OF DIGESTIBLE CONSTITUENTS PER 100 G	GRAM	16.00	15.20	0.00	
ENERGY VALUE (AVERAGE) OF THE DIGESTIBLE FRACTION PER 100 G EDIBLE PORTION	KJOULE	295	604	0.0	899
	(KCAL)	70	144	0.0	215

WASTE PERCENTAGE AVERAGE 5.0 MINIMUM 2.0 MAXIMUM 7.0

CONSTITUENTS	DIM	AV	VARIATION			AVR	NUTR. DENS.		MOLPERC.
WATER	GRAM	62.20	55.00	-	69.20	59.09	GRAM/MJ	69.20	
PROTEIN	GRAM	16.50	15.80	-	18.30	15.68	GRAM/MJ	18.36	
FAT	GRAM	16.00	6.81	-	24.30	15.20	GRAM/MJ	17.80	
MINERALS	GRAM	2.75	2.22	-	3.71	2.61	GRAM/MJ	3.06	
SODIUM	GRAM	1.03	0.67	-	1.51	0.98	GRAM/MJ	1.15	
POTASSIUM	MILLI	98.00	78.00	-	108.00	93.10	MILLI/MJ	109.03	
MAGNESIUM	MILLI	12.00	9.00	-	18.00	11.40	MILLI/MJ	13.35	
CALCIUM	MILLI	38.00	28.00	-	56.00	36.10	MILLI/MJ	42.28	
PHOSPHORUS	MILLI	149.00	147.00	-	150.00	141.55	MILLI/MJ	165.77	
CHLORIDE	GRAM	1.52	0.97	-	2.31	1.44	GRAM/MJ	1.69	
IODIDE	MICRO	-	6.00	-	77.00				
VITAMIN A	MICRO	36.00	30.00	-	46.00	34.20	MICRO/MJ	40.05	
VITAMIN D	MICRO	13.00	-	-	-	12.35	MICRO/MJ	14.46	
VITAMIN B1	MICRO	50.00	-	-	-	47.50	MICRO/MJ	55.63	
VITAMIN B2	MILLI	0.21	0.17	-	0.25	0.20	MILLI/MJ	0.23	
VITAMIN B6	MILLI	0.15	0.12	-	0.17	0.14	MILLI/MJ	0.17	
ARGININE	GRAM	1.02	0.91	-	1.21	0.97	GRAM/MJ	1.13	
CYSTINE	GRAM	0.20	0.17	-	0.24	0.19	GRAM/MJ	0.22	
HISTIDINE	GRAM	0.44	0.40	-	0.51	0.42	GRAM/MJ	0.49	
ISOLEUCINE	GRAM	0.83	0.77	-	0.87	0.79	GRAM/MJ	0.92	
LEUCINE	GRAM	1.31	1.17	-	1.48	1.24	GRAM/MJ	1.46	
LYSINE	GRAM	1.51	1.29	-	1.64	1.43	GRAM/MJ	1.68	
METHIONINE	GRAM	0.49	0.44	-	0.58	0.47	GRAM/MJ	0.55	
PHENYLALANINE	GRAM	0.66	0.56	-	0.78	0.63	GRAM/MJ	0.73	
THREONINE	GRAM	0.79	0.72	-	0.88	0.75	GRAM/MJ	0.88	
TRYPTOPHAN	GRAM	0.17	0.14	-	0.24	0.16	GRAM/MJ	0.19	
TYROSINE	GRAM	0.51	0.44	-	0.61	0.48	GRAM/MJ	0.57	
VALINE	GRAM	0.92	0.83	-	1.02	0.87	GRAM/MJ	1.02	
ACETIC ACID	GRAM	1.94	1.68	-	2.30	1.84	GRAM/MJ	2.16	

KATFISCH (STEINBEISSER) GERÄUCHERT

CATFISH SMOKED

LOUP FUMÉ

		PROTEIN	FAT	CARBOHYDRATES	TOTAL
ENERGY VALUE (AVERAGE) PER 100 G EDIBLE PORTION	KJOULE	423	142	0.0	566
	(KCAL)	101	34	0.0	135
AMOUNT OF DIGESTIBLE CONSTITUENTS PER 100 G	GRAM	22.31	3.40	0.00	
ENERGY VALUE (AVERAGE) OF THE DIGESTIBLE FRACTION PER 100 G EDIBLE PORTION	KJOULE	411	135	0.0	546
	(KCAL)	98	32	0.0	130

WASTE PERCENTAGE AVERAGE 19 MINIMUM 15 MAXIMUM 23

CONSTITUENTS	DIM	AV	VARIATION	AVR	NUTR. DENS.		MOLPERC.
WATER	GRAM	71.20	69.50 - 72.10	57.67	GRAM/MJ	130.43	
PROTEIN	GRAM	23.00	21.60 - 24.40	18.63	GRAM/MJ	42.13	
FAT	GRAM	3.58	2.80 - 4.35	2.90	GRAM/MJ	6.56	
MINERALS	GRAM	1.76	0.81 - 2.99	1.43	GRAM/MJ	3.22	
SODIUM	MILLI	701.00	615.00 - 785.00	567.81	GRAM/MJ	1.28	
POTASSIUM	MILLI	409.00	371.00 - 440.00	331.29	MILLI/MJ	749.22	

KAVIAR
ECHT
(RUSSISCHER KAVIAR)

CAVIAR
REAL

CAVIAR
VÉRITABLE
(CAVIAR RUSSE)

		PROTEIN	FAT	CARBOHYDRATES	TOTAL
ENERGY VALUE (AVERAGE)	KJOULE	480	616	0.0	1097
PER 100 G	(KCAL)	115	147	0.0	262
EDIBLE PORTION					
AMOUNT OF DIGESTIBLE	GRAM	25.31	14.72	0.00	
CONSTITUENTS PER 100 G					
ENERGY VALUE (AVERAGE)	KJOULE	466	585	0.0	1051
OF THE DIGESTIBLE	(KCAL)	111	140	0.0	251
FRACTION PER 100 G					
EDIBLE PORTION					

WASTE PERCENTAGE AVERAGE 0.00

CONSTITUENTS	DIM	AV	VARIATION			AVR	NUTR.	DENS.	MOLPERC.
WATER	GRAM	47.10	36.60	-	50.30	47.10	GRAM/MJ	44.80	
PROTEIN	GRAM	26.10	25.40	-	26.90	26.10	GRAM/MJ	24.82	
FAT	GRAM	15.50	14.20	-	16.70	15.50	GRAM/MJ	14.74	
MINERALS	GRAM	6.73	5.36	-	8.09	6.73	GRAM/MJ	6.40	
SODIUM	GRAM	1.94	1.55	-	2.46	1.94	GRAM/MJ	1.85	
POTASSIUM	MILLI	164.00	154.00	-	173.00	164.00	MILLI/MJ	155.99	
CALCIUM	MILLI	51.00	23.00	-	71.00	51.00	MILLI/MJ	48.51	
MANGANESE	MICRO	50.00	-	-	-	50.00	MICRO/MJ	47.56	
IRON	MILLI	1.40	-	-	-	1.40	MILLI/MJ	1.33	
COBALT	MICRO	20.00	-	-	-	20.00	MICRO/MJ	19.02	
COPPER	MILLI	0.11	-	-	-	0.11	MILLI/MJ	0.10	
ZINC	MILLI	0.95	-	-	-	0.95	MILLI/MJ	0.90	
PHOSPHORUS	MILLI	300.00	-	-	-	300.00	MILLI/MJ	285.34	
VITAMIN A	MILLI	0.56	-	-	-	0.56	MILLI/MJ	0.53	
VITAMIN D	MICRO	5.87	-	-	-	5.87	MICRO/MJ	5.58	
ALANINE	GRAM	1.63	-	-	-	1.63	GRAM/MJ	1.55	
ARGININE	GRAM	1.77	-	-	-	1.77	GRAM/MJ	1.68	
ASPARTIC ACID	GRAM	2.29	-	-	-	2.29	GRAM/MJ	2.18	
GLUTAMIC ACID	GRAM	3.70	-	-	-	3.70	GRAM/MJ	3.52	
GLYCINE	GRAM	0.83	-	-	-	0.83	GRAM/MJ	0.79	
HISTIDINE	GRAM	0.82	-	-	-	0.82	GRAM/MJ	0.78	
ISOLEUCINE	GRAM	1.47	-	-	-	1.47	GRAM/MJ	1.40	
LEUCINE	GRAM	2.30	-	-	-	2.30	GRAM/MJ	2.19	
LYSINE	GRAM	2.07	-	-	-	2.07	GRAM/MJ	1.97	
METHIONINE	GRAM	0.80	-	-	-	0.80	GRAM/MJ	0.76	
PHENYLALANINE	GRAM	1.06	-	-	-	1.06	GRAM/MJ	1.01	
PROLINE	GRAM	1.15	-	-	-	1.15	GRAM/MJ	1.09	
SERINE	GRAM	2.00	-	-	-	2.00	GRAM/MJ	1.90	
THREONINE	GRAM	1.35	-	-	-	1.35	GRAM/MJ	1.28	
TYROSINE	GRAM	1.13	-	-	-	1.13	GRAM/MJ	1.07	
VALINE	GRAM	1.60	-	-	-	1.60	GRAM/MJ	1.52	
CHOLESTEROL	MILLI	300.00	-	-	-	300.00	MILLI/MJ	285.34	
TOTAL PURINES	MILLI	144.00 [1]	-	-	-	144.00	MILLI/MJ	136.96	

1 VALUE EXPRESSED IN MG URIC ACID/100G

KAVIAR-ERSATZ (DEUTSCHER KAVIAR) | CAVIAR SUBSTITUTE | SUCCÉDANÉRS DE CAVIAR PRODUIT DE REMPLACEMENT (CAVIAR ALLEMAND)

		PROTEIN	FAT	CARBOHYDRATES	TOTAL
ENERGY VALUE (AVERAGE)	KJOULE	258	256	0.0	514
PER 100 G	(KCAL)	62	61	0.0	123
EDIBLE PORTION					
AMOUNT OF DIGESTIBLE	GRAM	13.58	6.12	0.00	
CONSTITUENTS PER 100 G					
ENERGY VALUE (AVERAGE)	KJOULE	250	244	0.0	494
OF THE DIGESTIBLE	(KCAL)	60	58	0.0	118
FRACTION PER 100 G					
EDIBLE PORTION					

WASTE PERCENTAGE AVERAGE 0.00

CONSTITUENTS	DIM	AV	VARIATION	AVR	NUTR. DENS.		MOLPERC.
WATER	GRAM	71.20	70.10 - 72.90	71.20	GRAM/MJ	144.26	
PROTEIN	GRAM	14.00	13.60 - 14.30	14.00	GRAM/MJ	28.37	
FAT	GRAM	6.45	4.93 - 7.52	6.45	GRAM/MJ	13.07	
MINERALS	GRAM	5.52	4.50 - 6.54	5.52	GRAM/MJ	11.18	
SODIUM	GRAM	2.12	1.76 - 2.46	2.12	GRAM/MJ	4.30	
POTASSIUM	MILLI	101.00	70.00 - 154.00	101.00	MILLI/MJ	204.64	
CALCIUM	MILLI	51.00	23.00 - 71.00	51.00	MILLI/MJ	103.33	

KLIPPFISCH KLIPPFISH KLIPPFISH

		PROTEIN	FAT	CARBOHYDRATES	TOTAL
ENERGY VALUE (AVERAGE)	KJOULE	823	29	0.0	852
PER 100 G	(KCAL)	197	6.8	0.0	204
EDIBLE PORTION					
AMOUNT OF DIGESTIBLE	GRAM	43.35	0.68	0.00	
CONSTITUENTS PER 100 G					
ENERGY VALUE (AVERAGE)	KJOULE	798	27	0.0	825
OF THE DIGESTIBLE	(KCAL)	191	6.5	0.0	197
FRACTION PER 100 G					
EDIBLE PORTION					

WASTE PERCENTAGE AVERAGE 1.0

CONSTITUENTS	DIM	AV	VARIATION			AVR	NUTR. DENS.		MOLPERC.
WATER	GRAM	34.10	-	-	-	33.76	GRAM/MJ	41.31	
PROTEIN	GRAM	44.70	-	-	-	44.25	GRAM/MJ	54.15	
FAT	GRAM	0.72	-	-	-	0.71	GRAM/MJ	0.87	
MINERALS	GRAM	20.40	-	-	-	20.20	GRAM/MJ	24.72	
CALCIUM	MILLI	60.00	-	-	-	59.40	MILLI/MJ	72.69	
IRON	MILLI	1.60	-	-	-	1.58	MILLI/MJ	1.94	
PHOSPHORUS	MILLI	300.00	-	-	-	297.00	MILLI/MJ	363.46	
VITAMIN B1	MICRO	20.00	-	-	-	19.80	MICRO/MJ	24.23	
VITAMIN B2	MILLI	0.27	0.23	-	0.30	0.27	MILLI/MJ	0.33	
NICOTINAMIDE	MILLI	2.90	2.40	-	3.40	2.87	MILLI/MJ	3.51	
PANTOTHENIC ACID	MILLI	0.34	-	-	-	0.34	MILLI/MJ	0.41	
VITAMIN B12	MICRO	3.60	-	-	-	3.56	MICRO/MJ	4.36	
CYSTINE	GRAM	0.78	-	-	-	0.77	GRAM/MJ	0.94	
LYSINE	GRAM	4.49	-	-	-	4.45	GRAM/MJ	5.44	
METHIONINE	GRAM	1.34	-	-	-	1.33	GRAM/MJ	1.62	
TRYPTOPHAN	GRAM	0.67	-	-	-	0.66	GRAM/MJ	0.81	

KREBSFLEISCH	CRAYFISH MEAT	ECREVISSE
IN DOSEN	CANNED	EN BOÎTES
1	1	1

		PROTEIN	FAT	CARBOHYDRATES	TOTAL
ENERGY VALUE (AVERAGE)	KJOULE	331	66	0.0	398
PER 100 G	(KCAL)	79	16	0.0	95
EDIBLE PORTION					
AMOUNT OF DIGESTIBLE	GRAM	17.46	1.58	0.00	
CONSTITUENTS PER 100 G					
ENERGY VALUE (AVERAGE)	KJOULE	321	63	0.0	384
OF THE DIGESTIBLE	(KCAL)	77	15	0.0	92
FRACTION PER 100 G					
EDIBLE PORTION					

WASTE PERCENTAGE AVERAGE 0.00

CONSTITUENTS	DIM	AV	VARIATION			AVR	NUTR. DENS.		MOLPERC.
WATER	GRAM	78.40	76.60	-	80.10	78.40	GRAM/MJ	203.91	
PROTEIN	GRAM	18.00	16.30	-	20.40	18.00	GRAM/MJ	46.82	
FAT	GRAM	1.67	0.84	-	2.50	1.67	GRAM/MJ	4.34	
MINERALS	GRAM	1.95	1.80	-	2.09	1.95	GRAM/MJ	5.07	
SODIUM	MILLI	356.00	300.00	-	441.00	356.00	MILLI/MJ	925.90	
POTASSIUM	MILLI	296.00	236.00	-	322.00	296.00	MILLI/MJ	769.85	
CALCIUM	MILLI	45.00	-	-	-	45.00	MILLI/MJ	117.04	
IRON	MILLI	0.80	-	-	-	0.80	MILLI/MJ	2.08	
COPPER	MILLI	1.57	0.93	-	2.21	1.57	MILLI/MJ	4.08	
PHOSPHORUS	MILLI	180.00	-	-	-	180.00	MILLI/MJ	468.15	
VITAMIN B1	MILLI	0.14	0.08	-	0.18	0.14	MILLI/MJ	0.36	
VITAMIN B2	MICRO	51.00	16.00	-	84.00	51.00	MICRO/MJ	132.64	
NICOTINAMIDE	MILLI	1.64	1.37	-	1.90	1.64	MILLI/MJ	4.27	
PANTOTHENIC ACID	MILLI	0.49	-	-	-	0.49	MILLI/MJ	1.27	
BIOTIN	MICRO	4.60	-	-	-	4.60	MICRO/MJ	11.96	
FOLIC ACID	MICRO	0.30	-	-	-	0.30	MICRO/MJ	0.78	
VITAMIN B12	MICRO	0.46	-	-	-	0.46	MICRO/MJ	1.20	

1 TOTAL CANNED AMOUNT

LACHS (SALM) IN DOSEN 1	**SALMON** CANNED 1	**SAUMON** EN BOÎTES 1

		PROTEIN	FAT	CARBOHYDRATES	TOTAL
ENERGY VALUE (AVERAGE) PER 100 G EDIBLE PORTION	KJOULE	388	355	0.0	743
	(KCAL)	93	85	0.0	178
AMOUNT OF DIGESTIBLE CONSTITUENTS PER 100 G	GRAM	20.46	8.48	0.00	
ENERGY VALUE (AVERAGE) OF THE DIGESTIBLE FRACTION PER 100 G EDIBLE PORTION	KJOULE	377	337	0.0	714
	(KCAL)	90	81	0.0	171

WASTE PERCENTAGE AVERAGE 2.0

CONSTITUENTS	DIM	AV	VARIATION		AVR	NUTR.	DENS.	MOLPERC.
WATER	GRAM	67.00	64.70	- 69.20	65.66	GRAM/MJ	93.84	
PROTEIN	GRAM	21.10	19.70	- 22.70	20.68	GRAM/MJ	29.55	
FAT	GRAM	8.93	4.00	- 13.20	8.75	GRAM/MJ	12.51	
MINERALS	GRAM	2.53	2.30	- 2.90	2.48	GRAM/MJ	3.54	
SODIUM	MILLI	540.00	-	- -	529.20	MILLI/MJ	756.31	
POTASSIUM	MILLI	300.00	-	- -	294.00	MILLI/MJ	420.17	
CALCIUM	MILLI	185.00	154.00	- 216.00	181.30	MILLI/MJ	259.11	
IRON	MILLI	1.07	0.90	- 1.30	1.05	MILLI/MJ	1.50	
PHOSPHORUS	MILLI	292.00	283.00	- 305.00	286.16	MILLI/MJ	408.97	
VITAMIN A	MICRO	59.00	24.00	- 84.00	57.82	MICRO/MJ	82.63	
VITAMIN D	MICRO	11.50	4.70	- 19.00	11.27	MICRO/MJ	16.11	
VITAMIN B1	MICRO	30.00	18.00	- 52.00	29.40	MICRO/MJ	42.02	
VITAMIN B2	MILLI	0.17	0.14	- 0.20	0.17	MILLI/MJ	0.24	
NICOTINAMIDE	MILLI	6.81	5.63	- 7.40	6.67	MILLI/MJ	9.54	
PANTOTHENIC ACID	MILLI	0.58	0.57	- 0.58	0.57	MILLI/MJ	0.81	
VITAMIN B6	MILLI	0.45	-	- -	0.44	MILLI/MJ	0.63	
BIOTIN	MICRO	15.00	-	- -	14.70	MICRO/MJ	21.01	
FOLIC ACID	MICRO	38.00	26.00	- 50.00	37.24	MICRO/MJ	53.22	
VITAMIN B12	MICRO	4.50	2.07	- 6.89	4.41	MICRO/MJ	6.30	
VITAMIN C	-	0.00	-	- -	0.00			
ARGININE	GRAM	1.16	-	- -	1.14	GRAM/MJ	1.62	
HISTIDINE	GRAM	0.27	-	- -	0.26	GRAM/MJ	0.38	
ISOLEUCINE	GRAM	1.03	-	- -	1.01	GRAM/MJ	1.44	
LEUCINE	GRAM	1.54	-	- -	1.51	GRAM/MJ	2.16	
LYSINE	GRAM	1.20	-	- -	1.18	GRAM/MJ	1.68	
METHIONINE	GRAM	0.63	-	- -	0.62	GRAM/MJ	0.88	
THREONINE	GRAM	0.92	-	- -	0.90	GRAM/MJ	1.29	
TRYPTOPHAN	GRAM	0.19	-	- -	0.19	GRAM/MJ	0.27	
VALINE	GRAM	1.18	-	- -	1.16	GRAM/MJ	1.65	
CADAVERINE	MILLI	-	0.00	- 0.10				
PUTRESCINE	MILLI	0.35	-	- -	0.34	MILLI/MJ	0.49	

1 TOTAL CANNED AMOUNT

LACHS (SALM) IN ÖL	SALMON IN OIL	SAUMON À L'HUILE

		PROTEIN	FAT	CARBOHYDRATES	TOTAL
ENERGY VALUE (AVERAGE)	KJOULE	302	906	0.0	1208
PER 100 G	(KCAL)	72	217	0.0	289
EDIBLE PORTION					
AMOUNT OF DIGESTIBLE	GRAM	15.90	21.66	0.00	
CONSTITUENTS PER 100 G					
ENERGY VALUE (AVERAGE)	KJOULE	293	861	0.0	1154
OF THE DIGESTIBLE	(KCAL)	70	206	0.0	276
FRACTION PER 100 G					
EDIBLE PORTION					

WASTE PERCENTAGE AVERAGE 0.00

CONSTITUENTS	DIM	AV	VARIATION			AVR	NUTR. DENS.		MOLPERC.
WATER	GRAM	49.40	-	-	-	49.40	GRAM/MJ	42.81	
PROTEIN	GRAM	16.40	-	-	-	16.40	GRAM/MJ	14.21	
FAT	GRAM	22.80	-	-	-	22.80	GRAM/MJ	19.76	
MINERALS	GRAM	10.80	-	-	-	10.80	GRAM/MJ	9.36	
SODIUM	GRAM	4.07	-	-	-	4.07	GRAM/MJ	3.53	
POTASSIUM	MILLI	282.00	-	-	-	282.00	MILLI/MJ	244.41	
CHLORIDE	GRAM	6.30	-	-	-	6.30	GRAM/MJ	5.46	

MAKRELE	**MACKEREL**	**MAQUEREAU**
GERÄUCHERT	SMOKED	FUMÉ

		PROTEIN	FAT	CARBOHYDRATES	TOTAL
ENERGY VALUE (AVERAGE)	KJOULE	381	616	0.0	997
PER 100 G	(KCAL)	91	147	0.0	238
EDIBLE PORTION					
AMOUNT OF DIGESTIBLE	GRAM	20.07	14.72	0.00	
CONSTITUENTS PER 100 G					
ENERGY VALUE (AVERAGE)	KJOULE	370	585	0.0	955
OF THE DIGESTIBLE	(KCAL)	88	140	0.0	228
FRACTION PER 100 G					
EDIBLE PORTION					

WASTE PERCENTAGE AVERAGE 31 MINIMUM 30 MAXIMUM 31

CONSTITUENTS	DIM	AV	VARIATION		AVR	NUTR. DENS.		MOLPERC.
WATER	GRAM	62.30	58.90	- 63.80	42.99	GRAM/MJ	65.24	
PROTEIN	GRAM	20.70	18.50	- 21.80	14.28	GRAM/MJ	21.68	
FAT	GRAM	15.50	11.00	- 20.20	10.70	GRAM/MJ	16.23	
MINERALS	GRAM	1.32	1.15	- 1.52	0.91	GRAM/MJ	1.38	
SODIUM	MILLI	261.00	109.00	- 469.00	180.09	MILLI/MJ	273.32	
POTASSIUM	MILLI	275.00	215.00	- 396.00	189.75	MILLI/MJ	287.98	
CALCIUM	MILLI	5.00	-	- -	3.45	MILLI/MJ	5.24	
IRON	MILLI	1.20	-	- -	0.83	MILLI/MJ	1.26	
PHOSPHORUS	MILLI	240.00	-	- -	165.60	MILLI/MJ	251.33	
VITAMIN A	MICRO	60.00	-	- -	41.40	MICRO/MJ	62.83	
VITAMIN B1	MILLI	0.14	0.10	- 0.20	0.10	MILLI/MJ	0.15	
VITAMIN B2	MILLI	0.35	0.22	- 0.47	0.24	MILLI/MJ	0.37	
NICOTINAMIDE	MILLI	10.00	5.90	- 13.00	6.90	MILLI/MJ	10.47	
VITAMIN B6	MILLI	0.50	-	- -	0.35	MILLI/MJ	0.52	
TOTAL PURINES	MILLI	140.00 [1]	140.00	- 140.00	96.60	MILLI/MJ	146.61	
ADENINE	MILLI	10.00	10.00	- 10.00	6.90	MILLI/MJ	10.47	
GUANINE	MILLI	30.00	30.00	- 30.00	20.70	MILLI/MJ	31.42	
HYPOXANTHINE	MILLI	94.00	94.00	- 94.00	64.86	MILLI/MJ	98.44	

1 VALUE OF TOTAL PURINES IS EXPRESSSED IN MG URIC ACID/100G

MATJESHERING | MATJE CURED HERRING | MATJE HARENG

		PROTEIN	FAT	CARBOHYDRATES	TOTAL
ENERGY VALUE (AVERAGE)	KJOULE	295	898	0.0	1193
PER 100 G	(KCAL)	70	215	0.0	285
EDIBLE PORTION					
AMOUNT OF DIGESTIBLE	GRAM	15.52	21.47	0.00	
CONSTITUENTS PER 100 G					
ENERGY VALUE (AVERAGE)	KJOULE	286	853	0.0	1139
OF THE DIGESTIBLE	(KCAL)	68	204	0.0	272
FRACTION PER 100 G					
EDIBLE PORTION					

WASTE PERCENTAGE AVERAGE 0.00

CONSTITUENTS	DIM	AV	VARIATION		AVR	NUTR. DENS.		MOLPERC.
WATER	GRAM	54.40	-	- -	54.40	GRAM/MJ	47.76	
PROTEIN	GRAM	16.00	-	- -	16.00	GRAM/MJ	14.05	
FAT	GRAM	22.60	-	- -	22.60	GRAM/MJ	19.84	
MINERALS	GRAM	6.80	-	- -	6.80	GRAM/MJ	5.97	
SODIUM	GRAM	2.50	-	- -	2.50	GRAM/MJ	2.19	
MAGNESIUM	MILLI	35.00	-	- -	35.00	MILLI/MJ	30.73	
CALCIUM	MILLI	43.00	-	- -	43.00	MILLI/MJ	37.75	
IRON	MILLI	1.30	-	- -	1.30	MILLI/MJ	1.14	
COPPER	MILLI	0.44	-	- -	0.44	MILLI/MJ	0.39	
PHOSPHORUS	MILLI	200.00	-	- -	200.00	MILLI/MJ	175.58	
CHLORIDE	GRAM	3.90	-	- -	3.90	GRAM/MJ	3.42	
TOTAL PURINES	MILLI	219.00 1	219.00	- 219.00	219.00	MILLI/MJ	192.26	
ADENINE	MILLI	13.00	13.00	- 13.00	13.00	MILLI/MJ	11.41	
GUANINE	MILLI	35.00	35.00	- 35.00	35.00	MILLI/MJ	30.73	
HYPOXANTHINE	MILLI	133.00	133.00	- 133.00	133.00	MILLI/MJ	116.76	

1 VALUE OF TOTAL PURINES IS EXPRESSSED IN MG URIC ACID/100G

ROTBARSCH	REDFISH	SÉBASTE
GERÄUCHERT	(RED PERCH) SMOKED	FUMÉ

		PROTEIN	FAT	CARBOHYDRATES	TOTAL
ENERGY VALUE (AVERAGE)	KJOULE	438	219	0.0	657
PER 100 G	(KCAL)	105	52	0.0	157
EDIBLE PORTION					
AMOUNT OF DIGESTIBLE	GRAM	23.08	5.22	0.00	
CONSTITUENTS PER 100 G					
ENERGY VALUE (AVERAGE)	KJOULE	425	208	0.0	633
OF THE DIGESTIBLE	(KCAL)	102	50	0.0	151
FRACTION PER 100 G					
EDIBLE PORTION					

WASTE PERCENTAGE AVERAGE 27 MINIMUM 20 MAXIMUM 31

CONSTITUENTS	DIM	AV	VARIATION			AVR	NUTR.	DENS.	MOLPERC.
WATER	GRAM	68.40	61.00	-	72.40	49.93	GRAM/MJ	108.11	
PROTEIN	GRAM	23.80	22.80	-	25.80	17.37	GRAM/MJ	37.62	
FAT	GRAM	5.50	2.90	-	8.20	4.02	GRAM/MJ	8.69	
MINERALS	GRAM	2.09	1.89	-	2.36	1.53	GRAM/MJ	3.30	
SODIUM	GRAM	0.55	0.32	-	1.08	0.40	GRAM/MJ	0.87	
POTASSIUM	MILLI	367.00	282.00	-	445.00	267.91	MILLI/MJ	580.07	
CALCIUM	MILLI	25.00	-	-	-	18.25	MILLI/MJ	39.51	
IRON	MILLI	4.70	-	-	-	3.43	MILLI/MJ	7.43	
PHOSPHORUS	MILLI	230.00	-	-	-	167.90	MILLI/MJ	363.53	
IODIDE	MICRO	20.00	18.00	-	21.00	14.60	MICRO/MJ	31.61	

SALZHERING (PÖKELHERING) | PICKLED HERRING | HARENG SAUMURE

		PROTEIN	FAT	CARBOHYDRATES	TOTAL
ENERGY VALUE (AVERAGE) PER 100 G EDIBLE PORTION	KJOULE	365	612	0.0	977
	(KCAL)	87	146	0.0	233
AMOUNT OF DIGESTIBLE CONSTITUENTS PER 100 G	GRAM	19.20	14.63	0.00	
ENERGY VALUE (AVERAGE) OF THE DIGESTIBLE FRACTION PER 100 G EDIBLE PORTION	KJOULE	354	582	0.0	935
	(KCAL)	85	139	0.0	223

WASTE PERCENTAGE AVERAGE 57 MINIMUM 46 MAXIMUM 68

CONSTITUENTS	DIM	AV	VARIATION			AVR	NUTR. DENS.		MOLPERC.
WATER	GRAM	48.80	44.00	-	52.50	20.98	GRAM/MJ	52.19	
PROTEIN	GRAM	19.80	18.00	-	23.00	8.51	GRAM/MJ	21.17	
FAT	GRAM	15.40	11.00	-	25.00	6.62	GRAM/MJ	16.47	
MINERALS	GRAM	16.00	-	-	-	6.88	GRAM/MJ	17.11	
SODIUM	GRAM	5.93	5.31	-	6.29	2.55	GRAM/MJ	6.34	
POTASSIUM	MILLI	240.00	-	-	-	103.20	MILLI/MJ	256.66	
MAGNESIUM	MILLI	39.00	-	-	-	16.77	MILLI/MJ	41.71	
CALCIUM	MILLI	112.00	-	-	-	48.16	MILLI/MJ	119.77	
IRON	MILLI	20.00	-	-	-	8.60	MILLI/MJ	21.39	
PHOSPHORUS	MILLI	341.00	-	-	-	146.63	MILLI/MJ	364.67	
CHLORIDE	GRAM	9.17	8.19	-	9.71	3.94	GRAM/MJ	9.81	
VITAMIN A	MICRO	48.00	30.00	-	81.00	20.64	MICRO/MJ	51.33	
VITAMIN B1	MICRO	35.00	24.00	-	40.00	15.05	MICRO/MJ	37.43	
VITAMIN B2	MILLI	0.29	0.16	-	0.35	0.12	MILLI/MJ	0.31	
NICOTINAMIDE	MILLI	3.00	2.08	-	3.38	1.29	MILLI/MJ	3.21	
VITAMIN B6	MILLI	0.22	0.18	-	0.30	0.09	MILLI/MJ	0.24	
VITAMIN B12	MICRO	6.00	2.70	-	9.20	2.58	MICRO/MJ	6.42	
VITAMIN C	-	0.00	-	-	-	0.00			
ARGININE	GRAM	1.12	-	-	-	0.48	GRAM/MJ	1.20	
CYSTINE	GRAM	0.27	-	-	-	0.12	GRAM/MJ	0.29	
ISOLEUCINE	GRAM	1.00	-	-	-	0.43	GRAM/MJ	1.07	
LEUCINE	GRAM	1.50	-	-	-	0.65	GRAM/MJ	1.60	
LYSINE	GRAM	1.74	-	-	-	0.75	GRAM/MJ	1.86	
METHIONINE	GRAM	0.58	-	-	-	0.25	GRAM/MJ	0.62	
PHENYLALANINE	GRAM	0.74	-	-	-	0.32	GRAM/MJ	0.79	
THREONINE	GRAM	0.86	-	-	-	0.37	GRAM/MJ	0.92	
TRYPTOPHAN	GRAM	0.20	-	-	-	0.09	GRAM/MJ	0.21	
TYROSINE	GRAM	0.54	-	-	-	0.23	GRAM/MJ	0.58	
VALINE	GRAM	1.06	-	-	-	0.46	GRAM/MJ	1.13	
TYRAMINE	MILLI	-	0.00	-	300.00				

SARDINEN	SARDINES	SARDINES
IN ÖL	IN OIL	À L'HUILE
1	1	1

		PROTEIN	FAT	CARBOHYDRATES	TOTAL
ENERGY VALUE (AVERAGE)	KJOULE	444	552	0.0	996
PER 100 G	(KCAL)	106	132	0.0	238
EDIBLE PORTION					
AMOUNT OF DIGESTIBLE	GRAM	23.37	13.20	0.00	
CONSTITUENTS PER 100 G					
ENERGY VALUE (AVERAGE)	KJOULE	430	525	0.0	955
OF THE DIGESTIBLE	(KCAL)	103	125	0.0	228
FRACTION PER 100 G					
EDIBLE PORTION					

WASTE PERCENTAGE AVERAGE 0.00

CONSTITUENTS	DIM	AV	VARIATION			AVR	NUTR. DENS.		MOLPERC.
WATER	GRAM	55.60	50.00	-	64.50	55.60	GRAM/MJ	58.21	
PROTEIN	GRAM	24.10	23.00	-	25.70	24.10	GRAM/MJ	25.23	
FAT	GRAM	13.90	8.30	-	14.40	13.90	GRAM/MJ	14.55	
MINERALS	GRAM	3.62	3.04	-	4.12	3.62	GRAM/MJ	3.79	
SODIUM	MILLI	505.00	395.00	-	614.00	505.00	MILLI/MJ	528.67	
POTASSIUM	MILLI	397.00	393.00	-	400.00	397.00	MILLI/MJ	415.61	
CALCIUM	MILLI	330.00	174.00	-	460.00	330.00	MILLI/MJ	345.47	
IRON	MILLI	2.70	2.16	-	3.58	2.70	MILLI/MJ	2.83	
COPPER	MICRO	40.00	-	-	-	40.00	MICRO/MJ	41.87	
PHOSPHORUS	MILLI	430.00	290.00	-	590.00	430.00	MILLI/MJ	450.15	
CHLORIDE	GRAM	1.07	0.80	-	1.56	1.07	GRAM/MJ	1.12	
FLUORIDE	MILLI	1.60	1.10	-	4.50	1.60	MILLI/MJ	1.67	
IODIDE	MICRO	24.00	13.00	-	54.00	24.00	MICRO/MJ	25.12	
VITAMIN A	MICRO	58.00	18.00	-	90.00	58.00	MICRO/MJ	60.72	
VITAMIN B1	MICRO	40.00	20.00	-	70.00	40.00	MICRO/MJ	41.87	
VITAMIN B2	MILLI	0.30	0.14	-	0.38	0.30	MILLI/MJ	0.31	
NICOTINAMIDE	MILLI	6.48	3.88	-	10.50	6.48	MILLI/MJ	6.78	
VITAMIN B6	MILLI	0.22	0.16	-	0.24	0.22	MILLI/MJ	0.23	
FOLIC ACID	MICRO	16.00	-	-	-	16.00	MICRO/MJ	16.75	
VITAMIN B12	MICRO	0.16	0.14	-	0.18	0.16	MICRO/MJ	0.17	
VITAMIN C	-	0.00	-	-	-	0.00			
ARGININE	GRAM	1.33	-	-	-	1.33	GRAM/MJ	1.39	
CYSTINE	GRAM	0.24	0.20	-	0.36	0.24	GRAM/MJ	0.25	
HISTIDINE	GRAM	0.53	-	-	-	0.53	GRAM/MJ	0.55	
ISOLEUCINE	GRAM	1.12	1.11	-	1.13	1.12	GRAM/MJ	1.17	

1 ONLY THE DRAINED SOLIDS

CONSTITUENTS	DIM	AV	VARIATION			AVR	NUTR. DENS.		MOLPERC.
LEUCINE	GRAM	1.91	1.74	-	2.08	1.91	GRAM/MJ	2.00	
LYSINE	GRAM	2.24	1.90	-	2.57	2.24	GRAM/MJ	2.34	
METHIONINE	GRAM	0.74	0.65	-	0.81	0.74	GRAM/MJ	0.77	
PHENYLALANINE	GRAM	0.89	0.87	-	0.91	0.89	GRAM/MJ	0.93	
THREONINE	GRAM	1.13	1.09	-	1.14	1.13	GRAM/MJ	1.18	
TRYPTOPHAN	GRAM	0.21	0.19	-	0.23	0.21	GRAM/MJ	0.22	
TYROSINE	GRAM	0.73	-	-	-	0.73	GRAM/MJ	0.76	
VALINE	GRAM	1.30	1.23	-	1.36	1.30	GRAM/MJ	1.36	
PALMITIC ACID	GRAM	1.95	1.95	-	1.95	1.95	GRAM/MJ	2.04	
STEARIC ACID	GRAM	0.27	0.27	-	0.27	0.27	GRAM/MJ	0.28	
PALMITOLEIC ACID	GRAM	1.52	1.52	-	1.52	1.52	GRAM/MJ	1.59	
OLEIC ACID	GRAM	2.71	2.71	-	2.71	2.71	GRAM/MJ	2.84	
EICOSENOIC ACID	GRAM	1.57	1.57	-	1.57	1.57	GRAM/MJ	1.64	
LINOLEIC ACID	GRAM	0.30	0.30	-	0.30	0.30	GRAM/MJ	0.31	
LINOLENIC ACID	GRAM	0.15	0.15	-	0.15	0.15	GRAM/MJ	0.16	
ARACHIDONIC ACID	MILLI	90.00	90.00	-	90.00	90.00	MILLI/MJ	94.22	
C20 : 5 N-3 FATTY ACID	GRAM	1.20	1.20	-	1.20	1.20	GRAM/MJ	1.26	
C22 : 5 N-3 FATTY ACID	GRAM	0.23	0.23	-	0.23	0.23	GRAM/MJ	0.24	
C22 : 6 N-3 FATTY ACID	GRAM	1.24	1.24	-	1.24	1.24	GRAM/MJ	1.30	
CHOLESTEROL	MILLI	140.00	130.00	-	160.00	140.00	MILLI/MJ	146.56	
TOTAL PURINES	MILLI	480.00 2	399.00	-	560.00	480.00	MILLI/MJ	502.49	

2 VALUE EXPRESSED IN MG URIC ACID/100G

SCHELLFISCH GERÄUCHERT — HADDOCK SMOKED — AIGLEFIN FUMÉ

		PROTEIN	FAT	CARBOHYDRATES	TOTAL
ENERGY VALUE (AVERAGE) PER 100 G EDIBLE PORTION	KJOULE	407	19	0.0	426
	(KCAL)	97	4.5	0.0	102
AMOUNT OF DIGESTIBLE CONSTITUENTS PER 100 G	GRAM	21.43	0.44	0.00	
ENERGY VALUE (AVERAGE) OF THE DIGESTIBLE FRACTION PER 100 G EDIBLE PORTION	KJOULE	395	18	0.0	412
	(KCAL)	94	4.2	0.0	99

WASTE PERCENTAGE AVERAGE 40 MINIMUM 37 MAXIMUM 44

CONSTITUENTS	DIM	AV	VARIATION			AVR	NUTR.	DENS.	MOLPERC.
WATER	GRAM	75.30	74.90	-	76.20	45.18	GRAM/MJ	182.59	
PROTEIN	GRAM	22.10	21.10	-	23.00	13.26	GRAM/MJ	53.59	
FAT	GRAM	0.47	0.20	-	0.73	0.28	GRAM/MJ	1.14	
MINERALS	GRAM	2.04	1.70	-	2.32	1.22	GRAM/MJ	4.95	
SODIUM	MILLI	557.00	392.00	-	672.00	334.20	GRAM/MJ	1.35	
POTASSIUM	MILLI	300.00	281.00	-	321.00	180.00	MILLI/MJ	727.46	
MAGNESIUM	MILLI	25.00	-	-	-	15.00	MILLI/MJ	60.62	
CALCIUM	MILLI	20.00	-	-	-	12.00	MILLI/MJ	48.50	
IRON	MILLI	1.00	-	-	-	0.60	MILLI/MJ	2.42	
PHOSPHORUS	MILLI	262.00	248.00	-	275.00	157.20	MILLI/MJ	635.31	
VITAMIN A	-	TRACES	-	-	-				
VITAMIN D	-	0.00	-	-	-	0.00			
VITAMIN B1	MICRO	50.00	-	-	-	30.00	MICRO/MJ	121.24	
VITAMIN B2	MILLI	0.10	-	-	-	0.06	MILLI/MJ	0.24	
NICOTINAMIDE	MILLI	2.50	-	-	-	1.50	MILLI/MJ	6.06	
VITAMIN C	-	TRACES	-	-	-				

SCHILLERLOCKEN "SCHILLERLOCKEN" "SCHILLERLOCKEN"

1 1 1

		PROTEIN	FAT	CARBOHYDRATES	TOTAL
ENERGY VALUE (AVERAGE)	KJOULE	392	958	0.0	1350
PER 100 G	(KCAL)	94	229	0.0	323
EDIBLE PORTION					
AMOUNT OF DIGESTIBLE	GRAM	20.66	22.89	0.00	
CONSTITUENTS PER 100 G					
ENERGY VALUE (AVERAGE)	KJOULE	380	910	0.0	1290
OF THE DIGESTIBLE	(KCAL)	91	218	0.0	308
FRACTION PER 100 G					
EDIBLE PORTION					

WASTE PERCENTAGE AVERAGE 0.00

CONSTITUENTS	DIM	AV	VARIATION	AVR	NUTR. DENS.		MOLPERC.
WATER	GRAM	52.50	49.50 - 60.50	52.50	GRAM/MJ	40.69	
PROTEIN	GRAM	21.30	18.70 - 28.70	21.30	GRAM/MJ	16.51	
FAT	GRAM	24.10	22.40 - 30.40	24.10	GRAM/MJ	18.68	
MINERALS	GRAM	2.07	1.66 - 2.45	2.07	GRAM/MJ	1.60	
SODIUM	MILLI	704.00	672.00 - 735.00	704.00	MILLI/MJ	545.57	
POTASSIUM	MILLI	219.00	212.00 - 225.00	219.00	MILLI/MJ	169.72	
MAGNESIUM	MILLI	28.00	21.00 - 36.00	28.00	MILLI/MJ	21.70	
CALCIUM	MILLI	18.00	12.00 - 21.00	18.00	MILLI/MJ	13.95	
IRON	MILLI	1.09	0.96 - 1.19	1.09	MILLI/MJ	0.84	
PHOSPHORUS	MILLI	230.00	177.00 - 284.00	230.00	MILLI/MJ	178.24	
MYRISTIC ACID	GRAM	0.78	0.78 - 0.78	0.78	GRAM/MJ	0.60	
PALMITIC ACID	GRAM	5.01	5.01 - 5.01	5.01	GRAM/MJ	3.88	
STEARIC ACID	GRAM	0.61	0.61 - 0.61	0.61	GRAM/MJ	0.47	
PALMITOLEIC ACID	GRAM	1.74	1.74 - 1.74	1.74	GRAM/MJ	1.35	
OLEIC ACID	GRAM	5.53	5.53 - 5.53	5.53	GRAM/MJ	4.29	
EICOSENOIC ACID	GRAM	2.45	2.45 - 2.45	2.45	GRAM/MJ	1.90	
LINOLEIC ACID	GRAM	0.39	0.39 - 0.39	0.39	GRAM/MJ	0.30	
ARACHIDONIC ACID	GRAM	0.59	0.59 - 0.59	0.59	GRAM/MJ	0.46	
C20 : 5 N-3 FATTY ACID	GRAM	1.34	1.34 - 1.34	1.34	GRAM/MJ	1.04	
C22 : 6 N-3 FATTY ACID	GRAM	2.26	2.26 - 2.26	2.26	GRAM/MJ	1.75	

1 SMOKED BELLY OF PICKED DOGFISH

SEEAAL	**EEL MARITIME**	**ANGUILLE DE MER**
GERÄUCHERT	SMOKED	FUMÉ

		PROTEIN	FAT	CARBOHYDRATES	TOTAL
ENERGY VALUE (AVERAGE)	KJOULE	480	279	0.0	760
PER 100 G	(KCAL)	115	67	0.0	182
EDIBLE PORTION					
AMOUNT OF DIGESTIBLE	GRAM	25.31	6.66	0.00	
CONSTITUENTS PER 100 G					
ENERGY VALUE (AVERAGE)	KJOULE	466	265	0.0	731
OF THE DIGESTIBLE	(KCAL)	111	63	0.0	175
FRACTION PER 100 G					
EDIBLE PORTION					

WASTE PERCENTAGE AVERAGE 20 MINIMUM 14 MAXIMUM 32

CONSTITUENTS	DIM	AV	VARIATION		AVR	NUTR. DENS.		MOLPERC.
WATER	GRAM	64.40	62.10	- 67.30	51.52	GRAM/MJ	88.08	
PROTEIN	GRAM	26.10	23.80	- 29.00	20.88	GRAM/MJ	35.70	
FAT	GRAM	7.02	3.87	- 9.46	5.62	GRAM/MJ	9.60	
MINERALS	GRAM	2.30	1.80	- 3.39	1.84	GRAM/MJ	3.15	
SODIUM	MILLI	626.00	414.00	- 788.00	500.80	MILLI/MJ	856.18	
POTASSIUM	MILLI	311.00	243.00	- 396.00	248.80	MILLI/MJ	425.35	
MAGNESIUM	MILLI	34.00	25.00	- 40.00	27.20	MILLI/MJ	46.50	
CALCIUM	MILLI	20.00	17.00	- 25.00	16.00	MILLI/MJ	27.35	
IRON	MILLI	0.84	0.71	- 0.96	0.67	MILLI/MJ	1.15	
PHOSPHORUS	MILLI	260.00	242.00	- 278.00	208.00	MILLI/MJ	355.60	
BROMINE	MICRO	535.00	506.00	- 564.00	428.00	MICRO/MJ	731.72	

SEELACHS GERÄUCHERT — SAITHE SMOKED — LIEU NOIR FUMÉ

		PROTEIN	FAT	CARBOHYDRATES	TOTAL
ENERGY VALUE (AVERAGE)	KJOULE	420	32	0.0	452
PER 100 G	(KCAL)	100	7.7	0.0	108
EDIBLE PORTION					
AMOUNT OF DIGESTIBLE	GRAM	22.11	0.00	0.00	
CONSTITUENTS PER 100 G					
ENERGY VALUE (AVERAGE)	KJOULE	407	0.0	0.0	407
OF THE DIGESTIBLE	(KCAL)	97	0.0	0.0	97
FRACTION PER 100 G					
EDIBLE PORTION					

WASTE PERCENTAGE AVERAGE 18 MINIMUM 14 MAXIMUM 25

CONSTITUENTS	DIM	AV	VARIATION	AVR	NUTR. DENS.		MOLPERC.
WATER	GRAM	73.60	71.90 - 74.50	60.35	GRAM/MJ	180.77	
PROTEIN	GRAM	22.80	22.20 - 23.30	18.70	GRAM/MJ	56.00	
FAT	GRAM	0.81	0.60 - 1.02	0.66	GRAM/MJ	1.99	
MINERALS	GRAM	2.58	2.16 - 3.47	2.12	GRAM/MJ	6.34	
SODIUM	GRAM	0.64	0.53 - 1.11	0.53	GRAM/MJ	1.59	
POTASSIUM	MILLI	398.00	320.00 - 457.00	326.36	MILLI/MJ	977.53	

SPROTTE GERÄUCHERT	**SPRAT** SMOKED	**ESPROT** FUMÉ

		PROTEIN	FAT	CARBOHYDRATES	TOTAL
ENERGY VALUE (AVERAGE)	KJOULE	357	731	0.0	1089
PER 100 G EDIBLE PORTION	(KCAL)	85	175	0.0	260
AMOUNT OF DIGESTIBLE CONSTITUENTS PER 100 G	GRAM	18.81	17.48	0.00	
ENERGY VALUE (AVERAGE)	KJOULE	346	695	0.0	1041
OF THE DIGESTIBLE FRACTION PER 100 G EDIBLE PORTION	(KCAL)	83	166	0.0	249

WASTE PERCENTAGE AVERAGE 41 MINIMUM 40 MAXIMUM 42

CONSTITUENTS	DIM	AV	VARIATION		AVR	NUTR. DENS.		MOLPERC.
WATER	GRAM	60.70	59.80	- 61.70	35.81	GRAM/MJ	58.30	
PROTEIN	GRAM	19.40	16.70	- 22.00	11.45	GRAM/MJ	18.63	
FAT	GRAM	18.40	16.60	- 20.00	10.86	GRAM/MJ	17.67	
SODIUM	GRAM	0.78	0.47	- 1.10	0.46	GRAM/MJ	0.75	
POTASSIUM	MILLI	590.00	535.00	- 645.00	348.10	MILLI/MJ	566.64	
CALCIUM	GRAM	1.70	0.62	- 2.78	1.00	GRAM/MJ	1.63	
VITAMIN A	MILLI	0.15	0.03	- 0.24	0.09	MILLI/MJ	0.14	
VITAMIN D	MICRO	32.00	30.00	- 33.00	18.88	MICRO/MJ	30.73	
VITAMIN B1	MICRO	25.00	20.00	- 30.00	14.75	MICRO/MJ	24.01	
VITAMIN B2	MILLI	0.40	-	- -	0.24	MILLI/MJ	0.38	
NICOTINAMIDE	MILLI	4.00	3.00	- 5.00	2.36	MILLI/MJ	3.84	
VITAMIN C	MILLI	1.00	-	- -	0.59	MILLI/MJ	0.96	
TOTAL PURINES	MILLI	804.00 1	804.00	- 804.00	474.36	MILLI/MJ	772.17	
ADENINE	MILLI	20.00	20.00	- 20.00	11.80	MILLI/MJ	19.21	
GUANINE	MILLI	473.00	473.00	- 473.00	279.07	MILLI/MJ	454.27	
XANTHINE	MILLI	16.00	16.00	- 16.00	9.44	MILLI/MJ	15.37	
HYPOXANTHINE	MILLI	191.00	191.00	- 191.00	112.69	MILLI/MJ	183.44	

1 VALUE OF TOTAL PURINES IS EXPRESSSED IN MG URIC ACID/100G

STOCKFISCH | STOCKFISH | STOCKFISH

		PROTEIN	FAT	CARBOHYDRATES	TOTAL
ENERGY VALUE (AVERAGE)	KJOULE	458	99	0.0	1557
PER 100 G	(KCAL)	348	24	0.0	372
EDIBLE PORTION					
AMOUNT OF DIGESTIBLE	GRAM	76.82	2.37	0.00	
CONSTITUENTS PER 100 G					
ENERGY VALUE (AVERAGE)	KJOULE	414	94	0.0	1509
OF THE DIGESTIBLE	(KCAL)	338	23	0.0	361
FRACTION PER 100 G					
EDIBLE PORTION					

WASTE PERCENTAGE AVERAGE 36

CONSTITUENTS	DIM	AV	VARIATION			AVR	NUTR. DENS.		MOLPERC.
WATER	GRAM	15.20	15.00	-	15.40	9.73	GRAM/MJ	10.07	
PROTEIN	GRAM	79.20	78.50	-	80.00	50.69	GRAM/MJ	52.50	
FAT	GRAM	2.50	2.00	-	3.00	1.60	GRAM/MJ	1.66	
MINERALS	GRAM	3.10	-	-	-	1.98	GRAM/MJ	2.05	
SODIUM	MILLI	500.00	-	-	-	320.00	MILLI/MJ	331.41	
POTASSIUM	GRAM	1.50	-	-	-	0.96	GRAM/MJ	0.99	
CALCIUM	MILLI	60.00	-	-	-	38.40	MILLI/MJ	39.77	
IRON	MILLI	4.30	3.60	-	5.00	2.75	MILLI/MJ	2.85	
PHOSPHORUS	MILLI	450.00	-	-	-	288.00	MILLI/MJ	298.27	
FLUORIDE	MILLI	0.50	-	-	-	0.32	MILLI/MJ	0.33	
VITAMIN A	MICRO	-	0.00	-	45.00				
VITAMIN B1	MICRO	87.00	26.00	-	130.00	55.68	MICRO/MJ	57.67	
VITAMIN B2	MILLI	0.11	0.07	-	0.13	0.07	MILLI/MJ	0.07	
NICOTINAMIDE	MILLI	3.50	1.56	-	5.00	2.24	MILLI/MJ	2.32	
VITAMIN B6	MILLI	0.20	0.16	-	0.31	0.13	MILLI/MJ	0.13	
VITAMIN B12	MICRO	1.00	0.70	-	1.30	0.64	MICRO/MJ	0.66	
VITAMIN C	-	0.00	-	-	-	0.00			
ARGININE	GRAM	4.46	-	-	-	2.85	GRAM/MJ	2.96	
CYSTINE	GRAM	1.06	-	-	-	0.68	GRAM/MJ	0.70	
ISOLEUCINE	GRAM	4.02	-	-	-	2.57	GRAM/MJ	2.66	
LEUCINE	GRAM	5.99	-	-	-	3.83	GRAM/MJ	3.97	
LYSINE	GRAM	6.94	-	-	-	4.44	GRAM/MJ	4.60	
METHIONINE	GRAM	2.31	-	-	-	1.48	GRAM/MJ	1.53	
PHENYLALANINE	GRAM	2.94	-	-	-	1.88	GRAM/MJ	1.95	
THREONINE	GRAM	3.43	-	-	-	2.20	GRAM/MJ	2.27	
TRYPTOPHAN	GRAM	0.79	-	-	-	0.51	GRAM/MJ	0.52	
TYROSINE	GRAM	2.14	-	-	-	1.37	GRAM/MJ	1.42	
VALINE	GRAM	4.22	-	-	-	2.70	GRAM/MJ	2.80	

THUNFISCH IN ÖL	**TUNA** IN OIL	**THON** À L'HUILE
1	1	1

		PROTEIN	FAT	CARBOHYDRATES	TOTAL
ENERGY VALUE (AVERAGE)	KJOULE	438	831	0.0	1269
PER 100 G	(KCAL)	105	199	0.0	303
EDIBLE PORTION					
AMOUNT OF DIGESTIBLE	GRAM	23.08	19.85	0.00	
CONSTITUENTS PER 100 G					
ENERGY VALUE (AVERAGE)	KJOULE	425	789	0.0	1214
OF THE DIGESTIBLE	(KCAL)	102	189	0.0	290
FRACTION PER 100 G					
EDIBLE PORTION					

WASTE PERCENTAGE AVERAGE 0.00

CONSTITUENTS	DIM	AV	VARIATION		AVR	NUTR.	DENS.	MOLPERC.
WATER	GRAM	52.50	-	-	52.50	GRAM/MJ	43.24	
PROTEIN	GRAM	23.80	-	-	23.80	GRAM/MJ	19.60	
FAT	GRAM	20.90	-	-	20.90	GRAM/MJ	17.21	
MINERALS	GRAM	2.30	-	-	2.30	GRAM/MJ	1.89	
SODIUM	MILLI	361.00	297.00	484.00	361.00	MILLI/MJ	297.31	
POTASSIUM	MILLI	343.00	300.00	424.00	343.00	MILLI/MJ	282.49	
CALCIUM	MILLI	7.00	-	-	7.00	MILLI/MJ	5.77	
IRON	MILLI	1.20	-	-	1.20	MILLI/MJ	0.99	
PHOSPHORUS	MILLI	294.00	-	-	294.00	MILLI/MJ	242.13	
IODIDE	MICRO	53.00	16.00	90.00	53.00	MICRO/MJ	43.65	
VITAMIN A	MILLI	0.37	0.12	0.83	0.37	MILLI/MJ	0.30	
VITAMIN B1	MICRO	50.00	22.00	65.00	50.00	MICRO/MJ	41.18	
VITAMIN B2	MICRO	60.00	53.00	77.00	60.00	MICRO/MJ	49.42	
NICOTINAMIDE	MILLI	10.80	-	-	10.80	MILLI/MJ	8.89	
VITAMIN B6	MILLI	0.25	0.20	0.31	0.25	MILLI/MJ	0.21	
VITAMIN B12	MICRO	1.30	-	-	1.30	MICRO/MJ	1.07	
VITAMIN C	-	0.00	-	-	0.00			
ARGININE	GRAM	1.24	-	-	1.24	GRAM/MJ	1.02	
HISTIDINE	GRAM	1.40	-	-	1.40	GRAM/MJ	1.15	
ISOLEUCINE	GRAM	1.17	-	-	1.17	GRAM/MJ	0.96	
LEUCINE	GRAM	1.74	-	-	1.74	GRAM/MJ	1.43	
LYSINE	GRAM	1.98	-	-	1.98	GRAM/MJ	1.63	
METHIONINE	GRAM	0.67	-	-	0.67	GRAM/MJ	0.55	
PHENYLALANINE	GRAM	0.88	-	-	0.88	GRAM/MJ	0.72	
THREONINE	GRAM	1.02	-	-	1.02	GRAM/MJ	0.84	
TRYPTOPHAN	GRAM	0.21	-	-	0.21	GRAM/MJ	0.17	
VALINE	GRAM	1.26	-	-	1.26	GRAM/MJ	1.04	
TRYPTAMINE	MILLI	0.10	-	-	0.10	MILLI/MJ	0.09	
TYRAMINE	MICRO	72.00	-	-	72.00	MICRO/MJ	59.30	
CHOLESTEROL	MILLI	32.30	-	-	32.30	MILLI/MJ	26.60	
TOTAL PURINES	MILLI	290.00 [2]	-	-	290.00	MILLI/MJ	238.84	

1 TOTAL CANNED AMOUNT

2 VALUE EXPRESSED IN MG URIC ACID/100G

Getreide

Cereals
Céréales

Getreide und Buchweizen

CEREALS AND BUCKWHEAT **CÉRÉALES ET SARRASIN**

BUCHWEIZEN — BUCKWHEAT — SARRASIN

GESCHÄLTES KORN — SHUCKED CORN — GRAINS DÉCORTIQUÉS

FAGOPYRUM ESCULENTUM MÖNCH

		PROTEIN	FAT	CARBOHYDRATES	TOTAL
ENERGY VALUE (AVERAGE) PER 100 G EDIBLE PORTION	KJOULE	173	67	1188	1428
	(KCAL)	41	16	284	341
AMOUNT OF DIGESTIBLE CONSTITUENTS PER 100 G	GRAM	7.07	1.55	70.98	
ENERGY VALUE (AVERAGE) OF THE DIGESTIBLE FRACTION PER 100 G EDIBLE PORTION	KJOULE	135	61	1188	1383
	(KCAL)	32	14	284	331

WASTE PERCENTAGE AVERAGE 0.00

CONSTITUENTS	DIM	AV		VARIATION			AVR	NUTR. DENS.		MOLPERC.
WATER	GRAM	12.80		-	-	-	12.80	GRAM/MJ	9.25	
PROTEIN	GRAM	9.07		8.26	-	10.21	9.07	GRAM/MJ	6.56	
FAT	GRAM	1.73		1.40	-	2.00	1.73	GRAM/MJ	1.25	
AVAILABLE CARBOHYDR.	GRAM	70.98	1	-	-	-	70.98	GRAM/MJ	51.32	
TOTAL DIETARY FIBRE	GRAM	3.70	2	-	-	-	3.70	GRAM/MJ	2.67	
MINERALS	GRAM	1.72		1.60	-	1.86	1.72	GRAM/MJ	1.24	
POTASSIUM	MILLI	324.00		-	-	-	324.00	MILLI/MJ	234.24	
MAGNESIUM	MILLI	85.00		80.00	-	89.00	85.00	MILLI/MJ	61.45	
CALCIUM	MILLI	21.00		11.00	-	30.00	21.00	MILLI/MJ	15.18	
IRON	MILLI	3.20		-	-	-	3.20	MILLI/MJ	2.31	
NICKEL	MILLI	0.10		0.08	-	0.12	0.10	MILLI/MJ	0.07	
MOLYBDENUM	MILLI	0.48		-	-	-	0.49	MILLI/MJ	0.35	
PHOSPHORUS	MILLI	254.00		-	-	-	254.00	MILLI/MJ	183.63	
CHLORIDE	MILLI	12.00		-	-	-	12.00	MILLI/MJ	8.68	
BORON	MILLI	0.68		0.53	-	0.75	0.68	MILLI/MJ	0.49	
SELENIUM	MICRO	18.00		-	-	-	18.00	MICRO/MJ	13.01	
TOTAL TOCOPHEROLS	MILLI	3.70		-	-	-	3.70	MILLI/MJ	2.67	
VITAMIN B1	MILLI	0.24		-	-	-	0.24	MILLI/MJ	0.17	
VITAMIN B2	MILLI	0.15		-	-	-	0.15	MILLI/MJ	0.11	
NICOTINAMIDE	MILLI	2.90		-	-	-	2.90	MILLI/MJ	2.10	
PANTOTHENIC ACID	MILLI	1.20		1.10	-	1.30	1.20	MILLI/MJ	0.87	
VITAMIN C	-	0.00		-	-	-	0.00			

1 ESTIMATED BY THE DIFFERENCE METHOD
100 - (WATER + PROTEIN + FAT + MINERALS + TOTAL DIETARY FIBRE)

2 METHOD OF MEUSER, SUCKOW AND KULIKOWSKI ("BERLINER METHODE")

CONSTITUENTS	DIM	AV	VARIATION			AVR	NUTR. DENS.		MOLPERC.
ALANINE	GRAM	0.56	-	-	-	0.56	GRAM/MJ	0.40	7.6
ARGININE	GRAM	0.97	0.84	-	1.05	0.97	GRAM/MJ	0.70	6.7
ASPARTIC ACID	GRAM	0.97	-	-	-	0.97	GRAM/MJ	0.70	8.8
CYSTINE	GRAM	0.22	0.20	-	0.24	0.22	GRAM/MJ	0.16	1.1
GLUTAMIC ACID	GRAM	1.88	-	-	-	1.88	GRAM/MJ	1.36	15.5
GLYCINE	GRAM	0.79	-	-	-	0.79	GRAM/MJ	0.57	12.7
HISTIDINE	GRAM	0.22	-	-	-	0.22	GRAM/MJ	0.16	1.7
ISOLEUCINE	GRAM	0.49	-	-	-	0.49	GRAM/MJ	0.35	4.5
LEUCINE	GRAM	0.66	0.58	-	0.74	0.66	GRAM/MJ	0.48	6.1
LYSINE	GRAM	0.58	0.40	-	0.66	0.58	GRAM/MJ	0.42	4.8
METHIONINE	GRAM	0.19	0.16	-	0.21	0.19	GRAM/MJ	0.14	1.5
PHENYLALANINE	GRAM	0.41	-	-	-	0.41	GRAM/MJ	0.30	3.0
PROLINE	GRAM	0.48	-	-	-	0.48	GRAM/MJ	0.35	5.0
SERINE	GRAM	0.58	-	-	-	0.58	GRAM/MJ	0.42	6.7
THREONINE	GRAM	0.47	0.41	-	0.56	0.47	GRAM/MJ	0.34	4.8
TRYPTOPHAN	GRAM	0.17	-	-	-	0.17	GRAM/MJ	0.12	1.0
TYROSINE	GRAM	0.22	-	-	-	0.22	GRAM/MJ	0.16	1.5
VALINE	GRAM	0.66	0.53	-	0.74	0.66	GRAM/MJ	0.48	6.8
MYRISTIC ACID	MILLI	10.00	-	-	-	10.00	MILLI/MJ	7.23	
PALMITIC ACID	GRAM	0.26	-	-	-	0.26	GRAM/MJ	0.19	
STEARIC ACID	MILLI	40.00	-	-	-	40.00	MILLI/MJ	28.92	
ARACHIDIC ACID	MILLI	40.00	-	-	-	40.00	MILLI/MJ	28.92	
PALMITOLEIC ACID	-	0.00	-	-	-	0.00			
OLEIC ACID	GRAM	0.58	-	-	-	0.58	GRAM/MJ	0.42	
LINOLEIC ACID	GRAM	0.53	-	-	-	0.53	GRAM/MJ	0.38	
LINOLENIC ACID	MILLI	80.00	-	-	-	80.00	MILLI/MJ	57.84	
TOTAL STEROLS	MILLI	198.00	-	-	-	198.00	MILLI/MJ	143.15	
CAMPESTEROL	MILLI	20.00	-	-	-	20.00	MILLI/MJ	14.46	
BETA-SITOSTEROL	MILLI	164.00	-	-	-	164.00	MILLI/MJ	118.57	
STIGMASTEROL	MILLI	8.00	-	-	-	8.00	MILLI/MJ	5.78	
DIETARY FIBRE, WAT.SOL.	GRAM	1.60	-	-	-	1.60	GRAM/MJ	1.16	
DIETARY FIBRE, WAT.INS.	GRAM	2.10	-	-	-	2.10	GRAM/MJ	1.52	

BUCHWEIZENGRÜTZE BUCKWHEAT GROATS GRUAU DE SARRASIN

		PROTEIN	FAT	CARBOHYDRATES	TOTAL
ENERGY VALUE (AVERAGE) PER 100 G EDIBLE PORTION	KJOULE	144	60	1216	1419
	(KCAL)	34	14	291	339
AMOUNT OF DIGESTIBLE CONSTITUENTS PER 100 G	GRAM	5.57	1.39	72.63	
ENERGY VALUE (AVERAGE) OF THE DIGESTIBLE FRACTION PER 100 G EDIBLE PORTION	KJOULE	106	54	1216	1376
	(KCAL)	25	13	291	329

WASTE PERCENTAGE AVERAGE 0.00

CONSTITUENTS	DIM	AV		VARIATION			AVR	NUTR. DENS.		MOLPERC.
WATER	GRAM	13.20		12.70	-	14.00	13.20	GRAM/MJ	9.59	
PROTEIN	GRAM	7.54		4.64	-	9.47	7.54	GRAM/MJ	5.48	
FAT	GRAM	1.55		0.50	-	2.30	1.55	GRAM/MJ	1.13	
AVAILABLE CARBOHYDR.	GRAM	72.63	1	-	-	-	72.63	GRAM/MJ	52.78	
TOTAL DIETARY FIBRE	GRAM	3.22	2	-	-	-	3.22	GRAM/MJ	2.34	
MINERALS	GRAM	1.86		-	-	-	1.86	GRAM/MJ	1.35	
SODIUM	MILLI	1.00		0.00	-	2.00	1.00	MILLI/MJ	0.73	
POTASSIUM	MILLI	218.00		130.00	-	324.00	218.00	MILLI/MJ	158.43	
MAGNESIUM	MILLI	48.00		-	-	-	48.00	MILLI/MJ	34.88	
CALCIUM	MILLI	12.00		10.00	-	15.00	12.00	MILLI/MJ	8.72	
IRON	MILLI	2.00		-	-	-	2.00	MILLI/MJ	1.45	
COPPER	MILLI	0.70		-	-	-	0.70	MILLI/MJ	0.51	
PHOSPHORUS	MILLI	150.00		-	-	-	150.00	MILLI/MJ	109.01	
VITAMIN B1	MILLI	0.28		0.20	-	0.39	0.28	MILLI/MJ	0.20	
VITAMIN B2	MICRO	75.00		50.00	-	100.00	75.00	MICRO/MJ	54.50	
NICOTINAMIDE	MILLI	2.80		2.00	-	3.50	2.80	MILLI/MJ	2.03	
PANTOTHENIC ACID	MILLI	1.45		1.45	-	1.45	1.45	MILLI/MJ	1.05	
VITAMIN B6	MILLI	0.40		0.28	-	0.57	0.40	MILLI/MJ	0.29	
VITAMIN C	-	0.00		-	-	-	0.00			
ALANINE	GRAM	0.43		0.43	-	0.43	0.43	GRAM/MJ	0.31	8.4
ARGININE	GRAM	0.60		-	-	-	0.60	GRAM/MJ	0.44	6.0
ASPARTIC ACID	GRAM	0.78		0.78	-	0.78	0.78	GRAM/MJ	0.57	10.2
CYSTINE	GRAM	0.15		-	-	-	0.15	GRAM/MJ	0.11	1.1
GLUTAMIC ACID	GRAM	1.23		1.23	-	1.23	1.23	GRAM/MJ	0.89	14.6
GLYCINE	GRAM	0.55		0.55	-	0.55	0.55	GRAM/MJ	0.40	12.8

1 ESTIMATED BY THE DIFFERENCE METHOD
100 - (WATER + PROTEIN + FAT + MINERALS + TOTAL DIETARY FIBRE)

2 MODIFIED AOAC METHOD

CONSTITUENTS	DIM	AV	VARIATION			AVR	NUTR. DENS.		MOLPERC.
HISTIDINE	GRAM	0.17	-	-	-	0.17	GRAM/MJ	0.12	1.9
ISOLEUCINE	GRAM	0.29	-	-	-	0.29	GRAM/MJ	0.21	3.9
LEUCINE	GRAM	0.45	-	-	-	0.45	GRAM/MJ	0.33	6.0
LYSINE	GRAM	0.39	-	-	-	0.39	GRAM/MJ	0.28	4.7
METHIONINE	GRAM	0.13	-	-	-	0.13	GRAM/MJ	0.09	1.5
PHENYLALANINE	GRAM	0.29	-	-	-	0.29	GRAM/MJ	0.21	3.1
PROLINE	GRAM	0.33	0.33	-	0.33	0.33	GRAM/MJ	0.24	5.0
SERINE	GRAM	0.43	0.43	-	0.43	0.43	GRAM/MJ	0.31	7.2
THREONINE	GRAM	0.30	-	-	-	0.30	GRAM/MJ	0.22	4.4
TRYPTOPHAN	GRAM	0.11	-	-	-	0.11	GRAM/MJ	0.08	0.9
TYROSINE	GRAM	0.16	-	-	-	0.16	GRAM/MJ	0.12	1.5
VALINE	GRAM	0.45	-	-	-	0.45	GRAM/MJ	0.33	6.7
SUCROSE	GRAM	0.29	0.29	-	0.29	0.29	GRAM/MJ	0.21	
DIETARY FIBRE, WAT.SOL.	GRAM	1.39	-	-	-	1.39	GRAM/MJ	1.01	
DIETARY FIBRE, WAT.INS.	GRAM	1.83	-	-	-	1.83	GRAM/MJ	1.33	

BUCHWEIZENVOLL-MEHL — BUCKWHEAT FLOUR WHOLE MEAL — FARINE DE SARRASIN ENTIÈRE

		PROTEIN	FAT	CARBOHYDRATES	TOTAL
ENERGY VALUE (AVERAGE) PER 100 G EDIBLE PORTION	KJOULE	207	105	1184	1496
	(KCAL)	49	25	283	358
AMOUNT OF DIGESTIBLE CONSTITUENTS PER 100 G	GRAM	8.47	2.43	70.74	
ENERGY VALUE (AVERAGE) OF THE DIGESTIBLE FRACTION PER 100 G EDIBLE PORTION	KJOULE	161	95	1184	1440
	(KCAL)	39	23	283	344

WASTE PERCENTAGE AVERAGE 0.00

CONSTITUENTS	DIM	AV	VARIATION			AVR	NUTR. DENS.		MOLPERC.
WATER	GRAM	14.10	13.40	-	19.40	14.10	GRAM/MJ	9.79	
PROTEIN	GRAM	10.86	10.39	-	11.04	10.86	GRAM/MJ	7.54	
FAT	GRAM	2.71	2.40	-	2.80	2.71	GRAM/MJ	1.88	
AVAILABLE CARBOHYDR.	GRAM	70.74 1	-	-	-	70.74	GRAM/MJ	49.12	
MINERALS	GRAM	1.59	1.37	-	1.67	1.59	GRAM/MJ	1.10	
SODIUM	MILLI	1.00	-	-	-	1.00	MILLI/MJ	0.69	
POTASSIUM	MILLI	680.00	-	-	-	680.00	MILLI/MJ	472.20	
CALCIUM	MILLI	33.00	-	-	-	33.00	MILLI/MJ	22.92	
IRON	MILLI	2.20	1.80	-	2.80	2.20	MILLI/MJ	1.53	
MOLYBDENUM	MICRO	8.00	-	-	-	8.00	MICRO/MJ	5.56	
PHOSPHORUS	MILLI	263.00	179.00	-	347.00	263.00	MILLI/MJ	182.63	
IODIDE	MICRO	2.50	-	-	-	2.50	MICRO/MJ	1.74	
VITAMIN B1	MILLI	0.58	-	-	-	0.58	MILLI/MJ	0.40	
VITAMIN B2	MILLI	0.15	-	-	-	0.15	MILLI/MJ	0.10	
NICOTINAMIDE	MILLI	2.90	-	-	-	2.90	MILLI/MJ	2.01	
PANTOTHENIC ACID	MILLI	1.45	1.36	-	1.51	1.45	MILLI/MJ	1.01	
ALANINE	GRAM	0.61	-	-	-	0.61	GRAM/MJ	0.42	7.0
ARGININE	GRAM	1.22	0.74	-	1.41	1.22	GRAM/MJ	0.85	7.2
ASPARTIC ACID	GRAM	1.26	-	-	-	1.26	GRAM/MJ	0.87	9.7
CYSTINE	GRAM	0.23	0.15	-	0.34	0.23	GRAM/MJ	0.16	1.0
GLUTAMIC ACID	GRAM	2.14	-	-	-	2.14	GRAM/MJ	1.49	14.9
GLYCINE	GRAM	0.83	-	-	-	0.83	GRAM/MJ	0.58	11.3
HISTIDINE	GRAM	0.28	0.27	-	0.30	0.28	GRAM/MJ	0.19	1.8
ISOLEUCINE	GRAM	0.52	0.43	-	0.55	0.52	GRAM/MJ	0.36	4.1
LEUCINE	GRAM	0.80	0.71	-	0.86	0.80	GRAM/MJ	0.56	6.2
LYSINE	GRAM	0.77	0.72	-	0.77	0.77	GRAM/MJ	0.53	5.4
METHIONINE	GRAM	0.24	0.22	-	0.25	0.24	GRAM/MJ	0.17	1.6
PHENYLALANINE	GRAM	0.52	0.45	-	0.55	0.52	GRAM/MJ	0.36	3.2
PROLINE	GRAM	0.76	-	-	-	0.76	GRAM/MJ	0.53	6.8
SERINE	GRAM	0.66	-	-	-	0.66	GRAM/MJ	0.46	6.4
THREONINE	GRAM	0.54	0.49	-	0.58	0.54	GRAM/MJ	0.37	4.6
TRYPTOPHAN	GRAM	0.19	0.13	-	0.22	0.19	GRAM/MJ	0.13	1.0
TYROSINE	GRAM	0.25	0.21	-	0.27	0.25	GRAM/MJ	0.17	1.4
VALINE	GRAM	0.73	0.42	-	0.78	0.73	GRAM/MJ	0.51	6.4
STARCH	GRAM	67.15	-	-	-	67.15	GRAM/MJ	46.63	

1 ESTIMATED BY THE DIFFERENCE METHOD
100 - (WATER + PROTEIN + FAT + MINERALS)

GERSTE / BARLEY / ORGE

ENTSPELZT / WITHOUT HUSK / DÉGLUMÉS

GANZES KORN / WHOLE GRAIN / GRAINS ENTIERS

HORDEUM VULGARE L.

		PROTEIN	FAT	CARBOHYDRATES	TOTAL
ENERGY VALUE (AVERAGE) PER 100 G EDIBLE PORTION	KJOULE (KCAL)	187 45	82 20	1060 253	1329 318
AMOUNT OF DIGESTIBLE CONSTITUENTS PER 100 G	GRAM	7.67	1.89	63.31	
ENERGY VALUE (AVERAGE) OF THE DIGESTIBLE FRACTION PER 100 G EDIBLE PORTION	KJOULE (KCAL)	146 35	74 18	1060 253	1279 306

WASTE PERCENTAGE AVERAGE 0.00

CONSTITUENTS	DIM	AV		VARIATION			AVR	NUTR. DENS.		MOLPERC.
WATER	GRAM	11.70		10.50	-	12.80	11.70	GRAM/MJ	9.15	
PROTEIN	GRAM	9.84		9.00	-	10.49	9.84	GRAM/MJ	7.69	
FAT	GRAM	2.10		1.80	-	2.25	2.10	GRAM/MJ	1.64	
AVAILABLE CARBOHYDR.	GRAM	63.31	1	-	-	-	63.31	GRAM/MJ	49.49	
TOTAL DIETARY FIBRE	GRAM	9.80	2	-	-	-	9.80	GRAM/MJ	7.66	
MINERALS	GRAM	2.25		2.15	-	2.39	2.25	GRAM/MJ	1.76	
SODIUM	MILLI	18.00		6.00	-	29.00	18.00	MILLI/MJ	14.07	
POTASSIUM	MILLI	444.00		371.00	-	521.00	444.00	MILLI/MJ	347.09	
MAGNESIUM	MILLI	114.00		110.00	-	130.00	114.00	MILLI/MJ	89.12	
CALCIUM	MILLI	38.00		33.00	-	42.00	38.00	MILLI/MJ	29.71	
MANGANESE	MILLI	1.65		1.50	-	1.80	1.65	MILLI/MJ	1.29	
IRON	MILLI	2.80		2.00	-	3.60	2.80	MILLI/MJ	2.19	
COBALT	MICRO	6.80		2.50	-	10.00	6.80	MICRO/MJ	5.32	
COPPER	MILLI	0.30		0.10	-	0.50	0.30	MILLI/MJ	0.23	
ZINC	MILLI	3.10		2.60	-	4.40	3.10	MILLI/MJ	2.42	
NICKEL	MICRO	50.00		10.00	-	110.00	50.00	MICRO/MJ	39.09	
CHROMIUM	MICRO	13.00		1.90	-	23.00	13.00	MICRO/MJ	10.16	
MOLYBDENUM	MICRO	43.00		32.00	-	52.00	43.00	MICRO/MJ	33.61	
PHOSPHORUS	MILLI	342.00		-	-	-	342.00	MILLI/MJ	267.35	
CHLORIDE	MILLI	23.00		-	-	-	23.00	MILLI/MJ	17.98	
FLUORIDE	MILLI	0.12		0.04	-	0.50	0.12	MILLI/MJ	0.09	
IODIDE	MICRO	7.00		-	-	-	7.00	MICRO/MJ	5.47	
BORON	MILLI	0.45		0.27	-	0.65	0.46	MILLI/MJ	0.36	
SELENIUM	MICRO	-	3	0.20	-	24.00				
SILICON	MILLI	188.00		160.00	-	210.00	188.00	MILLI/MJ	146.97	
CAROTENE	MICRO	1.00		-	-	-	1.00	MICRO/MJ	0.78	

1 ESTIMATED BY THE DIFFERENCE METHOD
100 - (WATER + PROTEIN + FAT + MINERALS + TOTAL DIETARY FIBRE)

2 METHOD OF MEUSER, SUCKOW AND KULIKOWSKI ("BERLINER METHODE")

3 GREAT REGIONAL DIFFERENCES: USA: 26 - 60 UG/100 G
MIDDLE AMERICA: 400 - 2000 UG/100 G
SCANDINAVIAN COUNTRIES: 0,2 - 1,0 UG/100 G

CONSTITUENTS	DIM	AV	VARIATION			AVR	NUTR. DENS.		MOLPERC.
VITAMIN E ACTIVITY	MILLI	0.65	0.55	-	0.77	0.65	MILLI/MJ	0.51	
TOTAL TOCOPHEROLS	MILLI	2.20	2.00	-	2.70	2.20	MILLI/MJ	1.72	
ALPHA-TOCOPHEROL	MILLI	0.31	0.20	-	0.41	0.31	MILLI/MJ	0.24	
BETA-TOCOPHEROL	MICRO	30.00	20.00	-	40.00	30.00	MICRO/MJ	23.45	
GAMMA-TOCOPHEROL	MICRO	40.00	30.00	-	60.00	40.00	MICRO/MJ	31.27	
DELTA-TOCOPHEROL	MICRO	10.00	-	-	-	10.00	MICRO/MJ	7.82	
ALPHA-TOCOTRIENOL	MILLI	1.10	-	-	-	1.10	MILLI/MJ	0.86	
BETA-TOCOTRIENOL	MILLI	0.27	0.24	-	0.30	0.27	MILLI/MJ	0.21	
GAMMA-TOCOTRIENOL	MILLI	0.47	0.20	-	0.74	0.47	MILLI/MJ	0.37	
VITAMIN B1	MILLI	0.43	0.29	-	0.50	0.43	MILLI/MJ	0.34	
VITAMIN B2	MILLI	0.18	0.10	-	0.30	0.18	MILLI/MJ	0.14	
NICOTINAMIDE	MILLI	4.80	1.50	-	7.04	4.80	MILLI/MJ	3.75	
PANTOTHENIC ACID	MILLI	0.68	0.35	-	1.00	0.68	MILLI/MJ	0.53	
VITAMIN B6	MILLI	0.56	-	-	-	0.56	MILLI/MJ	0.44	
FOLIC ACID	MICRO	65.00	-	-	-	65.00	MICRO/MJ	50.81	
VITAMIN C	-	0.00	-	-	-	0.00			
ALANINE	GRAM	0.56	0.54	-	0.57	0.56	GRAM/MJ	0.44	7.2
ARGININE	GRAM	0.56	0.54	-	0.58	0.56	GRAM/MJ	0.44	3.7
ASPARTIC ACID	GRAM	0.68	0.64	-	0.73	0.68	GRAM/MJ	0.53	5.8
CYSTINE	GRAM	0.22	0.15	-	0.27	0.22	GRAM/MJ	0.17	1.0
GLUTAMIC ACID	GRAM	2.81	2.77	-	2.98	2.81	GRAM/MJ	2.20	21.8
GLYCINE	GRAM	0.54	0.49	-	0.59	0.54	GRAM/MJ	0.42	8.2
HISTIDINE	GRAM	0.21	0.19	-	0.25	0.21	GRAM/MJ	0.16	1.5
ISOLEUCINE	GRAM	0.46	0.39	-	0.50	0.46	GRAM/MJ	0.36	4.0
LEUCINE	GRAM	0.80	0.76	-	0.85	0.80	GRAM/MJ	0.63	7.0
LYSINE	GRAM	0.38	0.31	-	0.44	0.38	GRAM/MJ	0.30	3.0
METHIONINE	GRAM	0.18	0.14	-	0.21	0.18	GRAM/MJ	0.14	1.4
PHENYLALANINE	GRAM	0.59	0.52	-	0.63	0.59	GRAM/MJ	0.46	4.1
PROLINE	GRAM	1.26	1.05	-	1.43	1.26	GRAM/MJ	0.98	12.5
SERINE	GRAM	0.54	0.49	-	0.63	0.54	GRAM/MJ	0.42	5.9
THREONINE	GRAM	0.43	0.41	-	0.44	0.43	GRAM/MJ	0.34	4.1
TRYPTOPHAN	GRAM	0.15	0.12	-	0.19	0.15	GRAM/MJ	0.12	0.8
TYROSINE	GRAM	0.39	0.33	-	0.45	0.39	GRAM/MJ	0.30	2.5
VALINE	GRAM	0.58	0.49	-	0.66	0.58	GRAM/MJ	0.45	5.6
GLUCOSE	GRAM	0.10	-	-	-	0.10	GRAM/MJ	0.08	
FRUCTOSE	GRAM	0.10	-	-	-	0.10	GRAM/MJ	0.08	
SUCROSE	GRAM	0.99	0.51	-	1.49	0.99	GRAM/MJ	0.77	
RAFFINOSE	GRAM	0.25	-	-	-	0.25	GRAM/MJ	0.20	
MALTOSE	GRAM	0.52	-	-	-	0.52	GRAM/MJ	0.41	
STARCH	GRAM	64.30	56.00	-	64.70	64.30	GRAM/MJ	50.27	
PHYTIC ACID	GRAM	1.07	0.97	-	1.16	1.07	GRAM/MJ	0.84	
MYRISTIC ACID	MILLI	40.00	-	-	-	40.00	MILLI/MJ	31.27	
PALMITIC ACID	GRAM	0.45	0.42	-	0.48	0.45	GRAM/MJ	0.35	
STEARIC ACID	MILLI	40.00	-	-	-	40.00	MILLI/MJ	31.27	
PALMITOLEIC ACID	MILLI	20.00	-	-	-	20.00	MILLI/MJ	15.63	
OLEIC ACID	GRAM	0.23	0.18	-	0.28	0.23	GRAM/MJ	0.18	
LINOLEIC ACID	GRAM	1.15	1.12	-	1.18	1.15	GRAM/MJ	0.90	
LINOLENIC ACID	GRAM	0.11	0.08	-	0.14	0.11	GRAM/MJ	0.09	
TOTAL PHOSPHOLIPIDS	MILLI	354.00	-	-	-	354.00	MILLI/MJ	276.73	
PHOSPHATIDYLCHOLINE	MILLI	282.00	-	-	-	282.00	MILLI/MJ	220.45	
PHOSPHATIDYLETHANOLAMINE	MILLI	30.00	-	-	-	30.00	MILLI/MJ	23.45	
PHOSPHATIDYLSERINE	MILLI	20.00	-	-	-	20.00	MILLI/MJ	15.63	
PHOSPHATIDYLINOSITOL	MILLI	6.00	-	-	-	6.00	MILLI/MJ	4.69	
DIETARY FIBRE, WAT.SOL.	GRAM	1.70	-	-	-	1.70	GRAM/MJ	1.33	
DIETARY FIBRE, WAT.INS.	GRAM	8.10	-	-	-	8.10	GRAM/MJ	6.33	

GERSTENGRAUPEN PEARL BARLEY ORGE MONDÉ

		PROTEIN	FAT	CARBOHYDRATES	TOTAL
ENERGY VALUE (AVERAGE) PER 100 G EDIBLE PORTION	KJOULE	184	53	1188	1424
	(KCAL)	44	13	284	340
AMOUNT OF DIGESTIBLE CONSTITUENTS PER 100 G	GRAM	7.52	1.22	70.96	
ENERGY VALUE (AVERAGE) OF THE DIGESTIBLE FRACTION PER 100 G EDIBLE PORTION	KJOULE	143	48	1188	1379
	(KCAL)	34	11	284	329

WASTE PERCENTAGE AVERAGE 0.00

CONSTITUENTS	DIM	AV		VARIATION		AVR	NUTR. DENS.		MOLPERC.
WATER	GRAM	12.20		11.10 - 13.30		12.20	GRAM/MJ	8.85	
PROTEIN	GRAM	9.65		7.24 - 11.14		9.65	GRAM/MJ	7.00	
FAT	GRAM	1.36		1.00 - 2.70		1.36	GRAM/MJ	0.99	
AVAILABLE CARBOHYDR.	GRAM	70.96	1	- - -		70.96	GRAM/MJ	51.48	
TOTAL DIETARY FIBRE	GRAM	4.63	2	- - -		4.63	GRAM/MJ	3.36	
MINERALS	GRAM	1.20		0.90 - 1.70		1.20	GRAM/MJ	0.87	
SODIUM	MILLI	5.00		3.00 - 7.00		5.00	MILLI/MJ	3.63	
POTASSIUM	MILLI	190.00		152.00 - 229.00		190.00	MILLI/MJ	137.83	
CALCIUM	MILLI	14.00		11.00 - 16.00		14.00	MILLI/MJ	10.16	
IRON	MILLI	2.00		- - -		2.00	MILLI/MJ	1.45	
PHOSPHORUS	MILLI	189.00		- - -		189.00	MILLI/MJ	137.10	
FLUORIDE	MILLI	-		0.04 - 0.48					
VITAMIN B1	MICRO	92.00		50.00 - 110.00		92.00	MICRO/MJ	66.74	
VITAMIN B2	MICRO	80.00		- - -		80.00	MICRO/MJ	58.03	
NICOTINAMIDE	MILLI	3.10		- - -		3.10	MILLI/MJ	2.25	
PANTOTHENIC ACID	MILLI	0.50		0.49 - 0.50		0.50	MILLI/MJ	0.36	
VITAMIN B6	MILLI	0.22		- - -		0.22	MILLI/MJ	0.16	
FOLIC ACID	MICRO	20.00		- - -		20.00	MICRO/MJ	14.51	
VITAMIN C	-	0.00		- - -		0.00			
ARGININE	GRAM	0.49		- - -		0.49	GRAM/MJ	0.36	
CYSTINE	GRAM	0.20		- - -		0.20	GRAM/MJ	0.15	
HISTIDINE	GRAM	0.18		- - -		0.18	GRAM/MJ	0.13	
ISOLEUCINE	GRAM	0.41		- - -		0.41	GRAM/MJ	0.30	
LEUCINE	GRAM	0.66		- - -		0.66	GRAM/MJ	0.48	
LYSINE	GRAM	0.32		- - -		0.32	GRAM/MJ	0.23	

1 ESTIMATED BY THE DIFFERENCE METHOD
100 - (WATER + PROTEIN + FAT + MINERALS + TOTAL DIETARY FIBRE)

2 MODIFIED AOAC METHOD

CONSTITUENTS	DIM	AV	VARIATION			AVR	NUTR. DENS.		MOLPERC.
METHIONINE	GRAM	0.14	-	-	-	0.14	GRAM/MJ	0.10	
PHENYLALANINE	GRAM	0.49	-	-	-	0.49	GRAM/MJ	0.36	
THREONINE	GRAM	0.32	-	-	-	0.32	GRAM/MJ	0.23	
TRYPTOPHAN	GRAM	0.12	-	-	-	0.12	GRAM/MJ	0.09	
TYROSINE	GRAM	0.34	-	-	-	0.34	GRAM/MJ	0.25	
VALINE	GRAM	0.48	-	-	-	0.48	GRAM/MJ	0.35	
CELLULOSE	GRAM	1.01	-	-	-	1.01	GRAM/MJ	0.73	
DIETARY FIBRE, WAT.SOL.	GRAM	1.94	-	-	-	1.94	GRAM/MJ	1.41	
DIETARY FIBRE, WAT.INS.	GRAM	2.69	-	-	-	2.69	GRAM/MJ	1.95	

GERSTENGRÜTZE BARLEY GROATS GRUAU D'ORGE

		PROTEIN	FAT	CARBOHYDRATES	TOTAL
ENERGY VALUE (AVERAGE) PER 100 G EDIBLE PORTION	KJOULE	150	58	1106	1315
	(KCAL)	36	14	264	314
AMOUNT OF DIGESTIBLE CONSTITUENTS PER 100 G	GRAM	6.15	1.35	66.11	
ENERGY VALUE (AVERAGE) OF THE DIGESTIBLE FRACTION PER 100 G EDIBLE PORTION	KJOULE	117	53	1106	1276
	(KCAL)	28	13	264	305

WASTE PERCENTAGE AVERAGE 0.00

CONSTITUENTS	DIM	AV		VARIATION		AVR	NUTR. DENS.		MOLPERC.
WATER	GRAM	13.00		11.10 -	14.00	13.00	GRAM/MJ	10.19	
PROTEIN	GRAM	7.89		6.22 -	9.74	7.89	GRAM/MJ	6.18	
FAT	GRAM	1.50		1.00 -	2.00	1.50	GRAM/MJ	1.18	
AVAILABLE CARBOHYDR.	GRAM	66.11	1	- -	-	66.11	GRAM/MJ	51.81	
TOTAL DIETARY FIBRE	GRAM	10.30	2	10.30 -	10.30	10.30	GRAM/MJ	8.07	
MINERALS	GRAM	1.20		0.90 -	1.50	1.20	GRAM/MJ	0.94	
SODIUM	MILLI	3.00		- -	-	3.00	MILLI/MJ	2.35	
POTASSIUM	MILLI	160.00		- -	-	160.00	MILLI/MJ	125.38	
MAGNESIUM	MILLI	66.00		66.00 -	66.00	66.00	MILLI/MJ	51.72	
CALCIUM	MILLI	16.00		- -	-	16.00	MILLI/MJ	12.54	
MANGANESE	MILLI	1.27		1.27 -	1.27	1.27	MILLI/MJ	1.00	
IRON	MILLI	2.00		- -	-	2.00	MILLI/MJ	1.57	
COPPER	MILLI	0.12		0.12 -	0.12	0.12	MILLI/MJ	0.09	
ZINC	MILLI	1.28		1.28 -	1.28	1.28	MILLI/MJ	1.00	
CHROMIUM	MICRO	1.50		1.50 -	1.50	1.50	MICRO/MJ	1.18	
PHOSPHORUS	MILLI	189.00		- -	-	189.00	MILLI/MJ	148.11	
IODIDE	MICRO	1.00		1.00 -	1.00	1.00	MICRO/MJ	0.78	
SELENIUM	MICRO	1.00		1.00 -	1.00	1.00	MICRO/MJ	0.78	
VITAMIN B1	MILLI	0.20		0.12 -	0.29	0.20	MILLI/MJ	0.16	
VITAMIN B2	MICRO	80.00		- -	-	80.00	MICRO/MJ	62.69	
NICOTINAMIDE	MILLI	3.10		- -	-	3.10	MILLI/MJ	2.43	
PANTOTHENIC ACID	MILLI	0.49		0.49 -	0.49	0.49	MILLI/MJ	0.38	
VITAMIN B6	MILLI	0.29		0.29 -	0.29	0.29	MILLI/MJ	0.23	
FOLIC ACID	MICRO	19.00		19.00 -	19.00	19.00	MICRO/MJ	14.89	
VITAMIN C	-	0.00		- -	-	0.00			

1 ESTIMATED BY THE DIFFERENCE METHOD
100 - (WATER + PROTEIN + FAT + MINERALS + TOTAL DIETARY FIBRE)

2 METHOD OF SOUTHGATE AND ENGLYST

CONSTITUENTS	DIM	AV	VARIATION			AVR	NUTR. DENS.		MOLPERC.
ARGININE	GRAM	0.36	-	-	-	0.36	GRAM/MJ	0.28	
CYSTINE	GRAM	0.13	-	-	-	0.13	GRAM/MJ	0.10	
HISTIDINE	GRAM	0.14	-	-	-	0.14	GRAM/MJ	0.11	
ISOLEUCINE	GRAM	0.37	-	-	-	0.37	GRAM/MJ	0.29	
LEUCINE	GRAM	0.60	-	-	-	0.60	GRAM/MJ	0.47	
LYSINE	GRAM	0.19	-	-	-	0.19	GRAM/MJ	0.15	
METHIONINE	MILLI	80.00	-	-	-	80.00	MILLI/MJ	62.69	
PHENYLALANINE	GRAM	0.45	-	-	-	0.45	GRAM/MJ	0.35	
THREONINE	GRAM	0.29	-	-	-	0.29	GRAM/MJ	0.23	
TRYPTOPHAN	MILLI	90.00	-	-	-	90.00	MILLI/MJ	70.53	
TYROSINE	GRAM	0.29	-	-	-	0.29	GRAM/MJ	0.23	
VALINE	GRAM	0.41	-	-	-	0.41	GRAM/MJ	0.32	
STARCH	GRAM	60.30	60.30	-	60.30	60.30	GRAM/MJ	47.25	
PENTOSAN	GRAM	3.72	3.72	-	3.72	3.72	GRAM/MJ	2.92	
HEXOSAN	GRAM	4.80	4.80	-	4.80	4.80	GRAM/MJ	3.76	
CELLULOSE	GRAM	0.98	0.98	-	0.98	0.98	GRAM/MJ	0.77	
POLYURONIC ACID	GRAM	0.51	0.51	-	0.51	0.51	GRAM/MJ	0.40	

GRÜNKERN
(DINKEL, SPELZ)

UNRIPE SPELT GRAIN

GRAIN DE BLÉ VERT

TRITICUM SPELTA L.

		PROTEIN	FAT	CARBOHYDRATES	TOTAL
ENERGY VALUE (AVERAGE)	KJOULE	205	105	1059	1368
PER 100 G	(KCAL)	49	25	253	327
EDIBLE PORTION					
AMOUNT OF DIGESTIBLE	GRAM	9.14	2.43	63.25	
CONSTITUENTS PER 100 G					
ENERGY VALUE (AVERAGE)	KJOULE	174	95	1059	1327
OF THE DIGESTIBLE	(KCAL)	42	23	253	317
FRACTION PER 100 G					
EDIBLE PORTION					

WASTE PERCENTAGE AVERAGE 0.00

CONSTITUENTS	DIM	AV	VARIATION			AVR	NUTR. DENS.		MOLPERC.
WATER	GRAM	12.50	12.00	-	12.90	12.50	GRAM/MJ	9.42	
PROTEIN	GRAM	10.76	-	-	-	10.76	GRAM/MJ	8.11	
FAT	GRAM	2.70	-	-	-	2.70	GRAM/MJ	2.03	
AVAILABLE CARBOHYDR.	GRAM	63.25 1	-	-	-	63.25	GRAM/MJ	47.66	
TOTAL DIETARY FIBRE	GRAM	8.80 2	-	-	-	8.80	GRAM/MJ	6.63	
MINERALS	GRAM	1.99	1.90	-	2.08	1.99	GRAM/MJ	1.50	

1 ESTIMATED BY THE DIFFERENCE METHOD
100 - (WATER + PROTEIN + FAT + MINERALS + TOTAL DIETARY FIBRE)

2 METHOD OF MEUSER, SUCKOW AND KULIKOWSKI ("BERLINER METHODE")

CONSTITUENTS	DIM	AV	VARIATION			AVR	NUTR. DENS.		MOLPERC.
SODIUM	MILLI	2.80	0.90	-	6.50	2.80	MILLI/MJ	2.11	
POTASSIUM	MILLI	447.00	380.00	-	500.00	447.00	MILLI/MJ	336.79	
MAGNESIUM	MILLI	130.00	77.00	-	190.00	130.00	MILLI/MJ	97.95	
CALCIUM	MILLI	22.00	16.00	-	38.00	22.00	MILLI/MJ	16.58	
IRON	MILLI	4.20	2.80	-	6.20	4.20	MILLI/MJ	3.16	
COPPER	MILLI	0.26	0.13	-	0.54	0.26	MILLI/MJ	0.20	
PHOSPHORUS	MILLI	411.00	212.00	-	475.00	411.00	MILLI/MJ	309.67	
DIETARY FIBRE, WAT.SOL.	GRAM	3.40	-	-	-	3.40	GRAM/MJ	2.56	
DIETARY FIBRE, WAT.INS.	GRAM	5.40	-	-	-	5.40	GRAM/MJ	4.07	

GRÜNKERNMEHL — FLOUR OF UNRIPE SPELT GRAIN — FARINE DE BLÉ VERT

		PROTEIN	FAT	CARBOHYDRATES	TOTAL
ENERGY VALUE (AVERAGE)	KJOULE	184	78	1285	1547
PER 100 G EDIBLE PORTION	(KCAL)	44	19	307	370
AMOUNT OF DIGESTIBLE CONSTITUENTS PER 100 G	GRAM	8.00	1.80	76.80	
ENERGY VALUE (AVERAGE)	KJOULE	152	70	1285	1508
OF THE DIGESTIBLE FRACTION PER 100 G EDIBLE PORTION	(KCAL)	36	17	307	360

WASTE PERCENTAGE AVERAGE 0.00

CONSTITUENTS	DIM	AV	VARIATION			AVR	NUTR. DENS.		MOLPERC.
WATER	GRAM	10.00	9.50	-	11.00	10.00	GRAM/MJ	6.63	
PROTEIN	GRAM	9.65	8.26	-	11.04	9.65	GRAM/MJ	6.40	
FAT	GRAM	2.00	1.80	-	2.20	2.00	GRAM/MJ	1.33	
AVAILABLE CARBOHYDR.	GRAM	76.80 1	-	-	-	76.80	GRAM/MJ	50.93	
MINERALS	GRAM	1.55	1.40	-	1.70	1.55	GRAM/MJ	1.03	
SODIUM	MILLI	3.00	-	-	-	3.00	MILLI/MJ	1.99	
POTASSIUM	MILLI	349.00	-	-	-	349.00	MILLI/MJ	231.46	
CALCIUM	MILLI	20.00	-	-	-	20.00	MILLI/MJ	13.26	

1 ESTIMATED BY THE DIFFERENCE METHOD
100 - (WATER + PROTEIN + FAT + MINERALS)

HAFER
ENTSPELZT
GANZES KORN

OATS
WITHOUT HUSK
WHOLE GRAIN

AVOINE
DÉGLUMÉS
GRAINS ENTIERS

AVENA SATIVA L.

		PROTEIN	FAT	CARBOHYDRATES	TOTAL
ENERGY VALUE (AVERAGE)	KJOULE	223	276	1001	1499
PER 100 G EDIBLE PORTION	(KCAL)	53	66	239	358
AMOUNT OF DIGESTIBLE CONSTITUENTS PER 100 G	GRAM	8.88	6.38	59.80	
ENERGY VALUE (AVERAGE)	KJOULE	169	248	1001	1418
OF THE DIGESTIBLE FRACTION PER 100 G EDIBLE PORTION	(KCAL)	40	59	239	339

WASTE PERCENTAGE AVERAGE 0.00

CONSTITUENTS	DIM	AV	VARIATION		AVR	NUTR. DENS.		MOLPERC.
WATER	GRAM	13.00	10.90	- 15.00	13.00	GRAM/MJ	9.17	
PROTEIN	GRAM	11.69	11.04	- 12.25	11.69	GRAM/MJ	8.24	
FAT	GRAM	7.09	6.81	- 7.47	7.09	GRAM/MJ	5.00	
AVAILABLE CARBOHYDR.	GRAM	59.80 1	-	- -	59.80	GRAM/MJ	42.16	
TOTAL DIETARY FIBRE	GRAM	5.57 2	-	- -	5.57	GRAM/MJ	3.93	
MINERALS	GRAM	2.85	2.02	- 3.37	2.85	GRAM/MJ	2.01	
SODIUM	MILLI	8.40	3.10	- 12.00	8.40	MILLI/MJ	5.92	
POTASSIUM	MILLI	355.00	338.00	- 387.00	355.00	MILLI/MJ	250.31	
MAGNESIUM	MILLI	129.00	87.00	- 176.00	129.00	MILLI/MJ	90.96	
CALCIUM	MILLI	79.60	25.20	- 107.00	79.60	MILLI/MJ	56.13	
MANGANESE	MILLI	3.70	2.80	- 4.60	3.70	MILLI/MJ	2.61	
IRON	MILLI	5.80	4.50	- 7.00	5.80	MILLI/MJ	4.09	
COBALT	MICRO	8.50	2.20	- 20.00	8.50	MICRO/MJ	5.99	
COPPER	MILLI	0.47	0.23	- 0.70	0.47	MILLI/MJ	0.33	
ZINC	MILLI	4.50	3.10	- 4.90	4.50	MILLI/MJ	3.17	
NICKEL	MILLI	0.21	0.05	- 0.23	0.21	MILLI/MJ	0.15	
CHROMIUM	MICRO	13.10	2.40	- 24.00	13.10	MICRO/MJ	9.24	
MOLYBDENUM	MICRO	70.00	10.00	- 150.00	70.00	MICRO/MJ	49.36	
PHOSPHORUS	MILLI	342.00	332.00	- 352.00	342.00	MILLI/MJ	241.14	
CHLORIDE	MILLI	119.00	-	- -	119.00	MILLI/MJ	83.91	
FLUORIDE	MICRO	95.00	25.00	- 300.00	95.00	MICRO/MJ	66.98	
IODIDE	MICRO	6.00	-	- -	6.00	MICRO/MJ	4.23	
BORON	MILLI	-	0.39	- 0.75				
SELENIUM	MICRO	-	0.30	- 4.60				
SILICON	MILLI	425.00	340.00	- 580.00	425.00	MILLI/MJ	299.67	
VITAMIN E ACTIVITY	MILLI	0.84	0.71	- 0.99	0.84	MILLI/MJ	0.59	

1 ESTIMATED BY THE DIFFERENCE METHOD
100 - (WATER + PROTEIN + FAT + MINERALS + TOTAL DIETARY FIBRE)

2 METHOD OF MEUSER, SUCKOW AND KULIKOWSKI ("BERLINER METHODE")

CONSTITUENTS	DIM	AV	VARIATION			AVR	NUTR. DENS.		MOLPERC.
TOTAL TOCOPHEROLS	MILLI	1.80	1.55	-	2.15	1.80	MILLI/MJ	1.27	
ALPHA-TOCOPHEROL	MILLI	0.47	0.37	-	0.55	0.47	MILLI/MJ	0.33	
BETA-TOCOPHEROL	MICRO	80.00	70.00	-	90.00	80.00	MICRO/MJ	56.41	
ALPHA-TOCOTRIENOL	MILLI	1.10	1.00	-	1.30	1.10	MILLI/MJ	0.78	
BETA-TOCOTRIENOL	MILLI	0.17	0.10	-	0.20	0.17	MILLI/MJ	0.12	
VITAMIN K	MICRO	50.00	10.00	-	80.00	50.00	MICRO/MJ	35.25	
VITAMIN B1	MILLI	0.52	0.30	-	0.70	0.52	MILLI/MJ	0.37	
VITAMIN B2	MILLI	0.17	0.10	-	0.30	0.17	MILLI/MJ	0.12	
NICOTINAMIDE	MILLI	2.37	0.88	-	6.00	2.37	MILLI/MJ	1.67	
PANTOTHENIC ACID	MILLI	0.71	0.57	-	1.00	0.71	MILLI/MJ	0.50	
VITAMIN B6	MILLI	0.96	0.22	-	2.30	0.96	MILLI/MJ	0.68	
BIOTIN	MICRO	13.00	6.50	-	20.00	13.00	MICRO/MJ	9.17	
FOLIC ACID	MICRO	33.00	-	-	-	33.00	MICRO/MJ	23.27	
ALANINE	GRAM	0.72	0.69	-	0.75	0.72	GRAM/MJ	0.51	7.7
ARGININE	GRAM	0.85	0.54	-	0.97	0.85	GRAM/MJ	0.60	4.7
ASPARTIC ACID	GRAM	1.11	1.00	-	1.18	1.11	GRAM/MJ	0.78	7.9
CYSTINE	GRAM	0.32	0.17	-	0.38	0.32	GRAM/MJ	0.23	1.3
GLUTAMIC ACID	GRAM	2.90	2.58	-	3.06	2.90	GRAM/MJ	2.04	18.8
GLYCINE	GRAM	0.78	0.73	-	0.82	0.78	GRAM/MJ	0.55	9.9
HISTIDINE	GRAM	0.27	0.15	-	0.31	0.27	GRAM/MJ	0.19	1.7
ISOLEUCINE	GRAM	0.56	0.47	-	0.79	0.56	GRAM/MJ	0.39	4.1
LEUCINE	GRAM	1.02	0.71	-	1.20	1.02	GRAM/MJ	0.72	7.4
LYSINE	GRAM	0.55	0.25	-	0.69	0.55	GRAM/MJ	0.39	3.6
METHIONINE	GRAM	0.23	0.14	-	0.29	0.23	GRAM/MJ	0.16	1.5
PHENYLALANINE	GRAM	0.70	0.57	-	0.96	0.70	GRAM/MJ	0.49	4.0
PROLINE	GRAM	0.87	0.76	-	1.17	0.87	GRAM/MJ	0.61	7.2
SERINE	GRAM	0.74	0.66	-	0.85	0.74	GRAM/MJ	0.52	6.7
THREONINE	GRAM	0.49	0.30	-	0.60	0.49	GRAM/MJ	0.35	3.9
TRYPTOPHAN	GRAM	0.19	0.09	-	0.22	0.19	GRAM/MJ	0.13	0.9
TYROSINE	GRAM	0.45	0.19	-	0.59	0.45	GRAM/MJ	0.32	2.4
VALINE	GRAM	0.79	0.53	-	0.97	0.79	GRAM/MJ	0.56	6.4
SUCROSE	GRAM	1.05	-	-	-	1.05	GRAM/MJ	0.74	
RAFFINOSE	GRAM	0.25	-	-	-	0.25	GRAM/MJ	0.18	
MALTOSE	MILLI	30.00	-	-	-	30.00	MILLI/MJ	21.15	
GLUCODIFRUCTOSE	MILLI	90.00	-	-	-	90.00	MILLI/MJ	63.46	
STARCH	GRAM	56.58	55.90	-	60.00	56.58	GRAM/MJ	39.89	
PHYTIC ACID	GRAM	0.90	0.79	-	1.01	0.90	GRAM/MJ	0.63	
MYRISTIC ACID	MILLI	40.00	-	-	-	40.00	MILLI/MJ	28.20	
PALMITIC ACID	GRAM	1.28	-	-	-	1.28	GRAM/MJ	0.90	
STEARIC ACID	GRAM	0.11	-	-	-	0.11	GRAM/MJ	0.08	
PALMITOLEIC ACID	MILLI	54.00	-	-	-	54.00	MILLI/MJ	38.08	
OLEIC ACID	GRAM	2.46	-	-	-	2.46	GRAM/MJ	1.73	
LINOLEIC ACID	GRAM	2.74	-	-	-	2.74	GRAM/MJ	1.93	
LINOLENIC ACID	GRAM	0.12	-	-	-	0.12	GRAM/MJ	0.08	
DIETARY FIBRE,WAT.SOL.	GRAM	1.65	-	-	-	1.65	GRAM/MJ	1.16	
DIETARY FIBRE,WAT.INS.	GRAM	3.92	-	-	-	3.92	GRAM/MJ	2.76	

HAFERFLOCKEN ROLLED OATS FLOCONS D'AVOINE

		PROTEIN	FAT	CARBOHYDRATES	TOTAL
ENERGY VALUE (AVERAGE) PER 100 G EDIBLE PORTION	KJOULE	239	272	1059	1570
	(KCAL)	57	65	253	375
AMOUNT OF DIGESTIBLE CONSTITUENTS PER 100 G	GRAM	9.52	6.30	63.29	
ENERGY VALUE (AVERAGE) OF THE DIGESTIBLE FRACTION PER 100 G EDIBLE PORTION	KJOULE	181	245	1059	1486
	(KCAL)	43	59	253	355

WASTE PERCENTAGE AVERAGE 0.00

CONSTITUENTS	DIM	AV		VARIATION		AVR	NUTR. DENS.		MOLPERC.
WATER	GRAM	10.00		8.90 -	10.90	10.00	GRAM/MJ	6.73	
PROTEIN	GRAM	12.53		11.14 -	13.36	12.53	GRAM/MJ	8.43	
FAT	GRAM	7.00		6.30 -	8.50	7.00	GRAM/MJ	4.71	
AVAILABLE CARBOHYDR.	GRAM	63.29	1	- -	-	63.29	GRAM/MJ	42.60	
TOTAL DIETARY FIBRE	GRAM	5.43	2	- -	-	5.43	GRAM/MJ	3.65	
MINERALS	GRAM	1.75		1.65 -	1.90	1.75	GRAM/MJ	1.18	
SODIUM	MILLI	5.00		3.00 -	8.00	5.00	MILLI/MJ	3.37	
POTASSIUM	MILLI	335.00		290.00 -	368.00	335.00	MILLI/MJ	225.49	
MAGNESIUM	MILLI	139.00		113.00 -	150.00	139.00	MILLI/MJ	93.56	
CALCIUM	MILLI	54.00		50.00 -	56.00	54.00	MILLI/MJ	36.35	
MANGANESE	MILLI	4.90		4.00 -	5.80	4.90	MILLI/MJ	3.30	
IRON	MILLI	4.60		3.80 -	6.70	4.60	MILLI/MJ	3.10	
COBALT	MICRO	3.00		1.00 -	5.00	3.00	MICRO/MJ	2.02	
COPPER	MILLI	0.53		0.23 -	0.74	0.53	MILLI/MJ	0.36	
ZINC	MILLI	4.40		3.50 -	6.90	4.40	MILLI/MJ	2.96	
PHOSPHORUS	MILLI	391.00		380.00 -	416.00	391.00	MILLI/MJ	263.18	
CHLORIDE	MILLI	61.00		49.00 -	73.00	61.00	MILLI/MJ	41.06	
FLUORIDE	MICRO	37.00		30.00 -	45.00	37.00	MICRO/MJ	24.90	
IODIDE	MICRO	4.00		3.60 -	4.50	4.00	MICRO/MJ	2.69	
BORON	MILLI	0.20		- -	-	0.20	MILLI/MJ	0.13	
SELENIUM	MICRO	-		8.00 -	11.40				
BROMINE	MICRO	16.00		- -	-	16.00	MICRO/MJ	10.77	
VITAMIN B1	MILLI	0.59		0.40 -	0.68	0.59	MILLI/MJ	0.40	
VITAMIN B2	MILLI	0.15		0.10 -	0.19	0.15	MILLI/MJ	0.10	
NICOTINAMIDE	MILLI	1.00		0.90 -	1.10	1.00	MILLI/MJ	0.67	
PANTOTHENIC ACID	MILLI	1.09		0.92 -	1.41	1.09	MILLI/MJ	0.73	

1 ESTIMATED BY THE DIFFERENCE METHOD
100 - (WATER + PROTEIN + FAT + MINERALS + TOTAL DIETARY FIBRE)

2 MODIFIED AOAC METHOD

CONSTITUENTS	DIM	AV	VARIATION			AVR	NUTR. DENS.		MOLPERC.
VITAMIN B6	MILLI	0.16	0.12	-	0.20	0.16	MILLI/MJ	0.11	
BIOTIN	MICRO	20.00	-	-	-	20.00	MICRO/MJ	13.46	
FOLIC ACID	MICRO	24.00	16.00	-	26.00	24.00	MICRO/MJ	16.15	
VITAMIN C	-	0.00	-	-	-	0.00			
ALANINE	GRAM	0.79	0.73	-	0.84	0.79	GRAM/MJ	0.53	7.9
ARGININE	GRAM	0.87	0.73	-	1.01	0.87	GRAM/MJ	0.59	4.5
ASPARTIC ACID	GRAM	1.29	1.12	-	1.45	1.29	GRAM/MJ	0.87	8.7
CYSTINE	GRAM	0.39	0.22	-	0.50	0.39	GRAM/MJ	0.26	1.5
GLUTAMIC ACID	GRAM	3.08	2.97	-	3.18	3.08	GRAM/MJ	2.07	18.7
GLYCINE	GRAM	0.85	0.81	-	0.90	0.85	GRAM/MJ	0.57	10.1
HISTIDINE	GRAM	0.30	0.22	-	0.32	0.30	GRAM/MJ	0.20	1.7
ISOLEUCINE	GRAM	0.61	0.59	-	0.69	0.61	GRAM/MJ	0.41	4.2
LEUCINE	GRAM	1.13	1.07	-	1.17	1.13	GRAM/MJ	0.76	7.7
LYSINE	GRAM	0.50	0.47	-	0.57	0.50	GRAM/MJ	0.34	3.1
METHIONINE	GRAM	0.24	0.21	-	0.28	0.24	GRAM/MJ	0.16	1.4
PHENYLALANINE	GRAM	0.78	0.69	-	0.87	0.78	GRAM/MJ	0.53	4.2
PROLINE	GRAM	0.84	0.82	-	0.87	0.84	GRAM/MJ	0.57	6.5
SERINE	GRAM	0.71	0.71	-	0.72	0.71	GRAM/MJ	0.48	6.0
THREONINE	GRAM	0.53	0.47	-	0.57	0.53	GRAM/MJ	0.36	4.0
TRYPTOPHAN	GRAM	0.19	0.18	-	0.22	0.19	GRAM/MJ	0.13	0.8
TYROSINE	GRAM	0.57	0.49	-	0.69	0.57	GRAM/MJ	0.38	2.8
VALINE	GRAM	0.81	0.76	-	0.89	0.81	GRAM/MJ	0.55	6.2
SUCROSE	GRAM	0.70	-	-	-	0.70	GRAM/MJ	0.47	
STARCH	GRAM	60.50	58.50	-	61.80	60.50	GRAM/MJ	40.72	
CELLULOSE	GRAM	0.80	-	-	-	0.80	GRAM/MJ	0.54	
MYRISTIC ACID	MILLI	20.00	-	-	-	20.00	MILLI/MJ	13.46	
PALMITIC ACID	GRAM	1.21	-	-	-	1.21	GRAM/MJ	0.81	
STEARIC ACID	GRAM	0.10	-	-	-	0.10	GRAM/MJ	0.07	
ARACHIDIC ACID	MILLI	40.00	-	-	-	40.00	MILLI/MJ	26.92	
PALMITOLEIC ACID	MILLI	20.00	-	-	-	20.00	MILLI/MJ	13.46	
OLEIC ACID	GRAM	2.60	-	-	-	2.60	GRAM/MJ	1.75	
LINOLEIC ACID	GRAM	2.60	2.50	-	2.70	2.60	GRAM/MJ	1.75	
LINOLENIC ACID	GRAM	0.10	0.08	-	0.12	0.10	GRAM/MJ	0.07	
DIETARY FIBRE, WAT.SOL.	GRAM	1.75	-	-	-	1.75	GRAM/MJ	1.18	
DIETARY FIBRE, WAT.INS.	GRAM	3.68	-	-	-	3.68	GRAM/MJ	2.48	

HAFERGRÜTZE GROATS GRUAU D'AVOINE

		PROTEIN	FAT	CARBOHYDRATES	TOTAL
ENERGY VALUE (AVERAGE) PER 100 G EDIBLE PORTION	KJOULE	246	227	1167	1640
	(KCAL)	59	54	279	392
AMOUNT OF DIGESTIBLE CONSTITUENTS PER 100 G	GRAM	9.80	5.25	69.73	
ENERGY VALUE (AVERAGE) OF THE DIGESTIBLE FRACTION PER 100 G EDIBLE PORTION	KJOULE	187	205	1167	1558
	(KCAL)	45	49	279	372

WASTE PERCENTAGE AVERAGE 0.00

CONSTITUENTS	DIM	AV	VARIATION		AVR	NUTR. DENS.		MOLPERC.
WATER	GRAM	9.53	9.17	9.70	9.53	GRAM/MJ	6.12	
PROTEIN	GRAM	12.90	10.58	15.13	12.90	GRAM/MJ	8.28	
FAT	GRAM	5.84	4.80	6.48	5.84	GRAM/MJ	3.75	
AVAILABLE CARBOHYDR.	GRAM	69.73 1	-	-	69.73	GRAM/MJ	44.75	
MINERALS	GRAM	2.00	1.65	2.27	2.00	GRAM/MJ	1.28	
SODIUM	MILLI	6.33	6.33	7.24	6.33	MILLI/MJ	4.06	
POTASSIUM	MILLI	308.00	295.00	320.00	308.00	MILLI/MJ	197.67	
MAGNESIUM	MILLI	71.00	64.00	77.00	71.00	MILLI/MJ	45.57	
CALCIUM	MILLI	67.00	47.00	87.00	67.00	MILLI/MJ	43.00	
IRON	MILLI	3.89	3.80	4.07	3.89	MILLI/MJ	2.50	
COBALT	MICRO	1.45	0.90	2.71	1.45	MICRO/MJ	0.93	
PHOSPHORUS	MILLI	349.00	-	-	349.00	MILLI/MJ	223.98	
CHLORIDE	MILLI	88.00	-	-	88.00	MILLI/MJ	56.48	
FLUORIDE	MICRO	30.00	19.00	44.00	30.00	MICRO/MJ	19.25	
IODIDE	MICRO	4.52	4.52	5.43	4.52	MICRO/MJ	2.90	
VITAMIN B1	MILLI	0.60	-	-	0.60	MILLI/MJ	0.39	
VITAMIN B2	MILLI	0.22	-	-	0.22	MILLI/MJ	0.14	
ARGININE	GRAM	0.79	0.54	0.97	0.79	GRAM/MJ	0.51	
CYSTINE	GRAM	0.26	0.17	0.39	0.26	GRAM/MJ	0.17	
HISTIDINE	GRAM	0.22	0.16	0.31	0.22	GRAM/MJ	0.14	
ISOLEUCINE	GRAM	0.62	0.54	0.76	0.62	GRAM/MJ	0.40	
LEUCINE	GRAM	0.90	0.69	1.15	0.90	GRAM/MJ	0.58	
LYSINE	GRAM	0.45	0.24	0.66	0.45	GRAM/MJ	0.29	
METHIONINE	GRAM	0.18	0.13	0.30	0.18	GRAM/MJ	0.12	
PHENYLALANINE	GRAM	0.64	0.54	0.93	0.64	GRAM/MJ	0.41	
THREONINE	GRAM	0.40	0.27	0.46	0.40	GRAM/MJ	0.26	
TRYPTOPHAN	GRAM	0.16	0.07	0.20	0.16	GRAM/MJ	0.10	
TYROSINE	GRAM	0.45	0.19	0.58	0.45	GRAM/MJ	0.29	
VALINE	GRAM	0.71	0.49	0.90	0.71	GRAM/MJ	0.46	
LINOLEIC ACID	GRAM	2.40	-	-	2.40	GRAM/MJ	1.54	
LINOLENIC ACID	MILLI	67.00	-	-	67.00	MILLI/MJ	43.00	

1 ESTIMATED BY THE DIFFERENCE METHOD
100 - (WATER + PROTEIN + FAT + MINERALS)

HAFERMEHL OAT MEAL FARINE D'AVOINE

		PROTEIN	FAT	CARBOHYDRATES	TOTAL
ENERGY VALUE (AVERAGE)	KJOULE	263	278	1136	1677
PER 100 G	(KCAL)	63	66	271	401
EDIBLE PORTION					
AMOUNT OF DIGESTIBLE	GRAM	10.51	6.43	67.87	
CONSTITUENTS PER 100 G					
ENERGY VALUE (AVERAGE)	KJOULE	200	250	1136	1586
OF THE DIGESTIBLE	(KCAL)	48	60	271	379
FRACTION PER 100 G					
EDIBLE PORTION					

WASTE PERCENTAGE AVERAGE 0.00

CONSTITUENTS	DIM	AV	VARIATION	AVR	NUTR. DENS.	MOLPERC.
WATER	GRAM	9.35	8.30 - 10.00	9.35	GRAM/MJ 5.89	
PROTEIN	GRAM	13.83	13.18 - 14.76	13.83	GRAM/MJ 8.72	
FAT	GRAM	7.15	6.70 - 7.50	7.15	GRAM/MJ 4.51	
AVAILABLE CARBOHYDR.	GRAM	67.87 1	- - -	67.87	GRAM/MJ 42.78	
MINERALS	GRAM	1.80	1.65 - 1.90	1.80	GRAM/MJ 1.13	
SODIUM	MILLI	6.00	4.00 - 7.00	6.00	MILLI/MJ 3.78	
POTASSIUM	MILLI	268.00	253.00 - 282.00	268.00	MILLI/MJ 168.94	
MAGNESIUM	MILLI	131.00	131.00 - 132.00	131.00	MILLI/MJ 82.58	
CALCIUM	MILLI	54.70	32.00 - 71.50	54.70	MILLI/MJ 34.48	
IRON	MILLI	4.20	3.80 - 4.50	4.20	MILLI/MJ 2.65	
CHROMIUM	MICRO	6.00	- - -	6.00	MICRO/MJ 3.78	
PHOSPHORUS	MILLI	405.00	- - -	405.00	MILLI/MJ 255.30	
IODIDE	MICRO	4.20	- - -	4.20	MICRO/MJ 2.65	
BORON	MILLI	0.13	- - -	0.13	MILLI/MJ 0.08	
VITAMIN B1	MILLI	0.56	0.50 - 0.63	0.56	MILLI/MJ 0.35	
VITAMIN B2	MILLI	0.12	0.10 - 0.14	0.12	MILLI/MJ 0.08	
NICOTINAMIDE	MILLI	0.93	0.80 - 1.00	0.93	MILLI/MJ 0.59	
VITAMIN B6	MILLI	0.20	0.17 - 0.23	0.20	MILLI/MJ 0.13	
VITAMIN C	-	0.00	- - -	0.00		
ALANINE	GRAM	0.88	0.84 - 0.91	0.88	GRAM/MJ 0.55	7.9
ARGININE	GRAM	0.95	0.79 - 1.01	0.95	GRAM/MJ 0.60	4.4
ASPARTIC ACID	GRAM	1.44	1.30 - 1.58	1.44	GRAM/MJ 0.91	8.7
CYSTINE	GRAM	0.39	0.32 - 0.42	0.39	GRAM/MJ 0.25	1.3
GLUTAMIC ACID	GRAM	3.44	3.41 - 3.48	3.44	GRAM/MJ 2.17	18.8
GLYCINE	GRAM	0.96	0.92 - 0.99	0.96	GRAM/MJ 0.61	10.3
HISTIDINE	GRAM	0.32	0.27 - 0.33	0.32	GRAM/MJ 0.20	1.7
ISOLEUCINE	GRAM	0.70	0.65 - 0.82	0.70	GRAM/MJ 0.44	4.3
LEUCINE	GRAM	1.24	1.20 - 1.31	1.24	GRAM/MJ 0.78	7.6
LYSINE	GRAM	0.58	0.51 - 0.64	0.58	GRAM/MJ 0.37	3.2
METHIONINE	GRAM	0.28	0.23 - 0.31	0.28	GRAM/MJ 0.18	1.5
PHENYLALANINE	GRAM	0.85	0.82 - 0.95	0.85	GRAM/MJ 0.54	4.1
PROLINE	GRAM	0.93	0.90 - 0.95	0.93	GRAM/MJ 0.59	6.5

1 ESTIMATED BY THE DIFFERENCE METHOD
100 - (WATER + PROTEIN + FAT + MINERALS)

CONSTITUENTS	DIM	AV	VARIATION			AVR	NUTR. DENS.		MOLPERC.
SERINE	GRAM	0.81	0.76	-	0.84	0.81	GRAM/MJ	0.51	6.2
THREONINE	GRAM	0.56	0.53	-	0.60	0.56	GRAM/MJ	0.35	3.8
TRYPTOPHAN	GRAM	0.20	0.19	-	0.24	0.20	GRAM/MJ	0.13	0.8
TYROSINE	GRAM	0.58	0.54	-	0.65	0.58	GRAM/MJ	0.37	2.6
VALINE	GRAM	0.92	0.89	-	0.97	0.92	GRAM/MJ	0.58	6.3
GLUCOSE	MILLI	70.00	-	-	-	70.00	MILLI/MJ	44.13	
FRUCTOSE	MILLI	30.00	20.00	-	50.00	30.00	MILLI/MJ	18.91	
SUCROSE	GRAM	0.49	0.40	-	0.63	0.49	GRAM/MJ	0.31	
RAFFINOSE	GRAM	0.20	0.16	-	0.26	0.20	GRAM/MJ	0.13	
MALTOSE	MILLI	20.00	10.00	-	30.00	20.00	MILLI/MJ	12.61	
STACHYOSE	MILLI	80.00	-	-	-	80.00	MILLI/MJ	50.43	
MYRISTIC ACID	MILLI	20.00	-	-	-	20.00	MILLI/MJ	12.61	
PALMITIC ACID	GRAM	1.21	-	-	-	1.21	GRAM/MJ	0.76	
STEARIC ACID	GRAM	0.10	-	-	-	0.10	GRAM/MJ	0.06	
ARACHIDIC ACID	MILLI	40.00	-	-	-	40.00	MILLI/MJ	25.21	
PALMITOLEIC ACID	MILLI	20.00	-	-	-	20.00	MILLI/MJ	12.61	
OLEIC ACID	GRAM	2.60	-	-	-	2.60	GRAM/MJ	1.64	
LINOLEIC ACID	GRAM	2.93	-	-	-	2.93	GRAM/MJ	1.85	
LINOLENIC ACID	MILLI	82.00	-	-	-	82.00	MILLI/MJ	51.69	

HIRSE
GESCHÄLTES KORN

MILLET
SHUCKED CORN

MILLET
GRAINS DÉCORTIQUÉS

PANICUM MILIACEUM L.

		PROTEIN	FAT	CARBOHYDRATES	TOTAL
ENERGY VALUE (AVERAGE)	KJOULE	187	152	1151	1490
PER 100 G	(KCAL)	45	36	275	356
EDIBLE PORTION					
AMOUNT OF DIGESTIBLE	GRAM	8.36	3.51	68.76	
CONSTITUENTS PER 100 G					
ENERGY VALUE (AVERAGE)	KJOULE	159	137	1151	1447
OF THE DIGESTIBLE	(KCAL)	38	33	275	346
FRACTION PER 100 G					
EDIBLE PORTION					

WASTE PERCENTAGE AVERAGE 0.00

CONSTITUENTS	DIM	AV		VARIATION			AVR	NUTR. DENS.		MOLPERC.
WATER	GRAM	12.10		11.00	-	13.60	12.10	GRAM/MJ	8.36	
PROTEIN	GRAM	9.84		9.28	-	10.58	9.84	GRAM/MJ	6.80	
FAT	GRAM	3.90		2.00	-	5.00	3.90	GRAM/MJ	2.70	
AVAILABLE CARBOHYDR.	GRAM	68.76	1	-	-	-	68.76	GRAM/MJ	47.53	
TOTAL DIETARY FIBRE	GRAM	3.80	2	-	-	-	3.80	GRAM/MJ	2.63	
MINERALS	GRAM	1.60		0.75	-	2.40	1.60	GRAM/MJ	1.11	

1 ESTIMATED BY THE DIFFERENCE METHOD
100 - (WATER + PROTEIN + FAT + MINERALS + TOTAL DIETARY FIBRE)

2 METHOD OF MEUSER, SUCKOW AND KULIKOWSKI ("BERLINER METHODE")

CONSTITUENTS	DIM	AV	VARIATION			AVR	NUTR. DENS.		MOLPERC.
SODIUM	MILLI	3.00	-	-	-	3.00	MILLI/MJ	2.07	
POTASSIUM	MILLI	-	150.00	-	280.00				
MAGNESIUM	MILLI	170.00	-	-	-	170.00	MILLI/MJ	117.52	
CALCIUM	MILLI	-	0.00	-	50.00				
MANGANESE	MILLI	1.90	-	-	-	1.90	MILLI/MJ	1.31	
IRON	MILLI	9.00	-	-	-	9.00	MILLI/MJ	6.22	
COPPER	MILLI	0.85	-	-	-	0.85	MILLI/MJ	0.59	
ZINC	MILLI	1.80	-	-	-	1.80	MILLI/MJ	1.24	
PHOSPHORUS	MILLI	310.00	-	-	-	310.00	MILLI/MJ	214.30	
CHLORIDE	MILLI	15.00	-	-	-	15.00	MILLI/MJ	10.37	
FLUORIDE	MICRO	50.00	20.00	-	80.00	50.00	MICRO/MJ	34.56	
IODIDE	MICRO	2.50	-	-	-	2.50	MICRO/MJ	1.73	
BORON	MILLI	0.52	0.23	-	0.80	0.52	MILLI/MJ	0.36	
VITAMIN E ACTIVITY	MICRO	70.00	-	-	-	70.00	MICRO/MJ	48.39	
TOTAL TOCOPHEROLS	MILLI	1.70	-	-	-	1.70	MILLI/MJ	1.18	
BETA-TOCOPHEROL	MICRO	40.00	-	-	-	40.00	MICRO/MJ	27.65	
DELTA-TOCOPHEROL	MILLI	0.31	-	-	-	0.31	MILLI/MJ	0.21	
BETA-TOCOTRIENOL	MILLI	1.30	-	-	-	1.30	MILLI/MJ	0.90	
VITAMIN B1	MILLI	0.26	0.20	-	0.31	0.26	MILLI/MJ	0.18	
VITAMIN B2	MILLI	0.14	-	-	-	0.14	MILLI/MJ	0.10	
NICOTINAMIDE	MILLI	1.80	-	-	-	1.80	MILLI/MJ	1.24	
VITAMIN B6	MILLI	0.75	0.70	-	0.80	0.75	MILLI/MJ	0.52	
ALANINE	GRAM	1.34	-	-	-	1.34	GRAM/MJ	0.93	14.9
ARGININE	GRAM	0.37	0.32	-	0.50	0.37	GRAM/MJ	0.26	2.1
ASPARTIC ACID	GRAM	0.64	-	-	-	0.64	GRAM/MJ	0.44	4.8
CYSTINE	GRAM	0.15	0.12	-	0.17	0.15	GRAM/MJ	0.10	0.6
GLUTAMIC ACID	GRAM	2.24	-	-	-	2.24	GRAM/MJ	1.55	15.1
GLYCINE	GRAM	0.33	-	-	-	0.33	GRAM/MJ	0.23	4.4
HISTIDINE	GRAM	0.19	-	-	-	0.19	GRAM/MJ	0.13	1.2
ISOLEUCINE	GRAM	0.55	0.33	-	0.86	0.55	GRAM/MJ	0.38	4.2
LEUCINE	GRAM	1.35	1.14	-	2.25	1.35	GRAM/MJ	0.93	10.2
LYSINE	GRAM	0.28	0.21	-	0.45	0.28	GRAM/MJ	0.19	1.9
METHIONINE	GRAM	0.25	0.17	-	0.31	0.25	GRAM/MJ	0.17	1.7
PHENYLALANINE	GRAM	0.46	0.36	-	0.53	0.46	GRAM/MJ	0.32	2.8
PROLINE	GRAM	1.09	-	-	-	1.09	GRAM/MJ	0.75	9.4
SERINE	GRAM	1.68	-	-	-	1.68	GRAM/MJ	1.16	15.9
THREONINE	GRAM	0.42	0.26	-	0.62	0.42	GRAM/MJ	0.29	3.5
TRYPTOPHAN	GRAM	0.18	0.13	-	0.25	0.18	GRAM/MJ	0.12	0.9
TYROSINE	GRAM	0.26	-	-	-	0.26	GRAM/MJ	0.18	1.4
VALINE	GRAM	0.61	0.45	-	0.80	0.61	GRAM/MJ	0.42	5.2
SUCROSE	GRAM	1.45	1.20	-	1.60	1.45	GRAM/MJ	1.00	
RAFFINOSE	GRAM	0.63	0.57	-	0.72	0.63	GRAM/MJ	0.44	
STACHYOSE	MILLI	82.00	50.00	-	110.00	82.00	MILLI/MJ	56.69	
STARCH	GRAM	60.00	58.10	-	60.95	60.00	GRAM/MJ	41.48	
CELLULOSE	GRAM	2.10	-	-	-	2.10	GRAM/MJ	1.45	
PALMITIC ACID	GRAM	0.76	0.60	-	0.93	0.76	GRAM/MJ	0.53	
STEARIC ACID	GRAM	0.19	0.07	-	0.30	0.19	GRAM/MJ	0.13	
ARACHIDIC ACID	MILLI	37.00	-	-	-	37.00	MILLI/MJ	25.58	
OLEIC ACID	GRAM	0.93	0.71	-	1.16	0.93	GRAM/MJ	0.64	
LINOLEIC ACID	GRAM	1.77	1.49	-	2.05	1.77	GRAM/MJ	1.22	
LINOLENIC ACID	GRAM	0.13	0.07	-	0.19	0.13	GRAM/MJ	0.09	
DIETARY FIBRE, WAT.SOL.	GRAM	1.40	-	-	-	1.40	GRAM/MJ	0.97	
DIETARY FIBRE, WAT.INS.	GRAM	2.40	-	-	-	2.40	GRAM/MJ	1.66	

MAIS
GANZES KORN

MAIZE
WHOLE GRAIN

MAÏS
GRAINS ENTIERS

ZEA MAYS L.

		PROTEIN	FAT	CARBOHYDRATES	TOTAL
ENERGY VALUE (AVERAGE) PER 100 G EDIBLE PORTION	KJOULE	163	148	1082	1393
	(KCAL)	39	35	259	333
AMOUNT OF DIGESTIBLE CONSTITUENTS PER 100 G	GRAM	5.12	3.42	64.66	
ENERGY VALUE (AVERAGE) OF THE DIGESTIBLE FRACTION PER 100 G EDIBLE PORTION	KJOULE	98	133	1082	1313
	(KCAL)	23	32	259	314

WASTE PERCENTAGE AVERAGE 0.00

CONSTITUENTS	DIM	AV	VARIATION			AVR	NUTR. DENS.		MOLPERC.
WATER	GRAM	12.50	12.00	-	13.20	12.50	GRAM/MJ	9.52	
PROTEIN	GRAM	8.54	7.61	-	9.84	8.54	GRAM/MJ	6.51	
FAT	GRAM	3.80	3.20	-	4.30	3.80	GRAM/MJ	2.89	
AVAILABLE CARBOHYDR.	GRAM	64.66 1	-	-	-	64.66	GRAM/MJ	49.25	
TOTAL DIETARY FIBRE	GRAM	9.20 2	-	-	-	9.20	GRAM/MJ	7.01	
MINERALS	GRAM	1.30	1.12	-	1.51	1.30	GRAM/MJ	0.99	
SODIUM	MILLI	6.00	1.00	-	10.00	6.00	MILLI/MJ	4.57	
POTASSIUM	MILLI	330.00	310.00	-	350.00	330.00	MILLI/MJ	251.38	
MAGNESIUM	MILLI	120.00	-	-	-	120.00	MILLI/MJ	91.41	
CALCIUM	MILLI	15.00	10.00	-	19.00	15.00	MILLI/MJ	11.43	
MANGANESE	MILLI	0.48	0.15	-	0.80	0.48	MILLI/MJ	0.37	
IRON	MILLI	-	0.50	-	2.40				
COBALT	MICRO	-	0.20	-	8.00				
COPPER	MILLI	-	0.07	-	0.25				
ZINC	MILLI	2.50	-	-	-	2.50	MILLI/MJ	1.90	
NICKEL	MILLI	0.12	0.04	-	0.14	0.13	MILLI/MJ	0.10	
CHROMIUM	MICRO	32.00	27.00	-	37.00	32.00	MICRO/MJ	24.38	
MOLYBDENUM	MICRO	55.00	50.00	-	58.00	55.00	MICRO/MJ	41.90	
PHOSPHORUS	MILLI	256.00	-	-	-	256.00	MILLI/MJ	195.01	
CHLORIDE	MILLI	12.00	-	-	-	12.00	MILLI/MJ	9.14	
FLUORIDE	MICRO	62.00	-	-	-	62.00	MICRO/MJ	47.23	
IODIDE	MICRO	2.60	2.50	-	2.70	2.60	MICRO/MJ	1.98	
BORON	MILLI	0.15	0.07	-	0.22	0.15	MILLI/MJ	0.11	
SELENIUM	MICRO	-	0.40	-	9.30				
CAROTENE	MILLI	0.37	0.09	-	0.60	0.37	MILLI/MJ	0.28	
VITAMIN E ACTIVITY	MILLI	1.95	0.36	-	2.70	1.95	MILLI/MJ	1.49	

1 ESTIMATED BY THE DIFFERENCE METHOD
100 - (WATER + PROTEIN + FAT + MINERALS + TOTAL DIETARY FIBRE)

2 METHOD OF MEUSER, SUCKOW AND KULIKOWSKI ("BERLINER METHODE")

CONSTITUENTS	DIM	AV	VARIATION			AVR	NUTR.	DENS.	MOLPERC.
TOTAL TOCOPHEROLS	MILLI	6.60	2.00	-	9.90	6.60	MILLI/MJ	5.03	
ALPHA-TOCOPHEROL	MILLI	1.50	0.20	-	1.90	1.50	MILLI/MJ	1.14	
GAMMA-TOCOPHEROL	MILLI	4.40	1.50	-	6.60	4.40	MILLI/MJ	3.35	
ALPHA-TOCOTRIENOL	MILLI	0.23	0.03	-	0.56	0.23	MILLI/MJ	0.18	
GAMMA-TOCOTRIENOL	MILLI	0.48	0.28	-	0.86	0.48	MILLI/MJ	0.37	
VITAMIN K	MICRO	40.00	-	-	-	40.00	MICRO/MJ	30.47	
VITAMIN B1	MILLI	0.36	0.20	-	0.60	0.36	MILLI/MJ	0.27	
VITAMIN B2	MILLI	0.20	0.10	-	0.24	0.20	MILLI/MJ	0.15	
NICOTINAMIDE	MILLI	1.50	1.00	-	2.00	1.50	MILLI/MJ	1.14	
PANTOTHENIC ACID	MILLI	0.65	0.60	-	0.70	0.65	MILLI/MJ	0.50	
VITAMIN B6	MILLI	0.40	-	-	-	0.40	MILLI/MJ	0.30	
BIOTIN	MICRO	6.00	-	-	-	6.00	MICRO/MJ	4.57	
FOLIC ACID	MICRO	26.00	20.00	-	40.00	26.00	MICRO/MJ	19.81	
VITAMIN C	-	0.00	-	-	-	0.00			
ALANINE	GRAM	0.79	0.77	-	0.83	0.79	GRAM/MJ	0.60	11.3
ARGININE	GRAM	0.42	0.19	-	0.56	0.42	GRAM/MJ	0.32	3.1
ASPARTIC ACID	GRAM	0.62	0.59	-	0.63	0.62	GRAM/MJ	0.47	5.9
CYSTINE	GRAM	0.14	0.07	-	0.28	0.14	GRAM/MJ	0.11	0.7
GLUTAMIC ACID	GRAM	1.78	1.74	-	1.88	1.78	GRAM/MJ	1.36	15.4
GLYCINE	GRAM	0.43	0.43	-	0.44	0.43	GRAM/MJ	0.33	7.3
HISTIDINE	GRAM	0.26	0.13	-	0.33	0.26	GRAM/MJ	0.20	2.1
ISOLEUCINE	GRAM	0.43	0.35	-	0.62	0.43	GRAM/MJ	0.33	4.2
LEUCINE	GRAM	1.22	0.91	-	2.11	1.22	GRAM/MJ	0.93	11.8
LYSINE	GRAM	0.29	0.04	-	0.48	0.29	GRAM/MJ	0.22	2.5
METHIONINE	GRAM	0.19	0.09	-	0.40	0.19	GRAM/MJ	0.14	1.6
PHENYLALANINE	GRAM	0.46	0.32	-	0.51	0.46	GRAM/MJ	0.35	3.5
PROLINE	GRAM	1.02	0.93	-	1.19	1.02	GRAM/MJ	0.78	11.3
SERINE	GRAM	0.52	0.50	-	0.53	0.52	GRAM/MJ	0.40	6.3
THREONINE	GRAM	0.39	0.32	-	0.51	0.39	GRAM/MJ	0.30	4.2
TRYPTOPHAN	MILLI	70.00	40.00	-	110.00	70.00	MILLI/MJ	53.32	0.4
TYROSINE	GRAM	0.38	0.19	-	0.69	0.38	GRAM/MJ	0.29	2.7
VALINE	GRAM	0.51	0.43	-	0.74	0.51	GRAM/MJ	0.39	5.5
GLUCOSE	GRAM	0.10	0.05	-	0.15	0.10	GRAM/MJ	0.08	
FRUCTOSE	MILLI	90.00	60.00	-	120.00	90.00	MILLI/MJ	68.56	
SUCROSE	GRAM	1.20	0.78	-	1.56	1.20	GRAM/MJ	0.91	
RAFFINOSE	GRAM	0.23	0.19	-	0.27	0.23	GRAM/MJ	0.18	
STARCH	GRAM	61.45	60.98	-	63.80	61.45	GRAM/MJ	46.81	
CELLULOSE	GRAM	2.20	-	-	-	2.20	GRAM/MJ	1.68	
PHYTIC ACID	GRAM	0.94	0.89	-	0.99	0.94	GRAM/MJ	0.72	
PALMITIC ACID	GRAM	0.47	0.25	-	0.69	0.47	GRAM/MJ	0.36	
STEARIC ACID	MILLI	90.00	36.00	-	145.00	90.00	MILLI/MJ	68.56	
ARACHIDIC ACID	MILLI	73.00	-	-	-	73.00	MILLI/MJ	55.61	
PALMITOLEIC ACID	MILLI	36.00	-	-	-	36.00	MILLI/MJ	27.42	
OLEIC ACID	GRAM	1.10	-	-	-	1.10	GRAM/MJ	0.84	
LINOLEIC ACID	GRAM	1.63	0.59	-	2.46	1.63	GRAM/MJ	1.24	
LINOLENIC ACID	MILLI	40.00	30.00	-	70.00	40.00	MILLI/MJ	30.47	
TOTAL STEROLS	MILLI	178.00	-	-	-	178.00	MILLI/MJ	135.59	
CAMPESTEROL	MILLI	32.00	-	-	-	32.00	MILLI/MJ	24.38	
BETA-SITOSTEROL	MILLI	120.00	-	-	-	120.00	MILLI/MJ	91.41	
STIGMASTEROL	MILLI	21.00	-	-	-	21.00	MILLI/MJ	16.00	
TOTAL PHOSPHOLIPIDS	MILLI	254.00	-	-	-	254.00	MILLI/MJ	193.48	
PHOSPHATIDYLCHOLINE	MILLI	186.00	-	-	-	186.00	MILLI/MJ	141.68	
PHOSPHATIDYLETHANOLAMI	MILLI	14.00	-	-	-	14.00	MILLI/MJ	10.66	
PHOSPHATIDYLINOSITOL	MILLI	14.00	-	-	-	14.00	MILLI/MJ	10.66	
DIETARY FIBRE,WAT.SOL.	GRAM	2.30	-	-	-	2.30	GRAM/MJ	1.75	
DIETARY FIBRE,WAT.INS.	GRAM	6.90	-	-	-	6.90	GRAM/MJ	5.26	

MAIS-FRÜHSTÜCKS-FLOCKEN (CORN FLAKES) — CORN FLAKES — FLOCONS DE MAÏS (CORN FLAKES)

		PROTEIN	FAT	CARBOHYDRATES	TOTAL
ENERGY VALUE (AVERAGE)	KJOULE	136	23	1333	1492
PER 100 G	(KCAL)	33	5.6	319	357
EDIBLE PORTION					
AMOUNT OF DIGESTIBLE	GRAM	5.43	0.54	79.65	
CONSTITUENTS PER 100 G					
ENERGY VALUE (AVERAGE)	KJOULE	103	21	1333	1457
OF THE DIGESTIBLE	(KCAL)	25	5.0	319	348
FRACTION PER 100 G					
EDIBLE PORTION					

WASTE PERCENTAGE AVERAGE 0.00

CONSTITUENTS	DIM	AV	VARIATION	AVR	NUTR. DENS.		MOLPERC.
WATER	GRAM	5.70	3.40 - 9.50	5.70	GRAM/MJ	3.91	
PROTEIN	GRAM	7.15	6.40 - 7.52	7.15	GRAM/MJ	4.91	
FAT	GRAM	0.60	0.40 - 0.80	0.60	GRAM/MJ	0.41	
AVAILABLE CARBOHYDR.	GRAM	79.65 [1]	- - -	79.65	GRAM/MJ	54.65	
TOTAL DIETARY FIBRE	GRAM	4.00 [2]	- - -	4.00	GRAM/MJ	2.74	
MINERALS	GRAM	2.90	- - -	2.90	GRAM/MJ	1.99	
SODIUM	GRAM	0.91	0.66 - 1.17	0.92	GRAM/MJ	0.63	
POTASSIUM	MILLI	139.00	106.00 - 160.00	139.00	MILLI/MJ	95.37	
MAGNESIUM	MILLI	14.00	- - -	14.00	MILLI/MJ	9.61	
CALCIUM	MILLI	13.00	10.00 - 17.00	13.00	MILLI/MJ	8.92	
MANGANESE	MICRO	50.00	- - -	50.00	MICRO/MJ	34.31	
IRON	MILLI	2.00	1.30 - 2.70	2.00	MILLI/MJ	1.37	
COPPER	MILLI	0.20	- - -	0.20	MILLI/MJ	0.14	
ZINC	MILLI	0.30	- - -	0.30	MILLI/MJ	0.21	
MOLYBDENUM	MICRO	8.00	- - -	8.00	MICRO/MJ	5.49	
PHOSPHORUS	MILLI	59.00	58.00 - 60.00	59.00	MILLI/MJ	40.48	
CHLORIDE	GRAM	1.80	- - -	1.80	GRAM/MJ	1.24	
IODIDE	MICRO	1.00	1.00 - 1.00	1.00	MICRO/MJ	0.69	
BORON	MICRO	20.00	- - -	20.00	MICRO/MJ	13.72	
SELENIUM	MICRO	2.60	2.40 - 2.80	2.60	MICRO/MJ	1.78	
VITAMIN B1	MICRO	60.00	20.00 - 100.00	60.00	MICRO/MJ	41.17	
VITAMIN B2	MICRO	60.00	60.00 - 60.00	60.00	MICRO/MJ	41.17	
NICOTINAMIDE	MILLI	1.40	1.10 - 1.60	1.40	MILLI/MJ	0.96	
PANTOTHENIC ACID	MILLI	0.17	0.16 - 0.20	0.17	MILLI/MJ	0.12	
VITAMIN B6	MICRO	70.00	66.00 - 74.00	70.00	MICRO/MJ	48.03	
FOLIC ACID	MICRO	5.70	4.20 - 7.10	5.70	MICRO/MJ	3.91	

1 ESTIMATED BY THE DIFFERENCE METHOD
100 - (WATER + PROTEIN + FAT + MINERALS + TOTAL DIETARY FIBRE)

2 METHOD OF MEUSER, SUCKOW AND KULIKOWSKI ("BERLINER METHODE")

CONSTITUENTS	DIM	AV	VARIATION			AVR	NUTR. DENS.		MOLPERC.
VITAMIN C	-	0.00	-	-	-	0.00			
ALANINE	GRAM	0.80	0.74	-	0.86	0.80	GRAM/MJ	0.55	12.5
ARGININE	GRAM	0.24	0.15	-	0.35	0.24	GRAM/MJ	0.16	1.9
ASPARTIC ACID	GRAM	0.54	0.50	-	0.57	0.54	GRAM/MJ	0.37	5.6
CYSTINE	GRAM	0.16	0.14	-	0.18	0.16	GRAM/MJ	0.11	0.9
GLUTAMIC ACID	GRAM	1.86	1.70	-	2.00	1.86	GRAM/MJ	1.28	17.6
GLYCINE	GRAM	0.34	0.31	-	0.38	0.34	GRAM/MJ	0.23	6.3
HISTIDINE	GRAM	0.24	-	-	-	0.24	GRAM/MJ	0.16	2.2
ISOLEUCINE	GRAM	0.33	0.32	-	0.34	0.33	GRAM/MJ	0.23	3.5
LEUCINE	GRAM	1.24	1.14	-	1.40	1.24	GRAM/MJ	0.85	13.2
LYSINE	GRAM	0.18	-	-	-	0.18	GRAM/MJ	0.12	1.7
METHIONINE	GRAM	0.17	0.15	-	0.18	0.17	GRAM/MJ	0.12	1.6
PHENYLALANINE	GRAM	0.43	0.39	-	0.46	0.43	GRAM/MJ	0.30	3.6
PROLINE	GRAM	0.97	0.83	-	1.11	0.97	GRAM/MJ	0.67	11.7
SERINE	GRAM	0.47	-	-	-	0.47	GRAM/MJ	0.32	6.2
THREONINE	GRAM	0.32	0.31	-	0.34	0.32	GRAM/MJ	0.22	3.7
TRYPTOPHAN	MILLI	50.00	-	-	-	50.00	MILLI/MJ	34.31	0.3
TYROSINE	GRAM	0.27	0.17	-	0.33	0.27	GRAM/MJ	0.19	2.1
VALINE	GRAM	0.44	-	-	-	0.44	GRAM/MJ	0.30	5.2
STARCH	GRAM	77.80	-	-	-	77.80	GRAM/MJ	53.38	
CELLULOSE	GRAM	2.42	-	-	-	2.42	GRAM/MJ	1.66	
DIETARY FIBRE,WAT.SOL.	GRAM	1.20	-	-	-	1.20	GRAM/MJ	0.82	
DIETARY FIBRE,WAT.INS.	GRAM	2.80	-	-	-	2.80	GRAM/MJ	1.92	

MAISMEHL — CORN FLOUR — FARINE DE MAÏS

		PROTEIN	FAT	CARBOHYDRATES	TOTAL
ENERGY VALUE (AVERAGE) PER 100 G EDIBLE PORTION	KJOULE	158	110	1267	1535
	(KCAL)	38	26	303	367
AMOUNT OF DIGESTIBLE CONSTITUENTS PER 100 G	GRAM	4.98	2.53	75.71	
ENERGY VALUE (AVERAGE) OF THE DIGESTIBLE FRACTION PER 100 G EDIBLE PORTION	KJOULE	95	99	1267	1461
	(KCAL)	23	24	303	349

WASTE PERCENTAGE AVERAGE 0.00

CONSTITUENTS	DIM	AV	VARIATION	AVR	NUTR. DENS.		MOLPERC.
WATER	GRAM	12.00	8.10 - 14.00	12.00	GRAM/MJ	8.21	
PROTEIN	GRAM	8.31	6.24 - 9.28	8.31	GRAM/MJ	5.69	
FAT	GRAM	2.82	1.56 - 3.90	2.82	GRAM/MJ	1.93	
AVAILABLE CARBOHYDR.	GRAM	75.71 1	- - -	75.71	GRAM/MJ	51.83	
MINERALS	GRAM	1.16	1.14 - 1.20	1.16	GRAM/MJ	0.79	
SODIUM	MILLI	0.70	- - -	0.70	MILLI/MJ	0.48	
POTASSIUM	MILLI	120.00	- - -	120.00	MILLI/MJ	82.15	
MAGNESIUM	MILLI	47.00	47.00 - 47.00	47.00	MILLI/MJ	32.18	
CALCIUM	MILLI	18.00	10.00 - 26.00	18.00	MILLI/MJ	12.32	
IRON	MILLI	2.40	- - -	2.40	MILLI/MJ	1.64	
PHOSPHORUS	MILLI	256.00	- - -	256.00	MILLI/MJ	175.25	
CAROTENE	MILLI	0.30	0.30 - 0.31	0.30	MILLI/MJ	0.21	
TOTAL TOCOPHEROLS	MILLI	1.30	- - -	1.30	MILLI/MJ	0.89	
VITAMIN B1	MILLI	0.44	0.38 - 0.49	0.44	MILLI/MJ	0.30	
VITAMIN B2	MILLI	0.13	0.11 - 0.17	0.13	MILLI/MJ	0.09	
NICOTINAMIDE	MILLI	1.93	1.70 - 2.10	1.93	MILLI/MJ	1.32	
PANTOTHENIC ACID	MILLI	0.55	0.50 - 0.59	0.55	MILLI/MJ	0.38	
VITAMIN B6	MICRO	60.00	- - -	60.00	MICRO/MJ	41.07	
BIOTIN	MICRO	6.60	- - -	6.60	MICRO/MJ	4.52	
FOLIC ACID	MICRO	10.10	- - -	10.10	MICRO/MJ	6.91	
VITAMIN C	-	0.00	- - -	0.00			
ARGININE	GRAM	0.30	- - -	0.30	GRAM/MJ	0.21	
CYSTINE	GRAM	0.11	- - -	0.11	GRAM/MJ	0.08	
HISTIDINE	GRAM	0.18	- - -	0.18	GRAM/MJ	0.12	
ISOLEUCINE	GRAM	0.38	- - -	0.38	GRAM/MJ	0.26	

1 ESTIMATED BY THE DIFFERENCE METHOD
100 - (WATER + PROTEIN + FAT + MINERALS)

CONSTITUENTS	DIM	AV	VARIATION			AVR	NUTR. DENS.		MOLPERC.
LEUCINE	GRAM	1.08	-	-	-	1.08	GRAM/MJ	0.74	
LYSINE	GRAM	0.24	-	-	-	0.24	GRAM/MJ	0.16	
METHIONINE	GRAM	0.16	-	-	-	0.16	GRAM/MJ	0.11	
PHENYLALANINE	GRAM	0.38	-	-	-	0.38	GRAM/MJ	0.26	
THREONINE	GRAM	0.33	-	-	-	0.33	GRAM/MJ	0.23	
TRYPTOPHAN	MILLI	50.00	-	-	-	50.00	MILLI/MJ	34.23	
TYROSINE	GRAM	0.51	-	-	-	0.51	GRAM/MJ	0.35	
VALINE	GRAM	0.43	-	-	-	0.43	GRAM/MJ	0.29	
STARCH	GRAM	71.60	-	-	-	71.60	GRAM/MJ	49.02	
MYRISTIC ACID	-	0.00	-	-	-	0.00			
PALMITIC ACID	GRAM	0.28	-	-	-	0.28	GRAM/MJ	0.19	
STEARIC ACID	MILLI	70.00	-	-	-	70.00	MILLI/MJ	47.92	
ARACHIDIC ACID	MILLI	10.00	-	-	-	10.00	MILLI/MJ	6.85	
PALMITOLEIC ACID	MILLI	10.00	-	-	-	10.00	MILLI/MJ	6.85	
OLEIC ACID	GRAM	0.87	-	-	-	0.87	GRAM/MJ	0.60	
LINOLEIC ACID	GRAM	1.41	-	-	-	1.41	GRAM/MJ	0.97	
LINOLENIC ACID	MILLI	25.00	-	-	-	25.00	MILLI/MJ	17.11	

REIS UNPOLIERT — RICE UNPOLISHED — RIZ NATUREL

ORYZA SATIVA L.

		PROTEIN	FAT	CARBOHYDRATES	TOTAL
ENERGY VALUE (AVERAGE)	KJOULE	137	86	1229	1452
PER 100 G	(KCAL)	33	20	294	347
EDIBLE PORTION					
AMOUNT OF DIGESTIBLE	GRAM	5.41	1.98	73.41	
CONSTITUENTS PER 100 G					
ENERGY VALUE (AVERAGE)	KJOULE	103	77	1229	1409
OF THE DIGESTIBLE	(KCAL)	25	18	294	337
FRACTION PER 100 G					
EDIBLE PORTION					

WASTE PERCENTAGE AVERAGE 0.00

CONSTITUENTS	DIM	AV		VARIATION			AVR	NUTR. DENS.		MOLPERC.
WATER	GRAM	13.10		9.60	-	14.10	13.10	GRAM/MJ	9.30	
PROTEIN	GRAM	7.22		7.02	-	7.31	7.22	GRAM/MJ	5.13	
FAT	GRAM	2.20		1.70	-	2.90	2.20	GRAM/MJ	1.56	
AVAILABLE CARBOHYDR.	GRAM	73.41	1	-	-	-	73.41	GRAM/MJ	52.11	
TOTAL DIETARY FIBRE	GRAM	2.87	2	-	-	-	2.87	GRAM/MJ	2.04	
MINERALS	GRAM	1.20		1.10	-	1.40	1.20	GRAM/MJ	0.85	
SODIUM	MILLI	10.00		-	-	-	10.00	MILLI/MJ	7.10	
POTASSIUM	MILLI	150.00		-	-	-	150.00	MILLI/MJ	106.48	
MAGNESIUM	MILLI	157.00		148.00	-	166.00	157.00	MILLI/MJ	111.45	
CALCIUM	MILLI	23.00		12.00	-	39.00	23.00	MILLI/MJ	16.33	
MANGANESE	MILLI	1.10		-	-	-	1.10	MILLI/MJ	0.78	
IRON	MILLI	2.60		2.00	-	3.10	2.60	MILLI/MJ	1.85	
COPPER	MILLI	0.24		-	-	-	0.24	MILLI/MJ	0.17	
ZINC	MILLI	1.40		0.80	-	2.00	1.40	MILLI/MJ	0.99	
MOLYBDENUM	MICRO	75.00		50.00	-	100.00	75.00	MICRO/MJ	53.24	
PHOSPHORUS	MILLI	325.00		290.00	-	383.00	325.00	MILLI/MJ	230.71	
FLUORIDE	MICRO	50.00		44.00	-	61.00	50.00	MICRO/MJ	35.49	
IODIDE	MICRO	2.20		-	-	-	2.20	MICRO/MJ	1.56	
BORON	MILLI	0.27		0.12	-	0.40	0.28	MILLI/MJ	0.20	
SELENIUM	MICRO	40.00		13.00	-	71.00	40.00	MICRO/MJ	28.39	
TOTAL TOCOPHEROLS	MILLI	0.76		0.37	-	1.10	0.76	MILLI/MJ	0.54	
ALPHA-TOCOPHEROL	MILLI	0.18		0.06	-	0.30	0.18	MILLI/MJ	0.13	
GAMMA-TOCOPHEROL	MILLI	0.19		0.07	-	0.30	0.19	MILLI/MJ	0.13	
DELTA-TOCOPHEROL	MICRO	30.00		20.00	-	40.00	30.00	MICRO/MJ	21.30	
GAMMA-TOCOTRIENOL	MILLI	0.36		0.22	-	0.50	0.36	MILLI/MJ	0.26	
VITAMIN B1	MILLI	0.41		0.32	-	0.51	0.41	MILLI/MJ	0.29	

1 ESTIMATED BY THE DIFFERENCE METHOD
100 - (WATER + PROTEIN + FAT + MINERALS + TOTAL DIETARY FIBRE)

2 MODIFIED AOAC METHOD

CONSTITUENTS	DIM	AV	VARIATION			AVR	NUTR. DENS.		MOLPERC.
VITAMIN B2	MICRO	91.00	50.00	-	130.00	91.00	MICRO/MJ	64.60	
NICOTINAMIDE	MILLI	5.20	4.60	-	5.80	5.20	MILLI/MJ	3.69	
PANTOTHENIC ACID	MILLI	1.70	-	-	-	1.70	MILLI/MJ	1.21	
VITAMIN B6	MILLI	0.67	0.35	-	1.00	0.68	MILLI/MJ	0.48	
BIOTIN	MICRO	12.00	-	-	-	12.00	MICRO/MJ	8.52	
FOLIC ACID	MICRO	16.00	-	-	-	16.00	MICRO/MJ	11.36	
VITAMIN C	-	0.00	-	-	-	0.00			
ALANINE	GRAM	0.55	0.49	-	0.58	0.55	GRAM/MJ	0.39	9.4
ARGININE	GRAM	0.60	0.44	-	0.91	0.60	GRAM/MJ	0.43	5.3
ASPARTIC ACID	GRAM	0.84	0.79	-	0.87	0.84	GRAM/MJ	0.60	9.6
CYSTINE	GRAM	0.10	0.06	-	0.19	0.10	GRAM/MJ	0.07	0.6
GLUTAMIC ACID	GRAM	1.64	1.52	-	1.76	1.64	GRAM/MJ	1.16	17.0
GLYCINE	GRAM	0.46	0.42	-	0.49	0.46	GRAM/MJ	0.33	9.3
HISTIDINE	GRAM	0.19	0.12	-	0.27	0.19	GRAM/MJ	0.13	1.9
ISOLEUCINE	GRAM	0.34	0.26	-	0.57	0.34	GRAM/MJ	0.24	4.0
LEUCINE	GRAM	0.69	0.50	-	0.93	0.69	GRAM/MJ	0.49	8.0
LYSINE	GRAM	0.30	0.10	-	0.42	0.30	GRAM/MJ	0.21	3.1
METHIONINE	GRAM	0.17	0.05	-	0.31	0.17	GRAM/MJ	0.12	1.7
PHENYLALANINE	GRAM	0.42	0.30	-	0.55	0.42	GRAM/MJ	0.30	3.9
PROLINE	GRAM	0.39	0.37	-	0.40	0.39	GRAM/MJ	0.28	5.2
SERINE	GRAM	0.47	0.41	-	0.50	0.47	GRAM/MJ	0.33	6.8
THREONINE	GRAM	0.33	0.19	-	0.62	0.33	GRAM/MJ	0.23	4.2
TRYPTOPHAN	MILLI	90.00	30.00	-	110.00	90.00	MILLI/MJ	63.89	0.7
TYROSINE	GRAM	0.32	0.21	-	0.47	0.32	GRAM/MJ	0.23	2.7
VALINE	GRAM	0.50	0.40	-	0.76	0.50	GRAM/MJ	0.35	6.5
SUCROSE	GRAM	0.60	-	-	-	0.60	GRAM/MJ	0.43	
STARCH	GRAM	72.70	-	-	-	72.70	GRAM/MJ	51.61	
PHYTIC ACID	GRAM	0.89	-	-	-	0.89	GRAM/MJ	0.63	
MYRISTIC ACID	MILLI	30.00	-	-	-	30.00	MILLI/MJ	21.30	
PALMITIC ACID	GRAM	0.54	-	-	-	0.54	GRAM/MJ	0.38	
STEARIC ACID	MILLI	40.00	-	-	-	40.00	MILLI/MJ	28.39	
PALMITOLEIC ACID	MILLI	10.00	-	-	-	10.00	MILLI/MJ	7.10	
OLEIC ACID	GRAM	0.54	-	-	-	0.54	GRAM/MJ	0.38	
LINOLEIC ACID	GRAM	0.78	-	-	-	0.78	GRAM/MJ	0.55	
LINOLENIC ACID	MILLI	30.00	-	-	-	30.00	MILLI/MJ	21.30	
TOTAL PHOSPHOLIPIDS	MILLI	89.00	-	-	-	89.00	MILLI/MJ	63.18	
PHOSPHATIDYLCHOLINE	MILLI	32.00	-	-	-	32.00	MILLI/MJ	22.72	
PHOSPHATIDYLETHANOLAMINE	MILLI	35.00	-	-	-	35.00	MILLI/MJ	24.85	
PHOSPHATIDYLSERINE	MILLI	3.00	-	-	-	3.00	MILLI/MJ	2.13	
DIETARY FIBRE, WAT.SOL.	GRAM	1.48	-	-	-	1.48	GRAM/MJ	1.05	
DIETARY FIBRE, WAT.INS.	GRAM	1.39	-	-	-	1.39	GRAM/MJ	0.99	

REIS
POLIERT

RICE
POLISHED

RIZ
POLI

ORYZA SATIVA L.

		PROTEIN	FAT	CARBOHYDRATES	TOTAL
ENERGY VALUE (AVERAGE) PER 100 G EDIBLE PORTION	KJOULE	130	24	1301	1455
	(KCAL)	31	5.8	311	348
AMOUNT OF DIGESTIBLE CONSTITUENTS PER 100 G	GRAM	5.73	0.55	77.73	
ENERGY VALUE (AVERAGE) OF THE DIGESTIBLE FRACTION PER 100 G EDIBLE PORTION	KJOULE	109	22	1301	1432
	(KCAL)	26	5.2	311	342

WASTE PERCENTAGE AVERAGE 0.00

CONSTITUENTS	DIM	AV	VARIATION		AVR	NUTR. DENS.		MOLPERC.
WATER	GRAM	12.90	12.60	- 13.80	12.90	GRAM/MJ	9.01	
PROTEIN	GRAM	6.83	6.53	- 7.31	6.83	GRAM/MJ	4.77	
FAT	GRAM	0.62	0.50	- 1.00	0.62	GRAM/MJ	0.43	
AVAILABLE CARBOHYDR.	GRAM	77.73 1	-	- -	77.73	GRAM/MJ	54.29	
TOTAL DIETARY FIBRE	GRAM	1.39 2	-	- -	1.39	GRAM/MJ	0.97	
MINERALS	GRAM	0.53	0.40	- 0.60	0.53	GRAM/MJ	0.37	
SODIUM	MILLI	6.00	1.00	- 10.00	6.00	MILLI/MJ	4.19	
POTASSIUM	MILLI	103.00	97.00	- 113.00	103.00	MILLI/MJ	71.94	
MAGNESIUM	MILLI	64.00	-	- -	64.00	MILLI/MJ	44.70	
CALCIUM	MILLI	6.00	3.00	- 10.00	6.00	MILLI/MJ	4.19	
MANGANESE	MILLI	2.00	1.00	- 3.00	2.00	MILLI/MJ	1.40	
IRON	MILLI	0.60	0.40	- 0.80	0.60	MILLI/MJ	0.42	
COBALT	MICRO	0.60	-	- -	0.60	MICRO/MJ	0.42	
COPPER	MILLI	0.13	-	- -	0.13	MILLI/MJ	0.09	
ZINC	MILLI	0.50	0.20	- 0.80	0.50	MILLI/MJ	0.35	
MOLYBDENUM	MICRO	80.00	40.00	- 110.00	80.00	MICRO/MJ	55.87	
PHOSPHORUS	MILLI	120.00	100.00	- 140.00	120.00	MILLI/MJ	83.81	
FLUORIDE	MICRO	50.00	10.00	- 80.00	50.00	MICRO/MJ	34.92	
IODIDE	MICRO	2.20	-	- -	2.20	MICRO/MJ	1.54	
BORON	MICRO	24.00	-	- -	24.00	MICRO/MJ	16.76	
SELENIUM	MICRO	40.00	10.00	- 70.00	40.00	MICRO/MJ	27.94	
TOTAL TOCOPHEROLS	MILLI	0.10	-	- -	0.10	MILLI/MJ	0.07	
VITAMIN B1	MICRO	60.00	40.00	- 93.00	60.00	MICRO/MJ	41.90	
VITAMIN B2	MICRO	32.00	22.00	- 55.00	32.00	MICRO/MJ	22.35	
NICOTINAMIDE	MILLI	1.30	1.00	- 1.50	1.30	MILLI/MJ	0.91	
PANTOTHENIC ACID	MILLI	0.63	0.62	- 0.64	0.63	MILLI/MJ	0.44	

1 ESTIMATED BY THE DIFFERENCE METHOD
100 - (WATER + PROTEIN + FAT + MINERALS + TOTAL DIETARY FIBRE)

2 METHOD OF MEUSER, SUCKOW AND KULIKOWSKI ("BERLINER METHODE")

CONSTITUENTS	DIM	AV	VARIATION			AVR	NUTR. DENS.		MOLPERC.
VITAMIN B6	MILLI	0.15	0.11	-	0.20	0.15	MILLI/MJ	0.10	
BIOTIN	MICRO	3.00	3.00	-	3.00	3.00	MICRO/MJ	2.10	
FOLIC ACID	MICRO	29.00	-	-	-	29.00	MICRO/MJ	20.25	
VITAMIN C	-	0.00	-	-	-	0.00			
ALANINE	GRAM	0.50	-	-	-	0.50	GRAM/MJ	0.35	9.1
ARGININE	GRAM	0.57	0.43	-	0.61	0.57	GRAM/MJ	0.40	5.3
ASPARTIC ACID	GRAM	0.78	0.77	-	0.80	0.78	GRAM/MJ	0.54	9.5
CYSTINE	GRAM	0.11	0.06	-	0.19	0.11	GRAM/MJ	0.08	0.7
GLUTAMIC ACID	GRAM	1.58	1.51	-	1.68	1.58	GRAM/MJ	1.10	17.4
GLYCINE	GRAM	0.41	0.38	-	0.44	0.41	GRAM/MJ	0.29	8.9
HISTIDINE	GRAM	0.17	0.13	-	0.26	0.17	GRAM/MJ	0.12	1.8
ISOLEUCINE	GRAM	0.34	0.24	-	0.56	0.34	GRAM/MJ	0.24	4.2
LEUCINE	GRAM	0.66	0.49	-	0.91	0.66	GRAM/MJ	0.46	8.2
LYSINE	GRAM	0.29	0.27	-	0.32	0.29	GRAM/MJ	0.20	3.2
METHIONINE	GRAM	0.17	0.05	-	0.29	0.17	GRAM/MJ	0.12	1.9
PHENYLALANINE	GRAM	0.39	0.29	-	0.53	0.39	GRAM/MJ	0.27	3.8
PROLINE	GRAM	0.42	0.39	-	0.44	0.42	GRAM/MJ	0.29	5.9
SERINE	GRAM	0.41	0.38	-	0.44	0.41	GRAM/MJ	0.29	6.3
THREONINE	GRAM	0.28	0.26	-	0.31	0.28	GRAM/MJ	0.20	3.8
TRYPTOPHAN	MILLI	90.00	30.00	-	120.00	90.00	MILLI/MJ	62.86	0.7
TYROSINE	GRAM	0.26	0.21	-	0.46	0.26	GRAM/MJ	0.18	2.3
VALINE	GRAM	0.49	0.39	-	0.74	0.49	GRAM/MJ	0.34	6.8
SUCROSE	GRAM	0.15	0.00	-	0.30	0.15	GRAM/MJ	0.10	
PENTOSAN	GRAM	0.35	0.30	-	0.40	0.35	GRAM/MJ	0.24	
PHYTIC ACID	GRAM	0.24	0.14	-	0.34	0.24	GRAM/MJ	0.17	
PALMITIC ACID	GRAM	0.11	-	-	-	0.12	GRAM/MJ	0.08	
STEARIC ACID	MILLI	12.00	-	-	-	12.00	MILLI/MJ	8.38	
OLEIC ACID	GRAM	0.22	-	-	-	0.23	GRAM/MJ	0.16	
LINOLEIC ACID	GRAM	0.22	-	-	-	0.22	GRAM/MJ	0.15	
LINOLENIC ACID	MILLI	12.00	-	-	-	12.00	MILLI/MJ	8.38	
DIETARY FIBRE, WAT.SOL.	GRAM	0.87	-	-	-	0.87	GRAM/MJ	0.61	
DIETARY FIBRE, WAT.INS.	GRAM	0.52	-	-	-	0.52	GRAM/MJ	0.36	

REIS	RICE	RIZ
POLIERT, GEKOCHT, ABGETROPFT	POLISHED, COOKED, DRAINED	POLI, CUITE, DEGOUTTÉ

		PROTEIN	FAT	CARBOHYDRATES	TOTAL
ENERGY VALUE (AVERAGE) PER 100 G EDIBLE PORTION	KJOULE	37	6.2	326	370
	(KCAL)	8.9	1.5	78	88
AMOUNT OF DIGESTIBLE CONSTITUENTS PER 100 G	GRAM	1.63	0.14	19.50	
ENERGY VALUE (AVERAGE) OF THE DIGESTIBLE FRACTION PER 100 G EDIBLE PORTION	KJOULE	31	5.6	326	363
	(KCAL)	7.5	1.3	78	87

WASTE PERCENTAGE AVERAGE 0.00

CONSTITUENTS	DIM	AV		VARIATION			AVR	NUTR. DENS.		MOLPERC.
WATER	GRAM	78.00		-	-	-	78.00	GRAM/MJ	214.79	
PROTEIN	GRAM	1.95		-	-	-	1.95	GRAM/MJ	5.37	
FAT	GRAM	0.16		-	-	-	0.16	GRAM/MJ	0.44	
AVAILABLE CARBOHYDR.	GRAM	19.50	1	-	-	-	19.50	GRAM/MJ	53.70	
MINERALS	GRAM	1.10		-	-	-	1.10	GRAM/MJ	3.03	
SODIUM	MILLI	448.00		-	-	-	448.00	GRAM/MJ	1.23	
POTASSIUM	MILLI	31.00		-	-	-	31.00	MILLI/MJ	85.37	
CALCIUM	MILLI	3.00		-	-	-	3.00	MILLI/MJ	8.26	
IRON	MILLI	0.10		-	-	-	0.10	MILLI/MJ	0.28	
PHOSPHORUS	MILLI	36.00		-	-	-	36.00	MILLI/MJ	99.14	
CHLORIDE	MILLI	681.00		-	-	-	681.00	GRAM/MJ	1.88	
VITAMIN B1	MICRO	20.00		-	-	-	20.00	MICRO/MJ	55.08	
VITAMIN B2	MICRO	10.00		-	-	-	10.00	MICRO/MJ	27.54	
NICOTINAMIDE	MILLI	0.32		-	-	-	0.32	MILLI/MJ	0.88	

1 CALCULATED FROM THE DRY PRODUCT

REISMEHL RICE FLOUR FARINE DE RIZ

		PROTEIN	FAT	CARBOHYDRATES	TOTAL
ENERGY VALUE (AVERAGE)	KJOULE	127	25	1333	1485
PER 100 G	(KCAL)	30	6.0	318	355
EDIBLE PORTION					
AMOUNT OF DIGESTIBLE	GRAM	5.61	0.58	79.62	
CONSTITUENTS PER 100 G					
ENERGY VALUE (AVERAGE)	KJOULE	107	23	1333	1462
OF THE DIGESTIBLE	(KCAL)	26	5.4	318	349
FRACTION PER 100 G					
EDIBLE PORTION					

WASTE PERCENTAGE AVERAGE 0.00

CONSTITUENTS	DIM	AV	VARIATION			AVR	NUTR.	DENS.	MOLPERC.
WATER	GRAM	12.50	-	-	-	12.50	GRAM/MJ	8.55	
PROTEIN	GRAM	6.68	6.22	-	6.96	6.68	GRAM/MJ	4.57	
FAT	GRAM	0.65	0.50	-	1.00	0.65	GRAM/MJ	0.44	
AVAILABLE CARBOHYDR.	GRAM	79.62 1	-	-	-	79.62	GRAM/MJ	54.46	
MINERALS	GRAM	0.55	0.50	-	0.58	0.55	GRAM/MJ	0.38	
SODIUM	MILLI	4.00	3.00	-	5.00	4.00	MILLI/MJ	2.74	
POTASSIUM	MILLI	104.00	100.00	-	111.00	104.00	MILLI/MJ	71.13	
MAGNESIUM	MILLI	23.00	-	-	-	23.00	MILLI/MJ	15.73	
CALCIUM	MILLI	7.00	2.00	-	10.00	7.00	MILLI/MJ	4.79	
MANGANESE	MILLI	0.60	0.60	-	0.60	0.60	MILLI/MJ	0.41	
IRON	MILLI	0.40	-	-	-	0.40	MILLI/MJ	0.27	
COPPER	MILLI	0.20	-	-	-	0.20	MILLI/MJ	0.14	
PHOSPHORUS	MILLI	90.00	79.00	-	100.00	90.00	MILLI/MJ	61.56	
CHLORIDE	MILLI	0.30	-	-	-	0.30	MILLI/MJ	0.21	
IODIDE	MICRO	1.00	1.00	-	1.00	1.00	MICRO/MJ	0.68	
VITAMIN B1	MICRO	60.00	40.00	-	80.00	60.00	MICRO/MJ	41.04	
VITAMIN B2	MICRO	30.00	-	-	-	30.00	MICRO/MJ	20.52	
NICOTINAMIDE	MILLI	1.40	1.00	-	2.00	1.40	MILLI/MJ	0.96	
VITAMIN B6	MILLI	0.20	0.19	-	0.22	0.20	MILLI/MJ	0.14	
FOLIC ACID	MICRO	10.00	10.00	-	10.00	10.00	MICRO/MJ	6.84	
ALANINE	GRAM	0.44	0.44	-	0.44	0.44	GRAM/MJ	0.30	9.0
ARGININE	GRAM	0.58	0.58	-	0.58	0.58	GRAM/MJ	0.40	6.0
ASPARTIC ACID	GRAM	0.67	0.67	-	0.67	0.67	GRAM/MJ	0.46	9.1
CYSTINE	MILLI	90.00	90.00	-	90.00	90.00	MILLI/MJ	61.56	0.7
GLUTAMIC ACID	GRAM	1.15	1.15	-	1.15	1.15	GRAM/MJ	0.79	14.2

1 ESTIMATED BY THE DIFFERENCE METHOD
100 - (WATER + PROTEIN + FAT + MINERALS)

CONSTITUENTS	DIM	AV	VARIATION			AVR	NUTR. DENS.		MOLPERC.
GLYCINE	GRAM	0.34	0.34	-	0.34	0.34	GRAM/MJ	0.23	8.2
HISTIDINE	GRAM	0.18	0.18	-	0.18	0.18	GRAM/MJ	0.12	2.1
ISOLEUCINE	GRAM	0.33	0.33	-	0.33	0.33	GRAM/MJ	0.23	4.6
LEUCINE	GRAM	0.62	0.62	-	0.62	0.62	GRAM/MJ	0.42	8.6
LYSINE	GRAM	0.28	0.28	-	0.28	0.28	GRAM/MJ	0.19	3.5
METHIONINE	GRAM	0.19	0.19	-	0.19	0.19	GRAM/MJ	0.13	2.3
PHENYLALANINE	GRAM	0.39	0.39	-	0.39	0.39	GRAM/MJ	0.27	4.3
PROLINE	GRAM	0.33	0.33	-	0.33	0.33	GRAM/MJ	0.23	5.2
SERINE	GRAM	0.41	0.41	-	0.41	0.41	GRAM/MJ	0.28	7.1
THREONINE	GRAM	0.24	0.24	-	0.24	0.24	GRAM/MJ	0.16	3.7
TRYPTOPHAN	GRAM	0.10	0.10	-	0.10	0.10	GRAM/MJ	0.07	0.9
TYROSINE	GRAM	0.28	0.28	-	0.28	0.28	GRAM/MJ	0.19	2.8
VALINE	GRAM	0.50	0.50	-	0.50	0.50	GRAM/MJ	0.34	7.8

ROGGEN
GANZES KORN

SECALE CEREALE L.

RYE
WHOLE GRAIN

SEIGLE
GRAINS ENTIERS

		PROTEIN	FAT	CARBOHYDRATES	TOTAL
ENERGY VALUE (AVERAGE) PER 100 G EDIBLE PORTION	KJOULE	168	66	1016	1250
	(KCAL)	40	16	243	299
AMOUNT OF DIGESTIBLE CONSTITUENTS PER 100 G	GRAM	5.90	1.53	60.73	
ENERGY VALUE (AVERAGE) OF THE DIGESTIBLE FRACTION PER 100 G EDIBLE PORTION	KJOULE	112	60	1016	1188
	(KCAL)	27	14	243	284

WASTE PERCENTAGE AVERAGE 0.00

CONSTITUENTS	DIM	AV		VARIATION		AVR	NUTR.	DENS.	MOLPERC.
WATER	GRAM	13.70		11.00 -	18.20	13.70	GRAM/MJ	11.53	
PROTEIN	GRAM	8.82		7.50 -	10.30	8.82	GRAM/MJ	7.42	
FAT	GRAM	1.70		1.60 -	2.60	1.70	GRAM/MJ	1.43	
AVAILABLE CARBOHYDR.	GRAM	60.73	1	- -	-	60.73	GRAM/MJ	51.10	
TOTAL DIETARY FIBRE	GRAM	13.15	2	- -	-	13.15	GRAM/MJ	11.07	
MINERALS	GRAM	1.90		1.80 -	1.90	1.90	GRAM/MJ	1.60	
SODIUM	MILLI	40.00		- -	-	40.00	MILLI/MJ	33.66	
POTASSIUM	MILLI	510.00		440.00 -	620.00	510.00	MILLI/MJ	429.14	
MAGNESIUM	MILLI	120.00		100.00 -	140.00	120.00	MILLI/MJ	100.98	
CALCIUM	MILLI	64.00		38.00 -	115.00	64.00	MILLI/MJ	53.85	
MANGANESE	MILLI	2.40		1.90 -	2.90	2.40	MILLI/MJ	2.02	
IRON	MILLI	4.60		3.50 -	10.00	4.60	MILLI/MJ	3.87	
COBALT	MICRO	3.10		0.70 -	18.00	3.10	MICRO/MJ	2.61	
COPPER	MILLI	0.50		0.40 -	0.60	0.50	MILLI/MJ	0.42	
ZINC	MILLI	1.30		- -	-	1.30	MILLI/MJ	1.09	
NICKEL	MILLI	-		0.02 -	0.27				
CHROMIUM	MICRO	25.00		4.00 -	32.00	25.00	MICRO/MJ	21.04	
MOLYBDENUM	MICRO	-		7.00 -	62.00				
PHOSPHORUS	MILLI	336.55		332.00 -	373.00	336.56	MILLI/MJ	283.20	
CHLORIDE	MILLI	20.00		- -	-	20.00	MILLI/MJ	16.83	
FLUORIDE	MILLI	0.15		0.04 -	0.71	0.15	MILLI/MJ	0.13	
IODIDE	MICRO	7.20		- -	-	7.20	MICRO/MJ	6.06	
BORON	MILLI	0.70		0.50 -	0.90	0.70	MILLI/MJ	0.59	
SELENIUM	MICRO	-		0.20 -	8.00				
SILICON	MILLI	9.00		7.00 -	12.00	9.00	MILLI/MJ	7.57	
VITAMIN E ACTIVITY	MILLI	1.95		1.70 -	2.20	1.95	MILLI/MJ	1.64	

1 ESTIMATED BY THE DIFFERENCE METHOD
100 - (WATER + PROTEIN + FAT + MINERALS + TOTAL DIETARY FIBRE)

2 METHOD OF MEUSER, SUCKOW AND KULIKOWSKI ("BERLINER METHODE")

CONSTITUENTS	DIM	AV	VARIATION			AVR	NUTR. DENS.		MOLPERC.
TOTAL TOCOPHEROLS	MILLI	4.50	3.70	-	5.30	4.50	MILLI/MJ	3.79	
ALPHA-TOCOPHEROL	MILLI	1.35	1.10	-	1.60	1.35	MILLI/MJ	1.14	
BETA-TOCOPHEROL	MILLI	0.36	0.33	-	0.40	0.36	MILLI/MJ	0.30	
ALPHA-TOCOTRIENOL	MILLI	1.40	1.30	-	1.50	1.40	MILLI/MJ	1.18	
BETA-TOCOTRIENOL	MILLI	0.90	0.80	-	1.00	0.90	MILLI/MJ	0.76	
VITAMIN B1	MILLI	0.35	0.16	-	0.76	0.35	MILLI/MJ	0.29	
VITAMIN B2	MILLI	0.17	0.13	-	0.22	0.17	MILLI/MJ	0.14	
NICOTINAMIDE	MILLI	1.81	0.78	-	3.00	1.81	MILLI/MJ	1.52	
PANTOTHENIC ACID	MILLI	1.50	1.00	-	2.00	1.50	MILLI/MJ	1.26	
VITAMIN B6	MILLI	0.29	-	-	-	0.29	MILLI/MJ	0.24	
BIOTIN	MICRO	5.00	-	-	-	5.00	MICRO/MJ	4.21	
FOLIC ACID	MICRO	42.00	-	-	-	42.00	MICRO/MJ	35.34	
VITAMIN C	-	0.00	-	-	-	0.00			
ALANINE	GRAM	0.52	0.47	-	0.53	0.52	GRAM/MJ	0.44	7.4
ARGININE	GRAM	0.49	0.45	-	0.53	0.49	GRAM/MJ	0.41	3.6
ASPARTIC ACID	GRAM	0.68	0.36	-	0.80	0.68	GRAM/MJ	0.57	6.5
CYSTINE	GRAM	0.19	0.06	-	0.29	0.19	GRAM/MJ	0.16	1.0
GLUTAMIC ACID	GRAM	2.57	2.26	-	2.84	2.57	GRAM/MJ	2.16	22.1
GLYCINE	GRAM	0.50	0.45	-	0.55	0.50	GRAM/MJ	0.42	8.4
HISTIDINE	GRAM	0.19	0.13	-	0.22	0.19	GRAM/MJ	0.16	1.5
ISOLEUCINE	GRAM	0.39	0.33	-	0.50	0.39	GRAM/MJ	0.33	3.8
LEUCINE	GRAM	0.67	0.61	-	0.79	0.67	GRAM/MJ	0.56	6.5
LYSINE	GRAM	0.40	0.32	-	0.48	0.40	GRAM/MJ	0.34	3.5
METHIONINE	GRAM	0.14	0.06	-	0.24	0.14	GRAM/MJ	0.12	1.2
PHENYLALANINE	GRAM	0.47	0.41	-	0.60	0.47	GRAM/MJ	0.40	3.6
PROLINE	GRAM	1.25	0.22	-	1.70	1.25	GRAM/MJ	1.05	13.7
SERINE	GRAM	0.45	0.43	-	0.50	0.45	GRAM/MJ	0.38	5.4
THREONINE	GRAM	0.36	0.26	-	0.46	0.36	GRAM/MJ	0.30	3.8
TRYPTOPHAN	GRAM	0.11	0.06	-	0.20	0.11	GRAM/MJ	0.09	0.7
TYROSINE	GRAM	0.23	0.13	-	0.50	0.23	GRAM/MJ	0.19	1.6
VALINE	GRAM	0.53	0.49	-	0.55	0.53	GRAM/MJ	0.45	5.7
GLUCOSE	MILLI	50.00	-	-	-	50.00	MILLI/MJ	42.07	
FRUCTOSE	MILLI	50.00	-	-	-	50.00	MILLI/MJ	42.07	
SUCROSE	GRAM	0.79	-	-	-	0.79	GRAM/MJ	0.66	
STARCH	GRAM	52.42	49.10	-	52.70	52.42	GRAM/MJ	44.11	
CELLULOSE	GRAM	2.20	-	-	-	2.20	GRAM/MJ	1.85	
PHYTIC ACID	GRAM	0.97	-	-	-	0.97	GRAM/MJ	0.82	
PALMITIC ACID	GRAM	0.29	-	-	-	0.29	GRAM/MJ	0.24	
STEARIC ACID	MILLI	20.00	-	-	-	20.00	MILLI/MJ	16.83	
PALMITOLEIC ACID	MILLI	60.00	-	-	-	60.00	MILLI/MJ	50.49	
OLEIC ACID	GRAM	0.41	-	-	-	0.41	GRAM/MJ	0.34	
LINOLEIC ACID	GRAM	0.75	-	-	-	0.75	GRAM/MJ	0.63	
LINOLENIC ACID	MILLI	65.00	-	-	-	65.00	MILLI/MJ	54.69	
TOTAL PURINES	MILLI	47.00 [3]	47.00	-	47.00	47.00	MILLI/MJ	39.55	
ADENINE	MILLI	16.00	16.00	-	16.00	16.00	MILLI/MJ	13.46	
GUANINE	MILLI	24.00	24.00	-	24.00	24.00	MILLI/MJ	20.20	
DIETARY FIBRE, WAT.SOL.	GRAM	4.70	-	-	-	4.70	GRAM/MJ	3.95	
DIETARY FIBRE, WAT.INS.	GRAM	8.45	-	-	-	8.45	GRAM/MJ	7.11	

3 VALUE OF TOTAL PURINES IS EXPRESSSED IN MG URIC ACID/100G

ROGGENMEHL TYPE 815 | RYE FLOUR TYPE 815 | FARINE DE SEIGLE TYPE 815

		PROTEIN	FAT	CARBOHYDRATES	TOTAL
ENERGY VALUE (AVERAGE)	KJOULE	123	40	1189	1351
PER 100 G EDIBLE PORTION	(KCAL)	29	9.6	284	323
AMOUNT OF DIGESTIBLE CONSTITUENTS PER 100 G	GRAM	4.83	0.92	71.03	
ENERGY VALUE (AVERAGE)	KJOULE	92	36	1189	1317
OF THE DIGESTIBLE FRACTION PER 100 G EDIBLE PORTION	(KCAL)	22	8.6	284	315

WASTE PERCENTAGE AVERAGE 0.00

CONSTITUENTS	DIM	AV		VARIATION			AVR	NUTR. DENS.		MOLPERC.
WATER	GRAM	14.30		12.30	-	16.00	14.30	GRAM/MJ	10.86	
PROTEIN	GRAM	6.44		-	-	-	6.44	GRAM/MJ	4.89	
FAT	GRAM	1.03		0.91	-	1.25	1.03	GRAM/MJ	0.78	
AVAILABLE CARBOHYDR.	GRAM	71.03	1	-	-	-	71.03	GRAM/MJ	53.94	
TOTAL DIETARY FIBRE	GRAM	6.50	2	-	-	-	6.50	GRAM/MJ	4.94	
MINERALS	GRAM	0.70		0.68	-	0.75	0.70	GRAM/MJ	0.53	
SODIUM	MILLI	1.00		-	-	-	1.00	MILLI/MJ	0.76	
POTASSIUM	MILLI	170.00		156.00	-	181.00	170.00	MILLI/MJ	129.10	
MAGNESIUM	MILLI	26.00		-	-	-	26.00	MILLI/MJ	19.75	
CALCIUM	MILLI	22.00		20.00	-	25.00	22.00	MILLI/MJ	16.71	
IRON	MILLI	2.10		1.73	-	2.37	2.10	MILLI/MJ	1.59	
COPPER	MILLI	0.20		-	-	-	0.20	MILLI/MJ	0.15	
ZINC	MILLI	0.77		0.68	-	0.86	0.77	MILLI/MJ	0.58	
MOLYBDENUM	MILLI	-		0.03	-	0.16				
PHOSPHORUS	MILLI	125.66		121.00	-	135.00	125.67	MILLI/MJ	95.44	
CHLORIDE	MILLI	32.00		-	-	-	32.00	MILLI/MJ	24.30	
TOTAL TOCOPHEROLS	MILLI	2.60		-	-	-	2.60	MILLI/MJ	1.97	
VITAMIN B1	MILLI	0.18		0.15	-	0.20	0.18	MILLI/MJ	0.14	
VITAMIN B2	MICRO	92.00		60.00	-	140.00	92.00	MICRO/MJ	69.87	
NICOTINAMIDE	MILLI	0.60		0.46	-	0.71	0.60	MILLI/MJ	0.46	
VITAMIN B6	MILLI	0.11		-	-	-	0.11	MILLI/MJ	0.08	
FOLIC ACID	MICRO	15.00		-	-	-	15.00	MICRO/MJ	11.39	
ALANINE	GRAM	0.40		-	-	-	0.40	GRAM/MJ	0.30	7.7
ARGININE	GRAM	0.32		0.23	-	0.41	0.32	GRAM/MJ	0.24	3.2
ASPARTIC ACID	GRAM	0.58		-	-	-	0.58	GRAM/MJ	0.44	7.5

1 ESTIMATED BY THE DIFFERENCE METHOD
100 - (WATER + PROTEIN + FAT + MINERALS + TOTAL DIETARY FIBRE)

2 METHOD OF MEUSER, SUCKOW AND KULIKOWSKI ("BERLINER METHODE")

CONSTITUENTS	DIM	AV	VARIATION			AVR	NUTR. DENS.		MOLPERC.
CYSTINE	GRAM	0.13	0.04	-	0.20	0.13	GRAM/MJ	0.10	0.9
GLUTAMIC ACID	GRAM	1.92	-	-	-	1.92	GRAM/MJ	1.46	22.4
GLYCINE	GRAM	0.40	-	-	-	0.40	GRAM/MJ	0.30	9.1
HISTIDINE	GRAM	0.16	0.12	-	0.18	0.16	GRAM/MJ	0.12	1.8
ISOLEUCINE	GRAM	0.29	0.26	-	0.33	0.29	GRAM/MJ	0.22	3.8
LEUCINE	GRAM	0.50	0.43	-	0.54	0.50	GRAM/MJ	0.38	6.5
LYSINE	GRAM	0.26	0.24	-	0.31	0.26	GRAM/MJ	0.20	3.1
METHIONINE	GRAM	0.10	0.05	-	0.16	0.10	GRAM/MJ	0.08	1.2
PHENYLALANINE	GRAM	0.34	0.27	-	0.42	0.34	GRAM/MJ	0.26	3.5
PROLINE	GRAM	0.78	-	-	-	0.78	GRAM/MJ	0.59	11.6
SERINE	GRAM	0.33	-	-	-	0.33	GRAM/MJ	0.25	5.4
THREONINE	GRAM	0.29	0.17	-	0.32	0.29	GRAM/MJ	0.22	4.2
TRYPTOPHAN	MILLI	70.00	40.00	-	140.00	70.00	MILLI/MJ	53.16	0.6
TYROSINE	GRAM	0.21	0.09	-	0.35	0.21	GRAM/MJ	0.16	2.0
VALINE	GRAM	0.38	0.29	-	0.43	0.38	GRAM/MJ	0.29	5.6
STARCH	GRAM	64.07	-	-	-	64.07	GRAM/MJ	48.66	
DIETARY FIBRE,WAT.SOL.	GRAM	2.60	-	-	-	2.60	GRAM/MJ	1.97	
DIETARY FIBRE,WAT.INS.	GRAM	3.90	-	-	-	3.90	GRAM/MJ	2.96	

ROGGENMEHL TYPE 997 | RYE FLOUR TYPE 997 | FARINE DE SEIGLE TYPE 997

		PROTEIN	FAT	CARBOHYDRATES	TOTAL
ENERGY VALUE (AVERAGE)	KJOULE	131	44	1137	1312
PER 100 G EDIBLE PORTION	(KCAL)	31	11	272	314
AMOUNT OF DIGESTIBLE CONSTITUENTS PER 100 G	GRAM	4.87	1.02	67.93	
ENERGY VALUE (AVERAGE)	KJOULE	93	40	1137	1270
OF THE DIGESTIBLE FRACTION PER 100 G EDIBLE PORTION	(KCAL)	22	9.5	272	303

WASTE PERCENTAGE AVERAGE 0.00

CONSTITUENTS	DIM	AV		VARIATION			AVR	NUTR. DENS.		MOLPERC.
WATER	GRAM	14.60		14.10	-	15.00	14.60	GRAM/MJ	11.50	
PROTEIN	GRAM	6.86		6.40	-	7.37	6.86	GRAM/MJ	5.40	
FAT	GRAM	1.14		0.90	-	1.41	1.14	GRAM/MJ	0.90	
AVAILABLE CARBOHYDR.	GRAM	67.93	1	-	-	-	67.93	GRAM/MJ	53.51	
TOTAL DIETARY FIBRE	GRAM	8.62	2	-	-	-	8.62	GRAM/MJ	6.79	
MINERALS	GRAM	0.85		0.81	-	0.91	0.85	GRAM/MJ	0.67	

1 ESTIMATED BY THE DIFFERENCE METHOD
100 - (WATER + PROTEIN + FAT + MINERALS + TOTAL DIETARY FIBRE)

2 MODIFIED AOAC METHOD

CONSTITUENTS	DIM	AV	VARIATION			AVR	NUTR. DENS.		MOLPERC.
SODIUM	MILLI	1.00	-	-	-	1.00	MILLI/MJ	0.79	
POTASSIUM	MILLI	240.00	-	-	-	240.00	MILLI/MJ	189.05	
CALCIUM	MILLI	31.00	-	-	-	31.00	MILLI/MJ	24.42	
IRON	MILLI	2.20	1.89	-	2.50	2.20	MILLI/MJ	1.73	
ZINC	MILLI	1.02	0.90	-	1.14	1.02	MILLI/MJ	0.80	
MOLYBDENUM	MILLI	-	0.03	-	0.16				
PHOSPHORUS	MILLI	180.00	180.00	-	180.00	180.00	MILLI/MJ	141.79	
TOTAL TOCOPHEROLS	MILLI	3.40	-	-	-	3.40	MILLI/MJ	2.68	
VITAMIN B1	MILLI	0.19	0.17	-	0.20	0.19	MILLI/MJ	0.15	
VITAMIN B2	MILLI	0.11	-	-	-	0.11	MILLI/MJ	0.09	
NICOTINAMIDE	MILLI	0.80	-	-	-	0.80	MILLI/MJ	0.63	
ALANINE	GRAM	0.41	-	-	-	0.41	GRAM/MJ	0.32	7.4
ARGININE	GRAM	0.33	0.24	-	0.42	0.33	GRAM/MJ	0.26	3.0
ASPARTIC ACID	GRAM	0.61	-	-	-	0.61	GRAM/MJ	0.48	7.4
CYSTINE	GRAM	0.14	0.04	-	0.21	0.14	GRAM/MJ	0.11	0.9
GLUTAMIC ACID	GRAM	2.05	-	-	-	2.05	GRAM/MJ	1.61	22.4
GLYCINE	GRAM	0.43	-	-	-	0.43	GRAM/MJ	0.34	9.2
HISTIDINE	GRAM	0.18	0.13	-	0.19	0.18	GRAM/MJ	0.14	1.9
ISOLEUCINE	GRAM	0.32	0.28	-	0.34	0.32	GRAM/MJ	0.25	3.9
LEUCINE	GRAM	0.54	0.46	-	0.64	0.54	GRAM/MJ	0.43	6.6
LYSINE	GRAM	0.28	0.26	-	0.32	0.28	GRAM/MJ	0.22	3.1
METHIONINE	GRAM	0.12	0.05	-	0.17	0.12	GRAM/MJ	0.09	1.3
PHENYLALANINE	GRAM	0.36	0.29	-	0.46	0.36	GRAM/MJ	0.28	3.5
PROLINE	GRAM	0.84	-	-	-	0.84	GRAM/MJ	0.66	11.7
SERINE	GRAM	0.34	-	-	-	0.34	GRAM/MJ	0.27	5.2
THREONINE	GRAM	0.31	0.19	-	0.34	0.31	GRAM/MJ	0.24	4.2
TRYPTOPHAN	MILLI	70.00	40.00	-	150.00	70.00	MILLI/MJ	55.14	0.6
TYROSINE	GRAM	0.22	0.10	-	0.35	0.22	GRAM/MJ	0.17	2.0
VALINE	GRAM	0.41	0.31	-	0.45	0.41	GRAM/MJ	0.32	5.6
STARCH	GRAM	62.78	61.74	-	70.30	62.78	GRAM/MJ	49.45	
DIETARY FIBRE, WAT.SOL.	GRAM	4.01	-	-	-	4.01	GRAM/MJ	3.16	
DIETARY FIBRE, WAT.INS.	GRAM	4.61	-	-	-	4.61	GRAM/MJ	3.63	

ROGGENMEHL TYPE 1150 | RYE FLOUR TYPE 1150 | FARINE DE SEIGLE TYPE 1150

		PROTEIN	FAT	CARBOHYDRATES	TOTAL
ENERGY VALUE (AVERAGE)	KJOULE	158	51	1135	1344
PER 100 G	(KCAL)	38	12	271	321
EDIBLE PORTION					
AMOUNT OF DIGESTIBLE	GRAM	5.90	1.17	67.81	
CONSTITUENTS PER 100 G					
ENERGY VALUE (AVERAGE)	KJOULE	112	46	1135	1293
OF THE DIGESTIBLE	(KCAL)	27	11	271	309
FRACTION PER 100 G					
EDIBLE PORTION					

WASTE PERCENTAGE AVERAGE 0.00

CONSTITUENTS	DIM	AV	VARIATION			AVR	NUTR. DENS.		MOLPERC.
WATER	GRAM	13.60	-	-	-	13.60	GRAM/MJ	10.52	
PROTEIN	GRAM	8.31	6.18	-	10.02	8.31	GRAM/MJ	6.43	
FAT	GRAM	1.30	-	-	-	1.30	GRAM/MJ	1.01	
AVAILABLE CARBOHYDR.	GRAM	67.81 1	-	-	-	67.81	GRAM/MJ	52.46	
TOTAL DIETARY FIBRE	GRAM	8.00 2	-	-	-	8.00	GRAM/MJ	6.19	
MINERALS	GRAM	0.98	0.93	-	1.06	0.98	GRAM/MJ	0.76	
SODIUM	MILLI	1.00	0.00	-	2.00	1.00	MILLI/MJ	0.77	
POTASSIUM	MILLI	297.00	283.00	-	330.00	297.00	MILLI/MJ	229.75	
MAGNESIUM	MILLI	67.00	-	-	-	67.00	MILLI/MJ	51.83	
CALCIUM	MILLI	20.00	17.00	-	23.00	20.00	MILLI/MJ	15.47	
IRON	MILLI	2.42	-	-	-	2.42	MILLI/MJ	1.87	
COPPER	MILLI	0.80	-	-	-	0.80	MILLI/MJ	0.62	
MOLYBDENUM	MILLI	-	0.03	-	0.16				
PHOSPHORUS	MILLI	196.00	196.00	-	196.00	196.00	MILLI/MJ	151.62	
VITAMIN B1	MILLI	0.22	0.20	-	0.26	0.22	MILLI/MJ	0.17	
VITAMIN B2	MILLI	0.10	0.08	-	0.11	0.10	MILLI/MJ	0.08	
NICOTINAMIDE	MILLI	1.15	0.69	-	1.60	1.15	MILLI/MJ	0.89	
ALANINE	GRAM	0.49	-	-	-	0.49	GRAM/MJ	0.38	7.3
ARGININE	GRAM	0.42	0.32	-	0.53	0.42	GRAM/MJ	0.32	3.2
ASPARTIC ACID	GRAM	0.75	-	-	-	0.75	GRAM/MJ	0.58	7.5
CYSTINE	GRAM	0.17	0.05	-	0.27	0.17	GRAM/MJ	0.13	0.9
GLUTAMIC ACID	GRAM	2.50	-	-	-	2.50	GRAM/MJ	1.93	22.6
GLYCINE	GRAM	0.51	-	-	-	0.51	GRAM/MJ	0.39	9.0
HISTIDINE	GRAM	0.20	0.15	-	0.23	0.20	GRAM/MJ	0.15	1.7
ISOLEUCINE	GRAM	0.39	0.34	-	0.43	0.39	GRAM/MJ	0.30	4.0

1 ESTIMATED BY THE DIFFERENCE METHOD
100 - (WATER + PROTEIN + FAT + MINERALS + TOTAL DIETARY FIBRE)

2 METHOD OF MEUSER, SUCKOW AND KULIKOWSKI ("BERLINER METHODE")

CONSTITUENTS	DIM	AV	VARIATION			AVR	NUTR. DENS.		MOLPERC.
LEUCINE	GRAM	0.62	0.55	-	0.70	0.62	GRAM/MJ	0.48	6.3
LYSINE	GRAM	0.35	0.32	-	0.39	0.35	GRAM/MJ	0.27	3.2
METHIONINE	GRAM	0.14	0.06	-	0.21	0.14	GRAM/MJ	0.11	1.2
PHENYLALANINE	GRAM	0.43	0.35	-	0.53	0.43	GRAM/MJ	0.33	3.5
PROLINE	GRAM	1.01	-	-	-	1.01	GRAM/MJ	0.78	11.7
SERINE	GRAM	0.43	-	-	-	0.43	GRAM/MJ	0.33	5.4
THREONINE	GRAM	0.36	0.20	-	0.42	0.36	GRAM/MJ	0.28	4.0
TRYPTOPHAN	MILLI	90.00	50.00	-	190.00	90.00	MILLI/MJ	69.62	0.6
TYROSINE	GRAM	0.29	0.12	-	0.45	0.29	GRAM/MJ	0.22	2.1
VALINE	GRAM	0.49	0.38	-	0.54	0.49	GRAM/MJ	0.38	5.6
STARCH	GRAM	61.58	59.30	-	68.40	61.58	GRAM/MJ	47.64	
DIETARY FIBRE,WAT.SOL.	GRAM	3.10	-	-	-	3.10	GRAM/MJ	2.40	
DIETARY FIBRE,WAT.INS.	GRAM	4.90	-	-	-	4.90	GRAM/MJ	3.79	

ROGGENMEHL TYPE 1370

RYE FLOUR TYPE 1370

FARINE DE SEIGLE TYPE 1370

		PROTEIN	FAT	CARBOHYDRATES	TOTAL
ENERGY VALUE (AVERAGE)	KJOULE	157	55	1117	1329
PER 100 G	(KCAL)	38	13	267	318
EDIBLE PORTION					
AMOUNT OF DIGESTIBLE	GRAM	5.37	1.27	66.72	
CONSTITUENTS PER 100 G					
ENERGY VALUE (AVERAGE)	KJOULE	102	50	1117	1269
OF THE DIGESTIBLE	(KCAL)	24	12	267	303
FRACTION PER 100 G					
EDIBLE PORTION					

WASTE PERCENTAGE AVERAGE 0.00

CONSTITUENTS	DIM	AV		VARIATION			AVR	NUTR. DENS.		MOLPERC.
WATER	GRAM	13.40		10.60	-	15.10	13.40	GRAM/MJ	10.56	
PROTEIN	GRAM	8.27		6.63	-	10.86	8.27	GRAM/MJ	6.52	
FAT	GRAM	1.42		1.26	-	1.50	1.42	GRAM/MJ	1.12	
AVAILABLE CARBOHYDR.	GRAM	66.72	1	-	-	-	66.72	GRAM/MJ	52.59	
TOTAL DIETARY FIBRE	GRAM	9.00	2	-	-	-	9.00	GRAM/MJ	7.09	
MINERALS	GRAM	1.19		1.13	-	1.26	1.19	GRAM/MJ	0.94	
SODIUM	MILLI	1.00		-	-	-	1.00	MILLI/MJ	0.79	
POTASSIUM	MILLI	303.00		300.00	-	305.00	303.00	MILLI/MJ	238.83	
MAGNESIUM	MILLI	46.80		-	-	-	46.80	MILLI/MJ	36.89	
CALCIUM	MILLI	31.10		22.00	-	43.30	31.10	MILLI/MJ	24.51	
IRON	MILLI	2.60		-	-	-	2.60	MILLI/MJ	2.05	
MOLYBDENUM	MILLI	-		0.03	-	0.16				
PHOSPHORUS	MILLI	170.00		-	-	-	170.00	MILLI/MJ	134.00	
VITAMIN B1	MILLI	0.30		0.15	-	0.44	0.30	MILLI/MJ	0.24	
VITAMIN B2	MILLI	0.13		-	-	-	0.13	MILLI/MJ	0.10	
NICOTINAMIDE	MILLI	1.60		0.77	-	2.40	1.60	MILLI/MJ	1.26	
VITAMIN C	-	0.00		-	-	-	0.00			
ALANINE	GRAM	0.49		-	-	-	0.49	GRAM/MJ	0.39	7.3
ARGININE	GRAM	0.42		0.32	-	0.53	0.42	GRAM/MJ	0.33	3.2
ASPARTIC ACID	GRAM	0.75		-	-	-	0.75	GRAM/MJ	0.59	7.5
CYSTINE	GRAM	0.17		0.05	-	0.27	0.17	GRAM/MJ	0.13	0.9
GLUTAMIC ACID	GRAM	2.51		-	-	-	2.51	GRAM/MJ	1.98	22.7
GLYCINE	GRAM	0.51		-	-	-	0.51	GRAM/MJ	0.40	9.0
HISTIDINE	GRAM	0.20		0.15	-	0.23	0.20	GRAM/MJ	0.16	1.7
ISOLEUCINE	GRAM	0.39		0.34	-	0.43	0.39	GRAM/MJ	0.31	4.0

1 ESTIMATED BY THE DIFFERENCE METHOD
100 - (WATER + PROTEIN + FAT + MINERALS + TOTAL DIETARY FIBRE)

2 METHOD OF MEUSER, SUCKOW AND KULIKOWSKI ("BERLINER METHODE")

CONSTITUENTS	DIM	AV	VARIATION			AVR	NUTR. DENS.		MOLPERC.
LEUCINE	GRAM	0.62	0.55	-	0.70	0.62	GRAM/MJ	0.49	6.3
LYSINE	GRAM	0.35	0.32	-	0.39	0.35	GRAM/MJ	0.28	3.2
METHIONINE	GRAM	0.14	0.06	-	0.21	0.14	GRAM/MJ	0.11	1.2
PHENYLALANINE	GRAM	0.43	0.35	-	0.53	0.43	GRAM/MJ	0.34	3.5
PROLINE	GRAM	1.01	-	-	-	1.01	GRAM/MJ	0.80	11.7
SERINE	GRAM	0.43	-	-	-	0.43	GRAM/MJ	0.34	5.4
THREONINE	GRAM	0.36	0.20	-	0.42	0.36	GRAM/MJ	0.28	4.0
TRYPTOPHAN	MILLI	90.00	50.00	-	190.00	90.00	MILLI/MJ	70.94	0.6
TYROSINE	GRAM	0.29	0.12	-	0.45	0.29	GRAM/MJ	0.23	2.1
VALINE	GRAM	0.49	0.38	-	0.54	0.49	GRAM/MJ	0.39	5.6
STARCH	GRAM	60.91	58.90	-	67.00	60.91	GRAM/MJ	48.01	
DIETARY FIBRE, WAT.SOL.	GRAM	3.30	-	-	-	3.30	GRAM/MJ	2.60	
DIETARY FIBRE, WAT.INS.	GRAM	5.70	-	-	-	5.70	GRAM/MJ	4.49	

ROGGENMEHL
TYPE 1800

RYE FLOUR
TYPE 1800

FARINE DE SEIGLE
TYPE 1800

		PROTEIN	FAT	CARBOHYDRATES	TOTAL
ENERGY VALUE (AVERAGE)	KJOULE	191	58	987	1236
PER 100 G EDIBLE PORTION	(KCAL)	46	14	236	295
AMOUNT OF DIGESTIBLE CONSTITUENTS PER 100 G	GRAM	6.51	1.35	58.97	
ENERGY VALUE (AVERAGE)	KJOULE	124	53	987	1163
OF THE DIGESTIBLE FRACTION PER 100 G EDIBLE PORTION	(KCAL)	30	13	236	278

WASTE PERCENTAGE AVERAGE 0.00

CONSTITUENTS	DIM	AV		VARIATION			AVR	NUTR. DENS.		MOLPERC.
WATER	GRAM	14.30		14.00	-	14.50	14.30	GRAM/MJ	12.29	
PROTEIN	GRAM	10.02		8.27	-	10.95	10.02	GRAM/MJ	8.61	
FAT	GRAM	1.50		1.40	-	1.60	1.50	GRAM/MJ	1.29	
AVAILABLE CARBOHYDR.	GRAM	58.97	1	-	-	-	58.97	GRAM/MJ	50.69	
TOTAL DIETARY FIBRE	GRAM	13.67	2	-	-	-	13.67	GRAM/MJ	11.75	
MINERALS	GRAM	1.54		1.41	-	1.71	1.54	GRAM/MJ	1.32	

1 ESTIMATED BY THE DIFFERENCE METHOD
100 - (WATER + PROTEIN + FAT + MINERALS + TOTAL DIETARY FIBRE)

2 MODIFIED AOAC METHOD

CONSTITUENTS	DIM	AV	VARIATION			AVR	NUTR. DENS.		MOLPERC.
SODIUM	MILLI	2.00	-	-	-	2.00	MILLI/MJ	1.72	
POTASSIUM	MILLI	439.00	-	-	-	439.00	MILLI/MJ	377.33	
MAGNESIUM	MILLI	83.00	-	-	-	83.00	MILLI/MJ	71.34	
CALCIUM	MILLI	23.00	-	-	-	23.00	MILLI/MJ	19.77	
MOLYBDENUM	MILLI	-	0.03	-	0.16				
PHOSPHORUS	MILLI	326.00	308.00	-	362.00	326.00	MILLI/MJ	280.20	
CHLORIDE	MILLI	73.00	-	-	-	73.00	MILLI/MJ	62.74	
VITAMIN B1	MILLI	0.30	-	-	-	0.30	MILLI/MJ	0.26	
VITAMIN B2	MILLI	0.14	-	-	-	0.14	MILLI/MJ	0.12	
NICOTINAMIDE	MILLI	1.90	0.91	-	3.10	1.90	MILLI/MJ	1.63	
VITAMIN C	-	0.00	-	-	-	0.00			
ALANINE	GRAM	0.58	-	-	-	0.58	GRAM/MJ	0.50	7.2
ARGININE	GRAM	0.54	0.36	-	0.64	0.54	GRAM/MJ	0.46	3.4
ASPARTIC ACID	GRAM	0.90	-	-	-	0.90	GRAM/MJ	0.77	7.5
CYSTINE	GRAM	0.21	0.06	-	0.30	0.21	GRAM/MJ	0.18	1.0
GLUTAMIC ACID	GRAM	2.99	-	-	-	2.99	GRAM/MJ	2.57	22.4
GLYCINE	GRAM	0.62	-	-	-	0.62	GRAM/MJ	0.53	9.1
HISTIDINE	GRAM	0.25	0.19	-	0.26	0.25	GRAM/MJ	0.21	1.8
ISOLEUCINE	GRAM	0.45	0.41	-	0.53	0.45	GRAM/MJ	0.39	3.8
LEUCINE	GRAM	0.78	0.66	-	0.84	0.78	GRAM/MJ	0.67	6.6
LYSINE	GRAM	0.42	0.37	-	0.47	0.42	GRAM/MJ	0.36	3.2
METHIONINE	GRAM	0.17	0.06	-	0.25	0.17	GRAM/MJ	0.15	1.3
PHENYLALANINE	GRAM	0.51	0.43	-	0.65	0.51	GRAM/MJ	0.44	3.4
PROLINE	GRAM	1.21	-	-	-	1.21	GRAM/MJ	1.04	11.6
SERINE	GRAM	0.58	-	-	-	0.58	GRAM/MJ	0.50	6.1
THREONINE	GRAM	0.43	0.26	-	0.49	0.43	GRAM/MJ	0.37	4.0
TRYPTOPHAN	GRAM	0.11	0.06	-	0.22	0.11	GRAM/MJ	0.09	0.6
TYROSINE	GRAM	0.27	0.15	-	0.55	0.27	GRAM/MJ	0.23	1.6
VALINE	GRAM	0.60	0.47	-	0.56	0.60	GRAM/MJ	0.52	5.6
DIETARY FIBRE, WAT.SOL.	GRAM	3.81	-	-	-	3.81	GRAM/MJ	3.27	
DIETARY FIBRE, WAT.INS.	GRAM	9.86	-	-	-	9.86	GRAM/MJ	8.47	

ROGGENKEIME RYE GERM GERMES DE SEIGLE

		PROTEIN	FAT	CARBOHYDRATES	TOTAL
ENERGY VALUE (AVERAGE)	KJOULE	742	436	546	1724
PER 100 G	(KCAL)	177	104	131	412
EDIBLE PORTION					
AMOUNT OF DIGESTIBLE	GRAM	25.33	10.08	32.65	
CONSTITUENTS PER 100 G					
ENERGY VALUE (AVERAGE)	KJOULE	482	392	546	1421
OF THE DIGESTIBLE	(KCAL)	115	94	131	340
FRACTION PER 100 G					
EDIBLE PORTION					

WASTE PERCENTAGE AVERAGE 0.00

CONSTITUENTS	DIM	AV	VARIATION			AVR	NUTR. DENS.		MOLPERC.
WATER	GRAM	12.00	-	-	-	12.00	GRAM/MJ	8.44	
PROTEIN	GRAM	38.98	36.47	-	41.48	38.98	GRAM/MJ	27.43	
FAT	GRAM	11.20	10.50	-	12.00	11.20	GRAM/MJ	7.88	
AVAILABLE CARBOHYDR.	GRAM	32.65 1	-	-	-	32.65	GRAM/MJ	22.98	
MINERALS	GRAM	5.17	4.80	-	5.54	5.17	GRAM/MJ	3.64	
COPPER	MILLI	0.75	0.70	-	0.80	0.75	MILLI/MJ	0.53	
ZINC	MILLI	20.80	20.50	-	21.50	20.80	MILLI/MJ	14.64	
VITAMIN B1	MILLI	1.00	0.60	-	1.50	1.00	MILLI/MJ	0.70	
VITAMIN B2	MILLI	0.84	0.63	-	1.00	0.84	MILLI/MJ	0.59	
NICOTINAMIDE	MILLI	2.30	-	-	-	2.30	MILLI/MJ	1.62	
PANTOTHENIC ACID	MILLI	0.80	-	-	-	0.80	MILLI/MJ	0.56	
VITAMIN C	-	0.00	-	-	-	0.00			
ARGININE	GRAM	3.04	-	-	-	3.04	GRAM/MJ	2.14	
CYSTINE	GRAM	0.35	-	-	-	0.35	GRAM/MJ	0.25	
HISTIDINE	GRAM	1.24	-	-	-	1.24	GRAM/MJ	0.87	
ISOLEUCINE	GRAM	1.55	-	-	-	1.55	GRAM/MJ	1.09	
LEUCINE	GRAM	2.77	-	-	-	2.77	GRAM/MJ	1.95	
LYSINE	GRAM	2.13	-	-	-	2.13	GRAM/MJ	1.50	
METHIONINE	GRAM	0.62	-	-	-	0.62	GRAM/MJ	0.44	
PHENYLALANINE	GRAM	1.30	-	-	-	1.30	GRAM/MJ	0.91	
THREONINE	GRAM	1.67	-	-	-	1.67	GRAM/MJ	1.18	
TRYPTOPHAN	GRAM	0.39	-	-	-	0.39	GRAM/MJ	0.27	
TYROSINE	GRAM	0.92	-	-	-	0.92	GRAM/MJ	0.65	
VALINE	GRAM	2.13	-	-	-	2.13	GRAM/MJ	1.50	
TOTAL PHOSPHOLIPIDS	MILLI	726.00	-	-	-	726.00	MILLI/MJ	510.91	
PHOSPHATIDYLCHOLINE	MILLI	235.00	-	-	-	235.00	MILLI/MJ	165.38	
PHOSPHATIDYLETHANOLAMINE	MILLI	90.00	-	-	-	90.00	MILLI/MJ	63.34	
PHOSPHATIDYLSERINE	MILLI	156.00	-	-	-	156.00	MILLI/MJ	109.78	

1 ESTIMATED BY THE DIFFERENCE METHOD
100 - (WATER + PROTEIN + FAT + MINERALS)

SORGHUM (MOHRENHIRSE) | SORGHUM | SORGHO

SORGHUM BICOLOR L. (MOENCH) (SY)

		PROTEIN	FAT	CARBOHYDRATES	TOTAL
ENERGY VALUE (AVERAGE)	KJOULE	196	125	1106	1426
PER 100 G EDIBLE PORTION	(KCAL)	47	30	264	341
AMOUNT OF DIGESTIBLE CONSTITUENTS PER 100 G	GRAM	8.75	2.88	66.07	
ENERGY VALUE (AVERAGE)	KJOULE	167	112	1106	1384
OF THE DIGESTIBLE FRACTION PER 100 G EDIBLE PORTION	(KCAL)	40	27	264	331

WASTE PERCENTAGE AVERAGE 0.00

CONSTITUENTS	DIM	AV		VARIATION			AVR	NUTR.	DENS.	MOLPERC.
WATER	GRAM	11.40		4.60	-	18.00	11.40	GRAM/MJ	8.23	
PROTEIN	GRAM	10.30		6.77	-	17.54	10.30	GRAM/MJ	7.44	
FAT	GRAM	3.20		0.10	-	5.80	3.20	GRAM/MJ	2.31	
AVAILABLE CARBOHYDR.	GRAM	66.07	1	-	-	-	66.07	GRAM/MJ	47.72	
TOTAL DIETARY FIBRE	GRAM	7.28	2	6.70	-	8.09	7.28	GRAM/MJ	5.26	
MINERALS	GRAM	1.75		1.10	-	4.50	1.75	GRAM/MJ	1.26	
CALCIUM	MILLI	26.00		14.00	-	61.00	26.00	MILLI/MJ	18.78	
IRON	MILLI	2.70		2.10	-	34.50	2.70	MILLI/MJ	1.95	
COPPER	MILLI	0.37		0.26	-	0.69	0.37	MILLI/MJ	0.27	
ZINC	MILLI	2.00		1.50	-	2.90	2.00	MILLI/MJ	1.44	
NICKEL	MILLI	0.17		0.00	-	1.20	0.17	MILLI/MJ	0.12	
CHROMIUM	MILLI	-		0.00	-	2.60				
MOLYBDENUM	MILLI	0.17		0.00	-	0.17	0.17	MILLI/MJ	0.12	
VANADIUM	MICRO	80.00		0.00	-	170.00	80.00	MICRO/MJ	57.78	
PHOSPHORUS	MILLI	330.00		180.00	-	440.00	330.00	MILLI/MJ	238.36	
CAROTENE	MILLI	10.00		0.00	-	30.00	10.00	MILLI/MJ	7.22	
VITAMIN E ACTIVITY	MILLI	0.17		-	-	-	0.17	MILLI/MJ	0.12	
TOTAL TOCOPHEROLS	MILLI	1.10		-	-	-	1.10	MILLI/MJ	0.79	
ALPHA-TOCOPHEROL	MICRO	70.00		-	-	-	70.00	MICRO/MJ	50.56	
GAMMA-TOCOPHEROL	MILLI	1.00		-	-	-	1.00	MILLI/MJ	0.72	
VITAMIN B1	MILLI	0.34		0.17	-	0.62	0.34	MILLI/MJ	0.25	
VITAMIN B2	MILLI	0.15		0.05	-	0.68	0.15	MILLI/MJ	0.11	
NICOTINAMIDE	MILLI	3.30		1.30	-	5.10	3.30	MILLI/MJ	2.38	
VITAMIN C	-	TRACES		0.00	-	12.00				
ALANINE	GRAM	0.88		-	-	-	0.88	GRAM/MJ	0.64	11.1

1 ESTIMATED BY THE DIFFERENCE METHOD
100 - (WATER + PROTEIN + FAT + MINERALS + TOTAL DIETARY FIBRE)

2 MODIFIED AOAC METHOD

CONSTITUENTS	DIM	AV	VARIATION			AVR	NUTR. DENS.		MOLPERC.
ARGININE	GRAM	0.38	-	-	-	0.38	GRAM/MJ	0.27	2.4
ASPARTIC ACID	GRAM	0.71	-	-	-	0.71	GRAM/MJ	0.51	6.0
CYSTINE	GRAM	0.10	-	-	-	0.10	GRAM/MJ	0.07	0.5
GLUTAMIC ACID	GRAM	2.29	-	-	-	2.29	GRAM/MJ	1.65	17.5
GLYCINE	GRAM	0.43	-	-	-	0.43	GRAM/MJ	0.31	6.4
HISTIDINE	GRAM	0.22	-	-	-	0.22	GRAM/MJ	0.16	1.6
ISOLEUCINE	GRAM	0.58	-	-	-	0.58	GRAM/MJ	0.42	5.0
LEUCINE	GRAM	1.36	-	-	-	1.36	GRAM/MJ	0.98	11.6
LYSINE	GRAM	0.26	-	-	-	0.26	GRAM/MJ	0.19	2.0
METHIONINE	GRAM	0.20	-	-	-	0.20	GRAM/MJ	0.14	1.5
PHENYLALANINE	GRAM	0.44	-	-	-	0.44	GRAM/MJ	0.32	3.0
PROLINE	GRAM	1.55	-	-	-	1.55	GRAM/MJ	1.12	15.1
SERINE	GRAM	0.42	-	-	-	0.42	GRAM/MJ	0.30	4.5
THREONINE	GRAM	0.44	-	-	-	0.44	GRAM/MJ	0.32	4.1
TRYPTOPHAN	GRAM	0.11	-	-	-	0.11	GRAM/MJ	0.08	0.6
TYROSINE	GRAM	0.25	-	-	-	0.25	GRAM/MJ	0.18	1.5
VALINE	GRAM	0.58	-	-	-	0.58	GRAM/MJ	0.42	5.6
SUCROSE	GRAM	1.68	0.93	-	3.90	1.68	GRAM/MJ	1.21	
RAFFINOSE	GRAM	0.23	0.10	-	0.39	0.23	GRAM/MJ	0.17	
STACHYOSE	GRAM	0.10	0.05	-	0.21	0.10	GRAM/MJ	0.07	
CELLULOSE	GRAM	3.50	1.20	-	5.20	3.50	GRAM/MJ	2.53	
MYRISTIC ACID	MILLI	5.00	0.00	-	10.00	5.00	MILLI/MJ	3.61	
PALMITIC ACID	MILLI	340.00	310.00	-	350.00	340.00	MILLI/MJ	245.58	
STEARIC ACID	MILLI	90.00	70.00	-	110.00	90.00	MILLI/MJ	65.01	
ARACHIDIC ACID	MILLI	70.00	0.00	-	200.00	70.00	MILLI/MJ	50.56	
BEHENIC ACID	MILLI	-	0.00	-	110.00				
LIGNOCERIC ACID	MILLI	-	0.00	-	30.00				
PALMITOLEIC ACID	MILLI	25.00	10.00	-	30.00	25.00	MILLI/MJ	18.06	
OLEIC ACID	GRAM	0.99	-	-	-	0.99	GRAM/MJ	0.72	
LINOLEIC ACID	GRAM	1.01	-	-	-	1.01	GRAM/MJ	0.73	
LINOLENIC ACID	MILLI	70.00	-	-	-	70.00	MILLI/MJ	50.56	
DIETARY FIBRE,WAT.SOL.	GRAM	1.01	0.93	-	1.09	1.01	GRAM/MJ	0.73	
DIETARY FIBRE,WAT.INS.	GRAM	6.27	5.78	-	6.99	6.27	GRAM/MJ	4.53	

WEIZEN
GANZES KORN

WHEAT
WHOLE GRAIN

BLÉ
GRAINS ENTIERS

TRITICUM VULGARE VILL.

		PROTEIN	FAT	CARBOHYDRATES	TOTAL
ENERGY VALUE (AVERAGE)	KJOULE	223	78	1020	1322
PER 100 G	(KCAL)	53	19	244	316
EDIBLE PORTION					
AMOUNT OF DIGESTIBLE	GRAM	9.26	1.80	60.97	
CONSTITUENTS PER 100 G					
ENERGY VALUE (AVERAGE)	KJOULE	176	70	1020	1267
OF THE DIGESTIBLE	(KCAL)	42	17	244	303
FRACTION PER 100 G					
EDIBLE PORTION					

WASTE PERCENTAGE AVERAGE 0.00

CONSTITUENTS	DIM	AV		VARIATION			AVR	NUTR. DENS.		MOLPERC.
WATER	GRAM	13.20		12.80	-	13.50	13.20	GRAM/MJ	10.42	
PROTEIN	GRAM	11.73	1	10.20	-	13.21	11.73	GRAM/MJ	9.26	
FAT	GRAM	2.00		1.90	-	2.10	2.00	GRAM/MJ	1.58	
AVAILABLE CARBOHYDR.	GRAM	60.97	2	-	-	-	60.97	GRAM/MJ	48.13	
TOTAL DIETARY FIBRE	GRAM	10.30	3	-	-	-	10.30	GRAM/MJ	8.13	
MINERALS	GRAM	1.80		1.38	-	2.50	1.80	GRAM/MJ	1.42	
SODIUM	MILLI	7.80		6.60	-	9.00	7.80	MILLI/MJ	6.16	
POTASSIUM	MILLI	502.00		432.00	-	571.00	502.00	MILLI/MJ	396.26	
MAGNESIUM	MILLI	147.00		119.00	-	175.00	147.00	MILLI/MJ	116.04	
CALCIUM	MILLI	43.70		39.40	-	48.00	43.70	MILLI/MJ	34.50	
MANGANESE	MILLI	3.40		2.40	-	4.30	3.40	MILLI/MJ	2.68	
IRON	MILLI	3.30		3.10	-	3.50	3.30	MILLI/MJ	2.60	
COBALT	MICRO	2.00		0.50	-	9.00	2.00	MICRO/MJ	1.58	
COPPER	MILLI	0.63		0.48	-	0.78	0.63	MILLI/MJ	0.50	
ZINC	MILLI	4.10		2.20	-	10.00	4.10	MILLI/MJ	3.24	
NICKEL	MICRO	34.00		16.00	-	89.00	34.00	MICRO/MJ	26.84	
CHROMIUM	MICRO	3.00		2.00	-	175.00	3.00	MICRO/MJ	2.37	
MOLYBDENUM	MILLI	-		0.02	-	0.08				
VANADIUM	MICRO	-		2.00	-	230.00				
PHOSPHORUS	MILLI	344.42		341.00	-	406.00	344.42	MILLI/MJ	271.87	
CHLORIDE	MILLI	55.00		-	-	-	55.00	MILLI/MJ	43.41	
FLUORIDE	MICRO	90.00		10.00	-	400.00	90.00	MICRO/MJ	71.04	
IODIDE	MICRO	0.60		-	-	-	0.60	MICRO/MJ	0.47	
BORON	MILLI	-		0.20	-	0.73				
SELENIUM	MICRO	-	4	0.70	-	130.00				
SILICON	MILLI	8.00		5.00	-	19.00	8.00	MILLI/MJ	6.31	

1 VARIATION: LOW VALUE: MW SOFT WHEAT VARIETIES
HIGH VALUE: MW OF HARD WHEAT VARIETIES

2 ESTIMATED BY THE DIFFERENCE METHOD
100 - (WATER + PROTEIN + FAT + MINERALS + TOTAL DIETARY FIBRE)

3 METHOD OF MEUSER, SUCKOW AND KULIKOWSKI ("BERLINER METHODE")

4 GREAT REGIONAL DIFFERENCES: USA 5 - 100 UG/100 G
MIDDLE AMERICA: 100 - 3000 UG/100 G
SCANDINAVIAN COUNTRIES: 0,3 - 1,0 UG/100 G

CONSTITUENTS	DIM	AV	VARIATION			AVR	NUTR. DENS.		MOLPERC.
CAROTENE	MICRO	20.00	10.00	-	30.00	20.00	MICRO/MJ	15.79	
VITAMIN E ACTIVITY	MILLI	1.40	1.10	-	1.50	1.40	MILLI/MJ	1.11	
TOTAL TOCOPHEROLS	MILLI	4.30	3.00	-	4.90	4.30	MILLI/MJ	3.39	
ALPHA-TOCOPHEROL	MILLI	1.00	0.80	-	1.20	1.00	MILLI/MJ	0.79	
BETA-TOCOPHEROL	MILLI	0.38	0.28	-	54.00	0.38	MILLI/MJ	0.30	
ALPHA-TOCOTRIENOL	MILLI	0.27	0.12	-	0.48	0.27	MILLI/MJ	0.21	
BETA-TOCOTRIENOL	MILLI	2.40	1.60	-	2.90	2.40	MILLI/MJ	1.89	
VITAMIN K	MILLI	-	0.00	-	0.02				
VITAMIN B1	MILLI	0.48	0.14	-	1.08	0.48	MILLI/MJ	0.38	
VITAMIN B2	MILLI	0.14	0.05	-	0.31	0.14	MILLI/MJ	0.11	
NICOTINAMIDE	MILLI	5.10	2.20	-	11.10	5.10	MILLI/MJ	4.03	
PANTOTHENIC ACID	MILLI	1.18	0.91	-	1.75	1.18	MILLI/MJ	0.93	
VITAMIN B6	MILLI	0.44	0.21	-	0.70	0.44	MILLI/MJ	0.35	
BIOTIN	MICRO	6.00	2.00	-	11.00	6.00	MICRO/MJ	4.74	
FOLIC ACID	MICRO	49.00	16.00	-	78.00	49.00	MICRO/MJ	38.68	
VITAMIN C	MILLI	-	0.00	-	1.50				
ALANINE	GRAM	0.51	0.49	-	0.55	0.51	GRAM/MJ	0.40	5.4
ARGININE	GRAM	0.62	0.58	-	0.71	0.62	GRAM/MJ	0.49	3.4
ASPARTIC ACID	GRAM	0.70	0.69	-	0.73	0.70	GRAM/MJ	0.55	5.0
CYSTINE	GRAM	0.29	0.28	-	0.35	0.29	GRAM/MJ	0.23	1.1
GLUTAMIC ACID	GRAM	4.08	3.87	-	4.18	4.08	GRAM/MJ	3.22	26.1
GLYCINE	GRAM	0.72	0.63	-	0.93	0.72	GRAM/MJ	0.57	9.0
HISTIDINE	GRAM	0.28	0.26	-	0.31	0.28	GRAM/MJ	0.22	1.7
ISOLEUCINE	GRAM	0.54	0.47	-	0.58	0.54	GRAM/MJ	0.43	3.9
LEUCINE	GRAM	0.92	0.88	-	0.98	0.92	GRAM/MJ	0.73	6.6
LYSINE	GRAM	0.38	0.34	-	0.43	0.38	GRAM/MJ	0.30	2.4
METHIONINE	GRAM	0.22	0.19	-	0.27	0.22	GRAM/MJ	0.17	1.4
PHENYLALANINE	GRAM	0.64	0.62	-	0.65	0.64	GRAM/MJ	0.51	3.7
PROLINE	GRAM	1.56	1.45	-	1.77	1.56	GRAM/MJ	1.23	12.8
SERINE	GRAM	0.71	0.43	-	0.79	0.71	GRAM/MJ	0.56	6.4
THREONINE	GRAM	0.43	0.40	-	0.49	0.43	GRAM/MJ	0.34	3.4
TRYPTOPHAN	GRAM	0.15	0.14	-	0.16	0.15	GRAM/MJ	0.12	0.7
TYROSINE	GRAM	0.41	0.28	-	0.48	0.41	GRAM/MJ	0.32	2.1
VALINE	GRAM	0.62	0.61	-	0.65	0.62	GRAM/MJ	0.49	5.0
FRUCTOSE	MILLI	42.00	23.00	-	52.00	42.00	MILLI/MJ	33.15	
SUCROSE	GRAM	0.60	-	-	-	0.60	GRAM/MJ	0.47	
RAFFINOSE	GRAM	0.17	-	-	-	0.17	GRAM/MJ	0.13	
GLUCODIFRUCTOSE	GRAM	0.23	-	-	-	0.23	GRAM/MJ	0.18	
STARCH	GRAM	58.16	56.78	-	58.50	58.16	GRAM/MJ	45.91	
CELLULOSE	GRAM	2.80	-	-	-	2.80	GRAM/MJ	2.21	
PHYTIC ACID	GRAM	0.99	0.62	-	1.35	0.99	GRAM/MJ	0.78	
PALMITIC ACID	GRAM	0.39	0.33	-	0.76	0.39	GRAM/MJ	0.31	
STEARIC ACID	MILLI	29.00	19.00	-	38.00	29.00	MILLI/MJ	22.89	
PALMITOLEIC ACID	MILLI	29.00	19.00	-	38.00	29.00	MILLI/MJ	22.89	
OLEIC ACID	GRAM	0.28	0.15	-	0.40	0.28	GRAM/MJ	0.22	
LINOLEIC ACID	GRAM	1.10	1.05	-	1.15	1.10	GRAM/MJ	0.87	
LINOLENIC ACID	MILLI	76.00	57.00	-	96.00	76.00	MILLI/MJ	59.99	
TOTAL STEROLS	MILLI	69.00	-	-	-	69.00	MILLI/MJ	54.47	
CAMPESTEROL	MILLI	27.00	-	-	-	27.00	MILLI/MJ	21.31	
BETA-SITOSTEROL	MILLI	40.00	-	-	-	40.00	MILLI/MJ	31.57	
TOTAL PURINES	MILLI	40.00 [5]	40.00	-	40.00	40.00	MILLI/MJ	31.57	
ADENINE	MILLI	16.00	16.00	-	16.00	16.00	MILLI/MJ	12.63	
GUANINE	MILLI	19.00	19.00	-	19.00	19.00	MILLI/MJ	15.00	
TOTAL PHOSPHOLIPIDS	MILLI	711.00	-	-	-	711.00	MILLI/MJ	561.24	
PHOSPHATIDYLCHOLINE	MILLI	472.00	-	-	-	472.00	MILLI/MJ	372.58	
PHOSPHATIDYLETHANOLAMINE	MILLI	75.00	-	-	-	75.00	MILLI/MJ	59.20	
PHOSPHATIDYLINOSITOL	MILLI	51.00	-	-	-	51.00	MILLI/MJ	40.26	
DIETARY FIBRE,WAT.SOL.	GRAM	1.80	-	-	-	1.80	GRAM/MJ	1.42	
DIETARY FIBRE,WAT.INS.	GRAM	8.50	-	-	-	8.50	GRAM/MJ	6.71	

5 VALUE OF TOTAL PURINES IS EXPRESSSED IN MG URIC ACID/100G

WEIZENGRIESS WHEAT GRITS SEMOULE DE BLÉ

		PROTEIN	FAT	CARBOHYDRATES	TOTAL
ENERGY VALUE (AVERAGE) PER 100 G EDIBLE PORTION	KJOULE (KCAL)	182 43	31 7.3	1154 276	1367 327
AMOUNT OF DIGESTIBLE CONSTITUENTS PER 100 G	GRAM	8.50	0.71	68.96	
ENERGY VALUE (AVERAGE) OF THE DIGESTIBLE FRACTION PER 100 G EDIBLE PORTION	KJOULE (KCAL)	162 39	28 6.6	1154 276	1344 321

WASTE PERCENTAGE AVERAGE 0.00

CONSTITUENTS	DIM	AV		VARIATION			AVR	NUTR. DENS.		MOLPERC.
WATER	GRAM	13.10		12.20	-	14.00	13.10	GRAM/MJ	9.75	
PROTEIN	GRAM	9.56		8.58	-	10.21	9.56	GRAM/MJ	7.11	
FAT	GRAM	0.79		0.63	-	1.00	0.79	GRAM/MJ	0.59	
AVAILABLE CARBOHYDR.	GRAM	68.96	1	-	-	-	68.96	GRAM/MJ	51.32	
TOTAL DIETARY FIBRE	GRAM	7.12	2	-	-	-	7.12	GRAM/MJ	5.30	
MINERALS	GRAM	0.47		0.42	-	0.52	0.47	GRAM/MJ	0.35	
SODIUM	MILLI	1.00		0.00	-	2.00	1.00	MILLI/MJ	0.74	
POTASSIUM	MILLI	112.00		105.00	-	115.00	112.00	MILLI/MJ	83.35	
CALCIUM	MILLI	17.00		15.00	-	18.00	17.00	MILLI/MJ	12.65	
IRON	MILLI	1.00		-	-	-	1.00	MILLI/MJ	0.74	
CHLORIDE	MILLI	87.00		84.00	-	90.00	87.00	MILLI/MJ	64.74	
TOTAL TOCOPHEROLS	MILLI	1.80		-	-	-	1.80	MILLI/MJ	1.34	
VITAMIN B1	MILLI	0.12		0.05	-	0.21	0.12	MILLI/MJ	0.09	
VITAMIN B2	MICRO	38.00		30.00	-	45.00	38.00	MICRO/MJ	28.28	
NICOTINAMIDE	MILLI	1.30		0.90	-	2.00	1.30	MILLI/MJ	0.97	
VITAMIN B6	MICRO	85.00		77.00	-	94.00	85.00	MICRO/MJ	63.26	
REDUCING SUGAR	GRAM	0.32		-	-	-	0.32	GRAM/MJ	0.24	
NONREDUCING SUGAR	GRAM	1.53		-	-	-	1.53	GRAM/MJ	1.14	
DIETARY FIBRE,WAT.SOL.	GRAM	2.17		-	-	-	2.17	GRAM/MJ	1.61	
DIETARY FIBRE,WAT.INS.	GRAM	4.95		-	-	-	4.95	GRAM/MJ	3.68	

1 ESTIMATED BY THE DIFFERENCE METHOD
100 - (WATER + PROTEIN + FAT + MINERALS + TOTAL DIETARY FIBRE)

2 MODIFIED AOAC METHOD

WEIZENMEHL
TYPE 405

WHEAT FLOUR
TYPE 405

FARINE DE BLÉ
TYPE 405

		PROTEIN	FAT	CARBOHYDRATES	TOTAL
ENERGY VALUE (AVERAGE) PER 100 G EDIBLE PORTION	KJOULE	187	38	1187	1413
	(KCAL)	45	9.1	284	338
AMOUNT OF DIGESTIBLE CONSTITUENTS PER 100 G	GRAM	8.75	0.88	70.93	
ENERGY VALUE (AVERAGE) OF THE DIGESTIBLE FRACTION PER 100 G EDIBLE PORTION	KJOULE	167	34	1187	1388
	(KCAL)	40	8.2	284	332

WASTE PERCENTAGE AVERAGE 0.00

CONSTITUENTS	DIM	AV		VARIATION			AVR	NUTR. DENS.		MOLPERC.
WATER	GRAM	13.90		13.00	-	14.70	13.90	GRAM/MJ	10.01	
PROTEIN	GRAM	9.84		9.56	-	10.12	9.84	GRAM/MJ	7.09	
FAT	GRAM	0.98		0.90	-	1.03	0.98	GRAM/MJ	0.71	
AVAILABLE CARBOHYDR.	GRAM	70.93	1	-	-	-	70.93	GRAM/MJ	51.10	
TOTAL DIETARY FIBRE	GRAM	4.00	2	-	-	-	4.00	GRAM/MJ	2.88	
MINERALS	GRAM	0.35		0.33	-	0.38	0.35	GRAM/MJ	0.25	
SODIUM	MILLI	2.00		1.00	-	2.00	2.00	MILLI/MJ	1.44	
POTASSIUM	MILLI	108.00		102.00	-	113.00	108.00	MILLI/MJ	77.80	
CALCIUM	MILLI	15.00		13.00	-	16.00	15.00	MILLI/MJ	10.81	
MANGANESE	MILLI	0.74		-	-	-	0.74	MILLI/MJ	0.53	
IRON	MILLI	1.95		-	-	-	1.95	MILLI/MJ	1.40	
COPPER	MILLI	0.29		0.15	-	0.43	0.29	MILLI/MJ	0.21	
ZINC	MILLI	1.10		-	-	-	1.10	MILLI/MJ	0.79	
MOLYBDENUM	MICRO	45.00		25.00	-	64.00	45.00	MICRO/MJ	32.42	
PHOSPHORUS	MILLI	74.00		74.00	-	74.00	74.00	MILLI/MJ	53.31	
BORON	MICRO	59.00		-	-	-	59.00	MICRO/MJ	42.50	
SELENIUM	MICRO	19.00		-	-	-	19.00	MICRO/MJ	13.69	
VITAMIN B1	MICRO	60.00		0.00	-	60.00	60.00	MICRO/MJ	43.22	
VITAMIN B2	MICRO	30.00		0.00	-	30.00	30.00	MICRO/MJ	21.61	
NICOTINAMIDE	MILLI	0.70		-	-	-	0.70	MILLI/MJ	0.50	
PANTOTHENIC ACID	MILLI	0.21		0.20	-	0.22	0.21	MILLI/MJ	0.15	
VITAMIN B6	MILLI	0.18		0.15	-	0.20	0.18	MILLI/MJ	0.13	
BIOTIN	MICRO	1.50		1.00	-	2.00	1.50	MICRO/MJ	1.08	
FOLIC ACID	MICRO	10.00		-	-	-	10.00	MICRO/MJ	7.20	
ALANINE	GRAM	0.37		-	-	-	0.37	GRAM/MJ	0.27	4.8

1 ESTIMATED BY THE DIFFERENCE METHOD
100 - (WATER + PROTEIN + FAT + MINERALS + TOTAL DIETARY FIBRE)

2 METHOD OF MEUSER, SUCKOW AND KULIKOWSKI ("BERLINER METHODE")

CONSTITUENTS	DIM	AV	VARIATION			AVR	NUTR. DENS.		MOLPERC.
ARGININE	GRAM	0.43	0.40	-	0.52	0.43	GRAM/MJ	0.31	2.8
ASPARTIC ACID	GRAM	0.48	-	-	-	0.48	GRAM/MJ	0.35	4.1
CYSTINE	GRAM	0.24	0.18	-	0.26	0.24	GRAM/MJ	0.17	1.1
GLUTAMIC ACID	GRAM	3.66	-	-	-	3.66	GRAM/MJ	2.64	28.6
GLYCINE	GRAM	0.42	-	-	-	0.42	GRAM/MJ	0.30	6.4
HISTIDINE	GRAM	0.22	0.20	-	0.23	0.22	GRAM/MJ	0.16	1.6
ISOLEUCINE	GRAM	0.46	0.45	-	0.52	0.46	GRAM/MJ	0.33	4.0
LEUCINE	GRAM	0.82	0.80	-	0.87	0.82	GRAM/MJ	0.59	7.2
LYSINE	GRAM	0.24	0.21	-	0.25	0.24	GRAM/MJ	0.17	1.9
METHIONINE	GRAM	0.17	0.15	-	0.19	0.17	GRAM/MJ	0.12	1.3
PHENYLALANINE	GRAM	0.55	0.52	-	0.59	0.55	GRAM/MJ	0.40	3.8
PROLINE	GRAM	1.45	-	-	-	1.45	GRAM/MJ	1.04	14.5
SERINE	GRAM	0.66	-	-	-	0.66	GRAM/MJ	0.48	7.2
THREONINE	GRAM	0.32	0.30	-	0.32	0.32	GRAM/MJ	0.23	3.1
TRYPTOPHAN	GRAM	0.12	0.11	-	0.14	0.12	GRAM/MJ	0.09	0.7
TYROSINE	GRAM	0.32	0.27	-	0.40	0.32	GRAM/MJ	0.23	2.0
VALINE	GRAM	0.49	-	-	-	0.49	GRAM/MJ	0.35	4.8
GLUCOSE	MILLI	50.00	-	-	-	50.00	MILLI/MJ	36.02	
FRUCTOSE	MILLI	20.00	-	-	-	20.00	MILLI/MJ	14.41	
SUCROSE	GRAM	0.16	-	-	-	0.16	GRAM/MJ	0.12	
RAFFINOSE	MILLI	50.00	-	-	-	50.00	MILLI/MJ	36.02	
STARCH	GRAM	70.60	-	-	-	70.60	GRAM/MJ	50.86	
CELLULOSE	GRAM	0.60	-	-	-	0.60	GRAM/MJ	0.43	
DIETARY FIBRE,WAT.SOL.	GRAM	1.70	-	-	-	1.70	GRAM/MJ	1.22	
DIETARY FIBRE,WAT.INS.	GRAM	2.30	-	-	-	2.30	GRAM/MJ	1.66	

WEIZENMEHL	WHEAT FLOUR	FARINE DE BLÉ
TYPE 550	TYPE 550	TYPE 550

		PROTEIN	FAT	CARBOHYDRATES	TOTAL
ENERGY VALUE (AVERAGE) PER 100 G EDIBLE PORTION	KJOULE	187	44	1184	1416
	(KCAL)	45	11	283	338
AMOUNT OF DIGESTIBLE CONSTITUENTS PER 100 G	GRAM	8.75	1.01	70.76	
ENERGY VALUE (AVERAGE) OF THE DIGESTIBLE FRACTION PER 100 G EDIBLE PORTION	KJOULE	167	40	1184	1391
	(KCAL)	40	9.5	283	332

WASTE PERCENTAGE AVERAGE 0.00

CONSTITUENTS	DIM	AV		VARIATION			AVR	NUTR. DENS.		MOLPERC.
WATER	GRAM	13.70		12.10	-	15.00	13.70	GRAM/MJ	9.85	
PROTEIN	GRAM	9.84		9.28	-	10.49	9.84	GRAM/MJ	7.08	
FAT	GRAM	1.13		1.00	-	1.40	1.13	GRAM/MJ	0.81	
AVAILABLE CARBOHYDR.	GRAM	70.76	1	-	-	-	70.76	GRAM/MJ	50.89	
TOTAL DIETARY FIBRE	GRAM	4.10	2	-	-	-	4.10	GRAM/MJ	2.95	
MINERALS	GRAM	0.47		0.42	-	0.50	0.47	GRAM/MJ	0.34	
SODIUM	MILLI	3.00		1.00	-	4.00	3.00	MILLI/MJ	2.16	
POTASSIUM	MILLI	126.00		115.00	-	136.00	126.00	MILLI/MJ	90.61	
CALCIUM	MILLI	16.00		8.00	-	20.00	16.00	MILLI/MJ	11.51	
IRON	MILLI	1.10		0.70	-	1.50	1.10	MILLI/MJ	0.79	
COPPER	MILLI	0.29		0.13	-	0.43	0.29	MILLI/MJ	0.21	
MOLYBDENUM	MICRO	45.00		25.00	-	64.00	45.00	MICRO/MJ	32.36	
PHOSPHORUS	MILLI	113.17		95.00	-	114.00	113.17	MILLI/MJ	81.39	
BORON	MICRO	25.00		0.00	-	45.00	25.00	MICRO/MJ	17.98	
VITAMIN B1	MILLI	0.11		0.07	-	0.14	0.11	MILLI/MJ	0.08	
VITAMIN B2	MICRO	80.00		40.00	-	100.00	80.00	MICRO/MJ	57.53	
NICOTINAMIDE	MILLI	0.50		-	-	-	0.50	MILLI/MJ	0.36	
PANTOTHENIC ACID	MILLI	0.40		-	-	-	0.40	MILLI/MJ	0.29	
VITAMIN B6	MILLI	0.10		-	-	-	0.10	MILLI/MJ	0.07	
BIOTIN	MICRO	1.10		-	-	-	1.10	MICRO/MJ	0.79	
FOLIC ACID	MICRO	16.00		-	-	-	16.00	MICRO/MJ	11.51	
VITAMIN C	-	0.00		-	-	-	0.00			
ALANINE	GRAM	0.35		0.31	-	0.38	0.35	GRAM/MJ	0.25	4.5
ARGININE	GRAM	0.40		0.33	-	0.45	0.40	GRAM/MJ	0.29	2.7
ASPARTIC ACID	GRAM	0.46		0.45	-	0.51	0.46	GRAM/MJ	0.33	4.0
CYSTINE	GRAM	0.24		0.18	-	0.25	0.24	GRAM/MJ	0.17	1.2
GLUTAMIC ACID	GRAM	3.83		3.66	-	3.98	3.83	GRAM/MJ	2.75	30.1
GLYCINE	GRAM	0.46		0.43	-	0.49	0.46	GRAM/MJ	0.33	7.1
HISTIDINE	GRAM	0.21		0.20	-	0.23	0.21	GRAM/MJ	0.15	1.6
ISOLEUCINE	GRAM	0.43		0.40	-	0.50	0.43	GRAM/MJ	0.31	3.8

1 ESTIMATED BY THE DIFFERENCE METHOD
100 - (WATER + PROTEIN + FAT + MINERALS + TOTAL DIETARY FIBRE)

2 MODIFIED AOAC METHOD

CONSTITUENTS	DIM	AV	VARIATION			AVR	NUTR. DENS.		MOLPERC.
LEUCINE	GRAM	0.81	0.73	-	0.85	0.81	GRAM/MJ	0.58	7.1
LYSINE	GRAM	0.22	0.20	-	0.26	0.22	GRAM/MJ	0.16	1.7
METHIONINE	GRAM	0.17	0.15	-	0.19	0.17	GRAM/MJ	0.12	1.3
PHENYLALANINE	GRAM	0.54	0.51	-	0.58	0.54	GRAM/MJ	0.39	3.8
PROLINE	GRAM	1.41	1.36	-	1.49	1.41	GRAM/MJ	1.01	14.2
SERINE	GRAM	0.58	0.54	-	0.59	0.58	GRAM/MJ	0.42	6.4
THREONINE	GRAM	0.32	0.30	-	0.32	0.32	GRAM/MJ	0.23	3.1
TRYPTOPHAN	GRAM	0.12	0.10	-	0.13	0.12	GRAM/MJ	0.09	0.7
TYROSINE	GRAM	0.33	0.24	-	0.36	0.33	GRAM/MJ	0.24	2.1
VALINE	GRAM	0.48	0.46	-	0.54	0.48	GRAM/MJ	0.35	4.7
TOTAL SUGAR	GRAM	0.70	-	-	-	0.70	GRAM/MJ	0.50	
STARCH	GRAM	70.60	-	-	-	70.60	GRAM/MJ	50.77	
CELLULOSE	GRAM	0.60	-	-	-	0.60	GRAM/MJ	0.43	
DIETARY FIBRE,WAT.SOL.	GRAM	2.38	-	-	-	2.38	GRAM/MJ	1.71	
DIETARY FIBRE,WAT.INS.	GRAM	1.72	-	-	-	1.72	GRAM/MJ	1.24	

WEIZENMEHL
TYPE 630

WHEAT FLOUR
TYPE 630

FARINE DE BLÉ
TYPE 630

		PROTEIN	FAT	CARBOHYDRATES	TOTAL
ENERGY VALUE (AVERAGE) PER 100 G EDIBLE PORTION	KJOULE	201	60	1155	1417
	(KCAL)	48	14	276	339
AMOUNT OF DIGESTIBLE CONSTITUENTS PER 100 G	GRAM	9.41	1.38	69.04	
ENERGY VALUE (AVERAGE) OF THE DIGESTIBLE FRACTION PER 100 G EDIBLE PORTION	KJOULE	179	54	1155	1389
	(KCAL)	43	13	276	332

WASTE PERCENTAGE AVERAGE 0.00

CONSTITUENTS	DIM	AV		VARIATION			AVR	NUTR. DENS.		MOLPERC.
WATER	GRAM	14.20		13.00	-	14.80	14.20	GRAM/MJ	10.23	
PROTEIN	GRAM	10.58		10.21	-	11.69	10.58	GRAM/MJ	7.62	
FAT	GRAM	1.54		1.50	-	1.56	1.54	GRAM/MJ	1.11	
AVAILABLE CARBOHYDR.	GRAM	69.04	1	-	-	-	69.04	GRAM/MJ	49.72	
TOTAL DIETARY FIBRE	GRAM	4.10	2	-	-	-	4.10	GRAM/MJ	2.95	
MINERALS	GRAM	0.54		0.51	-	0.60	0.54	GRAM/MJ	0.39	

1 ESTIMATED BY THE DIFFERENCE METHOD
100 - (WATER + PROTEIN + FAT + MINERALS + TOTAL DIETARY FIBRE)

2 MODIFIED AOAC METHOD

CONSTITUENTS	DIM	AV	VARIATION			AVR	NUTR. DENS.		MOLPERC.
SODIUM	MILLI	3.25	2.50	-	4.00	3.25	MILLI/MJ	2.34	
POTASSIUM	MILLI	142.00	115.00	-	190.00	142.00	MILLI/MJ	102.26	
MAGNESIUM	MILLI	10.00	-	-	-	10.00	MILLI/MJ	7.20	
CALCIUM	MILLI	18.00	8.00	-	23.00	18.00	MILLI/MJ	12.96	
IRON	MILLI	0.80	0.70	-	1.80	0.80	MILLI/MJ	0.58	
COPPER	MILLI	0.29	0.13	-	0.43	0.29	MILLI/MJ	0.21	
MOLYBDENUM	MICRO	45.00	25.00	-	64.00	45.00	MICRO/MJ	32.41	
BORON	MICRO	25.00	0.00	-	45.00	25.00	MICRO/MJ	18.00	
VITAMIN B1	MILLI	0.12	0.07	-	0.19	0.12	MILLI/MJ	0.09	
VITAMIN B2	MICRO	50.00	40.00	-	80.00	50.00	MICRO/MJ	36.01	
NICOTINAMIDE	MILLI	0.84	0.77	-	1.00	0.84	MILLI/MJ	0.60	
PANTOTHENIC ACID	MILLI	0.30	-	-	-	0.30	MILLI/MJ	0.22	
VITAMIN B6	MILLI	0.22	0.20	-	0.22	0.22	MILLI/MJ	0.16	
BIOTIN	MICRO	2.00	-	-	-	2.00	MICRO/MJ	1.44	
FOLIC ACID	MICRO	17.00	-	-	-	17.00	MICRO/MJ	12.24	
VITAMIN C	-	0.00	-	-	-	0.00			
ALANINE	GRAM	0.40	-	-	-	0.40	GRAM/MJ	0.29	4.8
ARGININE	GRAM	0.43	0.41	-	0.50	0.43	GRAM/MJ	0.31	2.7
ASPARTIC ACID	GRAM	0.51	0.49	-	0.53	0.51	GRAM/MJ	0.37	4.1
CYSTINE	GRAM	0.29	-	-	-	0.29	GRAM/MJ	0.21	1.3
GLUTAMIC ACID	GRAM	4.04	3.90	-	4.18	4.04	GRAM/MJ	2.91	29.5
GLYCINE	GRAM	0.46	0.44	-	0.49	0.46	GRAM/MJ	0.33	6.6
HISTIDINE	GRAM	0.24	0.23	-	0.25	0.24	GRAM/MJ	0.17	1.7
ISOLEUCINE	GRAM	0.49	0.45	-	0.65	0.49	GRAM/MJ	0.35	4.0
LEUCINE	GRAM	0.86	0.85	-	0.97	0.86	GRAM/MJ	0.62	7.1
LYSINE	GRAM	0.26	0.22	-	0.32	0.26	GRAM/MJ	0.19	1.9
METHIONINE	GRAM	0.18	0.17	-	0.19	0.18	GRAM/MJ	0.13	1.3
PHENYLALANINE	GRAM	0.60	0.58	-	0.67	0.60	GRAM/MJ	0.43	3.9
PROLINE	GRAM	1.50	1.44	-	1.55	1.50	GRAM/MJ	1.08	14.0
SERINE	GRAM	0.64	0.59	-	0.77	0.64	GRAM/MJ	0.46	6.6
THREONINE	GRAM	0.35	0.34	-	0.37	0.35	GRAM/MJ	0.25	3.2
TRYPTOPHAN	GRAM	0.15	-	-	-	0.15	GRAM/MJ	0.11	0.8
TYROSINE	GRAM	0.29	0.27	-	0.31	0.29	GRAM/MJ	0.21	1.7
VALINE	GRAM	0.53	0.52	-	0.57	0.53	GRAM/MJ	0.38	4.9
GLUCOSE	MILLI	40.00	-	-	-	40.00	MILLI/MJ	28.81	
FRUCTOSE	MILLI	30.00	-	-	-	30.00	MILLI/MJ	21.60	
SUCROSE	GRAM	0.47	-	-	-	0.47	GRAM/MJ	0.34	
RAFFINOSE	GRAM	0.31	-	-	-	0.31	GRAM/MJ	0.22	
STARCH	GRAM	68.19	-	-	-	68.19	GRAM/MJ	49.11	

WEIZENMEHL / WHEAT FLOUR / FARINE DE BLÉ

TYPE 812 / TYPE 812 / TYPE 812

		PROTEIN	FAT	CARBOHYDRATES	TOTAL
ENERGY VALUE (AVERAGE) PER 100 G EDIBLE PORTION	KJOULE (KCAL)	224 54	51 12	1117 267	1392 333
AMOUNT OF DIGESTIBLE CONSTITUENTS PER 100 G	GRAM	10.02	1.17	66.74	
ENERGY VALUE (AVERAGE) OF THE DIGESTIBLE FRACTION PER 100 G EDIBLE PORTION	KJOULE (KCAL)	191 46	46 11	1117 267	1353 323

WASTE PERCENTAGE AVERAGE 0.00

CONSTITUENTS	DIM	AV	VARIATION			AVR	NUTR. DENS.		MOLPERC.
WATER	GRAM	14.70	13.70	-	15.50	14.70	GRAM/MJ	10.86	
PROTEIN	GRAM	11.79	11.14	-	12.44	11.79	GRAM/MJ	8.71	
FAT	GRAM	1.30	-	-	-	1.30	GRAM/MJ	0.96	
AVAILABLE CARBOHYDR.	GRAM	66.74 1	-	-	-	66.74	GRAM/MJ	49.32	
TOTAL DIETARY FIBRE	GRAM	4.78 2	-	-	-	4.78	GRAM/MJ	3.53	
MINERALS	GRAM	0.69	0.64	-	0.74	0.69	GRAM/MJ	0.51	
SODIUM	MILLI	3.00	2.00	-	5.00	3.00	MILLI/MJ	2.22	
POTASSIUM	MILLI	170.00	150.00	-	203.00	170.00	MILLI/MJ	125.62	
CALCIUM	MILLI	20.00	15.00	-	24.00	20.00	MILLI/MJ	14.78	
IRON	MILLI	1.70	1.30	-	2.00	1.70	MILLI/MJ	1.26	
COPPER	MILLI	0.29	0.13	-	0.43	0.29	MILLI/MJ	0.21	
MOLYBDENUM	MICRO	45.00	25.00	-	64.00	45.00	MICRO/MJ	33.25	
PHOSPHORUS	MILLI	161.00	130.00	-	191.00	161.00	MILLI/MJ	118.97	
BORON	MICRO	25.00	0.00	-	45.00	25.00	MICRO/MJ	18.47	
VITAMIN B1	MILLI	0.26	0.21	-	0.29	0.26	MILLI/MJ	0.19	
VITAMIN B2	MICRO	60.00	50.00	-	70.00	60.00	MICRO/MJ	44.34	
NICOTINAMIDE	MILLI	0.89	-	-	-	0.89	MILLI/MJ	0.66	
PANTOTHENIC ACID	MILLI	0.63	-	-	-	0.63	MILLI/MJ	0.47	
VITAMIN B6	MILLI	0.28	-	-	-	0.28	MILLI/MJ	0.21	
BIOTIN	MICRO	2.90	-	-	-	2.90	MICRO/MJ	2.14	
FOLIC ACID	MICRO	22.00	-	-	-	22.00	MICRO/MJ	16.26	
VITAMIN C	-	0.00	-	-	-	0.00			
ALANINE	GRAM	0.45	-	-	-	0.45	GRAM/MJ	0.33	4.9
ARGININE	GRAM	0.47	0.45	-	0.55	0.47	GRAM/MJ	0.35	2.6
ASPARTIC ACID	GRAM	0.57	0.56	-	0.58	0.57	GRAM/MJ	0.42	4.2

1 ESTIMATED BY THE DIFFERENCE METHOD
100 - (WATER + PROTEIN + FAT + MINERALS + TOTAL DIETARY FIBRE)

2 METHOD OF MEUSER, SUCKOW AND KULIKOWSKI ("BERLINER METHODE")

CONSTITUENTS	DIM	AV	VARIATION			AVR		NUTR. DENS.	MOLPERC.
CYSTINE	GRAM	0.35	-	-	-	0.35	GRAM/MJ	0.26	1.4
GLUTAMIC ACID	GRAM	4.50	4.34	-	4.65	4.50	GRAM/MJ	3.33	29.7
GLYCINE	GRAM	0.53	0.50	-	0.54	0.53	GRAM/MJ	0.39	6.8
HISTIDINE	GRAM	0.27	0.26	-	0.28	0.27	GRAM/MJ	0.20	1.7
ISOLEUCINE	GRAM	0.56	0.54	-	0.66	0.56	GRAM/MJ	0.41	4.1
LEUCINE	GRAM	0.96	0.95	-	0.97	0.96	GRAM/MJ	0.71	7.1
LYSINE	GRAM	0.29	0.25	-	0.35	0.29	GRAM/MJ	0.21	1.9
METHIONINE	GRAM	0.20	0.19	-	0.21	0.20	GRAM/MJ	0.15	1.3
PHENYLALANINE	GRAM	0.67	0.65	-	0.75	0.67	GRAM/MJ	0.50	3.9
PROLINE	GRAM	1.59	1.53	-	1.64	1.59	GRAM/MJ	1.17	13.4
SERINE	GRAM	0.71	0.67	-	0.78	0.71	GRAM/MJ	0.52	6.6
THREONINE	GRAM	0.38	0.37	-	0.42	0.38	GRAM/MJ	0.28	3.1
TRYPTOPHAN	GRAM	0.14	-	-	-	0.14	GRAM/MJ	0.10	0.7
TYROSINE	GRAM	0.33	0.30	-	0.35	0.33	GRAM/MJ	0.24	1.8
VALINE	GRAM	0.58	0.57	-	0.62	0.58	GRAM/MJ	0.43	4.8
GLUCOSE	MILLI	60.00	-	-	-	60.00	MILLI/MJ	44.34	
FRUCTOSE	MILLI	30.00	-	-	-	30.00	MILLI/MJ	22.17	
SUCROSE	GRAM	0.60	-	-	-	0.60	GRAM/MJ	0.44	
RAFFINOSE	GRAM	0.43	-	-	-	0.43	GRAM/MJ	0.32	
STARCH	GRAM	66.90	-	-	-	66.90	GRAM/MJ	49.44	

WEIZENMEHL TYPE 1050 — WHEAT FLOUR TYPE 1050 — FARINE DE BLÉ TYPE 1050

		PROTEIN	FAT	CARBOHYDRATES	TOTAL
ENERGY VALUE (AVERAGE) PER 100 G EDIBLE PORTION	KJOULE	214	68	1124	1406
	(KCAL)	51	16	269	336
AMOUNT OF DIGESTIBLE CONSTITUENTS PER 100 G	GRAM	9.32	1.57	67.19	
ENERGY VALUE (AVERAGE) OF THE DIGESTIBLE FRACTION PER 100 G EDIBLE PORTION	KJOULE	177	61	1124	1363
	(KCAL)	42	15	269	326

WASTE PERCENTAGE AVERAGE 0.00

CONSTITUENTS	DIM	AV		VARIATION			AVR	NUTR.	DENS.	MOLPERC.
WATER	GRAM	13.70		11.60	-	14.70	13.70	GRAM/MJ	10.05	
PROTEIN	GRAM	11.23		10.39	-	12.06	11.23	GRAM/MJ	8.24	
FAT	GRAM	1.75		1.60	-	1.90	1.75	GRAM/MJ	1.28	
AVAILABLE CARBOHYDR.	GRAM	67.19	1	-	-	-	67.19	GRAM/MJ	49.29	
TOTAL DIETARY FIBRE	GRAM	5.22	2	-	-	-	5.22	GRAM/MJ	3.83	
MINERALS	GRAM	0.91		0.86	-	0.99	0.91	GRAM/MJ	0.67	
SODIUM	MILLI	2.00		-	-	-	2.00	MILLI/MJ	1.47	
POTASSIUM	MILLI	203.00		-	-	-	203.00	MILLI/MJ	148.91	
MAGNESIUM	MILLI	53.00		-	-	-	53.00	MILLI/MJ	38.88	
CALCIUM	MILLI	14.00		10.00	-	18.00	14.00	MILLI/MJ	10.27	
IRON	MILLI	2.81		-	-	-	2.81	MILLI/MJ	2.06	
COPPER	MILLI	0.40		0.13	-	0.64	0.40	MILLI/MJ	0.29	
NICKEL	MICRO	28.00		18.00	-	38.00	28.00	MICRO/MJ	20.54	
MOLYBDENUM	MICRO	45.00		25.00	-	64.00	45.00	MICRO/MJ	33.01	
PHOSPHORUS	MILLI	208.00		202.00	-	232.00	208.00	MILLI/MJ	152.58	
BORON	MICRO	25.00		0.00	-	45.00	25.00	MICRO/MJ	18.34	
VITAMIN B1	MILLI	0.43		0.34	-	0.56	0.43	MILLI/MJ	0.32	
VITAMIN B2	MICRO	70.00		-	-	-	70.00	MICRO/MJ	51.35	
NICOTINAMIDE	MILLI	1.42		-	-	-	1.42	MILLI/MJ	1.04	
PANTOTHENIC ACID	MILLI	0.63		-	-	-	0.63	MILLI/MJ	0.46	
VITAMIN B6	MILLI	0.28		-	-	-	0.28	MILLI/MJ	0.21	
BIOTIN	MICRO	2.90		-	-	-	2.90	MICRO/MJ	2.13	
FOLIC ACID	MICRO	22.00		-	-	-	22.00	MICRO/MJ	16.14	
VITAMIN C	-	0.00		-	-	-	0.00			
ALANINE	GRAM	0.42		-	-	-	0.42	GRAM/MJ	0.31	4.8

1 ESTIMATED BY THE DIFFERENCE METHOD
100 - (WATER + PROTEIN + FAT + MINERALS + TOTAL DIETARY FIBRE)

2 MODIFIED AOAC METHOD

CONSTITUENTS	DIM	AV	VARIATION			AVR	NUTR. DENS.		MOLPERC.
ARGININE	GRAM	0.49	0.38	-	0.52	0.49	GRAM/MJ	0.36	2.8
ASPARTIC ACID	GRAM	0.54	0.49	-	0.57	0.54	GRAM/MJ	0.40	4.1
CYSTINE	GRAM	0.27	0.19	-	0.32	0.27	GRAM/MJ	0.20	1.1
GLUTAMIC ACID	GRAM	4.27	4.19	-	4.44	4.27	GRAM/MJ	3.13	29.4
GLYCINE	GRAM	0.50	0.49	-	0.54	0.50	GRAM/MJ	0.37	6.7
HISTIDINE	GRAM	0.24	0.21	-	0.27	0.24	GRAM/MJ	0.18	1.6
ISOLEUCINE	GRAM	0.48	0.43	-	0.61	0.48	GRAM/MJ	0.35	3.7
LEUCINE	GRAM	0.87	0.80	-	0.92	0.87	GRAM/MJ	0.64	6.7
LYSINE	GRAM	0.30	0.22	-	0.37	0.30	GRAM/MJ	0.22	2.1
METHIONINE	GRAM	0.20	0.19	-	0.21	0.20	GRAM/MJ	0.15	1.4
PHENYLALANINE	GRAM	0.63	0.58	-	0.71	0.63	GRAM/MJ	0.46	3.9
PROLINE	GRAM	1.61	1.57	-	1.66	1.61	GRAM/MJ	1.18	14.2
SERINE	GRAM	0.71	0.66	-	0.74	0.71	GRAM/MJ	0.52	6.8
THREONINE	GRAM	0.38	0.35	-	0.42	0.38	GRAM/MJ	0.28	3.2
TRYPTOPHAN	GRAM	0.13	-	-	-	0.13	GRAM/MJ	0.10	0.6
TYROSINE	GRAM	0.37	0.32	-	0.39	0.37	GRAM/MJ	0.27	2.1
VALINE	GRAM	0.56	0.48	-	0.58	0.56	GRAM/MJ	0.41	4.8
TOTAL SUGAR	GRAM	1.40	-	-	-	1.40	GRAM/MJ	1.03	
STARCH	GRAM	65.80	64.10	-	67.50	65.80	GRAM/MJ	48.27	
DIETARY FIBRE, WAT.SOL.	GRAM	2.11	-	-	-	2.11	GRAM/MJ	1.55	
DIETARY FIBRE, WAT.INS.	GRAM	3.11	-	-	-	3.11	GRAM/MJ	2.28	

WEIZENMEHL TYPE 1700 / WHEAT FLOUR TYPE 1700 / FARINE DE BLÉ TYPE 1700

		PROTEIN	FAT	CARBOHYDRATES	TOTAL
ENERGY VALUE (AVERAGE)	KJOULE	214	82	998	1294
PER 100 G	(KCAL)	51	20	239	309
EDIBLE PORTION					
AMOUNT OF DIGESTIBLE	GRAM	9.32	1.89	59.65	
CONSTITUENTS PER 100 G					
ENERGY VALUE (AVERAGE)	KJOULE	177	74	998	1249
OF THE DIGESTIBLE	(KCAL)	42	18	239	299
FRACTION PER 100 G					
EDIBLE PORTION					

WASTE PERCENTAGE AVERAGE 0.00

CONSTITUENTS	DIM	AV		VARIATION			AVR	NUTR. DENS.		MOLPERC.
WATER	GRAM	12.60		12.00	-	13.30	12.60	GRAM/MJ	10.09	
PROTEIN	GRAM	11.23		10.58	-	12.34	11.23	GRAM/MJ	8.99	
FAT	GRAM	2.10		2.00	-	2.30	2.10	GRAM/MJ	1.68	
AVAILABLE CARBOHYDR.	GRAM	59.65	1	-	-	-	59.65	GRAM/MJ	47.75	
TOTAL DIETARY FIBRE	GRAM	12.93	2	-	-	-	12.93	GRAM/MJ	10.35	
MINERALS	GRAM	1.49		1.40	-	1.66	1.49	GRAM/MJ	1.19	
SODIUM	MILLI	2.00		-	-	-	2.00	MILLI/MJ	1.60	
POTASSIUM	MILLI	290.00		-	-	-	290.00	MILLI/MJ	232.13	
CALCIUM	MILLI	41.00		-	-	-	41.00	MILLI/MJ	32.82	
IRON	MILLI	3.30		-	-	-	3.30	MILLI/MJ	2.64	
COPPER	MILLI	0.40		0.13	-	0.60	0.40	MILLI/MJ	0.32	
ZINC	MILLI	1.30		-	-	-	1.30	MILLI/MJ	1.04	
MOLYBDENUM	MICRO	45.00		25.00	-	64.00	45.00	MICRO/MJ	36.02	
PHOSPHORUS	MILLI	372.00		-	-	-	372.00	MILLI/MJ	297.77	
IODIDE	MICRO	-		3.00	-	14.00				
BORON	MICRO	25.00		0.00	-	45.00	25.00	MICRO/MJ	20.01	
VITAMIN B1	MILLI	0.47		0.39	-	0.55	0.47	MILLI/MJ	0.38	
VITAMIN B2	MILLI	0.17		0.12	-	0.20	0.17	MILLI/MJ	0.14	
NICOTINAMIDE	MILLI	5.00		4.20	-	6.50	5.00	MILLI/MJ	4.00	
PANTOTHENIC ACID	MILLI	1.20		-	-	-	1.20	MILLI/MJ	0.96	
VITAMIN B6	MILLI	0.46		-	-	-	0.46	MILLI/MJ	0.37	
BIOTIN	MICRO	8.30		-	-	-	8.30	MICRO/MJ	6.64	
FOLIC ACID	MICRO	50.00		-	-	-	50.00	MICRO/MJ	40.02	
VITAMIN C	-	0.00		-	-	-	0.00			
ALANINE	GRAM	0.49		0.45	-	0.52	0.49	GRAM/MJ	0.39	5.5

1 ESTIMATED BY THE DIFFERENCE METHOD
100 - (WATER + PROTEIN + FAT + MINERALS + TOTAL DIETARY FIBRE)

2 MODIFIED AOAC METHOD

CONSTITUENTS	DIM	AV	VARIATION			AVR	NUTR. DENS.		MOLPERC.
ARGININE	GRAM	0.60	0.58	-	0.67	0.60	GRAM/MJ	0.48	3.4
ASPARTIC ACID	GRAM	0.66	-	-	-	0.66	GRAM/MJ	0.53	4.9
CYSTINE	GRAM	0.25	0.16	-	0.32	0.25	GRAM/MJ	0.20	1.0
GLUTAMIC ACID	GRAM	3.75	3.60	-	3.91	3.75	GRAM/MJ	3.00	25.4
GLYCINE	GRAM	0.63	-	-	-	0.63	GRAM/MJ	0.50	8.4
HISTIDINE	GRAM	0.25	0.24	-	0.26	0.25	GRAM/MJ	0.20	1.6
ISOLEUCINE	GRAM	0.52	0.45	-	0.55	0.52	GRAM/MJ	0.42	4.0
LEUCINE	GRAM	0.86	0.84	-	0.92	0.86	GRAM/MJ	0.69	6.5
LYSINE	GRAM	0.35	0.32	-	0.36	0.35	GRAM/MJ	0.28	2.4
METHIONINE	GRAM	0.21	0.19	-	0.24	0.21	GRAM/MJ	0.17	1.4
PHENYLALANINE	GRAM	0.59	0.58	-	0.60	0.59	GRAM/MJ	0.47	3.6
PROLINE	GRAM	1.57	1.45	-	1.69	1.57	GRAM/MJ	1.26	13.6
SERINE	GRAM	0.74	-	-	-	0.74	GRAM/MJ	0.59	7.0
THREONINE	GRAM	0.39	0.36	-	0.45	0.39	GRAM/MJ	0.31	3.3
TRYPTOPHAN	GRAM	0.15	0.14	-	0.17	0.15	GRAM/MJ	0.12	0.7
TYROSINE	GRAM	0.37	0.26	-	0.45	0.37	GRAM/MJ	0.30	2.0
VALINE	GRAM	0.60	0.59	-	0.61	0.60	GRAM/MJ	0.48	5.1
STARCH	GRAM	58.70	-	-	-	58.70	GRAM/MJ	46.99	
CELLULOSE	GRAM	2.50	-	-	-	2.50	GRAM/MJ	2.00	
DIETARY FIBRE,WAT.SOL.	GRAM	4.37	-	-	-	4.37	GRAM/MJ	3.50	
DIETARY FIBRE,WAT.INS.	GRAM	8.56	-	-	-	8.56	GRAM/MJ	6.85	

WEIZENKEIME — WHEAT GERM — GERMES DE BLÉ

		PROTEIN	FAT	CARBOHYDRATES	TOTAL
ENERGY VALUE (AVERAGE)	KJOULE	506	358	395	1259
PER 100 G EDIBLE PORTION	(KCAL)	121	86	94	301
AMOUNT OF DIGESTIBLE CONSTITUENTS PER 100 G	GRAM	21.01	8.28	23.58	
ENERGY VALUE (AVERAGE)	KJOULE	400	322	395	1117
OF THE DIGESTIBLE FRACTION PER 100 G EDIBLE PORTION	(KCAL)	96	77	94	267

WASTE PERCENTAGE AVERAGE 0.00

CONSTITUENTS	DIM	AV		VARIATION			AVR	NUTR. DENS.		MOLPERC.
WATER	GRAM	11.70		11.00	-	13.00	11.70	GRAM/MJ	10.48	
PROTEIN	GRAM	26.60		-	-	-	26.60	GRAM/MJ	23.82	
FAT	GRAM	9.20		6.50	-	10.40	9.20	GRAM/MJ	8.24	
AVAILABLE CARBOHYDR.	GRAM	23.58	1	-	-	-	23.58	GRAM/MJ	21.11	
TOTAL DIETARY FIBRE	GRAM	24.72	2	-	-	-	24.72	GRAM/MJ	22.13	
MINERALS	GRAM	4.20		2.90	-	4.80	4.20	GRAM/MJ	3.76	
SODIUM	MILLI	5.00		2.00	-	9.00	5.00	MILLI/MJ	4.48	
POTASSIUM	MILLI	837.00		780.00	-	951.00	837.00	MILLI/MJ	749.42	
MAGNESIUM	MILLI	250.00		140.00	-	360.00	250.00	MILLI/MJ	223.84	
CALCIUM	MILLI	69.00		38.00	-	84.00	69.00	MILLI/MJ	61.78	
MANGANESE	MILLI	9.30		-	-	-	9.30	MILLI/MJ	8.33	
IRON	MILLI	8.10		-	-	-	8.10	MILLI/MJ	7.25	
COBALT	MICRO	1.70		-	-	-	1.70	MICRO/MJ	1.52	
COPPER	MILLI	0.95		0.70	-	1.20	0.95	MILLI/MJ	0.85	
ZINC	MILLI	12.00		10.00	-	14.00	12.00	MILLI/MJ	10.74	
MOLYBDENUM	MILLI	0.10		0.07	-	0.13	0.10	MILLI/MJ	0.09	
PHOSPHORUS	GRAM	1.10		-	-	-	1.10	GRAM/MJ	0.98	
BORON	MILLI	1.65		-	-	-	1.65	MILLI/MJ	1.48	
SELENIUM	MILLI	0.11		-	-	-	0.11	MILLI/MJ	0.10	
ALPHA-TOCOPHEROL	MILLI	11.70		-	-	-	11.70	MILLI/MJ	10.48	
VITAMIN K	MILLI	0.35		-	-	-	0.35	MILLI/MJ	0.31	
VITAMIN B1	MILLI	2.01		0.79	-	2.70	2.01	MILLI/MJ	1.80	
VITAMIN B2	MILLI	0.72		0.57	-	0.80	0.72	MILLI/MJ	0.64	
NICOTINAMIDE	MILLI	4.52		4.00	-	5.30	4.52	MILLI/MJ	4.05	
PANTOTHENIC ACID	MILLI	1.00		-	-	-	1.00	MILLI/MJ	0.90	
VITAMIN B6	MILLI	3.30		3.10	-	3.50	3.30	MILLI/MJ	2.95	

1 ESTIMATED BY THE DIFFERENCE METHOD
100 - (WATER + PROTEIN + FAT + MINERALS + TOTAL DIETARY FIBRE)

2 MODIFIED AOAC METHOD

CONSTITUENTS	DIM	AV	VARIATION			AVR	NUTR. DENS.		MOLPERC.
BIOTIN	MICRO	17.00	-	-	-	17.00	MICRO/MJ	15.22	
FOLIC ACID	MILLI	0.52	0.33	-	0.70	0.52	MILLI/MJ	0.47	
VITAMIN C	-	0.00	-	-	-	0.00			
ALANINE	GRAM	2.14	-	-	-	2.14	GRAM/MJ	1.92	9.8
ARGININE	GRAM	2.31	2.26	-	2.34	2.31	GRAM/MJ	2.07	5.4
ASPARTIC ACID	GRAM	2.79	-	-	-	2.79	GRAM/MJ	2.50	8.6
CYSTINE	GRAM	0.46	0.35	-	0.59	0.46	GRAM/MJ	0.41	0.8
GLUTAMIC ACID	GRAM	5.25	-	-	-	5.25	GRAM/MJ	4.70	14.6
GLYCINE	GRAM	2.16	-	-	-	2.16	GRAM/MJ	1.93	11.8
HISTIDINE	GRAM	0.84	0.82	-	0.85	0.84	GRAM/MJ	0.75	2.2
ISOLEUCINE	GRAM	1.32	1.09	-	1.54	1.32	GRAM/MJ	1.18	4.1
LEUCINE	GRAM	2.17	2.10	-	2.24	2.17	GRAM/MJ	1.94	6.8
LYSINE	GRAM	1.90	1.89	-	1.93	1.90	GRAM/MJ	1.70	5.3
METHIONINE	GRAM	0.56	0.52	-	0.58	0.56	GRAM/MJ	0.50	1.5
PHENYLALANINE	GRAM	1.18	1.16	-	1.20	1.18	GRAM/MJ	1.06	2.9
PROLINE	GRAM	1.71	-	-	-	1.71	GRAM/MJ	1.53	6.1
SERINE	GRAM	1.52	-	-	-	1.52	GRAM/MJ	1.36	5.9
THREONINE	GRAM	1.55	1.30	-	1.79	1.55	GRAM/MJ	1.39	5.3
TRYPTOPHAN	GRAM	0.33	-	-	-	0.33	GRAM/MJ	0.30	0.7
TYROSINE	GRAM	1.01	0.90	-	1.11	1.01	GRAM/MJ	0.90	2.3
VALINE	GRAM	1.68	1.44	-	1.70	1.68	GRAM/MJ	1.50	5.9
CADAVERINE	MILLI	-	2.00	-	23.00				
PUTRESCINE	MILLI	-	1.20	-	14.00				
SPERMIDINE	MILLI	-	8.00	-	31.00				
SPERMINE	MILLI	-	2.00	-	14.00				
GLUCOSE	GRAM	0.70	-	-	-	0.70	GRAM/MJ	0.63	
FRUCTOSE	GRAM	0.50	-	-	-	0.50	GRAM/MJ	0.45	
SUCROSE	GRAM	14.00	-	-	-	14.00	GRAM/MJ	12.54	
RAFFINOSE	GRAM	9.50	-	-	-	9.50	GRAM/MJ	8.51	
PALMITIC ACID	MILLI	10.00	-	-	-	10.00	MILLI/MJ	8.95	
STEARIC ACID	GRAM	1.81	-	-	-	1.81	GRAM/MJ	1.62	
ARACHIDIC ACID	MILLI	60.00	-	-	-	60.00	MILLI/MJ	53.72	
PALMITOLEIC ACID	MILLI	40.00	-	-	-	40.00	MILLI/MJ	35.81	
OLEIC ACID	GRAM	1.54	-	-	-	1.54	GRAM/MJ	1.38	
LINOLEIC ACID	GRAM	4.40	-	-	-	4.40	GRAM/MJ	3.94	
LINOLENIC ACID	GRAM	0.59	-	-	-	0.59	GRAM/MJ	0.53	
DIETARY FIBRE,WAT.SOL.	GRAM	6.09	-	-	-	6.09	GRAM/MJ	5.45	
DIETARY FIBRE,WAT.INS.	GRAM	18.63	-	-	-	18.63	GRAM/MJ	16.68	

WEIZENKLEIE
SPEISEKLEIE

WHEAT BRAN

SON DE BLÉ

		PROTEIN	FAT	CARBOHYDRATES	TOTAL
ENERGY VALUE (AVERAGE) PER 100 G EDIBLE PORTION	KJOULE	283	181	342	806
	(KCAL)	68	43	82	193
AMOUNT OF DIGESTIBLE CONSTITUENTS PER 100 G	GRAM	5.94	4.18	20.45	
ENERGY VALUE (AVERAGE) OF THE DIGESTIBLE FRACTION PER 100 G EDIBLE PORTION	KJOULE	113	163	342	618
	(KCAL)	27	39	82	148

WASTE PERCENTAGE AVERAGE 0.00

CONSTITUENTS	DIM	AV		VARIATION			AVR	NUTR. DENS.		MOLPERC.
WATER	GRAM	11.50		11.00	-	12.00	11.50	GRAM/MJ	18.60	
PROTEIN	GRAM	14.85		12.90	-	16.98	14.85	GRAM/MJ	24.02	
FAT	GRAM	4.65		4.00	-	5.38	4.65	GRAM/MJ	7.52	
AVAILABLE CARBOHYDR.	GRAM	20.45	1	-	-	-	20.45	GRAM/MJ	33.08	
TOTAL DIETARY FIBRE	GRAM	42.40	2	-	-	-	42.40	GRAM/MJ	68.59	
MINERALS	GRAM	6.15		5.30	-	7.59	6.15	GRAM/MJ	9.95	
SODIUM	MILLI	2.00		-	-	-	2.00	MILLI/MJ	3.24	
POTASSIUM	GRAM	1.39		1.38	-	1.40	1.39	GRAM/MJ	2.25	
MAGNESIUM	MILLI	590.00		-	-	-	590.00	MILLI/MJ	954.42	
CALCIUM	MILLI	43.00		-	-	-	43.00	MILLI/MJ	69.56	
MANGANESE	MILLI	3.70		-	-	-	3.70	MILLI/MJ	5.99	
IRON	MILLI	3.58		2.95	-	4.20	3.58	MILLI/MJ	5.79	
COPPER	MILLI	1.55		-	-	-	1.55	MILLI/MJ	2.51	
ZINC	MILLI	13.30		-	-	-	13.30	MILLI/MJ	21.51	
CHROMIUM	MICRO	5.00		2.00	-	13.00	5.00	MICRO/MJ	8.09	
PHOSPHORUS	GRAM	1.28		1.24	-	1.30	1.29	GRAM/MJ	2.08	
SELENIUM	MILLI	-		0.06	-	0.13				
CAROTENE	MICRO	5.00		2.00	-	10.00	5.00	MICRO/MJ	8.09	
VITAMIN E ACTIVITY	MILLI	2.80		-	-	-	2.80	MILLI/MJ	4.53	
TOTAL TOCOPHEROLS	MILLI	9.80		-	-	-	9.80	MILLI/MJ	15.85	
ALPHA-TOCOPHEROL	MILLI	1.70		-	-	-	1.70	MILLI/MJ	2.75	
BETA-TOCOPHEROL	MILLI	1.10		-	-	-	1.10	MILLI/MJ	1.78	
ALPHA-TOCOTRIENOL	MILLI	1.10		-	-	-	1.10	MILLI/MJ	1.78	
BETA-TOCOTRIENOL	MILLI	5.90		5.30	-	6.40	5.90	MILLI/MJ	9.54	
VITAMIN K	MICRO	80.00		-	-	-	80.00	MICRO/MJ	129.41	
VITAMIN B1	MILLI	0.65		0.55	-	0.80	0.65	MILLI/MJ	1.05	

1 ESTIMATED BY THE DIFFERENCE METHOD
100 - (WATER + PROTEIN + FAT + MINERALS + TOTAL DIETARY FIBRE)

2 METHOD OF SOUTHGATE AND ENGLYST

CONSTITUENTS	DIM	AV	VARIATION			AVR	NUTR. DENS.		MOLPERC.
VITAMIN B2	MILLI	0.51	0.21	-	0.80	0.51	MILLI/MJ	0.83	
NICOTINAMIDE	MILLI	17.70	14.20	-	22.00	17.70	MILLI/MJ	28.63	
PANTOTHENIC ACID	MILLI	2.50	2.30	-	2.70	2.50	MILLI/MJ	4.04	
VITAMIN B6	MILLI	2.50	-	-	-	2.50	MILLI/MJ	4.04	
BIOTIN	MICRO	44.00	-	-	-	44.00	MICRO/MJ	71.18	
FOLIC ACID	MILLI	0.40	-	-	-	0.40	MILLI/MJ	0.65	
VITAMIN C	-	0.00	-	-	-	0.00			
ALANINE	GRAM	1.07	-	-	-	1.07	GRAM/MJ	1.73	8.8
ARGININE	GRAM	1.23	1.17	-	1.31	1.23	GRAM/MJ	1.99	5.2
ASPARTIC ACID	GRAM	1.46	-	-	-	1.46	GRAM/MJ	2.36	8.1
CYSTINE	GRAM	0.39	0.32	-	0.45	0.39	GRAM/MJ	0.63	1.2
GLUTAMIC ACID	GRAM	3.59	-	-	-	3.59	GRAM/MJ	5.81	17.9
GLYCINE	GRAM	1.32	-	-	-	1.32	GRAM/MJ	2.14	12.9
HISTIDINE	GRAM	0.44	0.33	-	0.53	0.44	GRAM/MJ	0.71	2.1
ISOLEUCINE	GRAM	0.77	0.59	-	0.82	0.77	GRAM/MJ	1.25	4.3
LEUCINE	GRAM	1.12	1.06	-	1.18	1.12	GRAM/MJ	1.81	6.3
LYSINE	GRAM	0.72	0.69	-	0.74	0.72	GRAM/MJ	1.16	3.6
METHIONINE	GRAM	0.25	0.21	-	0.29	0.25	GRAM/MJ	0.40	1.2
PHENYLALANINE	GRAM	0.65	0.54	-	0.73	0.65	GRAM/MJ	1.05	2.9
PROLINE	GRAM	1.14	-	-	-	1.14	GRAM/MJ	1.84	7.3
SERINE	GRAM	0.90	-	-	-	0.90	GRAM/MJ	1.46	6.3
THREONINE	GRAM	0.59	0.53	-	0.65	0.59	GRAM/MJ	0.95	3.6
TRYPTOPHAN	GRAM	0.25	-	-	-	0.25	GRAM/MJ	0.40	0.9
TYROSINE	GRAM	0.46	0.38	-	0.54	0.46	GRAM/MJ	0.74	1.9
VALINE	GRAM	0.88	0.84	-	0.91	0.88	GRAM/MJ	1.42	5.5
FERULIC ACID	MILLI	3.40	-	-	-	3.40	MILLI/MJ	5.50	
PARA-COUMARIC ACID	MILLI	4.00	-	-	-	4.00	MILLI/MJ	6.47	
GLUCOSE	MILLI	90.00	-	-	-	90.00	MILLI/MJ	145.59	
FRUCTOSE	MILLI	50.00	-	-	-	50.00	MILLI/MJ	80.88	
SUCROSE	GRAM	1.75	-	-	-	1.75	GRAM/MJ	2.83	
RAFFINOSE	GRAM	1.30	-	-	-	1.30	GRAM/MJ	2.10	
MALTOSE	MILLI	30.00	-	-	-	30.00	MILLI/MJ	48.53	
STACHYOSE	MILLI	40.00	-	-	-	40.00	MILLI/MJ	64.71	
STARCH	GRAM	13.43	8.74	-	15.70	13.43	GRAM/MJ	21.73	
CELLULOSE	GRAM	9.00	-	-	-	9.00	GRAM/MJ	14.56	
MYRISTIC ACID	MILLI	10.00	-	-	-	10.00	MILLI/MJ	16.18	
PALMITIC ACID	GRAM	0.69	-	-	-	0.69	GRAM/MJ	1.12	
STEARIC ACID	MILLI	40.00	-	-	-	40.00	MILLI/MJ	64.71	
PALMITOLEIC ACID	MILLI	20.00	-	-	-	20.00	MILLI/MJ	32.35	
OLEIC ACID	GRAM	0.71	-	-	-	0.71	GRAM/MJ	1.15	
LINOLEIC ACID	GRAM	2.20	-	-	-	2.20	GRAM/MJ	3.56	
LINOLENIC ACID	GRAM	0.16	-	-	-	0.16	GRAM/MJ	0.26	
TOTAL STEROLS	MILLI	121.00	-	-	-	121.00	MILLI/MJ	195.74	
DIETARY FIBRE,WAT.SOL.	GRAM	2.05	-	-	-	2.05	GRAM/MJ	3.32	
DIETARY FIBRE,WAT.INS.	GRAM	40.30	-	-	-	40.30	GRAM/MJ	65.19	

Brot und Kleingebäck

BREAD AND ROLLS **PAIN ET PETITS PAINS**

BRÖTCHEN (SEMMELN) BREAD ROLLS PETIT PAIN

		PROTEIN	FAT	CARBOHYDRATES	TOTAL
ENERGY VALUE (AVERAGE) PER 100 G EDIBLE PORTION	KJOULE	158	73	831	1062
	(KCAL)	38	17	199	254
AMOUNT OF DIGESTIBLE CONSTITUENTS PER 100 G	GRAM	7.39	1.68	49.64	
ENERGY VALUE (AVERAGE) OF THE DIGESTIBLE FRACTION PER 100 G EDIBLE PORTION	KJOULE	141	65	831	1037
	(KCAL)	34	16	199	248

WASTE PERCENTAGE AVERAGE 0.00

CONSTITUENTS	DIM	AV		VARIATION			AVR	NUTR. DENS.		MOLPERC.
WATER	GRAM	35.40		32.90	-	38.50	35.40	GRAM/MJ	34.13	
PROTEIN	GRAM	8.31		7.70	-	9.21	8.31	GRAM/MJ	8.01	
FAT	GRAM	1.87		1.10	-	2.60	1.87	GRAM/MJ	1.80	
AVAILABLE CARBOHYDR.	GRAM	49.64	1	-	-	-	49.64	GRAM/MJ	47.87	
TOTAL DIETARY FIBRE	GRAM	3.03	2	-	-	-	3.03	GRAM/MJ	2.92	
MINERALS	GRAM	1.75		0.90	-	2.60	1.75	GRAM/MJ	1.69	
SODIUM	MILLI	553.00		486.00	-	600.00	553.00	MILLI/MJ	533.24	
POTASSIUM	MILLI	110.00		100.00	-	219.00	110.00	MILLI/MJ	106.07	
MAGNESIUM	MILLI	30.00		24.00	-	41.00	30.00	MILLI/MJ	28.93	
CALCIUM	MILLI	27.00		21.00	-	49.00	27.00	MILLI/MJ	26.04	
MANGANESE	MILLI	0.89		0.27	-	1.20	0.89	MILLI/MJ	0.86	
IRON	MILLI	1.20		0.40	-	2.30	1.20	MILLI/MJ	1.16	
COPPER	MILLI	0.26		0.16	-	0.47	0.26	MILLI/MJ	0.25	
ZINC	MILLI	1.10		-	-	-	1.10	MILLI/MJ	1.06	
PHOSPHORUS	MILLI	102.00		89.00	-	122.00	102.00	MILLI/MJ	98.35	
CHLORIDE	MILLI	656.00		450.00	-	862.00	656.00	MILLI/MJ	632.56	
FLUORIDE	MICRO	1.10		-	-	-	1.10	MICRO/MJ	1.06	

1 ESTIMATED BY THE DIFFERENCE METHOD
100 - (WATER + PROTEIN + FAT + MINERALS + TOTAL DIETARY FIBRE)

2 METHOD OF MEUSER, SUCKOW AND KULIKOWSKI ("BERLINER METHODE")

CONSTITUENTS	DIM	AV	VARIATION			AVR	NUTR.	DENS.	MOLPERC.
TOTAL TOCOPHEROLS	MILLI	0.39	0.11	-	0.66	0.39	MILLI/MJ	0.38	
VITAMIN B1	MICRO	98.00	42.00	-	180.00	98.00	MICRO/MJ	94.50	
VITAMIN B2	MICRO	34.00	20.00	-	50.00	34.00	MICRO/MJ	32.78	
NICOTINAMIDE	MILLI	1.10	0.80	-	1.70	1.10	MILLI/MJ	1.06	
PANTOTHENIC ACID	MILLI	0.50	0.46	-	0.53	0.50	MILLI/MJ	0.48	
VITAMIN B6	MICRO	40.00	30.00	-	50.00	40.00	MICRO/MJ	38.57	
BIOTIN	MICRO	1.00	-	-	-	1.00	MICRO/MJ	0.96	
FOLIC ACID	MICRO	27.00	25.00	-	29.00	27.00	MICRO/MJ	26.04	
VITAMIN C	-	0.00	-	-	-	0.00			
ALANINE	GRAM	0.32	-	-	-	0.32	GRAM/MJ	0.31	4.7
ARGININE	GRAM	0.29	-	-	-	0.29	GRAM/MJ	0.28	2.2
ASPARTIC ACID	GRAM	0.22	-	-	-	0.22	GRAM/MJ	0.21	2.2
CYSTINE	MILLI	50.00	-	-	-	50.00	MILLI/MJ	48.21	0.3
GLUTAMIC ACID	GRAM	2.74	-	-	-	2.74	GRAM/MJ	2.64	24.6
GLYCINE	GRAM	0.58	-	-	-	0.58	GRAM/MJ	0.56	10.2
HISTIDINE	GRAM	0.18	-	-	-	0.18	GRAM/MJ	0.17	1.5
ISOLEUCINE	GRAM	0.48	0.36	-	0.60	0.48	GRAM/MJ	0.46	4.8
LEUCINE	GRAM	0.76	0.63	-	0.87	0.76	GRAM/MJ	0.73	7.6
LYSINE	GRAM	0.19	0.16	-	0.27	0.19	GRAM/MJ	0.18	1.7
METHIONINE	GRAM	0.13	0.08	-	0.17	0.13	GRAM/MJ	0.13	1.1
PHENYLALANINE	GRAM	0.44	0.33	-	0.57	0.44	GRAM/MJ	0.42	3.5
PROLINE	GRAM	1.53	-	-	-	1.53	GRAM/MJ	1.48	17.5
SERINE	GRAM	0.50	-	-	-	0.50	GRAM/MJ	0.48	6.3
THREONINE	GRAM	0.32	0.26	-	0.37	0.32	GRAM/MJ	0.31	3.5
TRYPTOPHAN	MILLI	70.00	40.00	-	90.00	70.00	MILLI/MJ	67.50	0.5
TYROSINE	GRAM	0.30	0.24	-	0.36	0.30	GRAM/MJ	0.29	2.2
VALINE	GRAM	0.49	0.44	-	0.61	0.49	GRAM/MJ	0.47	5.5
GLUCOSE	GRAM	0.16	-	-	-	0.16	GRAM/MJ	0.15	
FRUCTOSE	GRAM	0.21	-	-	-	0.21	GRAM/MJ	0.20	
SUCROSE	MILLI	30.00	-	-	-	30.00	MILLI/MJ	28.93	
STARCH	GRAM	45.03	-	-	-	45.03	GRAM/MJ	43.42	
CELLULOSE	GRAM	0.41	-	-	-	0.41	GRAM/MJ	0.40	
TOTAL PURINES	MILLI	21.00 [3]	-	-	-	21.00	MILLI/MJ	20.25	
DIETARY FIBRE,WAT.SOL.	GRAM	1.68	-	-	-	1.68	GRAM/MJ	1.62	
DIETARY FIBRE,WAT.INS.	GRAM	1.35	-	-	-	1.35	GRAM/MJ	1.30	
ETHANOL	MILLI	20.00	-	-	-	20.00	MILLI/MJ	19.29	

3 VALUE EXPRESSED IN MG URIC ACID/100G

GRAHAMBROT (WEIZENSCHROTBROT) — BREAD GRAHAM (SHREDDED WHEAT BREAD) — PAIN GRAHAM (PAIN DE BLÉ ÉGRUGRÉ)

		PROTEIN	FAT	CARBOHYDRATES	TOTAL
ENERGY VALUE (AVERAGE)	KJOULE	148	39	728	915
PER 100 G EDIBLE PORTION	(KCAL)	35	9.3	174	219
AMOUNT OF DIGESTIBLE CONSTITUENTS PER 100 G	GRAM	6.16	0.90	43.50	
ENERGY VALUE (AVERAGE)	KJOULE	117	35	728	880
OF THE DIGESTIBLE FRACTION PER 100 G EDIBLE PORTION	(KCAL)	28	8.4	174	210

WASTE PERCENTAGE AVERAGE 0.00

CONSTITUENTS	DIM	AV		VARIATION		AVR	NUTR. DENS.		MOLPERC.
WATER	GRAM	39.70		38.50 -	42.20	39.70	GRAM/MJ	45.10	
PROTEIN	GRAM	7.80		6.22 -	9.00	7.80	GRAM/MJ	8.86	
FAT	GRAM	1.00		0.60 -	1.40	1.00	GRAM/MJ	1.14	
AVAILABLE CARBOHYDR.	GRAM	43.50	1	- -	-	43.50	GRAM/MJ	49.41	
TOTAL DIETARY FIBRE	GRAM	6.40	2	- -	-	6.40	GRAM/MJ	7.27	
MINERALS	GRAM	1.60		1.52 -	1.60	1.60	GRAM/MJ	1.82	
SODIUM	MILLI	430.00		352.00 -	506.00	430.00	MILLI/MJ	488.45	
POTASSIUM	MILLI	209.00		189.00 -	230.00	209.00	MILLI/MJ	237.41	
MAGNESIUM	MILLI	42.00		40.00 -	44.00	42.00	MILLI/MJ	47.71	
CALCIUM	MILLI	-		23.00 -	61.00				
IRON	MILLI	1.60		- -	-	1.60	MILLI/MJ	1.82	
CHROMIUM	MICRO	5.00		2.00 -	13.00	5.00	MICRO/MJ	5.68	
PHOSPHORUS	MILLI	244.57		176.00 -	256.00	244.57	MILLI/MJ	277.81	
CHLORIDE	MILLI	378.00		377.00 -	380.00	378.00	MILLI/MJ	429.38	
ALPHA-TOCOTRIENOL	MILLI	1.10		- -	-	1.10	MILLI/MJ	1.25	
VITAMIN B1	MILLI	0.21		0.16 -	0.27	0.21	MILLI/MJ	0.24	
VITAMIN B2	MILLI	0.11		0.09 -	0.12	0.11	MILLI/MJ	0.12	
NICOTINAMIDE	MILLI	2.50		2.36 -	2.75	2.50	MILLI/MJ	2.84	
PANTOTHENIC ACID	MILLI	0.79		- -	-	0.79	MILLI/MJ	0.90	
VITAMIN B6	MILLI	0.24		0.17 -	0.28	0.24	MILLI/MJ	0.27	
BIOTIN	MICRO	1.74		- -	-	1.74	MICRO/MJ	1.98	
FOLIC ACID	MICRO	30.00		- -	-	30.00	MICRO/MJ	34.08	
ALANINE	GRAM	0.33		- -	-	0.33	GRAM/MJ	0.37	5.4
ARGININE	GRAM	0.32		- -	-	0.32	GRAM/MJ	0.36	2.7
ASPARTIC ACID	GRAM	0.43		- -	-	0.43	GRAM/MJ	0.49	4.7

1 ESTIMATED BY THE DIFFERENCE METHOD
100 - (WATER + PROTEIN + FAT + MINERALS + TOTAL DIETARY FIBRE)

2 METHOD OF MEUSER, SUCKOW AND KULIKOWSKI ("BERLINER METHODE")

CONSTITUENTS	DIM	AV	VARIATION			AVR	NUTR. DENS.		MOLPERC.
CYSTINE	MILLI	60.00	-	-	-	60.00	MILLI/MJ	68.16	0.4
GLUTAMIC ACID	GRAM	2.66	-	-	-	2.66	GRAM/MJ	3.02	26.2
GLYCINE	GRAM	0.32	-	-	-	0.32	GRAM/MJ	0.36	6.2
HISTIDINE	GRAM	0.19	-	-	-	0.19	GRAM/MJ	0.22	1.8
ISOLEUCINE	GRAM	0.43	-	-	-	0.43	GRAM/MJ	0.49	4.7
LEUCINE	GRAM	0.73	-	-	-	0.73	GRAM/MJ	0.83	8.1
LYSINE	GRAM	0.20	-	-	-	0.20	GRAM/MJ	0.23	2.0
METHIONINE	MILLI	50.00	-	-	-	50.00	MILLI/MJ	56.80	0.5
PHENYLALANINE	GRAM	0.68	-	-	-	0.68	GRAM/MJ	0.77	6.0
PROLINE	GRAM	1.15	-	-	-	1.15	GRAM/MJ	1.31	14.5
SERINE	GRAM	0.47	-	-	-	0.47	GRAM/MJ	0.53	6.5
THREONINE	GRAM	0.22	-	-	-	0.22	GRAM/MJ	0.25	2.7
TRYPTOPHAN	MILLI	60.00	-	-	-	60.00	MILLI/MJ	68.16	0.4
TYROSINE	GRAM	0.33	-	-	-	0.33	GRAM/MJ	0.37	2.6
VALINE	GRAM	0.40	-	-	-	0.40	GRAM/MJ	0.45	4.9
GLUCOSE	GRAM	1.00	-	-	-	1.00	GRAM/MJ	1.14	
FRUCTOSE	GRAM	0.74	-	-	-	0.74	GRAM/MJ	0.84	
SUCROSE	GRAM	0.16	-	-	-	0.16	GRAM/MJ	0.18	
STARCH	GRAM	35.33	-	-	-	35.33	GRAM/MJ	40.13	
OLEIC ACID	GRAM	0.31	-	-	-	0.31	GRAM/MJ	0.35	
LINOLEIC ACID	GRAM	0.42	-	-	-	0.42	GRAM/MJ	0.48	
LINOLENIC ACID	MICRO	30.00	-	-	-	30.00	MICRO/MJ	34.08	
DIETARY FIBRE, WAT.SOL.	GRAM	1.30	-	-	-	1.30	GRAM/MJ	1.48	
DIETARY FIBRE, WAT.INS.	GRAM	5.10	-	-	-	5.10	GRAM/MJ	5.79	

KNÄCKEBROT | CRISPBREAD | PAIN COMPLET CROQUANT

		PROTEIN	FAT	CARBOHYDRATES	TOTAL
ENERGY VALUE (AVERAGE) PER 100 G EDIBLE PORTION	KJOULE	178	54	1093	1326
	(KCAL)	43	13	261	317
AMOUNT OF DIGESTIBLE CONSTITUENTS PER 100 G	GRAM	6.27	1.26	65.33	
ENERGY VALUE (AVERAGE) OF THE DIGESTIBLE FRACTION PER 100 G EDIBLE PORTION	KJOULE	120	49	1093	1262
	(KCAL)	29	12	261	302

WASTE PERCENTAGE AVERAGE 0.00

CONSTITUENTS	DIM	AV		VARIATION			AVR	NUTR. DENS.		MOLPERC.
WATER	GRAM	7.00		5.50	-	8.00	7.00	GRAM/MJ	5.55	
PROTEIN	GRAM	9.37		7.42	-	10.58	9.37	GRAM/MJ	7.43	
FAT	GRAM	1.40		0.60	-	2.00	1.40	GRAM/MJ	1.11	
AVAILABLE CARBOHYDR.	GRAM	65.33	1	-	-	-	65.33	GRAM/MJ	51.77	
TOTAL DIETARY FIBRE	GRAM	14.60	2	-	-	-	14.60	GRAM/MJ	11.57	
MINERALS	GRAM	2.30		-	-	-	2.30	GRAM/MJ	1.82	
SODIUM	MILLI	463.00		339.00	-	709.00	463.00	MILLI/MJ	366.91	
POTASSIUM	MILLI	436.00		410.00	-	461.00	436.00	MILLI/MJ	345.51	
MAGNESIUM	MILLI	68.00		-	-	-	68.00	MILLI/MJ	53.89	
CALCIUM	MILLI	55.00		48.00	-	66.00	55.00	MILLI/MJ	43.58	
IRON	MILLI	4.70		4.40	-	5.00	4.70	MILLI/MJ	3.72	
COPPER	MILLI	0.40		-	-	-	0.40	MILLI/MJ	0.32	
ZINC	MILLI	3.10		-	-	-	3.10	MILLI/MJ	2.46	
PHOSPHORUS	MILLI	300.85		298.00	-	318.00	300.86	MILLI/MJ	238.42	
CHLORIDE	MILLI	370.00		-	-	-	370.00	MILLI/MJ	293.21	
VITAMIN B1	MILLI	0.20		0.15	-	0.28	0.20	MILLI/MJ	0.16	
VITAMIN B2	MILLI	0.18		0.16	-	0.20	0.18	MILLI/MJ	0.14	
NICOTINAMIDE	MILLI	1.10		0.91	-	1.26	1.10	MILLI/MJ	0.87	
PANTOTHENIC ACID	MILLI	1.10		-	-	-	1.10	MILLI/MJ	0.87	
VITAMIN B6	MILLI	0.30		0.22	-	0.39	0.30	MILLI/MJ	0.24	
BIOTIN	MICRO	7.00		-	-	-	7.00	MICRO/MJ	5.55	
FOLIC ACID	MICRO	40.00		-	-	-	40.00	MICRO/MJ	31.70	
VITAMIN C	-	0.00		-	-	-	0.00			
ARGININE	GRAM	0.45		-	-	-	0.45	GRAM/MJ	0.36	
CYSTINE	GRAM	0.19		-	-	-	0.19	GRAM/MJ	0.15	

1 ESTIMATED BY THE DIFFERENCE METHOD
100 - (WATER + PROTEIN + FAT + MINERALS + TOTAL DIETARY FIBRE)

2 METHOD OF MEUSER, SUCKOW AND KULIKOWSKI ("BERLINER METHODE")

CONSTITUENTS	DIM	AV		VARIATION		AVR	NUTR. DENS.		MOLPERC.
HISTIDINE	GRAM	0.21		-	-	-	0.21	GRAM/MJ	0.17	
ISOLEUCINE	GRAM	0.40		-	-	-	0.40	GRAM/MJ	0.32	
LEUCINE	GRAM	0.63		-	-	-	0.63	GRAM/MJ	0.50	
LYSINE	GRAM	0.38		-	-	-	0.38	GRAM/MJ	0.30	
METHIONINE	GRAM	0.14		-	-	-	0.14	GRAM/MJ	0.11	
PHENYLALANINE	GRAM	0.45		-	-	-	0.45	GRAM/MJ	0.36	
THREONINE	GRAM	0.34		-	-	-	0.34	GRAM/MJ	0.27	
TRYPTOPHAN	GRAM	0.10		-	-	-	0.10	GRAM/MJ	0.08	
TYROSINE	GRAM	0.31		-	-	-	0.31	GRAM/MJ	0.25	
VALINE	GRAM	0.49		-	-	-	0.49	GRAM/MJ	0.39	
TOTAL PURINES	MILLI	60.00	3	-	-	-	60.00	MILLI/MJ	47.55	
DIETARY FIBRE, WAT.SOL.	GRAM	4.60		-	-	-	4.60	GRAM/MJ	3.65	
DIETARY FIBRE, WAT.INS.	GRAM	10.00		-	-	-	10.00	GRAM/MJ	7.92	

3 VALUE EXPRESSED IN MG URIC ACID/100G

PUMPERNICKEL PUMPERNICKEL PUMPERNICKEL

		PROTEIN	FAT	CARBOHYDRATES	TOTAL
ENERGY VALUE (AVERAGE)	KJOULE	130	36	688	854
PER 100 G	(KCAL)	31	8.6	164	204
EDIBLE PORTION					
AMOUNT OF DIGESTIBLE	GRAM	5.39	0.82	41.09	
CONSTITUENTS PER 100 G					
ENERGY VALUE (AVERAGE)	KJOULE	103	32	688	823
OF THE DIGESTIBLE	(KCAL)	25	7.7	164	197
FRACTION PER 100 G					
EDIBLE PORTION					

WASTE PERCENTAGE AVERAGE 0.00

CONSTITUENTS	DIM	AV	VARIATION			AVR	NUTR. DENS.		MOLPERC.
WATER	GRAM	40.10	37.20	-	44.00	40.10	GRAM/MJ	48.75	
PROTEIN	GRAM	6.83	5.60	-	8.10	6.83	GRAM/MJ	8.30	
FAT	GRAM	0.92	0.70	-	1.16	0.92	GRAM/MJ	1.12	
AVAILABLE CARBOHYDR.	GRAM	41.09 1	-	-	-	41.09	GRAM/MJ	49.95	
TOTAL DIETARY FIBRE	GRAM	9.76 2	-	-	-	9.76	GRAM/MJ	11.86	
MINERALS	GRAM	1.30	1.21	-	1.40	1.30	GRAM/MJ	1.58	
SODIUM	MILLI	370.00	335.00	-	404.00	370.00	MILLI/MJ	449.78	
POTASSIUM	MILLI	338.00	304.00	-	371.00	338.00	MILLI/MJ	410.88	
MAGNESIUM	MILLI	80.00	-	-	-	80.00	MILLI/MJ	97.25	
CALCIUM	MILLI	55.00	43.00	-	61.00	55.00	MILLI/MJ	66.86	
PHOSPHORUS	MILLI	147.00	116.00	-	210.00	147.00	MILLI/MJ	178.70	
VITAMIN B1	MICRO	50.00	-	-	-	50.00	MICRO/MJ	60.78	
PHYTIC ACID	GRAM	0.10	-	-	-	0.10	GRAM/MJ	0.12	
DIETARY FIBRE, WAT.SOL.	GRAM	3.23	-	-	-	3.23	GRAM/MJ	3.93	
DIETARY FIBRE, WAT.INS.	GRAM	6.53	-	-	-	6.53	GRAM/MJ	7.94	

1 ESTIMATED BY THE DIFFERENCE METHOD
100 - (WATER + PROTEIN + FAT + MINERALS + TOTAL DIETARY FIBRE)

2 MODIFIED AOAC METHOD

ROGGENBROT RYE BREAD PAIN DE SEIGLE

		PROTEIN	FAT	CARBOHYDRATES	TOTAL
ENERGY VALUE (AVERAGE)	KJOULE	118	39	796	953
PER 100 G	(KCAL)	28	9.3	190	228
EDIBLE PORTION					
AMOUNT OF DIGESTIBLE	GRAM	4.41	0.90	47.55	
CONSTITUENTS PER 100 G					
ENERGY VALUE (AVERAGE)	KJOULE	84	35	796	915
OF THE DIGESTIBLE	(KCAL)	20	8.4	190	219
FRACTION PER 100 G					
EDIBLE PORTION					

WASTE PERCENTAGE AVERAGE 0.00

CONSTITUENTS	DIM	AV		VARIATION		AVR	NUTR.	DENS.	MOLPERC.
WATER	GRAM	38.10		35.10 -	42.90	38.10	GRAM/MJ	41.64	
PROTEIN	GRAM	6.22		5.01 -	7.61	6.22	GRAM/MJ	6.80	
FAT	GRAM	1.00		0.80 -	1.10	1.00	GRAM/MJ	1.09	
AVAILABLE CARBOHYDR.	GRAM	47.55	1	- -	-	47.55	GRAM/MJ	51.97	
TOTAL DIETARY FIBRE	GRAM	5.50	2	4.10 -	7.60	5.50	GRAM/MJ	6.01	
MINERALS	GRAM	1.63		1.40 -	2.20	1.63	GRAM/MJ	1.78	
SODIUM	MILLI	552.00		467.00 -	652.00	552.00	MILLI/MJ	603.35	
POTASSIUM	MILLI	169.00		117.00 -	233.00	169.00	MILLI/MJ	184.72	
MAGNESIUM	MILLI	35.00		21.00 -	60.00	35.00	MILLI/MJ	38.26	
CALCIUM	MILLI	29.00		25.00 -	36.00	29.00	MILLI/MJ	31.70	
MANGANESE	MILLI	0.92		0.40 -	1.50	0.92	MILLI/MJ	1.01	
IRON	MILLI	2.50		1.50 -	3.70	2.50	MILLI/MJ	2.73	
COBALT	MICRO	1.80		1.30 -	2.40	1.80	MICRO/MJ	1.97	
COPPER	MILLI	0.27		0.22 -	0.35	0.27	MILLI/MJ	0.30	
ZINC	MILLI	0.86		0.59 -	1.20	0.86	MILLI/MJ	0.94	
NICKEL	MICRO	26.00		12.00 -	44.00	26.00	MICRO/MJ	28.42	
CHROMIUM	MICRO	8.50		5.00 -	12.00	8.50	MICRO/MJ	9.29	
MOLYBDENUM	MICRO	50.00		- -	-	50.00	MICRO/MJ	54.65	
PHOSPHORUS	MILLI	118.57		115.00 -	140.00	118.57	MILLI/MJ	129.60	
CHLORIDE	MILLI	670.00		660.00 -	690.00	670.00	MILLI/MJ	732.33	
FLUORIDE	MICRO	13.00		- -	-	13.00	MICRO/MJ	14.21	
IODIDE	MICRO	8.50		- -	-	8.50	MICRO/MJ	9.29	
BORON	MICRO	80.00		40.00 -	130.00	80.00	MICRO/MJ	87.44	
CAROTENE	-	0.00		- -	-	0.00			
VITAMIN B1	MILLI	0.18		0.14 -	0.26	0.18	MILLI/MJ	0.20	
VITAMIN B2	MILLI	0.11		0.07 -	0.15	0.11	MILLI/MJ	0.12	

1 ESTIMATED BY THE DIFFERENCE METHOD
100 - (WATER + PROTEIN + FAT + MINERALS + TOTAL DIETARY FIBRE)

2 METHOD OF MEUSER, SUCKOW AND KULIKOWSKI ("BERLINER METHODE")

CONSTITUENTS	DIM	AV	VARIATION			AVR	NUTR. DENS.		MOLPERC.
NICOTINAMIDE	MILLI	0.92	0.50	-	1.40	0.92	MILLI/MJ	1.01	
PANTOTHENIC ACID	MILLI	0.47	0.45	-	0.52	0.47	MILLI/MJ	0.51	
FOLIC ACID	MICRO	16.00	-	-	-	16.00	MICRO/MJ	17.49	
VITAMIN C	-	0.00	-	-	-	0.00			
ALANINE	GRAM	0.30	0.30	-	0.32	0.30	GRAM/MJ	0.33	6.2
ARGININE	GRAM	0.42	0.40	-	0.44	0.42	GRAM/MJ	0.46	4.4
ASPARTIC ACID	GRAM	0.48	0.44	-	0.50	0.48	GRAM/MJ	0.52	6.6
CYSTINE	GRAM	0.13	0.12	-	0.14	0.13	GRAM/MJ	0.14	1.0
GLUTAMIC ACID	GRAM	1.92	1.88	-	1.95	1.92	GRAM/MJ	2.10	23.9
GLYCINE	GRAM	0.32	0.32	-	0.33	0.32	GRAM/MJ	0.35	7.8
HISTIDINE	GRAM	0.24	0.23	-	0.25	0.24	GRAM/MJ	0.26	2.8
ISOLEUCINE	GRAM	0.26	0.24	-	0.29	0.26	GRAM/MJ	0.28	3.6
LEUCINE	GRAM	0.47	0.47	-	0.50	0.47	GRAM/MJ	0.51	6.6
LYSINE	GRAM	0.30	0.28	-	0.32	0.30	GRAM/MJ	0.33	3.8
METHIONINE	MILLI	60.00	50.00	-	70.00	60.00	MILLI/MJ	65.58	0.7
PHENYLALANINE	GRAM	0.35	0.34	-	0.36	0.35	GRAM/MJ	0.38	3.9
PROLINE	GRAM	0.72	0.69	-	0.75	0.72	GRAM/MJ	0.79	11.5
SERINE	GRAM	0.35	0.32	-	0.40	0.35	GRAM/MJ	0.38	6.1
THREONINE	GRAM	0.25	0.23	-	0.26	0.25	GRAM/MJ	0.27	3.8
TRYPTOPHAN	MILLI	50.00	-	-	-	50.00	MILLI/MJ	54.65	0.4
TYROSINE	GRAM	0.17	0.12	-	0.19	0.17	GRAM/MJ	0.19	1.7
VALINE	GRAM	0.33	0.32	-	0.35	0.33	GRAM/MJ	0.36	5.2
GLUCOSE	GRAM	0.52	-	-	-	0.52	GRAM/MJ	0.57	
FRUCTOSE	GRAM	0.38	-	-	-	0.38	GRAM/MJ	0.42	
SUCROSE	GRAM	0.61	-	-	-	0.61	GRAM/MJ	0.67	
MALTOSE	GRAM	2.30	-	-	-	2.30	GRAM/MJ	2.51	
STARCH	GRAM	44.00	-	-	-	44.00	GRAM/MJ	48.09	
CELLULOSE	GRAM	1.00	-	-	-	1.00	GRAM/MJ	1.09	
PHYTIC ACID	GRAM	0.25	-	-	-	0.25	GRAM/MJ	0.27	
DIETARY FIBRE, WAT.SOL.	GRAM	2.20	1.70	-	2.60	2.20	GRAM/MJ	2.40	
DIETARY FIBRE, WAT.INS.	GRAM	3.30	2.40	-	5.00	3.30	GRAM/MJ	3.61	

ROGGENMISCHBROT RYE AND WHEAT BREAD PAIN BIS DE SEIGLE

		PROTEIN	FAT	CARBOHYDRATES	TOTAL
ENERGY VALUE (AVERAGE)	KJOULE	122	43	759	924
PER 100 G	(KCAL)	29	10	181	221
EDIBLE PORTION					
AMOUNT OF DIGESTIBLE	GRAM	4.54	0.99	45.36	
CONSTITUENTS PER 100 G					
ENERGY VALUE (AVERAGE)	KJOULE	87	39	759	884
OF THE DIGESTIBLE	(KCAL)	21	9.2	181	211
FRACTION PER 100 G					
EDIBLE PORTION					

WASTE PERCENTAGE AVERAGE 0.00

CONSTITUENTS	DIM	AV		VARIATION			AVR	NUTR. DENS.		MOLPERC.
WATER	GRAM	39.10		-	-	-	39.10	GRAM/MJ	44.22	
PROTEIN	GRAM	6.40		5.75	-	6.68	6.40	GRAM/MJ	7.24	
FAT	GRAM	1.10		0.90	-	1.38	1.10	GRAM/MJ	1.24	
AVAILABLE CARBOHYDR.	GRAM	45.36	1	-	-	-	45.36	GRAM/MJ	51.30	
TOTAL DIETARY FIBRE	GRAM	6.20	2	-	-	-	6.20	GRAM/MJ	7.01	
MINERALS	GRAM	1.84		1.77	-	1.91	1.84	GRAM/MJ	2.08	
SODIUM	MILLI	537.00		322.00	-	768.00	537.00	MILLI/MJ	607.35	
POTASSIUM	MILLI	185.00		133.00	-	242.00	185.00	MILLI/MJ	209.24	
CALCIUM	MILLI	23.00		19.00	-	30.00	23.00	MILLI/MJ	26.01	
IRON	MILLI	2.40		2.00	-	2.80	2.40	MILLI/MJ	2.71	
PHOSPHORUS	MILLI	135.85		128.00	-	183.00	135.86	MILLI/MJ	153.65	
CHLORIDE	GRAM	1.03		0.97	-	1.08	1.03	GRAM/MJ	1.16	
VITAMIN B1	MILLI	0.17		0.14	-	0.20	0.17	MILLI/MJ	0.19	
VITAMIN B2	MICRO	79.00		70.00	-	86.00	79.00	MICRO/MJ	89.35	
NICOTINAMIDE	MILLI	0.96		0.64	-	1.49	0.96	MILLI/MJ	1.09	
PANTOTHENIC ACID	MILLI	0.26		0.12	-	0.49	0.26	MILLI/MJ	0.29	
VITAMIN B6	MILLI	0.12		0.08	-	0.18	0.12	MILLI/MJ	0.14	
VITAMIN C	-	0.00		-	-	-	0.00			
ALANINE	GRAM	0.29		0.21	-	0.30	0.29	GRAM/MJ	0.33	5.8
ARGININE	GRAM	0.41		0.40	-	0.42	0.41	GRAM/MJ	0.46	4.2
ASPARTIC ACID	GRAM	0.45		0.44	-	0.45	0.45	GRAM/MJ	0.51	6.0
CYSTINE	GRAM	0.15		0.13	-	0.16	0.15	GRAM/MJ	0.17	1.1
GLUTAMIC ACID	GRAM	2.15		2.11	-	2.19	2.15	GRAM/MJ	2.43	26.1
GLYCINE	GRAM	0.31		0.28	-	0.32	0.31	GRAM/MJ	0.35	7.4
HISTIDINE	GRAM	0.24		0.22	-	0.26	0.24	GRAM/MJ	0.27	2.8

1 ESTIMATED BY THE DIFFERENCE METHOD
100 - (WATER + PROTEIN + FAT + MINERALS + TOTAL DIETARY FIBRE)

2 MODIFIED AOAC METHOD

CONSTITUENTS	DIM	AV	VARIATION			AVR	NUTR. DENS.		MOLPERC.
ISOLEUCINE	GRAM	0.28	0.27	-	0.28	0.28	GRAM/MJ	0.32	3.8
LEUCINE	GRAM	0.51	0.49	-	0.54	0.51	GRAM/MJ	0.58	6.9
LYSINE	GRAM	0.27	0.26	-	0.30	0.27	GRAM/MJ	0.31	3.3
METHIONINE	MILLI	60.00	20.00	-	80.00	60.00	MILLI/MJ	67.86	0.7
PHENYLALANINE	GRAM	0.36	0.35	-	0.38	0.36	GRAM/MJ	0.41	3.9
PROLINE	GRAM	0.74	0.71	-	0.79	0.74	GRAM/MJ	0.84	11.5
SERINE	GRAM	0.34	0.32	-	0.36	0.34	GRAM/MJ	0.38	5.8
THREONINE	GRAM	0.25	0.23	-	0.26	0.25	GRAM/MJ	0.28	3.7
TRYPTOPHAN	MILLI	60.00	-	-	-	60.00	MILLI/MJ	67.86	0.5
TYROSINE	GRAM	0.13	0.08	-	0.16	0.13	GRAM/MJ	0.15	1.3
VALINE	GRAM	0.34	0.32	-	0.36	0.34	GRAM/MJ	0.38	5.2
GLUCOSE	GRAM	0.14	-	-	-	0.14	GRAM/MJ	0.16	
FRUCTOSE	GRAM	0.45	-	-	-	0.45	GRAM/MJ	0.51	
SUCROSE	GRAM	0.11	-	-	-	0.11	GRAM/MJ	0.12	
STARCH	GRAM	38.73	38.50	-	40.30	38.73	GRAM/MJ	43.80	
DIETARY FIBRE,WAT.SOL.	GRAM	2.41	-	-	-	2.41	GRAM/MJ	2.73	
DIETARY FIBRE,WAT.INS.	GRAM	3.79	-	-	-	3.79	GRAM/MJ	4.29	

ROGGENMISCHBROT MIT WEIZENKLEIE
RYE AND WHEAT BREAD WITH WHEAT BRAN
PAIN BIS DE SEIGLE AVEC SON DE BLÉ

		PROTEIN	FAT	CARBOHYDRATES	TOTAL
ENERGY VALUE (AVERAGE)	KJOULE	111	58	719	889
PER 100 G EDIBLE PORTION	(KCAL)	27	14	172	213
AMOUNT OF DIGESTIBLE CONSTITUENTS PER 100 G	GRAM	4.38	1.35	42.99	
ENERGY VALUE (AVERAGE)	KJOULE	84	53	719	856
OF THE DIGESTIBLE FRACTION PER 100 G EDIBLE PORTION	(KCAL)	20	13	172	204

WASTE PERCENTAGE AVERAGE 0.00

CONSTITUENTS	DIM	AV		VARIATION		AVR	NUTR. DENS.		MOLPERC.
WATER	GRAM	39.80		-	- -	39.80	GRAM/MJ	46.52	
PROTEIN	GRAM	5.85		-	- -	5.85	GRAM/MJ	6.84	
FAT	GRAM	1.50		-	- -	1.50	GRAM/MJ	1.75	
AVAILABLE CARBOHYDR.	GRAM	42.99	1	-	- -	42.99	GRAM/MJ	50.25	
TOTAL DIETARY FIBRE	GRAM	7.36	2	-	- -	7.36	GRAM/MJ	8.60	
MINERALS	GRAM	2.50		-	- -	2.50	GRAM/MJ	2.92	
SODIUM	MILLI	511.00		404.00	- 632.00	511.00	MILLI/MJ	597.29	
POTASSIUM	MILLI	223.00		183.00	- 272.00	223.00	MILLI/MJ	260.66	
DIETARY FIBRE, WAT.SOL.	GRAM	2.56		-	- -	2.56	GRAM/MJ	2.99	
DIETARY FIBRE, WAT.INS.	GRAM	4.80		-	- -	4.80	GRAM/MJ	5.61	

1 ESTIMATED BY THE DIFFERENCE METHOD
100 - (WATER + PROTEIN + FAT + MINERALS + TOTAL DIETARY FIBRE)

2 METHOD OF MEUSER, SUCKOW AND KULIKOWSKI ("BERLINER METHODE")

ROGGENVOLLKORN-BROT | RYE WHOLE-MEAL BREAD | PAIN COMPLET DE SEIGLE

		PROTEIN	FAT	CARBOHYDRATES	TOTAL
ENERGY VALUE (AVERAGE) PER 100 G EDIBLE PORTION	KJOULE	129	47	683	858
	(KCAL)	31	11	163	205
AMOUNT OF DIGESTIBLE CONSTITUENTS PER 100 G	GRAM	4.53	1.08	40.79	
ENERGY VALUE (AVERAGE) OF THE DIGESTIBLE FRACTION PER 100 G EDIBLE PORTION	KJOULE	86	42	683	811
	(KCAL)	21	10	163	194

WASTE PERCENTAGE AVERAGE 0.00

CONSTITUENTS	DIM	AV		VARIATION		AVR	NUTR. DENS.		MOLPERC.
WATER	GRAM	42.00		40.00 - 44.00		42.00	GRAM/MJ	51.79	
PROTEIN	GRAM	6.77		5.10 - 7.24		6.77	GRAM/MJ	8.35	
FAT	GRAM	1.20		1.10 - 1.30		1.20	GRAM/MJ	1.48	
AVAILABLE CARBOHYDR.	GRAM	40.79	1	- - -		40.79	GRAM/MJ	50.29	
TOTAL DIETARY FIBRE	GRAM	7.74	2	- - -		7.74	GRAM/MJ	9.54	
MINERALS	GRAM	1.50		- - -		1.50	GRAM/MJ	1.85	
SODIUM	MILLI	527.00		329.00 - 720.00		527.00	MILLI/MJ	649.79	
POTASSIUM	MILLI	291.00		269.00 - 309.00		291.00	MILLI/MJ	358.80	
CALCIUM	MILLI	43.00		35.00 - 50.00		43.00	MILLI/MJ	53.02	
IRON	MILLI	3.30		2.50 - 4.20		3.30	MILLI/MJ	4.07	
COPPER	MILLI	0.68		- - -		0.68	MILLI/MJ	0.84	
PHOSPHORUS	MILLI	197.71		194.00 - 220.00		197.71	MILLI/MJ	243.78	
TOTAL TOCOPHEROLS	MILLI	2.40		- - -		2.40	MILLI/MJ	2.96	
VITAMIN B1	MILLI	0.18		0.11 - 0.54		0.18	MILLI/MJ	0.22	
VITAMIN B2	MILLI	0.15		0.14 - 0.18		0.15	MILLI/MJ	0.18	
NICOTINAMIDE	MILLI	0.56		- - -		0.56	MILLI/MJ	0.69	
ALANINE	GRAM	0.32		0.29 - 0.33		0.32	GRAM/MJ	0.39	5.9
ARGININE	GRAM	0.41		0.36 - 0.47		0.41	GRAM/MJ	0.51	3.9
ASPARTIC ACID	GRAM	0.50		0.44 - 0.56		0.50	GRAM/MJ	0.62	6.2
CYSTINE	GRAM	0.14		0.13 - 0.15		0.14	GRAM/MJ	0.17	1.0
GLUTAMIC ACID	GRAM	2.10		2.05 - 2.12		2.10	GRAM/MJ	2.59	23.6
GLYCINE	GRAM	0.35		0.34 - 0.35		0.35	GRAM/MJ	0.43	7.7
HISTIDINE	GRAM	0.22		0.18 - 0.27		0.22	GRAM/MJ	0.27	2.3
ISOLEUCINE	GRAM	0.32		0.26 - 0.33		0.32	GRAM/MJ	0.39	4.0
LEUCINE	GRAM	0.53		0.52 - 0.55		0.53	GRAM/MJ	0.65	6.7

1 ESTIMATED BY THE DIFFERENCE METHOD
100 - (WATER + PROTEIN + FAT + MINERALS + TOTAL DIETARY FIBRE)

2 MODIFIED AOAC METHOD

CONSTITUENTS	DIM	AV	VARIATION			AVR	NUTR. DENS.		MOLPERC.
LYSINE	GRAM	0.32	0.30	-	0.36	0.32	GRAM/MJ	0.39	3.6
METHIONINE	GRAM	0.10	0.06	-	0.13	0.10	GRAM/MJ	0.12	1.1
PHENYLALANINE	GRAM	0.37	0.35	-	0.40	0.37	GRAM/MJ	0.46	3.7
PROLINE	GRAM	0.84	0.79	-	0.85	0.84	GRAM/MJ	1.04	12.1
SERINE	GRAM	0.38	0.34	-	0.44	0.38	GRAM/MJ	0.47	6.0
THREONINE	GRAM	0.29	0.25	-	0.30	0.29	GRAM/MJ	0.36	4.0
TRYPTOPHAN	MILLI	60.00	-	-	-	60.00	MILLI/MJ	73.98	0.5
TYROSINE	GRAM	0.20	0.13	-	0.25	0.20	GRAM/MJ	0.25	1.8
VALINE	GRAM	0.40	0.31	-	0.42	0.40	GRAM/MJ	0.49	5.7
GLUCOSE	GRAM	0.72	-	-	-	0.72	GRAM/MJ	0.89	
FRUCTOSE	GRAM	1.06	-	-	-	1.06	GRAM/MJ	1.31	
SUCROSE	GRAM	0.22	-	-	-	0.22	GRAM/MJ	0.27	
STARCH	GRAM	33.33	33.17	-	34.30	33.33	GRAM/MJ	41.10	
DIETARY FIBRE, WAT.SOL.	GRAM	2.76	-	-	-	2.76	GRAM/MJ	3.40	
DIETARY FIBRE, WAT.INS.	GRAM	4.98	-	-	-	4.98	GRAM/MJ	6.14	

WEIZEN(MEHL)BROT (WEISSBROT) | WHEAT (FLOUR) BREAD (WHITE BREAD) | PAIN DE BLÉ

		PROTEIN	FAT	CARBOHYDRATES	TOTAL
ENERGY VALUE (AVERAGE) PER 100 G EDIBLE PORTION	KJOULE	145	47	800	992
	(KCAL)	35	11	191	237
AMOUNT OF DIGESTIBLE CONSTITUENTS PER 100 G	GRAM	6.77	1.08	47.83	
ENERGY VALUE (AVERAGE) OF THE DIGESTIBLE FRACTION PER 100 G EDIBLE PORTION	KJOULE	129	42	800	971
	(KCAL)	31	10	191	232

WASTE PERCENTAGE AVERAGE 0.00

CONSTITUENTS	DIM	AV		VARIATION			AVR	NUTR. DENS.		MOLPERC.
WATER	GRAM	38.30		33.80	-	41.60	38.30	GRAM/MJ	39.43	
PROTEIN	GRAM	7.61		7.33	-	8.72	7.61	GRAM/MJ	7.83	
FAT	GRAM	1.20		0.90	-	2.30	1.20	GRAM/MJ	1.24	
AVAILABLE CARBOHYDR.	GRAM	47.83	1	-	-	-	47.83	GRAM/MJ	49.24	
TOTAL DIETARY FIBRE	GRAM	3.46	2	-	-	-	3.46	GRAM/MJ	3.56	
MINERALS	GRAM	1.60		1.30	-	1.90	1.60	GRAM/MJ	1.65	
SODIUM	MILLI	540.00		404.00	-	722.00	540.00	MILLI/MJ	555.87	
POTASSIUM	MILLI	132.00		94.00	-	175.00	132.00	MILLI/MJ	135.88	
MAGNESIUM	MILLI	24.00		-	-	-	24.00	MILLI/MJ	24.71	
CALCIUM	MILLI	58.00		27.00	-	88.00	58.00	MILLI/MJ	59.70	
MANGANESE	MILLI	0.60		0.20	-	1.00	0.60	MILLI/MJ	0.62	
IRON	MILLI	0.95		0.40	-	1.50	0.95	MILLI/MJ	0.98	
COBALT	MICRO	2.20		-	-	-	2.20	MICRO/MJ	2.26	
COPPER	MILLI	0.22		0.10	-	0.34	0.22	MILLI/MJ	0.23	
ZINC	MILLI	0.50		0.20	-	0.80	0.50	MILLI/MJ	0.51	
NICKEL	MICRO	13.00		-	-	-	13.00	MICRO/MJ	13.38	
CHROMIUM	MICRO	37.00		14.00	-	60.00	37.00	MICRO/MJ	38.09	
MOLYBDENUM	MICRO	21.00		-	-	-	21.00	MICRO/MJ	21.62	
VANADIUM	MICRO	7.00		-	-	-	7.00	MICRO/MJ	7.21	
PHOSPHORUS	MILLI	87.28		87.00	-	89.00	87.29	MILLI/MJ	89.85	
CHLORIDE	MILLI	450.00		-	-	-	450.00	MILLI/MJ	463.23	
FLUORIDE	MICRO	80.00		60.00	-	110.00	80.00	MICRO/MJ	82.35	
IODIDE	MICRO	5.80		-	-	-	5.80	MICRO/MJ	5.97	
BORON	MICRO	90.00		70.00	-	100.00	90.00	MICRO/MJ	92.65	
SELENIUM	MICRO	28.00		-	-	-	28.00	MICRO/MJ	28.82	
TOTAL TOCOPHEROLS	MILLI	-		0.05	-	2.20				

1 ESTIMATED BY THE DIFFERENCE METHOD
100 - (WATER + PROTEIN + FAT + MINERALS + TOTAL DIETARY FIBRE)

2 METHOD OF MEUSER, SUCKOW AND KULIKOWSKI ("BERLINER METHODE")

CONSTITUENTS	DIM	AV	VARIATION			AVR	NUTR. DENS.		MOLPERC.
VITAMIN B1	MICRO	86.00	42.00	-	140.00	86.00	MICRO/MJ	88.53	
VITAMIN B2	MICRO	60.00	24.00	-	120.00	60.00	MICRO/MJ	61.76	
NICOTINAMIDE	MILLI	0.85	0.50	-	1.50	0.85	MILLI/MJ	0.87	
PANTOTHENIC ACID	MILLI	0.69	-	-	-	0.69	MILLI/MJ	0.71	
VITAMIN B6	MICRO	40.00	30.00	-	50.00	40.00	MICRO/MJ	41.18	
BIOTIN	MICRO	2.90	-	-	-	2.90	MICRO/MJ	2.99	
FOLIC ACID	MICRO	15.00	-	-	-	15.00	MICRO/MJ	15.44	
ALANINE	GRAM	0.24	-	-	-	0.24	GRAM/MJ	0.25	4.1
ARGININE	GRAM	0.31	-	-	-	0.31	GRAM/MJ	0.32	2.7
ASPARTIC ACID	GRAM	0.39	-	-	-	0.39	GRAM/MJ	0.40	4.5
CYSTINE	GRAM	0.18	-	-	-	0.18	GRAM/MJ	0.19	1.1
GLUTAMIC ACID	GRAM	3.15	-	-	-	3.15	GRAM/MJ	3.24	32.5
GLYCINE	GRAM	0.29	-	-	-	0.29	GRAM/MJ	0.30	5.9
HISTIDINE	GRAM	0.18	-	-	-	0.18	GRAM/MJ	0.19	1.8
ISOLEUCINE	GRAM	0.38	-	-	-	0.38	GRAM/MJ	0.39	4.4
LEUCINE	GRAM	0.59	-	-	-	0.59	GRAM/MJ	0.61	6.8
LYSINE	GRAM	0.20	-	-	-	0.20	GRAM/MJ	0.21	2.1
METHIONINE	GRAM	0.13	-	-	-	0.13	GRAM/MJ	0.13	1.3
PHENYLALANINE	GRAM	0.42	-	-	-	0.42	GRAM/MJ	0.43	3.9
PROLINE	GRAM	0.96	-	-	-	0.96	GRAM/MJ	0.99	12.7
SERINE	GRAM	0.39	-	-	-	0.39	GRAM/MJ	0.40	5.6
THREONINE	GRAM	0.25	-	-	-	0.25	GRAM/MJ	0.26	3.2
TRYPTOPHAN	MILLI	80.00	-	-	-	80.00	MILLI/MJ	82.35	0.6
TYROSINE	GRAM	0.21	-	-	-	0.21	GRAM/MJ	0.22	1.8
VALINE	GRAM	0.39	-	-	-	0.39	GRAM/MJ	0.40	5.1
TOTAL SUGAR	GRAM	1.80	-	-	-	1.80	GRAM/MJ	1.85	
STARCH	GRAM	41.77	-	-	-	41.77	GRAM/MJ	43.00	
CELLULOSE	GRAM	0.20	-	-	-	0.20	GRAM/MJ	0.21	
PHYTIC ACID	MILLI	20.00	-	-	-	20.00	MILLI/MJ	20.59	
TOTAL PURINES	MILLI	15.00 [3]	-	-	-	15.00	MILLI/MJ	15.44	
DIETARY FIBRE,WAT.SOL.	GRAM	1.67	-	-	-	1.67	GRAM/MJ	1.72	
DIETARY FIBRE,WAT.INS.	GRAM	1.79	-	-	-	1.79	GRAM/MJ	1.84	
ETHANOL	GRAM	0.17	-	-	-	0.17	GRAM/MJ	0.17	

3 VALUE EXPRESSED IN MG URIC ACID/100G

WEIZENMISCHBROT WHEAT AND RYE BREAD PAIN BIS DE BLÉ

		PROTEIN	FAT	CARBOHYDRATES	TOTAL
ENERGY VALUE (AVERAGE) PER 100 G EDIBLE PORTION	KJOULE (KCAL)	119 28	43 10	837 200	999 239
AMOUNT OF DIGESTIBLE CONSTITUENTS PER 100 G	GRAM	5.17	0.99	49.99	
ENERGY VALUE (AVERAGE) OF THE DIGESTIBLE FRACTION PER 100 G EDIBLE PORTION	KJOULE (KCAL)	99 24	39 9.3	837 200	974 233

WASTE PERCENTAGE AVERAGE 0.00

CONSTITUENTS	DIM	AV		VARIATION	AVR	NUTR. DENS.		MOLPERC.
WATER	GRAM	37.60		35.90 - 41.00	37.60	GRAM/MJ	38.60	
PROTEIN	GRAM	6.24		5.75 - 6.73	6.24	GRAM/MJ	6.41	
FAT	GRAM	1.11		0.80 - 1.37	1.11	GRAM/MJ	1.14	
AVAILABLE CARBOHYDR.	GRAM	49.99	1	- - -	49.99	GRAM/MJ	51.32	
TOTAL DIETARY FIBRE	GRAM	3.52	2	- - -	3.52	GRAM/MJ	3.61	
MINERALS	GRAM	1.54		0.96 - 2.23	1.54	GRAM/MJ	1.58	
SODIUM	MILLI	553.00		395.00 - 729.00	553.00	MILLI/MJ	567.70	
POTASSIUM	MILLI	177.00		126.00 - 258.00	177.00	MILLI/MJ	181.71	
CALCIUM	MILLI	17.00		16.00 - 18.00	17.00	MILLI/MJ	17.45	
IRON	MILLI	1.70		1.60 - 1.80	1.70	MILLI/MJ	1.75	
COPPER	MILLI	0.18		0.13 - 0.23	0.18	MILLI/MJ	0.18	
ZINC	MILLI	3.50		2.00 - 5.00	3.50	MILLI/MJ	3.59	
PHOSPHORUS	MILLI	127.28		111.00 - 130.00	127.29	MILLI/MJ	130.67	
CHLORIDE	GRAM	0.95		0.90 - 1.01	0.95	GRAM/MJ	0.98	
VITAMIN B1	MILLI	0.14		0.13 - 0.16	0.14	MILLI/MJ	0.14	
VITAMIN B2	MICRO	73.00		60.00 - 85.00	73.00	MICRO/MJ	74.94	
NICOTINAMIDE	MILLI	1.20		0.87 - 1.50	1.20	MILLI/MJ	1.23	
PANTOTHENIC ACID	MILLI	0.25		- - -	0.25	MILLI/MJ	0.26	
VITAMIN B6	MICRO	94.00		63.00 - 120.00	94.00	MICRO/MJ	96.50	
VITAMIN C	-	0.00		- - -	0.00			
ALANINE	GRAM	0.27		0.26 - 0.28	0.27	GRAM/MJ	0.28	5.5
ARGININE	GRAM	0.37		0.34 - 0.38	0.37	GRAM/MJ	0.38	3.9
ASPARTIC ACID	GRAM	0.40		0.37 - 0.43	0.40	GRAM/MJ	0.41	5.5
CYSTINE	GRAM	0.14		0.13 - 0.15	0.14	GRAM/MJ	0.14	1.1
GLUTAMIC ACID	GRAM	2.24		2.17 - 2.33	2.24	GRAM/MJ	2.30	27.7

1 ESTIMATED BY THE DIFFERENCE METHOD
100 - (WATER + PROTEIN + FAT + MINERALS + TOTAL DIETARY FIBRE)

2 MODIFIED AOAC METHOD

CONSTITUENTS	DIM	AV	VARIATION			AVR	NUTR. DENS.		MOLPERC.
GLYCINE	GRAM	0.30	0.28	-	0.32	0.30	GRAM/MJ	0.31	7.3
HISTIDINE	GRAM	0.23	0.21	-	0.25	0.23	GRAM/MJ	0.24	2.7
ISOLEUCINE	GRAM	0.27	0.26	-	0.28	0.27	GRAM/MJ	0.28	3.7
LEUCINE	GRAM	0.51	0.41	-	0.53	0.51	GRAM/MJ	0.52	7.1
LYSINE	GRAM	0.24	0.22	-	0.26	0.24	GRAM/MJ	0.25	3.0
METHIONINE	MILLI	50.00	20.00	-	60.00	50.00	MILLI/MJ	51.33	0.6
PHENYLALANINE	GRAM	0.35	0.34	-	0.37	0.35	GRAM/MJ	0.36	3.9
PROLINE	GRAM	0.74	0.69	-	0.77	0.74	GRAM/MJ	0.76	11.7
SERINE	GRAM	0.37	0.35	-	0.38	0.37	GRAM/MJ	0.38	6.4
THREONINE	GRAM	0.23	0.22	-	0.24	0.23	GRAM/MJ	0.24	3.5
TRYPTOPHAN	MILLI	60.00	50.00	-	80.00	60.00	MILLI/MJ	61.60	0.5
TYROSINE	GRAM	0.12	0.08	-	0.18	0.12	GRAM/MJ	0.12	1.2
VALINE	GRAM	0.32	0.31	-	0.33	0.32	GRAM/MJ	0.33	5.0
GLUCOSE	GRAM	0.18	-	-	-	0.18	GRAM/MJ	0.18	
FRUCTOSE	GRAM	0.42	-	-	-	0.42	GRAM/MJ	0.43	
SUCROSE	MILLI	50.00	-	-	-	50.00	MILLI/MJ	51.33	
STARCH	GRAM	41.94	-	-	-	41.94	GRAM/MJ	43.06	
DIETARY FIBRE, WAT.SOL.	GRAM	1.75	-	-	-	1.75	GRAM/MJ	1.80	
DIETARY FIBRE, WAT.INS.	GRAM	1.78	-	-	-	1.78	GRAM/MJ	1.83	

WEIZENTOASTBROT WHEAT TOAST BREAD PAIN DE BLÉ POUR TOAST

		PROTEIN	FAT	CARBOHYDRATES	TOTAL
ENERGY VALUE (AVERAGE)	KJOULE	131	171	805	1107
PER 100 G	(KCAL)	31	41	192	265
EDIBLE PORTION					
AMOUNT OF DIGESTIBLE	GRAM	5.83	3.96	48.12	
CONSTITUENTS PER 100 G					
ENERGY VALUE (AVERAGE)	KJOULE	111	154	805	1071
OF THE DIGESTIBLE	(KCAL)	27	37	192	256
FRACTION PER 100 G					
EDIBLE PORTION					

WASTE PERCENTAGE AVERAGE 0.00

CONSTITUENTS	DIM	AV		VARIATION		AVR	NUTR. DENS.		MOLPERC.
WATER	GRAM	35.10		-	- -	35.10	GRAM/MJ	32.79	
PROTEIN	GRAM	6.87		-	- -	6.87	GRAM/MJ	6.42	
FAT	GRAM	4.40		-	- -	4.40	GRAM/MJ	4.11	
AVAILABLE CARBOHYDR.	GRAM	48.12	1	-	- -	48.12	GRAM/MJ	44.95	
TOTAL DIETARY FIBRE	GRAM	3.57	2	-	- -	3.57	GRAM/MJ	3.33	
MINERALS	GRAM	1.94		-	- -	1.94	GRAM/MJ	1.81	
SODIUM	MILLI	551.00		446.00	- 662.00	551.00	MILLI/MJ	514.67	
POTASSIUM	MILLI	160.00		126.00	- 204.00	160.00	MILLI/MJ	149.45	
PHOSPHORUS	MILLI	92.00		92.00	- 92.00	92.00	MILLI/MJ	85.93	
DIETARY FIBRE, WAT.SOL.	GRAM	1.56		-	- -	1.56	GRAM/MJ	1.46	
DIETARY FIBRE, WAT.INS.	GRAM	2.01		-	- -	2.01	GRAM/MJ	1.88	

1 ESTIMATED BY THE DIFFERENCE METHOD
100 - (WATER + PROTEIN + FAT + MINERALS + TOTAL DIETARY FIBRE)

2 MODIFIED AOAC METHOD

WEIZENVOLLKORN-BROT / WHEAT WHOLE-MEAL BREAD / PAIN COMPLET DE BLÉ

		PROTEIN	FAT	CARBOHYDRATES	TOTAL
ENERGY VALUE (AVERAGE) PER 100 G EDIBLE PORTION	KJOULE	133	33	693	860
	(KCAL)	32	8.0	166	205
AMOUNT OF DIGESTIBLE CONSTITUENTS PER 100 G	GRAM	5.53	0.77	41.39	
ENERGY VALUE (AVERAGE) OF THE DIGESTIBLE FRACTION PER 100 G EDIBLE PORTION	KJOULE	105	30	693	828
	(KCAL)	25	7.2	166	198

WASTE PERCENTAGE AVERAGE 0.00

CONSTITUENTS	DIM	AV		VARIATION		AVR	NUTR. DENS.		MOLPERC.
WATER	GRAM	41.70		40.00 -	44.00	41.70	GRAM/MJ	50.35	
PROTEIN	GRAM	7.01		6.50 -	7.52	7.01	GRAM/MJ	8.46	
FAT	GRAM	0.86		0.72 -	1.00	0.86	GRAM/MJ	1.04	
AVAILABLE CARBOHYDR.	GRAM	41.39	1	- -	-	41.39	GRAM/MJ	49.97	
TOTAL DIETARY FIBRE	GRAM	7.52	2	- -	-	7.52	GRAM/MJ	9.08	
MINERALS	GRAM	1.52		- -	-	1.52	GRAM/MJ	1.84	
SODIUM	MILLI	380.00		- -	-	380.00	MILLI/MJ	458.80	
POTASSIUM	MILLI	270.00		- -	-	270.00	MILLI/MJ	325.99	
MAGNESIUM	MILLI	92.00		- -	-	92.00	MILLI/MJ	111.08	
CALCIUM	MILLI	63.00		30.00 -	95.00	63.00	MILLI/MJ	76.06	
MANGANESE	MILLI	2.30		- -	-	2.30	MILLI/MJ	2.78	
IRON	MILLI	2.00		- -	-	2.00	MILLI/MJ	2.41	
COPPER	MILLI	0.42		- -	-	0.42	MILLI/MJ	0.51	
ZINC	MILLI	2.10		- -	-	2.10	MILLI/MJ	2.54	
NICKEL	MILLI	0.13		- -	-	0.13	MILLI/MJ	0.16	
CHROMIUM	MICRO	49.00		0.00 -	60.00	49.00	MICRO/MJ	59.16	
MOLYBDENUM	MICRO	31.00		30.00 -	32.00	31.00	MICRO/MJ	37.43	
PHOSPHORUS	MILLI	195.57		184.00 -	265.00	195.57	MILLI/MJ	236.13	
BORON	MICRO	48.00		- -	-	48.00	MICRO/MJ	57.95	
SELENIUM	MICRO	55.00		41.00 -	68.00	55.00	MICRO/MJ	66.41	
VITAMIN B1	MILLI	0.25		0.16 -	0.45	0.25	MILLI/MJ	0.30	
VITAMIN B2	MILLI	0.15		0.15 -	0.16	0.15	MILLI/MJ	0.18	
NICOTINAMIDE	MILLI	3.30		3.00 -	3.50	3.30	MILLI/MJ	3.98	
PANTOTHENIC ACID	MILLI	0.65		0.57 -	0.80	0.65	MILLI/MJ	0.78	
VITAMIN B6	MILLI	0.36		- -	-	0.36	MILLI/MJ	0.43	
BIOTIN	MICRO	3.50		2.00 -	5.00	3.50	MICRO/MJ	4.23	

1 ESTIMATED BY THE DIFFERENCE METHOD
100 - (WATER + PROTEIN + FAT + MINERALS + TOTAL DIETARY FIBRE)

2 MODIFIED AOAC METHOD

CONSTITUENTS	DIM	AV	VARIATION			AVR	NUTR. DENS.		MOLPERC.
FOLIC ACID	MICRO	60.00	-	-	-	60.00	MICRO/MJ	72.44	
VITAMIN C	-	0.00	-	-	-	0.00			
ALANINE	GRAM	0.32	0.28	-	0.32	0.32	GRAM/MJ	0.39	5.7
ARGININE	GRAM	0.36	0.20	-	0.43	0.36	GRAM/MJ	0.43	3.3
ASPARTIC ACID	GRAM	0.42	-	-	-	0.42	GRAM/MJ	0.51	5.0
CYSTINE	GRAM	0.16	0.10	-	0.20	0.16	GRAM/MJ	0.19	1.1
GLUTAMIC ACID	GRAM	2.38	2.28	-	2.48	2.38	GRAM/MJ	2.87	25.8
GLYCINE	GRAM	0.39	-	-	-	0.39	GRAM/MJ	0.47	8.3
HISTIDINE	GRAM	0.15	0.10	-	0.20	0.15	GRAM/MJ	0.18	1.5
ISOLEUCINE	GRAM	0.32	0.23	-	0.38	0.32	GRAM/MJ	0.39	3.9
LEUCINE	GRAM	0.53	0.45	-	0.58	0.53	GRAM/MJ	0.64	6.4
LYSINE	GRAM	0.20	0.17	-	0.26	0.20	GRAM/MJ	0.24	2.2
METHIONINE	GRAM	0.14	0.06	-	0.21	0.14	GRAM/MJ	0.17	1.5
PHENYLALANINE	GRAM	0.36	0.28	-	0.44	0.36	GRAM/MJ	0.43	3.5
PROLINE	GRAM	0.99	0.91	-	1.07	0.99	GRAM/MJ	1.20	13.7
SERINE	GRAM	0.48	-	-	-	0.48	GRAM/MJ	0.58	7.3
THREONINE	GRAM	0.24	0.14	-	0.30	0.24	GRAM/MJ	0.29	3.2
TRYPTOPHAN	GRAM	0.10	0.06	-	0.11	0.10	GRAM/MJ	0.12	0.8
TYROSINE	GRAM	0.22	0.17	-	0.36	0.22	GRAM/MJ	0.27	1.9
VALINE	GRAM	0.36	0.18	-	0.40	0.36	GRAM/MJ	0.43	4.9
CELLULOSE	GRAM	1.20	-	-	-	1.20	GRAM/MJ	1.45	
PHYTIC ACID	GRAM	0.33	-	-	-	0.33	GRAM/MJ	0.40	
DIETARY FIBRE,WAT.SOL.	GRAM	1.57	-	-	-	1.57	GRAM/MJ	1.90	
DIETARY FIBRE,WAT.INS.	GRAM	5.95	-	-	-	5.95	GRAM/MJ	7.18	

Teigwaren

PASTRY | PÂTES

EIERTEIGWAREN (NUDELN, MAKKARONI, SPAGHETTI ETC.)

PASTA MADE WITH EGGS (NOODELS, MACCARONI, SPAGHETTI ETC.)

PÂTES AUX OEUFS (NOUILLES, MACCARONI, SPAGHETTI ETC.)

		PROTEIN	FAT	CARBOHYDRATES	TOTAL
ENERGY VALUE (AVERAGE)	KJOULE	235	108	1170	1513
PER 100 G	(KCAL)	56	26	280	362
EDIBLE PORTION					
AMOUNT OF DIGESTIBLE	GRAM	10.61	2.50	69.91	
CONSTITUENTS PER 100 G					
ENERGY VALUE (AVERAGE)	KJOULE	202	97	1170	1469
OF THE DIGESTIBLE	(KCAL)	48	23	280	351
FRACTION PER 100 G					
EDIBLE PORTION					

WASTE PERCENTAGE AVERAGE 0.00

CONSTITUENTS	DIM	AV	VARIATION			AVR	NUTR. DENS.		MOLPERC.
WATER	GRAM	10.70	9.60	-	12.50	10.70	GRAM/MJ	7.28	
PROTEIN	GRAM	12.34	10.86	-	14.38	12.34	GRAM/MJ	8.40	
FAT	GRAM	2.78	2.02	-	3.50	2.78	GRAM/MJ	1.89	
AVAILABLE CARBOHYDR.	GRAM	69.91 1	-	-	-	69.91	GRAM/MJ	47.58	
TOTAL DIETARY FIBRE	GRAM	3.38 2	-	-	-	3.38	GRAM/MJ	2.30	
MINERALS	GRAM	0.89	0.71	-	1.14	0.89	GRAM/MJ	0.61	
SODIUM	MILLI	17.00	10.00	-	25.60	17.00	MILLI/MJ	11.57	
POTASSIUM	MILLI	164.00	136.00	-	217.00	164.00	MILLI/MJ	111.61	
MAGNESIUM	MILLI	67.00	57.00	-	74.00	67.00	MILLI/MJ	45.60	
CALCIUM	MILLI	27.00	21.00	-	36.00	27.00	MILLI/MJ	18.37	
MANGANESE	MILLI	0.73	0.50	-	1.10	0.73	MILLI/MJ	0.50	
IRON	MILLI	1.60	1.00	-	2.10	1.60	MILLI/MJ	1.09	
COBALT	MICRO	2.20	1.40	-	2.90	2.20	MICRO/MJ	1.50	
COPPER	MILLI	0.15	0.07	-	0.28	0.15	MILLI/MJ	0.10	
ZINC	MILLI	1.60	1.00	-	2.20	1.60	MILLI/MJ	1.09	
MOLYBDENUM	MICRO	49.00	46.00	-	51.00	49.00	MICRO/MJ	33.35	
PHOSPHORUS	MILLI	191.00	152.00	-	214.00	191.00	MILLI/MJ	129.99	
CHLORIDE	MILLI	31.40	-	-	-	31.40	MILLI/MJ	21.37	
FLUORIDE	MICRO	80.00	73.00	-	87.00	80.00	MICRO/MJ	54.44	
BORON	MICRO	75.00	63.00	-	105.00	75.00	MICRO/MJ	51.04	
SELENIUM	MICRO	65.00	58.00	-	66.00	65.00	MICRO/MJ	44.24	

1 ESTIMATED BY THE DIFFERENCE METHOD
100 - (WATER + PROTEIN + FAT + MINERALS + TOTAL DIETARY FIBRE)

2 METHOD OF MEUSER, SUCKOW AND KULIKOWSKI ("BERLINER METHODE")

CONSTITUENTS	DIM	AV	VARIATION			AVR	NUTR. DENS.		MOLPERC.
VITAMIN A	MICRO	63.00	60.00	-	66.00	63.00	MICRO/MJ	42.87	
VITAMIN D	-	0.00	-	-	-	0.00			
TOTAL TOCOPHEROLS	MILLI	0.30	-	-	-	0.30	MILLI/MJ	0.20	
VITAMIN B1	MILLI	0.17	0.13	-	0.24	0.17	MILLI/MJ	0.12	
VITAMIN B2	MICRO	73.00	40.00	-	110.00	73.00	MICRO/MJ	49.68	
NICOTINAMIDE	MILLI	1.90	1.10	-	2.30	1.90	MILLI/MJ	1.29	
PANTOTHENIC ACID	MILLI	0.30	-	-	-	0.30	MILLI/MJ	0.20	
VITAMIN B6	MICRO	60.00	-	-	-	60.00	MICRO/MJ	40.83	
BIOTIN	MICRO	1.00	-	-	-	1.00	MICRO/MJ	0.68	
FOLIC ACID	MICRO	11.00	-	-	-	11.00	MICRO/MJ	7.49	
VITAMIN C	-	0.00	-	-	-	0.00			
ALANINE	GRAM	0.47	0.45	-	0.50	0.47	GRAM/MJ	0.32	4.9
ARGININE	GRAM	0.71	0.66	-	0.76	0.71	GRAM/MJ	0.48	3.8
ASPARTIC ACID	GRAM	0.76	0.68	-	0.82	0.76	GRAM/MJ	0.52	5.3
CYSTINE	GRAM	0.31	0.24	-	0.33	0.31	GRAM/MJ	0.21	1.2
GLUTAMIC ACID	GRAM	4.49	4.27	-	4.76	4.49	GRAM/MJ	3.06	28.4
GLYCINE	GRAM	0.52	0.47	-	0.55	0.52	GRAM/MJ	0.35	6.4
HISTIDINE	GRAM	0.46	0.43	-	0.55	0.46	GRAM/MJ	0.31	2.8
ISOLEUCINE	GRAM	0.53	0.50	-	0.59	0.53	GRAM/MJ	0.36	3.8
LEUCINE	GRAM	1.01	0.97	-	1.07	1.01	GRAM/MJ	0.69	7.2
LYSINE	GRAM	0.45	0.42	-	0.47	0.45	GRAM/MJ	0.31	2.9
METHIONINE	GRAM	0.10	0.06	-	0.20	0.10	GRAM/MJ	0.07	0.6
PHENYLALANINE	GRAM	0.64	0.61	-	0.68	0.64	GRAM/MJ	0.44	3.6
PROLINE	GRAM	1.49	1.41	-	1.54	1.49	GRAM/MJ	1.01	12.1
SERINE	GRAM	0.74	0.69	-	0.78	0.74	GRAM/MJ	0.50	6.6
THREONINE	GRAM	0.42	0.39	-	0.45	0.42	GRAM/MJ	0.29	3.3
TRYPTOPHAN	MILLI	80.00	60.00	-	110.00	80.00	MILLI/MJ	54.44	0.4
TYROSINE	GRAM	0.42	0.37	-	0.45	0.42	GRAM/MJ	0.29	2.2
VALINE	GRAM	0.59	0.57	-	0.65	0.59	GRAM/MJ	0.40	4.7
REDUCING SUGAR	GRAM	1.95	-	-	-	1.95	GRAM/MJ	1.33	
NONREDUCING SUGAR	GRAM	1.70	-	-	-	1.70	GRAM/MJ	1.16	
STARCH	GRAM	64.47	63.10	-	64.70	64.47	GRAM/MJ	43.88	
LINOLEIC ACID	GRAM	0.83	0.80	-	0.85	0.83	GRAM/MJ	0.56	
LINOLENIC ACID	MILLI	76.00	67.00	-	83.00	76.00	MILLI/MJ	51.72	
CHOLESTEROL	MILLI	93.80	68.70	-	105.00	93.80	MILLI/MJ	63.84	
TOTAL PURINES	MILLI	40.00 3	-	-	-	40.00	MILLI/MJ	27.22	
DIETARY FIBRE, WAT.SOL.	GRAM	2.40	-	-	-	2.40	GRAM/MJ	1.63	
DIETARY FIBRE, WAT.INS.	GRAM	0.98	-	-	-	0.98	GRAM/MJ	0.67	

3 VALUE EXPRESSED IN MG URIC ACID/100G

EIERTEIGWAREN	PASTA	PÂTES AUX OEUFS
NUDELN GEKOCHT, ABGETROPFT	MADE WITH EGGS NOODELS COOKED, DRAINED	NOUILLES CUITE, DEGOUTÉE

		PROTEIN	FAT	CARBOHYDRATES	TOTAL
ENERGY VALUE (AVERAGE)	KJOULE	76	113	305	493
PER 100 G EDIBLE PORTION	(KCAL)	18	27	73	118
AMOUNT OF DIGESTIBLE CONSTITUENTS PER 100 G	GRAM	3.39	2.61	18.20	
ENERGY VALUE (AVERAGE)	KJOULE	65	102	305	471
OF THE DIGESTIBLE FRACTION PER 100 G EDIBLE PORTION	(KCAL)	15	24	73	113

WASTE PERCENTAGE AVERAGE 0.00

CONSTITUENTS	DIM	AV		VARIATION			AVR	NUTR.	DENS.	MOLPERC.
WATER	GRAM	76.80		-	-	-	76.80	GRAM/MJ	163.15	
PROTEIN	GRAM	3.99		-	-	-	3.99	GRAM/MJ	8.48	
FAT	GRAM	2.90		-	-	-	2.90	GRAM/MJ	6.16	
AVAILABLE CARBOHYDR.	GRAM	18.20	1	-	-	-	18.20	GRAM/MJ	38.66	
MINERALS	GRAM	0.78		-	-	-	0.78	GRAM/MJ	1.66	
POTASSIUM	MILLI	53.00		-	-	-	53.00	MILLI/MJ	112.59	
CALCIUM	MILLI	9.00		-	-	-	9.00	MILLI/MJ	19.12	
IRON	MILLI	0.80		-	-	-	0.80	MILLI/MJ	1.70	
PHOSPHORUS	MILLI	62.00		-	-	-	62.00	MILLI/MJ	131.71	
VITAMIN B1	MICRO	6.00		-	-	-	6.00	MICRO/MJ	12.75	
VITAMIN B2	MICRO	4.00		-	-	-	4.00	MICRO/MJ	8.50	
NICOTINAMIDE	MILLI	0.40		-	-	-	0.40	MILLI/MJ	0.85	

1 CALCULATED FROM THE DRY PRODUCT

Fein- und Dauerbackwaren

CAKES PÂTISSERIE

KEKS (BUTTERKEKS, HARTKEKS) BISCUIT (COOKIES, KEKS) BISCUITS (PETIT-BEURRE)

		PROTEIN	FAT	CARBOHYDRATES	TOTAL
ENERGY VALUE (AVERAGE)	KJOULE	144	428	1250	1822
PER 100 G	(KCAL)	34	102	299	435
EDIBLE PORTION					
AMOUNT OF DIGESTIBLE	GRAM	6.72	9.90	74.69	
CONSTITUENTS PER 100 G					
ENERGY VALUE (AVERAGE)	KJOULE	128	385	1250	1763
OF THE DIGESTIBLE	(KCAL)	31	92	299	421
FRACTION PER 100 G					
EDIBLE PORTION					

WASTE PERCENTAGE AVERAGE 0.00

CONSTITUENTS	DIM	AV		VARIATION		AVR	NUTR. DENS.		MOLPERC.
WATER	GRAM	1.99		0.81	- 3.51	1.99	GRAM/MJ	1.13	
PROTEIN	GRAM	7.56		6.46	- 8.32	7.56	GRAM/MJ	4.29	
FAT	GRAM	11.00		8.00	- 17.20	11.00	GRAM/MJ	6.24	
AVAILABLE CARBOHYDR.	GRAM	74.69	1	-	- -	74.69	GRAM/MJ	42.36	
TOTAL DIETARY FIBRE	GRAM	3.33	2	-	- -	3.33	GRAM/MJ	1.89	
MINERALS	GRAM	1.43		0.85	- 1.62	1.43	GRAM/MJ	0.81	
SODIUM	MILLI	387.00		210.00	- 509.00	387.00	MILLI/MJ	219.47	
POTASSIUM	MILLI	139.00		94.00	- 172.00	139.00	MILLI/MJ	78.83	
MAGNESIUM	MILLI	23.00		9.40	- 50.00	23.00	MILLI/MJ	13.04	
CALCIUM	MILLI	47.00		33.00	- 71.00	47.00	MILLI/MJ	26.65	
IRON	MILLI	1.77		0.90	- 4.08	1.77	MILLI/MJ	1.00	
PHOSPHORUS	MILLI	122.33		109.00	- 129.00	122.33	MILLI/MJ	69.38	
REDUCING SUGAR	GRAM	2.80		-	- -	2.80	GRAM/MJ	1.59	
SUCROSE	GRAM	20.00		18.00	- 22.00	20.00	GRAM/MJ	11.34	
DIETARY FIBRE, WAT.SOL.	GRAM	2.45		-	- -	2.45	GRAM/MJ	1.39	
DIETARY FIBRE, WAT.INS.	GRAM	0.88		-	- -	0.88	GRAM/MJ	0.50	

1 ESTIMATED BY THE DIFFERENCE METHOD
100 - (WATER + PROTEIN + FAT + MINERALS + TOTAL DIETARY FIBRE)

2 METHOD OF MEUSER, SUCKOW AND KULIKOWSKI ("BERLINER METHODE")

SALZSTANGEN (SALZBREZELN) ALS DAUERGEBÄCK | SALT STICKS (SALT CRACKERS, PRETZELS) | STIXI AU SEL

		PROTEIN	FAT	CARBOHYDRATES	TOTAL
ENERGY VALUE (AVERAGE) PER 100 G EDIBLE PORTION	KJOULE	171	19	1272	1463
	(KCAL)	41	4.7	304	350
AMOUNT OF DIGESTIBLE CONSTITUENTS PER 100 G	GRAM	8.01	0.45	76.00	
ENERGY VALUE (AVERAGE) OF THE DIGESTIBLE FRACTION PER 100 G EDIBLE PORTION	KJOULE	152	18	1272	1442
	(KCAL)	36	4.2	304	345

WASTE PERCENTAGE AVERAGE 0.00

CONSTITUENTS	DIM	AV		VARIATION			AVR	NUTR.	DENS.	MOLPERC.
WATER	GRAM	9.00		8.00	-	10.00	9.00	GRAM/MJ	6.24	
PROTEIN	GRAM	9.00		8.17	-	9.84	9.00	GRAM/MJ	6.24	
FAT	GRAM	0.50		-	-	-	0.50	GRAM/MJ	0.35	
AVAILABLE CARBOHYDR.	GRAM	76.00	1	-	-	-	76.00	GRAM/MJ	52.71	
MINERALS	GRAM	5.50		-	-	-	5.50	GRAM/MJ	3.81	
SODIUM	GRAM	1.79		0.78	-	3.68	1.79	GRAM/MJ	1.24	
POTASSIUM	MILLI	124.00		110.00	-	135.00	124.00	MILLI/MJ	86.00	
CALCIUM	MILLI	147.00		45.00	-	263.00	147.00	MILLI/MJ	101.95	
IRON	MILLI	0.70		-	-	-	0.70	MILLI/MJ	0.49	
VITAMIN B1	MICRO	10.00		-	-	-	10.00	MICRO/MJ	6.94	
VITAMIN B2	MICRO	40.00		-	-	-	40.00	MICRO/MJ	27.74	
NICOTINAMIDE	MILLI	0.70		-	-	-	0.70	MILLI/MJ	0.49	
VITAMIN C	-	0.00		-	-	-	0.00			

1 ESTIMATED BY THE DIFFERENCE METHOD 100 - (WATER + PROTEIN + FAT + MINERALS)

ZWIEBACK EIFREI	RUSK (CRACKER, ZWIEBACK, EGGLES)	BISCOTTES SANS OEUFS

		PROTEIN	FAT	CARBOHYDRATES	TOTAL
ENERGY VALUE (AVERAGE) PER 100 G EDIBLE PORTION	KJOULE	175	167	1224	1566
	(KCAL)	42	40	292	374
AMOUNT OF DIGESTIBLE CONSTITUENTS PER 100 G	GRAM	8.17	3.87	73.11	
ENERGY VALUE (AVERAGE) OF THE DIGESTIBLE FRACTION PER 100 G EDIBLE PORTION	KJOULE	156	151	1224	1530
	(KCAL)	37	36	292	366

WASTE PERCENTAGE AVERAGE 0.00

CONSTITUENTS	DIM	AV		VARIATION			AVR	NUTR. DENS.		MOLPERC.
WATER	GRAM	8.50		7.00	-	9.50	8.50	GRAM/MJ	5.56	
PROTEIN	GRAM	9.19		8.35	-	10.12	9.19	GRAM/MJ	6.01	
FAT	GRAM	4.30		2.00	-	8.23	4.30	GRAM/MJ	2.81	
AVAILABLE CARBOHYDR.	GRAM	73.11	1	-	-	-	73.11	GRAM/MJ	47.79	
TOTAL DIETARY FIBRE	GRAM	3.50	2	-	-	-	3.50	GRAM/MJ	2.29	
MINERALS	GRAM	1.40		-	-	-	1.40	GRAM/MJ	0.92	
SODIUM	MILLI	263.00		250.00	-	275.00	263.00	MILLI/MJ	171.91	
POTASSIUM	MILLI	160.00		150.00	-	175.00	160.00	MILLI/MJ	104.58	
MAGNESIUM	MILLI	16.00		-	-	-	16.00	MILLI/MJ	10.46	
CALCIUM	MILLI	42.00		37.00	-	47.00	42.00	MILLI/MJ	27.45	
IRON	MILLI	1.50		-	-	-	1.50	MILLI/MJ	0.98	
PHOSPHORUS	MILLI	132.00		120.00	-	138.00	132.00	MILLI/MJ	86.28	
NICOTINAMIDE	MILLI	1.30		1.09	-	1.56	1.30	MILLI/MJ	0.85	
VITAMIN B6	MICRO	90.00		83.00	-	100.00	90.00	MICRO/MJ	58.83	
ALANINE	GRAM	0.37		-	-	-	0.37	GRAM/MJ	0.24	5.1
ARGININE	GRAM	0.42		-	-	-	0.42	GRAM/MJ	0.27	3.0
ASPARTIC ACID	GRAM	0.81		-	-	-	0.81	GRAM/MJ	0.53	7.5
CYSTINE	GRAM	0.20		-	-	-	0.20	GRAM/MJ	0.13	1.0
GLUTAMIC ACID	GRAM	3.04		-	-	-	3.04	GRAM/MJ	1.99	25.3
GLYCINE	GRAM	0.49		-	-	-	0.49	GRAM/MJ	0.32	8.0
HISTIDINE	GRAM	0.16		-	-	-	0.16	GRAM/MJ	0.10	1.3
ISOLEUCINE	GRAM	0.49		-	-	-	0.49	GRAM/MJ	0.32	4.6
LEUCINE	GRAM	0.82		-	-	-	0.82	GRAM/MJ	0.54	7.7
LYSINE	GRAM	0.21		-	-	-	0.21	GRAM/MJ	0.14	1.8
METHIONINE	GRAM	0.14		-	-	-	0.14	GRAM/MJ	0.09	1.2

1 ESTIMATED BY THE DIFFERENCE METHOD
100 - (WATER + PROTEIN + FAT + MINERALS + TOTAL DIETARY FIBRE)

2 METHOD OF MEUSER, SUCKOW AND KULIKOWSKI ("BERLINER METHODE")

CONSTITUENTS	DIM	AV	VARIATION			AVR	NUTR. DENS.		MOLPERC.
PHENYLALANINE	GRAM	0.52	-	-	-	0.52	GRAM/MJ	0.34	3.9
PROLINE	GRAM	1.29	-	-	-	1.29	GRAM/MJ	0.84	13.7
SERINE	GRAM	0.40	-	-	-	0.40	GRAM/MJ	0.26	4.7
THREONINE	GRAM	0.32	-	-	-	0.32	GRAM/MJ	0.21	3.3
TRYPTOPHAN	GRAM	0.10	-	-	-	0.10	GRAM/MJ	0.07	0.6
TYROSINE	GRAM	0.39	-	-	-	0.39	GRAM/MJ	0.25	2.6
VALINE	GRAM	0.47	-	-	-	0.47	GRAM/MJ	0.31	4.9
STARCH	GRAM	53.70	-	-	-	53.70	GRAM/MJ	35.10	
DIETARY FIBRE,WAT.SOL.	GRAM	2.00	-	-	-	2.00	GRAM/MJ	1.31	
DIETARY FIBRE,WAT.INS.	GRAM	1.50	-	-	-	1.50	GRAM/MJ	0.98	

Stärken

STARCHES AMIDON

KARTOFFELSTÄRKE POTATO STARCH FÉCULE DE POMME DE TERRE

		PROTEIN	FAT	CARBOHYDRATES	TOTAL
ENERGY VALUE (AVERAGE)	KJOULE	8.9	3.9	1391	1404
PER 100 G	(KCAL)	2.1	0.9	332	335
EDIBLE PORTION					
AMOUNT OF DIGESTIBLE	GRAM	0.42	0.09	83.10	
CONSTITUENTS PER 100 G					
ENERGY VALUE (AVERAGE)	KJOULE	6.6	3.5	1391	1401
OF THE DIGESTIBLE	(KCAL)	1.6	0.8	332	335
FRACTION PER 100 G					
EDIBLE PORTION					

WASTE PERCENTAGE AVERAGE 0.00

CONSTITUENTS	DIM	AV	VARIATION			AVR	NUTR. DENS.		MOLPERC.
WATER	GRAM	15.50	12.00	-	17.80	15.50	GRAM/MJ	11.06	
PROTEIN	GRAM	0.57	0.08	-	0.90	0.57	GRAM/MJ	0.41	
FAT	GRAM	0.10	0.05	-	0.20	0.10	GRAM/MJ	0.07	
AVAILABLE CARBOHYDR.	GRAM	83.10	-	-	-	83.10	GRAM/MJ	59.32	
MINERALS	GRAM	0.39	0.24	-	0.57	0.39	GRAM/MJ	0.28	
SODIUM	MILLI	7.60	5.00	-	10.00	7.60	MILLI/MJ	5.43	
POTASSIUM	MILLI	15.00	-	-	-	15.00	MILLI/MJ	10.71	
MAGNESIUM	MILLI	6.00	6.00	-	6.00	6.00	MILLI/MJ	4.28	
CALCIUM	MILLI	35.00	30.00	-	41.00	35.00	MILLI/MJ	24.98	
MANGANESE	MILLI	0.10	0.10	-	0.10	0.10	MILLI/MJ	0.07	
IRON	MILLI	1.80	-	-	-	1.80	MILLI/MJ	1.28	
COPPER	MILLI	0.13	0.13	-	0.13	0.13	MILLI/MJ	0.09	
ZINC	MILLI	0.15	0.15	-	0.15	0.15	MILLI/MJ	0.11	
PHOSPHORUS	MILLI	6.67	6.11	-	7.70	6.67	MILLI/MJ	4.76	
CHLORIDE	MILLI	4.75	4.50	-	5.00	4.75	MILLI/MJ	3.39	
IODIDE	MICRO	1.00	1.00	-	1.00	1.00	MICRO/MJ	0.71	
VITAMIN B6	MICRO	10.00	10.00	-	10.00	10.00	MICRO/MJ	7.14	
VITAMIN C	-	0.00	-	-	-	0.00			
STARCH	GRAM	83.10	80.90	-	85.40	83.10	GRAM/MJ	59.32	

MAISSTÄRKE CORN STARCH AMIDON DE MAÏS

		PROTEIN	FAT	CARBOHYDRATES	TOTAL
ENERGY VALUE (AVERAGE)	KJOULE	7.6	3.1	1438	1448
PER 100 G	(KCAL)	1.8	0.7	344	346
EDIBLE PORTION					
AMOUNT OF DIGESTIBLE	GRAM	0.24	0.07	85.90	
CONSTITUENTS PER 100 G					
ENERGY VALUE (AVERAGE)	KJOULE	4.6	2.8	1438	1445
OF THE DIGESTIBLE	(KCAL)	1.1	0.7	344	345
FRACTION PER 100 G					
EDIBLE PORTION					

WASTE PERCENTAGE AVERAGE 0.00

CONSTITUENTS	DIM	AV	VARIATION			AVR	NUTR.	DENS.	MOLPERC.
WATER	GRAM	12.60	11.00	-	14.00	12.60	GRAM/MJ	8.72	
PROTEIN	GRAM	0.40	0.24	-	0.63	0.40	GRAM/MJ	0.28	
FAT	MILLI	80.00	0.00	-	200.00	80.00	MILLI/MJ	55.36	
AVAILABLE CARBOHYDR.	GRAM	85.90	-	-	-	85.90	GRAM/MJ	59.45	
MINERALS	GRAM	0.22	0.06	-	0.30	0.22	GRAM/MJ	0.15	
SODIUM	MILLI	3.00	2.00	-	4.00	3.00	MILLI/MJ	2.08	
POTASSIUM	MILLI	7.00	3.00	-	10.00	7.00	MILLI/MJ	4.84	
MAGNESIUM	MILLI	2.00	-	-	-	2.00	MILLI/MJ	1.38	
MANGANESE	MILLI	1.00	1.00	-	1.00	1.00	MILLI/MJ	0.69	
IRON	MILLI	0.50	-	-	-	0.50	MILLI/MJ	0.35	
PHOSPHORUS	MILLI	30.00	-	-	-	30.00	MILLI/MJ	20.76	
CHLORIDE	MILLI	6.00	-	-	-	6.00	MILLI/MJ	4.15	
IODIDE	MICRO	2.50	-	-	-	2.50	MICRO/MJ	1.73	
VITAMIN B1	MICRO	-	0.00	-	1.00				
VITAMIN B2	MICRO	8.00	4.00	-	12.00	8.00	MICRO/MJ	5.54	
NICOTINAMIDE	MICRO	30.00	20.00	-	50.00	30.00	MICRO/MJ	20.76	
VITAMIN B6	MICRO	5.00	-	-	-	5.00	MICRO/MJ	3.46	
VITAMIN C	-	0.00	-	-	-	0.00			
STARCH	GRAM	85.90	-	-	-	85.90	GRAM/MJ	59.45	

REISSTÄRKE RICE STARCH AMIDON DE RIZ

		PROTEIN	FAT	CARBOHYDRATES	TOTAL
ENERGY VALUE (AVERAGE) PER 100 G EDIBLE PORTION	KJOULE (KCAL)	14 3.5	0.0 0.0	1423 340	1437 343
AMOUNT OF DIGESTIBLE CONSTITUENTS PER 100 G	GRAM	0.63	0.00	85.00	
ENERGY VALUE (AVERAGE) OF THE DIGESTIBLE FRACTION PER 100 G EDIBLE PORTION	KJOULE (KCAL)	12 2.9	0.0 0.0	1423 340	1435 343

WASTE PERCENTAGE AVERAGE 0.00

CONSTITUENTS	DIM	AV	VARIATION	AVR	NUTR. DENS.		MOLPERC.
WATER	GRAM	13.80	13.60 - 14.00	13.80	GRAM/MJ	9.62	
PROTEIN	GRAM	0.76	- - -	0.76	GRAM/MJ	0.53	
FAT	-	0.00	0.00 - 1.00	0.00			
AVAILABLE CARBOHYDR.	GRAM	85.00	84.00 - 85.00	85.00	GRAM/MJ	59.25	
MINERALS	GRAM	0.44	0.30 - 0.50	0.44	GRAM/MJ	0.31	
SODIUM	MILLI	61.00	- - -	61.00	MILLI/MJ	42.52	
POTASSIUM	MILLI	8.00	- - -	8.00	MILLI/MJ	5.58	
MAGNESIUM	MILLI	20.00	- - -	20.00	MILLI/MJ	13.94	
CALCIUM	MILLI	20.00	17.00 - 24.00	20.00	MILLI/MJ	13.94	
STARCH	GRAM	85.00	- - -	85.00	GRAM/MJ	59.25	

TAPIOKASTÄRKE TAPIOCA STARCH AMIDON DE TAPIOCA

		PROTEIN	FAT	CARBOHYDRATES	TOTAL
ENERGY VALUE (AVERAGE)	KJOULE	9.1	9.3	1421	1439
PER 100 G	(KCAL)	2.2	2.2	340	344
EDIBLE PORTION					
AMOUNT OF DIGESTIBLE	GRAM	0.42	0.21	84.90	
CONSTITUENTS PER 100 G					
ENERGY VALUE (AVERAGE)	KJOULE	6.7	8.4	1421	1436
OF THE DIGESTIBLE	(KCAL)	1.6	2.0	340	343
FRACTION PER 100 G					
EDIBLE PORTION					

WASTE PERCENTAGE AVERAGE 0.00

CONSTITUENTS	DIM	AV	VARIATION			AVR	NUTR. DENS.		MOLPERC.
WATER	GRAM	12.60	11.00	-	15.00	12.60	GRAM/MJ	8.77	
PROTEIN	GRAM	0.58	0.05	-	0.87	0.58	GRAM/MJ	0.40	
FAT	GRAM	0.24	0.20	-	0.30	0.24	GRAM/MJ	0.17	
AVAILABLE CARBOHYDR.	GRAM	84.90	-	-	-	84.90	GRAM/MJ	59.12	
MINERALS	GRAM	0.29	0.20	-	0.38	0.29	GRAM/MJ	0.20	
SODIUM	MILLI	4.00	-	-	-	4.00	MILLI/MJ	2.79	
POTASSIUM	MILLI	20.00	-	-	-	20.00	MILLI/MJ	13.93	
MAGNESIUM	MILLI	3.00	2.00	-	3.00	3.00	MILLI/MJ	2.09	
CALCIUM	MILLI	12.00	-	-	-	12.00	MILLI/MJ	8.36	
MANGANESE	MILLI	-	0.20	-	1.00				
IRON	MILLI	1.00	-	-	-	1.00	MILLI/MJ	0.70	
COPPER	MILLI	-	0.05	-	0.20				
PHOSPHORUS	MILLI	12.00	7.00	-	27.00	12.00	MILLI/MJ	8.36	
VITAMIN B1	-	0.00	-	-	-	0.00			
VITAMIN B2	MILLI	0.10	-	-	-	0.10	MILLI/MJ	0.07	
NICOTINAMIDE	-	0.00	-	-	-	0.00			
VITAMIN C	-	0.00	-	-	-	0.00			
STARCH	GRAM	84.90	-	-	-	84.90	GRAM/MJ	59.12	

WEIZENSTÄRKE WHEAT STARCH AMIDON DE BLÉ

		PROTEIN	FAT	CARBOHYDRATES	TOTAL
ENERGY VALUE (AVERAGE) PER 100 G EDIBLE PORTION	KJOULE	7.6	5.4	1441	1454
	(KCAL)	1.8	1.3	344	348
AMOUNT OF DIGESTIBLE CONSTITUENTS PER 100 G	GRAM	0.35	0.12	86.10	
ENERGY VALUE (AVERAGE) OF THE DIGESTIBLE FRACTION PER 100 G EDIBLE PORTION	KJOULE	6.8	4.9	1441	1453
	(KCAL)	1.6	1.2	344	347

WASTE PERCENTAGE AVERAGE 0.00

CONSTITUENTS	DIM	AV	VARIATION			AVR	NUTR. DENS.		MOLPERC.
WATER	GRAM	12.30	10.20	-	14.20	12.30	GRAM/MJ	8.47	
PROTEIN	GRAM	0.40	0.21	-	0.50	0.40	GRAM/MJ	0.28	
FAT	GRAM	0.14	0.03	-	0.20	0.14	GRAM/MJ	0.10	
AVAILABLE CARBOHYDR.	GRAM	86.10	-	-	-	86.10	GRAM/MJ	59.27	
MINERALS	GRAM	0.20	0.10	-	0.30	0.20	GRAM/MJ	0.14	
SODIUM	MILLI	2.00	-	-	-	2.00	MILLI/MJ	1.38	
POTASSIUM	MILLI	16.00	-	-	-	16.00	MILLI/MJ	11.01	
CALCIUM	-	0.00	-	-	-	0.00			
IRON	-	0.00	-	-	-	0.00			
VITAMIN B1	-	0.00	-	-	-	0.00			
VITAMIN B2	-	0.00	-	-	-	0.00			
PANTOTHENIC ACID	-	0.00	-	-	-	0.00			
VITAMIN C	-	0.00	-	-	-	0.00			
STARCH	GRAM	82.50	-	-	-	82.50	GRAM/MJ	56.79	

Gemüse

Vegetables
Légumes

Wurzel- und Knollengemüse

ROOTS AND TUBERS **RACINES ET TUBERCULES**

BATATE (SÜSSKARTOFFEL) — SWEET POTATO — PATATE DOUCE

IPOMOEA BATATAS POIR.

		PROTEIN	FAT	CARBOHYDRATES	TOTAL
ENERGY VALUE (AVERAGE)	KJOULE	26	23	524	573
PER 100 G EDIBLE PORTION	(KCAL)	6.1	5.6	125	137
AMOUNT OF DIGESTIBLE CONSTITUENTS PER 100 G	GRAM	1.20	0.54	31.32	
ENERGY VALUE (AVERAGE)	KJOULE	19	21	524	564
OF THE DIGESTIBLE FRACTION PER 100 G EDIBLE PORTION	(KCAL)	4.5	5.0	125	135

WASTE PERCENTAGE AVERAGE 19 MINIMUM 14 MAXIMUM 23

CONSTITUENTS	DIM	AV	VARIATION		AVR	NUTR. DENS.		MOLPERC.
WATER	GRAM	69.20	66.10	71.10	56.05	GRAM/MJ	122.67	
PROTEIN	GRAM	1.63	1.57	2.08	1.32	GRAM/MJ	2.89	
FAT	GRAM	0.60	0.40	1.00	0.49	GRAM/MJ	1.06	
AVAILABLE CARBOHYDR.	GRAM	31.32	-	-	25.37	GRAM/MJ	55.52	
TOTAL DIETARY FIBRE	GRAM	7.80 1	-	-	6.32	GRAM/MJ	13.83	
MINERALS	GRAM	1.12	1.03	1.19	0.91	GRAM/MJ	1.99	
SODIUM	MILLI	4.00	-	-	3.24	MILLI/MJ	7.09	
POTASSIUM	MILLI	413.00	229.00	530.00	334.53	MILLI/MJ	732.13	
MAGNESIUM	MILLI	25.00	15.00	28.00	20.25	MILLI/MJ	44.32	
CALCIUM	MILLI	35.00	30.00	49.00	28.35	MILLI/MJ	62.04	
IRON	MILLI	0.85	0.70	1.00	0.69	MILLI/MJ	1.51	
COPPER	MILLI	0.16	-	-	0.13	MILLI/MJ	0.28	
PHOSPHORUS	MILLI	45.00	21.00	65.00	36.45	MILLI/MJ	79.77	
CHLORIDE	MILLI	31.00	-	-	25.11	MILLI/MJ	54.95	
IODIDE	MICRO	2.40	-	-	1.94	MICRO/MJ	4.25	
SELENIUM	MICRO	0.65	0.60	0.70	0.53	MICRO/MJ	1.15	

1 METHOD OF SOUTHGATE

CONSTITUENTS	DIM	AV	VARIATION			AVR	NUTR. DENS.		MOLPERC.
CAROTENE	MILLI	-	0.30	-	4.62 2				
VITAMIN B1	MICRO	64.00	30.00	-	100.00	51.84	MICRO/MJ	113.45	
VITAMIN B2	MICRO	50.00	40.00	-	55.00	40.50	MICRO/MJ	88.64	
NICOTINAMIDE	MILLI	0.60	-	-	-	0.49	MILLI/MJ	1.06	
PANTOTHENIC ACID	MILLI	0.83	0.79	-	0.87	0.67	MILLI/MJ	1.47	
VITAMIN B6	MILLI	0.27	-	-	-	0.22	MILLI/MJ	0.48	
BIOTIN	MICRO	4.30	-	-	-	3.48	MICRO/MJ	7.62	
FOLIC ACID	MICRO	12.00	6.00	-	19.00	9.72	MICRO/MJ	21.27	
VITAMIN C	MILLI	30.00	7.00	-	58.00	24.30	MILLI/MJ	53.18	
ARGININE	MILLI	65.00	47.00	-	82.00	52.65	MILLI/MJ	115.23	
CYSTINE	MILLI	25.00	-	-	-	20.25	MILLI/MJ	44.32	
HISTIDINE	MILLI	29.00	21.00	-	36.00	23.49	MILLI/MJ	51.41	
ISOLEUCINE	MILLI	68.00	59.00	-	76.00	55.08	MILLI/MJ	120.54	
LEUCINE	MILLI	84.00	78.00	-	90.00	68.04	MILLI/MJ	148.91	
LYSINE	MILLI	66.00	58.00	-	74.00	53.46	MILLI/MJ	117.00	
METHIONINE	MILLI	28.00	-	-	-	22.68	MILLI/MJ	49.64	
PHENYLALANINE	MILLI	79.00	70.00	-	87.00	63.99	MILLI/MJ	140.04	
THREONINE	MILLI	68.00	62.00	-	74.00	55.08	MILLI/MJ	120.54	
TRYPTOPHAN	MILLI	28.00	27.00	-	29.00	22.68	MILLI/MJ	49.64	
TYROSINE	MILLI	71.00	-	-	-	57.51	MILLI/MJ	125.86	
VALINE	GRAM	0.11	0.09	-	0.12	0.09	GRAM/MJ	0.19	
SALICYLIC ACID	MILLI	0.49	0.49	-	0.49	0.40	MILLI/MJ	0.87	
GLUCOSE	GRAM	0.70	0.45	-	0.95	0.57	GRAM/MJ	1.24	
FRUCTOSE	GRAM	0.41	0.41	-	0.41	0.33	GRAM/MJ	0.73	
SUCROSE	GRAM	10.75	4.61	-	16.90	8.71	GRAM/MJ	19.07	
STARCH	GRAM	19.46	19.46	-	19.46	15.76	GRAM/MJ	34.50	
CELLULOSE	GRAM	1.90	-	-	-	1.54	GRAM/MJ	3.37	
DIETARY FIBRE,WAT.SOL.	GRAM	4.40	-	-	-	3.56	GRAM/MJ	7.80	
DIETARY FIBRE,WAT.INS.	GRAM	3.40	-	-	-	2.75	GRAM/MJ	6.03	

2 GREAT VARIATION: LOW VALUE VALID FOR THE PALE-YELLOW VARIETY, HIGH VALUE VALID FOR THE DARK YELLOW VARIETY

CASSAVE	**CASSAVA**	**MANIOC**
KNOLLEN (MANIOK, TAPIOKA) MANIHOT UTILISSIMA, POHL 1	TUBER 1	TUBERCULE 1

		PROTEIN	FAT	CARBOHYDRATES	TOTAL
ENERGY VALUE (AVERAGE)	KJOULE	16	8.9	537	561
PER 100 G	(KCAL)	3.8	2.1	128	134
EDIBLE PORTION					
AMOUNT OF DIGESTIBLE	GRAM	0.74	0.20	32.07	
CONSTITUENTS PER 100 G					
ENERGY VALUE (AVERAGE)	KJOULE	12	8.1	537	556
OF THE DIGESTIBLE	(KCAL)	2.8	1.9	128	133
FRACTION PER 100 G					
EDIBLE PORTION					

WASTE PERCENTAGE AVERAGE 26 MINIMUM 16 MAXIMUM 40

CONSTITUENTS	DIM	AV	VARIATION			AVR	NUTR. DENS.		MOLPERC.
WATER	GRAM	63.10	60.00	-	65.50	46.69	GRAM/MJ	113.41	
PROTEIN	GRAM	1.00	0.70	-	1.20	0.74	GRAM/MJ	1.80	
FAT	GRAM	0.23	0.20	-	0.35	0.17	GRAM/MJ	0.41	
AVAILABLE CARBOHYDR.	GRAM	32.07 2	-	-	-	23.73	GRAM/MJ	57.64	
TOTAL DIETARY FIBRE	GRAM	2.90 3	-	-	-	2.15	GRAM/MJ	5.21	
MINERALS	GRAM	0.70	0.40	-	0.90	0.52	GRAM/MJ	1.26	
SODIUM	MILLI	2.00	-	-	-	1.48	MILLI/MJ	3.59	
POTASSIUM	MILLI	394.00	-	-	-	291.56	MILLI/MJ	708.14	
CALCIUM	MILLI	37.00	25.00	-	68.00	27.38	MILLI/MJ	66.50	
IRON	MILLI	1.20	0.80	-	1.90	0.89	MILLI/MJ	2.16	
PHOSPHORUS	MILLI	38.00	32.00	-	42.00	28.12	MILLI/MJ	68.30	
CAROTENE	MICRO	30.00	5.00	-	35.00	22.20	MICRO/MJ	53.92	
VITAMIN B1	MICRO	60.00	40.00	-	70.00	44.40	MICRO/MJ	107.84	
VITAMIN B2	MICRO	30.00	20.00	-	50.00	22.20	MICRO/MJ	53.92	
NICOTINAMIDE	MILLI	0.60	-	-	-	0.44	MILLI/MJ	1.08	
VITAMIN C	MILLI	30.00	25.00	-	34.00	22.20	MILLI/MJ	53.92	
DIETARY FIBRE, WAT.SOL.	GRAM	0.69	-	-	-	0.51	GRAM/MJ	1.24	
DIETARY FIBRE, WAT.INS.	GRAM	2.20	-	-	-	1.63	GRAM/MJ	3.95	

1 THE RAW PRODUCT IS POISONOUS DUE TO THE PRESENCE OF LINAMARIN (= PHASEOLUNATINE), A HYDROCYANIC CONTAINING GLYCOSIDE (55 MG/100G WHOLE ROOT, 250 MG/100G ROOT CORTEX).

2 ESTIMATED BY THE DIFFERENCE METHOD 100 - (WATER + PROTEIN + FAT + MINERALS + TOTAL DIETARY FIBRE)

3 METHOD OF SOUTHGATE

KARTOFFEL POTATO POMME DE TERRE

SOLANUM TUBEROSUM L.

		PROTEIN	FAT	CARBOHYDRATES	TOTAL
ENERGY VALUE (AVERAGE)	KJOULE	32	4.3	258	294
PER 100 G	(KCAL)	7.7	1.0	62	70
EDIBLE PORTION					
AMOUNT OF DIGESTIBLE	GRAM	1.50	0.09	15.40	
CONSTITUENTS PER 100 G					
ENERGY VALUE (AVERAGE)	KJOULE	24	3.9	258	285
OF THE DIGESTIBLE	(KCAL)	5.7	0.9	62	68
FRACTION PER 100 G					
EDIBLE PORTION					

WASTE PERCENTAGE AVERAGE 20 MINIMUM 15 MAXIMUM 25

CONSTITUENTS	DIM	AV	VARIATION		AVR	NUTR. DENS.		MOLPERC.
WATER	GRAM	77.80	73.80	- 81.80	62.24	GRAM/MJ	272.72	
PROTEIN	GRAM	2.04	1.42	- 2.93	1.63	GRAM/MJ	7.15	
FAT	GRAM	0.11	0.04	- 0.17	0.09	GRAM/MJ	0.39	
AVAILABLE CARBOHYDR.	GRAM	15.40	-	- -	12.32	GRAM/MJ	53.98	
TOTAL DIETARY FIBRE	GRAM	2.51 [1]	1.00	- 2.60	2.01	GRAM/MJ	8.80	
MINERALS	GRAM	1.02	0.60	- 1.30	0.82	GRAM/MJ	3.58	

CONSTITUENTS	DIM	AV	VARIATION		AVR	NUTR. DENS.	
SODIUM	MILLI	3.20	0.80	- 5.20	2.56	MILLI/MJ	11.22
POTASSIUM	MILLI	443.00	340.00	- 600.00	354.40	GRAM/MJ	1.55
MAGNESIUM	MILLI	25.00	17.00	- 32.00	20.00	MILLI/MJ	87.64
CALCIUM	MILLI	9.50	6.40	- 14.00	7.60	MILLI/MJ	33.30
MANGANESE	MILLI	0.15	0.10	- 0.25	0.12	MILLI/MJ	0.53
IRON	MILLI	0.80	0.44	- 1.50	0.64	MILLI/MJ	2.80
COBALT	MICRO	1.30	0.80	- 1.60	1.04	MICRO/MJ	4.56
COPPER	MILLI	0.15	0.08	- 0.23	0.12	MILLI/MJ	0.53
ZINC	MILLI	0.27	0.12	- 0.49	0.22	MILLI/MJ	0.95
NICKEL	MICRO	26.00	5.00	- 56.00	20.80	MICRO/MJ	91.14
CHROMIUM	MICRO	33.00	0.00	- 65.00	26.40	MICRO/MJ	115.68
MOLYBDENUM	MICRO	-	5.00	- 86.00			
VANADIUM	MICRO	-	1.00	- 149.00 [2]			
PHOSPHORUS	MILLI	50.00	35.00	- 79.00	40.00	MILLI/MJ	175.27
CHLORIDE	MILLI	45.00	26.00	- 59.00	36.00	MILLI/MJ	157.74
FLUORIDE	MICRO	9.70	4.00	- 21.00	7.76	MICRO/MJ	34.00
IODIDE	MICRO	3.80	2.30	- 5.80	3.04	MICRO/MJ	13.32
BORON	MILLI	0.10	0.05	- 0.15	0.08	MILLI/MJ	0.35
SELENIUM	MICRO	-	4.00	- 20.00			
NITRATE	MILLI	13.00 [3]	0.00	- 100.00	10.40	MILLI/MJ	45.57

1 METHOD OF SOUTHGATE AND ENGLYST

2 GREAT REGIONAL DIFFERENCES:
E.G. USSR: 1 UG/100G; USA: 150 UG/100G

3 STONG DEPENDENCE ON ENVIROMENTAL CONDITIONS

CONSTITUENTS	DIM	AV	VARIATION			AVR	NUTR. DENS.		MOLPERC.
CAROTENE	MICRO	10.00	2.00	-	20.00	8.00	MICRO/MJ	35.05	
VITAMIN E ACTIVITY	MICRO	60.00	-	-	-	48.00	MICRO/MJ	210.33	
ALPHA-TOCOPHEROL	MICRO	60.00	-	-	-	48.00	MICRO/MJ	210.33	
VITAMIN K	MICRO	50.00	20.00	-	80.00	40.00	MICRO/MJ	175.27	
VITAMIN B1	MILLI	0.11	0.07	-	0.14	0.09	MILLI/MJ	0.39	
VITAMIN B2	MICRO	47.00	30.00	-	75.00	37.60	MICRO/MJ	164.75	
NICOTINAMIDE	MILLI	1.22	0.80	-	1.60	0.98	MILLI/MJ	4.28	
PANTOTHENIC ACID	MILLI	0.40	0.30	-	0.50	0.32	MILLI/MJ	1.40	
VITAMIN B6	MILLI	0.21	0.14	-	0.27	0.17	MILLI/MJ	0.74	
BIOTIN	MICRO	0.40	0.10	-	0.60	0.32	MICRO/MJ	1.40	
FOLIC ACID	MICRO	7.00	4.00	-	19.00	5.60	MICRO/MJ	24.54	
VITAMIN B12	-	0.00	-	-	-	0.00			
VITAMIN C	MILLI	17.00	10.00	-	40.00	13.60	MILLI/MJ	59.59	
ALANINE	GRAM	0.11	-	-	-	0.09	GRAM/MJ	0.39	6.8
ARGININE	GRAM	0.12	-	-	-	0.10	GRAM/MJ	0.42	3.8
ASPARTIC ACID	GRAM	0.43	0.31	-	0.50	0.34	GRAM/MJ	1.51	17.8
CYSTINE	MILLI	20.00	-	-	-	16.00	MILLI/MJ	70.11	0.5
GLUTAMIC ACID	GRAM	0.46	0.41	-	0.48	0.37	GRAM/MJ	1.61	17.3
GLYCINE	GRAM	0.12	0.09	-	0.13	0.10	GRAM/MJ	0.42	8.8
HISTIDINE	MILLI	40.00	30.00	-	50.00	32.00	MILLI/MJ	140.22	1.4
ISOLEUCINE	GRAM	0.10	0.09	-	0.12	0.08	GRAM/MJ	0.35	4.2
LEUCINE	GRAM	0.14	0.12	-	0.16	0.11	GRAM/MJ	0.49	5.9
LYSINE	GRAM	0.13	0.11	-	0.14	0.10	GRAM/MJ	0.46	4.9
METHIONINE	MILLI	30.00	-	-	-	24.00	MILLI/MJ	105.16	1.1
PHENYLALANINE	GRAM	0.10	0.09	-	0.11	0.08	GRAM/MJ	0.35	3.3
PROLINE	GRAM	0.11	0.08	-	0.12	0.09	GRAM/MJ	0.39	5.3
SERINE	GRAM	0.10	0.08	-	0.11	0.08	GRAM/MJ	0.35	5.3
THREONINE	MILLI	90.00	80.00	-	110.00	72.00	MILLI/MJ	315.49	4.2
TRYPTOPHAN	MILLI	30.00	20.00	-	40.00	24.00	MILLI/MJ	105.16	0.8
TYROSINE	MILLI	80.00	70.00	-	90.00	64.00	MILLI/MJ	280.43	2.4
VALINE	GRAM	0.13	0.12	-	0.14	0.10	GRAM/MJ	0.46	6.1
MALIC ACID	MILLI	91.60	79.50	-	103.80	73.28	MILLI/MJ	321.10	
CITRIC ACID	MILLI	520.00	142.00	-	650.00	416.00	GRAM/MJ	1.82	
OXALIC ACID TOTAL	-	0.00	-	-	-	0.00			
SUCCINIC ACID	MILLI	3.70	-	-	-	2.96	MILLI/MJ	12.97	
GLUCONIC ACID	MILLI	5.40	4.70	-	6.20	4.32	MILLI/MJ	18.93	
SALICYLIC ACID	MILLI	0.12	0.12	-	0.12	0.10	MILLI/MJ	0.42	
GLUCOSE	GRAM	0.24	0.03	-	0.70	0.19	GRAM/MJ	0.84	
FRUCTOSE	GRAM	0.17	0.03	-	0.53	0.14	GRAM/MJ	0.60	
SUCROSE	GRAM	0.30	0.01	-	0.80	0.24	GRAM/MJ	1.05	
STARCH	GRAM	14.10 [4]	11.60	-	20.00	11.28	GRAM/MJ	49.43	
PENTOSAN	GRAM	0.26	-	-	-	0.21	GRAM/MJ	0.91	
HEXOSAN	GRAM	0.93	-	-	-	0.74	GRAM/MJ	3.26	
CELLULOSE	GRAM	0.89	-	-	-	0.71	GRAM/MJ	3.12	
POLYURONIC ACID	GRAM	0.46	-	-	-	0.37	GRAM/MJ	1.61	
MYRISTIC ACID	MILLI	0.40	-	-	-	0.32	MILLI/MJ	1.40	
PALMITIC ACID	MILLI	18.87	18.00	-	19.00	15.10	MILLI/MJ	66.16	
STEARIC ACID	MILLI	4.28	3.50	-	4.40	3.43	MILLI/MJ	15.03	
ARACHIDIC ACID	MILLI	1.10	-	-	-	0.88	MILLI/MJ	3.86	
BEHENIC ACID	MILLI	0.70	-	-	-	0.56	MILLI/MJ	2.45	
PALMITOLEIC ACID	MILLI	0.51	0.30	-	2.00	0.41	MILLI/MJ	1.80	
OLEIC ACID	MILLI	1.75	1.50	-	3.50	1.40	MILLI/MJ	6.13	
LINOLEIC ACID	MILLI	32.12	26.00	-	33.00	25.70	MILLI/MJ	112.61	
LINOLENIC ACID	MILLI	22.75	21.00	-	35.00	18.20	MILLI/MJ	79.75	

4 MEAN VALUE: VALID FOR THE EDIBLE VARIETIES
STARCH CONTENT HIGHER THAN 15 %: VARIETIES FOR INDUSTRIAL
APPLICATION

CONSTITUENTS	DIM	AV	VARIATION			AVR	NUTR. DENS.		MOLPERC.
TOTAL STEROLS	MILLI	5.00	-	-	-	4.00	MILLI/MJ	17.53	
BETA-SITOSTEROL	MILLI	3.00	-	-	-	2.40	MILLI/MJ	10.52	
STIGMASTEROL	MILLI	1.00	-	-	-	0.80	MILLI/MJ	3.51	
TOTAL PURINES	MILLI	5.00 5	-	-	-	4.00	MILLI/MJ	17.53	
TOTAL PHOSPHOLIPIDS	MILLI	56.00	-	-	-	44.80	MILLI/MJ	196.30	
PHOSPHATIDYLCHOLINE	MILLI	28.00	-	-	-	22.40	MILLI/MJ	98.15	
PHOSPHATIDYLETHANOLAMINE	MILLI	16.00	-	-	-	12.80	MILLI/MJ	56.09	
PHOSPHATIDYLSERINE	MILLI	1.00	-	-	-	0.80	MILLI/MJ	3.51	
PHOSPHATIDYLINOSITOL	MILLI	9.00	-	-	-	7.20	MILLI/MJ	31.55	
DIETARY FIBRE, WAT.SOL.	GRAM	0.61	-	-	-	0.49	GRAM/MJ	2.14	
DIETARY FIBRE, WAT.INS.	GRAM	1.90	-	-	-	1.52	GRAM/MJ	6.66	

5 VALUE EXPRESSED IN MG URIC ACID/100G

KARTOFFELFLOCKEN (KARTOFFELPÜREE) TROCKENPRODUKT — POTATO FLAKES DRIED — FLOCONS DE POMMES DE TERRE (PURÉE) PRODUIT SÉCHÉ

		PROTEIN	FAT	CARBOHYDRATES	TOTAL
ENERGY VALUE (AVERAGE)	KJOULE	135	22	1188	1345
PER 100 G	(KCAL)	32	5.2	284	321
EDIBLE PORTION					
AMOUNT OF DIGESTIBLE	GRAM	6.36	0.50	71.00	
CONSTITUENTS PER 100 G					
ENERGY VALUE (AVERAGE)	KJOULE	100	20	1188	1308
OF THE DIGESTIBLE	(KCAL)	24	4.7	284	313
FRACTION PER 100 G					
EDIBLE PORTION					

WASTE PERCENTAGE AVERAGE 0.00

CONSTITUENTS	DIM	AV	VARIATION			AVR	NUTR. DENS.		MOLPERC.
WATER	GRAM	6.95	5.63	-	8.20	6.95	GRAM/MJ	5.31	
PROTEIN	GRAM	8.60	6.96	-	9.80	8.60	GRAM/MJ	6.58	
FAT	GRAM	0.56	0.26	-	0.80	0.56	GRAM/MJ	0.43	
AVAILABLE CARBOHYDR.	GRAM	71.00	-	-	-	71.00	GRAM/MJ	54.29	
MINERALS	GRAM	3.01	2.43	-	3.80	3.01	GRAM/MJ	2.30	
SODIUM	MILLI	160.00	120.00	-	200.00	160.00	MILLI/MJ	122.35	
POTASSIUM	GRAM	1.15	0.98	-	1.44	1.15	GRAM/MJ	0.88	
CALCIUM	MILLI	30.00	-	-	-	30.00	MILLI/MJ	22.94	
PHOSPHORUS	MILLI	310.00	244.00	-	349.00	310.00	MILLI/MJ	237.05	
ALANINE	GRAM	0.35	0.31	-	0.39	0.35	GRAM/MJ	0.27	5.3
ARGININE	GRAM	0.61	0.53	-	0.73	0.61	GRAM/MJ	0.47	4.7
ASPARTIC ACID	GRAM	1.83	1.61	-	2.11	1.83	GRAM/MJ	1.40	18.4
CYSTINE	GRAM	0.14	0.09	-	0.18	0.14	GRAM/MJ	0.11	0.8
GLUTAMIC ACID	GRAM	1.89	1.57	-	2.12	1.89	GRAM/MJ	1.45	17.2
GLYCINE	GRAM	0.34	0.26	-	0.39	0.34	GRAM/MJ	0.26	6.1
HISTIDINE	GRAM	0.29	0.26	-	0.35	0.29	GRAM/MJ	0.22	2.5
ISOLEUCINE	GRAM	0.33	0.26	-	0.35	0.33	GRAM/MJ	0.25	3.4
LEUCINE	GRAM	0.59	0.44	-	0.69	0.59	GRAM/MJ	0.45	6.0
LYSINE	GRAM	0.69	0.61	-	0.76	0.69	GRAM/MJ	0.53	6.3
METHIONINE	MILLI	80.00	60.00	-	90.00	80.00	MILLI/MJ	61.18	0.7
PHENYLALANINE	GRAM	0.42	0.37	-	0.44	0.42	GRAM/MJ	0.32	3.4
PROLINE	GRAM	0.47	0.30	-	0.58	0.47	GRAM/MJ	0.36	5.5
SERINE	GRAM	0.46	0.45	-	0.48	0.46	GRAM/MJ	0.35	5.9
THREONINE	GRAM	0.36	0.34	-	0.40	0.36	GRAM/MJ	0.28	4.1
TRYPTOPHAN	MILLI	70.00	-	-	-	70.00	MILLI/MJ	53.53	0.5
TYROSINE	GRAM	0.50	0.33	-	0.63	0.50	GRAM/MJ	0.38	3.7
VALINE	GRAM	0.49	0.46	-	0.52	0.49	GRAM/MJ	0.37	5.6
STARCH	GRAM	70.70	-	-	-	70.70	GRAM/MJ	54.06	
DEXTRIN	GRAM	0.30	-	-	-	0.30	GRAM/MJ	0.23	
PECTIN	GRAM	0.84	-	-	-	0.84	GRAM/MJ	0.64	

KARTOFFEL	**POTATO**	**POMME DE TERRE**
GEBACKEN, MIT SCHALE	BAKED, WITH SKIN	CUITE, AVEC PELURE

		PROTEIN	FAT	CARBOHYDRATES	TOTAL
ENERGY VALUE (AVERAGE)	KJOULE	40	4.3	316	360
PER 100 G	(KCAL)	9.5	1.0	76	86
EDIBLE PORTION					
AMOUNT OF DIGESTIBLE	GRAM	1.87	0.09	18.90	
CONSTITUENTS PER 100 G					
ENERGY VALUE (AVERAGE)	KJOULE	29	3.9	316	350
OF THE DIGESTIBLE	(KCAL)	7.0	0.9	76	84
FRACTION PER 100 G					
EDIBLE PORTION					

WASTE PERCENTAGE AVERAGE 13

CONSTITUENTS	DIM	AV		VARIATION			AVR	NUTR.	DENS.	MOLPERC.
WATER	GRAM	73.20		-	-	-	63.68	GRAM/MJ	209.35	
PROTEIN	GRAM	2.54	1	-	-	-	2.21	GRAM/MJ	7.26	
FAT	GRAM	0.11	1	-	-	-	0.10	GRAM/MJ	0.31	
AVAILABLE CARBOHYDR.	GRAM	18.90	1	-	-	-	16.44	GRAM/MJ	54.05	
TOTAL DIETARY FIBRE	GRAM	3.10		-	-	-	2.70	GRAM/MJ	8.87	
MINERALS	GRAM	1.20		-	-	-	1.04	GRAM/MJ	3.43	
SODIUM	MILLI	4.00		-	-	-	3.48	MILLI/MJ	11.44	
POTASSIUM	MILLI	547.00		-	-	-	475.89	GRAM/MJ	1.56	
CALCIUM	MILLI	12.00		-	-	-	10.44	MILLI/MJ	34.32	
IRON	MILLI	0.93		-	-	-	0.81	MILLI/MJ	2.66	
PHOSPHORUS	MILLI	61.00		-	-	-	53.07	MILLI/MJ	174.46	
VITAMIN B1	MILLI	0.11		-	-	-	0.10	MILLI/MJ	0.31	
VITAMIN B2	MICRO	50.00		-	-	-	43.50	MICRO/MJ	143.00	
NICOTINAMIDE	MILLI	1.40		-	-	-	1.22	MILLI/MJ	4.00	
VITAMIN C	MILLI	17.00		-	-	-	14.79	MILLI/MJ	48.62	

1 CALCULATED FROM THE FRESH PRODUCT

KARTOFFEL / POTATO / POMMES DE TERRE

GEKOCHT, MIT SCHALE / COOKED, WITH SKIN / BOUILLIE, AVEC PELURE

		PROTEIN	FAT	CARBOHYDRATES	TOTAL
ENERGY VALUE (AVERAGE)	KJOULE	32	4.3	258	294
PER 100 G EDIBLE PORTION	(KCAL)	7.7	1.0	62	70
AMOUNT OF DIGESTIBLE CONSTITUENTS PER 100 G	GRAM	1.50	0.09	15.40	
ENERGY VALUE (AVERAGE)	KJOULE	24	3.9	258	285
OF THE DIGESTIBLE FRACTION PER 100 G EDIBLE PORTION	(KCAL)	5.7	0.9	62	68

WASTE PERCENTAGE AVERAGE 13

CONSTITUENTS	DIM	AV	VARIATION	AVR	NUTR. DENS.		MOLPERC.
WATER	GRAM	77.80	- - -	67.69	GRAM/MJ	272.72	
PROTEIN	GRAM	2.04	- - -	1.77	GRAM/MJ	7.15	
FAT	GRAM	0.11	- - -	0.10	GRAM/MJ	0.39	
AVAILABLE CARBOHYDR.	GRAM	15.40	- - -	13.40	GRAM/MJ	53.98	
TOTAL DIETARY FIBRE	GRAM	2.50	- - -	2.18	GRAM/MJ	8.76	
MINERALS	GRAM	1.00	- - -	0.87	GRAM/MJ	3.51	
SODIUM	MILLI	3.00	- - -	2.61	MILLI/MJ	10.52	
POTASSIUM	MILLI	443.00	- - -	385.41	GRAM/MJ	1.55	
CALCIUM	MILLI	10.00	- - -	8.70	MILLI/MJ	35.05	
IRON	MILLI	0.80	- - -	0.70	MILLI/MJ	2.80	
PHOSPHORUS	MILLI	50.00	- - -	43.50	MILLI/MJ	175.27	
VITAMIN B1	MILLI	0.10	- - -	0.09	MILLI/MJ	0.35	
VITAMIN B2	MICRO	50.00	- - -	43.50	MICRO/MJ	175.27	
NICOTINAMIDE	MILLI	1.20	- - -	1.04	MILLI/MJ	4.21	
VITAMIN C	MILLI	14.00	- - -	12.18	MILLI/MJ	49.08	

KARTOFFELKNÖDEL (KARTOFFELKLÖSSE) GEKOCHT TROCKENPRODUKT

POTATO-DUMPLING BOILED DRIED

BOULETTES DE POMMES DE TERRE CUITES PRODUIT SÉCHÉ

		PROTEIN	FAT	CARBOHYDRATES	TOTAL
ENERGY VALUE (AVERAGE)	KJOULE	112	53	1230	1395
PER 100 G EDIBLE PORTION	(KCAL)	27	13	294	334
AMOUNT OF DIGESTIBLE CONSTITUENTS PER 100 G	GRAM	5.28	1.23	73.50	
ENERGY VALUE (AVERAGE)	KJOULE	83	48	1230	1361
OF THE DIGESTIBLE FRACTION PER 100 G EDIBLE PORTION	(KCAL)	20	11	294	325

WASTE PERCENTAGE AVERAGE 0.00

CONSTITUENTS	DIM	AV	VARIATION			AVR	NUTR.	DENS.	MOLPERC.
WATER	GRAM	9.74	9.39	-	10.10	9.74	GRAM/MJ	7.16	
PROTEIN	GRAM	7.14	6.54	-	7.74	7.14	GRAM/MJ	5.25	
FAT	GRAM	1.37	1.16	-	1.58	1.37	GRAM/MJ	1.01	
AVAILABLE CARBOHYDR.	GRAM	73.50 1	-	-	-	73.50	GRAM/MJ	54.01	
MINERALS	GRAM	5.21	4.71	-	5.71	5.21	GRAM/MJ	3.83	
SODIUM	GRAM	1.18	1.16	-	1.24	1.18	GRAM/MJ	0.87	
PHOSPHORUS	MILLI	240.00	214.00	-	271.00	240.00	MILLI/MJ	176.34	
CHLORIDE	GRAM	1.82	1.66	-	1.99	1.82	GRAM/MJ	1.34	
ALANINE	GRAM	0.31	0.28	-	0.34	0.31	GRAM/MJ	0.23	5.6
ARGININE	GRAM	0.55	0.48	-	0.66	0.55	GRAM/MJ	0.40	5.1
ASPARTIC ACID	GRAM	1.71	1.58	-	1.81	1.71	GRAM/MJ	1.26	20.8
CYSTINE	GRAM	0.11	0.07	-	0.14	0.11	GRAM/MJ	0.08	0.7
GLUTAMIC ACID	GRAM	1.41	1.19	-	1.59	1.41	GRAM/MJ	1.04	15.5
GLYCINE	GRAM	0.29	0.25	-	0.33	0.29	GRAM/MJ	0.21	6.2
HISTIDINE	GRAM	0.20	0.15	-	0.24	0.20	GRAM/MJ	0.15	2.1
ISOLEUCINE	GRAM	0.30	0.25	-	0.43	0.30	GRAM/MJ	0.22	3.7
LEUCINE	GRAM	0.47	0.39	-	0.55	0.47	GRAM/MJ	0.35	5.8
LYSINE	GRAM	0.57	0.53	-	0.61	0.57	GRAM/MJ	0.42	6.3
METHIONINE	MILLI	50.00	20.00	-	100.00	50.00	MILLI/MJ	36.74	0.5
PHENYLALANINE	GRAM	0.34	0.32	-	0.36	0.34	GRAM/MJ	0.25	3.3
PROLINE	GRAM	0.30	0.28	-	0.41	0.30	GRAM/MJ	0.22	4.2
SERINE	GRAM	0.35	0.31	-	0.40	0.35	GRAM/MJ	0.26	5.4
THREONINE	GRAM	0.32	0.29	-	0.33	0.32	GRAM/MJ	0.24	4.3
TRYPTOPHAN	MILLI	70.00	60.00	-	110.00	70.00	MILLI/MJ	51.43	0.6
TYROSINE	GRAM	0.47	0.37	-	0.57	0.47	GRAM/MJ	0.35	4.2
VALINE	GRAM	0.40	0.38	-	0.43	0.40	GRAM/MJ	0.29	5.5
DIETARY FIBRE,WAT.INS.	GRAM	3.00	-	-	-	3.00	GRAM/MJ	2.20	

1 CALCULATED BY THE DIFFERENCE USING THE VALUE OF WATER INSOLUBLE DIETARY FIBRE

KARTOFFELKNÖDEL ROH TROCKENPRODUKT

POTATO-DUMPLING RAW DRIED

BOULETTES DE POMMES DE TERRE CRUES PRODUIT SÉCHÉ

		PROTEIN	FAT	CARBOHYDRATES	TOTAL
ENERGY VALUE (AVERAGE) PER 100 G EDIBLE PORTION	KJOULE	89	9.7	1289	1388
	(KCAL)	21	2.3	308	332
AMOUNT OF DIGESTIBLE CONSTITUENTS PER 100 G	GRAM	4.21	0.22	77.00	
ENERGY VALUE (AVERAGE) OF THE DIGESTIBLE FRACTION PER 100 G EDIBLE PORTION	KJOULE	66	8.8	1289	1364
	(KCAL)	16	2.1	308	326

WASTE PERCENTAGE AVERAGE 0.00

CONSTITUENTS	DIM	AV	VARIATION			AVR	NUTR. DENS.		MOLPERC.
WATER	GRAM	9.26	8.24	-	10.30	9.26	GRAM/MJ	6.79	
PROTEIN	GRAM	5.70	4.90	-	6.60	5.70	GRAM/MJ	4.18	
FAT	GRAM	0.25	0.20	-	0.31	0.25	GRAM/MJ	0.18	
AVAILABLE CARBOHYDR.	GRAM	77.00 1	-	-	-	77.00	GRAM/MJ	56.47	
MINERALS	GRAM	4.79	4.20	-	5.37	4.79	GRAM/MJ	3.51	
SODIUM	GRAM	1.26	1.17	-	1.42	1.26	GRAM/MJ	0.92	
POTASSIUM	MILLI	749.00	-	-	-	749.00	MILLI/MJ	549.28	
MAGNESIUM	MILLI	45.00	-	-	-	45.00	MILLI/MJ	33.00	
CALCIUM	MILLI	40.00	27.00	-	59.00	40.00	MILLI/MJ	29.33	
MANGANESE	MILLI	0.39	-	-	-	0.39	MILLI/MJ	0.29	
ZINC	MILLI	1.05	-	-	-	1.05	MILLI/MJ	0.77	
PHOSPHORUS	MILLI	170.00	160.00	-	183.00	170.00	MILLI/MJ	124.67	
CHLORIDE	GRAM	1.78	1.70	-	1.85	1.78	GRAM/MJ	1.31	
VITAMIN B1	MILLI	0.14	-	-	-	0.14	MILLI/MJ	0.10	
ALANINE	GRAM	0.25	0.23	-	0.27	0.25	GRAM/MJ	0.18	5.6
ARGININE	GRAM	0.40	0.36	-	0.43	0.40	GRAM/MJ	0.29	4.6
ASPARTIC ACID	GRAM	1.40	1.30	-	1.56	1.40	GRAM/MJ	1.03	21.0
CYSTINE	GRAM	0.12	0.08	-	0.15	0.12	GRAM/MJ	0.09	1.0
GLUTAMIC ACID	GRAM	1.17	1.00	-	1.30	1.17	GRAM/MJ	0.86	15.8
GLYCINE	GRAM	0.26	0.23	-	0.34	0.26	GRAM/MJ	0.19	6.9
HISTIDINE	GRAM	0.17	0.14	-	0.19	0.17	GRAM/MJ	0.12	2.2
ISOLEUCINE	GRAM	0.23	0.14	-	0.29	0.23	GRAM/MJ	0.17	3.5
LEUCINE	GRAM	0.37	0.24	-	0.47	0.37	GRAM/MJ	0.27	5.6
LYSINE	GRAM	0.47	0.43	-	0.56	0.47	GRAM/MJ	0.34	6.4
METHIONINE	MILLI	50.00	-	-	-	50.00	MILLI/MJ	36.67	0.7
PHENYLALANINE	GRAM	0.28	0.23	-	0.34	0.28	GRAM/MJ	0.21	3.4
PROLINE	GRAM	0.25	0.17	-	0.30	0.25	GRAM/MJ	0.18	4.3
SERINE	GRAM	0.28	0.27	-	0.30	0.28	GRAM/MJ	0.21	5.3
THREONINE	GRAM	0.25	0.23	-	0.27	0.25	GRAM/MJ	0.18	4.2
TRYPTOPHAN	MILLI	80.00	-	-	-	80.00	MILLI/MJ	58.67	0.8
TYROSINE	GRAM	0.29	0.20	-	0.36	0.29	GRAM/MJ	0.21	3.2
VALINE	GRAM	0.33	0.31	-	0.34	0.33	GRAM/MJ	0.24	5.6
DIETARY FIBRE, WAT.INS.	GRAM	3.00	-	-	-	3.00	GRAM/MJ	2.20	

1 CALCULATED BY THE DIFFERENCE USING THE VALUE OF WATER INSOLUBLE DIETARY FIBRE

KARTOFFEL-KROKETTEN
TROCKENPRODUKT

POTATO CROQUETTES
DRIED

CROQUETTES DE POMMES DE TERRE
PRODUIT SÉCHÉ

		PROTEIN	FAT	CARBOHYDRATES	TOTAL
ENERGY VALUE (AVERAGE)	KJOULE	127	63	1260	1450
PER 100 G	(KCAL)	30	15	301	347
EDIBLE PORTION					
AMOUNT OF DIGESTIBLE	GRAM	5.99	1.44	75.30	
CONSTITUENTS PER 100 G					
ENERGY VALUE (AVERAGE)	KJOULE	94	56	1260	1411
OF THE DIGESTIBLE	(KCAL)	22	13	301	337
FRACTION PER 100 G					
EDIBLE PORTION					

WASTE PERCENTAGE AVERAGE 0.00

CONSTITUENTS	DIM	AV	VARIATION			AVR	NUTR. DENS.		MOLPERC.
WATER	GRAM	8.16	7.03	-	11.30	8.16	GRAM/MJ	5.78	
PROTEIN	GRAM	8.10	7.04	-	9.73	8.10	GRAM/MJ	5.74	
FAT	GRAM	1.61	0.82	-	2.55	1.61	GRAM/MJ	1.14	
AVAILABLE CARBOHYDR.	GRAM	75.30 1	-	-	-	75.30	GRAM/MJ	53.38	
MINERALS	GRAM	4.81	4.68	-	5.93	4.81	GRAM/MJ	3.41	
SODIUM	GRAM	1.38	0.92	-	1.51	1.38	GRAM/MJ	0.98	
POTASSIUM	MILLI	936.00	-	-	-	936.00	MILLI/MJ	663.52	
MAGNESIUM	MILLI	56.00	-	-	-	56.00	MILLI/MJ	39.70	
CALCIUM	MILLI	52.00	-	-	-	52.00	MILLI/MJ	36.86	
MANGANESE	MILLI	0.57	-	-	-	0.57	MILLI/MJ	0.40	
ZINC	MILLI	2.41	-	-	-	2.41	MILLI/MJ	1.71	
PHOSPHORUS	MILLI	267.00	255.00	-	305.00	267.00	MILLI/MJ	189.27	
CHLORIDE	GRAM	1.96	1.54	-	2.08	1.96	GRAM/MJ	1.39	
ALANINE	GRAM	0.33	0.23	-	0.35	0.33	GRAM/MJ	0.23	5.4
ARGININE	GRAM	0.63	0.60	-	0.67	0.63	GRAM/MJ	0.45	5.3
ASPARTIC ACID	GRAM	1.95	1.85	-	2.07	1.95	GRAM/MJ	1.38	21.4
CYSTINE	GRAM	0.16	0.08	-	0.21	0.16	GRAM/MJ	0.11	1.0
GLUTAMIC ACID	GRAM	1.57	1.29	-	1.94	1.57	GRAM/MJ	1.11	15.6
GLYCINE	GRAM	0.22	0.21	-	0.23	0.22	GRAM/MJ	0.16	4.3
HISTIDINE	GRAM	0.22	0.21	-	0.23	0.22	GRAM/MJ	0.16	2.1
ISOLEUCINE	GRAM	0.32	0.31	-	0.33	0.32	GRAM/MJ	0.23	3.6
LEUCINE	GRAM	0.53	0.42	-	0.62	0.53	GRAM/MJ	0.38	5.9
LYSINE	GRAM	0.67	0.63	-	0.68	0.67	GRAM/MJ	0.47	6.7
METHIONINE	MILLI	50.00	-	-	-	50.00	MILLI/MJ	35.44	0.5
PHENYLALANINE	GRAM	0.43	0.40	-	0.44	0.43	GRAM/MJ	0.30	3.8
PROLINE	GRAM	0.40	0.32	-	0.54	0.40	GRAM/MJ	0.28	5.1
SERINE	GRAM	0.38	0.36	-	0.42	0.38	GRAM/MJ	0.27	5.3
THREONINE	GRAM	0.34	0.32	-	0.36	0.34	GRAM/MJ	0.24	4.2
TRYPTOPHAN	MILLI	90.00	-	-	-	90.00	MILLI/MJ	63.80	0.6
TYROSINE	GRAM	0.44	0.32	-	0.54	0.44	GRAM/MJ	0.31	3.5
VALINE	GRAM	0.48	-	-	-	0.48	GRAM/MJ	0.34	6.0
DIETARY FIBRE,WAT.INS.	GRAM	2.00	-	-	-	2.00	GRAM/MJ	1.42	

1 CALCULATED BY THE DIFFERENCE USING THE VALUE OF WATER INSOLUBLE DIETARY FIBRE

KARTOFFELPUFFER (REIBEKUCHEN) TROCKENPRODUKT

POTATO PANCAKES DRIED

CRÊPES DE POMMES DE TERRE PRODUIT SÉCHÉ

		PROTEIN	FAT	CARBOHYDRATES	TOTAL
ENERGY VALUE (AVERAGE)	KJOULE	100	21	1257	1377
PER 100 G	(KCAL)	24	4.9	300	329
EDIBLE PORTION					
AMOUNT OF DIGESTIBLE	GRAM	4.69	0.47	75.10	
CONSTITUENTS PER 100 G					
ENERGY VALUE (AVERAGE)	KJOULE	74	19	1257	1349
OF THE DIGESTIBLE	(KCAL)	18	4.4	300	322
FRACTION PER 100 G					
EDIBLE PORTION					

WASTE PERCENTAGE AVERAGE 0.00

CONSTITUENTS	DIM	AV	VARIATION			AVR	NUTR. DENS.		MOLPERC.
WATER	GRAM	9.40	8.00	-	11.00	9.40	GRAM/MJ	6.97	
PROTEIN	GRAM	6.35	5.26	-	8.40	6.35	GRAM/MJ	4.71	
FAT	GRAM	0.53	0.15	-	0.83	0.53	GRAM/MJ	0.39	
AVAILABLE CARBOHYDR.	GRAM	75.10 1	-	-	-	75.10	GRAM/MJ	55.66	
MINERALS	GRAM	4.61	4.10	-	4.98	4.61	GRAM/MJ	3.42	
PHOSPHORUS	MILLI	192.00	179.00	-	201.00	192.00	MILLI/MJ	142.31	
ALANINE	GRAM	0.27	0.20	-	0.28	0.27	GRAM/MJ	0.20	
ARGININE	GRAM	0.55	0.47	-	0.78	0.55	GRAM/MJ	0.41	
ASPARTIC ACID	GRAM	1.47	1.14	-	1.80	1.47	GRAM/MJ	1.09	
CYSTINE	GRAM	0.13	0.08	-	0.19	0.13	GRAM/MJ	0.10	
GLUTAMIC ACID	GRAM	1.34	1.14	-	1.49	1.34	GRAM/MJ	0.99	
GLYCINE	GRAM	0.26	0.23	-	0.30	0.26	GRAM/MJ	0.19	
HISTIDINE	GRAM	0.19	0.18	-	0.20	0.19	GRAM/MJ	0.14	
ISOLEUCINE	GRAM	0.25	0.21	-	0.34	0.25	GRAM/MJ	0.19	
LEUCINE	GRAM	0.41	0.36	-	0.48	0.41	GRAM/MJ	0.30	
LYSINE	GRAM	0.49	0.46	-	0.52	0.49	GRAM/MJ	0.36	
METHIONINE	-	TRACES	-	-	-				
PHENYLALANINE	GRAM	0.31	0.25	-	0.35	0.31	GRAM/MJ	0.23	
THREONINE	GRAM	0.25	0.23	-	0.27	0.25	GRAM/MJ	0.19	
TRYPTOPHAN	-	0.00	-	-	-	0.00			
TYROSINE	GRAM	0.38	0.29	-	0.44	0.38	GRAM/MJ	0.28	
VALINE	GRAM	0.34	0.29	-	0.43	0.34	GRAM/MJ	0.25	
DIETARY FIBRE,WAT.INS.	GRAM	4.00	-	-	-	4.00	GRAM/MJ	2.96	

1 CALCULATED BY THE DIFFERENCE USING THE VALUE OF WATER INSOLUBLE DIETARY FIBRE

KARTOFFELSCHEIBEN (KARTOFFELCHIPS) ÖLGERÖSTET, GESALZEN

POTATO SLICES (POTATO CRISPS) FRIED IN OIL, SALTED

TRANCHES DE POMMES DE TERRE (CHIPS) FRITES À L'HUILE

		PROTEIN	FAT	CARBOHYDRATES	TOTAL
ENERGY VALUE (AVERAGE)	KJOULE	86	1533	679	2299
PER 100 G EDIBLE PORTION	(KCAL)	21	366	162	549
AMOUNT OF DIGESTIBLE CONSTITUENTS PER 100 G	GRAM	4.05	35.46	40.60	
ENERGY VALUE (AVERAGE)	KJOULE	64	1380	679	2123
OF THE DIGESTIBLE FRACTION PER 100 G EDIBLE PORTION	(KCAL)	15	330	162	507

WASTE PERCENTAGE AVERAGE 0.00

CONSTITUENTS	DIM	AV	VARIATION			AVR	NUTR.	DENS.	MOLPERC.
WATER	GRAM	2.30	1.66	-	2.95	2.30	GRAM/MJ	1.08	
PROTEIN	GRAM	5.48	5.00	-	5.84	5.48	GRAM/MJ	2.58	
FAT	GRAM	39.40	36.10	-	42.50	39.40	GRAM/MJ	18.56	
AVAILABLE CARBOHYDR.	GRAM	40.60 [1]	-	-	-	40.60	GRAM/MJ	19.12	
MINERALS	GRAM	3.54	2.73	-	4.25	3.54	GRAM/MJ	1.67	
SODIUM	MILLI	450.00	240.00	-	630.00	450.00	MILLI/MJ	211.97	
POTASSIUM	GRAM	1.00	-	-	-	1.00	GRAM/MJ	0.47	
MAGNESIUM	MILLI	64.00	-	-	-	64.00	MILLI/MJ	30.15	
CALCIUM	MILLI	52.00	-	-	-	52.00	MILLI/MJ	24.49	
MANGANESE	MILLI	0.46	-	-	-	0.46	MILLI/MJ	0.22	
IRON	MILLI	2.30	-	-	-	2.30	MILLI/MJ	1.08	
COPPER	MILLI	0.73	-	-	-	0.73	MILLI/MJ	0.34	
ZINC	MILLI	1.60	-	-	-	1.60	MILLI/MJ	0.75	
PHOSPHORUS	MILLI	147.00	131.00	-	162.00	147.00	MILLI/MJ	69.24	
CHLORIDE	MILLI	700.00	360.00	-	970.00	700.00	MILLI/MJ	329.74	
CAROTENE	MICRO	60.00	-	-	-	60.00	MICRO/MJ	28.26	
VITAMIN B1	MILLI	0.22	0.19	-	0.24	0.22	MILLI/MJ	0.10	
VITAMIN B2	MILLI	0.10	0.10	-	0.11	0.10	MILLI/MJ	0.05	
NICOTINAMIDE	MILLI	3.40	1.20	-	4.50	3.40	MILLI/MJ	1.60	
VITAMIN C	MILLI	8.00	0.00	-	13.00	8.00	MILLI/MJ	3.77	
ARGININE	GRAM	0.31	0.30	-	0.32	0.31	GRAM/MJ	0.15	
CYSTINE	MILLI	40.00	30.00	-	50.00	40.00	MILLI/MJ	18.84	
HISTIDINE	GRAM	0.15	0.14	-	0.15	0.15	GRAM/MJ	0.07	
ISOLEUCINE	GRAM	0.20	0.19	-	0.23	0.20	GRAM/MJ	0.09	
LEUCINE	GRAM	0.35	0.32	-	0.40	0.35	GRAM/MJ	0.16	

[1] CALCULATED FROM THE FAT FREE SOLID CONTENT OF RAW POTATOES

CONSTITUENTS	DIM	AV	VARIATION			AVR	NUTR. DENS.		MOLPERC.
LYSINE	GRAM	0.40	0.36	-	0.43	0.40	GRAM/MJ	0.19	
METHIONINE	-	TRACES	-	-	-				
PHENYLALANINE	GRAM	0.25	0.23	-	0.30	0.25	GRAM/MJ	0.12	
THREONINE	GRAM	0.21	0.19	-	0.25	0.21	GRAM/MJ	0.10	
TRYPTOPHAN	MILLI	80.00	-	-	-	80.00	MILLI/MJ	37.68	
TYROSINE	GRAM	0.27	0.22	-	0.30	0.27	GRAM/MJ	0.13	
VALINE	GRAM	0.28	0.25	-	0.31	0.28	GRAM/MJ	0.13	
INVERT SUGAR	GRAM	1.41	0.84	-	1.99	1.41	GRAM/MJ	0.66	
SUCROSE	GRAM	1.22	0.98	-	1.46	1.22	GRAM/MJ	0.57	

KARTOFFEL-STÄBCHEN
(KARTOFFELSTICKS)
ÖLGERÖSTET, GESALZEN

POTATO STICKS
FRIED IN OIL, SALTED

STICKS
DE POMMES DE TERRE
FRITES À L'HUILE, AVEC SEL

		PROTEIN	FAT	CARBOHYDRATES	TOTAL
ENERGY VALUE (AVERAGE)	KJOULE	102	1226	772	2099
PER 100 G	(KCAL)	24	293	184	502
EDIBLE PORTION					
AMOUNT OF DIGESTIBLE	GRAM	4.81	28.35	46.10	
CONSTITUENTS PER 100 G					
ENERGY VALUE (AVERAGE)	KJOULE	75	1103	772	1950
OF THE DIGESTIBLE	(KCAL)	18	264	184	466
FRACTION PER 100 G					
EDIBLE PORTION					

WASTE PERCENTAGE AVERAGE 0.00

CONSTITUENTS	DIM	AV	VARIATION			AVR	NUTR. DENS.		MOLPERC.
WATER	GRAM	2.33	1.59	-	3.38	2.33	GRAM/MJ	1.19	
PROTEIN	GRAM	6.50	5.70	-	7.72	6.50	GRAM/MJ	3.33	
FAT	GRAM	31.50	23.50	-	37.30	31.50	GRAM/MJ	16.15	
AVAILABLE CARBOHYDR.	GRAM	46.10 1	-	-	-	46.10	GRAM/MJ	23.64	
MINERALS	GRAM	4.61	3.89	-	5.57	4.61	GRAM/MJ	2.36	

1 CALCULATED FROM THE FAT FREE SOLID CONTENT OF RAW POTATOES

CONSTITUENTS	DIM	AV	VARIATION			AVR	NUTR.	DENS.	MOLPERC.
SODIUM	MILLI	720.00	640.00	-	800.00	720.00	MILLI/MJ	369.21	
POTASSIUM	GRAM	1.16	-	-	-	1.16	GRAM/MJ	0.59	
MAGNESIUM	MILLI	74.00	-	-	-	74.00	MILLI/MJ	37.95	
CALCIUM	MILLI	60.00	-	-	-	60.00	MILLI/MJ	30.77	
MANGANESE	MILLI	0.53	-	-	-	0.53	MILLI/MJ	0.27	
IRON	MILLI	2.60	-	-	-	2.60	MILLI/MJ	1.33	
COPPER	MILLI	0.84	-	-	-	0.84	MILLI/MJ	0.43	
ZINC	MILLI	1.80	-	-	-	1.80	MILLI/MJ	0.92	
PHOSPHORUS	MILLI	169.00	151.00	-	186.00	169.00	MILLI/MJ	86.66	
CHLORIDE	GRAM	1.08	0.96	-	1.20	1.08	GRAM/MJ	0.55	
CAROTENE	MICRO	60.00	-	-	-	60.00	MICRO/MJ	30.77	
VITAMIN B1	MILLI	0.23	-	-	-	0.23	MILLI/MJ	0.12	
VITAMIN B2	MILLI	0.10	-	-	-	0.10	MILLI/MJ	0.05	
NICOTINAMIDE	MILLI	3.40	-	-	-	3.40	MILLI/MJ	1.74	
VITAMIN C	MILLI	8.00	-	-	-	8.00	MILLI/MJ	4.10	
ALANINE	GRAM	0.34	0.24	-	0.42	0.34	GRAM/MJ	0.17	
ARGININE	GRAM	0.45	0.42	-	0.46	0.45	GRAM/MJ	0.23	
ASPARTIC ACID	GRAM	1.44	0.92	-	2.00	1.44	GRAM/MJ	0.74	
CYSTINE	GRAM	0.13	0.10	-	0.16	0.13	GRAM/MJ	0.07	
GLUTAMIC ACID	GRAM	1.28	1.17	-	1.44	1.28	GRAM/MJ	0.66	
GLYCINE	GRAM	0.33	0.31	-	0.38	0.33	GRAM/MJ	0.17	
HISTIDINE	GRAM	0.21	0.20	-	0.22	0.21	GRAM/MJ	0.11	
ISOLEUCINE	GRAM	0.29	0.26	-	0.32	0.29	GRAM/MJ	0.15	
LEUCINE	GRAM	0.51	0.46	-	0.56	0.51	GRAM/MJ	0.26	
LYSINE	GRAM	0.57	0.52	-	0.61	0.57	GRAM/MJ	0.29	
METHIONINE	-	TRACES	-	-	-				
PHENYLALANINE	GRAM	0.35	0.33	-	0.42	0.35	GRAM/MJ	0.18	
PROLINE	GRAM	0.31	0.29	-	0.33	0.31	GRAM/MJ	0.16	
SERINE	GRAM	0.37	0.32	-	0.43	0.37	GRAM/MJ	0.19	
THREONINE	GRAM	0.31	0.27	-	0.35	0.31	GRAM/MJ	0.16	
TRYPTOPHAN	GRAM	0.10	-	-	-	0.10	GRAM/MJ	0.05	
TYROSINE	GRAM	0.38	0.33	-	0.42	0.38	GRAM/MJ	0.19	
VALINE	GRAM	0.13	0.10	-	0.16	0.13	GRAM/MJ	0.07	
INVERT SUGAR	GRAM	0.93	0.68	-	1.18	0.93	GRAM/MJ	0.48	
SUCROSE	GRAM	1.02	0.81	-	1.23	1.02	GRAM/MJ	0.52	

KARTOFFELSUPPE
TROCKENPRODUKT

POTATO SOUP
DRIED

POTAGE AUX
POMMES DE TERRE
PRODUIT SÉCHÉ

		PROTEIN	FAT	CARBOHYDRATES	TOTAL
ENERGY VALUE (AVERAGE)	KJOULE	114	144	1188	1446
PER 100 G	(KCAL)	27	34	284	346
EDIBLE PORTION					
AMOUNT OF DIGESTIBLE	GRAM	5.35	3.33	71.01	
CONSTITUENTS PER 100 G					
ENERGY VALUE (AVERAGE)	KJOULE	84	130	1188	1402
OF THE DIGESTIBLE	(KCAL)	20	31	284	335
FRACTION PER 100 G					
EDIBLE PORTION					

WASTE PERCENTAGE AVERAGE 0.00

CONSTITUENTS	DIM	AV	VARIATION			AVR	NUTR. DENS.		MOLPERC.
WATER	GRAM	6.55	6.15	-	6.99	6.55	GRAM/MJ	4.67	
PROTEIN	GRAM	7.24	6.75	-	7.72	7.24	GRAM/MJ	5.16	
FAT	GRAM	3.70	2.20	-	5.68	3.70	GRAM/MJ	2.64	
AVAILABLE CARBOHYDR.	GRAM	71.01 1	-	-	-	71.01	GRAM/MJ	50.65	
MINERALS	GRAM	11.50	10.80	-	12.40	11.50	GRAM/MJ	8.20	
ALANINE	GRAM	0.27	0.24	-	0.35	0.27	GRAM/MJ	0.19	4.9
ARGININE	GRAM	0.46	0.41	-	0.50	0.46	GRAM/MJ	0.33	4.3
ASPARTIC ACID	GRAM	1.47	1.37	-	1.53	1.47	GRAM/MJ	1.05	17.9
CYSTINE	GRAM	0.11	0.06	-	0.16	0.11	GRAM/MJ	0.08	0.7
GLUTAMIC ACID	GRAM	2.43	2.38	-	2.47	2.43	GRAM/MJ	1.73	26.7
GLYCINE	GRAM	0.25	0.22	-	0.28	0.25	GRAM/MJ	0.18	5.4
HISTIDINE	GRAM	0.20	0.17	-	0.22	0.20	GRAM/MJ	0.14	2.1
ISOLEUCINE	GRAM	0.27	0.22	-	0.29	0.27	GRAM/MJ	0.19	3.3
LEUCINE	GRAM	0.43	0.31	-	0.53	0.43	GRAM/MJ	0.31	5.3
LYSINE	GRAM	0.51	0.46	-	0.59	0.51	GRAM/MJ	0.36	5.6
METHIONINE	MILLI	30.00	-	-	-	30.00	MILLI/MJ	21.40	0.3
PHENYLALANINE	GRAM	0.31	0.25	-	0.34	0.31	GRAM/MJ	0.22	3.0
PROLINE	GRAM	0.25	0.22	-	0.27	0.25	GRAM/MJ	0.18	3.5
SERINE	GRAM	0.29	0.25	-	0.32	0.29	GRAM/MJ	0.21	4.5
THREONINE	GRAM	0.28	0.23	-	0.30	0.28	GRAM/MJ	0.20	3.8
TRYPTOPHAN	MILLI	90.00	-	-	-	90.00	MILLI/MJ	64.19	0.7
TYROSINE	GRAM	0.34	0.23	-	0.41	0.34	GRAM/MJ	0.24	3.0
VALINE	GRAM	0.35	0.30	-	0.41	0.35	GRAM/MJ	0.25	4.8

1 ESTIMATED BY THE DIFFERENCE METHOD
100 - (WATER + PROTEIN + FAT + MINERALS)

POMMES FRITES	POMMES FRITES	POMMES FRITES
VERZEHRSFERTIG	READY TO EAT	PRÊT À MANGER
UNGESALZEN	WITHOUT SALT	SANS SEL

		PROTEIN	FAT	CARBOHYDRATES	TOTAL
ENERGY VALUE (AVERAGE)	KJOULE	66	564	597	1228
PER 100 G	(KCAL)	16	135	143	293
EDIBLE PORTION					
AMOUNT OF DIGESTIBLE	GRAM	3.10	13.05	35.70	
CONSTITUENTS PER 100 G					
ENERGY VALUE (AVERAGE)	KJOULE	49	508	597	1154
OF THE DIGESTIBLE	(KCAL)	12	121	143	276
FRACTION PER 100 G					
EDIBLE PORTION					

WASTE PERCENTAGE AVERAGE 0.00

CONSTITUENTS	DIM	AV		VARIATION		AVR	NUTR. DENS.		MOLPERC.
WATER	GRAM	43.60	-	-	-	43.60	GRAM/MJ	37.78	
PROTEIN	GRAM	4.20	-	-	-	4.20	GRAM/MJ	3.64	
FAT	GRAM	14.50	-	-	-	14.50	GRAM/MJ	12.56	
AVAILABLE CARBOHYDR.	GRAM	35.70 1	-	-	-	35.70	GRAM/MJ	30.94	
MINERALS	GRAM	2.00	-	-	-	2.00	GRAM/MJ	1.73	
SODIUM	MILLI	6.00	-	-	-	6.00	MILLI/MJ	5.20	
POTASSIUM	MILLI	926.00	-	-	-	926.00	MILLI/MJ	802.40	
CALCIUM	MILLI	20.00	-	-	-	20.00	MILLI/MJ	17.33	
IRON	MILLI	1.70	-	-	-	1.70	MILLI/MJ	1.47	
PHOSPHORUS	MILLI	105.00	-	-	-	105.00	MILLI/MJ	90.99	
VITAMIN B1	MILLI	0.14	-	-	-	0.14	MILLI/MJ	0.12	
VITAMIN B2	MICRO	90.00	-	-	-	90.00	MICRO/MJ	77.99	
NICOTINAMIDE	MILLI	2.50	-	-	-	2.50	MILLI/MJ	2.17	
VITAMIN C	MILLI	28.00	-	-	-	28.00	MILLI/MJ	24.26	

1 ESTIMATED BY THE DIFFERENCE METHOD
100 - (WATER + PROTEIN + FAT + MINERALS)

KOHLRABI KOHLRABI CHOU-RAVE

BRASSICA OLERACEA L.
VAR. GONGYLODES L.

		PROTEIN	FAT	CARBOHYDRATES	TOTAL
ENERGY VALUE (AVERAGE) PER 100 G EDIBLE PORTION	KJOULE	30	3.9	64	99
	(KCAL)	7.3	0.9	15	24
AMOUNT OF DIGESTIBLE CONSTITUENTS PER 100 G	GRAM	1.26	0.09	3.85	
ENERGY VALUE (AVERAGE) OF THE DIGESTIBLE FRACTION PER 100 G EDIBLE PORTION	KJOULE	20	3.5	64	88
	(KCAL)	4.7	0.8	15	21

WASTE PERCENTAGE AVERAGE 34 MINIMUM 20 MAXIMUM 46

CONSTITUENTS	DIM	AV	VARIATION			AVR	NUTR. DENS.		MOLPERC.
WATER	GRAM	91.60	90.00	-	92.70	60.46	KILO/MJ	1.04	
PROTEIN	GRAM	1.94	1.60	-	2.30	1.28	GRAM/MJ	22.12	
FAT	GRAM	0.10	-	-	-	0.07	GRAM/MJ	1.14	
AVAILABLE CARBOHYDR.	GRAM	3.85	3.23	-	4.31	2.54	GRAM/MJ	43.89	
TOTAL DIETARY FIBRE	GRAM	1.44 [1]	-	-	-	0.95	GRAM/MJ	16.42	
MINERALS	GRAM	0.95	0.89	-	1.00	0.63	GRAM/MJ	10.83	
SODIUM	MILLI	32.00	10.00	-	56.00	21.12	MILLI/MJ	364.79	
POTASSIUM	MILLI	380.00	340.00	-	500.00	250.80	GRAM/MJ	4.33	
MAGNESIUM	MILLI	43.00	32.00	-	48.00	28.38	MILLI/MJ	490.19	
CALCIUM	MILLI	68.00	43.00	-	90.00	44.88	MILLI/MJ	775.19	
MANGANESE	MILLI	0.13	0.08	-	0.20	0.09	MILLI/MJ	1.48	
IRON	MILLI	0.90	0.60	-	1.20	0.59	MILLI/MJ	10.26	
COBALT	MICRO	3.50	1.00	-	6.00	2.31	MICRO/MJ	39.90	
COPPER	MILLI	0.12	0.09	-	0.14	0.08	MILLI/MJ	1.37	
ZINC	MILLI	0.26	-	-	-	0.17	MILLI/MJ	2.96	
CHROMIUM	-	0.00	-	-	-	0.00			
PHOSPHORUS	MILLI	49.70	44.20	-	55.00	32.80	MILLI/MJ	566.57	
CHLORIDE	MILLI	57.00	-	-	-	37.62	MILLI/MJ	649.79	
IODIDE	MICRO	1.40	-	-	-	0.92	MICRO/MJ	15.96	
SELENIUM	MICRO	-	8.00	-	167.00				
NITRATE	MILLI	192.00	36.00	-	438.00	126.72	GRAM/MJ	2.19	
CAROTENE	MILLI	0.20	0.10	-	0.45	0.13	MILLI/MJ	2.28	
VITAMIN B1	MICRO	48.00	30.00	-	60.00	31.68	MICRO/MJ	547.19	
VITAMIN B2	MICRO	46.00	30.00	-	55.00	30.36	MICRO/MJ	524.39	
NICOTINAMIDE	MILLI	1.80	1.04	-	2.62	1.19	MILLI/MJ	20.52	
PANTOTHENIC ACID	MILLI	0.10	-	-	-	0.07	MILLI/MJ	1.14	

1 METHOD OF MEUSER, SUCKOW AND KULIKOWSKI ("BERLINER METHODE")

CONSTITUENTS	DIM	AV	VARIATION			AVR	NUTR. DENS.		MOLPERC.
VITAMIN B6	MILLI	0.12	0.10	-	0.14	0.08	MILLI/MJ	1.37	
BIOTIN	MICRO	2.70	2.70	-	2.70	1.78	MICRO/MJ	30.78	
VITAMIN C	MILLI	63.30	41.00	-	92.00	41.78	MILLI/MJ	721.61	
ARGININE	GRAM	0.12	0.11	-	0.13	0.08	GRAM/MJ	1.37	
HISTIDINE	MILLI	37.00	33.00	-	41.00	24.42	MILLI/MJ	421.79	
ISOLEUCINE	MILLI	89.00	81.00	-	97.00	58.74	GRAM/MJ	1.01	
LEUCINE	MILLI	78.00	63.00	-	93.00	51.48	MILLI/MJ	889.19	
LYSINE	MILLI	64.00	56.00	-	72.00	42.24	MILLI/MJ	729.59	
METHIONINE	MILLI	16.00	13.00	-	19.00	10.56	MILLI/MJ	182.40	
PHENYLALANINE	MILLI	45.00	29.00	-	61.00	29.70	MILLI/MJ	512.99	
THREONINE	MILLI	56.00	53.00	-	59.00	36.96	MILLI/MJ	638.39	
TRYPTOPHAN	MILLI	12.00	-	-	-	7.92	MILLI/MJ	136.80	
VALINE	MILLI	58.00	51.00	-	65.00	38.28	MILLI/MJ	661.19	
CITRIC ACID	MILLI	155.00	149.00	-	160.00	102.30	GRAM/MJ	1.77	
OXALIC ACID TOTAL	MILLI	2.80	-	-	-	1.85	MILLI/MJ	31.92	
OXALIC ACID SOLUBLE	MILLI	2.20	-	-	-	1.45	MILLI/MJ	25.08	
GLUCOSE	GRAM	1.38	1.33	-	1.49	0.91	GRAM/MJ	15.79	
FRUCTOSE	GRAM	1.23	1.23	-	1.37	0.81	GRAM/MJ	14.06	
SUCROSE	GRAM	1.08	0.52	-	1.29	0.72	GRAM/MJ	12.38	
MYRISTIC ACID	MILLI	0.20	-	-	-	0.13	MILLI/MJ	2.28	
PALMITIC ACID	MILLI	23.50	22.00	-	25.00	15.51	MILLI/MJ	267.90	
STEARIC ACID	MILLI	3.35	2.30	-	4.40	2.21	MILLI/MJ	38.19	
ARACHIDIC ACID	MILLI	1.00	-	-	-	0.66	MILLI/MJ	11.40	
PALMITOLEIC ACID	MILLI	2.30	1.90	-	2.70	1.52	MILLI/MJ	26.22	
OLEIC ACID	MILLI	11.50	5.00	-	18.00	7.59	MILLI/MJ	131.10	
LINOLEIC ACID	MILLI	24.00	21.00	-	27.00	15.84	MILLI/MJ	273.60	
LINOLENIC ACID	MILLI	47.00	44.00	-	50.00	31.02	MILLI/MJ	535.79	
TOTAL PURINES	MILLI	16.00 2	11.00	-	20.00	10.56	MILLI/MJ	182.40	
DIETARY FIBRE,WAT.SOL.	GRAM	0.48	-	-	-	0.32	GRAM/MJ	5.47	
DIETARY FIBRE,WAT.INS.	GRAM	0.96	-	-	-	0.63	GRAM/MJ	10.94	

2 VALUE EXPRESSED IN MG URIC ACID/100G

KOHLRÜBE (STECKRÜBE, WRUCKE, DOTSCHE) BRASSICA NAPUS VAR. NAPOBRASSICA MILL.

SWEDE (TURNIP, RAPE, RUTABAGA)

CHOU-NAVET

		PROTEIN	FAT	CARBOHYDRATES	TOTAL
ENERGY VALUE (AVERAGE) PER 100 G EDIBLE PORTION	KJOULE (KCAL)	18 4.4	6.2 1.5	14 3.4	39 9.2
AMOUNT OF DIGESTIBLE CONSTITUENTS PER 100 G	GRAM	0.85	0.14	0.85	
ENERGY VALUE (AVERAGE) OF THE DIGESTIBLE FRACTION PER 100 G EDIBLE PORTION	KJOULE (KCAL)	13 3.2	5.6 1.3	14 3.4	33 8.0

WASTE PERCENTAGE AVERAGE 17 MINIMUM 15 MAXIMUM 29

CONSTITUENTS	DIM	AV	VARIATION			AVR	NUTR. DENS.		MOLPERC.
WATER	GRAM	89.30	88.90	-	90.00	74.12	KILO/MJ	2.68	
PROTEIN	GRAM	1.16	1.00	-	1.39	0.96	GRAM/MJ	34.84	
FAT	GRAM	0.16	0.10	-	0.20	0.13	GRAM/MJ	4.81	
AVAILABLE CARBOHYDR.	GRAM	0.85	-	-	-	0.71	GRAM/MJ	25.53	
MINERALS	GRAM	0.77	0.74	-	0.80	0.64	GRAM/MJ	23.13	
SODIUM	MILLI	10.00	5.00	-	15.00	8.30	MILLI/MJ	300.33	
POTASSIUM	MILLI	227.00	200.00	-	227.00	188.41	GRAM/MJ	6.82	
MAGNESIUM	MILLI	11.00	-	-	-	9.13	MILLI/MJ	330.36	
CALCIUM	MILLI	47.50	40.00	-	55.00	39.43	GRAM/MJ	1.43	
MANGANESE	MICRO	40.00	-	-	-	33.20	MILLI/MJ	1.20	
IRON	MILLI	0.45	0.40	-	0.50	0.37	MILLI/MJ	13.51	
COPPER	MICRO	80.00	-	-	-	66.40	MILLI/MJ	2.40	
ZINC	MICRO	80.00	-	-	-	66.40	MILLI/MJ	2.40	
PHOSPHORUS	MILLI	31.00	-	-	-	25.73	MILLI/MJ	931.01	
CHLORIDE	MILLI	31.00	-	-	-	25.73	MILLI/MJ	931.01	
FLUORIDE	MICRO	30.00	-	-	-	24.90	MICRO/MJ	900.98	
IODIDE	MICRO	4.00	-	-	-	3.32	MICRO/MJ	120.13	
BORON	MILLI	5.00	2.50	-	7.00	4.15	MILLI/MJ	150.16	
CAROTENE	MICRO	99.00	-	-	-	82.17	MILLI/MJ	2.97	
VITAMIN B1	MICRO	50.00	30.00	-	70.00	41.50	MILLI/MJ	1.50	
VITAMIN B2	MICRO	58.00	35.00	-	58.00	48.14	MILLI/MJ	1.74	
NICOTINAMIDE	MILLI	0.85	0.80	-	0.85	0.71	MILLI/MJ	25.53	
PANTOTHENIC ACID	MILLI	0.11	-	-	-	0.09	MILLI/MJ	3.30	
VITAMIN B6	MILLI	0.20	-	-	-	0.17	MILLI/MJ	6.01	
BIOTIN	MICRO	0.10	-	-	-	0.08	MICRO/MJ	3.00	
FOLIC ACID	MICRO	27.00	-	-	-	22.41	MICRO/MJ	810.88	
VITAMIN C	MILLI	33.00	30.00	-	36.00	27.39	MILLI/MJ	991.08	
OXALIC ACID TOTAL	-	0.00	-	-	-	0.00			
SALICYLIC ACID	MILLI	0.16	0.16	-	0.16	0.13	MILLI/MJ	4.81	

CONSTITUENTS	DIM	AV	VARIATION			AVR	NUTR. DENS.		MOLPERC.
GLUCOSE	GRAM	0.30	0.30	-	0.30	0.25	GRAM/MJ	9.01	
FRUCTOSE	GRAM	0.55	0.55	-	0.55	0.46	GRAM/MJ	16.52	
CELLULOSE	GRAM	0.77	-	-	-	0.64	GRAM/MJ	23.13	
HEMICELLULOSE	GRAM	0.68	-	-	-	0.56	GRAM/MJ	20.42	
MYRISTIC ACID	MILLI	0.50	-	-	-	0.42	MILLI/MJ	15.02	
PALMITIC ACID	MILLI	20.00	-	-	-	16.60	MILLI/MJ	600.65	
STEARIC ACID	MILLI	1.00	-	-	-	0.83	MILLI/MJ	30.03	
PALMITOLEIC ACID	MILLI	3.10	-	-	-	2.57	MILLI/MJ	93.10	
OLEIC ACID	MILLI	0.80	-	-	-	0.66	MILLI/MJ	24.03	
EICOSENOIC ACID	MILLI	0.30	-	-	-	0.25	MILLI/MJ	9.01	
LINOLEIC ACID	MILLI	28.00	-	-	-	23.24	MILLI/MJ	840.91	
LINOLENIC ACID	MILLI	69.00	-	-	-	57.27	GRAM/MJ	2.07	

MEERRETTICH HORSE-RADISH RAIFORT

COCHLEARIA ARMORACIA L.

		PROTEIN	FAT	CARBOHYDRATES	TOTAL
ENERGY VALUE (AVERAGE) PER 100 G EDIBLE PORTION	KJOULE	44	12	207	262
	(KCAL)	11	2.8	49	63
AMOUNT OF DIGESTIBLE CONSTITUENTS PER 100 G	GRAM	2.07	0.27	12.35	
ENERGY VALUE (AVERAGE) OF THE DIGESTIBLE FRACTION PER 100 G EDIBLE PORTION	KJOULE	33	11	207	250
	(KCAL)	7.8	2.5	49	60

WASTE PERCENTAGE AVERAGE 47

CONSTITUENTS	DIM	AV	VARIATION			AVR	NUTR. DENS.		MOLPERC.
WATER	GRAM	76.60	-	-	-	40.60	GRAM/MJ	306.76	
PROTEIN	GRAM	2.80	2.00	-	3.73	1.48	GRAM/MJ	11.21	
FAT	GRAM	0.30	0.30	-	0.31	0.16	GRAM/MJ	1.20	
AVAILABLE CARBOHYDR.	GRAM	12.35	-	-	-	6.55	GRAM/MJ	49.46	
MINERALS	GRAM	2.20	-	-	-	1.17	GRAM/MJ	8.81	

CONSTITUENTS	DIM	AV	VARIATION			AVR	NUTR. DENS.		MOLPERC.
SODIUM	MILLI	9.00	1.00	-	32.00	4.77	MILLI/MJ	36.04	
POTASSIUM	MILLI	554.00	420.00	-	686.00	293.62	GRAM/MJ	2.22	
MAGNESIUM	MILLI	33.00	29.00	-	37.00	17.49	MILLI/MJ	132.16	
CALCIUM	MILLI	105.00	92.00	-	127.00	55.65	MILLI/MJ	420.50	
MANGANESE	MILLI	0.46	0.42	-	0.49	0.24	MILLI/MJ	1.84	
IRON	MILLI	1.40	-	-	-	0.74	MILLI/MJ	5.61	
COBALT	MICRO	1.00	0.50	-	2.00	0.53	MICRO/MJ	4.00	
COPPER	MILLI	0.14	-	-	-	0.07	MILLI/MJ	0.56	
ZINC	MILLI	1.40	1.30	-	1.40	0.74	MILLI/MJ	5.61	
NICKEL	MICRO	30.00	20.00	-	30.00	15.90	MICRO/MJ	120.14	
CHROMIUM	MICRO	3.00	2.00	-	4.00	1.59	MICRO/MJ	12.01	
PHOSPHORUS	MILLI	65.30	56.70	-	70.70	34.61	MILLI/MJ	261.51	
CHLORIDE	MILLI	18.00	16.00	-	19.00	9.54	MILLI/MJ	72.08	
IODIDE	MICRO	1.00	1.00	-	1.00	0.53	MICRO/MJ	4.00	
SELENIUM	MICRO	0.20	0.10	-	0.30	0.11	MICRO/MJ	0.80	
SILICON	MILLI	0.50	-	-	-	0.27	MILLI/MJ	2.00	
CAROTENE	MICRO	20.00	13.00	-	28.00	10.60	MICRO/MJ	80.09	
VITAMIN B1	MILLI	0.14	0.03	-	0.21	0.07	MILLI/MJ	0.56	
VITAMIN B2	MILLI	0.11	0.08	-	0.15	0.06	MILLI/MJ	0.44	
NICOTINAMIDE	MILLI	0.60	0.40	-	0.70	0.32	MILLI/MJ	2.40	
VITAMIN B6	MILLI	0.18	0.13	-	0.27	0.10	MILLI/MJ	0.72	
VITAMIN C	MILLI	114.00	90.00	-	260.00	60.42	MILLI/MJ	456.54	
ALANINE	MILLI	49.00	49.00	-	49.00	25.97	MILLI/MJ	196.23	4.1
ARGININE	MILLI	607.00	607.00	-	607.00	321.71	GRAM/MJ	2.43	26.2
ASPARTIC ACID	MILLI	132.00	132.00	-	132.00	69.96	MILLI/MJ	528.62	7.5
CYSTINE	MILLI	3.50	3.50	-	3.50	1.86	MILLI/MJ	14.02	0.1
GLUTAMIC ACID	MILLI	246.00	246.00	-	246.00	130.38	MILLI/MJ	985.16	12.6
GLYCINE	MILLI	49.00	49.00	-	49.00	25.97	MILLI/MJ	196.23	4.9
HISTIDINE	MILLI	49.00	49.00	-	49.00	25.97	MILLI/MJ	196.23	2.4
ISOLEUCINE	MILLI	59.00	59.00	-	59.00	31.27	MILLI/MJ	236.28	3.4
LEUCINE	MILLI	72.00	72.00	-	72.00	38.16	MILLI/MJ	288.34	4.1
LYSINE	MILLI	81.00	81.00	-	81.00	42.93	MILLI/MJ	324.38	4.2
METHIONINE	MILLI	15.00	15.00	-	15.00	7.95	MILLI/MJ	60.07	0.8
PHENYLALANINE	MILLI	45.00	45.00	-	45.00	23.85	MILLI/MJ	180.21	2.1
PROLINE	MILLI	185.00	185.00	-	185.00	98.05	MILLI/MJ	740.87	12.1
SERINE	MILLI	68.00	68.00	-	68.00	36.04	MILLI/MJ	272.32	4.9
THREONINE	MILLI	54.00	54.00	-	54.00	28.62	MILLI/MJ	216.25	3.4
TRYPTOPHAN	MILLI	16.00	16.00	-	16.00	8.48	MILLI/MJ	64.08	0.6
TYROSINE	MILLI	32.00	32.00	-	32.00	16.96	MILLI/MJ	128.15	1.3
VALINE	MILLI	85.00	85.00	-	85.00	45.05	MILLI/MJ	340.40	5.5
MALIC ACID	MILLI	680.00	-	-	-	360.40	GRAM/MJ	2.72	
GLUCOSE	GRAM	1.40	1.40	-	1.40	0.74	GRAM/MJ	5.61	
FRUCTOSE	GRAM	0.13	0.13	-	0.13	0.07	GRAM/MJ	0.52	
SUCROSE	GRAM	6.72	6.72	-	6.72	3.56	GRAM/MJ	26.91	
STARCH	GRAM	3.42	3.42	-	3.42	1.81	GRAM/MJ	13.70	
PALMITIC ACID	MILLI	44.00	44.00	-	44.00	23.32	MILLI/MJ	176.21	
STEARIC ACID	MILLI	3.80	3.80	-	3.80	2.01	MILLI/MJ	15.22	
PALMITOLEIC ACID	MILLI	2.10	2.10	-	2.10	1.11	MILLI/MJ	8.41	
OLEIC ACID	MILLI	27.00	27.00	-	27.00	14.31	MILLI/MJ	108.13	
LINOLEIC ACID	MILLI	51.00	51.00	-	51.00	27.03	MILLI/MJ	204.24	
LINOLENIC ACID	MILLI	116.00	116.00	-	116.00	61.48	MILLI/MJ	464.55	

MÖHRE (KAROTTE, MOHRRÜBE) CARROT CAROTTE

DAUCUS CAROTA L.

		PROTEIN	FAT	CARBOHYDRATES	TOTAL
ENERGY VALUE (AVERAGE)	KJOULE	15	7.8	83	106
PER 100 G EDIBLE PORTION	(KCAL)	3.7	1.9	20	25
AMOUNT OF DIGESTIBLE CONSTITUENTS PER 100 G	GRAM	0.72	0.18	4.93	
ENERGY VALUE (AVERAGE)	KJOULE	11	7.0	83	101
OF THE DIGESTIBLE FRACTION PER 100 G EDIBLE PORTION	(KCAL)	2.7	1.7	20	24

WASTE PERCENTAGE AVERAGE 19 MINIMUM 10 MAXIMUM 25

CONSTITUENTS	DIM	AV		VARIATION		AVR	NUTR. DENS.		MOLPERC.
WATER	GRAM	88.20		87.50 -	92.10	71.44	GRAM/MJ	874.21	
PROTEIN	GRAM	0.98		0.70 -	1.20	0.79	GRAM/MJ	9.71	
FAT	GRAM	0.20		0.10 -	0.30	0.16	GRAM/MJ	1.98	
AVAILABLE CARBOHYDR.	GRAM	4.93		3.56 -	8.21	3.99	GRAM/MJ	48.86	
TOTAL DIETARY FIBRE	GRAM	3.43	1	3.40 -	3.70	2.78	GRAM/MJ	34.00	
MINERALS	GRAM	0.86		0.66 -	1.02	0.70	GRAM/MJ	8.52	
SODIUM	MILLI	60.00		32.00 -	83.00	48.60	MILLI/MJ	594.70	
POTASSIUM	MILLI	290.00		201.00 -	346.00	234.90	GRAM/MJ	2.87	
MAGNESIUM	MILLI	18.00		15.00 -	24.00	14.58	MILLI/MJ	178.41	
CALCIUM	MILLI	41.00		25.00 -	52.00	33.21	MILLI/MJ	406.38	
MANGANESE	MILLI	0.21		0.12 -	0.36	0.17	MILLI/MJ	2.08	
IRON	MILLI	2.10		0.40 -	4.80	1.70	MILLI/MJ	20.81	
COBALT	MICRO	2.00		- -	-	1.62	MICRO/MJ	19.82	
COPPER	MILLI	0.13		0.08 -	0.28	0.11	MILLI/MJ	1.29	
ZINC	MILLI	0.64		0.18 -	2.10	0.52	MILLI/MJ	6.34	
NICKEL	MICRO	25.00		6.00 -	53.00	20.25	MICRO/MJ	247.79	
CHROMIUM	MICRO	5.00		2.00 -	8.00	4.05	MICRO/MJ	49.56	
MOLYBDENUM	MICRO	8.00		- -	-	6.48	MICRO/MJ	79.29	
VANADIUM	MICRO	10.00		- -	-	8.10	MICRO/MJ	99.12	
PHOSPHORUS	MILLI	35.00		30.00 -	44.00	28.35	MILLI/MJ	346.91	
CHLORIDE	MILLI	61.00		39.00 -	75.00	49.41	MILLI/MJ	604.61	
FLUORIDE	MICRO	27.00		16.00 -	38.00	21.87	MICRO/MJ	267.62	
IODIDE	MICRO	15.00		13.00 -	17.00	12.15	MICRO/MJ	148.68	
BORON	MILLI	0.31		0.18 -	0.42	0.25	MILLI/MJ	3.09	
SELENIUM	MICRO	0.20	2	0.00 -	2.00	0.16	MICRO/MJ	1.98	
BROMINE	MICRO	147.00		- -	-	119.07	MILLI/MJ	1.46	
NITRATE	MILLI	50.00		9.00 -	110.00	40.50	MILLI/MJ	495.58	

1 METHOD OF SCHWEIZER AND WUERSCH

2 AMOUNTS, DETERMINED IN THE FEDERAL REPUBLIC OF GERMANY

CONSTITUENTS	DIM	AV	VARIATION			AVR	NUTR. DENS.		MOLPERC.
CAROTENE	MILLI	12.00	6.00	-	21.00	9.72	MILLI/MJ	118.94	
VITAMIN E ACTIVITY	MILLI	0.60	-	-	-	0.49	MILLI/MJ	5.95	
TOTAL TOCOPHEROLS	MILLI	0.68	-	-	-	0.55	MILLI/MJ	6.74	
ALPHA-TOCOPHEROL	MILLI	0.60	-	-	-	0.49	MILLI/MJ	5.95	
BETA-TOCOPHEROL	MICRO	30.00	-	-	-	24.30	MICRO/MJ	297.35	
ALPHA-TOCOTRIENOL	MICRO	50.00	-	-	-	40.50	MICRO/MJ	495.58	
VITAMIN K	MICRO	80.00	-	-	-	64.80	MICRO/MJ	792.94	
VITAMIN B1	MICRO	69.00	50.00	-	100.00	55.89	MICRO/MJ	683.91	
VITAMIN B2	MICRO	53.00	30.00	-	80.00	42.93	MICRO/MJ	525.32	
NICOTINAMIDE	MILLI	0.58	0.40	-	1.00	0.47	MILLI/MJ	5.75	
PANTOTHENIC ACID	MILLI	0.27	0.20	-	1.00	0.22	MILLI/MJ	2.68	
VITAMIN B6	MICRO	93.00	65.00	-	120.00	75.33	MICRO/MJ	921.79	
BIOTIN	MICRO	5.00	2.00	-	7.00	4.05	MICRO/MJ	49.56	
FOLIC ACID	MICRO	8.00	-	-	-	6.48	MICRO/MJ	79.29	
VITAMIN C	MICRO	7.00	-	-	-	5.67	MICRO/MJ	69.38	
ARGININE	MILLI	41.00	35.00	-	46.00	33.21	MILLI/MJ	406.38	
CYSTINE	MILLI	13.00	2.00	-	24.00	10.53	MILLI/MJ	128.85	
HISTIDINE	MILLI	15.00	7.00	-	19.00	12.15	MILLI/MJ	148.68	
ISOLEUCINE	MILLI	43.00	38.00	-	47.00	34.83	MILLI/MJ	426.20	
LEUCINE	MILLI	42.00	35.00	-	54.00	34.02	MILLI/MJ	416.29	
LYSINE	MILLI	47.00	40.00	-	56.00	38.07	MILLI/MJ	465.85	
METHIONINE	MILLI	8.00	4.00	-	10.00	6.48	MILLI/MJ	79.29	
PHENYLALANINE	MILLI	31.00	28.00	-	35.00	25.11	MILLI/MJ	307.26	
THREONINE	MILLI	36.00	33.00	-	39.00	29.16	MILLI/MJ	356.82	
TRYPTOPHAN	MILLI	10.00	8.00	-	12.00	8.10	MILLI/MJ	99.12	
TYROSINE	MILLI	16.00	12.00	-	21.00	12.96	MILLI/MJ	158.59	
VALINE	MILLI	40.00	38.00	-	47.00	32.40	MILLI/MJ	396.47	
MALIC ACID	MILLI	293.00	276.00	-	309.00	237.33	GRAM/MJ	2.90	
CITRIC ACID	MILLI	51.00	47.00	-	55.00	41.31	MILLI/MJ	505.50	
OXALIC ACID TOTAL	MILLI	6.10	-	-	-	4.94	MILLI/MJ	60.46	
OXALIC ACID SOLUBLE	MILLI	1.50	-	-	-	1.22	MILLI/MJ	14.87	
QUINIC ACID	MILLI	51.00	42.00	-	60.00	41.31	MILLI/MJ	505.50	
FERULIC ACID	MILLI	1.50	1.00	-	2.00	1.22	MILLI/MJ	14.87	
CAFFEIC ACID	MILLI	6.00	2.00	-	10.00	4.86	MILLI/MJ	59.47	
PARA-COUMARIC ACID	MILLI	0.10	0.00	-	0.20	0.08	MILLI/MJ	0.99	
SUCCINIC ACID	MILLI	5.00	0.00	-	10.00	4.05	MILLI/MJ	49.56	
FUMARIC ACID	MILLI	7.00	5.00	-	8.00	5.67	MILLI/MJ	69.38	
SALICYLIC ACID	MILLI	0.23	0.23	-	0.23	0.19	MILLI/MJ	2.28	
GLUCOSE	GRAM	1.40	0.84	-	1.71	1.13	GRAM/MJ	13.89	
FRUCTOSE	GRAM	1.29	0.84	-	1.96	1.05	GRAM/MJ	12.84	
SUCROSE	GRAM	1.90	1.55	-	4.17	1.55	GRAM/MJ	18.91	
PECTIN	GRAM	1.35	-	-	-	1.09	GRAM/MJ	13.38	
PENTOSAN	GRAM	0.32	-	-	-	0.26	GRAM/MJ	3.17	
HEXOSAN	GRAM	0.64	-	-	-	0.52	GRAM/MJ	6.34	
CELLULOSE	GRAM	0.95	-	-	-	0.77	GRAM/MJ	9.42	
INOSITOL	MILLI	26.50	26.00	-	27.00	21.47	MILLI/MJ	262.66	
MANNITOL	GRAM	0.16	-	-	-	0.13	GRAM/MJ	1.59	
MYRISTIC ACID	MILLI	0.30	-	-	-	0.24	MILLI/MJ	2.97	
PALMITIC ACID	MILLI	35.00	32.00	-	38.00	28.35	MILLI/MJ	346.91	
STEARIC ACID	MILLI	2.55	2.10	-	3.00	2.07	MILLI/MJ	25.27	
ARACHIDIC ACID	MILLI	0.60	-	-	-	0.49	MILLI/MJ	5.95	
OLEIC ACID	MILLI	3.20	3.00	-	3.40	2.59	MILLI/MJ	31.72	
EICOSENOIC ACID	MILLI	0.60	-	-	-	0.49	MILLI/MJ	5.95	
LINOLEIC ACID	MILLI	104.50	104.00	-	105.00	84.65	GRAM/MJ	1.04	

CONSTITUENTS	DIM	AV	VARIATION			AVR	NUTR. DENS.		MOLPERC.
LINOLENIC ACID	MILLI	12.30	9.60	-	15.00	9.96	MILLI/MJ	121.91	
TOTAL STEROLS	MILLI	12.00	-	-	-	9.72	MILLI/MJ	118.94	
CAMPESTEROL	MILLI	1.00	-	-	-	0.81	MILLI/MJ	9.91	
BETA-SITOSTEROL	MILLI	7.00	-	-	-	5.67	MILLI/MJ	69.38	
STIGMASTEROL	MILLI	3.00	-	-	-	2.43	MILLI/MJ	29.74	
TOTAL PURINES	MILLI	23.00	3 20.00	-	25.00	18.63	MILLI/MJ	227.97	
TOTAL PHOSPHOLIPIDS	MILLI	39.00	-	-	-	31.59	MILLI/MJ	386.56	
PHOSPHATIDYLCHOLINE	MILLI	16.00	-	-	-	12.96	MILLI/MJ	158.59	
PHOSPHATIDYLETHANOLAMINE	MILLI	11.00	-	-	-	8.91	MILLI/MJ	109.03	
PHOSPHATIDYLSERINE	MILLI	2.00	-	-	-	1.62	MILLI/MJ	19.82	
PHOSPHATIDYLINOSITOL	MILLI	4.00	-	-	-	3.24	MILLI/MJ	39.65	
DIETARY FIBRE, WAT.SOL.	GRAM	1.51	-	-	-	1.22	GRAM/MJ	14.97	
DIETARY FIBRE, WAT.INS.	GRAM	1.92	-	-	-	1.56	GRAM/MJ	19.03	

3 VALUE EXPRESSED IN MG URIC ACID/100G

MÖHREN
GEKOCHT, ABGETROPFT

CARROTS
BOILED, DRAINED

CAROTTES
BOUILLIES, DEGOUTTÉES

		PROTEIN	FAT	CARBOHYDRATES	TOTAL
ENERGY VALUE (AVERAGE)	KJOULE	13	7.8	57	77
PER 100 G	(KCAL)	3.0	1.9	14	18
EDIBLE PORTION					
AMOUNT OF DIGESTIBLE	GRAM	0.52	0.18	3.40	
CONSTITUENTS PER 100 G					
ENERGY VALUE (AVERAGE)	KJOULE	8.2	7.0	57	72
OF THE DIGESTIBLE	(KCAL)	2.0	1.7	14	17
FRACTION PER 100 G					
EDIBLE PORTION					

WASTE PERCENTAGE AVERAGE 0.00

CONSTITUENTS	DIM	AV	VARIATION			AVR	NUTR. DENS.		MOLPERC.
WATER	GRAM	91.20	-	-	-	91.20	KILO/MJ	1.27	
PROTEIN	GRAM	0.80	-	-	-	0.80	GRAM/MJ	11.10	
FAT	GRAM	0.20	-	-	-	0.20	GRAM/MJ	2.78	
AVAILABLE CARBOHYDR.	GRAM	3.40	3.24	-	3.55	3.40	GRAM/MJ	47.18	
MINERALS	GRAM	0.65	-	-	-	0.65	GRAM/MJ	9.02	
SODIUM	MILLI	42.00	-	-	-	42.00	MILLI/MJ	582.81	
POTASSIUM	MILLI	189.00	-	-	-	189.00	GRAM/MJ	2.62	
CALCIUM	MILLI	37.00	-	-	-	37.00	MILLI/MJ	513.42	
IRON	MILLI	0.57	-	-	-	0.57	MILLI/MJ	7.91	
PHOSPHORUS	MILLI	30.00	-	-	-	30.00	MILLI/MJ	416.29	
VITAMIN B1	MICRO	60.00	-	-	-	60.00	MICRO/MJ	832.58	
VITAMIN B2	MICRO	30.00	-	-	-	30.00	MICRO/MJ	416.29	
NICOTINAMIDE	MILLI	0.48	-	-	-	0.48	MILLI/MJ	6.66	
VITAMIN C	MILLI	5.00	-	-	-	5.00	MILLI/MJ	69.38	
MALIC ACID	GRAM	0.22	0.21	-	0.23	0.22	GRAM/MJ	3.05	
CITRIC ACID	MILLI	40.00	-	-	-	40.00	MILLI/MJ	555.05	
GLUCOSE	GRAM	1.06	1.04	-	1.08	1.06	GRAM/MJ	14.71	
FRUCTOSE	GRAM	0.94	0.92	-	0.95	0.94	GRAM/MJ	13.04	
SUCROSE	GRAM	1.14	1.03	-	1.25	1.14	GRAM/MJ	15.82	

MÖHREN	CARROTS	CAROTTES
GETROCKNET	DRIED	SÉCHÉS

		PROTEIN	FAT	CARBOHYDRATES	TOTAL
ENERGY VALUE (AVERAGE)	KJOULE	107	56	633	797
PER 100 G	(KCAL)	26	13	151	190
EDIBLE PORTION					
AMOUNT OF DIGESTIBLE	GRAM	4.43	1.30	37.85	
CONSTITUENTS PER 100 G					
ENERGY VALUE (AVERAGE)	KJOULE	70	51	633	754
OF THE DIGESTIBLE	(KCAL)	17	12	151	180
FRACTION PER 100 G					
EDIBLE PORTION					

WASTE PERCENTAGE AVERAGE 0.00

CONSTITUENTS	DIM	AV		VARIATION		AVR	NUTR. DENS.		MOLPERC.
WATER	GRAM	9.40	4.00	-	14.60	9.40	GRAM/MJ	12.47	
PROTEIN	GRAM	6.82	4.00	-	9.30	6.82	GRAM/MJ	9.05	
FAT	GRAM	1.45	1.40	-	1.50	1.45	GRAM/MJ	1.92	
AVAILABLE CARBOHYDR.	GRAM	37.85 1	-	-	-	37.85	GRAM/MJ	50.21	
TOTAL DIETARY FIBRE	GRAM	26.30 1	-	-	-	26.30	GRAM/MJ	34.89	
MINERALS	GRAM	5.55	5.32	-	6.00	5.55	GRAM/MJ	7.36	
SODIUM	MILLI	495.00	205.00	-	756.00	495.00	MILLI/MJ	656.68	
POTASSIUM	GRAM	2.64	2.11	-	2.92	2.64	GRAM/MJ	3.50	
CALCIUM	MILLI	256.00	242.00	-	293.00	256.00	MILLI/MJ	339.62	
IRON	MILLI	4.70	2.30	-	5.90	4.70	MILLI/MJ	6.24	
PHOSPHORUS	MILLI	103.00	102.00	-	104.00	103.00	MILLI/MJ	136.64	
CAROTENE	MILLI	93.30	36.00	-	135.00	93.30	MILLI/MJ	123.77	
VITAMIN B1	MILLI	0.36	0.29	-	0.48	0.36	MILLI/MJ	0.48	
VITAMIN B2	MILLI	0.32	0.28	-	0.37	0.32	MILLI/MJ	0.42	
NICOTINAMIDE	MILLI	3.40	3.00	-	4.00	3.40	MILLI/MJ	4.51	
VITAMIN C	MILLI	19.00	11.00	-	34.00	19.00	MILLI/MJ	25.21	

1 CALCULATED FROM THE FRESH PRODUCT

MÖHREN IN DOSEN 1 — CARROTS CANNED 1 — CAROTTES EN BOÎTES 1

		PROTEIN	FAT	CARBOHYDRATES	TOTAL
ENERGY VALUE (AVERAGE)	KJOULE	9.1	13	60	82
PER 100 G	(KCAL)	2.2	3.1	14	20
EDIBLE PORTION					
AMOUNT OF DIGESTIBLE	GRAM	0.42	0.29	3.59	
CONSTITUENTS PER 100 G					
ENERGY VALUE (AVERAGE)	KJOULE	6.7	12	60	78
OF THE DIGESTIBLE	(KCAL)	1.6	2.8	14	19
FRACTION PER 100 G					
EDIBLE PORTION					

WASTE PERCENTAGE AVERAGE 0.00

CONSTITUENTS	DIM	AV	VARIATION			AVR	NUTR. DENS.		MOLPERC.
WATER	GRAM	91.40	90.00	-	92.20	91.40	KILO/MJ	1.17	
PROTEIN	GRAM	0.58	0.50	-	0.70	0.58	GRAM/MJ	7.40	
FAT	GRAM	0.33	0.20	-	0.40	0.33	GRAM/MJ	4.21	
AVAILABLE CARBOHYDR.	GRAM	3.59 2	-	-	-	3.59	GRAM/MJ	45.81	
MINERALS	GRAM	0.91	0.90	-	0.91	0.91	GRAM/MJ	11.61	
SODIUM	MILLI	61.00	13.00	-	211.00	61.00	MILLI/MJ	778.33	
POTASSIUM	MILLI	140.00	74.00	-	189.00	140.00	GRAM/MJ	1.79	
CALCIUM	MILLI	24.00	10.00	-	49.00	24.00	MILLI/MJ	306.23	
MANGANESE	MILLI	-	0.05	-	0.20				
IRON	MILLI	0.65	0.60	-	0.70	0.65	MILLI/MJ	8.29	
ZINC	MILLI	-	0.20	-	0.80				
PHOSPHORUS	MILLI	22.00	20.00	-	24.00	22.00	MILLI/MJ	280.71	
FLUORIDE	MICRO	20.00	-	-	-	20.00	MICRO/MJ	255.19	
SELENIUM	MICRO	1.30	-	-	-	1.30	MICRO/MJ	16.59	
CAROTENE	MILLI	7.28	7.20	-	7.35	7.28	MILLI/MJ	92.89	
VITAMIN B1	MICRO	24.00	20.00	-	30.00	24.00	MICRO/MJ	306.23	
VITAMIN B2	MICRO	20.00	-	-	-	20.00	MICRO/MJ	255.19	
NICOTINAMIDE	MILLI	0.30	-	-	-	0.30	MILLI/MJ	3.83	
VITAMIN B6	MICRO	22.00	-	-	-	22.00	MICRO/MJ	280.71	
BIOTIN	MICRO	1.50	-	-	-	1.50	MICRO/MJ	19.14	
VITAMIN C	MILLI	2.50	2.00	-	3.00	2.50	MILLI/MJ	31.90	
ARGININE	MILLI	20.00	8.00	-	32.00	20.00	MILLI/MJ	255.19	
CYSTINE	MILLI	14.00	-	-	-	14.00	MILLI/MJ	178.63	
HISTIDINE	MILLI	8.00	4.00	-	11.00	8.00	MILLI/MJ	102.08	
ISOLEUCINE	MILLI	22.00	16.00	-	28.00	22.00	MILLI/MJ	280.71	
LEUCINE	MILLI	31.00	27.00	-	37.00	31.00	MILLI/MJ	395.54	
LYSINE	MILLI	25.00	13.00	-	35.00	25.00	MILLI/MJ	318.99	
METHIONINE	MILLI	5.00	1.00	-	8.00	5.00	MILLI/MJ	63.80	
PHENYLALANINE	MILLI	20.00	15.00	-	30.00	20.00	MILLI/MJ	255.19	
THREONINE	MILLI	21.00	16.00	-	76.00	21.00	MILLI/MJ	267.95	
TRYPTOPHAN	MILLI	4.00	2.00	-	7.00	4.00	MILLI/MJ	51.04	
TYROSINE	MILLI	9.00	7.00	-	12.00	9.00	MILLI/MJ	114.84	
VALINE	MILLI	27.00	19.00	-	39.00	27.00	MILLI/MJ	344.51	

1 TOTAL CONTENT (SOLID AND LIQUID)

2 CALCULATED FROM THE RAW PRODUCT

MÖHRENSAFT (KAROTTENSAFT) — CARROT JUICE — JUS DE CAROTTES

		PROTEIN	FAT	CARBOHYDRATES	TOTAL
ENERGY VALUE (AVERAGE)	KJOULE	9.9	0.0	82	91
PER 100 G	(KCAL)	2.4	0.0	19	22
EDIBLE PORTION					
AMOUNT OF DIGESTIBLE	GRAM	0.40	0.00	4.87	
CONSTITUENTS PER 100 G					
ENERGY VALUE (AVERAGE)	KJOULE	6.4	0.0	82	88
OF THE DIGESTIBLE	(KCAL)	1.5	0.0	19	21
FRACTION PER 100 G					
EDIBLE PORTION					

WASTE PERCENTAGE AVERAGE 0.00

CONSTITUENTS	DIM	AV	VARIATION			AVR	NUTR. DENS.		MOLPERC.
WATER	GRAM	92.70	-	-	-	92.70	KILO/MJ	1.05	
PROTEIN	GRAM	0.63	0.46	-	0.78	0.63	GRAM/MJ	7.16	
AVAILABLE CARBOHYDR.	GRAM	4.87	-	-	-	4.87	GRAM/MJ	55.39	
MINERALS	GRAM	0.67	0.50	-	0.78	0.67	GRAM/MJ	7.62	
SODIUM	MILLI	52.00	33.00	-	85.00	52.00	MILLI/MJ	591.38	
POTASSIUM	MILLI	219.00	182.00	-	273.00	219.00	GRAM/MJ	2.49	
CALCIUM	MILLI	27.00	13.00	-	48.00	27.00	MILLI/MJ	307.06	
PHOSPHORUS	MILLI	31.00	24.00	-	35.00	31.00	MILLI/MJ	352.56	
CHLORIDE	MILLI	41.00	29.00	-	53.00	41.00	MILLI/MJ	466.28	
NITRITE	MICRO	40.00	0.00	-	340.00	40.00	MICRO/MJ	454.91	
NITRATE	MILLI	23.00	1.60	-	84.00	23.00	MILLI/MJ	261.57	
CAROTENE	MILLI	2.62	-	-	-	2.62	MILLI/MJ	29.80	
VITAMIN C	MILLI	3.78	1.70	-	5.60	3.78	MILLI/MJ	42.99	
TOTAL ACIDS	GRAM	0.12	0.11	-	0.13	0.12	GRAM/MJ	1.36	
INVERT SUGAR	GRAM	2.33	2.03	-	2.63	2.33	GRAM/MJ	26.50	
SUCROSE	GRAM	2.42	-	-	-	2.42	GRAM/MJ	27.52	
EXTRACT	GRAM	8.97	8.30	-	9.64	8.97	GRAM/MJ	102.01	

PASTINAKE PARSNIP PANAIS

PASTINACA EUSATIVA L.

		PROTEIN	FAT	CARBOHYDRATES	TOTAL
ENERGY VALUE (AVERAGE) PER 100 G EDIBLE PORTION	KJOULE	21	17	57	94
	(KCAL)	4.9	4.0	14	22
AMOUNT OF DIGESTIBLE CONSTITUENTS PER 100 G	GRAM	0.96	0.38	3.39	
ENERGY VALUE (AVERAGE) OF THE DIGESTIBLE FRACTION PER 100 G EDIBLE PORTION	KJOULE	15	15	57	87
	(KCAL)	3.6	3.6	14	21

WASTE PERCENTAGE AVERAGE 26 MINIMUM 22 MAXIMUM 30

CONSTITUENTS	DIM	AV	VARIATION			AVR	NUTR. DENS.		MOLPERC.
WATER	GRAM	80.20	78.60	-	82.50	59.35	GRAM/MJ	921.80	
PROTEIN	GRAM	1.31	0.90	-	1.70	0.97	GRAM/MJ	15.06	
FAT	GRAM	0.43	0.30	-	0.53	0.32	GRAM/MJ	4.94	
AVAILABLE CARBOHYDR.	GRAM	3.39	-	-	-	2.51	GRAM/MJ	38.96	
MINERALS	GRAM	1.18	1.14	-	1.20	0.87	GRAM/MJ	13.56	
SODIUM	MILLI	8.00	7.00	-	9.00	5.92	MILLI/MJ	91.95	
POTASSIUM	MILLI	469.00	342.00	-	740.00	347.06	GRAM/MJ	5.39	
MAGNESIUM	MILLI	22.00	21.00	-	22.00	16.28	MILLI/MJ	252.86	
CALCIUM	MILLI	51.00	37.00	-	57.00	37.74	MILLI/MJ	586.18	
MANGANESE	MILLI	0.40	0.25	-	0.63	0.30	MILLI/MJ	4.60	
IRON	MILLI	0.62	0.50	-	0.70	0.46	MILLI/MJ	7.13	
COPPER	MILLI	0.10	-	-	-	0.07	MILLI/MJ	1.15	
ZINC	MILLI	0.85	0.49	-	1.40	0.63	MILLI/MJ	9.77	
NICKEL	MICRO	20.00	4.00	-	30.00	14.80	MICRO/MJ	229.88	
CHROMIUM	MICRO	13.00	-	-	-	9.62	MICRO/MJ	149.42	
PHOSPHORUS	MILLI	73.00	61.00	-	80.00	54.02	MILLI/MJ	839.05	
FLUORIDE	MICRO	10.00	-	-	-	7.40	MICRO/MJ	114.94	
IODIDE	MICRO	3.60	-	-	-	2.66	MICRO/MJ	41.38	
SILICON	MILLI	0.50	-	-	-	0.37	MILLI/MJ	5.75	
CAROTENE	MICRO	20.00	8.00	-	40.00	14.80	MICRO/MJ	229.88	
VITAMIN E ACTIVITY	MILLI	1.00	-	-	-	0.74	MILLI/MJ	11.49	
ALPHA-TOCOPHEROL	MILLI	1.00	-	-	-	0.74	MILLI/MJ	11.49	
VITAMIN B1	MICRO	80.00	50.00	-	100.00	59.20	MICRO/MJ	919.50	
VITAMIN B2	MILLI	0.13	0.08	-	0.18	0.10	MILLI/MJ	1.49	
NICOTINAMIDE	MILLI	0.94	0.10	-	3.20	0.70	MILLI/MJ	10.80	
PANTOTHENIC ACID	MILLI	0.50	-	-	-	0.37	MILLI/MJ	5.75	

CONSTITUENTS	DIM	AV	VARIATION			AVR	NUTR. DENS.		MOLPERC.
VITAMIN B6	MILLI	0.11	0.08	-	0.15	0.08	MILLI/MJ	1.26	
BIOTIN	MICRO	0.10	-	-	-	0.07	MICRO/MJ	1.15	
FOLIC ACID	MICRO	59.00	59.00	-	59.00	43.66	MICRO/MJ	678.13	
VITAMIN C	MILLI	18.00	6.00	-	32.00	13.32	MILLI/MJ	206.89	
ALANINE	MILLI	67.00	67.00	-	67.00	49.58	MILLI/MJ	770.08	9.4
ARGININE	MILLI	139.00	139.00	-	139.00	102.86	GRAM/MJ	1.60	10.0
ASPARTIC ACID	MILLI	131.00	131.00	-	131.00	96.94	GRAM/MJ	1.51	12.3
CYSTINE	MILLI	4.00	4.00	-	4.00	2.96	MILLI/MJ	45.98	0.2
GLUTAMIC ACID	MILLI	123.00	123.00	-	123.00	91.02	GRAM/MJ	1.41	10.5
GLYCINE	MILLI	42.00	42.00	-	42.00	31.08	MILLI/MJ	482.74	7.0
HISTIDINE	MILLI	21.00	21.00	-	21.00	15.54	MILLI/MJ	241.37	1.7
ISOLEUCINE	MILLI	52.00	52.00	-	52.00	38.48	MILLI/MJ	597.68	5.0
LEUCINE	MILLI	70.00	70.00	-	70.00	51.80	MILLI/MJ	804.56	6.7
LYSINE	MILLI	78.00	78.00	-	78.00	57.72	MILLI/MJ	896.52	6.7
METHIONINE	MILLI	16.00	16.00	-	16.00	11.84	MILLI/MJ	183.90	1.3
PHENYLALANINE	MILLI	45.00	45.00	-	45.00	33.30	MILLI/MJ	517.22	3.4
PROLINE	MILLI	44.00	44.00	-	44.00	32.56	MILLI/MJ	505.73	4.8
SERINE	MILLI	48.00	48.00	-	48.00	35.52	MILLI/MJ	551.70	5.7
THREONINE	MILLI	49.00	49.00	-	49.00	36.26	MILLI/MJ	563.20	5.2
TRYPTOPHAN	MILLI	14.00	14.00	-	14.00	10.36	MILLI/MJ	160.91	0.9
TYROSINE	MILLI	26.00	26.00	-	26.00	19.24	MILLI/MJ	298.84	1.8
VALINE	MILLI	70.00	70.00	-	70.00	51.80	MILLI/MJ	804.56	7.5
MALIC ACID	MILLI	350.00	-	-	-	259.00	GRAM/MJ	4.02	
CITRIC ACID	MILLI	130.00	-	-	-	96.20	GRAM/MJ	1.49	
SALICYLIC ACID	MILLI	0.45	0.45	-	0.45	0.33	MILLI/MJ	5.17	
GLUCOSE	GRAM	0.18	0.17	-	0.19	0.14	GRAM/MJ	2.13	
FRUCTOSE	GRAM	0.16	0.14	-	0.23	0.12	GRAM/MJ	1.87	
SUCROSE	GRAM	2.57	2.49	-	2.81	1.90	GRAM/MJ	29.54	
RAFFINOSE	GRAM	0.63	-	-	-	0.47	GRAM/MJ	7.24	
PALMITIC ACID	MILLI	62.00	62.00	-	62.00	45.88	MILLI/MJ	712.61	
STEARIC ACID	MILLI	2.80	2.80	-	2.80	2.07	MILLI/MJ	32.18	
OLEIC ACID	MILLI	24.00	24.00	-	24.00	17.76	MILLI/MJ	275.85	
LINOLEIC ACID	MILLI	68.00	68.00	-	68.00	50.32	MILLI/MJ	781.58	
LINOLENIC ACID	MILLI	21.00	21.00	-	21.00	15.54	MILLI/MJ	241.37	

PETERSILIE
WURZEL
PETROSELINUM SATIVUM
HOFFMANN

PARSLEY
ROOT

PERSIL
RACINE

		PROTEIN	FAT	CARBOHYDRATES	TOTAL
ENERGY VALUE (AVERAGE) PER 100 G EDIBLE PORTION	KJOULE	45	18	91	155
	(KCAL)	11	4.4	22	37
AMOUNT OF DIGESTIBLE CONSTITUENTS PER 100 G	GRAM	2.13	0.42	5.44	
ENERGY VALUE (AVERAGE) OF THE DIGESTIBLE FRACTION PER 100 G EDIBLE PORTION	KJOULE	33	16	91	141
	(KCAL)	8.0	3.9	22	34

WASTE PERCENTAGE AVERAGE 39 MINIMUM 28 MAXIMUM 45

CONSTITUENTS	DIM	AV	VARIATION	AVR	NUTR. DENS.		MOLPERC.
WATER	GRAM	88.00	86.50 - 90.20	53.68	GRAM/MJ	624.37	
PROTEIN	GRAM	2.88	2.11 - 3.36	1.76	GRAM/MJ	20.43	
FAT	GRAM	0.47	- - -	0.29	GRAM/MJ	3.33	
AVAILABLE CARBOHYDR.	GRAM	5.44	- - -	3.32	GRAM/MJ	38.60	
MINERALS	GRAM	1.62	1.51 - 1.74	0.99	GRAM/MJ	11.49	
SODIUM	MILLI	12.00	12.00 - 12.00	7.32	MILLI/MJ	85.14	
POTASSIUM	MILLI	399.00	399.00 - 399.00	243.39	GRAM/MJ	2.83	
MAGNESIUM	MILLI	26.00	26.00 - 26.00	15.86	MILLI/MJ	184.47	
CALCIUM	MILLI	39.00	39.00 - 39.00	23.79	MILLI/MJ	276.71	
IRON	MILLI	0.85	0.85 - 0.85	0.52	MILLI/MJ	6.03	
PHOSPHORUS	MILLI	56.60	49.90 - 69.20	34.53	MILLI/MJ	401.58	
BORON	MILLI	-	0.36 - 0.54				
NITRATE	MILLI	120.00	1.00 - 469.00	73.20	MILLI/MJ	851.42	
CAROTENE	MICRO	10.00	10.00 - 65.00	6.10	MICRO/MJ	70.95	
VITAMIN B1	MILLI	0.10	0.08 - 0.14	0.06	MILLI/MJ	0.71	
VITAMIN B2	MICRO	86.00	85.00 - 88.00	52.46	MICRO/MJ	610.18	
NICOTINAMIDE	MILLI	2.00	1.35 - 2.80	1.22	MILLI/MJ	14.19	
VITAMIN B6	MILLI	0.23	0.18 - 0.33	0.14	MILLI/MJ	1.63	
VITAMIN C	MILLI	41.00	24.00 - 54.00	25.01	MILLI/MJ	290.90	
ALANINE	MILLI	98.00	98.00 - 98.00	59.78	MILLI/MJ	695.32	
ARGININE	MILLI	104.00	104.00 - 104.00	63.44	MILLI/MJ	737.89	
ASPARTIC ACID	MILLI	301.00	301.00 - 301.00	183.61	GRAM/MJ	2.14	
CYSTINE	MILLI	9.00	9.00 - 9.00	5.49	MILLI/MJ	63.86	
GLUTAMIC ACID	MILLI	764.00	764.00 - 764.00	466.04	GRAM/MJ	5.42	
GLYCINE	MILLI	80.00	80.00 - 80.00	48.80	MILLI/MJ	567.61	

CONSTITUENTS	DIM	AV	VARIATION			AVR	NUTR. DENS.		MOLPERC.
ISOLEUCINE	MILLI	89.00	89.00	-	89.00	54.29	MILLI/MJ	631.47	
LEUCINE	MILLI	117.00	117.00	-	117.00	71.37	MILLI/MJ	830.13	
LYSINE	MILLI	113.00	113.00	-	113.00	68.93	MILLI/MJ	801.75	
METHIONINE	MILLI	29.00	29.00	-	29.00	17.69	MILLI/MJ	205.76	
PHENYLALANINE	MILLI	68.00	68.00	-	68.00	41.48	MILLI/MJ	482.47	
PROLINE	MILLI	288.00	288.00	-	288.00	175.68	GRAM/MJ	2.04	
THREONINE	MILLI	80.00	80.00	-	80.00	48.80	MILLI/MJ	567.61	
TRYPTOPHAN	MILLI	28.00	28.00	-	28.00	17.08	MILLI/MJ	198.66	
TYROSINE	MILLI	36.00	36.00	-	36.00	21.96	MILLI/MJ	255.42	
VALINE	MILLI	138.00	138.00	-	138.00	84.18	MILLI/MJ	979.13	
GLUCOSE	GRAM	0.28	0.00	-	0.56	0.17	GRAM/MJ	1.99	
FRUCTOSE	GRAM	0.33	0.00	-	0.66	0.20	GRAM/MJ	2.34	
SUCROSE	GRAM	4.83	3.90	-	5.76	2.95	GRAM/MJ	34.27	
RAFFINOSE	GRAM	0.31	-	-	-	0.19	GRAM/MJ	2.20	
MANNITOL	GRAM	0.82	-	-	-	0.50	GRAM/MJ	5.82	
MYRISTIC ACID	MILLI	0.40	-	-	-	0.24	MILLI/MJ	2.84	
PALMITIC ACID	MILLI	72.00	70.00	-	74.00	43.92	MILLI/MJ	510.85	
STEARIC ACID	MILLI	2.10	1.90	-	2.30	1.28	MILLI/MJ	14.90	
ARACHIDIC ACID	MILLI	1.10	-	-	-	0.67	MILLI/MJ	7.80	
PALMITOLEIC ACID	MILLI	1.10	-	-	-	0.67	MILLI/MJ	7.80	
OLEIC ACID	MILLI	28.50	15.00	-	42.00	17.39	MILLI/MJ	202.21	
EICOSENOIC ACID	MILLI	2.30	-	-	-	1.40	MILLI/MJ	16.32	
LINOLEIC ACID	MILLI	220.50	209.00	-	232.00	134.51	GRAM/MJ	1.56	
LINOLENIC ACID	MILLI	31.50	31.00	-	32.00	19.22	MILLI/MJ	223.50	
DIETARY FIBRE, WAT.INS.	GRAM	4.00	-	-	-	2.44	GRAM/MJ	28.38	

RADIESCHEN RADISHES PETIT RADIS

RAPHANUS SATIVUS L.
VAR. RADICULA PERS.

		PROTEIN	FAT	CARBOHYDRATES	TOTAL
ENERGY VALUE (AVERAGE)	KJOULE	16	5.4	37	59
PER 100 G	(KCAL)	3.9	1.3	8.9	14
EDIBLE PORTION					
AMOUNT OF DIGESTIBLE	GRAM	0.77	0.12	2.22	
CONSTITUENTS PER 100 G					
ENERGY VALUE (AVERAGE)	KJOULE	12	4.9	37	54
OF THE DIGESTIBLE	(KCAL)	2.9	1.2	8.9	13
FRACTION PER 100 G					
EDIBLE PORTION					

WASTE PERCENTAGE AVERAGE 37 MINIMUM 20 MAXIMUM 46

CONSTITUENTS	DIM	AV	VARIATION			AVR	NUTR. DENS.		MOLPERC.
WATER	GRAM	94.40	93.30	-	95.50	59.47	KILO/MJ	1.74	
PROTEIN	GRAM	1.05	0.84	-	1.20	0.66	GRAM/MJ	19.36	
FAT	GRAM	0.14	0.10	-	0.20	0.09	GRAM/MJ	2.58	
AVAILABLE CARBOHYDR.	GRAM	2.22	1.55	-	2.53	1.40	GRAM/MJ	40.92	
TOTAL DIETARY FIBRE	GRAM	1.50 1	-	-	-	0.95	GRAM/MJ	27.65	
MINERALS	GRAM	0.90	0.79	-	1.00	0.57	GRAM/MJ	16.59	
SODIUM	MILLI	17.00	9.00	-	25.00	10.71	MILLI/MJ	313.38	
POTASSIUM	MILLI	255.00	250.00	-	260.00	160.65	GRAM/MJ	4.70	
MAGNESIUM	MILLI	8.00	-	-	-	5.04	MILLI/MJ	147.47	
CALCIUM	MILLI	34.00	30.00	-	37.00	21.42	MILLI/MJ	626.75	
MANGANESE	MILLI	-	0.05	-	0.20				
IRON	MILLI	1.50	1.00	-	2.00	0.95	MILLI/MJ	27.65	
COPPER	MILLI	0.15	0.15	-	0.16	0.09	MILLI/MJ	2.77	
ZINC	MILLI	0.16	-	-	-	0.10	MILLI/MJ	2.95	
NICKEL	MICRO	8.00	-	-	-	5.04	MICRO/MJ	147.47	
VANADIUM	MICRO	-	5.00	-	300.00 2				
PHOSPHORUS	MILLI	26.40	21.30	-	31.00	16.63	MILLI/MJ	486.66	
CHLORIDE	MILLI	44.00	-	-	-	27.72	MILLI/MJ	811.09	
FLUORIDE	MICRO	70.00	60.00	-	80.00	44.10	MILLI/MJ	1.29	
IODIDE	MICRO	8.00	-	-	-	5.04	MICRO/MJ	147.47	
BORON	MILLI	0.10	0.06	-	0.18	0.06	MILLI/MJ	1.84	
NITRATE	MILLI	220.00	8.00	-	453.00	138.60	GRAM/MJ	4.06	
CAROTENE	MICRO	23.00	12.00	-	30.00	14.49	MICRO/MJ	423.98	
VITAMIN B1	MICRO	33.00	18.00	-	50.00	20.79	MICRO/MJ	608.32	
VITAMIN B2	MICRO	30.00	20.00	-	53.00	18.90	MICRO/MJ	553.02	
NICOTINAMIDE	MILLI	0.25	0.20	-	0.30	0.16	MILLI/MJ	4.61	

1 METHOD OF SOUTHGATE AND ENGLYST

2 STRONG REGIONAL DEPENDENCE

CONSTITUENTS	DIM	AV	VARIATION			AVR	NUTR. DENS.		MOLPERC.
PANTOTHENIC ACID	MILLI	0.18	0.18	-	0.19	0.11	MILLI/MJ	3.32	
VITAMIN B6	MICRO	60.00	45.00	-	77.00	37.80	MILLI/MJ	1.11	
FOLIC ACID	MICRO	24.00	-	-	-	15.12	MICRO/MJ	442.41	
VITAMIN C	MILLI	29.00	22.00	-	50.00	18.27	MILLI/MJ	534.58	
CITRIC ACID	MILLI	100.00	3.00	-	200.00	63.00	GRAM/MJ	1.84	
OXALIC ACID TOTAL	-	0.00	-	-	-	0.00			
SALICYLIC ACID	MILLI	1.24	1.24	-	1.24	0.78	MILLI/MJ	22.86	
GLUCOSE	GRAM	1.29	1.00	-	1.37	0.82	GRAM/MJ	23.85	
FRUCTOSE	GRAM	0.71	0.44	-	0.73	0.45	GRAM/MJ	13.22	
SUCROSE	GRAM	0.12	0.11	-	0.23	0.08	GRAM/MJ	2.25	
PENTOSAN	GRAM	0.16	-	-	-	0.10	GRAM/MJ	2.95	
HEXOSAN	GRAM	0.23	-	-	-	0.14	GRAM/MJ	4.24	
CELLULOSE	GRAM	0.70	-	-	-	0.44	GRAM/MJ	12.90	
POLYURONIC ACID	GRAM	0.40	-	-	-	0.25	GRAM/MJ	7.37	
PALMITIC ACID	MILLI	28.00	28.00	-	28.00	17.64	MILLI/MJ	516.15	
STEARIC ACID	MILLI	4.60	4.60	-	4.60	2.90	MILLI/MJ	84.80	
PALMITOLEIC ACID	MILLI	1.50	1.50	-	1.50	0.95	MILLI/MJ	27.65	
OLEIC ACID	MILLI	15.00	15.00	-	15.00	9.45	MILLI/MJ	276.51	
LINOLEIC ACID	MILLI	9.20	9.20	-	9.20	5.80	MILLI/MJ	169.59	
LINOLENIC ACID	MILLI	46.00	46.00	-	46.00	28.98	MILLI/MJ	847.96	
TOTAL PURINES	MILLI	15.00 [3]	-	-	-	9.45	MILLI/MJ	276.51	
DIETARY FIBRE,WAT.SOL.	GRAM	0.19	-	-	-	0.12	GRAM/MJ	3.50	
DIETARY FIBRE,WAT.INS.	GRAM	1.31	-	-	-	0.83	GRAM/MJ	24.15	

3 VALUE EXPRESSED IN MG URIC ACID/100G

RETTICH RADISH RADIS

RAPHANUS SATIVUS L.
VAR. NIGER KERNER

		PROTEIN	FAT	CARBOHYDRATES	TOTAL
ENERGY VALUE (AVERAGE)	KJOULE	16	5.8	31	54
PER 100 G	(KCAL)	3.9	1.4	7.5	13
EDIBLE PORTION					
AMOUNT OF DIGESTIBLE	GRAM	0.77	0.13	1.88	
CONSTITUENTS PER 100 G					
ENERGY VALUE (AVERAGE)	KJOULE	12	5.3	31	49
OF THE DIGESTIBLE	(KCAL)	2.9	1.3	7.5	12
FRACTION PER 100 G					
EDIBLE PORTION					

WASTE PERCENTAGE AVERAGE 24

CONSTITUENTS	DIM	AV	VARIATION			AVR	NUTR. DENS.		MOLPERC.
WATER	GRAM	93.50	92.70	-	94.30	71.06	KILO/MJ	1.91	
PROTEIN	GRAM	1.05	0.90	-	1.10	0.80	GRAM/MJ	21.47	
FAT	GRAM	0.15	0.10	-	0.20	0.11	GRAM/MJ	3.07	
AVAILABLE CARBOHYDR.	GRAM	1.88	-	-	-	1.43	GRAM/MJ	38.44	
TOTAL DIETARY FIBRE	GRAM	1.16 [1]	-	-	-	0.88	GRAM/MJ	23.72	
MINERALS	GRAM	0.75	0.70	-	0.80	0.57	GRAM/MJ	15.33	
SODIUM	MILLI	18.00	-	-	-	13.68	MILLI/MJ	368.04	
POTASSIUM	MILLI	322.00	-	-	-	244.72	GRAM/MJ	6.58	
MAGNESIUM	MILLI	15.00	-	-	-	11.40	MILLI/MJ	306.70	
CALCIUM	MILLI	33.00	30.00	-	35.00	25.08	MILLI/MJ	674.74	
MANGANESE	MICRO	50.00	-	-	-	38.00	MILLI/MJ	1.02	
IRON	MILLI	0.80	0.60	-	1.00	0.61	MILLI/MJ	16.36	
COPPER	MILLI	0.13	-	-	-	0.10	MILLI/MJ	2.66	
ZINC	MILLI	0.20	-	-	-	0.15	MILLI/MJ	4.09	
NICKEL	MICRO	8.00	1.00	-	15.00	6.08	MICRO/MJ	163.57	
CHROMIUM	MICRO	1.00	-	-	-	0.76	MICRO/MJ	20.45	
VANADIUM	MICRO	5.00	3.00	-	7.00	3.80	MICRO/MJ	102.23	
PHOSPHORUS	MILLI	29.00	26.00	-	31.00	22.04	MILLI/MJ	592.95	
CHLORIDE	MILLI	19.00	-	-	-	14.44	MILLI/MJ	388.49	
IODIDE	MICRO	8.00	-	-	-	6.08	MICRO/MJ	163.57	
BORON	MILLI	2.08	1.63	-	2.53	1.58	MILLI/MJ	42.53	
SELENIUM	MICRO	1.90	1.90	-	30.00	1.44	MICRO/MJ	38.85	
SILICON	MILLI	2.00	1.00	-	3.00	1.52	MILLI/MJ	40.89	
NITRATE	MILLI	259.00	30.00	-	496.00	196.84	GRAM/MJ	5.30	
CAROTENE	MICRO	6.00	-	-	-	4.56	MICRO/MJ	122.68	
VITAMIN B1	MICRO	30.00	-	-	-	22.80	MICRO/MJ	613.40	

1. METHOD OF MEUSER, SUCKOW AND KULIKOWSKI ("BERLINER METHODE")

CONSTITUENTS	DIM	AV	VARIATION			AVR	NUTR. DENS.		MOLPERC.
VITAMIN B2	MICRO	30.00	20.00	-	30.00	22.80	MICRO/MJ	613.40	
NICOTINAMIDE	MILLI	0.40	0.30	-	0.40	0.30	MILLI/MJ	8.18	
PANTOTHENIC ACID	MILLI	0.18	-	-	-	0.14	MILLI/MJ	3.68	
VITAMIN B6	MICRO	60.00	-	-	-	45.60	MILLI/MJ	1.23	
FOLIC ACID	MICRO	24.00	-	-	-	18.24	MICRO/MJ	490.72	
VITAMIN C	MILLI	27.00	20.00	-	30.00	20.52	MILLI/MJ	552.06	
OXALIC ACID TOTAL	-	0.00	-	-	-	0.00			
GLUCOSE	GRAM	1.15	0.64	-	1.33	0.88	GRAM/MJ	23.70	
FRUCTOSE	GRAM	0.60	0.39	-	0.68	0.46	GRAM/MJ	12.35	
SUCROSE	GRAM	0.13	0.07	-	0.15	0.10	GRAM/MJ	2.66	
PALMITIC ACID	MILLI	29.00	29.00	-	29.00	22.04	MILLI/MJ	592.95	
STEARIC ACID	MILLI	3.60	3.60	-	3.60	2.74	MILLI/MJ	73.61	
OLEIC ACID	MILLI	15.00	15.00	-	15.00	11.40	MILLI/MJ	306.70	
LINOLEIC ACID	MILLI	18.00	18.00	-	18.00	13.68	MILLI/MJ	368.04	
LINOLENIC ACID	MILLI	55.00	55.00	-	55.00	41.80	GRAM/MJ	1.12	
TOTAL STEROLS	MILLI	11.00	-	-	-	8.36	MILLI/MJ	224.91	
CAMPESTEROL	MILLI	5.00	-	-	-	3.80	MILLI/MJ	102.23	
BETA-SITOSTEROL	MILLI	6.00	-	-	-	4.56	MILLI/MJ	122.68	
TOTAL PURINES	MILLI	15.00 [2]	-	-	-	11.40	MILLI/MJ	306.70	
DIETARY FIBRE,WAT.SOL.	GRAM	0.27	-	-	-	0.21	GRAM/MJ	5.52	
DIETARY FIBRE,WAT.INS.	GRAM	0.89	-	-	-	0.68	GRAM/MJ	18.20	

2 VALUE EXPRESSED IN MG URIC ACID/100G

ROTE RÜBE (ROTE BETE) — BEETROOT — BETTERAVE ROUGE

BETA VULGARIS L. VAR. CRUENTA ALEF.

		PROTEIN	FAT	CARBOHYDRATES	TOTAL
ENERGY VALUE (AVERAGE)	KJOULE	24	3.9	144	172
PER 100 G	(KCAL)	5.7	0.9	34	41
EDIBLE PORTION					
AMOUNT OF DIGESTIBLE	GRAM	1.13	0.09	8.61	
CONSTITUENTS PER 100 G					
ENERGY VALUE (AVERAGE)	KJOULE	18	3.5	144	165
OF THE DIGESTIBLE	(KCAL)	4.2	0.8	34	40
FRACTION PER 100 G					
EDIBLE PORTION					

WASTE PERCENTAGE AVERAGE 22 MINIMUM 20 MAXIMUM 25

CONSTITUENTS	DIM	AV		VARIATION		AVR	NUTR. DENS.		MOLPERC.
WATER	GRAM	88.80		82.90 -	91.70	69.26	GRAM/MJ	537.00	
PROTEIN	GRAM	1.53		1.12 -	2.00	1.19	GRAM/MJ	9.25	
FAT	GRAM	0.10		0.10 -	0.20	0.08	GRAM/MJ	0.60	
AVAILABLE CARBOHYDR.	GRAM	8.61		- -	-	6.72	GRAM/MJ	52.07	
TOTAL DIETARY FIBRE	GRAM	2.53	1	- -	-	1.97	GRAM/MJ	15.30	
MINERALS	GRAM	1.00		0.77 -	1.10	0.78	GRAM/MJ	6.05	
SODIUM	MILLI	58.00		28.00 -	86.00	45.24	MILLI/MJ	350.74	
POTASSIUM	MILLI	336.00		321.00 -	350.00	262.08	GRAM/MJ	2.03	
MAGNESIUM	MILLI	25.00		2.00 -	40.00	19.50	MILLI/MJ	151.18	
CALCIUM	MILLI	29.00		26.00 -	40.00	22.62	MILLI/MJ	175.37	
MANGANESE	MILLI	1.00		0.50 -	1.40	0.78	MILLI/MJ	6.05	
IRON	MILLI	0.93		0.70 -	1.00	0.73	MILLI/MJ	5.62	
COBALT	MICRO	0.70		0.50 -	0.90	0.55	MICRO/MJ	4.23	
COPPER	MILLI	0.19		- -	-	0.15	MILLI/MJ	1.15	
ZINC	MILLI	0.59		0.28 -	0.90	0.46	MILLI/MJ	3.57	
NICKEL	MICRO	10.00		1.00 -	30.00	7.80	MICRO/MJ	60.47	
CHROMIUM	MICRO	1.00		0.10 -	4.00	0.78	MICRO/MJ	6.05	
PHOSPHORUS	MILLI	45.00		30.00 -	66.00	35.10	MILLI/MJ	272.13	
CHLORIDE	MILLI	82.00		56.00 -	100.00	63.96	MILLI/MJ	495.88	
FLUORIDE	MICRO	20.00		- -	-	15.60	MICRO/MJ	120.95	
IODIDE	MICRO	0.40		0.40 -	0.40	0.31	MICRO/MJ	2.42	
BORON	MILLI	2.10		1.00 -	3.45	1.64	MILLI/MJ	12.70	
SELENIUM	MICRO	0.80	2	0.80 -	22.00	0.62	MICRO/MJ	4.84	
SILICON	MILLI	0.40		0.10 -	1.00	0.31	MILLI/MJ	2.42	
NITRATE	MILLI	195.00	3	18.00 -	536.00	152.10	GRAM/MJ	1.18	
CAROTENE	MICRO	11.00		10.00 -	12.00	8.58	MICRO/MJ	66.52	

1 METHOD OF SOUTHGATE AND ENGLYST

2 AMOUNTS, DETERMINED IN THE FEDERAL REPUBLIC OF GERMANY

3 THE AMOUNTS ARE DEPENDENT ON THE COMPOSITION OF THE SOIL, ON THE CLIMATE AND ON THE TIME OF THE HARVEST

CONSTITUENTS	DIM	AV	VARIATION			AVR	NUTR. DENS.		MOLPERC.
VITAMIN E ACTIVITY	MICRO	30.00	-	-	-	23.40	MICRO/MJ	181.42	
ALPHA-TOCOPHEROL	MICRO	30.00	-	-	-	23.40	MICRO/MJ	181.42	
VITAMIN B1	MICRO	22.00	20.00	-	25.00	17.16	MICRO/MJ	133.04	
VITAMIN B2	MICRO	42.00	30.00	-	45.00	32.76	MICRO/MJ	253.99	
NICOTINAMIDE	MILLI	0.23	0.10	-	0.40	0.18	MILLI/MJ	1.39	
PANTOTHENIC ACID	MILLI	0.13	0.13	-	0.13	0.10	MILLI/MJ	0.79	
VITAMIN B6	MICRO	50.00	-	-	-	39.00	MICRO/MJ	302.36	
FOLIC ACID	MICRO	93.00	-	-	-	72.54	MICRO/MJ	562.40	
VITAMIN C	MILLI	10.00	8.00	-	13.00	7.80	MILLI/MJ	60.47	
ARGININE	MILLI	27.00	-	-	-	21.06	MILLI/MJ	163.28	
HISTIDINE	MILLI	21.00	-	-	-	16.38	MILLI/MJ	126.99	
ISOLEUCINE	MILLI	49.00	-	-	-	38.22	MILLI/MJ	296.32	
LEUCINE	MILLI	53.00	-	-	-	41.34	MILLI/MJ	320.51	
LYSINE	MILLI	82.00	-	-	-	63.96	MILLI/MJ	495.88	
METHIONINE	MILLI	5.00	-	-	-	3.90	MILLI/MJ	30.24	
PHENYLALANINE	MILLI	26.00	-	-	-	20.28	MILLI/MJ	157.23	
THREONINE	MILLI	33.00	-	-	-	25.74	MILLI/MJ	199.56	
TRYPTOPHAN	MILLI	13.00	-	-	-	10.14	MILLI/MJ	78.61	
VALINE	MILLI	47.00	-	-	-	36.66	MILLI/MJ	284.22	
MALIC ACID	MILLI	37.00	-	-	-	28.86	MILLI/MJ	223.75	
CITRIC ACID	MILLI	195.00	-	-	-	152.10	GRAM/MJ	1.18	
OXALIC ACID TOTAL	MILLI	181.00	89.00	-	327.00	141.18	GRAM/MJ	1.09	
OXALIC ACID SOLUBLE	MILLI	116.00	56.00	-	209.00	90.48	MILLI/MJ	701.49	
FERULIC ACID	MILLI	6.00	2.00	-	10.00	4.68	MILLI/MJ	36.28	
CAFFEIC ACID	MILLI	0.50	0.00	-	1.00	0.39	MILLI/MJ	3.02	
PARA-COUMARIC ACID	MILLI	0.50	0.00	-	1.00	0.39	MILLI/MJ	3.02	
SUCCINIC ACID	MILLI	10.00	-	-	-	7.80	MILLI/MJ	60.47	
FUMARIC ACID	MILLI	2.00	-	-	-	1.56	MILLI/MJ	12.09	
GLYCOLIC ACID	MILLI	23.00	-	-	-	17.94	MILLI/MJ	139.09	
SALICYLIC ACID	MILLI	0.18	0.18	-	0.18	0.14	MILLI/MJ	1.09	
GLUCOSE	GRAM	0.27	0.07	-	0.38	0.21	GRAM/MJ	1.66	
FRUCTOSE	GRAM	0.25	0.05	-	0.35	0.20	GRAM/MJ	1.52	
SUCROSE	GRAM	7.85	6.12	-	8.88	6.13	GRAM/MJ	47.52	
PENTOSAN	GRAM	1.10	-	-	-	0.86	GRAM/MJ	6.65	
HEXOSAN	GRAM	0.50	-	-	-	0.39	GRAM/MJ	3.02	
CELLULOSE	GRAM	0.80	-	-	-	0.62	GRAM/MJ	4.84	
POLYURONIC ACID	GRAM	0.20	-	-	-	0.16	GRAM/MJ	1.21	
PALMITIC ACID	MILLI	18.00	18.00	-	18.00	14.04	MILLI/MJ	108.85	
STEARIC ACID	MICRO	80.00	80.00	-	80.00	62.40	MICRO/MJ	483.78	
OLEIC ACID	MILLI	9.90	9.90	-	9.90	7.72	MILLI/MJ	59.87	
LINOLEIC ACID	MILLI	41.00	41.00	-	41.00	31.98	MILLI/MJ	247.94	
LINOLENIC ACID	MILLI	8.20	8.20	-	8.20	6.40	MILLI/MJ	49.59	
TOTAL STEROLS	MILLI	25.00	-	-	-	19.50	MILLI/MJ	151.18	
TOTAL PURINES	MILLI	15.00 [4]	-	-	-	11.70	MILLI/MJ	90.71	
DIETARY FIBRE, WAT.SOL.	GRAM	0.48	-	-	-	0.37	GRAM/MJ	2.90	
DIETARY FIBRE, WAT.INS.	GRAM	2.05	-	-	-	1.60	GRAM/MJ	12.40	

4 VALUE EXPRESSED IN MG URIC ACID/100G

ROTE-RÜBEN-SAFT — BEETROOT JUICE — JUS DE BETTERAVES ROUGES

		PROTEIN	FAT	CARBOHYDRATES	TOTAL
ENERGY VALUE (AVERAGE) PER 100 G EDIBLE PORTION	KJOULE	17	0.0	136	153
	(KCAL)	4.1	0.0	32	36
AMOUNT OF DIGESTIBLE CONSTITUENTS PER 100 G	GRAM	0.79	0.00	8.11	
ENERGY VALUE (AVERAGE) OF THE DIGESTIBLE FRACTION PER 100 G EDIBLE PORTION	KJOULE	13	0.0	136	148
	(KCAL)	3.0	0.0	32	35

WASTE PERCENTAGE AVERAGE 0.00

CONSTITUENTS	DIM	AV	VARIATION			AVR	NUTR. DENS.		MOLPERC.
WATER	GRAM	88.40	-	-	-	88.40	GRAM/MJ	596.22	
PROTEIN	GRAM	1.08	-	-	-	1.08	GRAM/MJ	7.28	
AVAILABLE CARBOHYDR.	GRAM	8.11	-	-	-	8.11	GRAM/MJ	54.70	
MINERALS	GRAM	0.99	-	-	-	0.99	GRAM/MJ	6.68	
SODIUM	MILLI	200.00	-	-	-	200.00	GRAM/MJ	1.35	
POTASSIUM	MILLI	242.00	-	-	-	242.00	GRAM/MJ	1.63	
PHOSPHORUS	MILLI	28.80	-	-	-	28.80	MILLI/MJ	194.24	
NITRITE	MILLI	0.64	0.45	-	1.80	0.64	MILLI/MJ	4.32	
NITRATE	MILLI	169.00	97.00	-	217.00	169.00	GRAM/MJ	1.14	
VITAMIN C	MILLI	2.90	-	-	-	2.90	MILLI/MJ	19.56	
CITRIC ACID	GRAM	0.12	-	-	-	0.12	GRAM/MJ	0.81	
INVERT SUGAR	GRAM	1.89	-	-	-	1.89	GRAM/MJ	12.75	
SUCROSE	GRAM	6.10	-	-	-	6.10	GRAM/MJ	41.14	

SCHWARZWURZEL — VIPER'S GRASS (BLACK SALSIFY) — SALSIFIS NOIR (SCORSONÈRE)

SCORZONERA HISPANICA L.

		PROTEIN	FAT	CARBOHYDRATES	TOTAL
ENERGY VALUE (AVERAGE) PER 100 G EDIBLE PORTION	KJOULE	22	17	27	66
	(KCAL)	5.2	4.0	6.5	16
AMOUNT OF DIGESTIBLE CONSTITUENTS PER 100 G	GRAM	1.02	0.38	1.63	
ENERGY VALUE (AVERAGE) OF THE DIGESTIBLE FRACTION PER 100 G EDIBLE PORTION	KJOULE	16	15	27	58
	(KCAL)	3.9	3.6	6.5	14

WASTE PERCENTAGE AVERAGE 44

CONSTITUENTS	DIM	AV	VARIATION	AVR	NUTR.	DENS.	MOLPERC.
WATER	GRAM	78.60	77.50 - 80.40	44.02	KILO/MJ	1.34	
PROTEIN	GRAM	1.39	1.00 - 2.51	0.78	GRAM/MJ	23.77	
FAT	GRAM	0.43	0.23 - 0.50	0.24	GRAM/MJ	7.35	
AVAILABLE CARBOHYDR.	GRAM	1.63	- - -	0.91	GRAM/MJ	27.87	
TOTAL DIETARY FIBRE	GRAM	16.96 1	- - -	9.50	GRAM/MJ	290.03	
MINERALS	GRAM	0.99	- - -	0.55	GRAM/MJ	16.93	
SODIUM	MILLI	5.00	- - -	2.80	MILLI/MJ	85.50	
POTASSIUM	MILLI	320.00	240.00 - 400.00	179.20	GRAM/MJ	5.47	
MAGNESIUM	MILLI	23.00	- - -	12.88	MILLI/MJ	393.32	
CALCIUM	MILLI	53.00	46.00 - 60.00	29.68	MILLI/MJ	906.34	
MANGANESE	MILLI	0.41	- - -	0.23	MILLI/MJ	7.01	
IRON	MILLI	3.30	1.50 - 5.00	1.85	MILLI/MJ	56.43	
COPPER	MILLI	0.30	- - -	0.17	MILLI/MJ	5.13	
ZINC	MILLI	0.22	- - -	0.12	MILLI/MJ	3.76	
PHOSPHORUS	MILLI	75.70	50.00 - 105.00	42.39	GRAM/MJ	1.29	
CHLORIDE	MILLI	31.00	- - -	17.36	MILLI/MJ	530.12	
BORON	MILLI	0.18	0.18 - 0.19	0.10	MILLI/MJ	3.16	
NITRATE	MILLI	31.00	17.00 - 40.00	17.36	MILLI/MJ	530.12	
CAROTENE	MICRO	20.00	19.00 - 20.00	11.20	MICRO/MJ	342.01	
VITAMIN B1	MILLI	0.11	0.07 - 0.15	0.06	MILLI/MJ	1.88	
VITAMIN B2	MICRO	35.00	20.00 - 50.00	19.60	MICRO/MJ	598.53	
NICOTINAMIDE	MILLI	0.35	0.30 - 0.40	0.20	MILLI/MJ	5.99	
VITAMIN C	MILLI	4.00	3.00 - 5.00	2.24	MILLI/MJ	68.40	
OXALIC ACID SOLUBLE	-	0.00	- - -	0.00			
FERULIC ACID	MILLI	0.50	0.00 - 1.00	0.28	MILLI/MJ	8.55	

1 ESTIMATED BY THE DIFFERENCE METHOD:
100 - (WATER + PROTEIN + FAT + AVAILABLE CARBOHYDRATES + MINERALS)

A LARGE PART OF THE CARBOHYDRATE AMOUNT IS INULIN, WHICH IS NOT DIGESTED BY THE ENZYME SYSTEM OF THE HUMAN INTESTINE

CONSTITUENTS	DIM	AV	VARIATION			AVR	NUTR. DENS.		MOLPERC.
CAFFEIC ACID	MILLI	9.50	5.00	-	14.00	5.32	MILLI/MJ	162.46	
PARA-COUMARIC ACID	MILLI	7.00	5.00	-	9.00	3.92	MILLI/MJ	119.71	
GLUCOSE	-	TRACES	TRACES	-					
FRUCTOSE	MILLI	70.00	70.00	-	70.00	39.20	GRAM/MJ	1.20	
SUCROSE	GRAM	1.56	1.56	-	1.56	0.87	GRAM/MJ	26.68	
RAFFINOSE	GRAM	1.56	-	-	-	0.87	GRAM/MJ	26.68	
STACHYOSE	GRAM	1.12	-	-	-	0.63	GRAM/MJ	19.15	
MANNITOL	MILLI	70.00	-	-	-	39.20	GRAM/MJ	1.20	

SCHWARZWURZEL
GEKOCHT, ABGETROPFT

VIPER'S GRASS
(BLACK SALSIFY)
BOILED, DRAINED

SALSIFIS NOIR
(SCORSONÈRE)
BOUILLI, DEGOUTTÉ

		PROTEIN	FAT	CARBOHYDRATES	TOTAL
ENERGY VALUE (AVERAGE)	KJOULE	20	16	33	69
PER 100 G EDIBLE PORTION	(KCAL)	4.9	3.7	8.0	17
AMOUNT OF DIGESTIBLE CONSTITUENTS PER 100 G	GRAM	0.84	0.36	2.00	
ENERGY VALUE (AVERAGE)	KJOULE	13	14	33	61
OF THE DIGESTIBLE FRACTION PER 100 G EDIBLE PORTION	(KCAL)	3.2	3.3	8.0	15

WASTE PERCENTAGE AVERAGE 0.00

CONSTITUENTS	DIM	AV	VARIATION			AVR	NUTR. DENS.		MOLPERC.
WATER	GRAM	82.10	-	-	-	82.10	KILO/MJ	1.35	
PROTEIN	GRAM	1.30	-	-	-	1.30	GRAM/MJ	21.40	
FAT	GRAM	0.40	-	-	-	0.40	GRAM/MJ	6.59	
AVAILABLE CARBOHYDR.	GRAM	2.00	-	-	-	2.00	GRAM/MJ	32.93	
MINERALS	GRAM	0.80	-	-	-	0.80	GRAM/MJ	13.17	
SODIUM	MILLI	4.00	-	-	-	4.00	MILLI/MJ	65.86	
POTASSIUM	MILLI	224.00	-	-	-	224.00	GRAM/MJ	3.69	
CALCIUM	MILLI	47.00	-	-	-	47.00	MILLI/MJ	773.81	
IRON	MILLI	2.90	-	-	-	2.90	MILLI/MJ	47.75	
PHOSPHORUS	MILLI	61.00	-	-	-	61.00	GRAM/MJ	1.00	
VITAMIN B1	MICRO	80.00	-	-	-	80.00	MILLI/MJ	1.32	
VITAMIN B2	MICRO	40.00	-	-	-	40.00	MICRO/MJ	658.57	
NICOTINAMIDE	MILLI	0.23	-	-	-	0.23	MILLI/MJ	3.79	
VITAMIN C	MILLI	3.00	-	-	-	3.00	MILLI/MJ	49.39	
GLUCOSE	MILLI	70.00	-	-	-	70.00	GRAM/MJ	1.15	
FRUCTOSE	GRAM	1.93	-	-	-	1.93	GRAM/MJ	31.78	

SELLERIE KNOLLE — CELERIAC — CÉLÉRI-RAVE

APIUM GRAVEOLENS L.
VAR. RAPACEUM

		PROTEIN	FAT	CARBOHYDRATES	TOTAL
ENERGY VALUE (AVERAGE)	KJOULE	24	13	38	75
PER 100 G	(KCAL)	5.8	3.1	9.0	18
EDIBLE PORTION					
AMOUNT OF DIGESTIBLE	GRAM	1.14	0.29	2.25	
CONSTITUENTS PER 100 G					
ENERGY VALUE (AVERAGE)	KJOULE	18	12	38	67
OF THE DIGESTIBLE	(KCAL)	4.3	2.8	9.0	16
FRACTION PER 100 G					
EDIBLE PORTION					

WASTE PERCENTAGE AVERAGE 27 MINIMUM 14 MAXIMUM 37

CONSTITUENTS	DIM	AV		VARIATION		AVR	NUTR. DENS.		MOLPERC.
WATER	GRAM	88.60	87.30	-	90.50	64.68	KILO/MJ	1.32	
PROTEIN	GRAM	1.55	1.20	-	2.00	1.13	GRAM/MJ	23.06	
FAT	GRAM	0.33	0.20	-	0.46	0.24	GRAM/MJ	4.91	
AVAILABLE CARBOHYDR.	GRAM	2.25	-	-	-	1.64	GRAM/MJ	33.48	
TOTAL DIETARY FIBRE	GRAM	4.23 1	-	-	-	3.09	GRAM/MJ	62.94	
MINERALS	GRAM	0.94	0.91	-	0.97	0.69	GRAM/MJ	13.99	
SODIUM	MILLI	77.00	75.00	-	79.00	56.21	GRAM/MJ	1.15	
POTASSIUM	MILLI	321.00	276.00	-	350.00	234.33	GRAM/MJ	4.78	
MAGNESIUM	MILLI	9.30	-	-	-	6.79	MILLI/MJ	138.37	
CALCIUM	MILLI	68.00	35.00	-	88.00	49.64	GRAM/MJ	1.01	
MANGANESE	MILLI	0.15	0.14	-	0.16	0.11	MILLI/MJ	2.23	
IRON	MILLI	0.53	0.09	-	1.00	0.39	MILLI/MJ	7.89	
COPPER	MICRO	20.00	10.00	-	20.00	14.60	MICRO/MJ	297.58	
ZINC	MILLI	0.31	-	-	-	0.23	MILLI/MJ	4.61	
NICKEL	MICRO	5.00	4.00	-	10.00	3.65	MICRO/MJ	74.39	
CHROMIUM	MICRO	0.50	0.20	-	0.50	0.37	MICRO/MJ	7.44	
MOLYBDENUM	MICRO	2.00	-	-	-	1.46	MICRO/MJ	29.76	
PHOSPHORUS	MILLI	80.00	52.00	-	99.00	58.40	GRAM/MJ	1.19	
CHLORIDE	MILLI	150.00	-	-	-	109.50	GRAM/MJ	2.23	
FLUORIDE	MICRO	14.00	-	-	-	10.22	MICRO/MJ	208.31	
IODIDE	MICRO	2.83	0.20	-	5.45	2.07	MICRO/MJ	42.11	
BORON	MILLI	1.06	0.59	-	1.57	0.77	MILLI/MJ	15.77	
SELENIUM	MICRO	1.10	1.10	-	10.00	0.80	MICRO/MJ	16.37	
NITRATE	MILLI	98.00	7.00	-	364.00	71.54	GRAM/MJ	1.46	
CAROTENE	MICRO	15.00	10.00	-	30.00	10.95	MICRO/MJ	223.18	
VITAMIN K	MILLI	0.10	-	-	-	0.07	MILLI/MJ	1.49	

1 METHOD OF SOUTHGATE AND ENGLYST

CONSTITUENTS	DIM	AV	VARIATION			AVR	NUTR. DENS.		MOLPERC.
VITAMIN B1	MICRO	36.00	25.00	-	50.00	26.28	MICRO/MJ	535.64	
VITAMIN B2	MICRO	70.00	35.00	-	140.00	51.10	MILLI/MJ	1.04	
NICOTINAMIDE	MILLI	0.90	0.80	-	1.00	0.66	MILLI/MJ	13.39	
PANTOTHENIC ACID	MILLI	0.51	0.51	-	0.51	0.37	MILLI/MJ	7.59	
VITAMIN B6	MILLI	0.20	0.20	-	0.21	0.15	MILLI/MJ	2.98	
FOLIC ACID	MICRO	7.00	-	-	-	5.11	MICRO/MJ	104.15	
VITAMIN C	MILLI	8.25	7.00	-	11.00	6.02	MILLI/MJ	122.75	
ALANINE	MILLI	86.00	86.00	-	86.00	62.78	GRAM/MJ	1.28	10.7
ARGININE	MILLI	44.00	44.00	-	44.00	32.12	MILLI/MJ	654.67	2.8
ASPARTIC ACID	MILLI	164.00	164.00	-	164.00	119.72	GRAM/MJ	2.44	13.7
CYSTINE	MILLI	4.00	4.00	-	4.00	2.92	MILLI/MJ	59.52	0.2
GLUTAMIC ACID	MILLI	283.00	283.00	-	283.00	206.59	GRAM/MJ	4.21	21.3
GLYCINE	MILLI	47.00	47.00	-	47.00	34.31	MILLI/MJ	699.31	6.9
HISTIDINE	MILLI	24.00	24.00	-	24.00	17.52	MILLI/MJ	357.09	1.7
ISOLEUCINE	MILLI	48.00	48.00	-	48.00	35.04	MILLI/MJ	714.19	4.1
LEUCINE	MILLI	75.00	75.00	-	75.00	54.75	GRAM/MJ	1.12	6.3
LYSINE	MILLI	74.00	74.00	-	74.00	54.02	GRAM/MJ	1.10	5.6
METHIONINE	MILLI	18.00	18.00	-	18.00	13.14	MILLI/MJ	267.82	1.3
PHENYLALANINE	MILLI	47.00	47.00	-	47.00	34.31	MILLI/MJ	699.31	3.2
PROLINE	MILLI	40.00	40.00	-	40.00	29.20	MILLI/MJ	595.16	3.9
SERINE	MILLI	49.00	49.00	-	49.00	35.77	MILLI/MJ	729.07	5.2
THREONINE	MILLI	44.00	44.00	-	44.00	32.12	MILLI/MJ	654.67	4.1
TRYPTOPHAN	MILLI	12.00	12.00	-	12.00	8.76	MILLI/MJ	178.55	0.7
TYROSINE	MILLI	25.00	25.00	-	25.00	18.25	MILLI/MJ	371.97	1.5
VALINE	MILLI	73.00	73.00	-	73.00	53.29	GRAM/MJ	1.09	6.9
OXALIC ACID TOTAL	MILLI	6.80	-	-	-	4.96	MILLI/MJ	101.18	
OXALIC ACID SOLUBLE	MILLI	2.10	-	-	-	1.53	MILLI/MJ	31.25	
FERULIC ACID	MILLI	5.00	3.00	-	6.00	3.65	MILLI/MJ	74.39	
CAFFEIC ACID	MILLI	11.00	8.00	-	13.00	8.03	MILLI/MJ	163.67	
PARA-COUMARIC ACID	MILLI	0.20	-	-	-	0.15	MILLI/MJ	2.98	
FRUCTOSE	GRAM	0.10	0.10	-	0.10	0.07	GRAM/MJ	1.49	
SUCROSE	GRAM	1.71	-	-	-	1.25	GRAM/MJ	25.44	
STARCH	GRAM	0.44	0.44	-	0.44	0.32	GRAM/MJ	6.55	
PENTOSAN	GRAM	0.98	-	-	-	0.72	GRAM/MJ	14.58	
HEXOSAN	GRAM	0.77	-	-	-	0.56	GRAM/MJ	11.46	
CELLULOSE	GRAM	1.40	-	-	-	1.02	GRAM/MJ	20.83	
POLYURONIC ACID	GRAM	1.13	-	-	-	0.82	GRAM/MJ	16.81	
PALMITIC ACID	MILLI	63.00	63.00	-	63.00	45.99	MILLI/MJ	937.37	
STEARIC ACID	MILLI	3.50	3.50	-	3.50	2.56	MILLI/MJ	52.08	
OLEIC ACID	MILLI	13.00	13.00	-	13.00	9.49	MILLI/MJ	193.43	
LINOLEIC ACID	MILLI	156.00	156.00	-	156.00	113.88	GRAM/MJ	2.32	
LINOLENIC ACID	MILLI	17.00	17.00	-	17.00	12.41	MILLI/MJ	252.94	
TOTAL PURINES	MILLI	30.00 2	-	-	-	21.90	MILLI/MJ	446.37	
DIETARY FIBRE,WAT.SOL.	GRAM	0.55	-	-	-	0.40	GRAM/MJ	8.18	
DIETARY FIBRE,WAT.INS.	GRAM	3.68	-	-	-	2.69	GRAM/MJ	54.75	

2 VALUE EXPRESSED IN MG URIC ACID/100G

TARO (WASSERBROTWURZEL) | **TARO** (COCO-YAM) | **TARO** (COLOCASIE)

COLOCASIA ESCULENTA L. SCHOTT, VAR. ANTIQUORUM

		PROTEIN	FAT	CARBOHYDRATES	TOTAL
ENERGY VALUE (AVERAGE)	KJOULE	31	9.7	402	443
PER 100 G	(KCAL)	7.5	2.3	96	106
EDIBLE PORTION					
AMOUNT OF DIGESTIBLE	GRAM	1.48	0.22	24.00	
CONSTITUENTS PER 100 G					
ENERGY VALUE (AVERAGE)	KJOULE	23	8.8	402	434
OF THE DIGESTIBLE	(KCAL)	5.6	2.1	96	104
FRACTION PER 100 G					
EDIBLE PORTION					

WASTE PERCENTAGE AVERAGE 16 MINIMUM 14 MAXIMUM 20

CONSTITUENTS	DIM	AV	VARIATION			AVR	NUTR. DENS.		MOLPERC.
WATER	GRAM	72.00	66.70	-	75.40	60.48	GRAM/MJ	166.04	
PROTEIN	GRAM	2.00	1.70	-	2.20	1.68	GRAM/MJ	4.61	
FAT	GRAM	0.25	0.10	-	0.40	0.21	GRAM/MJ	0.58	
AVAILABLE CARBOHYDR.	GRAM	24.00	-	-	-	20.16	GRAM/MJ	55.35	
TOTAL DIETARY FIBRE	GRAM	3.80 1	-	-	-	3.19	GRAM/MJ	8.76	
MINERALS	GRAM	1.00	0.90	-	1.20	0.84	GRAM/MJ	2.31	
SODIUM	MILLI	5.00	1.00	-	10.00	4.20	MILLI/MJ	11.53	
POTASSIUM	MILLI	433.00	340.00	-	510.00	363.72	MILLI/MJ	998.52	
MAGNESIUM	MILLI	31.00	28.00	-	34.00	26.04	MILLI/MJ	71.49	
CALCIUM	MILLI	31.00	19.00	-	51.00	26.04	MILLI/MJ	71.49	
IRON	MILLI	1.10	0.60	-	1.70	0.92	MILLI/MJ	2.54	
ZINC	MILLI	1.70	0.20	-	6.30	1.43	MILLI/MJ	3.92	
PHOSPHORUS	MILLI	61.00	60.00	-	62.00	51.24	MILLI/MJ	140.67	
CAROTENE	MICRO	10.00	-	-	-	8.40	MICRO/MJ	23.06	
VITAMIN B1	MILLI	0.12	0.10	-	0.15	0.10	MILLI/MJ	0.28	
VITAMIN B2	MICRO	35.00	20.00	-	40.00	29.40	MICRO/MJ	80.71	
NICOTINAMIDE	MILLI	1.00	0.80	-	1.10	0.84	MILLI/MJ	2.31	
VITAMIN C	MILLI	6.00	4.00	-	8.00	5.04	MILLI/MJ	13.84	
MALIC ACID	MILLI	175.00	80.00	-	380.00	147.00	MILLI/MJ	403.56	
CITRIC ACID	MILLI	68.00	30.00	-	100.00	57.12	MILLI/MJ	156.81	
OXALIC ACID	MILLI	36.00	20.00	-	60.00	30.24	MILLI/MJ	83.02	
GLUCOSE	GRAM	0.40	0.20	-	1.00	0.34	GRAM/MJ	0.92	
FRUCTOSE	GRAM	0.40	0.20	-	0.70	0.34	GRAM/MJ	0.92	
SUCROSE	GRAM	0.10	0.00	-	0.50	0.08	GRAM/MJ	0.23	
MALTOSE	GRAM	0.10	0.00	-	0.20	0.08	GRAM/MJ	0.23	
STARCH	GRAM	23.00	15.00	-	26.00	19.32	GRAM/MJ	53.04	
DIETARY FIBRE, WAT.SOL.	GRAM	1.70	-	-	-	1.43	GRAM/MJ	3.92	
DIETARY FIBRE, WAT.INS.	GRAM	2.10	-	-	-	1.76	GRAM/MJ	4.84	

1 METHOD OF SOUTHGATE

TOPINAMBUR (ERDARTISCHOCKE) — TOPINAMBOUR (JERUSALEM-ARTICHOKE) — TOPINAMBOUR

HELLIANTHUS TUBEROSUS L.

		PROTEIN	FAT	CARBOHYDRATES	TOTAL
ENERGY VALUE (AVERAGE)	KJOULE	38	16	67	121
PER 100 G	(KCAL)	9.2	3.8	16	29
EDIBLE PORTION					
AMOUNT OF DIGESTIBLE	GRAM	1.80	0.36	4.00	
CONSTITUENTS PER 100 G					
ENERGY VALUE (AVERAGE)	KJOULE	28	14	67	110
OF THE DIGESTIBLE	(KCAL)	6.8	3.4	16	26
FRACTION PER 100 G					
EDIBLE PORTION					

WASTE PERCENTAGE AVERAGE 31

CONSTITUENTS	DIM	AV	VARIATION	AVR	NUTR. DENS.		MOLPERC.
WATER	GRAM	78.90	67.00 - 81.30	54.44	GRAM/MJ	719.68	
PROTEIN	GRAM	2.44	1.85 - 3.17	1.68	GRAM/MJ	22.26	
FAT	GRAM	0.41	0.10 - 0.72	0.28	GRAM/MJ	3.74	
AVAILABLE CARBOHYDR.	GRAM	4.00 [1]	- - -	2.76	GRAM/MJ	36.49	
TOTAL DIETARY FIBRE	GRAM	12.51	- - -	8.63	GRAM/MJ	114.11	
MINERALS	GRAM	1.74	1.10 - 2.84	1.20	GRAM/MJ	15.87	
POTASSIUM	MILLI	478.00	428.00 - 524.00	329.82	GRAM/MJ	4.36	
MAGNESIUM	MILLI	20.00	11.00 - 24.00	13.80	MILLI/MJ	182.43	
CALCIUM	MILLI	10.00	6.00 - 14.00	6.90	MILLI/MJ	91.21	
IRON	MILLI	3.70	3.40 - 4.00	2.55	MILLI/MJ	33.75	
PHOSPHORUS	MILLI	78.00	- - -	53.82	MILLI/MJ	711.47	
CAROTENE	MICRO	12.00	- - -	8.28	MICRO/MJ	109.46	
VITAMIN B1	MILLI	0.20	- - -	0.14	MILLI/MJ	1.82	
VITAMIN B2	MICRO	60.00	- - -	41.40	MICRO/MJ	547.28	
NICOTINAMIDE	MILLI	1.30	- - -	0.90	MILLI/MJ	11.86	
VITAMIN C	MILLI	4.00	- - -	2.76	MILLI/MJ	36.49	
MALIC ACID	MILLI	200.00	180.00 - 220.00	138.00	GRAM/MJ	1.82	
CITRIC ACID	MILLI	235.00	225.00 - 250.00	162.15	GRAM/MJ	2.14	
SUCCINIC ACID	MILLI	7.00	- - -	4.83	MILLI/MJ	63.85	
FUMARIC ACID	MILLI	12.00	- - -	8.28	MILLI/MJ	109.46	
SUCROSE	GRAM	4.00	3.00 - 5.00	2.76	GRAM/MJ	36.49	
PALMITIC ACID	MILLI	90.00	90.00 - 90.00	62.10	MILLI/MJ	820.93	
STEARIC ACID	MILLI	5.10	5.10 - 5.10	3.52	MILLI/MJ	46.52	
OLEIC ACID	MILLI	7.70	7.70 - 7.70	5.31	MILLI/MJ	70.23	
LINOLEIC ACID	MILLI	166.00	166.00 - 166.00	114.54	GRAM/MJ	1.51	
LINOLENIC ACID	MILLI	43.00	43.00 - 43.00	29.67	MILLI/MJ	392.22	

1 A LARGE PART OF CARBOHYDRATE IN THE FRESH TUBER IS INULIN, WHICH IS NOT DIGESTED BY THE ENZYME SYSTEM OF THE HUMAN INTESTINE

WEISSE RÜBE (WASSERRÜBE, HERBSTRÜBE) — **TURNIP** — **NAVET**

BRASSICA RAPA VAR. RAPIFERA METZGER

		PROTEIN	FAT	CARBOHYDRATES	TOTAL
ENERGY VALUE (AVERAGE) PER 100 G EDIBLE PORTION	KJOULE	16	8.6	78	102
	(KCAL)	3.7	2.0	19	24
AMOUNT OF DIGESTIBLE CONSTITUENTS PER 100 G	GRAM	0.73	0.19	4.66	
ENERGY VALUE (AVERAGE) OF THE DIGESTIBLE FRACTION PER 100 G EDIBLE PORTION	KJOULE	11	7.7	78	97
	(KCAL)	2.7	1.8	19	23

WASTE PERCENTAGE AVERAGE 31 MINIMUM 27 MAXIMUM 34

CONSTITUENTS	DIM	AV	VARIATION			AVR	NUTR. DENS.		MOLPERC.
WATER	GRAM	90.50	89.50	-	91.30	62.45	GRAM/MJ	931.18	
PROTEIN	GRAM	0.99	0.80	-	1.11	0.68	GRAM/MJ	10.19	
FAT	GRAM	0.22	0.20	-	0.24	0.15	GRAM/MJ	2.26	
AVAILABLE CARBOHYDR.	GRAM	4.66	-	-	-	3.22	GRAM/MJ	47.95	
MINERALS	GRAM	0.73	0.70	-	0.76	0.50	GRAM/MJ	7.51	
SODIUM	MILLI	58.00	-	-	-	40.02	MILLI/MJ	596.78	
POTASSIUM	MILLI	238.00	-	-	-	164.22	GRAM/MJ	2.45	
MAGNESIUM	MILLI	7.40	-	-	-	5.11	MILLI/MJ	76.14	
CALCIUM	MILLI	49.40	40.00	-	58.80	34.09	MILLI/MJ	508.29	
MANGANESE	MICRO	68.00	-	-	-	46.92	MICRO/MJ	699.67	
IRON	MILLI	0.44	0.37	-	0.50	0.30	MILLI/MJ	4.53	
COPPER	MICRO	70.00	-	-	-	48.30	MICRO/MJ	720.25	
ZINC	MILLI	0.23	-	-	-	0.16	MILLI/MJ	2.37	
NICKEL	MICRO	1.00	-	-	-	0.69	MICRO/MJ	10.29	
CHROMIUM	MICRO	0.50	-	-	-	0.35	MICRO/MJ	5.14	
MOLYBDENUM	MICRO	10.00	-	-	-	6.90	MICRO/MJ	102.89	
PHOSPHORUS	MILLI	30.80	27.50	-	34.00	21.25	MILLI/MJ	316.91	
CHLORIDE	MILLI	70.00	-	-	-	48.30	MILLI/MJ	720.25	
IODIDE	MICRO	7.50	-	-	-	5.18	MICRO/MJ	77.17	
SELENIUM	MICRO	-	3.00	-	27.00				
SILICON	MILLI	12.00	-	-	-	8.28	MILLI/MJ	123.47	
CAROTENE	MICRO	62.00	0.00	-	190.00	42.78	MICRO/MJ	637.93	
VITAMIN B1	MICRO	40.00	30.00	-	50.00	27.60	MICRO/MJ	411.57	
VITAMIN B2	MICRO	51.00	40.00	-	70.00	35.19	MICRO/MJ	524.75	
NICOTINAMIDE	MILLI	0.67	0.50	-	0.92	0.46	MILLI/MJ	6.89	
PANTOTHENIC ACID	MILLI	0.20	0.19	-	0.23	0.14	MILLI/MJ	2.06	

CONSTITUENTS	DIM	AV	VARIATION			AVR	NUTR. DENS.		MOLPERC.
VITAMIN B6	MICRO	80.00	78.00	-	82.00	55.20	MICRO/MJ	823.14	
BIOTIN	MICRO	2.00	-	-	-	1.38	MICRO/MJ	20.58	
FOLIC ACID	MICRO	20.00	-	-	-	13.80	MICRO/MJ	205.79	
VITAMIN C	MILLI	20.00	10.00	-	28.00	13.80	MILLI/MJ	205.79	
ISOLEUCINE	MILLI	18.00	-	-	-	12.42	MILLI/MJ	185.21	
LYSINE	MILLI	51.00	-	-	-	35.19	MILLI/MJ	524.75	
METHIONINE	MILLI	11.00	-	-	-	7.59	MILLI/MJ	113.18	
PHENYLALANINE	MILLI	18.00	-	-	-	12.42	MILLI/MJ	185.21	
TYROSINE	MILLI	35.00	-	-	-	24.15	MILLI/MJ	360.12	
GLUCOSE	GRAM	1.92	1.92	-	1.92	1.32	GRAM/MJ	19.76	
FRUCTOSE	GRAM	1.51	1.51	-	1.51	1.04	GRAM/MJ	15.54	
SUCROSE	GRAM	0.54	0.54	-	0.54	0.37	GRAM/MJ	5.56	
STARCH	GRAM	0.69	-	-	-	0.48	GRAM/MJ	7.10	
HEMICELLULOSE	GRAM	1.45	-	-	-	1.00	GRAM/MJ	14.92	

YAM KNOLLE	**YAM** TUBER	**IGNAME** TUBERCULE

DIOSCOREA SPP.

		PROTEIN	FAT	CARBOHYDRATES	TOTAL
ENERGY VALUE (AVERAGE) PER 100 G EDIBLE PORTION	KJOULE	31	5.1	375	411
	(KCAL)	7.5	1.2	90	98
AMOUNT OF DIGESTIBLE CONSTITUENTS PER 100 G	GRAM	1.48	0.11	22.40	
ENERGY VALUE (AVERAGE) OF THE DIGESTIBLE FRACTION PER 100 G EDIBLE PORTION	KJOULE	23	4.6	375	403
	(KCAL)	5.6	1.1	90	96

WASTE PERCENTAGE AVERAGE 16 MINIMUM 13 MAXIMUM 22

CONSTITUENTS	DIM	AV		VARIATION		AVR	NUTR. DENS.		MOLPERC.
WATER	GRAM	71.20		67.30 -	76.40	59.81	GRAM/MJ	176.82	
PROTEIN	GRAM	2.00		1.50 -	3.00	1.68	GRAM/MJ	4.97	
FAT	GRAM	0.13		0.10 -	0.20	0.11	GRAM/MJ	0.32	
AVAILABLE CARBOHYDR.	GRAM	22.40	1	- -	-	18.82	GRAM/MJ	55.63	
TOTAL DIETARY FIBRE	GRAM	3.30	2	- -	-	2.77	GRAM/MJ	8.20	
MINERALS	GRAM	1.00		0.60 -	1.50	0.84	GRAM/MJ	2.48	
SODIUM	MILLI	10.00		8.00 -	12.00	8.40	MILLI/MJ	24.83	
POTASSIUM	MILLI	393.00		294.00 -	600.00	330.12	MILLI/MJ	976.01	
CALCIUM	MILLI	25.00		10.00 -	40.00	21.00	MILLI/MJ	62.09	
IRON	MILLI	0.90		0.30 -	1.10	0.76	MILLI/MJ	2.24	
PHOSPHORUS	MILLI	44.00		35.00 -	60.00	36.96	MILLI/MJ	109.27	
CAROTENE	MICRO	10.00		- -	-	8.40	MICRO/MJ	24.83	
VITAMIN B1	MICRO	90.00		50.00 -	100.00	75.60	MICRO/MJ	223.51	
VITAMIN B2	MICRO	30.00		10.00 -	40.00	25.20	MICRO/MJ	74.50	
NICOTINAMIDE	MILLI	0.60		0.40 -	0.80	0.50	MILLI/MJ	1.49	
VITAMIN C	MILLI	10.00		4.00 -	18.00	8.40	MILLI/MJ	24.83	
LAURIC ACID	MILLI	1.80		- -	-	1.51	MILLI/MJ	4.47	
MYRISTIC ACID	MILLI	1.00		- -	-	0.84	MILLI/MJ	2.48	
PALMITIC ACID	MILLI	41.00		- -	-	34.44	MILLI/MJ	101.82	
ARACHIDIC ACID	MILLI	2.20		- -	-	1.85	MILLI/MJ	5.46	
PALMITOLEIC ACID	MILLI	2.40		- -	-	2.02	MILLI/MJ	5.96	
OLEIC ACID	MILLI	58.40		- -	-	49.06	MILLI/MJ	145.04	
LINOLEIC ACID	MILLI	1.20		- -	-	1.01	MILLI/MJ	2.98	
DIETARY FIBRE,WAT.SOL.	GRAM	0.70		- -	-	0.59	GRAM/MJ	1.74	
DIETARY FIBRE,WAT.INS.	GRAM	2.60		- -	-	2.18	GRAM/MJ	6.46	

1 ESTIMATED BY THE DIFFERENCE METHOD
100 - (WATER + PROTEIN + FAT + MINERALS + TOTAL DIETARY FIBRE)

2 METHOD OF SOUTHGATE

Blatt-, Stengel- und Blütengemüse

LEAVES, STEMS AND FLOWERS FEUILLES, TIGES ET FLEURS

ARTISCHOCKE ARTICHOKE ARTICHAUT

CYNARA SCOLYMUS L.

		PROTEIN	FAT	CARBOHYDRATES	TOTAL
ENERGY VALUE (AVERAGE)	KJOULE	38	4.7	49	91
PER 100 G	(KCAL)	9.0	1.1	12	22
EDIBLE PORTION					
AMOUNT OF DIGESTIBLE	GRAM	1.56	0.10	2.90	
CONSTITUENTS PER 100 G					
ENERGY VALUE (AVERAGE)	KJOULE	24	4.2	49	77
OF THE DIGESTIBLE	(KCAL)	5.9	1.0	12	18
FRACTION PER 100 G					
EDIBLE PORTION					

WASTE PERCENTAGE AVERAGE 52

CONSTITUENTS	DIM	AV	VARIATION			AVR	NUTR. DENS.		MOLPERC.
WATER	GRAM	82.50	79.60	-	83.70	39.60	KILO/MJ	1.07	
PROTEIN	GRAM	2.40	1.80	-	3.00	1.15	GRAM/MJ	31.08	
FAT	GRAM	0.12	0.08	-	0.20	0.06	GRAM/MJ	1.55	
AVAILABLE CARBOHYDR.	GRAM	2.90	-	-	-	1.39	GRAM/MJ	37.56	
TOTAL DIETARY FIBRE	GRAM	10.79 [1]	-	-	-	5.18	GRAM/MJ	139.74	
MINERALS	GRAM	1.29	1.07	-	1.51	0.62	GRAM/MJ	16.71	
SODIUM	MILLI	47.00	43.00	-	56.00	22.56	MILLI/MJ	608.70	
POTASSIUM	MILLI	353.00	200.00	-	430.00	169.44	GRAM/MJ	4.57	
MAGNESIUM	MILLI	26.00	25.00	-	27.00	12.48	MILLI/MJ	336.73	
CALCIUM	MILLI	53.00	40.00	-	70.00	25.44	MILLI/MJ	686.41	
MANGANESE	MILLI	0.38	-	-	-	0.18	MILLI/MJ	4.92	
IRON	MILLI	1.50	1.00	-	1.90	0.72	MILLI/MJ	19.43	
COPPER	MILLI	0.32	-	-	-	0.15	MILLI/MJ	4.14	
PHOSPHORUS	MILLI	130.00	90.00	-	170.00	62.40	GRAM/MJ	1.68	
CHLORIDE	MILLI	40.00	22.00	-	57.00	19.20	MILLI/MJ	518.05	
BORON	MILLI	0.35	0.24	-	0.49	0.17	MILLI/MJ	4.53	

[1] ESTIMATED BY THE DIFFERENCE METHOD:
100 - (WATER + PROTEIN + FAT + AVAILABLE CARBOHYDRATES + MINERALS)
A LARGE PART OF THE CARBOHYDRATE AMOUNT IS INULIN, WHICH IS
NOT DIGESTED BY THE ENZYME SYSTEM OF THE HUMAN INTESTINE

CONSTITUENTS	DIM	AV	VARIATION			AVR	NUTR. DENS.		MOLPERC.
CAROTENE	MILLI	0.10	0.06	-	0.12	0.05	MILLI/MJ	1.30	
VITAMIN E ACTIVITY	MILLI	0.19	-	-	-	0.09	MILLI/MJ	2.46	
ALPHA-TOCOPHEROL	MILLI	0.19	-	-	-	0.09	MILLI/MJ	2.46	
VITAMIN B1	MILLI	0.14	0.08	-	0.18	0.07	MILLI/MJ	1.81	
VITAMIN B2	MICRO	12.00	4.00	-	30.00	5.76	MICRO/MJ	155.41	
NICOTINAMIDE	MILLI	0.90	0.10	-	1.60	0.43	MILLI/MJ	11.66	
VITAMIN C	MILLI	7.60	5.00	-	11.00	3.65	MILLI/MJ	98.43	
MALIC ACID	MILLI	170.00	-	-	-	81.60	GRAM/MJ	2.20	
CITRIC ACID	MILLI	100.00	-	-	-	48.00	GRAM/MJ	1.30	
OXALIC ACID TOTAL	MILLI	8.80	-	-	-	4.22	MILLI/MJ	113.97	
OXALIC ACID SOLUBLE	MILLI	8.30	-	-	-	3.98	MILLI/MJ	107.49	
GLUCOSE	GRAM	0.76	0.76	-	0.76	0.36	GRAM/MJ	9.84	
FRUCTOSE	GRAM	1.73	1.73	-	1.73	0.83	GRAM/MJ	22.41	
SUCROSE	GRAM	0.14	0.14	-	0.14	0.07	GRAM/MJ	1.81	

BLEICHSELLERIE · CELERY · CÉLERI EN BRANCHES

APIUM GRAVEOLENS VAR. DULCE MILLER

BLEICHSELLERIE [1] · CELERY [1] · CÉLERI EN BRANCHES [1]

		PROTEIN	FAT	CARBOHYDRATES	TOTAL
ENERGY VALUE (AVERAGE)	KJOULE	19	7.5	36	63
PER 100 G EDIBLE PORTION	(KCAL)	4.5	1.8	8.7	15
AMOUNT OF DIGESTIBLE CONSTITUENTS PER 100 G	GRAM	0.78	0.18	2.18	
ENERGY VALUE (AVERAGE)	KJOULE	12	6.8	36	56
OF THE DIGESTIBLE FRACTION PER 100 G EDIBLE PORTION	(KCAL)	2.9	1.6	8.7	13

WASTE PERCENTAGE AVERAGE 37

CONSTITUENTS	DIM	AV	VARIATION			AVR	NUTR. DENS.		MOLPERC.
WATER	GRAM	92.90	92.00	-	93.70	58.53	KILO/MJ	1.67	
PROTEIN	GRAM	1.20	1.00	-	1.30	0.76	GRAM/MJ	21.62	
FAT	GRAM	0.20	0.10	-	0.20	0.13	GRAM/MJ	3.60	
AVAILABLE CARBOHYDR.	GRAM	2.18	-	-	-	1.37	GRAM/MJ	39.28	
MINERALS	GRAM	1.10	-	-	-	0.69	GRAM/MJ	19.82	

1 THESE ARE THE PULPY STEMS OF THE LEAVES, WHICH REMAINED PALE BY BINDING AND COVERING

CONSTITUENTS	DIM	AV	VARIATION			AVR	NUTR. DENS.		MOLPERC.
SODIUM	MILLI	132.00	113.00	-	150.00	83.16	GRAM/MJ	2.38	
POTASSIUM	MILLI	344.00	287.00	-	400.00	216.72	GRAM/MJ	6.20	
MAGNESIUM	MILLI	12.00	-	-	-	7.56	MILLI/MJ	216.21	
CALCIUM	MILLI	80.00	50.00	-	100.00	50.40	GRAM/MJ	1.44	
IRON	MILLI	0.50	-	-	-	0.32	MILLI/MJ	9.01	
COPPER	MILLI	0.12	0.12	-	0.12	0.08	MILLI/MJ	2.16	
ZINC	MILLI	0.11	0.11	-	0.11	0.07	MILLI/MJ	1.98	
PHOSPHORUS	MILLI	48.00	40.00	-	65.00	30.24	MILLI/MJ	864.85	
IODIDE	MICRO	0.10	0.10	-	0.10	0.06	MICRO/MJ	1.80	
NITRATE	MILLI	223.00	5.00	-	527.00	140.49	GRAM/MJ	4.02	
CAROTENE	MICRO	17.00	15.00	-	18.00	10.71	MICRO/MJ	306.30	
VITAMIN B1	MICRO	48.00	30.00	-	75.00	30.24	MICRO/MJ	864.85	
VITAMIN B2	MICRO	76.00	40.00	-	150.00	47.88	MILLI/MJ	1.37	
NICOTINAMIDE	MILLI	0.55	0.45	-	0.80	0.35	MILLI/MJ	9.91	
PANTOTHENIC ACID	MILLI	0.43	-	-	-	0.27	MILLI/MJ	7.75	
VITAMIN B6	MICRO	90.00	77.00	-	97.00	56.70	MILLI/MJ	1.62	
BIOTIN	MICRO	0.10	0.10	-	0.10	0.06	MICRO/MJ	1.80	
FOLIC ACID	MICRO	7.00	-	-	-	4.41	MICRO/MJ	126.12	
VITAMIN C	MILLI	7.00	-	-	-	4.41	MILLI/MJ	126.12	
ARGININE	MILLI	100.00	-	-	-	63.00	GRAM/MJ	1.80	
CYSTINE	MILLI	6.00	-	-	-	3.78	MILLI/MJ	108.11	
HISTIDINE	MILLI	44.00	-	-	-	27.72	MILLI/MJ	792.78	
LYSINE	MILLI	19.00	-	-	-	11.97	MILLI/MJ	342.34	
METHIONINE	MILLI	14.00	-	-	-	8.82	MILLI/MJ	252.25	
PHENYLALANINE	MILLI	69.00	-	-	-	43.47	GRAM/MJ	1.24	
TRYPTOPHAN	MILLI	11.00	-	-	-	6.93	MILLI/MJ	198.20	
TYROSINE	MILLI	31.00	-	-	-	19.53	MILLI/MJ	558.55	
VALINE	MILLI	130.00	-	-	-	81.90	GRAM/MJ	2.34	
GLUCOSE	-	0.00		-		0.00			
FRUCTOSE	GRAM	0.10	0.10	-	0.10	0.06	GRAM/MJ	1.80	
SUCROSE	GRAM	2.08	2.08	-	2.08	1.31	GRAM/MJ	37.48	
MANNITOL	GRAM	0.11	-	-	-	0.07	GRAM/MJ	1.98	

BLUMENKOHL CAULIFLOWER CHOU-FLEUR

BRASSICA OLERACEA L.
VAR. BOTRYTIS

		PROTEIN	FAT	CARBOHYDRATES	TOTAL
ENERGY VALUE (AVERAGE)	KJOULE	39	11	43	92
PER 100 G EDIBLE PORTION	(KCAL)	9.2	2.6	10	22
AMOUNT OF DIGESTIBLE CONSTITUENTS PER 100 G	GRAM	1.59	0.25	2.54	
ENERGY VALUE (AVERAGE)	KJOULE	25	9.8	43	77
OF THE DIGESTIBLE FRACTION PER 100 G EDIBLE PORTION	(KCAL)	6.0	2.3	10	18

WASTE PERCENTAGE AVERAGE 38 MINIMUM 25 MAXIMUM 55

CONSTITUENTS	DIM	AV		VARIATION		AVR	NUTR. DENS.		MOLPERC.
WATER	GRAM	91.60	90.90	-	93.00	56.79	KILO/MJ	1.18	
PROTEIN	GRAM	2.46	2.00	-	2.72	1.53	GRAM/MJ	31.78	
FAT	GRAM	0.28	0.20	-	0.31	0.17	GRAM/MJ	3.62	
AVAILABLE CARBOHYDR.	GRAM	2.54	-	-	-	1.57	GRAM/MJ	32.82	
TOTAL DIETARY FIBRE	GRAM	2.94 1	-	-	-	1.82	GRAM/MJ	37.98	
MINERALS	GRAM	0.82	0.80	-	0.83	0.51	GRAM/MJ	10.59	
SODIUM	MILLI	16.00	10.00	-	24.00	9.92	MILLI/MJ	206.71	
POTASSIUM	MILLI	328.00	313.00	-	350.00	203.36	GRAM/MJ	4.24	
MAGNESIUM	MILLI	17.00	-	-	-	10.54	MILLI/MJ	219.63	
CALCIUM	MILLI	20.00	17.00	-	22.00	12.40	MILLI/MJ	258.39	
MANGANESE	MILLI	0.17	-	-	-	0.11	MILLI/MJ	2.20	
IRON	MILLI	0.63	0.30	-	1.10	0.39	MILLI/MJ	8.14	
COBALT	MICRO	-	0.10	-	1.00				
COPPER	MILLI	0.14	-	-	-	0.09	MILLI/MJ	1.81	
ZINC	MILLI	0.23	-	-	-	0.14	MILLI/MJ	2.97	
NICKEL	MICRO	30.00	3.00	-	100.00	18.60	MICRO/MJ	387.58	
CHROMIUM	MICRO	2.00	-	-	-	1.24	MICRO/MJ	25.84	
MOLYBDENUM	MICRO	-	0.00	-	10.00				
PHOSPHORUS	MILLI	54.00	30.00	-	72.00	33.48	MILLI/MJ	697.64	
CHLORIDE	MILLI	29.00	-	-	-	17.98	MILLI/MJ	374.66	
FLUORIDE	MICRO	12.00	-	-	-	7.44	MICRO/MJ	155.03	
IODIDE	MICRO	0.12	-	-	-	0.07	MICRO/MJ	1.55	
BORON	MILLI	0.15	0.12	-	0.18	0.09	MILLI/MJ	1.94	
SELENIUM	MICRO	3.20 2	0.60	-	16.00	1.98	MICRO/MJ	41.34	
SILICON	MILLI	0.50	-	-	-	0.31	MILLI/MJ	6.46	
NITRATE	MILLI	42.00	4.00	-	103.00	26.04	MILLI/MJ	542.61	

1 METHOD OF SOUTHGATE AND ENGLYST

2 AMOUNTS, DETERMINED IN THE FEDERAL REPUBLIC OF GERMANY

CONSTITUENTS	DIM	AV	VARIATION			AVR	NUTR. DENS.		MOLPERC.
CAROTENE	MICRO	33.00	18.00	-	54.00	20.46	MICRO/MJ	426.34	
VITAMIN E ACTIVITY	MICRO	30.00	-	-	-	18.60	MICRO/MJ	387.58	
TOTAL TOCOPHEROLS	MICRO	90.00	-	-	-	55.80	MILLI/MJ	1.16	
ALPHA-TOCOPHEROL	MICRO	30.00	-	-	-	18.60	MICRO/MJ	387.58	
GAMMA-TOCOPHEROL	MICRO	50.00	-	-	-	31.00	MICRO/MJ	645.97	
DELTA-TOCOPHEROL	MICRO	10.00	-	-	-	6.20	MICRO/MJ	129.19	
VITAMIN K	MILLI	0.30	-	-	-	0.19	MILLI/MJ	3.88	
VITAMIN B1	MILLI	0.11	0.05	-	0.15	0.07	MILLI/MJ	1.42	
VITAMIN B2	MILLI	0.10	0.07	-	0.13	0.06	MILLI/MJ	1.29	
NICOTINAMIDE	MILLI	0.60	0.53	-	0.75	0.37	MILLI/MJ	7.75	
PANTOTHENIC ACID	MILLI	1.01	0.83	-	1.19	0.63	MILLI/MJ	13.05	
VITAMIN B6	MILLI	0.20	0.15	-	0.30	0.12	MILLI/MJ	2.58	
BIOTIN	MICRO	1.50	-	-	-	0.93	MICRO/MJ	19.38	
FOLIC ACID	MICRO	55.00	-	-	-	34.10	MICRO/MJ	710.56	
VITAMIN C	MILLI	73.00	57.00	-	124.00	45.26	MILLI/MJ	943.11	
ARGININE	GRAM	0.11	0.09	-	0.12	0.07	GRAM/MJ	1.42	
HISTIDINE	MILLI	49.00	43.00	-	54.00	30.38	MILLI/MJ	633.05	
ISOLEUCINE	GRAM	0.11	0.10	-	0.11	0.07	GRAM/MJ	1.42	
LEUCINE	GRAM	0.17	0.13	-	0.18	0.11	GRAM/MJ	2.20	
LYSINE	GRAM	0.14	0.13	-	0.15	0.09	GRAM/MJ	1.81	
METHIONINE	MILLI	48.00	13.00	-	69.00	29.76	MILLI/MJ	620.13	
PHENYLALANINE	MILLI	77.00	42.00	-	91.00	47.74	MILLI/MJ	994.79	
THREONINE	GRAM	0.11	0.07	-	0.13	0.07	GRAM/MJ	1.42	
TRYPTOPHAN	MILLI	34.00	22.00	-	44.00	21.08	MILLI/MJ	439.26	
TYROSINE	MILLI	35.00	-	-	-	21.70	MILLI/MJ	452.18	
VALINE	GRAM	0.15	0.13	-	0.18	0.09	GRAM/MJ	1.94	
OXALIC ACID SOLUBLE	MILLI	4.30	-	-	-	2.67	MILLI/MJ	55.55	
FERULIC ACID	MILLI	0.20	-	-	-	0.12	MILLI/MJ	2.58	
CAFFEIC ACID	MILLI	0.90	-	-	-	0.56	MILLI/MJ	11.63	
PARA-COUMARIC ACID	MILLI	3.50	-	-	-	2.17	MILLI/MJ	45.22	
SALICYLIC ACID	MILLI	0.16	0.16	-	0.16	0.10	MILLI/MJ	2.07	
GLUCOSE	GRAM	1.10	0.51	-	1.18	0.68	GRAM/MJ	14.24	
FRUCTOSE	GRAM	0.98	0.63	-	1.51	0.61	GRAM/MJ	12.75	
SUCROSE	GRAM	0.17	0.07	-	0.48	0.11	GRAM/MJ	2.25	
STARCH	GRAM	0.29	0.29	-	0.29	0.18	GRAM/MJ	3.75	
PENTOSAN	GRAM	0.54	-	-	-	0.33	GRAM/MJ	6.98	
HEXOSAN	GRAM	0.39	-	-	-	0.24	GRAM/MJ	5.04	
CELLULOSE	GRAM	1.12	-	-	-	0.69	GRAM/MJ	14.47	
POLYURONIC ACID	GRAM	0.89	-	-	-	0.55	GRAM/MJ	11.50	
XYLITOL	MILLI	26.00	-	-	-	16.12	MILLI/MJ	335.90	
MYRISTIC ACID	MILLI	0.50	-	-	-	0.31	MILLI/MJ	6.46	
PALMITIC ACID	MILLI	39.00	-	-	-	24.18	MILLI/MJ	503.85	
STEARIC ACID	MILLI	1.80	-	-	-	1.12	MILLI/MJ	23.25	
ARACHIDIC ACID	MILLI	1.10	-	-	-	0.68	MILLI/MJ	14.21	
PALMITOLEIC ACID	MILLI	3.80	-	-	-	2.36	MILLI/MJ	49.09	
OLEIC ACID	MILLI	1.80	-	-	-	1.12	MILLI/MJ	23.25	
EICOSENOIC ACID	MILLI	1.10	-	-	-	0.68	MILLI/MJ	14.21	
LINOLEIC ACID	MILLI	29.00	-	-	-	17.98	MILLI/MJ	374.66	
LINOLENIC ACID	MILLI	109.00	-	-	-	67.58	GRAM/MJ	1.41	
TOTAL STEROLS	MILLI	18.00	-	-	-	11.16	MILLI/MJ	232.55	
CAMPESTEROL	MILLI	3.00	-	-	-	1.86	MILLI/MJ	38.76	
BETA-SITOSTEROL	MILLI	12.00	-	-	-	7.44	MILLI/MJ	155.03	
STIGMASTEROL	MILLI	2.00	-	-	-	1.24	MILLI/MJ	25.84	
TOTAL PURINES	MILLI	25.00 3	-	-	-	15.50	MILLI/MJ	322.98	
DIETARY FIBRE,WAT.SOL.	GRAM	0.49	-	-	-	0.30	GRAM/MJ	6.33	
DIETARY FIBRE,WAT.INS.	GRAM	2.43	-	-	-	1.51	GRAM/MJ	31.39	

3 VALUE EXPRESSED IN MG URIC ACID/100G

BLUMENKOHL GEKOCHT, ABGETROPFT	**CAULIFLOWER** BOILED, DRAINED	**CHOU-FLEUR** BOUILLI, DEGOUTTÉ

		PROTEIN	FAT	CARBOHYDRATES	TOTAL
ENERGY VALUE (AVERAGE)	KJOULE	31	11	34	76
PER 100 G	(KCAL)	7.5	2.6	8.1	18
EDIBLE PORTION					
AMOUNT OF DIGESTIBLE	GRAM	1.30	0.25	2.03	
CONSTITUENTS PER 100 G					
ENERGY VALUE (AVERAGE)	KJOULE	20	9.8	34	64
OF THE DIGESTIBLE	(KCAL)	4.9	2.3	8.1	15
FRACTION PER 100 G					
EDIBLE PORTION					

WASTE PERCENTAGE AVERAGE 0.00

CONSTITUENTS	DIM	AV	VARIATION			AVR	NUTR. DENS.		MOLPERC.
WATER	GRAM	93.40	-	-	-	93.40	KILO/MJ	1.46	
PROTEIN	GRAM	2.00	-	-	-	2.00	GRAM/MJ	31.16	
FAT	GRAM	0.28	-	-	-	0.28	GRAM/MJ	4.36	
AVAILABLE CARBOHYDR.	GRAM	2.03	-	-	-	2.03	GRAM/MJ	31.63	
MINERALS	GRAM	0.55	-	-	-	0.55	GRAM/MJ	8.57	
SODIUM	MILLI	11.00	-	-	-	11.00	MILLI/MJ	171.40	
POTASSIUM	MILLI	229.00	-	-	-	229.00	GRAM/MJ	3.57	
CALCIUM	MILLI	17.00	-	-	-	17.00	MILLI/MJ	264.89	
IRON	MILLI	0.40	-	-	-	0.40	MILLI/MJ	6.23	
PHOSPHORUS	MILLI	41.00	-	-	-	41.00	MILLI/MJ	638.86	
VITAMIN B1	MICRO	90.00	-	-	-	90.00	MILLI/MJ	1.40	
VITAMIN B2	MICRO	80.00	-	-	-	80.00	MILLI/MJ	1.25	
NICOTINAMIDE	MILLI	0.51	-	-	-	0.51	MILLI/MJ	7.95	
VITAMIN C	MILLI	49.00	-	-	-	49.00	MILLI/MJ	763.52	
GLUCOSE	GRAM	0.88	-	-	-	0.88	GRAM/MJ	13.71	
FRUCTOSE	GRAM	0.76	-	-	-	0.76	GRAM/MJ	11.84	
SUCROSE	GRAM	0.16	-	-	-	0.16	GRAM/MJ	2.49	
STARCH	GRAM	0.23	-	-	-	0.23	GRAM/MJ	3.58	

BROCCOLI BROCCOLI BROCCOLI

BRASSICA OLERACEA L.
VAR. ITALICA PLENCK

		PROTEIN	FAT	CARBOHYDRATES	TOTAL
ENERGY VALUE (AVERAGE) PER 100 G EDIBLE PORTION	KJOULE (KCAL)	52 12	7.8 1.9	47 11	107 26
AMOUNT OF DIGESTIBLE CONSTITUENTS PER 100 G	GRAM	2.14	0.18	2.82	
ENERGY VALUE (AVERAGE) OF THE DIGESTIBLE FRACTION PER 100 G EDIBLE PORTION	KJOULE (KCAL)	34 8.0	7.0 1.7	47 11	88 21

WASTE PERCENTAGE AVERAGE 39 MINIMUM 23 MAXIMUM 55

CONSTITUENTS	DIM	AV	VARIATION			AVR	NUTR. DENS.		MOLPERC.
WATER	GRAM	89.70	88.70	-	90.40	54.72	KILO/MJ	1.02	
PROTEIN	GRAM	3.30	-	-	-	2.01	GRAM/MJ	37.56	
FAT	GRAM	0.20	-	-	-	0.12	GRAM/MJ	2.28	
AVAILABLE CARBOHYDR.	GRAM	2.82	-	-	-	1.72	GRAM/MJ	32.10	
TOTAL DIETARY FIBRE	GRAM	3.00 [1]	-	-	-	1.83	GRAM/MJ	34.15	
MINERALS	GRAM	1.10	-	-	-	0.67	GRAM/MJ	12.52	
SODIUM	MILLI	13.00	-	-	-	7.93	MILLI/MJ	147.97	
POTASSIUM	MILLI	464.00	368.00	-	560.00	283.04	GRAM/MJ	5.28	
MAGNESIUM	MILLI	24.00	22.00	-	28.00	14.64	MILLI/MJ	273.18	
CALCIUM	MILLI	105.00	80.00	-	130.00	64.05	GRAM/MJ	1.20	
MANGANESE	MILLI	0.26	-	-	-	0.16	MILLI/MJ	2.96	
IRON	MILLI	1.30	-	-	-	0.79	MILLI/MJ	14.80	
COBALT	MICRO	5.00	2.00	-	7.00	3.05	MICRO/MJ	56.91	
COPPER	MILLI	0.20	-	-	-	0.12	MILLI/MJ	2.28	
ZINC	MILLI	0.94	0.46	-	1.30	0.57	MILLI/MJ	10.70	
NICKEL	MICRO	50.00	30.00	-	80.00	30.50	MICRO/MJ	569.12	
CHROMIUM	MICRO	1.00	0.50	-	2.00	0.61	MICRO/MJ	11.38	
PHOSPHORUS	MILLI	82.00	76.00	-	87.00	50.02	MILLI/MJ	933.36	
CHLORIDE	MILLI	78.00	-	-	-	47.58	MILLI/MJ	887.83	
FLUORIDE	MICRO	10.00	3.00	-	10.00	6.10	MICRO/MJ	113.82	
IODIDE	MICRO	15.00	-	-	-	9.15	MICRO/MJ	170.74	
BORON	MILLI	0.16	-	-	-	0.10	MILLI/MJ	1.82	
NITRATE	MILLI	71.00	13.00	-	127.00	43.31	MILLI/MJ	808.15	
CAROTENE	MILLI	1.90	0.83	-	2.40	1.16	MILLI/MJ	21.63	
VITAMIN E ACTIVITY	MILLI	0.47	-	-	-	0.29	MILLI/MJ	5.35	
ALPHA-TOCOPHEROL	MILLI	0.46	-	-	-	0.28	MILLI/MJ	5.24	

1 METHOD OF MEUSER, SUCKOW AND KULIKOWSKI ("BERLINER METHODE")

CONSTITUENTS	DIM	AV	VARIATION			AVR	NUTR. DENS.		MOLPERC.
VITAMIN K	MILLI	0.13	0.06	-	0.20	0.08	MILLI/MJ	1.48	
VITAMIN B1	MICRO	95.00	90.00	-	100.00	57.95	MILLI/MJ	1.08	
VITAMIN B2	MILLI	0.21	-	-	-	0.13	MILLI/MJ	2.39	
NICOTINAMIDE	MILLI	1.00	0.90	-	1.10	0.61	MILLI/MJ	11.38	
PANTOTHENIC ACID	MILLI	1.29	1.06	-	1.53	0.79	MILLI/MJ	14.68	
VITAMIN B6	MILLI	0.17	-	-	-	0.10	MILLI/MJ	1.94	
BIOTIN	MICRO	0.50	0.50	-	0.50	0.31	MICRO/MJ	5.69	
FOLIC ACID	MICRO	33.00	27.00	-	42.00	20.13	MICRO/MJ	375.62	
VITAMIN C	MILLI	114.00	110.00	-	118.00	69.54	GRAM/MJ	1.30	
ARGININE	GRAM	0.19	-	-	-	0.12	GRAM/MJ	2.16	
HISTIDINE	MILLI	63.00	-	-	-	38.43	MILLI/MJ	717.09	
ISOLEUCINE	GRAM	0.13	-	-	-	0.08	GRAM/MJ	1.48	
LEUCINE	GRAM	0.16	-	-	-	0.10	GRAM/MJ	1.82	
LYSINE	GRAM	0.15	-	-	-	0.09	GRAM/MJ	1.71	
METHIONINE	MILLI	50.00	-	-	-	30.50	MILLI/MJ	569.12	
PHENYLALANINE	GRAM	0.12	-	-	-	0.07	GRAM/MJ	1.37	
THREONINE	GRAM	0.12	-	-	-	0.07	GRAM/MJ	1.37	
TRYPTOPHAN	MILLI	37.00	-	-	-	22.57	MILLI/MJ	421.15	
VALINE	GRAM	0.17	-	-	-	0.10	GRAM/MJ	1.94	
MALIC ACID	MILLI	120.00	-	-	-	73.20	GRAM/MJ	1.37	
CITRIC ACID	MILLI	210.00	-	-	-	128.10	GRAM/MJ	2.39	
FERULIC ACID	MILLI	1.30	-	-	-	0.79	MILLI/MJ	14.80	
CAFFEIC ACID	MILLI	0.80	-	-	-	0.49	MILLI/MJ	9.11	
PARA-COUMARIC ACID	MILLI	1.30	-	-	-	0.79	MILLI/MJ	14.80	
SALICYLIC ACID	MILLI	0.65	0.65	-	0.65	0.40	MILLI/MJ	7.40	
GLUCOSE	GRAM	0.94	0.63	-	1.11	0.57	GRAM/MJ	10.72	
FRUCTOSE	GRAM	1.03	0.58	-	1.22	0.63	GRAM/MJ	11.81	
SUCROSE	GRAM	0.52	0.37	-	1.04	0.32	GRAM/MJ	5.98	
CELLULOSE	GRAM	1.46	-	-	-	0.89	GRAM/MJ	16.62	
DIETARY FIBRE, WAT.SOL.	GRAM	1.30	-	-	-	0.79	GRAM/MJ	14.80	
DIETARY FIBRE, WAT.INS.	GRAM	1.70	-	-	-	1.04	GRAM/MJ	19.35	

BROCCOLI
GEKOCHT, ABGETROPFT

BROCCOLI
BOILED, DRAINED

BROCCOLI
BOUILLI, DEGOUTTÉ

		PROTEIN	FAT	CARBOHYDRATES	TOTAL
ENERGY VALUE (AVERAGE)	KJOULE	44	7.8	38	90
PER 100 G	(KCAL)	11	1.9	9.2	22
EDIBLE PORTION					
AMOUNT OF DIGESTIBLE	GRAM	1.82	0.18	2.30	
CONSTITUENTS PER 100 G					
ENERGY VALUE (AVERAGE)	KJOULE	29	7.0	38	74
OF THE DIGESTIBLE	(KCAL)	6.8	1.7	9.2	18
FRACTION PER 100 G					
EDIBLE PORTION					

WASTE PERCENTAGE AVERAGE 0.00

CONSTITUENTS	DIM	AV	VARIATION			AVR	NUTR. DENS.		MOLPERC.
WATER	GRAM	91.90	-	-	-	91.90	KILO/MJ	1.24	
PROTEIN	GRAM	2.80	-	-	-	2.80	GRAM/MJ	37.81	
FAT	GRAM	0.20	-	-	-	0.20	GRAM/MJ	2.70	
AVAILABLE CARBOHYDR.	GRAM	2.30	-	-	-	2.30	GRAM/MJ	31.06	
MINERALS	GRAM	0.80	-	-	-	0.80	GRAM/MJ	10.80	
SODIUM	MILLI	9.00	-	-	-	9.00	MILLI/MJ	121.54	
POTASSIUM	MILLI	324.00	-	-	-	324.00	GRAM/MJ	4.38	
CALCIUM	MILLI	87.00	-	-	-	87.00	GRAM/MJ	1.17	
IRON	MILLI	0.90	-	-	-	0.90	MILLI/MJ	12.15	
PHOSPHORUS	MILLI	65.00	-	-	-	65.00	MILLI/MJ	877.75	
VITAMIN B1	MICRO	90.00	-	-	-	90.00	MILLI/MJ	1.22	
VITAMIN B2	MILLI	0.18	-	-	-	0.18	MILLI/MJ	2.43	
NICOTINAMIDE	MILLI	0.90	-	-	-	0.90	MILLI/MJ	12.15	
VITAMIN C	MILLI	90.00	-	-	-	90.00	GRAM/MJ	1.22	
MALIC ACID	GRAM	0.10	-	-	-	0.10	GRAM/MJ	1.35	
CITRIC ACID	GRAM	0.17	-	-	-	0.17	GRAM/MJ	2.30	
GLUCOSE	GRAM	0.75	-	-	-	0.75	GRAM/MJ	10.13	
FRUCTOSE	GRAM	0.80	-	-	-	0.80	GRAM/MJ	10.80	
SUCROSE	GRAM	0.48	-	-	-	0.48	GRAM/MJ	6.48	

BRUNNENKRESSE — WATER CRESS — CRESSON DE FONTAINE

NASTURTIUM OFFICIN. R. BR.

		PROTEIN	FAT	CARBOHYDRATES	TOTAL
ENERGY VALUE (AVERAGE) PER 100 G EDIBLE PORTION	KJOULE	25	12	34	71
	(KCAL)	6.0	2.8	8.1	17
AMOUNT OF DIGESTIBLE CONSTITUENTS PER 100 G	GRAM	1.04	0.27	2.03	
ENERGY VALUE (AVERAGE) OF THE DIGESTIBLE FRACTION PER 100 G EDIBLE PORTION	KJOULE	16	11	34	61
	(KCAL)	3.9	2.5	8.1	15

WASTE PERCENTAGE AVERAGE 0.00

CONSTITUENTS	DIM	AV	VARIATION	AVR	NUTR. DENS.		MOLPERC.
WATER	GRAM	93.50	92.30 - 95.00	93.50	KILO/MJ	1.54	
PROTEIN	GRAM	1.60	0.70 - 2.00	1.60	GRAM/MJ	26.32	
FAT	GRAM	0.30	0.10 - 0.50	0.30	GRAM/MJ	4.93	
AVAILABLE CARBOHYDR.	GRAM	2.03 1	- - -	2.03	GRAM/MJ	33.39	
TOTAL DIETARY FIBRE	GRAM	1.47 2	- - -	1.47	GRAM/MJ	24.18	
MINERALS	GRAM	1.10	- - -	1.10	GRAM/MJ	18.09	
SODIUM	MILLI	12.00	7.00 - 17.00	12.00	MILLI/MJ	197.38	
POTASSIUM	MILLI	276.00	238.00 - 301.00	276.00	GRAM/MJ	4.54	
MAGNESIUM	MILLI	34.00	28.00 - 40.00	34.00	MILLI/MJ	559.23	
CALCIUM	MILLI	180.00	132.00 - 195.00	180.00	GRAM/MJ	2.96	
IRON	MILLI	3.14	2.00 - 7.20	3.14	MILLI/MJ	51.65	
PHOSPHORUS	MILLI	63.60	46.00 - 75.00	63.60	GRAM/MJ	1.05	
CHLORIDE	MILLI	109.00	- - -	109.00	GRAM/MJ	1.79	
BORON	MILLI	0.11	- - -	0.11	MILLI/MJ	1.81	
CAROTENE	MILLI	2.68	1.20 - 4.00	2.68	MILLI/MJ	44.08	
VITAMIN B1	MICRO	85.00	66.00 - 110.00	85.00	MILLI/MJ	1.40	
VITAMIN B2	MILLI	0.17	0.10 - 0.27	0.17	MILLI/MJ	2.80	
NICOTINAMIDE	MILLI	0.65	0.14 - 1.00	0.65	MILLI/MJ	10.69	
VITAMIN C	MILLI	51.00	25.00 - 77.00	51.00	MILLI/MJ	838.85	
ARGININE	MILLI	50.00	- - -	50.00	MILLI/MJ	822.40	
HISTIDINE	MILLI	32.00	- - -	32.00	MILLI/MJ	526.34	
ISOLEUCINE	MILLI	72.00	- - -	72.00	GRAM/MJ	1.18	
LEUCINE	MILLI	120.00	- - -	120.00	GRAM/MJ	1.97	
LYSINE	MILLI	9.00	- - -	9.00	MILLI/MJ	148.03	
PHENYLALANINE	MILLI	58.00	- - -	58.00	MILLI/MJ	953.98	
THREONINE	MILLI	79.00	- - -	79.00	GRAM/MJ	1.30	
TRYPTOPHAN	MILLI	26.00	- - -	26.00	MILLI/MJ	427.65	
TYROSINE	MILLI	34.00	- - -	34.00	MILLI/MJ	559.23	
VALINE	MILLI	79.00	- - -	79.00	GRAM/MJ	1.30	
SALICYLIC ACID	MILLI	0.84	0.84 - 0.84	0.84	MILLI/MJ	13.82	
DIETARY FIBRE, WAT.SOL.	GRAM	0.66	- - -	0.66	GRAM/MJ	10.86	
DIETARY FIBRE, WAT.INS.	GRAM	0.81	- - -	0.81	GRAM/MJ	13.32	

1 ESTIMATED BY THE DIFFERENCE METHOD
100 - (WATER + PROTEIN + FAT + MINERALS + TOTAL DIETARY FIBRE)

2 METHOD OF ENGLYST

CHICOREE CHICORY CHICORÉE

CICHORIUM INTIBUS L.

		PROTEIN	FAT	CARBOHYDRATES	TOTAL
ENERGY VALUE (AVERAGE) PER 100 G EDIBLE PORTION	KJOULE (KCAL)	20 4.9	7.0 1.7	39 9.3	66 16
AMOUNT OF DIGESTIBLE CONSTITUENTS PER 100 G	GRAM	0.84	0.16	2.32	
ENERGY VALUE (AVERAGE) OF THE DIGESTIBLE FRACTION PER 100 G EDIBLE PORTION	KJOULE (KCAL)	13 3.2	6.3 1.5	39 9.3	58 14

WASTE PERCENTAGE AVERAGE 11 MINIMUM 9.0 MAXIMUM 14

CONSTITUENTS	DIM	AV	VARIATION			AVR	NUTR. DENS.		MOLPERC.
WATER	GRAM	94.40	93.10	-	96.20	84.02	KILO/MJ	1.62	
PROTEIN	GRAM	1.30	0.80	-	1.70	1.16	GRAM/MJ	22.26	
FAT	GRAM	0.18	0.10	-	0.30	0.16	GRAM/MJ	3.08	
AVAILABLE CARBOHYDR.	GRAM	2.32	-	-	-	2.06	GRAM/MJ	39.73	
TOTAL DIETARY FIBRE	GRAM	1.26 1	-	-	-	1.12	GRAM/MJ	21.58	
MINERALS	GRAM	1.00	-	-	-	0.89	GRAM/MJ	17.13	
SODIUM	MILLI	4.43	1.00	-	7.30	3.94	MILLI/MJ	75.87	
POTASSIUM	MILLI	192.00	182.00	-	200.00	170.88	GRAM/MJ	3.29	
MAGNESIUM	MILLI	12.90	12.60	-	13.00	11.48	MILLI/MJ	220.93	
CALCIUM	MILLI	25.60	15.00	-	49.00	22.78	MILLI/MJ	438.44	
MANGANESE	MILLI	0.30	-	-	-	0.27	MILLI/MJ	5.14	
IRON	MILLI	0.74	0.50	-	1.10	0.66	MILLI/MJ	12.67	
COPPER	MILLI	0.14	-	-	-	0.12	MILLI/MJ	2.40	
ZINC	MILLI	0.19	-	-	-	0.17	MILLI/MJ	3.25	
PHOSPHORUS	MILLI	26.00	20.00	-	28.00	23.14	MILLI/MJ	445.29	
CHLORIDE	MILLI	25.00	-	-	-	22.25	MILLI/MJ	428.16	
BORON	MILLI	2.06	-	-	-	1.83	MILLI/MJ	35.28	
NITRATE	MILLI	15.00	8.00	-	21.00	13.35	MILLI/MJ	256.90	
CAROTENE	MILLI	1.29	0.60	-	2.16	1.15	MILLI/MJ	22.09	
VITAMIN B1	MICRO	51.00	40.00	-	85.00	45.39	MICRO/MJ	873.45	
VITAMIN B2	MICRO	33.00	30.00	-	35.00	29.37	MICRO/MJ	565.17	
NICOTINAMIDE	MILLI	0.24	0.20	-	0.30	0.21	MILLI/MJ	4.11	
VITAMIN B6	MICRO	50.00	-	-	-	44.50	MICRO/MJ	856.32	
BIOTIN	MICRO	4.80	4.80	-	4.80	4.27	MICRO/MJ	82.21	
FOLIC ACID	MICRO	52.00	-	-	-	46.28	MICRO/MJ	890.58	
VITAMIN C	MILLI	10.20	5.00	-	15.00	9.08	MILLI/MJ	174.69	

1 METHOD OF MEUSER, SUCKOW AND KULIKOWSKI ("BERLINER METHODE")

CONSTITUENTS	DIM	AV	VARIATION			AVR	NUTR. DENS.		MOLPERC.
CYSTINE	MILLI	5.00	-	-	-	4.45	MILLI/MJ	85.63	
HISTIDINE	MILLI	20.00	-	-	-	17.80	MILLI/MJ	342.53	
LYSINE	MILLI	42.00	-	-	-	37.38	MILLI/MJ	719.31	
METHIONINE	MILLI	13.00	-	-	-	11.57	MILLI/MJ	222.64	
TRYPTOPHAN	MILLI	20.00	-	-	-	17.80	MILLI/MJ	342.53	
TYROSINE	MILLI	33.00	-	-	-	29.37	MILLI/MJ	565.17	
OXALIC ACID	MILLI	27.30	-	-	-	24.30	MILLI/MJ	467.55	
SALICYLIC ACID	MILLI	1.00	1.00	-	1.00	0.89	MILLI/MJ	17.13	
GLUCOSE	GRAM	1.11	0.11	-	1.74	0.99	GRAM/MJ	19.13	
FRUCTOSE	GRAM	0.68	0.12	-	1.03	0.61	GRAM/MJ	11.70	
SUCROSE	GRAM	0.53	0.23	-	0.87	0.48	GRAM/MJ	9.20	
RAFFINOSE	GRAM	1.15	-	-	-	1.02	GRAM/MJ	19.70	
STACHYOSE	GRAM	0.34	-	-	-	0.30	GRAM/MJ	5.82	
PALMITIC ACID	MILLI	29.00	29.00	-	29.00	25.81	MILLI/MJ	496.67	
STEARIC ACID	MILLI	1.50	1.50	-	1.50	1.34	MILLI/MJ	25.69	
OLEIC ACID	MILLI	4.50	4.50	-	4.50	4.01	MILLI/MJ	77.07	
LINOLEIC ACID	MILLI	73.00	73.00	-	73.00	64.97	GRAM/MJ	1.25	
LINOLENIC ACID	MILLI	29.00	29.00	-	29.00	25.81	MILLI/MJ	496.67	
DIETARY FIBRE, WAT.SOL.	GRAM	0.37	-	-	-	0.33	GRAM/MJ	6.34	
DIETARY FIBRE, WAT.INS.	GRAM	0.89	-	-	-	0.79	GRAM/MJ	15.24	

CHINAKOHL CHINESE LEAVES CHOU DE CHINE

BRASSICA CHINENSIS JUSLEN.

		PROTEIN	FAT	CARBOHYDRATES	TOTAL
ENERGY VALUE (AVERAGE) PER 100 G EDIBLE PORTION	KJOULE (KCAL)	19 4.5	12 2.8	22 5.4	53 13
AMOUNT OF DIGESTIBLE CONSTITUENTS PER 100 G	GRAM	0.77	0.27	1.34	
ENERGY VALUE (AVERAGE) OF THE DIGESTIBLE FRACTION PER 100 G EDIBLE PORTION	KJOULE (KCAL)	12 2.9	11 2.5	22 5.4	45 11

WASTE PERCENTAGE AVERAGE 21 MINIMUM 13 MAXIMUM 28

CONSTITUENTS	DIM	AV	VARIATION	AVR	NUTR. DENS.		MOLPERC.
WATER	GRAM	95.40	- - -	75.37	KILO/MJ	2.12	
PROTEIN	GRAM	1.19	1.17 - 1.20	0.94	GRAM/MJ	26.40	
FAT	GRAM	0.30	- - -	0.24	GRAM/MJ	6.66	
AVAILABLE CARBOHYDR.	GRAM	1.34	- - -	1.06	GRAM/MJ	29.73	
TOTAL DIETARY FIBRE	GRAM	1.70 1	- - -	1.34	GRAM/MJ	37.72	
MINERALS	GRAM	0.65	0.59 - 0.70	0.51	GRAM/MJ	14.42	
SODIUM	MILLI	7.00	6.00 - 8.00	5.53	MILLI/MJ	155.32	
POTASSIUM	MILLI	202.00	190.00 - 211.00	159.58	GRAM/MJ	4.48	
MAGNESIUM	MILLI	11.00	11.00 - 12.00	8.69	MILLI/MJ	244.07	
CALCIUM	MILLI	40.00	36.00 - 43.00	31.60	MILLI/MJ	887.54	
MANGANESE	MILLI	0.28	0.24 - 0.33	0.22	MILLI/MJ	6.21	
IRON	MILLI	0.60	0.30 - 0.90	0.47	MILLI/MJ	13.31	
COPPER	MICRO	20.00	- - -	15.80	MICRO/MJ	443.77	
ZINC	MILLI	0.34	- - -	0.27	MILLI/MJ	7.54	
PHOSPHORUS	MILLI	30.00	18.00 - 41.00	23.70	MILLI/MJ	665.65	
FLUORIDE	MICRO	15.00	- - -	11.85	MICRO/MJ	332.83	
IODIDE	MICRO	0.30	0.30 - 0.30	0.24	MICRO/MJ	6.66	
NITRATE	MILLI	112.00	20.00 - 261.00	88.48	GRAM/MJ	2.49	
CAROTENE	MICRO	78.00	- - -	61.62	MILLI/MJ	1.73	
VITAMIN B1	MICRO	30.00	- - -	23.70	MICRO/MJ	665.65	
VITAMIN B2	MICRO	40.00	- - -	31.60	MICRO/MJ	887.54	
NICOTINAMIDE	MILLI	0.40	- - -	0.32	MILLI/MJ	8.88	
PANTOTHENIC ACID	MILLI	0.20	0.20 - 0.20	0.16	MILLI/MJ	4.44	
VITAMIN B6	MILLI	0.16	0.16 - 0.16	0.13	MILLI/MJ	3.55	
FOLIC ACID	MICRO	83.00	83.00 - 83.00	65.57	MILLI/MJ	1.84	
VITAMIN C	MILLI	36.00	31.00 - 40.00	28.44	MILLI/MJ	798.78	

1 METHOD OF MEUSER, SUCKOW AND KULIKOWSKI ("BERLINER METHODE")

CONSTITUENTS	DIM	AV	VARIATION			AVR	NUTR. DENS.		MOLPERC.
ARGININE	MILLI	80.00	76.00	-	84.00	63.20	GRAM/MJ	1.78	
CYSTINE	MILLI	10.00	-	-	-	7.90	MILLI/MJ	221.88	
HISTIDINE	MILLI	26.00	-	-	-	20.54	MILLI/MJ	576.90	
LYSINE	MILLI	58.00	57.00	-	59.00	45.82	GRAM/MJ	1.29	
METHIONINE	MILLI	32.00	28.00	-	36.00	25.28	MILLI/MJ	710.03	
PHENYLALANINE	MILLI	47.00	-	-	-	37.13	GRAM/MJ	1.04	
THREONINE	MILLI	52.00	51.00	-	53.00	41.08	GRAM/MJ	1.15	
TRYPTOPHAN	MILLI	20.00	20.00	-	21.00	15.80	MILLI/MJ	443.77	
TYROSINE	MILLI	39.00	38.00	-	39.00	30.81	MILLI/MJ	865.35	
VALINE	MILLI	70.00	59.00	-	80.00	55.30	GRAM/MJ	1.55	
GLUCOSE	GRAM	0.75	0.75	-	0.75	0.59	GRAM/MJ	16.64	
FRUCTOSE	GRAM	0.59	0.59	-	0.59	0.47	GRAM/MJ	13.09	
DIETARY FIBRE,WAT.SOL.	GRAM	0.17	-	-	-	0.13	GRAM/MJ	3.77	
DIETARY FIBRE,WAT.INS.	GRAM	1.50	-	-	-	1.19	GRAM/MJ	33.28	

ENDIVIE (ESCARIOL) — ENDIVE — ENDIVE

CICHORIUM ENDIVIA L.

		PROTEIN	FAT	CARBOHYDRATES	TOTAL
ENERGY VALUE (AVERAGE) PER 100 G EDIBLE PORTION	KJOULE	27	7.8	5.0	40
	(KCAL)	6.6	1.9	1.2	9.6
AMOUNT OF DIGESTIBLE CONSTITUENTS PER 100 G	GRAM	1.13	0.18	0.30	
ENERGY VALUE (AVERAGE) OF THE DIGESTIBLE FRACTION PER 100 G EDIBLE PORTION	KJOULE	18	7.0	5.0	30
	(KCAL)	4.3	1.7	1.2	7.1

WASTE PERCENTAGE AVERAGE 23 MINIMUM 10 MAXIMUM 44

CONSTITUENTS	DIM	AV	VARIATION		AVR	NUTR. DENS.		MOLPERC.
WATER	GRAM	94.30	93.30	95.00	72.61	KILO/MJ	3.16	
PROTEIN	GRAM	1.75	1.60	2.00	1.35	GRAM/MJ	58.58	
FAT	GRAM	0.20	-	-	0.15	GRAM/MJ	6.70	
AVAILABLE CARBOHYDR.	GRAM	0.30	-	-	0.23	GRAM/MJ	10.04	
TOTAL DIETARY FIBRE	GRAM	1.53 [1]	-	-	1.18	GRAM/MJ	51.22	
MINERALS	GRAM	0.90	-	-	0.69	GRAM/MJ	30.13	
SODIUM	MILLI	53.00	18.00	90.00	40.81	GRAM/MJ	1.77	
POTASSIUM	MILLI	346.00	300.00	400.00	266.42	GRAM/MJ	11.58	
MAGNESIUM	MILLI	10.00	-	-	7.70	MILLI/MJ	334.76	
CALCIUM	MILLI	54.00	20.00	79.00	41.58	GRAM/MJ	1.81	
MANGANESE	MILLI	0.22	-	-	0.17	MILLI/MJ	7.36	
IRON	MILLI	1.40	1.00	1.70	1.08	MILLI/MJ	46.87	
COPPER	MILLI	0.10	-	-	0.08	MILLI/MJ	3.35	
ZINC	MILLI	0.34	-	-	0.26	MILLI/MJ	11.38	
MOLYBDENUM	MICRO	4.00	-	-	3.08	MICRO/MJ	133.90	
PHOSPHORUS	MILLI	54.30	37.00	70.00	41.81	GRAM/MJ	1.82	
CHLORIDE	MILLI	71.00	-	-	54.67	GRAM/MJ	2.38	
IODIDE	MICRO	6.40	-	-	4.93	MICRO/MJ	214.25	
BORON	MILLI	-	0.05	0.29				
SELENIUM	MICRO	13.00	-	-	10.01	MICRO/MJ	435.19	
NITRATE	MILLI	106.00	7.00	259.00	81.62	GRAM/MJ	3.55	
CAROTENE	MILLI	1.14	0.72	1.80	0.88	MILLI/MJ	38.16	
VITAMIN B1	MICRO	52.00	30.00	100.00	40.04	MILLI/MJ	1.74	
VITAMIN B2	MILLI	0.12	0.06	0.20	0.09	MILLI/MJ	4.02	
NICOTINAMIDE	MILLI	0.41	0.20	0.80	0.32	MILLI/MJ	13.73	
FOLIC ACID	MICRO	49.00	-	-	37.73	MILLI/MJ	1.64	

1 METHOD OF MEUSER, SUCKOW AND KULIKOWSKI ("BERLINER METHODE")

CONSTITUENTS	DIM	AV		VARIATION		AVR	NUTR. DENS.		MOLPERC.
VITAMIN C	MILLI	9.40	5.00	-	13.00	7.24	MILLI/MJ	314.67	
OXALIC ACID TOTAL	MILLI	2.50	-	-	-	1.93	MILLI/MJ	83.69	
OXALIC ACID SOLUBLE	-	0.00	-	-	-	0.00			
FERULIC ACID	MICRO	50.00	0.00	-	100.00	38.50	MILLI/MJ	1.67	
CAFFEIC ACID	MILLI	51.50	16.00	-	87.00	39.66	GRAM/MJ	1.72	
SALICYLIC ACID	MILLI	1.90	1.90	-	1.90	1.46	MILLI/MJ	63.60	
GLUCOSE	MILLI	70.00	70.00	-	70.00	53.90	GRAM/MJ	2.34	
FRUCTOSE	GRAM	0.16	0.16	-	0.16	0.12	GRAM/MJ	5.36	
SUCROSE	MILLI	70.00	70.00	-	70.00	53.90	GRAM/MJ	2.34	
TOTAL PURINES	MILLI	20.00 2	-	-	-	15.40	MILLI/MJ	669.52	
DIETARY FIBRE, WAT.SOL.	GRAM	0.37	-	-	-	0.28	GRAM/MJ	12.39	
DIETARY FIBRE, WAT.INS.	GRAM	1.16	-	-	-	0.89	GRAM/MJ	38.83	

2 VALUE EXPRESSED IN MG URIC ACID/100G

FELDSALAT (RAPUNZEL) — LAMB'S LETTUCE — MÂCHE

VALERIANELLA OLITORIA L.

		PROTEIN	FAT	CARBOHYDRATES	TOTAL
ENERGY VALUE (AVERAGE) PER 100 G EDIBLE PORTION	KJOULE	29	14	12	55
	(KCAL)	6.9	3.3	2.8	13
AMOUNT OF DIGESTIBLE CONSTITUENTS PER 100 G	GRAM	1.19	0.32	0.70	
ENERGY VALUE (AVERAGE) OF THE DIGESTIBLE FRACTION PER 100 G EDIBLE PORTION	KJOULE	19	13	12	43
	(KCAL)	4.5	3.0	2.8	10

WASTE PERCENTAGE AVERAGE 0.00

CONSTITUENTS	DIM	AV	VARIATION	AVR	NUTR. DENS.		MOLPERC.
WATER	GRAM	93.40	86.90 - 94.80	90.60	KILO/MJ	2.17	
PROTEIN	GRAM	1.84	1.20 - 2.14	1.78	GRAM/MJ	42.70	
FAT	GRAM	0.36	0.20 - 0.50	0.35	GRAM/MJ	8.36	
AVAILABLE CARBOHYDR.	GRAM	0.70	- - -	0.68	GRAM/MJ	16.25	
TOTAL DIETARY FIBRE	GRAM	1.52 1	- - -	1.47	GRAM/MJ	35.28	
MINERALS	GRAM	0.80	- - -	0.78	GRAM/MJ	18.57	
SODIUM	MILLI	4.00	3.00 - 6.00	3.88	MILLI/MJ	92.83	
POTASSIUM	MILLI	421.00	404.00 - 436.00	408.37	GRAM/MJ	9.77	
MAGNESIUM	MILLI	13.00	- - -	12.61	MILLI/MJ	301.71	
CALCIUM	MILLI	35.00	24.00 - 40.00	33.95	MILLI/MJ	812.30	
IRON	MILLI	2.00	- - -	1.94	MILLI/MJ	46.42	
COPPER	MILLI	0.11	0.04 - 0.19	0.11	MILLI/MJ	2.55	
ZINC	MILLI	0.54	- - -	0.52	MILLI/MJ	12.53	
PHOSPHORUS	MILLI	49.00	48.00 - 50.00	47.53	GRAM/MJ	1.14	
CHLORIDE	MILLI	70.00	- - -	67.90	GRAM/MJ	1.62	
IODIDE	MICRO	-	5.00 - 62.00				
NITRATE	MILLI	219.00	18.00 - 433.00	212.43	GRAM/MJ	5.08	
CAROTENE	MILLI	3.90	0.90 - 7.50	3.78	MILLI/MJ	90.51	
VITAMIN E ACTIVITY	MILLI	0.60	0.40 - 0.80	0.58	MILLI/MJ	13.93	
ALPHA-TOCOPHEROL	MILLI	0.60	- - -	0.58	MILLI/MJ	13.93	
VITAMIN B1	MICRO	65.00	40.00 - 80.00	63.05	MILLI/MJ	1.51	
VITAMIN B2	MICRO	80.00	70.00 - 100.00	77.60	MILLI/MJ	1.86	
NICOTINAMIDE	MILLI	0.38	0.20 - 0.40	0.37	MILLI/MJ	8.82	
VITAMIN B6	MILLI	0.25	0.20 - 0.30	0.24	MILLI/MJ	5.80	
VITAMIN C	MILLI	35.00	15.00 - 60.00	33.95	MILLI/MJ	812.30	

1 METHOD OF SOUTHGATE AND ENGLYST

CONSTITUENTS	DIM	AV	VARIATION			AVR	NUTR. DENS.		MOLPERC.
ARGININE	MILLI	88.00	71.00	-	110.00	85.36	GRAM/MJ	2.04	
HISTIDINE	MILLI	35.00	31.00	-	39.00	33.95	MILLI/MJ	812.30	
ISOLEUCINE	GRAM	0.13	0.10	-	0.15	0.13	GRAM/MJ	3.02	
LEUCINE	GRAM	0.14	0.13	-	0.16	0.14	GRAM/MJ	3.25	
LYSINE	GRAM	0.11	0.10	-	0.13	0.11	GRAM/MJ	2.55	
METHIONINE	MILLI	11.00	5.00	-	17.00	10.67	MILLI/MJ	255.29	
PHENYLALANINE	MILLI	96.00	86.00	-	110.00	93.12	GRAM/MJ	2.23	
THREONINE	MILLI	85.00	74.00	-	95.00	82.45	GRAM/MJ	1.97	
TRYPTOPHAN	MILLI	20.00	12.00	-	29.00	19.40	MILLI/MJ	464.17	
VALINE	GRAM	0.11	0.10	-	0.12	0.11	GRAM/MJ	2.55	
OXALIC ACID TOTAL	-	0.00	-	-	-	0.00			
CAFFEIC ACID	MILLI	40.00	36.00	-	44.00	38.80	MILLI/MJ	928.34	
GLUCOSE	GRAM	0.47	0.47	-	0.47	0.46	GRAM/MJ	10.91	
FRUCTOSE	GRAM	0.20	0.20	-	0.20	0.19	GRAM/MJ	4.64	
SUCROSE	MILLI	30.00	30.00	-	30.00	29.10	MILLI/MJ	696.25	
TOTAL PURINES	MILLI	45.00 2	-	-	-	43.65	GRAM/MJ	1.04	
DIETARY FIBRE, WAT.SOL.	GRAM	0.15	-	-	-	0.15	GRAM/MJ	3.48	
DIETARY FIBRE, WAT.INS.	GRAM	1.37	-	-	-	1.33	GRAM/MJ	31.80	

2 VALUE EXPRESSED IN MG URIC ACID/100G

FENCHEL FENNEL LEAVES FENOUILLE

(BOLOGNESER FENCHEL)

FOENICULUM VULGARE
F. DULCE MILL.

		PROTEIN	FAT	CARBOHYDRATES	TOTAL
ENERGY VALUE (AVERAGE)	KJOULE	38	12	47	97
PER 100 G	(KCAL)	9.1	2.8	11	23
EDIBLE PORTION					
AMOUNT OF DIGESTIBLE	GRAM	1.57	0.27	2.82	
CONSTITUENTS PER 100 G					
ENERGY VALUE (AVERAGE)	KJOULE	25	11	47	82
OF THE DIGESTIBLE	(KCAL)	5.9	2.5	11	20
FRACTION PER 100 G					
EDIBLE PORTION					

WASTE PERCENTAGE AVERAGE 7.0

CONSTITUENTS	DIM	AV		VARIATION			AVR	NUTR. DENS.		MOLPERC.
WATER	GRAM	86.00		81.90	-	90.00	79.98	KILO/MJ	1.04	
PROTEIN	GRAM	2.43		1.90	-	2.80	2.26	GRAM/MJ	29.46	
FAT	GRAM	0.30		0.20	-	0.40	0.28	GRAM/MJ	3.64	
AVAILABLE CARBOHYDR.	GRAM	2.82		-	-	-	2.62	GRAM/MJ	34.19	
TOTAL DIETARY FIBRE	GRAM	3.33	1	-	-	-	3.10	GRAM/MJ	40.37	
MINERALS	GRAM	1.70		-	-	-	1.58	GRAM/MJ	20.61	
SODIUM	MILLI	86.00		52.00	-	120.00	79.98	GRAM/MJ	1.04	
POTASSIUM	MILLI	494.00		339.00	-	612.00	459.42	GRAM/MJ	5.99	
MAGNESIUM	MILLI	49.00		-	-	-	45.57	MILLI/MJ	594.06	
CALCIUM	MILLI	109.00		100.00	-	117.00	101.37	GRAM/MJ	1.32	
IRON	MILLI	2.70		-	-	-	2.51	MILLI/MJ	32.73	
COPPER	MICRO	60.00		-	-	-	55.80	MICRO/MJ	727.41	
ZINC	MILLI	0.25		-	-	-	0.23	MILLI/MJ	3.03	
PHOSPHORUS	MILLI	51.00		-	-	-	47.43	MILLI/MJ	618.30	
NITRATE	MILLI	127.00		30.00	-	420.00	118.11	GRAM/MJ	1.54	
CAROTENE	MILLI	4.70		2.10	-	7.80	4.37	MILLI/MJ	56.98	
VITAMIN B1	MILLI	0.23		0.10	-	0.35	0.21	MILLI/MJ	2.79	
VITAMIN B2	MILLI	0.11		0.02	-	0.20	0.10	MILLI/MJ	1.33	
NICOTINAMIDE	MILLI	0.20		-	-	-	0.19	MILLI/MJ	2.42	
PANTOTHENIC ACID	MILLI	0.25		0.20	-	0.30	0.23	MILLI/MJ	3.03	
VITAMIN B6	MILLI	0.10		-	-	-	0.09	MILLI/MJ	1.21	
BIOTIN	MICRO	2.50		-	-	-	2.33	MICRO/MJ	30.31	
FOLIC ACID	MILLI	0.10		0.09	-	0.10	0.09	MILLI/MJ	1.21	
VITAMIN C	MILLI	93.00		60.00	-	120.00	86.49	GRAM/MJ	1.13	
OXALIC ACID TOTAL	MILLI	5.00		-	-	-	4.65	MILLI/MJ	60.62	
GLUCOSE	GRAM	1.25		1.12	-	1.38	1.17	GRAM/MJ	15.25	
FRUCTOSE	GRAM	1.06		1.04	-	1.27	0.99	GRAM/MJ	12.85	
SUCROSE	GRAM	0.51		0.18	-	2.57	0.48	GRAM/MJ	6.27	
TOTAL PURINES	MILLI	10.00	2	-	-	-	9.30	MILLI/MJ	121.24	
DIETARY FIBRE, WAT.SOL.	GRAM	0.83		-	-	-	0.77	GRAM/MJ	10.06	
DIETARY FIBRE, WAT.INS.	GRAM	2.50		-	-	-	2.33	GRAM/MJ	30.31	

1 METHOD OF MEUSER, SUCKOW AND KULIKOWSKI ("BERLINER METHODE")

2 VALUE EXPRESSED IN MG URIC ACID/100G

GARTENKRESSE CRESS CRESSON ALÉNOIS

LEPIDIUM SATIVUM L.

		PROTEIN	FAT	CARBOHYDRATES	TOTAL
ENERGY VALUE (AVERAGE)	KJOULE	66	54	30	150
PER 100 G	(KCAL)	16	13	7.1	36
EDIBLE PORTION					
AMOUNT OF DIGESTIBLE	GRAM	2.73	1.26	1.78	
CONSTITUENTS PER 100 G					
ENERGY VALUE (AVERAGE)	KJOULE	43	49	30	122
OF THE DIGESTIBLE	(KCAL)	10	12	7.1	29
FRACTION PER 100 G					
EDIBLE PORTION					

WASTE PERCENTAGE AVERAGE 37 MINIMUM 29 MAXIMUM 46

CONSTITUENTS	DIM	AV	VARIATION			AVR	NUTR.	DENS.	MOLPERC.
WATER	GRAM	87.20	-	-	-	54.94	GRAM/MJ	716.80	
PROTEIN	GRAM	4.20	-	-	-	2.65	GRAM/MJ	34.52	
FAT	GRAM	1.40	-	-	-	0.88	GRAM/MJ	11.51	
AVAILABLE CARBOHYDR.	GRAM	1.78 1	-	-	-	1.12	GRAM/MJ	14.63	
TOTAL DIETARY FIBRE	GRAM	3.52 2	-	-	-	2.22	GRAM/MJ	28.94	
MINERALS	GRAM	1.90	-	-	-	1.20	GRAM/MJ	15.62	
SODIUM	MILLI	5.00	2.00	-	9.00	3.15	MILLI/MJ	41.10	
POTASSIUM	MILLI	550.00	455.00	-	665.00	346.50	GRAM/MJ	4.52	
CALCIUM	MILLI	214.00	211.00	-	217.00	134.82	GRAM/MJ	1.76	
IRON	MILLI	2.90	-	-	-	1.83	MILLI/MJ	23.84	
MOLYBDENUM	MICRO	10.00	-	-	-	6.30	MICRO/MJ	82.20	
PHOSPHORUS	MILLI	38.00	-	-	-	23.94	MILLI/MJ	312.37	
FLUORIDE	MICRO	24.00	-	-	-	15.12	MICRO/MJ	197.28	
NITRATE	MILLI	245.00	63.00	-	463.00	154.35	GRAM/MJ	2.01	
CAROTENE	MILLI	2.19	1.78	-	2.60	1.38	MILLI/MJ	18.00	
VITAMIN E ACTIVITY	MILLI	0.70	-	-	-	0.44	MILLI/MJ	5.75	
TOTAL TOCOPHEROLS	MILLI	3.60	-	-	-	2.27	MILLI/MJ	29.59	
ALPHA-TOCOPHEROL	MILLI	0.70	-	-	-	0.44	MILLI/MJ	5.75	
VITAMIN B1	MILLI	0.15	0.11	-	0.18	0.09	MILLI/MJ	1.23	
VITAMIN B2	MILLI	0.19	0.17	-	0.20	0.12	MILLI/MJ	1.56	
NICOTINAMIDE	MILLI	1.75	1.00	-	2.50	1.10	MILLI/MJ	14.39	
VITAMIN B6	MILLI	0.30	0.21	-	0.37	0.19	MILLI/MJ	2.47	
VITAMIN C	MILLI	59.00	35.00	-	87.00	37.17	MILLI/MJ	484.99	
PALMITIC ACID	MILLI	87.00	87.00	-	87.00	54.81	MILLI/MJ	715.16	
STEARIC ACID	MILLI	6.00	6.00	-	6.00	3.78	MILLI/MJ	49.32	
PALMITOLEIC ACID	MILLI	10.00	10.00	-	10.00	6.30	MILLI/MJ	82.20	
OLEIC ACID	MILLI	4.00	4.00	-	4.00	2.52	MILLI/MJ	32.88	
LINOLEIC ACID	MILLI	96.00	96.00	-	96.00	60.48	MILLI/MJ	789.14	
LINOLENIC ACID	MILLI	290.00	290.00	-	290.00	182.70	GRAM/MJ	2.38	
DIETARY FIBRE, WAT.SOL.	GRAM	1.48	-	-	-	0.93	GRAM/MJ	12.17	
DIETARY FIBRE, WAT.INS.	GRAM	2.04	-	-	-	1.29	GRAM/MJ	16.77	

1 ESTIMATED BY THE DIFFERENCE METHOD
100 - (WATER + PROTEIN + FAT + MINERALS + TOTAL DIETARY FIBRE)

2 METHOD OF ENGLYST

GRÜNKOHL (BRAUNKOHL) | KALE | CHOU VERT

BRASSICA OLERACEA L.
VAR. ACEPHALA D. C.

		PROTEIN	FAT	CARBOHYDRATES	TOTAL
ENERGY VALUE (AVERAGE)	KJOULE	67	35	50	152
PER 100 G	(KCAL)	16	8.4	12	36
EDIBLE PORTION					
AMOUNT OF DIGESTIBLE	GRAM	2.79	0.81	2.97	
CONSTITUENTS PER 100 G					
ENERGY VALUE (AVERAGE)	KJOULE	44	32	50	125
OF THE DIGESTIBLE	(KCAL)	10	7.5	12	30
FRACTION PER 100 G					
EDIBLE PORTION					

WASTE PERCENTAGE AVERAGE 49 MINIMUM 30 MAXIMUM 57

CONSTITUENTS	DIM	AV	VARIATION		AVR	NUTR. DENS.		MOLPERC.
WATER	GRAM	86.30	80.50	- 86.60	44.01	GRAM/MJ	689.97	
PROTEIN	GRAM	4.30	3.90	- 4.90	2.19	GRAM/MJ	34.38	
FAT	GRAM	0.90	0.60	- 1.30	0.46	GRAM/MJ	7.20	
AVAILABLE CARBOHYDR.	GRAM	2.97	1.66	- 4.28	1.51	GRAM/MJ	23.75	
TOTAL DIETARY FIBRE	GRAM	4.20 1	-	- -	2.14	GRAM/MJ	33.58	
MINERALS	GRAM	1.70	-	- -	0.87	GRAM/MJ	13.59	
SODIUM	MILLI	42.00	23.00	- 56.00	21.42	MILLI/MJ	335.79	
POTASSIUM	MILLI	490.00	401.00	- 553.00	249.90	GRAM/MJ	3.92	
MAGNESIUM	MILLI	31.00	-	- -	15.81	MILLI/MJ	247.85	
CALCIUM	MILLI	212.00	110.00	- 287.00	108.12	GRAM/MJ	1.69	
MANGANESE	MILLI	0.55	-	- -	0.28	MILLI/MJ	4.40	
IRON	MILLI	1.90	1.00	- 2.20	0.97	MILLI/MJ	15.19	
COPPER	MICRO	90.00	-	- -	45.90	MICRO/MJ	719.55	
ZINC	MILLI	0.33	-	- -	0.17	MILLI/MJ	2.64	
NICKEL	MICRO	30.00	-	- -	15.30	MICRO/MJ	239.85	
CHROMIUM	MICRO	10.00	-	- -	5.10	MICRO/MJ	79.95	
MOLYBDENUM	MICRO	4.00	2.00	- 6.00	2.04	MICRO/MJ	31.98	
PHOSPHORUS	MILLI	87.00	58.00	- 114.00	44.37	MILLI/MJ	695.57	
CHLORIDE	MILLI	60.00	-	- -	30.60	MILLI/MJ	479.70	
FLUORIDE	MICRO	20.00	10.00	- 40.00	10.20	MICRO/MJ	159.90	
IODIDE	MICRO	12.00	-	- -	6.12	MICRO/MJ	95.94	
BORON	MILLI	0.24	0.11	- 0.37	0.12	MILLI/MJ	1.92	
SELENIUM	MICRO	2.30	-	- -	1.17	MICRO/MJ	18.39	
NITRATE	MILLI	101.00	1.00	- 174.00	51.51	MILLI/MJ	807.50	
CAROTENE	MILLI	4.10	0.90	- 7.58	2.09	MILLI/MJ	32.78	
VITAMIN E ACTIVITY	MILLI	1.70	-	- -	0.87	MILLI/MJ	13.59	

1 METHOD OF SOUTHGATE AND ENGLYST

CONSTITUENTS	DIM	AV	VARIATION	AVR	NUTR.	DENS.	MOLPERC.
ALPHA-TOCOPHEROL	MILLI	1.70	- - -	0.87	MILLI/MJ	13.59	
VITAMIN B1	MILLI	0.10	0.02 - 0.21	0.05	MILLI/MJ	0.80	
VITAMIN B2	MILLI	0.25	0.20 - 0.42	0.13	MILLI/MJ	2.00	
NICOTINAMIDE	MILLI	2.10	1.50 - 4.60	1.07	MILLI/MJ	16.79	
PANTOTHENIC ACID	MILLI	-	0.10 - 1.40				
VITAMIN B6	MILLI	0.25	0.10 - 0.46	0.13	MILLI/MJ	2.00	
BIOTIN	MICRO	0.50	- - -	0.26	MICRO/MJ	4.00	
FOLIC ACID	MICRO	60.00	- - -	30.60	MICRO/MJ	479.70	
VITAMIN C	MILLI	105.00	60.00 - 392.00	53.55	MILLI/MJ	839.48	
ARGININE	GRAM	0.30	0.25 - 0.32	0.15	GRAM/MJ	2.40	
CYSTINE	MILLI	69.00	- - -	35.19	MILLI/MJ	551.66	
HISTIDINE	GRAM	0.10	0.09 - 0.10	0.05	GRAM/MJ	0.80	
ISOLEUCINE	GRAM	0.14	0.06 - 0.28	0.07	GRAM/MJ	1.12	
LEUCINE	GRAM	0.25	0.16 - 0.32	0.13	GRAM/MJ	2.00	
LYSINE	GRAM	0.24	0.20 - 0.24	0.12	GRAM/MJ	1.92	
METHIONINE	MILLI	52.00	47.00 - 53.00	26.52	MILLI/MJ	415.74	
PHENYLALANINE	GRAM	0.14	0.10 - 0.18	0.07	GRAM/MJ	1.12	
THREONINE	GRAM	0.13	0.06 - 0.19	0.07	GRAM/MJ	1.04	
TRYPTOPHAN	MILLI	64.00	50.00 - 72.00	32.64	MILLI/MJ	511.68	
TYROSINE	GRAM	0.18	- - -	0.09	GRAM/MJ	1.44	
VALINE	GRAM	0.23	0.16 - 0.27	0.12	GRAM/MJ	1.84	
MALIC ACID	MILLI	215.00	50.00 - 380.00	109.65	GRAM/MJ	1.72	
CITRIC ACID	MILLI	220.00	90.00 - 350.00	112.20	GRAM/MJ	1.76	
OXALIC ACID TOTAL	MILLI	7.50	- - -	3.83	MILLI/MJ	59.96	
OXALIC ACID SOLUBLE	MILLI	5.30	- - -	2.70	MILLI/MJ	42.37	
FERULIC ACID	MILLI	20.00	13.00 - 28.00	10.20	MILLI/MJ	159.90	
CAFFEIC ACID	MILLI	21.00	13.00 - 30.00	10.71	MILLI/MJ	167.90	
PARA-COUMARIC ACID	MILLI	1.50	2.00 - 3.00	0.77	MILLI/MJ	11.99	
SUCCINIC ACID	MILLI	32.00	- - -	16.32	MILLI/MJ	255.84	
FUMARIC ACID	MILLI	22.00	- - -	11.22	MILLI/MJ	175.89	
GLUCOSE	GRAM	0.61	0.38 - 0.85	0.31	GRAM/MJ	4.92	
FRUCTOSE	GRAM	0.92	0.29 - 1.55	0.47	GRAM/MJ	7.36	
SUCROSE	GRAM	1.00	0.85 - 1.15	0.51	GRAM/MJ	8.00	
PENTOSAN	GRAM	0.92	- - -	0.47	GRAM/MJ	7.36	
HEXOSAN	GRAM	0.72	- - -	0.37	GRAM/MJ	5.76	
CELLULOSE	GRAM	1.47	- - -	0.75	GRAM/MJ	11.75	
POLYURONIC ACID	GRAM	1.06	- - -	0.54	GRAM/MJ	8.47	
MYRISTIC ACID	MILLI	1.40	- - -	0.71	MILLI/MJ	11.19	
PALMITIC ACID	MILLI	94.50	94.00 - 95.00	48.20	MILLI/MJ	755.53	
STEARIC ACID	MILLI	8.60	- - -	4.39	MILLI/MJ	68.76	
ARACHIDIC ACID	MILLI	2.20	- - -	1.12	MILLI/MJ	17.59	
PALMITOLEIC ACID	MILLI	11.95	2.90 - 21.00	6.09	MILLI/MJ	95.54	
OLEIC ACID	MILLI	5.00	- - -	2.55	MILLI/MJ	39.98	
EICOSENOIC ACID	MILLI	2.20	- - -	1.12	MILLI/MJ	17.59	
LINOLEIC ACID	MILLI	130.00	127.00 - 133.00	66.30	GRAM/MJ	1.04	
LINOLENIC ACID	MILLI	354.50	349.00 - 360.00	180.80	GRAM/MJ	2.83	
TOTAL PURINES	MILLI	25.00 [2]	20.00 - 30.00	12.75	MILLI/MJ	199.88	

[2] VALUE EXPRESSED IN MG URIC ACID/100G

KNOBLAUCH GARLIC AIL

ALLIUM SATIVUM L.

		PROTEIN	FAT	CARBOHYDRATES	TOTAL
ENERGY VALUE (AVERAGE) PER 100 G EDIBLE PORTION	KJOULE	95	4.7	475	575
	(KCAL)	23	1.1	114	137
AMOUNT OF DIGESTIBLE CONSTITUENTS PER 100 G	GRAM	3.93	0.10	28.41	
ENERGY VALUE (AVERAGE) OF THE DIGESTIBLE FRACTION PER 100 G EDIBLE PORTION	KJOULE	62	4.2	475	541
	(KCAL)	15	1.0	114	129

WASTE PERCENTAGE AVERAGE 12 MINIMUM 8.0 MAXIMUM 16

CONSTITUENTS	DIM	AV	VARIATION			AVR	NUTR. DENS.		MOLPERC.
WATER	GRAM	64.00	63.00	-	64.60	56.32	GRAM/MJ	118.22	
PROTEIN	GRAM	6.05	5.30	-	6.76	5.32	GRAM/MJ	11.18	
FAT	GRAM	0.12	0.06	-	0.20	0.11	GRAM/MJ	0.22	
AVAILABLE CARBOHYDR.	GRAM	28.41 1	-	-	-	25.00	GRAM/MJ	52.48	
MINERALS	GRAM	1.42	1.40	-	1.44	1.25	GRAM/MJ	2.62	
CALCIUM	MILLI	38.00	-	-	-	33.44	MILLI/MJ	70.19	
MANGANESE	MILLI	0.46	-	-	-	0.40	MILLI/MJ	0.85	
IRON	MILLI	1.40	-	-	-	1.23	MILLI/MJ	2.59	
COPPER	MILLI	0.26	-	-	-	0.23	MILLI/MJ	0.48	
ZINC	MILLI	1.00	-	-	-	0.88	MILLI/MJ	1.85	
NICKEL	MICRO	10.00	-	-	-	8.80	MICRO/MJ	18.47	
MOLYBDENUM	MICRO	70.00	-	-	-	61.60	MICRO/MJ	129.30	
PHOSPHORUS	MILLI	134.00	-	-	-	117.92	MILLI/MJ	247.52	
CHLORIDE	MILLI	30.00	-	-	-	26.40	MILLI/MJ	55.41	
IODIDE	MICRO	2.70	-	-	-	2.38	MICRO/MJ	4.99	
BORON	MILLI	0.44	0.34	-	0.63	0.39	MILLI/MJ	0.81	
SELENIUM	MICRO	20.00	14.00	-	28.00	17.60	MICRO/MJ	36.94	
VITAMIN E ACTIVITY	MICRO	11.00	-	-	-	9.68	MICRO/MJ	20.32	
TOTAL TOCOPHEROLS	MILLI	0.10	-	-	-	0.09	MILLI/MJ	0.18	
ALPHA-TOCOPHEROL	MICRO	10.00	-	-	-	8.80	MICRO/MJ	18.47	
DELTA-TOCOPHEROL	MICRO	90.00	-	-	-	79.20	MICRO/MJ	166.24	
VITAMIN B1	MILLI	0.20	0.18	-	0.21	0.18	MILLI/MJ	0.37	
VITAMIN B2	MICRO	80.00	-	-	-	70.40	MICRO/MJ	147.77	
NICOTINAMIDE	MILLI	0.60	-	-	-	0.53	MILLI/MJ	1.11	
VITAMIN C	MILLI	14.00	9.00	-	18.00	12.32	MILLI/MJ	25.86	
SALICYLIC ACID	MILLI	0.10	0.10	-	0.10	0.09	MILLI/MJ	0.18	
LAURIC ACID	MILLI	0.50	0.50	-	0.50	0.44	MILLI/MJ	0.92	
PALMITIC ACID	MILLI	24.00	24.00	-	24.00	21.12	MILLI/MJ	44.33	
STEARIC ACID	-	TRACES	TRACES	-					
OLEIC ACID	MILLI	3.00	3.00	-	3.00	2.64	MILLI/MJ	5.54	
LINOLEIC ACID	MILLI	62.00	62.00	-	62.00	54.56	MILLI/MJ	114.52	
LINOLENIC ACID	MILLI	5.50	5.50	-	5.50	4.84	MILLI/MJ	10.16	

1 ESTIMATED BY THE DIFFERENCE METHOD
100 - (WATER + PROTEIN + FAT + MINERALS)

KOPFSALAT LETTUCE LAITUE

LACTUCA SATIVA L.

		PROTEIN	FAT	CARBOHYDRATES	TOTAL
ENERGY VALUE (AVERAGE)	KJOULE	20	8.6	18	47
PER 100 G	(KCAL)	4.7	2.0	4.4	11
EDIBLE PORTION					
AMOUNT OF DIGESTIBLE	GRAM	0.81	0.19	1.10	
CONSTITUENTS PER 100 G					
ENERGY VALUE (AVERAGE)	KJOULE	13	7.7	18	39
OF THE DIGESTIBLE	(KCAL)	3.0	1.8	4.4	9.3
FRACTION PER 100 G					
EDIBLE PORTION					

WASTE PERCENTAGE AVERAGE 32 MINIMUM 20 MAXIMUM 47

CONSTITUENTS	DIM	AV	VARIATION	AVR	NUTR. DENS.		MOLPERC.
WATER	GRAM	95.00	93.00 - 96.00	64.60	KILO/MJ	2.44	
PROTEIN	GRAM	1.25	0.80 - 1.63	0.85	GRAM/MJ	32.16	
FAT	GRAM	0.22	0.17 - 0.25	0.15	GRAM/MJ	5.66	
AVAILABLE CARBOHYDR.	GRAM	1.10	- - -	0.75	GRAM/MJ	28.31	
TOTAL DIETARY FIBRE	GRAM	1.52 1	- - -	1.03	GRAM/MJ	39.11	
MINERALS	GRAM	0.72	0.34 - 1.04	0.49	GRAM/MJ	18.53	
SODIUM	MILLI	10.00	5.00 - 14.00	6.80	MILLI/MJ	257.32	
POTASSIUM	MILLI	224.00	140.00 - 313.00	152.32	GRAM/MJ	5.76	
MAGNESIUM	MILLI	11.00	6.00 - 13.00	7.48	MILLI/MJ	283.05	
CALCIUM	MILLI	37.00	17.00 - 51.00	25.16	MILLI/MJ	952.08	
MANGANESE	MILLI	0.35	0.12 - 0.53	0.24	MILLI/MJ	9.01	
IRON	MILLI	1.10	0.50 - 2.00	0.75	MILLI/MJ	28.31	
COBALT	MICRO	5.40	1.80 - 12.00	3.67	MICRO/MJ	138.95	
COPPER	MICRO	54.00	30.00 - 78.00	36.72	MILLI/MJ	1.39	
ZINC	MILLI	0.22	0.16 - 0.35	0.15	MILLI/MJ	5.66	
NICKEL	MICRO	11.50	5.00 - 14.00	7.82	MICRO/MJ	295.92	
CHROMIUM	MICRO	14.00	7.00 - 21.00	9.52	MICRO/MJ	360.25	
MOLYBDENUM	MICRO	6.00	2.00 - 11.00	4.08	MICRO/MJ	154.39	
PHOSPHORUS	MILLI	33.00	19.00 - 57.00	22.44	MILLI/MJ	849.16	
CHLORIDE	MILLI	57.00	39.00 - 74.00	38.76	GRAM/MJ	1.47	
FLUORIDE	MICRO	32.00	28.00 - 38.00	21.76	MICRO/MJ	823.42	
IODIDE	MICRO	3.30	2.60 - 4.00	2.24	MICRO/MJ	84.92	
BORON	MICRO	82.00	30.00 - 90.00	55.76	MILLI/MJ	2.11	
SELENIUM	MICRO	0.75 2	0.40 - 10.00	0.51	MICRO/MJ	19.30	
SILICON	MILLI	2.00	1.00 - 4.00	1.36	MILLI/MJ	51.46	
NITRATE	MILLI	262.00	23.00 - 661.00	178.16	GRAM/MJ	6.74	

1 METHOD OF SOUTHGATE AND ENGLYST

2 AMOUNTS, DETERMINED IN THE FEDERAL REPUBLIC OF GERMANY

CONSTITUENTS	DIM	AV	VARIATION			AVR	NUTR. DENS.		MOLPERC.
CAROTENE	MILLI	0.79	0.16	-	1.60	0.54	MILLI/MJ	20.33	
VITAMIN E ACTIVITY	MILLI	0.44	-	-	-	0.30	MILLI/MJ	11.32	
ALPHA-TOCOPHEROL	MILLI	0.44	-	-	-	0.30	MILLI/MJ	11.32	
VITAMIN K	MILLI	0.20	-	-	-	0.14	MILLI/MJ	5.15	
VITAMIN B1	MICRO	62.00	40.00	-	80.00	42.16	MILLI/MJ	1.60	
VITAMIN B2	MICRO	78.00	60.00	-	100.00	53.04	MILLI/MJ	2.01	
NICOTINAMIDE	MILLI	0.32	0.20	-	0.50	0.22	MILLI/MJ	8.23	
PANTOTHENIC ACID	MILLI	0.11	-	-	-	0.07	MILLI/MJ	2.83	
VITAMIN B6	MICRO	55.00	36.00	-	75.00	37.40	MILLI/MJ	1.42	
BIOTIN	MICRO	1.90	0.70	-	3.10	1.29	MICRO/MJ	48.89	
FOLIC ACID	MICRO	37.00	-	-	-	25.16	MICRO/MJ	952.08	
VITAMIN B12	-	0.00	-	-	-	0.00			
VITAMIN C	MILLI	13.00	8.00	-	22.00	8.84	MILLI/MJ	334.52	
ARGININE	MILLI	62.00	56.00	-	69.00	42.16	GRAM/MJ	1.60	
HISTIDINE	MILLI	21.00	20.00	-	23.00	14.28	MILLI/MJ	540.37	
ISOLEUCINE	MILLI	70.00	48.00	-	83.00	47.60	GRAM/MJ	1.80	
LEUCINE	MILLI	77.00	75.00	-	79.00	52.36	GRAM/MJ	1.98	
LYSINE	MILLI	70.00	48.00	-	81.00	47.60	GRAM/MJ	1.80	
METHIONINE	MILLI	12.00	4.00	-	22.00	8.16	MILLI/MJ	308.78	
PHENYLALANINE	MILLI	54.00	46.00	-	64.00	36.72	GRAM/MJ	1.39	
THREONINE	MILLI	56.00	51.00	-	59.00	38.08	GRAM/MJ	1.44	
TRYPTOPHAN	MILLI	11.00	9.00	-	13.00	7.48	MILLI/MJ	283.05	
TYROSINE	MILLI	34.00	-	-	-	23.12	MILLI/MJ	874.89	
VALINE	MILLI	66.00	60.00	-	68.00	44.88	GRAM/MJ	1.70	
CITRIC ACID	MILLI	13.00	-	-	-	8.84	MILLI/MJ	334.52	
OXALIC ACID TOTAL	-	0.00	-	-	-	0.00			
FERULIC ACID	MICRO	50.00	0.00	-	100.00	34.00	MILLI/MJ	1.29	
CAFFEIC ACID	MILLI	38.00	16.00	-	60.00	25.84	MILLI/MJ	977.82	
GLUCOSE	GRAM	0.41	0.25	-	1.11	0.28	GRAM/MJ	10.65	
FRUCTOSE	GRAM	0.56	0.38	-	1.38	0.38	GRAM/MJ	14.41	
SUCROSE	GRAM	0.10	0.04	-	0.13	0.07	GRAM/MJ	2.80	
STARCH	MILLI	19.00	13.00	-	27.00	12.92	MILLI/MJ	488.91	
PENTOSAN	GRAM	0.18	-	-	-	0.12	GRAM/MJ	4.63	
HEXOSAN	GRAM	0.16	-	-	-	0.11	GRAM/MJ	4.12	
CELLULOSE	GRAM	0.76	-	-	-	0.52	GRAM/MJ	19.56	
POLYURONIC ACID	GRAM	0.42	-	-	-	0.29	GRAM/MJ	10.81	
PALMITIC ACID	MILLI	34.00	34.00	-	34.00	23.12	MILLI/MJ	874.89	
STEARIC ACID	MILLI	3.90	3.90	-	3.90	2.65	MILLI/MJ	100.35	
PALMITOLEIC ACID	MILLI	1.00	1.00	-	1.00	0.68	MILLI/MJ	25.73	
OLEIC ACID	MILLI	5.20	5.20	-	5.20	3.54	MILLI/MJ	133.81	
LINOLEIC ACID	MILLI	52.00	52.00	-	52.00	35.36	GRAM/MJ	1.34	
LINOLENIC ACID	MILLI	71.00	71.00	-	71.00	48.28	GRAM/MJ	1.83	
TOTAL STEROLS	MILLI	10.00	-	-	-	6.80	MILLI/MJ	257.32	
CAMPESTEROL	MILLI	1.00	-	-	-	0.68	MILLI/MJ	25.73	
BETA-SITOSTEROL	MILLI	5.00	-	-	-	3.40	MILLI/MJ	128.66	
STIGMASTEROL	MILLI	4.00	-	-	-	2.72	MILLI/MJ	102.93	
TOTAL PURINES	MILLI	20.00 [3]	10.00	-	30.00	13.60	MILLI/MJ	514.64	
DIETARY FIBRE,WAT.SOL.	GRAM	0.15	-	-	-	0.10	GRAM/MJ	3.86	
DIETARY FIBRE,WAT.INS.	GRAM	1.37	-	-	-	0.93	GRAM/MJ	35.25	

3 VALUE EXPRESSED IN MG URIC ACID/100G

LÖWENZAHN-BLÄTTER DANDELION LEAVES PISSENLIT FEUILLES

TARAXACUM OFFICIN. WEBER

		PROTEIN	FAT	CARBOHYDRATES	TOTAL
ENERGY VALUE (AVERAGE) PER 100 G EDIBLE PORTION	KJOULE	40	24	153	217
	(KCAL)	9.6	5.8	37	52
AMOUNT OF DIGESTIBLE CONSTITUENTS PER 100 G	GRAM	1.65	0.55	9.13	
ENERGY VALUE (AVERAGE) OF THE DIGESTIBLE FRACTION PER 100 G EDIBLE PORTION	KJOULE	26	22	153	201
	(KCAL)	6.2	5.2	37	48

WASTE PERCENTAGE AVERAGE 0.00

CONSTITUENTS	DIM	AV	VARIATION		AVR	NUTR. DENS.		MOLPERC.
WATER	GRAM	85.70	85.50	- 85.80	85.70	GRAM/MJ	427.39	
PROTEIN	GRAM	2.55	2.00	- 2.81	2.55	GRAM/MJ	12.72	
FAT	GRAM	0.62	0.40	- 0.70	0.62	GRAM/MJ	3.09	
AVAILABLE CARBOHYDR.	GRAM	9.13 1	-	- -	9.13	GRAM/MJ	45.53	
MINERALS	GRAM	2.00	1.99	- 2.00	2.00	GRAM/MJ	9.97	
SODIUM	MILLI	76.00	-	- -	76.00	MILLI/MJ	379.02	
POTASSIUM	MILLI	440.00	430.00	- 460.00	440.00	GRAM/MJ	2.19	
MAGNESIUM	MILLI	36.00	-	- -	36.00	MILLI/MJ	179.53	
CALCIUM	MILLI	158.00	100.00	- 187.00	158.00	MILLI/MJ	787.96	
MANGANESE	MILLI	0.34	-	- -	0.34	MILLI/MJ	1.70	
IRON	MILLI	3.10	-	- -	3.10	MILLI/MJ	15.46	
COPPER	MILLI	0.17	-	- -	0.17	MILLI/MJ	0.85	
ZINC	MILLI	1.20	-	- -	1.20	MILLI/MJ	5.98	
PHOSPHORUS	MILLI	70.00	-	- -	70.00	MILLI/MJ	349.10	
CHLORIDE	MILLI	100.00	99.00	- 100.00	100.00	MILLI/MJ	498.71	
CAROTENE	MILLI	7.90	7.20	- 8.20	7.90	MILLI/MJ	39.40	
VITAMIN E ACTIVITY	MILLI	2.50	-	- -	2.50	MILLI/MJ	12.47	
ALPHA-TOCOPHEROL	MILLI	2.50	-	- -	2.50	MILLI/MJ	12.47	
VITAMIN B1	MILLI	0.19	-	- -	0.19	MILLI/MJ	0.95	
VITAMIN B2	MILLI	0.17	0.14	- 0.23	0.17	MILLI/MJ	0.85	
NICOTINAMIDE	MILLI	0.80	-	- -	0.80	MILLI/MJ	3.99	
VITAMIN C	MILLI	30.00	20.00	- 36.00	30.00	MILLI/MJ	149.61	
OXALIC ACID	MILLI	24.60	-	- -	24.60	MILLI/MJ	122.68	

1 ESTIMATED BY THE DIFFERENCE METHOD
100 - (WATER + PROTEIN + FAT + MINERALS)

MANGOLD — MANGOLD — BETTE POIRÉE

BETA VULGARIS L.
VAR. CICLA L.

		PROTEIN	FAT	CARBOHYDRATES	TOTAL
ENERGY VALUE (AVERAGE)	KJOULE	33	11	12	56
PER 100 G	(KCAL)	8.0	2.6	2.8	13
EDIBLE PORTION					
AMOUNT OF DIGESTIBLE	GRAM	1.38	0.25	0.69	
CONSTITUENTS PER 100 G					
ENERGY VALUE (AVERAGE)	KJOULE	22	9.8	12	43
OF THE DIGESTIBLE	(KCAL)	5.2	2.3	2.8	10
FRACTION PER 100 G					
EDIBLE PORTION					

WASTE PERCENTAGE AVERAGE 19 MINIMUM 10 MAXIMUM 28

CONSTITUENTS	DIM	AV	VARIATION			AVR	NUTR. DENS.		MOLPERC.
WATER	GRAM	92.20	91.00	-	94.00	74.68	KILO/MJ	2.14	
PROTEIN	GRAM	2.13	1.40	-	2.60	1.73	GRAM/MJ	49.45	
FAT	GRAM	0.28	0.10	-	0.42	0.23	GRAM/MJ	6.50	
AVAILABLE CARBOHYDR.	GRAM	0.69	-	-	-	0.56	GRAM/MJ	16.02	
MINERALS	GRAM	1.68	0.20	-	2.20	1.36	GRAM/MJ	39.00	
SODIUM	MILLI	90.00	84.00	-	100.00	72.90	GRAM/MJ	2.09	
POTASSIUM	MILLI	376.00	349.00	-	400.00	304.56	GRAM/MJ	8.73	
CALCIUM	MILLI	103.00	100.00	-	105.00	83.43	GRAM/MJ	2.39	
IRON	MILLI	2.70	2.50	-	3.00	2.19	MILLI/MJ	62.68	
COPPER	MICRO	80.00	-	-	-	64.80	MILLI/MJ	1.86	
ZINC	MILLI	0.35	-	-	-	0.28	MILLI/MJ	8.13	
PHOSPHORUS	MILLI	39.00	36.00	-	45.00	31.59	MILLI/MJ	905.37	
NITRATE	MILLI	487.00 1	352.00	-	704.00	394.47	GRAM/MJ	11.31	
CAROTENE	MILLI	3.53	1.20	-	6.00	2.86	MILLI/MJ	81.95	
VITAMIN B1	MICRO	98.00	60.00	-	170.00	79.38	MILLI/MJ	2.28	
VITAMIN B2	MILLI	0.16	0.07	-	0.25	0.13	MILLI/MJ	3.71	
NICOTINAMIDE	MILLI	0.65	0.40	-	1.00	0.53	MILLI/MJ	15.09	
PANTOTHENIC ACID	MILLI	0.17	0.15	-	0.19	0.14	MILLI/MJ	3.95	
FOLIC ACID	MICRO	30.00	29.00	-	31.00	24.30	MICRO/MJ	696.44	
VITAMIN C	MILLI	39.00	-	-	-	31.59	MILLI/MJ	905.37	
ARGININE	MILLI	53.00	-	-	-	42.93	GRAM/MJ	1.23	
HISTIDINE	MILLI	27.00	-	-	-	21.87	MILLI/MJ	626.80	
ISOLEUCINE	MILLI	91.00	-	-	-	73.71	GRAM/MJ	2.11	
LEUCINE	MILLI	120.00	-	-	-	97.20	GRAM/MJ	2.79	
LYSINE	MILLI	84.00	-	-	-	68.04	GRAM/MJ	1.95	

1 THE AMOUNTS ARE VALID FOR THE CULTIVATION IN THE GLASSHOUSE
FOR OUTDOOR CULTIVATION, THEY ARE LOWER AND RANGE FROM 100- 220
MG/100G

CONSTITUENTS	DIM	AV	VARIATION			AVR	NUTR. DENS.		MOLPERC.
METHIONINE	MILLI	6.00	-	-	-	4.86	MILLI/MJ	139.29	
PHENYLALANINE	MILLI	70.00	-	-	-	56.70	GRAM/MJ	1.63	
THREONINE	MILLI	88.00	-	-	-	71.28	GRAM/MJ	2.04	
VALINE	MILLI	84.00	-	-	-	68.04	GRAM/MJ	1.95	
OXALIC ACID TOTAL	MILLI	650.00	-	-	-	526.50	GRAM/MJ	15.09	
GLUCOSE	GRAM	0.21	0.21	-	0.21	0.17	GRAM/MJ	4.88	
FRUCTOSE	GRAM	0.27	0.27	-	0.27	0.22	GRAM/MJ	6.27	
SUCROSE	GRAM	0.21	0.21	-	0.21	0.17	GRAM/MJ	4.88	

PETERSILIE
BLATT

PARSLEY
LEAVE

PERSIL
FEUILLE

PETROSELINUM SATIV. HOFFMANN

		PROTEIN	FAT	CARBOHYDRATES	TOTAL
ENERGY VALUE (AVERAGE)	KJOULE	70	14	22	105
PER 100 G	(KCAL)	17	3.3	5.2	25
EDIBLE PORTION					
AMOUNT OF DIGESTIBLE	GRAM	2.87	0.32	1.31	
CONSTITUENTS PER 100 G					
ENERGY VALUE (AVERAGE)	KJOULE	45	13	22	80
OF THE DIGESTIBLE	(KCAL)	11	3.0	5.2	19
FRACTION PER 100 G					
EDIBLE PORTION					

WASTE PERCENTAGE AVERAGE 40 MINIMUM 33 MAXIMUM 46

CONSTITUENTS	DIM	AV	VARIATION			AVR	NUTR. DENS.		MOLPERC.
WATER	GRAM	81.90	78.70	-	85.10	49.14	KILO/MJ	1.03	
PROTEIN	GRAM	4.43	3.66	-	5.20	2.66	GRAM/MJ	55.58	
FAT	GRAM	0.36	0.00	-	0.72	0.22	GRAM/MJ	4.52	
AVAILABLE CARBOHYDR.	GRAM	1.31	-	-	-	0.79	GRAM/MJ	16.43	
TOTAL DIETARY FIBRE	GRAM	4.25 1	-	-	-	2.55	GRAM/MJ	53.32	
MINERALS	GRAM	1.68	-	-	-	1.01	GRAM/MJ	21.08	

1 METHOD OF SOUTHGATE AND ENGLYST

CONSTITUENTS	DIM	AV	VARIATION			AVR	NUTR. DENS.		MOLPERC.
SODIUM	MILLI	33.00	30.00	-	33.00	19.80	MILLI/MJ	414.00	
POTASSIUM	GRAM	1.00	0.55	-	1.08	0.60	GRAM/MJ	12.55	
MAGNESIUM	MILLI	41.10	30.00	-	52.20	24.66	MILLI/MJ	515.61	
CALCIUM	MILLI	245.00	165.00	-	325.00	147.00	GRAM/MJ	3.07	
MANGANESE	MILLI	2.70	0.90	-	5.30	1.62	MILLI/MJ	33.87	
IRON	MILLI	5.50	2.50	-	8.00	3.30	MILLI/MJ	69.00	
COBALT	MICRO	1.00	0.10	-	3.00	0.60	MICRO/MJ	12.55	
COPPER	MILLI	0.52	-	-	-	0.31	MILLI/MJ	6.52	
ZINC	MILLI	0.90	-	-	-	0.54	MILLI/MJ	11.29	
NICKEL	MICRO	75.00	32.00	-	148.00	45.00	MICRO/MJ	940.90	
CHROMIUM	MICRO	7.00	6.00	-	10.00	4.20	MICRO/MJ	87.82	
PHOSPHORUS	MILLI	128.00	-	-	-	76.80	GRAM/MJ	1.61	
CHLORIDE	MILLI	156.00	-	-	-	93.60	GRAM/MJ	1.96	
FLUORIDE	MILLI	0.11	-	-	-	0.07	MILLI/MJ	1.38	
BORON	MILLI	0.54	0.41	-	0.76	0.32	MILLI/MJ	6.77	
SELENIUM	MILLI	-	0.00	-	0.11				
SILICON	MILLI	12.00	6.00	-	20.00	7.20	MILLI/MJ	150.54	
NITRATE	MILLI	212.00	10.00	-	564.00	127.20	GRAM/MJ	2.66	
CAROTENE	MILLI	7.25	6.50	-	8.00	4.35	MILLI/MJ	90.95	
VITAMIN K	MILLI	0.79	0.79	-	0.79	0.47	MILLI/MJ	9.91	
VITAMIN B1	MILLI	0.14	0.12	-	0.15	0.08	MILLI/MJ	1.76	
VITAMIN B2	MILLI	0.30	-	-	-	0.18	MILLI/MJ	3.76	
NICOTINAMIDE	MILLI	1.35	1.00	-	1.70	0.81	MILLI/MJ	16.94	
PANTOTHENIC ACID	MILLI	0.30	-	-	-	0.18	MILLI/MJ	3.76	
VITAMIN B6	MILLI	0.20	0.12	-	0.30	0.12	MILLI/MJ	2.51	
BIOTIN	MICRO	0.40	-	-	-	0.24	MICRO/MJ	5.02	
VITAMIN C	MILLI	166.00	150.00	-	182.00	99.60	GRAM/MJ	2.08	
ALANINE	MILLI	314.00	314.00	-	314.00	188.40	GRAM/MJ	3.94	
ARGININE	MILLI	172.00	172.00	-	172.00	103.20	GRAM/MJ	2.16	
ASPARTIC ACID	MILLI	486.00	486.00	-	486.00	291.60	GRAM/MJ	6.10	
CYSTINE	MILLI	20.00	20.00	-	20.00	12.00	MILLI/MJ	250.91	
GLUTAMIC ACID	MILLI	400.00	400.00	-	400.00	240.00	GRAM/MJ	5.02	
GLYCINE	MILLI	214.00	214.00	-	214.00	128.40	GRAM/MJ	2.68	
HISTIDINE	MILLI	100.00	100.00	-	100.00	60.00	GRAM/MJ	1.25	
ISOLEUCINE	MILLI	214.00	214.00	-	214.00	128.40	GRAM/MJ	2.68	
LEUCINE	MILLI	300.00	300.00	-	300.00	180.00	GRAM/MJ	3.76	
LYSINE	GRAM	0.28	-	-	-	0.17	GRAM/MJ	3.51	
METHIONINE	MILLI	21.00	-	-	-	12.60	MILLI/MJ	263.45	
PHENYLALANINE	MILLI	243.00	243.00	-	243.00	145.80	GRAM/MJ	3.05	
PROLINE	MILLI	429.00	429.00	-	429.00	257.40	GRAM/MJ	5.38	
SERINE	MILLI	200.00	200.00	-	200.00	120.00	GRAM/MJ	2.51	
THREONINE	MILLI	186.00	186.00	-	186.00	111.60	GRAM/MJ	2.33	
TRYPTOPHAN	MILLI	89.00	-	-	-	53.40	GRAM/MJ	1.12	
TYROSINE	MILLI	126.00	126.00	-	126.00	75.60	GRAM/MJ	1.58	
VALINE	MILLI	300.00	300.00	-	300.00	180.00	GRAM/MJ	3.76	
OXALIC ACID TOTAL	MILLI	-	0.00	-	10.00				
OXALIC ACID SOLUBLE	MILLI	-	0.00	-	8.90				
SALICYLIC ACID	MICRO	80.00	80.00	-	80.00	48.00	MILLI/MJ	1.00	
GLUCOSE	GRAM	0.53	0.53	-	0.53	0.32	GRAM/MJ	6.65	
FRUCTOSE	GRAM	0.32	0.32	-	0.32	0.19	GRAM/MJ	4.01	
STARCH	GRAM	0.46	0.46	-	0.46	0.28	GRAM/MJ	5.77	
PENTOSAN	GRAM	0.89	-	-	-	0.53	GRAM/MJ	11.17	
HEXOSAN	GRAM	0.89	-	-	-	0.53	GRAM/MJ	11.17	
CELLULOSE	GRAM	1.60	-	-	-	0.96	GRAM/MJ	20.07	
POLYURONIC ACID	GRAM	0.28	-	-	-	0.17	GRAM/MJ	3.51	
PALMITIC ACID	MILLI	41.00	41.00	-	41.00	24.60	MILLI/MJ	514.36	
STEARIC ACID	MILLI	1.90	1.90	-	1.90	1.14	MILLI/MJ	23.84	
PALMITOLEIC ACID	MILLI	7.20	7.20	-	7.20	4.32	MILLI/MJ	90.33	
LINOLEIC ACID	MILLI	72.00	72.00	-	72.00	43.20	MILLI/MJ	903.27	
LINOLENIC ACID	MILLI	120.00	120.00	-	120.00	72.00	GRAM/MJ	1.51	

PORREE (LAUCH) — LEEK — POIREAU

ALLIUM PORRUM L.

		PROTEIN	FAT	CARBOHYDRATES	TOTAL
ENERGY VALUE (AVERAGE)	KJOULE	35	13	53	102
PER 100 G	(KCAL)	8.4	3.2	13	24
EDIBLE PORTION					
AMOUNT OF DIGESTIBLE	GRAM	1.45	0.30	3.19	
CONSTITUENTS PER 100 G					
ENERGY VALUE (AVERAGE)	KJOULE	23	12	53	88
OF THE DIGESTIBLE	(KCAL)	5.5	2.8	13	21
FRACTION PER 100 G					
EDIBLE PORTION					

WASTE PERCENTAGE AVERAGE 42 MINIMUM 35 MAXIMUM 48

CONSTITUENTS	DIM	AV	VARIATION		AVR	NUTR. DENS.		MOLPERC.
WATER	GRAM	89.00	86.30	- 90.80	51.62	KILO/MJ	1.01	
PROTEIN	GRAM	2.24	2.00	- 2.50	1.30	GRAM/MJ	25.41	
FAT	GRAM	0.34	0.25	- 0.44	0.20	GRAM/MJ	3.86	
AVAILABLE CARBOHYDR.	GRAM	3.19	-	- -	1.85	GRAM/MJ	36.19	
TOTAL DIETARY FIBRE	GRAM	2.27 [1]	-	- -	1.32	GRAM/MJ	25.75	
MINERALS	GRAM	0.86	0.82	- 0.90	0.50	GRAM/MJ	9.76	
SODIUM	MILLI	5.00	-	- -	2.90	MILLI/MJ	56.73	
POTASSIUM	MILLI	225.00	200.00	- 250.00	130.50	GRAM/MJ	2.55	
MAGNESIUM	MILLI	18.00	14.00	- 21.00	10.44	MILLI/MJ	204.22	
CALCIUM	MILLI	87.00	52.00	- 114.00	50.46	MILLI/MJ	987.07	
MANGANESE	MILLI	0.19	0.12	- 0.23	0.11	MILLI/MJ	2.16	
IRON	MILLI	1.00	0.90	- 1.10	0.58	MILLI/MJ	11.35	
COBALT	MICRO	0.25	0.10	- 0.50	0.15	MICRO/MJ	2.84	
COPPER	MICRO	55.00	40.00	- 80.00	31.90	MICRO/MJ	624.01	
ZINC	MILLI	0.31	-	- -	0.18	MILLI/MJ	3.52	
NICKEL	MICRO	10.00	3.00	- 20.00	5.80	MICRO/MJ	113.46	
MOLYBDENUM	MICRO	10.00	2.00	- 20.00	5.80	MICRO/MJ	113.46	
PHOSPHORUS	MILLI	46.00	30.00	- 57.00	26.68	MILLI/MJ	521.90	
CHLORIDE	MILLI	24.00	-	- -	13.92	MILLI/MJ	272.30	
FLUORIDE	MICRO	10.00	3.00	- 30.00	5.80	MICRO/MJ	113.46	
IODIDE	MICRO	1.30	1.30	- 1.30	0.75	MICRO/MJ	14.75	
BORON	MILLI	0.28	0.15	- 0.42	0.16	MILLI/MJ	3.18	
SELENIUM	MILLI	-	0.00	- 0.01				
SILICON	MILLI	6.00	0.50	- 16.00	3.48	MILLI/MJ	68.07	
NITRATE	MILLI	51.00	4.00	- 448.00	29.58	MILLI/MJ	578.63	
CAROTENE	MILLI	-	0.02	- 0.70				

1 METHOD OF MEUSER, SUCKOW AND KULIKOWSKI ("BERLINER METHODE")

CONSTITUENTS	DIM	AV		VARIATION		AVR	NUTR. DENS.		MOLPERC.
VITAMIN E ACTIVITY	MILLI	0.90	-	-	-	0.52	MILLI/MJ	10.21	
ALPHA-TOCOPHEROL	MILLI	0.90	-	-	-	0.52	MILLI/MJ	10.21	
VITAMIN B1	MILLI	0.10	0.07	-	0.12	0.06	MILLI/MJ	1.13	
VITAMIN B2	MICRO	60.00	40.00	-	100.00	34.80	MICRO/MJ	680.74	
NICOTINAMIDE	MILLI	0.53	0.50	-	0.60	0.31	MILLI/MJ	6.01	
PANTOTHENIC ACID	MILLI	0.14	0.14	-	0.14	0.08	MILLI/MJ	1.59	
VITAMIN B6	MILLI	0.25	0.20	-	0.30	0.15	MILLI/MJ	2.84	
BIOTIN	MICRO	1.60	1.60	-	1.60	0.93	MICRO/MJ	18.15	
VITAMIN C	MILLI	30.00	17.00	-	78.00	17.40	MILLI/MJ	340.37	
ALANINE	MILLI	155.00	155.00	-	155.00	89.90	GRAM/MJ	1.76	12.5
ARGININE	MILLI	116.00	116.00	-	116.00	67.28	GRAM/MJ	1.32	4.8
ASPARTIC ACID	MILLI	193.00	193.00	-	193.00	111.94	GRAM/MJ	2.19	10.4
CYSTINE	MILLI	12.00	12.00	-	12.00	6.96	MILLI/MJ	136.15	0.4
GLUTAMIC ACID	MILLI	386.00	386.00	-	386.00	223.88	GRAM/MJ	4.38	18.8
GLYCINE	MILLI	75.00	75.00	-	75.00	43.50	MILLI/MJ	850.93	7.2
HISTIDINE	MILLI	39.00	39.00	-	39.00	22.62	MILLI/MJ	442.48	1.8
ISOLEUCINE	MILLI	77.00	77.00	-	77.00	44.66	MILLI/MJ	873.62	4.2
LEUCINE	MILLI	131.00	131.00	-	131.00	75.98	GRAM/MJ	1.49	7.2
LYSINE	MILLI	139.00	139.00	-	139.00	80.62	GRAM/MJ	1.58	6.8
METHIONINE	MILLI	29.00	29.00	-	29.00	16.82	MILLI/MJ	329.02	1.4
PHENYLALANINE	MILLI	63.00	63.00	-	63.00	36.54	MILLI/MJ	714.78	2.7
PROLINE	MILLI	67.00	67.00	-	67.00	38.86	MILLI/MJ	760.16	4.2
SERINE	MILLI	75.00	75.00	-	75.00	43.50	MILLI/MJ	850.93	5.1
THREONINE	MILLI	76.00	76.00	-	76.00	44.08	MILLI/MJ	862.27	4.6
TRYPTOPHAN	MILLI	22.00	22.00	-	22.00	12.76	MILLI/MJ	249.60	0.8
TYROSINE	MILLI	38.00	38.00	-	38.00	22.04	MILLI/MJ	431.14	1.5
VALINE	MILLI	93.00	93.00	-	93.00	53.94	GRAM/MJ	1.06	5.7
OXALIC ACID TOTAL	-	0.00	-	-	-	0.00			
SALICYLIC ACID	MICRO	80.00	80.00	-	80.00	46.40	MICRO/MJ	907.65	
GLUCOSE	GRAM	0.93	0.46	-	1.16	0.54	GRAM/MJ	10.62	
FRUCTOSE	GRAM	1.22	0.74	-	1.45	0.71	GRAM/MJ	13.91	
SUCROSE	GRAM	0.84	0.30	-	1.29	0.49	GRAM/MJ	9.60	
RAFFINOSE	GRAM	0.72	-	-	-	0.42	GRAM/MJ	8.17	
MALTOSE	MILLI	80.00	-	-	-	46.40	MILLI/MJ	907.65	
STARCH	GRAM	0.12	0.12	-	0.12	0.07	GRAM/MJ	1.36	
MANNITOL	GRAM	0.15	-	-	-	0.09	GRAM/MJ	1.70	
PALMITIC ACID	MILLI	74.00	74.00	-	74.00	42.92	MILLI/MJ	839.58	
STEARIC ACID	MILLI	3.70	3.70	-	3.70	2.15	MILLI/MJ	41.98	
OLEIC ACID	MILLI	13.00	13.00	-	13.00	7.54	MILLI/MJ	147.49	
LINOLEIC ACID	MILLI	139.00	139.00	-	139.00	80.62	GRAM/MJ	1.58	
LINOLENIC ACID	MILLI	37.00	37.00	-	37.00	21.46	MILLI/MJ	419.79	
TOTAL PURINES	MILLI	30.00 [2]	-	-	-	17.40	MILLI/MJ	340.37	
DIETARY FIBRE, WAT. SOL.	GRAM	0.52	-	-	-	0.30	GRAM/MJ	5.90	
DIETARY FIBRE, WAT. INS.	GRAM	1.75	-	-	-	1.02	GRAM/MJ	19.85	

2 VALUE EXPRESSED IN MG URIC ACID/100G

PORTULAK PURSLANE POURPIER

PORTULACA SATIVA
HAWORTH

		PROTEIN	FAT	CARBOHYDRATES	TOTAL
ENERGY VALUE (AVERAGE) PER 100 G EDIBLE PORTION	KJOULE	23	13	71	108
	(KCAL)	5.6	3.2	17	26
AMOUNT OF DIGESTIBLE CONSTITUENTS PER 100 G	GRAM	0.96	0.30	4.27	
ENERGY VALUE (AVERAGE) OF THE DIGESTIBLE FRACTION PER 100 G EDIBLE PORTION	KJOULE	15	12	71	98
	(KCAL)	3.6	2.8	17	24

WASTE PERCENTAGE AVERAGE 0.00

CONSTITUENTS	DIM	AV	VARIATION		AVR	NUTR. DENS.		MOLPERC.
WATER	GRAM	92.50	91.20	- 93.70	92.50	GRAM/MJ	939.44	
PROTEIN	GRAM	1.48	0.90	- 2.00	1.48	GRAM/MJ	15.03	
FAT	GRAM	0.34	0.20	- 0.40	0.34	GRAM/MJ	3.45	
AVAILABLE CARBOHYDR.	GRAM	4.27 [1]	-	- -	4.27	GRAM/MJ	43.37	
MINERALS	GRAM	1.41	1.24	- 1.60	1.41	GRAM/MJ	14.32	
SODIUM	MILLI	2.00	-	- -	2.00	MILLI/MJ	20.31	
POTASSIUM	MILLI	390.00	-	- -	390.00	GRAM/MJ	3.96	
MAGNESIUM	MILLI	151.00	-	- -	151.00	GRAM/MJ	1.53	
CALCIUM	MILLI	95.00	84.00	- 112.00	95.00	MILLI/MJ	964.83	
IRON	MILLI	3.60	-	- -	3.60	MILLI/MJ	36.56	
PHOSPHORUS	MILLI	35.00	32.00	- 39.00	35.00	MILLI/MJ	355.46	
NITRATE	MILLI	615.00	411.00	- 898.00	615.00	GRAM/MJ	6.25	
CAROTENE	MILLI	1.06	0.75	- 1.53	1.06	MILLI/MJ	10.77	
VITAMIN B1	MICRO	30.00	20.00	- 30.00	30.00	MICRO/MJ	304.68	
VITAMIN B2	MILLI	0.10	-	- -	0.10	MILLI/MJ	1.02	
NICOTINAMIDE	MILLI	0.50	-	- -	0.50	MILLI/MJ	5.08	
VITAMIN B6	MILLI	0.15	-	- -	0.15	MILLI/MJ	1.52	
VITAMIN C	MILLI	22.00	18.00	- 25.00	22.00	MILLI/MJ	223.43	
OXALIC ACID TOTAL	MILLI	-	218.00	- 650.00				
OXALIC ACID SOLUBLE	MILLI	74.00	-	- -	74.00	MILLI/MJ	751.55	

1 ESTIMATED BY THE DIFFERENCE METHOD
100 - (WATER + PROTEIN + FAT + MINERALS)

RHABARBER RHUBARB RHUBARBE

RHEUM UNDULATUM L.

		PROTEIN	FAT	CARBOHYDRATES	TOTAL
ENERGY VALUE (AVERAGE)	KJOULE	9.4	5.4	46	60
PER 100 G	(KCAL)	2.3	1.3	11	14
EDIBLE PORTION					
AMOUNT OF DIGESTIBLE	GRAM	0.39	0.12	2.72	
CONSTITUENTS PER 100 G					
ENERGY VALUE (AVERAGE)	KJOULE	6.1	4.9	46	57
OF THE DIGESTIBLE	(KCAL)	1.5	1.2	11	14
FRACTION PER 100 G					
EDIBLE PORTION					

WASTE PERCENTAGE AVERAGE 22 MINIMUM 14 MAXIMUM 30

CONSTITUENTS	DIM	AV	VARIATION			AVR	NUTR. DENS.		MOLPERC.
WATER	GRAM	94.50	93.00	-	95.10	73.71	KILO/MJ	1.67	
PROTEIN	GRAM	0.60	0.50	-	0.80	0.47	GRAM/MJ	10.61	
FAT	GRAM	0.14	0.10	-	0.20	0.11	GRAM/MJ	2.48	
AVAILABLE CARBOHYDR.	GRAM	2.72	1.92	-	4.39	2.12	GRAM/MJ	48.10	
TOTAL DIETARY FIBRE	GRAM	3.20 1	-	-	-	2.50	GRAM/MJ	56.59	
MINERALS	GRAM	0.64	0.44	-	0.93	0.50	GRAM/MJ	11.32	
SODIUM	MILLI	2.00	0.60	-	3.50	1.56	MILLI/MJ	35.37	
POTASSIUM	MILLI	270.00	200.00	-	320.00	210.60	GRAM/MJ	4.78	
MAGNESIUM	MILLI	13.00	11.00	-	16.00	10.14	MILLI/MJ	229.91	
CALCIUM	MILLI	52.00	20.00	-	96.00	40.56	MILLI/MJ	919.64	
MANGANESE	MILLI	0.13	0.05	-	0.20	0.10	MILLI/MJ	2.30	
IRON	MILLI	0.53	0.15	-	0.80	0.41	MILLI/MJ	9.37	
COBALT	MICRO	0.50	0.10	-	2.00	0.39	MICRO/MJ	8.84	
COPPER	MICRO	50.00	20.00	-	70.00	39.00	MICRO/MJ	884.27	
ZINC	MILLI	0.13	-	-	-	0.10	MILLI/MJ	2.30	
NICKEL	MICRO	10.00	1.00	-	20.00	7.80	MICRO/MJ	176.85	
CHROMIUM	MICRO	2.00	-	-	-	1.56	MICRO/MJ	35.37	
PHOSPHORUS	MILLI	24.00	17.00	-	30.00	18.72	MILLI/MJ	424.45	
CHLORIDE	MILLI	60.00	30.00	-	90.00	46.80	GRAM/MJ	1.06	
FLUORIDE	MICRO	40.00	-	-	-	31.20	MICRO/MJ	707.42	
IODIDE	MICRO	1.00	-	-	-	0.78	MICRO/MJ	17.69	
BORON	MILLI	-	0.04	-	0.10				
SILICON	MILLI	0.50	0.30	-	1.00	0.39	MILLI/MJ	8.84	
NITRATE	MILLI	215.00	71.00	-	545.00	167.70	GRAM/MJ	3.80	
CAROTENE	MICRO	70.00	43.00	-	120.00	54.60	MILLI/MJ	1.24	
VITAMIN B1	MICRO	27.00	10.00	-	50.00	21.06	MICRO/MJ	477.51	

1 METHOD OF SOUTHGATE AND ENGLYST

CONSTITUENTS	DIM	AV	VARIATION			AVR	NUTR. DENS.		MOLPERC.
VITAMIN B2	MICRO	30.00	22.00	-	41.00	23.40	MICRO/MJ	530.56	
NICOTINAMIDE	MILLI	0.25	0.19	-	0.38	0.20	MILLI/MJ	4.42	
PANTOTHENIC ACID	MICRO	84.00	79.00	-	87.00	65.52	MILLI/MJ	1.49	
VITAMIN B6	MICRO	35.00	25.00	-	64.00	27.30	MICRO/MJ	618.99	
FOLIC ACID	MICRO	2.50	1.40	-	3.50	1.95	MICRO/MJ	44.21	
VITAMIN B12	-	0.00	-	-	-	0.00			
VITAMIN C	MILLI	10.00	5.00	-	18.00	7.80	MILLI/MJ	176.85	
ALANINE	MILLI	18.00	18.00	-	18.00	14.04	MILLI/MJ	318.34	8.0
ARGININE	MILLI	20.00	20.00	-	20.00	15.60	MILLI/MJ	353.71	4.6
ASPARTIC ACID	MILLI	29.00	29.00	-	29.00	22.62	MILLI/MJ	512.88	8.6
CYSTINE	MILLI	1.00	1.00	-	1.00	0.78	MILLI/MJ	17.69	0.2
GLUTAMIC ACID	MILLI	48.00	48.00	-	48.00	37.44	MILLI/MJ	848.90	12.9
GLYCINE	MILLI	17.00	17.00	-	17.00	13.26	MILLI/MJ	300.65	9.0
HISTIDINE	MILLI	11.00	11.00	-	11.00	8.58	MILLI/MJ	194.54	2.8
ISOLEUCINE	MILLI	17.00	17.00	-	17.00	13.26	MILLI/MJ	300.65	5.1
LEUCINE	MILLI	24.00	24.00	-	24.00	18.72	MILLI/MJ	424.45	7.3
LYSINE	MILLI	25.00	25.00	-	25.00	19.50	MILLI/MJ	442.13	6.8
METHIONINE	MILLI	7.00	7.00	-	7.00	5.46	MILLI/MJ	123.80	1.9
PHENYLALANINE	MILLI	17.00	17.00	-	17.00	13.26	MILLI/MJ	300.65	4.1
PROLINE	MILLI	15.00	15.00	-	15.00	11.70	MILLI/MJ	265.28	5.2
SERINE	MILLI	17.00	17.00	-	17.00	13.26	MILLI/MJ	300.65	6.4
THREONINE	MILLI	14.00	14.00	-	14.00	10.92	MILLI/MJ	247.60	4.7
TRYPTOPHAN	MILLI	7.00	7.00	-	7.00	5.46	MILLI/MJ	123.80	1.4
TYROSINE	MILLI	11.00	11.00	-	11.00	8.58	MILLI/MJ	194.54	2.4
VALINE	MILLI	26.00	26.00	-	26.00	20.28	MILLI/MJ	459.82	8.8
MALIC ACID	GRAM	1.25	1.03	-	1.82	0.98	GRAM/MJ	22.11	
CITRIC ACID	MILLI	130.00	67.00	-	180.00	101.40	GRAM/MJ	2.30	
OXALIC ACID TOTAL	MILLI	460.00	290.00	-	640.00	358.80	GRAM/MJ	8.14	
OXALIC ACID SOLUBLE	MILLI	270.00	180.00	-	350.00	210.60	GRAM/MJ	4.78	
ACETIC ACID	MILLI	59.00	19.00	-	160.00	46.02	GRAM/MJ	1.04	
CAFFEIC ACID	MILLI	0.10	-	-	-	0.08	MILLI/MJ	1.77	
PARA-COUMARIC ACID	MILLI	0.50	0.00	-	1.00	0.39	MILLI/MJ	8.84	
PROTOCATECHUIC ACID	MILLI	1.00	0.50	-	1.50	0.78	MILLI/MJ	17.69	
GALLIC ACID	MILLI	3.00	2.00	-	4.00	2.34	MILLI/MJ	53.06	
GLUCOSE	GRAM	0.40	0.24	-	0.94	0.32	GRAM/MJ	7.18	
FRUCTOSE	GRAM	0.38	0.27	-	0.80	0.30	GRAM/MJ	6.86	
SUCROSE	GRAM	0.33	0.08	-	0.42	0.26	GRAM/MJ	5.92	
STARCH	GRAM	0.23	0.23	-	0.23	0.18	GRAM/MJ	4.07	
PENTOSAN	GRAM	0.35	-	-	-	0.27	GRAM/MJ	6.19	
HEXOSAN	GRAM	0.81	-	-	-	0.63	GRAM/MJ	14.33	
CELLULOSE	GRAM	1.31	-	-	-	1.02	GRAM/MJ	23.17	
POLYURONIC ACID	GRAM	0.60	-	-	-	0.47	GRAM/MJ	10.61	
PALMITIC ACID	MILLI	25.00	25.00	-	25.00	19.50	MILLI/MJ	442.13	
STEARIC ACID	MILLI	3.50	3.50	-	3.50	2.73	MILLI/MJ	61.90	
OLEIC ACID	MILLI	7.00	7.00	-	7.00	5.46	MILLI/MJ	123.80	
LINOLEIC ACID	MILLI	53.00	53.00	-	53.00	41.34	MILLI/MJ	937.33	
LINOLENIC ACID	MILLI	17.00	17.00	-	17.00	13.26	MILLI/MJ	300.65	
TOTAL PURINES	MILLI	10.00 [2]	-	-	-	7.80	MILLI/MJ	176.85	

2 VALUE EXPRESSED IN MG URIC ACID/100G

ROSENKOHL — BRUSSEL SPROUTS — CHOU DE BRUXELLES

BRASSICA OLERACEA L.
VAR. GEMMIFERA (DC.) THELL.

		PROTEIN	FAT	CARBOHYDRATES	TOTAL
ENERGY VALUE (AVERAGE) PER 100 G EDIBLE PORTION	KJOULE	70	13	63	146
	(KCAL)	17	3.2	15	35
AMOUNT OF DIGESTIBLE CONSTITUENTS PER 100 G	GRAM	2.89	0.30	3.77	
ENERGY VALUE (AVERAGE) OF THE DIGESTIBLE FRACTION PER 100 G EDIBLE PORTION	KJOULE	45	12	63	120
	(KCAL)	11	2.8	15	29

WASTE PERCENTAGE AVERAGE 22 MINIMUM 16 MAXIMUM 30

CONSTITUENTS	DIM	AV	VARIATION	AVR	NUTR. DENS.		MOLPERC.
WATER	GRAM	85.00	84.00 - 86.00	66.30	GRAM/MJ	706.07	
PROTEIN	GRAM	4.45	4.00 - 5.00	3.47	GRAM/MJ	36.96	
FAT	GRAM	0.34	- - -	0.27	GRAM/MJ	2.82	
AVAILABLE CARBOHYDR.	GRAM	3.77	- - -	2.94	GRAM/MJ	31.32	
TOTAL DIETARY FIBRE	GRAM	4.40 1	- - -	3.43	GRAM/MJ	36.55	
MINERALS	GRAM	1.40	1.30 - 1.51	1.09	GRAM/MJ	11.63	
SODIUM	MILLI	7.00	4.00 - 10.00	5.46	MILLI/MJ	58.15	
POTASSIUM	MILLI	411.00	375.00 - 500.00	320.58	GRAM/MJ	3.41	
MAGNESIUM	MILLI	22.00	20.00 - 30.00	17.16	MILLI/MJ	182.75	
CALCIUM	MILLI	31.00	25.00 - 40.00	24.18	MILLI/MJ	257.51	
MANGANESE	MILLI	0.26	0.25 - 0.27	0.20	MILLI/MJ	2.16	
IRON	MILLI	1.10	1.00 - 1.30	0.86	MILLI/MJ	9.14	
COBALT	MICRO	12.00	- - -	9.36	MICRO/MJ	99.68	
COPPER	MICRO	90.00	80.00 - 100.00	70.20	MICRO/MJ	747.60	
ZINC	MILLI	0.87	- - -	0.68	MILLI/MJ	7.23	
MOLYBDENUM	MICRO	7.00	- - -	5.46	MICRO/MJ	58.15	
PHOSPHORUS	MILLI	83.60	78.00 - 86.40	65.21	MILLI/MJ	694.44	
CHLORIDE	MILLI	40.00	- - -	31.20	MILLI/MJ	332.27	
IODIDE	MICRO	0.70	- - -	0.55	MICRO/MJ	5.81	
BORON	MILLI	0.27	- - -	0.21	MILLI/MJ	2.24	
SELENIUM	MICRO	18.00	- - -	14.04	MICRO/MJ	149.52	
NITRATE	MILLI	12.00	0.00 - 62.00	9.36	MILLI/MJ	99.68	
CAROTENE	MILLI	0.40	0.32 - 0.60	0.31	MILLI/MJ	3.32	
VITAMIN E ACTIVITY	MILLI	0.88	- - -	0.69	MILLI/MJ	7.31	
ALPHA-TOCOPHEROL	MILLI	0.88	- - -	0.69	MILLI/MJ	7.31	
VITAMIN K	MILLI	0.57	- - -	0.44	MILLI/MJ	4.73	

1 METHOD OF SOUTHGATE AND ENGLYST

CONSTITUENTS	DIM	AV	VARIATION			AVR	NUTR. DENS.		MOLPERC.
VITAMIN B2	MILLI	0.14	0.10	-	0.21	0.11	MILLI/MJ	1.16	
NICOTINAMIDE	MILLI	0.67	0.50	-	1.07	0.52	MILLI/MJ	5.57	
PANTOTHENIC ACID	MILLI	-	0.10	-	1.40				
VITAMIN B6	MILLI	0.28	0.22	-	0.30	0.22	MILLI/MJ	2.33	
BIOTIN	MICRO	0.40	-	-	-	0.31	MICRO/MJ	3.32	
FOLIC ACID	MICRO	78.00	-	-	-	60.84	MICRO/MJ	647.92	
VITAMIN B12	-	0.00	-	-	-	0.00			
VITAMIN C	MILLI	114.00	73.00	-	152.00	88.92	MILLI/MJ	946.96	
ARGININE	GRAM	0.28	0.24	-	0.30	0.22	GRAM/MJ	2.33	
HISTIDINE	GRAM	0.11	0.08	-	0.14	0.09	GRAM/MJ	0.91	
ISOLEUCINE	GRAM	0.21	0.17	-	0.27	0.16	GRAM/MJ	1.74	
LEUCINE	GRAM	0.23	0.16	-	0.34	0.18	GRAM/MJ	1.91	
LYSINE	GRAM	0.25	0.24	-	0.39	0.20	GRAM/MJ	2.08	
METHIONINE	MILLI	40.00	30.00	-	50.00	31.20	MILLI/MJ	332.27	
PHENYLALANINE	GRAM	0.15	0.11	-	0.17	0.12	GRAM/MJ	1.25	
THREONINE	GRAM	0.16	0.13	-	0.19	0.12	GRAM/MJ	1.33	
TRYPTOPHAN	MILLI	50.00	40.00	-	60.00	39.00	MILLI/MJ	415.33	
VALINE	GRAM	0.24	0.22	-	0.32	0.19	GRAM/MJ	1.99	
MALIC ACID	MILLI	200.00	-	-	-	156.00	GRAM/MJ	1.66	
CITRIC ACID	MILLI	350.00	240.00	-	450.00	273.00	GRAM/MJ	2.91	
OXALIC ACID TOTAL	MILLI	6.10	-	-	-	4.76	MILLI/MJ	50.67	
OXALIC ACID SOLUBLE	MILLI	5.80	-	-	-	4.52	MILLI/MJ	48.18	
SALICYLIC ACID	MICRO	70.00	70.00	-	70.00	54.60	MICRO/MJ	581.47	
GLUCOSE	GRAM	0.84	0.25	-	1.40	0.66	GRAM/MJ	7.01	
FRUCTOSE	GRAM	0.79	0.33	-	1.10	0.62	GRAM/MJ	6.58	
SUCROSE	GRAM	1.10	0.25	-	2.60	0.87	GRAM/MJ	9.21	
STARCH	GRAM	0.49	-	-	-	0.38	GRAM/MJ	4.07	
PENTOSAN	GRAM	1.20	-	-	-	0.94	GRAM/MJ	9.97	
HEXOSAN	GRAM	0.93	-	-	-	0.73	GRAM/MJ	7.73	
CELLULOSE	GRAM	1.28	-	-	-	1.00	GRAM/MJ	10.63	
POLYURONIC ACID	GRAM	0.90	-	-	-	0.70	GRAM/MJ	7.48	
MYRISTIC ACID	MILLI	1.40	-	-	-	1.09	MILLI/MJ	11.63	
PALMITIC ACID	MILLI	42.50	39.00	-	46.00	33.15	MILLI/MJ	353.03	
STEARIC ACID	MILLI	2.80	2.50	-	3.10	2.18	MILLI/MJ	23.26	
ARACHIDIC ACID	MILLI	4.40	-	-	-	3.43	MILLI/MJ	36.55	
PALMITOLEIC ACID	MILLI	1.40	-	-	-	1.09	MILLI/MJ	11.63	
OLEIC ACID	MILLI	5.40	1.60	-	9.20	4.21	MILLI/MJ	44.86	
EICOSENOIC ACID	MILLI	1.10	-	-	-	0.86	MILLI/MJ	9.14	
LINOLEIC ACID	MILLI	38.50	31.00	-	46.00	30.03	MILLI/MJ	319.81	
LINOLENIC ACID	MILLI	156.00	153.00	-	159.00	121.68	GRAM/MJ	1.30	
TOTAL STEROLS	MILLI	24.00	-	-	-	18.72	MILLI/MJ	199.36	
CAMPESTEROL	MILLI	6.00	-	-	-	4.68	MILLI/MJ	49.84	
BETA-SITOSTEROL	MILLI	17.00	-	-	-	13.26	MILLI/MJ	141.21	
TOTAL PURINES	MILLI	15.00 [2]	-	-	-	11.70	MILLI/MJ	124.60	

2 VALUE EXPRESSED IN MG URIC ACID/100G

ROSENKOHL / BRUSSEL SPROUTS / CHOU DE BRUXELLES

GEKOCHT, ABGETROPFT / BOILED, DRAINED / BOUILLI, DEGOUTTÉ

		PROTEIN	FAT	CARBOHYDRATES	TOTAL
ENERGY VALUE (AVERAGE)	KJOULE	60	20	48	128
PER 100 G EDIBLE PORTION	(KCAL)	14	4.8	12	31
AMOUNT OF DIGESTIBLE CONSTITUENTS PER 100 G	GRAM	2.47	0.45	2.88	
ENERGY VALUE (AVERAGE)	KJOULE	39	18	48	105
OF THE DIGESTIBLE FRACTION PER 100 G EDIBLE PORTION	(KCAL)	9.3	4.3	12	25

WASTE PERCENTAGE AVERAGE 0.00

CONSTITUENTS	DIM	AV	VARIATION	AVR	NUTR. DENS.		MOLPERC.
WATER	GRAM	87.90	- - -	87.90	GRAM/MJ	838.42	
PROTEIN	GRAM	3.80	- - -	3.80	GRAM/MJ	36.25	
FAT	GRAM	0.50	- - -	0.50	GRAM/MJ	4.77	
AVAILABLE CARBOHYDR.	GRAM	2.88	- - -	2.88	GRAM/MJ	27.47	
MINERALS	GRAM	0.90	- - -	0.90	GRAM/MJ	8.58	
SODIUM	MILLI	5.00	- - -	5.00	MILLI/MJ	47.69	
POTASSIUM	MILLI	288.00	- - -	288.00	GRAM/MJ	2.75	
CALCIUM	MILLI	27.00	- - -	27.00	MILLI/MJ	257.53	
IRON	MILLI	0.80	- - -	0.80	MILLI/MJ	7.63	
PHOSPHORUS	MILLI	76.00	- - -	76.00	MILLI/MJ	724.91	
VITAMIN B1	MICRO	90.00	- - -	90.00	MICRO/MJ	858.45	
VITAMIN B2	MILLI	0.12	- - -	0.12	MILLI/MJ	1.14	
NICOTINAMIDE	MILLI	0.59	- - -	0.59	MILLI/MJ	5.63	
VITAMIN C	MILLI	85.00	- - -	85.00	MILLI/MJ	810.75	
MALIC ACID	GRAM	0.16	- - -	0.16	GRAM/MJ	1.53	
CITRIC ACID	GRAM	0.28	- - -	0.28	GRAM/MJ	2.67	
GLUCOSE	GRAM	0.51	- - -	0.51	GRAM/MJ	4.86	
FRUCTOSE	GRAM	0.54	- - -	0.54	GRAM/MJ	5.15	
SUCROSE	GRAM	1.00	- - -	1.00	GRAM/MJ	9.54	
STARCH	GRAM	0.39	- - -	0.39	GRAM/MJ	3.72	

ROTKOHL (BLAUKRAUT) RED CABBAGE CHOU ROUGE

BRASSICA OLERACEA L.
V. CAPITATA L. F. RUBRA

		PROTEIN	FAT	CARBOHYDRATES	TOTAL
ENERGY VALUE (AVERAGE)	KJOULE	24	7.0	59	89
PER 100 G EDIBLE PORTION	(KCAL)	5.6	1.7	14	21
AMOUNT OF DIGESTIBLE CONSTITUENTS PER 100 G	GRAM	0.97	0.16	3.52	
ENERGY VALUE (AVERAGE)	KJOULE	15	6.3	59	81
OF THE DIGESTIBLE FRACTION PER 100 G EDIBLE PORTION	(KCAL)	3.7	1.5	14	19

WASTE PERCENTAGE AVERAGE 22 MINIMUM 8.0 MAXIMUM 42

CONSTITUENTS	DIM	AV	VARIATION			AVR	NUTR. DENS.		MOLPERC.
WATER	GRAM	91.80	91.50	-	92.40	71.60	KILO/MJ	1.14	
PROTEIN	GRAM	1.50	1.39	-	1.70	1.17	GRAM/MJ	18.63	
FAT	GRAM	0.18	0.10	-	0.20	0.14	GRAM/MJ	2.24	
AVAILABLE CARBOHYDR.	GRAM	3.52	-	-	-	2.75	GRAM/MJ	43.72	
TOTAL DIETARY FIBRE	GRAM	2.50 1	-	-	-	1.95	GRAM/MJ	31.05	
MINERALS	GRAM	0.67	0.50	-	0.80	0.52	GRAM/MJ	8.32	
SODIUM	MILLI	4.00	2.00	-	6.00	3.12	MILLI/MJ	49.68	
POTASSIUM	MILLI	266.00	245.00	-	302.00	207.48	GRAM/MJ	3.30	
MAGNESIUM	MILLI	18.00	17.00	-	18.00	14.04	MILLI/MJ	223.57	
CALCIUM	MILLI	35.00	29.00	-	46.00	27.30	MILLI/MJ	434.72	
MANGANESE	MILLI	0.10	-	-	-	0.08	MILLI/MJ	1.24	
IRON	MILLI	0.50	-	-	-	0.39	MILLI/MJ	6.21	
COBALT	MICRO	7.00	0.50	-	24.00	5.46	MICRO/MJ	86.94	
COPPER	MICRO	60.00	-	-	-	46.80	MICRO/MJ	745.23	
ZINC	MILLI	0.22	0.16	-	1.50	0.17	MILLI/MJ	2.73	
NICKEL	MICRO	24.00	-	-	-	18.72	MICRO/MJ	298.09	
CHROMIUM	MICRO	0.50	-	-	-	0.39	MICRO/MJ	6.21	
MOLYBDENUM	MILLI	0.12	-	-	-	0.10	MILLI/MJ	1.58	
PHOSPHORUS	MILLI	30.00	27.50	-	31.00	23.40	MILLI/MJ	372.61	
CHLORIDE	MILLI	100.00	-	-	-	78.00	GRAM/MJ	1.24	
FLUORIDE	MICRO	12.00	-	-	-	9.36	MICRO/MJ	149.05	
IODIDE	MICRO	5.20	-	-	-	4.06	MICRO/MJ	64.59	
BORON	MILLI	0.25	0.16	-	0.40	0.20	MILLI/MJ	3.11	
SELENIUM	MICRO	1.80	0.80	-	15.00	1.40	MICRO/MJ	22.36	
NITRATE	MILLI	28.00	7.00	-	268.00	21.84	MILLI/MJ	347.77	
CAROTENE	MICRO	30.00	11.00	-	48.00	23.40	MICRO/MJ	372.61	

1 METHOD OF SOUTHGATE AND ENGLYST

CONSTITUENTS	DIM	AV	VARIATION			AVR	NUTR. DENS.		MOLPERC.
VITAMIN E ACTIVITY	MILLI	1.70	0.20	-	7.00	1.33	MILLI/MJ	21.11	
ALPHA-TOCOPHEROL	MILLI	1.70	0.20	-	7.00	1.33	MILLI/MJ	21.11	
VITAMIN K	MILLI	-	0.01	-	3.00				
VITAMIN B1	MICRO	68.00	40.00	-	100.00	53.04	MICRO/MJ	844.59	
VITAMIN B2	MICRO	50.00	38.00	-	70.00	39.00	MICRO/MJ	621.02	
NICOTINAMIDE	MILLI	0.43	0.30	-	0.96	0.34	MILLI/MJ	5.34	
PANTOTHENIC ACID	MILLI	0.32	0.30	-	0.37	0.25	MILLI/MJ	3.97	
VITAMIN B6	MILLI	0.15	0.12	-	0.21	0.12	MILLI/MJ	1.86	
BIOTIN	MICRO	2.00	-	-	-	1.56	MICRO/MJ	24.84	
FOLIC ACID	MICRO	35.00	-	-	-	27.30	MICRO/MJ	434.72	
VITAMIN C	MILLI	50.00	40.00	-	72.00	39.00	MILLI/MJ	621.02	
ARGININE	MILLI	110.00	-	-	-	85.80	GRAM/MJ	1.37	
CYSTINE	MILLI	30.00	-	-	-	23.40	MILLI/MJ	372.61	
HISTIDINE	MILLI	27.00	-	-	-	21.06	MILLI/MJ	335.35	
ISOLEUCINE	MILLI	43.00	-	-	-	33.54	MILLI/MJ	534.08	
LEUCINE	MILLI	61.00	-	-	-	47.58	MILLI/MJ	757.65	
LYSINE	MILLI	71.00	-	-	-	55.38	MILLI/MJ	881.86	
METHIONINE	MILLI	14.00	-	-	-	10.92	MILLI/MJ	173.89	
PHENYLALANINE	MILLI	32.00	-	-	-	24.96	MILLI/MJ	397.46	
THREONINE	MILLI	42.00	-	-	-	32.76	MILLI/MJ	521.66	
TRYPTOPHAN	MILLI	12.00	-	-	-	9.36	MILLI/MJ	149.05	
VALINE	MILLI	46.00	-	-	-	35.88	MILLI/MJ	571.34	
OXALIC ACID TOTAL	MILLI	7.40	-	-	-	5.77	MILLI/MJ	91.91	
OXALIC ACID SOLUBLE	MILLI	3.00	-	-	-	2.34	MILLI/MJ	37.26	
FERULIC ACID	MILLI	5.00	3.00	-	8.00	3.90	MILLI/MJ	62.10	
CAFFEIC ACID	MILLI	1.20	1.00	-	1.50	0.94	MILLI/MJ	14.90	
PARA-COUMARIC ACID	MILLI	1.00	7.00	-	15.00	0.78	MILLI/MJ	12.42	
PROTOCATECHUIC ACID	MILLI	5.00	2.00	-	9.00	3.90	MILLI/MJ	62.10	
SALICYLIC ACID	MICRO	80.00	80.00	-	80.00	62.40	MICRO/MJ	993.64	
GLUCOSE	GRAM	1.67	1.18	-	1.93	1.31	GRAM/MJ	20.83	
FRUCTOSE	GRAM	1.27	0.84	-	1.67	1.00	GRAM/MJ	15.85	
SUCROSE	GRAM	0.51	0.15	-	0.88	0.40	GRAM/MJ	6.41	
STARCH	MILLI	72.00	72.00	-	72.00	56.16	MILLI/MJ	894.28	
PENTOSAN	GRAM	0.71	-	-	-	0.55	GRAM/MJ	8.82	
HEXOSAN	GRAM	0.38	-	-	-	0.30	GRAM/MJ	4.72	
CELLULOSE	GRAM	0.97	-	-	-	0.76	GRAM/MJ	12.05	
POLYURONIC ACID	GRAM	0.48	-	-	-	0.37	GRAM/MJ	5.96	
MYRISTIC ACID	MILLI	0.20	-	-	-	0.16	MILLI/MJ	2.48	
PALMITIC ACID	MILLI	29.50	27.00	-	32.00	23.01	MILLI/MJ	366.40	
STEARIC ACID	MILLI	2.35	2.00	-	2.70	1.83	MILLI/MJ	29.19	
ARACHIDIC ACID	MILLI	1.40	-	-	-	1.09	MILLI/MJ	17.39	
PALMITOLEIC ACID	MILLI	1.80	0.60	-	3.00	1.40	MILLI/MJ	22.36	
OLEIC ACID	MILLI	3.85	0.70	-	7.00	3.00	MILLI/MJ	47.82	
EICOSENOIC ACID	MILLI	0.90	-	-	-	0.70	MILLI/MJ	11.18	
LINOLEIC ACID	MILLI	54.00	43.00	-	65.00	42.12	MILLI/MJ	670.71	
LINOLENIC ACID	MILLI	43.50	30.00	-	57.00	33.93	MILLI/MJ	540.29	
TOTAL PURINES	MILLI	23.00 [2]	20.00	-	25.00	17.94	MILLI/MJ	285.67	

2 VALUE EXPRESSED IN MG URIC ACID/100G

SAUERKRAUT
ABGETROPFT

SAUERKRAUT
DRIPPED OFF

CHOUCROUTE
SANS JUS

		PROTEIN	FAT	CARBOHYDRATES	TOTAL
ENERGY VALUE (AVERAGE)	KJOULE	24	12	40	76
PER 100 G EDIBLE PORTION	(KCAL)	5.7	2.9	9.5	18
AMOUNT OF DIGESTIBLE CONSTITUENTS PER 100 G	GRAM	0.98	0.27	2.37	
ENERGY VALUE (AVERAGE)	KJOULE	16	11	40	66
OF THE DIGESTIBLE FRACTION PER 100 G EDIBLE PORTION	(KCAL)	3.7	2.6	9.5	16

WASTE PERCENTAGE AVERAGE 0.00

CONSTITUENTS	DIM	AV	VARIATION			AVR	NUTR.	DENS.	MOLPERC.
WATER	GRAM	90.70	88.00	-	92.00	90.70	KILO/MJ	1.37	
PROTEIN	GRAM	1.52	1.00	-	2.00	1.52	GRAM/MJ	23.02	
FAT	GRAM	0.31	0.20	-	0.54	0.31	GRAM/MJ	4.70	
AVAILABLE CARBOHYDR.	GRAM	2.37	-	-	-	2.37	GRAM/MJ	35.90	
TOTAL DIETARY FIBRE	GRAM	2.14 [1]	-	-	-	2.14	GRAM/MJ	32.41	
MINERALS	GRAM	2.35	1.40	-	4.00	2.35	GRAM/MJ	35.59	
SODIUM	MILLI	355.00	134.00	-	890.00	355.00	GRAM/MJ	5.38	
POTASSIUM	MILLI	288.00	140.00	-	475.00	288.00	GRAM/MJ	4.36	
MAGNESIUM	MILLI	14.00	7.00	-	24.00	14.00	MILLI/MJ	212.05	
CALCIUM	MILLI	48.00	36.00	-	57.00	48.00	MILLI/MJ	727.03	
MANGANESE	MILLI	0.14	0.10	-	0.20	0.14	MILLI/MJ	2.12	
IRON	MILLI	0.60	0.50	-	0.70	0.60	MILLI/MJ	9.09	
COBALT	MICRO	0.50	-	-	-	0.50	MICRO/MJ	7.57	
COPPER	MILLI	0.13	-	-	-	0.13	MILLI/MJ	1.97	
ZINC	MILLI	0.32	0.15	-	0.63	0.32	MILLI/MJ	4.85	
NICKEL	MICRO	5.00	2.00	-	10.00	5.00	MICRO/MJ	75.73	
CHROMIUM	MICRO	5.00	0.50	-	14.00	5.00	MICRO/MJ	75.73	
MOLYBDENUM	MICRO	10.00	-	-	-	10.00	MICRO/MJ	151.46	
PHOSPHORUS	MILLI	43.00	18.00	-	94.00	43.00	MILLI/MJ	651.30	
FLUORIDE	MICRO	45.00	10.00	-	90.00	45.00	MICRO/MJ	681.59	
CAROTENE	MICRO	18.00	0.00	-	30.00	18.00	MICRO/MJ	272.64	
VITAMIN K	MILLI	1.54	0.08	-	3.00	1.54	MILLI/MJ	23.33	
VITAMIN B1	MICRO	27.00	20.00	-	30.00	27.00	MICRO/MJ	408.95	
VITAMIN B2	MICRO	50.00	40.00	-	60.00	50.00	MICRO/MJ	757.32	
NICOTINAMIDE	MILLI	0.17	0.10	-	0.20	0.17	MILLI/MJ	2.57	
PANTOTHENIC ACID	MILLI	0.23	-	-	-	0.23	MILLI/MJ	3.48	

1 METHOD OF MEUSER, SUCKOW AND KULIKOWSKI ("BERLINER METHODE")

CONSTITUENTS	DIM	AV	VARIATION			AVR	NUTR. DENS.		MOLPERC.
VITAMIN B6	MILLI	0.21	-	-	-	0.21	MILLI/MJ	3.18	
FOLIC ACID	MICRO	19.00	-	-	-	19.00	MICRO/MJ	287.78	
VITAMIN C	MILLI	20.00	10.00	-	38.00	20.00	MILLI/MJ	302.93	
HISTAMINE	MILLI	6.95	0.90	-	13.00	6.95	MILLI/MJ	105.27	
TYRAMINE	MILLI	2.00	-	-	-	2.00	MILLI/MJ	30.29	
LACTIC ACID	GRAM	1.60	1.20	-	1.80	1.60	GRAM/MJ	24.23	
GLUCOSE	GRAM	0.42	0.42	-	0.42	0.42	GRAM/MJ	6.36	
FRUCTOSE	GRAM	0.21	0.21	-	0.21	0.21	GRAM/MJ	3.18	
SUCROSE	GRAM	0.14	0.14	-	0.14	0.14	GRAM/MJ	2.12	
TOTAL PURINES	MILLI	16.00 2	12.00	-	20.00	16.00	MILLI/MJ	242.34	
DIETARY FIBRE,WAT.SOL.	GRAM	0.84	-	-	-	0.84	GRAM/MJ	12.72	
DIETARY FIBRE,WAT.INS.	GRAM	1.30	-	-	-	1.30	GRAM/MJ	19.69	

2 VALUE EXPRESSED IN MG URIC ACID/100G

SCHNITTLAUCH CHIVES CIVETTE

ALLIUM SCHOENOPRASUM L.

		PROTEIN	FAT	CARBOHYDRATES	TOTAL
ENERGY VALUE (AVERAGE)	KJOULE	56	29	27	112
PER 100 G	(KCAL)	13	6.9	6.4	27
EDIBLE PORTION					
AMOUNT OF DIGESTIBLE	GRAM	2.32	0.66	1.59	
CONSTITUENTS PER 100 G					
ENERGY VALUE (AVERAGE)	KJOULE	37	26	27	89
OF THE DIGESTIBLE	(KCAL)	8.7	6.2	6.4	21
FRACTION PER 100 G					
EDIBLE PORTION					

WASTE PERCENTAGE AVERAGE 0.00

CONSTITUENTS	DIM	AV	VARIATION			AVR	NUTR. DENS.		MOLPERC.
WATER	GRAM	83.30	82.00	-	86.00	83.30	GRAM/MJ	935.58	
PROTEIN	GRAM	3.58	2.48	-	4.00	3.58	GRAM/MJ	40.21	
FAT	GRAM	0.74	0.60	-	0.90	0.74	GRAM/MJ	8.31	
AVAILABLE CARBOHYDR.	GRAM	1.59	-	-	-	1.59	GRAM/MJ	17.86	
MINERALS	GRAM	1.70	1.66	-	1.80	1.70	GRAM/MJ	19.09	

CONSTITUENTS	DIM	AV	VARIATION			AVR	NUTR. DENS.		MOLPERC.
SODIUM	MILLI	3.00	2.00	-	4.00	3.00	MILLI/MJ	33.69	
POTASSIUM	MILLI	434.00	388.00	-	479.00	434.00	GRAM/MJ	4.87	
MAGNESIUM	MILLI	44.00	32.00	-	55.00	44.00	MILLI/MJ	494.18	
CALCIUM	MILLI	129.00	119.00	-	146.00	129.00	GRAM/MJ	1.45	
IRON	MILLI	1.90	1.10	-	2.80	1.90	MILLI/MJ	21.34	
PHOSPHORUS	MILLI	75.00	57.00	-	111.00	75.00	MILLI/MJ	842.36	
CHLORIDE	MILLI	74.00	-	-	-	74.00	MILLI/MJ	831.13	
IODIDE	MICRO	4.20	4.20	-	4.20	4.20	MICRO/MJ	47.17	
CAROTENE	MILLI	0.30	-	-	-	0.30	MILLI/MJ	3.37	
VITAMIN K	MILLI	0.57	0.57	-	0.57	0.57	MILLI/MJ	6.40	
VITAMIN B1	MILLI	0.14	0.12	-	0.15	0.14	MILLI/MJ	1.57	
VITAMIN B2	MILLI	0.15	0.10	-	0.20	0.15	MILLI/MJ	1.68	
NICOTINAMIDE	MILLI	0.60	0.46	-	0.65	0.60	MILLI/MJ	6.74	
VITAMIN B6	MILLI	0.42	0.42	-	0.42	0.42	MILLI/MJ	4.72	
VITAMIN C	MILLI	47.00	40.00	-	60.00	47.00	MILLI/MJ	527.88	
OXALIC ACID TOTAL	-	0.00	-	-	-	0.00			
SALICYLIC ACID	MICRO	30.00	30.00	-	30.00	30.00	MICRO/MJ	336.94	
GLUCOSE	GRAM	0.65	0.65	-	0.65	0.65	GRAM/MJ	7.30	
FRUCTOSE	GRAM	0.76	0.76	-	0.76	0.76	GRAM/MJ	8.54	
SUCROSE	GRAM	0.18	0.18	-	0.18	0.18	GRAM/MJ	2.02	
PALMITIC ACID	MILLI	86.00	86.00	-	86.00	86.00	MILLI/MJ	965.90	
STEARIC ACID	MILLI	7.80	7.80	-	7.80	7.80	MILLI/MJ	87.61	
PALMITOLEIC ACID	MILLI	12.00	12.00	-	12.00	12.00	MILLI/MJ	134.78	
OLEIC ACID	MILLI	7.80	7.80	-	7.80	7.80	MILLI/MJ	87.61	
LINOLEIC ACID	MILLI	132.00	132.00	-	132.00	132.00	GRAM/MJ	1.48	
LINOLENIC ACID	MILLI	288.00	288.00	-	288.00	288.00	GRAM/MJ	3.23	

SOJASPROSSEN (SOJAKEIMLINGE) SOYBEAN SPROUTS POUSSES DE SOJA

		PROTEIN	FAT	CARBOHYDRATES	TOTAL
ENERGY VALUE (AVERAGE)	KJOULE	99	47	99	244
PER 100 G EDIBLE PORTION	(KCAL)	24	11	24	58
AMOUNT OF DIGESTIBLE CONSTITUENTS PER 100 G	GRAM	4.13	1.08	5.90	
ENERGY VALUE (AVERAGE)	KJOULE	77	42	99	218
OF THE DIGESTIBLE FRACTION PER 100 G EDIBLE PORTION	(KCAL)	18	10	24	52

WASTE PERCENTAGE AVERAGE 17

CONSTITUENTS	DIM	AV	VARIATION		AVR	NUTR. DENS.		MOLPERC.
WATER	GRAM	86.90	81.50	- 93.50	72.13	GRAM/MJ	399.11	
PROTEIN	GRAM	5.30	1.92	- 7.04	4.40	GRAM/MJ	24.34	
FAT	GRAM	1.20	0.50	- 1.80	1.00	GRAM/MJ	5.51	
AVAILABLE CARBOHYDR.	GRAM	5.90 1	-	- -	4.90	GRAM/MJ	27.10	
MINERALS	GRAM	0.70	0.40	- 1.00	0.58	GRAM/MJ	3.21	
SODIUM	MILLI	30.00	-	- -	24.90	MILLI/MJ	137.78	
POTASSIUM	MILLI	218.00	157.00	- 279.00	180.94	GRAM/MJ	1.00	
MAGNESIUM	MILLI	15.00	-	- -	12.45	MILLI/MJ	68.89	
CALCIUM	MILLI	42.00	32.00	- 52.00	34.86	MILLI/MJ	192.89	
IRON	MILLI	0.85	0.60	- 1.10	0.71	MILLI/MJ	3.90	
PHOSPHORUS	MILLI	58.00	-	- -	48.14	MILLI/MJ	266.38	
CAROTENE	MICRO	25.00	-	- -	20.75	MICRO/MJ	114.82	
VITAMIN B1	MILLI	0.19	-	- -	0.16	MILLI/MJ	0.87	
VITAMIN B2	MILLI	0.15	-	- -	0.12	MILLI/MJ	0.69	
NICOTINAMIDE	MILLI	1.80	0.80	- 2.70	1.49	MILLI/MJ	8.27	
VITAMIN C	MILLI	16.00	7.00	- 24.00	13.28	MILLI/MJ	73.48	

1 ESTIMATED BY THE DIFFERENCE METHOD
100 - (WATER + PROTEIN + FAT + MINERALS)

SPARGEL **ASPARAGUS** **ASPERGE**

ASPARAGUS OFFICINALIS L.

		PROTEIN	FAT	CARBOHYDRATES	TOTAL
ENERGY VALUE (AVERAGE) PER 100 G EDIBLE PORTION	KJOULE	30	5.4	37	72
	(KCAL)	7.1	1.3	8.8	17
AMOUNT OF DIGESTIBLE CONSTITUENTS PER 100 G	GRAM	1.40	0.12	2.19	
ENERGY VALUE (AVERAGE) OF THE DIGESTIBLE FRACTION PER 100 G EDIBLE PORTION	KJOULE	22	4.9	37	64
	(KCAL)	5.3	1.2	8.8	15

WASTE PERCENTAGE AVERAGE 26 MINIMUM 20 MAXIMUM 33

CONSTITUENTS	DIM	AV	VARIATION			AVR	NUTR. DENS.		MOLPERC.
WATER	GRAM	93.60	93.00	-	94.00	69.26	KILO/MJ	1.47	
PROTEIN	GRAM	1.90	1.50	-	2.20	1.41	GRAM/MJ	29.87	
FAT	GRAM	0.14	0.10	-	0.20	0.10	GRAM/MJ	2.20	
AVAILABLE CARBOHYDR.	GRAM	2.19	0.86	-	3.30	1.62	GRAM/MJ	34.43	
TOTAL DIETARY FIBRE	GRAM	1.47 [1]	-	-	-	1.09	GRAM/MJ	23.11	
MINERALS	GRAM	0.62	0.54	-	0.80	0.46	GRAM/MJ	9.75	
SODIUM	MILLI	4.00	2.00	-	5.00	2.96	MILLI/MJ	62.88	
POTASSIUM	MILLI	207.00	187.00	-	280.00	153.18	GRAM/MJ	3.25	
MAGNESIUM	MILLI	20.00	19.00	-	22.00	14.80	MILLI/MJ	314.39	
CALCIUM	MILLI	21.00	20.00	-	22.00	15.54	MILLI/MJ	330.11	
MANGANESE	MILLI	0.27	0.25	-	0.30	0.20	MILLI/MJ	4.24	
IRON	MILLI	1.00	0.90	-	1.10	0.74	MILLI/MJ	15.72	
COPPER	MILLI	0.15	0.14	-	0.16	0.11	MILLI/MJ	2.36	
ZINC	MILLI	0.50	0.20	-	0.80	0.37	MILLI/MJ	7.86	
PHOSPHORUS	MILLI	46.00	35.00	-	62.00	34.04	MILLI/MJ	723.10	
CHLORIDE	MILLI	53.00	-	-	-	39.22	MILLI/MJ	833.14	
FLUORIDE	MICRO	48.00	-	-	-	35.52	MICRO/MJ	754.54	
IODIDE	MICRO	7.00	4.00	-	10.00	5.18	MICRO/MJ	110.04	
BORON	MILLI	0.17	0.09	-	0.26	0.13	MILLI/MJ	2.67	
NITRATE	MILLI	66.00	1.30	-	70.00	48.84	GRAM/MJ	1.04	
CAROTENE	MICRO	30.00	-	-	-	22.20	MICRO/MJ	471.59	
VITAMIN E ACTIVITY	MILLI	2.00	-	-	-	1.48	MILLI/MJ	31.44	
TOTAL TOCOPHEROLS	MILLI	2.10	-	-	-	1.55	MILLI/MJ	33.01	
ALPHA-TOCOPHEROL	MILLI	2.00	-	-	-	1.48	MILLI/MJ	31.44	
BETA-TOCOPHEROL	MICRO	50.00	-	-	-	37.00	MICRO/MJ	785.98	
GAMMA-TOCOPHEROL	MICRO	70.00	-	-	-	51.80	MILLI/MJ	1.10	

1 METHOD OF SOUTHGATE AND ENGLYST

CONSTITUENTS	DIM	AV		VARIATION		AVR	NUTR. DENS.		MOLPERC.
VITAMIN K	MICRO	40.00		20.00 -	60.00	29.60	MICRO/MJ	628.78	
VITAMIN B1	MILLI	0.11		0.03 -	0.20	0.08	MILLI/MJ	1.73	
VITAMIN B2	MILLI	0.12		0.07 -	0.19	0.09	MILLI/MJ	1.89	
NICOTINAMIDE	MILLI	1.00		0.60 -	1.40	0.74	MILLI/MJ	15.72	
PANTOTHENIC ACID	MILLI	0.62		0.60 -	0.64	0.46	MILLI/MJ	9.75	
VITAMIN B6	MICRO	60.00		51.00 -	88.00	44.40	MICRO/MJ	943.18	
BIOTIN	MICRO	2.00		- -	-	1.48	MICRO/MJ	31.44	
FOLIC ACID	MICRO	86.00		58.00 -	120.00	63.64	MILLI/MJ	1.35	
VITAMIN C	MILLI	21.00		5.00 -	33.00	15.54	MILLI/MJ	330.11	
ARGININE	GRAM	0.11		0.10 -	0.11	0.08	GRAM/MJ	1.73	
HISTIDINE	MILLI	31.00		28.00 -	36.00	22.94	MILLI/MJ	487.31	
ISOLEUCINE	MILLI	69.00		68.00 -	70.00	51.06	GRAM/MJ	1.08	
LEUCINE	MILLI	83.00		63.00 -	100.00	61.42	GRAM/MJ	1.30	
LYSINE	MILLI	89.00		83.00 -	96.00	65.86	GRAM/MJ	1.40	
METHIONINE	MILLI	27.00		24.00 -	31.00	19.98	MILLI/MJ	424.43	
PHENYLALANINE	MILLI	60.00		55.00 -	64.00	44.40	MILLI/MJ	943.18	
THREONINE	MILLI	57.00		56.00 -	58.00	42.18	MILLI/MJ	896.02	
TRYPTOPHAN	MILLI	23.00		16.00 -	30.00	17.02	MILLI/MJ	361.55	
VALINE	MILLI	91.00		87.00 -	96.00	67.34	GRAM/MJ	1.43	
MALIC ACID	MILLI	95.00		70.00 -	120.00	70.30	GRAM/MJ	1.49	
CITRIC ACID	MILLI	60.00		30.00 -	90.00	44.40	MILLI/MJ	943.18	
OXALIC ACID TOTAL	-	0.00		- -	-	0.00			
QUINIC ACID	MILLI	39.00		- -	-	28.86	MILLI/MJ	613.06	
SUCCINIC ACID	MILLI	9.00		4.00 -	13.00	6.66	MILLI/MJ	141.48	
SALICYLIC ACID	MILLI	0.14		0.14 -	0.14	0.10	MILLI/MJ	2.20	
GLUCOSE	GRAM	0.80		0.29 -	1.10	0.60	GRAM/MJ	12.69	
FRUCTOSE	GRAM	0.99		0.44 -	1.40	0.73	GRAM/MJ	15.61	
SUCROSE	GRAM	0.24		0.03 -	0.59	0.18	GRAM/MJ	3.84	
PENTOSAN	GRAM	0.23		- -	-	0.17	GRAM/MJ	3.62	
HEXOSAN	GRAM	0.19		- -	-	0.14	GRAM/MJ	2.99	
CELLULOSE	GRAM	0.75		- -	-	0.56	GRAM/MJ	11.79	
POLYURONIC ACID	GRAM	0.21		- -	-	0.16	GRAM/MJ	3.30	
PALMITIC ACID	MILLI	31.00		31.00 -	31.00	22.94	MILLI/MJ	487.31	
STEARIC ACID	MILLI	1.10		1.10 -	1.10	0.81	MILLI/MJ	17.29	
PALMITOLEIC ACID	MILLI	0.80		0.80 -	0.80	0.59	MILLI/MJ	12.58	
OLEIC ACID	MILLI	2.20		2.20 -	2.20	1.63	MILLI/MJ	34.58	
LINOLEIC ACID	MILLI	70.00		70.00 -	70.00	51.80	GRAM/MJ	1.10	
LINOLENIC ACID	MILLI	6.00		6.00 -	6.00	4.44	MILLI/MJ	94.32	
TOTAL STEROLS	MILLI	24.00		- -	-	17.76	MILLI/MJ	377.27	
CAMPESTEROL	MILLI	1.00		- -	-	0.74	MILLI/MJ	15.72	
BETA-SITOSTEROL	MILLI	14.00		- -	-	10.36	MILLI/MJ	220.07	
STIGMASTEROL	MILLI	4.00		- -	-	2.96	MILLI/MJ	62.88	
TOTAL PURINES	MILLI	30.00	2	- -	-	22.20	MILLI/MJ	471.59	
DIETARY FIBRE,WAT.SOL.	GRAM	0.14		- -	-	0.10	GRAM/MJ	2.20	
DIETARY FIBRE,WAT.INS.	GRAM	1.33		- -	-	0.98	GRAM/MJ	20.91	

2 VALUE EXPRESSED IN MG URIC ACID/100G

SPARGEL	ASPARAGUS	ASPERGE
GEKOCHT, ABGETROPFT	BOILED, DRAINED	BOUILLI, DEGOUTTÉ

		PROTEIN	FAT	CARBOHYDRATES	TOTAL
ENERGY VALUE (AVERAGE)	KJOULE	27	5.4	20	52
PER 100 G EDIBLE PORTION	(KCAL)	6.4	1.3	4.8	13
AMOUNT OF DIGESTIBLE CONSTITUENTS PER 100 G	GRAM	1.10	0.12	1.21	
ENERGY VALUE (AVERAGE)	KJOULE	17	4.9	20	42
OF THE DIGESTIBLE FRACTION PER 100 G EDIBLE PORTION	(KCAL)	4.1	1.2	4.8	10

WASTE PERCENTAGE AVERAGE 0.00

CONSTITUENTS	DIM	AV	VARIATION			AVR	NUTR. DENS.		MOLPERC.
WATER	GRAM	95.50	-	-	-	95.50	KILO/MJ	2.25	
PROTEIN	GRAM	1.70	-	-	-	1.70	GRAM/MJ	40.01	
FAT	GRAM	0.14	-	-	-	0.14	GRAM/MJ	3.29	
AVAILABLE CARBOHYDR.	GRAM	1.21	-	-	-	1.21	GRAM/MJ	28.48	
MINERALS	GRAM	0.40	-	-	-	0.40	GRAM/MJ	9.41	
SODIUM	GRAM	2.00	-	-	-	2.00	GRAM/MJ	47.07	
POTASSIUM	MILLI	136.00	-	-	-	136.00	GRAM/MJ	3.20	
CALCIUM	MILLI	20.00	-	-	-	20.00	MILLI/MJ	470.69	
IRON	MILLI	0.60	-	-	-	0.60	MILLI/MJ	14.12	
PHOSPHORUS	MILLI	37.00	-	-	-	37.00	MILLI/MJ	870.78	
VITAMIN B1	MICRO	90.00	-	-	-	90.00	MILLI/MJ	2.12	
VITAMIN B2	MILLI	0.11	-	-	-	0.11	MILLI/MJ	2.59	
NICOTINAMIDE	MILLI	0.93	-	-	-	0.93	MILLI/MJ	21.89	
VITAMIN C	MILLI	16.00	-	-	-	16.00	MILLI/MJ	376.55	
MALIC ACID	MILLI	50.00	-	-	-	50.00	GRAM/MJ	1.18	
CITRIC ACID	MILLI	30.00	-	-	-	30.00	MILLI/MJ	706.03	
GLUCOSE	GRAM	0.42	-	-	-	0.42	GRAM/MJ	9.88	
FRUCTOSE	GRAM	0.56	-	-	-	0.56	GRAM/MJ	13.18	
SUCROSE	GRAM	0.15	-	-	-	0.15	GRAM/MJ	3.53	

SPARGEL IN DOSEN [1]

ASPARAGUS CANNED [1]

ASPERGES EN BOÎTES [1]

		PROTEIN	FAT	CARBOHYDRATES	TOTAL
ENERGY VALUE (AVERAGE)	KJOULE	30	12	17	59
PER 100 G	(KCAL)	7.1	2.8	4.1	14
EDIBLE PORTION					
AMOUNT OF DIGESTIBLE	GRAM	1.40	0.27	1.02	
CONSTITUENTS PER 100 G					
ENERGY VALUE (AVERAGE)	KJOULE	22	11	17	50
OF THE DIGESTIBLE	(KCAL)	5.3	2.5	4.1	12
FRACTION PER 100 G					
EDIBLE PORTION					

WASTE PERCENTAGE AVERAGE 0.00

CONSTITUENTS	DIM	AV	VARIATION			AVR	NUTR. DENS.		MOLPERC.
WATER	GRAM	93.50	93.30	-	93.60	93.50	KILO/MJ	1.88	
PROTEIN	GRAM	1.90	-	-	-	1.90	GRAM/MJ	38.28	
FAT	GRAM	0.30	-	-	-	0.30	GRAM/MJ	6.04	
AVAILABLE CARBOHYDR.	GRAM	1.02	-	-	-	1.02	GRAM/MJ	20.55	
TOTAL DIETARY FIBRE	GRAM	1.28	-	-	-	1.28	GRAM/MJ	25.79	
MINERALS	GRAM	1.38	1.30	-	1.46	1.38	GRAM/MJ	27.80	
SODIUM	MILLI	355.00	243.00	-	469.00	355.00	GRAM/MJ	7.15	
POTASSIUM	MILLI	104.00	63.00	-	152.00	104.00	GRAM/MJ	2.10	
MAGNESIUM	MILLI	6.00	6.00	-	6.00	6.00	MILLI/MJ	120.88	
CALCIUM	MILLI	17.00	10.00	-	20.00	17.00	MILLI/MJ	342.49	
MANGANESE	MILLI	0.12	0.12	-	0.12	0.12	MILLI/MJ	2.42	
IRON	MILLI	0.90	-	-	-	0.90	MILLI/MJ	18.13	
ZINC	MILLI	0.38	0.38	-	0.38	0.38	MILLI/MJ	7.66	
CHROMIUM	MICRO	3.40	3.40	-	3.40	3.40	MICRO/MJ	68.50	
PHOSPHORUS	MILLI	38.00	20.00	-	46.00	38.00	MILLI/MJ	765.56	
IODIDE	MICRO	0.90	0.90	-	0.90	0.90	MICRO/MJ	18.13	
CAROTENE	MILLI	0.35	0.33	-	0.36	0.35	MILLI/MJ	7.05	
VITAMIN B1	MICRO	60.00	50.00	-	70.00	60.00	MILLI/MJ	1.21	
VITAMIN B2	MICRO	80.00	60.00	-	100.00	80.00	MILLI/MJ	1.61	
NICOTINAMIDE	MILLI	0.80	0.70	-	0.90	0.80	MILLI/MJ	16.12	
PANTOTHENIC ACID	MILLI	0.13	-	-	-	0.13	MILLI/MJ	2.62	
VITAMIN B6	MICRO	30.00	-	-	-	30.00	MICRO/MJ	604.39	
BIOTIN	MICRO	1.70	-	-	-	1.70	MICRO/MJ	34.25	
FOLIC ACID	MICRO	55.00	55.00	-	55.00	55.00	MILLI/MJ	1.11	
VITAMIN C	MILLI	15.00	-	-	-	15.00	MILLI/MJ	302.19	

1 TOTAL CONTENT (SOLID AND LIQUID)

CONSTITUENTS	DIM	AV	VARIATION		AVR	NUTR. DENS.		MOLPERC.
ARGININE	MILLI	110.00	100.00	- 110.00	110.00	GRAM/MJ	2.22	
CYSTINE	MILLI	18.00	18.00	- 18.00	18.00	MILLI/MJ	362.63	
HISTIDINE	MILLI	31.00	28.00	- 36.00	31.00	MILLI/MJ	624.54	
ISOLEUCINE	MILLI	69.00	68.00	- 70.00	69.00	GRAM/MJ	1.39	
LEUCINE	MILLI	69.00	68.00	- 70.00	69.00	GRAM/MJ	1.39	
LYSINE	MILLI	89.00	83.00	- 96.00	89.00	GRAM/MJ	1.79	
METHIONINE	MILLI	27.00	24.00	- 31.00	27.00	MILLI/MJ	543.95	
PHENYLALANINE	MILLI	60.00	55.00	- 64.00	60.00	GRAM/MJ	1.21	
THREONINE	MILLI	57.00	56.00	- 58.00	57.00	GRAM/MJ	1.15	
TRYPTOPHAN	MILLI	23.00	16.00	- 30.00	23.00	MILLI/MJ	463.37	
TYROSINE	MILLI	40.00	40.00	- 40.00	40.00	MILLI/MJ	805.85	
VALINE	MILLI	91.00	87.00	- 96.00	91.00	GRAM/MJ	1.83	
GLUCOSE	GRAM	0.28	0.28	- 0.28	0.28	GRAM/MJ	5.64	
FRUCTOSE	GRAM	0.58	0.58	- 0.58	0.58	GRAM/MJ	11.68	
SUCROSE	GRAM	0.16	0.16	- 0.16	0.16	GRAM/MJ	3.22	
PENTOSAN	GRAM	0.20	0.20	- 0.20	0.20	GRAM/MJ	4.03	
HEXOSAN	GRAM	0.16	0.16	- 0.16	0.16	GRAM/MJ	3.22	
CELLULOSE	GRAM	0.65	0.65	- 0.65	0.65	GRAM/MJ	13.10	
POLYURONIC ACID	GRAM	0.18	0.18	- 0.18	0.18	GRAM/MJ	3.63	

SPINAT | SPINACH | ÉPINARD

SPINACIA OLERACEA L.

		PROTEIN	FAT	CARBOHYDRATES	TOTAL
ENERGY VALUE (AVERAGE)	KJOULE	40	12	10	61
PER 100 G	(KCAL)	9.5	2.8	2.4	15
EDIBLE PORTION					
AMOUNT OF DIGESTIBLE	GRAM	1.63	0.27	0.61	
CONSTITUENTS PER 100 G					
ENERGY VALUE (AVERAGE)	KJOULE	26	11	10	46
OF THE DIGESTIBLE	(KCAL)	6.1	2.5	2.4	11
FRACTION PER 100 G					
EDIBLE PORTION					

WASTE PERCENTAGE AVERAGE 15 MINIMUM 8.0 MAXIMUM 22

CONSTITUENTS	DIM	AV		VARIATION			AVR	NUTR.	DENS.	MOLPERC.
WATER	GRAM	91.60		89.00	-	93.70	77.86	KILO/MJ	1.97	
PROTEIN	GRAM	2.52		2.00	-	3.20	2.14	GRAM/MJ	54.29	
FAT	GRAM	0.30		0.20	-	0.41	0.26	GRAM/MJ	6.46	
AVAILABLE CARBOHYDR.	GRAM	0.61		-	-	-	0.52	GRAM/MJ	13.14	
TOTAL DIETARY FIBRE	GRAM	1.84	1	-	-	-	1.56	GRAM/MJ	39.64	
MINERALS	GRAM	1.51		1.26	-	1.87	1.28	GRAM/MJ	32.53	
SODIUM	MILLI	65.00		40.00	-	86.00	55.25	GRAM/MJ	1.40	
POTASSIUM	MILLI	633.00		470.00	-	742.00	538.05	GRAM/MJ	13.64	
MAGNESIUM	MILLI	58.00		39.00	-	88.00	49.30	GRAM/MJ	1.25	
CALCIUM	MILLI	126.00		80.00	-	190.00	107.10	GRAM/MJ	2.71	
MANGANESE	MILLI	0.76		0.25	-	1.09	0.65	MILLI/MJ	16.37	
IRON	MILLI	4.10		2.80	-	6.60	3.49	MILLI/MJ	88.33	
COBALT	MICRO	1.90		0.50	-	4.60	1.62	MICRO/MJ	40.93	
COPPER	MILLI	0.12		0.07	-	0.20	0.10	MILLI/MJ	2.59	
ZINC	MILLI	0.50		0.22	-	0.79	0.43	MILLI/MJ	10.77	
NICKEL	MICRO	23.00		-	-	-	19.55	MICRO/MJ	495.53	
CHROMIUM	MICRO	5.00		-	-	-	4.25	MICRO/MJ	107.72	
MOLYBDENUM	MILLI	-		0.03	-	0.08				
PHOSPHORUS	MILLI	55.00		37.00	-	70.00	46.75	GRAM/MJ	1.18	
CHLORIDE	MILLI	54.00		32.00	-	76.00	45.90	GRAM/MJ	1.16	
FLUORIDE	MILLI	0.11		0.02	-	0.18	0.09	MILLI/MJ	2.37	
IODIDE	MICRO	12.00		4.00	-	20.00	10.20	MICRO/MJ	258.54	
BORON	MILLI	-		0.26	-	1.00				
SELENIUM	MICRO	1.70	2	1.70	-	18.00	1.45	MICRO/MJ	36.63	
BROMINE	MICRO	88.00		85.00	-	92.00	74.80	MILLI/MJ	1.90	
NITRATE	MILLI	166.00		2.00	-	670.00	141.10	GRAM/MJ	3.58	
CAROTENE	MILLI	4.20		3.00	-	5.70	3.57	MILLI/MJ	90.49	
VITAMIN E ACTIVITY	MILLI	1.60		-	-	-	1.36	MILLI/MJ	34.47	
TOTAL TOCOPHEROLS	MILLI	2.50		-	-	-	2.13	MILLI/MJ	53.86	
ALPHA-TOCOPHEROL	MILLI	1.60		-	-	-	1.36	MILLI/MJ	34.47	
GAMMA-TOCOPHEROL	MILLI	0.12		-	-	-	0.10	MILLI/MJ	2.59	
DELTA-TOCOPHEROL	MILLI	0.82		-	-	-	0.70	MILLI/MJ	17.67	
VITAMIN K	MILLI	0.35		-	-	-	0.30	MILLI/MJ	7.54	

1 METHOD OF SOUTHGATE AND ENGLYST

2 AMOUNTS, DETERMINED IN THE FEDERAL REPUBLIC OF GERMANY

CONSTITUENTS	DIM	AV	VARIATION			AVR	NUTR. DENS.		MOLPERC.
VITAMIN B1	MILLI	0.11	0.07	-	0.18	0.09	MILLI/MJ	2.37	
VITAMIN B2	MILLI	0.23	0.18	-	0.33	0.20	MILLI/MJ	4.96	
NICOTINAMIDE	MILLI	0.62	0.50	-	0.72	0.53	MILLI/MJ	13.36	
PANTOTHENIC ACID	MILLI	0.25	0.19	-	0.31	0.21	MILLI/MJ	5.39	
VITAMIN B6	MILLI	0.22	0.18	-	0.31	0.19	MILLI/MJ	4.74	
BIOTIN	MICRO	6.90	5.90	-	7.80	5.87	MICRO/MJ	148.66	
FOLIC ACID	MICRO	78.00	49.00	-	200.00	66.30	MILLI/MJ	1.68	
VITAMIN B12	-	0.00	-	-	-	0.00			
VITAMIN C	MILLI	52.00	15.00	-	120.00	44.20	GRAM/MJ	1.12	
ARGININE	GRAM	0.13	0.07	-	0.18	0.11	GRAM/MJ	2.80	
CYSTINE	MILLI	38.00	23.00	-	50.00	32.30	MILLI/MJ	818.70	
HISTIDINE	MILLI	53.00	28.00	-	81.00	45.05	GRAM/MJ	1.14	
ISOLEUCINE	GRAM	0.12	0.06	-	0.16	0.10	GRAM/MJ	2.59	
LEUCINE	GRAM	0.19	0.11	-	0.24	0.16	GRAM/MJ	4.09	
LYSINE	GRAM	0.16	0.08	-	0.23	0.14	GRAM/MJ	3.45	
METHIONINE	MILLI	43.00	17.00	-	58.00	36.55	MILLI/MJ	926.42	
PHENYLALANINE	GRAM	0.11	0.04	-	0.15	0.09	GRAM/MJ	2.37	
THREONINE	GRAM	0.11	0.06	-	0.17	0.09	GRAM/MJ	2.37	
TRYPTOPHAN	MILLI	41.00	27.00	-	56.00	34.85	MILLI/MJ	883.33	
TYROSINE	MILLI	80.00	51.00	-	110.00	68.00	GRAM/MJ	1.72	
VALINE	GRAM	0.14	0.08	-	0.21	0.12	GRAM/MJ	3.02	
MALIC ACID	MILLI	42.46	-	-	-	36.09	MILLI/MJ	914.79	
CITRIC ACID	MILLI	23.56	-	-	-	20.03	MILLI/MJ	507.59	
LACTIC ACID	MILLI	9.20	-	-	-	7.82	MILLI/MJ	198.21	
FORMIC ACID	MILLI	6.18	-	-	-	5.25	MILLI/MJ	133.15	
OXALIC ACID TOTAL	MILLI	442.00	-	-	-	375.70	GRAM/MJ	9.52	
OXALIC ACID SOLUBLE	MILLI	126.00	-	-	-	107.10	GRAM/MJ	2.71	
ACETIC ACID	MILLI	3.55	-	-	-	3.02	MILLI/MJ	76.48	
FERULIC ACID	MILLI	3.00	1.50	-	4.50	2.55	MILLI/MJ	64.63	
PARA-COUMARIC ACID	MILLI	10.00	7.00	-	13.00	8.50	MILLI/MJ	215.45	
SUCCINIC ACID	MILLI	12.00	-	-	-	10.20	MILLI/MJ	258.54	
FUMARIC ACID	MILLI	17.58	-	-	-	14.94	MILLI/MJ	378.76	
ALPHA-KETOGLUT.ACID	MILLI	19.00	-	-	-	16.15	MILLI/MJ	409.35	
SALICYLIC ACID	MILLI	0.58	0.58	-	0.58	0.49	MILLI/MJ	12.50	
GLUCOSE	GRAM	0.13	0.02	-	0.15	0.11	GRAM/MJ	2.82	
FRUCTOSE	GRAM	0.12	0.04	-	0.14	0.11	GRAM/MJ	2.78	
SUCROSE	GRAM	0.20	0.01	-	0.23	0.17	GRAM/MJ	4.35	
STARCH	MILLI	90.00	-	-	-	76.50	GRAM/MJ	1.94	
PENTOSAN	GRAM	0.31	-	-	-	0.26	GRAM/MJ	6.68	
HEXOSAN	GRAM	0.28	-	-	-	0.24	GRAM/MJ	6.03	
CELLULOSE	GRAM	0.74	-	-	-	0.63	GRAM/MJ	15.94	
POLYURONIC ACID	GRAM	0.51	-	-	-	0.43	GRAM/MJ	10.99	
PALMITIC ACID	MILLI	29.00	29.00	-	29.00	24.65	MILLI/MJ	624.80	
STEARIC ACID	MILLI	1.30	1.30	-	1.30	1.11	MILLI/MJ	28.01	
PALMITOLEIC ACID	MILLI	5.10	5.10	-	5.10	4.34	MILLI/MJ	109.88	
OLEIC ACID	MILLI	10.00	10.00	-	10.00	8.50	MILLI/MJ	215.45	
LINOLEIC ACID	MILLI	28.00	28.00	-	28.00	23.80	MILLI/MJ	603.25	
LINOLENIC ACID	MILLI	134.00	134.00	-	134.00	113.90	GRAM/MJ	2.89	
TOTAL STEROLS	MILLI	9.00	-	-	-	7.65	MILLI/MJ	193.90	
TOTAL PURINES	MILLI	70.00 [3]	-	-	-	59.50	GRAM/MJ	1.51	
TOTAL PHOSPHOLIPIDS	MILLI	157.00	-	-	-	133.45	GRAM/MJ	3.38	
PHOSPHATIDYLCHOLINE	MILLI	37.00	-	-	-	31.45	MILLI/MJ	797.15	
PHOSPHATIDYLETHANOLAMINE	MILLI	36.00	-	-	-	30.60	MILLI/MJ	775.61	
PHOSPHATIDYLINOSITOL	MILLI	11.00	-	-	-	9.35	MILLI/MJ	236.99	
DIETARY FIBRE,WAT.SOL.	GRAM	0.53	-	-	-	0.45	GRAM/MJ	11.42	
DIETARY FIBRE,WAT.INS.	GRAM	1.31	-	-	-	1.11	GRAM/MJ	28.22	

3 VALUE EXPRESSED IN MG URIC ACID/100G

SPINAT
GEKOCHT, ABGETROPFT

SPINACH
BOILED, DRAINED

ÉPINARD
BOUILLI, DEGOUTTÉ

		PROTEIN	FAT	CARBOHYDRATES	TOTAL
ENERGY VALUE (AVERAGE)	KJOULE	37	12	8.4	57
PER 100 G	(KCAL)	8.9	2.8	2.0	14
EDIBLE PORTION					
AMOUNT OF DIGESTIBLE	GRAM	1.53	0.27	0.50	
CONSTITUENTS PER 100 G					
ENERGY VALUE (AVERAGE)	KJOULE	24	11	8.4	43
OF THE DIGESTIBLE	(KCAL)	5.8	2.5	2.0	10
FRACTION PER 100 G					
EDIBLE PORTION					

WASTE PERCENTAGE AVERAGE 0.00

CONSTITUENTS	DIM	AV	VARIATION			AVR	NUTR. DENS.		MOLPERC.
WATER	GRAM	92.90	-	-	-	92.90	KILO/MJ	2.16	
PROTEIN	GRAM	2.36	-	-	-	2.36	GRAM/MJ	54.96	
FAT	GRAM	0.30	-	-	-	0.30	GRAM/MJ	6.99	
AVAILABLE CARBOHYDR.	GRAM	0.50	-	-	-	0.50	GRAM/MJ	11.64	
MINERALS	GRAM	1.10	-	-	-	1.10	GRAM/MJ	25.62	
SODIUM	MILLI	46.00	-	-	-	46.00	GRAM/MJ	1.07	
CALCIUM	MILLI	126.00	-	-	-	126.00	GRAM/MJ	2.93	
IRON	MILLI	2.90	-	-	-	2.90	MILLI/MJ	67.53	
PHOSPHORUS	MILLI	41.00	-	-	-	41.00	MILLI/MJ	954.77	
VITAMIN B1	MICRO	77.00	-	-	-	77.00	MILLI/MJ	1.79	
VITAMIN B2	MILLI	0.16	-	-	-	0.16	MILLI/MJ	3.73	
NICOTINAMIDE	MILLI	0.52	-	-	-	0.52	MILLI/MJ	12.11	
VITAMIN C	MILLI	29.00	-	-	-	29.00	MILLI/MJ	675.32	
MALIC ACID	MILLI	30.00	-	-	-	30.00	MILLI/MJ	698.61	
CITRIC ACID	MILLI	20.00	-	-	-	20.00	MILLI/MJ	465.74	
GLUCOSE	GRAM	0.10	-	-	-	0.10	GRAM/MJ	2.33	
FRUCTOSE	MILLI	90.00	-	-	-	90.00	GRAM/MJ	2.10	
SUCROSE	GRAM	0.19	-	-	-	0.19	GRAM/MJ	4.42	
STARCH	MILLI	70.00	-	-	-	70.00	GRAM/MJ	1.63	

SPINAT	SPINACH	ÉPINARD
IN DOSEN	CANNED	EN BOÎTES
1	1	1

		PROTEIN	FAT	CARBOHYDRATES	TOTAL
ENERGY VALUE (AVERAGE) PER 100 G EDIBLE PORTION	KJOULE	35	16	8.9	59
	(KCAL)	8.3	3.7	2.1	14
AMOUNT OF DIGESTIBLE CONSTITUENTS PER 100 G	GRAM	1.43	0.36	0.53	
ENERGY VALUE (AVERAGE) OF THE DIGESTIBLE FRACTION PER 100 G EDIBLE PORTION	KJOULE	22	14	8.9	45
	(KCAL)	5.4	3.3	2.1	11

WASTE PERCENTAGE AVERAGE 0.00

CONSTITUENTS	DIM	AV	VARIATION	AVR	NUTR.	DENS.	MOLPERC.
WATER	GRAM	92.70	92.30 - 93.00	92.70	KILO/MJ	2.05	
PROTEIN	GRAM	2.20	2.00 - 2.30	2.20	GRAM/MJ	48.55	
FAT	GRAM	0.40	- - -	0.40	GRAM/MJ	8.83	
AVAILABLE CARBOHYDR.	GRAM	0.53 2	- - -	0.53	GRAM/MJ	11.70	
MINERALS	GRAM	1.70	1.64 - 1.80	1.70	GRAM/MJ	37.52	
SODIUM	MILLI	170.00	18.00 - 308.00	170.00	GRAM/MJ	3.75	
POTASSIUM	MILLI	213.00	135.00 - 348.00	213.00	GRAM/MJ	4.70	
MAGNESIUM	MILLI	63.00	63.00 - 63.00	63.00	GRAM/MJ	1.39	
CALCIUM	MILLI	85.00	43.00 - 163.00	85.00	GRAM/MJ	1.88	
IRON	MILLI	-	0.80 - 10.90				
PHOSPHORUS	MILLI	30.00	11.00 - 44.00	30.00	MILLI/MJ	662.04	
IODIDE	MICRO	2.00	2.00 - 2.00	2.00	MICRO/MJ	44.14	
CAROTENE	MILLI	3.30	2.10 - 3.50	3.30	MILLI/MJ	72.82	
VITAMIN E ACTIVITY	MICRO	60.00	- - -	60.00	MILLI/MJ	1.32	
ALPHA-TOCOPHEROL	MICRO	60.00	- - -	60.00	MILLI/MJ	1.32	
VITAMIN K	MILLI	0.29	0.29 - 0.29	0.29	MILLI/MJ	6.40	
VITAMIN B1	MICRO	17.00	10.00 - 20.00	17.00	MICRO/MJ	375.15	
VITAMIN B2	MICRO	60.00	42.00 - 90.00	60.00	MILLI/MJ	1.32	
NICOTINAMIDE	MILLI	0.30	0.28 - 0.33	0.30	MILLI/MJ	6.62	
PANTOTHENIC ACID	MICRO	50.00	- - -	50.00	MILLI/MJ	1.10	
VITAMIN B6	MICRO	69.00	69.00 - 69.00	69.00	MILLI/MJ	1.52	
VITAMIN C	MILLI	14.00	13.00 - 14.00	14.00	MILLI/MJ	308.95	
ARGININE	MILLI	110.00	59.00 - 160.00	110.00	GRAM/MJ	2.43	
CYSTINE	MILLI	44.00	- - -	44.00	MILLI/MJ	970.98	
HISTIDINE	MILLI	47.00	24.00 - 70.00	47.00	GRAM/MJ	1.04	
ISOLEUCINE	MILLI	100.00	54.00 - 140.00	100.00	GRAM/MJ	2.21	
LEUCINE	MILLI	170.00	95.00 - 210.00	170.00	GRAM/MJ	3.75	
LYSINE	MILLI	140.00	67.00 - 200.00	140.00	GRAM/MJ	3.09	
METHIONINE	MILLI	38.00	14.00 - 50.00	38.00	MILLI/MJ	838.58	
PHENYLALANINE	MILLI	94.00	38.00 - 130.00	94.00	GRAM/MJ	2.07	
THREONINE	MILLI	97.00	52.00 - 150.00	97.00	GRAM/MJ	2.14	
TRYPTOPHAN	MILLI	36.00	24.00 - 49.00	36.00	MILLI/MJ	794.44	
TYROSINE	MILLI	70.00	45.00 - 95.00	70.00	GRAM/MJ	1.54	
VALINE	MILLI	120.00	66.00 - 180.00	120.00	GRAM/MJ	2.65	

1 TOTAL CONTENT (SOLID AND LIQUID)

2 CALCULATED FROM THE FRESH PRODUCT

SPINATSAFT SPINACH JUICE JUS D'ÉPINARD

		PROTEIN	FAT	CARBOHYDRATES	TOTAL
ENERGY VALUE (AVERAGE)	KJOULE	22	0.0	8.4	30
PER 100 G	(KCAL)	5.3	0.0	2.0	7.3
EDIBLE PORTION					
AMOUNT OF DIGESTIBLE	GRAM	1.03	0.00	0.50	
CONSTITUENTS PER 100 G					
ENERGY VALUE (AVERAGE)	KJOULE	16	0.0	8.4	25
OF THE DIGESTIBLE	(KCAL)	3.9	0.0	2.0	5.9
FRACTION PER 100 G					
EDIBLE PORTION					

WASTE PERCENTAGE AVERAGE 0.00

CONSTITUENTS	DIM	AV	VARIATION			AVR	NUTR. DENS.		MOLPERC.
WATER	GRAM	95.40	94.90	-	95.90	95.40	KILO/MJ	3.87	
PROTEIN	GRAM	1.40	1.03	-	1.76	1.40	GRAM/MJ	56.86	
AVAILABLE CARBOHYDR.	GRAM	0.50	-	-	-	0.50	GRAM/MJ	20.31	
MINERALS	GRAM	1.47	1.39	-	1.54	1.47	GRAM/MJ	59.70	
SODIUM	MILLI	73.00	-	-	-	73.00	GRAM/MJ	2.96	
POTASSIUM	MILLI	412.00	-	-	-	412.00	GRAM/MJ	16.73	
CALCIUM	MILLI	1.00	-	-	-	1.00	MILLI/MJ	40.61	
PHOSPHORUS	MILLI	43.90	40.60	-	47.10	43.90	GRAM/MJ	1.78	
VITAMIN C	MILLI	29.00	17.00	-	40.00	29.00	GRAM/MJ	1.18	
INVERT SUGAR	GRAM	0.50	0.39	-	0.61	0.50	GRAM/MJ	20.31	
SUCROSE	-	0.00	-	-	-	0.00			
EXTRACT	GRAM	3.67	3.41	-	3.92	3.67	GRAM/MJ	149.05	

WEISSKOHL (WEISSKRAUT) — WHITE CABBAGE — CHOU BLANC

BRASSICA OLERACEA L.
V. CAPITATA L. F. ALBA

		PROTEIN	FAT	CARBOHYDRATES	TOTAL
ENERGY VALUE (AVERAGE) PER 100 G EDIBLE PORTION	KJOULE	21	7.8	76	106
	(KCAL)	5.1	1.9	18	25
AMOUNT OF DIGESTIBLE CONSTITUENTS PER 100 G	GRAM	0.89	0.18	4.57	
ENERGY VALUE (AVERAGE) OF THE DIGESTIBLE FRACTION PER 100 G EDIBLE PORTION	KJOULE	14	7.0	76	97
	(KCAL)	3.3	1.7	18	23

WASTE PERCENTAGE AVERAGE 22 MINIMUM 8.0 MAXIMUM 42

CONSTITUENTS	DIM	AV	VARIATION			AVR	NUTR.	DENS.	MOLPERC.
WATER	GRAM	92.10	91.00	-	93.00	71.84	GRAM/MJ	945.01	
PROTEIN	GRAM	1.37	1.18	-	1.50	1.07	GRAM/MJ	14.06	
FAT	GRAM	0.20	0.10	-	0.20	0.16	GRAM/MJ	2.05	
AVAILABLE CARBOHYDR.	GRAM	4.57	3.03	-	5.27	3.56	GRAM/MJ	46.89	
TOTAL DIETARY FIBRE	GRAM	2.50 1	-	-	-	1.95	GRAM/MJ	25.65	
MINERALS	GRAM	0.59	0.37	-	0.80	0.46	GRAM/MJ	6.05	
SODIUM	MILLI	13.00	6.00	-	19.00	10.14	MILLI/MJ	133.39	
POTASSIUM	MILLI	227.00	177.00	-	250.00	177.06	GRAM/MJ	2.33	
MAGNESIUM	MILLI	23.00	-	-	-	17.94	MILLI/MJ	236.00	
CALCIUM	MILLI	46.00	17.00	-	76.00	35.88	MILLI/MJ	471.99	
MANGANESE	MILLI	0.10	-	-	-	0.08	MILLI/MJ	1.03	
IRON	MILLI	0.50	-	-	-	0.39	MILLI/MJ	5.13	
COBALT	MICRO	8.00	0.50	-	24.00	6.24	MICRO/MJ	82.09	
COPPER	MICRO	60.00	-	-	-	46.80	MICRO/MJ	615.64	
ZINC	MILLI	0.21	0.16	-	1.50	0.16	MILLI/MJ	2.15	
NICKEL	MICRO	23.00	14.00	-	32.00	17.94	MICRO/MJ	236.00	
CHROMIUM	MICRO	0.50	-	-	-	0.39	MICRO/MJ	5.13	
VANADIUM	MICRO	-	10.00	-	175.00 2				
PHOSPHORUS	MILLI	27.50	21.40	-	31.00	21.45	MILLI/MJ	282.17	
CHLORIDE	MILLI	37.00	-	-	-	28.86	MILLI/MJ	379.64	
FLUORIDE	MICRO	12.00	10.00	-	300.00	9.36	MICRO/MJ	123.13	
IODIDE	MICRO	5.20	-	-	-	4.06	MICRO/MJ	53.36	
BORON	MILLI	0.60	-	-	-	0.47	MILLI/MJ	6.16	
SELENIUM	MICRO	18.00	8.00	-	20.00	14.04	MICRO/MJ	184.69	
SILICON	MILLI	0.15	0.10	-	0.20	0.12	MILLI/MJ	1.54	
NITRATE	MILLI	107.00	1.00	-	323.00	83.46	GRAM/MJ	1.10	

1 METHOD OF SOUTHGATE AND ENGLYST

2 STRONG REGIONAL DEPENDENCE

CONSTITUENTS	DIM	AV	VARIATION			AVR	NUTR. DENS.		MOLPERC.
CAROTENE	MICRO	42.00	16.00	-	66.00	32.76	MICRO/MJ	430.95	
VITAMIN E ACTIVITY	MILLI	1.70	-	-	-	1.33	MILLI/MJ	17.44	
ALPHA-TOCOPHEROL	MILLI	1.70	-	-	-	1.33	MILLI/MJ	17.44	
VITAMIN K	MILLI	-	0.04	-	0.25				
VITAMIN B1	MICRO	48.00	32.00	-	60.00	37.44	MICRO/MJ	492.51	
VITAMIN B2	MICRO	43.00	31.00	-	50.00	33.54	MICRO/MJ	441.21	
NICOTINAMIDE	MILLI	0.32	0.20	-	0.56	0.25	MILLI/MJ	3.28	
PANTOTHENIC ACID	MILLI	0.26	0.24	-	0.29	0.20	MILLI/MJ	2.67	
VITAMIN B6	MILLI	0.11	0.09	-	0.14	0.09	MILLI/MJ	1.13	
BIOTIN	MICRO	3.07	0.10	-	3.50	2.40	MICRO/MJ	31.55	
FOLIC ACID	MICRO	79.00	57.00	-	95.00	61.62	MICRO/MJ	810.59	
VITAMIN C	MILLI	45.80	30.00	-	60.00	35.72	MILLI/MJ	469.94	
ARGININE	MILLI	100.00	100.00	-	110.00	78.00	GRAM/MJ	1.03	
CYSTINE	MILLI	27.00	22.00	-	33.00	21.06	MILLI/MJ	277.04	
HISTIDINE	MILLI	25.00	21.00	-	28.00	19.50	MILLI/MJ	256.52	
ISOLEUCINE	MILLI	39.00	36.00	-	43.00	30.42	MILLI/MJ	400.17	
LEUCINE	MILLI	56.00	51.00	-	61.00	43.68	MILLI/MJ	574.60	
LYSINE	MILLI	65.00	45.00	-	96.00	50.70	MILLI/MJ	666.94	
METHIONINE	MILLI	13.00	7.00	-	18.00	10.14	MILLI/MJ	133.39	
PHENYLALANINE	MILLI	30.00	22.00	-	44.00	23.40	MILLI/MJ	307.82	
THREONINE	MILLI	38.00	36.00	-	41.00	29.64	MILLI/MJ	389.91	
VALINE	MILLI	42.00	34.00	-	54.00	32.76	MILLI/MJ	430.95	
MALIC ACID	MILLI	330.00	60.00	-	600.00	257.40	GRAM/MJ	3.39	
CITRIC ACID	MILLI	100.00	50.00	-	150.00	78.00	GRAM/MJ	1.03	
OXALIC ACID TOTAL	MILLI	-	0.00	-	13.00				
QUINIC ACID	MILLI	4.10	-	-	-	3.20	MILLI/MJ	42.07	
SUCCINIC ACID	MILLI	-	1.10	-	130.00				
GLUCOSE	GRAM	2.04	1.56	-	2.11	1.60	GRAM/MJ	21.02	
FRUCTOSE	GRAM	1.76	1.26	-	2.07	1.37	GRAM/MJ	18.07	
SUCROSE	GRAM	0.31	0.10	-	0.34	0.25	GRAM/MJ	3.27	
STARCH	MILLI	30.00	-	-	-	23.40	MILLI/MJ	307.82	
PENTOSAN	GRAM	0.37	-	-	-	0.29	GRAM/MJ	3.80	
HEXOSAN	GRAM	0.27	-	-	-	0.21	GRAM/MJ	2.77	
CELLULOSE	GRAM	0.70	-	-	-	0.55	GRAM/MJ	7.18	
POLYURONIC ACID	GRAM	1.09	-	-	-	0.85	GRAM/MJ	11.18	
MYRISTIC ACID	MILLI	0.50	-	-	-	0.39	MILLI/MJ	5.13	
PALMITIC ACID	MILLI	29.50	23.00	-	36.00	23.01	MILLI/MJ	302.69	
STEARIC ACID	MILLI	2.90	1.80	-	4.00	2.26	MILLI/MJ	29.76	
ARACHIDIC ACID	MILLI	0.50	-	-	-	0.39	MILLI/MJ	5.13	
PALMITOLEIC ACID	MILLI	2.70	1.40	-	4.00	2.11	MILLI/MJ	27.70	
OLEIC ACID	MILLI	2.10	-	-	-	1.64	MILLI/MJ	21.55	
LINOLEIC ACID	MILLI	26.50	23.00	-	30.00	20.67	MILLI/MJ	271.91	
LINOLENIC ACID	MILLI	87.00	81.00	-	93.00	67.86	MILLI/MJ	892.68	
DIETARY FIBRE,WAT.SOL.	GRAM	0.89	-	-	-	0.69	GRAM/MJ	9.13	
DIETARY FIBRE,WAT.INS.	GRAM	1.58	-	-	-	1.23	GRAM/MJ	16.21	

WIRSINGKOHL (SAVOYERKOHL) — SAVOY CABBAGE — CHOU FRISÉ

BRASSICA OLERACEA L.
VAR. SABAUDA L.

		PROTEIN	FAT	CARBOHYDRATES	TOTAL
ENERGY VALUE (AVERAGE) PER 100 G EDIBLE PORTION	KJOULE	46	15	40	101
	(KCAL)	11	3.5	9.6	24
AMOUNT OF DIGESTIBLE CONSTITUENTS PER 100 G	GRAM	1.91	0.34	2.41	
ENERGY VALUE (AVERAGE) OF THE DIGESTIBLE FRACTION PER 100 G EDIBLE PORTION	KJOULE	30	13	40	84
	(KCAL)	7.2	3.2	9.6	20

WASTE PERCENTAGE AVERAGE 28 MINIMUM 18 MAXIMUM 40

CONSTITUENTS	DIM	AV		VARIATION			AVR	NUTR. DENS.		MOLPERC.
WATER	GRAM	90.00		88.00	-	91.80	64.80	KILO/MJ	1.07	
PROTEIN	GRAM	2.95		2.56	-	3.35	2.12	GRAM/MJ	35.23	
FAT	GRAM	0.38		0.27	-	0.60	0.27	GRAM/MJ	4.54	
AVAILABLE CARBOHYDR.	GRAM	2.41		-	-	-	1.74	GRAM/MJ	28.78	
TOTAL DIETARY FIBRE	GRAM	1.50	1	-	-	-	1.08	GRAM/MJ	17.92	
MINERALS	GRAM	1.10		0.96	-	1.20	0.79	GRAM/MJ	13.14	
SODIUM	MILLI	9.00		3.00	-	15.00	6.48	MILLI/MJ	107.49	
POTASSIUM	MILLI	282.00		237.00	-	350.00	203.04	GRAM/MJ	3.37	
MAGNESIUM	MILLI	12.00		-	-	-	8.64	MILLI/MJ	143.32	
CALCIUM	MILLI	47.00		41.00	-	55.00	33.84	MILLI/MJ	561.35	
MANGANESE	MILLI	0.20		-	-	-	0.14	MILLI/MJ	2.39	
IRON	MILLI	0.90		-	-	-	0.65	MILLI/MJ	10.75	
COBALT	MICRO	1.10		-	-	-	0.79	MICRO/MJ	13.14	
COPPER	MICRO	70.00		-	-	-	50.40	MICRO/MJ	836.05	
ZINC	MILLI	0.30		-	-	-	0.22	MILLI/MJ	3.58	
NICKEL	MILLI	0.16		-	-	-	0.12	MILLI/MJ	1.91	
PHOSPHORUS	MILLI	55.60		46.00	-	67.30	40.03	MILLI/MJ	664.06	
CHLORIDE	MILLI	22.00		-	-	-	15.84	MILLI/MJ	262.76	
BORON	MILLI	0.30		0.19	-	0.40	0.22	MILLI/MJ	3.58	
SELENIUM	MICRO	2.50	2	2.00	-	8.00	1.80	MICRO/MJ	29.86	
NITRATE	MILLI	48.00		5.00	-	165.00	34.56	MILLI/MJ	573.29	
CAROTENE	MICRO	39.00		-	-	-	28.08	MICRO/MJ	465.80	
VITAMIN E ACTIVITY	MILLI	2.50		2.00	-	3.00	1.80	MILLI/MJ	29.86	
ALPHA-TOCOPHEROL	MILLI	2.50		0.20	-	3.00	1.80	MILLI/MJ	29.86	
VITAMIN B1	MICRO	50.00		40.00	-	50.00	36.00	MICRO/MJ	597.18	
VITAMIN B2	MICRO	57.00		40.00	-	80.00	41.04	MICRO/MJ	680.78	

1 METHOD OF MEUSER, SUCKOW AND KULIKOWSKI ("BERLINER METHODE")

2 AMOUNTS, DETERMINED IN THE FEDERAL REPUBLIC OF GERMANY

CONSTITUENTS	DIM	AV	VARIATION			AVR	NUTR. DENS.		MOLPERC.
NICOTINAMIDE	MILLI	0.33	0.20	-	0.50	0.24	MILLI/MJ	3.94	
PANTOTHENIC ACID	MILLI	0.21	-	-	-	0.15	MILLI/MJ	2.51	
VITAMIN B6	MILLI	0.20	0.10	-	0.30	0.14	MILLI/MJ	2.39	
BIOTIN	MICRO	0.10	-	-	-	0.07	MICRO/MJ	1.19	
FOLIC ACID	MICRO	90.00	-	-	-	64.80	MILLI/MJ	1.07	
VITAMIN C	MILLI	45.00	40.00	-	50.00	32.40	MILLI/MJ	537.46	
ARGININE	GRAM	0.15	-	-	-	0.11	GRAM/MJ	1.79	
HISTIDINE	MILLI	47.00	-	-	-	33.84	MILLI/MJ	561.35	
ISOLEUCINE	GRAM	0.10	-	-	-	0.07	GRAM/MJ	1.19	
LEUCINE	GRAM	0.19	-	-	-	0.14	GRAM/MJ	2.27	
LYSINE	MILLI	92.00	-	-	-	66.24	GRAM/MJ	1.10	
METHIONINE	MILLI	27.00	-	-	-	19.44	MILLI/MJ	322.48	
PHENYLALANINE	GRAM	0.12	-	-	-	0.09	GRAM/MJ	1.43	
THREONINE	GRAM	0.11	-	-	-	0.08	GRAM/MJ	1.31	
TRYPTOPHAN	MILLI	32.00	-	-	-	23.04	MILLI/MJ	382.19	
VALINE	GRAM	0.14	-	-	-	0.10	GRAM/MJ	1.67	
OXALIC ACID TOTAL	MILLI	4.90	-	-	-	3.53	MILLI/MJ	58.52	
OXALIC ACID SOLUBLE	MILLI	3.30	-	-	-	2.38	MILLI/MJ	39.41	
CAFFEIC ACID	MILLI	1.00	0.50	-	2.00	0.72	MILLI/MJ	11.94	
PARA-COUMARIC ACID	MILLI	2.00	1.00	-	4.00	1.44	MILLI/MJ	23.89	
GLUCOSE	GRAM	0.81	0.81	-	0.81	0.58	GRAM/MJ	9.67	
FRUCTOSE	GRAM	0.90	0.90	-	0.90	0.65	GRAM/MJ	10.75	
SUCROSE	GRAM	0.70	0.70	-	0.70	0.50	GRAM/MJ	8.36	
TOTAL PURINES	MILLI	20.00 3	-	-	-	14.40	MILLI/MJ	238.87	
DIETARY FIBRE,WAT.SOL.	GRAM	0.65	-	-	-	0.47	GRAM/MJ	7.76	
DIETARY FIBRE,WAT.INS.	GRAM	0.84	-	-	-	0.60	GRAM/MJ	10.03	

3 VALUE EXPRESSED IN MG URIC ACID/100G

ZWIEBEL ONION OIGNON

ALLIUM CEPA L.

		PROTEIN	FAT	CARBOHYDRATES	TOTAL
ENERGY VALUE (AVERAGE) PER 100 G EDIBLE PORTION	KJOULE	20	9.7	97	126
	(KCAL)	4.7	2.3	23	30
AMOUNT OF DIGESTIBLE CONSTITUENTS PER 100 G	GRAM	0.92	0.22	5.79	
ENERGY VALUE (AVERAGE) OF THE DIGESTIBLE FRACTION PER 100 G EDIBLE PORTION	KJOULE	15	8.8	97	120
	(KCAL)	3.5	2.1	23	29

WASTE PERCENTAGE AVERAGE 8.0 MINIMUM 3.0 MAXIMUM 14

CONSTITUENTS	DIM	AV	VARIATION			AVR	NUTR. DENS.		MOLPERC.
WATER	GRAM	87.60	86.00	-	89.00	80.59	GRAM/MJ	728.97	
PROTEIN	GRAM	1.25	1.00	-	1.40	1.15	GRAM/MJ	10.40	
FAT	GRAM	0.25	0.10	-	0.40	0.23	GRAM/MJ	2.08	
AVAILABLE CARBOHYDR.	GRAM	5.79	-	-	-	5.33	GRAM/MJ	48.18	
TOTAL DIETARY FIBRE	GRAM	3.05 [1]	-	-	-	2.81	GRAM/MJ	25.38	
MINERALS	GRAM	0.59	0.57	-	0.60	0.54	GRAM/MJ	4.91	
SODIUM	MILLI	9.00	7.00	-	10.00	8.28	MILLI/MJ	74.89	
POTASSIUM	MILLI	175.00	149.00	-	200.00	161.00	GRAM/MJ	1.46	
MAGNESIUM	MILLI	11.00	9.00	-	16.00	10.12	MILLI/MJ	91.54	
CALCIUM	MILLI	31.00	30.00	-	32.00	28.52	MILLI/MJ	257.97	
MANGANESE	MILLI	0.23	0.05	-	0.36	0.21	MILLI/MJ	1.91	
IRON	MILLI	0.50	-	-	-	0.46	MILLI/MJ	4.16	
COBALT	MICRO	13.00	-	-	-	11.96	MICRO/MJ	108.18	
COPPER	MICRO	80.00	-	-	-	73.60	MICRO/MJ	665.73	
ZINC	MILLI	1.40	-	-	-	1.29	MILLI/MJ	11.65	
NICKEL	MICRO	21.00	6.00	-	35.00	19.32	MICRO/MJ	174.75	
CHROMIUM	MICRO	15.50	1.00	-	30.00	14.26	MICRO/MJ	128.98	
MOLYBDENUM	MILLI	-	0.01	-	0.05				
VANADIUM	MICRO	5.00	4.00	-	6.00	4.60	MICRO/MJ	41.61	
PHOSPHORUS	MILLI	42.00	40.00	-	44.00	38.64	MILLI/MJ	349.51	
FLUORIDE	MICRO	42.00	24.00	-	60.00	38.64	MICRO/MJ	349.51	
IODIDE	MICRO	2.00	-	-	-	1.84	MICRO/MJ	16.64	
BORON	MILLI	0.17	0.13	-	0.27	0.16	MILLI/MJ	1.41	
SELENIUM	MICRO	1.20	1.00	-	10.00	1.10	MICRO/MJ	9.99	
NITRATE	MILLI	20.00	0.00	-	225.00	18.40	MILLI/MJ	166.43	
CAROTENE	MICRO	30.00	10.00	-	49.00	27.60	MICRO/MJ	249.65	

1 METHOD OF SOUTHGATE AND ENGLYST

CONSTITUENTS	DIM	AV	VARIATION			AVR	NUTR. DENS.		MOLPERC.
VITAMIN E ACTIVITY	MILLI	0.14	-	-	-	0.13	MILLI/MJ	1.17	
TOTAL TOCOPHEROLS	MILLI	0.30	-	-	-	0.28	MILLI/MJ	2.50	
ALPHA-TOCOPHEROL	MILLI	0.12	-	-	-	0.11	MILLI/MJ	1.00	
DELTA-TOCOPHEROL	MILLI	0.18	-	-	-	0.17	MILLI/MJ	1.50	
VITAMIN B1	MICRO	33.00	30.00	-	40.00	30.36	MICRO/MJ	274.61	
VITAMIN B2	MICRO	28.00	20.00	-	40.00	25.76	MICRO/MJ	233.00	
NICOTINAMIDE	MILLI	0.20	0.11	-	0.32	0.18	MILLI/MJ	1.66	
PANTOTHENIC ACID	MILLI	0.17	-	-	-	0.16	MILLI/MJ	1.41	
VITAMIN B6	MILLI	0.13	0.10	-	0.17	0.12	MILLI/MJ	1.08	
BIOTIN	MICRO	3.50	-	-	-	3.22	MICRO/MJ	29.13	
FOLIC ACID	MICRO	7.00	6.10	-	8.10	6.44	MICRO/MJ	58.25	
VITAMIN C	MILLI	8.50	6.00	-	8.01	7.82	MILLI/MJ	70.73	
ARGININE	GRAM	0.16	-	-	-	0.15	GRAM/MJ	1.33	
HISTIDINE	MILLI	13.00	-	-	-	11.96	MILLI/MJ	108.18	
ISOLEUCINE	MILLI	19.00	-	-	-	17.48	MILLI/MJ	158.11	
LEUCINE	MILLI	33.00	-	-	-	30.36	MILLI/MJ	274.61	
LYSINE	MILLI	57.00	-	-	-	52.44	MILLI/MJ	474.33	
METHIONINE	MILLI	12.00	-	-	-	11.04	MILLI/MJ	99.86	
PHENYLALANINE	MILLI	35.00	-	-	-	32.20	MILLI/MJ	291.25	
THREONINE	MILLI	20.00	-	-	-	18.40	MILLI/MJ	166.43	
TRYPTOPHAN	MILLI	19.00	-	-	-	17.48	MILLI/MJ	158.11	
TYROSINE	MILLI	41.00	-	-	-	37.72	MILLI/MJ	341.18	
VALINE	MILLI	28.00	-	-	-	25.76	MILLI/MJ	233.00	
MALIC ACID	MILLI	170.00	-	-	-	156.40	GRAM/MJ	1.41	
CITRIC ACID	MILLI	20.00	-	-	-	18.40	MILLI/MJ	166.43	
OXALIC ACID TOTAL	MILLI	5.50	-	-	-	5.06	MILLI/MJ	45.77	
OXALIC ACID SOLUBLE	MILLI	3.90	-	-	-	3.59	MILLI/MJ	32.45	
SALICYLIC ACID	MILLI	0.16	0.16	-	0.16	0.15	MILLI/MJ	1.33	
GLUCOSE	GRAM	1.87	1.85	-	2.46	1.73	GRAM/MJ	15.64	
FRUCTOSE	GRAM	1.60	1.59	-	2.01	1.47	GRAM/MJ	13.34	
SUCROSE	GRAM	2.13	2.10	-	2.14	1.97	GRAM/MJ	17.80	
PENTOSAN	GRAM	0.16	-	-	-	0.15	GRAM/MJ	1.33	
HEXOSAN	GRAM	1.11	-	-	-	1.02	GRAM/MJ	9.24	
CELLULOSE	GRAM	0.86	-	-	-	0.79	GRAM/MJ	7.16	
POLYURONIC ACID	GRAM	0.85	-	-	-	0.78	GRAM/MJ	7.07	
PALMITIC ACID	MILLI	67.00	67.00	-	67.00	61.64	MILLI/MJ	557.54	
STEARIC ACID	MILLI	27.00	27.00	-	27.00	24.84	MILLI/MJ	224.68	
LINOLEIC ACID	MILLI	93.00	93.00	-	93.00	85.56	MILLI/MJ	773.91	
LINOLENIC ACID	MILLI	13.00	13.00	-	13.00	11.96	MILLI/MJ	108.18	
TOTAL STEROLS	MILLI	15.00	-	-	-	13.80	MILLI/MJ	124.82	
CAMPESTEROL	MILLI	1.00	-	-	-	0.92	MILLI/MJ	8.32	
BETA-SITOSTEROL	MILLI	12.00	-	-	-	11.04	MILLI/MJ	99.86	
TOTAL PURINES	MILLI	9.00 [2]	-	-	-	8.28	MILLI/MJ	74.89	
ALLYLISOCYANATE	MILLI	57.20	-	-	-	52.62	MILLI/MJ	475.99	
DIETARY FIBRE,WAT.SOL.	GRAM	0.91	-	-	-	0.84	GRAM/MJ	7.57	
DIETARY FIBRE,WAT.INS.	GRAM	2.14	-	-	-	1.97	GRAM/MJ	17.81	

2 VALUE EXPRESSED IN MG URIC ACID/100G

ZWIEBEL GETROCKNET / ONIONS DRIED / OIGNONS SÉCHÉS

		PROTEIN	FAT	CARBOHYDRATES	TOTAL
ENERGY VALUE (AVERAGE)	KJOULE	165	36	697	898
PER 100 G EDIBLE PORTION	(KCAL)	39	8.6	167	215
AMOUNT OF DIGESTIBLE CONSTITUENTS PER 100 G	GRAM	6.82	0.83	41.65	
ENERGY VALUE (AVERAGE)	KJOULE	107	33	697	837
OF THE DIGESTIBLE FRACTION PER 100 G EDIBLE PORTION	(KCAL)	26	7.8	167	200

WASTE PERCENTAGE AVERAGE 0.00

CONSTITUENTS	DIM	AV	VARIATION			AVR	NUTR. DENS.		MOLPERC.
WATER	GRAM	10.80	4.00	-	15.40	10.80	GRAM/MJ	12.91	
PROTEIN	GRAM	10.50	9.40	-	11.80	10.50	GRAM/MJ	12.55	
FAT	GRAM	0.93	0.70	-	1.10	0.93	GRAM/MJ	1.11	
AVAILABLE CARBOHYDR.	GRAM	41.65 1	-	-	-	41.65	GRAM/MJ	49.78	
TOTAL DIETARY FIBRE	GRAM	21.90 1	-	-	-	21.90	GRAM/MJ	26.17	
MINERALS	GRAM	4.20	3.90	-	4.50	4.20	GRAM/MJ	5.02	
SODIUM	MILLI	105.00	26.00	-	298.00	105.00	MILLI/MJ	125.49	
POTASSIUM	GRAM	1.04	0.79	-	1.34	1.04	GRAM/MJ	1.24	
CALCIUM	MILLI	162.00	158.00	-	168.00	162.00	MILLI/MJ	193.62	
IRON	MILLI	3.25	3.10	-	3.40	3.25	MILLI/MJ	3.88	
PHOSPHORUS	MILLI	243.00	200.00	-	273.00	243.00	MILLI/MJ	290.42	
CAROTENE	MILLI	0.26	0.00	-	0.78	0.26	MILLI/MJ	0.31	
VITAMIN B1	MILLI	0.26	0.23	-	0.30	0.26	MILLI/MJ	0.31	
VITAMIN B2	MILLI	0.18	0.15	-	0.20	0.18	MILLI/MJ	0.22	
NICOTINAMIDE	MILLI	1.07	0.70	-	1.10	1.07	MILLI/MJ	1.28	
PANTOTHENIC ACID	MILLI	1.05	-	-	-	1.05	MILLI/MJ	1.25	
VITAMIN B6	MILLI	0.50	-	-	-	0.50	MILLI/MJ	0.60	
BIOTIN	MICRO	28.00	-	-	-	28.00	MICRO/MJ	33.46	
FOLIC ACID	MILLI	0.11	-	-	-	0.11	MILLI/MJ	0.13	
VITAMIN C	MILLI	42.00	36.00	-	52.00	42.00	MILLI/MJ	50.20	

1 CALCULATED FROM THE FRESH PRODUCT

Gemüsefrüchte

VEGETABLE FRUITS **FRUITS DE LEGUMES**

AUBERGINE (EIERFRUCHT) | AUBERGINE (EGG PLANT) | AUBERGINE

SOLANUM MELONGENA L.

		PROTEIN	FAT	CARBOHYDRATES	TOTAL
ENERGY VALUE (AVERAGE) PER 100 G EDIBLE PORTION	KJOULE	19	7.0	45	71
	(KCAL)	4.7	1.7	11	17
AMOUNT OF DIGESTIBLE CONSTITUENTS PER 100 G	GRAM	0.80	0.16	2.66	
ENERGY VALUE (AVERAGE) OF THE DIGESTIBLE FRACTION PER 100 G EDIBLE PORTION	KJOULE	13	6.3	45	63
	(KCAL)	3.0	1.5	11	15

WASTE PERCENTAGE AVERAGE 17 MINIMUM 13 MAXIMUM 23

CONSTITUENTS	DIM	AV	VARIATION		AVR	NUTR. DENS.		MOLPERC.
WATER	GRAM	92.60	92.00	- 93.40	76.86	KILO/MJ	1.46	
PROTEIN	GRAM	1.24	0.70	- 2.30	1.03	GRAM/MJ	19.54	
FAT	GRAM	0.18	0.10	- 0.20	0.15	GRAM/MJ	2.84	
AVAILABLE CARBOHYDR.	GRAM	2.66	-	- -	2.21	GRAM/MJ	41.91	
TOTAL DIETARY FIBRE	GRAM	1.37 1	-	- -	1.14	GRAM/MJ	21.59	
MINERALS	GRAM	0.50	-	- -	0.42	GRAM/MJ	7.88	
SODIUM	MILLI	3.50	-	- -	2.91	MILLI/MJ	55.15	
POTASSIUM	MILLI	266.00	238.00	- 294.00	220.78	GRAM/MJ	4.19	
MAGNESIUM	MILLI	10.80	9.50	- 12.00	8.96	MILLI/MJ	170.17	
CALCIUM	MILLI	13.10	9.00	- 22.00	10.87	MILLI/MJ	206.40	
MANGANESE	MILLI	0.19	-	- -	0.16	MILLI/MJ	2.99	
IRON	MILLI	0.42	0.39	- 0.50	0.35	MILLI/MJ	6.62	
COPPER	MICRO	90.00	80.00	- 100.00	74.70	MILLI/MJ	1.42	
ZINC	MILLI	0.28	-	- -	0.23	MILLI/MJ	4.41	
NICKEL	MICRO	11.00	4.00	- 19.00	9.13	MICRO/MJ	173.32	
PHOSPHORUS	MILLI	21.40	12.00	- 31.00	17.76	MILLI/MJ	337.18	
CHLORIDE	MILLI	55.00	50.00	- 61.00	45.65	MILLI/MJ	866.59	
IODIDE	MICRO	0.80	-	- -	0.66	MICRO/MJ	12.60	
BORON	MILLI	0.19	0.09	- 0.29	0.16	MILLI/MJ	2.99	
NITRATE	MILLI	20.00	12.00	- 30.00	16.60	MILLI/MJ	315.12	

1 METHOD OF SOUTHGATE

CONSTITUENTS	DIM	AV	VARIATION			AVR	NUTR. DENS.		MOLPERC.
CAROTENE	MICRO	31.00	18.00	-	35.00	25.73	MICRO/MJ	488.44	
VITAMIN E ACTIVITY	MICRO	30.00	-	-	-	24.90	MICRO/MJ	472.68	
ALPHA-TOCOPHEROL	MICRO	30.00	-	-	-	24.90	MICRO/MJ	472.68	
VITAMIN B1	MICRO	40.00	30.00	-	60.00	33.20	MICRO/MJ	630.24	
VITAMIN B2	MICRO	50.00	40.00	-	60.00	41.50	MICRO/MJ	787.80	
NICOTINAMIDE	MILLI	0.60	0.50	-	0.80	0.50	MILLI/MJ	9.45	
PANTOTHENIC ACID	MILLI	0.23	-	-	-	0.19	MILLI/MJ	3.62	
VITAMIN B6	MICRO	90.00	-	-	-	74.70	MILLI/MJ	1.42	
FOLIC ACID	MICRO	31.00	-	-	-	25.73	MICRO/MJ	488.44	
VITAMIN C	MILLI	5.00	3.00	-	8.00	4.15	MILLI/MJ	78.78	
ARGININE	MILLI	42.00	-	-	-	34.86	MILLI/MJ	661.76	
HISTIDINE	MILLI	21.00	-	-	-	17.43	MILLI/MJ	330.88	
ISOLEUCINE	MILLI	63.00	-	-	-	52.29	MILLI/MJ	992.63	
LEUCINE	MILLI	77.00	-	-	-	63.91	GRAM/MJ	1.21	
LYSINE	MILLI	34.00	-	-	-	28.22	MILLI/MJ	535.71	
METHIONINE	MILLI	7.00	-	-	-	5.81	MILLI/MJ	110.29	
PHENYLALANINE	MILLI	54.00	-	-	-	44.82	MILLI/MJ	850.83	
THREONINE	MILLI	43.00	-	-	-	35.69	MILLI/MJ	677.51	
TRYPTOPHAN	MILLI	11.00	-	-	-	9.13	MILLI/MJ	173.32	
VALINE	MILLI	73.00	-	-	-	60.59	GRAM/MJ	1.15	
MALIC ACID	MILLI	170.00	-	-	-	141.10	GRAM/MJ	2.68	
CITRIC ACID	MILLI	10.00	-	-	-	8.30	MILLI/MJ	157.56	
OXALIC ACID TOTAL	MILLI	9.50	-	-	-	7.89	MILLI/MJ	149.68	
QUINIC ACID	MILLI	41.00	-	-	-	34.03	MILLI/MJ	646.00	
CAFFEIC ACID	MILLI	40.00	36.00	-	44.00	33.20	MILLI/MJ	630.24	
SALICYLIC ACID	MILLI	0.30	0.30	-	0.30	0.25	MILLI/MJ	4.73	
GLUCOSE	GRAM	1.03	0.73	-	1.57	0.86	GRAM/MJ	16.34	
FRUCTOSE	GRAM	1.03	0.78	-	1.57	0.85	GRAM/MJ	16.23	
SUCROSE	GRAM	0.20	0.11	-	0.28	0.17	GRAM/MJ	3.15	
STARCH	GRAM	0.22	0.22	-	0.22	0.18	GRAM/MJ	3.47	
CELLULOSE	GRAM	0.91	-	-	-	0.76	GRAM/MJ	14.34	
TOTAL STEROLS	MILLI	7.00	-	-	-	5.81	MILLI/MJ	110.29	
BETA-SITOSTEROL	MILLI	3.00	-	-	-	2.49	MILLI/MJ	47.27	
STIGMASTEROL	MILLI	2.00	-	-	-	1.66	MILLI/MJ	31.51	
DIETARY FIBRE, WAT.SOL.	MILLI	90.00	-	-	-	74.70	GRAM/MJ	1.42	
DIETARY FIBRE, WAT.INS.	GRAM	1.28	-	-	-	1.06	GRAM/MJ	20.17	

BOHNEN (SCHNITTBOHNEN) GRÜN	**BEANS** (STRING BEANS, FRENCH)	**HARICOTS** VERTS

PHASEOLUS VULGARIS L.

		PROTEIN	FAT	CARBOHYDRATES	TOTAL
ENERGY VALUE (AVERAGE) PER 100 G EDIBLE PORTION	KJOULE	44	9.3	89	142
	(KCAL)	11	2.2	21	34
AMOUNT OF DIGESTIBLE CONSTITUENTS PER 100 G	GRAM	1.86	0.21	5.29	
ENERGY VALUE (AVERAGE) OF THE DIGESTIBLE FRACTION PER 100 G EDIBLE PORTION	KJOULE	35	8.4	89	132
	(KCAL)	8.3	2.0	21	31

WASTE PERCENTAGE AVERAGE 6.0 MINIMUM 3.0 MAXIMUM 10

CONSTITUENTS	DIM	AV	VARIATION		AVR	NUTR. DENS.		MOLPERC.
WATER	GRAM	90.30	88.90	91.50	84.88	GRAM/MJ	685.92	
PROTEIN	GRAM	2.39	1.97	3.00	2.25	GRAM/MJ	18.15	
FAT	GRAM	0.24	0.19	0.40	0.23	GRAM/MJ	1.82	
AVAILABLE CARBOHYDR.	GRAM	5.29	3.81	6.22	4.97	GRAM/MJ	40.18	
TOTAL DIETARY FIBRE	GRAM	1.89 [1]	-	-	1.76	GRAM/MJ	14.32	
MINERALS	GRAM	0.72	0.66	0.80	0.68	GRAM/MJ	5.47	
SODIUM	MILLI	2.40	1.30	4.80	2.26	MILLI/MJ	18.23	
POTASSIUM	MILLI	248.00	200.00	300.00	233.12	GRAM/MJ	1.88	
MAGNESIUM	MILLI	25.00	16.00	31.00	23.50	MILLI/MJ	189.90	
CALCIUM	MILLI	57.00	40.00	82.00	53.58	MILLI/MJ	432.97	
IRON	MILLI	0.83	0.50	1.10	0.78	MILLI/MJ	6.30	
COBALT	MICRO	0.80	0.30	1.40	0.75	MICRO/MJ	6.08	
COPPER	MILLI	0.14	0.10	0.18	0.13	MILLI/MJ	1.06	
ZINC	MILLI	0.18	0.12	0.31	0.17	MILLI/MJ	1.37	
NICKEL	MICRO	10.00	5.00	21.00	9.40	MICRO/MJ	75.96	
MOLYBDENUM	MICRO	43.00	10.00	70.00	40.42	MICRO/MJ	326.63	
VANADIUM	MICRO	15.00	9.00	87.00	14.10	MICRO/MJ	113.94	
PHOSPHORUS	MILLI	37.80	23.00	44.00	35.53	MILLI/MJ	287.13	
CHLORIDE	MILLI	18.60	15.00	25.00	17.48	MILLI/MJ	141.29	
FLUORIDE	MICRO	12.00	11.00	15.00	11.28	MICRO/MJ	91.15	
IODIDE	MICRO	3.00	-	-	2.82	MICRO/MJ	22.79	
BORON	MILLI	0.15	0.06	0.25	0.15	MILLI/MJ	1.18	
SELENIUM	MICRO	0.60	0.00	2.00	0.56	MICRO/MJ	4.56	
SILICON	MILLI	10.00	8.00	12.00	9.40	MILLI/MJ	75.96	
NITRATE	MILLI	25.00	9.00	110.00	23.50	MILLI/MJ	189.90	
CAROTENE	MILLI	0.33	0.20	0.50	0.31	MILLI/MJ	2.51	

1 METHOD OF ENGLYST

CONSTITUENTS	DIM	AV	VARIATION			AVR	NUTR. DENS.		MOLPERC.
VITAMIN E ACTIVITY	MICRO	70.00	-	-	-	65.80	MICRO/MJ	531.72	
TOTAL TOCOPHEROLS	MILLI	0.28	-	-	-	0.26	MILLI/MJ	2.13	
ALPHA-TOCOPHEROL	MICRO	50.00	-	-	-	47.00	MICRO/MJ	379.80	
GAMMA-TOCOPHEROL	MILLI	0.23	-	-	-	0.22	MILLI/MJ	1.75	
VITAMIN K	MICRO	22.00	10.00	-	50.00	20.68	MICRO/MJ	167.11	
VITAMIN B1	MICRO	81.00	60.00	-	110.00	76.14	MICRO/MJ	615.28	
VITAMIN B2	MILLI	0.12	0.09	-	0.20	0.11	MILLI/MJ	0.91	
NICOTINAMIDE	MILLI	0.57	0.40	-	0.70	0.54	MILLI/MJ	4.33	
PANTOTHENIC ACID	MILLI	0.50	0.20	-	0.80	0.47	MILLI/MJ	3.80	
VITAMIN B6	MILLI	0.28	0.10	-	0.55	0.26	MILLI/MJ	2.13	
BIOTIN	MICRO	7.00	-	-	-	6.58	MICRO/MJ	53.17	
FOLIC ACID	MICRO	44.00	-	-	-	41.36	MICRO/MJ	334.23	
VITAMIN C	MILLI	20.00	10.00	-	27.40	18.80	MILLI/MJ	151.92	
ARGININE	GRAM	0.10	-	-	-	0.09	GRAM/MJ	0.76	
CYSTINE	MILLI	24.00	-	-	-	22.56	MILLI/MJ	182.31	
HISTIDINE	MILLI	49.00	45.00	-	53.00	46.06	MILLI/MJ	372.21	
ISOLEUCINE	GRAM	0.11	0.10	-	0.13	0.10	GRAM/MJ	0.84	
LEUCINE	GRAM	0.14	0.13	-	0.15	0.13	GRAM/MJ	1.06	
LYSINE	GRAM	0.14	0.13	-	0.16	0.13	GRAM/MJ	1.06	
METHIONINE	MILLI	34.00	28.00	-	38.00	31.96	MILLI/MJ	258.27	
PHENYLALANINE	MILLI	73.00	57.00	-	83.00	68.62	MILLI/MJ	554.51	
THREONINE	MILLI	93.00	91.00	-	98.00	87.42	MILLI/MJ	706.43	
TRYPTOPHAN	MILLI	27.00	22.00	-	33.00	25.38	MILLI/MJ	205.09	
TYROSINE	MILLI	50.00	26.00	-	75.00	47.00	MILLI/MJ	379.80	
VALINE	GRAM	0.13	0.11	-	0.15	0.12	GRAM/MJ	0.99	
MALIC ACID	MILLI	177.00	167.00	-	188.00	166.38	GRAM/MJ	1.34	
CITRIC ACID	MILLI	23.00	14.00	-	32.00	21.62	MILLI/MJ	174.71	
OXALIC ACID TOTAL	MILLI	43.50	-	-	-	40.89	MILLI/MJ	330.43	
OXALIC ACID SOLUBLE	MILLI	9.60	-	-	-	9.02	MILLI/MJ	72.92	
QUINIC ACID	MILLI	6.00	3.00	-	8.00	5.64	MILLI/MJ	45.58	
SUCCINIC ACID	MILLI	3.00	-	-	-	2.82	MILLI/MJ	22.79	
FUMARIC ACID	MILLI	1.00	-	-	-	0.94	MILLI/MJ	7.60	
ISOCITRIC ACID	MILLI	1.50	1.00	-	2.00	1.41	MILLI/MJ	11.39	
OXALOACETIC ACID	MILLI	2.00	1.00	-	3.00	1.88	MILLI/MJ	15.19	
ALPHA-KETOGLUT.ACID	MILLI	2.00	1.00	-	3.00	1.88	MILLI/MJ	15.19	
SALICYLIC ACID	MILLI	0.11	0.11	-	0.11	0.10	MILLI/MJ	0.84	
GLUCOSE	GRAM	0.96	0.45	-	1.57	0.90	GRAM/MJ	7.30	
FRUCTOSE	GRAM	1.31	0.68	-	1.35	1.23	GRAM/MJ	9.97	
SUCROSE	GRAM	0.42	0.25	-	0.53	0.40	GRAM/MJ	3.20	
VERBASCOSE	GRAM	1.60	0.60	-	3.10	1.50	GRAM/MJ	12.15	
STARCH	GRAM	2.40	2.25	-	2.55	2.26	GRAM/MJ	18.23	
PENTOSAN	GRAM	0.50	-	-	-	0.47	GRAM/MJ	3.80	
HEXOSAN	GRAM	1.39	-	-	-	1.31	GRAM/MJ	10.56	
CELLULOSE	GRAM	1.45	-	-	-	1.36	GRAM/MJ	11.01	
POLYURONIC ACID	GRAM	0.50	-	-	-	0.47	GRAM/MJ	3.80	
MYOINOSITOL	GRAM	0.20	-	-	-	0.19	GRAM/MJ	1.52	
PALMITIC ACID	MILLI	62.00	62.00	-	62.00	58.28	MILLI/MJ	470.95	
STEARIC ACID	MILLI	8.80	8.80	-	8.80	8.27	MILLI/MJ	66.85	
OLEIC ACID	MILLI	7.00	7.00	-	7.00	6.58	MILLI/MJ	53.17	
LINOLEIC ACID	MILLI	53.00	53.00	-	53.00	49.82	MILLI/MJ	402.59	
LINOLENIC ACID	MILLI	62.00	62.00	-	62.00	58.28	MILLI/MJ	470.95	
TOTAL PURINES	MILLI	25.00 [2]	20.00	-	30.00	23.50	MILLI/MJ	189.90	
DIETARY FIBRE,WAT.SOL.	GRAM	0.88	-	-	-	0.83	GRAM/MJ	6.68	
DIETARY FIBRE,WAT.INS.	GRAM	1.01	-	-	-	0.95	GRAM/MJ	7.67	

2 VALUE EXPRESSED IN MG URIC ACID/100G

BOHNEN	BEANS	HARICOTS
GRÜN	FRENCH	VERTS
IN DOSEN	CANNED	EN BOÎTES
1	1	1

		PROTEIN	FAT	CARBOHYDRATES	TOTAL
ENERGY VALUE (AVERAGE) PER 100 G EDIBLE PORTION	KJOULE	22	3.9	66	92
	(KCAL)	5.3	0.9	16	22
AMOUNT OF DIGESTIBLE CONSTITUENTS PER 100 G	GRAM	0.93	0.09	3.92	
ENERGY VALUE (AVERAGE) OF THE DIGESTIBLE FRACTION PER 100 G EDIBLE PORTION	KJOULE	17	3.5	66	87
	(KCAL)	4.2	0.8	16	21

WASTE PERCENTAGE AVERAGE 0.00

CONSTITUENTS	DIM	AV	VARIATION		AVR	NUTR. DENS.		MOLPERC.
WATER	GRAM	92.80	91.40	- 94.00	92.80	KILO/MJ	1.07	
PROTEIN	GRAM	1.20	1.00	- 1.90	1.20	GRAM/MJ	13.87	
FAT	GRAM	0.10	0.10	- 0.20	0.10	GRAM/MJ	1.16	
AVAILABLE CARBOHYDR.	GRAM	3.92 2	-	- -	3.92	GRAM/MJ	45.30	
MINERALS	GRAM	0.97	0.60	- 1.23	0.97	GRAM/MJ	11.21	
SODIUM	MILLI	275.00	162.00	- 410.00	275.00	GRAM/MJ	3.18	
POTASSIUM	MILLI	148.00	104.00	- 219.00	148.00	GRAM/MJ	1.71	
MAGNESIUM	MILLI	20.00	13.00	- 27.00	20.00	MILLI/MJ	231.12	
CALCIUM	MILLI	34.00	27.00	- 50.00	34.00	MILLI/MJ	392.91	
MANGANESE	MILLI	0.17	0.15	- 0.19	0.17	MILLI/MJ	1.96	
IRON	MILLI	1.30	1.10	- 1.40	1.30	MILLI/MJ	15.02	
COPPER	MILLI	0.23	0.22	- 0.23	0.23	MILLI/MJ	2.66	
NICKEL	MICRO	17.00	-	- -	17.00	MICRO/MJ	196.45	
PHOSPHORUS	MILLI	24.00	19.00	- 37.00	24.00	MILLI/MJ	277.35	
CHLORIDE	MILLI	307.00	239.00	- 374.00	307.00	GRAM/MJ	3.55	
IODIDE	MICRO	1.40	1.40	- 1.40	1.40	MICRO/MJ	16.18	
SELENIUM	MICRO	0.90	-	- -	0.90	MICRO/MJ	10.40	
CAROTENE	MILLI	0.20	0.10	- 0.25	0.20	MILLI/MJ	2.31	
VITAMIN E ACTIVITY	MICRO	50.00	-	- -	50.00	MICRO/MJ	577.81	
ALPHA-TOCOPHEROL	MICRO	50.00	-	- -	50.00	MICRO/MJ	577.81	
VITAMIN B1	MICRO	70.00	20.00	- 240.00	70.00	MICRO/MJ	808.93	
VITAMIN B2	MICRO	40.00	30.00	- 50.00	40.00	MICRO/MJ	462.24	
NICOTINAMIDE	MILLI	0.30	0.21	- 0.41	0.30	MILLI/MJ	3.47	
PANTOTHENIC ACID	MICRO	90.00	70.00	- 110.00	90.00	MILLI/MJ	1.04	
VITAMIN B6	MICRO	30.00	20.00	- 30.00	30.00	MICRO/MJ	346.68	
FOLIC ACID	MICRO	13.00	10.00	- 16.00	13.00	MICRO/MJ	150.23	

1 TOTAL CONTENT (SOLID AND LIQUID)

2 CALCULATED FROM THE FRESH PRODUCT

CONSTITUENTS	DIM	AV	VARIATION			AVR	NUTR. DENS.		MOLPERC.
VITAMIN C	MILLI	4.30	2.00	-	8.00	4.30	MILLI/MJ	49.69	
ARGININE	MILLI	51.00	49.00	-	53.00	51.00	MILLI/MJ	589.36	
CYSTINE	MILLI	10.00	8.00	-	12.00	10.00	MILLI/MJ	115.56	
HISTIDINE	MILLI	29.00	28.00	-	29.00	29.00	MILLI/MJ	335.13	
ISOLEUCINE	MILLI	45.00	41.00	-	49.00	45.00	MILLI/MJ	520.02	
LEUCINE	MILLI	83.00	82.00	-	84.00	83.00	MILLI/MJ	959.16	
LYSINE	MILLI	66.00	-	-	-	66.00	MILLI/MJ	762.70	
METHIONINE	MILLI	16.00	-	-	-	16.00	MILLI/MJ	184.90	
PHENYLALANINE	MILLI	51.00	47.00	-	55.00	51.00	MILLI/MJ	589.36	
THREONINE	MILLI	47.00	46.00	-	47.00	47.00	MILLI/MJ	543.14	
TRYPTOPHAN	MILLI	21.00	18.00	-	24.00	21.00	MILLI/MJ	242.68	
TYROSINE	MILLI	41.00	40.00	-	41.00	41.00	MILLI/MJ	473.80	
VALINE	MILLI	59.00	58.00	-	60.00	59.00	MILLI/MJ	681.81	

BOHNEN (BRECH- OD. SCHNITT-BOHNEN) GRÜN, GETROCKNET	BEANS FRENCH, DRIED	HARICOTS VERTS, SÉCHÉS

		PROTEIN	FAT	CARBOHYDRATES	TOTAL
ENERGY VALUE (AVERAGE)	KJOULE	325	56	793	1174
PER 100 G EDIBLE PORTION	(KCAL)	78	13	190	280
AMOUNT OF DIGESTIBLE CONSTITUENTS PER 100 G	GRAM	13.45	1.28	47.39	
ENERGY VALUE (AVERAGE)	KJOULE	211	50	793	1054
OF THE DIGESTIBLE FRACTION PER 100 G EDIBLE PORTION	(KCAL)	50	12	190	252

WASTE PERCENTAGE AVERAGE 0.00

CONSTITUENTS	DIM	AV		VARIATION			AVR	NUTR. DENS.		MOLPERC.
WATER	GRAM	13.10		11.20	-	16.60	13.10	GRAM/MJ	12.43	
PROTEIN	GRAM	20.70		18.90	-	21.90	20.70	GRAM/MJ	19.63	
FAT	GRAM	1.43		1.00	-	1.70	1.43	GRAM/MJ	1.36	
AVAILABLE CARBOHYDR.	GRAM	47.39	1	-	-	-	47.39	GRAM/MJ	44.95	
MINERALS	GRAM	4.40		3.40	-	5.80	4.40	GRAM/MJ	4.17	
SODIUM	GRAM	0.57		0.01	-	1.14	0.57	GRAM/MJ	0.54	
POTASSIUM	GRAM	1.77		1.15	-	2.60	1.77	GRAM/MJ	1.68	
CALCIUM	MILLI	197.00		100.00	-	329.00	197.00	MILLI/MJ	186.85	
IRON	MILLI	7.00		6.90	-	7.00	7.00	MILLI/MJ	6.64	
COBALT	MICRO	23.00		-	-	-	23.00	MICRO/MJ	21.82	
PHOSPHORUS	MILLI	419.00		400.00	-	437.00	419.00	MILLI/MJ	397.42	
CAROTENE	MILLI	1.50		0.00	-	2.90	1.50	MILLI/MJ	1.42	
VITAMIN E ACTIVITY	MILLI	0.51		-	-	-	0.51	MILLI/MJ	0.48	
ALPHA-TOCOPHEROL	MILLI	0.51		-	-	-	0.51	MILLI/MJ	0.48	
VITAMIN B1	MILLI	0.54		0.40	-	0.67	0.54	MILLI/MJ	0.51	
VITAMIN B2	MILLI	0.38		0.20	-	0.85	0.38	MILLI/MJ	0.36	
NICOTINAMIDE	MILLI	3.40		1.00	-	7.00	3.40	MILLI/MJ	3.22	
PANTOTHENIC ACID	MILLI	1.24		-	-	-	1.24	MILLI/MJ	1.18	
VITAMIN C	MILLI	24.00		0.00	-	38.00	24.00	MILLI/MJ	22.76	
TOTAL PURINES	MILLI	45.00	2	40.00	-	50.00	45.00	MILLI/MJ	42.68	

1 CALCULATED FROM THE FRESH PRODUCT

2 VALUE EXPRESSED IN MG URIC ACID/100G

GURKE CUCUMBER CONCOMBRE

CUCUMIS SATIVUS L.

		PROTEIN	FAT	CARBOHYDRATES	TOTAL
ENERGY VALUE (AVERAGE) PER 100 G EDIBLE PORTION	KJOULE	9.4	7.8	35	52
	(KCAL)	2.3	1.9	8.3	12
AMOUNT OF DIGESTIBLE CONSTITUENTS PER 100 G	GRAM	0.39	0.18	2.07	
ENERGY VALUE (AVERAGE) OF THE DIGESTIBLE FRACTION PER 100 G EDIBLE PORTION	KJOULE	6.1	7.0	35	48
	(KCAL)	1.5	1.7	8.3	11

WASTE PERCENTAGE AVERAGE 26 MINIMUM 14 MAXIMUM 42

CONSTITUENTS	DIM	AV	VARIATION			AVR	NUTR. DENS.		MOLPERC.
WATER	GRAM	96.80	96.10	-	97.30	71.63	KILO/MJ	2.03	
PROTEIN	GRAM	0.60	0.50	-	0.80	0.44	GRAM/MJ	12.56	
FAT	GRAM	0.20	0.05	-	0.30	0.15	GRAM/MJ	4.19	
AVAILABLE CARBOHYDR.	GRAM	2.07	-	-	-	1.53	GRAM/MJ	43.34	
TOTAL DIETARY FIBRE	GRAM	0.93 1	-	-	-	0.69	GRAM/MJ	19.47	
MINERALS	GRAM	0.60	0.40	-	0.89	0.44	GRAM/MJ	12.56	
SODIUM	MILLI	8.50	5.00	-	13.00	6.29	MILLI/MJ	177.95	
POTASSIUM	MILLI	141.00	67.00	-	200.00	104.34	GRAM/MJ	2.95	
MAGNESIUM	MILLI	8.00	-	-	-	5.92	MILLI/MJ	167.48	
CALCIUM	MILLI	15.00	10.00	-	20.00	11.10	MILLI/MJ	314.03	
MANGANESE	MILLI	0.15	-	-	-	0.11	MILLI/MJ	3.14	
IRON	MILLI	0.50	-	-	-	0.37	MILLI/MJ	10.47	
COPPER	MICRO	90.00	-	-	-	66.60	MILLI/MJ	1.88	
ZINC	MILLI	0.16	-	-	-	0.12	MILLI/MJ	3.35	
NICKEL	MICRO	23.00	5.00	-	45.00	17.02	MICRO/MJ	481.51	
CHROMIUM	MICRO	3.00	1.00	-	5.00	2.22	MICRO/MJ	62.81	
MOLYBDENUM	MICRO	1.00	-	-	-	0.74	MICRO/MJ	20.94	
VANADIUM	MICRO	0.21	-	-	-	0.16	MICRO/MJ	4.40	
PHOSPHORUS	MILLI	23.00	17.60	-	30.00	17.02	MILLI/MJ	481.51	
CHLORIDE	MILLI	37.00	-	-	-	27.38	MILLI/MJ	774.60	
FLUORIDE	MICRO	20.00	-	-	-	14.80	MICRO/MJ	418.70	
IODIDE	MICRO	2.50	-	-	-	1.85	MICRO/MJ	52.34	
BORON	MILLI	3.63	1.81	-	4.65	2.69	MILLI/MJ	75.99	
SELENIUM	MILLI	-	0.00	-	0.06				
SILICON	MILLI	3.00	1.00	-	4.00	2.22	MILLI/MJ	62.81	
NITRATE	MILLI	19.00	2.00	-	56.00	14.06	MILLI/MJ	397.77	

1 METHOD OF SOUTHGATE AND ENGLYST

CONSTITUENTS	DIM	AV	VARIATION			AVR	NUTR. DENS.		MOLPERC.
CAROTENE	MILLI	0.17	0.12	-	0.29	0.13	MILLI/MJ	3.56	
VITAMIN E ACTIVITY	MILLI	0.10	-	-	-	0.07	MILLI/MJ	2.09	
TOTAL TOCOPHEROLS	MILLI	0.19	-	-	-	0.14	MILLI/MJ	3.98	
ALPHA-TOCOPHEROL	MICRO	90.00	-	-	-	66.60	MILLI/MJ	1.88	
VITAMIN K	MICRO	5.00	-	-	-	3.70	MICRO/MJ	104.68	
VITAMIN B1	MICRO	18.00	5.00	-	30.00	13.32	MICRO/MJ	376.83	
VITAMIN B2	MICRO	30.00	15.00	-	50.00	22.20	MICRO/MJ	628.05	
NICOTINAMIDE	MILLI	0.20	0.12	-	0.21	0.15	MILLI/MJ	4.19	
PANTOTHENIC ACID	MILLI	0.24	0.18	-	0.30	0.18	MILLI/MJ	5.02	
VITAMIN B6	MICRO	35.00	24.00	-	43.00	25.90	MICRO/MJ	732.73	
BIOTIN	MICRO	0.90	-	-	-	0.67	MICRO/MJ	18.84	
FOLIC ACID	MICRO	20.00	15.00	-	24.00	14.80	MICRO/MJ	418.70	
VITAMIN C	MILLI	8.00	6.00	-	11.00	5.92	MILLI/MJ	167.48	
ARGININE	MILLI	45.00	-	-	-	33.30	MILLI/MJ	942.08	
HISTIDINE	MILLI	8.00	-	-	-	5.92	MILLI/MJ	167.48	
ISOLEUCINE	MILLI	19.00	-	-	-	14.06	MILLI/MJ	397.77	
LEUCINE	MILLI	25.00	-	-	-	18.50	MILLI/MJ	523.38	
LYSINE	MILLI	26.00	24.00	-	28.00	19.24	MILLI/MJ	544.31	
METHIONINE	MILLI	6.00	3.00	-	9.00	4.44	MILLI/MJ	125.61	
PHENYLALANINE	MILLI	14.00	-	-	-	10.36	MILLI/MJ	293.09	
THREONINE	MILLI	16.00	-	-	-	11.84	MILLI/MJ	334.96	
TRYPTOPHAN	MILLI	4.00	4.00	-	5.00	2.96	MILLI/MJ	83.74	
VALINE	MILLI	21.00	-	-	-	15.54	MILLI/MJ	439.64	
MALIC ACID	MILLI	240.00	-	-	-	177.60	GRAM/MJ	5.02	
CITRIC ACID	MILLI	20.00	-	-	-	14.80	MILLI/MJ	418.70	
OXALIC ACID TOTAL	-	0.00	-	-	-	0.00			
SALICYLIC ACID	MILLI	0.78	0.78	-	0.78	0.58	MILLI/MJ	16.33	
GLUCOSE	GRAM	0.89	0.75	-	1.25	0.66	GRAM/MJ	18.70	
FRUCTOSE	GRAM	0.86	0.46	-	0.92	0.64	GRAM/MJ	18.07	
SUCROSE	MILLI	57.35	30.00	-	60.00	42.44	GRAM/MJ	1.20	
PENTOSAN	MILLI	70.00	-	-	-	51.80	GRAM/MJ	1.47	
HEXOSAN	GRAM	0.17	-	-	-	0.13	GRAM/MJ	3.56	
CELLULOSE	GRAM	0.39	-	-	-	0.29	GRAM/MJ	8.16	
POLYURONIC ACID	GRAM	0.27	-	-	-	0.20	GRAM/MJ	5.65	
MANNITOL	MILLI	70.00	-	-	-	51.80	GRAM/MJ	1.47	
LAURIC ACID	MILLI	0.30	0.30	-	0.30	0.22	MILLI/MJ	6.28	
MYRISTIC ACID	MILLI	1.30	1.30	-	1.30	0.96	MILLI/MJ	27.22	
PALMITIC ACID	MILLI	59.00	59.00	-	59.00	43.66	GRAM/MJ	1.24	
STEARIC ACID	MILLI	6.00	6.00	-	6.00	4.44	MILLI/MJ	125.61	
PALMITOLEIC ACID	MILLI	0.30	0.30	-	0.30	0.22	MILLI/MJ	6.28	
OLEIC ACID	MILLI	5.00	5.00	-	5.00	3.70	MILLI/MJ	104.68	
LINOLEIC ACID	MILLI	46.00	46.00	-	46.00	34.04	MILLI/MJ	963.02	
LINOLENIC ACID	MILLI	42.00	42.00	-	42.00	31.08	MILLI/MJ	879.27	
TOTAL STEROLS	MILLI	14.00	-	-	-	10.36	MILLI/MJ	293.09	
BETA-SITOSTEROL	MILLI	14.00	-	-	-	10.36	MILLI/MJ	293.09	
TOTAL PURINES	MILLI	8.00 2	-	-	-	5.92	MILLI/MJ	167.48	
DIETARY FIBRE,WAT.SOL.	GRAM	0.12	-	-	-	0.09	GRAM/MJ	2.51	
DIETARY FIBRE,WAT.INS.	GRAM	0.81	-	-	-	0.60	GRAM/MJ	16.96	

2 VALUE EXPRESSED IN MG URIC ACID/100G

GURKEN
(SALZGURKEN, SALZDILL-GURKEN)
MILCHSAUER

PICKLED CUCUMBER
(GHERKINS)

CONCOMBRES
CONSERVÉS

		PROTEIN	FAT	CARBOHYDRATES	TOTAL
ENERGY VALUE (AVERAGE) PER 100 G EDIBLE PORTION	KJOULE	16	5.8	64	86
	(KCAL)	3.9	1.4	15	21
AMOUNT OF DIGESTIBLE CONSTITUENTS PER 100 G	GRAM	0.67	0.13	3.82	
ENERGY VALUE (AVERAGE) OF THE DIGESTIBLE FRACTION PER 100 G EDIBLE PORTION	KJOULE	11	5.3	64	80
	(KCAL)	2.5	1.3	15	19

WASTE PERCENTAGE AVERAGE 0.00

CONSTITUENTS	DIM	AV	VARIATION			AVR	NUTR. DENS.		MOLPERC.
WATER	GRAM	90.70	-	-	-	90.70	KILO/MJ	1.14	
PROTEIN	GRAM	1.04	-	-	-	1.04	GRAM/MJ	13.03	
FAT	GRAM	0.15	-	-	-	0.15	GRAM/MJ	1.88	
AVAILABLE CARBOHYDR.	GRAM	3.82 1	-	-	-	3.82	GRAM/MJ	47.88	
MINERALS	GRAM	4.29	2.09	-	6.48	4.29	GRAM/MJ	53.77	
SODIUM	MILLI	960.00	-	-	-	960.00	GRAM/MJ	12.03	
CALCIUM	MILLI	30.00	-	-	-	30.00	MILLI/MJ	375.98	
IRON	MILLI	1.63	1.35	-	1.91	1.63	MILLI/MJ	20.43	
COPPER	MILLI	8.40	-	-	-	8.40	MILLI/MJ	105.28	
PHOSPHORUS	MILLI	30.00	-	-	-	30.00	MILLI/MJ	375.98	
CHLORIDE	GRAM	1.50	-	-	-	1.50	GRAM/MJ	18.80	
VITAMIN A	-	0.00	-	-	-	0.00			
VITAMIN B1	MICRO	3.00	-	-	-	3.00	MICRO/MJ	37.60	
VITAMIN B2	MICRO	22.00	-	-	-	22.00	MICRO/MJ	275.72	
VITAMIN C	MILLI	2.00	-	-	-	2.00	MILLI/MJ	25.07	
TOTAL ACIDS	GRAM	1.25	0.70	-	1.80	1.25	GRAM/MJ	15.67	

1 ESTIMATED BY THE DIFFERENCE METHOD
100 - (WATER + PROTEIN + FAT + MINERALS)

KÜRBIS PUMPKIN CITROUILLE

CUCURBITA PEPO L.

		PROTEIN	FAT	CARBOHYDRATES	TOTAL
ENERGY VALUE (AVERAGE)	KJOULE	17	5.1	80	102
PER 100 G	(KCAL)	4.1	1.2	19	24
EDIBLE PORTION					
AMOUNT OF DIGESTIBLE	GRAM	0.71	0.11	4.78	
CONSTITUENTS PER 100 G					
ENERGY VALUE (AVERAGE)	KJOULE	11	4.6	80	96
OF THE DIGESTIBLE	(KCAL)	2.7	1.1	19	23
FRACTION PER 100 G					
EDIBLE PORTION					

WASTE PERCENTAGE AVERAGE 30 MINIMUM 23 MAXIMUM 38

CONSTITUENTS	DIM	AV	VARIATION			AVR	NUTR. DENS.		MOLPERC.
WATER	GRAM	91.30	90.30	-	93.00	63.91	GRAM/MJ	953.34	
PROTEIN	GRAM	1.10	1.00	-	1.20	0.77	GRAM/MJ	11.49	
FAT	GRAM	0.13	0.10	-	0.20	0.09	GRAM/MJ	1.36	
AVAILABLE CARBOHYDR.	GRAM	4.78	-	-	-	3.35	GRAM/MJ	49.91	
MINERALS	GRAM	0.77	0.73	-	0.80	0.54	GRAM/MJ	8.04	
SODIUM	MILLI	1.10	-	-	-	0.77	MILLI/MJ	11.49	
POTASSIUM	MILLI	383.00	-	-	-	268.10	GRAM/MJ	4.00	
MAGNESIUM	MILLI	8.00	-	-	-	5.60	MILLI/MJ	83.53	
CALCIUM	MILLI	22.00	21.00	-	23.00	15.40	MILLI/MJ	229.72	
MANGANESE	MICRO	66.00	66.00	-	66.00	46.20	MICRO/MJ	689.16	
IRON	MILLI	0.80	-	-	-	0.56	MILLI/MJ	8.35	
COPPER	MICRO	80.00	-	-	-	56.00	MICRO/MJ	835.34	
ZINC	MILLI	0.20	-	-	-	0.14	MILLI/MJ	2.09	
CHROMIUM	MICRO	2.00	-	-	-	1.40	MICRO/MJ	20.88	
MOLYBDENUM	MILLI	-	0.00	-	0.02				
PHOSPHORUS	MILLI	44.00	-	-	-	30.80	MILLI/MJ	459.44	
CHLORIDE	MILLI	18.00	-	-	-	12.60	MILLI/MJ	187.95	
IODIDE	MICRO	1.40	-	-	-	0.98	MICRO/MJ	14.62	
BORON	MILLI	0.10	0.08	-	0.11	0.07	MILLI/MJ	1.04	
NITRATE	MILLI	68.00	42.00	-	100.00	47.60	MILLI/MJ	710.04	
CAROTENE	MILLI	1.96	1.80	-	2.04	1.37	MILLI/MJ	20.47	
VITAMIN E ACTIVITY	MILLI	1.00	-	-	-	0.70	MILLI/MJ	10.44	
ALPHA-TOCOPHEROL	MILLI	1.00	-	-	-	0.70	MILLI/MJ	10.44	
BETA-TOCOPHEROL	MILLI	0.14	-	-	-	0.10	MILLI/MJ	1.46	
VITAMIN B1	MICRO	47.00	40.00	-	50.00	32.90	MICRO/MJ	490.76	
VITAMIN B2	MICRO	65.00	55.00	-	80.00	45.50	MICRO/MJ	678.72	

CONSTITUENTS	DIM	AV	VARIATION			AVR	NUTR. DENS.		MOLPERC.
NICOTINAMIDE	MILLI	0.50	0.30	-	0.60	0.35	MILLI/MJ	5.22	
PANTOTHENIC ACID	MILLI	0.40	-	-	-	0.28	MILLI/MJ	4.18	
VITAMIN B6	MILLI	0.11	0.10	-	0.11	0.08	MILLI/MJ	1.15	
BIOTIN	MICRO	0.40	-	-	-	0.28	MICRO/MJ	4.18	
FOLIC ACID	MICRO	36.00	-	-	-	25.20	MICRO/MJ	375.90	
VITAMIN C	MILLI	12.00	8.00	-	20.00	8.40	MILLI/MJ	125.30	
ARGININE	MILLI	39.00	-	-	-	27.30	MILLI/MJ	407.23	
HISTIDINE	MILLI	17.00	-	-	-	11.90	MILLI/MJ	177.51	
ISOLEUCINE	MILLI	40.00	-	-	-	28.00	MILLI/MJ	417.67	
LEUCINE	MILLI	58.00	-	-	-	40.60	MILLI/MJ	605.62	
LYSINE	MILLI	53.00	-	-	-	37.10	MILLI/MJ	553.41	
METHIONINE	MILLI	10.00	-	-	-	7.00	MILLI/MJ	104.42	
PHENYLALANINE	MILLI	29.00	-	-	-	20.30	MILLI/MJ	302.81	
THREONINE	MILLI	26.00	-	-	-	18.20	MILLI/MJ	271.49	
TRYPTOPHAN	MILLI	15.00	-	-	-	10.50	MILLI/MJ	156.63	
TYROSINE	MILLI	15.00	-	-	-	10.50	MILLI/MJ	156.63	
VALINE	MILLI	41.00	-	-	-	28.70	MILLI/MJ	428.11	
MALIC ACID	MILLI	199.00	-	-	-	139.30	GRAM/MJ	2.08	
CITRIC ACID	MILLI	6.50	-	-	-	4.55	MILLI/MJ	67.87	
OXALIC ACID TOTAL	-	0.00	-	-	-	0.00			
PYRROLIDONE CARB.ACID	MILLI	7.80	-	-	-	5.46	MILLI/MJ	81.45	
SALICYLIC ACID	MILLI	0.12	0.12	-	0.12	0.08	MILLI/MJ	1.25	
GLUCOSE	GRAM	1.50	0.95	-	2.06	1.05	GRAM/MJ	15.71	
FRUCTOSE	GRAM	1.32	0.90	-	1.74	0.92	GRAM/MJ	13.78	
SUCROSE	GRAM	1.06	0.54	-	1.59	0.75	GRAM/MJ	11.12	
STARCH	GRAM	0.70	-	-	-	0.49	GRAM/MJ	7.31	
SORBITOL	-	0.00	-	-	-	0.00			
MANNITOL	MILLI	17.40	-	-	-	12.18	MILLI/MJ	181.69	
XYLITOL	MILLI	8.40	-	-	-	5.88	MILLI/MJ	87.71	
PALMITIC ACID	MILLI	34.00	34.00	-	34.00	23.80	MILLI/MJ	355.02	
OLEIC ACID	MILLI	2.00	2.00	-	2.00	1.40	MILLI/MJ	20.88	
LINOLEIC ACID	MILLI	23.00	23.00	-	23.00	16.10	MILLI/MJ	240.16	
LINOLENIC ACID	MILLI	41.00	41.00	-	41.00	28.70	MILLI/MJ	428.11	
TOTAL STEROLS	MILLI	12.00	-	-	-	8.40	MILLI/MJ	125.30	
BETA-SITOSTEROL	MILLI	12.00	-	-	-	8.40	MILLI/MJ	125.30	

PAPRIKAFRÜCHTE (PAPRIKASCHOTE) GREEN PEPPERS PAPRIKA

CAPSICUM ANNUUM

		PROTEIN	FAT	CARBOHYDRATES	TOTAL
ENERGY VALUE (AVERAGE) PER 100 G EDIBLE PORTION	KJOULE	18	13	54	85
	(KCAL)	4.4	3.1	13	20
AMOUNT OF DIGESTIBLE CONSTITUENTS PER 100 G	GRAM	0.76	0.29	3.21	
ENERGY VALUE (AVERAGE) OF THE DIGESTIBLE FRACTION PER 100 G EDIBLE PORTION	KJOULE	12	12	54	77
	(KCAL)	2.9	2.8	13	18

WASTE PERCENTAGE AVERAGE 23 MINIMUM 16 MAXIMUM 28

CONSTITUENTS	DIM	AV	VARIATION			AVR	NUTR. DENS.		MOLPERC.
WATER	GRAM	91.00	87.00	-	93.00	70.07	KILO/MJ	1.18	
PROTEIN	GRAM	1.17	0.70	-	1.90	0.90	GRAM/MJ	15.15	
FAT	GRAM	0.33	0.20	-	0.60	0.25	GRAM/MJ	4.27	
AVAILABLE CARBOHYDR.	GRAM	3.21	-	-	-	2.47	GRAM/MJ	41.57	
TOTAL DIETARY FIBRE	GRAM	1.97 [1]	-	-	-	1.52	GRAM/MJ	25.51	
MINERALS	GRAM	0.57	0.50	-	0.70	0.44	GRAM/MJ	7.38	
SODIUM	MILLI	1.75	0.50	-	3.00	1.35	MILLI/MJ	22.67	
POTASSIUM	MILLI	212.00	160.00	-	435.00	163.24	GRAM/MJ	2.75	
MAGNESIUM	MILLI	12.00	-	-	-	9.24	MILLI/MJ	155.42	
CALCIUM	MILLI	11.20	7.00	-	20.00	8.62	MILLI/MJ	145.06	
MANGANESE	MILLI	0.10	-	-	-	0.08	MILLI/MJ	1.30	
IRON	MILLI	0.75	0.40	-	1.70	0.58	MILLI/MJ	9.71	
COPPER	MILLI	0.10	-	-	-	0.08	MILLI/MJ	1.30	
ZINC	MILLI	0.18	-	-	-	0.14	MILLI/MJ	2.33	
PHOSPHORUS	MILLI	29.00	25.00	-	38.00	22.33	MILLI/MJ	375.59	
CHLORIDE	MILLI	19.00	-	-	-	14.63	MILLI/MJ	246.08	
IODIDE	MICRO	2.30	-	-	-	1.77	MICRO/MJ	29.79	
NITRATE	MILLI	12.00	0.00	-	35.00	9.24	MILLI/MJ	155.42	
CAROTENE	MILLI	0.20 [2]	0.06	-	1.00	0.15	MILLI/MJ	2.59	
VITAMIN E ACTIVITY	MILLI	3.10	-	-	-	2.39	MILLI/MJ	40.15	
ALPHA-TOCOPHEROL	MILLI	3.10	-	-	-	2.39	MILLI/MJ	40.15	
VITAMIN B1	MICRO	60.00	40.00	-	90.00	46.20	MICRO/MJ	777.09	
VITAMIN B2	MICRO	50.00	30.00	-	70.00	38.50	MICRO/MJ	647.57	
NICOTINAMIDE	MILLI	0.33	0.20	-	0.40	0.25	MILLI/MJ	4.27	
PANTOTHENIC ACID	MILLI	0.23	-	-	-	0.18	MILLI/MJ	2.98	
VITAMIN B6	MILLI	0.27	-	-	-	0.21	MILLI/MJ	3.50	

1 METHOD OF SOUTHGATE AND ENGLYST

2 VALID FOR RIPE FRUITS AS MARKETED (GREEN, GREEN-YELLOW OR PALE YELLOW). FULL RIPE FRUITS HAVE A CAROTENE CONTENT OF 4,1 (2,7 - 6,7) MG/100G

CONSTITUENTS	DIM	AV	VARIATION			AVR	NUTR. DENS.		MOLPERC.
FOLIC ACID	MICRO	17.50	16.00	-	19.00	13.48	MICRO/MJ	226.65	
VITAMIN C	MILLI	139.00	91.00	-	200.00	107.03	GRAM/MJ	1.80	
ARGININE	MILLI	23.00	-	-	-	17.71	MILLI/MJ	297.88	
HISTIDINE	MILLI	14.00	-	-	-	10.78	MILLI/MJ	181.32	
ISOLEUCINE	MILLI	45.00	-	-	-	34.65	MILLI/MJ	582.82	
LEUCINE	MILLI	45.00	-	-	-	34.65	MILLI/MJ	582.82	
LYSINE	MILLI	50.00	-	-	-	38.50	MILLI/MJ	647.57	
METHIONINE	MILLI	16.00	-	-	-	12.32	MILLI/MJ	207.22	
PHENYLALANINE	MILLI	54.00	-	-	-	41.58	MILLI/MJ	699.38	
THREONINE	MILLI	49.00	-	-	-	37.73	MILLI/MJ	634.62	
TRYPTOPHAN	MILLI	9.00	-	-	-	6.93	MILLI/MJ	116.56	
VALINE	MILLI	32.00	-	-	-	24.64	MILLI/MJ	414.45	
MALIC ACID	MILLI	60.00	-	-	-	46.20	MILLI/MJ	777.09	
CITRIC ACID	MILLI	262.00	-	-	-	201.74	GRAM/MJ	3.39	
OXALIC ACID TOTAL	MILLI	16.00	-	-	-	12.32	MILLI/MJ	207.22	
SALICYLIC ACID	MILLI	1.20	1.20	-	1.20	0.92	MILLI/MJ	15.54	
GLUCOSE	GRAM	1.37	1.11	-	1.41	1.06	GRAM/MJ	17.83	
FRUCTOSE	GRAM	1.25	0.87	-	1.90	0.96	GRAM/MJ	16.19	
SUCROSE	GRAM	0.14	0.09	-	0.26	0.11	GRAM/MJ	1.93	
STARCH	GRAM	0.13	0.13	-	0.13	0.10	GRAM/MJ	1.68	
PALMITIC ACID	MILLI	50.00	50.00	-	50.00	38.50	MILLI/MJ	647.57	
STEARIC ACID	MILLI	17.00	17.00	-	17.00	13.09	MILLI/MJ	220.17	
PALMITOLEIC ACID	MILLI	10.00	10.00	-	10.00	7.70	MILLI/MJ	129.51	
LINOLEIC ACID	MILLI	117.00	117.00	-	117.00	90.09	GRAM/MJ	1.52	
LINOLENIC ACID	MILLI	67.00	67.00	-	67.00	51.59	MILLI/MJ	867.75	
DIETARY FIBRE,WAT.SOL.	GRAM	0.27	-	-	-	0.21	GRAM/MJ	3.50	
DIETARY FIBRE,WAT.INS.	GRAM	1.70	-	-	-	1.31	GRAM/MJ	22.02	

SQUASH SQUASH, WINTER SQUASH

CUCURBITA MAXIMA

		PROTEIN	FAT	CARBOHYDRATES	TOTAL
ENERGY VALUE (AVERAGE) PER 100 G EDIBLE PORTION	KJOULE	22	7.8	77	107
	(KCAL)	5.3	1.9	18	25
AMOUNT OF DIGESTIBLE CONSTITUENTS PER 100 G	GRAM	0.91	0.18	4.59	
ENERGY VALUE (AVERAGE) OF THE DIGESTIBLE FRACTION PER 100 G EDIBLE PORTION	KJOULE	14	7.0	77	98
	(KCAL)	3.4	1.7	18	23

WASTE PERCENTAGE AVERAGE 26

CONSTITUENTS	DIM	AV	VARIATION			AVR	NUTR. DENS.		MOLPERC.
WATER	GRAM	88.70	-	-	-	65.64	GRAM/MJ	904.18	
PROTEIN	GRAM	1.40	-	-	-	1.04	GRAM/MJ	14.27	
FAT	GRAM	0.20	-	-	-	0.15	GRAM/MJ	2.04	
AVAILABLE CARBOHYDR.	GRAM	4.59	-	-	-	3.40	GRAM/MJ	46.79	
TOTAL DIETARY FIBRE	GRAM	0.78 1	-	-	-	0.58	GRAM/MJ	7.95	
MINERALS	GRAM	0.70	-	-	-	0.52	GRAM/MJ	7.14	
SODIUM	MILLI	7.00	-	-	-	5.18	MILLI/MJ	71.36	
POTASSIUM	MILLI	351.00	-	-	-	259.74	GRAM/MJ	3.58	
CALCIUM	MILLI	27.00	-	-	-	19.98	MILLI/MJ	275.23	
MANGANESE	MILLI	0.22	-	-	-	0.16	MILLI/MJ	2.24	
IRON	MILLI	0.77	-	-	-	0.57	MILLI/MJ	7.85	
COPPER	MILLI	0.10	-	-	-	0.07	MILLI/MJ	1.02	
PHOSPHORUS	MILLI	43.00	-	-	-	31.82	MILLI/MJ	438.33	
CAROTENE	MILLI	1.40	-	-	-	1.04	MILLI/MJ	14.27	
VITAMIN B1	MICRO	90.00	-	-	-	66.60	MICRO/MJ	917.43	
VITAMIN B2	MICRO	60.00	-	-	-	44.40	MICRO/MJ	611.62	
NICOTINAMIDE	MILLI	1.60	1.40	-	1.80	1.18	MILLI/MJ	16.31	
VITAMIN C	MILLI	14.00	-	-	-	10.36	MILLI/MJ	142.71	
SALICYLIC ACID	MILLI	0.63	0.63	-	0.63	0.47	MILLI/MJ	6.42	
GLUCOSE	GRAM	1.37	1.08	-	1.66	1.01	GRAM/MJ	13.97	
FRUCTOSE	GRAM	1.40	1.31	-	1.49	1.04	GRAM/MJ	14.27	
SUCROSE	GRAM	1.82	1.82	-	1.82	1.35	GRAM/MJ	18.55	
CELLULOSE	GRAM	0.55	-	-	-	0.41	GRAM/MJ	5.61	
DIETARY FIBRE,WAT.SOL.	MILLI	30.00	-	-	-	22.20	MILLI/MJ	305.81	
DIETARY FIBRE,WAT.INS.	GRAM	0.75	-	-	-	0.56	GRAM/MJ	7.65	

1 METHOD OF SOUTHGATE

TOMATE | TOMATO | TOMATE

LYCOPERSICUM ESCULENTUM MILL.

		PROTEIN	FAT	CARBOHYDRATES	TOTAL
ENERGY VALUE (AVERAGE) PER 100 G EDIBLE PORTION	KJOULE	15	8.2	58	81
	(KCAL)	3.6	2.0	14	19
AMOUNT OF DIGESTIBLE CONSTITUENTS PER 100 G	GRAM	0.61	0.18	3.45	
ENERGY VALUE (AVERAGE) OF THE DIGESTIBLE FRACTION PER 100 G EDIBLE PORTION	KJOULE	9.7	7.4	58	75
	(KCAL)	2.3	1.8	14	18

WASTE PERCENTAGE AVERAGE 0.00

CONSTITUENTS	DIM	AV		VARIATION			AVR	NUTR. DENS.		MOLPERC.
WATER	GRAM	94.20		93.40	-	95.20	90.43	KILO/MJ	1.26	
PROTEIN	GRAM	0.95		0.69	-	1.00	0.91	GRAM/MJ	12.70	
FAT	GRAM	0.21		0.20	-	0.30	0.20	GRAM/MJ	2.81	
AVAILABLE CARBOHYDR.	GRAM	3.45		2.43	-	5.20	3.31	GRAM/MJ	46.13	
TOTAL DIETARY FIBRE	GRAM	1.83	1	0.80	-	1.85	1.76	GRAM/MJ	24.47	
MINERALS	GRAM	0.61		0.60	-	0.61	0.59	GRAM/MJ	8.16	
SODIUM	MILLI	6.30		1.00	-	33.00	6.05	MILLI/MJ	84.24	
POTASSIUM	MILLI	297.00		92.00	-	376.00	285.12	GRAM/MJ	3.97	
MAGNESIUM	MILLI	20.00		5.00	-	20.00	19.20	MILLI/MJ	267.44	
CALCIUM	MILLI	14.00		4.00	-	21.00	13.44	MILLI/MJ	187.21	
MANGANESE	MILLI	0.14		0.04	-	0.30	0.13	MILLI/MJ	1.87	
IRON	MILLI	0.50		0.35	-	0.95	0.48	MILLI/MJ	6.69	
COBALT	MICRO	9.00		-	-	-	8.64	MICRO/MJ	120.35	
COPPER	MICRO	90.00		-	-	-	86.40	MILLI/MJ	1.20	
ZINC	MILLI	0.24		0.00	-	0.25	0.23	MILLI/MJ	3.21	
NICKEL	MICRO	23.00		3.00	-	62.00	22.08	MICRO/MJ	307.56	
CHROMIUM	MICRO	5.00		2.00	-	7.00	4.80	MICRO/MJ	66.86	
PHOSPHORUS	MILLI	26.00		8.00	-	53.00	24.96	MILLI/MJ	347.68	
CHLORIDE	MILLI	60.00		1.30	-	30.01	57.60	MILLI/MJ	802.33	
FLUORIDE	MICRO	24.00		-	-	-	23.04	MICRO/MJ	320.93	
IODIDE	MICRO	1.70		-	-	-	1.63	MICRO/MJ	22.73	
BORON	MILLI	0.11		0.03	-	0.21	0.11	MILLI/MJ	1.54	
SELENIUM	MICRO	0.60	2	0.50	-	10.00	0.58	MICRO/MJ	8.02	
NITRATE	MILLI	5.00	3	0.00	-	17.00	4.80	MILLI/MJ	66.86	
CAROTENE	MILLI	0.82		0.15	-	2.30	0.79	MILLI/MJ	10.97	
VITAMIN E ACTIVITY	MILLI	0.80		-	-	-	0.77	MILLI/MJ	10.70	

1 METHOD OF SOUTHGATE AND ENGLYST

2 AMOUNTS, DETERMINED IN THE FEDERAL REPUBLIC OF GERMANY

3 THE AMOUNTS ARE DEPENDENT ON THE N-CONTENT OF SOIL AND ON THE CLIMATE.

CONSTITUENTS	DIM	AV	VARIATION			AVR	NUTR. DENS.		MOLPERC.
TOTAL TOCOPHEROLS	MILLI	0.93	0.50	-	1.30	0.89	MILLI/MJ	12.44	
ALPHA-TOCOPHEROL	MILLI	0.80	0.34	-	1.20	0.77	MILLI/MJ	10.70	
GAMMA-TOCOPHEROL	MILLI	0.13	-	-	-	0.12	MILLI/MJ	1.74	
VITAMIN K	MICRO	8.00	6.00	-	11.00	7.68	MICRO/MJ	106.98	
VITAMIN B1	MICRO	57.00	20.00	-	80.00	54.72	MICRO/MJ	762.22	
VITAMIN B2	MICRO	35.00	20.00	-	50.00	33.60	MICRO/MJ	468.03	
NICOTINAMIDE	MILLI	0.53	0.30	-	0.85	0.51	MILLI/MJ	7.09	
PANTOTHENIC ACID	MILLI	0.31	0.28	-	0.34	0.30	MILLI/MJ	4.15	
VITAMIN B6	MILLI	0.10	0.07	-	0.15	0.10	MILLI/MJ	1.34	
BIOTIN	MICRO	4.00	-	-	-	3.84	MICRO/MJ	53.49	
FOLIC ACID	MICRO	39.00	-	-	-	37.44	MICRO/MJ	521.52	
VITAMIN C	MILLI	24.20	20.00	-	28.80	23.23	MILLI/MJ	323.61	
ALANINE	MILLI	26.00	26.00	-	26.00	24.96	MILLI/MJ	347.68	5.2
ARGININE	MILLI	18.00	18.00	-	18.00	17.28	MILLI/MJ	240.70	1.8
ASPARTIC ACID	MILLI	121.00	121.00	-	121.00	116.16	GRAM/MJ	1.62	16.2
CYSTINE	MILLI	1.00	1.00	-	1.00	0.96	MILLI/MJ	13.37	0.1
GLUTAMIC ACID	MILLI	337.00	337.00	-	337.00	323.52	GRAM/MJ	4.51	40.9
GLYCINE	MILLI	18.00	18.00	-	18.00	17.28	MILLI/MJ	240.70	4.3
HISTIDINE	MILLI	13.00	13.00	-	13.00	12.48	MILLI/MJ	173.84	1.5
ISOLEUCINE	MILLI	23.00	23.00	-	23.00	22.08	MILLI/MJ	307.56	3.1
LEUCINE	MILLI	30.00	30.00	-	30.00	28.80	MILLI/MJ	401.17	4.1
LYSINE	MILLI	29.00	29.00	-	29.00	27.84	MILLI/MJ	387.79	3.5
METHIONINE	MILLI	7.00	7.00	-	7.00	6.72	MILLI/MJ	93.61	0.8
PHENYLALANINE	MILLI	24.00	24.00	-	24.00	23.04	MILLI/MJ	320.93	2.6
PROLINE	MILLI	16.00	16.00	-	16.00	15.36	MILLI/MJ	213.96	2.5
SERINE	MILLI	28.00	28.00	-	28.00	26.88	MILLI/MJ	374.42	4.8
THREONINE	MILLI	23.00	23.00	-	23.00	22.08	MILLI/MJ	307.56	3.4
TRYPTOPHAN	MILLI	6.00	6.00	-	6.00	5.76	MILLI/MJ	80.23	0.5
TYROSINE	MILLI	12.00	12.00	-	12.00	11.52	MILLI/MJ	160.47	1.2
VALINE	MILLI	23.00	23.00	-	23.00	22.08	MILLI/MJ	307.56	3.5
HISTAMINE	MILLI	2.00	-	-	-	1.92	MILLI/MJ	26.74	
SEROTONINE	MILLI	1.20	-	-	-	1.15	MILLI/MJ	16.05	
TRYPTAMINE	MILLI	0.40	-	-	-	0.38	MILLI/MJ	5.35	
TYRAMINE	MILLI	0.40	-	-	-	0.38	MILLI/MJ	5.35	
MALIC ACID	MILLI	37.00	20.00	-	230.00	35.52	MILLI/MJ	494.77	
CITRIC ACID	GRAM	0.44	0.13	-	0.68	0.42	GRAM/MJ	5.88	
LACTIC ACID	MILLI	6.00	0.00	-	32.00	5.76	MILLI/MJ	80.23	
OXALIC ACID TOTAL	-	0.00	-	-	-	0.00			
ACETIC ACID	MILLI	8.00	0.00	-	17.00	7.68	MILLI/MJ	106.98	
CHLOROGENIC ACID	MILLI	9.70	-	-	-	9.31	MILLI/MJ	129.71	
QUINIC ACID	MILLI	8.10	-	-	-	7.78	MILLI/MJ	108.31	
FERULIC ACID	MILLI	0.70	1.00	-	2.00	0.67	MILLI/MJ	9.36	
MALONIC ACID	MILLI	-	0.00	-	112.00				
FUMARIC ACID	MILLI	1.60	-	-	-	1.54	MILLI/MJ	21.40	
PYRUVIC ACID	MILLI	0.19	0.12	-	0.26	0.18	MILLI/MJ	2.54	
OXALOACETIC ACID	MILLI	24.00	18.00	-	29.00	23.04	MILLI/MJ	320.93	
ALPHA-KETOGLUT.ACID	MILLI	-	1.40	-	8.10				
SALICYLIC ACID	MILLI	0.13	0.13	-	0.13	0.12	MILLI/MJ	1.74	
GLUCOSE	GRAM	1.21	0.90	-	1.62	1.17	GRAM/MJ	16.30	
FRUCTOSE	GRAM	1.54	1.29	-	1.70	1.48	GRAM/MJ	20.62	
SUCROSE	GRAM	0.13	0.01	-	0.14	0.13	GRAM/MJ	1.77	
STARCH	MILLI	80.00	80.00	-	80.00	76.80	GRAM/MJ	1.07	
PENTOSAN	GRAM	0.13	-	-	-	0.12	GRAM/MJ	1.74	
HEXOSAN	GRAM	0.22	-	-	-	0.21	GRAM/MJ	2.94	

CONSTITUENTS	DIM	AV		VARIATION			AVR	NUTR. DENS.		MOLPERC.
CELLULOSE	GRAM	0.70		-	-	-	0.67	GRAM/MJ	9.36	
POLYURONIC ACID	GRAM	0.45		0.45	-	0.45	0.43	GRAM/MJ	6.02	
MYOINOSITOL	MILLI	11.00		-	-	-	10.56	MILLI/MJ	147.09	
PALMITIC ACID	MILLI	32.00		32.00	-	32.00	30.72	MILLI/MJ	427.91	
STEARIC ACID	MILLI	5.00		5.00	-	5.00	4.80	MILLI/MJ	66.86	
PALMITOLEIC ACID	MILLI	2.00		2.00	-	2.00	1.92	MILLI/MJ	26.74	
OLEIC ACID	MILLI	23.00		23.00	-	23.00	22.08	MILLI/MJ	307.56	
LINOLEIC ACID	MILLI	91.00		91.00	-	91.00	87.36	GRAM/MJ	1.22	
LINOLENIC ACID	MILLI	9.00		9.00	-	9.00	8.64	MILLI/MJ	120.35	
TOTAL STEROLS	MILLI	7.00		-	-	-	6.72	MILLI/MJ	93.61	
CAMPESTEROL	MILLI	1.00		-	-	-	0.96	MILLI/MJ	13.37	
BETA-SITOSTEROL	MILLI	3.00		-	-	-	2.88	MILLI/MJ	40.12	
STIGMASTEROL	MILLI	3.50		-	-	-	3.36	MILLI/MJ	46.80	
TOTAL PURINES	MILLI	10.00	4	-	-	-	9.60	MILLI/MJ	133.72	
DIETARY FIBRE,WAT.SOL.	GRAM	0.14		-	-	-	0.13	GRAM/MJ	1.87	
DIETARY FIBRE,WAT.INS.	GRAM	1.69		-	-	-	1.62	GRAM/MJ	22.60	

4 VALUE EXPRESSED IN MG URIC ACID/100G

TOMATEN IN DOSEN [1] | TOMATOES CANNED [1] | TOMATES EN BOÎTES [1]

		PROTEIN	FAT	CARBOHYDRATES	TOTAL
ENERGY VALUE (AVERAGE) PER 100 G EDIBLE PORTION	KJOULE	18	7.8	61	87
	(KCAL)	4.3	1.9	15	21
AMOUNT OF DIGESTIBLE CONSTITUENTS PER 100 G	GRAM	0.85	0.18	3.63	
ENERGY VALUE (AVERAGE) OF THE DIGESTIBLE FRACTION PER 100 G EDIBLE PORTION	KJOULE	13	7.0	61	81
	(KCAL)	3.2	1.7	15	19

WASTE PERCENTAGE AVERAGE 0.00

CONSTITUENTS	DIM	AV	VARIATION			AVR	NUTR. DENS.		MOLPERC.
WATER	GRAM	93.90	93.60	-	94.20	93.90	KILO/MJ	1.16	
PROTEIN	GRAM	1.15	1.00	-	1.30	1.15	GRAM/MJ	14.18	
FAT	GRAM	0.20	-	-	-	0.20	GRAM/MJ	2.47	
AVAILABLE CARBOHYDR.	GRAM	3.63 [2]	-	-	-	3.63	GRAM/MJ	44.76	
MINERALS	GRAM	0.68	0.66	-	0.70	0.68	GRAM/MJ	8.38	
SODIUM	MILLI	9.00	1.00	-	22.00	9.00	MILLI/MJ	110.96	
POTASSIUM	MILLI	230.00	196.00	-	268.00	230.00	GRAM/MJ	2.84	
MAGNESIUM	MILLI	25.00	6.00	-	47.00	25.00	MILLI/MJ	308.23	
IRON	MILLI	0.20	-	-	-	0.20	MILLI/MJ	2.47	
PHOSPHORUS	MILLI	12.00	-	-	-	12.00	MILLI/MJ	147.95	
SELENIUM	MICRO	-	0.90	-	1.00				
CAROTENE	MILLI	0.61	-	-	-	0.61	MILLI/MJ	7.52	
VITAMIN B1	MICRO	60.00	-	-	-	60.00	MICRO/MJ	739.76	
VITAMIN B2	MICRO	29.00	27.00	-	30.00	29.00	MICRO/MJ	357.55	
NICOTINAMIDE	MILLI	0.70	0.69	-	0.70	0.70	MILLI/MJ	8.63	
VITAMIN C	MILLI	16.50	16.00	-	17.00	16.50	MILLI/MJ	203.43	
ALANINE	MILLI	31.00	31.00	-	31.00	31.00	MILLI/MJ	382.21	
ARGININE	MILLI	11.00	22.00	-	39.00	11.00	MILLI/MJ	135.62	
ASPARTIC ACID	MILLI	150.00	150.00	-	150.00	150.00	GRAM/MJ	1.85	
CYSTINE	MILLI	1.00	1.00	-	1.00	1.00	MILLI/MJ	12.33	
GLUTAMIC ACID	MILLI	403.00	403.00	-	403.00	403.00	GRAM/MJ	4.97	
GLYCINE	MILLI	22.00	22.00	-	22.00	22.00	MILLI/MJ	271.24	
HISTIDINE	MILLI	16.50	12.00	-	22.00	16.50	MILLI/MJ	203.43	
LYSINE	MILLI	41.50	28.00	-	57.00	41.50	MILLI/MJ	511.66	
METHIONINE	MILLI	7.50	2.00	-	11.00	7.50	MILLI/MJ	92.47	
PHENYLALANINE	MILLI	31.50	28.00	-	38.00	31.50	MILLI/MJ	388.37	
PROLINE	MILLI	21.00	21.00	-	21.00	21.00	MILLI/MJ	258.91	
SERINE	MILLI	33.00	33.00	-	33.00	33.00	MILLI/MJ	406.87	
THREONINE	MILLI	33.00	28.00	-	42.00	33.00	MILLI/MJ	406.87	
TRYPTOPHAN	MILLI	8.50	5.00	-	16.00	8.50	MILLI/MJ	104.80	
TYROSINE	MILLI	16.00	15.00	-	19.00	16.00	MILLI/MJ	197.27	
VALINE	MILLI	29.50	25.00	-	39.00	29.50	MILLI/MJ	363.71	

1 TOTAL CONTENT (SOLID AND LIQUID)

2 CALCULATED FROM THE FRESH PRODUCT

TOMATENMARK | TOMATO PUREE | CONCENTRÉ DE TOMATES

		PROTEIN	FAT	CARBOHYDRATES	TOTAL
ENERGY VALUE (AVERAGE) PER 100 G EDIBLE PORTION	KJOULE	36	19	159	215
	(KCAL)	8.6	4.7	38	51
AMOUNT OF DIGESTIBLE CONSTITUENTS PER 100 G	GRAM	1.70	0.45	9.50	
ENERGY VALUE (AVERAGE) OF THE DIGESTIBLE FRACTION PER 100 G EDIBLE PORTION	KJOULE	27	18	159	203
	(KCAL)	6.4	4.2	38	49

WASTE PERCENTAGE AVERAGE 0.00

CONSTITUENTS	DIM	AV	VARIATION			AVR	NUTR. DENS.		MOLPERC.
WATER	GRAM	86.00	-	-	-	86.00	GRAM/MJ	423.21	
PROTEIN	GRAM	2.30	-	-	-	2.30	GRAM/MJ	11.32	
FAT	GRAM	0.50	-	-	-	0.50	GRAM/MJ	2.46	
AVAILABLE CARBOHYDR.	GRAM	9.50 1	-	-	-	9.50	GRAM/MJ	46.75	
MINERALS	GRAM	1.70	-	-	-	1.70	GRAM/MJ	8.37	
SODIUM	MILLI	590.00	360.00	-	820.00	590.00	GRAM/MJ	2.90	
POTASSIUM	GRAM	1.16	0.67	-	1.44	1.16	GRAM/MJ	5.71	
MAGNESIUM	MILLI	32.00	32.00	-	32.00	32.00	MILLI/MJ	157.48	
CALCIUM	MILLI	60.00	41.00	-	80.00	60.00	MILLI/MJ	295.27	
IRON	MILLI	1.00	0.80	-	1.10	1.00	MILLI/MJ	4.92	
PHOSPHORUS	MILLI	34.00	30.00	-	37.00	34.00	MILLI/MJ	167.32	
IODIDE	MICRO	2.20	2.20	-	2.20	2.20	MICRO/MJ	10.83	
CAROTENE	MILLI	1.24	0.72	-	4.48	1.24	MILLI/MJ	6.10	
VITAMIN B1	MICRO	93.00	88.00	-	110.00	93.00	MICRO/MJ	457.66	
VITAMIN B2	MICRO	58.00	40.00	-	70.00	58.00	MICRO/MJ	285.42	
NICOTINAMIDE	MILLI	1.48	0.95	-	2.08	1.48	MILLI/MJ	7.28	
VITAMIN B6	MILLI	0.18	0.18	-	0.19	0.18	MILLI/MJ	0.89	
VITAMIN C	MILLI	9.00	0.00	-	16.00	9.00	MILLI/MJ	44.29	

1 ESTIMATED BY THE DIFFERENCE METHOD
100 - (WATER + PROTEIN + FAT + MINERALS)

TOMATENSAFT HANDELSWARE | TOMATO JUICE COMMERCIAL PRODUCT | JUS DE TOMATES PRODUIT DE VENTE

		PROTEIN	FAT	CARBOHYDRATES	TOTAL
ENERGY VALUE (AVERAGE) PER 100 G EDIBLE PORTION	KJOULE	13	1.8	57	71
	(KCAL)	3.0	0.4	14	17
AMOUNT OF DIGESTIBLE CONSTITUENTS PER 100 G	GRAM	0.64	0.04	3.40	
ENERGY VALUE (AVERAGE) OF THE DIGESTIBLE FRACTION PER 100 G EDIBLE PORTION	KJOULE	11	1.6	57	69
	(KCAL)	2.6	0.4	14	17

WASTE PERCENTAGE AVERAGE 0.00

CONSTITUENTS	DIM	AV	VARIATION		AVR	NUTR. DENS.		MOLPERC.
WATER	GRAM	94.10	93.50	- 95.00	94.10	KILO/MJ	1.36	
PROTEIN	GRAM	0.76	0.54	- 0.83	0.76	GRAM/MJ	10.98	
FAT	MILLI	46.00	40.00	- 52.00	46.00	MILLI/MJ	664.84	
AVAILABLE CARBOHYDR.	GRAM	3.40	-	- -	3.40	GRAM/MJ	49.14	
MINERALS	GRAM	0.63	0.45	- 0.90	0.63	GRAM/MJ	9.11	
SODIUM	MILLI	5.10	4.50	- 6.50	5.10	MILLI/MJ	73.71	
POTASSIUM	MILLI	236.00	179.00	- 332.00	236.00	GRAM/MJ	3.41	
MAGNESIUM	MILLI	9.50	7.00	- 10.80	9.50	MILLI/MJ	137.30	
CALCIUM	MILLI	15.00	9.20	- 19.00	15.00	MILLI/MJ	216.80	
MANGANESE	MICRO	7.70	2.40	- 13.00	7.70	MICRO/MJ	111.29	
IRON	MILLI	0.56	0.40	- 0.90	0.56	MILLI/MJ	8.09	
COPPER	MILLI	0.12	0.05	- 0.20	0.12	MILLI/MJ	1.73	
ZINC	MICRO	86.00	64.00	- 130.00	86.00	MILLI/MJ	1.24	
NICKEL	MICRO	5.00	-	- -	5.00	MICRO/MJ	72.27	
MOLYBDENUM	MICRO	-	0.00	- 4.40				
PHOSPHORUS	MILLI	16.00	12.00	- 20.00	16.00	MILLI/MJ	231.25	
BORON	MILLI	-	0.02	- 0.14				
CAROTENE	MILLI	0.54	0.41	- 0.63	0.54	MILLI/MJ	7.80	
VITAMIN B1	MICRO	56.00	50.00	- 75.00	56.00	MICRO/MJ	809.37	
VITAMIN B2	MICRO	26.00	20.00	- 30.00	26.00	MICRO/MJ	375.78	
NICOTINAMIDE	MILLI	0.72	0.50	- 0.80	0.72	MILLI/MJ	10.41	
PANTOTHENIC ACID	MILLI	0.20	-	- -	0.20	MILLI/MJ	2.89	
VITAMIN B6	MILLI	0.11	-	- -	0.11	MILLI/MJ	1.59	
BIOTIN	MICRO	-	1.00	- 4.00				
FOLIC ACID	MICRO	13.00	-	- -	13.00	MICRO/MJ	187.89	
VITAMIN C	MILLI	14.80	11.60	- 17.90	14.80	MILLI/MJ	213.90	

CONSTITUENTS	DIM	AV	VARIATION			AVR	NUTR. DENS.		MOLPERC.
ALANINE	MILLI	20.00	20.00	-	20.00	20.00	MILLI/MJ	289.06	5.0
ARGININE	MILLI	14.00	14.00	-	14.00	14.00	MILLI/MJ	202.34	1.8
ASPARTIC ACID	MILLI	101.00	101.00	-	101.00	101.00	GRAM/MJ	1.46	16.8
CYSTINE	MILLI	1.00	1.00	-	1.00	1.00	MILLI/MJ	14.45	0.1
GLUTAMIC ACID	MILLI	270.00	270.00	-	270.00	270.00	GRAM/MJ	3.90	40.5
GLYCINE	MILLI	14.00	14.00	-	14.00	14.00	MILLI/MJ	202.34	4.1
HISTIDINE	MILLI	11.00	11.00	-	11.00	11.00	MILLI/MJ	158.98	1.6
ISOLEUCINE	MILLI	19.00	19.00	-	19.00	19.00	MILLI/MJ	274.61	3.2
LEUCINE	MILLI	25.00	25.00	-	25.00	25.00	MILLI/MJ	361.33	4.2
LYSINE	MILLI	23.00	23.00	-	23.00	23.00	MILLI/MJ	332.42	3.5
METHIONINE	MILLI	5.00	5.00	-	5.00	5.00	MILLI/MJ	72.27	0.7
PHENYLALANINE	MILLI	19.00	19.00	-	19.00	19.00	MILLI/MJ	274.61	2.5
PROLINE	MILLI	14.00	14.00	-	14.00	14.00	MILLI/MJ	202.34	2.7
SERINE	MILLI	22.00	22.00	-	22.00	22.00	MILLI/MJ	317.97	4.6
THREONINE	MILLI	19.00	19.00	-	19.00	19.00	MILLI/MJ	274.61	3.5
TRYPTOPHAN	MILLI	4.00	4.00	-	4.00	4.00	MILLI/MJ	57.81	0.4
TYROSINE	MILLI	10.00	10.00	-	10.00	10.00	MILLI/MJ	144.53	1.2
VALINE	MILLI	19.00	19.00	-	19.00	19.00	MILLI/MJ	274.61	3.6
MALIC ACID	MILLI	41.00	36.00	-	45.00	41.00	MILLI/MJ	592.57	
CITRIC ACID	GRAM	0.44	0.37	-	0.51	0.44	GRAM/MJ	6.36	
SALICYLIC ACID	MILLI	0.13	0.13	-	0.14	0.13	MILLI/MJ	1.92	
GLUCOSE	GRAM	1.31	1.31	-	1.31	1.31	GRAM/MJ	18.93	
FRUCTOSE	GRAM	1.51	1.42	-	1.56	1.51	GRAM/MJ	21.87	
SUCROSE	-	0.00	-	-	-	0.00			
STARCH	GRAM	0.10	0.10	-	0.10	0.10	GRAM/MJ	1.45	

ZUCCHINI (SOMMER-SQUASH) — SQUASH, SUMMER — COURGE, COURGETTE

CUCURBITA PEPO L. CONVAR. GIROMONTIINA GREB.

		PROTEIN	FAT	CARBOHYDRATES	TOTAL
ENERGY VALUE (AVERAGE) PER 100 G EDIBLE PORTION	KJOULE	25	16	34	75
	(KCAL)	6.0	3.7	8.2	18
AMOUNT OF DIGESTIBLE CONSTITUENTS PER 100 G	GRAM	1.04	0.36	2.05	
ENERGY VALUE (AVERAGE) OF THE DIGESTIBLE FRACTION PER 100 G EDIBLE PORTION	KJOULE	16	14	34	65
	(KCAL)	3.9	3.3	8.2	15

WASTE PERCENTAGE AVERAGE 13

CONSTITUENTS	DIM	AV		VARIATION			AVR	NUTR. DENS.		MOLPERC.
WATER	GRAM	92.20		-	-	-	80.21	KILO/MJ	1.43	
PROTEIN	GRAM	1.60		-	-	-	1.39	GRAM/MJ	24.75	
FAT	GRAM	0.40		-	-	-	0.35	GRAM/MJ	6.19	
AVAILABLE CARBOHYDR.	GRAM	2.0[illegible]		-	-	-	1.78	GRAM/MJ	31.72	
TOTAL DIETARY FIBRE	GRAM	1.08	1	-	-	-	0.94	GRAM/MJ	16.71	
MINERALS	GRAM	0.70		-	-	-	0.61	GRAM/MJ	10.83	
CALCIUM	MILLI	30.00		-	-	-	26.10	MILLI/MJ	464.15	
MANGANESE	MILLI	0.14		-	-	-	0.12	MILLI/MJ	2.17	
IRON	MILLI	1.50		0.53	-	2.40	1.31	MILLI/MJ	23.21	
COPPER	MICRO	80.00		-	-	-	69.60	MILLI/MJ	1.24	
MOLYBDENUM	MICRO	12.00		-	-	-	10.44	MICRO/MJ	185.66	
PHOSPHORUS	MILLI	23.00		-	-	-	20.01	MILLI/MJ	355.85	
IODIDE	MICRO	2.30		-	-	-	2.00	MICRO/MJ	35.58	
CAROTENE	MILLI	0.35		-	-	-	0.30	MILLI/MJ	5.42	
VITAMIN B1	MILLI	0.50		-	-	-	0.44	MILLI/MJ	7.74	
VITAMIN B2	MICRO	90.00		-	-	-	78.30	MILLI/MJ	1.39	
NICOTINAMIDE	MILLI	0.40		-	-	-	0.35	MILLI/MJ	6.19	
VITAMIN C	MILLI	16.00		-	-	-	13.92	MILLI/MJ	247.55	
SALICYLIC ACID	MILLI	1.04		1.04	-	1.04	0.90	MILLI/MJ	16.09	
GLUCOSE	GRAM	0.90		0.85	-	1.08	0.79	GRAM/MJ	13.97	
FRUCTOSE	GRAM	1.02		0.98	-	1.15	0.89	GRAM/MJ	15.78	
SUCROSE	GRAM	0.13		0.13	-	0.13	0.11	GRAM/MJ	2.01	
DIETARY FIBRE,WAT.SOL.	GRAM	0.26		-	-	-	0.23	GRAM/MJ	4.02	
DIETARY FIBRE,WAT.INS.	GRAM	0.82		-	-	-	0.71	GRAM/MJ	12.69	

1 METHOD OF MEUSER, SUCKOW AND KULIKOWSKI ("BERLINER METHODE")

ZUCKERMAIS (SPEISEMAIS) — SWEETCORN — EPI DE MAÏS

ZEA MAYS L.

		PROTEIN	FAT	CARBOHYDRATES	TOTAL
ENERGY VALUE (AVERAGE) PER 100 G EDIBLE PORTION	KJOULE	62	48	264	374
	(KCAL)	15	11	63	89
AMOUNT OF DIGESTIBLE CONSTITUENTS PER 100 G	GRAM	1.96	1.10	15.76	
ENERGY VALUE (AVERAGE) OF THE DIGESTIBLE FRACTION PER 100 G EDIBLE PORTION	KJOULE	37	43	264	344
	(KCAL)	9.0	10	63	82

WASTE PERCENTAGE AVERAGE 63 MINIMUM 37 MAXIMUM 85

CONSTITUENTS	DIM	AV	VARIATION		AVR		NUTR. DENS.	MOLPERC.
WATER	GRAM	74.70	73.90	75.60	27.64	GRAM/MJ	216.96	
PROTEIN	GRAM	3.28	2.86	3.70	1.21	GRAM/MJ	9.53	
FAT	GRAM	1.23	-	-	0.46	GRAM/MJ	3.57	
AVAILABLE CARBOHYDR.	GRAM	15.76	-	-	5.83	GRAM/MJ	45.77	
MINERALS	GRAM	0.80	0.70	0.93	0.30	GRAM/MJ	2.32	
SODIUM	MILLI	0.30	0.15	0.50	0.11	MILLI/MJ	0.87	
POTASSIUM	MILLI	300.00	269.00	340.00	111.00	MILLI/MJ	871.33	
MAGNESIUM	MILLI	48.00	40.00	56.00	17.76	MILLI/MJ	139.41	
CALCIUM	MILLI	5.80	2.00	9.00	2.15	MILLI/MJ	16.85	
MANGANESE	MILLI	0.20	0.15	0.25	0.07	MILLI/MJ	0.58	
IRON	MILLI	0.55	0.34	0.72	0.20	MILLI/MJ	1.60	
COPPER	MICRO	60.00	50.00	80.00	22.20	MICRO/MJ	174.27	
ZINC	MILLI	1.00	1.00	1.00	0.37	MILLI/MJ	2.90	
PHOSPHORUS	MILLI	114.00	90.00	130.00	42.18	MILLI/MJ	331.11	
IODIDE	MICRO	3.30	-	-	1.22	MICRO/MJ	9.58	
BORON	MICRO	70.00	30.00	140.00	25.90	MICRO/MJ	203.31	
SELENIUM	MICRO	2.90	2.90	2.90	1.07	MICRO/MJ	8.42	
CAROTENE	MILLI	0.18	0.18	0.18	0.07	MILLI/MJ	0.52	
VITAMIN E ACTIVITY	MILLI	0.10	-	-	0.04	MILLI/MJ	0.29	
TOTAL TOCOPHEROLS	MILLI	0.62	-	-	0.23	MILLI/MJ	1.80	
ALPHA-TOCOPHEROL	MICRO	40.00	-	-	14.80	MICRO/MJ	116.18	
GAMMA-TOCOPHEROL	MILLI	0.16	-	-	0.06	MILLI/MJ	0.46	
ALPHA-TOCOTRIENOL	MILLI	0.12	-	-	0.04	MILLI/MJ	0.35	
GAMMA-TOCOTRIENOL	MILLI	0.30	-	-	0.11	MILLI/MJ	0.87	
VITAMIN K	MICRO	1.50	-	-	0.56	MICRO/MJ	4.36	
VITAMIN B1	MILLI	0.15	-	-	0.06	MILLI/MJ	0.44	

CONSTITUENTS	DIM	AV	VARIATION			AVR	NUTR. DENS.		MOLPERC.
VITAMIN B2	MILLI	0.12	-	-	-	0.04	MILLI/MJ	0.35	
NICOTINAMIDE	MILLI	1.70	-	-	-	0.63	MILLI/MJ	4.94	
PANTOTHENIC ACID	MILLI	0.89	-	-	-	0.33	MILLI/MJ	2.58	
VITAMIN B6	MILLI	0.22	-	-	-	0.08	MILLI/MJ	0.64	
FOLIC ACID	MICRO	43.00	-	-	-	15.91	MICRO/MJ	124.89	
VITAMIN C	MILLI	12.00	-	-	-	4.44	MILLI/MJ	34.85	
ARGININE	MILLI	160.00	-	-	-	59.20	MILLI/MJ	464.71	
HISTIDINE	MILLI	85.00	-	-	-	31.45	MILLI/MJ	246.88	
ISOLEUCINE	MILLI	130.00	-	-	-	48.10	MILLI/MJ	377.58	
LEUCINE	MILLI	350.00	-	-	-	129.50	GRAM/MJ	1.02	
LYSINE	MILLI	130.00	-	-	-	48.10	MILLI/MJ	377.58	
METHIONINE	MILLI	56.00	-	-	-	20.72	MILLI/MJ	162.65	
PHENYLALANINE	MILLI	200.00	-	-	-	74.00	MILLI/MJ	580.89	
THREONINE	MILLI	130.00	-	-	-	48.10	MILLI/MJ	377.58	
TRYPTOPHAN	MILLI	16.00	-	-	-	5.92	MILLI/MJ	46.47	
VALINE	MILLI	220.00	-	-	-	81.40	MILLI/MJ	638.98	
MALIC ACID	MILLI	29.00	-	-	-	10.73	MILLI/MJ	84.23	
CITRIC ACID	MILLI	21.00	-	-	-	7.77	MILLI/MJ	60.99	
QUINIC ACID	MILLI	3.50	-	-	-	1.30	MILLI/MJ	10.17	
SUCCINIC ACID	MILLI	7.70	-	-	-	2.85	MILLI/MJ	22.36	
PYRROLIDONE CARB.ACID	MILLI	6.40	-	-	-	2.37	MILLI/MJ	18.59	
SALICYLIC ACID	MILLI	0.13	0.13	-	0.13	0.05	MILLI/MJ	0.38	
GLUCOSE	GRAM	0.62	0.36	-	0.88	0.23	GRAM/MJ	1.80	
FRUCTOSE	GRAM	0.37	0.22	-	0.53	0.14	GRAM/MJ	1.09	
SUCROSE	GRAM	2.15 [1]	1.61	-	2.70	0.80	GRAM/MJ	6.26	
RAFFINOSE	GRAM	0.20	-	-	-	0.07	GRAM/MJ	0.58	
MALTOSE	GRAM	0.27	-	-	-	0.10	GRAM/MJ	0.78	
STARCH	GRAM	12.30	-	-	-	4.55	GRAM/MJ	35.72	
MANNITOL	GRAM	0.20	-	-	-	0.07	GRAM/MJ	0.58	

1 EXTRA SWEET VARIETIES: SUCROSE CONTENT 9,1 (8,4-9,8) G/100G

Hülsenfrüchte und Ölsamen

AUGENBOHNE
(KUHBOHNE)
SAMEN, TROCKEN
VIGNA UNGUICULATA (L.) WALP.
V. SINENSIS (L.) SAVI EX HASSK

COWPEA COMMON
SEED, DRY

DOLIQUE, NIEBE
(HARICOT INDIGENE)
GRAINE, SÈCHE

		PROTEIN	FAT	CARBOHYDRATES	TOTAL
ENERGY VALUE (AVERAGE)	KJOULE	438	54	697	1189
PER 100 G	(KCAL)	105	13	167	284
EDIBLE PORTION					
AMOUNT OF DIGESTIBLE	GRAM	18.33	1.26	41.66	
CONSTITUENTS PER 100 G					
ENERGY VALUE (AVERAGE)	KJOULE	341	49	697	1088
OF THE DIGESTIBLE	(KCAL)	82	12	167	260
FRACTION PER 100 G					
EDIBLE PORTION					

WASTE PERCENTAGE AVERAGE 0.00

CONSTITUENTS	DIM	AV	VARIATION			AVR	NUTR. DENS.		MOLPERC.
WATER	GRAM	11.20	10.00	-	12.30	11.20	GRAM/MJ	10.30	
PROTEIN	GRAM	23.50	22.00	-	26.00	23.50	GRAM/MJ	21.61	
FAT	GRAM	1.40	1.10	-	1.60	1.40	GRAM/MJ	1.29	
AVAILABLE CARBOHYDR.	GRAM	41.66	-	-	-	41.66	GRAM/MJ	38.31	
MINERALS	GRAM	3.50	3.20	-	4.00	3.50	GRAM/MJ	3.22	
SODIUM	MILLI	6.00	-	-	-	6.00	MILLI/MJ	5.52	
POTASSIUM	MILLI	688.00	-	-	-	688.00	MILLI/MJ	632.62	
CALCIUM	MILLI	101.00	90.00	-	110.00	101.00	MILLI/MJ	92.87	
IRON	MILLI	6.40	5.00	-	7.60	6.40	MILLI/MJ	5.88	
PHOSPHORUS	MILLI	400.00	382.00	-	416.00	400.00	MILLI/MJ	367.81	
CAROTENE	MICRO	12.00	10.00	-	13.00	12.00	MICRO/MJ	11.03	
VITAMIN B1	MILLI	0.80	0.60	-	0.90	0.80	MILLI/MJ	0.74	
VITAMIN B2	MILLI	0.16	0.09	-	0.22	0.16	MILLI/MJ	0.15	
NICOTINAMIDE	MILLI	2.80	2.00	-	4.00	2.80	MILLI/MJ	2.57	
VITAMIN C	MILLI	1.50	1.00	-	2.00	1.50	MILLI/MJ	1.38	

CONSTITUENTS	DIM	AV	VARIATION			AVR	NUTR. DENS.		MOLPERC.
ALANINE	GRAM	1.08	-	-	-	1.08	GRAM/MJ	0.99	6.0
ARGININE	GRAM	2.04	-	-	-	2.04	GRAM/MJ	1.88	5.8
ASPARTIC ACID	GRAM	3.17	-	-	-	3.17	GRAM/MJ	2.91	11.8
CYSTINE	GRAM	0.12	-	-	-	0.12	GRAM/MJ	0.11	0.2
GLUTAMIC ACID	GRAM	3.87	-	-	-	3.87	GRAM/MJ	3.56	13.0
GLYCINE	GRAM	1.08	-	-	-	1.08	GRAM/MJ	0.99	7.1
HISTIDINE	GRAM	1.18	-	-	-	1.18	GRAM/MJ	1.09	3.8
ISOLEUCINE	GRAM	1.66	-	-	-	1.66	GRAM/MJ	1.53	6.3
LEUCINE	GRAM	2.11	-	-	-	2.11	GRAM/MJ	1.94	8.0
LYSINE	GRAM	1.57	-	-	-	1.57	GRAM/MJ	1.44	5.3
METHIONINE	GRAM	0.33	-	-	-	0.33	GRAM/MJ	0.30	1.1
PHENYLALANINE	GRAM	1.30	-	-	-	1.30	GRAM/MJ	1.20	3.9
PROLINE	GRAM	1.67	-	-	-	1.67	GRAM/MJ	1.54	7.2
SERINE	GRAM	1.36	-	-	-	1.36	GRAM/MJ	1.25	6.4
THREONINE	GRAM	1.01	-	-	-	1.01	GRAM/MJ	0.93	4.2
TRYPTOPHAN	GRAM	0.33	-	-	-	0.33	GRAM/MJ	0.30	0.8
TYROSINE	GRAM	0.62	-	-	-	0.62	GRAM/MJ	0.57	1.7
VALINE	GRAM	1.74	-	-	-	1.74	GRAM/MJ	1.60	7.4
GLUCOSE	GRAM	0.18	0.18	-	0.18	0.18	GRAM/MJ	0.17	
FRUCTOSE	GRAM	0.36	0.36	-	0.36	0.36	GRAM/MJ	0.33	
SUCROSE	GRAM	1.32	0.81	-	1.97	1.33	GRAM/MJ	1.22	
RAFFINOSE	GRAM	1.46	0.63	-	2.56	1.47	GRAM/MJ	1.35	
STACHYOSE	GRAM	2.91	2.42	-	3.33	2.92	GRAM/MJ	2.68	
VERBASCOSE	GRAM	2.33	0.81	-	3.33	2.34	GRAM/MJ	2.15	
STARCH	GRAM	39.80	31.50	-	48.00	39.80	GRAM/MJ	36.60	
LAURIC ACID	MILLI	0.88	0.29	-	2.20	0.88	MILLI/MJ	0.81	
MYRISTIC ACID	MILLI	2.80	-	-	-	2.80	MILLI/MJ	2.57	
PALMITIC ACID	MILLI	351.00	-	-	-	351.00	MILLI/MJ	322.75	
STEARIC ACID	MILLI	76.00	-	-	-	76.00	MILLI/MJ	69.88	
ARACHIDIC ACID	MILLI	29.00	-	-	-	29.00	MILLI/MJ	26.67	
BEHENIC ACID	MILLI	79.00	-	-	-	79.00	MILLI/MJ	72.64	
OLEIC ACID	MILLI	112.00	-	-	-	112.00	MILLI/MJ	102.99	
LINOLEIC ACID	MILLI	443.00	-	-	-	443.00	MILLI/MJ	407.34	
LINOLENIC ACID	MILLI	262.00	-	-	-	262.00	MILLI/MJ	240.91	

BOHNE (GARTENBOHNE) SAMEN, WEISS, TROCKEN	**BEAN** (HARICOT) SEED, WHITE, DRY	**HARICOT** GRAINE, BLANCHE, SÈCHE

PHASEOLUS VULGARIS L.

		PROTEIN	FAT	CARBOHYDRATES	TOTAL
ENERGY VALUE (AVERAGE) PER 100 G EDIBLE PORTION	KJOULE	397	62	800	1259
	(KCAL)	95	15	191	301
AMOUNT OF DIGESTIBLE CONSTITUENTS PER 100 G	GRAM	16.61	1.44	47.80	
ENERGY VALUE (AVERAGE) OF THE DIGESTIBLE FRACTION PER 100 G EDIBLE PORTION	KJOULE	309	56	800	1165
	(KCAL)	74	13	191	279

WASTE PERCENTAGE AVERAGE 1.0

CONSTITUENTS	DIM	AV	VARIATION			AVR	NUTR.	DENS.	MOLPERC.
WATER	GRAM	11.60	11.20	-	12.00	11.48	GRAM/MJ	9.95	
PROTEIN	GRAM	21.30	20.00	-	24.30	21.09	GRAM/MJ	18.28	
FAT	GRAM	1.60	1.30	-	2.00	1.58	GRAM/MJ	1.37	
AVAILABLE CARBOHYDR.	GRAM	47.80	-	-	-	47.32	GRAM/MJ	41.02	
TOTAL DIETARY FIBRE	GRAM	17.00 1	-	-	-	16.83	GRAM/MJ	14.59	
MINERALS	GRAM	4.00	3.90	-	4.20	3.96	GRAM/MJ	3.43	
SODIUM	MILLI	2.00	1.00	-	18.00	1.98	MILLI/MJ	1.72	
POTASSIUM	GRAM	1.31	1.16	-	1.57	1.30	GRAM/MJ	1.12	
MAGNESIUM	MILLI	132.00	130.00	-	203.00	130.68	MILLI/MJ	113.27	
CALCIUM	MILLI	106.00	52.00	-	185.00	104.94	MILLI/MJ	90.96	
MANGANESE	MILLI	2.00	1.00	-	3.00	1.98	MILLI/MJ	1.72	
IRON	MILLI	6.10	2.90	-	7.10	6.04	MILLI/MJ	5.23	
COBALT	MICRO	8.00	-	-	-	7.92	MICRO/MJ	6.86	
COPPER	MILLI	0.80	0.60	-	1.20	0.79	MILLI/MJ	0.69	
ZINC	MILLI	2.80	1.70	-	5.70	2.77	MILLI/MJ	2.40	
NICKEL	MILLI	0.28	0.07	-	0.50	0.28	MILLI/MJ	0.24	
CHROMIUM	MICRO	20.00	20.00	-	20.00	19.80	MICRO/MJ	17.16	
MOLYBDENUM	MILLI	-	0.04	-	0.46				
PHOSPHORUS	MILLI	429.00	335.00	-	503.00	424.71	MILLI/MJ	368.13	
CHLORIDE	MILLI	47.00	-	-	-	46.53	MILLI/MJ	40.33	
FLUORIDE	MILLI	-	0.02	-	0.17				
IODIDE	MICRO	0.60	0.60	-	0.60	0.59	MICRO/MJ	0.51	
BORON	MILLI	0.43	0.35	-	0.51	0.43	MILLI/MJ	0.37	
SELENIUM	MICRO	22.00	-	-	-	21.78	MICRO/MJ	18.88	
CAROTENE	MILLI	0.40	0.18	-	0.72	0.40	MILLI/MJ	0.34	
VITAMIN E ACTIVITY	MILLI	0.21	-	-	-	0.21	MILLI/MJ	0.18	

1 METHOD OF MEUSER, SUCKOW AND KULIKOWSKI ("BERLINER METHODE")

CONSTITUENTS	DIM	AV	VARIATION			AVR	NUTR. DENS.		MOLPERC.
TOTAL TOCOPHEROLS	MILLI	2.10	-	-	-	2.08	MILLI/MJ	1.80	
ALPHA-TOCOPHEROL	-	TRACES	-	-	-				
GAMMA-TOCOPHEROL	MILLI	2.10	-	-	-	2.08	MILLI/MJ	1.80	
VITAMIN B1	MILLI	0.46	0.22	-	1.20	0.46	MILLI/MJ	0.39	
VITAMIN B2	MILLI	0.16	0.09	-	0.36	0.16	MILLI/MJ	0.14	
NICOTINAMIDE	MILLI	2.10	0.75	-	2.80	2.08	MILLI/MJ	1.80	
PANTOTHENIC ACID	MILLI	0.98	0.49	-	1.32	0.97	MILLI/MJ	0.84	
VITAMIN B6	MILLI	0.28	0.27	-	0.58	0.28	MILLI/MJ	0.24	
FOLIC ACID	MILLI	0.13	0.13	-	0.60	0.13	MILLI/MJ	0.11	
VITAMIN C	MILLI	2.50	0.00	-	5.00	2.48	MILLI/MJ	2.15	
ALANINE	GRAM	0.74	-	-	-	0.73	GRAM/MJ	0.64	4.5
ARGININE	GRAM	1.49	1.16	-	2.21	1.48	GRAM/MJ	1.28	4.6
ASPARTIC ACID	GRAM	2.45	-	-	-	2.43	GRAM/MJ	2.10	9.9
CYSTINE	GRAM	0.23	0.15	-	0.49	0.23	GRAM/MJ	0.20	0.5
GLUTAMIC ACID	GRAM	4.33	-	-	-	4.29	GRAM/MJ	3.72	15.8
GLYCINE	GRAM	0.95	-	-	-	0.94	GRAM/MJ	0.82	6.8
HISTIDINE	GRAM	0.70	0.37	-	0.90	0.69	GRAM/MJ	0.60	2.4
ISOLEUCINE	GRAM	1.49	1.13	-	1.76	1.48	GRAM/MJ	1.28	6.1
LEUCINE	GRAM	2.26	1.89	-	3.46	2.24	GRAM/MJ	1.94	9.3
LYSINE	GRAM	1.87	1.34	-	2.38	1.85	GRAM/MJ	1.60	6.9
METHIONINE	GRAM	0.26	0.14	-	0.48	0.26	GRAM/MJ	0.22	0.9
PHENYLALANINE	GRAM	1.40	0.84	-	1.86	1.39	GRAM/MJ	1.20	4.6
PROLINE	GRAM	0.98	-	-	-	0.97	GRAM/MJ	0.84	4.6
SERINE	GRAM	1.38	-	-	-	1.37	GRAM/MJ	1.18	7.1
THREONINE	GRAM	1.15	0.69	-	1.43	1.14	GRAM/MJ	0.99	5.2
TRYPTOPHAN	GRAM	0.23	0.11	-	0.41	0.23	GRAM/MJ	0.20	0.6
TYROSINE	GRAM	0.97	0.53	-	1.48	0.96	GRAM/MJ	0.83	2.9
VALINE	GRAM	1.63	1.29	-	1.91	1.61	GRAM/MJ	1.40	7.5
GLUCOSE	-	TRACES	-	-	-				
FRUCTOSE	-	TRACES	-	-	-				
SUCROSE	GRAM	1.58	-	-	-	1.56	GRAM/MJ	1.36	
RAFFINOSE	GRAM	0.39	0.32	-	0.46	0.39	GRAM/MJ	0.33	
STACHYOSE	GRAM	3.10	2.90	-	3.30	3.07	GRAM/MJ	2.66	
STARCH	GRAM	46.20	-	-	-	45.74	GRAM/MJ	39.64	
PECTIN	GRAM	3.10	2.90	-	3.80	3.07	GRAM/MJ	2.66	
CELLULOSE	GRAM	6.10	5.70	-	6.40	6.04	GRAM/MJ	5.23	
PHYTIC ACID	GRAM	0.80	0.55	-	1.05	0.79	GRAM/MJ	0.69	
PHOSPHATIDYLCHOLINE	MILLI	341.00	-	-	-	337.59	MILLI/MJ	292.62	
PHOSPHATIDYLETHANOLAMINE	MILLI	96.00	-	-	-	95.04	MILLI/MJ	82.38	
PHOSPHATIDYLSERINE	MILLI	107.00	-	-	-	105.93	MILLI/MJ	91.82	
DIETARY FIBRE,WAT.SOL.	GRAM	8.70	-	-	-	8.61	GRAM/MJ	7.47	
DIETARY FIBRE,WAT.INS.	GRAM	8.30	-	-	-	8.22	GRAM/MJ	7.12	

BOHNEN
SAMEN, WEISS
GEKOCHT

BEANS
SEED, WHITE
COOKED

HARICOTS
GRAINES, BLANCHES
CUITES

		PROTEIN	FAT	CARBOHYDRATES	TOTAL
ENERGY VALUE (AVERAGE)	KJOULE	139	23	241	403
PER 100 G EDIBLE PORTION	(KCAL)	33	5.6	58	96
AMOUNT OF DIGESTIBLE CONSTITUENTS PER 100 G	GRAM	5.81	0.54	14.40	
ENERGY VALUE (AVERAGE)	KJOULE	108	21	241	370
OF THE DIGESTIBLE FRACTION PER 100 G EDIBLE PORTION	(KCAL)	26	5.0	58	88

WASTE PERCENTAGE AVERAGE 0.00

CONSTITUENTS	DIM	AV	VARIATION			AVR	NUTR. DENS.		MOLPERC.
WATER	GRAM	73.40	-	-	-	73.40	GRAM/MJ	198.27	
PROTEIN	GRAM	7.45	-	-	-	7.45	GRAM/MJ	20.12	
FAT	GRAM	0.60	-	-	-	0.60	GRAM/MJ	1.62	
AVAILABLE CARBOHYDR.	GRAM	14.40 1	-	-	-	14.40	GRAM/MJ	38.90	
MINERALS	GRAM	1.40	-	-	-	1.40	GRAM/MJ	3.78	
SODIUM	MILLI	0.70	-	-	-	0.70	MILLI/MJ	1.89	
POTASSIUM	MILLI	455.00	-	-	-	455.00	GRAM/MJ	1.23	
CALCIUM	MILLI	36.00	-	-	-	36.00	MILLI/MJ	97.24	
IRON	MILLI	21.00	-	-	-	21.00	MILLI/MJ	56.73	
PHOSPHORUS	MILLI	149.00	-	-	-	149.00	MILLI/MJ	402.48	
VITAMIN B1	MILLI	0.10	-	-	-	0.10	MILLI/MJ	0.27	
VITAMIN B2	MICRO	50.00	-	-	-	50.00	MICRO/MJ	135.06	
NICOTINAMIDE	MILLI	0.61	-	-	-	0.61	MILLI/MJ	1.65	

1 CALCULATED FROM THE DRY PRODUCT

ERBSE
SCHOTE UND SAMEN, GRÜN

PISUM SATIVUM L.

PEA
POD AND SEED, GREEN

PETIT POIS
VERT, AVEC COSSE

		PROTEIN	FAT	CARBOHYDRATES	TOTAL
ENERGY VALUE (AVERAGE) PER 100 G EDIBLE PORTION	KJOULE	122	19	211	351
	(KCAL)	29	4.5	50	84
AMOUNT OF DIGESTIBLE CONSTITUENTS PER 100 G	GRAM	5.10	0.43	12.58	
ENERGY VALUE (AVERAGE) OF THE DIGESTIBLE FRACTION PER 100 G EDIBLE PORTION	KJOULE	95	17	211	322
	(KCAL)	23	4.0	50	77

WASTE PERCENTAGE AVERAGE 60 MINIMUM 55 MAXIMUM 75

CONSTITUENTS	DIM	AV	VARIATION			AVR	NUTR. DENS.		MOLPERC.
WATER	GRAM	77.30	74.30	-	78.90	30.92	GRAM/MJ	239.71	
PROTEIN	GRAM	6.55	5.85	-	7.40	2.62	GRAM/MJ	20.31	
FAT	GRAM	0.48	0.40	-	0.70	0.19	GRAM/MJ	1.49	
AVAILABLE CARBOHYDR.	GRAM	12.58	-	-	-	5.03	GRAM/MJ	39.01	
TOTAL DIETARY FIBRE	GRAM	4.25 1	-	-	-	1.70	GRAM/MJ	13.18	
MINERALS	GRAM	0.92	0.85	-	1.10	0.37	GRAM/MJ	2.85	
SODIUM	MILLI	2.00	0.50	-	4.00	0.80	MILLI/MJ	6.20	
POTASSIUM	MILLI	304.00	215.00	-	370.00	121.60	MILLI/MJ	942.72	
MAGNESIUM	MILLI	33.00	19.00	-	43.00	13.20	MILLI/MJ	102.33	
CALCIUM	MILLI	24.00	15.00	-	34.00	9.60	MILLI/MJ	74.43	
MANGANESE	MILLI	0.66	0.41	-	0.98	0.26	MILLI/MJ	2.05	
IRON	MILLI	1.84	1.65	-	2.10	0.74	MILLI/MJ	5.71	
COBALT	MICRO	3.00	-	-	-	1.20	MICRO/MJ	9.30	
COPPER	MILLI	0.38	0.20	-	0.59	0.15	MILLI/MJ	1.18	
ZINC	MILLI	1.03	1.03	-	1.03	0.41	MILLI/MJ	3.19	
CHROMIUM	MICRO	2.00	0.50	-	4.00	0.80	MICRO/MJ	6.20	
MOLYBDENUM	MICRO	70.00	20.00	-	100.00	28.00	MICRO/MJ	217.07	
PHOSPHORUS	MILLI	108.00	78.00	-	145.00	43.20	MILLI/MJ	334.91	
CHLORIDE	MILLI	40.00	28.00	-	52.00	16.00	MILLI/MJ	124.04	
FLUORIDE	MICRO	27.00	24.00	-	29.00	10.80	MICRO/MJ	83.73	
IODIDE	MICRO	4.20	2.30	-	6.00	1.68	MICRO/MJ	13.02	
BORON	MICRO	45.00	20.00	-	80.00	18.00	MICRO/MJ	139.55	
SELENIUM	MICRO	0.30	-	-	-	0.12	MICRO/MJ	0.93	
SILICON	MICRO	0.40	0.20	-	0.50	0.16	MICRO/MJ	1.24	
NITRATE	MILLI	3.00	0.00	-	11.00	1.20	MILLI/MJ	9.30	
CAROTENE	MILLI	0.38	0.18	-	0.55	0.15	MILLI/MJ	1.18	
VITAMIN E ACTIVITY	MILLI	0.39	-	-	-	0.16	MILLI/MJ	1.21	
TOTAL TOCOPHEROLS	MILLI	2.70	-	-	-	1.08	MILLI/MJ	8.37	
ALPHA-TOCOPHEROL	MILLI	0.13	-	-	-	0.05	MILLI/MJ	0.40	
GAMMA-TOCOPHEROL	MILLI	2.60	-	-	-	1.04	MILLI/MJ	8.06	
VITAMIN K	MICRO	22.00	7.00	-	36.00	8.80	MICRO/MJ	68.22	
VITAMIN B1	MILLI	0.30	0.17	-	0.40	0.12	MILLI/MJ	0.93	
VITAMIN B2	MILLI	0.16	0.12	-	0.19	0.06	MILLI/MJ	0.50	
NICOTINAMIDE	MILLI	2.38	1.91	-	2.70	0.95	MILLI/MJ	7.38	
PANTOTHENIC ACID	MILLI	0.72	0.65	-	0.82	0.29	MILLI/MJ	2.23	

1 METHOD OF SOUTHGATE AND ENGLYST

CONSTITUENTS	DIM	AV	VARIATION			AVR	NUTR. DENS.		MOLPERC.
VITAMIN B6	MILLI	0.16	0.13	-	0.18	0.06	MILLI/MJ	0.50	
BIOTIN	MICRO	5.30	3.50	-	7.00	2.12	MICRO/MJ	16.44	
FOLIC ACID	MICRO	33.00	10.00	-	50.00	13.20	MICRO/MJ	102.33	
VITAMIN C	MILLI	25.00	13.00	-	39.00	10.00	MILLI/MJ	77.53	
ALANINE	GRAM	0.14	-	-	-	0.06	GRAM/MJ	0.43	2.9
ARGININE	GRAM	1.06	0.95	-	1.34	0.42	GRAM/MJ	3.29	11.3
ASPARTIC ACID	GRAM	0.55	-	-	-	0.22	GRAM/MJ	1.71	7.7
CYSTINE	GRAM	0.13	0.10	-	0.16	0.05	GRAM/MJ	0.40	1.0
GLUTAMIC ACID	GRAM	0.99	-	-	-	0.40	GRAM/MJ	3.07	12.5
GLYCINE	GRAM	0.17	-	-	-	0.07	GRAM/MJ	0.53	4.2
HISTIDINE	GRAM	0.22	0.17	-	0.23	0.09	GRAM/MJ	0.68	2.6
ISOLEUCINE	GRAM	0.54	0.47	-	0.57	0.22	GRAM/MJ	1.67	7.7
LEUCINE	GRAM	0.67	0.54	-	0.81	0.27	GRAM/MJ	2.08	9.5
LYSINE	GRAM	0.61	0.52	-	0.68	0.24	GRAM/MJ	1.89	7.8
METHIONINE	GRAM	0.10	0.09	-	0.10	0.04	GRAM/MJ	0.31	1.2
PHENYLALANINE	GRAM	0.40	0.30	-	0.49	0.16	GRAM/MJ	1.24	4.5
PROLINE	GRAM	0.14	-	-	-	0.06	GRAM/MJ	0.43	2.3
SERINE	GRAM	0.28	-	-	-	0.11	GRAM/MJ	0.87	5.0
THREONINE	GRAM	0.45	0.42	-	0.52	0.18	GRAM/MJ	1.40	7.0
TRYPTOPHAN	GRAM	0.10	0.08	-	0.10	0.04	GRAM/MJ	0.31	0.9
TYROSINE	GRAM	0.35	0.27	-	0.46	0.14	GRAM/MJ	1.09	3.6
VALINE	GRAM	0.52	0.48	-	0.56	0.21	GRAM/MJ	1.61	8.3
MALIC ACID	MILLI	139.00	99.00	-	178.00	55.60	MILLI/MJ	431.05	
CITRIC ACID	MILLI	139.00	123.00	-	154.00	55.60	MILLI/MJ	431.05	
LACTIC ACID	-	TRACES	-	-	-				
OXALIC ACID TOTAL	-	0.00	-	-	-	0.00			
SUCCINIC ACID	MILLI	6.00	5.00	-	7.00	2.40	MILLI/MJ	18.61	
FUMARIC ACID	MILLI	14.00	6.00	-	21.00	5.60	MILLI/MJ	43.41	
GLYCOLIC ACID	MILLI	7.00	5.00	-	8.00	2.80	MILLI/MJ	21.71	
ISOCITRIC ACID	-	TRACES	-	-	-				
GALACTURONIC ACID	-	TRACES	-	-	-				
OXALOACETIC ACID	MILLI	2.00	1.00	-	3.00	0.80	MILLI/MJ	6.20	
ALPHA-KETOGLUT.ACID	MILLI	6.00	3.00	-	9.00	2.40	MILLI/MJ	18.61	
SALICYLIC ACID	MICRO	40.00	40.00	-	40.00	16.00	MICRO/MJ	124.04	
GLUCOSE	MILLI	88.00	62.00	-	220.00	35.20	MILLI/MJ	272.89	
FRUCTOSE	MILLI	65.00	19.00	-	190.00	26.00	MILLI/MJ	201.57	
SUCROSE	GRAM	1.15	0.89	-	1.41	0.46	GRAM/MJ	3.57	
STARCH	GRAM	11.00	-	-	-	4.40	GRAM/MJ	34.11	
PENTOSAN	GRAM	0.67	-	-	-	0.27	GRAM/MJ	2.08	
CELLULOSE	GRAM	2.09	-	-	-	0.84	GRAM/MJ	6.48	
POLYURONIC ACID	GRAM	0.22	-	-	-	0.09	GRAM/MJ	0.68	
PALMITIC ACID	MILLI	93.00	93.00	-	93.00	37.20	MILLI/MJ	288.40	
STEARIC ACID	MILLI	6.20	6.20	-	6.20	2.48	MILLI/MJ	19.23	
OLEIC ACID	MILLI	37.00	37.00	-	37.00	14.80	MILLI/MJ	114.74	
LINOLEIC ACID	MILLI	247.00	247.00	-	247.00	98.80	MILLI/MJ	765.96	
LINOLENIC ACID	MILLI	50.00	50.00	-	50.00	20.00	MILLI/MJ	155.05	
CAMPESTEROL	MILLI	10.00	-	-	-	4.00	MILLI/MJ	31.01	
BETA-SITOSTEROL	MILLI	106.00	-	-	-	42.40	MILLI/MJ	328.71	
STIGMASTEROL	MILLI	10.00	-	-	-	4.00	MILLI/MJ	31.01	
TOTAL PURINES	MILLI	84.00 [2]	-	-	-	33.60	MILLI/MJ	260.49	
TOTAL PHOSPHOLIPIDS	MILLI	345.00	-	-	-	138.00	GRAM/MJ	1.07	
PHOSPHATIDYLCHOLINE	MILLI	214.00	-	-	-	85.60	MILLI/MJ	663.62	
PHOSPHATIDYLETHANOLAMINE	MILLI	62.00	-	-	-	24.80	MILLI/MJ	192.26	
PHOSPHATIDYLINOSITOL	MILLI	70.00	-	-	-	28.00	MILLI/MJ	217.07	
DIETARY FIBRE, WAT.SOL.	GRAM	0.26	-	-	-	0.10	GRAM/MJ	0.81	
DIETARY FIBRE, WAT.INS.	GRAM	3.99	-	-	-	1.60	GRAM/MJ	12.37	

2 VALUE EXPRESSED IN MG URIC ACID/100G

ERBSEN	PEAS	PETITS POIS
SAMEN, GRÜN, GEKOCHT, ABGETROPFT	SEEDS, GREEN, BOILED, DRAINED	GRAINES, VERTES, BOUILLIES, DEGOUTTÉES

		PROTEIN	FAT	CARBOHYDRATES	TOTAL
ENERGY VALUE (AVERAGE) PER 100 G EDIBLE PORTION	KJOULE (KCAL)	104 25	19 4.7	181 43	304 73
AMOUNT OF DIGESTIBLE CONSTITUENTS PER 100 G	GRAM	4.36	0.45	10.80	
ENERGY VALUE (AVERAGE) OF THE DIGESTIBLE FRACTION PER 100 G EDIBLE PORTION	KJOULE (KCAL)	81 19	18 4.2	181 43	280 67

WASTE PERCENTAGE AVERAGE 0.00

CONSTITUENTS	DIM	AV	VARIATION			AVR	NUTR. DENS.		MOLPERC.
WATER	GRAM	80.40	-	-	-	80.40	GRAM/MJ	287.57	
PROTEIN	GRAM	5.60	-	-	-	5.60	GRAM/MJ	20.03	
FAT	GRAM	0.50	-	-	-	0.50	GRAM/MJ	1.79	
AVAILABLE CARBOHYDR.	GRAM	10.80 1	-	-	-	10.80	GRAM/MJ	38.63	
MINERALS	GRAM	0.70	-	-	-	0.70	GRAM/MJ	2.50	
SODIUM	MILLI	2.00	-	-	-	2.00	MILLI/MJ	7.15	
POTASSIUM	MILLI	213.00	-	-	-	213.00	MILLI/MJ	761.84	
CALCIUM	MILLI	22.00	-	-	-	22.00	MILLI/MJ	78.69	
IRON	MILLI	1.30	-	-	-	1.30	MILLI/MJ	4.65	
PHOSPHORUS	MILLI	91.00	-	-	-	91.00	MILLI/MJ	325.48	
VITAMIN B1	MILLI	0.23	-	-	-	0.23	MILLI/MJ	0.82	
VITAMIN B2	MILLI	0.16	-	-	-	0.16	MILLI/MJ	0.57	
NICOTINAMIDE	MILLI	2.00	-	-	-	2.00	MILLI/MJ	7.15	
VITAMIN C	MILLI	17.00	-	-	-	17.00	MILLI/MJ	60.80	

1 CALCULATED FROM THE FRESH PRODUCT

ERBSEN SAMEN, GRÜN IN DOSEN 1	**PEAS** SEEDS, GREEN CANNED 1	**PETITS POIS** GRAINES, VERTES EN BOÎTES 1

		PROTEIN	FAT	CARBOHYDRATES	TOTAL
ENERGY VALUE (AVERAGE)	KJOULE	67	14	147	227
PER 100 G	(KCAL)	16	3.3	35	54
EDIBLE PORTION					
AMOUNT OF DIGESTIBLE	GRAM	2.80	0.31	8.76	
CONSTITUENTS PER 100 G					
ENERGY VALUE (AVERAGE)	KJOULE	52	12	147	211
OF THE DIGESTIBLE	(KCAL)	12	2.9	35	50
FRACTION PER 100 G					
EDIBLE PORTION					

WASTE PERCENTAGE AVERAGE 0.00

CONSTITUENTS	DIM	AV	VARIATION			AVR	NUTR. DENS.		MOLPERC.
WATER	GRAM	84.20	82.00	-	87.80	84.20	GRAM/MJ	398.78	
PROTEIN	GRAM	3.60	3.40	-	4.05	3.60	GRAM/MJ	17.05	
FAT	GRAM	0.35	0.30	-	0.40	0.35	GRAM/MJ	1.66	
AVAILABLE CARBOHYDR.	GRAM	8.76 2	-	-	-	8.76	GRAM/MJ	41.49	
MINERALS	GRAM	1.13	1.07	-	1.21	1.13	GRAM/MJ	5.35	
SODIUM	MILLI	236.00	-	-	-	236.00	GRAM/MJ	1.12	
POTASSIUM	MILLI	99.00	71.00	-	148.00	99.00	MILLI/MJ	468.87	
MAGNESIUM	MILLI	20.00	12.00	-	29.00	20.00	MILLI/MJ	94.72	
CALCIUM	MILLI	20.00	19.00	-	22.00	20.00	MILLI/MJ	94.72	
MANGANESE	MILLI	0.20	0.19	-	0.21	0.20	MILLI/MJ	0.95	
IRON	MILLI	1.53	1.33	-	1.70	1.53	MILLI/MJ	7.25	
COPPER	MILLI	0.13	0.08	-	0.16	0.13	MILLI/MJ	0.62	
ZINC	MILLI	0.65	0.60	-	0.68	0.65	MILLI/MJ	3.08	
PHOSPHORUS	MILLI	62.00	58.00	-	66.00	62.00	MILLI/MJ	293.64	
IODIDE	MICRO	2.20	-	-	-	2.20	MICRO/MJ	10.42	
BORON	MILLI	0.16	-	-	-	0.16	MILLI/MJ	0.76	
CAROTENE	MILLI	0.26	0.22	-	0.30	0.26	MILLI/MJ	1.23	
VITAMIN B1	MILLI	0.10	0.09	-	0.11	0.10	MILLI/MJ	0.47	
VITAMIN B2	MICRO	55.00	50.00	-	60.00	55.00	MICRO/MJ	260.48	
NICOTINAMIDE	MILLI	0.88	0.70	-	1.00	0.88	MILLI/MJ	4.17	
PANTOTHENIC ACID	MILLI	0.11	0.09	-	0.14	0.11	MILLI/MJ	0.52	
VITAMIN B6	MICRO	46.00	37.00	-	66.00	46.00	MICRO/MJ	217.86	
BIOTIN	MICRO	1.50	-	-	-	1.50	MICRO/MJ	7.10	
FOLIC ACID	MICRO	12.00	8.00	-	17.00	12.00	MICRO/MJ	56.83	
VITAMIN C	MILLI	8.80	7.40	-	10.00	8.80	MILLI/MJ	41.68	

1 TOTAL CONTENT (SOLID AND LIQUID)

2 CALCULATED FROM THE FRESH PRODUCT

CONSTITUENTS	DIM	AV	VARIATION			AVR	NUTR. DENS.		MOLPERC.
ARGININE	GRAM	0.34	0.25	-	0.40	0.34	GRAM/MJ	1.61	
CYSTINE	MILLI	29.00	25.00	-	32.00	29.00	MILLI/MJ	137.35	
HISTIDINE	MILLI	64.00	54.00	-	76.00	64.00	MILLI/MJ	303.11	
ISOLEUCINE	GRAM	0.14	0.13	-	0.15	0.14	GRAM/MJ	0.66	
LEUCINE	GRAM	0.23	0.21	-	0.26	0.23	GRAM/MJ	1.09	
LYSINE	GRAM	0.23	0.18	-	0.33	0.23	GRAM/MJ	1.09	
METHIONINE	MILLI	30.00	25.00	-	36.00	30.00	MILLI/MJ	142.08	
PHENYLALANINE	GRAM	0.13	0.12	-	0.17	0.13	GRAM/MJ	0.62	
THREONINE	GRAM	0.15	0.12	-	0.20	0.15	GRAM/MJ	0.71	
TRYPTOPHAN	MILLI	36.00	22.00	-	50.00	36.00	MILLI/MJ	170.50	
TYROSINE	MILLI	100.00	86.00	-	120.00	100.00	MILLI/MJ	473.61	
VALINE	GRAM	0.16	0.16	-	0.17	0.16	GRAM/MJ	0.76	
GLUCOSE	MILLI	36.00	16.00	-	52.00	36.00	MILLI/MJ	170.50	
FRUCTOSE	MILLI	34.00	18.00	-	50.00	34.00	MILLI/MJ	161.03	
SUCROSE	GRAM	2.30	0.62	-	3.89	2.30	GRAM/MJ	10.89	

ERBSEN	PEAS	PETITS POIS
SAMEN, GRÜN IN DOSEN, ABGETROPFT	SEEDS, GREEN CANNED, DRAINED	GRAINES, VERTES, EN BOÎTES, DEGOUTTÉES
1	1	1

		PROTEIN	FAT	CARBOHYDRATES	TOTAL
ENERGY VALUE (AVERAGE) PER 100 G EDIBLE PORTION	KJOULE	99	17	199	315
	(KCAL)	24	4.0	48	75
AMOUNT OF DIGESTIBLE CONSTITUENTS PER 100 G	GRAM	4.05	0.38	11.91	
ENERGY VALUE (AVERAGE) OF THE DIGESTIBLE FRACTION PER 100 G EDIBLE PORTION	KJOULE	77	15	199	292
	(KCAL)	18	3.6	48	70

WASTE PERCENTAGE AVERAGE 0.00

CONSTITUENTS	DIM	AV		VARIATION		AVR	NUTR. DENS.		MOLPERC.
WATER	GRAM	78.50		72.70 -	83.10	78.50	GRAM/MJ	269.21	
PROTEIN	GRAM	5.20		4.20 -	5.90	5.20	GRAM/MJ	17.83	
FAT	GRAM	0.43		0.40 -	0.50	0.43	GRAM/MJ	1.47	
AVAILABLE CARBOHYDR.	GRAM	11.91	2	- -	-	11.91	GRAM/MJ	40.84	
MINERALS	GRAM	1.05		1.00 -	1.10	1.05	GRAM/MJ	3.60	
SODIUM	MILLI	211.00	3	157.00 -	270.00	211.00	MILLI/MJ	723.60	
POTASSIUM	MILLI	135.00		96.00 -	201.00	135.00	MILLI/MJ	462.96	
MAGNESIUM	MILLI	27.00		16.00 -	40.00	27.00	MILLI/MJ	92.59	
CALCIUM	MILLI	29.00		26.00 -	34.00	29.00	MILLI/MJ	99.45	
MANGANESE	MILLI	0.28		0.26 -	0.29	0.28	MILLI/MJ	0.96	
IRON	MILLI	1.72		1.48 -	1.90	1.72	MILLI/MJ	5.90	
COPPER	MILLI	0.18		0.11 -	0.22	0.18	MILLI/MJ	0.62	
ZINC	MILLI	0.25		0.25 -	0.25	0.25	MILLI/MJ	0.86	
NICKEL	MICRO	16.00		- -	-	16.00	MICRO/MJ	54.87	
PHOSPHORUS	MILLI	83.00		72.00 -	99.00	83.00	MILLI/MJ	284.64	
CHLORIDE	MILLI	332.00		267.00 -	410.00	332.00	GRAM/MJ	1.14	
IODIDE	MICRO	3.00		- -	-	3.00	MICRO/MJ	10.29	
BORON	MILLI	0.16		- -	-	0.16	MILLI/MJ	0.55	
CAROTENE	MILLI	0.36		0.30 -	0.41	0.36	MILLI/MJ	1.23	
VITAMIN B1	MILLI	0.12		0.09 -	0.17	0.12	MILLI/MJ	0.41	
VITAMIN B2	MICRO	67.00		60.00 -	80.00	67.00	MICRO/MJ	229.77	
NICOTINAMIDE	MILLI	0.98		0.80 -	1.20	0.98	MILLI/MJ	3.36	
PANTOTHENIC ACID	MILLI	0.15		0.12 -	0.19	0.15	MILLI/MJ	0.51	
VITAMIN B6	MICRO	63.00		50.00 -	90.00	63.00	MICRO/MJ	216.05	
BIOTIN	MICRO	2.00		- -	-	2.00	MICRO/MJ	6.86	
FOLIC ACID	MICRO	17.00		11.00 -	23.00	17.00	MICRO/MJ	58.30	

1 ONLY THE DRAINED SOLIDS

2 CALCULATED FROM THE FRESH PRODUCT

3 LOW SODIUM (DIETARY) PRODUCTS CONTAIN 1-2 MG SODIUM PER 100G

CONSTITUENTS	DIM	AV	VARIATION			AVR	NUTR. DENS.		MOLPERC.
VITAMIN C	MILLI	9.60	6.80	-	13.00	9.60	MILLI/MJ	32.92	
ARGININE	GRAM	0.50	0.36	-	0.57	0.50	GRAM/MJ	1.71	
CYSTINE	MILLI	42.00	36.00	-	47.00	42.00	MILLI/MJ	144.03	
HISTIDINE	MILLI	93.00	78.00	-	110.00	93.00	MILLI/MJ	318.93	
ISOLEUCINE	GRAM	0.21	0.19	-	0.22	0.21	GRAM/MJ	0.72	
LEUCINE	GRAM	0.33	0.31	-	0.37	0.33	GRAM/MJ	1.13	
LYSINE	GRAM	0.33	0.27	-	0.48	0.33	GRAM/MJ	1.13	
METHIONINE	MILLI	43.00	36.00	-	52.00	43.00	MILLI/MJ	147.46	
PHENYLALANINE	GRAM	0.19	0.17	-	0.24	0.19	GRAM/MJ	0.65	
THREONINE	GRAM	0.21	0.18	-	0.28	0.21	GRAM/MJ	0.72	
TRYPTOPHAN	MILLI	52.00	31.00	-	73.00	52.00	MILLI/MJ	178.33	
TYROSINE	GRAM	0.15	0.12	-	0.17	0.15	GRAM/MJ	0.51	
VALINE	GRAM	0.24	0.23	-	0.24	0.24	GRAM/MJ	0.82	
GLUCOSE	MILLI	27.00	14.00	-	41.00	27.00	MILLI/MJ	92.59	
FRUCTOSE	MILLI	30.00	18.00	-	39.00	30.00	MILLI/MJ	102.88	
SUCROSE	GRAM	1.90	0.88	-	2.79	1.90	GRAM/MJ	6.52	

ERBSE
SAMEN, TROCKEN

PEA
SEED, DRY

POIS JAUNE
GRAINE, SÈCHE

PISUM SATIVUM L.

		PROTEIN	FAT	CARBOHYDRATES	TOTAL
ENERGY VALUE (AVERAGE)	KJOULE	426	56	949	1431
PER 100 G	(KCAL)	102	13	227	342
EDIBLE PORTION					
AMOUNT OF DIGESTIBLE	GRAM	17.86	1.29	56.70	
CONSTITUENTS PER 100 G					
ENERGY VALUE (AVERAGE)	KJOULE	333	50	949	1332
OF THE DIGESTIBLE	(KCAL)	79	12	227	318
FRACTION PER 100 G					
EDIBLE PORTION					

WASTE PERCENTAGE AVERAGE 0.00

CONSTITUENTS	DIM	AV	VARIATION			AVR	NUTR. DENS.		MOLPERC.
WATER	GRAM	11.00	9.00	-	13.60	10.89	GRAM/MJ	8.26	
PROTEIN	GRAM	22.90	20.70	-	24.80	22.67	GRAM/MJ	17.19	
FAT	GRAM	1.44	1.00	-	2.11	1.43	GRAM/MJ	1.08	
AVAILABLE CARBOHYDR.	GRAM	56.70	-	-	-	56.13	GRAM/MJ	42.57	
TOTAL DIETARY FIBRE	GRAM	16.60 1	-	-	-	16.43	GRAM/MJ	12.46	
MINERALS	GRAM	2.68	2.42	-	3.10	2.65	GRAM/MJ	2.01	
SODIUM	MILLI	26.00	10.00	-	42.00	25.74	MILLI/MJ	19.52	
POTASSIUM	GRAM	0.93	0.83	-	1.04	0.93	GRAM/MJ	0.70	
MAGNESIUM	MILLI	116.00	93.00	-	140.00	114.84	MILLI/MJ	87.09	
CALCIUM	MILLI	51.00	32.00	-	84.00	50.49	MILLI/MJ	38.29	
MANGANESE	MILLI	1.30	0.90	-	2.00	1.29	MILLI/MJ	0.98	
IRON	MILLI	5.00	3.50	-	6.40	4.95	MILLI/MJ	3.75	
COBALT	MICRO	4.20	-	-	-	4.16	MICRO/MJ	3.15	
COPPER	MILLI	0.74	0.49	-	1.23	0.73	MILLI/MJ	0.56	
ZINC	MILLI	3.80	3.10	-	4.90	3.76	MILLI/MJ	2.85	
MOLYBDENUM	MICRO	70.00	10.00	-	260.00	69.30	MICRO/MJ	52.56	
PHOSPHORUS	MILLI	378.00	303.00	-	448.00	374.22	MILLI/MJ	283.80	
CHLORIDE	MILLI	55.00	44.00	-	66.00	54.45	MILLI/MJ	41.29	
FLUORIDE	MICRO	40.00	20.00	-	60.00	39.60	MICRO/MJ	30.03	
IODIDE	MICRO	14.00	-	-	-	13.86	MICRO/MJ	10.51	
BORON	MILLI	-	1.30	-	2.30				
SILICON	MILLI	3.00	2.00	-	5.00	2.97	MILLI/MJ	2.25	
VITAMIN A	-	0.00	-	-	-	0.00			
CAROTENE	MICRO	80.00	60.00	-	110.00	79.20	MICRO/MJ	60.06	
VITAMIN D	-	0.00	-	-	-	0.00			
VITAMIN B1	MILLI	0.76	0.50	-	1.10	0.75	MILLI/MJ	0.57	

1 METHOD OF MEUSER, SUCKOW AND KULIKOWSKI ("BERLINER METHODE")

CONSTITUENTS	DIM	AV	VARIATION			AVR	NUTR. DENS.		MOLPERC.
VITAMIN B2	MILLI	0.27	0.18	-	0.35	0.27	MILLI/MJ	0.20	
NICOTINAMIDE	MILLI	2.80	1.79	-	3.20	2.77	MILLI/MJ	2.10	
PANTOTHENIC ACID	MILLI	2.10	2.00	-	2.18	2.08	MILLI/MJ	1.58	
VITAMIN B6	MICRO	63.00	50.00	-	75.00	62.37	MICRO/MJ	47.30	
BIOTIN	MICRO	19.00	18.00	-	20.00	18.81	MICRO/MJ	14.27	
FOLIC ACID	MICRO	59.00	-	-	-	58.41	MICRO/MJ	44.30	
VITAMIN B12	-	0.00	-	-	-	0.00			
VITAMIN C	MILLI	1.60	0.50	-	2.00	1.58	MILLI/MJ	1.20	
ALANINE	GRAM	0.48	-	-	-	0.48	GRAM/MJ	0.36	2.9
ARGININE	GRAM	3.71	3.32	-	4.68	3.67	GRAM/MJ	2.79	11.4
ASPARTIC ACID	GRAM	1.92	-	-	-	1.90	GRAM/MJ	1.44	7.7
CYSTINE	GRAM	0.45	0.35	-	0.56	0.45	GRAM/MJ	0.34	1.0
GLUTAMIC ACID	GRAM	3.46	-	-	-	3.43	GRAM/MJ	2.60	12.5
GLYCINE	GRAM	0.59	-	-	-	0.58	GRAM/MJ	0.44	4.2
HISTIDINE	GRAM	0.77	0.59	-	0.80	0.76	GRAM/MJ	0.58	2.6
ISOLEUCINE	GRAM	1.88	1.64	-	1.99	1.86	GRAM/MJ	1.41	7.6
LEUCINE	GRAM	2.34	1.88	-	2.83	2.32	GRAM/MJ	1.76	9.5
LYSINE	GRAM	2.13	1.81	-	2.37	2.11	GRAM/MJ	1.60	7.8
METHIONINE	GRAM	0.35	0.31	-	0.35	0.35	GRAM/MJ	0.26	1.3
PHENYLALANINE	GRAM	1.39	1.05	-	1.71	1.38	GRAM/MJ	1.04	4.5
PROLINE	GRAM	0.49	-	-	-	0.49	GRAM/MJ	0.37	2.3
SERINE	GRAM	0.98	-	-	-	0.97	GRAM/MJ	0.74	5.0
THREONINE	GRAM	1.57	2.31	-	1.82	1.55	GRAM/MJ	1.18	7.0
TRYPTOPHAN	GRAM	0.35	0.28	-	0.35	0.35	GRAM/MJ	0.26	0.9
TYROSINE	GRAM	1.22	0.94	-	1.61	1.21	GRAM/MJ	0.92	3.6
VALINE	GRAM	1.82	1.68	-	1.96	1.80	GRAM/MJ	1.37	8.3
MALIC ACID	MILLI	80.00	-	-	-	79.20	MILLI/MJ	60.06	
CITRIC ACID	MILLI	-	290.00	-	800.00				
OXALIC ACID	MILLI	2.70	0.00	-	6.90	2.67	MILLI/MJ	2.03	
SUCROSE	GRAM	2.47	-	-	-	2.45	GRAM/MJ	1.85	
RAFFINOSE	GRAM	0.45	-	-	-	0.45	GRAM/MJ	0.34	
STACHYOSE	GRAM	1.48	-	-	-	1.47	GRAM/MJ	1.11	
VERBASCOSE	GRAM	0.92	-	-	-	0.91	GRAM/MJ	0.69	
STARCH	GRAM	54.10	53.50	-	54.70	53.56	GRAM/MJ	40.62	
PENTOSAN	GRAM	2.47	2.47	-	2.47	2.45	GRAM/MJ	1.85	
HEXOSAN	GRAM	11.20	11.20	-	11.20	11.09	GRAM/MJ	8.41	
CELLULOSE	GRAM	2.43	2.43	-	2.43	2.41	GRAM/MJ	1.82	
POLYURONIC ACID	GRAM	0.37	0.37	-	0.37	0.37	GRAM/MJ	0.28	
PALMITIC ACID	GRAM	0.23	0.23	-	0.23	0.23	GRAM/MJ	0.17	
STEARIC ACID	MILLI	30.00	30.00	-	30.00	29.70	MILLI/MJ	22.52	
OLEIC ACID	GRAM	0.11	0.11	-	0.11	0.11	GRAM/MJ	0.08	
LINOLEIC ACID	GRAM	0.63	0.60	-	0.65	0.63	GRAM/MJ	0.48	
LINOLENIC ACID	GRAM	0.14	0.13	-	0.17	0.14	GRAM/MJ	0.11	
DIETARY FIBRE, WAT.SOL.	GRAM	5.10	-	-	-	5.05	GRAM/MJ	3.83	
DIETARY FIBRE, WAT.INS.	GRAM	11.60	-	-	-	11.48	GRAM/MJ	8.71	

GOABOHNE (FLÜGEL-, MANILA-, PRINZESS-BOHNE) SAMEN, TROCKEN PSOPHOCARPUS TETRAGONOLOBUS DC	**WINGED BEAN** (GOABEAN) SEED, DRY	**POIS CARRÉ** (POIS AILE) GRAINE, SÈCHE

		PROTEIN	FAT	CARBOHYDRATES	TOTAL
ENERGY VALUE (AVERAGE) PER 100 G EDIBLE PORTION	KJOULE	616	630	515	1762
	(KCAL)	147	151	123	421
AMOUNT OF DIGESTIBLE CONSTITUENTS PER 100 G	GRAM	25.81	14.58	30.79	
ENERGY VALUE (AVERAGE) OF THE DIGESTIBLE FRACTION PER 100 G EDIBLE PORTION	KJOULE	481	567	515	1563
	(KCAL)	115	136	123	374

WASTE PERCENTAGE AVERAGE 0.00

CONSTITUENTS	DIM	AV	VARIATION		AVR	NUTR.	DENS.	MOLPERC.
WATER	GRAM	10.60	5.30	14.90	10.60	GRAM/MJ	6.78	
PROTEIN	GRAM	33.10	27.70	38.30	33.10	GRAM/MJ	21.17	
FAT	GRAM	16.20	15.10	17.00	16.20	GRAM/MJ	10.36	
AVAILABLE CARBOHYDR.	GRAM	30.79 1	-	-	30.79	GRAM/MJ	19.70	
MINERALS	GRAM	3.80	3.40	4.10	3.80	GRAM/MJ	2.43	
SODIUM	MILLI	50.00	40.00	60.00	50.00	MILLI/MJ	31.98	
POTASSIUM	GRAM	1.02	0.95	1.09	1.02	GRAM/MJ	0.65	
MAGNESIUM	MILLI	170.00	150.00	190.00	170.00	MILLI/MJ	108.74	
CALCIUM	MILLI	530.00	510.00	550.00	530.00	MILLI/MJ	339.02	
MANGANESE	MILLI	3.90	3.60	4.20	3.90	MILLI/MJ	2.49	
IRON	MILLI	14.50	11.20	17.80	14.50	MILLI/MJ	9.28	
COPPER	MILLI	3.50	3.20	3.80	3.50	MILLI/MJ	2.24	
ZINC	MILLI	4.60	4.50	4.70	4.60	MILLI/MJ	2.94	
PHOSPHORUS	MILLI	480.00	440.00	520.00	480.00	MILLI/MJ	307.04	
ALANINE	GRAM	1.72	1.54	1.84	1.72	GRAM/MJ	1.10	6.6
ARGININE	GRAM	2.51	2.37	2.95	2.51	GRAM/MJ	1.61	4.9
ASPARTIC ACID	GRAM	3.86	2.86	4.55	3.86	GRAM/MJ	2.47	9.9
CYSTINE	GRAM	0.68	0.54	0.76	0.68	GRAM/MJ	0.43	1.0
GLUTAMIC ACID	GRAM	5.77	5.48	6.08	5.77	GRAM/MJ	3.69	13.4
GLYCINE	GRAM	1.80	1.62	1.93	1.80	GRAM/MJ	1.15	8.2
HISTIDINE	GRAM	1.06	0.98	1.17	1.06	GRAM/MJ	0.68	2.3
ISOLEUCINE	GRAM	1.85	1.77	1.89	1.85	GRAM/MJ	1.18	4.8
LEUCINE	GRAM	3.21	3.07	3.29	3.21	GRAM/MJ	2.05	8.4
LYSINE	GRAM	2.93	2.77	3.15	2.93	GRAM/MJ	1.87	6.9
METHIONINE	GRAM	0.45	0.34	0.53	0.45	GRAM/MJ	0.29	1.0
PHENYLALANINE	GRAM	1.87	1.83	1.95	1.87	GRAM/MJ	1.20	3.9

1 DIFFERENCE METHOD: 100 - (WATER + PROTEIN + FAT + ASH + NONAVAILABLE OLIGOSACCHARIDES)

CONSTITUENTS	DIM	AV	VARIATION			AVR	NUTR. DENS.		MOLPERC.
PROLINE	GRAM	2.69	2.53	-	2.88	2.69	GRAM/MJ	1.72	8.0
SERINE	GRAM	1.97	1.93	-	2.02	1.97	GRAM/MJ	1.26	6.4
THREONINE	GRAM	1.64	1.64	-	1.68	1.64	GRAM/MJ	1.05	4.7
TRYPTOPHAN	GRAM	0.36	-	-	-	0.36	GRAM/MJ	0.23	0.6
TYROSINE	GRAM	1.66	1.19	-	1.93	1.66	GRAM/MJ	1.06	3.1
VALINE	GRAM	2.01	1.93	-	2.09	2.01	GRAM/MJ	1.29	5.9
SUCROSE	GRAM	6.60	5.64	-	8.17	6.60	GRAM/MJ	4.22	
RAFFINOSE	GRAM	1.51	1.15	-	1.98	1.51	GRAM/MJ	0.97	
STACHYOSE	GRAM	3.20	2.21	-	3.56	3.20	GRAM/MJ	2.05	
VERBASCOSE	GRAM	0.80	0.68	-	0.91	0.80	GRAM/MJ	0.51	
MYRISTIC ACID	MILLI	10.00	-	-	-	10.00	MILLI/MJ	6.40	
PALMITIC ACID	GRAM	1.56	-	-	-	1.56	GRAM/MJ	1.00	
STEARIC ACID	GRAM	0.89	-	-	-	0.89	GRAM/MJ	0.57	
ARACHIDIC ACID	GRAM	0.22	-	-	-	0.22	GRAM/MJ	0.14	
BEHENIC ACID	GRAM	2.54	-	-	-	2.54	GRAM/MJ	1.62	
LIGNOCERIC ACID	GRAM	0.64	0.56	-	0.78	0.64	GRAM/MJ	0.41	
OLEIC ACID	GRAM	5.79	-	-	-	5.79	GRAM/MJ	3.70	
C22 : 1 FATTY ACID	MILLI	29.00	10.00	-	46.00	29.00	MILLI/MJ	18.55	
LINOLEIC ACID	GRAM	3.51	-	-	-	3.51	GRAM/MJ	2.25	
LINOLENIC ACID	GRAM	0.24	-	-	-	0.24	GRAM/MJ	0.15	

KICHERERBSE
SAMEN, GRÜN

CHICK PEA
SEED, GREEN

POIS CHICHE
GRAINE, VERTE

CICER ARIETINUM L.

		PROTEIN	FAT	CARBOHYDRATES	TOTAL
ENERGY VALUE (AVERAGE) PER 100 G EDIBLE PORTION	KJOULE (KCAL)	140 33	105 25	355 85	600 143
AMOUNT OF DIGESTIBLE CONSTITUENTS PER 100 G	GRAM	5.85	2.43	21.20	
ENERGY VALUE (AVERAGE) OF THE DIGESTIBLE FRACTION PER 100 G EDIBLE PORTION	KJOULE (KCAL)	109 26	95 23	355 85	558 133

WASTE PERCENTAGE AVERAGE 0.00

CONSTITUENTS	DIM	AV	VARIATION			AVR	NUTR.	DENS.	MOLPERC.
WATER	GRAM	61.10	-	-	-	61.10	GRAM/MJ	109.44	
PROTEIN	GRAM	7.50	-	-	-	7.50	GRAM/MJ	13.43	
FAT	GRAM	2.70	-	-	-	2.70	GRAM/MJ	4.84	
AVAILABLE CARBOHYDR.	GRAM	21.20	-	-	-	21.20	GRAM/MJ	37.97	
MINERALS	GRAM	1.30	-	-	-	1.30	GRAM/MJ	2.33	
SODIUM	MILLI	19.00	-	-	-	19.00	MILLI/MJ	34.03	
POTASSIUM	MILLI	53.00	-	-	-	53.00	MILLI/MJ	94.93	
MAGNESIUM	MILLI	56.00	-	-	-	56.00	MILLI/MJ	100.31	
CALCIUM	MILLI	58.00	-	-	-	58.00	MILLI/MJ	103.89	
MANGANESE	MILLI	1.21	-	-	-	1.21	MILLI/MJ	2.17	
IRON	MILLI	2.95	-	-	-	2.95	MILLI/MJ	5.28	
COPPER	MILLI	0.34	-	-	-	0.34	MILLI/MJ	0.61	
ZINC	MILLI	1.24	-	-	-	1.24	MILLI/MJ	2.22	
SUCROSE	GRAM	0.44	0.37	-	0.49	0.44	GRAM/MJ	0.79	
RAFFINOSE	GRAM	0.29	0.25	-	0.32	0.29	GRAM/MJ	0.52	
STACHYOSE	GRAM	0.84	0.75	-	0.93	0.84	GRAM/MJ	1.50	
STARCH	GRAM	20.80	18.90	-	22.70	20.80	GRAM/MJ	37.26	
PHYTIC ACID	GRAM	0.28	-	-	-	0.28	GRAM/MJ	0.50	

KICHERERBSE
SAMEN, TROCKEN

CHICK PEA
SEED, DRY

POIS CHICHE
GRAINE, SÈCHE

CICER ARIETINUM L.

		PROTEIN	FAT	CARBOHYDRATES	TOTAL
ENERGY VALUE (AVERAGE) PER 100 G EDIBLE PORTION	KJOULE	369	132	813	1314
	(KCAL)	88	32	194	314
AMOUNT OF DIGESTIBLE CONSTITUENTS PER 100 G	GRAM	15.44	3.06	48.60	
ENERGY VALUE (AVERAGE) OF THE DIGESTIBLE FRACTION PER 100 G EDIBLE PORTION	KJOULE	288	119	813	1220
	(KCAL)	69	28	194	292

WASTE PERCENTAGE AVERAGE 0.00

CONSTITUENTS	DIM	AV	VARIATION			AVR	NUTR.	DENS.	MOLPERC.
WATER	GRAM	11.00	10.60	-	11.40	11.00	GRAM/MJ	9.02	
PROTEIN	GRAM	19.80	13.00	-	24.90	19.80	GRAM/MJ	16.23	
FAT	GRAM	3.40	1.60	-	5.18	3.40	GRAM/MJ	2.79	
AVAILABLE CARBOHYDR.	GRAM	48.60	-	-	-	48.60	GRAM/MJ	39.84	
TOTAL DIETARY FIBRE	GRAM	10.68 [1]	-	-	-	10.68	GRAM/MJ	8.75	
MINERALS	GRAM	2.72	2.25	-	3.00	2.72	GRAM/MJ	2.23	
SODIUM	MILLI	27.00	-	-	-	27.00	MILLI/MJ	22.13	
POTASSIUM	MILLI	580.00	-	-	-	580.00	MILLI/MJ	475.42	
MAGNESIUM	MILLI	108.00	-	-	-	108.00	MILLI/MJ	88.53	
CALCIUM	MILLI	110.00	90.00	-	149.00	110.00	MILLI/MJ	90.16	
IRON	MILLI	7.20	7.10	-	7.20	7.20	MILLI/MJ	5.90	
PHOSPHORUS	MILLI	428.00	375.00	-	480.00	428.00	MILLI/MJ	350.82	
CHLORIDE	MILLI	80.00	-	-	-	80.00	MILLI/MJ	65.57	
CAROTENE	MILLI	0.18	-	-	-	0.18	MILLI/MJ	0.15	
VITAMIN B1	MILLI	0.48	0.40	-	0.55	0.48	MILLI/MJ	0.39	
VITAMIN B2	MILLI	0.18	0.17	-	0.18	0.18	MILLI/MJ	0.15	
NICOTINAMIDE	MILLI	1.60	1.50	-	1.60	1.60	MILLI/MJ	1.31	
PANTOTHENIC ACID	MILLI	1.30	-	-	-	1.30	MILLI/MJ	1.07	
VITAMIN B6	MILLI	0.54	-	-	-	0.54	MILLI/MJ	0.44	
FOLIC ACID	MILLI	0.20	-	-	-	0.20	MILLI/MJ	0.16	
VITAMIN C	MILLI	4.00	2.00	-	5.00	4.00	MILLI/MJ	3.28	
ARGININE	GRAM	1.48	-	-	-	1.48	GRAM/MJ	1.21	
CYSTINE	GRAM	0.28	-	-	-	0.28	GRAM/MJ	0.23	
HISTIDINE	GRAM	0.53	-	-	-	0.53	GRAM/MJ	0.43	
ISOLEUCINE	GRAM	1.14	-	-	-	1.14	GRAM/MJ	0.93	

1 METHOD OF MEUSER, SUCKOW AND KULIKOWSKI ("BERLINER METHODE")

CONSTITUENTS	DIM	AV	VARIATION			AVR	NUTR. DENS.		MOLPERC.
LEUCINE	GRAM	1.46	-	-	-	1.46	GRAM/MJ	1.20	
LYSINE	GRAM	1.37	-	-	-	1.37	GRAM/MJ	1.12	
METHIONINE	GRAM	0.26	-	-	-	0.26	GRAM/MJ	0.21	
PHENYLALANINE	GRAM	0.96	-	-	-	0.96	GRAM/MJ	0.79	
THREONINE	GRAM	0.70	-	-	-	0.70	GRAM/MJ	0.57	
TRYPTOPHAN	GRAM	0.16	-	-	-	0.16	GRAM/MJ	0.13	
TYROSINE	GRAM	0.66	-	-	-	0.66	GRAM/MJ	0.54	
VALINE	GRAM	0.98	-	-	-	0.98	GRAM/MJ	0.80	
SUCROSE	GRAM	0.94	0.86	-	1.10	0.94	GRAM/MJ	0.77	
RAFFINOSE	GRAM	0.66	0.58	-	0.75	0.66	GRAM/MJ	0.54	
STACHYOSE	GRAM	1.90	1.70	-	2.10	1.90	GRAM/MJ	1.56	
STARCH	GRAM	47.70	43.20	-	52.10	47.70	GRAM/MJ	39.10	
DIETARY FIBRE,WAT.SOL.	GRAM	3.30	-	-	-	3.30	GRAM/MJ	2.70	
DIETARY FIBRE,WAT.INS.	GRAM	7.34	-	-	-	7.34	GRAM/MJ	6.02	

LEIN LEINSAMEN | LINSEED | GRAINE DE LIN

LINUM USITATISSIMUM

		PROTEIN	FAT	CARBOHYDRATES	TOTAL
ENERGY VALUE (AVERAGE) PER 100 G EDIBLE PORTION	KJOULE	454	1202	0.0	1657
	(KCAL)	109	287	0.0	396
AMOUNT OF DIGESTIBLE CONSTITUENTS PER 100 G	GRAM	19.03	27.81	0.00	
ENERGY VALUE (AVERAGE) OF THE DIGESTIBLE FRACTION PER 100 G EDIBLE PORTION	KJOULE	354	1082	0.0	1436
	(KCAL)	85	259	0.0	343

WASTE PERCENTAGE AVERAGE 0.00

CONSTITUENTS	DIM	AV	VARIATION	AVR	NUTR. DENS.		MOLPERC.
WATER	GRAM	6.10	6.10 - 6.10	6.10	GRAM/MJ	4.25	
PROTEIN	GRAM	24.40	24.40 - 24.40	24.40	GRAM/MJ	16.99	
FAT	GRAM	30.90	30.90 - 30.90	30.90	GRAM/MJ	21.51	
TOTAL DIETARY FIBRE	GRAM	38.60 1	- - -	38.60	GRAM/MJ	26.87	
CALCIUM	MILLI	198.00	198.00 - 198.00	198.00	MILLI/MJ	137.84	
IRON	MILLI	8.20	8.20 - 8.20	8.20	MILLI/MJ	5.71	
PHOSPHORUS	MILLI	662.00	662.00 - 662.00	662.00	MILLI/MJ	460.85	
VITAMIN B1	MILLI	0.17	0.17 - 0.17	0.17	MILLI/MJ	0.12	
VITAMIN B2	MILLI	0.16	0.16 - 0.16	0.16	MILLI/MJ	0.11	
NICOTINAMIDE	MILLI	1.40	1.40 - 1.40	1.40	MILLI/MJ	0.97	
DIETARY FIBRE, WAT.SOL.	GRAM	19.90	- - -	19.90	GRAM/MJ	13.85	
DIETARY FIBRE, WAT.INS.	GRAM	18.70	- - -	18.70	GRAM/MJ	13.02	

1 METHOD OF MEUSER, SUCKOW AND KULIKOWSKI ("BERLINER METHODE")

LIMABOHNE	LIMA BEAN	HARICOT DE LIMA
(MONDBOHNE, BUTTER-BOHNE) SAMEN, TROCKEN PHASEOLUS LUNATUS (LIMENSIS) L. [1]	(BUTTER BEAN) SEED, DRY [1]	(POIS DU CAP) GRAINE, SÈCHE [1]

		PROTEIN	FAT	CARBOHYDRATES	TOTAL
ENERGY VALUE (AVERAGE)	KJOULE	384	54	753	1191
PER 100 G	(KCAL)	92	13	180	285
EDIBLE PORTION					
AMOUNT OF DIGESTIBLE	GRAM	16.06	1.26	44.99	
CONSTITUENTS PER 100 G					
ENERGY VALUE (AVERAGE)	KJOULE	299	49	753	1101
OF THE DIGESTIBLE	(KCAL)	72	12	180	263
FRACTION PER 100 G					
EDIBLE PORTION					

WASTE PERCENTAGE AVERAGE 0.00

CONSTITUENTS	DIM	AV	VARIATION			AVR	NUTR.	DENS.	MOLPERC.
WATER	GRAM	11.50	8.70	-	13.20	11.50	GRAM/MJ	10.44	
PROTEIN	GRAM	20.60	19.00	-	22.40	20.60	GRAM/MJ	18.71	
FAT	GRAM	1.40	1.10	-	1.60	1.40	GRAM/MJ	1.27	
AVAILABLE CARBOHYDR.	GRAM	44.99	-	-	-	44.99	GRAM/MJ	40.86	
MINERALS	GRAM	3.70	3.00	-	5.10	3.70	GRAM/MJ	3.36	
SODIUM	MILLI	20.60	18.00	-	23.30	20.60	MILLI/MJ	18.71	
POTASSIUM	GRAM	1.75	1.67	-	1.87	1.75	GRAM/MJ	1.59	
MAGNESIUM	MILLI	201.00	165.00	-	278.00	201.00	MILLI/MJ	182.54	
CALCIUM	MILLI	91.00	74.00	-	116.00	91.00	MILLI/MJ	82.64	
MANGANESE	MILLI	1.95	1.65	-	2.25	1.95	MILLI/MJ	1.77	
IRON	MILLI	6.00	4.90	-	7.50	6.00	MILLI/MJ	5.45	
COPPER	MILLI	0.78	0.51	-	1.29	0.78	MILLI/MJ	0.71	
ZINC	MILLI	3.10	2.45	-	4.24	3.10	MILLI/MJ	2.82	
PHOSPHORUS	MILLI	348.00	240.00	-	407.00	348.00	MILLI/MJ	316.03	
VITAMIN B1	MILLI	0.41	0.33	-	0.45	0.41	MILLI/MJ	0.37	
VITAMIN B2	MILLI	0.17	0.13	-	0.21	0.17	MILLI/MJ	0.15	
NICOTINAMIDE	MILLI	2.00	1.40	-	2.50	2.00	MILLI/MJ	1.82	
VITAMIN C	MILLI	-	0.00	-	1.00				
ALANINE	GRAM	1.21	1.09	-	1.40	1.21	GRAM/MJ	1.10	7.4
ARGININE	GRAM	1.39	1.25	-	1.55	1.39	GRAM/MJ	1.26	4.4
ASPARTIC ACID	GRAM	2.92	2.79	-	3.02	2.92	GRAM/MJ	2.65	12.0
CYSTINE	GRAM	0.28	0.24	-	0.32	0.28	GRAM/MJ	0.25	0.6
GLUTAMIC ACID	GRAM	3.33	3.11	-	3.51	3.33	GRAM/MJ	3.02	12.4
GLYCINE	GRAM	1.28	1.10	-	1.41	1.28	GRAM/MJ	1.16	9.3
HISTIDINE	GRAM	0.68	0.60	-	0.72	0.68	GRAM/MJ	0.62	2.4

1 THE RAW PRODUCT IST POISONOUS DUE TO THE PRESENCE OF PHASEOLUNATINE, A HYDROCYANIC ACID CONTAINING GLYCOSIDE (10 - 300 MG HCN/100G

CONSTITUENTS	DIM	AV	VARIATION			AVR	NUTR. DENS.		MOLPERC.
ISOLEUCINE	GRAM	1.29	1.16	-	1.44	1.29	GRAM/MJ	1.17	5.4
LEUCINE	GRAM	1.90	1.73	-	2.01	1.90	GRAM/MJ	1.73	7.9
LYSINE	GRAM	1.47	1.45	-	1.51	1.47	GRAM/MJ	1.33	5.5
METHIONINE	GRAM	0.27	0.20	-	0.33	0.27	GRAM/MJ	0.25	1.0
PHENYLALANINE	GRAM	1.35	1.07	-	1.71	1.35	GRAM/MJ	1.23	4.5
PROLINE	GRAM	1.01	0.88	-	1.11	1.01	GRAM/MJ	0.92	4.8
SERINE	GRAM	1.52	1.23	-	1.83	1.52	GRAM/MJ	1.38	7.9
THREONINE	GRAM	0.93	0.80	-	1.00	0.93	GRAM/MJ	0.84	4.3
TRYPTOPHAN	GRAM	0.30	0.27	-	0.33	0.30	GRAM/MJ	0.27	0.8
TYROSINE	GRAM	0.85	0.78	-	1.00	0.85	GRAM/MJ	0.77	2.6
VALINE	GRAM	1.45	1.35	-	1.55	1.45	GRAM/MJ	1.32	6.8
GLUCOSE	MILLI	70.00	70.00	-	70.00	70.00	MILLI/MJ	63.57	
FRUCTOSE	GRAM	0.50	0.50	-	0.50	0.50	GRAM/MJ	0.45	
SUCROSE	GRAM	1.42	1.35	-	1.63	1.42	GRAM/MJ	1.29	
RAFFINOSE	GRAM	0.93	0.41	-	1.11	0.94	GRAM/MJ	0.85	
STACHYOSE	GRAM	2.44	2.44	-	2.44	2.44	GRAM/MJ	2.22	
VERBASCOSE	GRAM	0.27	0.27	-	0.27	0.27	GRAM/MJ	0.25	
STARCH	GRAM	43.00	43.00	-	43.00	43.00	GRAM/MJ	39.05	
LAURIC ACID	MILLI	0.55	-	-	-	0.55	MILLI/MJ	0.50	
MYRISTIC ACID	MILLI	5.40	-	-	-	5.40	MILLI/MJ	4.90	
PALMITIC ACID	MILLI	298.00	-	-	-	298.00	MILLI/MJ	270.63	
STEARIC ACID	MILLI	84.00	-	-	-	84.00	MILLI/MJ	76.28	
ARACHIDIC ACID	MILLI	22.00	-	-	-	22.00	MILLI/MJ	19.98	
BEHENIC ACID	MILLI	12.00	-	-	-	12.00	MILLI/MJ	10.90	
PALMITOLEIC ACID	MILLI	8.80	-	-	-	8.80	MILLI/MJ	7.99	
OLEIC ACID	MILLI	126.00	-	-	-	126.00	MILLI/MJ	114.43	
LINOLEIC ACID	MILLI	559.00	-	-	-	559.00	MILLI/MJ	507.65	
LINOLENIC ACID	MILLI	249.00	-	-	-	249.00	MILLI/MJ	226.13	

LINSE
SAMEN, TROCKEN

LENTIL
SEED, DRY

LENTILLES
GRAINE, SÈCHE

LENS ESCULENTA MOENCH

		PROTEIN	FAT	CARBOHYDRATES	TOTAL
ENERGY VALUE (AVERAGE) PER 100 G EDIBLE PORTION	KJOULE	438	54	869	1361
	(KCAL)	105	13	208	325
AMOUNT OF DIGESTIBLE CONSTITUENTS PER 100 G	GRAM	18.33	1.26	51.95	
ENERGY VALUE (AVERAGE) OF THE DIGESTIBLE FRACTION PER 100 G EDIBLE PORTION	KJOULE	341	49	869	1260
	(KCAL)	82	12	208	301

WASTE PERCENTAGE AVERAGE 0.00

CONSTITUENTS	DIM	AV		VARIATION			AVR	NUTR. DENS.		MOLPERC.
WATER	GRAM	11.80		11.20	-	12.30	11.80	GRAM/MJ	9.37	
PROTEIN	GRAM	23.50		21.50	-	26.00	23.50	GRAM/MJ	18.65	
FAT	GRAM	1.40		1.00	-	1.90	1.40	GRAM/MJ	1.11	
AVAILABLE CARBOHYDR.	GRAM	51.95		-	-	-	51.95	GRAM/MJ	41.24	
TOTAL DIETARY FIBRE	GRAM	10.60	1	-	-	-	10.60	GRAM/MJ	8.41	
MINERALS	GRAM	3.20		3.04	-	3.30	3.20	GRAM/MJ	2.54	
SODIUM	MILLI	4.00		3.00	-	5.00	4.00	MILLI/MJ	3.18	
POTASSIUM	GRAM	0.81		0.52	-	1.20	0.81	GRAM/MJ	0.64	
MAGNESIUM	MILLI	77.00		-	-	-	77.00	MILLI/MJ	61.12	
CALCIUM	MILLI	74.00		59.00	-	82.00	74.00	MILLI/MJ	58.74	
IRON	MILLI	6.90		5.00	-	10.00	6.90	MILLI/MJ	5.48	
COBALT	MICRO	35.00		-	-	-	35.00	MICRO/MJ	27.78	
COPPER	MILLI	0.66		-	-	-	0.66	MILLI/MJ	0.52	
ZINC	MILLI	5.00		2.00	-	9.00	5.00	MILLI/MJ	3.97	
NICKEL	MILLI	0.31		-	-	-	0.31	MILLI/MJ	0.25	
MOLYBDENUM	MILLI	-		0.07	-	0.19				
PHOSPHORUS	MILLI	412.00		400.00	-	423.00	412.00	MILLI/MJ	327.05	
CHLORIDE	MILLI	84.00		-	-	-	84.00	MILLI/MJ	66.68	
FLUORIDE	MICRO	26.00		-	-	-	26.00	MICRO/MJ	20.64	
BORON	MILLI	0.70		0.54	-	0.85	0.70	MILLI/MJ	0.56	
SELENIUM	MICRO	11.00		-	-	-	11.00	MICRO/MJ	8.73	
CAROTENE	MILLI	0.10		-	-	-	0.10	MILLI/MJ	0.08	
VITAMIN B1	MILLI	0.43		0.20	-	0.70	0.43	MILLI/MJ	0.34	
VITAMIN B2	MILLI	0.26		0.10	-	0.41	0.26	MILLI/MJ	0.21	
NICOTINAMIDE	MILLI	2.20		2.00	-	2.60	2.20	MILLI/MJ	1.75	
PANTOTHENIC ACID	MILLI	1.36		-	-	-	1.36	MILLI/MJ	1.08	

1 METHOD OF MEUSER, SUCKOW AND KULIKOWSKI ("BERLINER METHODE")

CONSTITUENTS	DIM	AV	VARIATION			AVR	NUTR. DENS.		MOLPERC.
VITAMIN B6	MILLI	0.60	-	-	-	0.60	MILLI/MJ	0.48	
FOLIC ACID	MICRO	35.00	-	-	-	35.00	MICRO/MJ	27.78	
VITAMIN C	MILLI	-	0.00	-	5.00				
ALANINE	GRAM	1.29	-	-	-	1.29	GRAM/MJ	1.02	7.1
ARGININE	GRAM	2.24	-	-	-	2.24	GRAM/MJ	1.78	6.3
ASPARTIC ACID	GRAM	3.16	-	-	-	3.16	GRAM/MJ	2.51	11.6
CYSTINE	GRAM	0.25	-	-	-	0.25	GRAM/MJ	0.20	0.5
GLUTAMIC ACID	GRAM	4.49	-	-	-	4.49	GRAM/MJ	3.56	14.9
GLYCINE	GRAM	1.30	-	-	-	1.30	GRAM/MJ	1.03	8.5
HISTIDINE	GRAM	0.71	-	-	-	0.71	GRAM/MJ	0.56	2.2
ISOLEUCINE	GRAM	1.19	-	-	-	1.19	GRAM/MJ	0.94	4.4
LEUCINE	GRAM	2.11	-	-	-	2.11	GRAM/MJ	1.67	7.9
LYSINE	GRAM	1.89	-	-	-	1.89	GRAM/MJ	1.50	6.3
METHIONINE	GRAM	0.22	-	-	-	0.22	GRAM/MJ	0.17	0.7
PHENYLALANINE	GRAM	1.40	-	-	-	1.40	GRAM/MJ	1.11	4.1
PROLINE	GRAM	1.22	-	-	-	1.22	GRAM/MJ	0.97	5.2
SERINE	GRAM	1.51	-	-	-	1.51	GRAM/MJ	1.20	7.0
THREONINE	GRAM	1.12	-	-	-	1.12	GRAM/MJ	0.89	4.6
TRYPTOPHAN	GRAM	0.25	-	-	-	0.25	GRAM/MJ	0.20	0.6
TYROSINE	GRAM	0.84	-	-	-	0.84	GRAM/MJ	0.67	2.3
VALINE	GRAM	1.39	-	-	-	1.39	GRAM/MJ	1.10	5.8
SUCROSE	GRAM	1.15	1.15	-	1.15	1.15	GRAM/MJ	0.91	
RAFFINOSE	GRAM	0.44	0.44	-	0.44	0.44	GRAM/MJ	0.35	
STACHYOSE	GRAM	1.94	1.94	-	1.94	1.94	GRAM/MJ	1.54	
STARCH	GRAM	50.80	-	-	-	50.80	GRAM/MJ	40.33	
TOTAL PURINES	MILLI	114.00 [2]	114.00	-	114.00	114.00	MILLI/MJ	90.49	
ADENINE	MILLI	41.00	41.00	-	41.00	41.00	MILLI/MJ	32.55	
GUANINE	MILLI	57.00	57.00	-	57.00	57.00	MILLI/MJ	45.25	
DIETARY FIBRE,WAT.SOL.	GRAM	3.90	-	-	-	3.90	GRAM/MJ	3.10	
DIETARY FIBRE,WAT.INS.	GRAM	6.70	-	-	-	6.70	GRAM/MJ	5.32	

2 VALUE OF TOTAL PURINES IS EXPRESSSED IN MG URIC ACID/100G

LINSEN	**LENTILS**	**LENTILLES**
SAMEN	SEEDS	GRAINES
GEKOCHT	COOKED	CUITES

		PROTEIN	FAT	CARBOHYDRATES	TOTAL
ENERGY VALUE (AVERAGE)	KJOULE	138	16	224	378
PER 100 G	(KCAL)	33	3.7	54	90
EDIBLE PORTION					
AMOUNT OF DIGESTIBLE	GRAM	5.77	0.36	13.40	
CONSTITUENTS PER 100 G					
ENERGY VALUE (AVERAGE)	KJOULE	107	14	224	346
OF THE DIGESTIBLE	(KCAL)	26	3.3	54	83
FRACTION PER 100 G					
EDIBLE PORTION					

WASTE PERCENTAGE AVERAGE 0.00

CONSTITUENTS	DIM	AV	VARIATION			AVR	NUTR. DENS.		MOLPERC.
WATER	GRAM	76.60	-	-	-	76.60	GRAM/MJ	221.55	
PROTEIN	GRAM	7.40	-	-	-	7.40	GRAM/MJ	21.40	
FAT	GRAM	0.40	-	-	-	0.40	GRAM/MJ	1.16	
AVAILABLE CARBOHYDR.	GRAM	13.40 1	-	-	-	13.40	GRAM/MJ	38.76	
MINERALS	GRAM	1.00	-	-	-	1.00	GRAM/MJ	2.89	
SODIUM	MILLI	1.00	-	-	-	1.00	MILLI/MJ	2.89	
POTASSIUM	MILLI	255.00	-	-	-	255.00	MILLI/MJ	737.55	
CALCIUM	MILLI	23.00	-	-	-	23.00	MILLI/MJ	66.52	
IRON	MILLI	2.10	-	-	-	2.10	MILLI/MJ	6.07	
PHOSPHORUS	MILLI	130.00	-	-	-	130.00	MILLI/MJ	376.01	
VITAMIN B1	MICRO	80.00	-	-	-	80.00	MICRO/MJ	231.39	
VITAMIN B2	MICRO	70.00	-	-	-	70.00	MICRO/MJ	202.47	
NICOTINAMIDE	MILLI	0.66	-	-	-	0.66	MILLI/MJ	1.91	

1 CALCULATED FROM THE DRY PRODUCT

MOHN	**POPPY SEED**	**PAVOT**
(SCHLAFMOHN)	SEED, DRY	(PAVOT SOMNIFÈRE)
SAMEN, TROCKEN		GRAINE, SÈCHE
PAPAVER SOMNIFERUM L.		

		PROTEIN	FAT	CARBOHYDRATES	TOTAL
ENERGY VALUE (AVERAGE)	KJOULE	376	1642	70	2088
PER 100 G	(KCAL)	90	392	17	499
EDIBLE PORTION					
AMOUNT OF DIGESTIBLE	GRAM	15.75	37.98	4.20	
CONSTITUENTS PER 100 G					
ENERGY VALUE (AVERAGE)	KJOULE	293	1478	70	1841
OF THE DIGESTIBLE	(KCAL)	70	353	17	440
FRACTION PER 100 G					
EDIBLE PORTION					

WASTE PERCENTAGE AVERAGE 0.00

CONSTITUENTS	DIM	AV		VARIATION			AVR	NUTR. DENS.		MOLPERC.
WATER	GRAM	6.10		5.20	-	8.20	6.10	GRAM/MJ	3.31	
PROTEIN	GRAM	20.20		16.30	-	22.70	20.20	GRAM/MJ	10.97	
FAT	GRAM	42.20		40.80	-	44.60	42.20	GRAM/MJ	22.92	
AVAILABLE CARBOHYDR.	GRAM	4.20	1	-	-	-	4.20	GRAM/MJ	2.28	
TOTAL DIETARY FIBRE	GRAM	20.50	2	-	-	-	20.50	GRAM/MJ	11.13	
MINERALS	GRAM	6.80		6.20	-	7.30	6.80	GRAM/MJ	3.69	
SODIUM	MILLI	21.00		21.00	-	21.00	21.00	MILLI/MJ	11.40	
POTASSIUM	MILLI	705.00		705.00	-	705.00	705.00	MILLI/MJ	382.84	
MAGNESIUM	MILLI	333.00		333.00	-	333.00	333.00	MILLI/MJ	180.83	
CALCIUM	GRAM	1.46		1.46	-	1.46	1.46	GRAM/MJ	0.79	
IRON	MILLI	9.50		9.50	-	9.50	9.50	MILLI/MJ	5.16	
ZINC	MILLI	10.00		10.00	-	10.00	10.00	MILLI/MJ	5.43	
PHOSPHORUS	MILLI	854.00		854.00	-	854.00	854.00	MILLI/MJ	463.75	
VITAMIN B1	MILLI	0.86		0.86	-	0.86	0.86	MILLI/MJ	0.47	
VITAMIN B2	MILLI	0.17		0.17	-	0.17	0.17	MILLI/MJ	0.09	
NICOTINAMIDE	MILLI	0.99		0.99	-	0.99	0.99	MILLI/MJ	0.54	
VITAMIN B6	MILLI	0.44		0.44	-	0.44	0.44	MILLI/MJ	0.24	
ALANINE	GRAM	1.39		-	-	-	1.39	GRAM/MJ	0.75	7.3
ARGININE	GRAM	2.83		-	-	-	2.83	GRAM/MJ	1.54	7.7
ASPARTIC ACID	GRAM	2.73		-	-	-	2.73	GRAM/MJ	1.48	9.7
CYSTINE	GRAM	0.51		-	-	-	0.51	GRAM/MJ	0.28	1.0
GLUTAMIC ACID	GRAM	5.78		-	-	-	5.78	GRAM/MJ	3.14	18.5
GLYCINE	GRAM	1.45		-	-	-	1.45	GRAM/MJ	0.79	9.1
HISTIDINE	GRAM	0.72		-	-	-	0.72	GRAM/MJ	0.39	2.2
ISOLEUCINE	GRAM	1.23		-	-	-	1.23	GRAM/MJ	0.67	4.4

1 ESTIMATED BY THE DIFFERENCE METHOD
100 - (WATER + PROTEIN + FAT + MINERALS + TOTAL DIETARY FIBRE)

2 METHOD OF MEUSER, SUCKOW AND KULIKOWSKI ("BERLINER METHODE")

CONSTITUENTS	DIM	AV	VARIATION			AVR	NUTR. DENS.		MOLPERC.
LEUCINE	GRAM	1.96	-	-	-	1.96	GRAM/MJ	1.06	7.0
LYSINE	GRAM	1.39	-	-	-	1.39	GRAM/MJ	0.75	4.5
METHIONINE	GRAM	0.43	-	-	-	0.43	GRAM/MJ	0.23	1.4
PHENYLALANINE	GRAM	1.10	-	-	-	1.10	GRAM/MJ	0.60	3.1
PROLINE	GRAM	1.48	-	-	-	1.48	GRAM/MJ	0.80	6.1
SERINE	GRAM	1.04	-	-	-	1.04	GRAM/MJ	0.56	4.7
THREONINE	GRAM	1.20	-	-	-	1.20	GRAM/MJ	0.65	4.7
TRYPTOPHAN	GRAM	0.38	-	-	-	0.38	GRAM/MJ	0.21	0.9
TYROSINE	GRAM	0.42	-	-	-	0.42	GRAM/MJ	0.23	1.1
VALINE	GRAM	1.67	-	-	-	1.67	GRAM/MJ	0.91	6.7
DIETARY FIBRE, WAT.SOL.	GRAM	2.00	-	-	-	2.00	GRAM/MJ	1.09	
DIETARY FIBRE, WAT.INS.	GRAM	18.50	-	-	-	18.50	GRAM/MJ	10.05	

MUNGBOHNE
SAMEN, TROCKEN

VIGNA RADIATA (L.) WILCZEK
PHASEOLUS AUREUS ROXB.

MUNG BEAN
(INDIAN BEAN, GREEN GRAM, GOLDEN GRAM)
SEED, DRY

AMBÉRIQUE
(HARICOT A GRAIN VERT)
GRAINE, SÈCHE

		PROTEIN	FAT	CARBOHYDRATES	TOTAL
ENERGY VALUE (AVERAGE) PER 100 G EDIBLE PORTION	KJOULE	447	43	778	1268
	(KCAL)	107	10	186	303
AMOUNT OF DIGESTIBLE CONSTITUENTS PER 100 G	GRAM	18.72	0.99	46.50	
ENERGY VALUE (AVERAGE) OF THE DIGESTIBLE FRACTION PER 100 G EDIBLE PORTION	KJOULE	349	39	778	1165
	(KCAL)	83	9.2	186	279

WASTE PERCENTAGE AVERAGE 0.00

CONSTITUENTS	DIM	AV	VARIATION			AVR	NUTR. DENS.		MOLPERC.
WATER	GRAM	10.00	7.20	-	12.00	10.00	GRAM/MJ	8.58	
PROTEIN	GRAM	24.00	22.00	-	26.20	24.00	GRAM/MJ	20.60	
FAT	GRAM	1.10	0.90	-	1.40	1.10	GRAM/MJ	0.94	
AVAILABLE CARBOHYDR.	GRAM	46.50	37.30	-	55.60	46.50	GRAM/MJ	39.90	
TOTAL DIETARY FIBRE	GRAM	10.08 1	-	-	-	10.08	GRAM/MJ	8.65	
MINERALS	GRAM	3.50	3.40	-	3.70	3.50	GRAM/MJ	3.00	

1 METHOD OF ENGLYST

CONSTITUENTS	DIM	AV	VARIATION			AVR	NUTR. DENS.		MOLPERC.
SODIUM	MILLI	6.00	-	-	-	6.00	MILLI/MJ	5.15	
POTASSIUM	GRAM	1.22	1.13	-	1.30	1.22	GRAM/MJ	1.05	
MAGNESIUM	MILLI	170.00	-	-	-	170.00	MILLI/MJ	145.89	
CALCIUM	MILLI	122.00	105.00	-	138.00	122.00	MILLI/MJ	104.69	
IRON	MILLI	6.90	5.50	-	8.00	6.90	MILLI/MJ	5.92	
PHOSPHORUS	MILLI	378.00	320.00	-	474.00	378.00	MILLI/MJ	324.38	
VITAMIN B1	MILLI	0.48	0.45	-	0.53	0.48	MILLI/MJ	0.41	
VITAMIN B2	MILLI	0.23	0.20	-	0.26	0.23	MILLI/MJ	0.20	
NICOTINAMIDE	MILLI	2.30	2.00	-	2.50	2.30	MILLI/MJ	1.97	
VITAMIN C	MILLI	-	0.00	-	4.00				
ALANINE	GRAM	1.30	1.08	-	1.40	1.30	GRAM/MJ	1.12	6.9
ARGININE	GRAM	1.90	1.76	-	1.96	1.90	GRAM/MJ	1.63	5.2
ASPARTIC ACID	GRAM	3.38	3.16	-	3.51	3.38	GRAM/MJ	2.90	12.0
CYSTINE	GRAM	0.27	0.14	-	0.39	0.27	GRAM/MJ	0.23	0.5
GLUTAMIC ACID	GRAM	4.81	3.96	-	5.51	4.81	GRAM/MJ	4.13	15.5
GLYCINE	GRAM	1.24	1.11	-	1.34	1.24	GRAM/MJ	1.06	7.8
HISTIDINE	GRAM	0.83	0.74	-	1.04	0.83	GRAM/MJ	0.71	2.5
ISOLEUCINE	GRAM	1.27	1.23	-	1.33	1.27	GRAM/MJ	1.09	4.6
LEUCINE	GRAM	2.22	2.09	-	2.31	2.22	GRAM/MJ	1.91	8.0
LYSINE	GRAM	1.95	1.90	-	2.03	1.95	GRAM/MJ	1.67	6.3
METHIONINE	GRAM	0.39	0.29	-	0.58	0.39	GRAM/MJ	0.33	1.2
PHENYLALANINE	GRAM	1.65	1.58	-	1.74	1.65	GRAM/MJ	1.42	4.7
PROLINE	GRAM	1.26	1.05	-	1.47	1.26	GRAM/MJ	1.08	5.2
SERINE	GRAM	1.45	1.22	-	1.57	1.45	GRAM/MJ	1.24	6.5
THREONINE	GRAM	1.02	0.95	-	1.15	1.02	GRAM/MJ	0.88	4.1
TRYPTOPHAN	GRAM	0.38	0.23	-	0.52	0.38	GRAM/MJ	0.33	0.9
TYROSINE	GRAM	0.80	0.75	-	0.84	0.80	GRAM/MJ	0.69	2.1
VALINE	GRAM	1.45	1.31	-	1.56	1.45	GRAM/MJ	1.24	5.9
SUCROSE	GRAM	1.20	0.30	-	2.00	1.20	GRAM/MJ	1.03	
RAFFINOSE	GRAM	1.50	0.30	-	2.60	1.50	GRAM/MJ	1.29	
STACHYOSE	GRAM	2.00	1.20	-	2.80	2.00	GRAM/MJ	1.72	
VERBASCOSE	GRAM	2.80	1.70	-	3.80	2.80	GRAM/MJ	2.40	
STARCH	GRAM	45.30	37.00	-	53.60	45.30	GRAM/MJ	38.87	
CELLULOSE	GRAM	5.00	-	-	-	5.00	GRAM/MJ	4.29	
DIETARY FIBRE,WAT.SOL.	GRAM	2.79	-	-	-	2.79	GRAM/MJ	2.39	
DIETARY FIBRE,WAT.INS.	GRAM	7.29	-	-	-	7.29	GRAM/MJ	6.26	

SESAM SAMEN, TROCKEN	**ORIENTAL SESAME** GINGELLY SEED, DRY	**SESAME** GRAINE, SÈCHE

SESAMUM INDICUM L.

		PROTEIN	FAT	CARBOHYDRATES	TOTAL
ENERGY VALUE (AVERAGE) PER 100 G EDIBLE PORTION	KJOULE	330	1961	179	2470
	(KCAL)	79	469	43	590
AMOUNT OF DIGESTIBLE CONSTITUENTS PER 100 G	GRAM	13.82	45.36	9.89	
ENERGY VALUE (AVERAGE) OF THE DIGESTIBLE FRACTION PER 100 G EDIBLE PORTION	KJOULE	257	1765	174	2196
	(KCAL)	62	422	42	525

WASTE PERCENTAGE AVERAGE 0.00

CONSTITUENTS	DIM	AV		VARIATION	AVR	NUTR. DENS.		MOLPERC.
WATER	GRAM	5.20		4.80 - 5.80	5.20	GRAM/MJ	2.37	
PROTEIN	GRAM	17.72		14.59 - 22.39	17.72	GRAM/MJ	8.07	
FAT	GRAM	50.40		48.40 - 52.80	50.40	GRAM/MJ	22.95	
AVAILABLE CARBOHYDR.	GRAM	10.20	1	- - -	10.20	GRAM/MJ	4.64	
TOTAL DIETARY FIBRE	GRAM	11.18	2	- - -	11.18	GRAM/MJ	5.09	
MINERALS	GRAM	5.30		4.40 - 6.20	5.30	GRAM/MJ	2.41	
SODIUM	MILLI	45.00		40.00 - 49.00	45.00	MILLI/MJ	20.49	
POTASSIUM	MILLI	458.00		407.00 - 508.00	458.00	MILLI/MJ	208.54	
MAGNESIUM	MILLI	347.00		- - -	347.00	MILLI/MJ	158.00	
CALCIUM	MILLI	783.00		750.00 - 816.00	783.00	MILLI/MJ	356.52	
IRON	MILLI	10.00		8.10 - 12.00	10.00	MILLI/MJ	4.55	
PHOSPHORUS	MILLI	607.00		600.00 - 614.00	607.00	MILLI/MJ	276.38	
ALANINE	GRAM	1.13		1.09 - 1.14	1.13	GRAM/MJ	0.51	7.8
ARGININE	GRAM	2.20		2.14 - 2.26	2.20	GRAM/MJ	1.00	7.8
ASPARTIC ACID	GRAM	1.37		1.09 - 1.63	1.37	GRAM/MJ	0.62	6.4
CYSTINE	GRAM	0.26		0.18 - 0.35	0.26	GRAM/MJ	0.12	0.7
GLUTAMIC ACID	GRAM	3.73		3.48 - 3.99	3.73	GRAM/MJ	1.70	15.7
GLYCINE	GRAM	1.48		1.39 - 1.55	1.48	GRAM/MJ	0.67	12.2
HISTIDINE	GRAM	0.49		0.46 - 0.51	0.49	GRAM/MJ	0.22	2.0
ISOLEUCINE	GRAM	0.93		- - -	0.93	GRAM/MJ	0.42	4.4
LEUCINE	GRAM	1.54		1.53 - 1.55	1.54	GRAM/MJ	0.70	7.3
LYSINE	GRAM	0.64		0.63 - 0.64	0.64	GRAM/MJ	0.29	2.7
METHIONINE	GRAM	0.64		- - -	0.64	GRAM/MJ	0.29	2.7
PHENYLALANINE	GRAM	1.25		- - -	1.25	GRAM/MJ	0.57	4.7
PROLINE	GRAM	1.13		0.99 - 1.25	1.13	GRAM/MJ	0.51	6.1
SERINE	GRAM	0.99		0.98 - 1.00	0.99	GRAM/MJ	0.45	5.8

1 ESTIMATED BY THE DIFFERENCE METHOD
100 - (WATER + PROTEIN + FAT + MINERALS + TOTAL DIETARY FIBRE)

2 METHOD OF MEUSER, SUCKOW AND KULIKOWSKI ("BERLINER METHODE")

CONSTITUENTS	DIM	AV	VARIATION			AVR	NUTR. DENS.		MOLPERC.
THREONINE	GRAM	0.91	0.86	-	0.96	0.91	GRAM/MJ	0.41	4.7
TRYPTOPHAN	GRAM	0.29	0.26	-	0.31	0.29	GRAM/MJ	0.13	0.9
TYROSINE	GRAM	0.72	0.67	-	0.78	0.72	GRAM/MJ	0.33	2.5
VALINE	GRAM	1.11	1.09	-	1.15	1.11	GRAM/MJ	0.51	5.9
SALICYLIC ACID	MILLI	0.23	0.23	-	0.23	0.23	MILLI/MJ	0.10	
PALMITIC ACID	GRAM	5.70	4.80	-	6.50	5.70	GRAM/MJ	2.60	
STEARIC ACID	GRAM	1.60	0.96	-	2.30	1.60	GRAM/MJ	0.73	
ARACHIDIC ACID	GRAM	0.25	-	-	-	0.25	GRAM/MJ	0.11	
OLEIC ACID	GRAM	19.90	19.50	-	20.30	19.90	GRAM/MJ	9.06	
LINOLEIC ACID	GRAM	18.70	15.90	-	21.40	18.70	GRAM/MJ	8.51	
LINOLENIC ACID	GRAM	0.67	0.00	-	0.86	0.67	GRAM/MJ	0.31	
DIETARY FIBRE,WAT.SOL.	GRAM	3.22	-	-	-	3.22	GRAM/MJ	1.47	
DIETARY FIBRE,WAT.INS.	GRAM	7.96	-	-	-	7.96	GRAM/MJ	3.62	

SOJABOHNE
SAMEN, TROCKEN

SOYA BEAN
SEED, DRY

SOJA
GRAINE, SÈCHE

GLYCINE HYSPIDA MAXIM.

		PROTEIN	FAT	CARBOHYDRATES	TOTAL
ENERGY VALUE (AVERAGE)	KJOULE	628	704	102	1434
PER 100 G	(KCAL)	150	168	24	343
EDIBLE PORTION					
AMOUNT OF DIGESTIBLE	GRAM	26.30	16.29	6.10	
CONSTITUENTS PER 100 G					
ENERGY VALUE (AVERAGE)	KJOULE	490	634	102	1226
OF THE DIGESTIBLE	(KCAL)	117	151	24	293
FRACTION PER 100 G					
EDIBLE PORTION					

WASTE PERCENTAGE AVERAGE 17 MINIMUM 16 MAXIMUM 18

CONSTITUENTS	DIM	AV	VARIATION			AVR	NUTR. DENS.		MOLPERC.
WATER	GRAM	8.50	7.50	-	10.10	7.06	GRAM/MJ	6.93	
PROTEIN	GRAM	33.73	31.08	-	36.56	28.00	GRAM/MJ	27.52	
FAT	GRAM	18.10	16.30	-	21.30	15.02	GRAM/MJ	14.77	
AVAILABLE CARBOHYDR.	GRAM	6.10	-	-	-	5.06	GRAM/MJ	4.98	
TOTAL DIETARY FIBRE	GRAM	15.18 1	-	-	-	12.60	GRAM/MJ	12.38	
MINERALS	GRAM	4.70	3.90	-	5.24	3.90	GRAM/MJ	3.83	
SODIUM	MILLI	4.00	-	-	-	3.32	MILLI/MJ	3.26	
POTASSIUM	GRAM	1.74	1.65	-	1.84	1.44	GRAM/MJ	1.42	
MAGNESIUM	MILLI	247.00	210.00	-	284.00	205.01	MILLI/MJ	201.50	
CALCIUM	MILLI	257.00	227.00	-	304.00	213.31	MILLI/MJ	209.66	
MANGANESE	MILLI	2.80	-	-	-	2.32	MILLI/MJ	2.28	
IRON	MILLI	8.59	7.32	-	10.00	7.13	MILLI/MJ	7.01	
COPPER	MILLI	0.11	-	-	-	0.09	MILLI/MJ	0.09	
ZINC	MILLI	1.00	-	-	-	0.83	MILLI/MJ	0.82	
NICKEL	MILLI	0.70	-	-	-	0.58	MILLI/MJ	0.57	
PHOSPHORUS	MILLI	591.00	551.00	-	641.00	490.53	MILLI/MJ	482.13	
CHLORIDE	MILLI	7.00	-	-	-	5.81	MILLI/MJ	5.71	
IODIDE	MICRO	6.30	-	-	-	5.23	MICRO/MJ	5.14	
SELENIUM	MICRO	60.00	48.00	-	71.00	49.80	MICRO/MJ	48.95	
CAROTENE	MILLI	0.38	0.34	-	0.40	0.32	MILLI/MJ	0.31	
VITAMIN E ACTIVITY	MILLI	1.50	-	-	-	1.25	MILLI/MJ	1.22	
TOTAL TOCOPHEROLS	MILLI	15.30	-	-	-	12.70	MILLI/MJ	12.48	
ALPHA-TOCOPHEROL	MILLI	0.64	-	-	-	0.53	MILLI/MJ	0.52	
GAMMA-TOCOPHEROL	MILLI	8.20	-	-	-	6.81	MILLI/MJ	6.69	
DELTA-TOCOPHEROL	MILLI	6.50	-	-	-	5.40	MILLI/MJ	5.30	
VITAMIN K	MILLI	0.19	-	-	-	0.16	MILLI/MJ	0.16	

1 METHOD OF ENGLYST

CONSTITUENTS	DIM	AV		VARIATION		AVR	NUTR. DENS.		MOLPERC.
VITAMIN B1	MILLI	0.99		0.30 - 1.44		0.82	MILLI/MJ	0.81	
VITAMIN B2	MILLI	0.52		0.25 - 1.30		0.43	MILLI/MJ	0.42	
NICOTINAMIDE	MILLI	2.51		1.00 - 4.90		2.08	MILLI/MJ	2.05	
PANTOTHENIC ACID	MILLI	1.92		0.80 - 2.75		1.59	MILLI/MJ	1.57	
VITAMIN B6	MILLI	1.19		- - -		0.99	MILLI/MJ	0.97	
BIOTIN	MICRO	60.00		- - -		49.80	MICRO/MJ	48.95	
FOLIC ACID	MILLI	0.23		- - -		0.19	MILLI/MJ	0.19	
ALANINE	GRAM	1.53		- - -		1.27	GRAM/MJ	1.25	6.5
ARGININE	GRAM	2.36		2.01 - 2.67		1.96	GRAM/MJ	1.93	5.1
ASPARTIC ACID	GRAM	3.99		- - -		3.31	GRAM/MJ	3.26	11.4
CYSTINE	GRAM	0.59		0.52 - 0.66		0.49	GRAM/MJ	0.48	0.9
GLUTAMIC ACID	GRAM	6.49		- - -		5.39	GRAM/MJ	5.29	16.7
GLYCINE	GRAM	1.42		- - -		1.18	GRAM/MJ	1.16	7.2
HISTIDINE	GRAM	0.83		0.78 - 0.88		0.69	GRAM/MJ	0.68	2.0
ISOLEUCINE	GRAM	1.78		1.58 - 1.98		1.48	GRAM/MJ	1.45	5.1
LEUCINE	GRAM	2.84		- - -		2.36	GRAM/MJ	2.32	8.2
LYSINE	GRAM	1.90		1.43 - 2.33		1.58	GRAM/MJ	1.55	4.9
METHIONINE	GRAM	0.58		0.49 - 0.68		0.48	GRAM/MJ	0.47	1.5
PHENYLALANINE	GRAM	1.97		1.83 - 2.15		1.64	GRAM/MJ	1.61	4.5
PROLINE	GRAM	1.82		- - -		1.51	GRAM/MJ	1.48	6.0
SERINE	GRAM	1.69		- - -		1.40	GRAM/MJ	1.38	6.1
THREONINE	GRAM	1.49		1.35 - 1.66		1.24	GRAM/MJ	1.22	4.7
TRYPTOPHAN	GRAM	0.45		0.40 - 0.51		0.37	GRAM/MJ	0.37	0.8
TYROSINE	GRAM	1.25		1.18 - 1.33		1.04	GRAM/MJ	1.02	2.6
VALINE	GRAM	1.76		1.42 - 1.94		1.46	GRAM/MJ	1.44	5.7
GLUCOSE	MILLI	5.00		- - -		4.15	MILLI/MJ	4.08	
SUCROSE	GRAM	6.10		3.30 - 9.90		5.06	GRAM/MJ	4.98	
RAFFINOSE	GRAM	1.10		- - -		0.91	GRAM/MJ	0.90	
ARABINOSE	MILLI	2.00		- - -		1.66	MILLI/MJ	1.63	
STACHYOSE	GRAM	3.70		- - -		3.07	GRAM/MJ	3.02	
PENTOSAN	GRAM	3.60		- - -		2.99	GRAM/MJ	2.94	
PHYTIC ACID	GRAM	1.25		1.00 - 1.50		1.04	GRAM/MJ	1.02	
LAURIC ACID	MILLI	10.00		10.00 - 10.00		8.30	MILLI/MJ	8.16	
MYRISTIC ACID	MILLI	30.00		30.00 - 30.00		24.90	MILLI/MJ	24.47	
PALMITIC ACID	GRAM	1.58		0.40 - 1.84		1.32	GRAM/MJ	1.29	
STEARIC ACID	GRAM	0.59		0.42 - 1.19		0.49	GRAM/MJ	0.49	
BEHENIC ACID	MILLI	10.00		10.00 - 10.00		8.30	MILLI/MJ	8.16	
OLEIC ACID	GRAM	3.79		3.61 - 5.32		3.15	GRAM/MJ	3.10	
EICOSENOIC ACID	MILLI	30.00		30.00 - 30.00		24.90	MILLI/MJ	24.47	
LINOLEIC ACID	GRAM	8.65		7.50 - 10.60		7.18	GRAM/MJ	7.06	
LINOLENIC ACID	GRAM	1.00		0.82 - 1.07		0.83	GRAM/MJ	0.82	
CAMPESTEROL	MILLI	23.00		- - -		19.09	MILLI/MJ	18.76	
BETA-SITOSTEROL	MILLI	90.00		- - -		74.70	MILLI/MJ	73.42	
STIGMASTEROL	MILLI	40.00		- - -		33.20	MILLI/MJ	32.63	
TOTAL PURINES	MILLI	380.00	2	- - -		315.40	MILLI/MJ	310.00	
TOTAL PHOSPHOLIPIDS	GRAM	1.78		- - -		1.48	GRAM/MJ	1.45	
PHOSPHATIDYLCHOLINE	MILLI	798.00		- - -		662.34	MILLI/MJ	651.00	
PHOSPHATIDYLETHANOLAMINE	MILLI	466.00		- - -		386.78	MILLI/MJ	380.16	
PHOSPHATIDYLINOSITOL	MILLI	250.00		- - -		207.50	MILLI/MJ	203.95	
DIETARY FIBRE,WAT.SOL.	GRAM	6.58		- - -		5.46	GRAM/MJ	5.37	
DIETARY FIBRE,WAT.INS.	GRAM	8.60		- - -		7.14	GRAM/MJ	7.02	

2 VALUE EXPRESSED IN MG URIC ACID/100G

SOJAMEHL VOLLFETT	**SOYA FLOUR** FULL FAT	**FARINE DE SOJA** GRASSE (FLOCONS DE SOJA)

		PROTEIN	FAT	CARBOHYDRATES	TOTAL
ENERGY VALUE (AVERAGE)	KJOULE	694	802	52	1548
PER 100 G	(KCAL)	166	192	12	370
EDIBLE PORTION					
AMOUNT OF DIGESTIBLE	GRAM	29.09	18.54	3.10	
CONSTITUENTS PER 100 G					
ENERGY VALUE (AVERAGE)	KJOULE	542	721	52	1315
OF THE DIGESTIBLE	(KCAL)	129	172	12	314
FRACTION PER 100 G					
EDIBLE PORTION					

WASTE PERCENTAGE AVERAGE 0.00

CONSTITUENTS	DIM	AV	VARIATION			AVR	NUTR. DENS.		MOLPERC.
WATER	GRAM	9.10	8.94	-	9.54	9.10	GRAM/MJ	6.92	
PROTEIN	GRAM	37.30	35.90	-	38.80	37.30	GRAM/MJ	28.37	
FAT	GRAM	20.60	19.80	-	22.10	20.60	GRAM/MJ	15.67	
AVAILABLE CARBOHYDR.	GRAM	3.10	-	-	-	3.10	GRAM/MJ	2.36	
TOTAL DIETARY FIBRE	GRAM	10.91 [1]	-	-	-	10.91	GRAM/MJ	8.30	
MINERALS	GRAM	4.40	4.08	-	4.60	4.40	GRAM/MJ	3.35	
SODIUM	MILLI	4.00	-	-	-	4.00	MILLI/MJ	3.04	
POTASSIUM	GRAM	1.87	1.83	-	1.90	1.87	GRAM/MJ	1.42	
MAGNESIUM	MILLI	247.00	-	-	-	247.00	MILLI/MJ	187.83	
CALCIUM	MILLI	195.00	-	-	-	195.00	MILLI/MJ	148.29	
MANGANESE	MILLI	4.00	-	-	-	4.00	MILLI/MJ	3.04	
IRON	MILLI	12.10	-	-	-	12.10	MILLI/MJ	9.20	
ZINC	MILLI	4.90	-	-	-	4.90	MILLI/MJ	3.73	
NICKEL	MILLI	0.41	-	-	-	0.41	MILLI/MJ	0.31	
MOLYBDENUM	MILLI	0.18	-	-	-	0.18	MILLI/MJ	0.14	
PHOSPHORUS	MILLI	553.00	-	-	-	553.00	MILLI/MJ	420.54	
CHLORIDE	MILLI	106.00	-	-	-	106.00	MILLI/MJ	80.61	
FLUORIDE	MILLI	0.11	-	-	-	0.11	MILLI/MJ	0.08	
IODIDE	MICRO	0.80	-	-	-	0.80	MICRO/MJ	0.61	
BORON	MILLI	0.30	-	-	-	0.30	MILLI/MJ	0.23	
CAROTENE	MICRO	84.00	-	-	-	84.00	MICRO/MJ	63.88	
VITAMIN E ACTIVITY	MILLI	1.50	-	-	-	1.50	MILLI/MJ	1.14	
TOTAL TOCOPHEROLS	MILLI	15.30	-	-	-	15.30	MILLI/MJ	11.64	
ALPHA-TOCOPHEROL	MILLI	0.64	-	-	-	0.64	MILLI/MJ	0.49	
GAMMA-TOCOPHEROL	MILLI	8.20	-	-	-	8.20	MILLI/MJ	6.24	
DELTA-TOCOPHEROL	MILLI	6.50	-	-	-	6.50	MILLI/MJ	4.94	

1 METHOD OF ENGLYST

CONSTITUENTS	DIM	AV	VARIATION			AVR	NUTR. DENS.		MOLPERC.
VITAMIN K	MILLI	0.20	-	-	-	0.20	MILLI/MJ	0.15	
VITAMIN B1	MILLI	0.77	-	-	-	0.77	MILLI/MJ	0.59	
VITAMIN B2	MILLI	0.28	-	-	-	0.28	MILLI/MJ	0.21	
NICOTINAMIDE	MILLI	2.20	-	-	-	2.20	MILLI/MJ	1.67	
PANTOTHENIC ACID	MILLI	1.80	-	-	-	1.80	MILLI/MJ	1.37	
VITAMIN B6	MILLI	-	0.80	-	9.50				
FOLIC ACID	MILLI	0.19	0.14	-	0.27	0.19	MILLI/MJ	0.14	
VITAMIN C	-	0.00	-	-	-	0.00			
ALANINE	GRAM	1.68	-	-	-	1.68	GRAM/MJ	1.28	6.2
ARGININE	GRAM	3.14	-	-	-	3.14	GRAM/MJ	2.39	5.9
ASPARTIC ACID	GRAM	4.48	-	-	-	4.48	GRAM/MJ	3.41	11.0
CYSTINE	GRAM	0.59	-	-	-	0.59	GRAM/MJ	0.45	0.8
GLUTAMIC ACID	GRAM	7.83	-	-	-	7.83	GRAM/MJ	5.95	17.5
GLYCINE	GRAM	1.68	-	-	-	1.68	GRAM/MJ	1.28	7.3
HISTIDINE	GRAM	0.95	-	-	-	0.95	GRAM/MJ	0.72	2.0
ISOLEUCINE	GRAM	1.90	-	-	-	1.90	GRAM/MJ	1.44	4.8
LEUCINE	GRAM	2.88	-	-	-	2.88	GRAM/MJ	2.19	7.2
LYSINE	GRAM	2.56	-	-	-	2.56	GRAM/MJ	1.95	5.7
METHIONINE	GRAM	0.58	-	-	-	0.58	GRAM/MJ	0.44	1.3
PHENYLALANINE	GRAM	1.86	-	-	-	1.86	GRAM/MJ	1.41	3.7
PROLINE	GRAM	2.34	-	-	-	2.34	GRAM/MJ	1.78	6.7
SERINE	GRAM	2.08	-	-	-	2.08	GRAM/MJ	1.58	6.5
THREONINE	GRAM	1.61	-	-	-	1.61	GRAM/MJ	1.22	4.4
TRYPTOPHAN	GRAM	0.48	-	-	-	0.48	GRAM/MJ	0.37	0.8
TYROSINE	GRAM	1.45	-	-	-	1.45	GRAM/MJ	1.10	2.6
VALINE	GRAM	1.97	-	-	-	1.97	GRAM/MJ	1.50	5.5
SUCROSE	GRAM	3.10	-	-	-	3.10	GRAM/MJ	2.36	
RAFFINOSE	GRAM	3.28	-	-	-	3.28	GRAM/MJ	2.49	
STACHYOSE	GRAM	3.68	-	-	-	3.68	GRAM/MJ	2.80	
POLYURONIC ACID	GRAM	4.80	4.00	-	5.60	4.80	GRAM/MJ	3.65	
PALMITIC ACID	GRAM	1.96	0.51	-	2.09	1.96	GRAM/MJ	1.49	
STEARIC ACID	GRAM	0.68	0.48	-	1.36	0.68	GRAM/MJ	0.52	
ARACHIDIC ACID	GRAM	-	0.10	-	0.48				
PALMITOLEIC ACID	MILLI	94.00	-	-	-	94.00	MILLI/MJ	71.48	
OLEIC ACID	GRAM	3.35	-	-	-	3.35	GRAM/MJ	2.55	
LINOLEIC ACID	GRAM	10.70	-	-	-	10.70	GRAM/MJ	8.14	
LINOLENIC ACID	GRAM	1.40	-	-	-	1.40	GRAM/MJ	1.06	
DIETARY FIBRE,WAT.SOL.	GRAM	5.09	-	-	-	5.09	GRAM/MJ	3.87	
DIETARY FIBRE,WAT.INS.	GRAM	5.82	-	-	-	5.82	GRAM/MJ	4.43	

SONNENBLUME / SUNFLOWER / TOURNESOL

SAMEN, TROCKEN — SEED, DRY — GRAINE, SÈCHE

HELIANTHUS ANNUUS L.

		PROTEIN	FAT	CARBOHYDRATES	TOTAL
ENERGY VALUE (AVERAGE)	KJOULE	418	1907	206	2531
PER 100 G EDIBLE PORTION	(KCAL)	100	456	49	605
AMOUNT OF DIGESTIBLE CONSTITUENTS PER 100 G	GRAM	17.52	44.10	12.33	
ENERGY VALUE (AVERAGE)	KJOULE	326	1716	206	2249
OF THE DIGESTIBLE FRACTION PER 100 G EDIBLE PORTION	(KCAL)	78	410	49	537

WASTE PERCENTAGE AVERAGE 48

CONSTITUENTS	DIM	AV		VARIATION			AVR	NUTR.	DENS.	MOLPERC.
WATER	GRAM	6.60		4.50	-	8.50	3.43	GRAM/MJ	2.94	
PROTEIN	GRAM	22.47		20.95	-	23.83	11.68	GRAM/MJ	9.99	
FAT	GRAM	49.00		36.00	-	56.40	25.48	GRAM/MJ	21.79	
AVAILABLE CARBOHYDR.	GRAM	12.33	1	-	-	-	6.41	GRAM/MJ	5.48	
TOTAL DIETARY FIBRE	GRAM	6.30	2	-	-	-	3.28	GRAM/MJ	2.80	
MINERALS	GRAM	3.30		2.00	-	5.00	1.72	GRAM/MJ	1.47	
SODIUM	MILLI	2.00		0.00	-	4.00	1.04	MILLI/MJ	0.89	
POTASSIUM	MILLI	725.00		633.00	-	815.00	377.00	MILLI/MJ	322.41	
MAGNESIUM	MILLI	420.00		370.00	-	470.00	218.40	MILLI/MJ	186.78	
CALCIUM	MILLI	98.00		59.00	-	142.00	50.96	MILLI/MJ	43.58	
MANGANESE	MILLI	2.40		1.40	-	4.10	1.25	MILLI/MJ	1.07	
IRON	MILLI	6.30		5.80	-	7.00	3.28	MILLI/MJ	2.80	
COPPER	MILLI	2.80		1.90	-	3.10	1.46	MILLI/MJ	1.25	
ZINC	MILLI	5.20		4.20	-	6.50	2.70	MILLI/MJ	2.31	
PHOSPHORUS	MILLI	618.00		486.00	-	735.00	321.36	MILLI/MJ	274.83	
VITAMIN B1	MILLI	1.90		-	-	-	0.99	MILLI/MJ	0.84	
VITAMIN B2	MILLI	0.14		0.07	-	0.20	0.07	MILLI/MJ	0.06	
NICOTINAMIDE	MILLI	4.10		2.40	-	5.80	2.13	MILLI/MJ	1.82	
VITAMIN C	-	0.00		-	-	-	0.00			
ALANINE	GRAM	1.29		1.08	-	1.48	0.67	GRAM/MJ	0.57	7.2
ARGININE	GRAM	2.20		2.09	-	2.31	1.14	GRAM/MJ	0.98	6.3
ASPARTIC ACID	GRAM	2.38		2.37	-	2.40	1.24	GRAM/MJ	1.06	8.9
CYSTINE	GRAM	0.39		0.25	-	0.53	0.20	GRAM/MJ	0.17	0.8
GLUTAMIC ACID	GRAM	6.40		5.94	-	6.86	3.33	GRAM/MJ	2.85	21.7
GLYCINE	GRAM	1.63		1.47	-	1.77	0.85	GRAM/MJ	0.72	10.8

1 ESTIMATED BY THE DIFFERENCE METHOD
100 - (WATER + PROTEIN + FAT + MINERALS + TOTAL DIETARY FIBRE)

2 METHOD OF MEUSER, SUCKOW AND KULIKOWSKI ("BERLINER METHODE")

CONSTITUENTS	DIM	AV	VARIATION			AVR	NUTR. DENS.		MOLPERC.
HISTIDINE	GRAM	0.63	0.61	-	0.64	0.33	GRAM/MJ	0.28	2.0
ISOLEUCINE	GRAM	1.37	1.15	-	1.59	0.71	GRAM/MJ	0.61	5.2
LEUCINE	GRAM	1.71	1.61	-	1.80	0.89	GRAM/MJ	0.76	6.5
LYSINE	GRAM	0.89	0.86	-	0.92	0.46	GRAM/MJ	0.40	3.0
METHIONINE	GRAM	0.49	0.40	-	0.58	0.25	GRAM/MJ	0.22	1.6
PHENYLALANINE	GRAM	1.26	1.17	-	1.34	0.66	GRAM/MJ	0.56	3.8
PROLINE	GRAM	1.07	0.92	-	1.21	0.56	GRAM/MJ	0.48	4.6
SERINE	GRAM	1.17	1.13	-	1.20	0.61	GRAM/MJ	0.52	5.6
THREONINE	GRAM	0.91	0.90	-	0.92	0.47	GRAM/MJ	0.40	3.8
TRYPTOPHAN	GRAM	0.31	0.29	-	0.33	0.16	GRAM/MJ	0.14	0.8
TYROSINE	GRAM	0.65	0.54	-	0.75	0.34	GRAM/MJ	0.29	1.8
VALINE	GRAM	1.26	1.18	-	1.34	0.66	GRAM/MJ	0.56	5.4
SALICYLIC ACID	MILLI	0.12	0.12	-	0.12	0.06	MILLI/MJ	0.05	
PALMITIC ACID	GRAM	3.14	-	-	-	1.63	GRAM/MJ	1.40	
STEARIC ACID	GRAM	2.12	-	-	-	1.10	GRAM/MJ	0.94	
ARACHIDIC ACID	GRAM	0.14	-	-	-	0.07	GRAM/MJ	0.06	
PALMITOLEIC ACID	MILLI	90.00	-	-	-	46.80	MILLI/MJ	40.02	
OLEIC ACID	GRAM	13.38	-	-	-	6.96	GRAM/MJ	5.95	
LINOLEIC ACID	GRAM	27.87	-	-	-	14.49	GRAM/MJ	12.39	
LINOLENIC ACID	MILLI	90.00	-	-	-	46.80	MILLI/MJ	40.02	
5-AVENASTEROL	MILLI	13.20	-	-	-	6.86	MILLI/MJ	5.87	
BRASSICASTEROL	MILLI	0.30	-	-	-	0.16	MILLI/MJ	0.13	
CAMPESTEROL	MILLI	13.90	-	-	-	7.23	MILLI/MJ	6.18	
BETA-SITOSTEROL	MILLI	95.10	-	-	-	49.45	MILLI/MJ	42.29	
STIGMASTEROL	MILLI	14.80	-	-	-	7.70	MILLI/MJ	6.58	
TOTAL PHOSPHOLIPIDS	MILLI	932.00	-	-	-	484.64	MILLI/MJ	414.47	
PHOSPHATIDYLCHOLINE	MILLI	329.00	-	-	-	171.08	MILLI/MJ	146.31	
PHOSPHATIDYLETHANOLAMINE	MILLI	121.00	-	-	-	62.92	MILLI/MJ	53.81	
PHOSPHATIDYLINOSITOL	MILLI	226.00	-	-	-	117.52	MILLI/MJ	100.50	
DIETARY FIBRE, WAT.SOL.	GRAM	2.50	-	-	-	1.30	GRAM/MJ	1.11	
DIETARY FIBRE, WAT.INS.	GRAM	3.80	-	-	-	1.98	GRAM/MJ	1.69	

SONNENBLUMEN-KERNMEHL | SUNFLOWER SEED FLOUR | FARINE DE TOURNESOL

		PROTEIN	FAT	CARBOHYDRATES	TOTAL
ENERGY VALUE (AVERAGE)	KJOULE	611	412	597	1621
PER 100 G	(KCAL)	146	99	143	388
EDIBLE PORTION					
AMOUNT OF DIGESTIBLE	GRAM	31.45	10.07	35.70	
CONSTITUENTS PER 100 G					
ENERGY VALUE (AVERAGE)	KJOULE	520	392	597	1509
OF THE DIGESTIBLE	(KCAL)	124	94	143	361
FRACTION PER 100 G					
EDIBLE PORTION					

WASTE PERCENTAGE AVERAGE 0.00

CONSTITUENTS	DIM	AV		VARIATION			AVR	NUTR. DENS.		MOLPERC.
WATER	GRAM	7.80		-	-	-	7.80	GRAM/MJ	5.17	
PROTEIN	GRAM	37.00		-	-	-	37.00	GRAM/MJ	24.52	
FAT	GRAM	10.60		-	-	-	10.60	GRAM/MJ	7.02	
AVAILABLE CARBOHYDR.	GRAM	35.70	1	-	-	-	35.70	GRAM/MJ	23.66	
TOTAL DIETARY FIBRE	GRAM	8.90	2	-	-	-	8.90	GRAM/MJ	5.90	
CALCIUM	MILLI	360.00		-	-	-	360.00	MILLI/MJ	238.56	
PHOSPHORUS	MILLI	600.00		-	-	-	600.00	MILLI/MJ	397.59	
VITAMIN B1	MILLI	1.50		-	-	-	1.50	MILLI/MJ	0.99	
VITAMIN B2	MILLI	0.47		-	-	-	0.47	MILLI/MJ	0.31	
NICOTINAMIDE	MILLI	31.20		-	-	-	31.20	MILLI/MJ	20.67	
ARGININE	GRAM	3.81		-	-	-	3.81	GRAM/MJ	2.52	
CYSTINE	GRAM	0.75		-	-	-	0.75	GRAM/MJ	0.50	
HISTIDINE	GRAM	0.94		-	-	-	0.94	GRAM/MJ	0.62	
ISOLEUCINE	GRAM	2.05		-	-	-	2.05	GRAM/MJ	1.36	
LEUCINE	GRAM	2.79		-	-	-	2.79	GRAM/MJ	1.85	
LYSINE	GRAM	1.40		-	-	-	1.40	GRAM/MJ	0.93	
METHIONINE	GRAM	0.71		-	-	-	0.71	GRAM/MJ	0.47	
PHENYLALANINE	GRAM	1.96		-	-	-	1.96	GRAM/MJ	1.30	
THREONINE	GRAM	1.47		-	-	-	1.47	GRAM/MJ	0.97	
TRYPTOPHAN	GRAM	0.55		-	-	-	0.55	GRAM/MJ	0.36	
TYROSINE	GRAM	1.04		-	-	-	1.04	GRAM/MJ	0.69	
VALINE	GRAM	2.18		-	-	-	2.18	GRAM/MJ	1.44	
PALMITIC ACID	GRAM	0.66		-	-	-	0.66	GRAM/MJ	0.44	
STEARIC ACID	GRAM	0.51		-	-	-	0.51	GRAM/MJ	0.34	
ARACHIDIC ACID	MILLI	64.00		-	-	-	64.00	MILLI/MJ	42.41	
PALMITOLEIC ACID	MILLI	25.00		-	-	-	25.00	MILLI/MJ	16.57	
OLEIC ACID	GRAM	2.30		-	-	-	2.30	GRAM/MJ	1.52	
LINOLEIC ACID	GRAM	6.30		-	-	-	6.30	GRAM/MJ	4.17	
LINOLENIC ACID	MILLI	50.00		-	-	-	50.00	MILLI/MJ	33.13	

1 ESTIMATED BY THE DIFFERENCE METHOD
100 - (WATER + PROTEIN + FAT + MINERALS + TOTAL DIETARY FIBRE)

2 METHOD OF MEUSER, SUCKOW AND KULIKOWSKI ("BERLINER METHODE")

STRAUCHERBSE (ERBSENBOHNE, TAUBEN-ERBSE) SAMEN, TROCKEN CAJANUS CAJAN (L.) HUTH. C. INDICUS SPRENG.	PIGEON PEA (RED GRAM) SEED, DRY	POIS D'ANGOLE (AMBREVADE) GRAINE, SÈCHE

		PROTEIN	FAT	CARBOHYDRATES	TOTAL
ENERGY VALUE (AVERAGE)	KJOULE	376	54	787	1217
PER 100 G EDIBLE PORTION	(KCAL)	90	13	188	291
AMOUNT OF DIGESTIBLE CONSTITUENTS PER 100 G	GRAM	15.75	1.26	47.00	
ENERGY VALUE (AVERAGE)	KJOULE	293	49	787	1129
OF THE DIGESTIBLE FRACTION PER 100 G EDIBLE PORTION	(KCAL)	70	12	188	270

WASTE PERCENTAGE AVERAGE 0.00

CONSTITUENTS	DIM	AV	VARIATION		AVR	NUTR. DENS.		MOLPERC.
WATER	GRAM	11.20	9.90	13.40	11.20	GRAM/MJ	9.92	
PROTEIN	GRAM	20.20	19.00	22.30	20.20	GRAM/MJ	17.89	
FAT	GRAM	1.40	1.20	1.70	1.40	GRAM/MJ	1.24	
AVAILABLE CARBOHYDR.	GRAM	47.00	-	-	47.00	GRAM/MJ	41.63	
MINERALS	GRAM	3.60	3.50	3.80	3.60	GRAM/MJ	3.19	
SODIUM	MILLI	26.00	-	-	26.00	MILLI/MJ	23.03	
POTASSIUM	MILLI	654.00	-	-	654.00	MILLI/MJ	579.29	
CALCIUM	MILLI	121.00	100.00	161.00	121.00	MILLI/MJ	107.18	
IRON	MILLI	5.00	-	-	5.00	MILLI/MJ	4.43	
PHOSPHORUS	MILLI	265.00	224.00	285.00	265.00	MILLI/MJ	234.73	
CAROTENE	MILLI	0.10	0.06	0.16	0.10	MILLI/MJ	0.09	
VITAMIN B1	MILLI	0.57	0.49	0.72	0.57	MILLI/MJ	0.50	
VITAMIN B2	MILLI	0.17	0.14	0.21	0.17	MILLI/MJ	0.15	
NICOTINAMIDE	MILLI	2.50	2.20	2.90	2.50	MILLI/MJ	2.21	
VITAMIN C	MILLI	-	0.00	2.00				
ALANINE	GRAM	1.12	-	-	1.12	GRAM/MJ	0.99	7.2
ARGININE	GRAM	1.39	-	-	1.39	GRAM/MJ	1.23	4.6
ASPARTIC ACID	GRAM	2.74	-	-	2.74	GRAM/MJ	2.43	11.8
CYSTINE	GRAM	0.15	-	-	0.15	GRAM/MJ	0.13	0.4
GLUTAMIC ACID	GRAM	5.21	-	-	5.21	GRAM/MJ	4.61	20.3
GLYCINE	GRAM	0.97	-	-	0.97	GRAM/MJ	0.86	7.4
HISTIDINE	GRAM	0.70	-	-	0.70	GRAM/MJ	0.62	2.6
ISOLEUCINE	GRAM	0.95	-	-	0.95	GRAM/MJ	0.84	4.2
LEUCINE	GRAM	1.88	-	-	1.88	GRAM/MJ	1.67	8.2
LYSINE	GRAM	1.50	-	-	1.50	GRAM/MJ	1.33	5.9

CONSTITUENTS	DIM	AV	VARIATION		AVR	NUTR. DENS.		MOLPERC.
METHIONINE	GRAM	0.15	-	-	0.15	GRAM/MJ	0.13	0.6
PHENYLALANINE	GRAM	1.74	-	-	1.74	GRAM/MJ	1.54	6.0
PROLINE	GRAM	1.04	-	-	1.04	GRAM/MJ	0.92	5.2
SERINE	GRAM	0.94	-	-	0.94	GRAM/MJ	0.83	5.1
THREONINE	GRAM	0.80	-	-	0.80	GRAM/MJ	0.71	3.9
TRYPTOPHAN	GRAM	0.12	-	-	0.12	GRAM/MJ	0.11	0.3
TYROSINE	GRAM	0.70	-	-	0.70	GRAM/MJ	0.62	2.2
VALINE	GRAM	0.83	-	-	0.83	GRAM/MJ	0.74	4.1
SUCROSE	GRAM	2.70	-	-	2.70	GRAM/MJ	2.39	
RAFFINOSE	GRAM	1.10	-	-	1.10	GRAM/MJ	0.97	
STACHYOSE	GRAM	2.80	-	-	2.80	GRAM/MJ	2.48	
VERBASCOSE	GRAM	4.10	-	-	4.10	GRAM/MJ	3.63	
STARCH	GRAM	44.30	40.40	48.20	44.30	GRAM/MJ	39.24	

URDBOHNE	BLACK GRAM	HARICOT MUNGO
(MUNGOBOHNE) SAMEN, TROCKEN PHASEOLUS MUNGO L. PH. RADIATUS ROXB.	(MUNGO BEAN) SEED, DRY	GRAINE, SÈCHE

		PROTEIN	FAT	CARBOHYDRATES	TOTAL
ENERGY VALUE (AVERAGE) PER 100 G EDIBLE PORTION	KJOULE	430	47	695	1171
	(KCAL)	103	11	166	280
AMOUNT OF DIGESTIBLE CONSTITUENTS PER 100 G	GRAM	18.01	1.08	41.50	
ENERGY VALUE (AVERAGE) OF THE DIGESTIBLE FRACTION PER 100 G EDIBLE PORTION	KJOULE	335	42	695	1072
	(KCAL)	80	10	166	256

WASTE PERCENTAGE AVERAGE 0.00

CONSTITUENTS	DIM	AV	VARIATION			AVR	NUTR. DENS.		MOLPERC.
WATER	GRAM	10.30	9.70	-	11.00	10.30	GRAM/MJ	9.61	
PROTEIN	GRAM	23.10	21.00	-	24.00	23.10	GRAM/MJ	21.55	
FAT	GRAM	1.20	1.00	-	1.60	1.20	GRAM/MJ	1.12	
AVAILABLE CARBOHYDR.	GRAM	41.50	33.50	-	49.40	41.50	GRAM/MJ	38.71	
MINERALS	GRAM	4.60	3.40	-	5.50	4.60	GRAM/MJ	4.29	
POTASSIUM	MILLI	171.00	-	-	-	171.00	MILLI/MJ	159.51	
MAGNESIUM	MILLI	243.00	-	-	-	243.00	MILLI/MJ	226.67	
CALCIUM	MILLI	123.00	110.00	-	148.00	123.00	MILLI/MJ	114.73	
IRON	MILLI	9.80	8.40	-	12.00	9.80	MILLI/MJ	9.14	
ZINC	MILLI	5.50	-	-	-	5.50	MILLI/MJ	5.13	
PHOSPHORUS	MILLI	365.00	348.00	-	382.00	365.00	MILLI/MJ	340.47	
CAROTENE	MICRO	36.00	20.00	-	45.00	36.00	MICRO/MJ	33.58	
TOTAL TOCOPHEROLS	MILLI	1.90	-	-	-	1.90	MILLI/MJ	1.77	
VITAMIN K	MILLI	0.13	-	-	-	0.13	MILLI/MJ	0.12	
VITAMIN B1	MILLI	0.49	0.41	-	0.58	0.49	MILLI/MJ	0.46	
VITAMIN B2	MILLI	0.23	0.20	-	0.27	0.23	MILLI/MJ	0.21	
NICOTINAMIDE	MILLI	2.30	-	-	-	2.30	MILLI/MJ	2.15	
PANTOTHENIC ACID	MILLI	3.50	-	-	-	3.50	MILLI/MJ	3.26	
VITAMIN B6	MILLI	-	0.20	-	1.10				
BIOTIN	MICRO	7.50	-	-	-	7.50	MICRO/MJ	7.00	
FOLIC ACID	MILLI	0.14	-	-	-	0.14	MILLI/MJ	0.13	
VITAMIN C	MILLI	3.00	TRACES	-	4.00	3.00	MILLI/MJ	2.80	
ALANINE	GRAM	0.99	-	-	-	0.99	GRAM/MJ	0.92	
ARGININE	GRAM	2.08	-	-	-	2.08	GRAM/MJ	1.94	
ASPARTIC ACID	GRAM	3.14	-	-	-	3.14	GRAM/MJ	2.93	

CONSTITUENTS	DIM	AV	VARIATION			AVR	NUTR. DENS.		MOLPERC.
CYSTINE	-	TRACES	-	-	-				
GLUTAMIC ACID	GRAM	4.53	-	-	-	4.53	GRAM/MJ	4.23	
GLYCINE	GRAM	0.95	-	-	-	0.95	GRAM/MJ	0.89	
HISTIDINE	GRAM	1.04	-	-	-	1.04	GRAM/MJ	0.97	
ISOLEUCINE	GRAM	2.17	-	-	-	2.17	GRAM/MJ	2.02	
LEUCINE	GRAM	1.32	-	-	-	1.32	GRAM/MJ	1.23	
LYSINE	GRAM	2.08	-	-	-	2.08	GRAM/MJ	1.94	
METHIONINE	GRAM	0.21	-	-	-	0.21	GRAM/MJ	0.20	
PHENYLALANINE	GRAM	1.57	-	-	-	1.57	GRAM/MJ	1.46	
PROLINE	GRAM	1.04	-	-	-	1.04	GRAM/MJ	0.97	
SERINE	GRAM	0.95	-	-	-	0.95	GRAM/MJ	0.89	
THREONINE	GRAM	0.81	-	-	-	0.81	GRAM/MJ	0.76	
TRYPTOPHAN	GRAM	1.34	-	-	-	1.34	GRAM/MJ	1.25	
TYROSINE	GRAM	0.88	-	-	-	0.88	GRAM/MJ	0.82	
VALINE	GRAM	1.46	-	-	-	1.46	GRAM/MJ	1.36	
SUCROSE	GRAM	1.40	1.30	-	1.50	1.40	GRAM/MJ	1.31	
RAFFINOSE	-	TRACES	-	-	-				
STACHYOSE	GRAM	0.72	0.55	-	0.89	0.72	GRAM/MJ	0.67	
VERBASCOSE	GRAM	3.60	3.40	-	3.70	3.60	GRAM/MJ	3.36	
STARCH	GRAM	40.10	32.20	-	47.90	40.10	GRAM/MJ	37.41	
PALMITIC ACID	GRAM	0.21	-	-	-	0.21	GRAM/MJ	0.20	
STEARIC ACID	MILLI	60.00	-	-	-	60.00	MILLI/MJ	55.97	
OLEIC ACID	GRAM	0.20	-	-	-	0.20	GRAM/MJ	0.19	
LINOLEIC ACID	GRAM	0.14	-	-	-	0.14	GRAM/MJ	0.13	
LINOLENIC ACID	GRAM	0.57	-	-	-	0.57	GRAM/MJ	0.53	

Pilze

MUSHROOMS CHAMPIGNONS

BIRKENPILZ ROUGH-STEMMED BOLETUS BOLET RUDE

LECCINUM SCABRUM FR. EX. BULL.

		PROTEIN	FAT	CARBOHYDRATES	TOTAL
ENERGY VALUE (AVERAGE)	KJOULE	39	24	0.0	63
PER 100 G EDIBLE PORTION	(KCAL)	9.3	5.8	0.0	15
AMOUNT OF DIGESTIBLE CONSTITUENTS PER 100 G	GRAM	2.48	0.55	0.00	
ENERGY VALUE (AVERAGE)	KJOULE	39	22	0.0	61
OF THE DIGESTIBLE FRACTION PER 100 G EDIBLE PORTION	(KCAL)	9.3	5.2	0.0	14

WASTE PERCENTAGE AVERAGE 30 MINIMUM 20 MAXIMUM 40

CONSTITUENTS	DIM	AV	VARIATION			AVR	NUTR. DENS.		MOLPERC.
WATER	GRAM	88.50	86.40	-	90.80	61.95	KILO/MJ	1.46	
PROTEIN	GRAM	2.48	1.16	-	3.58	1.74	GRAM/MJ	40.91	
FAT	GRAM	0.62	0.57	-	0.64	0.43	GRAM/MJ	10.23	
TOTAL DIETARY FIBRE	GRAM	7.30 1	-	-	-	5.11	GRAM/MJ	120.42	
MINERALS	GRAM	1.10	1.09	-	1.11	0.77	GRAM/MJ	18.14	
SODIUM	MILLI	2.00	-	-	-	1.40	MILLI/MJ	32.99	
POTASSIUM	MILLI	346.00	-	-	-	242.20	GRAM/MJ	5.71	
CALCIUM	MILLI	-	9.00	-	58.00				
MANGANESE	MILLI	0.74	-	-	-	0.52	MILLI/MJ	12.21	
IRON	MILLI	1.60	1.00	-	2.20	1.12	MILLI/MJ	26.39	
PHOSPHORUS	MILLI	115.00	-	-	-	80.50	GRAM/MJ	1.90	
VITAMIN B1	MILLI	0.10	-	-	-	0.07	MILLI/MJ	1.65	
VITAMIN B2	MILLI	0.44	-	-	-	0.31	MILLI/MJ	7.26	
NICOTINAMIDE	MILLI	4.90	-	-	-	3.43	MILLI/MJ	80.83	
VITAMIN C	MILLI	7.00	5.00	-	8.00	4.90	MILLI/MJ	115.47	

1 ESTIMATED BY THE DIFFERENCE METHOD
100 - (WATER + PROTEIN + FAT + MINERALS)

CONSTITUENTS	DIM	AV	VARIATION			AVR	NUTR. DENS.		MOLPERC.
ARGININE	MILLI	61.00	-	-	-	42.70	GRAM/MJ	1.01	
CYSTINE	MILLI	53.00	-	-	-	37.10	MILLI/MJ	874.25	
HISTIDINE	MILLI	16.00	-	-	-	11.20	MILLI/MJ	263.92	
ISOLEUCINE	MILLI	30.00	-	-	-	21.00	MILLI/MJ	494.86	
LEUCINE	MILLI	99.00	-	-	-	69.30	GRAM/MJ	1.63	
LYSINE	MILLI	41.00	-	-	-	28.70	MILLI/MJ	676.30	
METHIONINE	MILLI	3.00	-	-	-	2.10	MILLI/MJ	49.49	
PHENYLALANINE	MILLI	63.00	-	-	-	44.10	GRAM/MJ	1.04	
THREONINE	MILLI	81.00	-	-	-	56.70	GRAM/MJ	1.34	
TRYPTOPHAN	MILLI	18.00	-	-	-	12.60	MILLI/MJ	296.91	
TYROSINE	MILLI	66.00	-	-	-	46.20	GRAM/MJ	1.09	
VALINE	MILLI	49.00	-	-	-	34.30	MILLI/MJ	808.27	

BUTTERPILZ — RINGED BOLETUS — NONETTE VOILÉE

SUILLUS LUTEUS FR.

		PROTEIN	FAT	CARBOHYDRATES	TOTAL
ENERGY VALUE (AVERAGE) PER 100 G EDIBLE PORTION	KJOULE	27	14	5.0	46
	(KCAL)	6.4	3.3	1.2	11
AMOUNT OF DIGESTIBLE CONSTITUENTS PER 100 G	GRAM	1.70	0.32	0.30	
ENERGY VALUE (AVERAGE) OF THE DIGESTIBLE FRACTION PER 100 G EDIBLE PORTION	KJOULE	27	13	5.0	44
	(KCAL)	6.4	3.0	1.2	11

WASTE PERCENTAGE AVERAGE 20

CONSTITUENTS	DIM	AV		VARIATION			AVR	NUTR. DENS.		MOLPERC.
WATER	GRAM	91.10		87.40	-	93.40	72.88	KILO/MJ	2.06	
PROTEIN	GRAM	1.70		0.99	-	2.33	1.36	GRAM/MJ	38.37	
FAT	GRAM	0.36		0.27	-	0.45	0.29	GRAM/MJ	8.13	
AVAILABLE CARBOHYDR.	GRAM	0.30		-	-	-	0.24	GRAM/MJ	6.77	
TOTAL DIETARY FIBRE	GRAM	5.92	1	-	-	-	4.74	GRAM/MJ	133.63	
MINERALS	GRAM	0.62		0.45	-	0.80	0.50	GRAM/MJ	14.00	
POTASSIUM	MILLI	190.00		120.00	-	270.00	152.00	GRAM/MJ	4.29	
MAGNESIUM	MILLI	6.00		4.00	-	8.00	4.80	MILLI/MJ	135.44	
CALCIUM	MILLI	25.00		8.00	-	30.00	20.00	MILLI/MJ	564.32	
MANGANESE	MICRO	62.00		-	-	-	49.60	MILLI/MJ	1.40	
IRON	MILLI	1.28		-	-	-	1.02	MILLI/MJ	28.89	
VITAMIN C	MILLI	8.00		-	-	-	6.40	MILLI/MJ	180.58	
REDUCING SUGAR	GRAM	0.30		-	-	-	0.24	GRAM/MJ	6.77	
TREHALOSE	-	0.00		-	-	-	0.00			
MANNITOL	GRAM	1.44		1.39	-	1.50	1.15	GRAM/MJ	32.50	

1 ESTIMATED BY THE DIFFERENCE METHOD:
100 - (WATER + PROTEIN + FAT + AVAILABLE CARBOHYDRATES + MINERALS)

CHAMPIGNON (ZUCHTCHAMPIGNON) — MUSHROOM — CHAMPIGNON CULTIVÉ

AGARICUS BISPORUS (LANGE) SING.

		PROTEIN	FAT	CARBOHYDRATES	TOTAL
ENERGY VALUE (AVERAGE) PER 100 G EDIBLE PORTION	KJOULE	43	9.3	11	64
	(KCAL)	10	2.2	2.7	15
AMOUNT OF DIGESTIBLE CONSTITUENTS PER 100 G	GRAM	2.74	0.21	0.68	
ENERGY VALUE (AVERAGE) OF THE DIGESTIBLE FRACTION PER 100 G EDIBLE PORTION	KJOULE	43	8.4	11	63
	(KCAL)	10	2.0	2.7	15

WASTE PERCENTAGE AVERAGE 0.00

CONSTITUENTS	DIM	AV	VARIATION			AVR	NUTR. DENS.		MOLPERC.
WATER	GRAM	90.70	89.00	-	93.00	88.89	KILO/MJ	1.44	
PROTEIN	GRAM	2.74	2.21	-	3.27	2.69	GRAM/MJ	43.65	
FAT	GRAM	0.24	0.18	-	0.30	0.24	GRAM/MJ	3.82	
AVAILABLE CARBOHYDR.	GRAM	0.68	-	-	-	0.67	GRAM/MJ	10.83	
TOTAL DIETARY FIBRE	GRAM	1.92 1	-	-	-	1.88	GRAM/MJ	30.58	
MINERALS	GRAM	1.02	0.88	-	1.25	1.00	GRAM/MJ	16.25	
SODIUM	MILLI	8.00	4.00	-	15.00	7.84	MILLI/MJ	127.44	
POTASSIUM	MILLI	422.00	354.00	-	486.00	413.56	GRAM/MJ	6.72	
MAGNESIUM	MILLI	13.00	11.00	-	16.00	12.74	MILLI/MJ	207.09	
CALCIUM	MILLI	8.00	1.00	-	9.00	7.84	MILLI/MJ	127.44	
MANGANESE	MILLI	0.11	0.08	-	0.13	0.11	MILLI/MJ	1.75	
IRON	MILLI	1.26	0.73	-	1.95	1.23	MILLI/MJ	20.07	
COPPER	MILLI	0.40	0.14	-	0.64	0.39	MILLI/MJ	6.37	
ZINC	MILLI	0.39	0.28	-	0.50	0.38	MILLI/MJ	6.21	
NICKEL	MICRO	2.00	-	-	-	1.96	MICRO/MJ	31.86	
CHROMIUM	MICRO	7.00	1.00	-	14.00	6.86	MICRO/MJ	111.51	
PHOSPHORUS	MILLI	123.00	99.00	-	150.00	120.54	GRAM/MJ	1.96	
CHLORIDE	MILLI	67.00	25.00	-	90.00	65.66	GRAM/MJ	1.07	
FLUORIDE	MICRO	31.00	25.00	-	37.00	30.38	MICRO/MJ	493.82	
IODIDE	MICRO	18.00	-	-	-	17.64	MICRO/MJ	286.73	
BORON	MILLI	5.40	4.90	-	5.80	5.29	MILLI/MJ	86.02	
SELENIUM	MICRO	7.00	3.00	-	10.00	6.86	MICRO/MJ	111.51	
CAROTENE	MICRO	10.00	5.00	-	16.00	9.80	MICRO/MJ	159.30	
VITAMIN D	MICRO	1.94	0.50	-	3.75	1.90	MICRO/MJ	30.90	
VITAMIN E ACTIVITY	MILLI	0.11	-	-	-	0.11	MILLI/MJ	1.75	
TOTAL TOCOPHEROLS	MILLI	0.30	-	-	-	0.29	MILLI/MJ	4.78	

1 ESTIMATED BY THE DIFFERENCE METHOD:
100 - (WATER + PROTEIN + FAT + AVAILABLE CARBOHYDRATES + MINERALS)

CONSTITUENTS	DIM	AV	VARIATION			AVR	NUTR. DENS.		MOLPERC.
ALPHA-TOCOPHEROL	MICRO	80.00	-	-	-	78.40	MILLI/MJ	1.27	
ALPHA-TOCOTRIENOL	MILLI	0.12	-	-	-	0.12	MILLI/MJ	1.91	
VITAMIN K	MICRO	17.00	10.00	-	23.00	16.66	MICRO/MJ	270.80	
VITAMIN B1	MILLI	0.10	0.07	-	0.12	0.10	MILLI/MJ	1.59	
VITAMIN B2	MILLI	0.44	0.30	-	0.52	0.43	MILLI/MJ	7.01	
NICOTINAMIDE	MILLI	5.20	4.00	-	6.50	5.10	MILLI/MJ	82.83	
PANTOTHENIC ACID	MILLI	2.10	1.70	-	2.70	2.06	MILLI/MJ	33.45	
VITAMIN B6	MICRO	65.00	45.00	-	100.00	63.70	MILLI/MJ	1.04	
BIOTIN	MICRO	16.00	-	-	-	15.68	MICRO/MJ	254.87	
FOLIC ACID	MICRO	25.00	20.00	-	30.00	24.50	MICRO/MJ	398.24	
VITAMIN C	MILLI	4.90	3.00	-	9.00	4.80	MILLI/MJ	78.06	
ARGININE	MILLI	200.00	160.00	-	250.00	196.00	GRAM/MJ	3.19	
CYSTINE	MILLI	14.00	-	-	-	13.72	MILLI/MJ	223.02	
HISTIDINE	MILLI	57.00	37.00	-	77.00	55.86	MILLI/MJ	907.99	
ISOLEUCINE	MILLI	110.00	60.00	-	160.00	107.80	GRAM/MJ	1.75	
LEUCINE	MILLI	120.00	100.00	-	130.00	117.60	GRAM/MJ	1.91	
LYSINE	MILLI	170.00	120.00	-	260.00	166.60	GRAM/MJ	2.71	
METHIONINE	MILLI	23.00	13.00	-	36.00	22.54	MILLI/MJ	366.38	
PHENYLALANINE	MILLI	74.00	55.00	-	110.00	72.52	GRAM/MJ	1.18	
THREONINE	MILLI	87.00	71.00	-	110.00	85.26	GRAM/MJ	1.39	
TRYPTOPHAN	MILLI	24.00	19.00	-	28.00	23.52	MILLI/MJ	382.31	
TYROSINE	MILLI	66.00	53.00	-	78.00	64.68	GRAM/MJ	1.05	
VALINE	MILLI	90.00	70.00	-	130.00	88.20	GRAM/MJ	1.43	
MALIC ACID	MILLI	14.00	8.00	-	20.00	13.72	MILLI/MJ	223.02	
CITRIC ACID	MILLI	120.00	50.00	-	190.00	117.60	GRAM/MJ	1.91	
SUCCINIC ACID	MILLI	10.00	7.00	-	12.00	9.80	MILLI/MJ	159.30	
FUMARIC ACID	MILLI	12.00	-	-	-	11.76	MILLI/MJ	191.16	
SALICYLIC ACID	MILLI	0.24	0.24	-	0.24	0.24	MILLI/MJ	3.82	
GLUCOSE	GRAM	0.20	0.20	-	0.22	0.20	GRAM/MJ	3.27	
FRUCTOSE	GRAM	0.21	0.16	-	0.48	0.21	GRAM/MJ	3.39	
SUCROSE	GRAM	0.14	0.02	-	0.52	0.14	GRAM/MJ	2.29	
MANNITOL	GRAM	1.15	-	-	-	1.13	GRAM/MJ	18.32	
MYRISTIC ACID	MILLI	2.00	-	-	-	1.96	MILLI/MJ	31.86	
PALMITIC ACID	MILLI	32.00	-	-	-	31.36	MILLI/MJ	509.75	
STEARIC ACID	MILLI	5.00	-	-	-	4.90	MILLI/MJ	79.65	
LINOLEIC ACID	MILLI	149.00	-	-	-	146.02	GRAM/MJ	2.37	
TOTAL PURINES	MILLI	20.00 2	-	-	-	19.60	MILLI/MJ	318.59	
DIETARY FIBRE,WAT.SOL.	GRAM	0.37	-	-	-	0.36	GRAM/MJ	5.89	
DIETARY FIBRE,WAT.INS.	GRAM	1.55	-	-	-	1.52	GRAM/MJ	24.69	

2 VALUE EXPRESSED IN MG URIC ACID/100G

CHAMPIGNONS
IN DOSEN

1

MUSHROOMS
CANNED

1

CHAMPIGNONS
EN BOÎTES DE CONSERVES

1

		PROTEIN	FAT	CARBOHYDRATES	TOTAL
ENERGY VALUE (AVERAGE)	KJOULE	35	19	11	65
PER 100 G EDIBLE PORTION	(KCAL)	8.4	4.7	2.6	16
AMOUNT OF DIGESTIBLE CONSTITUENTS PER 100 G	GRAM	2.25	0.45	0.64	
ENERGY VALUE (AVERAGE)	KJOULE	35	18	11	64
OF THE DIGESTIBLE FRACTION PER 100 G EDIBLE PORTION	(KCAL)	8.4	4.2	2.6	15

WASTE PERCENTAGE AVERAGE 0.00

CONSTITUENTS	DIM	AV		VARIATION		AVR	NUTR. DENS.		MOLPERC.
WATER	GRAM	91.20	90.80	-	91.60	91.20	KILO/MJ	1.44	
PROTEIN	GRAM	2.25	1.95	-	2.52	2.25	GRAM/MJ	35.42	
FAT	GRAM	0.50	0.40	-	0.50	0.50	GRAM/MJ	7.87	
AVAILABLE CARBOHYDR.	GRAM	0.64 2	-	-	-	0.64	GRAM/MJ	10.07	
TOTAL DIETARY FIBRE	GRAM	7.02	-	-	-	7.02	GRAM/MJ	110.51	
MINERALS	GRAM	1.10	1.00	-	1.30	1.10	GRAM/MJ	17.32	
SODIUM	MILLI	319.00	247.00	-	391.00	319.00	GRAM/MJ	5.02	
POTASSIUM	MILLI	127.00	111.00	-	142.00	127.00	GRAM/MJ	2.00	
MAGNESIUM	MILLI	15.00	8.80	-	22.00	15.00	MILLI/MJ	236.13	
CALCIUM	MILLI	19.00	11.00	-	33.00	19.00	MILLI/MJ	299.10	
MANGANESE	MICRO	50.00	-	-	-	50.00	MICRO/MJ	787.11	
IRON	MILLI	0.80	0.70	-	0.80	0.80	MILLI/MJ	12.59	
COPPER	MILLI	0.48	0.24	-	0.71	0.48	MILLI/MJ	7.56	
PHOSPHORUS	MILLI	69.00	49.00	-	96.00	69.00	GRAM/MJ	1.09	
CHLORIDE	MILLI	400.00	-	-	-	400.00	GRAM/MJ	6.30	
BORON	MILLI	4.15	3.10	-	5.20	4.15	MILLI/MJ	65.33	
SELENIUM	MICRO	10.00	-	-	-	10.00	MICRO/MJ	157.42	
VITAMIN A	-	0.00	-	-	-	0.00			
CAROTENE	-	0.00	-	-	-	0.00			
VITAMIN B1	MICRO	20.00	10.00	-	50.00	20.00	MICRO/MJ	314.84	
VITAMIN B2	MILLI	0.19	0.11	-	0.25	0.19	MILLI/MJ	2.99	
NICOTINAMIDE	MILLI	1.22	0.62	-	1.75	1.22	MILLI/MJ	19.21	
PANTOTHENIC ACID	MILLI	0.80	-	-	-	0.80	MILLI/MJ	12.59	
VITAMIN B6	MICRO	60.00	-	-	-	60.00	MICRO/MJ	944.53	
VITAMIN C	MILLI	1.70	1.00	-	2.30	1.70	MILLI/MJ	26.76	

1 TOTAL CONTENT (SOLID AND LIQUID)

2 CALCULATED FROM THE FRESH PRODUCT

CONSTITUENTS	DIM	AV	VARIATION			AVR	NUTR. DENS.		MOLPERC.
ARGININE	GRAM	0.12	-	-	-	0.12	GRAM/MJ	1.89	
CYSTINE	MILLI	18.00	-	-	-	18.00	MILLI/MJ	283.36	
HISTIDINE	MILLI	46.00	-	-	-	46.00	MILLI/MJ	724.14	
ISOLEUCINE	MILLI	94.00	-	-	-	94.00	GRAM/MJ	1.48	
LEUCINE	GRAM	0.15	-	-	-	0.15	GRAM/MJ	2.36	
LYSINE	GRAM	0.14	-	-	-	0.14	GRAM/MJ	2.20	
METHIONINE	MILLI	35.00	-	-	-	35.00	MILLI/MJ	550.98	
PHENYLALANINE	MILLI	90.00	-	-	-	90.00	GRAM/MJ	1.42	
THREONINE	MILLI	94.00	-	-	-	94.00	GRAM/MJ	1.48	
TRYPTOPHAN	MILLI	37.00	-	-	-	37.00	MILLI/MJ	582.46	
TYROSINE	MILLI	76.00	-	-	-	76.00	GRAM/MJ	1.20	
VALINE	GRAM	0.11	-	-	-	0.11	GRAM/MJ	1.73	

HALLIMASCH HONEY MUSHROOM ARMILLAIRE DE MIEL

ARMILLARIELLA MELLEA FR.

		PROTEIN	FAT	CARBOHYDRATES	TOTAL
ENERGY VALUE (AVERAGE)	KJOULE	25	26	1.7	53
PER 100 G	(KCAL)	6.0	6.3	0.4	13
EDIBLE PORTION					
AMOUNT OF DIGESTIBLE	GRAM	1.60	0.61	0.10	
CONSTITUENTS PER 100 G					
ENERGY VALUE (AVERAGE)	KJOULE	25	24	1.7	51
OF THE DIGESTIBLE	(KCAL)	6.0	5.7	0.4	12
FRACTION PER 100 G					
EDIBLE PORTION					

WASTE PERCENTAGE AVERAGE 50

CONSTITUENTS	DIM	AV		VARIATION			AVR	NUTR. DENS.		MOLPERC.
WATER	GRAM	89.00		86.00	-	92.70	44.50	KILO/MJ	1.76	
PROTEIN	GRAM	1.60		1.19	-	2.02	0.80	GRAM/MJ	31.63	
FAT	GRAM	0.68		0.52	-	0.90	0.34	GRAM/MJ	13.44	
AVAILABLE CARBOHYDR.	GRAM	0.10		-	-	-	0.05	GRAM/MJ	1.98	
TOTAL DIETARY FIBRE	GRAM	7.59	1	-	-	-	3.80	GRAM/MJ	150.03	
MINERALS	GRAM	1.03		0.83	-	1.27	0.52	GRAM/MJ	20.36	
POTASSIUM	MILLI	440.00		380.00	-	530.00	220.00	GRAM/MJ	8.70	
MAGNESIUM	MILLI	12.50		8.00	-	17.00	6.25	MILLI/MJ	247.08	
CALCIUM	MILLI	7.00		4.00	-	10.00	3.50	MILLI/MJ	138.36	
MANGANESE	MILLI	0.16		-	-	-	0.08	MILLI/MJ	3.16	
IRON	MILLI	0.89		-	-	-	0.45	MILLI/MJ	17.59	
VITAMIN C	MILLI	5.00		3.00	-	7.00	2.50	MILLI/MJ	98.83	
INVERT SUGAR	GRAM	0.10		-	-	-	0.05	GRAM/MJ	1.98	
TREHALOSE	GRAM	0.10		-	-	-	0.05	GRAM/MJ	1.98	
MANNITOL	GRAM	1.09		-	-	-	0.55	GRAM/MJ	21.55	

1 ESTIMATED BY THE DIFFERENCE METHOD:
100 - (WATER + PROTEIN + FAT + AVAILABLE CARBOHYDRATES + MINERALS)

MORCHEL (SPEISEMORCHEL) — MOREL — MORILLE

MORCHELLA ESCULENTA PERS.

		PROTEIN	FAT	CARBOHYDRATES	TOTAL
ENERGY VALUE (AVERAGE)	KJOULE	26	12	0.0	38
PER 100 G EDIBLE PORTION	(KCAL)	6.2	3.0	0.0	9.2
AMOUNT OF DIGESTIBLE CONSTITUENTS PER 100 G	GRAM	1.66	0.28	0.00	
ENERGY VALUE (AVERAGE)	KJOULE	26	11	0.0	37
OF THE DIGESTIBLE FRACTION PER 100 G EDIBLE PORTION	(KCAL)	6.2	2.7	0.0	8.9

WASTE PERCENTAGE AVERAGE 17 MINIMUM 15 MAXIMUM 20

CONSTITUENTS	DIM	AV		VARIATION			AVR	NUTR. DENS.		MOLPERC.
WATER	GRAM	90.00		-	-	-	74.70	KILO/MJ	2.42	
PROTEIN	GRAM	1.66		1.32	-	2.19	1.38	GRAM/MJ	44.56	
FAT	GRAM	0.32		0.25	-	0.43	0.27	GRAM/MJ	8.59	
TOTAL DIETARY FIBRE	GRAM	7.01	1	-	-	-	5.82	GRAM/MJ	188.18	
MINERALS	GRAM	1.01		-	-	-	0.84	GRAM/MJ	27.11	
SODIUM	MILLI	2.00		-	-	-	1.66	MILLI/MJ	53.69	
POTASSIUM	MILLI	390.00		-	-	-	323.70	GRAM/MJ	10.47	
MAGNESIUM	MILLI	11.00		-	-	-	9.13	MILLI/MJ	295.29	
CALCIUM	MILLI	11.00		-	-	-	9.13	MILLI/MJ	295.29	
MANGANESE	MILLI	0.45		-	-	-	0.37	MILLI/MJ	12.08	
IRON	MILLI	1.20		-	-	-	1.00	MILLI/MJ	32.21	
PHOSPHORUS	MILLI	162.00		-	-	-	134.46	GRAM/MJ	4.35	
CHLORIDE	MILLI	8.00		-	-	-	6.64	MILLI/MJ	214.75	
VITAMIN D	MICRO	3.10		-	-	-	2.57	MICRO/MJ	83.22	
VITAMIN E ACTIVITY	MILLI	0.21		-	-	-	0.17	MILLI/MJ	5.64	
TOTAL TOCOPHEROLS	MILLI	0.63		-	-	-	0.52	MILLI/MJ	16.91	
ALPHA-TOCOPHEROL	MICRO	50.00		-	-	-	41.50	MILLI/MJ	1.34	
ALPHA-TOCOTRIENOL	MILLI	0.46		-	-	-	0.38	MILLI/MJ	12.35	
VITAMIN B1	MILLI	0.13		-	-	-	0.11	MILLI/MJ	3.49	
VITAMIN B2	MICRO	60.00		-	-	-	49.80	MILLI/MJ	1.61	
VITAMIN C	MILLI	5.00		-	-	-	4.15	MILLI/MJ	134.22	
TOTAL PURINES	MILLI	30.00	2	-	-	-	24.90	MILLI/MJ	805.33	

1 ESTIMATED BY THE DIFFERENCE METHOD 100 - (WATER + PROTEIN + FAT + MINERALS)

2 VALUE EXPRESSED IN MG URIC ACID/100G

PFIFFERLING (REHLING)	CHANTERELLE	CHANTERELLE (GIROLE)

CANTHARELLUS CIBARIUS FRIES

		PROTEIN	FAT	CARBOHYDRATES	TOTAL
ENERGY VALUE (AVERAGE)	KJOULE	24	19	2.8	46
PER 100 G EDIBLE PORTION	(KCAL)	5.7	4.6	0.7	11
AMOUNT OF DIGESTIBLE CONSTITUENTS PER 100 G	GRAM	1.52	0.44	0.17	
ENERGY VALUE (AVERAGE)	KJOULE	24	17	2.8	44
OF THE DIGESTIBLE FRACTION PER 100 G EDIBLE PORTION	(KCAL)	5.7	4.1	0.7	10

WASTE PERCENTAGE AVERAGE 39 MINIMUM 29 MAXIMUM 48

CONSTITUENTS	DIM	AV	VARIATION		AVR	NUTR. DENS.		MOLPERC.
WATER	GRAM	91.50	90.90	92.90	55.82	KILO/MJ	2.09	
PROTEIN	GRAM	1.52	-	-	0.93	GRAM/MJ	34.66	
FAT	GRAM	0.49	0.43	0.55	0.30	GRAM/MJ	11.17	
AVAILABLE CARBOHYDR.	GRAM	0.17	-	-	0.10	GRAM/MJ	3.88	
TOTAL DIETARY FIBRE	GRAM	5.55 [1]	-	-	3.39	GRAM/MJ	126.56	
MINERALS	GRAM	0.77	0.67	0.88	0.47	GRAM/MJ	17.56	
SODIUM	MILLI	3.00	1.00	7.00	1.83	MILLI/MJ	68.41	
POTASSIUM	MILLI	507.00	305.00	914.00	309.27	GRAM/MJ	11.56	
MAGNESIUM	MILLI	14.00	8.00	20.00	8.54	MILLI/MJ	319.24	
CALCIUM	MILLI	8.00	0.00	26.00	4.88	MILLI/MJ	182.42	
MANGANESE	MILLI	0.18	0.05	0.32	0.11	MILLI/MJ	4.10	
IRON	MILLI	6.50	-	-	3.97	MILLI/MJ	148.22	
COBALT	MICRO	3.00	2.00	4.00	1.83	MICRO/MJ	68.41	
COPPER	MILLI	-	0.20	1.00				
ZINC	MILLI	0.65	0.02	1.10	0.40	MILLI/MJ	14.82	
NICKEL	MICRO	10.00	5.00	10.00	6.10	MICRO/MJ	228.03	
CHROMIUM	MICRO	4.00	2.00	9.00	2.44	MICRO/MJ	91.21	
PHOSPHORUS	MILLI	44.20	-	-	26.96	GRAM/MJ	1.01	
FLUORIDE	MICRO	50.00	-	-	30.50	MILLI/MJ	1.14	
BORON	MICRO	35.00	21.00	63.00	21.35	MICRO/MJ	798.11	
SELENIUM	MICRO	0.40	0.30	0.50	0.24	MICRO/MJ	9.12	
VITAMIN D	MICRO	2.10	-	-	1.28	MICRO/MJ	47.89	
VITAMIN E ACTIVITY	MICRO	20.00	-	-	12.20	MICRO/MJ	456.06	
VITAMIN B1	MICRO	20.00	10.00	30.00	12.20	MICRO/MJ	456.06	
VITAMIN B2	MILLI	0.23	0.15	0.37	0.14	MILLI/MJ	5.24	
NICOTINAMIDE	MILLI	6.50	-	-	3.97	MILLI/MJ	148.22	

1 ESTIMATED BY THE DIFFERENCE METHOD:
100 - (WATER + PROTEIN + FAT + AVAILABLE CARBOHYDRATES + MINERALS)

CONSTITUENTS	DIM	AV		VARIATION		AVR	NUTR. DENS.		MOLPERC.
VITAMIN C	MILLI	6.00		2.00 - 10.00		3.66	MILLI/MJ	136.82	
ARGININE	MILLI	90.00		- - -		54.90	GRAM/MJ	2.05	
CYSTINE	MILLI	120.00		- - -		73.20	GRAM/MJ	2.74	
HISTIDINE	MILLI	30.00		- - -		18.30	MILLI/MJ	684.09	
ISOLEUCINE	MILLI	39.00		- - -		23.79	MILLI/MJ	889.32	
LEUCINE	MILLI	110.00		- - -		67.10	GRAM/MJ	2.51	
LYSINE	MILLI	39.00		- - -		23.79	MILLI/MJ	889.32	
METHIONINE	MILLI	8.00		- - -		4.88	MILLI/MJ	182.42	
PHENYLALANINE	MILLI	89.00		- - -		54.29	GRAM/MJ	2.03	
THREONINE	MILLI	130.00		- - -		79.30	GRAM/MJ	2.96	
TRYPTOPHAN	MILLI	48.00		- - -		29.28	GRAM/MJ	1.09	
TYROSINE	MILLI	85.00		- - -		51.85	GRAM/MJ	1.94	
VALINE	MILLI	62.00		- - -		37.82	GRAM/MJ	1.41	
GLUCOSE	MILLI	95.00		64.00 - 100.00		57.95	GRAM/MJ	2.17	
FRUCTOSE	MILLI	70.00		70.00 - 70.00		42.70	GRAM/MJ	1.60	
TREHALOSE	GRAM	1.01		1.01 - 1.01		0.62	GRAM/MJ	23.03	
MANNITOL	GRAM	0.39		0.39 - 0.39		0.24	GRAM/MJ	8.89	
TOTAL PURINES	MILLI	30.00	2	25.00 - 35.00		18.30	MILLI/MJ	684.09	

2 VALUE EXPRESSED IN MG URIC ACID/100G

PFIFFERLING GETROCKNET	CHANTERELLE DRIED	CHANTERELLE SÉCHÉE

		PROTEIN	FAT	CARBOHYDRATES	TOTAL
ENERGY VALUE (AVERAGE)	KJOULE	259	84	30	373
PER 100 G	(KCAL)	62	20	7.2	89
EDIBLE PORTION					
AMOUNT OF DIGESTIBLE	GRAM	16.50	1.93	1.80	
CONSTITUENTS PER 100 G					
ENERGY VALUE (AVERAGE)	KJOULE	259	75	30	364
OF THE DIGESTIBLE	(KCAL)	62	18	7.2	87
FRACTION PER 100 G					
EDIBLE PORTION					

WASTE PERCENTAGE AVERAGE 0.00

CONSTITUENTS	DIM	AV		VARIATION			AVR	NUTR. DENS.		MOLPERC.
WATER	GRAM	10.00		5.40	-	12.00	10.00	GRAM/MJ	27.45	
PROTEIN	GRAM	16.50		-	-	-	16.50	GRAM/MJ	45.29	
FAT	GRAM	2.15		-	-	-	2.15	GRAM/MJ	5.90	
AVAILABLE CARBOHYDR.	GRAM	1.80	1	-	-	-	1.80	GRAM/MJ	4.94	
TOTAL DIETARY FIBRE	GRAM	60.48	2	-	-	-	60.48	GRAM/MJ	166.02	
MINERALS	GRAM	9.07		8.50	-	9.64	9.07	GRAM/MJ	24.90	
SODIUM	MILLI	32.00		-	-	-	32.00	MILLI/MJ	87.84	
POTASSIUM	GRAM	5.37		-	-	-	5.37	GRAM/MJ	14.74	
CALCIUM	MILLI	85.00		-	-	-	85.00	MILLI/MJ	233.32	
IRON	MILLI	17.20		-	-	-	17.20	MILLI/MJ	47.21	
PHOSPHORUS	MILLI	581.00		-	-	-	581.00	GRAM/MJ	1.59	
VITAMIN C	MILLI	2.00		-	-	-	2.00	MILLI/MJ	5.49	
ARGININE	GRAM	0.98		-	-	-	0.98	GRAM/MJ	2.69	
CYSTINE	GRAM	1.30		-	-	-	1.30	GRAM/MJ	3.57	
HISTIDINE	GRAM	0.33		-	-	-	0.33	GRAM/MJ	0.91	
ISOLEUCINE	GRAM	0.42		-	-	-	0.42	GRAM/MJ	1.15	
LEUCINE	GRAM	1.20		-	-	-	1.20	GRAM/MJ	3.29	
LYSINE	GRAM	0.42		-	-	-	0.42	GRAM/MJ	1.15	
METHIONINE	MILLI	87.00		-	-	-	87.00	MILLI/MJ	238.81	
PHENYLALANINE	GRAM	0.97		-	-	-	0.97	GRAM/MJ	2.66	
THREONINE	GRAM	1.40		-	-	-	1.40	GRAM/MJ	3.84	
TRYPTOPHAN	GRAM	0.52		-	-	-	0.52	GRAM/MJ	1.43	
TYROSINE	GRAM	0.92		-	-	-	0.92	GRAM/MJ	2.53	
VALINE	GRAM	0.67		-	-	-	0.67	GRAM/MJ	1.84	
GLUCOSE	GRAM	0.31		0.31	-	0.31	0.31	GRAM/MJ	0.85	
TREHALOSE	GRAM	10.29		10.29	-	10.29	10.29	GRAM/MJ	28.25	
MANNITOL	GRAM	3.90		3.90	-	3.90	3.90	GRAM/MJ	10.71	

1 CALCULATED FROM THE FRESH PRODUCT

2 ESTIMATED BY THE DIFFERENCE METHOD:
100 - (WATER + PROTEIN + FAT + AVAILABLE CARBOHYDRATES + MINERALS)

PFIFFERLINGE	CHANTERELLES	CHANTERELLES
IN DOSEN	CANNED	EN BOÎTES DE CONSERVES
1	1	1

		PROTEIN	FAT	CARBOHYDRATES	TOTAL
ENERGY VALUE (AVERAGE)	KJOULE	22	27	3.3	52
PER 100 G	(KCAL)	5.2	6.4	0.8	12
EDIBLE PORTION					
AMOUNT OF DIGESTIBLE	GRAM	1.39	0.62	0.20	
CONSTITUENTS PER 100 G					
ENERGY VALUE (AVERAGE)	KJOULE	22	24	3.3	49
OF THE DIGESTIBLE	(KCAL)	5.2	5.8	0.8	12
FRACTION PER 100 G					
EDIBLE PORTION					

WASTE PERCENTAGE AVERAGE 0.00

CONSTITUENTS	DIM	AV	VARIATION			AVR	NUTR. DENS.		MOLPERC.
WATER	GRAM	90.10	87.60	-	92.70	90.10	KILO/MJ	1.83	
PROTEIN	GRAM	1.39	1.25	-	1.53	1.39	GRAM/MJ	28.18	
FAT	GRAM	0.69	0.50	-	0.95	0.69	GRAM/MJ	13.99	
AVAILABLE CARBOHYDR.	GRAM	0.20 2	-	-	-	0.20	GRAM/MJ	4.06	
TOTAL DIETARY FIBRE	GRAM	6.46 2	-	-	-	6.46	GRAM/MJ	130.98	
MINERALS	GRAM	0.80	0.77	-	0.82	0.80	GRAM/MJ	16.22	
SODIUM	MILLI	165.00	67.00	-	378.00	165.00	GRAM/MJ	3.35	
POTASSIUM	MILLI	155.00	81.00	-	276.00	155.00	GRAM/MJ	3.14	
MAGNESIUM	MILLI	6.10	-	-	-	6.10	MILLI/MJ	123.68	
CALCIUM	MILLI	4.70	-	-	-	4.70	MILLI/MJ	95.30	
IRON	MILLI	0.96	-	-	-	0.96	MILLI/MJ	19.46	
IODIDE	MICRO	3.30	3.30	-	3.30	3.30	MICRO/MJ	66.91	
CAROTENE	MILLI	1.30	1.30	-	1.30	1.30	MILLI/MJ	26.36	
VITAMIN B6	MICRO	40.00	40.00	-	40.00	40.00	MICRO/MJ	811.03	
VITAMIN C	MILLI	3.00	2.00	-	5.00	3.00	MILLI/MJ	60.83	

1 TOTAL CONTENT (SOLID AND LIQUID)

2 CALCULATED FROM THE FRESH PRODUCT

REIZKER ORANGE-AGARIC ORONGE

LACTARIUS DELICIOSUS FR. EX. L.

		PROTEIN	FAT	CARBOHYDRATES	TOTAL
ENERGY VALUE (AVERAGE) PER 100 G EDIBLE PORTION	KJOULE	30	26	1.7	57
	(KCAL)	7.1	6.2	0.4	14
AMOUNT OF DIGESTIBLE CONSTITUENTS PER 100 G	GRAM	1.32	0.60	0.10	
ENERGY VALUE (AVERAGE) OF THE DIGESTIBLE FRACTION PER 100 G EDIBLE PORTION	KJOULE	21	23	1.7	46
	(KCAL)	5.0	5.6	0.4	11

WASTE PERCENTAGE AVERAGE 39 MINIMUM 29 MAXIMUM 48

CONSTITUENTS	DIM	AV		VARIATION			AVR	NUTR.	DENS.	MOLPERC.
WATER	GRAM	89.80		88.80	-	91.90	54.78	KILO/MJ	1.96	
PROTEIN	GRAM	1.89		1.24	-	2.46	1.15	GRAM/MJ	41.18	
FAT	GRAM	0.67		0.50	-	0.83	0.41	GRAM/MJ	14.60	
AVAILABLE CARBOHYDR.	GRAM	0.10		-	-	-	0.06	GRAM/MJ	2.18	
TOTAL DIETARY FIBRE	GRAM	6.89	1	-	-	-	4.20	GRAM/MJ	150.13	
MINERALS	GRAM	0.65		0.50	-	0.78	0.40	GRAM/MJ	14.16	
SODIUM	MILLI	6.00		-	-	-	3.66	MILLI/MJ	130.73	
POTASSIUM	MILLI	310.00		300.00	-	320.00	189.10	GRAM/MJ	6.75	
MAGNESIUM	MILLI	8.00		-	-	-	4.88	MILLI/MJ	174.31	
CALCIUM	MILLI	6.00		-	-	-	3.66	MILLI/MJ	130.73	
MANGANESE	MILLI	0.30		-	-	-	0.18	MILLI/MJ	6.54	
IRON	MILLI	1.30		-	-	-	0.79	MILLI/MJ	28.33	
PHOSPHORUS	MILLI	74.00		-	-	-	45.14	GRAM/MJ	1.61	
CHLORIDE	MILLI	20.00		-	-	-	12.20	MILLI/MJ	435.78	
VITAMIN B1	MILLI	0.13		-	-	-	0.08	MILLI/MJ	2.83	
VITAMIN B2	MICRO	60.00		-	-	-	36.60	MILLI/MJ	1.31	
VITAMIN C	MILLI	6.00		3.00	-	10.00	3.66	MILLI/MJ	130.73	
ARGININE	MILLI	14.00		-	-	-	8.54	MILLI/MJ	305.04	
HISTIDINE	MILLI	18.00		-	-	-	10.98	MILLI/MJ	392.20	
ISOLEUCINE	MILLI	130.00		-	-	-	79.30	GRAM/MJ	2.83	
LEUCINE	MILLI	90.00		-	-	-	54.90	GRAM/MJ	1.96	
LYSINE	MILLI	57.00		-	-	-	34.77	GRAM/MJ	1.24	
METHIONINE	MILLI	14.00		-	-	-	8.54	MILLI/MJ	305.04	
PHENYLALANINE	MILLI	12.00		-	-	-	7.32	MILLI/MJ	261.47	
THREONINE	MILLI	100.00		-	-	-	61.00	GRAM/MJ	2.18	
TRYPTOPHAN	MILLI	4.00		-	-	-	2.44	MILLI/MJ	87.16	
VALINE	MILLI	75.00		-	-	-	45.75	GRAM/MJ	1.63	
REDUCING SUGAR	GRAM	0.10		-	-	-	0.06	GRAM/MJ	2.18	
TREHALOSE	-	0.00		-	-	-	0.00			
MANNITOL	GRAM	1.36		1.32	-	1.40	0.83	GRAM/MJ	29.63	

1 ESTIMATED BY THE DIFFERENCE METHOD:
100 - (WATER + PROTEIN + FAT + AVAILABLE CARBOHYDRATES + MINERALS)

ROTKAPPE RED BOLETUS „ROTKAPPE“

LECCINUM AURANTIACUM GRAY

		PROTEIN	FAT	CARBOHYDRATES	TOTAL
ENERGY VALUE (AVERAGE)	KJOULE	23	30	0.0	53
PER 100 G EDIBLE PORTION	(KCAL)	5.4	7.3	0.0	13
AMOUNT OF DIGESTIBLE CONSTITUENTS PER 100 G	GRAM	1.45	0.70	0.00	
ENERGY VALUE (AVERAGE)	KJOULE	23	27	0.0	50
OF THE DIGESTIBLE FRACTION PER 100 G EDIBLE PORTION	(KCAL)	5.4	6.5	0.0	12

WASTE PERCENTAGE AVERAGE 20

CONSTITUENTS	DIM	AV		VARIATION			AVR	NUTR. DENS.		MOLPERC.
WATER	GRAM	92.30		-	-	-	73.84	KILO/MJ	1.84	
PROTEIN	GRAM	1.45		-	-	-	1.16	GRAM/MJ	28.96	
FAT	GRAM	0.78		-	-	-	0.62	GRAM/MJ	15.58	
TOTAL DIETARY FIBRE	GRAM	4.72	1	-	-	-	3.78	GRAM/MJ	94.28	
MINERALS	GRAM	0.75		0.63	-	0.86	0.60	GRAM/MJ	14.98	
SODIUM	-	TRACES		-	-	-				
POTASSIUM	MILLI	314.00		-	-	-	251.20	GRAM/MJ	6.27	
MAGNESIUM	MILLI	9.00		8.00	-	10.00	7.20	MILLI/MJ	179.76	
CALCIUM	MILLI	30.00		15.00	-	45.00	24.00	MILLI/MJ	599.21	
FLUORIDE	MICRO	30.00		-	-	-	24.00	MICRO/MJ	599.21	
ARGININE	MILLI	140.00		-	-	-	112.00	GRAM/MJ	2.80	
CYSTINE	MILLI	93.00		-	-	-	74.40	GRAM/MJ	1.86	
HISTIDINE	MILLI	46.00		-	-	-	36.80	MILLI/MJ	918.78	
ISOLEUCINE	MILLI	30.00		-	-	-	24.00	MILLI/MJ	599.21	
LEUCINE	MILLI	110.00		-	-	-	88.00	GRAM/MJ	2.20	
LYSINE	MILLI	98.00		-	-	-	78.40	GRAM/MJ	1.96	
METHIONINE	MILLI	6.00		-	-	-	4.80	MILLI/MJ	119.84	
PHENYLALANINE	MILLI	59.00		-	-	-	47.20	GRAM/MJ	1.18	
THREONINE	MILLI	59.00		-	-	-	47.20	GRAM/MJ	1.18	
TRYPTOPHAN	MILLI	28.00		-	-	-	22.40	MILLI/MJ	559.26	
TYROSINE	MILLI	61.00		-	-	-	48.80	GRAM/MJ	1.22	
VALINE	MILLI	54.00		-	-	-	43.20	GRAM/MJ	1.08	

1 ESTIMATED BY THE DIFFERENCE METHOD
100 - (WATER + PROTEIN + FAT + MINERALS)

STEINPILZ YELLOW BOLETUS CÈPE

BOLETUS EDULIS BULL.

		PROTEIN	FAT	CARBOHYDRATES	TOTAL
ENERGY VALUE (AVERAGE)	KJOULE	43	16	8.9	68
PER 100 G	(KCAL)	10	3.7	2.1	16
EDIBLE PORTION					
AMOUNT OF DIGESTIBLE	GRAM	2.77	0.36	0.53	
CONSTITUENTS PER 100 G					
ENERGY VALUE (AVERAGE)	KJOULE	43	14	8.9	66
OF THE DIGESTIBLE	(KCAL)	10	3.3	2.1	16
FRACTION PER 100 G					
EDIBLE PORTION					

WASTE PERCENTAGE AVERAGE 20

CONSTITUENTS	DIM	AV		VARIATION			AVR	NUTR.	DENS.	MOLPERC.
WATER	GRAM	88.60		-	-	-	70.88	KILO/MJ	1.34	
PROTEIN	GRAM	2.77		-	-	-	2.22	GRAM/MJ	41.75	
FAT	GRAM	0.40		-	-	-	0.32	GRAM/MJ	6.03	
AVAILABLE CARBOHYDR.	GRAM	0.53		-	-	-	0.42	GRAM/MJ	7.99	
TOTAL DIETARY FIBRE	GRAM	6.89	1	-	-	-	5.51	GRAM/MJ	103.86	
MINERALS	GRAM	0.81		-	-	-	0.65	GRAM/MJ	12.21	
SODIUM	MILLI	6.00		3.70	-	9.00	4.80	MILLI/MJ	90.44	
POTASSIUM	MILLI	486.00		468.00	-	520.00	388.80	GRAM/MJ	7.33	
MAGNESIUM	MILLI	12.00		8.00	-	16.00	9.60	MILLI/MJ	180.89	
CALCIUM	MILLI	23.00		6.00	-	40.00	18.40	MILLI/MJ	346.70	
MANGANESE	MILLI	0.17		0.05	-	0.20	0.14	MILLI/MJ	2.56	
IRON	MILLI	1.00		-	-	-	0.80	MILLI/MJ	15.07	
COBALT	MICRO	0.50		0.20	-	1.00	0.40	MICRO/MJ	7.54	
COPPER	MILLI	0.23		-	-	-	0.18	MILLI/MJ	3.47	
ZINC	MILLI	0.70		0.20	-	1.10	0.56	MILLI/MJ	10.55	
NICKEL	MICRO	10.00		-	-	-	8.00	MICRO/MJ	150.74	
CHROMIUM	MICRO	5.00		3.00	-	7.00	4.00	MICRO/MJ	75.37	
PHOSPHORUS	MILLI	115.00		-	-	-	92.00	GRAM/MJ	1.73	
FLUORIDE	MICRO	63.00		-	-	-	50.40	MICRO/MJ	949.66	
SELENIUM	MILLI	0.10		0.06	-	0.14	0.08	MILLI/MJ	1.51	
SILICON	MILLI	4.00		-	-	-	3.20	MILLI/MJ	60.30	
VITAMIN D	MICRO	3.10		-	-	-	2.48	MICRO/MJ	46.73	
VITAMIN E ACTIVITY	MILLI	0.15		-	-	-	0.12	MILLI/MJ	2.26	
TOTAL TOCOPHEROLS	MILLI	0.60		-	-	-	0.48	MILLI/MJ	9.04	
ALPHA-TOCOPHEROL	MICRO	40.00		-	-	-	32.00	MICRO/MJ	602.96	
DELTA-TOCOPHEROL	MICRO	60.00		-	-	-	48.00	MICRO/MJ	904.44	

1 ESTIMATED BY THE DIFFERENCE METHOD:
100 - (WATER + PROTEIN + FAT + AVAILABLE CARBOHYDRATES + MINERALS)

CONSTITUENTS	DIM	AV	VARIATION			AVR	NUTR. DENS.		MOLPERC.
ALPHA-TOCOTRIENOL	MILLI	0.39	-	-	-	0.31	MILLI/MJ	5.88	
VITAMIN B1	MICRO	33.00	-	-	-	26.40	MICRO/MJ	497.44	
VITAMIN B2	MILLI	0.37	-	-	-	0.30	MILLI/MJ	5.58	
NICOTINAMIDE	MILLI	4.90	-	-	-	3.92	MILLI/MJ	73.86	
PANTOTHENIC ACID	MILLI	2.70	2.43	-	2.97	2.16	MILLI/MJ	40.70	
VITAMIN C	MILLI	2.50	-	-	-	2.00	MILLI/MJ	37.68	
ARGININE	GRAM	0.26	-	-	-	0.21	GRAM/MJ	3.92	
CYSTINE	GRAM	0.29	-	-	-	0.23	GRAM/MJ	4.37	
HISTIDINE	GRAM	0.22	-	-	-	0.18	GRAM/MJ	3.32	
ISOLEUCINE	MILLI	30.00	-	-	-	24.00	MILLI/MJ	452.22	
LEUCINE	GRAM	0.12	-	-	-	0.10	GRAM/MJ	1.81	
LYSINE	GRAM	0.19	-	-	-	0.15	GRAM/MJ	2.86	
METHIONINE	MILLI	58.00	-	-	-	46.40	MILLI/MJ	874.29	
PHENYLALANINE	GRAM	0.10	-	-	-	0.08	GRAM/MJ	1.51	
THREONINE	GRAM	0.11	-	-	-	0.09	GRAM/MJ	1.66	
TRYPTOPHAN	GRAM	0.21	-	-	-	0.17	GRAM/MJ	3.17	
TYROSINE	GRAM	0.12	-	-	-	0.10	GRAM/MJ	1.81	
VALINE	MILLI	78.00	-	-	-	62.40	GRAM/MJ	1.18	
GLUCOSE	GRAM	0.27	0.27	-	0.27	0.22	GRAM/MJ	4.07	
FRUCTOSE	GRAM	0.26	0.26	-	0.26	0.21	GRAM/MJ	3.92	
TOTAL PURINES	MILLI	50.00 [2]	-	-	-	40.00	MILLI/MJ	753.70	

2 VALUE EXPRESSED IN MG URIC ACID/100G

STEINPILZ GETROCKNET | **YELLOW BOLETUS** DRIED | **CÈPE** SÉCHÉ

		PROTEIN	FAT	CARBOHYDRATES	TOTAL
ENERGY VALUE (AVERAGE) PER 100 G EDIBLE PORTION	KJOULE (KCAL)	309 74	125 30	69 16	502 120
AMOUNT OF DIGESTIBLE CONSTITUENTS PER 100 G	GRAM	19.70	2.88	4.11	
ENERGY VALUE (AVERAGE) OF THE DIGESTIBLE FRACTION PER 100 G EDIBLE PORTION	KJOULE (KCAL)	309 74	112 27	69 16	490 117

WASTE PERCENTAGE AVERAGE 0.00

CONSTITUENTS	DIM	AV		VARIATION			AVR	NUTR.	DENS.	MOLPERC.
WATER	GRAM	11.60		-	-	-	11.60	GRAM/MJ	23.68	
PROTEIN	GRAM	19.70		-	-	-	19.70	GRAM/MJ	40.21	
FAT	GRAM	3.20		-	-	-	3.20	GRAM/MJ	6.53	
AVAILABLE CARBOHYDR.	GRAM	4.11	1	-	-	-	4.11	GRAM/MJ	8.39	
TOTAL DIETARY FIBRE	GRAM	55.29	2	-	-	-	55.29	GRAM/MJ	112.85	
MINERALS	GRAM	6.10		-	-	-	6.10	GRAM/MJ	12.45	
SODIUM	MILLI	14.00		8.00	-	24.00	14.00	MILLI/MJ	28.57	
POTASSIUM	GRAM	2.00		1.60	-	2.50	2.00	GRAM/MJ	4.08	
CALCIUM	MILLI	34.00		18.00	-	63.00	34.00	MILLI/MJ	69.40	
IRON	MILLI	8.40		-	-	-	8.40	MILLI/MJ	17.14	
PHOSPHORUS	MILLI	642.00		-	-	-	642.00	GRAM/MJ	1.31	
VITAMIN E ACTIVITY	MILLI	0.19		-	-	-	0.19	MILLI/MJ	0.39	
CYSTINE	GRAM	2.08		-	-	-	2.08	GRAM/MJ	4.25	
HISTIDINE	GRAM	1.59		-	-	-	1.59	GRAM/MJ	3.25	
ISOLEUCINE	GRAM	0.21		-	-	-	0.21	GRAM/MJ	0.43	
LEUCINE	GRAM	0.84		-	-	-	0.84	GRAM/MJ	1.71	
LYSINE	GRAM	1.35		-	-	-	1.35	GRAM/MJ	2.76	
METHIONINE	GRAM	0.42		-	-	-	0.42	GRAM/MJ	0.86	
PHENYLALANINE	GRAM	0.73		-	-	-	0.73	GRAM/MJ	1.49	
THREONINE	GRAM	0.75		-	-	-	0.75	GRAM/MJ	1.53	
TRYPTOPHAN	GRAM	1.46		-	-	-	1.46	GRAM/MJ	2.98	
TYROSINE	GRAM	0.86		-	-	-	0.86	GRAM/MJ	1.76	
VALINE	GRAM	0.56		-	-	-	0.56	GRAM/MJ	1.14	

1 CALCULATED FROM THE FRESH PRODUCT

2 ESTIMATED BY THE DIFFERENCE METHOD:
100 - (WATER + PROTEIN + FAT + AVAILABLE CARBOHYDRATES + MINERALS)

TRÜFFEL TRUFFLES TRUFFE

TUBER MELANOSPORUM VITT.

		PROTEIN	FAT	CARBOHYDRATES	TOTAL
ENERGY VALUE (AVERAGE)	KJOULE	87	20	0.0	107
PER 100 G	(KCAL)	21	4.7	0.0	25
EDIBLE PORTION					
AMOUNT OF DIGESTIBLE	GRAM	5.53	0.45	0.00	
CONSTITUENTS PER 100 G					
ENERGY VALUE (AVERAGE)	KJOULE	87	18	0.0	105
OF THE DIGESTIBLE	(KCAL)	21	4.3	0.0	25
FRACTION PER 100 G					
EDIBLE PORTION					

WASTE PERCENTAGE AVERAGE 0.00

CONSTITUENTS	DIM	AV	VARIATION			AVR	NUTR. DENS.		MOLPERC.
WATER	GRAM	75.50	74.00	-	77.10	75.50	GRAM/MJ	721.62	
PROTEIN	GRAM	5.53	5.07	-	6.00	5.53	GRAM/MJ	52.85	
FAT	GRAM	0.51	-	-	-	0.51	GRAM/MJ	4.87	
TOTAL DIETARY FIBRE	GRAM	16.54 1	-	-	-	16.54	GRAM/MJ	158.09	
MINERALS	GRAM	1.92	-	-	-	1.92	GRAM/MJ	18.35	
SODIUM	MILLI	77.00	-	-	-	77.00	MILLI/MJ	735.96	
POTASSIUM	MILLI	526.00	431.00	-	602.00	526.00	GRAM/MJ	5.03	
MAGNESIUM	MILLI	23.80	15.30	-	28.00	23.80	MILLI/MJ	227.48	
CALCIUM	MILLI	24.00	-	-	-	24.00	MILLI/MJ	229.39	
IRON	MILLI	3.50	-	-	-	3.50	MILLI/MJ	33.45	
PHOSPHORUS	MILLI	62.00	-	-	-	62.00	MILLI/MJ	592.59	
CHLORIDE	MILLI	27.70	18.00	-	39.00	27.70	MILLI/MJ	264.75	
ARGININE	GRAM	0.65	-	-	-	0.65	GRAM/MJ	6.21	
CYSTINE	GRAM	0.15	-	-	-	0.15	GRAM/MJ	1.43	
HISTIDINE	GRAM	0.10	-	-	-	0.10	GRAM/MJ	0.96	
ISOLEUCINE	GRAM	0.16	-	-	-	0.16	GRAM/MJ	1.53	
LEUCINE	GRAM	0.40	-	-	-	0.40	GRAM/MJ	3.82	
LYSINE	GRAM	0.49	-	-	-	0.49	GRAM/MJ	4.68	
PHENYLALANINE	GRAM	0.19	-	-	-	0.19	GRAM/MJ	1.82	
THREONINE	GRAM	0.38	-	-	-	0.38	GRAM/MJ	3.63	
TRYPTOPHAN	MILLI	20.00	-	-	-	20.00	MILLI/MJ	191.16	
TYROSINE	GRAM	0.18	-	-	-	0.18	GRAM/MJ	1.72	
VALINE	GRAM	0.25	-	-	-	0.25	GRAM/MJ	2.39	

1 ESTIMATED BY THE DIFFERENCE METHOD
100 - (WATER + PROTEIN + FAT + MINERALS)

Früchte

Fruits
Fruits

Kernobst

FRUITS WITH CORES **FRUITS À PEPINS**

APFEL APPLE POMME

MALUS SYLVESTRIS MILL.

		PROTEIN	FAT	CARBOHYDRATES	TOTAL
ENERGY VALUE (AVERAGE)	KJOULE	5.6	16	207	229
PER 100 G	(KCAL)	1.3	3.7	50	55
EDIBLE PORTION					
AMOUNT OF DIGESTIBLE	GRAM	0.28	0.36	12.39	
CONSTITUENTS PER 100 G					
ENERGY VALUE (AVERAGE)	KJOULE	4.8	14	207	226
OF THE DIGESTIBLE	(KCAL)	1.1	3.3	50	54
FRACTION PER 100 G					
EDIBLE PORTION					

WASTE PERCENTAGE AVERAGE 8.0 MINIMUM 2.0 MAXIMUM 30

CONSTITUENTS	DIM	AV		VARIATION			AVR	NUTR. DENS.		MOLPERC.
WATER	GRAM	85.30		80.40	-	90.00	78.48	GRAM/MJ	377.19	
PROTEIN	GRAM	0.34		0.20	-	0.45	0.31	GRAM/MJ	1.50	
FAT	GRAM	0.40		0.20	-	0.65	0.37	GRAM/MJ	1.77	
AVAILABLE CARBOHYDR.	GRAM	12.39		8.85	-	13.33	11.40	GRAM/MJ	54.79	
TOTAL DIETARY FIBRE	GRAM	2.30	1	2.10	-	2.50	2.12	GRAM/MJ	10.17	
MINERALS	GRAM	0.32		0.26	-	0.36	0.29	GRAM/MJ	1.42	
SODIUM	MILLI	3.00		1.50	-	4.00	2.76	MILLI/MJ	13.27	
POTASSIUM	MILLI	144.00		100.00	-	175.00	132.48	MILLI/MJ	636.76	
MAGNESIUM	MILLI	6.40		2.80	-	9.00	5.89	MILLI/MJ	28.30	
CALCIUM	MILLI	7.10		3.60	-	10.50	6.53	MILLI/MJ	31.40	
MANGANESE	MICRO	65.00		35.00	-	100.00	59.80	MICRO/MJ	287.43	
IRON	MILLI	0.48		0.26	-	0.85	0.44	MILLI/MJ	2.12	
COBALT	MICRO	10.00		-	-	-	9.20	MICRO/MJ	44.22	
COPPER	MILLI	0.10		0.05	-	0.16	0.09	MILLI/MJ	0.44	
ZINC	MILLI	0.12		0.04	-	0.22	0.11	MILLI/MJ	0.53	
NICKEL	MICRO	11.00		8.00	-	13.00	10.12	MICRO/MJ	48.64	
MOLYBDENUM	MICRO	0.15		-	-	-	0.14	MICRO/MJ	0.66	
VANADIUM	MICRO	3.50		0.90	-	6.10	3.22	MICRO/MJ	15.48	
PHOSPHORUS	MILLI	12.00		7.00	-	17.00	11.04	MILLI/MJ	53.06	
CHLORIDE	MILLI	2.20		1.50	-	4.00	2.02	MILLI/MJ	9.73	
FLUORIDE	MICRO	6.60		3.50	-	13.00	6.07	MICRO/MJ	29.19	
IODIDE	MICRO	1.60		0.10	-	2.00	1.47	MICRO/MJ	7.08	
BORON	MILLI	0.24		0.12	-	0.43	0.23	MILLI/MJ	1.08	

1 METHOD OF SCHWEIZER AND WUERSCH

CONSTITUENTS	DIM	AV	VARIATION	AVR	NUTR. DENS.		MOLPERC.
SELENIUM	MICRO	0.90	0.30 - 6.00	0.83	MICRO/MJ	3.98	
BROMINE	MICRO	20.00	- - -	18.40	MICRO/MJ	88.44	
SILICON	MILLI	0.50	0.10 - 1.00	0.46	MILLI/MJ	2.21	
CAROTENE	MICRO	47.00	28.00 - 70.00	43.24	MICRO/MJ	207.83	
VITAMIN E ACTIVITY	MILLI	0.49	- - -	0.45	MILLI/MJ	2.17	
ALPHA-TOCOPHEROL	MILLI	0.49	- - -	0.45	MILLI/MJ	2.17	
VITAMIN K	MICRO	-	0.00 - 5.00				
VITAMIN B1	MICRO	35.00	15.00 - 60.00	32.20	MICRO/MJ	154.77	
VITAMIN B2	MICRO	32.00	20.00 - 50.00	29.44	MICRO/MJ	141.50	
NICOTINAMIDE	MILLI	0.30	0.10 - 0.50	0.28	MILLI/MJ	1.33	
PANTOTHENIC ACID	MILLI	0.10	0.09 - 0.13	0.09	MILLI/MJ	0.44	
VITAMIN B6	MICRO	45.00	35.00 - 57.00	41.40	MICRO/MJ	198.99	
BIOTIN	MICRO	-	1.00 - 8.00				
FOLIC ACID	MICRO	6.50	5.00 - 8.00	5.98	MICRO/MJ	28.74	
VITAMIN C	MILLI	12.00	3.00 - 25.00	11.04	MILLI/MJ	53.06	
ALANINE	MILLI	15.00	15.00 - 15.00	13.80	MILLI/MJ	66.33	8.1
ARGININE	MILLI	8.00	8.00 - 8.00	7.36	MILLI/MJ	35.38	2.2
ASPARTIC ACID	MILLI	101.00	101.00 - 101.00	92.92	MILLI/MJ	446.62	36.3
CYSTINE	MILLI	1.00	1.00 - 1.00	0.92	MILLI/MJ	4.42	0.2
GLUTAMIC ACID	MILLI	25.00	25.00 - 25.00	23.00	MILLI/MJ	110.55	8.1
GLYCINE	MILLI	9.00	9.00 - 9.00	8.28	MILLI/MJ	39.80	5.7
HISTIDINE	MILLI	6.00	6.00 - 6.00	5.52	MILLI/MJ	26.53	1.9
ISOLEUCINE	MILLI	10.00	10.00 - 10.00	9.20	MILLI/MJ	44.22	3.6
LEUCINE	MILLI	16.00	16.00 - 16.00	14.72	MILLI/MJ	70.75	5.8
LYSINE	MILLI	15.00	15.00 - 15.00	13.80	MILLI/MJ	66.33	4.9
METHIONINE	MILLI	3.00	3.00 - 3.00	2.76	MILLI/MJ	13.27	1.0
PHENYLALANINE	MILLI	9.00	9.00 - 9.00	8.28	MILLI/MJ	39.80	2.6
PROLINE	MILLI	10.00	10.00 - 10.00	9.20	MILLI/MJ	44.22	4.2
SERINE	MILLI	12.00	12.00 - 12.00	11.04	MILLI/MJ	53.06	5.5
THREONINE	MILLI	8.00	8.00 - 8.00	7.36	MILLI/MJ	35.38	3.2
TRYPTOPHAN	MILLI	2.00	2.00 - 2.00	1.84	MILLI/MJ	8.84	0.5
TYROSINE	MILLI	5.00	5.00 - 5.00	4.60	MILLI/MJ	22.11	1.3
VALINE	MILLI	12.00	12.00 - 12.00	11.04	MILLI/MJ	53.06	4.9
MALIC ACID	GRAM	0.55	0.27 - 0.79	0.51	GRAM/MJ	2.43	
CITRIC ACID	MILLI	16.00	9.00 - 30.00	14.72	MILLI/MJ	70.75	
OXALIC ACID TOTAL	MILLI	0.50	0.30 - 0.70	0.46	MILLI/MJ	2.21	
CHLOROGENIC ACID	MILLI	7.80	- - -	7.18	MILLI/MJ	34.49	
FERULIC ACID	MILLI	0.60	- - -	0.55	MILLI/MJ	2.65	
CAFFEIC ACID	MILLI	8.00	4.00 - 11.00	7.36	MILLI/MJ	35.38	
PARA-COUMARIC ACID	MILLI	2.00	1.50 - 3.00	1.84	MILLI/MJ	8.84	
SALICYLIC ACID	MILLI	0.31	0.31 - 0.31	0.29	MILLI/MJ	1.37	
GLUCOSE	GRAM	2.21	1.40 - 2.35	2.04	GRAM/MJ	9.81	
FRUCTOSE	GRAM	6.04	5.52 - 6.40	5.56	GRAM/MJ	26.74	
SUCROSE	GRAM	2.47	0.54 - 2.58	2.27	GRAM/MJ	10.93	
STARCH	GRAM	0.60	0.60 - 0.60	0.55	GRAM/MJ	2.65	
PENTOSAN	GRAM	0.25	- - -	0.23	GRAM/MJ	1.11	
HEXOSAN	GRAM	0.12	- - -	0.11	GRAM/MJ	0.53	
CELLULOSE	GRAM	0.76	- - -	0.70	GRAM/MJ	3.36	
SORBITOL	GRAM	0.51	0.51 - 0.58	0.47	GRAM/MJ	2.27	
LAURIC ACID	MILLI	10.50	10.50 - 10.50	9.66	MILLI/MJ	46.43	
MYRISTIC ACID	MILLI	24.50	24.50 - 24.50	22.54	MILLI/MJ	108.34	
PALMITIC ACID	MILLI	125.66	64.00 - 156.50	115.61	MILLI/MJ	555.70	
STEARIC ACID	MILLI	45.00	12.00 - 61.50	41.40	MILLI/MJ	198.99	
ARACHIDIC ACID	MILLI	9.00	9.00 - 9.00	8.28	MILLI/MJ	39.80	
BEHENIC ACID	MILLI	13.00	13.00 - 13.00	11.96	MILLI/MJ	57.49	
OLEIC ACID	MILLI	11.00	10.50 - 12.00	10.12	MILLI/MJ	48.64	
EICOSENOIC ACID	MILLI	2.00	2.00 - 2.00	1.84	MILLI/MJ	8.84	

CONSTITUENTS	DIM	AV		VARIATION			AVR	NUTR. DENS.		MOLPERC.
LINOLEIC ACID	MILLI	174.00		174.00	-	174.00	160.08	MILLI/MJ	769.42	
LINOLENIC ACID	MILLI	46.00		46.00	-	46.00	42.32	MILLI/MJ	203.41	
TOTAL STEROLS	MILLI	12.00		-	-	-	11.04	MILLI/MJ	53.06	
CAMPESTEROL	MILLI	1.00		-	-	-	0.92	MILLI/MJ	4.42	
BETA-SITOSTEROL	MILLI	11.00		-	-	-	10.12	MILLI/MJ	48.64	
TOTAL PURINES	MILLI	3.00	2	2.00	-	5.00	2.76	MILLI/MJ	13.27	
DIETARY FIBRE,WAT.SOL.	GRAM	0.90		-	-	-	0.83	GRAM/MJ	3.98	
DIETARY FIBRE,WAT.INS.	GRAM	1.40		1.20	-	1.60	1.29	GRAM/MJ	6.19	

2 VALUE EXPRESSED IN MG URIC ACID/100G

APFEL GETROCKNET

APPLE DRIED

POMME SÈCHE

		PROTEIN	FAT	CARBOHYDRATES	TOTAL
ENERGY VALUE (AVERAGE)	KJOULE	23	63	1018	1104
PER 100 G EDIBLE PORTION	(KCAL)	5.4	15	243	264
AMOUNT OF DIGESTIBLE CONSTITUENTS PER 100 G	GRAM	1.16	1.46	60.81	
ENERGY VALUE (AVERAGE)	KJOULE	19	57	1018	1094
OF THE DIGESTIBLE FRACTION PER 100 G EDIBLE PORTION	(KCAL)	4.6	14	243	261

WASTE PERCENTAGE AVERAGE 0.00

CONSTITUENTS	DIM	AV		VARIATION			AVR	NUTR. DENS.		MOLPERC.
WATER	GRAM	26.70		20.00	-	32.00	26.70	GRAM/MJ	24.40	
PROTEIN	GRAM	1.37		1.30	-	1.42	1.37	GRAM/MJ	1.25	
FAT	GRAM	1.63		1.00	-	1.94	1.63	GRAM/MJ	1.49	
AVAILABLE CARBOHYDR.	GRAM	60.81	1	-	-	-	60.81	GRAM/MJ	55.58	
TOTAL DIETARY FIBRE	GRAM	8.03	2	-	-	-	8.03	GRAM/MJ	7.34	
MINERALS	GRAM	1.46		-	-	-	1.46	GRAM/MJ	1.33	

1 CALCULATED FROM THE FRESH PRODUCT

2 ESTIMATED BY THE DIFFERENCE METHOD:
100 - (WATER + PROTEIN + FAT + AVAILABLE CARBOHYDRATES + MINERALS)

CONSTITUENTS	DIM	AV	VARIATION			AVR	NUTR. DENS.		MOLPERC.
SODIUM	MILLI	10.00	5.00	-	18.00	10.00	MILLI/MJ	9.14	
POTASSIUM	MILLI	622.00	405.00	-	846.00	622.00	MILLI/MJ	568.53	
CALCIUM	MILLI	30.00	19.00	-	50.00	30.00	MILLI/MJ	27.42	
IRON	MILLI	1.20	1.00	-	1.40	1.20	MILLI/MJ	1.10	
PHOSPHORUS	MILLI	50.00	-	-	-	50.00	MILLI/MJ	45.70	
VITAMIN B1	MILLI	0.10	-	-	-	0.10	MILLI/MJ	0.09	
VITAMIN B2	MILLI	0.10	-	-	-	0.10	MILLI/MJ	0.09	
NICOTINAMIDE	MILLI	0.80	0.60	-	1.00	0.80	MILLI/MJ	0.73	
VITAMIN C	MILLI	12.00	-	-	-	12.00	MILLI/MJ	10.97	
ALANINE	MILLI	62.00	62.00	-	62.00	62.00	MILLI/MJ	56.67	7.7
ARGININE	MILLI	35.00	35.00	-	35.00	35.00	MILLI/MJ	31.99	2.2
ASPARTIC ACID	MILLI	431.00	431.00	-	431.00	431.00	MILLI/MJ	393.95	35.9
CYSTINE	MILLI	2.00	2.00	-	2.00	2.00	MILLI/MJ	1.83	0.1
GLUTAMIC ACID	MILLI	108.00	108.00	-	108.00	108.00	MILLI/MJ	98.72	8.1
GLYCINE	MILLI	39.00	39.00	-	39.00	39.00	MILLI/MJ	35.65	5.8
HISTIDINE	MILLI	22.00	22.00	-	22.00	22.00	MILLI/MJ	20.11	1.6
ISOLEUCINE	MILLI	44.00	44.00	-	44.00	44.00	MILLI/MJ	40.22	3.7
LEUCINE	MILLI	71.00	71.00	-	71.00	71.00	MILLI/MJ	64.90	6.0
LYSINE	MILLI	66.00	66.00	-	66.00	66.00	MILLI/MJ	60.33	5.0
METHIONINE	MILLI	13.00	13.00	-	13.00	13.00	MILLI/MJ	11.88	1.0
PHENYLALANINE	MILLI	39.00	39.00	-	39.00	39.00	MILLI/MJ	35.65	2.6
PROLINE	MILLI	44.00	44.00	-	44.00	44.00	MILLI/MJ	40.22	4.2
SERINE	MILLI	55.00	55.00	-	55.00	55.00	MILLI/MJ	50.27	5.8
THREONINE	MILLI	35.00	35.00	-	35.00	35.00	MILLI/MJ	31.99	3.3
TRYPTOPHAN	MILLI	12.00	12.00	-	12.00	12.00	MILLI/MJ	10.97	0.7
TYROSINE	MILLI	20.00	20.00	-	20.00	20.00	MILLI/MJ	18.28	1.2
VALINE	MILLI	53.00	53.00	-	53.00	53.00	MILLI/MJ	48.44	5.0
MALIC ACID	GRAM	2.74	-	-	-	2.74	GRAM/MJ	2.50	
CITRIC ACID	MILLI	80.00	-	-	-	80.00	MILLI/MJ	73.12	
GLUCOSE	GRAM	11.02	-	-	-	11.02	GRAM/MJ	10.07	
FRUCTOSE	GRAM	30.12	-	-	-	30.12	GRAM/MJ	27.53	
SUCROSE	GRAM	12.32	-	-	-	12.32	GRAM/MJ	11.26	
STARCH	GRAM	2.99	-	-	-	2.99	GRAM/MJ	2.73	
SORBITOL	GRAM	2.54	-	-	-	2.54	GRAM/MJ	2.32	

APFELMUS — APPLE PUREE — PURÉE DE POMMES

		PROTEIN	FAT	CARBOHYDRATES	TOTAL
ENERGY VALUE (AVERAGE) PER 100 G EDIBLE PORTION	KJOULE	3.6	3.9	321	329
	(KCAL)	0.9	0.9	77	79
AMOUNT OF DIGESTIBLE CONSTITUENTS PER 100 G	GRAM	0.18	0.09	19.20	
ENERGY VALUE (AVERAGE) OF THE DIGESTIBLE FRACTION PER 100 G EDIBLE PORTION	KJOULE	3.1	3.5	321	328
	(KCAL)	0.7	0.8	77	78

WASTE PERCENTAGE AVERAGE 0.00

CONSTITUENTS	DIM	AV	VARIATION		AVR	NUTR. DENS.		MOLPERC.
WATER	GRAM	77.90	73.10	- 80.30	77.90	GRAM/MJ	237.56	
PROTEIN	GRAM	0.22	0.20	- 0.30	0.22	GRAM/MJ	0.67	
FAT	GRAM	0.10	0.00	- 0.10	0.10	GRAM/MJ	0.30	
AVAILABLE CARBOHYDR.	GRAM	19.20	16.80	- 24.00	19.20	GRAM/MJ	58.55	
MINERALS	GRAM	0.18	0.15	- 0.20	0.18	GRAM/MJ	0.55	
SODIUM	MILLI	2.70	1.00	- 4.30	2.70	MILLI/MJ	8.23	
POTASSIUM	MILLI	114.00	92.00	- 131.00	114.00	MILLI/MJ	347.64	
MAGNESIUM	MILLI	9.80	3.40	- 16.70	9.80	MILLI/MJ	29.88	
CALCIUM	MILLI	4.40	2.20	- 6.00	4.40	MILLI/MJ	13.42	
MANGANESE	MICRO	30.00	-	- -	30.00	MICRO/MJ	91.48	
IRON	MILLI	0.30	0.20	- 1.60	0.30	MILLI/MJ	0.91	
COPPER	MILLI	-	0.03	- 0.26				
ZINC	MILLI	0.10	-	- -	0.10	MILLI/MJ	0.30	
PHOSPHORUS	MILLI	7.20	6.00	- 15.00	7.20	MILLI/MJ	21.96	
CHLORIDE	MILLI	5.00	-	- -	5.00	MILLI/MJ	15.25	
BORON	MILLI	0.19	-	- -	0.19	MILLI/MJ	0.58	
SELENIUM	MICRO	0.20	-	- -	0.20	MICRO/MJ	0.61	
CAROTENE	MICRO	36.00	-	- -	36.00	MICRO/MJ	109.78	
VITAMIN B1	MICRO	10.00	0.00	- 10.00	10.00	MICRO/MJ	30.49	
VITAMIN B2	MICRO	20.00	10.00	- 30.00	20.00	MICRO/MJ	60.99	
NICOTINAMIDE	MILLI	0.12	0.08	- 0.17	0.12	MILLI/MJ	0.37	
PANTOTHENIC ACID	MICRO	70.00	-	- -	70.00	MICRO/MJ	213.46	
VITAMIN B6	MICRO	30.00	-	- -	30.00	MICRO/MJ	91.48	
BIOTIN	MICRO	0.30	-	- -	0.30	MICRO/MJ	0.91	
FOLIC ACID	MICRO	4.00	-	- -	4.00	MICRO/MJ	12.20	
VITAMIN C	MILLI	2.00	1.00	- 2.90	2.00	MILLI/MJ	6.10	
GLUCOSE	GRAM	4.20	-	- -	4.20	GRAM/MJ	12.81	
FRUCTOSE	GRAM	7.50	-	- -	7.50	GRAM/MJ	22.87	
SUCROSE	GRAM	7.50	-	- -	7.50	GRAM/MJ	22.87	

BIRNE PEAR POIRE

PYRUS COMMUNIS L.

		PROTEIN	FAT	CARBOHYDRATES	TOTAL
ENERGY VALUE (AVERAGE) PER 100 G EDIBLE PORTION	KJOULE	7.8	11	212	231
	(KCAL)	1.9	2.7	51	55
AMOUNT OF DIGESTIBLE CONSTITUENTS PER 100 G	GRAM	0.39	0.26	12.66	
ENERGY VALUE (AVERAGE) OF THE DIGESTIBLE FRACTION PER 100 G EDIBLE PORTION	KJOULE	6.6	10	212	229
	(KCAL)	1.6	2.4	51	55

WASTE PERCENTAGE AVERAGE 7.0 MINIMUM 4.0 MAXIMUM 10

CONSTITUENTS	DIM	AV	VARIATION			AVR	NUTR.	DENS.	MOLPERC.
WATER	GRAM	84.30	82.00	-	87.40	78.40	GRAM/MJ	368.71	
PROTEIN	GRAM	0.47	0.27	-	0.70	0.44	GRAM/MJ	2.06	
FAT	GRAM	0.29	0.10	-	0.40	0.27	GRAM/MJ	1.27	
AVAILABLE CARBOHYDR.	GRAM	12.66	11.15	-	14.93	11.77	GRAM/MJ	55.37	
TOTAL DIETARY FIBRE	GRAM	2.80 [1]	-	-	-	2.60	GRAM/MJ	12.25	
MINERALS	GRAM	0.33	0.23	-	0.40	0.31	GRAM/MJ	1.44	
SODIUM	MILLI	2.10	1.00	-	3.00	1.95	MILLI/MJ	9.18	
POTASSIUM	MILLI	126.00	100.00	-	147.00	117.18	MILLI/MJ	551.09	
MAGNESIUM	MILLI	7.80	5.00	-	9.60	7.25	MILLI/MJ	34.12	
CALCIUM	MILLI	10.00	7.00	-	13.00	9.30	MILLI/MJ	43.74	
MANGANESE	MICRO	49.00	32.00	-	70.00	45.57	MICRO/MJ	214.31	
IRON	MILLI	0.26	0.19	-	0.30	0.24	MILLI/MJ	1.14	
COBALT	MICRO	15.00	2.00	-	32.00	13.95	MICRO/MJ	65.61	
COPPER	MICRO	90.00	60.00	-	120.00	83.70	MICRO/MJ	393.64	
ZINC	MILLI	0.23	0.16	-	0.32	0.21	MILLI/MJ	1.01	
NICKEL	MICRO	16.00	12.00	-	20.00	14.88	MICRO/MJ	69.98	
CHROMIUM	MICRO	2.00	-	-	-	1.86	MICRO/MJ	8.75	
VANADIUM	MICRO	2.50	1.00	-	5.00	2.33	MICRO/MJ	10.93	
PHOSPHORUS	MILLI	15.00	10.00	-	22.00	13.95	MILLI/MJ	65.61	
CHLORIDE	MILLI	2.00	1.00	-	4.00	1.86	MILLI/MJ	8.75	
FLUORIDE	MICRO	12.00	6.00	-	22.00	11.16	MICRO/MJ	52.49	
IODIDE	MICRO	1.50	1.00	-	2.00	1.40	MICRO/MJ	6.56	
BORON	MILLI	0.18	0.13	-	0.38	0.17	MILLI/MJ	0.80	
SELENIUM	MICRO	1.20	0.60	-	8.00	1.12	MICRO/MJ	5.25	
SILICON	MILLI	0.20	0.20	-	0.30	0.19	MILLI/MJ	0.87	
CAROTENE	MICRO	32.00	10.00	-	100.00	29.76	MICRO/MJ	139.96	

1 METHOD OF MEUSER, SUCKOW AND KULIKOWSKI ("BERLINER METHODE")

CONSTITUENTS	DIM	AV	VARIATION			AVR	NUTR. DENS.		MOLPERC.
VITAMIN E ACTIVITY	MILLI	0.43	0.36	-	0.50	0.40	MILLI/MJ	1.88	
ALPHA-TOCOPHEROL	MILLI	0.43	-	-	-	0.40	MILLI/MJ	1.88	
VITAMIN B1	MICRO	33.00	10.00	-	70.00	30.69	MICRO/MJ	144.33	
VITAMIN B2	MICRO	38.00	20.00	-	60.00	35.34	MICRO/MJ	166.20	
NICOTINAMIDE	MILLI	0.22	0.10	-	0.30	0.20	MILLI/MJ	0.96	
PANTOTHENIC ACID	MICRO	62.00	50.00	-	73.00	57.66	MICRO/MJ	271.17	
VITAMIN B6	MICRO	15.00	9.00	-	32.00	13.95	MICRO/MJ	65.61	
BIOTIN	MICRO	0.10	-	-	-	0.09	MICRO/MJ	0.44	
FOLIC ACID	MICRO	14.00	-	-	-	13.02	MICRO/MJ	61.23	
VITAMIN C	MILLI	4.60	2.00	-	9.90	4.28	MILLI/MJ	20.12	
MALIC ACID	MILLI	170.00	100.00	-	240.00	158.10	MILLI/MJ	743.54	
CITRIC ACID	MILLI	140.00	80.00	-	200.00	130.20	MILLI/MJ	612.33	
OXALIC ACID TOTAL	MILLI	6.20	-	-	-	5.77	MILLI/MJ	27.12	
OXALIC ACID SOLUBLE	MILLI	5.10	-	-	-	4.74	MILLI/MJ	22.31	
CHLOROGENIC ACID	MILLI	25.00	-	-	-	23.25	MILLI/MJ	109.34	
CAFFEIC ACID	MILLI	8.00	4.00	-	11.00	7.44	MILLI/MJ	34.99	
GLUCOSE	GRAM	1.66	1.50	-	1.69	1.55	GRAM/MJ	7.30	
FRUCTOSE	GRAM	6.72	6.07	-	7.70	6.26	GRAM/MJ	29.42	
SUCROSE	GRAM	1.80	1.24	-	2.50	1.68	GRAM/MJ	7.89	
PENTOSAN	GRAM	0.77	0.77	-	0.77	0.72	GRAM/MJ	3.37	
HEXOSAN	GRAM	0.40	0.40	-	0.40	0.37	GRAM/MJ	1.75	
CELLULOSE	GRAM	0.77	0.77	-	0.77	0.72	GRAM/MJ	3.37	
POLYURONIC ACID	GRAM	0.40	0.40	-	0.40	0.37	GRAM/MJ	1.75	
SORBITOL	GRAM	2.17	2.16	-	2.60	2.02	GRAM/MJ	9.49	
PALMITIC ACID	MILLI	43.00	43.00	-	43.00	39.99	MILLI/MJ	188.07	
STEARIC ACID	MILLI	11.00	11.00	-	11.00	10.23	MILLI/MJ	48.11	
PALMITOLEIC ACID	MILLI	4.00	4.00	-	4.00	3.72	MILLI/MJ	17.50	
OLEIC ACID	MILLI	29.00	29.00	-	29.00	26.97	MILLI/MJ	126.84	
LINOLEIC ACID	MILLI	108.00	108.00	-	108.00	100.44	MILLI/MJ	472.37	
LINOLENIC ACID	MILLI	36.00	36.00	-	36.00	33.48	MILLI/MJ	157.46	
TOTAL STEROLS	MILLI	8.00	-	-	-	7.44	MILLI/MJ	34.99	
BETA-SITOSTEROL	MILLI	7.00	-	-	-	6.51	MILLI/MJ	30.62	
TOTAL PURINES	MILLI	2.00 [2]	-	-	-	1.86	MILLI/MJ	8.75	
DIETARY FIBRE, WAT.SOL.	GRAM	0.60	-	-	-	0.56	GRAM/MJ	2.62	
DIETARY FIBRE, WAT.INS.	GRAM	2.20	-	-	-	2.05	GRAM/MJ	9.62	

2 VALUE EXPRESSED IN MG URIC ACID/100G

BIRNEN	**PEARS**	**POIRES**
IN DOSEN	CANNED	EN BOÎTES
1	1	1

		PROTEIN	FAT	CARBOHYDRATES	TOTAL
ENERGY VALUE (AVERAGE) PER 100 G EDIBLE PORTION	KJOULE	4.5	3.9	313	322
	(KCAL)	1.1	0.9	75	77
AMOUNT OF DIGESTIBLE CONSTITUENTS PER 100 G	GRAM	0.22	0.09	18.73	
ENERGY VALUE (AVERAGE) OF THE DIGESTIBLE FRACTION PER 100 G EDIBLE PORTION	KJOULE	3.8	3.5	313	321
	(KCAL)	0.9	0.8	75	77

WASTE PERCENTAGE AVERAGE 0.00

CONSTITUENTS	DIM	AV	VARIATION			AVR	NUTR.	DENS.	MOLPERC.
WATER	GRAM	80.70	78.30	-	84.30	80.70	GRAM/MJ	251.59	
PROTEIN	GRAM	0.27	0.17	-	0.40	0.27	GRAM/MJ	0.84	
FAT	GRAM	0.10	-	-	-	0.10	GRAM/MJ	0.31	
AVAILABLE CARBOHYDR.	GRAM	18.73 2	-	-	-	18.73	GRAM/MJ	58.39	
MINERALS	GRAM	0.20	-	-	-	0.20	GRAM/MJ	0.62	
SODIUM	MILLI	6.00	1.00	-	15.00	6.00	MILLI/MJ	18.71	
POTASSIUM	MILLI	66.00	52.00	-	90.00	66.00	MILLI/MJ	205.76	
MAGNESIUM	MILLI	4.00	3.00	-	6.00	4.00	MILLI/MJ	12.47	
CALCIUM	MILLI	7.00	5.00	-	9.00	7.00	MILLI/MJ	21.82	
MANGANESE	MICRO	20.00	-	-	-	20.00	MICRO/MJ	62.35	
IRON	MILLI	0.40	0.20	-	0.50	0.40	MILLI/MJ	1.25	
COPPER	MICRO	70.00	40.00	-	90.00	70.00	MICRO/MJ	218.23	
PHOSPHORUS	MILLI	8.00	5.00	-	10.00	8.00	MILLI/MJ	24.94	
CHLORIDE	MILLI	3.90	2.80	-	5.00	3.90	MILLI/MJ	12.16	
BORON	MILLI	0.15	-	-	-	0.15	MILLI/MJ	0.47	
SELENIUM	MICRO	0.20	-	-	-	0.20	MICRO/MJ	0.62	
CAROTENE	MICRO	10.00	0.00	-	10.00	10.00	MICRO/MJ	31.18	
VITAMIN B1	MICRO	10.00	0.00	-	10.00	10.00	MICRO/MJ	31.18	
VITAMIN B2	MICRO	20.00	10.00	-	30.00	20.00	MICRO/MJ	62.35	
NICOTINAMIDE	MILLI	0.13	0.07	-	0.20	0.13	MILLI/MJ	0.41	
PANTOTHENIC ACID	MICRO	20.00	-	-	-	20.00	MICRO/MJ	62.35	
VITAMIN B6	MICRO	10.00	-	-	-	10.00	MICRO/MJ	31.18	
FOLIC ACID	MICRO	5.90	-	-	-	5.90	MICRO/MJ	18.39	
VITAMIN C	MILLI	2.00	1.00	-	4.00	2.00	MILLI/MJ	6.24	

1 TOTAL CONTENT (SOLID AND LIQUID)

2 ESTIMATED BY THE DIFFERENCE METHOD
100 - (WATER + PROTEIN + FAT + MINERALS)

QUITTE QUINCE COING

CYDONIA VULGARIS PERSOON

		PROTEIN	FAT	CARBOHYDRATES	TOTAL
ENERGY VALUE (AVERAGE) PER 100 G EDIBLE PORTION	KJOULE	6.9	19	139	165
	(KCAL)	1.7	4.7	33	40
AMOUNT OF DIGESTIBLE CONSTITUENTS PER 100 G	GRAM	0.35	0.45	8.30	
ENERGY VALUE (AVERAGE) OF THE DIGESTIBLE FRACTION PER 100 G EDIBLE PORTION	KJOULE	5.9	18	139	162
	(KCAL)	1.4	4.2	33	39

WASTE PERCENTAGE AVERAGE 16

CONSTITUENTS	DIM	AV	VARIATION			AVR	NUTR. DENS.		MOLPERC.
WATER	GRAM	83.10	81.90	-	84.20	69.80	GRAM/MJ	511.96	
PROTEIN	GRAM	0.42	0.30	-	0.57	0.35	GRAM/MJ	2.59	
FAT	GRAM	0.50	0.20	-	0.90	0.42	GRAM/MJ	3.08	
AVAILABLE CARBOHYDR.	GRAM	8.30	-	-	-	6.97	GRAM/MJ	51.13	
MINERALS	GRAM	0.44	0.30	-	0.57	0.37	GRAM/MJ	2.71	
SODIUM	MILLI	2.00	0.50	-	3.20	1.68	MILLI/MJ	12.32	
POTASSIUM	MILLI	201.00	168.00	-	247.00	168.84	GRAM/MJ	1.24	
MAGNESIUM	MILLI	8.20	6.00	-	10.30	6.89	MILLI/MJ	50.52	
CALCIUM	MILLI	10.30	5.00	-	15.70	8.65	MILLI/MJ	63.46	
IRON	MILLI	0.60	0.30	-	1.28	0.50	MILLI/MJ	3.70	
COPPER	MILLI	0.13	-	-	-	0.11	MILLI/MJ	0.80	
PHOSPHORUS	MILLI	21.40	19.00	-	23.80	17.98	MILLI/MJ	131.84	
CHLORIDE	MILLI	1.70	1.54	-	1.90	1.43	MILLI/MJ	10.47	
FLUORIDE	MICRO	6.00	-	-	-	5.04	MICRO/MJ	36.96	
CAROTENE	MICRO	33.00	12.00	-	54.00	27.72	MICRO/MJ	203.30	
VITAMIN B1	MICRO	30.00	20.00	-	40.00	25.20	MICRO/MJ	184.82	
VITAMIN B2	MICRO	30.00	20.00	-	30.00	25.20	MICRO/MJ	184.82	
NICOTINAMIDE	MILLI	0.20	-	-	-	0.17	MILLI/MJ	1.23	
VITAMIN C	MILLI	13.00	12.00	-	13.00	10.92	MILLI/MJ	80.09	
MALIC ACID	GRAM	0.93	-	-	-	0.78	GRAM/MJ	5.73	
INVERT SUGAR	GRAM	6.68	-	-	-	5.61	GRAM/MJ	41.15	
SUCROSE	GRAM	0.64	-	-	-	0.54	GRAM/MJ	3.94	
PECTIN	GRAM	0.61	-	-	-	0.51	GRAM/MJ	3.76	

Steinobst

STONE FRUITS **FRUITS À NOYAU**

APRIKOSE APRICOT ABRICOT

PRUNUS ARMENIACA L.

		PROTEIN	FAT	CARBOHYDRATES	TOTAL
ENERGY VALUE (AVERAGE)	KJOULE	15	5.1	166	186
PER 100 G	(KCAL)	3.6	1.2	40	45
EDIBLE PORTION					
AMOUNT OF DIGESTIBLE	GRAM	0.76	0.11	9.94	
CONSTITUENTS PER 100 G					
ENERGY VALUE (AVERAGE)	KJOULE	13	4.6	166	184
OF THE DIGESTIBLE	(KCAL)	3.0	1.1	40	44
FRACTION PER 100 G					
EDIBLE PORTION					

WASTE PERCENTAGE AVERAGE 9.0 MINIMUM 6.0 MAXIMUM 11

CONSTITUENTS	DIM	AV	VARIATION			AVR	NUTR. DENS.		MOLPERC.
WATER	GRAM	85.30	82.70	-	89.30	77.62	GRAM/MJ	464.72	
PROTEIN	GRAM	0.90	0.80	-	1.10	0.82	GRAM/MJ	4.90	
FAT	GRAM	0.13	0.10	-	0.20	0.12	GRAM/MJ	0.71	
AVAILABLE CARBOHYDR.	GRAM	9.94	-	-	-	9.05	GRAM/MJ	54.15	
TOTAL DIETARY FIBRE	GRAM	2.02 1	-	-	-	1.84	GRAM/MJ	11.01	
MINERALS	GRAM	0.66	0.59	-	0.77	0.60	GRAM/MJ	3.60	
SODIUM	MILLI	2.00	0.60	-	6.10	1.82	MILLI/MJ	10.90	
POTASSIUM	MILLI	278.00	190.00	-	370.00	252.98	GRAM/MJ	1.51	
MAGNESIUM	MILLI	9.20	7.00	-	14.00	8.37	MILLI/MJ	50.12	
CALCIUM	MILLI	16.00	12.00	-	20.00	14.56	MILLI/MJ	87.17	
MANGANESE	MILLI	0.27	0.20	-	0.37	0.25	MILLI/MJ	1.47	
IRON	MILLI	0.65	0.49	-	0.85	0.59	MILLI/MJ	3.54	
COBALT	MICRO	1.90	1.10	-	3.00	1.73	MICRO/MJ	10.35	
COPPER	MILLI	0.15	0.11	-	0.20	0.14	MILLI/MJ	0.82	
ZINC	MICRO	70.00	40.00	-	100.00	63.70	MICRO/MJ	381.36	
NICKEL	MICRO	17.00	-	-	-	15.47	MICRO/MJ	92.62	
MOLYBDENUM	MICRO	14.00	-	-	-	12.74	MICRO/MJ	76.27	
PHOSPHORUS	MILLI	21.00	18.00	-	23.00	19.11	MILLI/MJ	114.41	
CHLORIDE	MILLI	1.00	0.80	-	2.00	0.91	MILLI/MJ	5.45	
FLUORIDE	MICRO	9.60	2.40	-	20.00	8.74	MICRO/MJ	52.30	
IODIDE	MICRO	0.50	-	-	-	0.46	MICRO/MJ	2.72	
BORON	MILLI	0.47	0.37	-	0.87	0.43	MILLI/MJ	2.59	

1 METHOD OF ENGLYST

CONSTITUENTS	DIM	AV	VARIATION			AVR	NUTR. DENS.		MOLPERC.
CAROTENE	MILLI	1.79	0.45	-	3.50	1.63	MILLI/MJ	9.75	
VITAMIN E ACTIVITY	MILLI	0.50	-	-	-	0.46	MILLI/MJ	2.72	
ALPHA-TOCOPHEROL	MILLI	0.50	-	-	-	0.46	MILLI/MJ	2.72	
VITAMIN B1	MICRO	40.00	30.00	-	60.00	36.40	MICRO/MJ	217.92	
VITAMIN B2	MICRO	53.00	30.00	-	90.00	48.23	MICRO/MJ	288.75	
NICOTINAMIDE	MILLI	0.77	0.70	-	0.80	0.70	MILLI/MJ	4.20	
PANTOTHENIC ACID	MILLI	0.29	0.27	-	0.30	0.26	MILLI/MJ	1.58	
VITAMIN B6	MICRO	70.00	-	-	-	63.70	MICRO/MJ	381.36	
FOLIC ACID	MICRO	3.60	2.50	-	4.70	3.28	MICRO/MJ	19.61	
VITAMIN B12	-	0.00	-	-	-	0.00			
VITAMIN C	MILLI	9.40	5.00	-	15.20	8.55	MILLI/MJ	51.21	
MALIC ACID	GRAM	1.00	0.70	-	1.30	0.91	GRAM/MJ	5.45	
CITRIC ACID	GRAM	0.40	0.14	-	0.70	0.36	GRAM/MJ	2.18	
OXALIC ACID TOTAL	MILLI	6.80	-	-	-	6.19	MILLI/MJ	37.05	
OXALIC ACID SOLUBLE	MILLI	3.40	-	-	-	3.09	MILLI/MJ	18.52	
CHLOROGENIC ACID	MILLI	7.50	-	-	-	6.83	MILLI/MJ	40.86	
QUINIC ACID	MILLI	7.00	-	-	-	6.37	MILLI/MJ	38.14	
MALONIC ACID	MILLI	10.00	-	-	-	9.10	MILLI/MJ	54.48	
SUCCINIC ACID	MILLI	10.00	-	-	-	9.10	MILLI/MJ	54.48	
OXALOACETIC ACID	-	TRACES	-	-	-				
SALICYLIC ACID	MILLI	2.58	2.58	-	2.58	2.35	MILLI/MJ	14.06	
GLUCOSE	GRAM	1.73	0.95	-	2.88	1.57	GRAM/MJ	9.43	
FRUCTOSE	GRAM	0.87	0.37	-	1.57	0.79	GRAM/MJ	4.74	
SUCROSE	GRAM	5.12	3.60	-	5.98	4.66	GRAM/MJ	27.89	
STARCH	-	TRACES	-	-	-				
PECTIN	GRAM	0.96	0.50	-	1.30	0.87	GRAM/MJ	5.23	
SORBITOL	GRAM	0.82	-	-	-	0.75	GRAM/MJ	4.47	
TOTAL STEROLS	MILLI	18.00	-	-	-	16.38	MILLI/MJ	98.07	
CAMPESTEROL	MILLI	1.00	-	-	-	0.91	MILLI/MJ	5.45	
BETA-SITOSTEROL	MILLI	16.00	-	-	-	14.56	MILLI/MJ	87.17	
DIETARY FIBRE,WAT.SOL.	GRAM	1.18	-	-	-	1.07	GRAM/MJ	6.43	
DIETARY FIBRE,WAT.INS.	GRAM	0.84	-	-	-	0.76	GRAM/MJ	4.58	

APRIKOSE	**APRICOT**	**ABRICOT**
GETROCKNET	DRIED	SEC

		PROTEIN	FAT	CARBOHYDRATES	TOTAL
ENERGY VALUE (AVERAGE) PER 100 G EDIBLE PORTION	KJOULE	83	19	933	1035
	(KCAL)	20	4.7	223	247
AMOUNT OF DIGESTIBLE CONSTITUENTS PER 100 G	GRAM	4.25	0.45	55.72	
ENERGY VALUE (AVERAGE) OF THE DIGESTIBLE FRACTION PER 100 G EDIBLE PORTION	KJOULE	70	18	933	1020
	(KCAL)	17	4.2	223	244

WASTE PERCENTAGE AVERAGE 0.00

CONSTITUENTS	DIM	AV	VARIATION			AVR	NUTR.	DENS.	MOLPERC.
WATER	GRAM	17.60	15.00	-	24.00	17.60	GRAM/MJ	17.25	
PROTEIN	GRAM	5.00	4.80	-	5.20	5.00	GRAM/MJ	4.90	
FAT	GRAM	0.50	0.40	-	0.50	0.50	GRAM/MJ	0.49	
AVAILABLE CARBOHYDR.	GRAM	55.72 [1]	-	-	-	55.72	GRAM/MJ	54.61	
TOTAL DIETARY FIBRE	GRAM	8.00	-	-	-	8.00	GRAM/MJ	7.84	
MINERALS	GRAM	3.50	3.10	-	3.80	3.50	GRAM/MJ	3.43	
SODIUM	MILLI	11.00	4.00	-	33.00	11.00	MILLI/MJ	10.78	
POTASSIUM	GRAM	1.37	0.98	-	1.70	1.37	GRAM/MJ	1.34	
MAGNESIUM	MILLI	50.00	38.00	-	76.00	50.00	MILLI/MJ	49.01	
CALCIUM	MILLI	82.00	67.00	-	92.00	82.00	MILLI/MJ	80.37	
MANGANESE	MILLI	1.50	1.10	-	2.00	1.50	MILLI/MJ	1.47	
IRON	MILLI	4.40	3.50	-	5.50	4.40	MILLI/MJ	4.31	
COBALT	MICRO	16.00	-	-	-	16.00	MICRO/MJ	15.68	
COPPER	MILLI	0.80	0.60	-	1.10	0.80	MILLI/MJ	0.78	
ZINC	MILLI	0.40	0.20	-	0.50	0.40	MILLI/MJ	0.39	
PHOSPHORUS	MILLI	114.00	108.00	-	120.00	114.00	MILLI/MJ	111.73	
CHLORIDE	MILLI	5.40	4.40	-	10.90	5.40	MILLI/MJ	5.29	
FLUORIDE	MICRO	50.00	10.00	-	100.00	50.00	MICRO/MJ	49.01	
IODIDE	MICRO	2.70	-	-	-	2.70	MICRO/MJ	2.65	
CAROTENE	MILLI	4.62	3.42	-	6.54	4.62	MILLI/MJ	4.53	
VITAMIN B1	MICRO	7.00	3.00	-	10.00	7.00	MICRO/MJ	6.86	
VITAMIN B2	MILLI	0.11	0.06	-	0.16	0.11	MILLI/MJ	0.11	
NICOTINAMIDE	MILLI	3.20	2.80	-	3.30	3.20	MILLI/MJ	3.14	
PANTOTHENIC ACID	MILLI	0.83	0.79	-	0.87	0.83	MILLI/MJ	0.81	
VITAMIN B6	MILLI	0.17	-	-	-	0.17	MILLI/MJ	0.17	
FOLIC ACID	MICRO	5.10	4.20	-	5.50	5.10	MICRO/MJ	5.00	

1 CALCULATED FROM THE FRESH PRODUCT

CONSTITUENTS	DIM	AV	VARIATION			AVR	NUTR. DENS.		MOLPERC.
VITAMIN C	MILLI	11.00	3.00	-	17.00	11.00	MILLI/MJ	10.78	
MALIC ACID	GRAM	5.61	-	-	-	5.61	GRAM/MJ	5.50	
CITRIC ACID	GRAM	2.24	-	-	-	2.24	GRAM/MJ	2.20	
GLUCOSE	GRAM	9.69	-	-	-	9.69	GRAM/MJ	9.50	
FRUCTOSE	GRAM	4.88	-	-	-	4.88	GRAM/MJ	4.78	
SUCROSE	GRAM	28.70	-	-	-	28.70	GRAM/MJ	28.13	
STARCH	-	TRACES	-	-	-				
PECTIN	GRAM	5.40	2.80	-	7.30	5.40	GRAM/MJ	5.29	
SORBITOL	GRAM	4.60	-	-	-	4.60	GRAM/MJ	4.51	
DIETARY FIBRE, WAT.SOL.	GRAM	4.30	-	-	-	4.30	GRAM/MJ	4.21	
DIETARY FIBRE, WAT.INS.	GRAM	3.70	-	-	-	3.70	GRAM/MJ	3.63	

APRIKOSE IN DOSEN	**APRICOTS** CANNED	**ABRICOTS** EN BOÎTES
1	1	1

		PROTEIN	FAT	CARBOHYDRATES	TOTAL
ENERGY VALUE (AVERAGE) PER 100 G EDIBLE PORTION	KJOULE	8.3	3.9	303	315
	(KCAL)	2.0	0.9	72	75
AMOUNT OF DIGESTIBLE CONSTITUENTS PER 100 G	GRAM	0.42	0.09	18.08	
ENERGY VALUE (AVERAGE) OF THE DIGESTIBLE FRACTION PER 100 G EDIBLE PORTION	KJOULE	7.0	3.5	303	313
	(KCAL)	1.7	0.8	72	75

WASTE PERCENTAGE AVERAGE 0.00

CONSTITUENTS	DIM	AV		VARIATION		AVR	NUTR.	DENS.	MOLPERC.
WATER	GRAM	80.90		77.90 -	83.90	80.90	GRAM/MJ	258.37	
PROTEIN	GRAM	0.50		0.50 -	0.60	0.50	GRAM/MJ	1.60	
FAT	GRAM	0.10		0.10 -	0.20	0.10	GRAM/MJ	0.32	
AVAILABLE CARBOHYDR.	GRAM	18.08	2	- -	-	18.08	GRAM/MJ	57.74	
MINERALS	GRAM	0.42		0.32 -	0.53	0.42	GRAM/MJ	1.34	
SODIUM	MILLI	13.00		1.00 -	25.00	13.00	MILLI/MJ	41.52	
POTASSIUM	MILLI	196.00		157.00 -	234.00	196.00	MILLI/MJ	625.97	
MAGNESIUM	MILLI	9.60		7.10 -	10.80	9.60	MILLI/MJ	30.66	
CALCIUM	MILLI	11.00		8.00 -	18.00	11.00	MILLI/MJ	35.13	
IRON	MILLI	0.70		0.30 -	1.20	0.70	MILLI/MJ	2.24	
COPPER	MICRO	50.00		- -	-	50.00	MICRO/MJ	159.69	
ZINC	MILLI	0.10		- -	-	0.10	MILLI/MJ	0.32	
PHOSPHORUS	MILLI	15.00		11.00 -	16.00	15.00	MILLI/MJ	47.91	
CHLORIDE	MILLI	2.00		- -	-	2.00	MILLI/MJ	6.39	
BORON	MILLI	0.58		- -	-	0.58	MILLI/MJ	1.85	
CAROTENE	MILLI	1.17		1.04 -	1.30	1.17	MILLI/MJ	3.74	
VITAMIN B1	MICRO	20.00		- -	-	20.00	MICRO/MJ	63.87	
VITAMIN B2	MICRO	20.00		- -	-	20.00	MICRO/MJ	63.87	
NICOTINAMIDE	MILLI	0.50		0.30 -	0.70	0.50	MILLI/MJ	1.60	
PANTOTHENIC ACID	MILLI	0.10		- -	-	0.10	MILLI/MJ	0.32	
VITAMIN B6	MICRO	50.00		- -	-	50.00	MICRO/MJ	159.69	
VITAMIN C	MILLI	4.00		- -	-	4.00	MILLI/MJ	12.77	
SALICYLIC ACID	MILLI	1.42		1.42 -	1.42	1.42	MILLI/MJ	4.54	

1 TOTAL CONTENT (SOLID AND LIQUID)

2 ESTIMATED BY THE DIFFERENCE METHOD
100 - (WATER + PROTEIN + FAT + MINERALS)

KIRSCHE SAUER | CHERRY (MORELLO) | GRIOTTE

PRUNUS CERASUS L.

		PROTEIN	FAT	CARBOHYDRATES	TOTAL
ENERGY VALUE (AVERAGE)	KJOULE	15	19	195	230
PER 100 G	(KCAL)	3.6	4.7	47	55
EDIBLE PORTION					
AMOUNT OF DIGESTIBLE	GRAM	0.76	0.45	11.67	
CONSTITUENTS PER 100 G					
ENERGY VALUE (AVERAGE)	KJOULE	13	18	195	225
OF THE DIGESTIBLE	(KCAL)	3.0	4.2	47	54
FRACTION PER 100 G					
EDIBLE PORTION					

WASTE PERCENTAGE AVERAGE 11 MINIMUM 6.0 MAXIMUM 18

CONSTITUENTS	DIM	AV	VARIATION			AVR	NUTR. DENS.		MOLPERC.
WATER	GRAM	84.80	-	-	-	75.47	GRAM/MJ	376.12	
PROTEIN	GRAM	0.90	0.80	-	0.90	0.80	GRAM/MJ	3.99	
FAT	GRAM	0.50	-	-	-	0.45	GRAM/MJ	2.22	
AVAILABLE CARBOHYDR.	GRAM	11.67	-	-	-	10.39	GRAM/MJ	51.76	
TOTAL DIETARY FIBRE	GRAM	1.04 1	-	-	-	0.93	GRAM/MJ	4.61	
MINERALS	GRAM	0.50	0.50	-	0.60	0.45	GRAM/MJ	2.22	
SODIUM	MILLI	2.00	-	-	-	1.78	MILLI/MJ	8.87	
POTASSIUM	MILLI	114.00	78.00	-	150.00	101.46	MILLI/MJ	505.63	
MAGNESIUM	MILLI	8.00	-	-	-	7.12	MILLI/MJ	35.48	
CALCIUM	MILLI	-	6.00	-	10.00				
IRON	MILLI	0.60	0.40	-	0.90	0.53	MILLI/MJ	2.66	
PHOSPHORUS	MILLI	19.00	-	-	-	16.91	MILLI/MJ	84.27	
CHLORIDE	MILLI	21.00	-	-	-	18.69	MILLI/MJ	93.14	
VITAMIN A	-	0.00	-	-	-	0.00			
CAROTENE	MILLI	0.30	-	-	-	0.27	MILLI/MJ	1.33	
VITAMIN E ACTIVITY	MILLI	0.13	-	-	-	0.12	MILLI/MJ	0.58	
ALPHA-TOCOPHEROL	MILLI	0.13	-	-	-	0.12	MILLI/MJ	0.58	
VITAMIN B1	MICRO	50.00	-	-	-	44.50	MICRO/MJ	221.77	
VITAMIN B2	MICRO	60.00	-	-	-	53.40	MICRO/MJ	266.12	
NICOTINAMIDE	MILLI	0.40	-	-	-	0.36	MILLI/MJ	1.77	
VITAMIN C	MILLI	12.00	-	-	-	10.68	MILLI/MJ	53.22	
MALIC ACID	GRAM	1.80	-	-	-	1.60	GRAM/MJ	7.98	
OXALIC ACID TOTAL	MILLI	4.70	-	-	-	4.18	MILLI/MJ	20.85	
OXALIC ACID SOLUBLE	MILLI	3.30	-	-	-	2.94	MILLI/MJ	14.64	
CHLOROGENIC ACID	MILLI	16.30	-	-	-	14.51	MILLI/MJ	72.30	
FERULIC ACID	MILLI	0.20	-	-	-	0.18	MILLI/MJ	0.89	
CAFFEIC ACID	MILLI	4.00	1.00	-	8.00	3.56	MILLI/MJ	17.74	
PARA-COUMARIC ACID	MILLI	13.00	10.00	-	16.00	11.57	MILLI/MJ	57.66	
GLUCOSE	GRAM	5.18	4.54	-	5.50	4.61	GRAM/MJ	22.98	
FRUCTOSE	GRAM	4.28	3.85	-	4.50	3.81	GRAM/MJ	19.00	
SUCROSE	GRAM	0.41	0.32	-	0.58	0.37	GRAM/MJ	1.85	
DIETARY FIBRE, WAT.SOL.	GRAM	0.57	-	-	-	0.51	GRAM/MJ	2.53	
DIETARY FIBRE, WAT.INS.	GRAM	0.47	-	-	-	0.42	GRAM/MJ	2.08	

1 METHOD OF MEUSER, SUCKOW AND KULIKOWSKI ("BERLINER METHODE")

Früchte Fruits Fruits

KIRSCHE — CHERRY — CERISE

SÜSS — SWEET

PRUNUS AVIUM L.

		PROTEIN	FAT	CARBOHYDRATES	TOTAL
ENERGY VALUE (AVERAGE) PER 100 G EDIBLE PORTION	KJOULE	15	12	238	265
	(KCAL)	3.6	2.9	57	63
AMOUNT OF DIGESTIBLE CONSTITUENTS PER 100 G	GRAM	0.76	0.27	14.21	
ENERGY VALUE (AVERAGE) OF THE DIGESTIBLE FRACTION PER 100 G EDIBLE PORTION	KJOULE	13	11	238	261
	(KCAL)	3.0	2.6	57	62

WASTE PERCENTAGE AVERAGE 12 MINIMUM 9.0 MAXIMUM 16

CONSTITUENTS	DIM	AV		VARIATION		AVR	NUTR. DENS.		MOLPERC.
WATER	GRAM	82.80		79.80 -	86.00	72.86	GRAM/MJ	316.86	
PROTEIN	GRAM	0.90		0.60 -	1.30	0.79	GRAM/MJ	3.44	
FAT	GRAM	0.31		0.12 -	0.50	0.27	GRAM/MJ	1.19	
AVAILABLE CARBOHYDR.	GRAM	14.21		10.38 -	17.26	12.50	GRAM/MJ	54.38	
TOTAL DIETARY FIBRE	GRAM	1.90	1	1.50 -	2.20	1.67	GRAM/MJ	7.27	
MINERALS	GRAM	0.49		0.40 -	0.60	0.43	GRAM/MJ	1.88	
SODIUM	MILLI	2.70		1.90 -	4.10	2.38	MILLI/MJ	10.33	
POTASSIUM	MILLI	229.00		162.00 -	305.00	201.52	MILLI/MJ	876.33	
MAGNESIUM	MILLI	11.00		10.00 -	14.00	9.68	MILLI/MJ	42.09	
CALCIUM	MILLI	17.00		8.00 -	24.00	14.96	MILLI/MJ	65.05	
MANGANESE	MICRO	63.00		30.00 -	95.00	55.44	MICRO/MJ	241.09	
IRON	MILLI	0.35		0.21 -	0.50	0.31	MILLI/MJ	1.34	
COBALT	MICRO	1.60		0.50 -	2.40	1.41	MICRO/MJ	6.12	
COPPER	MICRO	94.00		61.00 -	120.00	82.72	MICRO/MJ	359.72	
ZINC	MILLI	0.15		- -	-	0.13	MILLI/MJ	0.57	
NICKEL	MICRO	60.00		- -	-	52.80	MICRO/MJ	229.61	
CHROMIUM	MICRO	3.00		- -	-	2.64	MICRO/MJ	11.48	
PHOSPHORUS	MILLI	20.00		16.00 -	32.00	17.60	MILLI/MJ	76.54	
CHLORIDE	MILLI	3.00		- -	-	2.64	MILLI/MJ	11.48	
FLUORIDE	MICRO	18.00		10.00 -	25.00	15.84	MICRO/MJ	68.88	
IODIDE	MICRO	1.00		- -	-	0.88	MICRO/MJ	3.83	
BORON	MILLI	-		0.04 -	0.65				
CAROTENE	MICRO	84.00		35.00 -	190.00	73.92	MICRO/MJ	321.45	
VITAMIN E ACTIVITY	MILLI	0.13		- -	-	0.11	MILLI/MJ	0.50	
ALPHA-TOCOPHEROL	MILLI	0.13		- -	-	0.11	MILLI/MJ	0.50	
VITAMIN B1	MICRO	39.00		20.00 -	50.00	34.32	MICRO/MJ	149.24	

1 METHOD OF MEUSER, SUCKOW AND KULIKOWSKI ("BERLINER METHODE")

CONSTITUENTS	DIM	AV	VARIATION			AVR	NUTR. DENS.		MOLPERC.
VITAMIN B2	MICRO	42.00	25.00	-	60.00	36.96	MICRO/MJ	160.72	
NICOTINAMIDE	MILLI	0.27	0.15	-	0.40	0.24	MILLI/MJ	1.03	
PANTOTHENIC ACID	MILLI	0.19	0.12	-	0.26	0.17	MILLI/MJ	0.73	
VITAMIN B6	MICRO	45.00	31.00	-	56.00	39.60	MICRO/MJ	172.20	
BIOTIN	MICRO	0.40	-	-	-	0.35	MICRO/MJ	1.53	
FOLIC ACID	MICRO	6.00	5.20	-	6.70	5.28	MICRO/MJ	22.96	
VITAMIN B12	-	0.00	-	-	-	0.00			
VITAMIN C	MILLI	15.00	8.00	-	37.00	13.20	MILLI/MJ	57.40	
ALANINE	MILLI	24.00	24.00	-	24.00	21.12	MILLI/MJ	91.84	4.4
ARGININE	MILLI	14.00	14.00	-	14.00	12.32	MILLI/MJ	53.57	1.3
ASPARTIC ACID	MILLI	483.00	483.00	-	483.00	425.04	GRAM/MJ	1.85	59.8
CYSTINE	MILLI	3.00	3.00	-	3.00	2.64	MILLI/MJ	11.48	0.2
GLUTAMIC ACID	MILLI	31.00	31.00	-	31.00	27.28	MILLI/MJ	118.63	3.5
GLYCINE	MILLI	19.00	19.00	-	19.00	16.72	MILLI/MJ	72.71	4.2
HISTIDINE	MILLI	11.00	11.00	-	11.00	9.68	MILLI/MJ	42.09	1.2
ISOLEUCINE	MILLI	16.00	16.00	-	16.00	14.08	MILLI/MJ	61.23	2.0
LEUCINE	MILLI	23.00	23.00	-	23.00	20.24	MILLI/MJ	88.02	2.9
LYSINE	MILLI	31.00	31.00	-	31.00	27.28	MILLI/MJ	118.63	3.5
METHIONINE	MILLI	4.00	4.00	-	4.00	3.52	MILLI/MJ	15.31	0.4
PHENYLALANINE	MILLI	16.00	16.00	-	16.00	14.08	MILLI/MJ	61.23	1.6
PROLINE	MILLI	26.00	26.00	-	26.00	22.88	MILLI/MJ	99.50	3.7
SERINE	MILLI	26.00	26.00	-	26.00	22.88	MILLI/MJ	99.50	4.1
THREONINE	MILLI	18.00	18.00	-	18.00	15.84	MILLI/MJ	68.88	2.5
TRYPTOPHAN	MILLI	8.00	8.00	-	8.00	7.04	MILLI/MJ	30.61	0.6
TYROSINE	MILLI	10.00	10.00	-	10.00	8.80	MILLI/MJ	38.27	0.9
VALINE	MILLI	22.00	22.00	-	22.00	19.36	MILLI/MJ	84.19	3.1
MALIC ACID	GRAM	0.94	0.73	-	1.11	0.83	GRAM/MJ	3.60	
CITRIC ACID	MILLI	13.00	10.00	-	15.00	11.44	MILLI/MJ	49.75	
TARTARIC ACID	-	TRACES	-	-	-				
OXALIC ACID TOTAL	MILLI	7.20	-	-	-	6.34	MILLI/MJ	27.55	
OXALIC ACID SOLUBLE	MILLI	4.30	-	-	-	3.78	MILLI/MJ	16.46	
CHLOROGENIC ACID	MILLI	6.10	-	-	-	5.37	MILLI/MJ	23.34	
FERULIC ACID	MILLI	0.30	-	-	-	0.26	MILLI/MJ	1.15	
CAFFEIC ACID	MILLI	7.00	5.00	-	9.00	6.16	MILLI/MJ	26.79	
PARA-COUMARIC ACID	MILLI	9.00	4.00	-	13.00	7.92	MILLI/MJ	34.44	
SALICYLIC ACID	MILLI	0.85	0.85	-	0.85	0.75	MILLI/MJ	3.25	
GLUCOSE	GRAM	6.93	5.29	-	7.80	6.10	GRAM/MJ	26.53	
FRUCTOSE	GRAM	6.14	4.20	-	7.09	5.40	GRAM/MJ	23.50	
SUCROSE	GRAM	0.19	0.15	-	1.25	0.17	GRAM/MJ	0.74	
STARCH	-	0.00	-	-	-	0.00			
PECTIN	GRAM	0.36	0.28	-	0.45	0.32	GRAM/MJ	1.38	
MYRISTIC ACID	MILLI	1.70	1.70	-	1.70	1.50	MILLI/MJ	6.51	
PALMITIC ACID	MILLI	49.00	49.00	-	49.00	43.12	MILLI/MJ	187.51	
STEARIC ACID	MILLI	17.00	17.00	-	17.00	14.96	MILLI/MJ	65.05	
ARACHIDIC ACID	MILLI	2.20	2.20	-	2.20	1.94	MILLI/MJ	8.42	
PALMITOLEIC ACID	MILLI	1.00	1.00	-	1.00	0.88	MILLI/MJ	3.83	
OLEIC ACID	MILLI	83.00	83.00	-	83.00	73.04	MILLI/MJ	317.62	
LINOLEIC ACID	MILLI	47.00	47.00	-	47.00	41.36	MILLI/MJ	179.86	
LINOLENIC ACID	MILLI	46.00	46.00	-	46.00	40.48	MILLI/MJ	176.03	
BETA-SITOSTEROL	MILLI	12.00	-	-	-	10.56	MILLI/MJ	45.92	
DIETARY FIBRE,WAT.SOL.	GRAM	0.90	-	-	-	0.79	GRAM/MJ	3.44	
DIETARY FIBRE,WAT.INS.	GRAM	1.00	0.60	-	1.30	0.88	GRAM/MJ	3.83	

KIRSCHEN IN DOSEN [1]	**CHERRIES** CANNED [1]	**CERISES** EN BOÎTES [1]

		PROTEIN	FAT	CARBOHYDRATES	TOTAL
ENERGY VALUE (AVERAGE) PER 100 G EDIBLE PORTION	KJOULE (KCAL)	11 2.6	7.0 1.7	326 78	344 82
AMOUNT OF DIGESTIBLE CONSTITUENTS PER 100 G	GRAM	0.56	0.16	19.45	
ENERGY VALUE (AVERAGE) OF THE DIGESTIBLE FRACTION PER 100 G EDIBLE PORTION	KJOULE (KCAL)	9.4 2.2	6.3 1.5	326 78	341 82

WASTE PERCENTAGE AVERAGE 7.0 MINIMUM 6.0 MAXIMUM 8.0

CONSTITUENTS	DIM	AV		VARIATION		AVR	NUTR. DENS.		MOLPERC.
WATER	GRAM	79.20	78.20	-	84.20	73.66	GRAM/MJ	232.10	
PROTEIN	GRAM	0.67	0.43	-	0.90	0.62	GRAM/MJ	1.96	
FAT	GRAM	0.18	0.10	-	0.30	0.17	GRAM/MJ	0.53	
AVAILABLE CARBOHYDR.	GRAM	19.45 [2]	-	-	-	18.09	GRAM/MJ	57.00	
MINERALS	GRAM	0.50	0.40	-	0.60	0.47	GRAM/MJ	1.47	
SODIUM	MILLI	1.80	0.70	-	3.50	1.67	MILLI/MJ	5.28	
POTASSIUM	MILLI	135.00	124.00	-	146.00	125.55	MILLI/MJ	395.63	
MAGNESIUM	MILLI	21.00	-	-	-	19.53	MILLI/MJ	61.54	
CALCIUM	MILLI	12.00	11.00	-	14.00	11.16	MILLI/MJ	35.17	
MANGANESE	MICRO	50.00	-	-	-	46.50	MICRO/MJ	146.53	
IRON	MILLI	0.50	0.30	-	0.90	0.47	MILLI/MJ	1.47	
COPPER	MILLI	0.11	-	-	-	0.10	MILLI/MJ	0.32	
PHOSPHORUS	MILLI	14.00	12.00	-	16.00	13.02	MILLI/MJ	41.03	
CHLORIDE	MILLI	3.00	-	-	-	2.79	MILLI/MJ	8.79	
BORON	MILLI	0.57	-	-	-	0.53	MILLI/MJ	1.67	
CAROTENE	MILLI	0.41	0.20	-	0.52	0.38	MILLI/MJ	1.20	
VITAMIN B1	MICRO	30.00	20.00	-	30.00	27.90	MICRO/MJ	87.92	
VITAMIN B2	MICRO	20.00	10.00	-	20.00	18.60	MICRO/MJ	58.61	
NICOTINAMIDE	MILLI	0.18	0.14	-	0.20	0.17	MILLI/MJ	0.53	
PANTOTHENIC ACID	MILLI	0.10	0.09	-	0.10	0.09	MILLI/MJ	0.29	
VITAMIN B6	MICRO	10.00	-	-	-	9.30	MICRO/MJ	29.31	
FOLIC ACID	MICRO	11.00	-	-	-	10.23	MICRO/MJ	32.24	
VITAMIN C	MILLI	5.00	3.00	-	6.00	4.65	MILLI/MJ	14.65	

1 TOTAL CONTENT (SOLID AND LIQUID)

2 ESTIMATED BY THE DIFFERENCE METHOD
100 - (WATER + PROTEIN + FAT + MINERALS)

MIRABELLE MIRABELLE MIRABELLE

PRUNUS DOMESTICA L. SSP.
SYRIACIA B.

		PROTEIN	FAT	CARBOHYDRATES	TOTAL
ENERGY VALUE (AVERAGE)	KJOULE	12	7.8	249	269
PER 100 G	(KCAL)	2.9	1.9	60	64
EDIBLE PORTION					
AMOUNT OF DIGESTIBLE	GRAM	0.62	0.18	14.90	
CONSTITUENTS PER 100 G					
ENERGY VALUE (AVERAGE)	KJOULE	10	7.0	249	267
OF THE DIGESTIBLE	(KCAL)	2.5	1.7	60	64
FRACTION PER 100 G					
EDIBLE PORTION					

WASTE PERCENTAGE AVERAGE 6.0

CONSTITUENTS	DIM	AV	VARIATION			AVR	NUTR. DENS.		MOLPERC.
WATER	GRAM	82.40	80.70	-	85.70	77.46	GRAM/MJ	309.05	
PROTEIN	GRAM	0.73	0.70	-	0.80	0.69	GRAM/MJ	2.74	
FAT	GRAM	0.20	-	-	-	0.19	GRAM/MJ	0.75	
AVAILABLE CARBOHYDR.	GRAM	14.90	-	-	-	14.01	GRAM/MJ	55.88	
MINERALS	GRAM	0.46	0.36	-	0.52	0.43	GRAM/MJ	1.73	
SODIUM	MILLI	0.40	0.20	-	0.50	0.38	MILLI/MJ	1.50	
POTASSIUM	MILLI	230.00	204.00	-	256.00	216.20	MILLI/MJ	862.63	
MAGNESIUM	MILLI	15.00	-	-	-	14.10	MILLI/MJ	56.26	
CALCIUM	MILLI	12.00	7.00	-	17.00	11.28	MILLI/MJ	45.01	
IRON	MILLI	0.50	-	-	-	0.47	MILLI/MJ	1.88	
PHOSPHORUS	MILLI	33.00	20.00	-	41.00	31.02	MILLI/MJ	123.77	
CHLORIDE	MILLI	2.45	2.39	-	2.50	2.30	MILLI/MJ	9.19	
CAROTENE	MILLI	0.21	-	-	-	0.20	MILLI/MJ	0.79	
VITAMIN B1	MICRO	60.00	60.00	-	200.00	56.40	MICRO/MJ	225.03	
VITAMIN B2	MICRO	40.00	0.00	-	40.00	37.60	MICRO/MJ	150.02	
NICOTINAMIDE	MILLI	0.60	0.00	-	0.60	0.56	MILLI/MJ	2.25	
VITAMIN C	MILLI	7.20	3.00	-	14.00	6.77	MILLI/MJ	27.00	
MALIC ACID	GRAM	0.89	-	-	-	0.84	GRAM/MJ	3.34	
OXALIC ACID TOTAL	MILLI	10.70	-	-	-	10.06	MILLI/MJ	40.13	
OXALIC ACID SOLUBLE	MILLI	6.60	-	-	-	6.20	MILLI/MJ	24.75	
GLUCOSE	GRAM	5.10	-	-	-	4.79	GRAM/MJ	19.13	
FRUCTOSE	GRAM	4.30	-	-	-	4.04	GRAM/MJ	16.13	
SUCROSE	GRAM	4.60	-	-	-	4.32	GRAM/MJ	17.25	

PFIRSICH PEACH PÊCHE

PRUNUS PERSICA (L.) STOKES

		PROTEIN	FAT	CARBOHYDRATES	TOTAL
ENERGY VALUE (AVERAGE) PER 100 G EDIBLE PORTION	KJOULE	13	4.3	158	175
	(KCAL)	3.0	1.0	38	42
AMOUNT OF DIGESTIBLE CONSTITUENTS PER 100 G	GRAM	0.64	0.09	9.44	
ENERGY VALUE (AVERAGE) OF THE DIGESTIBLE FRACTION PER 100 G EDIBLE PORTION	KJOULE	11	3.9	158	173
	(KCAL)	2.6	0.9	38	41

WASTE PERCENTAGE AVERAGE 8.0 MINIMUM 6.0 MAXIMUM 10

CONSTITUENTS	DIM	AV	VARIATION		AVR	NUTR. DENS.		MOLPERC.
WATER	GRAM	87.50	86.20	- 89.10	80.50	GRAM/MJ	507.20	
PROTEIN	GRAM	0.76	0.50	- 1.13	0.70	GRAM/MJ	4.41	
FAT	GRAM	0.11	0.05	- 0.28	0.10	GRAM/MJ	0.64	
AVAILABLE CARBOHYDR.	GRAM	9.44	7.02	- 11.69	8.68	GRAM/MJ	54.72	
TOTAL DIETARY FIBRE	GRAM	1.68 1	-	- -	1.55	GRAM/MJ	9.74	
MINERALS	GRAM	0.45	0.34	- 0.50	0.41	GRAM/MJ	2.61	
SODIUM	MILLI	1.30	0.50	- 2.70	1.20	MILLI/MJ	7.54	
POTASSIUM	MILLI	205.00	160.00	- 259.00	188.60	GRAM/MJ	1.19	
MAGNESIUM	MILLI	9.20	7.50	- 11.00	8.46	MILLI/MJ	53.33	
CALCIUM	MILLI	7.80	4.80	- 9.00	7.18	MILLI/MJ	45.21	
MANGANESE	MILLI	0.11	-	- -	0.10	MILLI/MJ	0.64	
IRON	MILLI	0.48	0.32	- 0.60	0.44	MILLI/MJ	2.78	
COPPER	MICRO	50.00	41.00	- 60.00	46.00	MICRO/MJ	289.83	
ZINC	MICRO	20.00	-	- -	18.40	MICRO/MJ	115.93	
NICKEL	MICRO	40.00	30.00	- 50.00	36.80	MICRO/MJ	231.86	
CHROMIUM	MICRO	2.00	1.00	- 3.00	1.84	MICRO/MJ	11.59	
PHOSPHORUS	MILLI	23.00	17.00	- 35.00	21.16	MILLI/MJ	133.32	
CHLORIDE	MILLI	2.60	1.00	- 5.00	2.39	MILLI/MJ	15.07	
FLUORIDE	MICRO	21.00	-	- -	19.32	MICRO/MJ	121.73	
IODIDE	MICRO	1.00	-	- -	0.92	MICRO/MJ	5.80	
BORON	MILLI	7.00	5.00	- 9.00	6.44	MILLI/MJ	40.58	
SELENIUM	MICRO	0.40	-	- -	0.37	MICRO/MJ	2.32	
SILICON	MILLI	0.40	-	- -	0.37	MILLI/MJ	2.32	
CAROTENE	MILLI	0.44	0.21	- 0.80	0.40	MILLI/MJ	2.55	
VITAMIN B1	MICRO	27.00	20.00	- 40.00	24.84	MICRO/MJ	156.51	
VITAMIN B2	MICRO	51.00	25.00	- 65.00	46.92	MICRO/MJ	295.62	

1 METHOD OF ENGLYST

CONSTITUENTS	DIM	AV	VARIATION			AVR	NUTR. DENS.		MOLPERC.
NICOTINAMIDE	MILLI	0.85	0.50	-	1.00	0.78	MILLI/MJ	4.93	
PANTOTHENIC ACID	MILLI	0.14	0.12	-	0.15	0.13	MILLI/MJ	0.81	
VITAMIN B6	MICRO	26.00	20.00	-	32.00	23.92	MICRO/MJ	150.71	
BIOTIN	MICRO	1.90	1.70	-	2.00	1.75	MICRO/MJ	11.01	
FOLIC ACID	MICRO	2.70	1.80	-	4.00	2.48	MICRO/MJ	15.65	
VITAMIN B12	-	0.00	-	-	-	0.00			
VITAMIN C	MILLI	9.50	5.00	-	28.80	8.74	MILLI/MJ	55.07	
ALANINE	MILLI	39.00	39.00	-	39.00	35.88	MILLI/MJ	226.07	9.6
ARGININE	MILLI	17.00	17.00	-	17.00	15.64	MILLI/MJ	98.54	2.1
ASPARTIC ACID	MILLI	90.00	90.00	-	90.00	82.80	MILLI/MJ	521.69	14.8
CYSTINE	MILLI	9.00	9.00	-	9.00	8.28	MILLI/MJ	52.17	0.8
GLUTAMIC ACID	MILLI	139.00	139.00	-	139.00	127.88	MILLI/MJ	805.72	20.7
GLYCINE	MILLI	15.00	15.00	-	15.00	13.80	MILLI/MJ	86.95	4.4
HISTIDINE	MILLI	17.00	17.00	-	17.00	15.64	MILLI/MJ	98.54	2.4
ISOLEUCINE	MILLI	13.00	13.00	-	13.00	11.96	MILLI/MJ	75.36	2.2
LEUCINE	MILLI	28.00	28.00	-	28.00	25.76	MILLI/MJ	162.30	4.7
LYSINE	MILLI	29.00	29.00	-	29.00	26.68	MILLI/MJ	168.10	4.3
METHIONINE	MILLI	30.00	30.00	-	30.00	27.60	MILLI/MJ	173.90	4.4
PHENYLALANINE	MILLI	18.00	18.00	-	18.00	16.56	MILLI/MJ	104.34	2.4
PROLINE	MILLI	27.00	27.00	-	27.00	24.84	MILLI/MJ	156.51	5.1
SERINE	MILLI	33.00	33.00	-	33.00	30.36	MILLI/MJ	191.29	6.9
THREONINE	MILLI	27.00	27.00	-	27.00	24.84	MILLI/MJ	156.51	5.0
TRYPTOPHAN	MILLI	5.00	5.00	-	5.00	4.60	MILLI/MJ	28.98	0.5
TYROSINE	MILLI	20.00	20.00	-	20.00	18.40	MILLI/MJ	115.93	2.4
VALINE	MILLI	39.00	39.00	-	39.00	35.88	MILLI/MJ	226.07	7.3
MALIC ACID	GRAM	0.33	0.28	-	0.37	0.30	GRAM/MJ	1.91	
CITRIC ACID	GRAM	0.24	0.16	-	0.32	0.22	GRAM/MJ	1.39	
OXALIC ACID TOTAL	-	0.00	-	-	-	0.00			
CHLOROGENIC ACID	MILLI	23.20	-	-	-	21.34	MILLI/MJ	134.48	
SALICYLIC ACID	MILLI	0.58	0.58	-	0.58	0.53	MILLI/MJ	3.36	
GLUCOSE	GRAM	1.03	0.72	-	1.50	0.95	GRAM/MJ	5.99	
FRUCTOSE	GRAM	1.23	0.86	-	1.80	1.14	GRAM/MJ	7.15	
SUCROSE	GRAM	5.72	4.50	-	6.80	5.27	GRAM/MJ	33.20	
PECTIN	GRAM	0.54	0.35	-	0.80	0.50	GRAM/MJ	3.13	
PENTOSAN	GRAM	0.90	0.60	-	1.20	0.83	GRAM/MJ	5.22	
SORBITOL	GRAM	0.89	0.50	-	0.90	0.82	GRAM/MJ	5.16	
TOTAL STEROLS	MILLI	10.00	-	-	-	9.20	MILLI/MJ	57.97	
CAMPESTEROL	MILLI	1.00	-	-	-	0.92	MILLI/MJ	5.80	
BETA-SITOSTEROL	MILLI	6.00	-	-	-	5.52	MILLI/MJ	34.78	
STIGMASTEROL	MILLI	3.00	-	-	-	2.76	MILLI/MJ	17.39	
DIETARY FIBRE,WAT.SOL.	GRAM	0.88	-	-	-	0.81	GRAM/MJ	5.10	
DIETARY FIBRE,WAT.INS.	GRAM	0.80	-	-	-	0.74	GRAM/MJ	4.64	

PFIRSICH
GETROCKNET

PEACH
DRIED

PÊCHE
SÈCHE

		PROTEIN	FAT	CARBOHYDRATES	TOTAL
ENERGY VALUE (AVERAGE)	KJOULE	50	23	960	1033
PER 100 G	(KCAL)	12	5.6	230	247
EDIBLE PORTION					
AMOUNT OF DIGESTIBLE	GRAM	2.55	0.54	57.39	
CONSTITUENTS PER 100 G					
ENERGY VALUE (AVERAGE)	KJOULE	42	21	960	1024
OF THE DIGESTIBLE	(KCAL)	10	5.0	230	245
FRACTION PER 100 G					
EDIBLE PORTION					

WASTE PERCENTAGE AVERAGE 0.00

CONSTITUENTS	DIM	AV	VARIATION			AVR	NUTR. DENS.		MOLPERC.
WATER	GRAM	24.00	-	-	-	24.00	GRAM/MJ	23.45	
PROTEIN	GRAM	3.00	-	-	-	3.00	GRAM/MJ	2.93	
FAT	GRAM	0.60	-	-	-	0.60	GRAM/MJ	0.59	
AVAILABLE CARBOHYDR.	GRAM	57.39 1	-	-	-	57.39	GRAM/MJ	56.06	
TOTAL DIETARY FIBRE	GRAM	10.20 1	-	-	-	10.20	GRAM/MJ	9.96	
MINERALS	GRAM	3.00	-	-	-	3.00	GRAM/MJ	2.93	
SODIUM	MILLI	9.00	6.00	-	12.00	9.00	MILLI/MJ	8.79	
POTASSIUM	GRAM	1.34	1.10	-	1.58	1.34	GRAM/MJ	1.31	
MAGNESIUM	MILLI	54.00	-	-	-	54.00	MILLI/MJ	52.75	
CALCIUM	MILLI	44.00	37.00	-	57.00	44.00	MILLI/MJ	42.98	
IRON	MILLI	6.90	-	-	-	6.90	MILLI/MJ	6.74	
COPPER	MILLI	0.63	-	-	-	0.63	MILLI/MJ	0.62	
PHOSPHORUS	MILLI	126.00	-	-	-	126.00	MILLI/MJ	123.09	
CHLORIDE	MILLI	11.00	-	-	-	11.00	MILLI/MJ	10.75	
IODIDE	MICRO	1.00	-	-	-	1.00	MICRO/MJ	0.98	
CAROTENE	MILLI	0.50	0.36	-	0.69	0.50	MILLI/MJ	0.49	
VITAMIN B1	MICRO	10.00	-	-	-	10.00	MICRO/MJ	9.77	
VITAMIN B2	MILLI	0.14	0.07	-	0.20	0.14	MILLI/MJ	0.14	
NICOTINAMIDE	MILLI	3.30	1.20	-	5.40	3.30	MILLI/MJ	3.22	
VITAMIN B6	MILLI	0.15	0.10	-	0.22	0.15	MILLI/MJ	0.15	
VITAMIN C	MILLI	16.70	12.00	-	19.00	16.70	MILLI/MJ	16.31	
MALIC ACID	GRAM	2.01	-	-	-	2.01	GRAM/MJ	1.96	
CITRIC ACID	GRAM	1.45	-	-	-	1.45	GRAM/MJ	1.42	
GLUCOSE	GRAM	6.26	-	-	-	6.26	GRAM/MJ	6.12	
FRUCTOSE	GRAM	7.48	-	-	-	7.48	GRAM/MJ	7.31	
SUCROSE	GRAM	34.78	-	-	-	34.78	GRAM/MJ	33.98	
SORBITOL	GRAM	5.41	-	-	-	5.41	GRAM/MJ	5.29	

1 CALCULATED FROM THE FRESH PRODUCT

PFIRSICHE IN DOSEN[1] · PEACHES CANNED[1] · PECHES EN BOÎTES[1]

		PROTEIN	FAT	CARBOHYDRATES	TOTAL
ENERGY VALUE (AVERAGE)	KJOULE	6.9	2.3	276	285
PER 100 G	(KCAL)	1.7	0.6	66	68
EDIBLE PORTION					
AMOUNT OF DIGESTIBLE	GRAM	0.35	0.05	16.50	
CONSTITUENTS PER 100 G					
ENERGY VALUE (AVERAGE)	KJOULE	5.9	2.1	276	284
OF THE DIGESTIBLE	(KCAL)	1.4	0.5	66	68
FRACTION PER 100 G					
EDIBLE PORTION					

WASTE PERCENTAGE AVERAGE 0.00

CONSTITUENTS	DIM	AV	VARIATION			AVR	NUTR. DENS.		MOLPERC.
WATER	GRAM	81.70	78.20	-	84.20	81.70	GRAM/MJ	287.53	
PROTEIN	GRAM	0.42	0.40	-	0.50	0.42	GRAM/MJ	1.48	
FAT	MILLI	60.00	10.00	-	100.00	60.00	MILLI/MJ	211.16	
AVAILABLE CARBOHYDR.	GRAM	16.50	14.00	-	20.00	16.50	GRAM/MJ	58.07	
TOTAL DIETARY FIBRE	GRAM	1.10	-	-	-	1.10	GRAM/MJ	3.87	
MINERALS	GRAM	0.27	0.18	-	0.33	0.27	GRAM/MJ	0.95	
SODIUM	MILLI	2.50	1.40	-	5.00	2.50	MILLI/MJ	8.80	
POTASSIUM	MILLI	130.00	107.00	-	151.00	130.00	MILLI/MJ	457.51	
MAGNESIUM	MILLI	5.40	3.80	-	6.30	5.40	MILLI/MJ	19.00	
CALCIUM	MILLI	3.90	3.00	-	5.00	3.90	MILLI/MJ	13.73	
MANGANESE	MICRO	60.00	40.00	-	70.00	60.00	MICRO/MJ	211.16	
IRON	MILLI	0.30	0.19	-	0.36	0.30	MILLI/MJ	1.06	
COPPER	MICRO	33.00	25.00	-	41.00	33.00	MICRO/MJ	116.14	
PHOSPHORUS	MILLI	13.00	9.00	-	16.00	13.00	MILLI/MJ	45.75	
CHLORIDE	MILLI	1.60	0.60	-	4.20	1.60	MILLI/MJ	5.63	
FLUORIDE	MICRO	13.00	-	-	-	13.00	MICRO/MJ	45.75	
SELENIUM	MICRO	0.30	0.20	-	0.40	0.30	MICRO/MJ	1.06	
CAROTENE	MILLI	0.27	0.20	-	0.40	0.27	MILLI/MJ	0.95	
VITAMIN B1	MICRO	10.00	-	-	-	10.00	MICRO/MJ	35.19	
VITAMIN B2	MICRO	22.00	20.00	-	30.00	22.00	MICRO/MJ	77.43	
NICOTINAMIDE	MILLI	0.58	0.50	-	0.70	0.58	MILLI/MJ	2.04	
PANTOTHENIC ACID	MICRO	50.00	43.00	-	54.00	50.00	MICRO/MJ	175.97	
VITAMIN B6	MICRO	18.00	14.00	-	22.00	18.00	MICRO/MJ	63.35	
BIOTIN	MICRO	0.20	-	-	-	0.20	MICRO/MJ	0.70	
FOLIC ACID	MICRO	5.00	0.30	-	5.00	5.00	MICRO/MJ	17.60	
VITAMIN C	MILLI	4.00	3.00	-	6.10	4.00	MILLI/MJ	14.08	

1 TOTAL CONTENT (SOLID AND LIQUID)

CONSTITUENTS	DIM	AV	VARIATION			AVR	NUTR. DENS.		MOLPERC.
ALANINE	MILLI	82.00	82.00	-	82.00	82.00	MILLI/MJ	288.58	11.1
ARGININE	MILLI	40.00	40.00	-	40.00	40.00	MILLI/MJ	140.77	2.8
ASPARTIC ACID	MILLI	114.50	114.50	-	114.50	114.50	MILLI/MJ	402.96	10.4
CYSTINE	MILLI	3.00	3.00	-	3.00	3.00	MILLI/MJ	10.56	0.2
GLUTAMIC ACID	MILLI	96.00	96.00	-	96.00	96.00	MILLI/MJ	337.86	7.9
GLYCINE	MILLI	58.00	58.00	-	58.00	58.00	MILLI/MJ	204.12	9.3
HISTIDINE	MILLI	31.00	31.00	-	31.00	31.00	MILLI/MJ	109.10	2.4
ISOLEUCINE	MILLI	55.00	55.00	-	55.00	55.00	MILLI/MJ	193.56	5.1
LEUCINE	MILLI	82.00	82.00	-	82.00	82.00	MILLI/MJ	288.58	7.5
LYSINE	MILLI	65.00	65.00	-	65.00	65.00	MILLI/MJ	228.76	5.4
METHIONINE	MILLI	14.00	14.00	-	14.00	14.00	MILLI/MJ	49.27	1.1
PHENYLALANINE	MILLI	64.00	64.00	-	64.00	64.00	MILLI/MJ	225.24	4.7
PROLINE	MILLI	85.00	85.00	-	85.00	85.00	MILLI/MJ	299.14	8.9
SERINE	MILLI	70.00	70.00	-	70.00	70.00	MILLI/MJ	246.35	8.0
THREONINE	MILLI	49.00	49.00	-	49.00	49.00	MILLI/MJ	172.45	5.0
TRYPTOPHAN	MILLI	8.00	8.00	-	8.00	8.00	MILLI/MJ	28.15	0.5
TYROSINE	MILLI	24.00	24.00	-	24.00	24.00	MILLI/MJ	84.46	1.6
VALINE	MILLI	80.00	80.00	-	80.00	80.00	MILLI/MJ	281.55	8.2
MALIC ACID	GRAM	0.14	0.12	-	0.16	0.14	GRAM/MJ	0.49	
CITRIC ACID	GRAM	0.10	0.07	-	0.14	0.10	GRAM/MJ	0.35	
SALICYLIC ACID	MILLI	0.31	0.31	-	0.31	0.31	MILLI/MJ	1.09	
GLUCOSE	GRAM	3.60	3.10	-	4.30	3.60	GRAM/MJ	12.67	
FRUCTOSE	GRAM	3.40	2.90	-	4.10	3.40	GRAM/MJ	11.97	
SUCROSE	GRAM	9.20	5.80	-	13.90	9.20	GRAM/MJ	32.38	

PFLAUME PLUM PRUNE

PRUNUS DOMESTICA L.

		PROTEIN	FAT	CARBOHYDRATES	TOTAL
ENERGY VALUE (AVERAGE)	KJOULE	9.9	6.6	191	207
PER 100 G	(KCAL)	2.4	1.6	46	50
EDIBLE PORTION					
AMOUNT OF DIGESTIBLE	GRAM	0.51	0.15	11.41	
CONSTITUENTS PER 100 G					
ENERGY VALUE (AVERAGE)	KJOULE	8.4	6.0	191	205
OF THE DIGESTIBLE	(KCAL)	2.0	1.4	46	49
FRACTION PER 100 G					
EDIBLE PORTION					

WASTE PERCENTAGE AVERAGE 6.0 MINIMUM 4.0 MAXIMUM 10

CONSTITUENTS	DIM	AV		VARIATION		AVR	NUTR. DENS.		MOLPERC.
WATER	GRAM	83.70	78.70	-	87.90	78.68	GRAM/MJ	407.62	
PROTEIN	GRAM	0.60	0.50	-	0.80	0.56	GRAM/MJ	2.92	
FAT	GRAM	0.17	0.10	-	0.20	0.16	GRAM/MJ	0.83	
AVAILABLE CARBOHYDR.	GRAM	11.41	-	-	-	10.73	GRAM/MJ	55.57	
TOTAL DIETARY FIBRE	GRAM	1.70 [1]	-	-	-	1.60	GRAM/MJ	8.28	
MINERALS	GRAM	0.49	0.40	-	0.60	0.46	GRAM/MJ	2.39	
SODIUM	MILLI	1.70	0.60	-	2.30	1.60	MILLI/MJ	8.28	
POTASSIUM	MILLI	221.00	150.00	-	299.00	207.74	GRAM/MJ	1.08	
MAGNESIUM	MILLI	10.00	7.00	-	13.00	9.40	MILLI/MJ	48.70	
CALCIUM	MILLI	14.00	10.00	-	18.00	13.16	MILLI/MJ	68.18	
MANGANESE	MICRO	82.00	21.00	-	130.00	77.08	MICRO/MJ	399.34	
IRON	MILLI	0.44	0.30	-	0.54	0.41	MILLI/MJ	2.14	
COBALT	MICRO	0.90	0.40	-	1.70	0.85	MICRO/MJ	4.38	
COPPER	MICRO	93.00	60.00	-	150.00	87.42	MICRO/MJ	452.91	
ZINC	MICRO	70.00	30.00	-	100.00	65.80	MICRO/MJ	340.90	
NICKEL	MICRO	17.00	-	-	-	15.98	MICRO/MJ	82.79	
CHROMIUM	MICRO	2.00	-	-	-	1.88	MICRO/MJ	9.74	
MOLYBDENUM	MICRO	6.00	-	-	-	5.64	MICRO/MJ	29.22	
VANADIUM	MICRO	2.00	-	-	-	1.88	MICRO/MJ	9.74	
PHOSPHORUS	MILLI	18.00	15.00	-	26.00	16.92	MILLI/MJ	87.66	
CHLORIDE	MILLI	1.50	0.80	-	2.00	1.41	MILLI/MJ	7.30	
FLUORIDE	MICRO	1.80	1.60	-	2.00	1.69	MICRO/MJ	8.77	
IODIDE	MICRO	1.00	-	-	-	0.94	MICRO/MJ	4.87	
BORON	MILLI	0.34	0.13	-	0.64	0.32	MILLI/MJ	1.66	
SELENIUM	MICRO	0.15	0.10	-	0.20	0.14	MICRO/MJ	0.73	
SILICON	MILLI	0.40	0.20	-	1.00	0.38	MILLI/MJ	1.95	

1 METHOD OF MEUSER, SUCKOW AND KULIKOWSKI ("BERLINER METHODE")

CONSTITUENTS	DIM	AV	VARIATION			AVR	NUTR.	DENS.	MOLPERC.
VITAMIN A	-	0.00	-	-	-	0.00			
CAROTENE	MILLI	0.21	0.08	-	0.57	0.20	MILLI/MJ	1.02	
VITAMIN D	-	0.00	-	-	-	0.00			
VITAMIN E ACTIVITY	MILLI	0.80	0.28	-	1.30	0.75	MILLI/MJ	3.90	
VITAMIN B1	MICRO	72.00	20.00	-	120.00	67.68	MICRO/MJ	350.64	
VITAMIN B2	MICRO	43.00	25.00	-	70.00	40.42	MICRO/MJ	209.41	
NICOTINAMIDE	MILLI	0.44	0.25	-	0.60	0.41	MILLI/MJ	2.14	
PANTOTHENIC ACID	MILLI	0.18	0.13	-	0.23	0.17	MILLI/MJ	0.88	
VITAMIN B6	MICRO	45.00	31.00	-	67.00	42.30	MICRO/MJ	219.15	
BIOTIN	MICRO	0.10	-	-	-	0.09	MICRO/MJ	0.49	
FOLIC ACID	MICRO	2.00	1.80	-	2.20	1.88	MICRO/MJ	9.74	
VITAMIN B12	-	0.00	-	-	-	0.00			
VITAMIN C	MILLI	5.40	2.40	-	14.10	5.08	MILLI/MJ	26.30	
SEROTONINE	MILLI	-	0.00	-	1.00				
TRYPTAMINE	MILLI	-	0.00	-	0.50				
TYRAMINE	MILLI	-	0.00	-	0.60				
MALIC ACID	GRAM	1.22	0.82	-	1.99	1.15	GRAM/MJ	5.94	
CITRIC ACID	MILLI	34.00	23.00	-	55.00	31.96	MILLI/MJ	165.58	
OXALIC ACID TOTAL	MILLI	11.90	-	-	-	11.19	MILLI/MJ	57.95	
OXALIC ACID SOLUBLE	MILLI	6.00	-	-	-	5.64	MILLI/MJ	29.22	
CHLOROGENIC ACID	MILLI	9.00	-	-	-	8.46	MILLI/MJ	43.83	
FERULIC ACID	MILLI	0.90	-	-	-	0.85	MILLI/MJ	4.38	
CAFFEIC ACID	MILLI	14.00	8.00	-	22.00	13.16	MILLI/MJ	68.18	
PARA-COUMARIC ACID	MILLI	2.40	-	-	-	2.26	MILLI/MJ	11.69	
SALICYLIC ACID	MILLI	0.14	0.14	-	0.14	0.13	MILLI/MJ	0.68	
GLUCOSE	GRAM	3.36	3.36	-	3.36	3.16	GRAM/MJ	16.36	
FRUCTOSE	GRAM	2.01	2.01	-	2.01	1.89	GRAM/MJ	9.79	
SUCROSE	GRAM	3.38	3.38	-	3.38	3.18	GRAM/MJ	16.46	
PECTIN	GRAM	0.76	0.57	-	0.90	0.71	GRAM/MJ	3.70	
CELLULOSE	GRAM	0.23	-	-	-	0.22	GRAM/MJ	1.12	
SORBITOL	GRAM	1.41	1.41	-	1.41	1.33	GRAM/MJ	6.87	
PALMITIC ACID	MILLI	25.00	25.00	-	25.00	23.50	MILLI/MJ	121.75	
STEARIC ACID	MILLI	8.00	8.00	-	8.00	7.52	MILLI/MJ	38.96	
OLEIC ACID	MILLI	9.00	9.00	-	9.00	8.46	MILLI/MJ	43.83	
LINOLEIC ACID	MILLI	63.00	63.00	-	63.00	59.22	MILLI/MJ	306.81	
LINOLENIC ACID	MILLI	31.00	31.00	-	31.00	29.14	MILLI/MJ	150.97	
TOTAL STEROLS	MILLI	7.00	-	-	-	6.58	MILLI/MJ	34.09	
BETA-SITOSTEROL	MILLI	6.00	-	-	-	5.64	MILLI/MJ	29.22	
DIETARY FIBRE,WAT.SOL.	GRAM	0.85	-	-	-	0.80	GRAM/MJ	4.14	
DIETARY FIBRE,WAT.INS.	GRAM	0.85	-	-	-	0.80	GRAM/MJ	4.14	

PFLAUME GETROCKNET | PLUM DRIED | PRUNEAU

		PROTEIN	FAT	CARBOHYDRATES	TOTAL
ENERGY VALUE (AVERAGE)	KJOULE	38	23	890	952
PER 100 G	(KCAL)	9.1	5.6	213	227
EDIBLE PORTION					
AMOUNT OF DIGESTIBLE	GRAM	1.95	0.54	53.20	
CONSTITUENTS PER 100 G					
ENERGY VALUE (AVERAGE)	KJOULE	32	21	890	944
OF THE DIGESTIBLE	(KCAL)	7.7	5.0	213	226
FRACTION PER 100 G					
EDIBLE PORTION					

WASTE PERCENTAGE AVERAGE 15 MINIMUM 10 MAXIMUM 19

CONSTITUENTS	DIM	AV		VARIATION		AVR	NUTR. DENS.		MOLPERC.
WATER	GRAM	24.00	-	-	-	20.40	GRAM/MJ	25.43	
PROTEIN	GRAM	2.30	-	-	-	1.96	GRAM/MJ	2.44	
FAT	GRAM	0.60	-	-	-	0.51	GRAM/MJ	0.64	
AVAILABLE CARBOHYDR.	GRAM	53.20 [1]	-	-	-	45.22	GRAM/MJ	56.38	
TOTAL DIETARY FIBRE	GRAM	9.00 [2]	-	-	-	7.65	GRAM/MJ	9.54	
MINERALS	GRAM	2.10	-	-	-	1.79	GRAM/MJ	2.23	
SODIUM	MILLI	8.00	4.00	-	11.00	6.80	MILLI/MJ	8.48	
POTASSIUM	MILLI	824.00	721.00	-	926.00	700.40	MILLI/MJ	873.18	
MAGNESIUM	MILLI	27.00	-	-	-	22.95	MILLI/MJ	28.61	
CALCIUM	MILLI	41.00	30.00	-	54.00	34.85	MILLI/MJ	43.45	
IRON	MILLI	2.30	1.00	-	3.90	1.96	MILLI/MJ	2.44	
COPPER	MILLI	0.40	-	-	-	0.34	MILLI/MJ	0.42	
PHOSPHORUS	MILLI	73.00	60.00	-	85.00	62.05	MILLI/MJ	77.36	
CHLORIDE	MILLI	52.00	-	-	-	44.20	MILLI/MJ	55.10	
IODIDE	MICRO	1.00	-	-	-	0.85	MICRO/MJ	1.06	
CAROTENE	MILLI	0.67	0.20	-	1.20	0.57	MILLI/MJ	0.71	
VITAMIN B1	MILLI	0.15	0.06	-	0.30	0.13	MILLI/MJ	0.16	
VITAMIN B2	MILLI	0.12	0.06	-	0.16	0.10	MILLI/MJ	0.13	
NICOTINAMIDE	MILLI	1.73	1.70	-	1.96	1.47	MILLI/MJ	1.83	
PANTOTHENIC ACID	MILLI	0.46	-	-	-	0.39	MILLI/MJ	0.49	
VITAMIN B6	MILLI	0.15	0.12	-	0.18	0.13	MILLI/MJ	0.16	
FOLIC ACID	MICRO	4.00	-	-	-	3.40	MICRO/MJ	4.24	
VITAMIN C	MILLI	4.00	2.00	-	5.00	3.40	MILLI/MJ	4.24	
MALIC ACID	GRAM	5.69	-	-	-	4.84	GRAM/MJ	6.03	
CITRIC ACID	GRAM	0.14	-	-	-	0.12	GRAM/MJ	0.15	
GLUCOSE	GRAM	15.67	-	-	-	13.32	GRAM/MJ	16.61	
FRUCTOSE	GRAM	9.37	-	-	-	7.96	GRAM/MJ	9.93	
SUCROSE	GRAM	15.76	-	-	-	13.40	GRAM/MJ	16.70	
SORBITOL	GRAM	6.57	-	-	-	5.58	GRAM/MJ	6.96	
DIETARY FIBRE, WAT.SOL.	GRAM	4.90	-	-	-	4.17	GRAM/MJ	5.19	
DIETARY FIBRE, WAT.INS.	GRAM	4.10	-	-	-	3.49	GRAM/MJ	4.34	

1 CALCULATED FROM THE FRESH PRODUCT

2 METHOD OF MEUSER, SUCKOW AND KULIKOWSKI ("BERLINER METHODE")

PFLAUMEN IN DOSEN 1	**PLUMS** CANNED 1	**PRUNES** EN BOÎTES 1

		PROTEIN	FAT	CARBOHYDRATES	TOTAL
ENERGY VALUE (AVERAGE) PER 100 G EDIBLE PORTION	KJOULE (KCAL)	7.9 1.9	4.3 1.0	303 72	315 75
AMOUNT OF DIGESTIBLE CONSTITUENTS PER 100 G	GRAM	0.40	0.09	18.11	
ENERGY VALUE (AVERAGE) OF THE DIGESTIBLE FRACTION PER 100 G EDIBLE PORTION	KJOULE (KCAL)	6.7 1.6	3.9 0.9	303 72	314 75

WASTE PERCENTAGE AVERAGE 0.00

CONSTITUENTS	DIM	AV	VARIATION	AVR	NUTR. DENS.		MOLPERC.
WATER	GRAM	80.80	77.80 - 83.80	80.80	GRAM/MJ	257.58	
PROTEIN	GRAM	0.48	0.40 - 0.50	0.48	GRAM/MJ	1.53	
FAT	GRAM	0.11	0.10 - 0.14	0.11	GRAM/MJ	0.35	
AVAILABLE CARBOHYDR.	GRAM	18.11 2	- - -	18.11	GRAM/MJ	57.73	
MINERALS	GRAM	0.50	0.50 - 0.51	0.50	GRAM/MJ	1.59	
SODIUM	MILLI	11.50	5.00 - 18.00	11.50	MILLI/MJ	36.66	
POTASSIUM	MILLI	118.00	110.00 - 126.00	118.00	MILLI/MJ	376.17	
CALCIUM	MILLI	10.30	8.00 - 15.00	10.30	MILLI/MJ	32.84	
MANGANESE	MILLI	0.13	0.05 - 0.20	0.13	MILLI/MJ	0.41	
IRON	MILLI	1.10	1.00 - 1.20	1.10	MILLI/MJ	3.51	
PHOSPHORUS	MILLI	13.70	12.00 - 16.00	13.70	MILLI/MJ	43.67	
FLUORIDE	MICRO	21.00	- - -	21.00	MICRO/MJ	66.95	
CAROTENE	MILLI	0.14	- - -	0.14	MILLI/MJ	0.45	
VITAMIN B1	MICRO	28.00	23.00 - 30.00	28.00	MICRO/MJ	89.26	
VITAMIN B2	MICRO	27.00	20.00 - 30.00	27.00	MICRO/MJ	86.07	
NICOTINAMIDE	MILLI	0.39	0.36 - 0.40	0.39	MILLI/MJ	1.24	
VITAMIN C	MILLI	1.50	1.00 - 2.00	1.50	MILLI/MJ	4.78	

1 TOTAL CONTENT (SOLID AND LIQUID)

2 ESTIMATED BY THE DIFFERENCE METHOD
100 - (WATER + PROTEIN + FAT + MINERALS)

REINECLAUDE GREENGAGE REINE-CLAUDE

PRUNUS INSITITIA L.

		PROTEIN	FAT	CARBOHYDRATES	TOTAL
ENERGY VALUE (AVERAGE)	KJOULE	13	0.0	226	239
PER 100 G	(KCAL)	3.1	0.0	54	57
EDIBLE PORTION					
AMOUNT OF DIGESTIBLE	GRAM	0.67	0.00	13.52	
CONSTITUENTS PER 100 G					
ENERGY VALUE (AVERAGE)	KJOULE	11	0.0	226	237
OF THE DIGESTIBLE	(KCAL)	2.7	0.0	54	57
FRACTION PER 100 G					
EDIBLE PORTION					

WASTE PERCENTAGE AVERAGE 4.0 MINIMUM 2.0 MAXIMUM 5.0

CONSTITUENTS	DIM	AV		VARIATION		AVR	NUTR. DENS.		MOLPERC.
WATER	GRAM	80.70	78.20	-	81.90	77.47	GRAM/MJ	339.98	
PROTEIN	GRAM	0.79	0.76	-	0.80	0.76	GRAM/MJ	3.33	
AVAILABLE CARBOHYDR.	GRAM	13.52	-	-	-	12.98	GRAM/MJ	56.96	
TOTAL DIETARY FIBRE	GRAM	2.25 1	-	-	-	2.16	GRAM/MJ	9.48	
MINERALS	GRAM	0.60	-	-	-	0.58	GRAM/MJ	2.53	
SODIUM	MILLI	0.75	0.10	-	1.40	0.72	MILLI/MJ	3.16	
POTASSIUM	MILLI	243.00	144.00	-	305.00	233.28	GRAM/MJ	1.02	
MAGNESIUM	MILLI	9.60	7.70	-	11.50	9.22	MILLI/MJ	40.44	
CALCIUM	MILLI	13.30	4.00	-	19.10	12.77	MILLI/MJ	56.03	
IRON	MILLI	1.14	0.37	-	1.90	1.09	MILLI/MJ	4.80	
COPPER	MICRO	80.00	-	-	-	76.80	MICRO/MJ	337.03	
PHOSPHORUS	MILLI	24.80	22.60	-	27.00	23.81	MILLI/MJ	104.48	
CHLORIDE	MILLI	1.95	1.00	-	2.90	1.87	MILLI/MJ	8.22	
VITAMIN A	-	0.00	-	-	-	0.00			
CAROTENE	MILLI	0.18	-	-	-	0.17	MILLI/MJ	0.76	
VITAMIN C	MILLI	5.80	5.00	-	8.00	5.57	MILLI/MJ	24.43	
MALIC ACID	GRAM	1.25	-	-	-	1.20	GRAM/MJ	5.27	
GLUCOSE	GRAM	4.96	-	-	-	4.76	GRAM/MJ	20.90	
FRUCTOSE	GRAM	3.67	-	-	-	3.52	GRAM/MJ	15.46	
SUCROSE	GRAM	3.64	-	-	-	3.49	GRAM/MJ	15.33	
DIETARY FIBRE,WAT.SOL.	GRAM	1.38	-	-	-	1.32	GRAM/MJ	5.81	
DIETARY FIBRE,WAT.INS.	GRAM	0.87	-	-	-	0.84	GRAM/MJ	3.67	

1 METHOD OF ENGLYST

Beeren

BERRIES BAIES

BOYSENBEERE BOYSENBERRY BAIE DE BOYSEN

RUBUS-HYBRIDEN

		PROTEIN	FAT	CARBOHYDRATES	TOTAL
ENERGY VALUE (AVERAGE)	KJOULE	8.3	12	103	123
PER 100 G	(KCAL)	2.0	2.8	25	29
EDIBLE PORTION					
AMOUNT OF DIGESTIBLE	GRAM	0.42	0.27	6.18	
CONSTITUENTS PER 100 G					
ENERGY VALUE (AVERAGE)	KJOULE	7.0	11	103	121
OF THE DIGESTIBLE	(KCAL)	1.7	2.5	25	29
FRACTION PER 100 G					
EDIBLE PORTION					

WASTE PERCENTAGE AVERAGE 0.00

CONSTITUENTS	DIM	AV	VARIATION			AVR	NUTR. DENS.		MOLPERC.
WATER	GRAM	87.00	-	-	-	87.00	GRAM/MJ	719.26	
PROTEIN	GRAM	0.50	-	-	-	0.50	GRAM/MJ	4.13	
FAT	GRAM	0.30	-	-	-	0.30	GRAM/MJ	2.48	
AVAILABLE CARBOHYDR.	GRAM	6.18	-	-	-	6.18	GRAM/MJ	51.09	
MINERALS	GRAM	0.90	-	-	-	0.90	GRAM/MJ	7.44	
POTASSIUM	MILLI	150.00	-	-	-	150.00	GRAM/MJ	1.24	
MAGNESIUM	MILLI	18.00	-	-	-	18.00	MILLI/MJ	148.81	
CALCIUM	MILLI	25.00	-	-	-	25.00	MILLI/MJ	206.68	
IRON	MILLI	1.60	-	-	-	1.60	MILLI/MJ	13.23	
PHOSPHORUS	MILLI	24.00	-	-	-	24.00	MILLI/MJ	198.42	
CAROTENE	MILLI	0.10	-	-	-	0.10	MILLI/MJ	0.83	
VITAMIN B1	MICRO	20.00	-	-	-	20.00	MICRO/MJ	165.35	
VITAMIN B2	MILLI	0.13	-	-	-	0.13	MILLI/MJ	1.07	
NICOTINAMIDE	MILLI	1.00	-	-	-	1.00	MILLI/MJ	8.27	
VITAMIN C	MILLI	13.00	-	-	-	13.00	MILLI/MJ	107.47	
GLUCOSE	GRAM	2.48	2.48	-	2.48	2.48	GRAM/MJ	20.50	
FRUCTOSE	GRAM	3.70	3.70	-	3.70	3.70	GRAM/MJ	30.59	
SUCROSE	-	0.00		-		0.00			

BROMBEERE BLACKBERRY MÛRE

RUBUS FRUTICOSUS L.

		PROTEIN	FAT	CARBOHYDRATES	TOTAL
ENERGY VALUE (AVERAGE)	KJOULE	20	39	120	178
PER 100 G EDIBLE PORTION	(KCAL)	4.7	9.3	29	43
AMOUNT OF DIGESTIBLE CONSTITUENTS PER 100 G	GRAM	1.02	0.90	7.15	
ENERGY VALUE (AVERAGE)	KJOULE	17	35	120	172
OF THE DIGESTIBLE FRACTION PER 100 G EDIBLE PORTION	(KCAL)	4.0	8.4	29	41

WASTE PERCENTAGE AVERAGE 0.00

CONSTITUENTS	DIM	AV	VARIATION			AVR	NUTR. DENS.		MOLPERC.
WATER	GRAM	84.70	82.20	-	87.00	84.70	GRAM/MJ	493.76	
PROTEIN	GRAM	1.20	1.20	-	1.30	1.20	GRAM/MJ	7.00	
FAT	GRAM	1.00	-	-	-	1.00	GRAM/MJ	5.83	
AVAILABLE CARBOHYDR.	GRAM	7.15	5.38	-	8.13	7.15	GRAM/MJ	41.68	
TOTAL DIETARY FIBRE	GRAM	3.16 1	-	-	-	3.16	GRAM/MJ	18.42	
MINERALS	GRAM	0.51	0.50	-	0.52	0.51	GRAM/MJ	2.97	
SODIUM	MILLI	3.00	1.00	-	5.00	3.00	MILLI/MJ	17.49	
POTASSIUM	MILLI	189.00	179.00	-	200.00	189.00	GRAM/MJ	1.10	
MAGNESIUM	MILLI	30.00	-	-	-	30.00	MILLI/MJ	174.89	
CALCIUM	MILLI	44.00	25.00	-	63.00	44.00	MILLI/MJ	256.50	
MANGANESE	MILLI	0.59	-	-	-	0.59	MILLI/MJ	3.44	
IRON	MILLI	0.90	0.90	-	1.00	0.90	MILLI/MJ	5.25	
COPPER	MILLI	0.14	-	-	-	0.14	MILLI/MJ	0.82	
PHOSPHORUS	MILLI	30.00	25.00	-	32.00	30.00	MILLI/MJ	174.89	
IODIDE	MICRO	0.40	0.40	-	0.40	0.40	MICRO/MJ	2.33	
BORON	MICRO	87.00	51.00	-	137.00	87.00	MICRO/MJ	507.17	
CAROTENE	MILLI	0.27	0.12	-	0.48	0.27	MILLI/MJ	1.57	
VITAMIN E ACTIVITY	MILLI	0.72	-	-	-	0.72	MILLI/MJ	4.20	
TOTAL TOCOPHEROLS	MILLI	2.70	-	-	-	2.70	MILLI/MJ	15.74	
ALPHA-TOCOPHEROL	MILLI	0.60	-	-	-	0.60	MILLI/MJ	3.50	
GAMMA-TOCOPHEROL	MILLI	1.10	-	-	-	1.10	MILLI/MJ	6.41	
DELTA-TOCOPHEROL	MILLI	1.00	-	-	-	1.00	MILLI/MJ	5.83	
VITAMIN B1	MICRO	30.00	17.00	-	40.00	30.00	MICRO/MJ	174.89	
VITAMIN B2	MICRO	40.00	40.00	-	50.00	40.00	MICRO/MJ	233.18	
NICOTINAMIDE	MILLI	0.40	0.40	-	0.50	0.40	MILLI/MJ	2.33	
PANTOTHENIC ACID	MILLI	0.22	-	-	-	0.22	MILLI/MJ	1.28	

1 METHOD OF MEUSER, SUCKOW AND KULIKOWSKI ("BERLINER METHODE")

CONSTITUENTS	DIM	AV	VARIATION			AVR	NUTR. DENS.		MOLPERC.
VITAMIN B6	MICRO	50.00	40.00	-	57.00	50.00	MICRO/MJ	291.48	
VITAMIN C	MILLI	17.00	12.00	-	21.00	17.00	MILLI/MJ	99.10	
MALIC ACID	GRAM	0.90	0.86	-	0.95	0.90	GRAM/MJ	5.25	
CITRIC ACID	MILLI	18.00	15.00	-	21.00	18.00	MILLI/MJ	104.93	
OXALIC ACID TOTAL	MILLI	12.40	-	-	-	12.40	MILLI/MJ	72.29	
OXALIC ACID SOLUBLE	MILLI	6.80	-	-	-	6.80	MILLI/MJ	39.64	
FERULIC ACID	MILLI	0.80	0.60	-	1.20	0.80	MILLI/MJ	4.66	
CAFFEIC ACID	MILLI	2.40	1.00	-	3.40	2.40	MILLI/MJ	13.99	
PARA-COUMARIC ACID	MILLI	1.10	0.90	-	1.40	1.10	MILLI/MJ	6.41	
PROTOCATECHUIC ACID	MILLI	13.10	6.80	-	18.90	13.10	MILLI/MJ	76.37	
GALLIC ACID	MILLI	4.20	0.80	-	6.70	4.20	MILLI/MJ	24.48	
ISOCITRIC ACID	GRAM	0.81	-	-	-	0.81	GRAM/MJ	4.72	
GLUCOSE	GRAM	2.96	2.16	-	3.28	2.96	GRAM/MJ	17.26	
FRUCTOSE	GRAM	3.11	2.35	-	3.38	3.11	GRAM/MJ	18.15	
SUCROSE	GRAM	0.17	0.00	-	0.50	0.17	GRAM/MJ	0.99	
PECTIN	GRAM	0.48	0.32	-	0.63	0.48	GRAM/MJ	2.80	
DIETARY FIBRE,WAT.SOL.	GRAM	0.96	-	-	-	0.96	GRAM/MJ	5.60	
DIETARY FIBRE,WAT.INS.	GRAM	2.20	-	-	-	2.20	GRAM/MJ	12.83	

ERDBEERE
FRAGARIA-ARTEN

STRAWBERRY

FRAISE

		PROTEIN	FAT	CARBOHYDRATES	TOTAL
ENERGY VALUE (AVERAGE)	KJOULE	14	16	108	137
PER 100 G EDIBLE PORTION	(KCAL)	3.2	3.7	26	33
AMOUNT OF DIGESTIBLE CONSTITUENTS PER 100 G	GRAM	0.69	0.36	6.45	
ENERGY VALUE (AVERAGE)	KJOULE	12	14	108	133
OF THE DIGESTIBLE FRACTION PER 100 G EDIBLE PORTION	(KCAL)	2.8	3.3	26	32

WASTE PERCENTAGE AVERAGE 3.0 MINIMUM 2.0 MAXIMUM 5.0

CONSTITUENTS	DIM	AV		VARIATION		AVR	NUTR. DENS.		MOLPERC.
WATER	GRAM	89.50		84.10 -	92.40	86.82	GRAM/MJ	670.54	
PROTEIN	GRAM	0.82		0.23 -	1.18	0.80	GRAM/MJ	6.14	
FAT	GRAM	0.40		0.20 -	0.50	0.39	GRAM/MJ	3.00	
AVAILABLE CARBOHYDR.	GRAM	6.45		4.87 -	7.35	6.26	GRAM/MJ	48.32	
TOTAL DIETARY FIBRE	GRAM	2.00	1	1.90 -	2.10	1.94	GRAM/MJ	14.98	
MINERALS	GRAM	0.50		0.30 -	0.74	0.49	GRAM/MJ	3.75	
SODIUM	MILLI	2.50		0.50 -	5.00	2.43	MILLI/MJ	18.73	
POTASSIUM	MILLI	147.00		105.00 -	169.00	142.59	GRAM/MJ	1.10	
MAGNESIUM	MILLI	15.00		11.00 -	20.00	14.55	MILLI/MJ	112.38	
CALCIUM	MILLI	26.00		16.00 -	30.00	25.22	MILLI/MJ	194.79	
MANGANESE	MILLI	0.20		0.13 -	0.25	0.19	MILLI/MJ	1.50	
IRON	MILLI	0.96		0.80 -	1.31	0.93	MILLI/MJ	7.19	
COBALT	MICRO	2.30		1.60 -	3.00	2.23	MICRO/MJ	17.23	
COPPER	MILLI	0.12		0.05 -	0.17	0.12	MILLI/MJ	0.90	
ZINC	MILLI	0.12		0.09 -	0.14	0.12	MILLI/MJ	0.90	
NICKEL	MICRO	6.00		- -	-	5.82	MICRO/MJ	44.95	
CHROMIUM	MICRO	1.00		0.50 -	2.00	0.97	MICRO/MJ	7.49	
MOLYBDENUM	MICRO	9.00		- -	-	8.73	MICRO/MJ	67.43	
PHOSPHORUS	MILLI	29.00		24.00 -	38.00	28.13	MILLI/MJ	217.27	
CHLORIDE	MILLI	14.00		- -	-	13.58	MILLI/MJ	104.89	
FLUORIDE	MICRO	24.00		18.00 -	30.00	23.28	MICRO/MJ	179.81	
IODIDE	MICRO	1.00		- -	-	0.97	MICRO/MJ	7.49	
BORON	MICRO	88.00		63.00 -	113.00	85.36	MICRO/MJ	659.30	
SILICON	MILLI	2.00		1.00 -	3.00	1.94	MILLI/MJ	14.98	
CAROTENE	MICRO	49.00		36.00 -	64.00	47.53	MICRO/MJ	367.11	
VITAMIN E ACTIVITY	MILLI	0.12		- -	-	0.12	MILLI/MJ	0.90	

1 METHOD OF MEUSER, SUCKOW AND KULIKOWSKI ("BERLINER METHODE")

CONSTITUENTS	DIM	AV	VARIATION			AVR	NUTR. DENS.		MOLPERC.
ALPHA-TOCOPHEROL	MILLI	0.12	-	-	-	0.12	MILLI/MJ	0.90	
VITAMIN K	MICRO	13.00	-	-	-	12.61	MICRO/MJ	97.40	
VITAMIN B1	MICRO	31.00	20.00	-	40.00	30.07	MICRO/MJ	232.25	
VITAMIN B2	MICRO	54.00	30.00	-	70.00	52.38	MICRO/MJ	404.57	
NICOTINAMIDE	MILLI	0.51	0.19	-	1.11	0.49	MILLI/MJ	3.82	
PANTOTHENIC ACID	MILLI	0.30	0.29	-	0.32	0.29	MILLI/MJ	2.25	
VITAMIN B6	MICRO	60.00	49.00	-	90.00	58.20	MICRO/MJ	449.52	
BIOTIN	MICRO	4.00	-	-	-	3.88	MICRO/MJ	29.97	
FOLIC ACID	MICRO	16.00	-	-	-	15.52	MICRO/MJ	119.87	
VITAMIN C	MILLI	64.00	45.00	-	94.00	62.08	MILLI/MJ	479.49	
ALANINE	MILLI	44.00	-	-	-	42.68	MILLI/MJ	329.65	8.7
ARGININE	MILLI	37.00	-	-	-	35.89	MILLI/MJ	277.21	3.7
ASPARTIC ACID	MILLI	191.00	-	-	-	185.27	GRAM/MJ	1.43	25.3
CYSTINE	MILLI	7.00	-	-	-	6.79	MILLI/MJ	52.44	0.5
GLUTAMIC ACID	MILLI	126.00	-	-	-	122.22	MILLI/MJ	944.00	15.1
GLYCINE	MILLI	34.00	-	-	-	32.98	MILLI/MJ	254.73	8.0
HISTIDINE	MILLI	16.00	-	-	-	15.52	MILLI/MJ	119.87	1.8
ISOLEUCINE	MILLI	19.00	-	-	-	18.43	MILLI/MJ	142.35	2.6
LEUCINE	MILLI	44.00	-	-	-	42.68	MILLI/MJ	329.65	5.9
LYSINE	MILLI	34.00	-	-	-	32.98	MILLI/MJ	254.73	4.1
METHIONINE	MILLI	1.00	-	-	-	0.97	MILLI/MJ	7.49	0.1
PHENYLALANINE	MILLI	25.00	-	-	-	24.25	MILLI/MJ	187.30	2.7
PROLINE	MILLI	27.00	-	-	-	26.19	MILLI/MJ	202.29	4.1
SERINE	MILLI	33.00	-	-	-	32.01	MILLI/MJ	247.24	5.5
THREONINE	MILLI	26.00	-	-	-	25.22	MILLI/MJ	194.79	3.9
TRYPTOPHAN	MILLI	15.00	-	-	-	14.55	MILLI/MJ	112.38	1.3
TYROSINE	MILLI	29.00	-	-	-	28.13	MILLI/MJ	217.27	2.8
VALINE	MILLI	25.00	-	-	-	24.25	MILLI/MJ	187.30	3.8
MALIC ACID	GRAM	0.14	0.09	-	0.17	0.14	GRAM/MJ	1.05	
CITRIC ACID	GRAM	0.87	0.67	-	0.94	0.84	GRAM/MJ	6.52	
OXALIC ACID TOTAL	MILLI	15.80	-	-	-	15.33	MILLI/MJ	118.37	
OXALIC ACID SOLUBLE	MILLI	9.90	-	-	-	9.60	MILLI/MJ	74.17	
CHLOROGENIC ACID	MILLI	3.10	-	-	-	3.01	MILLI/MJ	23.23	
QUINIC ACID	MILLI	-	10.00	-	80.00				
PARA-COUMARIC ACID	MILLI	1.50	-	-	-	1.46	MILLI/MJ	11.24	
GALLIC ACID	MILLI	-	0.00	-	5.00				
SALICYLIC ACID	MILLI	1.40	1.40	-	1.40	1.36	MILLI/MJ	10.49	
GLUCOSE	GRAM	2.19	1.90	-	2.33	2.13	GRAM/MJ	16.42	
FRUCTOSE	GRAM	2.30	2.13	-	2.80	2.23	GRAM/MJ	17.25	
SUCROSE	GRAM	0.92	0.08	-	1.11	0.90	GRAM/MJ	6.94	
XYLOSE	MILLI	15.00	-	-	-	14.55	MILLI/MJ	112.38	
PECTIN	GRAM	0.81	0.50	-	1.36	0.79	GRAM/MJ	6.07	
CELLULOSE	GRAM	0.33	-	-	-	0.32	GRAM/MJ	2.47	
SORBITOL	MILLI	32.00	-	-	-	31.04	MILLI/MJ	239.75	
XYLITOL	MILLI	28.00	-	-	-	27.16	MILLI/MJ	209.78	
PALMITIC ACID	MILLI	36.00	36.00	-	36.00	34.92	MILLI/MJ	269.71	
OLEIC ACID	MILLI	60.00	60.00	-	60.00	58.20	MILLI/MJ	449.52	
LINOLEIC ACID	MILLI	132.00	132.00	-	132.00	128.04	MILLI/MJ	988.95	
LINOLENIC ACID	MILLI	112.00	112.00	-	112.00	108.64	MILLI/MJ	839.11	
TOTAL STEROLS	MILLI	12.00	-	-	-	11.64	MILLI/MJ	89.90	
CAMPESTEROL	-	TRACES	-	-	-				
BETA-SITOSTEROL	MILLI	10.00	-	-	-	9.70	MILLI/MJ	74.92	
STIGMASTEROL	-	TRACES	-	-	-				
TOTAL PURINES	MILLI	12.00 [2]	-	-	-	11.64	MILLI/MJ	89.90	
DIETARY FIBRE,WAT.SOL.	GRAM	0.50	0.20	-	0.80	0.49	GRAM/MJ	3.75	
DIETARY FIBRE,WAT.INS.	GRAM	1.50	1.10	-	1.90	1.46	GRAM/MJ	11.24	
TANNINS	GRAM	0.22	-	-	-	0.21	GRAM/MJ	1.65	

2 VALUE EXPRESSED IN MG URIC ACID/100G

ERDBEEREN IN DOSEN	**STRAWBERRIES** CANNED	**FRAISES** EN BOÎTES
1	1	1

		PROTEIN	FAT	CARBOHYDRATES	TOTAL
ENERGY VALUE (AVERAGE)	KJOULE	9.4	6.6	303	319
PER 100 G	(KCAL)	2.3	1.6	72	76
EDIBLE PORTION					
AMOUNT OF DIGESTIBLE	GRAM	0.48	0.15	18.10	
CONSTITUENTS PER 100 G					
ENERGY VALUE (AVERAGE)	KJOULE	8.0	6.0	303	317
OF THE DIGESTIBLE	(KCAL)	1.9	1.4	72	76
FRACTION PER 100 G					
EDIBLE PORTION					

WASTE PERCENTAGE AVERAGE 0.00

CONSTITUENTS	DIM	AV	VARIATION			AVR	NUTR. DENS.		MOLPERC.
WATER	GRAM	79.80	75.90	-	83.90	79.80	GRAM/MJ	251.83	
PROTEIN	GRAM	0.57	0.50	-	0.63	0.57	GRAM/MJ	1.80	
FAT	GRAM	0.17	0.14	-	0.20	0.17	GRAM/MJ	0.54	
AVAILABLE CARBOHYDR.	GRAM	18.10	14.00	-	22.00	18.10	GRAM/MJ	57.12	
MINERALS	GRAM	0.30	0.25	-	0.34	0.30	GRAM/MJ	0.95	
SODIUM	MILLI	3.60	2.00	-	5.10	3.60	MILLI/MJ	11.36	
POTASSIUM	MILLI	107.00	98.00	-	115.00	107.00	MILLI/MJ	337.66	
MAGNESIUM	MILLI	22.00	14.00	-	29.00	22.00	MILLI/MJ	69.43	
CALCIUM	MILLI	7.00	2.00	-	11.00	7.00	MILLI/MJ	22.09	
MANGANESE	MILLI	0.19	-	-	-	0.19	MILLI/MJ	0.60	
IRON	MILLI	1.90	0.80	-	2.90	1.90	MILLI/MJ	6.00	
COPPER	MICRO	40.00	-	-	-	40.00	MICRO/MJ	126.23	
ZINC	MILLI	0.20	-	-	-	0.20	MILLI/MJ	0.63	
PHOSPHORUS	MILLI	25.00	16.00	-	33.00	25.00	MILLI/MJ	78.89	
CHLORIDE	MILLI	5.00	-	-	-	5.00	MILLI/MJ	15.78	
VITAMIN B1	MICRO	10.00	0.00	-	10.00	10.00	MICRO/MJ	31.56	
VITAMIN B2	MICRO	30.00	20.00	-	40.00	30.00	MICRO/MJ	94.67	
NICOTINAMIDE	MILLI	0.30	0.17	-	0.39	0.30	MILLI/MJ	0.95	
PANTOTHENIC ACID	MILLI	0.21	-	-	-	0.21	MILLI/MJ	0.66	
VITAMIN B6	MICRO	30.00	-	-	-	30.00	MICRO/MJ	94.67	
BIOTIN	MICRO	1.00	-	-	-	1.00	MICRO/MJ	3.16	
FOLIC ACID	MICRO	12.00	-	-	-	12.00	MICRO/MJ	37.87	
VITAMIN C	MILLI	30.00	15.00	-	55.00	30.00	MILLI/MJ	94.67	
REDUCING SUGAR	GRAM	9.47	4.65	-	12.90	9.47	GRAM/MJ	29.88	
SUCROSE	GRAM	8.65	3.53	-	12.60	8.65	GRAM/MJ	27.30	

1 TOTAL CONTENT (SOLID AND LIQUID)

HEIDELBEERE (BLAUBEERE, BICKBEERE) — BILBERRY (BLUE BERRY, HUCKLE BERRY) — MYRTILLE

VACCINIUM MYRTILLUS L.

		PROTEIN	FAT	CARBOHYDRATES	TOTAL
ENERGY VALUE (AVERAGE) PER 100 G EDIBLE PORTION	KJOULE	9.9	23	123	156
	(KCAL)	2.4	5.6	29	37
AMOUNT OF DIGESTIBLE CONSTITUENTS PER 100 G	GRAM	0.51	0.54	7.36	
ENERGY VALUE (AVERAGE) OF THE DIGESTIBLE FRACTION PER 100 G EDIBLE PORTION	KJOULE	8.4	21	123	153
	(KCAL)	2.0	5.0	29	36

WASTE PERCENTAGE AVERAGE 0.00

CONSTITUENTS	DIM	AV		VARIATION			AVR	NUTR. DENS.		MOLPERC.
WATER	GRAM	84.61		-	-	-	82.07	GRAM/MJ	554.39	
PROTEIN	GRAM	0.60		0.60	-	0.78	0.58	GRAM/MJ	3.93	
FAT	GRAM	0.60		-	-	-	0.58	GRAM/MJ	3.93	
AVAILABLE CARBOHYDR.	GRAM	7.36		-	-	-	7.14	GRAM/MJ	48.23	
TOTAL DIETARY FIBRE	GRAM	4.90	1	-	-	-	4.75	GRAM/MJ	32.11	
MINERALS	GRAM	0.30		0.30	-	0.37	0.29	GRAM/MJ	1.97	
SODIUM	MILLI	1.00		0.30	-	2.00	0.97	MILLI/MJ	6.55	
POTASSIUM	MILLI	65.00		31.00	-	99.00	63.05	MILLI/MJ	425.90	
MAGNESIUM	MILLI	2.40		-	-	-	2.33	MILLI/MJ	15.73	
CALCIUM	MILLI	10.00		3.00	-	16.00	9.70	MILLI/MJ	65.52	
MANGANESE	MILLI	-		0.30	-	4.80				
IRON	MILLI	0.74		0.40	-	1.00	0.72	MILLI/MJ	4.85	
COPPER	MILLI	0.11		-	-	-	0.11	MILLI/MJ	0.72	
ZINC	MILLI	0.10		-	-	-	0.10	MILLI/MJ	0.66	
NICKEL	MICRO	10.00		-	-	-	9.70	MICRO/MJ	65.52	
VANADIUM	MICRO	0.16		-	-	-	0.16	MICRO/MJ	1.05	
PHOSPHORUS	MILLI	13.00		-	-	-	12.61	MILLI/MJ	85.18	
CHLORIDE	MILLI	5.00		-	-	-	4.85	MILLI/MJ	32.76	
FLUORIDE	MICRO	2.00		-	-	-	1.94	MICRO/MJ	13.10	
BORON	MICRO	50.00		32.00	-	67.00	48.50	MICRO/MJ	327.62	
SILICON	MICRO	5.00		-	-	-	4.85	MICRO/MJ	32.76	
CAROTENE	MILLI	0.13		0.06	-	0.17	0.13	MILLI/MJ	0.85	
VITAMIN B1	MICRO	20.00		20.00	-	30.00	19.40	MICRO/MJ	131.05	
VITAMIN B2	MICRO	20.00		20.00	-	30.00	19.40	MICRO/MJ	131.05	
NICOTINAMIDE	MILLI	0.40		0.30	-	0.40	0.39	MILLI/MJ	2.62	
PANTOTHENIC ACID	MILLI	0.16		-	-	-	0.16	MILLI/MJ	1.05	

1 METHOD OF MEUSER, SUCKOW AND KULIKOWSKI ("BERLINER METHODE")

CONSTITUENTS	DIM	AV	VARIATION			AVR	NUTR. DENS.		MOLPERC.
VITAMIN B6	MICRO	60.00	30.00	-	101.00	58.20	MICRO/MJ	393.14	
BIOTIN	MICRO	1.10	-	-	-	1.07	MICRO/MJ	7.21	
FOLIC ACID	MICRO	6.00	-	-	-	5.82	MICRO/MJ	39.31	
VITAMIN C	MILLI	22.00	10.00	-	44.00	21.34	MILLI/MJ	144.15	
MALIC ACID	MILLI	850.00	-	-	-	824.50	GRAM/MJ	5.57	
CITRIC ACID	MILLI	523.00	-	-	-	507.31	GRAM/MJ	3.43	
TARTARIC ACID	MILLI	0.22	-	-	-	0.21	MILLI/MJ	1.44	
OXALIC ACID	-	0.00	-	-	-	0.00			
CHLOROGENIC ACID	MILLI	121.00	-	-	-	117.37	MILLI/MJ	792.83	
QUINIC ACID	MILLI	33.00	-	-	-	32.01	MILLI/MJ	216.23	
GLUCOSE	GRAM	2.46	2.30	-	5.00	2.39	GRAM/MJ	16.15	
FRUCTOSE	GRAM	3.34	3.11	-	5.20	3.24	GRAM/MJ	21.92	
SUCROSE	GRAM	0.19	0.00	-	0.25	0.19	GRAM/MJ	1.29	
XYLOSE	MILLI	21.30	-	-	-	20.66	MILLI/MJ	139.56	
PENTOSAN	GRAM	1.08	0.76	-	1.28	1.05	GRAM/MJ	7.08	
SORBITOL	MILLI	4.30	-	-	-	4.17	MILLI/MJ	28.17	
XYLITOL	MILLI	2.10	-	-	-	2.04	MILLI/MJ	13.76	
DIETARY FIBRE,WAT.SOL.	GRAM	1.40	-	-	-	1.36	GRAM/MJ	9.17	
DIETARY FIBRE,WAT.INS.	GRAM	3.50	-	-	-	3.40	GRAM/MJ	22.93	
TANNINS	GRAM	0.22	-	-	-	0.21	GRAM/MJ	1.44	

HEIDELBEEREN (BLAUBEEREN) IN DOSEN [1]	BILBERRIES (BLUE BERRIES) CANNED [1]	MYRTILLES EN BOÎTES [1]

		PROTEIN	FAT	CARBOHYDRATES	TOTAL
ENERGY VALUE (AVERAGE) PER 100 G EDIBLE PORTION	KJOULE	15	23	306	344
	(KCAL)	3.5	5.6	73	82
AMOUNT OF DIGESTIBLE CONSTITUENTS PER 100 G	GRAM	0.74	0.54	18.30	
ENERGY VALUE (AVERAGE) OF THE DIGESTIBLE FRACTION PER 100 G EDIBLE PORTION	KJOULE	12	21	306	340
	(KCAL)	3.0	5.0	73	81

WASTE PERCENTAGE AVERAGE 0.00

CONSTITUENTS	DIM	AV	VARIATION			AVR	NUTR. DENS.		MOLPERC.
WATER	GRAM	76.80	72.40	-	82.30	76.80	GRAM/MJ	226.12	
PROTEIN	GRAM	0.88	0.86	-	0.90	0.88	GRAM/MJ	2.59	
FAT	GRAM	0.60	0.46	-	0.74	0.60	GRAM/MJ	1.77	
AVAILABLE CARBOHYDR.	GRAM	18.30	-	-	-	18.30	GRAM/MJ	53.88	
MINERALS	GRAM	0.25	0.19	-	0.30	0.25	GRAM/MJ	0.74	
SODIUM	MILLI	3.80	3.00	-	4.70	3.80	MILLI/MJ	11.19	
POTASSIUM	MILLI	59.00	51.00	-	66.00	59.00	MILLI/MJ	173.71	
MAGNESIUM	MILLI	3.60	2.20	-	5.00	3.60	MILLI/MJ	10.60	
CALCIUM	MILLI	12.00	11.00	-	13.00	12.00	MILLI/MJ	35.33	
MANGANESE	MILLI	1.90	-	-	-	1.90	MILLI/MJ	5.59	
IRON	MILLI	2.60	1.70	-	3.50	2.60	MILLI/MJ	7.66	
COPPER	MILLI	0.39	0.13	-	0.77	0.39	MILLI/MJ	1.15	
PHOSPHORUS	MILLI	16.00	14.00	-	17.00	16.00	MILLI/MJ	47.11	
CHLORIDE	MILLI	2.00	-	-	-	2.00	MILLI/MJ	5.89	
VITAMIN B1	MICRO	10.00	0.00	-	10.00	10.00	MICRO/MJ	29.44	
VITAMIN B2	MICRO	20.00	10.00	-	30.00	20.00	MICRO/MJ	58.89	
NICOTINAMIDE	MILLI	0.39	0.22	-	0.60	0.39	MILLI/MJ	1.15	
VITAMIN B6	MICRO	20.00	-	-	-	20.00	MICRO/MJ	58.89	
FOLIC ACID	MICRO	15.00	-	-	-	15.00	MICRO/MJ	44.16	
GLUCOSE	GRAM	8.34	5.36	-	10.60	8.34	GRAM/MJ	24.56	
FRUCTOSE	GRAM	9.04	5.92	-	12.40	9.04	GRAM/MJ	26.62	
SUCROSE	GRAM	0.93	0.66	-	1.47	0.93	GRAM/MJ	2.74	

1 TOTAL CONTENT (SOLID AND LIQUID)

HEIDELBEEREN	BILBERRIES	MYRTILLES
(BLAUBEEREN, DUNSTHEIDELBEEREN, DUNSTBLAUBEEREN) IN DOSEN, OHNE ZUCKERZUSATZ 1	(BLUE BERRIES, STEWED BILBERRIES) CANNED, UNSWEETENED 1	EN BOÎTES SANS SUCRE (À LA VAPEUR) 1

		PROTEIN	FAT	CARBOHYDRATES	TOTAL
ENERGY VALUE (AVERAGE)	KJOULE	6.6	16	80	102
PER 100 G	(KCAL)	1.6	3.7	19	24
EDIBLE PORTION					
AMOUNT OF DIGESTIBLE	GRAM	0.34	0.36	4.78	
CONSTITUENTS PER 100 G					
ENERGY VALUE (AVERAGE)	KJOULE	5.6	14	80	100
OF THE DIGESTIBLE	(KCAL)	1.3	3.3	19	24
FRACTION PER 100 G					
EDIBLE PORTION					

WASTE PERCENTAGE AVERAGE 0.00

CONSTITUENTS	DIM	AV		VARIATION	AVR	NUTR. DENS.		MOLPERC.
WATER	GRAM	90.00		- - -	90.00	GRAM/MJ	903.39	
PROTEIN	GRAM	0.40		- - -	0.40	GRAM/MJ	4.02	
FAT	GRAM	0.40		- - -	0.40	GRAM/MJ	4.02	
AVAILABLE CARBOHYDR.	GRAM	4.78	2	- - -	4.78	GRAM/MJ	47.98	
MINERALS	GRAM	0.20		- - -	0.20	GRAM/MJ	2.01	
CALCIUM	MILLI	11.00		- - -	11.00	MILLI/MJ	110.41	
PHOSPHORUS	MILLI	6.00		- - -	6.00	MILLI/MJ	60.23	
CAROTENE	MICRO	24.00		- - -	24.00	MICRO/MJ	240.90	
VITAMIN B1	MICRO	12.00		- - -	12.00	MICRO/MJ	120.45	
VITAMIN B2	MICRO	12.00		- - -	12.00	MICRO/MJ	120.45	
NICOTINAMIDE	MILLI	0.20		- - -	0.20	MILLI/MJ	2.01	
VITAMIN C	MILLI	12.00		9.00 - 14.00	12.00	MILLI/MJ	120.45	

1 TOTAL CONTENT (SOLID AND LIQUID)

2 CALCULATED FROM THE FRESH PRODUCT

HIMBEERE RASPBERRY FRAMBOISE

RUBUS IDAEUS L.

		PROTEIN	FAT	CARBOHYDRATES	TOTAL
ENERGY VALUE (AVERAGE) PER 100 G EDIBLE PORTION	KJOULE	21	12	116	149
	(KCAL)	5.1	2.8	28	36
AMOUNT OF DIGESTIBLE CONSTITUENTS PER 100 G	GRAM	1.10	0.27	6.92	
ENERGY VALUE (AVERAGE) OF THE DIGESTIBLE FRACTION PER 100 G EDIBLE PORTION	KJOULE	18	11	116	145
	(KCAL)	4.4	2.5	28	35

WASTE PERCENTAGE AVERAGE 0.00

CONSTITUENTS	DIM	AV		VARIATION		AVR	NUTR.	DENS.	MOLPERC.
WATER	GRAM	84.50	84.00	-	86.00	84.50	GRAM/MJ	584.45	
PROTEIN	GRAM	1.30	1.20	-	1.40	1.30	GRAM/MJ	8.99	
FAT	GRAM	0.30	0.20	-	0.40	0.30	GRAM/MJ	2.07	
AVAILABLE CARBOHYDR.	GRAM	6.92	4.14	-	10.34	6.92	GRAM/MJ	47.86	
TOTAL DIETARY FIBRE	GRAM	4.68 1	-	-	-	4.68	GRAM/MJ	32.37	
MINERALS	GRAM	0.51	0.37	-	0.58	0.51	GRAM/MJ	3.53	
SODIUM	MILLI	-	0.50	-	2.00				
POTASSIUM	MILLI	170.00	130.00	-	200.00	170.00	GRAM/MJ	1.18	
MAGNESIUM	MILLI	30.00	-	-	-	30.00	MILLI/MJ	207.50	
CALCIUM	MILLI	40.00	-	-	-	40.00	MILLI/MJ	276.66	
IRON	MILLI	1.00	0.90	-	1.00	1.00	MILLI/MJ	6.92	
COPPER	MILLI	0.14	-	-	-	0.14	MILLI/MJ	0.97	
ZINC	MILLI	0.53	0.53	-	0.53	0.53	MILLI/MJ	3.67	
CHROMIUM	MICRO	3.00	3.00	-	3.00	3.00	MICRO/MJ	20.75	
PHOSPHORUS	MILLI	44.00	37.00	-	50.00	44.00	MILLI/MJ	304.33	
IODIDE	MICRO	0.60	0.60	-	0.60	0.60	MICRO/MJ	4.15	
BORON	MICRO	98.00	71.00	-	125.00	98.00	MICRO/MJ	677.82	
CAROTENE	MICRO	80.00	-	-	-	80.00	MICRO/MJ	553.32	
VITAMIN E ACTIVITY	MILLI	0.48	-	-	-	0.48	MILLI/MJ	3.32	
TOTAL TOCOPHEROLS	MILLI	4.50	-	-	-	4.50	MILLI/MJ	31.12	
ALPHA-TOCOPHEROL	MILLI	0.30	-	-	-	0.30	MILLI/MJ	2.07	
GAMMA-TOCOPHEROL	MILLI	1.50	-	-	-	1.50	MILLI/MJ	10.37	
DELTA-TOCOPHEROL	MILLI	2.70	-	-	-	2.70	MILLI/MJ	18.67	
VITAMIN B1	MICRO	23.00	10.00	-	30.00	23.00	MICRO/MJ	159.08	
VITAMIN B2	MICRO	50.00	40.00	-	60.00	50.00	MICRO/MJ	345.83	
NICOTINAMIDE	MILLI	0.30	0.20	-	0.50	0.30	MILLI/MJ	2.07	

1 METHOD OF MEUSER, SUCKOW AND KULIKOWSKI ("BERLINER METHODE")

CONSTITUENTS	DIM	AV	VARIATION			AVR	NUTR. DENS.		MOLPERC.
PANTOTHENIC ACID	MILLI	0.30	-	-	-	0.30	MILLI/MJ	2.07	
VITAMIN B6	MICRO	75.00	60.00	-	90.00	75.00	MICRO/MJ	518.74	
VITAMIN C	MILLI	25.00	16.00	-	30.00	25.00	MILLI/MJ	172.91	
TYRAMINE	MILLI	-	1.00	-	9.00				
MALIC ACID	GRAM	0.40	0.00	-	0.80	0.40	GRAM/MJ	2.77	
CITRIC ACID	GRAM	1.72	1.06	-	2.48	1.72	GRAM/MJ	11.90	
OXALIC ACID TOTAL	MILLI	16.40	-	-	-	16.40	MILLI/MJ	113.43	
OXALIC ACID SOLUBLE	MILLI	11.30	-	-	-	11.30	MILLI/MJ	78.16	
CHLOROGENIC ACID	MILLI	3.10	-	-	-	3.10	MILLI/MJ	21.44	
QUINIC ACID	MILLI	15.00	-	-	-	15.00	MILLI/MJ	103.75	
FERULIC ACID	MILLI	1.00	0.30	-	1.70	1.00	MILLI/MJ	6.92	
CAFFEIC ACID	MILLI	0.80	0.60	-	1.00	0.80	MILLI/MJ	5.53	
PARA-COUMARIC ACID	MILLI	1.40	0.70	-	2.00	1.40	MILLI/MJ	9.68	
PROTOCATECHUIC ACID	MILLI	3.10	2.50	-	3.70	3.10	MILLI/MJ	21.44	
GALLIC ACID	MILLI	2.50	1.90	-	3.80	2.50	MILLI/MJ	17.29	
PARA-HYDROXYBENZ.ACID	MILLI	1.70	1.50	-	2.50	1.70	MILLI/MJ	11.76	
SALICYLIC ACID	MILLI	5.10	5.10	-	5.10	5.10	MILLI/MJ	35.27	
GLUCOSE	GRAM	1.78	1.13	-	2.63	1.79	GRAM/MJ	12.35	
FRUCTOSE	GRAM	2.05	1.83	-	2.92	2.05	GRAM/MJ	14.19	
SUCROSE	GRAM	0.96	0.12	-	1.51	0.97	GRAM/MJ	6.70	
XYLOSE	MILLI	12.80	-	-	-	12.80	MILLI/MJ	88.53	
PECTIN	GRAM	0.40	-	-	-	0.40	GRAM/MJ	2.77	
SORBITOL	MILLI	8.50	-	-	-	8.50	MILLI/MJ	58.79	
XYLITOL	MILLI	2.60	-	-	-	2.60	MILLI/MJ	17.98	
DIETARY FIBRE,WAT.SOL.	GRAM	0.98	-	-	-	0.98	GRAM/MJ	6.78	
DIETARY FIBRE,WAT.INS.	GRAM	3.70	-	-	-	3.70	GRAM/MJ	25.59	

HIMBEEREN	**RASPBERRIES**	**FRAMBOISES**
IN DOSEN	CANNED	EN BOÎTES
1	1	1

		PROTEIN	FAT	CARBOHYDRATES	TOTAL
ENERGY VALUE (AVERAGE) PER 100 G EDIBLE PORTION	KJOULE	12	10	338	359
	(KCAL)	2.8	2.4	81	86
AMOUNT OF DIGESTIBLE CONSTITUENTS PER 100 G	GRAM	0.59	0.23	20.18	
ENERGY VALUE (AVERAGE) OF THE DIGESTIBLE FRACTION PER 100 G EDIBLE PORTION	KJOULE	9.8	9.1	338	357
	(KCAL)	2.4	2.2	81	85

WASTE PERCENTAGE AVERAGE 0.00

CONSTITUENTS	DIM	AV	VARIATION			AVR	NUTR. DENS.		MOLPERC.
WATER	GRAM	78.60	74.60	-	82.60	78.60	GRAM/MJ	220.37	
PROTEIN	GRAM	0.70	0.60	-	0.80	0.70	GRAM/MJ	1.96	
FAT	GRAM	0.26	-	-	-	0.26	GRAM/MJ	0.73	
AVAILABLE CARBOHYDR.	GRAM	20.18 [2]	-	-	-	20.18	GRAM/MJ	56.58	
MINERALS	GRAM	0.26	0.22	-	0.30	0.26	GRAM/MJ	0.73	
SODIUM	MILLI	7.00	4.00	-	9.00	7.00	MILLI/MJ	19.63	
POTASSIUM	MILLI	92.00	78.00	-	105.00	92.00	MILLI/MJ	257.94	
MAGNESIUM	MILLI	13.00	11.00	-	15.00	13.00	MILLI/MJ	36.45	
CALCIUM	MILLI	18.00	14.00	-	21.00	18.00	MILLI/MJ	50.47	
MANGANESE	MILLI	0.12	-	-	-	0.12	MILLI/MJ	0.34	
IRON	MILLI	1.80	1.70	-	1.80	1.80	MILLI/MJ	5.05	
COPPER	MILLI	0.10	0.02	-	0.28	0.10	MILLI/MJ	0.28	
PHOSPHORUS	MILLI	13.00	11.00	-	14.00	13.00	MILLI/MJ	36.45	
CHLORIDE	MILLI	5.00	-	-	-	5.00	MILLI/MJ	14.02	
CAROTENE	MICRO	75.00	-	-	-	75.00	MICRO/MJ	210.28	
VITAMIN B1	MICRO	10.00	0.00	-	10.00	10.00	MICRO/MJ	28.04	
VITAMIN B2	MICRO	60.00	10.00	-	80.00	60.00	MICRO/MJ	168.22	
NICOTINAMIDE	MILLI	0.27	0.23	-	0.29	0.27	MILLI/MJ	0.76	
PANTOTHENIC ACID	MILLI	0.17	-	-	-	0.17	MILLI/MJ	0.48	
VITAMIN B6	MICRO	40.00	-	-	-	40.00	MICRO/MJ	112.15	
FOLIC ACID	MICRO	13.00	-	-	-	13.00	MICRO/MJ	36.45	
VITAMIN C	MILLI	5.00	2.00	-	9.00	5.00	MILLI/MJ	14.02	

1 TOTAL CONTENT (SOLID AND LIQUID)

2 ESTIMATED BY THE DIFFERENCE METHOD
100 - (WATER + PROTEIN + FAT + MINERALS)

JOHANNISBEERE ROT | RED-CURRANT | GROSEILLE ROUGE

RIBES RUBRUM L.

		PROTEIN	FAT	CARBOHYDRATES	TOTAL
ENERGY VALUE (AVERAGE)	KJOULE	19	7.8	125	151
PER 100 G	(KCAL)	4.5	1.9	30	36
EDIBLE PORTION					
AMOUNT OF DIGESTIBLE	GRAM	0.96	0.18	7.44	
CONSTITUENTS PER 100 G					
ENERGY VALUE (AVERAGE)	KJOULE	16	7.0	125	147
OF THE DIGESTIBLE	(KCAL)	3.8	1.7	30	35
FRACTION PER 100 G					
EDIBLE PORTION					

WASTE PERCENTAGE AVERAGE 2.0 MINIMUM 1.0 MAXIMUM 3.0

CONSTITUENTS	DIM	AV		VARIATION			AVR	NUTR. DENS.		MOLPERC.
WATER	GRAM	84.70		81.40	-	89.60	83.01	GRAM/MJ	574.65	
PROTEIN	GRAM	1.13		0.90	-	1.40	1.11	GRAM/MJ	7.67	
FAT	GRAM	0.20		0.10	-	0.40	0.20	GRAM/MJ	1.36	
AVAILABLE CARBOHYDR.	GRAM	7.44		4.82	-	9.04	7.29	GRAM/MJ	50.48	
TOTAL DIETARY FIBRE	GRAM	3.50	1	-	-	-	3.43	GRAM/MJ	23.75	
MINERALS	GRAM	0.63		0.51	-	0.71	0.62	GRAM/MJ	4.27	
SODIUM	MILLI	1.40		0.30	-	2.30	1.37	MILLI/MJ	9.50	
POTASSIUM	MILLI	238.00		156.00	-	278.00	233.24	GRAM/MJ	1.61	
MAGNESIUM	MILLI	13.00		8.00	-	17.00	12.74	MILLI/MJ	88.20	
CALCIUM	MILLI	29.00		17.00	-	37.00	28.42	MILLI/MJ	196.75	
MANGANESE	MILLI	0.60		0.20	-	1.00	0.59	MILLI/MJ	4.07	
IRON	MILLI	0.91		0.53	-	1.22	0.89	MILLI/MJ	6.17	
COPPER	MILLI	0.10		0.07	-	0.12	0.10	MILLI/MJ	0.68	
ZINC	MILLI	0.20		-	-	-	0.20	MILLI/MJ	1.36	
NICKEL	MICRO	5.00		3.00	-	10.00	4.90	MICRO/MJ	33.92	
CHROMIUM	MICRO	2.00		1.00	-	2.00	1.96	MICRO/MJ	13.57	
MOLYBDENUM	MICRO	10.00		-	-	-	9.80	MICRO/MJ	67.85	
PHOSPHORUS	MILLI	27.00		18.00	-	33.00	26.46	MILLI/MJ	183.18	
CHLORIDE	MILLI	14.00		-	-	-	13.72	MILLI/MJ	94.98	
FLUORIDE	MICRO	23.00		12.00	-	33.00	22.54	MICRO/MJ	156.04	
IODIDE	MICRO	1.00		-	-	-	0.98	MICRO/MJ	6.78	
BORON	MILLI	0.18		0.13	-	0.21	0.18	MILLI/MJ	1.22	
SELENIUM	MICRO	0.10		0.10	-	0.20	0.10	MICRO/MJ	0.68	
SILICON	MILLI	2.00		1.00	-	5.00	1.96	MILLI/MJ	13.57	
CAROTENE	MICRO	38.00		18.00	-	72.00	37.24	MICRO/MJ	257.81	
VITAMIN E ACTIVITY	MILLI	0.21		0.10	-	0.32	0.21	MILLI/MJ	1.42	

1 METHOD OF MEUSER, SUCKOW AND KULIKOWSKI ("BERLINER METHODE")

CONSTITUENTS	DIM	AV	VARIATION			AVR	NUTR. DENS.		MOLPERC.
ALPHA-TOCOPHEROL	MILLI	0.21	-	-	-	0.21	MILLI/MJ	1.42	
VITAMIN B1	MICRO	40.00	25.00	-	60.00	39.20	MICRO/MJ	271.38	
VITAMIN B2	MICRO	30.00	20.00	-	40.00	29.40	MICRO/MJ	203.54	
NICOTINAMIDE	MILLI	0.23	0.10	-	0.30	0.23	MILLI/MJ	1.56	
PANTOTHENIC ACID	MICRO	60.00	-	-	-	58.80	MICRO/MJ	407.07	
VITAMIN B6	MICRO	45.00	28.00	-	58.00	44.10	MICRO/MJ	305.30	
BIOTIN	MICRO	2.60	-	-	-	2.55	MICRO/MJ	17.64	
VITAMIN C	MILLI	36.00	26.00	-	47.00	35.28	MILLI/MJ	244.24	
MALIC ACID	GRAM	0.29	0.24	-	0.34	0.28	GRAM/MJ	1.97	
CITRIC ACID	GRAM	2.07	1.84	-	2.30	2.03	GRAM/MJ	14.04	
OXALIC ACID TOTAL	MILLI	9.90	-	-	-	9.70	MILLI/MJ	67.17	
OXALIC ACID SOLUBLE	MILLI	2.20	-	-	-	2.16	MILLI/MJ	14.93	
CHLOROGENIC ACID	MILLI	2.70	-	-	-	2.65	MILLI/MJ	18.32	
QUINIC ACID	MILLI	15.00	11.00	-	19.00	14.70	MILLI/MJ	101.77	
PARA-HYDROXYBENZ.ACID	MILLI	1.80	1.00	-	2.50	1.76	MILLI/MJ	12.21	
SALICYLIC ACID	MILLI	5.10	5.10	-	5.10	5.00	MILLI/MJ	34.60	
GLUCOSE	GRAM	2.13	0.73	-	2.90	2.09	GRAM/MJ	14.49	
FRUCTOSE	GRAM	2.57	1.87	-	3.00	2.52	GRAM/MJ	17.46	
SUCROSE	GRAM	0.38	0.14	-	0.50	0.38	GRAM/MJ	2.62	
PECTIN	GRAM	0.93	0.70	-	1.16	0.91	GRAM/MJ	6.31	
CELLULOSE	GRAM	0.88	-	-	-	0.86	GRAM/MJ	5.97	
PALMITIC ACID	MILLI	32.00	32.00	-	32.00	31.36	MILLI/MJ	217.11	
STEARIC ACID	MILLI	7.00	7.00	-	7.00	6.86	MILLI/MJ	47.49	
PALMITOLEIC ACID	MILLI	3.00	3.00	-	3.00	2.94	MILLI/MJ	20.35	
OLEIC ACID	MILLI	26.00	26.00	-	26.00	25.48	MILLI/MJ	176.40	
LINOLEIC ACID	MILLI	41.00	41.00	-	41.00	40.18	MILLI/MJ	278.17	
LINOLENIC ACID	MILLI	32.00	32.00	-	32.00	31.36	MILLI/MJ	217.11	
DIETARY FIBRE, WAT.SOL.	GRAM	0.50	0.30	-	0.60	0.49	GRAM/MJ	3.39	
DIETARY FIBRE, WAT.INS.	GRAM	3.00	-	-	-	2.94	GRAM/MJ	20.35	

JOHANNISBEERE SCHWARZ — BLACK-CURRANT — CASSIS

RIBES NIGRUM L.

		PROTEIN	FAT	CARBOHYDRATES	TOTAL
ENERGY VALUE (AVERAGE)	KJOULE	21	8.6	167	196
PER 100 G	(KCAL)	5.1	2.0	40	47
EDIBLE PORTION					
AMOUNT OF DIGESTIBLE	GRAM	1.08	0.19	9.96	
CONSTITUENTS PER 100 G					
ENERGY VALUE (AVERAGE)	KJOULE	18	7.7	167	192
OF THE DIGESTIBLE	(KCAL)	4.3	1.8	40	46
FRACTION PER 100 G					
EDIBLE PORTION					

WASTE PERCENTAGE AVERAGE 2.0 MINIMUM 2.0 MAXIMUM 3.0

CONSTITUENTS	DIM	AV		VARIATION			AVR	NUTR. DENS.		MOLPERC.
WATER	GRAM	81.30		77.40	-	84.70	79.67	GRAM/MJ	422.61	
PROTEIN	GRAM	1.28		0.90	-	1.70	1.25	GRAM/MJ	6.65	
FAT	GRAM	0.22		0.10	-	0.50	0.22	GRAM/MJ	1.14	
AVAILABLE CARBOHYDR.	GRAM	9.96		8.52	-	14.57	9.76	GRAM/MJ	51.77	
TOTAL DIETARY FIBRE	GRAM	6.80	1	-	-	-	6.66	GRAM/MJ	35.35	
MINERALS	GRAM	0.80		0.61	-	1.10	0.78	GRAM/MJ	4.16	
SODIUM	MILLI	1.50		0.30	-	2.70	1.47	MILLI/MJ	7.80	
POTASSIUM	MILLI	310.00		258.00	-	372.00	303.80	GRAM/MJ	1.61	
MAGNESIUM	MILLI	17.00		10.00	-	24.00	16.66	MILLI/MJ	88.37	
CALCIUM	MILLI	46.00		30.00	-	65.00	45.08	MILLI/MJ	239.11	
MANGANESE	MILLI	0.68		-	-	-	0.67	MILLI/MJ	3.53	
IRON	MILLI	1.29		0.90	-	2.20	1.26	MILLI/MJ	6.71	
COBALT	MICRO	0.25		0.20	-	0.50	0.25	MICRO/MJ	1.30	
COPPER	MILLI	0.11		0.07	-	0.14	0.11	MILLI/MJ	0.57	
ZINC	MILLI	0.18		0.15	-	0.20	0.18	MILLI/MJ	0.94	
NICKEL	MICRO	10.00		10.00	-	20.00	9.80	MICRO/MJ	51.98	
CHROMIUM	MICRO	3.00		1.00	-	5.00	2.94	MICRO/MJ	15.59	
PHOSPHORUS	MILLI	40.00		30.00	-	47.00	39.20	MILLI/MJ	207.93	
CHLORIDE	MILLI	15.00		-	-	-	14.70	MILLI/MJ	77.97	
FLUORIDE	MICRO	29.00		15.00	-	42.00	28.42	MICRO/MJ	150.75	
IODIDE	MICRO	1.00		-	-	-	0.98	MICRO/MJ	5.20	
BORON	MILLI	0.20		0.18	-	0.22	0.20	MILLI/MJ	1.04	
SILICON	MILLI	3.00		1.00	-	4.00	2.94	MILLI/MJ	15.59	
CAROTENE	MILLI	0.14		0.07	-	0.20	0.14	MILLI/MJ	0.73	
VITAMIN E ACTIVITY	MILLI	1.00		-	-	-	0.98	MILLI/MJ	5.20	
ALPHA-TOCOPHEROL	MILLI	1.00		-	-	-	0.98	MILLI/MJ	5.20	

1 METHOD OF SOUTHGATE

CONSTITUENTS	DIM	AV	VARIATION			AVR	NUTR. DENS.		MOLPERC.
VITAMIN B1	MICRO	51.00	25.00	-	80.00	49.98	MICRO/MJ	265.11	
VITAMIN B2	MICRO	44.00	25.00	-	60.00	43.12	MICRO/MJ	228.72	
NICOTINAMIDE	MILLI	0.28	0.20	-	0.32	0.27	MILLI/MJ	1.46	
PANTOTHENIC ACID	MILLI	0.40	-	-	-	0.39	MILLI/MJ	2.08	
VITAMIN B6	MICRO	80.00	60.00	-	100.00	78.40	MICRO/MJ	415.85	
BIOTIN	MICRO	2.40	-	-	-	2.35	MICRO/MJ	12.48	
VITAMIN B12	-	0.00	-	-	-	0.00			
VITAMIN C	MILLI	177.00	132.00	-	220.00	173.46	MILLI/MJ	920.07	
MALIC ACID	GRAM	0.41	0.35	-	0.44	0.40	GRAM/MJ	2.13	
CITRIC ACID	GRAM	2.88	2.48	-	3.11	2.82	GRAM/MJ	14.97	
QUINIC ACID	MILLI	-	21.00	-	48.00				
FERULIC ACID	MILLI	1.50	-	-	-	1.47	MILLI/MJ	7.80	
CAFFEIC ACID	MILLI	6.00	3.00	-	9.00	5.88	MILLI/MJ	31.19	
PARA-COUMARIC ACID	MILLI	9.00	6.00	-	12.00	8.82	MILLI/MJ	46.78	
PROTOCATECHUIC ACID	MILLI	2.80	1.50	-	4.00	2.74	MILLI/MJ	14.55	
SALICYLIC ACID	MILLI	3.10	3.10	-	3.10	3.04	MILLI/MJ	16.11	
GLUCOSE	GRAM	2.83	2.69	-	4.62	2.78	GRAM/MJ	14.76	
FRUCTOSE	GRAM	3.65	3.00	-	5.40	3.58	GRAM/MJ	19.01	
SUCROSE	GRAM	0.19	0.00	-	1.00	0.19	GRAM/MJ	1.00	
PECTIN	GRAM	1.70	-	-	-	1.67	GRAM/MJ	8.84	
PENTOSAN	GRAM	0.45	-	-	-	0.44	GRAM/MJ	2.34	
HEXOSAN	GRAM	1.27	-	-	-	1.24	GRAM/MJ	6.60	
CELLULOSE	GRAM	1.95	-	-	-	1.91	GRAM/MJ	10.14	
PALMITIC ACID	MILLI	31.00	31.00	-	31.00	30.38	MILLI/MJ	161.14	
STEARIC ACID	MILLI	3.00	3.00	-	3.00	2.94	MILLI/MJ	15.59	
OLEIC ACID	MILLI	12.00	12.00	-	12.00	11.76	MILLI/MJ	62.38	
LINOLEIC ACID	MILLI	69.00	69.00	-	69.00	67.62	MILLI/MJ	358.67	
LINOLENIC ACID	MILLI	34.00	34.00	-	34.00	33.32	MILLI/MJ	176.74	
DIETARY FIBRE, WAT.SOL.	GRAM	0.38	-	-	-	0.37	GRAM/MJ	1.98	
DIETARY FIBRE, WAT.INS.	GRAM	6.40	-	-	-	6.27	GRAM/MJ	33.27	

JOHANNISBEERE WEISS — WHITE-CURRANT — GROSEILLE BLANCHE

RIBES RUBRUM L.

		PROTEIN	FAT	CARBOHYDRATES	TOTAL
ENERGY VALUE (AVERAGE) PER 100 G EDIBLE PORTION	KJOULE	15	0.0	154	169
	(KCAL)	3.6	0.0	37	40
AMOUNT OF DIGESTIBLE CONSTITUENTS PER 100 G	GRAM	0.76	0.00	9.20	
ENERGY VALUE (AVERAGE) OF THE DIGESTIBLE FRACTION PER 100 G EDIBLE PORTION	KJOULE	13	0.0	154	167
	(KCAL)	3.0	0.0	37	40

WASTE PERCENTAGE AVERAGE 2.0 MINIMUM 2.0 MAXIMUM 3.0

CONSTITUENTS	DIM	AV	VARIATION			AVR	NUTR. DENS.		MOLPERC.
WATER	GRAM	84.20	81.20	-	88.00	82.52	GRAM/MJ	505.36	
PROTEIN	GRAM	0.90	0.40	-	1.30	0.88	GRAM/MJ	5.40	
AVAILABLE CARBOHYDR.	GRAM	9.20	-	-	-	9.02	GRAM/MJ	55.22	
SODIUM	MILLI	1.90	1.50	-	2.20	1.86	MILLI/MJ	11.40	
POTASSIUM	MILLI	268.00	250.00	-	291.00	262.64	GRAM/MJ	1.61	
MAGNESIUM	MILLI	8.80	4.90	-	12.70	8.62	MILLI/MJ	52.82	
CALCIUM	MILLI	30.10	22.40	-	38.00	29.50	MILLI/MJ	180.66	
IRON	MILLI	0.97	0.93	-	1.00	0.95	MILLI/MJ	5.82	
COPPER	MILLI	0.14	-	-	-	0.14	MILLI/MJ	0.84	
PHOSPHORUS	MILLI	23.10	11.30	-	30.00	22.64	MILLI/MJ	138.64	
CHLORIDE	MILLI	8.00	5.30	-	10.70	7.84	MILLI/MJ	48.02	
BORON	MILLI	1.31	-	-	-	1.28	MILLI/MJ	7.86	
CAROTENE	-	0.00	-	-	-	0.00			
VITAMIN B1	MICRO	80.00	-	-	-	78.40	MICRO/MJ	480.15	
VITAMIN B2	MICRO	20.00	-	-	-	19.60	MICRO/MJ	120.04	
NICOTINAMIDE	MILLI	0.20	-	-	-	0.20	MILLI/MJ	1.20	
VITAMIN C	MILLI	35.00	30.00	-	40.00	34.30	MILLI/MJ	210.07	
GALLIC ACID	MILLI	4.50	3.00	-	6.00	4.41	MILLI/MJ	27.01	
GLUCOSE	GRAM	3.10	-	-	-	3.04	GRAM/MJ	18.61	
FRUCTOSE	GRAM	3.00	-	-	-	2.94	GRAM/MJ	18.01	
SUCROSE	GRAM	0.58	-	-	-	0.57	GRAM/MJ	3.48	

MOOSBEERE (TORFBEERE) · SMALL-CRANBERRY (MOORBERRY, BOGBERRY) · CANNEBERGE

VACCINIUM OXYCOCCUS L.

		PROTEIN	FAT	CARBOHYDRATES	TOTAL
ENERGY VALUE (AVERAGE)	KJOULE	5.8	27	131	164
PER 100 G EDIBLE PORTION	(KCAL)	1.4	6.5	31	39
AMOUNT OF DIGESTIBLE CONSTITUENTS PER 100 G	GRAM	0.29	0.63	7.80	
ENERGY VALUE (AVERAGE)	KJOULE	4.9	25	131	160
OF THE DIGESTIBLE FRACTION PER 100 G EDIBLE PORTION	(KCAL)	1.2	5.9	31	38

WASTE PERCENTAGE AVERAGE 8.0 MINIMUM 4.0 MAXIMUM 13

CONSTITUENTS	DIM	AV	VARIATION			AVR	NUTR. DENS.		MOLPERC.
WATER	GRAM	87.40	87.00	-	88.00	80.41	GRAM/MJ	546.35	
PROTEIN	GRAM	0.35	0.26	-	0.40	0.32	GRAM/MJ	2.19	
FAT	GRAM	0.70	-	-	-	0.64	GRAM/MJ	4.38	
AVAILABLE CARBOHYDR.	GRAM	7.80	-	-	-	7.18	GRAM/MJ	48.76	
MINERALS	GRAM	0.24	0.20	-	0.27	0.22	GRAM/MJ	1.50	
SODIUM	MILLI	1.76	1.00	-	2.00	1.62	MILLI/MJ	11.00	
POTASSIUM	MILLI	90.00	64.00	-	119.00	82.80	MILLI/MJ	562.60	
MAGNESIUM	MILLI	7.20	6.00	-	8.40	6.62	MILLI/MJ	45.01	
CALCIUM	MILLI	14.40	14.00	-	14.70	13.25	MILLI/MJ	90.02	
IRON	MILLI	0.86	0.60	-	1.11	0.79	MILLI/MJ	5.38	
COPPER	MILLI	0.14	-	-	-	0.13	MILLI/MJ	0.88	
PHOSPHORUS	MILLI	9.87	7.40	-	11.20	9.08	MILLI/MJ	61.70	
CHLORIDE	MILLI	0.70	0.40	-	1.00	0.64	MILLI/MJ	4.38	
BORON	MILLI	1.26	1.08	-	1.45	1.16	MILLI/MJ	7.88	
CAROTENE	MICRO	22.00	20.00	-	24.00	20.24	MICRO/MJ	137.52	
VITAMIN D	-	0.00				0.00			
VITAMIN B1	MICRO	30.00	-	-	-	27.60	MICRO/MJ	187.53	
VITAMIN B2	MICRO	20.00	-	-	-	18.40	MICRO/MJ	125.02	
NICOTINAMIDE	MILLI	0.10	-	-	-	0.09	MILLI/MJ	0.63	
VITAMIN C	MILLI	11.00	10.00	-	12.00	10.12	MILLI/MJ	68.76	
TOTAL ACIDS	GRAM	3.89	-	-	-	3.58	GRAM/MJ	24.32	
INVERT SUGAR	GRAM	3.70	-	-	-	3.40	GRAM/MJ	23.13	
SUCROSE	GRAM	0.23	-	-	-	0.21	GRAM/MJ	1.44	

PREISELBEERE (KRONSBEERE) · MOUNTAIN-CRANBERRY (RED-BILBERRY) · AIRELLE ROUGE

VACCINIUM VITIS IDAEA L.

		PROTEIN	FAT	CARBOHYDRATES	TOTAL
ENERGY VALUE (AVERAGE) PER 100 G EDIBLE PORTION	KJOULE	4.6	21	127	152
	(KCAL)	1.1	4.9	30	36
AMOUNT OF DIGESTIBLE CONSTITUENTS PER 100 G	GRAM	0.23	0.47	7.57	
ENERGY VALUE (AVERAGE) OF THE DIGESTIBLE FRACTION PER 100 G EDIBLE PORTION	KJOULE	3.9	19	127	149
	(KCAL)	0.9	4.4	30	36

WASTE PERCENTAGE AVERAGE 6.0 MINIMUM 4.0 MAXIMUM 7.0

CONSTITUENTS	DIM	AV	VARIATION			AVR	NUTR. DENS.		MOLPERC.
WATER	GRAM	87.40	85.30	-	89.00	82.16	GRAM/MJ	585.85	
PROTEIN	GRAM	0.28	0.12	-	0.40	0.26	GRAM/MJ	1.88	
FAT	GRAM	0.53	0.40	-	0.70	0.50	GRAM/MJ	3.55	
AVAILABLE CARBOHYDR.	GRAM	7.57	-	-	-	7.12	GRAM/MJ	50.74	
TOTAL DIETARY FIBRE	GRAM	2.89 [1]	-	-	-	2.72	GRAM/MJ	19.37	
MINERALS	GRAM	0.26	0.25	-	0.32	0.24	GRAM/MJ	1.74	
SODIUM	MILLI	2.00	-	-	-	1.88	MILLI/MJ	13.41	
POTASSIUM	MILLI	71.70	53.00	-	87.00	67.40	MILLI/MJ	480.61	
MAGNESIUM	MILLI	5.50	-	-	-	5.17	MILLI/MJ	36.87	
CALCIUM	MILLI	14.00	13.00	-	15.00	13.16	MILLI/MJ	93.84	
MANGANESE	MILLI	-	0.20	-	2.70				
IRON	MILLI	0.50	0.40	-	0.60	0.47	MILLI/MJ	3.35	
COPPER	MILLI	0.26	0.12	-	0.40	0.24	MILLI/MJ	1.74	
ZINC	MILLI	0.25	-	-	-	0.24	MILLI/MJ	1.68	
NICKEL	MICRO	5.00	-	-	-	4.70	MICRO/MJ	33.52	
CHROMIUM	MICRO	1.00	-	-	-	0.94	MICRO/MJ	6.70	
MOLYBDENUM	MICRO	10.00	-	-	-	9.40	MICRO/MJ	67.03	
PHOSPHORUS	MILLI	9.70	8.00	-	11.00	9.12	MILLI/MJ	65.02	
CHLORIDE	MILLI	4.00	-	-	-	3.76	MILLI/MJ	26.81	
IODIDE	MICRO	5.00	-	-	-	4.70	MICRO/MJ	33.52	
BORON	MILLI	0.10	-	-	-	0.09	MILLI/MJ	0.67	
CAROTENE	MICRO	23.00	21.00	-	24.00	21.62	MICRO/MJ	154.17	
VITAMIN B1	MICRO	14.00	-	-	-	13.16	MICRO/MJ	93.84	
VITAMIN B2	MICRO	24.00	3.00	-	45.00	22.56	MICRO/MJ	160.87	
NICOTINAMIDE	MILLI	-	0.03	-	0.50				
PANTOTHENIC ACID	MILLI	-	0.03	-	0.22				

1 METHOD OF ENGLYST

CONSTITUENTS	DIM	AV	VARIATION			AVR	NUTR. DENS.		MOLPERC.
VITAMIN B6	MICRO	12.00	10.00	-	13.00	11.28	MICRO/MJ	80.44	
FOLIC ACID	MICRO	2.60	2.60	-	2.70	2.44	MICRO/MJ	17.43	
VITAMIN C	MILLI	12.00	11.00	-	20.00	11.28	MILLI/MJ	80.44	
MALIC ACID	GRAM	0.26	-	-	-	0.24	GRAM/MJ	1.74	
CITRIC ACID	GRAM	1.10	-	-	-	1.03	GRAM/MJ	7.37	
OXALIC ACID	-	0.00	-	-	-	0.00			
BENZOIC ACID	MILLI	65.00	-	-	-	61.10	MILLI/MJ	435.70	
GLUCOSE	GRAM	3.03	2.66	-	4.00	2.85	GRAM/MJ	20.31	
FRUCTOSE	GRAM	2.92	0.74	-	3.36	2.75	GRAM/MJ	19.63	
SUCROSE	GRAM	0.14	0.14	-	0.14	0.13	GRAM/MJ	0.94	
STARCH	GRAM	0.12	0.10	-	0.15	0.11	GRAM/MJ	0.80	
PECTIN	GRAM	1.20	-	-	-	1.13	GRAM/MJ	8.04	
DIETARY FIBRE,WAT.SOL.	GRAM	1.06	-	-	-	1.00	GRAM/MJ	7.11	
DIETARY FIBRE,WAT.INS.	GRAM	1.83	-	-	-	1.72	GRAM/MJ	12.27	

PREISELBEEREN
IN DOSEN

1

MOUNTAIN-CRANBERRIES
CANNED

1

AIRELLES
EN BOÎTES

1

		PROTEIN	FAT	CARBOHYDRATES	TOTAL
ENERGY VALUE (AVERAGE)	KJOULE	8.9	13	743	765
PER 100 G EDIBLE PORTION	(KCAL)	2.1	3.1	178	183
AMOUNT OF DIGESTIBLE CONSTITUENTS PER 100 G	GRAM	0.45	0.29	44.40	
ENERGY VALUE (AVERAGE)	KJOULE	7.6	12	743	762
OF THE DIGESTIBLE FRACTION PER 100 G EDIBLE PORTION	(KCAL)	1.8	2.8	178	182

WASTE PERCENTAGE AVERAGE 0.00

CONSTITUENTS	DIM	AV	VARIATION	AVR	NUTR. DENS.		MOLPERC.
WATER	GRAM	51.70	41.10 - 72.10	51.70	GRAM/MJ	67.83	
PROTEIN	GRAM	0.54	0.42 - 0.77	0.54	GRAM/MJ	0.71	
FAT	GRAM	0.33	0.26 - 0.38	0.33	GRAM/MJ	0.43	
AVAILABLE CARBOHYDR.	GRAM	44.40	24.00 - 55.00	44.40	GRAM/MJ	58.25	
MINERALS	GRAM	0.20	0.15 - 0.30	0.20	GRAM/MJ	0.26	
SODIUM	MILLI	16.00	5.90 - 25.00	16.00	MILLI/MJ	20.99	
POTASSIUM	MILLI	69.00	53.00 - 101.00	69.00	MILLI/MJ	90.52	
MAGNESIUM	MILLI	6.90	1.60 - 8.50	6.90	MILLI/MJ	9.05	
CALCIUM	MILLI	11.00	8.70 - 14.00	11.00	MILLI/MJ	14.43	
IRON	MILLI	2.72	2.05 - 4.76	2.72	MILLI/MJ	3.57	
PHOSPHORUS	MILLI	10.00	8.30 - 16.00	10.00	MILLI/MJ	13.12	
GLUCOSE	GRAM	21.00	19.70 - 21.90	21.00	GRAM/MJ	27.55	
FRUCTOSE	GRAM	19.80	18.60 - 21.20	19.80	GRAM/MJ	25.98	
SUCROSE	GRAM	3.64	0.70 - 7.76	3.64	GRAM/MJ	4.78	

1 TOTAL CONTENT (SOLID AND LIQUID)

PREISELBEEREN
IN DOSEN, OHNE ZUCKERZUSATZ
(DUNSTPREISELBEEREN)
1

MOUNTAIN-CRANBERRIES
CANNED, UNSWEETENED
(STEWED CRANBERRIES)
1

AIRELLES
EN BOÎTES
SANS SUCRE (À LA VAPEUR)
1

		PROTEIN	FAT	CARBOHYDRATES	TOTAL
ENERGY VALUE (AVERAGE)	KJOULE	11	23	109	143
PER 100 G EDIBLE PORTION	(KCAL)	2.7	5.5	26	34
AMOUNT OF DIGESTIBLE CONSTITUENTS PER 100 G	GRAM	0.57	0.53	6.50	
ENERGY VALUE (AVERAGE)	KJOULE	9.6	21	109	139
OF THE DIGESTIBLE FRACTION PER 100 G EDIBLE PORTION	(KCAL)	2.3	4.9	26	33

WASTE PERCENTAGE AVERAGE 0.00

CONSTITUENTS	DIM	AV	VARIATION			AVR	NUTR.	DENS.	MOLPERC.
WATER	GRAM	87.60	-	-	-	87.60	GRAM/MJ	630.22	
PROTEIN	GRAM	0.68	0.54	-	0.96	0.68	GRAM/MJ	4.89	
FAT	GRAM	0.59	0.48	-	0.78	0.59	GRAM/MJ	4.24	
AVAILABLE CARBOHYDR.	GRAM	6.50 2	-	-	-	6.50	GRAM/MJ	46.76	
MINERALS	GRAM	0.22	0.19	-	0.28	0.22	GRAM/MJ	1.58	
SODIUM	MILLI	9.00	8.00	-	10.00	9.00	MILLI/MJ	64.75	
POTASSIUM	MILLI	72.00	51.00	-	99.00	72.00	MILLI/MJ	517.99	
MAGNESIUM	MILLI	12.00	-	-	-	12.00	MILLI/MJ	86.33	
CALCIUM	MILLI	13.00	11.00	-	14.00	13.00	MILLI/MJ	93.53	
IRON	MILLI	1.50	1.40	-	1.60	1.50	MILLI/MJ	10.79	
PHOSPHORUS	MILLI	14.00	12.00	-	15.00	14.00	MILLI/MJ	100.72	
BORON	MILLI	0.28	-	-	-	0.28	MILLI/MJ	2.01	

1 TOTAL CONTENT (SOLID AND LIQUID)

2 CALCULATED FROM THE FRESH PRODUCT

STACHELBEERE GOOSEBERRY GROSEILLE À MAQUEREAU

RIBES GROSSULARIA L.

		PROTEIN	FAT	CARBOHYDRATES	TOTAL
ENERGY VALUE (AVERAGE) PER 100 G EDIBLE PORTION	KJOULE	13	5.8	142	161
	(KCAL)	3.2	1.4	34	39
AMOUNT OF DIGESTIBLE CONSTITUENTS PER 100 G	GRAM	0.68	0.13	8.49	
ENERGY VALUE (AVERAGE) OF THE DIGESTIBLE FRACTION PER 100 G EDIBLE PORTION	KJOULE	11	5.3	142	159
	(KCAL)	2.7	1.3	34	38

WASTE PERCENTAGE AVERAGE 2.0

CONSTITUENTS	DIM	AV	VARIATION			AVR	NUTR. DENS.		MOLPERC.
WATER	GRAM	87.30	85.10	-	88.90	85.55	GRAM/MJ	550.51	
PROTEIN	GRAM	0.80	0.70	-	0.91	0.78	GRAM/MJ	5.04	
FAT	GRAM	0.15	0.10	-	0.20	0.15	GRAM/MJ	0.95	
AVAILABLE CARBOHYDR.	GRAM	8.49	-	-	-	8.32	GRAM/MJ	53.54	
TOTAL DIETARY FIBRE	GRAM	2.95	-	-	-	2.89	GRAM/MJ	18.60	
MINERALS	GRAM	0.45	0.40	-	0.49	0.44	GRAM/MJ	2.84	
SODIUM	MILLI	1.60	1.20	-	2.00	1.57	MILLI/MJ	10.09	
POTASSIUM	MILLI	203.00	170.00	-	230.00	198.94	GRAM/MJ	1.28	
MAGNESIUM	MILLI	15.00	12.00	-	19.00	14.70	MILLI/MJ	94.59	
CALCIUM	MILLI	29.00	22.00	-	34.00	28.42	MILLI/MJ	182.87	
MANGANESE	MICRO	40.00	-	-	-	39.20	MICRO/MJ	252.24	
IRON	MILLI	0.63	0.50	-	1.00	0.62	MILLI/MJ	3.97	
COPPER	MICRO	95.00	80.00	-	110.00	93.10	MICRO/MJ	599.07	
ZINC	MILLI	0.10	-	-	-	0.10	MILLI/MJ	0.63	
NICKEL	MICRO	3.00	0.40	-	5.00	2.94	MICRO/MJ	18.92	
CHROMIUM	MICRO	1.00	0.20	-	1.00	0.98	MICRO/MJ	6.31	
PHOSPHORUS	MILLI	30.00	28.00	-	33.00	29.40	MILLI/MJ	189.18	
CHLORIDE	MILLI	0.90	0.80	-	1.10	0.88	MILLI/MJ	5.68	
FLUORIDE	MICRO	11.00	-	-	-	10.78	MICRO/MJ	69.37	
IODIDE	MICRO	0.20	-	-	-	0.20	MICRO/MJ	1.26	
BORON	MILLI	0.14	0.12	-	0.18	0.14	MILLI/MJ	0.88	
SILICON	MILLI	0.30	0.10	-	0.50	0.29	MILLI/MJ	1.89	
CAROTENE	MILLI	0.21	0.17	-	0.24	0.21	MILLI/MJ	1.32	
VITAMIN E ACTIVITY	MILLI	0.37	-	-	-	0.36	MILLI/MJ	2.33	
ALPHA-TOCOPHEROL	MILLI	0.37	-	-	-	0.36	MILLI/MJ	2.33	
VITAMIN B1	MICRO	16.00	11.00	-	20.00	15.68	MICRO/MJ	100.90	

CONSTITUENTS	DIM	AV	VARIATION			AVR	NUTR.	DENS.	MOLPERC.
VITAMIN B2	MICRO	18.00	10.00	-	25.00	17.64	MICRO/MJ	113.51	
NICOTINAMIDE	MILLI	0.25	0.16	-	0.35	0.25	MILLI/MJ	1.58	
PANTOTHENIC ACID	MILLI	0.20	-	-	-	0.20	MILLI/MJ	1.26	
VITAMIN B6	MICRO	15.00	-	-	-	14.70	MICRO/MJ	94.59	
BIOTIN	MICRO	0.50	-	-	-	0.49	MICRO/MJ	3.15	
VITAMIN C	MILLI	35.00	30.00	-	48.00	34.30	MILLI/MJ	220.71	
MALIC ACID	GRAM	0.72	-	-	-	0.71	GRAM/MJ	4.54	
CITRIC ACID	GRAM	0.72	-	-	-	0.71	GRAM/MJ	4.54	
OXALIC ACID TOTAL	MILLI	19.30	-	-	-	18.91	MILLI/MJ	121.71	
OXALIC ACID SOLUBLE	MILLI	10.10	-	-	-	9.90	MILLI/MJ	63.69	
CHLOROGENIC ACID	-	0.00	-	-	-	0.00			
CAFFEIC ACID	MILLI	3.00	2.00	-	3.00	2.94	MILLI/MJ	18.92	
PARA-COUMARIC ACID	MILLI	2.00	2.00	-	3.00	1.96	MILLI/MJ	12.61	
SHIKIMIC ACID	GRAM	0.12	-	-	-	0.12	GRAM/MJ	0.76	
GLUCOSE	GRAM	3.01	2.89	-	4.40	2.96	GRAM/MJ	19.02	
FRUCTOSE	GRAM	3.33	3.26	-	4.10	3.26	GRAM/MJ	21.00	
SUCROSE	GRAM	0.71	0.71	-	0.71	0.70	GRAM/MJ	4.48	
PECTIN	GRAM	0.62	-	-	-	0.61	GRAM/MJ	3.91	
PENTOSAN	GRAM	0.52	-	-	-	0.51	GRAM/MJ	3.28	
HEXOSAN	GRAM	0.39	-	-	-	0.38	GRAM/MJ	2.46	
CELLULOSE	GRAM	1.19	-	-	-	1.17	GRAM/MJ	7.50	
PALMITIC ACID	MILLI	16.00	16.00	-	16.00	15.68	MILLI/MJ	100.90	
STEARIC ACID	MILLI	2.00	2.00	-	2.00	1.96	MILLI/MJ	12.61	
PALMITOLEIC ACID	MILLI	1.00	1.00	-	1.00	0.98	MILLI/MJ	6.31	
OLEIC ACID	MILLI	14.00	14.00	-	14.00	13.72	MILLI/MJ	88.28	
LINOLEIC ACID	MILLI	39.00	39.00	-	39.00	38.22	MILLI/MJ	245.93	
LINOLENIC ACID	MILLI	28.00	28.00	-	28.00	27.44	MILLI/MJ	176.57	

WEINBEERE (WEINTRAUBE) — GRAPE — RAISIN

VITIS VINIFERA L.

		PROTEIN	FAT	CARBOHYDRATES	TOTAL
ENERGY VALUE (AVERAGE) PER 100 G EDIBLE PORTION	KJOULE	11	11	270	292
	(KCAL)	2.7	2.6	64	70
AMOUNT OF DIGESTIBLE CONSTITUENTS PER 100 G	GRAM	0.57	0.25	16.11	
ENERGY VALUE (AVERAGE) OF THE DIGESTIBLE FRACTION PER 100 G EDIBLE PORTION	KJOULE	9.6	9.8	270	289
	(KCAL)	2.3	2.3	64	69

WASTE PERCENTAGE AVERAGE 4.0 MINIMUM 2.0 MAXIMUM 6.0

CONSTITUENTS	DIM	AV	VARIATION			AVR	NUTR. DENS.		MOLPERC.
WATER	GRAM	81.10	77.30	-	83.60	77.86	GRAM/MJ	280.65	
PROTEIN	GRAM	0.68	0.50	-	0.90	0.65	GRAM/MJ	2.35	
FAT	GRAM	0.28	0.14	-	0.40	0.27	GRAM/MJ	0.97	
AVAILABLE CARBOHYDR.	GRAM	16.11	-	-	-	15.47	GRAM/MJ	55.75	
TOTAL DIETARY FIBRE	GRAM	1.62 [1]	-	-	-	1.56	GRAM/MJ	5.61	
MINERALS	GRAM	0.48	0.40	-	0.58	0.46	GRAM/MJ	1.66	
SODIUM	MILLI	1.90	0.50	-	3.00	1.82	MILLI/MJ	6.57	
POTASSIUM	MILLI	192.00	140.00	-	250.00	184.32	MILLI/MJ	664.42	
MAGNESIUM	MILLI	9.30	6.00	-	15.00	8.93	MILLI/MJ	32.18	
CALCIUM	MILLI	18.00	12.00	-	21.00	17.28	MILLI/MJ	62.29	
MANGANESE	MICRO	73.00	42.00	-	90.00	70.08	MICRO/MJ	252.62	
IRON	MILLI	0.51	0.30	-	0.70	0.49	MILLI/MJ	1.76	
COBALT	MICRO	1.40	0.90	-	1.90	1.34	MICRO/MJ	4.84	
COPPER	MICRO	61.00	35.00	-	80.00	58.56	MICRO/MJ	211.09	
ZINC	MICRO	82.00	35.00	-	110.00	78.72	MICRO/MJ	283.76	
NICKEL	MICRO	8.00	-	-	-	7.68	MICRO/MJ	27.68	
CHROMIUM	MICRO	2.00	-	-	-	1.92	MICRO/MJ	6.92	
PHOSPHORUS	MILLI	20.00	13.00	-	30.00	19.20	MILLI/MJ	69.21	
CHLORIDE	MILLI	2.00	-	-	-	1.92	MILLI/MJ	6.92	
FLUORIDE	MICRO	14.00	12.00	-	16.00	13.44	MICRO/MJ	48.45	
IODIDE	MICRO	0.70	-	-	-	0.67	MICRO/MJ	2.42	
SELENIUM	MICRO	2.80	0.00	-	20.00	2.69	MICRO/MJ	9.69	
SILICON	MILLI	0.30	0.10	-	0.50	0.29	MILLI/MJ	1.04	
VITAMIN A	-	0.00	-	-	-	0.00			
CAROTENE	MICRO	27.00	10.00	-	48.00	25.92	MICRO/MJ	93.43	
VITAMIN D	-	0.00	-	-	-	0.00			

1 METHOD OF MEUSER, SUCKOW AND KULIKOWSKI ("BERLINER METHODE")

CONSTITUENTS	DIM	AV	VARIATION	AVR	NUTR. DENS.		MOLPERC.
VITAMIN B1	MICRO	46.00	30.00 - 61.00	44.16	MICRO/MJ	159.18	
VITAMIN B2	MICRO	25.00	10.00 - 40.00	24.00	MICRO/MJ	86.51	
NICOTINAMIDE	MILLI	0.23	0.15 - 0.30	0.22	MILLI/MJ	0.80	
PANTOTHENIC ACID	MICRO	63.00	50.00 - 80.00	60.48	MICRO/MJ	218.01	
VITAMIN B6	MICRO	73.00	40.00 - 100.00	70.08	MICRO/MJ	252.62	
BIOTIN	MICRO	1.50	1.00 - 2.00	1.44	MICRO/MJ	5.19	
FOLIC ACID	MICRO	5.40	4.40 - 6.00	5.18	MICRO/MJ	18.69	
VITAMIN B12	-	0.00	- - -	0.00			
VITAMIN C	MILLI	4.20	2.00 - 7.40	4.03	MILLI/MJ	14.53	
CADAVERINE	MICRO	70.00	20.00 - 160.00	67.20	MICRO/MJ	242.24	
HISTAMINE	MICRO	6.00	- - -	5.76	MICRO/MJ	20.76	
PUTRESCINE	MILLI	0.11	0.04 - 0.21	0.11	MILLI/MJ	0.38	
SPERMIDINE	MILLI	0.34	0.25 - 0.41	0.33	MILLI/MJ	1.18	
SPERMINE	MILLI	0.77	0.59 - 0.88	0.74	MILLI/MJ	2.66	
MALIC ACID	MILLI	540.00	430.00 - 650.00	518.40	GRAM/MJ	1.87	
CITRIC ACID	MILLI	23.00	- - -	22.08	MILLI/MJ	79.59	
TARTARIC ACID	MILLI	530.00	390.00 - 670.00	508.80	GRAM/MJ	1.83	
OXALIC ACID TOTAL	MILLI	8.00	- - -	7.68	MILLI/MJ	27.68	
OXALIC ACID SOLUBLE	MILLI	3.30	- - -	3.17	MILLI/MJ	11.42	
CHLOROGENIC ACID	MILLI	12.50	- - -	12.00	MILLI/MJ	43.26	
SALICYLIC ACID	MILLI	1.40	1.40 - 1.40	1.34	MILLI/MJ	4.84	
GLUCOSE	GRAM	7.37	3.98 - 9.05	7.08	GRAM/MJ	25.53	
FRUCTOSE	GRAM	7.53	3.86 - 9.30	7.23	GRAM/MJ	26.07	
SUCROSE	GRAM	0.45	0.18 - 1.61	0.44	GRAM/MJ	1.57	
PECTIN	GRAM	0.28	0.20 - 0.35	0.27	GRAM/MJ	0.97	
SORBITOL	GRAM	0.20	0.20 - 0.20	0.19	GRAM/MJ	0.69	
MYRISTIC ACID	MILLI	1.10	1.10 - 1.10	1.06	MILLI/MJ	3.81	
C15 : 0 FATTY ACID	MILLI	0.90	0.90 - 0.90	0.86	MILLI/MJ	3.11	
PALMITIC ACID	MILLI	50.00	50.00 - 50.00	48.00	MILLI/MJ	173.03	
STEARIC ACID	MILLI	6.00	6.00 - 6.00	5.76	MILLI/MJ	20.76	
BEHENIC ACID	MILLI	4.00	4.00 - 4.00	3.84	MILLI/MJ	13.84	
PALMITOLEIC ACID	MILLI	1.50	1.50 - 1.50	1.44	MILLI/MJ	5.19	
OLEIC ACID	MILLI	13.00	13.00 - 13.00	12.48	MILLI/MJ	44.99	
LINOLEIC ACID	MILLI	111.00	111.00 - 111.00	106.56	MILLI/MJ	384.12	
LINOLENIC ACID	MILLI	36.00	36.00 - 36.00	34.56	MILLI/MJ	124.58	
TOTAL STEROLS	MILLI	4.00	- - -	3.84	MILLI/MJ	13.84	
BETA-SITOSTEROL	MILLI	3.00	- - -	2.88	MILLI/MJ	10.38	
DIETARY FIBRE,WAT.SOL.	GRAM	0.42	- - -	0.40	GRAM/MJ	1.45	
DIETARY FIBRE,WAT.INS.	GRAM	1.20	- - -	1.15	GRAM/MJ	4.15	

WEINBEERE (WEINTRAUBE) GETROCKNET (ROSINE)

GRAPE DRIED (RAISIN, SULTANA)

RAISIN SEC

		PROTEIN	FAT	CARBOHYDRATES	TOTAL
ENERGY VALUE (AVERAGE) PER 100 G EDIBLE PORTION	KJOULE	41	21	1108	1170
	(KCAL)	9.7	5.1	265	280
AMOUNT OF DIGESTIBLE CONSTITUENTS PER 100 G	GRAM	2.09	0.49	66.20	
ENERGY VALUE (AVERAGE) OF THE DIGESTIBLE FRACTION PER 100 G EDIBLE PORTION	KJOULE	35	19	1108	1162
	(KCAL)	8.3	4.6	265	278

WASTE PERCENTAGE AVERAGE 0.00

CONSTITUENTS	DIM	AV	VARIATION			AVR	NUTR. DENS.		MOLPERC.
WATER	GRAM	15.70	12.80	-	19.50	15.70	GRAM/MJ	13.51	
PROTEIN	GRAM	2.46	2.21	-	2.60	2.46	GRAM/MJ	2.12	
FAT	GRAM	0.55	-	-	-	0.55	GRAM/MJ	0.47	
AVAILABLE CARBOHYDR.	GRAM	66.20	-	-	-	66.20	GRAM/MJ	56.98	
TOTAL DIETARY FIBRE	GRAM	5.40 1	-	-	-	5.40	GRAM/MJ	4.65	
MINERALS	GRAM	2.02	1.87	-	2.16	2.02	GRAM/MJ	1.74	
SODIUM	MILLI	21.00	10.00	-	30.00	21.00	MILLI/MJ	18.08	
POTASSIUM	MILLI	782.00	639.00	-	875.00	782.00	MILLI/MJ	673.13	
MAGNESIUM	MILLI	15.00	-	-	-	15.00	MILLI/MJ	12.91	
CALCIUM	MILLI	31.00	-	-	-	31.00	MILLI/MJ	26.68	
IRON	MILLI	0.30	-	-	-	0.30	MILLI/MJ	0.26	
COPPER	MILLI	0.10	-	-	-	0.10	MILLI/MJ	0.09	
ZINC	MILLI	0.10	-	-	-	0.10	MILLI/MJ	0.09	
MOLYBDENUM	MICRO	10.50	8.00	-	13.00	10.50	MICRO/MJ	9.04	
PHOSPHORUS	MILLI	110.00	74.00	-	129.00	110.00	MILLI/MJ	94.69	
CHLORIDE	MILLI	10.00	-	-	-	10.00	MILLI/MJ	8.61	
BORON	MILLI	1.20	0.70	-	1.60	1.20	MILLI/MJ	1.03	
CAROTENE	MICRO	30.00	-	-	-	30.00	MICRO/MJ	25.82	
VITAMIN B1	MILLI	0.12	0.08	-	0.15	0.12	MILLI/MJ	0.10	
VITAMIN B2	MICRO	55.00	30.00	-	80.00	55.00	MICRO/MJ	47.34	
NICOTINAMIDE	MILLI	0.50	-	-	-	0.50	MILLI/MJ	0.43	
PANTOTHENIC ACID	MILLI	0.10	-	-	-	0.10	MILLI/MJ	0.09	
VITAMIN B6	MILLI	0.11	0.09	-	0.12	0.11	MILLI/MJ	0.09	
FOLIC ACID	MICRO	4.00	-	-	-	4.00	MICRO/MJ	3.44	
VITAMIN C	MILLI	1.00	1.00	-	2.00	1.00	MILLI/MJ	0.86	

1 METHOD OF MEUSER, SUCKOW AND KULIKOWSKI ("BERLINER METHODE")

CONSTITUENTS	DIM	AV	VARIATION			AVR	NUTR.	DENS.	MOLPERC.
ALANINE	MILLI	91.00	91.00	-	91.00	91.00	MILLI/MJ	78.33	10.5
ARGININE	MILLI	305.00	305.00	-	305.00	305.00	MILLI/MJ	262.54	14.9
ASPARTIC ACID	MILLI	87.00	87.00	-	87.00	87.00	MILLI/MJ	74.89	6.7
CYSTINE	MILLI	6.00	6.00	-	6.00	6.00	MILLI/MJ	5.16	0.3
GLUTAMIC ACID	MILLI	118.00	118.00	-	118.00	118.00	MILLI/MJ	101.57	8.3
GLYCINE	MILLI	63.00	63.00	-	63.00	63.00	MILLI/MJ	54.23	8.6
HISTIDINE	MILLI	51.00	51.00	-	51.00	51.00	MILLI/MJ	43.90	2.4
ISOLEUCINE	MILLI	47.00	47.00	-	47.00	47.00	MILLI/MJ	40.46	3.7
LEUCINE	MILLI	75.00	75.00	-	75.00	75.00	MILLI/MJ	64.56	5.9
LYSINE	MILLI	71.00	71.00	-	71.00	71.00	MILLI/MJ	61.12	4.0
METHIONINE	MILLI	13.00	13.00	-	13.00	13.00	MILLI/MJ	11.19	0.9
PHENYLALANINE	MILLI	47.00	47.00	-	47.00	47.00	MILLI/MJ	40.46	2.9
PROLINE	MILLI	157.00	157.00	-	157.00	157.00	MILLI/MJ	135.14	14.0
SERINE	MILLI	51.00	51.00	-	51.00	51.00	MILLI/MJ	43.90	5.0
THREONINE	MILLI	55.00	55.00	-	55.00	55.00	MILLI/MJ	47.34	4.8
TRYPTOPHAN	MILLI	5.00	5.00	-	5.00	5.00	MILLI/MJ	4.30	0.3
TYROSINE	MILLI	10.00	10.00	-	10.00	10.00	MILLI/MJ	8.61	0.6
VALINE	MILLI	71.00	71.00	-	71.00	71.00	MILLI/MJ	61.12	6.2
MALIC ACID	GRAM	2.30	-	-	-	2.30	GRAM/MJ	1.98	
TARTARIC ACID	GRAM	2.30	-	-	-	2.30	GRAM/MJ	1.98	
SALICYLIC ACID	MILLI	6.73	6.73	-	6.74	6.73	MILLI/MJ	5.80	
GLUCOSE	GRAM	31.20	26.60	-	36.70	31.20	GRAM/MJ	26.86	
FRUCTOSE	GRAM	31.60	28.30	-	36.30	31.60	GRAM/MJ	27.20	
SUCROSE	GRAM	1.10	0.53	-	1.73	1.10	GRAM/MJ	0.95	
DIETARY FIBRE, WAT.SOL.	GRAM	1.60	-	-	-	1.60	GRAM/MJ	1.38	
DIETARY FIBRE, WAT.INS.	GRAM	3.80	-	-	-	3.80	GRAM/MJ	3.27	

Wildfrüchte

EBERESCHENFRUCHT (VOGELBEERE, SÜSS) · ROWANBERRY SWEET · BAIE DE SORBIER DOUCE

SORBUS AUCUPARIA L. VAR. EDULIS (DIECK)

		PROTEIN	FAT	CARBOHYDRATES	TOTAL
ENERGY VALUE (AVERAGE)	KJOULE	25	0.0	340	365
PER 100 G EDIBLE PORTION	(KCAL)	5.9	0.0	81	87
AMOUNT OF DIGESTIBLE CONSTITUENTS PER 100 G	GRAM	1.27	0.00	20.32	
ENERGY VALUE (AVERAGE)	KJOULE	21	0.0	340	361
OF THE DIGESTIBLE FRACTION PER 100 G EDIBLE PORTION	(KCAL)	5.0	0.0	81	86

WASTE PERCENTAGE AVERAGE 50 MINIMUM 44 MAXIMUM 56

CONSTITUENTS	DIM	AV	VARIATION			AVR	NUTR. DENS.		MOLPERC.
WATER	GRAM	71.70	63.50	-	75.40	35.85	GRAM/MJ	198.53	
PROTEIN	GRAM	1.50	-	-	-	0.75	GRAM/MJ	4.15	
AVAILABLE CARBOHYDR.	GRAM	20.32	-	-	-	10.16	GRAM/MJ	56.27	
MINERALS	GRAM	0.68	0.50	-	0.80	0.34	GRAM/MJ	1.88	
SODIUM	-	TRACES	-	-	-				
POTASSIUM	MILLI	234.00	194.00	-	277.00	117.00	MILLI/MJ	647.94	
MAGNESIUM	MILLI	17.00	-	-	-	8.50	MILLI/MJ	47.07	
CALCIUM	MILLI	42.00	27.00	-	57.00	21.00	MILLI/MJ	116.30	
MANGANESE	MILLI	1.60	1.20	-	2.00	0.80	MILLI/MJ	4.43	
IRON	MILLI	1.50	0.70	-	2.00	0.75	MILLI/MJ	4.15	
COBALT	MICRO	1.00	-	-	-	0.50	MICRO/MJ	2.77	
COPPER	MICRO	90.00	80.00	-	110.00	45.00	MICRO/MJ	249.21	
ZINC	MILLI	0.26	-	-	-	0.13	MILLI/MJ	0.72	
NICKEL	MICRO	10.00	-	-	-	5.00	MICRO/MJ	27.69	
CHROMIUM	MICRO	3.00	2.00	-	5.00	1.50	MICRO/MJ	8.31	
PHOSPHORUS	MILLI	33.00	-	-	-	16.50	MILLI/MJ	91.38	
CHLORIDE	MILLI	5.00	-	-	-	2.50	MILLI/MJ	13.84	
FLUORIDE	MICRO	30.00	20.00	-	40.00	15.00	MICRO/MJ	83.07	
BORON	MILLI	0.53	0.42	-	0.62	0.27	MILLI/MJ	1.47	
SILICON	MILLI	3.00	2.00	-	5.00	1.50	MILLI/MJ	8.31	

CONSTITUENTS	DIM	AV	VARIATION			AVR	NUTR. DENS.		MOLPERC.
CAROTENE	MILLI	2.45	1.46	-	3.80	1.23	MILLI/MJ	6.78	
VITAMIN C	MILLI	98.00	78.00	-	117.00	49.00	MILLI/MJ	271.36	
TOTAL ACIDS	GRAM	2.37	1.60	-	2.42	1.19	GRAM/MJ	6.56	
PARASORBIC ACID	MILLI	10.00	-	-	-	5.00	MILLI/MJ	27.69	
REDUCING SUGAR	GRAM	9.30	6.10	-	13.10	4.65	GRAM/MJ	25.75	
SUCROSE	GRAM	0.15	0.00	-	0.30	0.08	GRAM/MJ	0.42	
SORBITOL	GRAM	8.50	5.00	-	12.00	4.25	GRAM/MJ	23.54	

HAGEBUTTE ROSE HIP, HAW BAIE D'ÉGLANTIER

ROSA CANINA L. UND VERWANDTE ARTEN

		PROTEIN	FAT	CARBOHYDRATES	TOTAL
ENERGY VALUE (AVERAGE)	KJOULE	59	0.0	323	383
PER 100 G	(KCAL)	14	0.0	77	91
EDIBLE PORTION					
AMOUNT OF DIGESTIBLE	GRAM	3.06	0.00	19.30	
CONSTITUENTS PER 100 G					
ENERGY VALUE (AVERAGE)	KJOULE	51	0.0	323	374
OF THE DIGESTIBLE	(KCAL)	12	0.0	77	89
FRACTION PER 100 G					
EDIBLE PORTION					

WASTE PERCENTAGE AVERAGE 35 MINIMUM 30 MAXIMUM 40

CONSTITUENTS	DIM	AV	VARIATION			AVR	NUTR. DENS.		MOLPERC.
WATER	GRAM	50.20	41.80	-	59.00	32.63	GRAM/MJ	134.38	
PROTEIN	GRAM	3.60	3.28	-	4.10	2.34	GRAM/MJ	9.64	
AVAILABLE CARBOHYDR.	GRAM	19.30	-	-	-	12.55	GRAM/MJ	51.66	
MINERALS	GRAM	2.60	1.90	-	3.30	1.69	GRAM/MJ	6.96	
SODIUM	MILLI	146.00	-	-	-	94.90	MILLI/MJ	390.82	
POTASSIUM	MILLI	291.00	-	-	-	189.15	MILLI/MJ	778.96	
MAGNESIUM	MILLI	104.00	-	-	-	67.60	MILLI/MJ	278.39	
CALCIUM	MILLI	257.00	-	-	-	167.05	MILLI/MJ	687.94	
MANGANESE	MILLI	1.20	-	-	-	0.78	MILLI/MJ	3.21	
IRON	MILLI	0.52	-	-	-	0.34	MILLI/MJ	1.39	
COBALT	MICRO	1.00	-	-	-	0.65	MICRO/MJ	2.68	
COPPER	MILLI	1.80	-	-	-	1.17	MILLI/MJ	4.82	
ZINC	MILLI	0.92	-	-	-	0.60	MILLI/MJ	2.46	
NICKEL	MICRO	40.00	-	-	-	26.00	MICRO/MJ	107.07	
CHROMIUM	MICRO	3.00	-	-	-	1.95	MICRO/MJ	8.03	
PHOSPHORUS	MILLI	258.00	-	-	-	167.70	MILLI/MJ	690.62	
CHLORIDE	MILLI	15.60	-	-	-	10.14	MILLI/MJ	41.76	
FLUORIDE	MICRO	60.00	-	-	-	39.00	MICRO/MJ	160.61	
IODIDE	MICRO	1.00	1.00	-	1.00	0.65	MICRO/MJ	2.68	
BORON	MILLI	0.88	-	-	-	0.57	MILLI/MJ	2.36	
SELENIUM	MICRO	0.20	-	-	-	0.13	MICRO/MJ	0.54	
SILICON	MILLI	1.00	-	-	-	0.65	MILLI/MJ	2.68	
CAROTENE	MILLI	-	3.60	-	6.00				
VITAMIN K	MILLI	-	0.08	-	0.10				
VITAMIN B1	MICRO	58.00	58.00	-	58.00	37.70	MICRO/MJ	155.26	
VITAMIN B2	MICRO	67.00	67.00	-	67.00	43.55	MICRO/MJ	179.35	
NICOTINAMIDE	MILLI	0.48	0.48	-	0.48	0.31	MILLI/MJ	1.28	
VITAMIN B6	MICRO	48.00	48.00	-	48.00	31.20	MICRO/MJ	128.49	
VITAMIN C	GRAM	1.25	0.25	-	2.90	0.81	GRAM/MJ	3.35	
MALIC ACID	GRAM	3.10	1.90	-	4.00	2.02	GRAM/MJ	8.30	
TARTARIC ACID	GRAM	-	0.70	-	2.60				
GLUCOSE	GRAM	7.30	7.30	-	7.30	4.75	GRAM/MJ	19.54	
FRUCTOSE	GRAM	7.30	7.30	-	7.30	4.75	GRAM/MJ	19.54	
SUCROSE	GRAM	1.60	0.00	-	2.10	1.04	GRAM/MJ	4.28	

HOLUNDERBEERE SCHWARZ — ELDERBERRY BLACK — BAIE DE SUREAU

HOLUNDERBEERE SCHWARZ

ELDERBERRY BLACK

BAIE DE SUREAU

SAMBUCUS NIGRA L.

		PROTEIN	FAT	CARBOHYDRATES	TOTAL
ENERGY VALUE (AVERAGE)	KJOULE	42	0.0	124	166
PER 100 G	(KCAL)	10.0	0.0	30	40
EDIBLE PORTION					
AMOUNT OF DIGESTIBLE	GRAM	2.15	0.00	7.40	
CONSTITUENTS PER 100 G					
ENERGY VALUE (AVERAGE)	KJOULE	36	0.0	124	159
OF THE DIGESTIBLE	(KCAL)	8.5	0.0	30	38
FRACTION PER 100 G					
EDIBLE PORTION					

WASTE PERCENTAGE AVERAGE 30

CONSTITUENTS	DIM	AV	VARIATION			AVR	NUTR.	DENS.	MOLPERC.
WATER	GRAM	80.90	80.40	-	81.80	56.63	GRAM/MJ	507.57	
PROTEIN	GRAM	2.53	2.50	-	2.54	1.77	GRAM/MJ	15.87	
FAT	GRAM	1.70	-	-	-	1.19	GRAM/MJ	10.67	
AVAILABLE CARBOHYDR.	GRAM	7.40	-	-	-	5.18	GRAM/MJ	46.43	
MINERALS	GRAM	0.69	0.58	-	0.84	0.48	GRAM/MJ	4.33	
SODIUM	MILLI	0.50	-	-	-	0.35	MILLI/MJ	3.14	
POTASSIUM	MILLI	305.00	-	-	-	213.50	GRAM/MJ	1.91	
CALCIUM	MILLI	35.00	-	-	-	24.50	MILLI/MJ	219.59	
PHOSPHORUS	MILLI	57.00	48.00	-	61.00	39.90	MILLI/MJ	357.62	
CAROTENE	MILLI	0.36	-	-	-	0.25	MILLI/MJ	2.26	
VITAMIN B1	MICRO	65.00	60.00	-	70.00	45.50	MICRO/MJ	407.81	
VITAMIN B2	MICRO	78.00	60.00	-	96.00	54.60	MICRO/MJ	489.37	
NICOTINAMIDE	MILLI	1.48	1.25	-	1.70	1.04	MILLI/MJ	9.29	
PANTOTHENIC ACID	MILLI	0.18	-	-	-	0.13	MILLI/MJ	1.13	
VITAMIN B6	MILLI	0.25	-	-	-	0.18	MILLI/MJ	1.57	
BIOTIN	MICRO	1.80	-	-	-	1.26	MICRO/MJ	11.29	
FOLIC ACID	MICRO	17.00	-	-	-	11.90	MICRO/MJ	106.66	
VITAMIN C	MILLI	18.00	10.00	-	29.00	12.60	MILLI/MJ	112.93	
TOTAL ACIDS	GRAM	0.91	0.42	-	1.40	0.64	GRAM/MJ	5.71	
INVERT SUGAR	GRAM	6.27	4.62	-	7.50	4.39	GRAM/MJ	39.34	
SUCROSE	GRAM	0.25	-	-	-	0.18	GRAM/MJ	1.57	
PALMITIC ACID	GRAM	0.10	0.10	-	0.10	0.07	GRAM/MJ	0.63	
STEARIC ACID	MILLI	30.00	30.00	-	30.00	21.00	MILLI/MJ	188.22	
OLEIC ACID	GRAM	0.11	0.11	-	0.11	0.08	GRAM/MJ	0.69	
LINOLEIC ACID	GRAM	0.61	0.61	-	0.61	0.43	GRAM/MJ	3.83	
LINOLENIC ACID	GRAM	0.51	0.51	-	0.51	0.36	GRAM/MJ	3.20	

KORNELKIRSCHE CORNELIAN CHERRY CORNOUILLE

DÜRLITZE

CORNUS MAS L.

		PROTEIN	FAT	CARBOHYDRATES	TOTAL
ENERGY VALUE (AVERAGE)	KJOULE	34	0	250	322
PER 100 G EDIBLE PORTION	(KCAL)	8.1	0.0	60	77
AMOUNT OF DIGESTIBLE CONSTITUENTS PER 100 G	GRAM	1.73	0.00	14.91	
ENERGY VALUE (AVERAGE)	KJOULE	29	0.0	250	278
OF THE DIGESTIBLE FRACTION PER 100 G EDIBLE PORTION	(KCAL)	6.8	0.0	60	66

WASTE PERCENTAGE AVERAGE 3.0

CONSTITUENTS	DIM	AV	VARIATION	AVR	NUTR. DENS.		MOLPERC.
WATER	GRAM	83.08	83.08 - 83.08	80.59	GRAM/MJ	298.64	
PROTEIN	GRAM	2.04	2.04 - 2.04	1.98	GRAM/MJ	7.33	
AVAILABLE CARBOHYDR.	GRAM	14.91	14.91 - 14.91	14.46	GRAM/MJ	53.60	
MINERALS	GRAM	0.73	0.73 - 0.73	0.71	GRAM/MJ	2.62	
VITAMIN C	MILLI	78.00	78.00 - 78.00	75.66	MILLI/MJ	280.38	
TOTAL ACIDS	GRAM	2.81	2.81 - 2.81	2.73	GRAM/MJ	10.10	
INVERT SUGAR	GRAM	12.43	12.43 - 12.43	12.06	GRAM/MJ	44.68	
SUCROSE	GRAM	0.30	0.30 - 0.30	0.29	GRAM/MJ	1.08	

SANDDORNBEERE SEA BUCKTHORN (SALLOW THORN) BAIE D'ARGOUSIER

HIPPOPHAE RHAMNOIDES L.

		PROTEIN	FAT	CARBOHYDRATES	TOTAL
ENERGY VALUE (AVERAGE)	KJOULE	23	276	88	387
PER 100 G EDIBLE PORTION	(KCAL)	5.6	66	21	93
AMOUNT OF DIGESTIBLE CONSTITUENTS PER 100 G	GRAM	1.20	6.39	5.24	
ENERGY VALUE (AVERAGE)	KJOULE	20	249	88	356
OF THE DIGESTIBLE FRACTION PER 100 G EDIBLE PORTION	(KCAL)	4.8	59	21	85

WASTE PERCENTAGE AVERAGE 40

CONSTITUENTS	DIM	AV	VARIATION			AVR	NUTR. DENS.		MOLPERC.
WATER	GRAM	82.60	81.20	-	85.60	49.56	GRAM/MJ	231.84	
PROTEIN	GRAM	1.42	1.20	-	2.09	0.85	GRAM/MJ	3.99	
FAT	GRAM	7.10	5.00	-	9.20	4.26	GRAM/MJ	19.93	
AVAILABLE CARBOHYDR.	GRAM	5.24	-	-	-	3.14	GRAM/MJ	14.71	
MINERALS	GRAM	0.45	0.28	-	0.50	0.27	GRAM/MJ	1.26	
SODIUM	MILLI	3.50	1.00	-	4.00	2.10	MILLI/MJ	9.82	
POTASSIUM	MILLI	133.00	59.00	-	207.00	79.80	MILLI/MJ	373.29	
MAGNESIUM	MILLI	30.00	-	-	-	18.00	MILLI/MJ	84.20	
CALCIUM	MILLI	42.00	12.00	-	72.00	25.20	MILLI/MJ	117.88	
IRON	MILLI	0.44	-	-	-	0.26	MILLI/MJ	1.23	
PHOSPHORUS	MILLI	8.60	5.80	-	9.50	5.16	MILLI/MJ	24.14	
CHLORIDE	MILLI	0.39	-	-	-	0.23	MILLI/MJ	1.09	
CAROTENE	MILLI	1.50	0.80	-	2.80	0.90	MILLI/MJ	4.21	
VITAMIN B1	MICRO	34.00	24.00	-	52.00	20.40	MICRO/MJ	95.43	
VITAMIN B2	MILLI	0.21	0.15	-	0.27	0.13	MILLI/MJ	0.59	
NICOTINAMIDE	MILLI	0.26	0.17	-	0.35	0.16	MILLI/MJ	0.73	
PANTOTHENIC ACID	MILLI	0.15	-	-	-	0.09	MILLI/MJ	0.42	
VITAMIN B6	MILLI	0.11	-	-	-	0.07	MILLI/MJ	0.31	
BIOTIN	MICRO	3.30	-	-	-	1.98	MICRO/MJ	9.26	
FOLIC ACID	MICRO	10.00	-	-	-	6.00	MICRO/MJ	28.07	
VITAMIN C	GRAM	0.45	0.10	-	1.20	0.27	GRAM/MJ	1.26	
TOTAL ACIDS	GRAM	1.95	1.50	-	2.85	1.17	GRAM/MJ	5.47	
INVERT SUGAR	GRAM	3.00	-	-	-	1.80	GRAM/MJ	8.42	
SUCROSE	GRAM	0.29	0.09	-	0.49	0.17	GRAM/MJ	0.81	
INOSITOL	MILLI	67.00	-	-	-	40.20	MILLI/MJ	188.05	

SCHLEHE BLACKTHORN PRUNELLE
SCHWARZDORNBEERE

		PROTEIN	FAT	CARBOHYDRATES	TOTAL
ENERGY VALUE (AVERAGE) PER 100 G EDIBLE PORTION	KJOULE	12	0	196	248
	(KCAL)	3.0	0.0	47	59
AMOUNT OF DIGESTIBLE CONSTITUENTS PER 100 G	GRAM	0.63	0.00	11.73	
ENERGY VALUE (AVERAGE) OF THE DIGESTIBLE FRACTION PER 100 G EDIBLE PORTION	KJOULE	11	0.0	196	207
	(KCAL)	2.5	0.0	47	49

WASTE PERCENTAGE AVERAGE 9.0

CONSTITUENTS	DIM	AV	VARIATION	AVR	NUTR. DENS.		MOLPERC.
WATER	GRAM	81.70	81.70 - 81.70	74.35	GRAM/MJ	394.97	
PROTEIN	GRAM	0.75	0.75 - 0.75	0.68	GRAM/MJ	3.63	
AVAILABLE CARBOHYDR.	GRAM	11.73	11.73 - 11.73	10.67	GRAM/MJ	56.71	
MINERALS	GRAM	1.05	1.05 - 1.05	0.96	GRAM/MJ	5.08	
TOTAL ACIDS	GRAM	3.09	3.09 - 3.09	2.81	GRAM/MJ	14.94	
INVERT SUGAR	GRAM	8.40	8.40 - 8.40	7.64	GRAM/MJ	40.61	
SUCROSE	GRAM	0.24	0.24 - 0.24	0.22	GRAM/MJ	1.16	

WEISSDORN HAWTHORN AUBEPINE

		PROTEIN	FAT	CARBOHYDRATES	TOTAL
ENERGY VALUE (AVERAGE) PER 100 G EDIBLE PORTION	KJOULE	28	0	183	250
	(KCAL)	6.6	0.0	44	60
AMOUNT OF DIGESTIBLE CONSTITUENTS PER 100 G	GRAM	1.42	0.00	10.95	
ENERGY VALUE (AVERAGE) OF THE DIGESTIBLE FRACTION PER 100 G EDIBLE PORTION	KJOULE	24	0.0	183	207
	(KCAL)	5.6	0.0	44	49

WASTE PERCENTAGE AVERAGE 21

CONSTITUENTS	DIM	AV	VARIATION	AVR	NUTR. DENS.	MOLPERC.
WATER	GRAM	81.77	81.77 - 81.77	64.60	GRAM/MJ 395.29	
PROTEIN	GRAM	1.68	1.68 - 1.68	1.33	GRAM/MJ 8.12	
AVAILABLE CARBOHYDR.	GRAM	10.95	10.95 - 10.95	8.65	GRAM/MJ 52.93	
MINERALS	GRAM	1.22	1.22 - 1.22	0.96	GRAM/MJ 5.90	
VITAMIN C	MILLI	13.00	13.00 - 13.00	10.27	MILLI/MJ 62.84	
TOTAL ACIDS	GRAM	5.22	5.22 - 5.22	4.12	GRAM/MJ 25.23	
INVERT SUGAR	GRAM	5.09	5.09 - 5.09	4.02	GRAM/MJ 24.61	
SUCROSE	GRAM	0.64	0.64 - 0.64	0.51	GRAM/MJ 3.09	

Exotische Früchte

EXOTIC FRUITS **FRUITS EXOTIQUES**

ACEROLA (WESTINDISCHE KIRSCHE) · ACEROLA (WEST. IND. CHERRY) · CERISE „ACEROLA“

MALPIGHIA PUNICIFOLIA L.

		PROTEIN	FAT	CARBOHYDRATES	TOTAL
ENERGY VALUE (AVERAGE) PER 100 G EDIBLE PORTION	KJOULE	3.5	8.9	60	72
	(KCAL)	0.8	2.1	14	17
AMOUNT OF DIGESTIBLE CONSTITUENTS PER 100 G	GRAM	0.17	0.20	3.57	
ENERGY VALUE (AVERAGE) OF THE DIGESTIBLE FRACTION PER 100 G EDIBLE PORTION	KJOULE	3.0	8.1	60	71
	(KCAL)	0.7	1.9	14	17

WASTE PERCENTAGE AVERAGE 19

CONSTITUENTS	DIM	AV	VARIATION			AVR	NUTR. DENS.		MOLPERC.
WATER	GRAM	89.20	81.90	-	93.50	72.25	KILO/MJ	1.26	
PROTEIN	GRAM	0.21	0.06	-	0.39	0.17	GRAM/MJ	2.97	
FAT	GRAM	0.23	0.09	-	0.34	0.19	GRAM/MJ	3.25	
AVAILABLE CARBOHYDR.	GRAM	3.57	-	-	-	2.89	GRAM/MJ	50.46	
MINERALS	GRAM	0.45	0.20	-	0.82	0.36	GRAM/MJ	6.36	
SODIUM	MILLI	2.70	-	-	-	2.19	MILLI/MJ	38.16	
POTASSIUM	MILLI	83.00	-	-	-	67.23	GRAM/MJ	1.17	
MAGNESIUM	MILLI	12.00	12.00	-	12.00	9.72	MILLI/MJ	169.61	
CALCIUM	MILLI	11.70	-	-	-	9.48	MILLI/MJ	165.37	
IRON	MILLI	0.24	-	-	-	0.19	MILLI/MJ	3.39	
PHOSPHORUS	MILLI	17.10	7.30	-	33.10	13.85	MILLI/MJ	241.69	
NITRATE	MILLI	0.50	0.50	-	0.50	0.41	MILLI/MJ	7.07	
CAROTENE	MILLI	0.17	0.11	-	0.24	0.14	MILLI/MJ	2.40	
VITAMIN B1	MICRO	20.00	10.00	-	30.00	16.20	MICRO/MJ	282.68	
VITAMIN B2	MICRO	73.00	50.00	-	90.00	59.13	MILLI/MJ	1.03	
NICOTINAMIDE	MILLI	0.41	-	-	-	0.33	MILLI/MJ	5.79	
PANTOTHENIC ACID	MILLI	0.33	0.33	-	0.33	0.27	MILLI/MJ	4.66	
VITAMIN B6	MICRO	8.70	-	-	-	7.05	MICRO/MJ	122.96	
BIOTIN	MICRO	2.50	2.50	-	2.50	2.03	MICRO/MJ	35.33	
VITAMIN C	GRAM	1.70	1.00	-	2.00	1.38	GRAM/MJ	24.03	
MALIC ACID	GRAM	0.90	0.90	-	0.90	0.73	GRAM/MJ	12.72	
CITRIC ACID	MILLI	6.00	6.00	-	6.00	4.86	MILLI/MJ	84.80	
ISOCITRIC ACID	MILLI	0.50	0.50	-	0.50	0.41	MILLI/MJ	7.07	
GLUCOSE	GRAM	1.20	1.20	-	1.20	0.97	GRAM/MJ	16.96	
FRUCTOSE	GRAM	1.46	1.46	-	1.46	1.18	GRAM/MJ	20.64	

AKEE
(AKIPFLAUME, AKINUSS)

BLIGHIA SAPIDA KOENIG
1

AKEE
1

RIS DE VEAU
1

		PROTEIN	FAT	CARBOHYDRATES	TOTAL
ENERGY VALUE (AVERAGE) PER 100 G EDIBLE PORTION	KJOULE	78	778	77	934
	(KCAL)	19	186	18	223
AMOUNT OF DIGESTIBLE CONSTITUENTS PER 100 G	GRAM	3.25	18.00	4.60	
ENERGY VALUE (AVERAGE) OF THE DIGESTIBLE FRACTION PER 100 G EDIBLE PORTION	KJOULE	51	700	77	828
	(KCAL)	12	167	18	198

WASTE PERCENTAGE AVERAGE 0.00

CONSTITUENTS	DIM	AV		VARIATION			AVR	NUTR. DENS.		MOLPERC.
WATER	GRAM	69.20		-	-	-	69.20	GRAM/MJ	83.54	
PROTEIN	GRAM	5.00		-	-	-	5.00	GRAM/MJ	6.04	
FAT	GRAM	20.00		-	-	-	20.00	GRAM/MJ	24.14	
AVAILABLE CARBOHYDR.	GRAM	4.60	1	-	-	-	4.60	GRAM/MJ	5.55	
MINERALS	GRAM	1.20		-	-	-	1.20	GRAM/MJ	1.45	
CALCIUM	MILLI	40.00		-	-	-	40.00	MILLI/MJ	48.29	
IRON	MILLI	2.70		-	-	-	2.70	MILLI/MJ	3.26	
PHOSPHORUS	MILLI	16.00		-	-	-	16.00	MILLI/MJ	19.31	
CAROTENE	MILLI	0.56		-	-	-	0.56	MILLI/MJ	0.68	
VITAMIN B1	MILLI	0.13		-	-	-	0.13	MILLI/MJ	0.16	
VITAMIN B2	MILLI	0.14		-	-	-	0.14	MILLI/MJ	0.17	
NICOTINAMIDE	MILLI	1.40		-	-	-	1.40	MILLI/MJ	1.69	
VITAMIN C	MILLI	26.00		-	-	-	26.00	MILLI/MJ	31.39	

1 ESTIMATED BY THE DIFFERENCE METHOD
100 - (WATER + PROTEIN + FAT + MINERALS)

ANANAS PINEAPPLE ANANAS

ANANAS SATIVUS SCHULT.

		PROTEIN	FAT	CARBOHYDRATES	TOTAL
ENERGY VALUE (AVERAGE)	KJOULE	7.6	5.8	220	233
PER 100 G	(KCAL)	1.8	1.4	52	56
EDIBLE PORTION					
AMOUNT OF DIGESTIBLE	GRAM	0.39	0.13	13.12	
CONSTITUENTS PER 100 G					
ENERGY VALUE (AVERAGE)	KJOULE	6.5	5.3	220	231
OF THE DIGESTIBLE	(KCAL)	1.5	1.3	52	55
FRACTION PER 100 G					
EDIBLE PORTION					

WASTE PERCENTAGE AVERAGE 46 MINIMUM 36 MAXIMUM 57

CONSTITUENTS	DIM	AV	VARIATION			AVR	NUTR. DENS.		MOLPERC.
WATER	GRAM	85.30	82.00	-	88.80	46.06	GRAM/MJ	368.80	
PROTEIN	GRAM	0.46	0.40	-	0.50	0.25	GRAM/MJ	1.99	
FAT	GRAM	0.15	0.00	-	0.25	0.08	GRAM/MJ	0.65	
AVAILABLE CARBOHYDR.	GRAM	13.12	-	-	-	7.08	GRAM/MJ	56.73	
TOTAL DIETARY FIBRE	GRAM	1.40 1	-	-	-	0.76	GRAM/MJ	6.05	
MINERALS	GRAM	0.39	0.32	-	0.50	0.21	GRAM/MJ	1.69	
SODIUM	MILLI	2.10	0.30	-	5.90	1.13	MILLI/MJ	9.08	
POTASSIUM	MILLI	173.00	123.00	-	250.00	93.42	MILLI/MJ	747.97	
MAGNESIUM	MILLI	17.00	13.00	-	22.00	9.18	MILLI/MJ	73.50	
CALCIUM	MILLI	16.00	12.00	-	18.00	8.64	MILLI/MJ	69.18	
MANGANESE	MILLI	0.11	-	-	-	0.06	MILLI/MJ	0.48	
IRON	MILLI	0.40	0.30	-	0.50	0.22	MILLI/MJ	1.73	
COPPER	MICRO	80.00	70.00	-	80.00	43.20	MICRO/MJ	345.88	
ZINC	MILLI	0.26	-	-	-	0.14	MILLI/MJ	1.12	
PHOSPHORUS	MILLI	9.00	7.00	-	11.00	4.86	MILLI/MJ	38.91	
CHLORIDE	MILLI	39.00	29.00	-	46.00	21.06	MILLI/MJ	168.62	
FLUORIDE	MICRO	14.00	-	-	-	7.56	MICRO/MJ	60.53	
IODIDE	MICRO	-	0.20	-	9.70				
SELENIUM	MICRO	0.55	0.50	-	0.60	0.30	MICRO/MJ	2.38	
CAROTENE	MICRO	60.00	42.00	-	90.00	32.40	MICRO/MJ	259.41	
VITAMIN E ACTIVITY	MILLI	0.10	-	-	-	0.05	MILLI/MJ	0.43	
ALPHA-TOCOPHEROL	MILLI	0.10	-	-	-	0.05	MILLI/MJ	0.43	
VITAMIN B1	MICRO	80.00	70.00	-	107.00	43.20	MICRO/MJ	345.88	
VITAMIN B2	MICRO	30.00	20.00	-	60.00	16.20	MICRO/MJ	129.71	
NICOTINAMIDE	MILLI	0.22	0.20	-	0.34	0.12	MILLI/MJ	0.95	
PANTOTHENIC ACID	MILLI	0.18	0.16	-	0.19	0.10	MILLI/MJ	0.78	

1 METHOD OF MEUSER, SUCKOW AND KULIKOWSKI ("BERLINER METHODE")

CONSTITUENTS	DIM	AV	VARIATION			AVR	NUTR.	DENS.	MOLPERC.
VITAMIN B6	MICRO	75.00	10.00	-	140.00	40.50	MICRO/MJ	324.27	
FOLIC ACID	MICRO	4.00	2.50	-	4.80	2.16	MICRO/MJ	17.29	
VITAMIN C	MILLI	19.00	10.00	-	25.00	10.26	MILLI/MJ	82.15	
SEROTONINE	MILLI	2.00	-	-	-	1.08	MILLI/MJ	8.65	
TOTAL ACIDS	GRAM	0.72	0.67	-	0.77	0.39	GRAM/MJ	3.11	
MALIC ACID	MILLI	94.00	87.00	-	100.00	50.76	MILLI/MJ	406.41	
CITRIC ACID	GRAM	0.63	0.58	-	0.67	0.34	GRAM/MJ	2.72	
SALICYLIC ACID	MILLI	2.10	2.10	-	2.10	1.13	MILLI/MJ	9.08	
GLUCOSE	GRAM	2.13	1.58	-	2.50	1.15	GRAM/MJ	9.21	
FRUCTOSE	GRAM	2.44	1.40	-	3.30	1.32	GRAM/MJ	10.55	
SUCROSE	GRAM	7.83	6.60	-	9.39	4.23	GRAM/MJ	33.85	
TOTAL STEROLS	MILLI	6.00	-	-	-	3.24	MILLI/MJ	25.94	
CAMPESTEROL	MILLI	1.00	-	-	-	0.54	MILLI/MJ	4.32	
BETA-SITOSTEROL	MILLI	4.00	-	-	-	2.16	MILLI/MJ	17.29	
DIETARY FIBRE,WAT.SOL.	GRAM	0.50	-	-	-	0.27	GRAM/MJ	2.16	
DIETARY FIBRE,WAT.INS.	GRAM	0.90	-	-	-	0.49	GRAM/MJ	3.89	

ANANAS IN DOSEN [1] | PINEAPPLE CANNED [1] | ANANAS EN BOÎTES [1]

		PROTEIN	FAT	CARBOHYDRATES	TOTAL
ENERGY VALUE (AVERAGE) PER 100 G EDIBLE PORTION	KJOULE	6.6	7.8	338	352
	(KCAL)	1.6	1.9	81	84
AMOUNT OF DIGESTIBLE CONSTITUENTS PER 100 G	GRAM	0.34	0.18	20.20	
ENERGY VALUE (AVERAGE) OF THE DIGESTIBLE FRACTION PER 100 G EDIBLE PORTION	KJOULE	5.6	7.0	338	351
	(KCAL)	1.3	1.7	81	84

WASTE PERCENTAGE AVERAGE 0.00

CONSTITUENTS	DIM	AV	VARIATION			AVR	NUTR. DENS.		MOLPERC.
WATER	GRAM	75.80	74.00	-	80.80	75.80	GRAM/MJ	216.15	
PROTEIN	GRAM	0.40	0.30	-	0.40	0.40	GRAM/MJ	1.14	
FAT	GRAM	0.20	0.10	-	0.20	0.20	GRAM/MJ	0.57	
AVAILABLE CARBOHYDR.	GRAM	20.20	-	-	-	20.20	GRAM/MJ	57.60	
TOTAL DIETARY FIBRE	GRAM	0.90 [2]	-	-	-	0.90	GRAM/MJ	2.57	
MINERALS	GRAM	0.40	-	-	-	0.40	GRAM/MJ	1.14	
SODIUM	MILLI	1.00	0.50	-	2.00	1.00	MILLI/MJ	2.85	
POTASSIUM	MILLI	75.00	57.00	-	140.00	75.00	MILLI/MJ	213.86	
MAGNESIUM	MILLI	8.00	-	-	-	8.00	MILLI/MJ	22.81	
CALCIUM	MILLI	13.00	11.00	-	15.00	13.00	MILLI/MJ	37.07	
IRON	MILLI	0.30	-	-	-	0.30	MILLI/MJ	0.86	
COPPER	MICRO	50.00	-	-	-	50.00	MICRO/MJ	142.58	
PHOSPHORUS	MILLI	7.00	5.00	-	8.00	7.00	MILLI/MJ	19.96	
CHLORIDE	MILLI	4.00	-	-	-	4.00	MILLI/MJ	11.41	
BORON	MICRO	70.00	55.00	-	85.00	70.00	MICRO/MJ	199.61	
SELENIUM	MICRO	1.00	0.80	-	1.20	1.00	MICRO/MJ	2.85	
CAROTENE	MICRO	40.00	30.00	-	70.00	40.00	MICRO/MJ	114.06	
VITAMIN B1	MICRO	70.00	35.00	-	80.00	70.00	MICRO/MJ	199.61	
VITAMIN B2	MICRO	20.00	-	-	-	20.00	MICRO/MJ	57.03	
NICOTINAMIDE	MILLI	0.20	-	-	-	0.20	MILLI/MJ	0.57	
PANTOTHENIC ACID	MICRO	30.00	-	-	-	30.00	MICRO/MJ	85.55	
VITAMIN B6	MICRO	70.00	-	-	-	70.00	MICRO/MJ	199.61	
FOLIC ACID	MICRO	2.00	-	-	-	2.00	MICRO/MJ	5.70	
VITAMIN C	MILLI	7.00	5.00	-	9.00	7.00	MILLI/MJ	19.96	
SALICYLIC ACID	MILLI	1.40	1.40	-	1.40	1.40	MILLI/MJ	3.99	
TOTAL SUGAR	GRAM	20.20	-	-	-	20.20	GRAM/MJ	57.60	

1 TOTAL CONTENT (SOLID AND LIQUID)

2 METHOD OF SOUTHGATE

APFELSINE (ORANGE) — ORANGE — ORANGE

CITRUS AURANTIUM L. SUBSP. SINENSIS L.

		PROTEIN	FAT	CARBOHYDRATES	TOTAL
ENERGY VALUE (AVERAGE) PER 100 G EDIBLE PORTION	KJOULE	17	7.8	154	178
	(KCAL)	4.0	1.9	37	43
AMOUNT OF DIGESTIBLE CONSTITUENTS PER 100 G	GRAM	0.85	0.18	9.19	
ENERGY VALUE (AVERAGE) OF THE DIGESTIBLE FRACTION PER 100 G EDIBLE PORTION	KJOULE	14	7.0	154	175
	(KCAL)	3.4	1.7	37	42

WASTE PERCENTAGE AVERAGE 28 MINIMUM 20 MAXIMUM 41

CONSTITUENTS	DIM	AV		VARIATION		AVR	NUTR. DENS.		MOLPERC.
WATER	GRAM	85.70		84.30 -	87.20	61.70	GRAM/MJ	490.12	
PROTEIN	GRAM	1.00		0.80 -	1.30	0.72	GRAM/MJ	5.72	
FAT	GRAM	0.20		0.10 -	0.37	0.14	GRAM/MJ	1.14	
AVAILABLE CARBOHYDR.	GRAM	9.19		8.46 -	10.20	6.62	GRAM/MJ	52.56	
TOTAL DIETARY FIBRE	GRAM	2.20	1	1.50 -	2.90	1.58	GRAM/MJ	12.58	
MINERALS	GRAM	0.48		0.38 -	0.57	0.35	GRAM/MJ	2.75	
SODIUM	MILLI	1.40		0.30 -	3.00	1.01	MILLI/MJ	8.01	
POTASSIUM	MILLI	177.00		150.00 -	206.00	127.44	GRAM/MJ	1.01	
MAGNESIUM	MILLI	14.00		11.00 -	18.00	10.08	MILLI/MJ	80.07	
CALCIUM	MILLI	42.00		33.00 -	58.00	30.24	MILLI/MJ	240.20	
MANGANESE	MICRO	29.00		24.00 -	35.00	20.88	MICRO/MJ	165.85	
IRON	MILLI	0.40		0.20 -	0.55	0.29	MILLI/MJ	2.29	
COBALT	MICRO	0.50		0.10 -	1.00	0.36	MICRO/MJ	2.86	
COPPER	MICRO	67.00		43.00 -	100.00	48.24	MICRO/MJ	383.17	
ZINC	MILLI	0.10		0.08 -	0.17	0.07	MILLI/MJ	0.57	
NICKEL	MICRO	10.00		1.00 -	17.00	7.20	MICRO/MJ	57.19	
CHROMIUM	MICRO	1.00		0.50 -	2.00	0.72	MICRO/MJ	5.72	
PHOSPHORUS	MILLI	23.00		20.00 -	26.00	16.56	MILLI/MJ	131.54	
CHLORIDE	MILLI	4.00		- -	-	2.88	MILLI/MJ	22.88	
FLUORIDE	MICRO	5.00		2.00 -	7.00	3.60	MICRO/MJ	28.60	
IODIDE	MICRO	2.10		- -	-	1.51	MICRO/MJ	12.01	
BORON	MILLI	0.18		0.10 -	0.25	0.13	MILLI/MJ	1.03	
SELENIUM	MICRO	3.50		1.00 -	24.00	2.52	MICRO/MJ	20.02	
BROMINE	MICRO	54.00		53.00 -	54.00	38.88	MICRO/MJ	308.83	
CAROTENE	MICRO	90.00		60.00 -	120.00	64.80	MICRO/MJ	514.71	
VITAMIN E ACTIVITY	MILLI	0.24		- -	-	0.17	MILLI/MJ	1.37	

1 METHOD OF MEUSER, SUCKOW AND KULIKOWSKI ("BERLINER METHODE")

CONSTITUENTS	DIM	AV	VARIATION			AVR	NUTR. DENS.		MOLPERC.
ALPHA-TOCOPHEROL	MILLI	0.24	-	-	-	0.17	MILLI/MJ	1.37	
VITAMIN K	MICRO	2.50	0.00	-	5.00	1.80	MICRO/MJ	14.30	
VITAMIN B1	MICRO	79.00	70.00	-	100.00	56.88	MICRO/MJ	451.80	
VITAMIN B2	MICRO	42.00	20.00	-	67.00	30.24	MICRO/MJ	240.20	
NICOTINAMIDE	MILLI	0.30	0.20	-	0.50	0.22	MILLI/MJ	1.72	
PANTOTHENIC ACID	MILLI	0.24	0.20	-	0.31	0.17	MILLI/MJ	1.37	
VITAMIN B6	MICRO	50.00	30.00	-	80.00	36.00	MICRO/MJ	285.95	
BIOTIN	MICRO	2.30	1.90	-	2.60	1.66	MICRO/MJ	13.15	
FOLIC ACID	MICRO	24.00	13.00	-	38.00	17.28	MICRO/MJ	137.26	
VITAMIN C	MILLI	50.00	39.00	-	65.00	36.00	MILLI/MJ	285.95	
ALANINE	MILLI	29.00	29.00	-	29.00	20.88	MILLI/MJ	165.85	5.5
ARGININE	MILLI	73.00	73.00	-	73.00	52.56	MILLI/MJ	417.49	7.0
ASPARTIC ACID	MILLI	122.00	122.00	-	122.00	87.84	MILLI/MJ	697.72	15.4
CYSTINE	MILLI	3.00	3.00	-	3.00	2.16	MILLI/MJ	17.16	0.2
GLUTAMIC ACID	MILLI	66.00	66.00	-	66.00	47.52	MILLI/MJ	377.45	7.5
GLYCINE	MILLI	23.00	23.00	-	23.00	16.56	MILLI/MJ	131.54	5.1
HISTIDINE	MILLI	12.00	12.00	-	12.00	8.64	MILLI/MJ	68.63	1.3
ISOLEUCINE	MILLI	20.00	20.00	-	20.00	14.40	MILLI/MJ	114.38	2.6
LEUCINE	MILLI	32.00	32.00	-	32.00	23.04	MILLI/MJ	183.01	4.1
LYSINE	MILLI	39.00	39.00	-	39.00	28.08	MILLI/MJ	223.04	4.5
METHIONINE	MILLI	8.00	8.00	-	8.00	5.76	MILLI/MJ	45.75	0.9
PHENYLALANINE	MILLI	20.00	20.00	-	20.00	14.40	MILLI/MJ	114.38	2.0
PROLINE	MILLI	189.00	189.00	-	189.00	136.08	GRAM/MJ	1.08	27.6
SERINE	MILLI	43.00	43.00	-	43.00	30.96	MILLI/MJ	245.92	6.9
THREONINE	MILLI	20.00	20.00	-	20.00	14.40	MILLI/MJ	114.38	2.8
TRYPTOPHAN	MILLI	7.00	7.00	-	7.00	5.04	MILLI/MJ	40.03	0.6
TYROSINE	MILLI	13.00	13.00	-	13.00	9.36	MILLI/MJ	74.35	1.2
VALINE	MILLI	33.00	33.00	-	33.00	23.76	MILLI/MJ	188.73	4.7
TRYPTAMINE	MILLI	0.10	-	-	-	0.07	MILLI/MJ	0.57	
TYRAMINE	MILLI	1.00	-	-	-	0.72	MILLI/MJ	5.72	
MALIC ACID	GRAM	0.16	0.13	-	0.19	0.12	GRAM/MJ	0.92	
CITRIC ACID	GRAM	1.06	0.93	-	1.31	0.76	GRAM/MJ	6.06	
FERULIC ACID	MILLI	1.00	-	-	-	0.72	MILLI/MJ	5.72	
CAFFEIC ACID	MILLI	5.00	-	-	-	3.60	MILLI/MJ	28.60	
PARA-COUMARIC ACID	MILLI	0.50	-	-	-	0.36	MILLI/MJ	2.86	
SALICYLIC ACID	MILLI	2.40	2.40	-	2.40	1.73	MILLI/MJ	13.73	
GLUCOSE	GRAM	2.23	1.81	-	2.44	1.61	GRAM/MJ	12.78	
FRUCTOSE	GRAM	2.52	2.38	-	3.03	1.82	GRAM/MJ	14.45	
SUCROSE	GRAM	3.22	3.21	-	3.23	2.32	GRAM/MJ	18.42	
PECTIN	GRAM	-	0.90	-	3.80				
PENTOSAN	GRAM	0.30	-	-	-	0.22	GRAM/MJ	1.72	
LAURIC ACID	MILLI	0.60	0.60	-	0.60	0.43	MILLI/MJ	3.43	
MYRISTIC ACID	MILLI	0.70	0.70	-	0.70	0.50	MILLI/MJ	4.00	
PALMITIC ACID	MILLI	20.66	14.00	-	34.00	14.88	MILLI/MJ	118.19	
STEARIC ACID	MILLI	2.80	1.20	-	6.00	2.02	MILLI/MJ	16.01	
PALMITOLEIC ACID	MILLI	9.06	6.00	-	10.60	6.53	MILLI/MJ	51.85	
OLEIC ACID	MILLI	43.73	37.00	-	47.10	31.49	MILLI/MJ	250.11	
EICOSENOIC ACID	MILLI	1.20	1.20	-	1.20	0.86	MILLI/MJ	6.86	
LINOLEIC ACID	MILLI	51.83	49.75	-	56.00	37.32	MILLI/MJ	296.43	
LINOLENIC ACID	MILLI	27.83	22.00	-	30.75	20.04	MILLI/MJ	159.18	
CAMPESTEROL	MILLI	4.00	-	-	-	2.88	MILLI/MJ	22.88	
BETA-SITOSTEROL	MILLI	17.00	-	-	-	12.24	MILLI/MJ	97.22	
STIGMASTEROL	MILLI	2.00	-	-	-	1.44	MILLI/MJ	11.44	
DIETARY FIBRE,WAT.SOL.	GRAM	1.30	0.90	-	1.70	0.94	GRAM/MJ	7.43	
DIETARY FIBRE,WAT.INS.	GRAM	0.90	0.60	-	1.20	0.65	GRAM/MJ	5.15	

AVOCADO AVOCADO AVOCAT

PERSEA GRATISSIMA GAERTN.

		PROTEIN	FAT	CARBOHYDRATES	TOTAL
ENERGY VALUE (AVERAGE) PER 100 G EDIBLE PORTION	KJOULE	31	914	6.7	953
	(KCAL)	7.5	219	1.6	228
AMOUNT OF DIGESTIBLE CONSTITUENTS PER 100 G	GRAM	1.61	21.15	0.40	
ENERGY VALUE (AVERAGE) OF THE DIGESTIBLE FRACTION PER 100 G EDIBLE PORTION	KJOULE	27	823	6.7	856
	(KCAL)	6.4	197	1.6	205

WASTE PERCENTAGE AVERAGE 25

CONSTITUENTS	DIM	AV	VARIATION			AVR	NUTR. DENS.		MOLPERC.
WATER	GRAM	68.00	58.70	-	82.30	51.00	GRAM/MJ	79.41	
PROTEIN	GRAM	1.90	1.14	-	4.39	1.43	GRAM/MJ	2.22	
FAT	GRAM	23.50	9.80	-	31.60	17.63	GRAM/MJ	27.44	
AVAILABLE CARBOHYDR.	GRAM	0.40	-	-	-	0.30	GRAM/MJ	0.47	
TOTAL DIETARY FIBRE	GRAM	3.30 [1]	-	-	-	2.48	GRAM/MJ	3.85	
MINERALS	GRAM	1.36	0.54	-	1.40	1.02	GRAM/MJ	1.59	
SODIUM	MILLI	3.00	-	-	-	2.25	MILLI/MJ	3.50	
POTASSIUM	MILLI	503.00	400.00	-	700.00	377.25	MILLI/MJ	587.37	
MAGNESIUM	MILLI	29.00	-	-	-	21.75	MILLI/MJ	33.86	
CALCIUM	MILLI	10.00	8.00	-	16.00	7.50	MILLI/MJ	11.68	
MANGANESE	-	TRACES	-	-	-				
IRON	MILLI	0.60	0.50	-	1.50	0.45	MILLI/MJ	0.70	
COPPER	MILLI	0.21	-	-	-	0.16	MILLI/MJ	0.25	
VANADIUM	MICRO	9.00	-	-	-	6.75	MICRO/MJ	10.51	
PHOSPHORUS	MILLI	38.00	30.00	-	50.00	28.50	MILLI/MJ	44.37	
CHLORIDE	MILLI	6.00	-	-	-	4.50	MILLI/MJ	7.01	
BORON	MILLI	0.95	0.54	-	1.34	0.72	MILLI/MJ	1.12	
CAROTENE	MICRO	72.00	60.00	-	87.00	54.00	MICRO/MJ	84.08	
VITAMIN E ACTIVITY	MILLI	1.30	-	-	-	0.98	MILLI/MJ	1.52	
ALPHA-TOCOPHEROL	MILLI	1.30	-	-	-	0.98	MILLI/MJ	1.52	
VITAMIN K	MICRO	8.00	-	-	-	6.00	MICRO/MJ	9.34	
VITAMIN B1	MICRO	80.00	60.00	-	100.00	60.00	MICRO/MJ	93.42	
VITAMIN B2	MILLI	0.15	0.13	-	0.17	0.11	MILLI/MJ	0.18	
NICOTINAMIDE	MILLI	1.10	1.00	-	2.40	0.83	MILLI/MJ	1.28	
PANTOTHENIC ACID	MILLI	1.10	-	-	-	0.83	MILLI/MJ	1.28	
VITAMIN B6	MILLI	0.53	-	-	-	0.40	MILLI/MJ	0.62	

1 METHOD OF SOUTHGATE

CONSTITUENTS	DIM	AV	VARIATION			AVR	NUTR. DENS.		MOLPERC.
BIOTIN	MICRO	10.00	-	-	-	7.50	MICRO/MJ	11.68	
FOLIC ACID	MICRO	30.00	-	-	-	22.50	MICRO/MJ	35.03	
VITAMIN C	MILLI	13.00	9.00	-	16.00	9.75	MILLI/MJ	15.18	
ALANINE	MILLI	175.00	-	-	-	131.25	MILLI/MJ	204.35	10.9
ARGININE	MILLI	60.00	-	-	-	45.00	MILLI/MJ	70.06	1.9
ASPARTIC ACID	MILLI	242.00	-	-	-	181.50	MILLI/MJ	282.59	10.1
CYSTINE	MILLI	6.00	-	-	-	4.50	MILLI/MJ	7.01	0.1
GLUTAMIC ACID	MILLI	284.00	-	-	-	213.00	MILLI/MJ	331.64	10.8
GLYCINE	MILLI	158.00	-	-	-	118.50	MILLI/MJ	184.50	11.7
HISTIDINE	MILLI	30.00	-	-	-	22.50	MILLI/MJ	35.03	1.1
ISOLEUCINE	MILLI	111.00	-	-	-	83.25	MILLI/MJ	129.62	4.7
LEUCINE	MILLI	197.00	-	-	-	147.75	MILLI/MJ	230.04	8.4
LYSINE	MILLI	155.00	-	-	-	116.25	MILLI/MJ	181.00	5.9
METHIONINE	MILLI	43.00	-	-	-	32.25	MILLI/MJ	50.21	1.6
PHENYLALANINE	MILLI	111.00	-	-	-	83.25	MILLI/MJ	129.62	3.7
PROLINE	MILLI	114.00	-	-	-	85.50	MILLI/MJ	133.12	5.5
SERINE	MILLI	124.00	-	-	-	93.00	MILLI/MJ	144.80	6.6
THREONINE	MILLI	119.00	-	-	-	89.25	MILLI/MJ	138.96	5.6
TRYPTOPHAN	MILLI	22.00	-	-	-	16.50	MILLI/MJ	25.69	0.6
TYROSINE	MILLI	73.00	-	-	-	54.75	MILLI/MJ	85.24	2.2
VALINE	MILLI	172.00	-	-	-	129.00	MILLI/MJ	200.85	8.2
GAMMA-AMINO-BUT.ACID	MILLI	5.00	-	-	-	3.75	MILLI/MJ	5.84	0.3
SEROTONINE	MILLI	1.00	-	-	-	0.75	MILLI/MJ	1.17	
TYRAMINE	MILLI	2.30	-	-	-	1.73	MILLI/MJ	2.69	
SALICYLIC ACID	MILLI	0.60	0.60	-	0.60	0.45	MILLI/MJ	0.70	
GLUCOSE	GRAM	0.10	0.10	-	0.10	0.08	GRAM/MJ	0.12	
FRUCTOSE	GRAM	0.20	0.20	-	0.20	0.15	GRAM/MJ	0.23	
SUCROSE	GRAM	0.10	0.10	-	0.10	0.08	GRAM/MJ	0.12	
PALMITIC ACID	GRAM	3.37	2.92	-	3.82	2.53	GRAM/MJ	3.94	
STEARIC ACID	-	TRACES	-	-	-				
ARACHIDIC ACID	-	TRACES	-	-	-				
PALMITOLEIC ACID	GRAM	0.91	0.67	-	1.15	0.68	GRAM/MJ	1.06	
OLEIC ACID	GRAM	15.60	15.10	-	16.20	11.70	GRAM/MJ	18.22	
LINOLEIC ACID	GRAM	1.97	1.41	-	2.54	1.48	GRAM/MJ	2.30	
LINOLENIC ACID	-	TRACES	-	-	-				
DIETARY FIBRE,WAT.SOL.	GRAM	1.30	-	-	-	0.98	GRAM/MJ	1.52	
DIETARY FIBRE,WAT.INS.	GRAM	2.00	-	-	-	1.50	GRAM/MJ	2.34	

BAMBUSSPROSSEN BAMBOO SHOOTS POUSSES DE BAMBOU

BAMBUSA VULGARIS SCHRAD.

		PROTEIN	FAT	CARBOHYDRATES	TOTAL
ENERGY VALUE (AVERAGE) PER 100 G EDIBLE PORTION	KJOULE	39	12	19	70
	(KCAL)	9.4	2.8	4.5	17
AMOUNT OF DIGESTIBLE CONSTITUENTS PER 100 G	GRAM	1.62	0.27	1.13	
ENERGY VALUE (AVERAGE) OF THE DIGESTIBLE FRACTION PER 100 G EDIBLE PORTION	KJOULE	25	11	19	55
	(KCAL)	6.1	2.5	4.5	13

WASTE PERCENTAGE AVERAGE 58 MINIMUM 44 MAXIMUM 71

CONSTITUENTS	DIM	AV	VARIATION			AVR	NUTR. DENS.		MOLPERC.
WATER	GRAM	91.00	-	-	-	38.22	KILO/MJ	1.66	
PROTEIN	GRAM	2.50	2.50	-	2.60	1.05	GRAM/MJ	45.53	
FAT	GRAM	0.30	-	-	-	0.13	GRAM/MJ	5.46	
AVAILABLE CARBOHYDR.	GRAM	1.13	1.13	-	1.13	0.47	GRAM/MJ	20.58	
MINERALS	GRAM	0.90	-	-	-	0.38	GRAM/MJ	16.39	
SODIUM	MILLI	6.00	-	-	-	2.52	MILLI/MJ	109.26	
POTASSIUM	MILLI	468.00	402.00	-	533.00	196.56	GRAM/MJ	8.52	
CALCIUM	MILLI	15.00	13.00	-	17.00	6.30	MILLI/MJ	273.15	
IRON	MILLI	0.70	0.50	-	0.90	0.29	MILLI/MJ	12.75	
PHOSPHORUS	MILLI	53.00	47.00	-	59.00	22.26	MILLI/MJ	965.15	
CAROTENE	MICRO	14.00	12.00	-	15.00	5.88	MICRO/MJ	254.94	
VITAMIN B1	MILLI	0.13	0.11	-	0.15	0.05	MILLI/MJ	2.37	
VITAMIN B2	MICRO	80.00	70.00	-	90.00	33.60	MILLI/MJ	1.46	
NICOTINAMIDE	MILLI	0.60	-	-	-	0.25	MILLI/MJ	10.93	
VITAMIN C	MILLI	6.50	4.00	-	9.00	2.73	MILLI/MJ	118.37	
ALANINE	MILLI	120.00	120.00	-	120.00	50.40	GRAM/MJ	2.19	7.4
ARGININE	MILLI	92.00	92.00	-	92.00	38.64	GRAM/MJ	1.68	2.9
ASPARTIC ACID	MILLI	400.00	400.00	-	400.00	168.00	GRAM/MJ	7.28	16.5
CYSTINE	MILLI	21.00	21.00	-	21.00	8.82	MILLI/MJ	382.42	0.5
GLUTAMIC ACID	MILLI	240.00	240.00	-	240.00	100.80	GRAM/MJ	4.37	9.0
GLYCINE	MILLI	84.00	84.00	-	84.00	35.28	GRAM/MJ	1.53	6.1
HISTIDINE	MILLI	40.00	40.00	-	40.00	16.80	MILLI/MJ	728.41	1.4
ISOLEUCINE	MILLI	84.00	84.00	-	84.00	35.28	GRAM/MJ	1.53	3.5
LEUCINE	MILLI	136.00	136.00	-	136.00	57.12	GRAM/MJ	2.48	5.7
LYSINE	MILLI	128.00	128.00	-	128.00	53.76	GRAM/MJ	2.33	4.8

CONSTITUENTS	DIM	AV	VARIATION			AVR	NUTR. DENS.		MOLPERC.
METHIONINE	MILLI	28.00	28.00	-	28.00	11.76	MILLI/MJ	509.89	1.0
PHENYLALANINE	MILLI	88.00	88.00	-	88.00	36.96	GRAM/MJ	1.60	2.9
PROLINE	MILLI	212.00	212.00	-	212.00	89.04	GRAM/MJ	3.86	10.1
SERINE	MILLI	124.00	124.00	-	124.00	52.08	GRAM/MJ	2.26	6.5
THREONINE	MILLI	84.00	84.00	-	84.00	35.28	GRAM/MJ	1.53	3.9
TRYPTOPHAN	MILLI	26.00	26.00	-	26.00	10.92	MILLI/MJ	473.47	0.7
TYROSINE	MILLI	400.00	400.00	-	400.00	168.00	GRAM/MJ	7.28	12.1
VALINE	MILLI	104.00	104.00	-	104.00	43.68	GRAM/MJ	1.89	4.9
MALIC ACID	MILLI	76.00	76.00	-	76.00	31.92	GRAM/MJ	1.38	
CITRIC ACID	MILLI	93.00	93.00	-	93.00	39.06	GRAM/MJ	1.69	
OXALIC ACID	MILLI	252.00	252.00	-	252.00	105.84	GRAM/MJ	4.59	
SUCCINIC ACID	MILLI	2.70	2.70	-	2.70	1.13	MILLI/MJ	49.17	
FUMARIC ACID	MILLI	6.00	6.00	-	6.00	2.52	MILLI/MJ	109.26	
GLUCOSE	GRAM	0.35	0.35	-	0.35	0.15	GRAM/MJ	6.37	
FRUCTOSE	GRAM	0.41	0.41	-	0.41	0.17	GRAM/MJ	7.47	
SUCROSE	GRAM	0.20	0.20	-	0.20	0.08	GRAM/MJ	3.64	
PALMITIC ACID	MILLI	55.00	55.00	-	55.00	23.10	GRAM/MJ	1.00	
OLEIC ACID	MILLI	11.00	11.00	-	11.00	4.62	MILLI/MJ	200.31	
LINOLEIC ACID	MILLI	120.00	120.00	-	120.00	50.40	GRAM/MJ	2.19	
LINOLENIC ACID	MILLI	52.00	52.00	-	52.00	21.84	MILLI/MJ	946.94	

BANANE BANANA BANANE

MUSA PARADISIACA L.

		PROTEIN	FAT	CARBOHYDRATES	TOTAL
ENERGY VALUE (AVERAGE) PER 100 G EDIBLE PORTION	KJOULE	19	7.0	358	384
	(KCAL)	4.5	1.7	86	92
AMOUNT OF DIGESTIBLE CONSTITUENTS PER 100 G	GRAM	0.97	0.16	21.39	
ENERGY VALUE (AVERAGE) OF THE DIGESTIBLE FRACTION PER 100 G EDIBLE PORTION	KJOULE	16	6.3	358	380
	(KCAL)	3.9	1.5	86	91

WASTE PERCENTAGE AVERAGE 33 MINIMUM 29 MAXIMUM 42

CONSTITUENTS	DIM	AV		VARIATION			AVR	NUTR.	DENS.	MOLPERC.
WATER	GRAM	73.90		70.70	-	76.50	49.51	GRAM/MJ	194.25	
PROTEIN	GRAM	1.15		1.05	-	1.30	0.77	GRAM/MJ	3.02	
FAT	GRAM	0.18		0.10	-	0.38	0.12	GRAM/MJ	0.47	
AVAILABLE CARBOHYDR.	GRAM	21.39		-	-	-	14.33	GRAM/MJ	56.22	
TOTAL DIETARY FIBRE	GRAM	2.00	1	1.50	-	2.50	1.34	GRAM/MJ	5.26	
MINERALS	GRAM	0.83		0.68	-	0.95	0.56	GRAM/MJ	2.18	
SODIUM	MILLI	1.00		0.30	-	2.00	0.67	MILLI/MJ	2.63	
POTASSIUM	MILLI	393.00		348.00	-	509.00	263.31	GRAM/MJ	1.03	
MAGNESIUM	MILLI	36.00		31.00	-	42.00	24.12	MILLI/MJ	94.63	
CALCIUM	MILLI	8.70		5.60	-	11.40	5.83	MILLI/MJ	22.87	
MANGANESE	MILLI	0.53		0.18	-	0.80	0.36	MILLI/MJ	1.39	
IRON	MILLI	0.55		0.41	-	0.70	0.37	MILLI/MJ	1.45	
COBALT	MICRO	-		0.25	-	5.00				
COPPER	MILLI	0.13		0.07	-	0.21	0.09	MILLI/MJ	0.34	
ZINC	MILLI	0.22		0.15	-	0.28	0.15	MILLI/MJ	0.58	
NICKEL	MICRO	34.00		-	-	-	22.78	MICRO/MJ	89.37	
CHROMIUM	MICRO	7.50		0.00	-	15.00	5.03	MICRO/MJ	19.71	
MOLYBDENUM	MICRO	3.00		-	-	-	2.01	MICRO/MJ	7.89	
VANADIUM	MICRO	6.00		-	-	-	4.02	MICRO/MJ	15.77	
PHOSPHORUS	MILLI	28.00		22.00	-	31.00	18.76	MILLI/MJ	73.60	
CHLORIDE	MILLI	109.00		79.00	-	125.00	73.03	MILLI/MJ	286.51	
FLUORIDE	MICRO	20.00		17.00	-	23.00	13.40	MICRO/MJ	52.57	
IODIDE	MICRO	2.80		-	-	-	1.88	MICRO/MJ	7.36	
BORON	MICRO	79.00		50.00	-	125.00	52.93	MICRO/MJ	207.65	
SELENIUM	MICRO	4.40		1.00	-	17.00	2.95	MICRO/MJ	11.57	
BROMINE	MICRO	28.00		-	-	-	18.76	MICRO/MJ	73.60	
SILICON	MILLI	8.00		7.00	-	9.00	5.36	MILLI/MJ	21.03	

1 METHOD OF MEUSER, SUCKOW AND KULIKOWSKI ("BERLINER METHODE")

CONSTITUENTS	DIM	AV	VARIATION			AVR	NUTR. DENS.		MOLPERC.
CAROTENE	MILLI	0.23	0.11	-	0.40	0.15	MILLI/MJ	0.60	
VITAMIN E ACTIVITY	MILLI	0.27	-	-	-	0.18	MILLI/MJ	0.71	
ALPHA-TOCOPHEROL	MILLI	0.27	-	-	-	0.18	MILLI/MJ	0.71	
VITAMIN B1	MICRO	44.00	30.00	-	50.00	29.48	MICRO/MJ	115.66	
VITAMIN B2	MICRO	57.00	47.00	-	80.00	38.19	MICRO/MJ	149.83	
NICOTINAMIDE	MILLI	0.65	0.56	-	0.70	0.44	MILLI/MJ	1.71	
PANTOTHENIC ACID	MILLI	0.23	0.18	-	0.31	0.15	MILLI/MJ	0.60	
VITAMIN B6	MILLI	0.37	0.30	-	0.50	0.25	MILLI/MJ	0.97	
BIOTIN	MICRO	5.50	4.00	-	8.00	3.69	MICRO/MJ	14.46	
FOLIC ACID	MICRO	20.00	10.00	-	30.00	13.40	MICRO/MJ	52.57	
VITAMIN C	MILLI	12.00	7.00	-	21.00	8.04	MILLI/MJ	31.54	
ALANINE	MILLI	46.00	46.00	-	46.00	30.82	MILLI/MJ	120.91	7.5
ARGININE	MILLI	54.00	54.00	-	54.00	36.18	MILLI/MJ	141.94	4.5
ASPARTIC ACID	MILLI	115.00	115.00	-	115.00	77.05	MILLI/MJ	302.28	12.5
CYSTINE	MILLI	2.00	2.00	-	2.00	1.34	MILLI/MJ	5.26	0.1
GLUTAMIC ACID	MILLI	105.00	105.00	-	105.00	70.35	MILLI/MJ	276.00	10.4
GLYCINE	MILLI	42.00	42.00	-	42.00	28.14	MILLI/MJ	110.40	8.1
HISTIDINE	MILLI	77.00	77.00	-	77.00	51.59	MILLI/MJ	202.40	7.2
ISOLEUCINE	MILLI	38.00	38.00	-	38.00	25.46	MILLI/MJ	99.88	4.2
LEUCINE	MILLI	85.00	85.00	-	85.00	56.95	MILLI/MJ	223.42	9.4
LYSINE	MILLI	57.00	57.00	-	57.00	38.19	MILLI/MJ	149.83	5.7
METHIONINE	MILLI	9.00	9.00	-	9.00	6.03	MILLI/MJ	23.66	0.9
PHENYLALANINE	MILLI	34.00	34.00	-	34.00	22.78	MILLI/MJ	89.37	3.0
PROLINE	MILLI	40.00	40.00	-	40.00	26.80	MILLI/MJ	105.14	5.0
SERINE	MILLI	49.00	49.00	-	49.00	32.83	MILLI/MJ	128.80	6.8
THREONINE	MILLI	38.00	38.00	-	38.00	25.46	MILLI/MJ	99.88	4.6
TRYPTOPHAN	MILLI	18.00	18.00	-	18.00	12.06	MILLI/MJ	47.31	1.3
TYROSINE	MILLI	21.00	21.00	-	21.00	14.07	MILLI/MJ	55.20	1.7
VALINE	MILLI	57.00	57.00	-	57.00	38.19	MILLI/MJ	149.83	7.1
DOPAMINE	MILLI	65.00	-	-	-	43.55	MILLI/MJ	170.85	
NORADRENALINE	MILLI	10.40	-	-	-	6.97	MILLI/MJ	27.34	
SEROTONINE	MILLI	7.70	-	-	-	5.16	MILLI/MJ	20.24	
TYRAMINE	MILLI	0.70	-	-	-	0.47	MILLI/MJ	1.84	
MALIC ACID	MILLI	360.00	240.00	-	500.00	241.20	MILLI/MJ	946.27	
CITRIC ACID	MILLI	270.00	150.00	-	390.00	180.90	MILLI/MJ	709.70	
ACETIC ACID	MILLI	7.80	6.00	-	9.60	5.23	MILLI/MJ	20.50	
GLUCOSE	GRAM	3.89	2.41	-	6.32	2.61	GRAM/MJ	10.23	
FRUCTOSE	GRAM	3.71	2.58	-	4.13	2.49	GRAM/MJ	9.77	
SUCROSE	GRAM	10.38	5.88	-	14.00	6.95	GRAM/MJ	27.28	
STARCH	GRAM	2.76	2.70	-	2.94	1.85	GRAM/MJ	7.25	
PECTIN	GRAM	0.60	0.50	-	0.70	0.40	GRAM/MJ	1.58	
CELLULOSE	GRAM	0.37	-	-	-	0.25	GRAM/MJ	0.97	
LAURIC ACID	MILLI	0.20	0.20	-	0.20	0.13	MILLI/MJ	0.53	
MYRISTIC ACID	MILLI	0.55	0.55	-	0.55	0.37	MILLI/MJ	1.45	
PALMITIC ACID	MILLI	52.00	52.00	-	52.00	34.84	MILLI/MJ	136.68	
STEARIC ACID	MILLI	3.00	3.00	-	3.00	2.01	MILLI/MJ	7.89	
ARACHIDIC ACID	MILLI	0.50	0.50	-	0.50	0.34	MILLI/MJ	1.31	
BEHENIC ACID	MILLI	0.50	0.50	-	0.50	0.34	MILLI/MJ	1.31	
PALMITOLEIC ACID	MILLI	4.50	4.50	-	4.50	3.02	MILLI/MJ	11.83	
OLEIC ACID	MILLI	14.00	14.00	-	14.00	9.38	MILLI/MJ	36.80	
EICOSENOIC ACID	MILLI	0.30	0.30	-	0.30	0.20	MILLI/MJ	0.79	
LINOLEIC ACID	MILLI	34.50	34.50	-	34.50	23.12	MILLI/MJ	90.68	
LINOLENIC ACID	MILLI	24.50	24.50	-	24.50	16.42	MILLI/MJ	64.40	
TOTAL STEROLS	MILLI	16.00	-	-	-	10.72	MILLI/MJ	42.06	
CAMPESTEROL	MILLI	2.00	-	-	-	1.34	MILLI/MJ	5.26	
BETA-SITOSTEROL	MILLI	11.00	-	-	-	7.37	MILLI/MJ	28.91	
STIGMASTEROL	MILLI	3.00	-	-	-	2.01	MILLI/MJ	7.89	
DIETARY FIBRE,WAT.SOL.	GRAM	0.60	0.50	-	0.80	0.40	GRAM/MJ	1.58	
DIETARY FIBRE,WAT.INS.	GRAM	1.40	1.00	-	1.80	0.94	GRAM/MJ	3.68	

BAUMTOMATE
(TAMARILLO)

CYPHOMONDRA BETACEA
(CAV.) SENDTN.

TOMATO TREE

CYPHOMANDE BÊTASSE
(TOMATE EN ARBRE)

		PROTEIN	FAT	CARBOHYDRATES	TOTAL
ENERGY VALUE (AVERAGE)	KJOULE	27	31	177	235
PER 100 G	(KCAL)	6.4	7.4	42	56
EDIBLE PORTION					
AMOUNT OF DIGESTIBLE	GRAM	1.10	0.72	10.60	
CONSTITUENTS PER 100 G					
ENERGY VALUE (AVERAGE)	KJOULE	17	28	177	223
OF THE DIGESTIBLE	(KCAL)	4.1	6.7	42	53
FRACTION PER 100 G					
EDIBLE PORTION					

WASTE PERCENTAGE AVERAGE 27

CONSTITUENTS	DIM	AV		VARIATION			AVR	NUTR. DENS.		MOLPERC.
WATER	GRAM	86.00		-	-	-	62.78	GRAM/MJ	386.07	
PROTEIN	GRAM	1.70		1.50	-	1.90	1.24	GRAM/MJ	7.63	
FAT	GRAM	0.80		0.30	-	1.20	0.58	GRAM/MJ	3.59	
AVAILABLE CARBOHYDR.	GRAM	10.60	1	-	-	-	7.74	GRAM/MJ	47.59	
MINERALS	GRAM	0.90		0.80	-	1.00	0.66	GRAM/MJ	4.04	
POTASSIUM	MILLI	320.00		-	-	-	233.60	GRAM/MJ	1.44	
MAGNESIUM	MILLI	21.00		-	-	-	15.33	MILLI/MJ	94.27	
CALCIUM	MILLI	12.00		11.00	-	13.00	8.76	MILLI/MJ	53.87	
IRON	MILLI	0.70		0.60	-	0.80	0.51	MILLI/MJ	3.14	
PHOSPHORUS	MILLI	32.00		24.00	-	39.00	23.36	MILLI/MJ	143.66	
CAROTENE	MILLI	1.30		0.46	-	2.10	0.95	MILLI/MJ	5.84	
VITAMIN B1	MICRO	80.00		40.00	-	120.00	58.40	MICRO/MJ	359.14	
VITAMIN B2	MICRO	40.00		-	-	-	29.20	MICRO/MJ	179.57	
NICOTINAMIDE	MILLI	1.10		1.00	-	1.20	0.80	MILLI/MJ	4.94	
VITAMIN C	MILLI	24.00		17.00	-	30.00	17.52	MILLI/MJ	107.74	

1 ESTIMATED BY THE DIFFERENCE METHOD
100 - (WATER + PROTEIN + FAT + MINERALS)

BROTFRUCHT BREADFRUIT ARBRE À PAIN

ARTOCARPUS ALTILIS
(PARKINS.) FO.

		PROTEIN	FAT	CARBOHYDRATES	TOTAL
ENERGY VALUE (AVERAGE)	KJOULE	25	12	423	460
PER 100 G	(KCAL)	5.9	2.8	101	110
EDIBLE PORTION					
AMOUNT OF DIGESTIBLE	GRAM	1.27	0.27	25.30	
CONSTITUENTS PER 100 G					
ENERGY VALUE (AVERAGE)	KJOULE	21	11	423	455
OF THE DIGESTIBLE	(KCAL)	5.0	2.5	101	109
FRACTION PER 100 G					
EDIBLE PORTION					

WASTE PERCENTAGE AVERAGE 20

CONSTITUENTS	DIM	AV	VARIATION			AVR	NUTR. DENS.		MOLPERC.
WATER	GRAM	72.00	66.00	-	75.00	57.60	GRAM/MJ	158.24	
PROTEIN	GRAM	1.50	1.30	-	1.70	1.20	GRAM/MJ	3.30	
FAT	GRAM	0.30	-	-	-	0.24	GRAM/MJ	0.66	
AVAILABLE CARBOHYDR.	GRAM	25.30 1	-	-	-	20.24	GRAM/MJ	55.60	
MINERALS	GRAM	0.90	0.80	-	1.00	0.72	GRAM/MJ	1.98	
SODIUM	MILLI	13.00	-	-	-	10.40	MILLI/MJ	28.57	
POTASSIUM	MILLI	422.00	396.00	-	440.00	337.60	MILLI/MJ	927.48	
CALCIUM	MILLI	31.00	29.00	-	33.00	24.80	MILLI/MJ	68.13	
IRON	MILLI	1.00	0.70	-	1.20	0.80	MILLI/MJ	2.20	
PHOSPHORUS	MILLI	36.00	32.00	-	40.00	28.80	MILLI/MJ	79.12	
CAROTENE	MICRO	20.00	-	-	-	16.00	MICRO/MJ	43.96	
VITAMIN B1	MILLI	0.10	0.08	-	0.11	0.08	MILLI/MJ	0.22	
VITAMIN B2	MICRO	40.00	30.00	-	60.00	32.00	MICRO/MJ	87.91	
NICOTINAMIDE	MILLI	1.00	0.90	-	1.20	0.80	MILLI/MJ	2.20	
VITAMIN C	MILLI	21.00	12.00	-	29.00	16.80	MILLI/MJ	46.15	
STARCH	GRAM	16.80	-	-	-	13.44	GRAM/MJ	36.92	

1 ESTIMATED BY THE DIFFERENCE METHOD
100 - (WATER + PROTEIN + FAT + MINERALS)

CARISSA (NATAL PFLAUME)	**CARISSA** (NATAL PLUM)	**CARISSA** (PRUNE DE NATAL)

CARISSA GRANDIFLORA DC.

		PROTEIN	FAT	CARBOHYDRATES	TOTAL
ENERGY VALUE (AVERAGE) PER 100 G EDIBLE PORTION	KJOULE	7.4	43	277	327
	(KCAL)	1.8	10	66	78
AMOUNT OF DIGESTIBLE CONSTITUENTS PER 100 G	GRAM	0.38	0.99	16.55	
ENERGY VALUE (AVERAGE) OF THE DIGESTIBLE FRACTION PER 100 G EDIBLE PORTION	KJOULE	6.3	39	277	322
	(KCAL)	1.5	9.2	66	77

WASTE PERCENTAGE AVERAGE 13

CONSTITUENTS	DIM	AV		VARIATION			AVR	NUTR.	DENS.	MOLPERC.
WATER	GRAM	81.50		81.00	-	81.90	70.91	GRAM/MJ	253.24	
PROTEIN	GRAM	0.45		0.40	-	0.50	0.39	GRAM/MJ	1.40	
FAT	GRAM	1.10		0.90	-	1.30	0.96	GRAM/MJ	3.42	
AVAILABLE CARBOHYDR.	GRAM	16.55	1	-	-	-	14.40	GRAM/MJ	51.43	
MINERALS	GRAM	0.40		-	-	-	0.35	GRAM/MJ	1.24	
CALCIUM	MILLI	14.00		11.00	-	17.00	12.18	MILLI/MJ	43.50	
IRON	MILLI	1.30		-	-	-	1.13	MILLI/MJ	4.04	
PHOSPHORUS	MILLI	9.00		7.00	-	11.00	7.83	MILLI/MJ	27.97	
CAROTENE	MICRO	23.00		20.00	-	25.00	20.01	MICRO/MJ	71.47	
VITAMIN B1	MICRO	40.00		-	-	-	34.80	MICRO/MJ	124.29	
VITAMIN B2	MICRO	60.00		-	-	-	52.20	MICRO/MJ	186.44	
NICOTINAMIDE	MILLI	0.20		-	-	-	0.17	MILLI/MJ	0.62	
VITAMIN C	MILLI	47.00		38.00	-	56.00	40.89	MILLI/MJ	146.04	

1 ESTIMATED BY THE DIFFERENCE METHOD
100 - (WATER + PROTEIN + FAT + MINERALS)

CASHEW-APFEL CASHEW FRUIT POMME DE CAJOU

ANACARDIUM OCCIDENTALE L.

		PROTEIN	FAT	CARBOHYDRATES	TOTAL
ENERGY VALUE (AVERAGE)	KJOULE	17	27	180	224
PER 100 G EDIBLE PORTION	(KCAL)	4.0	6.5	43	54
AMOUNT OF DIGESTIBLE CONSTITUENTS PER 100 G	GRAM	0.85	0.63	10.76	
ENERGY VALUE (AVERAGE)	KJOULE	14	25	180	219
OF THE DIGESTIBLE FRACTION PER 100 G EDIBLE PORTION	(KCAL)	3.4	5.9	43	52

WASTE PERCENTAGE AVERAGE 0.00

CONSTITUENTS	DIM	AV	VARIATION			AVR	NUTR. DENS.		MOLPERC.
WATER	GRAM	85.60	-	-	-	85.60	GRAM/MJ	391.51	
PROTEIN	GRAM	1.00	-	-	-	1.00	GRAM/MJ	4.57	
FAT	GRAM	0.70	-	-	-	0.70	GRAM/MJ	3.20	
AVAILABLE CARBOHYDR.	GRAM	10.76	-	-	-	10.76	GRAM/MJ	49.21	
MINERALS	GRAM	0.40	-	-	-	0.40	GRAM/MJ	1.83	
SODIUM	MILLI	6.00	6.00	-	6.00	6.00	MILLI/MJ	27.44	
POTASSIUM	MILLI	150.00	150.00	-	150.00	150.00	MILLI/MJ	686.06	
MAGNESIUM	MILLI	10.00	10.00	-	10.00	10.00	MILLI/MJ	45.74	
CALCIUM	MILLI	12.00	-	-	-	12.00	MILLI/MJ	54.88	
IRON	MILLI	0.90	0.50	-	1.40	0.90	MILLI/MJ	4.12	
PHOSPHORUS	MILLI	28.00	10.00	-	45.00	28.00	MILLI/MJ	128.06	
CHLORIDE	MILLI	6.00	6.00	-	6.00	6.00	MILLI/MJ	27.44	
CAROTENE	MILLI	0.76	-	-	-	0.76	MILLI/MJ	3.48	
VITAMIN B1	MICRO	30.00	-	-	-	30.00	MICRO/MJ	137.21	
VITAMIN B2	MILLI	-	0.02	-	0.20				
NICOTINAMIDE	MILLI	0.22	0.13	-	0.30	0.22	MILLI/MJ	1.01	
PANTOTHENIC ACID	MILLI	0.11	0.11	-	0.11	0.11	MILLI/MJ	0.50	
BIOTIN	MICRO	1.50	1.50	-	1.50	1.50	MICRO/MJ	6.86	
VITAMIN C	MILLI	252.00	150.00	-	400.00	252.00	GRAM/MJ	1.15	
MALIC ACID	GRAM	0.24	0.24	-	0.24	0.24	GRAM/MJ	1.10	
CITRIC ACID	MILLI	20.00	20.00	-	20.00	20.00	MILLI/MJ	91.47	
ISOCITRIC ACID	MILLI	0.60	0.60	-	0.60	0.60	MILLI/MJ	2.74	
GLUCOSE	GRAM	5.35	5.35	-	5.35	5.35	GRAM/MJ	24.47	
FRUCTOSE	GRAM	5.17	5.17	-	5.17	5.17	GRAM/MJ	23.65	

CHAYOTE (SCHUSCHU) — CHAYOTE — CHOUCHOU

SECHIUM EDULE (JACQ.)

		PROTEIN	FAT	CARBOHYDRATES	TOTAL
ENERGY VALUE (AVERAGE) PER 100 G EDIBLE PORTION	KJOULE	12	3.9	95	110
	(KCAL)	2.8	0.9	23	26
AMOUNT OF DIGESTIBLE CONSTITUENTS PER 100 G	GRAM	0.48	0.09	5.65	
ENERGY VALUE (AVERAGE) OF THE DIGESTIBLE FRACTION PER 100 G EDIBLE PORTION	KJOULE	7.6	3.5	95	106
	(KCAL)	1.8	0.8	23	25

WASTE PERCENTAGE AVERAGE 23

CONSTITUENTS	DIM	AV		VARIATION			AVR	NUTR.	DENS.	MOLPERC.
WATER	GRAM	93.20		92.00	-	94.30	71.76	GRAM/MJ	881.66	
PROTEIN	GRAM	0.75		0.70	-	0.80	0.58	GRAM/MJ	7.09	
FAT	GRAM	0.10		-	-	-	0.08	GRAM/MJ	0.95	
AVAILABLE CARBOHYDR.	GRAM	5.65	1	-	-	-	4.35	GRAM/MJ	53.45	
MINERALS	GRAM	0.30		-	-	-	0.23	GRAM/MJ	2.84	
SODIUM	MILLI	2.00		-	-	-	1.54	MILLI/MJ	18.92	
POTASSIUM	MILLI	108.00		-	-	-	83.16	GRAM/MJ	1.02	
CALCIUM	MILLI	13.50		10.00	-	17.00	10.40	MILLI/MJ	127.71	
IRON	MILLI	0.40		-	-	-	0.31	MILLI/MJ	3.78	
PHOSPHORUS	MILLI	14.00		-	-	-	10.78	MILLI/MJ	132.44	
CAROTENE	MICRO	15.00		-	-	-	11.55	MICRO/MJ	141.90	
VITAMIN B1	MICRO	30.00		10.00	-	50.00	23.10	MICRO/MJ	283.80	
VITAMIN B2	MICRO	35.00		20.00	-	50.00	26.95	MICRO/MJ	331.10	
NICOTINAMIDE	MILLI	0.45		0.40	-	0.50	0.35	MILLI/MJ	4.26	
VITAMIN C	MILLI	17.00		14.00	-	20.00	13.09	MILLI/MJ	160.82	
SALICYLIC ACID	MICRO	10.00		10.00	-	10.00	7.70	MICRO/MJ	94.60	

1 ESTIMATED BY THE DIFFERENCE METHOD
100 - (WATER + PROTEIN + FAT + MINERALS)

CHERIMOYA CHERIMOYA CHÉRIMOLE

ANNONA CHERIMOLA MILL.

		PROTEIN	FAT	CARBOHYDRATES	TOTAL
ENERGY VALUE (AVERAGE)	KJOULE	25	12	224	261
PER 100 G	(KCAL)	5.9	2.8	54	62
EDIBLE PORTION					
AMOUNT OF DIGESTIBLE	GRAM	1.27	0.27	13.40	
CONSTITUENTS PER 100 G					
ENERGY VALUE (AVERAGE)	KJOULE	21	11	224	256
OF THE DIGESTIBLE	(KCAL)	5.0	2.5	54	61
FRACTION PER 100 G					
EDIBLE PORTION					

WASTE PERCENTAGE AVERAGE 35

CONSTITUENTS	DIM	AV	VARIATION			AVR	NUTR. DENS.		MOLPERC.
WATER	GRAM	74.10	68.50	-	80.20	48.17	GRAM/MJ	289.63	
PROTEIN	GRAM	1.50	0.90	-	2.10	0.98	GRAM/MJ	5.86	
FAT	GRAM	0.30	0.10	-	0.40	0.20	GRAM/MJ	1.17	
AVAILABLE CARBOHYDR.	GRAM	13.40	-	-	-	8.71	GRAM/MJ	52.38	
MINERALS	GRAM	0.80	0.70	-	0.90	0.52	GRAM/MJ	3.13	
CALCIUM	MILLI	13.00	8.00	-	23.00	8.45	MILLI/MJ	50.81	
IRON	MILLI	0.40	0.20	-	0.60	0.26	MILLI/MJ	1.56	
PHOSPHORUS	MILLI	32.00	24.00	-	40.00	20.80	MILLI/MJ	125.08	
VITAMIN B1	MICRO	90.00	40.00	-	130.00	58.50	MICRO/MJ	351.78	
VITAMIN B2	MILLI	0.11	-	-	-	0.07	MILLI/MJ	0.43	
NICOTINAMIDE	MILLI	1.10	0.80	-	1.30	0.72	MILLI/MJ	4.30	
VITAMIN C	MILLI	15.00	9.00	-	24.00	9.75	MILLI/MJ	58.63	
REDUCING SUGAR	GRAM	8.90	-	-	-	5.79	GRAM/MJ	34.79	
NONREDUCING SUGAR	GRAM	4.50	-	-	-	2.93	GRAM/MJ	17.59	

DATTEL GETROCKNET	**DATE** DRIED	**DATTE** SÈCHE

PHOENIX DACTYLIFERA L.

		PROTEIN	FAT	CARBOHYDRATES	TOTAL
ENERGY VALUE (AVERAGE) PER 100 G EDIBLE PORTION	KJOULE	31	21	1110	1161
	(KCAL)	7.3	4.9	265	278
AMOUNT OF DIGESTIBLE CONSTITUENTS PER 100 G	GRAM	1.57	0.47	66.32	
ENERGY VALUE (AVERAGE) OF THE DIGESTIBLE FRACTION PER 100 G EDIBLE PORTION	KJOULE	26	19	1110	1154
	(KCAL)	6.2	4.4	265	276

WASTE PERCENTAGE AVERAGE 13 MINIMUM 7.0 MAXIMUM 18

CONSTITUENTS	DIM	AV	VARIATION			AVR	NUTR. DENS.		MOLPERC.
WATER	GRAM	20.20	18.50	-	22.00	17.57	GRAM/MJ	17.50	
PROTEIN	GRAM	1.85	1.60	-	2.20	1.61	GRAM/MJ	1.60	
FAT	GRAM	0.53	0.40	-	0.60	0.46	GRAM/MJ	0.46	
AVAILABLE CARBOHYDR.	GRAM	66.32	-	-	-	57.70	GRAM/MJ	57.45	
TOTAL DIETARY FIBRE	GRAM	9.20 [1]	-	-	-	8.00	GRAM/MJ	7.97	
MINERALS	GRAM	1.82	1.80	-	1.83	1.58	GRAM/MJ	1.58	
SODIUM	MILLI	35.00	11.00	-	48.00	30.45	MILLI/MJ	30.32	
POTASSIUM	MILLI	650.00	540.00	-	754.00	565.50	MILLI/MJ	563.02	
MAGNESIUM	MILLI	50.00	-	-	-	43.50	MILLI/MJ	43.31	
CALCIUM	MILLI	63.00	46.00	-	72.00	54.81	MILLI/MJ	54.57	
MANGANESE	MILLI	0.15	-	-	-	0.13	MILLI/MJ	0.13	
IRON	MILLI	1.90	1.50	-	2.10	1.65	MILLI/MJ	1.65	
COPPER	MILLI	0.33	0.20	-	0.40	0.29	MILLI/MJ	0.29	
ZINC	MILLI	0.34	-	-	-	0.30	MILLI/MJ	0.29	
CHROMIUM	MICRO	29.00	-	-	-	25.23	MICRO/MJ	25.12	
PHOSPHORUS	MILLI	57.00	48.00	-	62.00	49.59	MILLI/MJ	49.37	
CHLORIDE	MILLI	117.00	108.00	-	125.00	101.79	MILLI/MJ	101.34	
CAROTENE	MICRO	28.00	20.00	-	36.00	24.36	MICRO/MJ	24.25	
VITAMIN B1	MICRO	36.00	17.00	-	60.00	31.32	MICRO/MJ	31.18	
VITAMIN B2	MICRO	73.00	30.00	-	100.00	63.51	MICRO/MJ	63.23	
NICOTINAMIDE	MILLI	1.90	1.20	-	2.40	1.65	MILLI/MJ	1.65	
PANTOTHENIC ACID	MILLI	0.80	-	-	-	0.70	MILLI/MJ	0.69	
VITAMIN B6	MILLI	0.13	0.10	-	0.17	0.11	MILLI/MJ	0.11	
FOLIC ACID	MICRO	21.00	-	-	-	18.27	MICRO/MJ	18.19	
VITAMIN C	MILLI	3.00	-	-	-	2.61	MILLI/MJ	2.60	

1 METHOD OF SCHWEIZER AND WUERSCH

CONSTITUENTS	DIM	AV		VARIATION			AVR	NUTR. DENS.		MOLPERC.
ARGININE	MILLI	40.00		-	-	-	34.80	MILLI/MJ	34.65	
HISTIDINE	MILLI	40.00		-	-	-	34.80	MILLI/MJ	34.65	
ISOLEUCINE	MILLI	60.00		-	-	-	52.20	MILLI/MJ	51.97	
LEUCINE	MILLI	62.00		-	-	-	53.94	MILLI/MJ	53.70	
LYSINE	MILLI	44.00		-	-	-	38.28	MILLI/MJ	38.11	
METHIONINE	MILLI	22.00		-	-	-	19.14	MILLI/MJ	19.06	
PHENYLALANINE	MILLI	51.00		-	-	-	44.37	MILLI/MJ	44.18	
THREONINE	MILLI	49.00		-	-	-	42.63	MILLI/MJ	42.44	
TRYPTOPHAN	MILLI	49.00		-	-	-	42.63	MILLI/MJ	42.44	
VALINE	MILLI	76.00		-	-	-	66.12	MILLI/MJ	65.83	
SEROTONINE	MILLI	0.80		-	-	-	0.70	MILLI/MJ	0.69	
MALIC ACID	GRAM	1.26		-	-	-	1.10	GRAM/MJ	1.09	
SALICYLIC ACID	MILLI	4.50		4.50	-	4.50	3.92	MILLI/MJ	3.90	
GLUCOSE	GRAM	25.02		15.88	-	29.59	21.77	GRAM/MJ	21.67	
FRUCTOSE	GRAM	24.91		15.97	-	29.39	21.68	GRAM/MJ	21.58	
SUCROSE	GRAM	13.78		6.89	-	27.58	11.99	GRAM/MJ	11.94	
SORBITOL	GRAM	1.35		1.35	-	1.35	1.17	GRAM/MJ	1.17	
TOTAL PURINES	MILLI	15.00	2	-	-	-	13.05	MILLI/MJ	12.99	
DIETARY FIBRE,WAT.SOL.	GRAM	2.30		-	-	-	2.00	GRAM/MJ	1.99	
DIETARY FIBRE,WAT.INS.	GRAM	6.90		-	-	-	6.00	GRAM/MJ	5.98	

Früchte Fruits Fruits

2 VALUE EXPRESSED IN MG URIC ACID/100G

DURIAN — DURIAN (CIVET) — DURIO

DURIO ZIBETHINUS MURR.

		PROTEIN	FAT	CARBOHYDRATES	TOTAL
ENERGY VALUE (AVERAGE) PER 100 G EDIBLE PORTION	KJOULE	45	70	477	592
	(KCAL)	11	17	114	141
AMOUNT OF DIGESTIBLE CONSTITUENTS PER 100 G	GRAM	2.29	1.62	28.50	
ENERGY VALUE (AVERAGE) OF THE DIGESTIBLE FRACTION PER 100 G EDIBLE PORTION	KJOULE	38	63	477	578
	(KCAL)	9.1	15	114	138

WASTE PERCENTAGE AVERAGE 75

CONSTITUENTS	DIM	AV	VARIATION			AVR	NUTR. DENS.		MOLPERC.
WATER	GRAM	61.50	56.30	-	66.80	15.38	GRAM/MJ	106.41	
PROTEIN	GRAM	2.70	2.00	-	3.20	0.68	GRAM/MJ	4.67	
FAT	GRAM	1.80	1.00	-	2.90	0.45	GRAM/MJ	3.11	
AVAILABLE CARBOHYDR.	GRAM	28.50 [1]	-	-	-	7.13	GRAM/MJ	49.31	
TOTAL DIETARY FIBRE	GRAM	4.40	-	-	-	1.10	GRAM/MJ	7.61	
MINERALS	GRAM	1.10	0.80	-	1.50	0.28	GRAM/MJ	1.90	
SODIUM	MILLI	1.00	-	-	-	0.25	MILLI/MJ	1.73	
POTASSIUM	MILLI	601.00	-	-	-	150.25	GRAM/MJ	1.04	
CALCIUM	MILLI	12.00	5.30	-	20.00	3.00	MILLI/MJ	20.76	
IRON	MILLI	1.00	0.80	-	1.10	0.25	MILLI/MJ	1.73	
PHOSPHORUS	MILLI	45.00	28.00	-	63.00	11.25	MILLI/MJ	77.86	
CAROTENE	MICRO	15.00	10.00	-	20.00	3.75	MILLI/MJ	25.95	
VITAMIN B1	MILLI	0.45	0.27	-	0.67	0.11	MILLI/MJ	0.78	
VITAMIN B2	MILLI	0.35	0.17	-	0.53	0.09	MILLI/MJ	0.61	
NICOTINAMIDE	MILLI	1.20	1.10	-	1.20	0.30	MILLI/MJ	2.08	
VITAMIN C	MILLI	42.00	32.00	-	58.00	10.50	MILLI/MJ	72.67	
CELLULOSE	GRAM	1.40	-	-	-	0.35	GRAM/MJ	2.42	

1 ESTIMATED BY THE DIFFERENCE METHOD
100 - (WATER + PROTEIN + FAT + MINERALS)

FEIGE　　FIG　　FIGUE

FICUS CARICA L.

		PROTEIN	FAT	CARBOHYDRATES	TOTAL
ENERGY VALUE (AVERAGE) PER 100 G EDIBLE PORTION	KJOULE	21	19	216	257
	(KCAL)	5.1	4.7	52	61
AMOUNT OF DIGESTIBLE CONSTITUENTS PER 100 G	GRAM	1.10	0.45	12.90	
ENERGY VALUE (AVERAGE) OF THE DIGESTIBLE FRACTION PER 100 G EDIBLE PORTION	KJOULE	18	18	216	252
	(KCAL)	4.4	4.2	52	60

WASTE PERCENTAGE AVERAGE 0.00

CONSTITUENTS	DIM	AV	VARIATION			AVR	NUTR. DENS.		MOLPERC.
WATER	GRAM	80.20	78.00	-	83.50	80.20	GRAM/MJ	318.68	
PROTEIN	GRAM	1.30	1.10	-	1.40	1.30	GRAM/MJ	5.17	
FAT	GRAM	0.50	0.40	-	0.50	0.50	GRAM/MJ	1.99	
AVAILABLE CARBOHYDR.	GRAM	12.90	-	-	-	12.90	GRAM/MJ	51.26	
TOTAL DIETARY FIBRE	GRAM	2.04 [1]	-	-	-	2.04	GRAM/MJ	8.11	
MINERALS	GRAM	0.70	0.60	-	0.80	0.70	GRAM/MJ	2.78	
SODIUM	MILLI	2.00	-	-	-	2.00	MILLI/MJ	7.95	
POTASSIUM	MILLI	240.00	190.00	-	290.00	240.00	MILLI/MJ	953.64	
MAGNESIUM	MILLI	20.00	-	-	-	20.00	MILLI/MJ	79.47	
CALCIUM	MILLI	54.00	-	-	-	54.00	MILLI/MJ	214.57	
IRON	MILLI	0.60	-	-	-	0.60	MILLI/MJ	2.38	
COPPER	MICRO	70.00	-	-	-	70.00	MICRO/MJ	278.15	
ZINC	MILLI	0.25	0.10	-	0.40	0.25	MILLI/MJ	0.99	
PHOSPHORUS	MILLI	32.00	-	-	-	32.00	MILLI/MJ	127.15	
CHLORIDE	MILLI	18.00	-	-	-	18.00	MILLI/MJ	71.52	
FLUORIDE	MICRO	20.00	-	-	-	20.00	MICRO/MJ	79.47	
IODIDE	MICRO	1.50	-	-	-	1.50	MICRO/MJ	5.96	
BORON	MILLI	0.13	0.08	-	0.18	0.13	MILLI/MJ	0.52	
CAROTENE	MICRO	48.00	-	-	-	48.00	MICRO/MJ	190.73	
VITAMIN B1	MICRO	46.00	31.00	-	50.00	46.00	MICRO/MJ	182.78	
VITAMIN B2	MICRO	50.00	-	-	-	50.00	MICRO/MJ	198.68	
NICOTINAMIDE	MILLI	0.42	0.29	-	0.50	0.43	MILLI/MJ	1.70	
PANTOTHENIC ACID	MILLI	0.30	-	-	-	0.30	MILLI/MJ	1.19	
VITAMIN B6	MILLI	0.11	-	-	-	0.11	MILLI/MJ	0.44	
FOLIC ACID	MICRO	6.70	4.90	-	8.80	6.70	MICRO/MJ	26.62	
VITAMIN C	MILLI	2.74	0.66	-	3.30	2.74	MILLI/MJ	10.89	

1 METHOD OF MEUSER, SUCKOW AND KULIKOWSKI ("BERLINER METHODE")

CONSTITUENTS	DIM	AV	VARIATION			AVR	NUTR. DENS.		MOLPERC.
SEROTONINE	MILLI	1.29	-	-	-	1.29	MILLI/MJ	5.13	
SALICYLIC ACID	MILLI	0.18	0.18	-	0.18	0.18	MILLI/MJ	0.72	
INVERT SUGAR	GRAM	12.90	10.20	-	15.60	12.90	GRAM/MJ	51.26	
TOTAL STEROLS	MILLI	31.00	-	-	-	31.00	MILLI/MJ	123.18	
CAMPESTEROL	MILLI	1.00	-	-	-	1.00	MILLI/MJ	3.97	
BETA-SITOSTEROL	MILLI	27.00	-	-	-	27.00	MILLI/MJ	107.28	
STIGMASTEROL	MILLI	3.00	-	-	-	3.00	MILLI/MJ	11.92	
DIETARY FIBRE, WAT.SOL.	GRAM	1.21	-	-	-	1.21	GRAM/MJ	4.81	
DIETARY FIBRE, WAT.INS.	GRAM	0.83	-	-	-	0.83	GRAM/MJ	3.30	

FEIGE GETROCKNET / FIG DRIED / FIGUE SÈCHE

		PROTEIN	FAT	CARBOHYDRATES	TOTAL
ENERGY VALUE (AVERAGE) PER 100 G EDIBLE PORTION	KJOULE	59	51	904	1013
	(KCAL)	14	12	216	242
AMOUNT OF DIGESTIBLE CONSTITUENTS PER 100 G	GRAM	3.00	1.17	54.00	
ENERGY VALUE (AVERAGE) OF THE DIGESTIBLE FRACTION PER 100 G EDIBLE PORTION	KJOULE	50	46	904	999
	(KCAL)	12	11	216	239

WASTE PERCENTAGE AVERAGE 1.0

CONSTITUENTS	DIM	AV	VARIATION			AVR	NUTR. DENS.		MOLPERC.
WATER	GRAM	24.60	23.40	-	26.10	24.35	GRAM/MJ	24.62	
PROTEIN	GRAM	3.54	3.00	-	4.00	3.50	GRAM/MJ	3.54	
FAT	GRAM	1.30	1.20	-	1.39	1.29	GRAM/MJ	1.30	
AVAILABLE CARBOHYDR.	GRAM	54.00	-	-	-	53.46	GRAM/MJ	54.05	
TOTAL DIETARY FIBRE	GRAM	9.60 [1]	-	-	-	9.50	GRAM/MJ	9.61	
MINERALS	GRAM	2.38	2.25	-	2.50	2.36	GRAM/MJ	2.38	

1 METHOD OF MEUSER, SUCKOW AND KULIKOWSKI ("BERLINER METHODE")

CONSTITUENTS	DIM	AV	VARIATION			AVR	NUTR. DENS.		MOLPERC.
SODIUM	MILLI	40.00	34.00	-	46.00	39.60	MILLI/MJ	40.04	
POTASSIUM	MILLI	850.00	780.00	-	970.00	841.50	MILLI/MJ	850.85	
MAGNESIUM	MILLI	70.00	-	-	-	69.30	MILLI/MJ	70.07	
CALCIUM	MILLI	193.00	186.00	-	200.00	191.07	MILLI/MJ	193.19	
MANGANESE	MILLI	0.35	-	-	-	0.35	MILLI/MJ	0.35	
IRON	MILLI	3.30	3.00	-	4.00	3.27	MILLI/MJ	3.30	
COPPER	MILLI	0.38	0.35	-	0.40	0.38	MILLI/MJ	0.38	
PHOSPHORUS	MILLI	108.00	100.00	-	113.00	106.92	MILLI/MJ	108.11	
CHLORIDE	MILLI	43.00	-	-	-	42.57	MILLI/MJ	43.04	
IODIDE	MICRO	4.00	-	-	-	3.96	MICRO/MJ	4.00	
BORON	MILLI	0.71	-	-	-	0.70	MILLI/MJ	0.71	
CAROTENE	MICRO	51.00	30.00	-	100.00	50.49	MICRO/MJ	51.05	
VITAMIN B1	MILLI	0.12	0.09	-	0.16	0.12	MILLI/MJ	0.12	
VITAMIN B2	MICRO	85.00	50.00	-	110.00	84.15	MICRO/MJ	85.09	
NICOTINAMIDE	MILLI	1.15	0.75	-	1.70	1.14	MILLI/MJ	1.15	
PANTOTHENIC ACID	MILLI	0.39	0.35	-	0.41	0.39	MILLI/MJ	0.39	
VITAMIN B6	MILLI	0.12	-	-	-	0.12	MILLI/MJ	0.12	
FOLIC ACID	MICRO	14.00	11.00	-	18.00	13.86	MICRO/MJ	14.01	
VITAMIN C	MILLI	2.50	0.00	-	5.00	2.48	MILLI/MJ	2.50	
MALIC ACID	GRAM	1.08	0.82	-	1.33	1.07	GRAM/MJ	1.08	
SALICYLIC ACID	MILLI	0.64	0.64	-	0.64	0.63	MILLI/MJ	0.64	
TOTAL SUGAR	GRAM	52.90	-	-	-	52.37	GRAM/MJ	52.95	
DIETARY FIBRE,WAT.SOL.	GRAM	1.90	-	-	-	1.88	GRAM/MJ	1.90	
DIETARY FIBRE,WAT.INS.	GRAM	7.70	-	-	-	7.62	GRAM/MJ	7.71	

GRANATAPFEL POMEGRANATE GRENADE

PUNICA GRANATUM L.

		PROTEIN	FAT	CARBOHYDRATES	TOTAL
ENERGY VALUE (AVERAGE) PER 100 G EDIBLE PORTION	KJOULE	12	23	279	314
	(KCAL)	2.8	5.6	67	75
AMOUNT OF DIGESTIBLE CONSTITUENTS PER 100 G	GRAM	0.59	0.54	16.70	
ENERGY VALUE (AVERAGE) OF THE DIGESTIBLE FRACTION PER 100 G EDIBLE PORTION	KJOULE	9.8	21	279	310
	(KCAL)	2.4	5.0	67	74

WASTE PERCENTAGE AVERAGE 65

CONSTITUENTS	DIM	AV		VARIATION			AVR	NUTR.	DENS.	MOLPERC.
WATER	GRAM	82.50		80.00	-	84.20	28.88	GRAM/MJ	265.84	
PROTEIN	GRAM	0.70		0.30	-	1.00	0.25	GRAM/MJ	2.26	
FAT	GRAM	0.60		-	-	-	0.21	GRAM/MJ	1.93	
AVAILABLE CARBOHYDR.	GRAM	16.70		-	-	-	5.85	GRAM/MJ	53.81	
TOTAL DIETARY FIBRE	GRAM	3.10	1	-	-	-	1.09	GRAM/MJ	9.99	
MINERALS	GRAM	0.70		0.30	-	0.80	0.25	GRAM/MJ	2.26	
SODIUM	MILLI	7.00		-	-	-	2.45	MILLI/MJ	22.56	
POTASSIUM	MILLI	290.00		200.00	-	380.00	101.50	MILLI/MJ	934.47	
MAGNESIUM	MILLI	3.00		-	-	-	1.05	MILLI/MJ	9.67	
CALCIUM	MILLI	8.00		3.00	-	13.00	2.80	MILLI/MJ	25.78	
IRON	MILLI	0.50		0.20	-	0.70	0.18	MILLI/MJ	1.61	
PHOSPHORUS	MILLI	17.00		10.00	-	23.00	5.95	MILLI/MJ	54.78	
CAROTENE	MICRO	40.00		-	-	-	14.00	MICRO/MJ	128.89	
VITAMIN B1	MICRO	50.00		20.00	-	70.00	17.50	MICRO/MJ	161.12	
VITAMIN B2	MICRO	20.00		10.00	-	30.00	7.00	MICRO/MJ	64.45	
NICOTINAMIDE	MILLI	0.30		-	-	-	0.11	MILLI/MJ	0.97	
VITAMIN C	MILLI	7.00		5.00	-	20.00	2.45	MILLI/MJ	22.56	
MALIC ACID	GRAM	0.10		-	-	-	0.04	GRAM/MJ	0.32	
CITRIC ACID	GRAM	0.50		-	-	-	0.18	GRAM/MJ	1.61	
SALICYLIC ACID	MICRO	70.00		70.00	-	70.00	24.50	MICRO/MJ	225.56	
GLUCOSE	GRAM	7.20		-	-	-	2.52	GRAM/MJ	23.20	
FRUCTOSE	GRAM	7.90		-	-	-	2.77	GRAM/MJ	25.46	
SUCROSE	GRAM	1.00		-	-	-	0.35	GRAM/MJ	3.22	
TOTAL STEROLS	MILLI	17.00		-	-	-	5.95	MILLI/MJ	54.78	
CAMPESTEROL	-	TRACES		-	-	-				
BETA-SITOSTEROL	MILLI	16.00		-	-	-	5.60	MILLI/MJ	51.56	
DIETARY FIBRE, WAT.SOL.	GRAM	0.63		-	-	-	0.22	GRAM/MJ	2.03	
DIETARY FIBRE, WAT.INS.	GRAM	2.47		-	-	-	0.86	GRAM/MJ	7.96	

1 METHOD OF ENGLYST

GRAPEFRUIT (PAMPELMUSE) | GRAPEFRUIT | PAMPLEMOUSSE

CITRUS DECUMANA L.

		PROTEIN	FAT	CARBOHYDRATES	TOTAL
ENERGY VALUE (AVERAGE)	KJOULE	9.9	5.8	150	166
PER 100 G	(KCAL)	2.4	1.4	36	40
EDIBLE PORTION					
AMOUNT OF DIGESTIBLE	GRAM	0.51	0.13	8.95	
CONSTITUENTS PER 100 G					
ENERGY VALUE (AVERAGE)	KJOULE	8.4	5.3	150	163
OF THE DIGESTIBLE	(KCAL)	2.0	1.3	36	39
FRACTION PER 100 G					
EDIBLE PORTION					

WASTE PERCENTAGE AVERAGE 34 MINIMUM 27 MAXIMUM 43

CONSTITUENTS	DIM	AV		VARIATION			AVR	NUTR. DENS.		MOLPERC.
WATER	GRAM	89.00		86.00	-	91.00	58.74	GRAM/MJ	544.45	
PROTEIN	GRAM	0.60		0.50	-	0.80	0.40	GRAM/MJ	3.67	
FAT	GRAM	0.15		0.10	-	0.20	0.10	GRAM/MJ	0.92	
AVAILABLE CARBOHYDR.	GRAM	8.95		5.93	-	11.51	5.91	GRAM/MJ	54.75	
TOTAL DIETARY FIBRE	GRAM	0.58	1	-	-	-	0.38	GRAM/MJ	3.55	
MINERALS	GRAM	0.35		0.27	-	0.43	0.23	GRAM/MJ	2.14	
SODIUM	MILLI	1.60		1.00	-	2.00	1.06	MILLI/MJ	9.79	
POTASSIUM	MILLI	180.00		125.00	-	234.00	118.80	GRAM/MJ	1.10	
MAGNESIUM	MILLI	10.00		8.00	-	12.00	6.60	MILLI/MJ	61.17	
CALCIUM	MILLI	18.00		14.00	-	23.00	11.88	MILLI/MJ	110.11	
MANGANESE	MICRO	13.00		10.00	-	20.00	8.58	MICRO/MJ	79.53	
IRON	MILLI	0.34		0.26	-	0.50	0.22	MILLI/MJ	2.08	
COPPER	MICRO	40.00		30.00	-	60.00	26.40	MICRO/MJ	244.69	
ZINC	MILLI	0.17		0.10	-	0.20	0.11	MILLI/MJ	1.04	
NICKEL	MICRO	10.00		4.00	-	30.00	6.60	MICRO/MJ	61.17	
CHROMIUM	MICRO	1.00		0.20	-	2.00	0.66	MICRO/MJ	6.12	
PHOSPHORUS	MILLI	17.00		13.00	-	20.00	11.22	MILLI/MJ	104.00	
CHLORIDE	MILLI	2.30		1.30	-	3.00	1.52	MILLI/MJ	14.07	
FLUORIDE	MICRO	24.00		12.00	-	36.00	15.84	MICRO/MJ	146.82	
IODIDE	MICRO	1.30		-	-	-	0.86	MICRO/MJ	7.95	
BORON	MILLI	0.15		0.12	-	0.23	0.10	MILLI/MJ	0.92	
SELENIUM	MICRO	0.20		0.10	-	0.30	0.13	MICRO/MJ	1.22	
CAROTENE	MICRO	15.00		10.00	-	20.00	9.90	MICRO/MJ	91.76	
VITAMIN E ACTIVITY	MILLI	0.25		-	-	-	0.17	MILLI/MJ	1.53	
ALPHA-TOCOPHEROL	MILLI	0.25		-	-	-	0.17	MILLI/MJ	1.53	
VITAMIN B1	MICRO	48.00		31.00	-	70.00	31.68	MICRO/MJ	293.63	

1 METHOD OF MEUSER, SUCKOW AND KULIKOWSKI ("BERLINER METHODE")

CONSTITUENTS	DIM	AV	VARIATION			AVR	NUTR. DENS.		MOLPERC.
VITAMIN B2	MICRO	24.00	10.00	-	40.00	15.84	MICRO/MJ	146.82	
NICOTINAMIDE	MILLI	0.24	0.13	-	0.41	0.16	MILLI/MJ	1.47	
PANTOTHENIC ACID	MILLI	0.25	0.18	-	0.30	0.17	MILLI/MJ	1.53	
VITAMIN B6	MICRO	28.00	20.00	-	45.00	18.48	MICRO/MJ	171.29	
BIOTIN	MICRO	0.35	0.30	-	0.40	0.23	MICRO/MJ	2.14	
FOLIC ACID	MICRO	11.00	-	-	-	7.26	MICRO/MJ	67.29	
VITAMIN C	MILLI	44.00	38.00	-	55.00	29.04	MILLI/MJ	269.16	
ALANINE	MILLI	20.00	-	-	-	13.20	MILLI/MJ	122.35	7.4
ARGININE	MILLI	40.00	-	-	-	26.40	MILLI/MJ	244.69	7.6
ASPARTIC ACID	MILLI	105.00	-	-	-	69.30	MILLI/MJ	642.32	26.0
CYSTINE	MILLI	2.00	-	-	-	1.32	MILLI/MJ	12.23	0.3
GLUTAMIC ACID	MILLI	42.00	-	-	-	27.72	MILLI/MJ	256.93	9.4
GLYCINE	MILLI	10.00	-	-	-	6.60	MILLI/MJ	61.17	4.4
HISTIDINE	MILLI	6.00	-	-	-	3.96	MILLI/MJ	36.70	1.3
ISOLEUCINE	MILLI	10.00	-	-	-	6.60	MILLI/MJ	61.17	2.5
LEUCINE	MILLI	15.00	-	-	-	9.90	MILLI/MJ	91.76	3.8
LYSINE	MILLI	19.00	-	-	-	12.54	MILLI/MJ	116.23	4.3
METHIONINE	MILLI	3.00	-	-	-	1.98	MILLI/MJ	18.35	0.7
PHENYLALANINE	MILLI	10.00	-	-	-	6.60	MILLI/MJ	61.17	2.0
PROLINE	MILLI	47.00	-	-	-	31.02	MILLI/MJ	287.52	13.5
SERINE	MILLI	26.00	-	-	-	17.16	MILLI/MJ	159.05	8.1
THREONINE	MILLI	11.00	-	-	-	7.26	MILLI/MJ	67.29	3.0
TRYPTOPHAN	MILLI	4.00	-	-	-	2.64	MILLI/MJ	24.47	0.6
TYROSINE	MILLI	7.00	-	-	-	4.62	MILLI/MJ	42.82	1.3
VALINE	MILLI	14.00	-	-	-	9.24	MILLI/MJ	85.64	3.9
MALIC ACID	GRAM	0.18	0.05	-	0.31	0.12	GRAM/MJ	1.10	
CITRIC ACID	GRAM	1.37	1.25	-	1.46	0.90	GRAM/MJ	8.38	
QUINIC ACID	-	TRACES	-	-	-				
FERULIC ACID	MILLI	3.00	-	-	-	1.98	MILLI/MJ	18.35	
CAFFEIC ACID	MILLI	4.00	-	-	-	2.64	MILLI/MJ	24.47	
PARA-COUMARIC ACID	-	TRACES	-	-	-				
SALICYLIC ACID	MILLI	0.68	0.68	-	0.68	0.45	MILLI/MJ	4.16	
GLUCOSE	GRAM	2.38	2.14	-	2.81	1.57	GRAM/MJ	14.57	
FRUCTOSE	GRAM	2.09	0.79	-	2.38	1.38	GRAM/MJ	12.82	
SUCROSE	GRAM	2.93	1.70	-	4.55	1.93	GRAM/MJ	17.93	
PALMITIC ACID	MILLI	33.00	33.00	-	33.00	21.78	MILLI/MJ	201.87	
STEARIC ACID	MILLI	5.00	5.00	-	5.00	3.30	MILLI/MJ	30.59	
PALMITOLEIC ACID	MILLI	3.00	3.00	-	3.00	1.98	MILLI/MJ	18.35	
OLEIC ACID	MILLI	24.00	24.00	-	24.00	15.84	MILLI/MJ	146.82	
LINOLEIC ACID	MILLI	42.00	42.00	-	42.00	27.72	MILLI/MJ	256.93	
LINOLENIC ACID	MILLI	12.00	12.00	-	12.00	7.92	MILLI/MJ	73.41	
TOTAL STEROLS	MILLI	17.00	-	-	-	11.22	MILLI/MJ	104.00	
CAMPESTEROL	MILLI	2.00	-	-	-	1.32	MILLI/MJ	12.23	
BETA-SITOSTEROL	MILLI	13.00	-	-	-	8.58	MILLI/MJ	79.53	
STIGMASTEROL	MILLI	2.00	-	-	-	1.32	MILLI/MJ	12.23	
DIETARY FIBRE, WAT.SOL.	GRAM	0.30	-	-	-	0.20	GRAM/MJ	1.84	
DIETARY FIBRE, WAT.INS.	GRAM	0.28	-	-	-	0.18	GRAM/MJ	1.71	

GUAVE (GUAJAVE) · GUAVA · GOYAVE

PSIDIUM GUAYAVA L.

		PROTEIN	FAT	CARBOHYDRATES	TOTAL
ENERGY VALUE (AVERAGE)	KJOULE	15	19	112	146
PER 100 G EDIBLE PORTION	(KCAL)	3.6	4.7	27	35
AMOUNT OF DIGESTIBLE CONSTITUENTS PER 100 G	GRAM	0.76	0.45	6.70	
ENERGY VALUE (AVERAGE)	KJOULE	13	18	112	142
OF THE DIGESTIBLE FRACTION PER 100 G EDIBLE PORTION	(KCAL)	3.0	4.2	27	34

WASTE PERCENTAGE AVERAGE 11 MINIMUM 2.0 MAXIMUM 20

CONSTITUENTS	DIM	AV	VARIATION			AVR	NUTR. DENS.		MOLPERC.
WATER	GRAM	83.50	81.00	-	86.00	74.32	GRAM/MJ	586.85	
PROTEIN	GRAM	0.90	0.80	-	1.00	0.80	GRAM/MJ	6.33	
FAT	GRAM	0.50	0.40	-	0.60	0.45	GRAM/MJ	3.51	
AVAILABLE CARBOHYDR.	GRAM	6.70	-	-	-	5.96	GRAM/MJ	47.09	
TOTAL DIETARY FIBRE	GRAM	5.20 [1]	-	-	-	4.63	GRAM/MJ	36.55	
MINERALS	GRAM	0.68	0.60	-	0.75	0.61	GRAM/MJ	4.78	
SODIUM	MILLI	4.00	-	-	-	3.56	MILLI/MJ	28.11	
POTASSIUM	MILLI	290.00	289.00	-	291.00	258.10	GRAM/MJ	2.04	
MAGNESIUM	MILLI	13.00	-	-	-	11.57	MILLI/MJ	91.37	
CALCIUM	MILLI	17.00	15.00	-	23.00	15.13	MILLI/MJ	119.48	
IRON	MILLI	0.75	-	-	-	0.67	MILLI/MJ	5.27	
ZINC	MILLI	0.90	0.70	-	1.00	0.80	MILLI/MJ	6.33	
PHOSPHORUS	MILLI	31.00	24.00	-	42.00	27.59	MILLI/MJ	217.87	
CAROTENE	MILLI	0.22	0.08	-	0.40	0.20	MILLI/MJ	1.55	
VITAMIN B1	MICRO	30.00	20.00	-	50.00	26.70	MICRO/MJ	210.85	
VITAMIN B2	MICRO	40.00	40.00	-	50.00	35.60	MICRO/MJ	281.13	
NICOTINAMIDE	MILLI	1.10	0.50	-	1.20	0.98	MILLI/MJ	7.73	
VITAMIN C	MILLI	273.00	132.00	-	450.00	242.97	GRAM/MJ	1.92	
MALIC ACID	MILLI	325.00	182.00	-	469.00	289.25	GRAM/MJ	2.28	
CITRIC ACID	MILLI	537.00	532.00	-	541.00	477.93	GRAM/MJ	3.77	
LACTIC ACID	MILLI	19.00	12.00	-	25.00	16.91	MILLI/MJ	133.54	
GLUCOSE	GRAM	2.08	-	-	-	1.85	GRAM/MJ	14.62	
FRUCTOSE	GRAM	3.43	-	-	-	3.05	GRAM/MJ	24.11	
SUCROSE	GRAM	0.31	-	-	-	0.28	GRAM/MJ	2.18	
CELLULOSE	GRAM	1.40	-	-	-	1.25	GRAM/MJ	9.84	

1 METHOD OF SOUTHGATE

JABOTIKABA JABOTICABA JABOTICABA

MYRCIARIA CAULIFLORA (DC.)

		PROTEIN	FAT	CARBOHYDRATES	TOTAL
ENERGY VALUE (AVERAGE) PER 100 G EDIBLE PORTION	KJOULE	8.3	70	231	309
	(KCAL)	2.0	17	55	74
AMOUNT OF DIGESTIBLE CONSTITUENTS PER 100 G	GRAM	0.42	1.62	13.80	
ENERGY VALUE (AVERAGE) OF THE DIGESTIBLE FRACTION PER 100 G EDIBLE PORTION	KJOULE	7.0	63	231	301
	(KCAL)	1.7	15	55	72

WASTE PERCENTAGE AVERAGE 20

CONSTITUENTS	DIM	AV	VARIATION	AVR	NUTR. DENS.		MOLPERC.
WATER	GRAM	83.50	- - -	66.80	GRAM/MJ	277.39	
PROTEIN	GRAM	0.50	- - -	0.40	GRAM/MJ	1.66	
FAT	GRAM	1.80	- - -	1.44	GRAM/MJ	5.98	
AVAILABLE CARBOHYDR.	GRAM	13.80 [1]	- - -	11.04	GRAM/MJ	45.84	
MINERALS	GRAM	0.40	- - -	0.32	GRAM/MJ	1.33	
CALCIUM	MILLI	61.00	- - -	48.80	MILLI/MJ	202.65	
IRON	MILLI	0.20	- - -	0.16	MILLI/MJ	0.66	
PHOSPHORUS	MILLI	17.00	- - -	13.60	MILLI/MJ	56.48	
VITAMIN B1	MICRO	10.00	- - -	8.00	MICRO/MJ	33.22	
VITAMIN B2	MICRO	20.00	- - -	16.00	MICRO/MJ	66.44	
NICOTINAMIDE	MILLI	0.30	- - -	0.24	MILLI/MJ	1.00	
VITAMIN C	MILLI	17.00	- - -	13.60	MILLI/MJ	56.48	

1 ESTIMATED BY THE DIFFERENCE METHOD
100 - (WATER + PROTEIN + FAT + MINERALS)

JACKFRUCHT JACKFRUIT JAQUIER

ARTOCARPUS HETEROPHYLLA
LAM.

		PROTEIN	FAT	CARBOHYDRATES	TOTAL
ENERGY VALUE (AVERAGE)	KJOULE	18	18	256	292
PER 100 G	(KCAL)	4.3	4.2	61	70
EDIBLE PORTION					
AMOUNT OF DIGESTIBLE	GRAM	0.93	0.40	15.31	
CONSTITUENTS PER 100 G					
ENERGY VALUE (AVERAGE)	KJOULE	15	16	256	287
OF THE DIGESTIBLE	(KCAL)	3.7	3.8	61	69
FRACTION PER 100 G					
EDIBLE PORTION					

WASTE PERCENTAGE AVERAGE 0.00

CONSTITUENTS	DIM	AV		VARIATION			AVR	NUTR. DENS.		MOLPERC.
WATER	GRAM	74.55		73.10	-	76.00	74.55	GRAM/MJ	259.36	
PROTEIN	GRAM	1.10		0.60	-	1.70	1.10	GRAM/MJ	3.83	
FAT	GRAM	0.45		0.30	-	0.60	0.45	GRAM/MJ	1.57	
AVAILABLE CARBOHYDR.	GRAM	15.31		-	-	-	15.31	GRAM/MJ	53.26	
TOTAL DIETARY FIBRE	GRAM	4.15	1	-	-	-	4.15	GRAM/MJ	14.44	
MINERALS	GRAM	1.00		0.50	-	1.40	1.00	GRAM/MJ	3.48	
SODIUM	MILLI	2.00		-	-	-	2.00	MILLI/MJ	6.96	
POTASSIUM	MILLI	407.00		-	-	-	407.00	GRAM/MJ	1.42	
CALCIUM	MILLI	27.00		-	-	-	27.00	MILLI/MJ	93.93	
IRON	MILLI	0.60		-	-	-	0.60	MILLI/MJ	2.09	
PHOSPHORUS	MILLI	38.00		-	-	-	38.00	MILLI/MJ	132.20	
CAROTENE	MILLI	0.23		-	-	-	0.24	MILLI/MJ	0.82	
VITAMIN B1	MICRO	30.00		-	-	-	30.00	MICRO/MJ	104.37	
VITAMIN B2	MILLI	0.11		-	-	-	0.11	MILLI/MJ	0.38	
NICOTINAMIDE	MILLI	0.60		0.40	-	0.70	0.60	MILLI/MJ	2.09	
VITAMIN C	MILLI	9.00		-	-	-	9.00	MILLI/MJ	31.31	
GLUCOSE	GRAM	6.00		-	-	-	6.00	GRAM/MJ	20.87	
FRUCTOSE	GRAM	1.70		-	-	-	1.70	GRAM/MJ	5.91	
SUCROSE	GRAM	6.90		-	-	-	6.90	GRAM/MJ	24.01	
STARCH	GRAM	0.71		-	-	-	0.71	GRAM/MJ	2.47	
CELLULOSE	GRAM	1.30		-	-	-	1.30	GRAM/MJ	4.52	

1 METHOD OF SOUTHGATE

JAPANISCHE MISPEL (WOLLMISPEL, LOQUATE) · LOQUAT (JAPANESE MEDLAR) · NÈFLE DU JAPON (BIBACE)

ERIOBOTRYA JAPONICA (THUNB.) L.

		PROTEIN	FAT	CARBOHYDRATES	TOTAL
ENERGY VALUE (AVERAGE) PER 100 G EDIBLE PORTION	KJOULE	9.1	7.8	197	214
	(KCAL)	2.2	1.9	47	51
AMOUNT OF DIGESTIBLE CONSTITUENTS PER 100 G	GRAM	0.46	0.18	11.77	
ENERGY VALUE (AVERAGE) OF THE DIGESTIBLE FRACTION PER 100 G EDIBLE PORTION	KJOULE	7.7	7.0	197	212
	(KCAL)	1.8	1.7	47	51

WASTE PERCENTAGE AVERAGE 37

CONSTITUENTS	DIM	AV		VARIATION			AVR	NUTR. DENS.		MOLPERC.
WATER	GRAM	87.00		84.70	-	90.30	54.81	GRAM/MJ	410.93	
PROTEIN	GRAM	0.55		0.35	-	0.80	0.35	GRAM/MJ	2.60	
FAT	GRAM	0.20		-	-	-	0.13	GRAM/MJ	0.94	
AVAILABLE CARBOHYDR.	GRAM	11.77	1	-	-	-	7.42	GRAM/MJ	55.59	
MINERALS	GRAM	0.48		0.46	-	0.50	0.30	GRAM/MJ	2.27	
SODIUM	MILLI	4.00		-	-	-	2.52	MILLI/MJ	18.89	
POTASSIUM	MILLI	263.00		210.00	-	315.00	165.69	GRAM/MJ	1.24	
MAGNESIUM	MILLI	10.00		-	-	-	6.30	MILLI/MJ	47.23	
CALCIUM	MILLI	19.00		18.00	-	20.00	11.97	MILLI/MJ	89.74	
IRON	MILLI	0.30		0.20	-	0.40	0.19	MILLI/MJ	1.42	
PHOSPHORUS	MILLI	23.00		20.00	-	25.00	14.49	MILLI/MJ	108.64	
CAROTENE	MILLI	0.80		0.40	-	1.10	0.50	MILLI/MJ	3.78	
VITAMIN B1	MICRO	20.00		-	-	-	12.60	MICRO/MJ	94.47	
VITAMIN B2	MICRO	30.00		20.00	-	40.00	18.90	MICRO/MJ	141.70	
NICOTINAMIDE	MILLI	0.20		-	-	-	0.13	MILLI/MJ	0.94	
VITAMIN C	MILLI	4.00		-	-	-	2.52	MILLI/MJ	18.89	
SUCROSE	GRAM	3.70		2.40	-	4.90	2.33	GRAM/MJ	17.48	

1 ESTIMATED BY THE DIFFERENCE METHOD
100 - (WATER + PROTEIN + FAT + MINERALS)

JUJUBE
(CHINESISCHE DATTEL, INDISCHE BRUSTBEERE)

JUJUBE, COMMON
(CHINESE DATE)

JUJUBE, COMMUN

ZIZIPHUS JUJUBA MILL.

		PROTEIN	FAT	CARBOHYDRATES	TOTAL
ENERGY VALUE (AVERAGE)	KJOULE	23	12	403	438
PER 100 G	(KCAL)	5.5	2.8	96	105
EDIBLE PORTION					
AMOUNT OF DIGESTIBLE	GRAM	1.19	0.27	24.10	
CONSTITUENTS PER 100 G					
ENERGY VALUE (AVERAGE)	KJOULE	20	11	403	434
OF THE DIGESTIBLE	(KCAL)	4.7	2.5	96	104
FRACTION PER 100 G					
EDIBLE PORTION					

WASTE PERCENTAGE AVERAGE 6.0

CONSTITUENTS	DIM	AV		VARIATION			AVR	NUTR. DENS.		MOLPERC.
WATER	GRAM	73.50		70.00	-	76.90	69.09	GRAM/MJ	169.55	
PROTEIN	GRAM	1.40		1.20	-	1.60	1.32	GRAM/MJ	3.23	
FAT	GRAM	0.30		0.20	-	0.40	0.28	GRAM/MJ	0.69	
AVAILABLE CARBOHYDR.	GRAM	24.10	1	-	-	-	22.65	GRAM/MJ	55.59	
MINERALS	GRAM	0.70		0.60	-	0.80	0.66	GRAM/MJ	1.61	
SODIUM	MILLI	3.00		-	-	-	2.82	MILLI/MJ	6.92	
POTASSIUM	MILLI	278.00		-	-	-	261.32	MILLI/MJ	641.28	
CALCIUM	MILLI	33.00		29.00	-	49.00	31.02	MILLI/MJ	76.12	
IRON	MILLI	0.80		-	-	-	0.75	MILLI/MJ	1.85	
CAROTENE	MICRO	10.00		-	-	-	9.40	MICRO/MJ	23.07	
VITAMIN B1	MICRO	30.00		20.00	-	30.00	28.20	MICRO/MJ	69.20	
VITAMIN B2	MICRO	40.00		30.00	-	40.00	37.60	MICRO/MJ	92.27	
NICOTINAMIDE	MILLI	0.80		0.70	-	0.90	0.75	MILLI/MJ	1.85	
VITAMIN C	MILLI	58.00		46.00	-	70.00	54.52	MILLI/MJ	133.79	

1 ESTIMATED BY THE DIFFERENCE METHOD
100 - (WATER + PROTEIN + FAT + MINERALS)

KAKI (KAKIPFLAUME, CHINESISCHE QUITTE, JAPANISCHE PERSIMONE) DIOSPYROS KAKI THUNB.	**PERSIMMON** (KAKIPLUM)	**KAKI**

		PROTEIN	FAT	CARBOHYDRATES	TOTAL
ENERGY VALUE (AVERAGE) PER 100 G EDIBLE PORTION	KJOULE	11	12	268	290
	(KCAL)	2.5	2.8	64	69
AMOUNT OF DIGESTIBLE CONSTITUENTS PER 100 G	GRAM	0.54	0.27	16.00	
ENERGY VALUE (AVERAGE) OF THE DIGESTIBLE FRACTION PER 100 G EDIBLE PORTION	KJOULE	9.0	11	268	287
	(KCAL)	2.1	2.5	64	69

WASTE PERCENTAGE AVERAGE 13 MINIMUM 10 MAXIMUM 17

CONSTITUENTS	DIM	AV	VARIATION			AVR	NUTR.	DENS.	MOLPERC.
WATER	GRAM	81.00	78.70	-	83.70	70.47	GRAM/MJ	281.96	
PROTEIN	GRAM	0.64	0.50	-	0.75	0.56	GRAM/MJ	2.23	
FAT	GRAM	0.30	0.20	-	0.30	0.26	GRAM/MJ	1.04	
AVAILABLE CARBOHYDR.	GRAM	16.00	-	-	-	13.92	GRAM/MJ	55.70	
MINERALS	GRAM	0.67	0.51	-	0.90	0.58	GRAM/MJ	2.33	
SODIUM	MILLI	4.00	2.00	-	6.00	3.48	MILLI/MJ	13.92	
POTASSIUM	MILLI	170.00	130.00	-	176.00	147.90	MILLI/MJ	591.77	
MAGNESIUM	MILLI	8.00	-	-	-	6.96	MILLI/MJ	27.85	
CALCIUM	MILLI	8.00	6.00	-	14.00	6.96	MILLI/MJ	27.85	
IRON	MILLI	0.37	0.30	-	0.40	0.32	MILLI/MJ	1.29	
PHOSPHORUS	MILLI	25.00	20.00	-	26.00	21.75	MILLI/MJ	87.03	
CAROTENE	MILLI	1.60	1.30	-	1.90	1.39	MILLI/MJ	5.57	
VITAMIN B1	MICRO	24.00	14.00	-	30.00	20.88	MICRO/MJ	83.54	
VITAMIN B2	MICRO	30.00	20.00	-	40.00	26.10	MICRO/MJ	104.43	
NICOTINAMIDE	MILLI	0.23	0.10	-	0.33	0.20	MILLI/MJ	0.80	
VITAMIN C	MILLI	16.00	6.00	-	50.00	13.92	MILLI/MJ	55.70	
TOTAL ACIDS	GRAM	0.27	0.19	-	0.34	0.23	GRAM/MJ	0.94	
SALICYLIC ACID	MILLI	0.18	0.18	-	0.18	0.16	MILLI/MJ	0.63	
GLUCOSE	GRAM	7.00	-	-	-	6.09	GRAM/MJ	24.37	
FRUCTOSE	GRAM	8.00	-	-	--	6.96	GRAM/MJ	27.85	
SUCROSE	GRAM	1.00	-	-	-	0.87	GRAM/MJ	3.48	
PECTIN	GRAM	0.47	0.21	-	0.73	0.41	GRAM/MJ	1.64	
PALMITIC ACID	MILLI	49.00	49.00	-	49.00	42.63	MILLI/MJ	170.57	
PALMITOLEIC ACID	MILLI	36.00	36.00	-	36.00	31.32	MILLI/MJ	125.32	
OLEIC ACID	MILLI	58.00	58.00	-	58.00	50.46	MILLI/MJ	201.90	
LINOLEIC ACID	MILLI	62.00	62.00	-	62.00	53.94	MILLI/MJ	215.82	

KAPSTACHELBEERE CAPE GOOSBERRY COQUERET DU PEROU
(PHYSALISFRUCHT,
ANANASKIRSCHE)
PHYSALIS PERUVIANA L.
(PH.) EDU

		PROTEIN	FAT	CARBOHYDRATES	TOTAL
ENERGY VALUE (AVERAGE) PER 100 G EDIBLE PORTION	KJOULE	38	43	223	303
	(KCAL)	9.1	10	53	73
AMOUNT OF DIGESTIBLE CONSTITUENTS PER 100 G	GRAM	1.95	0.99	13.30	
ENERGY VALUE (AVERAGE) OF THE DIGESTIBLE FRACTION PER 100 G EDIBLE PORTION	KJOULE	32	39	223	293
	(KCAL)	7.7	9.2	53	70

WASTE PERCENTAGE AVERAGE 6.0

CONSTITUENTS	DIM	AV		VARIATION	AVR	NUTR. DENS.		MOLPERC.
WATER	GRAM	82.50		80.00 - 85.00	77.55	GRAM/MJ	281.17	
PROTEIN	GRAM	2.30		1.90 - 2.70	2.16	GRAM/MJ	7.84	
FAT	GRAM	1.10		0.70 - 1.30	1.03	GRAM/MJ	3.75	
AVAILABLE CARBOHYDR.	GRAM	13.30	1	- - -	12.50	GRAM/MJ	45.33	
MINERALS	GRAM	0.80		- - -	0.75	GRAM/MJ	2.73	
CALCIUM	MILLI	12.00		10.00 - 14.00	11.28	MILLI/MJ	40.90	
IRON	MILLI	1.30		1.10 - 1.50	1.22	MILLI/MJ	4.43	
PHOSPHORUS	MILLI	39.00		- - -	36.66	MILLI/MJ	132.91	
CAROTENE	MILLI	0.90		0.20 - 1.70	0.85	MILLI/MJ	3.07	
VITAMIN B1	MICRO	60.00		10.00 - 100.00	56.40	MICRO/MJ	204.48	
VITAMIN B2	MICRO	40.00		- - -	37.60	MICRO/MJ	136.32	
NICOTINAMIDE	MILLI	2.00		1.10 - 2.80	1.88	MILLI/MJ	6.82	
VITAMIN C	MILLI	28.00		10.00 - 40.00	26.32	MILLI/MJ	95.43	

1 ESTIMATED BY THE DIFFERENCE METHOD
100 - (WATER + PROTEIN + FAT + MINERALS)

KARAMBOLE (BAUMSTACHELBEERE) — CARAMBOLA — CARAMBOLE

AVERRHOA CARAMBOLA L.

		PROTEIN	FAT	CARBOHYDRATES	TOTAL
ENERGY VALUE (AVERAGE)	KJOULE	20	19	59	98
PER 100 G	(KCAL)	4.7	4.7	14	23
EDIBLE PORTION					
AMOUNT OF DIGESTIBLE	GRAM	1.02	0.45	3.50	
CONSTITUENTS PER 100 G					
ENERGY VALUE (AVERAGE)	KJOULE	17	18	59	93
OF THE DIGESTIBLE	(KCAL)	4.0	4.2	14	22
FRACTION PER 100 G					
EDIBLE PORTION					

WASTE PERCENTAGE AVERAGE 15

CONSTITUENTS	DIM	AV	VARIATION			AVR	NUTR. DENS.		MOLPERC.
WATER	GRAM	91.20	90.00	-	92.30	77.52	GRAM/MJ	981.24	
PROTEIN	GRAM	1.20	0.30	-	1.60	1.02	GRAM/MJ	12.91	
FAT	GRAM	0.50	0.40	-	0.50	0.43	GRAM/MJ	5.38	
AVAILABLE CARBOHYDR.	GRAM	3.50	-	-	-	2.98	GRAM/MJ	37.66	
MINERALS	GRAM	0.40	0.30	-	0.40	0.34	GRAM/MJ	4.30	
SODIUM	MILLI	2.00	-	-	-	1.70	MILLI/MJ	21.52	
POTASSIUM	MILLI	184.00	181.00	-	190.00	156.40	GRAM/MJ	1.98	
CALCIUM	MILLI	6.00	4.00	-	8.00	5.10	MILLI/MJ	64.56	
IRON	MILLI	0.90	-	-	-	0.77	MILLI/MJ	9.68	
PHOSPHORUS	MILLI	16.00	15.00	-	17.00	13.60	MILLI/MJ	172.15	
CAROTENE	MICRO	90.00	20.00	-	160.00	76.50	MICRO/MJ	968.33	
VITAMIN B1	MICRO	50.00	40.00	-	50.00	42.50	MICRO/MJ	537.96	
VITAMIN B2	MICRO	30.00	20.00	-	40.00	25.50	MICRO/MJ	322.78	
NICOTINAMIDE	MILLI	0.40	0.30	-	0.40	0.34	MILLI/MJ	4.30	
VITAMIN C	MILLI	36.00	35.00	-	38.00	30.60	MILLI/MJ	387.33	
OXALIC ACID	GRAM	-	0.04	-	0.68				
GLUCOSE	GRAM	1.60	-	-	-	1.36	GRAM/MJ	17.21	
FRUCTOSE	GRAM	1.20	-	-	-	1.02	GRAM/MJ	12.91	
SUCROSE	GRAM	0.70	-	-	-	0.60	GRAM/MJ	7.53	

KIWI
(CHINESISCHE STACHEL-BEERE)
ACTINIDIA CHINENSIS PLANCH.

KIWI FRUIT
(CHINESE GOOSEBERRY, STRAWBERRY PEACH)

KIWI

		PROTEIN	FAT	CARBOHYDRATES	TOTAL
ENERGY VALUE (AVERAGE) PER 100 G EDIBLE PORTION	KJOULE (KCAL)	17 4.0	25 5.9	180 43	221 53
AMOUNT OF DIGESTIBLE CONSTITUENTS PER 100 G	GRAM	0.85	0.56	10.77	
ENERGY VALUE (AVERAGE) OF THE DIGESTIBLE FRACTION PER 100 G EDIBLE PORTION	KJOULE (KCAL)	14 3.4	22 5.3	180 43	216 52

WASTE PERCENTAGE AVERAGE 13 MINIMUM 5.0 MAXIMUM 20

CONSTITUENTS	DIM	AV	VARIATION			AVR	NUTR. DENS.		MOLPERC.
WATER	GRAM	83.80	83.50	-	84.10	72.91	GRAM/MJ	387.32	
PROTEIN	GRAM	1.00	0.60	-	1.60	0.87	GRAM/MJ	4.62	
FAT	GRAM	0.63	0.46	-	0.80	0.55	GRAM/MJ	2.91	
AVAILABLE CARBOHYDR.	GRAM	10.77	9.21	-	12.35	9.37	GRAM/MJ	49.78	
TOTAL DIETARY FIBRE	GRAM	3.90 [1]	-	-	-	3.39	GRAM/MJ	18.03	
MINERALS	GRAM	0.72	0.70	-	0.74	0.63	GRAM/MJ	3.33	
SODIUM	MILLI	4.10	1.80	-	4.70	3.57	MILLI/MJ	18.95	
POTASSIUM	MILLI	295.00	260.00	-	430.00	256.65	GRAM/MJ	1.36	
MAGNESIUM	MILLI	23.80	14.00	-	27.00	20.71	MILLI/MJ	110.00	
CALCIUM	MILLI	38.00	25.00	-	52.00	33.06	MILLI/MJ	175.64	
IRON	MILLI	0.80	0.30	-	1.60	0.70	MILLI/MJ	3.70	
PHOSPHORUS	MILLI	31.00	26.00	-	67.00	26.97	MILLI/MJ	143.28	
CHLORIDE	MILLI	65.50	39.00	-	89.00	56.99	MILLI/MJ	302.74	
CAROTENE	MILLI	0.37	0.04	-	0.74	0.32	MILLI/MJ	1.71	
VITAMIN B1	MICRO	17.00	9.00	-	21.00	14.79	MICRO/MJ	78.57	
VITAMIN B2	MICRO	50.00	-	-	-	43.50	MICRO/MJ	231.10	
NICOTINAMIDE	MILLI	0.41	0.15	-	0.59	0.36	MILLI/MJ	1.90	
VITAMIN C	MILLI	71.00	17.00	-	295.00	61.77	MILLI/MJ	328.16	
MALIC ACID	MILLI	500.00	470.00	-	530.00	435.00	GRAM/MJ	2.31	
CITRIC ACID	GRAM	0.99	0.98	-	1.01	0.87	GRAM/MJ	4.60	
OXALIC ACID	-	TRACES	-	-	-				
QUINIC ACID	GRAM	0.96	0.92	-	1.00	0.84	GRAM/MJ	4.44	
SUCCINIC ACID	-	TRACES	-	-	-				
FUMARIC ACID	-	TRACES	-	-	-				
OXALOACETIC ACID	-	TRACES	-	-	-				
SALICYLIC ACID	MILLI	0.32	0.32	-	0.32	0.28	MILLI/MJ	1.48	
GLUCOSE	GRAM	4.49	3.98	-	5.00	3.91	GRAM/MJ	20.75	
FRUCTOSE	GRAM	3.54	2.68	-	4.40	3.08	GRAM/MJ	16.36	
SUCROSE	GRAM	1.25	1.10	-	1.41	1.09	GRAM/MJ	5.80	
PECTIN	GRAM	0.74	0.39	-	1.13	0.64	GRAM/MJ	3.42	
PENTOSAN	GRAM	0.61	-	-	-	0.53	GRAM/MJ	2.82	
DIETARY FIBRE,WAT.SOL.	GRAM	1.50	-	-	-	1.31	GRAM/MJ	6.93	
DIETARY FIBRE,WAT.INS.	GRAM	2.40	-	-	-	2.09	GRAM/MJ	11.09	

1 METHOD OF SCHWEIZER AND WUERSCH

KUMQUAT
(ZWERGPOMERANZE, KINOTO)
FORTUNELLA MARGARITA (LOUR.) S

OVAL KUMQUAT
(NAGAMI)

KUMQUAT

		PROTEIN	FAT	CARBOHYDRATES	TOTAL
ENERGY VALUE (AVERAGE) PER 100 G EDIBLE PORTION	KJOULE	11	12	244	267
	(KCAL)	2.6	2.8	58	64
AMOUNT OF DIGESTIBLE CONSTITUENTS PER 100 G	GRAM	0.55	0.27	14.60	
ENERGY VALUE (AVERAGE) OF THE DIGESTIBLE FRACTION PER 100 G EDIBLE PORTION	KJOULE	9.1	11	244	264
	(KCAL)	2.2	2.5	58	63

WASTE PERCENTAGE AVERAGE 2.0

CONSTITUENTS	DIM	AV	VARIATION		AVR	NUTR. DENS.		MOLPERC.
WATER	GRAM	83.90	81.00	86.70	82.22	GRAM/MJ	317.82	
PROTEIN	GRAM	0.65	0.40	0.90	0.64	GRAM/MJ	2.46	
FAT	GRAM	0.30	0.10	0.50	0.29	GRAM/MJ	1.14	
AVAILABLE CARBOHYDR.	GRAM	14.60 [1]	-	-	14.31	GRAM/MJ	55.31	
MINERALS	GRAM	0.55	0.50	0.60	0.54	GRAM/MJ	2.08	
SODIUM	MILLI	111.00	-	-	108.78	MILLI/MJ	420.48	
POTASSIUM	MILLI	198.00	156.00	240.00	194.04	MILLI/MJ	750.05	
CALCIUM	MILLI	16.00	-	-	15.68	MILLI/MJ	60.61	
IRON	MILLI	0.60	0.40	0.80	0.59	MILLI/MJ	2.27	
PHOSPHORUS	MILLI	44.00	23.00	65.00	43.12	MILLI/MJ	166.68	
CAROTENE	MILLI	0.21	0.03	0.40	0.21	MILLI/MJ	0.80	
VITAMIN B1	MICRO	85.00	80.00	90.00	83.30	MICRO/MJ	321.99	
VITAMIN B2	MICRO	80.00	60.00	100.00	78.40	MICRO/MJ	303.05	
NICOTINAMIDE	MILLI	0.50	-	-	0.49	MILLI/MJ	1.89	
VITAMIN C	MILLI	38.00	36.00	40.00	37.24	MILLI/MJ	143.95	

[1] ESTIMATED BY THE DIFFERENCE METHOD
100 - (WATER + PROTEIN + FAT + MINERALS)

LIMONE, LIMETTE / LIME / LIME

CITRUS AURANTIFOLIA (CHRISTM.)

		PROTEIN	FAT	CARBOHYDRATES	TOTAL
ENERGY VALUE (AVERAGE)	KJOULE	8.3	93	32	133
PER 100 G	(KCAL)	2.0	22	7.6	32
EDIBLE PORTION					
AMOUNT OF DIGESTIBLE	GRAM	0.42	2.16	1.90	
CONSTITUENTS PER 100 G					
ENERGY VALUE (AVERAGE)	KJOULE	7.0	84	32	123
OF THE DIGESTIBLE	(KCAL)	1.7	20	7.6	29
FRACTION PER 100 G					
EDIBLE PORTION					

WASTE PERCENTAGE AVERAGE 23

CONSTITUENTS	DIM	AV	VARIATION			AVR	NUTR. DENS.		MOLPERC.
WATER	GRAM	91.00	-	-	-	70.07	GRAM/MJ	740.62	
PROTEIN	GRAM	0.50	-	-	-	0.39	GRAM/MJ	4.07	
FAT	GRAM	2.40	-	-	-	1.85	GRAM/MJ	19.53	
AVAILABLE CARBOHYDR.	GRAM	1.90	-	-	-	1.46	GRAM/MJ	15.46	
MINERALS	GRAM	0.20	-	-	-	0.15	GRAM/MJ	1.63	
SODIUM	MILLI	2.00	-	-	-	1.54	MILLI/MJ	16.28	
POTASSIUM	MILLI	82.00	-	-	-	63.14	MILLI/MJ	667.37	
CALCIUM	MILLI	13.00	-	-	-	10.01	MILLI/MJ	105.80	
IRON	MILLI	0.20	-	-	-	0.15	MILLI/MJ	1.63	
PHOSPHORUS	MILLI	11.00	-	-	-	8.47	MILLI/MJ	89.53	
CAROTENE	MICRO	10.00	-	-	-	7.70	MICRO/MJ	81.39	
VITAMIN B1	MICRO	28.00	9.00	-	44.00	21.56	MICRO/MJ	227.88	
VITAMIN B2	MICRO	20.00	-	-	-	15.40	MICRO/MJ	162.77	
NICOTINAMIDE	MILLI	0.17	0.10	-	0.34	0.13	MILLI/MJ	1.38	
VITAMIN C	MILLI	43.50	37.50	-	46.00	33.50	MILLI/MJ	354.03	
GLUCOSE	GRAM	0.80	-	-	-	0.62	GRAM/MJ	6.51	
FRUCTOSE	GRAM	0.80	-	-	-	0.62	GRAM/MJ	6.51	
SUCROSE	GRAM	0.30	-	-	-	0.23	GRAM/MJ	2.44	

LITCHI (LITSCHIPFLAUME, CHINESISCHE HASELNUSS)

LITCHI (LYCHEE)

LITCHI

LITCHI CHINENSIS SONN.

		PROTEIN	FAT	CARBOHYDRATES	TOTAL
ENERGY VALUE (AVERAGE) PER 100 G EDIBLE PORTION	KJOULE (KCAL)	15 3.6	12 2.8	285 68	311 74
AMOUNT OF DIGESTIBLE CONSTITUENTS PER 100 G	GRAM	0.76	0.27	17.00	
ENERGY VALUE (AVERAGE) OF THE DIGESTIBLE FRACTION PER 100 G EDIBLE PORTION	KJOULE (KCAL)	13 3.0	11 2.5	285 68	308 74

WASTE PERCENTAGE AVERAGE 37 MINIMUM 33 MAXIMUM 44

CONSTITUENTS	DIM	AV	VARIATION			AVR	NUTR. DENS.		MOLPERC.
WATER	GRAM	81.20	80.10	-	84.00	51.16	GRAM/MJ	263.93	
PROTEIN	GRAM	0.90	0.80	-	1.20	0.57	GRAM/MJ	2.93	
FAT	GRAM	0.30	0.20	-	0.40	0.19	GRAM/MJ	0.98	
AVAILABLE CARBOHYDR.	GRAM	17.00	-	-	-	10.71	GRAM/MJ	55.26	
TOTAL DIETARY FIBRE	GRAM	1.60	-	-	-	1.01	GRAM/MJ	5.20	
MINERALS	GRAM	0.45	0.40	-	0.53	0.28	GRAM/MJ	1.46	
SODIUM	MILLI	2.50	1.00	-	3.00	1.58	MILLI/MJ	8.13	
POTASSIUM	MILLI	182.00	168.00	-	228.00	114.66	MILLI/MJ	591.56	
CALCIUM	MILLI	9.30	6.30	-	13.00	5.86	MILLI/MJ	30.23	
IRON	MILLI	0.35	0.30	-	0.40	0.22	MILLI/MJ	1.14	
PHOSPHORUS	MILLI	33.00	22.00	-	42.00	20.79	MILLI/MJ	107.26	
VITAMIN B1	MICRO	50.00	20.00	-	130.00	31.50	MICRO/MJ	162.52	
VITAMIN B2	MICRO	50.00	40.00	-	60.00	31.50	MICRO/MJ	162.52	
NICOTINAMIDE	MILLI	0.53	0.30	-	0.99	0.33	MILLI/MJ	1.72	
VITAMIN C	MILLI	39.20	25.00	-	50.00	24.70	MILLI/MJ	127.41	
MALIC ACID	MILLI	239.00	-	-	-	150.57	MILLI/MJ	776.83	
CITRIC ACID	MILLI	16.00	-	-	-	10.08	MILLI/MJ	52.01	
LACTIC ACID	MILLI	1.80	-	-	-	1.13	MILLI/MJ	5.85	
MALONIC ACID	MILLI	1.04	-	-	-	0.66	MILLI/MJ	3.38	
GLUTARIC ACID	MILLI	2.64	-	-	-	1.66	MILLI/MJ	8.58	
SUCCINIC ACID	MILLI	2.40	-	-	-	1.51	MILLI/MJ	7.80	
GLUCOSE	GRAM	5.00	-	-	-	3.15	GRAM/MJ	16.25	
FRUCTOSE	GRAM	3.20	-	-	-	2.02	GRAM/MJ	10.40	
SUCROSE	GRAM	8.60	-	-	-	5.42	GRAM/MJ	27.95	
CELLULOSE	GRAM	0.16	-	-	-	0.10	GRAM/MJ	0.52	

LONGAN (DRACHENAUGE) | LONGAN | LONGAN

NEPHELIUM LONGANUM (LAM.) CAM.
EUPHORIA LONGANA LAM.

		PROTEIN	FAT	CARBOHYDRATES	TOTAL
ENERGY VALUE (AVERAGE)	KJOULE	17	29	153	199
PER 100 G EDIBLE PORTION	(KCAL)	4.0	7.0	37	47
AMOUNT OF DIGESTIBLE CONSTITUENTS PER 100 G	GRAM	0.85	0.67	9.14	
ENERGY VALUE (AVERAGE)	KJOULE	14	26	153	193
OF THE DIGESTIBLE FRACTION PER 100 G EDIBLE PORTION	(KCAL)	3.4	6.3	37	46

WASTE PERCENTAGE AVERAGE 50

CONSTITUENTS	DIM	AV	VARIATION		AVR	NUTR. DENS.		MOLPERC.
WATER	GRAM	81.50	81.00	- 82.00	40.75	GRAM/MJ	421.67	
PROTEIN	GRAM	1.00	-	- -	0.50	GRAM/MJ	5.17	
FAT	GRAM	0.75	0.10	- 1.40	0.38	GRAM/MJ	3.88	
AVAILABLE CARBOHYDR.	GRAM	9.14	-	- -	4.57	GRAM/MJ	47.29	
TOTAL DIETARY FIBRE	GRAM	1.10 [1]	-	- -	0.55	GRAM/MJ	5.69	
MINERALS	GRAM	0.85	0.70	- 1.00	0.43	GRAM/MJ	4.40	
CALCIUM	MILLI	16.50	10.00	- 23.00	8.25	MILLI/MJ	85.37	
IRON	MILLI	0.80	0.40	- 1.20	0.40	MILLI/MJ	4.14	
PHOSPHORUS	MILLI	39.00	36.00	- 42.00	19.50	MILLI/MJ	201.78	
VITAMIN B1	MICRO	30.00	-	- -	15.00	MICRO/MJ	155.22	
VITAMIN B2	MILLI	0.14	-	- -	0.07	MILLI/MJ	0.72	
NICOTINAMIDE	MILLI	0.30	-	- -	0.15	MILLI/MJ	1.55	
VITAMIN C	MILLI	56.00	-	- -	28.00	MILLI/MJ	289.74	
MALIC ACID	GRAM	0.49	-	- -	0.25	GRAM/MJ	2.54	
CITRIC ACID	MILLI	50.00	-	- -	25.00	MILLI/MJ	258.69	
GLUCOSE	GRAM	2.30	-	- -	1.15	GRAM/MJ	11.90	
FRUCTOSE	GRAM	3.20	-	- -	1.60	GRAM/MJ	16.56	
SUCROSE	GRAM	3.10	-	- -	1.55	GRAM/MJ	16.04	
CELLULOSE	GRAM	0.28	-	- -	0.14	GRAM/MJ	1.45	

1 METHOD OF SOUTHGATE

MAMMEY-APFEL (MAMMIAPFEL) | MAMEY (MAMMIAPPLE) | POMME DE MAMMEY

MAMMEA AMERICANA L.

		PROTEIN	FAT	CARBOHYDRATES	TOTAL
ENERGY VALUE (AVERAGE)	KJOULE	8.3	14	207	229
PER 100 G	(KCAL)	2.0	3.3	49	55
EDIBLE PORTION					
AMOUNT OF DIGESTIBLE	GRAM	0.42	0.31	12.35	
CONSTITUENTS PER 100 G					
ENERGY VALUE (AVERAGE)	KJOULE	7.0	12	207	226
OF THE DIGESTIBLE	(KCAL)	1.7	2.9	49	54
FRACTION PER 100 G					
EDIBLE PORTION					

WASTE PERCENTAGE AVERAGE 38

CONSTITUENTS	DIM	AV		VARIATION			AVR	NUTR.	DENS.	MOLPERC.
WATER	GRAM	86.50		-	-	-	53.63	GRAM/MJ	382.79	
PROTEIN	GRAM	0.50		-	-	-	0.31	GRAM/MJ	2.21	
FAT	GRAM	0.35		0.20	-	0.50	0.22	GRAM/MJ	1.55	
AVAILABLE CARBOHYDR.	GRAM	12.35	1	-	-	-	7.66	GRAM/MJ	54.65	
MINERALS	GRAM	0.30		-	-	-	0.19	GRAM/MJ	1.33	
CALCIUM	MILLI	11.00		-	-	-	6.82	MILLI/MJ	48.68	
IRON	MILLI	0.60		0.50	-	0.70	0.37	MILLI/MJ	2.66	
PHOSPHORUS	MILLI	11.00		-	-	-	6.82	MILLI/MJ	48.68	
CAROTENE	MILLI	0.12		0.10	-	0.14	0.07	MILLI/MJ	0.53	
VITAMIN B1	MICRO	20.00		-	-	-	12.40	MICRO/MJ	88.51	
VITAMIN B2	MICRO	40.00		-	-	-	24.80	MICRO/MJ	177.01	
NICOTINAMIDE	MILLI	0.40		-	-	-	0.25	MILLI/MJ	1.77	
VITAMIN C	MILLI	14.00		4.00	-	22.00	8.68	MILLI/MJ	61.95	

1 ESTIMATED BY THE DIFFERENCE METHOD
100 - (WATER + PROTEIN + FAT + MINERALS)

MANDARINE | MANDARIN (TANGERINE) | MANDARINE

CITRUS NOBILIS LOUREIRO

		PROTEIN	FAT	CARBOHYDRATES	TOTAL
ENERGY VALUE (AVERAGE) PER 100 G EDIBLE PORTION	KJOULE	12	12	169	192
	(KCAL)	2.8	2.8	40	46
AMOUNT OF DIGESTIBLE CONSTITUENTS PER 100 G	GRAM	0.59	0.27	10.10	
ENERGY VALUE (AVERAGE) OF THE DIGESTIBLE FRACTION PER 100 G EDIBLE PORTION	KJOULE	9.8	11	169	189
	(KCAL)	2.4	2.5	40	45

WASTE PERCENTAGE AVERAGE 35

CONSTITUENTS	DIM	AV	VARIATION			AVR	NUTR. DENS.		MOLPERC.
WATER	GRAM	86.70	86.00	-	87.30	56.36	GRAM/MJ	457.83	
PROTEIN	GRAM	0.70	0.50	-	0.80	0.46	GRAM/MJ	3.70	
FAT	GRAM	0.30	-	-	-	0.20	GRAM/MJ	1.58	
AVAILABLE CARBOHYDR.	GRAM	10.10	-	-	-	6.57	GRAM/MJ	53.33	
MINERALS	GRAM	0.70	-	-	-	0.46	GRAM/MJ	3.70	
SODIUM	MILLI	1.20	0.40	-	2.00	0.78	MILLI/MJ	6.34	
POTASSIUM	MILLI	210.00	171.00	-	250.00	136.50	GRAM/MJ	1.11	
MAGNESIUM	MILLI	11.00	-	-	-	7.15	MILLI/MJ	58.09	
CALCIUM	MILLI	33.00	26.00	-	40.00	21.45	MILLI/MJ	174.26	
MANGANESE	MICRO	40.00	-	-	-	26.00	MICRO/MJ	211.22	
IRON	MILLI	0.30	-	-	-	0.20	MILLI/MJ	1.58	
COPPER	MICRO	90.00	-	-	-	58.50	MICRO/MJ	475.25	
ZINC	MICRO	80.00	-	-	-	52.00	MICRO/MJ	422.45	
NICKEL	MICRO	3.00	1.00	-	4.00	1.95	MICRO/MJ	15.84	
CHROMIUM	MICRO	1.00	-	-	-	0.65	MICRO/MJ	5.28	
PHOSPHORUS	MILLI	20.00	-	-	-	13.00	MILLI/MJ	105.61	
CHLORIDE	MILLI	4.00	2.00	-	6.00	2.60	MILLI/MJ	21.12	
FLUORIDE	MICRO	10.00	-	-	-	6.50	MICRO/MJ	52.81	
IODIDE	MICRO	0.80	-	-	-	0.52	MICRO/MJ	4.22	
BORON	MICRO	90.00	5.00	-	140.00	58.50	MICRO/MJ	475.25	
SELENIUM	MICRO	17.00	13.00	-	21.00	11.05	MICRO/MJ	89.77	
SILICON	MILLI	3.00	-	-	-	1.95	MILLI/MJ	15.84	
CAROTENE	MILLI	0.34	0.24	-	0.53	0.22	MILLI/MJ	1.80	
VITAMIN B1	MICRO	60.00	60.00	-	70.00	39.00	MICRO/MJ	316.83	
VITAMIN B2	MICRO	30.00	-	-	-	19.50	MICRO/MJ	158.42	
NICOTINAMIDE	MILLI	0.20	-	-	-	0.13	MILLI/MJ	1.06	

CONSTITUENTS	DIM	AV	VARIATION			AVR	NUTR. DENS.		MOLPERC.
VITAMIN B6	MICRO	23.00	-	-	-	14.95	MICRO/MJ	121.45	
BIOTIN	MICRO	0.45	-	-	-	0.29	MICRO/MJ	2.38	
FOLIC ACID	MICRO	7.00	-	-	-	4.55	MICRO/MJ	36.96	
VITAMIN C	MILLI	30.00	29.00	-	31.00	19.50	MILLI/MJ	158.42	
ALANINE	MILLI	43.00	-	-	-	27.95	MILLI/MJ	227.07	10.1
ARGININE	MILLI	44.00	-	-	-	28.60	MILLI/MJ	232.35	5.3
ASPARTIC ACID	MILLI	96.00	-	-	-	62.40	MILLI/MJ	506.94	15.1
CYSTINE	MILLI	9.00	-	-	-	5.85	MILLI/MJ	47.53	0.8
GLUTAMIC ACID	MILLI	86.00	-	-	-	55.90	MILLI/MJ	454.13	12.3
GLYCINE	MILLI	70.00	-	-	-	45.50	MILLI/MJ	369.64	19.6
HISTIDINE	MILLI	11.00	-	-	-	7.15	MILLI/MJ	58.09	1.5
ISOLEUCINE	MILLI	19.00	-	-	-	12.35	MILLI/MJ	100.33	3.0
LEUCINE	MILLI	18.00	-	-	-	11.70	MILLI/MJ	95.05	2.9
LYSINE	MILLI	36.00	-	-	-	23.40	MILLI/MJ	190.10	5.2
METHIONINE	MILLI	11.00	-	-	-	7.15	MILLI/MJ	58.09	1.5
PHENYLALANINE	MILLI	25.00	-	-	-	16.25	MILLI/MJ	132.01	3.2
PROLINE	MILLI	39.00	-	-	-	25.35	MILLI/MJ	205.94	7.1
SERINE	MILLI	19.00	-	-	-	12.35	MILLI/MJ	100.33	3.8
THREONINE	MILLI	11.00	-	-	-	7.15	MILLI/MJ	58.09	1.9
TRYPTOPHAN	MILLI	4.00	-	-	-	2.60	MILLI/MJ	21.12	0.4
TYROSINE	MILLI	14.00	-	-	-	9.10	MILLI/MJ	73.93	1.6
VALINE	MILLI	26.00	-	-	-	16.90	MILLI/MJ	137.30	4.7
SALICYLIC ACID	MILLI	0.56	0.56	-	0.56	0.36	MILLI/MJ	2.96	
GLUCOSE	GRAM	1.70	-	-	-	1.11	GRAM/MJ	8.98	
FRUCTOSE	GRAM	1.30	-	-	-	0.85	GRAM/MJ	6.86	
SUCROSE	GRAM	7.10	-	-	-	4.62	GRAM/MJ	37.49	

MANGO MANGO MANGUE

MAGNIFERA INDICA L.

		PROTEIN	FAT	CARBOHYDRATES	TOTAL
ENERGY VALUE (AVERAGE)	KJOULE	9.9	18	214	242
PER 100 G	(KCAL)	2.4	4.2	51	58
EDIBLE PORTION					
AMOUNT OF DIGESTIBLE	GRAM	0.51	0.40	12.80	
CONSTITUENTS PER 100 G					
ENERGY VALUE (AVERAGE)	KJOULE	8.4	16	214	238
OF THE DIGESTIBLE	(KCAL)	2.0	3.8	51	57
FRACTION PER 100 G					
EDIBLE PORTION					

WASTE PERCENTAGE AVERAGE 31 MINIMUM 28 MAXIMUM 34

CONSTITUENTS	DIM	AV	VARIATION		AVR	NUTR. DENS.		MOLPERC.
WATER	GRAM	82.00	80.50	83.00	56.58	GRAM/MJ	343.95	
PROTEIN	GRAM	0.60	0.40	0.80	0.41	GRAM/MJ	2.52	
FAT	GRAM	0.45	0.20	0.60	0.31	GRAM/MJ	1.89	
AVAILABLE CARBOHYDR.	GRAM	12.80	-	-	8.83	GRAM/MJ	53.69	
TOTAL DIETARY FIBRE	GRAM	1.70 1	-	-	1.17	GRAM/MJ	7.13	
MINERALS	GRAM	0.50	0.30	0.60	0.35	GRAM/MJ	2.10	
SODIUM	MILLI	5.00	3.00	7.00	3.45	MILLI/MJ	20.97	
POTASSIUM	MILLI	190.00	168.00	214.00	131.10	MILLI/MJ	796.95	
MAGNESIUM	MILLI	18.00	-	-	12.42	MILLI/MJ	75.50	
CALCIUM	MILLI	12.00	10.00	20.00	8.28	MILLI/MJ	50.33	
IRON	MILLI	0.40	0.30	0.50	0.28	MILLI/MJ	1.68	
PHOSPHORUS	MILLI	13.00	10.00	17.00	8.97	MILLI/MJ	54.53	
IODIDE	MICRO	1.60	-	-	1.10	MICRO/MJ	6.71	
BORON	MICRO	48.00	32.00	64.00	33.12	MICRO/MJ	201.34	
CAROTENE	MILLI	2.77	1.88	3.40	1.91	MILLI/MJ	11.62	
VITAMIN E ACTIVITY	MILLI	1.00	-	-	0.69	MILLI/MJ	4.19	
ALPHA-TOCOPHEROL	MILLI	1.00	-	-	0.69	MILLI/MJ	4.19	
VITAMIN B1	MICRO	45.00	20.00	80.00	31.05	MICRO/MJ	188.75	
VITAMIN B2	MICRO	50.00	40.00	80.00	34.50	MICRO/MJ	209.72	
NICOTINAMIDE	MILLI	0.70	0.30	1.20	0.48	MILLI/MJ	2.94	
FOLIC ACID	MICRO	36.00	-	-	24.84	MICRO/MJ	151.00	
VITAMIN C	MILLI	38.70	30.00	55.00	26.70	MILLI/MJ	162.33	
MALIC ACID	MILLI	74.00	-	-	51.06	MILLI/MJ	310.39	
CITRIC ACID	MILLI	296.00	265.00	327.00	204.24	GRAM/MJ	1.24	
TARTARIC ACID	MILLI	81.00	-	-	55.89	MILLI/MJ	339.75	

1 METHOD OF SCHWEIZER AND WUERSCH

CONSTITUENTS	DIM	AV	VARIATION			AVR	NUTR. DENS.		MOLPERC.
OXALIC ACID	MILLI	36.00	-	-	-	24.84	MILLI/MJ	151.00	
SUCCINIC ACID	MILLI	40.00	-	-	-	27.60	MILLI/MJ	167.78	
GLYCOLIC ACID	MILLI	61.00	-	-	-	42.09	MILLI/MJ	255.86	
SALICYLIC ACID	MILLI	0.11	0.11	-	0.11	0.08	MILLI/MJ	0.46	
GLUCOSE	GRAM	0.85	0.50	-	2.00	0.59	GRAM/MJ	3.57	
FRUCTOSE	GRAM	2.60	2.00	-	3.50	1.79	GRAM/MJ	10.91	
SUCROSE	GRAM	9.00	7.00	-	11.00	6.21	GRAM/MJ	37.75	
CELLULOSE	GRAM	1.71	-	-	-	1.18	GRAM/MJ	7.17	
LAURIC ACID	MILLI	1.10	1.10	-	1.10	0.76	MILLI/MJ	4.61	
MYRISTIC ACID	MILLI	10.00	10.00	-	10.00	6.90	MILLI/MJ	41.94	
PALMITIC ACID	MILLI	85.00	85.00	-	85.00	58.65	MILLI/MJ	356.53	
STEARIC ACID	MILLI	5.00	5.00	-	5.00	3.45	MILLI/MJ	20.97	
PALMITOLEIC ACID	MILLI	95.00	95.00	-	95.00	65.55	MILLI/MJ	398.48	
OLEIC ACID	MILLI	87.00	87.00	-	87.00	60.03	MILLI/MJ	364.92	
LINOLEIC ACID	MILLI	9.00	9.00	-	9.00	6.21	MILLI/MJ	37.75	
LINOLENIC ACID	MILLI	67.00	67.00	-	67.00	46.23	MILLI/MJ	281.03	
DIETARY FIBRE, WAT.SOL.	GRAM	0.63	-	-	-	0.43	GRAM/MJ	2.64	
DIETARY FIBRE, WAT.INS.	GRAM	1.07	-	-	-	0.74	GRAM/MJ	4.49	

MANGOSTANE — MANGOSTEEN — MANGOUSTAN DU MALABAR

GARCINIA MANGOSTANA L.

		PROTEIN	FAT	CARBOHYDRATES	TOTAL
ENERGY VALUE (AVERAGE)	KJOULE	9.9	23	266	299
PER 100 G	(KCAL)	2.4	5.6	64	72
EDIBLE PORTION					
AMOUNT OF DIGESTIBLE	GRAM	0.51	0.54	15.90	
CONSTITUENTS PER 100 G					
ENERGY VALUE (AVERAGE)	KJOULE	8.4	21	266	296
OF THE DIGESTIBLE	(KCAL)	2.0	5.0	64	71
FRACTION PER 100 G					
EDIBLE PORTION					

WASTE PERCENTAGE AVERAGE 71

CONSTITUENTS	DIM	AV		VARIATION	AVR	NUTR. DENS.		MOLPERC.
WATER	GRAM	81.30		79.70 - 84.30	23.58	GRAM/MJ	275.09	
PROTEIN	GRAM	0.60		0.50 - 0.70	0.17	GRAM/MJ	2.03	
FAT	GRAM	0.60		0.30 - 0.80	0.17	GRAM/MJ	2.03	
AVAILABLE CARBOHYDR.	GRAM	15.90	1	- - -	4.61	GRAM/MJ	53.80	
TOTAL DIETARY FIBRE	GRAM	1.40	2	- - -	0.41	GRAM/MJ	4.74	
MINERALS	GRAM	0.20		- - -	0.06	GRAM/MJ	0.68	
SODIUM	MILLI	1.00		- - -	0.29	MILLI/MJ	3.38	
CALCIUM	MILLI	15.00		10.00 - 18.00	4.35	MILLI/MJ	50.75	
IRON	MILLI	0.40		0.30 - 0.50	0.12	MILLI/MJ	1.35	
PHOSPHORUS	MILLI	11.00		- - -	3.19	MILLI/MJ	37.22	
VITAMIN B1	MILLI	0.50		0.30 - 0.60	0.15	MILLI/MJ	1.69	
VITAMIN B2	MICRO	15.00		10.00 - 20.00	4.35	MICRO/MJ	50.75	
NICOTINAMIDE	MILLI	-		0.04 - 0.60				
VITAMIN C	MILLI	2.70		2.00 - 4.00	0.78	MILLI/MJ	9.14	
CELLULOSE	GRAM	0.31		- - -	0.09	GRAM/MJ	1.05	

1 ESTIMATED BY THE DIFFERENCE METHOD
100 - (WATER + PROTEIN + FAT + MINERALS + TOTAL DIETARY FIBRE)

2 METHOD OF SOUTHGATE

NARANJILLA NARANJILLA NARANJILLA

(LULO, QUITO-ORANGE)

SOLANUM QUITOENSE LAM.

		PROTEIN	FAT	CARBOHYDRATES	TOTAL
ENERGY VALUE (AVERAGE) PER 100 G EDIBLE PORTION	KJOULE	16	5.8	163	184
	(KCAL)	3.8	1.4	39	44
AMOUNT OF DIGESTIBLE CONSTITUENTS PER 100 G	GRAM	0.65	0.13	9.71	
ENERGY VALUE (AVERAGE) OF THE DIGESTIBLE FRACTION PER 100 G EDIBLE PORTION	KJOULE	10	5.3	163	178
	(KCAL)	2.4	1.3	39	43

WASTE PERCENTAGE AVERAGE 0.00

CONSTITUENTS	DIM	AV		VARIATION		AVR	NUTR. DENS.		MOLPERC.
WATER	GRAM	88.50	85.80	-	90.00	88.50	GRAM/MJ	497.31	
PROTEIN	GRAM	1.00	0.90	-	1.30	1.00	GRAM/MJ	5.62	
FAT	GRAM	0.15	0.10	-	0.24	0.15	GRAM/MJ	0.84	
AVAILABLE CARBOHYDR.	GRAM	9.71 1	-	-	-	9.71	GRAM/MJ	54.56	
MINERALS	GRAM	0.64	0.60	-	0.75	0.64	GRAM/MJ	3.60	
CALCIUM	MILLI	14.00	-	-	-	14.00	MILLI/MJ	78.67	
IRON	MILLI	0.50	-	-	-	0.50	MILLI/MJ	2.81	
PHOSPHORUS	MILLI	50.00	41.00	-	78.00	50.00	MILLI/MJ	280.97	
CAROTENE	MILLI	0.13	0.07	-	0.23	0.13	MILLI/MJ	0.73	
VITAMIN B1	MICRO	60.00	-	-	-	60.00	MICRO/MJ	337.16	
VITAMIN B2	MICRO	40.00	-	-	-	40.00	MICRO/MJ	224.77	
NICOTINAMIDE	MILLI	1.50	1.20	-	1.80	1.50	MILLI/MJ	8.43	
VITAMIN C	MILLI	67.00	31.00	-	78.00	67.00	MILLI/MJ	376.49	

1 ESTIMATED BY THE DIFFERENCE METHOD
100 - (WATER + PROTEIN + FAT + MINERALS)

OKRA
(GOMBO, EIBISCH)

LADY'S FINGER

GOMBO
(BONNET GREC)

ABELMOSCHUS ESCULENTUS
(L.) MOENCH.

		PROTEIN	FAT	CARBOHYDRATES	TOTAL
ENERGY VALUE (AVERAGE)	KJOULE	33	7.8	37	78
PER 100 G	(KCAL)	7.9	1.9	8.8	19
EDIBLE PORTION					
AMOUNT OF DIGESTIBLE	GRAM	1.36	0.18	2.20	
CONSTITUENTS PER 100 G					
ENERGY VALUE (AVERAGE)	KJOULE	21	7.0	37	65
OF THE DIGESTIBLE	(KCAL)	5.1	1.7	8.8	16
FRACTION PER 100 G					
EDIBLE PORTION					

WASTE PERCENTAGE AVERAGE 15

CONSTITUENTS	DIM	AV		VARIATION			AVR	NUTR. DENS.		MOLPERC.
WATER	GRAM	88.60		-	-	-	75.31	KILO/MJ	1.36	
PROTEIN	GRAM	2.10		-	-	-	1.79	GRAM/MJ	32.19	
FAT	GRAM	0.20		-	-	-	0.17	GRAM/MJ	3.07	
AVAILABLE CARBOHYDR.	GRAM	2.20		-	-	-	1.87	GRAM/MJ	33.72	
TOTAL DIETARY FIBRE	GRAM	4.90	1	-	-	-	4.17	GRAM/MJ	75.11	
MINERALS	GRAM	0.90		-	-	-	0.77	GRAM/MJ	13.80	
SODIUM	MILLI	3.00		-	-	-	2.55	MILLI/MJ	45.98	
POTASSIUM	MILLI	285.00		-	-	-	242.25	GRAM/MJ	4.37	
MAGNESIUM	MILLI	60.00		-	-	-	51.00	MILLI/MJ	919.68	
CALCIUM	MILLI	84.00		-	-	-	71.40	GRAM/MJ	1.29	
IRON	MILLI	1.20		-	-	-	1.02	MILLI/MJ	18.39	
PHOSPHORUS	MILLI	75.00		60.00	-	90.00	63.75	GRAM/MJ	1.15	
IODIDE	MICRO	5.60		-	-	-	4.76	MICRO/MJ	85.84	
CAROTENE	MILLI	0.14		0.09	-	0.19	0.12	MILLI/MJ	2.15	
VITAMIN B1	MICRO	70.00		-	-	-	59.50	MILLI/MJ	1.07	
VITAMIN B2	MICRO	80.00		-	-	-	68.00	MILLI/MJ	1.23	
NICOTINAMIDE	MILLI	0.80		0.60	-	1.00	0.68	MILLI/MJ	12.26	
VITAMIN C	MILLI	36.00		25.00	-	47.00	30.60	MILLI/MJ	551.81	
GLUCOSE	GRAM	0.70		0.60	-	0.80	0.60	GRAM/MJ	10.73	
FRUCTOSE	GRAM	0.80		0.60	-	1.10	0.68	GRAM/MJ	12.26	
SUCROSE	GRAM	0.70		0.50	-	0.90	0.60	GRAM/MJ	10.73	
CELLULOSE	GRAM	0.98		-	-	-	0.83	GRAM/MJ	15.02	
TOTAL STEROLS	MILLI	24.00		-	-	-	20.40	MILLI/MJ	367.87	
CAMPESTEROL	MILLI	3.00		-	-	-	2.55	MILLI/MJ	45.98	
BETA-SITOSTEROL	MILLI	15.00		-	-	-	12.75	MILLI/MJ	229.92	
STIGMASTEROL	MILLI	6.00		-	-	-	5.10	MILLI/MJ	91.97	

1 METHOD OF SOUTHGATE

OLIVE	**OLIVE**	**OLIVE VERTE**
GRÜN	GREEN	SAUMURÉE
MARINIERT	MARINATED	

		PROTEIN	FAT	CARBOHYDRATES	TOTAL
ENERGY VALUE (AVERAGE) PER 100 G EDIBLE PORTION	KJOULE	23	541	29	593
	(KCAL)	5.5	129	7.0	142
AMOUNT OF DIGESTIBLE CONSTITUENTS PER 100 G	GRAM	1.17	12.51	1.76	
ENERGY VALUE (AVERAGE) OF THE DIGESTIBLE FRACTION PER 100 G EDIBLE PORTION	KJOULE	19	487	29	536
	(KCAL)	4.6	116	7.0	128

WASTE PERCENTAGE AVERAGE 20

CONSTITUENTS	DIM	AV		VARIATION			AVR	NUTR.	DENS.	MOLPERC.
WATER	GRAM	74.80		74.30	-	75.20	59.84	GRAM/MJ	139.65	
PROTEIN	GRAM	1.38		1.25	-	1.50	1.10	GRAM/MJ	2.58	
FAT	GRAM	13.90		13.50	-	14.30	11.12	GRAM/MJ	25.95	
AVAILABLE CARBOHYDR.	GRAM	1.76	1	-	-	-	1.41	GRAM/MJ	3.29	
TOTAL DIETARY FIBRE	GRAM	2.36	2	-	-	-	1.89	GRAM/MJ	4.41	
MINERALS	GRAM	5.80		-	-	-	4.64	GRAM/MJ	10.83	
SODIUM	GRAM	2.10		0.98	-	2.93	1.68	GRAM/MJ	3.92	
POTASSIUM	MILLI	43.00		23.00	-	55.00	34.40	MILLI/MJ	80.28	
MAGNESIUM	MILLI	19.00		19.00	-	19.00	15.20	MILLI/MJ	35.47	
CALCIUM	MILLI	96.00		80.00	-	126.00	76.80	MILLI/MJ	179.23	
MANGANESE	MICRO	58.00		58.00	-	58.00	46.40	MICRO/MJ	108.29	
IRON	MILLI	1.80		1.60	-	2.00	1.44	MILLI/MJ	3.36	
COPPER	MILLI	0.27		0.27	-	0.27	0.22	MILLI/MJ	0.50	
PHOSPHORUS	MILLI	17.00		-	-	-	13.60	MILLI/MJ	31.74	
BORON	MILLI	0.19		0.18	-	0.20	0.15	MILLI/MJ	0.35	
CAROTENE	MILLI	0.21		0.21	-	0.21	0.17	MILLI/MJ	0.39	
VITAMIN B1	MICRO	30.00		-	-	-	24.00	MICRO/MJ	56.01	
VITAMIN B2	MICRO	80.00		-	-	-	64.00	MICRO/MJ	149.36	
NICOTINAMIDE	MILLI	0.50		-	-	-	0.40	MILLI/MJ	0.93	
PANTOTHENIC ACID	MILLI	-		0.15	-	0.97				
VITAMIN B6	MICRO	23.00		23.00	-	23.00	18.40	MICRO/MJ	42.94	
VITAMIN C	-	0.00		-	-	-	0.00			
SALICYLIC ACID	MILLI	0.34		0.34	-	0.34	0.27	MILLI/MJ	0.63	
PALMITIC ACID	GRAM	1.50		-	-	-	1.20	GRAM/MJ	2.80	
STEARIC ACID	GRAM	0.33		-	-	-	0.26	GRAM/MJ	0.62	
PALMITOLEIC ACID	GRAM	0.21		-	-	-	0.17	GRAM/MJ	0.39	
OLEIC ACID	GRAM	10.00		8.39	-	11.10	8.00	GRAM/MJ	18.67	
LINOLEIC ACID	GRAM	1.12		0.49	-	1.78	0.90	GRAM/MJ	2.09	
LINOLENIC ACID	GRAM	0.13		-	-	-	0.10	GRAM/MJ	0.24	
DIETARY FIBRE,WAT.SOL.	GRAM	0.25		-	-	-	0.20	GRAM/MJ	0.47	
DIETARY FIBRE,WAT.INS.	GRAM	2.11		-	-	-	1.69	GRAM/MJ	3.94	

1 ESTIMATED BY THE DIFFERENCE METHOD
100 - (WATER + PROTEIN + FAT + MINERALS + TOTAL DIETARY FIBRE)

2 METHOD OF ENGLYST

OPUNTIE (KAKTUSBIRNE, KAKTUSFEIGE, KAKTUSAPFEL) OPUNTIA FICUS-INDICA (L.) MILL.

PRICKLY PEAR

FIGUE DE BARBARIE

		PROTEIN	FAT	CARBOHYDRATES	TOTAL
ENERGY VALUE (AVERAGE)	KJOULE	17	16	119	151
PER 100 G	(KCAL)	4.0	3.7	28	36
EDIBLE PORTION					
AMOUNT OF DIGESTIBLE	GRAM	0.85	0.36	7.10	
CONSTITUENTS PER 100 G					
ENERGY VALUE (AVERAGE)	KJOULE	14	14	119	147
OF THE DIGESTIBLE	(KCAL)	3.4	3.3	28	35
FRACTION PER 100 G					
EDIBLE PORTION					

WASTE PERCENTAGE AVERAGE 45

CONSTITUENTS	DIM	AV	VARIATION			AVR	NUTR. DENS.		MOLPERC.
WATER	GRAM	86.40	83.00	-	89.70	47.52	GRAM/MJ	588.23	
PROTEIN	GRAM	1.00	-	-	-	0.55	GRAM/MJ	6.81	
FAT	GRAM	0.40	-	-	-	0.22	GRAM/MJ	2.72	
AVAILABLE CARBOHYDR.	GRAM	7.10	-	-	-	3.91	GRAM/MJ	48.34	
MINERALS	GRAM	0.30	-	-	-	0.17	GRAM/MJ	2.04	
POTASSIUM	MILLI	90.00	-	-	-	49.50	MILLI/MJ	612.74	
CALCIUM	MILLI	28.00	24.00	-	31.00	15.40	MILLI/MJ	190.63	
IRON	MILLI	0.30	-	-	-	0.17	MILLI/MJ	2.04	
PHOSPHORUS	MILLI	27.00	-	-	-	14.85	MILLI/MJ	183.82	
CAROTENE	MICRO	40.00	-	-	-	22.00	MICRO/MJ	272.33	
VITAMIN B1	MICRO	18.00	6.00	-	30.00	9.90	MICRO/MJ	122.55	
VITAMIN B2	MICRO	30.00	-	-	-	16.50	MICRO/MJ	204.25	
NICOTINAMIDE	MILLI	0.38	0.30	-	0.45	0.21	MILLI/MJ	2.59	
VITAMIN C	MILLI	23.00	17.00	-	42.00	12.65	MILLI/MJ	156.59	
MALIC ACID	MILLI	23.00	-	-	-	12.65	MILLI/MJ	156.59	
CITRIC ACID	MILLI	62.00	-	-	-	34.10	MILLI/MJ	422.11	
QUINIC ACID	MILLI	19.00	-	-	-	10.45	MILLI/MJ	129.36	
SHIKIMIC ACID	MILLI	3.00	-	-	-	1.65	MILLI/MJ	20.42	
GLUCOSE	GRAM	6.50	-	-	-	3.58	GRAM/MJ	44.25	
FRUCTOSE	GRAM	0.60	-	-	-	0.33	GRAM/MJ	4.08	
SUCROSE	-	0.00	-	-	-	0.00			

PAPAYA (BAUMMELONE, MAMMAO)	**PAPAYA** (PAWPAW)	**PAPAYE**

CARICA PAPAYA L.

		PROTEIN	FAT	CARBOHYDRATES	TOTAL
ENERGY VALUE (AVERAGE)	KJOULE	8.6	3.5	40	52
PER 100 G	(KCAL)	2.1	0.8	9.6	12
EDIBLE PORTION					
AMOUNT OF DIGESTIBLE	GRAM	0.44	0.08	2.40	
CONSTITUENTS PER 100 G					
ENERGY VALUE (AVERAGE)	KJOULE	7.3	3.2	40	51
OF THE DIGESTIBLE	(KCAL)	1.7	0.8	9.6	12
FRACTION PER 100 G					
EDIBLE PORTION					

WASTE PERCENTAGE AVERAGE 28 MINIMUM 22 MAXIMUM 34

CONSTITUENTS	DIM	AV	VARIATION			AVR	NUTR.	DENS.	MOLPERC.
WATER	GRAM	87.90	86.80	-	90.00	63.29	KILO/MJ	1.74	
PROTEIN	GRAM	0.52	0.36	-	0.60	0.37	GRAM/MJ	10.27	
FAT	MILLI	90.00	60.00	-	100.00	64.80	GRAM/MJ	1.78	
AVAILABLE CARBOHYDR.	GRAM	2.40	-	-	-	1.73	GRAM/MJ	47.41	
TOTAL DIETARY FIBRE	GRAM	1.90 [1]	-	-	-	1.37	GRAM/MJ	37.53	
MINERALS	GRAM	0.55	0.50	-	0.60	0.40	GRAM/MJ	10.86	
SODIUM	MILLI	3.40	3.10	-	4.00	2.45	MILLI/MJ	67.16	
POTASSIUM	MILLI	211.00	178.00	-	234.00	151.92	GRAM/MJ	4.17	
MAGNESIUM	MILLI	40.50	40.00	-	41.00	29.16	MILLI/MJ	800.03	
CALCIUM	MILLI	20.70	11.00	-	29.90	14.90	MILLI/MJ	408.90	
MANGANESE	MICRO	30.00	-	-	-	21.60	MICRO/MJ	592.62	
IRON	MILLI	0.42	0.19	-	0.70	0.30	MILLI/MJ	8.30	
COPPER	MICRO	28.00	-	-	-	20.16	MICRO/MJ	553.11	
ZINC	MILLI	0.12	-	-	-	0.09	MILLI/MJ	2.37	
MOLYBDENUM	MICRO	25.00	-	-	-	18.00	MICRO/MJ	493.85	
PHOSPHORUS	MILLI	16.40	11.60	-	22.00	11.81	MILLI/MJ	323.96	
BORON	MILLI	0.16	-	-	-	0.12	MILLI/MJ	3.16	
CAROTENE	MILLI	0.56	0.15	-	1.05	0.40	MILLI/MJ	11.06	
VITAMIN B1	MICRO	30.00	24.00	-	40.00	21.60	MICRO/MJ	592.62	
VITAMIN B2	MICRO	39.00	30.00	-	50.00	28.08	MICRO/MJ	770.40	
NICOTINAMIDE	MILLI	0.30	0.20	-	0.40	0.22	MILLI/MJ	5.93	
VITAMIN C	MILLI	82.00	50.00	-	130.00	59.04	GRAM/MJ	1.62	
SEROTONINE	MILLI	1.50	1.00	-	2.00	1.08	MILLI/MJ	29.63	
MALIC ACID	MILLI	29.00	27.00	-	31.00	20.88	MILLI/MJ	572.86	
CITRIC ACID	MILLI	30.80	28.80	-	33.00	22.18	MILLI/MJ	608.42	
ALPHA-KETOGLUT.ACID	MILLI	0.30	-	-	-	0.22	MILLI/MJ	5.93	
SALICYLIC ACID	MICRO	80.00	80.00	-	80.00	57.60	MILLI/MJ	1.58	
GLUCOSE	GRAM	0.99	-	-	-	0.71	GRAM/MJ	19.56	
FRUCTOSE	GRAM	0.33	-	-	-	0.24	GRAM/MJ	6.52	
SUCROSE	GRAM	1.01	-	-	-	0.73	GRAM/MJ	19.95	
DIETARY FIBRE,WAT.SOL.	GRAM	0.79	-	-	-	0.57	GRAM/MJ	15.61	
DIETARY FIBRE,WAT.INS.	GRAM	1.11	-	-	-	0.80	GRAM/MJ	21.93	

1 METHOD OF SOUTHGATE

PASSIONSFRUCHT GRANADILLA (PASSION-FRUIT) GRENADILLE

PASSIFLORA EDULIS SIMS.

		PROTEIN	FAT	CARBOHYDRATES	TOTAL
ENERGY VALUE (AVERAGE)	KJOULE	40	16	225	280
PER 100 G	(KCAL)	9.5	3.7	54	67
EDIBLE PORTION					
AMOUNT OF DIGESTIBLE	GRAM	2.04	0.36	13.44	
CONSTITUENTS PER 100 G					
ENERGY VALUE (AVERAGE)	KJOULE	34	14	225	273
OF THE DIGESTIBLE	(KCAL)	8.1	3.3	54	65
FRACTION PER 100 G					
EDIBLE PORTION					

WASTE PERCENTAGE AVERAGE 39

CONSTITUENTS	DIM	AV	VARIATION			AVR	NUTR. DENS.		MOLPERC.
WATER	GRAM	75.80	73.30	-	79.00	46.24	GRAM/MJ	278.01	
PROTEIN	GRAM	2.40	2.20	-	2.80	1.46	GRAM/MJ	8.80	
FAT	GRAM	0.40	0.20	-	0.70	0.24	GRAM/MJ	1.47	
AVAILABLE CARBOHYDR.	GRAM	13.44	-	-	-	8.20	GRAM/MJ	49.29	
TOTAL DIETARY FIBRE	GRAM	1.45 [1]	-	-	-	0.88	GRAM/MJ	5.32	
MINERALS	GRAM	0.90	0.80	-	1.00	0.55	GRAM/MJ	3.30	
POTASSIUM	MILLI	340.00	348.00	-	350.00	207.40	GRAM/MJ	1.25	
CALCIUM	MILLI	17.00	11.00	-	25.00	10.37	MILLI/MJ	62.35	
IRON	MILLI	1.30	1.10	-	1.60	0.79	MILLI/MJ	4.77	
PHOSPHORUS	MILLI	57.00	48.00	-	64.00	34.77	MILLI/MJ	209.06	
CAROTENE	MILLI	-	0.01	-	0.42				
VITAMIN B1	MICRO	20.00	TRACES	-	40.00	12.20	MICRO/MJ	73.35	
VITAMIN B2	MILLI	0.10	-	-	-	0.06	MILLI/MJ	0.37	
NICOTINAMIDE	MILLI	2.10	1.40	-	2.80	1.28	MILLI/MJ	7.70	
VITAMIN C	MILLI	24.00	17.00	-	40.00	14.64	MILLI/MJ	88.02	
MALIC ACID	GRAM	0.65	-	-	-	0.40	GRAM/MJ	2.38	
CITRIC ACID	GRAM	3.25	-	-	-	1.98	GRAM/MJ	11.92	
SALICYLIC ACID	MILLI	0.14	0.14	-	0.14	0.09	MILLI/MJ	0.51	
GLUCOSE	GRAM	3.64	-	-	-	2.22	GRAM/MJ	13.35	
FRUCTOSE	GRAM	2.81	-	-	-	1.71	GRAM/MJ	10.31	
SUCROSE	GRAM	3.09	-	-	-	1.88	GRAM/MJ	11.33	
DIETARY FIBRE, WAT.SOL.	GRAM	0.72	-	-	-	0.44	GRAM/MJ	2.64	
DIETARY FIBRE, WAT.INS.	GRAM	0.73	-	-	-	0.45	GRAM/MJ	2.68	

1 METHOD OF MEUSER, SUCKOW AND KULIKOWSKI ("BERLINER METHODE")

RAMBUTAN RAMBUTAN RAMBOUTAN (LITCHI CHEVELU)

NEPHELIUM LAPPACEUM L.

		PROTEIN	FAT	CARBOHYDRATES	TOTAL
ENERGY VALUE (AVERAGE)	KJOULE	17	3.9	251	271
PER 100 G	(KCAL)	4.0	0.9	60	65
EDIBLE PORTION					
AMOUNT OF DIGESTIBLE	GRAM	0.85	0.09	15.00	
CONSTITUENTS PER 100 G					
ENERGY VALUE (AVERAGE)	KJOULE	14	3.5	251	269
OF THE DIGESTIBLE	(KCAL)	3.4	0.8	60	64
FRACTION PER 100 G					
EDIBLE PORTION					

WASTE PERCENTAGE AVERAGE 56

CONSTITUENTS	DIM	AV		VARIATION			AVR	NUTR. DENS.	MOLPERC.
WATER	GRAM	82.00		-	-	-	36.08	GRAM/MJ	305.30
PROTEIN	GRAM	1.00		-	-	-	0.44	GRAM/MJ	3.72
FAT	GRAM	0.10		-	-	-	0.04	GRAM/MJ	0.37
AVAILABLE CARBOHYDR.	GRAM	15.00	1	-	-	-	6.60	GRAM/MJ	55.85
TOTAL DIETARY FIBRE	GRAM	1.50	2	-	-	-	0.66	GRAM/MJ	5.58
MINERALS	GRAM	0.40		-	-	-	0.18	GRAM/MJ	1.49
SODIUM	MILLI	1.00		-	-	-	0.44	MILLI/MJ	3.72
POTASSIUM	MILLI	64.00		-	-	-	28.16	MILLI/MJ	238.28
CALCIUM	MILLI	20.00		-	-	-	8.80	MILLI/MJ	74.46
IRON	MILLI	1.90		-	-	-	0.84	MILLI/MJ	7.07
PHOSPHORUS	MILLI	15.00		-	-	-	6.60	MILLI/MJ	55.85
CAROTENE	-	0.00		-	-	-	0.00		
VITAMIN B1	MICRO	10.00		-	-	-	4.40	MICRO/MJ	37.23
VITAMIN B2	MICRO	60.00		-	-	-	26.40	MICRO/MJ	223.39
NICOTINAMIDE	MILLI	0.40		-	-	-	0.18	MILLI/MJ	1.49
VITAMIN C	MILLI	53.00		-	-	-	23.32	MILLI/MJ	197.33
CELLULOSE	GRAM	0.36		-	-	-	0.16	GRAM/MJ	1.34

1 ESTIMATED BY THE DIFFERENCE METHOD
100 - (WATER + PROTEIN + FAT + MINERALS + TOTAL DIETARY FIBRE)

2 METHOD OF SOUTHGATE

ROSENAPFEL
(JAMBOSE)
SYZYGIUM JAMBOS (L.)
ALSTON
EUGENIA JAMBOS L.

ROSE APPLE

POMME-ROSE

		PROTEIN	FAT	CARBOHYDRATES	TOTAL
ENERGY VALUE (AVERAGE) PER 100 G EDIBLE PORTION	KJOULE	9.9	12	113	135
	(KCAL)	2.4	2.8	27	32
AMOUNT OF DIGESTIBLE CONSTITUENTS PER 100 G	GRAM	0.51	0.27	6.77	
ENERGY VALUE (AVERAGE) OF THE DIGESTIBLE FRACTION PER 100 G EDIBLE PORTION	KJOULE	8.4	11	113	132
	(KCAL)	2.0	2.5	27	32

WASTE PERCENTAGE AVERAGE 34

CONSTITUENTS	DIM	AV	VARIATION			AVR	NUTR.	DENS.	MOLPERC.
WATER	GRAM	85.00	-	-	-	56.10	GRAM/MJ	642.78	
PROTEIN	GRAM	0.60	-	-	-	0.40	GRAM/MJ	4.54	
FAT	GRAM	0.30	-	-	-	0.20	GRAM/MJ	2.27	
AVAILABLE CARBOHYDR.	GRAM	6.77	-	-	-	4.47	GRAM/MJ	51.20	
MINERALS	GRAM	0.40	-	-	-	0.26	GRAM/MJ	3.02	
CALCIUM	MILLI	20.00	10.00	-	29.00	13.20	MILLI/MJ	151.24	
IRON	MILLI	0.90	0.50	-	1.20	0.59	MILLI/MJ	6.81	
PHOSPHORUS	MILLI	16.00	-	-	-	10.56	MILLI/MJ	120.99	
CAROTENE	MILLI	0.10	-	-	-	0.07	MILLI/MJ	0.76	
VITAMIN B1	MICRO	20.00	-	-	-	13.20	MICRO/MJ	151.24	
VITAMIN B2	MICRO	30.00	-	-	-	19.80	MICRO/MJ	226.86	
NICOTINAMIDE	MILLI	0.80	-	-	-	0.53	MILLI/MJ	6.05	
VITAMIN C	MILLI	22.00	-	-	-	14.52	MILLI/MJ	166.37	
GLUCOSE	GRAM	3.00	-	-	-	1.98	GRAM/MJ	22.69	
FRUCTOSE	GRAM	1.96	-	-	-	1.29	GRAM/MJ	14.82	
SUCROSE	GRAM	1.81	-	-	-	1.19	GRAM/MJ	13.69	

SAPODILLE SAPODILLA SAPOTIER

(BREIAPFEL)
MANILKARA ZAPOTA L.
VAN ROYEN
(SYN. ACHRAS ZAPOTA L.)

		PROTEIN	FAT	CARBOHYDRATES	TOTAL
ENERGY VALUE (AVERAGE) PER 100 G EDIBLE PORTION	KJOULE	8.3	35	318	361
	(KCAL)	2.0	8.4	76	86
AMOUNT OF DIGESTIBLE CONSTITUENTS PER 100 G	GRAM	0.42	0.81	19.00	
ENERGY VALUE (AVERAGE) OF THE DIGESTIBLE FRACTION PER 100 G EDIBLE PORTION	KJOULE	7.0	32	318	357
	(KCAL)	1.7	7.5	76	85

WASTE PERCENTAGE AVERAGE 20

CONSTITUENTS	DIM	AV		VARIATION			AVR	NUTR. DENS.		MOLPERC.
WATER	GRAM	74.00		-	-	-	59.20	GRAM/MJ	207.56	
PROTEIN	GRAM	0.50		0.40	-	0.50	0.40	GRAM/MJ	1.40	
FAT	GRAM	0.90		0.70	-	1.10	0.72	GRAM/MJ	2.52	
AVAILABLE CARBOHYDR.	GRAM	19.00		-	-	-	15.20	GRAM/MJ	53.29	
TOTAL DIETARY FIBRE	GRAM	5.10	1	-	-	-	4.08	GRAM/MJ	14.30	
MINERALS	GRAM	0.50		-	-	-	0.40	GRAM/MJ	1.40	
POTASSIUM	MILLI	187.00		181.00	-	193.00	149.60	MILLI/MJ	524.51	
CALCIUM	MILLI	24.00		21.00	-	27.00	19.20	MILLI/MJ	67.32	
IRON	MILLI	0.70		-	-	-	0.56	MILLI/MJ	1.96	
PHOSPHORUS	MILLI	12.00		-	-	-	9.60	MILLI/MJ	33.66	
CAROTENE	MICRO	40.00		-	-	-	32.00	MICRO/MJ	112.19	
VITAMIN B1	MICRO	20.00		-	-	-	16.00	MICRO/MJ	56.10	
VITAMIN B2	MICRO	20.00		-	-	-	16.00	MICRO/MJ	56.10	
NICOTINAMIDE	MILLI	0.20		-	-	-	0.16	MILLI/MJ	0.56	
VITAMIN C	MILLI	12.00		-	-	-	9.60	MILLI/MJ	33.66	
GLUCOSE	GRAM	6.00		-	-	-	4.80	GRAM/MJ	16.83	
FRUCTOSE	GRAM	6.00		-	-	-	4.80	GRAM/MJ	16.83	
SUCROSE	GRAM	7.00		-	-	-	5.60	GRAM/MJ	19.63	
DIETARY FIBRE,WAT.SOL.	GRAM	0.93		-	-	-	0.74	GRAM/MJ	2.61	
DIETARY FIBRE,WAT.INS.	GRAM	4.20		-	-	-	3.36	GRAM/MJ	11.78	

1 METHOD OF SOUTHGATE

SAPOTE SAPOTE SAPOTE

CALOCARPUM SAPOTA (JACQ.)

		PROTEIN	FAT	CARBOHYDRATES	TOTAL
ENERGY VALUE (AVERAGE)	KJOULE	23	19	349	392
PER 100 G	(KCAL)	5.5	4.7	84	94
EDIBLE PORTION					
AMOUNT OF DIGESTIBLE	GRAM	1.19	0.45	20.88	
CONSTITUENTS PER 100 G					
ENERGY VALUE (AVERAGE)	KJOULE	20	18	349	387
OF THE DIGESTIBLE	(KCAL)	4.7	4.2	84	92
FRACTION PER 100 G					
EDIBLE PORTION					

WASTE PERCENTAGE AVERAGE 26

CONSTITUENTS	DIM	AV	VARIATION			AVR	NUTR. DENS.		MOLPERC.
WATER	GRAM	67.60	65.00	-	70.20	50.02	GRAM/MJ	174.85	
PROTEIN	GRAM	1.40	1.00	-	1.80	1.04	GRAM/MJ	3.62	
FAT	GRAM	0.50	-	-	-	0.37	GRAM/MJ	1.29	
AVAILABLE CARBOHYDR.	GRAM	20.88	-	-	-	15.45	GRAM/MJ	54.01	
MINERALS	GRAM	0.90	0.70	-	1.10	0.67	GRAM/MJ	2.33	
SODIUM	MILLI	6.00	-	-	-	4.44	MILLI/MJ	15.52	
POTASSIUM	MILLI	226.00	-	-	-	167.24	MILLI/MJ	584.55	
CALCIUM	MILLI	31.00	22.00	-	39.00	22.94	MILLI/MJ	80.18	
IRON	MILLI	0.90	-	-	-	0.67	MILLI/MJ	2.33	
PHOSPHORUS	MILLI	21.00	14.00	-	28.00	15.54	MILLI/MJ	54.32	
CAROTENE	MICRO	60.00	-	-	-	44.40	MICRO/MJ	155.19	
VITAMIN B1	MICRO	40.00	10.00	-	60.00	29.60	MICRO/MJ	103.46	
VITAMIN B2	MICRO	20.00	-	-	-	14.80	MICRO/MJ	51.73	
NICOTINAMIDE	MILLI	1.60	1.40	-	1.80	1.18	MILLI/MJ	4.14	
VITAMIN C	MILLI	23.00	-	-	-	17.02	MILLI/MJ	59.49	
GLUCOSE	GRAM	3.37	-	-	-	2.49	GRAM/MJ	8.72	
FRUCTOSE	GRAM	1.38	-	-	-	1.02	GRAM/MJ	3.57	
SUCROSE	GRAM	16.13	-	-	-	11.94	GRAM/MJ	41.72	

TAMARINDE (SAUERDATTEL) | TAMARIND | TAMARIN

TAMARINDUS INDICA L.

		PROTEIN	FAT	CARBOHYDRATES	TOTAL
ENERGY VALUE (AVERAGE) PER 100 G EDIBLE PORTION	KJOULE	38	7.8	949	995
	(KCAL)	9.1	1.9	227	238
AMOUNT OF DIGESTIBLE CONSTITUENTS PER 100 G	GRAM	1.95	0.18	56.70	
ENERGY VALUE (AVERAGE) OF THE DIGESTIBLE FRACTION PER 100 G EDIBLE PORTION	KJOULE	32	7.0	949	988
	(KCAL)	7.7	1.7	227	236

WASTE PERCENTAGE AVERAGE 0.00

CONSTITUENTS	DIM	AV		VARIATION			AVR	NUTR. DENS.		MOLPERC.
WATER	GRAM	38.70		-	-	-	38.70	GRAM/MJ	39.16	
PROTEIN	GRAM	2.30		-	-	-	2.30	GRAM/MJ	2.33	
FAT	GRAM	0.20		-	-	-	0.20	GRAM/MJ	0.20	
AVAILABLE CARBOHYDR.	GRAM	56.70	1	-	-	-	56.70	GRAM/MJ	57.37	
MINERALS	GRAM	2.10		-	-	-	2.10	GRAM/MJ	2.12	
SODIUM	MILLI	3.00		-	-	-	3.00	MILLI/MJ	3.04	
POTASSIUM	MILLI	570.00		-	-	-	570.00	MILLI/MJ	576.78	
CALCIUM	MILLI	81.00		-	-	-	81.00	MILLI/MJ	81.96	
IRON	MILLI	1.30		-	-	-	1.30	MILLI/MJ	1.32	
PHOSPHORUS	MILLI	86.00		-	-	-	86.00	MILLI/MJ	87.02	
CAROTENE	MICRO	10.00		-	-	-	10.00	MICRO/MJ	10.12	
VITAMIN B1	MILLI	0.30		-	-	-	0.30	MILLI/MJ	0.30	
VITAMIN B2	MICRO	80.00		-	-	-	80.00	MICRO/MJ	80.95	
NICOTINAMIDE	MILLI	1.10		-	-	-	1.10	MILLI/MJ	1.11	
VITAMIN C	MILLI	3.00		-	-	-	3.00	MILLI/MJ	3.04	

1 ESTIMATED BY THE DIFFERENCE METHOD
100 - (WATER + PROTEIN + FAT + MINERALS)

WASSERKASTANIE (SUMPFSIMSE, SÜSS)

MATAI (CHINESE WATER CHESTNUT)

MARRON D'EAU

ELEOCHARIS DULCIS TRIN.

		PROTEIN	FAT	CARBOHYDRATES	TOTAL
ENERGY VALUE (AVERAGE) PER 100 G EDIBLE PORTION	KJOULE	22	12	234	268
	(KCAL)	5.3	2.8	56	64
AMOUNT OF DIGESTIBLE CONSTITUENTS PER 100 G	GRAM	1.03	0.27	14.00	
ENERGY VALUE (AVERAGE) OF THE DIGESTIBLE FRACTION PER 100 G EDIBLE PORTION	KJOULE	16	11	234	261
	(KCAL)	3.9	2.5	56	62

WASTE PERCENTAGE AVERAGE 27

CONSTITUENTS	DIM	AV		VARIATION			AVR	NUTR. DENS.		MOLPERC.
WATER	GRAM	79.60		78.00	-	81.10	58.11	GRAM/MJ	304.91	
PROTEIN	GRAM	1.40		-	-	-	1.02	GRAM/MJ	5.36	
FAT	GRAM	0.30		-	-	-	0.22	GRAM/MJ	1.15	
AVAILABLE CARBOHYDR.	GRAM	14.00	1	-	-	-	10.22	GRAM/MJ	53.63	
TOTAL DIETARY FIBRE	GRAM	3.60	2	-	-	-	2.63	GRAM/MJ	13.79	
MINERALS	GRAM	1.10		-	-	-	0.80	GRAM/MJ	4.21	
SODIUM	MILLI	10.00		-	-	-	7.30	MILLI/MJ	38.30	
POTASSIUM	MILLI	490.00		481.00	-	500.00	357.70	GRAM/MJ	1.88	
CALCIUM	MILLI	5.00		-	-	-	3.65	MILLI/MJ	19.15	
IRON	MILLI	0.70		-	-	-	0.51	MILLI/MJ	2.68	
PHOSPHORUS	MILLI	77.00		-	-	-	56.21	MILLI/MJ	294.95	
VITAMIN B1	MILLI	-		0.03	-	0.14				
VITAMIN B2	MILLI	-		0.02	-	0.20				
NICOTINAMIDE	MILLI	1.00		-	-	-	0.73	MILLI/MJ	3.83	
VITAMIN C	MILLI	5.00		4.00	-	6.00	3.65	MILLI/MJ	19.15	
CELLULOSE	GRAM	0.70		-	-	-	0.51	GRAM/MJ	2.68	

1 ESTIMATED BY THE DIFFERENCE METHOD
100 - (WATER + PROTEIN + FAT + MINERALS + TOTAL DIETARY FIBRE)

2 METHOD OF SOUTHGATE

WASSERMELONE WATERMELON PASTÈQUE

CITRULLUS LANATUS

		PROTEIN	FAT	CARBOHYDRATES	TOTAL
ENERGY VALUE (AVERAGE)	KJOULE	9.9	7.8	139	156
PER 100 G EDIBLE PORTION	(KCAL)	2.4	1.9	33	37
AMOUNT OF DIGESTIBLE CONSTITUENTS PER 100 G	GRAM	0.51	0.18	8.28	
ENERGY VALUE (AVERAGE)	KJOULE	8.4	7.0	139	154
OF THE DIGESTIBLE FRACTION PER 100 G EDIBLE PORTION	(KCAL)	2.0	1.7	33	37

WASTE PERCENTAGE AVERAGE 56 MINIMUM 40 MAXIMUM 71

CONSTITUENTS	DIM	AV	VARIATION	AVR	NUTR.	DENS.	MOLPERC.
WATER	GRAM	93.20	92.10 - 94.20	41.01	GRAM/MJ	605.17	
PROTEIN	GRAM	0.60	0.50 - 0.80	0.26	GRAM/MJ	3.90	
FAT	GRAM	0.20	- - -	0.09	GRAM/MJ	1.30	
AVAILABLE CARBOHYDR.	GRAM	8.28	- - -	3.64	GRAM/MJ	53.76	
TOTAL DIETARY FIBRE	GRAM	0.24 [1]	- - -	0.11	GRAM/MJ	1.56	
MINERALS	GRAM	0.40	0.30 - 0.52	0.18	GRAM/MJ	2.60	
SODIUM	MILLI	0.50	0.30 - 1.00	0.22	MILLI/MJ	3.25	
POTASSIUM	MILLI	158.00	116.00 - 200.00	69.52	GRAM/MJ	1.03	
MAGNESIUM	MILLI	2.90	- - -	1.28	MILLI/MJ	18.83	
CALCIUM	MILLI	10.50	7.00 - 14.00	4.62	MILLI/MJ	68.18	
MANGANESE	MILLI	-	0.02 - 0.20				
IRON	MILLI	0.40	0.20 - 0.80	0.18	MILLI/MJ	2.60	
COPPER	MICRO	70.00	- - -	30.80	MICRO/MJ	454.53	
ZINC	MILLI	0.10	- - -	0.04	MILLI/MJ	0.65	
NICKEL	MICRO	14.00	7.00 - 24.00	6.16	MICRO/MJ	90.91	
PHOSPHORUS	MILLI	11.00	10.00 - 12.00	4.84	MILLI/MJ	71.43	
CHLORIDE	MILLI	8.30	- - -	3.65	MILLI/MJ	53.89	
FLUORIDE	MICRO	11.00	- - -	4.84	MICRO/MJ	71.43	
IODIDE	MICRO	1.00	- - -	0.44	MICRO/MJ	6.49	
CAROTENE	MILLI	0.20	0.05 - 0.35	0.09	MILLI/MJ	1.30	
VITAMIN B1	MICRO	45.00	40.00 - 50.00	19.80	MICRO/MJ	292.19	
VITAMIN B2	MICRO	50.00	- - -	22.00	MICRO/MJ	324.66	
NICOTINAMIDE	MILLI	0.15	0.10 - 0.20	0.07	MILLI/MJ	0.97	
PANTOTHENIC ACID	MILLI	1.60	- - -	0.70	MILLI/MJ	10.39	
VITAMIN B6	MICRO	70.00	- - -	30.80	MICRO/MJ	454.53	
FOLIC ACID	MICRO	5.00	3.00 - 8.00	2.20	MICRO/MJ	32.47	

1 METHOD OF MEUSER, SUCKOW AND KULIKOWSKI ("BERLINER METHODE")

CONSTITUENTS	DIM	AV	VARIATION			AVR	NUTR. DENS.		MOLPERC.
VITAMIN C	MILLI	6.00	-	-	-	2.64	MILLI/MJ	38.96	
SALICYLIC ACID	MILLI	0.48	0.48	-	0.48	0.21	MILLI/MJ	3.12	
GLUCOSE	GRAM	2.02	1.81	-	2.06	0.89	GRAM/MJ	13.14	
FRUCTOSE	GRAM	3.91	3.54	-	3.98	1.72	GRAM/MJ	25.43	
SUCROSE	GRAM	2.35	2.35	-	2.35	1.03	GRAM/MJ	15.26	
PECTIN	MILLI	95.00	-	-	-	41.80	MILLI/MJ	616.86	
PALMITIC ACID	MILLI	45.00	45.00	-	45.00	19.80	MILLI/MJ	292.19	
STEARIC ACID	MILLI	9.00	9.00	-	9.00	3.96	MILLI/MJ	58.44	
PALMITOLEIC ACID	MILLI	4.00	4.00	-	4.00	1.76	MILLI/MJ	25.97	
OLEIC ACID	MILLI	22.00	22.00	-	22.00	9.68	MILLI/MJ	142.85	
LINOLEIC ACID	MILLI	27.00	27.00	-	27.00	11.88	MILLI/MJ	175.32	
LINOLENIC ACID	MILLI	40.00	40.00	-	40.00	17.60	MILLI/MJ	259.73	
TOTAL STEROLS	MILLI	2.00	-	-	-	0.88	MILLI/MJ	12.99	
BETA-SITOSTEROL	MILLI	1.00	-	-	-	0.44	MILLI/MJ	6.49	
DIETARY FIBRE,WAT.SOL.	GRAM	0.12	-	-	-	0.05	GRAM/MJ	0.78	
DIETARY FIBRE,WAT.INS.	GRAM	0.12	-	-	-	0.05	GRAM/MJ	0.78	

ZITRONE LEMON CITRON

CITRUS MEDICA L.

		PROTEIN	FAT	CARBOHYDRATES	TOTAL
ENERGY VALUE (AVERAGE)	KJOULE	12	23	135	170
PER 100 G	(KCAL)	2.8	5.6	32	41
EDIBLE PORTION					
AMOUNT OF DIGESTIBLE	GRAM	0.59	0.54	8.08	
CONSTITUENTS PER 100 G					
ENERGY VALUE (AVERAGE)	KJOULE	9.8	21	135	166
OF THE DIGESTIBLE	(KCAL)	2.4	5.0	32	40
FRACTION PER 100 G					
EDIBLE PORTION					

WASTE PERCENTAGE AVERAGE 36 MINIMUM 20 MAXIMUM 53

CONSTITUENTS	DIM	AV	VARIATION			AVR	NUTR.	DENS.	MOLPERC.
WATER	GRAM	90.20	89.30	-	91.00	57.73	GRAM/MJ	543.14	
PROTEIN	GRAM	0.70	0.30	-	0.90	0.45	GRAM/MJ	4.22	
FAT	GRAM	0.60	-	-	-	0.38	GRAM/MJ	3.61	
AVAILABLE CARBOHYDR.	GRAM	8.08	-	-	-	5.17	GRAM/MJ	48.65	
MINERALS	GRAM	0.50	-	-	-	0.32	GRAM/MJ	3.01	
SODIUM	MILLI	2.70	2.00	-	3.00	1.73	MILLI/MJ	16.26	
POTASSIUM	MILLI	149.00	148.00	-	150.00	95.36	MILLI/MJ	897.20	
MAGNESIUM	MILLI	28.00	-	-	-	17.92	MILLI/MJ	168.60	
CALCIUM	MILLI	11.00	10.00	-	40.00	7.04	MILLI/MJ	66.24	
MANGANESE	MICRO	30.00	0.00	-	50.00	19.20	MICRO/MJ	180.64	
IRON	MILLI	0.45	0.10	-	0.60	0.29	MILLI/MJ	2.71	
COPPER	MILLI	0.35	0.30	-	0.40	0.22	MILLI/MJ	2.11	
ZINC	MILLI	0.12	0.03	-	0.20	0.08	MILLI/MJ	0.72	
NICKEL	MICRO	16.00	-	-	-	10.24	MICRO/MJ	96.34	
PHOSPHORUS	MILLI	16.00	10.00	-	22.00	10.24	MILLI/MJ	96.34	
CHLORIDE	MILLI	4.50	-	-	-	2.88	MILLI/MJ	27.10	
FLUORIDE	MICRO	10.00	2.80	-	17.40	6.40	MICRO/MJ	60.21	
IODIDE	MICRO	0.50	-	-	-	0.32	MICRO/MJ	3.01	
BORON	MILLI	0.17	0.14	-	0.21	0.11	MILLI/MJ	1.05	
SELENIUM	MICRO	1.20	1.20	-	12.00	0.77	MICRO/MJ	7.23	
BROMINE	MICRO	38.00	-	-	-	24.32	MICRO/MJ	228.82	
CAROTENE	MICRO	15.00	11.00	-	18.00	9.60	MICRO/MJ	90.32	
VITAMIN B1	MICRO	51.00	34.00	-	60.00	32.64	MICRO/MJ	307.10	
VITAMIN B2	MICRO	20.00	10.00	-	34.00	12.80	MICRO/MJ	120.43	
NICOTINAMIDE	MILLI	0.17	0.10	-	0.23	0.11	MILLI/MJ	1.02	
PANTOTHENIC ACID	MILLI	0.27	0.26	-	0.27	0.17	MILLI/MJ	1.63	

CONSTITUENTS	DIM	AV	VARIATION			AVR	NUTR. DENS.		MOLPERC.
VITAMIN B6	MICRO	60.00	45.00	-	100.00	38.40	MICRO/MJ	361.29	
FOLIC ACID	MICRO	6.30	4.20	-	10.20	4.03	MICRO/MJ	37.94	
VITAMIN C	MILLI	53.00	35.00	-	62.00	33.92	MILLI/MJ	319.14	
ALANINE	MILLI	41.00	41.00	-	41.00	26.24	MILLI/MJ	246.88	10.0
ARGININE	MILLI	42.00	42.00	-	42.00	26.88	MILLI/MJ	252.90	5.2
ASPARTIC ACID	MILLI	96.00	96.00	-	96.00	61.44	MILLI/MJ	578.06	15.7
CYSTINE	MILLI	9.00	9.00	-	9.00	5.76	MILLI/MJ	54.19	0.8
GLUTAMIC ACID	MILLI	80.00	80.00	-	80.00	51.20	MILLI/MJ	481.72	11.8
GLYCINE	MILLI	67.00	67.00	-	67.00	42.88	MILLI/MJ	403.44	19.4
HISTIDINE	MILLI	10.00	10.00	-	10.00	6.40	MILLI/MJ	60.21	1.4
ISOLEUCINE	MILLI	19.00	19.00	-	19.00	12.16	MILLI/MJ	114.41	3.2
LEUCINE	MILLI	18.00	18.00	-	18.00	11.52	MILLI/MJ	108.39	3.0
LYSINE	MILLI	35.00	35.00	-	35.00	22.40	MILLI/MJ	210.75	5.2
METHIONINE	MILLI	10.00	10.00	-	10.00	6.40	MILLI/MJ	60.21	1.5
PHENYLALANINE	MILLI	25.00	25.00	-	25.00	16.00	MILLI/MJ	150.54	3.3
PROLINE	MILLI	37.00	37.00	-	37.00	23.68	MILLI/MJ	222.79	7.0
SERINE	MILLI	19.00	19.00	-	19.00	12.16	MILLI/MJ	114.41	3.9
THREONINE	MILLI	10.00	10.00	-	10.00	6.40	MILLI/MJ	60.21	1.8
TRYPTOPHAN	MILLI	4.00	4.00	-	4.00	2.56	MILLI/MJ	24.09	0.4
TYROSINE	MILLI	14.00	14.00	-	14.00	8.96	MILLI/MJ	84.30	1.7
VALINE	MILLI	25.00	25.00	-	25.00	16.00	MILLI/MJ	150.54	4.6
CITRIC ACID	GRAM	4.92	3.50	-	7.20	3.15	GRAM/MJ	29.63	
FERULIC ACID	MILLI	1.40	-	-	-	0.90	MILLI/MJ	8.43	
CAFFEIC ACID	MILLI	2.10	-	-	-	1.34	MILLI/MJ	12.65	
PARA-COUMARIC ACID	MILLI	0.60	-	-	-	0.38	MILLI/MJ	3.61	
SALICYLIC ACID	MILLI	0.18	0.18	-	0.18	0.12	MILLI/MJ	1.08	
GLUCOSE	GRAM	1.40	-	-	-	0.90	GRAM/MJ	8.43	
FRUCTOSE	GRAM	1.35	-	-	-	0.86	GRAM/MJ	8.13	
SUCROSE	GRAM	0.41	-	-	-	0.26	GRAM/MJ	2.47	
TOTAL STEROLS	MILLI	12.00	-	-	-	7.68	MILLI/MJ	72.26	
CAMPESTEROL	MILLI	2.00	-	-	-	1.28	MILLI/MJ	12.04	
BETA-SITOSTEROL	MILLI	8.00	-	-	-	5.12	MILLI/MJ	48.17	
STIGMASTEROL	MILLI	1.00	-	-	-	0.64	MILLI/MJ	6.02	

ZUCKERMELONE (HONIGMELONE) — MUSKMELON — CANTALOUP (MELON)

CUCUMIS MELO L.

		PROTEIN	FAT	CARBOHYDRATES	TOTAL
ENERGY VALUE (AVERAGE) PER 100 G EDIBLE PORTION	KJOULE	15	3.9	208	226
	(KCAL)	3.6	0.9	50	54
AMOUNT OF DIGESTIBLE CONSTITUENTS PER 100 G	GRAM	0.76	0.09	12.40	
ENERGY VALUE (AVERAGE) OF THE DIGESTIBLE FRACTION PER 100 G EDIBLE PORTION	KJOULE	13	3.5	208	224
	(KCAL)	3.0	0.8	50	53

WASTE PERCENTAGE AVERAGE 20

CONSTITUENTS	DIM	AV	VARIATION		AVR	NUTR. DENS.		MOLPERC.
WATER	GRAM	87.00	-	-	69.60	GRAM/MJ	388.96	
PROTEIN	GRAM	0.90	-	-	0.72	GRAM/MJ	4.02	
FAT	GRAM	0.10	-	-	0.08	GRAM/MJ	0.45	
AVAILABLE CARBOHYDR.	GRAM	12.40	-	-	9.92	GRAM/MJ	55.44	
TOTAL DIETARY FIBRE	GRAM	0.98 [1]	-	-	0.78	GRAM/MJ	4.38	
MINERALS	GRAM	0.40	-	-	0.32	GRAM/MJ	1.79	
SODIUM	MILLI	20.00	-	-	16.00	MILLI/MJ	89.42	
POTASSIUM	MILLI	330.00	-	-	264.00	GRAM/MJ	1.48	
MAGNESIUM	MILLI	10.00	-	-	8.00	MILLI/MJ	44.71	
CALCIUM	MILLI	6.00	-	-	4.80	MILLI/MJ	26.83	
IRON	MILLI	0.20	-	-	0.16	MILLI/MJ	0.89	
COPPER	MICRO	85.00	85.00	85.00	68.00	MICRO/MJ	380.02	
ZINC	MILLI	0.20	0.20	0.20	0.16	MILLI/MJ	0.89	
MOLYBDENUM	MICRO	34.00	34.00	34.00	27.20	MICRO/MJ	152.01	
PHOSPHORUS	MILLI	21.00	14.00	28.00	16.80	MILLI/MJ	93.89	
IODIDE	MICRO	0.70	0.70	0.70	0.56	MICRO/MJ	3.13	
BORON	MILLI	0.11	0.03	0.24	0.09	MILLI/MJ	0.50	
CAROTENE	MILLI	1.75	-	-	1.40	MILLI/MJ	7.82	
VITAMIN E ACTIVITY	MILLI	0.14	-	-	0.11	MILLI/MJ	0.63	
TOTAL TOCOPHEROLS	MILLI	0.31	-	-	0.25	MILLI/MJ	1.39	
ALPHA-TOCOPHEROL	MILLI	0.14	-	-	0.11	MILLI/MJ	0.63	
VITAMIN B1	MICRO	60.00	-	-	48.00	MICRO/MJ	268.25	
VITAMIN B2	MICRO	20.00	-	-	16.00	MICRO/MJ	89.42	
NICOTINAMIDE	MILLI	0.60	-	-	0.48	MILLI/MJ	2.68	
FOLIC ACID	MICRO	30.00	-	-	24.00	MICRO/MJ	134.13	
VITAMIN C	MILLI	32.00	-	-	25.60	MILLI/MJ	143.07	

1 METHOD OF MEUSER, SUCKOW AND KULIKOWSKI ("BERLINER METHODE")

CONSTITUENTS	DIM	AV	VARIATION			AVR	NUTR. DENS.		MOLPERC.
LYSINE	MILLI	24.00	24.00	-	24.00	19.20	MILLI/MJ	107.30	
PHENYLALANINE	MILLI	28.00	28.00	-	28.00	22.40	MILLI/MJ	125.18	
TRYPTOPHAN	MILLI	5.00	5.00	-	5.00	4.00	MILLI/MJ	22.35	
MALIC ACID	MILLI	-	0.00	-	50.00				
CITRIC ACID	MILLI	75.00	0.00	-	150.00	60.00	MILLI/MJ	335.31	
OXALIC ACID TOTAL	-	0.00	-	-	-	0.00			
GLUCOSE	GRAM	1.60	-	-	-	1.28	GRAM/MJ	7.15	
FRUCTOSE	GRAM	1.30	-	-	-	1.04	GRAM/MJ	5.81	
SUCROSE	GRAM	9.50	-	-	-	7.60	GRAM/MJ	42.47	
PECTIN	GRAM	0.30	0.24	-	0.35	0.24	GRAM/MJ	1.34	
PALMITIC ACID	MILLI	22.00	22.00	-	22.00	17.60	MILLI/MJ	98.36	
STEARIC ACID	MILLI	4.00	4.00	-	4.00	3.20	MILLI/MJ	17.88	
PALMITOLEIC ACID	MILLI	2.00	2.00	-	2.00	1.60	MILLI/MJ	8.94	
OLEIC ACID	MILLI	11.00	11.00	-	11.00	8.80	MILLI/MJ	49.18	
LINOLEIC ACID	MILLI	13.00	13.00	-	13.00	10.40	MILLI/MJ	58.12	
LINOLENIC ACID	MILLI	2.00	2.00	-	2.00	1.60	MILLI/MJ	8.94	
TOTAL STEROLS	MILLI	10.00	-	-	-	8.00	MILLI/MJ	44.71	
BETA-SITOSTEROL	MILLI	8.00	-	-	-	6.40	MILLI/MJ	35.77	
DIETARY FIBRE,WAT.SOL.	GRAM	0.29	-	-	-	0.23	GRAM/MJ	1.30	
DIETARY FIBRE,WAT.INS.	GRAM	0.69	-	-	-	0.55	GRAM/MJ	3.08	

Schalenfrüchte

NUTS **FRUITS À COQUES**

CASHEW NUSS — CASHEW NUT — NOIX CASHEW

(KASCHUNUSS, INDISCHE MANDEL, ACAJOUNUSS)
ANACARDIUM OCCIDENTALE L.

		PROTEIN	FAT	CARBOHYDRATES	TOTAL
ENERGY VALUE (AVERAGE) PER 100 G EDIBLE PORTION	KJOULE	326	1642	510	2478
	(KCAL)	78	392	122	592
AMOUNT OF DIGESTIBLE CONSTITUENTS PER 100 G	GRAM	13.65	37.98	30.50	
ENERGY VALUE (AVERAGE) OF THE DIGESTIBLE FRACTION PER 100 G EDIBLE PORTION	KJOULE	254	1478	510	2242
	(KCAL)	61	353	122	536

WASTE PERCENTAGE AVERAGE 0.00

CONSTITUENTS	DIM	AV		VARIATION			AVR	NUTR. DENS.		MOLPERC.
WATER	GRAM	4.00		2.70	-	5.20	4.00	GRAM/MJ	1.78	
PROTEIN	GRAM	17.50		-	-	-	17.50	GRAM/MJ	7.80	
FAT	GRAM	42.20		34.20	-	47.40	42.20	GRAM/MJ	18.82	
AVAILABLE CARBOHYDR.	GRAM	30.50	1	-	-	-	30.50	GRAM/MJ	13.60	
TOTAL DIETARY FIBRE	GRAM	2.90	2	-	-	-	2.90	GRAM/MJ	1.29	
MINERALS	GRAM	2.90		2.60	-	3.10	2.90	GRAM/MJ	1.29	
SODIUM	MILLI	14.00		13.00	-	15.00	14.00	MILLI/MJ	6.24	
POTASSIUM	MILLI	552.00		464.00	-	633.00	552.00	MILLI/MJ	246.16	
MAGNESIUM	MILLI	267.00		-	-	-	267.00	MILLI/MJ	119.07	
CALCIUM	MILLI	31.00		24.00	-	38.00	31.00	MILLI/MJ	13.82	
MANGANESE	MILLI	0.84		-	-	-	0.84	MILLI/MJ	0.37	
IRON	MILLI	2.80		1.80	-	3.80	2.80	MILLI/MJ	1.25	
COBALT	MICRO	10.00		-	-	-	10.00	MICRO/MJ	4.46	
COPPER	MILLI	3.70		-	-	-	3.70	MILLI/MJ	1.65	
ZINC	MILLI	4.80		-	-	-	4.80	MILLI/MJ	2.14	
NICKEL	MILLI	0.50		-	-	-	0.50	MILLI/MJ	0.22	
MOLYBDENUM	MICRO	10.00		-	-	-	10.00	MICRO/MJ	4.46	
PHOSPHORUS	MILLI	373.00		-	-	-	373.00	MILLI/MJ	166.34	
CHLORIDE	MILLI	18.40		-	-	-	18.40	MILLI/MJ	8.21	
FLUORIDE	MILLI	0.14		-	-	-	0.14	MILLI/MJ	0.06	
IODIDE	MICRO	10.00		-	-	-	10.00	MICRO/MJ	4.46	

1 ESTIMATED BY THE DIFFERENCE METHOD
100 - (WATER + PROTEIN + FAT + MINERALS + TOTAL DIETARY FIBRE)

2 METHOD OF MEUSER, SUCKOW AND KULIKOWSKI ("BERLINER METHODE")

CONSTITUENTS	DIM	AV	VARIATION			AVR	NUTR. DENS.		MOLPERC.
CAROTENE	MICRO	60.00	-	-	-	60.00	MICRO/MJ	26.76	
VITAMIN E ACTIVITY	MILLI	0.80	-	-	-	0.80	MILLI/MJ	0.36	
TOTAL TOCOPHEROLS	MILLI	5.70	-	-	-	5.70	MILLI/MJ	2.54	
ALPHA-TOCOPHEROL	MILLI	0.26	-	-	-	0.26	MILLI/MJ	0.12	
GAMMA-TOCOPHEROL	MILLI	5.20	-	-	-	5.20	MILLI/MJ	2.32	
DELTA-TOCOPHEROL	MILLI	0.23	-	-	-	0.23	MILLI/MJ	0.10	
VITAMIN B1	MILLI	0.63	0.43	-	0.85	0.63	MILLI/MJ	0.28	
VITAMIN B2	MILLI	0.26	0.20	-	0.32	0.26	MILLI/MJ	0.12	
NICOTINAMIDE	MILLI	2.00	1.80	-	2.10	2.00	MILLI/MJ	0.89	
PANTOTHENIC ACID	MILLI	1.20	1.10	-	1.20	1.20	MILLI/MJ	0.54	
ARGININE	GRAM	1.98	-	-	-	1.98	GRAM/MJ	0.88	
CYSTINE	GRAM	0.50	-	-	-	0.50	GRAM/MJ	0.22	
HISTIDINE	GRAM	0.39	-	-	-	0.39	GRAM/MJ	0.17	
ISOLEUCINE	GRAM	1.16	-	-	-	1.16	GRAM/MJ	0.52	
LEUCINE	GRAM	1.44	-	-	-	1.44	GRAM/MJ	0.64	
LYSINE	GRAM	0.75	-	-	-	0.75	GRAM/MJ	0.33	
METHIONINE	GRAM	0.33	-	-	-	0.33	GRAM/MJ	0.15	
PHENYLALANINE	GRAM	0.90	-	-	-	0.90	GRAM/MJ	0.40	
THREONINE	GRAM	0.70	-	-	-	0.70	GRAM/MJ	0.31	
TRYPTOPHAN	GRAM	0.45	-	-	-	0.45	GRAM/MJ	0.20	
TYROSINE	GRAM	0.68	-	-	-	0.68	GRAM/MJ	0.30	
VALINE	GRAM	1.51	-	-	-	1.51	GRAM/MJ	0.67	
SALICYLIC ACID	MICRO	70.00	70.00	-	70.00	70.00	MICRO/MJ	31.22	
CAPRYLIC ACID	GRAM	0.11	-	-	-	0.11	GRAM/MJ	0.05	
CAPRIC ACID	GRAM	0.11	-	-	-	0.11	GRAM/MJ	0.05	
LAURIC ACID	GRAM	0.76	-	-	-	0.76	GRAM/MJ	0.34	
MYRISTIC ACID	GRAM	0.19	-	-	-	0.19	GRAM/MJ	0.08	
PALMITIC ACID	GRAM	3.95	-	-	-	3.95	GRAM/MJ	1.76	
STEARIC ACID	GRAM	2.80	-	-	-	2.80	GRAM/MJ	1.25	
ARACHIDIC ACID	GRAM	0.54	-	-	-	0.54	GRAM/MJ	0.24	
PALMITOLEIC ACID	GRAM	0.22	-	-	-	0.22	GRAM/MJ	0.10	
OLEIC ACID	GRAM	24.20	-	-	-	24.20	GRAM/MJ	10.79	
LINOLEIC ACID	GRAM	6.70	3.00	-	6.70	6.70	GRAM/MJ	2.99	
LINOLENIC ACID	GRAM	0.15	0.15	-	0.17	0.15	GRAM/MJ	0.07	
TOTAL STEROLS	MILLI	158.00	-	-	-	158.00	MILLI/MJ	70.46	
CAMPESTEROL	MILLI	13.00	-	-	-	13.00	MILLI/MJ	5.80	
BETA-SITOSTEROL	MILLI	130.00	-	-	-	130.00	MILLI/MJ	57.97	
STIGMASTEROL	-	TRACES	-	-	-				
DIETARY FIBRE,WAT.SOL.	GRAM	1.60	-	-	-	1.60	GRAM/MJ	0.71	
DIETARY FIBRE,WAT.INS.	GRAM	1.30	-	-	-	1.30	GRAM/MJ	0.58	

EDELKASTANIE (MARONE) · SWEET CHESTNUT (SPANISH CHESTNUT) · CHÂTAIGNE (MARRON)

CASTANEA VESCA GAERTN.

		PROTEIN	FAT	CARBOHYDRATES	TOTAL
ENERGY VALUE (AVERAGE) PER 100 G EDIBLE PORTION	KJOULE	46	74	690	810
	(KCAL)	11	18	165	194
AMOUNT OF DIGESTIBLE CONSTITUENTS PER 100 G	GRAM	1.93	1.71	41.20	
ENERGY VALUE (AVERAGE) OF THE DIGESTIBLE FRACTION PER 100 G EDIBLE PORTION	KJOULE	36	67	690	792
	(KCAL)	8.6	16	165	189

WASTE PERCENTAGE AVERAGE 20 MINIMUM 11 MAXIMUM 28

CONSTITUENTS	DIM	AV	VARIATION			AVR	NUTR. DENS.		MOLPERC.
WATER	GRAM	50.10	47.00	-	53.20	40.08	GRAM/MJ	63.25	
PROTEIN	GRAM	2.48	1.95	-	2.88	1.98	GRAM/MJ	3.13	
FAT	GRAM	1.90	1.50	-	2.70	1.52	GRAM/MJ	2.40	
AVAILABLE CARBOHYDR.	GRAM	41.20	-	-	-	32.96	GRAM/MJ	52.02	
MINERALS	GRAM	1.18	1.00	-	1.43	0.94	GRAM/MJ	1.49	
SODIUM	MILLI	1.50	1.00	-	2.00	1.20	MILLI/MJ	1.89	
POTASSIUM	MILLI	707.00	705.00	-	708.00	565.60	MILLI/MJ	892.59	
MAGNESIUM	MILLI	45.00	33.00	-	55.00	36.00	MILLI/MJ	56.81	
CALCIUM	MILLI	33.00	29.00	-	35.00	26.40	MILLI/MJ	41.66	
MANGANESE	MILLI	0.75	-	-	-	0.60	MILLI/MJ	0.95	
IRON	MILLI	1.32	0.89	-	1.70	1.06	MILLI/MJ	1.67	
COPPER	MILLI	0.23	-	-	-	0.18	MILLI/MJ	0.29	
PHOSPHORUS	MILLI	87.00	74.00	-	96.00	69.60	MILLI/MJ	109.84	
CHLORIDE	MILLI	13.00	11.00	-	15.00	10.40	MILLI/MJ	16.41	
IODIDE	MICRO	0.10	0.10	-	0.10	0.08	MICRO/MJ	0.13	
CAROTENE	MICRO	24.00	0.00	-	24.00	19.20	MICRO/MJ	30.30	
VITAMIN E ACTIVITY	MILLI	1.20	-	-	-	0.96	MILLI/MJ	1.52	
TOTAL TOCOPHEROLS	MILLI	7.50	-	-	-	6.00	MILLI/MJ	9.47	
ALPHA-TOCOPHEROL	MILLI	0.50	-	-	-	0.40	MILLI/MJ	0.63	
GAMMA-TOCOPHEROL	MILLI	7.00	-	-	-	5.60	MILLI/MJ	8.84	
VITAMIN B1	MILLI	0.20	0.14	-	0.26	0.16	MILLI/MJ	0.25	
VITAMIN B2	MILLI	0.21	0.13	-	0.25	0.17	MILLI/MJ	0.27	
NICOTINAMIDE	MILLI	0.87	0.50	-	1.20	0.70	MILLI/MJ	1.10	
PANTOTHENIC ACID	MILLI	0.50	-	-	-	0.40	MILLI/MJ	0.63	
VITAMIN B6	MILLI	0.35	-	-	-	0.28	MILLI/MJ	0.44	
BIOTIN	MICRO	1.50	-	-	-	1.20	MICRO/MJ	1.89	

CONSTITUENTS	DIM	AV	VARIATION			AVR	NUTR. DENS.		MOLPERC.
VITAMIN C	MILLI	27.00	0.00	-	60.00	21.60	MILLI/MJ	34.09	
ARGININE	GRAM	0.20	-	-	-	0.16	GRAM/MJ	0.25	
CYSTINE	MILLI	30.00	-	-	-	24.00	MILLI/MJ	37.88	
HISTIDINE	MILLI	80.00	-	-	-	64.00	MILLI/MJ	101.00	
LYSINE	GRAM	0.15	-	-	-	0.12	GRAM/MJ	0.19	
METHIONINE	MILLI	40.00	-	-	-	32.00	MILLI/MJ	50.50	
PHENYLALANINE	GRAM	0.11	-	-	-	0.09	GRAM/MJ	0.14	
THREONINE	GRAM	0.14	-	-	-	0.11	GRAM/MJ	0.18	
TRYPTOPHAN	MILLI	30.00	-	-	-	24.00	MILLI/MJ	37.88	
TYROSINE	GRAM	0.14	-	-	-	0.11	GRAM/MJ	0.18	
VALINE	GRAM	0.16	-	-	-	0.13	GRAM/MJ	0.20	
SUCROSE	GRAM	13.90	-	-	-	11.12	GRAM/MJ	17.55	
STARCH	GRAM	27.30	-	-	-	21.84	GRAM/MJ	34.47	
TOTAL STEROLS	MILLI	22.00	-	-	-	17.60	MILLI/MJ	27.78	
CAMPESTEROL	MILLI	2.00	-	-	-	1.60	MILLI/MJ	2.53	
BETA-SITOSTEROL	MILLI	18.00	-	-	-	14.40	MILLI/MJ	22.73	
STIGMASTEROL	MILLI	2.00	-	-	-	1.60	MILLI/MJ	2.53	

ERDNUSS PEANUT CACAHUÈTE

ARACHIS HYPOGAEA L.

		PROTEIN	FAT	CARBOHYDRATES	TOTAL
ENERGY VALUE (AVERAGE)	KJOULE	470	1872	203	2545
PER 100 G EDIBLE PORTION	(KCAL)	112	447	49	608
AMOUNT OF DIGESTIBLE CONSTITUENTS PER 100 G	GRAM	19.69	43.29	12.15	
ENERGY VALUE (AVERAGE)	KJOULE	367	1684	203	2255
OF THE DIGESTIBLE FRACTION PER 100 G EDIBLE PORTION	(KCAL)	88	403	49	539

WASTE PERCENTAGE AVERAGE 0.00

CONSTITUENTS	DIM	AV		VARIATION			AVR	NUTR. DENS.		MOLPERC.
WATER	GRAM	5.21		4.50	-	6.40	5.21	GRAM/MJ	2.31	
PROTEIN	GRAM	25.25		23.79	-	26.70	25.25	GRAM/MJ	11.20	
FAT	GRAM	48.10		46.90	-	49.20	48.10	GRAM/MJ	21.34	
AVAILABLE CARBOHYDR.	GRAM	12.15	1	-	-	-	12.15	GRAM/MJ	5.39	
TOTAL DIETARY FIBRE	GRAM	7.10	2	-	-	-	7.10	GRAM/MJ	3.15	
MINERALS	GRAM	2.22		1.92	-	2.53	2.22	GRAM/MJ	0.98	
SODIUM	MILLI	5.20		3.90	-	6.30	5.20	MILLI/MJ	2.31	
POTASSIUM	MILLI	706.00		648.00	-	785.00	706.00	MILLI/MJ	313.15	
MAGNESIUM	MILLI	163.00		127.00	-	215.00	163.00	MILLI/MJ	72.30	
CALCIUM	MILLI	59.00		50.00	-	73.00	59.00	MILLI/MJ	26.17	
MANGANESE	MILLI	1.13		0.69	-	1.57	1.13	MILLI/MJ	0.50	
IRON	MILLI	2.11		1.90	-	2.48	2.11	MILLI/MJ	0.94	
COBALT	MICRO	34.00		30.00	-	37.00	34.00	MICRO/MJ	15.08	
COPPER	MILLI	0.55		0.27	-	0.85	0.55	MILLI/MJ	0.24	
ZINC	MILLI	3.07		2.90	-	3.24	3.07	MILLI/MJ	1.36	
NICKEL	MILLI	0.16		-	-	-	0.16	MILLI/MJ	0.07	
CHROMIUM	MICRO	8.00		-	-	-	8.00	MICRO/MJ	3.55	
MOLYBDENUM	MICRO	25.00		-	-	-	25.00	MICRO/MJ	11.09	
PHOSPHORUS	MILLI	372.00		294.00	-	455.00	372.00	MILLI/MJ	165.00	
CHLORIDE	MILLI	7.00		-	-	-	7.00	MILLI/MJ	3.10	
FLUORIDE	MILLI	0.13		0.12	-	0.14	0.13	MILLI/MJ	0.06	
IODIDE	MICRO	13.00		6.00	-	20.00	13.00	MICRO/MJ	5.77	
BORON	MILLI	1.20		0.53	-	1.80	1.20	MILLI/MJ	0.53	
SELENIUM	MICRO	2.00		-	-	-	2.00	MICRO/MJ	0.89	
CAROTENE	MICRO	11.00		0.00	-	18.00	11.00	MICRO/MJ	4.88	
VITAMIN E ACTIVITY	MILLI	9.10		-	-	-	9.10	MILLI/MJ	4.04	

1 ESTIMATED BY THE DIFFERENCE METHOD
100 - (WATER + PROTEIN + FAT + MINERALS + TOTAL DIETARY FIBRE)

2 METHOD OF SOUTHGATE

CONSTITUENTS	DIM	AV	VARIATION			AVR	NUTR. DENS.		MOLPERC.
TOTAL TOCOPHEROLS	MILLI	16.30	-	-	-	16.30	MILLI/MJ	7.23	
ALPHA-TOCOPHEROL	MILLI	8.30	-	-	-	8.30	MILLI/MJ	3.68	
GAMMA-TOCOPHEROL	MILLI	8.00	-	-	-	8.00	MILLI/MJ	3.55	
VITAMIN B1	MILLI	0.90	0.76	-	1.03	0.90	MILLI/MJ	0.40	
VITAMIN B2	MILLI	0.15	0.10	-	0.28	0.15	MILLI/MJ	0.07	
NICOTINAMIDE	MILLI	15.30	9.80	-	20.00	15.30	MILLI/MJ	6.79	
PANTOTHENIC ACID	MILLI	2.60	2.50	-	2.70	2.60	MILLI/MJ	1.15	
VITAMIN B6	MILLI	0.30	-	-	-	0.30	MILLI/MJ	0.13	
BIOTIN	MICRO	34.00	-	-	-	34.00	MICRO/MJ	15.08	
FOLIC ACID	MICRO	53.00	44.00	-	65.00	53.00	MICRO/MJ	23.51	
VITAMIN C	-	0.00	-	-	-	0.00			
ALANINE	GRAM	0.81	-	-	-	0.81	GRAM/MJ	0.36	4.1
ARGININE	GRAM	3.46	3.20	-	3.89	3.46	GRAM/MJ	1.53	9.0
ASPARTIC ACID	GRAM	3.31	-	-	-	3.31	GRAM/MJ	1.47	11.2
CYSTINE	GRAM	0.43	0.38	-	0.51	0.43	GRAM/MJ	0.19	0.8
GLUTAMIC ACID	GRAM	5.63	-	-	-	5.63	GRAM/MJ	2.50	17.3
GLYCINE	GRAM	1.64	-	-	-	1.64	GRAM/MJ	0.73	9.9
HISTIDINE	GRAM	0.71	0.60	-	0.76	0.71	GRAM/MJ	0.31	2.1
ISOLEUCINE	GRAM	1.23	1.05	-	1.45	1.23	GRAM/MJ	0.55	4.2
LEUCINE	GRAM	2.03	1.84	-	2.18	2.03	GRAM/MJ	0.90	7.0
LYSINE	GRAM	1.10	0.98	-	1.19	1.10	GRAM/MJ	0.49	3.4
METHIONINE	GRAM	0.31	0.25	-	0.40	0.31	GRAM/MJ	0.14	0.9
PHENYLALANINE	GRAM	1.54	-	-	-	1.54	GRAM/MJ	0.68	4.2
PROLINE	GRAM	1.43	-	-	-	1.43	GRAM/MJ	0.63	5.6
SERINE	GRAM	1.83	-	-	-	1.83	GRAM/MJ	0.81	7.9
THREONINE	GRAM	0.85	0.76	-	0.95	0.85	GRAM/MJ	0.38	3.2
TRYPTOPHAN	GRAM	0.32	0.28	-	0.34	0.32	GRAM/MJ	0.14	0.7
TYROSINE	GRAM	1.19	1.12	-	1.25	1.19	GRAM/MJ	0.53	3.0
VALINE	GRAM	1.45	1.30	-	1.74	1.45	GRAM/MJ	0.64	5.6
SALICYLIC ACID	MILLI	1.12	1.12	-	1.12	1.12	MILLI/MJ	0.50	
TOTAL SUGAR	GRAM	3.10	-	-	-	3.10	GRAM/MJ	1.38	
STARCH	GRAM	5.50	-	-	-	5.50	GRAM/MJ	2.44	
CELLULOSE	GRAM	1.30	-	-	-	1.30	GRAM/MJ	0.58	
CAPRYLIC ACID	-	0.00	-	-	-	0.00			
CAPRIC ACID	-	0.00	-	-	-	0.00			
LAURIC ACID	-	0.00	-	-	-	0.00			
MYRISTIC ACID	-	0.00	-	-	-	0.00			
PALMITIC ACID	GRAM	5.10	-	-	-	5.10	GRAM/MJ	2.26	
STEARIC ACID	GRAM	1.30	-	-	-	1.30	GRAM/MJ	0.58	
ARACHIDIC ACID	GRAM	0.56	-	-	-	0.56	GRAM/MJ	0.25	
PALMITOLEIC ACID	-	0.00	-	-	-	0.00			
OLEIC ACID	GRAM	22.10	17.00	-	25.30	22.10	GRAM/MJ	9.80	
LINOLEIC ACID	GRAM	13.90	11.50	-	18.90	13.90	GRAM/MJ	6.17	
LINOLENIC ACID	GRAM	0.53	-	-	-	0.53	GRAM/MJ	0.24	
TOTAL STEROLS	MILLI	220.00	-	-	-	220.00	MILLI/MJ	97.58	
CAMPESTEROL	MILLI	24.00	-	-	-	24.00	MILLI/MJ	10.65	
BETA-SITOSTEROL	MILLI	142.00	-	-	-	142.00	MILLI/MJ	62.98	
STIGMASTEROL	MILLI	23.00	-	-	-	23.00	MILLI/MJ	10.20	
TOTAL PURINES	MILLI	90.00 [3]	-	-	-	90.00	MILLI/MJ	39.92	
TOTAL PHOSPHOLIPIDS	MILLI	615.00	-	-	-	615.00	MILLI/MJ	272.79	
PHOSPHATIDYLCHOLINE	MILLI	268.00	-	-	-	268.00	MILLI/MJ	118.87	
PHOSPHATIDYLETHANOLAMINE	MILLI	50.00	-	-	-	50.00	MILLI/MJ	22.18	
PHOSPHATIDYLINOSITOL	MILLI	149.00	-	-	-	149.00	MILLI/MJ	66.09	
DIETARY FIBRE, WAT.SOL.	GRAM	0.77	-	-	-	0.77	GRAM/MJ	0.34	
DIETARY FIBRE, WAT.INS.	GRAM	6.30	-	-	-	6.30	GRAM/MJ	2.79	

3 VALUE EXPRESSED IN MG URIC ACID/100G

ERDNUSS GERÖSTET	**PEANUT** ROAST	**CACAHUÈTE** GRILLÉE, ECALÉE

		PROTEIN	FAT	CARBOHYDRATES	TOTAL
ENERGY VALUE (AVERAGE)	KJOULE	477	1922	224	2624
PER 100 G EDIBLE PORTION	(KCAL)	114	459	54	627
AMOUNT OF DIGESTIBLE CONSTITUENTS PER 100 G	GRAM	19.99	44.46	13.40	
ENERGY VALUE (AVERAGE)	KJOULE	372	1730	224	2326
OF THE DIGESTIBLE FRACTION PER 100 G EDIBLE PORTION	(KCAL)	89	413	54	556

WASTE PERCENTAGE AVERAGE 0.00

CONSTITUENTS	DIM	AV	VARIATION			AVR	NUTR. DENS.		MOLPERC.
WATER	GRAM	1.56	1.02	-	2.20	1.56	GRAM/MJ	0.67	
PROTEIN	GRAM	25.63	23.79	-	27.19	25.63	GRAM/MJ	11.02	
FAT	GRAM	49.40	48.10	-	50.90	49.40	GRAM/MJ	21.23	
AVAILABLE CARBOHYDR.	GRAM	13.40	-	-	-	13.40	GRAM/MJ	5.76	
TOTAL DIETARY FIBRE	GRAM	7.40	-	-	-	7.40	GRAM/MJ	3.18	
MINERALS	GRAM	2.61	2.25	-	2.96	2.61	GRAM/MJ	1.12	
SODIUM	MILLI	5.70	4.30	-	6.90	5.70	MILLI/MJ	2.45	
POTASSIUM	MILLI	777.00	713.00	-	864.00	777.00	MILLI/MJ	333.98	
MAGNESIUM	MILLI	182.00	144.00	-	237.00	182.00	MILLI/MJ	78.23	
CALCIUM	MILLI	65.00	55.00	-	80.00	65.00	MILLI/MJ	27.94	
MANGANESE	MILLI	1.24	0.76	-	1.73	1.24	MILLI/MJ	0.53	
IRON	MILLI	2.32	2.09	-	2.73	2.32	MILLI/MJ	1.00	
COBALT	MICRO	37.00	33.00	-	41.00	37.00	MICRO/MJ	15.90	
COPPER	MILLI	0.61	0.30	-	0.94	0.61	MILLI/MJ	0.26	
ZINC	MILLI	3.38	3.19	-	3.56	3.38	MILLI/MJ	1.45	
PHOSPHORUS	MILLI	409.00	323.00	-	501.00	409.00	MILLI/MJ	175.80	
CHLORIDE	MILLI	8.00	-	-	-	8.00	MILLI/MJ	3.44	
FLUORIDE	MILLI	0.14	0.13	-	0.15	0.14	MILLI/MJ	0.06	
IODIDE	MICRO	14.00	7.00	-	22.00	14.00	MICRO/MJ	6.02	
BORON	MILLI	1.70	-	-	-	1.70	MILLI/MJ	0.73	
SELENIUM	MILLI	-	0.00	-	0.04				
VITAMIN E ACTIVITY	MILLI	8.80	-	-	-	8.80	MILLI/MJ	3.78	
TOTAL TOCOPHEROLS	MILLI	16.00	-	-	-	16.00	MILLI/MJ	6.88	
ALPHA-TOCOPHEROL	MILLI	8.00	-	-	-	8.00	MILLI/MJ	3.44	
GAMMA-TOCOPHEROL	MILLI	8.00	-	-	-	8.00	MILLI/MJ	3.44	
VITAMIN B1	MILLI	0.25	0.17	-	0.32	0.25	MILLI/MJ	0.11	

CONSTITUENTS	DIM	AV	VARIATION			AVR	NUTR. DENS.		MOLPERC.
VITAMIN B2	MILLI	0.14	0.09	-	0.21	0.14	MILLI/MJ	0.06	
NICOTINAMIDE	MILLI	14.30	9.60	-	17.20	14.30	MILLI/MJ	6.15	
PANTOTHENIC ACID	MILLI	2.14	1.62	-	2.53	2.14	MILLI/MJ	0.92	
VITAMIN B6	MILLI	0.40	-	-	-	0.40	MILLI/MJ	0.17	
VITAMIN B12	-	0.00	-	-	-	0.00			
VITAMIN C	-	0.00	-	-	-	0.00			
ALANINE	GRAM	0.81	-	-	-	0.81	GRAM/MJ	0.35	4.1
ARGININE	GRAM	3.46	3.20	-	3.89	3.46	GRAM/MJ	1.49	9.0
ASPARTIC ACID	GRAM	3.31	-	-	-	3.31	GRAM/MJ	1.42	11.2
CYSTINE	GRAM	0.43	0.38	-	0.51	0.43	GRAM/MJ	0.18	0.8
GLUTAMIC ACID	GRAM	5.63	-	-	-	5.63	GRAM/MJ	2.42	17.3
GLYCINE	GRAM	1.64	-	-	-	1.64	GRAM/MJ	0.70	9.9
HISTIDINE	GRAM	0.71	0.60	-	0.76	0.71	GRAM/MJ	0.31	2.1
ISOLEUCINE	GRAM	1.23	1.05	-	1.45	1.23	GRAM/MJ	0.53	4.2
LEUCINE	GRAM	2.03	1.84	-	2.18	2.03	GRAM/MJ	0.87	7.0
LYSINE	GRAM	1.10	0.98	-	1.19	1.10	GRAM/MJ	0.47	3.4
METHIONINE	GRAM	0.31	0.25	-	0.40	0.31	GRAM/MJ	0.13	0.9
PHENYLALANINE	GRAM	1.54	-	-	-	1.54	GRAM/MJ	0.66	4.2
PROLINE	GRAM	1.43	-	-	-	1.43	GRAM/MJ	0.61	5.6
SERINE	GRAM	1.83	-	-	-	1.83	GRAM/MJ	0.79	7.9
THREONINE	GRAM	0.85	0.76	-	0.95	0.85	GRAM/MJ	0.37	3.2
TRYPTOPHAN	GRAM	0.32	0.28	-	0.34	0.32	GRAM/MJ	0.14	0.7
TYROSINE	GRAM	1.19	1.12	-	1.25	1.19	GRAM/MJ	0.51	3.0
VALINE	GRAM	1.45	1.30	-	1.74	1.45	GRAM/MJ	0.62	5.6
TOTAL SUGAR	GRAM	3.20	-	-	-	3.20	GRAM/MJ	1.38	
STARCH	GRAM	5.70	-	-	-	5.70	GRAM/MJ	2.45	
LINOLEIC ACID	GRAM	13.80	10.50	-	16.00	13.80	GRAM/MJ	5.93	
LINOLENIC ACID	GRAM	-	TRACES	-	0.54				

HASELNUSS HAZELNUT (COB) NOISETTE

CORYLUS AVELLANA L.

		PROTEIN	FAT	CARBOHYDRATES	TOTAL
ENERGY VALUE (AVERAGE) PER 100 G EDIBLE PORTION	KJOULE	223	2397	190	2810
	(KCAL)	53	573	45	672
AMOUNT OF DIGESTIBLE CONSTITUENTS PER 100 G	GRAM	9.32	55.44	11.36	
ENERGY VALUE (AVERAGE) OF THE DIGESTIBLE FRACTION PER 100 G EDIBLE PORTION	KJOULE	174	2157	190	2521
	(KCAL)	42	516	45	603

WASTE PERCENTAGE AVERAGE 58 MINIMUM 54 MAXIMUM 65

CONSTITUENTS	DIM	AV		VARIATION			AVR	NUTR. DENS.		MOLPERC.
WATER	GRAM	5.24		3.94	-	7.11	2.20	GRAM/MJ	2.08	
PROTEIN	GRAM	11.96		10.68	-	13.57	5.02	GRAM/MJ	4.74	
FAT	GRAM	61.60		58.00	-	62.70	25.87	GRAM/MJ	24.43	
AVAILABLE CARBOHYDR.	GRAM	11.36	1	-	-	-	4.77	GRAM/MJ	4.51	
TOTAL DIETARY FIBRE	GRAM	7.40	2	-	-	-	3.11	GRAM/MJ	2.94	
MINERALS	GRAM	2.44		2.30	-	2.50	1.02	GRAM/MJ	0.97	
SODIUM	MILLI	2.00		0.60	-	3.00	0.84	MILLI/MJ	0.79	
POTASSIUM	MILLI	636.00		560.00	-	750.00	267.12	MILLI/MJ	252.28	
MAGNESIUM	MILLI	156.00		140.00	-	184.00	65.52	MILLI/MJ	61.88	
CALCIUM	MILLI	226.00		200.00	-	250.00	94.92	MILLI/MJ	89.65	
MANGANESE	MILLI	5.70		4.20	-	10.00	2.39	MILLI/MJ	2.26	
IRON	MILLI	3.80		3.00	-	4.50	1.60	MILLI/MJ	1.51	
COBALT	MICRO	12.00		-	-	-	5.04	MICRO/MJ	4.76	
COPPER	MILLI	1.28		1.20	-	1.35	0.54	MILLI/MJ	0.51	
ZINC	MILLI	1.87		1.30	-	2.44	0.79	MILLI/MJ	0.74	
NICKEL	MILLI	0.12		-	-	-	0.05	MILLI/MJ	0.05	
CHROMIUM	MICRO	14.00		8.00	-	29.00	5.88	MICRO/MJ	5.55	
PHOSPHORUS	MILLI	333.00		300.00	-	355.00	139.86	MILLI/MJ	132.09	
CHLORIDE	MILLI	10.00		-	-	-	4.20	MILLI/MJ	3.97	
FLUORIDE	MICRO	17.00		3.00	-	30.00	7.14	MICRO/MJ	6.74	
IODIDE	MICRO	1.50		-	-	-	0.63	MICRO/MJ	0.59	
BORON	MILLI	2.15		-	-	-	0.90	MILLI/MJ	0.85	
SELENIUM	MICRO	2.00		-	-	-	0.84	MICRO/MJ	0.79	
SILICON	MILLI	10.00		-	-	-	4.20	MILLI/MJ	3.97	
CAROTENE	MICRO	29.00		20.00	-	60.00	12.18	MICRO/MJ	11.50	
VITAMIN E ACTIVITY	MILLI	26.20		-	-	-	11.00	MILLI/MJ	10.39	

1 ESTIMATED BY THE DIFFERENCE METHOD
100 - (WATER + PROTEIN + FAT + MINERALS + TOTAL DIETARY FIBRE)

2 METHOD OF MEUSER, SUCKOW AND KULIKOWSKI ("BERLINER METHODE")

CONSTITUENTS	DIM	AV	VARIATION			AVR	NUTR. DENS.		MOLPERC.
TOTAL TOCOPHEROLS	MILLI	28.00	-	-	-	11.76	MILLI/MJ	11.11	
ALPHA-TOCOPHEROL	MILLI	26.10	-	-	-	10.96	MILLI/MJ	10.35	
GAMMA-TOCOPHEROL	MILLI	1.90	-	-	-	0.80	MILLI/MJ	0.75	
VITAMIN B1	MILLI	0.39	0.19	-	0.47	0.16	MILLI/MJ	0.15	
VITAMIN B2	MILLI	0.21	0.16	-	0.26	0.09	MILLI/MJ	0.08	
NICOTINAMIDE	MILLI	1.35	0.90	-	2.00	0.57	MILLI/MJ	0.54	
PANTOTHENIC ACID	MILLI	1.15	1.05	-	1.18	0.48	MILLI/MJ	0.46	
VITAMIN B6	MILLI	0.45	0.36	-	0.54	0.19	MILLI/MJ	0.18	
FOLIC ACID	MICRO	71.00	56.00	-	89.00	29.82	MICRO/MJ	28.16	
VITAMIN C	MILLI	3.00	TRACES	-	7.00	1.26	MILLI/MJ	1.19	
ARGININE	GRAM	2.03	1.92	-	2.11	0.85	GRAM/MJ	0.81	
CYSTINE	GRAM	0.19	0.06	-	0.36	0.08	GRAM/MJ	0.08	
HISTIDINE	GRAM	0.28	0.25	-	0.31	0.12	GRAM/MJ	0.11	
ISOLEUCINE	GRAM	0.77	0.64	-	0.88	0.32	GRAM/MJ	0.31	
LEUCINE	GRAM	0.89	0.87	-	0.91	0.37	GRAM/MJ	0.35	
LYSINE	GRAM	0.38	0.34	-	0.42	0.16	GRAM/MJ	0.15	
METHIONINE	GRAM	0.14	0.11	-	0.17	0.06	GRAM/MJ	0.06	
PHENYLALANINE	GRAM	0.51	0.50	-	0.53	0.21	GRAM/MJ	0.20	
THREONINE	GRAM	0.40	0.37	-	0.42	0.17	GRAM/MJ	0.16	
TRYPTOPHAN	GRAM	0.20	0.16	-	0.28	0.08	GRAM/MJ	0.08	
TYROSINE	GRAM	0.47	0.41	-	0.53	0.20	GRAM/MJ	0.19	
VALINE	GRAM	0.87	0.83	-	0.90	0.37	GRAM/MJ	0.35	
SALICYLIC ACID	MILLI	0.14	0.14	-	0.14	0.06	MILLI/MJ	0.06	
REDUCING SUGAR	GRAM	0.48	-	-	-	0.20	GRAM/MJ	0.19	
NONREDUCING SUGAR	GRAM	4.14	-	-	-	1.74	GRAM/MJ	1.64	
CAPRYLIC ACID	-	0.00	-	-	-	0.00			
CAPRIC ACID	-	0.00	-	-	-	0.00			
LAURIC ACID	-	0.00	-	-	-	0.00			
MYRISTIC ACID	GRAM	0.15	-	-	-	0.06	GRAM/MJ	0.06	
PALMITIC ACID	GRAM	3.00	-	-	-	1.26	GRAM/MJ	1.19	
STEARIC ACID	GRAM	1.10	-	-	-	0.46	GRAM/MJ	0.44	
ARACHIDIC ACID	GRAM	0.11	-	-	-	0.05	GRAM/MJ	0.04	
PALMITOLEIC ACID	GRAM	0.16	-	-	-	0.07	GRAM/MJ	0.06	
OLEIC ACID	GRAM	47.40	-	-	-	19.91	GRAM/MJ	18.80	
LINOLEIC ACID	GRAM	6.30	5.90	-	9.90	2.65	GRAM/MJ	2.50	
LINOLENIC ACID	GRAM	0.15	-	-	-	0.06	GRAM/MJ	0.06	
TOTAL PURINES	MILLI	27.00 [3]	-	-	-	11.34	MILLI/MJ	10.71	
DIETARY FIBRE,WAT.SOL.	GRAM	2.80	-	-	-	1.18	GRAM/MJ	1.11	
DIETARY FIBRE,WAT.INS.	GRAM	4.60	-	-	-	1.93	GRAM/MJ	1.82	

3 VALUE EXPRESSED IN MG URIC ACID/100G

KOKOSNUSS COCONUT NOIX DE COCOS

COCOS NUCIFERA L.

		PROTEIN	FAT	CARBOHYDRATES	TOTAL
ENERGY VALUE (AVERAGE) PER 100 G EDIBLE PORTION	KJOULE	73	1420	80	1574
	(KCAL)	17	339	19	376
AMOUNT OF DIGESTIBLE CONSTITUENTS PER 100 G	GRAM	3.05	32.85	4.80	
ENERGY VALUE (AVERAGE) OF THE DIGESTIBLE FRACTION PER 100 G EDIBLE PORTION	KJOULE	57	1278	80	1415
	(KCAL)	14	306	19	338

WASTE PERCENTAGE AVERAGE 27 MINIMUM 19 MAXIMUM 37

CONSTITUENTS	DIM	AV	VARIATION	AVR	NUTR.	DENS.	MOLPERC.
WATER	GRAM	44.80	38.00 - 46.90	32.70	GRAM/MJ	31.65	
PROTEIN	GRAM	3.92	3.40 - 4.20	2.86	GRAM/MJ	2.77	
FAT	GRAM	36.50	34.00 - 40.00	26.65	GRAM/MJ	25.79	
AVAILABLE CARBOHYDR.	GRAM	4.80	- - -	3.50	GRAM/MJ	3.39	
TOTAL DIETARY FIBRE	GRAM	9.00 1	- - -	6.57	GRAM/MJ	6.36	
MINERALS	GRAM	1.18	1.00 - 1.50	0.86	GRAM/MJ	0.83	
SODIUM	MILLI	35.00	25.00 - 42.00	25.55	MILLI/MJ	24.73	
POTASSIUM	MILLI	379.00	363.00 - 400.00	276.67	MILLI/MJ	267.75	
MAGNESIUM	MILLI	39.00	- - -	28.47	MILLI/MJ	27.55	
CALCIUM	MILLI	20.00	19.00 - 21.00	14.60	MILLI/MJ	14.13	
MANGANESE	MILLI	1.31	- - -	0.96	MILLI/MJ	0.93	
IRON	MILLI	2.25	2.00 - 2.70	1.64	MILLI/MJ	1.59	
COPPER	MILLI	-	0.30 - 7.00				
ZINC	MILLI	0.50	- - -	0.37	MILLI/MJ	0.35	
MOLYBDENUM	MICRO	25.00	- - -	18.25	MICRO/MJ	17.66	
PHOSPHORUS	MILLI	94.00	85.00 - 100.00	68.62	MILLI/MJ	66.41	
CHLORIDE	MILLI	122.00	- - -	89.06	MILLI/MJ	86.19	
IODIDE	MICRO	1.20	- - -	0.88	MICRO/MJ	0.85	
SELENIUM	MILLI	0.81	- - -	0.59	MILLI/MJ	0.57	
VITAMIN E ACTIVITY	MILLI	0.73	- - -	0.53	MILLI/MJ	0.52	
TOTAL TOCOPHEROLS	MILLI	0.95	- - -	0.69	MILLI/MJ	0.67	
ALPHA-TOCOPHEROL	MILLI	0.70	- - -	0.51	MILLI/MJ	0.49	
GAMMA-TOCOPHEROL	MILLI	0.25	- - -	0.18	MILLI/MJ	0.18	
VITAMIN B1	MICRO	61.00	25.00 - 100.00	44.53	MICRO/MJ	43.09	
VITAMIN B2	MICRO	8.00	5.00 - 10.00	5.84	MICRO/MJ	5.65	
NICOTINAMIDE	MILLI	0.38	0.20 - 0.60	0.28	MILLI/MJ	0.27	

1 METHOD OF SCHWEIZER AND WUERSCH

CONSTITUENTS	DIM	AV	VARIATION			AVR	NUTR. DENS.		MOLPERC.
PANTOTHENIC ACID	MILLI	0.20	-	-	-	0.15	MILLI/MJ	0.14	
VITAMIN B6	MICRO	60.00	-	-	-	43.80	MICRO/MJ	42.39	
FOLIC ACID	MICRO	30.00	-	-	-	21.90	MICRO/MJ	21.19	
VITAMIN C	MILLI	2.00	-	-	-	1.46	MILLI/MJ	1.41	
ARGININE	GRAM	0.49	0.41	-	0.56	0.36	GRAM/MJ	0.35	
CYSTINE	MILLI	71.00	-	-	-	51.83	MILLI/MJ	50.16	
HISTIDINE	MILLI	71.00	62.00	-	80.00	51.83	MILLI/MJ	50.16	
ISOLEUCINE	GRAM	0.20	0.18	-	0.21	0.15	GRAM/MJ	0.14	
LEUCINE	GRAM	0.31	-	-	-	0.23	GRAM/MJ	0.22	
LYSINE	GRAM	0.15	0.12	-	0.18	0.11	GRAM/MJ	0.11	
METHIONINE	MILLI	70.00	59.00	-	82.00	51.10	MILLI/MJ	49.45	
PHENYLALANINE	GRAM	0.18	0.15	-	0.20	0.13	GRAM/MJ	0.13	
THREONINE	GRAM	0.13	0.12	-	0.15	0.09	GRAM/MJ	0.09	
TRYPTOPHAN	MILLI	39.00	38.00	-	39.00	28.47	MILLI/MJ	27.55	
TYROSINE	GRAM	0.12	-	-	-	0.09	GRAM/MJ	0.08	
VALINE	GRAM	0.22	0.19	-	0.24	0.16	GRAM/MJ	0.16	
REDUCING SUGAR	GRAM	0.12	-	-	-	0.09	GRAM/MJ	0.08	
NONREDUCING SUGAR	GRAM	4.66	-	-	-	3.40	GRAM/MJ	3.29	
CAPRYLIC ACID	GRAM	2.80	-	-	-	2.04	GRAM/MJ	1.98	
CAPRIC ACID	GRAM	2.20	-	-	-	1.61	GRAM/MJ	1.55	
LAURIC ACID	GRAM	16.20	-	-	-	11.83	GRAM/MJ	11.44	
MYRISTIC ACID	GRAM	6.10	-	-	-	4.45	GRAM/MJ	4.31	
PALMITIC ACID	GRAM	3.10	-	-	-	2.26	GRAM/MJ	2.19	
STEARIC ACID	GRAM	1.10	-	-	-	0.80	GRAM/MJ	0.78	
ARACHIDIC ACID	GRAM	0.34	-	-	-	0.25	GRAM/MJ	0.24	
PALMITOLEIC ACID	GRAM	0.13	-	-	-	0.09	GRAM/MJ	0.09	
OLEIC ACID	GRAM	2.10	-	-	-	1.53	GRAM/MJ	1.48	
LINOLEIC ACID	GRAM	0.68	-	-	-	0.50	GRAM/MJ	0.48	
LINOLENIC ACID	-	0.00	-	-	-	0.00			
TOTAL STEROLS	MILLI	47.00	-	-	-	34.31	MILLI/MJ	33.20	
CAMPESTEROL	MILLI	3.00	-	-	-	2.19	MILLI/MJ	2.12	
BETA-SITOSTEROL	MILLI	27.00	-	-	-	19.71	MILLI/MJ	19.07	
STIGMASTEROL	MILLI	7.00	-	-	-	5.11	MILLI/MJ	4.95	
DIETARY FIBRE,WAT.SOL.	GRAM	2.10	-	-	-	1.53	GRAM/MJ	1.48	
DIETARY FIBRE,WAT.INS.	GRAM	6.90	-	-	-	5.04	GRAM/MJ	4.87	

KOKOSNUSSMILCH COCONUT MILK LAIT DE COCO

		PROTEIN	FAT	CARBOHYDRATES	TOTAL
ENERGY VALUE (AVERAGE) PER 100 G EDIBLE PORTION	KJOULE	4.1	16	23	42
	(KCAL)	1.0	3.7	5.4	10
AMOUNT OF DIGESTIBLE CONSTITUENTS PER 100 G	GRAM	0.21	0.36	1.36	
ENERGY VALUE (AVERAGE) OF THE DIGESTIBLE FRACTION PER 100 G EDIBLE PORTION	KJOULE	3.5	14	23	40
	(KCAL)	0.8	3.3	5.4	9.6

WASTE PERCENTAGE AVERAGE 0.00

CONSTITUENTS	DIM	AV	VARIATION			AVR	NUTR. DENS.		MOLPERC.
WATER	GRAM	94.40	93.60	-	95.30	94.40	KILO/MJ	2.34	
PROTEIN	GRAM	0.25	-	-	-	0.25	GRAM/MJ	6.21	
FAT	GRAM	0.40	-	-	-	0.40	GRAM/MJ	9.93	
AVAILABLE CARBOHYDR.	GRAM	1.36	-	-	-	1.36	GRAM/MJ	33.76	
MINERALS	GRAM	0.66	0.62	-	0.70	0.66	GRAM/MJ	16.38	
SODIUM	MILLI	47.00	15.00	-	79.00	47.00	GRAM/MJ	1.17	
POTASSIUM	MILLI	282.00	251.00	-	312.00	282.00	GRAM/MJ	7.00	
MAGNESIUM	MILLI	30.00	-	-	-	30.00	MILLI/MJ	744.77	
CALCIUM	MILLI	27.00	24.00	-	29.00	27.00	MILLI/MJ	670.29	
IRON	MILLI	0.10	-	-	-	0.10	MILLI/MJ	2.48	
COPPER	MILLI	0.40	-	-	-	0.40	MILLI/MJ	9.93	
PHOSPHORUS	MILLI	33.00	29.00	-	37.00	33.00	MILLI/MJ	819.25	
CHLORIDE	MILLI	183.00	-	-	-	183.00	GRAM/MJ	4.54	
VITAMIN B1	-	TRACES	-	-	-				
VITAMIN B2	-	TRACES	-	-	-				
NICOTINAMIDE	MILLI	0.10	-	-	-	0.10	MILLI/MJ	2.48	
PANTOTHENIC ACID	MICRO	50.00	-	-	-	50.00	MILLI/MJ	1.24	
VITAMIN B6	MICRO	30.00	-	-	-	30.00	MICRO/MJ	744.77	
VITAMIN C	MILLI	2.00	-	-	-	2.00	MILLI/MJ	49.65	
REDUCING SUGAR	MILLI	80.00	-	-	-	80.00	GRAM/MJ	1.99	
SUCROSE	GRAM	1.28	-	-	-	1.28	GRAM/MJ	31.78	

KOLANUSS COLA NUT NOIX DE COLA

COLA ACUMINATA

		PROTEIN	FAT	CARBOHYDRATES	TOTAL
ENERGY VALUE (AVERAGE) PER 100 G EDIBLE PORTION	KJOULE (KCAL)	104 25	68 16	783 187	955 228
AMOUNT OF DIGESTIBLE CONSTITUENTS PER 100 G	GRAM	5.33	1.57	46.80	
ENERGY VALUE (AVERAGE) OF THE DIGESTIBLE FRACTION PER 100 G EDIBLE PORTION	KJOULE (KCAL)	88 21	61 15	783 187	933 223

WASTE PERCENTAGE AVERAGE 0.00

CONSTITUENTS	DIM	AV	VARIATION			AVR	NUTR. DENS.		MOLPERC.
WATER	GRAM	11.75	11.50	-	12.00	11.75	GRAM/MJ	12.60	
PROTEIN	GRAM	6.28	4.92	-	7.63	6.28	GRAM/MJ	6.73	
FAT	GRAM	1.75	1.50	-	2.00	1.75	GRAM/MJ	1.88	
AVAILABLE CARBOHYDR.	GRAM	46.80	-	-	-	46.80	GRAM/MJ	50.17	
MINERALS	GRAM	2.90	-	-	-	2.90	GRAM/MJ	3.11	
CALCIUM	MILLI	108.00	-	-	-	108.00	MILLI/MJ	115.79	
IRON	MILLI	6.00	-	-	-	6.00	MILLI/MJ	6.43	
PHOSPHORUS	MILLI	76.00	-	-	-	76.00	MILLI/MJ	81.48	
VITAMIN B1	MICRO	60.00	-	-	-	60.00	MICRO/MJ	64.33	
SUCROSE	GRAM	2.75	-	-	-	2.75	GRAM/MJ	2.95	
STARCH	GRAM	44.00	-	-	-	44.00	GRAM/MJ	47.17	

MACADAMIANUSS — MACADAMIA NUT — MACADAMIA

(AUSTRALNUSS)
MACADAMIA TERNIFOLIA F. V. MUE.
(M. INTEGRIFOLIA MAIDEN ET BET.)

(QUEENSLAND NUT)

		PROTEIN	FAT	CARBOHYDRATES	TOTAL
ENERGY VALUE (AVERAGE) PER 100 G EDIBLE PORTION	KJOULE	139	2841	258	3238
	(KCAL)	33	679	62	774
AMOUNT OF DIGESTIBLE CONSTITUENTS PER 100 G	GRAM	5.81	65.70	15.44	
ENERGY VALUE (AVERAGE) OF THE DIGESTIBLE FRACTION PER 100 G EDIBLE PORTION	KJOULE	108	2556	258	2923
	(KCAL)	26	611	62	699

WASTE PERCENTAGE AVERAGE 34 MINIMUM 24 MAXIMUM 42

CONSTITUENTS	DIM	AV	VARIATION		AVR	NUTR. DENS.		MOLPERC.
WATER	GRAM	2.50	2.00	- 3.00	1.65	GRAM/MJ	0.86	
PROTEIN	GRAM	7.46	6.61	- 8.48	4.92	GRAM/MJ	2.55	
FAT	GRAM	73.00	71.60	- 76.00	48.18	GRAM/MJ	24.97	
AVAILABLE CARBOHYDR.	GRAM	15.44 [1]	-	- -	10.19	GRAM/MJ	5.28	
MINERALS	GRAM	1.60	1.40	- 1.70	1.06	GRAM/MJ	0.55	
POTASSIUM	MILLI	265.00	-	- -	174.90	MILLI/MJ	90.65	
CALCIUM	MILLI	51.00	48.00	- 53.00	33.66	MILLI/MJ	17.45	
IRON	MILLI	0.20	-	- -	0.13	MILLI/MJ	0.07	
PHOSPHORUS	MILLI	201.00	161.00	- 241.00	132.66	MILLI/MJ	68.76	
VITAMIN B1	MILLI	0.28	0.22	- 3.40	0.18	MILLI/MJ	0.10	
VITAMIN B2	MILLI	0.12	-	- -	0.08	MILLI/MJ	0.04	
NICOTINAMIDE	MILLI	1.50	1.30	- 1.60	0.99	MILLI/MJ	0.51	
VITAMIN C	-	0.00	-	- -	0.00			
SALICYLIC ACID	MILLI	0.52	0.52	- 0.52	0.34	MILLI/MJ	0.18	
BUTYRIC ACID	GRAM	0.49	-	- -	0.32	GRAM/MJ	0.17	
LAURIC ACID	-	TRACES	-	- -				
PALMITIC ACID	GRAM	5.70	-	- -	3.76	GRAM/MJ	1.95	
STEARIC ACID	GRAM	1.90	-	- -	1.25	GRAM/MJ	0.65	
PALMITOLEIC ACID	GRAM	14.20	-	- -	9.37	GRAM/MJ	4.86	
OLEIC ACID	GRAM	43.10	-	- -	28.45	GRAM/MJ	14.74	
EICOSENOIC ACID	GRAM	1.70	1.50	- 1.80	1.12	GRAM/MJ	0.58	
LINOLEIC ACID	GRAM	1.30	-	- -	0.86	GRAM/MJ	0.44	

1 ESTIMATED BY THE DIFFERENCE METHOD
100 - (WATER + PROTEIN + FAT + MINERALS)

MANDEL / ALMOND / AMANDE

MANDEL	ALMOND	AMANDE
SÜSS	SWEET	DOUCE

AMYGDALUS COMMUNIS BUNGE

		PROTEIN	FAT	CARBOHYDRATES	TOTAL
ENERGY VALUE (AVERAGE)	KJOULE	349	2105	152	2606
PER 100 G EDIBLE PORTION	(KCAL)	83	503	36	623
AMOUNT OF DIGESTIBLE CONSTITUENTS PER 100 G	GRAM	14.60	48.69	9.08	
ENERGY VALUE (AVERAGE)	KJOULE	272	1895	152	2318
OF THE DIGESTIBLE FRACTION PER 100 G EDIBLE PORTION	(KCAL)	65	453	36	554

WASTE PERCENTAGE AVERAGE 49 MINIMUM 44 MAXIMUM 53

CONSTITUENTS	DIM	AV		VARIATION			AVR	NUTR. DENS.		MOLPERC.
WATER	GRAM	5.65		4.70	-	6.27	2.88	GRAM/MJ	2.44	
PROTEIN	GRAM	18.72		18.11	-	19.44	9.55	GRAM/MJ	8.07	
FAT	GRAM	54.10		53.20	-	55.00	27.59	GRAM/MJ	23.33	
AVAILABLE CARBOHYDR.	GRAM	9.08	1	-	-	-	4.63	GRAM/MJ	3.92	
TOTAL DIETARY FIBRE	GRAM	9.80	2	-	-	-	5.00	GRAM/MJ	4.23	
MINERALS	GRAM	2.65		2.30	-	3.00	1.35	GRAM/MJ	1.14	
SODIUM	MILLI	-		5.00	-	40.00				
POTASSIUM	MILLI	835.00		764.00	-	900.00	425.85	MILLI/MJ	360.16	
MAGNESIUM	MILLI	170.00		-	-	-	86.70	MILLI/MJ	73.33	
CALCIUM	MILLI	252.00		250.00	-	254.00	128.52	MILLI/MJ	108.70	
MANGANESE	MILLI	1.90		-	-	-	0.97	MILLI/MJ	0.82	
IRON	MILLI	4.13		4.00	-	4.40	2.11	MILLI/MJ	1.78	
COPPER	MILLI	0.85		0.14	-	1.20	0.43	MILLI/MJ	0.37	
ZINC	MILLI	2.10		-	-	-	1.07	MILLI/MJ	0.91	
NICKEL	MILLI	0.13		-	-	-	0.07	MILLI/MJ	0.06	
CHROMIUM	MICRO	12.00		10.00	-	19.00	6.12	MICRO/MJ	5.18	
PHOSPHORUS	MILLI	454.00		437.00	-	475.00	231.54	MILLI/MJ	195.82	
CHLORIDE	MILLI	40.00		-	-	-	20.40	MILLI/MJ	17.25	
FLUORIDE	MICRO	90.00		-	-	-	45.90	MICRO/MJ	38.82	
IODIDE	MICRO	2.00		-	-	-	1.02	MICRO/MJ	0.86	
BORON	MILLI	1.40		5.00	-	2.30	0.71	MILLI/MJ	0.60	
SELENIUM	MICRO	2.00		0.50	-	6.00	1.02	MICRO/MJ	0.86	
CAROTENE	MILLI	0.12		0.08	-	0.17	0.06	MILLI/MJ	0.05	
TOTAL TOCOPHEROLS	MILLI	26.10		-	-	-	13.31	MILLI/MJ	11.26	
ALPHA-TOCOPHEROL	MILLI	24.80		-	-	-	12.65	MILLI/MJ	10.70	
BETA-TOCOPHEROL	MILLI	0.15		-	-	-	0.08	MILLI/MJ	0.06	

1 ESTIMATED BY THE DIFFERENCE METHOD
100 - (WATER + PROTEIN + FAT + MINERALS + TOTAL DIETARY FIBRE)

2 METHOD OF MEUSER, SUCKOW AND KULIKOWSKI ("BERLINER METHODE")

CONSTITUENTS	DIM	AV	VARIATION			AVR	NUTR. DENS.		MOLPERC.
GAMMA-TOCOPHEROL	MILLI	1.10	-	-	-	0.56	MILLI/MJ	0.47	
VITAMIN B1	MILLI	0.22	0.11	-	0.28	0.11	MILLI/MJ	0.09	
VITAMIN B2	MILLI	0.62	0.35	-	0.80	0.32	MILLI/MJ	0.27	
NICOTINAMIDE	MILLI	4.18	3.50	-	4.60	2.13	MILLI/MJ	1.80	
PANTOTHENIC ACID	MILLI	0.58	0.53	-	0.64	0.30	MILLI/MJ	0.25	
VITAMIN B6	MICRO	60.00	-	-	-	30.60	MICRO/MJ	25.88	
BIOTIN	MICRO	-	0.40	-	20.00				
FOLIC ACID	MICRO	45.00	36.00	-	58.00	22.95	MICRO/MJ	19.41	
VITAMIN C	MILLI	-	0.80	-	6.50				
ARGININE	GRAM	2.75	-	-	-	1.40	GRAM/MJ	1.19	
CYSTINE	GRAM	0.38	-	-	-	0.19	GRAM/MJ	0.16	
HISTIDINE	GRAM	0.52	-	-	-	0.27	GRAM/MJ	0.22	
ISOLEUCINE	GRAM	0.88	-	-	-	0.45	GRAM/MJ	0.38	
LEUCINE	GRAM	1.46	-	-	-	0.74	GRAM/MJ	0.63	
LYSINE	GRAM	0.58	-	-	-	0.30	GRAM/MJ	0.25	
METHIONINE	GRAM	0.27	-	-	-	0.14	GRAM/MJ	0.12	
PHENYLALANINE	GRAM	1.16	-	-	-	0.59	GRAM/MJ	0.50	
THREONINE	GRAM	0.61	-	-	-	0.31	GRAM/MJ	0.26	
TRYPTOPHAN	GRAM	0.17	-	-	-	0.09	GRAM/MJ	0.07	
TYROSINE	GRAM	0.62	-	-	-	0.32	GRAM/MJ	0.27	
VALINE	GRAM	1.14	-	-	-	0.58	GRAM/MJ	0.49	
SALICYLIC ACID	MILLI	3.00	3.00	-	3.00	1.53	MILLI/MJ	1.29	
CAPRYLIC ACID	-	0.00	-	-	-	0.00			
CAPRIC ACID	-	0.00	-	-	-	0.00			
LAURIC ACID	-	0.00	-	-	-	0.00			
MYRISTIC ACID	MILLI	50.00	-	-	-	25.50	MILLI/MJ	21.57	
PALMITIC ACID	GRAM	3.26	-	-	-	1.66	GRAM/MJ	1.41	
STEARIC ACID	GRAM	0.90	-	-	-	0.46	GRAM/MJ	0.39	
ARACHIDIC ACID	GRAM	0.10	-	-	-	0.05	GRAM/MJ	0.04	
PALMITOLEIC ACID	GRAM	0.34	-	-	-	0.17	GRAM/MJ	0.15	
OLEIC ACID	GRAM	36.50	-	-	-	18.62	GRAM/MJ	15.74	
LINOLEIC ACID	GRAM	9.86	-	-	-	5.03	GRAM/MJ	4.25	
LINOLENIC ACID	GRAM	0.26	-	-	-	0.13	GRAM/MJ	0.11	
TOTAL STEROLS	MILLI	143.00	-	-	-	72.93	MILLI/MJ	61.68	
CAMPESTEROL	MILLI	5.00	-	-	-	2.55	MILLI/MJ	2.16	
BETA-SITOSTEROL	MILLI	122.00	-	-	-	62.22	MILLI/MJ	52.62	
STIGMASTEROL	MILLI	3.00	-	-	-	1.53	MILLI/MJ	1.29	
TOTAL PURINES	MILLI	30.00 [3]	-	-	-	15.30	MILLI/MJ	12.94	
DIETARY FIBRE,WAT.SOL.	GRAM	3.30	-	-	-	1.68	GRAM/MJ	1.42	
DIETARY FIBRE,WAT.INS.	GRAM	6.50	-	-	-	3.32	GRAM/MJ	2.80	

3 VALUE EXPRESSED IN MG URIC ACID/100G

PARANUSS | BRAZIL NUT | NOIX DU BRÉSIL

BERTHOLLETIA EXCELSA H. B. K.

		PROTEIN	FAT	CARBOHYDRATES	TOTAL
ENERGY VALUE (AVERAGE)	KJOULE	253	2599	61	2913
PER 100 G EDIBLE PORTION	(KCAL)	60	621	15	696
AMOUNT OF DIGESTIBLE CONSTITUENTS PER 100 G	GRAM	10.60	60.12	3.64	
ENERGY VALUE (AVERAGE)	KJOULE	197	2339	61	2598
OF THE DIGESTIBLE FRACTION PER 100 G EDIBLE PORTION	(KCAL)	47	559	15	621

WASTE PERCENTAGE AVERAGE 51 MINIMUM 50 MAXIMUM 52

CONSTITUENTS	DIM	AV		VARIATION			AVR	NUTR. DENS.		MOLPERC.
WATER	GRAM	5.62		5.30	-	5.94	2.75	GRAM/MJ	2.16	
PROTEIN	GRAM	13.59		13.11	-	13.98	6.66	GRAM/MJ	5.23	
FAT	GRAM	66.80		65.90	-	67.70	32.73	GRAM/MJ	25.72	
AVAILABLE CARBOHYDR.	GRAM	3.64	1	-	-	-	1.78	GRAM/MJ	1.40	
TOTAL DIETARY FIBRE	GRAM	6.70	2	-	-	-	3.28	GRAM/MJ	2.58	
MINERALS	GRAM	3.65		3.40	-	3.89	1.79	GRAM/MJ	1.41	
SODIUM	MILLI	2.00		1.00	-	2.00	0.98	MILLI/MJ	0.77	
POTASSIUM	MILLI	644.00		610.00	-	677.00	315.56	MILLI/MJ	247.92	
MAGNESIUM	MILLI	160.00		-	-	-	78.40	MILLI/MJ	61.59	
CALCIUM	MILLI	132.00		78.00	-	186.00	64.68	MILLI/MJ	50.82	
MANGANESE	MILLI	0.60		0.30	-	0.90	0.29	MILLI/MJ	0.23	
IRON	MILLI	3.40		-	-	-	1.67	MILLI/MJ	1.31	
COBALT	MICRO	-		3.00	-	160.00				
COPPER	MILLI	1.30		-	-	-	0.64	MILLI/MJ	0.50	
ZINC	MILLI	4.00		-	-	-	1.96	MILLI/MJ	1.54	
CHROMIUM	MILLI	0.10		-	-	-	0.05	MILLI/MJ	0.04	
PHOSPHORUS	MILLI	674.00		655.00	-	693.00	330.26	MILLI/MJ	259.47	
IODIDE	NANO	50.00		50.00	-	50.00	24.50	NANO/MJ	19.25	
SELENIUM	MILLI	0.10		-	-	-	0.05	MILLI/MJ	0.04	
CAROTENE	-	TRACES		-	-	-				
VITAMIN E ACTIVITY	MILLI	7.60		-	-	-	3.72	MILLI/MJ	2.93	
TOTAL TOCOPHEROLS	MILLI	17.50		-	-	-	8.58	MILLI/MJ	6.74	
ALPHA-TOCOPHEROL	MILLI	6.50		-	-	-	3.19	MILLI/MJ	2.50	
GAMMA-TOCOPHEROL	MILLI	11.00		-	-	-	5.39	MILLI/MJ	4.23	
VITAMIN B1	MILLI	1.00		0.86	-	1.13	0.49	MILLI/MJ	0.38	
VITAMIN B2	MICRO	35.00		-	-	-	17.15	MICRO/MJ	13.47	

1 ESTIMATED BY THE DIFFERENCE METHOD
100 - (WATER + PROTEIN + FAT + MINERALS + TOTAL DIETARY FIBRE)

2 METHOD OF MEUSER, SUCKOW AND KULIKOWSKI ("BERLINER METHODE")

CONSTITUENTS	DIM	AV	VARIATION			AVR	NUTR. DENS.		MOLPERC.
NICOTINAMIDE	MILLI	0.20	-	-	-	0.10	MILLI/MJ	0.08	
PANTOTHENIC ACID	MILLI	0.23	0.21	-	0.24	0.11	MILLI/MJ	0.09	
VITAMIN B6	MILLI	0.11	-	-	-	0.05	MILLI/MJ	0.04	
FOLIC ACID	MICRO	39.00	35.00	-	46.00	19.11	MICRO/MJ	15.01	
VITAMIN C	MILLI	0.70	-	-	-	0.34	MILLI/MJ	0.27	
ARGININE	GRAM	2.12	-	-	-	1.04	GRAM/MJ	0.82	
CYSTINE	GRAM	0.48	-	-	-	0.24	GRAM/MJ	0.18	
HISTIDINE	GRAM	0.35	-	-	-	0.17	GRAM/MJ	0.13	
ISOLEUCINE	GRAM	0.56	-	-	-	0.27	GRAM/MJ	0.22	
LEUCINE	GRAM	1.07	-	-	-	0.52	GRAM/MJ	0.41	
LYSINE	GRAM	0.42	-	-	-	0.21	GRAM/MJ	0.16	
METHIONINE	GRAM	0.89	-	-	-	0.44	GRAM/MJ	0.34	
PHENYLALANINE	GRAM	0.58	-	-	-	0.28	GRAM/MJ	0.22	
THREONINE	GRAM	0.40	-	-	-	0.20	GRAM/MJ	0.15	
TRYPTOPHAN	GRAM	0.17	-	-	-	0.08	GRAM/MJ	0.07	
TYROSINE	GRAM	0.46	-	-	-	0.23	GRAM/MJ	0.18	
VALINE	GRAM	0.78	-	-	-	0.38	GRAM/MJ	0.30	
SALICYLIC ACID	MILLI	0.46	0.46	-	0.46	0.23	MILLI/MJ	0.18	
TOTAL SUGAR	GRAM	2.34	-	-	-	1.15	GRAM/MJ	0.90	
CAPRYLIC ACID	-	0.00	-	-	-	0.00			
CAPRIC ACID	-	0.00	-	-	-	0.00			
LAURIC ACID	-	0.00	-	-	-	0.00			
MYRISTIC ACID	GRAM	0.15	-	-	-	0.07	GRAM/MJ	0.06	
PALMITIC ACID	GRAM	10.00	-	-	-	4.90	GRAM/MJ	3.85	
STEARIC ACID	GRAM	6.90	-	-	-	3.38	GRAM/MJ	2.66	
ARACHIDIC ACID	-	0.00	-	-	-	0.00			
PALMITOLEIC ACID	GRAM	0.28	-	-	-	0.14	GRAM/MJ	0.11	
OLEIC ACID	GRAM	21.70	-	-	-	10.63	GRAM/MJ	8.35	
LINOLEIC ACID	GRAM	24.90	-	-	-	12.20	GRAM/MJ	9.59	
LINOLENIC ACID	-	0.00	-	-	-	0.00			
DIETARY FIBRE, WAT.SOL.	GRAM	1.40	-	-	-	0.69	GRAM/MJ	0.54	
DIETARY FIBRE, WAT.INS.	GRAM	5.30	-	-	-	2.60	GRAM/MJ	2.04	

PEKANNUSS | PECAN NUT | PECANIER

CARYA ILLINOINENSIS

		PROTEIN	FAT	CARBOHYDRATES	TOTAL
ENERGY VALUE (AVERAGE) PER 100 G EDIBLE PORTION	KJOULE	154	2802	233	3188
	(KCAL)	37	670	56	762
AMOUNT OF DIGESTIBLE CONSTITUENTS PER 100 G	GRAM	7.90	64.80	13.90	
ENERGY VALUE (AVERAGE) OF THE DIGESTIBLE FRACTION PER 100 G EDIBLE PORTION	KJOULE	131	2521	233	2885
	(KCAL)	31	603	56	689

WASTE PERCENTAGE AVERAGE 53

CONSTITUENTS	DIM	AV		VARIATION			AVR	NUTR. DENS.		MOLPERC.
WATER	GRAM	3.20		-	-	-	1.50	GRAM/MJ	1.11	
PROTEIN	GRAM	9.30		-	-	-	4.37	GRAM/MJ	3.22	
FAT	GRAM	72.00		-	-	-	33.84	GRAM/MJ	24.96	
AVAILABLE CARBOHYDR.	GRAM	13.90	1	-	-	-	6.53	GRAM/MJ	4.82	
MINERALS	GRAM	1.60		-	-	-	0.75	GRAM/MJ	0.55	
SODIUM	MILLI	3.00		-	-	-	1.41	MILLI/MJ	1.04	
POTASSIUM	MILLI	604.00		-	-	-	283.88	MILLI/MJ	209.38	
MAGNESIUM	MILLI	142.00		-	-	-	66.74	MILLI/MJ	49.22	
CALCIUM	MILLI	73.00		-	-	-	34.31	MILLI/MJ	25.31	
MANGANESE	MILLI	3.50		-	-	-	1.65	MILLI/MJ	1.21	
IRON	MILLI	2.40		-	-	-	1.13	MILLI/MJ	0.83	
NICKEL	MILLI	1.50		0.90	-	2.00	0.71	MILLI/MJ	0.52	
PHOSPHORUS	MILLI	290.00		-	-	-	136.30	MILLI/MJ	100.53	
BORON	MILLI	0.76		-	-	-	0.36	MILLI/MJ	0.26	
SELENIUM	MICRO	3.00		-	-	-	1.41	MICRO/MJ	1.04	
CAROTENE	MICRO	80.00		-	-	-	37.60	MICRO/MJ	27.73	
VITAMIN E ACTIVITY	MILLI	3.10		-	-	-	1.46	MILLI/MJ	1.07	
TOTAL TOCOPHEROLS	MILLI	19.80		-	-	-	9.31	MILLI/MJ	6.86	
ALPHA-TOCOPHEROL	MILLI	1.20		-	-	-	0.56	MILLI/MJ	0.42	
GAMMA-TOCOPHEROL	MILLI	18.60		-	-	-	8.74	MILLI/MJ	6.45	
VITAMIN B1	MILLI	0.86		-	-	-	0.40	MILLI/MJ	0.30	
VITAMIN B2	MILLI	0.13		-	-	-	0.06	MILLI/MJ	0.05	
NICOTINAMIDE	MILLI	2.00		-	-	-	0.94	MILLI/MJ	0.69	
VITAMIN C	MILLI	2.00		-	-	-	0.94	MILLI/MJ	0.69	
SALICYLIC ACID	MILLI	0.12		0.12	-	0.12	0.06	MILLI/MJ	0.04	

1 ESTIMATED BY THE DIFFERENCE METHOD
100 - (WATER + PROTEIN + FAT + MINERALS)

CONSTITUENTS	DIM	AV	VARIATION			AVR	NUTR. DENS.		MOLPERC.
CAPRYLIC ACID	-	0.00	-	-	-	0.00			
CAPRIC ACID	-	0.00	-	-	-	0.00			
LAURIC ACID	-	0.00	-	-	-	0.00			
MYRISTIC ACID	-	0.00	-	-	-	0.00			
PALMITIC ACID	GRAM	4.20	-	-	-	1.97	GRAM/MJ	1.46	
STEARIC ACID	GRAM	1.90	-	-	-	0.89	GRAM/MJ	0.66	
ARACHIDIC ACID	-	0.00	-	-	-	0.00			
PALMITOLEIC ACID	GRAM	0.21	-	-	-	0.10	GRAM/MJ	0.07	
OLEIC ACID	GRAM	42.60	-	-	-	20.02	GRAM/MJ	14.77	
LINOLEIC ACID	GRAM	16.90	-	-	-	7.94	GRAM/MJ	5.86	
LINOLENIC ACID	GRAM	0.85	-	-	-	0.40	GRAM/MJ	0.29	
CAMPESTEROL	MILLI	4.00	-	-	-	1.88	MILLI/MJ	1.39	
BETA-SITOSTEROL	MILLI	88.00	-	-	-	41.36	MILLI/MJ	30.51	
STIGMASTEROL	MILLI	2.00	-	-	-	0.94	MILLI/MJ	0.69	

PISTAZIE | PISTACHIO | PISTACHE

(GRÜNE MANDEL, PISTAZIEN-MANDEL)

PISTACIA VERA L.

		PROTEIN	FAT	CARBOHYDRATES	TOTAL
ENERGY VALUE (AVERAGE) PER 100 G EDIBLE PORTION	KJOULE	328	2008	262	2598
	(KCAL)	78	480	63	621
AMOUNT OF DIGESTIBLE CONSTITUENTS PER 100 G	GRAM	13.75	46.44	15.66	
ENERGY VALUE (AVERAGE) OF THE DIGESTIBLE FRACTION PER 100 G EDIBLE PORTION	KJOULE	256	1807	262	2325
	(KCAL)	61	432	63	556

WASTE PERCENTAGE AVERAGE 47 MINIMUM 44 MAXIMUM 50

CONSTITUENTS	DIM	AV		VARIATION			AVR	NUTR. DENS.		MOLPERC.
WATER	GRAM	5.90		4.00	-	8.00	3.13	GRAM/MJ	2.54	
PROTEIN	GRAM	17.64		16.28	-	19.16	9.35	GRAM/MJ	7.59	
FAT	GRAM	51.60		43.40	-	58.90	27.35	GRAM/MJ	22.19	
AVAILABLE CARBOHYDR.	GRAM	15.66	1	-	-	-	8.30	GRAM/MJ	6.73	
TOTAL DIETARY FIBRE	GRAM	6.50	2	-	-	-	3.45	GRAM/MJ	2.80	
MINERALS	GRAM	2.70		2.40	-	3.10	1.43	GRAM/MJ	1.16	
POTASSIUM	GRAM	1.02		0.97	-	1.06	0.54	GRAM/MJ	0.44	
MAGNESIUM	MILLI	158.00		-	-	-	83.74	MILLI/MJ	67.95	
CALCIUM	MILLI	136.00		131.00	-	140.00	72.08	MILLI/MJ	58.49	
IRON	MILLI	7.30		-	-	-	3.87	MILLI/MJ	3.14	
NICKEL	MICRO	80.00		-	-	-	42.40	MICRO/MJ	34.40	
PHOSPHORUS	MILLI	500.00		-	-	-	265.00	MILLI/MJ	215.03	
SELENIUM	MILLI	0.45		-	-	-	0.24	MILLI/MJ	0.19	
CAROTENE	MILLI	0.15		0.06	-	0.25	0.08	MILLI/MJ	0.06	
VITAMIN E ACTIVITY	MILLI	5.20		-	-	-	2.76	MILLI/MJ	2.24	
ALPHA-TOCOPHEROL	MILLI	5.20		-	-	-	2.76	MILLI/MJ	2.24	
VITAMIN B1	MILLI	0.69		0.67	-	0.70	0.37	MILLI/MJ	0.30	
VITAMIN B2	MILLI	0.20		-	-	-	0.11	MILLI/MJ	0.09	
NICOTINAMIDE	MILLI	1.45		1.40	-	1.50	0.77	MILLI/MJ	0.62	
FOLIC ACID	MICRO	58.00		-	-	-	30.74	MICRO/MJ	24.94	
VITAMIN C	MILLI	7.00		0.00	-	14.00	3.71	MILLI/MJ	3.01	
SALICYLIC ACID	MILLI	0.55		0.55	-	0.55	0.29	MILLI/MJ	0.24	
CAPRYLIC ACID	-	0.00		-	-	-	0.00			
CAPRIC ACID	-	0.00		-	-	-	0.00			

1 ESTIMATED BY THE DIFFERENCE METHOD
100 - (WATER + PROTEIN + FAT + MINERALS + TOTAL DIETARY FIBRE)

2 METHOD OF MEUSER, SUCKOW AND KULIKOWSKI ("BERLINER METHODE")

CONSTITUENTS	DIM	AV	VARIATION			AVR	NUTR. DENS.		MOLPERC.
LAURIC ACID	MILLI	50.00	-	-	-	26.50	MILLI/MJ	21.50	
MYRISTIC ACID	GRAM	0.10	-	-	-	0.05	GRAM/MJ	0.04	
PALMITIC ACID	GRAM	6.00	-	-	-	3.18	GRAM/MJ	2.58	
STEARIC ACID	GRAM	0.68	-	-	-	0.36	GRAM/MJ	0.29	
ARACHIDIC ACID	GRAM	0.27	-	-	-	0.14	GRAM/MJ	0.12	
PALMITOLEIC ACID	GRAM	0.28	-	-	-	0.15	GRAM/MJ	0.12	
OLEIC ACID	GRAM	34.60	-	-	-	18.34	GRAM/MJ	14.88	
LINOLEIC ACID	GRAM	6.50	6.50	-	10.30	3.45	GRAM/MJ	2.80	
LINOLENIC ACID	GRAM	0.27	-	-	-	0.14	GRAM/MJ	0.12	
TOTAL STEROLS	MILLI	108.00	-	-	-	57.24	MILLI/MJ	46.45	
CAMPESTEROL	MILLI	6.00	-	-	-	3.18	MILLI/MJ	2.58	
BETA-SITOSTEROL	MILLI	90.00	-	-	-	47.70	MILLI/MJ	38.70	
STIGMASTEROL	MILLI	2.00	-	-	-	1.06	MILLI/MJ	0.86	
DIETARY FIBRE,WAT.SOL.	GRAM	3.00	-	-	-	1.59	GRAM/MJ	1.29	
DIETARY FIBRE,WAT.INS.	GRAM	3.50	-	-	-	1.86	GRAM/MJ	1.51	

WALNUSS WALNUT NOIX

JUGLANS REGIA L.

		PROTEIN	FAT	CARBOHYDRATES	TOTAL
ENERGY VALUE (AVERAGE)	KJOULE	268	2432	203	2903
PER 100 G	(KCAL)	64	581	49	694
EDIBLE PORTION					
AMOUNT OF DIGESTIBLE	GRAM	11.23	56.25	12.14	
CONSTITUENTS PER 100 G					
ENERGY VALUE (AVERAGE)	KJOULE	209	2189	203	2601
OF THE DIGESTIBLE	(KCAL)	50	523	49	622
FRACTION PER 100 G					
EDIBLE PORTION					

WASTE PERCENTAGE AVERAGE 57 MINIMUM 51 MAXIMUM 65

CONSTITUENTS	DIM	AV		VARIATION			AVR	NUTR. DENS.		MOLPERC.
WATER	GRAM	4.38		3.30	-	7.18	1.88	GRAM/MJ	1.68	
PROTEIN	GRAM	14.40		13.60	-	15.60	6.19	GRAM/MJ	5.54	
FAT	GRAM	62.50		56.00	-	68.10	26.88	GRAM/MJ	24.03	
AVAILABLE CARBOHYDR.	GRAM	12.14	1	-	-	-	5.22	GRAM/MJ	4.67	
TOTAL DIETARY FIBRE	GRAM	4.60	2	-	-	-	1.98	GRAM/MJ	1.77	
MINERALS	GRAM	1.98		1.65	-	2.40	0.85	GRAM/MJ	0.76	
SODIUM	MILLI	2.40		1.00	-	4.00	1.03	MILLI/MJ	0.92	
POTASSIUM	MILLI	544.00		440.00	-	700.00	233.92	MILLI/MJ	209.15	
MAGNESIUM	MILLI	129.00		92.00	-	144.00	55.47	MILLI/MJ	49.60	
CALCIUM	MILLI	87.00		60.00	-	100.00	37.41	MILLI/MJ	33.45	
MANGANESE	MILLI	1.97		0.75	-	3.21	0.85	MILLI/MJ	0.76	
IRON	MILLI	2.50		2.00	-	3.10	1.08	MILLI/MJ	0.96	
COBALT	MICRO	9.50		5.00	-	15.00	4.09	MICRO/MJ	3.65	
COPPER	MILLI	0.88		0.31	-	1.40	0.38	MILLI/MJ	0.34	
ZINC	MILLI	2.70		2.00	-	3.20	1.16	MILLI/MJ	1.04	
NICKEL	MILLI	0.13		-	-	-	0.06	MILLI/MJ	0.05	
PHOSPHORUS	MILLI	409.00		310.00	-	510.00	175.87	MILLI/MJ	157.24	
CHLORIDE	MILLI	23.00		-	-	-	9.89	MILLI/MJ	8.84	
FLUORIDE	MILLI	0.68		-	-	-	0.29	MILLI/MJ	0.26	
IODIDE	MICRO	3.00		-	-	-	1.29	MICRO/MJ	1.15	
BORON	MILLI	0.76		0.42	-	1.27	0.33	MILLI/MJ	0.29	
CAROTENE	MICRO	48.00		18.00	-	78.00	20.64	MICRO/MJ	18.45	
VITAMIN E ACTIVITY	MILLI	6.20		-	-	-	2.67	MILLI/MJ	2.38	
TOTAL TOCOPHEROLS	MILLI	45.60		-	-	-	19.61	MILLI/MJ	17.53	
ALPHA-TOCOPHEROL	MILLI	1.90		-	-	-	0.82	MILLI/MJ	0.73	
GAMMA-TOCOPHEROL	MILLI	41.40		-	-	-	17.80	MILLI/MJ	15.92	

1 ESTIMATED BY THE DIFFERENCE METHOD
100 - (WATER + PROTEIN + FAT + MINERALS + TOTAL DIETARY FIBRE)

2 METHOD OF MEUSER, SUCKOW AND KULIKOWSKI ("BERLINER METHODE")

CONSTITUENTS	DIM	AV	VARIATION			AVR	NUTR. DENS.		MOLPERC.
DELTA-TOCOPHEROL	MILLI	0.20	-	-	-	0.09	MILLI/MJ	0.08	
VITAMIN B1	MILLI	0.34	0.20	-	0.44	0.15	MILLI/MJ	0.13	
VITAMIN B2	MILLI	0.12	0.06	-	0.16	0.05	MILLI/MJ	0.05	
NICOTINAMIDE	MILLI	1.00	0.60	-	1.20	0.43	MILLI/MJ	0.38	
PANTOTHENIC ACID	MILLI	0.82	0.70	-	0.96	0.35	MILLI/MJ	0.32	
VITAMIN B6	MILLI	0.87	0.74	-	1.00	0.37	MILLI/MJ	0.33	
BIOTIN	MICRO	-	2.00	-	37.00				
FOLIC ACID	MICRO	77.00	68.00	-	87.00	33.11	MICRO/MJ	29.60	
VITAMIN C	MILLI	2.60	TRACES	-	4.50	1.12	MILLI/MJ	1.00	
ALANINE	GRAM	0.90	-	-	-	0.39	GRAM/MJ	0.35	7.2
ARGININE	GRAM	2.09	1.87	-	2.20	0.90	GRAM/MJ	0.80	8.5
ASPARTIC ACID	GRAM	2.12	-	-	-	0.91	GRAM/MJ	0.82	11.3
CYSTINE	GRAM	0.25	0.19	-	0.31	0.11	GRAM/MJ	0.10	0.7
GLUTAMIC ACID	GRAM	3.96	-	-	-	1.70	GRAM/MJ	1.52	19.2
GLYCINE	GRAM	1.03	-	-	-	0.44	GRAM/MJ	0.40	9.8
HISTIDINE	GRAM	0.36	0.33	-	0.39	0.15	GRAM/MJ	0.14	1.7
ISOLEUCINE	GRAM	0.67	0.62	-	0.74	0.29	GRAM/MJ	0.26	3.6
LEUCINE	GRAM	1.14	0.99	-	1.27	0.49	GRAM/MJ	0.44	6.2
LYSINE	GRAM	0.44	0.26	-	0.72	0.19	GRAM/MJ	0.17	2.1
METHIONINE	GRAM	0.22	0.13	-	0.29	0.09	GRAM/MJ	0.08	1.0
PHENYLALANINE	GRAM	0.66	0.59	-	0.74	0.28	GRAM/MJ	0.25	2.8
PROLINE	GRAM	1.14	-	-	-	0.49	GRAM/MJ	0.44	7.0
SERINE	GRAM	1.14	-	-	-	0.49	GRAM/MJ	0.44	7.7
THREONINE	GRAM	0.54	0.46	-	0.66	0.23	GRAM/MJ	0.21	3.2
TRYPTOPHAN	GRAM	0.17	0.14	-	0.19	0.07	GRAM/MJ	0.07	0.6
TYROSINE	GRAM	0.64	0.56	-	0.71	0.28	GRAM/MJ	0.25	2.5
VALINE	GRAM	0.77	0.62	-	0.91	0.33	GRAM/MJ	0.30	4.7
SALICYLIC ACID	MILLI	0.30	0.30	-	0.30	0.13	MILLI/MJ	0.12	
CAPRYLIC ACID	-	0.00	-	-	-	0.00			
CAPRIC ACID	-	0.00	-	-	-	0.00			
LAURIC ACID	-	0.00	-	-	-	0.00			
MYRISTIC ACID	GRAM	0.69	-	-	-	0.30	GRAM/MJ	0.27	
PALMITIC ACID	GRAM	4.40	-	-	-	1.89	GRAM/MJ	1.69	
STEARIC ACID	GRAM	1.30	-	-	-	0.56	GRAM/MJ	0.50	
ARACHIDIC ACID	GRAM	0.42	-	-	-	0.18	GRAM/MJ	0.16	
PALMITOLEIC ACID	GRAM	0.20	-	-	-	0.09	GRAM/MJ	0.08	
OLEIC ACID	GRAM	9.60	-	-	-	4.13	GRAM/MJ	3.69	
LINOLEIC ACID	GRAM	34.10	29.90	-	47.20	14.66	GRAM/MJ	13.11	
LINOLENIC ACID	GRAM	6.80	2.10	-	8.50	2.92	GRAM/MJ	2.61	
ARACHIDONIC ACID	GRAM	0.59	-	-	-	0.25	GRAM/MJ	0.23	
TOTAL STEROLS	MILLI	108.00	-	-	-	46.44	MILLI/MJ	41.52	
CAMPESTEROL	MILLI	6.00	-	-	-	2.58	MILLI/MJ	2.31	
BETA-SITOSTEROL	MILLI	87.00	-	-	-	37.41	MILLI/MJ	33.45	
TOTAL PURINES	MILLI	23.00 [3]	-	-	-	9.89	MILLI/MJ	8.84	
DIETARY FIBRE,WAT.SOL.	GRAM	2.10	-	-	-	0.90	GRAM/MJ	0.81	
DIETARY FIBRE,WAT.INS.	GRAM	2.50	-	-	-	1.08	GRAM/MJ	0.96	

3 VALUE EXPRESSED IN MG URIC ACID/100G

Obst- und Beeren-Säfte

Juices from fruits and berries
Jus de fruits et de baies

Obst- und Beeren-Säfte
Juices from fruits and berries
Jus de fruits et de baies

ANANASSAFT
IN DOSEN

PINAPPLE JUICE
CANNED

JUS D'ANANAS
EN BOÎTES

		PROTEIN	FAT	CARBOHYDRATES	TOTAL
ENERGY VALUE (AVERAGE)	KJOULE	6.6	3.9	202	213
PER 100 G	(KCAL)	1.6	0.9	48	51
EDIBLE PORTION					
AMOUNT OF DIGESTIBLE	GRAM	0.34	0.09	12.08	
CONSTITUENTS PER 100 G					
ENERGY VALUE (AVERAGE)	KJOULE	5.6	3.5	202	211
OF THE DIGESTIBLE	(KCAL)	1.3	0.8	48	51
FRACTION PER 100 G					
EDIBLE PORTION					

WASTE PERCENTAGE AVERAGE 0.00

CONSTITUENTS	DIM	AV	VARIATION			AVR	NUTR. DENS.		MOLPERC.
WATER	GRAM	86.10	-	-	-	86.10	GRAM/MJ	407.49	
PROTEIN	GRAM	0.40	-	-	-	0.40	GRAM/MJ	1.89	
FAT	GRAM	0.10	-	-	-	0.10	GRAM/MJ	0.47	
AVAILABLE CARBOHYDR.	GRAM	12.08	-	-	-	12.08	GRAM/MJ	57.17	
MINERALS	GRAM	0.20	-	-	-	0.20	GRAM/MJ	0.95	
SODIUM	MILLI	1.00	-	-	-	1.00	MILLI/MJ	4.73	
POTASSIUM	MILLI	140.00	-	-	-	140.00	MILLI/MJ	662.59	
MAGNESIUM	MILLI	12.00	-	-	-	12.00	MILLI/MJ	56.79	
CALCIUM	MILLI	12.00	-	-	-	12.00	MILLI/MJ	56.79	
IRON	MILLI	0.70	-	-	-	0.70	MILLI/MJ	3.31	
COPPER	MICRO	90.00	-	-	-	90.00	MICRO/MJ	425.95	
PHOSPHORUS	MILLI	10.00	-	-	-	10.00	MILLI/MJ	47.33	
CHLORIDE	MILLI	38.00	-	-	-	38.00	MILLI/MJ	179.85	
BORON	MILLI	0.85	0.06	-	0.11	0.85	MILLI/MJ	4.02	
VITAMIN B1	MICRO	50.00	-	-	-	50.00	MICRO/MJ	236.64	
VITAMIN B2	MICRO	20.00	-	-	-	20.00	MICRO/MJ	94.66	
NICOTINAMIDE	MILLI	0.20	-	-	-	0.20	MILLI/MJ	0.95	
PANTOTHENIC ACID	MILLI	0.10	-	-	-	0.10	MILLI/MJ	0.47	
VITAMIN B6	MILLI	0.10	-	-	-	0.10	MILLI/MJ	0.47	
FOLIC ACID	MICRO	2.00	-	-	-	2.00	MICRO/MJ	9.47	
VITAMIN C	MILLI	8.00	-	-	-	8.00	MILLI/MJ	37.86	
SEROTONINE	MILLI	3.00	2.50	-	3.50	3.00	MILLI/MJ	14.20	
MALIC ACID	GRAM	0.13	0.13	-	0.13	0.13	GRAM/MJ	0.62	
CITRIC ACID	GRAM	0.59	0.59	-	0.59	0.59	GRAM/MJ	2.79	
SALICYLIC ACID	MILLI	0.16	0.16	-	0.16	0.16	MILLI/MJ	0.76	
GLUCOSE	GRAM	3.52	3.52	-	3.52	3.52	GRAM/MJ	16.66	
FRUCTOSE	GRAM	3.61	3.61	-	3.61	3.61	GRAM/MJ	17.09	
SUCROSE	GRAM	4.23	4.23	-	4.23	4.23	GRAM/MJ	20.02	

APFELSAFT
HANDELSWARE

APPLE JUICE
COMMERCIAL PRODUCT

JUS DE POMMES
PRODUIT DE VENTE

		PROTEIN	FAT	CARBOHYDRATES	TOTAL
ENERGY VALUE (AVERAGE)	KJOULE	1.2	0.0	197	199
PER 100 G	(KCAL)	0.3	0.0	47	47
EDIBLE PORTION					
AMOUNT OF DIGESTIBLE	GRAM	0.05	0.00	11.80	
CONSTITUENTS PER 100 G					
ENERGY VALUE (AVERAGE)	KJOULE	1.0	0.0	197	198
OF THE DIGESTIBLE	(KCAL)	0.2	0.0	47	47
FRACTION PER 100 G					
EDIBLE PORTION					

WASTE PERCENTAGE AVERAGE 0.00

CONSTITUENTS	DIM	AV	VARIATION			AVR	NUTR. DENS.		MOLPERC.
WATER	GRAM	88.10	86.20	-	90.30	88.10	GRAM/MJ	443.90	
PROTEIN	MILLI	70.00	60.00	-	80.00	70.00	MILLI/MJ	352.70	
FAT	-	0.00	-	-	-	0.00			
AVAILABLE CARBOHYDR.	GRAM	11.80	-	-	-	11.80	GRAM/MJ	59.46	
MINERALS	GRAM	0.27	0.22	-	0.33	0.27	GRAM/MJ	1.36	
SODIUM	MILLI	2.20	0.60	-	5.00	2.20	MILLI/MJ	11.08	
POTASSIUM	MILLI	116.00	90.00	-	131.00	116.00	MILLI/MJ	584.48	
MAGNESIUM	MILLI	4.20	3.00	-	5.40	4.20	MILLI/MJ	21.16	
CALCIUM	MILLI	6.90	4.00	-	8.80	6.90	MILLI/MJ	34.77	
MANGANESE	MILLI	0.12	0.05	-	0.21	0.12	MILLI/MJ	0.60	
IRON	MILLI	0.26	0.10	-	0.39	0.26	MILLI/MJ	1.31	
COBALT	MICRO	1.00	1.00	-	2.00	1.00	MICRO/MJ	5.04	
COPPER	MICRO	59.00	23.00	-	105.00	59.00	MICRO/MJ	297.28	
ZINC	MILLI	0.12	0.07	-	0.19	0.12	MILLI/MJ	0.60	
NICKEL	MILLI	-	5.00	-	55.00				
CHROMIUM	MICRO	3.00	0.00	-	6.00	3.00	MICRO/MJ	15.12	
MOLYBDENUM	MICRO	-	0.00	-	1.60				
PHOSPHORUS	MILLI	7.00	5.20	-	8.80	7.00	MILLI/MJ	35.27	
CHLORIDE	MILLI	0.30	TRACES	-	0.60	0.30	MILLI/MJ	1.51	
FLUORIDE	MICRO	10.00	6.00	-	15.00	10.00	MICRO/MJ	50.39	
IODIDE	MICRO	1.00	-	-	-	1.00	MICRO/MJ	5.04	
BORON	MILLI	0.12	0.01	-	0.19	0.12	MILLI/MJ	0.60	
SILICON	MILLI	1.00	-	-	-	1.00	MILLI/MJ	5.04	
CAROTENE	MICRO	45.00	30.00	-	60.00	45.00	MICRO/MJ	226.74	
VITAMIN B1	MICRO	20.00	6.00	-	40.00	20.00	MICRO/MJ	100.77	
VITAMIN B2	MICRO	25.00	10.00	-	50.00	25.00	MICRO/MJ	125.96	

CONSTITUENTS	DIM	AV	VARIATION			AVR	NUTR. DENS.		MOLPERC.
NICOTINAMIDE	MILLI	0.30	0.10	-	0.50	0.30	MILLI/MJ	1.51	
PANTOTHENIC ACID	MICRO	55.00	20.00	-	100.00	55.00	MICRO/MJ	277.12	
VITAMIN B6	MICRO	50.00	30.00	-	70.00	50.00	MICRO/MJ	251.93	
BIOTIN	MICRO	1.00	0.40	-	2.00	1.00	MICRO/MJ	5.04	
FOLIC ACID	MICRO	3.10	1.50	-	4.00	3.10	MICRO/MJ	15.62	
VITAMIN C	MILLI	1.40	0.70	-	2.00	1.40	MILLI/MJ	7.05	
ALANINE	MILLI	3.00	3.00	-	3.00	3.00	MILLI/MJ	15.12	
ARGININE	MILLI	2.00	2.00	-	2.00	2.00	MILLI/MJ	10.08	
ASPARTIC ACID	MILLI	22.00	22.00	-	22.00	22.00	MILLI/MJ	110.85	
GLUTAMIC ACID	MILLI	7.00	7.00	-	7.00	7.00	MILLI/MJ	35.27	
GLYCINE	MILLI	2.00	2.00	-	2.00	2.00	MILLI/MJ	10.08	
HISTIDINE	MILLI	1.00	1.00	-	1.00	1.00	MILLI/MJ	5.04	
ISOLEUCINE	MILLI	2.00	2.00	-	2.00	2.00	MILLI/MJ	10.08	
LEUCINE	MILLI	4.00	4.00	-	4.00	4.00	MILLI/MJ	20.15	
LYSINE	MILLI	5.00	5.00	-	5.00	5.00	MILLI/MJ	25.19	
METHIONINE	MILLI	1.00	1.00	-	1.00	1.00	MILLI/MJ	5.04	
PHENYLALANINE	MILLI	2.00	2.00	-	2.00	2.00	MILLI/MJ	10.08	
PROLINE	MILLI	2.00	2.00	-	2.00	2.00	MILLI/MJ	10.08	
SERINE	MILLI	3.00	3.00	-	3.00	3.00	MILLI/MJ	15.12	
THREONINE	MILLI	2.00	2.00	-	2.00	2.00	MILLI/MJ	10.08	
TRYPTOPHAN	MILLI	1.00	1.00	-	1.00	1.00	MILLI/MJ	5.04	
TYROSINE	MILLI	1.00	1.00	-	1.00	1.00	MILLI/MJ	5.04	
VALINE	MILLI	3.00	3.00	-	3.00	3.00	MILLI/MJ	15.12	
VOLATILE ACID	MILLI	15.00	4.00	-	25.00	15.00	MILLI/MJ	75.58	
MALIC ACID	MILLI	740.00	450.00	-	970.00	740.00	GRAM/MJ	3.73	
CITRIC ACID	MILLI	9.00	5.00	-	13.00	9.00	MILLI/MJ	45.35	
LACTIC ACID	MILLI	17.00	5.00	-	28.00	17.00	MILLI/MJ	85.66	
FORMIC ACID	MILLI	2.30	-	-	-	2.30	MILLI/MJ	11.59	
ISOCITRIC ACID	MILLI	0.80	-	-	-	0.80	MILLI/MJ	4.03	
SALICYLIC ACID	MILLI	0.19	0.19	-	0.19	0.19	MILLI/MJ	0.96	
GLUCOSE	GRAM	2.40	1.70	-	3.00	2.40	GRAM/MJ	12.09	
FRUCTOSE	GRAM	6.40	5.10	-	7.70	6.40	GRAM/MJ	32.25	
SUCROSE	GRAM	1.70	1.20	-	2.30	1.70	GRAM/MJ	8.57	
PECTIN	MILLI	32.00	2.00	-	76.00	32.00	MILLI/MJ	161.23	
SORBITOL	GRAM	0.56	0.30	-	0.90	0.56	GRAM/MJ	2.82	
ETHANOL	GRAM	0.14	0.05	-	0.27	0.14	GRAM/MJ	0.71	

APFELSINENSAFT	ORANGE JUICE	JUS D'ORANGES
(ORANGENSAFT)	FRESH	FRAIS
FRISCH GEPRESST	ORIGINAL	ORIGINAL
MUTTERSAFT		

		PROTEIN	FAT	CARBOHYDRATES	TOTAL
ENERGY VALUE (AVERAGE)	KJOULE	11	7.0	178	196
PER 100 G EDIBLE PORTION	(KCAL)	2.6	1.7	43	47
AMOUNT OF DIGESTIBLE CONSTITUENTS PER 100 G	GRAM	0.55	0.16	10.66	
ENERGY VALUE (AVERAGE)	KJOULE	9.1	6.3	178	194
OF THE DIGESTIBLE FRACTION PER 100 G EDIBLE PORTION	(KCAL)	2.2	1.5	43	46

WASTE PERCENTAGE AVERAGE 0.00

CONSTITUENTS	DIM	AV	VARIATION			AVR	NUTR.	DENS.	MOLPERC.
WATER	GRAM	88.10	86.70	-	89.40	88.10	GRAM/MJ	454.50	
PROTEIN	GRAM	0.65	0.49	-	0.80	0.65	GRAM/MJ	3.35	
FAT	GRAM	0.18	0.08	-	0.22	0.18	GRAM/MJ	0.93	
AVAILABLE CARBOHYDR.	GRAM	10.66	-	-	-	10.66	GRAM/MJ	54.99	
MINERALS	GRAM	0.37	0.28	-	0.45	0.37	GRAM/MJ	1.91	
SODIUM	MILLI	1.00	0.50	-	1.60	1.00	MILLI/MJ	5.16	
POTASSIUM	MILLI	157.00	106.00	-	194.00	157.00	MILLI/MJ	809.94	
MAGNESIUM	MILLI	12.00	10.00	-	13.00	12.00	MILLI/MJ	61.91	
CALCIUM	MILLI	11.00	8.00	-	15.00	11.00	MILLI/MJ	56.75	
MANGANESE	MICRO	30.00	-	-	-	30.00	MICRO/MJ	154.77	
IRON	MILLI	0.20	0.10	-	0.35	0.20	MILLI/MJ	1.03	
COPPER	MICRO	80.00	-	-	-	80.00	MICRO/MJ	412.71	
ZINC	MICRO	42.00	30.00	-	60.00	42.00	MICRO/MJ	216.67	
NICKEL	MICRO	1.00	1.00	-	2.00	1.00	MICRO/MJ	5.16	
CHROMIUM	MICRO	1.00	0.60	-	1.00	1.00	MICRO/MJ	5.16	
MOLYBDENUM	MICRO	79.00	-	-	-	79.00	MICRO/MJ	407.55	
PHOSPHORUS	MILLI	15.00	10.00	-	19.00	15.00	MILLI/MJ	77.38	
CHLORIDE	MILLI	3.80	2.10	-	5.70	3.80	MILLI/MJ	19.60	
FLUORIDE	MICRO	0.90	0.00	-	3.60	0.90	MICRO/MJ	4.64	
IODIDE	MICRO	1.00	-	-	-	1.00	MICRO/MJ	5.16	
BORON	MILLI	0.10	-	-	-	0.10	MILLI/MJ	0.52	
SELENIUM	MICRO	6.00	-	-	-	6.00	MICRO/MJ	30.95	
SILICON	MILLI	1.00	-	-	-	1.00	MILLI/MJ	5.16	
CAROTENE	MICRO	70.00	20.00	-	120.00	70.00	MICRO/MJ	361.12	
VITAMIN B1	MICRO	95.00	80.00	-	120.00	95.00	MICRO/MJ	490.09	
VITAMIN B2	MICRO	30.00	28.00	-	32.00	30.00	MICRO/MJ	154.77	

CONSTITUENTS	DIM	AV	VARIATION			AVR	NUTR. DENS.		MOLPERC.
NICOTINAMIDE	MILLI	0.29	0.20	-	0.40	0.29	MILLI/MJ	1.50	
PANTOTHENIC ACID	MILLI	0.23	0.21	-	0.25	0.23	MILLI/MJ	1.19	
VITAMIN B6	MICRO	50.00	20.00	-	80.00	50.00	MICRO/MJ	257.94	
BIOTIN	MICRO	1.40	0.80	-	2.00	1.40	MICRO/MJ	7.22	
FOLIC ACID	MICRO	41.00	22.00	-	66.00	41.00	MICRO/MJ	211.51	
VITAMIN C	MILLI	54.00	42.00	-	69.00	54.00	MILLI/MJ	278.58	
VOLATILE ACID	MILLI	13.00	5.00	-	18.00	13.00	MILLI/MJ	67.07	
MALIC ACID	GRAM	0.17	0.13	-	0.20	0.17	GRAM/MJ	0.88	
CITRIC ACID	GRAM	1.09	0.96	-	1.35	1.09	GRAM/MJ	5.62	
GLUCOSE	GRAM	2.30	1.70	-	3.40	2.30	GRAM/MJ	11.87	
FRUCTOSE	GRAM	2.80	2.30	-	3.60	2.80	GRAM/MJ	14.44	
SUCROSE	GRAM	4.30	3.30	-	5.00	4.30	GRAM/MJ	22.18	
PECTIN	MILLI	86.00	57.00	-	120.00	86.00	MILLI/MJ	443.66	
INOSITOL	GRAM	0.19	0.18	-	0.20	0.19	GRAM/MJ	0.98	
CHOLINE	MILLI	7.20	-	-	-	7.20	MILLI/MJ	37.14	

APFELSINENSAFT (ORANGENSAFT) UNGESÜSST HANDELSWARE	ORANGE JUICE UNSWEETENED COMMERCIAL PRODUCT	JUS D'ORANGES SANS SUCRE PRODUIT DE VENTE

		PROTEIN	FAT	CARBOHYDRATES	TOTAL
ENERGY VALUE (AVERAGE)	KJOULE	11	8.9	170	190
PER 100 G EDIBLE PORTION	(KCAL)	2.6	2.1	41	45
AMOUNT OF DIGESTIBLE CONSTITUENTS PER 100 G	GRAM	0.55	0.20	10.16	
ENERGY VALUE (AVERAGE)	KJOULE	9.1	8.1	170	187
OF THE DIGESTIBLE FRACTION PER 100 G EDIBLE PORTION	(KCAL)	2.2	1.9	41	45

WASTE PERCENTAGE AVERAGE 0.00

CONSTITUENTS	DIM	AV	VARIATION			AVR	NUTR. DENS.		MOLPERC.
WATER	GRAM	87.70	85.70	-	89.20	87.70	GRAM/MJ	468.42	
PROTEIN	GRAM	0.65	0.44	-	1.00	0.65	GRAM/MJ	3.47	
FAT	GRAM	0.23	0.06	-	0.85	0.23	GRAM/MJ	1.23	
AVAILABLE CARBOHYDR.	GRAM	10.16	-	-	-	10.16	GRAM/MJ	54.27	
MINERALS	GRAM	0.38	0.33	-	0.43	0.38	GRAM/MJ	2.03	
SODIUM	MILLI	1.40	0.70	-	2.30	1.40	MILLI/MJ	7.48	
POTASSIUM	MILLI	172.00	124.00	-	199.00	172.00	MILLI/MJ	918.69	
MAGNESIUM	MILLI	12.00	7.00	-	16.00	12.00	MILLI/MJ	64.09	
CALCIUM	MILLI	15.00	9.00	-	25.00	15.00	MILLI/MJ	80.12	
MANGANESE	MICRO	30.00	-	-	-	30.00	MICRO/MJ	160.24	
IRON	MILLI	0.27	0.11	-	0.40	0.27	MILLI/MJ	1.44	
COBALT	MICRO	9.00	-	-	-	9.00	MICRO/MJ	48.07	
COPPER	MICRO	57.00	16.00	-	95.00	57.00	MICRO/MJ	304.45	
ZINC	MILLI	0.12	0.06	-	0.20	0.12	MILLI/MJ	0.64	
CHROMIUM	MICRO	13.00	-	-	-	13.00	MICRO/MJ	69.44	
PHOSPHORUS	MILLI	16.00	14.00	-	19.00	16.00	MILLI/MJ	85.46	
BORON	MILLI	0.11	0.05	-	0.21	0.12	MILLI/MJ	0.61	
CAROTENE	MICRO	74.00	10.00	-	120.00	74.00	MICRO/MJ	395.25	
VITAMIN B1	MICRO	77.00	70.00	-	110.00	77.00	MICRO/MJ	411.27	
VITAMIN B2	MICRO	21.00	13.00	-	32.00	21.00	MICRO/MJ	112.17	
NICOTINAMIDE	MILLI	0.25	0.20	-	0.33	0.25	MILLI/MJ	1.34	
PANTOTHENIC ACID	MILLI	0.16	-	-	-	0.16	MILLI/MJ	0.85	
VITAMIN B6	MICRO	28.00	23.00	-	33.00	28.00	MICRO/MJ	149.55	
BIOTIN	MICRO	0.80	-	-	-	0.80	MICRO/MJ	4.27	
FOLIC ACID	MICRO	35.00	26.00	-	40.00	35.00	MICRO/MJ	186.94	
VITAMIN C	MILLI	44.00	32.00	-	53.00	44.00	MILLI/MJ	235.01	

CONSTITUENTS	DIM	AV	VARIATION			AVR	NUTR. DENS.		MOLPERC.
ALANINE	MILLI	8.50	-	-	-	8.50	MILLI/MJ	45.40	
ARGININE	MILLI	72.10	-	-	-	72.10	MILLI/MJ	385.10	
ASPARTIC ACID	MILLI	29.00	-	-	-	29.00	MILLI/MJ	154.90	
GLUTAMIC ACID	MILLI	11.30	-	-	-	11.30	MILLI/MJ	60.36	
GLYCINE	MILLI	1.30	-	-	-	1.30	MILLI/MJ	6.94	
HISTIDINE	MILLI	1.20	-	-	-	1.20	MILLI/MJ	6.41	
ISOLEUCINE	MILLI	0.60	-	-	-	0.60	MILLI/MJ	3.20	
LEUCINE	MILLI	0.50	-	-	-	0.50	MILLI/MJ	2.67	
LYSINE	MILLI	3.10	-	-	-	3.10	MILLI/MJ	16.56	
METHIONINE	MILLI	0.30	-	-	-	0.30	MILLI/MJ	1.60	
PHENYLALANINE	MILLI	3.00	-	-	-	3.00	MILLI/MJ	16.02	
PROLINE	MILLI	79.00	-	-	-	79.00	MILLI/MJ	421.96	
SERINE	MILLI	12.90	-	-	-	12.90	MILLI/MJ	68.90	
THREONINE	MILLI	1.80	-	-	-	1.80	MILLI/MJ	9.61	
TYROSINE	MILLI	1.10	-	-	-	1.10	MILLI/MJ	5.88	
VALINE	MILLI	1.70	-	-	-	1.70	MILLI/MJ	9.08	
MALIC ACID	GRAM	0.16	0.12	-	0.19	0.16	GRAM/MJ	0.85	
CITRIC ACID	GRAM	1.00	0.70	-	1.20	1.00	GRAM/MJ	5.34	
ISOCITRIC ACID	MILLI	8.80	4.40	-	17.40	8.80	MILLI/MJ	47.00	
GLUCOSE	GRAM	2.50	2.30	-	2.90	2.50	GRAM/MJ	13.35	
FRUCTOSE	GRAM	2.60	2.00	-	3.40	2.60	GRAM/MJ	13.89	
SUCROSE	GRAM	3.90	2.70	-	4.80	3.90	GRAM/MJ	20.83	
PECTIN	MILLI	54.00	28.00	-	83.00	54.00	MILLI/MJ	288.43	

APFELSINEN-DICKSAFT
(ORANGENDICKSAFT, ORANGENKONZENTRAT)

ORANGE JUICE
CONCENTRATE

JUS CONCENTRÉ D'ORANGES

		PROTEIN	FAT	CARBOHYDRATES	TOTAL
ENERGY VALUE (AVERAGE) PER 100 G EDIBLE PORTION	KJOULE	39	57	890	986
	(KCAL)	9.4	14	213	236
AMOUNT OF DIGESTIBLE CONSTITUENTS PER 100 G	GRAM	2.02	1.31	53.20	
ENERGY VALUE (AVERAGE) OF THE DIGESTIBLE FRACTION PER 100 G EDIBLE PORTION	KJOULE	33	51	890	975
	(KCAL)	8.0	12	213	233

WASTE PERCENTAGE AVERAGE 0.00

CONSTITUENTS	DIM	AV	VARIATION		AVR	NUTR. DENS.		MOLPERC.
WATER	GRAM	36.80	34.50	- 38.10	36.80	GRAM/MJ	37.75	
PROTEIN	GRAM	2.38	2.05	- 2.71	2.38	GRAM/MJ	2.44	
FAT	GRAM	1.46	-	- -	1.46	GRAM/MJ	1.50	
AVAILABLE CARBOHYDR.	GRAM	53.20	-	- -	53.20	GRAM/MJ	54.57	
MINERALS	GRAM	2.23	1.51	- 2.94	2.23	GRAM/MJ	2.29	
SODIUM	MILLI	43.00	7.00	- 67.00	43.00	MILLI/MJ	44.11	
POTASSIUM	MILLI	674.00	433.00	- 873.00	674.00	MILLI/MJ	691.34	
MAGNESIUM	MILLI	83.00	28.00	- 137.00	83.00	MILLI/MJ	85.14	
CALCIUM	MILLI	34.00	17.00	- 50.00	34.00	MILLI/MJ	34.87	
CHLORIDE	MILLI	68.00	40.00	- 96.00	68.00	MILLI/MJ	69.75	
CAROTENE	MILLI	-	0.30	- 1.80				
VITAMIN C	MILLI	225.00	112.00	- 364.00	225.00	MILLI/MJ	230.79	
TOTAL ACIDS	GRAM	6.11	3.92	- 8.29	6.11	GRAM/MJ	6.27	
INVERT SUGAR	GRAM	32.30	28.50	- 36.10	32.30	GRAM/MJ	33.13	
SUCROSE	GRAM	14.80	8.76	- 20.90	14.80	GRAM/MJ	15.18	

GRAPEFRUITSAFT FRISCH GEPRESST MUTTERSAFT

GRAPEFRUIT JUICE FRESH ORIGINAL

JUS DE PAMPLEMOUSSE FRAIS ORIGINAL

		PROTEIN	FAT	CARBOHYDRATES	TOTAL
ENERGY VALUE (AVERAGE) PER 100 G EDIBLE PORTION	KJOULE	9.1	3.9	144	157
	(KCAL)	2.2	0.9	34	38
AMOUNT OF DIGESTIBLE CONSTITUENTS PER 100 G	GRAM	0.46	0.09	8.60	
ENERGY VALUE (AVERAGE) OF THE DIGESTIBLE FRACTION PER 100 G EDIBLE PORTION	KJOULE	7.7	3.5	144	155
	(KCAL)	1.8	0.8	34	37

WASTE PERCENTAGE AVERAGE 0.00

CONSTITUENTS	DIM	AV	VARIATION			AVR	NUTR. DENS.		MOLPERC.
WATER	GRAM	89.70	88.30	-	91.50	89.70	GRAM/MJ	578.12	
PROTEIN	GRAM	0.55	0.40	-	0.69	0.55	GRAM/MJ	3.54	
FAT	GRAM	0.10	-	-	-	0.10	GRAM/MJ	0.64	
AVAILABLE CARBOHYDR.	GRAM	8.60	-	-	-	8.60	GRAM/MJ	55.43	
MINERALS	GRAM	0.33	0.25	-	0.44	0.33	GRAM/MJ	2.13	
SODIUM	MILLI	1.20	0.60	-	2.10	1.20	MILLI/MJ	7.73	
POTASSIUM	MILLI	142.00	127.00	-	162.00	142.00	MILLI/MJ	915.20	
MAGNESIUM	MILLI	9.00	7.00	-	12.00	9.00	MILLI/MJ	58.01	
CALCIUM	MILLI	9.60	8.00	-	11.80	9.60	MILLI/MJ	61.87	
MANGANESE	MICRO	8.00	-	-	-	8.00	MICRO/MJ	51.56	
IRON	MILLI	0.22	0.08	-	0.30	0.22	MILLI/MJ	1.42	
COPPER	MICRO	30.00	10.00	-	60.00	30.00	MICRO/MJ	193.35	
PHOSPHORUS	MILLI	14.00	13.00	-	17.00	14.00	MILLI/MJ	90.23	
CHLORIDE	MILLI	3.60	1.90	-	8.20	3.60	MILLI/MJ	23.20	
FLUORIDE	MICRO	15.00	15.00	-	16.00	15.00	MICRO/MJ	96.68	
IODIDE	MICRO	1.00	-	-	-	1.00	MICRO/MJ	6.45	
CAROTENE	MICRO	6.00	-	-	-	6.00	MICRO/MJ	38.67	
VITAMIN B1	MICRO	40.00	-	-	-	40.00	MICRO/MJ	257.80	
VITAMIN B2	MICRO	20.00	-	-	-	20.00	MICRO/MJ	128.90	
NICOTINAMIDE	MILLI	0.20	-	-	-	0.20	MILLI/MJ	1.29	
PANTOTHENIC ACID	MILLI	0.16	-	-	-	0.16	MILLI/MJ	1.03	
VITAMIN B6	MICRO	14.00	-	-	-	14.00	MICRO/MJ	90.23	
BIOTIN	MICRO	0.53	0.36	-	0.70	0.53	MICRO/MJ	3.42	
FOLIC ACID	MICRO	1.00	0.80	-	1.80	1.00	MICRO/MJ	6.45	
VITAMIN C	MILLI	43.00	34.00	-	54.00	43.00	MILLI/MJ	277.14	

CONSTITUENTS	DIM	AV	VARIATION			AVR	NUTR. DENS.		MOLPERC.
ALANINE	MILLI	21.00	21.00	-	21.00	21.00	MILLI/MJ	135.35	7.8
ARGININE	MILLI	40.00	40.00	-	40.00	40.00	MILLI/MJ	257.80	7.6
ASPARTIC ACID	MILLI	103.00	103.00	-	103.00	103.00	MILLI/MJ	663.84	25.7
CYSTINE	MILLI	2.00	2.00	-	2.00	2.00	MILLI/MJ	12.89	0.3
GLUTAMIC ACID	MILLI	41.00	41.00	-	41.00	41.00	MILLI/MJ	264.25	9.2
GLYCINE	MILLI	10.00	10.00	-	10.00	10.00	MILLI/MJ	64.45	4.4
HISTIDINE	MILLI	6.00	6.00	-	6.00	6.00	MILLI/MJ	38.67	1.3
ISOLEUCINE	MILLI	10.00	10.00	-	10.00	10.00	MILLI/MJ	64.45	2.5
LEUCINE	MILLI	15.00	15.00	-	15.00	15.00	MILLI/MJ	96.68	3.8
LYSINE	MILLI	19.00	19.00	-	19.00	19.00	MILLI/MJ	122.46	4.3
METHIONINE	MILLI	3.00	3.00	-	3.00	3.00	MILLI/MJ	19.34	0.7
PHENYLALANINE	MILLI	10.00	10.00	-	10.00	10.00	MILLI/MJ	64.45	2.0
PROLINE	MILLI	46.00	46.00	-	46.00	46.00	MILLI/MJ	296.47	13.2
SERINE	MILLI	26.00	26.00	-	26.00	26.00	MILLI/MJ	167.57	8.2
THREONINE	MILLI	10.00	10.00	-	10.00	10.00	MILLI/MJ	64.45	2.8
TRYPTOPHAN	MILLI	4.00	4.00	-	4.00	4.00	MILLI/MJ	25.78	0.6
TYROSINE	MILLI	7.00	7.00	-	7.00	7.00	MILLI/MJ	45.12	1.3
VALINE	MILLI	15.00	15.00	-	15.00	15.00	MILLI/MJ	96.68	4.2
VOLATILE ACID	MILLI	16.00	15.00	-	20.00	16.00	MILLI/MJ	103.12	
MALIC ACID	MILLI	44.00	20.00	-	60.00	44.00	MILLI/MJ	283.58	
CITRIC ACID	GRAM	1.39	1.16	-	1.74	1.39	GRAM/MJ	8.96	
ISOCITRIC ACID	MILLI	19.40	-	-	-	19.40	MILLI/MJ	125.03	
GLUCOSE	GRAM	2.40	1.90	-	3.10	2.40	GRAM/MJ	15.47	
FRUCTOSE	GRAM	2.30	1.20	-	3.10	2.30	GRAM/MJ	14.82	
SUCROSE	GRAM	2.50	1.50	-	3.40	2.50	GRAM/MJ	16.11	
PENTOSAN	GRAM	1.30	-	-	-	1.30	GRAM/MJ	8.38	

GRAPEFRUITSAFT
HANDELSWARE

GRAPEFRUIT JUICE
COMMERCIAL PRODUCT

JUS DE PAMPLEMOUSSE
PRODUIT DE VENTE

		PROTEIN	FAT	CARBOHYDRATES	TOTAL
ENERGY VALUE (AVERAGE)	KJOULE	8.8	3.9	189	202
PER 100 G	(KCAL)	2.1	0.9	45	48
EDIBLE PORTION					
AMOUNT OF DIGESTIBLE	GRAM	0.45	0.09	11.30	
CONSTITUENTS PER 100 G					
ENERGY VALUE (AVERAGE)	KJOULE	7.4	3.5	189	200
OF THE DIGESTIBLE	(KCAL)	1.8	0.8	45	48
FRACTION PER 100 G					
EDIBLE PORTION					

WASTE PERCENTAGE AVERAGE 0.00

CONSTITUENTS	DIM	AV	VARIATION	AVR	NUTR. DENS.		MOLPERC.
WATER	GRAM	88.50	86.70 - 89.40	88.50	GRAM/MJ	442.36	
PROTEIN	GRAM	0.53	0.46 - 0.59	0.53	GRAM/MJ	2.65	
FAT	GRAM	0.10	- - -	0.10	GRAM/MJ	0.50	
AVAILABLE CARBOHYDR.	GRAM	11.30	- - -	11.30	GRAM/MJ	56.48	
MINERALS	GRAM	0.37	0.30 - 0.45	0.37	GRAM/MJ	1.85	
SODIUM	MILLI	1.30	0.80 - 2.00	1.30	MILLI/MJ	6.50	
POTASSIUM	MILLI	149.00	127.00 - 179.00	149.00	MILLI/MJ	744.76	
MAGNESIUM	MILLI	8.00	7.70 - 11.00	8.00	MILLI/MJ	39.99	
CALCIUM	MILLI	9.30	8.00 - 12.00	9.30	MILLI/MJ	46.49	
MANGANESE	MICRO	10.00	- - -	10.00	MICRO/MJ	49.98	
IRON	MILLI	0.57	0.15 - 1.13	0.57	MILLI/MJ	2.85	
COPPER	MICRO	12.00	10.00 - 14.00	12.00	MICRO/MJ	59.98	
ZINC	MILLI	0.11	0.04 - 0.26	0.11	MILLI/MJ	0.55	
PHOSPHORUS	MILLI	13.00	8.00 - 16.00	13.00	MILLI/MJ	64.98	
CHLORIDE	MILLI	5.50	4.80 - 7.10	5.50	MILLI/MJ	27.49	
IODIDE	MICRO	1.00	- - -	1.00	MICRO/MJ	5.00	
BORON	MICRO	90.00	0.00 - 150.00	90.00	MICRO/MJ	449.86	
CAROTENE	MICRO	6.00	2.00 - 9.00	6.00	MICRO/MJ	29.99	
VITAMIN B1	MICRO	33.00	20.00 - 420.00	33.00	MICRO/MJ	164.95	
VITAMIN B2	MICRO	18.00	6.00 - 30.00	18.00	MICRO/MJ	89.97	
NICOTINAMIDE	MILLI	0.21	0.17 - 0.26	0.21	MILLI/MJ	1.05	
PANTOTHENIC ACID	MILLI	0.15	0.15 - 0.16	0.15	MILLI/MJ	0.75	
VITAMIN B6	MICRO	13.00	12.00 - 14.00	13.00	MICRO/MJ	64.98	
BIOTIN	MICRO	0.52	0.30 - 0.73	0.52	MICRO/MJ	2.60	
FOLIC ACID	MICRO	1.90	1.20 - 2.20	1.90	MICRO/MJ	9.50	
VITAMIN C	MILLI	36.00	31.00 - 43.00	36.00	MILLI/MJ	179.94	
MALIC ACID	MILLI	32.00	26.00 - 42.00	32.00	MILLI/MJ	159.95	
CITRIC ACID	GRAM	1.23	1.02 - 1.63	1.23	GRAM/MJ	6.15	
SALICYLIC ACID	MILLI	0.42	0.42 - 0.42	0.42	MILLI/MJ	2.10	
GLUCOSE	GRAM	4.30	2.60 - 5.60	4.30	GRAM/MJ	21.49	
FRUCTOSE	GRAM	4.20	2.70 - 5.00	4.20	GRAM/MJ	20.99	
SUCROSE	GRAM	1.60	1.10 - 2.10	1.60	GRAM/MJ	8.00	

HIMBEERSAFT
FRISCH GEPRESST
MUTTERSAFT

RASPBERRY JUICE
FRESH
ORIGINAL

JUS DE FRAMBOISES
FRAIS
ORIGINAL

		PROTEIN	FAT	CARBOHYDRATES	TOTAL
ENERGY VALUE (AVERAGE)	KJOULE	5.1	0.0	119	125
PER 100 G	(KCAL)	1.2	0.0	29	30
EDIBLE PORTION					
AMOUNT OF DIGESTIBLE	GRAM	0.26	0.00	7.14	
CONSTITUENTS PER 100 G					
ENERGY VALUE (AVERAGE)	KJOULE	4.4	0.0	119	124
OF THE DIGESTIBLE	(KCAL)	1.0	0.0	29	30
FRACTION PER 100 G					
EDIBLE PORTION					

WASTE PERCENTAGE AVERAGE 0.00

CONSTITUENTS	DIM	AV	VARIATION			AVR	NUTR.	DENS.	MOLPERC.
WATER	GRAM	89.40	88.00	-	91.00	89.40	GRAM/MJ	721.84	
PROTEIN	GRAM	0.31	0.20	-	0.40	0.31	GRAM/MJ	2.50	
FAT	-	0.00	-	-	-	0.00			
AVAILABLE CARBOHYDR.	GRAM	7.14	-	-	-	7.14	GRAM/MJ	57.65	
MINERALS	GRAM	0.45	0.41	-	0.52	0.45	GRAM/MJ	3.63	
SODIUM	MILLI	3.00	1.00	-	5.00	3.00	MILLI/MJ	24.22	
POTASSIUM	MILLI	153.00	131.00	-	178.00	153.00	GRAM/MJ	1.24	
MAGNESIUM	MILLI	16.00	15.00	-	16.00	16.00	MILLI/MJ	129.19	
CALCIUM	MILLI	18.00	6.00	-	27.00	18.00	MILLI/MJ	145.34	
MANGANESE	MILLI	1.70	0.90	-	2.80	1.70	MILLI/MJ	13.73	
IRON	MILLI	2.60	-	-	-	2.60	MILLI/MJ	20.99	
PHOSPHORUS	MILLI	13.00	10.00	-	16.00	13.00	MILLI/MJ	104.97	
CHLORIDE	MILLI	5.00	-	-	-	5.00	MILLI/MJ	40.37	
NITRATE	MILLI	-	0.00	-	1.00				
PHOSPHATE	MILLI	43.00	28.00	-	72.00	43.00	MILLI/MJ	347.19	
SULPHATE	MILLI	13.00	4.00	-	19.00	13.00	MILLI/MJ	104.97	
CAROTENE	MICRO	60.00	-	-	-	60.00	MICRO/MJ	484.46	
VITAMIN B1	MICRO	30.00	-	-	-	30.00	MICRO/MJ	242.23	
VITAMIN C	MILLI	25.00	12.00	-	44.00	25.00	MILLI/MJ	201.86	
MALIC ACID	MILLI	46.00	-	-	-	46.00	MILLI/MJ	371.42	
CITRIC ACID	GRAM	1.61	1.20	-	2.10	1.61	GRAM/MJ	13.00	
ISOCITRIC ACID	MILLI	11.70	4.50	-	20.00	11.70	MILLI/MJ	94.47	
GLUCOSE	GRAM	2.40	-	-	-	2.40	GRAM/MJ	19.38	
FRUCTOSE	GRAM	3.08	-	-	-	3.08	GRAM/MJ	24.87	
SORBITOL	-	0.00	-	-	-	0.00			
EXTRACT SUGAR FREE	GRAM	4.40	3.20	-	5.00	4.40	GRAM/MJ	35.53	

HIMBEERSIRUP RASPBERRY SYRUP SIROP DE FRAMBOISES

		PROTEIN	FAT	CARBOHYDRATES	TOTAL
ENERGY VALUE (AVERAGE)	KJOULE	0	0.0	1101	1118
PER 100 G	(KCAL)	0.0	0.0	263	267
EDIBLE PORTION					
AMOUNT OF DIGESTIBLE	GRAM	0.00	0.00	65.80	
CONSTITUENTS PER 100 G					
ENERGY VALUE (AVERAGE)	KJOULE	0.0	0.0	1101	1101
OF THE DIGESTIBLE	(KCAL)	0.0	0.0	263	263
FRACTION PER 100 G					
EDIBLE PORTION					

WASTE PERCENTAGE AVERAGE 0.00

CONSTITUENTS	DIM	AV	VARIATION			AVR	NUTR. DENS.		MOLPERC.
WATER	GRAM	31.30	30.60	-	32.10	31.30	GRAM/MJ	28.42	
PROTEIN	-	TRACES	-	-	-				
FAT	-	0.00	-	-	-	0.00			
AVAILABLE CARBOHYDR.	GRAM	65.80	-	-	-	65.80	GRAM/MJ	59.75	
MINERALS	GRAM	0.16	0.13	-	0.21	0.16	GRAM/MJ	0.15	
SODIUM	MILLI	1.60	0.60	-	4.20	1.60	MILLI/MJ	1.45	
POTASSIUM	MILLI	90.00	63.00	-	122.00	90.00	MILLI/MJ	81.73	
MAGNESIUM	MILLI	7.00	6.00	-	8.00	7.00	MILLI/MJ	6.36	
CALCIUM	MILLI	16.00	11.00	-	20.00	16.00	MILLI/MJ	14.53	
IRON	MILLI	2.00	-	-	-	2.00	MILLI/MJ	1.82	
COPPER	MILLI	1.00	-	-	-	1.00	MILLI/MJ	0.91	
PHOSPHORUS	MILLI	15.00	10.00	-	18.00	15.00	MILLI/MJ	13.62	
CHLORIDE	MILLI	22.00	-	-	-	22.00	MILLI/MJ	19.98	
VITAMIN B1	MICRO	60.00	-	-	-	60.00	MICRO/MJ	54.48	
VITAMIN B2	MICRO	3.00	-	-	-	3.00	MICRO/MJ	2.72	
NICOTINAMIDE	MILLI	0.16	-	-	-	0.16	MILLI/MJ	0.15	
PANTOTHENIC ACID	MICRO	30.00	-	-	-	30.00	MICRO/MJ	27.24	
VITAMIN B6	MICRO	32.00	-	-	-	32.00	MICRO/MJ	29.06	
BIOTIN	MICRO	-	1.00	-	2.00				
VITAMIN C	MILLI	16.00	12.00	-	21.00	16.00	MILLI/MJ	14.53	
REDUCING SUGAR	GRAM	43.40	22.30	-	58.90	43.40	GRAM/MJ	39.41	
SUCROSE	GRAM	22.40	7.20	-	41.60	22.40	GRAM/MJ	20.34	
INOSITOL	MILLI	8.00	-	-	-	8.00	MILLI/MJ	7.26	

HOLUNDERBEER-SAFT
MUTTERSAFT

ELDERBERRY JUICE
ORIGINAL

JUS DE BAIES DE SUREAU
ORIGINAL

		PROTEIN	FAT	CARBOHYDRATES	TOTAL
ENERGY VALUE (AVERAGE)	KJOULE	33	0.0	126	159
PER 100 G EDIBLE PORTION	(KCAL)	8.0	0.0	30	38
AMOUNT OF DIGESTIBLE CONSTITUENTS PER 100 G	GRAM	1.71	0.00	7.50	
ENERGY VALUE (AVERAGE)	KJOULE	28	0.0	126	154
OF THE DIGESTIBLE FRACTION PER 100 G EDIBLE PORTION	(KCAL)	6.8	0.0	30	37

WASTE PERCENTAGE AVERAGE 0.00

CONSTITUENTS	DIM	AV	VARIATION			AVR	NUTR. DENS.		MOLPERC.
WATER	GRAM	86.50	-	-	-	86.50	GRAM/MJ	562.07	
PROTEIN	GRAM	2.02	1.60	-	2.25	2.02	GRAM/MJ	13.13	
AVAILABLE CARBOHYDR.	GRAM	7.50 1	-	-	-	7.50	GRAM/MJ	48.73	
MINERALS	GRAM	0.80	0.28	-	1.03	0.80	GRAM/MJ	5.20	
SODIUM	MILLI	0.50	-	-	-	0.50	MILLI/MJ	3.25	
POTASSIUM	MILLI	288.00	-	-	-	288.00	GRAM/MJ	1.87	
CALCIUM	MILLI	5.00	-	-	-	5.00	MILLI/MJ	32.49	
PHOSPHORUS	MILLI	49.00	34.00	-	66.00	49.00	MILLI/MJ	318.40	
VITAMIN B1	MICRO	30.00	-	-	-	30.00	MICRO/MJ	194.94	
VITAMIN B2	MICRO	60.00	-	-	-	60.00	MICRO/MJ	389.87	
NICOTINAMIDE	MILLI	0.43	-	-	-	0.43	MILLI/MJ	2.79	
PANTOTHENIC ACID	MILLI	0.21	-	-	-	0.21	MILLI/MJ	1.36	
VITAMIN B6	MICRO	90.00	-	-	-	90.00	MICRO/MJ	584.81	
BIOTIN	MICRO	0.70	-	-	-	0.70	MICRO/MJ	4.55	
FOLIC ACID	MICRO	6.00	-	-	-	6.00	MICRO/MJ	38.99	
VITAMIN C	MILLI	26.00	-	-	-	26.00	MILLI/MJ	168.94	
TOTAL ACIDS	GRAM	0.99	0.55	-	1.30	0.99	GRAM/MJ	6.43	
SUCROSE	MILLI	60.00	50.00	-	90.00	60.00	MILLI/MJ	389.87	
SORBITOL	MILLI	22.00	17.00	-	28.00	22.00	MILLI/MJ	142.95	

1 VALUE FROM THE RAW PRODUCT

JOHANNISBEER-NEKTAR
ROT
HANDELSWARE

RED CURRANT NECTAR
COMMERCIAL PRODUCT

NECTAR DE GROSSEILLES ROUGES
PRODUIT DE VENTE

		PROTEIN	FAT	CARBOHYDRATES	TOTAL
ENERGY VALUE (AVERAGE) PER 100 G EDIBLE PORTION	KJOULE	6.6	0.0	221	228
	(KCAL)	1.6	0.0	53	54
AMOUNT OF DIGESTIBLE CONSTITUENTS PER 100 G	GRAM	0.34	0.00	13.21	
ENERGY VALUE (AVERAGE) OF THE DIGESTIBLE FRACTION PER 100 G EDIBLE PORTION	KJOULE	5.6	0.0	221	227
	(KCAL)	1.3	0.0	53	54

WASTE PERCENTAGE AVERAGE 0.00

CONSTITUENTS	DIM	AV	VARIATION			AVR	NUTR. DENS.		MOLPERC.
WATER	GRAM	85.60	83.30	-	87.40	85.60	GRAM/MJ	377.59	
PROTEIN	GRAM	0.40	-	-	-	0.40	GRAM/MJ	1.76	
FAT	-	TRACES	-	-	-				
AVAILABLE CARBOHYDR.	GRAM	13.21	12.26	-	14.64	13.21	GRAM/MJ	58.27	
MINERALS	GRAM	0.29	0.22	-	0.35	0.29	GRAM/MJ	1.28	
SODIUM	-	TRACES	-	-	-				
POTASSIUM	MILLI	110.00	-	-	-	110.00	MILLI/MJ	485.22	
CALCIUM	MILLI	7.00	-	-	-	7.00	MILLI/MJ	30.88	
IRON	MILLI	0.34	-	-	-	0.34	MILLI/MJ	1.50	
COPPER	MICRO	20.00	-	-	-	20.00	MICRO/MJ	88.22	
PHOSPHORUS	MILLI	7.10	6.40	-	7.80	7.10	MILLI/MJ	31.32	
CAROTENE	MICRO	24.00	-	-	-	24.00	MICRO/MJ	105.87	
VITAMIN B1	MICRO	2.00	-	-	-	2.00	MICRO/MJ	8.82	
VITAMIN B2	MICRO	2.00	-	-	-	2.00	MICRO/MJ	8.82	
PANTOTHENIC ACID	-	TRACES	-	-	-				
VITAMIN B6	-	TRACES	-	-	-				
BIOTIN	-	TRACES	-	-	-				
FOLIC ACID	-	TRACES	-	-	-				
VITAMIN C	MILLI	6.00	3.00	-	10.00	6.00	MILLI/MJ	26.47	
VOLATILE ACID	MILLI	12.00	5.00	-	25.00	12.00	MILLI/MJ	52.93	
MALIC ACID	GRAM	0.63	0.63	-	0.65	0.64	GRAM/MJ	2.81	
CITRIC ACID	GRAM	0.20	0.19	-	0.22	0.21	GRAM/MJ	0.92	
LACTIC ACID	MILLI	6.00	4.00	-	7.00	6.00	MILLI/MJ	26.47	
GLUCOSE	GRAM	2.65	2.34	-	2.84	2.66	GRAM/MJ	11.72	
FRUCTOSE	GRAM	2.87	2.53	-	3.07	2.87	GRAM/MJ	12.68	
SUCROSE	GRAM	6.83	6.54	-	7.34	6.83	GRAM/MJ	30.13	
SORBITOL	MILLI	28.00	28.00	-	28.00	28.00	MILLI/MJ	123.51	
ETHANOL	GRAM	0.14	0.07	-	0.27	0.14	GRAM/MJ	0.62	

JOHANNISBEER-NEKTAR
SCHWARZ
HANDELSWARE

BLACK CURRANT NECTAR
COMMERCIAL PRODUCT

NECTAR DE CASSIS
PRODUIT DE VENTE

		PROTEIN	FAT	CARBOHYDRATES	TOTAL
ENERGY VALUE (AVERAGE) PER 100 G EDIBLE PORTION	KJOULE	7.1	0.0	224	232
	(KCAL)	1.7	0.0	54	55
AMOUNT OF DIGESTIBLE CONSTITUENTS PER 100 G	GRAM	0.36	0.00	13.41	
ENERGY VALUE (AVERAGE) OF THE DIGESTIBLE FRACTION PER 100 G EDIBLE PORTION	KJOULE	6.0	0.0	224	230
	(KCAL)	1.4	0.0	54	55

WASTE PERCENTAGE AVERAGE 0.00

CONSTITUENTS	DIM	AV	VARIATION			AVR	NUTR. DENS.		MOLPERC.
WATER	GRAM	85.30	83.10	-	87.30	85.30	GRAM/MJ	370.11	
PROTEIN	GRAM	0.43	-	-	-	0.43	GRAM/MJ	1.87	
FAT	-	TRACES	-	-	-				
AVAILABLE CARBOHYDR.	GRAM	13.41	12.69	-	13.98	13.41	GRAM/MJ	58.19	
MINERALS	GRAM	0.27	0.24	-	0.34	0.27	GRAM/MJ	1.17	
SODIUM	MILLI	5.00	2.00	-	7.00	5.00	MILLI/MJ	21.69	
POTASSIUM	MILLI	98.00	86.00	-	110.00	98.00	MILLI/MJ	425.22	
CALCIUM	MILLI	15.00	14.00	-	15.00	15.00	MILLI/MJ	65.08	
IRON	MILLI	0.30	0.27	-	0.34	0.30	MILLI/MJ	1.30	
COPPER	MICRO	20.00	-	-	-	20.00	MICRO/MJ	86.78	
PHOSPHORUS	MILLI	10.00	-	-	-	10.00	MILLI/MJ	43.39	
CAROTENE	MICRO	24.00	-	-	-	24.00	MICRO/MJ	104.13	
VITAMIN B1	MICRO	5.00	-	-	-	5.00	MICRO/MJ	21.69	
VITAMIN B2	MICRO	2.00	-	-	-	2.00	MICRO/MJ	8.68	
NICOTINAMIDE	MICRO	30.00	-	-	-	30.00	MICRO/MJ	130.17	
PANTOTHENIC ACID	-	TRACES	-	-	-				
VITAMIN B6	-	TRACES	-	-	-				
BIOTIN	-	TRACES	-	-	-				
FOLIC ACID	-	TRACES	-	-	-				
VITAMIN C	MILLI	30.00	20.00	-	48.00	30.00	MILLI/MJ	130.17	
MALIC ACID	MILLI	79.43	61.00	-	90.00	79.43	MILLI/MJ	344.65	
CITRIC ACID	GRAM	0.85	0.76	-	0.89	0.85	GRAM/MJ	3.70	
LACTIC ACID	MILLI	5.00	2.00	-	11.00	5.00	MILLI/MJ	21.69	
ISOCITRIC ACID	MILLI	5.40	3.90	-	7.10	5.40	MILLI/MJ	23.43	
GLUCOSE	GRAM	4.54	4.49	-	4.96	4.55	GRAM/MJ	19.74	
FRUCTOSE	GRAM	4.64	4.54	-	4.66	4.65	GRAM/MJ	20.15	
SUCROSE	GRAM	3.28	2.83	-	3.35	3.29	GRAM/MJ	14.25	
SORBITOL	MILLI	22.10	7.00	-	28.00	22.10	MILLI/MJ	95.90	
ETHANOL	GRAM	0.20	0.18	-	0.21	0.20	GRAM/MJ	0.87	

MANDARINENSAFT
FRISCH GEPRESST
MUTTERSAFT

MANDARIN JUICE
FRESH
ORIGINAL

JUS DE MANDARINE
FRAIS
ORIGINAL

		PROTEIN	FAT	CARBOHYDRATES	TOTAL
ENERGY VALUE (AVERAGE) PER 100 G EDIBLE PORTION	KJOULE	15	12	169	195
	(KCAL)	3.6	2.8	40	47
AMOUNT OF DIGESTIBLE CONSTITUENTS PER 100 G	GRAM	0.76	0.27	10.09	
ENERGY VALUE (AVERAGE) OF THE DIGESTIBLE FRACTION PER 100 G EDIBLE PORTION	KJOULE	13	11	169	192
	(KCAL)	3.0	2.5	40	46

WASTE PERCENTAGE AVERAGE 0.00

CONSTITUENTS	DIM	AV	VARIATION			AVR	NUTR. DENS.		MOLPERC.
WATER	GRAM	89.20	-	-	-	89.20	GRAM/MJ	464.55	
PROTEIN	GRAM	0.90	-	-	-	0.90	GRAM/MJ	4.69	
FAT	GRAM	0.30	-	-	-	0.30	GRAM/MJ	1.56	
AVAILABLE CARBOHYDR.	GRAM	10.09	-	-	-	10.09	GRAM/MJ	52.55	
MINERALS	GRAM	0.31	0.23	-	0.40	0.31	GRAM/MJ	1.61	
CALCIUM	MILLI	19.00	-	-	-	19.00	MILLI/MJ	98.95	
IRON	MILLI	0.20	-	-	-	0.20	MILLI/MJ	1.04	
PHOSPHORUS	MILLI	16.00	-	-	-	16.00	MILLI/MJ	83.33	
CAROTENE	MILLI	0.25	-	-	-	0.25	MILLI/MJ	1.30	
VITAMIN B1	MICRO	70.00	-	-	-	70.00	MICRO/MJ	364.55	
VITAMIN B2	MICRO	30.00	-	-	-	30.00	MICRO/MJ	156.24	
NICOTINAMIDE	MILLI	0.20	-	-	-	0.20	MILLI/MJ	1.04	
VITAMIN C	MILLI	32.00	31.00	-	32.00	32.00	MILLI/MJ	166.65	
CITRIC ACID	GRAM	0.50	0.36	-	0.76	0.50	GRAM/MJ	2.60	
GLUCOSE	GRAM	1.55	1.55	-	1.55	1.55	GRAM/MJ	8.07	
FRUCTOSE	GRAM	3.02	3.02	-	3.02	3.02	GRAM/MJ	15.73	
SUCROSE	GRAM	5.02	5.02	-	5.02	5.02	GRAM/MJ	26.14	

PASSIONSFRUCHT-SAFT
FRISCH
MUTTERSAFT

GRANADILLA JUICE
FRESH
ORIGINAL

JUS DE GRENADILLE
FRAIS
ORIGINAL

		PROTEIN	FAT	CARBOHYDRATES	TOTAL
ENERGY VALUE (AVERAGE)	KJOULE	13	16	226	255
PER 100 G	(KCAL)	3.2	3.7	54	61
EDIBLE PORTION					
AMOUNT OF DIGESTIBLE	GRAM	0.68	0.36	13.50	
CONSTITUENTS PER 100 G					
ENERGY VALUE (AVERAGE)	KJOULE	11	14	226	251
OF THE DIGESTIBLE	(KCAL)	2.7	3.3	54	60
FRACTION PER 100 G					
EDIBLE PORTION					

WASTE PERCENTAGE AVERAGE 0.00

CONSTITUENTS	DIM	AV	VARIATION			AVR	NUTR.	DENS.	MOLPERC.
WATER	GRAM	82.00	77.00	-	85.00	82.00	GRAM/MJ	326.46	
PROTEIN	GRAM	0.80	0.60	-	1.20	0.80	GRAM/MJ	3.18	
FAT	GRAM	0.40	0.20	-	0.60	0.40	GRAM/MJ	1.59	
AVAILABLE CARBOHYDR.	GRAM	13.50	-	-	-	13.50	GRAM/MJ	53.75	
MINERALS	GRAM	0.50	0.36	-	0.77	0.50	GRAM/MJ	1.99	
SODIUM	MILLI	0.80	0.60	-	1.00	0.80	MILLI/MJ	3.18	
POTASSIUM	MILLI	215.00	204.00	-	227.00	215.00	MILLI/MJ	855.95	
CALCIUM	MILLI	8.60	4.00	-	14.00	8.60	MILLI/MJ	34.24	
IRON	MILLI	0.30	-	-	-	0.30	MILLI/MJ	1.19	
PHOSPHORUS	MILLI	20.00	8.00	-	60.00	20.00	MILLI/MJ	79.62	
CAROTENE	MILLI	1.50	1.30	-	1.70	1.50	MILLI/MJ	5.97	
VITAMIN B1	MICRO	20.00	-	-	-	20.00	MICRO/MJ	79.62	
VITAMIN B2	MILLI	0.11	-	-	-	0.11	MILLI/MJ	0.44	
NICOTINAMIDE	MILLI	2.00	-	-	-	2.00	MILLI/MJ	7.96	
VITAMIN C	MILLI	30.00	20.00	-	40.00	30.00	MILLI/MJ	119.44	
SEROTONINE	MILLI	2.50	1.00	-	4.00	2.50	MILLI/MJ	9.95	
TOTAL ACIDS	GRAM	3.45	3.20	-	3.70	3.45	GRAM/MJ	13.74	
GLUCOSE	GRAM	3.77	3.59	-	3.94	3.77	GRAM/MJ	15.01	
FRUCTOSE	GRAM	3.14	3.04	-	3.24	3.14	GRAM/MJ	12.50	
SUCROSE	GRAM	3.13	2.30	-	7.90	3.13	GRAM/MJ	12.46	
PECTIN	GRAM	0.23	0.20	-	0.25	0.23	GRAM/MJ	0.92	

SANDDORNBEEREN-SAFT | SEA BUCKTHORN JUICE | JUS DE BAIES D'ARGOUSIER

		PROTEIN	FAT	CARBOHYDRATES	TOTAL
ENERGY VALUE (AVERAGE)	KJOULE	16	91	80	187
PER 100 G	(KCAL)	3.7	22	19	45
EDIBLE PORTION					
AMOUNT OF DIGESTIBLE	GRAM	0.79	2.09	4.80	
CONSTITUENTS PER 100 G					
ENERGY VALUE (AVERAGE)	KJOULE	13	82	80	175
OF THE DIGESTIBLE	(KCAL)	3.2	20	19	42
FRACTION PER 100 G					
EDIBLE PORTION					

WASTE PERCENTAGE AVERAGE 0.00

CONSTITUENTS	DIM	AV	VARIATION			AVR	NUTR. DENS.		MOLPERC.
WATER	GRAM	91.50	-	-	-	91.50	GRAM/MJ	522.46	
PROTEIN	GRAM	0.94	0.93	-	0.96	0.94	GRAM/MJ	5.37	
FAT	GRAM	2.33	2.00	-	2.93	2.33	GRAM/MJ	13.30	
AVAILABLE CARBOHYDR.	GRAM	4.80 [1]	-	-	-	4.80	GRAM/MJ	27.41	
MINERALS	GRAM	0.46	0.36	-	0.56	0.46	GRAM/MJ	2.63	
SODIUM	MILLI	6.00	-	-	-	6.00	MILLI/MJ	34.26	
POTASSIUM	MILLI	209.00	-	-	-	209.00	GRAM/MJ	1.19	
CALCIUM	MILLI	9.00	-	-	-	9.00	MILLI/MJ	51.39	
PHOSPHORUS	MILLI	-	10.00	-	140.00				
VITAMIN C	MILLI	266.00	111.00	-	664.00	266.00	GRAM/MJ	1.52	
MALIC ACID	GRAM	3.57	2.95	-	4.15	3.57	GRAM/MJ	20.38	
INVERT SUGAR	GRAM	-	0.09	-	1.07				
SUCROSE	GRAM	-	0.00	-	0.15				

1 ESTIMATED BY THE DIFFERENCE METHOD
100 - (WATER + PROTEIN + FAT + MINERALS)

SAUERKIRSCHSAFT
MUTTERSAFT

CHERRY MORELLO JUICE
ORIGINAL

JUS DE GRIOTTE
ORIGINAL

		PROTEIN	FAT	CARBOHYDRATES	TOTAL
ENERGY VALUE (AVERAGE)	KJOULE	5.0	0	233	276
PER 100 G	(KCAL)	1.2	0.0	56	66
EDIBLE PORTION					
AMOUNT OF DIGESTIBLE	GRAM	0.25	0.00	13.90	
CONSTITUENTS PER 100 G					
ENERGY VALUE (AVERAGE)	KJOULE	4.2	0.0	233	237
OF THE DIGESTIBLE	(KCAL)	1.0	0.0	56	57
FRACTION PER 100 G					
EDIBLE PORTION					

WASTE PERCENTAGE AVERAGE 0.00

CONSTITUENTS	DIM	AV	VARIATION	AVR	NUTR. DENS.		MOLPERC.
WATER	GRAM	85.60	- - -	85.60	GRAM/MJ	361.42	
PROTEIN	GRAM	0.30	0.20 - 0.40	0.30	GRAM/MJ	1.27	
AVAILABLE CARBOHYDR.	GRAM	13.90	11.50 - 17.00	13.90	GRAM/MJ	58.69	
MINERALS	GRAM	0.50	0.40 - 0.70	0.50	GRAM/MJ	2.11	
SODIUM	MILLI	1.00	0.80 - 1.20	1.00	MILLI/MJ	4.22	
POTASSIUM	MILLI	201.00	177.00 - 352.00	201.00	MILLI/MJ	848.66	
MAGNESIUM	MILLI	13.00	12.00 - 15.00	13.00	MILLI/MJ	54.89	
CALCIUM	MILLI	15.00	14.00 - 18.00	15.00	MILLI/MJ	63.33	
PHOSPHORUS	MILLI	17.00	15.00 - 25.00	17.00	MILLI/MJ	71.78	
MALIC ACID	GRAM	2.10	1.35 - 2.80	2.10	GRAM/MJ	8.87	
CITRIC ACID	MILLI	8.00	5.00 - 11.00	8.00	MILLI/MJ	33.78	
ISOCITRIC ACID	MILLI	3.50	3.00 - 5.00	3.50	MILLI/MJ	14.78	
GLUCOSE	GRAM	6.50	5.70 - 7.70	6.50	GRAM/MJ	27.44	
FRUCTOSE	GRAM	5.30	4.40 - 6.50	5.30	GRAM/MJ	22.38	
SUCROSE	-	0.00	- - -	0.00			

TRAUBENSAFT
HANDELSWARE

GRAPE JUICE
COMMERCIAL PRODUCT

JUS DE RAISINS
PRODUIT DE VENTE

		PROTEIN	FAT	CARBOHYDRATES	TOTAL
ENERGY VALUE (AVERAGE)	KJOULE	3.5	0.0	284	287
PER 100 G	(KCAL)	0.8	0.0	68	69
EDIBLE PORTION					
AMOUNT OF DIGESTIBLE	GRAM	0.17	0.00	16.97	
CONSTITUENTS PER 100 G					
ENERGY VALUE (AVERAGE)	KJOULE	3.0	0.0	284	287
OF THE DIGESTIBLE	(KCAL)	0.7	0.0	68	69
FRACTION PER 100 G					
EDIBLE PORTION					

WASTE PERCENTAGE AVERAGE 0.00

CONSTITUENTS	DIM	AV	VARIATION			AVR	NUTR. DENS.		MOLPERC.
WATER	GRAM	81.90	80.30	-	83.50	81.90	GRAM/MJ	285.41	
PROTEIN	GRAM	0.21	0.12	-	0.30	0.21	GRAM/MJ	0.73	
FAT	-	0.00	-	-	-	0.00			
AVAILABLE CARBOHYDR.	GRAM	16.97	-	-	-	16.97	GRAM/MJ	59.14	
MINERALS	GRAM	0.33	0.25	-	0.42	0.33	GRAM/MJ	1.15	
SODIUM	MILLI	2.60	0.90	-	4.30	2.60	MILLI/MJ	9.06	
POTASSIUM	MILLI	148.00	115.00	-	198.00	148.00	MILLI/MJ	515.75	
MAGNESIUM	MILLI	8.80	4.00	-	12.00	8.80	MILLI/MJ	30.67	
CALCIUM	MILLI	13.00	8.00	-	24.00	13.00	MILLI/MJ	45.30	
MANGANESE	MICRO	50.00	29.00	-	62.00	50.00	MICRO/MJ	174.24	
IRON	MILLI	0.43	0.30	-	0.64	0.43	MILLI/MJ	1.50	
COBALT	MICRO	1.00	0.60	-	1.30	1.00	MICRO/MJ	3.48	
COPPER	MICRO	48.00	9.00	-	100.00	48.00	MICRO/MJ	167.27	
ZINC	MICRO	40.00	-	-	-	40.00	MICRO/MJ	139.39	
NICKEL	MICRO	4.40	-	-	-	4.40	MICRO/MJ	15.33	
CHROMIUM	MICRO	3.00	0.00	-	6.00	3.00	MICRO/MJ	10.45	
MOLYBDENUM	MICRO	4.50	-	-	-	4.50	MICRO/MJ	15.68	
PHOSPHORUS	MILLI	12.00	7.00	-	23.00	12.00	MILLI/MJ	41.82	
CHLORIDE	MILLI	3.30	2.00	-	5.80	3.30	MILLI/MJ	11.50	
FLUORIDE	MICRO	10.00	8.00	-	11.00	10.00	MICRO/MJ	34.85	
IODIDE	MICRO	0.48	-	-	-	0.48	MICRO/MJ	1.67	
BORON	MILLI	0.39	0.23	-	0.49	0.40	MILLI/MJ	1.38	
SELENIUM	MICRO	4.00	-	-	-	4.00	MICRO/MJ	13.94	
CAROTENE	MILLI	-	0.00	-	0.04				
VITAMIN B1	MICRO	31.00	12.00	-	40.00	31.00	MICRO/MJ	108.03	
VITAMIN B2	MICRO	16.00	3.00	-	50.00	16.00	MICRO/MJ	55.76	

CONSTITUENTS	DIM	AV	VARIATION			AVR	NUTR. DENS.		MOLPERC.
NICOTINAMIDE	MILLI	0.18	0.14	-	0.21	0.18	MILLI/MJ	0.63	
PANTOTHENIC ACID	MICRO	49.00	38.00	-	74.00	49.00	MICRO/MJ	170.76	
VITAMIN B6	MICRO	22.00	11.00	-	39.00	22.00	MICRO/MJ	76.67	
BIOTIN	MICRO	1.20	1.00	-	1.60	1.20	MICRO/MJ	4.18	
FOLIC ACID	MICRO	-	0.20	-	3.00				
VITAMIN C	MILLI	1.70	0.80	-	2.90	1.70	MILLI/MJ	5.92	
CADAVERINE	MICRO	9.00	4.00	-	16.00	9.00	MICRO/MJ	31.36	
HISTAMINE	-	TRACES	-	-	-				
PUTRESCINE	MICRO	70.00	30.00	-	120.00	70.00	MICRO/MJ	243.94	
SPERMIDINE	MILLI	0.27	0.24	-	0.30	0.27	MILLI/MJ	0.94	
SPERMINE	MICRO	43.00	-	-	-	43.00	MICRO/MJ	149.85	
MALIC ACID	MILLI	350.00	280.00	-	470.00	350.00	GRAM/MJ	1.22	
CITRIC ACID	MILLI	22.00	15.00	-	39.00	22.00	MILLI/MJ	76.67	
TARTARIC ACID	MILLI	450.00	380.00	-	530.00	450.00	GRAM/MJ	1.57	
LACTIC ACID	MILLI	8.00	2.00	-	17.00	8.00	MILLI/MJ	27.88	
ISOCITRIC ACID	MILLI	5.60	-	-	-	5.60	MILLI/MJ	19.51	
SALICYLIC ACID	MILLI	0.53	0.53	-	0.53	0.53	MILLI/MJ	1.85	
GLUCOSE	GRAM	8.10	-	-	-	8.10	GRAM/MJ	28.23	
FRUCTOSE	GRAM	8.30	-	-	-	8.30	GRAM/MJ	28.92	
SUCROSE	GRAM	0.20	-	-	-	0.20	GRAM/MJ	0.70	
PECTIN	GRAM	0.20	0.03	-	0.37	0.20	GRAM/MJ	0.70	
ETHANOL	GRAM	0.17	0.01	-	0.37	0.17	GRAM/MJ	0.59	

ZITRONENSAFT	LEMON JUICE	JUS DE CITRON
FRISCH GEPRESST	FRESH	FRAIS
MUTTERSAFT	ORIGINAL	ORIGINAL

		PROTEIN	FAT	CARBOHYDRATES	TOTAL
ENERGY VALUE (AVERAGE)	KJOULE	6.6	3.9	120	130
PER 100 G	(KCAL)	1.6	0.9	29	31
EDIBLE PORTION					
AMOUNT OF DIGESTIBLE	GRAM	0.34	0.09	7.16	
CONSTITUENTS PER 100 G					
ENERGY VALUE (AVERAGE)	KJOULE	5.6	3.5	120	129
OF THE DIGESTIBLE	(KCAL)	1.3	0.8	29	31
FRACTION PER 100 G					
EDIBLE PORTION					

WASTE PERCENTAGE AVERAGE 0.00

CONSTITUENTS	DIM	AV	VARIATION		AVR	NUTR. DENS.		MOLPERC.
WATER	GRAM	91.00	89.70	- 92.80	91.00	GRAM/MJ	705.70	
PROTEIN	GRAM	0.40	0.30	- 0.50	0.40	GRAM/MJ	3.10	
FAT	GRAM	0.10	0.00	- 0.20	0.10	GRAM/MJ	0.78	
AVAILABLE CARBOHYDR.	GRAM	7.16	-	- -	7.16	GRAM/MJ	55.53	
MINERALS	GRAM	0.34	0.30	- 0.43	0.34	GRAM/MJ	2.64	
SODIUM	MILLI	1.00	0.50	- 1.50	1.00	MILLI/MJ	7.75	
POTASSIUM	MILLI	138.00	126.00	- 146.00	138.00	GRAM/MJ	1.07	
MAGNESIUM	MILLI	10.00	8.20	- 11.00	10.00	MILLI/MJ	77.55	
CALCIUM	MILLI	11.00	5.60	- 17.00	11.00	MILLI/MJ	85.30	
MANGANESE	MILLI	-	0.00	- 0.05				
IRON	MILLI	0.14	0.10	- 0.20	0.14	MILLI/MJ	1.09	
COPPER	MILLI	0.20	-	- -	0.20	MILLI/MJ	1.55	
ZINC	MILLI	-	0.03	- 0.20				
PHOSPHORUS	MILLI	11.00	9.20	- 14.00	11.00	MILLI/MJ	85.30	
CHLORIDE	MILLI	5.40	3.60	- 8.50	5.40	MILLI/MJ	41.88	
IODIDE	MICRO	5.20	-	- -	5.20	MICRO/MJ	40.33	
BORON	MICRO	24.00	21.00	- 27.00	24.00	MICRO/MJ	186.12	
CAROTENE	MICRO	46.00	43.00	- 50.00	46.00	MICRO/MJ	356.72	
VITAMIN B1	MICRO	40.00	30.00	- 60.00	40.00	MICRO/MJ	310.20	
VITAMIN B2	MICRO	10.00	-	- -	10.00	MICRO/MJ	77.55	
NICOTINAMIDE	MILLI	0.10	-	- -	0.10	MILLI/MJ	0.78	
PANTOTHENIC ACID	MILLI	0.10	0.10	- 0.10	0.10	MILLI/MJ	0.78	
VITAMIN B6	MICRO	52.00	52.00	- 52.00	52.00	MICRO/MJ	403.25	
BIOTIN	MICRO	0.30	0.30	- 0.30	0.30	MICRO/MJ	2.33	
FOLIC ACID	MICRO	0.90	-	- -	0.90	MICRO/MJ	6.98	
VITAMIN C	MILLI	53.00	46.00	- 62.00	53.00	MILLI/MJ	411.01	
MALIC ACID	GRAM	0.25	-	- -	0.25	GRAM/MJ	1.94	
CITRIC ACID	GRAM	4.50	-	- -	4.50	GRAM/MJ	34.90	
GLUCOSE	GRAM	1.00	0.52	- 1.49	1.00	GRAM/MJ	7.76	
FRUCTOSE	GRAM	1.02	0.80	- 1.44	1.03	GRAM/MJ	7.98	
SUCROSE	GRAM	0.39	0.20	- 0.51	0.40	GRAM/MJ	3.09	
INOSITOL	MILLI	66.00	-	- -	66.00	MILLI/MJ	511.82	

Obst- und Beeren-Marmeladen

Jams from fruits and berries
Confitures de fruits et de baies

APFELGELEE APPLE JELLY GELÉE DE POMMES

		PROTEIN	FAT	CARBOHYDRATES	TOTAL
ENERGY VALUE (AVERAGE)	KJOULE	0.0	0.0		1086
PER 100 G	(KCAL)	0.0	0.0		259
EDIBLE PORTION					
AMOUNT OF DIGESTIBLE	GRAM	0.00	0.00	64.87	
CONSTITUENTS PER 100 G					
ENERGY VALUE (AVERAGE)	KJOULE	0.0	0.0		1086
OF THE DIGESTIBLE	(KCAL)	0.0	0.0		259
FRACTION PER 100 G					
EDIBLE PORTION					

WASTE PERCENTAGE AVERAGE 0.00

CONSTITUENTS	DIM	AV	VARIATION	AVR	NUTR. DENS.	MOLPERC.
WATER	GRAM	35.00	- - -	35.00		
AVAILABLE CARBOHYDR.	GRAM	64.87 1	- - -	64.87		
MINERALS	GRAM	0.13	0.10 - 0.16	0.13		
SODIUM	MILLI	15.00	12.00 - 17.00	15.00		
POTASSIUM	MILLI	49.00	40.00 - 61.00	49.00		
CALCIUM	MILLI	10.00	8.00 - 11.00	10.00		
PHOSPHORUS	MILLI	3.30	2.80 - 3.70	3.30		
EXTRACT	GRAM	65.00	- - -	65.00		

1 ESTIMATED BY THE DIFFERENCE METHOD
100 - (WATER + MINERALS)

APFELSINEN-KONFITÜRE
(ORANGENKONFITÜRE)

ORANGE JAM

CONFITURE D'ORANGES

		PROTEIN	FAT	CARBOHYDRATES	TOTAL
ENERGY VALUE (AVERAGE) PER 100 G EDIBLE PORTION	KJOULE	5.8	0.0	1011	1017
	(KCAL)	1.4	0.0	242	243
AMOUNT OF DIGESTIBLE CONSTITUENTS PER 100 G	GRAM	0.29	0.00	60.40	
ENERGY VALUE (AVERAGE) OF THE DIGESTIBLE FRACTION PER 100 G EDIBLE PORTION	KJOULE	4.9	0.0	1011	1016
	(KCAL)	1.2	0.0	242	243

WASTE PERCENTAGE AVERAGE 0.00

CONSTITUENTS	DIM	AV	VARIATION			AVR	NUTR.	DENS.	MOLPERC.
WATER	GRAM	31.20	28.00	-	34.70	31.20	GRAM/MJ	30.72	
PROTEIN	GRAM	0.35	-	-	-	0.35	GRAM/MJ	0.34	
AVAILABLE CARBOHYDR.	GRAM	60.40	-	-	-	60.40	GRAM/MJ	59.46	
MINERALS	GRAM	0.14	0.09	-	0.22	0.14	GRAM/MJ	0.14	
SODIUM	MILLI	11.00	2.00	-	30.00	11.00	MILLI/MJ	10.83	
POTASSIUM	MILLI	53.00	35.00	-	84.00	53.00	MILLI/MJ	52.18	
CALCIUM	MILLI	32.00	25.00	-	44.00	32.00	MILLI/MJ	31.50	
PHOSPHORUS	MILLI	4.50	2.90	-	6.90	4.50	MILLI/MJ	4.43	
BORON	MICRO	40.00	5.00	-	65.00	40.00	MICRO/MJ	39.38	
VITAMIN C	MILLI	4.00	1.70	-	7.00	4.00	MILLI/MJ	3.94	
CITRIC ACID	GRAM	0.52	0.27	-	0.78	0.52	GRAM/MJ	0.51	
INVERT SUGAR	GRAM	43.90	37.60	-	46.00	43.90	GRAM/MJ	43.22	
SUCROSE	GRAM	16.00	-	-	-	16.00	GRAM/MJ	15.75	
EXTRACT WATER SOLUBLE	GRAM	68.00	63.80	-	72.90	68.00	GRAM/MJ	66.94	

APRIKOSEN-KONFITÜRE | APRICOT JAM | CONFITURE D'ABRICOTS

		PROTEIN	FAT	CARBOHYDRATES	TOTAL
ENERGY VALUE (AVERAGE)	KJOULE	6.8	0.0	1038	1044
PER 100 G	(KCAL)	1.6	0.0	248	250
EDIBLE PORTION					
AMOUNT OF DIGESTIBLE	GRAM	0.34	0.00	62.00	
CONSTITUENTS PER 100 G					
ENERGY VALUE (AVERAGE)	KJOULE	5.8	0.0	1038	1043
OF THE DIGESTIBLE	(KCAL)	1.4	0.0	248	249
FRACTION PER 100 G					
EDIBLE PORTION					

WASTE PERCENTAGE AVERAGE 0.00

CONSTITUENTS	DIM	AV	VARIATION	AVR	NUTR. DENS.		MOLPERC.
WATER	GRAM	33.00	30.00 - 35.00	33.00	GRAM/MJ	31.63	
PROTEIN	GRAM	0.41	0.36 - 0.45	0.41	GRAM/MJ	0.39	
FAT	-	0.00	- - -	0.00			
AVAILABLE CARBOHYDR.	GRAM	62.00	- - -	62.00	GRAM/MJ	59.42	
MINERALS	GRAM	0.36	0.28 - 0.47	0.36	GRAM/MJ	0.35	
POTASSIUM	MILLI	104.00	95.00 - 113.00	104.00	MILLI/MJ	99.67	
CALCIUM	MILLI	8.00	6.00 - 10.00	8.00	MILLI/MJ	7.67	
PHOSPHORUS	MILLI	11.00	6.00 - 14.00	11.00	MILLI/MJ	10.54	
VITAMIN B1	MICRO	10.00	- - -	10.00	MICRO/MJ	9.58	
VITAMIN B2	MICRO	20.00	- - -	20.00	MICRO/MJ	19.17	
NICOTINAMIDE	MILLI	-	0.50 - 1.00				
TOTAL ACIDS	GRAM	0.71	0.45 - 0.90	0.71	GRAM/MJ	0.68	
INVERT SUGAR	GRAM	33.10	24.20 - 42.10	33.10	GRAM/MJ	31.72	
SUCROSE	GRAM	28.20	14.90 - 36.90	28.20	GRAM/MJ	27.03	
PECTIN	GRAM	0.50	- - -	0.50	GRAM/MJ	0.48	
EXTRACT WATER SOLUBLE	GRAM	66.20	62.30 - 69.80	66.20	GRAM/MJ	63.45	

BROMBEER-KONFITÜRE | BLACKBERRY JAM | CONFITURE DE MÛRES

		PROTEIN	FAT	CARBOHYDRATES	TOTAL
ENERGY VALUE (AVERAGE) PER 100 G EDIBLE PORTION	KJOULE	8.8	0.0	982	991
	(KCAL)	2.1	0.0	235	237
AMOUNT OF DIGESTIBLE CONSTITUENTS PER 100 G	GRAM	0.45	0.00	58.70	
ENERGY VALUE (AVERAGE) OF THE DIGESTIBLE FRACTION PER 100 G EDIBLE PORTION	KJOULE	7.4	0.0	982	990
	(KCAL)	1.8	0.0	235	237

WASTE PERCENTAGE AVERAGE 0.00

CONSTITUENTS	DIM	AV	VARIATION			AVR	NUTR. DENS.		MOLPERC.
WATER	GRAM	32.20	31.40	-	33.00	32.20	GRAM/MJ	32.53	
PROTEIN	GRAM	0.53	-	-	-	0.53	GRAM/MJ	0.54	
FAT	-	0.00	-	-	-	0.00			
AVAILABLE CARBOHYDR.	GRAM	58.70	-	-	-	58.70	GRAM/MJ	59.30	
MINERALS	GRAM	0.30	0.28	-	0.33	0.30	GRAM/MJ	0.30	
PHOSPHORUS	MILLI	14.00	12.00	-	15.00	14.00	MILLI/MJ	14.14	
TOTAL ACIDS	GRAM	0.65	0.51	-	0.90	0.65	GRAM/MJ	0.66	
INVERT SUGAR	GRAM	34.20	32.20	-	36.10	34.20	GRAM/MJ	34.55	
SUCROSE	GRAM	23.80	23.60	-	24.00	23.80	GRAM/MJ	24.04	
PECTIN	GRAM	0.32	-	-	-	0.32	GRAM/MJ	0.32	
EXTRACT WATER SOLUBLE	GRAM	66.10	64.00	-	66.90	66.10	GRAM/MJ	66.78	

ERDBEER-KONFITÜRE

STRAWBERRY JAM

CONFITURE DE FRAISES

		PROTEIN	FAT	CARBOHYDRATES	TOTAL
ENERGY VALUE (AVERAGE)	KJOULE	6.6	0.0	974	981
PER 100 G	(KCAL)	1.6	0.0	233	234
EDIBLE PORTION					
AMOUNT OF DIGESTIBLE	GRAM	0.34	0.00	58.20	
CONSTITUENTS PER 100 G					
ENERGY VALUE (AVERAGE)	KJOULE	5.6	0.0	974	980
OF THE DIGESTIBLE	(KCAL)	1.3	0.0	233	234
FRACTION PER 100 G					
EDIBLE PORTION					

WASTE PERCENTAGE AVERAGE 0.00

CONSTITUENTS	DIM	AV	VARIATION			AVR	NUTR. DENS.		MOLPERC.
WATER	GRAM	33.30	30.60	-	34.40	33.30	GRAM/MJ	33.99	
PROTEIN	GRAM	0.40	0.27	-	0.59	0.40	GRAM/MJ	0.41	
FAT	-	0.00	-	-	-	0.00			
AVAILABLE CARBOHYDR.	GRAM	58.20	-	-	-	58.20	GRAM/MJ	59.41	
MINERALS	GRAM	0.30	0.22	-	0.39	0.30	GRAM/MJ	0.31	
POTASSIUM	MILLI	62.00	53.00	-	71.00	62.00	MILLI/MJ	63.29	
CALCIUM	MILLI	9.50	8.00	-	11.00	9.50	MILLI/MJ	9.70	
PHOSPHORUS	MILLI	10.00	9.00	-	11.00	10.00	MILLI/MJ	10.21	
BORON	MICRO	85.00	-	-	-	85.00	MICRO/MJ	86.77	
VITAMIN B1	MICRO	10.00	-	-	-	10.00	MICRO/MJ	10.21	
VITAMIN B2	MICRO	10.00	-	-	-	10.00	MICRO/MJ	10.21	
NICOTINAMIDE	MILLI	0.30	-	-	-	0.30	MILLI/MJ	0.31	
VITAMIN C	MILLI	9.30	7.10	-	12.30	9.30	MILLI/MJ	9.49	
TOTAL ACIDS	GRAM	0.49	0.44	-	0.54	0.49	GRAM/MJ	0.50	
INVERT SUGAR	GRAM	32.90	24.10	-	41.70	32.90	GRAM/MJ	33.58	
SUCROSE	GRAM	24.80	16.10	-	31.50	24.80	GRAM/MJ	25.32	
PECTIN	GRAM	0.36	-	-	-	0.36	GRAM/MJ	0.37	
EXTRACT WATER SOLUBLE	GRAM	65.70	64.00	-	68.00	65.70	GRAM/MJ	67.06	

HAGEBUTTEN-MARMELADE / ROSE HIP JAM / CONFITURES D'ÉGLANTES

		PROTEIN	FAT	CARBOHYDRATES	TOTAL
ENERGY VALUE (AVERAGE) PER 100 G EDIBLE PORTION	KJOULE	7.8	0.0	1038	1045
	(KCAL)	1.9	0.0	248	250
AMOUNT OF DIGESTIBLE CONSTITUENTS PER 100 G	GRAM	0.39	0.00	62.00	
ENERGY VALUE (AVERAGE) OF THE DIGESTIBLE FRACTION PER 100 G EDIBLE PORTION	KJOULE	6.6	0.0	1038	1044
	(KCAL)	1.6	0.0	248	250

WASTE PERCENTAGE AVERAGE 0.00

CONSTITUENTS	DIM	AV	VARIATION			AVR	NUTR. DENS.		MOLPERC.
WATER	GRAM	34.50	34.00	-	35.00	34.50	GRAM/MJ	33.04	
PROTEIN	GRAM	0.47	0.44	-	0.50	0.47	GRAM/MJ	0.45	
AVAILABLE CARBOHYDR.	GRAM	62.00 1	-	-	-	62.00	GRAM/MJ	59.37	
MINERALS	GRAM	0.60	0.51	-	0.68	0.60	GRAM/MJ	0.57	
SODIUM	MILLI	5.00	3.00	-	6.00	5.00	MILLI/MJ	4.79	
POTASSIUM	MILLI	165.00	100.00	-	230.00	165.00	MILLI/MJ	158.01	
CALCIUM	MILLI	71.00	49.00	-	93.00	71.00	MILLI/MJ	67.99	
VITAMIN C	MILLI	51.00	10.00	-	140.00	51.00	MILLI/MJ	48.84	
MALIC ACID	GRAM	0.64	0.48	-	0.80	0.64	GRAM/MJ	0.61	
EXTRACT WATER SOLUBLE	GRAM	62.00	60.00	-	63.60	62.00	GRAM/MJ	59.37	

1 AMOUNT OF EXTRACT

HEIDELBEER-KONFITÜRE | BILBERRY JAM | CONFITURE DE MYRTILLES

		PROTEIN	FAT	CARBOHYDRATES	TOTAL
ENERGY VALUE (AVERAGE)	KJOULE	4.5	0.0	1011	1015
PER 100 G	(KCAL)	1.1	0.0	242	243
EDIBLE PORTION					
AMOUNT OF DIGESTIBLE	GRAM	0.22	0.00	60.40	
CONSTITUENTS PER 100 G					
ENERGY VALUE (AVERAGE)	KJOULE	3.8	0.0	1011	1015
OF THE DIGESTIBLE	(KCAL)	0.9	0.0	242	243
FRACTION PER 100 G					
EDIBLE PORTION					

WASTE PERCENTAGE AVERAGE 0.00

CONSTITUENTS	DIM	AV	VARIATION			AVR	NUTR. DENS.		MOLPERC.
WATER	GRAM	30.20	27.90	-	32.40	30.20	GRAM/MJ	29.76	
PROTEIN	GRAM	0.27	0.27	-	0.35	0.27	GRAM/MJ	0.27	
FAT	-	0.00	-	-	-	0.00			
AVAILABLE CARBOHYDR.	GRAM	60.40	-	-	-	60.40	GRAM/MJ	59.53	
MINERALS	GRAM	0.22	0.21	-	0.22	0.22	GRAM/MJ	0.22	
POTASSIUM	MILLI	64.00	-	-	-	64.00	MILLI/MJ	63.08	
PHOSPHORUS	MILLI	14.00	-	-	-	14.00	MILLI/MJ	13.80	
TOTAL ACIDS	GRAM	0.78	0.77	-	0.79	0.78	GRAM/MJ	0.77	
INVERT SUGAR	GRAM	48.20	45.60	-	51.40	48.20	GRAM/MJ	47.50	
SUCROSE	GRAM	11.50	10.90	-	12.10	11.50	GRAM/MJ	11.33	
PECTIN	GRAM	0.23	-	-	-	0.23	GRAM/MJ	0.23	
EXTRACT WATER SOLUBLE	GRAM	67.00	66.20	-	69.80	67.00	GRAM/MJ	66.03	

HIMBEERGELEE · RASPBERRY JELLY · GELÉE DE FRAMBOISES

		PROTEIN	FAT	CARBOHYDRATES	TOTAL
ENERGY VALUE (AVERAGE)	KJOULE	0.0	0.0		1084
PER 100 G	(KCAL)	0.0	0.0		259
EDIBLE PORTION					
AMOUNT OF DIGESTIBLE	GRAM	0.00	0.00	64.78	
CONSTITUENTS PER 100 G					
ENERGY VALUE (AVERAGE)	KJOULE	0.0	0.0		1084
OF THE DIGESTIBLE	(KCAL)	0.0	0.0		259
FRACTION PER 100 G					
EDIBLE PORTION					

WASTE PERCENTAGE AVERAGE 0.00

CONSTITUENTS	DIM	AV	VARIATION	AVR	NUTR. DENS.	MOLPERC.
WATER	GRAM	35.00	- - -	35.00		
AVAILABLE CARBOHYDR.	GRAM	64.78 1	- - -	64.78		
MINERALS	GRAM	0.22	0.15 - 0.31	0.22		
POTASSIUM	MILLI	72.00	46.00 - 99.00	72.00		
PHOSPHORUS	MILLI	4.80	3.00 - 6.50	4.80		
EXTRACT	GRAM	65.00	- - -	65.00		

1 ESTIMATED BY THE DIFFERENCE METHOD
100 - (WATER + MINERALS)

HIMBEER-KONFITÜRE

RASPBERRY JAM

CONFITURE DE FRAMBOISES

		PROTEIN	FAT	CARBOHYDRATES	TOTAL
ENERGY VALUE (AVERAGE)	KJOULE	9.9	0.0	1026	1036
PER 100 G	(KCAL)	2.4	0.0	245	248
EDIBLE PORTION					
AMOUNT OF DIGESTIBLE	GRAM	0.51	0.00	61.30	
CONSTITUENTS PER 100 G					
ENERGY VALUE (AVERAGE)	KJOULE	8.4	0.0	1026	1034
OF THE DIGESTIBLE	(KCAL)	2.0	0.0	245	247
FRACTION PER 100 G					
EDIBLE PORTION					

WASTE PERCENTAGE AVERAGE 0.00

CONSTITUENTS	DIM	AV	VARIATION	AVR	NUTR. DENS.		MOLPERC.
WATER	GRAM	28.70	28.50 - 30.20	28.70	GRAM/MJ	27.75	
PROTEIN	GRAM	0.60	0.54 - 0.63	0.60	GRAM/MJ	0.58	
FAT	-	0.00	- - -	0.00			
AVAILABLE CARBOHYDR.	GRAM	61.30	- - -	61.30	GRAM/MJ	59.26	
MINERALS	GRAM	0.30	0.29 - 0.31	0.30	GRAM/MJ	0.29	
PHOSPHORUS	MILLI	16.00	13.00 - 20.00	16.00	MILLI/MJ	15.47	
VITAMIN C	MILLI	2.70	0.33 - 10.60	2.70	MILLI/MJ	2.61	
TOTAL ACIDS	GRAM	0.88	0.85 - 0.90	0.88	GRAM/MJ	0.85	
INVERT SUGAR	GRAM	35.70	31.20 - 40.10	35.70	GRAM/MJ	34.51	
SUCROSE	GRAM	24.70	18.10 - 31.30	24.70	GRAM/MJ	23.88	
PECTIN	GRAM	0.37	- - -	0.37	GRAM/MJ	0.36	
EXTRACT WATER SOLUBLE	GRAM	67.80	67.20 - 67.90	67.80	GRAM/MJ	65.55	

JOHANNISBEER-GELEE
ROT

RED-CURRANT JELLY

GELÉE DE GROSEILLES ROUGES

		PROTEIN	FAT	CARBOHYDRATES	TOTAL
ENERGY VALUE (AVERAGE) PER 100 G EDIBLE PORTION	KJOULE	0.0	0.0		1109
	(KCAL)	0.0	0.0		265
AMOUNT OF DIGESTIBLE CONSTITUENTS PER 100 G	GRAM	0.00	0.00	66.30	
ENERGY VALUE (AVERAGE) OF THE DIGESTIBLE FRACTION PER 100 G EDIBLE PORTION	KJOULE	0.0	0.0		1109
	(KCAL)	0.0	0.0		265

WASTE PERCENTAGE AVERAGE 0.00

CONSTITUENTS	DIM	AV	VARIATION	AVR	NUTR. DENS.	MOLPERC.
WATER	GRAM	33.70	30.40 - 39.00	33.70		
AVAILABLE CARBOHYDR.	GRAM	66.30 [1]	- - -	66.30		
SODIUM	MILLI	4.00	- - -	4.00		
POTASSIUM	MILLI	80.00	- - -	80.00		
CALCIUM	MILLI	6.00	- - -	6.00		
EXTRACT	GRAM	66.30	61.00 - 69.60	66.30		

[1] ESTIMATED BY THE DIFFERENCE METHOD
100 - WATER

JOHANNISBEER-KONFITÜRE ROT

RED-CURRANT JAM

CONFITURE DE GROSEILLES ROUGES

		PROTEIN	FAT	CARBOHYDRATES	TOTAL
ENERGY VALUE (AVERAGE)	KJOULE	8.3	0.0	982	990
PER 100 G	(KCAL)	2.0	0.0	235	237
EDIBLE PORTION					
AMOUNT OF DIGESTIBLE	GRAM	0.42	0.00	58.65	
CONSTITUENTS PER 100 G					
ENERGY VALUE (AVERAGE)	KJOULE	7.0	0.0	982	989
OF THE DIGESTIBLE	(KCAL)	1.7	0.0	235	236
FRACTION PER 100 G					
EDIBLE PORTION					

WASTE PERCENTAGE AVERAGE 0.00

CONSTITUENTS	DIM	AV	VARIATION	AVR	NUTR. DENS.		MOLPERC.
WATER	GRAM	33.20	31.50 - 33.40	33.20	GRAM/MJ	33.58	
PROTEIN	GRAM	0.50	0.45 - 0.59	0.50	GRAM/MJ	0.51	
FAT	-	0.00	- - -	0.00			
AVAILABLE CARBOHYDR.	GRAM	58.65	- - -	58.65	GRAM/MJ	59.33	
MINERALS	GRAM	0.34	0.33 - 0.35	0.34	GRAM/MJ	0.34	
VITAMIN C	MILLI	20.60	14.30 - 24.00	20.60	MILLI/MJ	20.84	
TOTAL ACIDS	GRAM	0.95	- - -	0.95	GRAM/MJ	0.96	
INVERT SUGAR	GRAM	47.60	- - -	47.60	GRAM/MJ	48.15	
SUCROSE	GRAM	10.10	- - -	10.10	GRAM/MJ	10.22	
PECTIN	GRAM	0.43	- - -	0.43	GRAM/MJ	0.43	
EXTRACT WATER SOLUBLE	GRAM	66.10	64.60 - 66.30	66.10	GRAM/MJ	66.86	

KIRSCH-KONFITÜRE | CHERRY JAM | CONFITURE DE CERISES

		PROTEIN	FAT	CARBOHYDRATES	TOTAL
ENERGY VALUE (AVERAGE)	KJOULE	6.6	0.0	1041	1048
PER 100 G	(KCAL)	1.6	0.0	249	250
EDIBLE PORTION					
AMOUNT OF DIGESTIBLE	GRAM	0.34	0.00	62.20	
CONSTITUENTS PER 100 G					
ENERGY VALUE (AVERAGE)	KJOULE	5.6	0.0	1041	1047
OF THE DIGESTIBLE	(KCAL)	1.3	0.0	249	250
FRACTION PER 100 G					
EDIBLE PORTION					

WASTE PERCENTAGE AVERAGE 0.00

CONSTITUENTS	DIM	AV	VARIATION			AVR	NUTR. DENS.		MOLPERC.
WATER	GRAM	32.50	24.60	-	33.70	32.50	GRAM/MJ	31.05	
PROTEIN	GRAM	0.40	0.36	-	0.41	0.40	GRAM/MJ	0.38	
FAT	-	0.00	-	-	-	0.00			
AVAILABLE CARBOHYDR.	GRAM	62.20	-	-	-	62.20	GRAM/MJ	59.43	
MINERALS	GRAM	0.38	0.34	-	0.50	0.38	GRAM/MJ	0.36	
POTASSIUM	MILLI	90.00	-	-	-	90.00	MILLI/MJ	85.99	
CALCIUM	MILLI	9.00	7.00	-	11.00	9.00	MILLI/MJ	8.60	
PHOSPHORUS	MILLI	9.00	7.00	-	11.00	9.00	MILLI/MJ	8.60	
CAROTENE	-	0.00	-	-	-	0.00			
VITAMIN C	MILLI	1.20	0.00	-	2.90	1.20	MILLI/MJ	1.15	
TOTAL ACIDS	GRAM	0.55	0.37	-	0.72	0.55	GRAM/MJ	0.53	
INVERT SUGAR	GRAM	32.50	24.20	-	42.40	32.50	GRAM/MJ	31.05	
SUCROSE	GRAM	29.10	22.40	-	33.80	29.10	GRAM/MJ	27.80	
PECTIN	GRAM	0.42	-	-	-	0.42	GRAM/MJ	0.40	
EXTRACT WATER SOLUBLE	GRAM	66.90	65.70	-	74.80	66.90	GRAM/MJ	63.92	

PFLAUMEN-KONFITÜRE

PLUM JAM

CONFITURE DE PRUNES

		PROTEIN	FAT	CARBOHYDRATES	TOTAL
ENERGY VALUE (AVERAGE)	KJOULE	5.3	0.0	1004	1009
PER 100 G	(KCAL)	1.3	0.0	240	241
EDIBLE PORTION					
AMOUNT OF DIGESTIBLE	GRAM	0.27	0.00	60.00	
CONSTITUENTS PER 100 G					
ENERGY VALUE (AVERAGE)	KJOULE	4.5	0.0	1004	1009
OF THE DIGESTIBLE	(KCAL)	1.1	0.0	240	241
FRACTION PER 100 G					
EDIBLE PORTION					

WASTE PERCENTAGE AVERAGE 0.00

CONSTITUENTS	DIM	AV	VARIATION			AVR	NUTR. DENS.		MOLPERC.
WATER	GRAM	31.10	25.30	-	34.60	31.10	GRAM/MJ	30.83	
PROTEIN	GRAM	0.32	-	-	-	0.32	GRAM/MJ	0.32	
AVAILABLE CARBOHYDR.	GRAM	60.00	-	-	-	60.00	GRAM/MJ	59.49	
MINERALS	GRAM	0.24	0.24	-	0.25	0.24	GRAM/MJ	0.24	
PHOSPHORUS	MILLI	8.60	6.90	-	10.30	8.60	MILLI/MJ	8.53	
TOTAL ACIDS	GRAM	0.42	0.31	-	0.53	0.42	GRAM/MJ	0.42	
INVERT SUGAR	GRAM	33.50	32.10	-	34.80	33.50	GRAM/MJ	33.21	
SUCROSE	GRAM	26.10	25.40	-	26.80	26.10	GRAM/MJ	25.88	
PECTIN	GRAM	0.41	-	-	-	0.41	GRAM/MJ	0.41	
EXTRACT WATER SOLUBLE	GRAM	67.80	64.40	-	73.70	67.80	GRAM/MJ	67.22	

QUITTEN-KONFITÜRE | QUINCE JAM | CONFITURE DE COINGS

		PROTEIN	FAT	CARBOHYDRATES	TOTAL
ENERGY VALUE (AVERAGE) PER 100 G EDIBLE PORTION	KJOULE	3.1	0.0	984	987
	(KCAL)	0.8	0.0	235	236
AMOUNT OF DIGESTIBLE CONSTITUENTS PER 100 G	GRAM	0.16	0.00	58.80	
ENERGY VALUE (AVERAGE) OF THE DIGESTIBLE FRACTION PER 100 G EDIBLE PORTION	KJOULE	2.7	0.0	984	987
	(KCAL)	0.6	0.0	235	236

WASTE PERCENTAGE AVERAGE 0.00

CONSTITUENTS	DIM	AV	VARIATION			AVR	NUTR. DENS.	MOLPERC.
WATER	GRAM	34.50	33.40	-	35.00	34.50	GRAM/MJ	34.96
PROTEIN	GRAM	0.19	0.14	-	0.26	0.19	GRAM/MJ	0.19
FAT	-	0.00	-	-	-	0.00		
AVAILABLE CARBOHYDR.	GRAM	58.80	-	-	-	58.80	GRAM/MJ	59.59
MINERALS	GRAM	0.23	0.22	-	0.24	0.23	GRAM/MJ	0.23
PHOSPHORUS	MILLI	8.50	-	-	-	8.50	MILLI/MJ	8.61
TOTAL ACIDS	GRAM	0.52	0.41	-	0.64	0.52	GRAM/MJ	0.53
INVERT SUGAR	GRAM	33.90	23.40	-	44.30	33.90	GRAM/MJ	34.36
SUCROSE	GRAM	24.40	17.60	-	31.20	24.40	GRAM/MJ	24.73
EXTRACT WATER SOLUBLE	GRAM	64.40	63.60	-	65.20	64.40	GRAM/MJ	65.27

Honig, Zucker, Süßwaren

Honey, sugar, sweets
Miel, sucre, confiserie

HONIG (BLÜTENHONIG) | HONEY | MIEL (DE FLEURS)

		PROTEIN	FAT	CARBOHYDRATES	TOTAL
ENERGY VALUE (AVERAGE)	KJOULE	6.3	0.0	1257	1263
PER 100 G	(KCAL)	1.5	0.0	300	302
EDIBLE PORTION					
AMOUNT OF DIGESTIBLE	GRAM	0.31	0.00	75.10	
CONSTITUENTS PER 100 G					
ENERGY VALUE (AVERAGE)	KJOULE	5.3	0.0	1257	1262
OF THE DIGESTIBLE	(KCAL)	1.3	0.0	300	302
FRACTION PER 100 G					
EDIBLE PORTION					

WASTE PERCENTAGE AVERAGE 0.00

CONSTITUENTS	DIM	AV	VARIATION		AVR	NUTR. DENS.		MOLPERC.
WATER	GRAM	18.60	17.50	20.60	18.60	GRAM/MJ	14.74	
PROTEIN	GRAM	0.38	0.30	0.50	0.38	GRAM/MJ	0.30	
FAT	-	0.00	-	-	0.00			
AVAILABLE CARBOHYDR.	GRAM	75.10	-	-	75.10	GRAM/MJ	59.50	
MINERALS	GRAM	0.22	0.20	0.24	0.22	GRAM/MJ	0.17	
SODIUM	MILLI	7.40	4.80	10.00	7.40	MILLI/MJ	5.86	
POTASSIUM	MILLI	47.00	43.00	50.00	47.00	MILLI/MJ	37.24	
MAGNESIUM	MILLI	5.50	-	-	5.50	MILLI/MJ	4.36	
CALCIUM	MILLI	4.50	3.60	5.00	4.50	MILLI/MJ	3.57	
MANGANESE	MICRO	30.00	-	-	30.00	MICRO/MJ	23.77	
IRON	MILLI	1.30	0.90	2.00	1.30	MILLI/MJ	1.03	
COPPER	MICRO	90.00	-	-	90.00	MICRO/MJ	71.31	
ZINC	MILLI	0.35	0.35	0.35	0.35	MILLI/MJ	0.28	
NICKEL	MICRO	5.50	3.00	8.00	5.50	MICRO/MJ	4.36	
CHROMIUM	MICRO	29.00	-	-	29.00	MICRO/MJ	22.98	
PHOSPHORUS	MILLI	18.00	16.00	20.00	18.00	MILLI/MJ	14.26	
IODIDE	MICRO	0.50	0.50	0.50	0.50	MICRO/MJ	0.40	
BORON	MILLI	0.35	-	-	0.35	MILLI/MJ	0.28	
CAROTENE	-	0.00	-	-	0.00			
VITAMIN K	MICRO	25.00	-	-	25.00	MICRO/MJ	19.81	
VITAMIN B1	MICRO	3.00	2.00	4.00	3.00	MICRO/MJ	2.38	
VITAMIN B2	MICRO	50.00	20.00	100.00	50.00	MICRO/MJ	39.61	
NICOTINAMIDE	MILLI	0.13	0.10	0.20	0.13	MILLI/MJ	0.10	
PANTOTHENIC ACID	MICRO	70.00	70.00	70.00	70.00	MICRO/MJ	55.46	
VITAMIN B6	MILLI	0.16	0.16	0.16	0.16	MILLI/MJ	0.13	
VITAMIN C	MILLI	2.40	1.00	4.00	2.40	MILLI/MJ	1.90	
SALICYLIC ACID	MILLI	6.28	6.28	6.29	6.29	MILLI/MJ	4.98	
GLUCOSE	GRAM	33.90	26.30	39.80	33.90	GRAM/MJ	26.86	
FRUCTOSE	GRAM	38.80	35.90	42.10	38.80	GRAM/MJ	30.74	
SUCROSE	GRAM	2.37	1.71	2.99	2.37	GRAM/MJ	1.88	
STARCH	-	0.00	-	-	0.00			

INVERTZUCKER-CREME
KUNSTHONIG

ARTIFICIAL HONEY

MIEL ARTIFICIEL

		PROTEIN	FAT	CARBOHYDRATES	TOTAL
ENERGY VALUE (AVERAGE)	KJOULE	2.5	0.0	1381	1383
PER 100 G	(KCAL)	0.6	0.0	330	331
EDIBLE PORTION					
AMOUNT OF DIGESTIBLE	GRAM	0.13	0.00	82.50	
CONSTITUENTS PER 100 G					
ENERGY VALUE (AVERAGE)	KJOULE	2.2	0.0	1381	1383
OF THE DIGESTIBLE	(KCAL)	0.5	0.0	330	331
FRACTION PER 100 G					
EDIBLE PORTION					

WASTE PERCENTAGE AVERAGE 0.00

CONSTITUENTS	DIM	AV	VARIATION			AVR	NUTR.	DENS.	MOLPERC.
WATER	GRAM	17.20	15.80	-	21.70	17.20	GRAM/MJ	12.44	
PROTEIN	GRAM	0.15	-	-	-	0.15	GRAM/MJ	0.11	
FAT	-	0.00	-	-	-	0.00			
AVAILABLE CARBOHYDR.	GRAM	82.50	-	-	-	82.50	GRAM/MJ	59.66	
MINERALS	GRAM	0.16	0.12	-	0.23	0.16	GRAM/MJ	0.12	
SODIUM	MILLI	19.00	8.00	-	29.00	19.00	MILLI/MJ	13.74	
POTASSIUM	MILLI	4.50	4.00	-	5.00	4.50	MILLI/MJ	3.25	
MAGNESIUM	-	0.00	-	-	-	0.00			
CALCIUM	MILLI	5.50	5.00	-	6.00	5.50	MILLI/MJ	3.98	
IRON	-	0.00	-	-	-	0.00			
VITAMIN A	-	0.00	-	-	-	0.00			
CAROTENE	-	0.00	-	-	-	0.00			
VITAMIN D	-	0.00	-	-	-	0.00			
VITAMIN E ACTIVITY	-	0.00	-	-	-	0.00			
VITAMIN K	-	0.00	-	-	-	0.00			
VITAMIN B1	-	0.00	-	-	-	0.00			
VITAMIN B2	-	0.00	-	-	-	0.00			
NICOTINAMIDE	-	0.00	-	-	-	0.00			
VITAMIN C	-	0.00	-	-	-	0.00			
FORMIC ACID	MILLI	73.30	71.10	-	73.60	73.30	MILLI/MJ	53.00	
GLUCOSE	GRAM	41.10	36.40	-	46.30	41.10	GRAM/MJ	29.72	
FRUCTOSE	GRAM	36.10	34.40	-	37.40	36.10	GRAM/MJ	26.10	
SUCROSE	GRAM	5.30	4.50	-	6.24	5.30	GRAM/MJ	3.83	

ZUCKER (ROHRZUCKER, RÜBENZUCKER)

SUGAR (CANE-SUGAR, BEET-SUGAR)

SUCRE (SUCRE DE CANNE, SUCRE DE BETTERAVES)

		PROTEIN	FAT	CARBOHYDRATES	TOTAL
ENERGY VALUE (AVERAGE)	KJOULE	0.0	0.0	1670	1670
PER 100 G EDIBLE PORTION	(KCAL)	0.0	0.0	399	399
AMOUNT OF DIGESTIBLE CONSTITUENTS PER 100 G	GRAM	0.00	0.00	99.80	
ENERGY VALUE (AVERAGE)	KJOULE	0.0	0.0	1670	1670
OF THE DIGESTIBLE FRACTION PER 100 G EDIBLE PORTION	(KCAL)	0.0	0.0	399	399

WASTE PERCENTAGE AVERAGE 0.00

CONSTITUENTS	DIM	AV	VARIATION		AVR	NUTR. DENS.		MOLPERC.
WATER	MILLI	50.00	30.00	- 100.00	50.00	MILLI/MJ	29.94	
PROTEIN	-	0.00	-	- -	0.00			
FAT	-	0.00	-	- -	0.00			
AVAILABLE CARBOHYDR.	GRAM	99.80	99.60	- 99.90	99.80	GRAM/MJ	59.75	
MINERALS	MILLI	40.00	10.00	- 50.00	40.00	MILLI/MJ	23.95	
SODIUM	MILLI	0.30	0.00	- 1.50	0.30	MILLI/MJ	0.18	
POTASSIUM	MILLI	2.20	0.00	- 6.00	2.20	MILLI/MJ	1.32	
MAGNESIUM	MILLI	0.20	0.00	- 0.30	0.20	MILLI/MJ	0.12	
CALCIUM	MILLI	0.60	0.00	- 2.00	0.60	MILLI/MJ	0.36	
MANGANESE	MICRO	10.00	-	- -	10.00	MICRO/MJ	5.99	
IRON	MILLI	0.29	0.16	- 0.52	0.29	MILLI/MJ	0.17	
MOLYBDENUM	MILLI	-	0.00	- 0.02				
PHOSPHORUS	MILLI	0.30	0.00	- 0.50	0.30	MILLI/MJ	0.18	
CHLORIDE	MILLI	1.50	-	- -	1.50	MILLI/MJ	0.90	
SELENIUM	MICRO	-	0.00	- 0.30				
VITAMIN A	-	0.00	-	- -	0.00			
CAROTENE	-	0.00	-	- -	0.00			
VITAMIN D	-	0.00	-	- -	0.00			
VITAMIN E ACTIVITY	-	0.00	-	- -	0.00			
VITAMIN K	-	0.00	-	- -	0.00			
VITAMIN B1	-	0.00	-	- -	0.00			
VITAMIN B2	-	0.00	-	- -	0.00			
NICOTINAMIDE	-	0.00	-	- -	0.00			
PANTOTHENIC ACID	-	0.00	-	- -	0.00			
VITAMIN B6	-	0.00	-	- -	0.00			
BIOTIN	-	0.00	-	- -	0.00			
FOLIC ACID	-	0.00	-	- -	0.00			
VITAMIN B12	-	0.00	-	- -	0.00			
VITAMIN C	-	0.00	-	- -	0.00			
REDUCING SUGAR	MILLI	10.00	0.00	- 50.00	10.00	MILLI/MJ	5.99	
SUCROSE	GRAM	99.80	-	- -	99.80	GRAM/MJ	59.75	
RAFFINOSE	MILLI	40.00	10.00	- 100.00	40.00	MILLI/MJ	23.95	

ROHZUCKER
AUS ZUCKERROHR
(BRAUNER ZUCKER)

CANE-SUGAR
UNREFINED

SUCRE BRUN
(DE CANNE À SUCRE)

		PROTEIN	FAT	CARBOHYDRATES	TOTAL
ENERGY VALUE (AVERAGE) PER 100 G EDIBLE PORTION	KJOULE	0.0	0.0	1630	1630
	(KCAL)	0.0	0.0	390	390
AMOUNT OF DIGESTIBLE CONSTITUENTS PER 100 G	GRAM	0.00	0.00	97.40	
ENERGY VALUE (AVERAGE) OF THE DIGESTIBLE FRACTION PER 100 G EDIBLE PORTION	KJOULE	0.0	0.0	1630	1630
	(KCAL)	0.0	0.0	390	390

WASTE PERCENTAGE AVERAGE 0.00

CONSTITUENTS	DIM	AV	VARIATION			AVR	NUTR.	DENS.	MOLPERC.
WATER	GRAM	0.58	0.36	-	0.90	0.58	GRAM/MJ	0.36	
PROTEIN	-	0.00	-	-	-	0.00			
FAT	-	0.00	-	-	-	0.00			
AVAILABLE CARBOHYDR.	GRAM	97.40	-	-	-	97.40	GRAM/MJ	59.75	
MINERALS	GRAM	0.45	0.35	-	0.79	0.45	GRAM/MJ	0.28	
SODIUM	MILLI	2.00	0.50	-	10.00	2.00	MILLI/MJ	1.23	
POTASSIUM	MILLI	90.00	50.00	-	110.00	90.00	MILLI/MJ	55.21	
MAGNESIUM	MILLI	14.00	10.00	-	19.00	14.00	MILLI/MJ	8.59	
CALCIUM	MILLI	55.00	20.00	-	80.00	55.00	MILLI/MJ	33.74	
IRON	MILLI	-	1.00	-	8.00				
MOLYBDENUM	-	0.00	-	-	-	0.00			
PHOSPHORUS	MILLI	24.00	-	-	-	24.00	MILLI/MJ	14.72	
SELENIUM	MICRO	-	0.00	-	1.20				
VITAMIN B1	MICRO	6.00	0.00	-	12.00	6.00	MICRO/MJ	3.68	
VITAMIN B2	MICRO	6.00	3.00	-	8.00	6.00	MICRO/MJ	3.68	
NICOTINAMIDE	MICRO	30.00	20.00	-	30.00	30.00	MICRO/MJ	18.40	
PANTOTHENIC ACID	MILLI	-	0.01	-	0.04				
VITAMIN C	MILLI	0.70	0.20	-	1.20	0.70	MILLI/MJ	0.43	
REDUCING SUGAR	GRAM	0.70	0.50	-	0.90	0.70	GRAM/MJ	0.43	
SUCROSE	GRAM	96.70	96.00	-	98.00	96.70	GRAM/MJ	59.32	

ROHZUCKER
AUS ZUCKERRÜBEN
(BRAUNER ZUCKER)

BEET-SUGAR
UNREFINED

SUCRE BRUN
(DE BETTERAVES SUCRIÈRES)

		PROTEIN	FAT	CARBOHYDRATES	TOTAL
ENERGY VALUE (AVERAGE) PER 100 G EDIBLE PORTION	KJOULE	0.0	0.0	1613	1613
	(KCAL)	0.0	0.0	386	386
AMOUNT OF DIGESTIBLE CONSTITUENTS PER 100 G	GRAM	0.00	0.00	96.40	
ENERGY VALUE (AVERAGE) OF THE DIGESTIBLE FRACTION PER 100 G EDIBLE PORTION	KJOULE	0.0	0.0	1613	1613
	(KCAL)	0.0	0.0	386	386

WASTE PERCENTAGE AVERAGE 0.00

CONSTITUENTS	DIM	AV	VARIATION			AVR	NUTR. DENS.		MOLPERC.
WATER	GRAM	1.40	0.70	-	1.90	1.40	GRAM/MJ	0.87	
FAT	-	0.00	-	-	-	0.00			
AVAILABLE CARBOHYDR.	GRAM	96.40	-	-	-	96.40	GRAM/MJ	59.75	
MINERALS	GRAM	0.80	0.60	-	1.10	0.80	GRAM/MJ	0.50	
SODIUM	MILLI	35.00	15.00	-	73.00	35.00	MILLI/MJ	21.69	
POTASSIUM	MILLI	240.00	200.00	-	280.00	240.00	MILLI/MJ	148.76	
CALCIUM	MILLI	8.50	4.00	-	15.00	8.50	MILLI/MJ	5.27	
IRON	MILLI	6.00	-	-	-	6.00	MILLI/MJ	3.72	
CHLORIDE	MILLI	40.00	-	-	-	40.00	MILLI/MJ	24.79	
SELENIUM	MICRO	-	0.00	-	1.20				
REDUCING SUGAR	GRAM	-	0.00	-	0.09				
SUCROSE	GRAM	96.40	-	-	-	96.40	GRAM/MJ	59.75	

FONDANT FONDANT FONDANT

		PROTEIN	FAT	CARBOHYDRATES	TOTAL
ENERGY VALUE (AVERAGE) PER 100 G EDIBLE PORTION	KJOULE	0.0	0.0	1473	1473
	(KCAL)	0.0	0.0	352	352
AMOUNT OF DIGESTIBLE CONSTITUENTS PER 100 G	GRAM	0.00	0.00	88.00	
ENERGY VALUE (AVERAGE) OF THE DIGESTIBLE FRACTION PER 100 G EDIBLE PORTION	KJOULE	0.0	0.0	1473	1473
	(KCAL)	0.0	0.0	352	352

WASTE PERCENTAGE AVERAGE 0.00

CONSTITUENTS	DIM	AV	VARIATION			AVR	NUTR. DENS.		MOLPERC.
WATER	GRAM	11.00	8.00	-	15.00	11.00	GRAM/MJ	7.47	
PROTEIN	-	0.00	-	-	-	0.00			
FAT	-	0.00	-	-	-	0.00			
AVAILABLE CARBOHYDR.	GRAM	88.00 1	-	-	-	88.00	GRAM/MJ	59.75	
MINERALS	GRAM	1.00	-	-	-	1.00	GRAM/MJ	0.68	
SODIUM	MILLI	17.00	5.00	-	46.00	17.00	MILLI/MJ	11.54	
POTASSIUM	MILLI	2.00	0.50	-	3.00	2.00	MILLI/MJ	1.36	
CALCIUM	MILLI	3.00	1.00	-	6.00	3.00	MILLI/MJ	2.04	
IRON	-	0.00	-	-	-	0.00			
PHOSPHORUS	-	0.00	-	-	-	0.00			
VITAMIN A	-	0.00	-	-	-	0.00			
VITAMIN B1	-	0.00	-	-	-	0.00			
VITAMIN B2	-	0.00	-	-	-	0.00			
NICOTINAMIDE	-	0.00	-	-	-	0.00			
VITAMIN C	-	0.00	-	-	-	0.00			
SUCROSE	GRAM	73.00	65.00	-	80.00	73.00	GRAM/MJ	49.57	
STARCH	GRAM	1.43	-	-	-	1.43	GRAM/MJ	0.97	

1 ESTIMATED BY THE DIFFERENCE METHOD
100 - (WATER + PROTEIN + FAT + MINERALS)

MARZIPAN MARZIPAN MASSEPAIN

		PROTEIN	FAT	CARBOHYDRATES	TOTAL
ENERGY VALUE (AVERAGE) PER 100 G EDIBLE PORTION	KJOULE	151	969	961	2081
	(KCAL)	36	232	230	497
AMOUNT OF DIGESTIBLE CONSTITUENTS PER 100 G	GRAM	6.24	22.41	57.44	
ENERGY VALUE (AVERAGE) OF THE DIGESTIBLE FRACTION PER 100 G EDIBLE PORTION	KJOULE	117	872	961	1951
	(KCAL)	28	208	230	466

WASTE PERCENTAGE AVERAGE 0.00

CONSTITUENTS	DIM	AV	VARIATION	AVR	NUTR. DENS.		MOLPERC.
WATER	GRAM	8.83	7.00 - 10.00	8.83	GRAM/MJ	4.53	
PROTEIN	GRAM	8.00	- - -	8.00	GRAM/MJ	4.10	
FAT	GRAM	24.90	- - -	24.90	GRAM/MJ	12.76	
AVAILABLE CARBOHYDR.	GRAM	57.44 [1]	- - -	57.44	GRAM/MJ	29.44	
MINERALS	GRAM	0.83	0.70 - 1.00	0.83	GRAM/MJ	0.43	
SODIUM	MILLI	5.00	2.00 - 16.00	5.00	MILLI/MJ	2.56	
POTASSIUM	MILLI	209.00	117.00 - 329.00	209.00	MILLI/MJ	107.14	
MAGNESIUM	MILLI	120.00	- - -	120.00	MILLI/MJ	61.51	
CALCIUM	MILLI	43.00	27.00 - 73.00	43.00	MILLI/MJ	22.04	
IRON	MILLI	2.00	- - -	2.00	MILLI/MJ	1.03	
COPPER	MICRO	80.00	- - -	80.00	MICRO/MJ	41.01	
ZINC	MILLI	1.50	- - -	1.50	MILLI/MJ	0.77	
PHOSPHORUS	MILLI	220.00	- - -	220.00	MILLI/MJ	112.77	
CHLORIDE	MILLI	13.00	- - -	13.00	MILLI/MJ	6.66	
VITAMIN B1	MILLI	0.10	0.09 - 0.12	0.10	MILLI/MJ	0.05	
VITAMIN B2	MILLI	0.45	- - -	0.45	MILLI/MJ	0.23	
NICOTINAMIDE	MILLI	1.38	- - -	1.38	MILLI/MJ	0.71	
PANTOTHENIC ACID	MILLI	0.35	- - -	0.35	MILLI/MJ	0.18	
VITAMIN B6	MICRO	60.00	- - -	60.00	MICRO/MJ	30.76	
BIOTIN	MICRO	2.00	- - -	2.00	MICRO/MJ	1.03	
VITAMIN C	MILLI	2.00	- - -	2.00	MILLI/MJ	1.03	
TOTAL SUGAR	GRAM	49.00	- - -	49.00	GRAM/MJ	25.12	

[1] ESTIMATED BY THE DIFFERENCE METHOD
100 - (WATER + PROTEIN + FAT + MINERALS)

NUSS-NOUGAT-CREME / NUT-NOUGAT-CREAM / NOIX-NOUGAT-CRÈME

		PROTEIN	FAT	CARBOHYDRATES	TOTAL
ENERGY VALUE (AVERAGE)	KJOULE	72	1218	977	2267
PER 100 G	(KCAL)	17	291	233	542
EDIBLE PORTION					
AMOUNT OF DIGESTIBLE	GRAM	4.32	28.16	58.36	
CONSTITUENTS PER 100 G					
ENERGY VALUE (AVERAGE)	KJOULE	72	1096	977	2145
OF THE DIGESTIBLE	(KCAL)	17	262	233	513
FRACTION PER 100 G					
EDIBLE PORTION					

WASTE PERCENTAGE AVERAGE 0.00

CONSTITUENTS	DIM	AV	VARIATION			AVR	NUTR.	DENS.	MOLPERC.
WATER	GRAM	0.59	0.59	-	0.59	0.59	GRAM/MJ	0.28	
PROTEIN	GRAM	4.32	4.32	-	4.32	4.32	GRAM/MJ	2.01	
FAT	GRAM	31.29	31.29	-	31.29	31.29	GRAM/MJ	14.59	
AVAILABLE CARBOHYDR.	GRAM	58.36	-	-	-	58.36	GRAM/MJ	27.21	
MINERALS	GRAM	0.87	0.87	-	0.87	0.87	GRAM/MJ	0.41	
CALCIUM	MILLI	13.00	13.00	-	13.00	13.00	MILLI/MJ	6.06	
IRON	MILLI	3.50	3.50	-	3.50	3.50	MILLI/MJ	1.63	
VITAMIN A	MICRO	31.00	31.00	-	31.00	31.00	MICRO/MJ	14.45	
VITAMIN B1	MILLI	0.23	0.23	-	0.23	0.23	MILLI/MJ	0.11	
VITAMIN B2	MILLI	0.12	0.12	-	0.12	0.12	MILLI/MJ	0.06	
VITAMIN B6	MILLI	0.69	0.69	-	0.69	0.69	MILLI/MJ	0.32	
SUCROSE	GRAM	55.69	55.69	-	55.69	55.69	GRAM/MJ	25.97	
LACTOSE	GRAM	1.92	1.92	-	1.92	1.92	GRAM/MJ	0.90	
STARCH	GRAM	0.75	0.75	-	0.75	0.75	GRAM/MJ	0.35	

Speiseeis

Ice cream
Glaces

EISCREME ICE CREAM CRÈME GLACÉE

		PROTEIN	FAT	CARBOHYDRATES	TOTAL
ENERGY VALUE (AVERAGE)	KJOULE	72	453	351	876
PER 100 G	(KCAL)	17	108	84	209
EDIBLE PORTION					
AMOUNT OF DIGESTIBLE	GRAM	3.78	11.11	21.00	
CONSTITUENTS PER 100 G					
ENERGY VALUE (AVERAGE)	KJOULE	70	430	351	851
OF THE DIGESTIBLE	(KCAL)	17	103	84	203
FRACTION PER 100 G					
EDIBLE PORTION					

WASTE PERCENTAGE AVERAGE 0.00

CONSTITUENTS	DIM	AV	VARIATION		AVR	NUTR. DENS.		MOLPERC.
WATER	GRAM	63.40	63.00	- 65.00	63.40	GRAM/MJ	74.48	
PROTEIN	GRAM	3.90	3.85	- 4.44	3.90	GRAM/MJ	4.58	
FAT	GRAM	11.70	10.00	- 13.00	11.70	GRAM/MJ	13.74	
AVAILABLE CARBOHYDR.	GRAM	21.00	-	- -	21.00	GRAM/MJ	24.67	
MINERALS	GRAM	0.80	-	- -	0.80	GRAM/MJ	0.94	
SODIUM	MILLI	110.00	100.00	- 120.00	110.00	MILLI/MJ	129.22	
POTASSIUM	MILLI	99.00	90.00	- 108.00	99.00	MILLI/MJ	116.30	
CALCIUM	MILLI	140.00	112.00	- 188.00	140.00	MILLI/MJ	164.46	
IRON	MILLI	0.14	0.12	- 0.16	0.14	MILLI/MJ	0.16	
PHOSPHORUS	MILLI	117.00	96.00	- 167.00	117.00	MILLI/MJ	137.44	
VITAMIN A	MILLI	0.13	0.07	- 0.16	0.13	MILLI/MJ	0.15	
VITAMIN B1	MICRO	39.00	38.00	- 42.00	39.00	MICRO/MJ	45.81	
VITAMIN B2	MILLI	0.25	0.24	- 0.29	0.25	MILLI/MJ	0.29	
NICOTINAMIDE	MICRO	98.00	-	- -	98.00	MICRO/MJ	115.12	
VITAMIN C	MILLI	-	0.00	- 1.50				
SUCROSE	GRAM	14.30	13.00	- 15.00	14.30	GRAM/MJ	16.80	
LACTOSE	GRAM	6.67	-	- -	6.67	GRAM/MJ	7.84	

FRUCHTEIS — ICE CREAM SUNDAE — GLACE AUX FRUITS

		PROTEIN	FAT	CARBOHYDRATES	TOTAL
ENERGY VALUE (AVERAGE)	KJOULE	28	70	487	584
PER 100 G	(KCAL)	6.6	17	116	140
EDIBLE PORTION					
AMOUNT OF DIGESTIBLE	GRAM	1.45	1.71	29.10	
CONSTITUENTS PER 100 G					
ENERGY VALUE (AVERAGE)	KJOULE	27	66	487	580
OF THE DIGESTIBLE	(KCAL)	6.4	16	116	139
FRACTION PER 100 G					
EDIBLE PORTION					

WASTE PERCENTAGE AVERAGE 0.00

CONSTITUENTS	DIM	AV	VARIATION	AVR	NUTR.	DENS.	MOLPERC.
WATER	GRAM	67.60	64.00 - 70.00	67.60	GRAM/MJ	116.55	
PROTEIN	GRAM	1.50	1.00 - 2.30	1.50	GRAM/MJ	2.59	
FAT	GRAM	1.80	1.00 - 3.00	1.80	GRAM/MJ	3.10	
AVAILABLE CARBOHYDR.	GRAM	29.10 1	- - -	29.10	GRAM/MJ	50.17	
SODIUM	MILLI	20.00	10.00 - 32.00	20.00	MILLI/MJ	34.48	
POTASSIUM	MILLI	38.00	22.00 - 54.00	38.00	MILLI/MJ	65.52	
CALCIUM	MILLI	41.00	16.00 - 75.00	41.00	MILLI/MJ	70.69	
IRON	MILLI	0.40	0.19 - 0.77	0.40	MILLI/MJ	0.69	
PHOSPHORUS	MILLI	26.00	13.00 - 39.00	26.00	MILLI/MJ	44.83	

1 ESTIMATED BY THE DIFFERENC METHOD:
100 - (WATER + PROTEIN + FAT)

Alkoholhaltige Getränke

Alcoholic beverages
Boissons alcooliques

NÄHRBIER NUTRIENT BEER BIÈRE NUTRITIVE

		PROTEIN	FAT	CARBOHYDRATES	TOTAL
ENERGY VALUE (AVERAGE) PER 100 G EDIBLE PORTION	KJOULE [1]	9.5	0.0	144	
	(KCAL)	2.3	0.0	34	
AMOUNT OF DIGESTIBLE CONSTITUENTS PER 100 G	GRAM	0.39	0.00	8.60	
ENERGY VALUE (AVERAGE) OF THE DIGESTIBLE FRACTION PER 100 G EDIBLE PORTION	KJOULE	7.4	0.0	144	
	(KCAL)	1.8	0.0	34	

WASTE PERCENTAGE AVERAGE 0.00

CONSTITUENTS	DIM	AV	VARIATION			AVR	NUTR. DENS.	MOLPERC.
WATER	GRAM	87.80	-	-	-	87.80		
PROTEIN	GRAM	0.50	0.40	-	0.60	0.50		
FAT	-	0.00	-	-	-	0.00		
AVAILABLE CARBOHYDR.	GRAM	8.60	-	-	-	8.60		
SODIUM	MILLI	1.00	-	-	-	1.00		
POTASSIUM	MILLI	47.00	-	-	-	47.00		
MAGNESIUM	MILLI	5.50	-	-	-	5.50		
CALCIUM	MILLI	1.00	-	-	-	1.00		
IRON	MILLI	0.21	-	-	-	0.21		
COPPER	MICRO	90.00	-	-	-	90.00		
PHOSPHORUS	MILLI	5.30	-	-	-	5.30		
HISTAMINE	MILLI	0.67	-	-	-	0.67		
GLUCOSE	GRAM	0.46	0.20	-	0.90	0.46		
FRUCTOSE	GRAM	0.25	0.17	-	0.38	0.25		
SUCROSE	MILLI	40.00	-	-	-	40.00		
MALTOSE	GRAM	4.95	4.24	-	5.41	4.95		
ETHANOL	GRAM	1.28	0.75	-	1.70	1.28		
EXTRACT	GRAM	10.90	10.20	-	11.30	10.90		

1 THE TOTAL ENERGY VALUE AMOUNTS TO 217 KJ (52 KCAL) AND WAS CALCULATED FROM EXTRACT AND ETHANOL

VOLLBIER DUNKEL	**REAL BEER** DARK	**BIÈRE ENTIÈRE** BRUNE

		PROTEIN	FAT	CARBOHYDRATES	TOTAL
ENERGY VALUE (AVERAGE) PER 100 G EDIBLE PORTION	KJOULE [1]	7.6	0.0	47	
	(KCAL)	1.8	0.0	11	
AMOUNT OF DIGESTIBLE CONSTITUENTS PER 100 G	GRAM	0.31	0.00	2.80	
ENERGY VALUE (AVERAGE) OF THE DIGESTIBLE FRACTION PER 100 G EDIBLE PORTION	KJOULE	5.9	0.0	47	
	(KCAL)	1.4	0.0	11	

WASTE PERCENTAGE AVERAGE 0.00

CONSTITUENTS	DIM	AV	VARIATION			AVR	NUTR. DENS.	MOLPERC.
WATER	GRAM	91.10	-	-	-	91.10		
PROTEIN	GRAM	0.40	-	-	-	0.40		
FAT	-	0.00	-	-	-	0.00		
AVAILABLE CARBOHYDR.	GRAM	2.80	-	-	-	2.80		
SODIUM	MILLI	3.00	0.30	-	7.00	3.00		
POTASSIUM	MILLI	50.00	40.00	-	57.00	50.00		
CALCIUM	MILLI	3.00	2.00	-	4.00	3.00		
PHOSPHORUS	MILLI	25.00	-	-	-	25.00		
SELENIUM	MICRO	19.00	-	-	-	19.00		
CAROTENE	MICRO	5.40	-	-	-	5.40		
VITAMIN B1	MICRO	4.00	3.00	-	6.00	4.00		
VITAMIN B2	MICRO	30.00	20.00	-	40.00	30.00		
NICOTINAMIDE	MILLI	0.88	0.65	-	1.10	0.88		
PANTOTHENIC ACID	MICRO	80.00	50.00	-	110.00	80.00		
VITAMIN B6	MICRO	50.00	30.00	-	80.00	50.00		
BIOTIN	MICRO	0.50	0.30	-	0.80	0.50		
TYRAMINE	MILLI	-	0.18	-	1.20			
STARCH	-	0.00	-	-	-	0.00		
DEXTRIN	GRAM	2.80	2.20	-	3.50	2.80		
ETHANOL	GRAM	3.50	2.90	-	4.10	3.50		
EXTRACT	GRAM	5.40	4.20	-	6.70	5.40		

[1] THE TOTAL ENERGY VALUE AMOUNTS TO 194 KJ (46 KCAL) AND WAS CALCULATED FROM EXTRACT AND ETHANOL

VOLLBIER
HELL

REAL BEER
LIGHT

BIÈRE ENTIÈRE
BLONDE

		PROTEIN	FAT	CARBOHYDRATES	TOTAL
ENERGY VALUE (AVERAGE) PER 100 G EDIBLE PORTION	KJOULE 1	9.5	0.0	49	
	(KCAL)	2.3	0.0	12	
AMOUNT OF DIGESTIBLE CONSTITUENTS PER 100 G	GRAM	0.39	0.00	2.90	
ENERGY VALUE (AVERAGE) OF THE DIGESTIBLE FRACTION PER 100 G EDIBLE PORTION	KJOULE	7.4	0.0	49	
	(KCAL)	1.8	0.0	12	

WASTE PERCENTAGE AVERAGE 0.00

CONSTITUENTS	DIM	AV	VARIATION			AVR	NUTR. DENS.	MOLPERC.
WATER	GRAM	90.60	90.00	-	91.10	90.60		
PROTEIN	GRAM	0.50	-	-	-	0.50		
FAT	-	0.00	-	-	-	0.00		
AVAILABLE CARBOHYDR.	GRAM	2.90	-	-	-	2.90		
MINERALS	GRAM	0.20	-	-	-	0.20		
SODIUM	MILLI	5.00	2.00	-	6.00	5.00		
POTASSIUM	MILLI	38.00	32.00	-	44.00	38.00		
MAGNESIUM	MILLI	9.00	7.00	-	12.00	9.00		
CALCIUM	MILLI	4.00	2.00	-	5.00	4.00		
MANGANESE	MICRO	29.00	200.00	-	300.00	29.00		
IRON	MICRO	42.00	10.00	-	200.00	42.00		
COPPER	MICRO	40.00	20.00	-	50.00	40.00		
ZINC	MICRO	20.00	10.00	-	20.00	20.00		
NICKEL	MICRO	-	1.00	-	200.00			
CHROMIUM	MICRO	0.70	0.20	-	2.00	0.70		
MOLYBDENUM	MICRO	0.20	0.10	-	0.30	0.20		
PHOSPHORUS	MILLI	28.00	25.00	-	30.00	28.00		
CHLORIDE	MILLI	35.00	-	-	-	35.00		
FLUORIDE	MICRO	50.00	20.00	-	80.00	50.00		
IODIDE	MICRO	0.70	0.50	-	0.90	0.70		
BORON	MICRO	5.00	0.00	-	17.00	5.00		
SELENIUM	MICRO	-	0.10	-	19.00			
SILICON	MILLI	6.00	-	-	-	6.00		
VITAMIN B1	MICRO	4.00	3.00	-	6.00	4.00		
VITAMIN B2	MICRO	30.00	20.00	-	40.00	30.00		
NICOTINAMIDE	MILLI	0.88	0.65	-	1.10	0.88		

1 THE TOTAL ENERGY VALUE AMOUNTS TO 188 KJ (45 KCAL) AND WAS CALCULATED FROM EXTRACT AND ETHANOL

CONSTITUENTS	DIM	AV	VARIATION			AVR	NUTR. DENS.	MOLPERC.
PANTOTHENIC ACID	MICRO	80.00	50.00	-	110.00	80.00		
VITAMIN B6	MICRO	50.00	30.00	-	80.00	50.00		
BIOTIN	MICRO	0.50	0.30	-	0.80	0.50		
FOLIC ACID	MICRO	5.00	4.00	-	8.00	5.00		
HISTAMINE	MILLI	0.63	0.29	-	1.10	0.63		
TYRAMINE	MILLI	-	0.18	-	1.20			
MALIC ACID	MILLI	5.50	-	-	-	5.50		
CITRIC ACID	MILLI	11.90	-	-	-	11.90		
LACTIC ACID	MILLI	3.20	-	-	-	3.20		
OXALIC ACID	MILLI	1.60	0.60	-	2.70	1.60		
SUCCINIC ACID	MILLI	2.20	-	-	-	2.20		
FUMARIC ACID	MILLI	1.20	-	-	-	1.20		
GLYCOLIC ACID	MILLI	1.80	-	-	-	1.80		
GALACTURONIC ACID	MILLI	0.25	-	-	-	0.25		
PYRUVIC ACID	MILLI	4.80	-	-	-	4.80		
GLUCONIC ACID	MILLI	4.90	-	-	-	4.90		
GLUCOSE	MILLI	13.30	-	-	-	13.30		
MALTOSE	MILLI	187.50	-	-	-	187.50		
ARABINOSE	MILLI	7.20	-	-	-	7.20		
GALACTOSE	MILLI	1.40	-	-	-	1.40		
STARCH	-	0.00	-	-	-	0.00		
DEXTRIN	GRAM	2.50	1.80	-	3.40	2.50		
SORBITOL	MILLI	1.70	-	-	-	1.70		
MANNITOL	MILLI	2.00	-	-	-	2.00		
ETHANOL	GRAM	3.61	2.95	-	4.59	3.61		
EXTRACT	GRAM	4.80	3.50	-	6.60	4.80		
GLYCEROL	MILLI	200.00	-	-	-	200.00		

WEISSBIER | PALE BEER | BIÈRE BLANCHE

		PROTEIN	FAT	CARBOHYDRATES	TOTAL
ENERGY VALUE (AVERAGE)	KJOULE 1	4.8	0.0	47	
PER 100 G	(KCAL)	1.1	0.0	11	
EDIBLE PORTION					
AMOUNT OF DIGESTIBLE	GRAM	0.19	0.00	2.80	
CONSTITUENTS PER 100 G					
ENERGY VALUE (AVERAGE)	KJOULE	3.8	0.0	47	
OF THE DIGESTIBLE	(KCAL)	0.9	0.0	11	
FRACTION PER 100 G					
EDIBLE PORTION					

WASTE PERCENTAGE AVERAGE 0.00

CONSTITUENTS	DIM	AV	VARIATION			AVR	NUTR. DENS.	MOLPERC.
WATER	GRAM	93.70	-	-	-	93.70		
PROTEIN	GRAM	0.25	-	-	-	0.25		
AVAILABLE CARBOHYDR.	GRAM	2.80	-	-	-	2.80		
MINERALS	GRAM	0.14	-	-	-	0.14		
SODIUM	MILLI	4.00	3.00	-	4.00	4.00		
POTASSIUM	MILLI	21.00	19.00	-	25.00	21.00		
CALCIUM	MILLI	1.00	0.00	-	2.00	1.00		
PHOSPHORUS	MILLI	13.00	-	-	-	13.00		
SELENIUM	MICRO	19.00	-	-	-	19.00		
HISTAMINE	MILLI	0.46	0.40	-	0.53	0.46		
MALIC ACID	MILLI	4.60	-	-	-	4.60		
CITRIC ACID	MILLI	17.40	-	-	-	17.40		
LACTIC ACID	MILLI	18.10	-	-	-	18.10		
OXALIC ACID	MILLI	1.50	-	-	-	1.50		
SUCCINIC ACID	MILLI	1.60	-	-	-	1.60		
FUMARIC ACID	MILLI	1.20	-	-	-	1.20		
GLYCOLIC ACID	MILLI	1.50	-	-	-	1.50		
PYRUVIC ACID	MILLI	4.00	-	-	-	4.00		
GLUCONIC ACID	MILLI	1.70	-	-	-	1.70		
GLUCOSE	MILLI	21.50	-	-	-	21.50		
FRUCTOSE	MILLI	11.20	-	-	-	11.20		
MALTOSE	MILLI	207.00	-	-	-	207.00		
ARABINOSE	MILLI	4.20	-	-	-	4.20		
XYLOSE	MILLI	1.40	-	-	-	1.40		
GALACTOSE	MILLI	1.90	-	-	-	1.90		
DEXTRIN	GRAM	2.60	1.70	-	3.20	2.60		
SORBITOL	MILLI	2.30	-	-	-	2.30		
MANNITOL	MILLI	2.10	-	-	-	2.10		
ETHANOL	GRAM	3.53	3.07	-	3.99	3.53		
EXTRACT	GRAM	5.02	4.12	-	6.20	5.02		
GLYCEROL	MILLI	201.00	-	-	-	201.00		

1 THE TOTAL ENERGY VALUE AMOUNTS TO 189 KJ (45 KCAL) AND WAS CALCULATED FROM EXTRACT AND ETHANOL

APFELWEIN

APPLE WINE
CIDER

CIDRE

		PROTEIN	FAT	CARBOHYDRATES	TOTAL
ENERGY VALUE (AVERAGE)	KJOULE	[1] 0	0.0	0.0	
PER 100 G	(KCAL)	0.0	0.0	0.0	
EDIBLE PORTION					
AMOUNT OF DIGESTIBLE	GRAM	0.00	0.00	0.00	
CONSTITUENTS PER 100 G					
ENERGY VALUE (AVERAGE)	KJOULE	0.0	0.0	0.0	
OF THE DIGESTIBLE	(KCAL)	0.0	0.0	0.0	
FRACTION PER 100 G					
EDIBLE PORTION					

WASTE PERCENTAGE AVERAGE 0.00

CONSTITUENTS	DIM	AV	VARIATION			AVR	NUTR. DENS.	MOLPERC.
WATER	GRAM	92.40	-	-	-	92.40		
MINERALS	GRAM	0.29	0.23	-	0.33	0.29		
TOTAL ACIDS	GRAM	0.53	0.38	-	0.67	0.53		
VOLATILE ACID	MILLI	65.00	18.00	-	100.00	65.00		
INVERT SUGAR	GRAM	0.15	0.06	-	0.38	0.15		
ETHANOL	GRAM	4.99	3.97	-	5.27	4.99		
EXTRACT	GRAM	2.63	2.38	-	2.90	2.63		
TANNINS	GRAM	0.12	0.04	-	0.54	0.12		
GLYCEROL	GRAM	0.41	0.33	-	0.47	0.41		

[1] THE TOTAL ENERGY VALUE AMOUNTS TO 192 KJ (46 KCAL) AND WAS CALCULATED FROM EXTRACT AND ETHANOL

DESSERTWEINE (PORTWEIN, SHERRY, WERMUT)

WINES, SPECIAL (PORT, SHERRY, VERMOUTH)

VINES, DESSERT (PORT, SHERRY, ABSINTH)

		PROTEIN	FAT	CARBOHYDRATES	TOTAL
ENERGY VALUE (AVERAGE)	KJOULE [1]	0.0	0.0	0.0	0.0
PER 100 G EDIBLE PORTION	(KCAL)	0.0	0.0	0.0	0.0
AMOUNT OF DIGESTIBLE CONSTITUENTS PER 100 G	GRAM	0.00	0.00	0.00	
ENERGY VALUE (AVERAGE)	KJOULE	0.0	0.0	0.0	0.0
OF THE DIGESTIBLE FRACTION PER 100 G EDIBLE PORTION	(KCAL)	0.0	0.0	0.0	0.0

WASTE PERCENTAGE AVERAGE 0.00

CONSTITUENTS	DIM	AV	VARIATION			AVR	NUTR. DENS.	MOLPERC.
WATER	GRAM	80.90	-	-	-	80.90		
PROTEIN	GRAM	0.16	0.10	-	0.30	0.16		
FAT	-	0.00	-	-	-	0.00		
AVAILABLE CARBOHYDR.	GRAM	-	1.40	-	22.00			
MINERALS	GRAM	0.28	0.17	-	0.41	0.28		
SODIUM	MILLI	-	4.00	-	18.00			
POTASSIUM	MILLI	-	40.00	-	110.00			
MAGNESIUM	MILLI	9.00	-	-	-	9.00		
CALCIUM	MILLI	7.00	-	-	-	7.00		
IRON	MILLI	0.39	-	-	-	0.39		
COPPER	MILLI	10.00	-	-	-	10.00		
ZINC	MILLI	0.27	-	-	-	0.27		
CHLORIDE	MILLI	10.00	-	-	-	10.00		
VITAMIN B2	MICRO	10.00	-	-	-	10.00		
NICOTINAMIDE	MICRO	50.00	-	-	-	50.00		
VITAMIN B6	MICRO	10.00	-	-	-	10.00		
SALICYLIC ACID	MILLI	0.50	0.50	-	0.51	0.50		
TOTAL SUGAR	GRAM	-	0.62	-	21.80			
ETHANOL	GRAM	14.80	12.20	-	16.70	14.80		
EXTRACT	GRAM	-	3.00	-	23.80			
GLYCEROL	GRAM	0.52	0.23	-	0.96	0.52		

1 THE TOTAL ENERGY VALUE AMOUNTS TO 414 - 895 KJ (99-214 KCAL) AND WAS CALCULATED FROM EXTRACT AND ETHANOL

ROTWEIN
LEICHTE QUALITÄT

RED WINE
LIGHT QUALITY

VIN ROUGE
QUALITÉ LÉGÈRE

		PROTEIN	FAT	CARBOHYDRATES	TOTAL
ENERGY VALUE (AVERAGE) PER 100 G EDIBLE PORTION	KJOULE	3.6 [1]	0.0	0.0	
	(KCAL)	0.9	0.0	0.0	
AMOUNT OF DIGESTIBLE CONSTITUENTS PER 100 G	GRAM	0.18	0.00	0.00	
ENERGY VALUE (AVERAGE) OF THE DIGESTIBLE FRACTION PER 100 G EDIBLE PORTION	KJOULE	3.1	0.0	0.0	
	(KCAL)	0.7	0.0	0.0	

WASTE PERCENTAGE AVERAGE 0.00

CONSTITUENTS	DIM	AV	VARIATION			AVR	NUTR. DENS.	MOLPERC.
WATER	GRAM	89.80	87.90	-	91.40	89.80		
PROTEIN	GRAM	0.22	0.11	-	0.45	0.22		
FAT	-	0.00	-	-	-	0.00		
MINERALS	GRAM	0.27	0.20	-	0.41	0.27		
SODIUM	MILLI	3.00	1.00	-	7.00	3.00		
POTASSIUM	MILLI	102.00	82.00	-	116.00	102.00		
MAGNESIUM	MILLI	8.30	6.50	-	11.00	8.30		
CALCIUM	MILLI	7.00	5.00	-	10.00	7.00		
MANGANESE	MILLI	-	0.00	-	0.20			
IRON	MILLI	0.85	-	-	-	0.85		
COBALT	MICRO	-	0.00	-	1.20			
COPPER	MILLI	0.10	-	-	-	0.10		
ZINC	MILLI	0.18	0.10	-	0.30	0.18		
NICKEL	MICRO	6.30	-	-	-	6.30		
MOLYBDENUM	MICRO	0.30	-	-	-	0.30		
PHOSPHORUS	MILLI	10.00	-	-	-	10.00		
FLUORIDE	MICRO	-	6.00	-	40.00			
IODIDE	MILLI	-	0.01	-	0.06			
BORON	MILLI	0.47	0.28	-	0.66	0.48		
VITAMIN B1	-	TRACES	-	-	-			
VITAMIN B2	-	TRACES	-	-	-			
VITAMIN B6	MICRO	20.00	-	-	-	20.00		
VITAMIN C	MILLI	1.80	-	-	-	1.80		
CADAVERINE	MICRO	40.00	0.00	-	300.00	40.00		
HISTAMINE	MILLI	-	0.00	-	1.50			

1 THE TOTAL ENERGY VALUE AMOUNTS TO 272 KJ (65 KCAL) AND WAS CALCULATED FROM EXTRACT AND ETHANOL

CONSTITUENTS	DIM	AV	VARIATION		AVR	NUTR. DENS.	MOLPERC.
BETA-PHENYLETHYLAMINE	MILLI	-	0.00	- 0.50			
PUTRESCINE	MILLI	0.76	0.07	- 2.40	0.76		
TYRAMINE	MILLI	-	0.00	- 2.00			
MALIC ACID	GRAM	0.51	0.51	- 0.51	0.51		
TARTARIC ACID	GRAM	0.10	0.10	- 0.10	0.10		
LACTIC ACID	GRAM	0.24	0.24	- 0.24	0.24		
FORMIC ACID	MILLI	15.00	15.00	- 15.00	15.00		
OXALIC ACID	MILLI	1.00	1.00	- 1.00	1.00		
ACETIC ACID	MILLI	77.00	36.00	- 117.00	77.00		
SUCCINIC ACID	MILLI	26.00	26.00	- 26.00	26.00		
GLUCONIC ACID	MILLI	35.00	35.00	- 35.00	35.00		
SALICYLIC ACID	MILLI	0.61	0.61	- 0.62	0.62		
GLUCOSE	GRAM	0.31	0.31	- 0.31	0.31		
FRUCTOSE	GRAM	0.25	0.25	- 0.25	0.25		
ARABINOSE	MILLI	54.00	54.00	- 54.00	54.00		
XYLOSE	MILLI	10.00	10.00	- 10.00	10.00		
GALACTOSE	MILLI	15.00	15.00	- 15.00	15.00		
SORBITOL	MILLI	6.00	6.00	- 6.00	6.00		
BUTYLENE GLYCOL	MILLI	60.00	40.00	- 70.00	60.00		
ETHANOL	GRAM	7.85	6.98	- 8.90	7.85		
EXTRACT	GRAM	2.40	1.65	- 3.20	2.40		
GLYCEROL	GRAM	0.80	0.80	- 0.80	0.80		

ROTWEIN
SCHWERE QUALITÄT

RED WINE
HEAVY QUALITY

VIN ROUGE
QUALITÉ LOURDE

		PROTEIN	FAT	CARBOHYDRATES	TOTAL
ENERGY VALUE (AVERAGE) PER 100 G EDIBLE PORTION	KJOULE [1]	3.6	0.0	0.0	
	(KCAL)	0.9	0.0	0.0	
AMOUNT OF DIGESTIBLE CONSTITUENTS PER 100 G	GRAM	0.18	0.00	0.00	
ENERGY VALUE (AVERAGE) OF THE DIGESTIBLE FRACTION PER 100 G EDIBLE PORTION	KJOULE	3.1	0.0	0.0	
	(KCAL)	0.7	0.0	0.0	

WASTE PERCENTAGE AVERAGE 0.00

CONSTITUENTS	DIM	AV	VARIATION			AVR	NUTR. DENS.	MOLPERC.
WATER	GRAM	88.00	86.60	-	89.10	88.00		
PROTEIN	GRAM	0.22	0.11	-	0.45	0.22		
FAT	-	0.00	-	-	-	0.00		
MINERALS	GRAM	0.27	0.22	-	0.35	0.27		
SODIUM	MILLI	4.00	2.00	-	14.00	4.00		
POTASSIUM	MILLI	93.00	75.00	-	112.00	93.00		
MAGNESIUM	MILLI	8.30	6.50	-	11.00	8.30		
CALCIUM	MILLI	7.60	3.40	-	14.00	7.60		
MANGANESE	MILLI	0.20	0.08	-	0.33	0.20		
IRON	MILLI	0.71	0.59	-	1.00	0.71		
COBALT	MICRO	0.70	-	-	-	0.70		
COPPER	MICRO	40.00	10.00	-	80.00	40.00		
ZINC	MILLI	0.13	0.03	-	0.26	0.13		
NICKEL	MICRO	5.00	3.00	-	120.00	5.00		
CHROMIUM	MICRO	1.00	-	-	-	1.00		
MOLYBDENUM	MICRO	1.40	0.70	-	2.00	1.40		
PHOSPHORUS	MILLI	28.00	15.00	-	40.00	28.00		
CHLORIDE	MILLI	9.90	1.80	-	39.00	9.90		
FLUORIDE	MICRO	12.00	0.40	-	20.00	12.00		
SILICON	MILLI	1.00	-	-	-	1.00		
VITAMIN B1	-	0.00	-	-	-	0.00		
VITAMIN B2	MICRO	20.00	-	-	-	20.00		
NICOTINAMIDE	MILLI	0.10	-	-	-	0.10		
PANTOTHENIC ACID	MILLI	0.20	-	-	-	0.20		
VITAMIN B6	MICRO	20.00	-	-	-	20.00		
BIOTIN	MICRO	1.40	-	-	-	1.40		

1 THE TOTAL ENERGY VALUE AMOUNTS TO 326 KJ (78 KCAL) AND WAS CALCULATED FROM EXTRACT AND ETHANOL

CONSTITUENTS	DIM	AV	VARIATION			AVR	NUTR. DENS.	MOLPERC.
FOLIC ACID	MICRO	0.20	-	-	-	0.20		
HISTAMINE	MILLI	-	0.00	-	1.50			
BETA-PHENYLETHYLAMINE	MILLI	-	0.00	-	0.30			
TYRAMINE	MILLI	-	0.00	-	2.00			
MALIC ACID	MILLI	23.00	20.00	-	25.00	23.00		
TARTARIC ACID	MILLI	150.00	119.00	-	180.00	150.00		
LACTIC ACID	MILLI	225.00	213.00	-	230.00	225.00		
ACETIC ACID	MILLI	77.00	36.00	-	117.00	77.00		
TOTAL SUGAR	GRAM	0.30	-	-	-	0.30		
BUTYLENE GLYCOL	MILLI	90.00	40.00	-	120.00	90.00		
ETHANOL	GRAM	9.50	8.90	-	10.30	9.50		
EXTRACT	GRAM	2.53	2.03	-	3.07	2.53		
GLYCEROL	GRAM	0.75	0.67	-	0.91	0.75		

SEKT WEISS (DEUTSCHER SCHAUMWEIN)	**SEKT** WHITE (GERMAN CHAMPAGNE)	**VIN MOUSSEUX** (ALLEMAND)

		PROTEIN	FAT	CARBOHYDRATES	TOTAL
ENERGY VALUE (AVERAGE) PER 100 G EDIBLE PORTION	KJOULE (KCAL)	2.6 [1] 0.6	0.0 0.0	0.0 0.0	
AMOUNT OF DIGESTIBLE CONSTITUENTS PER 100 G	GRAM	0.13	0.00	0.00	
ENERGY VALUE (AVERAGE) OF THE DIGESTIBLE FRACTION PER 100 G EDIBLE PORTION	KJOULE (KCAL)	2.3 0.5	0.0 0.0	0.0 0.0	

WASTE PERCENTAGE AVERAGE 0.00

CONSTITUENTS	DIM	AV	VARIATION			AVR	NUTR. DENS.	MOLPERC.
WATER	GRAM	86.00	84.30	-	87.70	86.00		
PROTEIN	GRAM	0.16	0.12	-	0.28	0.16		
FAT	-	0.00	-	-	-	0.00		
MINERALS	GRAM	0.19	0.17	-	0.23	0.19		
HISTAMINE	MILLI	0.65	0.51	-	0.78	0.65		
TOTAL ACIDS	GRAM	0.73	0.63	-	0.80	0.73		
VOLATILE ACID	MILLI	60.00	40.00	-	70.00	60.00		
TARTARIC ACID	GRAM	0.23	0.19	-	0.30	0.23		
ETHANOL	GRAM	8.90	8.38	-	9.33	8.90		
EXTRACT	GRAM	5.14	3.88	-	6.42	5.14		
GLYCEROL	GRAM	0.66	0.57	-	0.80	0.66		

1 THE TOTAL ENERGY VALUE AMOUNTS TO 351 KJ (84 KCAL) AND WAS CALCULATED FROM EXTRACT AND ETHANOL

WEISSWEIN
MITTLERE QUALITÄT

WHITE WINE
MIDDLE QUALITY

VIN BLANC
QUALITÉ MOYENNE

		PROTEIN	FAT	CARBOHYDRATES	TOTAL
ENERGY VALUE (AVERAGE) PER 100 G EDIBLE PORTION	KJOULE 1	2.5	0.0	0.0	
	(KCAL)	0.6	0.0	0.0	
AMOUNT OF DIGESTIBLE CONSTITUENTS PER 100 G	GRAM	0.13	0.00	0.00	
ENERGY VALUE (AVERAGE) OF THE DIGESTIBLE FRACTION PER 100 G EDIBLE PORTION	KJOULE	2.2	0.0	0.0	
	(KCAL)	0.5	0.0	0.0	

WASTE PERCENTAGE AVERAGE 0.00

CONSTITUENTS	DIM	AV	VARIATION			AVR	NUTR. DENS.	MOLPERC.
WATER	GRAM	89.00	87.50	-	90.40	89.00		
PROTEIN	GRAM	0.15	0.08	-	0.29	0.15		
FAT	-	0.00	-	-	-	0.00		
MINERALS	GRAM	0.24	0.16	-	0.41	0.24		
SODIUM	MILLI	2.00	0.50	-	4.00	2.00		
POTASSIUM	MILLI	82.00	66.00	-	92.00	82.00		
MAGNESIUM	MILLI	10.00	6.00	-	15.00	10.00		
CALCIUM	MILLI	9.00	8.00	-	11.00	9.00		
MANGANESE	MILLI	0.14	0.00	-	0.30	0.14		
IRON	MILLI	0.60	0.40	-	1.00	0.60		
COBALT	MICRO	-	0.00	-	1.20			
COPPER	MICRO	70.00	40.00	-	100.00	70.00		
ZINC	MILLI	0.23	0.10	-	0.34	0.23		
NICKEL	MICRO	6.30	-	-	-	6.30		
MOLYBDENUM	MICRO	-	0.20	-	1.70			
PHOSPHORUS	MILLI	15.00	10.00	-	20.00	15.00		
CHLORIDE	MILLI	5.00	2.00	-	8.00	5.00		
FLUORIDE	MICRO	30.00	20.00	-	50.00	30.00		
IODIDE	MILLI	-	0.01	-	0.06			
BORON	MILLI	0.25	0.19	-	0.31	0.25		
VITAMIN B1	-	TRACES	-	-	-			
VITAMIN B2	MICRO	10.00	-	-	-	10.00		
NICOTINAMIDE	MILLI	0.10	-	-	-	0.10		
PANTOTHENIC ACID	MICRO	22.00	-	-	-	22.00		
VITAMIN B6	MICRO	20.00	-	-	-	20.00		
BIOTIN	MICRO	0.50	-	-	-	0.50		

1 THE TOTAL ENERGY VALUE AMOUNTS TO 293 KJ (70 KCAL) AND WAS CALCULATED FROM EXTRACT AND ETHANOL

CONSTITUENTS	DIM	AV		VARIATION			AVR	NUTR. DENS.	MOLPERC.
FOLIC ACID	MICRO	0.30		-	-	-	0.30		
CADAVERINE	MILLI	0.14		0.00	-	0.28	0.14		
HISTAMINE	MILLI	-		0.03	-	0.50			
BETA-PHENYLETHYLAMINE	MILLI	-		0.00	-	0.38			
PUTRESCINE	MILLI	0.12		0.00	-	0.17	0.12		
SPERMINE	MILLI	-		0.00	-	0.08			
TYRAMINE	MILLI	-		0.00	-	0.30			
MALIC ACID	GRAM	0.35		0.29	-	0.39	0.36		
CITRIC ACID	MILLI	14.00		14.00	-	14.00	14.00		
TARTARIC ACID	GRAM	0.14		0.13	-	0.15	0.14		
LACTIC ACID	GRAM	0.17		0.17	-	0.17	0.17		
FORMIC ACID	MILLI	22.00		20.00	-	24.00	22.00		
OXALIC ACID	MILLI	1.66		1.00	-	2.00	1.67		
ACETIC ACID	MILLI	30.00		17.00	-	53.00	30.00		
SUCCINIC ACID	MILLI	21.33		19.00	-	24.00	21.33		
GLUCONIC ACID	MILLI	5.50		5.00	-	6.00	5.50		
SALICYLIC ACID	MILLI	0.63		0.63	-	0.64	0.63		
GLUCOSE	GRAM	0.38	2	0.38	-	0.38	0.38		
FRUCTOSE	GRAM	0.41		0.41	-	0.41	0.41		
ARABINOSE	MILLI	64.50		52.00	-	77.00	64.50		
XYLOSE	MILLI	9.33		6.00	-	13.00	9.33		
GALACTOSE	MILLI	10.66		6.00	-	15.00	10.67		
SORBITOL	MILLI	5.00		4.00	-	6.00	5.00		
BUTYLENE GLYCOL	MILLI	42.00		37.00	-	46.00	42.00		
ETHANOL	GRAM	8.40		7.40	-	9.10	8.40		
EXTRACT	GRAM	2.60		2.20	-	3.40	2.60		
GLYCEROL	GRAM	0.70		0.69	-	0.71	0.70		

2 SUGAR CONTENT OF DECORATED WINES. THE SUGAR CONTENT OF WINES WHICH ARE FERMENTED TOTALLY IS LOWER

WEINBRAND

BRANDY
COGNAC

COGNAC

		PROTEIN	FAT	CARBOHYDRATES	TOTAL
ENERGY VALUE (AVERAGE) PER 100 G EDIBLE PORTION	KJOULE [1]	0.0	0.0	0.0	0.0
	(KCAL)	0.0	0.0	0.0	0.0
AMOUNT OF DIGESTIBLE CONSTITUENTS PER 100 G	GRAM	0.00	0.00	0.00	
ENERGY VALUE (AVERAGE) OF THE DIGESTIBLE FRACTION PER 100 G EDIBLE PORTION	KJOULE	0.0	0.0	0.0	0.0
	(KCAL)	0.0	0.0	0.0	0.0

WASTE PERCENTAGE AVERAGE 0.00

CONSTITUENTS	DIM	AV	VARIATION	AVR	NUTR. DENS.	MOLPERC.
WATER	GRAM	64.90	61.70 - 68.90	64.90		
MINERALS	MILLI	26.00	13.00 - 39.00	26.00		
SODIUM	MILLI	2.40	0.60 - 4.80	2.40		
POTASSIUM	MILLI	1.60	1.20 - 2.00	1.60		
ETHANOL	GRAM	33.10	30.10 - 35.00	33.10		
EXTRACT	GRAM	1.98	1.00 - 3.25	1.98		

1 THE TOTAL ENERGY VALUE AMOUNTS TO 1016 KJ (243 KCAL) AND WAS CALCULATED FROM EXTRACT AND ETHANOL

WHISKY WHISKY WHISKY

		PROTEIN	FAT	CARBOHYDRATES	TOTAL
ENERGY VALUE (AVERAGE)	KJOULE [1]	0.0	0.0	0.0	0.0
PER 100 G	(KCAL)	0.0	0.0	0.0	0.0
EDIBLE PORTION					
AMOUNT OF DIGESTIBLE	GRAM	0.00	0.00	0.00	
CONSTITUENTS PER 100 G					
ENERGY VALUE (AVERAGE)	KJOULE	0.0	0.0	0.0	0.0
OF THE DIGESTIBLE	(KCAL)	0.0	0.0	0.0	0.0
FRACTION PER 100 G					
EDIBLE PORTION					

WASTE PERCENTAGE AVERAGE 0.00

CONSTITUENTS	DIM	AV	VARIATION	AVR	NUTR. DENS.	MOLPERC.
WATER	GRAM	64.70	64.50 - 65.00	64.70		
MINERALS	MILLI	5.00	4.00 - 7.00	5.00		
SODIUM	MILLI	0.14	0.10 - 0.30	0.14		
POTASSIUM	MILLI	2.80	2.60 - 3.00	2.80		
MAGNESIUM	MILLI	0.25	0.19 - 0.28	0.25		
CALCIUM	MILLI	1.50	1.10 - 1.70	1.50		
ETHANOL	GRAM	35.20	34.90 - 35.40	35.20		
EXTRACT	GRAM	0.11	0.09 - 0.12	0.11		

1 THE TOTAL ENERGY VALUE AMOUNTS IS 1046 KJ (250 KCAL) AND WAS CALCULATED FROM EXTRACT AND ETHANOL

Erfrischungsgetränke

Soft drinks
Boissons rafraîchissantes

COLA-GETRÄNKE COLA DRINKS BOISSONS COLA

		PROTEIN	FAT	CARBOHYDRATES	TOTAL
ENERGY VALUE (AVERAGE)	KJOULE	0.0	0.0	176	176
PER 100 G	(KCAL)	0.0	0.0	42	42
EDIBLE PORTION					
AMOUNT OF DIGESTIBLE	GRAM	0.00	0.00	10.50	
CONSTITUENTS PER 100 G					
ENERGY VALUE (AVERAGE)	KJOULE	0.0	0.0	176	176
OF THE DIGESTIBLE	(KCAL)	0.0	0.0	42	42
FRACTION PER 100 G					
EDIBLE PORTION					

WASTE PERCENTAGE AVERAGE 0.00

CONSTITUENTS	DIM	AV	VARIATION			AVR	NUTR. DENS.		MOLPERC.
WATER	GRAM	88.00	-	-	-	88.00	GRAM/MJ	500.77	
PROTEIN	-	TRACES	-	-	-				
AVAILABLE CARBOHYDR.	GRAM	10.50	-	-	-	10.50	GRAM/MJ	59.75	
SODIUM	MILLI	6.00	3.00	-	10.00	6.00	MILLI/MJ	34.14	
POTASSIUM	MILLI	1.00	0.00	-	3.00	1.00	MILLI/MJ	5.69	
MAGNESIUM	MILLI	1.00	-	-	-	1.00	MILLI/MJ	5.69	
CALCIUM	MILLI	4.00	0.00	-	9.00	4.00	MILLI/MJ	22.76	
COPPER	MICRO	30.00	-	-	-	30.00	MICRO/MJ	170.72	
PHOSPHORUS	MILLI	14.70	0.00	-	24.00	14.70	MILLI/MJ	83.65	
CHLORIDE	MILLI	10.00	-	-	-	10.00	MILLI/MJ	56.91	
VITAMIN A	-	0.00	-	-	-	0.00			
VITAMIN B1	-	0.00	-	-	-	0.00			
VITAMIN B2	-	0.00	-	-	-	0.00			
NICOTINAMIDE	-	0.00	-	-	-	0.00			
VITAMIN C	-	0.00	-	-	-	0.00			
TOTAL SUGAR	GRAM	10.50	-	-	-	10.50	GRAM/MJ	59.75	

Kakao

Cocoa
Cacao

KAKAOPULVER SCHWACH ENTÖLT

COCOA POWDER OIL PARTIALLY REMOVED

CACAO EN POUDRE LÉGÈREMENT DÉSHUILÉ

		PROTEIN	FAT	CARBOHYDRATES	TOTAL
ENERGY VALUE (AVERAGE) PER 100 G EDIBLE PORTION	KJOULE (KCAL)	360 86	953 228	181 43	1495 357
AMOUNT OF DIGESTIBLE CONSTITUENTS PER 100 G	GRAM	8.31	22.05	10.84	
ENERGY VALUE (AVERAGE) OF THE DIGESTIBLE FRACTION PER 100 G EDIBLE PORTION	KJOULE (KCAL)	151 36	858 205	181 43	1191 285

WASTE PERCENTAGE AVERAGE 0.00

CONSTITUENTS	DIM	AV		VARIATION			AVR	NUTR. DENS.		MOLPERC.
WATER	GRAM	5.60		3.90	-	6.65	5.60	GRAM/MJ	4.70	
PROTEIN	GRAM	19.80		17.00	-	22.00	19.80	GRAM/MJ	16.63	
FAT	GRAM	24.50		20.00	-	28.40	24.50	GRAM/MJ	20.58	
AVAILABLE CARBOHYDR.	GRAM	10.84		-	-	-	10.84	GRAM/MJ	9.10	
TOTAL DIETARY FIBRE	GRAM	37.73	1	-	-	-	37.73	GRAM/MJ	31.69	
MINERALS	GRAM	6.53		5.00	-	8.00	6.53	GRAM/MJ	5.48	
SODIUM	MILLI	17.00		0.50	-	27.00	17.00	MILLI/MJ	14.28	
POTASSIUM	GRAM	1.92		0.98	-	3.20	1.92	GRAM/MJ	1.61	
MAGNESIUM	MILLI	414.00		370.00	-	457.00	414.00	MILLI/MJ	347.68	
CALCIUM	MILLI	114.00		76.00	-	140.00	114.00	MILLI/MJ	95.74	
IRON	MILLI	12.50		11.60	-	15.00	12.50	MILLI/MJ	10.50	
COPPER	MILLI	3.90		-	-	-	3.90	MILLI/MJ	3.28	
ZINC	MILLI	3.50		2.00	-	5.00	3.50	MILLI/MJ	2.94	
NICKEL	MILLI	1.23		-	-	-	1.23	MILLI/MJ	1.03	
CHROMIUM	MICRO	60.00		-	-	-	60.00	MICRO/MJ	50.39	
MOLYBDENUM	MICRO	73.00		-	-	-	73.00	MICRO/MJ	61.31	
PHOSPHORUS	MILLI	656.00		600.00	-	712.00	656.00	MILLI/MJ	550.91	
CHLORIDE	MILLI	32.00		-	-	-	32.00	MILLI/MJ	26.87	
FLUORIDE	MILLI	0.12		0.05	-	0.20	0.12	MILLI/MJ	0.10	
BORON	MILLI	2.20		-	-	-	2.20	MILLI/MJ	1.85	
VITAMIN B1	MILLI	0.13		-	-	-	0.13	MILLI/MJ	0.11	
VITAMIN B2	MILLI	0.40		-	-	-	0.40	MILLI/MJ	0.34	
NICOTINAMIDE	MILLI	2.70		-	-	-	2.70	MILLI/MJ	2.27	
PANTOTHENIC ACID	MILLI	1.10		-	-	-	1.10	MILLI/MJ	0.92	
VITAMIN B6	MILLI	0.14		-	-	-	0.14	MILLI/MJ	0.12	
BIOTIN	MICRO	20.00		-	-	-	20.00	MICRO/MJ	16.80	
FOLIC ACID	MICRO	38.00		-	-	-	38.00	MICRO/MJ	31.91	
VITAMIN C	-	0.00		-	-	-	0.00			
OXALIC ACID	MILLI	470.00		450.00	-	480.00	470.00	MILLI/MJ	394.70	
REDUCING SUGAR	GRAM	2.10		-	-	-	2.10	GRAM/MJ	1.76	
NONREDUCING SUGAR	GRAM	0.14		-	-	-	0.14	GRAM/MJ	0.12	
STARCH	GRAM	8.60		-	-	-	8.60	GRAM/MJ	7.22	
THEOBROMINE	GRAM	2.30		1.90	-	2.70	2.30	GRAM/MJ	1.93	

1 ESTIMATED BY THE DIFFERENCE METHOD:
100 - (WATER + PROTEIN + FAT + AVAILABLE CARBOHYDRATES + MINERALS)

SCHOKOLADE MILCHFREI 1	**CHOCOLATE** MILK FREE 1	**CHOCOLAT** SANS LAIT 1

		PROTEIN	FAT	CARBOHYDRATES	TOTAL
ENERGY VALUE (AVERAGE) PER 100 G EDIBLE PORTION	KJOULE	96	1167	787	2050
	(KCAL)	23	279	188	490
AMOUNT OF DIGESTIBLE CONSTITUENTS PER 100 G	GRAM	2.22	27.00	47.00	
ENERGY VALUE (AVERAGE) OF THE DIGESTIBLE FRACTION PER 100 G EDIBLE PORTION	KJOULE	41	1051	787	1878
	(KCAL)	9.7	251	188	449

WASTE PERCENTAGE AVERAGE 0.00

CONSTITUENTS	DIM	AV		VARIATION		AVR	NUTR.	DENS.	MOLPERC.
WATER	GRAM	0.90	0.60	-	1.40	0.90	GRAM/MJ	0.48	
PROTEIN	GRAM	5.30	3.50	-	7.10	5.30	GRAM/MJ	2.82	
FAT	GRAM	30.00	28.70	-	33.50	30.00	GRAM/MJ	15.98	
AVAILABLE CARBOHYDR.	GRAM	47.00	-	-	-	47.00	GRAM/MJ	25.03	
TOTAL DIETARY FIBRE	GRAM	15.60 [2]	-	-	-	15.60	GRAM/MJ	8.31	
MINERALS	GRAM	1.20	0.80	-	1.60	1.20	GRAM/MJ	0.64	
SODIUM	MILLI	19.00	-	-	-	19.00	MILLI/MJ	10.12	
POTASSIUM	MILLI	397.00	-	-	-	397.00	MILLI/MJ	211.43	
MAGNESIUM	MILLI	100.00	-	-	-	100.00	MILLI/MJ	53.26	
CALCIUM	MILLI	63.00	-	-	-	63.00	MILLI/MJ	33.55	
IRON	MILLI	3.20	2.50	-	4.40	3.20	MILLI/MJ	1.70	
COPPER	MILLI	-	1.10	-	2.70				
ZINC	MILLI	2.00	-	-	-	2.00	MILLI/MJ	1.07	
NICKEL	MILLI	0.26	-	-	-	0.26	MILLI/MJ	0.14	
PHOSPHORUS	MILLI	287.00	-	-	-	287.00	MILLI/MJ	152.85	
CHLORIDE	MILLI	100.00	-	-	-	100.00	MILLI/MJ	53.26	
FLUORIDE	MICRO	50.00	-	-	-	50.00	MICRO/MJ	26.63	
IODIDE	MICRO	5.50	-	-	-	5.50	MICRO/MJ	2.93	
VITAMIN B1	MICRO	40.00	30.00	-	50.00	40.00	MICRO/MJ	21.30	
VITAMIN B2	MILLI	0.13	0.08	-	0.17	0.13	MILLI/MJ	0.07	
NICOTINAMIDE	MILLI	0.86	0.57	-	1.10	0.86	MILLI/MJ	0.46	
PANTOTHENIC ACID	MILLI	0.35	0.23	-	0.46	0.35	MILLI/MJ	0.19	
VITAMIN B6	MICRO	50.00	30.00	-	60.00	50.00	MICRO/MJ	26.63	
BIOTIN	MICRO	6.00	4.00	-	8.00	6.00	MICRO/MJ	3.20	
FOLIC ACID	MICRO	10.00	-	-	-	10.00	MICRO/MJ	5.33	
VITAMIN C	-	0.00	-	-	-	0.00			
SUCROSE	GRAM	47.00	44.00	-	60.00	47.00	GRAM/MJ	25.03	
THEOBROMINE	GRAM	0.63	0.40	-	0.80	0.63	GRAM/MJ	0.34	

1 THE AVERAGE VALUES ARE IN ACCORDANCE WITH THE RECIPE OF HALF-BITTER-CHOCOLATE (45 % COCOA, 5 % COCOABUTTER, 50 % SUGAR)
VARIATION: HALF-SWEET: 30 % COCOA, 12 % COCOABUTTER, 58 % SUGAR
BITTER: 60 % COCOA, 40 % SUGAR

2 ESTIMATED BY THE DIFFERENCE METHOD:
100 - (WATER + PROTEIN + FAT + AVAILABLE CARBOHYDRATES + MINERALS)

SCHOKOLADE / CHOCOLATE / CHOCOLAT

MILCHSCHOKOLADE / MILK CHOCOLATE / CHOCOLAT AU LAIT

		PROTEIN	FAT	CARBOHYDRATES	TOTAL
ENERGY VALUE (AVERAGE) PER 100 G EDIBLE PORTION	KJOULE	169	1226	905	2300
	(KCAL)	40	293	216	550
AMOUNT OF DIGESTIBLE CONSTITUENTS PER 100 G	GRAM	7.17	28.35	54.10	
ENERGY VALUE (AVERAGE) OF THE DIGESTIBLE FRACTION PER 100 G EDIBLE PORTION	KJOULE	132	1103	905	2140
	(KCAL)	31	264	216	511

WASTE PERCENTAGE AVERAGE 0.00

CONSTITUENTS	DIM	AV	VARIATION			AVR	NUTR. DENS.		MOLPERC.
WATER	GRAM	1.40	1.20	-	1.50	1.40	GRAM/MJ	0.65	
PROTEIN	GRAM	9.20	6.80	-	11.60	9.20	GRAM/MJ	4.30	
FAT	GRAM	31.50	29.40	-	33.60	31.50	GRAM/MJ	14.72	
AVAILABLE CARBOHYDR.	GRAM	54.10	46.00	-	62.30	54.10	GRAM/MJ	25.28	
MINERALS	GRAM	2.20	1.50	-	2.80	2.20	GRAM/MJ	1.03	
SODIUM	MILLI	58.00	45.00	-	80.00	58.00	MILLI/MJ	27.10	
POTASSIUM	MILLI	471.00	441.00	-	500.00	471.00	MILLI/MJ	220.09	
MAGNESIUM	MILLI	86.00	62.00	-	104.00	86.00	MILLI/MJ	40.19	
CALCIUM	MILLI	214.00	182.00	-	257.00	214.00	MILLI/MJ	100.00	
IRON	MILLI	2.30	1.50	-	4.00	2.30	MILLI/MJ	1.07	
COBALT	MICRO	6.00	4.00	-	10.00	6.00	MICRO/MJ	2.80	
COPPER	MILLI	1.30	0.27	-	4.00	1.30	MILLI/MJ	0.61	
ZINC	MILLI	2.00	-	-	-	2.00	MILLI/MJ	0.93	
NICKEL	MILLI	0.15	0.04	-	0.22	0.15	MILLI/MJ	0.07	
CHROMIUM	MICRO	6.00	4.00	-	10.00	6.00	MICRO/MJ	2.80	
PHOSPHORUS	MILLI	242.00	200.00	-	283.00	242.00	MILLI/MJ	113.08	
CHLORIDE	MILLI	131.00	-	-	-	131.00	MILLI/MJ	61.21	
FLUORIDE	MICRO	50.00	-	-	-	50.00	MICRO/MJ	23.36	
IODIDE	MICRO	5.50	-	-	-	5.50	MICRO/MJ	2.57	
BORON	MICRO	10.00	-	-	-	10.00	MICRO/MJ	4.67	
SELENIUM	MICRO	2.00	1.00	-	3.00	2.00	MICRO/MJ	0.93	
SILICON	MILLI	1.00	-	-	-	1.00	MILLI/MJ	0.47	
VITAMIN A	MICRO	53.00	45.00	-	60.00	53.00	MICRO/MJ	24.77	
CAROTENE	MICRO	34.00	27.00	-	40.00	34.00	MICRO/MJ	15.89	
VITAMIN B1	MILLI	0.11	0.08	-	0.14	0.11	MILLI/MJ	0.05	
VITAMIN B2	MILLI	0.37	0.18	-	0.45	0.37	MILLI/MJ	0.17	
NICOTINAMIDE	MILLI	0.46	0.32	-	0.59	0.46	MILLI/MJ	0.21	
PANTOTHENIC ACID	MILLI	0.90	0.68	-	1.10	0.90	MILLI/MJ	0.42	
VITAMIN B6	MILLI	0.11	0.08	-	0.13	0.11	MILLI/MJ	0.05	
BIOTIN	MICRO	3.00	-	-	-	3.00	MICRO/MJ	1.40	
FOLIC ACID	MICRO	10.00	-	-	-	10.00	MICRO/MJ	4.67	
VITAMIN C	-	0.00	-	-	-	0.00			
SUCROSE	GRAM	44.60	38.40	-	50.90	44.60	GRAM/MJ	20.84	
LACTOSE	GRAM	9.50	7.60	-	11.40	9.50	GRAM/MJ	4.44	
THEOBROMINE	GRAM	0.18	0.15	-	0.20	0.18	GRAM/MJ	0.08	

Kaffee und Tee

Coffee and tea
Café et thé

KAFFEE
GRÜN
(ROHKAFFEE)

COFFEE
GREEN
(COFFEE UNPROCESSED)

CAFÉ
VERT

		PROTEIN	FAT	CARBOHYDRATES	TOTAL
ENERGY VALUE (AVERAGE)	KJOULE	204	510	112	826
PER 100 G	(KCAL)	49	122	27	197
EDIBLE PORTION					
AMOUNT OF DIGESTIBLE	GRAM	4.70	11.79	6.70	
CONSTITUENTS PER 100 G					
ENERGY VALUE (AVERAGE)	KJOULE	86	459	112	657
OF THE DIGESTIBLE	(KCAL)	20	110	27	157
FRACTION PER 100 G					
EDIBLE PORTION					

WASTE PERCENTAGE AVERAGE 0.00

CONSTITUENTS	DIM	AV	VARIATION			AVR	NUTR. DENS.		MOLPERC.
WATER	GRAM	10.20	7.49	-	11.30	10.20	GRAM/MJ	15.54	
PROTEIN	GRAM	11.20	9.36	-	12.60	11.20	GRAM/MJ	17.06	
FAT	GRAM	13.10	11.60	-	16.50	13.10	GRAM/MJ	19.95	
AVAILABLE CARBOHYDR.	GRAM	6.70	-	-	-	6.70	GRAM/MJ	10.21	
TOTAL DIETARY FIBRE	GRAM	54.80 1	-	-	-	54.80	GRAM/MJ	83.47	
MINERALS	GRAM	4.00	3.80	-	4.70	4.00	GRAM/MJ	6.09	
SODIUM	MILLI	2.44	0.88	-	4.00	2.44	MILLI/MJ	3.72	
POTASSIUM	GRAM	1.86	1.71	-	2.08	1.86	GRAM/MJ	2.83	
MAGNESIUM	MILLI	220.00	200.00	-	300.00	220.00	MILLI/MJ	335.11	
CALCIUM	MILLI	160.00	120.00	-	180.00	160.00	MILLI/MJ	243.71	
IRON	MILLI	20.00	14.00	-	28.00	20.00	MILLI/MJ	30.46	
PHOSPHORUS	MILLI	160.00	130.00	-	170.00	160.00	MILLI/MJ	243.71	
CHLORIDE	MILLI	27.00	18.00	-	44.00	27.00	MILLI/MJ	41.13	
VITAMIN B1	MILLI	0.21	-	-	-	0.21	MILLI/MJ	0.32	
VITAMIN B2	MILLI	0.23	-	-	-	0.23	MILLI/MJ	0.35	
NICOTINAMIDE	MILLI	2.20	-	-	-	2.20	MILLI/MJ	3.35	
PANTOTHENIC ACID	MILLI	1.00	-	-	-	1.00	MILLI/MJ	1.52	
VITAMIN B6	MILLI	0.14	-	-	-	0.14	MILLI/MJ	0.21	
FOLIC ACID	MICRO	20.00	-	-	-	20.00	MICRO/MJ	30.46	
CHLOROGENIC ACID	GRAM	3.01	1.07	-	4.95	3.01	GRAM/MJ	4.58	
SUCROSE	GRAM	6.70	5.00	-	7.67	6.70	GRAM/MJ	10.22	
COFFEINE	GRAM	1.30	0.80	-	1.59	1.30	GRAM/MJ	1.98	

1 ESTIMATED BY THE DIFFERENCE METHOD
100 - (WATER + PROTEIN + FAT + AVAILABLE CARBOHYDRATES + MINERALS)

KAFFEE GERÖSTET | COFFEE ROAST | CAFÉ TORRÉFIÉ

		PROTEIN	FAT	CARBOHYDRATES	TOTAL
ENERGY VALUE (AVERAGE) PER 100 G EDIBLE PORTION	KJOULE	242	498	25	765
	(KCAL)	58	119	6.0	183
AMOUNT OF DIGESTIBLE CONSTITUENTS PER 100 G	GRAM	5.58	11.52	1.50	
ENERGY VALUE (AVERAGE) OF THE DIGESTIBLE FRACTION PER 100 G EDIBLE PORTION	KJOULE	102	448	25	575
	(KCAL)	24	107	6.0	137

WASTE PERCENTAGE AVERAGE 0.00

CONSTITUENTS	DIM	AV	VARIATION			AVR	NUTR.	DENS.	MOLPERC.
WATER	GRAM	2.75	0.36	-	5.00	2.75	GRAM/MJ	4.78	
PROTEIN	GRAM	13.30	12.00	-	14.70	13.30	GRAM/MJ	23.13	
FAT	GRAM	12.80	8.30	-	14.80	12.80	GRAM/MJ	22.26	
AVAILABLE CARBOHYDR.	GRAM	1.50	-	-	-	1.50	GRAM/MJ	2.61	
TOTAL DIETARY FIBRE	GRAM	65.52 [1]	-	-	-	65.52	GRAM/MJ	113.94	
MINERALS	GRAM	4.13	3.15	-	5.17	4.13	GRAM/MJ	7.18	
SODIUM	MILLI	3.60	1.40	-	8.90	3.60	MILLI/MJ	6.26	
POTASSIUM	GRAM	1.73	1.51	-	2.00	1.73	GRAM/MJ	3.01	
MAGNESIUM	MILLI	201.00	162.00	-	240.00	201.00	MILLI/MJ	349.55	
CALCIUM	MILLI	146.00	68.00	-	227.00	146.00	MILLI/MJ	253.90	
IRON	MILLI	16.80	4.70	-	28.90	16.80	MILLI/MJ	29.22	
COPPER	MILLI	3.00	-	-	-	3.00	MILLI/MJ	5.22	
NICKEL	MICRO	77.00	58.00	-	98.00	77.00	MICRO/MJ	133.91	
PHOSPHORUS	MILLI	192.00	145.00	-	234.00	192.00	MILLI/MJ	333.90	
CHLORIDE	MILLI	19.00	-	-	-	19.00	MILLI/MJ	33.04	
FLUORIDE	MICRO	90.00	20.00	-	160.00	90.00	MICRO/MJ	156.51	
IODIDE	MICRO	8.00	-	-	-	8.00	MICRO/MJ	13.91	
BORON	MILLI	1.37	1.32	-	1.40	1.37	MILLI/MJ	2.38	
SELENIUM	MICRO	-	0.00	-	25.00				
VITAMIN B1	MICRO	70.00	-	-	-	70.00	MICRO/MJ	121.73	
VITAMIN B2	MILLI	0.18	0.05	-	0.30	0.18	MILLI/MJ	0.31	
NICOTINAMIDE	MILLI	13.80	9.33	-	17.00	13.80	MILLI/MJ	24.00	
PANTOTHENIC ACID	MILLI	0.23	-	-	-	0.23	MILLI/MJ	0.40	
VITAMIN B6	MICRO	1.00	-	-	-	1.00	MICRO/MJ	1.74	
FOLIC ACID	MICRO	22.00	-	-	-	22.00	MICRO/MJ	38.26	
CHLOROGENIC ACID	GRAM	4.11	3.29	-	4.89	4.11	GRAM/MJ	7.15	
SUCROSE	GRAM	1.50	0.41	-	2.83	1.50	GRAM/MJ	2.61	
COFFEINE	GRAM	1.28	0.86	-	2.19	1.28	GRAM/MJ	2.23	

1 ESTIMATED BY THE DIFFERENCE METHOD:
100 - (WATER + PROTEIN + FAT + AVAILABLE CARBOHYDRATES + MINERALS)

KAFFEE-EXTRAKT-PULVER | COFFEE EXTRACT POWDER | CAFÉ EN POUDRE, EXTRAIT

		PROTEIN	FAT	CARBOHYDRATES	TOTAL
ENERGY VALUE (AVERAGE) PER 100 G EDIBLE PORTION	KJOULE	204	0.0	137	341
	(KCAL)	49	0.0	33	82
AMOUNT OF DIGESTIBLE CONSTITUENTS PER 100 G	GRAM	4.70	0.00	8.20	
ENERGY VALUE (AVERAGE) OF THE DIGESTIBLE FRACTION PER 100 G EDIBLE PORTION	KJOULE	86	0.0	137	223
	(KCAL)	20	0.0	33	53

WASTE PERCENTAGE AVERAGE 0.00

CONSTITUENTS	DIM	AV	VARIATION			AVR	NUTR. DENS.		MOLPERC.
WATER	GRAM	-	1.00	-	6.00				
PROTEIN	GRAM	11.20	9.30	-	13.10	11.20	GRAM/MJ	50.26	
AVAILABLE CARBOHYDR.	GRAM	8.20	-	-	-	8.20	GRAM/MJ	36.80	
MINERALS	GRAM	-	7.60	-	14.60				
SODIUM	MILLI	58.00	28.00	-	88.00	58.00	MILLI/MJ	260.26	
POTASSIUM	GRAM	4.38	4.36	-	4.40	4.38	GRAM/MJ	19.65	
MAGNESIUM	MILLI	390.00	-	-	-	390.00	GRAM/MJ	1.75	
CALCIUM	MILLI	168.00	106.00	-	235.00	168.00	MILLI/MJ	753.87	
IRON	MILLI	4.40	-	-	-	4.40	MILLI/MJ	19.74	
COPPER	MICRO	50.00	-	-	-	50.00	MICRO/MJ	224.37	
ZINC	MILLI	0.50	-	-	-	0.50	MILLI/MJ	2.24	
NICKEL	MICRO	96.00	61.00	-	130.00	96.00	MICRO/MJ	430.78	
CHLORIDE	MILLI	50.00	-	-	-	50.00	MILLI/MJ	224.37	
VITAMIN B2	MILLI	0.11	-	-	-	0.11	MILLI/MJ	0.49	
NICOTINAMIDE	MILLI	22.00	-	-	-	22.00	MILLI/MJ	98.72	
PANTOTHENIC ACID	MILLI	0.40	-	-	-	0.40	MILLI/MJ	1.79	
VITAMIN B6	MICRO	30.00	-	-	-	30.00	MICRO/MJ	134.62	
SALICYLIC ACID	MILLI	0.46	0.46	-	0.47	0.47	MILLI/MJ	2.09	
REDUCING SUGAR	GRAM	8.20	-	-	-	8.20	GRAM/MJ	36.80	

ZICHORIENKAFFEE CHICORY COFFEE CAFÉ EN CHICOREE

		PROTEIN	FAT	CARBOHYDRATES	TOTAL
ENERGY VALUE (AVERAGE)	KJOULE	107	62	0.0	169
PER 100 G	(KCAL)	26	15	0.0	40
EDIBLE PORTION					
AMOUNT OF DIGESTIBLE CONSTITUENTS PER 100 G	GRAM	5.03	1.44	0.00	
ENERGY VALUE (AVERAGE)	KJOULE	79	56	0.0	135
OF THE DIGESTIBLE	(KCAL)	19	13	0.0	32
FRACTION PER 100 G EDIBLE PORTION					

WASTE PERCENTAGE AVERAGE 0.00

CONSTITUENTS	DIM	AV	VARIATION			AVR	NUTR.	DENS.	MOLPERC.
WATER	GRAM	13.30	-	-	-	13.30	GRAM/MJ	98.53	
PROTEIN	GRAM	6.80	-	-	-	6.80	GRAM/MJ	50.38	
FAT	GRAM	1.60	-	-	-	1.60	GRAM/MJ	11.85	
TOTAL DIETARY FIBRE	GRAM	73.90 [1]	-	-	-	73.90	GRAM/MJ	547.47	
MINERALS	GRAM	4.40	-	-	-	4.40	GRAM/MJ	32.60	

1 ESTIMATED BY THE DIFFERENCE METHOD
100 - (WATER + PROTEIN + FAT + MINERALS)

TEE	TEA	THÉ
(SCHWARZER TEE)	(BLACK TEA)	(THÉ NOIR)
CAMELLIA SINENSIS (L.) KUNTZE		

		PROTEIN	FAT	CARBOHYDRATES	TOTAL
ENERGY VALUE (AVERAGE)	KJOULE	473	198	13	685
PER 100 G EDIBLE PORTION	(KCAL)	113	47	3.2	164
AMOUNT OF DIGESTIBLE CONSTITUENTS PER 100 G	GRAM	10.92	4.59	0.80	
ENERGY VALUE (AVERAGE)	KJOULE	199	179	13	391
OF THE DIGESTIBLE FRACTION PER 100 G EDIBLE PORTION	(KCAL)	48	43	3.2	93

WASTE PERCENTAGE AVERAGE 0.00

CONSTITUENTS	DIM	AV		VARIATION			AVR	NUTR. DENS.		MOLPERC.
WATER	GRAM	7.90		7.40	-	8.50	7.90	GRAM/MJ	20.22	
PROTEIN	GRAM	26.00		24.00	-	30.00	26.00	GRAM/MJ	66.54	
FAT	GRAM	5.10		2.00	-	8.20	5.10	GRAM/MJ	13.05	
AVAILABLE CARBOHYDR.	GRAM	0.80		-	-	-	0.80	GRAM/MJ	2.05	
TOTAL DIETARY FIBRE	GRAM	54.60	1	-	-	-	54.60	GRAM/MJ	139.74	
MINERALS	GRAM	5.60		5.10	-	7.20	5.60	GRAM/MJ	14.33	
SODIUM	MILLI	6.00		4.00	-	7.00	6.00	MILLI/MJ	15.36	
POTASSIUM	GRAM	1.79		1.77	-	1.80	1.79	GRAM/MJ	4.58	
MAGNESIUM	MILLI	184.00		-	-	-	184.00	MILLI/MJ	470.90	
CALCIUM	MILLI	302.00		289.00	-	314.00	302.00	MILLI/MJ	772.89	
MANGANESE	MILLI	73.40		-	-	-	73.40	MILLI/MJ	187.85	
IRON	MILLI	17.20		-	-	-	17.20	MILLI/MJ	44.02	
COBALT	MILLI	0.10		-	-	-	0.10	MILLI/MJ	0.26	
COPPER	MILLI	2.78		-	-	-	2.78	MILLI/MJ	7.11	
ZINC	MILLI	3.02		-	-	-	3.02	MILLI/MJ	7.73	
NICKEL	MILLI	0.65		0.51	-	0.76	0.65	MILLI/MJ	1.66	
CHROMIUM	MILLI	0.11		-	-	-	0.11	MILLI/MJ	0.28	
MOLYBDENUM	MICRO	13.00		13.00	-	13.00	13.00	MICRO/MJ	33.27	
PHOSPHORUS	MILLI	314.00		-	-	-	314.00	MILLI/MJ	803.60	
FLUORIDE	MILLI	9.50	2	9.20	-	10.00	9.50	MILLI/MJ	24.31	
IODIDE	MICRO	8.00		-	-	-	8.00	MICRO/MJ	20.47	
BORON	MILLI	1.59		1.26	-	1.90	1.59	MILLI/MJ	4.07	
SELENIUM	MICRO	-		0.00	-	6.00				
VITAMIN B1	MICRO	25.00		20.00	-	30.00	25.00	MICRO/MJ	63.98	
VITAMIN B2	MILLI	0.95		0.93	-	0.97	0.95	MILLI/MJ	2.43	
NICOTINAMIDE	MILLI	7.70		7.70	-	7.70	7.70	MILLI/MJ	19.71	

1 ESTIMATED BY THE DIFFERENCE METHOD:
100 - (WATER + PROTEIN + FAT + AVAILABLE CARBOHYDRATES + MINERALS)

2 77 % OF THE FLUORINE CONTENT DISSOLVES ON INFUSION WITH BOILING WATER

CONSTITUENTS	DIM	AV	VARIATION			AVR	NUTR. DENS.		MOLPERC.
PANTOTHENIC ACID	MILLI	1.30	1.30	-	1.30	1.30	MILLI/MJ	3.33	
VITAMIN B6	MILLI	0.31	0.31	-	0.31	0.31	MILLI/MJ	0.79	
VITAMIN C	-	0.00	-	-	-	0.00			
SALICYLIC ACID	MILLI	3.69	3.69	-	3.70	3.70	MILLI/MJ	9.47	
GLUCOSE	GRAM	0.10	-	-	-	0.10	GRAM/MJ	0.26	
FRUCTOSE	GRAM	0.70	-	-	-	0.70	GRAM/MJ	1.79	
TANNINS	GRAM	10.90	5.00	-	12.50	10.90	GRAM/MJ	27.90	

Hefe

Yeast
Levure

BÄCKERHEFE
GEPRESST

BAKER'S YEAST
COMPRESSED

LEVURE DE BOULANGER
COMPRIMÉE

		PROTEIN	FAT	CARBOHYDRATES	TOTAL
ENERGY VALUE (AVERAGE) PER 100 G EDIBLE PORTION	KJOULE	262	47	0.0	309
	(KCAL)	63	11	0.0	74
AMOUNT OF DIGESTIBLE CONSTITUENTS PER 100 G	GRAM	13.36	1.08	0.00	
ENERGY VALUE (AVERAGE) OF THE DIGESTIBLE FRACTION PER 100 G EDIBLE PORTION	KJOULE	210	42	0.0	252
	(KCAL)	50	10	0.0	60

WASTE PERCENTAGE AVERAGE 0.00

CONSTITUENTS	DIM	AV	VARIATION			AVR	NUTR. DENS.		MOLPERC.
WATER	GRAM	73.00	-	-	-	73.00	GRAM/MJ	290.09	
PROTEIN	GRAM	16.70	-	-	-	16.70	GRAM/MJ	66.36	
FAT	GRAM	1.20	-	-	-	1.20	GRAM/MJ	4.77	
MINERALS	GRAM	2.10	-	-	-	2.10	GRAM/MJ	8.35	
SODIUM	MILLI	34.00	26.00	-	46.00	34.00	MILLI/MJ	135.11	
POTASSIUM	MILLI	649.00	627.00	-	670.00	649.00	GRAM/MJ	2.58	
MAGNESIUM	MILLI	28.00	28.00	-	28.00	28.00	MILLI/MJ	111.27	
CALCIUM	MILLI	28.00	25.00	-	30.00	28.00	MILLI/MJ	111.27	
MANGANESE	MILLI	0.28	0.28	-	0.28	0.28	MILLI/MJ	1.11	
IRON	MILLI	4.90	-	-	-	4.90	MILLI/MJ	19.47	
COPPER	MILLI	0.14	0.14	-	0.14	0.14	MILLI/MJ	0.56	
ZINC	MILLI	4.05	4.05	-	4.05	4.05	MILLI/MJ	16.09	
MOLYBDENUM	MICRO	7.00	5.00	-	9.00	7.00	MICRO/MJ	27.82	
PHOSPHORUS	MILLI	605.00	-	-	-	605.00	GRAM/MJ	2.40	
IODIDE	MICRO	0.50	0.50	-	0.50	0.50	MICRO/MJ	1.99	
BORON	MILLI	4.60	2.50	-	6.70	4.60	MILLI/MJ	18.28	
SELENIUM	MICRO	0.90	0.90	-	0.90	0.90	MICRO/MJ	3.58	
VITAMIN B1	MILLI	1.43	0.44	-	2.40	1.43	MILLI/MJ	5.68	
VITAMIN B2	MILLI	2.31	0.68	-	3.00	2.31	MILLI/MJ	9.18	
NICOTINAMIDE	MILLI	17.40	7.56	-	40.00	17.40	MILLI/MJ	69.15	
PANTOTHENIC ACID	MILLI	3.46	1.50	-	7.02	3.46	MILLI/MJ	13.75	
VITAMIN B6	MILLI	0.81	0.43	-	1.51	0.81	MILLI/MJ	3.22	
BIOTIN	MICRO	33.00	16.00	-	49.00	33.00	MICRO/MJ	131.14	
FOLIC ACID	MILLI	1.02	0.40	-	2.16	1.02	MILLI/MJ	4.05	
VITAMIN C	-	0.00	-	-	-	0.00			

CONSTITUENTS	DIM	AV	VARIATION		AVR	NUTR. DENS.		MOLPERC.
ARGININE	GRAM	0.73	0.68	- 0.78	0.73	GRAM/MJ	2.90	
CYSTINE	GRAM	0.14	0.13	- 0.15	0.14	GRAM/MJ	0.56	
HISTIDINE	GRAM	0.40	0.34	- 0.45	0.40	GRAM/MJ	1.59	
ISOLEUCINE	GRAM	0.89	0.83	- 0.94	0.89	GRAM/MJ	3.54	
LEUCINE	GRAM	1.30	1.13	- 1.46	1.30	GRAM/MJ	5.17	
LYSINE	GRAM	1.23	1.16	- 1.29	1.23	GRAM/MJ	4.89	
METHIONINE	GRAM	0.29	0.25	- 0.32	0.29	GRAM/MJ	1.15	
PHENYLALANINE	GRAM	0.77	0.65	- 0.94	0.77	GRAM/MJ	3.06	
THREONINE	GRAM	0.82	0.80	- 0.83	0.82	GRAM/MJ	3.26	
TRYPTOPHAN	GRAM	0.15	0.08	- 0.20	0.15	GRAM/MJ	0.60	
TYROSINE	GRAM	0.67	0.59	- 0.74	0.67	GRAM/MJ	2.66	
VALINE	GRAM	1.00	0.93	- 1.07	1.00	GRAM/MJ	3.97	
TOTAL PURINES	MILLI	680.00 1	680.00	- 680.00	680.00	GRAM/MJ	2.70	
ADENINE	MILLI	229.00	229.00	- 229.00	229.00	MILLI/MJ	910.02	
GUANINE	MILLI	206.00	206.00	- 206.00	206.00	MILLI/MJ	818.62	
XANTHINE	MILLI	10.00	10.00	- 10.00	10.00	MILLI/MJ	39.74	
HYPOXANTHINE	MILLI	126.00	126.00	- 126.00	126.00	MILLI/MJ	500.71	

1 VALUE OF TOTAL PURINES IS EXPRESSSED IN MG URIC ACID/100G

BIERHEFE GETROCKNET | **BREWER'S YEAST** DRIED | **LEVURE DE BIERE** SÈCHE

		PROTEIN	FAT	CARBOHYDRATES	TOTAL
ENERGY VALUE (AVERAGE)	KJOULE	752	163	0.0	915
PER 100 G EDIBLE PORTION	(KCAL)	180	39	0.0	219
AMOUNT OF DIGESTIBLE CONSTITUENTS PER 100 G	GRAM	38.32	3.78	0.00	
ENERGY VALUE (AVERAGE)	KJOULE	601	147	0.0	748
OF THE DIGESTIBLE FRACTION PER 100 G EDIBLE PORTION	(KCAL)	144	35	0.0	179

WASTE PERCENTAGE AVERAGE 0.00

CONSTITUENTS	DIM	AV	VARIATION			AVR	NUTR. DENS.		MOLPERC.
WATER	GRAM	6.00	5.00	-	7.00	6.00	GRAM/MJ	8.02	
PROTEIN	GRAM	47.90	46.10	-	50.00	47.90	GRAM/MJ	64.01	
FAT	GRAM	4.20	-	-	-	4.20	GRAM/MJ	5.61	
MINERALS	GRAM	7.90	-	-	-	7.90	GRAM/MJ	10.56	
SODIUM	MILLI	77.00	54.00	-	100.00	77.00	MILLI/MJ	102.90	
POTASSIUM	GRAM	1.41	1.32	-	1.50	1.41	GRAM/MJ	1.88	
CALCIUM	MILLI	50.00	10.00	-	80.00	50.00	MILLI/MJ	66.82	
MANGANESE	MILLI	0.53	-	-	-	0.53	MILLI/MJ	0.71	
IRON	MILLI	17.60	13.80	-	20.00	17.60	MILLI/MJ	23.52	
COPPER	MILLI	3.32	-	-	-	3.32	MILLI/MJ	4.44	
ZINC	MILLI	8.00	-	-	-	8.00	MILLI/MJ	10.69	
PHOSPHORUS	MILLI	1.90	-	-	-	1.90	MILLI/MJ	2.54	
IODIDE	MICRO	4.00	-	-	-	4.00	MICRO/MJ	5.35	
SELENIUM	MICRO	-	8.00	-	90.00				
VITAMIN B1	MILLI	12.00	5.00	-	23.50	12.00	MILLI/MJ	16.04	
VITAMIN B2	MILLI	3.77	1.59	-	5.45	3.77	MILLI/MJ	5.04	
NICOTINAMIDE	MILLI	44.80	28.80	-	59.20	44.80	MILLI/MJ	59.87	
PANTOTHENIC ACID	MILLI	7.21	0.94	-	18.90	7.21	MILLI/MJ	9.63	
VITAMIN B6	MILLI	4.41	1.52	-	9.40	4.41	MILLI/MJ	5.89	
BIOTIN	MICRO	115.00	-	-	-	115.00	MICRO/MJ	153.68	
FOLIC ACID	MILLI	3.17	1.80	-	5.50	3.17	MILLI/MJ	4.24	
ARGININE	GRAM	2.18	2.01	-	2.34	2.18	GRAM/MJ	2.91	
CYSTINE	GRAM	0.57	0.56	-	0.57	0.57	GRAM/MJ	0.76	
HISTIDINE	GRAM	1.30	-	-	-	1.30	GRAM/MJ	1.74	
ISOLEUCINE	GRAM	2.49	-	-	-	2.49	GRAM/MJ	3.33	
LEUCINE	GRAM	3.35	-	-	-	3.35	GRAM/MJ	4.48	
LYSINE	GRAM	3.54	3.43	-	3.65	3.54	GRAM/MJ	4.73	
METHIONINE	GRAM	0.93	0.87	-	0.98	0.93	GRAM/MJ	1.24	
PHENYLALANINE	GRAM	2.17	1.98	-	2.35	2.17	GRAM/MJ	2.90	
THREONINE	GRAM	2.56	2.44	-	2.68	2.56	GRAM/MJ	3.42	
TRYPTOPHAN	GRAM	0.74	-	-	-	0.74	GRAM/MJ	0.99	
TYROSINE	GRAM	1.76	1.53	-	1.98	1.76	GRAM/MJ	2.35	
VALINE	GRAM	2.73	2.62	-	2.83	2.73	GRAM/MJ	3.65	
TOTAL PURINES	GRAM	1 1.81	-	-	-	1.81	GRAM/MJ	2.42	

1 VALUE EXPRESSED IN MG URIC ACID/100G

Würzmittel

Flavourings
Condiments

SUPPENWÜRFEL SOUPE CUBE CUBES À POTAGE

		PROTEIN	FAT	CARBOHYDRATES	TOTAL
ENERGY VALUE (AVERAGE) PER 100 G EDIBLE PORTION	KJOULE	291	867	0.0	1157
	(KCAL)	70	207	0.0	277
AMOUNT OF DIGESTIBLE CONSTITUENTS PER 100 G	GRAM	15.32	20.71	0.00	
ENERGY VALUE (AVERAGE) OF THE DIGESTIBLE FRACTION PER 100 G EDIBLE PORTION	KJOULE	282	823	0.0	1105
	(KCAL)	67	197	0.0	264

WASTE PERCENTAGE AVERAGE 0.00

CONSTITUENTS	DIM	AV	VARIATION	AVR	NUTR. DENS.		MOLPERC.
WATER	GRAM	4.84	2.89 - 8.27	4.84	GRAM/MJ	4.38	
PROTEIN	GRAM	15.80	13.40 - 19.80	15.80	GRAM/MJ	14.29	
FAT	GRAM	21.80	17.60 - 26.20	21.80	GRAM/MJ	19.72	
MINERALS	GRAM	43.90	38.30 - 46.30	43.90	GRAM/MJ	39.72	
FORMIC ACID	GRAM	0.10	0.02 - 0.29	0.10	GRAM/MJ	0.09	

Mayonnaise

Mayonnaise
Mayonnaise

Mayonnaise
Mayonnaise
Mayonnaise

MAYONNAISE FETTREICH	MAYONNAISE HIGH-FAT CONTENT	MAYONNAISE RICHE EN MATIÈRE GRASSE

		PROTEIN	FAT	CARBOHYDRATES	TOTAL
ENERGY VALUE (AVERAGE)	KJOULE	28	3210	0.0	3238
PER 100 G EDIBLE PORTION	(KCAL)	6.7	767	0.0	774
AMOUNT OF DIGESTIBLE CONSTITUENTS PER 100 G	GRAM	1.44	78.37	0.00	
ENERGY VALUE (AVERAGE)	KJOULE	27	3050	0.0	3077
OF THE DIGESTIBLE FRACTION PER 100 G EDIBLE PORTION	(KCAL)	6.5	729	0.0	735

WASTE PERCENTAGE AVERAGE 0.00

CONSTITUENTS	DIM	AV	VARIATION			AVR	NUTR. DENS.		MOLPERC.
WATER	GRAM	13.00	11.60	-	16.00	13.00	GRAM/MJ	4.23	
PROTEIN	GRAM	1.49	1.25	-	1.85	1.49	GRAM/MJ	0.48	
FAT	GRAM	82.50	78.00	-	86.60	82.50	GRAM/MJ	26.81	
MINERALS	GRAM	0.98	0.89	-	1.06	0.98	GRAM/MJ	0.32	
SODIUM	MILLI	481.00	362.00	-	600.00	481.00	MILLI/MJ	156.33	
POTASSIUM	MILLI	18.00	17.00	-	19.00	18.00	MILLI/MJ	5.85	
MAGNESIUM	MILLI	23.00	19.00	-	26.00	23.00	MILLI/MJ	7.48	
IRON	MILLI	1.00	-	-	-	1.00	MILLI/MJ	0.33	
PHOSPHORUS	MILLI	60.00	-	-	-	60.00	MILLI/MJ	19.50	
VITAMIN B1	MICRO	40.00	20.00	-	50.00	40.00	MICRO/MJ	13.00	
VITAMIN B2	MICRO	40.00	20.00	-	50.00	40.00	MICRO/MJ	13.00	
NICOTINAMIDE	MILLI	0.20	-	-	-	0.20	MILLI/MJ	0.07	
VITAMIN C	-	0.00	-	-	-	0.00			

Appendix I

Composition of the meat portions of animals for slaughter based on the recommendations of the German Society of Agriculture (Deutsche Landwirtschaftsgesellschaft, DLG).

Anhang I

Zusammensetzung der Fleischteile von Schlachttieren in grobgeweblicher Zerlegung nach der Empfehlung der Deutschen Landwirtschaftsgesellschaft (DLG).

Appendice I

Composition des pieces de viande de bétail decoupeés selon des recommandations de la Societé d'agriculture allemande (Deutsche Landwirtschaftsgesellschaft, DLG).

Page; Seite; Page

Rindfleisch; Beef; Viande de Boeuf

Meat portion	Water (g/100 g edible portion)	Protein (g/100 g edible portion)	Fat (g/100 g edible portion)	Minerals (g/100 g edible portion)	Cholesterol (mg/100 g edible portion)	Total purins* (mg/100 g edible portion)	kcal/100g edible portion	kJ/100g edible portion
Brust; Brisket; Poitrine								
Commercial class								
R1+2	65.7	19.6	13.7	1.07	70	110	202	854
R3	56.7	17.4	25.0	0.92	70	90	295	1246
R4+5	52.4	16.3	30.4	0.84	80	80	339	1432
Bug; Shoulder; Paleron								
Commercial class								
R1+2	72.1	19.5	7.6	1.13	60	120	146	620
R3	68.8	18.8	11.7	1.07	60	110	181	764
R4+5	65.9	18.2	15.2	1.03	70	110	210	887
Fehlrippe; Short loin; Entrecôte								
Commercial class								
R1+2	71.3	21.2	6.5	1.14	70	120	143	607
R3	67.6	20.2	11.2	1.08	60	120	182	769
R4+5	64.7	19.5	14.9	1.03	70	110	212	898
Filet; Fillet; Filet								
Commercial class								
R1+2	68.0	20.0	11.2	1.10	60	120	181	766
R3	63.7	18.9	16.6	1.02	70	110	225	952
R4+5	61.4	18.4	19.4	0.98	70	110	248	1050

Rindfleisch; Beef; Viande de Boeuf

Meat portion	Water	Protein	Fat	Minerals	Cholesterol	Total purins*	kcal/100g edible portion	kJ/100g edible portion
	g/100 g edible portion				mg/100 g edible portion			
Fleischdünnung; Flank, Belly; Ventre								
Commercial class								
R1+2	62.6	18.4	18.1	1.03	60	110	237	1001
R3	51.0	15.0	33.1	0.84	60	90	358	1513
R4+5	43.3	13.1	41.8	0.74	60	80	429	1811
Kamm; Chuck; Collet								
Commercial class								
R1+2	72.0	21.2	5.8	1.18	70	120	137	581
R3	69.4	20.6	9.1	1.13	60	120	164	696
R4	68.8	20.4	9.8	1.12	60	120	170	719
Keule; Leg; Culotte								
Commercial class								
R1+2	71.0	21.3	6.8	1.13	70	120	146	621
R3	68.9	20.7	9.6	1.09	60	120	169	717
R4+5	66.7	20.2	12.3	1.07	60	110	192	811
Roastbeef; Roastbeef; Rosbif								
Commercial class								
R1+2	73.6	21.6	3.75	1.21	60	120	120	510
R3	66.4	19.8	12.8	1.09	70	110	194	823
R4+5	62.2	18.8	18.1	1.01	70	100	238	1007
Spannrippe; Rib; Côte								
Commercial class								
R1+2	70.4	20.8	7.8	1.15	60	120	153	650
R3	62.2	18.8	18.1	1.01	70	100	238	1007
R4+5	55.9	17.6	26.1	0.91	70	90	305	1291

Schweinefleisch; Pork; Viande de Porc

Meat portion	Water (g/100 g edible portion)	Protein (g/100 g edible portion)	Fat (g/100 g edible portion)	Minerals (g/100 g edible portion)	Cholesterol (mg/100 g edible portion)	Total purins* (mg/100 g edible portion)	kcal/100g edible portion	kJ/100g edible portion
Bauch; Belly; Ventre								
Commercial class								
E	51.0	15.3	32.6	0.92	60	110	355	1499
U	46.4	14.0	38.55	0.91	60	100	403	1703
R	41.6	12.6	44.7	0.88	60	80	453	1913
Bug (ohne subkutanes Fett); Shoulder (without subcutaneous fat); Epaule (sans graisse souscutan)								
Commercial class								
E	69.4	20.4	8.8	0.98	70	150	161	681
U	68.3	20.1	10.3	0.96	70	150	173	733
R	67.1	19.7	12.0	0.95	70	145	187	791
Bug (mit subkutanem Fett); Shoulder (with subcutaneous fat); Epaule (avec graisse souscutan)								
Commercial class								
E	60.9	18.0	19.85	0.94	70	130	251	1060
U	58.0	17.1	23.7	0.91	70	120	285	1191
R	55.0	16.3	27.5	0.89	75	115	313	1322
Eisbein, vorne (Vorderhaxe); Hip-Bone (foreleg); Jambonneau de devant								
Commercial class								
E	63.35	18.6	16.9	0.90	70	130	227	958
U	61.6	18.0	19.15	0.88	70	130	244	1034
R	59.8	17.5	21.5	0.85	70	120	264	1115
Eisbein, hinten (Hinterhaxe); Hip-Bone (hind leg); Jambonneau de derrière								
Commercial class								
E	61.25	17.9	19.6	0.87	70	130	248	1049
U	59.5	17.4	21.9	0.85	70	120	267	1128
R	57.4	16.7	24.7	0.82	70	115	289	1223

Schweinefleisch; Pork; Viande de Porc

Meat portion	Water	Protein	Fat	Minerals	Cholesterol	Total purins*	kcal/100g edible portion	kJ/100g edible portion
	g/100 g edible portion				mg/100 g edible portion			
Filet; Fillet; Filet								
Commercial class								
E	69.4	20.4	8.9	0.98	70	150	162	685
U	68.6	20.2	9.9	0.97	70	150	170	720
R	67.6	19.9	11.2	0.96	70	145	180	764
Kamm; Chuck; Collet								
Commercial class								
E	67.4	19.8	11.5	0.95	70	145	183	774
U	65.8	19.3	13.4	0.93	70	140	198	837
R	64.2	18.8	15.8	0.91	70	135	217	920
Kotelett; Chop; Côtelette								
Commercial class								
E	69.3	20.4	9.1	0.97	70	150	164	693
U	67.6	19.9	11.2	0.96	70	145	180	764
R	65.7	19.3	13.8	0.93	70	140	201	853
Schinken (ohne subkutanes Fett); Leg (without subcutaneous fat); Culotte (sans graisse souscutan)								
Commercial class								
E	71.9	21.2	5.6	1.02	70	160	135	573
U	71.3	21.0	6.4	1.01	70	160	142	600
R	70.6	20.8	7.3	0.99	70	150	148	631
Schinken (mit subkutanem Fett); Leg (with subcutaneous fat); Culotte (avec graisse souscutan)								
Commercial class								
E	61.9	18.1	18.5	0.96	65	130	239	1011
U	58.8	16.4	22.2	0.94	70	125	265	1122
R	55.4	16.4	27.1	0.91	70	120	310	1309
Wamme; Dewlap; Fanon								
Commercial class								
E	42.3	12.8	43.9	0.88	60	90	446	1886
U	38.6	11.7	48.7	0.86	60	80	485	2050
R	35.2	10.8	53.1	0.84	60	70	521	2201

Mastlammfleisch; Lamb; Viande d'agneau

Meat portion	g/100 g edible portion Water	Protein	Fat	Minerals	mg/100 g edible portion Cholesterol	Total purins*	kcal/100g edible portion	kJ/100g edible portion
Brust; Brisket; Poitrine								
Commercial class								
I g	66.35	18.6	14.3	1.09	70	130	203	860
I m	56.1	15.3	28.0	0.92	70	90	313	1324
I s	53.0	14.3	32.1	0.87	70	80	346	1463
Bug; Shoulder; Epaule								
Commercial class								
I g	66.4	18.6	14.3	1.09	70	130	203	860
I m	64.1	18.0	17.3	1.05	70	120	228	963
I s	61.9	17.2	20.3	1.01	70	110	252	1064
Kamm + Hals; Chuck; Collet								
Commercial class								
I g	69.2	19.5	10.6	1.14	70	130	173	734
I m	64.4	18.0	16.9	1.06	80	120	224	948
I s	62.0	17.2	20.1	1.02	70	110	250	1056
Keule; Leg; Gigot								
Commercial class								
I g	68.8	19.4	11.0	1.13	70	130	177	748
I m	65.9	18.5	14.9	1.07	70	120	208	881
I s	64.9	18.2	16.1	1.06	70	120	218	921
Kotelett + Filet; Chop + Fillet; Côtelette + Filet								
Commercial class								
I g	66.4	18.6	14.2	1.09	70	130	202	856
I m	59.5	16.4	23.5	0.97	70	100	277	1172
I s	56.5	15.4	27.4	0.93	70	100	308	1303

Hammelfleisch; Mutton; Viande de Mouton

Meat portion	g/100 g edible portion Water	Protein	Fat	Minerals	mg/100 g edible portion Cholesterol	Total purins*	kcal/100g edible portion	kJ/100g edible portion
Brust; Brisket; Poitrine	48.00	12.00	37.00	-	-	-	381	1610
Bug; Shoulder; Epaule	58.00	15.60	25.00	0.80	-	-	287	1215
Keule; Leg; Gigot	64.00	18.00	18.00	0.90	70.0	150.0	234	990
Kotelett; Chop; Côtelette	52.00	14.90	32.00	0.80	-	-	348	1469
Lende; Sirloin; Contrefilet	66.70	18.70	13.20	1.20	-	-	194	820

Kalbfleisch; Veal; Viande de Veau

Meat portion	Water	Protein	Fat	Minerals	Cholesterol	Total purins*	kcal/100g edible portion	kJ/100g edible portion
	g/100 g edible portion				mg/100 g edible portion			
Bauch; Belly; Ventre								
Commercial class								
R1+2	71.1	20.0	7.46	1.11	70	140	147	623
R3	64.3	18.4	15.9	0.99	70	120	217	917
R4+5	60.7	17.6	20.4	0.94	80	110	254	1074
Bug; Shoulder; Epaule								
Commercial class								
R1+2	74.9	20.9	2.63	1.17	70	150	107	455
R3	70.4	19.8	8.22	1.09	70	140	153	649
R4+5	68.9	19.5	10.1	1.07	70	140	169	715
Filet; Fillet; Filet								
Commercial class								
R1+2	70.9	19.95	7.63	1.10	70	140	149	629
R3	69.9	19.7	8.85	1.08	70	140	158	671
R4+5	69.3	19.6	9.66	1.08	70	140	165	700
Hals; Neck; Collet								
Commercial class								
R1+2	76.4	21.2	2.66	1.19	70	150	109	462
R3	73.4	20.5	4.17	1.14	80	150	120	507
R4+5	72.7	20.4	5.45				131	554
Haxe (Hinterhaxe); Knuckle; Jarret de Veau								
Commercial class								
R1+2	75.6	21.0	1.83	1.17	70	160	101	427
R3	72.3	20.3	5.94	1.12	70	150	135	571
R4+5	71.5	20.1	6.87	1.12	70	140	142	603

Kalbfleisch; veal; Viande de Veau

Meat portion	Water	Protein	Fat	Minerals	Cholesterol	Total purins*	kcal/100g edible portion	kJ/100g edible portion
	g/100 g edible portion				mg/100 g edible portion			
Keule; Leg; Membre de Veau								
Commercial class								
R1+2	74.2	20.7	3.56	1.15	70	150	115	487
R3	72.0	20.2	6.31	1.12	70	150	138	583
R4+5	70.9	19.95	7.60	1.11	70	140	148	628
Kotelett; Chop; Côtelette								
Commercial class								
R1+2	74.9	20.9	2.62	1.17	70	150	107	455
R3	71.4	20.1	6.99	1.12	70	140	143	607
R4+5	68.7	19.4	10.4	1.07	70	140	171	725

* values expressed in mg uric acid /100 g.

Calculation of energy according to the regulation of the Federal Republic of Germany (Nährwertkennzeichnungsverordnung: BGBL 1977, pp. 2569).
Energy conversion factors: Protein: 4 kcal, 17 kJ/g; Fat: 9 kcal, 38kJ/g.

Appendix II
Free amino acids in fruits and fruit juices

Anhang II
Freie Aminosäuren in Früchten und Fruchtsäften

Appendice II
Aminoacides libres en fruits et en jus de fruits

Foods:		Free amino acids	Alanine	Arginine	Aspartic acid	Cystine	Glutamic acid	Glycine	Histidine	Isoleucine	Leucine	Lysine
Tomate	1		3,20	4,30	43,80	-	187,00	1,50	4,50	3,60	5,00	6,40
Tomato	2		2,80-	2,30-	29,00-	-	69,00-	1,10-	3,30-	3,00-	2,30-	4,00-
Tomate			3,40	6,60	53,00	-	252,00	2,30	5,70	4,60	9,70	10,90
Erdbeere	1		6,90	0,50	10,50	0,10	10,30	0,70	0,20	0,40	0,70	0,50
Strawberry	2		1,20-	0,40-	10,40-	-	1,80-	0,10-	-	0,00-	0,00-	0,00-
Fraise			26,50	0,70	28,60		44,40	5,00		1,25	7,60	1,70
Holunderbeere	1		20,30	8,60	15,90	-	17,30	0,30	9,90	8,30	58,10	1,90
Elderberry black	2		4,20-	5,80-	10,90-	-	13,80-	0,20-	6,90-	4,90-	36,70-	1,30-
Baie de sureau			48,80	14,80	21,50		22,70	0,40	13,80	12,90	85,40	2,90
Apfelsine	1		6,70	10,30	38,80	-	16,70	3,10	-	-	0,90	3,70
Orange	2		4,00-	0,00-	15,20-	-	7,00-	0,00-	-	-	0,00-	0,00-
Orange			10,00	20,00	90,00		80,00	10,00			1,40	4,30
Banane	1		4,10	2,82	11,70	0,50	22,00	1,90	16,40	0,70	7,30	2,90
Banana												
Banane												
Grapefruit	1		8,10	10,80	62,50	-	17,80	2,90	-	-	0,40	2,00
Grapefruit	2		4,00-	4,00-	30,00-	-	8,00-	1,00-	-	-	0,00-	0,00-
Pamplemousse			24,00	20,00	87,10		50,00	6,00			1,00	3,50
Mandarine	1		8,80	11,30	40,20	-	10,80	3,90	-	-	0,30	1,00
Mandarin	2		-	2,00-	25,00-	-	6,00-	0,00-	-	-	0,00-	0,00-
Mandanrine				20,00	70,00		15,00	15,00			1,00	2,00
Mango	1		9,70	19,70	0,20	0,10	1,90	0,70	0,60	0,20	0,70	0,80
Mango	2		4,50-	13,10-	0,12-	-	1,70-	0,50-	0,30-	-	0,20-	0,70-
Mangue			15,00	26,30	0,33		2,10	0,90	0,80		1,30	0,80
Zitrone	1		10,10	-	57,00	-	34,90	2,60	-	0,40	1,20	-
Lemon	2		7,70-	0,00-	31,20-	-	13,30-	0,00-	-	-	0,00-	0,00-
Citron			16,00	8,00	120,00		84,00	6,00			4,00	0,90
Apfelsinensaft	1		8,50	72,10	29,00	-	11,30	1,30	1,20	0,60	0,50	3,10
Orange juice												
Jus d'orange												
Himbeersaft	1		32,20	-	8,70	-	14,00	1,70	2,10	1,70	1,70	4,80
Rasberry juice	2		10,50-	-	3,90-	-	6,70-	0,80-	0,70-	0,80-	1,20-	2,70-
Jus de framboises			45,70		13,90		17,60	2,10	3,30	3,40	2,00	6,80
Passionsfruchtsaft	1		17,80	-	86,50	-	55,90	1,50	5,50	2,60	3,30	4,60
Granadilla juice	2		8,90-	-	39,90-	-	29,40-	0,70-	2,20-	1,30-	1,30-	1,80-
Jus de granadille			40,10		159,70		80,90	3,80	8,80	6,60	6,60	10,90
Sauerkirschsaft	1		traces	-	11,00	-	3,20	0,30	0,70	0,50	-	-
Cherry morello juice	2		-	-	5,00-	-	2,00-	0,20-	0,40-	0,20-	-	-
Jus de griotte					13,00		3,80	0,40	1,50	1,00		

mg/100g 1 = average value (AV) 2 = variation

Appendix II
Free amino acids in fruits and fruit juices

Anhang II
Freie Aminosäuren in Früchten und Fruchtsäften

Appendice II
Aminoacides libres en fruits et en jus de fruits

Foods:	Methionine	Phenylalanine	Proline	Serine	Threonine	Tryptophan	Tyrosine	Valine	Gamma-amino-but.a	Asparagine	Glutamine
Tomate	0,90	8,60	1,10	6,60	6,50	-	3,90	3,40	9,00	-	-
Tomato	0,50-	7,20-	0,60-	5,70-	-	-	1,50-	0,80-	-	-	-
Tomate	1,70	11,40	1,60	7,40			3,90	7,40			
Erdbeere	0,10	0,40	0,70	3,10	1,80	0,80	0,40	0,60	3,00	-	-
Strawberry		0,00-	0,00-	1,20-	0,00-	0,40-	0,00-	0,10-	0,10-	-	-
Fraise		1,20	1,30	5,20	4,40	1,20	1,50	2,10	9,70		
Holunderbeere	1,70	48,80	4,10	9,20	3,10	8,20	109,50	15,60	1,50	7,00	5,60
Elderberry black	1,30-	37,50-	1,50-	5,60-	1,80-	5,60-	84,00-	8,70-	1,30-	1,50-	0,50-
Baie de sureau	2,30	57,30	8,50	15,60	5,10	10,40	131,70	26,30	1,60	17,70	14,60
Apfelsine	1,00	3,30	35,00	9,10	4,00	4,70	2,40	1,40	18,10	-	-
Orange	0,00-	0,00-	7,00-	5,00-	0,00-	0,00-	0,00-	0,00-	7,70-	-	-
Orange	4,00	9,50	65,90	12,60	10,00	10,00	6,00	4,00	31,20		
Banane	0,18	1,00	2,30	3,00	1,80	-	1,90	5,40	11,20	12,40	0,02
Banana											
Banane											
Grapefruit	-	1,80	14,90	6,60	6,30	1,30	2,30	1,10	-	-	-
Grapefruit	-	0,00-	0,00-	4,00-	0,00-	0,00-	0,00-	0,00-	-	-	-
Pamplemousse		10,20	22,00	9,40	11,00	4,00	10,00	3,00			
Mandarine	2,00	2,90	2,50	10,50	6,60	3,50	6,10	8,30	-	-	-
Mandarin	0,00-	0,00-	0,00-	8,00-	0,00-	0,00-	2,00-	1,00-	-	-	-
Mandanrine	4,00	4,50	6,00	12,00	9,00	9,00	15,00	20,00			
Mango	-	0,40	1,40	1,40	1,30	-	0,40	1,00	5,20	0,67	-
Mango	-	0,10-	1,40-	1,30-	1,00-	-	0,30-	0,50-	0,10	-	-
Mangue		1,00	1,50	1,50	1,50		0,60	1,50	10,30		
Zitrone	-	2,20	-	12,60	5,20	-	-	2,10	5,10	-	-
Lemon	-	0,00-	0,00-	-	0,00-	-	0,00-	0,90-	-	-	-
Citron		6,00	34,10		18,00		12,00	6,00			
Apfelsinensaft	0,30	3,00	79,00	12,90	1,80	-	1,10	1,70	-	-	-
Orange juice											
Jus d'orange											
Himbeersaft	0,50	0,80	5,00	18,20	5,40	-	1,00	4,50	11,50	88,50	29,00
Rasberry juice	0,00-	0,30-	2,50-	5,10-	1,80-	-	0,40-	1,90-	2,30-	6,40-	5,50-
Jus de framboises	1,00	1,50	15,00	29,80	9,40		2,20	6,10	15,10	181,30	60,00
Passionsfruchtsaft	-	6,60	-	31,50	1,80	-	-	4,70	25,80	-	-
Granadilla juice	0,00-	3,30-	0,10-	14,70-	0,90-	-	0,00-	2,30-	15,50-	0,00-	0,00-
Jus de granadille	0,80	13,20	1,30	52,60	2,90		5,40	9,40	41,20	0,03	0,02
Sauerkirschsaft	-	0,30	4,30	1,90	1,20	-	0,20	-	16,00	-	-
Cherry morello juice	-	0,20-	1,60-	1,30-	0,20-	-	0,10-	-	-	-	-
Jus de griotte		0,40	11,70	2,40	1,70		0,30				

mg/100g 1 = average value (AV) 2 = variation

INDEX

A

B

C

D

E

F

G

H

I

J

K

L

M

N

O

P

Q

R

S

T

U

V

W

Y

Z

REGISTER

A

B

C

D

E

T

U

V

W

REGISTRE

A

B

C

D

E

F

G

H

R

S

T

V

W

Y

Concordance

between the Food Code of the Federal Republic of Germany (1) and the number of the page, where the corresponding food is to be found.

Konkordanz

zwischen dem Warencode der Bundesrepublik Deutschland (1) und der Zahl der Seite, auf welcher das entsprechende Lebensmittel zu finden ist.

Concordance

entre le code alimentaire officiel de la République d'Allemagne fédérale (1) et le numero de la page ou l'on trouve l'aliment en question.

(1) Warencode für die amtliche Lebensmittelüberwachung, Verzehrserhebungen und Fremdstoffberechnungen von Peter Weigert und Hubertus Klein Bundesgesundheitsamt, Berlin 1984.

Code	Seite
130409	162
130410	164
130411	168
130413	157
130414	167
130415	166
130416	150
130417	170
130418	158
130420	169
130421	156
130423	161
130504	173
130508	171
130512	174
130601	175
140212	555
150101	485
150102	449
150201	472
150301	444
150401	451
150501	459
150601	465
150603	467
150701	439
150801	457
160102	474
160103	475
160104	477
160105	479
160112	488
160113	490
160114	491
160115	493
160116	495
160125	456
160126	463
160127	470
160128	442
160130	450
160201	487
160301	480
160303	497
160401	533
160402	535
160403	532
160605	461
160701	499
160702	482
160801	501
160906	453
160908	448
160909	446
160911	455
160912	441
170101	517
170105	505
170106	522
170201	510
170204	515
170301	519
170305	512
170306	514
170401	521
170701	509
170702	507
171101	503
181007	527
181401	529
181800	528
200101	997
220200	524
230101	686
230103	691
230105	676
230110	674
230117	696
230118	713
230201	704
230203	706
230402	699
230403	693
230404	708
230408	702
230501	872
230502	891
230503	876
230504	887
230505	878
230506	885
230507	868
230508	883
230512	889
230600	870
230701	874
230801	880
240101	542
240301	547
240306	551
240308	550
240410	556
240502	545
240601	549
240603	548
240701	552
240702	553
240801	531
240901	541
240902	539
240903	585
240905	588
240906	584
241002	534
250101	612
250102	605
250105	599
250106	603
250107	601
250108	614
250109	623
250110	626
250111	642
250112	609
250113	644
250114	637
250115	590
250117	616
250118	629
250119	608
250119	598
250120	615
250122	618
250201	595
250202	557
250203	592
250204	589
250205	632
250206	611
250208	646
250210	812
250212	607
250213	631
250301	664
250302	661
250304	851
250305	656
250306	659
250307	862
250307	866
250308	649
250309	671
250310	672
250312	651
250313	679
250401	562
250403	582
250404	560
250405	575
250406	573
250407	559
250408	580
250409	577
250411	569
250412	571
260105	640
260401	628
260501	625
260506	639
260602	635
260803	648
261001	594
261003	597
261005	634
261101	667
261109	684
261109	682
261110	653
261117	668
261401	658
261409	852
261508	681
261603	567
261803	566
262001	565
262003	581
262601	669
262602	568
262603	579
262706	641
270101	718
270206	728
270208	722
270301	717
270307	730
270308	729
270309	715
270501	724
270801	733
270803	723
280101	720
280602	727
280818	732
280823	726
290101	797
290102	769
290103	776
290104	767
290105	800
290106	779
290107	781
290108	783
290109	789
290110	791
290112	785
290113	795
290114	784
290115	772
290116	798
290117	766
290201	737
290202	742
290203	745
290301	765
290302	755
290303	756
290304	746
290305	761
290307	752
290308	751
290309	810
290401	808
290402	845
290404	864
290405	829
290501	805
290502	814
290503	825
290504	855
290505	836
290506	854
290507	849
290509	847
290510	828
290511	831
290512	842
290513	839
290514	621
290517	819
290918	853
300101	787
300101	788
300103	771
300104	774
300104	775
300105	778
300304	793
300802	741
300803	744
301001	739
301503	759
301504	750
301505	764
301507	754
301701	758
301702	748
301703	763
302801	807
303002	826
303003	822
310101	915
310104	906
310109	908
310110	909
310111	910
310112	913
310404	907
310601	896
311103	914
311601	903
311601	905
311602	917
311603	898
311603	900
311604	911
311903	902
312101	895
312104	912
320900	969
360502	955
360601	952
360602	953
360900	951
361405	989
370103	966
390101	939
390206	940
390207	941
400000	937
400500	938
410201	925
410202	929
410203	924
410204	931
410208	927
410401	928
410403	930
410601	934
410701	921
411003	923
411004	932
411005	933
411404	922
420200	948
420500	947
430302	942
431602	943
431803	944
440101	974
440701	975
450401	973
460101	979
460201	980
460301	981
460504	982
470301	983
531200	993
560202	987
560900	250